Lecture Notes in Computer Science 5878

Commenced Publication in 1973
Founding and Former Series Editors:
Gerhard Goos, Juris Hartmanis, and Jan van Leeuwen

Yingfei Dong Ding-Zhu Du Oscar Ibarra (Eds.)

Algorithms and Computation

20th International Symposium, ISAAC 2009
Honolulu, Hawaii, USA, December 16-18, 2009
Proceedings

Springer

Volume Editors

Yingfei Dong
University of Hawaii
Deptartment of Electrical and Computer Engineering
Holmes Hall 483
2540 Dole Street
Honolulu, HI 96822, USA
E-mail: yingfei@hawaii.edu

Ding-Zhu Du
University of Texas at Dallas
Department of Computer Science
Richardson, TX 75080, USA
E-mail: dzdu@utdallas.edu

Oscar Ibarra
University of California
Department of Computer Science
Santa Barbara, CA 93106, USA
E-mail: ibarra@cs.ucsb.edu

Library of Congress Control Number: 2009939433

CR Subject Classification (1998): F.2, G.1, G.2, G.3, G.4, I.2.2, I.3, D.2

LNCS Sublibrary: SL 1 – Theoretical Computer Science and General Issues

ISSN 0302-9743
ISBN-10 3-642-10630-7 Springer Berlin Heidelberg New York
ISBN-13 978-3-642-10630-9 Springer Berlin Heidelberg New York

springer.com

Printed in Germany

Typesetting: Camera-ready by author, data conversion by Scientific Publishing Services, Chennai, India
Printed on acid-free paper SPIN: 12805785 06/3180 5 4 3 2 1 0

Preface

The papers in this volume were presented at the 20th Annual International Symposium on Algorithms and Computation, held December 16–18, 2009, in Hawaii, USA. In response to the Call-for-Papers, 279 papers were submitted. Each paper received at least three reviews by either Program Committee members or experts selected by Program Committee members. In all, 120 papers were selected based on the review reports and are included in this volume. We wish to thank all who submitted papers for consideration and all Program Committee members and reviewers for their excellent and hard work. To our invited speakers, we wish to thank you for sharing your expertise. Finally, we wish to thank our colleagues who contributed to the success of the symposium and the sponsors for their assistance and support.

December 2009

Yingfei Dong
Ding-Zhu Du
Oscar Ibarra

Organization

Program Committee Co-chairs

Yingfei Dong	University of Hawaii, USA
Ding-Zhu Du	University of Texas at Dallas, USA
Oscar Ibarra	University of California, Santa Barbara, USA

Advisory Committee Chair

Takeshi Tokuyama	Tohoku University, Japan

Advisory Committee

Tetsuo Asano	Japan Advanced Institute of Science and Technology, Japan
Kyung-Yong Chwa	Korea Advanced Institute of Science and Technology, Republic of Korea
Francis Y.L. Chin	The University of Hong Kong, Hong Kong
Ding-Zhu Du	The University of Texas at Dallas, USA
Peter Eades	The University of Sydney, Australia
Wen Lian Hsu	Academia Sinica, Taiwan
Der Tsai Lee	Academia Sinica, Taiwan
Takao Nishizeki	Tohoku University, Japan

Local Arrangements Chair

Yingfei Dong	University of Hawaii, USA

Publicity, Webmaster, and Online Conference System (OCS) Co-chairs

Xiaofeng Gao	University of Texas at Dallas, USA
Donghyun Kim	University of Texas at Dallas, USA

Financial Co-chairs

Jiaofei Zhong	University of Texas at Dallas, USA
Feng Zou	University of Texas at Dallas, USA

Program Committee

Alberto Apostolico	Georgia Insititute of Technology, USA
Giorgio Ausiello	Università di Roma, Italy
Wolfgang Bein	University of Nevada, USA
Jin-Yi Cai	University of Wisconsin, USA
Zhipeng Cai	Mississippi State University, USA
Zhenfu Cao	Shanghai Jiaotong University, China
Wei Chen	Microsoft Research Asia, China
Maggie Cheng	Missouri University of Science and Technology, USA
Siu-Wing Cheng	Hong Kong University of Science and Technology, Hong Kong
Yongxi Cheng	University of Alberta, Canada
Marek Chrobak	University of California, Riverside, USA
Kyung-Yong Chwa	Korea Advanced Institute of Science and Technology, Republic of Korea
Bhaskar Dasgupta	University of Illinois at Chicago, USA
Narsingh Deo	University of Central Florida, USA
Omer Egecioglu	University of California, Santa Barbara, USA
Juraj Hromkovic	ETH Zentrum, Switzerland
Wenlian Hsu	Academia Sinica, Taiwan
Xiaodong Hu	Chinese Academy of Sciences, China
Kazuo Iwama	Kyoto University, Japan
Ravi Janardan	University of Minnesota, USA
Tao Jiang	University of California, Riverside, USA
Liying Kang	Shanghai University, China
Samir Khuller	University of Maryland, USA
Angsheng Li	Chinese Academy of Sciences, China
Shlomo Moran	Technion, USA
Koji Nakano	Hiroshima University, Japan
Hung Ngo	University of Buffalo, US
Mitsunori Ogihara	University of Miami, USA
E.K. Park	National Science Foundation, USA
Suneeta Ramaswami	Rutgers University, USA
Sanjay Ranka	University of Florida, USA
Sanguthevar Rajasekaran	University of Connecticut, USA
Balasubramanian Ravikumar	Sonoma State University, USA
Paul Spirakis	University of Patras, Greece

Kazuo Sugihara	University of Hawaii, USA
Xiaoming Sun	Tsinghua University, China
Subhash Suri	University of California, Santa Barbara, USA
My T. Thai	University of Florida, USA
Takeaki Uno	NII, Japan
Jie Wang	University of Massachusetts Lowell, USA
Lusheng Wang	City University of Hong Kong, Hong Kong
Wei Wang	Xi'an Jiaotong University, China
Weili Wu	University of Texas at Dallas, USA
Yibo Xue	Tsinghua University, China
Hsu-Chun Yen	National Taiwan University, Taiwan
Wenan Zang	Hong Kong University, Hong Kong
Zhao Zhang	Xingjiang University, China
Hong Zhu	Huadong Normal University, China

Referees

Abdullah Arslan
Sang Won Bae
Gill Barequet
Doina Bein
Said Bettayeb
Hans-J. Boeckenhauer
Vincenzo Bonifaci
Peter Brucker
Tian-Ming Bu
Saverio Caminiti
B. Charron-Bost
Zhi-Zhong Chen
Ho-Lin Chen
Yinjie Chen
Bhadrachalam Chitturi
Ovidiu Daescu
Mirela Damian
Thomas Erlebach
Zheng Fang
Paolo Ferragina
Robin Flatland
Hervé Fournier
Paolo Giulio Franciosa
Rajiv Gandhi
Xiaofeng Gao
Bill Gauvin
Mordecai Golin
Prosenjit Gupta
Xin Han
Rafi Hassin
Tomio Hirata
Junwei Huang
Keiko Imai
Volkan Isler
Naoyuki Kamiyama
Donghyun Kim
Sangjin Kim
Carl Kingsford
Dennis Komm
Richard Kralovic
Zbyněk Křivka
Yokesh Kumar
Luigi Laura
Ronny Lempel
Jian Li
Chun-Cheng Lin
Hsueh-I Lu
Pinyan Lu
Joe Mitchell
Shuichi Miyazaki
Tobias Moemke
Linda Morales
Pat Morin
Hiroki Morizumi
Hiroshi Nagamochi
Harumichi Nishimura
Yoshio Okamoto
Joseph O'Rourke
Sheung-Hung Poon
Iris Reinbacher
Kunihiko Sadakane
Yan Shi
Yi Shi
Charles Shields Jr.
Marcelo Siqueira
Sagi Snir
Bettina Speckmann
Andreas Sprock
William Steiger
Monika Steinova
Hal Sudborough
Hisao Tamaki
Suguru Tamaki
Ramki Thurimella
Takeshi Tokuyama
Shi-Chun Tsai
Ryuhei Uehara
Antoine Vigneron
Xiang Wan
Yajun Wang
Zhong Wang
James Willson
Yulai Xie
Jinhui Xu

Bo Yan
Jie Yang
Sivan Yogev
Neal Young
Tian-Li Yu
Shmuel Zaks
Guochuan Zhang
Wei Zhang
Jiaofei Zhong
Binhai Zhu
Feng Zou

Sponsoring Institutions

The University of Hawaii, USA
The University of Texas at Dallas, USA

Table of Contents

Bubblesort and Juggling Sequences

Ronald L. Graham

Department of Computer Science and Engineering
University of California, San Diego, USA
graham@ucsd.edu

Abstract. In this talk I will describe some recent results concerning the connection between the bubblesort sorting algorithm and certain integer sequences used to analyze various juggling patterns. The analysis leads to new results on the joint distribution of the descent and maximum drop statistics of a permutation, as well as a new class of identities for the classical Eulerian numbers.

Y. Dong, D.-Z. Du, and O. Ibarra (Eds.): ISAAC 2009, LNCS 5878, p. 1, 2009.

A Proof of the Molecular Conjecture

Naoki Katoh

Department of Architecture and Architectural Engineering
Kyoto University, Kyoto Daigaku Katsura, Nishikyo-ku, Kyoto 615-8540, Japan
naoki@archi.kyoto-u.ac.jp

Abstract. To identify flexible/rigid region in a protein is one of the central issues in the field of molecular biology as this could provide insight into its function and a means to predict possible changes of structural flexibility by the environmental factors such as temperature and pH. Several methods were developed for this purpose.

One of standard methods is to model the protein as a geometric graph embedded in $\mathbb{R}^3$ by regarding an atom as a vertex and a bond between atoms as an edge with a fixed length. It analyzes the protein's rigidity by using the theory of structural rigidity. Algorithms such as 3D pebble game were developed for this. Computer software like FIRST (Floppy Inclusions and Rigid Substructure Topography) uses this algorithm. These software are used widely and successfully now. Algorithms used by FIRST and other programs rely their correctness upon the theory of structural rigidity. From the mathematical point of view, however, the correctness proof is incomplete because it relies on the so called "Molecular Conjecture" which has been a long standing open problem over twenty-five years in the field of combinatorial rigidity. In the past years, however, in spite of the absence of the rigorous proof of the Molecular Conjecture, empirical data have been accumulated that support the conjecture. Recently, we were able to settle the Molecular Conjecture affirmatively in $\mathbb{R}^3$ and in higher dimensions and provide the theoretical validity of the algorithms behind such software as FIRST, FRODA, etc.

A d-dimensional body-and-hinge framework is a structure consisting of rigid bodies connected by hinges in d-dimensional space. The generic infinitesimal rigidity of a body-and-hinge framework has been characterized in terms of the underlying multigraph independently by Tay and Whiteley as follows: A multigraph G can be realized as an infinitesimally rigid body-and-hinge framework by mapping each vertex to a body and each edge to a hinge if and only if $\left(\binom{d+1}{2}-1\right)G$ contains $\binom{d+1}{2}$ edge-disjoint spanning trees, where $\left(\binom{d+1}{2}-1\right)G$ is the graph obtained from G by replacing each edge by $\left(\binom{d+1}{2}-1\right)$ parallel edges. In 1984 they jointly posed a question about whether their combinatorial characterization can be further applied to a nongeneric case. Specifically, they conjectured that G can be realized as an infinitesimally rigid bofy-and-hinge framework if and only if G can be realized as that with the additional "hinge-coplanar" property, i.e., all the hinges incident to each

Y. Dong, D.-Z. Du, and O. Ibarra (Eds.): ISAAC 2009, LNCS 5878, pp. 2–3, 2009.

body are contained in a common hyperplane. This is the definition of "the Molecular Conjecture".

In this talk, we will first introduce main topics in the field of combinatorial rigidity and then give a brief overview of the proof of the moleclar conjecture.

This is a joint work with Shin-ichi Tanigawa.

Exact Algorithms for Dominating Clique Problems
(Extended Abstract)

N. Bourgeois[1], F. Della Croce[2], B. Escoffier[1], and V.Th. Paschos[1]

[1] LAMSADE, CNRS and Université Paris-Dauphine, France
{bourgeois,escoffier,paschos}@lamsade.dauphine.fr
[2] D.A.I., Politecnico di Torino, Italy
federico.dellacroce@@polito.it

Abstract. We handle in this paper three dominating clique problems, namely, the decision problem itself when one asks if there exists a dominating clique in a graph G and two optimization versions where one asks for a maximum- and a minimum-size dominating clique, if any. For the three problems we propose optimal algorithms with provably worst-case upper bounds improving existing ones by (D. Kratsch and M. Liedloff, *An exact algorithm for the minimum dominating clique problem*, Theoretical Computer Science 385(1-3), pp. 226–240, 2007). We then settle all the three problems in sparse and dense graphs also providing improved upper running time bounds.

1 Introduction

Given a graph $G(V, E)$, a dominating clique is a clique which is also a dominating set for G. Determining whether there exists a dominating clique or not is known to be **NP**-hard ([1]) and so are the two related problems MIN DOMINATING CLIQUE and MAX DOMINATING CLIQUE, where we are asked for a dominating clique of minimum and of maximum size, respectively. These problems can of course be solved by enumerating all the subsets of V. So, an interesting problem is to devise algorithms able to optimally solve the three problems EXISTING DOMINATING CLIQUE, MIN DOMINATING CLIQUE and MAX DOMINATING CLIQUE within time $O(2^{c|V|}p(|V|))$, where c is a constant lower than 1 and p some polynomial function. Notice that, compared to the slightest improvement of c, p is non relevant. So, from now on, we use notation $O^*(2^{c|V|})$ in order to omit polynomial factors.

Regarding EXISTING DOMINATING CLIQUE, trivial $O^*(2^{|V|})$ bound has been initially broken by [2] down to $O^*(3^{|V|/3}) = O^*(1.443^{|V|})$ using a result by [3], namely that the number of maximal (for inclusion) independent sets in a graph is at most $3^{|V|/3}$. Recently, [4] have proposed a branching algorithm that, according to a measure and conquer analysis [5], solves MIN DOMINATING CLIQUE with polynomial space and running time $O^*(1.3387^{|V|})$, and another one that requires $O^*(1.3234^{|V|})$ time and space. Naturally, these algorithms also solve EXISTING DOMINATING CLIQUE.

Y. Dong, D.-Z. Du, and O. Ibarra (Eds.): ISAAC 2009, LNCS 5878, pp. 4–13, 2009.

In this paper, we first devise several branching algorithms for EXISTING DOMINATING CLIQUE (Section 2), MAX DOMINATING CLIQUE (Section 3) and MIN DOMINATING CLIQUE (Section 4), achieving improved running times respect to the existing results mentioned above. Our algorithms work in polynomial space, but also provide interesting bounds using memorization technique to the case where an exponential memory space is allowed. The following table summarizes our results in Sections 2, 3 and 4.

	Former result	Our result
EXISTING DOMINATING CLIQUE	$O^*(1.3387^{\|V\|})$	$O^*(1.2740^{\|V\|})$
- exponential space allowed	$O^*(1.3234^{\|V\|})$	$O^*(1.2556^{\|V\|})$
MAX DOMINATING CLIQUE	$O^*(1.4423^{\|V\|})$	$O^*(1.3196^{\|V\|})$
- exponential space allowed		$O^*(1.2937^{\|V\|})$
MIN DOMINATING CLIQUE	$O^*(1.3387^{\|V\|})$	$O^*(1.3248^{\|V\|})$
- exponential space allowed	$O^*(1.3234^{\|V\|})$	$O^*(1.298^{\|V\|})$

In Section 5, we restrict ourselves to sparse and dense graphs, and produce parameterized algorithms depending on minimum, maximum and average degree. For instance, we show in Section 5 that if $n - \delta = o(n)$, or if the density of the graph $D(G) = 1 - o(1)$, MIN DOMINATING CLIQUE can be solved in subexponential time and MAX DOMINATING CLIQUE within roughly the same running time as MAX CLIQUE. On the other hand, if $D(G) = o(1)$ both MIN and MAX DOMINATING CLIQUE can be also solved in subexponential time.

It is easy to see that no polynomial time approximation algorithm can exist for MIN DOMINATING CLIQUE and MAX DOMINATING CLIQUE, since any such algorithm should first solve EXISTING DOMINATING CLIQUE that is **NP**-complete. For this reason, in Section 6, we present some approximation results for MIN DOMINATING CLIQUE and MAX DOMINATING CLIQUE by exponential time algorithms that run faster than the corresponding exact algorithms for these problems. The approximation ratio of an algorithm A for an **NP**-hard problem Π is defined by the ratio between the value of an optimal solution for Π over the value of a solution computed by A.

In what follows, given a graph $G(V, E)$ and a vertex $v \in V$, the neighborhood $N(v)$ of v is the set of vertices that are adjacent to v and the set $N[v] = N(v) \cup \{v\}$ is called the closed neighborhood of v. For the degree of v, we use the notation $d(v) = |N(v)|$; the anti-degree of v is the quantity $\bar{d}(v) = |V \setminus N[v]|$. For any set $H \subset V$, we write $N_H(v) = N(v) \cap H$, $d_H(v) = |N_H(v)|$ and $\bar{d}_H(v) = |H \setminus N_H[v]|$; $G[H]$ is the subgraph induced by H in G. Finally, for $v \in V$, we set $f(v) = 1 - |N(v)|/|V|$. For simplicity, we set $n = |V|$ and $m = |E|$. For any function $\mathcal{T}$, $T(n)$ stands for the maximum running time the algorithm requires to compute $\mathcal{T}$ on a graph containing at most n vertices.

Due to limits in paper's length some of the results are given without proofs that can be found in [6].

2 Existing Dominating Clique

2.1 A Tight $O^*(2^{2n/5})$ Branching Algorithm

We use the same notations as in [4]: S is the set of vertices we have added to the solution, D is the set of vertices we have discarded, $A = \cap_{s \in S} N(s) \setminus D$ is the set of vertices still available and $F = V \setminus (\bigcup_{s \in S} N(s))$ is the set of free vertices, i.e., the set of vertices that still remain to be dominated; $T(S, D, A, F)$ is a boolean function that returns TRUE if and only if there exists a dominating clique in G contained in A and with no vertex from D. As pointed out in [4], for MIN DOMINATING CLIQUE, it is equivalent to solve EXISTING DOMINATING CLIQUE or to determine if there exists $v \in V$ such that $T(\{v\}, \emptyset, N(v), V \setminus N[v]) =$ TRUE.

We devise in this section, an improved algorithm, called EDC that computes $T(\{v\}, \emptyset, N(v), V \setminus N[v])$ (for each vertex v) as follows:

1. if $F = \emptyset$, then $T(S, D, A, F) =$ TRUE;
2. if $\exists u \in F$, such that $d_A(u) = 0$, then $T(S, D, A, F) =$ FALSE;
3. if $\exists u \in A$, such that $d_F(u) = 0$, then $T(S, D, A, F) = T(S, D \cup \{u\}, A \setminus \{u\}, F)$
4. otherwise, fix $u \in A$ such that $d_A(u)$ is maximal and set $T(S, D, A, F) = \bigvee_{w \in A \setminus N_A(u)} T(S \cup \{w\}, D \cup (A \setminus N_A[w]) \cup N_F(w), N_A(w), F \setminus N_F(w))$.

Rules 1, 2 and 3 are straightforward, so we only need to prove correctness of the branching rule. If we add some vertex u to S, we must discard from A any vertex that is not neighbor of u. On the other hand, assume that every vertex in $A \setminus N(u)$ have been discarded. Then, any vertex still available is in $N_A(u)$, so we can safely add u to S. Hence, in any case we take one vertex from $A \setminus N_A(u)$, leading to the recurrence of the branching rule. Notice that, because of this insertion of u in the last case, the clique computed by Algorithm EDC may be not a minimum dominating clique.

Proposition 1. *Algorithm EDC decides whether there exists a dominating clique, or not, with running time $O^*(2^{2n/5}) = O^*(1.3196^n)$. This bound is tight.*

Proof. We simply count the number of vertices in the remaining graph having $n = |A \cup F|$ vertices. Fix $\delta = \min_{a \in A}\{|A \setminus N_A(a)|\} = |A \setminus N_A(u)|$. By definition of u we have: $T(n) \leqslant \sum_{w \in A \setminus N_A(u)} T(|N_A(w)| + |F| - d_F(w)) \leqslant \delta \times T(|A| - \delta + |F| - 1) \leqslant \delta \times T(n - \delta - 1)$. Then, a simple recursion shows that $T(n) \leqslant \max_{\delta \in \mathbb{N}}\{\delta^{n/(\delta+1)}\} = 4^{n/5}$.

Tightness is shown in the following graph. Consider a collection G_p of p stars of size 5 (1 center plus 4 outer vertices). Set A is the set of the outer vertices, and set F is the set of the centers. Any outer vertex u is adjacent to any other outer vertex in any other star; so, $d_A(v) = 4p - 4$. When the algorithm branches on any of these outer vertices, it has to solve EXISTING DOMINATING CLIQUE on four identical copies of G_{p-1} and, in this case $T(n) = \Omega(4^{n/5})$. □

2.2 Improvement of EDC

We now refine algorithm EDC in such a way that when branching on a vertex u (of maximum degree $d_A(u)$), instead of removing for each of the δ branches $\delta + 1$

vertices, we remove in the worst case $\delta + 2$ vertices in $\delta - 1$ branches, and $\delta + 1$ vertices only in one branch. To do this, we say that a vertex w is *good* if: (i) either $|A \setminus N_A(w)| \geqslant \delta + 1$, or (ii) w is adjacent to at least two vertices in F, or (iii) w is adjacent to only one vertex $v \in F$ and this vertex v has another neighbor w' in $N(w)$. Otherwise, we say that w is *bad*.

Suppose first that there exists a vertex u such that any vertex in $A \setminus N_A[u]$ is good. We branch on u, as in EDC. For each $w \in A \setminus N_A[u]$, in the branch we take w, we remove at least $\delta + 2$ vertices. Indeed, in the two first cases this is trivial. In the third one, if we take w we can safely discard w' (and we also remove $\delta + 2$ vertices) since, otherwise, if we take w' then w does not cover new elements so w is useless (this case being handled in some other branch).

Assume now there exist at least two non adjacent bad vertices u and u'. Suppose first that u is adjacent to v in F and u' is adjacent to $v' \neq v$. Let $u_1, u_2, \ldots, u_\delta$ the vertices in $A \setminus N_A[u]$ (with $u_\delta = u'$). When branching on u_1, we consider the cases of taking u_1 or not, u_2 or not, ... Then, in the branch where all the other vertices in $A \setminus N_A[u]$ but $u' = u_\delta$ have been discarded, v has degree 1 and then we have to take u in the solution (without branching on u'). In this way, we get $\delta - 1$ branches where we remove $\delta + 1$ vertices.

Suppose now $v' = v$. Then, in the branch where all the other vertices in $A \setminus N_A[u]$ but u' have been discarded, $d(v) = 2$ (since u is bad); so, we have to take either u or u', but it is not interesting to take u' (take u instead). So we can keep u in the solution (without branching on u'). In this way, we get also $\delta - 1$ branches where we remove $\delta + 1$ vertices.

In all, we get either $T(n) \leqslant (\delta - 1)T(n - \delta - 2) + T(n - \delta - 1)$, or $T(n) \leqslant (\delta - 1)T(n - \delta - 1)$. The worst case of these recurrence relations is $T(n) \leqslant 3T(n - 6) + T(n - 5)$, leading to the following proposition.

Proposition 2. *Algorithm EDC' decides whether there exists a dominating clique, or not, with running time $O^*(2^{0.35n}) = O^*(1.2740^n)$.*

2.3 Memorization: Trading Space for Time

The principle of memorization, as it has been explained for example in [7], is quite simple. Before running the algorithm on the main graph, we run it on every induced subgraph of size at most αn and store the results in a table (note that, in the algorithms, available vertices never become free (or vice-versa), hence the number of subproblems of size αn to consider is $\binom{n}{\alpha n}$). Now, we run the main algorithm until the remaining graph has size αn or less; then a polynomial-time query in the storage table allows us to conclude. Thus, the total running time and space is $O^*\left(\max\left\{\binom{n}{\alpha n}, 1.2740^{(1-\alpha)n}\right\}\right)$. This value is minimal for α verifying $1.2740^{1-\alpha} = 1/(\alpha^\alpha(1-\alpha)^{1-\alpha})$, leading to $T(n) = O^*(1.2556^n)$.

3 Max Dominating Clique

The function $T(S, D, A, F)$ is now an integer function that returns the cardinality of a maximum dominating clique in G contained in set A and with no vertex from set D, if any, and $-\infty$ otherwise. Once again, solving MAX DOMINATING CLIQUE is equivalent to finding $\max_{v \in V}\{T(\{v\}, \emptyset, N(v), V \setminus N[v])\}$. Note that now we cannot discard as in Algorithm EDC any vertex that has degree 0 in F, because we could be led to a solution that is not a maximum one; however, we claim that we can compute a solution with the same running time by the following algorithm called MDC1 that works as follows:

1. if $A = F = \emptyset$, then $T(S, D, A, F) = |S|$;
2. if $\exists u \in F$, such that $d_A(u) = 0$, then $T(S, D, A, F) = -\infty$;
3. fix $w \in A$ such that $d_A(w)$ is maximum, and $d_F(w)$ is maximum among vertices of maximum d_A. If there exists $u \in A \setminus N[w]$, such that $d_F(u) = 0$, then $T(S, D, A, F) = \max_{u' \in A \setminus (N_A(w) \cup \{u\})}\{T(S \cup \{u'\}, D \cup (A \setminus N_A[u']) \cup N_F(u'), N_A(u'), F \setminus N_F(u'))\}$;
   otherwise, $T(S, D, A, F) = \max_{u \in A \setminus N_A(w)}\{T(S \cup \{u\}, D \cup (A \setminus N_A[u]) \cup N_F(u), N_A(u), F \setminus N_F(u))\}$.

To prove correctness, just consider the following three facts: (i) if we add some vertex to S, we must discard any vertices from A that are not its neighbors; (ii) if every vertices in $A \setminus N(w)$ have been discarded, then any vertex still available is in $N_A(w)$, so we can safely add w to S; (iii) if every vertex in $A \setminus N(w)$ but u has been discarded, then only one among u and w may be added; furthermore, if $d_F(u) = 0$, we can safely discard u and add w to S.

Proposition 3. *Algorithm MDC computes a maximum-size dominating clique, if any, with running time $O^*(2^{2n/5})$. This bound is tight. Using memorization, Algorithm's MDC runs in time and space $O^*(1.2937^n)$.*

4 Min Dominating Clique

The function $T(S, D, A, F)$ is now an integer function that returns the cardinality of a minimum dominating clique in G contained in set A and with no vertex from set D, if any, and ∞ otherwise. Once again, solving MIN DOMINATING CLIQUE is equivalent to finding $\min_{v \in V}\{T(\{v\}, \emptyset, N(v), V \setminus N[v])\}$. We denote in this section by $d_A(v)$ the degree of v within the set A of available vertices and by $\bar{d_A}(v)$ the anti-degree of v always within set A. The following remarks hold.

Remark 1. Each vertex $j \in A$ is adjacent to at least one vertex $i \in F$, or else j can be discarded as it cannot be part of the dominating clique. □

Remark 2. There exists at least one vertex $j \in A$ such that $\bar{d_A}(j) \geqslant 1$, or else a pure set covering problem where the set F corresponds to the universe U of elements and the set A corresponds to the collection S of the (nonempty) subsets of U and the aim is to determine a minimum cardinality sub-collection $S' \subseteq S$ which covers U. This set covering problem is known to be solvable to optimality in $O^*(1.2301^{|A|+|F|})$ time ([5]). But, as $|A| + |F| \leqslant n$ this is not superior to $O^*(1.2301^n)$ time. □

Remark 3. Each vertex $i \in F$ is adjacent to at least three vertices $j, k, l \in A$, or else a low exponential complexity immediately holds in the worst case. Indeed, if $i \in F$ is adjacent only to the vertex $j \in A$, then no branch occurs, j is included in S dominating i which is removed from F; alternatively, $i \in F$ is adjacent to vertices $j, k \in A$ and either j or k must be included in S to dominate i. Then: if $d_F(j) \geqslant 2$, either j is included in S and at least 3 vertices are removed, or j is discarded, k is included and i is covered, *i.e.*, 3 vertices are removed. If $d_F(k) = d_F(k) = 1$, $\bar{d_A}(j) \geqslant 1$ holds, or else k could be discarded without branching. But then, in both cases at least three vertices are fixed leading to $T(n) \leqslant 2T(n-3)$ corresponding to $O^*(1.2599)$ time. □

We claim that we can solve MIN DOMINATING CLIQUE in time $O^*(1.3248^n)$ by the following algorithm called `MINDC1`:

1. if $A = F = \emptyset$, then $T(S, D, A, F) = |S|$;
2. else, if $\exists i \in F$, such that $d_A(i) = 0$, then $T(S, D, A, F) = \infty$;
3. else, if $\forall j \in A\ \bar{d_A}(j) = 0$, then solve the problem as a minimum set covering problem according to Remark 2 with complexity not superior to $O^*(1.2301^n)$;
4. else, if $\exists i \in F$, such that $2 \geqslant d_A(i) \geqslant 1$, then branch on the vertices adjacent to i according to Remark 3 with complexity not superior to $O^*(1.2599^n)$;
5. else, if $\forall i \in F\ d_A(i) \geqslant 3$, then select $j \in A$ such that $\bar{d_A}(j)$ is maximum and branch according to the following exhaustive cases
   (a) $\bar{d_A}(j) \geqslant 1$ with vertex $j \in A$ adjacent to only one vertex $i \in F$ with $d_A(i) = 3$.
   (b) $\bar{d_A}(j) \geqslant 1$ with vertex $j \in A$ adjacent to only one vertex $i \in F$ with $d_A(i) \geqslant 4$.
   (c) $\bar{d_A}(j) = 1$ with vertex $j \in A$ non adjacent to vertex $k \in A$ and adjacent to exactly two vertices $h, i \in F$ with $d_A(i) \geqslant d_A(h) = 3$.
   (d) $\bar{d_A}(j) = 1$ with vertex $j \in A$ non adjacent to vertex $k \in A$ and adjacent to exactly two vertices $h, i \in F$ with $d_A(i) \geqslant d_A(h) = 4$.
   (e) $\bar{d_A}(j) = 1$ with vertex $j \in A$ non adjacent to vertex $k \in A$ and adjacent to exactly two vertices $h, i \in F$ with $d_A(i) \geqslant d_A(h) \geqslant 5$.
   (f) $\bar{d_A}(j) = 1$ with vertex $j \in A$ non adjacent to vertex $k \in A$ and adjacent to at least three vertices $g, h, i \in F$.
   (g) $\bar{d_A}(j) \geqslant 2$ with vertex $j \in A$ non adjacent to vertices $k, m \in A$ and adjacent to at least two vertices $h, i \in F$.

Proposition 4. *Algorithm `MINDC` computes a minimum-size dominating clique, if any, with running time $O^*(1.3248^n)$. Using memorization, it runs in time and space $O^*(1.2980^n)$.*

5 Dense and Sparse Graphs

5.1 Graphs of Fixed Maximal or Minimal Degree

Notice that if a graph has maximum degree Δ, all the maximal cliques can be computed with running time $O^*(3^{\Delta/3})$ and all the cliques with running time $O^*(2^{\Delta})$. In particular, dominating clique problems are polynomial if Δ is finite and subexponential if $\Delta = o(n)$.

In the case of high minimum degree δ, this does not remain true. Indeed, MAX CLIQUE easily reduces to MAX DOMINATING CLIQUE by adding to the graph instance G a new vertex adjacent to any vertex in G, and MAX CLIQUE is well known to be **NP**-hard even in graphs of minimum degree $n-4$ (MAX INDEPENDENT SET being **NP**-hard in graphs of maximum degree 3,[1]). Then, even in graphs with minimum degree $\delta \geqslant n - \overline{\delta}$ for some constant $\overline{\delta}$ MAX DOMINATING CLIQUE is not polynomial (if **P**$\neq$ **NP**). However, there are some interesting results.

Let us begin with a preliminary remark: in a graph, for any $v \in V$, if there exists some dominating clique K containing v, then there exists a dominating clique $K' \subseteq K$ with $|K'| \leqslant |V \setminus N(v)|$. Clique K' could be, for instance, obtained by simply taking one neighbor in K for any vertex in $V \setminus N(v)$.

Then, by definition, for any $v \in V$, $nf(v) \leqslant n - \delta$. But thanks to the previous remark, if there exists a dominating clique, then there exists a dominating clique of size at most $nf(v)$. Thus, EXISTING, and MIN DOMINATING CLIQUE can be computed with running time $O^*\left(\binom{n}{n-\delta}\right)$. In particular, EXISTING, and MIN DOMINATING CLIQUE are polynomial if $n - \delta$ is finite and subexponential if $n - \delta = o(n)$.

We now exhibit Algorithm `MDC2` for MAX DOMINATING CLIQUE:

1. for every $v \in V$, form the collection $\mathcal{S}(v)$ of every subset of $N[v]$ of size at most $nf(v)$ that is a dominating clique;
2. for every $S \in \mathcal{S}(v)$, compute a maximum clique in $G[\bigcap_{s \in S} N[s]]$;
3. return the maximum-size clique among those computed in Step 2.

`MDC2` optimally solves for MAX DOMINATING CLIQUE in any graph and has running time $O^*\left(c^n \binom{n}{n-\delta}\right)$ if the algorithm used to solve MAX CLIQUE has complexity $O^*(c^n)$ (proofs omitted). In particular, MAX DOMINATING CLIQUE can be computed with the same exponential bound on running time as MAX CLIQUE in graphs such that $n - \delta = o(n)$; this bound cannot be improved (thanks to the reduction from MAX CLIQUE mentioned above).

Note that the best worst-case complexity bound for MAX CLIQUE is, to our knowledge, the $O^*(1.1889^n)$ bound claimed by [8] in his unpublished technical report, or the $O^*(1.2210^n)$ algorithm by [9], that is the best published result.

5.2 Graphs of Small Average Degree

More generally, the density of a graph can be defined as $D(G) = 2m/(n(n-1)) = d/(n-1)$ where d is the average degree of the graph. It can be easily seen that, if $D(G) = o(1)$, the size of a maximum clique is $o(n)$ and then we can enumerate all cliques in subexponential time.

We now focus ourselves on the case where $D(G) \notin o(1)$ but is bounded above by some constant $\lambda \in [0, 1]$. In this case, average degree is at most $\lambda(n-1)$, hence:

$$m \leqslant \frac{\lambda n(n-1)}{2} \leqslant \frac{\lambda n^2}{2} \qquad (1)$$

By enumeration of subsets of size at most $\sqrt{D(G)}n$, we trivially find if there is a dominating clique (and return the minimal and the maximal one) with running time $O^*\left(\binom{n}{\sqrt{\lambda}n}\right)$. This is only interesting for rather small values of λ; for example, if $\sqrt{\lambda} = 1/20$, running time is $O^*(1.22^n)$.

As far as EXISTING DOMINATING CLIQUE and MAX DOMINATING CLIQUE are concerned (not MIN DOMINATING CLIQUE), we can greatly improve this result. Notice that, for any dominating clique K we have the following inequality $m \geqslant |K|(|K|-1)/2 + \sum_{v \in K} d_{V \setminus K}(v)$.

Assume first that there exists a vertex $v \in K$ such that $d_{V \setminus K}(v) < \mu n + 1/2$, for a given μ whose value will be fixed later. Then $n(1 - f(v)) < \mu n + 1/2 + |K|$. In [10] it is established that in a graph of size n, the cliques of size k or less (where $n/k \geqslant 3$) can be enumerated within time $O^*((n/k)^k)$. Thus, we can enumerate all maximal cliques containing v with running time $T(n) = O^*\left(\left(\frac{\mu n + \frac{1}{2} + |K|}{|K|}\right)^{|K|}\right) = O^*\left(2^{|K| \log_2(1 + \mu n/|K|)}\right)$. Since this function is increasing with K and, according to (1), $|K| \leqslant \sqrt{2m} + 1 \leqslant \sqrt{\lambda}n + 1$, this leads to $T(n) = O^*\left(2^{\sqrt{\lambda}n \log_2(1 + \mu/\sqrt{\lambda})}\right)$.

On the other hand, if for any $v \in K$, $d_{V \setminus K}(v) \geqslant \mu n + 1/2$, using also (1), we get: $m \geqslant \frac{|K|(|K|-1)}{2} + |K|\left(\mu n + \frac{1}{2}\right) = \frac{|K|^2}{2} + |K|\mu n \Longrightarrow |K| \leqslant (\sqrt{\lambda + \mu^2} - \mu)n$. In this case, running time for enumerating all small subsets containing v is bounded above by $T(n) = O^*\left(\left(\frac{n}{(\sqrt{\lambda + \mu^2} - \mu)n}\right)^{(\sqrt{\lambda + \mu^2} - \mu)n}\right)$. We finally fix μ to its best value, *i.e.* when the two running times are equal.

In the following table, bounds on running time of the above method vs. running time of exhaustive search (both depending on parameter λ) are given.

$\sqrt{\lambda}$	1/4	1/6	1/8	1/10	1/20	1/50
Optimal μ	0.326	0.248	0.200	0.169	0.097	0.045
Running time of our algorithm	1.233^n	1.164^n	1.127^n	1.104^n	1.056^n	1.046^n
Running time of exhaustive search	1.755^n	1.570^n	1.458^n	1.385^n	1.220^n	1.104^n

5.3 Graphs of High Average Degree

In this section, we deal with graphs of density $D(G) \geqslant 3/4$. As previously, it is easy to see that in such graphs MAX DOMINATING CLIQUE is harder than MAX CLIQUE. Let us now define Algorithms `mDC1` and `MDC3` for MIN DOMINATING CLIQUE and MAX DOMINATING CLIQUE, respectively.

The former, `mDC1`, works as follows: fix $\epsilon = \sqrt{1 - D(G)}$, compute any subset of size at most ϵn and return a smallest one that is a dominating clique if any.

On the other hand, Algorithm `MDC3` works as follows:

1. fix $\epsilon = \sqrt{1 - D(G)}$ and compute any subset of size at most ϵn; let K_0 be a largest one that is a dominating clique, if any;
2. for every $v \in V$ such that $nf(v) \leqslant \epsilon n$, form the collection $\mathcal{S}(v)$ of every subset of size at most $nf(v)$ that is a dominating clique;

3. for every $S \in \mathcal{S}(v)$, compute a maximum clique in $G[\bigcap_{s \in S} N[s]]$;
4. let K_1 be a clique of maximum size among those computed in Step 3; output $\max\{K_0, K_1\}$.

Algorithm `mDC1` solves MIN DOMINATING CLIQUE in subexponential running time, while Algorithm `MDC3` solves MAX DOMINATING CLIQUE within the same running time as the MAX CLIQUE-algorithm called in Step 3, with a subexponential multiplicative factor (proofs omitted). Again, no improvement in the exponential basis of the running time seems possible thanks to the reduction from MAX CLIQUE to MAX DOMINATING CLIQUE mentioned above.

6 Moderately Exponential Approximation

Since any approximation algorithm for MIN DOMINATING CLIQUE or MAX DOMINATING CLIQUE solves EXISTING DOMINATING CLIQUE that is **NP**-complete, it is impossible to devise any polynomial approximation algorithm for MIN- and MAX DOMINATING CLIQUE. Thus, it seems interesting to see if it is possible to compute solutions for the two optimization problems that guarantee a good approximation ratio with moderately exponential running time; interesting running times of approximation algorithms lie between the best known complexity for solving EXISTING DOMINATING CLIQUE ($O^*(1.2740)$ with our algorithm) and the best known complexity for solving the corresponding optimization problem. We propose in what follows two algorithms `mMOD` and `MMOD` that do this for MIN DOMINATING CLIQUE and MAX DOMINATING CLIQUE, respectively. Obviously, in what follows we assume that we handle graphs admitting dominating cliques.

Algorithm `mMOD`(ρ) for MIN DOMINATING CLIQUE works as follows: run algorithm `EDC'` and let K_0 be the dominating clique computed; compute all the subsets of V whose size is at most n/ρ; let K_1 be a dominating clique of minimum size among them; if none is found, then set $K_1 = V$; return $\text{argmin}\{|K_0|, |K_1|\}$.

Proposition 5. *If $G(V, E)$ has a dominating clique then, for any ρ, $2 \leqslant \rho \leqslant 15.24$, it is possible to compute a ρ-approximation to* MIN DOMINATING CLIQUE *with polynomial space and running time $O^*\left(\binom{n}{n/\rho}\right)$. This is faster than the exact (polynomial space) algorithm for $\rho \geqslant 11.71$. If exponential space is allowed, then, for any ρ, $2 \leqslant \rho \leqslant 16.60$, it is possible to compute a ρ-approximation to* MIN DOMINATING CLIQUE *with running time and space $O^*\left(\binom{n}{n/\rho}\right)$. This is faster than the exact (exponential space) algorithm for $\rho \geqslant 12.40$.*

Unfortunately, this strategy does not work as far as MAX DOMINATING CLIQUE is concerned. Indeed, consider the following instance: $K = (k_i)_{i \leqslant n/3}$ is a clique and $S(s_i)_{i \leqslant n/3}$, $T(t_i)_{i \leqslant n/3}$ are two independent sets. Add an edge for each pair (k_i, s_i), (k_i, t_i) and (s_i, t_i). All the dominating cliques but K have size exactly 3. Thus, searching for all small and/or large cliques leads to an unbounded approximation ratio. However, it is possible to get some approximation result if we observe that *algorithm `EDC'` is able not only to find a dominating clique, but also, given a subset S, to find a dominating clique containing S.*

Based upon this remark we devise Algorithm MMOD(ρ) for MAX DOMINATING CLIQUE that works as follows: compute all the subsets K of V whose size is at most n/ρ; if K is a clique, then find $T(K, N[K] \setminus \bigcap_{v \in K} N[v], \bigcap_{v \in K} N(v), V \setminus N[K])$, according to algorithm EDC' (where $N[K] = \cup_{v \in K} N[v]$); return the largest dominating clique found, denoted by K.

Proposition 6. *If $G(V,E)$ has a dominating clique, then, for any $\rho \geqslant 2$ it is possible to compute a ρ-approximation to* MAX DOMINATING CLIQUE *with polynomial space and running time: $O^*\left(\binom{n}{n/\rho} 1.2740^{n(1-1/\rho)}\right)$. This is faster than the exact (polynomial space) algorithm for $\rho \geqslant 168$. If exponential space is allowed, then, for any $\rho \geqslant 2$ it is possible to compute a ρ-approximation to* MAX DOMINATING CLIQUE *with running time and space $O^*\left(\binom{n}{n/\rho} 1.2556^{n(1-1/\rho)}\right)$. This is faster than the exact (exponential space) algorithm for $\rho \geqslant 204$.*

References

1. Garey, M.R., Johnson, D.S.: Computers and intractability. A guide to the theory of NP-completeness. W. H. Freeman, San Francisco (1979)
2. Johnson, D.S., Yannakakis, M., Papadimitriou, C.H.: On generating all maximal independent sets. Inform. Process. Lett. 27, 119–123 (1988)
3. Moon, J.W., Moser, L.: On cliques in graphs. Israel J. of Mathematics 3, 23–28 (1965)
4. Kratsch, D., Liedloff, M.: An exact algorithm for the minimum dominating clique problem. Theoret. Comput. Sci. 385, 226–240 (2007)
5. Fomin, F., Grandoni, F., Kratsch, D.: Measure and conquer: Domination – A case study. In: Caires, L., Italiano, G.F., Monteiro, L., Palamidessi, C., Yung, M. (eds.) ICALP 2005. LNCS, vol. 3580, pp. 191–203. Springer, Heidelberg (2005)
6. Bourgeois, N., Della Croce, F., Escoffier, B., Paschos, V.T.: Exact algorithms for dominating clique problems. Cahier du LAMSADE 285, LAMSADE, Université Paris-Dauphine (2009), http://www.lamsade.dauphine.fr/cahiers/PDF/cahierLamsade285.pdf
7. Fomin, F., Grandoni, F., Kratsch, D.: Some new techniques in design and analysis of exact (exponential) algorithms. Bulletin of the European Association for Theoretical Computer Science 87, 47–77 (2005)
8. Robson, J.M.: Finding a maximum independent set in time $O(2^{n/4})$. Technical Report 1251-01, LaBRI, Université de Bordeaux I (2001)
9. Fomin, F., Grandoni, F., Kratsch, D.: Measure and conquer: a simple $O(2^{0.288n})$ independent set algorithm. In: Proc. Symposium on Discrete Algorithms, SODA 2006, pp. 18–25 (2006)
10. Byskov, J.: Enumerating maximal independent sets with applications to graph colouring. Oper. Res. Lett. 32, 547–556 (2004)

Enumerating Stereoisomers of Tree Structured Molecules Using Dynamic Programming*

Tomoki Imada[1], Shunsuke Ota[1], Hiroshi Nagamochi[1], and Tatsuya Akutsu[2]

[1] Department of Applied Mathematics and Physics, Graduate School of Informatics, Kyoto University, Japan
{imada,ota,nag}@amp.i.kyoto-u.ac.jp
[2] Bioinformatics Center, Institute for Chemical Research, Kyoto University
takutsu@kuicr.kyoto-u.ac.jp

Abstract. Exhaustive enumeration of stereoisomers is one of the most fundamental and important problems in chemoinformatics. In this paper we deal with chemical compounds composed of carbon, hydrogen, oxygen and nitrogen atoms whose graphical structures are tree-like graphs, and consider stereoisomers caused only by asymmetry around carbon atoms. We introduce a mathematical representation for stereoisomers, and propose a dynamic programming algorithm of generating all stereoisomers without duplication. The algorithm first computes the number of stereoisomers of a given tree with n vertices in $O(n)$ time and space. Then the algorithm constructs each stereoisomer by backtracking the process of computing the numbers of stereoisomers in $O(n)$ space and in $O(n)$ time per stereoisomer. The latter result is achieved by a fast bijection algorithm for combinations of distinct integers.

1 Introduction

Exhaustive enumeration of isomers/stereoisomers is one of the most fundamental and important problems in chemoinformatics because it plays core roles in structure elucidation and molecular design [5]. Since Cayley studied enumeration of alkanes in the 19th century [2], extensive studies have been done, which include Pólya's seminal work on counting the number of isomers using group theory [9,10]. Two chemical compounds with the same isomer may have different three-dimensional configurations due to asymmetry around carbon atoms and many other structural asymmetries. Stereoisomers often exhibit different chemical properties, and synthesis of a specific stereoisomer remains a challenging issue in chemistry. In this paper, we consider stereoisomers caused only by asymmetry around carbon atoms. Such stereoisomers might be further divided into more detailed classes according to their three-dimensional conformations and stabilities. However, if the combinatorial structures based on asymmetry around carbon atoms are different, then the stereoisomers are considered different in any definition. Then stereoisomers caused only by asymmetry around

* This work was partially supported by Grant-in-Aid for Scientific Research from the Ministry of Education, Culture, Sports, Science and Technology of Japan.

Y. Dong, D.-Z. Du, and O. Ibarra (Eds.): ISAAC 2009, LNCS 5878, pp. 14–23, 2009.

carbon atoms are fundamental and practically important. As to enumeration of such stereoisomers, several methods have been proposed [1,3,12], which mostly follow the work by Nourse [8]. Given a chemical compound with m carbon atoms, these methods first create a list of all 2^m combinations of the two choices of asymmetries around each carbon atom, and remove each set S of combinations that represent the same stereoisomer leaving one of them as their representative. Although such a set S of combinations can be constructed in $O(|S|m)$ time by a method on permutation groups called the *configuration groups*, the time and space complexity of the entire algorithm is $\Omega(2^m)$, which is always exponential even if the number of stereoisomers is any small.

In this paper, we focus on *tree structured molecules* (i.e., acyclic molecules) and develop algorithms for enumerating stereoisomers with guaranteed computational complexity. Differently from the existing approaches based on configuration groups, we use *dynamic programming*. For this, we treat a given tree structured molecule as a tree rooted at its structural center, and derive recursive formulas for the numbers of stereoisomers of rooted subtrees. However, it is nontrivial to represent stereoisomers with a mathematically consistent form, without which such recursive formulas cannot be derived. The main contribution of this paper is to give a mathematical representation for stereoisomers by introducing a new notion, "orientation of carbon circuits," and to design a dynamic programming algorithm that counts the total number K of stereoisomers of a given tree based on the derived recursive formulas and a traceback algorithm that constructs the k-th stereoisomer of the tree for each $k = 1, 2, \ldots, K$, by identifying the stereoisomer of each subtree corresponding to the k-th stereoisomer. In this paper, we assume that each of the four arithmetic operations can be done in constant time. Our algorithm counts the number K of stereoisomers in $O(n)$ time and enumerates all K stereoisomers without duplication in $O(n)$ time per stereoisomer, where n is the number of atoms in a given tree. Our algorithms are optimal provided that each stereoisomer needs to be output explicitly in $O(n)$ time. In particular, the latter result is achieved by a new bijection algorithm, which is required as a subroutine of the traceback algorithm. More specifically we show that, given integers $p \in \{1, 2, 3, 4\}$ and $n \geq p$, there is an $O(1)$ time algorithm that delivers the k-th set from the $\binom{n}{p}$ sets of p distinct integers $k_1, k_2, \ldots, k_p \in [1, n]$. We also conducted computational experiments to evaluate the practical computation time of the proposed algorithm.

2 Preliminary and Problem Formulation

Problem definition. In this paper, we deal with the problem defined as follows.

Input: A tree-like chemical graph whose vertex set V consists of carbon, hydrogen, oxygen and nitrogen atoms. A *vertex-number* $n : V \to \{1, 2, \ldots, |V|\}$, by which each vertex is numbered from 1 to $|V|$.
Output: All the "stereoisomers" that can be generated by asymmetric carbon atoms and double bonds between two adjacent carbon atoms.

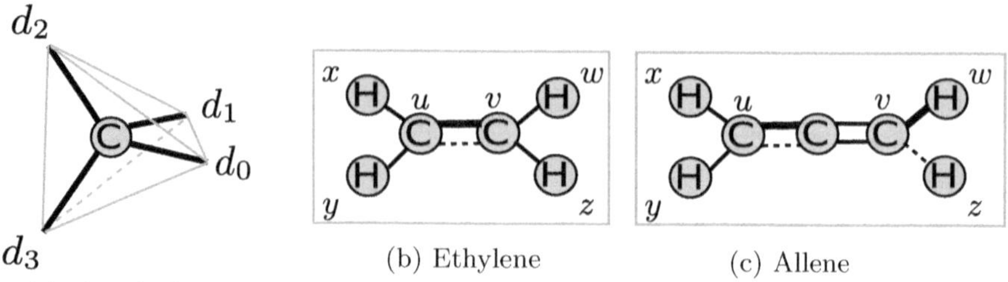

(a) Tetrahedron (b) Ethylene (c) Allene

Fig. 1. (a) The four directions d_0, d_1, d_2 and d_3 around a carbon atom in the three-dimensional space. (b) and (c) Configurations around a chain of double bonds between two carbon atoms u and v. The rectangle shows the plane that contains the left two hydrogen atoms x and y. Thick lines indicate edges on the front side of the plane and dashed lines indicate edges on the back side of the plane.

Our first task is to define "stereoisomers" by a mathematically consistent form, which will be given in Section 3.

We denote a given chemical graph by $G = (V, E)$ with a vertex set V and an edge set E. The vertex set V is partitioned into $V_C = \{v \mid v \text{ is a carbon atom}\}$, $V_H = \{v \mid v \text{ is a hydrogen atom}\}$, $V_O = \{v \mid v \text{ is an oxygen atom}\}$ and $V_N = \{v \mid v \text{ is a nitrogen atom}\}$. We denote $|V| = n$. Multiple edges are treated as one single edges and the edge set E is partitioned into $E_1 = \{e \mid e \text{ is a single bond}\}$, $E_2 = \{e \mid e \text{ is a double bond}\}$ and $E_3 = \{e \mid e \text{ is a triple bond}\}$.

The three-dimensional structure around a carbon atom forms a regular tetrahedron, as shown in Fig. 1(a), where d_0, d_1, d_2 and d_3 represent the *directions* along the four edges incident to the carbon atom. We define the *configuration* around a carbon atom v as a correspondence between the edges incident to v and d_i $(i = 0, 1, 2, 3)$, where we do not distinguish two correspondences which result in the same stereoisomorphic (stereochemically isomorphic) compounds. For example, there is only one configuration around v if v is adjacent to an atom by a triple bond. Informally, we consider that there are two different configurations around v only when one of the following cases occurs:

(i) v is adjacent to four different substructures;
(ii) v is adjacent to a substructure T_1 with a double bond and two different substructures T_2 and T_3 with single bonds, and T_1 is not symmetric along the double bond; and
(iii) v is adjacent to two substructures T_1 and T_2 with double bonds, and T_i, $i = 1, 2$ is not symmetric along the double bond.

The exact relationship between configurations and stereoisomers is given in the full version [6].

We here show our assumption on the three-dimensional structure of a chain of double bonds between two carbon atoms u and v such that u is adjacent to two atoms x and y by single bonds and v is adjacent to two atoms w and z by single bonds, as shown in Fig. 1(b) and (c). For the number k of double bonds between u and v, we assume that

- x, y, w and z are on the same plane when k is odd; and
- x, y, w and z are not on the same plane when k is even.

For example, Fig. 1(b) and (c) illustrate the chain of double bonds of ethylene ($k = 1$) and allene ($k = 2$), respectively.

Isomorphism of tree-like graphs. Let $V(G)$ and $E(G)$ denote the sets of vertices and edges of a multigraph G, respectively. Two chemical graphs G_u and G_v with roots $u \in V(G_u)$ and $v \in V(G_v)$ are *rooted-isomorphic*, which is denoted by $G_u \underset{r}{\approx} G_v$, if they admit an isomorphism such that u corresponds to v, and each vertex in G_u corresponds to a vertex in G_v of the same type of atoms.

Every tree admits a structurally unique vertex or edge, called the *unicentroid* and *bicentroid*, respectively [7]. The *root* of a given tree is defined by the vertex/vertices in its *unicentroid* or *bicentroid*. For every non-root vertex $u \in V$, the *parent* of u is defined to be the vertex v adjacent to u which is nearer to the root than u. For each vertex $v \in V$, $Ch(v)$ denotes the set of the children of vertex v, and the *rooted tree* T_v is defined to be the tree induced by v and all descendants of v.

For each subtree T_v, let $\sigma(v, T_v)$ ($\sigma(v)$ for short) denote the *signature* of the subtree T_v, that is a non-negative integer satisfying a property that $\sigma(v, T_v) = \sigma(u, T_u)$ if and only if $T_v \underset{r}{\approx} T_u$. It is known that signatures of all rooted subtrees of a given tree can be computed in linear time [4].

3 Definition of Stereoisomer

This section gives a formal definition of *stereoisomers*, considering difference of configurations.

Definition of representations and stereoisomorphism. To define stereoisomers of G, we first introduce a label $l(v)$ for each carbon atom $v \in V_{\mathrm{C}}$, where $l(v)$ takes one of $+$, $-$, *cis*, *trans* and nil (nil means that v has a unique configuration around v). As will be shown, labels *cis* and *trans* do not always correspond to chemical terms cis and trans. We define the total order among these labels by "$+$" > "$-$" > "*cis*" > "*trans*" > "nil." For every vertex $v \in V_{\mathrm{O}} \cup V_{\mathrm{N}} \cup V_{\mathrm{H}}$, define $l(v) = $ nil.

We next introduce a *representation* I of G as a set of pairs of vertex-number $n(v)$ and label $l(v)$ over all vertices $v \in V$. That is,

$$I = \{(n(v), l(v)) \mid v \in V\}.$$

Let $\mathcal{R}(G)$ denote the set of all representations I of G, where $|\mathcal{R}(G)| = 5^{|V_{\mathrm{C}}|}$ holds. Similarly, for each vertex $v \in V$, we define a *representation* I_v of the rooted subtree T_v as

$$I_v = \{(n(u), l(u)) \mid u \in V(T_v)\}.$$

5 Traceback Algorithm for Enumeration

The second phase of our algorithm, called *Output phase*, constructs proper representations for stereoisomers by using $f^*(G)$, $f(v), g(v)$ and $h(v)$ for all non-unicentroid vertices v. For $i = 1, 2, \ldots, f^*(G)$, we output the proper representation for the i-th stereoisomer of G by backtracking the computation process of Counting phase. When we compute the k-th rooted-stereoisomer of T_v, we detect the corresponding $l(v)$ and calculate k_u for every child u of v, and we trace the computation process recursively to the leaves of G.

We do not maintain any table of (rooted) stereoisomers during Counting phase. However, Output phase needs to find for a given k the k-th combination of numbers k_u of all children of u. To design an $O(1)$ time algorithm for finding a desired combination of such numbers k_u, we define bijections between a set of tuples and combinations of the elements in tuples.

Definition 8. *For positive integers $M_1, M_2, \ldots, M_p$, define*

$$D(M_1, M_2, \ldots, M_p) := \{[k_1, k_2, \ldots, k_p] \mid k_i \in \{1, 2, \ldots, M_i\},\ i = 1, 2, \ldots, p\}.$$

Let $D(; M_1, M_2, \ldots, M_p)$ denote a bijection between the set $\{1, 2, \ldots, M_1 M_2 \cdots M_p\}$ of integers and $D(M_1, M_2, \ldots, M_p)$. Let $D(k; M_1, M_2, \ldots, M_p)$ denote the k-th tuple $[k_1, k_2, \ldots, k_p] \in D(M_1, M_2, \ldots, M_p)$ in the bijection.

Definition 9. *For positive integers n and p, define the set $C_{n,p}$ of tuples by*

$$C_{n,p} := \{[k_1, k_2, \ldots, k_p] \in [1, n]^p \mid k_j \neq k_{j'},\ 1 \leq j < j' \leq p\}.$$

Let $C_{n,p}()$ denote a bijection between the set $\{1, 2, \ldots, \binom{n}{p}\}$ of integers and $C_{n,p}$. Let $C_{n,p}(k)$ denote the tuple $[k_1, k_2, \ldots, k_p] \in C_{n,p}$ corresponding to $k \in \{1, 2, \ldots, \binom{n}{p}\}$.

We have shown that there exist bijections $D(; M_1, M_2, \ldots, M_p)$ and $C_{n,p}()$ such that we can compute $D(k; M_1, M_2, \ldots, M_p)$ in $O(p)$ time for any integer $k \in \{1, 2, \ldots, M_1 M_2 \cdots M_p\}$ and we can compute $C_{n,p}(k)$ in $O(1)$ time for any integer $k \in \{1, 2, \ldots, \binom{n}{p}\}$ for $p \in \{1, 2, 3, 4\}$. See the full version [6] for the detail.

Here we show an example of computation process of traceback procedure. We consider the case when $v \in V_C$ and v has three children. Let x, y and w be three children of v (see Fig. 2(c)). Similarly to Counting phase, we consider the case when $T_x \underset{r}{\approx} T_y$ and $T_x \underset{r}{\not\approx} T_w$ hold. If $g(v) < k \leq g(v) + h(v)$ holds, then we set $\hat{k} = k - g(v)$ and choose the $\hat{k}$-th rooted-stereoisomer of $\{I_v \mid l(v) = +$ (i.e., v is an asymmetric carbon atom)$\}$ as the k-th rooted-stereoisomer. Then we set $l(v) := +$ and $[k', k_w] := D(\hat{k}; \binom{f(x)}{2}, f(w)), [k_x, k_y] := C_{f(x),2}(k')$. The details of computation processes for all the other cases are given in the full version [6] and the following theorem holds.

Theorem 10. *For a tree-like chemical graph $G = (V, E)$ with $|V| = n$, Output phase enumerates all the stereoisomers $I \in \mathcal{I}(G)$ without duplication in $O(n)$ space and in $O(n)$ time per stereoisomer.*

6 Concluding Remarks

In this paper, we introduced a mathematical representation of stereoisomers of tree-like chemical graphs, and designed a dynamic programming algorithm for generating stereoisomers. We implemented our algorithm and conducted some experiments to evaluate the practical performance. For a compound with compositional formula $C_{25}O_{24}H_{52}$, Counting phase took less than 0.01 sec., and Output phase took 39.39 sec. (on a PC with 1.2GHz CPU) for enumerating 8,388,608 stereoisomers. We observed that the computation time of Output phase increases linearly to the number of stereoisomers, and that the number of stereoisomers $f^*(G)$ is identical with that computed by Razinger et al. [11].

It is left as future work to extend our algorithm to a wider class of graphs, such as outerplanar graphs. It is also important to visualize the output representations.

References

1. Abe, H., Hayasaka, H., Miyashita, Y., Sasaki, S.: Generation of Stereoisomeric Structures Using Topological Information Alone. J. Chem. Inf. Comput. Sci. 24, 216–219 (1984)
2. Cayley, A.: On the Mathematical Theory of Isomers. Phil. Mag. Ser. 47, 444–446 (1874)
3. Contreras, M.L., Alvarez, J., Guajardo, D., Rozas, R.: Algorithm for Exhaustive and Nonredundant Organic Stereoisomer Generation. J. Chem. Inf. Model 46, 2288–2298 (2006)
4. Dinitz, Y., Itai, A., Rodeh, M.: On an Algorithm of Zemlyachenko for Subtree Isomorphism. Information Processing Letters 70, 141–146 (1999)
5. Faulon, J.-L., Visco Jr., D.P., Roe, D.: Enumerating Molecules. Reviews in Computational Chemistry 21, 209–286 (2005)
6. Imada, T., Ota, S., Nagamochi, H., Akutsu, T.: Enumerating Stereoisomers of Tree Structured Molecules Using Dynamic Programming. Technical report 2009-015, Graduate School of Informatics, Kyoto University (2009), `http://www-or.amp.i.kyoto-u.ac.jp/members/nag/Technicalreport/TR2009-015.pdf`
7. Jordan, C.: Sur Les Assemblages De Lignes. J. Reine Angew. Math. 70, 185–190 (1869)
8. Nourse, J.G.: The Configuration Symmetry Group and its Application to Stereoisomer Generation, Specification, and Enumeration. J. Am. Chem. Soc. 101, 1210–1216 (1979)
9. Pólya, G.: Kombinatorische Anzahlbestimmungen für Gruppen, Graphen, un chemische Verbindungen. Acta Math. 68, 145–253 (1937)
10. Pólya, G., Read, R.C.: Combinatorial Enumeration of Groups, Graphs, and Chemical Compounds. Springer, New York (1987)
11. Razinger, M., Balasubramanian, K., Perdih, M., Munk, M.E.: Stereoisomer Generation in Computer-Enhanced Structure Elucidation. J. Chem. Inf. Comput. Sci. 33, 812–825 (1993)
12. Wieland, T., Kerber, A., Laue, R.: Principles of the Generation of Constitutional and Configurational Isomers. J. Chem. Inf. Comput. Sci. 36, 413–419 (1996)

Exact Algorithms for the Bottleneck Steiner Tree Problem*
(Extended Abstract)

Sang Won Bae[1], Sunghee Choi[2], Chunseok Lee[2], and Shin-ichi Tanigawa[3]

[1] Department of Computer Science and Engineering, POSTECH, Pohang, Korea
swbae@postech.ac.kr
[2] Division of Computer Science, KAIST, Daejeon, Korea
{sunghee, stonecold}@tclab.kaist.ac.kr
[3] Department of Architecture and Architectural Engineering, Kyoto University, Kyoto, Japan
is.tanigawa@archi.kyoto-u.ac.jp

Abstract. Given n terminals in the plane $\mathbb{R}^2$ and a positive integer k, the bottleneck Steiner tree problem is to find k Steiner points in $\mathbb{R}^2$ so that the longest edge length of the resulting Steiner tree is minimized. In this paper, we study this problem in any L_p metric. We present the first fixed-parameter tractable algorithm running in $O(f(k) \cdot n^2 \log n)$ time for the L_1 and the L_∞ metrics, and the first exact algorithm for any other L_p metric with $1 < p < \infty$ whose time complexity is $O(f(k) \cdot (n^k + n \log n))$, where $f(k)$ is a function dependent only on k. Note that prior to this paper there was no known exact algorithm even for the L_2 metric, and our algorithms take a polynomial time in n for fixed k.

1 Introduction

We consider the following problem, called the *bottleneck Steiner tree problem* (BST).

Problem 1 (BST). Given n points, called *terminals*, in the plane and a positive integer k, find a Steiner tree spanning all terminals and at most k Steiner points that minimizes the length of its longest edge.

Such a Steiner tree with the length of the longest edge minimized is called a *bottleneck Steiner tree* (also known as a *min-max Steiner tree*).

Unlike the classical Steiner tree problem where the sum of the edge lengths of the Steiner tree is minimized, this problem asks a Steiner tree where the maximum of the edge lengths is minimized and the Steiner points in the resulting tree can be chosen in the whole plane $\mathbb{R}^2$.

The bottleneck Steiner tree problem (shortly BST) has been studied with many applications, like VLSI layout [4], multi-facility location, and wireless communication network design [9]. Previous work on BST considered the Euclidean plane (L_2) or the

* Work by S.W.Bae was supported by the Brain Korea 21 Project. Work by C.Lee and S.Choi was supported by the Korea Science and Engineering Fouundation (KOSEF) grant funded by the Korea Government (MEST) (No.R01-2007-000-20865-0). Work by S.Tanigawa was supported by Grant-in-Aid for JSPS Research Fellowship for Young Scientists.

Y. Dong, D.-Z. Du, and O. Ibarra (Eds.): ISAAC 2009, LNCS 5878, pp. 24–33, 2009.

rectilinear plane (L_1). BST is known to be NP-hard to approximate within ratio $\sqrt{2}$ in the L_2 metric or ratio 2 in the L_1 metric [9]. For the L_1 metric, the matching upper bound was known by Wang and Du [9] and for the L_2 metric, the best known upper bounds on approximation ratio is 1.866 by Wang and Li [10]. For the special case of this problem where there should be no edge connecting any two Steiner points in the optimal solution, Li et al. [7] present a $(\sqrt{2} + \epsilon)$-factor approximation algorithm and inapproximability within ratio $\sqrt{2}$.

Also, there has been effort on devising an exact algorithm for finding the locations of k Steiner points. Many researchers considered the variation of BST where a topology $\mathcal{T}$ of Steiner trees is fixed, called the bottleneck Steiner tree with fixed topology (BST-FT); the resulting Steiner tree should have the same topology as a given tree $\mathcal{T}$ [8,5]. Once we have an exact algorithm for this problem BST-FT, we can find an exact solution to BST by enumerating all valid topology trees.

For the L_1 (or the L_∞) metric, Ganley and Salowe [5] presented an $O((n+k)^2)$ time exact algorithm for BST-FT, which leads to an exact algorithm for BST with running time $O((n+k)!(n+k)^2)$. The situation for the L_2 metric (or the Euclidean metric) is much worse. There was no known exact algorithm for BST in the L_2 metric but few partial results: Sarrafzadeh and Wong [8] presented an $O((n+k)\log(n+k))$ time algorithm for the decision version of BST-FT. More recently, Bae et al. [3] presented exact algorithms for $k \leq 2$; $O(n \log n)$ time for $k = 1$ and $O(n^2)$ time for $k = 2$.

The aim of this paper is to devise exact algorithms for BST in any L_p metric, which run fast for small fixed k. We present two exact algorithms for BST: (1) an $O(f(k) \cdot n^2 \log n)$-time algorithm for the L_1 or the L_∞ metric and (2) an $O(f(k) \cdot (n^k + n \log n))$-time algorithm for the L_p metric with $1 < p < \infty$, where $f(k)$ is a function dependent on k only. Note that both algorithms run in time polynomial in n for fixed k. The algorithm for the L_1 and the L_∞ metrics is a *fixed-parameter algorithm* with respect to parameter k. A fixed-parameter algorithm is one that has time complexity of the form $f(k) \cdot n^{O(1)}$. Thus, we show that BST in the L_1 metric is fixed-parameter tractable with respect to parameter k. On the other hand, we failed to achieve fixed-parameter tractability for the other case but it is worth noting that our algorithm for the L_p metric with $1 < p < \infty$ is the first exact algorithm for BST.

2 Properties of Bottleneck Steiner Trees

Let $P \subset \mathbb{R}^2$ be a set of n points; we call each point in P a *terminal*. A *bottleneck spanning tree* of P is a spanning tree of P such that the length of a longest edge is minimized. We call the length of a longest edge in a bottleneck spanning tree of P the *bottleneck of the set* P, denoted by $b(P)$. Note that $b(P)$ is dependent only on the set P, and not on a particular spanning tree of the points P. A *minimum spanning tree* $MST(P)$ of given points P is a spanning tree of P with minimum sum of the edge lengths. Obviously, $MST(P)$ is a bottleneck spanning tree of P. Throughout this section, the underlying distance function is assumed to be the L_p metric for any $1 \leq p \leq \infty$, and $d(a, b)$ denotes the L_p-distance between two points $a, b \in \mathbb{R}^2$.

Since computing a minimum or bottleneck spanning tree of a given set can be done easily, the bottleneck Steiner tree problem can be viewed as finding a set Q of k optimal the maximum edge length in the resulting Steiner tree with the same topology as $\mathcal{T}_0$ is at most λ. Once we obtain an efficient decision algorithm, we do a binary search on *critical values* of λ to find the smallest λ^* that yields the positive answer "YES". Thus, λ^* is the optimal bottleneck value for a given abstract topology $\mathcal{T}_0$.

We should mention that the optimal objective value λ^* may be determined by $|e_{c+1}|$, the longest edge that is left from $MST(P)$. We, however, consider only the edges incident to any Steiner point and optimize their lengths; afterwards, we can simply compare the obtained value to $|e_{c+1}|$. Thus, in the remaining of this paper, we assume that λ^* is always determined by the length of an edge incident to a Steiner point.

3.1 Decision Algorithm

Our algorithm can be seen as an extension of the algorithm by Ganley and Salowe [5]; namely, from concrete topologies to abstract topologies. We choose an arbitrary Steiner vertex of $\mathcal{T}_0$ as a root, and consider $\mathcal{T}_0$ as a rooted tree. Let $C(v)$ be the set of children of a vertex v in $\mathcal{T}_0$. For each vertex v of $\mathcal{T}_0$, we shall associate a region $R_\lambda(v)$ as follows. If $v = t_i$, representing a subtree T_i of $MST(P)$, then $R_\lambda(v)$ is the set of terminals contained in T_i. Otherwise, if v is a Steiner vertex, $R_\lambda(v) := \bigcap_{u \in C(v)} R_\lambda(u) \oplus B_\lambda$, where B_λ denotes the L_∞-ball (i.e., a square) of radius λ centered at the origin and $\oplus$ denotes the Minkowski sum operation. We compute the regions $R_\lambda(v)$ in a bottom-up manner starting from the leaves, and answer "YES" if $R_\lambda(v) \neq \emptyset$ for all vertices v of $\mathcal{T}_0$; otherwise, report "NO". The correctness of the algorithm is straightforward.

To compute the regions $R_\lambda(v)$, we shall show that $R_\lambda(v)$ can be simply described in terms of squares centered at the terminals in P. We denote by $B(p, r)$ the L_∞-ball of radius r centered at point $p \in \mathbb{R}^2$. Also, for two vertices u and v in $\mathcal{T}_0$, let $\delta_{u,v}$ be the length of the path between u and v (i.e. the number of edges) in $\mathcal{T}_0$, and $L(v)$ be the set of leaves which are descendants of v in $\mathcal{T}_0$.

Lemma 5. *Let v be an internal vertex of a given abstract topology $\mathcal{T}_0$. Suppose $R_\lambda(u) \neq \emptyset$ for all descendants u of v in $\mathcal{T}_0$. Then, we have $R_\lambda(v) = \bigcap_{t_i \in L(v)} \bigcup_{p \in T_i} B(p, \lambda \cdot \delta_{t_i,v})$.*

Lemma 5 tells us the complexity of $R_\lambda(v)$ as well as how to compute it.

Lemma 6. *The decision version of BST-FT-ST can be solved in $O(kn^2)$ time.*

3.2 Optimization Algorithm

We shall show that the exact solution of BST-FT-ST can be obtained by $O(\log n)$ implementations of the decision algorithm. Following is a key lemma for our purpose.

Lemma 7. *Let λ^* be the smallest real number that the decision algorithm reports YES. Then, there exists a Steiner vertex s_i such that the area of $R_{\lambda^*}(s_i)$ is zero.*

For any Steiner vertex s_i and two terminals $p \in T_j$ and $p' \in T_{j'}$, consider the value λ such that $B(p, \lambda \cdot \delta_{s_i,t_j})$ touches $B(p', \lambda \cdot \delta_{s_i,t_{j'}})$. We call such λ a *critical value.* Lemma 7 implies that there is s_i such that we have two touching squares $B(p, \lambda^* \cdot \delta_{s_i,t_j})$ and $B(p', \lambda^* \cdot \delta_{s_i,t_{j'}})$ where $p \in T_j$ and $p' \in T_{j'}$; thus, λ^* is also a critical value.

Lemma 8. *BST-FT-ST in the L_∞ metric can be solved in $O(kn^2 \log n)$ time.*

In order to solve BST, we enumerate all possible abstract topologies and solve BST-FT-ST for each. The number of abstract topologies is at most $O((7k+1)^{7k-1})$ by Lemma 4 and Theorem 1. Finally, we obtain a fixed-parameter algorithm for BST in the L_1 or the L_∞ metric.

Theorem 2. *Given n terminals and a positive integer k, a bottleneck Steiner tree with k Steiner points in the L_1 or the L_∞ metric can be exactly computed in $O((7k+1)^{7k} \cdot n^2 \log n)$ time.*

4 Exact Algorithm for the Euclidean Metric and the L_p Metric

In this section we present an algorithm that finds an exact solution to BST-FT-ST in the L_2 (the Euclidean) metric, leading to the first exact algorithm for BST in the Euclidean metric. Throughout this section, $d(a,b)$ denotes the Euclidean distance between two points a and b. Also, we assume that every leaf of the given abstract topology $\mathcal{T}_0$ is a terminal vertex as in Section 3. Note that our algorithm works for any L_p metric by Theorem 1 while we mainly discuss bottleneck Steiner trees in the Euclidean metric.

As Ganley and Salowe [5] pointed out earlier, it seems much harder to find an exact solution to BST in the Euclidean metric (or any L_p metric) than in the L_1 metric. One could try a similar approach as done in Section 3 but would immediately face with a difficulty; a nice description of the region $R_\lambda(v)$ like Lemma 5 is hardly obtained. Indeed, that is one of the reasons why no exact algorithm has been discovered yet for the Euclidean case. This also causes another difficulty in collecting critical values λ among which the optimal objective value λ^* can be found. To overcome all difficulty, we make full use of the properties of bottleneck Steiner trees revealed in Section 2, introducing some new concepts; determinators and primary clusters.

4.1 Determinators and Primary Clusters

Consider an optimal placement $Q = \{q_1, \ldots, q_k\}$ of k Steiner points for a given abstract topology $\mathcal{T}_0$. Recall that the corresponding Steiner tree T^* is supposed to satisfy **BST1–4**. Let r_i be the length of the longest edge incident to q_i in T^* and B_i be the disk centered at q_i with radius r_i. By **BST2**, each B_i has two or three points on its boundary, which are among the neighbors of q_i in T^*. Note that a disk is said to be *determined by three points* or *by two diametral points* if the three points lie on the boundary of the disk or if the two points define the diameter of the disk, respectively. We call the three or two points determining a disk the *determinators* of the disk or of its center. Note that if s_i is adjacent to t_j in $\mathcal{T}_0$ we can assume that, in the resulting Steiner tree, q_i is adjacent to the closest terminal in subtree T_j.

Let D_i be the set of determinators of q_i in T^*. D_i consists of two or three points in $P \cup Q$. In the case where there are more than three points on the boundary of B_i, we arbitrarily choose three of them. Then, the following is an immediate observation.

Lemma 9. *If $q_i \in D_j$ in T^*, then $r_i \geq r_j$.*

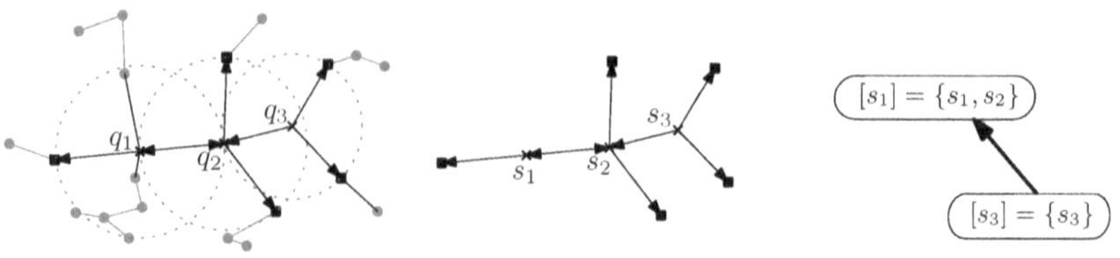

(a) BST T^* (b) concrete topology $\mathcal{T}$ (c) DAG G on $[S]$

Fig. 2. (a) A bottleneck Steiner tree T^* for abstract topology $\mathcal{T}_0$ in Figure 1; The arrows indicate the determinators D_i of each Steiner points and terminals that are determinators of some q_i are depicted as small solid squares. Dotted circles indicate B_i whose radius is r_i. (b) An induced concrete topology $\mathcal{T}$ taking only terminals that are determinators, $D_i \cap P$, with determinator information; An optimal placement of Steiner points is determined by these selected terminals only. Here, we have a primary cluster $[s_1] = \{s_1, s_2\}$. See $r_1 = r_2 \geq r_3$ in (a). (c) Directed acyclic graph G on $[S]$ based on the determinator relation on $\mathcal{T}$ induces a partial ordering on $[S]$.

Based on the sets of determinators, D_i, we can infer an ordering on the Steiner vertices. First, we build an equivalence relation $\equiv$ on $S = \{s_1, \ldots, s_k\}$; $s_i \equiv s_j$ if and only if $q_i \in D_j$ and $q_j \in D_i$. We denote by $[s_i]$ the equivalence class that includes s_i, and let $[S] := \{[s_i] \mid s_i \in S\}$. Now, we construct a directed graph G on $[S]$ in which there is a directed edge from $[s_i]$ to $[s_j]$ if and only if there are two indices i' and j' with $s_{i'} \in [s_i]$ and $s_{j'} \in [s_j]$ such that $q_{j'} \in D_{i'}$. Obviously, G is acyclic. We denote by $([S], \preceq)$ a partial ordered set induced by G. Lemma 9 implies that if $s_j \in [s_i]$, then $r_i = r_j$, and if $[s_i] \preceq [s_j]$ then $r_{i'} \leq r_{j'}$ for any i', j' with $s_{i'} \in [s_i]$ and $s_{j'} \in [s_j]$. Let $\mathcal{M}$ be the set of all the maximal elements in $[S]$ with respect to $\preceq$. More precisely, $\mathcal{M} = \{[s_i] \in [S] \mid$ there is no other $[s_j] \in [S]$ with $[s_i] \preceq [s_j]\}$. We call each subset of Steiner vertices in $\mathcal{M}$ a *primary cluster*. See Figure 2 for more illustrative explanation.

Since the optimal objective value λ^* for given abstract topology $\mathcal{T}_0$ is determined as $\max r_i$ (as assumed in the beginning of Section 3), it is obvious that λ^* is indeed determined by a primary cluster in $\mathcal{M}$.

Lemma 10. *Assume that all symbols and relations are induced from a bottleneck Steiner tree as above. Then, there always exists a primary cluster* $[s_i] \in \mathcal{M}$ *such that* $\lambda^* = r_i$.

4.2 Algorithm

Our algorithm enumerates all possible combinations of determinators $(D_1, \ldots, D_k)$ for a given abstract topology $\mathcal{T}_0$. By a fixed combination of determinators, the equivalence relation $(S, \equiv)$ and the partial ordering $([S], \preceq)$ are inferred as above. At last, we will be able to collect critical values by handling each primary cluster.

Thus, our algorithm for BST-FT-ST in the L_2 metric is summarized as follows:

(1) Enumerate all possible combinations of determinators D_i for abstract topology $\mathcal{T}_0$.
(2) For a fixed combination of the determinators D_i, build a concrete topology $\mathcal{T}$.
(3) For each primary cluster induced by the D_i, collect critical values.
(4) Do a binary search on the collected critical values to find the optimal value λ^*.

From now on, we describe each step of the algorithm in more details.

(1) Enumeration of all possible combinations of determinators For a given abstract topology $\mathcal{T}_0$, a rough calculation on the number of combinations of determinators gives us $20^k n^{3k}$: (i) we choose two or three neighbors from at most five neighbors of each s_i in $\mathcal{T}_0$ and, (ii) choose one terminal from at most n terminals in T_j if t_j was chosen for s_i at (i).

In order to prune a number of unnecessary combinations of determinators, we bring a known geometric structure, namely, the *farthest color Voronoi diagram.* The farthest color Voronoi diagram is a generalization of the standard farthest-neighbor Voronoi diagram to colored point sets [6,1]: Given a collection $\mathcal{C} = \{P_1, \ldots, P_l\}$ of l sets of colored points, define the *distance to a color* i for $1 \leq i \leq l$ as $d_i(x) := \min_{p \in P_i} d(x, p)$. Then, the farthest color Voronoi diagram $FCVD(\mathcal{C})$ is a decomposition of $\mathbb{R}^2$ into Voronoi regions VR_i for subset P_i, defined to be the set $\{x \in \mathbb{R}^2 \mid d_i(x) > d_j(x), 1 \leq j \leq l, j \neq i\}$. Each edge of $FCVD(\mathcal{C})$ is the set of points that have the same set of two farthest colors; each vertex is a point that has three or more farthest colors. Each Voronoi region VR_i is again refined into cells σ_p for each $p \in P_i$ such that $\sigma_p = \{x \in VR_i \mid d(x, p) = d_i(x) = \min_{q \in P_i} d(x, q)\}$. We regard $FCVD(\mathcal{C})$ as this refined diagram. Note that each cell, edge, or vertex of $FCVD(\mathcal{C})$ is determined by (thus, associated with) one, two, or three points in $\bigcup_i P_i$. For more details about the farthest color Voronoi diagram, we refer to Huttenlocher et al. [6] and Abellanas et al. [1]. We now introduce an interesting relation between the farthest color Voronoi diagram and the determinators.

Lemma 11. *Given the determinators D_i of q_i, let $m := |D_i \cap P|$, the number of terminals that are determinators of q_i. If $m > 0$, then q_i lies on the $(3 - m)$-face ϕ_i of $FCVD(\mathcal{C}_i)$ determined by $D_i \cap P$, where $\mathcal{C}_i = \{V(T_j) \mid t_j$ a neighbor of s_i in $\mathcal{T}_0\}$ and $V(T_j) \subseteq P$ denotes the vertex set of T_j.*

Thus, as a preprocessing, we compute $FCVD(\mathcal{C}_i)$ for each Steiner vertex s_i, where $\mathcal{C}_i$ is defined as in Lemma 11. Since we have at most 5 subsets of P in $\mathcal{C}_i$, the total combinatorial complexity of all $FCVD(\mathcal{C}_i)$ is $O(n)$ and $O(n \log n)$ time is sufficient to build them for any L_p metric with $1 < p < \infty$ [6].

To enumerate combinations of determinators, we first choose one face ϕ_i (of any dimension) from $FCVD(\mathcal{C}_i)$ to fix $D_i \cap P$. Then, for each edge between two Steiner vertices s_i, s_j in $\mathcal{T}_0$, we have four possibilities to fix $D_i \setminus P$; $q_i \in D_j$ or $q_i \notin D_j$, and $q_j \in D_i$ or $q_j \notin D_i$. Thus, we have $(O(n))^k$ combinations of $D_i \cap P$ and 4^k combinations of $D_i \setminus P$. Hence, we have at most $(O(n))^k \cdot 4^k = (O(n))^k$ possible combinations of the determinators in total.

(2) Building a corresponding concrete topology At this step, we are given the determinators D_i for each Steiner point. The D_i induce a concrete topology $\mathcal{T}$ from the abstract topology $\mathcal{T}_0$ by choosing the terminals appeared in any of the D_i. Note that $\mathcal{T}$ consists of at most $3k$ terminals and k Steiner points since D_i contains at most three terminals. We should mention that when building $\mathcal{T}$ from $\mathcal{T}_0$, we drop some edges, which are not related to the D_i. Since an optimal placement of Steiner points is determined only by the D_i, these dropped edges do not matter for getting critical values at Step (3).

(3) Collecting critical values In order to collect critical values λ, we search each primary cluster $[s_i] \in \mathcal{M}$. By Lemma 10, a primary cluster $[s_i]$ among $\mathcal{M}$ causes the longest edge in the resulting Steiner tree T^*; that is, $\lambda^* = r_i = r_j$ for any $s_j \in [s_i]$. Recall that the placement q_j for each $s_j \in [s_i]$ is determined by its determinators D_j. Thus, to precompute (candidates of) the value λ^*, we take a subtopology $\mathcal{T}_i$ of $\mathcal{T}$, which is an induced subtree by $[s_i]$ and their determinators.

In $\mathcal{T}_i$, every leaf is a terminal in P and every internal vertex is a Steiner vertex with degree 2 or 3. Observe that any degree-2 Steiner point is located at the midpoint of its two neighbors by **BST2**. Thus, degree-2 Steiner points are rather easy to find once we know an optimal placement of all degree-3 Steiner points. We modify $\mathcal{T}_i$ into a weighted tree $\mathcal{T}_i'$ by the following operations: (1) Initially, assign weight 1 to every edge of $\mathcal{T}_i$. (2) Whenever there is a degree-2 Steiner vertex, we remove it from $\mathcal{T}_i$ and its two incident edges are merged into one with summed weight.

Now, let m be the number of Steiner vertices in $\mathcal{T}_i'$, and $w(e)$ be the weight of an edge e of $\mathcal{T}_i'$. Then, $\mathcal{T}_i'$ consists of exactly $2m + 1$ edges and m internal vertices (all Steiner). Without loss of generality, we assume that $s_1, \ldots, s_m$ are the Steiner vertices in $\mathcal{T}_i'$. For each edge e of $\mathcal{T}_i'$, we assign a function $h_e : (\mathbb{R}^2)^m \to \mathbb{R}$ defined to be

$$h_e((q_1, \ldots, q_m)) := \frac{1}{w(e)}\{\text{the length of } e \text{ when } s_j \text{ is placed at } q_j \in \mathbb{R}^2 \text{ for each } j\}.$$

Let $q_1^*, \ldots, q_m^*$ be an optimal placement of m Steiner points. Then, at $(q_1^*, \ldots, q_m^*)$, we have equality $h_e((q_1^*, \ldots, q_m^*)) = h_{e'}((q_1^*, \ldots, q_m^*))$ for any pair of two edges e, e' of $\mathcal{T}_i'$, by Lemma 9.

The function h_e indeed returns the distance between two points, so it defines an algebraic surface of degree 2 in $\mathbb{R}^{2m+1}$, which is the graph of h_e. Furthermore, at every point $(q_1, \ldots, q_m) \in (\mathbb{R}^2)^m$ that satisfies all the equalities, the $2m+1$ surfaces meet all together at a point $(q_1, \ldots, q_m, \lambda) \in \mathbb{R}^{2m+1}$; this point is a vertex (0-face) of the lower (or upper) envelope of the $2m+1$ surfaces. Fortunately, all 0-faces of the lower envelope of algebraic surfaces in high dimension can be computed by Agarwal et al. [2]; in our case, it costs $O((2m + 1)^{2m+\epsilon})$ time for any positive ϵ. Once we find all the 0-faces of the lower envelope, we have at most $O((2m + 1)^{2m+\epsilon})$ critical values by taking the height (the $(2m + 1)$-st coordinate) of each 0-face of the lower envelope.

We do this procedure for every primary cluster, collecting $O((2k + 1)^{2k+\epsilon})$ critical values in total in the same time bound.

(4) Binary search on critical values At this step, we do a binary search on the collected critical values using a decision algorithm running on the concrete topology $\mathcal{T}$. We make use of a modified version of the decision algorithm by Sarrafzadeh and Wong [8]. We choose an arbitrary internal vertex of $\mathcal{T}$ as a root and consider $\mathcal{T}$ as a rooted tree.

By Lemma 11, every Steiner point q_i must lie in the face ϕ_i chosen at Step (1). We thus redefine the region $R_\lambda(s_i)$ for each Steiner vertex as follows:

$$R_\lambda(s_i) := \phi_i \cap \bigcap_{u \in C(s_i)} (R_\lambda(u) \oplus B_\lambda),$$

where $C(s_i)$ is the set of children of s_i in $\mathcal{T}$ and B_λ is a unit disk centered at the origin.

If ϕ_i is a segment or a point, this additional intersection with ϕ_i does not increase the complexity. For easy analysis of the case where ϕ_i is a two-dimensional face, we triangulate each two-dimensional face of $FCVD(\mathcal{C}_i)$ at Step (1) and take ϕ_i as a triangle. Since the triangulation does not increase the complexity of $FCVD(\mathcal{C}_i)$, we still have $O(n)$ number of vertices, edges, and triangular faces in $FCVD(\mathcal{C}_i)$. Also, computing $R_\lambda(s_i)$ is done without additional cost; ϕ_i is either a point, a segment, or a triangle. Since $\mathcal{T}$ consists of $O(k)$ vertices, our modified decision algorithm runs in $O(k \log k)$ time, based on the analysis by Sarrafzadeh and Wong [8].

Hence, for each concrete topology $\mathcal{T}$, we can find an optimal objective value λ^* in $O(k \log k \cdot \log((2k+1)^{2k+\epsilon})) = O(k^2 \log^2 k)$ time. To see the total time complexity, initially we spend $O(n \log n)$ time to compute $FCVD(\mathcal{C}_i)$, Steps (3)–(4) take $O((2k+1)^{2k+\epsilon})$ time, and we repeat Steps (3)–(4) $(O(n))^k$ times; $O((2k+1)^{2k+\epsilon} \cdot (O(n))^k + n \log n)$ time to solve BST-FT-ST in total. Combining this with Lemma 4, we finally conclude:

Theorem 3. *Given n terminals and a positive integer k, a bottleneck Steiner tree with k Steiner points in the L_p metric with $1 < p < \infty$ can be exactly computed in $O(f(k) \cdot (n^k + n \log n))$ time, where $f(k) = O((5k+1)^{5k-1}(2k+1)^{2k+\epsilon}2^{O(k)}) = O(k^{7k} \cdot 2^{O(k)})$.*

Remark that our algorithm only finds an optimal placement Q of k Steiner points. To obtain a bottleneck Steiner tree, it suffices to compute the minimum spanning tree $MST(P \cup Q)$ for points $P \cup Q$.

References

1. Abellanas, M., Hurtado, F., Icking, C., Klein, R., Langetepe, E., Ma, L., Sacristan, V.: The farthest color Voronoi diagram and related problems. In: Proc. 17th European Workshop Comput. Geom., pp. 113–116 (2001)
2. Agarwal, P.K., Aronov, B., Sharir, M.: Computing envelopes in four dimensions with applications. SIAM J. Comput. 26(6), 1714–1732 (1997)
3. Bae, S.W., Lee, C., Choi, S.: On exact solutions to the Euclidean bottleneck Steiner tree problem. In: Das, S., Uehara, R. (eds.) WALCOM 2009. LNCS, vol. 5431, pp. 105–116. Springer, Heidelberg (2009)
4. Chiang, C., Sarrafzadeh, M., Wong, C.: A powerful global router: based on Steiner min-max trees. In: Proc. IEEE Int. Conf. CAD, pp. 2–5 (1989)
5. Ganley, J.L., Salowe, J.S.: Optimal and approximate bottleneck Steiner trees. Oper. Res. Lett. 19, 217–224 (1996)
6. Huttenlocher, D.P., Kedem, K., Sharir, M.: The upper envelope of Voronoi surfaces and its applications. Discrete Comput. Geom. 9, 267–291 (1993)
7. Li, Z.-M., Zhu, D.-M., Ma, S.-H.: Approximation algorithm for bottleneck Steiner tree problem in the Euclidean plane. J. Comput. Sci. Tech. 19(6), 791–794 (2004)
8. Sarrafzadeh, M., Wong, C.: Bottleneck Steiner trees in the plane. IEEE Trans. Comput. 41(3), 370–374 (1992)
9. Wang, L., Du, D.-Z.: Approximations for a bottleneck Steiner tree problem. Algorithmica 32, 554–561 (2002)
10. Wang, L., Li, Z.: An approximation algorithm for a bottleneck k-Steiner tree problem in the Euclidean plane. Inform. Process. Lett. 81, 151–156 (2002)

Exact Algorithms for Set Multicover and Multiset Multicover Problems

Qiang-Sheng Hua[1], Dongxiao Yu[1], Francis C.M. Lau[1], and Yuexuan Wang[2]

[1] Department of Computer Science, The University of Hong Kong, Pokfulam, Hong Kong, China
qshua@cs.hku.hk, dxyu@cs.hku.hk, fcmlau@cs.hku.hk
[2] Institute for Theoretical Computer Science, Tsinghua University, Beijing, 100084, China
wangyuexuan@tsinghua.edu.cn

Abstract. Given a universe N containing n elements and a collection of multisets or sets over N, the multiset multicover (MSMC) or the set multicover (SMC) problem is to cover all elements at least a number of times as specified in their coverage requirements with the minimum number of multisets or sets. In this paper, we give various exact algorithms for these two problems, with or without constraints on the number of times a multiset or set may be picked. First, we can exactly solve the MSMC without multiplicity constraints problem in $O(((b+1)(c+1))^n)$ time where b and c ($c \leq b$ and $b \geq 2$) respectively are the maximum coverage requirement and the maximum number of times that each element can appear in a multiset. To our knowledge, this is the first known exact algorithm for the MSMC without multiplicity constraints problem. Second, we can solve the SMC without multiplicity constraints problem in $O((b+2)^n)$ time. Compared with the two recent results in [Hua et al., Set Multi-covering via inclusion-exclusion, Theoretical Computer Science, 410(38-40):3882-3892 (2009)] and [Nederlof, J.: Inclusion Exclusion for hard problems. Master Thesis. Utrecht University, The Netherlands (2008)], we have given the fastest exact algorithm for the SMC without multiplicity constraints problem. Finally, we give the first known exact algorithm for the MSMC or the SMC with multiplicity constraints problem in $O((b+1)^n|F|)$ time and $O((b+1)^n|F|)$ space where $|F|$ denotes the total number of given multisets or sets over N.

1 Introduction

In this paper, we study exact algorithms for the set multicover(SMC) and the multiset multicover(MSMC) problems. In the set multicover problem, we are given a universe N of n elements and a family of sets $F = \{S_1, \cdots, S_{|F|}\}$ where each S_i is a subset of N, and we need to find a minimum cardinality sub-family $F' \subseteq F$ such that each element $i \in N$ is covered b_i integral number of times. In the multiset multicover problem, we are given a collection of multisets instead of a collection of sets. Here a multiset S_i contains a specified number of copies of each element $i \in N$. Note that in order to minimize the total number of picked sets or multisets,

Y. Dong, D.-Z. Du, and O. Ibarra (Eds.): ISAAC 2009, LNCS 5878, pp. 34–44, 2009.

each set or multiset can be chosen a number of times. Here if we further require that each set or multiset can be chosen at most a specified number of times, the SMC or the MSMC problem becomes the SMC or the MSMC with multiplicity constraints problem. Much attention has been given to approximation algorithms for these problems in the past several decades [11,10,3,4,5]. Besides approximate algorithms, recently there has also been some effort in understanding how fast we can exactly solve these covering problems.

By using the inclusion-exclusion principle and a fast zeta transform technique, Björklund et al. [1] have shown that the set cover problem can be exactly solved in $O^*(2^n)$ time using $O^*(2^n)$ space. Here, using the $O^*(f(n))$ notation we omit a $(\log f(n))^{O(1)}$ factor. Based on this observation, they proposed a family of exact algorithms for the set partitioning problems which improve all the previous algorithms. Later on they showed that similar faster algorithms can also be obtained by using the so called fast subset convolution [2]. Very recently, Hua et al. and Nederlof have independently given their exact algorithms for the set multicover problem in [6] and [9], respectively. In [6], we show that the set multicover problem can be exactly solved in $O^*((2b)^n)$ time using $O^*((b+1)^n)$ space or in $O^*(2^{O(bn^2)})$ time with polynomial space. In [9], based on a novel counting formulation, the set multicover problem can be solved in $O((b+1)^n|F_{mc}|)$ time and polynomial space. Here $|F_{mc}|$ means the total number of given sets. Although this result greatly outperforms the polynomial space exact algorithm given in [6], as discussed in [8], the algorithm given in [6] can also exactly count the number of set multicovers that satisfy the coverage requirements. We are not aware of any known exact algorithms for solving the multiset multicover, the set multicover with multiplicity constraints and the multiset multicover with multiplicity constraints problems. Some key notations and their definitions are given in Table 1.

Table 1. Some key notations and definitions

Notations	Definitions
$N = \{1, \cdots, n\}$	The universe set.
F_{ms} (F_{mc}) (F_{sc})	A collection of (multi)sets in a multiset multicover (set multicover) (set cover) instance.
$B = (b_1, \cdots, b_n)$	The positive integral coverage requirement vector indicating that each element i must be covered at least b_i times and $\mathbf{b} = \max_{i \in N}(b_i)$.
$C = (F_{ms}(S, i))$	The vector indicating the number of times that each element $i \in N$ appears in each multiset $S \in F_{ms}$ and $\mathbf{c} = \max_{i \in N, S \in F_{ms}}(F_{ms}(S, i))$. We assume $c \le b$ and $b \ge 2$.
$c_k(F_{sc})$	The number of k-tuples $< s_1, \cdots, s_k >$ over F_{sc} such that the union of the sets $\bigcup_{i=1}^{k} s_i$ without removing duplicate elements satisfies the coverage requirements.

10. Rajagopalan, S., Vazirani, V.V.: Primal-dual RNC approximation algorithms for set cover and covering integer programs. SIAM Journal on Computing 28(2), 525–540 (1998)
11. Vazirani, V.V.: Approximation Algorithms. Springer, Berlin (2003)
12. Amzallag, D., Engelberg, R., Naor, J., Raz, D.: Capacitated Cell Planning of 4G Cellular Networks (Technical Report CS-2008-04). Computer Science Department, Technion, Haifa 32000, Israel (2008)

Practical Discrete Unit Disk Cover Using an Exact Line-Separable Algorithm*

Francisco Claude[1], Reza Dorrigiv[1], Stephane Durocher[2], Robert Fraser[1], Alejandro López-Ortiz[1,**], and Alejandro Salinger[1]

[1] Cheriton School of Computer Science, University of Waterloo, Waterloo, Canada
{fclaude,rdorrigiv,r3fraser,alopez-o,ajsalinger}@cs.uwaterloo.ca
[2] Department of Computer Science, University of Manitoba, Winnipeg, Canada
durocher@cs.umanitoba.ca

Abstract. Given m unit disks and n points in the plane, the discrete unit disk cover problem is to select a minimum subset of the disks to cover the points. This problem is NP-hard [11] and the best previous practical solution is a 38-approximation algorithm by Carmi et al. [4]. We first consider the line-separable discrete unit disk cover problem (the set of disk centres can be separated from the set of points by a line) for which we present an $O(m^2n)$-time algorithm that finds an exact solution. Combining our line-separable algorithm with techniques from the algorithm of Carmi et al. [4] results in an $O(m^2n^4)$ time 22-approximate solution to the discrete unit disk cover problem.

1 Introduction

Recent interest in specific geometric set cover problems is partly motivated by applications in wireless networking. In particular, when wireless clients and servers are modelled as points in the plane and the range of wireless transmission is assumed to be constant (say one unit), the resulting region of wireless communication is a disk of unit radius centred on the point representing the corresponding wireless transmitting device. Under this model, sender a successfully transmits a wireless message to receiver b if and only if point b is covered by the unit disk centred at point a. This model applies more generally to a variety of facility location problems for which the Euclidean distance between clients and facilities cannot exceed a given radius, and clients and candidate facility locations are represented by discrete sets of points. Examples include (1) selecting locations for wireless servers (e.g., gateways) from a set of candidate locations to cover a set of wireless clients, (2) positioning a fleet of water bombers at airports such that every active forest fire is within a given maximum distance of a water bomber, (3) selecting a set of weather radar antennae to cover a set of cities, and (4)

* Funding for this project was provided by the NSERC Strategic Grant on Optimal Data Structures for Organization and Retrieval of Spatial Data.

** Part of this work took place while the fifth author was on sabbatical at the Max-Planck-Institut für Informatik in Saarbrücken, Germany.

Y. Dong, D.-Z. Du, and O. Ibarra (Eds.): ISAAC 2009, LNCS 5878, pp. 45–54, 2009.

selecting locations for anti-ballistic defenses from a set of candidate locations to cover strategic sites. These problems can be modelled by the *discrete unit disk cover problem* (DUDC), whose definition is: Given sets P of m points and Q of n points in the plane (candidate facilities and clients, respectively), find a set $P' \subseteq P$ (facilities) of minimum cardinality such that Disk(P') covers Q, where Disk(A) denotes the set of unit disks centred on points in set A. In this work, we consider the *line-separable discrete unit disk cover* (LSDUDC), where P and Q are separated by a line l.

The DUDC problem is NP-hard [11]. In a recent result, Carmi et al. [4] describe a polynomial-time 38-approximate solution, improving on earlier 108-approximate [6] and 72-approximate solutions [16]. We present an $O(m^2n)$-time algorithm that returns an exact solution to the LSDUDC problem, as well as a thorough proof of correctness of the technique. By combining the LSDUDC algorithm with techniques from the algorithm of Carmi et al. [4], we present a 22-approximation algorithm to the DUDC problem, improving on the best previous practical polynomial-time approximation factor of 38.

1.1 Related Work

Line-Separable Discrete Unit Disk Cover. A solution to the LSDUDC problem was independently discovered and published by [3, Lemma 1], where they propose a dynamic programming algorithm with a time bound of $O(m^2n)$ but whose correctness is not straightforward nor is it formally argued. This paper presents an alternative algorithm together with a proof of correctness. Both algorithms follow natural approaches, yet a full proof of correctness is not immediate.

ε-nets for Geometric Hitting Problems. Using ϵ-nets, Mustafa and Ray [15,14] have recently presented a $(1+\epsilon)$-approximation to the DUDC problem. Their algorithm runs in $O(m^{2(c/\varepsilon)^2+1}n)$ time, where $c \leq 4\gamma$ [14]. Their γ value can be bounded from above by $2\sqrt{2}$ [8,12]. The fastest operation of this algorithm is obtained by setting $\varepsilon = 1$ for a 2-approximation, and this will run in $O(m^{2\cdot(8\sqrt{2})^2+1}n) = O(m^{257}n)$ time in the worst case. Clearly, this algorithm will not be practical for large values of m.

Minimum Geometric Disk Cover. In the minimum geometric disk cover problem, the input consists of a set of points in the plane, and the problem is to find a set of unit disks of minimum cardinality whose union covers the points. Unlike our problem, disk centres are not constrained to be selected from a given discrete set, but rather may be centred at arbitrary points in the plane. Again, this problem is NP-hard [7,17] and has a PTAS solution [9]. Of course the problem can be generalized further: see [5] for a discussion of geometric set cover problems.

Discrete k-Centre. Also related is the discrete Euclidean k-centre problem: given a set P of m points in the plane, a set Q of n points in the plane, and an integer k, find a set of k disks centred on points in P whose union covers Q such that the radius of the largest disk is minimized. Observe that set Q has a discrete unit disk cover consisting of k disks centred on points in P if and only if Q has a discrete k-centre centred on points in P with radius at most one. This problem is NP-hard if k is an input variable [2]. When k is fixed, Hwang et al. [10] give a $m^{O(\sqrt{k})}$-time algorithm, and Agarwal and Procopiuc [1] give an $m^{O(k^{1-1/d})}$-time algorithm for points in $\mathbb{R}^d$.

2 Overview of the Algorithm

In this section we describe a polynomial-time algorithm for the line-separable discrete unit disk cover (LSDUDC) problem and prove its correctness. Details of the algorithm and its running time will be discussed in Section 3. Recall that we have two sets $P = \{p_1, p_2, \ldots, p_m\}$ and $Q = \{q_1, q_2, \ldots, q_n\}$ of points in the plane that are separated by a line l. We want to find a subset $P' \subseteq P$ of minimum cardinality such that all points of Q are covered by unit disks centred at the points of P'. An instance of the problem is shown in Figure 1. Without loss of generality we assume that l is a horizontal line and points of P are above l. Let d_i denote the unit disk that is centred at p_i, for $i \in \{1, 2, \ldots, m\}$, and let D denote the set of these disks. We use p_i and d_i interchangeably, e.g., our solution can be considered both as a set of points (a subset of P) and as a set of disks. Further, when we discuss the intersection of a line with a disk, we are referring to the intersection of the line with the boundary of the disk.

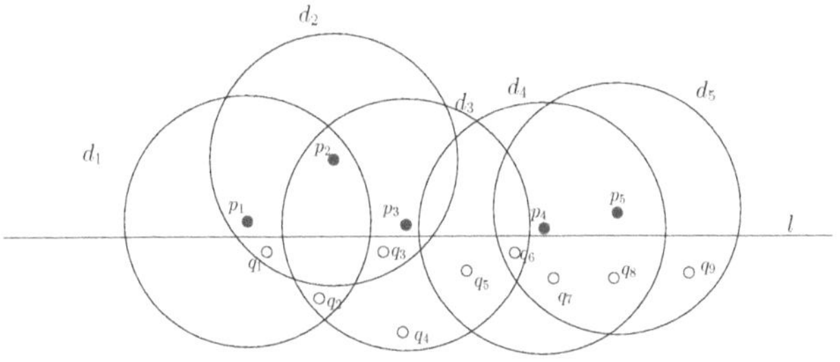

Fig. 1. An instance of the line-separable discrete unit disk cover problem

During the execution of our algorithm, it may be determined that a disk $d \in D$ should be added to the solution or that it is not relevant for the remainder of the computation of the solution set. When this occurs, we remove disk d from the problem. Similarly, we remove a point $q \in Q$ if this point is not relevant for the remainder of the computation (i.e., point q is covered by a disk in the partial solution being constructed). Our algorithm relies on the following three observations:

1. If a disk d_1 covers no points from Q, we remove it.
2. If a disk d_1 is *dominated* by a disk d_2, then we can remove d_1 from the problem instance. Disk d_2 dominates d_1 if it covers all points of Q covered by d_1. If two disks cover the same subset of points from Q, we designate the dominating disk as that whose left intersection with l is rightmost.

3. If a point $q_1 \in Q$ is only covered by a disk d_1, then d_1 must be part of the solution. We also remove d_1 together with all points of Q covered by d_1.

These three observations give us three SIMPLIFICATION rules. The idea is to apply these rules to as many disks as possible and simplify the problem. For example, consider the problem instance shown in Figure 1. Initially no disk dominates another, thus we cannot apply the second rule. Disk d_3 is the only disk that covers q_4 and, similarly, disk d_5 is the only disk that covers q_9. Thus we add d_3 and d_5 to the (initially empty) solution and remove them together with the points that are covered by them, namely $\{q_2, q_3, q_4, q_5, q_6, q_7, q_8, q_9\}$. Now disk d_4 covers no point and can be removed. There is only one remaining point (q_1) and it is covered by the two remaining disks (d_1 and d_2). According to our convention, d_1 is dominated by d_2 and is removed. Now d_2 is the only disk covering q_1. We add d_2 to the solution and remove d_2 and q_1. No disks or points remain and we are done. Thus the SIMPLIFICATION rules suffice for this instance and give an optimal solution $\{d_2, d_3, d_5\}$. This example also illustrates that an optimal solution is not necessarily unique, as $\{d_1, d_3, d_5\}$ is also an optimal solution. In general, however, these SIMPLIFICATION rules do not suffice to obtain an optimal solution.Referring to Fig. 1, if given only disks d_1, d_2 and d_3 and points q_1, q_2 and q_3, then no point $q \in Q$ is covered by only one disk and no disk dominates any other one.

We augment the SIMPLIFICATION rules with a simple greedy step to solve the problem. We rename the disks so that the left intersection of d_i with l is to the left of the left intersection of d_{i+1} with l. We say that d_i precedes d_{i+1} in the ordering (the disks in Figure 1 follow this ordering). This combined algorithm, GREEDY, works by first applying the SIMPLIFICATION rules as many times as possible. Next we find the first remaining disk in the left-to-right order, say d_j. We add d_j to our solution and remove d_j from D and all points covered by d_j from Q. We apply the SIMPLIFICATION rules followed by the greedy step repeatedly until all disks have been removed. Since we remove at least one disk at each greedy step, the algorithm terminates after at most m iterations. See Algorithm 1 for the corresponding pseudocode.

Algorithm 1. GREEDY (D, Q)

```
D ←sortLeftToRight(D) //sort in increasing order of left intersection with l
S ← ∅
while D ≠ ∅ do do
  SIMPLIFICATION (D, Q, S) //SIMPLIFICATION possibly modifies D, Q and S
  d_ℓ ← leftmost disk in D
  S ← S ∪ {d_ℓ}
  D ← D \ d_ℓ
  Q' ← {q ∈ Q | q is contained in d_ℓ}
  Q ← Q \ Q'
end while
return S
```

2.1 Correctness of Greedy

We now prove the correctness of the algorithm by proving that Greedy gives a minimum LSDUDC solution. Assume for the sake of contradiction that there is an algorithm Opt that gives a cover with fewer disks than Greedy. Let d_1 be the first disk in the ordering that is selected by Greedy but not by Opt. Let C be the set of points in Q that are covered by d_1 (we consider only the remaining points and disks, i.e., those that have not been removed by the algorithm). First assume that C is covered by a single disk d_0 in the solution of Opt. Since d_1 is not removed in the Simplification step, it is not dominated by any other disk. Thus the only possibility is that d_0 and d_1 cover exactly the same set of (remaining) points (i.e., set C) and d_0 precedes d_1 in the ordering. In this case, we replace d_0 with d_1 in Opt, pushing the first difference between the solution of Greedy and Opt to the right. Otherwise, C is covered by at least two disks in the solution of Opt. Let d_2 and d_3 be two disks in the solution of Opt such that each of them cover a strict subset of C. Without loss of generality assume that d_2 precedes d_3 in the ordering. We prove that $d_1 \cup d_3$ covers all points of Q covered by $d_2 \cup d_3$.

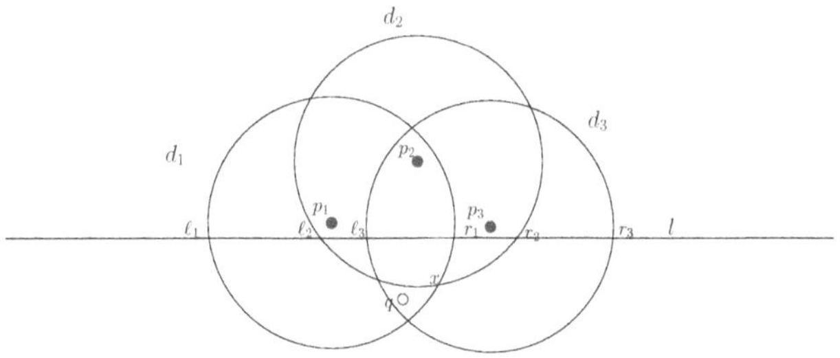

Fig. 2. Proof of correctness of Greedy. If d_1 is the first disk selected by Greedy and not by Opt, then Opt must have d_2 and d_3 in its solution.

Let ℓ_i and r_i denote the respective left and right intersection points of the boundary of the unit disk d_i with the line l, for $i \in \{1, 2, 3\}$. If d_2 precedes d_1 in the ordering, d_1 dominates d_2 (otherwise, Greedy would select d_2 and not d_1 at this step). In this case we replace d_2 with d_1 in Opt, pushing the difference between the two algorithms to the right. Hence we are left with the case in which d_1 precedes d_2 and d_2 precedes d_3 in the ordering. Thus the points are ordered $\ell_1, \ell_2, \ell_3, r_1, r_2, r_3$ along line l (see Figure 2). Note that we cannot have a pair of disks nested below l, otherwise the nested disk is dominated by the other. Furthermore, we know that $(d_1 \cap d_3) \setminus d_2 \neq \emptyset$. Let $R = d_2 \setminus d_1$. It suffices to prove that R is completely contained in d_3.

Proposition 1. *Region R is contained in disk d_3.*

Proof. Since points r_1 and r_2 both lie between ℓ_3 and r_3 on line l, both points r_1 and r_2 are in disk d_3. Let x denote the rightmost point of the intersection of the boundaries of disks d_1 and d_2. Observe that x lies on the boundary of region $(d_1 \cap d_3) \setminus d_2$. Consequently, $x \in d_3$. Since the boundary of R consists of arcs of unit disks joining the points x, r_1, and r_2, it follows that R is contained in the 1-hull of $\{x, r_1, r_2\}$, where the 1-hull of $\{x, r_1, r_2\}$ is the intersection of all unit disks that contain $\{x, r_1, r_2\}$, denoted 1-$H(\{x, r_1, r_2\})$. Since $\{x, r_1, r_2\} \subseteq d_3$, it follows that $R \subseteq$ 1-$H(\{x, r_1, r_2\}) \subseteq d_3$. □

Thus by removing d_2 from the solution of OPT and adding d_1 to it we will have a feasible solution with the same number of disks. This pushes the first difference between the solution of GREEDY and OPT to the right. By continuing this argument we can prove that the solution returned by GREEDY uses the same number of disks as OPT and therefore GREEDY is an optimal algorithm.

3 Implementation Details and Analysis

We construct a graph $G = (V, E)$, where each node $v_i \in V$ corresponds to disk d_i for $i \in \{1, \ldots, m\}$ (recall that d_i is the ith disk sorted according to its left intersection with l). We also associate a counter c_{v_i} to each node v_i that stores the number of points in Q contained in disk d_i that have not yet been covered by the algorithm. Similarly, we associate with each edge $e = (v_{i_1}, v_{i_2})$ a counter c_e that represents the number of points contained in $d_{i_1} \cap d_{i_2}$. This graph can be constructed in $O(m^2 n)$ time by checking which points are contained in the intersection of each pair of disks, adding the corresponding edges, and updating the node and edge counters. The algorithm GREEDY-GRAPH starts by traversing the nodes in order $v_1, v_2, \ldots, v_m$. At each node v_i, there are three possible cases: (1) The counter c_{v_i} is 0; in this case d_i does not contain any points or is dominated by a set of disks that has already been added to the solution. This disk will not be in the solution set, so we can ignore this node and continue with the next one. This is analogous to the first SIMPLIFICATION rule. (2) There is an edge $e = (v_i, v_k)$, $k > i$, such that $c_e = c_{v_i}$; in this case we know that d_i is dominated by disk d_k. Again, we ignore this node and continue. This corresponds to an application of the second SIMPLIFICATION rule. (3) Every edge $e = (v_i, v_k)$, $k > i$, satisfies $c_e < c_{v_i}$; that means that disk d_i is not dominated by any disk to its right. In this case we add d_i to the solution set and we eliminate all remaining points contained by this disk from the graph. We continue with the next node in the graph. Note that this is an application of the third rule of SIMPLIFICATION and the greedy step.

In order to identify the appropriate case above we traverse the adjacency list of each node we visit. This requires $O(m)$ time in the worst case. When a disk is added to the solution in the third case, all points contained in the disk must be eliminated. Consider the elimination of a point p in disk d_i. Let $N(v_i) = \{v_k \mid c_{(v_i, v_k)} > 0\}$. For all $v_k \in N(v_i)$, we decrease c_{v_k} and $c_{(v_i, v_k)}$ by one. In addition, for each pair of elements $\{v_{k_1}, v_{k_2}\} \subseteq N(v_i)$, we check whether the point is contained by both disks, and if this is case we decrease $c_{(v_{k_1}, v_{k_2})}$ by one. This can take at most $O(m^2)$ time per point, thus the time required for eliminating all points is bounded by $O(m^2n)$ time. Since the time required to construct the graph is $O(m^2n)$, the overall process takes $O(m^2n)$ time.

3.1 Correctness of Greedy-Graph

We now demonstrate that the Greedy-Graph algorithm is optimal by showing that the set of disks returned by this algorithm has the same cardinality as that returned by the Greedy algorithm presented in Section 2.

Lemma 1. *If S is the disk cover returned by* Greedy-Graph, *and S' is the disk cover returned by* Greedy, *then $|S| = |S'|$.*

Proof. Assume for the sake of contradiction that $|S| \neq |S'|$. Recall that Greedy is optimal, therefore $|S'|$ and $|S|$ can only differ if $|S| > |S'|$. Let d_1 be the first disk in the left-to-right order that is present in the solution of Greedy-Graph, and not in the solution of Greedy. At some point during its execution, Greedy must have decided to discard disk d_i. The only mechanisms in Greedy for discarding disks are the first and second Simplification rules. Recall that the first rule removes a disk if it contains no points, and the second rule discards a disk if it is dominated by some other disk. We now show that for any of the following possible events, Greedy-Graph will discard the same disk d_1.

- **Empty** - Suppose d_1 contains no points. In this case, Greedy-Graph will find that $c_{v_1} = 0$. Therefore, d_1 will be discarded by Case 1, in contradiction to our assumption.
- **Dominance (right)** - Now suppose d_1 is dominated by some disk to the right, d_r. In this case, we will encounter d_1 first during our walk, and we will have that $c_{v_1} = c_{(v_1,v_r)}$. Therefore, Greedy-Graph will remove d_1 by Rule 2, in contradiction to our assumption.
- **Dominance (left)** - Suppose d_1 is dominated by some disk to the left, d_ℓ. In this case, we will have encountered d_ℓ first during our walk. There are two possible cases in this scenario:
  (i) If $c_{v_\ell} > c_{(v_\ell,v_k)}$ for all d_k, d_ℓ is added to S by Rule 3 of Greedy-Graph. All points covered by d_ℓ are removed, leaving no points covered by d_1. This is now an instance of the Empty case.
  (ii) Otherwise, $c_{v_\ell} = c_{(v_\ell,v_k)}$ for some d_k. This means that d_ℓ is dominated by d_k. Greedy-Graph would discard d_ℓ by Rule 3. By transitivity, d_k also dominates d_1. If d_k is to the right of d_1, then this is now an instance of Dominance (right), and thus we reach a contradiction. If d_k is to the left of d_1, then this is again an instance of Dominance (left), so we apply this same argument recursively. The recursion stops either when we reach an instance of Dominance (right) or case (i) of Dominance (left). ⊔

We have shown that the solution of Greedy-Graph has the same cardinality as the solution of Greedy, and since Greedy is optimal, so is Greedy-Graph.

Theorem 1. *Given sets P of m points and Q of n points in the plane, where P and Q can be separated by a line l, LSDUDC can be solved in $O(m^2n)$ time.*

4 Approximate Discrete Unit Disk Cover

We now show that our algorithm for the line-separable discrete unit disk cover (LSDUDC) problem leads to a 22-approximation algorithm for the discrete unit disk cover (DUDC) problem. The approximation algorithm is based on a suitable adaptation of the 38-approximation algorithm of Carmi et al. [4].

For simplicity, we use the notation and assumptions of [4]. In that work, the DUDC problem is supported by a variant of the LSDUDC problem: suppose are given a set of disks $D = L \cup U$. The disks in U are centred above a line l while the disks in L are centred below l. We are also given a set of points Q covered by U. The goal is to obtain the subset G of D of smallest cardinality such that every point in Q is covered by a disk in G.

Note that our line-separable algorithm does not immediately result in a straightforward improvement to the approximation factor of the algorithm of Carmi et al. Their proof of correctness uses the fact that their 2-approximation to the LSDUDC problem consists of disks forming the lower boundary of U, where the lower boundary is the union of all disk boundary arc segments below l not contained in other disks. This is not necessarily the case in our solution.

Instead, we first solve the LSDUDC problem optimally using our algorithm on the set of disks U to obtain a disk set H and then use the greedy *minimum assisted cover* algorithm (see Carmi et al. [4, §2] for the formal definition) over the sets H and L to obtain an improved solution E. Now we wish to compare the cardinality of E with that of the global minimum disk cover G.

Consider the upper and lower components of the solutions E and G, i.e., $E_U = E \cap U$, $E_L = E \cap L$, $G_U = G \cap U$, and $G_L = G \cap L$. Note that $|G| \leq |E|$ since G is the global minimum. Similarly, since E is the minimum assisted cover based on H, it follows that $|E| = |E_U| + |E_L| \leq |H/G_L| + |G_L|$, where H/G_L is the smallest subset of H that forms an assisted cover with G_L.

Now we will show that $2|G_U| \geq |H/G_L|$. Given a disk d in G_U, there are two cases: either d lies above the lower boundary of H/G_L, i.e., d is contained in the union of all the disks in H/G_L, or d contains one or more arc segments of the lower boundary of H/G_L. In the first case, Carmi et al. show that at most two disks in H/G_L suffice to cover d and, hence, for every such disk in the global optimum solution G there are most two disks in H/G_L. In the second case, let V denote the subset of disks that have lower boundary segments that are contained in d. The set of arc segments of the disks in V consists, from left to right, of a partially-covered arc segment of the lower boundary, zero or more fully-covered arc segments, and a partially-covered arc segment. Let W denote the disks whose arcs are partially covered together with d. W dominates V and hence there is at most one arc of the lower boundary fully contained in d; otherwise replacing V with W results in a cover of smaller cardinality, deriving a contradiction, since $V \subset H$, and H is the optimal LSDUDC solution[1]. Furthermore, observe that the partially-covered arc disks must contain points not contained in the fully-covered disk; otherwise they can also be eliminated while reducing the

[1] Recall that all disks in V and U are centred above l, and all points in Q are below l.

cardinality of the cover. As those disks contain other points, each of the disks is partially covered by at least one other disk in G. We arbitrarily associate each disk covered more than once to its leftmost disk in G. Thus, of the (at most) three disks in V, at most two are associated to d. In sum, in either case each disk in G_U has at most two associated disks in H/G_L from which it follows that $2|G_U| \geq |H/G_L|$. Hence, $2|G| = 2(|G_U| + |G_L|) \geq 2|G_U| + |G_L| \geq |H/G_L| + |G_L| \geq |E_U| + |E_L| = |E|$, which gives the approximation factor of two as desired. Carmi et al. [4] prove that any disk can be used in up to eight applications of the assisted LSDUDC algorithm, for which they have a 4-approximation. These operations, followed by a 6-approximation for any remaining disks results in an $8 \times 4 + 6 = 38$-approximation for the general DUDC problem. As we have shown that our technique provides a 2-approximation for the assisted LSDUDC problem, we now have an approximation ratio of $8 \times 2 + 6 = 22$ for DUDC.

4.1 Algorithm Analysis

There are essentially two main components to the algorithm for solving DUDC by Carmi et al. [4]. First, they apply a grid of size $3/2 \times 3/2$ to the input data. Our LSDUDC algorithm supplemented by their assisting disk technique is run on all grid lines. Note that the number of relevant grid lines is $O(n)$. Our technique runs in $O(m^2 n)$, and the assisting disk operation is easily implementable in $O(mn)$, so the running time of the first component is dominated by our step.

The second major component to their technique is finding the 6-approximation for the DUDC of all disk centres and points contained in each of the $3/2 \times 3/2$ squares of the grid. Their technique is based on the application of a subset of nine properties depending on where the disk centres are located. First, they determine whether a solution exists using one or two centres by brute force, which is easily done in $O(m^2 n)$ time. The determination of which properties may be applied can be done in $O(m)$ time, and there are only two expensive steps that may be used in any of the procedures, each of which may only be used a constant number of times. First is the assisted LSDUDC technique, whose running time is $O(m^2 n)$, as we just discussed. The second technique that may be required is to determine the optimal disk cover of a set of points using centres contained in one of the $1/2 \times 1/2$ squares, which can be solved in $O(m^2 n^4)$ time using the technique presented in [13]. The centre of each disk can only be contained in one square, and so this operation is never performed twice for any given disk. Therefore, the complete DUDC algorithm achieves worst-case performance when all of the disk centres in the plane are confined to a single $1/2 \times 1/2$ square, so that the $O(m^2 n^4)$ operation is performed over the entire data set.

5 Conclusions

This paper presents a polynomial-time algorithm that returns an exact solution to the LSDUDC problem, as well as a proof of correctness of the approach. This algorithm for the line-separable problem allows us to improve the approximation algorithm of Carmi et al. [4], resulting in a 22-approximate solution to the general DUDC problem, which runs in $O(m^2n^4)$ time in the worst case.

Theorem 2. *Given sets P of m points and Q of n points in the plane, we can compute a 22-approximation of the DUDC problem in $O(m^2n^4)$ time in the worst case.*

Acknowledgements. The authors wish to thank Paz Carmi for sharing his insights and discussing details of his results on the discrete unit disk cover problem [4]. In addition, the authors acknowledge Sariel Har-Peled with whom a preliminary problem was discussed that inspired our examination of the disk cover problem.

References

1. Agarwal, P., Procopiuc, C.: Exact and approximation algorithms for clustering. Alg. 33, 201–226 (2002)
2. Agarwal, P., Sharir, M.: Efficient algorithms for geometric optimization. ACM Comp. Surv. 30, 412–458 (1998)
3. Ambühl, C., Erlebach, T., Mihal'ák, M., Nunkesser, M.: Constant-factor approximation for minimum-weight (connected) dominating sets in unit disk graphs. In: Proc. of Approx., Rand., and Comb. Opt. Alg. and Tech., pp. 3–14 (2006)
4. Carmi, P., Katz, M., Lev-Tov, N.: Covering points by unit disks of fixed location. In: Proc. Int'l Symp. on Alg. and Comp., pp. 644–655 (2007)
5. Clarkson, K., Varadarajan, K.: Improved approximation algorithms for geometric set cover. Disc. and Comp. Geom. 37(1), 43–58 (2007)
6. Călinescu, G., Măndoiu, I., Wan, P.J., Zelikovsky, A.: Selecting forwarding neighbours in wireless ad hoc networks. Mob. Net. and Appl. 9(2), 101–111 (2004)
7. Fowler, R., Paterson, M., Tanimoto, S.: Optimal packing and covering in the plane are NP-complete. Inf. Proc. Lett. 12(3), 133–137 (1981)
8. Frederickson, G.: Fast algorithms for shortest paths in planar graphs, with applications. SIAM J. on Comp. 16(6), 1004–1022 (1987)
9. Hochbaum, D., Maass, W.: Approximation schemes for covering and packing problems in image processing and VLSI. J. ACM 32, 130–136 (1985)
10. Hwang, R., Lee, R., Chang, R.: The generalized searching over separators strategy to solve some NP-hard problems in subexponential time. Alg. 9, 398–423 (1993)
11. Johnson, D.: The NP-completeness column: An ongoing guide. J. of Alg. 3(2), 182–195 (1982)
12. Koutis, I., Miller, G.: A linear work, $o(n^{1/6})$ time, parallel algorithm for solving planar Laplacians. In: Proc. Symp. Disc. Alg., pp. 1002–1011 (2007)
13. Lev-Tov, N.: Algorithms for Geometric Optimization Problems in Wireless Networks. Ph.D. thesis, Weizmann Institute of Science (2005)
14. Mustafa, N., Ray, S.: Improved results on geometric hitting set problems (2009), `http://www.mpi-inf.mpg.de/~saurabh/Papers/Hitting-Sets.pdf`
15. Mustafa, N., Ray, S.: PTAS for geometric hitting set problems via local search. In: Proc. Symp. on Comp. Geom., pp. 17–22 (2009)
16. Narayanappa, S., Voytechovsky, P.: An improved approximation factor for the unit disk covering problem. In: Proc. Can. Conf. Comp. Geom. (2006)
17. Supowit, K.: Topics in Computational Geometry. Ph.D. thesis, University of Illinois at Urbana-Champaign (1981)

Divide-and-Conquer Algorithms for Partitioning Hypergraphs and Submodular Systems*

Kazumasa Okumoto[1], Takuro Fukunaga[2], and Hiroshi Nagamochi[2]

[1] Graduate School of Economics, University of Tokyo, Japan
ee096048@mail.ecc.u-tokyo.ac.jp

[2] Department of Applied Mathematics and Physics, Graduate School of Informatics, Kyoto University, Japan
{takuro,nag}@amp.i.kyoto-u.ac.jp

Abstract. The submodular system k-partition problem is a problem of partitioning a given finite set V into k non-empty subsets $V_1, V_2, \ldots, V_k$ so that $\sum_{i=1}^{k} f(V_i)$ is minimized where f is a non-negative submodular function on V, and k is a fixed integer. This problem contains the hypergraph k-cut problem. In this paper, we design the first exact algorithm for $k = 3$ and approximation algorithms for $k \geq 4$. We also analyze the approximation factor for the hypergraph k-cut problem.

1 Introduction

A set function $f : 2^V \to \Re$ on a finite set V is called *submodular* if it satisfies $f(X) + f(Y) \geq f(X \cap Y) + f(X \cup Y)$ for every pair of sets $X, Y \subseteq V$ [1,2]. Moreover, f is called *symmetric* if $f(X) = f(V \setminus X)$ for every $X \subseteq V$, and *non-negative* if $f(X) \geq 0$ for every $X \subseteq V$. A *submodular system* is defined as a pair (V, f) of a finite set V and a submodular function f on V.

Let k be an integer at least 2. A k-partition P_k of (V, f) is defined as a partition of V into k non-empty sets $V_1, V_2, \ldots, V_k \subseteq V$, i.e., $V_i \neq \emptyset$ for $i \in \{1, 2, \ldots, k\}$, $V_i \cap V_j = \emptyset$ for $1 \leq i < j \leq k$, and $\cup_{i=1}^{k} V_i = V$. We denote the partition by $[V_1, V_2, \ldots, V_k]$. The cost of P_k, denoted by $f(P_k)$ or $f(V_1, V_2, \ldots, V_k)$, is defined as $\sum_{i=1}^{k} f(V_i)$. In this paper, we consider the *submodular system k-partition problem*, that asks to find a minimum cost k-partition of given submodular system (V, f). Throughout this paper, it is supposed that f is non-negative and given as an oracle which returns $f(X)$ for $X \subseteq V$.

Submodularity often plays an essential role in studies on the connectivity of graphs. In fact, the submodular system k-partition problem generalizes the *graph k-cut problem* and *hypergraph k-cut problem*. For a graph or hypergraph $G = (V, E)$ with weight $w : E \to \Re$, a k-cut is defined as a set of edges whose removal divides G into at least k connected components. The k-cut problem asks to find a minimum weight k-cut of the given graph or hypergraph.

* This work was partially supported by Grant-in-Aid for Scientific Research from the Ministry of Education, Culture, Sports, Science and Technology of Japan.

Y. Dong, D.-Z. Du, and O. Ibarra (Eds.): ISAAC 2009, LNCS 5878, pp. 55–64, 2009.

For a graph G and $X \subseteq V$, define $f(X)$ as $\sum_{e \in \delta(X)} w(e)$ where $\delta(X)$ denotes the set of edges joining vertices in X and $V \setminus X$. It is known that this set function is symmetric and submodular. For a k-partition $[V_1, V_2, \ldots, V_k]$ of V, $\sum_{i=1}^{k} f(V_i)$ is exactly twice the weight of a k-cut disconnecting $V_1, V_2, \ldots, V_k$ each other. Hence the k-cut problem of undirected graphs can be formulated as the k-partition problem with symmetric submodular systems.

For hypergraphs, the cut function defined above does not formulate the k-cut problem as the submodular k-partition problem. However, by defining submodular functions appropriately, it is possible to formulate the hypergraph k-cut problem as the submodular system k-partition problem (see Section 2). Note that the submodular functions used for this is not symmetric.

The graph k-cut problem is one of the fundamental problems in combinatorial optimization. It is closely related to the reliability of networks, and has many applications, for example, to the traveling salesperson problem, VLSI design, and evolutionary tree construction [5,12]. Goldschmidt and Hochbaum [7] showed that the problem is NP-hard when k is not fixed. For fixed k, they presented a polynomial-time algorithm. Its running time is $O(n^{k^2} T(n, m))$ where $T(n, m)$ is time for computing max-flow in a graph consisting of n vertices and m edges. Note that $T(n, m)$ is known to be $O(mn \log(n^2/m))$ [6]. After their work, many polynomial-time algorithms for fixed k are obtained. An algorithm due to Kamidoi, Yoshida and Nagamochi [9] runs in $O(n^{4k/(1-1.71/\sqrt{k})-34} T(n, m))$. An algorithm due to Xiao [17] runs in $O(n^{4k - \log k})$. An algorithm due to Thorup [14] runs in $\tilde{O}(n^{2k})$. In addition, Karger and Stein [10] gave a random algorithm running in $O(n^{2(k-1)} \log^3 n)$.

For the hypergraph k-cut problem, Xiao [16] designed a polynomial-time algorithm to the case of $k = 3$. However, no polynomial-time algorithm is known when k is a fixed integer larger than 3. With regards to approximation algorithms, Zhao, Nagamochi and Ibaraki [18] gave an algorithm achieving the approximation factor $(1 - \frac{2}{k}) \min\{k, d_{\max}\}$ with the result due to Xiao [16], where $d_{\max}$ denotes the maximum size of hyperedges. Moreover, with the reduction to the terminal k-vertex cut problem in bipartite graphs (see Section 2), we can apply the LP-rounding algorithm due to Garg, Vazirani and Yannakakis [4]. It achieves the approximation factor $(2 - \frac{2}{k})$.

Little is known about the submodular k-partition problem. Queyranne [13] gave a $(2 - \frac{2}{k})$-approximation algorithm for the problem with symmetric submodular functions. Zhao, Nagamochi and Ibaraki [18] presented a $(k - 1)$-approximation algorithm for the problem with non-negative submodular functions.

Besides the reduction from the hypergraph k-cut problem to the submodular system k-partition problem and the terminal k-vertex cut problem, our contribution is to design algorithms for the submodular system k-partition problem. For $k = 3$, we present a polynomial-time algorithm for $k = 3$. For $k \geq 4$, we give approximation algorithms. We also discuss the approximation factor of the algorithms for the hypergraph k-cut problem. Table 1 summarizes these approximation factors. Our algorithms perform well especially for small k. For hypergraph

Table 1. Comparison of approximation factors by the previous best and our algorithms

	hypergraph k-cut problem		submodular system k-partition problem	
$k = 3$	1	(Xiao [16])	2	(Zhao et al. [18])
			1	(This paper)
$k = 4$	1.5	(Garg et al. [4])	3	(Zhao et al. [18])
	4/3	(This paper)	1.5	(This paper)
$k = 5$	1.6	(Garg et al. [4])	4	(Zhao et al. [18])
	5/3	(This paper)	2	(This paper)
$k \geq 6$	$2 - 2/k$	(Garg et al. [4])	$k - 1$	(Zhao et al. [18])
	$k/2 - 1$	(This paper)	$k + 1 - 2\sqrt{k-1}$	(This paper)

k-cut problem with $k \geq 5$, our algorithms present worse approximation factor than the algorithm due to Garg, Vazirani and Yannakakis [4] for the terminal k-vertex cut problem. However, our algorithms have an advantage since they do not need to solve the linear programming problems.

The key of our algorithms is *uncrossing operation*, which has been applied to many problems related to submodular functions. In this paper, we prove a theorem on uncrossing k-partitions and 2-partitions satisfying some conditions (Theorem 4). This theorem is originally proven by Xiao [16] for hypergraphs. We reveal in this paper that his result essentially relies only on submodularity of the cut function in hypergraphs.

The rest of this paper is organized as follows. Section 2 introduces notations. It also describes the reduction of the hypergraph k-cut problem to the terminal k-vertex cut problem in bipartite graphs, and shows that the submodular system k-partition problem contains the hypergraph k-cut problem. Section 3 proves several properties of k-partitions. Section 4 gives an approximation algorithm for the submodular system 4-partition problem. It also analyzes the approximation factor of the algorithm for the hypergraph 4-cut problem. Several proofs and the detail on our result for $k \geq 5$ are not included in this article due to the space limitation. They will appear in the full version of this paper.

2 Preliminaries

2.1 Notations

In this paper, $\Re$ and $\Re_+$ stand for the sets of reals and non-negative reals, respectively. For a submodular system (V, f), we denote $|V|$ by n. Each $v \in V$ is called a *vertex*. The complement of a subset X of V is denoted by $\overline{X}$ (i.e., $\overline{X} = V \setminus X$).

Let $[V_1, V_2, \ldots, V_k]$ be a k-partition of a submodular system (V, f). Each V_i, $i \in \{1, 2, \ldots, k\}$ is called a *component* of the partition. A k-partition is called *h-size* if its all components contain at least h vertices. We note that a minimum

h-size k-partition stands for a partition of minimum cost among all the h-size k-partitions, and an h-size minimum k-partition stands for an h-size partition of minimum cost among all the k-partitions.

For $U \subset V$, define a set function f^U on U such that $f^U(X) = f(X)$ for all $X \subseteq U$. Notice that if f is submodular, then f^U is submodular. We call a submodular system (U, f^U) *subsystem of* (V, f) *induced by* U.

For $U \subset V$, define V_U as a set obtained from V by replacing U with a new single vertex u (i.e., $V_U = (V \setminus U) \cup \{u\}$). Moreover, define f_U as a set function on V_U such that for every $X \subseteq V_U$, $f_U(X) = f(X)$ if $u \notin X$ and $f_U(X) = f((X \setminus \{u\}) \cup U)$ otherwise. It is easy to check that if f is submodular, then f_U is submodular. We call this operation *shrinking* U *into* u.

We say that a 2-partition $[X, \overline{X}]$ *crosses* a k-partition $[Y_1, Y_2, \ldots, Y_k]$ if both $X \setminus Y_i$ and $\overline{X} \setminus Y_i$ are non-empty for all $i \in \{1, 2, \ldots, k\}$. This means that, if $[X, \overline{X}]$ does not cross $[Y_1, Y_2, \ldots, Y_k]$, then there exists some $i \in \{1, 2, \ldots, k\}$ such that $X \subseteq Y_i$ or $\overline{X} \subseteq Y_i$. If $k = 2$, then $[X, \overline{X}]$ crosses $[Y_1, Y_2]$ whenever none of $X \cap Y_1$, $X \cap Y_2$, $\overline{X} \cap Y_1$ and $\overline{X} \cap Y_2$ is non-empty.

Suppose that a k-partition $[V_1, V_2, \ldots, V_k]$ and a 2-partition $[X, \overline{X}]$ of (V, f) satisfies $X \subseteq V_i$ with some $i \in \{1, 2, \ldots, k\}$. Then (V_X, f_X) has a k-partition $[V_1', V_2', \ldots, V_k']$ that satisfies $f_X(V_1', V_2', \ldots, V_k') = f(V_1, V_2, \ldots, V_k)$; I.e, each k-partition of (V, f) whose some component contains X corresponds to a k-partition of (V_X, f_X). The contrary also holds. Throughout this paper, we do not distinguish such a k-partition of (V, f) with the corresponding one of (V_X, f_X).

For a finite set V, a hyperedge is defined as a subset of V. A hypergraph is defined as a pair (V, E) of V and a set E of hyperedges on V. Let $U_1, U_2, \ldots, U_\ell$ be disjoint subsets of V. Then $\delta(U_1, U_2, \ldots, U_\ell)$ denotes the set of edges in E that intersect at least two of $U_1, U_2, \ldots, U_\ell$. Moreover, $\delta_{\rm in}(U_1, U_2, \ldots, U_\ell)$ denotes the set of edges in $\delta(U_1, U_2, \ldots, U_\ell)$ that do not intersect $V \setminus (U_1 \cup U_2 \cup \cdots \cup U_\ell)$. When a weight $w : E \to \Re_+$ is given, we define $w(U_1, U_2, \ldots, U_i) = \sum_{e \in \delta(U_1, U_2, \ldots, U_\ell)} w(e)$, and $w_{\rm in}(U_1, U_2, \ldots, U_\ell) = \sum_{e \in \delta_{\rm in}(U_1, U_2, \ldots, U_\ell)} w(e)$.

The cost of a k-cut for a hypergraph G can be represented by $w(V_1, V_2, \ldots, V_k)$ where $V_1, V_2, \ldots, V_k$ represent vertex sets of connected components after removing the k-cut. Hence the hypergraph k-cut problem can be regarded as a problem of computing a k-partition $[V_1, V_2, \ldots, V_k]$ of V minimizing $w(V_1, V_2, \ldots, V_k)$. Hence we denote a k-cut by a k-partition of V in this paper.

Given two vertices s and t, an (s, t)*-partition* is a 2-partition such that s and t are in different components. In this paper, time complexities of algorithms are evaluated by the times of computing minimum (s, t)-partitions. A minimum (s, t)-partition of a submodular system (V, f) can be computed by minimizing a (asymmetric) submodular function. See [8] for recent algorithmic development of the submodular function minimization. In hypergraphs (V, E), minimum (s, t)-partitions can be computed by finding a maximum flow in digraphs with $|V| + 2|E|$ vertices and $|E| + 2\sum_{e \in E} |e|$ edges [11].

2.2 Reduction of the Hypergraph k-Cut Problem

In the *terminal k-vertex cut problem*, an undirected graph (V, E) with vertex weight $w : V \to \Re$ and k terminals $t_1, t_2, \ldots, t_k \in V$ are given. A *vertex cut* is defined as a subset U of V whose removal disconnects given terminals each other. The objective of the problem is to find a vertex cut U minimizing $\sum_{v \in U} w(v)$. First, let us observe that the hypergraph k-cut problem can be reduced to the terminal k-vertex cut problem with bipartite graphs.

Let $G = (V, E)$ be a hypergraph with weight $w : E \to \Re_+$. Let V_E be the set of vertices corresponding to E while $v_e \in V_E$ denotes the vertex corresponding to $e \in E$. Define E' as a set of edges on $V \cup V_E$ such that E' contains an edge joining $u \in V$ and $v_e \in V_E$ if and only if the hyperedge e contains u in G. Then the bipartite graph $B = (V \cup V_E, E')$ can be defined from G. Define a vertex-weight $w' : V \cup V_E$ such that $w'(v) = +\infty$ for $v \in V$ and $w'(v_e) = w(e)$ for $v_e \in V_E$.

Choosing k terminals from V, solve the terminal k-vertex cut problem with B and w'. Obtained minimum vertex cut U contains no vertex in V by the definition of w'. Let $F \subseteq E$ be the set of hyperedges corresponding to vertices in U. Then F is a k-cut of hypergraph G because removing F from G disconnects the k vertices in V chosen as terminals in the terminal k-vertex cut problem each other. Moreover, $\sum_{e \in F} w(e) = \sum_{v \in U} w'(v)$ holds. Therefore, by solving the terminal k-vertex cut problem with choosing every set of k vertices in V as terminals, we can solve the hypergraph k-cut problem.

Notice that this reduction preserves the approximation factor. Hence with a $(2 - \frac{2}{k})$-approximation algorithm [4] to the terminal k-vertex cut problem, we obtain the following.

Theorem 1. *The hypergraph k-cut problem can be approximated within a factor of $2 - \frac{2}{k}$ if k is fixed.* □

We can reduce the hypergraph k-cut problem to the submodular k-partition problem as follows. For each hyperedge $e \in E$, define the *head* denoted by $h(e)$ as an arbitrary vertex v contained by e. Moreover, define a set function $f_G : 2^V \to \Re$ so that $f_G(X) = \sum\{w(e) \mid e \in E, h(e) \in X, e \setminus X \neq \emptyset\}$ for $X \subseteq V$. It is easy to prove that function f_G is submodular. A k-cut $[V_1, V_2, \ldots, V_k]$ of the hypergraph $G = (V, E)$ satisfies $w(V_1, V_2, \ldots, V_k) = f_G(V_1, V_2, \ldots, V_k)$. Therefore, we obtain the following.

Theorem 2. *The hypergraph k-cut problem is contained by the submodular system k-partition problem.* □

We note that the submodular function defined from the hypergraph is not symmetric. This is the difference between the graph k-cut problem and the hypergraph k-cut problem.

3 Basic Property of Partitions

In this section, we prove two important properties of k-partitions.

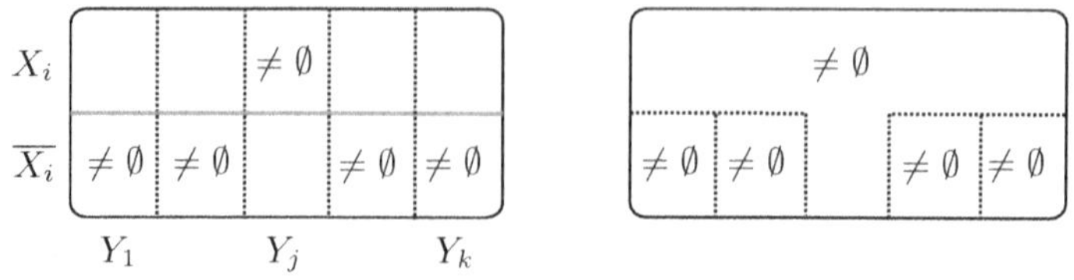

Fig. 1. P, P_k (left) and P'_k (right) in Theorem 4

Theorem 3. *Let $[X_1, X_2, \ldots, X_k]$ be a minimum k-partition of a submodular system (V, f). Moreover, let $X = \cup_{j=1}^{h} X_{i_j}$ for $\{i_1, i_2, \ldots, i_h\} \subseteq \{1, 2, \ldots, k\}$. Then $[X_{i_1}, X_{i_2}, \ldots, X_{i_h}]$ is a minimum h-partition of (X, f^X).* □

Theorem 4. *Let $P = [X_1, X_2]$ and $P_k = [Y_1, Y_2, \ldots, Y_k]$ be 2- and k-partitions of a submodular system (V, f), respectively. Let Z_{ij} denote $X_i \cap Y_j$ for each $i \in \{1, 2\}$ and $j \in \{1, 2, \ldots, k\}$. If some pair of $i \in \{1, 2\}$ and $j \in \{1, 2, \ldots, k\}$ satisfies $Z_{i'j'} \neq \emptyset$ for all $i' \neq i$ and $j' \neq j$, and $f(Z_{ij}, \overline{Z_{ij}}) \geq f(P)$, then*

$$P'_k = [Y_1 \setminus X_i, \ldots, Y_{j-1} \setminus X_i, X_i \cup Y_j, Y_{j+1} \setminus X_i, \ldots, Y_k \setminus X_i]$$

is a k-partition of (V, f) that satisfies $f(P'_k) \leq f(P_k)$. □

Theorems 3 and 4 have been already proven in [16] for hypergraphs. In particular, Theorem 4 is important because it tells that when a minimum k-partition crosses a 2-partition and they satisfies the conditions, we can uncross them by finding another minimum k-partition.

The algorithm for the hypergraph 3-cut problem in [16] can be extended to the submodular system 3-partition problem. We do not explain the detail because it is straightforward by Theorems 3 and 4.

Theorem 5. *A minimum 3-partition of a submodular system is computable by $O(n^3)$ (s,t)-partition computations.* □

4 Approximation Algorithm for 4-Partition Problem

4.1 Submodular System 4-Partition Problem

In this section, we consider the submodular 4-partition problem with non-negative submodular functions. As a consequence of Theorem 4, we obtain the following theorem. The proof is omitted due to the space limitation.

Theorem 6. *Let $P = [X_1, X_2]$ be a minimum 3-size 2-partition of a submodular system (V, f) with $|V| \geq 17$. Then (V, f) has a minimum 4-partition $P_4 = [Y_1, Y_2, Y_3, Y_4]$ satisfying one of the following conditions up to changing indices of components in P and P_4:*

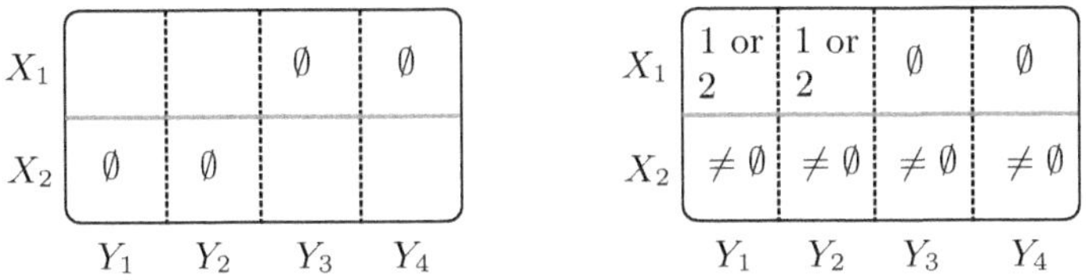

Fig. 2. P and P_4 in cases (ii) and (iii) of Theorem 6

(i) *P_4 is not 3-size;*
(ii) $Y_1 \cup Y_2 = X_1$ *and* $Y_3 \cup Y_4 = X_2$*;*
(iii) $1 \leq |X_1 \cap Y_1| \leq |X_1 \cap Y_2| \leq 2$, $X_1 \cap Y_3 = X_1 \cap Y_4 = \emptyset$, *and* $X_2 \cap Y_j \neq \emptyset$ *for* $j \in \{1, 2, 3, 4\}$*;*
(iv) *P does not cross P_4.* □

Now we give our algorithm. When $|V| \leq 16$, it is possible to compute an optimal solution in constant time. Hence let us examine the case with $|V| \geq 17$. Theorem 6 shows that there exists an optimal solution satisfying one of the four conditions with a minimum 3-size 2-partition P. We note that a minimum 3-size 2-partition is computable. The algorithm presented by Vazirani and Yannakakis [15] enumerates all the 2-partitions in the order of non-decreasing costs and with the delay between two successive outputs at most $O(n)$ minimum (s,t)-partition computations. Since the number of 2-partitions not being 3-size is $O(n^2)$, a minimum 3-size 2-partition is found after at most $O(n^3)$ minimum (s,t)-partition computations.

First, let us examine the case where (i) is satisfied, i.e., there exists a minimum 4-partition $[U, V_1, V_2, V_3]$ with $U \subseteq V$, $1 \leq |U| \leq 2$, and $V_1, V_2, V_3 \subseteq V \setminus U$. By Theorem 3, $[V_1, V_2, V_3]$ is a minimum 3-partition of the subsystem induced by $V \setminus U$. Such a 3-partition can be found in $O(n^3)$ (s,t)-partition computations by Theorem 5. By executing this for every $U \subseteq V$ with $1 \leq |U| \leq 2$, it is possible to compute an optimal solution in this case. The number of (s,t)-partition computations for this is $O(n^5)$.

Next, let us consider the case where (ii) is satisfied. By Theorem 3, $[Y_1, Y_2]$ is a minimum 2-partition of the subsystem (X_1, f^{X_1}), and $[Y_3, Y_4]$ is a minimum 2-partition of the subsystem (X_2, f^{X_2}). Hence an optimal solution can be found by computing minimum 2-partitions twice, which needs $O(n)$ minimum (s,t)-partition computations.

When (iv) is satisfied, we can recursively apply our algorithm to (V_{X_1}, f_{X_1}) and (V_{X_2}, f_{X_2}). Hence the remaining case is when (iii) is satisfied. In contrast with the cases above, we do not know how to compute an optimal solution. However, we can compute a 1.5-approximate solution in this case, or find a 2-size 2-partition not crossing an optimal solution.

Lemma 1. *Suppose that P and P_4 in Theorem 6 satisfy (iii), and that P crosses any minimum 4-partition of (V, f). Moreover, assume that there exists a* *minimum 2-partition of* (X_1, f^{X_1}) *different from* $[Z_{1,1}, Z_{1,2}]$. *Then a minimum 3-partition* $[R_1, R_2, R_3]$ *of* (X_1, f^{X_1}) *satisfies*

$$f(X_2, R_1, R_2, R_3) \leq 1.5f(P_4).$$

□

We describe the entire of our algorithm below.

Algorithm MIN4PT(V,f)
Input: A submodular system (V, f)
Output: A 4-partition of (V, f)
Step 1: Initialize the solution S with an arbitrary 4-partition of (V, f).
Step 2: If $|V| \leq 16$, then enumerate all 4-partitions of V, and terminate with outputting a minimum 4-partition among them.
Step 3: For every $U \subseteq V$ with $1 \leq |U| \leq 2$, compute a minimum 3-partition $\{V_1, V_2, V_3\}$ of $(V \setminus U, f^{V \setminus U})$ by Theorem 5, and $S \leftarrow \min\{S, [U, V_1, V_2, V_3]\}$. (This step is for (i) and (iii).)
Step 4: Compute a minimum 3-size 2-partition $[X_1, X_2]$ of (V, f) such that $|X_1| \leq |X_2|$.
Step 5: Compute a minimum 2-partition $[Q_1, Q_2]$ of (X_1, f^{X_1}) and a minimum 2-partition $[Q_3, Q_4]$ of (X_2, f^{X_2}). Then $S \leftarrow \min\{S, [Q_1, Q_2, Q_3, Q_4]\}$. (This step is for (ii).)
Step 6: If $|X_1| \geq 5$, then set $S \leftarrow \min\{S, \text{MIN4PT}(V_{X_1}, f_{X_1}), \text{MIN4PT}(V_{X_2}, f_{X_2})\}$. (This step is for (iv).)
Step 7: If $|X_1| \leq 4$, then compute a minimum 2-partition $[A, B]$ with $|A| \geq |B|$, and set $S \leftarrow \min\{S, \text{MIN4PT}(V_A, f_A), \text{MIN4PT}(V_{X_2}, f_{X_2})\}$. (This step is for (iii) and (iv).)
Step 8: Terminate with outputting S.

Theorem 7. *Algorithm* MIN4PT *computes a* 1.5*-approximate solution of the submodular system* 4*-partition problem by computing minimum* (s, t)*-partitions* $O(n^6)$ *times.*

Proof. First, let us prove the 1.5-approximability of the algorithm by the induction on n. When $|V| \leq 16$, Step 2 of the algorithm computes an optimal solution. Below, we examine the cases where the conditions discussed in Theorem 6 are satisfied by the minimum 3-size 2-partition $[X_1, X_2]$ computed in the algorithm and some minimum 4-partition. When (i) of Theorem 6 is satisfied, Steps 3 of the algorithm finds an optimal solution as we have already seen. When (ii) of Theorem 6 is satisfied, Step 5 of the algorithm finds an optimal solution. Then the remaining case is when (iii) or (iv) of Theorem 6 holds.

Consider the case where (iv) of Theorem 6 is satisfied. In this case, shrinking X_1 or X_2 preserves an optimal solution. Hence when $|X_5| \geq 5$, $\text{MIN4PT}(V_{X_1}, f_{X_1})$ or $\text{MIN4PT}(V_{X_2}, f_{X_2})$ in Step 6 returns a 1.5-approximate solution by the inductive hypothesis. When $|X_5| \leq 4$, $\text{MIN4PT}(V_A, f_A)$ or $\text{MIN4PT}(V_{X_2}, f_{X_2})$ in Step 7 returns a 1.5-approximate solution by the inductive hypothesis (Notice that $A \subseteq X_1$).

Consider the case where (iii) of Theorem 6 is satisfied. In this case, $|X_1| \leq 4$ holds. Lemma 1 implies that $A = Z_{1,1}$, $A = Z_{1,2}$, or $[R_1, R_2, R_3, X_2]$ is a 1.5-approximate solution where $[R_1, R_2, R_3]$ is a minimum 3-partition of (X_1, f^{X_1}). If $A = Z_{1,1}$ or $A = Z_{1,2}$ holds, then MIN4PT(V_A, f_A) returns a 1.5-approximate solution by the inductive hypothesis. Otherwise, Step 3 computes a 1.5-approximate solution since each of R_i, $i = 1, 2, 3$ contains at most two elements.

Next, let us consider the times of computing minimum (s,t)-partitions. Let $x = |X_1|$. Since $[X_1, X_2]$ is 3-size, $3 \leq x \leq n-3$. Moreover, $|V_{X_1}| = n-x+1$ and $|V_{X_2}| = x+1$ hold. Define $D(n)$ as the number of (s,t)-partition computations by the algorithm. By the discussion above, all operations in Steps 1 to 6 can be done by $O(n^5)$ (s,t)-computations. If $|X_1| \geq 5$, then the algorithm executes the operations in Step 6. In this case, we have

$$D(n) \leq D(x+1) + D(n-x+1) + O(n^5). \tag{1}$$

If $|X_1| \leq 4$, then the algorithm executes the operations in Step 7 instead of those in Step 6. In this case, we have

$$D(n) \leq D(x+1) + D(n-|B_1|+1) + O(n^5) \leq D(5) + D(n-1) + O(n^5). \tag{2}$$

It is easy to verify that $D(n) = O(n^6)$ satisfies both of (1) and (2). □

4.2 Hypergraph 4-Cut Problem

Since the hypergraph 4-cut problem is contained by the submodular system 4-partition problem as seen in Theorem 2, Algorithm MIN4PT works with the same approximation factor. In what follows, we show that Algorithm MIN4PT achieves better approximation factor for this problem.

Lemma 2. *Assume that the submodular system (V, f) is constructed from a hypergraph $G = (V, E)$ with weight $w : E \to \Re_+$ as described in Section 2.2, and that P and P_4 in Theorem 6 satisfy (iii). Moreover, suppose that P crosses any minimum 4-partition, and that there exists a minimum 2-partition of G^{X_1} different from $[Z_{1,1}, Z_{1,2}]$. Then a minimum 3-partition $[R_1, R_2, R_3]$ of (X_1, f^{X_1}) satisfies $w(R_1, R_2, R_3, X_2) \leq \frac{4}{3} f(P_4)$.*

By replacing Lemma 1 with Lemma 2, we have the following.

Theorem 8. *Algorithm* MIN4PT *achieves approximation factor 4/3 for the hypergraph 4-cut problem.* □

5 Concluding Remarks

In this paper, we have presented algorithms for the submodular k-partition problem and the hypergraph k-cut problem. In spite of the progress made by these, we do not know whether the hypergraph k-cut problem and the submodular system k-partition problem are polynomial-time solvable or NP-hard for fixed $k \geq 4$. Recently Fukunaga [3] showed that the hypergraph k-cut problem is polynomial-time solvable when both k and $\max_{e \in E} |e|$ are fixed. Still the other cases remain open. This is a challenging future work.

References

1. Frank, A.: Applications of Submodular Functions. Surveys in Combinatorics. Cambridge London Mathematical Society Lecture Notes Series, vol. 187, pp. 85–136 (1993)
2. Fujishige, S.: Submodular Function and Optimization. North-Holland, Amsterdam (1991)
3. Fukunaga, T.: Computing Minimum Multiway Cuts in Hypergraphs from Hypertree Packings. Technical report, Department of Applied Mathematics and Physics, Graduate School of Informatics, Kyoto University (2009)
4. Garg, N., Vazirani, V.V., Yannakakis, M.: Multiway Cuts in Node Weighted Graphs. J. Algorithms 50, 49–61 (2004)
5. Gasieniec, L., Jansson, J., Lingas, A., Óstlin, A.: On the Complexity of Constructing Evolutionary Trees. J. Comb. Opt. 3, 183–197 (1999)
6. Goldberg, A.V., Tarjan, R.E.: A New Approach to the Maximum Flow Problem. J. ACM 35, 921–940 (1988)
7. Goldschmidt, O., Hochbaum, D.: A Polynomial Algorithm for the k-cut Problem for Fixed k. Math. Operations Research 19, 24–37 (1994)
8. Iwata, S.: Submodular Function Minimization. Math. Prog. 112, 45–64 (2008)
9. Kamidoi, Y., Yoshida, N., Nagamochi, H.: A Deterministic Algorithm for Finding All Minimum k-way Cuts. SIAM J. Comp. 36, 1329–1341 (2006)
10. Karger, D.R., Stein, C.: A New Approach to the Minimum Cut Problem. J. ACM 43, 601–640 (1996)
11. Lawler, E.L.: Cutsets and Partitions of Hypergraphs. Networks 3, 275–285 (1973)
12. Nagamochi, H.: Algorithms for the Minimum Partitioning Problems in Graphs. IEICE Trans. Inf. Sys. J86-D-1, 53–68 (2003)
13. Queyranne, M.: On Optimum Size-constrained Set Partitions. In: Proceedings of AUSSOIS (1999)
14. Thorup, M.: Minimum k-way Cuts via Deterministic Greedy Tree Packing. In: 40th Annual ACM Symposium on Theory of Computing, pp. 159–166 (2008)
15. Vazirani, V.V., Yannakakis, M.: Suboptimal Cuts: Their Enumeration, Weight and Number. In: Kuich, W. (ed.) ICALP 1992. LNCS, vol. 623, pp. 366–377. Springer, Heidelberg (1992)
16. Xiao, M.: Finding Minimum 3-way Cuts in Hypergraphs. In: Agrawal, M., Du, D.-Z., Duan, Z., Li, A. (eds.) TAMC 2008. LNCS, vol. 4978, pp. 270–281. Springer, Heidelberg (2008)
17. Xiao, M.: An Improved Divide-and-Conquer Algorithm for Finding All Minimum k-Way Cuts. In: Hong, S.-H., Nagamochi, H., Fukunaga, T. (eds.) ISAAC 2008. LNCS, vol. 5369, pp. 208–219. Springer, Heidelberg (2008)
18. Zhao, L., Nagamochi, H., Ibaraki, T.: A Unified Framework for Approximating Multiway Partition Problems. In: Eades, P., Takaoka, T. (eds.) ISAAC 2001. LNCS, vol. 2223, pp. 682–694. Springer, Heidelberg (2001)

On Protein Structure Alignment under Distance Constraint

Shuai Cheng Li[1] and Yen Kaow Ng[2]

[1] David R. Cheriton School of Computer Science, University of Waterloo, Waterloo ON N2L 3G1 Canada
scli@cs.uwaterloo.ca
[2] Department of Computer and Information Sciences, Tokyo University of Agriculture and Technology, 2-24-16 Naka-cho, Koganei, Tokyo 184-8588, Japan
kalngyk@cc.tuat.ac.jp

Abstract. In this paper we study the protein structure comparison problem where each protein is modeled as a sequence of 3D points, and a contact edge is placed between every two of these points that are sufficiently close. Given two proteins represented this way, our problem is to find a subset of points from each protein, and a bijective matching of points between these two subsets, with the objective of maximizing either (A) the size of the subsets (LCP problem), or (B) the number of edges that exist simultaneously in both subsets (CMO problem), under the requirement that only points within a specified proximity can be matched. It is known that the general CMO problem (without the proximity requirement) is hard to approximate. However, with the proximity requirement, it is known that if a minimum inter-residue distance is imposed on the input, approximate solutions can be efficiently obtained. In this paper we mainly show that the CMO problem under these conditions: (1) is NP-hard, but (2) allows a PTAS. The rest of this paper shows algorithms for the LCP problem which improves on known results.

1 Introduction

The molecular shape of a protein typically determines the protein's biological mechanism. Hence, proteins with similar 3D structures can be expected to have similar functions. This allows one to predict the functions of a protein base on its structural resemblance to proteins of known functions. The incentive of making such predictions has resulted in very substantial development of approaches, algorithms, and software tools for evaluating the similarity between 3D protein structures, under the name of Protein Structure Alignment [18,17]. As a fundamental problem in bioinformatics, many heuristic algorithms have been proposed for this problem [2,3,6,8,10,11,13,16,20]. These systems generate good alignments in practice. However, relatively few theoretical studies specific to the problem have been made [1,7,12,15,21].

Due to differences in approaches to the problem, proteins have been modeled in many different ways. In this paper we consider both the cases of treating

Y. Dong, D.-Z. Du, and O. Ibarra (Eds.): ISAAC 2009, LNCS 5878, pp. 65–76, 2009.

proteins as sequences of 3D points *and* as contact maps. In modeling a protein as a sequence of 3D points, each point represents the (relative) coordinate of a residue in the protein. In this case one is to find a rigid transformation to superimpose two proteins such that there is a good positional correspondence between the proteins' residues. We consider the case where the quality of the correspondence is evaluated by the number of (disjoint) pairs of points, each from one of the two proteins, that are within a given proximity ϵ of each other. We call the problem of maximizing this number of pairs of points the ***largest common point set (LCP) problem under bottleneck distance***. This problem is known to be exactly solvable in $O(n^{32.5})$ time [4], where n is the length of the protein sequence. Akutsu [1] showed that there is an approximation algorithm of $O(n^{8.5})$ runtime which returns a solution that optimizes the target parameter but does not fulfill the proximity requirement ϵ strictly. More precisely, a pair of points in that solution may be up to 8ϵ apart. Chakraborty and Biswas [7] showed an improved algorithm with each pair of points at most 2ϵ apart, at the same time complexity. The present work improves this runtime to $O(n^8)$. When two properties of protein structures (namely minimum inter-residue distance and globular shape) are taken into account, this runtime can be shown to be $O(n^{6.5})$.

The main contribution of this paper is in the variant of the problem which models proteins as contact maps. In modeling a protein as a contact map, each residue is taken to be a vertex in a graph, where an edge (called a *contact edge*) exists between two vertices if and only if the vertices are no further than an allowed distance (e.g. 5Å) apart. If two proteins are similar, their contact maps tend to be similar. Hence to compare two proteins under such a model one typically looks for large common subgraphs of the contact maps of the two proteins. The residues' positions are typically not considered under such a model except for the purpose of creating edges. Nevertheless, a more realistic model where matched vertices has to be no further than a threshold distance apart in the protein molecules has been suggested [21]. We call the resultant problem ***contact map overlap (CMO) problem with distance constraint***. This threshold distance makes a difference when there is a requirement for any two residues in a protein to be at least some fixed distance apart. Under normal conditions, this restricts every residue to be matchable to only constantly many residues in a relatively small bounded space. An interesting consequence of this is that, while the CMO problem is NP-hard and hard to approximate in the general case [12,21], under the conditions, approximation solutions of a score at least $1-(c_1\varepsilon+c_2)$ factor of the optimal (for some constants c_1, c_2 and a runtime dependent factor ε) can be obtained in polynomial time ([21], Theorem 4.1). In this paper we show that even under these conditions, the CMO problem with distance constraint (1) is NP-hard, but (2) allows a PTAS (that is, the optimal score can be approximated to within any factor, which is not possible with the known method due to the factor c_2).

2 Preliminaries

We first provide some background on protein structures. A protein structure can be modeled as a finite, ordered sequence of 3D points. That is, each residue

($C\alpha$ atom) in the protein is represented by a coordinate which encodes the residue's relative position with respect to other residues in the protein. Hence a protein structure P of n 3D points is written as $(p_1, p_2, \ldots, p_n)$, where each $p_i \in \mathbb{R}^3$. We will consider two characteristics of protein structures in this paper:

Fact 1. The distance between any two residues in a protein structure cannot be too small due to steric clashes. (In fact, the distance between any two non-consecutive points is no less than $4\mathring{A}$, and the distance between any two consecutive points is about $3.8\mathring{A}$.) Throughout this paper we let D_l denote the minimum inter-residue distance in any given protein. A consequence of this is that, if two residues p_i, p_j are considered matchable iff $||p_i - p_j|| \leq d$ for some fixed $d \in \mathbb{R}$, then every residue is matchable with at most $O((1 + \frac{2d}{D_l})^3)$ residues [21].

Fact 2. The points in any protein structure P is known to be bounded within a sphere of radius R_P, where $R_P = O(n)$ for general proteins, and $R_P = cn^{1/3}$ for some constant c for globular proteins [15]. Furthermore, it can be assumed that the points are uniformly distributed in space. We now state our problems.

LCP PROBLEM UNDER BOTTLENECK DISTANCE	
Input:	sequences $P = (p_1, \ldots, p_n)$, $Q = (q_1, \ldots, q_m)$ and distance threshold $D_c \in \mathbb{R}$. Without loss of generality assume $m \geq n$.
Output:	(i) subsets $P' \subseteq P$, $Q' \subseteq Q$, $\|P'\| = \|Q'\|$, (ii) bijection $f : P' \mapsto Q'$, and (iii) rigid transformation (rotation and translation) t, fulfilling the following conditions: (A) $\max_{p \in P'} \|\|t(p) - f(p)\|\| \leq D_c$, (B) the score $S = \|P'\|$ is maximized.

We refer to f as an *alignment*. An alignment can be *sequential* or *non-sequential*: an alignment is *sequential* iff for any two points p_{i_1}, $p_{i_2} \in P'$, where the corresponding $f(p_{i_1}) = q_{j_1}$ and $f(p_{i_2}) = q_{j_2}$, we have $i_1 < i_2$ iff $j_1 < j_2$. Otherwise the alignment is *non-sequential*. The LCP problem which requires alignments to be sequential is said to be *sequential*, otherwise it is *non-sequential*. We let $\boldsymbol{P}$, $\boldsymbol{Q}$, $\boldsymbol{f}$, $\boldsymbol{T}$, $\boldsymbol{S}$ denote an optimal P', Q', f, t, S, respectively.

A protein can also be modeled as a *contact map graph*. A *contact* is a pair of points in a protein that are no more than a given distance apart. Throughout this paper we let D_u denote this distance. In order to form contact edges, $D_u \geq D_l$. A *contact map graph* of a protein consists of the residues (i.e., vertices) and their contacts (i.e., edges). Each vertex v is also associated with a 3D point pos(v) indicating the residue's relative position from other points in the protein. In this paper we consider the following problem using contact maps.

CMO PROBLEM WITH DISTANCE CONSTRAINT	
Input:	contact maps $G_1 = (V_1, E_1)$, $G_2 = (V_2, E_2)$ and distance threshold $D_c \in \mathbb{R}$. Without loss of generality assume $\|V_2\| \geq \|V_1\|$.
Output:	(i) subsets $V_1' \subseteq V_1$, $V_2' \subseteq V_2$, $\|V_1'\| = \|V_2'\|$, (ii) bijection $M : V_1' \mapsto V_2'$, and (iii) rigid transformation (rotation and translation) t, fulfilling the following conditions: (A) $\max_{v \in V_1'} \|\|t(\text{pos}(v)) - \text{pos}(M(v))\|\| \leq D_c$, (B) $S = \|\{(v, u) \in E_1' \mid (M(v), M(u)) \in E_2'\}\|$ is maximized.

We refer to M as an *alignment*. We consider vertices in V_1, V_2 to be ordered. We let $\boldsymbol{V_1}$, $\boldsymbol{V_2}$, $\boldsymbol{M}$, $\boldsymbol{T}$, $\boldsymbol{S}$ denote an optimal V_1, V_2, M, t, S, respectively.

Approximate Solutions. We show algorithms that output approximations to solutions for the problems above. The approximate solutions would: (1) fulfill Condition (A) for a slightly larger distance threshold, i.e. $(1+\epsilon)D_c$ for some $\epsilon \in \mathbb{R}$; (2) have a score at or above $r\boldsymbol{S}$ for some $r \in \mathbb{R}$, $r \leq 1$. We call such a solution an *(ϵ, r)-approximation.* The algorithms we show have runtime complexities that depend on $1/\epsilon$ and r, i.e. discrepencies from the optimal due to both (1) and (2) can be made arbitrarily small at the expense of higher runtimes.

Other notations. We let $\Delta_\epsilon = (1 + \frac{2(1+\epsilon)D_c}{D_l})^3$. Hence every residue in P can be matched to at most $O(\Delta_\epsilon)$ residues in Q. For any point p and transformation t, $t(p)$ denotes the point obtained by transforming p with t. For a set of points P, $t(P)$ denotes the set $\{t(p) \mid p \in P\}$. For $r \in \mathbb{R}$, the *r-sphere* of a point p is the sphere of radius r centered at p.

3 Hardness of the CMO Problem with Distance Constraint

As mentioned, it is known that approximate solutions for the non-sequential CMO problem with distance constraint can be obtained efficiently if $\frac{\max\{2D_c, D_u\}}{D_l}$ is bounded below a constant [21]. In fact a PTAS exists (see Theorem 5). We first show that even in such a case, the problem is NP-hard.

Theorem 1. *The non-sequential CMO problem with distance constraint is NP-hard, even when $\frac{\max\{2D_c, D_u\}}{D_l}$ is bounded below a constant.*

Proof. (Sketch) We use a reduction from the planar 1-in-3-SAT problem [9]. Planarity of the problem allows us to construct a geometrical representation of the 1-in-3-SAT problem where none of the legs cross, to be used as input to the CMO problem. We assume $D_c > D_u$ throughout this proof. As stated earlier, $D_u \geq D_l$. We make no other assumption regarding D_c, D_u, and D_l. That is, $\frac{\max\{2D_c, D_u\}}{D_l}$ can be set to a small constant.

Given an input formula we will construct two sequences of 3D points P and Q as input to the CMO problem. We assume that the optimal solution will have the points in exactly the positions we place them in our construction. For each clause we construct some points for both P and Q, where the contact edges form chain-like structures which we call *chains*. Points constructed for different clauses are separated by more than D_c apart, hence they form no contact edges except for a few end-points which we will explain later (in Fig 6). Each point in a chain form contact edges only with its two (one in the case of end-points) immediate neighbors. For each clause, we construct six such chains for Q, and three such chains for P. Each chain of the same clause has the same number of points, η say. A cyclic structure formed using three additional points (called *pivot points*) connects the six chains for Q through their end-points. One additional pivot point connects the end-points of the three chains of P.

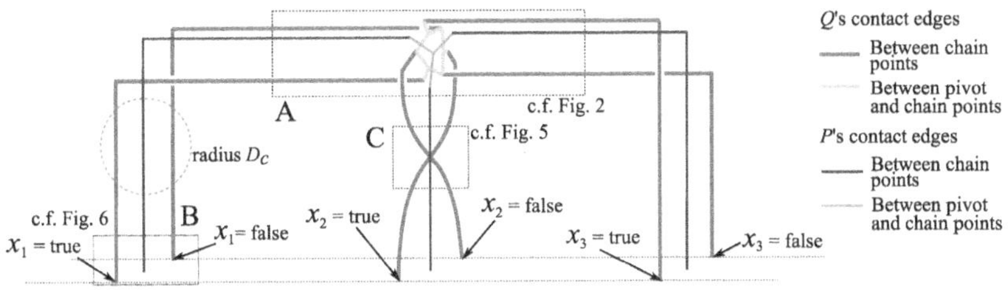

Fig. 1. Overall view of a construction for the clause $(x_1 \lor x_2 \lor x_3)$

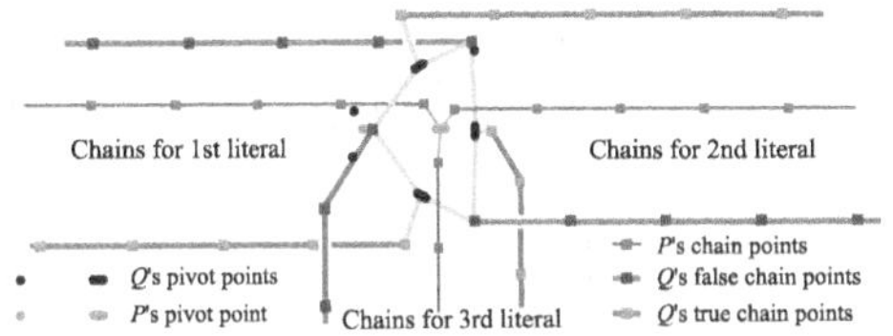

Fig. 2. Chains connected by pivot points (P not drawn accurately to allow clearer viewing)

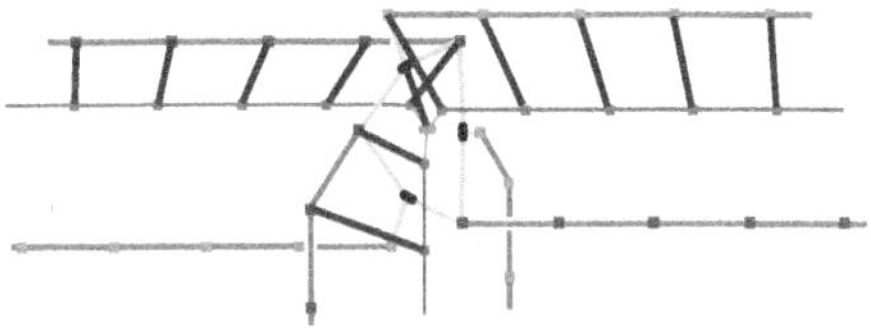

Fig. 3. An optimal mapping

Fig 1 shows how we would construct a structure for the clause $(x_1 \lor x_2 \lor x_3)$. Note that in the points constructed for a single clause no contact edge exists besides those that form the chains, and the connections between the end-points of the chains to the pivot points. In Fig 2 we show the details at the pivot points.

In the construction of a clause each literal is represented using two chains for Q and one chain for P. The CMO problem asks for a mapping between P and Q which maximizes the number of corresponding contact edges in P and Q. To achieve this maximization the chain for P must be completely mapped to only one of the two chains for Q (i.e. giving $\eta - 1$ corresponding contact edges). We will let the literal be assigned true or false depending on which of the two chains for Q we find P's chain mapped to in a CMO output. We call a Q chain for true assignment *true chain*, and a Q chain for false assignment *false chain*.

We now explain the pivot structure, which is constructed to ensure that exactly one literal in the clause will be assigned true. Fig 2 shows the contact edges near the structure. Each false chain is in contact with two pivot points; each true chain is in contact with one. The end-points of the chains for P need to be more than distance D_u from each other, but each within distance D_c from the points in Q that they are mapped to in the optimal mappings. Fig 4 shows a possible arrangement.

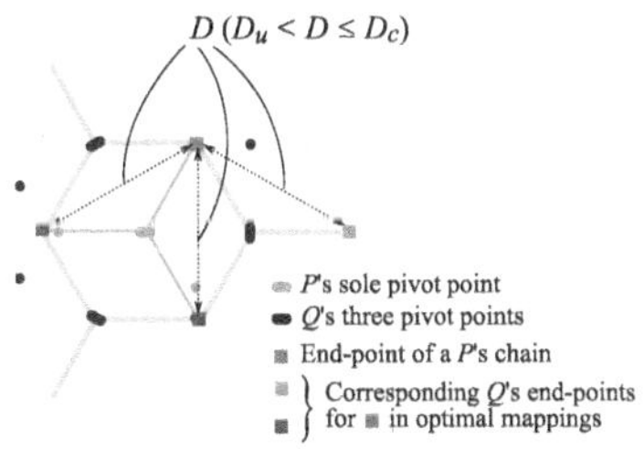

Fig. 4. Details at the pivots

There are exactly three mappings between P and Q which maximizes the number of corresponding contact edges, and they all map the single pivot point of P to one of the three pivot points of Q. Each mapping gives a total of $3(1 + \eta)$ corresponding contact edges. Fig 3 shows one such mapping. In all three optimal mappings, exactly one literal in the clause is assigned true.

The points at the other end of the chains (i.e. the end not connected to a pivot point) are collected along two lines — one for all the chains representing true assignment to a variable and the other for false assignment. (A true chain does not imply a true assignment to a variable, due to negation. That is, a literal of a negated variable should have its true assignment chains at the line for false assignment, and vice versa.) Such positioning may require the two Q chains for a literal to swap position, which can be done through arrangements as shown in Fig 5.

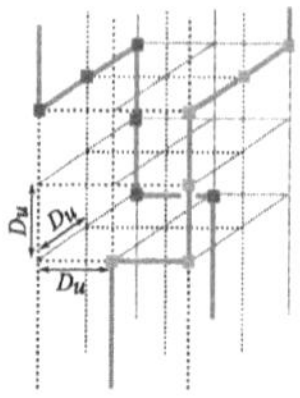

Fig. 5.

Chains for the same variable from different clauses are placed at close proximity to each other, with a distance of D_u between their consecutive end-points. This is shown in Fig 6. Planarity of the problem allows such a construction of the chains, which allows us to arrange it such that only the end-points form contact edges. Fig 7 shows the arrangement. Note that under the arrangement, points in P can be mapped to their respective points in Q.

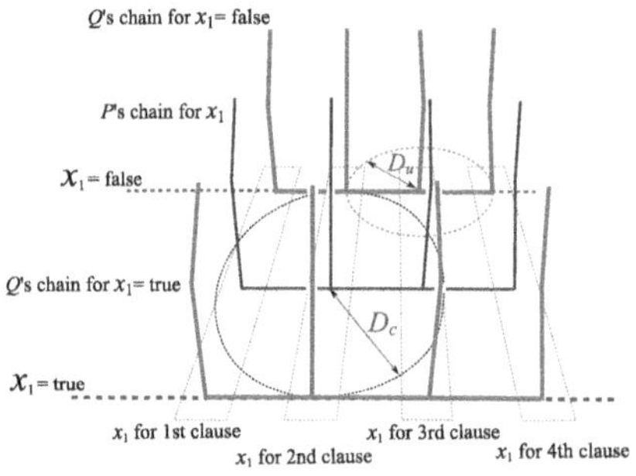

Fig. 6. Construction for a variable that appears in four clauses

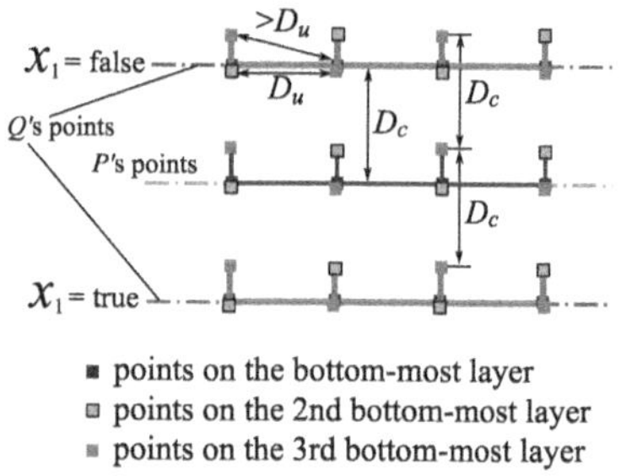

Fig. 7. View from top of Fig 6

For any variable, an optimal mapping would have either all its corresponding chains in P mapped to their corresponding chains in Q which represent true, or to their corresponding chains in Q which represent false. This is because such a mapping increases the number of corresponding contact edges through the additional edges at the bottom-most layer, that is, by exactly the number of occurrences of the variable (in the 1-in-3-SAT formula) minus one.

Hence an optimal solution for the CMO problem with distance constraint results in a specific number of corresponding contact edges, which is computable from the number of clauses and variable occurrences in the 1-in-3-SAT formula. A solution with this optimal score for the CMO problem corresponds to whether there is a set of variable assignments that fulfills the 1-in-3-SAT formula.

We now look at the number of points needed for constructing P and Q so far. Let α be the minimal number of points needed to construct a single clause, then it can be shown that the number of points needed to construct an input of n clauses is of order $O(n^2\alpha)$.

Finally, in order to make the positions of the above points immutable in an optimal solution in the CMO problem, we can add additional points on P and Q — the additional points have only one optimal mapping, and this mapping produces more corresponding edges than the total number of edges in the above construction. Such can be achieved with, for example, points that form a "grid" on the X-Y plane. These points can be placed sufficiently far away from the points in the above construction to avoid forming contact edges with the construction. It is clear that $O(n^2\alpha)$ points suffice to construct these "grid" points.

The polynomial reduction witnesses the hardness of the problem. ∎

4 Approximation Algorithms

We use the strategy in [19] to find a suitable rigid transformation to superimpose the two protein structures. The following results are from therein.

Definition 1. *Given a finite set of points P and two points p, $p' \in P$, we write $\langle [p]p' \rangle_P$ iff p' is the furthest point from p among all the points in P. We say that p and p' is a* radial pair *just in case either $\langle [p]p' \rangle_P$ or $\langle [p']p \rangle_P$.*

Lemma 1 ([19]). *Given a set of points P, rigid transformations $\boldsymbol{T}$, T, and radial pair p_1, $p_2 \in P$, if $||\boldsymbol{T}(p_1) - T(p_1)|| \leq \delta$ and $||\boldsymbol{T}(p_2) - T(p_2)|| \leq \delta$, then there exists a rotation R about the axis through the points $T(p_1)$ and $T(p_2)$, such that $\forall p \in P$, $||R(T(p)) - \boldsymbol{T}(p)|| \leq 3\delta$.*

Lemma 2 ([19]). *If it is known that p_i, $p_j \in \boldsymbol{P}$ is a radial pair, and that $\boldsymbol{f}(p_i) = q_k$ and $\boldsymbol{f}(p_j) = q_l$ for q_k, $q_l \in \boldsymbol{Q}$, then one only needs to search among $O(1/\epsilon^5)$ transformations to find the transformation T in Lemma 1.*

Lemma 3. *Given that T, p_1, p_2 in Lemma 1 is known, there are at most $O(mn)$ rotations to evaluate to find R of Lemma 1 in the general case, and at most $O(nm^{1/3})$ rotations to evaluate in the case of globular proteins. These rotations can be discovered in $O(mn)$ time.*

4.1 PTAS for LCP Problem under Bottleneck Distance

We now discuss how to evaluate a rotation in Lemma 3 for the LCP problem under bottleneck distance. Given that the points are fixed in position, this evaluation can be done by constructing a bipartite graph $G(P \cup Q, E)$ where $(u, v) \subset E$ iff $u \in P, v \in Q$ and $||u - v|| \leq (1+\epsilon)D_c$, and find the maximum bipartite matching of G. Constructing a bipartite matching in the general case takes $O(nm)$ time, but for a globular protein this construction can be done in $O(n\Delta_\epsilon)$ time, since by arranging points in Q into cells of size $D_c \times D_c \times D_c$ all the points in Q within distance $(1 + \epsilon)D_c$ of any point in P can be found in $O(\Delta_\epsilon)$ time.

It is clear that the same bipartite graph is constructed for every rotation within the same rotation interval. Furthermore, the bipartite graph for one rotation interval can be constructed in $O(1)$ time from that of an earlier rotation interval, since they differ by only a single edge. In order to know which edge is added or removed in a subsequently rotation interval, we sort the $O(mn)$ (resp. $O(mn^{1/3})$) entry/exit points.

The maximum bipartite matching differs by at most an edge for any two consecutive intervals. This allows us to use an algorithm for bipartite matching which reuse results obtained for an interval in the matching of the next interval.

Lemma 4 ([5]). *The bipartite matching problem can be solved with time complexity $O((|M| - |M_0|)|E|)$, where M is a maximum matching, and M_0 is some initial matching.*

By the algorithm a maximum matching M can be computed in $O(|E|)$ time from a matching M_0 from the earlier rotation interval, that is, $O(nm)$ in the general case and $O(n\Delta_\epsilon)$ in the case of globular proteins. Hence in total,

1. Discovering the entry/exit points of the intervals take $O(nm)$ time. Sorting them takes $O(nm \log nm)$ time, or $O(nm^{1/3} \log nm^{1/3})$ for globular proteins.
2. Construction of the bipartite graph for the first rotation interval takes $O(nm)$ time, or $O(n\Delta_\epsilon)$ time for proteins.
3. An initial match for the first rotation interval requires $O(m^{2.5})$ time [14].
4. The remaining $O(nm)$ rotation intervals each takes $O(1)$ time for input modification, and $O(|E|) = O(nm)$ time to find new matching. For globular proteins, $O(nm^{1/3})$ rotation intervals of $O(1)$ time for input modification and $O(n\Delta_\epsilon)$ time for new matching. In total,

Lemma 5. *If a rotation axis is specified, the non-sequential LCP problem under bottleneck distance can be solved in $O(m^{2.5} + n^2m^2)$ time in the general case, and $O(m^{2.5} + n^2m^{1/3}\Delta_\epsilon)$ time for globular proteins.*

We do not know which pair is a radial pair of $\boldsymbol{P}$, nor do we know their matching points in Q. For this reason we exhaustively search all the possible m^2n^2 combinations of pairs of points in P and Q. By Lemma 2, each combination results in $O(1/\epsilon^5)$ possible matches. By Lemma 5 we have the following.

Theorem 2. *There is an algorithm of time complexity $O((n^2m^{4.5} + n^4m^4)/\epsilon^5)$ in the general case, or $O((n^2m^{4.5} + n^4m^{2.\bar{3}}\Delta_\epsilon)/\epsilon^5)$ in the case of globular proteins, that outputs an $(\epsilon, 1)$-approximate solution to the non-sequential LCP problem under bottleneck distance.*

This improves the currently known best result of $O(n^{8.5})$ to $O(n^8)$ (letting $n = m$ and $\epsilon = 1$) for the general case [7]. When D_c/D_l is reasonably small, the runtime dependency on m, n in the case of globular proteins can be further improved to $O(n^3m^{2.\bar{3}})$, if we relax the approximation ratio. To do so we use the strategy for approximation as in Section 3.2 of [21]. The strategy divides the points spatially, forming smaller, mutually independent sub-cases which are each solved with bipartite matching exactly, and then merged to form the solution.

Let k, W_x, W_y, R_j be as defined in Section 3.2 of [21]. Let $D = 2D_c$ (i.e. instead of $D = \max\{2D_c, D_u\}$ as in [21], since there is no D_u for the LCP problem).

However, instead of partitioning in 1 axis and obtaining blocks of size $W_x \times W_y \times D$, we partition along all 3 axes to obtain blocks of size D^3, and we let k along the x axis, y axis, and z axis be k_x, k_y, k_z respectively. We revise the definition of R_j to adapt to such a partitioning. Instead of k we now have $k_xk_yk_z$ partitioning schemes. Since there are now $(k_x-1)(k_y-1)(k_z-1)$ blocks in each R_j, the number of residues in each R_j, $|R_j| = O(k_xk_yk_z(\frac{D}{D_l})^3)$ (since each block contains at most $(\frac{D}{D_l})^3$ residues). Bipartite matching for the initial rotation interval takes $O((|R_j|\Delta_\epsilon)^{2.5})$ time for each R_j [14]. To compute for all $k_xk_yk_z$ partitioning schemes and for all $\#R_j$ partitions of R_j in each partitioning scheme (i.e. $n = \#R_j|R_j|$) then takes $O(\#R_j(k_xk_yk_z)(|R_j|\Delta_\epsilon)^{2.5})$ $= O(n(k_xk_yk_z)^{2.5}(\frac{D}{D_l})^{4.5}{\Delta_\epsilon}^{2.5})$ time. The remaining $O(nm^{1/3})$ rotation intervals each requires recomputation for matching at most a single R_j to Q. For all $k_xk_yk_z$ partitioning schemes this takes $O(nm^{1/3}k_xk_yk_z|E|) = O(nm^{1/3}k_xk_yk_z|R_j|\Delta_\epsilon)$ $= O(nm^{1/3}(k_xk_yk_z)^2(\frac{D}{D_l})^3\Delta_\epsilon)$ time, using the algorithm in Lemma 4. A similar analysis as in Section 3.2 of [21] will show that the method leads to a $(1-\frac{4}{k_x})(1-\frac{4}{k_y})(1-\frac{4}{k_z})$ approximation. Hence,

Theorem 3. *In the case of globular proteins, there is an algorithm of*

$$O(n^3(m^2(k_xk_yk_z)^{2.5}(\tfrac{2D_c}{D_l})^{4.5}{\Delta_\epsilon}^{2.5} + m^{2.\bar{3}}(k_xk_yk_z)^2(\tfrac{2D_c}{D_l})^3\Delta_\epsilon)/\epsilon^5)$$

time complexity that outputs an $(\epsilon, 1-4(\frac{1}{k_x}+\frac{1}{k_y}+\frac{1}{k_z}))$*-approximate solution to the non-sequential LCP problem under bottleneck distance.*

For the sequential LCP problem under bottleneck distance, instead of bipartite matching, straight-forward dynamic programming can be used to find the maximum number of matches. We let $f(\eta,\mu)$ where $1 \le \eta \le n$ and $1 \le \mu \le m$, denote the optimal number of matches for the subsequences $(p_\eta, p_{\eta+1}, \ldots, p_n)$ and $(q_\mu, q_{\mu+1}, \ldots, q_m)$.

$$f(\eta,\mu) = \max \left\{ \begin{array}{ll} f(\eta+1,\mu) & ||p_{\eta+1}-q_\mu|| \le (1+\epsilon)D_c \\ & \quad (p_\eta \text{ is not matched in this case}) \\ f(\eta,\mu+1) & ||p_\eta - q_{\mu+1}|| \le (1+\epsilon)D_c \\ & \quad (q_\mu \text{ is not matched in this case}) \\ f(\eta+1,\mu+x+1)+1 & \mu \le \mu+x \le m \wedge ||p_\eta - q_{\mu+x}|| \le (1+\epsilon)D_c \\ f(\eta+y+1,\mu+1)+1 & \eta \le \eta+y \le n \wedge ||p_{\eta+y} - q_\mu|| \le (1+\epsilon)D_c \end{array} \right\}$$

The number of $f(\eta,\mu)$ values to compute in this dynamic programming is of $O(nm)$ in the general case and $O(n|\Delta_\epsilon|)$ for globular proteins. Note that by using the bipartite graph we can easily search for points within $(1+\epsilon)D_c$ apart. For each $f(\eta,\mu)$, there are $O(m)$ values from which to find a maximum in the general case, and $O(\Delta_\epsilon)$ values for globular proteins. Hence, the runtime complexity for a single rotation interval is $O(nm^2)$ in the general case, and $O(n{\Delta_\epsilon}^2)$ in the case of globular proteins. In total we have the following.

Theorem 4. *There is an algorithm of time complexity* $O(n^4m^5/\epsilon^5)$ *in the general case, and* $O(n^4\ m^{2.3\bar{3}}{\Delta_\epsilon}^2/\epsilon^5)$ *in the case of globular proteins, that outputs* latter algorithm runs in polynomial time if there exists an APTAS for 2-Dim Fractional Bin Packing, a question that was open before this paper.

Our results: The main result of this paper is a technical lemma on the structure of packings of items into a bin. Roughly speaking, the lemma concerns a packing into a bin of a set of items that can be partitioned into three subsets, namely "fat and tall", "fat and low", and "thin and tall", and for which the number of widths of the "fat and low" items as well as the number of heights of the "thin and tall" items is bounded by a constant. The (fairly complex) formal statement is:

Lemma 1 (Structural lemma). *Consider a set of items (rectangles) that fits into a bin (unit square) of the form $L \cup O \cup V$, where $w_i \geq \varepsilon$ for $i \in L \cup O$; $h_i \geq \varepsilon$ for $i \in L \cup V$; and the number of distinct widths of the items in O and heights of the items in V is at most d, where ε and d are given constants. Let $\overline{w}_1, \overline{w}_2 \ldots$ be the distinct widths of the items in O; $h(\overline{w}_j)$ be the total height of the items having width $\overline{w}_j$; $\overline{h}_1, \overline{h}_2 \ldots$ be the distinct heights of the items in V; and $b(\overline{h}_j)$ be the total width of the items having height $\overline{h}_j$. Then, there exists a constant $f(d, \varepsilon)$ such that, for any $\delta > 0$, the following set of rectangles fits into a unit square: the items in L plus, for $j = 1, 2, \ldots$, a set of rectangles of width $\overline{w}_j$ and height δ for a total height at least $h(\overline{w}_j) - \delta f(d, \varepsilon)$, and a set of rectangles of height $\overline{h}_j$ and width δ for a total width at least $b(\overline{h}_j) - \delta f(d, \varepsilon)$.*

By using this lemma, we are able to prove the following theorem, that shows that 2-Dim Geometric Knapsack has a PTAS if the range of the profit/area ratios, namely $\max_{i \in I}(p_i/a_i)/\min_{i \in I}(p_i/a_i)$, is bounded from above by a constant. Note that, by possibly scaling the profits, this is equivalent to saying that there exists a constant r such that $p_i/a_i \in [1, r]$ for $i \in I$.

Theorem 1. *For any fixed $r \geq 1$, there exists a PTAS for 2-Dim Geometric Knapsack with and without rotations restricted to instances I for which $p_i/a_i \in [1, r]$ for $i \in I$.*

As a corollary, we get a PTAS for the problem of maximizing the area occupied in a bin, whose existence was open so far.

Corollary 1. *There exists a PTAS for the special case of 2-Dim Geometric Knapsack with and without rotations in which $p_i = w_i \cdot h_i$ for $i \in I$.*

Although the straightforward column generation (or dual separation) problem for the customary LP formulation of 2-Dim Fractional Bin Packing is a general 2-Dim Geometric Knapsack, to which Theorem 1 does not apply, we show that the column generation problem for a closely-related variant can be solved near-optimally. By the well known connection between approximate separation and optimization [18,19,31], this implies:

Theorem 2. *There exists an APTAS for 2-Dim Fractional Bin Packing with and without rotations.*

As mentioned above, the results in [4,5] along with Theorem 2 imply:

Corollary 2 ([4,5]). *For any fixed $\varepsilon > 0$, there exists a polynomial-time approximation algorithm for 2-Dim Bin Packing without rotations with approximation guarantee arbitrarily close to $1 + \ln \Pi_\infty = 1.525\ldots$.*

For the case without rotations, the bins can be assumed to be unit squares without loss of generality. For the case with rotations, our results hold also for the case in which the bins can be arbitrary rectangles, and we address the case of unit squares only for simplicity of presentation. For the sake of readability, in the coming sections we present the above results in reverse order, which corresponds to increasing technical difficulty. For a full discription of the proof of the structural lemma we refer to the full version.

Next-fit decreasing height: Throughout the paper, we will extensively use the *next-fit decreasing height* (NFDH) procedure introduced by [13]:

Observation 1 ([13]) *Consider a set of items I and its packing into bins by NFDH, letting m be the number of these bins and, for $j = 1, \ldots, m$, S_j be the subset of items packed into the j-th bin. The following hold:*

(i) if $m > 1$ and $w_i \geq h_i$ for $i \in I$, then the area $a(S_1) \geq 1/4$;
(ii) if $m > 2$, then $\max\{a(S_1), a(S_2)\} \geq 1/4$;
(iii) if $m > 1$, then $a(S_1) \geq (1 - \max_{i \in S_1} w_i) \cdot (1 - \max_{i \in S_1} h_i)$.

2 An APTAS for 2-Dim Fractional Bin Packing

In this section we prove Theorem 2. It is well known that 2-Dim Bin Packing can be formulated as the Set Covering problem in which the set I of items has to be covered by *configurations* from the collection $\mathcal{C} \subseteq 2^I$, where each configuration $C \in \mathcal{C}$ corresponds to a set of items that can be packed into a bin. The associated 2-Dim Fractional Bin Packing is the continuous relaxation of this Set Covering problem:

$$\min\{\sum_{C \in \mathcal{C}} x_C : \sum_{C \ni i} x_C \geq 1 \; (i \in I), \; x_C \geq 0 \; (C \in \mathcal{C})\}. \tag{1}$$

The dual of this LP is given by:

$$\max\{\sum_{i \in I} \pi_i : \sum_{i \in C} \pi_i \leq 1 \; (C \in \mathcal{C}), \; \pi_i \geq 0 \; (i \in I)\}. \tag{2}$$

The well known connection between approximate separation and optimization for (1) reads:

Theorem 3 ([18,19,31]). *There exists a PTAS for (1) if, for any $\varepsilon > 0$, there exists a polynomial-time algorithm that, given $(\pi_i^*) \in \mathbb{R}_+^{|I|}$ such that $\max_{C \in \mathcal{C}} \sum_{i \in C} \pi_i^* \geq 1 + \varepsilon$, finds a configuration $C^* \in \mathcal{C}$ such that $\sum_{i \in C^*} \pi_i^* > 1$.*

Note that a PTAS for the 2-Dim Geometric Knapsack associated with the items in I in which the item profits correspond to the dual values π_i^* would suffice in Theorem 3. Since the existence of such a PTAS remains open, we now introduce a variant of (1) that, on the one hand, is almost equivalent to the original problem and, on the other, has a dual separation problem that fulfils the requirements of Theorem 1. The definition of this variant and its properties is the novelty of this section. The variant is simply obtained by imposing a bound of $4a_i$ on each dual variable π_i:

$$\max\{\sum_{i \in I} \pi_i : \sum_{i \in C} \pi_i \leq 1 \; (C \in \mathcal{C}), \; 0 \leq \pi_i \leq 4a_i \; (i \in I)\}, \tag{3}$$

which corresponds to the primal problem with the additional variables y_i:

$$\min\{\sum_{C\in\mathcal{C}} x_C + \sum_{i\in I} 4a_i y_i : \sum_{C\ni i} x_C + y_i \geq 1\ (i\in I),\ x_C, y_i \geq 0\ (C\in\mathcal{C}, i\in I)\}. \quad (4)$$

Lemma 2. *Given any solution of (4) of value z^*, one can obtain in polynomial time a solution of (1) of value at most z^*+1 for the case with rotations and z^*+2 for the case without rotations.*

Proof. Consider a solution (x^*_C, y^*_i) of (4). Let $I^* := \{i \in I : y^*_i > 0\}$ be the set of items associated with a positive y-component in the solution. We pack the items in I^* into bins by NFDH. If at least one of these bins contains a subset of items $S^* \subseteq I^*$ such that $a(S^*) \geq 1/4$, we do the following. We let $\alpha := \min_{i\in S^*} y^*_i$, and define the new solution of (4) in which y^*_i is decreased by α for $i \in S^*$ and x_{S^*} is increased by α. It is immediate to verify that the new solution is feasible, and not worse than the previous one since $\sum_{i\in S^*} 4a_i\alpha = 4\alpha a(S^*) \geq \alpha$.
We repeat the procedure above until no bin packed by NFDH with the items in I^* has area occupied at least $1/4$. In this case, by Observation 1(ii), for the case without rotations we have that NFDH packs the items in I^* into at most two bins, associated with, say, subsets S^*_1 and S^*_2. At this point, we define the new solution of (4) in which y^*_i is set to 0 for $i \in S^*_1 \cup S^*_2$ and $x_{S^*_1}, x_{S^*_2}$ are increased to value 1. This solution is feasible also for (1) (by neglecting the y variables) and has a value which is larger than the previous one by at most 2. On the other hand, for the case with rotations, by Observation 1(i) NFDH packs the items in I^* into one bin, and the reasoning is analogous. Note that the number of iterations of the above procedure is at most $|I|$ as, in each iteration, at least one y^*_i is decreased from a positive value to 0.

Lemma 3. *There exists a PTAS for (4) with and without item rotations.*

Proof. By the counterpart of Theorem 3 for (4), for any $\varepsilon > 0$ we need a polynomial-time algorithm that, given $(\pi^*_i) \in [0, 4a_i]^{|I|}$ such that $\max_{C\in\mathcal{C}} \sum_{i\in C} \pi^*_i \geq 1+\varepsilon$, finds a configuration $C^* \in \mathcal{C}$ such that $\sum_{i\in C^*} \pi^*_i > 1$. In other words, if the 2-Dim Geometric Knapsack associated with the items in I having profits π^*_i satisfies $\text{opt}_{2KP}(I) \geq 1+\varepsilon$, we want a solution of the problem of value > 1. Letting $\sigma := \varepsilon/3$, we first remove all the items $i \in I$ such that $\pi^*_i \leq \sigma a_i$, whose overall contribution to $\text{opt}_{2KP}(I)$ is at most σ. For the items left, the range of the profit/area ratios is $[\sigma, 4]$, i.e., it becomes $[1, 4/\sigma]$ after scaling. Then, we apply the PTAS of Theorem 1 with internal precision σ where now $r = 4/\sigma$. The solution found by this PTAS, after scaling profits back to their original values, has value at least $(1-\sigma)\,(\text{opt}_{2KP}(I) - \sigma) \geq (1-\sigma)(1+\varepsilon-\sigma) > 1$.

3 A PTAS for 2-Dim Geometric Knapsack

In this section we prove Theorem 1. Recall that we are assuming $p_i/a_i \in [1, r]$ for $i \in I$, where r is a constant. For simplicity, we will assume that r is integer. By Observation 1(ii), items for a total area at least $\min\{a(I), 1\}/4$ can be packed into the bin. Together with $p_i/a_i \geq 1$ for $i \in I$ this implies $\text{opt}_{2KP}(I) \geq \min\{a(I), 1\}/4$.

Let $\bar{\varepsilon} < 1/2$ denote the accuracy required. Letting $\delta < \bar{\varepsilon}^2$ be a suitable constant threshold specified below, we distinguish the case in which $a(I) \geq \delta$, for which we apply the algorithm described below, from the case in which $a(I) < \delta$. In this second case, if rotations are allowed all the items are packed into the bin by NFDH, by Observation 1(i). On the other hand, for the case without rotations handling instances in which the overall area $a(I)$ of the items is very small may be tricky. In fact, for the case in which $a(I) < \delta$, we adopt a completely different method illustrated in the full version.

Description of the main algorithm: We first illustrate the case without rotations, as it is more complex. Let $\varepsilon > 0$ denote an internal accuracy parameter, assuming for simplicity that $1/\varepsilon$ is integer. We will show how to find in polynomial time a solution of value at least $(1-\alpha\varepsilon)\,\text{opt}_{2\text{KP}}(I) - \varepsilon$, where $\alpha > 2$ is a suitable constant (independent of ε). Note that this yields a PTAS for the case in which $a(I) > \delta$, implying $\text{opt}_{2\text{KP}}(I) > \delta$ (recalling $\delta < \overline{\varepsilon}^2 < 1/4$), by setting for instance $\varepsilon := (\overline{\varepsilon}\delta)/\alpha$.

Size classification: Let $\varepsilon_0 := 1$ and, for $j = 0, \ldots, 2/\varepsilon$, ε_{j+1} be a suitable constant, depending on $\varepsilon, r, \varepsilon_j$, to be specified later, such that $\varepsilon_{j+1} < \varepsilon^2\varepsilon_j$. Let $I_j \subseteq I$ denote the subset of items that have width or height in the interval $(\varepsilon_{j+1}, \varepsilon_j]$. We apply the method that follows for all values $m = 0, \ldots, 2/\varepsilon$, and take the best solution produced. We neglect the items in I_m (i.e., we find a solution in which none of these items is packed) and partition the rectangles in $I \setminus I_m$ as follows: Let L (large) denote the rectangles having both height and width $> \varepsilon_m$; O (horizontal) denote the rectangles having width $> \varepsilon_m$ and height $\leq \varepsilon_{m+1}$; V (vertical) denote the rectangles having height $> \varepsilon_m$ and width $\leq \varepsilon_{m+1}$; S (small) denote the rectangles having both height and width $\leq \varepsilon_{m+1}$.

Rounding the items in O and V: In order to apply Lemma 1, we modify the widths of the items in O (resp., the heights of the items in V) so that there are only a constant number of distinct widths (resp., heights). In this phase we allow the items in O to be sliced horizontally (resp., the items in V to be sliced vertically) so as to be able to form subsets whose total height (resp., width) is exactly a given value. At the end of the algorithm, we will pack the items in O and V with their original sizes and without slicing them.

We partition the items in O into groups O_{jk} for which the width and the profit/area ratio is approximately the same, as follows:

$$O_{jk} := \{i \in O : w_i \in ((1-\varepsilon)^j, (1-\varepsilon)^{j-1}],\ p_i/a_i \in (r(1-\varepsilon)^k, r(1-\varepsilon)^{k-1}]\}.$$

Note that we have to consider $j = 1, \ldots, \lceil(\log \varepsilon_m)/(\log(1-\varepsilon))\rceil$, as $w_i \in (\varepsilon_m, 1]$, and $k = 1, \ldots, \lceil(\log 1/r)/(\log(1-\varepsilon))\rceil + 1$, as $p_i/a_i \in [1, r]$. This implies that the total number of groups is at most

$$g_m := \lceil(\log \varepsilon_m)/(\log(1-\varepsilon))\rceil \cdot (\lceil(\log 1/r)/(\log(1-\varepsilon))\rceil + 1) \tag{5}$$

For simplicity, we redefine (decrease) the profits of the items in each group O_{jk} so that their profit/*height* ratio is equal to $r(1-\varepsilon)^{j+k}$, i.e., the profit of any (slice of) item in O_{jk} having height h is given by $r(1-\varepsilon)^{j+k} \cdot h$. Given that items in O_{jk} can be sliced, this implies that it is better to pack the items in O_{jk} with smallest width. Analogously, we redefine the profits of the items in each group V_{jk} so that their profit/*width* ratio is equal to $r(1-\varepsilon)^{j+k}$.

For each group O_{jk}, if $h(O_{jk}) > 1/(1-\varepsilon)^j$, we keep only the items with the smallest width for a total height equal to $1/(1-\varepsilon)^j$. Accordingly, in the reminder of this section we will assume $h(O_{jk}) \leq 1/(1-\varepsilon)^j$. Then, we consider the items in increasing order of widths, and define rg_m/ε subgroups $O_{jk1}, O_{jk2}, \ldots$ of consecutive items, so that the total height of the items in each subgroup $O_{jk\ell}$ is $h(O_{jk})\varepsilon/(rg_m)$. Note that the overall number of subgroups of items in O is at most rg_m^2/ε.

For each subgroup $O_{jk\ell}$, we define the *increased width* $\overline{w}_{jk\ell}$ of all items as the largest original width of an item in the subgroup. Finally, for each subgroup $O_{jk\ell}$ we further slice the items into $\lfloor h(O_{jk})\varepsilon/(rg_m\delta_m)\rfloor$ identical slices of width $\overline{w}_{jk\ell}$ and height δ_m, where $\delta_m \in [(\varepsilon_{m+1}/\varepsilon), \varepsilon_m]$ is a suitable constant, depending on $\varepsilon, r, \varepsilon_m$, to be specified later. The possible residual slice of height $< \delta_m$ is neglected.

The rounding procedure for the items in V is analogous, leading to at most rg_m^2/ε distinct *increased heights* and, for each increased height $\overline{h}_{jk\ell}$, to identical slices of height $\overline{h}_{jk\ell}$ and width δ_m. After having defined the slices as above, we consider these slices as single items that cannot be sliced further. Overall, this leaves us with a modified instance I' with the items in L and the items corresponding to slices from O and V. Note that $|I'| \leq |I|$ as $\delta_m > \varepsilon_{m+1}$.

Enumeration of the solutions for I': We enumerate all 2-Dim Geometric Knapsack solutions associated with I' as these are polynomially many. Specifically, since the area of each item in I' is at least $\delta_m\varepsilon_m$, only the $O(|I|^{1/(\delta_m\varepsilon_m)})$ subsets with at most $1/(\delta_m\varepsilon_m)$ items may be fit into the bin. Moreover, we can test in constant time if each of these subsets indeed fits into the bin, since we can assume that the bottom left corner of each item is placed into the bin at some (x, y) position which is an integer linear combination of the widths and heights of the items in the subset, and therefore we have $O(2^{2/(\delta_m\varepsilon_m)})$ possible positions for each item. For each solution for I', and the associated packing into the bin, we pack the small items in S and the original items in O and V by the greedy procedure of the next section. Among the solutions defined in this way, we keep the best one.

Converting the solution for I' into one for I: We use the empty spaces left in the bin by the items in I' to pack the items in S, and the space occupied by the slices of items in O and V to pack the original items in O and V. All (original) items in O, V and S are *unpacked* at the beginning of this phase. In order to pack the items in S, we draw horizontal and vertical lines through the coordinates of each corner of the items in I', and let the *cells* be the rectangles that are empty among those defined by these lines. We consider the cells one by one (in an arbitrary order) and, for each cell C, having area a_C, we consider the unpacked items in S in decreasing order of profit/area ratios and define a subset R by selecting the first items until condition $a(R) \geq a_C - 2\varepsilon_{m+1}$ is satisfied. We pack all the items in R by NFDH into the cell, given that they fit as we now show. Indeed, by Observation 1(iii), letting w_C and h_C be the width and height of the cell, after scaling all small item widths by $1/w_C$ and all item heights by $1/h_C$, we have that the area of any subset of items in S packed by NFDH in the cell, in case some items are unpacked, is at least $(1-\varepsilon_{m+1}/w_C)\cdot(1-\varepsilon_{m+1}/h_C)\cdot(w_C\cdot h_C) \geq w_C\cdot h_C - 2\varepsilon_{m+1} = a_C - 2\varepsilon_{m+1}$.

As to the items in O, for each group O_{jk}, we consider the slices of width $\overline{w}_{jk\ell}$ and height δ_m in increasing order of widths (i.e., by increasing ℓ). For each such slice, we consider the unpacked (original) items in O_{jk} in increasing order of widths, and define a subset R by selecting the first items until condition $h(R) \geq \delta_m - \varepsilon_{m+1}$ is satisfied. We pack all items in R in the slice (noting that they clearly fit). Note that the order in which we consider slices and items guarantees that we never run out of items. We do the same for the items in V.

The case with rotations and proof of approximation guarantee: For a full description of the case with rotations and the following lemma we refer to the full version.

Lemma 4. *By defining* $\varepsilon_0 := 1$ *and, for each* $m = 0, \ldots, 2/\varepsilon$, $\delta_m := \varepsilon^2/(2r^2 g_m^2 f(rg_m^2/\varepsilon, \varepsilon_m))$ *and* $\varepsilon_{m+1} := \varepsilon/(2r(2/(\delta_m \varepsilon_m) + 1)^2)$, *the value of the 2-Dim Geometric Knapsack solution produced by the algorithm above is at least* $(1 - 13\varepsilon)\,\mathrm{opt}_{\mathrm{2KP}}(I) - \varepsilon$, *where* g_m *is defined by (5) and* $f(\cdot,\cdot)$ *is the constant in the statement of Lemma 1.*

References

1. Baker, B.S., Coffman Jr., E.G., Rivest, R.L.: Orthogonal packing in two dimensions. SIAM Journal on Computing 9, 846–855 (1980)
2. Baker, B.S., Schwartz, J.S.: Shelf algorithms for two-dimensional packing problems. SIAM Journal on Computing 12, 508–525 (1983)
3. Bansal, N., Correa, J., Kenyon, C., Sviridenko, M.: Bin packing in multiple dimensions: inapproximability results and approximation schemes. Mathematics of Operations Research 31, 31–49 (2006)
4. Bansal, N., Caprara, A., Sviridenko, M.: Improved approximation algorithms for multidimensional bin packing problems. In: Proceedings of the 47-th Annual IEEE Symposium on Foundations of Computer Science (FOCS 2006), pp. 697–708 (2006)
5. Bansal, N., Caprara, A., Sviridenko, M.: A New Approximation Method for Set Covering Problems with Applications to Multidimensional Bin Packing. In: SICOMP (to appear)
6. Bansal, N., Lodi, A., Sviridenko, M.: A tale of two dimensional bin packing. In: Proceedings of the 46-th Annual IEEE Symposium on Foundations of Computer Science (FOCS 2005), pp. 657–666 (2005)
7. Bansal, N., Sviridenko, M.: Two-dimensional bin packing with one dimensional resource augmentation. Discrete Optimization 4, 143–153 (2007)
8. Caprara, A.: Packing 2-dimensional bins in harmony. In: Proceedings of the 43-rd Annual IEEE Symposium on Foundations of Computer Science (FOCS 2002), pp. 490–499 (2002)
9. Caprara, A., Lodi, A., Monaci, M.: Fast approximation schemes for two-stage, two-dimensional bin packing. Mathematics of Operations Research 30, 150–172 (2005)
10. Caprara, A., Monaci, M.: On the 2-dimensional knapsack problem. Operations Research Letters 32, 5–14 (2004)
11. Caprara, A., Monaci, M.: Bidimensional packing by bilinear programming. Mathematical Programming 118, 75–108 (2009)
12. Chung, F.R.K., Garey, M.R., Johnson, D.S.: On packing two-dimensional bins. SIAM Journal on Algebraic and Discrete Methods 3, 66–76 (1982)
13. Coffman Jr., E.G., Garey, M.R., Johnson, D.S., Tarjan, R.E.: Performance bounds for level-oriented two-dimensional packing algorithms. SIAM Journal on Computing 9, 808–826 (1980)
14. Correa, J.R.: Resource augmentation in two-dimensional packing with orthogonal rotations. Operations Research Letters 34, 85–93 (2006)

15. Fernandez de la Vega, W., Lueker, G.S.: Bin packing can be solved within $1 + \varepsilon$ in linear time. Combinatorica 1, 349–355 (1981)
16. Fishkin, A.V., Gerber, O., Jansen, K.: On efficient weighted rectangle packing with large resources. In: Deng, X., Du, D.-Z. (eds.) ISAAC 2005. LNCS, vol. 3827, pp. 1039–1050. Springer, Heidelberg (2005)
17. Fishkin, A.V., Gerber, O., Jansen, K., Solis-Oba, R.: Packing weighted rectangles into a square. In: Jedrzejowicz, J., Szepietowski, A. (eds.) MFCS 2005. LNCS, vol. 3618, pp. 352–363. Springer, Heidelberg (2005)
18. Grigoriadis, M.D., Khachiyan, L.G., Porkolab, L., Villavicencio, J.: Approximate max-min resource sharing for structured concave optimization. SIAM Journal on Optimization 11, 1081–1091 (2001)
19. Grötschel, M., Lovsz, L., Schrijver, A.: Geometric algorithms and combinatorial optimization. Springer, Berlin (1988)
20. Harren, R., van Stee, R.: Improved absolute approximation ratios for two-dimensional packing problems. In: Naor, S. (ed.) APPROX 2009, pp. 177–189 (2009)
21. Ibarra, O.H., Kim, C.E.: Fast approximation algorithms for the knapsack and subset sum problems. Journal of the ACM 22, 463–468 (1975)
22. Jansen, K., Prädel, L.: An Approximation Algorithm for Two-Dimensional Strip-Packing with Absolute Performance Bound $\frac{7}{4} + \varepsilon$. Unpublished Manuscript
23. Jansen, K., Prädel, L., Schwarz, U.M.: A 2-approximation for 2D Bin Packing. In: WADS 2009. LNCS, vol. 5664, pp. 399–410. Springer, Heidelberg (2009)
24. Jansen, K., Solis-Oba, R.: New approximability results for 2-dimensional packing problems. In: Kučera, L., Kučera, A. (eds.) MFCS 2007. LNCS, vol. 4708, pp. 103–114. Springer, Heidelberg (2007)
25. Jansen, K., Solis-Oba, R.: A polynomial time approximation scheme for the square packing problem. In: Lodi, A., Panconesi, A., Rinaldi, G. (eds.) IPCO 2008. LNCS, vol. 5035, pp. 184–198. Springer, Heidelberg (2008)
26. Jansen, K., van Stee, R.: On strip packing with rotations. In: Proceedings of the 37-th Annual ACM Symposium on the Theory of Computing (STOC 2005), pp. 755–761 (2005)
27. Jansen, K., Zhang, G.: Maximizing the total profit of rectangles packed into a rectangle. Algorithmica 47, 323–342 (2007)
28. Kenyon, C., Rémila, E.: A near-optimal solution to a two-dimensional cutting stock problem. Mathematics of Operations Research 25, 645–656 (2000)
29. Karmarkar, N., Karp, R.M.: An efficient approximation scheme for the one-dimensional bin-packing problem. In: Proceedings of the 23rd Annual IEEE Symposium on Foundations of Computer Science (FOCS 1982), pp. 312–320 (1982)
30. Lee, C.C., Lee, D.T.: A simple on-line bin packing algorithm. Journal of the ACM 32, 562–572 (1985)
31. Plotkin, S.A., Shmoys, D., Tardos, E.: Fast approximation algorithms for fractional packing and covering problems. Mathematics of Operations Research 20, 257–301 (1995)
32. Sahni, S.: Approximate algorithms for the 0/1 knapsack problem. Journal of the ACM 22, 115–124 (1975)
33. Steinberg, A.: A strip-packing algorithm with absolute performance bound 2. SIAM Journal on Computing 26, 401–409 (1997)

Max-Coloring Paths: Tight Bounds and Extensions

Telikepalli Kavitha[1,*] and Julián Mestre[2]

[1] Indian Institute of Science, Bangalore, India
kavitha@csa.iisc.ernet.in
[2] Max-Plank-Institut für Informatik, Saarbrücken, Germany
jmestre@mpi-inf.mpg.de

Abstract. The max-coloring problem is to compute a legal coloring of the vertices of a graph $G = (V, E)$ with a non-negative weight function w on V such that $\sum_{i=1}^{k} \max_{v \in C_i} w(v_i)$ is minimized, where $C_1, \ldots, C_k$ are the various color classes. Max-coloring general graphs is as hard as the classical vertex coloring problem, a special case where vertices have unit weight. In fact, in some cases it can even be harder: for example, no polynomial time algorithm is known for max-coloring trees. In this paper we consider the problem of max-coloring paths and its generalization, max-coloring a broad class of trees and show it can be solved in time $O(|V|+$time for sorting the vertex weights). When vertex weights belong to $\mathbb{R}$, we show a matching lower bound of $\Omega(|V| \log |V|)$ in the algebraic computation tree model.

1 Introduction

The *max-coloring problem* takes as input a graph $G = (V, E)$ with a weight function $w : V \to \mathbb{N}$. The problem is to compute a legal coloring of V such that $\sum_{i=1}^{k} \max_{v \in C_i} w(v_i)$ is minimized, where $C_1, \ldots, C_k$ are the various color classes. When all the weights are 1, this problem reduces to the classical problem of determining a legal coloring of V using the least number of colors.

The max-coloring problem arises in the problem of partitioning a set of n jobs into batches, where all jobs in a batch start at the same time, and the batch is completed when its last job finishes. The length of a batch is the length of the longest job in this batch. The graph G has each job as a vertex and each edge in G captures a conflict between a pair of jobs, that is, these jobs cannot be processed in the same batch. Thus a valid schedule is a legal coloring of G, since each color class consists of jobs that are conflict-free and can be processed together as one batch. The goal is to minimize the makespan of the batch schedule, which is the same as computing a max-coloring of G.

Since classical coloring is a special case of max-coloring (when all weights are 1), the problem is hard for general graphs. Unlike the classical coloring

* Work done as part of the DST-MPG partner group "Efficient Graph Algorithms" at IISc Bangalore.

Y. Dong, D.-Z. Du, and O. Ibarra (Eds.): ISAAC 2009, LNCS 5878, pp. 87–96, 2009.

problem, max-coloring bipartite graphs is APX-hard [8]. Even max-coloring of trees is non-trivial: there is an exact algorithm with running time $O(n^{\log n})$ where n is the number of vertices, but only a PTAS is known if we restrict ourselves to polynomial time algorithms [8]. Constant factor approximation algorithms and NP-hardness results for various classes of perfect graphs were given in [2].

The focus of this paper is on max-coloring paths and the more general problem of max-coloring *skinny trees*, these are trees where the set of vertices of degree at least 3 form an independent set. It is easy to show that at most 3 color classes are needed to optimally max-color a path (see, e.g., [1]). This simple observation immediately yields an $O(n^4)$ time algorithm [5]. This was subsequently improved to $O(n^2)$ by Escoffier *et al.* [3]. More recently, Halldórsson and Shachnai [6] gave a recursive $O(n \log n)$ time algorithm.

In this paper we show algorithms for max-coloring paths and skinny trees that take time $O(n + S(n))$, where $S(n)$ is the time it takes to sort the vertex weights. Since the vertex weights are integers, when they are polynomially bounded, our algorithms take $O(n)$ time using radix sort to sort the vertex weights. More generally, using the randomized algorithm of Han and Thorup for integer sorting [7] makes the expected running time of our algorithms $O(n\sqrt{\log \log n})$. Our algorithms can be easily adapted to give $1 + \epsilon$ approximation of the optimal max-coloring in time $O(n + 1/\epsilon)$.

We next consider the problem of max-coloring a path with *real* vertex weights and show a lower bound of $\Omega(n \log n)$ in the algebraic computation tree model. Thus the complexity of max-coloring a path with real vertex weights is $\Theta(n \log n)$.

Organization of the paper. Though our result for skinny trees implies our result for paths, we first present our algorithm for max-coloring paths as this algorithm and analysis are simpler. This is done in Section 2. In Section 3 we present our algorithm to max-color skinny trees. Section 4 contains our approximation algorithm for this problem. Section 5 contains our lower bound result.

2 Max-Coloring a Path in Time for Sorting the Vertex Weights

Our input is a path $(v_1, v_2, \ldots, v_n)$ with a weight function $w : \{v_1, \ldots, v_n\} \to \mathbb{N}$. Since the degree of each vertex is at most 2, it is immediate that optimally max-coloring a path requires at most three colors: call them RED, BLUE, and GREEN. Let the weight of a color class be the weight of a heaviest vertex colored by that color. We assume without loss of generality that the color class RED has the heaviest weight, followed by the color class BLUE and then the color class GREEN. The problem of max-coloring a path is the problem of determining the weights of the BLUE and GREEN color classes so that their sum is minimized, since the color class RED has weight w_{max}, the weight of the heaviest vertex. Any coloring always has the form $[w_{max}, \beta, \gamma]$, where the first coordinate denotes the weight of the RED color class, the second coordinate β denotes the weight of the BLUE color class and the third coordinate γ the weight of the GREEN color class.

Algorithm 1. MAX-COLORING-PATH $(\langle v_1, \ldots, v_n \rangle, w)$

$b_0 \leftarrow \max_{1 \leq j \leq n-1} \min\{w(v_j), w(v_{j+1})\}$
Scan the path from left to right identifying nodes $u_0, \ldots, u_r$ s.t. $w(u_j) > b_0$. These vertices break the path into smaller subpaths $\rho_1, \ldots, \rho_r$.
Store the path decomposition $\langle u_0, (p_1, z_1), u_1, \ldots, (p_r, z_r), u_r \rangle$ as a linked list, where p_j is the parity of ρ_j and z_j is a least weight vertex in ρ_j.
$g_0 \leftarrow \max_{1 \leq j \leq r} w(z_i)$
$Q \leftarrow$ queue containing the vertices z_i in non-increasing order of $w(z_i)$.
for $i = 1, \ldots, r$ **do**
    $z \leftarrow$ pop first vertex from Q
    $r \leftarrow$ lighter u-vertex adjacent to z's subpath
    if r is the first or last vertex in the path decomposition **then**
        update decomposition: remove r and its adjacent subpath
    else if r appears in between two subpaths ρ' and ρ'' **then**
        update decomposition: remove r and merge ρ', r, ρ'' into single subpath
        set the parity of this merged path to 1+ sum of parities of ρ' and ρ''
    if z's subpath had odd parity **then**
        $b_i = b_{i-1}$.
    else if z's subpath had even parity **then**
        $b_i \leftarrow \max\{w(r), b_{i-1}\}$
    $g_i \leftarrow$ weight of the next z-vertex in Q
return best coloring among the candidates $[w_{\max}, b_j, g_j]$ for $j = 0, \ldots, r$.

Fig. 1. Pseudocode of our algorithm for max-coloring a path

The algorithm is formally presented in Fig. 1.

We need the following lemmas to prove the correctness of our algorithm. The proof of Lemma 1 is omitted here due to lack of space.

Lemma 1. *Each candidate coloring* $[w_{max}, b_j, g_j]$ *the algorithm produces is valid.*

Lemma 2. *Let* $i < r$ *be such that* $b_i < b_{i+1}$*. Then for every valid coloring* $[w_{max}, \beta, \gamma]$ *if* $\beta < b_{i+1}$ *then* $\gamma \geq g_i$.

Proof. Consider the $i+1$st iteration of the algorithm and let z be the vertex removed from the queue (recall that $w(z) = g_i$). Since $b_{i+1} > b_i$ it must be the case that the subpath of z has even parity and that the lightest u-neighbor of z has weight b_{i+1}. It follows that any valid coloring $[w_{max}, \beta, \gamma]$ such that $\beta < b_{i+1}$ must color both u-neighbors of z RED; however, since the subpath has even parity we must also color at least one vertex in the subpath GREEN and so $\gamma \geq g_i$. □

Everything is in place to prove the correctness of our algorithm.

Theorem 1. *Algorithm* MAX-COLORING-PATH *outputs a feasible optimal max-coloring.*

Proof. First, we note that from Lemma 1 it follows that the solution output is a valid coloring. To argue that it is optimal, we compare its cost to any other valid coloring $[w_{max}, \beta, \gamma]$. If $\beta \geq b_r$ then our algorithm considers the candidate solution $[w_{\max}, b_r, g_r]$ where $g_r = 0$ (at the last iteration the queue is empty), so the solution output is at least as good as $[w_{max}, \beta, \gamma]$. Otherwise, there exists $0 \leq i < r$ such that $b_i \leq \beta < b_{i+1}$ (note that for any valid coloring $\beta \geq b_0$ since no two adjacent vertices can be colored RED.). Therefore, by Lemma 2 we have $\gamma \geq g_i$ and since the algorithm considers the candidate solution $[w_{max}, b_i, g_i]$, the solution output is at least as good as $[w_{max}, \beta, \gamma]$. It follows that our algorithm outputs an optimal solution. □

Running time. The initial path decomposition and the first coloring can be computed in $O(n)$ time. Building the queue Q requires that we sort the weights of the z_j vertices. After that, in each iteration we update the path by removing a subpath or we merge two adjacent subpaths. We can carry out the update in $O(1)$ time by keeping along with the vertices in Q, a pointer to the subpath they belong to in the current path decomposition. Each iteration removes one u-vertex from our path decomposition, so the total running time of the for loop is $O(r)$. Hence, other than the sorting done in Line 5, our algorithm runs in $O(n)$ time. The sorting time is our most expensive step. In the RAM model, this step can be carried out in $O(n\sqrt{\log \log n})$ expected time using the algorithm of Han and Thorup [7] for integer sorting. In fact, this sorting step is more efficient in many cases, for example when the weights are polynomially bounded in n.

Theorem 2. *The running time of our algorithm is $O(n + S(r))$, where $S(r)$ is the time it takes to sort the weights of the vertices $z_1, \ldots, z_r$ used in the initial path decomposition.*

3 Extension to Skinny Trees

In this section we present our algorithm for max-coloring a tree $T = (V, E)$ having the property that the set of vertices of degree at least 3 forms an independent set. We call such a tree *skinny*. As we will see next, any such tree has the interesting property that it always has an optimal coloring with at most 3 colors. Two notable examples of trees falling in this class are paths (every vertex has degree is at most 2) and spiders (there is a single vertex with degree 3 or more).

Lemma 3. *Every skinny tree T has an optimal max-coloring using at most 3 colors.*

Proof. Suppose a fourth color is used, where colors RED $\geq$ BLUE $\geq$ GREEN denote the three heaviest color classes. Consider any vertex x that is colored by a color different from these 3 colors. If this is a degree ≤ 2 vertex, then one of RED, BLUE, GREEN is missing from its neighborhood. Hence we can change the color of x to this missing color and this results in a valid coloring and the value to be optimized does not increase by this change of color. Hence after we perform

Algorithm 2. MAX-COLORING-SKINNY-TREE(T, w)

1. $b_0 \leftarrow g_0 \leftarrow \max_{(u,v)\in T} \min\{w(u), w(v)\}$
2. $H \leftarrow \{v \in V \mid w(v) > b_0\}$
3. $L \leftarrow \{v \in V \mid w(v) \leq b_0\}$
4. $Q \leftarrow$ list $z_1, \ldots, z_{|L|}$ of vertices in L in non-increasing order of weight
5. **for** z_i in Q **do**
6. transfer z_i from L to H
7. let C be the connected component of H containing z_i
8. $[\alpha, \beta] \leftarrow$ optimally 2-color C
9. $b_i \leftarrow \max\{\beta, b_{i-1}\}$
10. $g_i \leftarrow \max_{v\in L} w(v)$
11. record $[w_{\max}, b_i, g_i]$ as a candidate solution
12. **return** best coloring among the candidate solutions

Fig. 2. Pseudocode of our algorithm for max-coloring a skinny tree

this step for every such x of degree at most 2, only vertices of degree 3 or higher may be colored by a color other than RED, BLUE, or GREEN.

Let y be such a vertex. If y has no GREEN-colored neighbors, then we can change the color of y to GREEN. So let us assume that y has a neighbor v that is colored GREEN. By the structure of T, we know that the degree of v is at most 2. Since one neighbor of v is y, there is at most another neighbor of v; hence one of RED, BLUE is missing from v's neighborhood – thus we can change the color of v from GREEN to this missing color in RED, BLUE. The value to be optimized does not increase by this change of color. Repeating this step for every neighbor of y that is colored GREEN allows us to eliminate GREEN from y's neighborhood; now y can be colored GREEN. Hence it follows that the optimal max-coloring on such a graph uses at most 3 colors. □

From Lemma 3, an $O(n^3)$ time algorithm follows easily: Pemmaraju and Raman [8] gave an $O(kn)$ time algorithm that given a tree T and a tuple $[W_1, \ldots, W_k]$ determines if there is a feasible coloring $[W_1, \ldots, W_k]$ of T; since there are $O(n^2)$ possible triples $[w_{max}, \beta, \gamma]$ for 3-colorings, we get an $O(n^3)$ time algorithm. We improve on this by showing an almost linear time algorithm for max-coloring skinny trees. We first describe the algorithm and then proceed to show its correctness. The implementation details will be described later. The complete pseudocode for our algorithm is given in Fig. 2.

As we did before, we look at each edge and determine a lower bound b_0 for BLUE as $b_0 = \max_{(u,v)\in T} \min\{w(u), w(v)\}$. Since for any edge (u, v), both u and v cannot be colored RED, it follows that b_0 is a lower bound for BLUE. We partition the vertices into a set of heavy nodes $H = \{u \in T \mid w(u) > b_0\}$ and a set of light nodes $L = \{u \in T \mid w(u) \leq b_0\}$.

In the ith iteration we transfer a node $u \in L$ of maximum weight to H. Then we optimally color with RED and BLUE the nodes in H to obtain a coloring with cost $[\alpha, \beta]$. We extend this into a proper 3-coloring by coloring the remaining vertices of L with RED, BLUE, and GREEN. The cost of this coloring is at most $[w_{max}, b_i, g_i]$ where $b_i = \max\{\beta, b_{i-1}\}$ and $g_i = \max_{u \in L} w(u)$. We record this coloring as a candidate solution and at the end of the algorithm (when $L = \emptyset$) we return the best candidate solution thus produced.

3.1 Correctness

The correctness of the algorithm relies on the following key lemmas.

Lemma 4. *Let H be a subset of the vertices of a skinny tree T. Then every coloring of H with* RED *and* BLUE *can be extended to the whole tree using an extra color* GREEN.

Proof. First color with GREEN all the vertices in $T \setminus H$ whose degree is 3 or more. The vertices that remain uncolored form a collection of paths whose endpoints have degree 2 or 1 in the original tree. These left-over paths can be easily colored without creating any conflicts. □

Lemma 5. *Let i be such that $b_i < b_{i+1}$ and $[w_{max}, b, g]$ be a valid coloring of T with $b < b_{i+1}$. Then we must have $g \geq g_i$.*

Proof. Let H and L be the set of heavy and light vertices after the ith iteration is executed and let z_{i+1} be the vertex in L to be transferred to H in the $i+1$st iteration. Recall that $g_i = max_{v \in L} w(v) = w(z_{i+1})$.

The vertices in H form a number of connected components in T. If the vertex z_{i+1} is adjacent to none of these components then the cost of 2-coloring $H + z_{i+1}$ cannot be more than 2-coloring H. Therefore, adding z_{i+1} must enlarge a single component or merge several components. At any rate the cost of 2-coloring this new component must be $[\alpha, b_{i+1}]$—the cost of coloring the remaining components does not change with the update. Notice that this coloring is optimal for the component. Since the coloring $[w_{max}, b, g]$ from the lemma statement has $b < b_{i+1}$, it must be the case that at least one vertex in the component is colored green. Since all nodes in the component have weight greater or equal than g_i, it follows that $g \geq g_i$. □

Theorem 3. *The algorithm* MAX-COLORING-SKINNY-TREES *outputs a feasible optimal max-coloring.*

Proof. First we note that by Lemma 4, every candidate solution is feasible and thus the algorithm outputs a feasible solution.

Let $[w_{max}, b_r, g_r]$ be the last coloring produced by the algorithm. For each valid 3-coloring $[w_{max}, b, g]$ where $b < b_r$ it follows from Lemma 5 and the fact that $b \geq b_0$ that our algorithm finds an algorithm just as good. If $b \geq b_r$ then the last coloring produced is at least as good since $g_r = 0$ (at this point $L = \emptyset$). □

Implementation details. We will show that except for Line 4, the algorithm runs in linear time. In fact, we only need to argue that the time spent on Lines 7 and 8 is linear as it is straightforward to implement the remaining lines in linear time.

The details are omitted here due to lack of space. The main ideas are:

(1) The connected components of H can be kept track of via a union-find data structure. The union operations follow the underlying tree structure and for precisely this setup there exists a data structure that can execute both operations in $O(1)$ amortized time [4]. This means that we can execute Line 7 in $O(\deg(z_i))$ amortized time, where z_i is the vertex being processed.

(2) Line 8 can be implemented in the same time bound by maintaining for each connected component of H the root of the component and the two possible 2-colorings of the component: one that colors the root of the component RED and another that colors the root of the component BLUE. We call these the red and blue options and we store their costs. When a vertex z_i is transferred from L to H, it can be shown how to compute the new root of the newly formed component and how to compute the costs of the red/blue options in $O(\deg(z_i))$ amortized time.

We conclude this section with Theorem 4.

Theorem 4. *The algorithm* MAX-COLORING-SKINNY-TREES *can be implemented to run in $O(n + S(n))$ time, where $S(n)$ is the time it takes to sort the vertices by weight.*

4 A $(1 + \epsilon)$-Approximation Algorithm in $O(n + 1/\epsilon)$ Time

Except for the sorting step, the rest of our algorithm for skinny trees runs in $O(n)$ time. In this section we exploit this fact to get an algorithm that computes a $(1 + \epsilon)$-approximation of the optimal max-coloring in $O(n + 1/\epsilon)$ time.

Theorem 5. *There is an $O(n+1/\epsilon)$ time $(1+\epsilon)$ approximation for max-coloring a skinny tree.*

The idea is very simple. Given a skinny tree and a weight function w and a real number $\epsilon > 0$, we create another weight function w' so that for every vertex v_i in the tree

$$w'(v_i) = \left\lfloor \frac{3\, w(v_i)}{\epsilon\, w_{max}} \right\rfloor .$$

The weights under w' are integers in the range $[0, \lceil \frac{3}{\epsilon} \rceil]$. Therefore, by Theorem 2 we can find an optimal max-coloring in $O(n + \frac{1}{\epsilon})$ time using bucket sort for sorting the weights of the z_j vertices. We claim that this coloring is a $1 + \epsilon$ approximation for the original weights w. Let OPT be the optimal coloring under w and $w(\text{OPT})$ be its cost; similarly, let OPT$'$ be the optimal coloring under w' (the one output by our algorithm) and $w'(\text{OPT})$ be its cost.

We have $w(\text{OPT}') \leq (w'(\text{OPT}') + 3)\ \frac{\epsilon\, w_{max}}{3}$, since there are only three color classes. Further, this quantity is bounded by $(w'(\text{OPT}) + 3)\ \frac{\epsilon\, w_{max}}{3}$ since OPT' is optimal for w'. This in turn is bounded by $w(\text{OPT}) + \epsilon\, w_{max}$ from the definition of w'. Finally, since $w(\text{OPT}) \geq w_{max}$, we have $w(\text{OPT}) + \epsilon\, w_{max} \leq (1+\epsilon)\, w(\text{OPT})$.

5 Lower Bounds on Max-Coloring a Path with Real Vertex Weights

Our algorithm for max-coloring a path took time $O(n + S(r))$, where $S(r)$ is the time it takes to sort the weights of a subset of r vertices. The max-coloring problem was defined for integral vertex weights, since this problem was motivated by scheduling problems, where the time to process each job is integral. However the max-coloring problem can be naturally extended to real vertex weights also. When the vertex weights are real, our algorithm takes $O(n \log n)$ time (this is the time needed for sorting n real numbers). We will now show that this is the optimal running time for max-coloring a path in the real algebraic computation tree model. In this model the operations $\{+, -, \times, \div, \sqrt{}, = 0, \geq 0\}$ on reals can be performed at unit cost.

Our lower bounding method works via a problem that we call the *max-difference problem*. In the max-difference problem, we are given a set $\{y_1, \ldots, y_n\}$ of real numbers and let $\langle y_{\pi(1)}, \ldots, y_{\pi(n)} \rangle$ be these elements sorted in nondecreasing order. The problem is to determine the maximum value of $y_{\pi(i+1)} - y_{\pi(i)}$ for $1 \leq i \leq n-1$.

5.1 The Reduction from Max-Difference to Max-Coloring a Path

We will show an $O(n)$ time reduction from the max-difference problem to the max-coloring a path problem. Given a set $\{y_1, \ldots, y_n\}$ of n real numbers, let us determine the maximum and minimum of $y_1, \ldots, y_n$ in $O(n)$ time. Assume that y_1 is the minimum and y_n is the maximum. We form an instance of the max-coloring problem on a path as follows.

The path $(v_1, v_2, \ldots, v_N)$ consists of $N = 5(n-1) + 2$ vertices. The leftmost vertex v_1 has weight $y_n + y_1 - \epsilon$, where $\epsilon > 0$ is a small constant. The next vertex v_2 has weight $2y_n$; recall that the largest value among the y_i's is y_n and this is the largest vertex weight here. We can assume that v_2 (vertex with the maximum weight) is colored RED. Let the next heaviest color class be BLUE. Note that since v_1 is adjacent to v_2, the BLUE color class has weight at least $y_n + y_1 - \epsilon$.

The remaining vertices $(v_3, \ldots, v_N)$ can be split into $n-1$ blocks, each with 5 vertices. Let us call the 5 vertices of the ith block $z_i, u_i, x_{i_1}, x_{i_2}, u_i'$, where z_i has weight 0, vertices u_i and u_i' have weight $y_n + y_i$ and vertices x_{i_1} and x_{i_2} have weight $y_n - y_i$. Note that if both u_i and u_i' are colored RED, then one of x_{i_1}, x_{i_2} has to be colored GREEN. The vertices z_i of weight 0 in each block allow the blocks to be independent of each other.

There is a coloring of the path where the BLUE color class has weight $y_n + y_1 - \epsilon$ (the least possible weight for BLUE). Since $\epsilon > 0$, this forces all the $2(n-1)$

vertices of weights in $\{y_n + y_1, \ldots, y_n + y_{n-1}\}$ to be colored RED in the initial coloring. This thereby forces certain vertices with weights in $\{y_n - y_1, \ldots, y_n - y_{n-1}\}$ to be colored GREEN. So the GREEN color class has weight $y_n - y_1$ here (since y_1 is the minimum among the y_i's, $y_n - y_1$ is the largest among the $y_n - y_i$'s). Thus the sum of the BLUE and GREEN color classes in the above coloring is $y_n + y_1 - \epsilon + y_n - y_1 = 2y_n - \epsilon$. Here we only compute the sum of BLUE and GREEN color classes to compare colorings since the RED always has weight $2y_n$.

If we want a lower weight for the GREEN color class, then we need to increase the weight of the BLUE color class. Suppose we want the weight of the GREEN color class to go down to $y_n - y_{\pi(k)}$; this means that all the vertices with weights in $\{y_n - y_1, y_n - y_{\pi(2)}, \ldots, y_n - y_{\pi(k-1)}\}$ are colored RED or BLUE. The following lemma holds because we have 2 vertices of weight $y_n - y_{\pi(k-1)}$ sandwiched between 2 vertices of weight $y_n + y_{\pi(k-1)}$.

Lemma 6. *For all the vertices of weight $y_n - y_{\pi(k-1)}$ to be colored* non-GREEN, *a vertex of weight $y_n + y_{\pi(k-1)}$ should be colored* BLUE.

It is easy to see that with $y_n + y_{\pi(k-1)}$ as the heaviest BLUE vertex weight, we are free to color vertices with weights in $\{y_n + y_{\pi(1)}, y_n + y_{\pi(2)}, \ldots, y_n + y_{\pi(k-2)}\}$ BLUE, thus we can make all vertices with weights in $\{y_n - y_{\pi(1)}, y_n - y_{\pi(2)}, \ldots, y_n - y_{\pi(k-1)}\}$ *non*-GREEN. Thus the GREEN color class has weight $y_n - y_{\pi(k)}$ now. This makes the sum of BLUE and GREEN color classes in such a coloring $y_n + y_{\pi(k-1)} + y_n - y_{\pi(k)}$.

We will take ϵ to be a very small value, let $\epsilon < (y_n - y_1)/(n-1)$, the average difference. Then the best (least weight) coloring among all the candidates for BLUE+GREEN color class weights which are: $2y_n - \epsilon, 2y_n + y_{\pi(1)} - y_{\pi(2)}, \ldots, 2y_n + y_{\pi(k-1)} - y_{\pi(k)}, \ldots, 2y_n + y_{\pi(n-1)} - y_{\pi(n)}$ is that $2y_n + y_{\pi(i-1)} - y_{\pi(i)}$, where "$y_{\pi(i-1)} - y_{\pi(i)}$" is the *most negative*, that is, $y_{\pi(i)} - y_{\pi(i-1)}$ is the largest.

Thus solving the max-coloring a path problem solves the max-difference problem. An $O(t(n))$ running time for max-coloring a path implies an $O(t(n) + n)$ algorithm, that is, an $O(t(n))$ algorithm for the maximum difference problem, since $t(n) \geq n$ for the max-coloring problem.

5.2 A Lower Bound for the Max-Difference Problem

In this section we will show the following: in the algebraic computation tree model, if there exists an $O(t(n))$ algorithm for the max-difference problem, then there exists an $O(t(n) + n)$ algorithm for the following problem: given $\{x_1, \ldots, x_n\} \in \mathbb{R}^n$, do these numbers in sorted increasing order form a *non-trivial arithmetic progression*? That is, this problem "*Is-it-an-AP*" asks if there a $d > 0$ such that

$$\{x_1, \ldots, x_n\} = \{\min, \min + d, \min + 2d, \ldots, \min + (n-1)d\},$$

where min is the minimum among $x_1, \ldots, x_n$. Then we will show an $\Omega(n \log n)$ lower bound for *Is-it-an-AP*.

The $O(n)$ time reduction from "Is-it-an-AP" to "max-difference" is simple: we compute the max-difference of $x_1, \ldots, x_n$, call this d. Then we compute the following values $z_0, \ldots, z_{n-1}$, where $z_0 = \min$ and $z_i = z_{i-1} + d$ for $1 \leq i \leq n-1$, (min is the minimum among $x_1, \ldots, x_n$). It is easy to see that $\{x_1, \ldots, x_n\}$ forms a *yes* instance of "Is-it-an-AP" if and only if $\sum_{i=1}^{n} x_i = \sum_{i=0}^{n-1} z_i$ and $d \neq 0$.

Lower bound on "Is-it-an-AP". Our lower bounding method is absolutely the same as in the Element Distinctness problem. The *yes* instances of this problem form at least $n!$ connected components (the same reasoning as in Element Distinctness), so it follows from Ben-Or's result that any algebraic computation tree solving the *Is-it-an-AP* problem has depth $\Omega(\log n! - n)$, which is $\Omega(n \log n)$.

Theorem 6. *The complexity of max-coloring a path with real vertex weights is $\Theta(n \log n)$ in the algebraic computation tree model.*

Acknowledgments. Thanks to Magnús Halldórsson and Rajiv Raman for helpful discussions.

References

1. Brucker, P.: Scheduling Algorithms, 4th edn. Springer, Heidelberg (2004)
2. Epstein, L., Levin, A.: On the max-coloring problem. In: Kaklamanis, C., Skutella, M. (eds.) WAOA 2007. LNCS, vol. 4927, pp. 142–155. Springer, Heidelberg (2008)
3. Escoffier, B., Monnot, J., Paschos, V.T.: Weighted coloring: further complexity and approximability results. Inf. Process. Lett. 97(3), 98–103 (2006)
4. Gabow, H.N., Tarjan, R.E.: A linear-time algorithm for a special case of disjoint set union. J. Comput. Syst. Sci. 30(2), 209–221 (1985)
5. Guan, D.J., Zhu, X.: A coloring problem for weighted graphs. Inf. Process. Lett. 61(2), 77–81 (1997)
6. Halldórsson, M.M., Shachnai, H.: Batch coloring flat graphs and thin. In: Gudmundsson, J. (ed.) SWAT 2008. LNCS, vol. 5124, pp. 198–209. Springer, Heidelberg (2008)
7. Han, Y., Thorup, M.: Integer sorting in $O(n\sqrt{\log \log n})$ expected time and linear space. In: Proceedings of the 43th Annual IEEE Symposium on Foundations of Computer Science, pp. 135–144 (2002)
8. Pemmaraju, S.V., Raman, R.: Approximation algorithms for the max-coloring problem. In: Caires, L., Italiano, G.F., Monteiro, L., Palamidessi, C., Yung, M. (eds.) ICALP 2005. LNCS, vol. 3580, pp. 1064–1075. Springer, Heidelberg (2005)

Fréchet Distance Problems in Weighted Regions*

Yam Ki Cheung and Ovidiu Daescu

Department of Computer Science,
University of Texas at Dallas,
Richardson , TX 75080, USA
{ykcheung,daescu}@utdallas.edu

Abstract. We discuss two versions of the Fréchet distance problem in weighted planar subdivisions. In the first one, the distance between two points is the weighted length of the line segment joining the points. In the second one, the distance between two points is the length of the shortest path between the points. In both cases we give algorithms for finding a $(1 + \epsilon)$-factor approximation of the Fréchet distance between two polygonal curves. We also consider the Fréchet distance between two polygonal curves among polyhedral obstacles in $\mathcal{R}^3$ ($1/\infty$ weighted region problem) and present a $(1 + \epsilon)$-factor approximation algorithm.

1 Introduction

Measuring similarity between curves is a fundamental problem that appears in various applications, including computer graphics and computer vision, pattern recognition, robotics, and structural biology. One common choice for measuring the similarity between curves is the *Fréchet distance*, introduced by Fréchet in 1906 [16]. The traditional (continuous) Fréchet distance δ_F for two parametric curves P,Q: $[0,1] \to \mathcal{R}^d$ is defined as

$$\delta_F(P,Q) = \inf_{\alpha,\beta:[0,1]\to[0,1]} \sup_{r\in[0,1]} S(P(\alpha(r)), Q(\beta(r))),$$

where α and β range over all continuous non-decreasing functions with $\alpha(0) = \beta(0) = 0$ and $\alpha(1) = \beta(1) = 1$, S is a distance metric between points, and $d > 0$ is the dimension of the problem.

The Fréchet distance is described intuitively by a man walking a dog on a leash. The man follows a curve (path), and the dog follows another path. Both can control their speed, but backtracking is not allowed. The Fréchet distance between the curves is the length of the shortest leash that is sufficient for the man and the dog to walk their paths from start to end.

A reparameterization is a continuous, non-decreasing surjective function. The pair α, β of reparametrizations define how the end points of the leash, i.e. $P(\alpha(r))$ and $Q(\beta(r))$, sweep along their respective curves. We say that α, β is a *matching* between P and Q or α, β define a *monotone walk* from leash $P(0)Q(0)$ to leash

* This work was supported in part by NSF award CCF-0635013.

Y. Dong, D.-Z. Du, and O. Ibarra (Eds.): ISAAC 2009, LNCS 5878, pp. 97–111, 2009.

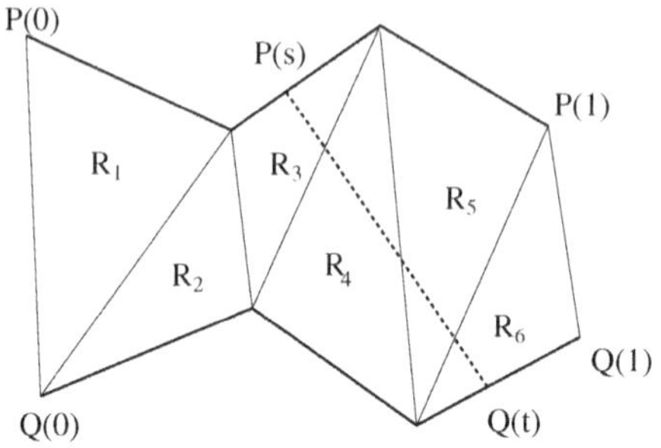

Fig. 1. An example of the Fréchet distance problem in weighted regions

$P(1)Q(1)$. In this paper, we use s and t to denote $\alpha(r)$ and $\beta(r)$, respectively. We assume P and Q are polygonal, and denote by $\delta(P) = (u_1, u_2, \ldots, u_p)$ and $\delta(Q) = (v_1, v_2, \ldots, v_q)$ the sequences of vertices of P and Q, respectively.

The *discrete Fréchet distance*, introduced by Either and Mannila [15], considers only the vertices of the (polygonal) curves. A coupling L between $\delta(P)$ and $\delta(Q)$ is defined as

$$(u_{a_1}, v_{b_1}), (u_{a_2}, v_{b_2}), \ldots, (u_{a_m}, v_{b_m}),$$

such that $a_1 = 1$, $b_1 = 1$, $a_m = p$, $b_m = q$, and for $i = 1, 2, \ldots, m$, we have $a_{i+1} = a_i$ or $a_{i+1} = a_i + 1$, and $b_{i+1} = b_i$ or $b_{i+1} = b_i + 1$. In other words, L is an order preserving pairing of vertices in P and Q. Note that each vertex can appear in L more than once. The discrete Fréchet distance is defined as

$$\delta_{dF}(P, Q) = \inf_{L} \max_{i=1,2,\ldots,m} S(u_{a_i}, v_{b_i}).$$

Let $R = \{R_1, R_2, \ldots, R_{n'}\}$ be a subdivision of the plane with a total of n vertices, with each region $R_i \in R$ having associated a positive integer weight w_i. The length of a path π that stays within a region R_i of R is defined as $w_i|\pi|$, where $|\pi|$ is the Euclidean length of π. The length of a path in R is the sum of the lengths of the subpaths within each region of R.

Let $P(s)Q(t)$ be the leash with endpoints $P(s)$ and $Q(t)$. Let $R_i(P(s)Q(t))$ be the Euclidean length of the line segment with endpoints $P(s)$ and $Q(t)$ within the region R_i. We define the distance between two points $P(s)$ and $Q(t)$ on P and Q, respectively, as (a) $S(P(s), Q(t)) = \sum_{i=1}^{n'} w_i * R_i(P(s)Q(t))$ or (b) $S(P(s), Q(t))$= the length of the shortest (weighted) path from $P(s)$ to $Q(t)$.

In this paper, we study the *Fréchet distance problem in weighted regions* in the plane: Given a weighted subdivision R and two parameterized polygonal chains P and Q in R, find the Fréchet distance between P and Q, where the distance between two points $P(s)$ and $Q(t)$ on P and Q, respectively, is defined either as in (a) or as in (b) above.

Without loss of generality, we assume R is triangulated and P and Q lie on boundaries of weighted regions (see Fig. 1 as an illustration). We also discuss a special case in $\mathcal{R}^3$, where the weights are either 1 or ∞ (i.e., free space and obstacles).

Motivation. The motivation for studying such measurement comes from path/travel planning. Let us consider the following scenario. Suppose during a military operation, there are two teams of military units traveling on separate paths. However, before reaching their own destinations, the two teams want to maintain constant radio communications so that in case of any emergency one team can rescue the other one in time. The question is how should these two teams schedule their trips such that at any given time, the radio signal received by one team from the other is stronger than a threshold, or what is the optimal traveling schedule for each team such that minimum signal strength received is maximized.

We use the *Beer-Lambert law* to model the decay (of the intensity) of an EM wave traveling through a region. Beer-Lambert law states that there is a logarithmic dependence between the *transmission* T of the EM wave and the product of the *absorption coefficient* associated with the region α and the distance l traveled by the EM wave in the region, i.e. $T = \frac{I}{I_o} = e^{-\alpha l}$, where I_o and I are the initial intensity (or power) of the EM wave and the intensity after the path, respectively [21]. If the EM wave travels through a set of regions $R = \{R_1, R_2, \ldots, R_k\}$, the transmission T can be written as $T = e^{\sum_{i=1}^{k} \alpha_i l_i}$, where α_i and l_i are the absorption coefficient associated with R_i and the distance traveled by the EM wave in R_i, respectively. The decay of the wave intensity can also be expressed in terms of the *absorbance* A which is defined as $A = -\log_{10} \frac{I}{I_0} = \sum_{i=1}^{k} \alpha_i l_i$. Notice that if we treat R as a weighted subdivision, where the weight of each region R_i is its absorption coefficient α_i, A is exactly the weighted length of the path traveled by the EM wave. Hence, this scheduling problem can be reduced to the Fréchet distance problem in weighted regions.

Previous Work. The Fréchet distance and its variants attracted considerable attention in literature. Most previous work assumes an unweighted environment and can be divided into two categories, depending on the distance metric used. In the first category, the distance between two points is the Euclidean distance. In other words, the leash is always a line segment, as in case (a) above. Bending of the leash is not allowed. Fréchet distances in this category are also referred to as non-geodesic Fréchet distances. Alt and Godau [4] give a fundamental study on the computational properties of the Fréchet distance. They present an algorithm that computes the exact Fréchet distance between two polygonal curves in $O(pq \log pq)$ time, where p and q are the number of vertices of P and Q, respectively. Either and Mannila [15] show how to find the discrete Fréchet distance between two polygonal curves. Rote [19] gives algorithms for finding the Fréchet distance between piecewise smooth curves. Buchin et al. [6] study the Fréchet distance between polygons. Wang et al. [7] presents the first exact, polynomial-time algorithm to compute a partial matching between two polygonal curves via Fréchet distance. In the second category, the distance between two points is the geodesic distance. Fréchet distances in this category are also referred to as geodesic Fréchet distances. The leash is allowed to bend to achieve to the minimum length. Maheshwari and Yi [17] study the case in which the curves lie on a convex polyhedron. Cook and Wenk [14] show how to compute the Fréchet distance between two curves when the leash is constrained inside a polygon or when polygonal obstacles are present between curves. Chambers et al. [8] give a polynomial-time algorithm to compute the *homotopic* Fréchet distance between two polygonal curves in the presence of obstacles. The homotopic Fréchet distance is the Fréchet distance with an additional continuity requirement: the leash is not allowed to switch discontinuously, e.g. jump over obstacles (unless the leash is long enough).

Weighted region shortest path problems have been investigated in computational geometry for about two decades. Using Snell's refraction low and the continuous Dijkstra algorithm, Mitchell and Papadimitriou [18] present an $O(n^8 \log \frac{nN'\rho}{\epsilon})$ time algorithm, where N' is the largest integer coordinate of vertices and ρ is the ratio of the maximum weight to the minimum weight. Aleksandrov et al. [2,3] provide two logarithmic discretization schemes that place Steiner points along edges or bisectors of angles of regions in R, forming a geometric progression. The placement of the Steiner points depends on an input parameter $\epsilon > 0$ and the geometry of the subdivision. The $(1+\epsilon)$-approximation algorithms in [2,3] take $O(\frac{n}{\epsilon}(\frac{1}{\sqrt{\epsilon}} + \log n) \log \frac{1}{\epsilon})$ and $O(\frac{n}{\sqrt{\epsilon}} \log \frac{n}{\epsilon} \log \frac{1}{\epsilon})$ time, respectively. Sun and Reif [20] give an algorithm, called BUSHWHACK, which constructs a discrete graph G by placing Steiner points along edges of the subdivision. By exploiting the geometric property of an optimal path, BUSHWHACK computes an approximate path more efficiently as it accesses only a subgraph of G. Combined with the logarithmic discretization scheme introduced in [2], BUSHWHACK takes $O(\frac{n}{\epsilon}(\log \frac{1}{\epsilon} + \log n) \log \frac{1}{\epsilon})$ time. Very recently, Aleksandrov et al. [1] gave a query algorithm that can find an ϵ-approximate shortest path between any two points in $O(\bar{q})$ time, where $\bar{q}$ is a query time parameter. The preprocessing time of this algorithm is $O(\frac{(g+1)n^2}{\epsilon^{2/3}\bar{q}} \log \frac{n}{\epsilon} \log^4 \frac{1}{\epsilon})$, where g is the genus of the discrete graph constructed by the discretization scheme. Cheng et al. [10] give an algorithm to approximate optimal paths in anisotropic regions, which is a generalized case of weighted regions. Their algorithm takes $O(\frac{\rho^2 \log \rho}{\epsilon^2} n^3 \log(\frac{\rho n}{\rho}))$ time, where $\rho \geq 1$ is a constant such that the convex distance function of any region contains a concentric Euclidean disk with radius $1/\rho$. In weighted regions, the time complexity of the algorithm is improved to $O(\frac{\rho \log \rho}{\epsilon} n^3 \log(\frac{\rho n}{\epsilon}))$ time, where ρ is the ratio of the maximum weight to the minimum weight. Very recently, Cheng et al. [11] also provided a query version of this algorithm that gives an approximate optimal path from a fixed source (in an anisotropic subdivision) in $O(\log \frac{\rho n}{\epsilon})$ time. The preprocessing time is $O(\frac{\rho^2 n^4}{\epsilon^2}(\log \frac{\rho n}{\epsilon})^2)$.

Our results. Let $\epsilon > 0$ be a positive constant given as part of the input. We also assume $\epsilon < 1$ at times. We have the following results: (1) For weighted regions in the plane, with $S(P(s), Q(t)) = \sum_{i=1}^{n'} w_i * R_i(P(s)Q(t))$ (case (a) above), we present a $(1+\epsilon)$-approximation algorithm that takes $O(pqN^4 \log(pqN))$ time, $N = O(C(R)(\frac{n}{\epsilon}(\log \frac{1}{\epsilon} + \log n) \log \frac{1}{\epsilon})$ is the total number of Steiner points used and $C(R)$ is a constant associated with the geometry of the subdivision R; (2) For weighted regions in the plane, with $S(P(s), Q(t))$=the length of the shortest (weighted) path from $P(s)$ to $Q(t)$, we present a $(1+\epsilon)$-approximation algorithm that takes $O(C(R)^2 \frac{pq}{\epsilon^2}(\log^4 \frac{1}{\epsilon})\bar{q} + \frac{(g+1)n^2}{\epsilon^{2/3}\bar{q}} \log \frac{n}{\epsilon} \log^4 \frac{1}{\epsilon})$ time, where $\bar{q}$ is a query time parameter related to computing shortest path and g is the genus of the graph constructed by the discretization scheme. (3) We give a $(1+\epsilon)$-approximation algorithm for finding the Fréchet distance between P and Q among polyhedral obstacles in $\mathcal{R}^3$ (1 and ∞ weights). To the best of our knowledge this is the first result for the $\mathcal{R}^3$ problem. The algorithm takes $O(C(R)^2 pq(1/\epsilon^2) \log^4(1/\epsilon)(n^2\lambda(n) \log(n/\epsilon)/\epsilon^4 + n^2 \log(n\gamma) \log(n \log \gamma)))$ time, where γ is the ratio of the length of the longest obstacle edge to the Euclidean distance between the two points, and $\lambda(n)$ is a very slowly-growing function related to the inverse of the Ackermann's function.

2 Preliminaries and Definitions

We denote by $P(s)Q(t)$ the leash (line segment, also called link, or shortest path, depending of context) from $P(s)$ to $Q(t)$, where $s, t \in [0, 1]$.

Alt and Godau [4] introduced the *free space diagram* to solve the decision version of the non-geodesic Fréchet distance problem (unweighted case): Given two polygonal curves P and Q, and a positive constant Δ, determine if $\delta_F(P, Q) < \Delta$. A free space diagram is a $[0, 1] \times [0, 1]$ parameter space such that each point (s, t) in the parameter space corresponds to a leash with end points $P(s)$ and $Q(t)$. A *free space cell* $C \subseteq [0, 1]^2$ is defined by two line segments, one from each curve. C corresponds to the leashes with endpoints on the two segments. The *free space* in the parameter space is defined as $\{(s, t) : S(P(s), Q(t)) < \Delta\}$. That is, the free space represents all links shorter than Δ. A *monotone path* is a path monotone along both coordinate axes. There is a one-to-one correspondence between all possible matchings of P and Q and all monotone paths from $(0, 0)$ to $(1, 1)$ in the free space diagram. Hence, $\delta_F(P, Q) < \Delta$ if and only if there exists a monotone path in the free space from $(0, 0)$ to $(1, 1)$.

3 The Line Segment Leash

In this section, we discuss the case of the line segment leash. That is, $S(P(s), Q(t)) = \sum_{i=1}^{n'} w_i * R_i(P(s)Q(t))$. We first briefly discuss a pseudo polynomial exact algorithm for solving the decision version of this problem. Then, we address the optimization version of this problem directly and give a polynomial time approximation algorithm.

3.1 An Exact Algorithm for the Decision Problem

In this section, we give an exact algorithm extending Alt and Godau's algorithm [4] to solve the decision version of the problem. Recall that the free space diagram introduced by Alt and Godau is a $[0,1] \times [0,1]$ parameter space such that each point (s,t) in the parameter space corresponds to a leash with end points $P(s)$ and $Q(t)$. The free space diagram can be decomposed into $O(pq)$ free space cells such that each cell C corresponds to the leashes with endpoints on two segments, one from each curve, where p and q are the number of vertices of P and Q, respectively. To determine if $\delta_F(P,Q) < \Delta$ for some positive constant Δ, we proceed as follows:

1. Partition each cell C into regions such that each region corresponds to leashes intersecting the same sequence of edges in R.
2. For each region, find the free space with respect to the positive constant Δ by solving an $O(n)$ order bivariate polynomial system with $O(n)$ equations and inequalities.
3. Determine if there exists a monotone path in the free space from $(0,0)$ to $(1,1)$.

The partition in step 1 is an arrangement of $O(n)$ curves of the form $st + c_1 s + c_2 t + c_3 = 0$, where c_1, c_2, and c_3 are constants. In the worst case, the total number of regions in the parameter space is (pqn^2). Each region can be defined mathematically by $O(n)$ inequalities. Once the free space in each region is computed, step 3 can be solved by decomposing the parameter space and constructing a directed shortest path graph. We will address steps 1 and 3 in detail in later sections. Both steps can be solved in polynomial time. Next, we will show that Step 2 takes pseudo polynomial time, which dominates the complexity of this algorithm.

For each region, let $f(s,t)$ be the weighted length of the leash $P(s)Q(t)$ in terms of s and t, let and $\{e_1, e_2, \ldots, e_k\}$ be the sequence of edges intersected by the leash. Note that $k = O(n)$ and e_1 and e_k are the two segments in P and Q, respectively. Following [9], the weighted length of leashes within the same region have the same functional expression, i.e. $S(P(s), Q(t)) = \sqrt{1+m^2}\sum_{i=1}^{k-1} w_i \left|\frac{p_{i+1}-p}{m-m_{i+1}} - \frac{p_i - p}{m-m_i}\right| = \sqrt{1+m^2}\sum_{i=1}^{k} \frac{a_i p + b_i}{m-m_i}$, where a_i, b_i are constants, w_i is the weight of the region bounded by edges e_i and e_{i+1}, m, p are the slope and intercept of the leash, respectively, and m_i, p_i are the slope and intercept of $e_i \in \{e_1, e_2, \ldots, e_k\}$, respectively. Let the end points of e_1 be $P(s_1) = (x_1, y_1)$ and $P(s_2) = (x_2, y_2)$, respectively, and the end points of e_k be $Q(t_1) = (x_3, y_3)$ and $Q(t_2) = (x_4, y_4)$, respectively. For any $s \in [s_1, s_2]$, we have

$$P(s) = (x_1 + (x_2 - x_1)(s - s_1)/(s_2 - s_1), y_1 + (y_2 - y_1)(s - s_1)/(s_2 - s_1)),$$

and similarly for any $t \in [t_1, t_2]$, we have

$$Q(t) = (x_3 + (x_4 - x_3)(t - t_1)/(t_2 - t_1), y_3 + (y_4 - y_3)(t - t_1)/(t_2 - t_1)).$$

Expressing m and p in terms of s and t , we obtain that

$$m = \frac{c_1 s + c_2 t + c_3}{d_1 s + d_2 t + d_3}$$

and

$$p = \frac{c_1' st + c_2' s + c_3' t + c_4'}{d_1 s + d_2 t + d_3},$$

where $c_1, c_2, c_3, c_1', c_2', c_3', c_4', d_1, d_2$, and d_3 are constants. It follows that

$$f(s,t) = \sqrt{1 + (\frac{c_1 s + c_2 t + c_3}{d_1 s + d_2 t + d_3})^2} \sum_{i=1}^{k} \frac{a_i(c_1' st + c_2' s + c_3' t + c_4')/(d_1 s + d_2 t + d_3) + b_i}{(c_1 s + c_2 t + c_3)/(d_1 s + d_2 t + d_3) - m_i}.$$

We can find the boundary of free space in this region by solving the following equation system: $f(s,t) = \Delta$ subject to $O(n)$ inequalities that define the region.

Since there are $O(n)$ fractional terms in $f(s,t)$, we can convert this equation system into a bivariate polynomial system of degree $O(n)$ which consists of one polynomial equation and $O(n)$ inequalities and requires pseudo polynomial time to solve [13].

Next, we are going propose an approximation algorithm for the optimization version of this problem.

3.2 Discretization Using Steiner Points

We first discretize the problem by placing Steiner points in R extending the discretization scheme given in [2]. Let E be the set of all edges in R. Let V be the set of vertices in R. For any point v on an edge in E, let $E(v)$ be the set of edges incident to v and let $d(v)$ be the minimum distance between v and edges in $E \setminus E(v)$. For each edge $e \in E$, let $d(e) = \sup\{d(v) | v \in e\}$ and let v_e be the point on e so that $d(v_e) = d(e)$. For each $v \in V$, the vertex radius for v is defined as $r(v) = \frac{\epsilon B}{n w_{max}(v)}$, where ϵ is a positive real number defining the quality of the approximation, B is a lower bound on $\delta_F(P, Q)$, and $w_{max}(v)$ is the maximum weight among all weighted regions incident to v. The disk of radius $r(v)$ centered at v defines the vertex-vicinity of v. B can be computed in $O(pq \log(pq))$ time using the standard (unweighted case) continuous Fréchet distance algorithm described in [4], where p and q are the number of vertices of P and Q, respectively.

For each edge $e = v_1 v_2$ in E, we place Steiner points $v_{i,1}, v_{i,2}, \ldots, v_{i,k_i}$ outside the vertex-vicinity of v_i, for $i = 1, 2$, such that $|v_i v_{i,1}| = r(v_i)$, $|v_{i,j} v_{i,j+1}| = \epsilon d(v_{i,j})$, for $j = 1, 2, \ldots, k_i - 1$, and $v_{i,k_i} = v_e$. It follows from [2] that the number of Steiner points placed on an edge is $O(C(e) 1/\epsilon \log 1/\epsilon)$, where $C(e) = O(\frac{|e|}{d(e)} \log \frac{|e|}{\sqrt{r(v_1) r(v_2)}})$. Let N denote the total number of Steiner points and vertices of R. Since we set $r(v) = \frac{\epsilon B}{n w_{max}(v)}$ for each $v \in R$, we have $N = O(C(R)(\frac{n}{\epsilon}(\log \frac{1}{\epsilon} + \log n) \log \frac{1}{\epsilon}))$, where $C(R) = \max_{e \in R}(\frac{|e| \log |e| w_{max}(e)}{d(e) B})$ is a constant associated with the geometry of R, and $w_{max}(e)$ is the maximum weight among all weighted regions incident to the end points of e.

We refer to a line segment bounded by two consecutive Steiner points on an edge of R as a *Steiner edge*. An *hourglass* is the union of all leashes (line segments) intersecting the same sequence of Steiner edges in the same order. Let

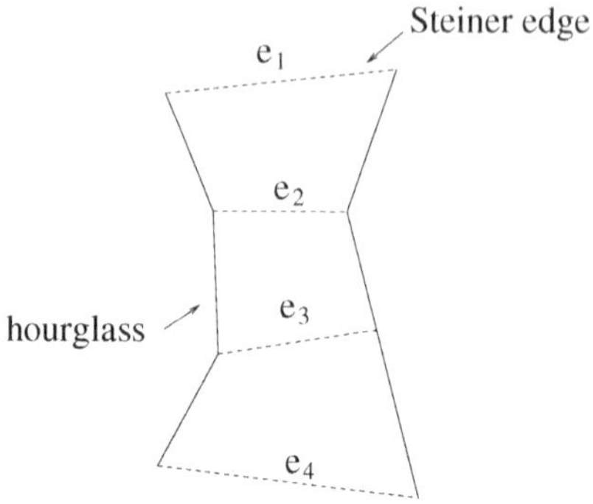

Fig. 2. An hourglass is the union of leashes intersecting with the same set of Steiner edges

H be an hourglass defined by a sequence of Steiner edges $\{e_1, e_2, \ldots, e_k\}$, where $e_1 \in P$ and $e_k \in Q$ (See Fig. 2).

Lemma 1. *Let l and l' be two segments in H. Then, $S(l) \leq (1+2\epsilon)S(l') + 2\epsilon B$.*

Proof. Let $R_{i_1}, R_{i_2}, \ldots, R_{i_{k-1}}$ be the weighted regions in H, such that R_{i_j} is between e_j and e_{j+1}. Recall that for a region R_k of R, $R_k(l)$ denotes the Euclidean length of l within R_k. We have

$$S(l) = \sum_{j=1}^{k-1} w_{i_j} R_{i_j}(l) \leq \sum_{j=1}^{k-1} w_{i_j}(R_{i_j}(l') + e_j + e_{j+1}),$$

where w_{i_j} is the positive weight associated with R_{i_j}.

If a Steiner edge e_j is outside of any vertex-vicinity, $|e_j| \leq \epsilon R_{i_{j-1}}(l')$ and $|e_j| \leq \epsilon R_{i_j}(l')$, due to the placement of the Steiner points. If e_j is incident to a vertex v, then $|e_j| = r(v) = (\epsilon B)/(n w_{max}(v))$. One segment can intersect at most $O(n)$ Steiner edges inside vertex vicinities. The result follows. □

Let l_H be an arbitrary segment in H, with endpoints on P and Q. Let α and β be two reparametrizations that define a matching between P and Q. Let $J = \{H_1, H_2, \ldots, H_k\}$ be the set of hourglasses that are traversed by the leash $P(\alpha(r))Q(\beta(r))$. For an hourglass $H \in J$, let $I_H = \{e | e = P(\alpha(r))Q(\beta(r)), r \in [0,1], e \in H\}$. Let $H(\alpha, \beta)$ be the segment in I_H with the largest weighted length. That is, $S(H(\alpha,\beta)) = \max_{e \in I_H} S(e)$. Let $\delta(\alpha, \beta) = \sup_{r \in [0,1]} S(P(\alpha(r)), Q(\beta(r)))$.

Lemma 2. $|\max_{H \in J} S(l_H) - \delta(\alpha, \beta)| \leq 4\epsilon\delta(\alpha, \beta)$.

Proof. Omitted.

We say that $\delta(J) = \max_{H \in J} S(l_H)$ is a 4ϵ-approximation of $\delta(\alpha, \beta)$. Given a sequence of hourglasses $J = \{H_1, H_2, \ldots, H_k\}$, we call J *legal* if there exist two reparmetrizations α, β that define a leash traversing the same sequence of hourglasses as J.

Lemma 3. *We can find a value $\delta'_F(P,Q)$ such that $|\delta'_F(P,Q) - \delta_F(P,Q)| \leq 4\epsilon\delta_F(P,Q)$, that is, $\delta'_F(P,Q)$ is a 4ϵ-approximation of $\delta_F(P,Q)$.*

Proof. (Sketch) Let $\hat{\alpha}$,$\hat{\beta}$ be the optimal reparametrizations that give the Fréchet distance between P and Q. Let $\hat{J}$ be the sequence of hourglasses traversed by the leash defined by α and β. Applying *Lemma* 2, we have,

$$|\delta(\hat{J}) - \delta(\hat{\alpha},\hat{\beta})| = |\delta(\hat{J}) - \delta_F(P,Q)| \leq 4\epsilon\delta(\hat{\alpha},\hat{\beta}) = 4\epsilon\delta_F(P,Q).$$

Then, there is a legal sequence of hourglasses that gives a 4ϵ-approximation of $\delta_F(P,Q)$. □

3.3 Fréchet Distance between Two Segments

In this section, we study a special case of the problem where each curve consists of one line segment. Similar to Alt and Godau's algorithm, we attack the problem in the parameter space $D = [0,1]^2$, where a leash $P(s)Q(t)$ is associated with a point $(s,t) \in D$. We refer to $(s,t) \in D$ as the *dual point* of the leash $P(s)Q(t)$.

Lemma 4. *All leashes through a Steiner point v correspond to a curve in D, with equation $C_v : st + c_1 s + c_2 t + c_3 = 0$, where c_1, c_2, and c_3 are constants.*

Proof. Let $v = (a,b)$, $P(0) = (x_1,y_1)$, $P(1) = (x_2,y_2)$, $Q(0) = (x_3,y_3)$, $Q(1) = (x_4,y_4)$. We have

$$P(s) = (x_1 + (x_2 - x_1)s, y_1 + (y_2 - y_1)s),$$

$$Q(t) = (x_3 + (x_4 - x_3)t, y_3 + (y_4 - y_3)t),$$

and $P(s)Q(t) : y - (y_1 + (y_2 - y_1)s) = \frac{y_1+(y_2-y_1)s-(y_3+(y_4-y_3)t)}{x_1+(x_2-x_1)s-(x_3+(x_4-x_3)t)}(x - (x_1 + (x_2 - x_1)s))$.

Since $P(s)Q(t)$ passes through v, we obtain that

$$(b - y_1 - (y_2 - y_1)s)(x_1 - x_3 + (x_2 - x_1)s - (x_4 - x_3)t)) = (a - x_1 - (x_2 - x_1)s)(y_1 - y_3 + (y_2 - y_1)s - (y_4 - y_3)t),$$

and thus we have

$$C_v : st + c_1 s + c_2 t + c_3 = 0,$$

where c_1, c_2, and c_3 are constants. □

We call the curve C_v the *dual curve* of Steiner point v. C_v is continuous and monotone along both (s and t) axes. Obviously, v lies on a leash if and only if the dual point of the leash in D lies on C_v. Next we define the relative position of two leashes with respect to a Steiner point v. Given two leashes $P(s)Q(t)$ and $P(s')Q(t')$, we say they are on the *same side* of v if and only if there exists two continuous functions $\alpha', \beta' : [0,1] \to [0,1]$, such that $\alpha'(0) = s$, $\alpha'(1) = s'$, $\beta'(0) = t$, $\beta'(1) = t'$ and $v \notin P(\alpha'(r)Q(\beta'(r)), \forall r \in [0,1]$. Note that α', β' are not necessarily monotone. Intuitively, two leashes are on the same side of v, if one leash can reach the other one by sweeping its end points on their respective curves without crossing v.

Lemma 5. *C_v divides D into two partitions, each partition corresponding to all links on the same side of point v.*

Proof. Omitted. □

We partition D by the dual curves of the Steiner points. Using the algorithm in [5], the partition can be computed in $O(N \log N + k)$ time and $O(N + k)$ space, where $N = O(C(R)(\frac{n}{\epsilon}(\log \frac{1}{\epsilon} + \log n) \log \frac{1}{\epsilon}))$ is the total number of Steiner points and k is the number cells, which is $O(N^2)$ in worst case. Let A denote the partition.

Lemma 6. *In A, each cell corresponds to an hourglass.*

Proof. If two points $(s, t), (s', t')$ belong to the same cell, then there exists a walk between $P(s)Q(t)$ and $P(s')Q(t')$ that does not cross any Steiner point, i.e. $P(s)Q(t)$ and $P(s')Q(t')$ intersect with the same set of Steiner edges. If $(s, t), (s', t')$ belong to different cells, then any walk between $P(s)Q(t)$ and $P(s')Q(t')$ must cross at least one Steiner point. That is, $P(s)Q(t), P(s')Q(t')$ do not intersect with the same set of Steiner edges. □

The Fréchet distance between P and Q can then be approximated as follows:
1. Place Steiner points as described previously.
2. Partition D by dual curves of all Steiner points.
3. For each cell Z, choose an arbitrary leash l and assign $S(l)$ as the weight of the cell.
4. Find a monotone path T in D, from its left bottom corner, $(0, 0)$, to its top right corner, $(1, 1)$, such that the maximum weight of the cells traversed by T is minimized.

Since the sequence of cells traversed by a monotone path in D corresponds to a legal sequence of hourglasses traversed by a leash following a match of P and Q, and vice versa, by Lemma 3 the maximum weight of the cells traversed by T is a 4ϵ-approximation of the Fréchet distance between P and Q.

3.4 Finding an Optimal Path in D

We define the cost of a path between two points in D as the maximum weight of the regions traversed by the path. We decompose D by extending a horizontal as well as a vertical line from every vertex until it reaches the boundary of D. See Fig. 3 for an illustration. Let the new subdivision be D'.

Lemma 7. *Given an edge e in D', let p be a point on e and let T_p be an arbitrary monotone path from $(0, 0)$, i.e. the bottom left corner of D', to p. Then, there exists a monotone path from $(0, 0)$ to any point on e with the same cost of T_p.*

Proof. Omitted. □

Thus, given an optimal monotone path T_p from $(0, 0)$ to an arbitrary point $p \in e$, we can construct a monotone path from $(0, 0)$ to any point in e, which has the same cost as T_p and is also optimal. To find the optimal monotone path from $(0,0)$ to $(1,1)$, we construct a directed graph G having as vertices the (open) edges as well as the vertices of D'. In G, a direct edge is added from a node v_1 to a node v_2 if, in D', v_1 and v_2 share a common cell and there exists a monotone path from v_1 to v_2. The sought path can be found by running Dijkstra's algorithm on G.

Time complexity: The complexity of D' is N^4, where $N = O(C(R)(\frac{n}{\epsilon}(\log \frac{1}{\epsilon} + \log n) \log \frac{1}{\epsilon}))$ is the number of Steiner points used, since each cell in D' has $O(1)$ edges and vertices, and thus each cell contributes $O(1)$ edges to G. Then, G has $O(N^4)$ vertices and $O(N^4)$ edges and Dijkstra's algorithm takes $O(N^4 \log N)$ time.

3.5 Fréchet Distance between Two Polygonal Chains

We extend the algorithm above to approximate the Fréchet distance between two polygonal chains P and Q. Recall that p and q are the number of vertices of P and Q, respectively. The parameter space D can be divided into pq subspaces, which are equivalent to the free space cells in the free space diagram, such that each subspace corresponds to all leashes bounded by the same two segments, one from each chain. The Fréchet distance $\delta_F(P,Q)$ can be approximated by finding the optimal path from the bottom left corner to the top right corner of D. The complexity of D is $O(pqN^2)$ and the complexity of the decomposed parameter space D' is $O(p^2q^2N^4)$, where $N = O(C(R)(\frac{n}{\epsilon}(\log \frac{1}{\epsilon} + \log n) \log \frac{1}{\epsilon}))$ is the total number of Steiner points. It takes $O(p^2q^2N^4 \log(pqN))$ time to approximate $\delta_F(P,Q)$.

4 Geodesic Fréchet Distance

In this section we study two versions of the geodesic Fréchet distance problem. We do not require the leash to be homotopic, i.e. the leash is allowed to sweep discontinuously without penalty. For example, the leash can pass through or jump over obstacles.

4.1 Geodesic Fréchet Distance in Weighted Regions in $\mathcal{R}^2$

Recall that the cost (weighted length) of a path in R is defined as the weighted sum of its Euclidean lengths within each region of R. A geodesic path between two points in R is a path between those points that has minimum cost. Here, the distance between two points in R is the cost of the geodesic path between those points.

We prove the geodesic Fréchet distance can be approximated by the discrete geodesic Fréchet distance. Given two polygonal curves P, Q, to approximate $\delta_F(P,Q)$, we need to add additional vertices to P and Q. We follow a similar approach as in [2], except that we define the vertex radius of a vertex v in P or Q as $r(v) = \frac{\epsilon B}{w_{max}(v)}$, where ϵ is a positive real number defining the quality

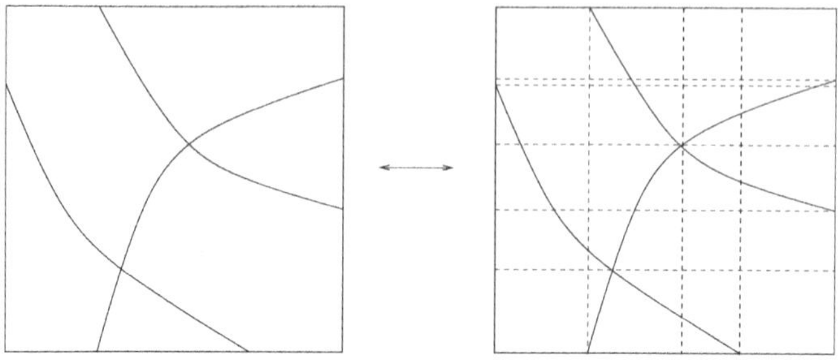

D D'

Fig. 3. An example of the decomposition of D

of the approximation, B is a lower bound on $\delta_F(P,Q)$, and $w_{max}(v)$ is the maximum weight among all weighted regions incident to v. B can be computed in $O(pq\log(pq))$ time using the standard (unweighted case) continuous Fréchet distance algorithm described in [4], where p and q are the number of vertices of P and Q, respectively. Recall that the disk of radius $r(v)$ centered at v defines the vertex-vicinity of v. New vertices are placed on edges of P nd Q forming a geometric progression with ratios depending on ϵ and on the geometry of R (see Section 3.1 for details). The total number of additional vertices introduced on each edge is $O(C(R)\frac{1}{\epsilon}\log^2\frac{1}{\epsilon})$, where $C(R)$ is a constant associated with geometry of R. Let P' and Q' be the new polygonal chains and let $p' = O(C(R)\frac{p}{\epsilon}\log^2\frac{1}{\epsilon})$ and $q' = O(C(R)\frac{q}{\epsilon}\log^2\frac{1}{\epsilon})$ be the number of vertices of P' and Q', respectively.

Lemma 8. *The discrete geodesic Fréchet distance between P' and Q' gives an ϵ-approximation of the geodesic Fréchet distance between P and Q, i.e. $(1-\epsilon)\delta_F(P,Q) \le \delta_{dF}(P',Q') \le (1+\epsilon)\delta_F(P,Q)$.*

Proof. Omitted. □

Instead of finding an exact geodesic path between two points we find an approximate path. Let $\delta'_{dF}(P',Q')$ be the approximate discrete Fréchet distance computed by replacing the exact shortest path algorithm by the approximation algorithm, which gives an ϵ-approximation of the geodesic distance between points. We have $\delta'_{dF}(P',Q') \le (1+\epsilon)\delta_d F(P,Q) \le (1+\epsilon)^2\delta_F(P,Q) \le (1+3\epsilon)\delta_F(P,Q)$, assuming that $\epsilon \le 1$, and similarly $\delta'_{dF}(P',Q') \ge (1-3\epsilon)\delta_F(P,Q)$. Thus, we obtain $|\delta'_{dF}(P',Q') - \delta_F(P,Q)| \le 3\epsilon\delta_F(P,Q)$. If a non-query based approximation algorithm is used, $\delta'_{dF}(P',Q')$ can be computed in time $O(p'q'T(n,\epsilon)) = O(C(R)^2\frac{pq}{\epsilon^2}(\log^4\frac{1}{\epsilon})T(n,\epsilon))$, where $T(n,\epsilon)$ is the time to compute an approximate shortest path between two points. If a query-based approximation algorithm is used, then $\delta'_{dF}(P',Q')$ can be computed in $O(p'q'QUERY(n,\epsilon) + PRE(n,\epsilon)) = O(C(R)^2\frac{pq}{\epsilon^2}(\log^4\frac{1}{\epsilon})QUERY(n,\epsilon) + PRE(n,\epsilon))$, where $QUERY(n,\epsilon)$ is the query time and $PRE(n,\epsilon)$ is the preprocessing time of the algorithm. For example, if we use the algorithm proposed by Aleksandrov et. al. [1], the algorithm takes $O(C(R)^2 \frac{pq}{\epsilon^2}(\log^4 \frac{1}{\epsilon})\bar{q} + \frac{(g+1)n^2}{\epsilon^{2/3}\bar{q}} \log \frac{n}{\epsilon} \log^4 \frac{1}{\epsilon})$ time, where $\bar{q}$ is a query time parameter and g is the genus of the graph constructed by the discretization scheme.

4.2 Geodesic Fréchet Distance in $\mathcal{R}^3$ with Obstacles

In this subsection, we briefly discuss the geodesic Fréchet distance problem in 1 or ∞ weighted regions in $\mathcal{R}^3$ (that is, among obstacles in $\mathcal{R}^3$). Let R be a weighted subdivision in $\mathcal{R}^3$ with a total of n vertices. The weight of each region $R_i \in R$ is either 1 or ∞. Given two polygonal curves P and Q in R, we want to find the Fréchet distance between P and Q, where the distance between two points in R is defined as the length of the geodesic path between those points, i.e. the length of the shortest obstacle-avoiding path.

We set $r(v) = \epsilon B$ and add additional vertices on P and Q as described in Section 4.1. Let P' and Q' be the new curves.

Lemma 9. *The discrete Fréchet distance between P' and Q' gives an ϵ-approximation of the Fréchet distance between P and Q, i.e. $(1-\epsilon)\delta_F(P,Q) \leq \delta_{dF}(P',Q') \leq (1+\epsilon)\delta_F(P,Q)$.*

Proof. Similar to the proof of *Lemma* 8. □

Let $\delta'_{dF}(P',Q')$ be the discrete Fréchet distance computed by a shortest path approximation algorithm, which gives an ϵ-approximation of the shortest path. We have

$$|\delta'_{dF}(P',Q') - \delta_F(P,Q)| \leq 3\epsilon\delta_F(P,Q).$$

$\delta'_{dF}(P',Q')$ can be computed in $O(C(R)^2 pq(1/\epsilon^2)\log^4(1/\epsilon)T(n,\epsilon)$ time, where $C(R)$ is a constant depending on the geometry of the problem and $T(n,\epsilon)$ is the time to approximate the shortest path between two points in R. For example, we can use the approximation algorithm given by Clarkson [12], which takes $O(n^2\lambda(n)\log(n/\epsilon)/\epsilon^4 + n^2\log(n\gamma)\log(n\log\gamma))$ time, where γ is the ratio of the length of the longest obstacle edge to the Euclidean distance between the two points, and $\lambda(n)$ is a very slowly-growing function related to the inverse of the Ackermann's function. Thus, our algorithm takes $O(C(R)^2 pq(1/\epsilon^2)\log^4(1/\epsilon)$ $(n^2\lambda(n)\log(n/\epsilon)/\epsilon^4 + n^2\log(n\gamma)\log(n\log\gamma)))$ time.

5 Conclusion

In this paper, we discussed three versions of the Fréchet distance problem in weighted regions and presented an approximation algorithm for each version. First, we discussed the non-geodesic Fréchet distance problem in planar weighted regions. We showed that we can approximate the Fréchet distance by using a parameter space, D, where each leash is associated with a point in D, and constructing a discrete graph G from D. We then discussed two geodesic Fréchet distance problems, in planar weighted regions and in 1 or ∞ weighted regions in $\mathcal{R}^3$, and showed that in both cases, by adding additional vertices to the polygonal curves, the discrete Fréchet distance can be used to approximate the continuous Fréchet distance.

References

1. Aleksandrov, L., Djidjev, H.N., Guo, H., Maheshwari, A., Nussbaum, D., Sack, J.: Approximate shortest path queries on weighted polyhedral surfaces. In: Královič, R., Urzyczyn, P. (eds.) MFCS 2006. LNCS, vol. 4162, pp. 98–109. Springer, Heidelberg (2006)
2. Aleksandrov, L., Maheshwari, A.A., Sack, J.R.: Approximation algorithms for geometric shortest path problems. In: Proc. 32nd Annual ACM Symposium on Theory of Computing, pp. 286–295 (2000)
3. Aleksandrov, L., Maheshwari, A., Sack, J.: Determining approximate shortest paths on weighted polyhedral surfaces. Journal of the ACM 52(1), 25–53 (2005)
4. Alt, H., Godau, M.: Computation the Fréchet distance between two polygonal curves. International Journal of Computational Geometry and Applications 5, 75–91 (1995)
5. Amato, N.M., Goodrich, M.T., Ramos, E.A.: Computing the arrangement of curve segments: Divide-and-conquer algorithms via sampling. In: Proc. 11th Annual CAM-SIAM Symposium on Discrete Algorithms, pp. 705–706 (2000)
6. Buchin, K., Buchin, M., Wenk, C.: Computing the Fréchet distance between simple polygons in polynomial time. In: Proc. 22nd Symposium on Computational Geometry, pp. 80–87 (2006)
7. Buchin, K., Buchin, M., Wang, Y.: Exact partial curve matching under the Fréchet distance. In: Proc. ACM-SIAM Symposium on Discrete Algorithms (to appear, 2009)
8. Chambers, E.W., De Verdière,C. É., Erickson, J., Lazard, S., Lazarus, F., Thite, S.: Walking your dog in the woods in polynomial time. In: Proc. 24th Annual ACM Symposium on Computational Geometry, pp. 101–109 (2008)
9. Chen, D.Z., Daescu, O., Hu, X., Wu, X., Xu, J.: Determining an optimal penetration among weighted regions in two and three dimensions. J. combinat. Optim. 5(1), 59–79 (2001)
10. Cheng, S.W., Na, H.S., Vigneron, A., Wang, Y.: Approximate shortest paths in anisotropic regions. In: Proc. 18th annual ACM-SIAM symposium on Discrete algorithms, pp. 766–774 (2007)
11. Cheng, S.W., Na, H.S., Vigneron, A., Wang, Y.: Querying approximate shortest paths in anisotropic regions. In: Proc. 23rd annual symposium on Computational geometry, pp. 84–91 (2005)
12. Clarkson, K.L.: Approximation algorithms for shortest path motion planning. In: Proc. 19th Annual ACM Symposium on Theory of Computing, pp. 56–65 (1987)
13. Collins, G.E.: Quantifier elimination for real closed fields by cylindrical algebraic decomposition. In: Brakhage, H. (ed.) GI-Fachtagung 1975. LNCS, vol. 33, pp. 134–183. Springer, Heidelberg (1975)
14. Cook IV, A.F., Wenk, C.: Geodesic Fréchet distance inside a simple polygon. In: Proc. 25th International Symposium on Theoretical Aspects of Computer Science, pp. 193–204 (2008)

15. Eiter, T., Mannila, H.: Computing discrete Fréchet distance. Technical Report CD-TR 94/64, Information Systems Department, Technical University of Vienna (1994)
16. Fréchet, M.: Sur quelques points du calcul fonctionnel. Rendiconti del Circolo Mathematico di Palermo 22, 1–74 (1906)
17. Maheshwari, A., Yi, J.: On computing Fréchet distance of two paths on a convex polyhedron. In: Proc. 21st European Workshop on computational Geometry, pp. 41–44 (2005)
18. Mitchell, J.S.B., Papadimitriou, C.H.: The weighted region problem: Finding shortest paths through a weighted planer subdivision. Journal of the ACM 38(1), 18–73 (1991)
19. Rote, G.: Computing the Fréchet distance between piecewise smooth curves. Technical Report, ECGTR-241108-01 (2005)
20. Sun, Z., Reif, J.H.: On finding approximate optimal path in weighted regions. Journal of Algorithms 58(1), 1–32 (2006)
21. Beer-Lambert law. Wikipedia, http://en.wikipedia.org/wiki/BeerLambert_law

The Complexity of Solving Stochastic Games on Graphs*

Daniel Andersson and Peter Bro Miltersen

Department of Computer Science, Aarhus University, Denmark

Abstract. We consider some well-known families of two-player zero-sum perfect-information stochastic games played on finite directed graphs. The families include stochastic parity games, stochastic mean payoff games, and simple stochastic games. We show that the tasks of solving games in each of these classes (quantitiatively or strategically) are all polynomial time equivalent. In addition, we exhibit a linear time algorithm that given a simple stochastic game *and* the values of all positions of that game, computes a pair of optimal strategies.

1 Introduction

We consider some well-known families of two-player zero-sum perfect-information stochastic games played on finite directed graphs.

Simple stochastic games were introduced to the algorithms and complexity community by Condon [5], who was motivated by the study of randomized Turing machine models. A simple stochastic game is given by a finite directed graph $G = (V, E)$, with the set of vertices V also called *positions* and the set of arcs E also called *actions*. There is a partition of the positions into V_1 (positions belonging to Player 1), V_2 (positions belonging to Player 2), V_R (coin-toss positions), and a special terminal position **1**. Positions of V_R have exactly two outgoing arcs, the terminal position **1** has none, while all positions in V_1, V_2 have at least one outgoing arc. Between moves, a pebble is resting at some vertex u. If u belongs to a player, this player should strategically pick an outgoing arc from u and move the pebble along this edge to another vertex. If u is a vertex in V_R, nature picks an outgoing arc from u uniformly at random and moves the pebble along this arc. The objective of the game for Player 1 is to reach **1** and should play so as to maximize his probability of doing so. The objective for Player 2 is to prevent Player 1 from reaching **1**.

Stochastic terminal-payoff games is the natural generalization of Condon's simple stochastic games where we allow each vertex in V_R to have more than two outgoing arcs and an arbitrary rational valued probability distribution on these, and several terminals with different payoffs, positive or negative. In a

* Work supported by *Center for Algorithmic Game Theory*, funded by the Carlsberg Foundation. A large fraction of the results of this paper appeared in a preprint [10], co-authored by Vladimir Gurvich and the second author of this paper. Vladimir Gurvich's contributions to that preprint will appear elsewhere.

Y. Dong, D.-Z. Du, and O. Ibarra (Eds.): ISAAC 2009, LNCS 5878, pp. 112–121, 2009.

stochastic terminal-payoff game, the *outcome* of the game is the payoff of the terminal reached, with the outcome being 0 if play is infinite. The objective for Player 1 is to maximize the expected outcome while the objective for Player 2 is to minimize it.

Stochastic parity games were introduced by Chatterjee, Jurdziński, and Henzinger at SODA'04 [4] and further studied in [3,2]. They are a natural generalization of the non-stochastic parity games of McNaughton [15], the latter having central importance in the computer-aided verification community, as solving them is equivalent to model checking the μ-calculus [6]. As for the case of simple stochastic games, a stochastic parity game is given by a directed graph $G = (V, E)$ with the same partition of the vertices into V_1, V_2 and V_R as for the case of simple stochastic games and terminal-payoff games. Also, a pebble is moved from vertex to vertex under the same rules as in these games. However, now, the players are not concerned with reaching or avoiding a particular position. Instead, each vertex is assigned an integral *priority*. The play continues forever. If the highest priority that appears infinitely often during play is odd, Player 1 wins the game; if it is even, Player 2 wins the game.

Stochastic mean-payoff games and the related **stochastic discounted-payoff games** were first studied in the game theory community by Gillette [7] as the perfect information special case of the stochastic games of Shapley [16]. A stochastic mean-payoff or discounted-payoff game G is given by a finite set of positions V, partitioned into V_1 (positions belonging to Player 1) and V_2 (positions belonging to Player 2). To each position u is associated a finite set of possible actions. To each such action is associated a real-valued *reward* and a probability distribution on positions. At any point in time of play, the game is in a particular position i. The player to move chooses an action strategically and the corresponding reward is paid by Player 2 to Player 1. Then, nature chooses the next position at random according to the probability distribution associated with the action. The play continues forever and the sum of rewards may therefore be unbounded. Nevertheless, one can associate a finite payoff to the players in spite of this, in more ways than one (so G is not just one game, but really a family of games): For a stochastic discounted-payoff game, we fix a *discount factor* $\beta \in (0, 1)$ and define the outcome of the play (the payoff to Player 1) to be $\sum_{i=0}^{\infty} \beta^i r_i$ where r_i is the reward incurred at stage i of the game. We shall denote the resulting game G_β. For a stochastic mean-payoff game we define the outcome of the play (the payoff to Player 1) to be the *limiting average* payoff $\liminf_{n\to\infty}(\sum_{i=0}^{n} r_i)/(n + 1)$. We shall denote the resulting game G_1. A natural restriction of stochastic mean-payoff games is to *deterministic* transitions (i.e., all probability distributions put all probability mass on one position). This class of games has been studied in the computer science literature under the names of cyclic games [9] and mean-payoff games [17]. We shall refer to them as deterministic mean-payoff games in this paper.

A *strategy* for a game is a (possibly randomized) procedure for selecting which arc or action to take, given the history of the play so far. A *positional strategy* is the special case of this where the choice is deterministic and only depends on the

Theorem 2 means that algorithms that efficiently and quantitatively solve simple stochastic games satisfying certain conditions (such as the algorithm of Gimbert and Horn [8], which is efficient when the number of coin-toss vertices is small) can also be used to strategically solve those games, in essentially the same time bound. We leave as an open problem whether a similar algorithm (even a polynomial-time one) can be obtained for stochastic mean-payoff games.

2 Proof of Theorem 1

Figure 1 shows a minimal set of reductions needed to establish all equivalences. We first enumerate a number of trivial and known reductions and afterwards fill in the remaining "gaps". For all games considered here, it is well-known that quantitatively solving them reduces to strategically solving them: Once the strategies have been fixed, a complete analysis of the resulting finite-state random process can be obtained using linear algebra and the theory of Markov chains [12]. Also, for the case of stochastic discounted-payoff games, the converse reduction is also easy: A strategy is optimal if and only if in each position it chooses an action that maximizes the sum of the reward obtained by this action and the discounted expected value of the next position. Thus, $\textsc{Discounted}^{\mathrm{s}}_{\mathrm{b}} \preceq \textsc{Discounted}^{\mathrm{q}}_{\mathrm{b}}$. Of course, each problem with unary input encoding trivially reduces to the corresponding binary version. Also, it is obvious that stochastic terminal-payoff games generalize simple stochastic games, and the only numbers that appear are 0, 1 and $\frac{1}{2}$, so $\textsc{Simple}^{\mathrm{q}} \preceq \textsc{Terminal}^{\mathrm{q}}_{\mathrm{u}}$ and $\textsc{Simple}^{\mathrm{s}} \preceq \textsc{Terminal}^{\mathrm{s}}_{\mathrm{u}}$. Quantitatively solving simple stochastic games easily reduces to quantitatively solving stochastic parity games, as was noted in the original work on stochastic parity games [4]. Also, Chatterjee and Henzinger [2] show that strategically solving stochastic parity games reduces to strategically solving stochastic mean-payoff games. Thus, to "complete the picture" and establish all equivalences, we only have to show: $\textsc{Mean}^{\mathrm{s}}_{\mathrm{b}} \preceq \textsc{Discounted}^{\mathrm{s}}_{\mathrm{b}}$, $\textsc{Discounted}^{\mathrm{q}}_{\mathrm{b}} \preceq \textsc{Simple}^{\mathrm{q}}$, $\textsc{Terminal}^{\mathrm{s}}_{\mathrm{b}} \preceq \textsc{Mean}^{\mathrm{s}}_{\mathrm{b}}$ and $\textsc{Terminal}^{\mathrm{q}}_{\mathrm{u}} \preceq \textsc{Mean}^{\mathrm{q}}_{\mathrm{u}}$. These reductions are provided by the following lemmas.

Lemma 1. *Let G be a stochastic discounted-payoff/mean-payoff game with n positions and all transition probabilities and rewards being fractions with integral numerators and denominators, all of absolute value at most M. Let $\beta^* = 1 - ((n!)^2 2^{2n+3} M^{2n^2})^{-1}$ and let $\beta \in [\beta^*, 1)$. Then, any optimal positional strategy (for either player) in the discounted-payoff game G_β is also an optimal strategy in the mean-payoff game G_1.*

Proof. The fact that *some* β^* with the desired property exists is explicit in the proof of Theorem 1 of Liggett and Lippman [13]. We assume familiarity with that proof in the following. Here, we derive a concrete value for β^*. From the proof of Liggett and Lippman, we have that for x^* to be an optimal positional strategy (for Player 1) in G_1, it is sufficient to be an optimal positional strategy in G_β for all values of β sufficiently close to 1, i.e., to satisfy the

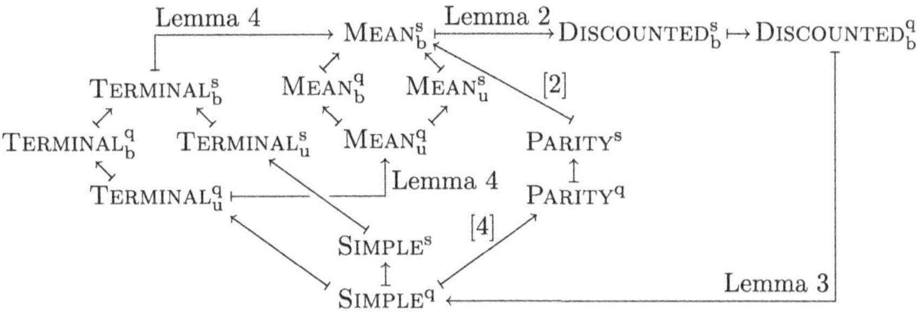

Fig. 1. Reductions used in the proof of Theorem 1

inequalities $\min_{y \in S'_2} u^i_\beta(x^*, y) \ge \max_{x \in S'_1} \min_{y \in S'_2} u^i_\beta(x, y)$ for all positions i and for all values of β sufficiently close to 1, where S'_1 (S'_2) is the set of positional, strategies for Player 1 (2) and u^i_β is the expected payoff when game starts in position i and the discount factor is β. Similarly, for y^* to be an optimal positional strategy (for Player 2) in G_1, it is sufficient to be an optimal positional strategy in G_β for all values of β sufficiently close to 1, i.e., to satisfy the inequalities $\max_{x \in S'_1} u^i_\beta(x, y^*) \le \min_{y \in S'_2} \max_{x \in S'_1} u^i_\beta(x, y)$. So, we can prove the lemma by showing that for all positions i and *all* positional strategies x, y, x', y', the sign of $u^i_\beta(x, y) - u^i_\beta(x', y')$ is the same for all $\beta \ge \beta^*$. For fixed strategies x, y we have that $v_i = u^i_\beta(x, y)$ is the expected total reward in a *discounted Markov process* and is therefore given by (see [12])

$$v = (I - \beta Q)^{-1} r, \tag{3}$$

where v is the vector of $u_\beta(x, y)$ values, one for each position, Q is the matrix of transition probabilities and r is the vector of rewards (note that for *fixed* positional strategies x, y, rewards can be assigned to positions in the natural way). Let $\gamma = 1 - \beta$. Then, (3) is a system of linear equations in the unknowns v, where each coefficient is of the form $a_{ij}\gamma + b_{ij}$ where a_{ij}, b_{ij} are rational numbers with numerators with absolute value bounded by $2M$ and with denominators with absolute value bounded by M. By multiplying the equations with all denominators, we can in fact assume that a_{ij}, b_{ij} are integers of absolute value less than $2M^n$. Solving the equations using Cramer's rule, we may write an entry of v as a quotient between determinants of $n \times n$ matrices containing terms of the form $a_{ij}\gamma + b_{ij}$. The determinant of such a matrix is a polynomial in γ of degree n with the coefficient of each term being of absolute value at most $n!(2M^n)^n = n! 2^n M^{n^2}$. We denote these two polynomials p_1, p_2. Arguing similarly about $u_\beta(x', y')$ and deriving corresponding polynomials p_3, p_4, we have that $u^i_\beta(x, y) - u^i_\beta(x', y') \ge 0$ is equivalent to $p_1(\gamma)/p_2(\gamma) - p_3(\gamma)/p_4(\gamma) \ge 0$, i.e., $p_1(\gamma)p_4(\gamma) - p_3(\gamma)p_2(\gamma) \ge 0$. Letting $q(\gamma) = p_1(\gamma)p_4(\gamma) - p_3(\gamma)p_2(\gamma)$, we have that q is a polynomial in γ, with integer coefficients, all of absolute value at most $R = 2(n!)^2 2^{2n} M^{2n^2}$. Since $1 - \beta^* < 1/(2R)$, the sign of $q(\gamma)$ is the same for all $\gamma \le 1 - \beta^*$, i.e., for all $\beta \ge \beta^*$. This completes the proof.

Lemma 2. $\text{MEAN}_{\text{b}}^{\text{s}} \preceq \text{DISCOUNTED}_{\text{b}}^{\text{s}}$.

Proof. This follows immediately from Lemma 1 by observing that the binary representation of the number $\beta^* = 1 - ((n!)^2 2^{2n+3} M^{2n^2})^{-1}$ has length polynomial in the size of the representation of the given game.

Lemma 3. $\text{DISCOUNTED}_{\text{b}}^{\text{q}} \preceq \text{SIMPLE}^{\text{q}}$.

Proof. Zwick and Paterson [17] considered solving *deterministic* discounted-payoff games, i.e., games where the action taken deterministically determines the transition taken and reduced these to solving simple stochastic games. It is natural to try to generalize their reduction so that it also works for stochastic discounted-payoff games. We find that such reduction indeed works, even though the correctness proof of Zwick and Paterson has to be modified slightly compared to their proof. The details follow.

The redution proceeds in two steps: First we reduce to stochastic terminal-payoff games with 0/1 payoffs, and then to simple stochastic games. We are given as input a stochastic discounted-payoff game G with discount factor β and must first produce a stochastic terminal-payoff game G' whose values can be used to construct the values for the stochastic discounted-payoff game G_β. First, we affinely scale and translate all rewards of G so that they are in the interval $[0, 1]$. This does not influence the optimal strategies, and all values are transformed accordingly. Vertices of G' include all positions of G (belonging to the same player in G' as in G), and, in addition, a random vertex $w_{u,A}$ for each possible action A of each position u of G. We also add terminals $\mathbf{0}$ and $\mathbf{1}$. We construct the arcs of G' by adding, for each (position,action) pair (u, A) the "gadget" indicated in Figure 2. To be precise, if the action has reward r and leads to positions $v_1, v_2, \ldots, v_k$ with probability weights $p_1, p_2, \ldots, p_k$, we include in G' an arc from u to $w_{u,A}$, arcs from $w_{u,A}$ to $v_1, \ldots, v_k$ with probability weights $(1 - \beta)p_1, \ldots, (1 - \beta)p_k$, an arc from $w_{u,A}$ to $\mathbf{0}$ with probability weight $\beta(1 - r)$ and finally an arc from $w_{u,A}$ to the terminal $\mathbf{1}$ with probability weight βr.

There is clearly a one-to-one correspondence between strategies in G and in G'. To see the correspondance between values, fix a strategy profile and consider any play. By construction, if the expected reward of the play in G is h, the probability that the play in G' reaches $\mathbf{1}$ is exactly βh.

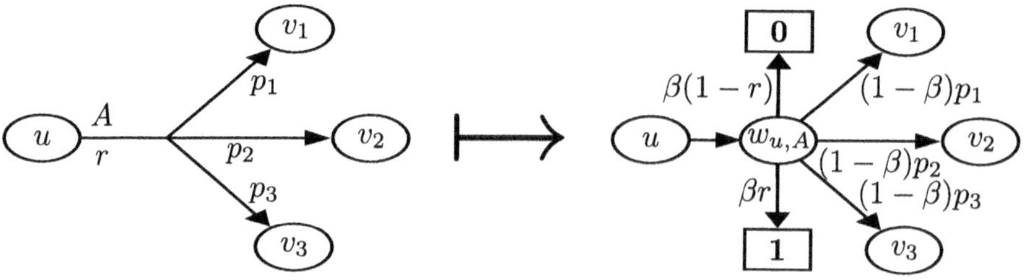

Fig. 2. Reducing discounted-payoff games to terminal-payoff games

The second step is identical to that of Zwick and Paterson [17]. They describe how arbitrary rational probability distributions can be implemented using a polynomial number of coin-toss vertices. Thus, we can transform G' into an equivalent simple stochastic game.

Lemma 4. $\text{TERMINAL}^{\text{s}}_{\text{b}} \preceq \text{MEAN}^{\text{s}}_{\text{b}}$ *and* $\text{TERMINAL}^{\text{q}}_{\text{u}} \preceq \text{MEAN}^{\text{q}}_{\text{u}}$.

Proof. We are given a stochastic terminal-payoff game G and must construct a stochastic mean-payoff game G'. Positions of G' will coincide with vertices of G, with the positions of G' including the terminals. Positions u belonging to a player in G belongs to the same player in G'. For each outgoing arc of u, we add an action in G' with reward 0, and with a deterministic transition to the endpoint of the arc of G. Random vertices of G can be assigned to either player in G', but he will only be given a single "dummy choice": If the random vertex has arcs to v_1 and v_2, we add a single action in G' with reward 0 and transitions into v_1, v_2, both with probability weight 1/2. Each terminal can be assigned to either player in G', but again he will be given only a dummy choice: We add a single action with reward equal to the payoff of the terminal and with a transition back into the same terminal with probability weight 1.

There is clearly a one-to-one correspondence between positional strategies in G and strategies in G'. To see the correspondence between values, fix a strategy profile for the two players and consider play starting from some vertex. By construction, if the probability of the play reaching some particular terminal in G is q, then the probability of the play reaching the corresponding self-loop in G' is also q. Thus, the expected reward is the same.

3 Proof of Theorem 2

In this section, we consider a given stochastic terminal-payoff game. Our goal is an algorithm for computing optimal strategies, given the values of all vertices. To simplify the presentation, we will focus on strategies for Player 1. Because of symmetry, there is no loss of generality. In this section, "strategy" means " positional strategy". An arc (u, v) is called *safe* if $\text{val}(u) = \text{val}(v)$. A *safe strategy* is a strategy that only uses safe arcs. A strategy x for Player 1 is called *stopping* if for any strategy y for Player 2 and any vertex v with positive value, there is a non-zero probability that the play $P(x, y, v)$ reaches a terminal.

It is not hard to see that any optimal strategy for Player 1 must be safe and stopping, so these two conditions are necessary for optimality. We will now show that they are also sufficient.

Lemma 5. *If a strategy is safe and stopping, then it is also optimal.*

Proof. Let x be any safe and stopping strategy for Player 1, let y be an arbitrary strategy for Player 2, and let v_0 be an arbitrary vertex. Consider the play $P(x, y, v_0)$. Denote by $q_i(v)$ the probability that after i steps, the play is at v. Since x is safe,

$$\forall i: \quad \text{val}(v_0) \leq \sum_{v \in V} \text{val}(v) q_i(v) \leq \sum_{v \in T \cup V^+} \text{val}(v) q_i(v), \tag{4}$$

where T denotes the set of terminal vertices and V^+ denotes the set of non-terminal vertices with positive value. Since x is stopping,

$$\forall v \in V^+ : \quad \lim_{i \to \infty} q_i(v) = 0. \tag{5}$$

Finally, note that

$$\begin{aligned} u^{v_0}(x,y) &= \sum_{t \in T} \mathrm{val}(t) \lim_{i \to \infty} q_i(t) = \sum_{v \in T \cup V^+} \mathrm{val}(v) \lim_{i \to \infty} q_i(v) && \text{(by (5))} \\ &= \lim_{i \to \infty} \sum_{v \in T \cup V^+} \mathrm{val}(v) q_i(v) \geq \mathrm{val}(v_0). && \text{(by (4))} \end{aligned}$$

Therefore, x is optimal.

Using this characterization of optimality, strategy recovery can be reduced to strategically solving a simple stochastic game without random vertices.

Theorem 2. *Given a stochastic terminal-payoff game and the values of all its vertices, optimal positional strategies can be computed in linear time.*

Proof. Construct from the given game G a simple stochastic game G' as follows:

1. Merge all terminals into one.
2. Remove all outgoing non-safe arcs from the vertices of Player 1.
3. Transfer ownership of all random vertices to Player 1.

Compute an optimal strategy x' for Player 1 in G' using linear-time retrograde analysis [1]. Let x be the interpretation of x' as a strategy in G obtained by restricting x' to the vertices of Player 1 in G. By construction, from any starting vertex v, Player 1 can ensure reaching the terminal in G' if and only if x ensures a non-zero probability of reaching a terminal in G. Let v be any vertex with positive value in G. Player 1 has a safe strategy that ensures a non-zero probability of reaching a terminal from v, specifically, the optimal strategy. This implies that there is a corresponding strategy for Player 1 in G' that ensures reaching the terminal from v. It follows that x is stopping, and it is safe by construction. Therefore, by Lemma 5, it is optimal.

Acknowledgements. We would like to thank Vladmir Gurvich for fruitful collaboration and Florian Horn for helpful comments.

References

1. Andersson, D., Hansen, K.A., Miltersen, P.B., Sørensen, T.B.: Deterministic graphical games revisited. In: Beckmann, A., Dimitracopoulos, C., Löwe, B. (eds.) CiE 2008. LNCS, vol. 5028, pp. 1–10. Springer, Heidelberg (2008)
2. Chatterjee, K., Henzinger, T.A.: Reduction of stochastic parity to stochastic mean-payoff games. Inf. Process. Lett. 106(1), 1–7 (2008)

3. Chatterjee, K., Jurdziński, M., Henzinger, T.: Simple stochastic parity games. In: Baaz, M., Makowsky, J.A. (eds.) CSL 2003. LNCS, vol. 2803, pp. 100–113. Springer, Heidelberg (2003)
4. Chatterjee, K., Jurdziński, M., Henzinger, T.A.: Quantitative stochastic parity games. In: SODA 2004: Proceedings of the fifteenth annual ACM-SIAM symposium on Discrete algorithms, pp. 121–130 (2004)
5. Condon, A.: The complexity of stochastic games. Information and Computation 96, 203–224 (1992)
6. Emerson, E.A., Jutla, C.S., Sistla, A.P.: On model checking for the mu-calculus and its fragments. Theoretical Computer Science 258(1-2), 491–522 (2001)
7. Gillette, D.: Stochastic games with zero stop probabilities. In: Contributions to the Theory of Games III. Annals of Math. Studies, vol. 39, pp. 179–187. Princeton University Press, Princeton (1957)
8. Gimbert, H., Horn, F.: Simple Stochastic Games with Few Random Vertices are Easy to Solve. In: Amadio, R.M. (ed.) FOSSACS 2008. LNCS, vol. 4962, pp. 5–19. Springer, Heidelberg (2008)
9. Gurvich, V., Karzanov, A., Khachiyan, L.: Cyclic games and an algorithm to find minimax cycle means in directed graphs. USSR Computational Mathematics and Mathematical Physics 28, 85–91 (1988)
10. Gurvich, V., Miltersen, P.B.: On the computational complexity of solving stochastic mean-payoff games. arXiv:0812.0486v1 [cs.GT] (2008)
11. Halman, N.: Simple stochastic games, parity games, mean payoff games and discounted payoff games are all LP-type problems. Algorithmica 49(1), 37–50 (2007)
12. Howard, R.A.: Dynamic Programming and Markov Processes. MIT Press, Cambridge (1960)
13. Liggett, T.M., Lippman, S.A.: Stochastic games with perfect information and time average payoff. SIAM Review 11(4), 604–607 (1969)
14. Littman, M.L.: Algorithms for sequential decision making. PhD thesis, Brown University, Department of Computer Science (1996)
15. McNaughton, R.: Infinite games played on finite graphs. An. Pure and Applied Logic 65, 149–184 (1993)
16. Shapley, L.: Stochastic games. Proc. Nat. Acad. Science 39, 1095–1100 (1953)
17. Zwick, U., Paterson, M.: The complexity of mean payoff games on graphs. Theor. Comput. Sci. 158(1-2), 343–359 (1996)

Computational Complexity of Cast Puzzles

Chuzo Iwamoto[1], Kento Sasaki[1], Kenji Nishio[2], and Kenichi Morita[1]

[1] Hiroshima University, Graduate School of Engineering
Higashi-Hiroshima, 739-8527 Japan
chuzo@hiroshima-u.ac.jp
[2] Sharp Corporation, Osaka, 545-8522 Japan

Abstract. A disentanglement puzzle consists of mechanically interlinked pieces, and the puzzle is solved by disentangling one piece from another set of pieces. A cast puzzle is a type of disentanglement puzzle, where each piece is a zinc die-casting alloy. In this paper, we consider the generalized cast puzzle problem whose input is the layout of a finite number of pieces (polyhedrons) in the 3-dimensional Euclidean space. For every integer $k \geq 0$, we present a polynomial-time transformation from an arbitrary k-exponential-space Turing machine M and its input x to a cast puzzle c_1 of size k-exponential in $|x|$ such that M accepts x if and only if c_1 is solvable. Here, the layout of c_1 is encoded as a string of length polynomial (even if c_1 has size k-exponential). Therefore, the cast puzzle problem of size k-exponential is k-EXPSPACE-hard for every integer $k \geq 0$. We also present a polynomial-time transformation from an arbitrary instance f of the SAT problem to a cast puzzle c_2 such that f is satisfiable if and only if c_2 is solvable.

1 Introduction

Disentanglement puzzles are one of the most fundamental and popular playthings. They consist of mechanically interlinked pieces, and a puzzle is solved by disentangling one piece from another set of pieces. Disentanglement puzzles are classified into two categories, wire puzzles and cast puzzles. A wire puzzle consists of two or more entangled stiff wires. Wires may or may not be closed loops, and they have complex shapes. Normally, wire puzzles are solved by disentangling one piece from another set of pieces without cutting or bending the wires. On the other hand, pieces of cast puzzles are zinc die-casting alloys. So, people who take them in hands feel their solidness and heaviness. See the official site [1] for the cast puzzles, produced by Hanayama Co.,Ltd. In this site, cast puzzles are classified into six levels (i.e., Easiest, Easy, Medium, Fairly hard, Hard, and Very hard). However, from the point of view of a computer scientist, they should be classified according to computational complexities (i.e., P, NP, PSPACE, EXPTIME, 2-EXPTIME, 2-EXPSPACE, and so on).

We will define a cast puzzle as a set of simple polyhedrons. We consider the generalized cast puzzle problem whose input is the layout of a finite number of pieces (polyhedrons) in the 3-dimensional Euclidean space. Since all cast puzzles in this paper are constructed by polynomial-time transformations, the layout of

Y. Dong, D.-Z. Du, and O. Ibarra (Eds.): ISAAC 2009, LNCS 5878, pp. 122–131, 2009.

a cast puzzle is encoded as a string of length polynomial (even if the cast puzzle has size k-exponential).

In this paper, for every integer $k \geq 0$, we present a polynomial-time transformation from an arbitrary k-exponential-space Turing machine M and its input x to a cast puzzle c_1 of size k-exponential in $|x|$ such that M accepts x if and only if c_1 is solvable. Therefore, the cast puzzle problem of size k-exponential is k-EXPSPACE-hard for every integer $k \geq 0$. We also present a polynomial-time transformation from an arbitrary instance f of the SAT problem to a cast puzzle c_2 such that f is satisfiable if and only if c_2 is solvable.

Flake and Baum proved that some PSPACE-complete problem can be reduced to the Rush hour problem (a sliding block puzzle on a board) [4]. From this result, it is not difficult to construct a PSPACE-hard cast puzzle by making sure that the pieces cannot use the third direction. Note that the k-EXPSPACE-hardness for a set of cast puzzles implies the NP-hardness and PSPACE-hardness for the *same* set of cast puzzles. Our NP-hard cast puzzle, transformed from the SAT problem, can be solved *by hand* in $n+1$ steps with n guesses. On the other hand, a PSPACE-hard cast puzzle constructed by the idea of [4] cannot be solved in polynomial steps with polynomial guesses unless NP = PSPACE.

There have been a huge amount of literatures on computational complexities of games and puzzles. For example, Tetris [2], Solitaire [6], Minesweeper [9], $(n \times n)$-extension of the 15-puzzle [12], and Sokoban (a transport puzzle in a maze) [3] are known to be NP-hard. As for higher complexity classes, Othello [8] is known to be PSPACE-hard; Chess [5] and Go [13] are EXPTIME-hard.

2 Definitions and Main Results

In our model, all pieces are defined as simple polyhedrons in the 3-dimensional Euclidean space E^3. The definitions of polygons and polyhedrons are mostly from [11]. In E^2, a *polygon* is defined by a finite set of segments such that every segment extreme is shared by exactly two edges and no subset of edges has the same property. The segments are the *edges* and their extremes are the *vertices* of the polygon. In E^3, a *polyhedron* is defined by a set of plane polygons such that every edge of a polygon is shared by exactly one other polygon (adjacent polygon) and no subset of polygons has the same property. The vertices and the edges of the polygons are the *vertices* and the *edges* of the polyhedron, and the polygons are the *facet* of the polyhedron. A polyhedron is said to be *simple* if there is no pair of non-adjacent facets sharing a point. A simple polyhedron partitions the space into two disjoint domains, the *interior* (bounded) and the *exterior* (unbounded). In this paper, the term polyhedron is used to denote the union of the boundary and of the interior.

A *cast puzzle* is a finite set of simple polyhedrons, called *pieces*, embedded in E^3. One of the pieces is called the *target piece*. The input of the *cast puzzle problem* is the layout of a finite number of pieces. A cast puzzle is said to be *solvable* if the target piece can be disentangled from another set of pieces without deforming, whittling, or breaking the pieces. In this paper, we assume that, for any piece p, no vertex of any other piece is in the interior of p. A vertex of a piece may touch the surface of another piece. The surface of any piece is frictionless.

The input of the cast puzzle problem must be represented by a string on the input tape of a Turing machine (TM). So, we consider a cast puzzle such that each vertex of any piece has integral coordinates. The layout of polyhedrons can be represented as a string as follows. Suppose P_0 is a polyhedron such that one of P_0's vertices is on the origin of coordinates. Polyhedron P_0 is represented by a set of P_0's faces $f_0, f_1, \ldots, f_{m-1}$, where one of f_0's vertices is on the origin of coordinates. Each face f_i is represented by a sequence of f_i's vertices. Let $\langle P_0, (x_1, y_1, z_1)\rangle$ denote the polyhedron embedded in E^3, obtained by *translating* P_0 from (x, y, z) to $(x', y', z') = (x, y, z) + (x_1, y_1, z_1)$ (which describes a transformation where each point is subject to a fixed displacement (x_1, y_1, z_1)). For example, let P'_0 be a polyhedron fitted inside a unit cube. Set $S = \{\langle P'_0, (2i, 0, 0)\rangle \mid 0 \leq i \leq h(n') - 1\}$ represents a sequence of $h(n')$ polyhedrons placed at regular intervals, where n' is an integer, and h is a function of n'. Note that such well-regulated polyhedrons can be encoded as a string *code*(S) over $\{0, 1\}$ of length $O(\log n')$ when polyhedron P'_0 and function h can be encoded as strings of length constant.

Let n be the length of the string representing the layout of a cast puzzle. The cast puzzle is said to have *size* $s(n)$ if the convex hull of all pieces is fitted inside a cuboid of size $s_1(n) \times s_2(n) \times s_3(n)$ such that $s_1(n) + s_2(n) + s_3(n) \leq s(n)$.

Let $g(k, p(n)) = 2^{g(k-1, p(n))}$ and $g(0, p(n)) = p(n)$, where $p(n)$ is an arbitrary polynomial function. A TM is said to be *k-exponential-space bounded* if, for every accepted input x of length n, M halts with an accepting state in $g(k, p(n))$ space. The cast puzzle problem of size k-exponential is said to be *k-EXPSPACE-hard* if there is a polynomial-time transformation from an arbitrary k-exponential-space TM M and its input x to a cast puzzle c_1 of size k-exponential in $|x|$ such that M accepts x if and only if c_1 is solvable. Since any cast puzzle in this paper is constructed by a polynomial-time transformation, the layout is encoded as a string of length polynomial (even if the cast puzzle has size k-exponential).

Theorem 1. *For every integer $k \geq 0$, there is a polynomial-time transformation from an arbitrary k-exponential-space TM M and its input x to a cast puzzle c_1 of size k-exponential in $|x|$ such that M accepts x if and only if c_1 is solvable.*

Corollary 1. *For every integer $k \geq 0$, the cast puzzle problem of size k-exponential is k-EXPSPACE-hard.*

Theorem 2. *There is a polynomial-time transformation from an arbitrary instance f of the SAT problem to a cast puzzle c_2 such that f is satisfiable if and only if c_2 is solvable.*

The proof of Theorem 1 is given in Section 3. For the proof of Theorem 2, see an unpublished manuscript [7] available as a pdf file.

3 Transformations from TMs to Cast Puzzles

In this section, we will prove Theorem 1. We show a polynomial-time transformation from a k-exponential-space TM and its input to the layout of a cast puzzle.

The definition of a TM is mostly from [10]. A one-tape two-symbol TM is a system defined by $M = (Q, \Sigma, q_0, q_{m-1}, R)$, where $Q = \{q_0, q_1, \dots, q_{m-1}\}$ is a finite set of states, $\Sigma = \{0, 1\}$ is a set of tape symbols, q_0 (resp. q_{m-1}) is the unique initial (resp. final) state, and R is a set of transition rules, where $R \subseteq ((Q - \{q_{m-1}\}) \times \Sigma \times \Sigma \times Q) \cup ((Q - \{q_{m-1}\}) \times \{/\} \times \{-1, +1\} \times Q)$. Each transition rule in R is of the form $[q_i, a, a', q_{i'}]$ or $[q_i, /, d, q_{i'}]$, where $q_i \in Q - \{q_{m-1}\}$, $q_{i'} \in Q$; $a, a' \in \Sigma$; and $d \in \{-1, +1\}$. Rule $[q_i, a, a', q_{i'}]$ means that if M reads symbol a in state q_i, then M writes a' and enters state $q_{i'}$. Rule $[q_i, /, d, q_{i'}]$ means that if M is in state q_i, then M moves the head to the right (resp. left) when $d = +1$ (resp. $d = -1$) and enters state $q_{i'}$. Let $\alpha_1 = [q_i, a, a', q_{i'}]$ and $\alpha_2 = [q_j, b, b', q_{j'}]$ be two transition rules in R. We say that α_1 and α_2 *overlap in domain* if ($q_i = q_j$ and $a = b$) or ($q_i = q_j$ and ($a = /$ or $b = /$)). A rule α is said to be *deterministic* in R if there is no other rule in R with which α overlaps in domain. A TM M is called deterministic if every rule in R is deterministic. In the rest of this paper, all TMs are one-tape two-symbol deterministic TMs.

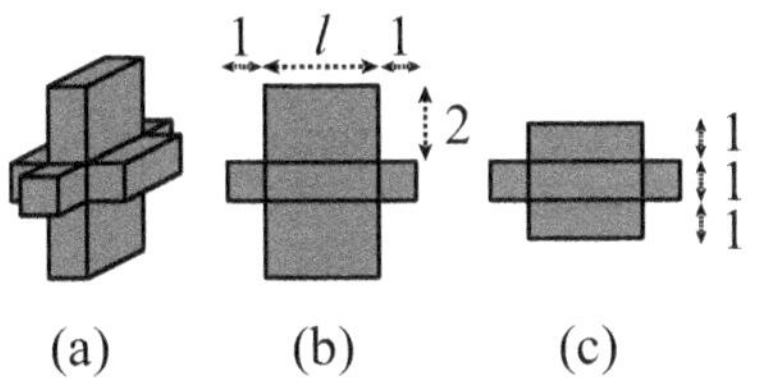

Fig. 1. (a) A target piece. (b) Lateral view. (c) Top view.

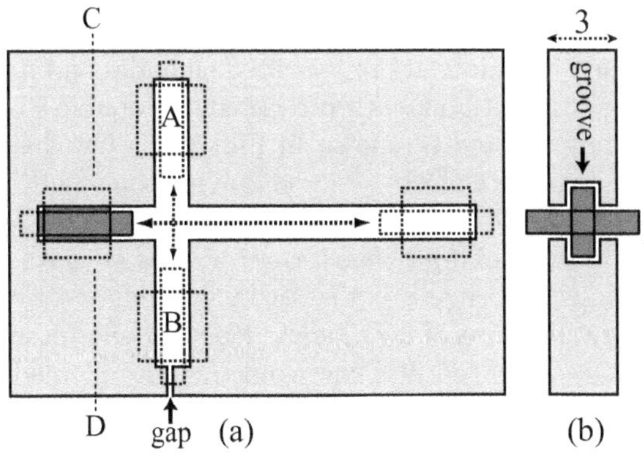

Fig. 2. (a) A board with a cross-shaped hole. (b) A section view between C and D.

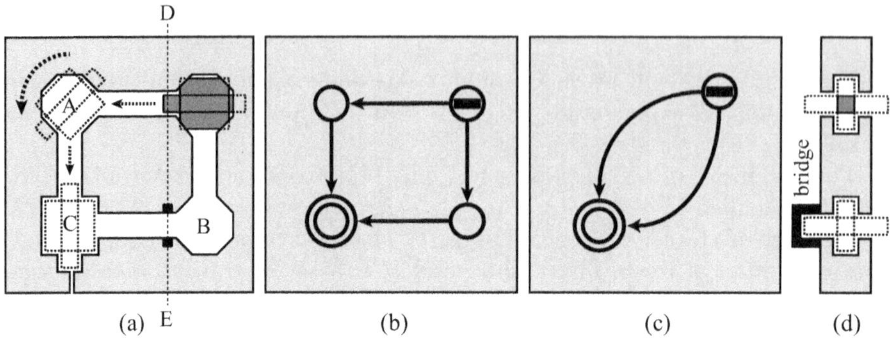

Fig. 3. (a) A board having corridors, octagonal and quadrilateral spaces. (b) Diagrammatic sketch of (a). (c) Unimportant octagonal spaces are omitted. (d) A bridge.

We first construct the target piece (see Fig. 1). It is composed of a square pole of size $1 \times 1 \times (l+2)$, two rectangular poles of size $1 \times 2 \times l$, and two square poles of size $1 \times 1 \times l$. The length l will be fixed later. The four poles of length l are welded to the pole of length $l+2$ so that they form a single polyhedron.

Consider a 3-thick board (see Fig. 2). The board has a cross-shaped hole. There is a groove on the inside wall of the hole (see Fig. 2(b)) so that the target piece cannot be taken out from the board. This cross-shaped hole is composed of horizontal and vertical *corridors*. In the horizontal corridor, the target piece can move to the right, and it can go back to the original position. The vertical corridor connects positions A and B. It should be noted that the target piece cannot move to position A or B from the current position. (Strictly speaking, this is not a cross-shaped *hole* because there is a narrow gap which connects the hole to the exterior so that the board is a polyhedron.)

Consider a board shown in Fig. 3(a), which has four corridors, three octagonal space, and one quadrilateral space. Suppose that the target piece is in the right-upper octagonal space. The target piece can move to position A, take a 90-degree turn using the octagonal space, and the target piece can reach position C. At this position, we can take out the target piece from the board. In the following, such octagonal spaces and corridors are represented as nodes and arcs (see Fig. 3(b)). A quadrilateral space (position C) is represented as a double circle. (Unimportant nodes are sometimes omitted as shown in Fig. 3(c).) On this board, there is a *bridge* so that the two pieces form a single polyhedron (see Fig. 3(d)).

We illustrate the transformation from a k-exponential-space TM M to a cast puzzle. There is a polynomial $p(n)$ such that M uses at most $g(k, p(n))$ cells on its tape. Each tape cell c_i, $1 \leq i \leq g(k, p(n))$, is simulated by a *block* shown in Figs. 4(a) and 4(b). The size of each block depends only on the number of M's states. All blocks are arranged in a row, and they are pierced by two L-shaped square poles, which are welded to another square pole (see Fig. 4(c)). These square poles form a single polyhedron. (A groove in Fig. 4(b) is explained later.)

Let $Q = \{q_0, q_1, \ldots, q_{m-1}\}$ be the set of M's states. Recall that tape symbols are 0 and 1 only. Each block has $2m$ holes, say, $h(q_0, 0), h(q_0, 1)$;

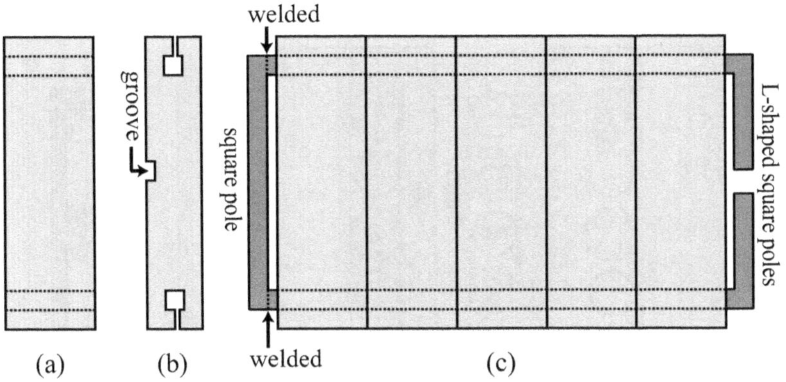

Fig. 4. (a) A block simulating a cell c_i of M. (b) Lateral view. (c) Blocks are pierced by two L-shaped square poles, which are welded to another square pole.

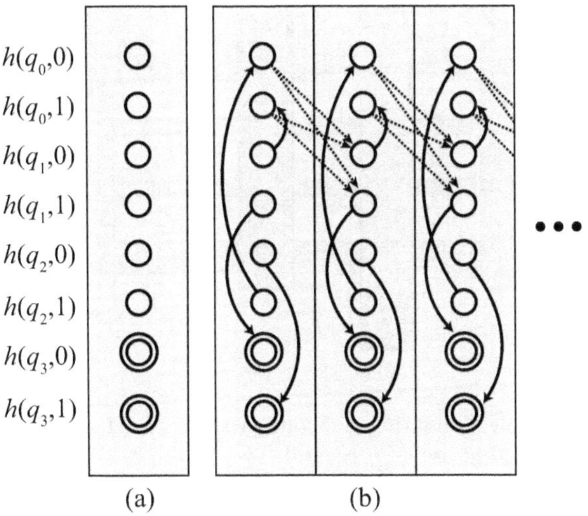

Fig. 5. (a) A block has a hole $h(q_i, a)$ for every pair of $q_i \in Q$ and $a \in \Sigma$. (b) Solid (resp. dotted) arcs correspond to transition rules of the form $[q_i, a, a', q_{i'}]$ (resp. $[q_i, /, d, q_{i'}]$).

$h(q_1, 0), h(q_1, 1); \cdots ; h(q_{m-1}, 0), h(q_{m-1}, 1)$ (see Fig. 5(a) for a four-state TM). For every transition rule of the form $[q_i, a, a', q_{i'}]$, $a \in \Sigma$, we add an arc from node $h(q_i, a)$ to $h(q_{i'}, a')$ (see solid arcs in Fig. 5(b)). Furthermore, for every transition rule of the form $[q_i, /, d, q_{i'}]$, $d \in \{-1, +1\}$, we add 2×2 dotted arcs from nodes $h(q_i, 0), h(q_i, 1)$ in the current block to $h(q_{i'}, 0), h(q_{i'}, 1)$ in the right (resp. left) adjacent block if $d = +1$ (resp. $d = -1$) (see dotted arcs in Fig. 5(b)). All blocks are the same polyhedron except for the first and last blocks. The first (resp. last) block does not have arcs which is to or from the left (resp. right) adjacent block.

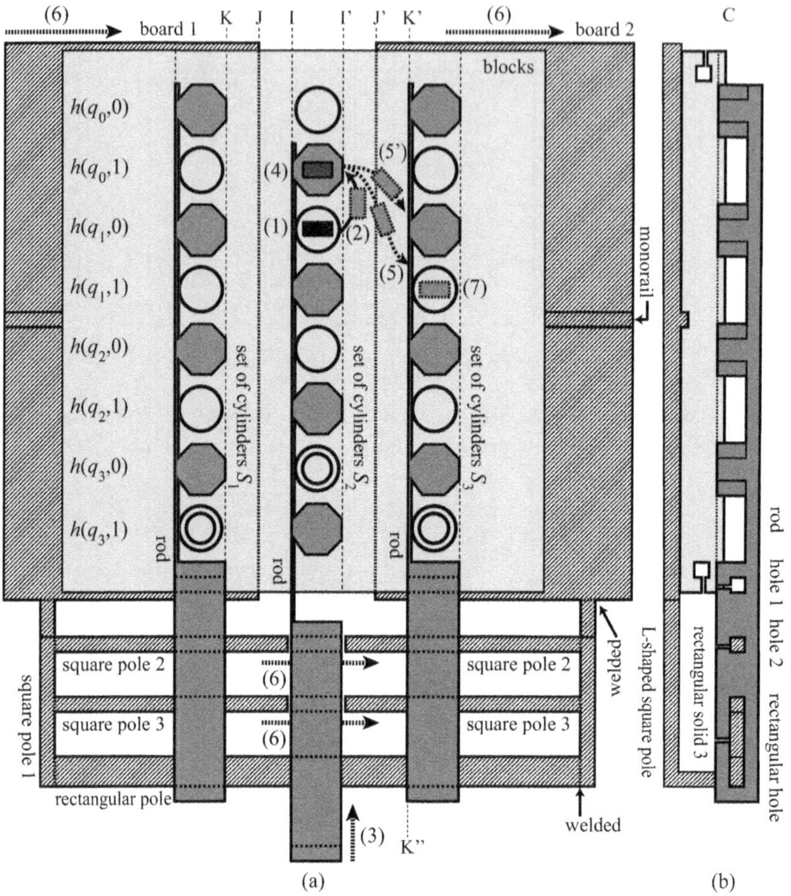

Fig. 6. (a) A cast puzzle simulating the configuration of a TM in Fig. 8. (b) A lateral view from the cross section between K′ and K″.

Fig. 5(b) is constructed according to a four-state deterministic TM $M = (Q, \Sigma, q_0, q_3, R)$ such that $Q = \{q_0, q_1, q_2, q_3\}$, $\Sigma = \{0, 1\}$, and

$$R = \{[q_0, /, +1, q_1], [q_1, 0, 1, q_0], [q_1, 1, 0, q_3], [q_2, 0, 1, q_3], [q_2, 1, 0, q_0]\}. \quad (1)$$

Fig. 6 illustrates a cast puzzle simulating a configuration of TM M. It is composed of a sequence of $g(k, p(n))$ blocks (see also Fig. 5(b)), two boards 1,2 behind them, and $g(k, p(n))$ sets of cylinders $S_1, S_2, \cdots, S_{g(k,p(n))}$. (Due to space limitation, Fig. 6 contains only three blocks, say, blocks 1,2,3, and three sets of cylinders S_1, S_2, S_3.) Fig. 6 corresponds the configuration shown in Fig. 8(a).

Fig. 6(b) is a lateral view from the cross section between K′ and K″. Each of boards 1,2 is welded to an L-shaped square pole (see Fig. 6(b)). The other end of the L-shaped square pole is welded to square pole 1, which is further welded

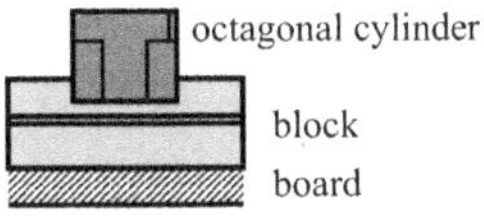

Fig. 7. A lateral view of a block and a cylinder from point C of Fig. 6(b)

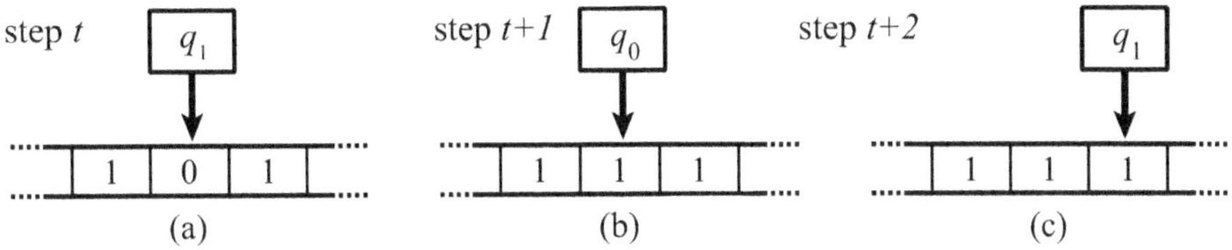

Fig. 8. Configurations of a TM at steps t, $t+1$, and $t+2$

to square poles 2,3 and a rectangular pole. It should be noted that boards 1,2 are connected via two L-shaped square poles and a rectangular pole; if board 1 is moved to the right, then board 2 is also moved to the right simultaneously (see (6) in Fig. 6(a)).

On block i, there is a set of cylinders S_i, where S_i contains m cylinders. Those m cylinders are connected by a rod so that they cover all holes in either $\{h(q_j, 0) \mid 0 \leq j \leq m-1\}$ or $\{h(q_j, 1) \mid 0 \leq j \leq m-1\}$. The other end of the rod is connected to rectangular solid i, which has two (square) holes 1,2 and one rectangular hole (see Fig. 6(b)). Square poles 2,3 pierce either holes 1,2 or hole 2 and the rectangular hole. If square poles 2,3 pierce holes 1,2, then the set of cylinders close holes $\{h(q_j, 1) \mid 0 \leq j \leq m-1\}$ (and holes $\{h(q_j, 0) \mid 0 \leq j \leq m-1\}$ are opened; see S_2 in Fig. 6(a)), which implies the tape cell has symbol 0. Similarly, if square poles 2,3 pierce hole 2 and the rectangular hole, then the tape cell has symbol 1 (see S_1, S_3). There is a gap between the left square poles 2,3 and the right square poles 2,3. The gap distance is the same as the width of each rectangular solid. Therefore, there exists at most one set of cylinders which we can move vertically.

The rectangular pole always pieces all rectangular solids. We dig a ditch on every block (see Fig. 7) and we fit the corresponding set of cylinders into the ditch so that they cannot move in the horizontal direction. On each board, there is a monorail which is fitted in the groove of each block so that the block does not move in the vertical direction.

Since the target piece has height 5 (see Fig. 1(b)), the top and bottom of the target piece protrude outside the block (see Fig. 2(b)). Therefore, the target piece must always be between boards 1 and 2. If the target piece reaches one of the "accepting" holes $h(q_{m-1}, 0), h(q_{m-1}, 1)$, then we can take it out from the block.

Consider a configuration of the TM M at step t (see Fig. 8(a)). In Fig. 6(a), the target piece is in the hole $h(q_1, 0)$ in the second block, and holes $\{h(q_j, 0) \mid 0 \leq j \leq m-1\}$ (resp. $\{h(q_j, 1) \mid 0 \leq j \leq m-1\}$) are opened in the second block

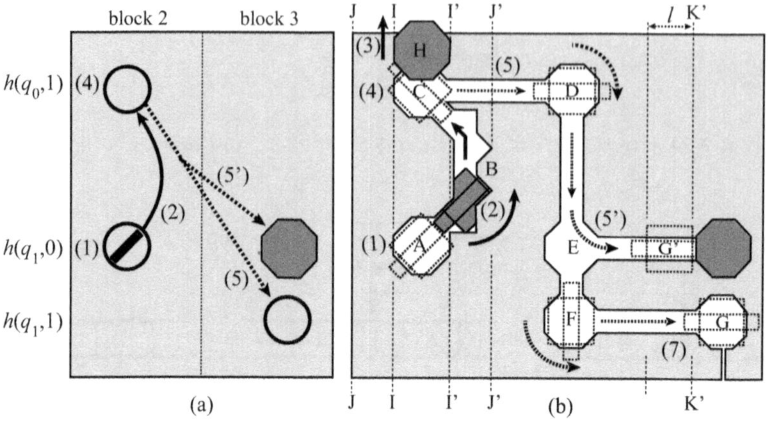

Fig. 9. (a) Solid arc from $h(q_1, 0)$ to $h(q_0, 1)$, and dotted arcs from $h(q_0, 1)$ to $h(q_1, 0)$ and $h(q_1, 1)$. (b) Detailed drawing of (a).

(resp. in the first and third blocks). From equation (1), there is a solid arc from hole $h(q_1, 0)$ to hole $h(q_0, 1)$ (see Figs. 6(a) and 9(a)), and there is a pair of dotted arcs from $h(q_0, 1)$ to $h(q_1, 0)$ and $h(q_1, 1)$ in the right adjacent block, where the hole $h(q_1, 0)$ is closed. Fig. 9(b) is the detailed drawing of Fig. 9(a).

(1) Suppose that the target piece is in the hole $h(q_1, 0)$. (2) We can carry the target piece forward until it reaches position B (see Fig. 9(b)). At this position, we can take a 45-degree turn. (3) During the evacuation of the target piece, we can move the set of cylinders S_2 upward. (In Fig. 9(b), cylinder H is just moving upward.) Since rectangular solid 2 is in the gap of square poles 2,3 (see Fig. 6(a)), we can freely move S_2 vertically (as far as the rectangular pole in the rectangular hole permits). (4) After the movement of cylinders, the hole $h(q_0, 1)$ at position C is opened, and $h(q_1, 0)$ at position A is closed. We can carry the target piece to position C. At this time, the edges of boards 1,2 are on borderlines J,J$'$, respectively. (5) We can carry the target piece to position D by moving board 2 to the right simultaneously. (6) Square poles 2,3 are connected to boards 1,2; thus the square hole 2 and the rectangular hole of rectangular solid 2 are pierced by square poles 2 and 3, respectively (see Fig. 6(a)). (7) Then we take a 90-degree turn at D, a downward movement from D to F, and a 90-degree turn at F. Again, we carry the target piece to position G by moving board 2 to the right. Finally, we move board 2 to the right so that cylinders S_3 are placed in the middle between boards 1 and 2, by which square poles 2,3 are pulled out of rectangular solid 3. Now the set of cylinders S_3 can move vertically, and the target piece can also move according to the solid arc in the third block.

The target piece cannot reach hole $h(q_1, 0)$ via dotted arc (5$'$) because of the following reason. In general, the set of cylinders S_i can move vertically if and only if the distance between I and J is the same as the distance between I$'$ and J$'$ (see Fig. 6(a)). Now we fix the length l of the target piece so that l is strictly longer than the distance between I and J. For such an l, the target piece gets stuck between the cylinder on $h(q_1, 0)$ and board 1 (see position G′ in Fig. 9(b)).

Initially, the first n sets of cylinders $S_1, S_2, \ldots, S_n$ are placed according to the input string $x \in \{0,1\}^n$, and the remaining $S_{n+1}, S_{n+2}, \ldots$ are placed so that holes $h(q_j, 0)$ are opened on every block. The target piece is initially in the hole $h(q_0, 0)$ or $h(q_0, 1)$ in the first block. The layout of such polyhedrons can be represented by a string of length polynomial in n. Since TM M is deterministic, the accepting computation is a sequence of configurations in which the last configuration is accepting. Such a sequence belongs to a directed *accepting tree* such that (i) every node is a configuration, and (ii) the root node is the unique accepting configuration in the tree. Therefore, M accepts input x if and only if the initial configuration of M belongs to an accepting tree. Moving the target piece in our cast puzzle corresponds to walking round nodes in the tree. The root node of this tree is an accepting configuration if and only if the target piece can be taken out of a block. By this construction, the TM M accepts x if and only if the cast puzzle is solvable. This completes the proof of Theorem 1.

References

1. `http://www.castpuzzle.net/english/index.html`
2. Demaine, E.D., Hohenberger, S., Lieben-Nowell, D.: Tetris is hard, even to approximate. In: Warnow, T.J., Zhu, B. (eds.) COCOON 2003. LNCS, vol. 2697, pp. 351–363. Springer, Heidelberg (2003)
3. Dor, D., Zwick, U.: Sokoban and other motion planning problems. Computational Geometry: Theory and Applications 13(4), 215–228 (1999)
4. Flake, G.W., Baum, E.B.: Rush Hour is PSPACE-complete, or why you should generously tip parking lot attendants. Theoret. Comput. Sci. 270(1/2), 895–911 (2002)
5. Fraenkel, A.S., Lichtenstein, D.: Computing a perfect strategy for $n \times n$ chess requires time exponential in n. J. Combinatorial Theory, Ser. A 31(2), 199–214 (1981)
6. Helmert, M.: Complexity results for standard benchmark domains in planning. Artificial Intelligence 143(2), 219–262 (2003)
7. Iwamoto, C., Nishio, K.: A polynomial-time transformation of SAT-instances to cast puzzles. An unpublished manuscript, `http://home.hiroshima-u.ac.jp/chuzo/IN09.pdf`
8. Iwata, S., Kasai, T.: The Othello game on an $n \times n$ board is PSPACE-complete. Theoret. Comput. Sci. 123(2), 329–340 (1994)
9. Kaye, R.: Minesweeper is NP-complete. Math. Intelligencer 22(2), 9–15 (2000)
10. Morita, K., Shirasaki, A., Gono, Y.: A 1-tape 2-symbol reversible Turing machine. Trans. IEICE E72(3), 223–228 (1989)
11. Preparata, F.P., Shamos, M.I.: Computational Geometry: An Introduction. Springer, New York (1985)
12. Ratner, D., Warmuth, M.: Finding a shortest solution for the $(N \times N)$-extension of the 15-puzzle is intractable. J. Symbolic Computation 10, 111–137 (1990)
13. Robson, J.M.: The complexity of Go. In: Proc. 9th World Computer Congress on Information Processing, pp. 413–417 (1983)

New Bounds on the Average Distance from the Fermat-Weber Center of a Planar Convex Body

Adrian Dumitrescu[1,*] and Csaba D. Tóth[2,**]

[1] University of Wisconsin–Milwaukee
ad@cs.uwm.edu
[2] University of Calgary and Tufts University
cdtoth@ucalgary.ca

Abstract. The Fermat-Weber center of a planar body Q is a point in the plane from which the average distance to the points in Q is minimal. We first show that for any convex body Q in the plane, the average distance from the Fermat-Weber center of Q to the points of Q is larger than $\frac{1}{6} \cdot \Delta(Q)$, where $\Delta(Q)$ is the diameter of Q. This proves a conjecture of Carmi, Har-Peled and Katz. From the other direction, we prove that the same average distance is at most $\frac{2(4-\sqrt{3})}{13} \cdot \Delta(Q) < 0.3490 \cdot \Delta(Q)$. The new bound substantially improves the previous bound of $\frac{2}{3\sqrt{3}} \cdot \Delta(Q) \approx 0.3849 \cdot \Delta(Q)$ due to Abu-Affash and Katz, and brings us closer to the conjectured value of $\frac{1}{3} \cdot \Delta(Q)$. We also confirm the upper bound conjecture for centrally symmetric planar convex bodies.

1 Introduction

The Fermat-Weber center of a measurable planar set Q with positive area is a point in the plane that minimizes the average distance to the points in Q. Such a point is the ideal location for a base station (e.g., fire station or a supply station) serving the region Q, assuming the region has uniform density. Given a measurable set Q with positive area and a point p in the plane, let $\mu_Q(p)$ be the average distance between p and the points in Q, namely,

$$\mu_Q(p) = \frac{\int_{q \in Q} \text{dist}(p, q)\, \mathrm{d}q}{\text{area}(Q)},$$

where $\text{dist}(p, q)$ is the Euclidean distance between p and q. Let FW_Q be the Fermat-Weber center of Q, and write $\mu_Q^* = \min\{\mu_Q(p) : p \in \mathbb{R}^2\} = \mu_Q(FW_Q)$.

Carmi, Har-Peled and Katz [3] showed that there exists a constant $c > 0$ such that $\mu_Q^* \geq c \cdot \Delta(Q)$ holds for any convex body Q, where $\Delta(Q)$ denotes the diameter of Q. The convexity is necessary, since it is easy to construct nonconvex regions where the average distance from the Fermat-Weber center is arbitrarily small compared to the diameter. Of course the opposite inequality $\mu_Q^* \leq c' \cdot \Delta(Q)$ holds for any body Q (convexity is not required), since we can trivially take $c' = 1$.

* Supported in part by NSF CAREER grant CCF-0444188.
** Supported in part by NSERC grant RGPIN 35586.

Y. Dong, D.-Z. Du, and O. Ibarra (Eds.): ISAAC 2009, LNCS 5878, pp. 132–141, 2009.

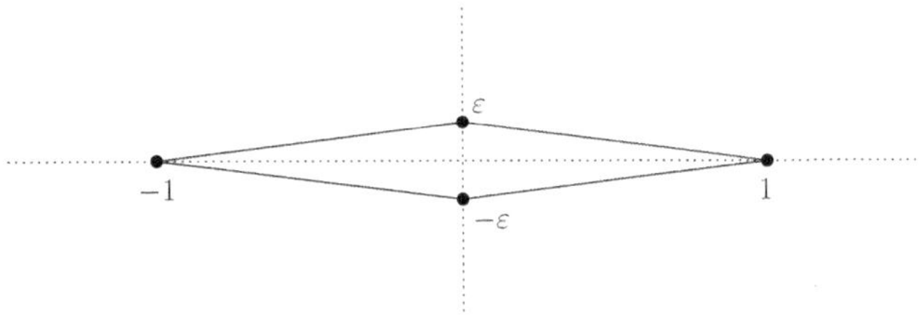

Fig. 1. A flat rhombus P_ε, with $\lim_{\varepsilon \to 0} \mu^*_{P_\varepsilon}/\Delta(P_\varepsilon) = \frac{1}{6}$

Let c_1 denote the infimum, and c_2 denote the supremum of $\mu^*_Q/\Delta(Q)$ over all convex bodies Q in the plane. Carmi, Har-Peled and Katz [3] conjectured that $c_1 = \frac{1}{6}$ and $c_2 = \frac{1}{3}$. Moreover, they conjectured that the supremum c_2 is attained for a circular disk D, where $\mu^*_D = \frac{1}{3} \cdot \Delta(D)$. They also proved that $\frac{1}{7} \leq c_1 \leq \frac{1}{6}$. The inequality $c_1 \leq \frac{1}{6}$ is given by an infinite sequence of rhombi, P_ε, where one diagonal has some fixed length, say 2, and the other diagonal tends to zero; see Fig. 1. By symmetry, the Fermat-Weber center of a rhombus is its center of symmetry, and one can verify that $\mu^*_{P_\varepsilon}/\Delta(P_\varepsilon)$ tends to $\frac{1}{6}$. The lower bound for c_1 has been recently further improved by Abu-Affash and Katz from $\frac{1}{7}$ to $\frac{4}{25}$ [1]. Here we establish that $c_1 = \frac{1}{6}$ and thereby confirm the first of the two conjectures of Carmi, Har-Peled and Katz.

Regarding the second conjecture, recently Abu-Affash and Katz proved that $c_2 \leq \frac{2}{3\sqrt{3}} = 0.3849\ldots$. Here we further improve this bound and bring it closer to the conjectured value of $\frac{1}{3}$. Finally, we also confirm the upper bound conjecture for centrally symmetric convex bodies Q.

Our main results are summarized in the following two theorems:

Theorem 1. *For any convex body Q in the plane, we have $\mu^*_Q > \frac{1}{6} \cdot \Delta(Q)$.*

Theorem 2. *For any convex body Q in the plane, we have*

$$\mu^*_Q \leq \frac{2(4-\sqrt{3})}{13} \cdot \Delta(Q) < 0.3490 \cdot \Delta(Q).$$

Moreover, if Q is centrally symmetric, then $\mu^*_Q \leq \frac{1}{3} \cdot \Delta(Q)$.

Remarks. **1.** The average distance from a point p in the plane can be defined analogously for finite point sets and for rectifiable curves. Observe that for a line segment I (a one-dimensional convex set), we would have $\mu^*_I/\Delta(I) = \frac{1}{4}$. It might be interesting to note that while the thin rhombi mentioned above tend in the limit to a line segment, the value of the limit $\mu^*_{P_\varepsilon}/\Delta(P_\varepsilon)$ equals $\frac{1}{6}$, not $\frac{1}{4}$.

2. In some applications, the cost of serving a location q from a facility at point p is $\mathrm{dist}^\kappa(p,q)$ for some exponent $\kappa \geq 1$, rather than $\mathrm{dist}(p,q)$. We can define $\mu^\kappa_Q(p) = \left(\int_{q \in Q} \mathrm{dist}^\kappa(p,q)\, \mathrm{d}q\right)/\mathrm{area}(Q)$ and $\mu^{\kappa *}_Q = \inf\{\mu^\kappa_Q(p) : p \in \mathbb{R}^2\}$, which is invariant under congruence. The ratio $\mu^{\kappa *}_Q/\Delta^\kappa(Q)$ is also invariant under similarity. The proof of Theorem 1 carries over for this variant and shows that $\mu^{\kappa *}_Q/\Delta^\kappa(Q) > \frac{1}{(\kappa+2)2^\kappa}$ for any convex body Q, and $\lim_{\varepsilon \to 0} \mu_{P_\varepsilon}^{\kappa *}/2^\kappa = \frac{1}{(\kappa+2)2^\kappa}$. For the upper bound, the picture is not so clear: $\mu_Q^*/\Delta(Q)$ is conjectured to be maximal for the circular disk, however, there is a $\kappa \geq 1$ such that $\mu_Q^{\kappa *}/\Delta^\kappa(Q)$ cannot be maximal for the disk. In particular, if D is a disk of diameter 2 and R is a convex body of diameter 2 whose smallest enclosing circle has diameter more than 2 (e.g., a regular or a Rouleaux triangle of diameter 2), then $\mu_D^{\kappa *} < \mu_R^{\kappa *}$, for a sufficiently large $\kappa > 1$. Let o be an arbitrary point in the plane, and let D be centered at o. Then $\int_{q \in D} \text{dist}^\kappa(o,q)\ dq = \int_0^{2\pi} \int_0^1 r^\kappa \cdot r\ dr\ d\theta = \frac{2\pi}{\kappa+2}$, and so $\lim_{\kappa \to \infty} \mu_D^{\kappa *} \leq \lim_{\kappa \to \infty} \frac{2}{\kappa+2} = 0$. On the other hand, for any region R' lying outside of D and for any $\kappa \geq 1$, we have $\int_{q \in R'} \text{dist}^\kappa(o,q)\ dq \geq \text{area}(R') > 0$. If $R' = R \setminus D$ is the part of R lying outside D, then $\lim_{\kappa \to \infty} \mu_R^{\kappa *} \geq \text{area}(R')/\pi > 0$.

Related work. Fekete, Mitchell, and Weinbrecht [7] studied a continuous version of the problem for polygons with holes, where the distance between two points is measured by the L_1 geodesic distance. A related question on Fermat-Weber centers in a discrete setting deals with stars and Steiner stars [5,6]. The reader can find more information on other variants of the Fermat-Weber problem in [4,10].

2 Lower Bound: Proof of Theorem 1

In a nutshell the proof goes as follows. Given a convex body Q, we take its Steiner symmetrization with respect to a supporting line of a diameter segment cd, followed by another Steiner symmetrization with respect to the perpendicular bisector of cd. The two Steiner symmetrizations preserve the area and the diameter, and do not increase the average distance from the corresponding Fermat-Weber centers. In the final step, we prove that the inequality holds for a convex body with two orthogonal symmetry axes.

Steiner symmetrization with respect to an axis. Steiner symmetrization of a convex figure Q with respect to an axis (line) ℓ consists in replacing Q by a new figure $S(Q,\ell)$ with symmetry axis ℓ by means of the following construction: Each chord of Q orthogonal to ℓ is displaced along its line to a new position where it is symmetric with respect to ℓ, see [11, pp. 64]. The resulting figure $S(Q,\ell)$ is also convex, and obviously has the same area as Q.

A body Q is x-monotone if the intersection of Q with every vertical line is either empty or is connected (that is, a point or a line segment). Every x-monotone body Q is bounded by the graphs of some functions $f : [a,b] \to \mathbb{R}$ and $g : [a,b] \to \mathbb{R}$ such that $g(x) \leq f(x)$ for all $x \in [a,b]$. The Steiner symmetrization with respect to the x-axis ℓ_x transforms Q into an x-monotone body $S(Q,\ell_x)$ bounded by the functions $\frac{1}{2}(f(x) - g(x))$ and $\frac{1}{2}(g(x) - f(x))$ for $x \in [a,b]$. As noted earlier, $\text{area}(S(Q,\ell_x)) = \text{area}(Q)$. The next two lemmas do not require the convexity of Q.

Lemma 1. *Let Q be an x-monotone body in the plane with a diameter parallel or orthogonal to the x-axis, then $\Delta(Q) = \Delta(S(Q,\ell_x))$.*

Proof. Let $Q' = S(Q,\ell_x)$. If Q has a diameter parallel to the x-axis, then the diameter is $[(a,c),(b,c)]$, with a value $c \in \mathbb{R}$, $g(a) = c = f(a)$ and $g(b) = c = f(b)$. That is,

$\Delta(Q) = b - a$. In this case, the diameter of Q' is at least $b - a$, since both points $(a, 0)$ and $(b, 0)$ are in Q'. If Q has a diameter orthogonal to the x-axis, then the diameter is $[(x_0, f(x_0)), (x_0, g(x_0))]$ for some $x_0 \in [a, b]$, and $\Delta(Q) = f(x_0) - g(x_0)$. In this case, the diameter of Q' is at least $f(x_0) - g(x_0)$, since both points $(x_0, \frac{1}{2}(f(x_0) - g(x_0)))$ and $(x_0, \frac{1}{2}(g(x_0) - f(x_0)))$ are in Q'. Therefore, we have $\Delta(Q') \geq \Delta(Q)$.

Let A_1 and A_2 be two points on the boundary of Q' such that $\Delta(Q') = \mathrm{dist}(A_1, A_2)$. Since Q' is symmetric to the x-axis, points A_1 and A_2 cannot both be on the upper (resp., lower) boundary of Q'. Assume w.l.o.g. that $A_1 = (x_1, \frac{1}{2}(f(x_1) - g(x_1)))$ and $A_2 = (x_2, \frac{1}{2}(g(x_2) - f(x_2)))$ for some $a \leq x_1, x_2 \leq b$.

$$\Delta(Q') = \mathrm{dist}(A_1, A_2) = \sqrt{(x_2 - x_1)^2 + \left(\frac{f(x_1) + f(x_2) - g(x_1) - g(x_2)}{2}\right)^2}.$$

Now consider the following two point pairs in Q. The distance between $B_1 = (x_1, f(x_1))$ and $B_2 = (x_2, g(x_2))$ is $\mathrm{dist}(B_1, B_2) = \sqrt{(x_2 - x_1)^2 + (f(x_1) - g(x_2))^2}$. Similarly, the distance between $C_1 = (x_1, g(x_1))$ and $C_2 = (x_2, f(x_2))$ is $\mathrm{dist}(C_1, C_2) = \sqrt{(x_2 - x_1)^2 + (g(x_1) - f(x_2))^2}$. Using the inequality between the arithmetic and quadratic means, we have

$$\left(\frac{f(x_1) + f(x_2) - g(x_1) - g(x_2)}{2}\right)^2 \leq \frac{(f(x_1) - g(x_2))^2 + (g(x_1) - f(x_2))^2}{2}.$$

This implies that $\mathrm{dist}(A_1, A_2) \leq \max(\mathrm{dist}(B_1, B_2), \mathrm{dist}(C_1, C_2))$, and so $\Delta(Q') \leq \Delta(Q)$. We conclude that $\Delta(Q) = \Delta(S(Q, \ell_x))$. □

Lemma 2. *If Q is an x-monotone body in the plane, then $\mu^*_Q \geq \mu^*_{S(Q,\ell_x)}$.*

Proof. If (x_0, y_0) is the Fermat-Weber center of Q, then

$$\mu^*_Q = \frac{\int_a^b \int_{g(x)}^{f(x)} \sqrt{(x - x_0)^2 + (y - y_0)^2} \,\mathrm{d}y \,\mathrm{d}x}{\mathrm{area}(Q)}.$$

Observe that $\int_{g(x)}^{f(x)} \sqrt{(x - x_0)^2 + (y - y_0)^2} \,\mathrm{d}y$ is the integral of the distances of the points in a line segment of length $f(x) - g(x)$ from a point at distance $|x - x_0|$ from the supporting line of the segment. This integral is minimal if the point is on the orthogonal bisector of the segment. That is, we have

$$\begin{aligned}\int_{g(x)}^{f(x)} \sqrt{(x - x_0)^2 + (y - y_0)^2} \,\mathrm{d}y &\geq \int_{g(x)}^{f(x)} \sqrt{(x - x_0)^2 + \left(y - \frac{f(x) - g(x)}{2}\right)^2} \,dy \\ &= \int_{\frac{1}{2}(g(x) - f(x))}^{\frac{1}{2}(f(x) - g(x))} \sqrt{(x - x_0)^2 + y^2} \,\mathrm{d}y.\end{aligned}$$

Therefore, we conclude that

$$\begin{aligned}\mu^*_Q &= \frac{\int_a^b \int_{g(x)}^{f(x)} \sqrt{(x - x_0)^2 + (y - y_0)^2} \,\mathrm{d}y \,\mathrm{d}x}{\mathrm{area}(Q)} \\ &\geq \frac{\int_a^b \int_{\frac{1}{2}(g(x) - f(x))}^{\frac{1}{2}(f(x) - g(x))} \sqrt{(x - x_0)^2 + y^2} \,\mathrm{d}y \,\mathrm{d}x}{\mathrm{area}(S(Q, x))} = \mu_{S(Q,\ell_x)}((x_0, 0)) \geq \mu^*_{S(Q,\ell_x)}.\end{aligned}$$ □

Triangles. We next consider right triangles of a special kind, lying in the first quadrant, and show that the average distance from the origin to their points is larger than $\frac{1}{3}$.

Lemma 3. *Let T a right triangle in the first quadrant based on the x-axis, with vertices $(a, 0)$, (a, b), and $(1, 0)$, where $0 \le a < 1$, and $b > 0$. Then $\mu_T(o) > \frac{1}{3}$.*

Proof. We use the simple fact that the x-coordinate of a point is a lower bound to the distance from the origin.

$$\begin{aligned}
\mu_T(o) &= \frac{\int_a^1 \left(\int_0^{b(1-x)/(1-a)} \sqrt{x^2+y^2}\, dy\right) dx}{b(1-a)/2} > \frac{\int_a^1 \left(\int_0^{b(1-x)/(1-a)} x\, dy\right) dx}{b(1-a)/2} \\
&= \frac{\frac{b}{1-a}\int_a^1 x(1-x)\, dx}{b(1-a)/2} = \frac{2}{(1-a)^2}\left(\frac{x^2}{2} - \frac{x^3}{3}\right)\Big|_a^1 \\
&= \frac{2}{(1-a)^2} \cdot \frac{(2a^3 - 3a^2 + 1)}{6} = \frac{2}{(1-a)^2} \cdot \frac{(1-a)(1+a-2a^2)}{6} \\
&= \frac{1}{(1-a)} \cdot \frac{(1+a-2a^2)}{3} > \frac{1}{3}.
\end{aligned}$$

The last inequality in the chain follows from $a < 1$. The inequality in the lemma is strict, since $\sqrt{x^2+y^2} > x$ for all points above the x-axis. □

Corollary 1. *Let P be any rhombus. Then $\mu_P^* > \frac{1}{6} \cdot \Delta(P)$.*

Proof. Without loss of generality, we may assume that P is symmetric with respect to both the x- and the y-axis. Let us denote the vertices of P by $(-1, 0)$, $(1, 0)$, $(0, -b)$, and $(0, b)$, where $b \le 1$. We have $\Delta(P) = 2$. By symmetry, μ_P^* equals the average distance between the origin $(0, 0)$ and the points in one of the four congruent right triangles forming P. Consider the triangle T in the first quadrant. By Lemma 3 (with $a = 0$), we have $\mu_P^* = \mu_T(o) > \frac{1}{3}$. Since $\Delta(P) = 2$, we have $\mu_P^* > \frac{1}{6} \cdot \Delta(P)$, as desired. □

Lemma 4. *Let T be a triangle in the first quadrant with a vertical side on the line $x = a$, where $0 \le a < 1$, and a third vertex at $(1, 0)$. Then $\mu_T(o) > \frac{1}{3}$.*

Proof. Refer to Fig. 2(ii). Let U be a right triangle obtained from T by translating each vertical chord of T down until its lower endpoint is on the x-axis. Note that $\text{area}(T) = \text{area}(U)$. Observe also that the average distance from the origin decreases in this transformation, namely $\mu_T(o) \ge \mu_U(o)$. By Lemma 3, we have $\mu_U(o) > \frac{1}{3}$, and so $\mu_T(o) > \frac{1}{3}$, as desired. □

We now have all necessary ingredients to prove Theorem 1.

Proof of Theorem 1. Refer to Fig. 2. Let Q be a convex body in the plane, and let $c, d \in Q$ be two points at $\Delta(Q)$ distance apart. We may assume that $c = (-1, 0)$ and $d = (1, 0)$, by a similarity transformation if necessary, so that $\Delta(Q) = 2$ (the ratio $\mu_Q^*/\Delta(Q)$ is invariant under similarities). Apply a Steiner symmetrization with respect to the x-axis, and then a second Steiner symmetrization with respect to the y-axis. The

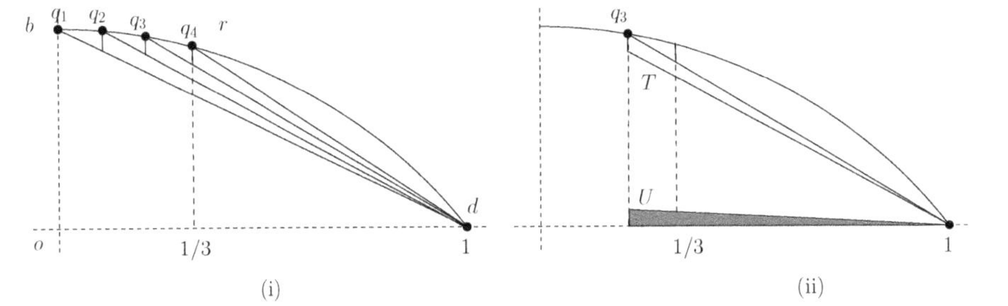

Fig. 2. (i) The subdivision of Q_1 for $n = 3$. Here $o = (0, 0)$, $q_1 = b = (0, h)$, $q_4 = r$, $d = (1, 0)$. (ii) Transformation in the proof of Lemma 4.

resulting body $Q' = S(S(Q, \ell_x), \ell_y)$ is convex, and it is symmetric with respect to both coordinate axes. We have $\Delta(Q') = \Delta(Q) = 2$ by Lemma 1, and in fact $c, d \in Q'$. We also have $\mu^*_{Q'} \le \mu^*_Q$ by Lemma 2.

Let Q_1 be the part of Q' lying in the first quadrant: $Q_1 = \{(x, y) \in Q' : x, y \ge 0\}$. By symmetry, $FW_{Q'} = o$ and we have $\mu^*_{Q'} = \mu_{Q'}(o) = \mu_{Q_1}(o)$. Let γ be the portion of the boundary of Q' lying in the first quadrant, between points $b = (0, h)$, with $0 < h \le 1$, and $d = (1, 0)$. For any two points $p, q \in \gamma$ along γ, denote by $\gamma(p, q)$ the portion of γ between p and q. Let r be the intersection point of γ and the vertical line $x = \frac{1}{3}$.

For a positive integer n, subdivide Q_1 into at most $2n + 2$ pieces as follows. Choose $n + 1$ points $b = q_1, q_2 \ldots, q_{n+1} = r$ along $\gamma(b, r)$ such that q_i is the intersection of γ and the vertical line $x = (i - 1)/3n$. Connect each of the $n + 1$ points to d by a straight line segment. These segments subdivide Q_1 into $n + 2$ pieces: the right triangle $T_0 = \Delta bod$; a convex body Q_0 bounded by rd and $\gamma(r, d)$; and n curvilinear triangles $\Delta q_i d q_{i+1}$ for $i = 1, 2, \ldots, n$. For simplicity, we assume that neither Q_0, nor any of the curvilinear triangles are degenerate; otherwise they can be safely ignored (they do not contribute to the value of $\mu^*_{Q'}$). Subdivide each curvilinear triangle $\Delta q_i d q_{i+1}$ along the vertical line through q_{i+1} into a small curvilinear triangle S_i on the left and a triangle T_i incident to point d on the right. The resulting subdivision has $2n + 2$ pieces, under the nondegeneracy assumption.

By Lemma 3, we have $\mu_{T_0}(o) > \frac{1}{3}$. Observe that the difference $\mu_{T_0}(o) - \frac{1}{3}$ does not depend on n, and let $\delta = \mu_{T_0}(o) - \frac{1}{3}$. By Lemma 4, we also have $\mu_{T_i}(o) > \frac{1}{3}$, for each $i = 1, 2, \ldots, n$. Since every point in Q_0 is at distance at least $\frac{1}{3}$ from the origin, we also have $\mu_{Q_0}(o) \ge \frac{1}{3}$.

For the n curvilinear triangles S_i, $i = 1, 2, \ldots, n$, we use the trivial lower bound $\mu_{S_i}(o) \ge 0$. We now show that their total area $s_n = \sum_{i=1}^{n} \text{area}(S_i)$ tends to 0 if n goes to infinity. Recall that the y-coordinates of the points q_i are at most 1, and their x-coordinates are at most $\frac{1}{3}$. This implies that the slope of every line $q_i d$, $i = 1, 2, \ldots, n+1$, is in the interval $[-3/2, 0]$. Therefore, S_i is contained in a right triangle bounded by a horizontal line through q_i, a vertical line through q_{i+1}, and the line $q_i d$. The area of this triangle is at most $\frac{1}{2}(\frac{1}{3n} \cdot (\frac{3}{2} \cdot \frac{1}{3n})) = 1/(12n^2)$. That is, $s_n = \sum_{i=1}^{n} \text{area}(S_i) \le 1/12n$. In particular, $s_n \le \delta \cdot \text{area}(T_0)$ for a sufficiently large n. Then we can write

$$\mu_{Q_1}(o) = \frac{\int_{p\in Q_1} \text{dist}(op)\,\text{d}p}{\text{area}(Q_1)} \geq \frac{\mu_{Q_0}(o)\cdot \text{area}(Q_0) + \sum_{i=0}^{n} \mu_{T_i}(o)\cdot \text{area}(T_i)}{\text{area}(Q_1)}$$
$$\geq \frac{\frac{1}{3}(\text{area}(Q_1) - s_n) + \delta \cdot \text{area}(T_0)}{\text{area}(Q_1)} \geq \frac{1}{3} + \frac{2\delta \cdot \text{area}(T_0)}{3\cdot \text{area}(Q_1)} > \frac{1}{3}.$$

This concludes the proof of Theorem 1. □

Remark. A finite triangulation, followed by taking the limit suffices to prove the slightly weaker, non-strict inequality: $\mu_Q^* \geq \frac{1}{6}\cdot \Delta(Q)$.

3 Upper Bounds: Proof of Theorem 2

Let Q be a planar convex body. Let ∂Q denote the boundary of Q, and let $\text{int}(Q)$ denote the interior of Q. Let Ω be the smallest circle enclosing Q, and let o be the center of Ω. Write $a = \frac{2(4-\sqrt{3})}{13}$, and $D = \Delta(Q)$. By the convexity of Q, $o \in Q$, as observed in [1]. Moreover, Abu-Affash and Katz [1] have shown that the average distance from o to the points in Q satisfies

$$\mu_Q(o) \leq \frac{2}{3\sqrt{3}}\cdot \Delta(Q) < 0.3850 \cdot \Delta(Q).$$

Here we further refine their analysis and derive a better upper bound on the average distance from o:

$$\mu_Q(o) \leq \frac{2(4-\sqrt{3})}{13}\cdot \Delta(Q) < 0.3490 \cdot \Delta(Q).$$

Since the average distance from the Fermat-Weber center of Q is not larger than that from o, we immediately get the same upper bound on c_2. We need the next simple lemma established in [1]. Its proof follows from the definition of average distance.

Lemma 5. [1]. *Let Q_1, Q_2 be two (not necessarily convex) disjoint bodies in the plane, and p be a point in the plane. Then $\mu_{(Q_1\cup Q_2)}(p) \leq \max(\mu_{Q_1}(p), \mu_{Q_2}(p))$.*

By induction, Lemma 5 yields:

Lemma 6. *Let $Q_1, Q_2, \ldots, Q_n$ be n (not necessarily convex) pairwise disjoint bodies in the plane, and p be a point in the plane. Then*

$$\mu_{(Q_1\cup\ldots\cup Q_n)}(p) \leq \max(\mu_{Q_1}(p), \ldots \mu_{Q_n}(p)).$$

We also need the following classical result of Jung [9] (also available in [8]).

Theorem 3. (Jung [9]). *Let S be a set of diameter $\Delta(S)$ in the plane. Then S is contained in a circle of radius $\frac{1}{\sqrt{3}}\cdot \Delta(S)$.*

Lastly, observe that the average distance from the center of a circular sector of radius r and center angle α to the points in the sector is

$$\frac{\int_0^r \alpha x^2 \, \mathrm{d}x}{\int_0^r \alpha x \, \mathrm{d}x} = \frac{\alpha r^3/3}{\alpha r^2/2} = \frac{2r}{3}. \tag{1}$$

We now proceed with the proof of Theorem 2. We distinguish two cases: Case 1: $o \in \partial Q$, and Case 2: $o \in \text{int}(Q)$. Assume that we are in the first (easy) case. Then Q is contained in a halfdisk Θ of Ω (with the same diameter D), and o is the midpoint of the this diameter. Then by (1), it follows that $\mu_Q(o) \le \frac{1}{3} \cdot D$.

Next we assume that we are in the second case, with o lying in the interior of Q. Let $\varepsilon > 0$ be sufficiently small. For a large positive integer n, subdivide Ω into n congruent circular double sectors (wedges) $W_1, \ldots, W_n$, symmetric about o (the center of Ω), where each sector subtends an angle $\alpha = \pi/n$. Consider a double sector $W_i = U_i \cup V_i$, where U_i and V_i are circular sectors of Ω. Let $X_i \subseteq U_i$, and $Y_i \subseteq V_i$ be two minimal circular sectors centered at o and containing $U_i \cap Q$, and $V_i \cap Q$, respectively: $U_i \cap Q \subseteq X_i$, and $V_i \cap Q \subseteq Y_i$. Let x_i and y_i be the radii of X_i and Y_i, respectively. Let $X_i' \subseteq X_i$, and $Y_i' \subseteq Y_i$ be two circular subsectors of radii $(1-\varepsilon)x_i$ and $(1-\varepsilon)y_i$, respectively. Since $o \in \text{int}(Q)$, we can select $n = n(Q, \varepsilon)$ large enough, so that for each $1 \le i \le n$, the subsectors X_i' and Y_i' are nonempty and entirely contained in Q. That is, for every i, we have

$$X_i' \cup Y_i' \subseteq W_i \cap Q \subseteq X_i \cup Y_i. \tag{2}$$

It is enough to show that for any double sector $W = W_i$, we have

$$\lim_{\varepsilon \to 0} \mu_{(W \cap Q)}(o) \le aD,$$

since then, Lemma 6 (with W_i being the n pairwise disjoint regions) will imply that $\mu_Q(o) \le aD$, concluding the proof of Theorem 2. For simplicity, write $x = x_i$, and $y = y_i$. Obviously the diameter of $W \cap Q$ is at most D, hence $x + y \le D$. We can assume w. l. o. g. that $y \le x$, so by Theorem 3 we also have $x \le \frac{1}{\sqrt{3}} \cdot D$. For easy reference, we summarize our constraints:

$$0 < y \le x \le \frac{1}{\sqrt{3}} \cdot D \qquad \text{and} \qquad x + y \le D. \tag{3}$$

By the definition of average distance, we can write

$$\begin{aligned} \mu_{(W \cap Q)}(o) &= \frac{\int_{p \in (W \cap Q)} \text{dist}(op) \, \mathrm{d}p}{\text{area}(W \cap Q)} \\ &\le \frac{\alpha \cdot \frac{x^2}{2} \cdot \frac{2x}{3} + \alpha \cdot \frac{y^2}{2} \cdot \frac{2y}{3}}{\alpha(1-\varepsilon)^2 \cdot \left(\frac{x^2}{2} + \frac{y^2}{2}\right)} \\ &= \frac{2}{3} \cdot \frac{x^3 + y^3}{(1-\varepsilon)^2 \cdot (x^2 + y^2)}. \end{aligned} \tag{4}$$

Let

$$f(x, y) = \frac{2}{3} \cdot \frac{x^3 + y^3}{x^2 + y^2}, \quad \text{and} \quad g(x, y, \varepsilon) = \frac{2}{3} \cdot \frac{x^3 + y^3}{(1-\varepsilon)^2 \cdot (x^2 + y^2)}.$$

Clearly for any feasible pair (x, y), we have

$$\lim_{\varepsilon \to 0} g(x, y, \varepsilon) = f(x, y).$$

It is a straightforward calculus exercise to show that under the constraints listed in (3), $f(x, y)$ is maximized for:

$$x_0 = \frac{D}{\sqrt{3}}, \text{ and } y_0 = \left(1 - \frac{1}{\sqrt{3}}\right) D.$$

By substituting these values for x and y, and letting ε tend to zero in (4), we obtain

$$\mu_Q(o) \leq \lim_{\varepsilon \to 0} g(x_0, y_0, \varepsilon) = f(x_0, y_0) = \frac{2(4 - \sqrt{3})}{13} \cdot D,$$

as required.

Assume now that Q is centrally symmetric with respect to point q. We repeat the same argument. It is enough to observe that: (i) the center of Ω coincides with q, that is, $o = q$; and (ii) $x = y \leq \frac{1}{2} \cdot D$ for any double sector W. The average distance calculation yields now

$$\mu_{(W \cap Q)}(o) \leq \frac{2x^3}{3(1-\varepsilon)^2 \cdot x^2} = \frac{2x}{3(1-\varepsilon)^2} \leq \frac{D}{3(1-\varepsilon)^2},$$

and by taking the limit when ε tends to zero, we obtain

$$\mu_Q(o) \leq \frac{D}{3},$$

as required. The proof of Theorem 2 is now complete.

4 Applications

1. Carmi, Har-Peled and Katz [3] showed that given a convex polygon Q with n vertices, and a parameter $\varepsilon > 0$, one can compute an ε-approximate Fermat-Weber center $q \in Q$ in $O(n + 1/\varepsilon^4)$ time such that $\mu_Q(q) \leq (1+\varepsilon)\mu_Q^*$. Abu-Affash and Katz [1] gave a simple $O(n)$-time algorithm for computing the center q of the smallest circle enclosing Q, and showed that q approximates the Fermat-Weber center of Q, with $\mu_Q(q) \leq \frac{25}{6\sqrt{3}}\mu_Q^*$. Our Theorems 1 and 2, combined with their analysis, improves the approximation ratio to about 2.09:

$$\mu_Q(q) \leq \frac{12(4 - \sqrt{3})}{13} \mu_Q^*.$$

2. The value of the constant c_1 (i.e., the infimum of $\mu_Q^*/\Delta(Q)$ over all convex bodies Q in the plane) plays a key role in the following load balancing problem introduced by Aronov, Carmi and Katz [2]. We are given a convex body D and m points $p_1, p_2, \ldots, p_m$ representing *facilities* in the interior of D. Subdivide D into m convex regions, $R_1, R_2, \ldots, R_m$, of equal area such that $\sum_{i=1}^{m} \mu_{p_i}(R_i)$ is minimal. Here

$\mu_{p_i}(R_i)$ is the *cost* associated with facility p_i, which may be interpreted as the average travel time from the facility to any location in its designated region, each of which has the same area. One of the main results in [2] is a $(8+\sqrt{2\pi})$-factor approximation in the case that D is an $n_1 \times n_2$ rectangle for some integers $n_1, n_2 \in \mathbb{N}$. This basic approximation bound is then used for several other cases, e.g., subdividing a convex fat domain D into m convex regions R_i.

By substituting $c_1 = \frac{1}{6}$ (Theorem 1) into the analysis in [2], the upper bound for the approximation ratio improves from $8 + \sqrt{2\pi} \approx 10.5067$ to $7 + \sqrt{2\pi} \approx 9.5067$. It can be further improved by optimizing another parameter used in their calculation. Let S be a unit square and let $s \in S$ be an arbitrary point in the square. Aronov et al. [2] used the upper bound $\mu_S(s) \leq \frac{2}{3}\sqrt{2} \approx .9429$. It is clear that $\max_{s\in S} \mu_S(s)$ is attained if s is a vertex of S. The average distance of S from such a vertex, say v, is $\mu_S(v) = \frac{1}{3}\left(\sqrt{2} + \ln(1+\sqrt{2})\right) \approx .7652$, and so $\mu_S(s) \leq \frac{1}{3}\left(\sqrt{2} + \ln(1+\sqrt{2})\right)$, for any $s \in S$. With these improvements, the upper bound on the approximation ratio becomes $7 + \frac{\sqrt{\pi}}{2}\left(\sqrt{2} + \ln(1+\sqrt{2})\right) \approx 9.0344$.

Acknowledgement. We are grateful to Alex Rand for stimulating discussions.

References

1. Abu-Affash, A.K., Katz, M.J.: Improved bounds on the average distance to the Fermat-Weber center of a convex object. Inf. Proc. Letts. 109(6), 329–333 (2009)
2. Aronov, B., Carmi, P., Katz, M.J.: Minimum-cost load-balancing partitions. Algorithmica 54(3), 318–336 (2009)
3. Carmi, P., Har-Peled, S., Katz, M.: On the Fermat-Weber center of a convex body. Comput. Geom. Theory Appl. 32, 188–195 (2005)
4. Drezner, Z., Klamroth, K., Schöbel, A., Wesolowsky, G.O.: The Weber problem. In: Hamacher, H.W., Drezner, Z. (eds.) Facility Location: Applications And Theory, pp. 1–36. Springer, Berlin (2002)
5. Dumitrescu, A., Tóth, Cs.D., Xu, G.: On stars and Steiner stars. Discrete Optimization 6(3), 324–332 (2009)
6. Fekete, S., Meijer, H.: On minimum stars and maximum matchings. Discrete & Computational Geometry 23, 389–407 (2000)
7. Fekete, S., Mitchell, J.S.B., Weinbrecht, K.: On the continuous Weber and k-median problems. In: Proc. 16th ACM Sympos. Comput. Geom., pp. 70–79. ACM Press, New York (2000)
8. Hadwiger, H., Debrunner, H., Klee, V.: Combinatorial Geometry in the Plane. Holt, Rinehart and Winston, New York (1964)
9. Jung, H.W.E.: Über der kleinsten Kreis, der eine ebene Figur einschließt. J. Angew. Math. 137, 310–313 (1910)
10. Wesolowsky, G.: The Weber problem: History and perspective. Location Science 1, 5–23 (1993)
11. Yaglom, I.M., Boltyanski, V.G.: Convex Figures. Holt, Rinehart and Winston, New York (1961)

Reconstructing Numbers from Pairwise Function Values

Shiteng Chen[1], Zhiyi Huang[2], and Sampath Kannan[2]

[1] Tsinghua University, Beijing 100084, China
cst100@gmail.com
[2] University of Pennsylvania, Philadelphia PA 19104, USA
{hzhiyi,kannan}@cis.upenn.edu

Abstract. The turnpike problem is one of the few "natural" problems that are neither known to be NP-complete nor solvable by efficient algorithms. We seek to study this problem in a more general setting. We consider the generalized problem which tries to resolve set $A = \{a_1, a_2, \cdots, a_n\}$ from pairwise function values $\{f(a_i, a_j) | 1 \leq i, j \leq n\}$ for a given bivariate function f. We call this problem the NUMBER RECONSTRUCTION problem. Our results include efficient algorithms when f is monotone and non-trivial bounds on the number of solutions when f is the sum. We also generalize previous backtracking and algebraic algorithms for the turnpike problem such that they work for the family of anti-monotone functions and linear-decomposable functions. Finally, we propose an efficient algorithm for the string reconstruction problem, which is related to an approach to protein reconstruction.

1 Introduction

Given a set of n numbers, it is easy to compute their pairwise distances. The reverse problem of reconstructing all possible sets of n numbers for a given unordered set of $\binom{n}{2}$ distances is known as the *turnpike problem*. This problem dates back to the origins of X-ray crystallography in the 1930's [11,12,13] and later arises in restriction site mapping of DNA, where it is known as the *Partial Digest Problem*, and in the area of computational geometry [15].

Rosenblatt and Seymour [14] studied a related concept called *homometric sets*. Two sets are homometric if they provide the same unordered set of pairwise distances. They introduced a generating function technique and proved two sets are homometric if and only if their generating functions have a certain relationship. We will give more details in Section 2.

Based on the generating function technique, Skiena, Lemke and Smith [8] studied the number of different solutions for a turnpike instance. Let $H(n)$ denote the largest number of different homometric sets of size n. They proved that there exist infinitely many n such that $H(n) \geq n^{\alpha}/2$, where $\alpha \approx 0.8107144$, and for any n, $H(n) \leq n^{\beta}/2$, where $\beta \approx 1.2324827$. Hence, the number of solutions is bounded by polynomials of the input size. They proved that the turnpike problem in arbitrary dimension is strongly NP-complete. Skiena et al.

Y. Dong, D.-Z. Du, and O. Ibarra (Eds.): ISAAC 2009, LNCS 5878, pp. 142–152, 2009.

also proposed two non-trivial algorithms for solving the problem. The first one is a combinatorial algorithm, known as the backtracking algorithm, that solves any turnpike instance in $\mathcal{O}(2^n n \log n)$ time. Later, Zhang [19] showed that the backtracking algorithm indeed requires exponential time in the worst case, and Skiena et al. [8] proved that the backtracking algorithm works reasonably well on average if the input is drawn from a certain distribution. The second algorithm is a pseudopolynomial algorithm based on the generating function technique and polynomial factorization. The running time of this algorithm is polynomial in the maximum distance.

While hardness of the original turnpike problem remains open, there have been many successes in the study of variants. The Double Digest Problem [7] and the Simplified Partial Digest Problem [1,2], originated from other methods for reconstructing the restriction site locations of enzymes from DNA fragments and are both known to be NP-complete. Cieliebak et al. [4,3] studied four types of error that the turnpike problem may possibly encounter in real experiments. They proved that solving the turnpike problem with any of these errors is NP-complete. Pandurangan and Ramesh [10] explored a variant known as labeled partial digest. In this variant, both ends are labeled. So in addition to the pairwise distances, we are also given the set of lengths of segments at least one of whose endpoints is one of the two ends. They proposed an efficient and robust algorithm for this problem that runs in $\mathcal{O}(n^4)$ time and tolerates an absolute error up to $\mathcal{O}(\min_i d_i/n)$, where $\{d_i\}$ are the lengths of segments.

In this paper, we consider the more general NUMBER RECONSTRUCTION problem. Suppose there are n unknown integers $a_1, a_2, \cdots, a_n$ and a bivariate function f. Given an unordered set of n^2 function values $f(a_i, a_j)$, $1 \leq i, j \leq n$, the goal is to reconstruct $a_1, a_2, \cdots, a_n$ from the given set of values. We will use REC_f to denote this problem. The turnpike problem is clearly the special case when f is the difference function. We consider the following questions for a given function f. Can we solve REC_f efficiently? Can we solve REC_f uniquely? If not, what is the upper and lower bound on the number of possible solutions for any given instance? We seek to study this problem for a large family of functions f as an intermediate step toward resolving the complexity of turnpike problem.

Since the function values $f(a_i, a_i)$ are trivially known in the turnpike case, we are also interested in a variant where we are only given an unordered set of $n(n-1)$ function values $f(a_i, a_j)$, $1 \leq i \neq j \leq n$. Again, the goal is to find efficient algorithms reconstructing $a_1, a_2, \cdots, a_n$ from the given values as well as bounding the number of solutions. We will use REC^*_f to denote this problem. We call this setting the *incomplete information setting* and refer to the original setting as the *full information setting*.

Our Contribution. We study the case when f is the sum and more generally when f is monotone. In this case, we give efficient algorithms for both the full information setting and the incomplete information setting. Furthermore, we show non-trivial bounds on the number of solutions for the case when f is the sum in incomplete information setting. We then generalize our algorithm such that it can even solve the case when f is a multi-variate function. We also generalize previous turnpike algorithms to more general families of functions f, namely anti-monotone or linear-decomposable functions. Finally, we resolve an open problem proposed in [6] by giving an efficient algorithm for the string reconstruction problem (known as general reconstruction in [6]) that is related to a new approach to protein reconstruction. Our algorithm relies on a reduction to the turnpike problem and polynomial factorization.

2 Preliminaries and Notations

2.1 Homometric Sets

In this paper, we abuse the notion of *homometric* and say that two sets of integers $\{a_1, a_2, \cdots, a_n\}$ and $\{b_1, b_2, \cdots, b_n\}$ are homometric if they give the same set of pairwise function values, that is, $\{f(a_i, a_j) | 1 \leq i, j \leq n\} = \{f(b_i, b_j) | 1 \leq i, j \leq n\}$. If two homometric sets are *distinct*, then we say they are *incongruent homometric sets*. Here the meaning of distinct varies for different functions f. For example, in the turnpike problem we say two sets are distinct if and only if they are not the same under shifting and mirroring since these two operations trivially preserve the set of pairwise differences. For the case when f is the sum function, two sets are distinct simply means the sets are not equal.

2.2 Generating Function Technique

We will use the following *generating functions*. Given a set of integers $A = \{a_1, a_2, \cdots, a_n\}$, let $P_A(x)$ denote the generating function $\sum_{i=1}^{n} x^{a_i}$. Let D denote the set of pairwise differences of $a_1, a_2, \cdots, a_n$, that is, $\{a_i - a_j | 1 \leq i, j \leq n\}$. We get that $P_D(x) = P_A(x)P_A(1/x)$. Suppose we now consider f to be the sum. Let S denote the set of pairwise sums of $a_1, a_2, \cdots, a_n$. We get that $P_S(x) = P_A^2(x)$. Rosenblatt and Seymour [14] proved that two sets A, B are homometric with respect to the turnpike problem if and only if there exists two polynomials Q, R and integer u such that $P_A(x) = Q(x)R(x)$ and $P_B(X) = x^u Q(x)R(1/x)$.

In this paper, we will use the generating function technique to study the number of solutions when f is the sum function. We also show that the generating function technique and polynomial factorization indeed capture the structure of NUMBER RECONSTRUCTION for a large family of functions f.

2.3 Measures of Polynomials

We will use the following measures of polynomials. Suppose $P(x) = c_0 + c_1 x + \cdots + c_d x^d$ is a polynomial whose roots are $\alpha_1, \alpha_2, \cdots, \alpha_n$. The Mahler Measure [18] of this polynomial $M(P)$ is $M(P) = c_d \prod_{i=1}^{m} \max\{1, |\alpha_i|\}$. The L_2 norm of the polynomial P is $L_2(P) = (\sum_{i=0}^{d} c_i^2)^{1/2}$. We have that $M(P_1 P_2) = M(P_1)M(P_2)$ and $M(P) \leq L_2(P)$ for any polynomial P.

2.4 Some Notation

With a little abuse of notation, we use $f(A)$ to denote the set $\{f(a_i, a_j) | 1 \leq i, j \leq n\}$ for a given bivariate function f and a set $A = \{a_1, a_2, \cdots, a_n\}$. We shall use $f^*(A)$ to denote the set $\{f(a_i, a_j) | 1 \leq i \neq j \leq n\}$. The notations can be naturally generalized to multi-variate functions f.

We let DIFF denote the difference function and let ADD denote the sum function. We will use $H_f(n)$ to denote the maximum number of incongruent homometric sets for any instance of NUMBER RECONSTRUCTION problem with function f and n integers in the full information setting. We define $H_f^*(n)$ similarly for the incomplete information setting.

We say a function $f(x, y)$ is *monotone* if and only if for all $x \geq x'$, $y \geq y'$, $f(x, y) \geq f(x', y')$. We say a function $f(x, y)$ is *anti-monotone* if and only if for all $x \geq x'$, $y \leq y'$, $f(x, y) \geq f(x', y')$ (Or $f(x, y) \leq f(x', y')$ for all $x \geq x'$ and $y \leq y'$. But they are equivalent if one switches the two dimensions of f. Thus we can without loss of generality discuss only the former case). Strict monotonicity and strict anti-monotonicity can be defined straightforwardly.

We say a function $f(x, y)$ is *linear decomposable* if there exist univariate functions h and g such that $f(x, y) = h(x) + g(y)$. We note that the turnpike problem and the special case of sum are both linear decomposable.

We use $\mathsf{P}^{\mathcal{O}}$ to denote the set of problems which can be solved in polynomial time given access to oracle $\mathcal{O}$.

3 Reconstructing Numbers from Pairwise Sums

3.1 Full Information Setting

Suppose f is sum and we consider the NUMBER RECONSTRUCTION problem in the full information setting. In this setting, we can reconstruct $a_1, a_2, \cdots, a_n$ both uniquely and efficiently as illustrated in Algorithm 1.

Algorithm 1. Sum Function, Full Information

1: Sort $S = \{a_i + a_j | 1 \leq i, j \leq n\}$.
2: **for all** $1 \leq i \leq n$ **do**
3: Solve a_i from the equation $a_1 + a_i = \max_{s \in S} s$.
4: **if** $\{a_i + a_j | 1 \leq j \leq i\} \subseteq S$ **then**
5: Let $S = S \setminus \{a_i + a_j | 1 \leq j < i\}$.
6: **else**
7: No solution for this instance.
8: **end if**
9: **end for**

The key observation is that, if we assume without loss of generality that $a_1 \geq a_2 \geq \cdots \geq a_n$, then $a_1 + a_{i+1}$ is the largest pairwise sum in S once we have solved for $a_1, a_2, \cdots, a_i$ and removed the set $\{a_j + a_k | 1 \leq j, k \leq i\}$ from the set S of pairwise sums, and at the first step $a_1 + a_1$ is the largest pairwise sum. It follows from the algorithm that the solution is unique (if exists) for any given instance. The running time of the algorithm is $\mathcal{O}(n^2)$ since steps 4-6 in the algorithm require at most $\mathcal{O}(n)$ time.

3.2 Incomplete Information Setting

Now we consider reconstructing numbers from pairwise sums in the incomplete information setting. We can also find all solutions (if one exists) efficiently in this setting as in Algorithm 2. However, there may be multiple solutions for a single instance. For example, $\{6,3,2,1\}$ and $\{5,4,3,0\}$ are both valid solutions to the instance $\{9,8,7,5,4,3\}$.

The correctness of the algorithm is based on the following observations. Since $f(x,y)$ is monotone increasing in both dimensions, we have $f(a_1,a_2) \geq f(a_1,a_3) \geq \cdots \geq f(a_1,a_n)$ and $f(a_2,a_3) \geq f(a_i,a_j)$ for any $i,j \geq 2$ if we assume without loss of generality that $a_1 \geq a_2 \geq \cdots \geq a_n$. So there exists some $3 \leq k \leq n$ such that $f(a_1,a_2) \geq f(a_1,a_3) \geq \cdots \geq f(a_1,a_k) \geq f(a_2,a_3)$ are the k largest pairwise sums. Hence we can guess the correct value of k and then solve the values of $a_1,a_2,\cdots,a_k$ and finally resolve the value of $a_{k+1},\cdots,a_n$ one by one. The running time of this algorithm if $\mathcal{O}(n^3)$.

Algorithm 2. Sum Function, Incomplete Information

Sort the set $S = \{a_i + a_j : 1 \leq i < j \leq n, i \neq j\}$.
Suppose $S = \{s_1, s_2, \cdots, s_N\}$ such that $s_1 \geq s_2 \geq \cdots \geq s_N$ and $N = \binom{n}{2}$.
for all $3 \leq k \leq n$ **do**
  Solve $a_1, a_2, \cdots, a_k$ from equations $a_1 + a_i = s_{i-1}$, $2 \leq i \leq k$, and $a_2 + a_3 = s_k$.
  for all $k+1 \leq \ell \leq n$ **do**
    if $\{a_i + a_j : 1 \leq i < j < \ell\} \subseteq S$ **then**
      Let $S^* = S \setminus \{a_i + a_j : 1 \leq i < j < \ell\}$.
      Solve a_ℓ from the equation $a_1 + a_\ell = \max_{s \in S^*} s$.
    else
      No solution for the current value of k.
    end if
  end for
end for

Now let us consider the upper and lower bound on the number of solutions.

Theorem 1. *The following facts hold: (1) For any* $n > 2$, $H^*_{\text{ADD}}(n) \leq n-2$. *(2) For any* $n > 2$ *and* n *is a power of* 2, $H^*_{\text{ADD}}(n) \geq 2$.

Proof. The first part of the theorem directly follows from Algorithm 2. Now we prove by induction that $H^*_{\text{ADD}}(2^k) \geq 2$ for any $k \geq 2$. The base case is true because $\{6,3,2,1\}$ and $\{5,4,3,0\}$ are incongruent homometric sets. Suppose $H^*_{\text{ADD}}(2^{k-1}) \geq 2$ for some $k > 2$. There exists incongruent homometric sets $A = \{a_1, a_2, \cdots, a_m\}$ and $B = \{b_1, b_2, \cdots, b_m\}$ such that $m = |A| = |B| = 2^{k-1}$. Let u be a large number such that $u > |a_i - b_j|$ for any $1 \leq i,j \leq m$. Consider $C = (A+u) \cup B$ and $D = (B+u) \cup A$, where we use $X + u$ to denote the set $\{x + u | x \in X\}$. It is easy to verify that C and D are incongruent homometric sets and $|C| = |D| = 2^k$. □

Theorem 2. *The number of incongruent homometric sets $H^*_{\text{ADD}}(n)$ is larger than 1 if and only if n is a power of 2.*

Proof. ($\Leftarrow$) The statement is trivially true when $n = 1, 2$. If n is a power of 2 and $n > 2$, by Theorem 1 we know that $H^*_{\text{ADD}}(n) \geq 2$.

($\Rightarrow$) Consider the following algebraic approach. Recall that given a set $C = \{c_1, c_2, \cdots, c_n\}$, $P_C(x)$ is the generating function $x^{c_1} + x^{c_2} + \cdots + x^{c_n}$. Suppose $A = \{a_1, a_2, \cdots, a_n\}$ and $B = \{b_1, b_2, \cdots, b_n\}$ are two incongruent homometric sets. We use S to denote the set of pairwise sums $\{a_i + a_j | 1 \leq i \neq j \leq n\}$ $= \{b_i + b_j | 1 \leq i \neq j \leq n\}$. We have $P_S(x) = \sum_{i \neq j} x^{a_i + a_j} = 2 \sum_{i<j} x^{a_i + a_j} = P_A(x)^2 - P_A(x^2)$ and $P_S(x) = \sum_{i \neq} x^{b_i + b_j} = 2 \sum_{i<j} x^{b_i + b_j} = P_B(x)^2 - P_B(x^2)$.

So we get that $P_A(x)^2 - P_A(x^2) = P_B(x)^2 - P_B(x^2)$. Hence we have $P_A(x^2) - P_B(x^2) = P_A(x)^2 - P_B(x)^2 = [P_A(x) - P_B(x)][P_A(x) + P_B(x)]$. Note that $P_A(1) = P_B(1) = n$ and hence $P_A(1) - P_B(1) = 0$. We have that $(x - 1)$ divides $P_A(x) - P_B(x)$. Assume that $(x - 1)^k | [P_A(x) - P_B(x)]$ and $(x - 1)^{k+1} \not| [P_A(x) - P_B(x)]$, then we may assume that $P_A(x) - P_B(x) = (x - 1)^k h(x)$ and $h(1) \neq 1$. We have $(x^2 - 1)^k h(x^2) = (x - 1)^k h(x) [P_A(x) + P_B(x)]$. Therefore $(x + 1)^k h(x^2) = h(x) [P_A(x) + P_B(x)]$. Note that $P_A(1) + P_B(1) = 2n$ and $h(1) \neq 0$. Let $x = 1$ and we have $n = 2^{k-1}$. □

4 The General Number Reconstruction Problem

In this section, we generalize Algorithm 1 and 2, the backtracking algorithm, and the polynomial factorization approach for more general NUMBER RECONSTRUCTION problems.

4.1 Monotone Functions

Note that in Algorithm 1 and 2 the key observation only rely on the fact that sum is a monotone function. We now generalize the idea in these two algorithms and propose analogous oracle algorithms for general monotone functions.

Incomplete Information Setting. Let us first consider the incomplete information setting. We shall discuss the case that f is symmetric and the case that f is asymmetric separately here. We also note that algorithms that resolve both cases simultaneously can be easily achieved by assuming a more powerful oracle and properly merging the algorithms we propose.

Now we study the case when f is symmetric. In this case, In order to reconstruct the numbers from the given values, we need to be able to answer some basic queries. We shall use $\mathcal{S}_f$ to denote an oracle that can answer the following two types of queries: (1) Given $f(x, y)$ and x, solve y; (2) Given $f(x, y)$, $f(y, z)$, $f(z, x)$, solve x, y, z. Assuming one can resolve the above queries efficiently is reasonable because of the following fact.

Lemma 1. *If $f(x, y)$ is strictly monotone, then: (1) Given f and x, there is a unique y (if exists) such that $f(x, y) = f$; (2) Given f_1, f_2 and f_3, there is a unique triplet x, y, z (if exists) such that $f(x, y) = f_1, f(y, z) = f_2, f(z, x) = f_3$.*

Proof. We prove both statement by contradiction. Suppose there exists two distinct y_1 and y_2 such that $f(x, y_1) = f$, and $f(x, y_2) = f$. Without loss of generality, we may assume that $y_1 > y_2$. From the strictly monotonicity we get that $f = f(x, y_1) > f(x, y_2) = f$, a contradiction. Now suppose there exist two distinct triples x_1, y_1, z_1 and x_2, y_2, z_2 such that $f(x_1, y_1) = f(x_2, y_2) = f_1$, $f(y_1, z_1) = f(y_2, z_2) = f_2$, $f(z_1, x_1) = f(z_2, x_2) = f_3$. Since two triples are distinct, without loss of generality we assume $x_1 \neq x_2$ and thus we may further assume $x_1 > x_2$. If $y_1 \geq y_2$ then from the strictly monotonicity we get $f(x_1, y_1) > f(x_2, y_2)$, which contradicts our assumption. So $y_1 < y_2$. Similarly $z_1 < z_2$. But now we get $f_2 = f(y_1, z_1) < f(y_2, z_2) = f_2$, a contradiction. □

Theorem 3. *Suppose f is a symmetric and monotone function, then we have* $\text{REC}_f^* \in \mathsf{P}^{\mathcal{S}_f}$ *and* $H_f^*(n) \leq n - 2$ *for all* $n > 2$.

This theorem follows from Algorithm 3. A similar idea can be used for asymmetric monotone functions. In this case, we need strict monotonicity of function f. Again, we need to assume that some basic queries can be solved efficiently. But the second type of query is different from the symmetric case. The oracle $\mathcal{A}_f$ answers two types of queries: (1) Given $f(x, y)$ and x, solve y; or given $f(x, y)$ and y, solve x; (2) Given $f(x, y)$ and $f(y, x)$, solve x and y.

Algorithm 3. Symmetric Monotone Functions, Incomplete Information

Sort the set $S = \{f(a_i, a_j) | 1 \leq i < j \leq n\}$.
Suppose $S = \{s_1, s_2, \cdots, s_N\}$ such that $s_1 \geq s_2 \geq \cdots \geq s_N$ and $N = \binom{n}{2}$.
for all $3 \leq i \leq n$ **do**
  Solve a_1, a_2, a_3 given $f(a_1, a_2) = s_1$, $f(a_1, a_3) = s_2$, and $f(a_2, a_3) = s_i$.
  for all $4 \leq j \leq i$ **do**
    Solve a_j given $f(a_1, a_j) = s_{j-1}$ and a_1.
  end for
  for all $i < j \leq n$ **do**
    if $\{f(a_k, a_\ell) | 1 \leq k < \ell < j\} \subseteq S$ **then**
      Let $S^* = S \setminus \{f(a_k, a_\ell) | 1 \leq k < \ell < j\}$.
      Solve a_j given $f(a_1, a_j) = \max_{s \in S^*} s$ and a_1.
    else
      No solution for this value i, go to next i.
    end if
  end for
end for

One may notice that the second type of query may be unsolvable in some instances even if we assume strict monotonicity. However, if we consider the base case when $n = 2$, it is of exactly the same form as the second type of queries. So it is reasonable to assume one can efficiently solve at least the base case.

We shall defer the proof of the following theorem to the full version.

Theorem 4. *Suppose f is an asymmetric and monotone function, then we have* $\text{REC}_f^* \in \mathsf{P}^{\mathcal{A}_f}$ *and* $H_f^*(n) \leq 2n-2$ *for all* $n>2$.

Therefore, the NUMBER RECONSTRUCTION problem with any monotone function can be solved in polynomial time under some reasonable assumptions. Now we turn to the case with monotone functions in full information setting.

Full Information Setting. In the full information setting, we propose Algorithm 4 and we need an oracle $\mathcal{F}_f$ that can answer two types of queries: (1) Given $f(x,x)$, solve x; (2) Given $f(x,y)$ and x, solve y; or given $f(x,y)$ and y, solve x. Again, we argue that these two types of queries are reasonable because the solution uniquely exists if we assume strict monotonicity.

Theorem 5. *Suppose f is an monotone function, then we have* $\text{REC}_f \in \mathsf{P}^{\mathcal{F}_f}$ *and* $H_f(n) = 1$ *for all* n.

Remark 1. We note that the above algorithms can be easily modified to solve the multi-variate version of NUMBER RECONSTRUCTION problems, in which case we want to resolve set $A = \{a_1, a_2, \cdots, a_n\}$ from the function values $\{f(a_{i_1}, a_{i_2}, \cdots, a_{i_k}) | 1 \leq i_1, i_2, \cdots, i_k \leq n\}$. We can define the problem for incomplete information setting similarly.

4.2 Anti-monotone Functions

Recall that we say a function $f(x,y)$ is *anti-monotone* if and only if for any $x \geq x'$, $y \leq y'$, $f(x,y) \geq f(x',y')$. Thus assuming that $a_1 \geq a_2 \geq \cdots \geq a_n$, we have the following: (1) $f(a_1,a_n) = \max_{i,j} f(a_i,a_j)$; (2) Given a subset $A' = \{a_1, \cdots, a_i, a_j \cdots, a_n\} \subseteq A = \{a_1, a_2, \cdots, a_n\}$ such that $i < j-1$, $\max\{f(a_1,a_{j-1}), f(a_{i+1},a_n)\} = \max\{f(a_k,a_\ell) | a_k \notin A' \vee a_\ell \notin A'\}$. Given these two properties, we have that the backtracking algorithm solves the NUMBER RECONSTRUCTION problem for any anti-monotone functions in $\mathcal{O}(2^n n \log n)$ time.

4.3 Linear Decomposable Functions

Now we consider decomposable functions $f(x,y)$ of the form $f(x,y) = h(x) + g(y)$. Let F denote that set $\{f(a_i,a_j) | 1 \leq i,j \leq n\}$ and let H and G denote the sets $\{h(a_i) | 1 \leq i \leq n\}$ and $\{g(a_j) | 1 \leq j \leq n\}$ respectively. We have that $P_F(x) = \sum_{1 \leq i,j \leq n} x^{f(a_i,a_j)} = \sum_{1 \leq i,j \leq n} x^{h(a_i)+g(a_j)} = P_H(x) P_G(x)$.

One can solving the NUMBER RECONSTRUCTION problem in two steps: (1) factorize a polynomial P_F of n^2 terms (we count the same term multiple times if the coefficient is larger than one) into the product of two polynomials P'_H and P'_G of n terms each (2) check whether $P_H(x) = x^u P'_H(x)$ and $P_G(x) = x^{-u} P'_G(x)$ give a feasible solution for each u. Suppose $P_F = x^{f_1} + x^{f_2} + \cdots + x^{f_{n^2}}$ such that $f_1 \geq f_2 \geq \cdots \geq f_{n^2}$. A naive way of factorizing is the following algorithm which capture the spirit of the backtracking algorithm:

Algorithm 4. Monotone Functions, Full Information

Sort the set $S = \{f(a_i, a_j) | 1 \leq i, j \leq n\}$.
for all $1 \leq i \leq n$ **do**
  Solve a_i given $f(a_1, a_i) = \max_{s \in S} s$.
  if $f(a_i, a_1) \notin S$ **then**
    Solve a_i given $f(a_i, a_1) = \max_{s \in S} s$.
  end if
  if $\{f(a_i, a_i)\} \cup \{f(a_i, a_j) | 1 \leq j < i\} \cup \{f(a_j, a_i) | 1 \leq j < i\} \subseteq S$ **then**
    Let $S^* = S \setminus (\{f(a_i, a_i)\} \cup \{f(a_i, a_j) | 1 \leq j < i\} \cup \{f(a_j, a_i) | 1 \leq j < i\})$.
  else
    No solution for this value i, go to next i.
  end if
end for

- Without loss of generality, let $P'_H(x) = x^{f_1}$ and $P'_G(x) = 1$ initially.
- Guess whether P'_H or P'_G contribute the next term x^{f_i} of highest degree in $P_F(x) - P'_H(x)P'_G(x)$. In the first case, let $P'_H(x) = P'_H(x) + x^{f_i}$. In the latter case, let $P'_G(x) = P'_G(x) + x^{f_i - f_1}$.
- Repeat until $P_F(x) = P'_H(x)P'_G(x)$. If any contradiction is found (some negative term in the polynomial $P_F(x) - P'_H(x)P'_G(x)$) then backtrack.

It is clear that the above algorithm is another version of backtracking algorithm. So the Number Reconstruction problem for any decomposable function can be solved in $\mathcal{O}(2^n n \log n)$ time.

The polynomial factorization approach which obtains pseudopolynomial algorithm for the turnpike problem can also be applied here. However, it is not clear that this approach still gives pseudopolynomial running time. The polynomial factorization approach proceeds in two steps: (1) factorize the polynomial P_F into the product of some irreducible polynomials and (2) for each feasible partition of these irreducible polynomials into two subsets, let P_H be the product of polynomials in one subset and let P_G be the product of polynomials in the other subset, then check if P_H and P_G give a feasible solution. Factorizing the polynomial P_F can be done in polynomial time (in $D = \max_{1 \leq i,j \leq n} f(a_i, a_j)$).

Further Discussion on the Factoring Approach. To bound the running time of this approach we still need to bound the number of irreducible factors one need to consider when finding feasible P_H and P_G. For the turnpike problem one only need to consider non-reciprocal factors and Smyth [17] proved that $M(P) \geq M(x^3 - x + 1) \approx 1.324$ for any non-reciprocal polynomial P. Note that $M(P_F) \leq L_2(P_F) = n$. We get that the number of non-reciprocal irreducible factors of P_F is bounded by $\mathcal{O}(\log n)$. However, we need to consider all irreducible factors for the general Number Reconstruction problem. Under Lehmer's conjecture on Mahler Measure Problem [9] that $M(P) \geq M(x^{10} + x^9 - x^7 - x^6 - x^5 - x^4 - x^3 + x + 1) \approx 1.176$, we can bound the number of non-cyclotomic factors by $\mathcal{O}(\log n)$. If we can further bound the number of cyclotomic factors of P_F by $\mathcal{O}(\log D)$, then the factoring approach solves the Number Reconstruction problem in pseudopolynomial time for linear decomposable function f.

5 Reconstructing Strings

Here we consider a string reconstruction problem that arises in reconstructing protein sequences [6]. Suppose there is an alphabet Σ and a string $s \in \Sigma^n$. The *profile* of a string is a vector consisting of the number of occurrences of each symbol in Σ. Given an unordered set of $\binom{n+1}{2}$ profiles, one for each substring of s, the goal is to reconstruct the string s.

5.1 An Efficient Algorithm for Binary Alphabet

We first consider the case when the alphabet is the binary set $B = \{\sigma_0, \sigma_1\}$. It is clear that this problem is closely related to the turnpike problem in the following sense. If we let the symbol σ_0 represent a segment of length 0 and let the symbol σ_1 represent a segment of length 1. We can easily translate the given set of number of σ_0's and σ_1's in the substrings into the set of $\binom{n+1}{2}$ pairwise differences of $n+1$ integers. Moreover, the largest pairwise difference is at most n. We note that not every turnpike solution correspond to a feasible solution for the string reconstruction instance. We can verify the turnpike solutions in polynomial time since the number of solution is polynomial in input size. An alternative approach is to let σ_0 and σ_1 represent segments of length 1 and $n+1$ respectively. In this approach, all turnpike solutions lead to a valid string for the original problem. Therefore, we can solve Rec_B in polynomial time.

5.2 General Alphabet

For the string reconstruction problem with general alphabet, we can reconstruct the string bit by bit. Given a general alphabet Σ, let $k = \lceil \log |\Sigma| \rceil$. Then without loss of generality, we may assume that $\Sigma \subseteq B^k$. So we reconstruct the ith bit of each symbol in the string using the above algorithm for binary alphabet for $1 \leq i \leq k$. Therefore we can solve Rec_Σ in time polynomial in n and $\log |\Sigma|$. We note that Das et al. [6] proposed a different reduction from general alphabet to binary alphabet. Our reduction improves the dependence on $|\Sigma|$.

Acknowledgement. We would like to thank Alon Orlitsky for suggesting to us the string reconstruction problem.

References

1. Abrams, Z., Chen, H.L.: The Simplified Partial Digest Problem: Hardness and a Probabilistic Analysis. In: RECOMB Satellite Meeting on DNA Sequencing Technologies and Computation (2004)
2. Blazewicz, J., Formanowicz, P., Kasprzak, M., Jaroszewski, M., Markiewicz, W.T.: Construction of DNA restriction maps based on a simplified experiment (2001)
3. Cieliebak, M., Eidenbenz, S.: Measurement errors make the partial digest problem NP-hard. In: Farach-Colton, M. (ed.) LATIN 2004. LNCS, vol. 2976, pp. 379–390. Springer, Heidelberg (2004)

4. Cieliebak, M., Eidenbenz, S., Penna, P.: Noisy data make the partial digest problem NP-hard. In: Benson, G., Page, R.D.M. (eds.) WABI 2003. LNCS (LNBI), vol. 2812, pp. 111–123. Springer, Heidelberg (2003)
5. Dakic, T.: On the turnpike problem. PhD thesis, Simon Fraser University (2000)
6. Das, H., Orlitsky, A., Santhanam, N.: Order from disorder. In: Information Theory and Applications Workshop (2009)
7. Goldstein, L., Waterman, M.S.: Mapping DNA by stochastic relaxation. Advances in Applied Mathematics 8(2), 194–207 (1987)
8. Lemke, P., Skiena, S.S., Smith, W.D.: Reconstructing sets from interpoint distances. Discrete and computational geometry: The Goodman-Pollack Festschrift 25, 597–631
9. O'Bryant, K., Weisstein, E.: Lehmer's Mahler measure problem. MathWorld–A Wolfram Web Resource
10. Pandurangan, G., Ramesh, H.: The restriction mapping problem revisited. Journal of Computer and System Sciences 65(3), 526–544 (2002)
11. Patterson, A.L.: A direct method for the determination of the components of interatomic distances in crystals. Zeitschr. Krist. 90, 517–542 (1935)
12. Patterson, A.L.: Ambiguities in the X-ray analysis of crystal structures. Phys. Review 65, 195–201 (1944)
13. Piccard, S.: Sur les Ensembles de Distances des Ensembles de Point d'un Espace Euclidean. Mem. Univ. Neuchatel 13 (1939)
14. Rosenblatt, J., Seymour, P.D.: The structure of homometric sets. SIAM Journal on Algebraic and Discrete Methods 3, 343 (1982)
15. Shamos, M.I.: Problems in computational geometry. Unpublished manuscript, Carnegie Mellon University, Pittsburgh, PA (1977)
16. Skiena, S.S., Sundaram, G.: A partial digest approach to restriction site mapping. Bulletin of Mathematical Biology 56(2), 275–294 (1994)
17. Smyth, C.J.: On the product of the conjugates outside the unit circle of an algebraic integer. Bulletin of the London Mathematical Society 3(2), 169 (1971)
18. Smyth, C.J.: The Mahler measure of algebraic numbers: a survey. Number Theory and Polynomials, 322 (2008)
19. Zhang, Z.: An exponential example for a partial digest mapping algorithm. Journal of Computational Biology 1(3), 235–239 (1994)

Hilbert's Thirteenth Problem and Circuit Complexity

Kristoffer Arnsfelt Hansen[1], Oded Lachish[2], and Peter Bro Miltersen[1]

[1] Department of Computer Science, Aarhus University
{arnsfelt,bromille}@cs.au.dk
[2] Department of Computer Science, University of Warwick
O.Lachish@warwick.ac.uk

Abstract. We study the following question, communicated to us by Miklós Ajtai: Can all explicit (e.g., polynomial time computable) functions $f : (\{0,1\}^w)^3 \to \{0,1\}^w$ be computed by word circuits of constant size? A word circuit is an acyclic circuit where each wire holds a word (i.e., an element of $\{0,1\}^w$) and each gate G computes some binary operation $g_G : (\{0,1\}^w)^2 \to \{0,1\}^w$, defined for all word lengths w. We present an explicit function so that its w'th slice for any $w \geq 8$ cannot be computed by word circuits with at most 4 gates. Also, we formally relate Ajtai's question to open problems concerning ACC^0 circuits.

1 Introduction

A *word* is a bit string of length w, where w is a parameter called the *word length.* We define a *word circuit* to be an acyclic circuit where each wire holds a word and each gate G computes some binary operation $g_G : (\{0,1\}^w)^2 \to \{0,1\}^w$, defined for all word lengths w. A word circuit with k input wires computes a function $g : (\{0,1\}^w)^k \to \{0,1\}^w$ in the natural way. Consider some ternary operation on words $f : (\{0,1\}^w)^3 \to \{0,1\}^w$ defined for all word length $w = 1,2,3,\ldots$. We say that f is *decomposable* if f can be computed by a constant size word circuit. Otherwise, f is called non-decomposable. For instance, identifying $\{0,1\}^w$ with $\{0,1,2,\ldots,2^w-1\}$, the function $f(x,y,z) = x+y+z \bmod 2^w$ is decomposable, as $f(x,y,z) = g(g(x,y),z)$, where $g(x,y) = x+y \bmod 2^w$. It is much harder to give examples of natural functions that are provably non-decomposable. However, a simple counting argument (for details, see Section 2) shows that functions $f : (\{0,1\}^w)^3 \to \{0,1\}^w$ *exist* so that any word circuit computing the slice of the function corresponding to word length w has size at least roughly 2^w. In particular, such an f is non-decomposable. Standard arguments translate this non-constructive existence result into a function computable in EXPSPACE with this property. The main question we investigate is the following: *Are all polynomial time computable functions $f : (\{0,1\}^w)^3 \to \{0,1\}^w$ decomposable?* This question was communicated to us by Miklós Ajtai (personal communication). We believe that the question should be resolved in the negative. That is, we conjecture the existence of ternary polynomial time computable functions that can not be expressed using a constant number of binary functions. The present

Y. Dong, D.-Z. Du, and O. Ibarra (Eds.): ISAAC 2009, LNCS 5878, pp. 153–162, 2009.

paper leaves the question unresolved, but we prove some weaker statements and present reductions suggesting that the question might be difficult to resolve. Before presenting these, we briefly explain how the question is motivated. The question can be viewed as a discrete analogue of Hilbert's 13th problem. In fact, according to personal communication with Miklós Ajtai, a version of the question (where "computable in polynomial time" was replaced with the less tangible "natural") has circulated in the combinatorics community with this motivation. Also, the question is arguably one of the simplest questions one can ask about the word circuit model. The study of word circuits in general may be motivated by their relationship to *network coding*. Concretely, Adler *et al.* considers the *transposition* problem $t : (\{0,1\}^w)^w \to (\{0,1\}^w)^w$ where $t(M) = M^T$ when the input and output are interpreted as matrices. They conjecture that any "oblivious I/O machine" computing t requires $\Omega(w \log w)$ I/O's and show that this conjecture is implied by the central "undirected k-pairs conjecture" of network coding. It is easy to see that their conjecture is *exactly* equivalent to conjecturing that any word circuit computing t has size $\Omega(w \log w)$. It seems likely (though we admit to having no formal argument for this) that one should first understand apparently very simple questions about word circuits such as our main question concerning 3-input functions, before one can make progress on questions such as the undirected k-pairs conjecture.

Our results are the following. We show (in Section 2) that there exists a function $f : (\{0,1\}^w)^3 \to \{0,1\}^w$ that requires word circuits of size $(1 - o(1))2^w$. We match this with an upper bound: All functions $f : (\{0,1\}^w)^3 \to \{0,1\}^w$ can be computed by a word circuit of size $(2 + o(1))2^w$. As our main technical result, we show (in Section 3) that there exists an explicit (polynomial time computable) function $F : (\{0,1\}^w)^3 \to \{0,1\}^w$ so that for all $w \geq 8$, the w'th slice of F cannot be computed by a word circuit of size 4. Finally, we show (in Section 4) that resolving our main question in the negative (i..e, proving non-constant lower bounds on word circuit size for ternary functions) is likely to be somewhat hard, by relating it to open problems in *Boolean* circuit complexity. More precisely, while non-linear bounds on the number of *wires* of ACC^0 circuits are known for explicit functions [2], no such lower bounds are currently known for the number of *gates* for explicit functions, even multi-output functions. We show that if a Boolean function $f : \{0,1\}^{3w} \to \{0,1\}^w$ has an ACC^0 circuit of size (i.e., number of gates) $O(w)$, then f, viewed as a function mapping three words to one word, is decomposable. Thus, resolving our main question in the negative would lead to new ACC^0 lower bounds.

2 The Complexity of the Hardest Ternary Function

Proposition 1. *There exists a function* $f : (\{0,1\}^w)^3 \to \{0,1\}^w$ *so that no word circuit of size smaller than* $(1 - o(1))2^w$ *computes* f.

Proof. We phrase the standard counting argument using Kolmogorov complexity lingo for readability: A word circuit of size s can be described using $s(2\log_2(s + 3) + w2^{2w})$ bits (the first term accounts for describing the wiring of inputs to the gates, the second for describing the binary function of the gate). A random ternary function has no description shorter than $w2^{3w}$ bits. Thus, if s is an upper bound on the word circuit size of all functions, we have

$$s(2\log_2(s+3) + w2^{2w}) \geq w2^{3w}$$

or $s \geq (1 - o(1))2^w$.

This lower bound is matched within a constant factor by the following upper bound.

Theorem 1. *Every function $f : (\{0,1\}^w)^3 \to \{0,1\}^w$ is computed by a word circuit of size at most $(2 + o(1))2^w$.*

Proof. Let the names of the three input variables be x, y, z. Partition the set $\{0,1\}^w \times \{0,1\}^w$ into $m = \lceil \frac{2^{2w}}{2^w - 1} \rceil$ sets $P_1, \ldots, P_m$, each of size at most $2^w - 1$. For all i, fix injections $\pi_i : P_i \to \{0,1\}^w \setminus \{0^w\}$, and define gates $g_1, \ldots, g_m$ and $h_1, \ldots, h_m$ as follows, letting g_0 denote x.

$$h_i(y,z) = \begin{cases} \pi_i(y,z) & \text{if} \quad (y,z) \in P_i \\ 0^w & \text{if} \quad (y,z) \notin P_i \end{cases}$$

$$g_i(g_{i-1}, h_i) = \begin{cases} f(g_{i-1}, \pi_i^{-1}(h_i)) & \text{if} \quad h_i \neq 0^w \\ g_{i-1} & \text{if} \quad h_i = 0^w \end{cases}$$

The output of the circuit is the gate g_m. It is readily verified that the circuit computes the function f correctly, and the size of the circuit is $2m = (2+o(1))2^w$.

We consider it an interesting open problem to get bounds tight within a low order term, similar to the bounds known for Boolean circuits. Note that the word circuit constructed in our upper bound proof is actually a *formula*. Thus, there is at most a difference of a factor of roughly two between the maximum possible word formula size and the maximum possible word circuit size. Again, this contrasts the case of Boolean circuits and formulas, where the hardest n-bit function has circuit size $\approx 2^n/n$ and formula size $\approx 2^n/\log n$ by results of Lupanov [3,4] (see, e.g., Wegener [5] for an exposition).

3 An Explicit Lower Bound

In this section, we prove our main technical result:

Theorem 2. *There exists a polytime computable function $F : (\{0,1\}^w)^3 \to \{0,1\}^w$ so that for all $w \geq 8$, the w'th slice of F cannot be computed by a word circuit of size 4.*

The section is organized as follows. First we present some structural theorems about word circuits in general and word circuits of size at most 4 in particular. Next, we define the function F and establish certain properties of F. Finally, we combine the two groups of statements to complete the proof that no word circuit of size at most 4 computes F.

Definition 1. Let C be a word circuit. We say a gate g is *redundant* if either of the following holds.

1. One of the inputs of g is a gate g^* and the other input of g is an input of g^*.
2. The inputs of g are gates g_1 and g_2 that have the exact same set of inputs.

A word circuit is called *redundancy free* if it does not contain any redundant gates.

Claim 3. *Let C be a word circuit that contains a redundant gate. There exists a redundancy free circuit C^* of size at most the size of C that computes the same function C does.*

Proof. We replace a redundant gate g of the first kind, computing $g(u, g^*(u, v))$ with a gate g' computing $g'(u, v) = g(u, g^*(u, v))$. We replace a redundant gate g of the second kind, computing $g(g_1(u, v), g_2(u, v))$ with a gate g' computing $g'(u, v) = g(g_1(u, v), g_2(u, v))$. If C turns out to be again redundant, we repeat this transformation. This process is easily seen to eventually terminate as each transformation structurally simplifies the circuit.

Definition 2. Let C be a word circuit. If there exists a variable in C that is connected to the output of the circuit by a unique path, we say C is *weak*. We also refer to such a variable as a *weak* variable. In general, a weak circuit may have more than one weak variable. In such a case we pick one of them arbitrarily to be *the* weak variable. A gate in C is called a *weak* gate if it is on the path connecting the weak variable to the output. Each one of the weak gates has an input that is not on the path connecting the weak variable to the output. We refer to this input as the *control input* of the gate.

Lemma 1. *Let C be a redundancy free word circuit of size exactly 4, that computes a function that depends on all 3 inputs. Then C is weak and satisfies one of the following.*

1. *There exist functions Y_1, Y_2 on the non-weak inputs so that the control input of each weak gate is one of Y_1, Y_2.*
2. *The control input of all but at most one of the weak gates is one of the non-weak input variables.*

Proof. Let C be a redundancy free word circuit of size 4. Since every gate takes 2 inputs the circuit contains 8 wires. Except for the output gate, all gates as well as the input variables must have an outgoing wire, thus accounting for 6 of the wires. If an internal gate has 3 outgoing wires, the last wires are accounted for and we must have a redundancy of the first kind. We are left with the following 3 cases.

1. There is an input variable that has 3 outgoing wires. In this case all gates have 1 outgoing wire and hence there is a unique path from the other 2 input variables to the output gate.

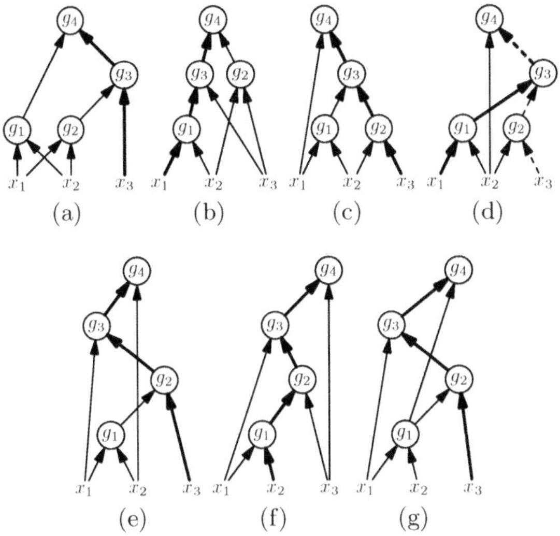

Fig. 1. All redundancy free circuits of size 4. The paths from weak variables are bold/dashed.

2. There are two input variables that have 2 outgoing wires. Here again all gates have 1 outgoing wire, hence there is a unique path from the last input variable to the output gate.
3. There is one input variable that has 2 outgoing wires and a gate g having 2 outgoing wires. If one of the other input variables does not have a path to g, then it will have a unique path to the output. If both of the other input variables had a path to g, then they had to be directly connected to g, as otherwise the circuit would be redundant. But then all of the last 3 gates would depend only on g and the third input variables, meaning that the circuit would be redundant.

In all cases we identify a weak variable, and hence C must be weak. The gates that are not on the path from the weak input variable to the output must depend solely on the other 2 input variables. We can have either 2 or 1 of these. When C contains 2 non-weak gates, both of the weak gates must take these as inputs and we have the first case of the statement. When C has a 1 non-weak gate, it is either connected to 2 or 1 weak gates in which case we have the first or the second case of the statement, respectively.

The above proof identifies several classes of circuits of size more than 4 that are not weak. For instance all 3 input variables may have 2 outgoing wires. All the redundancy free circuits of size 4 (up to relabeling) are shown in Figure 1. The first case of Lemma 1 occurs for circuits (a) and (g), whereas the second case occurs for the remaining circuits.

Remark. From here on, we assume $w \geq 8$.

Definition 3. We define $F : (\{0,1\}^w)^3 \to \{0,1\}^w$ as follows

1. If $\alpha_i = \alpha_j = 0^w$ for distinct i,j then $F(\alpha_0,\alpha_1,\alpha_2) = 0^w$.
2. If $\alpha_i = 0^w$ and $\alpha_j, \alpha_k \neq 0^w$ for distinct i,j,k then $F(\alpha_0,\alpha_1,\alpha_2) = \alpha_j^\ell \cdot \alpha_k^{-\ell}$, where $\ell = 1$ if $k - j \equiv 1 \pmod 3$ and $\ell = -1$ otherwise. Here, the multiplication is over $\mathrm{GF}(2^w)$, i.e., we identify $\{0,1\}^w$ with $\mathrm{GF}(2^w)$.
3. If $\alpha_0,\alpha_1,\alpha_2$ are distinct and all different from 0 then if there exists $\ell \in \{0,1,2\}$ such that $\alpha_0+\alpha_1+\alpha_2 \in \{8\ell+1,\ldots,8\ell+8\}$ we set $F(\alpha_0,\alpha_1,\alpha_2) = \alpha_\ell$ and otherwise $F(\alpha_0,\alpha_1,\alpha_2) = \alpha_0+\alpha_1+\alpha_2$. Here, the addition is modulo 2^w, i.e., we identify $\{0,1\}^w$ with $\mathbf{Z}/2^w\mathbf{Z}$.
4. If $\alpha_0,\alpha_1,\alpha_2$ are not 0 and $\alpha_i = \alpha_j$ then if $\alpha_k < \alpha_i$ we set $F(\alpha_0,\alpha_1,\alpha_2) = 1$ and otherwise $F(\alpha_0,\alpha_1,\alpha_2) = 1 + \alpha_k - \alpha_i$. Here, the arithmetic and the inequalities are defined by identifying $\{0,1\}^w$ with $\{0,1,2,\ldots,2^w-1\}$.

Before analyzing the properties of F, we introduce some useful concepts. We let the variables of F be denoted X_0, X_1, X_2. For $\alpha,\beta,\gamma \in \{0,1\}^w$ and distinct $i,j,k \in \{0,1,2\}$ we let $F_{\restriction X_i=\alpha,X_j=\beta}(\gamma)$ be the value of F when $X_i = \alpha, X_j = \beta$ and $X_k = \gamma$. For $\alpha,\beta \in \{0,1\}^w$ and distinct $i,j \in \{0,1,2\}$ we let $F_{\restriction X_i=\alpha,X_j=\beta}$ be the function we get from F over the last variable X_k when $X_i = \alpha, X_j = \beta$. For any function T and S that is a subset of T's domain we let $T(S) = \{T(\gamma) : \gamma \in S\}$. When S is the domain of T we abuse notation and write $|T|$ instead of $|T(S)|$. Let $T : \{0,1\}^w \to \{0,1\}^w$. For $\alpha \in \{0,1\}^w$, let $T^{-1}(\alpha) = \{\beta : T(\beta) = \alpha\}$. We let $M(T)$ be the (2^w+1)-tuple $M(T)_\ell = |\{\alpha : |T^{-1}(\alpha)| = \ell\}|$ for $\ell \in \{0,\ldots,2^w\}$ and we let $\mathrm{Mask}(F,i,j) = \{M(F_{\restriction X_i=\alpha,X_j=\beta}) : \alpha,\beta \in \{0,1\}^w\}$. Also, F may be replaced by a circuit C in all the above notation.

Claim 4. *For every distinct $i,j \in \{0,1,2\}$ and distinct $\alpha,\beta \in \{1,\ldots,2^w-1\}$ the following are satisfied:*

- $|F_{\restriction X_i=0,X_j=0}| = 1$.
- $|F_{\restriction X_i=\alpha,X_j=0}| = 2^w$.
- $|F_{\restriction X_i=\alpha,X_j=\beta}| \geq 2^w - 26$.
- $|F_{\restriction X_i=\alpha,X_j=\alpha}| = 2^w - \alpha$.
- *Let S be the set of all (γ,δ), where $\gamma,\delta \in \{0,1\}^w$ and $|F_{\restriction X_i=\gamma,X_j=\delta}| < 2^w - 2^5$. Then, $|S| < 2^w$.*

Proof. The first two equalities follow from the definition of F. Let S be the set of all $\gamma \in \{0,1\}^w$ satisfying $\gamma \notin \{0,\alpha,\beta\}$ and $\alpha+\beta+\gamma \notin \{1,\ldots,24\}$. Obviously, $|S| \geq 2^w - 26$. By the definition of F for every $\gamma \in S$ we have that $F_{\restriction X_i=\alpha,X_j=\beta}(\gamma) = \alpha+\beta+\gamma$ and hence $|F_{\restriction X_i=\alpha,X_j=\beta}| \geq |F_{\restriction X_i=\alpha,X_j=\beta}(S)| \geq 2^w - 26$. By Condition 4 of Definition 3 we have that $|F_{\restriction X_i=\alpha,X_j=\alpha}| = 2^w - \alpha$ for every $\alpha \in \{1,\ldots,2^w-1\}$.

From this, we also have that $|F_{\restriction X_i=\gamma,X_j=\delta}| < 2^w - 2^5$ only if $\gamma = \delta$ and either $\gamma = 0$ or $\gamma > 2^5$. Hence the set of all (γ,δ),where $\gamma,\delta \in \{0,1\}^w$ and $|F_{\restriction X_i=\gamma,X_j=\delta}| < 2^w - 2^5$, has less than 2^w members.

Claim 5. $|\mathrm{Mask}(F,i,j)| > 2^w$ *for every distinct $i,j \in \{0,1,2\}$.*

Proof. To prove the claim we show that Mask(F, i, j) contains more than 2^w different tuples. Condition 1 of Definition 3 asserts that $M(F_{\restriction X_i=0,X_j=0})$ is zero on all coordinates except for $M(F_{\restriction X_i=0,X_j=0})_{2^w} = 1$. Condition 2 asserts that $M(F_{\restriction X_i=0,X_j=1})$ is zero on all coordinates except for $M(F_{\restriction X_i=0,X_j=1})_1 = 2^w$. Condition 4 asserts that for every $\alpha \in \{1, \ldots, 2^w - 1\}$ we have $M(F_{\restriction X_i=\alpha,X_j=\alpha})$ is zero on all coordinates except for $M(F_{\restriction X_i=\alpha,X_j=\alpha})_1$ which is equal to $2^w - \alpha - 1$ and $M(F_{\restriction X_i=\alpha,X_j=\alpha})_{\alpha+1}$ which is 1. This constitutes at least 2^w different tuples. We next show that there is at least one more tuple that we have not described yet.

For every $\alpha \in \{8i + 1, \ldots, 8i + 8\}$ we have that $F_{\restriction X_i=1,X_j=2}(\alpha) = 1$ and for every $\alpha \in \{8j + 1, \ldots, 8j + 8\}$ we have that $F_{\restriction X_i=1,X_j=2}(\alpha) = 2$. Consequently, one of the following is true: there exists a coordinate $k > 1$ such that $M(F_{\restriction X_i=1,X_j=2})_k > 1$, or, there exists two distinct coordinates $k_1, k_2 > 1$ such that $M(F_{\restriction X_i=1,X_j=2})_{k_1} \geq 1$ and $M(F_{\restriction X_i=1,X_j=2})_{k_2} \geq 1$. Note that in both cases $M(F_{\restriction X_i=1,X_j=2})$ is different from all the other tuples mentioned above.

Claim 6. *$F_{\restriction X_i=\alpha,X_j=\beta} \neq F_{\restriction X_i=\gamma,X_j=\delta}$ for every distinct $i, j \in \{0, 1, 2\}$ and $\alpha, \beta, \gamma, \delta \in \{0, 1\}^w$ such that $(\alpha, \beta) \neq (\gamma, \delta)$.*

Proof. Let i, j be distinct members of $\{0, 1, 2\}$ and let $\alpha, \beta, \gamma, \delta \in \{0, 1\}^w$ such that $(\alpha, \beta) \neq (\gamma, \delta)$. Let $\ell = 1$ if $i - j \equiv 1 \pmod 3$ and otherwise $\ell = -1$. We prove the claim by showing that there exists κ such that $F_{\restriction X_i=\alpha,X_j=\beta}(\kappa) \neq F_{\restriction X_i=\gamma,X_j=\delta}(\kappa)$. This directly implies the claim.

1. Assume that $\alpha = \beta = 0$. Consequently at least one of γ, δ differs from 0. Hence, by Condition 4 and Condition 2 of Definition 3 we get that $F_{\restriction X_i=\gamma,X_j=\delta}(\max\{\gamma, \delta\}) = 1$. By Condition 1 of Definition 3 we get that $F_{\restriction X_i=\alpha,X_j=\beta}(\max\{\gamma, \delta\}) = 0$.
2. Assume that exactly one of α, β is different from 0 and at least one of γ, δ is different from 0 (due to symmetry the case that $\gamma = \delta = 0$ was dealt with previously). Without loss of generality assume $\alpha = 0$ and $\beta \neq 0$.
   - Assume that $\gamma = 0$ and $\delta \neq 0$. Thus, $\beta \neq \delta$ and hence Condition 2 of Definition 3 asserts that $F_{\restriction X_i=\gamma,X_j=\delta}(\beta) \neq 1$ and $F_{\restriction X_i=\alpha,X_j=\beta}(\beta) = 1$.
   - Assume that $\gamma \neq 0$ and $\delta = 0$ and $\beta \neq \gamma^{-1}$. Then by Condition 2 of Definition 3 we get that $F_{\restriction X_i=\alpha,X_j=\beta}(1) = \beta^\ell \neq \gamma^{-\ell} = F_{\restriction X_i=\gamma,X_j=\delta}(1)$.
   - Assume that $\gamma \neq 0$ and $\delta = 0$ and $\beta = \gamma^{-1} = \eta$. Let κ be such that $\kappa \neq \kappa^{-1}$. Now by Condition 2 of Definition 3 we get that $F_{\restriction X_i=\alpha,X_j=\beta}(\kappa) = \eta^\ell \kappa^\ell \neq \eta^\ell \kappa^{-\ell} = F_{\restriction X_i=\gamma,X_j=\delta}(\kappa)$.
   - Assume that $\gamma, \delta \notin \{0\}$. Thus $\gamma^\ell \delta^{-\ell} \neq 0$ and hence Condition 2 of Definition 3 asserts that $F_{\restriction X_i=\gamma,X_j=\delta}(0) \neq 0$. Condition 1 of Definition 3 asserts that $F_{\restriction X_i=\alpha,X_j=\beta}(0) = 0$.
3. Assume that $\alpha = \beta \neq 0$ and $\gamma, \delta \notin \{0\}$ (due to symmetry the case that at least one of γ, δ is 0 was dealt with previously).
   - Assume that $\gamma = \delta$. Thus $\alpha \neq \gamma$ and hence Condition 4 of Definition 3 implies that $F_{\restriction X_i=\gamma,X_j=\delta}(\max\{\alpha, \gamma\}) \neq F_{\restriction X_i=\alpha,X_j=\beta}(\max\{\alpha, \gamma\})$, since it asserts that one side of the inequality is 1 and the other side is strictly greater than 1.

  - Assume that $\gamma \neq \delta$. Then Condition 2 of Definition 3 implies that $F_{\restriction X_i=\gamma, X_j=\delta}(0) \neq 1$ and $F_{\restriction X_i=\alpha, X_j=\beta}(0) = 1$.
4. Assume that $\alpha \neq \beta$, $\gamma \neq \delta$ and $\alpha, \beta, \gamma, \delta \notin \{0\}$ (All other cases have already been dealt with).
  - Assume $\alpha+\beta = \gamma+\delta$. This implies that $\alpha \neq \gamma$. Choose $\kappa \notin \{0, \alpha, \beta, \gamma, \delta\}$ such that $\alpha + \beta + \kappa \in \{8i+1, \ldots, 8i+8\}$. Then by Condition 3 of Definition 3 we have that $F_{\restriction X_i=\alpha, X_j=\beta}(\kappa) = \alpha \neq \gamma = F_{\restriction X_i=\gamma, X_j=\delta}(\kappa)$.
  - Assume $\alpha + \beta \neq \gamma + \delta$. Then as $2^w > 54$ there exists κ such that Condition 3 of Definition 3 asserts that $F_{\restriction X_i=\alpha, X_j=\beta}(\kappa) = \alpha + \beta + \kappa \neq \gamma + \delta + \kappa = F_{\restriction X_i=\gamma, X_j=\delta}(\kappa)$.

Lemma 2. *Let C be a weak circuit for which there exist functions Y_1, Y_2 such that the control input of each weak gate is equivalent to one of Y_1, Y_2. If C computes F, then C has depth at least 2^{w-5}.*

Proof. Let the input variables be X_0, X_1, X_2 and let the weak variable be X_0. Assume for the sake of contradiction that C has depth strictly less than 2^{w-5}.

For each $i \in \{1, 2\}$ let G_i be the set of all weak gates g such that Y_i is equivalent to the control input of g. Observe that Claim 4 asserts that $|F_{\restriction X_1=0, X_2=0}| = 1$. Hence as the depth of C is less than 2^{w-5} we conclude that there exists $i \in \{1, 2\}$ and a gate $g \in G_i$ such that $|g_{\restriction Y_i(0,0)}| < 2^w - 2^5$. Let S be the set of all (α, β) such that $Y_i(\alpha, \beta) = Y_i(0, 0)$. Now as $|g_{\restriction Y_i(0,0)}| < 2^w - 2^5$ we have that $|C_{\restriction X_1=\alpha, X_2=\beta}| < 2^w - 2^5$ for every $(\alpha, \beta) \in S$. Hence, $|S| < 2^w$ by Claim 4.

We now show that we can also derive that $|S| = 2^w$. That is, we get the required contradiction. Let $\alpha, \beta, \gamma, \delta \in \{0, 1\}^w$ be such that $(\alpha, \beta) \neq (\gamma, \delta)$. By Claim 6 we have that $F_{\restriction X_1=\alpha, X_2=\beta} \neq F_{\restriction X_1=\gamma, X_2=\delta}$ and thus $(Y_1(\alpha), Y_2(\beta)) \neq (Y_1(\gamma), Y_2(\delta))$. Thus, $(Y_1(X_1, X_2), Y_2(X_1, X_2))$ is a bijection of (X_1, X_2) and therefore $|S| = 2^w$.

Lemma 3. *Let C be a weak circuit such that the control input of all but at most one of the weak gates is equivalent to one of the non-weak variables. Then, C does not compute F.*

Proof. Let the input variables be X_0, X_1, X_2 and let the weak variable be X_0. Assume for the sake of contradiction that C computes F. Thus, $|\mathrm{Mask}(C, 1, 2)| > 2^w$ by Claim 5. We shall get the required contradiction by showing that we can also derive that $|\mathrm{Mask}(C, 1, 2)| \leq 2^w$.

By Claim 4 we have that $|C_{\restriction X_1=\alpha, X_2=\beta}| = 2^w$ if either $\alpha = 0$ or $\beta = 0$ but not both. Thus, $|g_{\restriction Y=\alpha}| = 2^w$ for every weak gate g whose control input Y is equivalent to one of X_1, X_2. Assume there does not exist a weak gate g whose control input Y is neither equivalent to X_1 nor to X_2. Then, $|\mathrm{Mask}(C, 1, 2)| = 1$ because the output of C is a permutation of X_0 regardless for any values assigned to X_1, X_2. Assume that t is a weak gate, whose control input Y is neither equivalent to X_1 nor to X_2. Recall that t can be the only such gate and hence $|\mathrm{Mask}(C, 1, 2)| = |\{M(g_{\restriction Y=\alpha}) : \alpha \in \{0, 1\}^w\}| \leq 2^w$.

From Lemma 1, Lemma 2 and Lemma 3 we conclude:

Corollary 1. *Any redundancy free word circuit that computes F for any fixed $w \geq 8$ must be of size at least* 5.

Finally by applying Claim 3 we get the statement of the main Theorem.

4 Converting ACC^0 Circuits to Word Circuits

Our simulation of ACC^0 circuits by word circuits can be regarded as a simple application of a technique from the word RAM literature [1]: *word parallelism*, i.e., the technique of using a gate operating on words as a miniature vector processor. The technique works for ACC^0 circuits as the intermediate results needed in the simulation of such a circuit have bounded bit-size and can therefore be compactly represented within a word. In contrast, for slightly more powerful circuit classes (such as TC^0), we do not know a simulation that converts non-trivial circuits into constant size word circuits.

Lemma 4. *Let k and m be positive integers so that $km \leq w$. Let $+^{(1)}, +^{(2)}, \ldots, +^{(m)}$ be associative and commutative operations on $K = \{0,1\}^k$. For some constant a, embed the domain K^m into $\{0,1\}^w$ and the domain K^{ma} into $(\{0,1\}^w)^a$ by identifying consecutive blocks of bits with elements of K. Let sets $S_1, \ldots, S_m \subseteq \{1, 2, \ldots, ma\}$ be given. Consider a function $g : K^{ma} \to K^m$ defined in the following way: $g(x_1, x_2, \ldots, x_{ma}) = (\sum_{i \in S_1}^{(1)} x_i, \ldots, \sum_{i \in S_m}^{(m)} x_i)$ where $\sum^{(j)}$ is summation with respect to $+^{(j)}$. Then g has a word circuit of size $a - 1$.*

Proof. The word circuit is a tree (i.e., a formula) consisting of gates computing pointwise addition of two elements of K^m, with the j'th entries added with respect to the operation $+^{(j)}$. At the bottom of the tree we have for each segment of inputs $x_i, x_{i+1}, \ldots, x_{i+m-1}$ corresponding to a word a unary gate mapping this word into a word representing an element of K^m whose j'th entry is $\sum_{\ell \in \{i,i+1,\ldots,i+m-1\} \cap S_j}^{(j)} x_\ell$. These unary gates feed into the pointwise addition gates of the tree. The correctness of the construction is immediate. To make the size of the circuit exactly $a - 1$, we merge the unary gates with their immediate successors in the circuit.

Theorem 7. *Let a family of ACC^0 circuits computing a family of functions $g : \{0,1\}^{3w} \to \{0,1\}^w, w \geq 2k$ be given so that the w'th circuit contains aw gates, has depth d and so that all counting gates count modulo some integer less than 2^k. Then, viewed as a function mapping three words to one, g is decomposable and has a word circuit of size less than $(d(d+1)/2)(2ka+1)^2 + 4ka + 2$.*

Proof. Let $m = \lfloor w/k \rfloor$. Divide the circuit into d layers so that each layer gets inputs from (possibly all) previous layers. Layer 0 is the layer of input bits. Add dummy gates so that each layer contains exactly $\ell = m\lceil aw/m \rceil$ gates. The *expanded* representation of the bits computed by a layer is the string of words defined as follows: Divide the bits computed by the gates in the layer into

$\ell/m = \lceil aw/m \rceil \leq aw/m + 1 \leq 2ka + 1$ words. This gives us m bits per word, allowing us to embed each bit into the domain $\{0,1\}^k$, representing 0 as 0^k and 1 and 1^k.

The expanded representation of layer 0 is easily seen to be computable by a word circuit of size at most $2ka + 1$ (one gate computes each word in the representation). By Lemma 4, assuming the computation of previous layers has been handled, the computation of one of the words of the expanded representation of layer j can be done by a word circuit of size at most $j(2ka+1) - 1$. That is, the computation of the entire j'th layer can be simulated by a word circuit of size less than $j(2ka + 1)^2$. So, the entire ACC^0 circuit can be simulated by a word circuit of size less than $\frac{1}{2}d(d+1)(2ka+1)^2$. Finally, a simple word circuit of size at most $2ka + 1$ converts the expanded representation of the output bits of the output layer into a single word.

This almost finishes the proof, up to one subtle point which was not relevant in this paper until now and hence ignored: To be decomposable, a function much be computed on its entire infinite domain $\cup_w(\{0,1\}^w)^3$ by *one* word circuit, not just by a family of constant sized ones. That is, we should worry whether we are using differently structured constant sized circuits for different word lengths. However, by inspecting the construction, we find that the structure of the circuit is in fact exactly the same for every word length. Indeed, this was our reason for adding the seemingly wasteful dummy gates.

Acknowledgments

We thank Miklós Ajtai for introducing us to the problem considered in this paper, for explaining its motivation and for helpful discussions during the course of this work.

References

1. Hagerup, T.: Sorting and searching on the word RAM. In: Meinel, C., Morvan, M. (eds.) STACS 1998. LNCS, vol. 1373, pp. 366–398. Springer, Heidelberg (1998)
2. Koucký, M., Pudlák, P., Thérien, D.: Bounded-depth circuits: separating wires from gates. In: STOC 2005: Proceedings of the 37th annual ACM symposium on Theory of computing, pp. 257–265. ACM, New York (2005)
3. Lupanov, O.B.: The synthesis of contact circuits. Dokl. Akad. Nauk SSSR (N.S.) 119, 23–26 (1958)
4. Lupanov, O.B.: Complexity of formula realization of functions of logical algebra. Prob. Kibernetki 3, 782–811 (1962)
5. Wegener, I.: The complexity of Boolean functions. John Wiley & Sons, Inc., New York (1987)

Interval Stabbing Problems in Small Integer Ranges*

Jens M. Schmidt

Freie Universität, Berlin, Germany
jens.schmidt@inf.fu-berlin.de

Abstract. Given a set I of n intervals, a *stabbing query* consists of a point q and asks for all intervals in I that contain q. The *Interval Stabbing Problem* is to find a data structure that can handle stabbing queries efficiently. We propose a new, simple and optimal approach for different kinds of interval stabbing problems in a static setting where the query points and interval ends are in $\{1, \ldots, O(n)\}$.

Keywords: interval stabbing, interval intersection, static, discrete, point enclosure.

1 Introduction

Interval stabbing, also known as the one-dimensional *point enclosure problem* is one of the most fundamental problems in computational geometry and has been studied for decades. Let l_a be the left endpoint and r_a be the right endpoint of an interval a. We address the following static setting:

Let I be a given set of n intervals with l_a, $r_a \in Q := \{1, ..., O(n)\}$ for every $a \in I$. An interval $a \in I$ is *stabbed* by a point $q \in Q$ if $q \in a$. We want to construct simple and lightweight data structures that answer the following queries on I efficiently:

1. *Interval Stabbing Problem*: Given a query point $q \in Q$, report all intervals in I that are stabbed by q.
2. *Interval Intersection Problem*: Given a query interval $[l_q, r_q]$ with $l_q, r_q \in Q$, report all intervals $i \in I$ with $[l_i, r_i] \cap [l_q, r_q] \neq \emptyset$.
3. *Interval Cover Problem*: Given an interval $q \in I$, report all intervals in I that contain the interval q.
4. *Multiple Query Problems*: These problems extend each of the problems 1-3 by allowing multiple queries $q_1 < \ldots < q_t, \forall i : q_i \in Q$, at the same time. The query points have to be given as a sorted list while the output consists of the intervals that are stabbed by at least one q_i (without double occurences).

We demand in addition that the intervals in each output are reported in lexicographical order. In general, queries do not admit a worst-case running time better

* This research was supported by the Deutsche Forschungsgemeinschaft within the research training group "Methods for Discrete Structures" (GRK 1408).

Y. Dong, D.-Z. Du, and O. Ibarra (Eds.): ISAAC 2009, LNCS 5878, pp. 163–172, 2009.

than $O(n)$, since the output itself can be that large. But for many inputs the output will be much smaller. Therefore, it is reasonable to consider the *output-sensitive complexity* for queries, where the running time is given with respect to the input size and the output size k. We assume the uniform cost model, thus k is the number of intervals in the output. Clearly, every data structure needs to store all intervals and, thus, needs at least $\Omega(n)$ space and preprocessing time to be built. The query time is at least $\Omega(1+k)$ (or $\Omega(t+k)$ for problems of type 4), the 1 (or t) coming from queries in Q that are not covered by any interval.

We will only focus on solutions that reach that bounds, i. e., that solve problems 1-4 in asymptotic optimal space and time. Therefore common interval data structures like *interval trees* [6,8], *segment trees* [4] and *priority trees* [9] cease to apply, as each of them needs a preprocessing time of $\Omega(n \log n)$ and query times of at least $\Omega(\log n + k)$. Alstrup, Brodal and Rauhe [1] describe a data structure, based on results in [7], that can be used for solving the problems optimally. The idea is to interpret every interval $a \in I$ as a point (l_a, r_a) in the integer grid $n \times n$ and then model the given problem by *three-sided range queries* in this grid, i. e., by rectangular range queries with one side going to ∞ or $-\infty$. Each three-sided range query can be performed in time $O(1 + k)$ by computing iteratively *nearest common ancestors* in a cartesian tree as shown by Gabow, Bentley and Tarjan [7]. However, this procedure seems far too involved for the type of problems we look at and comes with a significant implementation overhead.

The essence of this paper is a direct, new approach that solves all problems optimally and does not rely on computing nearest common ancestors, thus has considerably less overhead. Problems 1 and 2 can as well be solved by the *filtering search* data structure due to Chazelle [5] in the same asymptotic time and space requirements. However, filtering search does not solve Problems 3 and 4 and experiments show that our data structure performs faster than filtering search in practice. That can be explained with the lower number of comparisons needed for one query in the theoretical worst case: Our data structure needs $3k$ point-to-q comparisons instead of $8k$ comparisons for Chazelle's data structure.

We assume all given intervals a to be closed, but, if necessary, open and half open intervals may be easily modeled by increasing l_a and/or decreasing r_a by one in advance. Let l_a and r_a be the *endpoints* (or shorter *ends*) of an interval $a \in I$ and let l_a be the *left endpoint* and r_a be the *right endpoint.*

If the query range Q is not $\{1, ..., O(n)\}$ there are techniques that reduce problems to work within a small integer range [1,7]. E. g., any universe can be reduced to the integer range $\{2, ..., 4n\}$ by first sorting the $\leq 2n$ interval ends and then assigning to each one two times its rank. This leaves a gap between every pair of consecutive interval ends. Then a binary search transforms any stabbing query $q \in Q$ to a query in $\{2, ..., 4n\}$, reflecting its relative position in Q, either at an interval end or a gap. This goes along with a blow-up of the preprocessing time to $O(n \log n)$ and query time to $O(\log(n) + k)$ for problems 1-3 and to $O(\min(t \log(n), n) + k)$ for problems of type 4. If the model of computation is the unit-cost word RAM and all query points fit in a constant number of words, much faster algorithms for sorting and predecessor searching of query points can be applied (Andersson et al. [2], Beame and Fitch [3], although these results are not needed for the restricted universe we consider here.

2 The Data Structure

We identify intervals with their left and right endpoints and sort all intervals according to the *lexicographic order* $< \subseteq N \times N$, i.e., for two intervals a and b holds $a < b$ if $l_a < l_b$ or $(l_a = l_b \wedge r_a \leq r_b)$. The computation time of this lexicographic list is $O(n)$ by using (stable) bucket sort for the right endpoints followed by a bucket sort for the left endpoints, since all interval ends are by definition in Q.

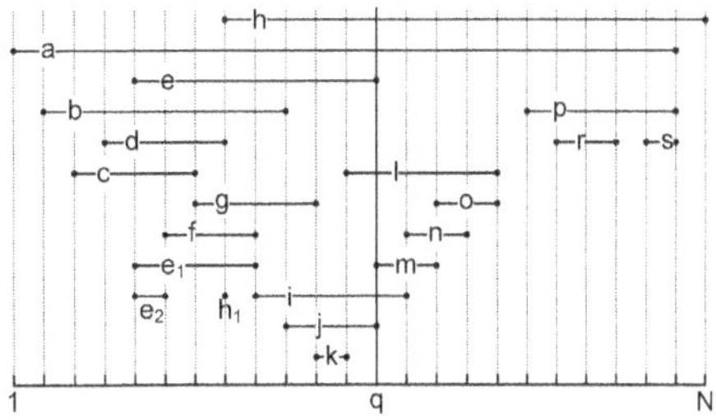

Fig. 1. The intervals e_1, e_2 and h_1 are removed from I in advance, because $Smaller(e) = \{e_1, e_2\}$ and $Smaller(h) = \{h_1\}$. Only intervals a and b cover d and $Parent(d) = b$. Moreover, c overlaps d, e, f, g and e_1 but d does not overlap c.

Intervals that share right endpoints will integrate well in our data structure, thus the frequently used input transformation to intervals with completely distinct ends is not necessary. To get rid of intervals sharing their left endpoint l (for every l), we apply the following preprocessing: All the intervals with left endpoint l, except one such longest interval a, are stored in a list called *Smaller*(a) (see Figure 1). These lists are sorted by length in descending order, get a link to a, and every element in them is removed from I (i.e., from now on I does not contain intervals in *Smaller*(a) and n is redefined to $|I|$ afterwards). This establishes $<$ to be a *strict total order* on I relying only on left endpoints. Later, a simple trick will deal with the omitted intervals *Smaller*(a).

Two intervals $a, b \in I$ *intersect* if $a \cap b \neq \emptyset$. Otherwise, they are called *disjoint*. We say that interval a *overlaps* interval b if $l_a < l_b \leq r_a < r_b$. Moreover, let a be *covered* by b (and b *cover* a) if $a \subseteq b$. Let the *rightmost* interval in a non-empty subset of I be the interval with the maximal left endpoint. Note that this is well-defined as the left endpoint is unique in I. Then *Parent*(a) is defined as the rightmost among all intervals that cover a (see Figure 1). If a is not covered by any interval, *Parent*(a) $:= \emptyset$.

Proposition 1. *For two intervals $a, b \in I$ with $a < b$ exactly one of the following statements holds:*

- *a and b are disjoint*
- *a is covering b*
- *a overlaps b.*

We attach each interval a to *Parent*(a), yielding a forest F with intervals as nodes and the *Parent*-function as edges. Let $root_i$ denote the root of a maximal tree T_i in F. We construct a spanning tree $S = (V, E)$ by augmenting the forest with a special dummy node *root* (representing the interval Q) and attaching the roots of all trees T_i to it (see Figure 2).

Let S be ordered by sorting the children of each node according to their left endpoints. The children of a node $v \in V$ are stored in a doubly linked list, denoted by *Children*(v). Every entry in *Children*(v) is a *sibling* of each other entry. We call the sibling immediately to the left (right) of an entry the *left sibling* (*right sibling*). In a tree, a node w is an *ancestor* of a node v, if w is contained in the path from v to the root (including the node v).

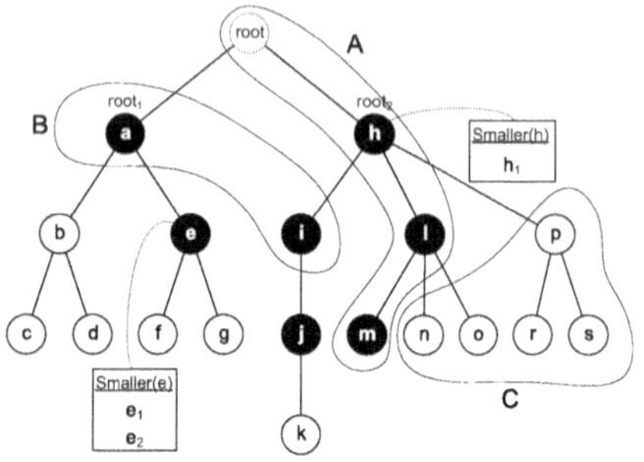

Fig. 2. The spanning tree S of the example in Figure 1. Black nodes indicate intervals stabbed by q.

3 The Interval Stabbing Problem

We show how to solve Problem 1 using the spanning tree S and extend this result later to problems 2-4. First, imagine that all pairs of intervals in I would either be disjoint or cover each other. In this restricted case it suffices to precompute the rightmost stabbed interval *Start*(q) for every $q \in Q$, if it exists. If a query q arises, let T_s be the tree in F that contains *Start*(q) and P be the path from *Start*(q) to $root_s$ in T_s. Then we can get all k stabbed intervals by traversing P in time $O(1+k)$, since *Start*(q) must be the smallest stabbed interval and all other stabbed intervals have to be ancestors of it in T_s.

However, in general intervals may overlap and stabbed ones can even be contained in different trees of F. We partition $V(S)$ into four classes subject to P (see Figure 2). A node $v \in V(S)$ is in class

- A, if v is in P or the dummy node
- B, if v has a sibling w in P with $l_w > l_v$
- C, if $l_v > q$
- D, otherwise

Lemma 1. *For every* $v \in V(S)$ *the stabbed children of* v *are adjacent in* Children(v).

Proof. We assume to the contrary that there is at least one child $b \in I$ that is not stabbed between two stabbed children a and c. Since siblings cannot cover each other and a and c cannot be disjoint a must overlap c. Then $a \cap c$ contains the query point and b is stabbed as well, since $l_b < l_c$ and $r_b > r_a$. □

Given a query point q, we first show how to obtain all stabbed intervals in the sets A, B and C efficiently with a traversal starting at *Start*(q). If *Start*(q) is not stabbed, no interval can be stabbed and the query time is $O(1)$. Otherwise, all intervals in A must contain q and we traverse them. The stabbed intervals in B can then be easily computed with Lemma 1 by iteratively traversing to the left sibling for each node in P until the list ends or a node was not stabbed. No interval in C can be stabbed because their left endpoints are greater than q by definition, so only class D remains.

Lemma 2. *Every stabbed node $v \in D$ has a stabbed ancestor in B.*

Proof. With v all ancestors of v are stabbed and at least one of them is contained in A, since the dummy node is in A. Let w be the ancestor that is not in A but has a parent z in A. If $z \neq \mathit{Start}(q)$ then w is a stabbed sibling left of a node in P and therefore in B and the claim follows. Otherwise, $z = \mathit{Start}(q)$ contradicts $v \in D$, since all intervals in the subtree of z have a greater left endpoint than z has. □

Lemma 3. *If $v \in D$ has a right sibling w and is stabbed, w is stabbed as well.*

Proof. According to Lemma 2, there is a stabbed ancestor of v and w in B. Then the right sibling z of this ancestor exists, is either in A or B and is stabbed. By construction of the spanning tree $l_v < l_w < l_z$ must hold and the query point q is in $v \cap z$. Since $v \cap z \subseteq w$, the point q has to stab w as well. □

Lemmas 2 and 3 lead immediately to a recursive characterization of all stabbed nodes in D. Let $U(v)$ for a node $v \in D$ be the sequence of nodes from v to the first node in B where each successor is the right sibling, if it exists, and otherwise the parent.

Corollary 1. *The node $v \in D$ is stabbed if and only if all nodes of $U(v)$ are stabbed.*

Corollary 1 allows us to compute all stabbed nodes in D by traversing paths back from stabbed nodes in B.

Definition 1. *The* rightmost path *$R(v)$ of a node $v \in V(S)$ is empty if v has no left sibling or its left sibling w is not stabbed. Otherwise, $R(v)$ is the path from w to the rightmost stabbed node in the subtree of w in S.*

Note that $R(v)$ contains only stabbed intervals and can be constructed by iteratively taking the last child, starting with w. We are now in a position to compute all stabbed nodes by traversing P from the bottom up and recursively computing and traversing $R(v)$ from the bottom-up for each visited node v (see Algorithm 1). All stabbed nodes in A and B are found, since the computation of rightmost paths considers left children and continues with them at some point, if they are stabbed. The same holds for stabbed nodes in D, since Corollary 1 ensures that all stabbed nodes in C are reachable by a sequence of rightmost paths that start with a stabbed node in B.

We can find all k stabbed intervals in $O(1 + k)$ time, because checking an interval to be stabbed by q, computing $\mathit{Start}(q)$ and traversing to the parent, left sibling or last child can be done in constant time.

Algorithm 1. Traverse ($v \in V(S)$, stack $O, q \in Q$)

Push v to stack O ▷ for output purposes
while next interval w in *Smaller*(v) exists and $q \in w$ **do**
  Push w to stack O
Compute the rightmost path $R(v)$
for all nodes w in $R(v)$ (from the bottom up) **do**
  Traverse(w, O, q)

It only remains to show how to deal with the intervals in the *Smaller*-lists and ensure that the output is sorted in lexicographic order. Each time we reach a node v with *Smaller*(v) $\neq \emptyset$, we traverse that list until the end or the first non-stabbed node was found. This way we do not visit intervals that are not stabbed and, thus, preserve the running time of $O(1+k)$ for each query. We use the following lemma to verify that the output is sorted in lexicographic order.

Lemma 4. *A preorder traversal on root returns all intervals of S sorted by their left endpoints.*

Proof. All children of a node $v \in V(S)$ are sorted and have left endpoints strictly greater than l_v for $v \neq root$. Let w be the right sibling of v. Then, due to the definition of *Parent*, every interval in the subtree on v has a left endpoint of strictly less than l_w. Recursively collecting the actual node and traversing the children from left to right returns the intervals sorted by their left endpoints. □

The traversal of S starts with the stabbed interval *Start*(q) that has the maximal left endpoint and visits subsequent intervals containing q in a postorder traversal that prefers right children to left children. As this postorder reverses the preorder traversal and the output of the preorder traversal is sorted in inverse lexicographic order with Lemma 4, we need to reverse the order of intervals found. This is done by using a stack (see Algorithm 2).

All preprocessing steps, i. e., computing the *Parent* and *Start* pointers can be done with one sweep line procedure in time $O(n)$ by maintaining a list of stabbed intervals for each $q \in Q$ (see the pseudocode description in Algorithm 3). For each stabbed interval v of a query, we check at most three subsequent intervals on containing q, the left sibling of v, the last child of v and the successor in

Algorithm 2. Stabbing query ($q \in Q$)

Stack $O = \emptyset$ ▷ O for output purposes
if *Start*(q) $= \emptyset$ **then** STOP ▷ q stabs no interval
Compute the path P from *Start*(q) to $root_s$
for all nodes v in $V(P)$ (from the bottom up) **do**
  Traverse(v, O, q)
while $O \neq \emptyset$ **do** ▷ reverse list of stabbed intervals
  Append *pop*(O) to output

Algorithm 3. Preprocessing

List $L = \emptyset$; create dummy node *root*
$\forall q \in Q, a \in I$: create pointers to intervals *Start*(q) and *Parent*(a)
$\forall q \in Q$: compute lists *Smaller*(q); update I
for all $a \in I$ in lexicographic order **do** ▷ build event structure
    Append a to *Event*(r_a) ▷ event list of intervals on r_a
    Append a to *Event*(l_a)
for $q = 1$ to N **do** ▷ sweep line
    If $L \neq \emptyset$, store the last element in L as *Start*(q), else *Start*(q) := *NULL*
    for all intervals $a \in$ *Event*(q) in reverse order **do**
        if $l_a = q$ **then**
            Append a to L and save a link to its position
        else ▷ $r_a = q$
            if a has a predecessor b in L **then**
                Store b as *Parent*(a) and append a to *Children*(b)
            else
                Store *root* as *Parent*(a) and append a to *Children*(*root*)
            Remove a from L

Smaller(v). However, we need only to compare the right endpoints of those intervals with q, since Lemma 4 ensures that $l_v \leq q$ holds.

Theorem 1. *All k intervals stabbed by a query point q can be found sorted in lexicographic order in query time $O(1 + k)$ and with at most $3k$ comparisons with q ($2k$ comparisons if all left endpoints in I are pairwise distinct). The preprocessing time and space requirement is $O(n)$.*

4 Variants of the Problem

We discuss the problems 2-4. The *Interval Intersection Problem* differs from the *Interval Stabbing Problem* only in having a query interval $[l_q, r_q]$ instead of a query point. Let an interval be stabbed if its intersection with $[l_q, r_q]$ is non-empty. Then Lemmas 1, 2, 3 and Corollary 1 remain valid and we can still use the data structure of the *Interval Stabbing Problem* to get all stabbed intervals. The traversal starts with the intervals containing r_q, recurses to their rightmost paths and stops at intervals that are not stabbed. These lie with Lemma 4 completely left of l_q and are not part of the output. Testing a visited interval a on being stabbed can be done with one comparison by checking $r_a \geq l_q$, leading to a $O(1 + k)$ query time in total.

For the *Interval Cover Problem* an interval $q \in I$ is given. We set *Start*(q) $= q$, because there is no interval with a higher left endpoint covering q. Since ancestors cover q if one of their children does, we can build the path P and partition $V(S)$ subject to P as in the *Interval Stabbing Problem.* When we replace the property *stabbed* with *covering* q on intervals, the Lemmas 1, 2, 3 and Corollary 1 still hold. Every visited interval a can be tested on covering q by checking $r_a \geq r_q$, which gives a query time of $O(1 + k)$.

We show that the *Multiple Interval Stabbing Problem* in 4 allows for a query time of $O(t + k)$, the other problems in 4 can then be solved using the same technique. If every interval in the output would be stabbed by only one value q_i, $1 \leq i \leq t$, the problem could be solved in time $O(t + k)$ by applying the queries $q_1, \ldots, q_t$ subsequently. In general that is not the case and we have to ensure that the traversals of different query points do not both visit a node.

Assume that we start with the traversal of the rightmost query point q_t and compute recursively rightmost paths. Then with Lemma 4 the sequence of left endpoints of visited intervals is strictly monotone decreasing. For every visited interval a we check in advance if $l_a \leq q_{t-1}$ holds and if so, replace the current query point q_t with the maximal query point $q_j \geq l_a$, $j < t$. If now a or any subsequent interval is stabbed by q_t, it will also be stabbed by q_j and we can perform all comparisons with q_j instead of q_t. If a traversal of q_i, $i > 1$, ends without switching to q_{i-1} we invoke the traversal on the next query point q_{i-1}. Since the list $q_1, \ldots, q_t$ is ordered, the additional expense to update the query point is bounded by t constant time comparisons, which gives a total query time of $O(t + k)$.

Corollary 2. *The k intervals in the output of problems 2 and 3 can be found sorted in lexicographic order in query time $O(1 + k)$ and with at most $3k$ comparisons with q ($2k$ comparisons if all left endpoints in I are pairwise distinct). For problems of type 4 a query can be done in time $O(t + k)$ with at most $4k$ comparisons with values in $\{q_1, \ldots, q_t\}$. For all problems the preprocessing time and space requirement is $O(n)$.*

5 Experimental Analysis

We implemented Chazelle's data structure [5] with various window sizes ($\delta = 1.2$, 1.5, 2, 3, 5) for the *Interval Stabbing Problem* and compared the running times to our approach on Problem 1. However, $\delta = 2$ gave the best results in both preprocessing and query times and we will focus on that parameter, since $\delta < 2$ did not lead to observable better query times but to a considerably worse preprocessing time instead (see Figure 3). Both data structures use identical representations for intervals, lists and stacks and work under the same conditions as much as possible. All tests are performed on a 1.86 GHz CPU and 2GB RAM using the MS compiler 9.0 with optimization level O2. The source code is available online: `http://page.mi.fu-berlin.de/jeschmid/pub`.

The input consists of various n from 10000 to 1000000, $Q := \{1, \ldots, 5n\}$ (other constants than 5 led to similar results) and either *random intervals* with uniformly distributed interval ends in Q or *short random intervals*. *Short random intervals* have an exponentially distributed length with expected value 1000, while their left endpoints and all query points are uniformly distributed on Q.

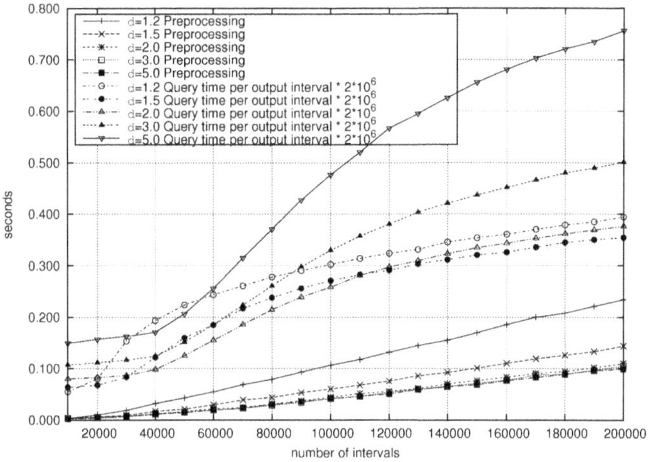

Fig. 3. Different values for the window size δ in Chazelle's data structure (short random intervals).

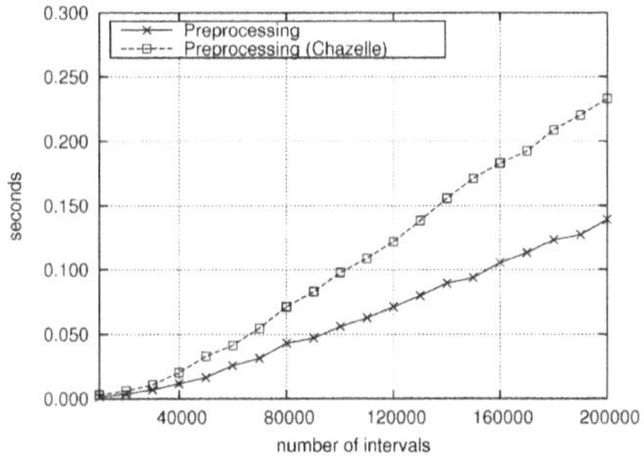

(a) Preprocessing (random intervals)

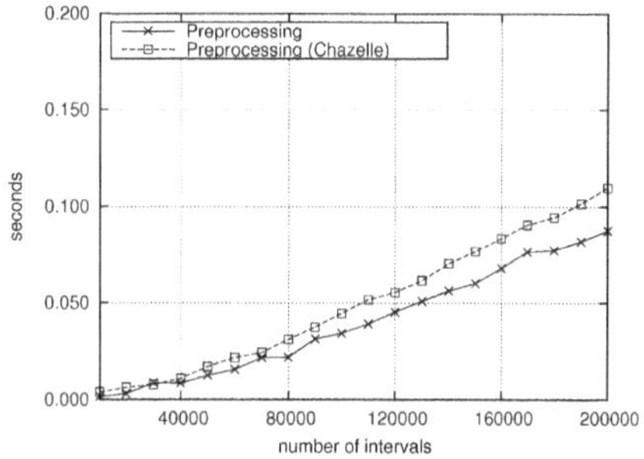

(b) Preprocessing (short random intervals)

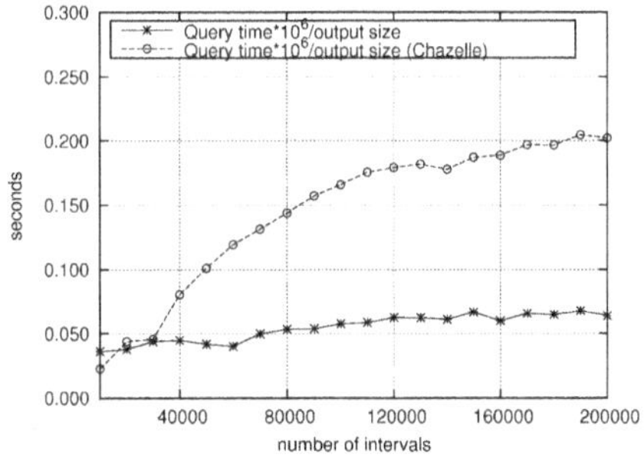

(c) Query times (random intervals)

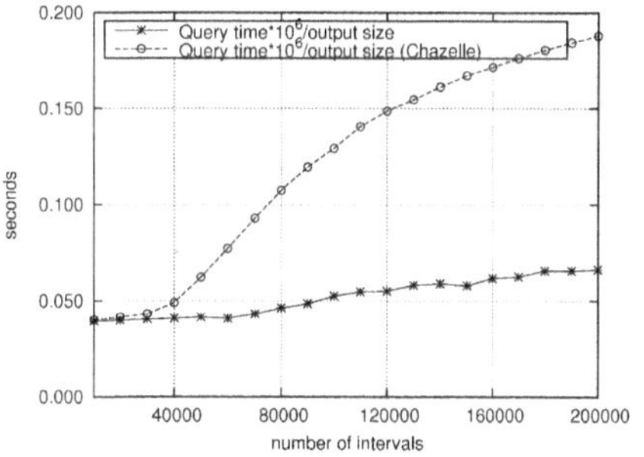

(d) Query times (short random intervals)

Fig. 4. Comparison of preprocessing and query times

The exponential distribution is generated with inverse transform sampling on a uniform distribution. Both query and preprocessing times are averaged over 20 instances for each n with up to 10000 queries per instance.

The preprocessing times of both data structures in practice reflect the theoretical linear bound of $\Theta(n)$, except for small n (see Figures 4(a) and 4(b)). In both figures, our approach performs faster, although on short random intervals the advantage is marginal. Since query times are primarily dependent on the output length, we measure the average computation time needed for one interval in the output. Theoretically, each query time should be constant, although the memory hierarchy can increase the time in practice when n grows. For large n, the query times of our data structure are significantly faster than Chazelle's for both input types (see Figures 4(c) and 4(d)).

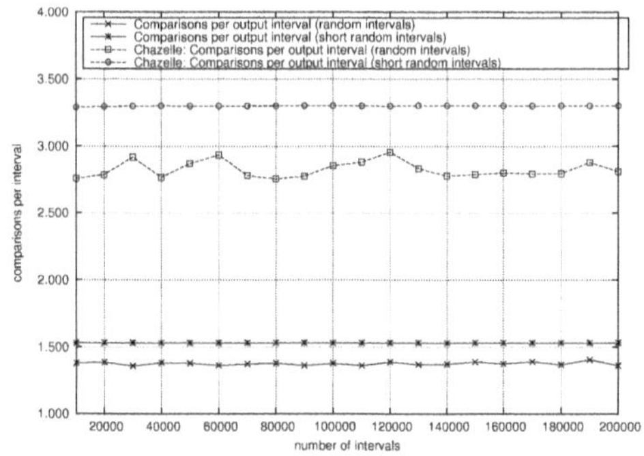

Fig. 5. Point-to-q comparisons

Figure 5 shows how many point comparisons with q are made on average for each interval in the output. Both data structures need about half of the comparisons of the theoretical worst case (3 point-to-q comparisons for our data structure, 8 point-to-q comparisons for Chazelle's data structure).

References

1. Alstrup, S., Brodal, G.S., Rauhe, T.: New data structures for orthogonal range searching. In: FOCS 2000, pp. 198–207 (2000)
2. Andersson, A., Hagerup, T., Nilsson, S., Raman, R.: Sorting in linear time? In: STOC 1995, pp. 427–436 (1995)
3. Beame, P., Fich, F.E.: Optimal bounds for the predecessor problem. In: STOC 1999: Proceedings of the 31st annual ACM symposium on Theory of computing, pp. 295–304. ACM, New York (1999)
4. Bentley, J.L.: Solutions to Klee's rectangle problems. Tech. report, Carnegie-Mellon Univ., Pittsburgh, PA (1977)
5. Chazelle, B.: Filtering search: A new approach to query answering. SIAM J. Comput. 15(3), 703–724 (1986)
6. Edelsbrunner, H.: Dynamic data structures for orthogonal intersection queries. Tech. Report F59, Inst. Informationsverarb, Tech. Univ. Graz (1980)
7. Gabow, H.N., Bentley, J.L., Tarjan, R.E.: Scaling and related techniques for geometry problems. In: STOC 1984: Proceedings of the 16th annual ACM symposium on Theory of computing, pp. 135–143 (1984)
8. McCreight, E.M.: Efficient algorithms for enumerating intersecting intervals and rectangles. Tech. Report CSL-80-9, Xerox Palo Alto Res. Center, CA (1980)
9. McCreight, E.M.: Priority search trees. SIAM Journal on Computing 14(2), 257–276 (1985)

Online Sorted Range Reporting

Gerth Stølting Brodal[1], Rolf Fagerberg[2], Mark Greve[1], and Alejandro López-Ortiz[3]

[1] MADALGO*, Dept. of Computer Science, Aarhus University, Denmark
{gerth,mgreve}@cs.au.dk
[2] Dept. of Math. and Computer Science, University of Southern Denmark
rolf@imada.sdu.dk
[3] School of Computer Science, University of Waterloo, Canada
alopez-o@uwaterloo.ca

Abstract. We study the following one-dimensional range reporting problem: On an array A of n elements, support queries that given two indices $i \leq j$ and an integer k report the k smallest elements in the sub-array $A[i..j]$ in sorted order. We present a data structure in the RAM model supporting such queries in optimal $O(k)$ time. The structure uses $O(n)$ words of space and can be constructed in $O(n \log n)$ time. The data structure can be extended to solve the online version of the problem, where the elements in $A[i..j]$ are reported one-by-one in sorted order, in $O(1)$ worst-case time per element. The problem is motivated by (and is a generalization of) a problem with applications in search engines: On a tree where leaves have associated rank values, report the highest ranked leaves in a given subtree. Finally, the problem studied generalizes the classic range minimum query (RMQ) problem on arrays.

1 Introduction

In information retrieval, the basic query types are exact word matches, and combinations such as intersections of these. Besides exact word matches, search engines may also support more advanced query types like prefix matches on words, general pattern matching on words, and phrase matches. Many efficient solutions for these involve string tree structures such as tries and suffix trees, with query algorithms returning nodes of the tree. The leaves in the subtree of the returned node then represent the answer to the query, e.g. as pointers to documents.

An important part of any search engine is the ranking of the returned documents. Often, a significant element of this ranking is a query-independent pre-calculated rank of each document, with PageRank [1] being the canonical example. In the further processing of the answer to a tree search, where it is merged with results of other searches, it is beneficial to return the answer set ordered by the pre-calculated rank, and even better if it is possible to generate an increasing prefix of this ordered set on demand. In short, we would like a

* Center for Massive Data Algorithmics, a Center of the Danish National Research Foundation.

Y. Dong, D.-Z. Du, and O. Ibarra (Eds.): ISAAC 2009, LNCS 5878, pp. 173–182, 2009.

functionality similar to storing at each node in the tree a list of the leaves in its subtree sorted by their pre-calculated rank, but without the prohibitive space cost incurred by this solution.

Motivated by the above example, we consider the following set of problems, listed in order of increasing generality. For each problem, an array $A[0..n-1]$ of n numbers is given, and the task is to preprocess A into a space-efficient data structure that efficiently supports the query stated. A query consists of two indices $i \leq j$, and if applicable also an integer k.

Sorted range reporting:
Report the elements in $A[i..j]$ in sorted order.
Sorted range selection:
Report the k smallest elements in $A[i..j]$ in sorted order.
Online sorted range reporting:
Report the elements in $A[i..j]$ one-by-one in sorted order.

Note that if the leaves of a tree are numbered during a depth-first traversal, the leaves of any subtree form a consecutive segment of the numbering. By placing leaf number i at entry $A[i]$, and annotating each node of the tree by the maximal and minimal leaf number in its subtree, we see that the three problems above generalize our motivating problems on trees. The aim of this paper is to present linear space data structures with optimal query bounds for each of these three problems. We remark that the two last problems above also form a generalization of the well-studied range minimum query (RMQ) problem [2]. The RMQ problem is to preprocess an array A such that given two indices $i \leq j$, the minimum element in $A[i..j]$ can be returned efficiently.

Contributions: We present data structures to support sorted range reporting queries in $O(j-i+1)$ time, sorted range selection queries in $O(k)$ time, and online sorted range reporting queries in worst case $O(1)$ time per element reported. For all problems the solutions take $O(n)$ words of space and can be constructed in $O(n \log n)$ time. We assume a unit-cost RAM whose operations include addition, subtraction, bitwise AND, OR, XOR, and left and right shifting and multiplication. Multiplication is not crucial to our constructions and can be avoided by the use of table lookup. By w we denote the word length in bits, and assume that $w \geq \log n$.

Outline: In the remainder of this section, we give definitions and simple constructions used extensively in the solution of all three problems. In Section 2, we give a simple solution to the sorted range reporting problem which illustrates some of the ideas used in our more involved solution of the sorted range selection problem. In Section 3, we present the main result of the paper which is our solution to the sorted range selection problem. Building on the solution to the previous problem, we give a solution to the online sorted range reporting problem in Section 4.

We now give some definitions and simple results used by our constructions.

Essential for our constructions to achieve overall linear space, is the standard trick of packing multiple equally sized items into words.

Lemma 1. *Let X be a two-dimensional array of size $s \times t$ where each entry $X[i][j]$ consists of $b \leq w$ bits for $0 \leq i < s$ and $0 \leq j < t$. We can store this using $O(stb + w)$ bits, and access entries of X in $O(1)$ time.*

We often need an array of s secondary arrays, where the secondary arrays have variable lengths. For this, we simply pad these to have the length of the longest secondary array. For our applications we also need to be able to determine the length of a secondary array. Since a secondary array has length at most t, we need $\lfloor \log t \rfloor + 1$ bits to represent the length. By packing the length of each secondary array as in Lemma 1, we get a total space usage of $O(stb+s \log t+w) = O(stb+w)$ bits of space, and lookups can still be performed in $O(1)$ time. We summarize this in a lemma.

Lemma 2. *Let X be an array of s secondary arrays, where each secondary array $X[i]$ contains up to t elements, and each entry $X[i][j]$ in a secondary array $X[i]$ takes up $b \leq w$ bits. We can store this using $O(stb + w)$ bits of space and access an entry or length of a secondary array in $O(1)$ time.*

Complete binary trees: Throughout this paper $\mathcal{T}$ will denote a complete binary tree with n leaves where n is a power of two. We number the nodes as in binary heaps: the root has index 1, and an internal node with index x has left child $2x$, right child $2x + 1$ and parent $\lfloor x/2 \rfloor$. Below, we identify a node by its number.

We let $\mathcal{T}_u$ denote the subtree of $\mathcal{T}$ rooted at node u, and $h(\mathcal{T}_u)$ the *height* of the subtree $\mathcal{T}_u$, with heights of leaves defined to be 0. The height of a node $h(u)$ is defined to be $h(\mathcal{T}_u)$, and *level* ℓ of $\mathcal{T}$ is defined to be the set of nodes with height ℓ. The height $h(u)$ of a node u can be found in $O(1)$ time as $h(\mathcal{T}) - d(u)$, where $d(u)$ is the *depth* of node u. The depth $d(u)$ of a node u can be found by computing the index of the most significant bit set in u (the root has depth 0). This can be done in $O(1)$ time and space using multiplications [3], or in $O(1)$ time and $O(n)$ space without multiplications, by using a lookup table mapping every integer from $0 \ldots 2n - 1$ to the index of its most significant bit set.

To navigate efficiently in $\mathcal{T}$, we explain a few additional operations that can be performed in $O(1)$ time. First we define $\text{anc}(u, \ell)$ as the ℓ'th ancestor of the node u, where $\text{anc}(u, 0) = u$ and $\text{anc}(u, \ell) = \text{parent}(\text{anc}(u, \ell - 1))$. Note that $\text{anc}(u, \ell)$ can be computed in $O(1)$ time by right shifting u ℓ times. Finally, we note that we can find the leftmost leaf in a subtree $\mathcal{T}_u$ in $O(1)$ time by left shifting u $h(u)$ times. Similarly we can find the rightmost leaf in a subtree $\mathcal{T}_u$ in $O(1)$ time by left shifting u $h(u)$ times, and setting the bits shifted to 1 using bitwise OR.

LCA queries in complete binary trees: An important component in our construction is finding the lowest common ancestor (LCA) of two nodes at the same depth in a complete binary tree in $O(1)$ time. This is just $\text{anc}(u, \ell)$ where ℓ is the index of the most significant bit set in the word u XOR v.

Selection in sorted arrays: The following theorem due to Frederickson and Johnson [4] is essential to our construction in Section 3.4.

Theorem 1. *Given m sorted arrays, we can find the overall k smallest elements in time $O(m + k)$.*

Proof. By Theorem 1 in [4] with $p = \min(m, k)$ we can find the k'th smallest element in time $O(m + p\log(k/p)) = O(m + k)$ time. When we have the k'th smallest element x, we can go through each of the m sorted arrays and select elements that are $\leq x$ until we have collected k elements or exhausted all lists. This takes time $O(m + k)$. □

2 Sorted Range Reporting

In this section, we give a simple solution to the sorted range reporting problem with query time $O(j - i + 1)$. The solution introduces the concept of *local rank labellings* of elements, and shows how to combine this with radix sorting to answer sorted range reporting queries. These two basic techniques will be used in a similar way in the solution of the more general sorted range selection problem in Section 3.

We construct local rank labellings for each r in $0 \ldots \lceil \log\log n \rceil$ as follows (the rank of an element x in a set X is defined as $|\{y \in X \mid y < x\}|$). For each r, the input array is divided into $\lceil n/2^{2^r} \rceil$ consecutive subarrays each of size 2^{2^r} (except possibly the last subarray), and for each element $A[x]$ the r'th local rank labelling is defined as its rank in the subarray $A[\lfloor x/2^{2^r} \rfloor 2^{2^r} .. (\lfloor x/2^{2^r} \rfloor + 1)2^{2^r} - 1]$. Thus, the r'th local rank for an element $A[x]$ consists of 2^r bits. Using Lemma 1 we can store all local rank labels of length 2^r using space $O(n2^r + w)$ bits. For all $\lceil \log\log n \rceil$ local rank labellings, the total number of bits used is $O(w \log\log n + n \log n) = O(nw)$ bits. All local rank labellings can be built in $O(n \log n)$ time while performing mergesort on A. The r'th structure is built by writing out the sorted lists, when we reach level 2^r. Given a query for $k = j - i + 1$ elements, we find the r for which $2^{2^{r-1}} < k \leq 2^{2^r}$. Since each subarray in the r'th local rank labelling contains 2^{2^r} elements, we know that i and j are either in the same or in two consecutive subarrays. If i and j are in consecutive subarrays, we compute the start index of the subarray where the index j belongs, i.e. $x = \lfloor j/2^{2^r} \rfloor 2^{2^r}$. We then radix sort the elements in $A[i..x-1]$ using the local rank labels of length 2^r. This can be done in $O(k)$ time using two passes by dividing the 2^r bits into two parts of 2^{r-1} bits each, since $2^{2^{r-1}} < k$. Similarly we radix sort the elements from $A[x..j]$ using the labels of length 2^r in $O(k)$ time. Finally, we merge these two sorted sequences in $O(k)$ time, and return the k smallest elements. If i and j are in the same subarray, we just radix sort $A[i..j]$.

3 Sorted Range Selection

Before presenting our solution to the sorted range selection problem we note that if we do not require the output of a query to be sorted, it is possible to get a conceptually simple solution with $O(k)$ query time using $O(n)$ preprocessing time and space. First build a data structure to support range minimum queries in $O(1)$ time using $O(n)$ preprocessing time and space [2]. Given a query on $A[i..j]$ with parameter k, we lazily build the Cartesian tree [5] for the subarray $A[i..j]$ using range minimum queries. The Cartesian tree is defined recursively by choosing the root to be the minimum element in $A[i..j]$, say $A[x]$, and recursively constructing its left subtree using $A[i..x-1]$ and its right subtree using $A[x+1..j]$. By observing that the Cartesian tree is a heap-ordered binary tree storing the elements of $A[i..j]$, we can use the heap selection algorithm of Frederickson [6] to select the k smallest elements in $O(k)$ time. Thus, we can find the k smallest elements in $A[i..j]$ in *unsorted* order in $O(k)$ time.

In the remainder of this section, we present our data structure for the sorted range selection problem. The data structure supports queries in $O(k)$ time, uses $O(n)$ words of space and can be constructed in $O(n \log n)$ time. When answering a query we choose to have our algorithm return the indices of the elements of the output, and not the actual elements. Our solution consists of two separate data structures for the cases where $k \leq \lfloor \log n/(2 \log\log n)^2 \rfloor$ and $k > \lfloor \log n/(2 \log\log n)^2 \rfloor$. The data structures are described in Sections 3.3 and 3.4 respectively. In Sections 3.1 and 3.2, we present simple techniques used by both data structures. In Section 3.1, we show how to decompose sorted range selection queries into a constant number of smaller ranges, and in Section 3.2 we show how to answer a subset of the queries from the decomposition by precomputing the answers. In Section 3.5, we describe how to build our data structures.

3.1 Decomposition of Queries

For both data structures described in Sections 3.3 and 3.4, we consider a complete binary tree $\mathcal{T}$ with the leaves storing the input array A. We assume without loss of generality that the size n of A is a power of two. Given an index i for $0 \leq i < n$ into A we denote the corresponding leaf in $\mathcal{T}$ as $\text{leaf}[i] = n + i$. For a node x in $\mathcal{T}$, we define the canonical subset C_x as the leaves in $\mathcal{T}_x$. For a query range $A[i..j]$, we let $u = \text{leaf}[i]$, $v = \text{leaf}[j]$, and $w = \text{LCA}(u, v)$. On the two paths from the two leaves to their LCA w we get at most $2 \log n$ disjoint canonical subsets, whose union represents all elements in $A[i..j]$, see Figure 1(a).

For a node x we define the sets $\mathcal{R}(x, \ell)$ (and $\mathcal{L}(x, \ell)$) as the union of the canonical subsets of nodes rooted at the right (left for $\mathcal{L}$) children of the nodes on

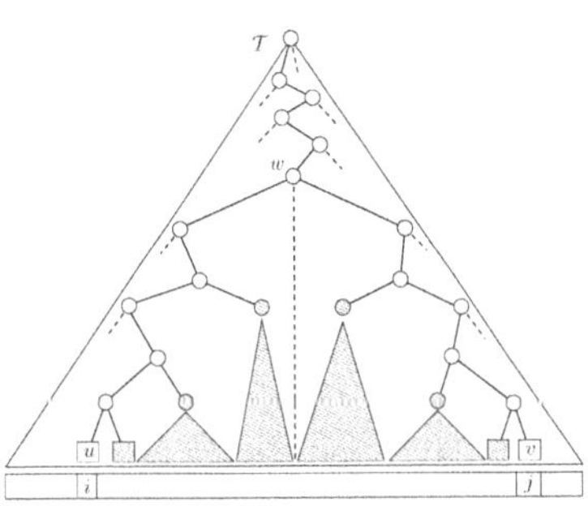

(a) The shaded parts and the leaves u and v cover the query range.

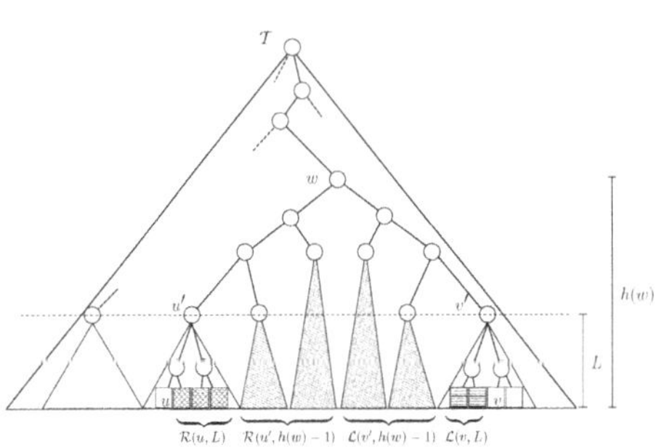

(b) Decomposition of a query range into four smaller ranges by cutting the tree at level L.

Fig. 1. Query decomposition

the path from x to the ancestor of x at level ℓ, but excluding the canonical subsets of nodes that are on this path, see Figure 1(a). Using the definition of the sets $\mathcal{R}$ and $\mathcal{L}$, we see that the set of leaves strictly between leaves u and v is equal to $\mathcal{R}(u, h(w)-1) \cup \mathcal{L}(v, h(w)-1)$. In particular, we will decompose queries as shown in Figure 1(b). Assume L is a fixed level in $\mathcal{T}$, and that the LCA w is at a level $> L$. Define the ancestors $u' = \text{anc}(u, L)$ and $v' = \text{anc}(v, L)$ of u and v at level L. We observe that the query range, i.e. the set of leaves strictly between leaves u and v can be represented as $\mathcal{R}(u, L) \cup \mathcal{R}(u', h(w)-1) \cup \mathcal{L}(v', h(w)-1) \cup \mathcal{L}(v, L)$. In the case that the LCA w is below or at level L, the set of leaves strictly between u and v is equal to $\mathcal{R}(u, h(w)-1) \cup \mathcal{L}(v, h(w)-1)$.

Hence to answer a sorted range selection query on k elements, we need only find the k smallest elements in sorted order of each of these at most four sets, and then select the k overall smallest elements in sorted order (including the leaves u and v). Assuming we have a sorted list over the k smallest elements for each set, this can be done in $O(k)$ time by merging the sorted lists (including u and v), and extracting the k smallest of the merged list. Thus, assuming we have a procedure for finding the k smallest elements in each set in $O(k)$ time, we obtain a general procedure for sorted range queries in $O(k)$ time.

The above decomposition motivates the definition of *bottom* and *top* queries relative to a fixed level L. A *bottom* query on k elements is the computation of the k smallest elements in sorted order in $\mathcal{R}(x, \ell)$ (or $\mathcal{L}(x, \ell)$) where x is a leaf and $\ell \leq L$. A *top* query on k elements is the computation of the k smallest elements in sorted order in $\mathcal{R}(x, \ell)$ (or $\mathcal{L}(x, \ell)$) where x is a node at level L. From now on we only state the level L where we cut $\mathcal{T}$, and then discuss how to answer bottom and top queries in $O(k)$ time, i.e. implicitly assuming that we use the procedure described in this section to decompose the original query, and obtain the final result from the answers to the smaller queries.

3.2 Precomputing Answers to Queries

We now describe a simple solution that can be used to answer a subset of possible queries, where a query is the computation of the k smallest elements in sorted order of $\mathcal{R}(x, \ell)$ or $\mathcal{L}(x, \ell)$ for some node x and a level ℓ, where $\ell \geq h(x)$. The solution works by precomputing answers to queries. We apply this solution later on to solve some of the cases that we split a sorted range selection query into.

Let x be a fixed node, and let y and K be fixed integer thresholds. We now describe how to support queries for the k smallest elements in sorted order of $\mathcal{R}(x, \ell)$ (or $\mathcal{L}(x, \ell)$) where $h(x) \leq \ell \leq y$ and $k \leq K$. We precompute the answer to all queries that satisfy the constraints set forth by K and y by storing two arrays R_x and L_x for the node x. In $R_x[\ell]$, we store the indices of the K smallest leaves in sorted order of $\mathcal{R}(x, \ell)$. The array L_x is defined symmetrically. We summarize this solution in a lemma, where we also discuss the space usage and how to represent indices of leaves.

Lemma 3. *For a fixed node x and fixed parameters y and K, where $y \geq h(x)$, we can store R_x and L_x using $O(Ky^2 + w)$ bits of space. Queries for the k smallest* *elements in sorted order in $\mathcal{R}(x,\ell)$ (or $\mathcal{L}(x,\ell)$) can be supported in time $O(k)$ provided $k \le K$ and $h(x) \le \ell \le y$.*

Proof. By storing indices relative to the index of the rightmost leaf in $\mathcal{T}_x$, we only need to store y bits per element in R_x and L_x. We can store the two arrays R_x and L_x with a space usage of $O(Ky^2+w)$ bits using Lemma 2. When reading an entry, we can add the index of the rightmost leaf in $\mathcal{T}_x$ in $O(1)$ time. The k smallest elements in $\mathcal{R}(x,\ell)$ can be reported by returning the k first entries in $R_x[\ell]$ (and similarly for $L_x[\ell]$). □

3.3 Solution for $k \le \lfloor \log n/(2\log\log n)^2 \rfloor$

In this section, we show how to answer queries for $k \le \lfloor \log n/(2\log\log n)^2 \rfloor$. Having discussed how to decompose a query into bottom and top queries in Section 3.1, and how to answer queries by storing precomputed answers in Section 3.2, this case is now simple to explain.

Theorem 2. *For $k \le \lfloor \log n/(2\log\log n)^2 \rfloor$, we can answer sorted range selection queries in $O(k)$ time using $O(n)$ words of space.*

Proof. We cut $\mathcal{T}$ at level $2\lfloor \log\log n \rfloor$. A bottom query is solved using the construction in Lemma 3 with $K = \lfloor \log n/(2\log\log n)^2 \rfloor$ and $y = 2\lfloor \log\log n \rfloor$. The choice of parameters is justified by the fact that we cut $\mathcal{T}$ at level $2\lfloor \log\log n \rfloor$, and by assumption $k \le \lfloor \log n/(2\log\log n)^2 \rfloor$. As a bottom query can be on any of the n leaves, we must store arrays L_x and R_x for each leaf as described in Lemma 3. All R_x structures are stored in one single array which is indexed by a leaf x. Using Lemma 3 the space usage for all R_x becomes $O(n(w + \lfloor \log n/(2\log\log n)^2 \rfloor (2\lfloor \log\log n \rfloor)^2) = O(n(w+\log n)) = O(nw)$ bits (and similarly for L_x). For the top query, we for all nodes x at level $2\lfloor \log\log n \rfloor$ use the same construction with $K = \lfloor \log n/(2\log\log n)^2 \rfloor$ and $y = \log n$. As we only have $n/2^{2\lfloor \log\log n \rfloor} = \Theta(n/(\log n)^2)$ nodes at level $2\lfloor \log\log n \rfloor$, the space usage becomes $O(\frac{n}{(\log n)^2}(w + \lfloor \log n/(2\log\log n)^2 \rfloor (\log n)^2)) = O(n(w+\log n)) = O(nw)$ bits (as before we store all the R_x structures in one single array, which is indexed by a node x, and similarly for L_x). For both query types the $O(k)$ time bound follows from Lemma 3. □

3.4 Solution for $k > \lfloor \log n/(2\log\log n)^2 \rfloor$

In this case, we build $O(\log\log n)$ different structures each handling some range of the query parameter k. The r'th structure is used to answer queries for $2^{2^r} < k \le 2^{2^{r+1}}$. Note that no structure is required for r satisfying $2^{2^{r+1}} \le \lfloor \log n/(2\log\log n)^2 \rfloor$ since this is handled by the case $k \le \lfloor \log n/(2\log\log n)^2 \rfloor$.

The r'th structure uses $O(w+n(2^r+w/2^r))$ bits of space, and supports sorted range selection queries in $O(2^{2^r}+k)$ time for $k \le 2^{2^{r+1}}$. The total space usage of the $O(\log\log n)$ structures becomes $O(w\log\log n + n\log n + nw)$ bits, i.e. $O(n)$

words, since $r \le \lceil \log\log n \rceil$. Given a sorted range selection query, we find the right structure. This can be done in done in $o(k)$ time. Finally, we query the r'th structure in $O(2^{2^r} + k) = O(k)$ time, since $2^{2^r} \le k$.

In the r'th structure, we cut $\mathcal{T}$ at level 2^r and again at level $7 \cdot 2^r$. By generalizing the idea of decomposing queries as explained in Section 3.1, we split the original sorted range selection query into three types of queries, namely *bottom*, *middle* and *top* queries. We define u' as the ancestor of u at level 2^r and u'' as the ancestor of u at level $7 \cdot 2^r$. We define v' and v'' in the same way for v. When the level of $w = \mathrm{LCA}(u, v)$ is at a level $> 7 \cdot 2^r$, we see that the query range (i.e. all the leaves strictly between the leaves u and v) is equal to $\mathcal{R}(u, 2^r) \cup \mathcal{R}(u', 7 \cdot 2^r) \cup \mathcal{R}(u'', h(w) - 1) \cup \mathcal{L}(v'', h(w) - 1) \cup \mathcal{L}(v', 7 \cdot 2^r) \cup \mathcal{L}(v, 2^r)$. In the case that w is below or at level $7 \cdot 2^r$, we can use the decomposition as in Section 3.1. In the following we focus on describing how to support each type of query in $O(2^{2^r} + k)$ time.

Bottom query: A bottom query is a query on a leaf u for $\mathcal{R}(u, \ell)$ (or $\mathcal{L}(u, \ell)$) where $\ell \le 2^r$. For all nodes x at level 2^r, we store an array S_x containing the canonical subset C_x in sorted order. Using Lemma 1 we can store the S_x arrays for all x using $O(n2^r + w)$ bits as each leaf can be indexed with 2^r bits (relative to the leftmost leaf in $\mathcal{T}_x$). Now, to answer a bottom query we make a linear pass through the array $S_{\mathrm{anc}(u,2^r)}$ discarding elements that are not within the query range. We stop once we have k elements, or we have no more elements left in the array. This takes $O(2^{2^r} + k)$ time.

Top query: A top query is a query on a node x at level $7 \cdot 2^r$ for $\mathcal{R}(x, \ell)$ (or $\mathcal{L}(x, \ell)$) where $7 \cdot 2^r < \ell \le \log n$. We use the construction in Lemma 3 with $K = 2^{2^{r+1}}$ and $y = \log n$. We have $n/(2^{7 \cdot 2^r})$ nodes at level $7 \cdot 2^r$, so to store all structures at this level the total number of bits of space used is

$$O\left(\frac{n}{2^{7 \cdot 2^r}}(w + 2^{2^{r+1}}(\log n)^2)\right) = O\left(n\frac{w}{2^r} + \frac{n}{2^{5 \cdot 2^r}}(\log n)^2\right)$$

$$= O\left(n\frac{w}{2^r} + \frac{n}{\lfloor \log n/(2\log\log n)^2 \rfloor^{5/2}}(\log n)^2\right) = O\left(n\frac{w}{2^r}\right) ,$$

where we used that $\lfloor \log n/(2\log\log n)^2 \rfloor < k \le 2^{2^{r+1}}$. By Lemma 3 a top query takes $O(k)$ time.

Middle query: A middle query is a query on a node z at level 2^r for $\mathcal{R}(z, \ell)$ (or $\mathcal{L}(z, \ell)$) with $2^r < \ell \le 7 \cdot 2^r$. For all nodes x at level 2^r, let $\min_x = \min C_x$. The idea in answering middle queries is as follows. Suppose we could find the nodes at level 2^r corresponding to the up to k smallest $\min_x$ values within the query range. To answer a middle query, we would only need to extract the k overall smallest elements from the up to k corresponding sorted S_x arrays of the nodes, we just found. The insight is that both subproblems mentioned can be solved using Theorem 1 as the key part. Once we have the k smallest elements in the middle query range, all that remains is to sort them.

We describe a solution in line with the above idea. For each node x at levels 2^r to $7 \cdot 2^r$, we have a sorted array $\mathcal{M}_x^r$ of all nodes x' at level 2^r in $\mathcal{T}_x$ sorted with respect to the $\min_{x'}$ values. To store the $\mathcal{M}_x^r$ arrays for all x, the space required is $O(\frac{n}{2^{2^r}} \cdot 6 \cdot 2^r) = O(\frac{n}{2^r})$ words (i.e. $O(n\frac{w}{2^r}$ bits), since we have $\frac{n}{2^{2^r}}$ nodes at level 2^r, and each such node will appear $7 \cdot 2^r - 2^r = 6 \cdot 2^r$ times in an $\mathcal{M}_x^r$ array (and to store the index of a node we use a word).

To answer a middle query for the k smallest elements in $\mathcal{R}(z, \ell)$, we walk $\ell - 2^r$ levels up from z while collecting the $\mathcal{M}_x^r$ arrays for the nodes x whose canonical subset is a part of the query range (at most $6 \cdot 2^r$ arrays since we collect at most one per level). Using Theorem 1 we select the k smallest elements from the $O(2^r)$ sorted arrays in $O(2^r + k) = O(k)$ time (note that there may not be k elements to select, so in reality we select up to k elements). This gives us the k smallest $\min_{x'}$ values of the nodes $x'_1, x'_2, \ldots, x'_k$ at level 2^r that are within the query range. Finally, we select the k overall smallest elements of the sorted arrays $S_{x'_1}, S_{x'_2}, \ldots, S_{x'_k}$ in $O(k)$ time using Theorem 1. This gives us the k smallest elements of $\mathcal{R}(z, \ell)$, but not in sorted order. We now show how to sort these elements in $O(k)$ time. For every leaf u, we store its local rank relative to $C_{u''}$, where u'' is the the ancestor of u at level $7 \cdot 2^r$. Since each subtree $\mathcal{T}_{u''}$ contains $2^{7 \cdot 2^r}$ leaves, we need $7 \cdot 2^r$ bits to index a leaf (relative to the leftmost leaf in $\mathcal{T}_{u''}$). We store all local rank labels of length $7 \cdot 2^r$ in a single array, and using Lemma 1 the space usage becomes $O(n2^r + w)$ bits. Given $O(k)$ leaves from C_x for a node x at level $7 \cdot 2^r$, we can use the local rank labellings of the leaves of length $7 \cdot 2^r$ bits to radix sort them in $O(k)$ time (for the analysis we use that $2^{2^r} < k$). This completes how to support queries.

3.5 Construction

In this section, we show how to build the data structures in Sections 3.3 and 3.4 in $O(n \log n)$ time using $O(n)$ extra words of space. The structures to be created for node x are a subset of the possible structures S_x, $\mathcal{M}_x^r$, $R_x[\ell]$, $L_x[\ell]$ (where ℓ is a level above x), and the local rank labellings. In total, the structures to be created store $O(n\frac{\log n}{\log\log n})$ elements which is dominated by the number of elements stored in the R_x and L_x structures for all leaves in Section 3.3. The general idea in the construction is to perform mergesort bottom up on $\mathcal{T}$ (level-by-level) starting at the leaves. The time spent on mergesort is $O(n \log n)$, and we use $O(n)$ words of space for the mergesort as we only store the sorted lists for the current and previous level. Note that when visiting a node x during mergesort the set C_x has been sorted, i.e. we have computed the array S_x. The structures S_x and $\mathcal{M}_x^r$ will be constructed while visiting x during the traversal of $\mathcal{T}$, while $R_x[\ell]$ and $L_x[\ell]$ will be constructed at the ancestor of x at level ℓ. As soon as a set has been computed, we store it in the data structure, possibly in a packed manner. For the structures in Section 3.3, when visiting a node x at level $\ell \leq 2\lfloor \log\log n \rfloor$ we compute for each leaf z in the right subtree of x the structure $R_z[\ell] = R_z[\ell - 1]$ (where $R_z[0] = \emptyset$), and the structure $L_z[\ell]$ containing the (up to) $\lfloor \log n/(2\log\log n)^2 \rfloor$ smallest elements in sorted order of $L_z[\ell - 1] \cup S_{2x}$. Both structures can be computed in time $O(\lfloor \log n/(2\log\log n)^2 \rfloor)$. Symmetrically, we compute the same structures for all leaves z in the left subtree of x. In the case that x is at level $\ell > 2\lfloor \log\log n \rfloor$, we compute for each node z at level $2\lfloor \log\log n \rfloor$

in the right subtree of x the structure $R_z[\ell] = R_z[\ell-1]$ (where $R_z[2\lfloor\log\log n\rfloor] = \emptyset$), and the structure $L_z[\ell]$ containing the $\lfloor \log n/(2\log\log n)^2 \rfloor$ smallest elements in sorted order of $L_z[\ell-1] \cup S_{2x}$. Both structures can be computed in time $O(\lfloor \log n/(2\log\log n)^2 \rfloor)$. Symmetrically, we compute the same structures for all nodes z at level $2\lfloor\log\log n\rfloor$ in the left subtree of x. For the structures in Section 3.4, when visiting a node x we first decide in $O(\log\log n)$ time if we need to compute any structures at x for any r. In the case that x is a node at level 2^r, we store $S_x = C_x$ and $\mathcal{M}_x^r = \min C_x$. For x at level $2^r < \ell \leq 7 \cdot 2^r$ we store $\mathcal{M}_x^r = \mathcal{M}_{2x}^r \cup \mathcal{M}_{2x+1}^r$. This can be computed in time linear in the size of $\mathcal{M}_x^r$. In the case that x is a node at level $7 \cdot 2^r$, we store the local rank labelling for each leaf in $\mathcal{T}_x$ using the sorted C_x list. For x at level $\ell > 7 \cdot 2^r$, we compute for each z at level $7 \cdot 2^r$ in the right subtree of x the structure $R_z[\ell] = R_z[\ell-1]$ (where $R_z[7 \cdot 2^r] = \emptyset$), and the structure $L_z[\ell]$ containing the $2^{2^{r+1}}$ smallest elements in sorted order of $L_z[\ell-1] \cup S_{2x}$. Both structures can be computed in time $O(2^{2^{r+1}})$. Symmetrically, we compute the same structures for all nodes z at level $7 \cdot 2^r$ in the left subtree of x. Since all structures can be computed in time linear in the size and that we have $O(n\frac{\log n}{\log\log n})$ elements in total, the overall construction time becomes $O(n \log n)$.

4 Online Sorted Range Reporting

We now describe how to extend the solution for the sorted range selection problem from Section 3 to a solution for the online sorted range reporting problem. We solve the problem by performing a sequence of sorted range selection queries Q_y with indices i and j and $k = 2^y$ for $y = 0, 1, 2, \ldots$. The initial query to the range $A[i..j]$ is Q_0. Each time we report an element from the current query Q_y, we spend $O(1)$ time building part of the next query Q_{y+1} so that when we have exhausted Q_y, we will have finished building Q_{y+1}. Since we report the 2^{y-1} largest elements in Q_y (the 2^{y-1} smallest are reported for $Q_0, Q_1, \ldots, Q_{y-1}$), we can distribute the $O(2^{y+1})$ computation time of Q_{y+1} over the 2^{y-1} reportings from Q_y. Hence the query time becomes $O(1)$ worst-case per element reported.

References

1. Page, L., Brin, S., Motwani, R., Winograd, T.: The PageRank citation ranking: Bringing order to the web. Technical Report 1999-66, Stanford InfoLab (1999)
2. Harel, D., Tarjan, R.E.: Fast algorithms for finding nearest common ancestors. SIAM Journal on Computing 13(2), 338–355 (1984)
3. Fredman, M.L., Willard, D.E.: Surpassing the information theoretic bound with fusion trees. J. Comput. Syst. Sci. 47(3), 424–436 (1993)
4. Frederickson, G.N., Johnson, D.B.: The complexity of selection and ranking in $X+Y$ and matrices with sorted columns. J. Comput. Syst. Sci. 24(2), 197–208 (1982)
5. Vuillemin, J.: A unifying look at data structures. CACM 23(4), 229–239 (1980)
6. Frederickson, G.N.: An optimal algorithm for selection in a min-heap. Inf. Comput. 104(2), 197–214 (1993)

Data Structures for Approximate Orthogonal Range Counting

Yakov Nekrich

Dept. of Computer Science
University of Bonn
yasha@cs.uni-bonn.de

Abstract. We present new data structures for approximately counting the number of points in an orthogonal range. There is a deterministic linear space data structure that supports updates in $O(1)$ time and approximates the number of elements in a 1-D range up to an additive term $k^{1/c}$ in $O(\log\log U \cdot \log\log n)$ time, where k is the number of elements in the answer, U is the size of the universe and c is an arbitrary fixed constant. We can estimate the number of points in a two-dimensional orthogonal range up to an additive term k^{ρ} in $O(\log\log U + (1/\rho)\log\log n)$ time for any $\rho > 0$. We can estimate the number of points in a three-dimensional orthogonal range up to an additive term k^{ρ} in $O(\log\log U + (\log\log n)^3 + (3^v)\log\log n)$ time for $v = \log\frac{1}{\rho}/\log\frac{3}{2} + 2$.

1 Introduction

Range reporting and range counting are two variants of the range searching problem. In the range counting problem, the data structure returns the number of points in an arbitrary query range. In the range reporting problem the data structure reports all points in the query range. Both variants were studied extensively and in many cases we know the matching upper and lower bounds for those problems for dimension $d \leq 4$. Answering an orthogonal range counting query takes more time than answering the orthogonal range reporting query in the same dimension. This gap cannot be closed because of the lower bounds for the range counting queries: while range reporting queries can be answered in constant time in one dimension and in almost-constant time in two and three dimensions (if the universe size is not too big)[1], range counting queries take super-constant time in one dimension and poly-logarithmic time in two and three dimensions.

Approximate range counting queries help us bridge the gap between range reporting and counting: instead of exactly counting the number of points (elements) in the query range, the data structure provides a good estimation. There are data structures that approximate the number of points in a one-dimensional interval [4,19] or in a halfspace [7], [15], [2], [8] up to a constant

[1] For simplicity, we consider only emptiness queries. In other words, we ignore the time needed to output the points in the answer: if range reporting data structure supports queries in $O(f(n) + k)$ time, we simply say that the query time is $O(f(n))$.

Y. Dong, D.-Z. Du, and O. Ibarra (Eds.): ISAAC 2009, LNCS 5878, pp. 183–192, 2009.

factor: given a query Q, the data structure returns the number k' such that $(1-\varepsilon)k \leq k' \leq (1+\varepsilon)k$, where k is the exact number of points in the answer and ε is an arbitrarily small positive constant. In this paper we consider the following new variant of approximate range counting: If k is the number of points in the answer, the answer to a query Q is an integer k' such that $k - \varepsilon k^{\alpha} \leq k' \leq k + \varepsilon k^{\alpha}$ for some constant $\alpha < 1$. Thus we obtain better estimation for the number of points in the answer for large (superconstant) values of k. On the other hand, if the range Q is empty, then $k' = 0$. We present data structures that approximate the number of points in a d-dimensional orthogonal range for $d = 2, 3$. We also describe a dynamic one-dimensional data structure.

Dynamic 1-D Data Structure. A static data structure that answers 1-D reporting queries in $O(1)$ time is described in [4]. In [4] the authors also describe a static data structure that approximates the number of points in a 1-D range up to an arbitrary constant factor in constant time. Pătraşcu and Demaine [24] show that any dynamic data structure with polylogarithmic update time needs $\Omega(\log n/\log\log U)$ time to answer an exact range counting query; henceforth U denotes the size of the universe. The dynamic randomized data structure of Mortensen [19] supports approximate range counting queries in $O(1)$ time and updates in $O(\log^{\varepsilon} U)$ time; see [19] for other trade-offs between query and update times. In this paper we present a new result on approximate range counting in 1-D:

- There is a deterministic data structure that can answer one-dimensional approximate range counting queries using the best known data structure for predecessor queries, i.e. dynamic data structure supports range reporting queries in $O(dpred(n, U))$ time, where $dpred(n, U)$ is the time to answer a predecessor query in the dynamic setting; currently $dpred(n, U) = O(\min(\log\log U \cdot \log\log n, \sqrt{\log n/\log\log n}))$ [6]. We show that we can approximate the number of points in the query range up to an additive factor $k^{1/c}$, where k is the number of points in the answer and c is an arbitrary constant, in $O(dpred(n, U))$ time. We thus significantly improve the precision of the estimation; the query time is still much less than the lower bound for the exact counting queries in the dynamic scenario.

Using the standard techniques, we can extend the results for one-dimensional approximate range counting to an arbitrary constant dimension d. There is a data structure that approximates the number of points in a d-dimensional range up to an additive term k^c for any $c > 0$ in $O(\log\log n(\log n/\log\log n)^{d-1})$ time and supports updates in $O(\log^{d-1+\varepsilon} n)$ time. For comparison, the fastest known dynamic data structure [18] supports emptiness queries in $O((\log n/\log\log n)^{d-1})$ time. Dynamic data structures are described in section 2.

Approximate Range Counting in 2-D and 3-D. We match or almost match the best upper bounds for 2-D and 3-D emptiness queries. Best data structures for exact range counting in 2-D and 3-D support queries in $O(\log n/\log\log n)$ and $O((\log n/\log\log n)^2)$ time respectively [14].

– If all point coordinates do not exceed n, we can approximate the number of points in a two-dimensional query rectangle up to an additive term k^ρ for an arbitrary parameter ρ, $0 < \rho < 1$, in $O((1/\rho)\log\log n)$ time.
– If all point coordinates do not exceed n, we can approximate the number of points in three-dimensional query rectangle up to an additive term k^ρ in $O((\log\log n)^3 + (3^v)\log\log n)$ time for an arbitrary parameter ρ, $0 < \rho < 1$, and $v = \log\frac{1}{\rho}/\log\frac{3}{2} + 2$.

The parameter ρ is not fixed in advance, i.e. the same data structures can be used for answering queries with arbitrary precision. If point coordinates are arbitrary integers, then the query time of the above data structures increases by an additive term $O(\min(\log\log U, \sqrt{\log n/\log\log n}))$. Data structure for range counting in 2-D and 3-D are described in section 3. In section 3.1 we describe space-efficient variants of two- and three-dimensional data structures that estimate the number of points in a range up to an additive error k^c for some fixed constant c.

Our results for approximate range counting queries are valid in the word RAM model. Throughout this paper ε denotes an arbitrarily small constant.

2 Dynamic Approximate Range Counting

We show that in the dynamic scenario answering one-dimensional counting queries with an additive error $k^{1/c}$ can be performed as efficiently as answering predecessor queries. The best known deterministic data structure supports one-dimensional emptiness queries in $O(dpred(n,U))$ time, where $dpred(n,U) = \min(\sqrt{\log n/\log\log n}, \log\log U \cdot \log\log n)$ is the time needed to answer a predecessor query in dynamic scenario [5], [6].

Theorem 1. *For any fixed constant $c > 1$, there exists a linear space data structure that supports approximate range counting queries with additive error $k^{1/c}$ in $O(dpred(n,U))$ time, deletions in $O(\log\log n)$ amortized time, and insertions in $O(dpred(n,U))$ amortized time.*

Proof: First we observe that if the query interval contains less than $(\log\log n)^c$ points for an arbitrary constant c, $k = |P \cap [a,b]| \le (\log\log n)^c$, then we can use a simple modification of the standard binary tree solution: the set P is divided into groups of $(\log\log n)^c$ consecutive elements, i.e., $|G_i| = (\log\log n)^c$ and every element in G_i is smaller than any element in G_{i+1}. Using a dynamic data structure for predecessor queries we can find in $O((dpred(n,U))$ time the successor a' of a in P and the predecessor b' of b in P. If a and b belong to the same group G_i, then we can count elements in $[a,b]$ in $O(\log\log\log n)$ time using the standard binary range tree solution. If a' and b' belong to two consecutive groups G_i and G_{i+1}, then we count the number of elements $e \in G_i$, $e \ge a$, and the number of elements $e' \in G_{i+1}$, $e' \le b$. If a' belongs to a group G_i and b' belongs to a group G_j so that $j > i+1$, then $[a,b]$ contains more than $(\log\log n)^c$ elements. We also assume w.l.o.g. that $c > 2$.

We maintain the exponential tree [5], [6] for the set P. The root node has $\Theta(n^{1/c})$ children, so that each child node contains between $n^{(c-1)/c}/2$ and

$2n^{(c-1)/c}$ points from P. In a general case, if a node v contains n_v points of P, then node v has $\Theta(n_v^{1/c})$ children, so that each child contains between $n_v^{(c-1)/c}/2$ and $2n_v^{(c-1)/c}$ points from P. The exponential tree can be maintained as described in [5], so that insertions and deletions are supported in $O(\log\log n)$ time. Additionally in every node v we store the approximate number of elements in any consecutive sequence of children of v, denoted by $c_v(i,j)$: for any $i < j$, $n_{v_i} + n_{v_{i+1}} + \ldots + n_{v_j} - n_v^{3/c}/2 \le c_v(i,j) \le n_{v_i} + n_{v_{i+1}} + \ldots + n_{v_j} + n_v^{3/c}/2$. When $n_v^{3/c}/2$ elements are inserted into a node v or deleted from v, we set $c_v(i,j) = n_{v_i} + n_{v_{i+1}} + \ldots + n_{v_j}$ for all $i < j$. Recomputing $c_v(i,j)$ for a node v takes $O(n_v^{2/c})$ time. Since insertion or deletion results in incrementing or decrementing the value of n_v in $O(\log\log n)$ nodes v, recomputing $c_v(i,j)$ incurs an amortized cost $O(\log\log n)$. Thus amortized cost of a delete operation is $O(\log\log n)$. When we insert a new point, we also have to find its position in the exponential tree; therefore an insertion takes $O(dpred(n,U))$ time.

We store $O(n_v^{2/c})$ auxiliary values in each node v; hence, we can show that the space usage is $O(n)$ in exactly the same way as in [5,6].

Given an interval $[a,b]$, we find $b' = \text{pred}(b,P)$ and $a' = \text{succ}(a,P)$ and identify the leaves of the exponential tree in which they are stored. The lowest common ancestor q of those leaves can be found in $O(\log\log n)$ time because the height of the tree is $O(\log\log n)$. If a' and b' are stored in the i-th and the j-th children of q and $i+1 < j$, then all elements stored in $q_{i+1}, \ldots, q_{j-1}$ belong to $[a,b]$ and we initialize a variable *count* to $c_v(i+1, j-1)$. Otherwise *count* is set to 0. Then, we traverse the path from q to a' and in every visited node v we increment *count* by $c_v(i_v+1, r_v)$, such that a' is in the i_v-th child of v, and r_v is the total number of v's children. Finally, we traverse the path from q to b' and in every visited node v we increment *count* by $c_v(1, i_v - 1)$, such that b' is in the i_v-th child of v, Suppose that the variable *count* was incremented by $s_v > 0$ when a node v was visited. Let k_v be the exact number of elements in all children of v whose ranges are entirely contained in v. Then, $k_v - n_v^{3/c} \le s_v \le k_v + n_v^{3/c}$. Since $k_v \ge n_v^{(c-1)/c}$, $k_v - k_v^{3/(c-1)} \le s_v \le k_v + k_v^{3/(c-1)}$. Clearly, the total number of points equals to the sum of k_v for all visited nodes v. The search procedure visits less than $c_h \log\log n$ nodes for a constant c_h. Hence, $k - k^{3/(c-1)} \log\log n \le count \le k + k^{3/(c-1)} \log\log n$ for $k = |P \cap [a,b]|$. Since $\log\log n \le k^{1/(c-1)}$, $k - k^{4/(c-1)} \le count \le k + k^{4/(c-1)}$. We obtain the result of the Theorem by replacing c with $c' = \max(5c, 5)$ in the above proof. □

Our dynamic data structure can be extended to d dimensions using the standard range tree [10].

Theorem 2. *For any fixed constant $c > 1$, there exists a data structure that supports d-dimensional approximate range counting queries with additive error $k^{1/c}$ in $O(\log\log n(\log n/\log\log n)^{d-1})$ time and updates in $O(\log^{d-1+\varepsilon} n)$ amortized time.*

Proof: This result can be obtained by combining the standard range tree technique (node degree in a range tree is $O(\log^{\varepsilon'} n)$ for an appropriate constant $\varepsilon' = \varepsilon/(d-1)$) with the data structure for one-dimensional approximate range counting of Theorem 1. Details will be given in the full version of this paper. □

3 Approximate Range Counting in 2-D and 3-D

A point p dominates a point q if each coordinate of p is greater than or equal to the corresponding coordinate of q. The goal of the (approximate) dominance counting query is to (approximately) count the number of points in P that dominate q. The dominance query is equivalent to the orthogonal range query with a restriction that query range Q is a product of half-open intervals. We start this section with a description of the data structure that estimates the number of points in the answer to a 2-D dominance query up to a constant factor. We can obtain a data structure for general orthogonal range counting queries using a standard technique. Then, we show that queries can be answered with higher precision without increasing the query time. Finally, we describe a data structure for approximate range counting in 3-D. For simplicity, we only consider the case when all point coordinates are bounded by n. We can obtain the results for the case of arbitrarily large point coordinates by a standard reduction to rank space technique [13]: the space usage remains linear and the query time increases by $pred(n, U)$ - the time needed to answer a static predecessor query.

Theorem 3. *There exists a linear space data structure that answers approximate two-dimensional dominance range counting queries on $n \times n$ grid in $O(\log \log n)$ time.*

A t-approximate boundary, introduced by Vengroff and Vitter [26] is a polyline $\mathcal{M}$ consisting of $O(n/t)$ axis-parallel segments that partitions the space[2], so that every point $\mathcal{M}$ is dominated by at most $2t$ and at least t points of P. This notion can be straightforwardly extended to a t_α-boundary $\mathcal{M}_\alpha$: $\mathcal{M}_\alpha$ partitions the space into two parts, and every point $\mathcal{M}_\alpha$ is dominated by at most $\alpha \cdot t$ and at least t points of P. We can construct a t_α-boundary with the same algorithm as in [26]. Let p be a point with coordinates $(0, 0)$. We move p in the positive x direction until p is dominated by at most αt points. Then, we repeat the following steps until the x-coordinate of p equals to 0: a) move p in $+y$ direction as long as p is dominated by more than t points of P b) move p in the $-x$ direction until p is dominated by αt points of P. The path traced by p is a t_α-boundary. *Inward corners* are formed when we move p in $+y$ direction, i.e. inward corners mark the beginning of step a) resp. the end of step b). Inward corners of $\mathcal{M}$ have a property that no point of $\mathcal{M}$ is strictly dominated by an inward corner and for every point $m \in \mathcal{M}$ that is not an inward corner, there is an inward corner m_i dominated by m. There are $O(n/t)$ inward corners in a t_α-approximate boundary because for every inward corner $c = (c_x, c_y)$ there are $(\alpha - 1)t$ points that dominate c and

[2] In this section we assume that all points have positive coordinates.

do not dominate inward corners whose x-coordinates are larger than c_x. Our data structure consists of $\log_\alpha n$ t_α-approximate boundaries $\mathcal{M}_1, \mathcal{M}_2, \ldots, \mathcal{M}_s$ such that $\mathcal{M}_i$ is an α^i-approximate boundary of P, i.e. every point on $\mathcal{M}_i$ is dominated by at least α^i and at most α^{i+1} points of P. If a point $p \in \mathcal{M}_i$ is dominated by a query point q, then q is dominated by at most α^{i+1} points of P. If q dominates a point on $\mathcal{M}_i$, then it also dominates an inward corner of $\mathcal{M}_i$. Hence, we can estimate the number of points that dominate q up to a constant α by finding the minimal index j such that q dominates an inward corner of $\mathcal{M}_j$. Since q is dominated by a point of $\mathcal{M}_{j-1}$, q is dominated by $k \geq \alpha^{j-1}$ points of P. On the other hand, $k \leq \alpha^{j+1}$ because a point of $\mathcal{M}_j$ is dominated by q.

We can store inward corners of all boundaries $\mathcal{M}_i$ in a linear space data structure so that for any point q the minimal index j, such that some point on $\mathcal{M}_j$ is dominated by q, can be found in $O(\log\log n)$ time. We denote by $\mathrm{pred}_x(a, S)$ the point $p = (p_x, p_y) \in S$, such that $p_x = \mathrm{pred}(a, S_x)$ where S_x is the set of x-coordinates of all points in S. For simplicity, we sometimes do not distinguish between a boundary $\mathcal{M}_i$ and the set of its inward corners. Let $q = (q_x, q_y)$. Let $c_i = (c_x, c_y)$ be the inward corner on a boundary $\mathcal{M}_i$ whose x-coordinate c_x precedes q_x, $c_i = \mathrm{pred}_x(q_x, \mathcal{M}_i)$. For any other inward corner $c'_i = (c'_x, c'_y)$ on $\mathcal{M}_i$, $c'_y > c_y$ if and only if $c'_x < c_x$ because the y-coordinates of inward corners decrease monotonously as their x-coordinates increase. Hence, q dominates a point on $\mathcal{M}_i$ if and only if $q_y \geq c_y$. Thus given a query point q, it suffices to identify the minimal index j, such that the y-coordinate of the inward corner $c_j \in \mathcal{M}_j$ that precedes q_x is smaller than or equal to q_y. The x-axis is subdivided into intervals of size $\log n$. For each interval I_s the list L_s contains indexes of boundaries $\mathcal{M}_i$ such that the x-coordinate of at least one inward corner of $\mathcal{M}_i$ belongs to I_s. For a query point q with $q_x \in I_s$ and for every $j \in L_s$, we can find the inward corner preceding q_x with respect to its x-coordinate, $\mathrm{pred}_x(q_x, \mathcal{M}_j)$, in $O(1)$ time because x-coordinates of all relevant inward corners belong to an interval of size $\log n$. Hence, we can find the minimal index $j_s \in L_s$, such that q dominates a point on $\mathcal{M}_{j_s}$ in $O(\log\log n)$ time by binary search among indexes in L_s. For the left bound a_s of an interval $I_s = [a_s, b_s]$ and for all indexes $j = 1, \ldots, \log_\alpha n$, the list A_s contains the inward corner c_j, such that $c_j = \mathrm{pred}_x(a_s, \mathcal{M}_j)$. By binary search in A_s we can find the minimal j_a such that q dominates the inward corner $c_{j_a} \in A_s$. Clearly $j = \min(j_a, j_s)$ is the minimal index of a boundary dominated by q.

Theorem 4. *There exists a $O(n \log^2 n)$ space data structure that supports two-dimensional approximate range counting queries on $n \times n$ grid in $O(\log\log n)$ time.*

The next Lemma will enable us to obtain a better estimation of the number of points.

Lemma 1. *There exists a $O(n \log n)$ space data structure that supports two-dimensional approximate range counting queries on $n \times n$ grid with an additive error n^ρ in $O((1/\rho)\log\log n)$ time for any ρ, $0 < \rho < 1$.*

Proof: We divide the grid into x-slabs $X_i = [x_{i-1}, x_i] \times [1, n]$ and y-slabs $Y_j = [1, n] \times [y_{j-1}, y_j]$, so that each slab contains $n^{1/2}$ points. For every point (x_i, y_j), $0 \le i, j, \le n^{1/2}$ we store the number of points in P that dominate it. There is also a recursively defined data structure for each slab. The total space usage is $s(n) = O(n) + 2n^{1/2}s(n^{1/2})$ and $s(n) = O(n \log n)$.

We can easily obtain an approximation with additive error $2n^{1/2}$ using the first level data structure: for a query $q = (q_x, q_y)$ we identify the indexes i and j, such that $x_{i-1} \le q_x \le x_i$ and $y_{j-1} \le q_y \le y_j$, i.e. we identify the x-slab X_i and the y-slab Y_j that contain q. Indexes i and j can be found in $O(\log \log n)$ time. Let $c(x, y)$ be the number of points that dominate a point $p = (x, y)$; let $c(x, y, X_i)$ $(c(x, y, Y_j))$ be the number of points in the slab X_i (Y_j) that dominate $p = (x, y)$. Then $c(q_x, q_y) = c(x_i, y_j) + c(x_i, q_y, Y_j) + c(q_x, q_y, X_i)$. Since $c(x_i, q_y, Y_j) \le n^{1/2}$ and $c(q_x, q_y, X_i) \le n^{1/2}$, the value of $c(x_i, y_j)$ is an approximation of $c(q_x, q_y)$ with an additive error $2n^{1/2}$. Using recursive data structures for slabs X_i and Y_j we can estimate $c(q_x, q_y, X_i)$ and $c(x_i, q_y, Y_j)$ with an additive error $2n^{1/4}$ and estimate $c(q_x, q_y)$ with an additive error $4n^{1/4}$. If the recursion depth is v (i.e. if we apply recursion v times), then the total number of recursive calls is $O(2^v)$ and we obtain in $O((2^v) \log \log n)$ time an approximation with additive error $2^v \cdot n^{1/2^v}$ for any positive integer v.

We set recursion depth $v = \lceil \log(1/\rho) \rceil + 2$. Then, $v + (1/2^v) \log n \le (\rho/4) \log n + \log(1/\rho) = (\rho/4 + \frac{\log(1/\rho)}{\log n}) \log n < \rho \log n$. Hence, $n^\rho > 2^v n^{1/2^v}$. Therefore, if recursion depth is set to v, then our data structure provides an answer with additive error n^ρ. □

Theorem 5. *There exists a $O(n \log^2 n)$ space data structure that supports two-dimensional dominance counting queries on $n \times n$ grid with an additive error k^ρ for an arbitrary parameter ρ, $0 < \rho < 1$, in $O((1/\rho) \log \log n)$ time.*
There exists a $O(n \log^4 n)$ space data structure that supports two-dimensional range counting queries on $n \times n$ grid with an additive error k^ρ for an arbitrary parameter ρ, $0 < \rho < 1$, in $O((1/\rho) \log \log n)$ time.

Proof: As in Theorem 3 we construct t-boundaries $\mathcal{M}_1, \ldots, \mathcal{M}_{\log n}$, such that M_i is a 2^i-approximate boundary, i.e. each point on $\mathcal{M}_i$ is dominated by at least 2^i and at most 2^{2i} points of P. For each inward corner $c_{i,j}$ of every M_j, we store a data structure $D_{i,j}$ that contains all points that dominate $c_{i,j}$ and supports approximate counting queries as described in Lemma 1. For a fixed j, there are $O(\frac{n}{2^j})$ data structures $D_{i,j}$, and each $D_{i,j}$ contains $O(2^j)$ points. Hence, all data structures $D_{i,j}$ use $O(n \log^2 n)$ space.

As described in Theorem 4, we can find in $O(\log \log n)$ time the minimal index j, such that $\mathcal{M}_j$ is dominated by the query point q and an inward corner $c_{i,j} \in \mathcal{M}_j$ dominated by q. Then, we use the data structure $D_{i,j}$ to obtain a better approximation. Since $D_{i,j}$ contains $O(k)$ points, by Lemma 1 $D_{i,j}$ estimates the number of points that dominate q with an additive error k^ρ in $O((1/\rho) \log \log n)$ time. We can extend the result for dominance counting to the general three-dimensional counting using the standard technique from range reporting [12,25]; see also the proof of Theorem 4. □

Lemma 2. *There exists a* $O(n \log^3 n)$ *space data structure that supports three-dimensional approximate range counting queries on* $n \times n \times n$ *grid with an additive error* n^ρ *in* $O(3^v \log\log n)$ *time for any* ρ, $0 < \rho < 1$, *and for* $v = \log\frac{1}{\rho}/\log\frac{3}{2} + 2$.

Proof: We divide the grid into x-, y-, and z-slabs, $X_i = [x_{i-1}, x_i] \times [1, n] \times [1, n]$, $Y_j = [1, n] \times [y_{j-1}, y_j] \times [1, n]$, $Z_d = [1, n] \times [1, n] \times [z_{d-1}, z_d]$, so that each slab contains $n^{2/3}$ points. For each point (x_i, y_j, z_d) we store the number of points in P that dominate it. There is also a recursively defined data structure for each slab. The total space usage is $s(n) = O(n) + 3n^{1/3}s(n^{2/3})$ and $s(n) = O(n \log^3 n)$.

For a query $q = (q_x, q_y, q_z)$ we identify the x-, y-, and z-slabs X_i, Y_j, and Z_d that contain q. By the same argument as in Lemma 1, the number of points that dominate (x_i, y_j, z_d) differs from the number of points that dominate q by at most $3n^{2/3}$. We can estimate the number of points that dominate q and belong to one of the slabs X_i, Y_j, and Z_d using recursively defined data structures. If the recursion depth is v, then we obtain in $O(3^v \log\log n)$ time an approximation with additive error $3^v \cdot n^{(2/3)^v}$ for any positive integer v. The result of the Lemma follows if we set $v = \log\frac{1}{\rho}/\log\frac{3}{2} + 2$. □

Theorem 6. *There exists a* $O(n \log^4 n)$ *space data structure that supports approximate dominance range counting queries on* $n \times n \times n$ *grid with an additive error* k^ρ *in* $O((\log\log n)^3 + 3^v \log\log n)$ *time for any* ρ, $0 < \rho < 1$, *and for* $v = \log\frac{1}{\rho}/\log\frac{3}{2} + 2$.

There exists a $O(n \log^7 n)$ *space data structure that supports approximate range counting queries on* $n \times n \times n$ *grid with an additive error* k^ρ *in* $O((\log\log n)^3 + 3^v \log\log n)$ *time for any* ρ, $0 < \rho < 1$, *and for* $v = \log\frac{1}{\rho}/\log\frac{3}{2} + 2$.

Proof: Instead of counting points that dominate q we count points dominated by q. Both types of queries are equivalent. Hence, the data structure of Lemma 2 can be used to approximately count points dominated by q.

A downward corner of a point p consists of all points dominated by p. We define an approximate t-level as a set of downward corners $\mathcal{L}$, such that (1) any point p that dominates at most t points of P is contained in some $r \in \mathcal{L}$ (2) any downward corner $r \in \mathcal{L}$ contains at most $\alpha \cdot t$ points of P. Afshani [1] showed that for an arbitrary constant α there exists an approximate t-level of size $O(\frac{n}{t})$. We can assume that no $r \in \mathcal{L}$ dominates $r' \in \mathcal{L}$ in an approximate t-level $\mathcal{L}$: if r dominates r', then the downward corner r' can be removed from $\mathcal{L}$. Identifying an inward corner $r \in \mathcal{L}$ that dominates a query point q (or answering that no $r \in \mathcal{L}$ dominates q) is equivalent to answering a point location query in a rectangular planar subdivision [26,21] and takes $O((\log\log n)^2)$ time.

Our data structure consists of approximate levels $\mathcal{M}_1, \mathcal{M}_2, \ldots, \mathcal{M}_{\log n}$, such that $\mathcal{M}_i$ is a 2^i-approximate level and the constant α is chosen to be 2. For every downward corner $r_{i,j} \in \mathcal{M}_j$, we store all points dominated by $r_{i,j}$ in a data structure $D_{i,j}$; $D_{i,j}$ contains $O(2^j)$ points and supports counting queries with additive error $O(2^{\rho j})$ by Lemma 2. All data structures $D_{i,j}$ use $O(n \log^4 n)$ space.

We can find a minimal j, such that $\mathcal{M}_j$ dominates q in $O((\log\log n)^3)$ time by binary search. Let $r_{i,j}$ be the downward corner that dominates q. We can use the data structure $D_{i,j}$ to estimate the number of points that are dominated by q with an additive error k^{ρ}; by Lemma 2 this takes $O(3^v \log\log n)$ time for $v = \log \frac{1}{\rho} / \log \frac{3}{2} + 2$.

We can extend the result for dominance counting to the general three-dimensional counting using the standard technique [12,25]; see also the proof of Theorem 4. □

3.1 Space-Efficient Approximate Range Counting in 2-D and 3-D

If we are interested in counting with an additive error k^c for some predefined constant $c > 0$, then the space usage can be significantly reduced. The two-dimensional data structure uses $O(n \log^2 n)$ space ($O(n)$ space for dominance counting), and the three-dimensional data structure uses $O(n \log^3 n)$ space ($O(n)$ space for dominance counting). The main idea of our improvement is that in the construction of Lemma 1 (resp. Lemma 2) each slab contains $n^{1/2+\varepsilon}$ points ($n^{2/3+\varepsilon}$ points) for some $\varepsilon > 0$ and there is a constant number of recursion levels. Proofs of Theorem 7 and Theorem 8 can be found in the full version of this paper [22].

Theorem 7. *For any fixed constant $c < 1$, there exists a $O(n)$ space data structure that supports two-dimensional dominance counting queries on $n \times n$ grid with an additive error k^c in $O(\log\log n)$ time.*

For any fixed constant $c < 1$, there exists a $O(n \log^2 n)$ space data structure that supports two-dimensional range counting queries on $n \times n$ grid with an additive error k^c in $O(\log\log n)$ time.

Theorem 8. *For any fixed constant $c < 1$, there exists a $O(n)$ space data structure that supports approximate dominance range counting queries on $n \times n \times n$ grid with an additive error k^c in $O((\log\log n)^3)$ time.*

For any fixed constant $c < 1$, there exists a $O(n \log^4 n)$ space data structure that supports approximate range counting queries on $n \times n \times n$ grid with an additive error k^c in $O((\log\log n)^3)$ time.

Acknowledgment

We would like to thank an anonymous reviewer of the previous version of this paper for stimulating suggestions that helped us improve some of our results.

References

1. Afshani, P.: On Dominance Reporting in 3D. In: Halperin, D., Mehlhorn, K. (eds.) ESA 2008. LNCS, vol. 5193, pp. 41–51. Springer, Heidelberg (2008)
2. Afshani, P., Chan, T.M.: On Approximate Range Counting and Depth. In: Proc. SoCG 2007, pp. 337–343 (2007)
3. Alstrup, S., Brodal, G.S., Rauhe, T.: New Data Structures for Orthogonal Range Searching. In: Proc. FOCS, pp. 198–207 (2000)

4. Alstrup, S., Brodal, G.S., Rauhe, T.: Optimal Static Range Reporting in One Dimension. In: Proc. STOC 2001, pp. 476–482 (2001)
5. Andersson, A.: Faster Deterministic Sorting and Searching in Linear Space. In: Proc. FOCS 1996, pp. 135–141 (1996)
6. Andersson, A., Thorup, M.: Dynamic Ordered Sets with Exponential Search Trees. J. ACM (JACM) 54(3), 13 (2007)
7. Aronov, B., Har-Peled, S.: On Approximating the Depth and Related Problems. SIAM J. Comput. 38(3), 899–921 (2008)
8. Aronov, B., Har-Peled, S., Sharir, M.: On Approximate Halfspace Range Counting and Relative Epsilon-Approximations. In: Proc. SoCG 2007, pp. 327–336 (2007)
9. Beame, P., Fich, F.E.: Optimal Bounds for the Predecessor Problem and Related Problems. J. Comput. Syst. Sci. 65(1), 38–72 (2002)
10. Bentley, J.L.: Multidimensional Divide-and-Conquer. Commun. ACM 23, 214–229 (1980)
11. de Berg, M., van Kreveld, M.J., Snoeyink, J.: Two- and Three-Dimensional Point Location in Rectangular Subdivisions. J. Algorithms 18(2), 256–277 (1995)
12. Chazelle, B., Guibas, L.J.: Fractional Cascading: I. A Data Structuring Technique. Algorithmica 1(2), 133–162 (1986)
13. Gabow, H., Bentley, J.L., Tarjan, R.E.: Scaling and Related Techniques for Geometry Problems. In: Proc. STOC 1984, pp. 135–143 (1984)
14. JaJa, J., Mortensen, C.W., Shi, Q.: Space-Efficient and Fast Algorithms for Multidimensional Dominance Reporting and Counting. In: Fleischer, R., Trippen, G. (eds.) ISAAC 2004. LNCS, vol. 3341, pp. 558–568. Springer, Heidelberg (2004)
15. Kaplan, H., Sharir, M.: Randomized Incremental Constructions of Three-dimensional Convex Hulls and Planar Voronoi Diagrams, and Approximate Range Counting. In: Proc. SODA 2006, pp. 484–493 (2006)
16. Matias, Y., Vitter, J.S., Young, N.E.: Approximate Data Structures with Applications. In: Proc. SODA 1994, pp. 187–194 (1994)
17. Miltersen, P.B., Nisan, N., Safra, S., Wigderson, A.: On Data Structures and Asymmetric Communication Complexity. J. Comput. Syst. Sci. 57(1), 37–49 (1998)
18. Mortensen, C.W.: Fully Dynamic Orthogonal Range Reporting on RAM. SIAM J. Comput. 35(6), 1494–1525 (2006)
19. Mortensen, C.W.: Data Structures for Orthogonal Intersection Searching and Other Problems, Ph.D. thesis (2006)
20. Mortensen, C.W., Pagh, R., Patrascu, M.: On Dynamic Range Reporting in One Dimension. In: Proc. STOC 2005, pp. 104–111 (2005)
21. Nekrich, Y.: A Data Structure for Multi-Dimensional Range Reporting. In: Proc. SoCG 2007, pp. 344–353 (2007)
22. Nekrich, Y.: Data Structures for Approximate Orthogonal Range Counting, arXiv:0906.2738 (2009)
23. Overmars, M.H.: Efficient Data Structures for Range Searching on a Grid. J. Algorithms 9(2), 254–275 (1988)
24. Patrascu, M., Demaine, E.D.: Logarithmic Lower Bounds in the Cell-Probe Model. SIAM J. Comput. 35(4), 932–963 (2006)
25. Subramanian, S., Ramaswamy, S.: The P-range Tree: A New Data Structure for Range Searching in Secondary Memory. In: Proc. SODA 1995, pp. 378–387 (1995)
26. Vengroff, D.E., Vitter, J.S.: Efficient 3-D Range Searching in External Memory. In: Proc. STOC 1996, pp. 192–201 (1996)

Dynamic 3-Sided Planar Range Queries with Expected Doubly Logarithmic Time

Gerth Stølting Brodal[1], Alexis C. Kaporis[2], Spyros Sioutas[3], Konstantinos Tsakalidis[1], and Kostas Tsichlas[4]

[1] MADALGO*, Department of Computer Science, Aarhus University, Denmark
{gerth,tsakalid}@madalgo.au.dk
[2] Computer Engineering and Informatics Department, University of Patras, Greece
kaporis@ceid.upatras.gr
[3] Department of Informatics, Ionian University, Corfu, Greece
sioutas@ionio.gr
[4] Department of Informatics, Aristotle University of Thessaloniki, Greece
tsichlas@csd.auth.gr

Abstract. We consider the problem of maintaining dynamically a set of points in the plane and supporting range queries of the type $[a, b] \times (-\infty, c]$. We assume that the inserted points have their x-coordinates drawn from a class of *smooth* distributions, whereas the y-coordinates are arbitrarily distributed. The points to be deleted are selected uniformly at random among the inserted points. For the RAM model, we present a linear space data structure that supports queries in $O(\log \log n + t)$ expected time with high probability and updates in $O(\log \log n)$ expected amortized time, where n is the number of points stored and t is the size of the output of the query. For the I/O model we support queries in $O(\log \log_B n + t/B)$ expected I/Os with high probability and updates in $O(\log_B \log n)$ expected amortized I/Os using linear space, where B is the disk block size. The data structures are deterministic and the expectation is with respect to the input distribution.

1 Introduction

We consider the dynamic 3-sided range reporting problem in the plane. That is, to design a data structure that supports insertions and deletions of points, and supports range reporting queries of the type $[a, b] \times (-\infty, c]$, i.e. report all points contained in the query rectangle with one side unbounded. The more general *orthogonal range searching* problem finds applications in databases and is used as a subroutine for solving general geometric problems. A survey can be found at [1]. In particular, multidimensional instances can be decomposed into two dimensional subproblems, where 3-sided queries are of major importance [2].

Previous results. In the internal memory, the most commonly used data structure for supporting 3-sided queries is the *priority search tree* of McCreight [3].

* Center for Massive Data Algorithmics, a Center of the Danish National Research Foundation.

Y. Dong, D.-Z. Du, and O. Ibarra (Eds.): ISAAC 2009, LNCS 5878, pp. 193–202, 2009.

It supports queries in $O(\log n + t)$ worst case time, insertions and deletions of points in $O(\log n)$ worst case time and uses linear space, where n is the number of points and t the size of the output of a query. It is a hybrid of a binary heap for the y-coordinates and of a balanced search tree for the x-coordinates.

In the RAM model, the only dynamic sublogarithmic bounds for this problem are due to Willard [4] who attains $O(\log n / \log\log n)$ worst case or $O(\sqrt{\log n})$ randomized update time and $O(\log n / \log\log n + t)$ query time using linear space. In the I/O model, Arge et al. [5] proposed an indexing scheme that consumes $O(n/B)$ space, supports updates in $O(\log_B n)$ amortized I/Os, and 3-sided range queries in $O(\log_B n + t/B)$ I/Os, where B denotes the block size. Both data structures pose no assumptions on the input distribution.

Our results. We consider the case where the x-coordinates of inserted points are drawn from a *smooth* probabilistic distribution, and the y-coordinates are arbitrarily distributed. Moreover, the deleted points are selected uniformly at random among the points in the data structure and queries can be adversarial. The assumption on the x-coordinates is broad enough to include distributions used in practice, such as uniform, regular and classes of non-uniform ones [6, 7].

We present two linear space data structures, for the RAM and the I/O model respectively. In the former model, we achieve a query time of $O(\log\log n + t)$ expected with high probability and update time of $O(\log\log n)$ expected amortized. In the latter model, the I/O complexity is $O(\log\log_B n + t/B)$ expected with high probability for the query and $O(\log_B \log n)$ expected amortized for the updates. In both cases, our data structures are deterministic and the expectation is derived from a probabilistic distribution of the x-coordinates, and an expected analysis of updates of points with respect to their y-coordinates.

2 Preliminaries

Weight Balanced Exponential Tree. The *exponential search tree* is a technique for converting static polynomial space search structures for ordered sets into fully-dynamic linear space data structures. It was introduced in [8, 9, 10] for searching and updating a dynamic set U of n integer keys in linear space and optimal $O(\sqrt{\log n / \log\log n})$ time in the RAM model. Effectively, to solve the dictionary problem, a doubly logarithmic height search tree is employed that stores static local search structures of size polynomial to the degree of the nodes.

Here we describe a variant of the exponential search tree that we dynamize using a rebalancing scheme relative to that of the *weight balanced search trees* [11]. In particular, a *weight balanced exponential tree* T on n points is a leaf-oriented rooted search tree where the degrees of the nodes increase double exponentially on a leaf-to-root path. All leaves have the same depth and reside on the lowest level of the tree (level zero). The *weight* of a subtree T_u rooted at node u is defined to be the number of its leaves. If u lies at level $i \geq 1$, the weight of T_u ranges within $\left[\frac{1}{2} \cdot w_i + 1, 2 \cdot w_i - 1\right]$, for a *weight parameter* $w_i = c_1^{c_2^i}$ and constants $c_2 > 1$ and $c_1 \geq 2^{3/(c_2-1)}$ (see Lem. 1). Note that $w_{i+1} = w_i^{c_2}$. The root does not need to satisfy the lower bound of this range. The tree has height $\Theta(\log_{c_2} \log_{c_1} n)$.

The insertion of a new leaf to the tree increases the weight of the nodes on the leaf-to-root path by one. This might cause some weights to exceed their range constraints ("overflow"). We *rebalance* the tree in order to revalidate the constraints by a leaf-to-root traversal, where we "split" each node that overflowed. An overflown node u at level i has weight $2w_i$. A split is performed by creating a new node v that is a sibling of u and redistributing the children of u among u and v such that each node acquires a weight within the allowed range. In particular, we scan the children of u, accumulating their weights until we exceed the value w_i, say at child x. Node u gets the scanned children and v gets the rest. Node x is assigned as a child to the node with the smallest weight. Processing the overflown nodes u bottom up guarantees that, during the split of u, its children satisfy their weight constraints.

The deletion of a leaf might cause the nodes on the leaf-to-root path to "underflow", i.e. a node u at level i reaches weight $\frac{1}{2}w_i$. By an upwards traversal of the path, we discover the underflown nodes. In order to revalidate their node constraints, each underflown node chooses a sibling node v to "merge" with. That is, we assign the children of u to v and delete u. Possibly, v needs to "split" again if its weight after the merge is more than $\frac{3}{2}w_i$ ("share"). In either case, the traversal continues upwards, which guarantees that the children of the underflown nodes satisfy their weight constraints. The following lemma, which is similar to [11, Lem. 9], holds.

Lemma 1. *After rebalancing a node u at level i, $\Omega(w_i)$ insertions or deletions need to be performed on T_u, for u to overflow or underflow again.*

Proof. A split, a merge or a share on a node u on level i yield nodes with weight in $\left[\frac{3}{4}w_i - w_{i-1}, \frac{3}{2}w_i + w_{i-1}\right]$. If we set $w_{i-1} \leq \frac{1}{8}w_i$, which always holds for $c_1 \geq 2^{3/(c_2-1)}$, this interval is always contained in $[\frac{5}{8}w_i, \frac{14}{8}w_i]$. □

Range Minimum Queries. The *range minimum query* (RMQ) problem asks to preprocess an array of size n such that, given an index range, one can report the position of the minimum element in the range. In [12] the RMQ problem is solved in $O(1)$ time using $O(n)$ space and preprocessing time.

Dynamic External Memory 3-sided Range Queries for $O(B^2)$ Points. [5, Lem. 1] A set of $K \leq B^2$ points can be stored in $O(K/B)$ blocks, so that 3-sided queries need $O(t/B + 1)$ I/Os and updates $O(1)$ I/Os, for output size t.

Smooth Distribution and Interpolation Search Structures. Informally, a distribution defined over an interval I is *smooth* if the probability density over any subinterval of I does not exceed a specific bound, however small this subinterval is (no "sharp peaks" exist).

Formally, given two functions f_1 and f_2, a density function $\mu = \mu[a,b](x)$ is (f_1, f_2)-*smooth* [13, 6] if there exists a constant β, such that for all c_1, c_2, c_3 where $a \leq c_1 < c_2 < c_3 \leq b$, and for all integers n and $\Delta = (c_3 - c_1)/f_1(n)$, it holds that $\int_{c_2-\Delta}^{c_2} \mu[c_1, c_3](x)dx \leq \frac{\beta \cdot f_2(n)}{n}$, when $\mu[c_1, c_3](x) = 0$ for $x < c_1$ or $x > c_3$, and $\mu[c_1, c_3](x) = \mu(x)/p$ for $c_1 \leq x \leq c_3$, where $p = \int_{c_1}^{c_3} \mu(x)dx$.

The *IS-tree* [13, 14] is a dynamic data structure based on interpolation search that consumes linear space and can be updated in $O(1)$ time when the update position is given. Furthermore, the elements can be searched in $O(\log^2 n)$ worst case time, or $O(\log\log n)$ time expected with high probability when they are drawn from an $(n^\alpha, n^{1/2})$-smooth distribution, for constant $1/2<\alpha<1$. Its externalization, the *ISB-tree* [15], consumes linear space and can be updated in $O(1)$ I/Os for given update position. It supports searches in $O(\log_B n)$ I/Os worst case, or $O(\log_B \log n)$ I/Os expected with high probability when the elements are drawn from an $(n/(\log\log n)^{1+\varepsilon}, n^{1/B})$-smooth distribution, for constant $\varepsilon>0$.

3 The Internal Memory Data Structure

Our internal memory construction for storing n points in the plane consists of an IS-tree storing the points in sorted order with respect to the x-coordinates. On the sorted points, we maintain a weight balanced exponential search tree T with $c_2 = 3/2$ and $c_1 = 2^6$. Thus its height is $\Theta(\log\log n)$. In order to use T as a priority search tree, we augment it as follows. The root stores the point with overall minimum y-coordinate. Points are assigned to nodes in a top-down manner, such that a node u stores the point with minimum y-coordinate among the points in T_u that is not already stored at an ancestor of u. Note that the point from a leaf of T can only be stored at an ancestor of the leaf and that the y-coordinates of the points stored at a leaf-to-root path are monotonically decreasing (*Min-Heap Property*). Finally, every node contains an RMQ-structure on the y-coordinates of the points in the children nodes and an array with pointers to the children nodes. Every point in a leaf can occur at most once in an internal node u and the RMQ-structure of u's parent. Since the space of the IS-tree is linear [13, 14], so is the total space.

3.1 Querying the Data Structure

Before we describe the query algorithm of the data structure, we will describe the query algorithm that finds all points with y-coordinate less than c in a subtree T_u. Let the query begin at an internal node u. At first we check if the y-coordinate of the point stored at u is smaller or equal to c (we call it a *member* of the query). If not we stop. Else, we identify the t_u children of u storing points with y-coordinate less than or equal to c, using the RMQ-structure of u. That is, we first query the whole array and then recurse on the two parts of the array partitioned by the index of the returned point. The recursion ends when the point found has y-coordinate larger than c (*non-member* point).

Lemma 2. *For an internal node u and value c, all points stored in T_u with y-coordinate $\leq c$ can be found in $O(t+1)$ time, when t points are reported.*

Proof. Querying the RMQ-structure at a node v that contains t_v member points will return at most t_v+1 non-member points. We only query the RMQ-structure of a node v if we have already reported its point as a member point. Summing over all visited nodes we get a total cost of $O\left(\sum_v (2t_v+1)\right)=O(t+1)$. □

In order to query the whole structure, we first process a 3-sided query $[a, b] \times (-\infty, c]$ by searching for a and b in the IS-tree. The two accessed leaves a, b of the IS-tree comprise leaves of T as well. We traverse T from a and b to the root. Let P_a (resp. P_b) be the root-to-leaf path for a (resp. b) in T and let $P_m = P_a \cap P_b$. During the traversal we also record the index of the traversed child. When we traverse a node u on the path $P_a - P_m$ (resp. $P_b - P_m$), the recorded index comprises the leftmost (resp. rightmost) margin of a query to the RMQ-structure of u. Thus all accessed children by the RMQ-query will be completely contained in the query's x-range $[a, b]$. Moreover, by Lem. 2 the RMQ-structure returns all member points in T_u.

For the lowest node in P_m, i.e. the lowest common ancestor (LCA) of a and b, we query the RMQ-structure for all subtrees contained completely within a and b. We don't execute RMQ-queries on the rest of the nodes of P_m, since they root subtrees that overlap the query's x-range. Instead, we merely check if the x- and y-coordinates of their stored point lies within the query. Since the paths P_m, $P_a - P_m$ and $P_b - P_m$ have length $O(\log \log n)$, the query time of T becomes $O(\log \log n + t)$. When the x-coordinates are smoothly distributed, the query to the IS-Tree takes $O(\log \log n)$ expected time with high probability [13]. Hence the total query time is $O(\log \log n + t)$ expected with high probability.

3.2 Inserting and Deleting Points

Before we describe the update algorithm of the data structure, we will first prove some properties of updating the points in T. Suppose that we decrease the y-value of a point p_u at node u to the value y'. Let v be the ancestor node of u highest in the tree with y-coordinate bigger than y'. We remove p_u from u. This creates an "empty slot" that has to be filled by the point of u's child with smallest y-coordinate. The same procedure has to be applied to the affected child, thus causing a *"bubble down"* of the empty slot until a node is reached with no points at its children. Next we replace v's point p_v with p_u (*swap*). We find the child of v that contains the leaf corresponding to p_v and swap its point with p_v. The procedure recurses on this child until an empty slot is found to place the last swapped out point (*"swap down"*). In case of increasing the y-value of a node the update to T is the same, except that p_u is now inserted at a node along the path from u to the leaf corresponding to p_u.

For every swap we will have to rebuild the RMQ-structures of the parents of the involved nodes, since the RMQ-structures are static data structures. This has a linear cost to the size of the RMQ-structure (Sect. 2).

Lemma 3. *Let i be the highest level where the point has been affected by an update. Rebuilding the RMQ-structures due to the update takes $O(w_i^{c_2-1})$ time.*

Proof. The executed "bubble down" and "swap down", along with the search for v, traverse at most two paths in T. We have to rebuild all the RMQ-structures that lie on the two v-to-leaf paths, as well as that of the parent of the top-most node of the two paths. The RMQ-structure of a node at level j

is proportional to its degree, namely $O(w_j/w_{j-1})$. Thus, the total time becomes $O(\sum_{j=1}^{i+1} w_j/w_{j-1}) = O(\sum_{j=0}^{i} w_j^{c_2-1}) = O(w_i^{c_2-1})$. □

To insert a point p, we first insert it in the IS-tree. This creates a new leaf in T, which might cause several of its ancestors to overflow. We split them as described in Sec. 2. For every split a new node is created that contains no point. This empty slot is filled by "bubbling down" as described above. Next, we search on the path to the root for the node that p should reside according to the Min-Heap Property and execute a "swap down", as described above. Finally, all affected RMQ-structures are rebuilt.

To delete point p, we first locate it in the IS-tree, which points out the corresponding leaf in T. By traversing the leaf-to-root path in T, we find the node in T that stores p. We delete the point from the node and "bubble down" the empty slot, as described above. Finally, we delete the leaf from T and rebalance T if required. Merging two nodes requires one point to be "swapped down" through the tree. In case of a share, we additionally "bubble down" the new empty slot. Finally we rebuild all affected RMQ-structures and update the IS-tree.

Analysis. We assume that the point to be deleted is selected uniformly at random among the points stored in the data structure. Moreover, we assume that the inserted points have their x-coordinates drawn independently at random from an $(n^\alpha, n^{1/2})$-smooth distribution for a constant $1/2<\alpha<1$, and that the y-coordinates are drawn from an arbitrary distribution. Searching and updating the IS-tree needs $O(\log\log n)$ expected with high probability [13, 14], under the same assumption for the x-coordinates.

Lemma 4. *Starting with an empty weight balanced exponential tree, the amortized time of rebalancing it due to insertions or deletions is $O(1)$.*

Proof. A sequence of n updates requires at most $O(n/w_i)$ rebalancings at level i (Lem. 1). Rebuilding the RMQ-structures after each rebalancing costs $O(w_i^{c_2-1})$ time (Lem. 3). Summing over all levels, the total time becomes $O(\sum_{i=1}^{height(T)} \frac{n}{w_i} \cdot w_i^{c_2-1}) = O(n \sum_{i=1}^{height(T)} w_i^{c_2-2}) = O(n)$, when $c_2<2$. □

Lemma 5. *The expected amortized time for inserting or deleting a point in a weight balanced exponential tree is $O(1)$.*

Proof. The insertion of a point creates a new leaf and thus T may rebalance, which by Lemma 4 costs $O(1)$ amortized time. Note that the shape of T only depends on the sequence of updates and the x-coordinates of the points that have been inserted. The shape of T is independent of the y-coordinates, but the assignment of points to the nodes of T follows uniquely from the y-coordinates, assuming all y-coordinates are distinct. Let u be the ancestor at level i of the leaf for the new point p. For any integer $k \geq 1$, the probability of p being inserted at u or an ancestor of u can be bounded by the probability that a point from a leaf of T_u is stored at the root down to the k-th ancestor of u plus the probability that the y-coordinate of p is among the k smallest y-coordinates of the leaves of T. The first probability is bounded by $\sum_{j=i+k}^{height(T)} \frac{2w_{j-1}}{\frac{1}{2}w_j}$, whereas the second probability is bounded by $k/\frac{1}{2}w_i$. It follows that p ends up at the i-th ancestor or higher with probability at most $O\Big(\sum_{j=i+k}^{height(T)} \frac{2w_{j-1}}{\frac{1}{2}w_j} + \frac{k}{\frac{1}{2}w_i}\Big) = O\Big(\sum_{j=i+k}^{height(T)} w_{j-1}^{1-c_2} + \frac{k}{w_i}\Big) = O\Big(w_{i+k-1}^{1-c_2} + \frac{k}{w_i}\Big) = O\Big(w_i^{(1-c_2)c_2^{k-1}} + \frac{k}{w_i}\Big) = O\Big(\frac{1}{w_i}\Big)$ for $c_2 = 3/2$ and $k = 3$. Thus the expected cost of "swapping down" p becomes $O\Big(\sum_{i=1}^{height(T)} \frac{1}{w_i} \cdot \frac{w_{i+1}}{w_i}\Big) = O\Big(\sum_{i=1}^{height(T)} w_i^{c_2-2}\Big) = O\Big(\sum_{i=1}^{height(T)} c_1^{(c_2-2)c_2^i}\Big) = O(1)$ for $c_2 < 2$.

A deletion results in "bubbling down" an empty slot, whose cost depends on the level of the node that contains it. Since the point to be deleted is selected uniformly at random and there are $O(n/w_i)$ points at level i, the probability that the deleted point is at level i is $O(1/w_i)$. Since the cost of an update at level i is $O(w_{i+1}/w_i)$, we get that the expected "bubble down" cost is $O\left(\sum_{i=1}^{height(T)} \frac{1}{w_i} \cdot \frac{w_{i+1}}{w_i}\right) = O(1)$ for $c_2 < 2$. □

Theorem 1. *In the RAM model, using $O(n)$ space, 3-sided queries can be supported in $O(\log\log n + t/B)$ expected time with high probability, and updates in $O(\log\log n)$ time expected amortized, given that the x-coordinates of the inserted points are drawn from an $(n^\alpha, n^{1/2})$-smooth distribution for constant $1/2<\alpha<1$, the y-coordinates from an arbitrary distribution, and that the deleted points are drawn uniformly at random among the stored points.*

4 The External Memory Data Structure

We now convert our internal memory into a solution for the I/O model. First we substitute the IS-tree with its variant in the I/O model, the ISB-Tree [15]. We implement every consecutive $\Theta(B^2)$ leaves of the ISB-Tree with the data structure of Arge et al. [5]. Each such structure constitutes a leaf of a weight balanced exponential tree T that we build on top of the $O(n/B^2)$ leaves.

In T every node now stores B points sorted by y-coordinate, such that the maximum y-coordinate of the points in a node is smaller than all the y-coordinates of the points of its children (Min-Heap Property). The B points with overall smallest y-coordinates are stored at the root. At a node u we store the B points from the leaves of T_u with smallest y-coordinates that are not stored at an ancestor of u. At the leaves we consider the B points with smallest y-coordinate among the remaining points in the leaf to comprise this list. Moreover, we define the weight parameter of a node at level i to be $w_i = B^{2\cdot(7/6)^i}$. Thus we get $w_{i+1} = w_i^{7/6}$, which yields a height of $\Theta(\log\log_B n)$. Let $d_i = \frac{w_i}{w_{i-1}} = w_i^{1/7}$ denote the *degree parameter* for level i. All nodes at level i have degree $O(d_i)$. Also every node stores an array that indexes the children according to their x-order.

We furthermore need a structure to identify the children with respect to their y-coordinates. We replace the RMQ-structure of the internal memory solution with a table. For every possible interval $[k, l]$ over the children of the node, we store in an entry of the table the points of the children that belong to this interval, sorted by y-coordinate. Since every node at level i has degree $O(d_i)$, there are $O(d_i^2)$ different intervals and for each interval we store $O(B \cdot d_i)$ points. Thus, the total size of this table is $O(B \cdot d_i^3)$ points or $O(d_i^3)$ disk blocks.

The ISB-Tree consumes $O(n/B)$ blocks [15]. Each of the $O(n/B^2)$ leaves of T contains B^2 points. Each of the n/w_i nodes at level i contains B points and a table with $O(B \cdot d_i^3)$ points. Thus, the total space is $O\left(n+\sum_{i=1}^{height(T)} n \cdot B \cdot d_i^3/w_i\right) = O\left(n+\sum_{i=1}^{height(T)} n \cdot B/(B^{2 \cdot \frac{7}{6}^i})^{\frac{4}{7}}\right) = O(n)$ points, i.e. $O(n/B)$ disk blocks.

4.1 Querying the Data Structure

The query is similar to the internal memory construction. First we access the ISB-Tree, spending $O(\log_B \log n)$ expected I/Os with high probability, given that the x-coordinates are smoothly distributed [15]. This points out the leaves of T that contain a, b. We perform a 3-sided range query at the two leaf structures. Next, we traverse upwards the leaf-to-root path P_a (resp. P_b) on T, while recording the index k (resp. l) of the traversed child in the table. That costs $\Theta(\log \log_B n)$ I/Os. At each node we report the points of the node that belong to the query range. For all nodes on $P_a - P_b$ and $P_b - P_a$ we query as follows: We access the table at the appropriate children range, recorded by the index k and l. These ranges are always $[k+1$,last child$]$ and $[0, l-1]$ for the node that lie on $P_a - P_b$ and $P_b - P_a$, respectively. The only node where we access a range $[k+1, l-1]$ is the LCA of the leaves that contain a and b. The recorded indices facilitate access to these entries in $O(1)$ I/Os. We scan the list of points sorted by y-coordinate, until we reach a point with y-coordinate bigger than c. All scanned points are reported. If the scan has reported all B elements of a child node, the query proceeds recursively to that child, since more member points may lie in its subtree. Note that for these recursive calls, we do not need to access the B points of a node v, since we accessed them in v's parent table. The table entries they access contain the complete range of children. If the recursion accesses a leaf, we execute a 3-sided query on it, with respect to a and b [5].

The list of B points in every node can be accessed in $O(1)$ I/Os. The construction of [5] allows us to load the B points with minimum y-coordinate in a leaf also in $O(1)$ I/Os. Thus, traversing P_a and P_b costs $\Theta(\log \log_B n)$ I/Os worst case. There are $O(\log \log_B n)$ nodes u on $P_a - P_m$ and $P_b - P_m$. The algorithm recurses on nodes that lie within the x-range. Since the table entries that we scan are sorted by y-coordinate, we access only points that belong to the answer. Thus, we can charge the scanning I/Os to the output. The algorithm recurses on all children nodes whose B points have been reported. The I/Os to access these children can be charged to their points reported by their parents, thus to the output. That allows us to access the child even if it contains only $o(B)$ member points to be reported. The same property holds also for the access to the leaves. Thus we can perform a query on a leaf in $O(t/B)$ I/Os. Summing up, the worst case query complexity of querying T is $O(\log \log_B n + \frac{t}{B})$ I/Os. Hence in total the query costs $O(\log \log_B n + \frac{t}{B})$ expected I/Os with high probability.

4.2 Inserting and Deleting Points

Insertions and deletions of points are in accordance with the internal solution. For the case of insertions, first we update the ISB-tree. This creates a new leaf in the ISB-tree that we also insert at the appropriate leaf of T in $O(1)$ I/Os [5]. This might cause some ancestors of the leaves to overflow. We split these nodes, as in the internal memory solution. For every split B empty slots "bubble down". Next, we update T with the new point. For the inserted point p we locate the highest ancestor node that contains a point with y-coordinate larger than p's. We insert p in the list of the node. This causes an excess point, namely the one with maximum y-coordinate among the B points stored in the node, to "swap down" towards the leaves. Next, we scan all affected tables to replace a single point with a new one.

In case of deletions, we search the ISB-tree for the deleted point, which points out the appropriate leaf of T. By traversing the leaf-to-root path and loading the list of B point, we find the point to be deleted. We remove the point from the list, which creates an empty slot that "bubbles down" T towards the leaves. Next we rebalance T as in the internal solution. For every merge we need to "swap down" the B largest excess points. For a share, we need to "bubble down" B empty slots. Next, we rebuild all affected tables and update the ISB-tree.

Analysis. Searching and updating the ISB-tree requires $O(\log_B \log n)$ expected I/Os with high probability, given that the x-coordinates are drawn from an $(n/(\log\log n)^{1+\varepsilon}, n^{1/B})$-smooth distribution, for constant $\varepsilon>0$ [15].

Lemma 6. *For every path corresponding to a "swap down" or a "bubble down" starting at level i, the cost of rebuilding the tables of the paths is $O(d_{i+1}^3)$ I/Os.*

Proof. Analogously to Lem. 3, a "swap down" or a "bubble down" traverse at most two paths in T. A table at level j costs $O(d_j^3)$ I/Os to be rebuilt, thus all tables on the paths need $O(\sum_{j=1}^{i+1} d_j^3) = O(d_{i+1}^3)$ I/Os. □

Lemma 7. *Starting with an empty external weight balanced exponential tree, the amortized I/Os for rebalancing it due to insertions or deletions is $O(1)$.*

Proof. We follow the proof of Lem. 4. Rebalancing a node at level i requires $O(d_{i+1}^3+B\cdot d_i^3)$ I/Os (Lem. 6), since we get B "swap downs" and "bubble downs" emanating from the node. The total I/O cost for a sequence of n updates is $O(\sum_{i=1}^{height(T)} \frac{n}{w_i}\cdot(d_{i+1}^3+B\cdot d_i^3))=O(n\cdot\sum_{i=1}^{height(T)} w_i^{-1/2}+B\cdot w_i^{-4/7})=O(n)$. □

Lemma 8. *The expected amortized I/Os for inserting or deleting a point in an external weight balanced exponential tree is $O(1)$.*

Proof. By similar arguments as in Lem. 5 and considering that a node contains B points, we bound the probability that point p ends up at the i-th ancestor or higher by $O(B/w_i)$. An update at level i costs $O(d_{i+1}^3)=O(w_i^{1/2})$ I/Os. Thus "swapping down" p costs $O(\sum_{i=1}^{height(T)} w_i^{1/2}\cdot\frac{B}{w_i})=O(1)$ expected I/Os. The same bound holds for deleting p, following similar arguments as in Lem. 5. □

Theorem 2. *In the I/O model, using $O(n/B)$ disk blocks, 3-sided queries can be supported in $O(\log\log_B n + t/B)$ expected I/Os with high probability, and updates in $O(\log_B \log n)$ I/Os expected amortized, given that the x-coordinates of the inserted points are drawn from an $(n/(\log\log n)^{1+\varepsilon}, n^{1/B})$-smooth distribution for a constant $\varepsilon > 0$, the y-coordinates from an arbitrary distribution, and that the deleted points are drawn uniformly at random among the stored points.*

References

[1] Agarwal, P., Erickson, J.: Geometric range rearching and its relatives. In: Chazelle, B., Goodman, J., Pollack, R. (eds.) Advances in Discrete and Computational Geometry. Contemporary Mathematics, pp. 1–56. American Mathematical Society Press (1999)

[2] Kanellakis, P.C., Ramaswamy, S., Vengroff, D.E., Vitter, J.S.: Indexing for data models with constraints and classes. In: Proc. ACM SIGACT-SIGMOD-SIGART PODS, pp. 233–243 (1993)

[3] McCreight, E.M.: Priority search trees. SIAM J. Comput. 14(2), 257–276 (1985)

[4] Willard, D.E.: Examining computational geometry, van emde boas trees, and hashing from the perspective of the fusion tree. SIAM J. Comput. 29(3), 1030–1049 (2000)

[5] Arge, L., Samoladas, V., Vitter, J.S.: On two-dimensional indexability and optimal range search indexing. In: Proc. ACM SIGMOD-SIGACT-SIGART PODS, pp. 346–357 (1999)

[6] Andersson, A., Mattsson, C.: Dynamic interpolation search in o(log log n) time. In: Lingas, A., Carlsson, S., Karlsson, R. (eds.) ICALP 1993. LNCS, vol. 700, pp. 15–27. Springer, Heidelberg (1993)

[7] Kaporis, A., Makris, C., Sioutas, S., Tsakalidis, A., Tsichlas, K., Zaroliagis, C.: Improved bounds for finger search on a RAM. In: Di Battista, G., Zwick, U. (eds.) ESA 2003. LNCS, vol. 2832, pp. 325–336. Springer, Heidelberg (2003)

[8] Andersson, A.: Faster deterministic sorting and searching in linear space. In: Proc. IEEE FOCS, pp. 135–141 (1996)

[9] Thorup, M.: Faster deterministic sorting and priority queues in linear space. In: Proc. ACM-SIAM SODA, pp. 550–555 (1998)

[10] Andersson, A., Thorup, M.: Dynamic ordered sets with exponential search trees. J. ACM 54(3), 13 (2007)

[11] Arge, L., Vitter, J.S.: Optimal dynamic interval management in external memory (extended abstract). In: Proc. IEEE FOCS, pp. 560–569 (1996)

[12] Harel, D., Tarjan, R.E.: Fast algorithms for finding nearest common ancestors. SIAM J. Comput. 13(2), 338–355 (1984)

[13] Mehlhorn, K., Tsakalidis, A.: Dynamic interpolation search. J. ACM 40(3), 621–634 (1993)

[14] Kaporis, A., Makris, C., Sioutas, S., Tsakalidis, A., Tsichlas, K., Zaroliagis, C.: Dynamic interpolation search revisited. In: Bugliesi, M., Preneel, B., Sassone, V., Wegener, I. (eds.) ICALP 2006. LNCS, vol. 4051, pp. 382–394. Springer, Heidelberg (2006)

[15] Kaporis, A.C., Makris, C., Mavritsakis, G., Sioutas, S., Tsakalidis, A.K., Tsichlas, K., Zaroliagis, C.D.: ISB-tree: A new indexing scheme with efficient expected behaviour. In: Deng, X., Du, D.-Z. (eds.) ISAAC 2005. LNCS, vol. 3827, pp. 318–327. Springer, Heidelberg (2005)

Untangled Monotonic Chains and Adaptive Range Search*

Diego Arroyuelo[1,**], Francisco Claude[2], Reza Dorrigiv[2], Stephane Durocher[3], Meng He[2], Alejandro López-Ortiz[2], J. Ian Munro[2], Patrick K. Nicholson[2], Alejandro Salinger[2], and Matthew Skala[2,4]

[1] Yahoo! Research Latin America, Chile
darroyue@dcc.uchile.cl

[2] Cheriton School of Computer Science, University of Waterloo, Waterloo, Ontario, Canada
{fclaude,rdorrigiv,sdurocher,mhe,alopez-o,imunro, p3nichol,ajsalinger,mskala}@cs.uwaterloo.ca

[3] Department of Computer Science, University of Manitoba, Winnipeg, Manitoba, Canada
durocher@cs.umanitoba.ca

[4] Department of Computer Science, University of Toronto, Toronto, Ontario, Canada
mskala@ansuz.sooke.bc.ca

Abstract. We present the first adaptive data structure for two-dimensional orthogonal range search. Our data structure is adaptive in the sense that it gives improved search performance for data with more inherent sortedness. Given n points on the plane, the linear-space data structure can answer range queries in $O(\log n + k + m)$ time, where m is the number of points in the output and k is the minimum number of monotonic chains into which the point set can be decomposed, which is $O(\sqrt{n})$ in the worst case. Our result matches the worst-case performance of other optimal-time linear-space data structures, or surpasses them when $k = o(\sqrt{n})$. Our data structure can also be made implicit, requiring no extra space beyond that of the data points themselves, in which case the query time becomes $O(k \log n + m)$. We present a novel algorithm of independent interest to decompose a point set into a minimum number of untangled, same-direction monotonic chains in $O(kn + n \log n)$ time.

1 Introduction

Applications in geographic information systems, among others, require structures that can store and retrieve spatial data efficiently in both space and time. In this work we describe a data structure and algorithm for two-dimensional orthogonal range search, which is a commonly-encountered spatial data retrieval

* Funding for this research was made possible by NSERC Discovery Grants, the Canada Research Chairs Program, and the NSERC Strategic Grant on Optimal Data Structures for Organization and Retrieval of Spatial Data.

** Much of this work took place while the first author was a visitor at the University of Waterloo.

Y. Dong, D.-Z. Du, and O. Ibarra (Eds.): ISAAC 2009, LNCS 5878, pp. 203–212, 2009.

problem. Our data structure is *adaptive*, giving improved query performance for data with more inherent sortedness; and can be *implicit*, requiring no added storage space beyond that of the data points themselves.

The problem of *two-dimensional orthogonal range search* can be defined as follows: let $P = \{p_1, p_2, \ldots, p_n\}$ be a set of n points in the plane, and let $r = [x_1, x_2] \times [y_1, y_2]$ be a *query range*. The orthogonal range search problem asks for all points $p_i \in P$ such that $x_1 \leq x(p_i) \leq x_2$ and $y_1 \leq y(p_i) \leq y_2$, where $x(p_i)$ and $y(p_i)$ denote the x and y coordinate values of point p_i, respectively. An orthogonal range search data structure preprocesses the set P in order to efficiently answer arbitrary range queries; a natural goal is to balance the conflicting objectives of minimizing both the space required by the data structure and the time required to answer queries.

Our basic data structure's worst-case query time is $O(k \log n + m)$, where n is the number of points in the point set; m the number of points in the output; and k the minimum number of monotonic chains into which the point set can be decomposed, which is $O(\sqrt{n})$ in the worst case. Applying fractional cascading [4] reduces the query time to $O(\log n + k + m)$ at the cost of implicitness.

We require that the monotonic chains should be untangled. That is, when successive vertices are connected by line segments, the chains should not intersect each other. This requirement does not increase the minimal number of chains. We present a novel algorithm for finding a minimal set of untangled chains (all monotonic in the same direction) in $O(kn + n \log n)$ time; this untangling algorithm is of independent interest.

2 Previous Work

Any set of n points can be split into some number k of chains in which the y coordinate is *monotonically* increasing or decreasing as the x coordinate increases. When all chains must be ascending (or all descending), the problem of finding a minimal chain decomposition is well-studied. With worst-case data the minimal number of chains may be $\Theta(n)$, even given a choice of ascending or descending chains. Supowit gives an algorithm for it with worst-case running time $\Theta(n \log n)$ [12], which is optimal [3]. If chains of both types are allowed, then minimizing the number of chains is NP-hard [5]. However, an algorithm of Fomin, Kratsch, and Novelli acheives a constant-factor approximation of the minimal number of chains in $O(n^3)$ time [6]. An algorithm of Yang, Chen, Lu, and Zheng generates a decomposition into at most $\lfloor \sqrt{2n + 1/4} - 1/2 \rfloor$ chains of both types (which is the minimal number for worst-case data) in $O(n^{3/2})$ time [14]. They do not prove any guaranteed approximation factor when the minimal number of chains is smaller than $O(\sqrt{n})$, but comment that in practical experiments their algorithm often achieves very close to the constant-factor approximation value.

The two-dimensional orthogonal range search problem has received considerable attention, and several efficient data structures exist. For instance, R-trees [7] are a multidimensional extension of B-trees. An R-tree is a height-balanced tree, where each tree node represents a region of the underlying space. Thus, the data structure divides the space with hierarchically nested (and possibly overlapping) minimum bounding rectangles. The search algorithm descends the tree, recursing into every subtree whose bounding rectangle overlaps the query. In the worst case a search could be forced to examine the entire tree in $O(n)$ time, even when the query rectangle is empty. However, R-trees are simple to implement, use linear space, tend to perform much better in practice than the theoretical worst case, and are popular as a result.

Range trees [9] support multidimensional range queries by generalizing balanced binary search trees to multiple dimensions. The data points are indexed along one dimension in a standard balanced binary search tree. At each node v of that tree, we collect all the descendants of v and store a new balanced binary search tree storing all those points indexed along the second dimension. A rectangle query descends the first tree to do a one-dimensional range search in $O(\log n)$ time, then searches along the other dimension for an overall time of $O(\log^2 n + m)$. More advanced techniques, like fractional cascading [4], allow the two-dimensional search time to be reduced to $O(\log n + m)$; and the technique can also be extended to higher dimensions at some cost in search time.

Alternative solutions exist that require linear space like R-trees but improve on the worst-case search time. Kanth shows that $O(\sqrt{n} + m)$ worst-case search time is optimal for non-replicating (or linear-space) data structures [8]. Bentley achieves it with kd-trees [2], which recursively divide a k-dimensional space with hyperplanes. Munro describes an *implicit* kd-tree, with optimal search time and no storage used beyond that of the points themselves [10]. Arge describes priority R-trees, or PR-trees [1], also with $O(\sqrt{n} + m)$ worst-case search time. In a recent result, Nekrich [11] presents a data structure that uses linear space with search time $O(\log n + m \log^{\epsilon} n)$, trading suboptimal performance in m for better performance in n. See Table 1 for a comparison of methods.

To summarize, R-trees are practical, but do not provide worst-case guarantees at search time, and range trees have an impractical $O(n \log n)$ space requirement. There are alternative solutions requiring linear space and providing better search time. However, none of these can profit from "easy" data. Here we present an *adaptive* data structure. When the data can be decomposed into a small number of monotonic chains, our search performance improves. If the number of

Table 1. Summary of orthogonal range query results; n is the number of points in the database, m is the number of points returned, and k is the number of chains

Data structure	Worst-case search time	Space
R-trees [7]	$O(n)$	$O(n)$
kd-trees [2,10]	$O(\sqrt{n} + m)$	implicit
PR-trees [1]	$O(\sqrt{n} + m)$	$O(n)$
Range trees [9]	$O(\log n + m)$	$O(n \log n)$
Nekrich [11]	$O(\log n + m \log^{\epsilon} n)$	$O(n)$
This paper	$O(\log n + k + m)$	$O(n)$
This paper	$O(k \log n + m)$	implicit

chains $k = o(\sqrt{n})$, we surpass the performance of optimal-time linear-space data structures [1,2,8,10].

3 Finding Untangled Chains

In the next section we describe an adaptive algorithm and data structure for two-dimensional orthogonal range search on data decomposed into a union of monotonic chains. The data structure performs better when there are fewer chains. Furthermore, we can search more efficiently by assuming that the chains are untangled: successive data points can be connected with line segments with no segments intersecting. That raises the question of how to find an optimal untangled chain decomposition, which we resolve in this section. To conserve space we omit proofs in this section, as they are mainly based on exhaustive case analysis.

Although our data structure asks for an optimal decomposition into chains with both ascending and descending monotonic chains allowed, it actually functions by splitting the points into the two directions as a preprocessing step and then considering the two directions separately; chains are only required to be untangled with respect to other chains of the same type. The untangling problem of interest to us, then, is how to decompose a set of points into a minimal number of untangled chains all in one direction (without loss of generality, descending). Also assume that points in the input set are in general position.

It is easy to see that removing a single tangle between two chains does not change the number of chains, so the minimum number of untangled chains is the same as the minimum number of possibly-tangled chains.

However, finding tangles to remove requires search, and each untangling move could introduce many new tangles, resulting in an expensive untangling procedure. Van Leeuwen and Schoone show that such a process must terminate after $O(n^3)$ moves [13]. They describe an $O(n^2)$ exhaustive search to find each tangle, for an overall time of $O(n^5)$. We describe an algorithm for finding a minimal number of chains in $O(n \log n + kn)$ time where k is the number of chains.

3.1 Untangling Monotonic Chains

Given two points $p_i, p_j \in P$, we say that the edge or line segment (p_i, p_j) is *valid* if $x(p_i) \leq x(p_j)$ and $y(p_j) \leq y(p_i)$. We also say that points p_i and p_j are *compatible* if (p_i, p_j) or (p_j, p_i) is valid. A *chain* is a sequence of edges $C = \{(p_1, p_2), (p_2, p_3), \ldots, (p_{n-1}, p_n)\}$ where each one is valid. We will often refer to a point $p \in C$ for some chain C, which means that p is an endpoint of some edge in C. A *sub-chain* S of C is a contiguous subset of the edges $\{(p_k, p_{k+1}), \ldots, (p_{k+\ell-1}, p_{k+\ell})\}$, where $k + \ell \leq n$. We call ℓ the length of S.

Supowit [12] proposed an algorithm, Algorithm 1, for decomposing points into a minimal number of possibly-intersecting same-direction monotonic chains. Let A be a chain and $\text{miny}(A) = \min\{y | (x, y) \in A\}$. Let $P = \{p_1, p_2, \ldots, p_n\}$ be the data points sorted by increasing x-coordinate.

Algorithm 1. Minimum number of descending chains

$S \leftarrow \emptyset$
for $i = 1 \ldots n$ **do**
  let $S' = \{A \in S, \text{miny}(A) \geq y(p_i)\}$
  if $S' \neq \emptyset$ **then**
    let $A_0 = \text{argmin}_A\{\text{miny}(A), A \in S'\}$
    append p_i to A_0
  else
    add p_i as a chain to S
return S

If an edge in one chain intersects an edge in another chain, we call the intersection a *tangle* and the chains *tangled* with each other. Let $\mathcal{L}(P)$ be the set of all valid edges and $\mathcal{L}^*(P)$ be the set of edges created by running Algorithm 1 on P. Then for any edge (p_i, p_j), define $H^+(p_i, p_j)$ to be the open half-plane bounded by the line through p_i and p_j and containing the point $(x(p_i) + 1, y(p_i) + 1)$, and $H^-(p_i, p_j)$ symmetrically. Now we can show that all tangles in the output of Algorithm 1 are of a special kind.

Definition 1. *Suppose we have two chains C_2 and C_1 with edges $(q_1, q_2) \in C_2$ and $(p_1, p_2), \ldots, (p_{\ell-1}, p_\ell) \in C_1$ such that $p_1 \in H^-(q_1, q_2)$, $p_\ell \in H^-(q_1, q_2)$, and $p_i \in H^+(q_1, q_2)$ for all $1 < i < \ell$. We call such a tangle a "valid"-tangle (abbreviated as v-tangle). Fig. 1 shows examples. We call (q_1, q_2) the upper part of the v-tangle, and $(p_1, p_2), \ldots, (p_{\ell-1}, p_\ell)$ the lower part.*

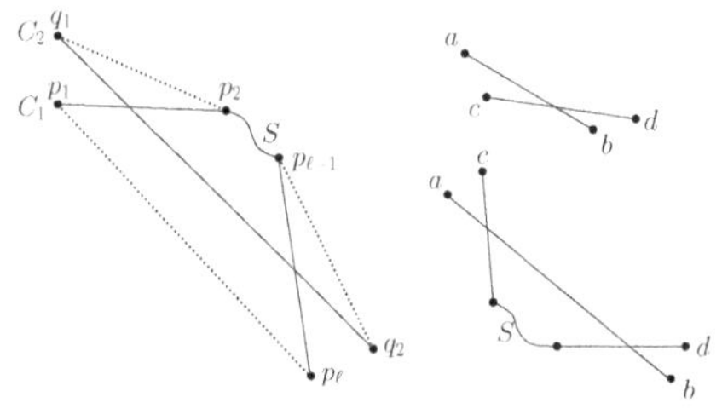

Fig. 1. (Left) Valid tangles (v-tangles) generated by Algorithm 1. (Right) Two examples of tangles that cannot be generated by Algorithm 1.

Lemma 1. *All tangles created by Algorithm 1 are v-tangles.*

Since only v-tangles are possible in the output of Algorithm 1, there is an intuitive ordering on the set of chains. Suppose we run Algorithm 1 on P and it generates k chains. We can create a set of k points $Q = \{q_1, \ldots, q_k\}$ such that $x(q_i) < x(q_{i+1})$, no two points in Q are compatible with each other, but every point in Q is compatible with every point in P, Then, if we execute Algorithm 1 again on $P \cup Q$, each q_i will be added to a single chain C_i, and we can order the chains based on these points. We will assume we have such a set at the beginning

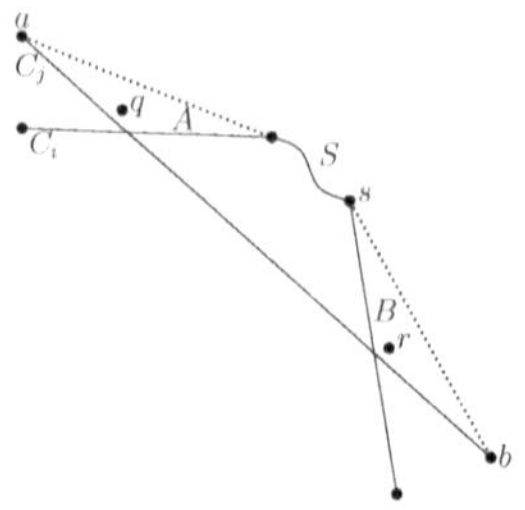

Fig. 2. Illustration of cases considered in Lemma 2

of the chains and another at the end in order to avoid special boundary cases. Thus, given two chains C_i and C_j, we can refer to C_j as the upper chain if $j > i$. The uppermost chain is C_k.

With this ordering in mind, we now discuss how to untangle a v-tangle. The following lemma illustrates that untangling a v-tangle does not create new tangles involving upper chains.

Remark 1. Given a v-tangle, as shown in Fig. 1, we can untangle it by using the dotted lines as edges. This is just moving S to be part of C_2. As we just explained, it does not matter how the points change and move around chains, chain C_i is the one that would contain q_i.

Lemma 2. *Consider two tangled chains C_i and C_j as in Fig. 2. By removing a v-tangle, where C_j is above C_i, we cannot generate new tangles involving chains above C_j.*

Consider Algorithm 2. Each iteration of the outer for loop ensures that chain C_i is not tangled with any chains below, $C_1, \ldots, C_{i-1}$.

Algorithm 2. Untangled-Chains(P)

```
1: Run Algorithm 1 on P to get chains C_1,...,C_k where C_k is the uppermost chain.
2: for i = k down to 1 do
3:    for j = i − 1 down to 1 do
4:       Find and untangle all v-tangles between C_i and C_j
```

To find the tangles we just traverse both chains in order of increasing x-coordinates of their points, so the process take time proportional to the sum of the lengths of the chains. Our method of untangling also has the following useful invariant properties.

Lemma 3. *Consider the set of points R in chains $C_1, \ldots, C_{i-1}$ after untangling $C_i, \ldots, C_k$. If we run Algorithm 1 with input R, the resulting set of chains is exactly $C_1, \ldots, C_{i-1}$.*

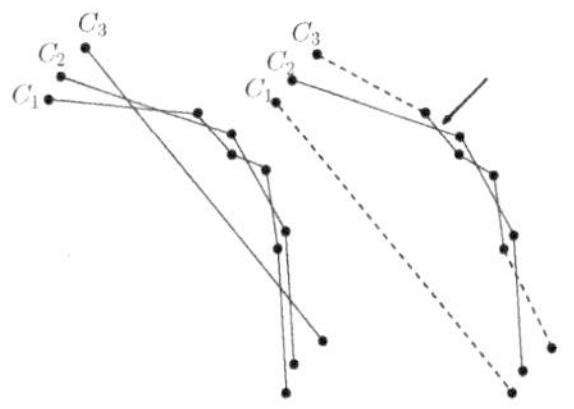

Fig. 3. Untangling chains in an arbitrary order may cause tangles which are not v-tangles. For example, untangling C_1 and C_3 results in such a situation. The arrow points to a new tangle that is not a v-tangle.

Lemma 4. *After we have untangled C_i with chains $C_{i-1}, \ldots, C_1$, no subsequent untangling operations occurring among chains $C_1, \ldots, C_{i-1}$ can cause a new tangle to form with C_i.*

The previous results allow for the possibility that during the untangling of C_i, we could (temporarily) create non-v-tangles involving C_i. In fact, such tangles are possible if the order in which the untangling is done is arbitrary, which can be seen in Fig. 3. However, since Algorithm 2 untangles the chains in descending order, this situation cannot occur.

What remains to be shown is that in the process of untangling the upper chain C_i from the chains $C_1, \ldots, C_{i-1}$, when untangling a v-tangle, any other v-tangles involving C_i either disappear or remain being a v-tangle.

Lemma 5. *Suppose a v-tangle between C_i and C_j is removed by Algorithm 2, where C_j is the upper chain. Any tangles between C_j and C_ℓ where $\ell < j$ may have been altered. However, the remaining tangles are still v-tangles.*

Now we state our main theorem about the untangling process:

Theorem 1. *A set of n points in the plane can be decomposed into a minimal set of chains without tangles in $O(n \log n + kn)$ time, where k is the number of chains.*

4 Adaptive Orthogonal Range Search

First, observe that if the data points form a single monotonic chain, then the answer to any query must be a contiguous interval of the ordered list of points, and we can find it with a binary search. We can store such a data set in $O(n)$ space and answer queries in $O(\log n + m)$ time, where n is the number of data points and m is the number of points returned by the query.

Now assume that as a preprocessing step the data points have been decomposed into a minimal number k of monotonic chains. A truly optimal decomposition would require solving an NP-hard problem, but we can come within a constant factor in $O(n^3)$ time with the algorithm of Fomin, Kratsch, and Novelli, and that is good enough to preserve the asymptotic search time of our data structure [6]. The $O(n^{3/2})$ partitioning algorithm of Yang, Chen, Lu, and Zheng offers no guarantee of a minimal decomposition, but appears to come close in practice and may be preferable in real applications [14]. In either case, once we have a decomposition of the data points into chains, we separate the ascending and descending chains, and treat the two directions separately, building a data structure for each and running every query on both.

The two-direction minimization algorithms are used only to decide for each point whether it will go into the ascending or descending structure. Having made that decision, we run the algorithm of the previous section to find a minimal set of untangled chains for each direction; doing so cannot increase the number of chains further.

Without loss of generality, we describe the data structure for descending chains here. The ascending case is symmetric. Let $\{C_1, C_2, \ldots, C_k\}$ be the set of untangled descending chains, and let ℓ_i be the length of C_i. Let $r = [x_1, x_2] \times [y_1, y_2]$ be the query range.

We first find the set of chains that intersect r. If we store the chains ordered from left to right as described in the previous section, we can find the first chain to pass above the point (x_1, y_1) and the last chain to pass below the point (x_2, y_2), and know that all chains intersecting the query range must be between those two chains in the ordering. Evaluating whether a point is above or below a chain can be accomplished by a simple binary search over the points in the chain in $O(\log n)$ time, so with two binary searches over the chains we can find the start and end of the range of chains that might intersect r, in $O(\log k \log n)$ time. Let $k' \leq k$ be the number of chains in that subset.

For each of the k' chains that might intersect r, we can do two more binary searches to find the start and end of the interval of data points within the chain, that are actually included in the query range. Note that because of the monotonicity of the chains, this must be a contiguous interval. The time to do these searches is $O(\log \ell_i)$ for each of the k' chains, and since $\sum \ell_i = n$, the time for this step is $O(k' \log(n/k'))$.

The number of points m returned by the query also places a lower bound on the running time simply because we must spend time writing them out. Adding up the times gives the following lemma:

Lemma 6. *Given a set of n points which can be decomposed into k monotonic chains, we can in $O(n^3)$ time construct a linear-space data structure answering two-dimensional orthogonal range search queries in $O(\log k \log n + k' \log(n/k') + m)$ time, where m is the number of points returned and $k' \leq k$ depends on the query.*

Observe that the above solution involves performing binary searches for the same keys in separate ordered lists. Thus, we can use the technique of fractional cascading [4] to speed up the query time and achieve the following result:

Theorem 2. *Given a set of n points which can be decomposed into k monotonic chains, we can in $O(n^3)$ time construct a linear-space data structure answering two-dimensional orthogonal range search queries in either $O(\log n + k + m)$ time* *or* $O(\log k \log n + k' + m)$ *time, where* m *is the number of points returned and* $k' \leq k$ *depends on the query.*

Proof. To check whether the query rectangle $[x_1, x_2] \times [y_1, y_2]$ intersects a given chain C_i, it is sufficient to perform binary searches on the list of x-coordinates (or y-coordinates) of the points on C_i using x_1 and x_2 (or y_1 and y_2) as search keys. This also finds which edge, if any, of C_i intersects each edge of the query rectangle. Therefore, we can report the points on C_i that are located in the query range in $O(\log n + k_i)$ time, where k_i is the number of such points.

Then to answer orthogonal range search queries using our data structure, we can perform two binary searches on the list of x-coordinates of the points on each chain, and two binary searches on the list of y-coordinates for each chain. Thus, we can store the sorted lists of x-coordinates and y-coordinates corresponding to the monotonically increasing chains separately, and use the technique of fractional cascading [4] to speed up the query time without increasing the asymptotic space cost of our data structure. We augment the data structure for the monotonically decreasing chains using the same approach. This yields a data structure of linear space that supports orthogonal range search in $O(\log n + k + m)$ time.

The other result in the theorem can be achieved by locating the start and the end of the range of chains that might intersect the query rectangle, and then using fractional cascading to compute the answer starting from the uppermost chain in this range. □

The $O(n^3)$ preprocessing time may be improved to $O(n^{3/2})$ (matching the untangling step) in practical cases when the partitioning algorithm of Yang, Chen, Lu, and Zheng gives acceptable results [14]. We can also make the data structure of Lemma 6 implicit:

Corollary 1. *A set of* n *points in the plane can be arranged as an array of* n *coordinate pairs so that any orthogonal range query over this point set can be answered in* $O(\log k \log n + k' \log(n/k') + m)$ *time with* $O(1)$ *working space.*

5 Conclusions

We have presented a new data structure for two-dimensional orthogonal range search that is adaptive to the minimum number of monotonic chains that the input points can be partitioned into. For data which is considered easy in this sense, our data structure outperforms existing alternatives, either in query time or space requirements. Furthermore, we show that our structure can be made implicit, requiring only constant space in addition to the space required to encode the input points.

As a contribution of independent interest, we show how to partition a set of two-dimensional points into a minimal number of untangled monotonic chains. This decomposition is a key element of our data structure, and could also be useful in other geometric applications.

References

1. Arge, L., de Berg, M., Haverkort, H.J., Yi, K.: The priority R-tree: A practically efficient and worst-case optimal R-tree. ACM Transactions on Algorithms 4(1) (2008)
2. Bentley, J.L.: Multidimensional binary search trees used for associative searching. Communications of the ACM 18(9), 509–517 (1975)
3. Bloniarz, P.A., Ravi, S.S.: An $\Omega(n \log n)$ lower bound for decomposing a set of points into chains. Information Processing Letters 31(6), 319–322 (1989)
4. Chazelle, B., Guibas, L.J.: Fractional cascading: I. a data structuring technique. Algorithmica 1(2), 133–162 (1986)
5. Di Stefano, G., Krause, S., Lübbecke, M.E., Zimmermann, U.T.: On minimum k-modal partitions of permutations. Journal of Discrete Algorithms 6(3), 381–392 (2008)
6. Fomin, F.V., Kratsch, D., Novelli, J.C.: Approximating minimum cocolorings. Information Processing Letters 84(5), 285–290 (2002)
7. Guttman, A.: *R*-trees: a dynamic index structure for spatial searching. SIGMOD Record (ACM Special Interest Group on Management of Data) 14(2), 47–57 (1984)
8. Kanth, K.V.R., Singh, A.: Optimal dynamic range searching in non-replicating index structures. In: Beeri, C., Bruneman, P. (eds.) ICDT 1999. LNCS, vol. 1540, pp. 257–276. Springer, Heidelberg (1999)
9. Lueker, G.S.: A data structure for orthogonal range queries. In: 19th Annual Symposium on Foundations of Computer Science (FOCS 1978), Long Beach, Ca., USA, pp. 28–34. IEEE Computer Society Press, Los Alamitos (1978)
10. Munro, J.I.: A multikey search problem. In: Proceedings of the 17th Allerton Conference on Communication, Control and Computing, University of Illinois, pp. 241–244 (1979)
11. Nekrich, Y.: Orthogonal range searching in linear and almost-linear space. Computational Geometry 42(4), 342–351 (2009)
12. Supowit, K.J.: Decomposing a set of points into chains, with applications to permutation and circle graphs. Information Processing Letters 21(5), 249–252 (1985)
13. van Leeuwen, J., Schoone, A.A.: Untangling a traveling salesman tour in the plane. In: Proceedings of the 7th Conference on Graphtheoretic Concepts in Computer Science (WG 1981), München, Germany, pp. 87–98. Hanser Verlag (1981)
14. Yang, B., Chen, J., Lu, E., Zheng, S.Q.: A comparative study of efficient algorithms for partitioning a sequence into monotone subsequences. In: Cai, J.-Y., Cooper, S.B., Zhu, H. (eds.) TAMC 2007. LNCS, vol. 4484, pp. 46–57. Springer, Heidelberg (2007)

Geodesic Spanners on Polyhedral Surfaces*

Sanjiv Kapoor and Xiang-Yang Li

Department of Computer Science, Illinois Institute of Technology, Chicago, IL, USA
kapoor@iit.edu, xli@cs.iit.edu

Abstract. In this paper we consider the problem of efficiently constructing geodesic t-spanners. We consider finding sparse spanners on the surface of a 3 dimensional polyhedron allowing for steiner vertices. If Steiner vertices are not allowed, then we establish lower bounds on the maximum node degree, depending on the spanning ratio t and also the total number of vertices of the polyhedron surface. We also consider the case of the surface of a convex polytope $\mathcal{P}$ with V vertices. Using its vertex set P and Steiner points, we can construct a t-spanner with a constant degree and weight $O(MST(U))$, where $MST(U)$ is the minimum spanning tree on the set U of vertices on convex polytope.

1 Introduction

Given an edge weighted graph $G = (V, E, w)$, where V is the set of vertices, E is the set of edges, and w is a weight function with $w(e)$ be the weight of an edge e, let $d_G(u, v)$ denote the shortest distance from node u to node v in graph G; let $\omega(G)$ be the sum of the edge weights of edges in G. A subgraph $H = (V, E', w)$ of G, where $E' \subseteq E$, is called a t-spanner of graph G, if for *any* pair of nodes u and v, $d_H(u, v) \leq t \cdot d_G(u, v)$. The minimum t such that H is a t-spanner of G is called the *stretch factor* of H with respect to G. An Euclidean graph is a graph where the weight of every edge (u, v) is the Euclidean distance $\|uv\|$ between its end-nodes. Given a geometric region Ω and a set V of points in Ω, a geodesic graph is a graph where the weight of each edge (u, v) with $u, v \in V$, is the *geodesic distance* from u to v in the region Ω.

For the case of geodesic spanners, our domain will be a 3-dimensional simplicial polygonal surface $\mathcal{P}$ that is formed of $m = O(n)$ triangles, and a set V of n nodes on the surface $\mathcal{P}$. While spanner construction has been well studied in general graphs and in Euclidean spaces, this is the first study of constructing geodesic spanners on a simplicial polygonal surface with some additional properties such as minimizing the node degree and/or total edge length. Notice that a $O(n^3 \log n)$ time construction of a t-spanner (without degree or weight bounds) for a polyhedral surface using Steiner points was implicitly studied in [10]. Another previous study of spanners involving geodesic distances in a 2-dimensional planar domain with obstacles can be found in [8]. Geodesic spanner graphs on the surface of the convex polyhedron $\mathcal{P}$ approximate the complete geodesic graph on a set of nodes, U on $\mathcal{P}$. The edges (u, v) have weight corresponding to the geodesic shortest distance between u and v. In order to construct a spanner,

* The research of authors is partially supported by NSF CNS-0916743, NSF CNS-0832120, National Natural Science Foundation of China under Grant No. 60828003, Hong Kong CERG under Grant PolyU-5232/07E, and Hong Kong RGC HKUST 6169/07.

Y. Dong, D.-Z. Du, and O. Ibarra (Eds.): ISAAC 2009, LNCS 5878, pp. 213–223, 2009.

we note that distances on the surface of a polyhedron could be stretched due to the folds of the surface. To capture this effect we define *geodesic dilation factor* to measure the difference between the distance on the geodesic surface and the Euclidean distance between two points. We consider both non-convex and convex three dimensional polytopes (simplicial 2-complex) in this paper.

A greedy algorithm has been used to construct spanners for various graphs [5,13,6,7, 3]. Peleg and Schaffer [12] showed that, for any $t > 1$, there exists a graph $G = (V, E)$ with $|V| = n$ and an edge weight, such that *any* t-spanner of this graph needs at least $n^{1+\frac{1}{t+2}}$ edges. Thus there are weighted graphs such that any t-spanner has weight at least $\Omega(n^{\frac{1}{t+2}}\omega(\mathrm{MST}))$ by letting the weight of each edge be 1. Althofer [1] proved that the greedy method produces a sparse t-spanner with at most $n^{1+\frac{2}{t-1}}$ edges; and Regev [14] showed that more precisely it has at most $n^{1+\frac{2}{\lfloor k \rfloor}} + n$ edges if $\lfloor k \rfloor$ is even, and at most $n^{1+\frac{2}{\lceil k \rceil}} + n$ edges if $\lfloor k \rfloor$ is odd. For arbitrary weighted graphs, Chandra *et al.* [3] showed that the greedy algorithm constructs a t-spanner of weight at most $(3 + \frac{16t}{\epsilon^2})n^{\frac{2+\epsilon}{t-1-\epsilon}} \cdot \omega(\mathrm{MST})$ for every $t > 1$ and any $\epsilon > 0$. Regev [14] proved that the t-spanner constructed by the greedy algorithm has weight at most $2e^2 \ln n \cdot n^{\frac{2}{t-1}} \cdot \omega(\mathrm{MST})$ when $t \in [3, 2\log n + 1]$, and has weight at most $(1 + \frac{4\log^2 n + 2\log n}{t+1-\log n}) \cdot \omega(\mathrm{MST})$ when $t > 2\log n + 1$, by studying the girth of the constructed t-spanner.

For geodesic spanners, we introduce geodesic cones and partition the space into geodesic cones. Using this space partition we get the following results: We develop an algorithm for computing a spanner graph for a set of nodes $U \subseteq P$ on a 3-D polyhedral surface $\mathcal{P}$. Here P is the set of vertices of $\mathcal{P}$. We construct a sparse t-spanner with Steiner vertices from P with $O(\gamma(\mathcal{P})n/\epsilon)$ edges in time $O(n^2/\epsilon + n^3)$, where $n = |P|$ and $\epsilon > 0$ is any small constant. Since this is the first result for constructing t-spanners for geodesic graphs, we have not attempted to optimize the time complexity. Here $\gamma(\mathcal{P})$ is the dilation factor (defined later) of the polyhedron. We also prove that, there is a polyhedral surface $\mathcal{P}$ and nodes' placement of U, such that the maximum node degree of any t-spanner, when no Steiner vertices are used in the spanner, is at least $\Omega(n^{1/t})$, for $t > 1$. When $t < 3$, we show by example that the maximum node degree of any t-spanner without using Steiner nodes for this example is at least $\gamma(\mathcal{P})$. Notice that in the worst case, $\gamma(\mathcal{P}) = \Theta(n)$. We also show that traditional greedy methods cannot build a t-spanner with degree bound $o(n)$, for any $t > 1$.

For a surface $\mathcal{P}$ of a convex polyhedron, by using Steiner vertices, we develop an algorithm to compute a t-spanner for a set of nodes U on the surface $\mathcal{P}$ with a constant maximum degree and weight $O(MST(U))$, where $MST(U)$ is the geodesic minimum spanning tree on the set U of nodes on convex polytopes, where the distance between two nodes is the geodesic distance between them.

2 Geodesic Spanners for Arbitrary Polyhedral Surfaces

2.1 Terminology

Assume that we are given a triangulated polyhedral surface $\mathcal{P}$ in three-dimension. Let P be the set of all vertices and $\mathcal{F}$ be the set of all faces of $\mathcal{P}$. Let U be a set of nodes on $\mathcal{P}$. In this paper, we always assume that $U \subseteq P$. For any two nodes $u, v \in U$, we let $\Pi_{\mathcal{P}}(u, v)$ be the shortest geodesic path on $\mathcal{P}$ between u and v; and let $\mathbf{d}_{\mathcal{P}}(u, v)$ denote the geodesic distance between u and v on $\mathcal{P}$. In general, given an edge weighted graph G and $u, v \in G$, let $\mathbf{d}_G(u, v)$ be the shortest distance between nodes u and v in G.

In this paper, we will focus on constructing spanners for the complete weighted graph $\mathcal{K}_{\mathcal{P}}(U, E)$ on the nodes in U with weight function, $\mathbf{d}_{\mathcal{P}} : E \rightarrow R^+$. A graph $H = (V, E)$, with $U \subseteq V$, is a geodesic t-spanner for U if $\mathbf{d}_H(u, v) \leq t \cdot \mathbf{d}_{\mathcal{P}}(u, v)$ for every pair of nodes $u, v \in U$. Here the weight of each edge xy in H is the geodesic distance $\mathbf{d}_{\mathcal{P}}(x, y)$ between x and y on the surface $\mathcal{P}$. The geodesic t-spanner H is said to use *Steiner vertices* if $U \subset V$. The geodesic t-spanner H is said to be on U (or without using Steiner vertices) if $V = U$. The geodesic t-spanner H is said to have size η if it has at most η nodes and edges. A *geodesic minimum spanning tree* of a set of vertices U on a surface $\mathcal{P}$ is the minimum spanning tree of graph $\mathcal{K}_{\mathcal{P}}(U, E)$.

Our algorithm to construct a geodesic t-spanner for $\mathcal{K}_{\mathcal{P}}(U, E)$ utilizes a new concept called *geodesic cones*. Given a point u and a region R, we let $P_u(R, d)$ be the set of all points p in region R that are at geodesic distance $\mathbf{d}_{\mathcal{P}}(u, p) = d$ from node u. For example, when R is a 2-D plane $\mathbb{R}^2$, then $P_u(R, d)$ is a circle centered at u with radius d. Geometric cones have been used widely to produce t-spanners in Euclidean space. An important property of the geometric cone $\mathcal{C}$, which was used in obtaining t-spanners, is that for any two points p_1, p_2 of $P_u(\mathcal{C}, d)$, $\|p_1 - p_2\| \leq \epsilon d$ for a small constant $0 < \epsilon < 1$. For example, for 2D geometric cones with angle θ, for any two points p_1, $p_2 \in P_u(\mathcal{C}, d)$, $\|p_1 - p_2\| \leq 2\sin(\frac{\theta}{2}) \cdot d$.

Definition 1 (pairwise-ϵ-neighbor property). *Given a surface $\mathcal{P}$ and the distance metric $\mathbf{d}_{\mathcal{P}}$, a set of points X is said to satisfy the pairwise-ϵ-neighbor property with respect to a node u if, for any two points $p_1, p_2 \in P_u(X, d)$, we have $\mathbf{d}_{\mathcal{P}}(p_1, p_2) \leq \epsilon \cdot d$.*

Given a surface $\mathcal{P}$, a set X of points on $\mathcal{P}$ is called an ϵ-geodesic cone *with respect to (w.r.t.) a node u, termed $\mathcal{C}(u)$, if the following two conditions are satisfied:*

1. *$P_u(X, d)$ satisfies the pairwise-ϵ-neighbor property with respect to a node u, and*
2. *if $x \in X$, then all points on a geodesic shortest path $\mathbf{d}_{\mathcal{P}}(u, x)$ are in X.*

Most of our results rely on constructing some special cone partition (or more precisely, cone covering) of space around every node in P. Notice that here a geodesic cone could be in a finite region. For any node u in the polyhedral surface $\mathcal{P}$, let $\mathcal{F}(u) = \{F_1(u), F_2(u), \cdots, F_p(u)\}$ be the p triangular faces incident onto node u. For each face $F_i(u) = \triangle xuy$ of $\mathcal{F}(u)$, let $\delta_i(u)$ be the radian value of the internal angle $\angle xuy$. Let $\delta(u) = \sum_{F_i(u) \in \mathcal{F}(u)} \delta_i(u)$.

Definition 2 (Dilation Factor). *The* dilation factor *$\gamma(u)$ of a node u in the polyhedral surface $\mathcal{P}$ is defined as $\gamma_{\mathcal{P}}(u) = \frac{\sum_i \delta_i(u)}{2\pi}$. We omit the subscript $\mathcal{P}$ if it is clear from the context. The* geodesic dilation factor, *$\gamma(\mathcal{P})$, of the polyhedral surface $\mathcal{P}$, is defined as $\gamma(\mathcal{P}) = \max_{u \in \mathcal{P}} \gamma(u)$.*

The dilation factor is a measure of the change in length of a "circle" on the surface of the polytope as compared to its length on a planar surface. When the surface $\mathcal{P}$ is planar, e.g. a single planar face, its geodesic dilation factor is $\gamma_{\mathcal{P}} = 1$. In general the dilation factor of a surface could be large and depends on the structure of the surface.

To determine geodesic shortest paths we will use the properties of these paths as defined in Mitchell et al. [11].

2.2 Constructing t-Spanner Using Steiner Vertices

In this section, given a polyhedral surface $\mathcal{P}$ with a set of vertices P, a set of nodes $U \subseteq P$ and a parameter $t > 1$, we first present a method to construct a t-spanner $H = (V, E)$ using some Steiner vertices, *i.e.*, $U \subseteq V$. Notice that here we will focus on the case $V \subseteq P$, *i.e.*, Steiner vertices must be a subset of P, where P is the set of vertices used to define the polyhedral surface.

If we can use arbitrary Steiner nodes, we can easily get a t-spanner with the maximum degree bound 3. Thus, in the rest of the paper, we always assume that the Steiner vertices in a t-spanner are restricted to the set P of vertices of the polytope. Observe that when we compute the shortest geodesic path between a pair of nodes u and v, the path found, often uses multiple line segments from different faces of the surface $\mathcal{P}$. We would like to clarify that the end-points of these segments are *not* considered as Steiner nodes in this paper, although they are not from P.

Our method is to partition the space near every vertex $u \in P$ by some ϵ-geodesic cones. Consider a vertex $u \in \mathcal{P}$ and all the triangular faces, $\mathcal{F}_{\mathcal{P}}(u) = \{F_1(u), F_2(u), \cdots, F_p(u)\}$, where $F_i(u)$ is a triangle $v_i u v_{i+1}$, where v_{p+1} is v_1. We define $\lceil \delta(u)/\epsilon \rceil$ cones $\mathcal{C}_{\mathcal{P}}(u) = \{C_1(u), C_2(u), \cdots, C_{\lceil \delta(u)/\epsilon \rceil}(u)\}$, where each of the cones has an angle at most $\epsilon < \pi/3$, where $t(\epsilon) = \frac{1}{1-2\sin\frac{\epsilon}{2}}$ is the spanning ratio that can be achieved by the first phase of our method. For the set of faces $\mathcal{F}(u)$, imagine that we cut the faces $F_1(u)$ and $F_p(u)$ along the segment uv_1 and "unfold" all faces in $\mathcal{F}(u)$ sequentially on $F_1(u)$, using successively the edges $uv_i, i = 1 \ldots p-1$. To construct the required cones we desire to construct rays with apex u on the faces $\mathcal{F}(u)$ such that when we unfold these faces, the angle between any two consecutive rays on the unfolded plane is at most ϵ. Using the unfolding, the cones in $\mathcal{C}(u)$ are produced in the unfolded 2-d space by dividing the surrounding unfolded region $\delta(u)$ (which could have angle arbitrarily larger than 2π for non-convex polyhedron, and smaller than 2π for convex polyhedron) using planar cones (sectors) with an angle at most ϵ, *i.e.*, a cone, when unfolded, has a shape of a sector. Observe that here a cone $C_i(u)$ may contain several triangular faces of $\mathcal{F}(u)$ inside. We can then fold the faces back and this will give us $\lceil \delta(u)/\epsilon \rceil$ rays: two consecutive rays define a cone. It is easy to show that (1) for any point x from $\mathcal{F}(u)$, $\mathbf{d}_{\mathcal{P}}(x, u) = \mathbf{d}(x, u)$; (2) for any two points x and y from $\mathcal{F}(u)$ that fall inside the same cone and $\mathbf{d}_{\mathcal{P}}(x, u) < \mathbf{d}_{\mathcal{P}}(y, u)$, we have $\mathbf{d}_{\mathcal{P}}(x, u) + t(\epsilon) \cdot \mathbf{d}_{\mathcal{P}}(x, y) \le t(\epsilon) \cdot \mathbf{d}_{\mathcal{P}}(y, u)$ for $t(\epsilon) = \frac{1}{1-2\sin\frac{\epsilon}{2}}$ when $\epsilon < \pi/3$.

Note that here a cone $C_i(u)$ only contains points from $\mathcal{F}(u)$ now. We later will show how to extend the cones to other triangular faces on $\mathcal{P}$ by propagating each cone.

We next present our method to construct a geodesic t-spanner, without using Steiner vertices, in phases:

1. **Phase 1: $t(\epsilon_1)$-Spanner Construction.** First, for every node $u \in U$, we construct a geodesic cone partition, $\Lambda_{\mathcal{P}}(u)$. The cone partition is achieved by a propagation method which develops the cones starting with $\lceil \frac{\delta(u)}{\epsilon_1} \rceil$ cones $\mathcal{C}_{\mathcal{P}}(u)$ on the faces $\mathcal{F}(u)$ containing u. The process is detailed in procedure $\mathbf{PropagateCones}(u)$

for each node $u \in U$. The process is also repeated for Steiner vertices S of the polyhedron encountered during the cone expansion from nodes in U, *i.e.*, we run procedure **PropagateCones**(u) for every node u of S and update S accordingly: adding the encountered node $v \in P$ to S if $v \notin U$. These new vertices are termed Steiner vertices, denoted as S. Let U' be the union of U and S.

For each ϵ-geodesic cone with apex node u, we add a (directed) edge uv if v is the closest node from P, *i.e.*, first encountered node in P. This phase results in a graph $H'(U \cup S, A')$ where A' is the set of edges added. We will prove that H' is a $t(\epsilon_1)$-spanner for U.

2. **Optional Phase 2: Further Degree Reduction.** The process described in this paragraph is used to *possibly* further reduce the node degree. Again, we partition the space around each node v, by cones of angle at most ϵ_2.

   Repeat the following step for each node $v \in U \cup S$. In $H'(U \cup S, A')$, for each node v, let $I(v)$ be the set of incoming neighbors of node v. For all nodes in $I(v)$, build a tree rooted at node v. Let $I_0(v)$ be the set of nodes that already has been processed. Initially, $I_0(v) = \{v\}$. A directed edge (u, v), where $u \in I(v)$, is retained if u is the closest node in some cone of node v, and we add u to the set $I_0(v)$. For each newly added node u in $I_0(v)$, recursively add directed edges xu where $x \notin I_0(v)$ is the closest node to u in some cone of u.
   Let the final structure be $H'(U \cup S, E')$.
3. **Optional Phase 3: Further Edge Reduction.** In this phase, edges in $H'(U \cup S, A')$ (or $H'(U \cup S, E')$) are pruned to create the graph $H(U \cup S, E)$ as follows: (1) Sort the edges A' in $H'(U \cup S, A')$ in decreasing order of the geodesic length, $e_1, e_2 \ldots e_{m'}$. Let $E = A'$. (2) Eliminate edge e_i from E if the edges in $E \setminus e_i$ provide a path of length at most $t \cdot \mathbf{d}_{\mathcal{P}}(u, v)$ for every pair of nodes u and v from U, where $t > 1$ is the spanning ratio. (3) Eliminate unnecessary Steiner vertices, where a Steiner vertex is unnecessary if it is not on the shortest path in H between any pair of nodes in U.

The construction of the Geodesic Cone partition is given in Algorithm 1.

Observe that, in Algorithm 1, clearly point $x(f(C, u))$ has only two choices: either it is some node v from P defining the original polyhedral surface, or it is inside some segment vw where uvw is a face incident on u. Note that even if cone C contains multiple triangular faces of $\mathcal{F}(u)$ inside, it is still possible that $x(f(C, u))$ is not one of the vertices in faces $\mathcal{F}(u)$. Additionally, in Algorithm 1, the sequence F_U of unfolding is not necessarily unique, and we have to test all necessary sequences of unfolding. After we unfold a triangle vwz, we should unfold both triangle vyz along the edge vz, and the triangle zwq along the edge zw to find the closest vertex from P that is inside the (extended) cone C. Here, either the vertex y or the vertex q, or both could be closer to u than the vertex z. The actual unfolding of faces will use the continuous Dijkstra method [11] as follows to define the **procedure** *Unfold*. For each cone C and each vertex u, we maintain an event heap: the event is the edge e of face f that has a point, denoted as $x(f, c)$, that is closest to the node u. When we unfold a face f represented by three vertices u, v, w along some edge, $e = (u, v)$, it will possibly introduce two new edges e_1 and e_2. We then add these two new edges to the heap based on the distance $\mathbf{d}_{\mathcal{P}}(e_i, u)$ to node u. We also add the distance to the vertex w. The top element of the

Algorithm 1. PropagateCones(u)

1: Let $\mathcal{C}_{\mathcal{P}}(u)$ be $\lceil \delta(u)/\epsilon_1 \rceil$ cones around u on the faces $f \in \mathcal{F}(u)$ containing u (each cone has an angle at most ϵ_1).
2: **for** each cone C in $\mathcal{C}_{\mathcal{P}}(u)$ **do**
3: For each face $f = uvw$ inside the cone C, let $uv'w'$ be the portion of the face that is completely contained inside C. Notice that here $v'w'$ could be a segment of vw. Let $x(f)$ be the point on the segment $v'w'$ that is *closest* from u, *i.e.*, $\mathbf{d}_{\mathcal{P}}(x(f), u) = \mathbf{d}(x(f), u) \leq \mathbf{d}(y, u)$ for any point y on the segment $v'w'$. Let $d = \min_{f \text{intersected by cone } C, f \in \mathcal{F}(u)} \mathbf{d}_{\mathcal{P}}(u, x(f))$. Let $f(C, u)$ (or simply f if no confusion) be the face that has the point $x(f)$ such that $\mathbf{d}_{\mathcal{P}}(u, x(f))$ is minimized among all faces intersected by C, *i.e.*, $\mathbf{d}_{\mathcal{P}}(u, x(f(C, u))) = d$.
4: **if** $x(f(C, u))$ is a vertex v from P **then**
5: Add v to S and add a *directed* edge (u, v) to the structure $H'(U \cup S, A')$.
6: If v is not marked *processed*, run procedure $\mathbf{PropagateCones(v)}$.
7: **else**
8: Let e be the edge that contains the point $x(f(C, u))$ and f be the face $f(C, u)$. Extend cone C across face f, adjacent along e and unfolded onto the sequence F_U of unfolded faces by **procedure** *Unfold*. Note: In this procedure we keep unfolding faces until the closest point to u among all points on segments defining polyhedral surface $\mathcal{P}$, say $x(C, d)$, is a vertex z from P. Then add an edge uz and update geodesic distance to z.
9: **end if**
10: **end for**
11: Mark node u processed.

heap is always the edge or vertex that is closest to the vertex u. Let e be the element in the top of the heap. We process the top event represented by e by unfolding the face adjacent to e as defined above. If the top element of the heap is a vertex v of P such that the distance $\mathbf{d}_{\mathcal{P}}(u, v) \leq \mathbf{d}_{\mathcal{P}}(u, e')$ for every element e' in the heap, the procedure returns v.

We can prove: By choosing $\epsilon_1 = \epsilon_2 = \epsilon$ for some small value ϵ, we have the following lemma.

Lemma 1. *The graph $H'(U \cup S, E')$ after degree-reduction procedure is a t-spanner with a maximum node degree $O(\min(n, (\gamma(\mathcal{P})/\epsilon)^2))$ where $t = (\frac{1}{1-2\sin(\epsilon/2)})^2$.*

Lemma 2. *Our algorithm constructs a t-spanner $H(U \cup S, E)$ in $O(n^2/\epsilon + n^3)$ time.*

We can construct an example surface and nodes placement (see Figure 1 for illustration) such that for some small enough t, a t-spanner (for some constant t) will have the maximum degree $\Omega(\gamma(\mathcal{P}))$ where $\gamma(\mathcal{P}) = \Theta(n)$ is the *dilation factor* of the surface. The basic idea of the example is as follows: There is a set U of n nodes $u_1, u_2, \cdots, u_n$. There are two triangular faces $u_i u_0 v_i$ and $v_i u_0 u_{i+1}$ between u_i and u_{i+1} such that the geodesic distance between u_i and u_{i+1} is larger than $(t-1) \cdot d + \eta$, for $t < 3$, $i \in [1, n-1]$, for a small constant $\eta > 0$. Actually, we can place these two triangular faces $u_i u_0 v_i$ and $v_i u_0 u_{i+1}$ such that the geodesic distance between u_i and u_{i+1} is $2d - \delta$ for any $0 \leq \delta < 2d$. A node u_0 at distance d from these nodes will then have to be connected directly to all these nodes to ensure that it is a t-spanner. Observe that, when $t \geq 3$, the preceding example does not imply that we have to connect u_0 with every node u_i, $i \geq 1$. This is because the geodesic distance between u_i and u_j is at most $\mathbf{d}_{\mathcal{P}}(u_i, u_0) + \mathbf{d}_{\mathcal{P}}(u_0, u_j) \leq 2d$. In this case, we can omit some edges $u_0 u_i$ without violating the t-spanner property for $t \geq 3$.

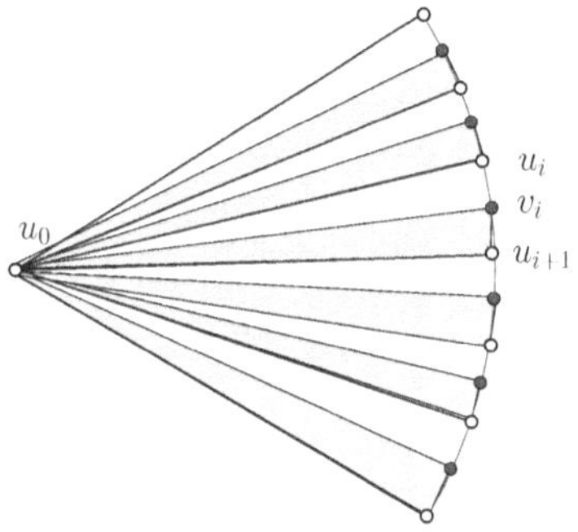

Fig. 1. An example of a surface and the set of nodes $U = \{u_0, u_1, \cdots, u_n\}$. Here $u_0 v_i$ defines a valley between $u_0 u_i$ and $u_0 u_{i+1}$, $d = \mathbf{d}_{\mathcal{P}}(u_i, u_0)$.

We further study reducing the weight of the structure. Chandra *et al.* [3] proved that for any metric space M, and every n-vertex complete graph G on this metric, if (1) there is an $O(g(n))$ time algorithm that builds a t-spanner for G with $O(f(n))$ edges, where $f(m)/2 \geq f(m/2)$ and $g(m)/2 \geq g(m/2)$ for any $m > 0$, and (2) there exists an $O(h(n))$ time algorithm that can build a spanning tree T for G with weight $O(1)\omega(\text{MST})$, then there exists an $O(\max(g(n), h(n), n \log n))$ time method which builds a $(t + \epsilon)$-spanner with $O(f(n))$ edges and weight $O(\frac{f(n)}{n} \log n)\omega(\text{MST})$. Notice that for geodesic metric, we have methods with $f(n) = O(\gamma(\mathcal{P})n)$ and $g(n) = O(n^3 + n^2/\epsilon)$. Our method for constructing a structure H' implies the following lemma.

Theorem 1. *A geodesic t-spanner can be constructed for any polyhedral surface* $\mathcal{P}$ *such that the total weight of the structure is* $O(\gamma(\mathcal{P}) \log n)\omega(\text{MST})$. *The construction requires* $g(n) = O(n^2/\epsilon)$ *time.*

Observe that with the optional degree-reduction phase 2, the running time of the method becomes $g(n) = O(n^3 + n^2/\epsilon)$. Notice that the method by Chandra [3] cannot preserve the degree bound of the final structure. We leave it as a future work to design a t-spanner structure with bounded degree $O(\gamma(\mathcal{P}))$, and total weight $O(\gamma(\mathcal{P}) \log n)\omega(\text{MST})$, or study whether it is possible to construct a t-spanner with weight $O(\gamma(\mathcal{P}) + \log n)\ \omega(\text{MST})$.

2.3 Geodesic Spanners without Using Steiner Vertices

We now study, given the polyhedral surface $\mathcal{P}$ (and its set of vertices P), a set of nodes $U \subseteq P$, and a number $t > 1$, how to construct a t-spanner $H = (U, E)$ without using Steiner vertices. Our objective is to construct a t-spanner with small node degree and small total edge weight. A more general question is following: given a complete weighted graph G with positive edge weights satisfying the triangular inequality, construct a t-spanner $H \subseteq G$ with small maximum degree and small total edge weight $\omega(H)$. Surprisingly, we could not find any results, except [9], in the literature that provide any degree bound on a t-spanner for an arbitrarily weighted graph.

We first show by example that, for any algorithm that constructs a t-spanner, there are inputs such that the constructed t-spanner will have a maximum degree at least $\Omega(n^{\frac{1}{t}})$ for nodes placed on a surface, where n is the size of U.

Lemma 3. *For any $t > 1$, there is a surface $\mathcal{P}$ on a set of nodes P, and a set of nodes $U \subseteq P$, such that the maximum node degree in any t-spanner $H = (U, E)$ without using Steiner vertices is at least $(\frac{n}{2})^{\frac{1}{t}}$, where $n = |U|$. For any $t > 1$, there is a surface $\mathcal{P}$ on a set of nodes P, and a set of nodes $U \subseteq P$, such that the weight of any t-spanner is at least $n^{\frac{1}{t+2}}/2$ times of MST.*

When $t < 3$, the placement the triangles $u_i u_0 v_i$, $1 \leq i \leq n-1$, and triangles $v_i u_0 u_{i+1}$, $1 \leq i \leq n-1$, ensures that the geodesic distance $\mathbf{d}_{\mathcal{P}}(u_i, u_{i+1})$ is $2\mathbf{d}_{\mathcal{P}}(u_0, u_i) - \delta$ (for small $0 < \delta < (3-t)\mathbf{d}_{\mathcal{P}}(u_0, u_i)$) and $\mathbf{d}_{\mathcal{P}}(u_0, u_i) = \mathbf{d}_{\mathcal{P}}(u_0, u_j)$ for $i \neq j$. Thus, we have to connect u_0 to every node u_i since $\mathbf{d}_{\mathcal{P}}(u_0, u_j) + \mathbf{d}_{\mathcal{P}}(u_j, u_i) \geq 3\mathbf{d}_{\mathcal{P}}(u_0, u_i) - \delta > t\mathbf{d}_{\mathcal{P}}(u_0, u_i)$ for every node u_j and $t < 3$.

Lemma 4. *For any t with $1 < t < 3$, there is a surface $\mathcal{P}$ on a set of nodes P, and a set of nodes $U \subseteq P$, such that the maximum node degree in any t-spanner $H = (U, E)$, without using Steiner vertices or all Steiner vertices are restricted to P, is at least $\gamma(\mathcal{P}) = \Theta(n)$, where $n = |U|$.*

Thus, generally, to get a t-spanner, which does not use any Steiner vertices or can only use Steiner vertices from P, with a maximum node degree $o(n)$, we must focus on $t \geq 3$. In this case, Lemma 3 shows that the maximum degree is at least $\Omega(n^{\frac{1}{t}})$.

3 Geodesic Spanners for Convex Polytopes

In this section, we study constructing geodesic spanners for a set of nodes on a convex polytope, and the distance is measured by geodesic distance. Our approach is to approximate a convex polytope by a constant number of 2D planar patches, similar to [4].

Let $\mathcal{P}$ be a convex polytope, with a set of polygonal faces $\mathcal{F}$. For any subset $F \subseteq \mathcal{F}$ of faces, let $\mathcal{N}(F) = \{N_f \mid f \in F\}$ be the set of normals to the faces where N_f is the normal to face f. Consider the angular representation of the normals: each normal N is represented by a pair (θ_N, ϕ_N) in the Spherical coordinates system, where $\theta_N, \phi_N \in [0, 2\pi]$ are the angle of the normal vector from the z-axis (called the colatitude or zenith) and the angle from the x-axis. The basic idea of our method for building the spanner is to partition the convex polygonal surface $\mathcal{P}$ into a constant number of convex patches such that each patch is almost flat (*i.e.*, the difference between the normals of any two faces in the patch is a small constant). Note that the patches we construct may overlap.

Definition 3. *A δ-patch of $\mathcal{P}$ is a set of faces, $F \subseteq \mathcal{F}$ such that (1) F forms a continuous region; (2) the patch is* flat, *i.e., $\Psi(F) = \sup_{f,g \in F} \max\left(|\theta_{N_f} - \theta_{N_g}|, |\phi_{N_f} - \phi_{N_g}|\right) \leq \delta$, i.e., the difference between any two normals is bounded by a constant.*

A δ-partition of $\mathcal{P}$, denoted as $\Delta_{\mathcal{P}}$, is a partition of the set of faces $\mathcal{F}$ such that each partition is a δ-patch. Here a δ-patch is not necessarily convex.

Definition 4. *A δ-planar projection, $\Xi(F)$ of a δ-patch F is the projection of points in F onto a plane, P, with normal N_P such that $N_P \in \mathcal{N}(F)$.*

A *linear convex patch* G is a connected closed subset of points with a piecewise linear boundary such that its δ-planar projection $\Xi(G)$ is convex.

Definition 5. *A convex extension $\mathcal{E}(F)$ of a δ-patch F is a minimal piece-wise linear convex patch, a collection of polygonal faces, that contains F with the property that $\Psi(\mathcal{E}(F)) - \Psi(F) \leq \epsilon$.*

It is not difficult to show that the following property holds for a convex surface $\mathcal{P}$.

Property 1. Low-distortion projection property: Let u and v be two points on a δ-patch, F. Then $\mathbf{d}_{\mathcal{P}}(u,v) \geq d(u,v) \geq \mathbf{d}_{\mathcal{P}}(u,v)/(1+2\cdot\delta)$ where $d(u,v)$ is the Euclidean distance between u and v on $\Xi(F)$ and $\mathbf{d}_{\mathcal{P}}(u,v)$ is the geodesic distance on F.

3.1 Algorithm

Our method for constructing a spanner for convex polytope is as follows:

1. Find a δ-partition, denoted as $\Delta_{\mathcal{P}} = \{F_1, F_2, \cdots F_p\}$, of $\mathcal{P}$. Here $F_i \subseteq \mathcal{F}$ is a subset of faces that form a δ-patch. Then we construct a convex-extension $\mathcal{E}(\Delta_{\mathcal{P}})$ as follows. For each δ-patch F_i in the δ-partition, we perform the following steps:
   (a) Find a δ-planar projection, $\Xi(F_i)$ of F_i to some plane with a normal $N \in N(F_i)$.
   (b) Find the convex hull $CH(\Xi(F_i))$ of $\Xi(F_i)$.
   (c) Find the inverse of the projection, i.e., find $\mathcal{E}(\Xi(F_i))$ such that its δ-planar projection is $CH(\Xi(F_i))$.
2. For every δ-patch in $\Delta_{\mathcal{P}}$ do the following
   (a) For each $u \in U_F$, construct a ϵ-cone partition of the surface as in the previous section 2. Let $\mathcal{C}(u)$ be the cone partition produced. Note that since the difference between normals is small, a simple method of projecting cone partitions of a plane suffices in this case.
   (b) Let U_F be the set of all nodes in F For node $u \in F$, let $\mathcal{I}(u)$ be the set of the intersection segments of all cones with the boundary $\partial\mathcal{E}(F)$. In each intersection region $\mathbf{C} \cap \partial\mathcal{E}(F)$ where $\mathbf{C} \in \mathcal{C}(u)$, we add a Steiner point if the cone $\mathbf{C}$ that created the region contains a shortest path from the apex, say u, of the cone to some other node $v \notin U_F$, *i.e.*, outside of the δ-patch. Note that this Steiner point is also added to the neighboring δ-patch, which the cone $\mathbf{C}$ intersects. Let the set of added Steiner points be $\mathcal{S}_F(u)$. Let the set of all Steiner points on F be $\mathcal{S}_F = \cup_{u \in U_F} \mathcal{S}_F(u)$.
   (c) Find a projection $\Xi(F)$ of F to some hyperplane perpendicular to the normal of a face in F.
   (d) Find an Euclidean t-spanner graph $H_F(\Xi(V_F), \Xi(A_F))$ of a constant maximum degree and of weight $O(MST(V_F))$ for the set of vertices, $\Xi(V_F)$, in $\Xi(F)$ where $V_F = U_F \cup \mathcal{S}_F$. $\Xi(A_F)$ is the set of edges created in the spanner graph. This can be done by several methods in the literature [2].

3. Let the spanner be $H(U, A) = \cup_{F \in \Delta_{\mathcal{P}}} H_F(V_F, A_F)$, where each edge (u, v) in A_F corresponds to an edge $(\Xi(u), \Xi(v))$ and is weighted by shortest geodesic distance between u and v.

It remains to determine a δ-partition and the convex extensions. Given the range of angles θ and ϕ we do the following

1. Partition the domain $[0, 2\pi]$ of θ and ϕ equally into $2\pi/\epsilon$ ranges of size ϵ, indexed by $(i, j), 1 \leq i, j \leq 2\pi/\epsilon$ which indicates that $\theta \in [(i-1) \cdot \epsilon, i \cdot \epsilon]$ and $\phi \in [(j-1) \cdot \epsilon, j \cdot \epsilon]$.
2. For every tuple of ranges (i, j) let $F = \cup f$ such that $\theta_{N_f} \in [(i-1) \cdot \epsilon, i \cdot \epsilon]$ and $\phi_{N_f} \in [(j-1) \cdot \epsilon, j \cdot \epsilon]$.

The preceding approach clearly creates a constant number of δ-patches because of the monotonicity of the normals for a convex polytope. The δ-patch F is obtained from the convex hull of points on F on the polyhedron $\mathcal{P}$. Note that our spanner uses Steiner points.

Theorem 2. *$H(V, A)$ is a t-spanner and requires $O((1/\epsilon)^2(n^2 \log n + T_E(n)))$ steps to constuct where $T_E(n)$ is the time required by any algorithm to compute a 2-dimensional t-spanner.*

Notice that $T_E(n) = O(n \log n)$ for any d-dimensional nodes. We also observe that our method of constructing t-spanner also works for any d-dimensional convex surface (where the cone must be small enough with an angle $O(\epsilon)$ with $(1+\epsilon)^d < t$).

References

1. Althofer, I., Das, G., Dopkin, D., Joseph, D., Soares, J.: On sparse spanners of weighted graphs. Discrete and Computational Geometry 9, 81–100 (1993)
2. Arya, S., Smid, M.: Efficient construction of a bounded degree spanner with low weight. Algorithmica 17, 33–54 (1997)
3. Chandra, B., Das, G., Narasimhan, G., Soares, J.: New sparseness results on graph spanners. International Journal of Computational Geometry and Applications 5, 125–144 (1995)
4. Dudley, R.: Metric entropy of some classes of sets with differentiable boundaries. J. Approximation Theory 10(3), 227–236 (1974)
5. Eppstein, D.: Spanning Trees and Spanners. In: Handbook of Computational Geometry, pp. 425–461. Elsevier Science, Amsterdam (1997)
6. Gudmundsson, J., Levcopoulos, C., Narasimhan, G.: Improved greedy algorithms for constructing sparse geometric spanners. In: Halldórsson, M.M. (ed.) SWAT 2000. LNCS, vol. 1851, pp. 314–327. Springer, Heidelberg (2000)
7. Gudmundsson, J., Levcopoulos, C., Narasimhan, G.: Fast greedy algorithms for constructing sparse geometric spanners. SIAM Journal on Computing 31(5) (2002)
8. Karavelas, M.I., Guibas, L.J.: Static and kinetic geometric spanners with applications. In: Proceeding of the Twelfth Annual Symposium on Discrete algorithms, pp. 168–176 (2001)
9. Kortsarz, G., Peleg, D.: Generating low-degree 2-spanners. In: Proceedings of the fifth annual ACM-SIAM symposium on Discrete algorithms (2000)

10. Lanthier, M., Maheshwari, A., Sack, R.J.: Approximating Shortest Paths on Weighted Polyhedral Surfaces. Algorithmica 30(4), 527–562 (2001)
11. Mitchell, J., Mount, D., Papadimitriou, C.: The discrete geodesic problem. SIAM Journal on Computing 16(4), 647–668 (1987)
12. Peleg, D., Schaffer, A.: Graph spanners. Journal of Graph Theory 13(1), 99–116 (1989)
13. Regev, H. The Weight of the Greedy Graph Spanner. Weizmann Institute of Science, Dept. of Applied Mathematics and Computer Science (1995)
14. Regev, H.: The weight of the greedy graph spanner. Tech. Rep. CS95-22, 1 (1995)

Approximating Points by a Piecewise Linear Function: I*

Danny Z. Chen and Haitao Wang**

Department of Computer Science and Engineering
University of Notre Dame, Notre Dame, IN 46556, USA
{dchen,hwang6}@nd.edu

Abstract. We study the problem of approximating a set of weighted planar points by a step function, and the problems of approximating non-weighted and weighted planar points by a (more general) piecewise linear function. We either improve the previously best-known results or give the first-known results for these problems. Our algorithms are based on interesting and nontrivial geometric techniques and data structures, which may find other applications. Further, we present the first-known results for the 3-D versions of the step function approximation problem.

1 Introduction

Approximating a set of points by a functional curve or surface in the d-D space is a fundamental topic in computational geometry. It finds applications in many areas, such as cartography, GIS, machine learning, image processing, database, etc. Different error metrics, constraints, and objective functions give rise to a large number of variations of the problem. For each variation, based on the optimization criteria, two problem versions, *min-#* and *min-ϵ*, are often considered in the literature. The definitions of the problems we study are given as follows.

Let $P = \{p_1, p_2, \ldots, p_n\}$ be the input point set, with $p_i = (x_i, y_i, z_i)$ (in the 2-D case, every $z_i = 0$). The *vertical distance* between any point $p_i \in P$ and an approximating functional curve (or surface) f is defined as $d(p_i, f) = |y_i - f(x_i)|$ in 2-D and $|z_i - f(x_i, y_i)|$ in 3-D. The *uniform metric* of error, also known as the L_∞ or *Chebychev* metric, is defined as $e(P, f) = \max_{p_i \in P} d(p_i, f)$. All problems in this paper use the uniform metric. The *size* of f is the total number of line segments in 2-D (or faces in 3-D) of f. Formally, the *min-#* and *min-ϵ* problem versions are defined as follows.

***min-#*:** Given an error tolerance $\epsilon \geq 0$, find an approximating function f under the specified constraints such that $e(P, f) \leq \epsilon$ and the size of f is minimized.

* This research was supported in part by NSF under Grants CCF-0515203 and CCF-0916606.

** Corresponding author. This author's work was also supported in part by a graduate fellowship from the Center for Applied Mathematics, University of Notre Dame.

Y. Dong, D.-Z. Du, and O. Ibarra (Eds.): ISAAC 2009, LNCS 5878, pp. 224–233, 2009.

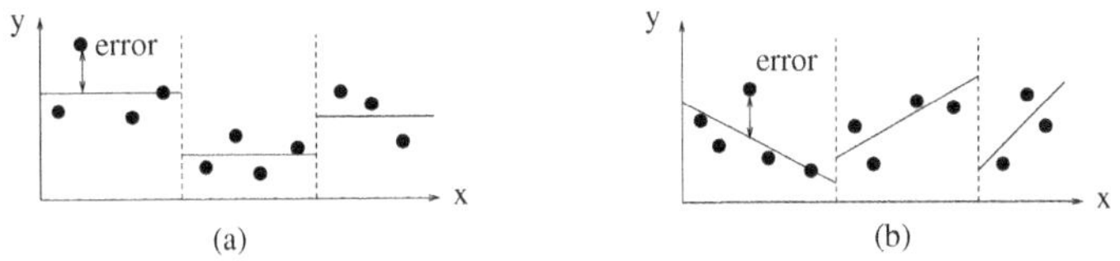

Fig. 1. (a) A step function. (b) A piecewise linear function.

***min-ε*:** Given an integer $k > 0$, find an approximating function f under the specified constraints such that the size of f is no bigger than k and the error $e(P, f)$ is minimized.

Depending on different constraints on f, there are several problem variations.

Planar points approximation by a step function. Given P in 2-D, the sought f is a step function, represented by a sequence of horizontal segments (see Fig. 1(a)). This problem is motivated by query optimizations and histogram constructions in database management systems [12,13,14,17,18]. In the paper, we use SF to denote this problem.

Planar points approximation by a piecewise linear function. Given P in 2-D, f is piecewise linear and any two consecutive line segments of f need not be joined (see Fig. 1(b)). This problem often arises in regression analysis and also in the histogram construction but in the steaming model [2]. Denote this problem by PF.

Weighted versions. Each point $p_i \in P$ has a weight $u_i \geq 0$ and $d(p_i, f)$ is defined as $u_i \cdot |y_i - f(x_i)|$. The weighed versions are motivated by applications with data of non-uniform significance [13]. Denote the weighted versions of SF and PF by WSF and WPF, respectively.

3-D versions. A step function in 3-D can be represented by a rectilinear surface consisting of rectangular faces parallel to the xy-plane, such that any line parallel to the z-axis intersects at most one such face. Denote the 3-D versions of SF and WSF by SF3 and WSF3, respectively.

The *min-#* versions of all above 2-D problems have been solved in linear time, and so has the *min-ε* version of SF. In this paper, we study the *min-ε* versions of the other three 2-D problems, i.e., WSF, PF, and WPF, as well as the *min-#* and *min-ε* versions of the two 3-D problems. By the way, our work on other extended problem variations is given in a companion paper [6].

To simplify the exposition, we assume all input points are in non-degenerate positions. Namely, no two points in P have the same x-coordinate in the 2-D problems and no two points in P have both the same x-coordinate and the same y-coordinate in the 3-D problems. In all 2-D problems, unless otherwise stated, we assume that the input points, $P = \{p_1, p_2, \ldots, p_n\}$, are already given sorted in increasing x-coordinates. We use P_{ij} $(i \leq j)$ to denote the subset of consecutive points $p_i, p_{i+1}, \ldots, p_j$ and use w_{ij} to denote the minimum error for approximating all points in P_{ij} by one line segment under the corresponding function constraints.

1.1 Related Work

The SF problem was studied in [7], which solved the *min-#* version in $O(n)$ time and gave an $O(n^2 \log n)$ time *min-ϵ* algorithm. Later on, the *min-ϵ* SF solution was improved in [13,22,26], and was eventually solved optimally in $O(n)$ time in a recent paper [9]. The algorithm in [9] is based on a more general formulation on the path partition problem [10]. The WSF *min-#* problem was solved in $O(n)$ time [18]. Its *min-ϵ* version was solved in $O(n \log n + k^2 \log^6 n)$ time [13] and $O(n \log^4 n)$ time [9], respectively; both these *min-ϵ* algorithms use a data structure developed in [13] to compute each w_{ij} in $O(\log^4 n)$ time, after $O(n \log n)$ time preprocessing.

The WPF *min-#* problem can be modeled as the *straight line fitting problem* (i.e., determining the minimum number of straight lines piercing each of n given data ranges), which was solved optimally in $O(n)$ time [24]. As a special case of WPF, the PF *min-#* problem can also be solved in linear time. No previous work specifically on the *min-ϵ* problems of either PF or WPF is found. However, for a problem related to PF, called the *polygonal fitting problem*, any two consecutive line segments of the approximating curve are required to be joined at their ends. This problem and its variations have been studied extensively. An $O(n)$ time *min-#* algorithm and an $O(n^2 \log n)$ time *min-ϵ* algorithm were first given [15]. The *min-ϵ* algorithm was improved to $O(n^2)$ time [27] and further to $O(n \log n)$ time [11]. The results on a more restricted case, in which the vertices of the approximating curve are constrained to be a subset of the input point set or the set of vertices of the input curve, can be found in [25].

SF3 and WSF3 can be modeled as certain variants of the rectangle tiling problem [1,19]. But we are not aware of any previous work on such tiling variants.

1.2 Our Contributions

We develop two high-level algorithmic frameworks for the *min-ϵ* versions of the three 2-D problems, WSF, PF, and WPF. Each framework, which consists of a set of computational components, is applied to every problem. For each individual problem, based on its specific geometric properties, we design different data structures and procedures for carrying out the major components of the corresponding algorithmic framework. Our data structures may be of independent interest and find other applications. Specifically, we solve the WSF *min-ϵ* problem in $O(\min\{n \log^2 n, n \log n + k^2 \log^2 \frac{n}{k} \log^2 n\})$ time, which improves the $O(n \log^4 n)$ time result in [9] and the $O(n \log n + k^2 \log^6 n)$ time result in [13]; for the PF and WPF *min-ϵ* problems, we give their first-known results: The two problems are solved in $O(\min\{n \log n, n + k^2 \log^2 \frac{n}{k} \log n \log\log n\})$ time and $O(\min\{n \log^3 n, n \log n + k^2 \log^2 \frac{n}{k} \log^3 n\})$ time, respectively.

For the 3-D problems, we prove that both the *min-#* and *min-ϵ* versions of SF3 and WSF3 are NP-hard. We also prove that the *min-ϵ* problems of both SF3 and WSF3 cannot be approximated with a factor smaller than 1.5 in polynomial time (unless P = NP). Furthermore, we present polynomial time 2-approximation *min-#* algorithms for SF3 and WSF3. Due to the space limit, our results on the 3-D problems are in the full paper [5].

2 Algorithmic Frameworks

In this section, we present the two high-level algorithmic frameworks $\mathbf{F_1}$ and $\mathbf{F_2}$.

The first framework $\mathbf{F_1}$ is a general formulation of the path partition algorithm [9,10]. Let $\theta(i,j)$ be a function for any two integers i and j with $1 \leq i \leq j \leq n$ such that $\theta(i,j) \geq 0$, and $\theta(i,j) = 0$ when $i = j$. Define the problem of MIN-MAX PARTITION(θ) as follows: Partition the (integer) interval $I = [1, 2, \ldots, n]$ into k subintervals $I_1, \ldots, I_k$, where each subinterval I_i consists of consecutive integers from $l_{i-1}+1$ to l_i (with $l_0 = 1$ and $l_k = n$), such that $\max_{1 \leq i \leq k} \theta(l_{i-1}+1, l_i)$ is minimized. Suppose θ has the following properties: (1) θ is non-decreasing, that is, $\theta(i,j) \leq \theta(i',j')$ for $1 \leq i' \leq i \leq j \leq j' \leq n$, and (2) after an $O(\pi(n))$ time preprocessing, for any integer pairs $i \leq j$, $\theta(i,j)$ can be computed in $q(n)$ time. Then based on the results in [9], the MIN-MAX PARTITION(θ) problem can be solved in $O(\pi(n) + n \cdot q(n))$ time. For each of the three 2-D problems, if we let $\theta(i,j) = w_{ij}$, then it is easy to see that $\theta(i,j) = 0$ when $i = j$ and θ is non-decreasing. Therefore, the following lemma holds.

Lemma 1. *For each of the three 2-D problems, if there is a data structure for queries w_{ij} with time bounds $O(\pi(n), q(n))$, then its min-ϵ version can be solved in $O(\pi(n) + n \cdot q(n))$ time.*

The second framework $\mathbf{F_2}$ relies on parametric search, in which we use an improved (after preprocessing) *min-#* algorithm as the "“master” algorithm. We give our improved *min-#* algorithm in Lemma 2, and then present $\mathbf{F_2}$. For the time bound below, note that $O(k \log \frac{n}{k}) \leq O(n)$ for any $k \leq n$.

Lemma 2. *For any min-# problem, if there is a data structure of $O(\pi(n), q(n))$ time for queries w_{ij}, then it is solvable in $O(q(n) \cdot k^* \log \frac{n}{k^*})$ time, where k^* is the size of the sought optimal approximating function for the given error ϵ.*

Proof. For the given error tolerance ϵ, by using the w_{ij} query as the "probing" mechanism, our *min-#* algorithm determines for any point p_i, the rightmost point p_j such that $i \leq j$ and $w_{ij} \leq \epsilon$ (i.e., the points from p_i to p_j can be approximated by one line segment under the specified constraints). Then the algorithm, starting at p_1, repeatedly finds the next line segment in a greedy fashion. If the i-th segment ($1 \leq i \leq k^*$) approximates n_i points, then computing that segment takes $O(q(n) \log n_i)$ time by using exponential search. Hence, the *min-#* algorithm runs in $O(q(n) \sum_{i=1}^{k^*} \log n_i)$ time, which is $O(q(n) \cdot k^* \log(\frac{n}{k^*}))$ due to the fact that $\sum_{i=1}^{k^*} \log n_i \leq k^* \log(\frac{n}{k^*})$.

Our parametric search uses the above *min-#* algorithm as the master algorithm. Given $k > 0$, our goal is to compute the smallest error ϵ^*. Suppose in the master *min-#* algorithm, the current line segment starts at p_i. When sweeping rightwards, for any encountered point p_j ($i < j$), we need to decide whether p_j can be approximated by the current line segment based on the size relation between w_{ij} and ϵ^* (e.g., $w_{ij} \leq \epsilon^*$ or not). Although the value of ϵ^* is not yet known, we can use an indirect way to determine this size relation, as follows.

Given an error ϵ, let $F(\epsilon)$ denote the size of an optimal function under the specified constraints to approximate P. The size relation between w_{ij} and ϵ^* can be decided by the values of k and $F(w_{ij})$. If $F(w_{ij}) > k$, then it must be $w_{ij} < \epsilon^*$. If $F(w_{ij}) \leq k$, however, then either $w_{ij} = \epsilon^*$ or $w_{ij} > \epsilon^*$ holds. If $w_{ij} = \epsilon^*$, then the algorithm can stop with $\epsilon^* = w_{ij}$; otherwise, p_j should not be approximated by the current line segment and a new segment should start at p_j. To determine whether $w_{ij} = \epsilon^*$ or $w_{ij} > \epsilon^*$, we use the lemma below with the proof omitted.

Lemma 3. *An error ϵ is ϵ^* if and only if $F(\epsilon) \leq k$ and $F(\epsilon - \delta) > k$ for any $\delta > 0$.*

Given an error ϵ, to compute the value of $F(\epsilon)$, we resort to our improved *min-#* algorithm in Lemma 2. To determine whether $F(\epsilon - \delta) > k$ for any $\delta > 0$, we only need to modify the *min-#* algorithm in Lemma 2 slightly: It computes an approximating function f with $e(P, f) < \epsilon$ instead of $e(P, f) \leq \epsilon$. Furthermore, since we only need to know the size relation between $F(\epsilon)$ (resp., $F(\epsilon - \delta)$) and k, both algorithms can stop whenever the current number of line segments is larger than k. Thus, after w_{ij} is computed in $O(q(n))$ time, we can determine the size relation between w_{ij} and the unknown ϵ^* in $O(q(n) \cdot k \log \frac{n}{k})$ time. As the analysis in Lemma 2, using exponential search, our *min-ϵ* algorithm takes $O(q(n)k^2 \log^2 \frac{n}{k})$ time to obtain the optimal solution.

Lemma 4. *For each of the three 2-D problems, if there is a data structure for the w_{ij} queries with time bounds $O(\pi(n), q(n))$, then its min-ϵ version can be solved in $O(\pi(n) + q(n)k^2 \log^2 \frac{n}{k})$ time.*

3 Data Structures for Major Computational Components

In this section, we present our data structures for the major components for each 2-D problem, i.e., the 2-D sublist LP queries for WSF, the 3-D sublist LP queries for WPF, and the vertical hull width queries for PF.

3.1 2-D Sublist LP Queries for Problem WSF

For WSF, the data structure for each w_{ij} query in [13] (also used in [9]) takes $O(\log^4 n)$ time after an $O(n \log n)$ time preprocessing. We develop an improved data structure with $O(n \log n, \log^2 n)$ time. As a result, both the *min-ϵ* algorithms in [9,13] are improved by a factor of $O(\log^2 n)$ time.

We model this query problem in a more "natural" way. Let each point $p_t = (x_t, y_t) \in P$ with a weight $u_t \geq 0$. For a point set P_{ij}, we first show that our goal is to find an approximating function $y = y^*$ such that $\max_{i \leq t \leq j}(u_t|y_t - y^*|)$ is minimized and that the sought w_{ij} is the minimized value of $\max_{i \leq t \leq j}(u_t|y_t - y^*|)$. Suppose we use a function $y = y'$ to approximate P_{ij} and ϵ is its error. Then for each point $p_t = (x_t, y_t)$ with weight u_t in P, we have two constraints: $u_t(y_t - y') \leq \epsilon$ and $-u_t(y_t - y') \leq \epsilon$. Consider a 2-D coordinate plane $\mathcal{P}$ with

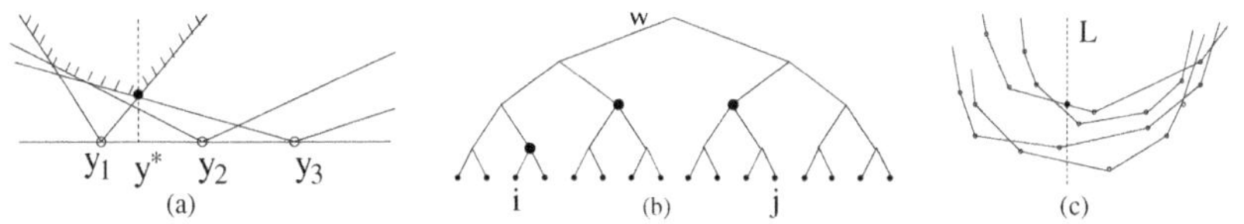

Fig. 2. (a) The black point is p^*. (b) A query (i, j). (c) L intersecting convex chains.

the y' value as its x-axis and the ϵ value as its y-axis. Then each point in P gives rise to two halfplanes in the plane $\mathcal{P}$. It is easy to see that if $p^* = (y^*, \epsilon^*)$ is the lowest point in the common intersection of the $2(i - j + 1)$ halfplanes for P_{ij}, then $y = y^*$ is the optimal approximating function and $w_{ij} = \epsilon^*$. In the example of Fig. 2(a), each point of P is associated with a "cone" that is bounded by two halflines (the parts below the x-axis are not drawn), i.e., defined by the two constraints for that point. Thus, we are considering the following problem: Given a set of upper halfplanes $H = \{h_i \mid 1 \leq i \leq n\}$ in the plane in an *arbitrary order*, compute the lowest point p^* in the common intersection of all halfplanes in $H_{ij} = \{h_t \mid i \leq t \leq j\}$ specified by a query $q(i, j)$. We call it the *2-D sublist LP query* problem. Clearly, answering such a 2-D sublist LP query also answers the query w_{ij} for WSF. An interesting technique, which we refer to as *binary search on sorted arrays*, is used repeatedly in this paper. The following result can be obtained by using similar techniques as in [21].

Lemma 5. *Given m arrays A_i, $1 \leq i \leq m$, each containing $O(n)$ elements in sorted order, a sought element δ in $A = \cup_{i=1}^{m} A_i$ can be determined in $O((m + T)\log(nm))$ time, where $O(T)$ is the time taken by one call to a decision procedure Π which, given any value a, can report $a \leq \delta$ or $a > \delta$.*

For ease of presentation, we first give a basic data structure in Lemma 6 with time bounds $O(n \log n, \log^3 n)$, and then apply the technique of fractional cascading [4] for further improvement.

Lemma 6. *After $O(n \log n)$ time preprocessing, any w_{ij} can be computed in $O(\log^3 n)$ time.*

Proof. For the given halfplane set H, we build a complete binary tree T whose i-th leaf stores the halfplane h_i. Each internal node v of T is associated with the (convex) common intersection chain C_v that bounds the common intersection of all halfplanes stored at the leaves of the subtree rooted at v. Each C_v is stored in an array A_v in increasing x-coordinates. Thus, in a bottom-up fashion, T can be built in $O(n \log n)$ time and space.

To answer a query $q(i, j)$, we first locate the lowest common ancestor (LCA) w of leaves i and j in T in $O(1)$ time [16]. By following the paths in T from w to i and to j, we can obtain $O(\log n)$ arrays representing $O(\log n)$ common intersection chains which together define the common intersection of all hyperplanes in H_{ij} (in Fig. 2(b), the arrays we need are stored at the black nodes). Denote the sets of these $O(\log n)$ chains and arrays by $C(i, j)$ and $A(i, j)$, respectively.

Since the elements in each array of $A(i,j)$ are in increasing x-coordinate order, to find p^*, we perform a binary search on these sorted arrays of $A(i,j)$ (for $l = O(\log n)$ chains), using Lemma 5. The decision procedure Π for this LP problem works as follows. For a given value $\overline{\delta}$, Π first computes, for every chain $C_t \in C(i,j)$ $(1 \leq t \leq l)$, the intersection point q_t of C_t and the vertical line L: $x = \overline{\delta}$. Using the point set $Q = \{q_1, \ldots, q_l\}$ and the local information around each point q_t on C_t, we can decide (as in [8]), in $O(l)$ time, whether p^* is at the highest point in Q, and if not, p^* lies on which side of the line L. Figure 2(c) gives an example in which p^* lies to the right of L. If a binary search is performed on each C_t to locate the point q_t, then computing the set Q takes $O(l \log n) = O(\log^2 n)$ time, which is the dominating time of the procedure Π.

Using the above procedure Π, p^* can be obtained from the chains C_t by applying Lemma 5. When the chains C_t have altogether $O(\log n)$ edges left, we simply use a linear time LP algorithm [8] to find p^*, in $O(\log n)$ time.

Because there are $O(\log n)$ sorted arrays A_t, $|A(i,j)| = O(n)$, and the time of the procedure Π is $O(T) = O(\log^2 n)$, computing p^* (and thus w_{ij}) takes $O((T + \log n)\log n) = O(\log^3 n)$ time.

To improve the query time, an immediate target is the $O(\log^2 n)$ time decision procedure Π on the $O(\log n)$ sorted chains C_t. Here, the dominating cost is at computing, for each chain C_t, its intersection with a vertical line L. A useful property is that the $O(\log n)$ chains C_t are organized along two ancestor-to-descendant paths in the tree T. This allows us to cast the computation of the intersections between L and the C_t's as an *iterative search problem* and thus handle it by the fractional cascading technique [4]. In the tree T of Lemma 6, a sorted array (for a C_t) is associated with each internal node. Using T as a *catalog graph* defined in [4] (of a total size $O(n \log n)$) with a *locally bounded degree* 3, by Theorem 2 in [4], we can build a data structure on T in $O(n \log n)$ time and space which enables us to search for the intersection between L and the first chain (say) C_1 in $O(\log n)$ time, and the subsequent intersections in $O(1)$ time per chain. Therefore, all intersections between L and the $O(\log n)$ C_t's are obtained in $O(\log n)$ time. Consequently, the time bound of the decision procedure Π becomes $O(T) = O(\log n)$. This gives us the following result.

Theorem 1. *The time complexity of our data structure for the 2-D sublist LP queries is $O(n \log n, \log^2 n)$.*

By combining Theorem 1 with Lemmas 1 and 4, we have the following result.

Theorem 2. *Given $k > 0$, the min-ϵ version of WSF can be solved in $O(n \log^2 n)$ time by the framework $\mathbf{F_1}$, and in $O(n \log n + k^2 \log^2 n \log^2 \frac{n}{k})$ time by the framework $\mathbf{F_2}$, respectively.*

3.2 3-D Sublist LP Queries for Problem WPF

For WPF, after a similar modeling as in the 2-D case, the problem of computing w_{ij}'s can be modeled to the following problem: Given a set of upper halfspaces,

$H = \{h_i \mid 1 \leq i \leq n\}$, in 3-D in an *arbitrary order*, a query $q(i, j)$ finds the lowest point p^* in the common intersection of all halfspaces in $H_{ij} = \{h_t \mid i \leq t \leq j\}$. We call this the *3-D sublist LP query* problem, and derive a data structure for answering each query in $O(\log^3 n)$ time with an $O(n \log n)$ time preprocessing. Note that this problem can be viewed an extension of the 2-D sublist LP query problem, but the techniques of the solution are much more sophisticated.

The first step of our preprocessing is to build a complete binary tree T for the halfspace set H, whose i-th leaf stores h_i and each internal node v stores a convex polyhedron S_v that bounds the common intersection of all halfspaces stored at the leaves of the subtree rooted at v. We use the doubly-connected-edge-list (DCEL) data structure [23] to represent each S_v. T is built in the bottom-up fashion. At an internal node v, from the two DCEL's representing the two polyhedra stored at v's two children, using Chazelle's algorithm [3], we compute the DCEL representing the common intersection of these two polyhedra in linear time. Since $|S_v| = O(m)$, where m is the number of leaves of the subtree rooted at v, all S_v's in T can be constructed in totally $O(n \log n)$ time. Further, to facilitate our query procedure, for S_v stored at each node v, in addition to its DCEL, we also represent it by Kirkpatrick's hierarchical data structure for point location [20], denoted by KH_v. We triangulate each face of S_v (and also KH_v) in $O(|S_v|)$ time. Note that each KH_v consists of $O(\log |S_v|)$ levels of triangular subdivisions [20]. Since every KH_v can be produced from S_v in $O(|S_v|)$ time [20], the total time for building the entire tree T is $O(n \log n)$.

Given a query $q(i, j)$, let w be the LCA of the leaves i and j in T. By following the paths from w to i and to j in T (see Fig. 2(b)), we obtain $O(\log n)$ nodes such that the common intersection C of the polyhedra stored at these nodes is the common intersection of all halfspaces in H_{ij}. Our goal is to find the lowest point p^* in the common intersection C of these $O(\log n)$ polyhedra.

Denote by S the set of these $O(\log n)$ polyhedra and let $h(n) = |S|$. For each polyhedron S_i in S, we also use S_i to denote its corresponding data structure KH_i. Let $g(n) = O(\log n)$. For every S_i, denote its j-th level subdivision by S_{ij}, $1 \leq j \leq g(n)$. Then $S_{i,g(n)} = S_i$ and $|S_{i1}| = O(1)$. Divide the level sequence of S_i into *level blocks* such that each block consists of $(\log \log n - \log \log \log n)/4$ consecutive levels. The number of level blocks is $l(n) = \frac{4g(n)}{\log \log n - \log \log \log n}$. For every S_i, denote each its block by BL_{ij}, $1 \leq j \leq l(n)$. Clearly, $S_{i1} \in BL_{i1}$ and $S_{i,g(n)} \in BL_{i,l(n)}$. It follows from [20] that every triangle at the lowest level (i.e., the level with the smallest index) of the j-th block can intersect at most $I(n) = d^{\frac{1}{4}(\log \log n - \log \log \log n)}$ triangles at the lowest level of the $(j+1)$-th block, where d is a constant whose value is upper bounded by 11. Thus $I(n) = O(2^{\log \log n - \log \log \log n}) = O(\frac{\log n}{\log \log n})$.

We briefly describe the basic idea of the query procedure, and leave the details in the full paper [5]. Starting from the first block BL_{i1} of every S_i, in each iteration, our idea is to find, at the lowest level of each block BL_{ij}, $O(1)$ triangles to one of which the sought point p^* belongs. Of course, we must do so without knowing where p^* actually locates. Here we say p^* *belongs to* a triangle t or t *contains* p^* if the projection of p^* is in the projection of t on the xy-plane.

After $O(l(n))$ iterations, for every S_i in S, we reach the last level $S_{i,g(n)}$, which is S_i itself, and obtain $O(1)$ triangles in S_i containing p^*. In other words, we find $O(\log n)$ triangles from all S_i's of S such that p^* is the lowest point in the common intersection of the halfspaces for these triangles. By applying the linear time 3-D LP algorithms [8], we then determine p^* in $O(\log n)$ time.

Theorem 3. *For the 3-D sublist LP query problem, there is a data structure that can answer each query in $O(\log^3 n)$ time with an $O(n \log n)$ time preprocessing.*

Theorem 4. *Given $k > 0$, the min-ϵ version of WPF can be solved in $O(n \log^3 n)$ time by the framework $\mathbf{F_1}$, and in $O(n \log n + k^2 \log^2 \frac{n}{k} \log^3 n)$ time by the framework $\mathbf{F_2}$, respectively.*

3.3 Vertical Hull Width Queries for Problem PF

Due to the space limit, we only give the following results and leave all the details in our full paper [5].

Theorem 5. *There exist data structures for the vertical hull width query problem with time bounds $O(n \log\log n, \log n)$ and $O(n, \log n \log\log n)$, respectively.*

Theorem 6. *Given $k > 0$, the min-ϵ version of PF can be solved in $O(n \log n)$ time by framework $\mathbf{F_1}$, and in $O(n + k^2 \log^2 \frac{n}{k} \log n \log\log n)$ time by $\mathbf{F_2}$.*

References

1. Berman, P., DasGupta, B., Muthukrishnan, S.: Slice and dice: A simple, improved approximate tiling recipe. In: Proc. of the 13th ACM-SIAM Symposium on Discrete Algorithms (SODA), pp. 455–464 (2002)
2. Buragohain, C., Shrivastava, N., Suri, S.: Space efficient streaming algorithms for the maximum error histogram. In: Proc. of the 23rd International Conference on Data Engineering (ICDE), pp. 1026–1035 (2007)
3. Chazelle, B.: An optimal algorithm for intersecting three-dimensional convex polyhedra. SIAM Journal on Computing 21(4), 671–696 (1992)
4. Chazelle, B., Guibas, L.: Fractional cascading: I. A data structuring technique. Algorithmica 1(1), 133–162 (1986)
5. Chen, D.Z., Wang, H.: Approximating points by a piecewise linear function (manuscript, 2009)
6. Chen, D.Z., Wang, H.: Approximating points by a piecewise linear function: II. Dealing with outliers. In: Proc. of the 20th International Symposium on Algorithms and Computation, ISAAC (2009)
7. Díaz-Bánez, J., Mesa, J.: Fitting rectilinear polygonal curves to a set of points in the plane. European Journal of Operational Research 130(1), 214–222 (2001)
8. Dyer, M.: Linear time algorithms for two- and three-variable linear programs. SIAM Journal on Computing 13(1), 31–45 (1984)
9. Fournier, H., Vigneron, A.: Fitting a step function to a point set. In: Halperin, D., Mehlhorn, K. (eds.) ESA 2008. LNCS, vol. 5193, pp. 442–453. Springer, Heidelberg (2008)

10. Frederickson, G.: Optimal algorithms for tree partitioning. In: Proc. of the 2nd Annual ACM-SIAM Symposium of Discrete Algorithms (SODA), pp. 168–177 (1991)
11. Goodrich, M.: Efficient piecewise-linear function approximation using the uniform metric. In: Proc. of the 10th Annual ACM Symposium on Computational Geometry, pp. 322–331 (1994)
12. Guha, S.: On the space—time of optimal, approximate and streaming algorithms for synopsis construction problems. The VLDB Journal – The International Journal on Very Large Data Bases 17(6), 1509–1535 (2008)
13. Guha, S., Shim, K.: A note on linear time algorithms for maximum error histograms. IEEE Transactions on Knowledge and Data Engineering 19(7), 993–997 (2007)
14. Guha, S., Shim, K., Woo, J.: Rehist: Relative error histogram construction algorithms. In: Proc. of the 30th International Conference on Very Large Data Bases (VLDB), pp. 300–311 (2004)
15. Hakimi, S., Schmeichel, E.: Fitting polygonal functions to a set of points in the plane. CVGIP: Graphical Models and Image Processing 53, 132–136 (1991)
16. Harel, D., Tarjan, R.E.: Fast algorithms for finding nearest common ancestors. SIAM Journal on Computing 13, 338–355 (1984)
17. Jagadish, H.V., Koudas, N., Muthukrishnan, S., Poosala, V., Sevcik, K.C., Suel, T.: Optimal histograms with quality gurantees. In: Proc. of the 24th International Conference on Very Large Data Bases (VLDB), pp. 275–286 (1998)
18. Karras, P., Sacharidis, D., Mamoulis, N.: Exploiting duality in summarization with deterministic guarantees. In: Proc. of the 13th ACM SIGKDD International Conference on Knowledge Discovery and Data Mining, pp. 380–389 (2007)
19. Khanna, S., Muthukrishnan, S., Paterson, M.: On approximating rectangle tiling and packing. In: Proc. of the 9th Annual ACM-SIAM Symposium on Discrete Algorithms (SODA), pp. 384–393 (1998)
20. Kirkpatrick, D.: Optimal search in planar subdivisions. SIAM Journal on Computing 12(1), 28–35 (1983)
21. Krizanc, D., Morin, P., Smid, M.: Range mode and range median queries on lists and trees. Nordic Journal of Computing 12(1), 1–17 (2005)
22. Mayster, Y., Lopez, M.: Approximating a set of points by a step function. Journal of Visual Communication & Image Representation 17, 1178–1189 (2006)
23. Muller, D., Preparata, F.: Finding the intersection of two convex polyhedra. Theoretical Computer Science 7, 217–236 (1978)
24. O'Rourke, J.: An on-line algorithm for fitting straight lines between data ranges. Communications of the ACM 24, 574–578 (1981)
25. Varadarajan, K.: Approximating monotone polygonal curves using the uniform metric. In: Proc. of the 12th Annual ACM Symposium on Computational Geometry, pp. 311–318 (1996)
26. Wang, D.: A new algorithm for fitting a rectilinear x-monotone curve to a set of points in the plane. Pattern Recognition Letters 23(1), 329–334 (2002)
27. Wang, D., Huang, N., Chao, H., Lee, R.: Plane sweep algorithms for the polynomial approximation problems with applications. In: Ng, K.W., Balasubramanian, N.V., Raghavan, P., Chin, F.Y.L. (eds.) ISAAC 1993. LNCS, vol. 762, pp. 515–522. Springer, Heidelberg (1993)

Approximating Points by a Piecewise Linear Function: II. Dealing with Outliers*

Danny Z. Chen and Haitao Wang**

Department of Computer Science and Engineering
University of Notre Dame, Notre Dame, IN 46556, USA
{dchen,hwang6}@nd.edu

Abstract. In this paper, we study the violation versions of the planar points approximation problems, which deal with outliers in the input points. We present efficient algorithms for both the step function and the more general piecewise linear function cases, and for both non-weighted and weighted points. Most of our results are first-known. Our algorithms are based on interesting and nontrivial geometric techniques and data structures, which may find other applications.

1 Introduction

Approximating a set of points by a functional curve in the plane is a fundamental topic in mathematics and computational geometry. It finds applications in many areas, such as cartography, geographic information systems, machine learning, image processing, database, etc. Different error metrics, constraints, and objective functions give rise to a large number of variations of the problem. For each variation, based on the optimization criteria, two problem versions, *min-#* and *min-ϵ*, are often considered in the literature. The formal definitions of the problems we study in this paper is give below.

Let $P = \{p_1, p_2, \ldots, p_n\}$ be the input point set in the plane, with $p_i = (x_i, y_i)$. The *vertical distance* between any point p_i and an approximating functional curve f is defined as $d(p_i, f) = |y_i - f(x_i)|$. The *uniform metric* of error, also known as the L_∞ or *Chebychev* metric, is defined as $e(P, f) = \max_{p_i \in P} d(p_i, f)$. All problems in this paper use the uniform metric. The *size* of f is the total number of line segments of f. The *min-#* and *min-ϵ* versions are defined below.

min-#: Given an error tolerance $\epsilon \geq 0$, find an approximating function f under the specified constraints such that $e(P, f) \leq \epsilon$ and the size of f is minimized.

min-ϵ: Given an integer $k > 0$, find an approximating function f under the specified constraints such that the size of f is no bigger than k and the error $e(P, f)$ is minimized.

* This research was supported in part by NSF under Grants CCF-0515203 and CCF-0916606.

** Corresponding author. The work of this author was also supported in part by a graduate fellowship from the Center for Applied Mathematics, University of Notre Dame.

Y. Dong, D.-Z. Du, and O. Ibarra (Eds.): ISAAC 2009, LNCS 5878, pp. 234–243, 2009.

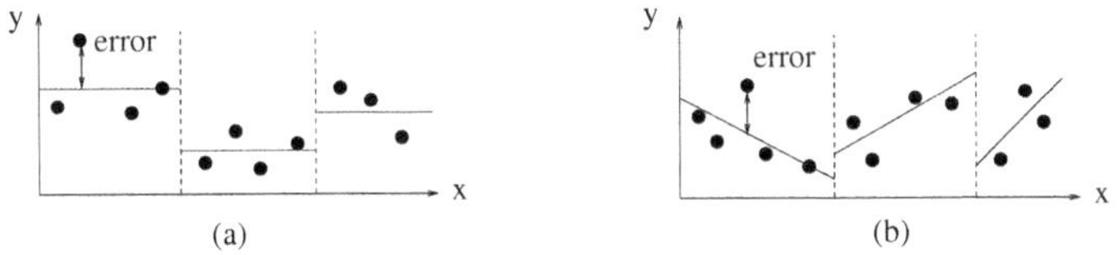

Fig. 1. (a) A step function. (b) A piecewise linear function.

Depending on different constraints on f, there are several problem variations.

Planar points approximation by a step function. Given P, the sought f is a step function, represented by a sequence of horizontal segments (see Fig. 1(a)). This problem is motivated by query optimizations and histogram constructions in database management systems [15,16,17,18,19]. In the paper, we use SF to denote this problem.

Planar points approximation by a piecewise linear function. Given P, f is piecewise linear and any two consecutive line segments of f need not be joined (see Fig. 1(b)). This problem often arises in regression analysis and also in the histogram construction but in the steaming model [3]. Denote this problem by PF.

Weighted versions. Each point $p_i \in P$ has a weight $u_i \geq 0$ and $d(p_i, f)$ is defined as $u_i \cdot |y_i - f(x_i)|$. The weighed versions are motivated by applications with data of non-uniform significance. Denote the weighted versions of SF and PF by WSF and WPF, respectively.

Violation versions. When approximating P with f, at most g points of P are allowed to violate the error tolerance and these points are called *outliers*. For example, if P' is the set of violation points with $|P'| \leq g$, then $e(P, f) = \max_{p_i \in P \setminus P'} d(p_i, f)$. This problem is motivated by applications in statistics, machine learning, data mining, databases, etc, where outliers must be reduced. Denote the violation versions of SF, PF, WSF, and WPF by VSF, VPF, VWSF, and VWPF, respectively.

All the non-violation problems, i.e., SF, WSF, PF, and WPF, are discussed in a companion paper [7]. In this paper, we study the *min-#* and *min-ϵ* versions of all the violation problems, i.e., VSF, VWSF, VPF, and VWPF.

To simplify the exposition, we assume all input points are in non-degenerate positions. Namely, no two points in P have the same x-coordinate in all problems. Unless otherwise stated, we assume that the input points, $P = \{p_1, p_2, \ldots, p_n\}$, are already given sorted in increasing x-coordinates. We use P_{ij} $(i \leq j)$ to denote the subset of consecutive points $p_i, p_{i+1}, \ldots, p_j$.

1.1 Related Work

The previous and related work on the *min-#* and *min-ϵ* versions of all non-violation problems, i.e., SF, WSF, PF, and WPF, is discussed in our companion paper [7]. The paper [7] also gives some new results on the above problems.

For the violation problems, VSF was studied in [11], with an $O(ng^2)$ time *min-#* algorithm and an $O(ng^2 \log n)$ time *min-ϵ* algorithm. The *min-#* algorithm is based on dynamic programming. Using an implicit matrix with sorted rows and sorted columns that contains all possible errors, its *min-ϵ* version was solved by applying the technique of binary search in such a matrix [12,13], which uses the *min-#* algorithm as a decision procedure. We are not aware of any previous work on other violation problems.

1.2 Our Contributions

We develop two high-level algorithmic frameworks, $\mathbf{F_1}$ and $\mathbf{F_2}$, for the *min-#* and *min-ϵ* versions of all four problems, i.e., VSF, VWSF, VPF, and VWPF. Each algorithmic framework, which consists of a set of computational components, is applied to every violation problem. For each individual problem, based on its specific geometric properties, we design different data structures and procedures for carrying out the major components in the corresponding framework. Some data structures may be of independent interest and find other applications.

The *min-#* algorithmic framework $\mathbf{F_1}$ solves the problems VWSF, VPF and VWPF in $O(ng^2)$, $O(ng^4 \log^2 n)$ and $O(ng^4 \log^2 n)$ time, respectively. Note that our VWSF *min-#* solution matches the one for VSF in [11]. The *min-ϵ* algorithmic framework $\mathbf{F_2}$ solves the problems VSF, VPF and VWPF in $O(ng^3 k \log(\log^* n))$, $O(ng^2 kW)$ and $O(ng^2 kW)$ time, respectively, where $W = (n \log g + g^3 n^\delta)$ for any constant $\delta > 0$. Thus, when $kg \log(\log^* n) = o(\log n)$ (e.g., when k and g are constant), our VSF *min-ϵ* algorithm is better than the one in [11]. For VWSF, due to its special geometric properties, we derive a more efficient *min-ϵ* algorithm than simply applying $\mathbf{F_2}$, with running time $O(n^2 + ng^2 \log n)$. Additionally, for the *min-ϵ* problem of VWPF (and also for VPF), based on geometric observations and the approach of binary search on sorted arrays [7], we give a new algorithm with time complexity $O(ngW \log n)$. Note that all our solutions are first-known for the corresponding problems except the VSF *min-ϵ* problem.

2 The *min-#* Algorithms

In this section, we discuss our *min-#* algorithms. We first give the high-level algorithmic framework $\mathbf{F_1}$, and then present the techniques for dealing with its major components for each problem.

2.1 High-Level Algorithmic Framework $\mathbf{F_1}$

Given an error tolerance ϵ and a number g of allowed violations, our framework $\mathbf{F_1}$ is based on dynamic programming. Let $N(i,t)$ $(1 \leq i \leq n, 0 \leq t \leq g)$ denote the minimum number of segments for approximating the points in P_{in} with at most t violations. Our goal is to compute $N(1,g)$. Let r_{iq} be the largest index, $r_{iq} \geq i$, such that all points from p_i to $p_{r_{iq}}$ can be approximated by one segment with q violations. Then $N(i,t) = 1 + \min_{0 \leq q \leq t}\{N(r_{iq}+1, t-q)\}$, i.e., for P_{in}, we use one segment with q violations to approximate the points from p_i to $p_{r_{iq}}$ and use $N(r_{iq}+1, t-q)$ segments to approximate the points in $P_{r_{iq}+1,n}$ with $t-q$ violations. The lemma below is based on the above dynamic programming.

Lemma 1. *For any problem, if all r_{iq}'s, $1 \leq i \leq n$, $0 \leq q \leq g$, can be computed in $O(T)$ time, then the min-# version can be solved in $O(T + ng^2)$ time.*

To compute all r_{iq}'s, we develop a high-level procedure which can be applied to all problems. First of all, note that for every q, $0 \leq q \leq g$, $r_{i-1,q} \leq r_{iq}$ holds for any $1 \leq i \leq n$. For each q, our algorithm computes all r_{iq}'s, $1 \leq i \leq n$, by scanning the points from left to right; after all r_{iq}'s are computed, we continue with $q+1$. For each q, suppose $r_{i-1,q}$ has been computed; to compute r_{iq}, we scan rightwards from the point p_j, $j = r_{i-1,q}+1$. For each p_j, let S_{ijq} denote the point set P_{ij} and call it a *feasible set* if all points in S_{ijq} can be approximated by one segment with q violations. Computing r_{iq} is to determine the largest feasible S_{ijq}. For each S_{ijq}, to determine efficiently whether it is a feasible set, we use a fully dynamic data structure that supports both deletions and insertions to maintain a point set. Initially (i.e., after $r_{i-1,q}$ is just computed), we let $j = r_{i-1,q}$ and $S_{ijq} = S_{i-1,j,q} \setminus \{p_{i-1}\}$, implying that p_{i-1} should be deleted from $S_{i-1,j,q}$. For each newly scanned point p_{j+1}, we insert it into S_{ijq}. After each insertion, we need to determine whether the new set $S_{i,j+1,q}$ is feasible. If it is feasible, then we continue with the next point; otherwise, we delete the newly inserted point p_{j+1} from the current set and S_{ijq} is the largest feasible set (i.e., $r_{iq} = j$). In summary, our dynamic data structure should support: (1) *Deletion* (delete a point from the current set); (2) *insertion* (insert a new point into the current set); (3) *feasibility test* (determine whether the current set is feasible).

To analyze the running time of the above framework, for each q, $0 \leq q \leq g$, when computing all r_{iq}'s for $1 \leq i \leq q$, we scan the points of P from left to right. It is easy to see that the number of deletions, insertions, and feasibility tests is totally $O(n)$. By Lemma 1, we have the following result.

Lemma 2. *For any problem, if there is a data structure that supports deletion, insertion, and feasibility test in $O(D)$, $O(I)$, and $O(F)$ time each, respectively, then all r_{iq}'s, $1 \leq i \leq n$, $0 \leq q \leq g$, can be computed in $O(ng(D+I+F))$ time, and its min-# is solvable in $O(ng(g+D+I+F))$ time.*

2.2 Problem VWSF

Given S_{ijq}, for any point $p_t = (x_t, y_t)$ in S_{ijq} with weight u_t, there is a y-interval $[y_t - \epsilon/u_t, y_t + \epsilon/u_t]$ such that p_t can be approximated if and only if the y-coordinate of the approximating segment is in that interval. We call the value $y_t - \epsilon/u_t$ (resp., $y_t + \epsilon/u_t$) the *lower end* (resp., *upper end*) of the y-interval for p_t, denoted by $le(p_t)$ (resp., $ue(p_t)$). Our data structure for S_{ijq} consists of four arrays U_{ij}, L_{ij}, IU_{ij}, and IL_{ij} (all indices start from 0), in addition to a range-minima and a range-maxima data structure [14]. Precisely, the array U_{ij} (resp., L_{ij}) stores the $q+1$ points with the smallest upper (resp., largest lower) ends in

S_{ijq} in increasing y-coordinates. Some points may be in both U_{ij} and L_{ij}. The array IU_{ij} (resp., IL_{ij}) stores the sorted indices of the points in U_{ij} (resp., L_{ij}). In addition, we build, in $O(n)$ time, a range-minima (resp., range-maxima) data structure on the upper (resp., lower) ends of all points in P.

Lemma 3. *For any point set S_{ijq}, the data structure can support insertion, deletion, and feasibility test in $O(q)$ time each.*

Proof. The feasibility test is hinged on the following claim: S_{ijq} is a feasible set if and only if there is an integer t, $0 \leq t \leq q$, such that the $(t+1)$-th smallest value $U_{ij}[t]$ of (the sorted) U_{ij} is no smaller than the $(t+1)$-th smallest value $L_{ij}[t]$ of L_{ij}. Refer to our full paper [6] for the proof of the claim.

To insert a new point p_{j+1} into S_{ijq}, if the number of elements in U_{ij} is less than $q+1$, then we simply insert p_{j+1} into the right place of U_{ij} (so that U_{ij} is still in sorted order). If not, we first determine whether there is a point in U_{ij} whose upper end is larger than that of p_{j+1}. If no, then p_{j+1} is not inserted. If yes, then we delete the point with the largest upper end (i.e., $U_{ij}[q]$) and insert p_{j+1} into the right place of U_{ij}. Whenever U_{ij} is updated, IU_{ij} is updated accordingly. We process the arrays L_{ij} and IL_{ij} in a similar way. Since the sizes of these arrays are at most $q+1$, an insertion takes $O(q)$ time.

To delete a point p_t from S_{ijq}, for the array U_{ij}, if $p_t \notin U_{ij}$, then nothing is done. Otherwise, we delete it from U_{ij} in $O(q)$ time. We also need to find (before removing p_t from U_{ij}) the point, say, p_z, in $S_{ijq} \setminus U_{ij}$ with the smallest upper end, and insert it to the right place of U_{ij}. To do so, we resort to the array IU_{ij} and the range-minima data structure. Recall that IU_{ij} contains the sorted indices of all points in U_{ij}. Thus by IU_{ij}, we can locate $q+2$ consecutive point intervals $[i, IU_{ij}[0])$, $\cdots$, $(IU_{ij}[a], IU_{ij}[a+1])$, $\cdots$, $(IU_{ij}[q], j]$, $0 \leq a \leq q-1$, such that the union of these intervals is $S_{ijq} \setminus U_{ij}$. Using the range-minima data structure for the upper ends of all points in P, we can obtain the point with the smallest upper end in each such point interval. Then the sought p_z is the one with the smallest upper end among the $q+2$ points obtained from these $q+2$ point intervals. Since each range-minima query takes $O(1)$ time, the total deletion time is $O(q)$. Whenever U_{ij} is updated, IU_{ij} is updated as well. The arrays L_{ij} and IL_{ij} can be processed in a similar way.

The above lemma, together with Lemma 2, leads to the following result.

Theorem 1. *There is a min-# algorithm for VWSF with running time $O(ng^2)$.*

2.3 Problems VPF and VWPF

Given $\epsilon > 0$, to determine whether S_{ijq} is a feasible set, we model the problem in the following way. Suppose we use a function $y = a \cdot x + b$ to approximate S_{ijq}. Each non-violation point $p_t = (x_t, y_t) \in S_{ijq}$ with weight u_t corresponds to two constraints $u_t(y_t - (ax_t + b)) \leq \epsilon$ and $u_t(y_t - (ax_t + b)) \geq -\epsilon$. Note that given any approximating function of S_{ijq}, for any point in S_{ijq}, the function can violate at most one constraint corresponding to the point. Therefore, we can determine whether S_{ijq} is a feasible set by the general 2-D LP with q violations [4,5,10,20], for which a and b are LP variables and the $2|S_{ijq}|$ constraints corresponding to all points in S_{ijq} are the LP constraints. The best-known deterministic time bounds for this LP problem are $O(n \log q + q^3 \log^2 n)$ [4] and $O(n \log n + nq)$ [10]. If we computes each r_{iq} separately, the running time for computing all $O(ng)$ r_{iq}'s is $\Omega(n^2 g)$. Since usually $g << n$, we have a more efficient solution below.

After a dynamic convex hull data structure is built, by the algorithm in [20], each feasibility test can be done in $O(q^3 \log^2 n)$ time. To handle each point insertion or deletion, we only need to update the convex hull. By using the dynamic data structure in [21] to maintain the convex hull of S_{ijq}, each insertion and deletion can be handled in $O(\log^2 n)$ time with $O(n \log n)$ time preprocessing. By Lemma 2, we have the following.

Theorem 2. *The min-# algorithms for VPF and VWPF run in* $O(ng^4 \log^2 n)$ *time.*

3 The *min-*ϵ Algorithms

In this section, we first give the algorithmic framework $\mathbf{F_2}$, and then present the techniques for its major components for each problem. Due to their special geometric properties, for VPF and VWPF, we come up with another algorithm. For VWSF, we have a more efficient algorithm than simply applying $\mathbf{F_2}$.

3.1 High-Level Algorithmic Framework $\mathbf{F_2}$

Our framework $\mathbf{F_2}$ is based on dynamic programming which involves three variables. To solve the problem efficiently, we find a way to transform the three-variable version to a set of two-variable cases whose matrices in question are proved to be totally monotone. Thus the linear time row minima algorithm in [1,2] is applicable to produce more efficient solutions.

Given $k > 0$ and $g > 0$, let $E(j,l,t)$ $(1 \leq j \leq n, 1 \leq l \leq k, 0 \leq t \leq g)$ denote the minimum error for approximating the points in P_{1j} using l segments with t violations. We have $E(j,l,t) = \min_{1<i\leq j, 0\leq q\leq t} \{\max\{E(i-1,l-1,t-q), w_{ijq}\}\}$, which means that we use $l-1$ segments to approximate $P_{1,i-1}$ with $t-q$ violations and use one segment to approximate P_{ij} with q violations. Our goal is to obtain $E(n,k,g)$. Suppose computing each w_{ijq} takes $O(W)$ time. A straightforward approach for this dynamic programming takes $O(kn^2g^2W)$ time.

We transform this 3-D dynamic programming problem to k 2-D sub-problems as follows. In computing $E(n,k,g)$, it the value of a variable is fixed, then it becomes a 2-D problem. More precisely, for each l, $1 < l \leq k$, we first compute the $O(ng)$ values of $E(j,l-1,t)$ for all $1 \leq j \leq n$ and $0 \leq t \leq g$, and then continue with computing $E(j,l,t)$ (for all $1 \leq j \leq n$ and $0 \leq t \leq g$). The reason for us to do so is that some geometric observations can be used to help handle the 2-D sub-problems more efficiently. To make this idea more clear and speedup the computation, we redefine the dependency relations among the $E(j, l, t)$'s into two parts, as follows.

Let $F(j, l, q, t) = \min_{1<i\leq j}\{\max\{E(i-1, l-1, t-q), w_{ijq}\}\}$. Thus $E(j, l, t) = \min_{0\leq q\leq t}\{F(j, l, q, t)\}$. Initially, we let $E(j, 1, t) = w_{1jt}$ for $1 \leq j \leq n$, $0 \leq t \leq g$ (note that, as to be shown below, the time for computing these $O(ng)$ values is much smaller than the total running time of the algorithm). For each $1 < l \leq k$, suppose the values $E(j, l-1, t)$ for all $1 \leq j \leq n, 0 \leq t \leq g$ have been obtained. To compute $E(j, l, t)$, for each t and for any $1 \leq j \leq n$, if $t = 0$, then it is easy to see that $E(j, l, 0) = F(j, l, 0, 0)$; else, $E(j, l, t) = \min_{0\leq q\leq t}\{F(j, l, q, t)\}$. For each q, $0 \leq q \leq t$, if the n values $F(j, l, q, t)$ for all $1 \leq j \leq n$ can be computed, say, in $O(F)$ time, then $E(j, l, t)$ for each t and all $1 \leq j \leq n$ can be computed in $O(g(F + n))$ time (note that $t \leq g$). Therefore, the total running time of the algorithm is $O(kg^2(n + F))$. Thus, our remaining task, for any two fixed values q and t with $0 \leq q \leq t$, is to derive an algorithm for computing $F(j, l, q, t)$ for all $1 \leq j \leq n$ in $O(F)$ time.

For two fixed values q and t, $0 \leq q \leq t$, let M be an $(n-1)\times(n-1)$ matrix such that each element $M(j, i) = \max\{E(i, l-1, t-q), w_{i+1,j+1,q}\}$ if $1 \leq i \leq j \leq n-1$ and $M(j, i) = +\infty$ otherwise. Clearly, for $j > 1$, the value of $F(j, l, q, t)$ is the row minima of the $(j-1)$-th row of M (when $j = 1$, $F(j, l, q, t) = 0$). Hence, the task of computing $F(j, l, q, t)$ for all $1 \leq j \leq n$ is reduced to finding all row minima of M. Here, M is represented *implicitly* (i.e., we compute only those elements of M that are actually needed by our algorithm). Suppose computing each w_{ijq} takes $O(W)$ time. Then a naive approach for finding all row minima of M takes $O(n^2W)$ time. A more efficient solution exploits the monotone property of the matrix M, as shown in the lemma below. A matrix is said to be *minimum monotone* if the minimum value in its i-th row lies below or to the right of the minimum value in its $(i-1)$-th row (if a row has several minima, then we take the rightmost one), and is *totally minimum monotone* if its every 2×2 submatrix is minimum monotone [1,2].

Lemma 4. *The matrix M is totally minimum monotone.*

Proof. Suppose M is not totally minimum monotone. Then there must exist a 2×2 submatrix $M[j, j + h_1; i - h_2, i]$ with $h_1 > 0$ and $h_2 > 0$, such that (1) $M(j, i - h_2) \geq M(j, i)$, and (2) $M(j + h_1, i - h_2) < M(j + h_1, i)$.

Note that $E(i-h_2, l-1, t-q) \leq E(i, l-1, t-q)$. We claim $E(i-h_2, l-1, t-q) \neq E(i, l-1, t-q)$ since otherwise, due to (2), it must be $w_{i-h_2+1,j+h_1+1,q} < w_{i+1,j+h_1+1,q}$, which cannot be true. Because $E(i-h_2, l-1, t-q) < E(i, l-1, t-q)$, due to (1), we have $w_{i-h_2+1,j+1,q} \geq E(i, l-1, t-q)$. Since $w_{i-h_2+1,j+h_1+1,q} \geq w_{i+1,j+h_1+1,q}$, due to (2), we have $w_{i-h_2+1,j+h_1+1,q} < E(i, l-1, t-q)$. Thus we can obtain $w_{i-h_2+1,j+1,q} > w_{i-h_2+1,j+h_1+1,q}$, a contradiction to the fact. Therefore, M is a totally minimum monotone matrix.

Based on Lemma 4, the linear time row minima algorithms in [1,2] can be applied to solve our row minima problem on M in $O(F) = O(nW)$ time (instead of $O(n^2W)$ time). Hence, we have the following result.

Lemma 5. *The min-ϵ version of any problem can be solved in $O(ng^2kW)$ time, where W is the time to compute each w_{ijq}.*

3.2 Problem VSF and q-Range-Minima Data Structure

For VSF, to compute w_{ijq}, suppose we have two sorted arrays $L[0\dots q]$ and $U[0\dots q]$, where L stores the y-coordinates of the lowest $q+1$ points in P_{ij} in increasing order and U stores the y-coordinates of the highest $q+1$ points in P_{ij} in decreasing order. Then it is easy to see that $w_{ijq} = \min_{0\le i\le q}\{U[i] - L[q-i]\}$. Thus the key issue is to obtain these two arrays L and U.

We design a data structure, called *q-range-minima*, on a given array $A[1\dots n]$ (not in any sorted order), which reports the q smallest elements, in sorted order, in the subarray $A[i\dots j]$ specified by a query (i, j). It is easy to see that if we build a g-range-minima data structure on the y-coordinates of all points in P, then the array L for any P_{ij} and q $(0 \le q \le g)$ can be obtained by a query. Likewise, we can define a q-range-maxima data structure similarly and use it to compute U. We only give the *q-range-minima* below and the q-range-maxima can be derived similarly. Clearly, the q-range-minima generalizes the range-minima [14], and may find more applications.

A straightforward method can answer each q-range-minima query in $O(q \log q)$ time by utilizing the range-minima data structure [14]. We give a better solution. Define $\phi(i, n) = \underbrace{\log\cdots\log}_{i\text{ times}} n$. The performance of our data structure is summarized in the next theorem, where $\Gamma(q, n, m) = \min\{i \ge 1 \mid \phi(i, n) \le O(q^m)\}$ with $m \ge 0$. Note that due to $q^m \ge 1$, we have $\Gamma(q, n, m) \le \log^* n$.

Theorem 3. *Given an array of n elements in arbitrary order, with $O(n \log q \cdot (\log q + q^m))$ time preprocessing, we can answer any q-range-minima query in $O(q \log(\Gamma(q, n, m)))$ time. Particularly, we answer each query in $O(q \log(\log^* n))$ time with $O(n \log^2 q)$ time preprocessing (by setting $m = 0$).*

Due to the space limit, the proof of the above theorem is in our full paper [6]. According to Theorem 3, by setting $m = 2$, we can compute each w_{ijq} for VSF in $O(g \log(\Gamma(g, n, 2)))$ time with an $O(ng^2 \log g)$ time preprocessing. By Lemma 5, we have the following result.

Theorem 4. *The VSF min-ϵ algorithm runs in $O(ng^3k \log(\Gamma(g, n, 2)))$ time.*

Note that when $g \ge \sqrt{\phi(O(1), n)}$ (e.g., $g \ge \sqrt{\log\log\log\log\log n}$), $\Gamma(g, n, 2) = O(1)$. When $gk \log(\Gamma(g, n, 2)) = o(\log n)$ (e.g., when k and g are constant), the result in Theorem 4 is better than the VSF *min-ϵ* algorithm in [11].

3.3 Problems VPF and VWPF

For VWPF, to compute each w_{ijq}, we need a geometric modeling. Suppose we use a linear function $y = a \cdot x + b$ to approximate all the points in P_{ij} with q

violations. For each point $p_t \in P_{ij}$ with weight u_t, we create two (upper) half-spaces $u_t(y_t-(ax_t+b)) \leq \epsilon$ and $-u_t(y_t-(ax_t+b)) \leq \epsilon$ in a 3-D coordinate space $\mathcal{S}$ with a as the x-axis, b as the y-axis, and ϵ as the z-axis. Denote by H the set of $2|P_{ij}|$ half-spaces created based on the points in P_{ij}. Let $p' = (a', b', \epsilon')$ be the lowest point (in terms of z-axis) in the common intersection of all half-spaces in H. Then, ϵ' is the minimum error to approximate P_{ij} by one segment without any violation and the corresponding approximation function is $y = a'x + b'$. Note that the any point higher than the xy plane in the space $\mathcal{S}$ cannot lie outside of both half-spaces created based on the same point in P_{ij}. Therefore, if $p^* = (a^*, b^*, \epsilon^*)$ is the lowest point in the common intersection of at least $|H|-q$ half-spaces in H (i.e., the lowest point on the q-level of the plane arrangement of H), then ϵ^* is the minimum error to approximate P_{ij} by one segment with q violations (i.e., $w_{ijq} = \epsilon^*$) and the corresponding approximation function is $y = a^*x + b^*$. To compute p^* is essentially the 3-D feasible LP with q violation problem [5,20]. By applying the algorithm in [20], we can compute each w_{ijq} in $O(n \log q + q^3 n^\delta)$ time for any constant $\delta > 0$. By lemma 5, $\mathbf{F_2}$ can solve the *min-*ϵ problem $O(ng^2k(n \log g + g^3n^\delta))$ time.

Based on geometric observations, we have another *min-*ϵ algorithm for VWPF, which is comparable to the framework $\mathbf{F_2}$. Let ϵ^* be the error of an optimal solution for the *min-*ϵ problem. Let S be the set of the values w_{ijq} for all $1 \leq i \leq j \leq n$ and $0 \leq q \leq g$. It must be $\epsilon^* \in S$. Let $A_{iq}[1 \ldots n]$ be an array, in which each element $A_{iq}[j] = w_{ijq}$ if $i \leq j$ and $A_{iq}[j] = 0$ otherwise. Then $S = \cup_{i=1}^{n} \cup_{q=0}^{g} A_{iq}$ and each A_{iq}, for any $1 \leq i \leq n$ and $0 \leq q \leq g$, is a sorted array. Given an error w_{ijq}, by using the *min-#* algorithm in Theorem 2, we can determine the size relation between ϵ^* and w_{ijq} in $O(ng^4 \log^2 n)$ time. There are $O(ng)$ sorted arrays of size $O(n)$ each. These sorted arrays can be represented *implicitly*. By the approach of binary search on sorted arrays [7], we can find ϵ^* in S in $O((ngW + ng^4 \log^2 n) \log n) = O(ngW \log n)$ time, where $W = O(n \log g + g^3 n^\delta)$. In summary, we have the following result.

Theorem 5. *The min-*ϵ *versions for VPF and VWPF can be solved in* $O(ngW \cdot \min\{kg, \log n\})$ *time, where* $W = O(n \log g + g^3 n^\delta)$ *for any constant* $\delta > 0$.

3.4 Problem VWSF

Due to the space limit, the proof of the following result is in [6].

Theorem 6. *The min-*ϵ *version of VWSF can be solved in* $O(n^2 + ng^2 \log n)$ *time.*

References

1. Aggarwal, A., Klawe, M., Moran, S., Shor, P., Wilbur, R.: Geometric applications of a matrix-searching algorithm. Algorithmica 2, 195–208 (1987)
2. Aggarwal, A., Park, J.: Notes on searching in multidimensional monotone arrays. In: Proc. of 29th Annual IEEE Symposium on Foundations of Computer Science (FOCS), pp. 497–512 (1988)

3. Buragohain, C., Shrivastava, N., Suri, S.: Space efficient streaming algorithms for the maximum error histogram. In: Proc. of the 23rd International Conference on Data Engineering (ICDE), pp. 1026–1035 (2007)
4. Chan, T.M.: Output-sensitive results on convex hulls and extreme points, and related problems. Discrete & Computational Geometry 16(3), 369–387 (1996)
5. Chan, T.M.: Low-dimensional linear programming with violations. In: Proc. of 43rd Annual IEEE Symposium on Foundations of Computer Science (FOCS), pp. 570–579 (2002)
6. Chen, D.Z., Wang, H.: Approximating points by a piecewise linear function (manuscript, 2009)
7. Chen, D.Z., Wang, H.: Approximating points by a piecewise linear function: I. In: Proc. of the 20th International Symposium on Algorithms and Computation, ISAAC (2009)
8. Cormen, T., Leiserson, C., Rivest, R., Stein, C.: Introduction to Algorithms, 2nd edn. MIT Press, Cambridge (2001)
9. Driscoll, J., Sarnak, N., Sleator, D., Tarjan, R.E.: Making data structures persistent. Journal of Computer and System Sciences 38(1), 86–124 (1989)
10. Everett, H., Robert, J.-M., van Kreveld, M.: An optimal algorithm for the ($\leq k$)-levels and with applications to separation and transversal problems. International Journal of Computational Geometry and Applications 6(3), 247–261 (1996)
11. Fournier, H., Vigneron, A.: Fitting a step function to a point set. In: Halperin, D., Mehlhorn, K. (eds.) ESA 2008. LNCS, vol. 5193, pp. 442–453. Springer, Heidelberg (2008)
12. Frederickson, G.: Optimal algorithms for tree partitioning. In: Proc. of the 2nd Annual ACM-SIAM Symposium of Discrete Algorithms (SODA), pp. 168–177 (1991)
13. Frederickson, G., Johnson, D.: Generalized selection and ranking: Sorted matrices. SIAM Journal on Computing 13(1), 14–30 (1984)
14. Gabow, H.N., Bentley, J., Tarjan, R.E.: Scaling and related techniques for geometry problems. In: Proc. of the 16th Annual ACM Symposium on Theory of Computing (STOC), pp. 135–143 (1984)
15. Guha, S.: On the space—time of optimal, approximate and streaming algorithms for synopsis construction problems. The VLDB Journal – The International Journal on Very Large Data Bases 17(6), 1509–1535 (2008)
16. Guha, S., Shim, K.: A note on linear time algorithms for maximum error histograms. IEEE Transactions on Knowledge and Data Engineering 19(7), 993–997 (2007)
17. Guha, S., Shim, K., Woo, J.: Rehist: Relative error histogram construction algorithms. In: Proc. of the 30th International Conference on Very Large Data Bases (VLDB), pp. 300–311 (2004)
18. Jagadish, H.V., Koudas, N., Muthukrishnan, S., Poosala, V., Sevcik, K.C., Suel, T.: Optimal histograms with quality gurantees. In: Proc. of the 24th International Conference on Very Large Data Bases (VLDB), pp. 275–286 (1998)
19. Karras, P., Sacharidis, D., Mamoulis, N.: Exploiting duality in summarization with deterministic guarantees. In: Proc. of the 13th ACM SIGKDD International Conference on Knowledge Discovery and Data Mining, pp. 380–389 (2007)
20. Matoušek, J.: On geometric optimization with few violated constraints. In: Proc. of the 10th Annual Symposium on Computational Geometry, pp. 312–321 (1994)
21. Overmars, M., van Leeuwen, J.: Maintenance of configurations in the plane. Journal of Computer System Sciences 23(2), 166–204 (1981)

Computing the Map of Geometric Minimal Cuts*

Jinhui Xu[1], Lei Xu[1], and Evanthia Papadopoulou[2]

[1] Department of Computer Science and Engineering
State University of New York at Buffalo
Buffalo, NY 14260, USA
{jinhui,lxu}@buffalo.edu

[2] Faculty of Informatics
Università della Svizzera italiana
Via Giuseppe Buffi 13
CH 6904 Lugano, Switzerland
evanthia.papadopoulou@unisi.ch

Abstract. In this paper we consider the problem of computing a map of geometric minimal cuts (called MGMC problem) induced by a planar rectilinear embedding of a subgraph $H = (V_H, E_H)$ of an input graph G. We first show that unlike the classic min-cut problem on graphs, the number of all rectilinear geometric minimal cuts is bounded by a low polynomial, $O(n^3)$. Our algorithm for identifying geometric minimum cuts runs in $O(n^3 \log n (\log\log n)^3)$ time in the worst case which can be reduced to $O(n \log n (\log\log n)^3)$ when the maximum size of the cut is bounded by a constant, where $n = |V_H|$. Once geometric minimal cuts are identified we show that the problem can be reduced to computing the L_∞ Hausdorff Voronoi diagram of axis aligned rectangles. We present the first output-sensitive algorithm to compute this diagram which runs in $O((N + K) \log^2 N \log\log N)$ time and $O(N \log^2 N)$ space, where N is the number of rectangles and K is the complexity of the diagram.

1 Introduction

In this paper, we consider the following problem (called Map of Geometric Minimal Cuts or MGMC problem): Given a graph $G = (V, E)$ and a planar embedding of a subgraph $H = (V_H, E_H)$ of G, compute a map $\mathcal{M}$ of the embedding plane P of H so that for every point $p \in P$, the cell in $\mathcal{M}$ containing p is associated with the "closest" *geometric cut* (in G) to p, where distance between a point p and a cut C is defined as the maximum distance between p and any individual element of C. A geometric cut C of G is a set of edges and vertices

* The research of the first two authors was supported in part by NSF through a CAREER Award CCF-0546509 and a grant IIS-0713489. The research of the third author was supported in part by an SNF project 200021_127137 and partially performed while affiliating with the IBM T.J. Watson Research center, Yorktown Heights, NY, USA, and the Athens University of Economics and Business, Greece.

Y. Dong, D.-Z. Du, and O. Ibarra (Eds.): ISAAC 2009, LNCS 5878, pp. 244–254, 2009.

in H that overlap a given geometric shape S in P and whose removal from G disconnects G. In this paper we consider the case where geometric cuts are induced by axis-aligned rectangles and distances are measured in the L_∞ metric. The main objective of the MGMC problem is to compute the map $\mathcal{M}$ of all geometric minimal (or canonical) cuts (the exact definition of geometric minimal cuts will be given in next section) of the planar embedding of H.

The MGMC problem is motivated by the VLSI critical area computation problem as explained in [6]. The critical area problem for various types of faults can be reduced to different variants of Voronoi diagrams that lead to accurate critical area extraction (see e.g. [11,10,4]). A VLSI net can be modeled as a graph $G = (V, E)$ with a subgraph embedded on every conducting layer. A subgraph $H = (V_H, E_H)$ on a layer X is vulnerable to random defects associated with layer X. Defects on layer X may create *cuts* on graph G that result in disconnecting the net N. The Voronoi framework for critical area extraction asks for a subdivision of layer X into regions that reveal for every point p the radius of the smallest disk centered at p inducing a cut of G. Open questions regarding the MGMC problem were posted in [6], some of which get addressed in this paper. The MGMC problem can also find applications in other networks, such as transportation networks.

In this paper we present a novel approach to solving the L_∞ MGMC problem for a rectilinear embedding of H based on a mix of geometric and graph algorithm techniques, that complements the one taken in [6]. All edges in H are assumed to be rectilinear in P. We first classify geometric cuts into two classes: 1-D cuts and 2-D cuts, and show that the number of all possible geometric 1-D and 2-D minimal cuts is $O(n^2)$ and $O(n^3)$ respectively. Directly computing the geometric minimal cuts could take $O(n^3(N+M))$ time, where $n = |V_H|$, $N = |V|$, and $M = |E|$. Based on interesting observations and dynamic connectivity data structures, we show that the worst case complexity can be reduced to $O(n^3 \log n (\log\log n)^3)$. We also consider the case in which the inducing rectangle of a cut has a constantly bounded edge length. For this case, we show that the time complexity of our algorithm can be significantly improved to $O(n \log n (\log\log n)^3)$ time. Once all geometric minimal cuts are identified, we show that the solution to the MGMC problem can be obtained by computing their Hausdorff Voronoi diagram.

We also revisit the plane sweep construction of the L_∞ Hausdorff Voronoi diagram of a set of rectangles. The Hausdorff Voronoi diagram of point clusters in the plane has been studied in [9,8,4,5,12,13]. We present the first output-sensitive algorithm which runs in $O((N + K)\log^2 N \log\log N)$ time and $O(N \log^2 N)$ space, where N is the number of rectangles and K is the complexity of the Hausdorff Voronoi diagram. One of the data structures used in the solution can also be used to speed up a query posted in [4] to achieve an optimal $O(N \log N)$ construction time for the case of N noncrossing rectangles, and an $O(N+M')\log N)$ algorithm in general, where M' is the number of *crossing* pairs of rectangles. K can be $\Theta(N + M')$ in the worst case.

2 Geometric Cuts

Let $G = (V, E)$ be the undirected graph in an MGMC problem and $H = (V_H, E_H)$ be its planar subgraph embedded in the plane P with $|V| = N$, $|E| = M$, $|V_H| = n$, and $|E_H| = m$. Due to the planarity of H, $m = O(n)$. In this paper, we assume that all edges in H are either horizontal or vertical straight line segments. A pair of vertices u and v in a graph is connected if there is a path in this graph from u to v. Otherwise, they are disconnected. A graph is connected if every pair of its distinct vertices is connected. Without loss of generality (WLOG), we assume that G is connected. A cut C of G is a subset of edges in G whose removal disconnects G. A cut C is minimal if removing any edge from C no longer forms a cut.

Definition 1. *Let R be a connected region in P, and $C = R \cap H$ be the set of edges in H intersected by R. C is called a geometric cut induced by R if the removal of C from G disconnects G.*

When there is no ambiguity of the region R, we often call the cut induced by R as a geometric cut for simplicity. For a given cut C, its *minimum inducing region* $R(C)$ is the minimum axis-aligned rectangle which intersects every edge of C For some geometric cut C, its $R(C)$ could be degenerated into a horizontal or vertical line segment, or even a single point. It is easy to see that if $R(C)$ is not degenerated, it is unique.

Definition 2. *A geometric cut C is called a 1-D geometric cut (or a 1-D cut) if $R(C)$ is a segment. If $R(C)$ is an axis-aligned rectangle, then C is called a 2-D geometric cut (or a 2-D cut).*

Definition 3. *A geometric cut C is a geometric minimal cut if the set of edges intersected by any rectangle shrinking from $R(C)$ is no longer a cut.*

Lemma 1. *Let C be any geometric minimal cut. If C is a 1-D cut, then each endpoint u of $R(C)$ is incident with either an endpoint of an edge e in H with the same orientation as $R(C)$ and $e \cap R(C) = u$ or an edge in H with different orientation as $R(C)$ (see Figure 1 (a)). If C is a 2-D cut, each bounding edge s of $R(C)$ is incident with either an endpoint v of an edge e in H of different orientation with s and $R(C) \cap e = v$ or an edge in H with the same orientation as s (see Figure 1(b); Note that s could be incident with multiple edges).*

3 Identifying Geometric Minimal Cuts

To compute the map $\mathcal{M}$ of geometric minimal cuts, we first identify all possible geometric minimal cuts and then construct a Hausdorff Voronoi diagram of their minimum inducing regions. There are three major problems: (1) How to find all 1-D minimal cuts; (2) How to find all 2-D minimal cuts; (3) How to efficiently construct the Hausdorff Voronoi diagram. This section focuses on problems (1) and (2). Next section deals with problem (3).

3.1 Computing 1-D Geometric Minimal Cuts

As discussed in last section, a 1-D geometric minimal cut can be induced by either horizontal or vertical segments. The following lemma upper bounds the total number of 1-D geometric minimal cuts.

Lemma 2. *There are $O(n^2)$ 1-D geometric minimal cuts in H.*

To compute the $O(n^2)$ 1-D geometric minimal cuts, the straightforward way will lead to a total of $O(n^3(N + M))$ time. To speed up the computation, we first simplify the graph G. We observe that the cuts involve only the edges in H. The connectivity in $G \setminus E_H$ will not be affected no matter which subset of edges in H is removed. To make use of this invariant, we first compute the connected components of $G \setminus E_H$. For each connected component CC, we contract it into a supernode v_{CC}. For each vertex $u \in H \cap CC$, we add an edge (u, v_{CC}). Let the resulting graph be $G' = (V', E')$ (called contracted graph). The following lemma gives some properties of this graph.

Lemma 3. *The number of vertices in G' is $|V'| = O(n)$ and the number of edges in G' is $|E'| = O(n)$. Furthermore, a subset of edges in H is a cut of G if and only if it is a cut of G'.*

From the above lemma, we know that the size of G' could be much smaller than G. Thus the time needed for answering a cut query is significantly reduced from $O(N + M)$ to $O(n)$.

Converting Cut Queries to Connectivity Queries: As discussed previously, to compute all 1-D geometric minimal cuts we have to check $O(n^3)$ possible subsets of edges in H. Many of them are quite similar (i.e., differ only by one or a small number of edges). Our main idea is to further decompose each cut queries into a set of connectivity queries with each connectivity query involving only one edge.

Connectivity Query: Most fully dynamic connectivity data structures support the following three operations: (1) $Insert(e)$, (2) $Delete(e)$, and (3) $Connectivity(u, v)$, where the $Insert(e)$ operation inserts edge e into G, the $Delete(e)$ operation removes edge e from G, and the $Connectivity(u, v)$ operation determines whether u and v are connected in the current graph G. Extensive research has been done on this problem and a number of results were obtained. In [2], Thorup *et al.* gave a simple and interesting solution for this problem which answers each connectivity query in $O(\log n)$ time and takes $O(\log^2 n)$ time for each update. Later Thorup gave a near optimal solution for this problem [7] which answers each connectivity query in $O(\log n / \log\log\log n)$ time and completes each insertion or deletion operation in $O(\log n(\log\log n)^3)$ expected amortized time.

In this paper we use the data structure in [7] for our problem. In practice (e.g., critical area computation), the simpler algorithm in [2] may be more practical. When the choice of the connectivity data structure is unclear, we use $MaxQU$ to represent the maximum time of the connectivity query and the update operation.

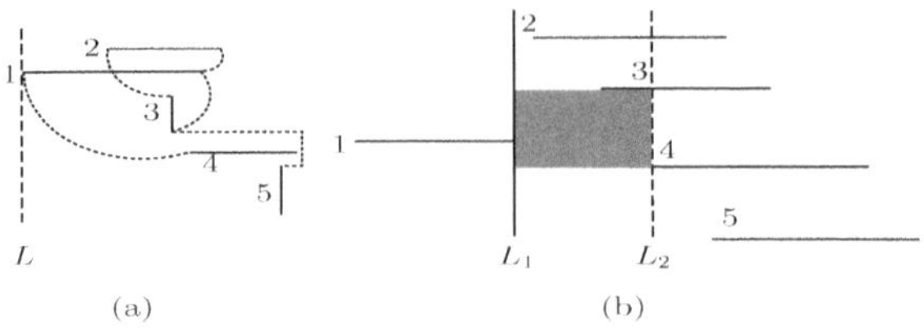

Fig. 3. (a) Plane sweep algorithm for 1-D cuts. (b) Double plane sweep algorithm for 2-D cuts.

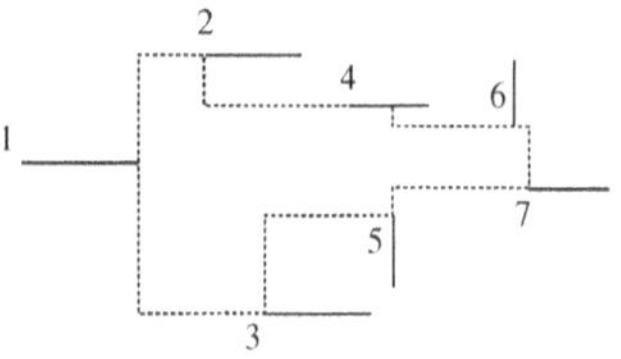

Fig. 4. The 2D algorithm searches the dotted region

4 Generating Map of Geometric Minimal Cuts

Given a set $\mathcal{C}$ of geometric minimal cuts of H, the Hausdorff Voronoi diagram of $\mathcal{C}$ is a partition of the embedding plane P of H into regions (or cells) so that the Hausdorff Voronoi cell of a cut $C \in \mathcal{C}$ is the union of all points whose Hausdorff distance to C is closer than to any other cut in $\mathcal{C}$.

In our MGMC problem, we have two types of objects, the minimum inducing regions of 1-D geometric minimal cuts and the minimum inducing regions of 2-D geometric minimal cuts. We are able to show that the Hausdorff Voronoi diagram of the two types of cuts can be constructed by propagating waves from the minimum inducing regions (details are omitted).

Lemma 5. *Let C be a 1-D or 2-D geometric minimal cut. At any moment, the wavefront of C is either empty or an axis-aligned rectangle. Furthermore, the wavefront in 3D is a facet cone apexed at a segment and with each facet forming a 45 degree angle with the xy plane.*

Lemma 6. *The Hausdorff Voronoi diagram can be obtained by projecting the lower envelope of the 3D facet cones to the xy plane.*

4.1 Properties and Plane Sweep Approach

Since every facet of a 3D facet cone forms a 45-degree angle with the xy plane and apexed at either a horizontal or vertical segment, the intersection of Q and a cone $\partial W(C)$ is either a V-shape curve (i.e., consisting of a 45-degree ray and a 135-degree ray on Q) or a U-shape curve (i.e., consisting of a 45-degree ray, a segment parallel to L, and a 135-degree ray).

The following difference fails the fortune's algorithm [1]. (1) When a cone is first encountered by Q, its corresponding initial V or U-shape curve may not necessarily be part of the beach line. (2) The initial V or U-shape curve may affect a number of curves in the beach line. (3) Once a V shape moves away from the beach line, it may re-appear in the beach line in a later time.

Definition 4. *A 3D facet cone $\partial W(C)$ is a U-cone (or V-cone) if its apex segment s_C is parallel to the y (or x) axis.*

First we consider U cones. Let $\partial W(C)$ be any U cone with apex segment s_C, and v_1 and v_2 be the two endpoints of s_C. When the sweep plane Q first encounters $\partial W(C)$, it introduces a U-shape curve C_u to Q. Let r_l, r_r, and s_m be the left and right rays and the middle segment of C_u respectively. Initially s_m is the apex segment s_C, and r_l and r_r are the two edges of facet cone. When Q (or L) moves, C_u grows and always maintains its U-shape.

Lemma 7. *Let $\partial W(C)$, C_u, r_l, r_r and s_m be defined as above. When Q moves in the direction of the x axis, C_u is always a U-shape curve. The supporting lines of r_l and r_r remain the same on Q, and the two endpoints of s_m (the fixed points of r_l and r_r) moves upwards in unit speed along the two supporting lines.*

For an arbitrary V cone $\partial W(C')$, let $s_{C'}$ be its apex segment, and v'_1 and v'_2 be its two endpoints (or left and right endpoints). When Q first touches $\partial W(C')$ at v'_1, it generates a V-shape curve C'_v. C'_v remains a V-shape curve before encountering v'_2. After that, C'_v becomes a U-shape curve.

Lemma 8. *Let r_l and r_r be the two rays of C'_v, and s_m be the middle segment of the U-shape curve C'_v after Q visiting v'_2. During the whole sweeping process, the supporting lines of r_l and r_r are fixed lines on Q. C'_v remains the same V-shape curve on Q before encountering v'_2. s_m moves upwards in unit speed along the supporting lines of r_l and r_r after Q encounters v'_2.*

Lemma 9. *Let $\partial W(C_1)$ be either a U or V cone and $\partial W(C_2)$ be a V cone with its left endpoint v_1 of s_{C_2} being inside of $\partial W(C_1)$ and its right endpoint v_2 being outside of $\partial W(C_1)$. If $\partial W(C_2)$ is not entirely contained by the union $\cup_{C_i \in \mathcal{C}; C_i \neq C_2} \partial W(C_i)$, the V-shape curve C_2 introduced by $\partial W(C_2)$ will be hidden by the beach line at the beginning and then becomes part of the beach line later. This is the only case in which a hidden U or V-shape curve could appear in the beach line.*

In the above lemma, the point v_i indicates that when Q sweeps it, the beach line is having a topological structure change. This indicates that in our problem there is a new type of event points.

Now we discuss our ideas for constructing the Hausdorff Voronoi diagram $HVD(\mathcal{C})$. First we consider the bisector of two rectangles (or cuts). Let C_1 and C_2 be two axis aligned rectangles in $\mathcal{C}$. The bisector of C_1 and C_2 is a line or a segment with two rays. In the latter case, each bisector contributes two vertices to the Hausdorff Voronoi diagram. Hence the Hausdorff Voronoi diagram consists of two types of vertices: (a) The intersection points of the bisectors and (b) the vertices of the bisectors.

Lemma 10. *Let $\mathcal{C}$ be a set of N rectangles. The edges of $HVD(\mathcal{C})$ are either segments or rays, and the vertices of the $HVD(\mathcal{C})$ are either the vertices of bisectors or the intersections of bisectors.*

To obtain a plane sweep algorithm, we need to design data structures to maintain the beach line and the event points. In our problem, the beach line is the lower envelope of the set of V and U-shape curves, and is a y-monotone polygonal

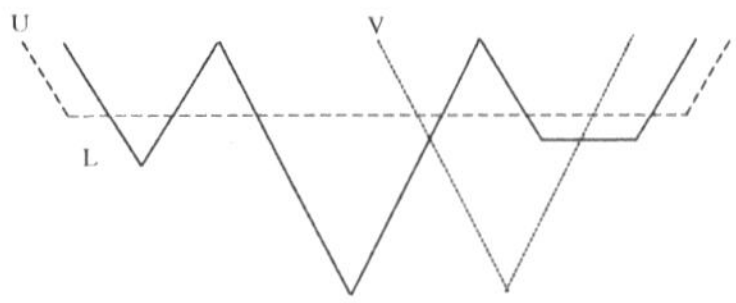

Fig. 5. Beach line L intersected by a U-shape or V-shape curve

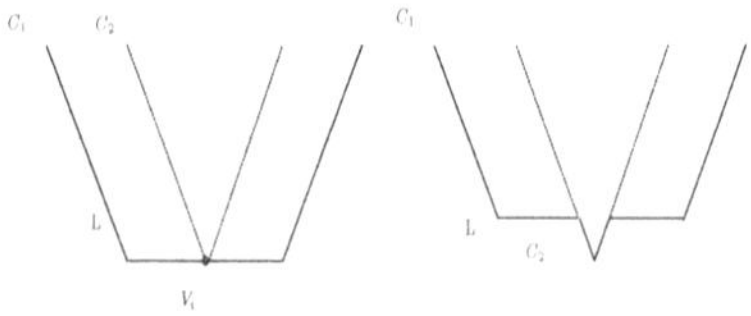

Fig. 6. (a) The bottom segment of C_1 coincides with V_i. (b) A hidden V-shape curve C_2 appears in the beach line.

curve. For non-disjoint 3D cones, the complexity of the beach line may not be linear in the number of the rectangles in $\mathcal{C}$. Figure 5(b) shows a newly generated U-shape curve intersecting the beach line a number of times and contributing multiple edges to the beach line. Consequently, the complexity of the $HVD(\mathcal{C})$ is not linear. The following lemma is a straightforward adaptation of Theorem 1 in [5] for the L_∞ metric.

Lemma 11. *The size K of the L_∞ Hausdorff Voronoi diagram of N rectangles is $O(N+M')$, where M' is the number of intersecting rectangle pairs. The bound is tight in the worst case.*

4.2 Events

For event points, we need to detect all events that cause the beach line to have topological structure changes. More specifically, we have to identify all the moments when a U or V-shape curve is inserted to or deleted from the beach line. There are two ways that a curve could appear in the beach line: (A) a newly generated U or V-shape curve becomes part of the beach line, *site events* and (B) a hidden V-shape curve appears in the beach line. For (B), it occurs when an unhidden portion of the bottom segment s_m of a U-shape curve C_1 moves upwards and encounters the apex point of a hidden V-shape curve C_2. We call this kind of events as V events (see Figure 6).There are also two ways for a curve or a portion of a curve to disappear from the beach line: (C) A curve (or part of the curve) is hidden by a newly generated curve, and (D) a U-shape curve (or part of the U-shape curve) moves out of the beach line.For (D), the disappearing U-shape curve C (or its unhidden portion) is caused the upwards movement of the bottom segment of C.Thus we have in total four types of events, site events, circle events, U events, and V events.

4.3 Data Structures and Algorithm

To construct $HVD(\mathcal{C})$, we use doubly-connected edge lists to store $HVD(\mathcal{C})$. We also need two data structures for the sweep line algorithm: an event queue and a sweep plane status structure representing the beach line.

The status structure for the beach line consists of three balanced binary search trees $\mathcal{T}$, $\mathcal{T}_{\pi/4}$, and $\mathcal{T}_{3\pi/4}$. $\mathcal{T}$ stores the y-monotone polygonal curve of the beach

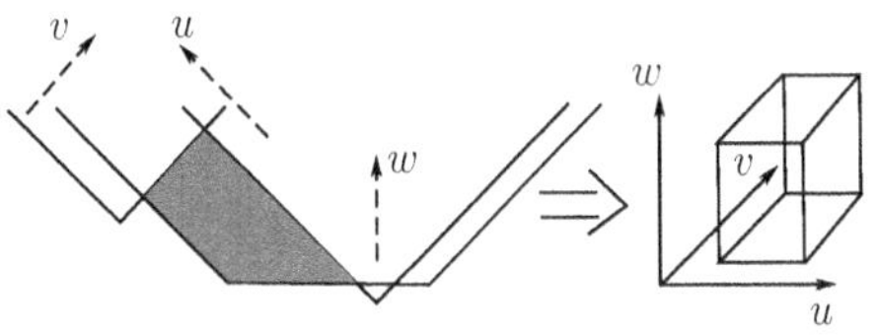

Fig. 7. A dominating region is converted to a 3D box in 3D space for MD

line. $\mathcal{T}_{\pi/4}$ is used to maintain the orders (along the norm direction) of the 45-degree rays of all U or V-shape curves which appear in the beach line. Similarly, $\mathcal{T}_{3\pi/4}$ maintains the orders of the 135-degree rays of U or V-shape curves which appear in the beach line. We can locate the positions in the beach line for the apex points of each newly encountered cone at a site event in $O(\log N)$ time, and update the beach line in $O(k \log N)$ time, where k is the number breakpoints destroyed and created after inserting the newly encountered U or V-shape curve into the beach line.

The event queue $\mathcal{Q}$ is implemented by a priority queue, where the priority of an event is the x coordinate of the corresponding event point.

To efficiently detect all possible V events, our main idea is not to maintain the arrangement, but rather to use the properties of V events to convert the problem into a query problem in 3D. Our idea is to process the apex points of all V-shape curves into a 3D dynamic range search tree data structure MD [3]. The three dimensions of MD are the orthogonal directions of the 45, 135 degrees lines in the sweep plane Q and the orthogonal direction of the bottom segment (or the y axis). In this way, we map the apex point of each hidden V-shape curve into a point and convert the dominating region (see the shaded region in Figure 7) of each unhidden portion of a U-shape curve into a 3D (possibly unbounded) box in the newly orthogonized space. By a 3D range query in MD, for each dominating region R we can find the closest apex point (among all hidden V-shape curves whose apex points fall in R) to the unhidden portion of the bottom segment of a U-shape curve. This takes $O(\log^2 N \log\log N)$ time [3].

With these data structures, we are able to show that all events can be efficiently handled (details are left for full paper).

Theorem 4. *The L_∞ Hausdorff Voronoi diagram $HVD(\mathcal{C})$ of a set $\mathcal{C}$ of axis aligned rectangles can be constructed by a plane sweep algorithm in $O((N + K)\log^2 N \log\log N)$ time, where $N = |\mathcal{C}|$ and K is the complexity of the Hausdorff Voronoi diagram.*

References

1. Fortune, S.: A sweepline algorithm for Voronoi diagrams. Algorithmica 2, 153–174 (1987)
2. Holm, J., Lichtenberg, K., Thorup, M.: Poly-logarithmic deterministic fully-dynamic algorithms for connectivity, minimum spanning tree, 2-edge, and biconnectivity. J. ACM 48(4), 723–760 (2001)

3. Mehlhorn, K., Näher, S.: Dynamic franctional cascading. Algorithmica 5, 215–241 (1990)
4. Papadopoulou, E.: Critical area computation for missing material defects in VLSI circuits. IEEE Transactions on Computer-Aided Design 20(5), 583–597 (2001)
5. Papadopoulou, E.: The Hausdorff Voronoi Diagram of Point Clusters in the Plane. Algorithmica 40, 63–82 (2004)
6. Papadopoulou, E.: Higher order Voronoi diagrams of segments for VLSI critical area extraction. In: Tokuyama, T. (ed.) ISAAC 2007. LNCS, vol. 4835, pp. 716–727. Springer, Heidelberg (2007)
7. Thorup, M.: Near-optimal fully-dynamic graph connectivity. In: STOC, pp. 343–350 (2000)
8. Abellanas, M., Hernandez, G., Klein, R., Neumann-Lara, V., Urrutia, J.: A Combinatorial Property of Convex Sets. Discrete Comput. Geometry 17, 307–318 (1997)
9. Edelsbrunner, H., Guibas, L.J., Sharir, M.: The upper envelope of piecewise linear functions: algorithms and applications. Discrete Comput. Geometry 4, 311–336 (1989)
10. Papadopoulou, E., Lee, D.T.: The L_∞ Voronoi Diagram of Segments and VLSI Applications. International Journal of Computational Geometry and Applications 11, 503–528 (2001)
11. Papadopoulou, E., Lee, D.T.: Critical Area computation via Voronoi diagrams. IEEE Transactions on Computer-Aided Design 18(4), 463–474 (1999)
12. Papadopoulou, E., Lee, D.T.: The Hausdorff Voronoi diagram of polygonal objects: A divide and conquer approach. International Journal of Computational Geometry and Applications 14(6), 421–452 (2004)
13. Dehne, F., Maheshwari, A., Taylor, R.: A Coarse Grained Parallel Algorithm for Hausdorff Voronoi Diagrams. In: Proc. 2006 International Conference on Parallel Processing, pp. 497–504 (2006)

On the Camera Placement Problem*

Rudolf Fleischer and Yihui Wang

Department of Computer Science and Engineering
IIPL, Fudan University, Shanghai, China
{rudolf,yihuiwang}@fudan.edu.cn

Abstract. We introduce a new probing problem: what is the minimum number of cameras at *fixed positions* necessary and sufficient to reconstruct *any* strictly convex polygon contained in a disk of radius 1 if cameras only see the silhouette of the polygon? The optimal number only depends on the largest angle α of the polygon. If no two camera tangents overlap, $\lceil \frac{3\pi}{\pi-\alpha} \rceil$ cameras are necessary and sufficient. Otherwise, approximately $\lceil \frac{4\pi}{\pi-\alpha} \rceil$ cameras are sufficient. Reconstruction only takes time linear in the number of cameras. We also give results for the 3D case.

1 Introduction

Geometric probing is concerned with determining or verifying the shape of an object by measurements, or *probes*. *Shape-based probing* is about determining the shape of an unknown object [16], while *model-based probing* is about identifying a given object from a known collection of objects [5]. *Finger probes* [2], *line probes* [10], *silhouette probes* [13], and *x-ray probes* [11] have been studied, for reconstruction of convex objects; non-convex objects can partially be reconstructed [1].

Silhouette probes assume a viewpoint at infinity, but they can also be defined for viewpoints closer to the object, in which case we speak of *camera probes* (e.g., see [9]). A camera probe is the silhouette of the object seen against a bright background. Note that we cannot see any details of the object in the interior of the silhouette, we only see its boundary edges. Now we ask:: how many cameras at fixed positions are necessary and sufficient to fully reconstruct any given object from its silhouettes? Note that we first fix the camera positions and then take silhouette probes of the object to be reconstructed.

Since a camera can only see a cone defined by the two tangents at the object, the *viewing cone* (see Fig. 1), we cannot reconstruct curved objects from a finite number of camera probes. Therefore, we only consider simple strictly convex polygons (or polyhedrons in $3D$). We assume the polygon is placed anywhere

* This work was supported by a grant from the National High Technology Research and Development Program of China (863 Program) (No. 2007AA01Z176) and the Shanghai Leading Academic Discipline Project (project number B114). The authors are ordered alphabetically by family name; otherwise, Y. Wang would be first author.

Y. Dong, D.-Z. Du, and O. Ibarra (Eds.): ISAAC 2009, LNCS 5878, pp. 255–264, 2009.

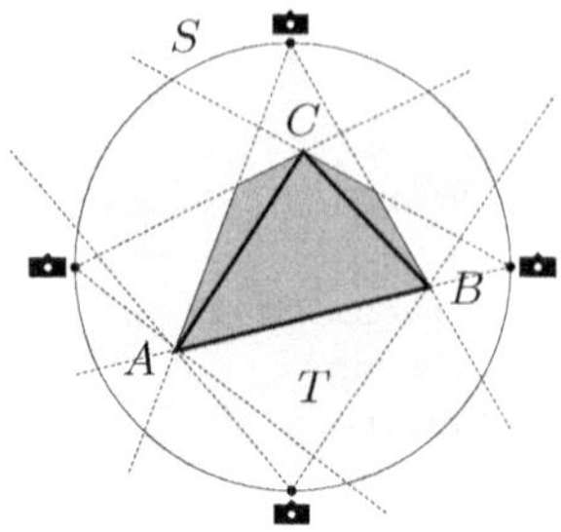

Fig. 1. The intersection of all camera cones (green) contains triangle ABC

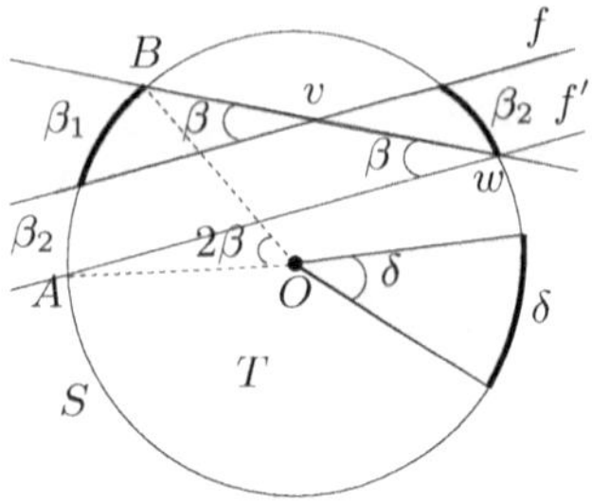

Fig. 2. Angles and cone arcs in a disk of radius 1, see Lemma 3

in a disk T of radius 1, and the cameras are positioned along the boundary of T. Each tangent of a viewing cone touches a strictly convex polygon in a vertex or edge. If a second tangent touches the same vertex, we can reconstruct its coordinates as the intersection point of the two tangents. Unfortunately, in a scene with many cameras and tangents, not every pair of tangents will intersect in a vertex of the polygon. The *camera placement problem* asks to determine the fixed positions for a minimum number of cameras around T such that we can reconstruct any strictly convex polygon in T, i.e., determine its number of vertices and compute their coordinates.

We need at least m cameras to reconstruct an m-gon from camera probes, because each camera can only see two vertex tangents of the m-gon, and we need two tangents at a vertex to compute its coordinates. One problem is that the cameras do not know whether their tangents touch at the same vertex. As it turns out, a stronger constraint comes from the largest angle α of the polygon. Under a weak general position assumption that no camera tangent to P contains another camera we show a tight bound of $\lceil \frac{3\pi}{\pi-\alpha} \rceil$, which is at least $\lceil \frac{3}{2}m \rceil$ in an m-gon (see Lemma 10), on the number of cameras needed to reconstruct any polygon with maximum angle α. If we want to reconstruct polygons in arbitrary position, then the bounds get slightly worse: $\lceil \frac{4\pi}{\pi-\alpha} \rceil$ evenly spaced cameras are sufficient to reconstruct any P. The distance between two cameras can be slightly increased in case the number of cameras is odd, but the exact value of the maximum distance can only be computed numerically.

For the three-dimensional version of the problem, we cannot hope to find tight bounds because we would have to solve quite irregular packing problems of cameras on a sphere, and such problems tend to be notoriously difficult [3]. Instead, we give some rough bounds.

There may be some applications for this research. For example, stereo-vision is concerned with the reconstruction of $3D$ objects from $2D$ data [17], which are often just silhouette or camera probes [8]. In computational medicine it is important to reconstruct the coronary arterial tree from angiograms (i.e., camera probes). Karl et al. [6] showed that three silhouettes are sufficient to reconstruct any ellipsoid, and artery reconstruction is very similar to ellipsoid reconstruction [7]. The gait recognition problem is the problem to identify a person from a database of gait sequences by watching the silhouette of the person walking around [15]. Determining the best walking route for snapshots from a singe camera is equivalent to determining the best camera positions for many cameras.

The camera placement problem differs from traditional probing problems in two ways: The camera positions are fixed a priori, while in previous works the position or direction of later probes usually depends on the outcome of earlier probes; and the optimal number of probes (camera positions) is not a function of the number of vertices of the polygon, but of the largest angle of the polygon.

2 Definitions

Let T be a disk of radius 1 in the plane with center O, and let S denote the circle bounding T; see Fig. 2. A cone with apex in T intersects S in a *cone arc*. We usually denote a cone arc by its length δ (note that a cone of angle δ centered at O cuts out a cone arc of length δ). We usually denote the complementary angle of α by $\beta = \pi - \alpha$. A double cone with apex v in T intersects S in two cone arcs.

Somewhere in T is placed a simple strictly convex m-gon P, but m is not known. Cameras can be placed anywhere on S. A *camera* at point A can only see the minimum cone with apex A containing P. That is, the camera only knows two tangents to P but no information about the vertices where the tangents touch P, or any other parts of P. The vertex touched by a tangent is *visible* to the camera, and the camera *sees* the vertex (though the camera does not know the coordinates of the vertex). If the tangent touches P in an edge of P, we say the camera *sees* the edge. Each vertex v of P induces a double cone containing P (by extending the two edges incident to v) and a *complementary double cone* not containing P (defined by the complementary angle at v), whose two complementary cone arcs define the *safe region* of v.

A vertex v of P can be *reconstructed* if we can determine its coordinates, i.e., at least two tangents meet in v and we can somehow infer that this intersection point must belong to P. P can be reconstructed if every vertex of P can be reconstructed. The *camera placement problem* is the problem of determining the minimum number of cameras and their placements on S such that any simple strictly convex polygon P in T with maximum angle α can be reconstructed.

Any cone in T of angle $\frac{\beta}{2}$ with apex on S has a cone arc of length β. In Fig. 2, w sees the arc AB from an angle of $\beta = \frac{1}{2}(\beta_1 + \beta_2)$. The next lemma generalizes this fact to any cone inside T.

Lemma 1. *Any double cone of angle β with apex somewhere in T has two cone arcs of total length 2β.* □

Corollary 2. *For any two non overlapping arcs β_1 and β_2 of total length $\beta_1 + \beta_2 = 2\beta$ there exists a double cone of angle β whose cone arcs are β_1 and β_2.* □

The following lemma is self-evident, see Fig. 2.

Lemma 3. *A camera can see vertex v of P if and only if it is placed in the safe region of v.* □

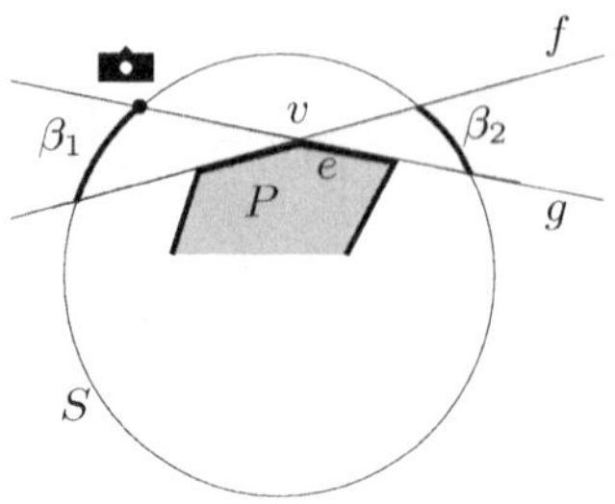

Fig. 3. The figure for Lemma 11

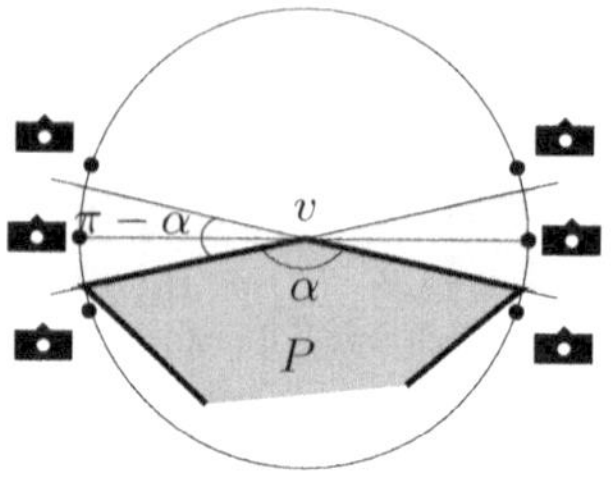

Fig. 4. Two antipodal cameras when k is even in Thm. 13

P is in *general position* if no vertex of P lies on a line segment connecting two cameras. We denote the intersection of all camera cones by P_C; clearly, $P \subseteq P_C$.

Lemma 4. *P_C is a polygon with at most $2m$ vertices.* □

Note that we do not need $P = P_C$ to reconstruct P if we assume general position and a sufficient number of cameras.

Lemma 5. *If three or more camera tangents to P intersect in a common point v and P is in general position, then v must be a vertex of P.* □

Let k_α denote the minimum number of cameras necessary to solve the camera placement problem for polygons of maximum angle α. Let α_k denote the largest angle such that any polygon with maximum angle α_k can be reconstructed by a set of k cameras. If we consider evenly spaced arrangements of cameras on S, then let γ_α denote the maximum arc length between two successive cameras such that any strictly convex n-gon of maximum angle α can be reconstructed. Whenever we speak about evenly spaced camera arrangements it is understood that exactly one pair of neighboring cameras may have a smaller distance. As we will see below, there is no advantage in placing cameras other than evenly spaced. Let $\hat{k}_\alpha$, $\hat{\alpha}_k$, and $\hat{\gamma}_\alpha$ denote these quantities for polygons in general position.

3 The Camera Placement Problem in 2D

We first consider the case of polygons in general position, and then extend the analysis to the case of arbitrary position. We only consider polygons with largest angle α, where $0 < \alpha < \pi$. Note that $\alpha \geq \frac{m-2}{m} \cdot \pi$ for m-gons. Intuitively, an optimal camera placement for polygons in general position should be evenly spaced, but this cannot easily be argued for the arbitrary position case.

Theorem 6. *For polygons in general position we have $\hat{k}_\alpha = \lceil \frac{3\pi}{\pi-\alpha} \rceil$, $\hat{\alpha}_k = \pi \cdot (1 - \frac{3}{k})$, and $\hat{\gamma}_\alpha = \frac{2}{3}(\pi - \alpha)$.*

Note that it is sufficient to prove the bounds for $\hat{k}_\alpha$ and $\hat{\gamma}_\alpha$, then the bound for $\hat{\alpha}_k$ follows immediately. Let $0 < \alpha < \pi$ be a given angle. Consider a camera placement consisting of k cameras such that every polygon P with largest angle α can be reconstructed. Let $M = \{c_0, \ldots, c_{k-1}\}$ be the positions (in polar coordinates) of the cameras in counterclockwise order around S. Let $dist(i,j)$ be the length of the arc in counterclockwise direction between c_i and c_j (all index operations are modulo k). The next lemma implies $\hat{\gamma}_\alpha \leq \frac{2}{3}(\pi - \alpha)$.

Lemma 7. *$dist(i, i+3) \leq 2\beta$, for all $i = 0, \ldots, k-1$.* □

Next we show $\hat{k}_\alpha \geq \lceil \frac{3\pi}{\pi-\alpha} \rceil$ (use $b = 3$ and $z = \frac{2}{3}\beta$).

Lemma 8. *Let $b \geq 1$. If an arrangement of k cameras satisfies $dist(i, i+b) \leq bz$, for all $i = 0, \ldots, k-1$, then there are at least $k \geq \lceil \frac{2\pi}{z} \rceil$ cameras.* □

Let $\mathcal{A}$ be an arrangement of evenly spaced cameras around S at distance $\frac{2}{3}(\pi-\alpha)$. The next lemma implies $\hat{\gamma}_\alpha \geq \frac{2}{3}(\pi - \alpha)$ and $\hat{k}_\alpha \leq \lceil \frac{3\pi}{\pi-\alpha} \rceil$.

Lemma 9. *Every vertex of P is visible to at least two cameras in $\mathcal{A}$.* □

Lemma 10. *At least one vertex of P is visible to three or more cameras in $\mathcal{A}$ and can therefore be reconstructed.*

Proof. Since $\alpha \geq \frac{m-2}{m} \cdot \pi$ for an m-gon P, $\mathcal{A}$ contains at least $k = \lceil \frac{3\pi}{\pi-\alpha} \rceil \geq \lceil \frac{3}{2}m \rceil$ cameras that see a total of at least $3m$ vertices (with multiplicity). □

Lemma 11. *If a camera in $\mathcal{A}$ sees an edge e of P, both endpoints of e are visible to at least three cameras and can therefore be reconstructed.* □

Now we will explain how we can reconstruct any polygon P of maximum angle α in general position. Assume P has $m \geq 3$ vertices (we do not need to know m to reconstruct P). By Lemma 4, P_C is at most a $2m$-gon. If P is in general position, then the vertices of P must be vertices of P_C. If a camera can see an edge e of P, we can reconstruct both endpoints by Lemma 11. We can reconstruct all vertices of P not incident to such an edge by choosing every second vertex from P_C, starting at an arbitrary endpoint of an edge visible to a camera. This is because every edge of P_C touches P and therefore it is not possible that two consecutive vertices of P_C are not vertices of P. If no such edge exists, then P_C must be exactly a $2m$-gon. Since we can reconstruct at least one vertex v of P by Lemma 10, we can reconstruct P by choosing every second vertex of P_C starting at v. This finishes the proof of Thm. 6. □

We now consider polygons in arbitrary position. Let P be an m-gon in T with largest angle at most α. If we use the camera arrangement $\mathcal{A}$ above, it is still true that every vertex is visible to at least two cameras. However, it may happen that a vertex v of P is not a vertex of P_C, because it may be lying somewhere in the middle of an edge of P_C. This can happen if v is lying exactly on the line segment connecting two cameras. In this case, we cannot reconstruct v from these two tangents.

To reconstruct v, we need a third camera that can see v. It is straightforward to generalize Lemmas 7 and 9 to the requirement that every double cone of angle

4 The Camera Placement Problem in 3D

Let P be a strictly convex polyhedron contained in a ball T of radius 1 with surface S. We assume that P is in general position, i.e., every tangent plane to P contains at most one camera. A camera can see a polygonal silhouette of P. Two planes intersecting in a line ℓ induce two double wedges with base ℓ. If ℓ passes through the interior of T, the intersection of each wedge with S is a *lune*. Correspondingly, a double wedge induces a *double lune*. The *angle* of a lune or double lune is the angle between its two defining planes. A lune is *spherical* if its base contains the center O of T.

A *cone* is defined as the intersection of three or more half-spaces induced by planes meeting in a single point, the *apex* of the cone. Since a plane intersects S in a circle on S (a spherical circle if the plane contains the center O), a cone with apex v inside T intersects S in a polygon Cap_v, called the *cap* of the cone. If v is a vertex of P, its incident faces induce two cones, one cone containing P and the other one its complementary cone. Let Cap_v and $\overline{Cap_v}$ denote the corresponding caps (see Fig. 6). Similarly as in the two-dimensional case, we define the *safe region* of v as $S - (Cap_v \cup \overline{Cap_v})$. If v is a vertex of P, then the planes supporting any two of the faces incident to v define two double lunes. As in the two-dimensional case, we call the double lune not containing P the *complementary double lune* at v. Note that the safe region of v is exactly the union of all complementary double lunes at v. Lemma 3 also holds in 3D, see Fig. 6. P is an *α-polyhedron* if the angle between any two faces (the *dihedral angle*) sharing a common edge of P is at most α.

Lemma 14. *If any double lune on S of angle at least $\pi - \alpha$ contains at least one camera in its interior, then any vertex of any α-polyhedron P in T is visible to at least two cameras.*

Proof. Let v be a vertex of P incident to $m \geq 3$ faces. Let $h_1, \ldots, h_m$ be the corresponding planes in cyclical order around v. If we consider the m complementary double lunes at v induced by h_i and h_{i+1}, for $i = 1, \ldots, m$ (where $h_{m+1} = h_1$), we see that they partially overlap, but no point in their union can be contained in all m double lunes. Since P is an α-polyhedron, every complementary double lune at v has angle at least $\pi - \alpha$ and therefore contains at least one camera, We conclude that the m complementary double lunes contain at least two cameras that can see v (by Lemma 3). □

The surface area of lunes does not behave like arc length of cones in 2D. There is no equivalent of Lemma 1 in 3D; the surface area of a double cone cap depends on the position of the apex within T.

We can then define P_C as the intersection of all half-spaces induced by tangent planes to P. Actually, we only need to consider those tangent planes that contain an edge of P. In contrast to the two-dimensional case (Lemma 4), P_C may have quadratic size. To reconstruct a vertex v of P, we must find three linearly independent planes tangent to P at v. The three-dimensional equivalent of Lemma 5 states that any intersection point of four or more camera tangent

Table 2. Numerically computed values of k_α

α	0.6799	0.9274	1.2311	1.3606	1.4615	1.5404	1.6650	2.1076	2.5028	2.6902
k_α	4	5	6	7	8	9	10	20	50	100

planes must already be a vertex of P. However, in connection with Lemma 14 we can get a better criterion for P being reconstructable.

Lemma 15. *We can reconstruct a vertex v of P in general position if it is visible to at least two cameras.*

Proof. Let v be a vertex of P. We need three linearly independent planes tangent to v to reconstruct the coordinates of v. Each camera can see v and either two edges of P incident to v or a line containing an edge incident to v. Thus we can construct v. □

To obtain good lower bounds for k_α, we would need good bounds on dense camera packing on a sphere such that any safe region of a vertex contains several camera. Unfortunately, even much simpler packing problems on the sphere, for example disk packing, are not fully understood (see, for example, [14]).

So let us turn to upper bounds for k_α. By Lemmas 14 and 15, it is sufficient to find a placement of k_α cameras on the unit sphere such that any double lune of angle at least $\pi - \alpha$ on S contains at least one camera, To get a rough bound on k_α, we consider the largest spherical cap contained in one of the two lunes of any double lune of angle at least $\pi-\alpha$; twice this radius is called the *width* of the lune. Clearly, the worst case is attained when this angle is equal to $\pi - \alpha$. The larger one of the two lunes in a double lune has smallest width if both lunes have the same width. It is not difficult to see that the double lune should be spherical to minimize this width. But then the radius r_α of the largest circle contained in the lune, i.e., half its width, is given by $r_\alpha = \sin \frac{\pi-\alpha}{2}$ (note that the surface area of a spherical lune of angle α is 2α). Thus, to satisfy Lemma 14, it suffices to find a camera placement on S such that any circle of radius r_α contains at least one camera. That is, we want to find a placement of the cameras with *covering radius* r_α, which is a well-studied (but not completely solved) problem. The optimal covering radius for n points on S is asymptotically $n^{-\frac{1}{2}}$. Hardin et al. [4] have made available placements of up to 130 points on the sphere minimizing the covering radius, and their numbers seem to indicate that $r_\alpha \approx \frac{130\pi}{180\sqrt{k_\alpha}}$. It follows, $k_\alpha \leq \left(\frac{130\cdot\pi}{180\cdot\cos\alpha/2}\right)^2$ for large α (in radians). Table 2 shows some of the values of the upper bound for k_α, computed using the values by Hardin et al. Note that these values are not tight.

We can get better bounds for small α. For example, four cameras are sufficient to reconstruct any α-polyhedron with $\alpha \leq \arccos(1/3) \approx 1.2309$. Let S be the circumscribing sphere of a regular tetrahedron C. We place the four cameras at the four vertices of C. Since C has a dihedral angle of $\arccos(1/3)$, any vertex of an α-polyhedron in T can be seen by at least two cameras if $\alpha \leq \arccos(1/3)$.

References

1. Boissonnat, J.-D., Yvinec, M.: Probing a scene of non-convex polyhedra. Algorithmica 8, 321–342 (1992)
2. Cole, R., Yap, C.K.: Shape from probing. Journal of Algorithms 8, 19–38 (1987)
3. Conway, J.H., Hardin, R.H., Sloane, J.A.: Packing lines, planes, etc.: packings in Grassmannian spaces. Experimental Mathematics 5 (1996)
4. Hardin, R.H., Sloane, N.J.A., Smith, W.D.: Spherical coverings (2008), `http://www.research.att.com/~njas/coverings/`
5. Joseph, E., Skiena, S.: Model-based probing strategies for convex polygons. Computational Geometry: Theory and Applications 2, 209–221 (1992)
6. Karl, W.C., Verghese, G.C., Willsky, A.S.: Reconstructing ellipsoids from projections. Computer Vision, Graphics, and Image Processing 56(2), 124–139 (1994)
7. Kayikcioglu, T., Gangal, A., Turhal, M.: Reconstructing coronary arterial segments from three projection boundaries. Pattern Recognition Letters 22(6-7), 611–624 (2001)
8. Laurentini, A.: How many $2D$ silhouettes does it take to reconstruct a $3D$ object? Computer Vision and Image Understanding 67(1), 81–87 (1997)
9. Lindenbaum, M., Bruckstein, A.: Reconstructing a convex polygon from binary perspective projections. Pattern Recognition 23(12), 1343–1350 (1990)
10. Lindenbaum, M., Bruckstein, A.: Reconstruction of polygonal sets by constrained and unconstrained double probing. Annals of Mathematics and Artificial Intelligence 4(3-4), 345–361 (1991)
11. Meijer, H., Skiena, S.S.: Reconstructing polygons from x-rays. Geometriae Dedicata 61(2), 191–204 (1996)
12. Mount, D.: Geometric intersection. In: Goodman, J.E., O'Rourke, J. (eds.) Handbook of Discrete and Computational Geometry, ch. 38, 2nd edn., pp. 859–879. CRC Press, New York (2004)
13. Rao, A.S., Goldberg, K.Y.: Shape from diameter: Recognizing polygonal parts with a parallel-jaw gripper. The International Journal of Robotics Research 13(1), 16–37 (1994)
14. Saff, E.B., Kuijlaars, A.B.J.: Distributing many points on a sphere. The Mathematical Intelligencer 19(1), 5–11 (1997)
15. Sarkar, S., Phillips, P.J., Liu, Z., Vega, I.R., Grother, P., Bowyer, K.W.: The humanid gait challenge problem: Data sets, performance, and analysis. IEEE Transactions on Pattern Analysis and Machine Intelligence 27(2), 162–177 (2005)
16. Skiena, S.S.: Problems in geometric probing. Algorithmica 4(4), 599–605 (1989)
17. Sormann, M., Bauer, J., Zach, C., Klaus, A., Karner, K.: VR modeler: from image sequences to 3D models. In: Proceedings of the 20th Spring Conference on Computer Graphics, pp. 148–156 (2004)

Graph Orientations with Set Connectivity Requirements*

Takuro Fukunaga

Department of Applied Mathematics and Physics,
Graduate School of Informatics, Kyoto University, Japan
takuro@amp.i.kyoto-u.ac.jp

Abstract. In an undirected or directed graph, the edge-connectivity between two disjoint vertex sets X and Y is defined as the minimum number of edges or arcs that should be removed for disconnecting all vertices in Y from those in X. In this paper, we discuss several conditions for a given undirected graph to have an orientation meeting the edge-connectivity requirements defined on some pairs of vertex sets.

1 Introduction

In this paper, we denote an undirected edge between vertices u and v by uv, and an arc from u to v by (u, v). An *orientation* $D = (V, A)$ of an undirected graph $G = (V, E)$ is a digraph obtained by replacing each edge $uv \in E$ with an arc (u, v) or (v, u). In an orientation problem, we are asked whether G has an orientation satisfying given connectivity demands. This is a basic problem in combinatorial optimization, and many beautiful results have been produced so far. The main purpose of this paper is to discuss possibility to extend those results with a general concept of the edge-connectivity.

Usually the edge-connectivity is defined on pairs of vertices. On the other hand, this paper deals with the edge-connectivity defined on pairs of vertex sets. Let X and Y be non-empty disjoint subsets of V, i.e., $X, Y \in 2^V$ and $X \cap Y = \emptyset$. We define the edge-connectivity $\lambda_G(X, Y)$ between X and Y in an undirected graph $G = (V, E)$ as $\min\{d_G(Z) \mid Z \in 2^V, X \subseteq Z \subseteq V - Y\}$ where $d_G(Z)$ stands for the number of edges joining vertices in Z and in $V - Z$. Equivalently $\lambda_G(X, Y)$ is the edge-connectivity $\lambda_{G'}(x, y)$ between two vertices x and y in the graph G' obtained by shrinking X and Y into single vertices x and y, respectively. For a digraph D, the arc-connectivity $\lambda_D(X, Y)$ from X and Y is defined as $\min\{\delta_D(Z) \mid Z \in 2^V, X \subseteq Z \subseteq V - Y\}$ where $\delta_D(Z)$ stands for the number of arcs from vertices in Z to those in $V - Z$. $\lambda_D(X, Y)$ is also defined as the arc-connectivity $\lambda_{D'}(x, y)$ from x to y in the digraph D' obtained by shrinking X and Y into single vertices x and y, respectively.

The connectivity between vertex sets is a useful notion in practice. For example, let X be a set of servers providing the same service in a communication

* This work was partially supported by Grant-in-Aid for Scientific Research from the Ministry of Education, Culture, Sports, Science and Technology of Japan.

Y. Dong, D.-Z. Du, and O. Ibarra (Eds.): ISAAC 2009, LNCS 5878, pp. 265–274, 2009.

network represented by an undirected graph $G = (V, E)$, and suppose that a vertex $v \in V - X$ represents a client of the service. Then $\lambda_G(\{v\}, X)$ stands for the minimum number of links which should be broken for disconnecting the client from all servers. By such motivation, several optimization problems defined by the edge-connectivity between vertex sets are considered (e.g., graph augmentation problem [4,5,6], source location problem [1], and minimum cost subgraph problem [3]).

In this paper, we discuss the existence of orientations that satisfies demands defined on given pairs of vertex sets. Our question is how large connectivity undirected graphs should have for obtaining its orientations satisfying the demands? This question can be formulated as finding the smallest C for which the following conjecture holds.

Conjecture 1. Let $G = (V, E)$ be an undirected graph, and $\{X_i, Y_i\}$ be pairs of disjoint subsets of V with demand $f_i \in \mathbb{Z}$ for $i \in \{1, \ldots, \ell\}$ ($\mathbb{Z}$ denotes the set of integers). If G satisfies $\lambda_G(X_i, Y_i) \geq C f_i$ for each $i \in \{1, \ldots, \ell\}$, then it has an orientation $D = (V, A)$ such that $\min\{\lambda_D(X_i, Y_i), \lambda_D(Y_i, X_i)\} \geq f_i$ for each $i \in \{1, \ldots, \ell\}$.

As examples such as cycles show, $C \geq 2$ is necessary for this conjecture to hold. With respects to the edge-connectivity between two vertices, Nash-Williams gave the following best possible result.

Theorem 1 (Nash-Williams [7]). *Let $f : \binom{V}{2} \to \mathbb{Z}$ be a demand function, where $\binom{V}{2}$ denotes the set of unordered pairs of vertices. Every undirected graph G has an orientation D such that $\lambda_D(u, v) \geq f(u, v)$ for each $u, v \in V$ if $\lambda_G(u, v) \geq 2f(u, v)$ for each $u, v \in V$.* □

Conjecture 1 is a natural extension of the theorem due to Nash-Williams [7].

One may consider that Conjecture 1 can be derived by applying Theorem 1 to the graph obtained by shrinking vertex sets in $\{X_i, Y_i \mid i = 1, \ldots, n\}$ into single vertices. We notice that this is not true by the following two reasons. While we are assuming $X_i \cap Y_i = \emptyset$ for $i = 1, \ldots, n$ in the conjecture, there are possibly intersecting sets belonging to different pairs, i.e., $X_i \cap X_j \neq \emptyset$, $Y_i \cap Y_j \neq \emptyset$ or $X_i \cap Y_j \neq \emptyset$ may hold for $i \neq j$. In addition, even if all sets in the given pairs are disjoint, shrinking a set makes influence on the edge-connectivity of other pairs. As an example, see Figure 1 illustrating a graph G with pairs $\{X_1, Y_1\}$, $\{X_2, Y_2\}$ and $\{X_3, Y_3\}$ of subsets of V, and G' obtained by shrinking X_1 and Y_1 into single vertices x_1 and y_1. Although the edge-connectivity of $\{X_2, Y_2\}$ is not changed by the shrinking ($\lambda_G(X_1, Y_1) = \lambda_{G'}(X_1, Y_1) = 4$), the edge-connectivity of $\{X_3, Y_3\}$ is ($\lambda_G(X_3, Y_3) = 4$ and $\lambda_{G'}(X_3, Y_3) = 5$).

In this paper, we prove Conjecture 1 with $C = \max\{2(|X_i| + |Y_i| - 1) \mid i = 1, \ldots, \ell\}$. We also give upper- and lower-bounds on the difference between $\min\{\lambda_D(X_i, Y_i), \lambda_D(Y_i, X_i)\}$ and $\lambda_G(X_i, Y_i)$ (Section 2). We also consider a special case where the demand is defined on the rooted edge-connectivity. Namely, G has the root $r \in V$ and $X_i = \{r\}$ for all $i \in \{1, \ldots, \ell\}$. In this case, Conjecture 1 can be stated as follows.

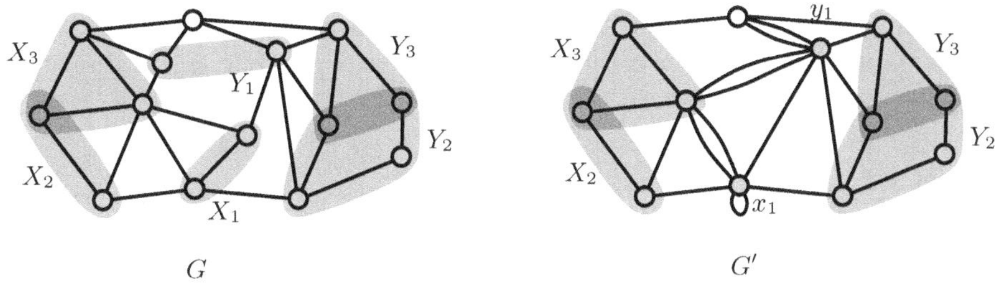

Fig. 1. An undirected graph and pairs of vertex subsets

Conjecture 2. Let $G = (V, E)$ be an undirected graph, $r \in V$, $X_i \subseteq V - r$ for $i \in \{1, \ldots, \ell\}$, and $f_i \in \mathbb{Z}$ for $i \in \{1, \ldots, \ell\}$. If G satisfies $\lambda_G(r, X_i) \geq Cf_i$ for each $i \in \{1, \ldots, \ell\}$, then it has an orientation $D = (V, A)$ such that $\lambda_D(r, X_i) \geq f_i$ for each $i \in \{1, \ldots, \ell\}$.

An affirmative answer to Conjecture 1 implies that to Conjecture 2 with the same ratio C. Moreover, $C = 2$ remains best possible also to Conjecture 2. In Section 3, we prove Conjecture 2 with $C = \ell$. Note that all the positive results for the conjectures are algorithmic. That is to say, if a given undirected graph satisfies the conditions specified in the conjectures, we can compute an orientation satisfying the demand.

In the remainder of this introduction, let us review the difficulty of our problems from the view points of demand functions. We say that a digraph $D = (V, A)$ *covers* a demand function $h : 2^V \to \mathbb{Z}$ if $\rho_D(X) \geq h(X)$ for all non-empty $X \subset V$ where $\rho_D(X)$ denotes the number of arcs from vertices in $V - X$ to those in X.

If subsets X and Y of V satisfy $X - Y, Y - X, X \cap Y, V - (X \cup Y) \neq \emptyset$, then X and Y are called *intersecting*. A set function $h : 2^V \to \mathbb{Z}$ is called *intersecting G-supermodular* if $h(X) + h(Y) \leq h(X \cup Y) + h(X \cap Y) + d_G(X, Y)$ holds for each intersecting $X, Y \in 2^V$ where $d_G(X, Y)$ denotes the number of edges in G joining vertices in $X - Y$ and in $Y - X$. If h satisfies $h(X) + h(Y) \leq h(X \cup Y) + h(X \cap Y)$ for each intersecting $X, Y \in 2^V$, then h is called *intersecting supermodular*.

The following theorem is due to Frank [2].

Theorem 2 (Frank [2]). *Let G be an undirected graph and h be an intersecting G-supermodular function (with possible negative values). There is an orientation of G covering h if and only if $d_G(\mathcal{P}) \geq \sum_{i=1}^{t} h(V_i)$ holds for every subpartition $\mathcal{P} = \{V_1, \ldots, V_t\}$ of V where $d_G(\mathcal{P})$ denotes the number of edges in G entering at least one member of $\mathcal{P}$.* □

This theorem is so general that it includes several known orientation theorems. However, our setting is not included by this because the edge-connectivity between vertex sets is not captured by intersecting G-supermodular functions. In Section 3, we observe that if Theorem 2 can be extended to skew-supermodular demand functions (defined in Section 3), then Conjecture 2 can be proven with $C = 2$.

This paper is organized as follows. Section 2 proves several results related to Conjecture 1. Section 3 discusses Conjecture 2. Section 4 concludes this paper by mentioning relationship between our results and previous works about tree packing theorems.

2 Orientation Preserving Local Edge-Connectivity

In [3], Fukunaga and Nagamochi gave a useful relationship between the edge-connectivity between vertices and that between vertex sets.

Lemma 1 (Fukunaga, Nagamochi [3]). *Let $\{X, Y\}$ be a pair of disjoint subsets of V. If $\lambda_G(X, Y) \geq k(|X| + |Y| - 1)$, then there exists a pair of vertices $x \in X$ and $y \in Y$ such that $\lambda_G(x, y) \geq k$.* □

From this fact, we can derive an answer to Conjecture 1 with $C = 2\max\{|X_i| + |Y_i| - 1 \mid 1 \leq i \leq \ell\}$ as follows.

Theorem 3. *Let $G = (V, E)$ be an undirected graph, and $\{X_1, Y_1\}, \ldots, \{X_\ell, Y_\ell\}$ be pairs of disjoint subsets of V with demand $f_i \in \mathbb{Z}$ for $i \in \{1, \ldots, \ell\}$. If G satisfies $\lambda_G(X_i, Y_i) \geq 2(|X_i| + |Y_i| - 1)f_i$ for each $i \in \{1, \ldots, \ell\}$, then it has an orientation D such that $\min\{\lambda_D(X_i, Y_i), \lambda_D(Y_i, X_i)\} \geq f_i$ for each $i \in \{1, \ldots, \ell\}$.*

Proof. Since $\lambda_G(X_i, Y_i) \geq 2(|X_i|+|Y_i|-1)f_i$ holds, there exists a pair of vertices $x_i \in X_i$ and $y_i \in Y_i$ such that $\lambda_G(x_i, y_i) \geq 2f_i$ by Lemma 1. The orientation D of G given by Theorem 1 satisfies $\min\{\lambda_D(x_i, y_i), \lambda_D(y_i, x_i)\} \geq f_i$ for each $i \in \{1, \ldots, \ell\}$. Since $\lambda_D(X_i, Y_i) \geq \lambda_D(x_i, y_i)$ and $\lambda_D(Y_i, X_i) \geq \lambda_D(y_i, x_i)$, D is a required orientation. □

It is easy to check that Conjecture 1 with $C = 2$ holds for Eulerian graph G. By this fact, we can give another bound following the approach taken by Nash-Williams [7] for proving Theorem 1. We let $O(G)$ stand for the number of vertices each of which has odd degree in G.

Theorem 4. *Let $G = (V, E)$ be an undirected graph, and $\{X_1, Y_1\}, \ldots, \{X_\ell, Y_\ell\}$ be pairs of disjoint subsets of V. Then G has an orientation D that satisfies $\lambda_D(X_i, Y_i) \geq \lceil \lambda_G(X_i, Y_i)/2 \rceil - O(G)/2$ and $\lambda_D(Y_i, X_i) \geq \lceil \lambda_G(X_i, Y_i)/2 \rceil - O(G)/2$ for each $i \in \{1, \ldots, \ell\}$.*

Proof. Observe that if $O(G) = 0$, then a digraph D obtained by orienting G along its Eulerian trail or path satisfies $\lambda_D(X_i, Y_i) \geq \lambda_G(X_i, Y_i)/2$ and $\lambda_D(Y_i, X_i) \geq \lambda_G(X_i, Y_i)/2$ for all $i \in \{1, \ldots, \ell\}$.

Let us consider the case where $O(G) > 0$. Augment G by adding a perfect matching M on the vertices of odd degrees in G. Then the obtained undirected graph $G+M$ is Eulerian (i.e., $O(G+M) = 0$). Hence $G+M$ has an orientation D' that satisfies $\lambda_{D'}(X_i, Y_i) \geq \lambda_{G+M}(X_i, Y_i)/2$ and $\lambda_{D'}(Y_i, X_i) \geq \lambda_{G+M}(X_i, Y_i)/2$ for all $i \in \{1, \ldots, \ell\}$ as mentioned above. Define D as the digraph obtained by removing arcs corresponding to M from D'. Then D is an orientation of

G. Since $|M| = O(G)/2$, $\lambda_D(X_i, Y_i) \geq \lambda_{D'}(X_i, Y_i) - O(G)/2$ and $\lambda_D(Y_i, X_i) \geq \lambda_{D'}(X_i, Y_i) - O(G)/2$ for all $i \in \{1, \ldots, \ell\}$. Since $G + M$ is Eulerian, every cut has even capacity. It means that $\lambda_{G+M}(X_i, Y_i)$ is even for all $i \in \{1, \ldots, \ell\}$. Thus $\lambda_{G+M}(X_i, Y_i) \geq 2\lceil \lambda_G(X_i, Y_i)/2 \rceil$ holds.

From these facts, we can derive

$$\begin{aligned}\lambda_D(X_i, Y_i) &\geq \lambda_{D'}(X_i, Y_i) - O(G)/2 \\ &\geq \lambda_{G+M}(X_i, Y_i)/2 - O(G)/2 \geq \lceil \lambda_G(X_i, Y_i)/2 \rceil - O(G)/2\end{aligned}$$

and

$$\begin{aligned}\lambda_D(Y_i, X_i) &\geq \lambda_{D'}(Y_i, X_i) - O(G)/2 \\ &\geq \lambda_{G+M}(X_i, Y_i)/2 - O(G)/2 \geq \lceil \lambda_G(X_i, Y_i)/2 \rceil - O(G)/2\end{aligned}$$

hold for all $i \in \{1, \ldots, \ell\}$. □

We also have a negative result for Conjecture 1.

Theorem 5. *Define $\{X_1, Y_1\}, \ldots, \{X_\ell, Y_\ell\}$ as all partitions of V into two non-empty subsets (i.e., $\{X_i \mid i = 1, \ldots, \ell\} = \{X \in 2^V \mid 0 < |X| < |V|/2\}$ and $Y_i = V - X_i$). Then G has no orientation D that satisfies*

$$\min\{\lambda_D(X_i, Y_i), \lambda_D(Y_i, X_i)\} > \lambda_G(X_i, Y_i)/2 - O(G)/4 \qquad (1)$$

for all $i \in \{1, \ldots, \ell\}$.

Proof. To contrary, suppose that G has an orientation D that satisfies (1) for all $i \in \{1, \ldots, \ell\}$.

Let us consider the case where at least $O(G)/2$ vertices in D have the in-degrees larger than the out-degrees Let X denote the set of those vertices in G, and $E(X)$ denote the set of edges in G whose both end vertices are in X. Then X satisfies

$$\rho_D(X) = \sum_{v \in X} \rho_D(v) - E(X) \geq \sum_{v \in X} (\delta_D(v) + 1) - E(X) \geq \delta_D(X) + O(G)/2.$$

On the other hand, $\rho_D(X) + \delta_D(X) = d_G(X)$. By these facts, $\delta_D(X) \leq d_G(X)/2 - O(G)/4$ holds. Hence we have $\lambda_D(X, V - X) = \delta_D(X) \leq d_G(X)/2 - O(G)/4 = \lambda_G(X, V - X)/2 - O(G)/4$, a contradiction.

If D has at least $O(G)/2$ vertices having the out-degrees larger than the in-degrees, then consider the digraph D' obtained by reversing all arcs in D. By applying the above argument to D', we have a contradiction also in this case. □

Theorem 5 implies that Conjecture 1 does not hold for $C < \min\{4/(2 - O(G)/\lambda_G(X_i, Y_i)) \mid 1 \leq i \leq \ell\}$ in general.

3 Orientation Preserving Rooted Edge-Connectivity

Let $G = (V, E)$ be an undirected graph with a root $r \in V$, vertex subsets $X_1, \ldots, X_\ell \subseteq V - r$ and demands $f_1, \ldots, f_\ell \in \mathbb{Z}$. In this section, we discuss sufficient conditions for $G = (V, E)$ to have an orientation D such that $\lambda_D(r, X_i) \geq f_i$ for all $i \in \{1, \ldots, \ell\}$. Since this connectivity demand is weaker than that discussed in Section 2, Theorems 4 and 5 also tell necessary conditions. In addition to these, we can obtain another condition as a benefit of weakening the demand.

Theorem 6. *Let $r \in V$ and $X_1, \ldots, X_\ell \subseteq V - r$. An undirected graph $G = (V, E)$ has an orientation $D = (V, A)$ such that $\lambda_D(r, X_i) \geq f_i$ for all $i \in \{1, \ldots, \ell\}$ if G satisfies $\lambda_G(r, X_i) \geq i f_i$ for all $i \in \{1, \ldots, \ell\}$.*

Proof. Let $M_i = (V, E_i, A_i)$, $i \in \{1, \ldots, \ell\}$ denote mixed graphs obtained from G by orienting some of its edges. We let G_i denote the undirected graph (V, E_i), and D_i denote the digraph (V, A_i). For proving Theorem 6, we show that it is possible to construct $M_1, \ldots, M_\ell$ inductively so that they satisfy

$$\lambda_{D_i}(r, X_j) \geq f_j \quad \text{for } j \in \{1, \ldots, i\} \tag{2}$$

and

$$\delta_{D_i}(Y) \leq i\rho_{D_i}(Y) \quad \text{for } X_j \subseteq Y \subseteq V - r \text{ with some } j \in \{i+1, \ldots, \ell\}. \tag{3}$$

Figure 2 shows an example of $M_1, \ldots, M_\ell$. Notice that a required orientation of G can be obtained from M_ℓ by orienting the edges in E_ℓ arbitrarily.

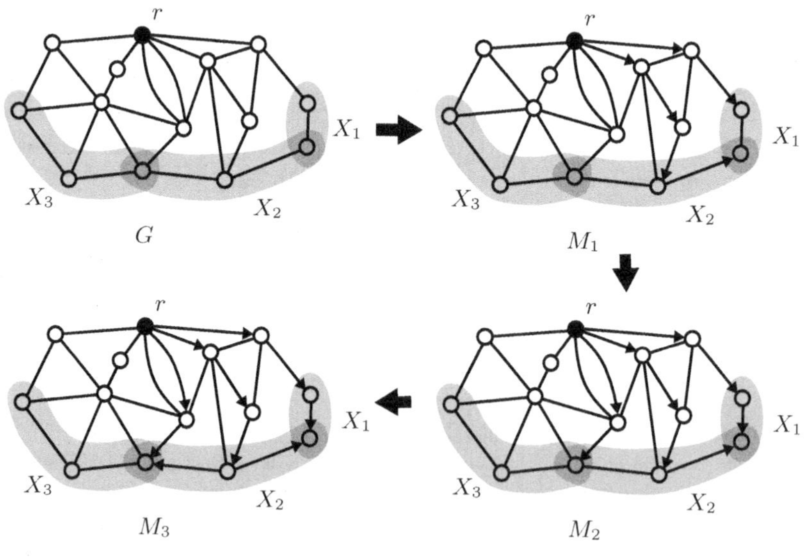

Fig. 2. Construction of M_i, $i \in \{1, \ldots, \ell\}$ from G

First, let us show how to construct M_1. Since $\lambda_G(r, X_1) \geq f_1$, G contains at least f_1 edge-disjoint paths between r and X_1. Orient the edges in those paths from r to X_1. Then the obtained mixed graph satisfies the conditions (2) and (3) for $i = 1$. In fact, the first condition $\lambda_{D_1}(r, X_1) \geq f_1$ obviously holds. The second condition $\delta_{D_1}(Y) \leq \rho_{D_1}(Y)$ also holds because each $e \in \delta_{D_1}(Y)$ is a part of a directed path from $r \notin Y$.

Now suppose that we have M_i for some $1 \leq i < \ell$. We show how to construct M_{i+1} from M_i. Let $M' = (V', E', A')$ be the mixed graph obtained from M_i by shrinking X_{i+1} into a single vertex x and deleting generated loops. We let G' and D' denote the undirected graph (V', E') and the digraph (V', A'), respectively. Define a set function $h : 2^{V'} \to \mathbb{Z}$ so that

$$h(Y) = \begin{cases} f_{i+1} - \rho_{D'}(Y) & \text{if } x \in Y \text{ and } r \notin Y, \\ -f_{i+1} - \rho_{D'}(Y) & \text{if } x \notin Y \text{ and } r \in Y, \\ -\rho_{D'}(Y) & \text{otherwise.} \end{cases}$$

Then we have a good property of h as follows.

Claim. Function h is intersecting supermodular.

Proof. It is easy to prove that $-\rho_{D'}$ is intersecting supermodular. Hence it suffices to show that $h' := h + \rho_{D'}$ is intersecting supermodular. Let $Y, Z \in 2^{V'}$ such that $Y \cap Z \neq \emptyset$. In the following, we observe that

$$h'(Y) + h'(Z) \leq h'(Y \cap Z) + h'(Y \cup Z) \tag{4}$$

always holds.

Let $h'(Y) = h'(Z) = f_{i+1}$. In this case, $x \in Y \cap Z, Y \cup Z$ and $r \notin Y \cap Z, Y \cup Z$ hold, which imply $h'(Y \cap Z) = h'(Y \cup Z) = f_{i+1}$. Hence (4) holds.

Let $h'(Y) = -f_{i+1}$ and $h'(Z) = f_{i+1}$. In this case, $h'(Y \cap Z) = 0$ because $x, r \notin Y \cap Z$, and $h'(Y \cup Z) = 0$ because $x, r \in Y \cup Z$. Hence (4) holds.

Let $h'(Y) = 0$ and $h'(Z) = k$. If $x, r \in Y$, then $x \in Y \cap Z$, $r \notin Y \cap Z$ and $x, r \in Y \cup Z$ hold, which implies that $h'(Y \cap Z) = f_{i+1}$ and $h'(Y \cup Z) = 0$. If $x, r \notin X$, then $x, r \notin Y \cap Z$, $x \in Y \cup Z$ and $r \notin Y \cup Z$ hold, which implies that $h'(Y \cap Z) = 0$ and $h'(Y \cup Z) = f_{i+1}$. In both cases, (4) holds.

Let $h'(Y) = 0$ and $h'(Z) = -f_{i+1}$. If $x, r \in Y$, then $x \notin Y \cap Z$, $r \in Y \cap Z$ and $x, r \in Y \cup Z$ hold, which implies that $h'(Y \cap Z) = -f_{i+1}$ and $h'(Y \cup Z) = 0$. If $x, r \notin Y$, then $x, r \notin Y \cap Z$, $x \notin Y \cup Z$ and $r \in Y \cup Z$ hold, which implies that $h'(Y \cap Z) = 0$ and $h'(Y \cup Z) = -f_{i+1}$. In both cases, (4) holds.

Finally, let $h'(Y) = h'(Z) = 0$. If $x, r \in Y, Z$, then $x, r \in Y \cap Z, Y \cup Z$, and hence $h'(Y \cap Z) = h'(Y \cup Z) = 0$ holds. If $x, r \in Y$ and $x, r \notin Z$, then $x, r \notin Y \cap Z$ and $x, r \in Y \cup Z$, and hence $h'(Y \cap Z) = h'(Y \cup Z) = 0$ holds. If $x, r \notin Y, Z$, then $x, r \notin Y \cap Z, Y \cup Z$, and hence $h'(Y \cap Z) = h'(Y \cup Z) = 0$ holds. In any cases, (4) holds. □

Recall that intersecting supermodular set functions on V' are intersecting G'-supermodular. Hence we can apply Theorem 2 for obtaining the following fact.

Claim. G' has an orientation covering h.

Proof. Let $Y \subseteq V'$. We first see that $d_{G'}(Y) \geq 2h(Y)$ holds. It suffices to consider the case where $x \in Y$ and $r \notin Y$. Notice that $d_G(Y) = d_{G'}(Y)+\rho_{D'}(Y)+\delta_{D'}(Y)$. Since $\lambda_G(r, X_{i+1}) \geq (i+1)f_{i+1}$, $d_G(Y) \geq (i+1)f_{i+1}$ holds by Menger's theorem. By these and condition (3), we have

$$d_{G'}(Y) + (i+1)\rho_{D'}(Y) \geq d_{G'}(Y) + \rho_{D'}(Y) + \delta_{D'}(Y) \geq (i+1)f_{i+1}.$$

Hence $d_{G'}(Y) \geq (i+1)(f_{i+1} - \rho_{D'}(Y)) = (i+1)h(Y) \geq 2h(Y)$.

Let $\mathcal{P} = \{V_1, \ldots, V_t\}$ be a subpartition of V'. It then satisfies $d_{G'}(\mathcal{P}) \geq \sum_{j=1}^{t} d_{G'}(V_j)/2 \geq \sum_{j=1}^{t} h(V_j)$. This means that G' satisfies the condition presented in Theorem 2. Therefore G' has an orientation covering h. □

Let D'' denote the orientation of G' covering h. Then $\lambda_{D'+D''}(r, x) \geq f_{i+1}$ since each $Y \subseteq V'$ with $x \in Y$ and $r \notin Y$ satisfies $\rho_{D'+D''}(Y) = \rho_{D'}(Y) + \rho_{D''}(Y) \geq \rho_{D'}(Y) + h(Y) \geq f_{i+1}$. Choose f_{i+1} edge-disjoint directed paths from r to x in $D' + D''$, and call them by $P_1, \ldots, P_{f_{i+1}}$, respectively. We denote the set of edges both in E_i and in P_ℓ by $E(P_\ell)$. Define M_{i+1} as the mixed graph obtained by orienting the edges in $E(P_\ell)$ with some $\ell \in \{1, \ldots, f_{i+1}\}$ from r to x. In the following, we see that the constructed M_{i+1} satisfies conditions (2) and (3).

For each $j \in \{1, \ldots, i\}$, $\lambda_{D_{i+1}}(r, X_j) \geq f_j$ holds because $A_i \subseteq A_{i+1}$ and M_i satisfies (2). Moreover, $\lambda_{D_{i+1}}(r, X_{i+1}) \geq f_{i+1}$ holds since D_{i+1} contains f_{i+1} edge-disjoint directed paths from r to X_{i+1}. Hence M_{i+1} satisfies (2).

Let $Y \in 2^V$ such that $X_j \subseteq Y \subseteq V - r$ with some $j \in \{i+2, \ldots, \ell\}$. Notice that each edge entering Y in $A_{i+1} - A_i$ is a part of a directed path from r to x in D'', where $r \notin Y$. Hence $\delta_{D_{i+1}}(Y) - \delta_{D_i}(Y) \leq \rho_{D_{i+1}}(Y)$ holds. By this fact and the assumption that M_i satisfies (3), it holds that

$$\delta_{D_{i+1}}(Y) = \delta_{D_{i+1}}(Y) - \delta_{D_i}(Y) + \delta_{D_i}(Y) \leq \rho_{D_{i+1}}(Y) + i\rho_{D_i}(Y) \leq (i+1)\rho_{D_{i+1}}(Y).$$

Therefore M_{i+1} satisfies (3). This completes the proof of Theorem 6. □

Combining Theorem 3, Theorem 6, and discussion at the beginning of this section presents an answer to Conjecture 2.

Corollary 1. *Let $G = (V, E)$ be an undirected graph, $r \in V$, $X_i \subseteq V - r$ for $i \in \{1, \ldots, \ell\}$, and $f_i \in \mathbb{Z}$ for $i \in \{1, \ldots, \ell\}$. Moreover, let $C = \min\{\ell, \max\{2|X_i| \mid i = 1, \ldots, \ell\}\}$. If G satisfies $\lambda_G(r, X_i) \geq Cf_i$ for each $i \in \{1, \ldots, \ell\}$, then it has an orientation $D = (V, A)$ such that $\lambda_D(r, X_i) \geq f_i$ for each $i \in \{1, \ldots, \ell\}$.* □

In the remainder of this section, we show that the orientation problem with arc-connectivity demands from r to X_i, $i \in \{1, \ldots, \ell\}$ can be reduced to the orientation problem in mixed graphs with rooted edge-connectivity demands. Given graph $G = (V, E)$ with $r \in V$ and $X_1, \ldots, X_\ell \subseteq V - r$, augment G with a new vertex x_i for each $i \in \{1, \ldots, \ell\}$ and f_i parallel arcs from each $v \in X_i$ to x_i. Let M denote the obtained mixed graph (see Figure 3). If we can orient the undirected edges in M so that the resultant digraph D satisfies $\lambda_D(r, x_i) \geq f_i$

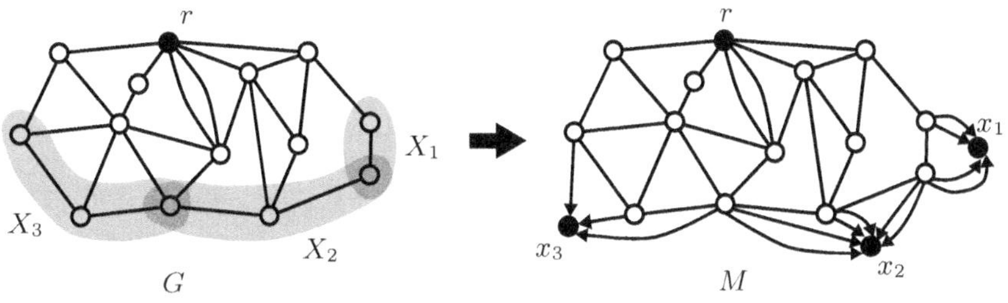

Fig. 3. Transformation from G to M

for $i \in \{1, \ldots, \ell\}$, then it gives an orientation D' of G such that $\lambda_{D'}(r, X_i) \geq f_i$ for $i \in \{1, \ldots, \ell\}$.

Unfortunately we do not know how to solve this orientation problem. If the connectivity demand is defined from r to x for all $x \in V - r$, then Theorem 2 gives a necessary and sufficient condition. However, in the above problem, the connectivity demand is defined only from r to x_i, $i \in \{1, \ldots, \ell\}$. This demand can not be formulated by G-supermodular functions, but by skew supermodular functions. If Theorem 2 can be extended to skew supermodular functions, then it gives a proof for Conjecture 2 with $C = 2$.

4 Concluding Remarks

As a concluding remark, let me mention relationship between rooted k-arc-connectivity and tree packings.

For an undirected graph $G = (V, E)$ with a root $r \in V$ and subsets X_i, $i \in \{1, 2, \ldots, \ell\}$ of $V - \{r\}$, a *group Steiner tree* is defined as a tree T in G that connects r and X_i, $i \in \{1, 2, \ldots, \ell\}$ each other. The *packing number* of group Steiner trees is defined as the maximum number of edge-disjoint group Steiner trees contained by G. Notice that if the packing number is at least k, then G has obviously an orientation D that satisfies $\lambda(r, X_i) \geq k$ for all $i \in \{1, \ldots, \ell\}$. We do not know whether its converse holds or not.

In [3], Fukunaga and Nagamochi have shown that the packing number is at least k if G satisfies $\lambda_G(r, X_i) \geq 2k|X_i|$ for all $i \in \{1, \ldots, \ell\}$. Based on this observation, they have presented an approximation algorithm for the minimum group Steiner tree problem, which is a problem of computing a minimum cost group Steiner tree[1]. The approximation factor of their algorithm heavily depends on the gap between the edge-connectivity in G and the packing number. Hence it is important to improve this gap.

Notice that the gap presented by Fukunaga and Nagamochi [3] coincides with Theorem 3. A natural question is whether Theorem 6 can be strengthen in order to obtain another gap between the edge-connectivity in G and the packing number. This question is formulated as the following conjecture.

[1] Fukunaga and Nagamochi [3] have actually solved a problem including the group Steiner tree problem by the above-mentioned approach.

Conjecture 3. Let $G = (V, E)$ be an undirected graph with a root $r \in V$ and subsets $X_1, \ldots, X_\ell \subseteq V - r$. If $\lambda_G(r, X_i) \geq \ell k$ for all $i \in \{1, \ldots, \ell\}$, then G contains k edge-disjoint group Steiner trees. □

Even if this conjecture is proven, it would not present an approximation algorithm better than known algorithms. Nevertheless the packing of group Steiner trees itself deserves attention.

References

1. Bárász, M., Beckder, J., Frank, A.: An Algorithm for Source Location in Directed Graphs. Operations Research Letters 33, 221–230 (2005)
2. Frank, A.: Orientations of Graphs and Submodular Flows. Conguressus Numerantium 113, 111–142 (1996)
3. Fukunaga, T., Nagamochi, H.: The Set Connector Problem in Graphs. In: Fischetti, M., Williamson, D.P. (eds.) IPCO 2007. LNCS, vol. 4513, pp. 484–498. Springer, Heidelberg (2007)
4. Ishii, T., Hagiwara, M.: Minimum Augmentation of Local Edge-connectivity between Vertices and Vertex Subsets in Undirected Graphs. Discrete Applied Mathematics 154, 2307–2329 (2006)
5. Ishii, T., Makino, K.: Augmenting Edge-connectivity between Vertex Subsets. In: The Australian Theory Symposium on Computing Theory (2008)
6. Ito, H.: Node-to-area Connectivity of Graphs. Transactions of the Institute of Electrical Engineers of Japan 11C, 463–469 (1994)
7. Nash-Williams, C.S.J.A.: On Orientations, Connectivity and Odd Vertex Pairings in Finite Graphs. Canadian Journal of Mathematics 12, 555–567 (1960)

A Linear Vertex Kernel for MAXIMUM INTERNAL SPANNING TREE

Fedor V. Fomin[1], Serge Gaspers[2], Saket Saurabh[1], and Stéphan Thomassé[2]

[1] Department of Informatics, University of Bergen, Norway
{fomin,saket}@ii.uib.no
[2] LIRMM – University of Montpellier 2, CNRS, France
{gaspers,thomasse}@lirmm.fr

Abstract. We present a polynomial time algorithm that for any graph G and integer $k \geq 0$, either finds a spanning tree with at least k internal vertices, or outputs a new graph G_R on at most $3k$ vertices and an integer k' such that G has a spanning tree with at least k internal vertices if and only if G_R has a spanning tree with at least k' internal vertices. In other words, we show that the MAXIMUM INTERNAL SPANNING TREE problem parameterized by the number of internal vertices k has a $3k$-vertex kernel. Our result is based on an innovative application of a classical min-max result about hypertrees in hypergraphs which states that "a hypergraph H contains a hypertree if and only if H is partition connected."

1 Introduction

In the MAXIMUM INTERNAL SPANNING TREE problem (MIST), we are given a graph G and the task is to find a spanning tree of G with a maximum number of internal vertices. MIST is a natural generalization of the HAMILTONIAN PATH problem because an n-vertex graph has a Hamiltonian path if and only if it has a spanning tree with $n-2$ internal vertices.

In this paper we study a parameterized version of MIST. Parameterized decision problems are defined by specifying the input (I), the parameter (k), and the question to be answered. A parameterized problem that can be solved in time $f(k)|I|^{O(1)}$, where f is a function of k alone is said to be *fixed parameter tractable* (FPT). The natural parameter k for MIST is the number of internal vertices in the spanning tree and the parameterized version of MIST, p-INTERNAL SPANNING TREE or p-IST for short, is for a given graph G and integer k, decide if G contains a spanning tree with at least k internal vertices. It follows from Robertson and Seymour's Graph Minors theory that p-IST is FPT [10]. Indeed, the property of not having a spanning tree with at least k internal vertices is closed under taking minors, and thus such graphs can be characterized by a finite set of forbidden minors. One of the consequences of the Graph Minors theory is that every graph property characterized by a finite set of forbidden minors is FPT, and thus p-IST is FPT. These arguments are however not constructive. The first constructive algorithm for p-IST is due to Prieto and Sloper [12] and has running time $2^{4k \log k} \cdot n^{O(1)}$. Recently this result was improved by Cohen et al. [2]

Y. Dong, D.-Z. Du, and O. Ibarra (Eds.): ISAAC 2009, LNCS 5878, pp. 275–282, 2009.

who solved a more general directed version of the problem in time $49.4^k \cdot n^{O(1)}$. In this paper we study p-IST from the kernelization viewpoint.

A parameterized problem is said to admit a *polynomial kernel* if there is a polynomial time algorithm (where the degree of the polynomial is independent of k), called a *kernelization* algorithm, that reduces the input instance to an instance whose size is bounded by a polynomial $p(k)$ in k, while preserving the answer. This reduced instance is called a $p(k)$ *kernel* for the problem. Let us remark that the instance size and the number of vertices in the instance may be different, and thus for bounding the number of vertices in the reduced graph, the term $p(k)$*-vertex kernel* is often used. While many problems on graphs are known to have polynomial kernels (parameterized by the solution size), there are not so many $O(k)$, or linear-vertex kernels known in the literature. Notable examples include a $2k$-vertex kernel for Vertex Cover [3], a k-vertex kernel for Set Splitting [6], and a $6k$-vertex kernel for Cluster Editing [5].

No linear-vertex kernel for p-IST was known prior to our work. Prieto and Sloper [11] provided an $O(k^3)$-kernel for the problem and then improved it to $O(k^2)$ in [12]. The main result of this paper is that p-IST has a $3k$-vertex kernel. The kernelization of Prieto and Sloper is based on the so-called "Crown Decomposition Method" [1]. Here, we use a different method, based on a min-max characterization of hypergraphs containing hypertrees by Frank et al. [4]. As a corollary of the new kernelization, we obtain an algorithm for solving p-IST running in time $8^k \cdot n^{O(1)}$.

The paper is organized as follows. In Section 2, we provide necessary definitions and facts about graphs and hypergraphs. In Section 3, we give the kernelization algorithm. Section 4 is devoted to the proof of the main combinatorial lemma, which is central to the correctness of the kernelization algorithm.

2 Preliminaries

2.1 Graphs

Let $G = (V, E)$ be an undirected simple graph with vertex set V and edge set E. For any nonempty subset $W \subseteq V$, the subgraph of G induced by W is denoted by $G[W]$. The *neighborhood* of a vertex v in G is $N_G(v) = \{u \in V : \{u, v\} \in E\}$, and for a vertex set $S \subseteq V$ we set $N_G(S) = \bigcup_{v \in S} N_G(v) \setminus S$. The degree of vertex v in G is $d_G(v) = |N_G(v)|$. Sometimes, when the graph is clear from the context, we omit the subscripts.

2.2 The Hypergraphic Matroid

Let $H = (V, E)$ be a hypergraph. A hyperedge $e \in E$ is a subset of V. A subset F of hyperedges is a *hyperforest* if $| \cup F'| \geq |F'| + 1$ for every subset F' of F, where $\cup F'$ denotes the union of vertices contained in the hyperedges of F'. This condition is also called the *strong Hall condition*, where *strong* stands for the extra plus one added to the usual Hall condition. A hyperforest with $|V| - 1$ edges is called a *hypertree*. Lorea proved (see [4] or [7]) that $\mathcal{M}_H = (E, \mathcal{F})$, where $\mathcal{F}$ consists of the hyperforests of H, is a matroid, called the *hypergraphic matroid*. Observe that these definitions are well-known when restricted to graphs.

Lovász proved (see [8]) that F is a hyperforest if and only if every hyperedge e of F can be shrunk into an edge e' (that is, $e' \subseteq e$ contains two vertices of e) in such a way that the set F' consisting of these contracted edges forms a forest in the usual sense, that is, forest of a graph. Observe that if F is a hypertree then its set of contracted edges F' forms a spanning tree on V.

The *border* of a partition $\mathcal{P} = \{V_1, \ldots, V_p\}$ of V is the set $\delta(\mathcal{P})$ of hyperedges of H which intersect at least two parts of $\mathcal{P}$. A hypergraph is *partition-connected* when $|\delta(\mathcal{P})| \geq |\mathcal{P}| - 1$ for every partition $\mathcal{P}$ of V. The following theorem can be found in [4, Corollary 2.6].

Theorem 1. *H contains a hypertree if and only if H is partition-connected.*

The proof of Theorem 1 can be turned into a polynomial time algorithm, that is, given a hypergraph $H = (V, E)$ we can either find a hypertree or find a partition $\mathcal{P}$ of V such that $|\delta(\mathcal{P})| < |\mathcal{P}| - 1$ in polynomial time. For the sake of completeness, we briefly mention a polynomial time algorithm to do this, though the running time may be easily improved. Recall that $\mathcal{M}_H = (E, \mathcal{F})$, where $\mathcal{F}$ consists of the hyperforests of H, is a matroid and hence we can construct a hypertree, if one exists, greedily. We start with an empty forest and iteratively try to grow our current hyperforest by adding new edges. When inspecting a new edge we either reject or accept it in our current hyperforest depending on whether by adding it we still have a hyperforest. The only question is to be able to test efficiently if a given collection of edges forms a hyperforest. In other words, we have to check if the strong Hall condition holds. This can be done in polynomial time by simply running the well-known polynomial time algorithm for testing the usual Hall condition for every subhypergraph $H \setminus v$, where v is a vertex and $H \setminus v$ is the hypergraph containing all hyperedges $e \setminus v$ for $e \in E$.

We can also find a contraction of the edges of a hypertree into a spanning tree in polynomial time. For this, consider any edge e of the hypertree with more than two vertices (if none exist, we already have our tree). By a result of Lovász [8] mentioned above, one of the vertices $v \in e$ can be deleted from e in such a way that we still have a hypertree. Hence we just find this vertex by checking the strong Hall condition for every choice of $e \setminus v$ where $v \in e$. This implies that we need to apply the algorithm to test the strong Hall condition at most $|V|$ times to obtain the desired spanning tree. Consequently, there exists a polynomial time algorithm which can find a contracted spanning tree out of a partition-connected hypergraph.

We now turn to the co-NP certificate, that is, we want to exhibit a partition $\mathcal{P}$ of V such that $|\delta(\mathcal{P})| < |\mathcal{P}| - 1$ when H is not partition-connected. The algorithm simply tries to contract every pair of vertices in $H = (V, E)$ and checks if the resulting hypergraph is partition-connected. When it is not, we contract the two vertices, and recurse. We stop when the resulting hypergraph H' is not partition-connected, and every contraction results in a partition-connected hypergraph. Observe then that if a partition $\mathcal{P}$ of H' is such that $|\delta(\mathcal{P})| < |\mathcal{P}| - 1$ and $\mathcal{P}$ has a part which is not a singleton, then contracting two vertices of this part results in a non partition-connected hypergraph. Hence, the singleton partition is the unique partition $\mathcal{P}$ of H' such that $|\delta(\mathcal{P})| < |\mathcal{P}| - 1$. This singleton partition corresponds to the partition of H which gives our co-NP certificate.

3 Kernelization Algorithm

Let $G = (V, E)$ be a connected graph on n vertices and $k \in \mathbb{N}$ be a parameter. In this section we describe an algorithm that takes G and k as an input, and in time polynomial in the size of G either solves p-IST, or produces a reduced graph G_R on at most $3k$ vertices and an integer $k' \leq k$, such that G has a spanning tree with at least k internal vertices if and only if G_R has a spanning tree with at least k' internal vertices. In other words, we show that p-IST has a $3k$-vertex kernel.

The algorithm is based on the following combinatorial lemma, which is interesting on its own. For two disjoint sets $X, Y \subseteq V$, we denote by $B(X, Y)$ the bipartite graph obtained from $G[X \cup Y]$ by removing all edges with both endpoints in X or Y.

Lemma 1. *If $n \geq 3$, and I is an independent set of G of cardinality at least $2n/3$, then there are nonempty subsets $S \subseteq V \setminus I$ and $L \subseteq I$ such that*

(i) $N(L) = S$, and
(ii) $B(S, L)$ has a spanning tree such that all vertices of S and $|S| - 1$ vertices of L are internal.

Moreover, given a graph on at least 3 vertices and an independent set of cardinality at least $2n/3$, such subsets can be found in time polynomial in the size of G.

The proof of Lemma 1 is postponed to Section 4. Now we give the description of the kernelization algorithm and use Lemma 1 to prove its correctness. The algorithm consists of the following reduction rules.

Rule 1. If $n \leq 3k$, then output graph G and stop. In this case G is a $3k$-vertex kernel. Otherwise proceed with Rule 2.

Rule 2. Choose an arbitrary vertex $v \in V$ and run a DFS (depth first search) from v. If the DFS tree T has at least k internal vertices, then the algorithm has found a solution and stops. Otherwise, because $n > 3k$, T has at least $2n/3 + 2$ leaves, and since all leaves but the root of the DFS tree are pairwise nonadjacent, the algorithm has found an independent set of G of cardinality at least $2n/3$. Proceed with Rule 3.

Rule 3 (reduction). Find nonempty subsets of vertices $S, L \subseteq V$ as in Lemma 1. Add a vertex v_S and make it adjacent to every vertex in $N(S) \setminus L$ and add a vertex v_L and make it adjacent to v_S. Finally, remove all vertices of $S \cup L$. Let $G_R = (V_R, E_R)$ be the new graph and $k' = k - 2|S| + 2$. Go to Rule 1 with $G := G_R$ and $k := k'$.

To prove the soundness of Rule 3, we need the following lemma. Here, S and L are as in Lemma 1. If T is a tree and X a vertex set, we denote by $\mathrm{i}_T(X)$ the number of vertices of X that are internal in T.

Lemma 2. *If G has a spanning tree with k internal vertices, then G has a spanning tree with at least k internal vertices in which all the vertices of S and exactly $|S|-1$ vertices of L are internal.*

Proof. Let T be a spanning tree of G with k internal vertices. Denote by F the forest obtained from T by removing all edges incident to L. Then, as long as 2 vertices of S are in the same connected component in F, remove an edge from F incident to one of these two vertices. Now, obtain the spanning tree T' by adding the edges of a spanning tree of $B(S, L)$ to F in which all vertices of S and $|S|-1$ vertices of L are internal (see Lemma 1). Clearly, all vertices of S and $|S|-1$ vertices of L are internal in T'. It remains to show that T' has at least as many internal vertices as T.

Let $U := V \setminus (S \cup L)$. Then, we have that $\mathrm{i}_T(L) \leq \sum_{u \in L} d_T(u) - |L|$ as every vertex in a tree has degree at least 1 and internal vertices have degree at least 2. We also have $\mathrm{i}_{T'}(U) \geq \mathrm{i}_T(U) - (|L| + |S| - 1 - \sum_{u \in L} d_T(u))$ as at most $|S| - 1 - (\sum_{u \in L} d_T(u) - |L|)$ edges incident to S are removed from F to separate $F \setminus L$ into $|S|$ connected components, one for each vertex of S. Thus,

$$\begin{aligned}
\mathrm{i}_{T'}(V) &= \mathrm{i}_{T'}(U) + \mathrm{i}_{T'}(S \cup L) \\
&\geq \mathrm{i}_T(U) - (|L| + |S| - 1 - \sum_{u \in L} d_T(u)) + \mathrm{i}_{T'}(S \cup L) \\
&= \mathrm{i}_T(U) + (\sum_{u \in L} d_T(u) - |L|) - |S| + 1 + \mathrm{i}_{T'}(S \cup L) \\
&\geq \mathrm{i}_T(U) + \mathrm{i}_T(L) - |S| + 1 + \mathrm{i}_{T'}(S \cup L) \\
&= \mathrm{i}_T(U) + \mathrm{i}_T(L) - (|S| - 1) + (|S| + |S| - 1) \\
&= \mathrm{i}_T(U) + \mathrm{i}_T(L) + |S| \\
&\geq \mathrm{i}_T(U) + \mathrm{i}_T(L) + \mathrm{i}_T(S) \\
&= \mathrm{i}_T(V).
\end{aligned}$$

This finishes the proof of the lemma. □

Lemma 3. *Rule 3 is sound, $|V_R| < |V|$, and $k' \leq k$.*

Proof. We claim first that the resulting graph $G_R = (V_R, E_R)$ has a spanning tree with at least $k' = k - 2|S| + 2$ internal vertices if and only if the original graph G has a spanning tree with at least k internal vertices. Indeed, assume G has a spanning tree with $\ell \geq k$ internal vertices. Then, let $B(S, L)$ be as in Lemma 1 and T be a spanning tree of G with ℓ internal vertices such that all vertices of S and $|S|-1$ vertices of L are internal (which exists by Lemma 2). Because $T[S \cup L]$ is connected, every two distinct vertices $u, v \in N_T(S) \setminus L$ are in different connected components of $T \setminus (L \cup S)$. But this means that the graph T' obtained from $T \setminus (L \cup S)$ by connecting v_S to all neighbors of S in $T \setminus (S \cup L)$

is also a tree in which the degree of every vertex in $N_G(S) \setminus L$ is unchanged. The graph T'' obtained from T' by adding v_L and connecting v_L to v_S is also a tree. Then T'' has exactly $\ell - 2|S| + 2$ internal vertices.

In the opposite direction, if G_R has a tree T'' with $\ell - 2|S| + 2$ internal vertices, then all neighbors of v_S in T'' are in different components of $T'' \setminus \{v_S\}$. By Lemma 1 we know that $B(S, L)$ has a spanning tree T_{SL} such that all the vertices of S and $|S|-1$ vertices of L are internal. We obtain a spanning tree T of G by considering the forest $T^* = T'' \setminus \{v_S, v_L\} \cup T_{SL}$ and adding edges between different components to make it connected. For each vertex $u \in N_{T''}(v_S) \setminus \{v_L\}$, add an edge uv to T^*, where uv is an edge of G and $v \in S$. By construction we know that such an edge always exists. Moreover, the degrees of the vertices in $N_G(S) \setminus L$ are the same in T as in T''. Thus T is a spanning tree with ℓ internal vertices.

Finally, as $|S| \geq 1$ and $|L \cup S| \geq 3$, we have that $|V_R| < |V|$ and $k' \leq k$. □

Thus Rule 3 compresses the graph and we conclude with the following theorem.

Theorem 2. p-IST *has a* $3k$*-vertex kernel.*

Corollary 1. p-IST *can be solved in time* $8^k \cdot n^{O(1)}$.

Proof. Obtain a $3k$-vertex kernel for the input graph G in polynomial time using Theorem 2 and run the $2^n n^{O(1)}$ time algorithm of Nederlof [9] on the kernel. □

4 Proof of Lemma 1

In this section we provide the postponed proof of Lemma 1. Let $G = (V, E)$ be a connected graph on n vertices, I be an independent set of G of cardinality at least $2n/3$ and $C := V \setminus I$.

Let Y be a subset of V. A subset $X \subseteq (V \setminus Y)$ has Y*-expansion* c, for some $c > 0$, if for each subset Z of X, $|N(Z) \cap Y| \geq c \cdot |Z|$. We first find an independent set $L \subseteq I$ whose neighborhood has L-expansion 2. For this, we need the following result.

Lemma 4 ([13]). *Let B be a nonempty bipartite graph with vertex bipartition (X, Y) with $|Y| \geq 2|X|$ and such that every vertex of Y has at least one neighbor in X. Then there exist nonempty subsets $X' \subseteq X$ and $Y' \subseteq Y$ such that the set of neighbors of Y' in B is exactly X', and such that X' has Y'-expansion 2. Moreover, such subsets X', Y' can be found in time polynomial in the size of B.*

By using Lemma 4, we find nonempty sets of vertices $S' \subseteq C$ and $L' \subseteq I$ such that $N(L') = S'$ and S' has L'-expansion 2.

Lemma 5. *Let $G = (V, E)$ be a connected graph on n vertices, I be an independent set of G of cardinality at least $2n/3$ and $C := V \setminus I$. Furthermore let $S' \subseteq C$ and $L' \subseteq I$ such that $N(L') = S'$ and S' has L'-expansion 2. Then there exist nonempty subsets $S \subseteq S'$ and $L \subseteq L'$ such that*

- *$B(S, L)$ has a spanning tree in which all the vertices of L have degree at most 2,*
- *S has L-expansion 2, and*
- *$N(L) = S$.*

Moreover, such sets S and L can be found in time polynomial in the size of G.

Proof. The proof is by induction on $|S'|$. If $|S'| = 1$, the lemma holds with $S := S'$ and $L := L'$. Let $H = (S', E')$ be the hypergraph with edge set $E' = \{N(v) \mid v \in L'\}$. If H contains a hypertree, then it has $|S'| - 1$ hyperedges and we can obtain a tree $T_{S'}$ on S' by contracting edges. We use this to find a subtree T' of $B(S', L')$ spanning S' as follows: for every edge $e = uv$ of $T_{S'}$ there exists a hyperedge corresponding to it and hence a unique vertex, say w, in L'; we delete the edge $e = uv$ from $T_{S'}$ and add the edges wu and wv to $T_{S'}$. Observe that the resulting subtree T' of $B(S', L')$ has the property that every vertex in T' which is in L' has degree 2 in it. Finally, we extend T' to a spanning tree of $B(S', L')$ by adding the remaining vertices of L' as pending vertices. All this can be done in polynomial time using the algorithm in Section 2.2. Thus S' and L' are the sets of vertices we are looking for. Otherwise, if H does not contain a hypertree, then H is not partition-connected by Theorem 1. Then we can find a partition $\mathcal{P} = \{P_1, P_2, \ldots, P_\ell\}$ of S' such that its border $\delta(\mathcal{P})$ contains at most $\ell - 2$ hyperedges of H in polynomial time. Let b_i be the number of hyperedges completely contained in P_i, where $1 \leq i \leq \ell$. Then there is $j, 1 \leq j \leq \ell$, such that $b_j \geq 2|P_j|$. Indeed, otherwise $|L'| \leq (\ell - 2) + \sum_{i=1}^{\ell}(2|P_i| - 1) < 2|S'|$, which contradicts the choice of L' and S' and the fact that S' has an L'-expansion 2. Let $X := P_j$ and $Y := \{w \in L' | \ N(w) \subseteq P_j\}$. We know that $|Y| \geq 2|X|$ and hence by Lemma 4 there exists a $S^* \subseteq X$ and $L^* \subseteq Y$ such that S^* has L^*-expansion 2 and $N(L^*) = S^*$. Thus, by the induction assumption, there exist $S \subseteq S^*$ and $L \subseteq L^*$ with the desired properties. □

Let S and L, be as in Lemma 5. We will prove in the following that there exists a spanning tree of $B(S, L)$ such that all the vertices of S and exactly $|S| - 1$ vertices of L are internal. Note that there cannot be more than $2|S| - 1$ internal vertices in a spanning tree of $B(S, L)$ without creating cycles. By Lemma 5, we know that there exists a spanning tree of $B(S, L)$ in which $|S| - 1$ vertices of L have degree exactly 2.

Consider the bipartite graph B_2 obtained from $B(S, L)$ by adding a copy S_c of S (each vertex in S has the same neighborhood as its copy in S_c and no vertex of S_c is adjacent to a vertex in S). As $|L| \geq |S \cup S_c|$ and each subset Z of $S \cup S_c$ has at least $|Z|$ neighbors in L, by Hall's theorem, there exists a matching in B_2 saturating $S \cup S_c$. This means that in $B(S, L)$, there exist two edge-disjoint matchings M_1 and M_2, both saturating S. We refer to the edges from $M_1 \cup M_2$ as the *favorite* edges.

Lemma 6. *$B(S, L)$ has a spanning tree T such that all the vertices of S and $|S| - 1$ vertices of L are internal in T.*

Proof. Let T be a spanning tree of $B(S, L)$ in which all vertices of L have degree at most 2, obtained using Lemma 5. As T is a tree, exactly $|S| - 1$ vertices of L have degree 2 in T. As long as a vertex $v \in S$ is not internal in T, add a favorite edge uv to T which was not yet in T ($u \in L$), and remove an appropriate edge from the tree which is incident to u so that T remains a spanning tree. Vertex v becomes internal and the degree of u in T remains unchanged. As u is only incident to one favorite edge, this rule increases the number of favorite edges in T even though it is possible that some other vertex in S would have become a leaf. We apply this rule until no longer possible. We know that this rule can only be applied at most $|S|$ times. In the end, all the vertices of S are internal and $|S| - 1$ vertices among L are internal as their degrees remain the same. □

To conclude with the proof of Lemma 1, we observe that $S \subseteq C$, $L \subseteq I$ and $N(L) = S$ by the construction of S and L, and by Lemma 6, $B(S, L)$ has a spanning tree in which all the vertices of S and $|S| - 1$ vertices of L are internal.

References

1. Abu-Khzam, F.N., Fellows, M.R., Langston, M.A., Suters, W.H.: Crown Structures for Vertex Cover Kernelization. Theory Comput. Syst. 41(3), 411–430 (2007)
2. Cohen, N., Fomin, F.V., Gutin, G., Kim, E.J., Saurabh, S., Yeo, A.: Algorithm for Finding k-Vertex Out-trees and its Application to k-Internal Out-branching Problem. In: Ngo, H.Q. (ed.) COCOON 2009. LNCS, vol. 5609, pp. 37–46. Springer, Heidelberg (2009)
3. Chen, J., Kanj, I.A., Jia, W.: Vertex Cover: Further observations and further improvements. J. Algorithms 41(2), 280–301 (2001)
4. Frank, A., Király, T., Kriesell, M.: On decomposing a hypergraph into k connected sub-hypergraphs. Discrete Appl. Math. 131, 373–383 (2003)
5. Guo, J.: A more effective linear kernelization for cluster editing. Theor. Comput. Sci. 410(8-10), 718–726 (2009)
6. Lokshtanov, D., Saurabh, S.: Even faster algorithm for Set Splitting!.To appear in the proceedings of IWPEC 2009 (2009)
7. Lorea, M.: Hypergraphes et matroides. Cahiers Centre Etud. Rech. Oper. 17, 289–291 (1975)
8. Lovász, L.: A generalization of König's theorem. Acta. Math. Acad. Sci. Hungar. 21, 443–446 (1970)
9. Nederlof, J.: Fast polynomial-space algorithms using Möbius inversion: Improving on Steiner Tree and related problems. In: Albers, S., Marchetti-Spaccamela, A., Matias, Y., Niko-letsea, S. (eds.) ICALP 2009. LNCS, vol. 5556, pp. 713–725. Springer, Heidelberg (2009)
10. Robertson, N., Seymour, P.D.: Graph minors-a survey. In: Anderson, I. (ed.) Surveys in Combinatorics, pp. 153–171. Cambridge Univ. Press, Cambridge (1985)
11. E. Prieto and C. Sloper. Either/or: Using vertex cover structure in designing FPT-algorithms—the case of k-internal spanning tree. In the proceedings of *WADS* 2003, volume 2748 of *LNCS*, pp. 465–483. Springer, 2003.
12. Prieto, E., Sloper, C.: Reducing to Independent Set Structure – the Case of k-Internal Spanning Tree. Nord. J. Comput. 12(3), 308–318 (2005)
13. Thomassé, S.: A quadratic kernel for feedback vertex set. In: The Proceedings of SODA 2009, pp. 115–119. SIAM, Philadelphia (2009)

Geometric Minimum Diameter Minimum Cost Spanning Tree Problem*

Dae Young Seo[1], D.T. Lee[2,**], and Tien-Ching Lin[2]

[1] Department of Computer Engineering, Korea Polytechnic University, Korea
seody@kpu.ac.kr
[2] Institute of Information Science, Academia Sinica, Taipei, Taiwan
{dtlee,kero}@iis.sinica.edu.tw
**Dept. of Computer Science and Information Engineering, National Taiwan University, Taipei, Taiwan

Abstract. In this paper we consider bi-criteria geometric optimization problems, in particular, the minimum diameter minimum cost spanning tree problem and the minimum radius minimum cost spanning tree problem for a set of points in the plane. The former problem is to construct a minimum diameter spanning tree among all possible minimum cost spanning trees, while the latter is to construct a minimum radius spanning tree among all possible minimum cost spanning trees. The graph-theoretic minimum diameter minimum cost spanning tree (MDMCST) problem and the minimum radius minimum cost spanning tree (MRM-CST) problem have been shown to be NP-hard. We will show that the geometric version of these two problems, GMDMCST problem and GMRMCST problem are also NP-hard. We also give two heuristic algorithms, one MCST-based and the other MDST-based for the GMDMCST problem and present some experimental results.

1 Introduction

Given a connected, undirected and weighted graph $G = (V, E)$ where each edge is associated with a nonnegative real number, referred to as the *cost*, a spanning tree of G whose total cost is minimum among all possible spanning trees is called a *minimum cost spanning tree* (MCST) of G or simply *minimum spanning tree* (MST). There are two algorithms commonly used, Prim's algorithm [8] and Kruskal's algorithm [7] for constructing a minimum spanning tree. Note that the MST may not be unique if two or more edges have the same cost. Given a tree T in which each edge has a cost representing the length of the edge, the *eccentricity* of a vertex v in T is defined to be the total length of the path from v to the farthest vertex u_v in T, denoted by $ecc(v)$. The longest path from vertex v to u_v in T whose $ecc(v)$ is the maximum among all v in T is referred to as the *diameter* of T, and the total cost of the diameter of T is denoted

* Research supported in part by the National Science Council under the Grants No. NSC-94-2213-E-001-004, NSC-95-2221-E-001-016-MY3, and NSC 94-2752-E-002-005-PAE, and by the Taiwan Information Security Center (TWISC), National Science Council under the Grant No. NSC94-3114-P-001-001-Y.

Y. Dong, D.-Z. Du, and O. Ibarra (Eds.): ISAAC 2009, LNCS 5878, pp. 283–292, 2009.

$D(T)$. Note that there exist at least two extreme vertices in T such that their eccentricity is the maximum. The vertex v that attains the minimum $ecc(v)$ for all v in T is called the *center* of T, and the minimum $ecc(v)$ is referred to as the *radius* of T, denoted $R(T)$. If the costs of edges of T are positive, then either we have a unique center vertex or at most two vertices for which the eccentricity is the minimum.

A spanning tree of a graph $G(V, E)$ that minimizes its diameter is called the *minimum diameter spanning tree*. The geometric version of this problem (GMDST) is defined as follows. Given a set P of n points in the plane in which the cost of an edge connecting any two points is the Euclidean distance between them, find a spanning tree of P such that its diameter is the minimum among all possible spanning trees. Given a graph $G(V, E)$ and a source vertex or *center* $v \in V$, the spanning tree that minimizes $ecc(v)$ is called the *minimum radius spanning tree*. The geometric version of this problem (GMRST) is defined similarly. In [5], Ho, et al. described an algorithm for finding the GMDST of a set of n points in $O(n^3)$ time. They proved that there always exists a GMDST such that it either has a *center point*, called *monopole*, and the rest of points are directly connected to the center point, or has a *center edge* with two points, called *dipole*, and the rest of the points are connected to one of the endpoints of the center edge. They further showed that the above results can be extended to any graph in which the edge costs satisfy the triangle inequality. In [3], Hassin and Tamir showed that the graph-theoretic version of the minimum diameter spanning tree (MDST) problem, where the edge costs do not necessarily satisfy the triangle inequality, is reducible to the *absolute 1-center problem* introduced by Hakimi [2]. The *absolute 1-center problem* can be solved in $O(mn + n^2 logn)$ time [6], where m and n denote respectively the numbers of edges and of vertices of G. Chan [10] improved the bound of GMDST problem in d-dimensional space R^d. He described a semi-online model that computes GMDST of an n-point set $P \subset R^d$ within $\tilde{O}(n^{3-\frac{1}{(d+1)(d/2+1)}})$ time by maintaining a dynamic data structure. (The $\tilde{O}$ notation hides factors that are $o(n^\epsilon)$ for any fixed $\epsilon > 0$.) In other words, the GMDST of an n-point set P in the plane can be found within $\tilde{O}(n^{17/6})$ time. The time bound of the GMDST problem is still very close to the cubic time.

In this paper we consider the geometric versions of these two related bi-criteria problems, the MDMCST problem and the MRMCST problem. The MDMCST problem is to construct a MDST among all possible MCSTs and the MRM-CST problem is to construct for a given center a MRST among all possible MCSTs. The graph-theoretic version of both problems have been shown to be NP-hard [4,5]. We will show that the geometric versions of both minimum diameter minimum cost spanning (GMDMCST) problem and minimum radius minimum cost spanning (GMRMCST) problem are NP-hard. We then give two heuristic algorithms, called MCST-based algorithm and MDST-based algorithm for the GMDMCST problem and present some experimental results.

The rest of the paper is organized as follows. In Section 2 we show that the GMDMCST and GMRMCST problems are NP-hard. In Section 3 we give MCST-based and MDST-based heuristic algorithms for the GMDMCST problem. In Section 4 we give the implementation results of these two kinds of heuristic algorithm on the *Algorithm Benchmark System* (ABS) [11] and conclude with an analysis and some remarks.

2 NP-Hardness of the GMDMCST and GMRMCST

Given a connected undirected graph $G(V, E)$ with nonnegative edge costs, we consider the set of all minimum cost spanning trees of G. In the following we will simply refer to minimum cost spanning tree as minimum spanning tree, when the cost measure is understood. In [9], Seo defined the *minimum spanning tree intersection graph* (MSTIG) which is the intersection of all the MST's, and the *minimum spanning tree union graph* (MSTUG) which is the union of all the MST's. The edges in MSTIG are called the *essential edges* and those in MSTUG but not in MSTIG are called the *optional edges.*

Given a set P of n points in the plane in which the cost of an edge connecting any two points is the Euclidean distance between them, Seo [9] gave an algorithm that runs in $O(n^2 \log n)$ time for finding the MSTIG(G) and MSTUG(G) and coloring all edges in G according the following coloring rule. The essential edges are colored blue, the optional edges are colored green and the edges not in the MSTUG are colored red. Thus all the edges of a graph G will be partitioned into three classes, each assigned a distinct color. The MSTUG will be a connected graph, and the MSTIG will be a forest unless the MST is unique. The subtrees in the forest of MSTIG that consist solely of essential edges are called *blue trees.*

In this section, we show that the decision version of the optimization GMRMCST problem, the so-called the *geometric bounded radius bounded cost spanning tree* (GBRBCST) problem, is NP-complete. We first show that the PARTITION problem, a well-known NP-complete problem [1], is polynomially reducible to the GBRBCST problem, and then show that the GBRBCST problem is polynomially reducible to the GBDBCST problem, and finally show that the GBRBCST and GBDBCST problems are respectively polynomially reducible to the GMRMCST and GMDMCST problems.

The PARTITION problem is defined as follows. Given a finite set W and a size $s(w) \in \mathbf{Z}^+$ for each $w \in W$, decide if there exists a subset $W' \subseteq W$ such that $\sum_{w \in W'} s(w) = \sum_{w \in W-W'} s(w)$?

The GBRBCST is defined as follows. Given a point set P in the plane, a center s and two positive values R and C, decide if there exists a Euclidean spanning tree T such that the distance, $ecc(s)$, from s to the farthest site along the tree is bounded above by R and that the total cost of the tree is bounded above by C?

Theorem 1. *The GBRBCST problem is NP-complete.*

Proof. GBRBCST is obviously in NP. For simplicity we assume that an instance of PARTITION is a multiset of positive integers (with repetition permitted) sorted in nondecreasing order, i.e., $w_1, w_2, \ldots, w_n$ where $w_i \leq w_{i+1}$ for $i = 1, ..., n-1$. We shall construct an instance of GBRBCST based on the input instance of PARTITION. The construction (see Fig. 1) is as follows. Let $ISqr^+(k)$

and $ISqr^-(k)$ be imaginary squares corresponding to w_k such that each side is of length $2w_k$. The top right (TR) corner of $ISqr^+(i)$ and the bottom left(BL) corner of $ISqr^+(i+1)$ are coincident, while BL($ISqr^-(i)$) and TR($ISqr^-(i+1)$) are coincident. Suppose that BL($ISqr^+(1)$) and TR($ISqr^-(1)$) are located at the origin s. For $k > 1$, BL($ISqr^+(k)$) = TR($ISqr^+(k-1)$) has coordinates $(2\sum_{i=1}^{k-1} w_i, 2\sum_{i=1}^{k-1} w_i)$, while TR($ISqr^-(k)$) = BL($ISqr^-(k-1)$) has coordinates $(-2\sum_{i=1}^{k-1} w_i, -2\sum_{i=1}^{k-1} w_i)$. The bottom and left sides of each $ISqr^+(k)$ or each $ISqr^-(k)$ are parallel to $x-$ and $y-$coordinate axes, respectively. Figure 2 shows a more detailed positioning of the points in a square of dimension $w \times w$. There are three types of points in our construction.

1. For each pair of imaginary squares $ISqr^+(k)$ and $ISqr^-(k)$, there is one basic TYPE1 point p at $(2\sum_{i=1}^{k} w_i, -2\sum_{i=1}^{k} w_i)$ shown in Fig. 2 on a *matching path* connecting $ISqr^+(k)$ and $ISqr^-(k)$, where the matching path consists of an upward vertical line from p to meet the right side of $ISqr^+(k)$ and a leftward horizontal line from p to meet the bottom side of $ISqr^-(k)$. Extraneous TYPE1 points on this matching path are added so that they equally divide the vertical and horizontal line segments $\overline{h,p}$ and $\overline{h',p}$ in Fig. 2, into $\lfloor \frac{(4\sum_{i=1}^{k} w_i)-w_k}{0.5w_k} \rfloor + 1$ edges, respectively. The divided edges will be of length strictly less than $0.5w_k$. Since w_i's are sorted in nondecreasing order, the number of TYPE1 extra points for each k is bounded by $O(n)$. Hence there will be $O(n^2)$ TYPE1 points.
2. Now consider TYPE2 points. For each ISqr, there will be 10 basic points (e.g. a,b,c,d,e,f,g,h,i and j for $ISqr^+(k)$ in Fig. 2) and 10 extra dividing points (e.g. in the middle of 5 edges (c,d),(d,e),(e,f),(g,j) and (i,j), 3 points equally dividing (a,b), and 2 points equally dividing (b,c), as shown). These dividing points are introduced to ensure that the distance of any divided segment in the corresponding ISqr is strictly less than $0.5w_k$. Similarly we have 20 points for $ISqr^-(k)$. Note that the distances between points c and h, between points h and i, and between points f and g in $ISqr^+(k)$ (or correspondingly those between points c' and h', between points h' and i', and between points f' and g' in $ISqr^-(k)$) are $0.5w_k$. Figure 2 shows the x and y-coordinates of basic TYPE2 points in $ISqr^+(k)$ and $ISqr^-(k)$; the subscript k is omitted from the figure. Note that points f and g are shifted so that the distance between points f and i is strictly greater than $0.5w_k$. Points f' and g' are shifted similarly.
3. Finally there are TYPE3 points, t_1, t_2, t_3, t_4 and other auxiliary points. Points t_1 and t_2 have x- and y- coordinates $((2+4.5\sqrt{2})\sum_{i=1}^{n} w_i, (2+4.5\sqrt{2})\sum_{i=1}^{n} w_i)$ and $(-(2+4.5\sqrt{2})\sum_{i=1}^{n} w_i, -(2+4.5\sqrt{2})\sum_{i=1}^{n} w_i)$, respectively. The distances between points j_n and t_1, and between points j'_n and t_2 are equal to $9\sum_{i=1}^{n} w_i$. Note that this distance is longer than the length of any matching path. The length of the matching path (h_k, p_k, h'_k) for $ISqr(k)$ is $8\sum_{i=1}^{k} w_i - 2w_k$. The shortest distance from s to t_1 (and to t_2) in a minimum spanning tree is $13\sum_{i=1}^{n} w_i$. Points t_3 and t_4 form an isosceles triangle with the apex s. The line segments $\overline{s,t_3}$ and $\overline{s,t_4}$ are of length $13.5\sum_{i=1}^{n} w_i$.

The slopes of $\overline{s,t_3}$ and $\overline{s,t_4}$ are -4 and -1/4, respectively. The auxiliary points on $\overline{s,t_3}$ are those with y-coordinates $(2\sum_{i=1}^{k} w_i)$ and $(2\sum_{i=1}^{k} w_i - w_k)$ for $1 \le k \le n$. On the other hand, the auxiliary points on $\overline{s,t_4}$ are those with x-coordinates $(-2\sum_{i=1}^{k} w_i)$ and $(-2\sum_{i=1}^{k} w_i + w_k)$. The slopes of $\overline{s,t_3}$ and $\overline{s,t_4}$ are selected so that the closest point from any auxiliary point must be s, t_3, t_4, or another auxiliary point on the same slope.

Let C be equal to the cost of the minimum spanning tree and R be equal to

$$13\sum_{i=1}^{n} w_i + \frac{1}{2}S = 13.5\sum_{i=1}^{n} w_i.$$

We now show that the instance of PARTITION has a solution if and only if the constructed instance of the GBRBCST problem has answer YES.

Let the solution be denoted by T^*. Note that all the edges shown as solid lines in Fig. 1 must belong to T^*, i.e., they are essential edges, and that those shown as dashed lines may or may not belong to T^*, i.e., they are optional edges. Let L_1 and L_2 denote the path length from s to t_1 and from s to t_2 in T^*, respectively. An imaginary square is said to be *straight*, *crooked*, or *bad* if it satisfies the following constraints. If both (c_ℓ, h_ℓ) and (h_ℓ, i_ℓ) of $ISqr^+(\ell)$ (or correspondingly (c'_ℓ, h'_ℓ) and (h'_ℓ, i'_ℓ) of $ISqr^-(\ell)$) are in T^*, then $ISqr^+(\ell)$ (or correspondingly $ISqr^-(\ell)$) is said to be *straight*. If only (f_ℓ, g_ℓ) of $ISqr^+(\ell)$ (or correspondingly (f'_ℓ, g'_ℓ) of $ISqr^-(\ell)$) is in T^*, then $ISqr^+(\ell)$ (resp. $ISqr^-(\ell)$) is said to be *crooked*. An imaginary square which is neither straight nor crooked is said to be *bad*. By Lemma 5, we can obtain from T^* a solution to the PARTITION problem by assigning w_ℓ to W' if $ISqr^+(\ell)$ is crooked and assigning w_ℓ to $W - W'$ if $ISqr^+(\ell)$ is straight, for $\ell = 1, 2, \ldots, n$. This completes the proof that the GBRBCST problem is NP-complete.

According to the construction of an instance of GBRBCST above, we have the following lemmas. Due to the page limit, some proofs are omited.

Lemma 1. $L_1 + L_2 \le 27\sum_{i=1}^{n} w_i$.

Lemma 2. *Each pair of imaginary squares $ISqr^+(\ell)$ and $ISqr^-(\ell)$ contributes $8w_\ell$ to $L_1 + L_2$ if both of them are straight,*
$9w_\ell$ to $L_1 + L_2$ if one of them is straight and the other one is crooked,
$10w_\ell$ to $L_1 + L_2$ if both of them are crooked, and
$\ge 14w_\ell$ $L_1 + L_2$ if at least one of them is bad.

Lemma 3. *For each pair of imaginary squares $ISqr^+(\ell)$ and $ISqr^-(\ell)$, at least two and at most four of the six optional edges (c_ℓ, h_ℓ), (h_ℓ, i_ℓ), (f_ℓ, g_ℓ), (c'_ℓ, h'_ℓ), (h'_ℓ, i'_ℓ) and (f'_ℓ, g'_ℓ) are in T^*.*

Lemma 4. *For each pair of imaginary squares $ISqr^+(\ell)$ and $ISqr^-(\ell)$, at most one of them is straight.*

Lemma 5. *For each pair of imaginary squares $ISqr^+(\ell)$ and $ISqr^-(\ell)$, exactly one of them is straight and the other one is crooked.*

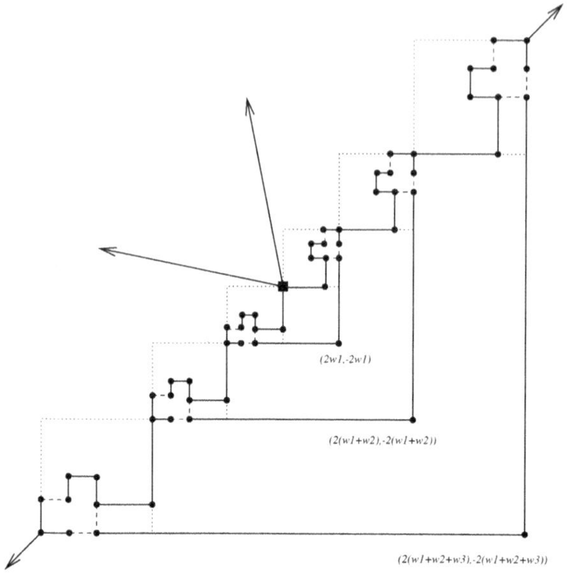

Fig. 1. MSTUG/MSTIG constructed from an instance of PARTITION

Lemma 6. *Given a graph G, suppose that the optional edges in MSTUG(G) are partitioned into l_{max} equivalence classes K_1, K_2, ..., $K_{l_{max}}$ such that the edges in the same class have an identical weight and the weight of an edge in K_i is strictly less than that in K_j if and only if $i < j$. Let $NO_i(T)$ denote the number of optional edges in K_i that are contained in the MST T. Then, for any pair of MST's T and T', $NO_i(T) = NO_i(T')$ for $1 \leq i \leq l_{max}$.*

Theorem 2. *The GBDBCST problem is NP-complete.*

Proof. Let $D = 2R$. Then the previous construction can be used to prove that the GBRBCST problem is polynomially reducible to the GBDBCST problem.

Theorem 3. *The GMRMCST and the GMDMCST problems are NP-hard.*

Proof. It is obvious that the GBRBCST problem and the GBDBCST problem are polynomially reducible to the GMRMCST problem and the GMDMCST problem, respectively.

3 Heuristic Algorithms for GMDMCST

In this section, we assume that MSTIG(G) and MSTUG(G) have been computed and will be used as the input of our heuristic algorithms [9]. Let BT_i denote a blue-tree in MSTIG(G). We define the *pseudo-center*, denoted $pc(BT_i)$, of BT_i to be the point on an edge or a vertex in BT_i such that the distances from the two extremes of the diameter $D(BT_i)$ to $pc(BT_i)$ are the same. It is trivial to see that $ecc(pc(BT_i)) = D(BT_i)/2$.

The crux of this problem lies in the selection of optional edges to be included in GMDMCST so that the resulting diameter of the MST is minimized. We shall adopt different strategies in our heuristic algorithms to select optional edges. The following properties and operations are needed for our heuristics. For each blue tree BT_i, we will maintain diameter $D(BT_i)$, pseudo-center $pc(BT_i)$, and the distance, denoted $\ell(v)$, from v to $pc(BT_i)$ for each vertex v in BT_i. When we select an optional edge $e = (u, v)$ concatenating two blue trees, BT_i and BT_j, we will create a new blue tree $BT_{i,j}$, which is the union of BT_i, BT_j and the optional edge e. $D(BT_{i,j}) = \max\{D(BT_i), D(BT_j), c(e) + \ell(u) + \ell(v) + ecc(pc(BT_i)) + ecc(pc(BT_j))\}$, where $c(e)$ denotes the cost of edge e.

We now start to develop our heuristic algorithms for GMDMCST. Both kinds of our heuristic algorithm for GMDMCST use the greedy method. The general idea is as follows: We greedily select an optional edge concatenating two blue trees so that the new blue tree has a minimum diameter until all the blue trees are connected to become a single MST. Note that we can easily modify our heuristic algorithm for GMDMCST to become a heuristic algorithm for GMRMCST by using a different greedy criterion which is to greedily select an optional edge concatenating two blue trees so that the new blue tree has a minimum radius.

3.1 The MST-Based Heuristic

In this subsection, we use the ideas of the Prim's and Kruskal's MST algorithm to develop our MST-based heuristic algorithm.

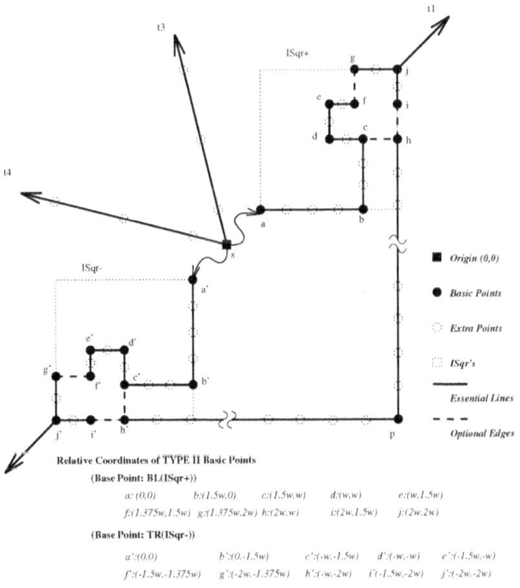

Fig. 2. The detailed picture for ISqr's in Figure 1

3.1.1 Prim-Like Heuristic

We start with a randomly selected blue tree BT_s among all blue trees in MSTIG(G) and consider the optional edges that are incident to BT_s with the minimum cost. Let $e = (u, v)$ be one of the optional edges with the minimum cost that is incident to BT_s, where $u \in BT_s$ and $v \in BT_j$ for some $j \neq s$.

$$D(BT_s) = \max\{D(BT_s), D(BT_j), c(e) + \ell(u) + \ell(v) + (D(BT_s) + D(BT_j))/2\}.$$

We greedily select among all optional edges with the minimum cost that are incident to BT_s the optional edge e that gives the minimum diameter according to the formula above. After selecting this optional edge e, we obtain a new blue tree BT_s by concatenating these two blue trees.

3.1.2 Kruskal-Like Heuristic

Let $e = (u, v)$ be one of the optional edges with the minimum cost, where $u \in BT_i$ and $v \in BT_j$. We will concatenate BT_i and BT_j to be a new blue tree for which $\delta(e)$ is minimum, where $\delta(e) = D(BT_{i,j}) - (D(BT_i) + D(BT_j))$.

3.2 The MDST-Based Heuristic

3.2.1 k-Center Heuristic

In [5], Ho et al. proved that there exists a GMDST of a set of n points which is monopolar or dipolar. The idea of k-center heuristic is to extend monopolar and dipolar spanning trees to k-polar. Let b be the total number of blue trees in MSTIG(G). If each combination contains k distinct blue trees in MSTIG(G), it has total $C(b, k)$ distinsct combinations. For each combination π of k blue trees $BT_{\pi(1)}, BT_{\pi(2)}, \ldots, BT_{\pi(k)}$, we will partition the set of remaining blue trees into k subsets $S_1, S_2, \ldots, S_k$ by some greedy criteria, connect each blue tree in S_i to $BT_{\pi(i)}$ for each $1 \leq i \leq k$ and then connect $BT_{\pi(1)}, BT_{\pi(2)}, \ldots, BT_{\pi(k)}$ by Kruskal-like heuristic to obtain a minimum spanning tree BT_π. The k-center heuristic will consider all combinations of k blue trees as possible k-poles, and select one such that its resulting minimum spanning tree has the minimum diameter.

Note that we can apply the Prim-like heuristic algorithm to implement the 1-center heuristic algorithm. We can implement the 1-center heuristic by calling the Prim-like heuristic b times, each starts with blue tree BT_i, where $1 \leq i \leq b$, and then select the MST with the minimum diameter. The k-center heuristic can also be viewed as a combination of Prim-like heuristic and Kruskal-like heuristic. For each combination π of k blue trees $BT_{\pi(1)}, BT_{\pi(2)}, \ldots, BT_{\pi(k)}$, we first connect all blue trees in S_i to $BT_{\pi(i)}$ by Prim-like heuristic and then connect blue trees $BT_{\pi(1)}, BT_{\pi(2)}, \ldots, BT_{\pi(k)}$ by Kruskal-like heuristic.

4 Implementation

The MST-based heuristic algorithm and the MDST-based heuristic algorithm have been implemented on the ABS. Tables 1 and 2 summarize the experimental

Table 1. The experimental results for MST-based heuristic algorithms

n	m	m_{opt}	m_{BT}	*MST_PRIM* avg. d.	*MST_PRIM* avg. t.	*MST_KRUSCAL* avg. d.	*MST_KRUSCAL* avg. t.
50	50	3.05	2.50	1265.028	0.0002	1263.837	0.0002
100	50	10.06	5.93	1437.457	0.0002	1431.074	0.0002
150	50	21.77	11.60	1509.909	0.0006	1496.996	0.0002
200	50	33.78	17.41	1584.962	0.0008	1555.767	0.0004
250	50	50.17	25.09	1646.061	0.0007	1596.836	0.0003
300	50	65.91	32.60	1639.306	0.0006	1608.231	0.0004
350	50	82.21	40.70	1693.195	0.0011	1631.054	0.0007
400	50	102.79	51.16	1655.356	0.0005	1598.080	0.0011

Table 2. The experimental results for MDST-based heuristic algorithms

n	m	m_{opt}	m_{BT}	*MDST_1C* avg. d.	*MDST_1C* avg. t.	*MDST_2C* avg. d.	*MDST_2C* avg. t.	*MDST_3C* avg. d.	*MDST_3C* avg. t.	*MDST_4C* avg. d.	*MDST_4C* avg. t.	*MDST_5C* avg. d.	*MDST_5C* avg. t.
50	50	3.05	2.50	1262.6	0.0001	1262.0	0.0001	1262.4	0.0001	1262.4	0.0001	1263.3	0.0002
100	50	10.06	5.93	1424.4	0.0006	1423.7	0.0011	1424.3	0.0011	1426.4	0.0015	1428.6	0.0018
150	50	21.77	11.60	1463.8	0.0009	1459.3	0.0044	1457.2	0.0135	1456.2	0.0381	1454.8	0.0847
200	50	33.78	17.41	1527.0	0.0016	1520.9	0.0142	1507.8	0.0785	1505.2	0.3383	1504.6	1.1756
250	50	50.17	25.09	1560.5	0.0048	1537.1	0.0446	1520.9	0.3563	1509.9	2.1918	1504.7	10.7995
300	50	65.91	32.60	1530.0	0.0077	1513.1	0.1066	1487.7	1.1158	1474.2	9.0257	1466.9	59.0691
350	50	82.21	40.70	1553.0	0.0127	1512.0	0.2253	1486.4	2.9002	1467.4	28.6890	1454.2	230.5542
400	50	102.79	51.16	1534.0	0.0217	1495.7	0.4741	1464.2	7.5041	1444.8	91.5174	1425.9	516.9337

results. Columns 1 - 2 indicate the total number of input vertices (n) and the size of grid ($m \times m$) respectively. Columns 3 - 4 contain the average number of optional edges (m_{opt}) and the average number of the blue trees (m_{BT}) for the input instances. The other columns contain the average diameter (avg. d.) and average running times (avg. t.) of the Prim-like heuristic algorithm (MST_PRIM), the Kruskal-like heuristic algorithm (MST_KRUSCAL) and the k-center heuristic algorithms (1-center (MST_1C), 2-center (MST_2C) , 3-center (MST_3C), 4-center (MST_4C), 5-center (MST_5C)), respectively. We generate the test instances with input size from $50, 100, 150, \ldots, 400$ vertices on a 50×50 grid. For each input size we randomly generate 1000 test instances and then execute all heuristic algorithms for every test instance. In general, we see the MDST-based heuristics are better than the MST-based heuristics, but they spend more time. As far as the MST-based heuristics are concerned, the Kruskal-like heuristic is better than the Prim-like heuristic. We also see that the 2-center heuristic obtains the smallest diameter for smaller input size 50 and 100, and the 5-center heuristic yields the smallest diameter spanning tree when the input size is larger than or equal to 150. It seems that the number of centers in an optimal MDMCST increases as the input size (or the size of blue trees) increases.

5 Conclusion

In this paper, we have considered two related bi-criteria problems GMDMCST and GMRMCST and shown that they both are NP-hard. We have presented two kinds of heuristic algorithms for the GMDMCST problem, called MST-based algorithm and MDST-based algorithm. The MST-based heuristic algorithm includes Prim-like heuristic algorithm and Kruskal-like heuristic algorithm. We have also introduced a k-center heuristic algorithm, which is an MDST-based heuristic algorithm. The time complexity of the Prim-like heuristic algorithm and the Kruskal-like heuristic algorithm both are $O(n^2 + m_{opt}m_{BT})$, where n is the number of vertices, m_{BT} is the number of blue trees and m_{opt} is the number of the optional edges. The time complexity of k-center heuristic algorithm is $O(n^k(n^2 + m_{opt}m_{BT} + m_{opt} \log m_{opt}))$.

So far, we have not been able to find any approximation algorithm for the GMDMCST. This is an interesting problem for further research. Our benchmark results give us a hint to find an approximation algorithm by approximating the number of centers in an optimal MDMCST.

References

1. Garey, M.R., Johnson, D.S.: Computers and Intractability: a guide to the theory of NP-completeness. Freeman, San Francisco (1979)
2. Hakimi, S.L.: Optimal locations of switching centers and medians of a graph. Operations Research 12, 405–459 (1964)
3. Hassin, R., Tamir, A.: On the minimum diameter spanning tree problem. Information Processing Letters 53, 109–111 (1995)
4. Ho, J.M.: Optimal Trees in Network Design, Ph.D Dissertation, Northwestern University (May 1989)
5. Ho, J.M., Lee, D.T., Chang, C.H., Wong, C.K.: Minimum Diameter Spanning Trees and Related Problems. SICOMP 20(5), 987–997 (1991)
6. Kariv, O., Hakimi, S.L.: An algorithmic approach to network location problems. I: The p-Centers. SIAM Journal on Applied Math. 37, 513–537 (1979)
7. Kruskal, J.B.: On the Shortest Spanning Subtree of a Graph and the Traveling Salesman Problem. Problem. Amer. Math. Soc. 7 (1956)
8. Prim, R.C.: Shortest Connection Networks and Some Generalizations. Bell System Tech. J. (1957)
9. Seo, D.Y.: On the Complexity of Bicriteria Spanning Tree Problems for a Set of Points in the Plane, Ph. D Dissertation, Northwestern University (1999)
10. Chan, T.M.: Semi-online maintenance of geometric optima and measures. SIAM Journal on Computing 32, 700–716 (2003)
11. Algorithm Benchmark System (ABS), `http://www.opencps.org`

On Shortest Disjoint Paths in Planar Graphs

Yusuke Kobayashi[1] and Christian Sommer[2]

[1] Department of Mathematical Informatics, Graduate School of Information Science and Technology, University of Tokyo, Tokyo 113-8656, Japan
Yusuke_Kobayashi@mist.i.u-tokyo.ac.jp

[2] Department of Computer Science, Graduate School of Information Science and Technology, University of Tokyo, and National Institute of Informatics, Tokyo, Japan
sommer@nii.ac.jp

Abstract. For a graph and a set of vertex pairs $\{(s_1, t_1), \ldots, (s_k, t_k)\}$, the k disjoint paths problem is to find k vertex-disjoint paths $P_1, \ldots, P_k$, where P_i is a path from s_i to t_i for each $i = 1, \ldots, k$. In the corresponding optimization problem, the shortest disjoint paths problem, the vertex-disjoint paths P_i have to be chosen such that a given objective function is minimized. We consider two different objectives, namely minimizing the total path length (minimum sum, or short: min-sum), and minimizing the length of the longest path (min-max), for $k = 2, 3$.

min-sum: We extend recent results by Colin de Verdière and Schrijver to prove that, for a planar graph and for terminals adjacent to at most two faces, the Min-Sum 2 Disjoint Paths Problem can be solved in polynomial time. We also prove that, for six terminals adjacent to one face in any order, the Min-Sum 3 Disjoint Paths Problem can be solved in polynomial time.

min-max: The Min-Max 2 Disjoint Paths Problem is known to be **NP**-hard for general graphs. We present an algorithm that solves the problem for graphs with tree-width 2 in polynomial time. We thus close the gap between easy and hard instances, since the problem is weakly **NP**-hard for graphs with tree-width at least 3.

1 Introduction

The *vertex-disjoint paths problem* is one of the classic problems in algorithmic graph theory and combinatorial optimization, and has many applications, for example in transportation networks, VLSI-design [5,14], or routing in networks [12,20]. The input of the vertex-disjoint paths problem is a graph $G = (V, E)$ and k pairs of vertices $(s_1, t_1), \ldots, (s_k, t_k)$, for which the algorithm has to find k pairwise vertex-disjoint paths connecting s_i and t_i, if they exist. Paths are called *vertex-disjoint* if they have no vertices in common (except, possibly, at the end points).

In the optimization version of the problem, we are interested in *short* vertex-disjoint paths. We may want to minimize the total length (minimum sum) or

Y. Dong, D.-Z. Du, and O. Ibarra (Eds.): ISAAC 2009, LNCS 5878, pp. 293–302, 2009.

the length of the longest path (min-max objective function). A more formal description of the problem is as follows.

Min-Sum k Disjoint Paths Problem (Min-Max k Disjoint Paths Problem)
Input: A graph $G = (V, E)$, k pairs of vertices $(s_1, t_1), (s_2, t_2), \ldots, (s_k, t_k)$ in G (which are sometimes called *terminals*), and a length function $l : E \to \mathbb{R}_+$.
Output: Vertex-disjoint paths $P_1, \ldots, P_k$ in G such that P_i is from s_i to t_i for $i = 1, 2, \ldots, k$, minimizing $\sum_{i=1}^{k} l(P_i)$ (or minimizing $\max_i l(P_i)$), where $l(P_i) = \sum_{e \in E(P_i)} l(e)$.

1.1 Related Work

If k is part of the input, the vertex-disjoint paths problem is one of Karp's **NP**-hard problems [8], and it remains **NP**-hard even if G is constrained to be planar [11]. If k is a fixed number, k pairwise vertex-disjoint paths can be found in polynomial time in directed planar graphs [17] and in directed acyclic graphs [4], whereas the problem in general directed graphs is **NP**-hard even if $k = 2$ [4]. It is known that the disjoint paths problem in undirected graphs is solvable in polynomial time when $k = 2$ [18,19,22]. Perhaps the biggest achievement in this area is Robertson and Seymour's polynomial-time algorithm for the problem in undirected graphs when k is fixed [15].

The optimization problem is considerably harder. The problem of finding disjoint paths minimizing the total length is wide open and only a few cases are known to be solvable in polynomial time (see also Table 1). First, finding k disjoint s-t paths (i.e., $s_1 = \cdots = s_k = s$ and $t_1 = \cdots = t_k = t$) with minimum total length (*min-sum*) is still possible in polynomial time, since it reduces to finding the standard minimum cost flow [21]. The min-sum problem is solvable in linear time for graphs with bounded tree-width [16]. For the following two cases, the min-sum problem can also be reduced to the minimum cost flow problem and can thus be solved in polynomial time:

- All sources (or sinks) coincide, that is, $s_1 = \cdots = s_k$ (or $t_1 = \cdots = t_k$, respectively).
- The graph is planar, all terminals are incident with a common face, and their cyclic order is $s_1, \ldots, s_k, t_k, \ldots, t_1$ (called *well-ordered*).

Another special case of the min-sum problem has recently been solved by Colin de Verdière and Schrijver [2]. They showed the following:

Theorem 1 (Colin de Verdière and Schrijver [2]). *If a given directed or undirected graph G is planar, all sources are incident to one face S, and all sinks are incident to another face $T \neq S$, then we can find k vertex-disjoint paths in G with minimum total length in $O(kn \log n)$ time.*

If the length of the longest path is to be minimized (*min-max*), the problem seems to be harder than the min-sum problem. The problem of finding two s-t paths minimizing the length of the longer path is **NP**-hard for an acyclic directed

Table 1. Results for the Min-Sum Disjoint Paths Problem

	Conditions	Complexity
$k = 2$	directed	**NP**-hard
	directed, planar, one face	OPEN
	undirected	OPEN
	undirected, planar, two faces	**P** (Theorem 2)
$k = 3$	undirected, planar, one face	**P** (Theorem 5)
k: fixed	undirected	OPEN
k: general	undirected	**NP**-hard
	$s_1 = \cdots = s_k$ and/or $t_1 = \cdots = t_k$	**P** (Min-cost flow)
	planar, one face, well-ordered	**P** (Min-cost flow)
	planar, $S \neq T$ faces	**P** [2]
	bounded tree-width	$O(n)$ [16]

Table 2. Results for the Min-Max Disjoint Paths Problem

	Conditions	Complexity
$k = 2$	directed, acyclic, $s_1 = s_2$, $t_1 = t_2$	**NP**-hard [7]
		pseudo-polynomial [10]
	directed, $s_1 = s_2$, $t_1 = t_2$	2-approx. [10]
	directed	strongly **NP**-hard [10]
	undirected, tree-width ≥ 3, planar	**NP**-hard ([24] and Theorem 6)
	undirected, tree-width ≤ 2	**P** (Theorem 7)

graph [7], but 2-approximable [10] using the *min-sum* version. Moreover, the problem is strongly **NP**-hard for general directed graphs when s_1, s_2, t_1, and t_2 are distinct [10]. For an overview, see Table 2.

Yet another variant of the objective function for the problem of finding two disjoint paths for one pair of terminals is the following: Before summing up the path lengths, the length of the longer path is multiplied by a factor $\alpha \in (0, 1)$, which parameterizes the cost function. $\alpha = 1$ would yield the min-sum variant and $\alpha = 0$ would yield the so-called 'min-min' variant, in which the length of the shorter path is to be minimized, which is **NP**-hard [23]. For $\alpha \in (0, 1)$ there is an approximation algorithm with ratio $\frac{1+\alpha}{2\alpha}$, which, for directed graphs, is claimed to be optimal unless the polynomial hierarchy collapses completely [25]. If the length of the shorter path is multiplied by $\alpha \in (0, 1)$, there is an approximation algorithm with ratio $\frac{2}{1+\alpha}$, which is claimed to be tight as well [24].

1.2 Contribution

We extend the min-sum results of Colin de Verdière and Schrijver [2] for undirected graphs and $k = 2$ as follows: the two disjoint faces F_1, F_2 may be 'mixed' such that

– s_1, s_2, t_1 are incident to F_1 and t_2 is incident to F_2 (Theorem 3),
– s_1, t_1 are incident to F_1 and s_2, t_2 are incident to F_2 (Theorem 4).

Our algorithms consist of non-trivial reductions to Theorem 1. By combining Theorem 1 with our new results, flow reductions, and trivially infeasible inputs, we obtain the following theorem.

Theorem 2. *Let $G = (V, E)$ be an undirected planar graph, and let F_1 and F_2 be its faces. If each terminal is on one of the boundaries of F_1 and F_2, then the Min-Sum 2 Disjoint Paths Problem in G is solvable in polynomial time.*

We also give a polynomial-time algorithm for the Min-Sum Disjoint Paths Problem when $k = 3$ and all terminals are incident to one face (Theorem 5). Our contribution is to give an algorithm for the case when the terminals are not well-ordered, by a non-trivial reduction to Theorem 1. For a summary, see Table 3.

Table 3. *min-sum* results in planar undirected graphs

	one face	two faces
$k = 2$	flow	Theorems 1, 3, 4
$k = 3$	flow, Theorem 5	OPEN

For the Min-Max 2 Disjoint Paths Problem, we draw the line between tractable and hard problems: We prove weak **NP**-hardness of the Min-Max 2 Disjoint Paths Problem for planar graphs with tree-width 3 using a reduction from the PARTITION problem (Theorem 6). We later learned that the reduction was used independently in almost the same manner in [24] already, without an explicit link to the tree-width and the min-max variant. For graphs with tree-width 2 (including series-parallel graphs and outer-planar graphs), we provide a polynomial-time algorithm (Theorem 7). The same algorithm also works for the Min-Min 2 Disjoint Paths Problem and the α-variants from [24,25].

For both the min-sum and the min-max versions and for the variants with cost functions parameterized by α as defined in [24,25], we give a pseudo-polynomial-time algorithm for graphs with bounded tree-width (Theorem 8). The algorithm runs in polynomial time for the min-sum objective function [16].

Due to space constraints, most proofs are given in [9].

2 Preliminaries

Let $G = (V, E)$ be an undirected graph with vertex set V and edge set E, and let $n = |V|$ denote the number of vertices. Since we consider vertex-disjoint paths, in what follows, we may assume that the graph has no multiple edges and no self-loops. An edge connecting $u, v \in V$ is denoted by uv, whereas (u, v) represents the arc from u to v in a directed graph. For a subgraph H of G, the vertex set and the edge set of H are denoted by $V(H)$ and $E(H)$, respectively.

Let $\delta(v)$ denote the set of edges incident to $v \in V$. For $U \subseteq V$, let $G[U]$ be the subgraph of $G = (V, E)$ induced by U, that is, its vertex set is U and its edge set consists of all edges in E with both ends in U. A graph G is *planar* if it can be embedded in a plane Σ such that no edges intersect, except at their end points. To simplify notation, we do not distinguish between a vertex of G and the point of Σ used in the embedding to represent the vertex, and we do not distinguish between an edge and the curve on Σ representing it. A *region* is a subset of Σ, and for a region R, let $G[[R]]$ denote the subgraph of G consisting of the vertices and the edges in R. For a face F of a planar graph, let ∂F denote the boundary of F. A planar graph is *outer-planar* if it allows for a planar embedding such that all its vertices are on the outer face. A *path* P, which is denoted by $P = (v_1, v_2, \dots, v_l)$, is a subgraph consisting of vertices $v_1, \dots, v_l$ and edges $e_1 = v_1v_2, \dots, e_{l-1} = v_{l-1}v_l$. When $v_1 = v_l$, it is called a *cycle*. A path (or a cycle) is *simple* if $v_i \neq v_j$ for distinct i, j (except for $v_1 = v_l$). For a simple path $P = (v_1, v_2, \dots, v_l)$, let $P^{[v_i, v_j]}$ denote the path $(v_i, v_{i+1}, \dots, v_j)$, and it is called a *subpath* of P. For a length function $l : E \to \mathbb{R}_+$, the length of a path P is denoted by $l(P)$, and for a pair of vertices $u, v \in V$, let $d_G(u, v)$ denote the length of a shortest path connecting u and v in G.

The tree-width of a graph was introduced by Halin [6], but it went unnoticed until it was rediscovered by Robertson and Seymour [13] and, independently, by Arnborg and Proskurowski [1]. The *tree-width* of a graph is defined as follows.

Definition 1. *Let G be a graph, T a tree and let $\mathcal{V} = \{V_t \subseteq V(G) \mid t \in V(T)\}$ be a family of vertex sets of G indexed by the vertices t of T. The pair $(T, \mathcal{V})$ is called a* tree-decomposition *of G if it satisfies the following three conditions:*

- $V(G) = \bigcup_{t \in T} V_t$
- *for every edge $e \in G$ there exists a $t \in T$ such that both ends of e lie in V_t*
- *If $t, t', t'' \in V(T)$ and t' lies on the path of T between t and t'', then $V_t \cap V_{t''} \subseteq V_{t'}$.*

The width *of $(T, \mathcal{V})$ is the number* $\max\{|V_t| - 1 \mid t \in T\}$ *and the* tree-width $\mathrm{tw}(G)$ *of G is the minimum width of any tree-decomposition of G.*

The tree-width is a good measure of the algorithmic tractability of graphs. It is known that a number of hard problems on graphs, such as "Hamiltonian cycle" and "chromatic number", can be solved efficiently when the given graph has small tree-width [1]. A graph has tree-width 1 if and only if it is a forest, and families of graphs with tree-width at most 2 include outer-planar graphs and series-parallel graphs.

3 Min-Sum Objective Function

In this section, we deal with the Min-Sum k Disjoint Paths Problem for $k = 2, 3$. To simplify the arguments, we use a perturbation technique such that all shortest paths are unique (see [3]). For $E = \{e_1, e_2, \dots, e_m\}$ and $l : E \to \mathbb{R}_+$, we use a new length function $l' : E \to \mathbb{R}_+$ defined by $l'(e_i) = l(e_i) + \varepsilon^i$ for each i, where ε

is an infinitely small positive number. Then, each path has a different length. In particular, the Min-Sum k Disjoint Paths Problem has a unique optimal solution if it has a feasible solution. In what follows, in this section, we simply denote the perturbed length function by l.

3.1 Min-Sum 2 Disjoint Paths Problem

In this section, we prove Theorem 2, which we restate here.

Theorem. *Let $G = (V, E)$ be an undirected planar graph, and let F_1 and F_2 be its faces. If every terminal is on $\partial F_1 \cup \partial F_2$, then the Min-Sum 2 Disjoint Paths Problem in G is solvable in polynomial time.*

Proof. The four terminals s_1, s_2, t_1, and t_2 may lie on two faces as follows:

- s_1, s_2, t_1, and t_2 are incident to F_1 (min-cost flow [21] or trivially infeasible),
- s_1, s_2 are incident to F_1 and t_1, t_2 are incident to F_2 (Theorem 1 due to [2]),
- s_1, s_2, t_1 are incident to F_1 and t_2 is incident to F_2 (Theorem 3), or
- s_1, t_1 are incident to F_1 and s_2, t_2 are incident to F_2 (Theorem 4).

The remaining cases (e.g. the case with s_2 alone on one face) are symmetric for undirected graphs. □

Theorem 3. *Let $G = (V, E)$ be an undirected planar graph, and F_1 and F_2 be its faces. If three terminals are on ∂F_1 and the remaining terminal is on ∂F_2, then the Min-Sum 2 Disjoint Paths Problem in G is solvable in $O(n^3 \log n)$ time.*

Proof. Let $s_1, s_2, t_1 \in \partial F_1$ and $t_2 \in \partial F_2$ be terminals. Let P be the shortest path connecting s_1 and t_2. The basic idea of the algorithm is to, for all pairs of vertices u, v on P, transform the original problem to an instance of the problem that can be solved using the algorithm by Colin de Verdière and Schrijver. The transformation is described in Lemma 2. In Lemma 1 we prove that the solution remains optimal.

Lemma 1. *Suppose that a pair of paths (P_1, P_2) is the unique optimal solution of the Min-Sum 2 Disjoint Paths Problem. Let u be the vertex in $V(P_1) \cap V(P)$ closest to t_2 in the ordering along P, and let v be the vertex in $V(P_2) \cap V(P^{[u,t_2]})$ closest to u in the ordering along P (Fig. 1). Then, $P^{[v,t_2]}$ is a subpath of P_2.*

Proof. Suppose that $P^{[v,t_2]}$ is not a subpath of P_2, and define $P_2' = P_2^{[s_2,v]} \cup P^{[v,t_2]}$. By definition of u and v, P_1 and P_2' are disjoint. Since P is the shortest path, every subpath $P^{[a,b]}$ is the shortest path between a and b, thus, $l(P^{[v,t_2]}) < l(P_2^{[v,t_2]})$, which implies that $l(P_2') < l(P_2)$. Here we use the fact that the shortest path is unique. Then, (P_1, P_2') is a shorter solution, which contradicts the optimality of (P_1, P_2). □

Lemma 2. *For distinct vertices u, v on P such that u is closer to s_1 than v, in $O(n \log n)$ time, we can either find two simple disjoint paths P_1 and P_2 minimizing the total length $l(P_1) + l(P_2)$ such that*

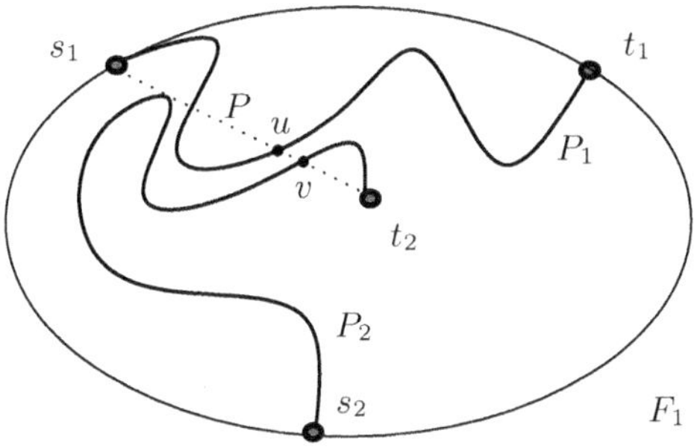

Fig. 1. Definitions of P, u, and v

1. P_i connects s_i and t_i for $i = 1, 2$,
2. $u \in V(P_1)$ and $V(P_1) \cap V(P) \subseteq V(P^{[s_1,u]})$, and
3. $P_2 \cap P^{[u,t_2]} = P^{[v,t_2]}$,

or conclude that such P_1 and P_2 do not exist.

Proof. Delete all vertices in $V(P^{[u,t_2]}) \setminus \{u, v, t_2\}$. This yields a graph G'. Note that u and v are on the boundary of the same face F' in G', because all internal vertices of $P^{[u,v]}$ have been removed.

We find three paths Q_1, Q_2, and Q_3 in G' minimizing the total length such that

- Q_1 connects s_1 and u, Q_2 connects t_1 and u, Q_3 connects s_2 and v,
- $V(Q_2) \cap V(Q_3) = V(Q_3) \cap V(Q_1) = \emptyset$, and $V(Q_1) \cap V(Q_2) = \{u\}$.

In order to apply Theorem 1, we divide u into two distinct vertices and construct a digraph as follows. Let $v_1, v_2, \ldots, v_p$ be the vertices in G' adjacent to u such that $v_1, v_p \in \partial F'$ and $uv_1, uv_2, \ldots, uv_p$ are incident to u in this order. Let $D_1 = (V_1, E_1)$ be the digraph obtained from $G - u$ by replacing each edge with two parallel arcs of opposite direction. Define a digraph $D_2 = (V_2, E_2)$ (see Fig. 2) by

$$V_2 = V_1 \cup \{w_1, w_2, \ldots, w_p, u_1, u_2\},$$

$$E_2 = E_1 \cup \bigcup_{i=1}^{p} \{(v_i, w_i)\} \cup \bigcup_{i=1}^{p-1} \{(w_i, w_{i+1}), (w_{i+1}, w_i)\} \cup \{(w_1, u_1), (w_p, u_2)\}.$$

Define a new length function $l' : E_2 \to \mathbb{R}_+$ as

$$l'(e) = \begin{cases} l(xy) & \text{if } e = (x, y) \text{ or } (y, x) \text{ for } xy \in E, \\ l(v_i u) & \text{if } e = (v_i, w_i), \\ 0 & \text{otherwise.} \end{cases}$$

By finding three disjoint paths Q_1', Q_2', Q_3' with minimum total length such that Q_1' is from s_1 to u_1 (or u_2, respectively), Q_2' is from t_1 to u_2 (or u_1, respectively),

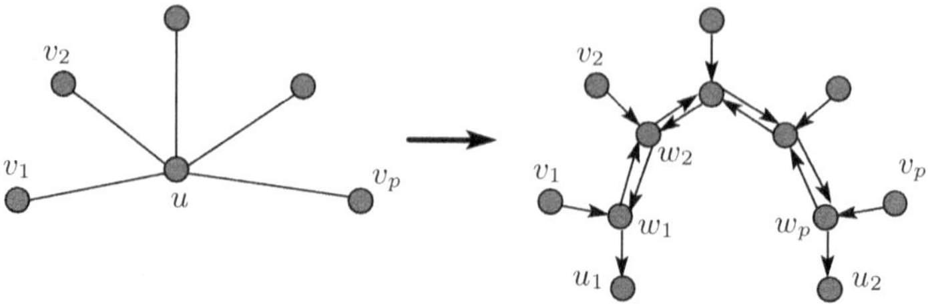

Fig. 2. Construction of D_2

and Q'_3 is from s_2 to v, we can obtain the desired paths Q_1, Q_2, and Q_3. This can be done in $O(n \log n)$ time by Theorem 1.

Then, $P_1 = Q_1 \cup Q_2$ and $P_2 = Q_3 \cup P^{[v,t_2]}$ are the desired disjoint paths in G. □

By Lemma 1, we can find the optimal solution of the Min-Sum 2 Disjoint Paths Problem by executing the procedure described in Lemma 2 for each pair of vertices u and v on the shortest path between s_1 and t_2. This concludes the proof of Theorem 3. □

Theorem 4. *Let $G = (V, E)$ be an undirected planar graph, and F_1 and F_2 be its faces. If $s_1, t_1 \in \partial F_1$ and $s_2, t_2 \in \partial F_2$ are terminals, then the Min-Sum 2 Disjoint Paths Problem in G is solvable in $O(n^3 \log n)$ time.*

3.2 Min-Sum 3 Disjoint Paths Problem

Theorem 5. *Let $G = (V, E)$ be an undirected planar graph and let F be its face. If all six terminals are on ∂F, then the Min-Sum 3 Disjoint Paths Problem in G is solvable in $O(n^4 \log n)$ time.*

4 Min-Max Objective Function

For graphs with tree-width 2 (including series-parallel graphs and outer-planar graphs), we provide a polynomial-time algorithm for the Min-Max 2 Disjoint Paths Problem. This closes the gap between tractable and hard instances due to the following theorem (and its proof in [24]).

Theorem 6 ([24]). *The Min-Max 2 Disjoint Paths Problem is (weakly)* **NP**-*hard for planar graphs with tree-width at least 3.*

Theorem 7. *The Min-Max 2 Disjoint Paths Problem for graphs with tree-width at most 2 can be solved in time $O(n^3)$.*

For fixed k, we also give a pseudo-polynomial-time algorithm for the Min-Max k Disjoint Paths Problem in bounded tree-width graphs. As shown in Theorem 6, the Min-Max k Disjoint Paths Problem is NP-hard even if $k = 2$ and the tree-width of the input graph is at most three, whereas the Min-Sum k Disjoint Paths Problem can be solved in polynomial time in bounded tree-width graphs [16]. Note that this technique also works for the weighted versions introduced in [24,25].

Theorem 8. *Let $G = (V, E)$ be a graph whose tree-width is bounded by a fixed constant, and let $\ell : E \to \mathbb{Z}_+$ be an integer-valued length function. Then, for fixed k, the Min-Max k Disjoint Paths Problem can be solved in time polynomial in $|V|$ and $\ell(E)$.*

Acknowledgement

The authors thank the anonymous reviewers for careful reading of the manuscript.

The first author is supported by JSPS Research Fellowships for Young Scientists. His work was partially supported by the Global COE Program "The research and training center for new development in mathematics", MEXT, Japan.

References

1. Arnborg, S., Proskurowski, A.: Linear time algorithms for NP-hard problems restricted to partial k-trees. Discrete Applied Mathematics 23(1), 11–24 (1989)
2. de Verdière, É.C., Schrijver, A.: Shortest vertex-disjoint two-face paths in planar graphs. In: STACS, pp. 181–192 (2008)
3. Edelsbrunner, H., Mücke, E.P.: Simulation of simplicity: a technique to cope with degenerate cases in geometric algorithms. ACM Transactions on Graphics 9(1), 66–104 (1990)
4. Fortune, S., Hopcroft, J.E., Wyllie, J.: The directed subgraph homeomorphism problem. Theoretical Computer Science 10, 111–121 (1980)
5. Frank, A.: Packing paths, cuts and circuits – a survey. In: Paths, Flows, and VLSI-Layout, pp. 49–100. Springer, Heidelberg (1990)
6. Halin, R.: S-functions for graphs. Journal of Geometry 8(1-2), 171–186 (1976)
7. Itai, A., Perl, Y., Shiloach, Y.: The complexity of finding maximum disjoint paths with length constraints. Networks 12, 277–286 (1981)
8. Karp, R.M.: On the computational complexity of combinatorial problems. Networks 5, 45–68 (1975)
9. Kobayashi, Y., Sommer, C.: On shortest disjoint paths in planar graphs. Technical Report METR 2009-38, Department of Mathematical Informatics, University of Tokyo (2009)
10. Li, C.-L., McCormick, S.T., Simchi-Levi, D.: The complexity of finding two disjoint paths with min-max objective function. Discrete Applied Mathematics 26(1), 105–115 (1990)
11. Lynch, J.F.: The equivalence of theorem proving and the interconnection problem. SIGDA Newsletter 5(3), 31–36 (1975)
12. Ogier, R.G., Rutenburg, V., Shacham, N.: Distributed algorithms for computing shortest pairs of disjoint paths. IEEE Transactions on Information Theory 39(2), 443–455 (1993)
13. Robertson, N., Seymour, P.D.: Graph minors. II. Algorithmic aspects of tree-width. Journal of Algorithms 7, 309–322 (1986)

14. Robertson, N., Seymour, P.D.: An outline of a disjoint paths algorithm. In: Paths, Flows, and VLSI-Layout, pp. 267–292. Springer, Heidelberg (1990)
15. Robertson, N., Seymour, P.D.: Graph minors. XIII. The disjoint paths problem. Journal of Combinatorial Theory, Series B 63(1), 65–110 (1995)
16. Scheffler, P.: A practical linear time algorithm for disjoint paths in graphs with bounded tree-width. Technical Report 396, Technische Universität Berlin (1994)
17. Schrijver, A.: Finding k disjoint paths in a directed planar graph. SIAM Journal on Computing 23(4), 780–788 (1994)
18. Seymour, P.D.: Disjoint paths in graphs. Discrete Mathematics 29, 293–309 (1980)
19. Shiloach, Y.: A polynomial solution to the undirected two paths problem. Journal of the ACM 27(3), 445–456 (1980)
20. Srinivas, A., Modiano, E.: Finding minimum energy disjoint paths in wireless ad-hoc networks. Wireless Networks 11(4), 401–417 (2005)
21. Tardos, É.: A strongly polynomial minimum cost circulation algorithm. Combinatorica 5(3), 247–256 (1985)
22. Thomassen, C.: 2-linked graphs. European Journal of Combinatorics 1, 371–378 (1980)
23. Yang, B., Zheng, S.-Q., Katukam, S.: Finding two disjoint paths in a network with min-min objective function. In: IASTED International Conference on Parallel and Distributed Computing and Systems, pp. 75–80 (2003)
24. Yang, B., Zheng, S.-Q., Lu, E.: Finding two disjoint paths in a network with normalized α^{-}-min-sum objective function. In: IASTED International Conference on Parallel and Distributed Computing and Systems, pp. 342–348 (2005)
25. Yang, B., Zheng, S.-Q., Lu, E.: Finding two disjoint paths in a network with normalized α^{+}-min-sum objective function. In: Deng, X., Du, D.-Z. (eds.) ISAAC 2005. LNCS, vol. 3827, pp. 954–963. Springer, Heidelberg (2005)

An Optimal Labeling for Node Connectivity*

Tai-Hsin Hsu and Hsueh-I Lu**

Department of Computer Science and Information Engineering
National Taiwan University
d92021@csie.ntu.edu.tw, hil@csie.ntu.edu.tw

Abstract. Given an n-node undirected simple graph G and a positive integer k, the *k-connectivity labeling problem* for G seeks short labels for the nodes of G such that whether any two nodes are k-connected in G can be determined merely by their labels. For $k = 1$, an optimal solution to the problem is to give each node in the same connected component of G a common $\lceil \log_2 n \rceil$-bit label, uniquely chosen for this connected component. For $k \geq 2$, Katz, Katz, Korman, and Peleg gave the first nontrivial solution to the problem, requiring $O(2^k \log n)$ bits per node. The best previously known solution, due to Korman, requires $O(k^2 \log n)$ bits per node. We give the first asymptotically optimal solution to the problem, requiring only $(2k - 1) \lceil \log_2 n \rceil$ bits per node, which matches a lower bound $\Omega(k \log n)$ proved by Katz, Katz, Korman, and Peleg.

1 Introduction

Let G be an n-node undirected simple graph. Let k be a positive integer. Let $V(G)$ consist of the nodes of G. For any node subset S of G, let $G \setminus S$ denote the subgraph of G induced by $V(G) \setminus S$. Let $|S|$ denote the cardinality of set S. Two nodes u and v are *k-connected* in G if u and v are connected in $G \setminus S$ for any node subset S of G with $\{u, v\} \cap S = \emptyset$ and $|S| \leq k - 1$. We study the *k-connectivity labeling problem* for G, which seeks a compact labeling L for the nodes of G such that whether any two nodes u and v are k-connected in G or not can be determined merely by the labels $L(u)$ and $L(v)$ of nodes u and v. For $k = 1$, an optimal solution to the problem is an L such that, for each node u in the i-th connected component of G, $L(u)$ is the $\lceil \log_2 n \rceil$-bit binary encoding of i. Clearly, nodes u and v are 1-connected in G if and only if $L(u) = L(v)$. For $k \geq 2$, Katz, Katz, Korman, and Peleg [1, 2] gave the first nontrivial solution, which requires $2^k \lceil \log_2 n \rceil$ (respectively, $(2^{k-1} + 1) \lceil \log_2 n \rceil$) bits per node for $k \geq 4$ (respectively, $2 \leq k \leq 3$). Using an encoding technique of Alstrup and Rauhe [3] for graphs with bounded arboricity, Korman [4, 5] gave the best previously known solution, which requires $(\frac{k(k-1)}{2} + 1) \lceil \log_2 n \rceil + O(\log^* n)$ bits per node. We give the first

* Research supported in part by NSC grants 97-2221-E-002-122 and 98-2221-E-002-079-MY3.

** Corresponding author. Web: www.csie.ntu.edu.tw/~hil. This author also holds joint appointments in the Graduate Institute of Networking and Multimedia and the Graduate Institute of Biomedical Electronics and Bioinformatics, National Taiwan University. Address: 1 Roosevelt Road, Section 4, Taipei 106, Taiwan, ROC.

Y. Dong, D.-Z. Du, and O. Ibarra (Eds.): ISAAC 2009, LNCS 5878, pp. 303–310, 2009.

known asymptotically optimal solution to the problem, as summarized in the following theorem.

Theorem 1. *For any positive integer k and any n-node graph G, there is a solution L to the k-connectivity labeling problem for G such that the label $L(u)$ of each node u in G has at most $(2k-1)\lceil\log_2 n\rceil$ bits. Moreover, the time required by our solution to determine whether any two nodes u and v are k-connected in G is linear in the sum of the numbers of bits in $L(u)$ and $L(v)$.*

The number of bits per node required by our solution matches a lower bound $\Omega(k \log n)$ proved by Katz et al. [1, 2]. The time required to determine whether two nodes u and v are k-connected in G by using $L(u)$ and $L(v)$ is also optimal.

1.1 Overview of Our Techniques

Let $H(G)$ denote the graph on $V(G)$ where any nodes u and v are adjacent in $H(G)$ if and only if u and v are k-connected in G. Adopting the same approach of all previous work [1, 2, 4, 5], our solution is actually an adjacency labeling L of $H(G)$, i.e., whether any two nodes u and v are adjacent in $H(G)$ can be determined merely by the labels $L(u)$ and $L(v)$ of u and v. A graph H is *closed under k-connectivity* if any two k-connected nodes u and v in H are adjacent in H. The following lemma ensures that it suffices to solve the adjacency labeling problem for a graph H that is closed under k-connectivity.

Lemma 1 (Katz, Katz, Korman, and Peleg [1, 2]). *For any graph G, we have that $H(G)$ is closed under k-connectivity.*

The rest of the paper assumes $k \geq 2$ and lets H be an n-node graph that is closed under k-connectivity. For any subgraph K of H, let $N(K)$ consist of the nodes u of $V(H)\setminus V(K)$ that is adjacent to some node v in $V(K)$. A subgraph K of graph H is a *clique* of H if K is complete, i.e., any two nodes of K are adjacent in K. A key step in our solution, to be shown in Lemma 2 of Section 2, is to identify a clique K of H with $|K| \geq 1$ and $|N(K)| \leq 2k-3$. The $(2k-1)\lceil\log_2 n\rceil$-bit label $L(u)$ of each node u of in K consists of the following three parts:

- A unique $\lceil\log_2 n\rceil$-bit node identifier $L_1(u)$ for u.
- A unique $\lceil\log_2 n\rceil$-bit clique identifier $L_2(u)$ for K.
- A set $L_3(u)$ of $2k-3$ node identifiers, one for each neighbor of u in $V(H) \setminus V(K)$.

After determining $L(u)$ for each node u in K, we delete $V(K)$ from H and repeat the above procedure of (a) identifying a clique K of H with $|K| \geq 1$ and $|N(K)| \leq 2k-3$ and (b) determining $L(u)$ for each node u in K until H is empty. We show, in Lemma 5 of Section 3, that nodes u and v are adjacent in H if and only if $L_2(u) = L_2(v)$, $L_1(v) \in L_3(u)$, or $L_1(u) \in L_3(v)$.

1.2 Related Work

Algorithms assigning informative labels to graph nodes are usually called *informative labeling schemes*. This paper and its previous work [2, 6] study the labels containing the information of node connectivity. Besides, there are various types of information such as adjacency [3, 7–9], distance [10–19], flow and edge connectivity [2, 3, 6], and queries on trees [20–31]. Other papers work on dynamic labeling schemes [32–35]. Gavoille and Peleg [36] had a comprehensive survey on labeling schemes.

The remainder of the paper is organized as follows. Section 2 shows how to identify a clique K of H with $|N(K)| \leq 2k-3$. Section 3 gives our adjacency labeling L for H and proves Theorem 1. Section 4 gives the conclusion.

2 Finding a Clique with a Bounded Number of Neighbors

This section proves the following lemma.

Lemma 2. *There is a clique K of H with $|V(K)| \geq 1$ and $|N(K)| \leq 2k-3$.*

2.1 Preliminaries

Lemma 3. *If S is a node subset of H, then $H \setminus S$ remains closed under k-connectivity.*

Proof. Let u and v be two nodes of $H \setminus S$ that are k-connected in $H \setminus S$. Clearly, u and v are k-connected in H. Since H is closed under k-connectivity, we know $(u, v) \in H$. Therefore, $(u, v) \in H \setminus S$. The lemma is proved. □

For any two non-adjacent nodes u and v of H, we say that (S, X, Y) is a (k, u, v)-*partition* of H if S, X, and Y are disjoint node subsets of H such that

- $S \cup X \cup Y = V(H)$,
- X contains exactly one of u and v,
- Y contains exactly one of u and v,
- $|S| \leq k-1$, and
- any node x of X is not adjacent to any node y of Y in H.

An illustration for a (k, u, v)-partition is shown in Figure 1.

Lemma 4. *For any two non-adjacent nodes u and v of H, there is a (k, u, v)-partition of H.*

Proof. Since H is closed under k-connectivity, $(u, v) \notin H$ implies that u and v are not k-connected in H. That is, there is a node set S of H with $|S| \leq k-1$ and $S \subseteq V(H) \setminus \{u, v\}$ such that u and v are not connected in $H \setminus S$. Let X consist of the nodes in the connected component of $H \setminus S$ that contains u. Let $Y = V(H) \setminus (X \cup S)$. One can verify that (S, X, Y) is a (k, u, v)-partition of H. □

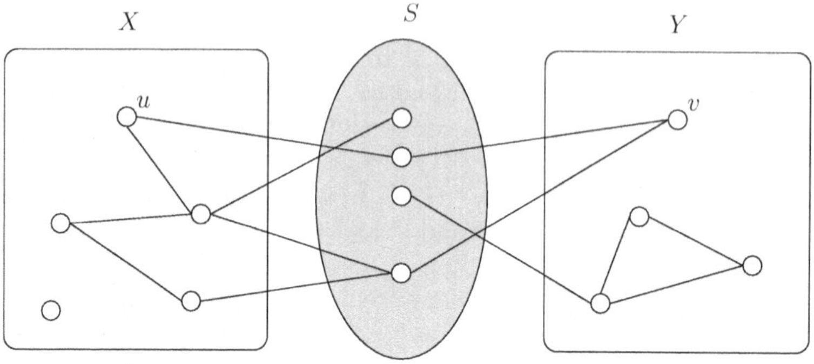

Fig. 1. An illustration for a (k, u, v)-partition

Algorithm 1

Input: A graph H that is closed under k-connectivity.
Output: A clique in H.

1: Let all nodes of H be white.
2: **while** the white nodes in H do not form a clique **do**
3: Let u and v be two non-adjacent white nodes of H.
4: Let (S, X, Y) be a (k, u, v)-partition of H such that the number of red nodes in X is no more than that in Y.
5: Let all nodes in S be red.
6: Let $H = H \setminus Y$.
7: **end while**
8: Return the clique formed by the white nodes in H.

2.2 Proving Lemma 2

Proof. We prove the lemma by showing that the output of Algorithm 1 satisfies the lemma. We first show that Algorithm 1 is well defined. Since the condition of the while-loop implies the existence of u and v, Step 3 is well defined. Observe that H can be changed only at Step 6, it follows from Lemma 3 that H remains closed under k-connectivity throughout the execution of Algorithm 1. By Lemma 4, Step 4 is well defined. Since Step 6 deletes at least one node from H, the while-loop terminates in at most n iterations.

Let K be the output of Algorithm 1. We next show $|K| \geq 1$. By Step 4, (S, X, Y) is a (k, u, v)-partition of H. Therefore, only one of u and v belongs to Y. That is to say, exactly one of u and v remains in H after Step 6 deletes Y from H. Since both u and v are white, H contains at least one white node at the end of each iteration of the while-loop.

It remains to show $|N(K)| \leq 2k-3$. By Step 4, (S, X, Y) is a (k, u, v)-partition of H, implying that any node in X is not adjacent to any node in Y. By Step 5, all white nodes belong to X at the end of each iteration. Therefore, at the end of

Algorithm 2

```
Let j = 0.
Let H_0 = H.
while H_j is not empty do
    Let K_j be a clique of H_j ensured by Lemma 2.
    Let H_{j+1} = H_j \ V(K_j).
    Let j = j + 1.
end while
```

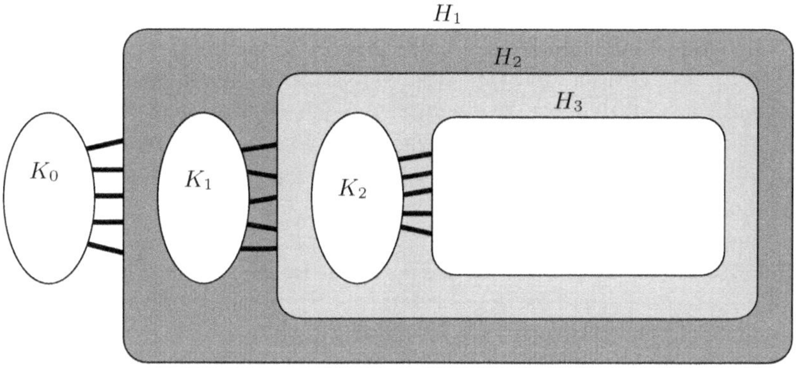

Fig. 2. An illustration for K_j and H_j. The whole graph H is H_0.

Algorithm 1, any node in K is not adjacent to any previously deleted nodes of H. Therefore, it suffices to prove that at the end of each iteration of the while-loop, the number of red nodes in H is at most $2k-3$. Since $k \geq 2$ implies $2k-3 \geq k-1$, it follows from the definition of (k, u, v)-partition of H that the statement holds if the while-loop executes for at most one iteration. Consider the situation that the while-loop executes for at least two iterations. Assume for contradiction that $i \geq 2$ is the smallest index such that the number of red nodes is at least $2k-2$ at the end of the i-th iteration. By $|S| \leq k-1$, as ensured by Lemma 4, Step 5 changes the color of at most $k-1$ white nodes to red. Therefore, the number of red nodes in X is at least $k-1$. Thus, the number of red nodes in Y is also no less than $k-1$. Since the colors of the nodes in $X \cup Y$ do not change in the i-th iteration, H has at least $2k-2$ red nodes at the end of the $(i-1)$-st iteration, contradicting the definition of i. The lemma is proved. □

3 Our Optimal Labeling

This section gives our adjacency labeling L for H. For each node u of H, let $L(u)$ be the concatenation of $L_1(u)$, $L_2(u)$, and $L_3(u)$ which are defined as follows. First of all, if u is the i-th node of H, then let $L_1(u)$ be the $\lceil \log_2 n \rceil$-bit binary representation of non-negative integer i. To define $L_2(u)$ and $L_3(u)$, we perform

Algorithm 2 to obtain H_j and K_j for each $j \geq 0$. An illustration is shown in Figure 2. For each node u in K_j, let $L_2(u)$ be the $\lceil \log_2 n \rceil$-bit binary representation of non-negative integer j. If $u \in V(K_j)$, let $L_3(u)$ be the concatenation of $L_1(v)$ for all nodes v in $V(H_j) \setminus V(K_j)$ that is adjacent to u in H_j. We write $L_1(v) \in L_3(u)$ to signify that $L_1(v)$ belongs to list $L_3(u)$.

Lemma 5

1. *For each node u of H, $L(u)$ has at most $(2k-1)\lceil \log_2 n \rceil$ bits.*
2. *For any two nodes u and v of H, we have that $(u, v) \in H$ if and only if $L_2(u) = L_2(v)$, $L_1(u) \in L_3(v)$, or $L_1(v) \in L_3(u)$.*

Proof. By Step 5 of Algorithm 2 and Lemma 3, each H_j is closed under k-connectivity. By Lemma 2, Step 4 of Algorithm 2 is well defined. Therefore, Algorithm 2 is well defined. By Lemma 2, each K_j obtained by Step 4 of Algorithm 2 is non-empty. Therefore, Algorithm 2 halts in at most n iterations of the while-loop. Observe that $V(K_j)$ with $j \geq 0$ form a partition of $V(H)$. That is, each node of H belongs to exactly one K_j.

Statement 1. Suppose that node u belongs to K_j. By Lemma 2, u has at most $2k-3$ neighbors in H_j. Therefore, $L_3(u)$ has at most $(2k-3)\lceil \log_2 n \rceil$ bits. Since each of $L_1(u)$ and $L_2(u)$ has $\lceil \log_2 n \rceil$ bits, the statement is proved.

Statement 2. The if-direction is straightforward from the definitions of $L(u)$ and $L(v)$. We show the other direction. Suppose that u and v are adjacent in H. Let i and j be the indices such that $u \in K_i$ and $v \in K_j$. If $i = j$, then $L_2(u) = L_2(v)$. If $i < j$, then $L_1(v) \in L_3(u)$. If $i > j$, then $L_1(u) \in L_3(v)$. The statement is proved. □

3.1 Proving Theorem 1

Proof. Straightforward by Lemmas 1 and 5. □

4 Concluding Remarks

This paper presents an asymptotically optimal solution to the k-connectivity labeling problem. It would be of interest to further reduce the number of bits per node, or to solve this problem on directed graphs.

References

1. Katz, M., Katz, N.A., Korman, A., Peleg, D.: Labeling schemes for flow and connectivity. In: Proceedings of the 13th Annual ACM-SIAM Symposium on Discrete Algorithms, pp. 927–936 (2002)
2. Katz, M., Katz, N.A., Korman, A., Peleg, D.: Labeling schemes for flow and connectivity. SIAM Journal on Computing 34(1), 23–40 (2005)
3. Alstrup, S., Rauhe, T.: Small induced-universal graphs and compact implicit graph representations. In: Proceedings of the 43rd Annual IEEE Symposium on Foundations of Computer Science, pp. 53–62 (2002)

4. Korman, A.: Labeling schemes for vertex connectivity. In: Arge, L., Cachin, C., Jurdziński, T., Tarlecki, A. (eds.) ICALP 2007. LNCS, vol. 4596, pp. 102–109. Springer, Heidelberg (2007)
5. Korman, A.: Labeling schemes for vertex connectivity. ACM Transactions on Algorithms (to appear)
6. Korman, A., Kutten, S.: Distributed verification of minimum spanning trees. Distributed Computing 20(4), 253–266 (2007)
7. Breuer, M.A.: Coding the vertexes of a graph. IEEE Transactions on Information Theory 12(2), 148–153 (1966)
8. Breuer, M.A., Folkman, J.: An unexpected result in coding the vertices of a graph. Journal of Mathematical Analysis and Applications 20(3), 583–600 (1967)
9. Kannan, S., Naor, M., Rudich, S.: Implicit representation of graphs. SIAM Journal on Discrete Mathematics 5(4), 596–603 (1992)
10. Peleg, D.: Proximity-preserving labeling schemes and their applications. In: Widmayer, P., Neyer, G., Eidenbenz, S. (eds.) WG 1999. LNCS, vol. 1665, pp. 30–41. Springer, Heidelberg (1999)
11. Katz, M., Katz, N.A., Peleg, D.: Distance labeling schemes for well-separated graph classes. Discrete Applied Mathematics 145(3), 384–402 (2005)
12. Gavoille, C., Peleg, D., Pérennes, S., Raz, R.: Distance labeling in graphs. Journal of Algorithms 53(1), 85–112 (2004)
13. Gavoille, C., Katz, M., Katz, N.A., Paul, C., Peleg, D.: Approximate distance labeling schemes. In: Meyer auf der Heide, F. (ed.) ESA 2001. LNCS, vol. 2161, pp. 476–487. Springer, Heidelberg (2001)
14. Kaplan, H., Milo, T.: Short and simple labels for small distances and other functions. In: Dehne, F., Sack, J.-R., Tamassia, R. (eds.) WADS 2001. LNCS, vol. 2125, pp. 246–257. Springer, Heidelberg (2001)
15. Cohen, E., Halperin, E., Kaplan, H., Zwick, U.: Reachability and distance queries via 2-hop labels. In: Proceedings of the 13th Annual ACM-SIAM Symposium on Discrete Algorithms, pp. 937–946 (2002)
16. Gavoille, C., Paul, C.: Distance labeling scheme and split decomposition. Discrete Mathematics 273(1-3), 115–130 (2003)
17. Thorup, M.: Compact oracles for reachability and approximate distances in planar digraphs. Journal of the ACM 51(6), 993–1024 (2004)
18. Alstrup, S., Bille, P., Rauhe, T.: Labeling schemes for small distances in trees. SIAM Journal on Discrete Mathematics 19(2), 448–462 (2005)
19. Korman, A., Peleg, D., Rodeh, Y.: Constructing labeling schemes through universal matrices. In: Asano, T. (ed.) ISAAC 2006. LNCS, vol. 4288, pp. 409–418. Springer, Heidelberg (2006)
20. Fraigniaud, P., Korman, A.: Compact ancestry labeling schemes for XML trees. In: Proceedings of the 21st Annual ACM-SIAM Symposium on Discrete Algorithms (to appear, 2010)
21. Cohen, R., Fraigniaud, P., Ilcinkas, D., Korman, A., Peleg, D.: Labeling schemes for tree representation. Algorithmica 53(1), 1–15 (2009)
22. Fischer, J.: Short labels for lowest common ancestors in trees. In: Fiat, A., Sanders, P. (eds.) ESA 2009. LNCS, vol. 5757, pp. 752–763. Springer, Heidelberg (2009)
23. Abiteboul, S., Kaplan, H., Milo, T.: Compact labeling schemes for ancestor queries. In: Proceedings of the 12th Annual ACM-SIAM Symposium on Discrete Algorithms, pp. 547–556 (2001)
24. Thorup, M., Zwick, U.: Compact routing schemes. In: Proceedings of the 13th Annual ACM Symposium on Parallel Algorithms and Architectures, pp. 1–10 (2001)

25. Fraigniaud, P., Gavoille, C.: Routing in trees. In: Orejas, F., Spirakis, P.G., van Leeuwen, J. (eds.) ICALP 2001. LNCS, vol. 2076, pp. 757–772. Springer, Heidelberg (2001)
26. Fraigniaud, P., Gavoille, C.: A space lower bound for routing in trees. In: Alt, H., Ferreira, A. (eds.) STACS 2002. LNCS, vol. 2285, pp. 65–75. Springer, Heidelberg (2002)
27. Alstrup, S., Rauhe, T.: Improved labeling scheme for ancestor queries. In: Proceedings of the 13th Annual ACM-SIAM Symposium on Discrete Algorithms, pp. 947–953 (2002)
28. Alstrup, S., Gavoille, C., Kaplan, H., Rauhe, T.: Nearest common ancestors: a survey and a new distributed algorithm. Theory of Computing Systems 37(3), 441–456 (2004)
29. Kaplan, H., Milo, T., Shabo, R.: A comparison of labeling schemes for ancestor queries. In: Proceedings of the 13th Annual ACM-SIAM Symposium on Discrete Algorithms, pp. 954–963 (2002)
30. Peleg, D.: Informative labeling schemes for graphs. Theoretical Computer Science 340(3), 577–593 (2005)
31. Kao, M.Y., Li, X.Y., Wang, W.: Average case analysis for tree labelling schemes. Theoretical Computer Science 378(3), 271–291 (2007)
32. Cohen, E., Kaplan, H., Milo, T.: Labeling dynamic XML trees. In: Proceedings of the 21st ACM SIGMOD-SIGACT-SIGART Symposium on Principles of Database Systems, pp. 271–281 (2002)
33. Korman, A., Peleg, D., Rodeh, Y.: Labeling schemes for dynamic tree networks. Theory of Computing Systems 37(1), 49–75 (2004)
34. Korman, A.: General compact labeling schemes for dynamic trees. Distributed Computing 20(3), 179–193 (2007)
35. Korman, A., Peleg, D.: Labeling schemes for weighted dynamic trees. Information and Computation 205(12), 1721–1740 (2007)
36. Gavoille, C., Peleg, D.: Compact and localized distributed data structures. Distributed Computing 16(2-3), 111–120 (2003)

SOFA: Strategyproof Online Frequency Allocation for Multihop Wireless Networks*

Ping Xu and Xiang-Yang Li

Department of Computer Science, Illinois Institute of Technology, Chicago, IL, 60616
pxu3@iit.edu, xli@cs.iit.edu

Abstract. A number of approaches, including cognitive radios, dynamic spectrum allocation, and spectrum auction, have been proposed and used to improve the spectrum usage. A natural characteristic of spectrum usage is that requests for spectrums often come in an online fashion. Thus, it is imperative to design efficient and effective *online* dynamic spectrum allocation methods. Another challenge is that the secondary users are often selfish and prefer to maximize their own benefits. In this paper, we address these two challenges by proposing **SOFA**, strategyproof online frequency allocation method. In our protocol, a frequency will be shared among a number of users, and secondary users are required to submit the spectrum bid α time slots before its usage. Upon receiving an online spectrum request, our protocol will decide whether to grant its exclusive usage, within γ time slots. Assume that existing spectrum usage can be preempted with some penalty. For various possible known information, we analytically prove that the competitive ratios of our methods are within small constant factors of the optimum online method. Furthermore, in our mechanisms, no selfish users will gain benefits by bidding lower than its willing payment.

1 Introduction

With the recent fast growing spectrum-based services and devices, the remaining spectrum available for future wireless services is being exhausted. The current fixed spectrum allocation scheme leads to significant spectrum *white spaces* where many allocated spectrum blocks are used only in certain geographical areas and/or in brief periods of time. A huge amount of precious spectrum, perfect for wireless communications that is worth billions of dollars, sit there silently.

Subleasing is widely regarded as a potential way to share spectrum. Previous studies (*e.g.*, [15, 20, 19]) mainly assume that the information of all requests is known before making allocation. This is true in some cases, but not true generally. In most applications, spectrum bidding requests often arrive online and the central authority (typically a primary user) needs to *quickly* make a decision whether the requests are granted or not. In this paper, we study this online model and propose online algorithms. We analytically study *competitive ratios* of our algorithms. The *competitive ratio* of online

* The research of authors is partially supported by NSF CNS-0832120, National Natural Science Foundation of China under Grant No. 60828003, the Natural Science Foundation of Zhejiang Province under Grant No.Z1080979, National Basic Research Program of China (973 Program) under grant No. 2010CB328100.

Y. Dong, D.-Z. Du, and O. Ibarra (Eds.): ISAAC 2009, LNCS 5878, pp. 311–320, 2009.

algorithm is defined as the ratio between its performance and the performance of the optimal offline algorithm for every possible input. To the best of our knowledge, we are the first to study online spectrum allocation with cancelation and preemption penalty.

The main contributions of this paper are as follows. We show that best competitive ratio achievable depends on penalty factor β and there are three regimes. For each regime, we design algorithm with competitive ratio that matches the upper bound asymptotically. We also design efficient auction mechanism. In our mechanism, to maximize its profit, no secondary user will bid lower than its actual valuation.

The rest of the paper is organized as follows. In Section 2, we define in detail the problems to be studied. In Section 3, we present upper bounds of any online methods. We then present our solutions in Section 4 and analytically prove the performance bounds. We present our mechanisms to deal with selfish users in Section 5. Finally we review the related work in Section 6 and conclude the paper in Section 7.

2 Preliminaries

2.1 Network Model

Consider a wireless network system consists of some primary users and a central authority who decides spectrum assignment on behalf of primary users. Some secondary users $\mathcal{V} = \{v_1, v_2, \cdots, v_n\}$ want to lease the right to use a channel in some region for some time period. We consider a simple scenario where only one channel is available.

Secondary users may reside at different geometry locations. Whether a secondary user's request conflict with others depends on their locations and time requirement. This location-dependent conflict can be modeled by a conflict graph $H = (\mathcal{V}, E)$, where two nodes v_i and v_j form an edge (v_i, v_j) if and only if they conflict with each other. We will first study a simple case where the conflict graph is a complete graph.Then we will show that our methods still have asymptotically optimum performance guarantees as long as the conflict graph is growth-bounded by a polynomial function. A graph H is *growth-bounded* by a function f, if for any node $v \in H$ and any integer $k > 0$, the number of independent nodes within k-hops of v is at most $f(k)$.

2.2 Problem Formulation

A user from $\mathcal{V} = \{v_1, v_2, \cdots, v_n\}$ could ask for the usage of spectrum at different time while time is slotted. Each request $\mathbf{e}_i = (v, b_i, a_i, s_i, t_i)$ is claimed by a secondary user v at time a_i, bids b_i for the usage of the channel from time s_i to time $s_i + t_i$. For most of our discussions we will omit the user v when it is clear from the context, or not needed in the notation. In other words, $\mathbf{e}_i = (b_i, a_i, s_i, t_i)$ denotes the ith request. Obviously, $a_i \le s_i$ for all requests, which means requests can only bid for the future usage of channel. Each user will ask for at most Δ timeslots. We call Δ the *time ratio*.

To improve the spectrum usage and revenue, the central authority can post a requirement that all bids must be submitted in advance of a certain time slots. In this paper, we assume that for every spectrum bid $\mathbf{e}_i$, $s_i - a_i \ge \alpha$ for a given value α. Hereafter, we call α the *advance factor*. Intuitively, a larger advance factor α will give more advantage to the central authority. We later will show that the performances of our methods do not depend on α as long as $\alpha \geq \gamma$. Our system also put some condition on the central authority. The central authority should make a decision within no more than γ time slots. We call γ *delay factor*. $\gamma \leq \alpha$ makes the system meaningful. If a request has been rejected, it will never be reconsidered and accepted later.

When a request is accepted, the secondary user who issued the request will be granted the usage of channel at the price of what he bids when the first price auction is used here. When current usage of channel is terminated, penalty should be paid to compensate the preemption. The penalty $\mu(b, \ell, t)$ is linear to the unfinished time $\ell \leq t$ of that request $\mathbf{e}(v, b, a, s, t)$, *i.e.*, $\mu(b, \ell, t) = \beta \frac{\ell}{t} b$ for a constant $\beta \geq 0$. Notice that the cancelation is modeled as $\ell = t$ if the spectrum usage was not started at all. Here the constant $\beta \geq 0$ is the *penalty factor*.

As we stated before, all requests arrive on the fly, thus any information about the future, *e.g.*, the distribution of future bids, the arrival time, the start time and the required time duration are unknown. The objective is to find an allocation which maximizes the total net profit, *i.e.*, the total profit minus the total penalties caused by preemption. As we will see later that the performance of our methods and the lower bounds on the performance is affected by penalty factor β, advance factor α and delay factor γ. For certain parameters β, α and γ, we call it (β, α, γ) problem in this paper.

3 Performance Upper Bounds

3.1 Upper Bounds for $(\beta = 1, \gamma, \alpha)$ Problem

We first show the performance lower bound when the penalty factor $\beta = 1$. There are two different cases. $\gamma = o(\Delta)$ or $\gamma = \Omega(\Delta)$.

1. When $\gamma = o(\Delta)$, *i.e.*, $\lim \frac{\gamma}{\Delta} = 0$, we first show that there is no online algorithm with competitive ratio more than $\sqrt[3]{2(\gamma + 1)}\Delta^{-\frac{1}{3}}$, then improve this bound to $\sqrt{2(\gamma + 1)}\Delta^{-\frac{1}{2}}$.
2. When $\gamma = \Omega(\Delta)$, we show that $(1, \alpha, \gamma)$ problem cannot have a competitive ratio $1 - \epsilon$ for an arbitrary $\epsilon > 0$.

Due to space limit, in this section and following sections, long proofs are omitted. To check the poofs, please see our technical report [21].

Theorem 1. *There is no online algorithm with a competitive ratio more than* $\sqrt[3]{2(\gamma + 1)}\Delta^{-\frac{1}{3}}$ *for* $(\beta = 1, \gamma, \alpha)$ *problem when* $\gamma = o(\Delta)$.

Theorem 2. *There is no online algorithm with competitive ratio more than* $\frac{1}{c}$ *for* $(\beta = 1, \gamma, \alpha)$ *problem, where* $\sqrt{2(\gamma + 1)}\Delta^{-\frac{1}{2}} < \frac{1}{c} \leq \sqrt[3]{2(\gamma + 1)}\Delta^{-\frac{1}{3}}$ *for* $(\beta = 1, \gamma, \alpha)$ *problem when* $\gamma = o(\Delta)$.

Theorem 3. *There is no online algorithm with a competitive ratio* $\geq 1 - \epsilon$ *for an arbitrary small* $\epsilon > 0$, *for* $(\beta = 1, \gamma, \gamma)$ *problem when* $\gamma = \Omega(\Delta)$.

All theorems in this section are proved by contradiction using adversary model. The adversary generates a set of requests in an online fashion based on the previous decision of online algorithm. The goal of the adversary is to make the competitive ratio of that online algorithm as bad as possible.

3.2 Upper Bounds for $(\beta > 1, \gamma, \alpha)$ Problem

For $(\beta > 1, \gamma, \alpha)$ problem, we show that there is no online algorithm with competitive ratio more than $O((\gamma+1)\Delta^{-1})$ when $\gamma = o(\Delta)$; no online algorithm with competitive ratio more than $1 - \epsilon$ for an arbitrary $\epsilon > 0$ when $\gamma = \Omega(\Delta)$.

Theorem 4. *There is no online algorithm with competitive ratio more than* $\frac{2\beta(\gamma+1)}{(\beta-1)^2}\Delta^{-1}$ *for* $(\beta > 1, \gamma, \alpha)$ *problem, when* $\gamma = o(\Delta)$.

From the analysis in this section, we can see that the performance lower bounds do not depend on the *advance factor* α. In other words, no matter how many time slots the secondary users claim their requests in advance, the theoretical lower bounds will not be improved asymptotically if the *delay factor* γ does not change.

4 Online Spectrum Allocation Methods

4.1 Asymptotically Optimal Method for $(\beta = 1, \gamma, \alpha)$ Problem

Let $\mathcal{R}_a(t)$ be all requests submitted before time t. Based on the processing delay requirement, we know that all requests in $\mathcal{R}_a(t)$ must be submitted during $[t-\gamma, t)$, and the requested starting time of these requests must be during $[t-\gamma+\alpha, t+\alpha]$. Among all requests in $\mathcal{R}_a(t)$, let $\mathcal{R}(t) \subseteq \mathcal{R}_a(t)$ be all requests whose starting times are during $[t, t+\gamma]$. Recall that $\gamma \leq \alpha$. Our method only make decisions at time t using the information from $\mathcal{R}(t)$, although a superset of requests $\mathcal{R}_a(t)$ is known. We will show that our method can achieve a competitive ratio that is already asymptotically optimum.

For the set of currently known requests $\mathcal{R}(t)$, we will find some subsets using dynamic programming to optimize some objective functions. Our method will then make decisions on whether to admit these subsets of requests under some conditions involving the currently running spectrum usages. We intorduce some notations first.

Definition 1. ***Candidate Requests Set:*** *A* strong candidate requests set *at time t, denoted as $\mathcal{C}_1(t)$, is a subset of requests from $\mathcal{R}(t)$ that has the largest total bids if $\mathcal{C}_1(t)$ is allowed to run without preemption from time $t-\gamma+\alpha$ to timeslots at most $t+\alpha+\Delta$. We abuse the notation little bit here by also letting $\mathcal{C}_1(t)$ denote the profit made by $\mathcal{C}_1(t)$.*

For set $\mathcal{C}_1(t)$, let $\mathcal{P}(\mathcal{C}_1(t), t')$ denote the profit made from $\mathcal{C}_1(t)$ if these requests are admitted and then possibly being preempted at a time-slot $t' \in [t-\gamma+\alpha, t+\alpha+\Delta]$.

A weak candidate requests set *at time t, denoted as $\mathcal{C}_2(t)$, is a subset of requests from $\mathcal{R}(t)$ that has the largest total bids if $\mathcal{C}_2(t)$ is allowed to run during time interval $[t-\gamma+\alpha, t+\alpha]$ (thus, these requests may be preempted by some requests started on time-slot $t+\alpha+1$). We abuse the notation little bit here by also letting $\mathcal{C}_2(t)$ denote the profit made by $\mathcal{C}_2(t)$.*

In the rest of paper, we always use $\mathcal{C}_1(t)$ ($\mathcal{C}_2(t)$, respectively) to denote the strong (weak, respectively) candidate requests set at time t. $\mathcal{C}_1(t)$ and $\mathcal{C}_2(t)$ can be solved by dynamic programming in $O(n^3)$ time where n is the total number of requests.

At each time t, algorithm $\mathcal{G}$ should decide whether a request that arrived at time slot $t-\gamma$ will be accepted immediately. Starting time of such requests are in the interval

$[t, t+\alpha-\gamma]$. Since $\gamma \leq \alpha$, at time t, we should know all requests whose start times are from t to $t+\gamma$. Algorithm $\mathcal{G}$ takes the following inputs: a constant parameter $c_1 > 1$, an adjustable control parameter $c_2 > 0$, delay requirement γ, advance factor α, time ratio Δ, $\mathcal{R}_a(t)$, $\mathcal{R}(t)$, $\mathcal{C}_1(t)$, and $\mathcal{C}_2(t)$. It works as follows.

Given $\mathcal{R}_a(t)$, if the channel will be *empty* at time $t-\gamma+\alpha$, we find the *strong candidate requests set* $\mathcal{C}_1(t)$ with the maximum overall profit. We accept the request in $\mathcal{C}_1(t)$ with starting time $t-\gamma+\alpha$, and we say that the channel is being used by candidate requests set $\mathcal{C}_1(t)$. In other words, here we treat $\mathcal{C}_1(t)$ as a large *virtual* spectrum request, although at current time slot t we only admit the first spectrum request from $\mathcal{C}_1(t)$ while leave the admission decisions on other requests in $\mathcal{C}_1(t)$ *pending*. Whether these pending "admitted" requests will be actually admitted depend on future coming requests. If future requests are better, we will preempt this virtual request $\mathcal{C}_1(t)$, thus, some of those pending admissions will not be issued at all.

If the channel will be used by a *weak candidate requests set* $\mathcal{C}_2(t)$ at time $t-\gamma+\alpha$ and this candidate requests set *weakly preempted* (exact definitions will be given later) some other candidate requests set before, all requests from $\mathcal{R}_a(t)$ submitted at time $t-\gamma$ will *not* be admitted. Otherwise, assume the request to be run at time $t-\gamma+\alpha$ is $\mathbf{e}_j$ from some virtual candidate requests $\mathcal{C}_1(t_1)$, we find the candidate requests set $\mathcal{C}_1(t)$. The first request $\mathbf{e}_i \in \mathcal{C}_1(t)$ such that $s_i = t-\gamma+\alpha$ will be accepted only if $\mathcal{C}_1(t) \geq c_1 \cdot \mathcal{C}_1(t_1)$. In other words, we use a strong request $\mathcal{C}_1(t)$ to replace another request $\mathcal{C}_1(t_1)$. We call it a *Strong-Preemption*. If strong-preemption cannot be applied, we find the candidate requests set $\mathcal{C}_2(t)$. The request $\mathbf{e}_i \in \mathcal{C}_2(t)$ such that $s_i = t-\gamma+\alpha$ will be accepted only if $\mathcal{C}_2(t)+\mathcal{P}(\mathcal{C}_1(t_1), t-\gamma+\alpha) \geq c_2 \cdot \mathcal{C}(t_1)$. In other words, we use a weak candidate requests set $\mathcal{C}_2(t)$. to replace another virtual weak candidate requests set $\mathcal{C}_1(t_1)$. We call it a *Weak-Preemption*. In this case, all requests in $\mathcal{C}_2(t)$ will be accepted and the last request will be terminated at time $t+\alpha+1$ automatically.

If the weak-preemption cannot be applied also, we accept the request in the previously used candidate requests set $\mathcal{C}(t_1)$, whose start time is $t-\gamma+\alpha$ (if there is any) or continue the request $\mathbf{e}_j$ that will continue run through the time slot $t-\gamma+\alpha$.

In following analysis, we show that algorithm $\mathcal{G}$ is asymptotically optimal if we choose constant c_1 and control parameter c_2 carefully. To analyze the performance of our method, we first give a definition *candidate sequence*.

Definition 2. ***Candidate Sequence:*** *An* candidate sequence *is a sequence of* candidate requests sets $\mathcal{C}_1(t_i)$, $\mathcal{C}_1(t_{i+1})$, $\cdots$, $\mathcal{C}_1(t_{j-1})$, $\mathcal{C}_2(t_j)$ *or* $\mathcal{C}_1(t_i)$, $\mathcal{C}_1(t_{i+1})$, $\cdots$, $\mathcal{C}_1(t_{j-1})$, $\mathcal{C}_1(t_j)$ *which satisfies all of following three conditions.*

1. $\mathcal{C}_1(t_i)$ *does not preempt another candidate requests set;*
2. $\mathcal{C}_1(t_{i+1})$ *strongly preempts* $\mathcal{C}_1(t_i)$, $\mathcal{C}_1(t_{i+2})$ *strongly preempts* $\mathcal{C}_1(t_{i+1})$, $\cdots$, $\mathcal{C}_1(t_{j-1})$ *strongly preempts* $\mathcal{C}_1(t_{j-2})$*;*
3. $\mathcal{C}_2(t_j)$ *weakly preempts* $\mathcal{C}_1(t_{j-1})$*; or* $\mathcal{C}_1(t_j)$ *strongly preempts* $\mathcal{C}_1(t_{j-1})$ *and is not preempted by another requests set.*

Here we use the indices of the first and last *candidate requests set* to denote a *candidate sequence, e.g.* $\mathcal{S}(t_i, t_j)$. According to the definition, we can decompose the solution of algorithm $\mathcal{G}$ into multiple *candidate sequences*. Notice that each spectrum request $\mathbf{e}$ will appear in exactly one *candidate sequence*. We use $\mathcal{G}(\mathcal{S}(t_i, t_j))$ to denote the profit made on *candidate sequence* $\mathcal{S}(t_i, t_j)$ by algorithm $\mathcal{G}$. And we use $\text{OPT}[t_i, t_j]$

Algorithm 1. Online Spectrum Allocation $\mathcal{G}$

Input: A constant parameter $c_1 > 1$, an adjustable control parameter $c_2 > 0$, γ, α, Δ, $\mathcal{R}_a(t)$, $\mathcal{R}(t)$, $\mathcal{C}_1(t)$, and $\mathcal{C}_2(t)$.
Current candidate requests set $\mathcal{C}$ from time $t' < t$. Here $\mathcal{C} = \mathcal{C}_1(t')$ if $\mathcal{C}_1(t')$ strongly preempted others, or $\mathcal{C} = \mathcal{C}_2(t')$ if $\mathcal{C}_2(t')$ strongly preempted others.
Output: whether requests submitted at time $t - \gamma$ will be admitted and new *current candidate requests set* $\mathcal{C}$.

if $\mathcal{C} = \mathcal{C}_2(t')$ **then**
  if $t - t' \geq \gamma$ **then**
    $\mathcal{C} = \emptyset$;
  else
    Accept request $\mathbf{e}_i \in \mathcal{C}_2(t)$ such that $s_i = t - \gamma + \alpha$.
if $\mathcal{C} = \mathcal{C}_1(t')$ or $\emptyset$ **then**
  if $\mathcal{C}_1(t) \geq c_1 \cdot \mathcal{C}_1(t')$ **then**
    $\mathcal{C} = \mathcal{C}_1(t)$;
    Accept request $\mathbf{e}_i \in \mathcal{C}_1(t)$ such that $s_i = t - \gamma + \alpha$.
  else if $\mathcal{C}_2(t) + \mathcal{P}(\mathcal{C}_1(t'), t) \geq c_2 \cdot \mathcal{C}_1(t)$ **then**
    $\mathcal{C} = \mathcal{C}_2(t)$;
    Accept request $\mathbf{e}_i \in \mathcal{C}_2(t)$ such that $s_i = t - \gamma + \alpha$.
  else
    Accept request $\mathbf{e}_i \in \mathcal{C}_1(t')$ such that $s_i = t - \gamma + \alpha$.

to denote the profit made by optimal offline algorithm on the requests whose starting times are in interval $[t_i, t_j]$.

Lemma 1. *For each* candidate sequence $\mathcal{S}(s_i, s_j)$ *in the solution given by algorithm* $\mathcal{G}$, *we have*

$$\mathcal{G}(\mathcal{S}(s_i, s_j)) \geq \min(c_1, c_2)\mathcal{C}_1(s_{j-1})$$

Proof. $\mathcal{C}_1(s_j)$ either strongly preempted $\mathcal{C}_1(s_{j-1})$ or weakly preempted $\mathcal{C}_1(s_{j-1})$. In first case, $\mathcal{G}$ makes at least $\mathcal{C}_1(s_j) \geq c_1 \cdot \mathcal{C}_1(s_{j-1}))$. otherwise, $\mathcal{G}$ makes at least $\mathcal{P}(\mathcal{C}_1(s_{i-1}), s_j) + \mathcal{C}_2(s_j) \geq c_2 \cdot \mathcal{C}_1(s_{j-1})$. So our lemma holds for either case.

Lemma 2. *For each* candidate sequence $\mathcal{S}(s_i, s_j)$ *in the solution given by algorithm* $\mathcal{G}$, *for each* $i \leq k < j$, *we have*

$$\mathrm{OPT}[s_k, s_{k+1}] \leq (c_1 + c_2 + \frac{c_2}{\sqrt{(\gamma+1)/\Delta}})\mathcal{C}_1(s_k)$$

Lemma 3. *For each* candidate sequence $\mathcal{S}(s_i, s_j)$ *in the solution given by algorithm* $\mathcal{G}$, *we have*

$$\mathrm{OPT}[s_i, s_j] \leq (c_1 + 1 + \frac{1}{c_1 - 1})(c_1 + c_2 + \frac{c_2}{\sqrt{(\gamma+1)/\Delta}})\mathcal{C}_1(s_{j-1})$$

Proof. Obviously $\mathrm{OPT}[s_i, s_j] = \sum_{k=i}^{j-1} \mathrm{OPT}[s_k, s_{k+1}]$. From Lemma 2, we have $\mathrm{OPT}[s_i, s_j] \leq (c_1 + c_2 + \frac{c_2}{\sqrt{(\gamma+1)/\Delta}}) \sum_{k=i}^{j-1} \mathcal{C}_1(s_k)$. Based on the condition of strong-preemption, we have $\mathcal{C}_1(s_k) \geq c_1 \cdot \mathcal{C}_1(s_{k-1})$ for all $i \leq k \leq j$. The lemma then follows.

Theorem 5. *Algorithm $\mathcal{G}$ is $\Theta(\sqrt{\gamma+1}\Delta^{-\frac{1}{2}})$-competitive when $\gamma = O(\Delta)$.*

Notice that algorithm $\mathcal{G}$ is $\Theta(1)$-competitive when $\gamma = \omega(\Delta)$, which is also asymptotically optimal. Let $n(t)$ be the cardinality of $\mathbf{e}_a(t)$. We also have following theorem.

Theorem 6. *Algorithm 1 takes $O(n(t)^3)$ to make decisions at a time instant t. Algorithm 1 takes $O((\gamma+\Delta)n^3)$ to make decisions on all n online requests.*

4.2 Asymptotically Optimal Method for $(\beta > 1, \gamma, \alpha)$ Problem

In this subsection, we propose a greedy algorithm $\mathcal{H}$ for $(\beta > 1, \gamma, \alpha)$ Problem, where $\mathcal{H}$ is asymptotically optimal.

Algorithm 2. Online Spectrum Allocation $\mathcal{H}$

Input: A constant parameter $c > 1+\beta$, γ, α, Δ, $\mathcal{R}_a(t)$, $\mathcal{R}(t)$, $\mathcal{C}_1(t)$. Previous *current candidate requests set* $\mathcal{C} = \mathcal{C}_1(t')$ where $t' < t$. Here $\mathcal{C}_1(t')$ may be empty.
Output: whether requests submitted at time $t-\gamma$ will be admitted and new *current candidate requests set* $\mathcal{C}$.

1: **if** $\mathcal{C}_1(t) \geq c \cdot \mathcal{C}_1(t')$ **then**
2: $\quad\mathcal{C} = \mathcal{C}_1(t)$;
3: $\quad$Accept request $\mathbf{e}_i \in \mathcal{C}_1(t)$ such that $a_i = t - \gamma$.
4: **else**
5: $\quad$Accept request $\mathbf{e}_i \in \mathcal{C}_1(t')$ such that $a_i = t - \gamma$.

Theorem 7. *Algorithm $\mathcal{H}$ is $\frac{(c-\beta-1)}{c^2}\frac{\gamma+1}{\Delta+\gamma+1}$ competitive.*

Theorem 8. *Method $\mathcal{H}$ is at least $\frac{a}{4(1+a)(1+\beta)}$-competitive (by choosing $c = 2(1+\beta)$), when $\gamma = a\Delta - 1$.*

4.3 General Conflict Graphs

Our algorithms can be easily extended for the case where the conflict graph has a bounded growth. The details are omitted due to space limit. Assume the bounded one-hop independent number of conflict graph H is λ. We have following theorem.

Theorem 9. *Algorithm $\mathcal{G}'$ is $\Theta(\sqrt{\gamma+1}\Delta^{-\frac{1}{2}})$-competitive for $\beta = 1$.*

5 Dealing with Selfish Users

When each secondary user declares his request, he may lie on the bid, and time requirement. We need to design rules such that each secondary user has incentives to declare his request truthfully. Each secondary user i has its own private information $\mathbf{t}_i$, including b_i, a_i, and t_i. Let $\mathbf{a}_i = (b'_i, a'_i, t'_i)$ be the value he will report.

For each vector of actions $\mathbf{a} = (\mathbf{a}_1, \mathbf{a}_2, \cdots, \mathbf{a}_n)$, a mechanism $M = (\mathcal{A}, P)$, computes a spectrum allocation $\mathcal{A}(a) = (\mathcal{A}_1(a), \mathcal{A}_2(a), \cdots, \mathcal{A}_n(a))$ and a payment vector $\mathbf{p}(a) = (\mathbf{p}_1(a), \mathbf{p}_2(a), \cdots, \mathbf{p}_n(a))$. Each user i will be allocated $\mathcal{A}_i(a)$ and be charged $\mathbf{p}_i(a)$.

Assume that no user will delay his/her spectrum request and a user will not lie about t_i and s_i. Consider a user i, assume the bids of all other users remain the same. Let $\underline{b}_i$ be the minimum bid that i has to bid to get admitted when its spectrum request is to be processed at γ timeslots later. Let $\overline{b}_i$ be the minimum bid that i has to bid to get admitted and not get preempted later. Clearly, $\underline{b}_i \leq \overline{b}_i$. $\underline{b}_i$ and $\overline{b}_i$ can be computed in polynomial time since all other bid values are known. Then the final profit of user i is

$$utility(i) = f \cdot b_i - \mathbf{p}_i + \mu(b'_i, \ell'_i, t_i).$$

Here if i is rejected, we have $f = 0$, $\mathbf{p}_i = 0$, and $\mu(b'_i, \ell'_i, t_i) = 0$; if i is admitted and later preempted, $f = 1 - \beta\frac{\ell'_i}{t_i}$ and $\mu(b'_i, \ell'_i, t_i) = \beta\frac{\ell'_i}{t_i}b'_i$, where ℓ'_i is the unserved time of its spectrum request; if i is admitted and not preempted, $f = 1$ and $\mu(b'_i, \ell'_i, t_i) = 0$. The payment $\mathbf{p}_i$ is always the upfront charge from i for being admitted.

It is a forklore result that the allocation method in a mechanism must have the monotone property. Here an allocation method $\mathcal{A}$ is monotone if a user i is granted the spectrum usage under $\mathcal{A}$ with a bid $\mathbf{e}_i = (b_i, a_i, t_i)$, then user i will still be granted under $\mathcal{A}$ if he increases b_i, and/or decreases t_i. First, our method (Algorithm 1) does have the monotone property. In our algorithm, we need to find strong candidate requests set, and weak candidate requests set by dynamic programming which can be shown as monotone. Thus, we can design a mechanism using our algorithms ($\mathcal{G}$ and $\mathcal{H}$) as allocation methods as follwoing

$$\begin{cases} \text{Use Algorithm } \mathcal{G} \text{ or } \mathcal{H} \text{ as allocation method} \\ \text{charge an admitted user } i \text{ a payment } \mathbf{p}_i = \underline{b}_i \end{cases}$$

Theorem 10. *In our mechanism using Algorithm $\mathcal{G}$ or $\mathcal{H}$ as the spectrum allocation method, to maximize its profit, every secondary user will not bid a price lower than its actual value.*

Observe that, in our scheme, the only scenario that a secondary user can gain benefit is when $b_i \in (\underline{b}_i, \overline{b}_i)$ and it bids a value $b'_i \in (b_i, \overline{b}_i)$.

6 Literature Reviews

The allocation of spectrums is essentially combinatorial allocation problem, which have been well studied [14, 1]. Yuan *et al.* [18] used time-spectrum block to model spectrum reservation in cognitive radio networks, and presented both centralized and distributed protocols. Li *et al.* [15] designed efficient methods and truthful mechanism for various dynamic spectrum assignment problems. Zhou *et al.* [19] propose a truthful and efficient dynamic spectrum auction system to serve many small players. In [20], Zhou and Zheng designed truthful double spectrum auctions where multiple parties can trade spectrum. All these results are based on offline models.

Our problem is also similar to online job scheduling problems. Various online scheduling problems focus on optimizing different objective functions. The most common objective function is *makespan*, which is the length of the schedule. Suppose that given m *identical* machines, jobs arrive one by one and no preemption is allowed. A number of results have been proved to improve the upper bounds [12, 7] and lower bounds [11]. Closing the gap between the best lower bound (1.88 [11]) and the upper bound (1.9201 [7]) is an open problem. All these results assume that preemption is not allowed and they focus on minimizing *makespan*. Many authors [16,8] also investigated the case where preemption is allowed without penalty. Online scheduling problem in which we pay penalty for rejecting jobs was first studied in [3] by Bartal *et al.* and improved later in [10] by Hoogeveen *et al.*.

For the model with deadline, it is usually impossible to finish all jobs. Thus, another model aims to maximize the profit. In 1991, Baruah *et al.* [4] proved that no online scheduling algorithm can make profit more than $\frac{1}{(1+\sqrt{D})^2}$ times the optimal. Koren *et al.* [13] gave an algorithm matching the lower bound [4]. Hoogeveen *et al.* [9] gave a $\frac{1}{2}$-competitive algorithm which maximizes the number of early jobs where preemption with no penalty is allowed. Chrobak *et al.* [5] gave a $\frac{2}{3}$-competitive algorithm which maximizes the number of jobs that have uniform length in the *preemption-restart* model.

The work that is most similar to our work is a recent result by Constantin *et al.* [6] in 2009. They proposed and studied a simple model for auctioning ad slot reservations in advance. A seller will display a set of slots at some point T in the future. Until T, bidders arrive sequentially and place a bid on the slots they are interested in. The seller must decide immediately whether or not to grant a reservation. Their model allows the seller to cancel at any time any reservation made earlier with a penalty proportional to the bid value. The major difference between our model and their model is that, in their model, they only auction a set of ad slots for a fixed time-slot T, while in our model, the bidders could bid the spectrum usage starting from any time-slot, and lasting for an arbitrary duration. We also consider preemption during the spectrum usage.

7 Conclusions

In this paper, we studied online spectrum allocation for wireless networks. For a number of variants, we designed efficient online scheduling algorithms and analytically showed that the competitive ratios of our methods are within small constant factors of the optimum. Especially, when γ is around the maximum requested time duration Δ, our algorithm results in a profit that is almost optimum. We also conducted extensive simulations to study the performances of our methods and our results show that they perform extremely well in practice.

We showed that no user will bid lower than its willing payment under our mechanism. We would like to study the Nash Equilibriums of our mechanism and investigate the price of anarchy of our mechanism. It remains open to design a mechanism in which every secondary user cannot gain benefit by bidding untruthfully. It is also interesting to extend our mechanism to deal the case when we know more information about requests, such as the distributions of bids, timeslots requested, and arrival times.

References

1. Archer, A., Papadimitriou, C., Talwar, K., Tardos, E.: An approximate truthful mechanism for combinatorial auctions with single parameter agents. Internet Mathematics 1(2), 129–150 (2004)
2. Archer, A., Tardos, E.: Truthful mechanisms for one-parameter agents. In: IEEE FOCS, p. 482 (2001)
3. Bartal, Y., Leonardi, S., Marchetti-Spaccamela, A., Sgall, J., Stougie, L.: Multiprocessor scheduling with rejection. In: ACM SODA, pp. 95–103 (1996)
4. Baruah, S., Koren, G., Mishra, B., Raghunathan, A., Rosier, L., Shasha, D.: On-line scheduling in the presence of overload. In: IEEE FOCS, pp. 100–110 (1991)
5. Chrobak, M., Jawor, W., Sgall, J., Tichy´, T.: Online scheduling of equal-length jobs: Randomization and restarts help. SIAM J. Comput. 36(6), 1709–1728 (2007)
6. Constantin, F., Feldman, J., Muthukrishnan, S., Pal, M.: An online mechanism for ad slot reservations with cancellations. In: ACM SODA, pp. 1265–1274 (2009)
7. Fleischer, R., Wahl, M.: On-line scheduling revisited. J. of Scheduling 3, 343–353 (2000)
8. Goemans, M.X., Wein, J.M., Williamson, D.P.: A 1.47-approximation algorithm for a preemptive single-machine scheduling problem. Operations Research Letters 26(4), 149–154 (2004)
9. Hoogeveen, H., Potts, C.N., Woeginger, G.J.: On-line scheduling on a single machine: maximizing the number of early jobs. Operations Research Letters 27(5), 193–197 (2000)
10. Hoogeveen, H., Skutella, M., Woeginger, G.J.: Preemptive scheduling with rejection. In: Paterson, M. (ed.) ESA 2000. LNCS, vol. 1879, pp. 268–277. Springer, Heidelberg (2000)
11. John, F., Rudin, I., Chandrasekaran, R.: Improved bounds for the online scheduling problem. SIAM J. Comput. 32(3), 717–735 (2003)
12. Karger, D.R., Phillips, S.J., Torng, E.: A better algorithm for an ancient scheduling problem. In: ACM SODA, pp. 132–140 (1994)
13. Koren, G., Shasha, D.: Dover: An optimal on-line scheduling algorithm for overloaded uniprocessor real-time systems. SIAM J. Comput. 24(2), 318–339 (1995)
14. Lehmann, D.J., O'Callaghan, L.I., Shoham, Y.: Truth revelation in approximately efficient combinatorial auctions. In: ACM Conf. on Electronic Commerce, pp. 96–102 (1999)
15. Li, X.-Y., Xu, P., Tang, S., Chu, X.: Spectrum bidding in wireless networks and related. In: Hu, X., Wang, J. (eds.) COCOON 2008. LNCS, vol. 5092, pp. 558–567. Springer, Heidelberg (2008)
16. Phillips, C.A., Stein, C., Wein, J.: Scheduling jobs that arrive over time (extended abstract). In: Sack, J.-R., Akl, S.G., Dehne, F., Santoro, N. (eds.) WADS 1995. LNCS, vol. 955, pp. 86–97. Springer, Heidelberg (1995)
17. Stine, J.A.: Spectrum management: The killer application of ad hoc and mesh networking. In: IEEE DySPAN (2005)
18. Yuan, Y., Bahl, P., Chandra, R., Moscibroda, T., Wu, Y.: Allocating dynamic time-spectrum blocks in cognitive radio networks. In: ACM MobiHoc, pp. 130–139 (2007)
19. Zhou, X., Gandhi, S., Suri, S., Zheng, H.: eBay in the Sky: strategy-proof wireless spectrum auctions. In: ACM MobiCom, pp. 2–13 (2008)
20. Zhou, X., Zheng, H.: Trust: A general framework for truthful double spectrum auctions. In: IEEE INFOCOM (2009)
21. Xu, P., Li, X.: SOFA: Strategyproof Online Frequency Allocation for Multihop Wireless Networks, http://www.cs.iit.edu/~xli/paper/spectrum-sofa.pdf

1-Bounded Space Algorithms for 2-Dimensional Bin Packing

Francis Y.L. Chin*, Hing-Fung Ting**, and Yong Zhang

Department of Computer Science, The University of Hong Kong, Hong Kong
{chin,hfting,yzhang}@cs.hku.hk

Abstract. In this paper, we study the bounded space variation, especially 1-bounded space, of 2-dimensional bin packing. A sequence of rectangular items arrive over time, and the following item arrives after the packing of the previous one. The height and width of each item are no more than 1, we need to pack these items into unit square bins of size 1×1 and our objective is to minimize the number of used bins. Once an item is packed into a square bin, the position of this item is fixed and it cannot be shifted within this bin. At any time, there is at most one active bin; the current unpacked item can be only packed into the active bin and the inactive bins (closed at some previous time) cannot be used for any future items. We first propose an online algorithm with a constant competitive ratio 12, then improve the competitive ratio to 8.84 by the some complicated analysis. Our results significantly improve the previous best known $O((\log \log m)^2)$-competitive algorithm[10], where m is the width of the square bin and the size of each item is $a \times b$, where a, b are integers no more than m. Furthermore, the lower bound for the competitive ratio is also improved to 2.5.

1 Introduction

Bin packing [1–5, 7–22] is one of the fundamental problems in theoretical computer science. The field of bin packing has been studied for more than thirty years and many problems still remain open.

In the bin packing problem, a sequence of items are packed into bins, the packed items in a bin do not overlap and all placed within a bin. The objective is to minimize the number of used bins. The size of each item should be no more than the size of a bin. Most of previous results have considered the *unbounded space* variation, i.e., each bin can be used for packing the coming item if this bin has enough space. In the *unbounded space* variation, each item can be packed into any bins. But in many applications, the number of active bins is always bounded by a constant, so, *bounded space* variation is valuable, too. In the *bounded space* variation, the current item can be only packed into the active bins. Once the active bins cannot accommodate the coming item, we have to close an active bin and a new bin will be opened and labeled with *active*.

* Supported by HK RGC grant HKU-7113/07E.
** Supported by HK RGC grant HKU-7171/08E.

Y. Dong, D.-Z. Du, and O. Ibarra (Eds.): ISAAC 2009, LNCS 5878, pp. 321–330, 2009.

It is an interesting problem that the number of active bin is only one (*1-bounded space*), i.e., each item can be packed either in the only active bin or in a new bin. This variant is motivated from grid computing, in which there is a cluster of computers arranged on a grid. Each job requests for a subgrid of $a \times b$ computers, whose orientation is not important (as long as a subgrid of $a \times b$ computers are assigned) and where the assignment in the grid cannot be changed. There is only one grid machine and the jobs come online. So we try to pack as many jobs as possible in this fixed size grid. When these jobs are finished, the grid can be used for another set of jobs. In this variant of bin packing problem, each item is rectangular in shape and its width and height are no more than 1. Items must be packed into square bins of size 1×1. Again, the target is to minimize the number of square bins used. For example, as shown in Fig. 1(a), there are 5 items to be packed into bins of size 1×1, item A, B, C and D are of the same size 0.75×0.25, while item E is of size 0.5×0.5. In the optimal solution, these five items can be packed into one bin (Fig. 1(b)), if we do not pack them carefully, the number of used bin may be more than one (Fig. 1(c)).

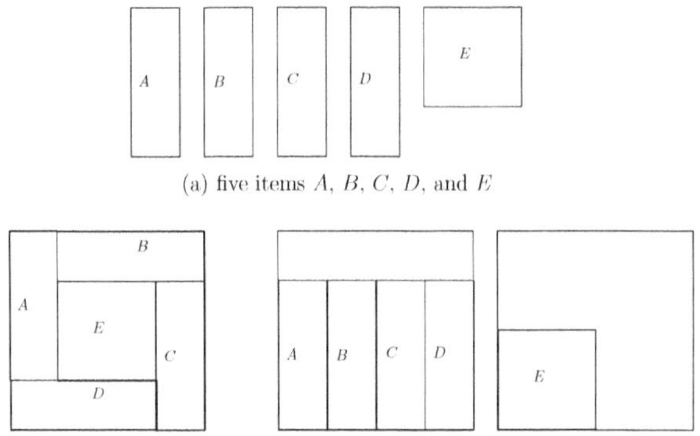

(a) five items A, B, C, D, and E

(b) optimal packing into one 4×4 grid (c) non-optimal packing into two 4×4 grids

Fig. 1. Example of optimal packing and non-optimal packing

The focus in this paper is the online variant of 1-bounded space 2-dimensional bin packing, where the items arrive over time, and when packing the current item, we have no information about the future items and the position of the packed items in the bin cannot be changed.

To evaluate an online algorithm for bin packing, we use the *asymptotic competitive ratio* defined as follows. Consider an online algorithm A and an optimal offline algorithm OPT. For any sequence S of items, let $A(S)$ be the cost (number of square bins used) incurred by algorithm A and let $OPT(S)$ be the corresponding optimal cost from OPT. Then the *asymptotic competitive ratio* for algorithm A is:

$$R_A^\infty = \lim_{k \to \infty} \sup_S \{ \frac{A(S)}{OPT(S)} | OPT(S) = k \}.$$

Bin packing has been well-studied. For the online one-dimensional case, Johnson et al. [14] showed that the First Fit algorithm (FF) has an asymptotic competitive ratio of 1.7. Yao [22] improved algorithm FF with a better upper bound of 5/3. Lee and Lee [15] introduced the class of Harmonic algorithms, and showed that an asymptotic competitive ratio of 1.63597 was achieved. Ramanan et al. [19] further improved the upper bound to 1.61217. The best known upper bound so far is from the Super Harmonic algorithm by Seiden [20] whose asymptotic competitive ratio is at most 1.58889. As for lower bound results, Yao [22] showed that no online algorithm has asymptotic competitive ratio less than 1.5. The best known lower bound to date is 1.54014 [21].

As for two-dimensional online bin packing, the best lower bound up to now is 1.907 [3], while the best known upper bound is 2.5545 [11]. We also briefly overview the offline results on two-dimensional bin packing. Chung et. al. [5] showed an approximation algorithm with an asymptotic performance ratio of 2.125. Caprara [4] improved the upper bound to 1.69103. Very recently Bansal et al. [2] derived a randomized algorithm with asymptotic performance ratio of at most 1.525. As for the offline lower bound results, Bansal et al. [1] showed that the two-dimensional bin packing problem does not admit any asymptotic polynomial time approximation scheme. For the special case where items are squares, there are also many results [8, 9, 12, 13, 16–18].

For bounded space online algorithms, Csirik and Johnson [6] presented an 1.7-competitive algorithm (K-Bounded Best Fit algorithms (BBF_K)) for one dimensional bin packing while the number of active bins $K \geq 2$; Epstein et al. [7] gave a 1.69103^d-competitive algorithm with $(2M-1)^d$ active bins, where $M \geq 1/(1-(1-\varepsilon)^{1/(d+2)})-1$, $\varepsilon > 0$ and d is the dimension of the bin packing problem. For the *1-bounded space* variant, Fujita [10] gave an $O((\log\log m)^2)$-competitive algorithm, where m is the width of the square bin and the size of each item is $a \times b$, where a, b are integers no more than m. He also proved that the competitive ratio for the 1-bounded space variant is at least 23/11.

The remainder of this paper is organized as follows. Section 2 gives some preliminary background of this problem, and the formal definition of the problem we address. In Section 3, we propose a 12-competitive algorithm, and in Section 4, we give a more complicated analysis and further improve the competitive ratio to 8.84. In Section 5, we show a lower bound of 2.5 for the competitive ratio of any algorithm for the 1-bounded space variant. Section 6 concludes this paper and gives some future research on this problem.

2 Preliminary

Let $S = \{r_1, r_2, ..., r_n\}$ denote the sequence of rectangle items, and let x_i and y_i be the width and height of item r_i, respectively, where x_i and y_i are positive and no more than 1. The size of square bin for packing these items is 1×1. The *occupied space (or occupation) of item* r_i is the product of x_i and y_i, i.e., $x_i y_i$, and the *occupied space (or occupation) of a packing in a bin* is the total occupied space of items in it. The bin packing problem is to pack all the items into bins without overlaps and the target is to minimize the number of bins used.

Fujita [10] proved that the competitive ratio for 1-bounded space version is $2m$ if rotation of items is not allowed. But an interesting problem is: can we have a better algorithm with a smaller competitive ratio if rotation of items is allowed? In this paper, we consider the 1-bounded space variant and rotation of items is allowed. Thus, we may assume that the width of each item is no less than its height, i.e., $x_i \geq y_i$.

In Fujita's algorithm [10], items are classified into three types $\{A : m/2 < x \leq m\}$, $\{B : m/\log_2 m < x \leq m/2\}$ and $\{C : 0 < x \leq m/\log_2 m\}$ (assuming $x \geq y$), and the square bin is partitioned into three disjoint parts. The packing strategies for each type are different and packing particular type of items in distinct part of a bin is based on the partition of the square bin. The worst case of the algorithm happens in handling the type B items, where the packing area is further partitioned into strips of equal height $m/\log\log m$. Items with height less than $m/\log\log m$ are packed within strips using a greedy approach. If the height of the newly arrived type B item is larger than $m/\log\log m$, the current active bin will be closed and a new bin will be opened for the unpacked item. Thus, the occupied space in the closed bin and the new created bin is at least $(m/\log\log m)^2$, which leads to an $O((\log\log m)^2)$-competitive algorithm. The packing of type A items is also based on a greedy approach, and the occupied space is at least $O(m^2/\log\log m)$. Thus, in the worst case, most area in the closed bin is wasted.

To efficiently use the area of each bin, we also classify the items into three types according to their width, and each bin is partitioned into two parts U and L. Items with larger or medium widths are packed into part U of the bin, and the narrow items are packed into part L of the bin. By setting the proper sizes of U and L, we can prove that the occupied space in each bin is never too small, and is at least a constant portion of the entire bin. The detailed description and analysis of our strategies will be given in Section 3 and 4.

3 Constant-Competitive Algorithm

Classify the rectangular items into three classes A, B and C, such that

$$\begin{aligned} A &= \{(x,y)|x \geq 1/2\}, \\ B &= \{(x,y)|1/6 \leq x < 1/2\}, and \\ C &= \{(x,y)|x < 1/6\}. \end{aligned}$$

For simplicity, let A-item denote an item belonging to class A, similarly for B-item and C-item.

Each square bin is partitioned into two parts: the upper part U and the lower part L, let u and l denote the heights of U and L respectively (Fig. 2).

In our packing strategy, A-items and B-items are packed into the upper part U, and C-items into the lower part L. If an item a cannot be packed into the corresponding part according to the strategy, we close the current active bin then open a new bin to pack item a.

Next, we will describe how to pack items into each part.

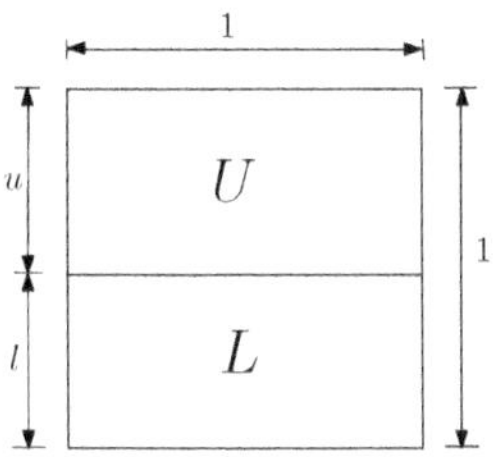

Fig. 2. Partition the bin into upper part U and lower part L

3.1 Packing Items in Upper Part U

Firstly, we consider the packing of A-items and B-items into the upper part U.

The A-items are packed by a top-down order in U, and the vertical symmetry axis of each item aligns with the vertical symmetry axis of the square bin. The B-items are packed by a bottom-up order in both left and right side of U while balancing the height of both sides, i.e., the new B-item is packed into the side of smaller height. If packing a new item leads to an overlap with other items, we close this bin and open a new square bin for this new item. For example, the current configuration of upper part U is shown in Fig. 3, the height of packed A-items is y, the left and right side of packed B-items are of height y_1 and y_2 respectively. Note that the width of each packed item in U is no less than its height, it is easy to show that the current occupied space in U is at least $(y/2 + y_1/6 + y_2/6)$. W.l.o.g., assume $y_1 \geq y_2$, we will try to pack a new B-item on the right side of U.

The following lemma shows that about 1/3 of the space in U of any inactive bins are occupied by the items if the inactive bins are due to the arrival of either A- or B- items. Intuitively, any horizontal strip of U is either covered by A-items or B-items. As the width of A-item $\geq 1/2$ and the total width of B-items is $2 \cdot 1/6 = 1/3$. So no less than 1/3 of the space will be covered by the A- or B-items. Note that, in the worst case, the B-items at the left and right side might not be of equal height, some of the horizontal strip can only be covered by one B-item and thus the occupied space can be strictly less than 1/3.

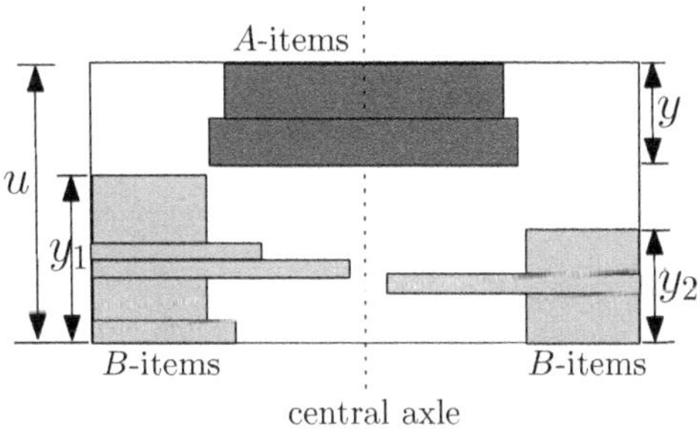

Fig. 3. Packing items into the upper part U

Lemma 1. *When the packing strategy cannot satisfy a newly arrived item a in the upper part U, the occupied space in U and the item a is at least $u/3 - 1/36$.*

The value $(u/3 - 1/36)$ is a lower bound of the occupied space of the packed items in U and the current unpacked item a. But we have to open a new bin for the unpacked item a, otherwise, overlap will appear. Thus, the average occupied space of the active bin and the bin to be opened next is at least half of the above value, i.e., $(u/6 - 1/72)$.

3.2 Packing Items in Lower Part L

Now we consider how to pack the C-items into the lower part L.

We further partition the C-items into subclasses C_0, C_1, C_2, ..., an item $x \times y$ belongs to subclass C_i if $2^{-i-1}/6 < x \leq 2^{-i}/6$. Let w_i denote the maximal possible width of items from subclass C_i. Thus, $w_0 = 1/6$, $w_1 = 1/12$, ... Based on this partition, the lower part L can be partitioned into columns of width w_i $(i > 0)$. Each item belonging to subclass C_j will be packed into a column of width w_j. Let f_l be the width of free space in right right part of L. When handling an item from subclass C_i, a new column of width w_i will be created if the existed columns of width w_i cannot satisfy this item. If the lower part L has not enough space to create a new column, i.e., $f_l < w_i$, we close the current active bin and open a new square bin for the current unpacked item.

For example, assume that the current configuration of the lower part L is shown in Fig. 4, there are three columns for subclass C_i and three columns for subclass C_j. If a new item of subclass C_i comes, it can be packed into the third column for C_i from left. If a new item of subclass C_j comes, the three columns for C_j are all full, we have to create a new column for C_j so as to allocate the new item, and the width of free space is decreased by w_i. If the free space cannot satisfy the creation of new columns, i.e., $f_l < w_i$, this square bin will be closed and a new one will be opened.

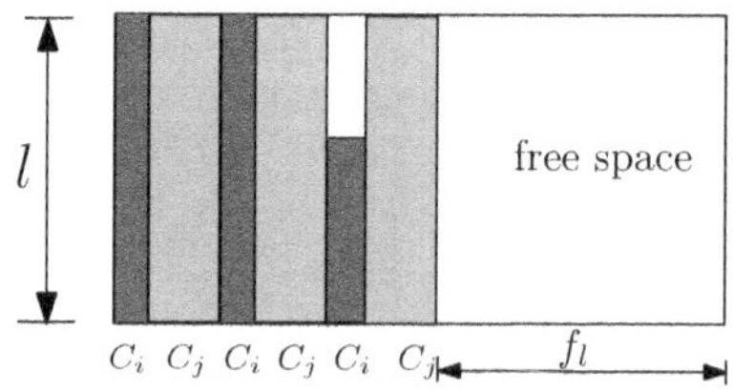

Fig. 4. Packing items into the lower part L

Lemma 2. *The occupied space in L is at least $(l/3 - 1/18)$ if L cannot accommodate a C-item.*

Based on the above two lemmas and note that $u + l = 1$, we can have the following conclusion.

Theorem 1. *The above described strategy is 12-competitive for 1-bounded space 2-dimensional bin packing when $u = 7/12$ and $l = 5/12$.*

Proof. From the above analysis, we have the following two facts:

(1) The average occupied space in U is at least $(u/6 - 1/72)$ if U cannot satisfy an item from class A or B. (Lemma 1)
(2) The average occupied space in L is at least $(l/3 - 1/18)$ if L cannot satisfy an item from class C. (Lemma 2)

Thus, the average occupied space of each bin is at least $\min\{(u/6-1/72),(l/3-1/18)\}$. Since $u+l=1$, $\min\{(u/6-1/72),(l/3-1/18)\} \geq 1/12$ when $l = 5/12$ and $u = 7/12$.

Therefore, the average occupied space of each bin is at least 1/12 of the total bin, and our strategy is 12-competitive. □

4 Further Improvement

In this section, we will give a more complicated analysis for the previous strategy and show that the algorithm is more competitive by modifying the values of u (the height of upper part) and l (the height of lower part).

For a given sequence of items, suppose the number of bins used by the packing strategy is n, let o_A^i, o_B^i and o_C^i be the occupied space of A-, B- and C- items in the i-th bin respectively. Define $\mathcal{U}$ as the set of bins which cannot satisfy the next A- or B- item. Define $\mathcal{L}$ as the set of bins which cannot satisfy the next C-item. It is easy to see that $\mathcal{U}$ and $\mathcal{L}$ are disjoint sets, and $|\mathcal{U}|+|\mathcal{L}|+1 = n$ since the last bin does not belong to either $\mathcal{U}$ or $\mathcal{L}$. The average occupation for all bins is $\sum_{i=1}^{n}(o_A^i+o_B^i+o_C^i)/n$. When n is very large, we can regard this average occupation as $\sum_{i=1}^{n}(o_A^i+o_B^i+o_C^i)/(|\mathcal{U}|+|\mathcal{L}|)$. It is easy to see that

$$\frac{\sum_{i=1}^{n}(o_A^i+o_B^i+o_C^i)}{|\mathcal{U}|+|\mathcal{L}|} \geq \min\{\frac{\sum_{i=1}^{n}(o_A^i+o_B^i)}{|\mathcal{U}|}, \frac{\sum_{i=1}^{n}(o_C^i)}{|\mathcal{L}|}\} \quad (1)$$

Now we compute the lower bounds for these two terms in the right part of above inequality.

Firstly, we consider the first term $\sum_{i=1}^{n}(o_A^i+o_B^i)/|\mathcal{U}|$ in inequality 1.

Let p_A^i and q_A^i be the disjoint occupations of A-items in the i-th bin, explicitly, p_A^i and q_A^i are the half occupied space of A-items in the i-th bin. For example, if there is an A-item of height h in a bin, p_A^i and q_A^i are at least $h/4$. Similarly, let p_B^i and q_B^i be the disjoint occupations of B-items in the i-th bin. Denote q_B^i be the half occupied space of either left side or right side with larger occupied space, and p_B^i be the remaining occupations of B-items in the i-th bin. For example, in the configuration shown in Fig. 3, if the occupied space in the right side with height y_2 is larger than the occupied space in the left side, q_B^i is equal to the value of half occupied space of the right side, which is at least $y_2/12$. Note that the side with higher height may not be the side with larger occupied space. Thus, we have

$$\frac{\sum_{i=1}^{n}(o_A^i + o_B^i)}{|\mathcal{U}|} = \frac{\sum_{i=1}^{n}(p_A^i + q_A^i + p_B^i + q_B^i)}{|\mathcal{U}|} \geq \min_{i\in\mathcal{U}}\{p_A^i + p_B^i + q_A^{i+1} + q_B^{i+1}\}$$

Similar to Section 3, we have the following lemma.

Lemma 3. $\min_{i\in\mathcal{U}}\{p_A^i + p_B^i + q_A^{i+1} + q_B^{i+1}\} \geq u/4 - 1/32$.

Then, we consider the second term $\sum_{k=1}^{n}(o_C^k)/|\mathcal{L}|$ in inequality (1).

Note that $\sum_{k=1}^{n}(o_C^k)/|\mathcal{L}| \geq \min_{k\in\mathcal{L}}\{o_C^k\}$, and we can compute the lower bound of o_C^k, as shown in Lemma 4.

Lemma 4. $o_C^k \geq l/3 - 1/36$ *for any* $k \in \mathcal{L}$.

Now we give the upper bound of competitive ratio of our strategy.

Theorem 2. *The strategy is 8.84-competitive for 1-bounded space 2-dimensional bin packing when* $u = 0.5773$ *and* $l = 0.4227$. *Furthermore, this strategy is tight.*

Proof. From Lemma 3 and Lemma 4, we have the following two facts:

(1) The first term of Inequality 1 is at least $u/4 - 1/32$. (Lemma 3)
(2) The second term of Inequality 1 is at least $l/3 - 1/36$. (Lemma 4)

Balancing these two values, we can achieve the balanced average occupation 0.1131 for all bins when $u = 0.5774$ and $l = 0.4226$. That means the average occupation in each bin is at least 0.1131 of the total bin. Thus, our strategy is 8.84-competitive.

For the worst case analysis of this strategy, consider the packing in upper part U. A sequence of items $(X, Y, X, Y, X, Y, ...)$ arrive over time. Item X belongs to class B and item Y belongs to class A. The size of X is 0.25×0.25, while the size of Y is $0.5 \times (0.3274 + \epsilon)$, where ϵ is a very small value. According to the strategy, X and Y cannot be packed into the same bin. Thus, the amortized occupation in each bin is $(0.25^2 + 0.5 \cdot (0.3274 + \epsilon))/2 = 0.1131 + 0.25\epsilon$. Therefore, we can say the strategy is tight. □

5 Lower Bound for 1-Bounded Space Algorithms

In this section, we will show that the lower bound of competitive ratio for any 1-bounded space algorithm is 2.5, which improves the previous bound 23/11.

Consider the input sequence of rectangle items:

$$(A_1, B_1, A_2, B_1, ...A_{2k+2}, B_1, X_1, X_2, X_3, k \cdot B_2, Y_1, Y_2, 2k \cdot C).$$

The size of item A_i is $(1/3 + a_i, 2/3 + x)$. The size of B_1 and B_2 are $(1/3 - \varepsilon, 1/3 - \varepsilon)$ and $(1/3 + 2\varepsilon, 1/3 + 2\varepsilon)$ respectively. The size of item C is $(1/3 + \varepsilon/2, 1/3 - x)$. The sizes of X_1, X_2, X_3, Y_1 and Y_2 are $(1/3 + \varepsilon - a_{2k+2}, 2/3 + x)$, $(1/3 - \varepsilon, 2/3 + \varepsilon + x)$, $(1/3 - x, 1)$, $(1/3 - 4\varepsilon, 1)$ and $(2/3 + 4\varepsilon, 1/3 - \varepsilon)$.

The sizes of these rectangle items must satisfy the following constraints.

1. $a_1 = \varepsilon_a$, $a_{2i} = (i+1)\varepsilon_a$, and $a_{2i+1} = -i\varepsilon_a - \delta_a$ for $i > 0$,
2. $\delta_a > 2\varepsilon > 0$
3. $\varepsilon_a > \delta_a + \varepsilon$,
4. $x > \varepsilon > 0$,
5. $a_i, \delta_a, \varepsilon, \varepsilon_a, x \ll 1$, i.e., compared with the size of the bin, the values of these parameters are very tiny.

Lemma 5. *For any online packing strategy, the number of used bins for the input sequence is at least* $2.5k + 2$.

Lemma 6. *The optimal solution for packing the above input sequence uses* $k+3$ *bins.*

From the above analysis, we have the following conclusion.

Theorem 1. *The lower bound of competitive ratio for 1-bounded space online algorithm is 2.5.*

6 Concluding Remarks

In this paper, we focus on the 1-bounded space 2-dimensional bin packing, and give a constant competitive algorithm, significantly improve the previous result. The lower bound of competitive ratio is also improved. In the following research, there are mainly three directions:

(1) A big gap still remains between the upper bound 8.84 and lower bound 2.5. Our future research will focus on how to close this gap.
(2) In previous results on bounded space variation, the number of active bins is always a big constant. We may study the variation with small number of active bins. For example, if the number of active bins is 2 or 3, can we significantly improve the competitive ratio?
(3) In online bin packing problem, the performance is often evaluated by comparing with the optimal offline algorithm, which does not consider the order of items arrive over time. It will make more sense if we compare with the optimal algorithm abiding by the order of arrival items.

References

1. Bansal, N., Correa, J.R., Kenyon, C., Sviridenko, M.: Bin Packing in Multiple Dimensions: In-approximability Results and Approximation Schemes. Mathematics of Operations Research 31(1), 31–49 (2006)
2. Bansal, N., Caprara, A., Sviridenko, M.: Improved approximation algorithm for multidimensional bin packing problems. In: FOCS 2006, pp. 697–708 (2006)
3. Blitz, D., van Vliet, A., Woeginger, G.J.: Lower bounds on the asymptotic worst-case ratio of on-line bin packing algorithms (1996) (unpublished manuscript)

4. Caprara, A.: Packing 2-dimensional bins in harmony. In: FOCS 2002, pp. 490–499 (2002)
5. Chung, F.R.K., Garey, M.R., Johnson, D.S.: On packing two-dimensional bins. SIAM J. Algebraic Discrete Methods 3(1), 66–76 (1982)
6. Csirik, J., Johnson, D.S.: Bounded Space On-Line Bin Packing: Best is Better than First. Algorithmica 31, 115–138 (2001)
7. Epstein, L., van Stee, R.: Optimal Online Algorithms for Multidimensional Packing Problems. SIAM Jouranl on Computing 35(2), 431–448 (2005)
8. Epstein, L., van Stee, R.: Online square and cube packing. Acta Inf. 41(9), 595–606 (2005)
9. Ferreira, C.E., Miyazawa, E.K., Wakabayashi, Y.: Packing squares into squares. Pesquisa Operacional 19, 223–237 (1999)
10. Fujita, S.: On-Line Grid-Packing with a Single Active Grid. In: Ibarra, O.H., Zhang, L. (eds.) COCOON 2002. LNCS, vol. 2387, pp. 476–483. Springer, Heidelberg (2002)
11. Han, X., Chin, F., Ting, H.-F., Zhang, G., Zhang, Y.: A New Upper Bound on 2D Online Bin Packing (manuscript)
12. Han, X., Iwama, K., Zhang, G.: Online removable square packing. Theory of Computing Systems 43(1), 38–55 (2008)
13. Januszewski, J., Lassak, M.: On-line packing sequences of cubes in the unit cube. Geometriae Dedicata 67, 285–293 (1997)
14. Johnson, D.S., Demers, A.J., Ullman, J.D., Garey, M.R., Graham, R.L.: Worst-Case performance bounds for simple one-dimensional packing algorithms. SIAM Journal on Computing 3(4), 299–325 (1974)
15. Lee, C.C., Lee, D.T.: A simple on-line bin packing algorithm. J. Assoc. Comput. Mach. 32, 562–572 (1985)
16. Leung, J.Y.-T., Tam, T.W., Wong, C.S., Young, G.H., Chin, F.Y.L.: Packing squares into a square. J. Parallel Distrib. Comput. 10, 271–275 (1990)
17. Kohayakawa, Y., Miyazawa, F.K., Raghavan, P., Wakabayashi, Y.: Multidimensionalcube packing. Algorithmica 40(3), 173–187 (2004)
18. Meir, A., Moser, L.: On packing of squares and cubes. Journal of Combinatorial Theory 5, 126–134 (1968)
19. Ramanan, P.V., Brown, D.J., Lee, C.C., Lee, D.T.: On-line bin packing in linear time. Journal of Algorithms 10, 305–326 (1989)
20. Seiden, S.S.: On the online bin packing problem. J. ACM 49, 640–671 (2002)
21. van Vliet, A.: An improved lower bound for on-line bin packing algorithms. Information Processing Letters 43, 277–284 (1992)
22. Yao, A.C.-C.: New Algorithms for Bin Packing. Journal of the ACM 27, 207–227 (1980)

On the Advice Complexity of Online Problems*
(Extended Abstract)

Hans-Joachim Böckenhauer[1], Dennis Komm[1], Rastislav Královič[2], Richard Královič[1], and Tobias Mömke[1]

[1] Department of Computer Science, ETH Zurich, Switzerland
{hjb,dennis.komm,richard.kralovic,tobias.moemke}@inf.ethz.ch
[2] Department of Computer Science, Comenius University, Bratislava, Slovakia
kralovic@dcs.fmph.uniba.sk

Abstract. In this paper, we investigate to what extent the solution quality of online algorithms can be improved by allowing the algorithm to extract a given amount of information about the input. We consider the recently introduced notion of *advice complexity* where the algorithm, in addition to being fed the requests one by one, has access to a tape of advice bits that were computed by some *oracle* function from the complete input. The advice complexity is the number of advice bits read. We introduce an improved model of advice complexity and investigate the connections of advice complexity to the competitive ratio of both deterministic and randomized online algorithms using the paging problem, job shop scheduling, and the routing problem on a line as sample problems. We provide both upper and lower bounds on the advice complexity of all three problems.

Our results for all of these problems show that very small advice (only three bits in the case of paging) already suffices to significantly improve over the best deterministic algorithm. Moreover, to achieve the same competitive ratio as any randomized online algorithm, a logarithmic number of advice bits is sufficient. On the other hand, to obtain optimality, much larger advice is necessary.

1 Introduction

Many problems such as routing, scheduling, or the paging problem work in so-called online environments and their algorithmic formulation and analysis demand a model in which an algorithm that deals with such a problem knows only a part of its input at any specific point during runtime. These problems are called *online problems* and the respective algorithms are called *online algorithms*. In such an online setting, an online algorithm A has a huge disadvantage compared to offline algorithms (i. e., algorithms knowing the whole input already at the beginning of their computation) since A has to make decisions at any time step i without knowing what the next chunk of input at time step $i+1$ will be.

* This work was partially supported by ETH grant TH 18 07-3, and APVV grant 0433-06.

Y. Dong, D.-Z. Du, and O. Ibarra (Eds.): ISAAC 2009, LNCS 5878, pp. 331–340, 2009.

ϕ is the content of the advice tape, i. e., an infinite binary sequence, and $I = (x_1, \ldots, x_n)$. Algorithm A *is* c-competitive with advice complexity $s(n)$ *if there exists a constant α such that, for every n and for each input sequence I of length at most n, there exists some ϕ such that $C(\mathtt{A}^\phi(I)) \le c \cdot C(\textsc{Opt}(I)) + \alpha$ and at most $s(n)$ bits of ϕ have been accessed during the computation of $\mathtt{A}^\phi(I)$. If $\alpha = 0$, then* A *is* strictly c-competitive *with advice complexity $s(n)$.*

For the ease of notation, for both randomized online algorithms and online algorithms with advice, if ϕ is clear from the context, we write $\mathtt{A}(I)$ instead of $\mathtt{A}^\phi(I)$. If A accesses b bits of the advice tape during some computation, we say that b *advice bits are communicated* to A, or that A *uses b bits of advice.* The advice complexity of A gives an upper bound on the number of communicated advice bits, depending on the size n of the input. Note that, if some randomized online algorithm achieves a competitive ratio of r using b random bits, the same competitive ratio r can be achieved by an online algorithm with advice using b advice bits. We use $\log(x)$ to denote the logarithm of x with base two.

For proving upper bounds, we sometimes use the following idea. The oracle needs to communicate some n-bit string to the algorithm, but this string always contains only few ones or few zeros, allowing for the following efficient encoding.

Lemma 1. *Consider an online algorithm* A *with advice, achieving competitive ratio r while using some n-bit advice string that contains at most n/t zeros or at least $n - n/t$ zeros, where $t \ge 2$ is a fixed constant. Then, it is possible to design an improved online algorithm* B *that knows the parameter t and achieves an advice complexity of*

$$s(n) = \min\left\{ n \log\left(t/(t-1)^{\frac{t-1}{t}} \right), \frac{n \log n}{t} \right\} + 3 \log n + \mathcal{O}(1).$$

3 Paging

To use computer memory as efficiently as possible, the well-known technique of *paging* is widely used. Formally, we define paging problem as follows.

Definition 2 (Paging Problem). *The input is a sequence of integers representing requests to logical pages $I = (x_1, \ldots, x_n)$, $x_i > 0$, $x_i \ne x_{i+1}$. An online algorithm* A *maintains a buffer (content of the physical memory) $B = \{b_1, \ldots, b_K\}$ of K integers, where K is a fixed constant known to* A*. Before processing the first request, the buffer gets initialized as $B = \{1, \ldots, K\}$. Upon receiving a request x_i, if $x_i \in B$, then $y_i = 0$. If $x_i \notin B$, then a* page fault *occurs, and the algorithm has to find some* victim *b_j, i. e., $B := B \setminus \{b_j\} \cup \{x_i\}$, and $y_i = b_j$. The cost of the solution $\mathcal{A} = \mathtt{A}(I)$ is the number of page faults, i. e., $C(\mathcal{A}) = |\{y_i : y_i > 0\}|$.*

The paging problem (or Paging for short) has been extensively studied (see [2]). It is known that every deterministic algorithm is at least K-competitive, and there exist K-competitive deterministic algorithms. A simple upper bound on the advice complexity of an optimal algorithm follows from [6, Lemma 3.1].

Fact 1. *For* PAGING, *there is an optimal online algorithm with advice complexity $n + K$ and a 1-competitive online algorithm with advice complexity n.*

A lower bound for the competitive ratio of any randomized paging algorithm was proven in [2]: every randomized algorithm is at least H_K-competitive, where H_K is the K-th harmonic number, i. e., $H_K = \sum_{i=1}^{K} \frac{1}{i}$. The lower bound is almost tightly matched by randomized *marking algorithm*[1]: the randomized marking algorithm is H_K-competitive, if the logical pages are chosen from a set of cardinality $K + 1$, and they are $2H_K$-competitive in general.

Constant Competitive Ratio. From Fact 1, we can see that using one bit of advice per request is sufficient to obtain an optimal paging algorithm. In the next theorem, we show that it is possible to obtain paging algorithms with good competitive ratios using smaller advice. More precisely, we show an upper bound on the advice complexity depending on the competitive ratio, indicating that using an amortized constant amount of bits per request $s(n)/n < 1$ is enough to achieve a constant competitive ratio.

Theorem 1. *For each constant $r \geq 1$, there exists an r-competitive algorithm with advice complexity $s(n) = n \log\left((r+1)/\left(r^{r/(r+1)}\right)\right) + 3 \log n + \mathcal{O}(1)$.*

The proof of Theorem 1 is presented in [3]. As a corollary, note that it is possible to obtain a constant competitive ratio with an amortized constant number of bits of advice per request. This number of bits can be arbitrarily close to 0 at the expense of a very high competitive ratio; on the other hand, a competitive ratio arbitrarily close to 1 is reachable with the number of bits per request approaching 1. More precisely, for each constant $r \geq 1$, there exists an r-competitive algorithm with advice complexity $s(n)$ such that $\lim_{n\to\infty}(s(n)/n) = \log\left((r+1)/\left(r^{r/(r+1)}\right)\right)$.

Next, we show a lower bound on the advice complexity required to obtain a constant competitive ratio, more precisely, we express the minimal number of advice bits per request required to obtain an r-competitive algorithm.

Theorem 2. *Let r be any constant such that $1 \leq r \leq 1.25$. For any paging algorithm* A *with advice complexity $s(n)$ and competitive ratio r it holds that $\frac{s(n)}{n} \geq \frac{1}{2K-2}\left[1 + \log(3-2r) - (2r-2)\log\left(\frac{1}{2r-2} - 1\right)\right] - \mathcal{O}\left(\frac{1}{n}\right)$. The constant of the $\mathcal{O}(1/n)$ depends on K and the parameters r, α of* A.

The core idea of the proof of Theorem 2 is to consider a certain (sufficiently large) class of inputs arranged into a complete tree. If the advice is small enough, there must be many inputs processed with the same advice. This makes it possible to prove that A must behave inefficiently on at least one of them, by exploiting the tree structure of the inputs. For a detailed proof, see [3].

---

[1] In general, a marking algorithm (see [2]) starts with all pages unmarked, and upon a hit marks the requested page. If there is no unmarked page upon a page fault, all pages in the memory are unmarked and a new phase begins.

To obtain an optimal algorithm, large advice is necessary. Similar arguments as in Theorem 2 yield a lower bound on the advice complexity of $\frac{s(n)}{n} \geq 1 - \frac{\log(K-1)+C}{4(K-1)} - \mathcal{O}\left(\frac{1}{n}\right)$, for some constant C, converging to 1 bit per request with growing cache size K. Thus, the upper bound from Fact 1 is tight for large K.

Small advice. We now analyze the case where the algorithm is allowed to use only a constant number of advice bits for the whole input. We show that, even in this very restricted case, the competitive ratio can be significantly improved with respect to the best deterministic algorithm. We start with upper bounds on the advice complexity, and complement them later with a lower bound.

Theorem 3. *Consider the class of inputs with a buffer of size K, and let $c < K$ be a power of 2. There is an algorithm with oracle size $\log c$ for* Paging *on this class of inputs with competitive ratio $r \leq 3\log c + (2(K+1)/c) + 1$.*

For the proof, we construct c deterministic marking algorithms. Their executions are naturally divided into phases, and the construction makes sure that, in each phase, the algorithms induce many different behaviors. A careful choice of page replacement strategies enables to use an accounting scheme to bound the overall number of faults occurring in all algorithms during a phase. The proof is finished by arguing that there must be an algorithm with at most $1/c$-th fraction of them, and comparing this amount with a trivial general lower bound. A detailed proof can be found in [3].

The presented results show that even very small advice can be used efficiently. For example, providing just two bits of advice per whole input instance reduces the competitive ratio from K (i. e., the best deterministically achievable ratio) to $K/2 + \mathcal{O}(1)$. Additionally, $\log K$ bits of advice (i. e., making c equal to the largest power of 2 smaller than K) can be used to achieve the competitive ratio $3\log K + \mathcal{O}(1)$ which is asymptotically equal (albeit the constant is worse) to the best possible ratio of $H_K = \sum_{i=1}^{K} \frac{1}{i}$ for a randomized algorithm without advice.

Next, we complement the result of Theorem 3 by presenting a lower bound on the competitive ratio for paging algorithms with constant advice complexity.

Theorem 4. *Consider the class of inputs with $K+1$ possible pages, where K is the size of the buffer, and let c be a power of 2. Any deterministic algorithm* $\mathtt{A}$ *with advice complexity $s(n) = \log c$ has competitive ratio at least K/c.*

The main proof idea is to consider all inputs of certain length arranged into a complete K-ary tree. Then, it is either possible to find an input with many faults in the beginning of the computation, or to select an input prefix such that some advice is not used for any extension of this prefix, and to continue by induction.

4 Disjoint Path Allocation

We consider a special type of network topology where the entities of a network are connected by one line (i. e., a bus network). The network is a path P and all

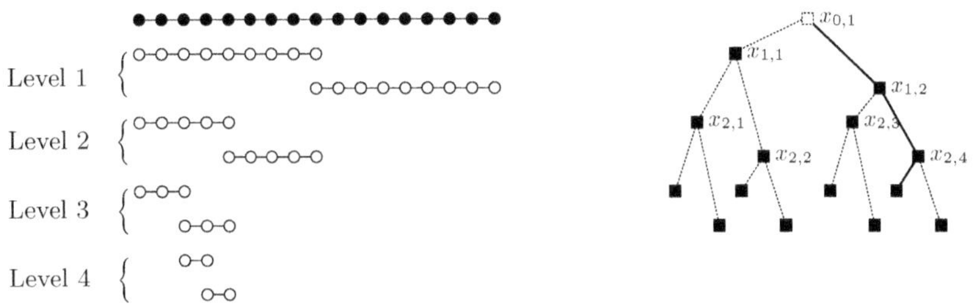

Fig. 1. A sample input from the class $\mathcal{I}_3$ and its representation as tree $\mathcal{T}$ of height h

connections in P have a capacity of one. For the *disjoint path allocation problem* (DisPathAlloc, described in [2]), additionally a set of subpaths of P is given. Each subpath $(v_i, \ldots, v_j)$ models a *call request*, i.e., a request to establish a permanent connection between the two endpoints v_i and v_j. If such a request is satisfied, no inner entity of the path is able to be part of any other call. Therefore, a *disjoint path allocation* is simply a set of edge-disjoint subpaths of P.

Definition 3 (Disjoint Path Allocation Problem). *Given a path* $P = (V, E)$, *where* $V = \{v_0, \ldots, v_L\}$ *is a set of entities, and a set* $\mathcal{P}$ *of subpaths of* P *where* $|\mathcal{P}| = n$, DisPathAlloc *is the problem of finding a maximum set* $\mathcal{P}' \subseteq \mathcal{P}$ *of edge-disjoint subpaths of* P.

Here, we deal with an environment in which the subpaths $P_1, \ldots, P_n \in \mathcal{P}$ arrive in an online fashion. We assume that $L = |V| - 1$ is known to the online algorithm in advance, but that this is not the case for n. For the ease of presentation, let (v_i, v_j) denote the path $(v_i, v_{i+1}, \ldots, v_j)$. The cost function $C(\mathtt{A}(I))$ denotes the number of requests satisfied by $\mathtt{A}$ on input I. Since DisPathAlloc is a maximization problem, an algorithm $\mathtt{A}$ solving it is c-competitive if $C(\textsc{Opt}(I)) \leq c \cdot C(\mathtt{A}(I)) + \alpha$ for some constant α and every input I and strictly c-competitive if $\alpha = 0$. The competitive ratio can be measured in terms of n as well as in terms of L. We focus on the former case, the latter is discussed in [3]. It is not difficult to check that any deterministic online algorithm $\mathtt{A}$ is no better than strictly $(n-1)$-competitive and strictly L-competitive. These bounds are tight, since they are reached by a simple greedy algorithm.

The Class $\mathcal{I}$ of Inputs. For some of the following proofs, we consider input instances as depicted in Fig. 1 for any even n. For every h, we define the class $\mathcal{I}_h$ (and $\mathcal{I} = \bigcup_{h \in \mathbb{N}} \mathcal{I}_h$) as follows. Every element of $\mathcal{I}_h$ consists of $h+1$ *levels*. Level 1 consists of two disjoint consecutive requests, splitting the line network into two parts of the same size. After that, two disjoint consecutive requests on level $i+1$ do the same with one of the intervals from level i. This is iterated until two requests of size 1 appear on level $h+1$. Fig. 1, for instance, shows an input from $\mathcal{I}_3$. It is obvious that any optimal algorithm satisfies exactly one request on each of the first h levels, allowing it to satisfy both requests on level $h+1$.

Clearly, we may represent an optimal solution, for any input sequence as described above, by a path from the root to a leaf in a complete binary tree $\mathcal{T}$

of height h with its root on a notional level 0 (see Fig. 1). The 2^h leaves of $\mathcal{T}$ represent the 2^h different inputs of the class $\mathcal{I}_h$. Let $\mathcal{OPT}$ denote a path corresponding to some input instance $\in \mathcal{I}_h$. We may say that an optimal algorithm Opt makes moves *according to this path.* For an arbitrary online algorithm A that satisfies one request on level i, we say that A makes the *correct* decision on this level if it also acts according to $\mathcal{OPT}$. However, if A satisfies one interval not according to this path $\mathcal{OPT}$ or satisfies both requests, we say that A makes a *wrong* decision on level i. In this case, we say that A is *out* (after level i). Moreover, for every i and an input instance $I_h \in \mathcal{I}_h$, let $C_i(\mathtt{A}(I_h))$ denote the overall number of requests satisfied by A up to level i. If A is out after level i, this means that, for every $j \geq i$, $C_j(\mathtt{A}(I_h)) = C_i(\mathtt{A}(I_h))$ (and obviously, therefore, $C(\mathtt{A}(I_h)) = C_i(\mathtt{A}(I_h)))$. Let $correct(\mathtt{A})$ [$wrong(\mathtt{A})$] denote the set of time steps in which A makes the correct [wrong] decision. Since making the wrong decision can only happen once (because the algorithm is out afterwards), the overall gain of A is $C(\mathtt{A}(I_h)) \leq |correct(\mathtt{A})| + 2 \cdot |wrong(\mathtt{A})| \leq |correct(\mathtt{A})| + 2$, which directly implies that, for an optimal algorithm Opt, $C(\textsc{Opt}(I_h)) = h + 2$.

Advice Complexity Bounds. DisPathAlloc is a hard online problem even for randomized algorithms. A lower bound of $\Omega(n)$ on the competitive ratio has been proven in [5, Theorem 9] and [3] independently. We now investigate how many advice bits are needed for an online algorithm A to yield optimality.

Theorem 5. *For any online algorithm with advice for* DisPathAlloc*, at least* $\frac{n+2}{2r} - 2$ *bits of advice are required to achieve a strict competitive ratio of* r*.*

Note that, by setting $r = 1$, we immediately get a lower bound of $b_{opt} = \frac{n-2}{2}$ advice bits for achieving an optimal solution. On the other hand, it is not difficult to construct an online algorithm with advice for DisPathAlloc whose advice complexity is only a factor of $\log n$ away from this lower bound for large r, and is asymptotically tight for constant r: Consider an algorithm A that reads one bit of advice per request, and satisfies the request if and only if this bit is one. Communicating an advice string with at most n/r ones is sufficient to achieve a competitive ratio r. Hence, Lemma 1 for $t := r$ implies the following theorem.

Theorem 6. *For every* r*, there exists an* r*-competitive online algorithm for* DisPathAlloc *with advice complexity*

$$s(n) = \min\left\{n \log\left(r/(r-1)^{(r-1)/r}\right), (n \log n)/r\right\} + 3 \log n + \mathcal{O}(1).$$

The previous theorem also shows that advice complexity $\mathcal{O}(\log^2 n)$ is sufficient to obtain a competitive ratio asymptotically better than any randomized algorithm is able to achieve.

5 Job Shop Scheduling

We consider a special case of job shop scheduling with two jobs consisting of n unit-length tasks each, which have to be processed on the same n machines, each task on a different one, but possibly in a different order within the two jobs. Two tasks requiring the same machine in one time step causes a delay in the schedule and the goal in this problem JSSCHEDULE is to minimize the number of these delays. This problem can be modelled as finding a shortest path in an $(n \times n)$-grid from the upper left to the lower right corner, where also diagonal steps are allowed, except for some grid cells which are blocked by obstacles modelling tasks competing for the same machine. For the details of the problem description, we refer to [3,8]. Full proofs of the results from this section can be found in [3].

Lemma 2. *For every instance of* JSSCHEDULE, *there exists an optimal solution $\mathcal{A}$ which, as long as it does not arrive at the right or bottom border of the grid, always moves diagonally if no obstacle is in its way.*

Consider the optimal solution for an arbitrary instance, and assume that it makes d diagonal moves, h moves to the right, and $v = h$ moves downwards. The cost of this solution is $d + 2h = n + h$, since $d + h = n$. According to [8], for every instance of JSSCHEDULE, there exists an optimal solution with cost at most $n + \lceil \sqrt{n} \rceil$ and therefore $h \leq \lceil \sqrt{n} \rceil$. Furthermore, using Lemma 2, it is easy to see that there exists an online algorithm A with advice needing at most $2h \leq 2\lceil \sqrt{n} \rceil$ advice bits: Algorithm A simply acts according to Lemma 2. Only at time steps where a diagonal move is not possible due to an obstacle, it has to read one bit from the advice tape indicating whether to move downwards or to the right. As a corollary, for every instance of JSSCHEDULE, there exists an optimal online algorithm A with advice complexity $s(n) = 2\lceil \sqrt{n} \rceil$. On the other hand, this upper bound is asymptotically tight, as claimed by the following lower bound on the number of advice bits necessary to compute an optimal solution.

Theorem 7. *Any online algorithm* A *with advice for* JSSCHEDULE *needs advice complexity $s(n) = \Omega(\sqrt{n})$ to achieve optimality.*

While an optimal solution of JSSCHEDULE requires large advice, much shorter advice is sufficient to achieve a solution with competitive ratio close to 1. In [8], a randomized algorithm for JSSCHEDULE with a competitive ratio of $\mathcal{O}(1 + 1/\sqrt{n})$ is presented which easily implies that there is an online algorithm for JSSCHEDULE with advice complexity $s(n) \leq 1 + \log(n)$ that achieves competitive ratio $\mathcal{O}(1 + 1/\sqrt{n})$.

6 Conclusion

Our results suggest the hypothesis that logarithmic advice complexity is sufficient to achieve the competitive ratio of the best randomized algorithm for *any* online problem. This claim, however, cannot be proven in full generality. In fact, we can construct an online problem where the number of required advice bits is as high as the number of random bits in order to keep up with randomized algorithms.

Consider the following problem, where the online algorithm has to guess a sequence of n bits. If all bits are guessed correctly, the algorithm is greatly rewarded, otherwise, the cost of the solution is zero. For this problem, we can construct a very simple randomized algorithm that has a positive number as expected gain whereas it is easy to show that any deterministic algorithm needs a linear number of advice bits for having a gain greater than zero. Unfortunately, this situation is quite artificial: if an algorithm with advice A outperforms a randomized algorithm R, then A is exponentially better than R, too. Nevertheless, this situation shows that, in general, the advice complexity is somehow orthogonal to randomization. It remains as an open problem to find more connections between these complexity measures, e. g., to somehow characterize classes of online problems where small advice is sufficient to keep up with randomized algorithms. Furthermore, it might be interesting to consider randomized online algorithms with advice.

Acknowledgments

The authors would like to thank Juraj Hromkovič for many helpful discussions on the topic of this paper and elaborating the model of advice complexity of online problems to the point we could start investigating the problems introduced.

References

1. Ben-David, S., Borodin, A.: A new measure for the study of on-line algorithms. Algorithmica 11(1), 73–91 (1994)
2. Borodin, A., El-Yaniv, R.: Online computation and competitive analysis. Cambridge University Press, New York (1998)
3. Böckenhauer, H.-J., Komm, D., Královič, R., Královič, R., Mömke, T.: Online Algorithms with Advice. Technical Report 614. ETH Zurich, Department of Computer Science (2009)
4. Brucker, P.: An efficient algorithm for the job-shop problem with two jobs. Computing 40(4), 353–359 (1988)
5. Caragiannis, I., Fishkin, A.V., Kaklamanis, C., Papaioannou, E.: Randomized online algorithms and lower bounds for computing large independent sets in disk graphs. Discrete Applied Mathematics 155(2), 119–136 (2007)
6. Dobrev, S., Královič, R., Pardubská, D.: Measuring the Problem-Relevant Information in Input. RAIRO-Theoretical Informatics and Applications 43(3), 585–613 (2009)
7. Emek, Y., Fraigniaud, P., Korman, A., Rosén, A.: Online Computation with Advice. In: Nikoletseas, S., Thomas, W. (eds.) ICALP 2009. LNCS, vol. 5555, pp. 427–438. Springer, Heidelberg (2009)
8. Hromkovič, J., Mömke, T., Steinhöfel, K., Widmayer, P.: Job shop scheduling with unit length tasks: bounds and algorithms. Algorithmic Operations Research 2(1), 1–14 (2007)
9. Sleator, D.D., Tarjan, R.E.: Amortized efficiency of list update and paging rules. Communications of the ACM 28(2), 202–208 (1985)

Online Knapsack Problems with Limited Cuts

Xin Han and Kazuhisa Makino

Department of Mathematical Informatics, Graduate School of Information and Technology, University of Tokyo, Tokyo, 113-8656, Japan
hanxin.mail@gmail.com, makino@mist.i.u-tokyo.ac.jp

Abstract. The (offline) maximization (resp., minimization) knapsack problem is given a set of items with weights and sizes, and the capacity of a knapsack, to maximize (resp., minimize) the total weight of selected items under the constraint that the total size of the selected items is at most (resp., at least) the capacity of the knapsack. In this paper, we study online maximization and minimization knapsack problems with limited cuts, in which 1) items are given one by one over time, i.e., after a decision is made on the current item, the next one is given, 2) items are allowed to be cut at most $k\,(\geq 1)$ times, and 3) items are allowed to be removed from the knapsack.

We obtain the following three results.

(i) For the maximization knapsack problem, we propose a $(k+1)/k$-competitive online algorithm, and show that it is the best possible, i.e., no online algorithm can have a competitive ratio less than $(k+1)/k$.
(ii) For the minimization knapsack problem, we show that no online algorithm can have a constant competitive ratio.
(iii) We extend the result (i) to the resource augmentation model, where an online algorithm is allowed to use a knapsack of capacity $m\,(> 1)$, while the optimal algorithm uses a unit capacity knapsack.

1 Introduction

The knapsack problem is one of the most classical and studied problems in combinatorial optimization and has a lot of applications in the real world [9]. The (classical) knapsack problem is given a set of items with weights and sizes, and the capacity value of a knapsack, to maximize the total weight of selected items in the knapsack satisfying the capacity constraint. This problem is also called the *maximization* knapsack problem (Max-Knapsack). Many kinds of variants and generalizations of the knapsack problem have been investigated so far [9]. Among them, the *minimization* knapsack problem (Min-Knapsack) is one of the most natural ones (see [1,2,3,4] and [9, pp. 412-413]), that is given a set of items associated with weights and sizes, and the size of a knapsack, to minimize the total weight of selected items that cover the knapsack. In this paper, we study online maximization and minimization knapsack problems with limited cuts, in which i) items are given one by one over time, i.e., after a decision is made on

Y. Dong, D.-Z. Du, and O. Ibarra (Eds.): ISAAC 2009, LNCS 5878, pp. 341–351, 2009.

the current item, the next one is given, ii) items are allowed to be cut at most k (≥ 1) times, and iii) items are allowed to be removed from the knapsack (but once they are removed, they cannot be used partially again).

Related work: It is well-known that offline Max-Knapsack and Min-Knapsack are both NP-hard, and admit fully polynomial time approximation schemes (FPTASs) [1,4,9]. As for the online maximization knapsack problem, it was first studied on average case analysis by Marchetti-Spaccamela and Vercellis [11]. They proposed a linear time approximation algorithm such that the expected difference between the optimal and the approximation solution value is $O(\log^{3/2} n)$ under the condition that the capacity of the knapsack grows proportionally to n, the number of items. Lueker [10] further improved the expected difference to $O(\log n)$ under a fairly general condition on the distribution. Iwama and Taketomi [7] studied the problem on worst case analysis. They obtained a 1.618-competitive algorithm for the online Max-Knapsack under the removable condition, if each item has its size equal to its profit. Here the removable condition means that it is allowed to remove some items in the knapsack in order to accept a new item. They also showed that this is the best possible by providing a lower bound 1.618 for this case. For the general case, Iwama and Zhang [8] showed that no algorithm for online Max-Knapsack has a bounded competitive ratio, even if the removal condition is allowed. Recently, Han and Makino [6] obtained an upper bound 8 and a lower bound 2 for minimization knapsack problem. Iwama and Zhang [8] presented the competitive ratio for the online Max-Knapsack problem with resource augmentation. Noga and Sarbua [12] studied an online partially fractional knapsack problem with resource augmentation, in which items are allowed to be cut at most once (i.e., $k = 1$), only before they are packed into the knapsack, whereas in our model items are allowed to be cut any time (but at most k (≥ 1) times). They gave an upper bound $2/m$ and proved the bound is the best possible, where $m \geq 1$ is the capacity of the knapsack used by online algorithms while the optimal offline algorithm uses a unit capacity knapsack. The online knapsack problem of using extra a bin and allowing to exchange items between two bins was studied by Horiyama, Iwama and Kawahara [5].

Our contributions: For the online maximization knapsack, we propose a simple greedy online algorithm, in which whenever a cut is necessary on an item, we almost cut off the fraction of size $\frac{1}{k+1}$ from the item. We show that our greedy algorithm is $\frac{k+1}{k}$-competitive, and it is the best possible by giving a lower bound of the cometitive ratio. We extend this result to to the model with resource augmentation. In the resource augmentation model, we show that the online maximization knapsack problem is $\max\{1, \frac{k+1}{m(k+1)-1}\}$-competitive (i.e., we present a $\max\{1, \frac{k+1}{m(k+1)-1}\}$-competitive algorithm and show that it is the best possible). When $k = 1$ and $1 \leq m < 2$, the competitive ratio $\frac{2}{2m-1}$ in our model is smaller than the ratio $\frac{2}{m}$ in the partial cut model discussed in [12]. This implies that our model for $k = 1$ is more powerful than the model given in [12].

Table 1. The current results on online Max-Knapsack with resource augmentation

Max-Knapsack	$k = 0$	$k = 1$ in the partial cut model	$k \geq 1$ in our model
Lower Bound	$1/(m-1)$ [8]	$2/m$ [12]	$\mathbf{(k+1)/(m(k+1)-1)}$
Upper Bound	$1/(m-1)$ [8]	$2/m$ [12]	$\mathbf{(k+1)/(m(k+1)-1)}$

For the minimization knapsack problem, we show that no online algorithm can have a constant competitive ratio, i.e., our cut condition does not help solving the problem.

Table 1 summarizes the current results on online Max-Knapsack with resource augmentation, where the bold letters represent the results obtained in this paper, $m\,(\geq 1)$ denotes the capacity used by online algorithms and k denotes the number of limited cuts to be allowed.

2 Preliminaries

In this section, we formally define our problems and review the basic concepts for the online algorithms.

Problem Max-Knapsack (resp., Min-Knapsack)

Input: A set of items $L = \{a_1, \ldots, a_n\}$ associated with weight $w : L \to \mathbb{R}_+$ and size $s : L \to \mathbb{R}_+$.

Output: A set of items $F \subseteq L$ that maximizes $w(F)$ subject to $s(F) \leq 1$ (resp., that minimizes $w(F)$ subject to $s(F) \geq 1$).

Here, for a set $U \subseteq L$, let $w(U) = \sum_{u \in U} w(u)$ and $s(U) = \sum_{u \in U} s(u)$, and we assume w.l.o.g. that the size of the knapsack is 1. The fractional version of the Max-Knapsack (resp., Min-Knapsack) is given as follows: $\max \sum_{u \in L} w(u)x(u)$ s.t. $\sum_{u \in L} s(u)x(u) \leq 1$ and $0 \leq x(u) \leq 1$ $(u \in L)$ (resp., $\min \sum_{u \in L} w(u)x(u)$ s.t. $\sum_{u \in L} s(u)x(u) \geq 1$ and $0 \leq x(u) \leq 1$ $(u \in L)$).

In our online model, the objective is the same with the offline version. But the input is given over time. Namely, the knapsack of size 1 is known beforehand, and after a decision is made on the current item a_t, the next one a_{t+1} is given. Besides this, our model satisfies the *removal* and *cut* conditions.

Removal condition: The items in the knapsack are allowed to be removed, where the items removed cannot be used again.

Cut condition: The current item and the items in the knapsack are allowed to be cut, where the part of the item cut off cannot be used again, and during the whole process, each item can be cut at most $k\,(\geq 1)$ times. Here k is a given positive integer.

By the cut condition, the knapsack keeps a set of fractional items, and hence our problems can be regared as the online fractional knapsack problems, rather than online 0-1 knapsack problems.

We analyze online algorithms by using one of the standards: the competitive ratio. Given an input sequence L and an online algorithm A, for the maximization problem, the *competitive ratio* of algorithm A is defined as follows:

$$R_A = \sup_L \frac{OPT(L)}{A(L)},$$

and for the minimization problem, the *competitive ratio* of algorithm A is defined as follows:

$$R_A = \sup_L \frac{A(L)}{OPT(L)},$$

where $OPT(L)$ and $A(L)$ denotes the weights obtained by an optimal algorithm and the algorithm A, respectively.

3 Online Maximization Knapsack with Limited Cuts

3.1 A Simple Greedy Algorithm A

The main ideas of our algorithm are as follows: when a new item is arrived, we apply a greedy algorithm to select items from the knapsack together with the new item. If the total size of the resulting items is greater than the capacity of the knapsack, then we cut the less efficient item, say b in the knapsack. Let $s(b)$ be the size of item b. The rule of cutting is below: if $s(b) > 1$ (the capacity of the knapsack) then we cut a fraction from b such that the remaining size $s(b)$ is exactly $\frac{k}{k+1}$, else cut a fraction of size $\min\{\frac{1}{k+1}, s(b)\}$ from item b. Then we repeatedly cut off item b, until the total size becomes at most the capacity of the knapsack.

Let $L = \{a_1, a_2, ..., a_n\}$ be the online input. Assume that items $a_1, ..., a_{i-1}$ have been dealt by our algorithm. Let B_{i-1} be the set of items in the knapsack. The execution of our algorithm on item a_i is the following.

Theorem 1. *The competitive ratio of algorithm A is $\frac{k+1}{k}$.*

Proof. It is not difficult to see that if an item originally has size at most 1 we never cut the item more than k item before it is totally removed, if an item originally has size larger than 1, this is also true since after the first cutting on the item the remaining size of the item is eactly $\frac{k}{k+1}$.

So next we need to prove that for all $1 \leq i \leq n$,

$$\frac{OPT(L_i)}{A(L_i)} \leq \frac{k+1}{k},$$

where $L_i = \{a_1, a_2, ..., a_i\}$ is the input just after time i, $OPT(L_i)$ and $A(L_i)$ are the total weights by an offline optimal and our online algorithms, respectively.

Just after time i, let B_i be the set of pieces in the knapsack. And let R_i be the set of pieces which have been discarded by algorithm A. Observe that algorithm

Algorithm: A
1. $B_i^{'} := B_{i-1} \cup \{a_i\}$, if $s(B_i^{'}) \leq 1$ then accept item a_i, else (a) Rename all the items in B_i' as $b_1, b_2, ...$ such that $w(b_1)/s(b_1) \geq w(b_2)/s(b_2) \geq \cdots$, where $w(b_j)$ and $s(b_j)$ respectively denote the weight and size of fractional item b_j. (b) Find a smallest index x such that $\sum_{h=1}^{x} s(b_h) > 1$, remove all the items with index larger than x in $B_i^{'}$. (c) If $\sum_{h=1}^{x-1} s(b_h) \geq \frac{k}{k+1}$, then remove item b_x. Else repeatedly chop off b_x by the following way until the total size in B_i' becomes at most 1: if $s(b_x) \leq 1$ chop off by a fraction of size $\frac{1}{k+1}$ from item b_x else chop off by a fraction such that the remaining size of item b_x is exactly $\frac{k}{k+1}$. 2. Update set B_i.

A always uses a greedy policy to select items. If both R_i and B_i are not empty, for any two pieces $q \in R_i$ and $p \in B_i$, we have

$$\frac{w(p)}{s(p)} \geq \frac{w(q)}{s(q)}, \tag{1}$$

where $w(p)$ (resp., $w(q)$) is the weight of fractional item p (resp., q) and $s(p)$ (resp., $s(q)$) is the size of fractional item p (resp., q). Moreover if R_i is not empty, we have

$$s(B_i) \geq \frac{k}{k+1}, \tag{2}$$

where $s(B_i)$ is the total size in set B_i.

If R_i is empty, then we have

$$A(L_i) = w(B_i) = OPT(L_i),$$

otherwise by (1) and (2), we immediately have

$$A(L_i) = w(B_i) \geq (1 - \frac{1}{k+1})OPT(L_i).$$

Hence this theorem holds. □

3.2 A Tight Lower Bound for the Competitive Ratio of the Maximum Knapsack

Surprisingly, the upper bound $\frac{1+k}{k}$ by online algorithm A is the best we can do, i.e., there exists no online algorithm with a competitive ratio less than $\frac{1+k}{k}$. We prove this in this subsection.

Assume there is an online algorithm with a competitive ratio c, which is less than $\frac{1+k}{k}$. Then the main ideas are as below:

1. We force the online algorithm to accept a large item with a low density and a unit size.
2. Sequentially, small items follow and their densities gradually increase, after some steps the online algorithm has to cut the large item to save space for small items otherwise its competitive ratio reaches to $\frac{1+k}{k}$; and more if the large item is cut then the size of the fraction cut off has to be less than $\frac{1}{k+1}$, otherwise the competitive ratio is as least $\frac{1+k}{k}$;
3. We keep doing the above operation k times, i.e., the online algorithm cuts the large item k times and every time a portion of size less than $\frac{1}{k+1}$ is cut off.
4. We finally continue to give new small items and increase their densities, then the online algorithm rejects the large item or does not, in both cases, we can prove that the online algorithm has a competitive ratio larger than c.

Theorem 2. *No online algorithm has a competitive ratio smaller than* $\frac{1+k}{k}$.

Proof. Assume there is an online algorithm A with a competitive ratio $c = \frac{1+k}{k+r} < \frac{1+k}{k}$, where $r > 0$. We prove that there is an input L such that $OPT(L)/A(L) > c$. The ideas to construct the list L are similar with the ones in [8].

In the input L, there are two kinds of sizes 1 and ϵ, i.e., *large* and *small*, where $\epsilon > 0$ is a sufficiently small and such that $\epsilon < \frac{r}{3(k+1)}$ and $\frac{1}{k\epsilon}$ is an integer. The input L is formed by phases. In phase 0, there is only a large item $(1, 1)$. For any $i > 0$, each phase i has $1/\epsilon$ items and each item has size ϵ and weight $\epsilon + i\epsilon^2$. Namely, the input L is below:

$$
\begin{array}{c}
(1,1) \\
(\epsilon+\epsilon^2,\epsilon),\ \ (\epsilon+\epsilon^2,\epsilon),\ldots,(\epsilon+\epsilon^2,\epsilon) \\
(\epsilon+2\epsilon^2,\epsilon),\ (\epsilon+2\epsilon^2,\epsilon),\ldots,(\epsilon+2\epsilon^2,\epsilon) \\
\cdots \\
(\epsilon+i\epsilon^2,\epsilon),\ (\epsilon+i\epsilon^2,\epsilon),\ldots,(\epsilon+i\epsilon^2,\epsilon) \\
\cdots
\end{array}
$$

Note that the information of the input L is gradually known to the online algorithm A and the input can stop at any step if the online algorithm performs poorly. Moreover online algorithm A does not know the future information of L and it can only use the information known so far.

We are going to prove that if there is a cut by the online algorithm A, then the size of the portion cut off is less than $1/(k+1)$. Let $OPT(i, j)$ be the optimal value just after the j-th item of phase i is given, where $1 \le j \le \frac{1}{\epsilon}$ and $i \ge 1$. It is not difficult to see

$$OPT(i,j) = (\frac{1}{\epsilon} - j)(\epsilon + (i-1)\epsilon^2) + j(\epsilon + i\epsilon^2) = 1 + \epsilon(i-1) + j\epsilon^2. \quad (3)$$

Lemma 1. *In order to achieve c-competitive, from phase* 0 *to phase* $\frac{1}{\epsilon k}$, *algorithm A has to cut the large item or discard it from the knapsack. If the algorithm*

A cuts the large item, then at the first cutting, the portion cut off has size smaller than $\frac{1}{k+1}$.

Proof. By (3), we have $OPT(i, \frac{1}{\epsilon}) = \frac{1+k}{k}$, where $i = \frac{1}{\epsilon k}$. If the algorithm A does not cut or discard the large item after phase $\frac{1}{\epsilon k}$, then any small item cannot be accepted in the knapsack. Hence the competitive ratio of algorithm A is at least $\frac{k+1}{k} > c$.

Assume that the online algorithm A makes the *first* cutting on the large item just after the j-th item of phase i is known, where $1 \leq j \leq \frac{1}{\epsilon}$ and $i \geq 0$. If $i = 0$, it is not difficult to see that algorithm A cannot cut off a portion with size at least $\frac{1}{k+1}$ at phase 0. If so, we stop this input at phase 0 and the online algorithm A has a competitive larger than c. Next we consider the case $i \geq 1$. By (3), during phase i the optimal value $OPT(L) \geq 1 + \epsilon(i-1)$. If the fraction cut off has size at least $\frac{1}{k+1}$ then we stop the input as this step and the weight $A(L)$ by online algorithm A is at most $1 - \frac{1}{k+1} + \epsilon + i\epsilon^2$. Hence the competitive ratio of algorithm A is at least

$$\frac{1+\epsilon(i-1)}{1-\frac{1}{k+1}+\epsilon+i\epsilon^2} > \frac{1}{\frac{k}{k+1}+2\epsilon} = \frac{k+1}{k+2\epsilon(k+1)} > \frac{k+1}{k+r},$$

where the first inequality holds from $\epsilon + i\epsilon^2 < 2\epsilon(1+\epsilon(i-1))$ for $\epsilon < 1$ and $i \geq 1$, the last one holds from $3\epsilon(k+1) < r$.

Hence this lemmas holds. □

Lemma 2. *Assume the large item has been cut* $j < k$ *times before phase* $i_0 \geq 0$ *and its remaining size in the knapsack is* $x \geq \frac{1}{k+1}$. *If there exists an integer* $i > i_0$ *such that*

$$\frac{OPT(i-1, \frac{1}{\epsilon})}{x + (1-x)\frac{\epsilon + i\epsilon^2}{\epsilon}} \geq \frac{k+1}{k},$$

then algorithm A has to cut the large item or discard it from the knapsack before phase i. *If the algorithm A makes its* $(j+1)$*-th cutting on the large item, then the size of the portion cut off is smaller than* $\frac{1}{k+1}$.

Proof. After the j-th cutting, if the online algorithm A does not cut the large item or discards it, then during phase i we have

$$A(L) \leq x + (1-x)\frac{\epsilon + i\epsilon^2}{\epsilon}.$$

Due to $OPT(L) > OPT(i-1, \frac{1}{\epsilon})$, then the competitive ratio of algorithm A is at least

$$\frac{OPT(i-1, \frac{1}{\epsilon})}{x + (1-x)\frac{\epsilon + i\epsilon^2}{\epsilon}} \geq \frac{k+1}{k}.$$

Assume that the online algorithm A cuts the large item at the $(j+1)$-th time during phase h, where $0 < h \leq i$. If the size of the fraction cut off is at least $\frac{1}{k+1}$, then we stop the input as this step. In this case, we have

$$OPT(L) > OPT(h-1, \frac{1}{\epsilon}) = 1 + (h-1)\epsilon,$$

and

$$A(L) \leq x - \frac{1}{k+1} + (1 - x + \epsilon)(1 + h\epsilon),$$

since the large item has size x and the space for small items is upper bounded by $(1 - x)$ before the cutting, and the input stops immediately after the cutting, hence there is at most one small item accepted after cutting, totally, the size of all the small items accepted is at most $(1 - x + \epsilon)$. We know each small item has its density at most $1 + h\epsilon$. So the competitive ratio of algorithm A is at least

$$\begin{aligned}\frac{1 + (h-1)\epsilon}{x - \frac{1}{k+1} + (1 - x + \epsilon)(1 + h\epsilon)} &\geq \frac{1 + (h-1)\epsilon}{x - \frac{1}{k+1} + (1+\epsilon)^2(1 + (h-1)\epsilon) - x(1 + h\epsilon)} \\ &\geq \frac{1}{-\frac{1}{k+1} + (1+\epsilon)^2} > \frac{1}{1 - \frac{1}{k+1} + 3\epsilon} = \frac{k+1}{k + 3\epsilon(k+1)} > \frac{k+1}{k+r} = c,\end{aligned}$$

where the first inequality holds from $(1 + h\epsilon) \leq (1 + \epsilon)(1 + (h-1)\epsilon)$ and the second inequality follows from $\frac{-1/(k+1) - xh\epsilon}{1 + (h-1)\epsilon} < -\frac{1}{k+1}$ for any $h > 0$, $\epsilon < 1$ and $x \geq \frac{1}{k+1}$.

Hence this lemma holds. □

Again, let x be the remaining size of the large item in the knapsack after the previous cutting. If $x > \frac{1}{k+1}$, there always exists an i such that the condition in Lemma 2 holds, i.e.,

$$\begin{aligned}\frac{OPT(i-1, \frac{1}{\epsilon})}{x + (1-x)\frac{\epsilon + i\epsilon^2}{\epsilon}} &\geq \frac{1 + (i-1)\epsilon}{x + (1-x)(1 + i\epsilon)} \quad \text{by (3)} \\ &= \frac{1 + (i-1)\epsilon}{1 + i\epsilon - x\epsilon} \geq \frac{k+1}{k},\end{aligned}$$

where the last inequality holds directly from $i \geq \frac{1}{\epsilon} \cdot \frac{1 + k\epsilon}{(k+1)x - 1}$.

Then by induction, we can see that the condition in Lemma 2 always holds before the large item has been cut k times.

By Lemmas 1 and 2, every time when the algorithm A cuts the large item, it cuts a portion of size less than $1/(k+1)$. Assume that the large item is discarded at size x. Therefore $x > 1 - k \times \frac{1}{k+1} = \frac{1}{k+1}$. Once the large item is discarded at phase $i > 0$, we stop the input L. At this step, $A(L) \leq (1 - x + \epsilon)(1 + i\epsilon)$ and $OPT(L) \geq 1 + (i-1)\epsilon$. Then the competitive ratio of algorithm A is at least

$$\begin{aligned}\frac{1 + \epsilon(i-1)}{(1 - x + \epsilon)(1 + i\epsilon)} &\geq \frac{1}{(1 - x + \epsilon)(1 + \epsilon)} \geq \frac{1}{(1 + 3\epsilon) - x(1 + \epsilon)} \\ &> \frac{k+1}{k + 3\epsilon(k+1)} > \frac{k+1}{k+r} = c,\end{aligned}$$

where the second inequality holds from $(1 + \epsilon)^2 \leq 1 + 3\epsilon$ and the third one holds from $x > \frac{1}{k+1}$ and $\epsilon > 0$.

After k times cutting, if the large item keeps staying in the knapsack after phase $i > \frac{1}{\epsilon}(\frac{1}{r}-1)$, we stop the input L just after phase i. Then by (3) $OPT(L) > \frac{1}{r}$ and $A(L)$ is at most $(1-x)OPT(L)+x$. Therefore, the competitive ratio after phase i is at least

$$\frac{OPT(L)}{(1-x)OPT(L)+x} \geq \frac{1}{(1-x)+x/OPT(L)} \geq \frac{1}{1-x(1-1/OPT(L))}$$
$$> \frac{1}{1-\frac{1-1/OPT(L)}{k+1}} = \frac{1+k}{k+1-(1-1/OPT(L))} > \frac{k+1}{k+r} = c,$$

where the third inequality follows from $x > \frac{1}{k+1}$ and $OPT(L) > 1$.

Hence, there exists an input L such that $OPT(L)/A(L) > c$, i.e., there is not an online algorithm with the competitive ratio strictly smaller than $\frac{k+1}{k}$. □

Remarks: In the model [12], the "cutting" is only allowed before packing, namely, when an item has been packed, it is not allow to cut it. In our model, there is not this restriction and we are allowed to cut items any time. So, our model is a generalization of the model in [12]. When $k = 1$, our upper and lower bounds are the same as the results in [12].

4 Resource Augmentation for the Online Maximization Knapsack with Limited Cuts

In this section, we study resource augmentation for the online maximization knapsack with limited cuts, in which the online algorithm uses a knapsack with capacity $m \geq 1$, while the offline algorithm uses a knapsack with capacity 1. We provide the competitive ratio in this model.

4.1 A Simple Greedy Algorithm

Let $L = \{a_1, a_2, ..., a_n\}$ be the online input. Assume that items $a_1, ..., a_{i-1}$ have been dealt by our algorithm. Let B_{i-1} be the set of items in the knapsack. The execution of our algorithm on item a_i is the following.

Observe that if there are some pieces of items discarded, then the total size in the knapsack is at least $m - \frac{1}{k+1}$. Due to the greedy police used in the above algorithm, we have the density of any item in the knapsack is not lower than the density of any item discarded. Then by the similar approach with Theorem 1, we have the following theorem.

Theorem 3. *The competitive ratio of algorithm B is* $\max\{1, \frac{k+1}{m(k+1)-1}\}$.

4.2 A Tight Lower Bound

In this subsection, we prove that the ratio $\max\{1, \frac{k+1}{m(k+1)-1}\}$ is the best possible ratio we can do, i.e., there is not an online algorithm with a competitive ratio strictly less than this ratio. The main ideas are the same with the model without resource augmentation. Here we only consider the non-trivial case $1 \leq m < \frac{k+2}{k+1}$ here. The details of the proof are in Appendix.

Algorithm: B for Resource Augmentation

1. $B_i' := B_{i-1} \cup \{a_i\}$, if $s(B_i') \leq m$ then accept item a_i, else
   (a) Rename all the items in B_i' as $b_1, b_2, ...$ such that $w(b_1)/s(b_1) \geq w(b_2)/s(b_2) \geq \cdots$, where $w(b_j)(s(b_j))$ is the weight(size) of item b_j.
   (b) Find a smallest index x such that $\sum_{h=1}^{x} s(b_h) > m$, remove all the items with index larger than x in B_i'.
   (c) If $\sum_{h=1}^{x-1} s(b_h) \geq m - \frac{1}{k+1}$, then remove item b_x. Else repeatedly chop off b_x by the following way until the total size in B_i' becomes at most m: if $s(b_x) \leq 1$ chop off by a fraction of size $\frac{1}{k+1}$ from item b_x else chop off by a fraction such that the remaining size of item b_x is exactly $\frac{k}{k+1}$.
2. Update set B_i.

Theorem 4. *No online algorithm has a competitive ratio smaller than* $\frac{k+1}{m(k+1)-1}$.

5 Online Minimization Knapsack with Limited Cuts

In this section, we consider the minimization version of knapsack problem, in which we are asked to select a subset of items to cover the knapsack such that the total weight of selected items is minimized.

Theorem 5. *No online algorithm has a constant competitive ratio for the minimization knapsack problem with limited cuts.*

References

1. Babat, L.G.: Linear functions on the N-dimensional unit cube. Dokl. AKad. Nauk SSSR 222, 761–762 (1975) (in Russian)
2. Csirik, J., Frenk, J.B.G., Labbé, M., Zhang, S.: Heuristics for the 0-1 Min-Knapsack problem. Acta Cybernetica 10(1-2), 15–20 (1991)
3. Güntzer, M.M., Jungnickel, D.: Approximate minimization algorithms for the 0/1 knapsack and subset-sum problem. Operations Research Letters 26, 55–66 (2000)
4. Gene, G., Levner, E.: Complexity of approximation algorithms for combinatorial problems: a survey. ACM SIGACT News 12(3), 52–65 (1980)
5. Horiyama, T., Iwama, K., Kawahara, J.: Finite-State Online Algorithms and Their Automated Competitive Analysis. In: Asano, T. (ed.) ISAAC 2006. LNCS, vol. 4288, pp. 71–80. Springer, Heidelberg (2006)
6. Han, X., Makino, K.: Online minimization knapsack problem. In: WAOA (to appear, 2009)
7. Iwama, K., Taketomi, S.: Removable online knapsack problems. In: Widmayer, P., Triguero, F., Morales, R., Hennessy, M., Eidenbenz, S., Conejo, R. (eds.) ICALP 2002. LNCS, vol. 2380, pp. 293–305. Springer, Heidelberg (2002)
8. Iwama, K., Zhang, G.: Optimal resource augmentations for online knapsack. In: Charikar, M., Jansen, K., Reingold, O., Rolim, J.D.P. (eds.) RANDOM 2007 and APPROX 2007. LNCS, vol. 4627, pp. 180–188. Springer, Heidelberg (2007)

9. Kellerer, H., Pferschy, U., Pisinger, D.: Knapsack Problems. Springer, Heidelberg (2004)
10. Lueker, G.S.: Average-case analysis of off-line and on-line knapsack problems. In: Proc. Sixth Annual ACM-SIAM SODA, pp. 179–188 (1995)
11. Marchetti-Spaccamela, A., Vercellis, C.: Stochastic on-line knapsack problems. Math. Programming 68(1, Ser. A), 73–104 (1995)
12. Noga, J., Sarbua, V.: An online partially fractional knapsack problem. In: ISPAN 2005, pp. 108–112 (2005)

Online Paging for Flash Memory Devices*

Annamária Kovács, Ulrich Meyer, Gabriel Moruz, and Andrei Negoescu

Institut für Informatik, Goethe-Universität Frankfurt am Main, Germany

Abstract. We propose a variation of online paging in two-level memory systems where pages in the fast cache get modified and therefore have to be explicitly written back to the slow memory upon evictions. For increased performance, up to α arbitrary pages can be moved from the cache to the slow memory within a single joint eviction, whereas fetching pages from the slow memory is still performed on a one-by-one basis. The main objective in this new α-paging scenario is to bound the number of evictions. After providing experimental evidence that α-paging can improve the performance of flash-memory devices in the context of translation layers we turn to the theoretical connections between α-paging and standard paging. We give lower bounds for deterministic and randomized α-paging algorithms. For deterministic algorithms, we show that an adaptation of LRU is strongly competitive, while for the randomized case we show that by adapting the classical Mark algorithm we get an algorithm with a competitive ratio larger than the lower bound by a multiplicative factor of approximately 1.7.

1 Introduction

In recent years flash memory is becoming increasingly popular as a viable storage support, especially for mobile computing. Flash memory devices are lighter, more shock-resistant, and consume less power than traditional hard-disks. For these reasons, flash memory is an appealing solution for end-user storage, partly even replacing traditional hard-disks. Motivated by the fact that, unlike traditional hard-disks, flash memory achieves the best performance when writes are done in blocks of size larger than the read block sizes [1], in this paper we consider paging algorithms for these devices. The key difference to traditional paging is that when a page fault occurs and the memory is full, instead of evicting only one page, up to α pages can be jointly evicted before the new page is loaded, for some fixed parameter $\alpha \geq 1$. The goal is to minimize the number of evictions.

Flash memory. Flash memory consists of an array of memory cells, divided into a number of *blocks* of α consecutive *pages*, where each page is a group of consecutive memory cells. Reading and writing are done on a page basis, but overwriting single pages is usually not possible. Instead, overwriting is done by *erasing a whole block* and then writing the new data. Since each block can sustain

* Partially supported by the DFG grant ME 3250/1-1, and by MADALGO – Center for Massive Data Algorithmics, a Center of the Danish National Research Foundation.

Y. Dong, D.-Z. Du, and O. Ibarra (Eds.): ISAAC 2009, LNCS 5878, pp. 352–361, 2009.

only a limited number of erase operations, typical flash memory devices include a wear-leveling mechanism that ensures an even usage of the blocks in time.

Because erase operations are slow, in applications that modify (i.e., overwrite) pages on disk in an unstructured way hardly any performance gain is obtained when replacing hard-disks by flash memory. Frequently, this problem can be resolved using an additional intermediate software-layer, i.e. different than the wear-leveling mechanism, that bundles up to α write requests (pages) and writes them jointly to new consecutive locations, thus exploiting the improved performance when writing in larger blocks. To subsequently find the respective data under their new locations on the flash-device, an internal-memory translation-table for page locations has to be maintained, too. Also, occasionally device space needs to be reclaimed by compressing blocks with respect to outdated pages. Such software translation-layers can be found both in algorithmic research (e.g. [1]) and commercial products (e.g. EasyCo's Managed Flash Technology [2]). Instead of actually transferring data blocks back and forth between main memory and flash device it is even more efficient to buffer as many blocks as possible in internal-memory. This is classically done using paging algorithms. Motivated by the asymmetry between reads and writes in flash devices, we adapt classical paging by having evictions done in groups of up to α pages.

Most other previous algorithmic works for flash memory focused on memory management and wear-leveling, i.e. another block re-mapping *within* the flash device to avoid a premature block wear-out, and flash-tailored file-systems (see e.g. [3] for an overview). Typically the software translation-layer with its write-page bundling has a positive effect with respect to efficiency. Recently, thorough benchmarks for flash memories were conducted [4,5], and based on their findings computational models exploiting the characteristics of these devices were proposed [1]. Other works use flash memory for model checking [6], route planning on mobile devices [7,8], or on flash-aware R-trees and dictionaries [9,10,11].

Paging algorithms. Online algorithms are not provided with the input in advance and therefore must serve input requests as they arrive. To measure the efficiency of such algorithms, Sleator and Tarjan [12] considered comparing their cost against the cost of an optimal offline algorithm, i.e. an algorithm that knows the input sequence in advance and processes it optimally. The resulting measure, denoted later *competitive ratio* [13], states that an online algorithm A is c-competitive if $A(\sigma) \leq c \cdot OPT(\sigma) + b$ for any input sequence σ, where b is a constant, and $A(\sigma)$ and $OPT(\sigma)$ are the costs of A and an optimal offline algorithm respectively (if A is randomized, $A(\sigma)$ is the expected cost of A).

Over the last decades, paging has been extensively studied in a variety of settings. In classical paging, we are provided with a two-level memory, a fast memory that can hold up to k pages and a disk that can store infinitely many pages. Given as input a sequence of pages, an algorithm must decide which pages to store in the memory so that it incurs as few page faults as possible, where a page fault occurs when some page does not reside in the memory when requested. In [12] it was proved that the competitive ratio of any deterministic algorithm is at least k, and that popular algorithms such as FIFO and LRU match this bound. Fiat et al. [14] proved a competitive ratio of at least H_k for any randomized paging algorithm, where $H_k = \sum_{i=1}^{k} 1/i$ is the k-th harmonic number. They gave an algorithm, denoted Mark, which is $(2H_k - 1)$-competitive. This bound was further improved in [15], where a H_k-competitive algorithm was proposed. More recently, Achlioptas et al. [16] gave another H_k-competitive algorithm which is more practical. For a detailed view on paging algorithms, we refer the interested reader to comprehensive surveys [17,18].

Our results. We propose α-paging as an adaptation of classical paging to improve the practical behavior of flash memory devices in the context of software translation layers like EasyCo's Managed Flash Technology. It is similar to classical paging, except that arbitrary sets of up to α pages are jointly evicted. Since in practice writes are typically more expensive than reads, we count the number of such joint evictions instead of page faults. More specifically, we are provided with a fast memory that can hold k pages and a slow memory which can hold infinitely many pages. The input consists of a sequence σ of pages to be served by the algorithm. For some request of page p, if it is not in the memory we say that a page fault occurs. Evicting pages from the fast to the slow memory is done in groups of at most α arbitrary pages. Therefore, each eviction increases the amount of free slots in the memory by up to α. As previously specified, the cost of the algorithm is given by the number of evictions performed. More generally, at any step, jointly evicting x pages costs $\lceil x/\alpha \rceil$.

We show that in our model it is easy to adapt classical paging algorithms, such as the optimal offline MIN [19], LRU, and Mark [14]. However, due to the fact that up to α pages are jointly evicted instead of only one, competitive ratios achieved by these algorithms are different and their analysis becomes significantly more involved. We prove lower bounds on the competitive ratio for randomized and deterministic online algorithms. In particular, the competitive ratios of deterministic and randomized algorithms cannot be smaller than k/α and $(H_{k+\alpha-1} - H_{\alpha-1})/(H_{2\alpha-1} - H_{\alpha-1})$ respectiviely, which are generalizations of the respective lower bounds for $\alpha = 1$. We show that, like in classical paging, our adaptation of LRU matches the deterministic lower bound. For our randomized version of Mark we prove that it achieves a competitive ratio of $((H_k - H_{2\alpha-1})/(H_{3\alpha-1} - H_{2\alpha-1})) + 3$. For large enough values of k and α this bound is by a factor of about 1.7 larger than the lower bound, whereas the classical Mark has a competitive ratio twice the lower bound.

2 α-Paging

We first give empirical results motivating α-paging, then we discuss generic properties of α-paging algorithms as generalizations of classical paging algorithms.

Motivation. We conduct experiments to demonstrate the practical relevance of writing in large blocks of data. We perform random writes in a very large array (about 1.5 the size of the memory) in two different settings. In the first one we

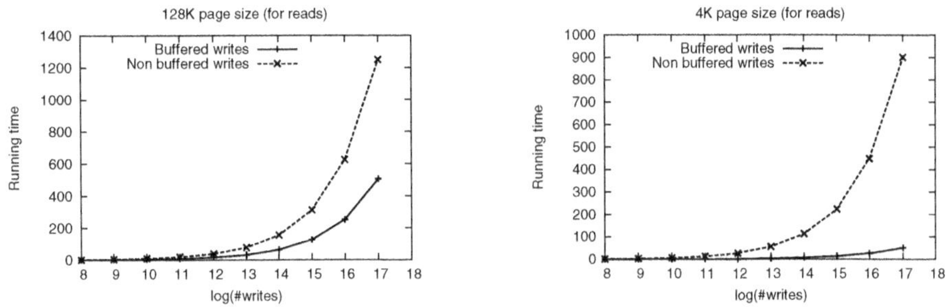

Fig. 1. Running times (in seconds) for buffered and non-buffered random writes on SSDs, when reading in blocks of 128 KB (left) and 4 KB(right)

write the modified page immediately, whereas in the second one we employ a translation layer, as in [1], which groups modified pages and writes them on the disk as a large block. This way, *any* pages can be grouped together in neighboring physical locations on the flash disk, regardless of the addresses from where they were loaded in memory. We measure the running time when varying the amount of random writes. Data is read in blocks of sizes 128KB and 4KB respectively, and written back to the flash disk in blocks of size 4MB when buffered. We note that for the disk used the best performance is achieved when the block size for reading is 128KB, and writing in blocks of 4MB also yields good performance.

The results of our experiments are shown in Figure 1. We note that in both cases writing large buffers, corresponding to evicting many pages at once, achieves significantly better performance than when writing data non-buffered. The improvements in running times are of about 250% and 1800% when the read-block size is 128KB and 4KB respectively. This confirms that evicting large blocks, i.e. groups of pages, yields significant performance improvements.

α-paging and classical paging. We note that α-paging is a generalization of classical paging. Every paging algorithm in the classical model is a valid α-paging algorithm and vice versa, by performing identical page replacements in both models. We denote by $A^\alpha(\sigma)$ the cost of some algorithm A when processing the request sequence σ in the α-paging model (note that $\alpha = 1$ corresponds to classical paging). Since an eviction in the α-model corresponds to at most α evictions in the classical model, we have that $A^\alpha(\sigma) \leq A^1(\sigma) \leq \alpha \cdot A^\alpha(\sigma)$. This inequality also holds for the cost of the optimal offline algorithm denoted by $OPT^\alpha(\sigma)$, which adapts its decisions to the value of α. Given a c-competitive online algorithm A in the α-model for fixed $\alpha \geq 1$, we obtain that A has a competitive ratio of at most $\alpha \cdot c$ in the classical model:

$$\frac{A^1(\sigma)}{OPT^1(\sigma)} \leq \frac{\alpha \cdot A^\alpha(\sigma)}{OPT^1(\sigma)} \leq \frac{\alpha \cdot A^\alpha(\sigma)}{OPT^\alpha(\sigma)} \leq \alpha c\ .$$

Lemma 1. *If c is a lower bound on the competitive ratio in classical paging then c/α is a lower bound on the competitive ratio in α-paging.*

We observe that, similarly to the case of classical paging, we can restrict ourselves to *lazy* algorithms. We call an α-paging algorithm *lazy*, if it performs an eviction only when the memory is full, and a page fault occurs; in this case it evicts at most α pages. Lemma 2 can be proved using a step-by-step modification of a general algorithm into a lazy one [20].

Lemma 2. *For any α-paging algorithm A, a lazy algorithm B exists such that $B^\alpha(\sigma) \leq A^\alpha(\sigma)$ for every input sequence σ.*

3 Lower Bounds

Recall that, for the competitive ratio of online paging algorithms, lower bounds of k and H_k were given in the deterministic [12] and randomized [14] settings, respectively. We generalize these bounds for α-paging algorithms. For deterministic α-paging, the result in Corollary 1 follows immediately from Lemma 1.

Corollary 1. *Every deterministic online algorithm for α-paging has a competitive ratio of at least k/α.*

For randomized α-paging, using the result in Lemma 1 yields a lower bound of H_k/α. In Lemma 3 this bound is significantly improved.

Lemma 3. *Every randomized online α-paging algorithm has a competitive ratio of at least $(H_{k+\alpha-1} - H_{\alpha-1})/(H_{2\alpha-1} - H_{\alpha-1})$.*

Proof. Let c_r be the claimed lower bound. By Yao's minimax principle for cost minimization problems [18], it suffices to prove that there exists a set of request sequences and a probability distribution over these inputs, such that the expected cost of any deterministic online algorithm is at least c_r times more than the expected cost of an optimal offline algorithm.

Consider input sequences that first request pages $(1, \dots, k+1)$ followed by n requests to pages in $\{1, \dots, k+\alpha\}$, drawn uniformly at random. For such inputs we prove [20] that the expected number of requests between two evictions is $(k+\alpha)(H_{2\alpha-1} - H_{\alpha-1})$ and $(k+\alpha)(H_{k+\alpha-1} - H_{\alpha-1})$ for any deterministic online algorithm and for optimal offline algorithms respectively. The proof follows.

4 Deterministic α-Paging

In this section we discuss deterministic α-paging algorithms. We give in Lemma 4 a lower bound on the number of evictions done by any offline algorithm.

Lemma 4. *Consider an arbitrary input sequence σ that we split into intervals $I_0, \dots, I_l$, so that I_j contains k pairwise distinct pages and is maximal with respect to this property, for all $j = 0, \dots, l-1$. Then any offline algorithm performs at least l evictions.*

Proof. Let A be an algorithm and $I_{j_1}, \ldots, I_{j_n}$ be all intervals where A performs no eviction. We first prove that for each pair $(I_{j_i}, I_{j_{i+1}})$ there exists an interval $I_{x'}$, with $j_i < x' < j_{i+1}$, such that A performs at least two evictions while processing $I_{x'}$. Assume that there exists some pair $(I_{j_i}, I_{j_{i+1}})$, such that each interval I_x performs one eviction, for all x, with $j_i < x < j_{i+1}$. Since I_{j_i} does no eviction, after its processing the memory is full and contains all pages requested in I_{j_i}. If some interval I_x starts with a full memory containing all pages in I_{x-1}, then the first page in I_x triggers an eviction, since by definition it is not requested in I_{x-1}. If there occurs no other eviction in I_x, after processing I_x the memory is full and contains all pages requested in I_x, since I_x contains k pairwise distinct pages. Therefore, if all I_x, with $j_i < x < j_{i+1}$, perform only one eviction, then the first request in $I_{j_{i+1}}$ causes an eviction, which is a contradiction. Therefore, between two intervals where A performs no evictions there exists one interval where it performs at least two evictions and this concludes the proof.

We propose an adaptation of the optimal offline algorithm MIN [19] from the classical paging, that we denote α-MIN, and prove that it achieves optimality also in α-paging. Upon a page request that is not in the memory, the MIN algorithm evicts the page whose first request occurs furthest away in the future. Similarly, upon a page fault when the memory is full, α-MIN evicts the α pages whose first requests occur furthest away in the future.

Lemma 5. *The α-MIN algorithm is optimal for α-paging.*

Due to space limitations, the proof of Lemma 5 is included in [20]. Like in the classic case, we modify an optimal algorithm step by step to eventually obtain α-MIN. Similarly to MIN, we adapt the classical LRU to the α-paging setting and obtain α-LRU which, when the memory is full and the requested page is not in memory, evicts the α least recently requested pages from the memory. We show that this algorithm achieves a competitive ratio of k/α, which is optimal.

Lemma 6. *α-LRU is k/α-competitive.*

Proof. We split the input sequence into consecutive intervals $(I_0, \ldots, I_l)$, each of them being maximal in requesting k pairwise distinct pages, except for I_l. We first prove that for each I_j, with $1 \leq j < l$, α-LRU performs at most k/α evictions. The first $k - \alpha$ pairwise distinct page requests do not evict any page previously requested in I_j, since there exist α pages requested less recently. These $k - \alpha$ pages cause at most $k/\alpha - 1$ evictions. If the remaining α pairwise distinct pages cause an eviction, then the memory contains only pages requested in I_j after this eviction, and thus no further eviction is possible in I_j. We conclude that while processing I_j α-LRU performs at most k/α evictions. Since by Lemma 4 any optimal offline algorithm performs amortized one eviction for I_j, we conclude that α-LRU is k/α-competitive.

5 Randomized α-Paging

We introduce an adaptation of the classical Mark algorithm, denoted α-Mark. Similarly to Mark, α-Mark keeps track of an interval splitting where each interval consists of exactly k pairwise distinct page requests. Additionally, α-Mark assigns priorities to pages, and pages are evicted based on these priorities.

α-Mark. Each page p is marked upon request. Assume that p causes a page fault, and the memory is full, containing x unmarked pages. If $x \geq 1$, then α-Mark evicts the subset of $\min\{x, \alpha\}$ unmarked pages with lowest priorities. In case $x = 0$, all pages get unmarked and are assigned k pairwise different priorities uniformly at random before choosing the pages to be evicted.

The analysis is based on interval splitting, where an interval ends just before all pages get unmarked. The key difference in analyzing the performance of Mark and α-Mark in such an interval is that the probability that some page p causes a page fault is determined solely by the input for classical Mark, whereas in the case of α-Mark it depends also on the random decisions before the request of p. Additionally, page faults can increase and decrease the probability of future page faults in an interval, such that an important simplifying assumption about the structure of the intervals (see Lemma 7) is not obvious like for classical Mark. We prove this assumption by using priorities instead of choosing a set of unmarked pages uniformly at random to be evicted.

Intervals. Suppose that the algorithm splits the input into the consecutive intervals $I_1, \cdots, I_l$, each containing k pairwise distinct pages, maximal with respect to this property. We call the pages requested in I_j but not in I_{j-1} *new pages* and denote their number by n_j. Pages requested in I_{j-1} are called *old pages.* In I_j exactly $o_j = k - n_j$ old pages are requested. When I_j starts, the memory contains all old pages, all of them are unmarked and have distinct priorities.

Lemma 7. *Let n_j be the number of pairwise distinct new pages requested in I_j. The expected number of evictions done by α-Mark in I_j does not decrease if we assume that:*

- *every page is requested only once in I_j;*
- *all n_j new pages are requested before all the old pages.*

Proof. The first claim follows immediately, since a second request of a page changes neither the number of evictions nor the state of the algorithm. To prove the second assumption we fix the priorities assigned to old pages at the beginning of each interval and obtain a deterministic algorithm D. It can be shown that D maximizes its cost if all new pages are requested before the old pages in I_j [20].

Expected cost of α-Mark. For some input σ, we bound the expected number of evictions done by α-Mark in an interval I_j, $(j \geq 1)$.

Lemma 8. *Consider an interval I_j, $j \geq 1$, in which n_j new pages are requested, and let $m_j = \lceil n_j/\alpha \rceil$. The expected number of evictions done by α-Mark in I_j is at most*

$$m_j + 1 + \frac{H_k - H_{n_j-1}}{H_{n_j+\alpha-1} - H_{n_j-1}} \, .$$

Proof. By Lemma 7, we assume that I_j consists of n_j requests of new pages followed by $o_j = k - n_j$ old page requests, and every page is requested only once in I_j. The new pages cause m_j evictions and all further evictions are caused by old pages.

We denote by $S(h)$ a state that occurs immediately after an eviction caused by an old page, where h is the number of old pages not yet requested (including the n_j old pages which are not requested at all) in I_j. Let y be the number of old pages not in memory for some $S(h)$. We have that $y = n_j + \alpha - 1$, since the memory contains n_j new pages, and $\alpha - 1$ free slots (as well as $k - h$ marked, and $h - n_j - \alpha + 1$ unmarked old pages).

For $S(h)$, let $(p_1, p_2, \ldots, p_h)$ be the old pages not yet requested in the order they appear in I_j; the pages not requested at all in I_j are at the end in arbitrary order. If p_i causes the next eviction then $S(h-i)$ occurs immediately after processing p_i. Let $E(h)$ be the expected number of evictions from state $S(h)$ until the end of I_j, and let $pb(i)$ be the probability that page p_i causes the next eviction. We have that:

$$n_j \leq h \leq n_j + \alpha - 1: \quad E(h) = 0 \,, \tag{1}$$

$$h > n_j + \alpha - 1: \quad E(h) = \sum_{i=\alpha}^{h-n_j} pb(i) \cdot (1 + E(h-i)) \,. \tag{2}$$

The base case is correct because there are $\alpha - 1$ empty locations in the memory and at most $\alpha - 1$ pages are requested until the end of I_j. For the recursive part we note that pages p_i, with $i < \alpha$, cannot cause an eviction because $\alpha - 1$ page faults must fill the memory before an eviction could take place. On the other hand, pages p_i, with $i > h - n_j$, are not requested at all in I_j and thus cannot cause evictions.

The next eviction is caused by p_i iff the y old pages not in memory at $S(h)$ contain p_i, moreover exactly $\alpha - 1$ pages from $(p_1, p_2, \ldots, p_{i-1})$, and $y - \alpha$ pages from $(p_{i+1}, \ldots, p_h)$. The total number of subsets of size y chosen from $(p_1, \ldots, p_h)$ with this property is $\binom{i-1}{\alpha-1} \cdot \binom{1}{1} \cdot \binom{h-i}{y-\alpha}$. Since all not yet requested pages have the same probability not to reside in memory, we get $pb(i) = \binom{i-1}{\alpha-1}\binom{h-i}{y-\alpha}/\binom{h}{y}$. For $h > n_j + \alpha - 1$, using $y = n_j + \alpha - 1$ and $\sum pb(i) \leq 1$, we obtain:

$$E(h) \leq 1 + \frac{1}{\binom{h}{n_j+\alpha-1}} \sum_{i=\alpha}^{h-n_j} \binom{i-1}{\alpha-1} \cdot \binom{h-i}{n_j-1} \cdot E(h-i) \,.$$

Surprisingly, we found [20] that the function $f(h)$, defined as

$$f(h) = \frac{H_h - H_{n_j-1}}{H_{n_j+\alpha-1} - H_{n_j-1}} \,,$$

satisfies nearly the same recurrence:

$$f(h) = 1 + \frac{1}{\binom{h}{n_j+\alpha-1}} \sum_{i=\alpha}^{h-n_j+1} \binom{i-1}{\alpha-1} \cdot \binom{h-i}{n_j-1} \cdot f(h-i) \,.$$

By induction it can be proven that $f(h) \geq E(h)$ for all h, $n_j \leq h \leq k-1$. Since the new pages in I_j cause m_j evictions, and the expected number of evictions caused by old pages is at most $1 + f(k-1) \leq 1 + f(k)$, the proof concludes.

Competitive ratio. Let σ be the input sequence, $(I_0, I_1, \ldots, I_l)$ the interval splitting, and consider n_j, the number of new pages requested in some interval I_j ($j \geq 1$), with $m_j = \lceil n_j/\alpha \rceil$.

Lemma 9. *Let $OPT(\sigma)$ be the number of evictions done by an optimal offline algorithm when processing σ. Then, $OPT(\sigma) \geq \max\left\{\frac{1}{2}\sum_{j=1}^{l} m_j, l\right\}$.*

Proof. By Lemma 4, OPT does at least l evictions. For some $j > 1$, the intervals I_j and I_{j-1} contain $k + n_j$ pairwise distinct pages, thus at least n_j of them are filled in memory slots resulted by evictions *during* I_{j-1} and I_j. Therefore, the number of evictions in these two intervals is at least m_j. We get the following:

$$OPT(\sigma) \geq \sum_{j\ odd, j\geq 1} m_j, \qquad OPT(\sigma) \geq \sum_{j\ even, j\geq 1} m_j\ .$$

Since $\max\{a, b\} \geq \frac{1}{2}(a+b)$, we obtain that $OPT(\sigma) \geq \frac{1}{2}\sum_{j\geq 1} m_j$.

Theorem 1. *The competitive ratio of α-Mark is at most*

$$3 + \frac{H_k - H_{2\alpha-1}}{H_{3\alpha-1} - H_{2\alpha-1}}\ .$$

Proof. Let M_j be the upper bound from Lemma 8 on the expected number of evictions done by α-Mark in interval I_j:

$$M_j = m_j + 1 + F(n_j), \qquad F(n_j) = \frac{H_k - H_{n_j-1}}{H_{n_j+\alpha-1} - H_{n_j-1}}\ .$$

We can assume w.l.o.g., that $n_j = m_j \cdot \alpha$, since it has no influence on the lower bound for OPT in Lemma 9, while in the upper bound for α-Mark M_j is increasing in n_j.

If $\frac{1}{2}\cdot\sum_{j=1}^{l} m_j < l$, then when incrementing some m_j the sum $\sum M_j$ increases. However, this does not affect the lower bound of OPT and thus we can assume that $\sum_{j=1}^{l} m_j \geq 2l$. Let $m = \sum_{j=1}^{l} m_j$. For fixed $m \geq 2l$, if the upper bound on the number of evictions done by α-Mark is maximal, then $m_j \geq 2$ for all j [20]. Denoting C the competitive ratio of α-Mark, we have:

$$C \leq \frac{\sum_{j=1}^{l} M_j}{\frac{1}{2}\sum_{j=1}^{l} m_j} \leq 3 + 2\frac{\sum_{j=1}^{l} F(m_j\cdot\alpha)}{\sum_{j=1}^{l} m_j} \leq 3 + 2\max_j\left\{\frac{F(m_j\cdot\alpha)}{m_j}\right\}\ .$$

Since $\frac{F(x\cdot\alpha)}{x}$ is decreasing in x, and $m_j \geq 2$, we get $C \leq 3 + \frac{H_k - H_{2\alpha-1}}{H_{3\alpha-1} - H_{2\alpha-1}}$.

By Lemma 3, for large enough k and α, approximating $H_x \approx \ln x + \gamma$, $\gamma \approx 0.57$, any randomized algorithm has a competitive ratio $C_{lb} \geq \ln(k/\alpha + 1)/\ln(2)$. Similarly, the competitive ratio of α-Mark satisfies $C \leq \ln(k/\alpha)/\ln(1.5)$, and thus the gap between the upper bound and the lower bound is approximately $\ln 2/\ln(1.5) \approx 1.7$, as opposed to a tight factor of 2 in the case of classical Mark.

References

1. Ajwani, D., Beckmann, A., Jacob, R., Meyer, U., Moruz, G.: On computational models for flash memory devices. In: Proc. 8th International Symposium on Experimental Algorithms, pp. 16–27 (2009)
2. EasyCo: Managed flash technology, `http://www.easyco.com/mft/`
3. Gal, E., Toledo, S.: Algorithms and data structures for flash memories. ACM Computing Surveys 37(2), 138–163 (2005)
4. Ajwani, D., Malinger, I., Meyer, U., Toledo, S.: Characterizing the performance of flash memory storage devices and its impact on algorithm design. In: McGeoch, C.C. (ed.) WEA 2008. LNCS, vol. 5038, pp. 208–219. Springer, Heidelberg (2008)
5. Bouganim, L., Jónsson, B.P., Bonnet, P.: uFLIP: Understanding Flash IO Patterns. In: Proc. 4th biennial conference on innovative data systems (CIDR) (2009)
6. Barnat, J., Brim, L., Edelkamp, S., Sulewski, D., Šimeček, P.: Can flash memory help in model checking? In: Proc. 13th International Workshop on Formal Methods for Industrial Critical Systems, pp. 159–174 (2008)
7. Goldberg, A.V., Werneck, R.: Computing point-to-point shortest paths from external memory. In: Proc. 7th Workshop on Algorithm Engineering and Experiments, pp. 26–40 (2005)
8. Sanders, P., Schultes, D., Vetter, C.: Mobile route planning. In: Halperin, D., Mehlhorn, K. (eds.) ESA 2008. LNCS, vol. 5193, pp. 732–743. Springer, Heidelberg (2008)
9. Li, Y., He, B., Luo, Q., Yi, K.: Tree indexing on flash disks. In: Proc. 25th International Conference on Data Engineering, 1303-1306 (2009)
10. Wu, C.H., Chang, L.P., Kuo, T.W.: An efficient R-tree implementation over flash-memory storage systems. In: Proc. 11th ACM International Symposium on Advances in Geographic Information Systems, pp. 17–24 (2003)
11. Wu, C.H., Kuo, T.W., Chang, L.P.: An efficient B-tree layer implementation for flash-memory storage systems. ACM Transactions on Embedded Computing Systems 6(3) (2007)
12. Sleator, D.D., Tarjan, R.E.: Amortized efficiency of list update and paging rules. Communications of the ACM 28(2), 202–208 (1985)
13. Karlin, A.R., Manasse, M.S., Rudolph, L., Sleator, D.D.: Competitive snoopy caching. Algorithmica 3, 77–119 (1988)
14. Fiat, A., Karp, R.M., Luby, M., McGeoch, L.A., Sleator, D.D., Young, N.E.: Competitive paging algorithms. Journal of Algorithms 12(4), 685–699 (1991)
15. McGeoch, L.A., Sleator, D.D.: A strongly competitive randomized paging algorithm. Algorithmica 6(6), 816–825 (1991)
16. Achlioptas, D., Chrobak, M., Noga, J.: Competitive analysis of randomized paging algorithms. Theoretical Computer Science 234(1-2), 203–218 (2000)
17. Albers, S.: Online algorithms: a survey. Mathematical Programming 97(1-2), 3–26 (2003)
18. Borodin, A., El-Yaniv, R.: Online computation and competitive anlysis. Cambridge University Press, Cambridge (1998)
19. Belady, L.A.: A study of replacement algorithms for virtual-storage computer. IBM Systems Journal 5(2), 78–101 (1966)
20. Kovács, A., Meyer, U., Moruz, G., Negoescu, A.: Online paging for flash memory devices. Technical report, Goethe-Universität Frankfurt am Main (2009)

Shifting Strategy for Geometric Graphs without Geometry*

Imran A. Pirwani

Department of Computing Science
University of Alberta
Edmonton, Alberta T6G 2E8, Canada
pirwani@cs.ualberta.ca

Abstract. We give a simple framework as an alternative to the celebrated shifting strategy of Hochbaum and Maass [J. ACM, 1985] which has yielded efficient algorithms with good approximation bounds for numerous optimization problems in low-dimensional Euclidean space. Our framework does not require the input graph/metric to have a geometric realization – it only requires that the input graph satisfy some weak property referred to as *growth boundedness*. We show how to obtain *polynomial time approximation schemes* (PTAS) for maximum (weighted) independent set problem on this graph class. Via a more sophisticated application of our framework, we also show how to obtain a PTAS for the maximum (weighted) independent set for intersection graphs of (low-dimensional) *fat objects* that are expressed without geometry.

1 Introduction

In their seminal work, Hochbaum and Maass [8] introduced a *divide-and-conquer* paradigm which is referred to as the *shifting strategy*. They applied their strategy to obtain PTAS for several covering and packing problems on instances in *low-dimensional Euclidean space* [8]. Simply put, the strategy first decomposes the original instance into several instances of "bounded width" [8]; each instance is amenable to "good", efficient solutions. The solutions to the instances are combined to obtain a feasible solution to the original instance, for the given decomposition. For any given $\varepsilon > 0$, the strategy considers $O(\frac{1}{\varepsilon^2})$ distinct decompositions and returns the best solution over the decompositions considered. It can be shown (depending on the problems they solve) that the cost of the returned solution is within a factor of $(1+\varepsilon)$ of an optimal one. The celebrated strategy has enjoyed success for a number of problems since their inception [6,5].

The shifting strategy [8] has required three ingredients: (**R1**) the input be expressed with geometry; (**R2**) the nature of the optimization problem be such that for some decomposition (into clusters of bounded diameter) of the input instance, the "interaction" between clusters be highly limited, that is, the problem be "separable" with respect to the objective; and (**R3**) the problem admit

* Supported by Alberta Ingenuity.

Y. Dong, D.-Z. Du, and O. Ibarra (Eds.): ISAAC 2009, LNCS 5878, pp. 362–371, 2009.

a reasonably good and efficient approximation for clusters in the decomposition. Note that (**R1**) pertains to the representation of the input, while (**R2**) and (**R3**) are properties of the objective function, and possibly of the input representation. This observation leads to this question which is the main motivation behind this work: how can one employ a strategy akin to the shifting strategy on combinatorial objects for which geometric representation exists, but is not part of the input? Our question is related to a line of research that bypasses the costly and potentially erroneous embedding step and attempts to solve the problems directly on the input metric/graph using special properties which are weaker than the properties of Euclidean spaces [11,10]; our work can be seen as being part of this line of research and is largely inspired by the work of Talwar [11] who gave a decomposition scheme for *doubling metrics* which he used to obtain approximation algorithms for various problems on such metrics.

Techniques and Results: We give a simple framework as an alternative to the shifting strategy [8]. Our framework does not require the input graph/metric to be geometric – it only requires that the input graph satisfy a weak property called *polynomially bounded growth* in literature [10]; intersection graphs of *fat* objects in low-dimensional Euclidean space satisfy this property. Graphs that have this property are called *growth bounded graphs*; we define this concept later. At the core of the framework is a probabilistic decomposition scheme that is an adaptation of Talwar's *split-tree decomposition* [11] but has some key differences: our decomposition scheme applies to metrics whose *doubling dimension* is not necessarily bounded. For example, the hop metric induced by growth bounded graphs does not have doubling dimension bounded by a constant [7]. Hence, an adaptation of Talwar's technique yielding good approximations for unrelated problems on unrelated graph classes is surprising.

Our scheme is akin to the shifting strategy in the sense that it has similar desirable properties for instances where (**R1**) can be relaxed without being required to relax (**R2**) and (**R3**). Our technique requires *very little randomness* and can be readily derandomized using the technique of *deterministic coin tossing* (see Talwar [11], for example). We only present the randomized variant.

As an application, in Section 2 we obtain a PTAS for the maximum weighted independent set problem on growth bounded graphs. We can also obtain PTAS for *minimum vertex cover*, *minimum dominating set* problems on these graphs, and a $(2 + \varepsilon)$ approximation for *minimum weighted clique partition* on UDGs; we defer a full treatment of these problems to a longer version of this article. These graphs are expressed in *standard form*, say as adjacency matrix.

In Section 3, we use this scheme to obtain a PTAS for the maximum weighted independent set problem on the intersection graph of low-dimensional *fat* objects[1] expressed without geometry, by way of a sophisticated use of the shifting strategy. Prior solutions for this problem have required use of geometry which were independently discovered by Erlebach et al. [5] and Chan [2]. Erlebach et al. [5] asked if it is possible to obtain a PTAS in the case when the geometric representation is *not* given. We answer this question in the affirmative; we

[1] For simplicity, we only present the case of disks in the Euclidean plane.

obtain a PTAS for the weighted independent set problem on low-dimensional ball graphs without the use of geometry. Instead of a geometric representation of the ball graph, we assume that some, albeit unknown, embedding exists in low-dimensional Euclidean space. We only require access to edge lengths and radii corresponding to balls centered at the input points with respect to some (unknown) realization. The UDG recognition problem is NP-complete even if all edge-lengths are known [1], and if the graph is a-priori known to be a UDG, Kuhn et al. [9] show that realizing the UDG is APX-hard. So, from a complexity theoretic perspective, having access to edge-lengths and radii with respect to an unknown realization is likely to be a significant weakening of the requirement that the input be a geometric representation. At the same time, a review of literature on wireless networks suggests that obtaining reasonable estimates on distances between nearby nodes is feasible (see [4,1], for example).

Related Work: Nieberg et al. invented techniques that yield PTAS for maximum (weighted) independent set, and minimum dominating set problems on growth bounded graphs, of which UDG is a subclass [10]. PTAS for such problems existed on subclasses of graphs e.g. intersection graphs of unit disks, fat objects, etc. [8,5,2], but they relied on geometric input. Until the work of Nieberg et al. [10], it was not known how to remove this reliance on geometry. Their techniques are not known to extend beyond growth bounded graphs [10].

Chan and Har-Peled [3] show that a simple *local search* on admissible objects, that is, a collection of geometric objects in the plane such that the *set difference* between any pair is connected, yields a PTAS for the *unweighted* case; they assume that the input is expressed in standard form. For the case of fat objects in low-dimensional space, they also show that their local search algorithm yields a PTAS [3]. Unfortunately, their technique does not generalize to the weighted case. They [3] also give a $O(1)$-approximation to the maximum weighted independent set problem on *pseudodisks* expressed in standard form.

Preliminaries: Let $G = (V, E)$, $|V| = n$, be a given graph. For any subgraph H of G, let $V(H)$ denote the set of vertices of H, but for simplicity, we refer to a vertex $u \in H$ to mean $u \in V(H)$. For any subgraph H of G and a pair of vertices $u, v \in H$, let $d_H(u, v)$ denote the length of the shortest path from u to v in H; if the graph is edge-weighted, $\ell : E \mapsto \mathbb{R}_{\geq 0}$, then $d_H(u, v)$ is the shortest weighted path between the two in H. When the context is clear, we simply refer to $d(u, v)$ to mean $d_H(u, v)$, for convenience. For a vertex $u \in H, r \geq 0$, let $B_H(u, r) = \{v \in H : d_H(u, v) \leq r\}$. When the context is clear, we refer to it as $B(u, r)$, for convenience. For a vertex $v \in G$ and $S \subset G$, let $d(v, S) = \min\{d(v, w) : w \in S\}$.

G is *f-growth bounded* if for every $v \in G$ and $r \geq 0$, every independent set of $B(v, r)$ has size at most $f(r)$ for function f which only depends on r. If $f(r)$ is a polynomial in r, then we say that G is a *growth bounded graph* (GBG).

For a graph G, an *r-cover* of V is a subset $S \subset V$ such that for any vertex $v \in V$, $d(v, S) \leq r$. A set S is called an *r-packing* if for any pair $u, v \in S, d(u, v) \geq 2r$. A set $S \subset V$ is called an *r-net* in V, if S is an r-cover and an $\frac{r}{2}$-packing. An r-net can be constructed greedily since any minimal r-cover is an r-net [11].

Let OPT denote an optimal solution and opt $= \mathbf{wt}$ (OPT), its value.

2 Random-Shifting Framework for GBGs

Here, we present the details of the framework which is analogous to the random shifting technique. Recall, the idea behind the grid-shifting strategy [8] is to first draw a crude partition of the Euclidean space via the use of a random grid to create a crude decomposition of the instance into multiple subproblems, from a distribution of crude partitions. The granularity of the grid depends inversely upon an input parameter $\varepsilon > 0$. This step is followed by solving the optimization problem on those subproblems exactly. The final step combines the solutions from the subproblems to yield a feasible solution to the original instance.

Without geometry, it suffices to draw a random decomposition of the instance which has roughly the same properties that a random grid-shifting has in the geometric setting. We seek a decomposition of the input instance into subproblems whose diameter is also inversely related to the input parameter, $\varepsilon > 0$.

Sample a random decomposition of the graph in the following manner: Pick an arbitrary $\frac{1}{\varepsilon}$-net, S of V, and choose a random permutation π over S. Also choose a radius, r, independently and uniformly at random from an interval of length $\Theta(\frac{1}{\varepsilon})$ such that $r \in \Theta(\frac{1}{\varepsilon})$. Vertices in S are processed in the order π to construct a random partition of G. Construct the ith. cluster to be the set of all remaining vertices that are at a distance at most r of s_i in the hop metric. Once the vertex set is partitioned into $\mathcal{C}$, solve the optimization problem $\mathcal{P}$ on each $C_i \in \mathcal{C}$. We describe the decomposition algorithm more precisely in Algorithm 1. The following lemma follows from the definition of r-net and that G is GBG.

Algorithm 1. Random-Shift(V, ε)

1: Let S denote a $\frac{1}{\varepsilon}$-net of V;
2: Let $\pi(S)$ denote a random permutation of points in S i.i.d;
3: Let r be a random number in $\{\lceil\frac{1}{2\varepsilon}\rceil, \lceil\frac{1}{2\varepsilon}\rceil + 1, \ldots, \lceil\frac{1}{\varepsilon}\rceil\}$ drawn uniformly;
4: $R \leftarrow V$; $\mathcal{C} \leftarrow \emptyset$;
5: **for all** $s_i \in \pi(S)$ **do**
6: $\quad C_i = \{u \in R : d(u, s_i) \leq r\}$;
7: $\quad \mathcal{C} \leftarrow \mathcal{C} \cup C_i$; $R \leftarrow R \setminus C_i$;
8: Solve problem $\mathcal{P}$ on every $C_v \in \mathcal{C}$;
9: Combine the solutions S_v for every C_v to get a feasible solution for G of $\mathcal{P}$;

Lemma 1. *$S \subset V$ be a r-net. For $u \in V$, $B(u, r)$ contains $r^{O(1)}$ points of S.*

2.1 PTAS for Maximum Weighted Independent Set on GBGs

Here, we are given a vertex weighted GBG. Let **wt**: $2^V \mapsto \mathbb{R}_{>0}$ denote the weight function. The objective is to find an independent set of vertices of maximum weight. This problem is NP-hard. We show how to obtain a PTAS by implementing steps "8" and "9" in Algorithm 1. The main idea mimics the PTAS which uses the random grid shifting strategy: (i) delete the subset of vertices that form the cluster boundaries; (ii) solve the subproblems optimally within each cluster, with the boundary vertices removed; and (iii), return the union of the solutions to the subproblems. Clearly, step (iii) yields a feasible solution for the original instance. We will show that in step (i), with reasonably high probability, only a small fraction of opt which depends only on ε, will be deleted. Since the diameter of each cluster in the decomposition is $O(1/\varepsilon)$, it follows, by definition of GBGs, that any optimal solution of any subproblem has size $O(1/\varepsilon^2)$; one can guess an optimal solution for the subproblem in $n^{O(1/\varepsilon^2)}$ time. This will show that step (iii) yields a solution of weight at least $(1+O(\varepsilon\cdot\log 1/\varepsilon))$ of opt, with reasonably high probability. In the next lemma, we bound the probability of a fixed vertex falling at the boundary of some cluster.

Lemma 2. *Probability that a fixed u lies at the boundary of $\mathcal{C}$ is $O(\varepsilon\cdot\log\frac{1}{\varepsilon})$.*

Proof. For u to lie at the boundary of the partition, $\mathcal{C}$, u has to lie at the boundary of some cluster. For u to lie at the boundary of a cluster, the corresponding cluster has to be at a distance at most $1/\varepsilon$ from u. According to Lemma 1, there are $O(1/\varepsilon)$ clusters of interest with respect to u; for other clusters, the probability of u lying at their boundaries is 0. Let t denote the number of such interesting clusters. Let $q_1, q_2, \ldots, q_t$ be the ordering of vertices in S according to increasing order of distance from u, breaking ties arbitrarily. We say that $\pi(s_i)$ *settles* u if i is the first index for which u belongs to C_i. Note that exactly one point in S settles u. We say that $\pi(s_i)$ *hits* u if $\pi(s_i)$ settles u, and the distance of u is exactly r from s_i. The probability that u is hit by $\mathcal{C}$ is:

$$\sum_i \mathbf{Pr}[\pi(s_i) \text{ hits } u] = \sum_j \mathbf{Pr}[q_j \text{ hits } u]$$

The event that q_j hits u requires the occurrence of two events: E_1, the event that r_j is exactly $d(u, q_j)$, and E_2, the event that q_j appears before $q_1, q_2, \ldots, q_{j-1}$ in the ordering $\pi(S)$. Using independence of E_1 and E_2,

$$\mathbf{Pr}[q_j \text{ hits } u] \leq \mathbf{Pr}[E_1]\cdot\mathbf{Pr}[E_2|E_1] = \mathbf{Pr}[E_1]\cdot\mathbf{Pr}[E_2] \leq \frac{1}{1/2\varepsilon}\cdot\frac{1}{j} = \frac{2\varepsilon}{j}$$

So, the probability that u is hit by $\mathcal{C}$ is, $\sum_j \mathbf{Pr}[q_j \text{ hits } u] \leq 2\varepsilon \sum_j \frac{1}{j} \leq c_0\varepsilon\cdot\log\frac{1}{\varepsilon}$

Applying Markov's inequality and union bound we see the following:

Corollary 1. *With constant probability, disks weighing $O(\varepsilon\log\frac{1}{\varepsilon}\cdot opt)$ are deleted.*

We independently sample $O(\log n)$ partitions, removing the vertices which belong to the boundary of each $\mathcal{C}$, solving the problem exactly on each of the clusters with the boundary removed, and take as the union of the solutions as the solution to the instance. Return, as the final answer, a heaviest solution over the $O(\log n)$ trials. We conclude this section with the following:

Theorem 1. *Given a vertex weight GBG in standard form and $\varepsilon > 0$, there is a randomized poly-time algorithm which constructs a maximum weighted independent set whose weight is at least a fraction $(1+\varepsilon\log\frac{1}{\varepsilon})$ of opt, w.h.p.*

3 Weighted Independent Set for Disk Graphs

We give a more sophisticated use of our random shifting strategy under slightly stronger assumptions on the input representation. We consider the problem of finding a maximum weighted independent set on a vertex weighted disk graph in the Euclidean plane expressed without geometry. We obtain a PTAS for the problem, which answers a question of Erlebach et al. [5]. Our main idea is that using a variant of the probabilistic decomposition scheme, one can mimic the algorithm of Erlebach et al. [5], even if the input does not come with a geometric representation of the graph. Instead, we assume that all edge-lengths, and disk radii $\mathbf{rad}: V \mapsto \mathbb{R}_{\geq 0}$ are known with respect to some unknown realization. As noted earlier, this is a significant weakening of the assumption that the input be expressed with geometry, that is, Euclidean coordinates and radii corresponding to disks. Given that our algorithm closely mimics theirs, it is worthwhile to see the algorithm in a geometric setting – we refer the reader to their work [5].

The discussion of our PTAS without geometry is split in two parts: (i) random shifting to construct a hierarchical decomposition of the original instance, and (ii) use of dynamic programming to solve the subproblems optimally. The random shifting is similar to the scheme described in Section 2: we choose a random permutation of the vertex set, and a random number r in the range $[1/2, 1]$, as the basis for the random partition. The approach is faithful to that of Erlebach et al. [5], rather than Chan's [2]. We use the terms disk/vertex interchangeably even though we do not have access to a realization. Details follow: We give some high-level idea before describing the details of `Partition` and `DeleteBoundary`. Let $r_m = \min_v\{\mathbf{rad}(v)\}$ and $L = \lfloor \log_\varepsilon r_m \rfloor$. Partition the disks into L levels by their radii: a disk u belongs to level l if $\varepsilon^{l+1} < \mathbf{rad}(u) \leq \varepsilon^l$; let $l(u)$ denote the level of u. Construct a level 0 clustering by first constructing a $\frac{1}{\varepsilon}$-net S_0 of the vertex set. Note that we are no longer using just a hop-metric – our graph is edge-weighted. Denote the weighted distance between u and v by $d(u, v)$. Process vertices in S_0 in the order as they appear in the random permutation, say, $(s_1^0, s_2^0, \ldots)$. Next, assign all (unassigned) vertices that are within (weighted) distance at most $r \cdot \frac{1}{\varepsilon}$ from s_1^0, to s_1^0. Then, assign all unassigned vertices at distance up to $r \cdot \frac{1}{\varepsilon}$ from s_2^0, to s_2^0, and so on, to construct a level 0 partition into clusters. Next, construct a finer partition (by a factor of ε) of vertices in each of the clusters by using a finer net, and so on, recursively, up to depth L. We give details of the recursive procedure, `Partition`, which returns a hierarchical partition of the input. Next, we create subproblems based on this

---

Algorithm 2. Random-Shift(V, ε)

1: $\mathcal{C} \leftarrow \emptyset$
2: Let $\pi(V)$ denote a random permutation i.i.d.
3: Let $r \in [1/2, 1)$ be a random number i.i.d.
4: $\mathcal{C} \leftarrow$ `Partition`$(V, \varepsilon, \pi, r, 0)$
5: `DeleteBoundary`$(\mathcal{C}, r, 0)$
6: `SolveSubproblemsOptimally`$(\mathcal{C}, r, 0)$

Algorithm 3. Partition$(V, \varepsilon, \pi, r, l)$

Let S be a $\left(\frac{1}{\varepsilon}\right)^{-l+1}$-net of V.
$R \leftarrow V$
for all $s_i \in \pi(S)$ **do**
  $C_i^l \leftarrow \{u \in R : d(u, s_i) \leq r \cdot \left(\frac{1}{\varepsilon}\right)^{-l+1}\}$
  $R \leftarrow R \setminus C_i^l$
  Call `Partition`$(C_i^l, \varepsilon, \pi, r, l+1)$ recursively so long as $l < L$.

partitioning by deleting points from the boundary of the clusters. This will result in subproblems, which we will solve optimally using dynamic programming. We decompose this hierarchical clustering by deleting vertices of level at least l that are "near" the cluster boundary corresponding to level l clusters. In this context, u being "near" the boundary means either (i) the sum of the distance u to its cluster center and $\mathbf{rad}\,(u)$ is more than the boundary defined by the cluster; or (ii) u has a neighbor in a different cluster with center s' and that $d(s', u)$ is more than the distance of s' to its cluster boundary, but $d(s', u) - \mathbf{rad}\,(u)$ is less than the distance of s' to its cluster boundary. This defines a probabilistic decomposition of the input. We give details of this procedure, `DeleteBoundary`, which deletes, on average, vertices that contribute at most an ε factor of an optimal (weighted) independent set because the chance of a fixed vertex getting deleted is small.

Algorithm 4. DeleteBoundary$(\mathcal{C}, r, l)$

for all level l clusters C_i^l **do**
  Let $B_i^l = \{u \in C_i^l : d(u, s_i) \leq r \cdot \left(\frac{1}{\varepsilon}\right)^{-l+1} < d(u, s_i) + \mathbf{rad}\,(u), l(u) \geq l\}$
  Let $B_j^l = \{u \in C_j^l : \exists s_i, d(u, s_i) - \mathbf{rad}\,(u) \leq r \cdot \left(\frac{1}{\varepsilon}\right)^{-l+1} < d(u, s_i), l(u) \geq l\}$
  for all $u \in B_i^l \cup B_j^l$ **do**
    if $u \in B_i^l$ **then**
      $C_i^l \leftarrow C_i^l \setminus u$
    else
      $C_j^l \leftarrow C_j^l \setminus u$
  Call `DeleteBoundary`$(\mathcal{C}, r, l+1)$ recursively so long as $l < L$.

Lemma 3. *For a fixed vertex u having level $l(u)$,* $\mathbf{Pr}\left[\begin{smallmatrix} u \textit{ is cut by} \\ \textit{DeleteBoundary}\end{smallmatrix}\right] \in O\left(\frac{\varepsilon}{1-\varepsilon} \cdot \log \frac{1}{\varepsilon}\right)$

Proof.

$$\mathbf{Pr}\left[\begin{smallmatrix} u \text{ is cut by} \\ \text{DeleteBoundary}\end{smallmatrix}\right] = \mathbf{Pr}\left[\bigvee_{l=0}^{l(u)} \begin{smallmatrix} \text{level l} \\ \text{cluster} \\ \text{cuts } u\end{smallmatrix}\right] \leq \sum_{l=0}^{l(u)} \mathbf{Pr}\left[\begin{smallmatrix} \text{level l} \\ \text{cluster} \\ \text{cuts } u\end{smallmatrix}\right] \quad (1)$$

Let us now bound the probability that a level l cluster cuts u. In order to show this, let us first reorder all the $\left(\frac{1}{\varepsilon}\right)^{-l+1}$-net vertices of cluster C_i^l, say S_i^l, according to their distance from u. Let this ordering be $q_1, q_2, \ldots$. So, the probability that u is cut by a level l cluster is, $\sum_i \mathbf{Pr}\left[\pi(s_i) \text{ cuts } u\right] = \sum_j \mathbf{Pr}\left[q_j \text{ cuts } u\right]$.

In order for q_j to cut u, two events must take place: (i) E_1, all closer $\left(\frac{1}{\varepsilon}\right)^{-l+1}$-net vertices appear later in permutation π, and (ii) E_2, $r \cdot \left(\frac{1}{\varepsilon}\right)^{-l+1}$ falls in the right range from q_j so as to cut the ball of radius $\mathbf{rad}(u)$ centered at u. Due to mutual independence of the two events, $\mathbf{Pr}\left[q_j \text{ cuts } u\right] \leq \frac{2 \cdot \varepsilon^{l(u)}}{\frac{1}{2 \cdot \varepsilon^{-l+1}}} \cdot \frac{1}{j} = \frac{4 \cdot \varepsilon^{l(u)-l+1}}{j}$.

So, $\mathbf{Pr}\begin{bmatrix}\text{level } l \\ \text{boundary} \\ \text{cuts } u\end{bmatrix} \leq \sum_j \mathbf{Pr}\left[q_j \text{ cuts } u\right] \leq c_0 \cdot \varepsilon^{l(u)-l+1} \cdot \log \frac{1}{\varepsilon}$, where, the last inequality follows from the fact that the maximum number of $\left(\frac{1}{\varepsilon}\right)^{-l+1}$-net vertices in C_i^l is $\left(\frac{1}{\varepsilon}\right)^{O(1)}$, and $c_0 \in O(1)$. This fact, combined with (1) implies that,

$$\mathbf{Pr}\begin{bmatrix}u \text{ is cut by} \\ \text{DeleteBoundary}\end{bmatrix} \leq c_0 \cdot \varepsilon^{l(u)+1} \cdot \log \frac{1}{\varepsilon} \cdot \sum_{l=0}^{l(u)} \left(\frac{1}{\varepsilon}\right)^l \leq c_0 \cdot \frac{\varepsilon}{1-\varepsilon} \cdot \log \frac{1}{\varepsilon}$$

Corollary 2. *For an instance of weight independent set on disk graphs, $0 < \varepsilon \leq 1/2$, with constant probability* `DeleteBoundary` *deletes weighing $O\left(\frac{\varepsilon}{1-\varepsilon} \cdot \log \frac{1}{\varepsilon} \cdot opt\right)$.*

We sample $O(\log n)$ random partitions, $\mathcal{C}$, deleting disks that hit the boundary of the hierarchical clusters, resulting in a number of independent subproblems. Assuming that the subproblems can be solved optimally, it follows that:

Theorem 2. *For every fixed $0 < \varepsilon \leq 1/2$, there is a randomized polynomial time algorithm which, given a vertex weighted disk graph (without geometry), a set of edge-lengths, and radii for the vertices, constructs a weighted independent set whose weight is at least a fraction $\left(1 + O(\frac{\varepsilon}{1-\varepsilon} \log \frac{1}{\varepsilon})\right)$ of opt, w.h.p.*

Solving Sub-problem Instances Optimally Using Dynamic Programming: Let $\mathcal{D}$ denote the set of disks that are obtained after `DeleteBoundary` deletes disks from the original instance. In this section, we show that maximum weighted independent set problem for $\mathcal{D}$ can be solved exactly in time polynomial in n by using edge-lengths and radii corresponding to the vertex set with respect to some (unknown) embedding, using dynamic programming. This will prove Theorem 2.

Call a level l cluster, C_i^l, an l-cluster. Call a l-cluster C_i^l *relevant* if it contains at least one level l vertex (disk) u, that is, $\varepsilon^{l+1} < \mathbf{rad}(u) \leq \varepsilon^l$. For a l-cluster C_i^l and l'-cluster $C_j^{l'}$, $l' > l$, say C_i^l *contains* $C_j^{l'}$ if before `DeleteBoundary` deletes disks that are cut by the boundary of clusters of levels higher than l, $C_j^{l'} \subset C_i^l$. For a relevant l-cluster C_i^l and a relevant l'-cluster $C_j^{l'}$, $l' > l$, call $C_j^{l'}$ a *child* or *child cluster* of C_i^l (and C_i^l is a *parent* of $C_j^{l'}$) if $C_j^{l'}$ is contained in C_i^l and if there is no relevant l''-cluster that is contained in C_i^l, and contains $C_j^{l'}$.

The hierarchical clustering obtained by deleting disks that hit cluster boundaries, satisfies two conditions summarized below. The following lemma is obvious.

Lemma 4. *For $0 \leq l < L$, every $(l+1)$-cluster is contained in some l-cluster.*

Lemma 5. *Let C_i^l be some l-cluster and I be a set of independent disks such that for every disk $u \in I, l(u) \leq l$, and u has a neighbor in C_i^l. There is a constant c_ε (depending only on ε) such that $|I| \leq c_\varepsilon$.*

Proof. By our assumption, C_i^l has a realization in the Euclidean plane. Therefore, C_i^l is fully contained in a disk of radius $O(\varepsilon^{l-1})$. Let us fix a realization and call the bounding disk D. The smallest radius disk that has a neighbor in C_i^l must have radius at least ε^{l+1} and must be centered at a distance $O(\varepsilon^{l+1})$ from D's boundary, with respect to the realization. So, the area occupied by any disk in I must be $\Omega(\varepsilon^{2(l+1)})$. Now, extend the radius of D by ε^{l+1} and call this larger disk D'. The area of D' is $O((\varepsilon^{l-1} + \varepsilon^{l+1})^2)$ and each disk of I occupies an area $\Omega(\varepsilon^{2(l+1)})$ of D'; the larger disks of I only occupy a larger area. By using standard packing arguments, it can be seen that $|I|$ is $O(\varepsilon^{-4})$.

For every relevant cluster C_i^l, denoted C for convenience, a table, $\texttt{Table}_C$, is maintained. Each entry corresponds to an independent set I of disks u such that $\mathbf{rad}(u) > \varepsilon^l$ (the large disks) which have a neighbor in C. By Lemma 5, $|I| \leq c_\varepsilon$. Hence, the combinatorial complexity of all subsets of independent large disks u that have a neighbor in C is polynomial in n. So, the size of $\texttt{Table}_C$ is polynomial in n. For every such set I, $\texttt{Table}_C(I)$ is a maximum weighted independent set of disks of level at least l that are also independent of I, and are contained in the l-cluster C. For a fixed choice I of independent disks of level less than l, we enumerate over all independent subsets X of disks of level l that are also independent of I and are contained in C. We then extend the union of the optimal solutions for child clusters contained in C for the subset of disks of $I \cup X$ that have neighbors in them by X. The best solution amongst all extensions is the solution for $\texttt{Table}_C$. We describe details of how to compute a particular table corresponding to a l-cluster C. Once the tables for all relevant clusters are computed, the final solution corresponding to $\mathcal{D}$ is the union over all clusters C that do not have a parent, of $\texttt{Table}_C(\emptyset)$. Next, we show that the solution returned is an optimal solution for $\mathcal{D}$, and that this solution is computed in time polynomial in n. The proof of the following lemma is similar to [5].

Algorithm 5. ComputeTable($l, \varepsilon, C, \mathcal{D}$)

$\texttt{Table}_C \leftarrow \emptyset$
Let R be the set of vertices (disks) in $\mathcal{D}$ having level at most l that have a neighbor in C
for all $J \subseteq R$, J is independent, $|J| \leq c_\varepsilon$ **do**
    Let X be the subset of J having level l
    for all C', C' is a child of C **do**
        Let I' be the set of disks in J that have a neighbor in C'
        $X \leftarrow X \cup \texttt{Table}_{C'}(I')$
    $I \leftarrow J \setminus X$
    if $\texttt{Table}_C(I)$ is undefined OR $\mathbf{wt}(X) > \mathbf{wt}(\texttt{Table}_C(I))$ **then**
        $\texttt{Table}_C(I) \leftarrow X$

Lemma 6. *For a relevant l-cluster C and any independent set I belonging to level less than l such that each have a neighbor in C, $I \cup \texttt{Table}_C(I)$ is independent and $\mathbf{wt}(\texttt{Table}_C(I))$ is most over all independent sets of C that are independent of I.*

Every disk in $\mathcal{D}$ is contained in some relevant cluster C that does not have a parent and that all such relevant clusters are independent of each other. So, the union of $\texttt{Table}_C(\emptyset)$ over all such clusters returns a heaviest independent set of $\mathcal{D}$. We conclude with the following lemma whose proof is similar to [5].

Lemma 7. *For any fixed $\varepsilon > 0$, the algorithm runs in time polynomial in n.*

4 Extensions and Concluding Remarks

Our PTAS can be extended to d-dimensional case. In addition, we do not specifically require that the objects be Euclidean balls – it suffices for the objects to be *fat*. We would require that radii corresponding to these objects refer to the radius of a minimum enclosing ball, for example, and the edge-lengths correspond to inter-point distances of the ball centers for objects that overlap.

Acknowledgments. We thank Matt Gibson, Sriram Pemmaraju, Mohammad Salavatipour, Zoya Svitkina, and Kasturi Varadarajan for helpful discussions.

References

1. Aspnes, J., Goldenberg, D., Yang, Y.: On the computational complexity of sensor network localization. In: Nikoletseas, S.E., Rolim, J.D.P. (eds.) ALGOSENSORS 2004. LNCS, vol. 3121, pp. 32–44. Springer, Heidelberg (2004)
2. Chan, T.M.: Polynomial-time approximation schemes for packing and piercing fat objects. J. Algorithms 46(2), 178–189 (2003)
3. Chan, T.M., Har-Peled, S.: Approximation algorithms for maximum indepen- dent set of pseudo-disks. In: SoCG (2009)
4. Chintalapudi, K., Govindan, R., Sukhatme, G., Dhariwal, A.: Ad-hoc localization using ranging and sectoring. In: INFOCOM (2004)
5. Erlebach, T., Jansen, K., Seidel, E.: Polynomial-time approximation schemes for geometric intersection graphs. SIAM J. Comput. 34(6), 1302–1323 (2005)
6. Erlebach, T., van Leeuwen, E.J.: Approximating geometric coverage problems. In: SODA, pp. 1267–1276 (2008)
7. Flury, R., Pemmaraju, S.V., Wattenhofer, R.: Greedy routing with bounded stretch. In: INFOCOM (2009)
8. Hochbaum, D.S., Maass, W.: Approximation schemes for covering and packing problems in image processing and VLSI. J. ACM 32(1), 130–136 (1985)
9. Kuhn, F., Moscibroda, T., O'Dell, R., Wattenhofer, M., Wattenhofer, R.: Virtual coordinates for ad hoc and sensor networks. To appear in Algorithmica
10. Nieberg, T., Hurink, J., Kern, W.: Approximation schemes for wireless networks. ACM Transactions on Algorithms 4(4), 1–17 (2008)
11. Talwar, K.: Bypassing the embedding: algorithms for low dimensional metrics. In: STOC, pp. 281–290. ACM, New York (2004)

Approximation Algorithms for Variable Voltage Processors: Min Energy, Max Throughput and Online Heuristics*

Minming Li

Department of Computer Science, City University of Hong Kong
minmli@cs.cityu.edu.hk

Abstract. Dynamic Voltage Scaling techniques allow the processor to set its speed dynamically in order to reduce energy consumption. It was shown that if the processor can run at arbitrary speeds and uses power s^α when running at speed s, the online heuristic AVR has a competitive ratio $(2\alpha)^\alpha/2$. In this paper we first study the online heuristics for the discrete model where the processor can only run at d given speeds. We propose a method to transform online heuristic AVR to an online heuristic for the discrete model and prove a competitive ratio $\frac{2^{\alpha-1}(\alpha-1)^{\alpha-1}(\delta^\alpha-1)^\alpha}{(\delta-1)(\delta^\alpha-\delta)^{\alpha-1}}+1$, where δ is the maximum ratio between adjacent non-zero speed levels. We also prove that the analysis holds for a class of heuristics that satisfy certain natural properties. We further study the throughput maximization problem when there is an upper bound for the maximum speed. We propose a greedy algorithm with running time $O(n^2 \log n)$ and prove that the output schedule is 3-approximation of the throughput and $\frac{(\alpha-1)^{\alpha-1}(3^\alpha-1)^\alpha}{2\alpha^\alpha(3^{\alpha-1}-1)^{\alpha-1}}$-approximation of the energy consumption.

1 Introduction

Energy efficiency currently becomes one of the major concerns in system designs. Portable electronic devices, which are typically powered by batteries, rely on energy efficient schedules to increase the battery lifetime; while non-portable systems need energy efficient schedules to reduce the operating cost. An important strategy to achieve energy saving is via *dynamic voltage scaling (DVS)*, which enables a processor to operate at a range of voltages and frequencies. Because energy consumption is at least a quadratic function of the voltage (hence CPU frenquency/speed), it saves energy by executing jobs as slowly as possible while still satisfying the required property like feasibility where all the jobs are required to be finished by their deadlines. The associated scheduling problem is referred to as min-energy feasibility DVS scheduling.

A theoretical study of min-energy feasibility DVS scheduling was initiated by Yao, Demers and Shenker[1]. They formulated the optimization problem and

* This work was supported by a grant from the Research Grants Council of the Hong Kong Special Administrative Region, China [Project No. CityU 116907].

Y. Dong, D.-Z. Du, and O. Ibarra (Eds.): ISAAC 2009, LNCS 5878, pp. 372–382, 2009.

gave an $O(n^3)$ algorithm YDS for computing the optimal schedule, which is later improved to $O(n^2 \log n)$ in [2]. In their formulation, each job j_i has an arrival time a_i, a workload R_i, and a deadline b_i. If job j_i is executed at speed s, then it needs R_i/s time to finish. They also assumed a speed to power function $P(s) = s^\alpha$, where $\alpha \geq 2$ is a system constant (usually $\alpha = 3$ if the cubic-root rule holds). Their model allows the processor speed to be set at any real value and therefore is referred to as the *continuous* model. An online heuristic AVR was proposed in [1] and proved to be $(2\alpha)^\alpha/2$-competitive in energy consumption. Bansal et.al. [3] further investigated the online heuristics for the model proposed in [1] and proved that the heuristic OA has a tight competitive ratio of α^α for all job sets. They also gave a new online heuristic which is the best possible heuristic for a wide range of power functions. Quan and Hu [4] considered scheduling jobs with fixed priorities and characterized the optimal schedule. Yun and Kim [5] later on showed the NP-hardness of computing the optimal schedule and also provided an FPTAS for this problem.

In practice, processors can run at only a finite number of preset speed levels. One can capture the discrete nature of the speed scale with a corresponding *discrete* model. Ishihara and Yasuura [6] initiated the research on discrete DVS problem and solved the case when the processor is only allowed to run at two different speeds. Kwon and Kim [7] extended that to the general discrete DVS model where the processor is allowed to run at speeds chosen from a finite speed set. They gave an algorithm for this problem based on the YDS algorithm in [1]. Recently, It was shown in [8] that the optimal schedule for the discrete model can be computed in $O(dn \log n)$ time where d is the number of speed levels. Compared to numerous analyses for online heuristics proposed for the continuous model, few works exist that design or analyze online heuristics for the discrete model which is more practical. We propose in this paper a systematic way to transform online heuristics for the continuous model to online heuristics for the discrete model with only a small increase in the competitive ratio.

Another practical constraint on the processor is the maximum speed limit, which sometimes makes it impossible for the processor to finish all jobs within their timing constraints. In this bounded speed model, a useful objective is to maximize the throughput of the whole system. In real time scheduling where the processor cannot change the execution speed, the throughput maximization problem where preemptions are not allowed is studied in [9][10][11]. [12] and [13] considered preemptive scheduling. It was shown in [12] that the best possible online scheduling algorithm is 4-competitive on throughput without considering energy and [13] proposed an online heuristic which is 14-competitive in throughput and O(1)-competitive in energy consumption. We also refer the readers to a recent survey on research in power/temperature management [14].

In this paper, we assume that preemption is allowed in job execution ,i.e., a job can be interrupted when being executed and resumed afterwards. We first investigate online heuristics for the discrete model where the allowed speed set $\{s_1, s_2, \ldots, s_d\}$ satisfies $s_1 > s_2 > \ldots > s_d > 0$. By suitably adjusting the speed used in AVR to adjacent allowed speed levels, we design an online heuristic AVR'_d

for the discrete model and prove it to be $(\frac{2^{\alpha-1}(\alpha-1)^{\alpha-1}(\delta^\alpha-1)^\alpha}{(\delta-1)(\delta^\alpha-\delta)^{\alpha-1}}+1)$-competitive where $\delta = \max_{1\leq i\leq d-1} \frac{s_i}{s_{i+1}}$. We also identify a class of heuristics which can be modified in a similar way from the continuous model to the discrete model with their competitive ratios increased by a factor of $\frac{(\alpha-1)^{\alpha-1}(\delta^\alpha-1)^\alpha}{\alpha^\alpha(\delta-1)(\delta^\alpha-\delta)^{\alpha-1}}$. The similar idea of adjusting speed to adjacent speed levels is widely known and proved to increase the competitive ratio by δ^α for throughput maximization algorithm proposed in [13]. Notice that our increased ratio is $\frac{(\delta+1)^2}{4\delta}$ when $\alpha = 2$ which improves upon δ^α almost by a factor of 4δ. Then we study the throughput maximization problem for the continuous model in the overloaded setting where the processor can only vary its speed between 0 and a maximum speed s_{max}. With this practical speed restriction, the processor sometimes can not finish all the jobs by their deadlines even if it always runs at s_{max}. We aim to find schedules which can minimize energy consumption while maximizing the throughput. The throughput is the total workload of the jobs completed by their deadlines. It is NP-hard to find schedules that maximize throughput since KNAPSACK is a special case when all deadlines are equal and all arrival times are equal. We propose a greedy algorithm to approximately maximize the throughput while also approximately minimizing the energy consumption. By using the properties of s-schedules defined in [8] and properties proved in the study of online heuristics for the discrete model, we prove that the greedy algorithm outputs a schedule which is 3-approximation of throughput and $\frac{(\alpha-1)^{\alpha-1}(3^\alpha-1)^\alpha}{2\alpha^\alpha(3^{\alpha-1}-1)^{\alpha-1}}$-approximation of energy consumption.

The remainder of the paper is organized as follows. We give the problem formulation and review some basic properties of the optimal schedules in Section 2. Section 3 describes a $(\frac{2^{\alpha-1}(\alpha-1)^{\alpha-1}(\delta^\alpha-1)^\alpha}{(\delta-1)(\delta^\alpha-\delta)^{\alpha-1}}+1)$-competitive online heuristic for the discrete model and extends the result to a class of online heuristics. We then study the offline throughput maximization problem in Section 4 and propose a greedy algorithm to compute a schedule that is 3-approximation in throughput and $\frac{(\alpha-1)^{\alpha-1}(3^\alpha-1)^\alpha}{2\alpha^\alpha(3^{\alpha-1}-1)^{\alpha-1}}$-approximation in energy consumption. Some concluding remarks are given in Section 5. Due to space limit, we omit some proofs in this version.

2 Models and Preliminaries

A job set $J = \{j_1, j_2, \ldots, j_n\}$ over $[0, 1]$ is given where each job j_k is characterized by three parameters: arrival time a_k, deadline b_k, and workload R_k. Here workload means the required number of CPU cycles. We also refer to $[a_k, b_k] \subseteq [0, 1]$ as the interval of j_k, and assume without loss of generality that $\cup_k[a_k, b_k] = [0, 1]$ (or J *spans* $[0, 1]$). A *schedule* S for J is a pair of functions $(s(t), job(t))$ which defines the processor speed and the job being executed at time t respectively. Both functions are assumed to be piecewise continuous with finitely many discontinuities. A *feasible* schedule must give each job its required workload between its arrival time and deadline with perhaps intermittent execution. We assume that the power P, or energy consumed per unit time, is $P(s) = s^\alpha$ $(\alpha \geq 2)$ where s is the processor speed. The total energy consumed by a schedule S is $E(S) = \int_0^1 P(s(t))dt$. The goal of the min-energy feasibility scheduling problem is to find a feasible schedule that minimizes $E(S)$ for any given job set J. We refer to this problem as the continuous *DVS* scheduling problem.

In the discrete version of the problem, we assume that the processer can run at d speed levels $s_1 > s_2 > \ldots > s_d$. The goal is to find a minimum-energy schedule for a job set using only these speeds. We refer to this problem as *Discrete DVS* scheduling problem if the highest speed s_1 is always fast enough to guarantee a feasible schedule for the given jobs.

For the continuous DVS scheduling problem, the optimal schedule S_{opt} is characterized by using the notion of a *critical interval* for J, which is an interval I in which a group of jobs must be scheduled at maximum constant speed $g(I)$ in any optimal schedule for J. The algorithm proceeds by identifying such a critical interval I, scheduling those 'critical' jobs at speed $g(I)$ over I, then constructing a subproblem for the remaining jobs and solving it recursively. The details are given below.

Definition 1. *For any interval $I \subseteq [0, 1]$, we use J_I to denote the subset of jobs in J whose intervals are completely contained in I. The intensity of an interval I is defined to be $g(I) = (\sum_{j_k \in J_I} R_k)/|I|$.*

An interval I^* achieving maximum $g(I)$ over all possible intervals I defines a critical interval for the current job set. It is known that the subset of jobs J_{I^*} can be feasibly scheduled at speed $g(I^*)$ over I^* by the earliest deadline first (EDF) principle. That is, at any time t, a job which is waiting to be executed and having earliest deadline will be executed during $[t, t + \epsilon]$. The interval I^* is then removed from $[0, 1]$; all the remaining job intervals $[a_k, b_k]$ are updated to reflect the removal, and the algorithm recurses. We denote the optimal schedule which guarantees feasibility and consumes minimum energy in the continuous model as *OPT*.

3 Online Heuristics for Discrete Model

In [1], two on-line heuristics AVR (Average Rate) and OA (Optimal Available) were introduced for the case that jobs arrive one after another. We define these two heuristics as follows.

At any time t, the heuristic AVR runs the earliest deadline job using the speed $\sum_{J(t)} \frac{R_k}{b_k - a_k}$ where $J(t)$ is the set of jobs with $a_k \leq t \leq b_k$. In other word, AVR always uses the speed equal to the summation of the average workload of all the available jobs.

At any time t, the heuristic OA always assume that the current available jobs are the only jobs to be scheduled. It calculates an optimal offline schedule for all these jobs and schedules them accordingly. Whenever a new job arrives, it will do a recalculation.

The heuristic AVR was shown to have a competitive ratio of at most $(2\alpha)^\alpha/2$ in [1]; recently a tight competitive ratio of α^α was proven for OA in [3]. Both of these results are for the continuous model where the speed can be set at an arbitrary value. In this paper, we design online heuristics for the discrete model where only a finite number of speeds are allowed for the processor but the highest speed is larger than the maximum speed needed by the online heuristic for the given job set in the continuous model.

We first consider a simple case. Suppose $s_1 > s > s_2$ and we have an execution interval of speed s which lasts for a time period of length t. Denote the energy consumption on this execution interval as E_c. Then, we adjust the speed s to s_1 and s_2 with appropriate proportion of time t_1 and t_2 in the execution interval so that the total workload executed remains the same, i.e., $st = s_1t_1 + s_2t_2$ and $t = t_1 + t_2$. We denote the energy consumption by the new schedule as E_d. We give an upper bound of E_d in Lemma 1.

Lemma 1. *If* $P(s) = s^\alpha$ *and* $s_1/s_2 \leq \delta$, *then* $E_d \leq \frac{(\alpha-1)^{\alpha-1}(\delta^\alpha-1)^\alpha}{\alpha^\alpha(\delta-1)(\delta^\alpha-\delta)^{\alpha-1}}E_c$.

Note that in order to make the lemma hold for all the adjustment, we implicitly need the assumption $s > s_d$ which means that s_d should be small enough. However, for $s < s_d$, if we round s to 0 and s_d, this part of energy will be less than the energy consumption of any schedule that finishes all the jobs using the given speed set observed by [13]. We define AVR_d as the schedule which adjusts each speed in AVR into proportioned execution of two adjacent speed levels. Notice that AVR_d is an offline schedule because online schedulers do not know when a new job will arrive and therefore do not know when the current piece of constant speed execution will end, which makes the accurate adjustment impossible. By the definition of AVR_d, we have the following property by Lemma 1.

Corollary 1. $E(AVR_d) \leq \frac{(\alpha-1)^{\alpha-1}(\delta^\alpha-1)^\alpha}{\alpha^\alpha(\delta-1)(\delta^\alpha-\delta)^{\alpha-1}}E(AVR) + E(OPT)$, *where* $\delta = \max_{1\leq i\leq d-1} \frac{s_i}{s_{i+1}}$.

We know that in AVR online heuristic, whenever an arrival time or a deadline is met, the scheduler will possibly adopt a different execution speed, i.e., it will recalculate the execution speed. The online heuristic we design for the discrete model tries to capture the same idea. The only difference is that some speeds are not allowed in the discrete model. Therefore, speed adjustment is needed to satisfy the speed requirement. We describe the online scheduling algorithm AVR'_d for the discrete model as follows. Whenever an arrival time or a deadline is met, the speed curve adopted by AVR is calculated. Then every speed is adjusted to adjacent speed levels according to the right proportion. For the contiguous time period which is originally assigned the same speed by AVR, the heuristic AVR'_d executes the higher speed portion first and then the lower. If some job's arrival at time a interrupts this proportioned schedule, the scheduler just recalculates the new speed using AVR for the remaining jobs in the job set J under AVR'_d. This remaining job set is denoted as $J_{a,remain}$. The arrival time of the job whose arrival time is earlier than a will be adjusted to a if it is not finished by a and the workload of the jobs will also be updated reflecting the previous execution. Notice that AVR'_d is different from AVR_d because it is not a simple offline adjustment of AVR. We prove the following lemma:

Lemma 2. $E(AVR'_d) \leq E(AVR_d)$.

We further combine Corollary 1 and Lemma 2 to derive the following theorem on the performance of AVR'_d when the power function is $P(s) = s^\alpha$.

Theorem 1. $E(AVR'_d) \leq (\frac{2^{\alpha-1}(\alpha-1)^{\alpha-1}(\delta^\alpha-1)^\alpha}{(\delta-1)(\delta^\alpha-\delta)^{\alpha-1}} + 1)E(OPT)$, *where* $\delta = \max_{1\leq i\leq d-1} \frac{s_i}{s_{i+1}}$.

Let $[0, a]$ be the first non-trivial stable interval and s be the speed calculated by the heuristic H at time 0 for $[0, a]$ where $s_i > s \geq s_{i+1}$. Define $J_{a,remain}$ similarly as above and let the speed functions of the heuristic H for J and $J_{a,remain}$ in $[a, 1]$ be s_H and s_{H_r} respectively. Lemma 2 can be extended to a class of online heuristics besides AVR which satisfy the following two properties.

1) Monotone Property: $s_H(t) \geq s_{H_r}(t)$ for $a \leq t \leq 1$.

2) Separation Property: there exists $b \leq 1$ where $s_{H_r}(t) \geq s_{i+1}$ for $a \leq t \leq b$ and $s_H(t) = s_{H_r}(t)$ for $b < t \leq 1$.

Therefore, we have the following theorem.

Theorem 2. *If H is a c-competitive online heuristic for the continuous model which satisfies Monotone Property and Separation Property, then the heuristic H'_d for the discrete model which adjusts the calculated speed to adjacent speed levels and executes the higher speed part first is* $(\frac{c(\alpha-1)^{\alpha-1}(\delta^\alpha-1)^\alpha}{\alpha^\alpha(\delta-1)(\delta^\alpha-\delta)^{\alpha-1}} + 1)$*-competitive.*

4 Minimizing Energy While Maximizing Throughput

In most systems, the processor can only vary its speed between 0 and a maximum speed s_{max}. Because the speed is bounded from above, the processor may be overloaded with jobs and no scheduling algorithms can finish all the jobs before their deadlines. In this section, we investigate the continuous model with a speed limit and investigate the offline throughput maximization problem in this overloaded setting where preemptions are allowed. Throughput is defined to be the total workload of the jobs completed by their deadlines. We propose a greedy algorithm which produces a schedule that is 3-approximation to the maximum throughput, while at the same time achieves $\frac{(\alpha-1)^{\alpha-1}(3^\alpha-1)^\alpha}{2\alpha^\alpha(3^{\alpha-1}-1)^{\alpha-1}}$-approximation with respect to energy consumption.

A tool called s-schedule was introduced in [8] which can provide useful information regarding the optimal speed function for J without explicitly computing it. The properties of the s-schedule are also crucial in our proof of approximation ratios for the greedy algorithm. For easy reference, we give the relevant definitions and properties below.

Definition 2. *For any constant s, the s-schedule for J is an EDF schedule which uses constant speed s in executing any jobs of J. It will give up a job when the deadline of the job has passed. In general, s-schedules may have idle periods or unfinished jobs.*

Definition 3. *In a schedule S, a maximal subinterval of* $[0, 1]$ *devoted to executing the same job* j_k *is called an execution interval for* j_k *(with respect to S). Denote by* $I_k(S)$ *the union of all execution intervals for* j_k *with respect to S. Execution intervals with respect to the s-schedule will be called s-execution intervals.*

It is easy to see that the s-schedule for n jobs contains at most $2n$ s-execution intervals, since the end of each execution interval (including an idle interval) corresponds to the moment when either a job is finished or a new job arrives. Also, the s-schedule can be computed in $O(n \log n)$ time by using a priority queue to keep all jobs currently available, prioritized by their deadlines. Denote the s-schedule of J by S_J. Next we prove a basic property of s-schedules.

Lemma 3. *Given a job set J, for any* $J' \subseteq J$*, we have* $|\bigcup_{j_k \in J'} I_k(S_J)| \leq |\bigcup_{j_k \in J'} I_k(S_{J'})|$.

In the following, we will use a slightly different result for s-partition introduced in [8] and summarize it into the following lemma.

Lemma 4. *The job set J can be partitioned into two job sets* $J^{\leq s}$ *and* $J^{>s}$ *in* $O(n \log n)$ *time, where jobs in* $J^{\leq s}$ *require speeds no larger than s in OPT while jobs in* $J^{>s}$ *require speeds larger than s in OPT.*

Recall that OPT is the min-energy schedule without speed limit. We use opt_T to denote the optimal schedule with speed upper bounded by s_{max}, and opt_T maximizes the throughput first and then minimizes the energy subject to this throughput. Given a job set, we first remove all the jobs whose average workload is more than s_{max} because these jobs will not contribute to the throughput even in opt_T. We denote the remaining job set as J. Then, we generate an s_{max}-schedule for J. By Lemma 4, we can identify critical intervals whose speeds are larger than s_{max}. We denote the union of those intervals as I. According to the way critical intervals are chosen, there is a schedule where each job j_k with $[a_k, b_k] \cap I \neq [a_k, b_k]$ can be finished by its deadline using execution speed at most s_{max} and is only executed outside I. We denote the set of jobs satisfying the above requirement as J_2. We know that all the jobs in J_2 can contribute to the throughput without affecting the feasibility of jobs in $J_1 = J \backslash J_2$. Therefore, we only consider jobs in J_1 in the following discussion.

Let $t = |I|$. We propose a greedy algorithm which achieves at least $\frac{1}{3}t \cdot s_{max}$ throughput within I as follows.

The intervals T_k^* is obtained from T_k in the following way. If $|T_k| = \frac{R_k}{s_{max}}$, then let T_k^* be T_k; otherwise, give all the time in T_k to T_k^* and then add more time to T_k^* in a backward manner starting from b_k until $|T_k^*| = \frac{R_k}{s_{max}}$. For example, if $T_k = [0, 1], [4, 5]$, $b_k = 6$ and $\frac{R_k}{s_{max}} = 4$, then $T_k^* = [0, 1], [3, 6]$.

Algorithm 1. Greedy Algorithm

1. $J^* = \emptyset$ and $J_h = J_1$.
2. $I = \cup_{j_i \in J_1} [a_i, b_i]$.

repeat

3. $J^* = J^* \cup j_k$, where $j_k \in J_h$ and $R_k \geq R_i$ for all $j_i \in J_h$.
4. Compute the s_{max}-schedule for J_h and denote the time intervals allocated to j_k as T_k.
5. Assign time intervals T_k^* to j_k (The way to choose T_k^* is explained below).
6. $I = I \backslash T_k^*$.
7. Update the job arrival times and deadlines after removing T_k^* from I (by treating time in T_k^* as non-existent).
8. Remove from J_h the job j_k and the jobs whose average workload are larger than s_{max} after the update.
9. Compute the s_{max}-partition for J_h on I using Lemma 4 and move the jobs in $J_h^{\leq s_{max}}$ from J_h to J^*.

until $J_h \neq \emptyset$.

10. Compute the optimal schedule S^* for $J^* \cup J_2$.

The set $J^* \cup J_2$ is the subset of jobs we choose to execute in the schedule. The feasibility of S^* using speeds at most s_{max} is guaranteed by the choice of jobs into J^*.

We can prove that jobs in J^* have a total workload at least $\frac{1}{3}t \cdot s_{max}$ by proving the following lemma.

Lemma 5. *Given a set J_h of n jobs, where the average workload of every job is no more than s_{max} and OPT always uses speeds higher than s_{max}, the total workload R of the jobs added to J^* in the first iteration of Algorithm 1 is at least $\frac{1}{3}(t-t')s_{max}$ where t represents the length of the union of jobs in J_h at the beginning of the iteration and t' represents the length of the union of jobs in J_h at the end of the iteration.*

Proof. First, it is easy to see that when Step 6 is finished, the s_{max}-schedule for $J_h \backslash j_k$ on I (we denote this schedule as S' in the following proof) runs at non-zero speed for a total period of $t - \frac{R_k}{s_{max}}$. We next prove that s_{max}-schedule for J_h computed in Step 9 runs at non-zero speed for a total period of at least $t - 3\frac{R_k}{s_{max}}$. Notice that in Step 8, if a job j_i is removed from J_h, then its interval $[a_i, b_i]$ should intersect with T_k^*, otherwise its average workload will not change after we remove T_k^* from I. Furthermore, $[a_i, b_i]$ should intersect with the last interval in T_k^*. Because if $b_i > b_k$, then $[a_i, b_i]$ has to intersect the last interval in T_k^*, otherwise it will not intersect T_k^*; if $b_i \leq b_k$, then $[a_i, b_i]$ only intersecting non-last intervals in T_k^* means that in the s_{max}-schedule for J_h (with j_k in), the job j_i can finish all its workload before the earliest intersection point; otherwise, the intersecting part with the non-last intervals in T_k^*, which is also intervals in T_k will not be used to execute j_k because j_i has a higher priority than j_k, therefore removing T_k^* will not make the average workload of j_i be larger than s_{max}. Furthermore, it is easy to see that $|[a_i, b_i] \backslash T_k^*| \leq \frac{R_i}{s_{max}} \leq \frac{R_k}{s_{max}}$. Based on the above analysis, we know that the jobs removed from J_h in Step 8 occupy a time period with length at most $2\frac{R_k}{s_{max}}$ (it is the time in $I\backslash T_k^*$ starting from both sides of the last interval of T_k^* with length at most $\frac{R_k}{s_{max}}$ on each side), which implies that the remaining jobs in J_h after Step 8 will be executed for a period of at least $t-3\frac{R_k}{s_{max}}$ in schedule S'. Therefore, s_{max}-schedule for J_h computed in Step 9 runs at non-zero speed for a total period of at least $t-3\frac{R_k}{s_{max}}$ by Lemma 3. In Step 9, by moving more jobs from J_h to J^*, we increase the workload in J^* by at least $t_r \cdot s_{max}$ where t_r is the execution time of those jobs in s_{max}-schedule for J_h computed in Step 9. Notice that the removal of jobs in Step 9 will not affect the s_{max}-schedule for the remaining jobs in J_h by the definition of critical intervals. From the above analysis, we have the following two relations: $t'+t_r \geq t-3\frac{R_k}{s_{max}}$ and $\frac{R}{s_{max}} \geq \frac{R_k}{s_{max}}+t_r$. The lemma then follows. □

We can interpret Lemma 5 in the following way. Picking the job j_k with the maximum R_k value will reduce the *useful* s-execution time by at most $3\frac{R_k}{s_{max}}$. We remark that T_k^* cannot be chosen arbitrarily within $[a_k, b_k]$. For example, if we choose T_k^* to be $[a_k, a_k+\frac{R_k}{s_{max}}]$, then picking the job j_k may reduce the useful s-execution time by $4\frac{R_k}{s_{max}}$ because the time in $I_k(S_{J_h})$ (s_{max}-execution intervals of j_k) may not overlap with any other job in J_h and therefore cannot be used to execute other jobs in the s_{max}-schedule.

Lemma 6. *The jobs in J^* have a total workload at least $\frac{1}{3}t \cdot s_{max}$ and can be feasibly scheduled using speeds at most s_{max}.*

The above lemma directly shows that Algorithm 1 is 3-approximation in throughput compared to opt_T because opt_T can at most select $t \cdot s_{max}$ workload from J_1 to execute. Next, we show that the energy consumption of S^* is at most $\frac{(\alpha-1)^{\alpha-1}(3^\alpha-1)^\alpha}{2\alpha^\alpha(3^{\alpha-1}-1)^{\alpha-1}}$ times that of opt_T.

Lemma 7. *The schedule S^* is $\frac{(\alpha-1)^{\alpha-1}(3^\alpha-1)^\alpha}{2\alpha^\alpha(3^{\alpha-1}-1)^{\alpha-1}}$-approximation of energy consumption compared to opt_T.*

We summarize the above results into the following theorem.

Theorem 3. *Algorithm 1 generates a schedule S^* in $O(n^2 \log n)$ time which is 3-approximation of throughput and $\frac{(\alpha-1)^{\alpha-1}(3^\alpha-1)^\alpha}{2\alpha^\alpha(3^{\alpha-1}-1)^{\alpha-1}}$-approximation of energy consumption compared to opt_T.*

The running time of Algorithm 1 is $O(n^2 \log n)$ because every step inside the loop takes $O(n \log n)$ time and the computation of S^* takes $O(n^2 \log n)$ time [2].

5 Conclusion

In this paper, we first investigate online heuristics for the discrete model. By suitably adjusting the speeds used in AVR to adjacent speed levels, we design an online heuristic AVR'_d for the discrete model and prove it to be

$(\frac{2^{\alpha-1}(\alpha-1)^{\alpha-1}(\delta^{\alpha}-1)^{\alpha}}{(\delta-1)(\delta^{\alpha}-\delta)^{\alpha-1}} + 1)$-competitive where $P(s) = s^{\alpha}$ and δ is the maximum ratio between adjacent non-zero speed levels. We also identify a class of heuristics which can be modified in a similar way from the continuous model to the discrete model with their competitive ratios increased by a constant factor $\frac{(\alpha-1)^{\alpha-1}(\delta^{\alpha}-1)^{\alpha}}{\alpha^{\alpha}(\delta-1)(\delta^{\alpha}-\delta)^{\alpha-1}}$ with an extra addition of 1. Then we study the throughput maximization problem for the preemptive continuous model in the overloaded setting where the maximum speed of the processor is upper bounded by s_{max}. We propose a greedy algorithm to approximately maximize throughput while also approximately minimizing the energy consumption. By using the properties of s-schedule defined in [8] and properties proved in the study of online heuristics for the discrete model, we prove that the greedy algorithm is 3-approximation of throughput and $\frac{(\alpha-1)^{\alpha-1}(3^{\alpha}-1)^{\alpha}}{2\alpha^{\alpha}(3^{\alpha-1}-1)^{\alpha-1}}$-approximation of energy consumption.

References

1. Yao, F., Demers, A., Shenker, S.: A Scheduling Model for Reduced CPU Energy. In: Proceedings of the 36th Annual Symposium on Foundations of Computer Science (FOCS), pp. 374–382 (1995)
2. Li, M., Yao, A.C., Yao, F.F.: Discrete and Continuous Min-Energy Schedules for Variable Voltage Processors. Proceedings of the National Academy of Sciences of USA 103, 3983–3987 (2006)
3. Bansal, N., Kimbrel, T., Pruhs, K.: Speed Scaling to Manage Energy and Temperature. Journal of the ACM 54(1), Article No. 3 (2007)
4. Quan, G., Hu, X.S.: Energy efficient fixed-priority scheduling for hard read-time systems. In: Proceedings of the 36th Conference on Design Automation, pp. 134–139 (1999)
5. Yun, H.S., Kim, J.: On Energy-Optimal Voltage Scheduling for Fixed-Priority Hard Real-Time Systems. ACM Transactions on Embedded Computing Systems 2(3), 393–430 (2003)
6. Ishihara, T., Yasuura, H.: Voltage Scheduling Problem for Dynamically Variable Voltage Processors. In: Proceedings of International Symposium on Low Power Electronics and Design, pp. 197–202 (1998)
7. Kwon, W., Kim, T.: Optimal Voltage Allocation Techniques for Dynamically Variable Voltage Processors. In: Proceedings of the 40th Conference on Design Automation, pp. 125–130 (2003)
8. Li, M., Yao, F.: An Efficient Algorithm for Computing Optimal Discrete Voltage Schedules. SIAM Journal on Computing 35(3), 658–671 (2005)
9. Garey, M.R., Johnson, D.S.: Two Processor scheduling with start times and deadlines. SIAM Journal on Computing 6(3), 416–426 (1977)
10. Spieksma, F.C.R.: On the approximability of an interval scheduling problem. Journal of Scheduling 2, 215–227 (1999)
11. Bar-Noy, A., Guha, S.: Approximating the throughput of multiple machines in real time scheduling. SIAM Journal on Computing 31(2), 331–352 (2001)

12. Koren, G., Shasha, D.: D^{over}: An optimal on-line scheduling algorithm for overloaded uniprocessor real-time systems. SIAM Journal on Computing 24(2), 318–339 (1995)
13. Chan, H.L., Chan, W.T., Lam, T.W., Lee, L.K., Mak, K.S., Wong, P.: Energy Efficient Online Deadline Scheduling. In: Proceedings of the 18th Annual ACM-SIAM Symposium on Discrete Algorithms (SODA), pp. 795–804 (2007)
14. Irani, S., Pruhs, K.: Online Algorithms: Algorithmic Problems in Power Management. ACM SIGACT News 36(2), 63–76 (2005)

Approximation Algorithms for Min-Max Path Cover Problems with Service Handling Time*

Zhou Xu and Liang Xu**

Department of Logistics and Maritime Studies, Faculty of Business
The Hong Kong Polytechnic University, Hong Kong
{lgtzx,08900198r}@polyu.edu.hk

Abstract. This paper presents improved approximation algorithms and inapproximability results for min-max path cover problems with service handling time, which have wide applications in practice when the latest service completion time for customers is critical. We study three variants of this problem, where paths must start (i) from a given depot, (ii) from any depot of a given set, and (iii) from any vertex of the given graph, respectively. For these three variants, we are able to achieve approximation ratios of 3, $(4 + \epsilon)$, and $(5 + \epsilon)$, respectively, for any $\epsilon > 0$. We have further shown that approximation ratios less than 4/3, 3/2, and 3/2 are impossible for them, respectively, unless NP = P.

Keywords: approximation algorithm, inapproximability, min-max vehicle routing, path covers.

1 Introduction

Consider a complete undirected graph $G = (V, E)$, where each vertex in V represents a customer to be served. For each edge $(u, v) \in E$, an edge weight $w(u, v)$ is given to represent the traveling time, which forms a metric. Given an integer $k > 0$, the min-max path cover problem (PCP) is to decide a path set $\mathbf{P} = \{P_1, P_2, ..., P_k\}$, for k vehicles to serve all vertices of V. Its objective is to minimize the latest service completion time of vertices, which can be represented by the maximum total edge weight of any path, i.e., $\max_{1 \le i \le k} w(P_i)$.

The min-max PCP has many applications in practice, especially when the latest service completion time of customers is critical, such as to deliver aid-suppliers through a humanitarian relief chain for a large-scale emergency [3], and to plan nurse service for patients [5]. However, most previous literatures on the min-max PCP [3,1] ignore the service handling times, which can be represented by vertex weights $h(v)$ for all $v \in V$. Since service handling times in many cases contribute to the latest service completion time significantly, $h(v)$ should be taken into considerations in the studies on the min-max PCP. We are thus motivated to study the min-max PCP with service handling time (PCPSHT).

* This research was supported by PolyU Grant A-PD0W.

** Corresponding author.

Y. Dong, D.-Z. Du, and O. Ibarra (Eds.): ISAAC 2009, LNCS 5878, pp. 383–392, 2009.

For any subgraph Q of G, we use $V(Q)$ and $E(Q)$ to represent the vertex and the edge sets of Q, respectively, and use $w(Q)$ and $h(Q)$ to denote the total edge weight and the total vertex weight of Q, respectively. For any path set $\mathbf{P}$, its latest service completion time, referred to as the cost of $\mathbf{P}$, can be represented by $cost(\mathbf{P})$ defined in below:

$$cost(\mathbf{P}) = \max_{P_i \in \mathbf{P}} \{w(P_i) + h(P_i)\} \tag{1}$$

Accordingly, the PCPSHT aims to minimize $cost(\mathbf{P})$, such that $\mathbf{P}$ contains k paths that cover each vertex in V for at least once. Since the classical traveling salesman path problem, as its special case (with $k = 1$), is NP-hard, the min-max PCPSHT is also NP-hard [7] for any $k \geq 1$.

In this paper, we design approximation algorithms and derive inapproximability results for the following three variants of the min-max PCPSHT.

- In the first case, the k vehicles must start from a given depot $r \in V$, which is so called a min-max PCPSHT with a single depot (PCPSHT-SD), whose instance is represented by (G, k, r, w, h).
- In the second case, the decision maker needs to choose depots from a given depot set $R \subseteq V$, for the k vehicles. Each vehicle must start from a depot in the depot set. We call this case as a min-max PCPSHT with a depot set (PCPSHT-DS), whose instance is represented by (G, k, R, w, h).
- In the third case, which is a special case of the second case, the depot set $R = V$. Thus, every vertex can be a depot of a vehicle, implying that the depot assignment to vehicles can be ignored. We call this case as a min-max PCPSHT with no depots (PCPSHT-ND), whose instance is represented by (G, k, w, h).

Related Work. As we have mentioned, most literatures on the min-max PCP ignore the service handling times. [3] derived a lower bound on the optimum objective value of the min-max PCP with a single depot. Based on this, a 4-approximation algorithm can be easily obtained. For the min-max PCP with no depots, [1] has devised a $(4 + \epsilon)$-approximation algorithm.

Some related literatures study the min-max tour (or tree) cover problems, where their feasible solution is a set of k tours (or trees) [2,8,9,11,10]. For example, [6] proposed an approximation algorithm for k traveling salesmen problem (k-TSP), which is equivalent to the min-max tour cover problem with a single depot. They applied the classical Christofides' heuristic [4] to solve the traveling salesman problem first, and then split the approximated solution almost evenly into k segments. By connecting the given depot to the two endpoints of each segment, respectively, they obtained a tour cover, which achieved an approximation ratio of $(5/2 - 1/k)$.

For the min-max tree cover problem, [5] devised a $(4 + \epsilon)$-approximation algorithm for its variant with a depot set. Although service handling times were ignored, an additional side constraint that forced each tree to be assigned to a distinct depot must be satisfied. By doubling edges of an approximated tree cover, one can directly obtain a $(8+\epsilon)$-approximation algorithm for the corresponding variant of the min-max PCP.

In contrast to approximation algorithms, inapproximability results for the min-max PCPSHT and its related problems are almost unknown.

Our Results. Our main results and their significance are the following:

1. We develop a 3-approximation algorithm for the min-max PCPSHT-SD, which improves the best approximation ratios even for its special case, the min-max PCP with a single depot, whose existing best ratio is 4 [3].
2. We develop a $(4+\epsilon)$-approximation algorithm for the min-max PCPSHT-ND, for any $\epsilon > 0$, which has achieved the best result even for its special case, the min-max PCP with no depots. Based on this algorithm, we develop a $(5+\epsilon)$-approximation algorithm for the min-max PCPSHT-DS.
3. We derive the first inapproximability results for the min-max PCPSHT.

Organization. In Section 2, a tour splitting procedure under revised edge weights is introduced. Based on this, we develop approximation algorithms for the PCPSHT-SD, the PCPSHT-ND, and the PCPSHT-DS, in Sections 3, 4, and 5, respectively, with analysis of their approximation ratios, and inapproximability results. Finally, some concluding remarks are provided in Section 6.

2 Preliminary

For any instance I of the three variants of the min-max PCPSHT, let $\mathbf{P}^* = \{P_1^*, \ldots, P_k^*\}$ denote its optimum solution. Thus the optimum objective value, denoted by $opt(I)$, can be represented by $opt(I) = cost(\mathbf{P}^*)$. We use T^* to denote the minimum spanning tree of the given graph G.

To develop approximation algorithms for three variants of the PCPSHT, we need to devise a tour splitting procedure first, as shown in Algorithm 1, which extends the work of [6] by taking service handling times into the consideration.

Algorithm 1 (Tour Splitting Procedure)
Input: *A tour C, a starting point $v_s \in V(C)$, an integer k, and a bound vector $\boldsymbol{B} = (B_1, \ldots, B_{k-1})$, where $B_i \le w'(C)$, for all $i \le k-1$.*
Output: *A set $\mathbf{S}$ of k segments of C.*

1. *For each edge $(u,v) \in E(C)$, revise the edge weight as $w'(u,v) = w(u,v) + h(u) + h(v)$. For each segment S of C, let $w'(S)$ denote its total revised edge weight, equal to $\sum_{e \in E(\mathbf{S})} w'(e)$.*
2. *From the starting point v_s, for each i, $1 \le i \le k-1$, find the last vertex $v_{\pi(i)}$ along the tour such that the segment $(v_s \ldots v_{\pi(i)})$ satisfies $w'(v_s \ldots v_{\pi(i)}) \le B_i$, and denote the next vertex along the tour as $v_{\pi'(i)}$.*
3. *Obtain a set $\mathcal{S}$ of k segments: $S_1 = (v_s \ldots v_{\pi(1)})$, ..., $S_i = (v_{\pi'(i-1)} \ldots v_{\pi(i)})$, ..., and $S_k = (v_{\pi'(k-1)} \ldots v_s)$.*

Algorithm 1 has a polynomial time complexity, and holds the following properties, which are used to derive approximation ratios for algorithms in Sections 3, 4, and 5. First, under the revised edge weights w', we have:

$$w'(C) = w(C) + 2h(C). \tag{2}$$

Secondly, for the segment set $\mathbf{S}$ returned by Algorithm 1, its $cost(\mathbf{S})$ (under w and h) has the following upper bound, as shown in Lemma 1.

Lemma 1. *For Algorithm 1, the segment set* $\mathbf{S}$ *that it returns satisfies:*

$$cost(\mathbf{S}) = \max_{1 \le i \le k} \{w(S_i) + h(S_i)\} \le \max_{1 \le i \le k} (B_i - B_{i-1}), \tag{3}$$

where $B_0 = 0$, $B_k = w'(C)$, *and* $\boldsymbol{B} = (B_1, ..., B_{k-1})$ *is the given bound vector.*

Proof. According to Algorithm 1, it can be verified that $w'(S_1) \le B_1 - B_0$ and $w'(S_k) \le w'(C) - B_{k-1}$. For $2 \le i \le k-1$, by $w'(v_s \dots v_{\pi(i)}) \le B_i$ and $B_{i-1} \le w'(v_s \dots v_{\pi'(i-1)})$, we have $w'(S_i) \le B_i - B_{i-1}$. Thus, for $1 \le i \le k$,

$$w'(S_i) \le B_i - B_{i-1}. \tag{4}$$

Moreover, for each S_i, where $1 \le i \le k$, let x_i and y_i denote the two endpoints of S_i. It can be observed that $w'(S_i) = w(S_i) + 2h(S_i) - h(x_i) - h(y_i)$, and that $h(x_i) + h(y_i) \le h(S_i)$. Thus, $w(S_i) + h(S_i) \le w'(S_i)$, implying $w(S_i) + h(S_i) \le B_i - B_{i-1}$ for $1 \le i \le k$ due to (4). From this, (3) can be obtained directly. □

3 PCPSHT-SD

3.1 Algorithm for the PCPSHT-SD

We develop Algorithm 2 for the PCPSHT-SD, which achieves an approximation ratio of 3 as shown in Theorem 1.

Algorithm 2 (PCPSHT-SD)
Input: *Instance* $I = (G, k, r, w, h)$ *of the PCPSHT-SD*
Output: *Feasible solution* $\mathbf{P}$ *to* I.

1. *Find a minimum spanning tree* T^* *of* G, *and double all edges of* T^* *to induce a tour* C *that covers all vertices of* V.
2. *Apply Algorithm 1 on* $(C, v_s, k, \boldsymbol{B})$ *to split* C *into* k *segments, where the starting point* v_s *equal to* r, *and the bound vector* $\boldsymbol{B}$ *has* $B_i = (i/k)w'(C)$ *for* $1 \le i \le k-1$.
3. *For* $i = 1, \dots, k$, *join* r *to one end endpoint of* S_i, *to construct a path* P_i. *Return the path set* $\mathbf{P} = \{P_1, \dots, P_k\}$.

To analyze the approximation ratio of Algorithm 2, we prove the following lower bound on $opt(I)$ for any PCPSHT-SD instance I.

Lemma 2. *For any instance $I = (G, k, r, w, h)$ of the PCPSHT-SD,*

$$opt(I) \geq max\{\frac{w(T^*) + h(G)}{k}, \max_{v \in V}\{h(r) + w(r, v)\}\}, \quad (5)$$

Proof. For the optimal solution $\mathbf{P}^*$ to instance I, since $\bigcup_{i=1}^{k} P_i^*$ spans all vertices of V, we have $\sum_{i=1}^{k} w(P_i^*) \geq w(T^*)$, which implies $\sum_{i=1}^{k}[w(P_i^*) + h(P_i^*)] \geq w(T^*) + h(G)$. Therefore, there must exist $P_i^* \in \mathbf{P}^*$ with $w(P_i^*) + h(P_i^*) \geq [w(T^*) + h(G)]/k$. Thus, $opt(I) \geq [w(T^*) + h(G)]/k$.

Let v^* denote a vertex with $h(r) + w(r, v^*) = \max_{v \in V}\{h(r) + w(r, v)\}$. Therefore, there must exist $P_j^* \in \mathbf{P}^*$ with $\{r, v^*\} \subseteq V(P_j^*)$. Since w is a metric, $w(P_j^*) + h(P_j^*) \geq w(r, v^*) + h(r) = \max_{v \in V}\{h(r) + w(r, v)\}$, which implies that $opt(I) \geq \max_{v \in V}\{h(r) + w(r, v)\}$. □

From Lemma 2, we can establish the following theorem, which implies that Algorithm 2 is a 3-approximation algorithm.

Theorem 1. *Algorithm 2 achieves an approximation ratio of* 3 *in polynomial time for the min-max PCPSHT-SD.*

Proof. It is easy to verify the polynomial time complexity for Algorithm 2. In the following, we prove its approximation ratio of 3.

Let $B_0 = 0$, and $B_k = w'(C)$. From Algorithm 2, by Lemma 1, we obtain $cost(\mathbf{S}) \leq w'(C)/k$. This implies that for $i = 1, \ldots, k$, the total weight of each path $P_i \in \mathcal{P}$, which connects r to S_i, must satisfy: $w(P_i) + h(P_i) \leq w'(C)/k + \max_{v \in V}\{h(r) + w(r, v)\}$.

Moreover, since $w(C) \leq 2w(T^*)$, by (2) we have $w'(C)/k \leq [2w(T^*)/k + 2h(G)]/k$. Thus, by Lemma 1, we obtain $cost(\mathbf{P}) \leq 3opt(I)$. Hence, Algorithm 2 achieves an approximation ratio of 3. □

3.2 Inapproximability Result for PCPSHT-SD

We establish an inapproximability result for PCPSHT-SD as follows.

Theorem 2. *Unless* NP = P, *there is no polynomial-time* $(4/3 - \epsilon)$*-approximation algorithm for the PCPSHT-SD, for any* $\epsilon > 0$, *even if* $h(v) = 0$ *for all* $v \in V$.

Proof. Suppose there exists a $(4/3 - \epsilon)$-approximation algorithm for $\epsilon > 0$ for the PCPSHT-SD with $h(v) = 0$ for all $v \in V$. We are going to show that this algorithm can be used to solve the following problem of three dimensional matching (3DM), a well known NP-complete problem, in polynomial time, which contradicts to NP $\neq$ P.

We first introduce the problem of 3DM [7]. Given a set $M \subseteq W \times X \times Y$ with $|M| = m$, where W, X and Y are disjoint sets with the same number of n elements. 3DM is to decide whether M contains an exact matching $M' \subseteq M$ such that $|M'| = n$ and no two elements of M' agree in any coordinate.

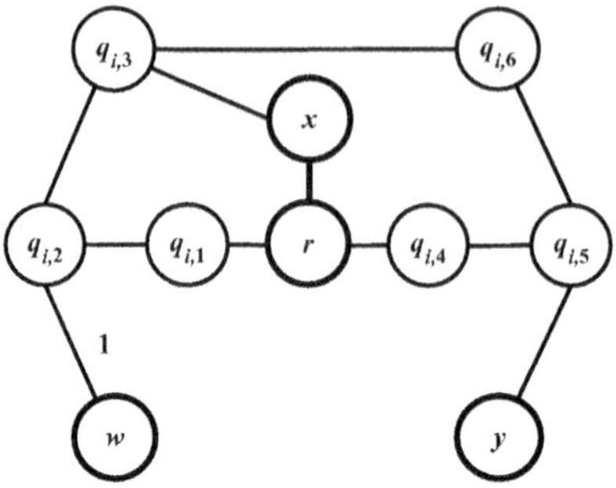

Fig. 1. A component for each tuple $M_i = (w, x, y) \in M$ used in transforming the 3DM to the PCPSHT-SD with handling times all equals to zeros

Given any 3DM instance, consider an instance $I = (G, k, r, w, h)$, defined as follows, of the PCPSHT-SD with handling times all equal to zeros. As shown in Figure 1, we use a local replacement to substitute any tuple $M_i = (w, x, y) \in M$, where $1 \leq i \leq m$, with a component of a complete graph $G = (V, E)$ with edge weights w. We define $V = \{r\} \cup W \cup X \cup Y \cup_{i=1}^{m} \{q_{i,j} : 1 \leq j \leq 6\}$ with $h(v) = 0$ for all $v \in V$. For each edge shown in the Figure 1, we set its edge weight to be 1, and for other edges of G, we set their edge weights all equal to 2. It is easy to verify that w forms a metric. Finally, we set $k = n + 2m$.

We are now going to show that the 3DM instance has an exact matching if and only if the $(4/3 - \epsilon)$-approximation algorithm for the PCPSHT-SD returns a feasible solution to instance I with cost at most 3.

On one hand, if the 3DM has an exact matching M', then a feasible solution $\mathbf{P}$ to I can be obtained by including three paths, $(rxq_{i,3}q_{i,6})$, $(rq_{i,1}q_{i,2}w)$, and $(rq_{i,4}q_{i,5}y)$, for each $M_i = (w, x, y) \in M'$, and including two paths, $(rq_{i,1}q_{i,2}q_{i,3})$, and $(rq_{i,4}q_{i,5}q_{i,6})$, for each $M_i = (w, x, y) \in M \backslash M'$. Thus, we have $|\mathbf{P}| = 3|M'| + 2(|M| - |M'|) = k$. Since M' is an exact matching for the 3DM instance, each vertex of $V \setminus \{r\}$ must be covered by a unique path in $\mathbf{P}$. Notice that each path in $\mathbf{P}$ contains the depot r, and has a total weight of 3. We know $\mathbf{P}$ is a feasible solution to instance I with cost exactly equal to 3. Thus, $opt(I) \leq 3$. Since the edge and vertex weights are integers, and since $\epsilon > 0$, the $(4/3-\epsilon)$-approximation algorithm must return a feasible solution to instance I with cost at most 3.

On the other hand, if the $(4/3-\epsilon)$-approximation algorithm returns a feasible solution $\mathcal{P}$, to instance I of the PCPSHT-SD, with $cost(\mathbf{P}) \leq 3$, then since the weight of each edge is either 1 or 2, each path of $\mathbf{P}$ must contain at most four vertices of V, and start from the depot r. Moreover, since $|V - \{r\}| = 6m + 3n$, and since $k = 2m+n$, each path of $\mathbf{P}$ must contain exact four vertices (including r) with each edge having weight equal to one, and no two paths of $\mathbf{P}$ can share the same vertex in $V - \{r\}$.

Therefore, for each $M_i = (w, x, y) \in M$, as shown in Figure 1, $q_{i,4}$ must be covered by either the path $(rq_{i,4}q_{i,5}y)$ or the path $(rq_{i,4}q_{i,5}q_{i,6})$, but not both. This implies that exact one of the two paths must be in $\mathbf{P}$. We therefore construct M' by including those $M_i = (w, x, y) \in M$ with $(rq_{i,4}q_{i,5}y) \in \mathbf{P}$, for $1 \leq i \leq m$.

To prove that M' is an exact matching for the 3DM, consider any element $M_i = (w, x, y) \in M$. If $M_i \in M'$, then $(rq_{i,4}q_{i,5}y) \in \mathbf{P}$, which implies that the path $(rxq_{i,3}q_{i,6})$ must be in $\mathbf{P}$ to cover $q_{i,6}$, and that the path $(rq_{i,1}q_{i,2}w)$ must be in $\mathbf{P}$ to cover $q_{i,1}$. Thus, since no two paths of $\mathbf{P}$ can share the same vertex in $V - \{r\}$, no elements in M other than M_i can match w, x, or y. Therefore, no two elements of M' can agree in any coordinate. Furthermore, if $M_i \in M \setminus M'$, then $(rq_{i,4}q_{i,5}q_{i,6}) \in \mathbf{P}$, which implies that the path $(rq_{i,1}q_{i,2}q_{i,3})$ must be in $\mathbf{P}$ to cover $q_{i,1}$ and $q_{i,3}$. Thus, we obtain $|\mathbf{P}| = 3|M'| + 2(|M| - |M'|)$, implying $|M'| = |\mathbf{P}| - 2|M| = k - 2m = n$. Therefore, M' is an exact matching. □

4 PCPSHT-ND

4.1 Algorithm for the PCPSHT-ND

Algorithm 3 for the PCPSHT-ND is also based on the Tour Splitting Procedure (Algorithm 1). Given a bound λ for $opt(I)$ of any PCPSHT-ND instance I, as shown in Lemma 3, Algorithm 3 either returns "λ is too small" (implying $\lambda < opt(I)$), or returns a feasible solution $\mathcal{P}$ with cost at most 4λ, in polynomial time. Since $opt(I)$ is bounded by the interval $[0, 2w(T^*) + h(G)]$, we can apply a binary search to obtain a closed value λ, such that Algorithm 3 will return a feasible solution with cost at most 4λ, and that for any $\epsilon > 0$, Algorithm 3 will return and guarantee that $(\lambda - \epsilon)$ is too low. Thus, a polynomial time $(4 + \epsilon)$-approximation algorithm can be obtained.

Algorithm 3 (PCPSHT-ND)
Input: *Instance* $I = (G, k, w, h)$ *of the PCPSHT-ND, and* $\lambda > 0$
Output: *Feasible solution* $\mathbf{P}$ *to* I.

1. *Remove all the edges with weights greater than* λ. *Let* G' *denote the remaining graph, and* $G_1, G_2, ..., G_m$ *denote the connected components of* G'.
2. *For* $1 \le i \le m$, *find a minimum spanning tree* T_i^* *of* G_i, *and double all edges of* T_i^* *to induce a tour* C_i *that covers all vertices of* G_i.
3. *For* $1 \le i \le m$, *apply Algorithm 1 on* $(C_i, s_i, k_i, \boldsymbol{B}_i)$, *where* s_i *is any vertex of* C_i, *and* k_i *equals* $\lceil w'(C_i)/4\lambda \rceil$, *and* $\boldsymbol{B}_i = (4\lambda, \ldots, 4i\lambda, \ldots, 4(k_i - 1)\lambda)$, *to split* C_i *into* k_i *segments denoted by* $S_{i,j}$ *for* $1 \le j \le k_i$.
4. *If* $\sum_{i=1}^{m} k_i > k$, *return "*λ *is too small". Otherwise, return a path set* $\mathbf{P} = \{S_{i,j} : 1 \le i \le m, 1 \le j \le k_i\}$.

Lemma 3. *Given any instance* I *of the PCPSHT-ND, and given any* $\lambda > 0$, *Algorithm 3 runs in polynomial time. If Algorithm 3 returns "*λ *is too small', then* $\lambda < opt(I)$; *otherwise, the path cover* $\mathbf{P}$ *that Algorithm 3 returns is a feasible solution to* I *with* $cost(\mathbf{P}) \le 4\lambda$.

Proof. The polynomial time complexity of Algorithm 3 is easy to be verified.

If Algorithm 3 returns "λ is too small", then $\sum_i k_i > k$. To prove $opt(I) > \lambda$ by contradiction, suppose $opt(I) \le \lambda$, which implies that there exists an optimal solution $\mathbf{P}^*$ to I, containing at most k paths, with $cost(\mathbf{P}^*) \le \lambda$. Thus there is no path in $\mathbf{P}^*$ that contains an edge whose two endpoints belong to two different connected components of G', where G' is the remaining graph, as defined in Step 1, obtained by removing edges with weights greater than λ from G. Hence, for $1 \le i \le m$, let k_i^* denote the number of paths in $\mathbf{P}^*$ that belong to the connected component G_i of G'. From these k_i^* paths, we can connect all vertices of G_i by adding at most $(k_i^* - 1)$ edges of G_i, whose edge weights must be less than or equal to λ. Since each path of $\mathbf{P}^*$ has a total weight less than or equal to λ, we obtain that $w(T_i^*) \le k_i^*\lambda + (k_i^* - 1)\lambda - h(T_i^*)$.

By $\lambda > 0$, we have $k_i^* \ge \lceil w(T_i^*) + h(T_i^*)]/(2\lambda)\rceil$. By Step 2 of Algorithm 3, $w(C_i) \le 2w(T_i^*)$, which implies $w'(C_i) \le 2[w(T_i^*) + h(T_i^*)]$ by (2). Hence,

$$k_i^* \ge \lceil w'(C_i)/(4\lambda)\rceil. \tag{6}$$

Since $\sum_{i=1}^m k_i > k$ and $k_i = \lceil w'(C_i)/(4\lambda)\rceil$, we obtain $\sum_{i=1}^m k_i^* > k$, implying $|\mathbf{P}^*| > k$, leading to a contradiction. Thus, $opt(I) \le \lambda$

Otherwise, Algorithm 3 returns a path set $\mathbf{P}$. According to Step 2 of Algorithm 3, we have $\cup_{i=1}^m V(C_i) = V$. According to Lemma 1, we have $\cup_{j=1}^{k_i} V(S_{i,j}) = C_i$ for $1 \le i \le m$. Thus, by Step 4 of Algorithm 3, $\mathbf{P}$ must cover all vertices of V. Furthermore, each path $S_{i,j} \in \mathbf{P}$, where $1 \le i \le m$ and $1 \le j \le k_i$ is a segment split from C_i by Algorithm 1 in Step 2. By Lemma 1, since $\boldsymbol{B}_i = (4\lambda, \ldots, 4i\lambda, \ldots, 4(k_i-1)\lambda)$, we obtain $w(S_{i,j}) + h(S_{i,j}) \le \max\{4\lambda, w'(C_i) - 4(k_i - 1)\lambda\}$. By the definition of k_i, $w'(C_i) - 4(k_i - 1)\lambda \le 4k_i\lambda - 4(k_i-1)\lambda \le 4\lambda$, which implies $w(S_{i,j}) + h(S_{i,j}) \le 4\lambda$. Hence, $cost(\mathbf{P}) \le 4\lambda$. □

From Lemma 3, we can directly obtain a $(4+\epsilon)$-approximation algorithm by a binary search, for any $\epsilon > 0$, due to the arguments in beginning of this section.

Theorem 3. *For any $\epsilon > 0$, there exists a polynomial time $(4+\epsilon)$-approximation algorithm for the PCPSHT-ND.*

4.2 Inapproximation Result for PCPSHT-ND

In the following content, theorem 4 presents the inapproximability result for the PCPSHT-ND:

Theorem 4. *Unless* NP $=$ P, *there is no polynomial time $(3/2 - \epsilon)$-approximation algorithm for the PCPSHT-ND, for any $\epsilon > 0$, even if $h(v) = 0$ for all $v \in V$.*

Proof. Suppose there exists a $(3/2 - \epsilon)$-approximation algorithm for $\epsilon > 0$ for the PCPSHT-ND with $h(v) = 0$ for all $v \in V$. Similar to the proof of Theorem 2, we are going to show that this algorithm can be used to solve the 3DM in polynomial time, which contradicts to NP $\neq$ P.

Given any 3DM instance (W, X, Y, M), let us consider an instance $I = (G, k, w, h)$, defined as follows, of the PCPSHT-ND with $h(v) = 0$ for all $v \in V$, as shown in Figure 2, for each edge shown in the Figure 2, we set its edge weight

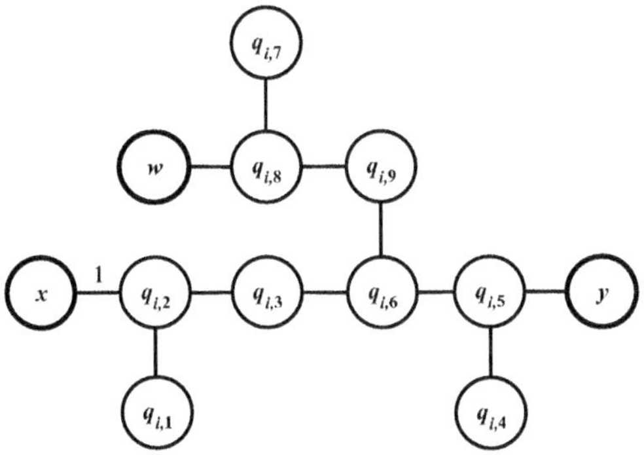

Fig. 2. A component for each tuple $M_i = (w, x, y) \in M$ used in transforming the 3DM to the PCPSHT-ND with handling times all equals to zeros

to be 1, and for other edges of G, we set their edge weights all equal to 2. It is easy to verify that w forms a metric. Finally, we set $k = n + 3m$.

We are now going to show that the 3DM instance has an exact matching if and only if the $(3/2 - \epsilon)$-approximation algorithm for the PCPSHT-ND returns a feasible solution to instance I with cost at most 2.

On one hand, if the 3DM instance has an exact matching M', then a feasible solution $\mathbf{P}$ to I can be obtained by including four paths, $(wq_{i,8}q_{i,7})$, $(yq_{i,5}q_{i,4})$, $(xq_{i,2}q_{i,1})$, and $(q_{i,3}q_{i,6}q_{i,9})$, for each $M_i = (w, x, y) \in M'$, and including three paths, $(q_{i,1}q_{i,2}q_{i,3})$, $(q_{i,4}q_{i,5}q_{i,6})$, and $(q_{i,7}q_{i,8}q_{i,9})$, for each $M_i = (w, x, y) \in M \backslash M'$. Similarly to Theorem 2, we can verify that the $(3/2 - \epsilon)$-approximation algorithm returns a feasible solution with cost at most 2.

On the other hand, if the $(3/2 - \epsilon)$-approximation algorithm returns a feasible solution $\mathbf{P}$, to instance I, with $cost(\mathbf{P}) \leq 2$, we can M' by including those $M_i = (w, x, y) \in M$ with $(yq_{i,5}q_{i,4}) \in \mathbf{P}$, for $1 \leq i \leq m$. Similarly to the proof of Theorem 2, it can be verified that such M' is an exact matching. □

5 PCPSHT-DS

For an instance I of the PCPSHT-DS, we can extent Algorithm 3 by connecting each path to its nearest depot in the depot set to obtain a feasible solution. It can be verified that, for any vertex v, $\min_{r \in R}\{w(r, v) + h(r)\} \leq opt(I)$. Similarly to Theorem 3, we can obtain that the approximation ratio of the algorithm to the PCPSHT-DS is $(5 + \epsilon)$, for any $\epsilon > 0$.

Since the PCPSHT-ND is a special instance of the PCPSHT-DS, the approximation hardness of the PCPSHT-DS is at least $3/2$.

6 Concluding Remarks

In this paper, we have presented improved approximation algorithms and inapproximability results for three variants of the min-max path cover problem with service handling time. Our future research may include to reduce the gap between approximation ratios and inapproximability results for different variants of the PCPSHT. An alternative direction is to develop approximation algorithms for problems similar to the PCPSHT, but with more complicated side constraints, such as the vehicle capacity constraint.

References

1. Arkin, E.M., Hassin, R., Levin, A.: Approximations for minimum and min-max vehicle routing problems. Journal of Algorithms 59(1), 1–18 (2006)
2. Berman, O., Averbakh, I.: (p-1)/(p+1)-approximate algorithms for p-travelling salesmen problems on a tree with minmax objective. Discrete Applied Mathematics 75, 201–216 (1997)
3. Campbell, A.M., Vandenbussche, D., Hermann, W.: Routing for relief efforts. Transportation Science 42(2), 127 (2008)
4. Christofides, N.: Worst-case analysis of a new heuristic for the travelling salesman problem. Technical Report 388, Graduate School of Industrial Adminsitration, Carnegie-Mellon University (1976)
5. Even, G., Garg, N., Könemann, J., Ravi, R., Sinha, A.: Minmax tree covers of graphs. Operations Research Letters 32(4), 309–315 (2004)
6. Frederickson, G.N.: Approximation algorithms for some routing problems. SIAM Joural on Computing 7(2) (1978)
7. Garey, M.R., Johnson, D.S.: Computers and intractability: a guide to the theory of NP-completeness. WH Freeman & Co., New York (1979)
8. Okada, K., Nagamochi, H.: Polynomial time 2-approximation algorithms for the minmax subtree cover problem. In: Ibaraki, T., Katoh, N., Ono, H. (eds.) ISAAC 2003. LNCS, vol. 2906, pp. 138–147. Springer, Heidelberg (2003)
9. Okada, K., Nagamochi, H.: A faster 2-approximation algorithm for the minmax p-travelling salesmen problem on a tree. Discrete Applied Mathematics 140, 103–114 (2004)
10. Okada, K., Nagamochi, H.: Approximating the minmax rooted-tree cover in a tree. Information Processing Letters 104, 173–178 (2007)
11. Nagamochi, H.: Approximating the minmax rooted-subtree cover problem. IEICE Transactoins on Fundamentals of Electronics, Communications and Computer Sciences E88-A(5) (2005)

Minimum Covering with Travel Cost*

Sándor P. Fekete[1], Joseph S.B. Mitchell[2,**], and Christiane Schmidt[1,***]

[1] Algorithms Group, Braunschweig Institute of Technology, Germany
{s.fekete,c.schmidt}@tu-bs.de
[2] Department of Applied Mathematics and Statistics, Stony Brook University
jsbm@ams.stonybrook.edu

Abstract. Given a polygon and a visibility range, the Myopic Watchman Problem with Discrete Vision (MWPDV) asks for a closed path P and a set of scan points S, such that (i) every point of the polygon is within visibility range of a scan point; and (ii) path length plus weighted sum of scan number along the tour is minimized. Alternatively, the bicriteria problem (ii') aims at minimizing both scan number and tour length. We consider both lawn mowing (in which tour and scan points may leave P) and milling (in which tour, scan points and visibility must stay within P) variants for the MWPDV; even for simple special cases, these problems are NP-hard.

We sketch a 2.5-approximation for rectilinear MWPDV milling in grid polygons with unit scan range; this holds for the bicriteria version, thus for any linear combination of travel cost and scan cost. For grid polygons and circular unit scan range, we describe a bicriteria 4-approximation. These results serve as stepping stones for the general case of circular scans with scan radius r and arbitrary polygons of feature size a, for which we extend the underlying ideas to a $\pi(\frac{r}{a} + \frac{r+1}{2})$ bicriteria approximation algorithm. Finally, we describe approximation schemes for MWPDV lawn mowing and milling of grid polygons, for fixed ratio between scan cost and travel cost.

1 Introduction

Covering a given polygonal region by a small set of disks or squares is a problem with many applications. Another classical problem is finding a short tour that visits a number of objects. Both of these aspects have been studied separately, with generalizations motivated by natural constraints.

In this paper, we study the combination of these problems, originally motivated by challenges from robotics, where accurate scanning requires a certain

---

* A 4-page abstract based on Sections 3 and 4 of this paper appeared in the informal and non-selective workshop "EuroCG", March 2009 [9].

** Partially supported by the National Science Foundation (CCF-0528209, CCF-0729019), Metron Aviation, and NASA Ames.

*** Supported by DFG Focus Program "Algorithm Engineering" (SPP 1307) project "RoboRithmics" (Fe 407/14-1).

Y. Dong, D.-Z. Du, and O. Ibarra (Eds.): ISAAC 2009, LNCS 5878, pp. 393–402, 2009.

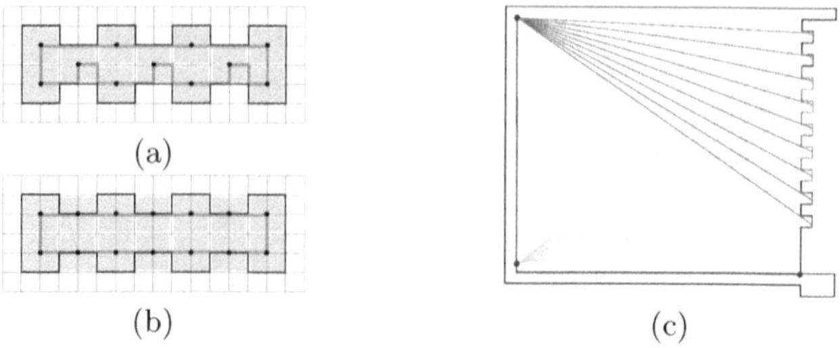

Fig. 1. (a) An MWPDV solution with a minimum number of scans; (b) an MWPDV solution with a minimum tour length. (c) A minimum guard cover may involve scan points that are not from an obvious set of candidate points.

amount of time for each scan; obviously, this is also the case for other surveillance tasks that combine changes of venue with stationary scanning. The crucial constraints are (a) a limited visibility range, and (b) the requirement to stop when scanning the environment, i.e., with vision only at discrete points. These constraints give rise to the *Myopic Watchman Problem with Discrete Vision* (MWPDV), the subject of this paper.

For a scan range that is not much bigger than the feature size of the polygon, the MWPDV combines two geometric problems that allow approximation schemes (minimum cover and TSP). This makes it tempting to assume that combining two approximation schemes will yield a polynomial-time approximation scheme (PTAS), e.g., by using a PTAS for minimum cover (Hochbaum and Maass [11]), then a PTAS for computing a tour on this solution. As can be seen from Figure 1 (a) and (b), this is not the case; moreover, an optimal solution depends on the relative weights of tour length and scan cost. This turns the task into a bicriteria problem; the example shows that there is no simultaneous PTAS for both aspects. As we will see in Sections 3 and 4, a different approach allows a simultaneous constant-factor approximation for both scan number and tour length, and thus of the combined cost. We show in Section 7, a more involved integrated guillotine approach allows a PTAS for combined cost in the case of a fixed ratio between scan cost and travel cost.

A different kind of difficulty is highlighted in Figure 1 (c): For a visibility range r that is large compared to the feature size a, it may be quite hard to determine a guard cover of small size. In fact, there is no known constant-factor approximation for minimum guard cover in general polygons; currently, the best result is an $O(\log OPT)$-approximation by Efrat and Har-Peled [7]. In addition, the optimal solution may change significantly with the relative weights between tour length: If tour length dominates the number of scans, an optimal tour can be forced to follow the row of niches on the right. We will show in Section 6 how to obtain a constant-factor approximation for bounded value $\frac{r}{a}$.

Related Work. Closely related to practical problems of searching with an autonomous robot is the classical theoretical problem of finding a *shortest watchman tour*; e.g., see [4,5]. Planning an optimal set of scan points (with unlimited visibility) is the *art gallery problem* [14]. Finally, visiting all grid points of a given set is a special case of the classical *Traveling Salesman Problem* (TSP); see [12]. Two generalizations of the TSP are the so-called *lawn mowing* and *milling problems*: Given a cutter of a certain shape, e.g., an axis-aligned square, the *milling problem* asks for a shortest tour along which the (center of the) cutter moves, such that the entire region is covered and the cutter stays inside the region at all times. Clearly, this takes care of the constraint of limited visibility, but it fails to account for discrete visibility. At this point, the best known approximation method for milling is a 2.5-approximation [2]. Related results for the TSP with neighborhoods (TSPN) include [6,13]; further variations arise from considering online scenarios, either with limited vision [3] or with discrete vision [10,8], but not both. Finally, [1] consider covering a set of points by a number of scans, and touring all scan points, with the objective function being a linear combination of scan cost and travel cost; however, the set to be scanned is discrete, and scan cost is a function of the scan radius, which may be small or large.

For an online watchman problem with unrestricted but discrete vision, Fekete and Schmidt [10] present a comprehensive study of the milling problem, including a strategy with constant competitive ratio for polygons of bounded feature size and with the assumption that each edge of the polygon is fully visibly from some scan point. For limited visibility range, Wagner et al. [15] discuss an online strategy that chooses an arbitrarily uncovered point on the boundary of the visibility circle and backtracks if no such point exists. For the cost they only consider the length of the path used between the scan points, scanning causes no cost. Then, they can give an upper bound on the cost as a ratio of total area to cover and squared radius.

Our Results. On the positive side, we give a 2.5-approximation for the case of grid polygons and a rectangular range of unit-range visibility, generalizing the 2.5-approximation by Arkin, Fekete, and Mitchell [2] for continuous milling. The underlying ideas form the basis for more general results: For circular scans of radius $r = 1$ and grid polygons we give a 4-approximation. Moreover, for circular scans of radius r and arbitrary polygons of feature size a, we extend the underlying ideas to a $\pi(\frac{r}{a} + \frac{r+1}{2})$-approximation algorithm. All these results also hold for the bicriteria versions, for which both scan cost and travel cost have to approximated simultaneously. Finally, we present a PTAS for MWPDV lawn mowing, and sketch a PTAS for MWPDV milling, both for the case of fixed ratio between scan cost and travel cost.

2 Notation and Preliminaries

We are given a polygon P. In general, P may be a polygon with holes; in Sections 3, 4 and 5, P is an axis-parallel polygon with integer coordinates.

Our robot, R, has discrete vision, i.e., it can perceive its environment when it stops at a point and performs a scan, which takes c time units. From a scan point p, only a ball of radius r is visible to R, either in L_∞- or L_2-metric. A set $\mathcal{S}$ of scan points *covers* the polygon P, if and only if for each point $q \in P$ there exists a scan point $p \in \mathcal{S}$ such that q sees p (i.e., $qp \subset P$) and $|qp| \leq r$.

We then define the *Myopic Watchman Problem with Discrete Vision* (MW-PDV) as follows: Our goal is to find a tour T and a set of scan points $\mathcal{S}(T)$ that covers P, such that the total travel and scan time is optimal, i.e., we minimize $t(T) = c \cdot |\mathcal{S}(T)| + L(T)$, where $L(T)$ is the length of tour T. Alternatively, we may consider the bicriteria problem, and aim for a simultaneous approximation of both scan number and tour length.

3 NP-Hardness

Even the simplest and extreme variants of MWPDV lawn mowing are still generalizations of NP-hard problems; proofs are omitted.

Theorem 1. *(1) The MWPDV is NP-hard, even for polyominoes and small or no scan cost, i.e.,* $c \ll 1$ *or* $c = 0$.

(2) The MWPDV is NP-hard, even for polyominoes and small or no travel cost, i.e., $c \gg 1$ *or* $t(T) = |\mathcal{S}|$.

4 Approximating Rectilinear MWPDV Milling for Rectangular Visibility Range

As a first step (and a warmup for more general cases), we sketch an approximation algorithm for rectilinear visibility range in rectilinear grid polygons.

Our approximation proceeds in two steps; details can be found in [9].

(I) Construct a set of scan points that is not larger than 2.5 times a minimum cardinality scan set.
(II) Construct a tour that contains all constructed scan points and that does not exceed 2.5 times the cost of an optimum milling tour.

First we describe how to construct a covering set of scan points:

1. Let S_{4e} be the "even quadruple" centers of all 2x2-squares that are fully contained in P, and which have two even coordinates.
2. Remove all 2x2-squares corresponding to S_{4e} from P; in the remaining polyomino P_{4e}, greedily pick a maximum disjoint set S_{4o} of "odd quadruple" 2x2-squares.
3. Remove all 2x2-squares corresponding to S_{4o} from P_{4e}; greedily pick a maximum disjoint set S_3 of "triple" 2x2-squares that cover 3 pixels each in the remaining polyomino $P_{4e,4o}$,
4. Remove all 2x2-squares corresponding to S_3 from $P_{4e,4o}$; in the remaining set $P_{4e,4o,3}$ of pixels, no three can be covered by the same scan. Considering edges between pixels that can be covered by the same scan, pick a minimum set of ("double" S_2 and "single" S_1) scans by computing a maximum matching.

Claim 1. The total number of scans is at most 2.5 times the size of a minimum cardinality scan set.

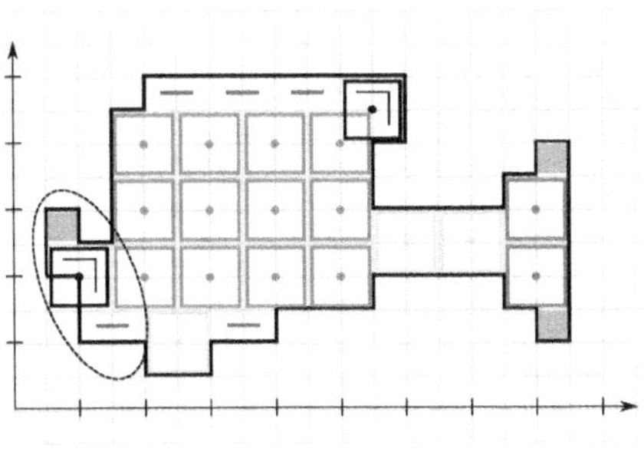

Fig. 2. An example for our approximation method: The set of "even quadruple"scans is shown in gray; the "odd quadruple" scans are light gray. A possible (greedy!) set of "triple" scan is shown in black, leaving the maximum matching (and the corresponding "double scans") shown in dark gray. The leftover single pixels are filled squares. The ellipse indicates a part that is covered by three scans instead of two: the triple scan with adjacent single and double scans could be covered by two triple scans.

Claim 2. All scan points lie on a 2.5-approximative milling tour.

The tour consists of (a) a "boundary" part following the contour of the polygon; (b) a "strip" part that covers the interior; (c) a "matching" part that allows an Eulerian tour. The cost for (a) and (b) is $L(T^*)$, while (c) can be bounded by $L(T^*)/2$. Proofs are omitted for lack of space.

Theorem 2. *A polyomino P allows a MWPDV with rectangular vision solution that contains at most 2.5 times the minimum number of scans necessary to scan the polygon, and has tour length at most 2.5 times the length of an optimum milling tour.*

5 Approximating Rectilinear MWPDV Milling for Circular Visibility Range

When considering a circular scan range, one additional difficulty are boundary effects of discrete scan points: While continuous vision allows simply sweeping a corridor of width $2r$, additional cleanup is required for the gaps left by discrete vision; this requires additional mathematical arguments.

We overlay the polyomino with a point grid as in Figure 3, left, i.e., a diagonal point grid with L_2-distance of $\sqrt{2}$. These are used as scan points; it is relatively straightforward to prove that this number is within a factor of 4 of the optimum number of scans.

For the movement between interior scan points and the boundary we use horizontal strips located on grid lines (and distance 1 to the boundary). As before, these are combined with a boundary tour. As strip ends do not fully extend to the boundary of the polygon, we link pairs of strips and connect them to the left boundary, other scan points are visited by paths of length 2 from the boundary. This can be achieved at the cost of one additional tour; see Figure 3, right, for an example.

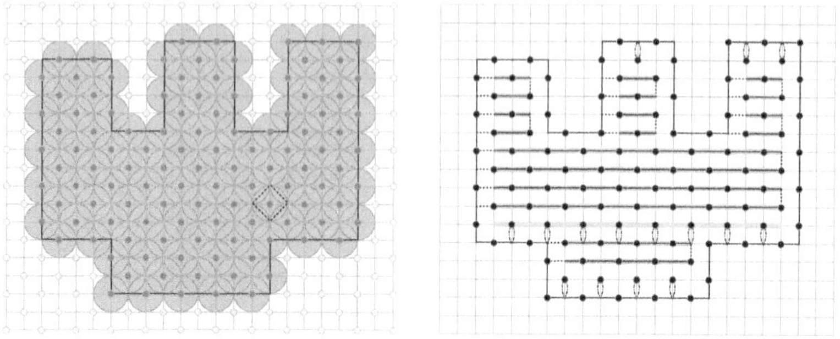

Fig. 3. Left: Point grid (light gray) with grid points within a polyomino (black) in dark gray. Circular visibility ranges of the grid points covering the plane, one square of side length $\sqrt{2}$ is indicated by a dashed line. Right: A polyomino P with the tour given by our strategy. Scan points are displayed in black. The horizontal strips of total length L_{strips} are indicated in bold light gray, the tour is in black for the links to the boundary and between strips (dotted) as well as for connections of points (dash-dotted) and parts located on the strips are indicated as a continuous line. (The rest of the tour runs on the boundary.)

Theorem 3. *A polyomino P allows a MWPDV solution for a circular visibility range with $r = 1$ that is 4-competitive.*

6 Approximating General MWPDV Milling for a Circular Visibility Range

In this section we discuss MWPDV milling for a circular visibility range r in general polygons. As discussed in Section 1, even the problem of minimum guard coverage has no known constant-factor approximation; therefore, we consider a bounded ratio r/a between visibility range and feature size.

Just as in the rectilinear case for a rectilinear scan range, our approximation proceeds in the two steps (I) and (II), see Section 4.

We start with a description of the second step, which will form the basis for the placement of scan points. Just as in the rectilinear case, we consider three parts.

(1) A "boundary" part: We use two "boundary tours" within distance of (at most) $\frac{1}{2}r$ and (at most) $\frac{3}{2}r$ to the boundary, $TR1$ and $TR2$ of length L_{TR1} and L_{TR2}, respectively. With $L_{\delta B}$ denoting the length of the boundary δB of B ($B \subset P$ is the inward offset region of all points within P that are feasible placements for the center of a milling cutter), we get:

$$L_{TR1} + L_{TR2} = 2 \cdot L_{\delta B} \leq 2 \cdot L(T^*) \tag{1}$$

(The length of the three tours differs at the vertices: drawing a line perpendicular there from $TR2$ to $TR1$ the Intercept Theorem shows that the

distance to the diagonal through the vertices of all tours on $TR1$ is twice as much as on the boundary tour with distance r to the boundary.)
The two "boundary" tours allow us to cover a corridor of width $2r$ with a bounded number of scans, while (1) enables us to bound the tour length in terms of the optimal length.

(2) A "strip" part: For the interior we use strips again: $P_{int} := P \backslash P_{\delta B}$—if nonempty—can be covered by a set of k_1 horizontal strips Σ_i^1. The y-coordinates of two strips differ by multiples of $2r$. We can consider another set of strips, Σ_i^2, shifted by r. Then, let $L_{str}^j = \sum_{i=1}^{k_j} L_{\Sigma_i^j}$. Similar to the argument for L_∞, we have $L_{str}^1 + L_{str}^2 \leq 2 \cdot L(T^*)$.

(3) A "matching" part: In order to combine the two "boundary parts" and the two sets of strips for a tour we add two more set of sections.
- The center lines of the strips have a distance of r to the boundary, thus they do not yet touch $TR1$. Consequently, we add $1/2r$ to each center line (on each end). For that purpose, we consider the matchings as defined above. (Consider the endpoints of strips on δB_i: every δB_i contains an even number of such endpoints. Hence, every δB_i is partitioned into two disjoint portions, $M_1(\delta B_i)$ and $M_2(\delta B_i)$. Using the shorter of these two $(M_*(\delta B_i))$ for every δB_i we obtain for the combined length, L_M: $L_M \leq L_{str}/2 \leq L(T^*)/2$.) Because two strips are at least a distance of r apart, the connection to $TR1$ costs less than $1/2 \cdot L_M \leq 1/2 \cdot L_{str}/2 \leq L(T^*)/4$.
- Moreover, we consider the above matchings defined on $TR1$ and insert the shorter sections of the disjoint parts, $(M_*^1(\delta B_i))$, for every δB_i. The Intercept Theorem in combination with the analogously defined sections on $TR2$ enables us to give an upper bound of $L_{M^1} \leq L_{str} \leq L(T^*)$.

Starting on some point on $TR1$, tracing the strips, and the inner "boundary" $TR2$ at once when passing it yields a tour; the above inequalities show that $L(T) \leq 21/4 \cdot L(T^*)$.

Now we only have to take care of (I), i.e., construct an appropriate set of scan points. For the "boundary" part we place scans with the center points located on $TR1$ and $TR2$ in distance $\sqrt{3} \cdot r$ if possible, but at corners we need to place scans, so the minimum width we are able to cover with the two scans (on both tours) is a. For the "strip" part the distance of scans is also $\sqrt{3} \cdot r$ on both strip sets, exactly the distance enabling us to cover a width of r.

It remains to consider the costs for the scans. We start with the inner part. Taking scans within a distance of $\sqrt{3} \cdot r$, we may need the length divided by this value, plus one scan. We only charge the first part to the strips, the (possible) additional scans are charged to the "boundary" part, as we have no minimum length of the strips. The optimum cannot cover more than πr^2 with one scan. Let $L_{str} = max(L_{str}^1, L_{str}^2)$:

$$|\mathcal{S}(T^*)| \geq \frac{L_{str}}{\pi r/2}, \quad |\mathcal{S}(T)| \leq \frac{2L_{str}}{\sqrt{3} \cdot r} \Rightarrow \frac{|\mathcal{S}(T)|}{|\mathcal{S}(T^*)|} \leq \frac{2L_{str}}{\sqrt{3} \cdot r} \cdot \frac{\pi r/2}{L_{str}} = \frac{\pi}{\sqrt{3}} \qquad (2)$$

Finally, we consider the "boundary". We assume $L_{\delta B} \geq 1$. So $|\mathcal{S}(T^*)| \geq \frac{L_{\delta B}}{\pi r/2}$. We may need to scan within a distance of a—on two strips—, need additional scans and have to charge the scans from the "strip" part, hence, this yields: $|\mathcal{S}(T)| \leq \frac{L_{\delta B}}{a/2} + 1 + \frac{L_{\delta B}}{r}$. Consequently, for $r \geq a$: $\frac{|\mathcal{S}(T)|}{|\mathcal{S}(T^*)|} \leq \frac{\pi r}{a} + \frac{\pi r}{2} + \frac{\pi}{2}$.

Theorem 4. *A polygon P allows a MWPDV solution that contains at most a cost of* $max(\frac{21}{4}, \frac{\pi r}{a} + \frac{\pi r}{2} + \frac{\pi}{2})$ *times the cost of an optimum MWPDV solution (for* $r \geq a$*).*

Note that Theorem 4 covers the case from Section 5; however, instead of the custom-built factor of 4 it yields a factor of $2 \cdot \pi$.

7 A PTAS for MWPDV Lawn Mowing

We describe here the following special case, which we generalize in the full paper. Consider a polyomino P (the "grass") that is to be "mowed" by a $k \times k$ square, M. At certain discrete set $\mathcal{S}(T)$ of positions of M along a tour T, the mower is activated (a "scan" is taken), causing all of the grass of P that lies below M at such a position to be mowed. For complete coverage, we require that P be contained in the union of $k \times k$ squares centered at points $\mathcal{S}(T)$. Between scan positions, the mower moves along the tour T.

In this "lawn mower" variant of the problem, the mower is not required to be fully inside P; the mower may extend outside P and move through the exterior of P, e.g., in order to reach different connected components of P. Since P may consist of singleton pixels, substantially separated, the problem is NP-hard even for $k = 1$, from TSP.

Here we describe a PTAS for the problem. We apply the m-guillotine method, with special care to handle the fact that we must have full coverage of P. Since the problem is closely related to the TSPN [6,13], we must address some of the similar difficulties in applying PTAS methods for the TSP: in particular, a mower centered on one side of a cut may be responsible to cover portions of P on the opposite side of the cut.

At the core of the method is a structure theorem, which shows that we can transform an arbitrary tour T, together with a set $\mathcal{S}(T)$ of scan points, into a tour and scan-point set, $(T_G, \mathcal{S}(T_G))$, that are m-guillotine in the following sense: the bounding box of the set of $k \times k$ squares centered at $\mathcal{S}(T)$ can be recursively partitioned into a rectangular subdivision by "m-perfect cuts". An axis-parallel cut line ℓ is m-perfect if its intersection with the tour has $O(m)$ connected components and its intersection with the union of $k \times k$ disks centered at scan points consists of $O(m)$ disks or "chains of disks" (meaning a set of disks whose centers lie equally spaced, at distance k, along a vertical/horizontal line).

The structure theorem is proved by showing the following lemma; the proof is omitted due to lack of space and can be found in the full version of the paper.

Lemma 1. *For any fixed* $m = \lceil 1/\epsilon \rceil$ *and any choice of* $(T, \mathcal{S}(T))$*, one can add a set of* doubled bridge segments*, of total length* $O(|T|/m)$*, to* T *and a set of* $O(|\mathcal{S}(T)|/m)$ bridging scans *to* $\mathcal{S}(T)$ *such that the resulting set,* $(T_G, \mathcal{S}(T_G))$*, is* m*-guillotine, with points* $\mathcal{S}(T_G)$ *on tour* T_G *and with* T_G *containing an Eulerian tour of* $\mathcal{S}(T_G)$*.*

The algorithm is based on dynamic programming to compute an optimal m-guillotine network. A subproblem is specified by a rectangle, R, with integer coordinates. The subproblem includes specification of *boundary information* for each of the four sides of R. The boundary information includes: (i) $O(m)$ integral points ("portals") where the tour is to cross the boundary, (ii) at most one (doubled) bridge and one disk-bridge (chain) per side of R, with each bridge having a parity (even or odd) specifying the parity of the number of connections to the bridge from within R, (iii) $O(m)$ scan positions (from $\mathcal{S}(T)$) such that a $k \times k$ square centered at each position intersects the corresponding side of R, (iv) a connection pattern, specifying which subsets of the portals/bridges are required to be connected within R. We summarize:

Theorem 5. *There is a PTAS for MWPDV lawn mowing of a (not necessarily connected) set of pixels by a $k \times k$ square.*

The Milling Variant. Our method does apply also to the "milling" variant of the MWPDV, in which the scans all must stay within the region P, provided that P is a simple rectilinear polygon. The details are rather involved and not included here. The main idea is this: Subproblems are defined, as before, by axis-aligned rectangles R. The difficulty now is that the restriction of R to P means that there may be many ($\Omega(n)$) vertical/horizontal chords of P along one side of R. We can ignore the boundary of P and construct an m-bridge (which we can "afford" to construct and charge off, by the same arguments as above) for T, but only the portions of such a bridge that lie inside P (and form chords of P) are traversable by our watchman. For each such chord, the subproblem must "know" if the chord is crossed by some edge of the tour, so that connections made inside R to a chord are not just made to a "dangling" component. We cannot afford to specify one bit per chord, as this would be $2^{\Omega(n)}$ information. However, in the case of a simple polygon P, no extra information must be specified to the subproblem – a chord is crossed by T if and only if the mower (scan) fits entirely inside the simple subpolygon on each side of the chord. Exploiting this fact, we are able to modify our PTAS to apply to MWPDV problem within a simple rectilinear polygon.

Theorem 6. *There is a PTAS for MWPDV milling of a simple rectilinear polygon by a $k \times k$ square.*

8 Conclusion

A number of open problems remain. Is it possible to remove the dependence on the ratio (r/a) of the approximation factor in our algorithm for general MWPDV milling? This would require a breakthrough for approximating minimum guard cover; a first step may be to achieve an approximation factor that depends on $\log(r/a)$ instead of (r/a).

For combined cost, we gave a PTAS for a lawn mowing variant, based on guillotine subdivisions. The PTAS extends to the milling case for simple rectilinear polygons. It is likely that the PTAS extends to other cases too (circular scan disks, Euclidean tour lengths), but the generalization to arbitrary domains with (many) holes seems particularly challenging. Our method makes use of a fixed ratio between scan cost and travel cost; as discussed in Figure 1, there is no PTAS for the bicriteria version.

References

1. Alt, H., Arkin, E.M., Brönnimann, H., Erickson, J., Fekete, S.P., Knauer, C., Lenchner, J., Mitchell, J.S.B., Whittlesey, K.: Minimum-cost coverage of point sets by disks. In: Symposium on Computational Geometry, pp. 449–458 (2006)
2. Arkin, E.M., Fekete, S.P., Mitchell, J.S.B.: Approximation algorithms for lawn mowing and milling. Comput. Geom. Theory Appl. 17(1-2), 25–50 (2000)
3. Bhattacharya, A., Ghosh, S.K., Sarkar, S.: Exploring an unknown polygonal environment with bounded visibility. In: Alexandrov, V.N., Dongarra, J., Juliano, B.A., Renner, R.S., Tan, C.J.K. (eds.) ICCS-ComputSci 2001. LNCS, vol. 2073, pp. 640–648. Springer, Heidelberg (2001)
4. Chin, W.-P., Ntafos, S.: Optimum watchman routes. In: Proc. 2nd ACM Symposium on Computational Geometry, vol. 28(1), pp. 39–44 (1988)
5. Chin, W.-P., Ntafos, S.C.: Shortest watchman routes in simple polygons. Discrete & Computational Geometry 6, 9–31 (1991)
6. Dumitrescu, A., Mitchell, J.S.B.: Approximation algorithms for TSP with neighborhoods in the plane. J. Algorithms 48(1), 135–159 (2003)
7. Efrat, A., Har-Peled, S.: Guarding galleries and terrains. Information Processing Letters 100(6), 238–245 (2006)
8. Fekete, S.P., Schmidt, C.: Polygon exploration with discrete vision. CoRR, abs/0807.2358 (2008)
9. Fekete, S.P., Schmidt, C.: Low-cost tours for nearsighted watchmen with discrete vision. In: 25th European Workshop on Comput.Geom., pp. 171–174 (2009)
10. Fekete, S.P., Schmidt, C.: Polygon exploration with time-discrete vision. Computational Geometry 43(2), 148–168 (2010)
11. Hochbaum, D.S., Maass, W.: Approximation schemes for covering and packing problems in image processing and vlsi. J. ACM 32(1), 130–136 (1985)
12. Itai, A., Papadimitriou, C.H., Szwarcfiter, J.L.: Hamilton paths in grid graphs. SIAM Journal on Computing 11(4), 676–686 (1982)
13. Mitchell, J.S.B.: A PTAS for TSP with neighborhoods among fat regions in the plane. In: Proc. 18th Annual ACM-SIAM Symposium on Discrete Algorithms (SODA 2007), pp. 11–18 (2007)
14. O'Rourke, J.: Art Gallery Theorems and Algorithms. Internat. Series of Monographs on Computer Science. Oxford University Press, New York (1987)
15. Wagner, I.A., Lindenbaum, M., Bruckstein, A.M.: MAC vs. PC: Determinism and randomness as complementary approaches to robotic exploration of continuous unknown domains. ROBRES: The International Journal of Robotics Research 19(1), 12–31 (2000)

Route-Enabling Graph Orientation Problems*

Takehiro Ito[1], Yuichiro Miyamoto[2],
Hirotaka Ono[3], Hisao Tamaki[4], and Ryuhei Uehara[5]

[1] Graduate School of Information Sciences, Tohoku University,
Aoba-yama 6-6-05, Sendai, 980-8579, Japan
takehiro@ecei.tohoku.ac.jp

[2] Faculty of Science and Technology, Sophia University,
Kioi-cho 7-1, Chiyoda-ku, Tokyo, 102-8554, Japan
miyamoto@sophia.ac.jp

[3] Graduate School of Information Science and Electrical Engineering,
Kyushu University, 744 Motooka, Nishi-ku, Fukuoka, 819-0395, Japan
ono@csce.kyushu-u.ac.jp

[4] School of Science and Technology, Meiji University,
Higashi-mita 1-1-1, Tama-ku, Kawasaki-shi, Kanagawa, 214-8571, Japan
tamaki@cs.meiji.ac.jp

[5] School of Information Science, JAIST,
Asahidai 1-1, Nomi, Ishikawa 923-1292, Japan
uehara@jaist.ac.jp

Abstract. Given an undirected and edge-weighted graph G together with a set of ordered vertex-pairs, called st-pairs, we consider the problems of finding an orientation of all edges in G: MIN-SUM ORIENTATION is to minimize the sum of the shortest directed distances between all st-pairs; and MIN-MAX ORIENTATION is to minimize the maximum shortest directed distance among all st-pairs. In this paper, we first show that both problems are strongly NP-hard for planar graphs even if all edge-weights are identical, and that both problems can be solved in polynomial time for cycles. We then consider the problems restricted to cacti, which form a graph class that contains trees and cycles but is a subclass of planar graphs. Then, MIN-SUM ORIENTATION is solvable in polynomial time, whereas MIN-MAX ORIENTATION remains NP-hard even for two st-pairs. However, based on LP-relaxation, we present a polynomial-time 2-approximation algorithm for MIN-MAX ORIENTATION. Finally, we give a fully polynomial-time approximation scheme (FPTAS) for MIN-MAX ORIENTATION on cacti if the number of st-pairs is a fixed constant.

1 Introduction

Consider the situation in which we wish to assign one-way restrictions to (narrow) aisles in a limited area, such as in an industrial factory, with keeping reachability between several sites. Since traffic jams rarely occur in industrial factories, the distances of routes between important sites are of great interest for the

* This work is partially supported by Grant-in-Aid for Scientific Research: 20650002, 20700003 and 21680001.

Y. Dong, D.-Z. Du, and O. Ibarra (Eds.): ISAAC 2009, LNCS 5878, pp. 403–412, 2009.

Fig. 1. (a) Solution for MIN-SUM ORIENTATION and (b) solution for MIN-MAX ORIENTATION

efficiency. This situation frequently appears in the context of the scheduling of automated guided vehicles without collision [5]. In this paper, we model the situation as graph orientation problems, in which we wish to find an orientation so that the distances of (directed) routes are not so long for given multiple st-pairs.

Let $G = (V, E)$ be an undirected graph together with an assignment of a non-negative integer, called the *weight* $\omega(e)$, to each edge e in G. Assume that we are given q ordered vertex-pairs (s_i, t_i), $1 \leq i \leq q$, called *st-pairs*. Then, an *orientation* of G is an assignment of exactly one direction to each edge in G so that there exists a directed (s_i, t_i)-path (i.e., a directed path from s_i to t_i) for every st-pair (s_i, t_i), $1 \leq i \leq q$. For an orientation $\boldsymbol{G}$ of G and an st-pair (s_i, t_i), we denote by $\omega(\boldsymbol{G}, s_i, t_i)$ the total weight of a shortest directed (s_i, t_i)-path on $\boldsymbol{G}$, that is, $\omega(\boldsymbol{G}, s_i, t_i) = \min\{\omega(P) \mid P \text{ is a directed } (s_i, t_i)\text{-path on } \boldsymbol{G}\}$ where $\omega(P)$ is the sum of weights of all edges in a path P.

We introduce two objective functions for orientations $\boldsymbol{G}$ of a graph G, and study the corresponding two minimization problems. The first objective is SUM-type, defined as follows: $g(\boldsymbol{G}) = \sum_{1 \leq i \leq q} \omega(\boldsymbol{G}, s_i, t_i)$. Its corresponding problem, called the MIN-SUM ORIENTATION problem, is to find an orientation $\boldsymbol{G}$ of G such that $g(\boldsymbol{G})$ is minimum; we denote by $g^*(G)$ the optimal value for G. The second objective is MAX-type, defined as follows: $h(\boldsymbol{G}) = \max\{\omega(\boldsymbol{G}, s_i, t_i) \mid 1 \leq i \leq q\}$. Its corresponding problem, called the MIN-MAX ORIENTATION problem, is to find an orientation $\boldsymbol{G}$ of G such that $h(\boldsymbol{G})$ is minimum; we denote by $h^*(G)$ the optimal value for G. For the sake of convenience, let $g^*(G) = +\infty$ and $h^*(G) = +\infty$ if G has no orientation for a given set of st-pairs. Clearly, both problems can be solved in polynomial time if we are given a single st-pair (s_1, t_1); in this case, we simply seek a shortest path between s_1 and t_1.

Figure 1 illustrates two orientations of the same graph G for the same set of st-pairs, where the weight $\omega(e)$ is attached to each edge e and the direction assigned to an edge is indicated by an arrow (but the direction is not indicated if the edge is not used in any shortest directed (s_i, t_i)-path, $1 \leq i \leq 3$). The orientation $\boldsymbol{G}$ in Fig.1(a) is an optimal solution for MIN-SUM ORIENTATION, where $g^*(G) = g(\boldsymbol{G}) = (1+6+8)+2+(6+5) = 28$. On the other hand, Fig.1(b) illustrates an optimal solution for MIN-MAX ORIENTATION, in which the st-pair (s_1, t_1) has the maximum distance; $h^*(G) = \max\{1+2+9,\ 4+3+1,\ 6+5\} = 12$.

Robbins [7] showed that every 2-edge-connected graph can be directed so that the resulting digraph is strongly connected. Therefore, a graph G has at least one orientation for any set of st-pairs if G is 2-edge-connected. Chvátal

Table 1. Summary of our results

	MIN-SUM ORIENTATION	MIN-MAX ORIENTATION
planar graphs	• strongly NP-hard • no $(2-\varepsilon)$-approximation	• strongly NP-hard • no $(2-\varepsilon)$-approximation
cacti	$O(nq^2)$	• NP-hard even for $q = 2$ • polynomial-time 2-approximation • FPTAS for a fixed constant q
cycles	$O(n+q^2)$	$O(n+q^2)$

and Thomassen [2] showed that it is NP-complete to determine whether a given unweighted graph can be directed so that the resulting digraph is strongly connected and whose (directed) diameter is 2. This implies that our MIN-MAX ORIENTATION is NP-hard in general. On the other hand, Hakimi *et al.* [4] proposed a quadratic algorithm for the problem of directing a 1-edge-connected graph so as to maximize the number of ordered vertex-pairs (x, y) having a directed (x, y)-path. The problem of [4] can be easily reduced to our MIN-SUM ORIENTATION.

In this paper, we mainly give the following three results. (Table1 summarizes our results, where n is the number of vertices in a graph.) The first is to show that both problems are strongly NP-hard for planar graphs even if all edge-weights are identical. We remark that the known result of [2] does not imply NP-completeness for planar graphs. The second is to show that both problems can be solved in polynomial time for cycles. By extending the algorithm for cycles, we show that MIN-SUM ORIENTATION is solvable in polynomial time for cacti, whereas MIN-MAX ORIENTATION remains NP-hard even for cacti with $q = 2$. (Cacti form a graph class that contains trees and cycles, but is a subclass of planar graphs.) The third is to give a fully polynomial-time approximation scheme (FPTAS) for MIN-MAX ORIENTATION on cacti if q is a fixed constant.

In addition, we give several results on the way to the three main results above. Firstly, our proof of strong NP-hardness implies that, for any constant $\varepsilon > 0$, both problems admit no polynomial-time $(2-\varepsilon)$-approximation algorithm unless P = NP. Secondly, in order to obtain a lower bound and an upper bound on $h^*(G)$ for a cactus G, we present a polynomial-time 2-approximation algorithm based on LP-relaxation; we remark that q is not required to be a fixed constant for this 2-approximation algorithm. We finally remark that our complexity analysis for MIN-MAX ORIENTATION on cacti is tight in some sense: the problem is in P if $q = 1$, and the problem for cacti cannot be strongly NP-hard if q is a fixed constant because our third result gives an FPTAS for the problem [6, p. 307].

2 Computational Hardness

In this section, we first show that our two problems are both strongly NP-hard for planar graphs, and then show that MIN-MAX ORIENTATION remains NP-hard even for cacti with $q = 2$.

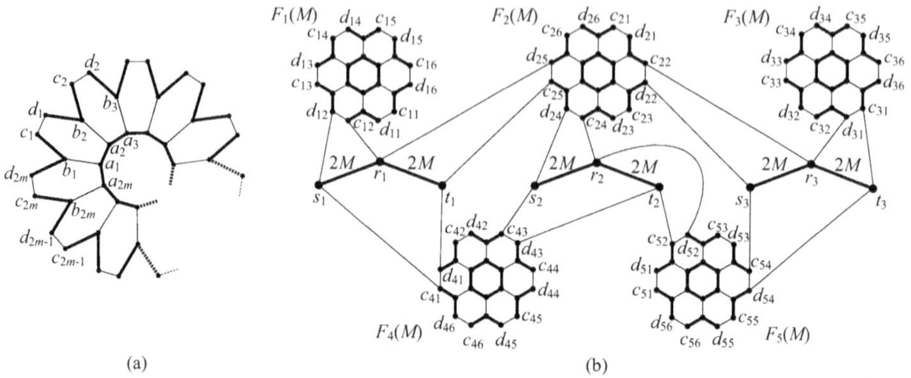

Fig. 2. (a) Flower gadget $F_i(M)$, and (b) planar graph G_ϕ corresponding to a Boolean formula ϕ with three clauses $c_1 = (u_1 \vee \bar{u}_2 \vee u_4)$, $c_2 = (u_2 \vee u_5 \vee u_4)$ and $c_3 = (u_2 \vee \bar{u}_3 \vee \bar{u}_5)$

Theorem 1. *Both* MIN-SUM ORIENTATION *and* MIN-MAX ORIENTATION *are strongly NP-hard for planar graphs even if all edge-weights are identical.*

Proof. We show that the PLANAR 3-SAT problem, which is known to be strongly NP-complete [3,6], can be reduced in polynomial time to MIN-MAX ORIENTATION. (The reduction to MIN-SUM ORIENTATION is similar.)

In PLANAR 3-SAT, we are given a Boolean formula ϕ in conjunctive normal form, say with set U of n variables $u_1, u_2, \ldots, u_n$ and set C of m clauses $c_1, c_2, \ldots, c_m$, such that each clause $c_j \in C$ contains exactly three literals and the following bipartite graph $B = (V', E')$ is planar: $V' = U \cup C$ and E' contains exactly those pairs $\{u_i, c_j\}$ such that either u_i or $\bar{u}_i$ appears in c_j. The PLANAR 3-SAT problem is to determine whether there is a satisfying truth assignment for ϕ. Given an instance of PLANAR 3-SAT, we construct the corresponding instance of MIN-MAX ORIENTATION. We first make a *flower gadget* $F_i(M)$ for each variable $u_i \in U$, and then construct the whole graph G_ϕ corresponding to ϕ.

We first define a flower gadget $F_i(M)$. Let M be a fixed constant (integer) such that $M \geq 3$. The flower gadget $F_i(M) = (V_i, E_i)$ for a variable $u_i \in U$ consists of $2m$ hexagonal elementary cycles, as illustrated in Fig.2(a). (Remember that m is the number of clauses in ϕ.) More precisely, $V_i = \{a_k, b_k, c_k, d_k \mid 1 \leq k \leq 2m\}$ and $E_i = \{\{a_{k+1}, a_k\}, \{a_k, b_k\}, \{b_k, c_k\}, \{c_k, d_k\}, \{d_k, b_{k+1}\} \mid 1 \leq k \leq 2m\}$, where $a_{2m+1} = a_1$ and $b_{2m+1} = b_1$. The edge-weights are defined as follows: for each k, $1 \leq k \leq 2m$, $\omega(\{a_{k+1}, a_k\}) = \omega(\{b_k, c_k\}) = \omega(\{d_k, b_{k+1}\}) = M$ and $\omega(\{a_k, b_k\}) = \omega(\{c_k, d_k\}) = 1$. (In Fig.2(a), the weight-M edges are depicted by thick lines.) Finally, we define the set ST_i of $12m$ st-pairs, as follows:

$$ST_i = \{(a_k, d_k), (d_k, a_k), (b_k, b_{k+1}), (b_{k+1}, b_k), (c_k, a_{k+1}), (a_{k+1}, c_k) \mid 1 \leq k \leq 2m\}.$$

For each k, $1 \leq k \leq 2m$, the kth hexagonal elementary cycle $a_k b_k c_k d_k b_{k+1} a_{k+1}$ is called the *kth petal* P_k; P_k is called an *odd petal* if k is odd, while is called an *even petal* if k is even. We call the edge $\{c_k, d_k\}$ in each petal P_k, $1 \leq k \leq 2m$, an *external edge* of P_k. For the sake of convenience, we fix the embedding of $F_i(M)$ such that the outer face consists of b_k, c_k, d_k, $1 \le k \le 2m$, which are placed in a clockwise direction, as illustrated in Fig.2(a).

It is easy to see that $F_i(M)$ has only two optimal orientations for ST_i: the one is to direct each odd petal in a clockwise direction and to direct each even petal in an anticlockwise direction; and the other is the reversed one. In the first optimal orientation, the external edges $\{c_k, d_k\}$ are directed from c_k to d_k in all odd petals P_k, while directed from d_k to c_k in all even petals; we call this optimal orientation of $F_i(M)$ a *true-orientation*, which corresponds to assigning TRUE to the variable u_i. On the other hand, the other optimal orientation of $F_i(M)$ is called a *false-orientation*, which corresponds to assigning FALSE to u_i. Clearly, $h^*(F_i(M)) = 2M + 1$.

We now construct the planar graph G_ϕ corresponding to the formula ϕ, as follows. We fix an embedding of the bipartite graph $B = (V', E')$ arbitrarily. For each variable u_i, $1 \le i \le n$, we replace it with a flower gadget $F_i(M)$. For each clause c_j, $1 \le j \le m$, we replace it with a path consisting of three vertices s_j, r_j, t_j; let $\omega(\{s_j, r_j\}) = \omega(\{r_j, t_j\}) = 2M$. We then connect flower gadgets $F_i(M)$, $1 \le i \le n$, with paths $s_j r_j t_j$, $1 \le j \le m$, as follows. For each clause c_j, $1 \le j \le m$, let l_{j1}, l_{j2}, l_{j3} be three literals in c_j, and assume without loss of generality that three flower gadgets corresponding to l_{j1}, l_{j2}, l_{j3} are placed in a clockwise direction around the path $s_j r_j t_j$ corresponding to c_j. Assume that l_{jk} is either u_i or $\bar{u}_i$. Then, we replace the edge of B joining variable u_i and clause c_j with a pair of weight-1 edges which, together with an external edge in $F_i(M)$, forms a path between two vertices chosen from $\{s_j, r_j, t_j\}$, according to the following rules (see Fig.2(b) as an example):

(i) The endpoints of this path are s_j and r_j if $k = 1$; r_j and t_j if $k = 2$; and s_j and t_j if $k = 3$.
(ii) The external edge is from an even petal if $l_{j1} = u_i$, $l_{j2} = u_i$, or $l_{j3} = \bar{u}_i$; while it is from an odd petal if $l_{j1} = \bar{u}_i$, $l_{j2} = \bar{u}_i$, or $l_{j3} = u_i$.
(iii) From the viewpoint of variable u_i, we choose a distinct external edge for each clause containing u_i, honoring the order of those clauses around u_i and thereby preserving the planarity of the embedding.

Finally, we replace each edge e in G_ϕ with a path of length $\omega(e)$ in which all edges are of weight 1. (Remember that M is a fixed constant.) Clearly, the resulting graph G_ϕ is planar, and can be constructed in polynomial time. The set of all st-pairs in this instance is defined as follows: $(\bigcup_{i=1}^{n} ST_i) \cup \{(s_j, t_j) \mid 1 \le j \le m\}$. Therefore, there are $(12mn+m)$ st-pairs in total. This completes the construction of the corresponding instance of MIN-MAX ORIENTATION.

Then, deciding whether $h^*(G_\phi) \le 2M + 3$ is equivalent to solving PLANAR 3-SAT for ϕ. (We omit the details due to the page limitation.) □

From our proof of Theorem 1, we immediately obtain the following corollary.

Corollary 1. *For any constant* $\varepsilon > 0$, *both* MIN-SUM ORIENTATION *and* MIN-MAX ORIENTATION *admit no polynomial-time* $(2 - \varepsilon)$*-approximation algorithm for planar graphs unless* P = NP.

A graph G is a *cactus* if every edge is part of at most one cycle in G [1,8]. Cacti form a subclass of planar graphs. However, we have the following theorem.

Theorem 2. MIN-MAX ORIENTATION *is NP-hard for cacti even if* $q = 2$.

3 Polynomial-Time Algorithms

The main result of this section is the following theorem.

Theorem 3. *Both* MIN-SUM ORIENTATION *and* MIN-MAX ORIENTATION *can be solved in time* $O(n + q^2)$ *for a cycle* C*, where* n *is the number of vertices in* C.

Proof. Suppose that we are given an edge-weighted cycle $C = (V, E)$ and q st-pairs (s_i, t_i), $1 \le i \le q$. Note that C has at least one orientation for any set of st-pairs: simply directing C in a clockwise direction.

For each st-pair (s_i, t_i), $1 \le i \le q$, let $\mathrm{cw}(i)$ be the set of all edges in the directed (s_i, t_i)-path when all edges in C are directed in a clockwise direction, and let $\mathrm{acw}(i)$ be the set of all edges in the directed (s_i, t_i)-path when all edges in C are directed in an anticlockwise direction. Clearly, for each i, $1 \le i \le q$, $\{\mathrm{cw}(i), \mathrm{acw}(i)\}$ is a partition of E, that is, $\mathrm{cw}(i) \cap \mathrm{acw}(i) = \emptyset$ and $\mathrm{cw}(i) \cup \mathrm{acw}(i) = E$. We introduce a $\{0, 1\}$-variable x_i for each st-pair (s_i, t_i), $1 \le i \le q$: if $x_i = 0$, then the edges in $\mathrm{cw}(i)$ are directed in a clockwise direction; if $x_i = 1$, then the edges in $\mathrm{acw}(i)$ are directed in an anticlockwise direction. For two st-pairs (s_i, t_i) and (s_j, t_j), it is easy to see that the two corresponding variables x_i and x_j have the following constraints (a)–(c):

(a) if $\mathrm{cw}(i) \cap \mathrm{acw}(j) \ne \emptyset$ and $\mathrm{acw}(i) \cap \mathrm{cw}(j) \ne \emptyset$, then $x_i = x_j$;
(b) if $\mathrm{cw}(i) \cap \mathrm{acw}(j) = \emptyset$ and $\mathrm{acw}(i) \cap \mathrm{cw}(j) \ne \emptyset$, then $x_i \le x_j$; and
(c) if $\mathrm{cw}(i) \cap \mathrm{acw}(j) \ne \emptyset$ and $\mathrm{acw}(i) \cap \mathrm{cw}(j) = \emptyset$, then $x_i \ge x_j$.

We now construct a *constraint graph* $\mathcal{C}$ in which each vertex v_i corresponds to an st-pair (s_i, t_i) and there is an edge between two vertices v_i and v_j if and only if $\mathrm{cw}(i) \cap \mathrm{acw}(j) \ne \emptyset$ and $\mathrm{acw}(i) \cap \mathrm{cw}(j) \ne \emptyset$, that is, the corresponding variables x_i and x_j have the constraint $x_i = x_j$. From an orientation of C, we can obtain an assignment of $\{0, 1\}$ to each variable x_k, $1 \le k \le q$; clearly, any two variables satisfy their constraint, and hence two variables x_i and x_j receive the same value if their corresponding vertices v_i and v_j are contained in the same connected component of $\mathcal{C}$.

Let $\mathcal{V} = \{V_1, V_2, \ldots, V_m\}$ be the partition of the vertex set of $\mathcal{C}$ such that each V_i, $1 \le i \le m$, forms a connected component of $\mathcal{C}$. Then, we define a relation "$\le$" on $\mathcal{V}$, as follows: $V_i \le V_j$ if and only if there exist two vertices $v_i \in V_i$ and $v_j \in V_j$ such that their corresponding variables x_i and x_j have the constraint $x_i \le x_j$. We show that $\mathcal{V}$ is totally ordered under the relation $\le$. (However, its proof is omitted from this extended abstract.) Then, for some index k, $1 \le k \le m$, we have $x_i = 0$ for all variables x_i whose corresponding vertices are contained in V_j with $V_j \le V_k$; otherwise $x_i = 1$. Therefore, both MIN-SUM ORIENTATION and MIN-MAX ORIENTATION can be reduced simply to finding such an appropriate index k on $\mathcal{V} = \{V_1, V_2, \ldots, V_m\}$. It is now easy to see that both problems can be solved in time $O(n + q^2)$. □

By extending Theorem 3, we can easily obtain the following theorem.

Theorem 4. MIN-SUM ORIENTATION *can be solved in time* $O(nq^2)$ *for a cactus* G*, where* n *is the number of vertices in* G.

4 FPTAS for MIN-MAX ORIENTATION on Cacti

In contrast to MIN-SUM ORIENTATION, MIN-MAX ORIENTATION remains NP-hard even for cacti with $q = 2$. However, in this section, we give an FPTAS for MIN-MAX ORIENTATION on cacti if q is a fixed constant.

In Section 4.1 we first present a polynomial-time 2-approximation algorithm based on LP-relaxation, which gives us both lower and upper bounds on $h^*(G)$ for a given cactus G. We then show in Section 4.2 that the problem can be solved in pseudo-polynomial time for cacti. In Section 4.3, we finally give our FPTAS based on the algorithm in Section 4.2 and using the lower and upper bounds on $h^*(G)$ obtained in Section 4.1.

It can be easily determined in time $O(nq)$ whether a given cactus $G = (V, E)$ has an orientation for the given set of st-pairs; we simply check the placements of st-pairs which pass through each bridge in G. Therefore, we may assume without loss of generality that G has at least one orientation, and hence $h^*(G) \neq +\infty$.

[Cactus and its underlay tree]
A cactus G can be represented by an *underlay tree* T, which is a rooted tree and can be easily obtained from G in a straightforward way. In the underlay tree T of G, each node represents either a bridge of G or an elementary cycle of G; and if there is an edge between nodes u and v of T, then bridges or cycles of G represented by u and v share exactly one vertex in G in common. (A similar idea can be found in [8, Theorem 11].) Each node v of T corresponds to a subgraph G_v of G induced by all bridges and cycles represented by the nodes that are descendants of v in T. Clearly, G_v is a cactus for each node v of T, and $G = G_r$ for the root r of T. It is easy to see that an underlay tree T of a given cactus G can be found in linear time, and hence we may assume that a cactus G and its underlay tree T are given. In Section 4.2, we solve MIN-MAX ORIENTATION by a dynamic programming approach based on the underlay tree T of G.

4.1 2-Approximation Algorithm Based on LP-Relaxation

In this subsection, we give the following theorem. It should be noted that the number q of st-pairs is not required to be a fixed constant in the theorem.

Theorem 5. *There is a polynomial-time 2-approximation algorithm for* MIN-MAX ORIENTATION *on cacti.*

For each st-pair (s_i, t_i), $1 \leq i \leq q$, let C_i be the set of elementary cycles represented by the nodes which are on the path from v_{s_i} to v_{t_i} in the underlay tree T of a given cactus G, where v_{s_i} and v_{t_i} are the nodes in T containing s_i and t_i, respectively. Let d_i be the sum of weights of bridges represented by the nodes which are on the path from v_{s_i} to v_{t_i} in T. Clearly, both C_i and d_i can be computed in time $O(nq)$ for all st-pairs (s_i, t_i), $1 \le i \le q$.

Consider the following two orientations of G: the one, denoted by $\boldsymbol{G}^a$, directs all elementary cycles in G in a clockwise direction; the other, denoted by $\boldsymbol{G}^b$, directs all elementary cycles in G in an anticlockwise direction. Clearly, both $\boldsymbol{G}^a$ and $\boldsymbol{G}^b$ are (feasible) orientations of G. For an st-pair (s_i, t_i), $1 \le i \le q$, and each elementary cycle $c \in C_i$, we denote by a_i^c and b_i^c the sums of weights of the edges which are contained in c and are in the directed (s_i, t_i)-paths on $\boldsymbol{G}^a$ and $\boldsymbol{G}^b$, respectively. For each elementary cycle c in G, we call an ordered index-pair (i, j), $1 \le i, j \le q$, a *conflicting pair on* c if the directed (s_i, t_i)-path on $\boldsymbol{G}^a$ and the directed (s_j, t_j)-path on $\boldsymbol{G}^b$ share at least one edge of c in common.

For an st-pair (s_i, t_i), $1 \le i \le q$, and each elementary cycle $c \in C_i$, we introduce two kinds of $\{0, 1\}$-variables x_i^c and y_i^c: if $x_i^c = 1$, then we direct edges of c so that there is a directed (s_i, t_i)-path which passes through c in a clockwise direction; if $y_i^c = 1$, then we direct edges of c so that there is a directed (s_i, t_i)-path which passes through c in an anticlockwise direction.

We are now ready to formulate MIN-MAX ORIENTATION for a cactus G.

$$\begin{aligned}
\text{minimize } & z && (1)\\
\text{subject to } & x_i^c + y_i^c = 1, \ \forall c \in C_i, \ i = 1, \ldots, q, && (2)\\
& x_i^c + y_j^c \le 1, \ \forall (i, j) \in \text{conflicting pairs on } c, \forall c \text{ in } G, && (3)\\
& d_i + \sum_{c \in C_i} (a_i^c x_i^c + b_i^c y_i^c) \le z, \ i = 1, \ldots, q, && (4)\\
& x_i^c, \ y_i^c \in \{0, 1\}, \ \forall c \in C_i, \ i = 1, \ldots, q. && (5)
\end{aligned}$$

Equations (2) and (3) ensure that there are directed (s_i, t_i)-paths for all st-pairs (s_i, t_i), $1 \le i \le q$. Therefore, according to the values of x_i^c and y_i^c, we can find an orientation $\boldsymbol{G}$ of G such that $h(\boldsymbol{G}) = z$. Thus, minimizing z in Eq. (1) is equivalent to computing $h^*(G)$ for G. Since the size of the above integer programming formulation is polynomial in n, its linear relaxation problem can be solved in polynomial time.

We now propose a polynomial-time 2-approximation algorithm for cacti. Our algorithm is very simple. We first solve the linear relaxation problem, and obtain a fractional solution $\bar{x}_i^c$ and $\bar{y}_i^c$, whose objective value is $\bar{z}$. Clearly, $h^*(G) \ge \bar{z}$ since $h^*(G)$ is the optimal value for the IP above. We then obtain an integer solution x_i^c and y_i^c by rounding the values of $\bar{x}_i^c$ and $\bar{y}_i^c$, as follows: $x_i^c = 1$ if $\bar{x}_i^c \ge 0.5$, otherwise $x_i^c = 0$; $y_i^c = 1$ if $\bar{y}_i^c > 0.5$, otherwise $y_i^c = 0$. Clearly, x_i^c and y_i^c satisfy Eqs. (2), (3) and (5), and hence x_i^c and y_i^c form a feasible solution for the IP above; we can thus obtain an orientation of G. Moreover, this algorithm clearly terminates in polynomial time. Therefore, it suffices to show that the approximation ratio of this algorithm is 2. Let z_A be the objective value for the solution x_i^c and y_i^c. Since $\bar{x}_i^c \ge \frac{1}{2} x_i^c$ and $\bar{y}_i^c \ge \frac{1}{2} y_i^c$, by Eq. (4) we have

$$h^*(G) \geq \bar{z} = \max\Big\{d_i + \sum_{c \in C_i} (a_i^c \bar{x}_i^c + b_i^c \bar{y}_i^c) \mid 1 \leq i \leq q\Big\}$$
$$\geq \frac{1}{2} \max\Big\{d_i + \sum_{c \in C_i} (a_i^c x_i^c + b_i^c y_i^c) \mid 1 \leq i \leq q\Big\} = \frac{1}{2} z_A. \quad (6)$$

4.2 Pseudo-polynomial-time Algorithm

From now on, assume that the number q of st-pairs is a fixed constant. The main result of this subsection is the following theorem.

Theorem 6. MIN-MAX ORIENTATION *can be solved in time* $O(nU^{2q})$ *for a cactus* G *if* q *is a fixed constant, where* U *is an arbitrary upper bound on* $h^*(G)$ *and* n *is the number of vertices in* G.

As the upper bound U on $h^*(G)$, we will employ the approximation value z_A obtained by the 2-approximation algorithm in Section 4.1; z_A can be computed in polynomial time.

Let $G = (V, E)$ be a given cactus, let v be a node of an underlay tree T of G, and let G_v be the subgraph of G for the node v. Then, G_v and $G \setminus G_v$ share exactly one vertex u in common. Consider an optimal orientation $\boldsymbol{G}$ of G. (Remember that G has at least one orientation for the given set of st-pairs.) Then, $\boldsymbol{G}$ naturally induces the "edge-direction" $\boldsymbol{G}_v$ of G_v, which is not always an orientation for the given set of st-pairs but satisfies the following four conditions: for each st-pair (s_i, t_i), $1 \leq i \leq q$,

(a) if both s_i and t_i are in G_v, then a shortest directed (s_i, t_i)-path on $\boldsymbol{G}$ is on $\boldsymbol{G}_v$ because $\boldsymbol{G}$ is optimal and all edge-weights are non-negative;
(b) if s_i is in G_v but t_i is in $G \setminus G_v$, then there is a directed (s_i, u)-path on $\boldsymbol{G}_v$;
(c) if s_i is in $G \setminus G_v$ but t_i is in G_v, then there is a directed (u, t_i)-path on $\boldsymbol{G}_v$; and
(d) if neither s_i nor t_i are in G_v, then $\boldsymbol{G}$ has a shortest directed (s_i, t_i)-path which contains no edge of $\boldsymbol{G}_v$.

For a q-tuple $(x_1, x_2, \ldots, x_q)$ of integers $0 \leq x_i \leq U$, $1 \leq i \leq q$, an edge-direction $\boldsymbol{G}_v$ of G_v is called an $(x_1, x_2, \ldots, x_q)$*-orientation* of G_v if the following three conditions (a)–(c) are satisfied: for each st-pair (s_i, t_i), $1 \leq i \leq q$,

(a) if both s_i and t_i are in G_v, then $\omega(\boldsymbol{G}_v, s_i, t_i) \leq x_i$;
(b) if s_i is in G_v but t_i is in $G \setminus G_v$, then $\omega(\boldsymbol{G}_v, s_i, u) \leq x_i$; and
(c) if s_i is in $G \setminus G_v$ but t_i is in G_v, then $\omega(\boldsymbol{G}_v, u, t_i) \leq x_i$.

We then define a set $F(G_v)$ of q-tuples, as follows:

$$F(G_v) = \{(x_1, x_2, \ldots, x_q) \mid G_v \text{ has an } (x_1, x_2, \ldots, x_q)\text{-orientation}\}.$$

Our algorithm computes $F(G_v)$ for each node v of T from the leaves to the root r of T by means of dynamic programming. Since $G = G_r$, we clearly have

$$h^*(G) = \min\left\{\max_{1 \leq i \leq q} x_i \mid (x_1, x_2, \ldots, x_q) \in F(G_r)\right\}. \quad (7)$$

Note that $F(G_r) \neq \emptyset$ since G has at least one orientation for the given set of *st*-pairs. Therefore, we can always compute $h^*(G)$ by Eq. (7). We omit the details of our pseudo-polynomial-time algorithm due to the page limitation.

4.3 FPTAS

We finally give the main result of this section, as in the following theorem.

Theorem 7. MIN-MAX ORIENTATION *admits a fully polynomial-time approximation scheme for cacti if q is a fixed constant.*

As a proof of Theorem 7, we give an algorithm to find an orientation $\boldsymbol{G}$ of a cactus G with $h(\boldsymbol{G}) < (1+\varepsilon)h^*(G)$ in time polynomial in both n and $1/\varepsilon$ for any real number $\varepsilon > 0$, where n is the number of vertices in G. Thus, our approximation value $h_A(G)$ for G is $h(\boldsymbol{G})$, and hence the error is bounded by $\varepsilon h^*(G)$, that is,

$$h_A(G) - h^*(G) = h(\boldsymbol{G}) - h^*(G) < \varepsilon h^*(G). \tag{8}$$

We now outline our algorithm and its analysis. We extend the ordinary "scaling and rounding" technique [9], and apply it to MIN-MAX ORIENTATION for a cactus $G = (V, E)$. For some scaling factor $\tau > 0$, let G_τ be the graph with the same vertex set V and edge set E as G, but the weight of each edge $e \in E$ is defined as follows: $\bar{\omega}(e) = \lceil \omega(e)/\tau \rceil$. We optimally solve MIN-MAX ORIENTATION for G_τ by using the pseudo-polynomial-time algorithm in Section 4.2. We take the optimal orientation $\boldsymbol{G}_\tau$ for G_τ as our approximation solution for G. Then, we can show that $h_A(G) - h^*(G) < \tau|E|$. Intuitively, this inequality holds because the error occurs at most τ at each edge in G_τ. By Eq. (6) and taking $\tau = \varepsilon z_A/2|E|$, we have Eq. (8). Since $h^*(G_\tau) \leq |E| + \frac{z_A}{\tau} = (1+\frac{2}{\varepsilon})|E|$, by Theorem 6 we can find the optimal orientation $\boldsymbol{G}_\tau$ for G_τ in time $O\Big(n\big(|E| + \frac{2|E|}{\varepsilon}\big)^{2q}\Big) = O\Big(\frac{n^{2q+1}}{\varepsilon^{2q}}\Big)$; since G is a cactus, $|E| = O(n)$.

References

1. Brandstädt, A., Le, V.B., Spinrad, J.P.: Graph Classes: A Survey. Society for Industrial and Applied Mathematics, Philadelphia (1999)
2. Chvátal, V., Thomassen, C.: Distances in orientations of graphs. J. Combinatorial Theory, Series B 24, 61–75 (1978)
3. Garey, M.R., Johnson, D.S.: Computers and Intractability: A Guide to the Theory of NP-Completeness. Freeman, San Francisco (1979)
4. Hakimi, S.L., Schmeichel, E.F., Young, N.E.: Orienting graphs to optimize reachability. Information Processing Letters 63, 229–235 (1997)
5. Lee, C.-Y., Lei, L., Pinedo, M.: Current trends in deterministic scheduling. Annals of Operations Research 70, 1–41 (1997)
6. Papadimitriou, C.H.: Computational Complexity. Addison-Wesley, Reading (1994)
7. Robbins, H.E.: A theorem on graphs with an application to a problem of traffic control. American Mathematical Monthly 46, 281–283 (1939)
8. Uehara, R., Uno, Y.: On computing longest paths in small graph classes. International Journal of Foundations of Computer Science 18, 911–930 (2007)
9. Vazirani, V.V.: Approximation Algorithms. Springer, Heidelberg (2001)

Complexity of Approximating the Vertex Centroid of a Polyhedron*

Khaled Elbassioni[1] and Hans Raj Tiwary[2]

[1] Max-Planck-Institut für Informatik, Saarbrücken, Germany
elbassio@mpi-inf.mpg.de
[2] Technische Universität Berlin Fakultät II: Institut für Mathematik, 10623 Berlin, Germany
tiwary@math.tu-berlin.de

Abstract. Let $\mathcal{P}$ be an $\mathcal{H}$-polytope in $\mathbb{R}^d$ with vertex set V. The vertex centroid is defined as the average of the vertices in V. We first prove that computing the vertex centroid of an $\mathcal{H}$-polytope, or even just checking whether it lies in a given halfspace, are #P-hard. We also consider the problem of approximating the vertex centroid by finding a point within an ϵ distance from it and prove this problem to be #P-easy by showing that given an oracle for counting the number of vertices of an $\mathcal{H}$-polytope, one can approximate the vertex centroid in polynomial time. We also show that any algorithm approximating the vertex centroid to *any* "sufficiently" non-trivial (for example constant) distance, can be used to construct a fully polynomial-time approximation scheme for approximating the centroid and also an output-sensitive polynomial algorithm for the Vertex Enumeration problem. Finally, we show that for unbounded polyhedra the vertex centroid can not be approximated to a distance of $d^{\frac{1}{2}-\delta}$ for any fixed constant $\delta > 0$.

1 Introduction

An intersection of a finite number of closed halfspaces in $\mathbb{R}^d$ defines a polyhedron. A polyhedron can also be represented as $conv(V) + cone(Y)$, the Minkowski sum of the convex hull of a finite set of points V and the cone of a finite set of rays. A bounded polyhedron is called a polytope. In what follows, we will discuss mostly polytopes for simplicity and refer to the unbounded case explicitly only towards the end.

Let $\mathcal{P}$ be an $\mathcal{H}$-polytope in $\mathbb{R}^d$ with vertex set V. Various notions try to capture the essence of a "center" of a polytope. Perhaps the most popular notion is that of the center of gravity of $\mathcal{P}$. Recently Rademacher proved that computing the center of gravity of a polytope is #P-hard [8]. The proof essentially relies on the fact that the center of gravity captures the volume of a polytope perfectly and that computing the volume of a polytope is #P-hard [4]. Note that, polynomial

* During part of this work the second author was supported by Graduiertenkolleg fellowship for PhD studies provided by Deutsche Forschungsgemeinschaft.

Y. Dong, D.-Z. Du, and O. Ibarra (Eds.): ISAAC 2009, LNCS 5878, pp. 413–422, 2009.

algorithms exist that approximate the volume of a polytope within any arbitrary factor [5]. It is also easy to see that the center of gravity can be approximated by simply sampling random points from the polytope, the number of samples depending polynomially on the desired approximation (See Algorithm 5.8 of [5]).

In this paper we study a variant of the notion of "center" defined as the centroid (average) of the vertices of P. Despite being quite a natural feature of polytopes, this variant seems to have received very little attention both from theoretical and computational perspectives. Throughout this paper we will refer to the vertex centroid just as centroid. The reader should note that in popular literature the word centroid refers more commonly to the center of gravity. We nevertheless use the same terminology for simplicity of language. Our motivation for studying the centroid stems from the fact that the centroid encodes the number of vertices of a polytope. As we will see, this also makes computing the centroid hard.

The parallels between centroid and the center of gravity of a polytope mimic the parallels between the volume and the number of vertices of a polytope. Computing the volume is #P-complete [4] but it can be approximated quite well [5]. Accordingly, the problem of computing the corresponding centroid is hard ([8], Theorem 1) but the volume centroid can be approximated quite well [5]. On the other hand computing the number of vertices is not only #P-complete [3,7], it can not be approximated within any factor polynomial in the number of facets and the dimension. As we will see in this paper, computing the vertex centroid of an $\mathcal{H}$-polytope exactly is #P-hard. Even approximating the vertex centroid for unbounded $\mathcal{H}$-polyhedra turns out to be NP-hard. We do not know the complexity of approximating the vertex centroid of an $\mathcal{H}$-polytope (bounded case).

The problem of enumerating vertices of an $\mathcal{H}$-polytope has been studied for a long time. However, in spite of years of research it is neither known to be hard nor is there an output sensitive polynomial algorithm for it. A problem that is polynomially equivalent to the Vertex Enumeration problem is to decide if a given list of vertices of an $\mathcal{H}$-polytope is complete [1]. In this paper we show that any algorithm that approximates the centroid of an arbitrary polytope to any "sufficiently" non-trivial distance can be used to obtain an output sensitive polynomial algorithm for the Vertex Enumeration problem.

The main results of this paper are the following:

(I) Computing the centroid of an $\mathcal{H}$-polytope is #P-hard, and it remains #P-hard even just to decide whether the centroid lies in a halfspace.

(II) Approximating the centroid of an $\mathcal{H}$-polytope is #P-easy.

(III) Any algorithm approximating the centroid of an arbitrary polytope within a distance $d^{\frac{1}{2}-\delta}$ for any fixed constant $\delta > 0$ can be used to obtain a fully polynomial-time approximation scheme for the centroid approximation problem and also an output sensitive polynomial algorithm for the Vertex Enumeration problem.

(IV) There is no polynomial algorithm that approximates the vertex centroid of an arbitrary $\mathcal{H}$-polyhedron within a distance $d^{\frac{1}{2}-\delta}$ for any fixed constant $\delta > 0$, unless $P = NP$.

The first two results in (I) follow easily from the hardness of counting the number of vertices of an $\mathcal{H}$-polytope. The next result is obtained by repeatedly slicing the given polytope, in a way somewhat similar to the one used to prove that computing the center of gravity is #P-hard [8]. The bootstrapping result in (III) is obtained by taking the product of the polytope with itself sufficiently many times. Using this result, and building on a construction in [6], we prove (IV). Namely, we use a modified version of the construction in [6] to show that it is NP-hard to approximate the centroid within a distance of $1/d$, then we use the result in (III) to bootstrap the hardness threshold to $d^{\frac{1}{2}-\delta}$ for any fixed constant $\delta > 0$.

We should remark that for the approximation of centroid, we only consider polytopes (and polyhedra) whose vertices lie inside a unit hypercube. To see how this assumption can easily be satisfied, notice that a halfspace h can be added to a polyhedron P such that $P \cap h$ is bounded and the vertices of P are preserved in $P \cap h$. Also, such a halfspace can be found in polynomial time from the inequalities defining P. Once we have a polytope in $\mathbb{R}^d$, solving $2d$ linear programs gives us the width along each coordinate axis. The polytope can be scaled by a factor depending on the width along each axis to obtain a polytope all whose vertices lie inside a unit hypercube. In case we started with a polyhedron P, the scaled counterpart of the halfspace h that was added can be thrown to get back a polyhedron that is a scaled version of P and all whose vertices lie inside the unit hypercube. In subsection 2.2 we provide further motivation for this assumption.

Since all the vertices of the polytope (or polyhedron) lie inside a unit hypercube, picking any arbitrary point from inside this hypercube yield a $d^{\frac{1}{2}}$-approximation of the vertex centroid. Thus, our last result above should be contrasted to the fact that approximating the vertex centroid within a distance of $d^{\frac{1}{2}}$ is trivial. Also, even though we discuss only polytopes *i.e.* bounded polyhedra in subsections 2.1 and 2.2, the results and the proofs are valid for the unbounded case as well. We discuss the unbounded case explicitly only in subsection 2.3.

2 Results

2.1 Exact Computation of the Centroid

The most natural computational question regarding the centroid of a polytope is whether we can compute the centroid efficiently. The problem is trivial if the input polytope is presented by its vertices. So we will assume that the polytope is presented by its facets. Perhaps not surprisingly, computing the centroid of an $\mathcal{H}$-polytope turns out be #P-hard. We prove this by showing that computing the centroid of an $\mathcal{H}$-polytope amounts to counting the vertices of the same polytope, a problem known to be #P-hard.

Proposition 1. *Given an $\mathcal{H}$-polytope $\mathcal{P} \subset \mathbb{R}^d$, it is #P-hard to compute its centroid $c(\mathcal{P})$.*

Proof. Embed $\mathcal{P}$ in $\mathbb{R}^{d+1}$ by putting a copy of $\mathcal{P}$ in the hyperplane $x_{d+1} = 1$ and making a pyramid with the base $\mathcal{P}$ and apex at the origin. Call this new polytope $\mathcal{Q}$. The facets of Q can be computed efficiently from the facets of P. Treating the direction of the positive x_{d+1}-axis as up, it is easy to see that the centroid of the new polytope lies at a height $1 - \frac{1}{n+1}$ if and only if the number of vertices of $\mathcal{P}$ is n. Thus any algorithm for computing the centroid can be run on $\mathcal{Q}$ and the number of vertices of $\mathcal{P}$ can be read off the $(d+1)$-st coordinate. □

Suppose, instead, that one does not want to compute the centroid exactly but is just interested in knowing whether the centroid lies to the left or to the right of a given arbitrary hyperplane. This problem turns out to be hard too, and it is not difficult to see why.

Proposition 2. *Given an $\mathcal{H}$-polytope $\mathcal{P} \subset \mathbb{R}^d$ and a hyperplane $h = \{a \cdot x = b\}$, it is #P-hard to decide whether $a \cdot c(\mathcal{P}) \leq b$.*

Proof. Consider the embedding and the direction pointing upwards as used in the proof of Proposition 1. Given an oracle answering sidedness queries for the centroid and any arbitrary hyperplane, one can perform a binary search on the height of the centroid and locate the exact height. The number of queries needed is only logarithmic in the number of vertices of $\mathcal{P}$, which is at most $O(\lfloor \frac{d}{2} \rfloor \log m)$ if $\mathcal{P}$ has m facets. □

2.2 Approximation of the Centroid

As stated before, even though computing the gravitational centroid of a polytope exactly is #P-hard, it can be approximated to any precision by random sampling. Now we consider the problem of similarly approximating the vertex centroid of an $\mathcal{H}$-polytope. Let $dist(x, y)$ denote the Euclidean distance between two points $x, y \in \mathbb{R}^d$. We are interested in the following problem:

Input: $\mathcal{H}$-polytope $P \subset \mathbb{R}^d$ and a real number $\epsilon > 0$.
Output: $p \in \mathbb{R}^d$ such that $dist(c(P), p) \leq \epsilon$.

We would like an algorithm for this problem that runs in time polynomial in the number of facets of P, the dimension d and $\frac{1}{\epsilon}$. Clearly, such an algorithm would be very useful because if such an algorithm is found then it can be used to test whether a polytope described by m facets has more than n vertices, in time polynomial in m, n and the dimension d of the polytope by setting $\epsilon < \frac{1}{2}\left(\frac{1}{n} - \frac{1}{n+1}\right)$ in the construction used in the proof of Theorem 1 This in turn would yield an algorithm that computes the number of vertices n of a d-dimensional polytope with m facets, in time polynomial in m, n and d. As stated before, a problem that is polynomially equivalent to the Vertex Enumeration problem is to decide if a given list of vertices of an $\mathcal{H}$-polytope is complete [1]. Clearly then, a polynomial-time approximation scheme for the centroid problem would yield an output-sensitive polynomial algorithm for the Vertex Enumeration problem.

Also, the problem of approximating the centroid is not so interesting if we allow polytopes that contain an arbitrarily large ball, since this would allow one to use an algorithm for approximating the centroid with *any* guarantee to obtain another algorithm with an arbitrary guarantee by simply scaling the input polytope appropriately, running the given algorithm and scaling back. So we will assume that the polytope is contained in a unit hypercube in $\mathbb{R}^d$.

Now we prove that the problem of approximating the centroid is #P-easy. We do this by showing that given an algorithm that computes the number of vertices of an arbitrary polytope (a #P-complete problem), one can compute the centroid to any desired precision by making a polynomial (in $\frac{1}{\epsilon}$, the number of facets and the dimension of the polytope) number of calls to this oracle. Notice that in the approximation problem at hand, we are required to find a point within a d-ball centered at the centroid of the polytope and of radius ϵ. We first modify the problem a bit by requiring to report a point that lies inside a hypercube, of side length 2ϵ, centered at the centroid of the polytope. (The hypercube has a clearly defined center of symmetry, namely its own vertex centroid.) To see why this does not essentially change the problem, note that the unit hypercube fits completely inside a d-ball with the same center and radius $\frac{\sqrt{d}}{2}$. We will call any point that is a valid output to this approximation problem, an ϵ-approximation of the centroid $c(P)$.

Given an $\mathcal{H}$-polytope P and a hyperplane $\{a \cdot x = b\}$ that intersects P in the relative interior and does not contain any vertex of P, define P_1 and P_2 as follows:

$$P_1 = P \cap \{x | a \cdot x < b\}, \quad P_2 = P \cap \{x | a \cdot x \geq b\}.$$

Let V_1 be the common vertices of P_1 and P, and V_2 be common vertices of P_2 and P. The following lemma gives a way to obtain the ϵ-approximation of the centroid of P from the ϵ-approximations of the centroids of V_1 and V_2.

Lemma 1. *Given P, V_1, V_2 defined as above, let n_1 and n_2 be the number of vertices in V_1 and V_2 respectively. If c_1 and c_2 are ϵ-approximations of the centroids of V_1 and V_2 respectively, then $c = \frac{n_1 c_1 + n_2 c_2}{n_1 + n_2}$ is an ϵ-approximation of the centroid c^* of P.*

Proof. Let c_{ij} be the j-th coordinate of c_i for $i \in \{1, 2\}$. Also, let c_i^* be the actual centroid of V_i with c_{ij}^* denoting the j-th coordinate of c_i^*. Since c_i approximates c_i^* within a hypercube of side-length 2ϵ, for each $j \in \{1, \cdots, d\}$ we have

$$c_{ij}^* - \epsilon \leq c_{ij} \leq c_{ij}^* + \epsilon.$$

Also, since c^* is the centroid of P,

$$c^* = \frac{n_1 c_1^* + n_2 c_2^*}{n_1 + n_2}.$$

Hence, for each coordinate c_j^* of c^* we have

$$\begin{aligned}
\tfrac{n_1(c_{1j}-\epsilon)+n_2(c_{2j}-\epsilon)}{n_1+n_2} &\leq c_j^* \leq \frac{n_1(c_{1j}+\epsilon)+n_2(c_{2j}+\epsilon)}{n_1+n_2} \\
\Rightarrow \quad \tfrac{n_1c_{1j}+n_2c_{2j}}{n_1+n_2} - \epsilon &\leq c_j^* \leq \frac{n_1c_{1j}+n_2c_{2j}}{n_1+n_2} + \epsilon \\
\Rightarrow \quad c_j - \epsilon &\leq c_j^* \leq c_j + \epsilon \\
\Rightarrow \quad c_j^* - \epsilon &\leq c_j \leq c_j^* + \epsilon.
\end{aligned}$$

□

Now to obtain an approximation of the centroid, we first slice the input polytope P from left to right (say, x_1 coordinate) into $\frac{1}{\epsilon}$ slices each of thickness at most ϵ. Using standard perturbation techniques we can ensure that any vertex of the input polytope does not lie on the left or right boundary of any slice. Any point in the interior of a slice gives us an ϵ-approximation of the x_1 coordinates of vertices of P that are contained in that slice. We can compute the number of vertices of P lying in this slice by subtracting the number of vertices on the boundary of the slice from the total number of vertices of the slice. This can be done using the oracle for vertex counting and then using the previous Lemma along with a slicing along each of the coordinate axes, we can obtain the centroid of P. Note that for slicing along one axis we only approximate that particular coordinate of the centroid and hence slicing along each of the axes is necessary and sufficient. Thus we have the following theorem:

Theorem 1. *Given a polytope P contained in the unit hypercube, an ϵ-approximation of the centroid of P can be computed by making a polynomial number of calls to an oracle for computing the number of vertices of a polytope.*

Now we present a bootstrapping theorem indicating that any "sufficiently" non-trivial approximation of the centroid can be used to obtain arbitrary approximations. For the notion of approximation let us revert back to the Euclidean distance function. Thus, any point x approximating the centroid c within a parameter ϵ satisfies $dist(x, c) \leq \epsilon$. As before we assume that the polytope $\mathcal{P}$ is contained in the unit hypercube. Since the polytope is thus contained in a hyperball with origin as its center and radius at most $\frac{\sqrt{d}}{2}$, any point inside $\mathcal{P}$ approximates the centroid within a factor $\sqrt{d}$. Before we make precise our notion of "sufficiently" non-trivial and present the bootstrapping theorem, some preliminaries are in order.

Lemma 2. *Suppose $(x, y), (u, u) \in \mathbb{R}^{2d}$, where $x, y, u \in \mathbb{R}^d$, then*

$$||u - \frac{x+y}{2}|| \leq \frac{||(u, u) - (x, y)||}{\sqrt{2}},$$

where $||\cdot||$ is the Euclidean norm.

The proof of the above lemma is easy and elementary, and hence we omit it here. Next, consider the product of two polytopes. Given d-dimensional polytopes $\mathcal{P}$, $\mathcal{Q}$

the product $\mathcal{P} \times \mathcal{Q}$ is defined as the set $\{(x, y)|x \in \mathcal{P}, y \in \mathcal{Q}\}$. The facet defining inequalities of the product of P, Q can be computed easily from the inequalities defining P and Q.

$$P = \{x|A_1 x \leq b_1\} \text{ and } Q = \{y|A_2 y \leq b_2\} \Rightarrow P \times Q = \{(x, y)|A_1 x \leq b_1, A_2 y \leq b_2\},$$

where $A_1 \in \mathbb{R}^{m_1 \times d_1}, A_2 \in \mathbb{R}^{m_2 \times d_2}, x \in \mathbb{R}^{d_1}, y \in \mathbb{R}^{d_2}, b_1 \in \mathbb{R}^{m_1 \times 1}, b_2 \in \mathbb{R}^{m_2 \times 1}$.

It is easy to see that the number of vertices of $\mathcal{P} \times \mathcal{Q}$ is the product of the number of vertices of $\mathcal{P}$ and that of $\mathcal{Q}$, and the number of facets of $\mathcal{P} \times \mathcal{Q}$ is the sum of the number of facets of $\mathcal{P}$ and that of $\mathcal{Q}$. Moreover, the dimension of $\mathcal{P} \times \mathcal{Q}$ is the sum of the dimensions of $\mathcal{P}$ and $\mathcal{Q}$.

Observation 1. *If c is the centroid of a polytope P then (c, c) is the centroid of $P \times P$.*

Suppose we are given an algorithm for finding ϵ-approximation of an arbitrary polytope contained in the unit hypercube. For example, for the simple algorithm that returns an arbitrary point inside the polytope, the approximation guarantee is $\frac{\sqrt{d}}{2}$. We consider similar algorithms whose approximation guarantee is a function of the ambient dimension of the polytope. Now suppose that for the given algorithm the approximation guarantee is $f(d)$. For some parameter k consider the k-fold product of P with itself $\overbrace{P \times \cdots \times P}^{k\ times}$, denoted by P^k. Using the given algorithm one can find the $f(kd)$ approximation of P^k and using Lemma 2 one can then find the $\frac{f(kd)}{\sqrt{k}}$-approximation of P. This gives us the following bootstrapping theorem:

Theorem 2. *Suppose we are given an algorithm that computes a $\frac{\sqrt{d}}{g(d)}$-approximation for any polytope contained in the unit hypercube in polynomial time, where $g(.)$ is an unbounded monotonically increasing function. Then, one can compute an ϵ-approximation in time polynomial in the size of the polytope and $g^{-1}(\frac{\sqrt{d}}{\epsilon})$.*

In particular, if we have an algorithm with $d^{\frac{1}{2}-\delta}$ approximation guarantee for finding the centroid of any polytope for some fixed constant $\delta > 0$, then this algorithm can be used to construct a fully polynomial-time approximation scheme for the general problem.

2.3 Approximating Centroid of a Polyhedron Is Hard

The reader should note that the analysis of subsections 2.1 and 2.2 remains valid even for the unbounded case (polyhedra). Even though we do not have any idea about the complexity of approximating the centroid of a polytope, now we show that for an arbitrary unbounded polyhedron the vertex centroid can not be $d^{\frac{1}{2}-\delta}$-approximated for any fixed constant $\delta > 0$ unless $P = NP$. To show this we first prove that for an $\mathcal{H}$-polyhedron $P \subset \mathbb{R}^d$ the vertex centroid of P can not be $\frac{1}{d}$-approximated in polynomial time unless $P = NP$. This together with Theorem 2 completes the proof for hardness of $d^{\frac{1}{2}-\delta}$-approximation of the centroid of an $\mathcal{H}$-polyhedron.

Our proof uses the construction from [6] and its slight modification in [2]. We give a sketch below. For completeness, we also give the complete construction in the appendix.

The proof goes as follows: Given a Boolean CNF formula ϕ, we construct a graph $G(\phi)$ such that $G(\phi)$ has a "long" negative cycle if and only if ϕ is satisfiable. For a given graph G we define a polyhedron $P(G)$ such that every negative cycle in G is a vertex of $P(G)$ and vice-versa. From the properties of the vertex centroid of this class of polyhedra, we then prove that for any formula ϕ, $\frac{1}{d}$-approximating the vertex centroid of $P(G(\phi))$ would reveal whether ϕ is satisfiable or not.

Graph of a CNF formula. Recall that the 3SAT problem is the following decision problem: Given a CNF Boolean formula $\phi = C_1 \wedge \cdots \wedge C_M$ on N literals $x_1, \cdots, x_N$ such that every clause C_i contains exactly 3 literals, is ϕ satisfiable?

Given a directed graph $G = (V, E)$ and a weight function $w : E \to \mathbb{R}$ on its arcs, a directed cycle will be called *short* if it has only two nodes and *long* otherwise. A cycle is *negative* if the total weight on its arcs is negative. The following result was established in [6] (see also [2]).

Lemma 3. *For any 3-CNF ϕ with m clauses we can obtain an arc weighted directed graph $G(\phi)$ with the following properties:*

(P1) $G(\phi)$ has $18m + 1$ edges;
(P2) $G(\phi)$ has $3m$ short negative cycles;
(P3) every negative cycle in $G(\phi)$ has total weight -1;
(P4) there is a distinguished arc e, such that every long negative cycle in $G(\phi)$ contains e; and
(P5) $G(\phi)$ has a long *negative cycle if and only if ϕ is satisfiable.*

The polyhedron of negative-weight flows of a graph. Given a directed graph $G = (V, E)$ and a weight function $w : E \to \mathbb{R}$ on its arcs, consider the following polyhedron:

$$P(G,w) = \left\{ y \in \mathbb{R}^E \;\middle|\; \begin{array}{lrcl} (F) & \displaystyle\sum_{v:(u,v)\in E} y_{uv} - \sum_{v:(v,u)\in E} y_{vu} & = & 0 \;\forall\; u \in V \\ (N) & \displaystyle\sum_{(u,v)\in E} w_{uv} y_{uv} & = & -1 \\ & y_{uv} \geq 0 & & \forall\; (u,v) \in E \end{array} \right\}.$$

If we think of $w_{u,v}$ as the cost/profit paid for edge (u, v) per unit of flow, then each point of $P(G, w)$ represents a *negative-weight circulation* in G, i.e., assigns a non-negative flow on the arcs, obeying the *conservation of flow* at each node of G, and such that total weight of the flow is strictly negative.

For a subset $X \subseteq E$, and a weight function $w : E \mapsto \mathbb{R}$, we denote by $w(X) = \sum_{e \in X} w_e$, the total weight of X. For $X \subseteq E$, we denote by $\chi(X) \in \{0,1\}^E$ the characteristic vector of X: $\chi_e(X) = 1$ if and only if $e \in X$, for $e \in E$. The following theorem states that the vertex set $\mathcal{V}(P(G,w))$ of $P(G,w)$ is in one-to-one correspondence with the negative cycles of the graph G.

Theorem 3 ([2]). *Let $G = (V,E)$ be a directed graph and $w : E \to \mathbb{R}$ be a real weight on the arcs. Then*

$$\mathcal{V}(P(G,w)) = \left\{ \frac{-1}{w(C)} \chi(C) : C \in \mathcal{C}^-(G,w) \right\}. \tag{1}$$

Since by $(P3)$, in the graph G arising from a 3-CNF formula, every negative cycle has weight exactly -1, Theorem 3 implies that the vertices of $P(G,w)$ are exactly the characteristic vectors of the negative cycles of G. By $(P5)$, finding whether G has any long negative cycle, *i.e.*, a negative cycle containing the distinguished arc e (cf. $(P4)$) is NP-complete. By $(P1)$ and $(P2)$, for a 3-CNF formula with m clauses the constructed graph G has $18m + 1$ arcs and $3m$ trivial short negative cycles. Consequently, the polyhedron $P(G,w)$ that is finally obtained has dimension $18m + 1$ and $3m$ trivial vertices corresponding to the short negative cycles of G.

Now, if there are no long negative cycles then the vertex centroid of $P(G,w)$ has value 0 in the coordinate corresponding to the edge e. For simplicity, we will refer to this coordinate axis as x_e. On the other hand, if there are $K \geq 1$ long negative cycles in G then in the centroid $x_e = \frac{K}{K+3m} \geq \frac{1}{3m+1}$. This implies that having an ϵ-approximation for the centroid of $P(G,w)$ for $\epsilon < \frac{1}{2(3m+1)}$ would reveal whether or not $P(G,w)$ has a non-trivial vertex and hence whether or not G has a long negative cycle. Thus we have the following theorem:

Theorem 4. *There is no polynomial algorithm that computes a $\frac{1}{d}$-approximation of the vertex centroid of an arbitrary $\mathcal{H}$-polyhedron $P \subset \mathbb{R}^d$, unless $P = NP$.*

An immediate consequence of Theorem 2 and Theorem 4 is that there is no polynomial algorithm that computes any "sufficiently non-trivial" approximation of the vertex centroid of an arbitrary $\mathcal{H}$-polyhedron unless $P = NP$. More formally,

Corollary 1. *There is no polynomial algorithm that $d^{\frac{1}{2}-\delta}$-approximates the centroid of an arbitrary d-dimensional $\mathcal{H}$-polyhedron for any fixed constant $\delta > 0$ unless $P = NP$.*

3 Open Problems

Although we can show that for unbounded polyhedra almost any non-trivial approximation of the vertex centroid is hard, we can not make a similar statement for the bounded case (*i.e. polytopes*). One interesting variant of Theorem 2 would be to consider a ball of radius r instead of a halfspace. If containment of vertex centroid in a ball of radius r can be decided in time polynomial in the number of inequalities defining the polytope, the dimension and r then one can perform a sort of random walk inside the polytope and approximate the centroid in polynomial time. We leave out the details of this random walk since we do not have a method to check containment inside a ball.

References

1. Avis, D., Bremner, D., Seidel, R.: How good are convex hull algorithms? Comput. Geom. 7, 265–301 (1997)
2. Boros, E., Elbassioni, K., Gurvich, V., Tiwary, H.R.: Characterization of the vertices and extreme directions of the negative cycle polyhedron and hardness of generating vertices of $0/1$-polyhedra. CoRR, abs/0801.3790 (2008)
3. Dyer, M.E.: The complexity of vertex enumeration methods. Mathematics of Operations Research 8(3), 381–402 (1983)
4. Dyer, M.E., Frieze, A.M.: On the complexity of computing the volume of a polyhedron. SIAM J. Comput. 17(5), 967–974 (1988)
5. Kannan, R., Lovász, L., Simonovits, M.: Random walks and an $o^*(n^5)$ volume algorithm for convex bodies. Random Structures and Algorithms 11(1), 1–50 (1998)
6. Khachiyan, L., Boros, E., Borys, K., Elbassioni, K.M., Gurvich, V.: Generating all vertices of a polyhedron is hard. In: SODA, pp. 758–765. ACM Press, New York (2006)
7. Linial, N.: Hard enumeration problems in geometry and combinatorics. SIAM J. Algebraic Discrete Methods 7(2), 331–335 (1986)
8. Rademacher, L.: Approximating the centroid is hard. In: Symposium on Computational Geometry, pp. 302–305 (2007)

Popular Matchings with Variable Job Capacities*

Telikepalli Kavitha and Meghana Nasre

Indian Institute of Science, Bangalore, India
{kavitha,meghana}@csa.iisc.ernet.in

Abstract. We consider the problem of matching people to jobs, where each person ranks a subset of jobs in an order of preference, possibly involving ties. There are several notions of optimality about how to best match each person to a job; in particular, *popularity* is a natural and appealing notion of optimality. However, popular matchings do not always provide an answer to the problem of determining an optimal matching since there are simple instances that do not admit popular matchings. This motivates the following extension of the popular matchings problem:

- Given a graph $G = (\mathcal{A} \cup \mathcal{J}, E)$ where $\mathcal{A}$ is the set of people and $\mathcal{J}$ is the set of jobs, and a list $\langle c_1, \ldots, c_{|\mathcal{J}|} \rangle$ denoting upper bounds on the capacities of each job, does there exist $(x_1, \ldots, x_{|\mathcal{J}|})$ such that setting the capacity of i-th job to x_i, where $1 \le x_i \le c_i$, for each i, enables the resulting graph to admit a popular matching.

In this paper we show that the above problem is NP-hard. We show that the problem is NP-hard even when each c_i is 1 or 2.

1 Introduction

In this paper we consider the problem of matching people to jobs, where each person ranks a subset of jobs in an order of preference possibly involving ties, that is, preference lists are *one-sided*. Our input is a bipartite graph $G = (\mathcal{A} \cup \mathcal{J}, E)$ where $\mathcal{A}$ is the set of people and $\mathcal{J}$ is the set of jobs, and $E = E_1 \dot\cup \cdots \dot\cup E_r$ is the set of edges, where E_t is the set of edges having rank t. For any $a \in \mathcal{A}$, we say a prefers job i to job j if the rank of edge (a, i) is smaller than the rank of edge (a, j). The goal is to come up with an *optimal* matching of people to jobs. Several notions of optimality like rank-maximality [7], maximum-utility, Pareto-optimality [1,3,13] have been studied in the literature for matchings with one-sided preferences. We consider the notion of popularity.

A person a *prefers* matching M to M' if (i) a is matched in M and unmatched in M', or (ii) a is matched in both M and M', and a prefers the job that it is matched to in M as compared to the job that it is matched to in M'.

* Work done as part of the DST-MPG partner group "Efficient Graph Algorithms" at IISc Bangalore.

Y. Dong, D.-Z. Du, and O. Ibarra (Eds.): ISAAC 2009, LNCS 5878, pp. 423–433, 2009.

Definition 1. *M is* more popular than *M', denoted by $M \succ M'$, if the number of people who prefer M to M' is higher than those that prefer M' to M. A matching M^* is* popular *if there is no matching that is more popular than M^*.*

Popular matchings were first introduced by Gardenfors [5] in the context of stable matchings. The notion of popularity is an attractive notion of optimality since it is based on *relative* ranking rather than the absolute ranks used by any person; also popular matchings can be considered stable in the sense that no majority vote of people can force a migration to another matching. Unfortunately there exist simple instances that do not admit any popular matching. Abraham et al. [2] designed efficient algorithms for determining if a given instance admits a popular matching and computing one, if it exists.

Our focus in this paper will be on instances that do not admit a popular matching. Intuitively, the absence of a popular matching in an instance is due to a small set of jobs being in too much demand by a large number of people. In such a case, we expect to improve the situation in terms of popularity by increasing the capacities of some jobs within certain bounds. This solution of increasing job capacities is appealing when the jobs represent items like books or DVDs and increasing capacities implies making copies of the relevant item.

Larger capacities need not always help. Note that it is not always the case that increasing the capacity of a job helps the cause that the resulting graph admits a popular matching. It was shown in [2] that any popular matching M is a maximum cardinality matching in the graph $G_1 = (\mathcal{A} \cup \mathcal{J}, E_1)$[1] and M matches every person to either one of her top choice jobs or to one of her most preferred jobs that is *non-critical* in G_1. A vertex u is critical in G_1 if every maximum cardinality matching in G_1 has to match u; otherwise, u is non-critical. Consider an instance where $\mathcal{A} = \{a_1, a_2, a_3, a_4, b\}$ and $\mathcal{J} = \{f_1, f_2, s_1, s_2, s_3, s_4\}$ and the preference lists of people are described in Figure 1. When the capacity of each job is 1, then the jobs f_1 and f_2 are critical in $(\mathcal{A} \cup \mathcal{J}, E_1)$ and $s_1, \ldots, s_4$ are non-critical in this graph. Thus s_i is the most preferred non-critical vertex in G_1 for person a_i, for $i = 1, \ldots, 4$ and the matching $M = \{(a_1, f_1), (b, f_2), (a_2, s_2), (a_3, s_3), (a_4, s_4)\}$ is a popular matching for this instance.

a_1	f_1	f_2	s_1
a_2	f_1	f_2	s_2
a_3	f_1	f_2	s_3
a_4	f_1	f_2	s_4
b	f_2		

Fig. 1.

a_1	f_1, f_1'	f_2, f_2'	s_1, s_1'
a_2	f_1, f_1'	f_2, f_2'	s_2, s_2'
a_3	f_1, f_1'	f_2, f_2'	s_3, s_3'
a_4	f_1, f_1'	f_2, f_2'	s_4, s_4'
b	f_2, f_2'		

Fig. 2.

[1] Note that only rank 1 edges, that is, edges (a, j) where j is a top choice job for a are included in G_1.

Consider the same problem where the capacity of each job is 2. Figure 2 shows the preference lists; we have explicitly included an identical copy j' of j to denote a capacity of 2 for each job j. The critical vertices in $(\mathcal{A} \cup \mathcal{J}, E_1)$ are now f_1 and f_1' while f_2 and f_2' are now *non-critical* in $(\mathcal{A} \cup \mathcal{J}, E_1)$. Thus f_2 and f_2' become the most preferred non-critical vertices for $a_1, \ldots, a_4$ - hence any popular matching has to match each of $a_1, \ldots, a_4$ to one of $\{f_1, f_1', f_2, f_2'\}$. Also, b has to be matched to f_2 or f_2' (since a popular matching is a maximum matching on rank 1 edges). Since there are 5 people and only 4 jobs that they can be matched to in any popular matching, there exists no popular matching now. Thus the instance shown in Figure 2 where each job capacity is 2 does not admit a popular matching while the instance in Figure 1 where each job capacity is 1 does.

The problem of fixing capacities. Given a graph $G = (\mathcal{A} \cup \mathcal{J}, E)$ where $\mathcal{J} = \{j_1, \ldots, j_{|\mathcal{J}|}\}$ and a list $\langle c_1, \ldots, c_{|\mathcal{J}|}\rangle$ of upper bounds on the job capacities, is there an $(x_1, \ldots, x_{|\mathcal{J}|})$ such that for each $i \in \{1, \ldots, |\mathcal{J}|\}$, setting the capacity of the i-th job to x_i, where $1 \leq x_i \leq c_i$, enables the resulting graph to admit a popular matching.

We assume that G does not admit a popular matching. Our problem is to determine if by fixing the capacities appropriately, the resulting graph admits a popular matching. We now define a special case of this problem which we call the 1-or-2 capacities problem.

The 1-or-2 capacities problem. In this case, each c_i is either 1 or 2. Note that when all the c_i's are 1, this is the standard popular matching problem. Thus the *1-or-2 capacities problem* is a generalization of the popular matching problem. Here we have a subset K of jobs whose capacities may be increased to 2, while the capacities of the remaining jobs should remain 1. The problem is to determine if by increasing the capacities of some elements in K from 1 to 2, we get a graph that admits a popular matching.

Related Work. Subsequent to the work on popular matching algorithms in [2], Manlove and Sng [10] generalized the algorithms of [2] to the case where each job j_i has an associated capacity c_i, the number of people that it can accommodate. Several other variants of the popular matchings problem were considered. Some of them include *weighted* popular matchings considered by Mestre [12] and random popular matchings considered by Mahdian [9]. In order to deal with the problem of the input instance not admitting a popular matching, the following extensions of popular matchings have been considered so far.

Least unpopular matching: The *unpopularity margin* of a matching M, call it $u(M)$, is $\max_{M'}$ |people who prefer M' to M| – |people who prefer M to M'|. The least unpopularity margin matching is that matching M with the least value of $u(M)$. McCutchen [11] showed that computing such a matching is NP-hard. In [6] Huang et al. gave efficient algorithms to compute matchings with bounded values of these unpopularity measures in certain graphs.

Mixed matchings: Very recently, Kavitha et al. [8] considered the problem of computing a probability distribution over matchings, also called a *mixed matching*, that is popular. It was shown that every instance admits a popular mixed matching and a polynomial time algorithm was given to compute such a probability distribution.

Our contributions. In this paper we consider the problem of fixing capacities described earlier. We show that this problem is NP-hard. In fact, we show that the 1-or-2 capacities problem is NP-hard.

2 Preliminaries

We first review the algorithmic characterization of popular matchings given in [2]. As was done in [2], it will be convenient to add a unique last item ℓ_a at the end of a's preference list for each person $a \in \mathcal{A}$. We will henceforth refer to this graph as $G = (\mathcal{A} \cup \mathcal{J}, E)$.

Recall from Section 1 that it was shown in [2] that any popular matching M is (i) a maximum cardinality matching in the graph $G_1 = (\mathcal{A} \cup \mathcal{J}, E_1)$, and (ii) M matches every person to either one of her top choice jobs or to one of her most preferred jobs that is *non-critical* in G_1.

A maximum matching M in a bipartite graph $G_1 = (\mathcal{A} \cup \mathcal{J}, E_1)$ has the following important properties: $M \cap E_1$ defines a partition of $\mathcal{A} \cup \mathcal{J}$ into three disjoint sets: $\mathcal{O}$, $\mathcal{U}$, and $\mathcal{E}$, where $\mathcal{O} \cup \mathcal{U}$ is the set of critical vertices and $\mathcal{E}$ is the set of non-critical vertices. A vertex $u \in \mathcal{E}$ if there is an *even* length alternating path in G_1 from an unmatched vertex to u. A vertex $u \in \mathcal{O}$ if there is an *odd* length alternating path in G_1 from an unmatched vertex to u. A vertex $u \in \mathcal{U}$, that is, it is *unreachable*, if there is no alternating path in G_1 from an unmatched vertex to u. The sets $\mathcal{E}, \mathcal{O}, \mathcal{U}$ are independent of M; the following definitions can be made:

Definition 2. *For each $a \in \mathcal{A}$, define $f(a)$ to be the elements of $\mathcal{O} \cup \mathcal{U}$ amongst a's top choice jobs. Define $s(a)$ to be the set of a's most-preferred jobs in $\mathcal{E}$.*

It was shown in [2] that in any popular matching, every person a has to be matched to a vertex in $f(a) \cup s(a)$. The algorithm for solving the popular matching problem is now straightforward: each $a \in \mathcal{A}$ determines the sets $f(a)$ and $s(a)$. An $\mathcal{A}$-perfect matching that is a maximum matching in G_1 and that matches each a to a job in $f(a) \cup s(a)$ needs to be determined.

3 The 1-or-2 Capacities Problem

Given a graph $G = (\mathcal{A} \cup \mathcal{J}, E)$ and a subset $K \subseteq \mathcal{J}$ of jobs whose capacity can be increased from 1 to 2, the problem is to determine if there exist capacities $\langle x_1, \ldots, x_{|\mathcal{J}|} \rangle$ where $x_t = 1$ for each job $j_t \in \mathcal{J} \setminus K$ and x_t is either 1 or 2 for each job $j_t \in K$ and with these capacities the resulting graph admits a popular matching.

To prove that this problem is NP-hard, we reduce the problem of monotone 1-in-3 SAT to the 1-or-2 capacities problem. Monotone 1-in-3 SAT is a variant of the 3-satisfiability problem (3SAT). Like 3SAT, the input instance is a collection of clauses, where each clause consists of exactly three variables and no variable appears in negated form. The monotone 1-in-3 SAT problem is to determine whether there exists a truth assignment to the variables so that each clause has exactly one true variable (and thus exactly two false variables). This problem is NP-hard [14]; in fact the variant of monotone 1-in-3 SAT where each variable occurs in at most 3 clauses is also NP-hard (refer to [4]). As with other reductions from the 3SAT problem, our reduction also involves designing small *gadgets* which build an instance of the 1-or-2 capacities problem. We first give an overview of our reduction and then describe these gadgets in detail.

3.1 Overview of the Reduction

Let I be an instance of the monotone 1-in-3 SAT problem with $\{X_1, X_2, \ldots, X_n\}$ being the set of variables and $\{C_1, C_2, \ldots, C_m\}$ being the set of clauses in I. We must construct from I an instance of the 1-or-2 capacities problem $G = (\mathcal{A} \cup \mathcal{J}, E)$ and a subset $K \subseteq \mathcal{J}$ of jobs whose capacities can be increased from 1 to 2. We want the following properties of our input $\langle G, K \rangle$:

(i) G (with each job capacity as 1) does not admit a popular matching.
(ii) By increasing the capacities of some jobs in K from 1 to 2 we get an instance that admits a popular matching *iff* there exists an assignment with exactly one variable in each clause of I set to true.

With these observations we design a gadget for every variable in I and another gadget for every clause in I. Each of these gadgets consists of a set of people and a set of jobs along with the preference lists of people. A subset of these jobs is *internal* to the gadget, that is, such jobs appear only on the preference lists of people within the gadget. In addition there will be jobs which are *public*, that is, such jobs appear on the preference lists of people across several gadgets.

Public jobs. The set of public jobs in our instance G is the set K of jobs whose capacities may be increased to 2. We now describe how we derive the set of public jobs from the instance I. For every occurrence of variable X_i in I, we have a public job in G. That is, if a variable X_i appears in clause C_t we have a job u_i^t in G. Note that since I is an instance of monotone 1-in-3 SAT, no variable appears in negated form in I.

We denote by $cap(j)$ the capacity status of job $j \in K$: $cap(j) = 1$ implies that capacity of job j is set to 1, while $cap(j) = 2$ implies that capacity of job j is set to 2. The value of $cap(u_i^t)$'s should capture the truth value of variable X_i, that is, $cap(u_i^t) = 1$ if $X_i = false$, whereas $cap(u_i^t) = 2$ if $X_i = true$.

For us to make the above translation of capacity status of $cap(u_i^t)$'s to the truth value of X_i, it has to be the case that for any i, all $cap(u_i^t)$'s have the same value. So if some $cap(u_i^t)$ is set to 2, we will need to set $cap(u_i^\ell) = 2$, for all ℓ where $u_i^\ell \in K$. Thus the set of jobs corresponding to the all occurrences of variable X_i should simultaneously have the same capacity status. To get such a control, for every variable X_i in the instance I, we introduce another public job u_i in K. The role of job u_i is to enforce the following:

(*) All jobs corresponding to the occurrences of variable X_i get the same $cap(\cdot)$ value as u_i, that is, $cap(u_i^t) = cap(u_i)$, for all t where $u_i^t \in K$.

It can be seen that, if this property is satisfied, the $cap(\cdot)$ status of u_i can be translated to the truth assignment for variable X_i in the instance I and such an assignment will be a consistent assignment to the variables of I. With this, we have completely described all the public jobs (elements of K) in our instance G. The set K, therefore, consists of $3m + n$ jobs as shown below.

$$K = \cup_{i=1}^{n}\{u_i\} \cup_{i,t} \{u_i^t : x_i \text{ appears in } C_t\} \quad (1)$$

Note that the instance I requires us to decide the true/false status of variables $\{X_1, \cdots X_n\}$. Similarly the instance G requires us to decide the capacity status for jobs in K. The non-triviality of the 1-or-2 capacities problem lies in the following: Let j be a job that is a unique rank-1 job for exactly one applicant in G, then j is critical in G restricted to rank-1 edges. Increasing the capacity of j by 1 makes j non-critical in the resulting graph restricted to rank-1 edges. This inturn may change the s-jobs of people in the resulting graph due to which the resulting graph may start admitting a popular matching.

The preference lists of people in our gadgets therefore ensure that every job in K is a unique rank-1 job for exactly one person in G. We now describe our gadgets - one corresponding to each clause and the other corresponding to each variable.

3.2 Gadget Corresponding to a Clause

Let $C_t = (X_{i_1} \vee X_{i_2} \vee X_{i_3})$ be a clause in I. Corresponding to C_t we have a clause gadget which we denote by G_{c_t}. The gadget G_{c_t} consists of a set $A_t = \{a_{t,1}, \ldots, a_{t,11}\}$ of 11 people and a set $D_t = \{p_1^t, p_2^t, q_0^t, q_1^t, q_2^t\}$ of 5 internal jobs. The public jobs that appear in the preference lists of people in G_{c_t} are $u_{i_1}, u_{i_2}, u_{i_3}, u_{i_1}^t, u_{i_2}^t$, and $u_{i_3}^t$.

Figures 3(a), 3(b) and 3(c) show the preference lists of the 11 people $a_1^t, \ldots, a_{11}^t$ associated with the clause C_t. Recall that we introduce a last resort job for each person to ensure that matchings are always $\mathcal{A}$-complete. The ℓ-jobs are these last resort jobs.

The preference lists are designed such that when each public job has capacity 1, then G_{c_t} does not admit a popular matching. Further, any new instance that admits a popular matching and is obtained by increasing capacities of public posts in G_{c_t} obeys the following two properties:

At least one of $u_{i_1}^t, u_{i_2}^t, u_{i_3}^t$ needs to have capacity 2. As seen in Figure 3(a), a_4^t and a_5^t have p_1^t as their top job and q_2^t as their second job – as q_2^t is nobody's top choice job, it follows that q_2^t is the most preferred non-critical job of a_4^t and

a_1^t	p_1^t	$u_{i_1}^t, u_{i_1}$	q_0^t, q_1^t	$\ell_{t,1}$
a_2^t	p_1^t	$u_{i_2}^t, u_{i_2}$	q_0^t, q_1^t	$\ell_{t,2}$
a_3^t	p_1^t	$u_{i_3}^t, u_{i_3}$	q_0^t, q_1^t	$\ell_{t,3}$
a_4^t	p_1^t	q_2^t	$\ell_{t,4}$	
a_5^t	p_1^t	q_2^t	$\ell_{t,5}$	

(a)

a_6^t	p_2^t	$u_{i_1}^t$	$\ell_{t,6}$
a_7^t	p_2^t	$u_{i_2}^t$	$\ell_{t,7}$
a_8^t	p_2^t	$u_{i_3}^t$	$\ell_{t,8}$

(b)

a_9^t	$u_{i_1}^t$	$\ell_{t,9}$
a_{10}^t	$u_{i_2}^t$	$\ell_{t,10}$
a_{11}^t	$u_{i_3}^t$	$\ell_{t,11}$

(c)

Fig. 3.

a_5^t. It is easy to see that in any popular matching one of a_4^t, a_5^t has to be matched to p_1^t and the other to q_2^t. The role of these 2 people is to ensure that a_1^t, a_2^t, a_3^t always get matched to jobs in $s(a_1^t), s(a_2^t), s(a_3^t)$, respectively.

Jobs $u_{i_1}^t, u_{i_2}^t, u_{i_3}^t$ appear as unique top jobs for a_9^t, a_{10}^t and a_{11}^t respectively (Figure 3(c)). Further, they do not appear as top jobs for any other person in G. The preference lists of people in the variable gadget ensure that jobs $u_{i_1}, u_{i_2}, u_{i_3}$ also appear as unique top jobs for exactly one person. Thus, with capacity 1, all these jobs remain *critical* on rank 1 edges. Hence $s(a_1^t) = s(a_2^t) = s(a_3^t) = \{q_0^t, q_1^t\}$. Thus these 3 people can be matched to only these 2 jobs in any popular matching, so there exists no popular matching when each job is capacity 1.

We will assume here that the capacities of each of $u_{i_1}, u_{i_2}, u_{i_3}$ is 1 since even if the capacity of some u_i is increased to 2, the gadget G_{x_i} corresponding to the variable X_i will be designed such that u_i with capacity 2 will be used up in G_{x_i}. Thus at least one of $u_{i_1}^t, u_{i_2}^t, u_{i_3}^t$ should have their capacity set to 2 for all of a_1^t, a_2^t, a_3^t to be matched to jobs in $s(a_1^t), s(a_2^t), s(a_3^t)$, respectively.

Exactly one of $u_{i_1}^t, u_{i_2}^t, u_{i_3}^t$ can have capacity 2. As seen in Figure 3(b) the 3 people a_6^t, a_7^t, a_8^t have the same top job p_2^t - when each of $u_{i_1}^t, u_{i_2}^t, u_{i_3}^t$ has its capacity set to 1, then one of these 3 people gets matched to p_2^t and the other 2 people get matched to their respective last resort jobs (most preferred non-critical job) in any popular matching. However we know that at least $u_{i_1}^t, u_{i_2}^t, u_{i_3}^t$ has its capacity set to 2 due to people $a_1^t, \ldots, a_5^t$.

If exactly one among these (say, $u_{i_1}^t$) becomes a non-critical job, then a_6^t gets matched to p_2^t while a_7^t and a_8^t get matched to their respective last resort jobs. If two or more among $u_{i_1}^t, u_{i_2}^t, u_{i_3}^t$ has its capacity set to 2, then while the corresponding people in a_1^t, a_2^t, a_3^t get matched to these jobs, we cannot match two or more among a_6^t, a_7^t, a_8^t to the single job p_2^t.

Thus the clause gadget G_{c_t} ensures that exactly one of the jobs $u_{i_1}^t, u_{i_2}^t, u_{i_3}^t$ has its capacity set to 2.

3.3 Gadget Corresponding to a Variable

Let X_i be a variable in I, then corresponding to X_i we construct a variable gadget G_{x_i}. The gadget G_{x_i} consists of a set B_i of 4 people $\{b_1^i, b_2^i, b_3^i, b_4^i\}$ in $\mathcal{A}$ and a set T_i of 2 internal jobs $\{p_i, q_i\}$. The public jobs associated with X_i

b_1^i	p_i	$u_i, u_i^{t_1}, u_i^{t_2}, u_i^{t_3}$	$\ell_{b_{i,1}}$
b_2^i	p_i	q_i	$\ell_{b_{i,2}}$
b_3^i	p_i	q_i	$\ell_{b_{i,3}}$
b_4^i	u_i	$\ell_{b_{i,4}}$	

Fig. 4.

are $u_i, u_i^{t_1}, u_i^{t_2}, u_i^{t_3}$ where t_1, t_2, t_3 are the clauses that the variable X_i occurs in. Note that we consider the restriction of monotone 1-in-3-SAT where each variable occurs in at most 3 clauses; therefore it suffices to consider at most 3 clauses in which X_i appears. Figure 4 shows the preference lists of people $b_1^i, b_2^i, b_3^i, b_4^i$.

The variable gadget for X_i will ensure that property ($*$) mentioned in Section 3.1 is enforced, that is, the $cap(u_i^t)$ values for all values of t are the same. As promised earlier, we have a person b_4^i who has job u_i as her top choice job. In any popular matching one of b_2^i, b_3^i has to be matched to p_i and the other to q_i. Thus b_1^i has to be matched to a job in $s(b_1^i)$ in any popular matching. Note that the jobs $u_i, u_i^{t_1}, u_i^{t_2}, u_i^{t_3}$ are tied as second choice jobs in the preference list of b_1^i. If each of these jobs has capacity 1, then $s(b_1^i) = \{\ell_{b_{i,1}}\}$; however if some of these jobs have their capacities set to 2, then $s(b_1^i)$ consists of such jobs.

3.4 Putting It All Together

As mentioned earlier, the graph G that we construct is the union of gadgets corresponding to variables as well as clauses. We define the sets $\mathcal{A}$ and $\mathcal{J}$ as: $\mathcal{A} = \cup_{t=1}^m A_t \cup_{i=1}^n B_i$ and $\mathcal{J} = \cup_{t=1}^m D_t \cup_{i=1}^n T_i \cup K$. The set K is the set of all public jobs as described earlier in Eqn. (1). The preference lists of the people are as shown in our gadgets. We note that every job $j \in K$ is a unique rank-1 job for exactly one person in $\mathcal{A}$. We now show how the variable and clause gadgets co-operate to enforce property ($*$) mentioned in Section 3.1.

Lemma 1. *If by setting the capacities of each job in K to either 1 or 2, there exists an instance that admits a popular matching, then $cap(u_i) = cap(u_i^t)$, for all t where $u_i^t \in K$.*

Proof. Let us assume that there exists a t_1 such that $u_i^{t_1} \in K$ and $cap(u_i^{t_1}) = 2$ and suppose $cap(u_i) = 1$. Recall that the job $u_i^{t_1}$ is in K due to the occurrence of variable X_i in clause C_{t_1} and $u_i^{t_1}$ is the unique top choice job of $a_l^{t_1}$ for some $l \in \{9, 10, 11\}$. This person always gets matched to $u_i^{t_1}$ in any popular matching (since such a matching has to be maximum on rank 1 edges). The job $u_i^{t_1}$ also appears on the preference lists of the following people:

(i)$a_k^{t_1}$ for some $k \in \{1, 2, 3\}$ (where X_i is the k-th variable in clause C_t) in the clause gadget $G_{c_{t_1}}$,
(ii) b_1^i in the variable gadget G_{x_i},
(iii) $a_l^{t_1}$ for some $l \in \{6, 7, 8\}$ in the clause gadget $G_{c_{t_1}}$.

Since $cap(u_i^{t_1}) = 2$, for all these people $u_i^{t_1}$ is an s-job. Further, by the design of our gadgets, while $a_l^{t_1}$ (for some $l \in \{6, 7, 8\}$) gets matched to her top choice job p_2^t in any popular matching, both $a_k^{t_1}$ and b_1^i have to be matched to their s-jobs. Clearly, both $a_k^{t_1}$ (for some $k \in \{1, 2, 3\}$) and b_1^i cannot be matched to $u_i^{t_1}$.

If $u_i^{t_1}$ is matched to b_1^i, then the appropriate $a_k^{t_1}$ does not have an s-job. Hence, assume $u_i^{t_1}$ is matched to $a_k^{t_1}$. Then our assumption that the resulting instance admits a popular matching implies that there exists a t_2 such that $u_i^{t_2}$ is in K and $cap(u_i^{t_2}) = 2$. Since $cap(u_i) = 1$, it is easy to see that a similar argument forbids the instance to admit an $\mathcal{A}$-complete matching restricted to f and s-jobs. This contradicts the lemma hypothesis or $cap(u_i) = 2$. Thus we have shown that if there exists t_1 such that $cap(u_i^{t_1}) = 2$, then $cap(u_i) = 2$.

We now show the other direction, that is, $cap(u_i) = 2$ implies that $cap(u_i^t) = 2$ for all t where $u_i^t \in K$. Assume for the sake of contradiction that $cap(u_i) = 2$ and there exists a t_1 such that $u_i^{t_1} \in K$ and $cap(u_i^{t_1}) = 1$. Note that the job $u_i^{t_1}$ was created because the variable X_i appears in clause C_{t_1}. Corresponding to clause C_{t_1}, we have a clause gadget $G_{c_{t_1}}$ in G. With $cap(u_i) = 2$, there is a person $a_k^{t_1}$ for some $k \in \{1, 2, 3\}$ (where X_i is the k-th variable in clause C_{t_1}) that treats u_i as an s-job.

But since b_1^i that belongs to the variable gadget also treats u_i as an s-job, there cannot be an $\mathcal{A}$-complete matching in which both b_1^i and $a_k^{t_1}$ are matched to u_i. Thus either the resulting graph does not admit a popular matching or $cap(u_i^{t_1}) = 2$. We therefore have $cap(u_i) = cap(u_i^t)$ for all t where $u_i^t \in K$. □

Thus if by setting the capacities of each job in K to 1 or 2, there exists an instance that admits a popular matching, then the capacity values of jobs in K always translate to a consistent truth assignment of the variables in I. Lemma 2 (proof omitted, refer to [15]) proves the correctness of our reduction.

Lemma 2. *There exists an instance that admits a popular matching by setting the capacities of each job in K to either 1 or 2* iff *there exists a 1-in-3 satisfying assignment for I.*

We can now conclude the following theorem.

Theorem 1. *The 1-or-2 capacities problem is NP-hard.*

Observe that our reduction constructed a graph G where preference lists included ties. This raises the question of the complexity of the 1-or-2 capacities problem when preference lists are *strict* (that is, no ties are allowed). Our gadgets can easily be modified to show that this problem is also NP-hard. We omit the proof of Corollary 1 that states our result for strict preference lists. Refer to [15] for the proof.

Theorem 2. *The 1-or-2 capacities problem is NP-hard for strict preference lists.*

A related problem: The difficulty of the problem of fixing capacities arises from the problem of deciding which jobs should be critical and which jobs can be non-critical in the resulting graph restricted to rank-1 edges. Consider a variant of the above problem where we only have to maintain an upper bound k on the *total* sum of increased capacities of all the jobs, rather an upper bound on the capacity of *each* job. That is, here we are given a graph $G = (\mathcal{A} \cup \mathcal{J}, E)$ and an integer $k \geq 0$ where we can increase the job capacity of the i-th job from 1 to $1 + \Delta_i$ for each i and the restriction is: $\sum_i \Delta_i \leq k$, in other words, the *sum of increases in capacities* is bounded by k. In this problem no job that is critical in G_1 needs to turn non-critical on rank 1 edges of the resulting graph (after appropriate capacity increases) and thus this problem becomes solvable in polynomial time. The details can be found in the full version of this paper [15].

Acknowledgment. We thank Sourav Chakraborty for discussions that motivated this work.

References

1. Abraham, D.J., Cechlárová, K., Manlove, D.F., Mehlhorn, K.: Pareto-optimality in house allocation problems. In: Fleischer, R., Trippen, G. (eds.) ISAAC 2004. LNCS, vol. 3341, pp. 3–15. Springer, Heidelberg (2004)
2. Abraham, D.J., Irving, R.W., Kavitha, T., Mehlhorn, K.: Popular matchings. SIAM Journal on Computing 37(4), 1030–1045 (2007)
3. Abdulkadiroğluand, A., Sönmez, T.: Random serial dictatorship and the core from random endowments in house allocation problems. Econometrica 66(3), 689–701 (1998)
4. Denman, R., Foster, S.: Using clausal graphs to determine the computational complexity of k-bounded positive one-in-three SAT. Discrete Applied Mathematics 157(7), 1655–1659 (2009)
5. Gardenfors, P.: Match making: assignments based on bilateral preferences. Behavioural Sciences 20, 166–173 (1975)
6. Huang, C.-C., Kavitha, T., Michail, D., Nasre, M.: Bounded unpopularity matchings. In: Gudmundsson, J. (ed.) SWAT 2008. LNCS, vol. 5124, pp. 127–137. Springer, Heidelberg (2008)
7. Irving, R.W., Kavitha, T., Mehlhorn, K., Michail, D., Paluch, K.: Rank-maximal matchings. ACM Transactions on Algorithms 2(4), 602–610 (2006)
8. Kavitha, T., Mestre, J., Nasre, M.: Popular Mixed Matchings. In: Albers, S., Marchetti-Spaccamela, A., Matias, Y., Niko-letsea, S. (eds.) ICALP 2009. LNCS, vol. 5556. Springer, Heidelberg (2009)
9. Mahdian, M.: Random popular matchings. In: Proceedings of the 8th ACM Conference on Electronic Commerce, pp. 238–242 (2006)
10. Manlove, D., Sng, C.: Popular matchings in the capacitated house allocation problem. In: Azar, Y., Erlebach, T. (eds.) ESA 2006. LNCS, vol. 4168, pp. 492–503. Springer, Heidelberg (2006)
11. McCutchen, R.M.: The least-unpopularity-factor and least-unpopularity-margin criteria for matching problems with one-sided preferences. In: Proceedings of the 15th Latin American Symposium on Theoretical Informatics, pp. 593–604 (2008)

12. Mestre, J.: Weighted popular matchings. In: Bugliesi, M., Preneel, B., Sassone, V., Wegener, I. (eds.) ICALP 2006. LNCS, vol. 4051, pp. 715–726. Springer, Heidelberg (2006)
13. Roth, A.E., Postlewaite, A.: Weak versus strong domination in a market with indivisible goods. Journal of Mathematical Economics 4, 131–137 (1977)
14. Schaefer, T.J.: The complexity of satisfiability problems. In: Proceedings of the 10th Annual ACM Symposium on Theory of Computing, pp. 216–226 (1978)
15. Kavitha, T., Nasre, M.: Popular matchings with variable job capacities. Indian Institute of Science. Technical Report. IISc-CSA-TR-2009-7

On the Tightness of the Buhrman-Cleve-Wigderson Simulation

Shengyu Zhang

Department of Computer Science and Engineering,
The Chinese University of Hong Kong
syzhang@cse.cuhk.edu.hk

Abstract. Buhrman, Cleve and Wigderson gave a general communication protocol for block-composed functions $f(g_1(x^1, y^1), \ldots, g_n(x^n, y^n))$ by simulating a decision tree computation for f [3]. It is also well-known that this simulation can be very inefficient for some functions f and $(g_1, \ldots, g_n)$. In this paper we show that the simulation is actually polynomially tight up to the choice of $(g_1, \ldots, g_n)$. This also implies that the classical and quantum communication complexities of certain block-composed functions are polynomially related.

1 Introduction

Decision tree complexity [4] and communication complexity [8] are two concrete models for studying computational complexity. In [3], Buhrman, Cleve and Wigderson gave a general method to design communication protocol for block-composed functions $f(g_1(x^1, y^1), \ldots, g_n(x^n, y^n))$. The basic idea is to simulate the decision tree computation for f, with queries of the i-th variable answered by a communication protocol for $g_i(x^i, y^i)$. In the language of complexity measures, this result gives

$$\mathsf{CC}(f(g_1, \ldots, g_n)) = \tilde{O}(\mathsf{DT}(f) \cdot \max_i \mathsf{CC}(g_i)) \tag{1}$$

where CC and DT are the communication complexity and the decision tree complexity, respectively. This simulation holds for all the models: deterministic, randomized and quantum[1].

It is also well known that the communication protocol by the above simulation can be very inefficient. For example, if f is the n-bit AND function and all g_i's are 2-bit AND function, then even the deterministic communication complexity of $f(g_1, \ldots, g_n)$ is just 1 bit, since Alice can compute and send the AND function of her bits. This is in sharp contrast to the decision tree complexity of the n-bit AND function f, which is $\Theta(n)$ in the randomized case and $\Theta(\sqrt{n})$ in the quantum case.

1 For the deterministic model, no error correction for each g_i is needed, so the $\tilde{O}$ can be changed to O.

Y. Dong, D.-Z. Du, and O. Ibarra (Eds.): ISAAC 2009, LNCS 5878, pp. 434–440, 2009.

Turning the simulation around, one can also get a lower bound method for the decision tree complexity by communication complexity. To lower bound $\mathsf{DT}(f)$, we can pick g_i's, then the communication complexity of $\mathsf{CC}(f(g_1,\ldots,g_n))/\max_i \mathsf{CC}(g_i)$ is a lower bound of $\mathsf{DT}(f)$. Actually in the same paper [3], they obtained the almost tight lower bound of $\tilde{\Omega}(n)$ for the quantum decision tree complexity of the Parity and Majority functions this way. But because of the counterexamples as shown in the last paragraph, it is not clear how tight this lower bound method can be in general.

In this paper, we will show that the simulation is polynomially tight, and actually this can be achieved by each g_i chosen only from $\{\wedge, \vee\}$, *i.e.* 2-bit AND and OR.

Theorem 1. *For Boolean functions f,*

$$\max_{g_i\in\{\wedge,\vee\}} \mathsf{R}^{\mathsf{CC}}(f(g_1,\ldots,g_n)) = \Omega(\mathsf{D}^{\mathsf{DT}}(f)^{1/3}), \tag{2}$$

$$\max_{g_i\in\{\wedge,\vee\}} \mathsf{Q}^{\mathsf{CC}}(f(g_1,\ldots,g_n)) = \Omega(\mathsf{D}^{\mathsf{DT}}(f)^{1/6}). \tag{3}$$

For monotone functions f, the bounds can be improved to the following. For two n-bit strings x and y, use $x \wedge_n y$ and $x \vee_n y$ to denote the bit-wise AND and OR of x and y, respectively. We drop the subscript when $n = 1$.

Theorem 2. *For monotone Boolean functions f,*

$$\max_{g\in\{\wedge_n,\vee_n\}} \mathsf{R}^{\mathsf{CC}}(f(g_1,\ldots,g_n)) = \Omega(\mathsf{D}^{\mathsf{DT}}(f)^{1/2}), \tag{4}$$

$$\max_{g\in\{\wedge_n,\vee_n\}} \mathsf{Q}^{\mathsf{CC}}(f(g_1,\ldots,g_n)) = \Omega(\mathsf{D}^{\mathsf{DT}}(f)^{1/4}). \tag{5}$$

Note that the improvement is two-fold: Besides the better bounds themselves, the range of inner function g is also restricted to $\{\wedge_n, \vee_n\}$. That is, we require all g_i's be the same; they are either all AND or all OR functions.

The bounds give the following corollary about the polynomial relation between quantum and classical communication complexity for composed functions.

Corollary 1. *For Boolean functions f,*

$$\max_{g_i\in\{\wedge,\vee\}} \mathsf{D}^{\mathsf{CC}}(f(g_1,\ldots,g_n)) = O\left(\max_{g_i\in\{\wedge,\vee\}} \mathsf{Q}^{\mathsf{CC}}(f(g_1,\ldots,g_n))^6\right). \tag{6}$$

If f is monotone, then

$$\max_{g\in\{\wedge_n,\vee_n\}} \mathsf{D}^{\mathsf{CC}}(f(g_1,\ldots,g_n)) = O\left(\max_{g\in\{\wedge_n,\vee_n\}} \mathsf{Q}^{\mathsf{CC}}(f(g_1,\ldots,g_n))^4\right). \tag{7}$$

Related work

1. After the results in the current paper being circulated at Institute for Quantum Computing at University of Waterloo and Centre for Quantum Technologies at National University of Singapore in May 2009, Sherstov posted a related paper [11], which does not have our Theorem 1 and Theorem 2, but shows the following result:

Theorem 3. *(Sherstov, [11]) For Boolean functions f,*

$$\max_{g\in\{\wedge_n,\vee_n\}} \mathsf{D}^{\mathsf{CC}}(f(g_1,\ldots,g_n)) = O\left(\max_{g\in\{\wedge_n,\vee_n\}} \mathsf{Q}^{\mathsf{CC}}(f(g_1,\ldots,g_n))^{12}\right). \quad (8)$$

The technical ingredient to achieve the above theorem is to observe that the function contains a subfunction g with size-2 block sensitivity $bs^2(g) \geq bs(f)$, and then use a theorem by Kenyon and Kutin [7] that $s(g) = \Omega(\sqrt{bs^2(g)})$.

2. A subsequent work jointly with Jain and Klauck [5] also uses the idea of embedding Disj (or its complement) in some other function, showing depth-independent lower bounds for communication complexity of read-once formulas composed with AND and OR.

2 Preliminaries

For an n-bit Boolean function $f : \{0,1\}^n \to \{0,1\}$, a deterministic query algorithm for f accesses the input x only by making queries in the form of "$x_i =?$". Each query has cost 1, and all the other computation between queries are free. A randomized query algorithm is the same except that the algorithm can toss coins to decide the next variable x_i to ask. The quantum query model, formally introduced in [2], has a working state in the form of $\sum_{i,a,z} \alpha_{i,a,z}|i,a,z\rangle$, where i ranges over $[n]$, a ranges over $\{0,1\}$ and z is the content in the working space. A quantum query on the input x corresponds to an oracle O_x, a unitary operation defined by

$$O_x\Big(\sum_{i,a,z} \alpha_{i,a,z}|i,a,z\rangle\Big) = \sum_{i,a,z} \alpha_{i,a,z}|i,a\oplus x_i,z\rangle \quad (9)$$

A T-query quantum query algorithm works as a sequence of operations

$$U_0 \to O_x \to U_1 \to O_x \to \cdots \to U_{T-1} \to O_x \to U_T \quad (10)$$

Here O_x is as defined above, and each U_t does not depend on the input x. In both randomized and quantum query models, we can allow a double-sided error probability of 1/3. The deterministic, randomized and quantum query complexities, denoted by $\mathsf{D}^{\mathsf{DT}}(f)$, $\mathsf{R}^{\mathsf{DT}}(f)$ and $\mathsf{Q}^{\mathsf{DT}}(f)$, are the minimum numbers of queries we need to make in order to compute the function by a deterministic, randomized and quantum query algorithm, respectively.

Communication complexity studies the minimum amount of communication that two or more parties need to compute a given function or a relation of their inputs. Since its inception in the seminal paper by Yao [13], communication complexity has been an important and widely studied research area, both because of the interesting and intriguing mathematics involved in its study, and also because of the fundamental connections it bears with many other areas in theoretical computer science. In the standard *two-party interactive* model, two parties Alice and Bob, each on receiving an input say $x \in \mathcal{X}$ and $y \in \mathcal{Y}$, respectively, sending messages back and forth to jointly compute a function f on input (x,y). Their computation and communication can be deterministic, randomized, and quantum. The deterministic, randomized, and quantum communication complexity of f, denoted by $\mathsf{D}^{\mathsf{CC}}(f)$, $\mathsf{R}^{\mathsf{CC}}(f)$ and $\mathsf{Q}^{\mathsf{CC}}(f)$, is the least number of bits (or qubits in the quantum case) needed to be transferred in the corresponding model, *s.t.* the protocol gives the correct answer with probability at least 2/3 for all inputs.

For a string $x \in \{0,1\}^n$ and a set $I \subseteq [n]$, the string $x^{(I)}$ is obtained from x by flipping all coordinates in I.

Definition 1 (block sensitivity). *The block sensitivity $bs(f,x)$ of f on x is the maximum number b such that there are disjoint sets $I_1, \ldots, I_b$ for which $f(x) \neq f(x^{I_j})$. The block sensitivity of f is $bs(f) = \max_x bs(f,x)$. For $z \in \{0,1\}$, the z-block sensitivity is $bs_z(f) = \max_{x:f(x)=z} bs(f,x)$.*

The block sensitivity is one of the complexity measures that are polynomially related to each other. In particular, it powered is known to be an upper bound of the deterministic decision tree complexity, as shown in the following theorem by Beals *et. al.* [2]. For many other complexity measures and their relations, we refer the reader to the excellent survey [4].

Theorem 4. *1.* $\mathsf{D}^{\mathsf{DT}}(f) = O(bs(f)^3)$.
2. For monotone functions f, $\mathsf{D}^{\mathsf{DT}}(f) = O(bs(f)^2)$.

The function $\mathsf{Disj}_n : \{0,1\}^n \times \{0,1\}^n$ is defined as follows: $\mathsf{Disj}_n(x,y) = \mathsf{OR}(x_1 \wedge y_1, \ldots, x_n \wedge y_n)$. The promise version of the problem, $\mathsf{PromiseDisj}_n$, is the same as Disj_n but with the promise that there is at most one i *s.t.* $x_i \wedge y_i = 1$. The randomized and quantum communication complexity for Disj_n and $\mathsf{PromiseDisj}_n$ are known as follows.

Theorem 5

$$\mathsf{R}^{\mathsf{CC}}(\mathsf{Disj}_n) \geq \mathsf{R}^{\mathsf{CC}}(\mathsf{PromiseDisj}_n) = \Omega(n), \tag{11}$$

$$\mathsf{Q}^{\mathsf{CC}}(\mathsf{Disj}_n) \geq \mathsf{Q}^{\mathsf{CC}}(\mathsf{PromiseDisj}_n) = \Omega(\sqrt{n}). \tag{12}$$

The original randomized lower bound [6,9,1] was for Disj instead of $\mathsf{PromiseDisj}$. But the same proof of [1] also carries to the same lower bound for $\mathsf{PromiseDisj}$. The original quantum lower bound [10,12] was also for Disj, but as mentioned in [11], the same method in [10] also applies to prove the same lower bound for $\mathsf{PromiseDisj}$. [11] also explicitly gives a proof for $\mathsf{Q}^{\mathsf{CC}}(\mathsf{PromiseDisj})$ by adapting the method in [12].

3 Lower Bounds for the Communication Complexity of Composed Functions

We will actually prove that

Lemma 1. *For Boolean functions f,*

$$\max_{g_i \in \{\wedge,\vee\}} \mathsf{R}^{\mathsf{CC}}(f(g_1, \ldots, g_n)) = \Omega(bs(f)), \tag{13}$$

$$\max_{g_i \in \{\wedge,\vee\}} \mathsf{Q}^{\mathsf{CC}}(f(g_1, \ldots, g_n)) = \Omega(\sqrt{bs(f)}). \tag{14}$$

If f is monotone, then

$$\max_{g\in\{\wedge_n,\vee_n\}} \mathsf{R}^{\mathsf{CC}}(f(g_1,\ldots,g_n)) = \Omega(bs(f)), \tag{15}$$

$$\max_{g\in\{\wedge_n,\vee_n\}} \mathsf{Q}^{\mathsf{CC}}(f(g_1,\ldots,g_n)) = \Omega(\sqrt{bs(f)}). \tag{16}$$

Proof. By the definition of block sensitivity, there are an input z and blocks $I_1,\ldots,I_b$, where $b = bs(f)$, *s.t.* f is sensitive on z at those blocks. That is,

$$f(z^{(I_1)}) = \ldots = f(z^{(I_b)}) \neq f(z). \tag{17}$$

We will define g_i's *s.t.* if there is a protocol for $f(g_1,\ldots,g_n)$, then there is a protocol for $\mathsf{PromiseDisj}_b$. The reduction is: on input $(x,y) \in \{0,1\}^b \times \{0,1\}^b$, we define an input $(x',y') \in \{0,1\}^n \times \{0,1\}^n$ for the function $f(g_1,\ldots,g_n)$ as follows.

1. For $i \notin \cup_{j=1}^b I_j$:

$$x'_i = y'_i = z_i,\ g_i = \wedge \tag{18}$$

2. For $i \in I_j$:

$$\begin{cases} x'_i = x_j,\ y'_i = y_i,\ g_i = \wedge, & \text{if } z_i = 0 \\ x'_i = \bar{x}_j,\ y'_i = \bar{y}_i,\ g_i = \vee, & \text{if } z_i = 1 \end{cases} \tag{19}$$

It is easy to see that for the first case, $g_i(x'_i,y'_i) = z_i \wedge z_i = z_i$. For the second case, if $z_i = 0$, then

$$g_i(x'_i,y'_i) = x_j \wedge y_j = (x_j \wedge y_j) \oplus z_i; \tag{20}$$

if $z_i = 1$, then

$$g_i(x'_i,y'_i) = \bar{x}_j \vee \bar{y}_j = \overline{x_j \wedge y_j} = (x_j \wedge y_j) \oplus z_i. \tag{21}$$

Thus for each $j = 1,2,\ldots,b$, it holds that

$$x_j \wedge y_j = 1 \Leftrightarrow g_i(x'_i,y'_i) = \bar{z}_i, \forall i \in I_j \tag{22}$$

Therefore,

$$\text{Distinguishing between } x \wedge y = 0 \text{ and } \exists \text{ unique } j,\ s.t.\ x_j \wedge y_j = 1 \tag{23}$$

$$\Leftrightarrow \text{Distinguishing between } g(x',y') = z \text{ and } \exists \text{ unique } j,\ s.t.\ g(x',y') = z^{(I_j)} \tag{24}$$

Now if we have a protocol to compute $f(g_1,\ldots,g_n)$, then we can use it to solve the problem in Eq. (24). By the equivalence, this also solves the problem in Eq. (23), *i.e.* the $\mathsf{PromiseDisj}_b$ problem. Since

$$\mathsf{R}^{\mathsf{CC}}(\mathsf{PromiseDisj}_b) = \Omega(b), \quad \mathsf{Q}^{\mathsf{CC}}(\mathsf{PromiseDisj}_b) = \Omega(\sqrt{b}) \tag{25}$$

we proved the conclusion for general Boolean function f.

If f is monotone, then observe that we can assume that each sensitive block I_j contains all 0's or all 1's. Actually, suppose $f(z) = 0$ and $I_j = I_{j,0} \uplus I_{j,1}$ where z on $I_{j,b}$ contains only bits equal to b. Then we can remove $I_{j,1}$ and let the $I_{j,0}$ be the new block. It is still disjoint with all other blocks, yet by monotonicity, it has $f(z^{(I_{j,0})}) \geq f(z^{(I_j)}) = 1$. In this way we can assume that for $f(z) = 0$, all sensitive blocks contains only 0's. Thus using similar reductions, we have

$$\mathsf{R}^{\mathsf{CC}}(f(\wedge, \ldots, \wedge)) = \Omega(\mathsf{R}^{\mathsf{CC}}(\mathsf{Disj}_{bs_0(f)})) = \Omega(bs_0(f)), \tag{26}$$

$$\mathsf{Q}^{\mathsf{CC}}(f(\wedge, \ldots, \wedge)) = \Omega(\mathsf{Q}^{\mathsf{CC}}(\mathsf{Disj}_{bs_0(f)})) = \Omega(\sqrt{bs_0(f)}). \tag{27}$$

Note that here because of the monotonicity, we can reduce the problem to Disj instead of PromiseDisj, though this does not give us any stronger bound. Similarly, we have

$$\mathsf{R}^{\mathsf{CC}}(f(\vee, \ldots, \vee)) = \Omega(\mathsf{R}^{\mathsf{CC}}(\mathsf{Disj}_{bs_1(f)})) = \Omega(bs_1(f)), \tag{28}$$

$$\mathsf{Q}^{\mathsf{CC}}(f(\vee, \ldots, \vee)) = \Omega(\mathsf{Q}^{\mathsf{CC}}(\mathsf{Disj}_{bs_1(f)})) = \Omega(\sqrt{bs_1(f)}). \tag{29}$$

Since $bs(f) = \max\{bs_0(f), bs_1(f)\}$, this finishes the proof of the lemma.

Theorem 1 and 2 follow from the above lemma and Theorem 4. Corollary 1 is also easy:

$$\max_{g_i \in \{\wedge,\vee\}} \mathsf{D}^{\mathsf{CC}}(f(g_1, \ldots, g_n)) \tag{30}$$

$$= O(\mathsf{D}^{\mathsf{DT}}(f)) \qquad \text{(by [3])} \tag{31}$$

$$= O\Big(\max_{g_i \in \{\wedge,\vee\}} \mathsf{Q}^{\mathsf{CC}}(f(g_1, \ldots, g_n))^6\Big) \qquad \text{(by Theorem 1)} \tag{32}$$

The monotone function case follows similarly.

Acknowledgments. The work is supported by the Hong Kong grant RGC-419309.

References

1. Bar-Yossef, Z., Jayram, T.S., Kumar, R., Sivakumar, D.: An information statistics approach to data stream and communication complexity. Journal of Computer and System Sciences 68(4), 702–732 (2004)
2. Beals, R., Buhrman, H., Cleve, R., Mosca, M., de Wolf, R.: Quantum lower bounds by polynomials. Journal of the ACM 48(4), 778–797 (2001)
3. Buhrman, H., Cleve, R., Wigderson, A.: Quantum vs. classical communication and computation. In: Proceedings of the Thirtieth Annual ACM Symposium on the Theory of Computing (STOC), pp. 63–68 (1998)
4. Buhrman, H., de Wolf, R.: Complexity measures and decision tree complexity: a survey. Theoretical Computer Science 288(1), 21–43 (2002)
5. Jain, R., Klauck, H., Zhang, S.: Depth-independent lower bounds on the communication complexity of read-once boolean formulas. arXiv:0908.4453 (2009)

6. Kalyanasundaram, B., Schintger, G.: The probabilistic communication complexity of set intersection. SIAM Journal on Discrete Mathematics 5(4), 545–557 (1992)
7. Kenyon, C., Kutin, S.: Sensitivity, block sensitivity, and l-block sensitivity of boolean functions. Information and Computation 189(1), 43–53 (2004)
8. Kushilevitz, E., Nisan, N.: Communication Complexity. Cambridge University Press, Cambridge (1997)
9. Razborov, A.: On the distributional complexity of disjointness. Theoretical Computer Science 106, 385–390 (1992)
10. Razborov, A.: Quantum communication complexity of symmetric predicates. Izvestiya: Mathematics 67(1), 145–159 (2003)
11. Sherstov, A.: On quantum-classical equivalence for composed communication problems. arXiv:quant-ph/0906.1399 (2009)
12. Sherstov, A.A.: The pattern matrix method for lower bounds on quantum communication. In: Proceedings of the 40th Annual ACM Symposium on the Theory of Computing (STOC), pp. 85–94 (2008)
13. Yao, A.: Some complexity questions related to distributive computing. In: Proceedings of the Eleventh Annual ACM Symposium on Theory of Computing (STOC), pp. 209–213 (1979)

Bounds on Contention Management Algorithms

Johannes Schneider and Roger Wattenhofer

Computer Engineering and Networks Laboratory, ETH Zurich, 8092 Zurich, Switzerland

Abstract. We present two new algorithms for contention management in transactional memory, the deterministic algorithm *CommitRounds* and the randomized algorithm *RandomizedRounds*. Our randomized algorithm is efficient: in some notorious problem instances (e.g., dining philosophers) it is exponentially faster than prior work from a worst case perspective. Both algorithms are (i) local and (ii) starvation-free. Our algorithms are local because they do not use global synchronization data structures (e.g., a shared counter), hence they do not introduce additional resource conflicts which eventually might limit scalability. Our algorithms are starvation-free because each transaction is guaranteed to complete. Prior work sometimes features either (i) or (ii), but not both. To analyze our algorithms (from a worst case perspective) we introduce a new measure of complexity that depends on the number of actual conflicts only. In addition, we show that even a non-constant approximation of the length of an optimal (shortest) schedule of a set of transactions is NP-hard – even if all transactions are known in advance and do not alter their resource requirements. Furthermore, in case the needed resources of a transaction varies over time, such that for a transaction the number of conflicting transactions increases by a factor k , the competitive ratio of any contention manager is $\Omega(k)$ for $k < \sqrt{m}$, where m denotes the number of cores.

1 Introduction

Designing and implementing concurrent programs is one of the biggest challenges a programmer can face. Transactional memory promises to resolve a couple of the difficulties by ensuring correctness and fast progress of computation at the same time. Transactions have been in use for database systems for a long time. They share several similarities with transactional memory. For instance, in case of a *conflict* (i.e. one transaction demanding a resource held by another) a transaction might get aborted and all the work done so far is lost, i.e. the values of all accessed variables will be restored (to the ones prior to the execution of the transaction).

The difficulty lies in making the right decision when conflicts arise. This task is done by so-called contention managers. They operate in a distributed fashion, that is to say, a separate instance of a contention manager is available for every thread, operating independently. If a transaction A stumbles upon a desired resource, held by another transaction B, it asks its contention manager for advice. We consider three choices for transaction A: (i) A might wait or help B, (ii) A

Y. Dong, D.-Z. Du, and O. Ibarra (Eds.): ISAAC 2009, LNCS 5878, pp. 441–451, 2009.

might abort B or (iii) abort itself. An abort wastes all computation of a transaction and might happen right before its completion. A waiting transaction blocks all other transactions trying to access any resource owned by it.

Our contributions are as follows: First, we show that even coarsely approximating the makespan of a schedule is a difficult task. (Informally, the makespan is the total time it takes to complete a set of transactions.) This holds even in the absence of an adversary. However, in case an adversary is able to modify resource requirements such that the number of conflicting transactions increases by a factor of k, the length of the schedule increases by a factor proportional to k. Second, we propose a complexity measure allowing more precise statements about the complexity of a contention management algorithm. Existing bounds on the makespan, for example, do not guarantee to be better than a sequential execution. However, we argue that since the complexity measure only depends on the number of (shared) resources overall, it does not capture the (local) nature of the problem well enough. In practice, the total number of (shared) resources may be large, though each single transaction might conflict with only a few other transactions. In other words, a lot of transactions can run in parallel, whereas the current measure only guarantees that one transaction runs at a time until commit. Third, we point out weaknesses of widely used contention managers. For instance, some algorithms schedule certain sets of transactions badly, while others require all transactions – also those facing no conflicts – to modify a global counter or access a global clock. Thus the amount of parallelism declines more and more with a growing number of cores. Fourth, we state and analyze two algorithms. Both refrain from using globally shared data. From a worst-case perspective, the randomized algorithm *RandomizedRounds* improves on existing contention managers drastically (exponentially) if for each transaction the number of conflicting transactions is small. In an extended version of this paper [14] we also show that to achieve a short makespan (from a worst case perspective) it is necessary to detect and handle all conflicts early, i.e. for every conflict a contention manager must have the possibility to abort any of the conflicting transactions.

2 Related Work

Transactional memory was introduced in the nineties [8,16]. In 2003 the FSTM system was proposed [6] and also the Dynamic STM (DSTM)[7] for dynamic data structures was described, which suggests the use of a contention manager as an independent module. After these milestones, a lot of systems have been proposed. An overview of design issues from a practical point of view can be found in [3].

Most proposed contention managers have been assessed by specific benchmarks only, and not analytically. A comparison of contention managers based on benchmarks can be found in [12,10]. The experiments yield best performance for randomized algorithms, which all leave a (small) chance for arbitrary large completion time. Apart from that, the choice of the best contention manager varies with the considered benchmark. Still, an algorithm called Polka [12] exhibits good overall performance for a variety of benchmarks and has been used successfully in various systems, e.g. [2,10]. In [10] an algorithm called SizeMatters is introduced, which gives higher priority to the transaction that has modified more (shared) memory. In an enhanced version of this paper [14] we show that from a worst-case perspective Polka and SizeMatters may perform exponentially worse than *RandomizedRounds*. In [4] the effects of selfishness among programmers on the makespan is investigated for various contention managers from a game theoretic perspective.

The first analysis of a contention manager named Greedy was given in [5], using time stamps to decide in favor of older transactions. Variants of time-stamping algorithms had been known previously (also in the field of STM [12]). However, [5] guaranteed that a transaction commits within bounded time and that the competitive ratio (i.e. the ratio of the makespan of the schedule defined by an online scheduler and by an optimal offline scheduler, knowing all transactions in advance) is $O(s^2)$, where s is the number of (shared) resources of all transactions *together*. The analysis was improved to $O(s)$ in [1]. In contrast to our contribution, access to a global clock or logical counter is needed for every transaction which clearly limits the possible parallelism with a growing number of cores. In [11] a scalable replacement for a global clock was presented using synchronized clocks. Unfortunately, these days most systems come without multiple clocks. Additionally, there are problems due to the drift of physical clocks.

Also in [1] a matching lower bound of $\Omega(s)$ for the competitive ratio of any (also randomized) algorithm is proven, where the adversary can alter resource requests of waiting transactions. We show that, more generally, if an adversary can reduce the possible parallelism (i.e., the number of concurrently running transactions) by a factor k, the competitive ratio is $\Omega(k)$ for deterministic algorithms and for randomized algorithms the expected ratio is $\Omega(\min\{k, \sqrt{m}\})$, where m is the number of cores. In the analysis of [1] an adversary can change the required resources such that instead of $\Omega(s)$ transactions only $O(1)$ can run in parallel, i.e. all of a sudden $\Omega(s)$ transactions write to the same resource. Though, indeed the needed resources of transactions do vary over time, we believe that the reduction in parallelism is rarely that high. Dynamic data structure such as (balanced) trees and lists usually do not vary from one extreme to the other. Therefore our lower bound directly incorporates the power of the adversary.

Furthermore, the complexity measure is not really satisfying, since the number of (shared) resources in total is not correlated well to the actual conflicting transactions an individual transaction potentially encounters. As a concrete example, consider the classical dining philosophers problem, where there are n unit length transactions sharing n resources, such that transaction T_i demands resource R_i as well as $R_{(i+1) \bmod n}$ exclusively. An optimal schedule finishes in constant time $O(1)$ by first executing all even transactions and afterwards all odd transactions. The best achievable bound by any scheduling algorithm using the number of shared resources as complexity measure is only $O(n)$. Furthermore, with our more local complexity measure, we prove that for a wide variety of scheduling tasks, the guarantee for algorithm Greedy is linearly worse, whereas our randomized algorithm *RandomizedRounds* is only a factor $\log n$ off the optimal, with high probability.

We relate the problem of contention management to coloring, where a large amount of distributed algorithms are available in different models of communication and for different graphs [9,13,15]. Our algorithm *RandomizedRounds* essentially computes a $O(\max\{\Delta, \log n\})$ coloring for a graph with maximum degree Δ.

For further related work in respect to online scheduling, transactional memory systems and coloring, see [14].

3 Model

A set of *transactions* $S_T := \{T_1, ..., T_n\}$ sharing up to s *resources* (such as memory cells) are executed on m *processors* $P_1, ..., P_m$.[1] For simplicity of the analysis we assume that a single processor runs one *thread* only, i.e., in total at most m threads are running concurrently. If a thread running on processor P_i creates transactions $T_0^{P_i}, T_1^{P_i}, T_2^{P_i}, ...$ one after the other, all of them are executed sequentially on the same processor, i.e., transaction $T_j^{P_i}$ is executed as soon as $T_{j-1}^{P_i}$ has completed, i.e. committed.

The *duration* of transaction T is denoted by t_T and refers to the time T executes until commit without contention (or equivalently, without interruption). The length of the longest transaction of a set S of transactions is denoted by $t_S^{max} := \max_{K \in S} t_K$. If an adversary can modify the duration of a transaction arbitrarily during the execution of the algorithm, the competitive ratio of any online algorithm is unbounded: Assume two transactions T_0 and T_1 face a conflict and an algorithm decides to let T_0 wait (or abort). The adversary could make the opposite decision and let T_0 proceed such that it commits at time t_0. Then it sets the execution time T_0 to infinity, i.e., $t_{T_0} = \infty$ after t_0. Since in the schedule produced by the online algorithm, transaction T_0 commits after t_0 its execution time is unbounded. Therefore, in the analysis we assume that t_T is fixed for all transactions T.[2] We consider an *oblivious adversary* that knows the (contention management) algorithm, but does not get to know the randomized choices of the algorithm before they take effect.

Each transaction consists of a sequence of operations. An operation can be a read or write access of a shared resource R or some arbitrary computation. A value written by a transaction T takes effect for other transactions only after T commits. A transaction either successfully finishes with a commit after executing all operations and acquiring all modified (written) resources or unsuccessfully with an abort anytime. A resource can be acquired either once it is used for the first time or at latest at commit time. A resource can be read in parallel

[1] Transactions are sometimes called jobs, and machines are sometimes called cores.

[2] In case the running time depends on the state/value of the resources and therefore the duration varied by a factor of c, the guarantees for our algorithms (see Section 6) would worsen only by the same factor c.

by arbitrarily many transactions. A read of transaction A of resource R is *visible*, if another transaction B accessing R after A is able to detect that A has already read R. We assume that conflicts that all reads are visible. In fact, we prove in [15] that systems with invisible readers can be very slow. To perform a write, a resource must be acquired exclusively. Only one transaction at a time can hold a resource exclusively. This leads to the following types of *conflicts*: (i) Read-Write: A transaction B tries to write to a resource that is read by another transaction A. (ii) Write-Write: A transaction tries to write to a resource that is already held exclusively (written) by another transaction, (iii) Write-Read: A transaction tries to read a resource that is already held exclusively (write) by another transaction. A contention manager comes into play if a conflict occurs and decides how to resolve the conflict. It can make a transaction wait (arbitrarily long), or abort, or assist the other transaction. We do not explicitly consider the third option. Helping requires that a transaction can be parallelized effectively itself, such that multiple processors can execute the same transaction in parallel with low coordination costs. In general, it is difficult to split a transaction into subtasks that can be executed in parallel. Consequently, state of the art systems do not employ helping. If a transaction gets aborted due to a conflict, it restores the values of all modified resources, frees its resources and restarts from scratch with its first operation. A transaction can request different resources in different executions or change the requested resource while waiting for another transaction.

We assume that a transaction notices a conflict once it actually occurs and a contention manager is called right away, i.e. *eagerly*.[3] A transaction keeps a resource locked until commit, i.e. *no early release*. By introducing additional writes in our examples, any transaction indeed cannot release its resources before commit.

A *schedule* shows for each processor P at any point in time whether it executes some transaction $T \in S_T$ or whether it is idle. The *makespan* of a schedule for a set of transactions S_T is defined as the duration from the start of the schedule until all transactions S_T have committed. We say a schedule for transactions S_T is *optimal*, if its makespan is minimum possible. We measure the quality of a contention manager in terms of the makespan. A contention manager is *optimal*, if it produces an optimal schedule for every set of transactions S_T.

4 Lower Bounds

Before elaborating on the problem complexity of contention management, we introduce some notation related to graph theory and scheduling. We show that even coarse approximations are NP-hard to compute. In [14] we give a lower bound of $\Omega(n)$ for the competitive ratio of algorithms Polka, SizeMatters and Greedy, which holds even if resource requirements remain the same over time.

[3] Even for "typical" cases neither eager nor lazy conflict handling consistently outperforms the other.

4.1 Notation

We use the notion of a *conflict graph* $G = (S, E)$ for a subset $S \subseteq S_T$ of transactions executing concurrently, and an edge between two conflicting transactions. The neighbors of transaction T in the conflict graph are denoted by N_T and represent all transactions that have a conflict with transaction T in G. The degree d_T of a transaction T in the graph corresponds to the number of neighbors in the graph, i.e., $d_T = |N_T|$. We have $d_T \leq |S| \leq \min\{m, n\}$, since at most m transactions can run in parallel, and since there are at most n transactions, i.e., $|S_T| = n$. The maximum degree Δ denotes the largest degree of a transaction, i.e., $\Delta := \max_{T \in S} d_T$. The term t_{N_T} denotes the total time it takes to execute all neighboring transactions of transaction T sequentially without contention, i.e., $t_{N_T} := \sum_{K \in N_T} t_K$. The time $t^+_{N_T}$ includes the execution of T, i.e., $t^+_{N_T} = t_{N_T} + t_T$. Note that the graph G is highly dynamic. It changes due to new or committed transactions or even after an abort of a transaction. Therefore, by d_T we refer to the maximum size of a neighborhood of transaction T that might arise in a conflict graph due to any sequence of aborts and commits. If the number of processors equals the number of transactions ($m = n$), all transactions can start concurrently. If, additionally, the resource requirements of transactions stay the same, then the maximum degree d_T can only decrease due to commits. However, if the resource demands of transactions are altered by an adversary, new conflicts might be introduced and d_T might increase up to $|S_T|$.

4.2 Problem Complexity

If an adversary is allowed to change resources after an abort, such that all restarted transactions require the same resource R, then for all aborted transactions T we can have $d_T = \min\{m, n\}$. This means that no algorithm can do better than a sequential execution (see lower bound in [1]).

We show that even if the adversary can only choose the initial conflict graph and does not influence it afterwards, it is computationally hard to get a reasonable approximation of an optimal schedule. Even, if the whole conflict graph is known and fixed, the best approximation of the schedule obtainable in polynomial time is exponentially worse than the optimal. The claim follows from a straight forward reduction to coloring. For the proof we refer to [14].

Theorem 1. *If the optimal schedule requires time k, it is NP-hard to compute a schedule of makespan less than $k^{\frac{\log k}{25}}$ (for sufficiently large constants), even if the conflict graph is known and transactions do not change their resource requirements.*

4.3 Power of the Adversary

We show that if the conflict graph can be modified, the competitive ratio is proportional to the possible change of a transaction's degree. Initially, a contention manager is not aware of any conflicts. Thus, it is likely to schedule (many) conflicting transactions. All transactions that faced a conflict (and aborted) change their resources on the next restart and require the same resource. Thus they must run sequentially. The contention manager might schedule transactions arbitrarily – in particular it might delay any transaction for an arbitrary amount of time (even before it executed the first time). The adversary has control of the initial transactions and can state how they are supposed to behave after an abort (i.e. if they should change their resource requirements). During the execution, it cannot alter its choices. Furthermore, we limit the power of the adversary as follows: Once the degree of a transaction T has increased by a factor of k, no new conflicts will be added for T, i.e. all initial proposals by the adversary for resource modifications augmenting the degree of T are ignored from then on.

The proof of the following theorem can be found in[14].

Theorem 2. *If the conflict graph can be modified by an oblivious adversary such that the degree of any transaction is increased by a factor of k, any deterministic contention manager has competitive ratio $\Omega(k)$ and any randomized has $\Omega(\min\{k, \sqrt{m}\})$.*

5 Algorithms

Our first algorithm *CommitRounds* (Section 5.1) gives assertions for the response time of individual transactions, i.e., how long a transaction needs to commit. Although we refrain from using global data and we can still give guarantees on the makespan, the result is not satisfying from a performance point of view, since the worst-case bound on the makespan is not better than a sequential execution. Therefore we derive a randomized algorithm *RandomizedRounds* (Section 5.2) with better performance.

Algorithm Commit Rounds (*CommitRounds*)

On conflict of transaction T^{P_i} with transaction T^{P_j}:
  $c_{P_i}^{max} := \max\{c_{P_i}^{max}, c_{P_j}^{max}\}$
  $c_{P_j}^{max} := c_{P_i}^{max}$
  if $c_{P_i} < c_{P_j} \vee (c_{P_i} = c_{P_j} \wedge P_i < P_j)$
    then Abort transaction T^{P_j}
    else Abort transaction T^{P_i}
  end if

After commit of transaction T^P:
  $c_P^{max} := c_P^{max} + 1$
  $c_P := c_P^{max}$

5.1 Deterministic Algorithm *CommitRounds*

The idea of the algorithm is to assign priorities to processors, i.e. a transaction T^P running on a processor P inherits P's priority, which stays the same until the transaction committed. When T commits, P's priority is altered, such that any transaction K having had a conflict with transaction T will have higher priority than all following transactions running on P. For the first execution of the first transaction on processor P_i, the variable $c_{P_i}^{max}$ and c_{P_i} are initialized with 0. We refer to the pseudocode of algorithm *CommitRounds* for details and to [14] for a more detailed textual description.

Algorithm Randomized Rounds (*RandomizedRounds*)

```
procedure Abort(transaction T, K)
  Abort transaction K
  K waits for T to commit or abort before restarting
end procedure

On (re)start of transaction T:
  x_T := random integer in [1, m]

On conflict of transaction T with transaction K:
  if x_T < x_K then Abort(T, K)
    else Abort(K, T)
  end if
```

5.2 Randomized Algorithm *RandomizedRounds*

For our randomized algorithm *RandomizedRounds* a transaction chooses a discrete number uniformly at random in the interval $[1, m]$ on start up and after every abort. In case of a conflict the transaction with the smaller random number proceeds and the other aborts. The routine Abort(transaction T, K) aborts transaction K. Moreover, K must hold off on restarting until T committed or aborted.

To incorporate priorities set by a user, a transaction simply has to modify the interval from which its random number is chosen. For example, choosing from $[1, \lfloor \frac{m}{2} \rfloor]$ instead of $[1, m]$ doubles the chance of succeeding in a round.[4]

6 Analysis

We study two classic efficiency measures of contention management algorithms, the makespan (the total time to complete a set of transactions) and the response time of the system (how long it takes for an individual transaction to commit).

[4] Any interval yields the same guarantees on the makespan as long as the number of distinct possible (random) values is at least m, i.e., the maximal number of parallel running jobs.

6.1 Deterministic Algorithm *CommitRounds*

Theorem 3. *Any transaction will commit after being in the system for a duration of at most* $2 \cdot m \cdot t_{S_T}^{max}$.

Proof. When transaction T^{P_i} runs and faces a conflict with a transaction T^{P_j} having lower priority than T^{P_i} i.e., $c_{P_i} < c_{P_j}$ or $c_{P_i} = c_{P_j}$ and also $P_i < P_j$, then T^{P_j} will lose against T^{P_i}. If not, transaction T^{P_j} will have $c_{P_j}^{max} \geq c_{P_i}^{max} \geq c_{P_i}$ after winning the conflict. Thus at latest after time $t_{S_T}^{max}$ one of the following two scenarios will have happened: The first is that T^{P_j} has committed and all transactions running on processor P_j later on will have $c_{P_j} > c_{P_j}^{max} \geq c_{P_i}^{max} \geq c_{P_i}$. The second is that T^{P_j} has had a conflict with another transaction T^{P_k} for which will also hold that $c_{P_k}^{max} \geq c_{P_i}^{max}$ after the conflict. Thus after time $t_{S_T}^{max}$ either a processor has got to know $c_{P_i}^{max}$ (or a larger value) or committed knowing $c_{P_i}^{max}$ (or a larger value). In the worst-case one processor after the other gets to know $c_{P_i}^{max}$ within time $t_{S_T}^{max}$, taking time at most $m \cdot t_{S_T}^{max}$ and then all transactions commit one after the other, yielding the bound of $2 \cdot m \cdot t_{S_T}^{max}$.

6.2 Randomized Algorithm *RandomizedRounds*

To analyze the response time, we use a complexity measure depending on local parameters, i.e., the neighborhood in the conflict graph (for definitions see Section 4.1).

Theorem 4. *The time span a transaction* T *needs from its first start until commit is* $O(d_T \cdot t_{N_T^+}^{max} \cdot \log n)$ *with probability* $1 - \frac{1}{n^2}$.

Proof. Consider an arbitrary conflict graph. The chance that for a transaction T no transaction $K \in N_T$ has the same random number given m discrete numbers are chosen from an interval $[1, m]$ is: $p(\nexists K \in N_T | x_K = x_T) = (1 - \frac{1}{m})^{d_T} \geq (1 - \frac{1}{m})^m \geq \frac{1}{e}$. We have $d_T \leq \min\{m, n\}$ (Section 4.1). The chances that x_T is at least as small as x_K of any transaction $K \in N_T$ is $\frac{1}{d_T+1}$. Thus the chance that x_T is smallest among all its neighbors is at least $\frac{1}{e \cdot (d_T+1)}$. If we conduct $y = 32 \cdot e \cdot (d_T + 1) \cdot \log n$ trials, each having success probability $\frac{1}{e \cdot (d_T+1)}$, then the probability that the number of successes X is less than $16 \cdot \log n$ becomes (using a Chernoff bound): $p(X < 16 \cdot \log n) < e^{-2 \cdot \log n} = \frac{1}{n^2}$

The duration of a trial, i.e., the time until T can pick a new random number, is at most the time until the first conflict occurs, i.e., the duration t_T plus the time T has to wait after losing a conflict, which is at most $t_{N_T}^{max}$. Thus the duration of a trial is bounded by $2 \cdot t_{N_T^+}^{max}$.

Theorem 5. *If* n *transactions* $S = \{T^{P_0}, ..., T^{P_n}\}$ *run on* n *processors, then the makespan of the schedule by algorithm* RandomizedRounds *is* $O(\max_{T \in S_T}(d_T \cdot t_{N_T^+}^{max}) \cdot \log n) + t_{last}$ *with probability* $1 - \frac{1}{n}$, *where* t_{last} *is the time, when the latest transaction started to execute.*

Proof. Once all transactions are executing, we can use Theorem 4 to show that $p(\exists K \in S$ finishing after $O(\max_{T\in S} d_T \cdot t^{max}_{N_T^+}) \cdot \log n) < \frac{1}{n}$. In the proof of Theorem 4, we showed that for any transaction T: $p(T$ finishes after $O(d_T \cdot t^{max}_{N_T^+} \cdot \log n) < \frac{1}{n^2}$. Since $O(d_T \cdot t^{max}_{N_T^+} \cdot \log n) \leq O(\max_{T\in S}(d_T \cdot t^{max}_{N_T^+}) \cdot \log n)$ we have $p(T$ finishes after $O(\max_{T\in S}(d_T \cdot t^{max}_{N_T^+}) \cdot \log n) < \frac{1}{n^2}$. The chance that no transaction out of all n transactions exceeds the bound of $O(\max_{T\in S}(d_T \cdot t^{max}_{N_T^+}) \cdot \log n)$ is $(1 - \frac{1}{n^2})^n \geq 1 - \frac{1}{n}$.

The theorem shows that if an adversary can increase the maximum degree d_T by a factor of k the running time also increases by the same factor. The bound still holds if an adversary can keep the degree constantly at d_T despite committing transactions. In practice, the degree might also be kept at the same level due to new transactions entering the system. In case, we do not allow any conflicts to be added to the initial conflict graph, the bound of Theorem 5 (and also the one of Theorem 4) can be improved to $O(\max_{T\in S_T}(\max\{d_T, \log n\} \cdot t^{max}_{N_T^+}))$, with an analogous derivation as in [15]. As explained in [14] the schedule corresponds then to a coloring using $O(\max\{\Delta, \log n\})$ colors.

Let us consider an example to get a better understanding of the bounds. Assume we have n transactions starting on n processors having equal length t. All transactions only need a constant amount of resources exclusively and each resource is only required by a constant number of transactions, i.e., d_T is a constant for all transactions T – as is the case in the dining philosophers problem mentioned in Section 2. Then the competitive ratio is $O(\log n)$, whereas it is $O(n)$ for the Greedy, Polka and SizeMatters algorithms as shown in [14].

References

1. Attiya, H., Epstein, L., Shachnai, H., Tamir, T.: Transactional contention management as a non-clairvoyant scheduling problem. In: PODC (2006)
2. Damron, P., Fedorova, A., Lev, Y., Luchangco, V., Moir, M., Nussbaum, D.: Hybrid transactional memory. In: ASPLOS (2006)
3. Dice, D., Shavit, N.: Understanding Tradeoffs in Software Transactional Memory. Symp. on Code Generation and Optimization (2007)
4. Eidenbenz, R., Wattenhofer, R.: Good Programming in Transactional Memory: Game Theory Meets Multicore Architecture. In: ISAAC (2009)
5. Guerraoui, R., Herlihy, M., Pochon, B.: Toward a theory of transactional contention managers. In: PODC (2005)
6. Harris, T., Fraser, K.: Language support for lightweight transactions. In: OOPSLA Conference (2003)
7. Herlihy, M., Luchangco, V., Moir, M., Scherer, W.: Software transactional memory for dynamic-sized data structures. In: PODC (2003)
8. Herlihy, M., Moss, J.: Transactional Memory: Architectural Support For Lock-free Data Structures. In: Symp. on Computer Architecture (1993)
9. Kuhn, F.: Weak Graph Coloring: Distributed Algorithms and Applications. In: SPAA (2009)

10. Ramadan, H., Rossbach, C., Porter, D., Hofmann, O., Bhandari, A., Witchel, E.: MetaTM/TxLinux: transactional memory for an operating system. In: Symp. on Computer Architecture (2007)
11. Riegel, T., Fetzer, C., Felber, P.: Time-based Transactional Memory with Scalable Time Bases. In: Parallel Algorithms and Architectures (2007)
12. Scherer, W., Scott, M.: Advanced contention management for dynamic software transactional memory. In: PODC (2005)
13. Schneider, J., Wattenhofer, R.: A Log-Star Distributed Maximal Independent Set Algorithm for Growth-Bounded Graphs. In: PODC (2008)
14. Schneider, J., Wattenhofer, R.: Bounds On Contention Management Algorithms. TIK Technical Report 311 (2009), `ftp://ftp.tik.ee.ethz.ch/pub/publications/TIK-Report-311.pdf`
15. Schneider, J., Wattenhofer, R.: Coloring Unstructured Wireless Multi-Hop Networks. In: PODC (2009)
16. Shavit, N., Touitou, D.: Software transactional memory. Distributed Computing 10 (1997)

Algorithmic Folding Complexity

Jean Cardinal[1], Erik D. Demaine[2], Martin L. Demaine[2], Shinji Imahori[3], Stefan Langerman[1,*], and Ryuhei Uehara[4]

[1] Université Libre de Bruxelles (ULB)
B-1050 Brussels, Belgium
{jcardin,slanger}@ulb.ac.be
[2] MIT
Cambridge, MA 02139, USA
{edemaine,mdemaine}@mit.edu
[3] University of Tokyo
Tokyo 113-8656, Japan
imahori@mist.i.u-tokyo.ac.jp
[4] Japan Advanced Institute of Science and Technology (JAIST)
Ishikawa 923-1292, Japan
uehara@jaist.ac.jp

Abstract. How do we most quickly fold a paper strip (modeled as a line) to obtain a desired mountain-valley pattern of equidistant creases (viewed as a binary string)? Define the *folding complexity* of a mountain-valley string as the minimum number of simple folds required to construct it. We show that the folding complexity of a length-n *uniform* string (all mountains or all valleys), and hence of a length-n *pleat* (alternating mountain/valley), is polylogarithmic in n. We also show that the maximum possible folding complexity of any string of length n is $O(n/\lg n)$, meeting a previously known lower bound.

1 Introduction

What is the best way to fold an origami model? Origamists around the world struggle with this problem daily, searching for clever, more accurate, or faster folding sequences and techniques. Many advanced origami models require substantial *precreasing* of a prescribed mountain–valley pattern (getting each crease folded slightly in the correct direction), and then folding all the creases at once. For example, in his instructional video for folding the MIT seal *Mens et Manus* in "three easy steps" [3], Brian Chan spends about three hours precreasing, then three hours folding those creases, and then four hours of artistic folding. The precreasing component is particularly tedious, leading us to a natural algorithmic problem of *optimal precreasing*: what is the fastest way to precrease a given mountain–valley pattern? Although the standard method of "fold one crease, unfold, repeat" is usually the most accurate, it might be possible to fold the paper along some of the desired creases to bring several other desired creases into alignment, and thereby precrease them all at once.

* Maître de recherches du F.R.S.-FNRS.

Y. Dong, D.-Z. Du, and O. Ibarra (Eds.): ISAAC 2009, LNCS 5878, pp. 452–461, 2009.

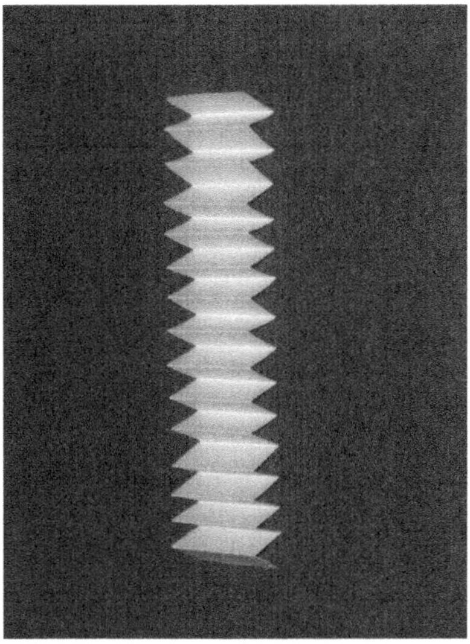

(a) How fast can we fold this?

(b) An origami angel with many pleats, folded by Takashi Hojyo (reproduced with his kind permission).

Fig. 1. Pleats

We focus here on a simple kind of one-dimensional precreasing, where the piece of paper is a long rectangular strip, which can be abstracted into a line segment, and the creases uniformly subdivide the strip. A mountain-valley pattern is then simply a binary string over the alphabet $\{M, V\}$ (M for mountain, V for valley), which we call a *mountain-valley string*. Of particular interest in origami is the *pleat*, which alternates $MVMVMV\cdots$; see Figure 1.

Our results. In this paper, we develop surprisingly efficient algorithms for precreasing a mountain-valley string, especially the pleat.

First, we show how to fold a uniform mountain–valley string $MMM\cdots$ of n mountain creases using just $O(\lg^{1+\sqrt{2}} n)$ simple fold operations. These operations fold only along desired creases, and the last direction that each crease gets folded is mountain. By combining two executions of this algorithm, we obtain the same bound for pleats. This folding is exponentially faster than both the standard folding and the best known folding of $O(n^{\varepsilon})$ folds [7]. From a complexity-theoretic perspective, this is the first polynomial-time algorithm for pleat folding, because the only input is the number n.

Second, we show how to fold an arbitrary mountain–valley string of n creases using just $O(n/\lg n)$ folds. This algorithm is the first to beat the straightforward $n-1$ upper bound by more than a constant factor, and is asymptotically optimal

[7]. We effectively exploit that every string has some redundancy in it, similar to how Lempel-Ziv can compress any string into $O(n/\lg n)$ block pointers.

Unfortunately, our algorithms are not about to revolutionize pleat folding or other practical paper precreasing, because they assume ideal zero-thickness paper. In reality, folding more than a few layers of paper leads to some inaccuracy in the creases, called *creep* in origami circles, and our algorithms require folding through $\Theta(n)$ layers. Nonetheless, our results lead the way for the development of practical algorithms that limit the number k of layers that can be folded through simultaneously, with speed increasing as k grows.

From an information-theory perspective, paper folding offers an intriguing new definition of the algorithmic complexity of a binary string. The *folding complexity* [7] of a mountain–valley string is the minimum number of folds needed to construct it. Similar to how Kolmogorov complexity compresses a string down to instructions for a Turing machine, folding complexity compresses a string down to instructions for an origamist. Unlike Kolmogorov complexity, however, folding complexity is computable, though its exact computational complexity (between P and EXPTIME) remains open. We lack a specific (deterministic) string whose folding complexity is asymptotically the maximum possible. (The pleat was an early candidate, now known to be far from the worst case.) Nonetheless, our results shed some light on the structure of this new measure.

Related work. Uehara [6] posed the problem we tackle here in August 2008. In March 2009, Ito, Kiyomi, Imahori, and Uehara [7] formalized the problem and made some partial progress. On the positive side, they showed how to fold any mountain–valley string using $\lfloor n/2 \rfloor + \lceil \lg(n+1) \rceil$ folds, a bound we improve on by a logarithmic factor; and they showed how to fold the uniform string and hence a pleat using $O(n^{\varepsilon})$ folds, for any $\varepsilon > 0$.[1] On the negative side, they showed that almost every mountain–valley string requires $\Omega(n/\lg n)$ folds using an information-theoretic argument. We tighten this lower bound to prove that a lead constant factor of 1 suffices, reasonably close to our asymptotically matching upper bound which has a lead constant factor of $4 + \varepsilon$, for any $\varepsilon > 0$.

About n different mountain–valley strings of length n can be folded using the absolute minimum number of folds, $\lceil \lg(n+1) \rceil$. These strings are called *paper folding sequences* and have been studied much earlier [8,4,1].

Model. We follow the folding and unfolding model of [7], which in turn is based on the simple-fold model of Arkin et al. [2] (see also [5, p. 225]).

The paper strip is a one-dimensional line with creases at every integer position. We are allowed to fold only at those positions, possibly many times, and the direction of a crease (in $\{M, V\}$) at the end of the algorithm is the one that was folded last. The goal is to achieve a particular string of Ms and Vs at the end. Folding has the effect of superimposing the layers of paper on the right and left of the crease; see Figure 2. The paper has zero thickness, and thus an arbitrary

[1] A somewhat more careful analysis shows that the same algorithm uses $2^{O(\sqrt{\lg n \lg \lg n})}$ folds.

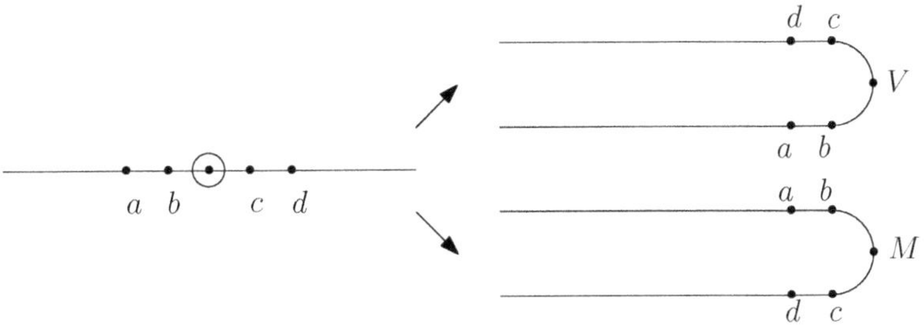

Fig. 2. Folding a mountain or a valley

number of layers can be folded simultaneously. Naturally, when many layers are folded, every other layer is reversed, and the direction of the crease for these layers is flipped. Finally we can, at any step, unfold the paper at zero cost, in the reverse order in which it was folded, and this does not change the directions of the creases. The complexity of the algorithm is the number of folds. (We do not count unfold operations, though doing so would increase the cost by only a constant factor.)

Several variants of this model arise from varying the allowed folding and unfolding operations. Clearly, a fold can be made simultaneously on several layers of the folded strip, allowing a form of parallelism. (Without this, we need n folds for any pattern.) We distinguish between two models for the folding operation.

All-layers fold model: We simultaneously fold all layers of paper under the crease point.

Some-layers fold model: We simultaneously fold the k successive layers immediately beneath the crease point, for any desired number k.

For upper bounds, we concentrate on the more-restrictive all-layers fold model, while for lower bounds, we use the more-flexible some-layers fold model. We also introduce the following three alternatives for the allowed unfolding operations.

All-unfold model: Once we decide to unfold, the paper is unfolded completely.

Reverse-unfold model: We can rewind any number of the last folds as far as we want.

General-unfold model: For a folded state s, we can obtain another folded state t by one general unfolding operation, provided s can be obtained from t by consecutive some-layers simple foldings.

The general-unfold model is the most realistic, but our algorithms require only the power of the reverse-unfold model, so we focus entirely on that model.

A folding algorithm is *end-free* if it can be applied to an infinite paper strip, viewed as a line instead of a line segment, with n equally spaced creases along it, without creasing anywhere else. We discuss end-free algorithms only, which is crucial for allowing us to apply an algorithm recursively (effectively ignoring any surrounding creases).

2 Folding Pleats

Every mountain-valley pattern with n creases requires at least $\lceil \lg(n+1) \rceil$ folds, just to get all the creases folded in some direction [7]. The purpose of this section is to show that the pleat pattern, while seemingly difficult to fold, has a fairly close upper bound. But first we need to show that the lower bound can be met for any n (not just of the form $2^k - 1$):

Lemma 1. *A string of n equally spaced creases (unspecified as mountain or valley) can be folded (and not unfolded) using at most $1 + \lceil \lg(n+1) \rceil$ simple folds in the end-free all-layers fold model.*

Proof. If the number of segments between creases, $n+1$, is 2^k for an integer k, then the folding is the obvious one: fold at the middle crease, k times. Otherwise, $n+1 = 2^k + r$ for some integers k and $0 \leq r < 2^k$. Valley-folding at the rth crease would place the initial r segments on top of the remaining 2^k segments, at which point we could apply the power-of-2 algorithm. But to make the folding end-free, we first mountain-fold at the $\lfloor r/2 \rfloor$th crease, so that these two creases form a zig-zag and the power-of-2 algorithm creases only where desired.

Theorem 1. *The folding complexity of a uniform string and of a pleat of length n in the all-layers fold, reverse-unfold model is $O(\lg^c n)$, where $c = 1 + \sqrt{2}$.*

Proof. Given an algorithm for folding a uniform sequence of n mountains, we can apply it twice to obtain the pleat sequence. We therefore concentrate on finding an algorithm for folding the uniform sequence.

Let $T(n)$ be a monotone upper bound on the number of required folds. Let $w := 2^k - 1$ for some value k. We suppose that $w < \sqrt{n}$. The exact value of w will be determined later.

The algorithm is decomposed into three steps.

Step 1. We apply Lemma 1 to fold the strip into $\lfloor n/w \rfloor$ layers of width w, at a cost of $\lg(n/w) + O(1)$ folds. Then we recurse on these layers, at a cost $T(w)$. After unfolding, we get sequences of w mountains separated by $w+2$ other creases (because half of the layers were reversed). The rest of the algorithm will work only with these creases; there are a final $n \bmod w$ creases which we will deal with at the end of the algorithm simply by recursing. The overall cost of this step (both preprocessing and postprocessing) is at most

$$\lg \frac{n}{w} + O(1) + 2T(w) = 2T(w) + O(\lg n). \tag{1}$$

Step 2. We apply Lemma 1 to a scaled version of the creases in order to align the mountain sequences (possible because the gap between the sequences, $w+2$, is odd). See Figure 3(a), where the lemma's folding is depicted as a zig-zag for simplicity. This folding costs $\lg \frac{n}{2w} + O(1)$ folds. The crease sequences within the layers are composed of three subsequences, the middle one being the aligned

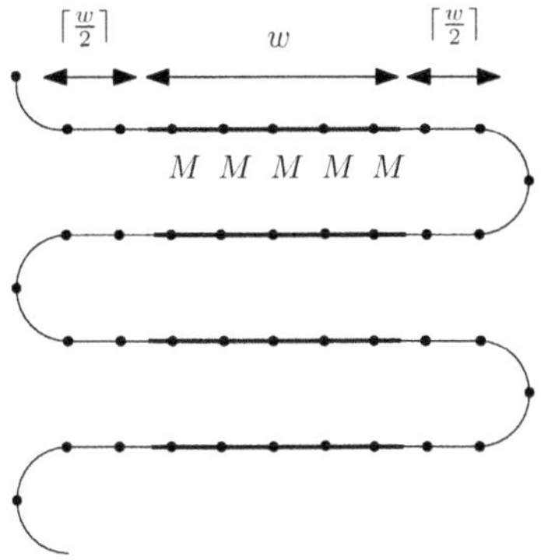

(a) Step 2: Extending the mountain sequences.

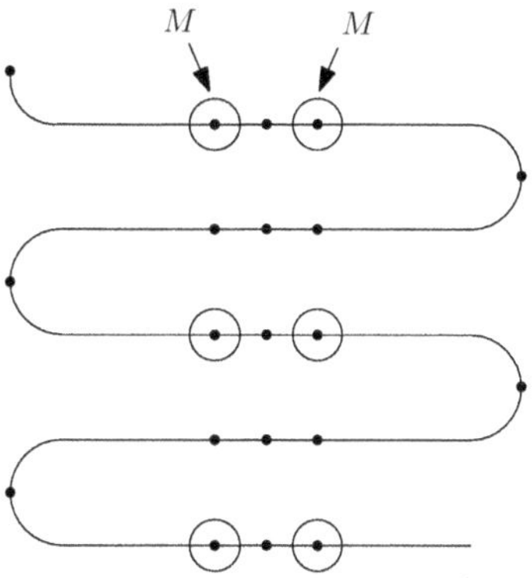

(b) Step 3: Matching triples. After folding a mountain on both sides, the circled creases become mountains.

Fig. 3. Folding pleats

mountains, of length w, and the two others being sequences of length $\lfloor(w+2)/2\rfloor = \lceil w/2\rceil = 2^{k-1}$. We call the two others *side* sequences.

We recurse again on the side sequences, folding mountains on the right, and valleys on the left. (Here we use the end-free property.) This folding costs $2T(\lceil w/2\rceil)$ folds, and has the effect of extending the lengths of the mountain sequences. Thus, after these recursive calls, we have mountain sequences of length $\lceil 3w/2\rceil$ separated by sequences of length $2^{k-1}+1$.

Next, we unfold everything and iterate these operations of aligning mountain sequences and recursing on both side sequences. This is possible because the distance between the mountain sequences remains odd. The width remains $2w$, so folding still costs $\lg \frac{n}{2w} + O(1)$, but the length of the side sequences decreases by a factor of 2 at every iteration. We iterate until the side sequences have length 1. At that point, we have mountain sequences of length $2w-3$ separated by triples of "junk" creases. Overall, the cost of this step is:

$$\lg w \cdot \left(\lg \frac{n}{2w} + O(1)\right) + 2\sum_i T\left(\left\lfloor \frac{w}{2^i}\right\rfloor\right) \leq \lg w \cdot \left(\lg \frac{n}{2w} + 2T(w)\right) + O(\lg n). \quad (2)$$

Step 3. At the end of the previous step, we are left with sequences of $2w-3$ mountains separated by three junk creases. We can again apply Lemma 1 to align the junk triples, (possible because the gap between the triples, $2w-3$, is odd). This folding costs $\lg \frac{n}{2w} + O(1)$ folds. Considering the aligned triples, we now fold a mountain on the left and right creases of the triples; see Figure 3(b). (More precisely, we fold a mountain on the left, then unfold once, then fold a mountain on the right, then unfold everything.) This procedure costs two folds, and creates mountains on the left and right creases of the layers that are not reversed. Hence, for half the triples, there is now only one junk crease left. We iterate the above steps on the remaining triples only, reducing by a factor of 2 the number of remaining triples at every step. This procedure remains possible because the remaining triples are separated by an odd number of creases, and the middle crease lies within one of the original triples. The overall cost is $\sum_i \left(\lg \frac{n}{2^i w} + O(1)\right)$.

When finished, we are left with all the middle creases of the original triples. By a simple scaling, this is a subproblem of size n/w that can be solved recursively. We therefore incur a final, additional cost of $T(n/w)$. Overall, the complexity of Step 3 is bounded by

$$T\left(\frac{n}{w}\right) + \sum_i \left(\lg \frac{n}{2^i w} + O(1)\right) \leq T\left(\frac{n}{w}\right) + \lg^2 \frac{n}{w} + O(\lg n). \tag{3}$$

Analysis. Summing the complexities of the three previous steps, we get

$$T(n) \leq (2 \lg w + 2)T(w) + \lg \frac{n}{w}\left(\lg w + \lg \frac{n}{w}\right) + T\left(\frac{n}{w}\right) + O(\lg n) \tag{4}$$

$$= (2 \lg w + 2)T(w) + O(\lg^2 n) + T\left(\frac{n}{w}\right). \tag{5}$$

We suppose that $T(n) < \lg^c n$ for some constant $c > 1$. We find a value for w such that the first term is also $O(\lg^2 n)$:

$$(2 \lg w + 2)T(w) < \lg^2 n \tag{6}$$

$$2 \lg^{c+1} w + 2 \lg^c w < \lg^2 n \tag{7}$$

$$w < 2^{b \cdot \lg^{\frac{2}{c+1}} n} \tag{8}$$

for some constant $b > 1$. The recurrence relation for $T(n)$ now becomes:

$$T(n) \leq T\left(\frac{n}{2^{b \cdot \lg^{\frac{2}{c+1}} n}}\right) + O(\lg^2 n). \tag{9}$$

This solves to:

$$T(n) = O(\lg^{\frac{3c+1}{c+1}} n). \tag{10}$$

(The proof is omitted in this version of the paper.)

But we made the hypothesis that $T(n) = O(\lg^c n)$, yielding

$$\frac{3c+1}{c+1} = c \tag{11}$$

$$c = 1 + \sqrt{2}. \tag{12}$$

□

3 Folding Arbitrary Sequences

Ito, Kiyomi, Imahori, and Uehara [7] proved that the folding complexity of a random sequence is $\Omega(n/\lg n)$ with high probability. We now refine this result.

Fig. 4. Partitioning the sequences in odd-length chunks. Note that every pair of chunks is separated by an odd number of creases.

Theorem 2 **([7]).** *The folding complexity of a random sequence of length n in the some-layers fold, reverse-unfold model is at least*

$$\frac{n}{3 + \lg n}$$

with high probability.

Proof. We use a counting argument. Suppose that we make k folds (and k unfolds). We note that, if a sequence of length n is obtained by less than k folds, we can also obtain the sequence using k folds exactly.

At each fold, there are two choices for the direction of folding (mountain or valley). There are at most n choices for the set of positions of the folding by considering the bottom-most layer of a valley folding or the top-most layer of a mountain folding. Thus we have $(2n)^k$ possibilities here.

At each unfolding operation, we may rewind some folding operations. We will rewind k folding operations in total. Moreover, for any step, the number of rewind operations so far cannot exceed the number of folding operations. Thus, there are at most $C(k)$ choices for the unfolding operations, where $C(k)$ is the kth Catalan number. Note that for large k, $C(k) < 4^k$.

Overall, the number of possible sequence of length n obtained by in at most k folds is bounded by $(2n)^k C(k)$. If we let $k = n/(3 + \lg n)$, then, for sufficiently large values of n, $(2n)^k C(k) < (2n)^k 4^k = (8n)^k = 2^n$.

Therefore, the number of possible sequences of length n by at most k folding (and unfolding) operations is less than the number of all sequences of length n. □

We give an upper bound that matches this lower bound up to a constant factor. Given an arbitrary sequence of length n, and an odd number $s \geq 3$, we divide the sequence into *chunks* of size s, each pair of successive chunks being separated by one crease. Note that, because s is odd, every pair of chunks is separated by an odd number of creases (see Figure 4). This will allow us to align any subset of chunks by folding them on top of each other.

Suppose there are at most k distinct patterns among such chunks, that is, the subsequence of creases in any chunk belongs to a set of at most k distinct sequences. We denote by $f(n, k, s)$ the worst-case complexity of folding such a sequence in the some-layers fold, reverse-unfold model.

Lemma 2.

$$f(n, k, s) \leq 4n/s + ks \lg n.$$

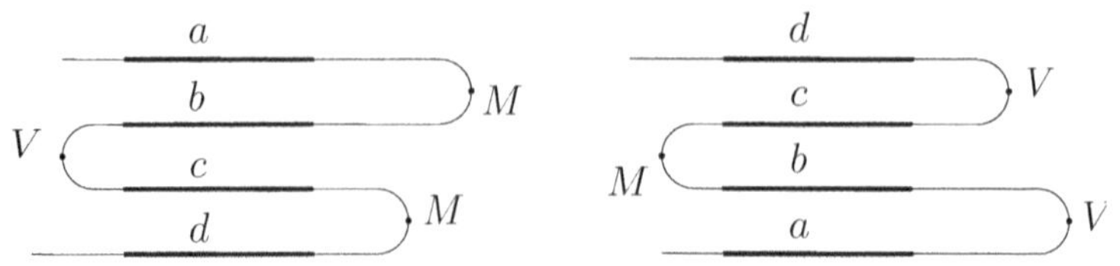

Fig. 5. Two ways to match chunks

Proof. Consider one kind of chunk, and suppose it appears t times. We can fold a zig-zag to match all the t chunks, so that we can fold them together. Note that this is always possible, because the distance between any two chunks is odd. Also note that folding the zig-zag involves some-layers folding operations. The number of required folds for the zig-zag is exactly $t - 1$.

When the chunks are folded, we have to fix the creases that may have been destroyed by the zig-zag. The zig-zag can be folded in at least two different ways (see Figure 5). By choosing one of them, we can ensure that at most one half of the zig-zag creases destroy previously folded creases. Hence the zig-zag costs at most $t - 1 + t/2 \leq 3t/2$ folds: $t - 1$ folds to create it, then $t/2$ folds to fix the creases that have been destroyed.

The folding of the chunks themselves costs s folds. But because they are mapped in alternating directions, only half of them are correctly folded. To fix this, we can recurse on the remaining half. The overall cost is therefore at most

$$(3t/2 + s) + (3t/4 + s) + (3t/8 + s) + \cdots + (1 + s) < 3t + s\lg(3t/2).$$

Now suppose that there are t_i occurrences of the chunk of type i. Note that $\sum_i t_i = n/s$. Thus repeating the steps above for each of the k kinds of chunks, we can fold all the chunks in

$$\sum_{i=1}^{k} \left(3t_i + s\lg\frac{3t_i}{2}\right) \leq 3n/s + ks\lg n$$

folds. Finally, we have to take care of the remaining creases separating the chunks. There are n/s of them. Folding them separately, we obtain

$$f(n,k,s) \leq 4n/s + ks\lg n. \qquad \square$$

This directly yields a $\Theta(n/\lg n)$ upper bound for folding arbitrary sequences.

Theorem 3. *The folding complexity of a binary sequence of length n in the some-layers fold, reverse-unfold model is at most*

$$(4+\varepsilon)\frac{n}{\lg n} + o\left(\frac{n}{\lg n}\right),$$

for any $\varepsilon > 0$.

Proof. Pick $s = (1 - \varepsilon)\lg n$, and $k = 2^s = n^{1-\varepsilon}$ in the formula of Lemma 2, with $\varepsilon := \varepsilon/(4 + \varepsilon)$. This yields:

$$\frac{4n}{(1-\varepsilon)\lg n} + (1-\varepsilon)n^{1-\varepsilon}\lg^2 n = (4+\varepsilon)\frac{n}{\lg n} + o\left(\frac{n}{\lg n}\right). \qquad \square$$

Note that the upper and lower bounds are within a factor 4 of each other. It remains open whether the same upper bound is possible in the all-layers fold model.

Acknowledgments

This work was initiated at the WAFOL'09 workshop in Brussels. We thank all the other participants, as well as Guy Louchard, for useful discussions.

References

1. Allouche, J.-P.: Sur la complexité des suites infinies. Bull. Belg. Math. Soc. 1, 133–143 (1994)
2. Arkin, E.M., Bender, M.A., Demaine, E.D., Demaine, M.L., Mitchell, J.S.B., Sethia, S., Skiena, S.S.: When can you fold a map? Comput. Geom. Theory Appl. 29(1), 23–46 (2004)
3. Chan, B.: The making of Mens et Manus (in origami), vol. 1 (March 2007), `http://techtv.mit.edu/collections/chosetec/videos/361-the-making-of-mens-et-manus-in-origami-vol-1`
4. Dekking, M., Mendès France, M., van der Poorten, A.J.: Folds! Math. Intell. 4, 130–138, 173–181, 190–195 (1982)
5. Demaine, E.D., O'Rourke, J.: Geometric Folding Algorithms. Cambridge University Press, Cambridge (2007)
6. Demaine, E.D., O'Rourke, J.: Open problems from CCCG 2008. In: Proc. 21st Canadian Conference on Computational Geometry, CCCG 2009 (to appear, 2009)
7. Ito, T., Kiyomi, M., Imahori, S., Uehara, R.: Complexity of pleat folding. In: Proc. 25th Workshop on Computational Geometry (EuroCG 2009), pp. 53–56 (2009)
8. Mendès France, M., van der Poorten, A.J.: Arithmetic and analytic properties of paper folding sequences. Bull. Austr. Math. Soc. 24, 123–131 (1981)

Min-Energy Scheduling for Aligned Jobs in Accelerate Model*

Weiwei Wu[1], Minming Li[2], and Enhong Chen[3]

[1] Department of Computer Science, University of Science and Technology of China
Department of Computer Science, City University of Hong Kong
USTC-CityU Joint Research Institute
wweiwei2@cityu.edu.hk
[2] Department of Computer Science, City University of Hong Kong
minmli@cs.cityu.edu.hk
[3] Department of Computer Science, University of Science and Technology of China
cheneh@ustc.edu.cn

Abstract. Dynamic voltage scaling technique provides the capability for processors to adjust the speed and control the energy consumption. We study the pessimistic accelerate model where the acceleration rate of the processor speed is at most K and jobs cannot be executed during the speed transition period. The objective is to find a min-energy (optimal) schedule that finishes every job within its deadline. The job set we study in this paper is aligned jobs where earlier released jobs have earlier deadlines. We start by investigating a special case where all jobs have common arrival time and design an $O(n^2)$ algorithm to compute the optimal schedule based on some nice properties of the optimal schedule. Then, we study the general aligned jobs and obtain an $O(n^2)$ algorithm to compute the optimal schedule by using the algorithm for the common arrival time case as a building block. Because our algorithm relies on the computation of the optimal schedule in the ideal model ($K = \infty$), in order to achieve $O(n^2)$ complexity, we improve the complexity of computing the optimal schedule in the ideal model for aligned jobs from the currently best known $O(n^2 \log n)$ to $O(n^2)$.

1 Introduction

Energy-efficiency has become the first-class design constraint besides the traditional time and space requirements. Portable devices (like laptops and PDAs) equipped with capacity limited batteries are popular in our daily life. Two facts make the energy problem more important. First, the battery capacity is increasing with a rate less than that of power consumption of the processors. Second,

* This work was supported in part by a grant from the Research Grants Council of the Hong Kong Special Administrative Region, China [Project No. CityU 117408], National Natural Science Foundation of China (grant no. 60775037) and the National High Technology Research and Development Program of China (863 Program) (grant no. 2009AA01Z123).

Y. Dong, D.-Z. Du, and O. Ibarra (Eds.): ISAAC 2009, LNCS 5878, pp. 462–472, 2009.

the accumulated heat due to energy consumption will reach a thermal wall and challenge the designers of electronic devices. It is found that, in the CMOS processors, the energy consumption can be saved by executing with a lower speed. Approximately, the speed is a cubic root of the power, which is famous as cube-root-rule. Dynamic voltage scaling (DVS) technique is widely adopted by modern processor manufactures, e.g., Intel, AMD, and IBM. It allows the processor to dynamically adjust its voltage/frequency to control the power consumption. The first theoretical study was initialed decades ago [14], where the authors make the standard generalization, a speed to power function $P(s) = s^\alpha$ ($\alpha \geq 1$). Usually, α is 2 or 3 according to the cube-root-rule of the processors. From then on, lots of studies are triggered in this field. It is usually formulated as a dual objective problem. That is, while conserving the energy, it also needs to satisfy some QoS metric. When all jobs are required to be completed before deadline, the metric is called *deadline feasibility*. There are also works trying to simultaneously minimize the response time of the jobs, namely, *flow*. A schedule consists of the *speed scaling policy* to determine what speed to run at time t and the *job selection policy* to decide which job to run at that time.

If the processor can run at arbitrary speeds, then based on how fast the voltage can be changed, there are two different models.

Ideal Model: It is assumed that the voltage/speed of the processor can be changed to any other value without any extra/physical cost or delay. This model provides an ideal fundamental benchmark and has been widely studied.

Accelerate Model: It is assumed that the voltage/speed change has some delay. In practice, the processor's acceleration capacity is limited. For example, in the low power ARM microprocessor system (lpARM) [5], the clock frequency transition takes approximately $25\mu s$ (1350 cycles) from 10MHz to 100MHz. Within this scope, there are two variations. In the *optimistic model*, the processor can execute jobs during the speed transition time, while in the *pessimistic model*, the execution of jobs in the transition time is not allowed [15].

1.1 Related Works

In recent years, there are many works on the impact of DVS technology. It is not practical to survey all of them, thus we just review the most related papers.

For the ideal model, Yao el. al. first studied the energy-efficient job scheduling to achieve deadline feasibility in their seminal paper [14]. They proposed an $O(n^3)$ time algorithm YDS to compute the optimal off-line schedule. Later on, the running time is improved to $O(n^2 \log n)$ in [12]. Another metric, the response time/flow, was examined in [13] with bounded energy consumption. It is first formulated as a linear single objective (energy+flow) optimization problem in [1]. This was then specifically studied in [4],[9],[6],[2],[3] el. al. under different assumptions. A good survey can be found in [8].

For the accelerate model, there are little theoretical studies to the best of our knowledge, except that the single task problem was studied in [7],[15]. In [7], they showed that the speed function which minimizes the energy is of some restricted shapes even when considering a single task. They also gave some empirical

Lemma 1. *There exists an optimal schedule, where the speed function is non-increasing and will accelerate as fast as possible, i.e., either* $|s'(t)| = K$ *or* $|s'(t)| = 0$*; and the jobs are completed in EDF (Earliest Deadline First) order with* J_i *executed in one speed and only in the lowest speed* $\min_{0 \le t \le d_i, s'(t)=0} s(t)$*. Furthermore, the finishing time* $\hat{t}$ *of each block (where* $\lim_{t \to \hat{t}^-} s'(t) = 0 \wedge \lim_{t \to \hat{t}^+} s'(t) = -K$*) is a tight deadline.*

Lemma 2. *In the optimal schedule,*

1) The first block is the interval $(0, d_t)$ *which maximizes* $\frac{\sum_{J \in \mathcal{J}_t} C(J)}{d_t}$ *where* $\mathcal{J}_t = \{J_j | d_j \le d_t]\}$ *and* $t \in \{1, \ldots, n\}$*, i.e. the maximum speed in the optimal schedule is* $s_1 = \max_i \frac{\sum_{J \in \mathcal{J}_i} C(J)}{d_i}$.

2) Suppose block j *has speed* s_j *and finishes at* J_{t_j}*'s deadline, then the speed in block* $j+1$ *is* $s_{j+1} = \max_t \frac{s_j - K(d_t - d_{t_j}) + \sqrt{(K(d_t - d_{t_j}) - s_j)^2 + 4K \sum_{i=t_j+1}^{t} C(J_i)}}{2}$ *where* $t \in \{t_j + 1, \ldots, n\}$.

Theorem 1. *The optimal schedule can be computed by Algorithm 1 at* $O(n^2)$*.*

Proof. Algorithm 1 is a direct implementation of Lemma 2. Steps 2-4 computes the first block. The two loops in Steps 6-10 computes the remaining blocks. By keeping the information of the summation on the computed jobs, the optimal schedule can be computed in $O(n^2)$ time.

Algorithm 1. *CRT_schedule*

1. $t = 0$;
2. $s_1 = \max_i \frac{\sum_{j=1}^{i} C(J_j)}{d_i}$;
3. $t = \arg\max_i s_1$;
4. Let the block with speed s_1 be $[0, d_t]$;
5. $m = 1$;

while $t < n$ **do**

   6. $s_{m+1} = \max_{t+1 \le i \le n} \frac{s_m - K(d_i - d_t) + \sqrt{(K(d_i - d_t) - s_m)^2 + 4K \sum_{j=t+1}^{i} C(J_j)}}{2}$;
   7. $t' = \arg\max_i s_{m+1}$;
   8. Let the block with speed s_{m+1} be $[d_t + (s_m - s_{m+1})/K, d_{t'}]$;
   9. $m = m + 1$;
   10. $t = t'$;

end while

---

4 Optimal Schedules for Aligned Jobs

In this section, we study the optimal schedule for general aligned jobs. Note that jobs with common arrival time is a special case of aligned jobs. We first extend some basic properties in Subsection 4.1. We will compute the optimal schedule for aligned jobs by adopting Algorithm 1 as a building block. We use OPT_K to denote the optimal schedule where K is the maximum acceleration rate.

In the ideal model, the acceleration rate is infinity $K = \infty$. We review the Algorithm YDS in [14] to compute OPT_∞. Let $w(t_1, t_2)$ denote the workload of the jobs that have release time at least t_1 and have deadline at most t_2, i.e. $w(t_1, t_2) = \sum_{I(J)\subseteq[t_1,t_2]} C(J)$. Define the intensity $Itt(t_1, t_2)$ of the time interval $[t_1, t_2]$ to be $w(t_1, t_2)/(t_2 - t_1)$. The algorithm tries every possible pair of arrival time and deadline to find an interval with largest intensity (called *critical interval*), schedule the jobs embedded in the critical interval and then repeatedly deal with the remaining jobs.

We first show that the optimal schedule for aligned jobs in the ideal model can be computed in $O(n^2)$ time.

Theorem 2. *The optimal schedule for aligned jobs in the ideal model can be computed in $O(n^2)$ time.*

Given a block $block_p$, we denote the corresponding interval as $[L(block_p), R(block_p)]$. We define *virtual canyon* to be a block with length 0. Next, we derive some properties of OPT_K.

4.1 Basic Properties

Among all the blocks, we define the block $[t_a, t_b]$ where $\lim_{t\to t_a^-} s'(t) = K \wedge \lim_{t\to t_a^+} s'(t) = 0$ and $\lim_{t\to t_b^-} s'(t) = 0 \wedge \lim_{t\to t_b^+} s'(t) = -K$ to be *peak*. Reversely, the block where $\lim_{t\to t_a^-} s'(t) = -K \wedge \lim_{t\to t_a^+} s'(t) = 0$ and $\lim_{t\to t_b^-} s'(t) = 0 \wedge \lim_{t\to t_b^+} s'(t) = K$ is called *canyon*.

We say $\hat{t}$ is *down-edge-time* if $\lim_{t\to \hat{t}^-} s'(t) = 0 \wedge \lim_{t\to \hat{t}^+} s'(t) = -K$ or $\lim_{t\to \hat{t}^-} s'(t) = K \wedge \lim_{t\to \hat{t}^+} s'(t) = 0$. For example, both the start time and finish time of a peak are down-edge-times.

We have the following lemma for OPT_K.

Lemma 3. *There is an optimal schedule, where the speed function will accelerate as fast as possible, i.e., either $|s'(t)| = K$ or $s'(t) = 0$, and every down-edge-time is either a tight deadline or a tight arrival time; and jobs are executed in EDF order; each job J is executed only in one block, and this block is the lowest one in interval $[r(J), d(J)]$.*

4.2 $O(n^2)$ Time Algorithm to Compute OPT_K

To find the optimal schedule, our method is to identify some special blocks belonging to OPT_K. After enough blocks are selected, the remaining interval of OPT_K can be easily computed. To be more specific, we compare OPT_K with schedule OPT_∞, which is the optimal schedule for the special case $K = \infty$, namely the ideal model. We observe that the block with the highest speed (we call it *global-peak*) of OPT_K can be computed first.

Algorithm 2. Computing a Monotone-interval

Input: OPT_∞, Schedule computed by YDS.
$[a, b]$, computed peak (it can be the global-peak or local-peak)
s_b, Starting speed at time b.
Output: $S_{[b,t_1]}$, a monotone-interval starting from b and its corresponding schedule.
/* Let $[t_L, t_R]$ be the current interval being handled. Let S be the the computed schedule for current interval $[t_L, t_R]$. s_{last} denotes the lowest speed in the computed S. $block_{p+1}$ is the first un-handled block in OPT_∞.*/
1. $s_{last} = s_b$; $t_L = t_R = b$; $S_{[b,t_1]} = \phi$; $S = \phi$; $p = 0$; $\overline{t} = b$.
2. In OPT_∞, index the blocks from the peak $[a, b]$ as $block_0, block_1, block_2, \ldots$ in the left to right order.
while $s_{last} \geq s(block_{p+1})$ **do**
  /*recover procedure*/
  if $S \neq \emptyset$ **then**
    3. For jobs that are executed in the lowest block of S, recover their arrival time/deadline to the original value.
    4. Reset s_{last} to be the speed of S in time $\overline{t}$;
  end if
  5. Select $block_i$ to be the block after t_R in OPT_∞ with i.e. $s(block_{p+1}) > \ldots > s(block_i)$ and $s(block_i) < s(block_{i+1})$; if such a block does not exist, then let $i = p + 1$; Reset $t_R = R(block_i)$.
  6. Set $p = i$;
  /*adjust procedure*/
  for every job with $I(J) \cap [t_L, t_R] \neq \phi$ **do**
    7. Adjust $r(J)$ to be $\max\{r(J), t_L\}$;
    8. Adjust $d(J)$ to be $\min\{d(J), t_R\}$;
    9. Backup the original value of $r(J)$ and $d(J)$;
  end for
  /*handle interval $[t_L, t_R]$ in OPT_∞*/
  10. Call **Algorithm 1** to compute a schedule S for jobs involved in Step 9 according to common arrival time t_L with starting speed s_{last}.
  11. If the $block_i$ found in Step 6 has speed 0, then we make S accelerate with rate $-K$ after the last time with positive speed and insert a virtual canyon at time t_R.
  12. Reset s_{last} to be the lowest positive speed in the computed S.
  if $s_{last} < s(block_{p+1})$ **then**
    13. $S_{[b,t_1]} = S_{[b,t_1]} \cup S$; Return $S_{[b,t_1]}$.
  else
    14. Let $\overline{t}$ be the finish time of the second lowest (including the virtual canyon inserted in Step 11) block in S.
    15. $S_{[b,t_1]} = S_{[b,t_1]} \cup$ (S restricted in interval $[t_L, \overline{t}]$).
    16. Reset $t_L = \overline{t}$.
  end if
end while

Lemma 4. *OPT_K executes the same as OPT_∞ in the first critical interval.*

After we have fixed the first block (global-peak) of OPT_K, a natural question is whether we can apply the same proof of Lemma 4 to select other blocks. For example, in the remaining interval of OPT_∞, does the block with maximum

Algorithm 3. Computing the Optimal Schedule between Two Adjacent Peaks

Input:
$[a_1, b_1], [a_2, b_2]$, the two adjacent peaks found in Algorithm 4.
$S_{[b_1,t_1]}, S_{[t_2,a_2]}$, the two schedules computed by Algorithm 2. Choose one of the intersection point as $\bar{t}$.
Output: schedule of OPT_K in interval $[b_1, a_2]$.
1. For each down-edge-time p on $s(t, S_{[b_1,t_1]})$ in $[b_1, \bar{t})$ or on $s(t, S_{[t_2,a_2]})$ in $(\bar{t}, a_2]$, let the point with the same speed on the other curve be p'. If there are more than one such point, let p' be the one minimizing $|pp'|$; if there are no such point, we do not consider line segment originating from p.
2. Sort all the segments pp' by increasing order of their speed (denoted by $Speed(p)$) into $p_1p'_1, p_2p'_2, \ldots, p_mp'_m$ (Duplicate segments are treated as one). The end points are relabeled so that p_i is always on $s(t, S_{[b_1,t_1]})$ and p'_i is always on $s(t, S_{[t_2,a_2]})$.
3. Find augment segment for each segment $p_ip'_i$ as follows. If p_i and p'_i are both down-edge-time, then the augment segment is $p_ip'_i$ itself; if p_i is a down-edge-time and p'_i is not, then the augment segment is p_ip' where p' is the closest down-edge-time on $s(t, S_{[t_2,a_2]})$ with respect to p'_i; the remaining case is similarly defined. We use $q_iq'_i$ to represent the augment segment of $p_ip'_i$.
for $i = 1$ to m **do**
  4. Let $C = \sum_{I(J)\cap[q_i,q'_i]\neq\emptyset} C(J)$.
  if $(\frac{C}{|p_ip'_i|} < Speed(p_i))$ **then**
    5. Let $S_{[\hat{t}_1,\hat{t}_2]}$ be the schedule that executes all jobs with $I(J) \cap [p_i, p'_i] \neq \phi$ with speed s in interval $[\hat{t}_1, \hat{t}_2]$.(The parameters can be calculated as $\hat{t}_1 = p_i + T$; $\hat{t}_2 = p'_i - T$; $s = Speed(p_i) - 2KT$; $T = \frac{Speed(p_i)+K|p_ip'_i| - \sqrt{(Speed(p_i)-K|p_ip'_i|)^2+4KC}}{4K}$)
    6. **break**;
  end if
end for
7. The optimal schedule in interval $[b_1, a_2]$ is $(S_{[b_1,\hat{t}_1]}$ restricted to $[b_1, \hat{t}_1]) \cup S_{[\hat{t}_1,\hat{t}_2]} \cup (S_{[t_2,a_2]}$ restricted to $[\hat{t}_2, a_2])$.

intensity have the same schedule as that of OPT_K? Although this is not true, we will show that some other blocks in OPT_∞ can be proved to be the same as OPT_K. The key observation is that by appropriately dividing the whole interval into two sub-intervals, the block with the maximum intensity inside one of the sub-intervals in OPT_∞ can be proved to be the same as OPT_K. Our partition of intervals is based on a *monotone-interval* defined below.

Definition 2. *Given a schedule, we define the sub-interval where the speed function/curve is strictly non-increasing or non-decreasing to be a monotone-interval.*

Since the speed in OPT_K outside the global-peak $[a, b]$ is at most $Itt(a, b)$, there exists a monotone-interval immediately after time b (non-increasing curve) and symmetrically before time a (non-decreasing curve). At time b and a, the speeds are respectively $s_b = Itt(a, b)$ and $s_a = Itt(a, b)$.

In the following, we will study a schedule $S_{[b,t_1]}$ (only specifying speeds in interval $[b, t_1]$) with monotone-interval $[b, t_1]$ (non-increasing speed with

Algorithm 4. Computing the Optimal Schedule for Aligned Jobs

Input: Aligned job set $\mathcal{J}$
Output: OPT_K
1. Compute OPT_∞.
2. Let the maximum intensity block in OPT_∞ be the global-peak in OPT_K .
3. Index the global-peak as an un-handled peak.
while there is a peak [L,R] un-handled **do**
  4. Let OPT_K execute jobs the same way as OPT_∞ in $[L, R]$.
  5. Call **Algorithm 2** to compute the monotone-interval starting from R (and also symmetrically a monotone-interval ending at L).
  6. If there are local-peaks in OPT_∞ in the un-handled interval on either side of the monotone-intervals, then index the local-peaks as un-handled peaks.
end while
7. Compute the OPT_K for all the intervals between adjacent peaks found in the previous while loop using Algorithm 3.

$s(b, S_{[b,t_1]}) = s(b, OPT_\infty))$. Suppose that t_1 is the first (earliest) intersection of the two curves $s(t, S_{[b,t_1]})$ and $s(t, OPT_\infty)$ with $\lim_{t \to t_1^+} s(t, OPT_\infty) > 0$. We will compare the speed curve of OPT_K with that of $S_{[b,t_1]}$.

Definition 3. *In interval $[b, t_1]$, we say t is a separation-time of OPT_K w.r.t $S_{[b,t_1]}$ if their speed curves totally overlap in interval $[b, t]$ and separate at $t + \Delta t$ where $\Delta t \to 0$.*

We can show that the schedule $S_{[b,t_1]}$ with non-increasing speed output by Algorithm 2 has the following property: let $[a_2, b_2]$ be the maximum intensity block in OPT_∞ among the remaining interval $[t_1, t_f]$, then OPT_K has the same schedule as OPT_∞ in interval $[a_2, b_2]$. Furthermore, Algorithm 2 runs in $O(n^2)$.

Among the un-handled interval (e.g. $[t_1, t_f]$), we define *local-peak* to be the peak which has the local maximal intensity in OPT_∞. The following lemma shows that the schedules OPT_K and OPT_∞ are the same in local-peaks.

Lemma 5. *The schedule of local-peaks in OPT_K is the same as OPT_∞.*

Note that there is a monotone-interval respectively before and after the computed global-peak or local-peak. We can repeatedly call Algorithm 2 (a symmetric version of Algorithm 2 can be used to compute a monotone-interval before a "peak") until no such peak exists in the un-handled intervals. Then the schedule of the remaining intervals (all intervals between the adjacent peaks computed in Algorithm 4) can be uniquely computed as shown in Lemma 6.

Lemma 6. *The schedule of OPT_K in intervals between two (local-)peaks found by Algorithm 4 can be computed by Algorithm 3. Notice that in this algorithm, "down-edge-time" means the corresponding point on the speed curve at the down-edge-time.*

Theorem 3. *Algorithm 4 computes OPT_K for aligned jobs in $O(n^2)$ time.*

5 Conclusion

In this paper, we study the energy-efficient dynamic voltage scaling problem and mainly focus on the pessimistic accelerate model and aligned jobs. All jobs are required to be completed before deadlines and the objective is to minimize the energy. We start by examining the properties for the special case where jobs are released at the same time. We show that the optimal schedule can be computed in $O(n^2)$. Based on this result, we study the general aligned jobs. The algorithm for jobs with common arrival time is adopted as an elementary procedure to compute the optimal schedule for general aligned jobs. By repeatedly computing heuristic schedules that is non-increasing, we fix some peaks of the optimal schedule first. This makes the optimal schedule in the remaining interval easier to compute. The complexity of the algorithm is $O(n^2)$ since we improve the computation of the optimal schedule for aligned jobs in the ideal model to $O(n^2)$.

References

1. Albers, S., Fujiwara, H.: Energy-efficient algorithms for flow time minimization. In: Durand, B., Thomas, W. (eds.) STACS 2006. LNCS, vol. 3884, pp. 621–633. Springer, Heidelberg (2006)
2. Bansal, N., Chan, H.L., Lam, T.W., Lee, L.K.: Scheduling for bounded speed processors. In: Aceto, L., Damgård, I., Goldberg, L.A., Halldórsson, M.M., Ingólfsdóttir, A., Walukiewicz, I. (eds.) ICALP 2008, Part I. LNCS, vol. 5125, pp. 409–420. Springer, Heidelberg (2008)
3. Bansal, N., Chan, H.-L., Pruhs, K.: Speed scaling with an arbitrary power function. In: Proceedings of the eighteenth annual ACM-SIAM symposium on Discrete algorithms (SODA), pp. 693–701 (2009)
4. Bansal, N., Pruhs, K., Stein, C.: Speed scaling for weighted flow time. In: Proceedings of the eighteenth annual ACM-SIAM symposium on Discrete algorithms (SODA), pp. 805–813 (2007)
5. Burd, T.D., Pering, T.A., Stratakos, A.J., Brodersen, R.W.: A dynamic voltage scaled microprocessor system. In: IEEE International Solid-State Circuits Conference, February 2000, pp. 294–295, 466 (2000)
6. Chan, H.-L., Edmonds, J., Lam, T.-W., Lee, L.-K., Marchetti-Spaccamela, A., Pruhs, K.: Nonclairvoyant speed scaling for fow and energy. In: STACS, Freiburg, Germany, pp. 409–420 (2009)
7. Hong, I., Qu, G., Potkonjak, M., Srivastava, M.B.: Synthesis Techniques for Low-Power Hard Real-Time Systems on Variable Voltage Processors. In: Proceedings of the IEEE Real-Time Systems Symposium (RTSS), pp. 178–187 (1998)
8. Irani, S., Pruhs, K.R.: Algorithmic problems in power management. SIGACT News 36(2), 63–76 (2005)
9. Lam, T.W., Lee, L.-K., To, I.K.-K., Wong, P.W.H.: Speed scaling functions for fow time scheduling based on active job count. In: Halperin, D., Mehlhorn, K. (eds.) ESA 2008. LNCS, vol. 5193, pp. 647–659. Springer, Heidelberg (2008)
10. Li, M., Liu, B.J., Yao, F.F.: Min-energy voltage allocation for tree-structured tasks. Journal of Combinatorial Optimization 11(3), 305–319 (2006)
11. Li, M., Yao, F.F.: An efficient algorithm for computing optimal discrete voltage schedules. SIAM J. on Computing 35, 658–671 (2005)

12. Li, M., Yao, A.C., Yao, F.F.: Discrete and continuous min-energy schedules for variable voltage processors. Proc. of the National Academy of Sciences USA (PNAS) 103, 3983–3987 (2006)
13. Pruhs, K., Uthaisombut, P., Woeginger, G.: Getting the best response for your erg. In: Scandanavian Workshop on Algorithms and Theory, pp. 14–25 (2004)
14. Yao, F., Demers, A., Shenker, S.: A scheduling model for reduced CPU energy. In: Proc. IEEE Symp. Foundations of Computer Science (FOCS), pp. 374–382 (1995)
15. Yuan, L., Qu, G.: Analysis of energy reduction on dynamic voltage scaling-enabled systems. IEEE Transactions on Computer-Aided Design of Integrated Circuits and Systems 24(12), 1827–1837 (2005)

Posi-modular Systems with Modulotone Requirements under Permutation Constraints

Toshimasa Ishii[1] and Kazuhisa Makino[2]

[1] Department of Information and Management Science, Otaru University of Commerce, Otaru 047-8501, Japan
ishii@res.otaru-uc.ac.jp

[2] Graduate School of Information Science and Technology, University of Tokyo, Tokyo 113-8656, Japan
makino@mist.i.u-tokyo.ac.jp

Abstract. Given a system (V, f, r) on a finite set V consisting of a posi-modular function $f : 2^V \to \mathbb{R}$ and a modulotone function $r : 2^V \to \mathbb{R}$, we consider the problem of finding a minimum set $R \subseteq V$ such that $f(X) \geq r(X)$ for all $X \subseteq V - R$. The problem, called the transversal problem, was introduced by Sakashita et al. [6] as a natural generalization of the source location problem and external network problem with edge-connectivity requirements in undirected graphs and hypergraphs.

By generalizing [8] for the source location problem, we show that the transversal problem can be solved by a simple greedy algorithm if r is π-monotone, where a modulotone function r is π-monotone if there exists a permutation π of V such that the function $p_r : V \times 2^V \to \mathbb{R}$ associated with r satisfies $p_r(u, W) \geq p_r(v, W)$ for all $W \subseteq V$ and $u, v \in V$ with $\pi(u) \geq \pi(v)$. Here we show that any modulotone function r can be characterized by p_r as $r(X) = \max\{p_r(v, W) \mid v \in X \subseteq V - W\}$.

We also show the structural properties on the minimal deficient sets $\mathcal{W}$ for the transversal problem for π-monotone function r, i.e., there exists a basic tree T for $\mathcal{W}$ such that $\pi(u) \leq \pi(v)$ for all arcs (u, v) in T, which, as a corollary, gives an alternative proof for the correctness of the greedy algorithm for the source location problem.

Furthermore, we show that a fractional version of the transversal problem can be solved by the algorithm similar to the one for the transversal problem.

1 Introduction

Given a system (V, f, r) on a finite set V consisting of a posi-modular function $f : 2^V \to \mathbb{R}$ and a modulotone function $r : 2^V \to \mathbb{R}$ with $f(\emptyset) \geq r(\emptyset)$, we consider the following problem:

$$\begin{aligned} &\text{Minimize } |R| \\ &\text{subject to } f(X) \geq r(X) \text{ for all } X \subseteq V - R \\ &\qquad\qquad R \subseteq V. \end{aligned} \tag{1.1}$$

Y. Dong, D.-Z. Du, and O. Ibarra (Eds.): ISAAC 2009, LNCS 5878, pp. 473–482, 2009.

Here $f(\emptyset) \geq r(\emptyset)$ is necessary for the problem to have a feasible solution. This problem was first introduced by Sakashita et al. [6] as a generalized framework of the source location problem and external network problem with edge-connectivity requirements in undirected graphs and hypergraphs [3,5,8]. They showed that the family of minimal *deficient* sets of (V, f, r) forms a tree hypergraph, and that conversely any tree hypergraph can be represented by minimal deficient sets of (V, f, r) for some posi-modular function f and some modulotone function r, where a set $X \subseteq V$ with $f(X) < r(X)$ is called deficient. Note that Problem (1.1) asks to find a minimum set hitting all deficient sets. By combining these results with properties shown in [2,3], it follows that Problem (1.1) can be solved in $O(|V|^3\rho(|V|))$ time, where $\rho(|V|)$ is the time required to check the feasibility (i.e., a given $R \subseteq V$ satisfies $f(X) \geq r(X)$ for all $X \subseteq V - R$), while it is still open whether the feasibility can be checked in polynomial time. They also gave a polynomial time algorithm for Problem (1.1) by utilizing a basic tree for the tree hypergraph, under the assumption that f is submodular and r is given by either $r(X) = \max\{d_1(v) \mid v \in X\}$ for a function $d_1 : V \to \mathbb{R}_+$ or $r(X) = \max\{d_2(u, v) \mid u \in X, v \in V - X\}$ for a function $d_2 : V \times V \to \mathbb{R}_+$. We here remark that these assumptions are necessary only for executing the algorithm in polynomial time. Both of the source location problem and external network problem satisfy these assumptions, and hence are polynomially solvable. On the other hand, it was shown by Tamura et al. [8] that the source location problem can be solved in polynomial time by a much simpler greedy algorithm without using any basic tree for the tree hypergraph.

Then natural questions arise: (i) is there some relationship between Sakashita et al.'s algorithm and Tamura et al.'s greedy one? (ii) if so, how can we characterize cases where such a greedy algorithm works? In this paper, we show that there exists a basic tree for the family of all minimal deficient sets for which Sakashita et al.'s algorithm can perform in the same way as Tamura et al.'s algorithm does. In other words, Sakashita et al.'s algorithm includes Tamura et al.'s one as its special case. Furthermore, we show that this relationship can be extended to Problem (1.1) in which a modulotone function r has a property called *π-monotonicity*.

The π-monotonicity of a modulotone function is defined as follows. An arbitrary modulotone function r can be characterized by using a function $p_r : V \times 2^V \to \mathbb{R}$, which is a slight generalization of similar properties shown in [4]. A modulotone function is called π-monotone if there exists a permutation π of V such that for all $u, v \in V$ and $W \subseteq V - \{u, v\}$, $\pi(u) \geq \pi(v)$ if and only if $p_r(u, W) \geq p_r(v, W)$. A modulotone function r in the above source location problem satisfies $r(X) = \max\{d_1(v) \mid v \in X\}$, $X \subseteq V$ for some function $d_1 : V \to \mathbb{R}_+$, and hence is π-monotone. Also, Problem (1.1) with a π-monotone modulotone function includes problems whose requirements are based on a function q on V; we will discuss these problems later in Subsection 3.2. We then show that if r is π-monotone, then there exists a tree hypergraph whose basic tree satisfies $\pi(u) \leq \pi(v)$ for each pair of u and its parent v. This interesting property enables that Sakashita et al.'s algorithm [6] can be executed in a simple greedy manner without computing any basic tree for the tree hypergraph.

Furthermore, we consider a fractional version of Problem (1.1):

$$\begin{array}{ll} \text{Minimize} & x(V) \\ \text{subject to} & f(X) + x(X) \geq r(X) \text{ for all } X \subseteq V \\ & x : V \to \mathbb{R}, \end{array} \tag{1.2}$$

where $x(X) = \sum_{v \in X} x(v)$ for all $X \subseteq V$. This problem can be regarded as a generalization of a capacitated type of the source location problem with edge-connectivity requirements in undirected graphs. Then we show that Sakashita et al.'s algorithm can be extended to this problem.

The rest of this paper is organized as follows. In Section 2, after giving basic definitions, we review properties and applications of Problem (1.1) shown in [6]. In Section 3, we define a π-monotonicity of a modulotone function. Furthermore, we show a structural property of minimal deficient sets of Problem (1.1) with a π-monotone modulotone function, which enables a greedy algorithm. Section 4 discusses Problem (1.2) as a fractional version of Problem (1.1).

2 Preliminaries

Let V be a finite set. For two sets $X, Y \subseteq V$, we say that X and Y *intersect* each other if $X \cap Y \neq \emptyset$, $X - Y \neq \emptyset$, and $Y - X \neq \emptyset$. For a family $\mathcal{E} \subseteq 2^V$, the hypergraph $(V, \mathcal{E})$ may be written as $\mathcal{E}$ simply. Let $V(\mathcal{E})$ denote the vertex set of a hypergraph $\mathcal{E}$. For a hypergraph $\mathcal{E}$, a subset $R \subseteq V$ is called a *transversal* (or *hitting set*) of $\mathcal{E}$ if $R \cap E \neq \emptyset$ for all $E \in \mathcal{E}$. A hypergraph $\mathcal{E}$ is called a *tree hypergraph* (or *hypertree*) if there exists a tree T with a vertex set V such that each hyperedge in $\mathcal{E}$ induces a subtree of T. We call such a tree T a *basic tree* for $\mathcal{E}$, and we may regard T as a rooted tree in describing algorithms. For a subset U of vertices in a tree T, $T[U]$ denotes the subgraph of T induced by U. For a vertex v in a rooted tree T, $T(v)$ denotes the subtree of T rooted at v.

2.1 Posi-modular Systems

In this subsection, let us review several properties about Problem (1.1) shown by Sakashita et al. [6]. A set function $f : 2^V \to \mathbb{R}$ is called *submodular* if

$$f(X) + f(Y) \geq f(X \cap Y) + f(X \cup Y) \tag{2.1}$$

for arbitrary two subsets X, Y of V, and *posi-modular* if

$$f(X) + f(Y) \geq f(X - Y) + f(Y - X) \tag{2.2}$$

for arbitrary two subsets X, Y of V. A set function $r : 2^V \to \mathbb{R}$ is called *modulotone* if for any nonempty subset X of V, there exists an element $v \in X$ such that all subsets Y of X with $Y \ni v$ satisfies $r(Y) \geq r(X)$.

Observe that Problem (1.1) is equivalent to that of asking to find a minimum transversal R of $\{X \subseteq V \mid f(X) < r(X)\}$. A set $X \subseteq V$ is called *deficient* if $f(X) < r(X)$. A deficient set X is called *minimal* if any proper subset Y of X is not deficient. We denote the family of all minimal deficient sets by $\mathcal{W}(f,r)$. It is known that the posi-modular systems have the following interesting property, where a *Sperner* family denotes a family of sets in V in which arbitrary two distinct sets E, E' satisfy neither $E \subseteq E'$ nor $E' \subseteq E$.

Theorem 1. [6] *A Sperner family $\mathcal{E} \subseteq 2^V$ is a tree hypergraph if and only if $\mathcal{E} = \mathcal{W}(f,r)$ holds for a posi-modular function $f : 2^V \to \mathbb{R}$ and a modulotone function $r : 2^V \to \mathbb{R}$.* □

By this theorem and properties observed in [2,3], it follows that Problem (1.1) can be solved in $O(|V|^3\rho(|V|))$ time, where $\rho(|V|)$ is the time required to check the feasibility, while it is still open whether the feasibility can be checked in polynomial time. On the other hand, we can solve Problem (1.1) more efficiently (more precisely, quadratically faster) by the following algorithm MinTransversal, under the assumption that f is submodular and r is given as

$$r(X) = \begin{cases} \max\{d_1(v) \mid v \in X\} & \text{if } X \neq \emptyset \\ 0 & \text{if } X = \emptyset, \end{cases} \tag{2.3}$$

for a function $d_1 : V \to \mathbb{R}_+$ or

$$r(X) = \begin{cases} \max\{d_2(u,v) \mid u \in X, v \in V - X\} & \text{if } X \neq \emptyset, V \\ 0 & \text{if } X = \emptyset \text{ or } V \end{cases} \tag{2.4}$$

for a function $d_2 : V \times V \to \mathbb{R}_+$. It is not difficult to observe that both functions defined as (2.3) and (2.4) are modulotone. Also, we remark that these assumptions are necessary only for executing this algorithm in polynomial time.

2.2 Applications of Problem (1.1)

We here introduce the source location problem and the external network problem in undirected graphs discussed as applications of Problem (1.1) in [6].

Algorithm 1. Algorithm MinTransversal[6]

Require: A posi-modular function $f : 2^V \to \mathbb{R}$, a modulotone function $r : 2^V \to \mathbb{R}$ with $f(\emptyset) \geq r(\emptyset)$.
Ensure: A minimum transversal R of $\mathcal{W}(f,r)$.
Compute a basic tree T for $\mathcal{W}(f,r)$.
Initialize $R := \emptyset$ and $U := V$.
while $U \neq \emptyset$ **do**
  Choose a leaf v of $T[U]$ and $U := U - \{v\}$.
  if $R \cup U$ is not a transversal **then**
    $R := R \cup \{v\}$.
  end if
end while
Output R as a solution.

Let $G = (V, E, c)$ be an undirected graph with a set V of vertices, a set E of edges, and a capacity function $c : E \to \mathbb{R}_+$. Suppose that each vertex $v \in V$ has a demand $d_1(v) \in \mathbb{R}_+$. The source location problem is defined as follows:

$$\begin{aligned} &\text{Minimize } |S| \\ &\text{subject to } \lambda_G(S, v) \geq d_1(v) \text{ for all } v \in V \\ &\qquad\qquad S \subseteq V, \end{aligned} \tag{2.5}$$

where $\lambda_G(S, v)$ denotes the maximum flow value (or edge-connectivity) between S and v in G, and we define $\lambda_G(S, v) = +\infty$ if $v \in S$. This problem has been studied as a location problem with requirements measured by a network flow amount or network connectivity [1,5,7,8].

In a multimedia network, a set S of some specified network nodes, such as the so-called mirror servers, may have functions of offering the same services for users. A user at a node v can use the service by communicating with at least one node $s \in S$ through a path between s and v. The edge-connectivity between S and v measures the robustness of the service against network link failures. Thus, location problems with such a fault-tolerancy can be formulated as the source location problem.

By the max-flow min-cut theorem, it is not difficult to see that the constraint of Problem (2.5) is equivalent to $u(X) \geq r(X)$ for all subsets X of $V - S$, where $u(X) = \sum\{c(u, v) \mid u \in X, v \in V - X, (u, v) \in E\}$ (i.e., u is a cut function in G) and r is given as (2.3). Since u is posi-modular, it follows that Problem (2.5) is a special case of Problem (1.1).

Given an undirected graph $G = (V, E, c)$ and a demand function $d_2 : V \times V \to \mathbb{R}_+$, the external network problem is given by:

$$\begin{aligned} &\text{Minimize } |S| \\ &\text{subject to } \lambda_{G/S}(u, v) \geq d_2(u, v) \text{ for all } u, v \in V \\ &\qquad\qquad S \subseteq V, \end{aligned} \tag{2.6}$$

where G/S denotes the graph obtained from G by contracting S into a single vertex s, and if $u \in S$, we define $\lambda_{G/S}(u, v) = \lambda_{G/S}(s, v)$. This problem has been studied as a problem of finding access points to some highly reliable external network while taking into account a network flow amount or connectivity [3].

In a communication network N, each pair of nodes may have some requirements measured by a network flow amount or connectivity. Suppose that we can use a highly reliable external network N' in which neither node nor link failures occurs. Then we can improve the reliability of N by adding access points to N'. The problem of asking to find a minimum set S of access points to N' in order to satisfy the connectivity requirements can be formulated as Problem (2.6).

Again by the max-flow min-cut theorem, we can see that the constraint of Problem (2.6) is equivalent to $u(X) \geq r(X)$ for all subsets X of $V - S$, where r is given as (2.4). Thus, Problem (2.6) is also a special case of Problem (1.1).

Furthermore, since a cut function u is submodular, both problems can be solved in polynomial time by Algorithm MinTransversal. In particular, for the source location problem, a much simpler greedy algorithm without using any

Algorithm 2. Algorithm MinSourceSet[8]

Require: An undirected graph $G = (V, E, c)$ and a demand function $d_1 : V \to \mathbb{R}_+$.
Ensure: A minimum set S satisfying $\lambda_G(S, v) \geq d_1(v)$ for all $v \in V$.

Order vertices of V such that $d_1(v_1) \leq \cdots \leq d_1(v_n)$.
Initialize $S := \emptyset$ and $U := V$.
for $j = 1$ to n **do**
  $U := U - \{v_j\}$.
  if $S \cup U$ is infeasible **then**
    $S := S \cup \{v_j\}$.
  end if
end for
Output S as a solution.

basic tree for the tree hypergraph was proposed [8]. This algorithm is described as Algorithm MinSourceSet.

3 Modulotone Function with π-Monotonicity

From the previous section, we can observe that as for Problem (2.5), if there exists a basic tree T for the family $\mathcal{W}(f, r)$ of minimal deficient sets such that $d_1(u) \leq d_1(v)$ holds for each pair of a vertex u and its parent v in T, then Algorithm MinTransversal can be executed in the same way as Algorithm MinSourceSet does; that is, in such cases we need not prepare any basic tree for the tree hypergraph. In this section, we will prove the existence of such a basic tree in a more general setting.

For this, we first characterize a modulotone function by using a function $p : V \times 2^V \to \mathbb{R}$ in Subsection 3.1. In Subsection 3.2, we define Problem (1.1) with a function r called π-monotone which is a generalization of Problem (2.5), discuss its applications, and prove the existence of basic trees for $\mathcal{W}(f, r)$ defined as above.

3.1 Characterization of a Modulotone Function

We here show that an arbitrary modulotone function can be characterized by using a function $p : V \times 2^V \to \mathbb{R}$. This is a slight generalization of similar properties observed in [4]. For a nonempty subset X of V and a function $p : V \times 2^V \to \mathbb{R}$, let

$$p^*(X) = \max\{p(v, U) \mid U \subseteq V, v \in X \subseteq V - U\}. \tag{3.1}$$

Lemma 1. (i) *Let $p : V \times 2^V \to \mathbb{R}$ be a function. Then, the set function $p^* : 2^V \to \mathbb{R}$ given as (3.1) is modulotone.*
(ii) *Let $p^* : 2^V \to \mathbb{R}$ be a modulotone function. Then, there exists a function $p : V \times 2^V \to \mathbb{R}$ that satisfies (3.1).* □

3.2 π-Monotonicity

For a modulotone function r, we denote by p_r a function $p : V \times 2^V \to \mathbb{R}$ such that r is given as (3.1). A modulotone function r is called *π-monotone* if there exist a function p_r and a permutation $\pi : V \to [|V|]$ of V such that for all $u, v \in V$ and $U \subseteq V - \{u, v\}$, $\pi(u) \geq \pi(v)$ if and only if $p_r(u, U) \geq p_r(v, U)$. In this section, we focus on Problem (1.1) in the case where r is π-monotone.

We first observe that the function r defined as (2.3) is π-monotone. Let $p_r(v, U) = d_1(v)$ for all $v \in V$ and $U \subseteq V$, and π be a permutation of V such that $\pi(u) \geq \pi(v)$ if and only if $d_1(u) \geq d_1(v)$ for each pair of two vertices u and v. Thus, r is clearly π-monotone. It follows that Problem (2.5) is a special case of Problem (1.1) with a π-monotone r.

For the function r defined as (2.4), if $d_2(u, v)$ is defined as a function of $(q(u), q(v))$ such as $q(u) + q(v)$ or $q(u)q(v)$ for a given function $q : V \to \mathbb{R}$, then we can observe that r is π-monotone. For example, it is natural to consider a situation where a user who pays more cost (or money) can communicate with a higher reliablity; $d_2(u, v)$ may be considered as a value proportional to $q(u)+q(v)$ where $q(u)$ is a payment of a user u. In another situation where each node u corresponds to a city whose population is $q(u)$, the reliablility requirement between two cities u and v may be assumed to be proportional to $q(u)q(v)$. In these settings, Problem (2.6) becomes a special case of Problem (1.1) with a π-monotone r.

On the other hand, we remark that even if r is given as (2.4), then r is not necessarily π-monotone. Consider p_r in the case where $V = \{v_1, v_2, v_3, v_4\}$, $d_2(v_1, v_2) = 1$, $d_2(v_3, v_4) = 2$, and $d_2(v_i, v_j) = 0$ otherwise. For $X_1 = \{v_1, v_3\}$, $p_r(v_1, U) \leq 1$ holds for all nonempty subsets U of $V - X_1$, since otherwise $r(v_1) > 2$, a contradiction. It follows by $r(X_1) = 2$ that $p_r(v_3, U') = 2$ for some $U' \subseteq V - X_1$. For $X_2 = \{v_1, v_3, v_4\}$, $p_r(v, V - X_2\ (= \{v_2\})) = 0$ for all $v \in \{v_3, v_4\}$ by $r(\{v_1, v_2\}) = 0$. It follows by $r(X_2) = 1$ that $p_r(v_1, V - X_2) = 1$. Thus, by $p_r(v_3, U') > p_r(v_1, U')$ and $p_r(v_3, V - X_2) < p_r(v_1, V - X_2)$, we can see that this r is not π-monotone.

In the rest of this subsection, we will show the following interesting structural property about $\mathcal{W}(f, r)$.

Theorem 2. *For a posi-modular function $f : 2^V \to \mathbb{R}$ and a π-monotone modulotone function $r : 2^V \to \mathbb{R}$, there exists a basic tree T for $\mathcal{E} = \mathcal{W}(f, r)$ (which is a tree hypergraph) such that for any pair of two vertices u and v in T,*

$$\text{if } u \text{ is a child of } v, \text{ then } \pi(u) \leq \pi(v). \tag{3.2}$$

This property enables us to execute Algorithm MinTransversal greedily based on π without any basic tree for $\mathcal{W}(f, r)$. Indeed, if we pick up all elements in V in nondecreasing order of their π-values, then it follows that we pick up a leaf of $T[U]$ for the current U in each iteration of the while loop of Algorithm MinTransversal. Also notice that this greedy procedure based on π is a generalization of Algorithm MinSourceSet.

Corollary 1. *If a modulotone function r is π-monotone, then Problem (1.1) can be solved in a greedy manner based on π.* □

Before proving this theorem, we show several preparatory lemmas. For a set $X \subseteq V$, let $\pi(X) = \max\{\pi(v) \mid v \in X\}$.

Lemma 2. *If W_1 and W_2 in $\mathcal{W}(f,r)$ satisfy $W_1 \cap W_2 \neq \emptyset$, then W_1 and W_2 intersect each other and we have $\pi(W_1 \cap W_2) > \pi(W_1 - W_2)$ or $\pi(W_1 \cap W_2) > \pi(W_2 - W_1)$.* □

Lemma 3. *Let $\mathcal{W} = \{W_1, W_2, \ldots, W_p\}$ be a family of sets in $\mathcal{W}(f,r)$ with $W_1 \cap W_2 \cap \ldots \cap W_p \neq \emptyset$. Then there exists a set $W_q \in \mathcal{W}$ such that all elements $w \in W_q$ with $\pi(w) = \pi(W_q)$ are contained in $W_1 \cap W_2 \cap \ldots \cap W_p$.* □

Proof of Theorem 2. Let $\mathcal{E}$ be a tree hypergraph with $\mathcal{E} = \mathcal{W}(f,r)$ and T_1 be its basic tree. Let v_r be a vertex with the maximum π-value (i.e., $\pi(v_r) = \max\{\pi(v) \mid v \in V\}$) and regard T_1 as a tree rooted at v_r. Assume that T_1 does not satisfy (3.2). Let u, v, and w be three vertices in T_1 such that $u = p_{T_1}(v)$, $v = p_{T_1}(w)$, $\pi(u) \geq \pi(v) < \pi(w)$, and $depth(w;T_1)$ is the minimum, where $p_T(x)$ denotes the parent of x in T, and $depth(x;T)$ denotes the length of the simple path connecting r and x in T rooted at r. Now, for a tree T rooted at r, define $F_T = \{(x, p_T(x)) \mid \pi(x) > \pi(p_T(x))\}$, and a potential function

$$\Phi(T) = depth(x^*_T;T) + \sum_{x:\ (x,p_T(x)) \in F_T} n(\pi(x) - \pi(p_T(x))),$$

where $n = |V|$ and x^*_T is a vertex x with $(x, p_T(x)) \in F_T$ such that $depth(x;T)$ is the minimum. Notice that if (3.2) is satisfied, $\Phi(T) = 0$, otherwise $\Phi(T) > 0$. Below, we will prove this theorem by showing the existence of a basic tree T' for $\mathcal{W}(f,r)$ such that $\Phi(T') < \Phi(T_1)$. Let $C(v)$ denote the set of all children of v other than w in T_1, and $\mathcal{W}^*$ denote the family of sets in $\mathcal{W}(f,r)$ containing v. Partition $\mathcal{W}^*$ into $\mathcal{X}_1 = \{X \in \mathcal{W}(f,r) \mid v, w \in X, u \notin X, C(v) \cap X \neq \emptyset\}$, $\mathcal{X}_2 = \{X \in \mathcal{W}(f,r) \mid v, w \in X, u \notin X, C(v) \cap X = \emptyset\}$, $\mathcal{Y}_1 = \{X \in \mathcal{W}(f,r) \mid u, v \in X, w \notin X, C(v) \cap X \neq \emptyset\}$, $\mathcal{Y}_2 = \{X \in \mathcal{W}(f,r) \mid u, v \in X, w \notin X, C(v) \cap X = \emptyset\}$, $\mathcal{Z}_1 = \{X \in \mathcal{W}(f,r) \mid u, v, w \in X, C(v) \cap X \neq \emptyset\}$, and $\mathcal{Z}_2 = \{X \in \mathcal{W}(f,r) \mid u, v, w \in X, C(v) \cap X = \emptyset\}$. Notice that every two sets in $\mathcal{W}^*$ intersect each other since every set is a minimal deficient set. There are the following three possible cases: (Case-1) $\mathcal{X}_1 \cup \mathcal{X}_2 = \emptyset$, (Case-2) $\mathcal{Y}_1 \cup \mathcal{Y}_2 = \emptyset$, and (Case-3) otherwise.

(Case-1) Let T_2 denote the tree from T_1 by deleting the edge (v, w) and adding a new edge connecting u and w (i.e., $p_{T_2}(w) := u$). T_2 is also a basic tree for $\mathcal{W}(f,r)$ because otherwise there exists a set $X \in \mathcal{W}(f,r)$ with $v, w \in X$ and $u \notin X$, contradicting $\mathcal{X}_1 \cup \mathcal{X}_2 = \emptyset$. Also, we can observe that $\Phi(T_2) < \Phi(T_1)$. Indeed, if $\pi(u) < \pi(w)$, then we have $\Phi(T_2) - \Phi(T_1) = depth(w;T_2) - depth(w;T_1) + n(-\pi(u) + \pi(v)) < 0$ because $x^*_{T_1} = x^*_{T_2} = w$, $depth(w;T_2) = depth(w;T_1) - 1$, and $\pi(u) \geq \pi(v)$. If $\pi(u) \geq \pi(w)$, then we have $\Phi(T_2) - \Phi(T_1) = depth(x^*_{T_2};T_2) - depth(w;T_1) + n(-\pi(u) + \pi(v)) < 0$ because $depth(x^*_{T_2};T_2) \leq n - 1$ and $\pi(u) > \pi(v)$ (by $\pi(u) \geq \pi(w) > \pi(v)$).

(Case-2) Let T_2 denote the tree from T_1 by deleting the edge (u, v), adding a new edge connecting u and w (i.e., $p_{T_2}(w) := u$), and making the parent of v the vertex w (i.e., $p_{T_2}(v) := w$). T_2 is also a basic tree for $\mathcal{W}(f, r)$ because otherwise there exists a set $X \in \mathcal{W}(f, r)$ with $u, v \in X$ and $w \notin X$, contradicting $\mathcal{Y}_1 \cup \mathcal{Y}_2 = \emptyset$.

Also, we can observe that $\Phi(T_2) < \Phi(T_1)$. Indeed, if $\pi(u) < \pi(w)$, then we have $\Phi(T_2) - \Phi(T_1) = depth(w; T_2) - depth(w; T_1) + n(-\pi(u) + \pi(v)) < 0$ because $x^*_{T_1} = x^*_{T_2} = w$, $depth(w; T_2) = depth(w; T_1) - 1$, $\pi(u) \geq \pi(v)$, and $(v, w) \notin F_{T_2}$. If $\pi(u) \geq \pi(w)$, then we have $\Phi(T_2) - \Phi(T_1) = depth(x^*_{T_2}; T_2) - depth(w; T_1) + n(-\pi(u) + \pi(v)) < 0$ because $depth(x^*_{T_2}; T_2) \leq n - 1$, $\pi(u) > \pi(v)$ (by $\pi(u) \geq \pi(w) > \pi(v)$), and $(v, w) \notin F_{T_2}$.

(Case-3) Omitted due to space limitation. □

In Algorithm MinSourceSet, if there exist two vertices u and v with $d_1(u) = d_1(v)$, then we can choose u before v or vice versa, depending on the sorting in line 1. The following corollary shows that there exists a basic tree for $\mathcal{W}(f, r)$ corresponding to each of these cases.

Corollary 2. *Let $\mathcal{E} = \mathcal{W}(f, r)$ be a tree hypergraph whose basic tree T satisfies (3.2), and (v, w) be an edge in T with $v = p_T(w)$ and $\pi(v) = \pi(w)$. Then there exists a basic tree T' for $\mathcal{W}(f, r)$ satisfying (3.2) such that $w = p_{T'}(v)$ or v and w have a common parent in T'.* □

Finally, we give a much simpler proof of the property that the above greedy algorithm based on π works. Notice that we scan all elements in nondecreasing order of their π-values. When we pick up an element v and $R \cup U - \{v\}$ is not a transversal of $\mathcal{W}(f, r)$ for the current transversal $R \cup U$, there exists a minimal deficient set $W_v \in \mathcal{W}(f, r)$ such that $W_v \cap U = \{v\}$. Then notice that all elements in W_v other than v have been scanned and deleted, and this implies that v has the maximum π-value among all elements in W_v; $\pi(v) = \pi(W_v)$. It follows that for each $v \in R$, there exists such a set W_v; let $\mathcal{W} = \{W_v \mid v \in R\}$. Then we can prove that every two sets in $\mathcal{W}$ are disjoint, which implies that R is a minimum transversal. Indeed, if two sets W_u and W_v satisfy $W_u \cap W_v \neq \emptyset$, then by $u \in W_u - W_v$, $\pi(u) = \pi(W_u)$, $v \in W_v - W_u$, and $\pi(v) = \pi(W_v)$, we have $\pi(W_u - W_v) = \pi(W_u)$ and $\pi(W_v - W_u) = \pi(W_v)$, contradicting Lemma 2.

4 Algorithm for Problem (1.2)

In this section, we consider Problem (1.2). This can be regarded as a generalization of a variant of the source location problem; given an undirected graph $G = (V, E, c)$ and a demand function $d_1 : V \to \mathbb{R}_+$, find an $x : V \to \mathbb{R}_+$ such that $x(X) + u(X) \geq \max\{d_1(v) \mid v \in X\}$ holds for all nonempty subsets X of V and $x(V)$ is the minimum. In applications of multimedia networks, we can locate a mirror server v with an arbitrarily finite capacity $x(v)$ and we want to minimize the total capacity $x(V)$ of servers to be located. Notice that in a setting discussed in Subsection 2.2, each server to be located has an infinite capacity.

Algorithm 3. Algorithm MinCover

Require: A posi-modular function $f : 2^V \to \mathbb{R}$, a modulotone function $r : 2^V \to \mathbb{R}$ with $f(\emptyset) \geq r(\emptyset)$.
Ensure: A minimum cover $x : V \to \mathbb{R}$ of $\mathcal{W}(f, r)$.
Compute a basic tree T for $\mathcal{W}(f, r)$.
Initialize $x(v) := 0$ and $x'(v) := \max\{r(W) \mid W \in \mathcal{W}(f, r)\}$ for all $v \in V$ and $U := V$.
while $U \neq \emptyset$ **do**
  Choose a leaf v of $T[U]$, $x'(v) := 0$, and $U := U - \{v\}$.
  if $x + x'$ is not a cover **then**
    $x(v) := \max\{r(W) - x(W) - f(W) \mid W \in \mathcal{W}(f, r)\}$.
  end if
end while
Output x as a solution.

For $f : 2^V \to \mathbb{R}$, $r : 2^V \to \mathbb{R}$, and $\mathcal{W}(f, r)$, $x : V \to \mathbb{R}$ is called a *cover* of $\mathcal{W}(f, r)$ if $x(X) + f(X) \geq r(X)$ for all $X \subseteq V$. Then we can find a minimum cover of $\mathcal{W}(f, r)$ by Algorithm MinCover, similar to Algorithm MinTransversal.

Lemma 4. *Algorithm* MinCover *finds a minimum cover x of $\mathcal{W}(f, r)$.* □

Finally, we remark that if f is submodular and r is given as (2.3) or (2.4), Algorithm MinCover can be implemented to run in the same complexity as Algorithm MinTransversal. Furthermore, similarly to the discussion in the previous section, if r is π-monotone, then we can execute it greedily based on π without any basic tree for $\mathcal{W}(f, r)$.

References

1. Arata, K., Iwata, S., Makino, K., Fujishige, S.: Locating sources to meet flow demands in undirected networks. Journal of Algorithms 42, 54–68 (2002)
2. van den Heuvel, J., Johnson, M.: Transversals of subtree hypergraphs and the source location problem in digraphs, CDAM Research Report, LSE-CDAM-2004-10, London School of Economics
3. van den Heuvel, J., Johnson, M.: The external network problem with edge- or arc-connectivity requirements. In: López-Ortiz, A., Hamel, A.M. (eds.) CAAN 2004. LNCS, vol. 3405, pp. 114–126. Springer, Heidelberg (2005)
4. Ishii, T., Makino, K.: Augmenting edge-connectivity between vertex subsets. In: Proceedings of the 15th Computing: The Australasian Theory Symposium, pp. 45–51 (2009)
5. Ito, H., Uehara, H., Yokoyama, M.: A faster and flexible algorithm for a location problem on undirected flow networks. IEICE Trans. E83-A, 704–712 (2000)
6. Sakashita, M., Makino, K., Nagamochi, H., Fujishige, S.: Minimum transversals in posi-modular systems. SIAM Journal on Discrete Mathematics 23, 858–871 (2009)
7. Tamura, H., Sengoku, M., Shinoda, S., Abe, T.: Some covering problems in location theory on flow networks. IEICE Trans. E75-A, 678–683 (1992)
8. Tamura, H., Sugawara, H., Sengoku, M., Shinoda, S.: Plural cover problem on undirected flow networks. IEICE Trans. J81-A, 863–869 (1998) (in Japanese)

Generalized Reduction to Compute Toric Ideals*

Deepanjan Kesh and Shashank K. Mehta

Indian Institute of Technology, Kanpur - 208016, India
{deepkesh,skmehta}@cse.iitk.ac.in

Abstract. Toric ideals have many applications including solving integer programs. Several algorithms for computing the toric ideal of an integer matrix are available in the literature. Since it is an NP hard problem the present approaches can only solve relatively small problems. We propose a new approach which improves upon a well known saturation technique.

1 Introduction

Let A be an integer matrix, and let $ker(A)$ be the lattice kernel of A, i.e., integer solutions of $Au = 0$. For any $u \in ker(A)$, let u_+ denote the vector such that $u_+[i] = u[i]$ if $u[i] > 0$ else $u_+[i] = 0$. Vector u_- is given by $u_+ - u$. For any positive integer vector $v \in \mathbb{N}^n$, monomial $x_1^{v[1]} x_2^{v[2]} \cdots x_n^{v[n]}$ is concisely denoted by x^v. The polynomial ideal generated by $\{x^{u_+} - x^{u_-} | u \in ker(A)\}$ is called the *toric ideal* of A and it is denoted by I_A. In this paper we address the problem of computing a generating set of I_A, which we loosely call the problem of computing a toric ideal.

This problem has some useful applications including solving integer programs [1,2,3], computing primitive partition identities [4] chapters 6 and 7, and solving scheduling problem [5] among a few others.

Suppose V is a lattice kernel basis, i.e., a basis of $ker(A)$ which generates the kernel vectors with integer coefficients. Let J_V be the ideal generated by $\{x^{u_+} - x^{u_-} | u \in V\}$. Then $I_A = J_V : (x_1 \ldots x_n)^\infty$ where the r.h.s. denotes the ideal $\{f \in k[\mathbf{x}] | x^\alpha \cdot f \in J_V \; for \; some \; x^\alpha\}$ is called the *saturation* of J_V. Thus the computation of a toric ideal has two steps: computation of lattice kernel basis and the saturation of J_V. The first step has a polynomial time solution by computing the Hermite normal form of A. Therefore the complex step is the saturation computation.

One of the most useful ideas in computational commutative algebra is Gröbner basis of an ideal. It has found many applications in computations related to ideal [6,7]. The first and the best known algorithm to compute a Gröbner basis is due to Büchberger [8]. It turns out that it is useful in the computation of the saturation of an ideal.

An early algorithm to compute I_A involved computation of a Gröbner basis in a ring of $n+d+1$ variables [4], where A is $n \times d$. It turned out to be too slow because Büchberger's algorithm is sensitive to the number of variables involved.

* This work was partly supported by Microsoft Research India through 2006 MSR India PhD Fellowship Program.

Y. Dong, D.-Z. Du, and O. Ibarra (Eds.): ISAAC 2009, LNCS 5878, pp. 483–492, 2009.

An algorithm, working in n variables, for saturation is due to Urbanke [9]. It transforms A to A' by negating some columns such that one of the rows has all non-negative entries. In this case $I_{A'} = J_{V'}$. Replacing one negated column at a time by the original one, it computes the toric ideal for the corresponding matrix from that of the previous matrix. Each step involves the computation of one Gröbner basis.

Another algorithm which also works in n variables is due to Sturmfels [10,4]. It computes the toric ideal iteratively, computing the saturation with x_i in the i-th iteration. Each iteration involves the computation of one Gröbner basis. The performances of the two algorithms are comparable, see [10] Bigatti et.al. [11] improved Sturmfels' algorithm.

Hemmecke and Malkin [12] presented an entirely new approach called *project and lift* . Let I_j denote the ideal I_A after setting $x_{j+1}, \ldots, x_n$ to 1. They begin by computing the Gröbner basis of I_1 and build their way up to I_A, one dimension at a time. In the i-th step they compute a certain grading vector r_i, and a Gröbner basis w.r.t. a term order based on r_i in the ring $k[x_1, \cdots, x_i]$.

We present an algorithm improving Sturmfels' algorithm. It requires the computation of a Gröbner basis in i-th iteration in $k[x_1, \cdots, x_i]$, i.e., in i variables. In our approach the basis is computed with respect to reverse lexicographic term order, which is known to be the most efficient among all term orders, [13,14]. Sturmfels' algorithm as well as our algorithm can compute saturation of any binomial ideal (ideal with at least one basis containing only binomials.)

Let $k[\mathbf{x}]$ denote the polynomial ring $k[x_1, \ldots, x_n]$ over a field k. Also, let $B \subset k[\mathbf{x}]$ be a set of polynomials. Then, we denote by $\langle B \rangle$, the ideal generated by the polynomials in B.

A total ordering $\prec$ on the monomials of $k[\mathbf{x}]$ is called a term order if it satisfies following properties: (i) it is a well-ordering (Artinian) and (ii) $x^\alpha \prec x^\beta \Rightarrow x^{\alpha+\gamma} \prec x^{\beta+\gamma}$ for all α, β, γ. Given a term order $\prec$ on $k[\mathbf{x}]$, we denote the leading term of a polynomial f by $in_\prec(f)$. A particular term order, *graded reverse lexicographic order* is frequently the most efficient ordering to compute Gröbner basis. Given an ordering on the variable-indices and a grading vector $\boldsymbol{d}$, the graded reverse lexicographic order $\prec_{\boldsymbol{d}}$ is given as follows: $x^\alpha \prec_{\boldsymbol{d}} x^\beta$ if $\alpha \cdot \boldsymbol{d} < \beta \cdot \boldsymbol{d}$ or if they are equal, then the least non-zero coordinate in $\alpha - \beta$ is negative. By $\prec_{\boldsymbol{d},i}$ we will represent any graded reverse lexicographic ordering with x_i as the least element.

Let I be an ideal in $k[\mathbf{x}]$ and $\prec$ be a term order. By $in_\prec(I)$ we denote the set $\langle in_\prec(f) | f \in I \rangle$, called the initial ideal if I. A basis of I, $G = \{f_1, \ldots, f_m\}$, is called a Gröbner basis if $in_\prec(I) = \langle in_\prec(f_1), \ldots, in_\prec(f_n) \rangle$.

Let f be a polynomial and B be a set of polynomials in $k[\mathbf{x}]$. If there is $f_i \in B$ and a term $c \cdot x^\alpha$ in f such that $in_\prec(f_i)$ divides $c \cdot x^\alpha$ and if the quotient is $c' \cdot x^\beta$ then $f \rightarrow f - c' \cdot x^\beta . f_i$ is called a *reduction* step. If a sequence of reductions lead to f', then we say that f is reduced to f' by B. If G is a Gröbner basis such that each $f \in G$ is irreducible by $G \setminus \{f\}$, then G is called *reduced* Gröbner basis. This basis is unique for a given ordering. We will denote it by $\mathcal{G}_\prec(I)$.

2 Surjective Ring Homomorphism

In this paper ϕ will denote a *surjective* ring homomorphism $k[\mathbf{x}] \to k[\mathbf{y}]$, where $k[\mathbf{x}]$ denotes $k[x_1 \dots, x_n]$ and $k[\mathbf{y}]$ denotes $k[y_1, \dots, y_m]$.

Definition 1. *Let $S \subseteq k[\mathbf{x}]$ be a set. Then, we define $\phi(S) = \{\phi(f) \mid f \in S\}$.*

Lemma 1. *Let $f_1, \dots, f_s \in k[\mathbf{x}]$. Then, $\phi(\langle f_1, \dots, f_s \rangle) = \langle \phi(f_1), \dots, \phi(f_s) \rangle$.*

Proof. Let $f' \in \phi(\langle f_1, \dots, f_s \rangle)$. Then, $\exists f \in \langle f_1, \dots, f_s \rangle$ such that $\phi(f) = f'$. Then $\exists g_1, \dots g_s \in k[\mathbf{x}]$ such that $f = \sum_i g_i f_i$. It implies $\phi(f) = \phi(\sum_i g_i f_i)$. Hence $f' = \sum_i \phi(g_i)\phi(f_i)$. Which implies $f' \in \langle \phi(f_1), \dots, \phi(f_s) \rangle$.

Conversely, let $f' \in \langle \phi(f_1), \dots, \phi(f_s) \rangle$. Then, $\exists g'_1, \dots, g'_s \in k[\mathbf{y}]$ such that $f' = \sum_i g'_i \phi(f_i)$. From surjectivity, $\exists g_1, \dots, g_s \in k[\mathbf{x}]$ such that $\phi(g_1) = g'_1, \dots, \phi(g_s) = g'_s$. So, $f' = \sum_i \phi(g_i)\phi(f_i) = \phi(\sum_i g_i f_i)$. Since $\sum_i g_i f_i \in \langle f_1, \cdots, f_s \rangle$, $f' \in \phi(\langle f_1, \cdots, f_s \rangle)$. ∎

Definition 2. *Kernel of a homomorphism ϕ is $ker(\phi) = \{f \in k[\mathbf{x}] \mid \phi(f) = 0\}$.*

From the first theorem of isomorphism, $k[\mathbf{x}]/ker(\phi)$ is isomorphic to $k[\mathbf{y}]$. We shall denote this isomorphism by Φ.

Definition 3. *Let S be a subset of $k[\mathbf{x}]$. Then $\phi^{-1}(S) = \{f \in k[\mathbf{x}] | \phi(f) \in S\}$.*

Observation 1. *Let J be an ideal in $k[\mathbf{y}]$. Then $\phi^{-1}(J)$ is an ideal. Also J, $\Phi^{-1}(J)$, and $\phi^{-1}(J)/ker(\phi)$ are isomorphic.*

Projections are examples of surjective homomorphisms which will be used in algorithms discussed in this paper.

Definition 4. *Let V be a set of variables and $V' \subset V$. Then the map $\phi : k[V] \to k[V \setminus V']$ is a said to be a projection map if $\phi(f) = f|_{x=1 \forall x \in V'}$. We shall denote the projections $k[\mathbf{x}] \to k[x_1, \dots, x_{i-1}, x_{i+1}, \dots, x_n]$, $k[x_i, \cdots, x_{i+j}, z] \to k[x_i, \cdots, x_{i+j}]$, and $k[\mathbf{x}] \to k[x_{i+1}, \dots, x_n]$ by π_i, π_z, and Π_i respectively.*

Observation 2. *π_i, π_z and Π_i are surjective ring homomorphisms.*

3 Homogeneous Polynomials and Saturation

3.1 Homogenization

Definition 5. *Let $f \in k[\mathbf{x}]$ and $\mathbf{d} \in \mathbb{N}^n$. We say f is homogeneous w.r.t. $\mathbf{d}$ if for all monomials $x^\alpha \in f$, $\mathbf{d} \cdot \alpha$ are equal. Vector $\mathbf{d}$ is called the grading vector.*

Let $\boldsymbol{d} \in \mathbb{N}^{n+1}$ be a 0/1 vector such that $d_{n+1} = 1$. Now we define a mapping $h_{\boldsymbol{d}} : k[\mathbf{x}] \to k[\mathbf{x}, z]$ such that $h_{\boldsymbol{d}}(f)$ is homogeneous w.r.t. $\boldsymbol{d}$ for every $f \in k[\mathbf{x}]$. Let $f = \sum_i c_i x^{\alpha_i} \in k[\mathbf{x}]$. Let $\alpha'_i \in \mathbb{N}^{n+1}$ such that its 1 to n components coincide with those of α_i and its $n+1$-st component is 0. Let $c_{i_0} x^{\alpha_{i_0}}$ be a term in f such that $\alpha'_{i_0} \cdot \boldsymbol{d} \geq \alpha'_i \cdot \boldsymbol{d}$ for all i. Let $\delta_i = (\alpha'_{i_0} - \alpha'_i) \cdot \boldsymbol{d}$. Define $h_{\boldsymbol{d}}(f) = \sum_i x^{\alpha_i} z^{\delta_i}$. We shall denote $h_{\boldsymbol{d}}(f)$ by $\tilde{f}$ when $\boldsymbol{d}$ is known from the context. Observe that $\pi_z(\tilde{f}) = f$. If $B = \{f_i\}_i$ is a set of polynomials of $k[\mathbf{x}]$, then by homogenization of B we would mean the set $\tilde{B}$ given by $\{\tilde{f}_i\}_i$.

3.2 Colon Ideals

Definition 6. *Let $J \subseteq k[\mathbf{x}]$ be an ideal. Then $J : x_i^\infty$ denotes the ideal $\{f \in k[\mathbf{x}] \mid x_i^a f \in J$ for some $a\}$. $(\dots(J : x_1^\infty)\dots) : x_i^\infty$ is equal to $J : (x_1 \cdots x_i)^\infty$ which is given by $\{f \in k[\mathbf{x}] \mid x^\alpha f \in J$ for some $\alpha \in \mathbb{N}^n\}$.*

In general the computation of $J : x_i^\infty$ is expensive, see section 4 in chapter 4 of [7]. But in a special case when J is a homogeneous ideal an efficient method to compute $J : x_i^\infty$ is known as described in the following theorem.

Notation. Let f be a polynomial and a be the largest integer such that x_j^a divides f, then we denote the quotient of the division by $f \div x_j^\infty$. If B be a set of polynomials, then $B : x_i^\infty$ denotes the set $\{f \div x_i^\infty | f \in B\}$.

Algorithm 2 computes $J : x_i^\infty$ for arbitrary ideal J using following lemmas..

Lemma 2 (lemma 12.1, [4]). *Let $J \subseteq k[\mathbf{x}]$ be a homogeneous ideal w.r.t. the grading vector $\mathbf{d}$. Also let $\mathcal{G}_{\prec_{\mathbf{d},j}}(J) = \{f_i\}_i$. Then $\{f_i \div x_j^\infty\}_i$ is a Gröbner basis of $J : x_j^\infty$.*

Lemma 3. *Let $J \subseteq k[\mathbf{x}]$ be any ideal. Then $\pi_i(J : x_j^\infty) = \pi_i(J) : x_j^\infty$.*

Data: A generating set, B, of an ideal $J \subseteq k[\mathbf{x}]$; An index i; A grading vector $\boldsymbol{d} \in \mathbb{N}^{n+1}$ be the vector with all components one.
Result: The Gröbner basis of $\langle B \rangle : x_i^\infty$

1 $\tilde{B} := \{\tilde{f} | f \in B\}$;
2 Compute $\mathcal{G}_{\prec_{\boldsymbol{d},i}}(\langle \tilde{B} \rangle)$;
3 Compute $B' = \{\ f \div x_i^\infty \mid f \in \mathcal{G}_{\prec_{\boldsymbol{d},i}}(\langle \tilde{B} \rangle)\ \}$;
4 **return** $\pi_z(B')$.

Algorithm 2. Computation of $\langle B \rangle : x_i^\infty$

4 Shadow Algorithms under a Surjective Homomorphism

Let I be an ideal in $k[\mathbf{x}]$. Lemma 1 shows that $\phi(I)$ is an ideal in $k[\mathbf{y}]$. In this section we show how to compute a basis B of I such that $\phi(B)$ is a Gröbner basis of $\phi(I)$.

Let α and β be two vectors in $\mathbb{N}^n$, and let $\alpha[i]$ and $\beta[i]$ denotes their i^{th} components. Then, $\alpha \vee \beta$ is the vector given by $(\alpha \vee \beta)[i] = max\{\alpha[i],\ \beta[i]\}$.

In this and the next section we will assume the existence of an oracle which computes any one member h of $\phi^{-1}(m)$ for any monomial $m \in k[\mathbf{y}]$. With an abuse of the notation we shall denote this by $h := \phi^{-1}(m)$ as a step in the algorithms given below.

Let $\prec$ denote a term order in $k[\mathbf{y}]$. Consider any $h_1, h_2 \in k[\mathbf{y}]$. Let $c_1 y^{\alpha_1} = in_\prec(h_1)$ and $c_2 y^{\alpha_2} = in_\prec(h_2)$. Also let $\beta_1 = (\alpha_1 \vee \alpha_2) - \alpha_1$ and $\beta_2 = (\alpha_1 \vee \alpha_2) - \alpha_2$. Then the S-polynomial of these polynomials is given by $S(h_1, h_2) = c_2 y^{\beta_1} h_1 - c_1 y^{\beta_2} h_2$. Observe that if $in_\prec(h_2)$ divides $in_\prec(h_1)$, then $S(h_1, h_2)$ is the reduction of h_1 by h_2. Algorithm 3 computes $g_1, g_2 \in k[\mathbf{x}]$ for given $f_1, f_2 \in k[\mathbf{x}]$ such that $\phi(g_1)\phi(f_1) - \phi(g_2)\phi(f_2) = S(\phi(f_1), \phi(f_2))$.

Observation 3. $(g_1, g_2) = \mathcal{A}(f_1, f_2, \phi, \prec) \Rightarrow \phi(g_1 f_1 - g_2 f_2) = S(\phi(f_1), \phi(f_2))$.

Data: $f_1, f_2 \in k[\mathbf{x}]$, a surjective ring homomorphism $\phi : k[\mathbf{x}] \to k[\mathbf{y}]$, a term order $\prec$ over $k[\mathbf{y}]$, an oracle that computes any one member of $\phi^{-1}(m)$ for any monomial m of $k[\mathbf{y}]$
Result: Two polynomials $g_1, g_2 \in k[\mathbf{x}]$ such that $\phi(f_1 g_1 - f_2 g_2) = S(\phi(f_1), \phi(f_2))$

Let $c_1 \cdot y^{\alpha_1} = in_\prec(\phi(f_1))$, $c_2 \cdot y^{\alpha_2} = in_\prec(\phi(f_2))$;
$\beta_1 := (\alpha_1 \vee \alpha_2) - \alpha_1$;
$\beta_2 := (\alpha_1 \vee \alpha_2) - \alpha_2$;
return $g_1 := \phi^{-1}(c_2 y^{\beta_1}); g_2 := \phi^{-1}(c_1 y^{\beta_2})$;

Algorithm 3. $\mathcal{A}(f_1, f_2, \phi, \prec)$: computation of g_1, g_2 for given f_1, f_2

4.1 Generalized Division Algorithm

Let $g, g_1, \cdots, g_s \in k[\mathbf{y}]$ and $\prec$ be a term order in $k[\mathbf{y}]$. Then $g = \sum_i q_i g_i + r$ is said to be a *standard* expression for g if (i) $in_\prec(q_i g_i) \preceq in_\prec(g)$ $\forall i$ and (ii) no monomial of r is divisible by $in_\prec(g_i)$ for any i, i.e., no monomial of r belongs to $\langle\{in_\prec(g_i) | 1 \leq i \leq s\}\rangle$. Standard expression generalizes the concept of division of a polynomial by another polynomial to the division of a polynomial by a set of polynomials. Here r is called the remainder and q_i are called the quotients of the division of g by $\{g_1, \cdots, g_s\}$. The algorithm to perform such a division is well known, see section 3 in chapter 2 of [7]. Let $f, f_1, \cdots, f_s \in k[\mathbf{x}]$. In Algorithm 4 we present a pseudo-division algorithm for f by $f_1, \cdots, f_s$ such that its image in $k[\mathbf{y}]$ gives a standard expression for $\phi(f)$ w.r.t. $\phi(f_1), \cdots, \phi(f_s)$.

Observe that the leading term of $\phi(p)$ strictly decreases after each pass of the while loop. Combining with this fact that $\prec$ is a well-ordering we observe that the algorithm terminates. Also observe that $\bar{f} \cdot f = \sum_j q_j f_j + r + p$ is an invariant of the loop. Thus we have the following claim.

Lemma 4. *Algorithm 4, SHADOW_DIV($f, f_1, \ldots, f_s, \phi, \prec$), terminates to give* $\bar{f} \cdot f = \sum_j q_j \cdot f_j + r$ *and* $\phi(f) = (1/\phi(\bar{f}))(\sum_j \phi(q_j)\phi(f_j) + \phi(r))$ *is a standard expression for* $\phi(f)$ *under* $\prec$*, where* $\phi(\bar{f})$ *is a non-zero constant.*

4.2 Büchberger's Algorithm with Generalized Division

Now we present Algorithm 5 to compute a basis of any ideal in $k[\mathbf{x}]$ such that the image of the basis under ϕ is a Gröbner basis of the image of the ideal.

Lemma 5. *Algorithm 5 terminates.*

Proof. We first consider the computation of C_1.

The algorithm iterates only if we detect that $B_{new} \neq B_{old}$. Let B_i denote the basis after the i-th iteration. So there must have been $f, g \in B_{i-1}$ such that the remainder, r, of division of $g_1 f - g_2 g$ by B_{i-1} is non-zero.

Then from Lemma 4, $in_\prec(r) \notin \langle in_\prec(\phi(B_{i-1}))\rangle$. Thus, $\langle in_\prec(\phi(B_0))\rangle \subsetneq \langle in_\prec(\phi(B_1))\rangle \subsetneq \langle in_\prec(\phi(B_2))\rangle \subsetneq \cdots$. $k[\mathbf{y}]$ is Noetherian hence this chain must be finite and consequently the algorithm must stop after finitely many iterations.

The termination of the second part is obvious. ∎

Data: $f \in k[\mathbf{x}]$; $\{f_1, \dots, f_s\} \subset k[\mathbf{x}]$; a surjective ring homomorphism, $\phi : k[\mathbf{x}] \longrightarrow k[\mathbf{y}]$; a term order $\prec$ over $k[\mathbf{y}]$; an oracle to compute one member of $\phi^{-1}(m)$ for any monomial m of $k[\mathbf{y}]$.

Result: $\bar{f}, q_1, \dots, q_s, r \in k[\mathbf{x}]$ such that $\bar{f}f = \sum_j q_j f_j + r$ and $\phi(\bar{f})\phi(f) = \sum_j \phi(q_j)\phi(f_j) + \phi(r)$ is a standard expression for $\phi(f)$ under $\prec$, where $\phi(\bar{f})$ is a constant

$\bar{f} := 1;\ q_1 := 0, \dots, q_s := 0; r := 0; p := f;$
while $\phi(p) \neq 0$ **do**
  if $\exists\ i$ *s.t.* $in_\prec(\phi(f_i))$ *divides* $in_\prec(\phi(p))$ **then**
    $(g_1, g_2) := \mathcal{A}(p, f_i, \phi, \prec)$; `/* ϕ(g1) is a constant */`
    $\bar{f} := \bar{f} * g_1;\ q_1 := q_1 * g_1, \dots, q_s := q_s * g_1;\ r := r * g_1;$
    $p := p * g_1 - f_i * g_2;$
    $q_i := q_i + g_2$
  end
  else
    $g_1 := \phi^{-1}(in_\prec(\phi(p)));$
    $r := r + g_1; p := p - g_1$ `/* ϕ(p) = 0 */`
  end
end
$r := r + p.$

Algorithm 4. SHADOW_DIV$(f, \{f_1, \dots, f_s\}, \phi, \prec)$

In the first part of the algorithm the remainder r is appended in the successive bases. But $r = g_1 f - g_2 g - \sum_i q_i f_i$ so it is already in the ideal before division, so appending it to the basis does not expand the ideal.

Lemma 6. $\langle B \rangle = \langle C_1 \rangle$.

Lemma 7. *$\phi(C_1)$ is the Gröbner basis of $\langle \phi(B) \rangle$. Further, $\phi(C_2)$ is the reduced Gröbner basis for $\langle \phi(B) \rangle$.*

Proof. It was pointed out after Algorithm 3 that $\phi(g_1 f_1 - g_2 f_2) = S(\phi(f_1), \phi(f_2))$. Upon termination, $\phi(r) = 0$ for all $f_1, f_2 \in B_{new}$. So from Lemma 4 the standard expression for the S-polynomial is $S(\phi(f_1), \phi(f_2)) = \sum_j \phi(q_j)\phi(f_j)$ for all pairs $\phi(f_1), \phi(f_2) \in \phi(B_{new})$. From Büchberger's criterion $\phi(B_{new})$ is a Gröbner basis, see [7] section 7 of chapter 2.

In the second part $\phi(r)$ is the result of the reduction of $\phi(f)$ by $\phi(B_{new}) \setminus \{f\}$. So upon termination, no polynomial in $\phi(B_{new})$ is reducible by the rest of the polynomials. Thus $\phi(B_{new})$ is the reduced Gröbner basis. ■

Lemma 8. *For every $f \in \langle B \rangle$ there exists $h \in \phi^{-1}(1)$ such that $h \cdot f \in \langle C_2 \rangle$.*

Proof. In the second part of SHADOW_BÜCH, let the successive bases after each reduction be $C_1 = B_0, B_1, \cdots, B_k = C_2$ such that $B_{i+1} = (B_i \cup \{r\}) \setminus \{f\}$ where r is the result of reduction of $f \in B_i$ by $B_i \setminus \{f\}$. We will show that for each j, if $g \in B_0$, then there exists $h \in \phi^{-1}(1)$such that $h \cdot g \in B_j$.

The claim is trivially true for $j = 1$. In order to prove the claim by induction let us assume that it holds for $j \leq i$.

Data: $B = \{ f_1, \ldots, f_s \} \subseteq k[\mathbf{x}]$; a surjective ring homomorphism $\phi : k[\mathbf{x}] \to k[\mathbf{y}]$, a term order, $\prec$, in $k[\mathbf{y}]$

Result: A subset $C_1 \subset k[\mathbf{x}]$ such that $\langle C_1 \rangle = \langle B \rangle$ and $\phi(C_1)$ is a Gröbner basis of $\phi(\langle B \rangle)$, a subset $C_2 \subset k[\mathbf{x}]$ such that $\phi(C_2)$ is the reduced Gröbner basis of $\phi(\langle B \rangle)$ and for each $f \in \langle B \rangle$ there is $h \in \phi^{-1}(1)$ such that $h \cdot f \in \langle C_2 \rangle$

$B_{new} := B$;
repeat
  $B_{old} := B_{new}$;
  for *each pair* $f_1, f_2 \in B_{new}$ *s.t.* $f_1 \neq f_2$ *and* $\phi(f_1) \neq 0, \phi(f_2) \neq 0$ **do**
    $(g_1, g_2) := \mathcal{A}(f_1, f_2, \phi, \prec)$;
    compute SHADOW_DIV$(g_1 f_1 - g_2 f_2, B_{new}, \phi, \prec)$;
    if $\phi(r) \neq 0$ **then**
      $B_{new} := B_{new} \cup \{r\}$
    end
  end
until $B_{new} = B_{old}$;
$C_1 := B_{new}; B_{old} := B_{new}$;
for *each* $f \in B_{old}$ **do**
  compute SHADOW_DIV$(f, B_{new} \setminus \{f\}, \phi, \prec)$;
  $B_{new} := B_{new} \setminus \{f\}$;
  if $r \neq 0$ **then**
    $B_{new} := B_{new} \cup \{r\}$;
  end
end
$C_2 := B_{new}$;

Algorithm 5. SHADOW_BÜCH$(B, \phi, \prec)$

We have $B_i \setminus \{f\} = B_{i+1} \setminus \{r\}$. From SHADOW_DIV algorithm we know that for some constant c there exists $\bar{f} \in \phi^{-1}(c)$ such that $\bar{f} \cdot f \in B_{i+1}$. Let $g \in \langle B_0 \rangle$ then from induction hypothesis $\exists \bar{g} \in \phi^{-1}(1)$ such that $\bar{g} \cdot g \in \langle B_i \rangle$. Thus $\bar{g} \cdot g = \sum_k q_k f_k$ where $f_k \in B_i$. If f does not occur in the sum then $\bar{g} \cdot g \in \langle B_{i+1} \rangle$. Otherwise, suppose $f_1 = f$. So $(1/c)\bar{f} \cdot \bar{g} \cdot g = (1/c) \sum_k q_k (\bar{f} f_k) \in B_{i+1}$. The desired $h = (1/c)\bar{f} \cdot \bar{g}$. ∎

Remark. If the computation of ϕ and ϕ^{-1} is $O(1)$, then SHADOW_BÜCH and Büchberger's algorithm have same time-complexity.

4.3 Projection Homomorphism and Binomial Ideal

From now onwards we shall restrict our consideration to only projection homomorphisms. In the following we shall use z to denote those x variables which are set to 1 by the projection homomorphism and the remaining variables will be denoted by symbol y. For example, if we are considering $\phi = \Pi_i$ then $x_1, \cdots, x_i$ will be denoted by $z_1, \cdots, z_i$ and $x_{i+1}, \cdots, x_n$ will be denoted by $y_1, \cdots, y_{n-i}$ The steps computing $\phi^{-1}()$ in $\mathcal{A}$ (algorithm 3) and SHADOW_DIV (algorithm 5) are described as follows.

In algorithm $\mathcal{A}$ for $j = 1$ and 2, f_j must contain a sub-polynomial of the form $h_j(z).y^{\alpha_j}$ such that $\phi(h_j(z)) = c_j$ and $in_{\prec}(\phi(f'_j))$ is strictly less than $in_{\prec}(\phi(f))$ where $f'_j = f_j - h_j(z).y^{\alpha_j}$. We define steps 4 and 5 as $g_1 := h_2(z)y^{\beta_1}$ and $g_2 := h_1(z)y^{\beta_2}$.

In algorithm *SHADOW_DIV* there exists a sub-polynomial $h(z)y^{\alpha}$ in $p(x)$ such that $in_{\prec}(\phi(p - h(z)y^{\alpha}))$ is strictly less than $in_{\prec}(\phi(p))$. We define step 10 as $g_1 := h(z)y^{\alpha}$.

Observation 4. *If ϕ is a projection homomorphism and f_1, f_2 are homogeneous w.r.t.* $\mathbf{d}$ *and $(g_1, g_2) = \mathcal{A}(f_1, f_2, \phi, \prec)$, then $g_1 f_1 - g_2 f_2$ is also homogeneous w.r.t.* $\mathbf{d}$.

We further restrict our discussion to binomial ideals. If a binomial $f = x^{\alpha_1} - x^{\alpha_2}$ is such that $\phi(f)$ is non-zero, then $\phi(x^{\alpha_1}) \neq \phi(x^{\alpha_2})$. Hence g_1, g_2 in steps 4,5 in $\mathcal{A}$ and g_1 in step 10 in SHADOW_BÜCH are all monomials.

Observation 5. *If ϕ is a projection, f_1 and f_2 are binomials and $(g_1, g_2) = \mathcal{A}(f_1, f_2, \phi, \prec)$, then $f_1 g_1 - f_2 g_2$ is the S-polynomial of f_1, f_2, hence it is also a binomial.*

Observation 6. *If ϕ is a projection and B is a set of binomials, then $\bar{f}$ computed by SHADOW_DIV$(f, B, \phi, \prec)$ is a monomial. Additionally, if f and each member of B is homogeneous, then so is the remainder r.*

In the notation for the variables in this section, $\bar{f}$ computed by SHADOW_DIV is z^{α} for some α. Using this fact in the proof of Lemma 8 we get the following lemma which is at the heart of algorithm proposed in the next section.

Lemma 9. *If ϕ is a projection, B is a set of binomials, and C_2 is computed by SHADOW_BÜCH$(B, \phi, \prec)$, then for each binomial $f \in \langle B \rangle$ there exists a monomial z^{α} such that $z^{\alpha} f \in \langle C_2 \rangle$.*

5 A Fast Algorithm for Toric Ideals

Let B be a finite set of binomials from $k[\mathbf{x}]$. In Algorithm 6 we present a new algorithm to compute $\langle B \rangle : (x_1 \ldots x_n)^{\infty}$ which is the key step in the computation of (generator set of) a toric ideal I_A given a matrix A.

To prove the correctness of Algorithm 6 let us assume that at the start of an iteration the value of B is B_{old} and at its end it is B_{new}. From Lemma 9 $\langle \tilde{C_2} \rangle : (x_1 \cdots x_i)^{\infty} = \langle \tilde{B} \rangle : (x_1 \cdots x_i)^{\infty}$ and from Lemma 7 $\Pi_i(\tilde{C_2})$ is the reduced Gröbner basis of $\langle \Pi(\tilde{B}) \rangle$. From Theorem 2 $\langle \tilde{C_2} \div x_{i+1}^{\infty} \rangle = \langle \tilde{C_2} \rangle : x_{i+1}^{\infty}$. Hence $\langle \tilde{C_2} \div x_{i+1}^{\infty} \rangle : (x_1 \cdots x_i)^{\infty} = (\langle \tilde{C_2} \rangle : (x_1 \cdots x_i)^{\infty}) : x_{i+1}^{\infty} = (\langle \tilde{B} \rangle : (x_1 \cdots x_i)^{\infty}) : x_{i+1}^{\infty} = (\langle \tilde{B} \rangle : (x_1 \cdots x_{i+1})^{\infty})$. Taking the projection π_y we get $\langle B_{new} \rangle : (x_1 \cdots x_i)^{\infty} = \langle B_{old} \rangle : (x_1 \cdots x_{i+1})^{\infty}$. Thus we have the correctness theorem.

Theorem 7. *Algorithm 6 correctly computes $\langle B \rangle : (x_1 \cdots x_n)^{\infty}$.*

Data: $B \subset k[\mathbf{x}]$, a finite set of binomials
Result: A generating set of $\langle B \rangle : (x_1 \dots x_n)^\infty$
for $i = n - 1$ *to* 0 **do**
  $\boldsymbol{d} := (0_1, \dots, 0_i, 1_{i+1}, \dots, 1_n, 1_y)$;
  $\tilde{B} = \{\tilde{f} | f \in B\}$;
  $\tilde{C_2} :=$ SHADOW_BÜCH$(\tilde{B}, \Pi_i, \prec_{\boldsymbol{d},i+1})$;
  $B := \pi_y(\tilde{C_2} \div x_{i+1}^\infty)$;
end
return B;

Algorithm 6. Computation of $I : (x_1 \cdots x_n)^\infty$ for a binomial ideal I

The advantage of the new algorithm is as follows. In this algorithm the dimension of the y-space is 1 in the first iteration, 2 in the second iteration, so on. Symbolically let $t(i)$ denote the time complexity of the Büchberger's algorithm in i variable problem. Then, as remarked after Lemma 8, the cost of the proposed algorithm is $\sum_{i=1}^{n} t(i)$ against the Sturmfels' algorithm's cost $n \cdot t(n)$.

6 Experimental Results

In this section we present the results on performance of the new algorithm and compare it with the existing algorithm by Sturmfels [10]. In these experiments we randomly generate binomials and compute $J_V : (x_1 \dots x_n)^\infty$. The programs were written in C. There are cases where one can ignore certain S-polynomial reduction in the Büchberger algorithm for Gröbner basis computation. There is a significant literature on criteria to select such S-polynomials. We only applied one such criterion, referred as *criterion tail* in Proposition 3.15 of [15] in the implementation of the new algorithm as well as to Sturmfels algorithm. Since every such criterion can be applied to both algorithms, we believe the performance gains shown here will remain same after the implementations are fully optimized.

Table 1.

Number of variables	Size of basis		Time taken (in sec.)		Speedup
	Initial	Final	Sturmfels	Proposed	
6	2	5	0.0	0.00	-
	4	51	0.001	0.00	-
8	4	186	0.12	0.02	6
	6	597	6.58	0.64	10.3
10	6	729	18.16	0.50	36.3
	8	357	2.68	0.29	9.2
12	6	423	4.04	0.27	14.9
	8	2695	822.12	27.21	30.2
14	10	1035	127.97	4.24	30.1

Table 1 shows performances of the two algorithms. Although only a few cases are shown in the table we ran an extensive experiment and in each and every case the proposed algorithm was faster. Also, as expected the performance ratio improves as the number of variables increase.

References

1. Conti, P., Traverso, C.: Buchberger algorithm and integer programming. In: AAECC, pp. 130–139 (1991)
2. Urbaniak, R., Weismantel, R., Ziegler, G.M.: A variant of the Buchberger algorithm for integer programming. SIAM J. Discret. Math. 10, 96–108 (1997)
3. Thomas, R.R., Weismantel, R.: Truncated Gröbner bases for integer programming. In: Engineering, Communication and Computing, pp. 241–257. Kluwer Academic Publishers, Dordrecht (1995)
4. Sturmfels, B.: Grobner Bases and Convex Polytopes. University Lecture Series, vol. 8. American Mathematical Society, Providence (1995)
5. Tayur, S.R., Thomas, R.R., Natraj, N.R.: An algebraic geometry algorithm for scheduling in the presence of setups and correlated demands. Mathematical Programming 69, 369–401 (1995)
6. Adams, W.W., Loustaunau, P.: An Introduction to Gröbner Bases. Graduate Studies in Mathematics, vol. 3. American Mathematical Society, Rhode Island (1994)
7. Cox, D.A., Little, J., O'Shea, D.: Ideals, Varieties, and Algorithms: An Introduction to Computational Algebraic Geometry and Commutative Algebra, 3/e (Undergraduate Texts in Mathematics). Springer, Secaucus (2007)
8. Buchberger, B.: A theoretical basis for the reduction of polynomials to canonical forms. SIGSAM Bull. 10(3), 19–29 (1976)
9. Biase, F.D., Urbanke, R.: An algorithm to calculate the kernel of certain polynomial ring homomorphisms. Experimental Mathematics 4, 227–234 (1995)
10. Hosten, S., Sturmfels, B.: Grin: An implementation of Gröbner bases for integer programming (1995)
11. Bigatti, A.M., Scala, R., Robbiano, L.: Computing toric ideals. J. Symb. Comput. 27, 351–365 (1999)
12. Hemmecke, R., Malkin, P.N.: Computing generating sets of lattice ideals and Markov bases of lattices. Journal of Symbolic Computation 44, 1463–1476 (2009)
13. Bayer, D., Stillman, M.: On the complexity of computing syzygies. J-SYMBOLIC-COMP 6, 135–148 (1988) (or 135–147??) Computational aspects of commutative algebra
14. Bayer, D., Stillman, M.: A theorem on refining division orders by the reverse lexicographic order. Duke Mathematical Journal 55, 321–328 (1987)
15. Bigatti, A.M., Scala, R., Robbiano, L.: Computing toric ideals. J. Symb. Comput. 27, 351–365 (1999)

Linear and Sublinear Time Algorithms for Basis of Abelian Groups*

Li Chen[1] and Bin Fu[2]

[1] Department of Computer Science, University of District of Columbia, Washington, DC 20008, USA
lchen@udc.edu

[2] Dept. of Computer Science, University of Texas - Pan American TX 78539, USA
binfu@cs.panam.edu

Abstract. It is well known that every finite abelian group G can be represented as a direct product of cyclic groups: $G \cong G_1 \times G_2 \times \cdots \times G_t$, where each G_i is a cyclic group of order p^j for some prime p and integer $j \geq 1$. If a_i generates the cyclic group of G_i, $i = 1, 2, \cdots, t$, then the elements $a_1, a_2, \cdots, a_t$ are called a basis of G. We show a randomized algorithm such that given a set of generators $M = \{x_1, \cdots, x_k\}$ for an abelian group G and the prime factorization of order $\text{ord}(x_i)$ $(i = 1, \cdots, k)$, it computes a basis of G in $O(|M|(\log n)^2 + \sum_{i=1}^{t} n_i p_i^{n_i/2})$ time, where $n = |G|$ has prime factorization $p_1^{n_1} p_2^{n_2} \cdots p_t^{n_t}$ (which is not a part of input). This generalizes Buchmann and Schmidt's algorithm that takes $O(|M|\sqrt{|G|})$ time. In another model, all elements in an abelian group are put into a list as a part of input. We obtain an $O(n)$ time deterministic algorithm and a subliner time randomized algorithm for computing a basis of an abelian group.

1 Introduction

Abelian groups are groups with commutative property. It is well known that a finite Abelian group can be decomposed to a direct product of cyclic groups with prime-power order (called cyclic p-groups) [9]. The set of generators with exactly one from each of those cyclic groups form a basis of the abelian group. Because a basis of an abelian group fully determines its structure, which is the nondecreasing orders of the elements in a basis, finding a basis is crucial in computing the general properties for abelian groups. The orders of all elements in a basis form the invariant structure of an abelian group. There is a long line of research about the algorithm for determining group isomorphism (e.g. [14,8,12,13,16,20,10,6,11]). Two abelian groups are isomorphic if and only if they have the same structure.

For finding a basis of abelian group, Chen [4] showed an $O(n^2)$ time algorithm for finding a basis of an abelian group G given all elements and size of G

* This research is supported in part by NSF Early Career Award 0845376.

Y. Dong, D.-Z. Du, and O. Ibarra (Eds.): ISAAC 2009, LNCS 5878, pp. 493–502, 2009.

as input. An abelian group is often represented by a set of generators in filed of computational group theory (e.g., [18]) as a set of generators costs a small amount memory. The algorithm for basis of abelian group with a set of generators as input was developed by Buchmann, et al [2], Teske [19], and Buchmann and Schmidt [3] with the fastest proven time $O(m\sqrt{|G|})$. The methods for computing the order for one element in a group also has connection with computing the abelian basis was also reported in [2,17].

We show a randomized algorithm such that given a set of generators $M = \{x_1, \cdots, x_k\}$ for an abelian group G and the prime factorization of order $\mathrm{ord}(x_i)$ ($i = 1, \cdots, k$), it computes a basis of G in $O(|M|(\log n)^2 + \sum_{i=1}^{t} n_i p_i^{n_i/2})$ time, where $n = |G|$ has prime factorization $p_1^{n_1} p_2^{n_2} \cdots p_t^{n_t}$ (which is not a part of input). This implies an algorithm such that given an abelian group G represented by a set of generators $M = \{x_1, \cdots, x_k\}$ without their orders information, it computes a basis of G in $O(|M|(\log n)^2 + \sum_{i=1}^{t} n_i p_i^{n_i/2} + (\sum_{i=1}^{t} \sqrt{\mathrm{ord}(x_i)}))$ time. This improves Buchmann and Schmidt's algorithm that takes $O(|M|\sqrt{|G|})$ time.

In the model of all elements in an abelian group being put into a list as a part of input, we derive an $O(\sum_{i=1}^{t} n_i \min(p_i^{n_i/2}, p_i^{n_i-1}) + \sum_{i=1}^{t} n_i \log n)$-time randomized algorithm to compute a basis of abelian group G of order n with factorization $n = p_1^{n_1} \cdots p_t^{n_t}$, which is also a part of the input. It implies an $O(n^{1/2} \sum_{i=1}^{t} n_i)$-time randomized algorithm to compute a basis of an abelian group G of order n. It also implies that if n is an integer in $\{1, 2, \cdots, m\} - G(m, c)$, then a basis of an abelian group of order n can be computed in $O((\log n)^{c+1})$-time, where c is any positive constant and $G(m, c)$ is a subset of the small fraction of integers in $\{1, 2, \cdots, m\}$ with $\frac{|G(m,c)|}{m} = O(\frac{1}{(\log m)^{c/2}})$ for every integer m. We show an algorithm such that given a set of generators $M = \{x_1, \cdots, x_k\}$ for an abelian group G and the prime factorizations of orders $\mathrm{ord}(x_i)$ ($i = 1, \cdots, k$), it computes a basis of G in $O(|M|(\sum_{i=1}^{t} p_i^{n_i/2}))$ time, where $n = |G|$ has prime factorization $p_1^{n_1} p_2^{n_2} \cdots p_t^{n_t}$ (which is not a part of input). We also obtain an $O(n)$-time deterministic algorithm for computing a basis of an abelian group with n elements. The existing algorithms need $O(n^2)$ time by Chen and $O(n^{1.5})$ time by Buchmann and Schmidt .

2 Notations

For two positive integers x and y, (x, y) represents the greatest common divisor (GCD) between them. For a set A, $|A|$ denotes the number of elements in A. For a real number x, $\lfloor x \rfloor$ is the largest integer $\leq x$ and $\lceil x \rceil$ is the smallest integer $\geq x$. For two integers x and y, $x|y$ means that $y = xc$ for some integer c.

A *group* is a nonempty set G with a binary operation "$\cdot$" that is closed in set G and satisfies the following properties (for simplicity, "ab" represents "$a \cdot b$"): 1)for every three elements a, b and c in G, $a(bc) = (ab)c$; 2)there exists an identity element $e \in G$ such that $ae = ea = a$ for every $a \in G$; 3)for every element $a \in G$, there exists $a^{-1} \in G$ with $aa^{-1} = a^{-1}a = e$. A group G is *finite* if G contains finite elements. Let e be the identity element of G, i.e. $ae = a$ for each $a \in G$.

For $a \in G$, $\text{ord}(a)$, the order of a, is the least integer k such that $a^k = e$. For $a \in G$, define $\langle a \rangle$ to be the subgroup of G generated by the element a (in other words, $\langle a \rangle = \{e, a, a^2, \cdots, a^{\text{ord}(a)-1}\}$). Let A and B be two subsets of group G, define $AB = A \cdot B = A \circ B = \{ab | a \in A \text{ and } b \in B\}$. We use $\cong$ to represent the isomorphism between two groups.

A group G is an *abelian* group if $ab = ba$ for every pair of elements $a, b \in G$. Assume that G is an abelian group with elements $g_1, g_2, \cdots, g_n$. For each element $g_i \in G$, it corresponds to an index i. According to the theory of abelian group, a finite abelian group G of n elements can be represented as $G = G(p_1^{n_1}) \circ G(p_2^{n_2}) \circ \cdots \circ G(p_t^{n_t}) \cong G(p_1^{n_1}) \times G(p_2^{n_2}) \times \cdots \times G(p_t^{n_t})$, where $n = p_1^{n_1} p_2^{n_2} \cdots p_t^{n_t}$, $p_1 < p_2 < \cdots < p_t$ are the prime factors of n, and $G(p_i^{n_i})$ is a subgroup of G with $p_i^{n_i}$ elements (see [9]). We also use the notation G_{p_i} to represent the subgroup of G with order $p_i^{n_i}$. Any abelian group G of order p^m can be represented by $G = G(p^{m_1}) \circ G(p^{m_2}) \circ \cdots \circ G(p^{m_k}) \cong G(p^{m_1}) \times G(p^{m_2}) \times \cdots \times G(p^{m_k})$, where $m = \sum_{i=1}^{k} m_i$ and $1 \le m_1 \le m_2 \le \cdots \le m_k$. Notice that each $G(p^{m_i})$ is a cyclic group.

For, $a_1, a_2, \cdots, a_k$ from the abelian group G, denote $\langle a_1, a_2, \cdots, a_k \rangle$ to be the set of all elements in G *generated* by $a_1, \cdots, a_k$. In other words, $\langle a_1, a_2, \cdots, a_k \rangle = \langle a_1 \rangle \langle a_2 \rangle \cdots \langle a_k \rangle$. An element $a \in G$ is *independent of* $a_1, a_2, \cdots, a_k$ in G if $a \neq e$ and $\langle a_1, a_2, \cdots, a_k \rangle \cap \langle a \rangle = \{e\}$. If $G = \langle a_1, a_2, \cdots, a_k \rangle$, then $\{a_1, a_2, \cdots, a_k\}$ is called a set of *generators* of G. If X is a set of elements in G, we also use $\langle X \rangle$ to represent the subgroup generated by set X.

The elements $a_1, a_2, \cdots, a_k$ from the abelian group G are *independent* if a_i is independent of $a_1, \cdots, a_{i-1}, a_{i+1}, \cdots, a_k$ for every i with $1 \le i \le k$. A basis of G is a set of independent elements $a_1, \cdots, a_k$ that can generate all elements of G (in other words, $G = \langle a_1, a_2, \cdots, a_k \rangle$).

3 Randomized Algorithm for Basis via Generators

An abelian group is often represented by a set of generators. The set of generators for a group is usually much less than the order of a group. It is important to find the algorithm for computing a basis of abelian group represented by a set of generators. The randomized algorithms in this paper belong to Monte Carlo algorithms [1], which have a small probability to output error results.

Let $B = \{b_1, \cdots, b_k\}$ be a set of basis for an abelian group G of size p^m (p is a prime) and assume that $\text{ord}(b_1) \le \text{ord}(b_2) \le \cdots \le \text{ord}(b_k)$. The *structure* of G is defined by $\langle \text{ord}(b_1), \text{ord}(b_2), \cdots, \text{ord}(b_k) \rangle$. We note that the structure of an abelian group is invariant, but its basis is not unique.

The theorem of Buchmann and Schmidt [3] is used in our algorithm for finding a basis of abelian group. The following Theorem 1 follows from Lemma 3.1 and Theorem 3.4 in [3].

Theorem 1 ([3]). *There exists an $O(m\sqrt{|G|})$ time algorithm such that given a set of generators of order m for an abelian group G of order p^t for some prime number p and intger $t \ge 1$, the algorithm returns a basis and the structure of G in $O(m\sqrt{|G|})$ steps.*

Theorem 2 ([3]). *There exists an algorithm such that given an element g of an abelian group G, it returns* $\text{ord}(g)$ *in* $O(\sqrt{\text{ord}(g)})$ *steps.*

Lemma 1. *Assume G is an abelian group of order n. We have the following two facts: 1) If $n = m_1 m_2$ with $(m_1, m_2) = 1$, $G' = \{a \in G | a^{m_1} = e\}$ and $G'' = \{a^{m_1} | a \in G\}$, then both G' and G'' are subgroups of G, $G = G' \circ G''$, $|G'| = m_1$ and $|G''| = m_2$. Furthermore, for every $a \in G$, if* $(\text{ord}(a), m_1) = 1$, *then $a \in G''$. 2)If $n = p_1^{n_1} p_2^{n_2} \cdots p_t^{n_t}$, then $G = G(p_1^{n_1}) \circ G(p_2^{n_2}) \circ \cdots \circ G(p_t^{n_t})$, where $G(p_i^{n_i}) = \{a \in G | a^{p_i^{n_i}} = e\}$ for $i = 1, \cdots, t$.*

Proof. It is easy to verify that G' is subgroup of G. Assume $a_1, \cdots, a_{s_1}, b_1, \cdots, b_{s_2}$ are the elements in a basis of G such that $\text{ord}(a_i) | m_1$ for $i = 1, \cdots, s_1$ and $\text{ord}(b_j) | m_2$ for $j = 1, \cdots, s_2$. It is easy to see that $a_i^{m_1} = e$ for $i = 1, \cdots, s_1$ and $b_j^{m_1} \neq e$ for $j = 1, \cdots, s_2$. For each b_j, $\langle b_j \rangle = \langle b_j^{m_1} \rangle$ since $(m_1, m_2) = 1$ and $\text{ord}(b_j) | m_2$. Assume that $x = a^{m_1}$ and $y = a'^{m_1}$. Both x and y belong to G''. Let's consider $xy = (aa')^{m_1}$. We still have $xy \in G''$. Thus, G'' is closed under multiplication. Since G'' is a subset of a finite group, G'' is a group. Therefore, G'' is a group generated by $b_1^{m_1}, \cdots, b_{s_2}^{m_1}$ that is the same as the group generated by $b_1, \cdots, b_{s_2}$. Therefore, G'' is of order m_2. On the other hand, G' has basis of elements $a_1, \cdots, a_{s_1}$ and is of order m_1. We also have that $G' \cap G'' = \{e\}$. It is easy to see that $G = G' \circ G''$. For $a \in G$ with $(\text{ord}(a), m_1) = 1$, $\langle a^{m_1} \rangle = \langle a \rangle$ and $a^{m_1} \neq e$. So, we have $a^{m_1} \in G''$, which implies that $a \in \langle a \rangle = \langle a^{m_1} \rangle \subseteq G''$. Part 2) follows from part 1).

Lemma 2. *Let $M = \{x_1, \cdots, x_k\}$ be a set of generators for an abelian group G. Assume that $|G| = n = p_1^{n_1} \cdots p_t^{n_t}$ is the prime factorization of the order of G. Let $m_i = \max\{t_i : p_i^{t_i} | \text{ord}(x_j)$ for some x_j in $M\}$ and $u_i = \prod_{v \neq i} p_v^{m_v}$ for $i = 1, \cdots, t$. Let $M_i = \{x_1^{u_i}, \cdots, x_k^{u_i}\}$. Then M_i is a set of generators for $G(p_i^{n_i})$.*

Proof. For each $x_j^{u_i} \in M_i$, we have $(x_j^{u_i})^{p_i^{n_i}} = e$. Therefore, all elements of M_i are in $G(p_i^{n_i})$ (by Lemma 1). Let g be an arbitrary element in $G(p_i^{n_i})$. By Lemma 1, $g^{p_i^{n_i}} = e$. Since M is a set of generators for G, let $g = x_1^{z_1} \cdots x_k^{z_k}$. Since the greatest common divisor $(u_i, p_i^{n_i}) = 1$, there exist two integers y_1 and y_2 such that $y_1 u_i + y_2 p_i^{n_i} = 1$. We have that

$$g = g^{y_1 u_i + y_2 p_i^{n_i}} = g^{y_1 u_i} g^{y_2 p_i^{n_i}} = g^{y_1 u_i} = (x_1^{z_1} \cdots x_k^{z_k})^{y_1 u_i} = (x_1^{u_i})^{z_1 y_1} \cdots (x_k^{u_i})^{z_k y_1}.$$

We just show that g can be generated by the elements in M_i. Therefore, M_i is a set of generator for $G(p_i^{n_i})$.

Let $X = \{x_1, \cdots, x_k\}$ be a set elements in a group G. Define a p-random product $x_1^{a_1} \cdots x_k^{a_k}$, where $a_1, \cdots, a_k$ are independent random integers in $[0, p-1]$.

Lemma 3. *Let G' be a proper subgroup of an abelian group $G = \langle x_1, \cdots, x_k \rangle$ of order p^m for some prime p. Let g be a p-random product of $\{x_1, \cdots, x_k\}$. Then* $\Pr(g \in G') \leq \frac{1}{p}$.

Proof. Since $G' \neq G$, let i be the least index such that $x_i \notin G'$. Consider $g = x_1^{a_1} \cdots x_{i-1}^{a_{i-1}} x_i^{a_i} x_{i+1}^{a_{i+1}} \cdots x_k^{a_k}$. Let $u = x_1^{a_1} \cdots x_{i-1}^{a_{i-1}}$ and $v = x_{i+1}^{a_{i+1}} \cdots x_k^{a_k}$. For any fixed u and v, there exists at most one integer $a_i \in [0, p-1]$ such that $ux_i^{a_i}v \in G'$. Assume that there exist $a_i' < a_i'' \in [0, p-1]$ such that $ux_i^{a_i'}v \in G'$ and $ux_i^{a_i''}v \in G'$. We have that $x_i^{a_i''-a_i'} \in G'$ since G is an abelian group. Let $\text{ord}(x_i) = p^s$. There exists an integer j such that $j(a_i'' - a_i') = 1 (\text{mod } p^s)$ since $a_i'' - a_i' \in (0, p-1]$. Clearly, $x_i^{a_i''-a_i'} \in G'$ implies $x_i = x_i^{j(a_i''-a_i')} \in G'$. A contradiction. Therefore, with probability at most $\frac{1}{p}$, $g \in G'$.

Lemma 4. *There exists a randomized algorithm such that given a set of generators $M = \{x_1, x_2, \cdots, x_k\}$ for a finite abelian p-group G, prime p, and integer $h \geq 1$, it computes a basis for G in $O(|M|hr\log p + (r+h)p^{r/2})$ time with probability at most p^{-h} to fail, where $|G| = p^r$ (which is not a part of input).*

Proof. We have the algorithm Randomly-Find-Basis-for-p-Group to find a basis for a p-group.

Algorithm Randomly-Find-Basis-for-p-Group

Input: prime p, a set of generators $x_1, \cdots, x_k$ of a finite abelian group G of order p^m (p^m is not a part of input), and a parameter h.

Output: a basis of G

Steps:

Let $A_0 = \{e\}$ (only contains the identity).

Let $B_0 = \{e\}$

Let $S_0 = \langle e \rangle$ (the structure for the group with one element).

$i = 0$.

Repeat

$i = i + 1$.

Generate h p-random products $a_1, \cdots, a_h$ of M.

Let $A_i = B_{i-1} \cup \{a_1, \cdots, a_h\}$.

Let B_i be a basis of $\langle A_i \rangle$ and S_i be the structure of $\langle A_i \rangle$ by the Algorithm in Theorem 1.

Until $S_i = S_{i-1}$.

Output B_{i-1} as a basis of G.

End of Algorithm

We prove that the algorithm has a small probability failing to return a basis of G. Assume that the subgroup $\langle A \rangle$ is not equal to G. By Lemma 3, for a p-random product g of M, the probability is at most $\frac{1}{p}$ that $g \in \langle A \rangle$. Therefore, for h p-random elements $a_1, \cdots, a_h$, the probability that all $a_1, \cdots, a_h$ are in $\langle A \rangle$ is at most p^{-h}. We have that the probability at most p^{-h} that the algorithm stops before returning a basis of G.

Each cycle in the loop of the algorithm is indexed by the variable i. Since G is of order p^r, the order $|\langle B_i \rangle|$ of subgroup $\langle B_i \rangle$ of G is p^{m_i} for some integer m_i. A basis of G contains at most r elements since $|G| = p^r$. Therefore, $|B_i| \leq r$. It takes $O(|M| \log p)$ time to generate one p-random product. The time spent in cycle i is $O(|M|h\log p+(|B_i|+h)\sqrt{|\langle B_i\rangle|})$. The loop is repeated at most r times since $\langle B_{i-1}\rangle \neq \langle B_i\rangle$. Assume the algorithm stops when $i=i_0$. The total time is $O(\sum_{i=1}^{i_0}(|M|h\log p+(|B_i|+h)\sqrt{|\langle B_i\rangle|})$. Since $\langle B_0\rangle \neq \langle B_1\rangle \neq \cdots \neq \langle B_{i_0}\rangle$, we have that $0=m_0<m_1<\cdots<m_{i_0}\leq r$. We have $\sum_{i=1}^{i_0}((|B_i|+h)\sqrt{|\langle B_i\rangle|})\leq \sum_{i=1}^{r}((r+h)\sqrt{p^i}=(r+h)\frac{(\sqrt{p})^{r+1}-1}{\sqrt{p}-1}$. The total time is $O(|M|hr\log p+(r+h)p^{r/2})$.

Theorem 3. *Let ϵ be a small constant greater than 0. Then there exists a randomized algorithm such that given a set of generators $M=\{x_1,x_2,\cdots,x_k\}$ for a finite abelian group G and the prime factorization for the order* $\mathrm{ord}(x_i)$ *of every $x_i (i=1,\cdots,k)$, it computes a basis for G in $O((|M|(\log n)^2+\sum_{i=1}^{t}n_ip_i^{n_i/2}))$ time and has probability at most ϵ to fail, where $n=|G|$ has prime factorization $p_1^{n_1}p_2^{n_2}\cdots p_t^{n_t}$ (which is not a part of input) with $p_1<p_2<\cdots<p_t$.*

Proof. Our algorithm to find a basis of G is decomposed into finding a basis of every p-group of G. The union of every basis among all p-subgroups of G is a basis of G. Let h be a constant such that $\frac{1}{(h-1)2^{h-1}}\leq\epsilon$.

Algorithm Randomly-Find-Basis-By-Generators

Input: a set of generators $x_1,\cdots,x_k$ of a finite abelian group G and the prime factorization for every $\mathrm{ord}(x_i)$ $(i=1,\cdots,k)$.

Output: a basis of G

Steps:

Let $p_1,\cdots,p_t$ be all of the prime numbers p_j with $p_j|\mathrm{ord}(x_i)$ for some i in $\{1,2,\cdots,k\}$.

For $i=1$ to t let $v_i=\max\{p_i^{t_i}: p_i^{t_i}|\mathrm{ord}(x_j)$ for some x_j in $M\}$.

Let $u=v_1v_2\cdots v_t$.

For $i=1$ to t let $u_i=\frac{u}{v_i}$.

For $i=1$ to t let $M_i=\{x_1^{u_i},\cdots,x_k^{u_i}\}$.

For $i=1$ to t

Let B_i be a basis of $\langle M_i\rangle$ by the Algorithm in Lemma 4 with input p_i, M_i, and h.

Output $B_1\cup B_2\cup\cdots\cup B_t$ as a basis of G.

End of Algorithm

By Lemma 2, M_i is a set of generator for G_{p_i}. By Lemma 4, the probability is at most p_i^{-h} that B_i is not a basis of G_{p_i}. The probability failing to output a basis of G is at most $\sum_{i=1}^{t}p_i^{-h}<\sum_{i=p_1}^{\infty}\frac{1}{i^h}\leq\int_{p_1}^{\infty}\frac{1}{x^h}d_x\leq\frac{1}{(h-1)p_1^{h-1}}\leq\epsilon$ since h is selected with $\frac{1}{(h-1)2^{h-1}}\leq\epsilon$. By Lemma 2, $B_1\cup B_2\cup\cdots\cup B_t$ is a basis of G.

Since the prime factorization of the order $\mathrm{ord}(x_i)$ for $i=1,\cdots,k$ is a part input, it takes $O(|M|t)$ time to compute one v_i. It takes $O(|M|t^2)=O(|M|(\log n)^2)$ time to compute $v_1,\cdots,v_t$. It takes $O(t)$ time to compute u and $u_1,\cdots,u_t$.

The time for computing each element in M_i is $O(\log n)$ since $u_i\leq n$ and computing the power function (x^n) takes $O(\log n)$ time. It takes $O(|M|\log n)$ time to generate one set M_i and $O(|M|t\log n)=O(|M|(\log n)^2)$ time to generate all $M_1, \cdots, M_t$. By Lemma 4, the computational time for computing each basis of $\langle M_i \rangle$ is $O(|M_i|n_i h + (n_i + h)p_i^{n_i/2})$. The total time is $O((|M|(\log n)^2 + (\sum_{i=1}^t n_i p_i^{n_i/2})))$ since h is a constant, $|M_i| = |M|$, and $\sum_{i=1}^t n_i = O(\log n)$.

The fastest-known fully proven deterministic algorithm for integer factorization is the Pollard-Strassen method, which is stated in Theorem 4.

Theorem 4 ([15,7]). *There exists an $2^{O((\log n)^{1/3}(\log\log n)^{2/3}))}$ time algorithm to factorize any integer n.*

We have Theorem 5 to compute a basis of an abelian group only given a set of generators. Some additional time is needed to compute the orders of elements among generators.

Theorem 5. *There exists a randomized algorithm such that given a set of generators $M = \{x_1, x_2, \cdots, x_k\}$ for a finite abelian group G of order n, it computes a basis for G in $O(|M|(\log n)^2 + \sum_{i=1}^t n_i p_i^{n_i/2} + \sum_{i=1}^t \sqrt{\mathrm{ord}(x_i)})$ time, where n has prime factorization $p_1^{n_1} p_2^{n_2} \cdots p_t^{n_t}$ (which is not a part of input).*

We have Theorem 6 to compute a basis of an abelian group only given a set of generators and their orders. Some additional time is needed to factorize the orders of elements among generators.

Theorem 6. *There exists a randomized algorithm such that given a set of generators $M = \{x_1, x_2, \cdots, x_k\}$ and their orders for a finite abelian group G of order n, it computes a basis for G in $O(|M|(\log n)^2 + \sum_{i=1}^t n_i p_i^{n_i/2} + |M| 2^{O((\log n)^{1/3}(\log\log n)^{2/3}))}$ time, where n has prime factorization $p_1^{n_1} p_2^{n_2} \cdots p_t^{n_t}$, which is not a part of input.*

4 Sublinear Time Algorithm with Entire Group as Input

In this section, we present a sublinear time randomized algorithm for finding a basis of a finite abelian group. The input contains a list that holds all the elements of an abelian group. We first show how to convert a random element from G to its subgroup $G(p_i^{n_i})$ in Lemma 5.

Lemma 5. *Let $n = p_1^{n_1} \cdots p_t^{n_t}$ and G be an abelian group of n elements. Assume $m_i = \frac{n}{p_i^{n_i}}$ for $i = 1, \cdots, t$. If a is a random element of G that with probability $\frac{1}{|G|}$, a is equal to b for each $b \in G$, then a^{m_i} is a random element of $G(p_i^{n_i})$, the subgroup of G with p^{n_i} elements, such that with probability $\frac{1}{p_i^{n_i}}$, a^{m_i} is b for any $b \in G(p_i^{n_i})$*

Proof. Let $b_{i,1}, b_{i,2}, \cdots, b_{i,k_i}$ form a basis of $G(p_i^{n_i})$, i.e. $G(p_i^{n_i}) = \langle b_{i,1} \rangle \circ \cdots \circ \langle b_{i,k_i} \rangle$. Assume a is a random element in G. Let $a = (\prod_{j=1}^{k_i} b_{i,j}^{c_{i,j}})a'$, where a' is an element in $\prod_{j \neq i} G(p_j^{n_j})$. For every two integers $x \neq y \in [0, p_i^{n_i} - 1]$, $m_i x \neq m_i y (\bmod\ p_i^{n_i})$ (Otherwise, $m_i x = m_i y (\bmod\ p_i^{n_i})$ implies $x = y$ because

$(m_i, p_i) = 1$). Thus, the list of numbers $m_i \cdot 0 (\bmod\ p_i^s), m_i \cdot 1 (\bmod\ p_i^s), \cdots, m_i(p_i^s - 1)(\bmod\ p_i^s)$ is a permutation of $0, 1, \cdots, p_i^s - 1$ for any integer $s \geq 1$. Thus, if $c_{i,j}$ is a random integer in the range $[0, \text{ord}(b_{i,j}) - 1]$ such that with probability $\frac{1}{\text{ord}(b_{i,j})}$, $c_{i,j} = c'$ for each $c' \in [0, \text{ord}(b_{i,j}) - 1]$, then the probability is also $\frac{1}{\text{ord}(b_{i,j})}$ that $m_i c_{i,j} = c'$ for each $c' \in [0, \text{ord}(b_{i,j}) - 1]$. Therefore, $a^{m_i} = ((\prod_{j=1}^{k_i} b_{i,j}^{c_{i,j}})a')^{m_i} = \prod_{j=1}^{k_i} b_{i,j}^{m_i c_{i,j}}$, which is a random element in $G(p_i^{n_i})$.

Lemma 6. *Let G be a group of order p^r. Then the probability is at most $\frac{2}{p^h \ln p}$ that a set of $r + 2h \log h + 9h$ random elements from G cannot generate G.*

Proof. For every subgroup G' of G, if $|G'| = p^s$, then the probability is p^{s-r} that a random element of G is in G'. We use this fact to construct a series of subgroups $G_0 = \langle e \rangle \subseteq G_1 \subseteq G_2 \subseteq \cdots \subseteq G_{r'}$ with $r' \leq r$. Each G_i is $\langle H_i \rangle$, where H_i is a set of random elements from G and we have the chain $H_0 = \{e\} \subset H_1 \subset H_2 \subset \cdots \subset H_{r'}$, which shows that H_{i+1} is extended from H_i by adding some additional random elements to H_i.

Let G_i be a subgroup generated by some elements $G_i = \langle H_i \rangle$. If $|G_i| = p^s \leq p^{r-h}$, then add one more random element to H_i to form H_{i+1}. With probability at most p^{s-r}, the new element is in G_i. Let a be the random element to be added to H_i. Therefore, $H_{i+1} = H_i \cup \{a\}$, $G_{i+1} = \langle H_{i+1} \rangle$, and the probability is at most p^{s-r} that $G_i = G_{i+1}$.

Now assume that $|G_i| > p^{r-h}$. We add new elements according to size of G_i. Let $|G_i| = p^s$. We have $r - s < h$ since $p^s = |G_i| > p^{r-h}$. We will construct at most $h - 1$ extensions (from $G_i = \langle H_i \rangle$ to $G_{i+1} = \langle H_{i+1} \rangle$). It is easy to see that $[1, h] \subseteq \cup_{k=0}^{\lfloor \log h \rfloor} (\frac{h}{2^{k+1}}, \frac{h}{2^k}]$. For $0 < r - s < h$, there exists an integer $k \in [0, \lfloor \ln h \rfloor]$ such that $r - s \in (\frac{h}{2^{k+1}}, \frac{h}{2^k}]$. If $r - s$ is in the range $(\frac{h}{2^k}, \frac{h}{2^{k+1}}]$, then in order to form H_{i+1}, we add $2 \cdot 2^{k+1}$ new random elements to H_i. Then the probability is at most $(p^{s-r})^{2^{k+2}} \leq (p^{-\frac{h}{2^{k+1}}})^{2^{k+2}} \leq \frac{1}{p^{2h}}$ that all of the $2 \cdot 2^{k+1}$ new elements are in G_i. Thus, with probability at most $\frac{1}{p^{2h}}$ that $G_i = G_{i+1}$.

Let i_0 be the least integer i with $|G_i| > p^{r-h}$. The number of random elements used in $H_{i_0 - 1}$ is at most $r - h$ since one element is increased from H_{i-1} to H_i for $i < i_0$.

Let $j = \lfloor \ln h \rfloor$. The number of integers in $(\frac{h}{2^{k+1}}, \frac{h}{2^k}]$ is at most $\frac{h}{2^k} - \frac{h}{2^{k+1}} + 1 = \frac{h}{2^{k+1}} + 1$. For $i \geq i_0$, H_{i+1} is increased by $2 \cdot 2^{k+1}$ new random elements from H_i, where $|G_i| = p^s$ with $r - s \in (\frac{h}{2^{k+1}}, \frac{h}{2^k}]$. For all extensions from H_i to H_{i+1} after $i \geq i_0$, we need at most $((h - \frac{h}{2} + 1) \cdot 4 + (\frac{h}{2} - \frac{h}{4} + 1) \cdot 8 + \cdots + (\frac{h}{2^j} - \frac{h}{2^{j+1}} + 1) \cdot 2 \cdot 2^{j+1}) = (\sum_{i=0}^{j} 2h + \sum_{i=0}^{j} 2^{i+2}) \leq 2h(\ln h + 1) + 8h = 2h \ln h + 10h$ elements. The total number of random elements used is at most $(r - h) + (2h \ln h + 10h) = r + 2h \ln h + 9h$.

The probability that $G_i = G_{i+1}$ for some $i < i_0$ is at most $\sum_{i=h}^{\infty} \frac{1}{p^i}$. The probability $G_i = G_{i+1}$ for some $i \geq i_0$ is at most $\frac{(h-1)}{p^{2h}}$. The probability that $r + 2h \ln h + 9h$ random elements of G are not generators for G is at most $\sum_{i=h}^{\infty} \frac{1}{p^i} + \frac{(h-1)}{p^{2h}} \leq 2 \sum_{i=h}^{\infty} \frac{1}{p^i} \leq 2 \int_h^{\infty} \frac{1}{p^x} d_x \leq \frac{2}{p^h \ln p}$.

Theorem 7. *Let h be an integer parameter. There exists a randomized algorithm such that given an abelian group G of order n with $n = p_1^{n_1} \cdots p_t^{n_t}$ with $p_1 < p_2 < \cdots < p_t$, the algorithm computes a basis of G in $O(\sum_{i=1}^{t}(n_i + h\log h)\min(p_i^{n_i/2}, p_i^{n_i-1}) + \sum_{i=1}^{t}(n_i + h\log h)\log n)$ running time and has probability at most $\frac{2}{(h-1)p_1^{h-1}\ln p_1}$ to fail.*

Proof. It takes $O(\log n)$ steps to compute a^{m_i} for an element $a \in G$, where $m_i = \frac{n}{p_i^{n_i}}$. Each random element of G can be converted into a random element of $G(p_i^{n_i})$ by Lemma 5. Each $G(p_i^{n_i})$ needs $O(n_i + h\log h)$ random elements to find a basis by Lemma 6. Each $G(p_i^{n_i})$ needs $O((n_i+h\log h)\log n)$ time to convert the $O(n_i + h\log h)$ random elements from G to $G(p_i^{n_i})$. It takes $O(\sum_{i=1}^{t}(n_i + h\log h)\log n))$ time to convert random elements of G into the random elements in all subgroups $G(p_i^{n_i})$ for $i = 1, \cdots, t$. For $n = p_1^{n_1}\cdots p_t^{n_t}$, $\sum_{i=1}^{t} n_i \log p_i = \log n$.

If $n_i = 1$, we just select an nonidentity element to be the basis for $G(p_i^{n_i})$. If $n_i > 1$, by Theorem 1, each $G(p_i^{n_i})$ needs $O((n_i + h\log h)p_i^{n_i/2})$ time to find a basis for $G(p_i^{n_i})$. The time spend for computing a basis of $G(p_i^{n_i})$ is $O((n_i+h\log h)\min(p_i^{n_i/2}, p_i^{n_i-1}))$. The sum of time for all $G(p_i^{n_i})$s to find basis is $O(\sum_{i=1}^{t}(n_i+h\log h)\min(p_i^{n_i/2}, p_i^{n_i-1}))$. The total time for the entire algorithm is equal to the time for generating random elements for k subgroups $G(p_i^{n_i})$ and the time for computing a basis of every $G(p_i^{n_i})$ $(i = 1, \cdots, t)$. Thus, the total time can be expressed as $O(\sum_{i=1}^{t}(n_i+h\log h)\min(p_i^{n_i/2}, p_i^{n_i-1})+\sum_{i=1}^{t}(n_i+h\log h)\log n)$.

By Lemma 6, the probability is at most $\frac{2}{p_i^h \ln p_i}$ that we cannot get a set of generators for $G(p_i^{n_i})$ by selecting $O(n_i + h\log h)$ random elements in $G(p_i^{n_i})$. The total probability to fail is $\sum_{i=1}^{t}\frac{2}{p_i^h\ln p_i} \le \frac{2}{\ln p_1}\sum_{i=1}^{t}\frac{1}{p_i^h} \le \frac{2}{\ln p_1}\int_{p_1}^{\infty}\frac{1}{x^h}d_x = \frac{2}{(h-1)p_1^{h-1}\ln p_1}$.

Definition 1. *For an integer n, define $F(n) = \max\{p^{i-1} : p^i|n,\ p^{i+1} \not| n,\ i \ge 1,$ and p is a prime $\}$. Define $G(m,c)$ to be the set of all integers n in $[1,m]$ with $F(n) \ge (\log n)^c$ and $H(m,c) = |G(m,c)|$.*

Theorem 8. $\frac{H(m,c)}{m} = O(\frac{1}{(\log m)^{c/2}})$ *for every constant $c > 0$.*

Theorem 9. *There exists a randomized algorithm such that if n is in $[1,m] - G(m,c)$, then a basis of an abelian group of order n whose prime factorization is also part of the input can be computed in $O((\log n)^{c+1})$-time, where $c \ge 1$ is an arbitrary constant and $G(m,c)$ is a subset of integers in $[1,m]$ with $\frac{|G(m,c)|}{m} = O(\frac{1}{(\log m)^{c/2}})$ for each integer m.*

We also develop deterministic algorithms to compute a basis of an abelian group. Our $O(n)$ time algorithm needs the results of Kavitha [10,11].

Theorem 10. *There is an $O(n)$ time algorithm for computing a basis of an abelian G group with n elements.*

Acknowledgements. We would like to thank Eric Allender, Hong Zhu, and Igor Shparlinski for their help and comments on an earlier version of this paper.

References

1. Babai, L.: Randomization in group algorithms: conceptual questions. In: Finklelstein, L., Kantor, W.M. (eds.) Groups and Computation. II, vol. 28, pp. 1–17 (1997)
2. Buchman, J., Jacobson Jr., M.J., Teske, E.: On some computational problems in finite abelian groups. Mathematics of Computation 66, 1663–1687 (1997)
3. Buchmann, J., Schmidt, A.: Computing the structure of a finite abelian group. Mathematics of Computation 74, 2017–2026 (2005)
4. Chen, L.: Algorithms and their complexity analysis for some problems in finite group. Journal of Sandong Normal University 2, 27–33 (1984) (in Chinese)
5. Cormen, T.H., Leiserson, C.E., Rivest, R.L., Stein, C.: Introduction to Algorithms, 2nd edn. The MIT Press, Cambridge (2001)
6. Garzon, M., Zalcstein, Y.: On isomorphism testing of a class of 2-nilpoten groups. Journal of Computer and System Sciences 42, 237–248 (1991)
7. Hardy, K., Muskat, J.B., Williams, K.S.: A deterministic algorithm for solving $n = fu^2 + gv^2$ in coprime integers u and v. Math. Comput. 55, 327–343 (1990)
8. Hoffmann, C.M.: Group-Theoretic Algorithms and Graph Isomorphism. Springer, Heidelberg (1982)
9. Hungerford, T.: Algebra. Springer, Heidelberg (1974)
10. Kavitha, T.: Efficient algorithms for abelian group isomorphism and related problems. In: Pandya, P.K., Radhakrishnan, J. (eds.) FSTTCS 2003. LNCS, vol. 2914, pp. 277–288. Springer, Heidelberg (2003)
11. Kavitha, T.: Linear time algorithms for abelian group isomorphism and related problem. Journal of Computer and System Sciences 73, 986–996 (2007)
12. Köbler, J., Schöning, U., Toran, J.: The Graph Isomorphism Problem: Its Structural Complexity. Birkhauser, Basel (1993)
13. Miller, G.L.: On the $n^{\log n}$ isomorphism technique. In: Proceedings of the tenth annual ACM symposium on theory of computing, pp. 128–142 (1978)
14. Miller, G.L.: Graph isomorphism, general remarks. Journal of Computer and System Sciences 18, 128–142 (1979)
15. Pomerance, C.: Analysis and comparison of some integer factorization algorithms. In: Lenstra, H.W., Tijdeman, R. (eds.) Computational Methods in Number Theory, Part 1, pp. 89–139. Mathematisch Centrum, Amsterdam (1982)
16. Savage, C.: An $O(n^2)$ algorithm for abelian group isomorphism. Technical report, North Carolina State University (January 1980)
17. Shanks, D.: Class number, a theory of factorization and genera. In: Proc. Symp. Pure Math., vol. 20, pp. 414–440 (1971)
18. Sims, C.: Computation with Finitely Presented Groups. Cambridge University Press, Cambridge (1994)
19. Teske, E.: A space efficient algorithm for group structure computation. Mathematics of Computation 67, 1637–1663 (1998)
20. Vikas, N.: An $O(n)$ algorithm for abelian p-group isomorphism and an $O(n \log n)$ algorithm for abelian group isomorphism. JCSS 53, 1–9 (1996)

Good Programming in Transactional Memory*
Game Theory Meets Multicore Architecture

Raphael Eidenbenz and Roger Wattenhofer

Computer Engineering and Networks Laboratory (TIK), ETH Zurich, Switzerland
{eidenbenz,wattenhofer}@tik.ee.ethz.ch

Abstract. In a multicore transactional memory (TM) system, concurrent execution threads interact and interfere with each other through shared memory. The less interference a program provokes the better for the system. However, as a programmer is primarily interested in optimizing her individual code's performance rather than the system's overall performance, she does not have a natural incentive to provoke as little interference as possible. Hence, a TM system must be designed compatible with good programming incentives (GPI), i.e., writing efficient code for the overall system coincides with writing code that optimizes an individual program's performance. We show that with most contention managers (CM) proposed in the literature so far, TM systems are not GPI compatible. We provide a generic framework for CMs that base their decisions on priorities and explain how to modify Timestamp-like CMs so as to feature GPI compatibility. In general, however, priority-based conflict resolution policies are prone to be exploited by selfish programmers. In contrast, a simple non-priority-based manager that resolves conflicts at random is GPI compatible.

1 Introduction

In traditional single core architecture, the performance of a computer program is usually measured in terms of space and time requirements. In multicore architecture, things are not so simple. Concurrency adds an incredible, almost unpredictable complexity to today's computers, as concurrent execution threads interact and interfere with each other. The paradigm of Transactional Memory (TM), proclaimed and implemented by Herlihy and Moss [6] in the 1990's, has emerged as a promising approach to keep the challenge of writing concurrent code manageable. Although today, TM is most-often associated with multithreading, its realm of application is much broader. It can for instance also be used in inter process communication where multiple threads in one or more processes exchange data. Or it can be used to manage concurrent access to system resources. Basically, the idea of TM can be employed to manage any situation where several tasks may concurrently access resources representable in memory. A TM system provides the possibility for programmers to wrap critical code that performs operations on shared memory into transactions. The system then guarantees an exclusive code execution such that no other code being currently processed interferes with the critical operations. To achieve this, TM systems employ a contention management policy. In *optimistic* contention management, transactional code is executed right away and modifications on

* A full version including all proofs, is available as TIK Report 310 at http://www.tik.ee.ethz.ch.

Y. Dong, D.-Z. Du, and O. Ibarra (Eds.): ISAAC 2009, LNCS 5878, pp. 503–513, 2009.

```
incRingCounters(Node start){
  var cur = start;
  atomic{
    while(cur.next!=start){
      c = cur.count;
      cur.count = c + 1;
      cur = cur.next; }}}
```

```
incRingCountersGP(Node start){
  var cur = start;
  while(cur.next!=start){
    atomic{
      c = cur.count;
      cur.count = c + 1;}
    cur = cur.next; }}
```

Fig. 1. Two variants of updating each node in a ring

shared resources take effect immediately. If another process, however, wants to access the same resource, a mechanism called contention manager (CM) resolves the conflict, i.e., it decides which transaction may continue and which must wait or abort. In case of an abort, all modifications done so far are undone. The aborted transaction will be restarted by the system until it is executed successfully. Thus, in multicore systems, the quality of a program must not only be judged in terms of space and (contention-free) time requirements, but also in terms of the amount of conflicts it provokes due to concurrent memory accesses.

Consider the example of a shared ring data structure. Let a ring consist of s nodes and let each node have a counter field as well as a pointer to the next node in the ring. Suppose a programmer wants to update each node in the ring. For the sake of simplicity we assume that she wants to increase each node's counter by one. Given a start node, her program accesses the current node, updates it and jumps to the next node until it ends up at the start node again. Since the ring is a shared data structure, node accesses must be wrapped into a transaction. We presume the programming language offers an **`atomic`** keyword for this purpose. The first method in Figure 1 (`incRingCounters`) is one way of implementing this task. It will have the desired effect. However, wrapping the entire while-loop into one transaction is not a very good solution, because by doing so, the update method keeps many nodes blocked although the update on these nodes is already done and the lock[1] is not needed anymore. A more desirable solution is to wrap each update in a separate transaction. This is achieved by a placement of the **`atomic`** block as in `incRingCountersGP` on the right in Figure 1. When there is no contention, i.e., no other transactions request access to any of the locked ring nodes, both `incRingCounters` and `incRingCountersGP` run equally fast[2] (cf. Figure 2). If there are interfering jobs, however, the affected transactions must compete for the resources whenever a conflict occurs. The defeated transaction then waits or aborts and hence system performance is lost. In our example, using `incRingCounters` instead of `incRingCountersGP` leads to many unnecessarily blocked resources and thereby increases the risk of conflicts with other program parts. In addition, if there is a conflict and the CM decides that the programmer's transaction must abort, then with `incRingCountersGP` only one modification needs to be undone, namely the update to the current node in the ring, whereas with `incRingCounters` all modifications back

[1] An optimistic, direct-update TM system "locks" a resource as soon as the transaction reads or writes it and releases it when committing or aborting. This is not to be confused with an explicit lock by the programmer. In TM, explicit locks are typically not supported.

[2] If we disregard locking overhead.

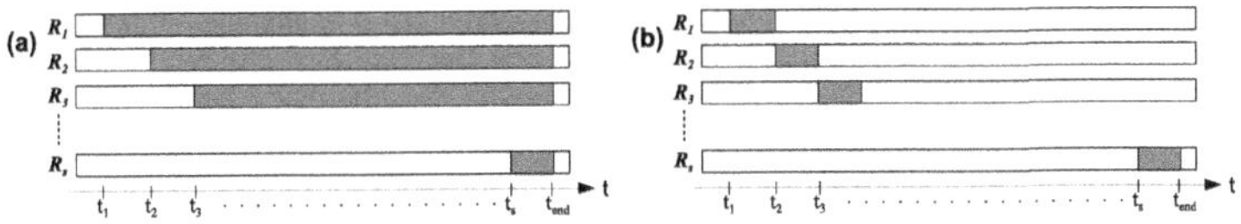

Fig. 2. Transactional allocation of ring nodes (a) by `incRingCounters` and (b) by `incRingCountersGP`

to the start node must be rolled back. In brief, employing `incRingCounters` causes an avoidable performance loss.

One might think that it is in the programmer's interest to choose the placement of atomic blocks as beneficial to the TM system as possible. The reasoning would be that by doing so she does not merely improve the system performance but the efficiency of her own piece of code as well. Unfortunately, in current TM systems, it is not necessarily true that if a thread is well designed—meaning that it avoids unnecessary accesses to shared data—it will also be executed faster. On the contrary, we will show that most CMs proposed so far privilege threads that incorporate long transactions rather than short ones. This is not a severe problem if there is no competition for the shared resources among the threads. Although in minor software projects all interfering threads might be programmed by the same developer, this is not the case in large software projects, where there are typically many developers involved, and code of different programmers will interfere with each other. Furthermore, we must not assume that all conflicting parties are primarily interested in keeping the contention low on the shared objects, especially if doing so slows down their own thread. It is rather realistic to assume that in many cases, a developer will push his threads' performance at the expense of other threads or even at the expense of the entire system's performance if the system does not prevent this option.To avoid this loss of efficiency, a multicore system must be designed such that the goal of achieving an optimal system performance is compatible with an individual programmer's goal of executing her code as fast as possible. This paper shows analytically that most CMs proposed in the literature so far lack such an incentive compatibility. As a practical proof of our findings, we implemented free-riding strategies in the TM library DSTM2[5] and tested them in several scenarios. These results can be found in [3].

2 Model

We use a model of a TM system with optimistic contention management, immediate conflict detection and direct update. As we do not want to restrict TM to the domain of multithreading, we will use the notion of jobs instead of threads to denote a set of transactions belonging together. In inter process communication, e.g., a job is rather a process than a thread. A job J_i consists of a sequence of *transactions* $T_{i1}, T_{i2}, \ldots, T_{ik}$. If J_i consists of only one transaction, we sometimes write T_i instead of T_{i1}. Transactions access shared *resources* R_i. At any point in time, we denote by n the number of running transactions in the system and by s the number of resources currently accessed. For the sake of simplicity, we consider all accesses as exclusive, thus, if two transactions both try to access resource R_i at the same time or if one has already locked R_i and the other desires access to R_i as well, they are in *conflict*. When a conflict occurs, a mechanism decides which transaction gains (or keeps) access of R_i and aborts the other competing transaction. Such a mechanism is called *contention manager (CM)*. We assume that once a transaction has accessed a resource, it keeps the exclusive access right until it either commits or aborts. We further assume that the time needed to detect a conflict, to decide which transaction wins and the time used to commit or start a transaction are negligible. We neither restrict the number of jobs running concurrently, nor do we impose any restrictions on the structure and length of transactions.[3] We say a job J_i *is running* if its first transaction T_{i1} has started and the last T_{ik} has not committed yet. Notice that in optimistic contention management, the starting time t_i of a job J_i and therewith the starting time t_{i1} of T_{i1} is not influenced by the CM, since it only reacts once a conflict occurs. The *environment* $\mathcal{E}$ is a potentially infinite set of tuples of a job and the time it enters the system, i.e., $\mathcal{E} = \{(J_0, t_0), (J_1, t_1), \ldots\}$. We assume that the state at a time t of a TM system managed by a deterministic CM is determined by the environment $\mathcal{E}$. The execution environment of a job J_i is then $\mathcal{E}_{-i} = \mathcal{E} \setminus \{(J_i, t_i)\}$.We further assume that once a job J_i is started at time t_i, any contained transaction T_{ij} accesses the same resources in each of its executions and for any resource, the time of its first access after a (re)start of T_{ij} remains the same in each execution. Once t_i is known, this allows a description of a contained transaction by a list of all resources accessed with their relative access time. E.g., $T_{ij} = (\{(R_1, t_1), \ldots, (R_k, t_k)\}, d_{ij})$ means that transaction T_{ij} accesses R_1 after t_1 time and so forth until it hopefully commits after d_{ij} time. The *contention-free execution time* $d_{ij}^{\mathcal{E}}$ is the time the system needs to execute T_{ij} if T_{ij} encounters no conflicts. The *job execution time* $d_i^{\mathcal{M},\mathcal{E}}$ is the time J_i's execution needs in a system managed by $\mathcal{M}$ in environment $\mathcal{E}$, i.e., the period from the time T_{i1} enters the system, t_{i1}, until the time T_{ik} commits. Similarly, $d_{ij}^{\mathcal{M},\mathcal{E}}$ denotes the execution time of transaction T_{ij} and $d^{\mathcal{M},\mathcal{E}}$ is the *makespan* of all jobs in $\mathcal{E}$, i.e., the time from $\min_i t_i$ until $\max_i(t_i + d_i^{\mathcal{M},\mathcal{E}})$. We denote by $\mathcal{M}^*$ an optimal offline CM. We presume $\mathcal{M}^*$ to know all future transactions. It can thus schedule all transactions optimally so as to minimize the makespan. We assume that the program code of each job is written by a different selfish developer and that there is competition among those developers. *Selfish* in this context means that the programmer only cares about how fast her job terminates. A developer is considered *rational*, i.e., she always acts so as to maximize her expected utility. This is, she minimizes her job's expected execution time. Further, we assume the developers to be *risk-averse* in the sense that they expect the worst case to happen, however they expect their job J_i to eventually terminate even if under certain environments, $\mathcal{M}$ does not terminate J_i. This assumption is inevitable since with many CMs, there exist (at least theoretically) execution environments $\mathcal{E}_{-i}$ which make J_i run forever. Thus a risk-averse developer could just as well twirl her thumbs instead of writing a piece of code without this assumption. In Lemma 1, the used notion of rationality will be further adapted in that we argue that delaying a transaction does not make sense if an arbitrary environment is assumed.

[3] That is why we do not address the problem of recognizing dead transactions and ignore heuristics included in CMs for this purpose.

3 Good Programming Incentives (GPI)

Our main goal is to design a multicore TM system that is as efficient as possible. As we may not assume programmers to write code so as to maximize the overall system performance but rather to optimize their individual job's runtime, we must design a system such that the goal of achieving an optimal system performance is compatible with an individual programmer's goal of executing her code as fast as possible. A first step in this direction is to determine the desired behavior, that is, we have to find the meaning of *good programming* in a TM system. We want to find out how a programmer should structure her code, or in particular, how she should place atomic blocks in order to optimize the overall efficiency of a TM system.

When a job accesses shared data structures it puts a load on the system. The insight gained by studying the example in the introduction is that the more resources a job locks and the longer it keeps those locks, the more potential conflicts it provokes. If the program logic does not require these locks, an unnecessary load is put on the system.

Fact 1. *Unnecessary locking of resources provokes a potential performance loss in a TM system.*

However the question remains whether *partitioning* a transaction into smaller transactions—even if this does not reduce the resource accesses—results in a better system performance. Consider an example where the program logic of a job J_1 requires exclusive access of resource R_1 for a period of d_1. One strategy for the programmer is to simply wrap all operations on R_1 into one transaction $T_1 = (\{(R_1, 0)\}, d_1)$. However, let the semantics also allow an execution of the code in two subsequent transactions $T_{11} = (\{(R_1, 0)\}, d_{11})$ and $T_{12} = (\{(R_1, 0)\}, d_{12})$ without losing consistency and without overhead, i.e., $d_{11} + d_{12} = d_1$. Figure 3 shows the execution of both strategic variants in a system managed by an optimal CM $\mathcal{M}^*$ in an execution environment $\mathcal{E}_{-1} = \{(J_2, 0)\}$ with only one concurrent job J_2. Both jobs J_1 and J_2 enter the system at time $t = 0$. Job J_2 consists of transactions T_{21} and T_{22} with $T_{21} = (\{(R_2, 0), (R_1, d_{21} - \delta_1)\}, d_{21})$ and $T_{22} = (\{(R_2, 0), (R_1, \delta_2)\}, d_{22})$. Furthermore, let $\delta_1 << d_1$ and $\delta_2 << d_1$. In situation **(a)**, the programmer does not partition T_1, $\mathcal{M}^*$ schedules T_1 first, at time $t = 0$, T_{21} at $t = \delta_1 + \delta_2$ and T_{22} at $t = d_1 + \delta_1$.

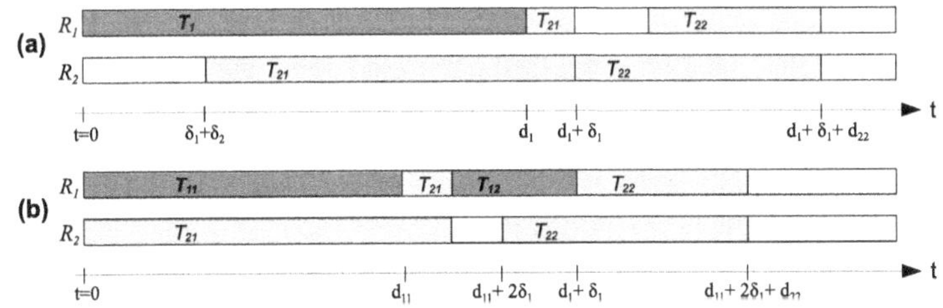

Fig. 3. Partitioning example. The picture depicts the optimal allocation of two resources R_1 and R_2 over time in two situations **(a)** and **(b)**. In **(a)**, the programmer of job J_1 does not partition T_1. In **(b)**, she partitions T_1 into T_{11} and T_{12}. The overall execution time is shorter in **(b)**, the individual execution of J_1, however, is faster in **(a)**.

This optimal schedule of T_1, T_{21} and T_{22} has a makespan of $d_1 + \delta_1 + d_{22}$. In situation **(b)**, the programmer partitions T_1 into T_{11} and T_{12}, an optimal CM schedules T_{11} and T_{21} concurrently at time $t = 0$, T_{12} at $t = d_{21} = d_{11} + \delta_1$ and T_{22} at $t = d_{11} + 2\delta_1$. This yields a makespan of $d_{11} + 2\delta_1 + d_{22} = d_1 + \delta_1 + d_{22} - \delta_2$. Thus, in the example of Figure 3, partitioning T_1 allows to schedule J_1 and J_2 by δ_2 faster. We can show that partitioning is beneficial in a system managed by $\mathcal{M}^*$ in general.

Theorem 1. *Let T_{ij1}, T_{ij2} be a valid partition of T_{ij}. Let J_i be a job containing T_{ij1}, T_{ij2} and J_i' the same job except it contains T_{ij} instead of T_{ij1}, T_{ij2}. A finer transaction granularity speeds up a transactional memory system managed by an optimal CM $\mathcal{M}^*$, i.e., $\forall\ \mathcal{E}_{-i}, t\ :\ d^{\mathcal{M}^*, \mathcal{E}_{-i} \cup \{(J_i,t)\}} \leq d^{\mathcal{M}^*, \mathcal{E}_{-i} \cup \{(J_i',t)\}}$ and $\exists\ \mathcal{E}_{-i}, t$ such that inequality holds.*

Proof (Sketch). Partitioning transactions only gives more freedom to $\mathcal{M}^*$. To be at least as fast with J_i as with J_i', $\mathcal{M}^*$ could execute T_{i2} right after T_{i1}. In some cases it might be even faster to schedule an intermediary transaction between T_{i1} and T_{i1}. □

Let us reconsider the example from Figure 3. We have seen that partitioning T_1 into T_{11} and T_{12} results in a smaller makespan. But what about the individual execution time of job J_1? In the unpartitioned execution, where J_1 only consists of T_1, J_1 terminates at time $t = d_1$. In the partitioned case, however, J_1 terminates at time $t = d_1 + \delta_1$. This means that partitioning a transaction speeds up the *overall* performance of a concurrent system managed by an optimal CM, but it possibly slows down an *individual* job. Thus, from a selfish programmer's point of view, it is not rational to simply make transactions as fine granular as possible. In fact, if a finer grained partitioning of transactions might result in a slower execution of a job, why should a selfish programmer make the effort of finding a transaction granularity as fine as possible?

Avoiding unnecessary locks and partitioning transactions whenever possible is beneficial to a TM system. We say a CM $\mathcal{M}$ *rewards partitioning* of transactions if in a system managed by $\mathcal{M}$, it is rational for a programmer to always partition a transaction whenever the program logic allows her to do so. Further, $\mathcal{M}$ *punishes unnecessary locking* if in a system managed by $\mathcal{M}$, it is rational for a programmer to never lock resources unnecessarily, i.e., she only locks a resource when required by the program logic. One can expect that, from a certain level of selfishness among developers, a CM which incentivizes these two crucial aspects of good programming, performs better than the best incentive incompatible CM. In the remainder, we are mainly concerned with the question of which CM policies fulfill the following property.

Property 1. A CM is *good programming incentive (GPI) compatible* if it rewards partitioning and punishes unnecessary locking.

As a remark, we would like to point out that the optimal CM $\mathcal{M}^*$ does not reward partitioning and hence is not GPI compatible (cf. Figure 3). If we assume developers act selfish then also a system managed by an optimal offline scheduler suffers a performance loss and a CM which offers incentives for good programming might be more efficient than $\mathcal{M}^*$. There is, however, an inherent loss due to the lack of collaboration, commonly known as *price of anarchy* (cf. [2,7]).

4 Priority-Based Contention Management (CM)

One key observation when analyzing the contention managers proposed in [1,4,8,9,10] is that most of them incorporate a mechanism that accumulates some sort of priority for a transaction. In the event of a conflict, the transaction with higher priority wins against the one with lower priority. Most often, priority is supposed to measure, in one way or another, the work already done by a transaction.The intuition behind this approach is that aborting old transactions discards more work already done and thus hurts the system efficiency more than discarding newer transactions. The proposed CMs base priority on a transaction's time in the system, the number of conflicts won, the number of aborts or the number of resources accessed. Definition 1 introduces a framework that comprises priority-based CMs. It allows us to classify priority-based CMs and to make generic statements about GPI compatibility of certain CM classes.

Definition 1. *A* priority-based *CM $\mathcal{M}$ associates with each job J_i a priority vector $\boldsymbol{\omega_i} \in \mathbb{R}^s$ where $\boldsymbol{\omega_i}[k]$ is J_i's priority on resource R_k. $\mathcal{M}$ resolves conflicts between two transactions $T_{ix} \in J_i$ and $T_{jy} \in J_j$ over resource R_k by aborting the transaction with lower priority, i.e., if $\boldsymbol{\omega_i}[k] \geq \boldsymbol{\omega_j}[k]$ then T_{ix} wins otherwise T_{iy} is aborted.*

In many CMs, all entries of the vector $\boldsymbol{\omega_i}$ are equal. In this case, we can also replace $\boldsymbol{\omega_i}$ by a scalar priority value $\omega_i \in \mathbb{R}$. We call such a CM *scalar-priority-based*. In the remainder we often use ω_i instead of $\boldsymbol{\omega_i}$, for the sake of simplicity, even if we are not talking about scalar-priority-based CMs only. Mostly, for a correct valuation of a job's competitiveness, absolute priority values are not relevant, but the relative value to other job priorities. A job J_i's *relative priority* vector $\tilde{\boldsymbol{\omega}}_{\boldsymbol{i}}$ is defined by $\tilde{\boldsymbol{\omega}}_{\boldsymbol{i}}[k] = \boldsymbol{\omega_i}[k] - \min_{i=1...n} \boldsymbol{\omega}_i[k]$, $\forall k = 1 \ldots s$. If the CM uses scalar priorities, J_i's relative priority $\tilde{\omega}_i$ is obtained by subtracting $\min_{i=1...n} \omega_i$ from the absolute priority ω_i. Since optimistic CMs feature a reactive nature it is best to consider the priority-building mechanism as event-driven. We find that the following *events* may occur for a transaction $T_{ij} \in J_i$ in a transactional memory system: A time step ($\mathcal{T}$); T_{ij} wins a conflict ($\mathcal{W}$); T_{ij} loses a conflict and is aborted ($\mathcal{A}$); T_{ij} successfully allocates a resource R_k ($\mathcal{R}_k$); T_{ij} commits ($\mathcal{C}$). The following two subtypes of priority-based CMs capture most contention management policies in the literature.

Definition 2. *A priority-based CM is* priority-accumulating *iff no event decreases a job's priority and there is at least one type of event which causes the priority to increase. A CM is* quasi-priority-accumulating *iff it is priority-accumulating w.r.t. events $\mathcal{T}$, $\mathcal{W}$, $\mathcal{A}$ and $\mathcal{R}$ but it resets J_i's priority when a transaction $T_{ij} \in J_i$ commits.*

As an example consider a Timestamp CM $\mathcal{M}_T$. $\mathcal{M}_T$ uses only events of type $\mathcal{T}$ and $\mathcal{C}$, i.e., in a time step dt after $T_{ij} \in J_i$ entered the system, ω_i is increased by $d\omega = \alpha dt, \alpha \in \mathbb{R}^+$ until $\mathcal{C}$ occurs, then reset to 0. J_i's scalar priority at time t, $t_{ij} < t \leq t_{ij} + d_{ij}^{\mathcal{M}_T,\mathcal{E}}$ is $\omega_i(t) = \int_{t_{ij}}^{t} \alpha dt = \alpha(t - t_{ij})$. Timestamp is quasi-priority-accumulating since a job's priority always increases and never decreases over time except it is reset when a contained transaction commits.

Waiting Lemma. We argue in this paragraph that delaying the execution of a job[4] is not rational with the assumption that the execution environment $\mathcal{E}_{-i}$ is arbitrary. This assumption implies that at any point in time, the history of the transactions does not hold any information about their future. Furthermore, we demand two restrictions on the CM's priority modification mechanism: (I.) An increase (or decrease) of ω_i never depends on ω_i's current value[5] or on any other job's priority value. (II.) In a period where no events occur except for time steps, all priorities ω_i increase by $\Delta\omega \geq 0$. Note that if $\Delta\omega$ is always 0, the priority is not based on time.

Lemma 1. *If $\mathcal{E}_{-i}$ is arbitrary, the strategy of waiting is irrational in a system managed by a priority-based CM $\mathcal{M}$ restricted by (I.–II.).*

Proof (Sketch). If a programmer delays a transaction by Δ, the adversary can preserve the environment and thus increase its execution time by Δ. □

Note that the assumption on $\mathcal{E}_{-i}$ being arbitrary naturally applies if the programmer has no information about the environment in which her program will be executed. Indeed, if the environment would be truly a worst case environment, the execution of job J_i would take forever. As with this assumption, starting a job would be completely pointless, we adapt our model of a risk-averse agent in that we let her suppose that a worst case environment yields a finite execution time. In practice, the programmer often has some information about the environment in which her programm will be deployed. Hence it might make sense to presume some structure of $\mathcal{E}_{-i}$. E.g., she could assume that lengths of locks follow a certain distribution, or that each resource has a given probability of being locked. In such cases waiting might not be irrational. In the following, we will sometimes argue that a CM is (not) GPI compatible by comparing two jobs J_i and J_i' where both are equal except for J_i' either locks a resource unnecessarily or does not partition a transaction although this would be semantically possible. We will show that in the same execution environment $\mathcal{E}_{-i}$, one job either perfoms faster or if it is slower, this is because it does not wait at a certain point in the execution. Since waiting is irrational, a developer will prefer this job even if it is not guaranteed to perform better in any environment.

Quasi-Priority-Accumulating CM. Quasi-priority-accumulating CMs increase a transaction's priority over time. Again, the intuition behind this approach is that, on one hand, aborting old transactions discards more work already done and thus hurts the system efficiency more than discarding newer transactions and on the other hand, any transaction will eventually have a priority high enough to win against all other competitors. This approach is legitimate. Although the former presupposes some structure of $\mathcal{E}$ and the latter is not automatically fulfilled, examples of quasi-priority-accumulating CMs showed to be useful in practice (cf. [9]). However, quasi-priority-accumulating CMs bear harmful potential. They incentivize programmers to not partition transactions and in some cases even to lock resources unnecessarily. Consider the case where a

[4] A programmer can implement waiting by executing code without allocating shared resources.

[5] E.g., rules such as "if ω_i is larger than 10 add 100" or "$\omega_i = 2\omega_i$" are prohibited. "$\omega_i = \omega_i + 2$" is permitted.

job has accumulated high priority on an resource R_i. It might be advisable for the job to keep locking R_i in order to maintain high priority. Although it does not need an exclusive access for the moment, maybe later on, the high priority will prevent an abort and thus save time. In fact, we can show that the entire class of quasi-priority-accumulating CMs is not GPI compatible.

Theorem 2. *Quasi-priority-acc. CMs restricted by (I.–II.) are not GPI compatible.*

Theorem 2 reflects the intuition, that if committing decreases an advantage in priority then there are cases where it is rational for a programmer not to commit and start a new transaction but to continue instead with the same transaction. Obviously, the opposite case is possible as well, namely that by not committing, the developer causes a conflict with a high priority transaction on a resource, which could have been freed if the transaction would have committed earlier, and thus is aborted. As in our model of a risk-averse programmer, she does not suppose any structure on $\mathcal{E}_{-i}$, she does not know which case is more likely to happen either and therefore has no preference among the two cases. She would probably just choose the strategy which is easier to implement. If we assumed, e.g., that a resource R_i is locked at time t with probability p by a transaction with priority x where both, p and x follow a certain probability distribution, then there would be a clear trade-off between executing a long transaction and therewith risking more conflicts and partitioning a transaction and thus losing priority.

Note that a similar proof can be used to show that no priority-based CM rewards partitioning unless it prevents the case where, after a commit of transaction $T_{ij} \in J_i$, the subsequent transaction $T_{i(j+1)} \in J_i$ starts with a lower priority than T_{ij} had just before committing. In fact, we can show that all priority-accumulating CMs proposed by [1,4,8,9,10] are not GPI compatible.

Corollary 1. *Polite, Greedy, Karma, Eruption, Kindergarten, Timestamp and Polka are not GPI compatible.*

Priority-Accumulating CM. The inherent problem of quasi-priority-accumulating mechanisms is not the fact that they accumulate priority over time but the fact that these priorities are reset when a transaction commits. Thus, by comitting early, a job loses its priority when starting a new transaction. One possibility to overcome this problem is to not reset ω_i when a transaction of J_i commits. With this trick, neither partitioning transactions nor letting resources go whenever they are not needed anymore resets the accumulated priority. We further need to ensure that two subsequent transactions of J_i are scheduled right after each other, because otherwise partitioning would result in a longer execution even in a contention-free environment. We denote this property of a CM as *gapless transaction scheduling*. If a CM $\mathcal{M}$ only modifies priorities on a certain event type $\mathcal{X}$, we say $\mathcal{M}$ *is based only on* $\mathcal{X}$*-events*.

Lemma 2. *Any priority-accumulating CM $\mathcal{M}$ which schedules transactions gapless and is based only on time ($\mathcal{T}$-events) is GPI compatible.*

Proof (Sketch). Unnecessary locking is punished since it can cause the transaction to abort and restart. Thus restarted, the transaction might be lucky and catch a better slot for execution. However, this is the same as waiting and hence irrational. Partitioning is rewarded since committing and restarting does not decrease priority. Furthermore, if a finer-grained job loses in a conflict, it has to redo less work. □

For instance, by simply not resetting a job J_i's priority when a contained transaction $T_{ij} \in J_i$ commits, we can make a Timestamp contention manager GPI compatible. Nevertheless, priority based CMs are generally dangerous in the sense that they bear a potential for programmers to cheat, i.e., to boost their job's priority at the expense of other jobs. E.g., consider a CM like Karma [8], where priority depends on the number of resources accessed. One way to gain high priority for a job would be to quickly access an unnecessarily large number of objects and thus become overly competitive. Or if priority is based on the number of aborts or the number of conflicts, a very smart programmer might use some dummy jobs which compete with the main job in such a way that they boost its priority. In fact, we can show that a large class of priority-accumulating CMs is not GPI compatible.

Theorem 3. *A priority-acc. CM $\mathcal{M}$ is* not GPI compatible *if one of the following holds:*

(i) *$\mathcal{M}$ increases a job's relative priority on $\mathcal{W}$-events (winning a conflict).*
(ii) *$\mathcal{M}$ increases relative priority on $\mathcal{R}$-events (having exclusive access of a resource).*
(iii) *$\mathcal{M}$ schedules transactions gapless and increases relative priorities on $\mathcal{C}$-events.*
(iv) *$\mathcal{M}$ restarts aborted transactions immediately and increases relative priorities on $\mathcal{A}$-events (aborting).*

5 Non-priority Based CM

One example of a CM which is not priority-based is Randomized (cf. [8]). To resolve conflicts, Randomized simply flips a coin in order to decide which competing transaction to abort. The advantage of this simple approach is that it bases decisions neither on information about a transaction's history nor on predictions about the future. This leaves programmers little possibility to boost their competitiveness.

Lemma 3. *Randomized is GPI compatible.*

Proof (Sketch). The proof works similarly to the proof of Lemma 2. Note that Lemma 1 does not apply here as Randomized is not priority-accumulating. However, to show that waiting is irrational also with Randomized is easy. An adversary can provoke the same conflicts for a transaction, if it is started immediately or if it is delayed for some time Δ. Since in any conflict, the probability of winning is the same, the expected runtime increases by Δ when the transaction is delayed. □

Employing such a simple Randomized CM is not a good solution although it rewards good programming. The probability $p_{success}$ that a transaction runs until commit decreases exponentially with the number of conflicts, i.e., $p_{success} \sim p^{|C|}$ where p is the probability of winning an individual conflict and C the set of conflicts. However, we see great potential for further developement of CMs based on randomization.

6 Conclusion and Future Work

While TM constitutes an inalienable convenience to programmers in concurrent environments, it does not automatically defuse the danger that selfish programmers might exploit a multicore system. GPI compatibility has to be addressed when designing a TM system. Priority-based CMs are prone to be corrupted unless they are based on time only. CMs not based on priority seem to feature incentive compatibility more naturally. We conjecture that by combining randomized conflict resolving with a time-based priority mechanism, chances of finding an efficient, GPI compatible CM are high.

References

1. Attiya, H., Epstein, L., Shachnai, H., Tamir, T.: Transactional contention management as a non-clairvoyant scheduling problem. In: PODC 2006: Proc. of the 25th annual ACM symposium on Principles of Distributed Computing, pp. 308–315 (2006)
2. Christodoulou, G., Koutsoupias, E.: The price of anarchy of finite congestion games. In: STOC 2005: Proc. of 37th annual symposium on Theory of computing, pp. 67–73 (2005)
3. Eidenbenz, R., Wattenhofer, R.: Brief announcement: Selfishness in transactional memory. In: SPAA 2009: Proc. of the 21st annual symposium on Parallelism in Algorithms and Architectures, pp. 41–42 (2009)
4. Guerraoui, R., Herlihy, M., Pochon, B.: Toward a theory of transactional contention managers. In: PODC 2005: Proc. of the 24th annual ACM symposium on Principles of Distributed Computing, pp. 258–264 (2005)
5. Herlihy, M., Luchangco, V., Moir, M.: A flexible framework for implementing software transactional memory. SIGPLAN Not. 41(10), 253–262 (2006)
6. Herlihy, M., Moss, J.E.B.: Transactional memory: architectural support for lock-free data structures. SIGARCH Comput. Archit. News 21(2), 289–300 (1993)
7. Roughgarden, T.: Selfish Routing and the Price of Anarchy. MIT Press, Cambridge (2005)
8. Scherer III, W.N., Scott, M.L.: Contention Management in Dynamic Software Transactional Memory. In: PODC Workshop on Concurrency and Synchronization in Java Programs (CSJP), St. John's, NL, Canada (July 2004)
9. Scherer III, W.N., Scott, M.L.: Advanced contention management for dynamic software transactional memory. In: PODC 2005: Proc. of the 24th annual ACM symposium on Principles of Distributed Computing, pp. 240–248 (2005)
10. Schneider, J., Wattenhofer, R.: Bounds On Contention Management Algorithms. In: Dong, Y., Du, D.-Z., Ibarra, O. (eds.) ISAAC 2009. LNCS, vol. 5878, pp. 441–451. Springer, Heidelberg (2009)

Induced Packing of Odd Cycles in a Planar Graph*

Petr A. Golovach[1], Marcin Kamiński[2,**],
Daniël Paulusma[3], and Dimitrios M. Thilikos[4]

[1] Department of Computer Science,
University of Bergen
Peter.Golovach@ii.uib.no
[2] Department of Computer Science,
Université Libre de Bruxelles
Marcin.Kaminski@ulb.ac.be
[3] Department of Computer Science,
University of Durham
Daniel.Paulusma@durham.ac.uk
[4] Department of Mathematics,
National and Kapodistrian University of Athens
sedthilk@math.uoa.gr

Abstract. An induced packing of odd cycles in a graph is a packing such that there is no edge in a graph between any two odd cycles in the packing. We prove that the problem is solvable in time $2^{\mathcal{O}(k^{3/2})} \cdot n^3 \log n$ when the input graph is planar. We also show that deciding if a graph has an induced packing of two odd cycles is NP-complete.

1 Introduction

We assume that the reader is familiar with notions of graph theory; for those not defined here, we refer to [6].

Packing graphs. Packing graphs is a classic field of graph theory with many results and many conjectures. Packing is finding (usually) vertex- or edge-disoint copies of graphs from some family (the guests graphs) into a fixed graph G (the host graph). There is a significant body of work on graph packing in the context of extremal combinatorics. The survey by H. Yap [21] presents many results on packing graphs into a complete graph.

Packing has also been studied from the algorithmic point of view. The goal is usually to find the maximum number of disjoint copies of a guest graph in the host graph. A *matching* in a graph is a packing of vertex disjoint K_2's and it can be solved in polynomial time by Edmond's algorithm [9].

* Supported by the project "Kapodistrias" (AΠ 02839/28.07.2008) of the National and Kapodistrian University of Athens (project code: 70/4/8757).
** Chargé de Recherches du F.R.S. – FNRS.

Y. Dong, D.-Z. Du, and O. Ibarra (Eds.): ISAAC 2009, LNCS 5878, pp. 514–523, 2009.

Packing cylces. The problem of finding the maximum number of vertex disjoint triangles in the input graph was proved to be NP-complete by Garey and Johnson in [11]. However, there is a randomized $(43/83 - \varepsilon)$-approximation algorithm, even for the weighted version of the problem presented by Hassin and Rubinstein [12], [13].

There is a large collection of results on edge-disjoint packing of cycles which has appliactions in genome rearrangement in computational biology. The problem is studied by Caprara et al. in [3] where the authors prove that it is APX-hard and can be approximated with the factor of $\mathcal{O}(\log n)$ by a greedy algorithm. The approximation factor was later improved by Krivelevich et al. to $\mathcal{O}(\log^{1/2} n)$ in [14]. Friggstad and Salavatipour showed that it is almost best possible [10]. They proved that to approximate the edge-disjoint cycle problem within ratio of $\mathcal{O}(\log^{1/2-\varepsilon} n)$ is quasi-NP-hard, for any constant $\varepsilon > 0$. They also note that the same results hold for packing vertex-disjoint cycles.

Packing odd cylces. The problem of packing odd cycles in a graph was studied by Bruce Reed in [18]. He was mainly concerned with Erdős-Pósa property for odd cycles. In the conclusion, he gives an argument that packing k odd cycles in a graph is NP-complete when k is a part of the input. He also points out that a consequence of his results on Erdős-Pósa property is a polynomial-time algorithm for packing odd cycles in a planar graph. Then he continues, "*As of the current writing, the author and P. Seymour believe they have a much more complicated algorithm for determining if a graph contains k vertex disjoint odd cycles, k fixed. However, the proof of this result is extremely complicated and may well never be written down.*" We are not aware of any materialization of this proof, or any other proof of this result. However, in the view of this claim packing k odd cycles in a graph can be done in polynomial-time when k is fixed. As we show in this paper, things are different when we require such cycles to be induced.

Even and odd holes. A hole is an induced cycle of length at least 4. Finding even and odd holes has been studied in the literature. The structure of graphs with no even hole is analyzed by Conforti et al. in [15], and in [16] they gave a polynomial-time algorithm for finding an even hole in a graph. Another algorithm, with a better running time, was proposed by Chudnovsky et al. in [17].

Despite a seeming similarity the complexity of detecting an odd hole in a graph has been a long standing open problem. The problem has been shown by Conforti et al. to be solvable in polynomial-time for graphs of bounded clique number [4]. Also, Bienstock has shown that the problem is NP-complete if the odd hole is required to contain a given vertex [1].

Our results. We are interested in induced packing of odd cycles. In this setting, odd cycles in the host graph are not only vertex-disjoint but also there is no edge in the host graph between two odd cycles in the packing. While packing of k odd cycles is polynomial, for any fixed k, as claimed in [18], we prove that *to decide if a graph contains an induced packing of* 2 *odd cycles is* NP-*complete.* Induced packing is then, not surprisingly, much harder than packing.

The two odd cycles in our hardness proof are in fact odd holes (and can be made arbitrarily long). While the problem of settling the complexity of detecting an odd hole is likely to be rather difficult to tackle, we show that to determine if a graph has two induced odd holes such that there is no edge between them is NP-complete.

A simple argument shows that induced packing of k odd cycles is NP-complete for planar graphs, if k is a part of the input. We prove that *when k is restricted to be a slowly growing function of the size of the graph, the induced packing of k odd cycles can be found in polynomial time.* Our strategy is to solve the problem by dynamic programming for graphs of small tree-width. If the tree-width is large, then the graph contains a large grid minor. In the model of the minor, we can either find an induced packing of k disjoint cycles, or a large bipartite graph. Our main technical result shows that in the latter case we can find an *irrelevant vertex* in the graph. The vertex is called irrelevant because one can remove it from the graph and be sure that the new graph has an induced packing of k odd cycles if and only if the original graph does.

2 Background

In this section, we gather some definitions and present results from the literature that we will use later. We consider graphs without loops and multiple edges and denote the number of vertices in the graph by n.

A graph is *chordal* if it does not contain an induced cycle of length ≥ 4. The *tree-width* of the graph G is the minimum size of the maximum clique minus 1, where the minimum is taken over all chordal supergraphs of G.

The $m \times m$ *grid* has all pairs (i, j) for $i, j = 0, 1, \ldots, m-1$ as the vertex set, and two vertices (i, j) and (i', j') are joined by an edge if and only if $|i-i'|+|j-j'| = 1$. A connected graph G contains H as a *minor* if H can be obtained from G by a sequence of vertex or edge deletions, and edge contractions (removing loops and multiple edges).

Here are two useful lemmas.

Lemma 1 ((6.2) in [19]). *Let $m \geq 1$ be an integer. Every planar graph with no $m \times m$ grid minor has tree-width $\leq 6m - 5$.*

Lemma 2 (from [2]). *The length of the side of the largest square grid minor in a planar graph can be approximated with the factor of 4 and the corresponding grid minor can be constructed in time $\mathcal{O}(n^2 \log n)$.*

Let G be a plane graph and f its outer face. For a cycle $C \subseteq G$ and a subgraph $H \subseteq G$, we say that H *lies inside* C if the boundary of f and H are contained in two different connected components of $G \setminus C$.

For two cycles C and Z, we define $\mu_Z(C)$ to be the number of connected components of $C \setminus Z$. We say that C *crosses* Z, for two cycles C and Z in a plane graph G, if there is a vertex of C inside Z.

Two subgraphs of a graph are called *mutually induced* if they are vertex disjoint and there is no vertex of one subgraph is adjacent to a vertex of the other. A set of subgraphs of a graph is mutually induced if any two subgraphs of the set are mutually induced.

A sequence of cycles $Z_1, \ldots, Z_q$ in a plane graph G is called *nested*, if there exist disks $\Delta_1, \ldots, \Delta_q$ such that for $i = 1, \ldots, q$, Z_i bounds Δ_i, and $\Delta_{i+1} \subseteq \Delta_i$, for $i = 1, \ldots, q-1$.

Now we are ready to formally define the problem we study here.

Problem k-INDUCED-PACKING-OF-ODD-CYCLES

Input: A planar graph G.
Output: A set of k mutually induced odd cycles in G if there exists one; NO otherwise.

The k-INDUCED-PACKING-OF-ODD-CYCLES problem is expressible in monadic second order logic. The seminal result of Courcelle [5] implies that for any class of graphs whose tree-width is bounded, there exists a linear-time algorithm solving the problem in this class of graphs. Even though the complexity is linear in n, the dependence on the parameter k is highly exponential. However, for our purposes, we can obtain a better dependence on the parameter using dynamic standard dynamic programming on tree decompositions. This approach can give an answer to the problem in $2^{O(w \cdot \log w)} \cdot n$ steps where w is the treewidth of the input graph. This running time can be further improved to $2^{O(w)} \cdot n$ for planar graphs using the technique of Catalan Structures developed by Bodlaender et al. [8] (see also [7] for a survey). Due to space restrictions, we omit the details (for the application of the same technique to similar problems see [20]). We conclude to the following lemma.

Lemma 3. k-INDUCED-PACKING-OF-ODD-CYCLES *is is solvable in time* $2^{O(w)}n$ *for graphs of tree-width at most* w.

3 Induced Packing of Odd Cycles

In this section, we present a combinatorial lemma on the existence of an irellevant vertex, and state our algorithm together with the proof of its correctness.

Irrelevant vertex

Lemma 4. *Let* k *be a positive integer,* G *a plane graph, and* Z_1 *a cycle in* G *such that the graph induced by the vertices of* Z_1 *and the vertices inside* Z_1 *is bipartite. Also, let* $Z_1, \ldots, Z_k$ *be a sequence of nested, mutually induced cycles in* G *and* v *a vertex inside* Z_k *such that it is not adjacent to any vertex of* Z_k*. Then,* G *has an induced packing of* k *odd cycles if and only if* $G \setminus v$ *does.*

Proof. The backward implication is clear. To prove the forward implication, let us assume that G has an induced packing of k odd cycles and let $\mathcal{C}$ be one for which $\sum_{C\in\mathcal{C}}\sum_{i=1,\ldots,k}\mu_{Z_i}(C)$ is minimum. We want to show that v is not contained in any cycle of $\mathcal{C}$, or adjacent to any vertex of any cycle in $\mathcal{C}$.

The relation of being inside defines a poset on $\mathcal{C}$ whose Hasse diagram $H_{\mathcal{C}}$ is a forest. We will work with $H_{\mathcal{C}}$ assuming that every tree in the forest is rooted; this can be done in a natural way. We define the *height* of a cycle in $\mathcal{C}$ to be 1 for the leaves of $H_{\mathcal{C}}$. For other cycles in $\mathcal{C}$, the height is the minimum of the height of its children in $H_{\mathcal{C}}$ plus 1. Notice that the depth of a tree in $T_{\mathcal{C}}$ is at most k and therefore the maximum height of a cycle in $\mathcal{C}$ is at most k.

Now it only remains to prove the following claim.

Claim. No cycle of height (at most) i from $\mathcal{C}$ crosses Z_i, for all $i = 1,\ldots,k$.

Let i be the smallest integer such that a cycle $C \in \mathcal{C}$ of height i crosses Z_i. Let Q the set of edges of C which are lie inside Z_i, and P be the set of edges of Z_i which lie inside C in the plane graph G. Notice that $(C \setminus Q) \cup P$ is a collection of cycles. Sets P and Q are disjoint and $P \cup Q$ is a collection of cycles every belonging to the bipartite graph and therefore even. Hence, the number of all edges in P and in Q is even. Since C is an odd cycle, the total length of cycles in $(C \setminus Q) \cup P$ is odd. Thus, there is an odd cycle $C' \in (C \setminus Q) \cup P$.

Cycle C' lies inside C in the plane graph G and there is no cycle inside C if $i = 1$, and no cycle inside C which would cross Z_{i-1}, for $i > 1$. Hence, C' is mutually induced with every cycle in $\mathcal{C} \setminus C$ and therefore $(\mathcal{C} \setminus C) \cup \{C'\}$ is an induced packing of k cycles in G. However, $\sum_{i=1,\ldots,k}\mu_{Z_i}(C') < \sum_{i=1,\ldots,k}\mu_{Z_i}(C)$; a contradiction with the choice of $\mathcal{C}$. □

The algorithm

Algorithm k-INDUCED-PACKING-OF-ODD-CYCLES

Input: A planar graph G.
Output: A collection of k mutually induced odd cycles in G if there exists one; NO otherwise.

1. **Run** the algorithm from Lemma 2 and construct a grid minor M.
2. If the side length of M is less than $\lceil\sqrt{k}\rceil(4k+2) - 1$, then **solve** the problem using the algorithm of Lemma 3 and **stop**.
3. Otherwise, **find** k mutually induced copies of a square grid of side length $4k + 1$ in M.
4. For every copy, **check** if the model of the copy in G is bipartite.
5. If models of all copies are non-bipartite, **return** an induced packing of k odd cycles and **stop**.
6. Otherwise, **construct** k mutually induced odd cycles in a bipartite copy H.
7. **Find** an irrelevant vertex v in H using Lemma 4.
8. **Run** the algorithm for $G \setminus v$.

Theorem 1. *Algorithm* k-INDUCED-PACKING-OF-ODD-CYCLES *is correct and runs in time* $2^{\mathcal{O}(k^{3/2})} n^3 \log n$.

Proof. Let us suppose that the input graph G contains an induced packing of k odd cycles.

If the algorithm from Lemma 2 in Step 1 finds a grid minor M of side length less than $\lceil\sqrt{k}\rceil(4k+2)-1$, then the largest grid minor is of side length less than $4\lceil\sqrt{k}\rceil(4k+2)-1$. This, from Lemma 1, means that the tree-width of the graph is bounded by a constant and the problem can be solved by Lemma 3. The induced packing of k odd cycles will be found in Step 2.

If the side length of M is at least $\lceil\sqrt{k}\rceil(4k+2)-1$, then M contains k mutually disjoint copies of $(4k+1)\times(4k+1)$ grid (Step 3). For each copy, we look at the graph induced by the union of branch sets in G corresponding to the vertices of the copy. If for every copy the graph is non-bipartite, there is an induced packing of k odd cylces (Step 4 & 5).

Otherwise, there is a copy H whose model is bipartite. In this copy, peeling off the $(4k+1)\times(4k+1)$ grid, we find k nested, mutually induced cycles $Z'_1,\ldots,Z'_k$. The central vertex of the grid v' is also mutually induced with the cycles. Notice that the model of cycle Z'_i in G, for all $i=1,\ldots,k$, contains a cycle passing through the all branch sets corresponding to the vertices of Z'_i. Therefore, we construct a collection of k nested, mutually induced cycles $Z_1,\ldots,Z_k$ (Step 6). These cycles, together with a vertex v from the model of v' in G (Step 7), satisfy conditions of Lemma 4. By the Lemma 4, the induced packing of k odd cycles will be found recursively (Step 8).

Notice that Step 8 of the algorithm will be executed at most n times and Step 3 is most time-consuming the algorithm, taking $\mathcal{O}(n^2\log n)$ time. Hence, the algorithm runs in time $\mathcal{O}(n^3\log n)$. □

Analyzing the proof of Theorem 1 we realize that if k is a slow-growing function of n, then k-INDUCED-PACKING-OF-ODD-CYCLES can be solved in polynomial-time.

Corollary 1. *If* $k = \mathcal{O}(\log^{2/3} n)$, *then* k-INDUCED-PACKING-OF-ODD-CYCLES *can be solved in polynomial time.*

4 Two Induced Disjoint Odd Cycles

Theorem 2. *It is* NP*-complete to decide whether a given graph* G *contains two mutually induced odd cycles.*

Proof. We reduce the well known NP-complete 3-SATISFIABILITY problem [11]. It is known that this problem remains NP-complete even for the case when each Boolean variable occurs at most two times in positive and at most two times in negations. We use this variant of the problem for our reduction. Let $x_1,\ldots,x_n$ be Boolean variables and let $C_1,\ldots,C_m$ be clauses of the given Boolean formula Φ in the conjunctive normal form. We construct a graph G as follows.

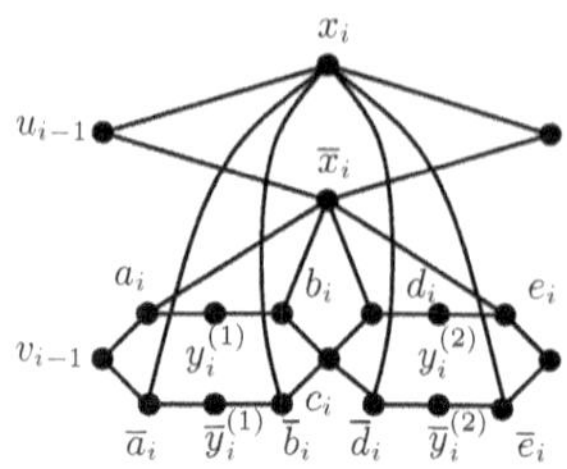

Fig. 1. First stage of construction of G

First, we introduce vertices $u_0, \ldots, u_n$ and vertices $v_0, \ldots, v_n$. For each $1 \leq i \leq n$, the following is done (see Fig. 1):

- Add vertices $x_i, \overline{x}_i$ and edges $u_{i-1}x_i$, x_iu_i, $u_{i-1}\overline{x}_i$ and $\overline{x}_iu_i$. Denote by P_i and $\overline{P}_i$ the paths $u_{i-1}x_iu_i$ and $u_{i-1}\overline{x}_iu_i$ respectively.
- Construct vertices $a_i, b_i, c_i, d_i, e_i, \overline{a}_i, \overline{b}_i, \overline{d}_i, \overline{e}_i$ and $y_i^{(1)}, y_i^{(2)}, \overline{y}_i^{(1)}, \overline{y}_i^{(2)}$, and then add edges $v_{i-1}a_i$, $a_iy_i^{(1)}$, $y_i^{(1)}b_i$, b_ic_i, c_id_i, $d_iy_i^{(2)}$, $y_i^{(2)}e_i$, e_iv_i, $v_{i-1}\overline{a}_i$, $\overline{a}_i\overline{y}_i^{(1)}$, $\overline{y}_i^{(1)}\overline{b}_i$, $\overline{b}_ic_i$, $c_i\overline{d}_i$, $\overline{d}_i\overline{y}_i^{(2)}$, $\overline{y}_i^{(2)}\overline{e}_i$ and $\overline{e}_iv_i$. Denote by Q_i the path $v_{i-1}a_iy_i^{(1)}b_ic_id_iy_i^{(2)}d_iv_i$ and let $\overline{Q}_i = v_{i-1}\overline{a}_i\overline{y}_i^{(1)}\overline{b}_ic_i\overline{d}_i\overline{y}_i^{(2)}\overline{d}_iv_i$.
- Add edges $x_i\overline{a}_i$, $x_i\overline{b}_i$, $x_i\overline{d}_i$, $x_i\overline{e}_i$, $\overline{x}_ia_i$, $\overline{x}_ib_i$, $\overline{x}_id_i$ and $\overline{x}_ie_i$.

Now we introduce vertices $w_0, \ldots, w_m$. For each $1 \leq j \leq m$, three vertices $z_j^{(1)}, z_j^{(2)}, z_j^{(3)}$ are constructed and joined by edges with w_{j-1} and w_j. We denote by $R_j^{(r)}$ the path $w_{j-1}z_j^{(r)}w_j$ for $r = 1, 2, 3$. Assume that the clause C_j contains literals l_1, l_2, l_3. For each literal l_r, $1 \leq r \leq 3$, the following is done (see Fig. 2):

- If $l_r = x_i$ for some $1 \leq i \leq n$, then the edge $x_iz_j^{(r)}$ is added, and also the vertex $z_j^{(r)}$ is joined by an edge with $y_i^{(1)}$ if l_r is the first occurrence of the literal x_i in the Boolean formula, and $z_j^{(r)}$ is joined with $y_i^{(2)}$ if l_r is the second occurrence of x_i.
- If $l_r = \overline{x}_i$ for some $1 \leq i \leq n$, then the edge $\overline{x}_iz_j^{(r)}$ is added, and also the vertex $z_j^{(r)}$ is joined by an edge with $\overline{y}_i^{(1)}$ if l_r is the first occurrence of the literal $\overline{x}_i$ in the Boolean formula, and $z_j^{(r)}$ is joined with $\overline{y}_i^{(2)}$ if l_r is the second occurrence of $\overline{x}_i$.

Finally, we add the edges u_0u_n and v_0w_0, introduce the vertex s and join the vertices v_n and w_m with s by edges.

We claim that Φ can be satisfied if and only if there are two disjoint induced odd cycles S_1 and S_2 in G.

Suppose that Φ can be satisfied, and variables $x_1, \ldots, x_n$ have corresponding truth assignment. We construct S_1 from paths P_i and $\overline{P_i}$ using them as segments. For each $1 \leq i \leq n$, we include P_i in the cycle if $x_i = false$ and $\overline{P}_i$ is included if $x_i = true$. The construction of S_1 is completed by adding the edge u_0u_n. The cycle S_2 is constructed by using paths $Q_i, \overline{Q}_i$ and $R_j^{(r)}$. For each $1 \leq i \leq n$, the

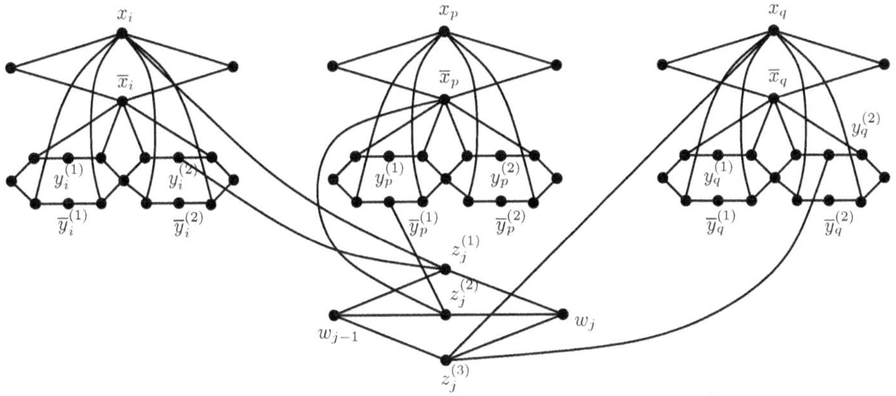

Fig. 2. Second stage of construction of G, the clause C_j contains literals x_i (second occurrence), $\overline{x}_p$ (first occurrence) and x_q (second occurrence)

path Q_i is included in S_2 if $x_i = false$ and $\overline{Q}_i$ is included if $x_i = true$. Now we consider clauses C_j for $1 \le j \le m$. Suppose that C_j contains literals l_1, l_2, l_3. Since $\Phi = true$, there is a literal $l_r = true$, and we include in S_2 the path $R_j^{(r)}$. Finally, the edge v_0w_0 and the path v_nsw_m is added to the cycle. It is easy to check that S_1 and S_2 are disjoint induced odd cycles, and they have no adjacent vertices.

Assume now that G contains two disjoint induced odd cycles S_1 and S_2. Notice that the graph obtained from G by removal of the edges u_0u_n and v_0w_0 is bipartite. Since S_1 and S_2 are odd, they have to include these edges. Suppose without loss of generality that u_0u_n is included in S_1 and v_0w_0 is included in S_2. We need now the following claim.

Claim. For any $1 \le i \le n$,

- either P_i or $\overline{P_i}$ is included in S_1 as a segment,
- if P_i is included in S_1 then Q_i is included in S_2, and if S_1 contains $\overline{P_i}$ then S_2 contains $\overline{Q}_i$.

For each $1 \le j \le n$, S_2 includes one of the paths $R_j^{(1)}, R_j^{(2)}, R_j^{(3)}$.

Proof (Proof of the Claim). First we prove that for any $1 \le i \le n$, either P_i or $\overline{P_i}$ is included in S_1 as a segment, and if P_i is included in S_1 then Q_i is included in S_2, and if S_1 contains $\overline{P_i}$ then S_2 contains $\overline{Q}_i$. Suppose that this claim holds for the lesser values of the parameter, i.e. S_1 contains the edge u_0u_n and either the path P_k or $\overline{P}_k$ for $1 \le k < i$, and similarly S_2 contains the edge v_0w_0 and either the path Q_k or $\overline{Q}_k$ for $1 < k < i$. Assume without loss of generality that if $i > 1$ then S_1 includes P_{i-1}. Since S_1 is an induced cycle, it contains either the edge $u_{i-1}x_i$ or the edge $u_{i-1}\overline{x}_i$. Assume using the symmetry of our construction that $u_{i-1}x_i$ is in S_1. Cycles S_1 and S_2 have no adjacent vertices. Therefore S_2 can not include edges $v_{i-1}\overline{e}_{i-1}$ (if $i > 1$) and $v_{i-1}\overline{a}_i$. Hence this cycle contains the edges $v_{i-1}a_i$ and $a_iy_i^{(1)}$. Notice that if $y_i^{(1)}$ is adjacent to some vertex $z_j^{(r)}$

then x_i is also adjacent to this vertex and S_2 can not include the edge $y_i^{(i)}z_j^{(r)}$. It means that S_2 contains the edges $y_i^{(1)}b_i$ and b_ic_i. By similar arguments we prove that S_2 includes the edges c_id_i, $d_iy_i^{(2)}$, $y_i^{(2)}e_i$ and e_iv_i. Therefore S_2 includes Q_i. Now we return to the cycle S_1. Since all vertices $z_j^{(r)}$ adjacent to x_i are adjacent either to $y_i^{(1)}$ or $y_i^{(2)}$, S_1 can not include edges $x_iz_j^{(r)}$. So, it contains the edge x_iu_i, and together with the fact that S_1 includes $u_{i-1}x_i$, it means that P_i is a segment of S_1.

Now we prove that for each $1 \le j \le n$, S_2 includes one of the paths $R_j^{(1)}, R_j^{(2)}, R_j^{(3)}$. Again suppose that this claim holds for the lesser values of the parameter, and S_2 contains the edge v_0w_0 and one of the paths $R_k^{(1)}, R_k^{(2)}, R_k^{(3)}$ for $1 \le k < j$. Since S_2 is an induced path, it includes one of the edges $w_{j-1}z_j^{(1)}$, $w_{j-1}z_j^{(2)}$, $w_{j-1}z_j^{(1)}$. Assume that the cycle contains $w_{j-1}z_j^{(3)}$. Suppose that $z_j^{(1)}$ is adjacent to some vertex x_i, and therefore to one of vertices $y_i^{(1)}$ or $y_i^{(2)}$ (say, the vertex $y_i^{(1)}$). Notice that in this case $\overline{x}_i$ is a vertex of S_1 by the first part of the claim. The cycle S_2 can not contain the edge $z_j^{(1)}x_i$ since x_i is adjacent to the vertex u_{i-1} which is included in S_1. If S_2 contains $z_j^{(1)}y_i^{(1)}$ then it should contain either the edge $y_i^{(1)}a_i$ or $y_i^{(1)}b_i$, but these vertices are adjacent to $\overline{x}_i$. Same arguments can be used for the case when $z_j^{(1)}$ is adjacent to some vertex $\overline{x}_i$. Hence S_2 includes the edge $z_j^{(1)}w_j$ and we conclude that $R_j^{(1)}$ is a segment of S_2.

Using this claim we assign values to the Boolean variables $x_1, \dots, x_n$: set $x_i = true$ if $\overline{P}_i$ is a segment of S_1 and $x_i = false$ if P_i is a segment of S_1. Consider clauses $C_1, \dots, C_m$. Suppose that the clause C_j contains literals l_1, l_2, l_3 which correspond to vertices $z_j^{(1)}, z_j^{(2)}, z_j^{(3)}$. The cycle S_2 contains one of the paths $R_j^{(1)}, R_j^{(2)}, R_j^{(3)}$, say the path $R_j^{(1)}$ which goes through $z_j^{(1)}$. This vertex is adjacent to one of the vertices x_i or $\overline{x}_i$, and this vertex is not included in S_1. It follows that by our truth assignment $l_i = true$. Since it holds for each $1 \le j \le m$, $\Phi = true$.

To conclude the proof of the theorem, it remains to note that G has $15n + 4m + 4$ vertices, and therefore can be constructed in polynomial time.

5 Conjecture

We conclude the paper with a conjecture. We have showed how to solve the problem in the class of planar graph but we believe that this can be generalized to larger classes.

Conjecture. k-Induced-Packing-Of-Odd-Cycles can be solved in polynomial time for any class of graphs in which genus is bounded.

References

1. Bienstock, D.: On the complexity of testing for odd holes and induced odd paths. Discrete Mathematics 90(1), 85–92 (1991)
2. Bodlaender, H.L., Grigoriev, A., Koster, A.M.C.A.: Treewidth lower bounds with brambles. Algorithmica 51, 81–98 (2008)
3. Caprara, A., Panconesi, A., Rizzi, R.: Packing cycles in undirected graphs. J. Algorithms 48, 239–256 (2003)
4. Conforti, M., Cornuéjols, G., Liu, X., Vuskovic, K., Zambelli, G.: Odd hole recognition in graphs of bounded clique size. SIAM J. Discrete Math. 20, 42–48 (2006)
5. Courcelle, B.: The monadic second-order logic of graphs. i. recognizable sets of finite graphs. Inf. Comput. 85, 12–75 (1990)
6. Diestel, R.: Graph Theory, Electronic Edition. Springer, Heidelberg (2005)
7. Dorn, F., Fomin, F.V., Thilikos, D.M.: Subexponential parameterized algorithms. Comp. Sci. Rev. 2, 29–39 (2008)
8. Dorn, F., Penninkx, E., Bodlaender, H.L., Fomin, F.V.: Efficient exact algorithms on planar graphs: exploiting sphere cut branch decompositions. In: Brodal, G.S., Leonardi, S. (eds.) ESA 2005. LNCS, vol. 3669, pp. 95–106. Springer, Heidelberg (2005)
9. Edmonds, J.: Paths, trees and flowers. Canadian Journal of Mathematics 17, 449–467 (1965)
10. Friggstad, Z., Salavatipour, M.R.: Approximability of packing disjoint cycles, pp. 304–315 (2007)
11. Garey, M.R., Johnson, D.S.: Computers and intractability. W. H. Freeman and Co., San Francisco (1979); A guide to the theory of NP-completeness, A Series of Books in the Mathematical Sciences
12. Hassin, R., Rubinstein, S.: An approximation algorithm for maximum triangle packing. Discrete Applied Mathematics 154, 971–979 (2006)
13. Hassin, R., Rubinstein, S.: Erratum to an approximation algorithm for maximum triangle packing. Discrete applied mathematics 154, 971–979 (2006); Discrete Applied Mathematics 154, 2620 (2006)
14. Krivelevich, M., Nutov, Z., Salavatipour, M.R., Yuster, J., Yuster, R.: Approximation algorithms and hardness results for cycle packing problems. ACM Transactions on Algorithms 3 (2007)
15. Conforti, A.K.M., Cornuéjols, G., Vušković, K.: Part i: Decomposition theorem. J. Graph Theory 39, 6–49 (2002)
16. Conforti, A.K.M., Cornuéjols, G., Vušković, K.: Part ii: Recognition algorithm. J. Graph Theory 40, 238–266 (2002)
17. Maria Chudnovsky, P.S., Kawarabayashi, K.-i.: Detecting even holes. J. Graph Theory 48, 85–111 (2005)
18. Reed, B.A.: Mangoes and blueberries. Combinatorica 19, 267–296 (1999)
19. Robertson, N., Seymour, P.D., Thomas, R.: Quickly excluding a planar graph. J. Comb. Theory, Ser. B 62, 323–348 (1994)
20. Sau, I., Thilikos, D.M.: Subexponential parameterized algorithms for bounded-degree connected subgraph problems on planar graphs. Electronic Notes in Discrete Mathematics 32, 59–66 (2009)
21. Yap, H.P.: Packing of graphs - a survey. Discrete Mathematics 72, 395–404 (1988)

On the Infinitesimal Rigidity of Bar-and-Slider Frameworks*

Naoki Katoh and Shin-ichi Tanigawa

Department of Architecture and Architectural Engineering, Kyoto University, Kyoto Daigaku Katsura, Nishikyo-ku, Kyoto 615-8540 Japan
{naoki,is.tanigawa}@archi.kyoto-u.ac.jp

Abstract. A bar-slider framework is a bar-joint framework a part of whose joints are constrained by using line-sliders. Such joints are allowed to move only along the sliders. Streinu and Theran proposed a combinatorial characterization of the infinitesimal rigidity of generic bar-slider frameworks in two dimensional space. In this paper we propose a generalization of their result. In particular, we prove that, even though the directions of the sliders are predetermined and degenerate, i.e., some sliders have the same direction, it is combinatorially decidable whether the framework is infinitesimally rigid or not. Also, in order to prove that, we present a new forest-partition theorem.

1 Introduction

A 2-dimensional *bar-joint framework* is defined as a pair $(G, \mathbf{p})$, where $G = (V, E)$ is a finite undirected graph having neither loops nor multiple edges and $\mathbf{p}$ is a mapping from V to $\mathbb{R}^2$, called a *joint configuration*. In a framework, each vertex and each edge are regarded as a universal joint and a rigid bar, respectively, and each joint is allowed to move continuously keeping the lengths of the bars. A framework is called *flexible* if it can be deformed by a continuous motion of joints, and otherwise *rigid*. A rigid framework is called *minimally rigid* if removing any bar results in a flexible framework.

A common strategy of dealing with the edge length constraints is to take the first order approximation, and this paper also focuses on this rigidity model, called the *infinitesimal rigidity*. The formal definition will be given in the next section. A joint configuration $\mathbf{p}$ is called *generic* if there is no special algebraic dependency between coordinates of $\mathbf{p}$. (This will be formally defined later.)

The celebrated Maxwell-Laman theorem [4] states that, if $\mathbf{p}$ is generic, $(G, \mathbf{p})$ is infinitesimally minimally rigid if and only if G satisfies $|E| = 2|V| - 3$ and $|F| \leq 2|V(F)| - 3$ for all nonempty $F \subseteq E$, where $V(F)$ denotes the set of vertices spanned by F. A graph satisfying *Laman's counting condition* is called a *minimally rigid graph* or a *Laman graph*. Instead of Laman's counting condition,

* The first author is supported by Grant-in-Aid for Scientific Research (B) and Grant-in-Aid for Scientific Research (C), JSPS. The second author is supported by Grant-in-Aid for JSPS Research Fellowships for Young Scientists.

Y. Dong, D.-Z. Du, and O. Ibarra (Eds.): ISAAC 2009, LNCS 5878, pp. 524–533, 2009.

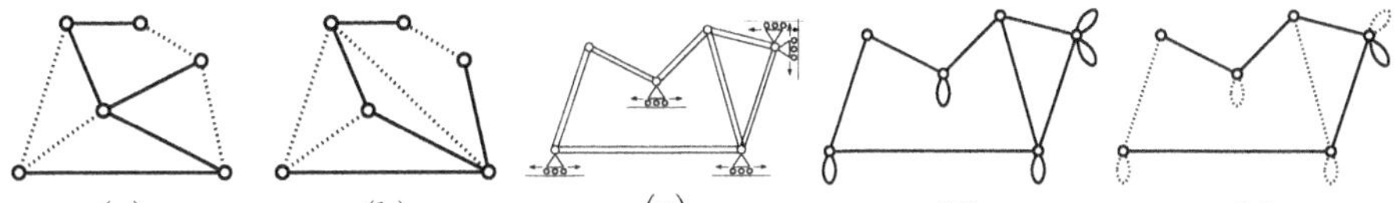

Fig. 1. The sets of bold and dotted edges represent a bipartition. (a) A proper 2forest-partition. (b) A non-proper 2forest-partition. (c) A bar-slider framework. (d) The underlying graph of (c). (e) A proper bipartition satisfying (P1) and (P2) of Proposition 1.

several equivalent characterizations are known, see e.g. [14]. Crapo [1] showed that G is minimally rigid if and only if $|E| = 2|V| - 3$ holds and E can be partitioned into two colored classes $\{R, B\}$ such that **(i)** each color forms a forest, and **(ii)** no subset $V' \subseteq V$ with $|V'| \geq 2$ induces two colored subtrees that span V' simultaneously. A bipartition of E into two forests is called *2forest partition*, while a bipartition satisfying the property (ii) is called *proper*. Hence, a bipartition satisfying both (i) and (ii) is said to be a *proper 2forest partition*[1] (see Figure 1(a)(b)).

Streinu and Theran [12] have extended Crapo's characterization to bar-slider frameworks in a natural way. A *bar-slider framework* is a bar-joint framework a part of whose joints are *constrained* by using *sliders*. As in [12,6], we shall handle each slider as a loop of a graph to extract the combinatorial aspect of frameworks. Let $G = (V, E)$ be an undirected graph that may have some loops, and let us denote the set of loops in $F \subseteq E$ by $L(F)$ and the set of loops incident to a vertex $u \in V$ by $\delta_{L(E)}(u)$. Then, a *bar-slider framework* is defined as a triple $(G, \mathbf{p}, \mathbf{d})$, where $\mathbf{p} : V \to \mathbb{R}^2$ is a joint configuration and $\mathbf{d} : L(E) \to \mathbb{R}^2$ represents a direction of each slider. Namely, for $u \in V$ and $e \in \delta_{L(E)}(u)$, $\{\mathbf{p}(u) + t\mathbf{d}(e) : t \in \mathbb{R}\}$ is a line representing a slider incident to $\mathbf{p}(u)$, and $\mathbf{p}(u)$ is allowed to move along this line (see Figures 1(c) and (d)).

A framework is *minimally rigid* if removing any bar or slider results in a framework that is not rigid. The following proposition is a result of [12].

Proposition 1. ([*12*]) *Let $G = (V, E)$ be an undirected graph. If E can be partitioned into two colored classes $\{R, B\}$ such that (i) $\{R \setminus L(R), B \setminus L(B)\}$ is a proper 2forest partition*[2] *of $E \setminus L(E)$, and (ii) each connected component of the graph (V, R) and (V, B) contains exactly one loop of its color, then there exist a joint configuration $\mathbf{p}$ and a direction mapping $\mathbf{d}$ such that $(G, \mathbf{p}, \mathbf{d})$ is infinitesimally rigid. In particular, each of $L(R)$ and $L(B)$ is realized as a slider parallel to the x-axis and the y-axis, respectively.*

Figure 1(e) shows an example of a partition of E satisfying (P1) and (P2). We refer to it as a *proper 2rooted-forest partition*. As a corollary of the algorithm by

[1] A partition of E into three trees such that each vertex is incident to exactly two of them is called a *3tree2 partition*, and Crapo's partition is usually called a *proper 3tree2 partition*. It is known that E admits a proper 3tree2 partition if and only if E admits a proper 2forest partition with $|E| = 2|V| - 3$.

[2] In [12], this is called an *induced-cut 2-forest* in order to emphasize the existence of a monochromatic cut in any induced subgraph.

Streinu and Theran [13] for checking the sparsity of a graph, it is known that E admits a proper 2rooted-forest partition if and only if

(L1) $|E| = 2|V|$,
(L2) $|F| \leq 2|V(F)| - 3$ for every nonempty $F \subseteq E \setminus L(E)$, and
(L3) $|F| \leq 2|V(F)|$ for every $F \subseteq E$.

In this paper, we will provide an extension of these results. Proposition 1 says that, if E admits a proper 2rooted-forest partition $\{R, B\}$, then each of $L(R)$ and $L(B)$ is realized as an x-slider and a y-slider, respectively. It is not however obvious whether a specified loop is realized as either x-or y-slider until we actually construct a partition specifically. In most practical situations, the directions of sliders are predetermined, and then we are not allowed to realize the predetermined x-slider as a y-slider or vice verse. This raises the following question. Given a set of joints connected by some bars as well as some sliders whose directions are specified and moreover some of which may have the same direction (e.g. x-direction or y-direction), we would like to decide whether it is rigid or flexible.

Notice that, even though a joint configuration is generic, $(G, \mathbf{p}, \mathbf{d})$ could be either rigid or flexible depending on $\mathbf{d}$. In this paper we will prove, however, that the generic rigidity does not actually depend on the specific values of $\mathbf{d}$, and it is combinatorially decidable whether a framework is rigid or not even though the directions of the sliders are "predetermined" and "degenerate".

To extract a combinatorial aspect of this problem, we shall consider a *loop-colored graph* $G_{\mathbf{c}} = (V, E, \mathbf{c})$, where $\mathbf{c}$ is a mapping from $L(E)$ to a finite set $\mathcal{C}$ of colors. Each color indicates a direction of a slider. We then redefine a bar-slider framework as a triple $(G_{\mathbf{c}}, \mathbf{p}, \mathbf{d})$, where $G_{\mathbf{c}}$ is a loop-colored graph, $\mathbf{p}$ is a joint configuration, and $\mathbf{d}$ is a *direction mapping* from $\mathcal{C}$ to $\mathbb{R}^2$ (not from $L(E)$) such that $\mathbf{d}(c)$ and $\mathbf{d}(c')$ are linearly independent for any pair of distinct colors c and c' in $\mathcal{C}$. A loop colored in $c \in \mathcal{C}$ is supposed to be realized as a slider with the direction $\mathbf{d}(c)$ (see Figure 2). A main result of this paper is stated as follows.

Theorem 1. *Let $G_{\mathbf{c}} = (V, E, \mathbf{c})$ be a loop-colored graph. Then, for any direction mapping $\mathbf{d}$ and for any generic joint-configuration $\mathbf{p}$, the bar-slider framework $(G_{\mathbf{c}}, \mathbf{d}, \mathbf{p})$ is infinitesimally rigid if and only if $G_{\mathbf{c}}$ satisfies (L1)$\sim$(L3) as well as*

(L4) *$|F| \leq 2|V(F)| - 1$ for any $F \subseteq E$ such that all loops of $L(F)$ are monochromatically colored.*

The necessity of Theorem 1 follows straightforwardly from the definition of the rigidity. In Section 3, we shall propose nontrivial forest-partition theorems (Theorem 2 and Theorem 3), which might be interesting results in their own right. In Section 4, we will provide a proof of the sufficiency of Theorem 1 based on the forest-partition theorem.

As for the related works of bar-slider frameworks, we should mention the *pinning problem* of a bar-joint framework in the plane. In this problem, given a bar-joint framework (having certain degree of freedom), we would like to stabilize it by *fixing* the positions of the smallest number of joints. Lovász [7] showed, as

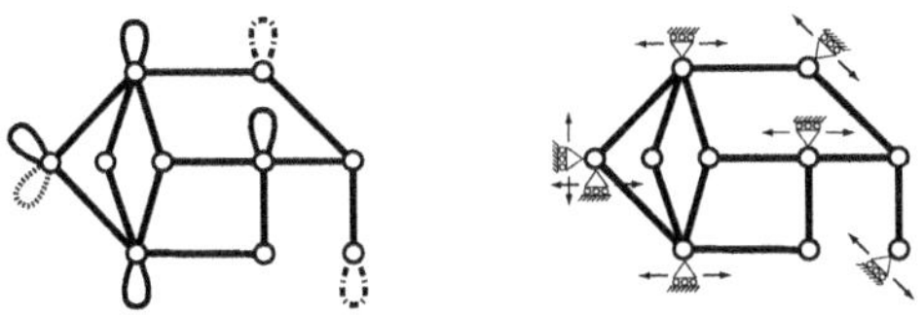

Fig. 2. A loop-colored graph $G_{\mathbf{c}}$ and a bar-slider framework $(G_{\mathbf{c}}, \mathbf{p}, \mathbf{d})$

an application of his matroid matching algorithm, that the pinning problem can be reduced to a 2-polymatroid matching problem, and is solvable in polynomial time. Fekete [2] provided a simpler min-max characterization of the optimal value in generic case. Fekete and Jordán further discussed in [3] the pinning problem from the view point of generic global rigidity. Also, Servatius, Shai and Whiteley [10] have presented a counting condition for the pinned bar-joint framework. Notice that pinning down a joint reduces the degree of freedom of a framework by at most two, while attaching a slider at some joint reduces the degree of freedom by at most one. In fact, attaching a slider seems easier to handle than pinning a joint and we thereby obtain a much clearer and extended combinatorial characterization of the rigidity of bar-slider frameworks.

2 Preliminaries

Matroids and Submodular Functions. We skip the definition, basic terminologies and fundamental properties of matroids (see e.g. [8]). Let E be a finite set. The function $f : 2^E \to \mathbb{R}$ is called *submodular* if $f(X) + f(Y) \geq f(X \cup Y) + f(X \cap Y)$ for any $X, Y \in 2^E$ and *nondecreasing* if $f(X) \leq f(Y)$ for any $X \subseteq Y \subseteq E$. Let $f : 2^E \to \mathbb{Z}$ be an integer-valued nondecreasing submodular function. It is known that f induces a matroid on E, denoted by $\mathcal{M}_f$, whose collection of independent sets is written by $\mathcal{I}(\mathcal{M}_f) = \{I \subseteq E : |I'| \leq f(I') \text{ for nonempty } I' \subseteq I\}$ see e.g., [8, Chapter 12]. For an edge set F, $\mathcal{P}(F)$ denotes the collection of all possible partitions $\{F_0, F_1, \dots, F_m\}$ of F for some integer m with $0 \leq m \leq |F|$ such that $F_i \neq \emptyset$ for each $i = 1, \dots, m$ (and F_0 may be empty). The following proposition provides an explicit formula expressing the rank function r_f of $\mathcal{M}_f$, which is in the form of the Dilworth truncation (restricted to a matroid), see e.g. [11, Chapter 48].

Proposition 2. *Let f be an integer-valued nondecreasing submodular function on E satisfying $f(F) \geq 0$ for every nonempty $F \subseteq E$. Then, for any nonempty $F \subseteq E$, the rank $r_f(F)$ of F in $\mathcal{M}_f$ is given by*

$$r_f(F) = \min_{\{F_0, \dots, F_m\} \in \mathcal{P}(F)} \{|F_0| + \sum_{i=1}^{m} f(F_i)\}. \tag{1}$$

Let us consider the matroid union $\mathcal{M}_f \vee \mathcal{M}_g$ of $\mathcal{M}_f$ and $\mathcal{M}_g$ induced by integer-valued nondecreasing submodular functions f and g on E. Pym and Perfect [9] showed that $\mathcal{M}_f \vee \mathcal{M}_g$ is the matroid induced by the submodular function

$f+g$, i.e. $\mathcal{M}_f \vee \mathcal{M}_g = \mathcal{M}_{f+g}$, if $f(F) \geq 0$ and $g(F) \geq 0$ hold for every $F \subseteq E$ including $\emptyset$. Whiteley further claimed $\mathcal{M}_f \vee \mathcal{M}_g = \mathcal{M}_{f+g}$ in [14] even in the case of $f(\emptyset) < 0$ or $g(\emptyset) < 0$. Although this statement is true for the union of the same matroids, say the union of graphic matroids, Jordán pointed out that this is not always true in general. We shall show below a sufficient condition for the statement to be true.

Lemma 1. *Let f and g be integer-valued nondecreasing submodular functions on E satisfying $f(F) \geq 0$ and $g(F) \geq 0$ for every nonempty $F \subseteq E$. Then, $\mathcal{M}_f \vee \mathcal{M}_g = \mathcal{M}_{f+g}$ holds if, for any $F \subseteq E$, there exists a partition $\{F_0, F_1, \ldots, F_m\} \in \mathcal{P}(F)$ that takes the minimum values of (1) for $r_f(F)$ and $r_g(F)$ simultaneously.*

Bar-joint Rigidity. Recall that a bar-joint framework is defined as a pair $(G, \mathbf{p})$ of a graph G and a joint configuration $\mathbf{p} : V \to \mathbb{R}^2$. An *infinitesimal motion of* $(G, \mathbf{p})$ is defined as an assignment $\mathbf{v} : V \to \mathbb{R}^2$ of a 2-dimensional vector for each joint $\mathbf{p}(v)$ such that

$$(\mathbf{p}(v) - \mathbf{p}(u)) \cdot (\mathbf{v}(v) - \mathbf{v}(u)) = 0 \qquad \text{for each } uv \in E. \tag{2}$$

We refer to (2) as the *length constraint* by the bar $\mathbf{p}(u)\mathbf{p}(v)$. Collecting (2) for all $e \in E$, we have a system of the $|E|$ equations on the unknown $\mathbf{v}(u), u \in V$. Hence $\mathbf{v}$ is an infinitesimal motion if and only if it is in the null space of the $|E| \times 2|V|$-matrix $R(G, \mathbf{p})$, so-called the *rigidity matrix* of $(G, \mathbf{p})$. If $|V| \geq 2$ and the rank of $R(G, \mathbf{p})$ is equal to $2|V| - 3$, the framework $(G, \mathbf{p})$ is said to be *infinitesimally rigid.* Equivalently, $(G, \mathbf{p})$ is infinitesimally rigid if and only if all solutions of $R(G, \mathbf{p})\mathbf{v} = 0$ are *trivial*, that is, (derivatives of) translations and rotations of the whole framework, see e.g. [14] for more details.

A configuration $\mathbf{p}$ is called *generic* with respect to G if the rank of the rigidity matrix $R(G, \mathbf{p})$ and those of all its row-induced submatrices have the maximum values taken over all configurations $\mathbf{p}$. Note that each minor of the rigidity matrix is written as a polynomial of coordinates of $\mathbf{p}$. If such a polynomial is not identically zero, then it takes nonzero value for almost all joint configurations. Namely, a set of generic joint configurations forms an open dense subset of the space of joint configurations. (see e.g. [14]). Therefore, in almost all cases, the rigidity of frameworks is completely determined by the underlying graphs.

Bar-slider Rigidity. Recall that a bar-slider framework is defined as a triple $(G_{\mathbf{c}}, \mathbf{p}, \mathbf{d})$, where $G_{\mathbf{c}}$ is a loop-colored graph, and $\mathbf{d}$ is a direction mapping from a set of colors to $\mathbb{R}^2$. A joint $\mathbf{p}(u)$ is constrained to be on the line $\{\mathbf{p}(u) + t\mathbf{d}(\mathbf{c}(e)) : t \in \mathbb{R}\}$ for each $e \in \delta_{L(E)}(u)$.

Rigidity of a bar-slider framework is defined in a similar way as in the case of a bar-joint framework, but it counts even trivial motions as its degree of freedom. $(G_{\mathbf{c}}, \mathbf{p}, \mathbf{d})$ is *rigid* if there exists no continuous motion of $\mathbf{p}$ which converts to a distinct framework under bar length constraints as well as slider constraints. Again, we shall consider the first-order rigidity of this concept, where an infinitesimal motion $\mathbf{v} : V \to \mathbb{R}^2$ of $(G_\mathbf{c}, \mathbf{p}, \mathbf{d})$ satisfies (2) as well as *direction constraints* written as

$$\mathbf{v}(u) \cdot \mathbf{d}(\mathbf{c}(e))^\perp = 0 \qquad \text{for each } u \in V \text{ and } e \in \delta_{L(E)}(u), \tag{3}$$

where $\mathbf{d}(\mathbf{c}(e))^\perp$ denotes a vector orthogonal to $\mathbf{d}(\mathbf{c}(e))$. As a result, taking new rows corresponding to the direction constraints (3) into account, we obtain the rigidity matrix $R(G_\mathbf{c}, \mathbf{p}, \mathbf{d})$, whose size becomes $|E| \times 2|V|$. It is called *infinitesimally rigid* if no infinitesimal motion exists (except for $\mathbf{0}$), equivalently the rank of $R(G_\mathbf{c}, \mathbf{p}, \mathbf{d})$ is equal to $2|V|$. A generic joint configuration $\mathbf{p}$ with respect to a given $G_\mathbf{c}$ and a given $\mathbf{d}$ is defined as in the case of bar-joint frameworks.

3 Combinatorial Results

Let $G_\mathbf{c} = (V, E, \mathbf{c})$ be a loop-colored graph, and $\{c_1, c_2, \ldots, c_k\}$ be the set of colors appearing in $G_\mathbf{c}$. By regarding each self-loop as a root, a subgraph $G' = (V', E')$ of $G_\mathbf{c}$ is said to be a *rooted-forest colored in* c_i if (i) $(V', E' \setminus L(E'))$ is a forest, (ii) E' does not contain any loop colored in c_j with $j \neq i$, and (iii) each connected component of G' contains exactly one loop colored in c_i. G' is further called a *spanning rooted-forest* colored in c_i if E' spans V.

We say that $G_\mathbf{c}$ satisfies the *strong counting condition* if it satisfies (L1)$\sim$(L4) given in the introduction. In this section we shall reveal properties of graphs satisfying the strong counting condition. In particular, we generalize a concept of the forest partitions to a partition $\mathcal{E} = \{E_1, E_2, \ldots, E_k\}$ of E into k components such that

(P1) each E_i induces a rooted-forest colored in c_i,
(P2) each vertex is spanned by exactly two components, i.e., $|\{i : \delta_E(v) \cap E_i \neq \emptyset\}| = 2$ holds for each $v \in V$.

We refer to each component of $\mathcal{E}$ as a *colored class*. If a partition of E satisfies these two conditions, we say that E (or G_c) admits a $(k,2)$*-rooted-forest partition.* As before, a $(k,2)$-rooted-forest partition is said to be *proper* if no two subtrees from $E_i \setminus L(E_i)$ and $E_j \setminus L(E_j)$ span the same set of vertices for any $1 \leq i, j \leq k$ with $i \neq j$. Figure 3 shows examples of $(k,2)$-rooted-forest partitions. The following forest partition theorem is our main combinatorial result.

Theorem 2. *Let $G_\mathbf{c}$ be a loop-colored graph, and let k be the number of colors used in $G_\mathbf{c}$. Then, $G_\mathbf{c}$ satisfies the strong counting condition if and only if it admits a proper $(k,2)$-rooted-forest partition.*

To prove Theorem 2, we define a counting condition weaker than (L2) as

(L2') $|F| \leq 2|V(F)| - 2$ for every nonempty $F \subseteq E \setminus L(E)$,

and let us refer to the set of the counting conditions (L1), (L2'), (L3) and (L4) as the *weak counting condition.* Although the detailed description is omitted, Theorem 2 easily follows from the next result.

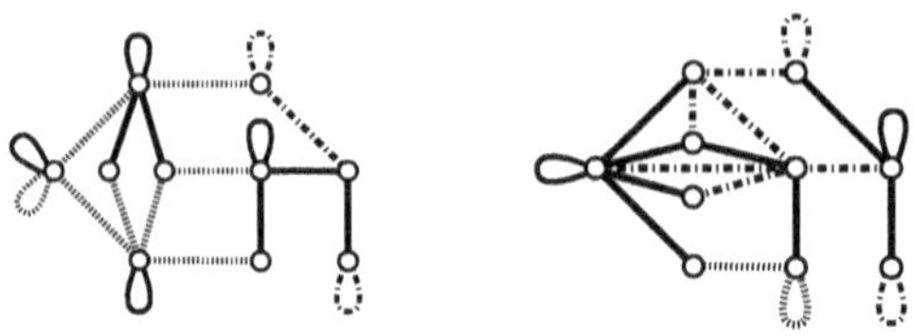

(a) (b)

Fig. 3. (a) A proper $(k, 2)$-rooted-forest partition for $k = 3$. (b) A non-proper $(k, 2)$-rooted-forest partition for $k = 3$.

Theorem 3. *Let $G_\mathbf{c}$ be a loop-colored graph, and let k be the number of colors used in $G_\mathbf{c}$. Then, $G_\mathbf{c}$ satisfies the weak counting condition if and only if it admits a $(k, 2)$-rooted-forest partition.*

Notice that, a set of edges satisfying the weak counting condition and the strong counting condition, respectively, is a base of the matroid induced by the integer-valued nondecreasing submodular functions $\mu : 2^E \to \mathbb{Z}$ and $\mu' : 2^E \to \mathbb{Z}$, respectively, defined as

$$\mu(F) = 2|V(F)| - 2 + \min\{\chi(F), 2\} \tag{4}$$

$$\mu'(F) = \begin{cases} 2|V(F)| - 3 & \text{if } L(F) = \emptyset \\ 2|V(F)| - 2 + \min\{\chi(F), 2\} & \text{otherwise,} \end{cases} \tag{5}$$

where $\chi(F)$ denotes the total number of colors appearing in $L(F)$. Therefore, Theorem 3 implies that a set of edges admitting a $(k, 2)$-rooted-forest partition is characterized in terms of a matroid as the well-known characterization of forest-partitions in terms of the union of graphic matroids. Also, Theorem 2 generalizes Crapo's characterization [1] of Laman graphs.

Proof of Theorem 3 for $k = 2$. We now consider a special case of Theorem 3 where the number of colors k is restricted to two. We prove that $(2, 2)$-rooted-forest partitions can be characterized in terms of the union of two matroids. For a vertex set V, let $K(V)$ denote the complete graph on V, and let $K^+(V)$ denotes the graph obtained from $K(V)$ by attaching two loops at each vertex. We simply denote by $K^+(V)$ the edge set of $K^+(V)$, if it is clear from the context. Throughout this restricted case (of $k = 2$), we assume that one of two loops incident to v in $K^+(V)$ is colored in red and the other one is colored in blue for each $v \in V$. For an edge set $F \subseteq K^+(V)$, let $L_r(F)$ and $L_b(F)$ denote the sets of loops colored in red and blue, respectively.

Let us first consider the following function $\tau_r : 2^{K^+(V)} \to \mathbb{Z}$; For $F \subseteq K^+(V)$,

$$\tau_r(F) = |V(F)| - 1 + \chi_r(F), \tag{6}$$

where $\chi_r(F)$ is defined as $\chi_r(F) = 0$ if $L_r(F) = \emptyset$ and otherwise $\chi_r(F) = 1$. Then, it is not difficult to see that τ_r is submodular. Also τ_r is nondecreasing, and hence it induces a matroid, denoted by $\mathcal{M}_{\tau_r}$, on $K^+(V)$. The functions τ_b and χ_b, and the matroid $\mathcal{M}_{\tau_b}$ (for the blue color) are symmetrically defined.

Lemma 2. *Let $c \in \{r,b\}$ and $\bar{c} \in \{r,b\} \setminus \{c\}$. An edge set F of $K^+(V)$ is a base of $\mathcal{M}_{\tau_c}$ if and only if it is a spanning rooted-forest colored in c.*

Let us consider how the independent sets of $\mathcal{M}_{\tau_r} \vee \mathcal{M}_{\tau_b}$ can be characterized in terms of the counting condition. To apply Lemma 1, we just need to show the following property.

Lemma 3. *Let $c \in \{r,b\}$ and let $F \subseteq K^+(V)$. Let m be the total number of connected components of $G[F]$ and $\{F_1,\ldots,F_m\}$ be a partition of F such that, for each $j = 1,\ldots,m$, F_j is the edge set of a connected component of $G[F]$. Also, let $F_0 = \emptyset$. Then, $\{F_0,F_1,\ldots,F_m\} \in \mathcal{P}(F)$ takes the minimum value of (1). Namely, the rank $r_{\tau_c}(F)$ can be described by $r_{\tau_c}(F) = \sum_{i=1}^{m} \tau_c(F_i)$.*

Combining Lemma 1 and Lemma 3, we obtain $\mathcal{M}_{\tau_r} \vee \mathcal{M}_{\tau_b} = \mathcal{M}_{\tau_r+\tau_b}$, and the following lemma easily follows.

Lemma 4. *Let $E \subseteq K^+(V)$. Then, E is a base of $\mathcal{M}_{\tau_r} \vee \mathcal{M}_{\tau_b}$ if and only if it satisfies the weak counting condition.*

We are now ready to show Theorem 3 for $k = 2$. By Lemma 2, an edge set is a base of $\mathcal{M}_{\tau_r} \vee \mathcal{M}_{\tau_b}$ if and only if it can be partitioned into E_r and E_b which are spanning rooted-forests colored in red and blue, respectively, and equivalently it admits a $(2,2)$-rooted-forest partition. Combining Lemma 4 with this fact, we conclude that an edge set admits a $(2,2)$-rooted-forest partition if and only if it satisfies the weak counting condition.

Due to the space limitation, we omit the proof of Theorem 3 for $k > 2$ in this extended abstract.

Other Combinatorial Results. In order to prove Theorem 1 we need one more combinatorial lemma related to the weak counting condition. We refer to a vertex v as a *(a,b)-vertex* in $G_{\mathbf{c}}$ if $\delta_{E \setminus L(E)}(v) = a$ and $\delta_{L(E)}(v) = b$. The following lemma claims the existence of small degree vertices.

Lemma 5. *Let $G_{\mathbf{c}} = (V,E,\mathbf{c})$ be a connected graph with $|V| > 1$ satisfying the weak counting condition. Suppose that there exists no $(1,1),(1,2),(2,0),(3,0)$-vertex. Then, for any $(k,2)$-rooted-forest partition $\mathcal{E}$ of $G_{\mathbf{c}}$, there exists a $(2,1)$-vertex v such that the two non-loop edges incident to v belong to distinct colored classes in $\mathcal{E}$.*

Remark. Each condition of the strong counting condition (and also the weak one) can be checked by using the so-called "pebble game" algorithm for checking the sparsity of graphs (see [5,13]). Since the pebble game works in $O(|V|^2)$ time, checking whether $G_{\mathbf{c}}$ satisfies the strong counting condition (also the weak one) can be done in $O(k|V|^2)$ time, where k is the number of colors. This can be improved to $O(|V|^2)$ time by just avoiding to start each pebble game from scratch. Therefore, assuming a generic joint configuration, we can decide in $O(|V|^2)$ time whether a bar-slider framework is rigid or not.

4 Infinitesimally Rigid Bar-Slider Frameworks

We now prove the following statement; for a given loop-colored graph $G_{\mathbf{c}}$ satisfying the strong counting condition and a direction mapping $\mathbf{d}$ on a set of colors, there exists a joint configuration $\mathbf{p}$ such that the bar-slider framework $(G_{\mathbf{c}}, \mathbf{p}, \mathbf{d})$ is infinitesimally rigid in the plane. Note that, if one particular realization $(G_{\mathbf{c}}, \mathbf{p}, \mathbf{d})$ is rigid, then $(G_{\mathbf{c}}, \mathbf{q}, \mathbf{d})$ becomes rigid for all generic joint configurations $\mathbf{q}$ as explained in Section 2, implying the nontrivial part (sufficiency) of Theorem 1.

The proof is done by induction on $|E \setminus L(E)|$ as follows. We convert $G_{\mathbf{c}}$ to a slightly smaller graph $G'_{\mathbf{c}'}$ with respect to $|E \setminus L(E)|$, and then prepare a rigid realization $(G'_{\mathbf{c}'}, \mathbf{p}, \mathbf{d})$ based on the induction hypothesis. Finally, we shall show that $(G_{\mathbf{c}}, \mathbf{p}, \mathbf{d})$ is infinitesimally rigid.

We omit the base case. Let us consider the case of $|E \setminus L(E)| > 0$. Applying Theorem 2, we have a proper $(k, 2)$-rooted-forest partition $\mathcal{E} = \{E_1, E_2, \ldots, E_k\}$ of $G_{\mathbf{c}}$, where $k = \chi(E)$. For the convenience of the description, let us assume that all edges of $G_{\mathbf{c}}$ (not only loops but also non-loop edges) are colored according to this partition $\mathcal{E}$ throughout the proof. Following Lemma 5, the proof is split into five cases depending on the existence of small degree vertices. Due to the space limitation, we shall omit the four cases, and show only the final case where $G_{\mathbf{c}}$ has no $(1,1), (1,2), (2,0), (3,0)$-vertex.

In this case, by Lemma 5, $G_{\mathbf{c}}$ has a $(2,1)$-vertex v such that the two non-loop edges incident to v belong to distinct colored classes in $\mathcal{E}$. Let e_1 be the loop attached to v with the color c_1, and let va and vb be the two non-loop edges incident to v. By condition (P2) of $\mathcal{E}$ and Lemma 5, we may assume that va and vb are colored in c_1 and c_2, respectively (see Figure 4).

We shall consider the graph $G'_{\mathbf{c}'}$ obtained from $G_{\mathbf{c}}$ by removing va, vb and then attaching new loops f_1 colored in c_1 to a and f_2 colored in c_2 to v, respectively (see Figure 4). Then, the coloring of the edge set induces a proper $(k, 2)$-rooted-forest partition of $G'_{\mathbf{c}'}$, implying that $G'_{\mathbf{c}'}$ satisfies the strong counting condition. By induction, there exists a rigid realization $(G'_{\mathbf{c}'}, \mathbf{p}, \mathbf{d})$. Since the joint $\mathbf{p}(v)$ is isolated, we may assume that $\mathbf{p}(v)$ is located such that $\mathbf{p}(v) - \mathbf{p}(a)$ is orthogonal to the direction $\mathbf{d}(c_1)$ (as shown in Figure 4).

We claim that $(G_{\mathbf{c}}, \mathbf{p}, \mathbf{d})$ is infinitesimally rigid. To see this, suppose that there exists a nonzero infinitesimal motion $\mathbf{v}$ for $(G_{\mathbf{c}}, \mathbf{p}, \mathbf{d})$. Due to the direction constraint by the slider associated with e_1, $\mathbf{v}(v)$ is a scalar multiple of $\mathbf{d}(c_1)$, and consequently $\mathbf{v}(a)$ is also a scalar multiple of $\mathbf{d}(c_1)$ by the length constraint of the bar $\mathbf{p}(v)\mathbf{p}(a)$. Let us define $\mathbf{v}' : V \to \mathbb{R}^2$ as $\mathbf{v}'(v) = 0$ and $\mathbf{v}'(u) = \mathbf{v}(u)$ for

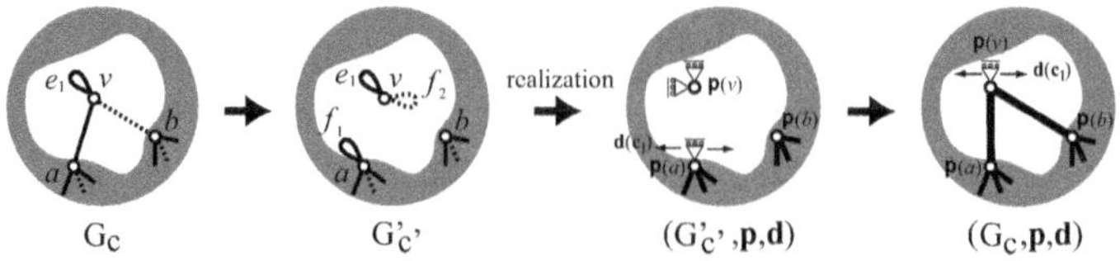

Fig. 4. $(2, 1)$-vertex

all $u \in V \setminus \{v\}$. Then, it is not difficult to see that $\mathbf{v}'$ satisfies all the constraints appearing in $(G'_{\mathbf{c}'}, \mathbf{p}, \mathbf{d})$ since $\mathbf{v}'$ satisfies the direction constraints by the sliders associated with f_1 and f_2, only which are not contained in $G_{\mathbf{c}}$. Therefore, all entries of $\mathbf{v}'$ must be zero because $(G'_{\mathbf{c}'}, \mathbf{p}, \mathbf{d})$ is rigid. Since $\mathbf{v}$ is nonzero, we obtain that $\mathbf{v}(v) \neq 0$ and $\mathbf{v}(u) = \mathbf{v}'(u) = 0$ for all $u \in V \setminus \{v\}$. However, $\mathbf{v}$ does not satisfy the length constraint (2) by the bar $\mathbf{p}(v)\mathbf{p}(b)$ which $(G_{\mathbf{c}}, \mathbf{p}, \mathbf{d})$ has. This contradicts that $\mathbf{v}$ is an infinitesimal motion of $(G_{\mathbf{c}}, \mathbf{p}, \mathbf{d})$. □

References

1. Crapo, H.: On the generic rigidity of plane frameworks. Technical Report 1278, Institute de recherche d'informatique et d'automatique (1988)
2. Fekete, Z.: Source location with rigidity and tree packing requirements. Operations Research Letters 34(6), 607–612 (2006)
3. Fekete, Z., Jordán, T.: Uniquely localizable networks with few anchors. In: Nikoletseas, S.E., Rolim, J.D.P. (eds.) ALGOSENSORS 2006. LNCS, vol. 4240, pp. 176–183. Springer, Heidelberg (2006)
4. Laman, G.: On graphs and rigidity of plane skeletal structures. Journal of Engineering Mathematics 4, 331–340 (1970)
5. Lee, A., Streinu, I.: Pebble game algorithms and sparse graphs. Discrete Mathematics 308(8), 1425–1437 (2008)
6. Lee, A., Streinu, I., Theran, L.: The slider-pinning problem. In: Proc. 19th Canadian Conference on Computational Geometry, CCCG 2007 (2007)
7. Lovász, L.: Matroid matching and some applications. J. Combinatorial Theory, Series (B) 28, 208–236 (1980)
8. Oxley, J.: Matroid Theory. Oxford University Press, Oxford (1992)
9. Pym, J.S., Perfect, H.: Submodular functions and independence structures. J. Math. Analysis Appl. 30, 1–31 (1970)
10. Servatius, B., Shai, O., Whiteley, W.: Combinatorial characterization of the assur graphs from engineering. arXiv:0801.252v1 (2008)
11. Schrijver, A.: Combinatorial Optimization: Polyhedra and Efficiency. Springer, Heidelberg (2003)
12. Streinu, I., Theran, L.: Combinatorial genericity and minimal rigidity. In: Proc. ACM Symposium on Computational Geometry (SoCG 2008), pp. 365–374 (2008)
13. Streinu, I., Theran, L.: Sparsity-certifying graph decompositions. Graphs and Combinatorics 25(2), 219–238 (2009)
14. Whiteley, W.: Matroids from discrete geometry. In: Bonin, J., Oxley, J., Servatius, B. (eds.) Matroid Theory. AMS Contemporary Mathematics, pp. 171–313 (1997)

Exploration of Periodically Varying Graphs

Paola Flocchini[1], Bernard Mans[2], and Nicola Santoro[3]

[1] University of Ottawa, Ottawa, Canada
flocchin@site.uottawa.ca
[2] Macquarie University, Sydney, Australia
bmans@science.mq.edu.au
[3] Carleton University, Ottawa, Canada
santoro@scs.carleton.ca

Abstract. We study the computability and complexity of the *exploration problem* in a class of highly dynamic graphs: *periodically varying* (PV) graphs, where the edges exist only at some (unknown) times defined by the periodic movements of carriers. These graphs naturally model highly dynamic infrastructure-less networks such as public transports with fixed timetables, low earth orbiting (LEO) satellite systems, security guards' tours, etc. We establish necessary conditions for the problem to be solved. We also derive lower bounds on the amount of time required in general, as well as for the PV graphs defined by restricted classes of carriers movements: simple routes, and circular routes. We then prove that the limitations on computability and complexity we have established are indeed tight. We do so constructively presenting two worst case optimal solution algorithms, one for anonymous systems, and one for those with distinct nodes ids.

1 Introduction

Graph exploration is a classical fundamental problem extensively studied since its initial formulation in 1951 by Shannon [10]. It has various applications in different areas, e.g, finding a path through a maze, or searching a computer network using a mobile software agent. In these cases, the environment to be explored is usually modelled as a (di)graph, where a single entity (called agent or robot) starting at a node of the graph, has to visit all the nodes and terminate within finite time. Different instances of the problem exist depending on a variety of factors (e.g., see [1,3,4,5,6]), but all these investigations assume that the graph to be explored is *connected*.

The connectivity assumption unfortunately does not hold for the new generation of networked environments that are highly dynamic and evolving in time. In these infrastructure-less networks, end-to-end multi-hop paths may not exist, and it is actually possible that, at every instant of time, the network is disconnected. However, communication routes may be available through time and mobility, and not only basic tasks like routing, but complex communication and computation services could still be performed. Almost all the existing work in this area focuses on the *routing* problem (e.g., [8,9,11,12]). No work exists on *exploration* of such networks, with the noticeable exception of the study of exploration by random walks [2]. The highly dynamic features of these networks can be described by means of *time-varying* graphs, that is

Y. Dong, D.-Z. Du, and O. Ibarra (Eds.): ISAAC 2009, LNCS 5878, pp. 534–543, 2009.

graphs where links between nodes exist only at some times (a priori unknown to the algorithm designer); thus, for example, the static graph defined by the set of edges existing at a given time might not be connected. Our research interest is on computability and complexity of the deterministic *exploration* of time-varying graphs.

In this paper, we start the investigation focusing on a particular class of time-varying graphs: the *periodically varying graphs* (PV graphs), where the edges of the graphs are defined by the periodic movements of some mobile entities, called carriers. This class models naturally infrastructure-less networks where mobile entities have fixed routes that they traverse regularly. Examples of such common settings are public transports with fixed timetables, low earth orbiting (LEO) satellite systems, security guards' tours, etc.; these networks have been investigated in the application/engineering community, with respect to routing and to the design of carriers' routes (e.g., see [9,12]). We view the system as composed of n sites and k carriers, each periodically moving among a subset of the sites. The routes of the carriers define the edges of the time-varying graph: a directed edge exists from node u to node v at time t only if there is a carrier that in its route moves from u to v at time t. In the system enters an explorer agent **a** that can ride with any carrier along its route, and it can switch to any carrier it meets while riding. Exploring a PV-graph is the process of **a** visiting all the nodes and exiting the system within finite time. We first investigate the computability of *PVG-Exploration* and establish necessary conditions for the problem to be solvable. We then consider the complexity of *PVG-Exploration* and establish lower bounds on the number of moves. We then prove that the limitations on computability and complexity established so far, are indeed tight. In fact we prove that all necessary conditions are also sufficient and all lower bounds on costs are tight. We do so constructively presenting worst case optimal solution algorithms, one for anonymous systems and one for those with ids. An added benefit is that the algorithms are rather simple and use a limited amount of memory. Due to space limitations, some proofs are omitted and are available in [7].

2 Model and Terminology

2.1 Periodically Varying Graphs

The system is composed of a set S of *sites*; depending on whether the sites have unique ids or no identifiers, the system will be said to be *with ids* or *anonymous*, respectively. In the system operates a set C of mobile entities called *carriers* moving among the sites; $|C| = k \leq n = |S|$. Each carrier c has a unique identifier $id(c)$ and an ordered sequence of sites $\pi(c) =< x_0, x_1, \ldots, x_{p(c)-1} >$, $x_i \in S$, called *route*; for any integer j we will denote by $\pi(c)[j]$ the component x_i of the route where $i = j \mod p(c)$, and $p(c)$ will be called the *period* of $\pi(c)$. A carrier $c \in C$ moves cyclically along its *route* $\pi(c)$: at time t, c will move from $\pi(c)[t]$ to $\pi(c)[t+1]$ where the indices are taken modulo $p(c)$. In the following, x_0 will be called the *starting* site of c, and the set $S(c) = \{x_0, x_1, \ldots, x_{p(c)-1}\}$, will be called the *domain* of c; clearly $|S(c)| \leq p(c)$. Each route $\pi(c) =< x_0, x_1, \ldots, x_{p(c)-1} >$ defines a directed edge-labelled multigraph $\boldsymbol{G}(c) = (S(c), \boldsymbol{E}(c))$, where $\boldsymbol{E}(c) = \{(x_i, x_{i+1}, i), 0 \leq i < p(c)\}$ and the operations on the indices are modulo $p(c)$. If $(x, y, t \mod p(c)) \in \boldsymbol{E}(c)$, we shall say that c *activates* the edge (x, y) at time t. A site $z \in S$ is the *meeting point* (or *connection*) of carriers a and b at time t if $\pi(a)[t] = \pi(b)[t] = z$; that is, there exist sites x and y such that, at time $t-1$, a activates the edge (x,z) and b activates the edge (y,z). A route $\pi(c) =< x_0, x_1, \ldots, x_{p(c)-1} >$ is *simple* if $\boldsymbol{G}(c)$ does not contain self loops nor multiple edges; that is $x_i \neq x_{i+1}$, for $0 \leq i < p(c)\}$, and if $(x,y,i),(x,y,j) \in \boldsymbol{E}(c)$ then $i=j$. A simple route $\pi(c)$ is *irredundant* (or *cyclic* if $\boldsymbol{G}(c)$ is either a simple cycle or a virtual cycle (i.e., a simple traversal of a tree). We shall denote by $R = \{\pi(c) : c \in C\}$ the set of all routes and by $p(R) = \text{Max}\{p(c) : c \in C\}$ the maximum *period* of the routes in R. The set R defines a directed edge-labelled multigraph $\boldsymbol{G}_R = (S, \boldsymbol{E})$, where $\boldsymbol{E} = \cup_{c \in C} \boldsymbol{E}(c)$, called *periodically varying graph* (or, shortly, *PV graph*). A *concrete walk* (or, simply, *walk*) σ in $\boldsymbol{G}_R$ is a (possibly infinite) ordered sequence σ =<$e_0, e_1, e_2 \ldots$> of edges in $\boldsymbol{E}$ where $e_i = (a_i, a_{i+1}, i) \in \boldsymbol{E}(c_i)$ for some $c_i \in C$, $0 \leq i$. To each route $\pi(c)$ in R corresponds an infinite concrete walk $\sigma(c)$ in $\boldsymbol{G}_R$ where $e_i = (\pi(c)[i], \pi(c)[i+1], i)$ for $i \geq 0$. A concrete walk σ is a *concrete cover* of $\boldsymbol{G}_R$ if it includes every site: $\cup_{0 \leq i \leq |\sigma|+1} \{a_i\} = S$. A set of routes R is *feasible* if there exists at least one concrete cover of $\boldsymbol{G}_R$ starting from any carrier. R is *homogeneous* if all routes have the same period: $\forall a, b \in C,\ p(a) = p(b)$; it is *heterogeneous* otherwise. R is *simple* (resp. *irredundant*) if every route $\pi(c) \in R$ is simple (resp., irredundant). With an abuse of notation, the above properties of R will be used also for $\boldsymbol{G}_R$; hence we will accordingly say that $\boldsymbol{G}_R$ is feasible (or homogeneous, simple, etc.). In the following, when no ambiguity arises, we will denote $p(R)$ simply as p, $\boldsymbol{G}_R$ simply as $\boldsymbol{G}$, and $(x, y, t \bmod p(c))$ simply as (x,y,t).

2.2 Exploring Agent and Traversal

In the system is injected an external computational entity **a** called *exploring agent*; the agent is injected at the starting site of some carrier at time $t=0$. The only two operations it can perform are: move with a carrier, switch carrier. Agent **a** can switch from carrier c to carrier c' at site y at time t iff it is riding with c at time t and both c and c' arrive at y at time t, that is: iff it is riding with c at time t and $\exists x, x' \in S$ such that $(x,y,t) \in E(c)$ and $(x',y,t) \in E(c')$. Agent **a** does not necessarily know n, k, nor $\boldsymbol{G}$; when at a site x at time t, **a** can however determine the identifier $id(c)$ of each carrier c that arrives at $x \in S$ at time t. The goal of **a** is to fully explore the system within finite time, that is to *visit* every site and terminate, exiting the system, within finite time, regardless of the starting position. We will call this problem *PVG-Exploration*.

An *exploration protocol* $\mathcal{A}$ is an algorithm that specifies the exploring agent's actions enabling it to traverse periodically varying graphs. More precisely, let *start*$(\boldsymbol{G}_R) = \{\pi(c)[0] : c \in C\}$ be the set of starting sites for a periodically varying graph $\boldsymbol{G}_R$, and let $C(t,x) = \{\pi(c)[t] = x : c \in C\}$, be the set of carriers that arrive at $x \in S$ at time $t \geq 0$. Initially, at time $t=0$, **a** is at a site $x \in start(\boldsymbol{G}_R)$. If **a** is at node y at time $t \geq 0$, $\mathcal{A}$ specifies *action* $\in C(t,x) \cup \{\text{halt}\}$: if *action* $= c \in C(t,x)$, **a** will move with c to $\pi(c)[t+1]$, traversing the edge $(x, \pi(c)[t+1], t)$; if *action*=halt, **a** will terminate the execution and exit the system. Hence the *execution* of $\mathcal{A}$ in $\boldsymbol{G}_R$ starting from injection site x uniquely defines the (possibly infinite) concrete walk $\xi(x) =< e_0, e_1, e_2, \cdots >$ of the edges traversed by **a** starting from x; the walk is infinite if **a** never executes *action*=halt, finite otherwise. Algorithm $\mathcal{A}$ *solves* the *PVG-Exploration* of $\boldsymbol{G}_R$ if $\forall x \in$*start*$(\boldsymbol{G}_R)$, $\xi(x)$ is a finite concrete cover of $\boldsymbol{G}_R$; that is, executing $\mathcal{A}$ in

G_R, **a** visits all sites of G_R and performs *action*=halt, regardless of the injection site $x \in$ *start*(G_R). Clearly, we have the following property.

Property 1. PVG-Exploration of G_R is possible only if R is feasible.

Hence, in the following, we will assume that R is *feasible* and restrict *PVG-Exploration* to the class of feasible periodically varying graphs. We will say that problem *PVG-Exploration* is *unsolvable* (in a class of PV graphs) if there is no deterministic exploration algorithm that solves the problem for all feasible PV graphs (in that class). The cost measure is the number of *moves* that the exploring agent **a** performs. Let $\mathcal{M}(G_R)$ denote the number of moves that need to be performed in the worst case by **a** to solve *PVG-Exploration* in feasible G_R. Given a class $\mathcal{G}$ of feasible graphs, let $\mathcal{M}$(G) be the largest $\mathcal{M}(G_R)$ over all $G_R \in \mathcal{G}$; and let $\mathcal{M}_{homo}(n, k)$ (resp. $\mathcal{M}_{hetero}(n, k)$) denote the largest $\mathcal{M}(G_R)$ in the class of all feasible homogeneous (resp. heterogeneous) PV graphs G_R with n sites and k carriers.

3 Computability and Lower Bounds

3.1 Knowledge and Solvability

The availability of a priori knowledge by **a** about the system has an immediate impact on the solvability of the problem *PVG-Exploration*. Consider first *anonymous* systems: the sites are indistinguishable to the exploring agent **a.** In this case, the problem is unsolvable if **a** has no knowledge of (an upper bound on) the system period.

Theorem 1. *Let the systems be* anonymous. PVG-Exploration *is unsolvable if* ***a*** *has no information on (an upper bound on) the system period. This result holds even if the systems are restricted to be* homogeneous, ***a*** *has unlimited memory and knows both* n *and* k.

Proof. By contradiction, let $\mathcal{A}$ solve *PVG-Exploration* in all anonymous feasible PV graphs without any information on (an upper bound on) the system period. Given n and k, let $S = \{x_0, \ldots, x_{n-1}\}$ be a set of n anonymous sites, and let π be an arbitrary sequence of elements of S such that all sites are included. Consider the homogeneous system where k carriers have exactly the same route π and let G be the corresponding graph. Without loss of generality, let x_0 be the starting site. Consider now the execution of $\mathcal{A}$ by **a** in G starting from x_0. Since $\mathcal{A}$ is correct, the walk $\xi(x_0)$ performed by **a** is a finite concrete cover; let m be its length. Furthermore, since all carriers have the same route, $\xi(x_0)$ is a prefix of the infinite walk $\sigma(c)$, performed by each carrier c; more precisely it consists of the first m edges of $\sigma(c)$. Let t_i denote the first time when x_i is visited in this execution; without loss of generality, let $t_i < t_{i+1}, 0 \le i < n-2$. Let π^* denote the sequence of sites in the order they are visited by **a** in the walk $\xi(x_0)$. Let α be the first $t_{n-2}+1$ sites of π^*, and β be the next $m+1-(t_{n-2}+1)$ sites (recall, m is the length of $\xi(x_0)$ and thus $m+1$ is that of π^*). Let γ be the sequence obtained from β by substituting each occurrence of x_{n-1} with x_{n-2}. Consider now the homogeneous system where all the k agents have the same route $\pi' =< \alpha, \gamma, \beta >$, and let G' be the corresponding graph. The execution of $\mathcal{A}$ in G' by **a** with injection site x_0 results (the y_j's); the distinguished sites are reached by the corresponding carriers simultaneously at time $t \equiv n - k \mod p$. The other $n - k - 1$ sites are common to all routes; however there is only a single meeting point in the system, x_0, and all carriers reach it simultaneously at time $t \equiv 0 \mod p$. More precisely, for all $1 \leq i \neq j \leq k$, c_i and c_j meet only at x_0; this will happen whenever $t \equiv 0 \mod p$. Let **a** start at x_0 at time $t = 0$. To visit y_i, **a** must hitch a ride on c_i; this can happen only at x_0 at time $t \equiv 0 \mod p$; in other words, until all y_i's are visited, **a** must traverse all k routes (otherwise it will not visit all distinguished sites) returning to x_0; only once the last distinguished site, say y_j has been visited, **a** can avoid returning to a_0. Each route, except the last, takes $2(n - k)$ moves; in the last, the agent can stop after only $n - k$ moves, for a total of $2k(n - k) - (n - k)$ moves. Since $k \leq \frac{n}{2}$, $2k(n - k) - (n - k) = 2nk - 2k^2 - n + k \geq (k - 1)\, n$.

Theorem 8. *Let the systems be* heterogeneous. *For any* $0 < \epsilon < 1$, $\frac{2}{\epsilon} \leq n$ *and* $2 \leq k \leq \epsilon\, n$, *there exists a feasible* irredundant *graph* $\mathbf{G}_R$ *with* n *sites and* k *carriers such that* $\mathcal{M}(\mathbf{G}_R) > \frac{1}{4}\,(1 - \epsilon)^2\, n^2\, (k - 2)$ *This result holds even if* **a** *knows* $\mathbf{G}_R$, n *and* k, *and has unlimited memory.*

4 Optimal Explorations

In this section we show that the limitations on computability and complexity presented in the previous section are tight. We do so constructively presenting worst case optimal solution algorithms. We introduce the notion of *meeting graph*, that will be useful in the description and analysis of our exploration algorithms. We will describe and analyze two exploration algorithms, one that does not require unique node identifiers (i.e., the PV graph could be anonymous), and one for the case when distinct site ids are available. The *meeting graph* of a PV graph $\mathbf{G}$ is the undirected graph $H(\mathbf{G}) = (C, E)$, where each node corresponds to one of the k carriers, and there is an edge between two nodes if there is at least a meeting point between the two corresponding carriers.

4.1 Exploration of Anonymous PV Graphs

We first consider the general problem of exploring any feasible periodically varying graph without making any assumption on the distinguishability of the nodes. By Theorem 1, under these conditions the problem is not solvable if an upper bound on the periods is not known to **a** (even if **a** has unbounded memory and knows n and k). We now prove that, if such a bound B is known, any feasible periodically varying graph can be explored even if the graph is anonymous, the system is heterogeneous, the routes are arbitrary, and n and k are unknown to **a.** The proof is constructive: we present a simple and efficient exploration algorithm for those conditions. Since the PV graph is anonymous and n and k are not known, to ensure that no node is left unvisited, the algorithm will have **a** explore all domains, according to a simple but effective strategy; the bound B will be used to determine termination. Let us now describe the algorithm, HITCH-A-RIDE. The exploration strategy used by the algorithm is best described as a *pre-order* traversal of a *spanning-tree* of the meeting graph H, where "visiting" a node of the meeting graph H really consists of riding with the carrier corresponding to that node for B' time units, where $B' = B$ if the set of routes is known to be homogeneous, $B' = B^2$ otherwise (the reason for this amount will be apparent later). More precisely, assume that agent **a** is riding with c for the first time; it will do so for B' time units keeping track of all new carriers encountered (list $Encounters$). By that time, **a** has not only visited the domain of c but, as we will show, **a** has encountered all carriers that can meet with c (i.e., all the neighbours of c in the meeting graph H). At this point **a** has "visited" c in H; it will then continue the traversal of H moving to an unvisited neighbour; this is done by **a** continuing to ride with c until a new carrier c' is encountered; c will become the "parent" of c'. If all neighbours of c in H have been visited, **a** will return to its "parent" in the traversal; this is done by **a** continuing the riding with c until its parent is encountered. The algorithm terminates when **a** returns to the starting carrier and the list $Encounters$ is empty.

Theorem 9. *Algorithm* HITCH-A-RIDE *correctly explores any feasible PV graph in finite time provided (an upper bound on) the size of largest route is known.*

Theorem 10. *The number of moves performed by* HITCH-A-RIDE *to traverse a feasible PV graph* **G** *is at most* $(3k-2)B'$.

Proof. Every time routine VISIT(c) is executed, **a** performs B' moves; since a visit is performed for each carrier, there will be a total of $k \cdot B'$ moves. Routine GO-TO-NEXT(c) is used to move from a carrier c to another c' having a meeting point in common. This is achieved by riding with c until c' is met; hence its execution costs at most B' moves. The routine is executed to move from a carrier to each of its "children", as well as to return to its "parent" in the post-order traversal of the spanning tree of H defined by the relation "parent-of". In other words, it will be executed precisely $2(k-1)$ times for a total cost of at most $2(k-1)B'$ moves. The theorem then follows.

The efficiency of Algorithm HITCH-A-RIDE clearly depends on the accuracy of the upperbound B on the size p of the longest route in the system, as large values of B affect the number of moves linearly in the case of homogeneous systems, and quadratically in the case of heterogeneous system. It is sufficient that the upperbound is linear in p for the algorithm to be *optimal*. From Theorem 10 and from Theorems 3-8 we have:

Theorem 11. *Let* $B = O(p)$*; then Algorithm* HITCH-A-RIDE *is worst-case optimal with respect to the amount of moves. This optimality holds even if (unknowingly) restricted to the class of feasible PV graphs* with ids, *and even if the class is further restricted to be* simple *or* circular *(anonymous or not).*

It is interesting to note that the amount of *memory* used by the algorithm is relatively small: $O(k \log k)$ bits are used to keep track of all the carriers and $O(\log B)$ bits to count up to B^2, for a total of $O(\log B + k \log k)$ bits.

4.2 Non-anonymous Systems

We now consider the case when the nodes have distinct Ids. By Theorem 2, either n or an upperbound on the system period must be available for the exploration to be

Parameterized Complexity of Arc-Weighted Directed Steiner Problems*

Jiong Guo[1,**], Rolf Niedermeier[2], and Ondřej Suchý[3,***]

[1] Universität des Saarlandes,
Campus E 1.4, D-66123 Saarbrücken, Germany
jguo@mmci.uni-saarland.de
[2] Institut für Informatik, Friedrich-Schiller-Universität Jena,
Ernst-Abbe-Platz 2, D-07743 Jena, Germany
rolf.niedermeier@uni-jena.de
[3] Department of Applied Mathematics and Institute for Theoretical Computer Science[†], Charles University,
Malostranské nám. 25, 118 00 Praha, Czech Republic
suchy@kam.mff.cuni.cz

Abstract. We initiate a systematic parameterized complexity study of three fundamental network design problems on arc-weighted directed graphs: DIRECTED STEINER TREE, STRONGLY CONNECTED STEINER SUBGRAPH, and DIRECTED STEINER NETWORK. We investigate their parameterized complexities with respect to the parameters "number of terminals", "an upper bound on the size of the connecting network", and the combination of both. We achieve several parameterized hardness as well as some fixed-parameter tractability results, in this way significantly extending previous results of Feldman and Ruhl [SIAM J. Comp. 2006].

1 Introduction

Steiner-type problems lie at the heart of network design and connectivity problems [9]. Roughly speaking, the task in these problems is to find in a given weighted graph a low-cost subgraph that satisfies prescribed connectivity requirements. Most of the corresponding optimization problems are NP-hard. Thus, there are numerous results on polynomial-time approximability [9]. By way of contrast, the study of the parameterized complexity of these problems is much less developed. Our work contributes new algorithmic and computational hardness results concerning the parameterized complexity of three fundamental types of Steiner problems in arc-weighted digraphs.

---

* Main work done while all authors were with Friedrich-Schiller-Universität Jena.

** Work partially supported by the Excellence Cluster on Multimodal Computing and Interaction (MMCI).

*** Work partially supported by the ERASMUS program and by the DFG, project NI 369/4 (PIAF) while visiting Friedrich-Schiller-Universität Jena (October 2008–March 2009) and by grant 201/05/H014 of the Czech Science Foundation.

† Supported by grant 1M0021620808 of the Czech Ministry of Education.

Y. Dong, D.-Z. Du, and O. Ibarra (Eds.): ISAAC 2009, LNCS 5878, pp. 544–553, 2009.

For the Directed Steiner Tree problem (DST), the task is to connect a distinguished root vertex by directed paths to a set of given terminals. For the Strongly Connected Steiner Subgraph problem (SCSS), the task is to connect all terminals among each other. Finally, for the Directed Steiner Network problem (DSN), the task is to connect given terminal vertex pairs. Refer to Section 2 for formal definitions. Obviously, DST and SCSS are special cases of DSN, whereas they are "incomparable" to each other. Note that, following standard modelling, we always assume the underlying directed graph to be complete; arcs that do not exist are modelled by assigning them the weight ∞. To achieve full modelling flexibility (including the cases where one wants to augment an already existing digraph), we sometimes also use arcs of weight 0 to represent already existing connection structure that comes for free. Allowing only arcs of weights 0 and 1 is studied and known in the literature as *augmentation problem*, and allowing only arc of weights 1 and ∞ models the case that one searches for a minimum-size subgraph. Thus, we distinguish between 0-DSN and DSN, indicating whether 0-weights are allowed or not (analogously, 0-SCSS, SCSS, 0-DST, DST). Moreover, we consider the (maximum) *ratio* r of arc weights to be the quotient of the maximum occurring arc weight and the minimum occurring arc weight, excluding 0-weights from consideration. If there are ∞-weight arcs, then we call this *unbounded ratio*. Clearly, a bounded ratio means that in principle every arc is a candidate for being part of the connecting subgraph. It is important to note that a higher ratio makes the problem harder as well as allowing arcs of weight 0 does. Some meaningful parameterizations of the considered Steiner problems are

- the parameter l that denotes the number of terminals to be connected,
- the weight p of the solution divided by the minimum arc weight $\min_W$ (again excluding 0), giving the parameter $p/\min_W$,[1] and
- the combined parameter $(l, p/\min_W)$.

Finally, note that a hardness result with respect to the combined parameter clearly means hardness results for each single parameter and, by way of contrast, a fixed-parameter tractability result for a single parameter trivially extends to the combined parameter.

Next, we discuss some known results for the introduced problems. Herein, n denotes the number of vertices and m the number of arcs of finite weight. In general terms, one may say that the considered problems are hard to approximate. For instance, it is known that DSN cannot be approximated to within a factor of $O(2^{\log^{1-\epsilon} n})$ for any fixed $\epsilon > 0$, unless NP $\subseteq$ TIME$(2^{\text{polylog}(n)})$ [4]. We refer to Kotsarz and Nutov [9] for a survey on the numerous approximation results for Steiner-type problems. By way of contrast, much less is known about the parameterized complexity of directed Steiner problems. The basic Steiner Tree problem in undirected graphs is known to be W[2]-complete with respect to the parameter "size of the Steiner tree" [5] whereas it is fixed-parameter tractable

[1] This parameter naturally reflects the number of arcs in the spanning subgraph by providing an upper bound on the number of (non-zero) arcs.

Table 1. Parameterized complexity results for (0-)DST, (0-)SCSS, and (0-)DSN. Herein, r denotes the ratio of the arc weights. For $r = 1$ the problems DST, SCSS, and DSN can be solved in a straightforward way in polynomial time.

Param.	l	$p/\min_W$	combined
DST	$r \geq 1$: FPT [3,6] $r = \infty$: no poly. kernel (Thm. 4)	$r \geq 1$: FPT [3,6] $r = \infty$: no poly. kernel (Thm. 4)	$r \geq 1$: FPT [3,6] $r = \infty$: no poly. kernel (Thm. 4)
0-DST	$r \geq 1$: FPT [3,6] $r = \infty$: no poly. kernel (Thm. 4)	$r \geq 1$: W[2]-h. [5]	$r \geq 1$: FPT [3,6] $r = \infty$: no poly. kernel (Thm. 4)
SCSS	$r \geq 9$: W[1]-h. (Thm. 1) $2 < r < 9$: open $r \leq 2$: FPT (Thm. 5)	$r \geq 9$: W[1]-h. (Thm. 1) $2 < r < 9$: open $r \leq 2$: FPT (Thm. 5)	$r \geq 9$: W[1]-h. (Thm. 1) $2 < r < 9$: open $r \leq 2$: FPT (Thm. 5)
0-SCSS	$r \geq 4$: W[1]-h. (Thm. 2) $1 < r < 4$: open $r = 1$: FPT (Thm. 6)	$r \geq 1$: W[2]-h. [1]	$r \geq 4$: W[1]-h. (Thm. 2) $1 < r < 4$: open $r = 1$: FPT (Thm. 6)
DSN	$r \geq 9$: W[1]-h. (Thm. 1) $1 < r < 9$: open	$r \geq 9$: W[1]-h. (Thm. 1) $1 < r < 9$: open	$r \geq 9$: W[1]-h. (Thm. 1) $1 < r < 9$: open
0-DSN	$r \geq 1$: W[1]-h. (Thm. 3)	$r \geq 1$: W[2]-h. [1]	$r \geq 1$: W[1]-h. (Thm. 3)

(FPT) with respect to the parameter "number of terminals" [6]. Both results also transfer to the directed case. In particular, the FPT-algorithm can also be used to solve 0-DST, yielding its fixed-parameter tractability with respect to the parameter "number of terminals". Moreover, since the W[2]-complete Set Cover problem (see [5]) can also be formulated as a special case of both 0-DST and 0-SCSS, it follows that 0-DST, 0-SCSS, and 0-DSN are W[2]-hard with respect to the parameter $p/\min_W$. Finally, Feldman and Ruhl [7] showed that 0-DSN can be solved in $O(mn^{4l-2} + n^{4l-1} \log n)$ time using their $O(mn^{2l-3} + n^{2l-2} \log n)$-time algorithm for 0-SCSS as a subprocedure. These algorithmic results directly lead to the question whether there are polynomial-time algorithms whose polynomial degree is independent of l. Thus, Feldman and Ruhl explicitly asked for the fixed-parameter tractability of (0-)DSN and (0-)SCSS.

We extend the above results by initiating a first systematic study of the parameterized complexity of the Steiner problems discussed above (also cf. Table 1). First, we observe that results of Arkin et al. [1] (seemingly Arkin et al. have been unaware of parameterized complexity issues) already answer Feldman and Ruhl's question, showing that 0-DSN is W[2]-hard with respect to the parameter $p/\min_W$. We significantly extend this result by showing W[1]-hardness results even with respect to the combined parameter $(l, p/\min_W)$ for all four types of problems (0-DSN, DSN, 0-SCSS, SCSS). Interestingly, for arc weight ratio r for SCSS (0-SCSS) we obtain W[1]-hardness with respect to the combined parameter only if $r \geq 9$ ($r \geq 4$), whereas we obtain fixed-parameter tractability for SCSS when $r \leq 2$ and mixed results for 0-SCSS for $r < 4$ (see Table 1 for details). Notably, also with respect to the combined parameter, 0-SCSS turns out to be fixed-parameter tractable for $r = 1$ whereas 0-DSN is W[1]-hard for $r = 1$. As a further negative result, we show that DST and 0-DST parameterized by the combined parameter have no polynomial-size problem kernel unless the polynomial hierarchy collapses to the third level. As indicated in Table 1, our work leaves some challenges for future research particularly concerning the parameterized complexity for small constant arc weight ratios.

Due to the lack of space, most details are deferred to the full paper.

2 Preliminaries and Formal Definitions

Let $\mathbb{N}$ be the set of natural numbers and let W be some subset of $\mathbb{N} \cup \{0, \infty\}$. If V is a set of vertices, $w : V \times V \to W$ a weight function[2], and $A \subseteq V \times V$ a set of arcs, then we define $w(A) := \sum_{a \in A} w(a)$. We study the following:

Directed Steiner Tree (DST)
Instance: A set of vertices V, a weight function $w : V \times V \to W$, a set $T \subseteq V$ of terminals ($l := |T|$), a root $s \in V$, and a weight bound $p \in \mathbb{N}$.
Question: Is there a set of arcs $A \subseteq V \times V$ of weight $w(A) \leq p$ such that in the digraph $D := (V, A)$ for every $t \in T$ there is a directed path from s to t?

Strongly Connected Steiner Subgraph (SCSS)
Instance: A set of vertices V, a weight function $w : V \times V \to W$, a set $S \subseteq V$ of terminals ($l := |S|$), and a weight bound $p \in \mathbb{N}$.
Question: Is there a set of arcs $A \subseteq V \times V$ of weight $w(A) \leq p$ such that in the digraph $D := (V, A)$ for every $s, t \in S$ there is a directed path from s to t?

Directed Steiner Network (DSN)
Instance: A set of vertices V, a weight function $w : V \times V \to W$, l pairs of vertices $(s_1, t_1), (s_2, t_2), \ldots, (s_l, t_l)$, and a weight bound $p \in \mathbb{N}$.
Question: Is there a set of arcs $A \subseteq V \times V$ of weight $w(A) \leq p$ such that in the digraph $D := (V, A)$ for every $1 \leq i \leq l$ there is a directed path from s_i to t_i?

We set $\min_W := \min(W \setminus \{0\})$ and $\max_W := \max(W \setminus \{\infty\})$. We use $\{u, v\}$ to denote the undirected edge between vertices u and v and use (u, v) to denote the arc directed from u to v. Moreover, in a directed graph $D = (V, A)$, we use *in-degree* (*out-degree*) of a vertex u to denote the number of vertices which have arcs directed to u (from u). A graph is *strongly connected* if between each pair of vertices u and v there is a path from u to v and a path from v to u.

A given parameterized problem is *fixed-parameter tractable (FPT)* with respect to a parameter k if there is an algorithm solving the problem in $f(k) \cdot n^{O(1)}$ time for some computable function f. A core tool in the development of parameterized algorithms is polynomial-time preprocessing by *data reduction rules*, often yielding a *problem kernel*. Herein, the goal is, given any problem instance G with parameter k, to transform it in polynomial time into a new instance G' with parameter k' such that the size of G' is bounded from above by some function only depending on k, $k' \leq k$, and (G, k) is a yes-instance if and only if (G', k')

[2] Observe that in this way we actually deal with complete digraphs in the sense that only arc weights are specified.

is a yes-instance. In analogy to the polynomial-time hierarchy, a hierarchy for parameterized complexity, called the *W-hierarchy*, has been defined. At the 0th level of this hierarchy lies the class FPT of fixed-parameter tractable problems. The class of all problems at the ith level of the W-hierarchy ($i > 0$) is denoted by W[i], W[1] is the basic class of fixed-parameter intractable problems. A parameterized problem Q is *FPT-reducible* to a parameterized problem Q' if there exists an algorithm of running time $f(k)|x|^{O(1)}$ that on an instance (x, k) of Q produces an instance $(x', g(k))$ of Q' such that (x, k) is a yes-instance of Q if and only if $(x', g(k))$ is a yes-instance of Q', where the functions f and g depend only on k. A parameterized problem Q is W[i]-hard if every problem in W[i] is FPT-reducible to Q. See [5,10] for more details.

3 Parameterized Hardness Results

In this section, we present the parameterized hardness results for DST, SCSS, and DSN. We start with SCSS. We provide an FPT-reduction from the W[1]-complete MULTI-COLORED CLIQUE (MCC) problem [8]. In MCC we are given an undirected graph that is properly colored by k colors and the question is whether there is a size-k clique in it taking exactly one vertex from each color class. The parameter is k.

Theorem 1. STRONGLY CONNECTED STEINER SUBGRAPH *with ratio at least* 9 *is W[1]-hard with respect to the combined parameter* $(l, p/\min_W)$.

Proof. (Sketch.) Let an undirected graph $G = (V, E)$, an integer $k \geq 1$, and the proper coloring $c : V \to \{1, \ldots, k\}$ form an instance of MCC. The high-level idea of the construction of the SCSS-instance (V', w, S, p) is as follows:

First, for every fixed arc weight ratio $r \geq 9$, we use only two weights for the arcs between the vertices in V', the cheap ones having weight $\min_W$ and the expensive ones having weight $\max_W$ (∞ if the ratio is unbounded) with $r = \max_W / \min_W$. It can be shown that there is always a solution using only cheap arcs. Thus, we consider in the following only such solutions.

Second, for each color i there is one terminal b_i that has only cheap arcs to and from the vertex gadgets representing the vertices in G of this color. Thus, the paths between b_i and other terminals, which consist of cheap arcs, have to pass through some arcs in some of these vertex gadgets. This corresponds to taking some vertex from this color class into the solution for the MCC-instance. A similar gadget is also used for every pair of distinct colors, representing the choice of the edge connecting the vertices of the appropriate colors.

Third, the vertex gadget for a vertex v of G consists of two vertices c_v and c'_v and a cheap arc (c_v, c'_v). This arc is the only cheap one leaving c_v and is also the only cheap arc entering c'_v. Taking this uniquely defined arc in solutions for the SCSS-instance represents the selection of the corresponding vertex into the solution for the MCC-instance. The edges of G are encoded in a similar way. Note, however, that every edge is encoded twice. Finally, the vertex and edge gadgets are connected by cheap arcs according to the incidence so that the selected edges are between the selected vertices.

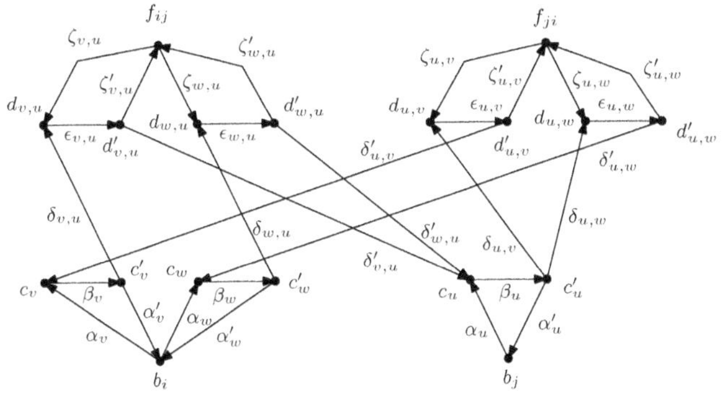

Fig. 1. Part of the construction from Theorem 1 with three vertices—v and w of color i and u of color j—and two edges $\{u, v\}$ and $\{u, w\}$. Only the arcs of $\mathcal{Y} \setminus \Gamma$ are drawn for simplicity. The gadget representing the choice of a vertex of color i is in the bottom left corner, the one for a vertex of color j in the bottom right, the edge selection is represented as a selection of an arc from color-i vertices to color-j vertices in top left and as a selection of an arc from color-j vertices to color-i vertices in top right. The gadgets are interconnected according to the incidence of the vertices and the edges.

Now let us present the construction more formally: We construct our instance (V', w, S, p) of SCSS as follows. The set of vertices V' consists of the following six vertex subsets (see Fig. 1):

$$B := \{b_i \mid 1 \leq i \leq k\}, \quad D := \{d_{u,v}, d_{v,u} \mid \{u, v\} \in E\},$$
$$C := \{c_v \mid v \in V\}, \quad D' := \{d'_{u,v}, d'_{v,u} \mid \{u, v\} \in E\},$$
$$C' := \{c'_v \mid v \in V\}, \quad F := \{f_{ij} \mid 1 \leq i, j \leq k, i \neq j\}.$$

The following arcs are given the weight $\min_W$, that is, they are cheap arcs (see Fig. 1):

$$\mathcal{A} := \{\alpha_v := (b_{c(v)}, c_v) \mid v \in V\},$$
$$\mathcal{A}' := \{\alpha'_v := (c'_v, b_{c(v)}) \mid v \in V\},$$
$$\mathcal{B} := \{\beta_v := (c_v, c'_v) \mid v \in V\},$$
$$\Gamma := \{\gamma_{u,v} := (c'_u, c_v) \mid u, v \in V\},$$
$$\mathcal{D} := \{\delta_{u,v} := (c'_u, d_{u,v}), \delta_{v,u} := (c'_v, d_{v,u}) \mid \{u, v\} \in E\},$$
$$\mathcal{D}' := \{\delta'_{u,v} := (d'_{u,v}, c_v), \delta'_{v,u} := (d'_{v,u}, c_u) \mid \{u, v\} \in E\},$$
$$\mathcal{H} := \{\epsilon_{u,v} := (d_{u,v}, d'_{u,v}), \epsilon_{v,u} := (d_{v,u}, d'_{v,u}) \mid \{u, v\} \in E\},$$
$$\mathcal{Z} := \{\zeta_{u,v} := (f_{c(u),c(v)}, d_{u,v}), \zeta_{v,u} := (f_{c(v),c(u)}, d_{v,u}) \mid \{u, v\} \in E\},$$
$$\mathcal{Z}' := \{\zeta'_{u,v} := (d'_{u,v}, f_{c(u),c(v)}), \zeta'_{v,u} := (d'_{v,u}, f_{c(v),c(u)}) \mid \{u, v\} \in E\},$$
$$\mathcal{Y} := \mathcal{A} \cup \mathcal{A}' \cup \mathcal{B} \cup \Gamma \cup \mathcal{D} \cup \mathcal{D}' \cup \mathcal{H} \cup \mathcal{Z} \cup \mathcal{Z}'.$$

All remaining arcs are set to be expensive arcs, that is, for $(x, y) \in ((V \times V) \setminus \mathcal{Y})$, $w((x, y)) := \max_W$ (∞ if the ratio is unbounded). The terminal set S is equal to $B \cup F$ and hence $l = |S| = |B| + |F| = k + k(k-1) = k^2$. Finally, we set $p := (3k + 5k(k-1)) \cdot \min_W$. It is clear that the instance is constructible in polynomial time and that both l and p/min_W only depend on the parameter k. We defer the correctness proof of the above reduction to the full paper. □

A similar reduction as above also works for 0-SCSS.

Theorem 2. 0-STRONGLY CONNECTED STEINER SUBGRAPH *with ratio at least 4 is W[1]-hard with respect to the combined parameter* $(l, p/\text{min}_W)$.

The reduction for 0-DSN with ratio 1 again is from MULTI-COLORED CLIQUE. However, with now only one non-zero weight allowed, more care, compared to the proofs of Theorems 1 and 2, is necessary to argue that only the arcs representing vertices and edges are needed.

Theorem 3. 0-DIRECTED STEINER NETWORK *is W[1]-hard with respect to the combined parameter* $(l, p/\text{min}_W)$ *even with ratio* 1.

Finally, we can use the technique introduced by Bodlaender et al. [2] to show that there is no polynomial-size problem kernel for DST.

Theorem 4. *There is no polynomial-size kernel for* DIRECTED STEINER TREE *with unbounded ratio with respect to the combined parameter* $(l, p/\text{min}_W)$ *unless the Polynomial Hierarchy collapses to the third level.*

4 Algorithmic Results

Here, we present two fixed-parameter algorithms for two variants of SCSS and 0-SCSS that restrict the allowed arc weight ratio, respectively. In addition, we indicate that these algorithms directly imply a significant running time improvement of the algorithm by Feldman and Ruhl [7] for these relevant cases.

Theorem 5. STRONGLY CONNECTED STEINER SUBGRAPH *with ratio at most* 2 *is solvable in* $O(l! \cdot n)$ *or* $O((p/\text{min}_W)! \cdot n)$ *time.*

Proof. We only consider the case that $p/(2\text{min}_W) < l \leq p/\text{min}_W$, since a Hamiltonian cycle over terminals gives a total weight at least $l \cdot \text{min}_W$ and at most $2l \cdot \text{min}_W$. This means that to strongly connect the terminals in S we need an arc set with a minimum weight at least $l \cdot \text{min}_W$ and a maximum weight at most $l \cdot \text{max}_W \leq 2l \cdot \text{min}_W$. Thus, $p \geq 2l \cdot \text{min}_W$ always gives yes-instances, while $p < l \cdot \text{min}_W$ gives no-instances. Therefore, the parameters l and p/min_W are linearly related and it suffices to show that the problem is solvable in $O(l! \cdot n)$ time. To this end, we claim that there is always a Hamiltonian cycle on S having the minimum total weight among all arc sets strongly connecting S. Since there are $l!$ Hamiltonian cycles on S and computing the total weight of a cycle can be done in $O(n)$ time, the theorem follows. The proof for the claim is omitted. □

Next, we consider the "augmentation case" of SCSS, namely, the case that there are only two weights and one of them is zero. In contrast to the W[1]-hardness of the augmentation case of DSN (cf. Theorem 3), for SCSS we achieve fixed-parameter tractability with respect to the number of terminals l.

Theorem 6. 0-Strongly Connected Steiner Subgraph *with ratio* 1 *is solvable in* $O(4^{l^2} + n^3)$ *time.*

Proof. First note that in this case we have only two weights 0 and $\min_W = \max_W$. Suppose that we are given an input instance of 0-SCSS consisting of V, w, S (where $|S| = l$), and p. If $p/\min_W \geq l$, then the answer is always yes, since we can connect all terminals to a cycle that costs at most $l \cdot \min_W$. So, for the rest of the proof we will assume that $p/\min_W < l$.

We provide four data reduction rules executable in $O(n^3)$ time that lead to a problem kernel with at most $2 \cdot 2^l + l$ vertices. Let $A_0 := \{a \in V \times V \mid w(a) = 0\}$. To simplify the presentation, the rules are described as modifications of the digraph $H := (V, A_0)$, which can be included in any solution. The vertices of $V \setminus S$ are called *non-terminals.* The rules are ordered and the next rule is always applied after the previous one cannot be applied any more. Later rules never produce a situation where an earlier rule could again be applied.

The instance produced by rules is a yes-instance if and only if the original instance is a yes-instance. For the first two rules and the fourth rule this is easy to check, reconnecting the arcs not in A_0 terminating in the removed vertices to the vertex from which the same vertices can be reached. Similarly for the arcs not in A_0 starting in the removed vertices. For the third rule a proof is given. Further details, as well as the running times of the rules, are defered to the full version of the paper.

Rule 1. *Contract strongly connected components into a single vertex.*

Rule 2. *For any non-terminal* $v \in V \setminus S$ *with both* $N^-(v) \neq \emptyset$ *and* $N^+(v) \neq \emptyset$, *where* $N^-(v)$ *is its in-neighborhood and* $N^+(v)$ *its out-neighborhood, delete* v *and connect its neighbors appropriately, that is, continue with the digraph* $H' := (V \setminus \{v\}, A_0 \setminus ((\{v\} \times N^+(v)) \cup (N^-(v) \times \{v\})) \cup (N^-(v) \times N^+(v)))$.

After this rule is exhaustively applied, there remain only sources, terminals, and sinks in the digraph.

Rule 3. *Delete all weight-0 arcs between two non-terminals.*

Claim. There is an optimal solution that uses no arc of weight 0 between two non-terminals.

Proof of Claim. Suppose, on the contrary, that each optimal solution uses some arc in $A_0 \cap ((V \setminus S) \times (V \setminus S))$. Let A be an optimal solution with the minimum number of such arcs and let $a := (x, y) \in A \cap A_0$ be such an arc, that is, $x, y \in V \setminus S$. Clearly, x is a source and y a sink in $H = (V, A_0)$. There is some arc in A ending in x and some arc in A starting in y since $A \setminus \{a\}$ is not a solution. We can assume that there is only one arc ending in x in A for the following reason: If $|N_A^-(x)| \geq 2$, then select an arbitrary $x' \in N_A^-(x)$ and replace the arcs ending in x (except (x', x)) by arcs ending in x' and get another optimal solution that satisfies this assumption. Let us call this unique arc (x', x). Similarly, we assume that there is a unique arc starting in y and we call it (y, y').

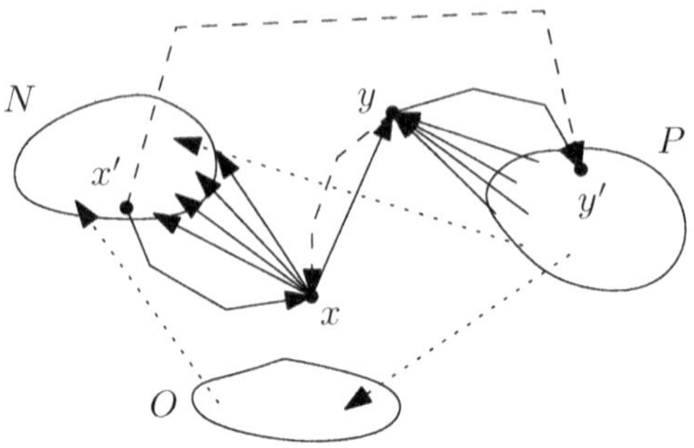

Fig. 2. Illustration to the proof of the claim in Theorem 6. Solid lines represent the sure connections in A. No other connections are possible in A, except for those drawn by dotted lines. Dashed lines represent the arcs in $A' \setminus A$.

Let V_0 denote the minimal set $S \subseteq V_0 \subseteq V$ such that $A \subseteq V_0 \times V_0$ and also assume that A is minimal in the sense that (V_0, A) is a strongly connected digraph. Now let us distinguish the following sets of vertices (see Fig. 2):

$$P := \{v \in V_0 \setminus \{y\} \mid \exists \text{ a path in } (V_0, A \setminus \{a\}) \text{ from } v \text{ to } y\},$$
$$N := \{v \in V_0 \setminus \{x\} \mid \exists \text{ a path in } (V_0, A \setminus \{a\}) \text{ from } x \text{ to } v\},$$
$$O := V_0 \setminus (P \cup N \cup \{x, y\}).$$

Observe that in $A \setminus \{a\}$ there is no path from any vertex in N to any vertex in P (in particular $N \cap P = \emptyset$), since otherwise there would be a path from x to y different from (x, y) and, thus, $A \setminus \{a\}$ would be a solution. There is also no path from O to P and from N to O according to the definition of N, P, and O. If N is empty, then $A' := (A \setminus \{(x', x), a\}) \cup \{(x', y)\}$ is a better solution, since $(V_0 \setminus \{x\}, A')$ is strongly connected, x is a non-terminal, and $w((x', x)) \geq w((x', y))$. Hence N is non-empty. Similarly, P is non-empty since otherwise $(A \setminus \{a, (y, y')\}) \cup \{(x, y')\}$ would be a better solution.

Since N is non-empty and (V_0, A) is strongly connected there must be some arc from some vertex in N to some vertex outside N. But, as we have shown, it can end neither in O nor in P nor in y. Hence it ends in x and thus $x' \in N$. Similarly, $y' \in P$. Now, let $A' := (A \setminus \{a, (x', x), (y, y')\}) \cup \{(x', y'), (y, x)\}$. It is straighforward to check that the digraph $H' := (V_0, A')$ is again strongly connected. Since $w((x', x)) + w((y, y')) = 2 \cdot \min_W \geq w((x', y')) + w((y, x))$ we have $w(A') \leq w(A)$. Thus, A' is an optimal solution which uses less arcs of weight 0 between two non-terminals—a contradiction. □

Rule 4. *If there are several non-terminals with the same neighborhood, then delete all of them except for one.*

Claim. If Rule 4 cannot be applied, then the digraph has at most $2 \cdot 2^l + l$ vertices.

Proof of Claim. (Sketch.) Each non-terminal is either source or sink and its (unique) neighborhood is formed only by terminals. □

To solve 0-SCSS on the reduced instance, try all possibilities to connect at most $p/\min_W$ sinks out of 2^l many to at most $p/\min_W$ sources out of 2^l many and check whether in the resulting digraph the terminals are mutually interconnected. This can be carried out in $O(\binom{2^l}{l} \cdot \binom{2^l}{l} \cdot l! \cdot 2^l \cdot l) = O(4^{l^2})$ time. Thus, 0-SCSS with ratio 1 can be solved in $O(4^{l^2} + n^3)$ time. □

The (0-)DSN algorithm developed by Feldman and Ruhl [7] uses an algorithm for (0-)SCSS as a subprocedure. Using the algorithms developed in the proofs of Theorems 5 and 6 in case of arc weight ratios 2 and 1, respectively, as the subprocedure, the running time of the overall algorithm of Feldman and Ruhl can be significantly improved by roughly halving the degree of its running time polynomial for the relevant case of these small arc weight ratios.

Corollary 1. Directed Steiner Network *with ratio* 2 *and* 0-Directed Steiner Network *with ratio* 1 *can be solved in* $O((2l)! \cdot n^{2l})$ *time and* $O(16^{l^2} n^{2l} + n^{2l+3})$ *time, respectively.*

References

1. Arkin, E.M., Hassin, R., Shahar, S.: Increasing digraph arc-connectivity by arc addition, reversal and complement. Discrete Appl. Math. 122(1-3), 13–22 (2002)
2. Bodlaender, H.L., Downey, R.G., Fellows, M.R., Hermelin, D.: On problems without polynomial kernels. In: Aceto, L., Damgård, I., Goldberg, L.A., Halldórsson, M.M., Ingólfsdóttir, A., Walukiewicz, I. (eds.) ICALP 2008, Part I. LNCS, vol. 5125, pp. 563–574. Springer, Heidelberg (2008); Long version to appear in J. Comput. Syst. Sci.
3. Ding, B., Yu, J.X., Wang, S., Qin, L., Zhang, X., Lin, X.: Finding top-k min-cost connected trees in databases. In: Proc. 23rd ICDE, pp. 836–845. IEEE, Los Alamitos (2007)
4. Dodis, Y., Khanna, S.: Design networks with bounded pairwise distance. In: Proc. 31th STOC, pp. 750–759. ACM, New York (1999)
5. Downey, R.G., Fellows, M.R.: Parameterized Complexity. Springer, Heidelberg (1999)
6. Dreyfus, S., Wagner, R.: The Steiner problem in graphs. Networks 1, 195–207 (1972)
7. Feldman, J., Ruhl, M.: The directed Steiner network problem is tractable for a constant number of terminals. SIAM J. Comput. 36(2), 543–561 (2006)
8. Fellows, M.R., Hermelin, D., Rosamond, F.A., Vialette, S.: On the parameterized complexity of multiple-interval graph problems. Theor. Comput. Sci. 410(1), 53–61 (2009)
9. Kortsarz, G., Nutov, Z.: Approximating minimum cost connectivity problems. In: Gonzalez, T.F. (ed.) Handbook of Approximation Algorithms and Metaheuristics. CRC, Boca Raton (2007)
10. Niedermeier, R.: Invitation to Fixed-Parameter Algorithms. Oxford University Press, Oxford (2006)

Worst Case Analysis for Pickup and Delivery Problems with Consecutive Pickups and Deliveries

Yoshitaka Nakao and Hiroshi Nagamochi

Department of Applied Mathematics and Physics
Graduate School of Informatics
Kyoto University
Yoshida Honmachi, Sakyo, Kyoto 606-8501, Japan
{yoshi,nag}@amp.i.kyoto-u.ac.jp

Abstract. The pickup and delivery problem (PDP) asks to find a set of routes with the minimum travel cost to serve a given set of requests. PDP is called the multi-trip PDP with consecutive pickups and deliveries (PDPCMT) if it has an additional requirement such that any vehicle which has begun a delivery action is not allowed to take pickup actions until all of the loads on the vehicle are delivered. In this paper, we are interested in how the least travel cost can be increased by the additional requirement, and examine the maximum ratio of the optimal value of PDPCMT to that of PDP over all instances with p requests. We show that the maximum ratio is bounded from above by $O(\log p)$ and from below by $\Omega(\log p/ \log\log p)$.

1 Introduction

Decreasing physical distribution cost is a fundamental subject for most manufacturing and distribution companies. The pickup and delivery problem (abbreviated as PDP) is a well studied physical distribution problem, and there exist several variations to PDP such as the vehicle routing problem (VRP) [10], the split delivery VRP (SDVRP) [1,4], PDP with transfer (PDPT) [6,7], PDP (resp., PDPT) with a constraint such that once a vehicle take a delivery action no pickup action is allowed during the same route (PDPC) (resp., (PDPTC)) [7], and so on.

Some theoretical analyses show how much travel cost can be increased by an additional requirement in routing problems [1,2,3,7]. Given an instance I and problem PRB, let p be the number of requests that need to be served, and let $opt_{\rm PRB}(I)$ denote the optimal cost to PRB with instance I. Archetti et al. [1] showed that $opt_{\rm VRP}(I) \leq 2opt_{\rm SDVRP}(I)$ holds for any I, and gave an instance where the bound is tight. Nakao and Nagamochi [7] studied the maximum cost that can be saved by introducing a transshipment point to PDP. They showed that $opt_{\rm PDPC}(I) \leq 6p^{1/2} \cdot opt_{\rm PDPTC}(I)$ holds for any instance I with p requests, while presenting an instance I' that satisfies $opt_{\rm PDPC}(I') = p^{1/4} \cdot opt_{\rm PDPTC}(I')$.

Y. Dong, D.-Z. Du, and O. Ibarra (Eds.): ISAAC 2009, LNCS 5878, pp. 554–563, 2009.

In this paper, we introduce a *multi-trip pickup and delivery problem with consecutive pickups and deliveries* (PDPCMT), which we encounter in many practical cases. We then analyze upper and lower bounds on the maximum travel cost that can be saved by regarding an instance I to PDP as that to PDPCMT. It is not difficult to obtain similar results between PDPT and PDPTCMT (PDPTCMT is an abbreviation of PDPCMT with transfer). Analyses for uncapacited cases with consecutive pickups and deliveries were reported in [2], while we consider capacitated cases.

After formulating PDP and PDPCMT in Section 2, we introduce a partition problem that asks to find a partition of vertices on a binary tree, and present a lower bound on the partition problem in Section 3. Section 4 shows an upper bound on the maximum ratio of the optimal cost of PDPCMT to that of PDP. Next, Section 5 presents an instance that gives a lower bound on the optimal cost by using the result of Section 3. Finally Section 6 makes concluding remarks.

2 Formulations of PDP and PDPCMT

This section formulates PDP and PDPCMT. Depots and customers to be visited by vehicles are represented by distance functions $d(u, v)$ for all ordered pairs of vertices u and v.

An instance consists of a set Q of *depots*, a set R of *requests*, and a vehicle capacity $c > 0$. Each request $r \in R$ consists of a *pickup action* r^+, a *delivery action* r^-, and a quantity $q(r)$ of loads for the request. That is, quantity $q(r)$ of loads is required to be picked up at a specified vertex where r^+ is taken and to be delivered to a specified vertex where r^- is taken. We may denote the vertex where request r is picked up (resp., delivered) by r^+ (resp., r^-), and call the vertex r^+ a *pickup point* (resp., r^- a *delivery point*). Let $A = \{r^+, r^- \mid r \in R\}$, i.e., the set of all actions for R.

Each vehicle must start from a depot in Q, and return to the same depot after serving some of the requests in R. A *route* is the requests assigned to a vehicle, and is represented by the sequence $\sigma = (a_0, a_1, a_2, \ldots, a_m, a_{m+1})$ of actions of the requests in the order that the vehicle serves, where $a_1, a_2, \ldots, a_m \in A$ (m is an even integer) and $a_0, a_{m+1} \in Q$. The travel cost of σ is defined by $cost(\sigma) = \sum_{0 \le i \le m} d(a_i, a_{i+1})$. A *solution* s is a set of routes that serves all requests in R, and its cost $cost(s)$ is defined by the sum of the travel costs of the routes σ in s. The objective is to find a minimum cost solution, and we assume that any number of vehicles is available. We start with defining PDP.

Pickup and Delivery Problem (PDP)
Input: An instance $I = (Q, R, c)$.
Output: A minimum cost solution that satisfies the following constraints:

(a) the total quantity of loads during a route does not exceed vehicle capacity c at any time;
(b) each request r is served by exactly one vehicle, and the actions r^+ and r^- are taken only once by the vehicle;

(c) (*coupling constraint*) actions r^+ and r^- of each request r must appear in the same route, and no request r is allowed to be temporarily dropped at any vertices; and
(d) (*precedence constraint*) For each request r, action r^+ must be taken before action r^-.

PDP is known to be NP-hard [9]. Many heuristics and metaheuristic algorithms have been developed for the problem [5,8,9]. We next define PDPCMT.

Multi-Trip PDP with Consecutive Pickups and Deliveries (PDPCMT)
Input: An instance $I = (Q, R, c)$.
Output: A minimum cost solution that satisfies (a),(b),(c),(d), and the following constraint:
(e) once a delivery action begins, pickup actions cannot be taken until all of the loads on the vehicle are delivered.

A route satisfying (e) is a sequence of subsequences such that first a vehicle with no loads on it takes pickup actions for some requests and takes delivery actions for these requests, which we call *trips*.

3 Lower Bound on a Partition Problem

This section derives a lower bound on a partition problem that asks to find a partition of vertices on a complete binary tree with the minimum cost.

Let h be a positive integer, and let $c = 2^h$, and let $\mathcal{T}_c = (V, E)$ denote a complete binary tree with 2^{c-1} leaves. Hence the number of vertices in V is $|V| = \sum_{i=1}^{c} 2^{i-1} = 2^c - 1$. For a vertex $v \in V$, let $L(v)$ denote the set of leaves that are the descendants of v (including v). For a subset $X \subseteq V$, let $L(X)$ denote the set $\cup_{v \in X} L(v)$. Fig 1 illustrates $\mathcal{T}_c$ with $c = 4$.

We now introduce a *partition problem of vertices on a binary tree* (PPBT). We call a partition $V_1, V_2, \ldots, V_n$ ($n \geq 1$) of V is *feasible* if it satisfies $V_1 \cup V_2 \cup \cdots \cup V_n = V$, $V_i \cap V_j = \emptyset$ for $1 \leq i < j \leq n$, and the number of vertices in each subset V_i is less than or equal to c. We call a feasible partition *a solution* to PPBT. The cost $cost_L(s)$ of a solution $s = \{V_1, V_2, \ldots, V_n\}$ is defined by the number of times that leaves appear in $L(X)$, $X \in s$, i.e., $cost_L(s) = \sum_{X \in s} |L(X)|$.

PPBT asks to find a solution s that minimizes $cost_L(s)$. This section proves the following theorem.

Theorem 1. *For a complete binary tree $\mathcal{T}_c$ with 2^{c-1} leaves, any solution s^* to PPBT satisfies $cost_L(s^*) \geq 2^{c-2} \cdot c/\log_2 c$. Furthermore there exists a solution $\hat{s}$ such that $cost_L(\hat{s}) = 2^{c-1} \cdot c/\log_2 c$.*

We introduce some notations on tree $\mathcal{T}_c$. We denote by $V[i]$, $i = 1, 2, \ldots, c$, the set of vertices on the i-th layer from the root vertex in $\mathcal{T}_c$. We denote set $V[c]$ of vertices on the c-th layer in $\mathcal{T}_c$ by $V[c] = \{v(0,1), v(1,2), v(2,3), \ldots, v(2^{c-1} - 1, 2^{c-1})\}$. For $i = 1, 2, \ldots, c$, set $V[i]$ consists of 2^{i-1} vertices and is denoted by

$$V[i] = \{v((j-1) \cdot 2^{c-i}, j \cdot 2^{c-i}) \mid j = 1, 2, \ldots, 2^{i-1}\}. \tag{1}$$

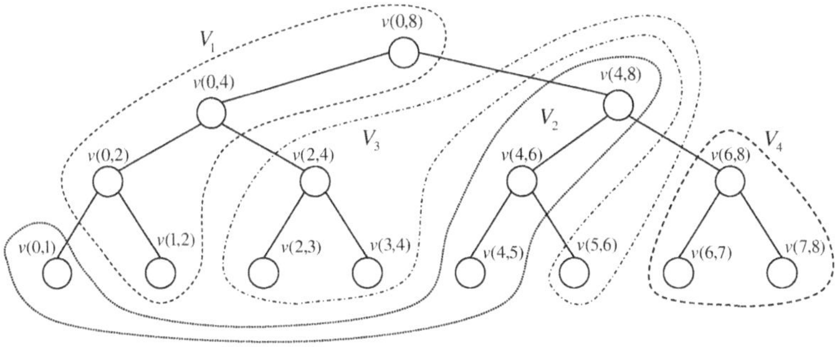

Fig. 1. An example of a solution $s = \{V_1 = \{v(0,8), v(0,4), v(0,2), v(1,2)\}, V_2 = \{v(0,1), v(4,5), v(4,6), v(4,8)\}, V_3 = \{v(2,4), v(2,3), v(3,4), v(5,6)\}, V_4 = \{v(6,8), v(6,7), v(7,8)\}$ to PPBT with $c = 4$

The set $V[i]$ satisfies

$$|L(V[i])| = |L(v(0, 2^{c-1}))| = 2^{c-1}. \tag{2}$$

Fig. 1 illustrates an example of a solution s to PPBT with $n = 4$. It hold $|L(V_1)| = 8$, $|L(V_2)| = 5$, $|L(V_3)| = 3$, and $|L(V_4)| = 2$.

We now introduce a certain solution $\hat{s}$ to PPBT, which has a systematic structure, and is used to analyze a lower bound on the optimal cost of PPBT.

Given a solution s to PPBT and a vertex $v \in V$, we denote by $V(v; s)$ the set of vertices in s to which v belongs. For a vertex $v_1 \in V(v; s) - \{v\}$, we say that v *is covered by* v_1 in s if v is a descendant vertex of v_1 on $\mathcal{T}_c$, i.e., it holds $L(v) \subset L(v_1)$. We denote by $Cover(v; s)$ the set of vertices by which v is covered in s. For example in Fig. 1, we have $Cover(v(1,2); s) = \{v(0,8), v(0,4), v(0,2)\}$, $Cover(v(2,3); s) = \{v(2,4)\}$, $Cover(v(4,8); s) = \emptyset$.

For a set X in a solution s, a vertex $v \in X$ is called a *head vertex* in X if X contains no vertex that covers v, i.e., $Cover(v; s) = \emptyset$. Let $Head(X)$ denote the set of head vertices in X. We easily see that for any set $X \in s$ it holds

$$\bigcup_{v \in Head(X)} L(v) = \bigcup_{v \in X} L(v) = L(X). \tag{3}$$

For the solution s shown in Fig. 1, we have $Head(V_1) = \{v(0,8)\}$, $Head(V_2) = \{v(0,1), v(4,8)\}$, $Head(V_3) = \{v(2,4), v(5,6)\}$, and $Head(V_4) = \{v(6,8)\}$.

Given a solution s, let $Head(s) = \bigcup_{X \in s} Head(X)$. By using the observation that no vertex appears as head vertices for two distinct sets in s, we easily derive

$$cost_L(s) = |L(Head(s))|. \tag{4}$$

We next introduce a solution $\hat{s}$ to PPBT. Recall that $c = 2^h$ holds for binary tree $\mathcal{T}_c$. We partition the vertex set V of $\mathcal{T}_c$ into subsets $V[k, j]$, $k = 1, \ldots, c/h$, $m = 1, \ldots, 2^{(k-1)h}$ such that each set $V[k, j]$ induces a complete binary subtree

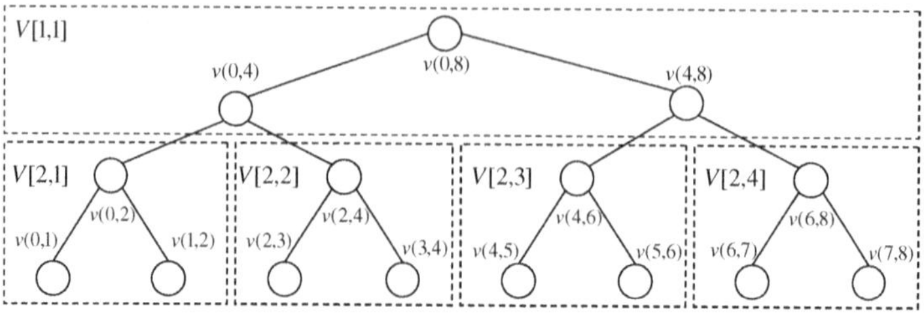

Fig. 2. A solution $\hat{s}$ with $c = 4$ and $h = 2$

with exactly $c - 1$ vertices from $\mathcal{T}_c$, as shown in Fig. 2. We then define solution $\hat{s}$ by

$$\hat{s} = \{V[k, j] \mid k = 1, 2, \ldots, c/h, j = 1, 2, \ldots, 2^{(k-1)h}\},$$

where the highest vertex in $V[k, j]$ is the unique head vertex in $V[k, j]$. Boxes with dashed lines in Fig. 2 show sets $V[k, j]$, $k = 1, 2$, $j = 1, 2, \ldots, 2^{2(k-1)}$, of vertices with $c = 4$ and $h = 2$. By using (2) and (4), we confirm that it holds

$$cost_L(\hat{s}) = |L(Head(\hat{s}))| = 2^{c-1} \cdot c/h = 2^{c-1} \cdot c/\log_2 c. \tag{5}$$

Let us analyze a lower bound on $cost_L(s^*)$ for a solution s^* to PPBT, by comparing s^* with solution $\hat{s}$. For this, we first partition the set V of vertices in $\mathcal{T}_c$ into five sets Z_1, Z_2, Z_2', Z_3, and Z_4 according to two solutions $\hat{s}$ and s^* as follows.

$$Z_1 = \{v \mid v \in Head(\hat{s}), v \in Head(s^*)\}, \tag{6}$$
$$Z_2 = \{v \mid v \in Head(\hat{s}), v \notin Head(s^*), Cover(v; s^*) \cap Head(\hat{s}) = \emptyset\}, \tag{7}$$
$$Z_2' = \{v \mid v \in Head(\hat{s}), v \notin Head(s^*), Cover(v; s^*) \cap Head(\hat{s}) \neq \emptyset\}, \tag{8}$$
$$Z_3 = \{v \mid v \notin Head(\hat{s}), v \in Head(s^*)\}, \tag{9}$$
$$Z_4 = \{v \mid v \notin Head(\hat{s}), v \notin Head(s^*)\}.$$

We obtain the following lemma.

Lemma 1. *The sets Z_1, Z_2, Z_2', and Z_3 of vertices defined in (6) - (9) satisfy*
(i) $|L(Z_1)| + |L(Z_2)| + |L(Z_2')| = 2^{c-1} \cdot c/h$;
(ii) $|L(Z_2)| \leq |L(Z_3)|$; *and*
(iii) $|L(Z_2')| \leq |L(Z_1)| + |L(Z_2)|$.

Proof: (i) The equality follows from $Z_1 \cup Z_2 \cup Z_2' = Head(s^*)$ and (5).
(ii) Let v_2 be an arbitrary vertex in Z_2. Then there exists a vertex $v_3 \in Z_3 \cap Cover(v_2; s^*)$, since v_2 is not a head vertex in $V(v_2; s^*)$, and is covered by a head vertex of $V(v_2; s^*)$ in s^*. Notice that it holds $L(v_2) \subset L(v_3)$ since v_2 is covered by v_3 in s^*. Hence $L(Z_2) \subseteq L(Z_3)$ holds.
(iii) Let $X \in s^*$ be an arbitrary set with $X \cap Z_2' \neq \emptyset$. For each vertex $u_2 \in X \cap Z_2'$, it holds $Cover(u_2; s^*) \cap Head(\hat{s}) \neq \emptyset$ by (8), and hence $Cover(u_2; s^*) \cap Head(\hat{s}) = Cover(u_2; s^*) \cap (Z_1 \cup Z_2 \cup Z_2') \neq \emptyset$. We choose a highest vertex

$v_1 \in Cover(u_2; s^*) \cap (Z_1 \cup Z_2 \cup Z_2')$ in $\mathcal{T}_c$. Then $v_1 \in Z_1 \cup Z_2$ holds, since $v_1 \in Z_2'$ would imply that there exists a vertex $v' \in Cover(v_1; s^*) \cap Head(\hat{s})$, which covers v_1 and also belongs to $Cover(u_2; s^*) \cap Head(\hat{s})$, a contradiction to the choice of v_1 (since v' is higher than v_1 in $\mathcal{T}_c$).

By the structure of $\hat{s}$, $v_1 \in Head(\hat{s})$ is the head vertex in a set $V[k_1, j_1] \in \hat{s}$ with $k_1 \in \{1, 2, \ldots, c/h - 1\}$ and $j_1 \in \{1, 2, \ldots, 2^{(k_1-1)h}\}$, and $u_2 \in Head(\hat{s})$ is the head vertex in a set $V[k_2, j_2] \in \hat{s}$ with $k_2 \in \{k_1 + 1, \ldots, c/h - 1\}$ and $j_2 \in \{1, 2, \ldots, 2^{(k_2-1)h}\}$. Since $k_2 \geq k_1 + 1$ and the number of layers in $\mathcal{T}_c$ for each set $V[k, j]$ is h, this implies

$$|L(u_2)| \leq |L(v_1)|/2^h.$$

Note that there may exist more than one vertex in $Z_2' \cap X$. By summing up the inequality over all vertices $u \in Z_2' \cap X$, we have

$$\sum_{u \in Z_2' \cap X} |L(u)| \leq \frac{|Z_2' \cap X|}{2^h} |L(v_1)| \leq |L(v_1)|,$$

where the second inequality follows from $|Z_2' \cap X| \leq c = 2^h$ by the feasibility of s^*. Note that $v_1 \in Cover(u_2; s^*) \subseteq V(u_2; s^*) = X$ holds. Therefore, by summing up the inequality over all sets in s^*, we have the lemma. □

We are ready to prove Theorem 1. By applying Lemma 1(ii), (iii) and (i), we have $cost_L(s^*) = |L(Head(s^*))| = |L(Z_1)| + |L(Z_3)| \geq |L(Z_1)| + |L(Z_2)| \geq \frac{1}{2}(|L(Z_1)| + |L(Z_2)| + |L(Z_2')|) = 2^{c-2} \cdot c/h$. This and (5) prove Theorem 1.

4 Upper Bound on Travel Cost of PDPCMT Solutions

This section derives an upper bound on the maximum ratio of the optimal cost of PDPCMT to that of PDP.

Given an instance I, we call a solution to PRB with instance I a *PRB solution* to I, and call a route in PRB solution with instance I a *PRB route* to I.

Given a solution s to PDPCMT or PDP, we denote by $T(s)$ the set of trips in s. Let $R(t)$ denote the set of requests that are served on trip t. Given a route σ, let $A(\sigma)$ represent the set of actions along σ.

Travel cost $d(j, j')$ from vertex j to j' for $j, j' \in A$ is a nonnegative real number, and in general asymmetric; i.e., $d(j, j') \neq d(j', j)$ may hold. Given a trip $t = (a_1, a_2, \ldots, a_n)$, let $cost(t) = \sum_{1 \leq i \leq n-1} d(a_i, a_{i+1})$. Given a solution s, let $cost(s) = \sum_{\sigma \in s} cost(\sigma)$, where σ is a route in s.

Given a PDPCMT route $\sigma = (a_0, t_1, t_2, \ldots, t_k, a_1)$ with $a_0, a_1 \in Q$ and trips $t_1, t_2, \ldots, t_k$, we define the *loaded travel cost* as the travel cost with loads, i.e., $cost_f(\sigma) = \sum_{i=1}^{k} cost(t_k)$. Given a PDPCMT solution s, let $cost_f(s) = \sum_{\sigma \in s} cost_f(\sigma)$.

Theorem 2. *Let $I = (Q, R, c)$ be an instance with $p \geq 2$ requests. Then it holds*

$$opt_{\text{PDPCMT}}(I) \leq \lceil \log_2 2p \rceil \cdot opt_{\text{PDP}}(I).$$

For this, we show how to convert a PDP solution to a PDPCMT solution.

Theorem 3. *Given a PDP route σ to an instance I, PDPCMT routes σ' with*

$$A(\sigma') = A(\sigma) \text{ and } cost(\sigma') \leq \lceil \log_2 2p' \rceil \cdot cost(\sigma)$$

can be constructed in polynomial time, where p' is the number of requests served in σ.

Theorem 2 is obtained by applying Theorem 3 to each route σ in an optimal PDP solution s to I.

In this section, we denote a given PDP route by $\sigma = (a_0, a_1, a_2, \ldots, a_m, a_{m+1})$, where $a_0, a_{m+1} \in Q$, $a_1, \ldots, a_m \in A$ and $m = 2p$. For notational simplicity, we assume that $m = 2^k$ holds for some integer $k \geq 1$ by introducing fictitious requests r with $q(r) = 0$ and $r^- = r^+$ on vertex a_h if necessary. We describe an algorithm that converts a PDP route $\sigma = (a_0, a_1, \ldots, a_m, a_{m+1})$ to k PDPCMT routes $\pi_1, \pi_2, \ldots, \pi_k$. Each route π_j visits the vertices in σ in the same order serving a set R_j of requests defined as follows.

We divide the set $A(1,1) = \{a_i \mid i = 1, 2, \ldots, 2^k\}$ of all actions in σ into two subsets $A(2,1) = \{a_i \mid i = 1, 2, \ldots, 2^{k-1}\}$ and $A(2,2) = \{a_i \mid i = 2^{k-1} + 1, 2^{k-1} + 2, \ldots, 2^k\}$ in the second level. By repeating this recursively, we define 2^{j-1} subsets at the j-th level to be

$$A(j,b) = \{a_i \mid i = (b-1)2^{k-j+1} + 1, (b-1)2^{k-j+1} + 2, \ldots, b2^{k-j+1}\}$$

for $b = 1, 2, \ldots, 2^{j-1}$, where $j = 1, 2, \ldots, k+1$. A request $r \in R$ is at the j-th level if j is the smallest level such that a subset $A(j,b)$ contains both actions r^+ and r^-. Let us denote the subsets of requests at the j-th level by

$$R(j,b) = \{r \in R \mid r^+ \in A(j+1, 2b-1), r^- \in A(j+1, 2b)\},$$

$j = 1, 2, \ldots, k$ and $b = 1, 2, \ldots, 2^{j-1}$. We assign to route π_j the requests in

$$R_j = R(j,1) \cup R(j,2) \cup \cdots \cup R(j, 2^{j-1}),$$

and let a vehicle serve the requests $R(j,1), R(j,2), \ldots, R(j, 2^{j-1})$ in this order along π_j. Note that it holds $cost(\pi_j) = cost(\sigma)$ if $R_j \neq \emptyset$, $cost(\pi_j) = 0$ otherwise, for each $j = 1, 2, \ldots, k$. The feasibility of routes $\pi_1, \pi_2, \ldots, \pi_k$ can be ensured. (The proof is omitted due to space limitation.)

We are now ready to prove Theorem 3. Since $m = 2p' = 2^k$, and k is an integer number, we have $k = \lceil 2p' \rceil$. Let σ' be a route where $\pi_1, \pi_2, \ldots, \pi_k$ appear in this order. Then it holds $A(\sigma') = A(\sigma)$. Thus we have $cost(\sigma') = \sum_{j=1}^{k} cost(\pi_j) \leq k \cdot cost(\sigma) = \lceil \log |2p'| \rceil \cdot cost(\sigma)$. Thus we have the theorem.

5 Lower Bound on Travel Cost of PDPCMT Solutions

This section proves that the upper bound in Theorem 2 is tight up to factor of $O(\log \log p)$ by identifying such an instance for any large p. Thus we prove the next result.

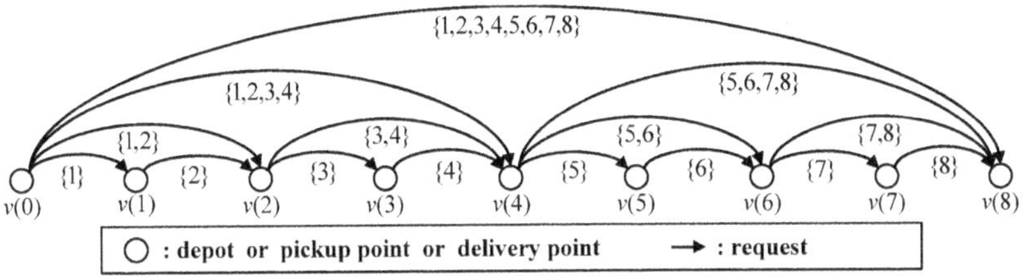

Fig. 3. An instance $G(p, \lambda)$ with $c = 4$, $h = 2$ and $p = 15$

Theorem 4. *For any integer $h \geq 1$, there exists an instance $I = (Q, R, c)$ with $p = 2^{2^h} - 1$ requests that satisfies*

$$opt_{\text{PDPCMT}}(I) \geq \frac{\log_2(p+1)}{4 \log_2 \log_2(p+1)} \cdot opt_{\text{PDP}}(I).$$

Such an instance I in Theorem 4, which we denote by $G(p, \lambda)$, is defined below. Given an integer $p = 2^{2^h} - 1$ $(h \geq 1)$ and a real $\lambda > 0$, let capacity be

$$c = 2^h = \log_2(p+1),$$

and let $V = \{v(0), v(1), v(2), \ldots, v(2^{c-1})\}$ be the set of vertices in $G(p, \lambda)$. We assume that all vertices in V are arranged uniformly on a horizontal straight line-segment of length λ from left to right as shown in Fig. 3. Hence $d(v(0), v(2^{c-1})) = \lambda$ and $d(v(i), v(i+1)) = \lambda/2^{c-1}$ for $i = 0, 1, \ldots, 2^{c-1} - 1$. A set of integers on each arc in Fig. 3 will be explained in the next subsection.

Let Q consist of a single depot $v(0)$, and the set R of requests in $G(p, \lambda)$ consist of the following c subsets $R[i]$, $i = 1, \ldots, c$, containing 2^{i-1} requests. Fig. 3 illustrates set R of requests in $G(p, \lambda)$ with $p = 15$. Let set $R[1]$ consist of a single request r with $r^+ = v(0)$, $r^- = v(2^{c-1})$. The i-th set $R[i]$, $i = 1, 2, \ldots, c$, consists of 2^{i-1} requests and expressed by

$$R[i] = \{r \mid (r^+, r^-) \in \{(v(0), v(2^{c-i})), (v(2^{c-i}), v(2 \cdot 2^{c-i})), (v(2 \cdot 2^{c-i}), v(3 \cdot 2^{c-i})), \ldots, (v(2^{c-1} - 2^{c-i}), v(2^{c-1}))\}\}.$$

Let $q(r) = 1$ for all $r \in R$.

We show that $G(p, \lambda)$ has a PDP route σ with travel cost 2λ, which is optimal since any vehicle must start from $v(0)$, visit $v(2^{c-1})$, and return to $v(0)$, requiring at least 2λ travel cost. Hence σ consists of the path from $v(0)$ to $v(2^{c-1})$ and the path from $v(2^{c-1})$ to $v(0)$. The total quantity of requests loaded on a vehicle is less than or equal to c any time on the path from $v(0)$ to $v(2^{c-1})$, since requests on a vehicle contains at most one request from each set $R[i]$, $i = 1, \ldots, c$.

Let us analyze a lower bound on the loaded travel cost of PDPCMT solutions by using Theorem 1. We analyze a lower bound on the loaded travel cost that does not depend on sequences in trips but are derived from travel cost between pickup points and delivery points of requests in the trips. We explain the cost in detail as follows.

We assign a set of integers to each request in R. Let $\varphi(r^+, r^-)$ denote a set of integers that are assigned to request $r = (r^+, r^-)$. We first assign a set of integers to requests in $R[c] = \{r \mid (r^+, r^-) \in \{(v(j-1), v(j)) \mid j = 1, 2, \ldots, 2^{c-1}\}\}$ by

$$\varphi(v(j-1), v(j)) = \{j\}, \quad j = 1, 2, \ldots, 2^{c-1}.$$

As shown in Fig. 3, the sets of integers that are assigned to the requests in $R - R[c]$ are defined by

$$\varphi(v(i), v(j)) = \varphi(v(i), v(i+1)) \cup \varphi(v(i+1), v(i+2)) \cup \cdots \cup \varphi(v(j-1), v(j)).$$

Given a trip t in a PDPCMT route, we define cost $cost_b(R(t))$ of a set $R(t)$ of requests that are served along t by

$$cost_b(R(t)) = |\bigcup_{r \in R(t)} \varphi(r^+, r^-)|.$$

The following lemma gives a lower bound on the loaded travel cost of a PDPCMT solution.

Lemma 2. *Given a PDPCMT solution s, the loaded travel cost $cost_f(s)$ of s satisfies*

$$cost_f(s) \geq \sum_{t \in T(s)} cost_b(R(t)) \cdot \lambda / 2^{c-1}.$$

Proof: For $i = 1, 2, \ldots, 2^{c-1}$, travel cost $d(v(i-1), v(i))$ along the segment from $v(i-1)$ to $v(i)$ is given by $d(v(i-1), v(i)) = \lambda/2^{c-1}$. It costs at least $cost_b(R(t)) \cdot \lambda/2^{c-1}$ to serve the requests in $R(t)$ since a vehicle that serves the requests in $R(t)$ is required to travel at least $cost_b(R(t))$ segments. Thus we have the lemma. □

We derive a lower bound on $cost_f(s)$ over all PDPCMT solution s as follows. For a PDPCMT solution s, let $R_1, R_2, \ldots, R_n$ of R be sets of requests such that each R_i corresponds to the set of requests that are served in a trip in s. Hence it holds $\sum_{r \in R_i} q(r) \leq c$ for $i = 1, 2, \ldots, n$. We show that minimizing $\sum_{i=1}^{n} cost_b(R_i)$ over all PDPCMT solution s is reduced to PPBT. For this, we associate request $r = (v((j-1) \cdot 2^{c-i}), v(j \cdot 2^{c-i}))$ in $G(p, \lambda)$ with vertex $v((j-1) \cdot 2^{c-i}, j \cdot 2^{c-i})$ on the binary tree $\mathcal{T}_c$ for PPBT. Then it holds

$$\varphi(v((j-1) \cdot 2^{c-i}), v(j \cdot 2^{c-i})) = |L(v((j-1) \cdot 2^{c-i}, j \cdot 2^{c-i}))|$$

for $i = 1, 2, \ldots, c$ and $j = 1, 2, \ldots, 2^{i-1}$. Let V_k be the set of vertices in $\mathcal{T}_c$ obtained from R_k in this way. If request $r = (v((j-1) \cdot 2^{c-i}), v(j \cdot 2^{c-i}))$ in $G(p, \lambda)$ belongs to R_k, then we let vertex $v((j-1) \cdot 2^{c-i}, j \cdot 2^{c-i})$ in V belong to V_k. Then $s^* = \{V_1, V_2, \ldots, V_n\}$ is a feasible partition in $\mathcal{T}_c$ since $|V_i| = |R_i| \leq c$ holds. Also $\sum_{t \in T(s)} cost_b(R(t)) = cost_L(s^*)$ holds. By Theorem 1, we have $cost_L(s^*) \geq 2^{c-2}c/h$. We are now ready to prove Theorem 4. We have

$$\begin{aligned} opt_{\mathrm{PDPCMT}}(G(p, \lambda)) &\geq cost_f(s) \geq \sum_{t \in T(s)} cost_b(R(t)) \cdot \lambda/2^{c-1} \\ &\geq 2^{c-2}c/h \cdot \lambda/2^{c-1} = \lambda \log_2(p+1)/(2 \log_2 \log_2(p+1)). \end{aligned}$$

Since $opt_{\mathrm{PDP}}(G(p, \lambda)) = 2\lambda$, this proves Theorem 4.

6 Conclusion

In this paper, we analyzed upper and lower bounds on the maximum travel cost that can be saved by regarding an instance to PDP as that to PDPCMT. We showed that $opt_{\mathrm{PDPCMT}}(I) \leq \lceil \log_2 2p \rceil \cdot opt_{\mathrm{PDP}}(I)$ holds for all instance I by constructing a method for converting a PDP solution to a PDPCMT solution. Thus we contributed to study on difference between travel cost of optimal solutions for several routing problems. In particular, we established a lower bound technique based on partitioning binary trees.

References

1. Archetti, C., Savelsbergh, M.W.P., Speranza, M.G.: Worst-case analysis for split delivery vehicle routing problems. Transportation Science 40, 226–234 (2006)
2. Arkin, E., Hassin, R., Klein, L.: Restricted delivery problems on a network. Networks 29, 205–216 (1997)
3. Charikar, M., Khuller, S., Raghavachari, B.: Algorithms for capacitated vehicle routing. In: STOC, pp. 349–358 (1998)
4. Dror, M., Trudeau, P.: Split delivery routing. Naval Research Logistics 37, 383–402 (1990)
5. Li, H., Lim, A.: A metaheuristic for the pickup and delivery problem with time windows. International Journal on Artificial Intelligence Tools 12, 173–186 (2003)
6. Mitrović-Minić, S., Laporte, G.: The pickup and delivery problem with time windows and transshipment. INFOR 44, 217–227 (2006)
7. Nakao, Y., Nagamochi, H.: Worst case analysis for pickup and delivery problems with transfer. IEICE Transaction 91-A(9), 2328–2334 (2008)
8. Nanry, W.P., Barnes, J.W.: Solving the pickup and delivery problem with time windows using reactive tabu search. Transportation Research Part B 34, 107–121 (2000)
9. Pankratz, G.: A grouping genetic algorithm for the pickup and delivery problem with time windows. OR Spectrum 27, 21–41 (2005)
10. Toth, P., Vigo, D.: The granular tabu search and its application to the vehicle-routing problem. INFORMS Journal on Computing 15, 333–346 (2003)

Minimum Cycle Bases of Weighted Outerplanar Graphs*

Tsung-Hao Liu and Hsueh-I. Lu**

Department of Computer Science and Information Engineering
National Taiwan University
r95122@csie.ntu.edu.tw, hil@csie.ntu.edu.tw

Abstract. We give the first known optimal algorithm that computes a minimum cycle basis for any weighted outerplanar graph. Specifically, for any n-node edge-weighted outerplanar graph G, we give an $O(n)$-time algorithm to obtain an $O(n)$-space compact representation $Z(\mathbb{C})$ for a minimum cycle basis $\mathbb{C}$ of G. Each cycle in $\mathbb{C}$ can be computed from $Z(\mathbb{C})$ in $O(1)$ time per edge. Our result works for directed and undirected outerplanar graphs G.

1 Introduction

Unless clearly specified otherwise, all graphs in the paper are undirected and have no multiple edges and self-loops. Let $G = (V, E)$ be an edge-weighted graph. Each cycle C of G can be represented by an incidence vector $\mathbf{x}$ with index set E such that e is an edge of C if and only if $\mathbf{x}_e = 1$. The *cycle space* of G is the vector space spanned by the incidence vectors of the cycles in G over $\mathbb{F}_2$. A *cycle basis* of G is a collection of cycles whose incidence vectors form a basis of the cycle space of G. A cycle basis of G with minimum weight is a *minimum cycle basis* (MCB) of G. An example is shown in Figure 1. In the present article, we study the problem of identifying an MCB for G, which finds applications in electrical engineering, biochemistry, structural engineering, surface reconstruction, and public transportation [1–3]. Kavitha, Liebchen, Mehlhorn, Michail, Rizzi, Ueckerdt, and Zweig [4] also presented an in-depth survey on the problem. The problem is NP-complete if G may contain negative cycles [4]. The rest of the paper assumes that G has nonnegative edge weights.

The cycle space of a graph is the direct sum of the cycle spaces of its 2-connected components. Therefore, without loss of generality, we assume that the input n-node m-edge graph G is 2-connected. One can verify that any MCB of G consists of $m - n + 1$ cycles. Horton [5] gave the first polynomial-time algorithm

* Research supported in part by NSC grants 97-2221-E-002-122 and 98-2221-E-002-079-MY3.

** Corresponding author. Web: www.csie.ntu.edu.tw/~hil. This author also holds joint appointments in the Graduate Institute of Networking and Multimedia and the Graduate Institute of Biomedical Electronics and Bioinformatics, National Taiwan University. Address: 1 Roosevelt Road, Section 4, Taipei 106, Taiwan, ROC.

Y. Dong, D.-Z. Du, and O. Ibarra (Eds.): ISAAC 2009, LNCS 5878, pp. 564–572, 2009.

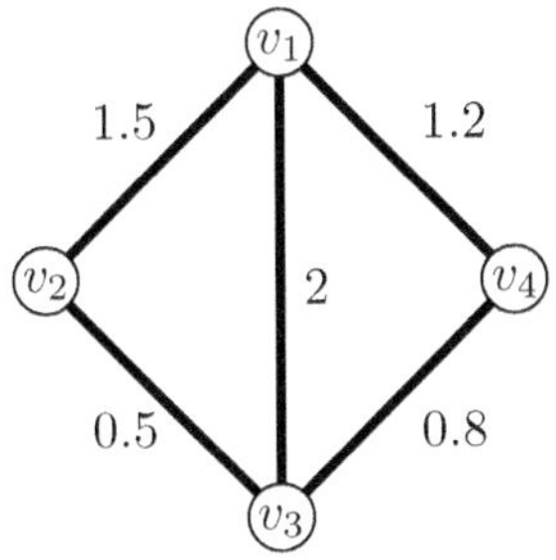

Fig. 1. An edge-weighted outerplanar graph G with three cycles: $\langle v_1, v_2, v_3 \rangle$, $\langle v_1, v_3, v_4 \rangle$, and $\langle v_1, v_2, v_3, v_4 \rangle$. Any two of them form an MCB of G.

for the problem, which runs in $O(m^3 n)$ time.[1] de Pina [8] presented a different algorithm to solve the problem in $O(m^3 + mn^2 \log n)$ time. (Recently, a combinatorial algorithm with the same time complexity was given by Berger, Gritzmann, and de Vries [9].) Golynski and Horton [10] then reduced the required time complexity of Horton's algorithm [5] to $O(m^\omega n)$ time, where ω can be any constant such that multiplying two $k \times k$ matrices can be done in $O(k^\omega)$ time.[2] Based on de Pina's approach [8], Kavitha, Mehlhorn, Michail, and Paluch [12–14] gave an $O(m^2 n + mn^2 \log n)$-time algorithm. Mehlhorn and Michail [1] then gave an approach which runs in $O(m^2 n / \log n + mn^2)$ time. Mehlhorn and Michail [15] also addressed the implementation issues of the above algorithms. The best currently known algorithm, due to Amaldi, Iuliano, Jurkiewicz, Mehlhorn, and Rizzi [16], runs in expected $O(m^\omega)$ time.

Algorithms for finding near minimum cycle bases of G are also investigated in the literature. For any $\epsilon > 0$, Kavitha, Mehlhorn, Michail, and Paluch [12, 13] gave a $(1+\epsilon)$-approximation algorithm that runs in $O(mn^\omega \epsilon^{-1} \log(\epsilon^{-1} W))$ time, where W is the largest edge weight in G. For any integer $k > 1$, Kavitha, Mehlhorn, and Michail [14, 17] presented a $(2k-1)$-approximation algorithm running in expected $O(kmn^{1+2/k} + mn^{(1+1/k)(\omega-1)})$ time or deterministic $O(n^{3+2/k})$ time. Moreover, Mehlhorn and Michail [1] gave a 2-approximation algorithm that runs in expected $O(m^2 \sqrt{n / \log n} + n^2 m + m^\omega)$ time and a $(2k-1)$-approximation algorithm, for any integer $k > 1$, that runs in $O(n^{3+2/k} / \log n + n^{3+1/k})$ time.

Various graph structures have been exploited for the MCB problem in the literature [16, 18–25]. Among which the results of Hartvigsen and Mardon [24] and Amaldi, Iuliano, Jurkiewicz, Mehlhorn, and Rizzi [16] seem to be the only algorithms that handle special graphs with edge weights. Specifically, Hartvigsen and Mardon [24] gave an $O(n^2 \log n)$-time algorithm for planar G, which was improved to $O(n^2)$ time by Amaldi, Iuliano, Jurkiewicz, Mehlhorn, and Rizzi [16], recently. In the present article, we focus on outerplanar G with edge weights. As shown in Figure 1, even if the edges of G have distinct weights, G may have

[1] Several researchers [6, 7] incorrectly claimed to solve the problem in polynomial time.

[2] Coppersmith and Winograd [11] proved that $\omega < 2.376$.

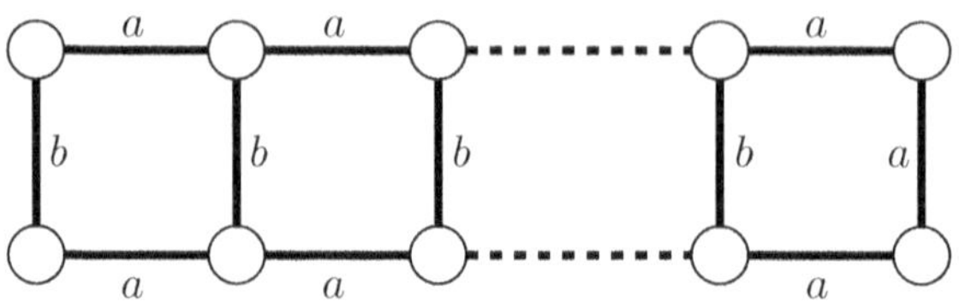

Fig. 2. If $b \gg a$, then this n-node outerplanar graph has a unique MCB with $\Theta(n^2)$ edges

more than one MCB. As summarized in the next theorem, we obtain an algorithm that is optimal in time and space. Our result extends the $O(n)$-time algorithm of Leydold and Stadler [25] for unweighted outerplanar G.

Theorem 1. *Given an n-node edge-weighted outerplanar graph G, an $O(n)$-space representation $Z(\mathbb{C})$ of a minimum cycle basis $\mathbb{C}$ of G can be computed from G in $O(n)$ time such that each cycle in $\mathbb{C}$ can be determined from $Z(\mathbb{C})$ in $O(1)$ time per edge.*

Although our result is presented only for undirected G, according to [16, Theorem 7], the MCB obtained by our algorithm found is also directed, weakly fundamental, totally unimodular, and integral [4]. Therefore, our result also works for directed outerplanar graph G.

The running time of our algorithm is linear in the size of input and output, but we have to point out that some G may have $\Theta(n^2)$ edges in its unique MCB. For instance, consider the case that G is as depicted in Figure 2 with $b \gg a$. Let $\ell = \frac{n}{2} - 1$. G has exactly ℓ edges with weight b. Every cycle of G contains at least one edge with weight b. Moreover, there are exactly ℓ cycles of G that contain exactly one edge with weight b. One can verify that these ℓ cycles, which contains $\Theta(n^2)$ edges, form the unique MCB of G.

The rest of the paper is organized as follows. Section 2 gives the basics. Section 3 gives our algorithm. Section 4 concludes the paper.

2 Basics

For the rest of the paper, let G be the input 2-connected outerplanar graph equipped with an outerplanar embedding. All nodes of G belong to the external boundary of G in the given embedding. An edge of G is *external* if it belongs to the external boundary of G. An edge of G is *internal* if it does not belong to the external boundary of G. For any subgraph H of G, let $w(H)$ denote the sum of edge weights in H.

2.1 Lex Short Cycles

For any set S, let $|S|$ denote the number of elements in S. If S is totally ordered, let $\min(S)$ denote the minimum element in S. Let the nodes of G form an

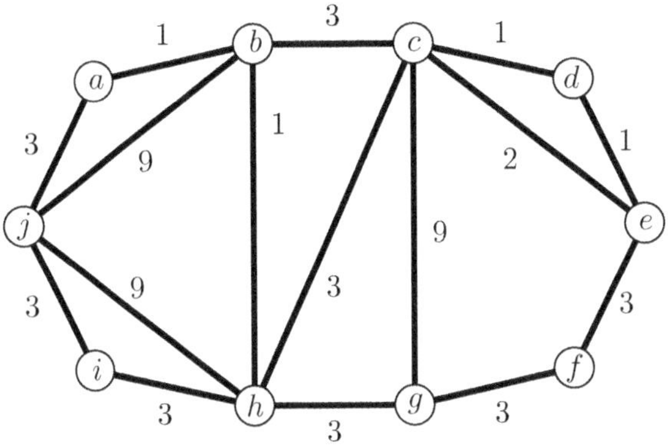

Fig. 3. An outerplanar graph with three tight edges (bh, ch, and ce) and three loose edges (bj, hj, and cg)

arbitrary but fixed total order. For any subgraph H of G, let $V(H)$ denote the node set of H and let $E(H)$ denote the edge set of H. For any nodes u and v of G, a *u-v path* of G is a path of G with endpoints u and v. A u-v path P of G is *lex shortest* [24] in G if P satisfies one of the following disjoint conditions for any u-v path P' of G other than P:

1. $w(P) < w(P')$,
2. $w(P) = w(P')$ and $|E(P)| < |E(P')|$,
3. $w(P) = w(P')$, $|E(P)| = |E(P')|$, and $\min(V(P) \setminus V(P')) < \min(V(P') \setminus V(P))$.

As noted by Hartvigsen and Mardon [24, Proposition 4.1], there is a unique lex shortest path of G between any two distinct nodes. Let $P(uv)$ denote the lex shortest u-v path of G. Clearly, $w(uv) \geq w(P(uv))$ holds for any edge uv of G. Moreover, $w(uv) = w(P(uv))$ if and only if $uv = P(uv)$.

Lemma 1 (Hartvigsen and Mardon [24, Proposition 4.3]). *It takes $O(n)$ time to compute an edge weight $\hat{w}$ from the original edge weight w such that $P(uv)$ is the unique shortest u-v path of G with respect to $\hat{w}$.*

A cycle C is *lex short* [24] in G if C contains $P(xy)$ for any two distinct nodes x and y in C.

Lemma 2 (Hartvigsen and Mardon [24, Proposition 4.5]). *There is an MCB $\mathbb{C}$ of G such that each cycle in $\mathbb{C}$ is lex short.*

2.2 Tight Cycles and Loose Cycles

An internal edge uv of G is *tight* if $w(uv) = w(P(uv))$. An internal edge uv of G is *loose* if $w(uv) > w(P(uv))$. For instance, if G is as shown in Figure 3, then G has three tight edges (bh, ch, and ce) and three loose edges (bj, jh, and cg). One can verify that $P(uv)$ does not contain any loose edge of G, since otherwise $P(uv)$ can be shortened by replacing that loose edge, say, xy with $P(xy)$.

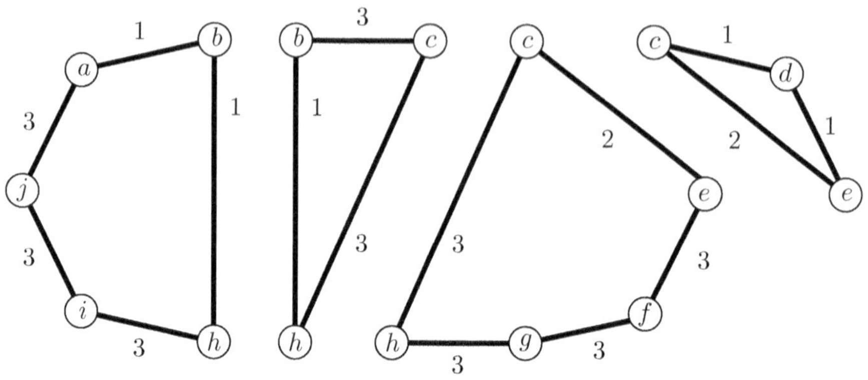

Fig. 4. The tight cycles of the outerplanar graph shown in Figure 3

Lemma 3. *It takes $O(n)$ time to classify the internal edges of G into tight and loose edges.*

Proof. We first obtain the weight function $\hat{w}$ ensured by Lemma 1 in $O(n)$ time. According to Frederickson and Janardan [26, Theorem 3.2], it takes $O(n)$ time to determine for all edges uv of G whether uv belongs to the shortest u-v path of G with respect to $\hat{w}$. If uv belongs to the shortest u-v path of G with respect to $\hat{w}$, then uv is tight; otherwise, uv is loose. Therefore, the lemma holds.

A cycle C of G is *tight* if

- each edge of C is either external or tight; and
- each edge interior of C is loose.

If G has k tight edges, then these tight edges and the external boundary of G form $k+1$ bounded regions, each of whose boundaries is a tight cycle. For instance, if G is as shown in Figure 3, then edges bh, ce and ch are the tight edges of G. Therefore, as shown in Figure 4, $\langle a,b,h,i,j\rangle$, $\langle b,c,h\rangle$, $\langle c,d,e\rangle$, and $\langle c,e,f,g,h\rangle$ are the tight cycles of G.

A cycle C of G is *loose* if C contains a loose edge uv such that $C = uv \cup P(uv)$. By the uniqueness of lex shortest u-v path and the fact that $P(uv)$ cannot contain any loose edge, we know that there cannot be two distinct loose cycles containing the same loose edge uv. Therefore, if G has ℓ loose edges, then G has exactly ℓ loose cycles. For instance, if G is as shown in Figure 3, then edges bj, cg, and hj are the loose edges. Therefore, as shown in Figure 5, $\langle a,b,j\rangle$, $\langle a,b,h,j\rangle$, and $\langle c,g,h\rangle$ are the loose cycles of G.

Let $\mathbb{T}$ consist of the tight cycles of G. Let $\mathbb{L}$ consist of the loose cycles of G. Clearly, $\mathbb{T}$ and $\mathbb{L}$ are disjoint. Let

$$\mathbb{C} = \mathbb{T} \cup \mathbb{L}.$$

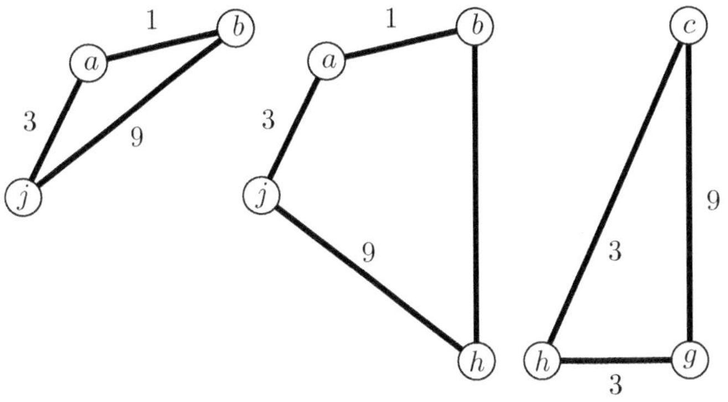

Fig. 5. The loose cycles of the outerplanar graph shown in Figure 3

Lemma 4. *$\mathbb{C}$ is an MCB of G. Moreover, $\mathbb{C}$ is exactly the set of lex short cycles of G.*

Proof. G has $m - n$ internal edges. Suppose that k of them are tight and the other $m - n - k$ are loose. Therefore, $\mathbb{C}$ consists of the $k + 1$ tight cycles and the $m - n - k$ loose cycles of G. We first prove that each lex short cycle C has to be a member of $\mathbb{C}$.

- For the case that C has no loose edge, i.e., C consists of the tight and the external edges, we show that C is tight by ensuring that any edge ab interior of C has to be loose. Edge ab, which is interior of C, cannot be an external edge. If ab is tight, then $P(ab) = ab$, a contradiction to the assumption that C is lex short.
- For the case that C contains some loose edge uv, it follows from the definition of lex short cycle that C contains $P(uv)$. Since uv is loose, $P(uv)$ does not contain uv. As a result, $C = uv \cup P(uv)$. Thus, C is a loose cycle.

It follows that G has at most $m - n + 1$ lex short cycles. Since each MCB of G has exactly $m - n + 1$ cycles, the lemma follows from Lemma 2.

3 Our Algorithm

Let H be the outerplane graph obtained from G by deleting its loose edges. It follows from Lemma 3 that H can be obtained in $O(n)$ time. The boundaries of the internal faces of H are the tight cycles. Therefore, the tight cycles of G can be obtained in $O(n)$ time. Let T_k denote the k-th cycle of $\mathbb{T}$. Let $T_k(i, j)$ denote the path of T_k starting from the i-th node of T_k to the j-th node of T_k in clockwise order, when $i \neq j$.

3.1 The Compact Representation for Tight Cycles

Let $Z(\mathbb{T})$ be an array of node lists, where the k-th list keeps the nodes of T_k in clockwise order. For instance, if G is as shown in Figure 3, then

$$Z(\mathbb{T}) = \langle T_1, T_2, T_3, T_4 \rangle, \tag{1}$$

where $T_1 = \langle c, d, e \rangle$, $T_2 = \langle c, e, f, g, h \rangle$, $T_3 = \langle b, c, h \rangle$, $T_4 = \langle a, b, h, i, j \rangle$.

Lemma 5

1. *$Z(\mathbb{T})$ takes $O(n)$ space.*
2. *It takes $O(n)$ time to compute $Z(\mathbb{T})$ from G.*
3. *For any indices i, j, and k, the path $T_k(i, j)$ can be determined from $Z(\mathbb{T})$ in $O(1)$ time per edge.*

Proof. The lemma follows immediately from the definition of $Z(\mathbb{T})$.

3.2 The Compact Representation for Loose Cycles

Let $Z(\mathbb{L})$ be an array of triples, one for each loose edge of G. Suppose that the y-th loose edge uv is enclosed by T_k. Let i and j be the indices such that the i-th (respectively, j-th) node of T_k in $Z(\mathbb{T})$ is u (respectively, v). By definition of tight cycle, $P(uv)$ is either $T_k(i, j)$ or $T_k(j, i)$. If $P(uv) = T_k(i, j)$, then let the y-th entry of $Z(\mathbb{L})$ be (i, j, k); otherwise let the y-th entry of $Z(\mathbb{L})$ be (j, i, k). For instance, if G is as shown in Figure 3 and $Z(\mathbb{T})$ is as shown in Equation (1), then

$$Z(\mathbb{L}) = \langle (4, 1, 2), (5, 2, 4), (5, 3, 4) \rangle.$$

Lemma 6

1. *$Z(\mathbb{L})$ takes $O(n)$ space.*
2. *It takes $O(n)$ time to compute $Z(\mathbb{L})$ from G.*
3. *For each index y, the y-th loose cycle of G can be obtained from $Z(\mathbb{T})$ and $Z(\mathbb{L})$ in $O(1)$ time per edge.*

Proof. The first statement is straightforward. The third statement follows immediately from the definition of $Z(\mathbb{L})$ and Lemma 5(3).

The rest of the proof proves the second statement. We first obtain the weight function $\hat{w}$ ensured by Lemma 1 in $O(n)$ time. For each index k, it takes time linear in the size of T_k to determine $\hat{w}(T_k)$ and $\hat{w}(T_k(1, t))$ for all indices t. Therefore, it takes $O(n)$ time to come up with a data structure, from which the value of $\hat{w}(T_k(i, j))$ for any indices i, j, ad k can be determined in $O(1)$ time.

For each index k, it takes time linear in the size of T_k, with the help of Lemma 3, to recognize the internal loose edges enclosed by T_k. Therefore, it takes $O(n)$ time to compute a data structure, from which it takes $O(1)$ time to determine for each index y the indices i, j, and k such that

- T_k encloses the y-th loose edge and
- the endpoints of the y-th loose edge are the i-th and j-th nodes of T_k.

As a result, for each index y, it takes $O(1)$ time to determine which one of $T_k(i, j)$ and $T_k(j, i)$ is the lex shortest path between the endpoints of the y-th loose edge. Specifically,

- if $\hat{w}(T_k(i, j)) < \hat{w}(T_k(j, i))$, then the y-th entry of $Z(\mathbb{L})$ is (i, j, k);
- if $\hat{w}(T_k(i, j)) > \hat{w}(T_k(j, i))$, then the y-th entry of $Z(\mathbb{L})$ is (j, i, k).

Thus, the second statement holds.

3.3 Proving Theorem 1

Proof. Let $Z(\mathbb{C})$ consist of $Z(\mathbb{T})$ and $Z(\mathbb{L})$. By Lemmas 5(1) and 6(1), $Z(\mathbb{C})$ takes $O(n)$ space. By Lemmas 5(2) and 6(2), $Z(\mathbb{C})$ can be computed from G in $O(n)$ time. By Lemmas 5(3) and 6(3), each cycle C in $\mathbb{C}$ can be computed from $Z(\mathbb{C})$ in $O(1)$ time per edge. It follows from Lemma 4 that the theorem is proved.

4 Concluding Remarks

Exploiting structures of larger classes of special graphs with edge weights for the MCB problems is an interesting direction for future work.

References

1. Mehlhorn, K., Michail, D.: Minimum cycle bases: Faster and simpler. ACM Transactions on Algorithms (accepted)
2. Liebchen, C., Rizzi, R.: Classes of cycle bases. Discrete Applied Mathematics 155(3), 337–355 (2007)
3. Mehlhorn, K.: Minimum cycle bases in graphs – algorithms and applications. In: Proceedings of the 32nd International Symposium on Mathematical Foundations of Computer Science, pp. 13–14 (2007)
4. Kavitha, T., Liebchen, C., Mehlhorn, K., Michail, D., Rizzi, R., Ueckerdt, T., Zweig, K.A.: Cycle bases in graphs characterization, algorithms, complexity, and applications. Computer Science Review (to appear)
5. Horton, J.D.: A polynomial-time algorithm to find the shortest cycle basis of a graph. SIAM Journal on Computing 16(2), 358–366 (1987)
6. Kolasinska, E.: On a minimum cycle basis of a graph. Zastosowania Matematyki 16, 631–639 (1980)
7. Hubicka, E., Syslo, M.M.: Minimal bases of cycle of a graph. In: Fiedler, M. (ed.) Recent Advances in Graph Theory, the 2nd Czech Symposium in Graph Theory, pp. 283–293 (1975)
8. de Pina, J.C.: Applications of Shortest Path Methods. PhD thesis, University of Amsterdam, Netherlands (1995)
9. Berger, F., Gritzmann, P., de Vries, S.: Minimum cycle bases for network graphs. Algorithmica 40, 51–62 (2004)
10. Golynski, A., Horton, J.D.: A polynomial time algorithm to find the minimum cycle basis of a regular matroid. In: Penttonen, M., Schmidt, E.M. (eds.) SWAT 2002. LNCS, vol. 2368, pp. 200–209. Springer, Heidelberg (2002)
11. Coppersmith, D., Winograd, S.: Matrix multiplication via arithmetic progressions. Journal of Symbolic Computation 9(3), 251–280 (1990)
12. Kavitha, T., Mehlhorn, K., Michail, D., Paluch, K.: A faster algorithm for minimum cycle basis of graphs. In: Díaz, J., Karhumäki, J., Lepistö, A., Sannella, D. (eds.) ICALP 2004. LNCS, vol. 3142, pp. 846–857. Springer, Heidelberg (2004)
13. Kavitha, T., Mehlhorn, K., Michail, D., Paluch, K.E.: An $\tilde{O}(m^2n)$ algorithm for minimum cycle basis of graphs. Algorithmica 52(3), 333–349 (2008)
14. Michail, D.: Minimum Cycle Basis. PhD thesis, Saarland University, Germany (2006)

15. Mehlhorn, K., Michail, D.: Implementing minimum cycle basis algorithms. Journal of Experimental Algorithmics 11, 2.5 (2006)
16. Amaldi, E., Iuliano, C., Jurkiewicz, T., Mehlhorn, K., Rizzi, R.: Breaking the $O(m^2n)$ barrier for minimum cycle bases. In: Proceedings of the the 17th European Symposium on Algorithms, pp. 301–312 (2009)
17. Kavitha, T., Mehlhorn, K., Michail, D.: New approximation algorithms for minimum cycles bases of graphs. In: Proceedings of the 24th Symposium on Theoretical Aspects of Computer Science, pp. 512–523 (2007)
18. Hammack, R.: Minimum cycle bases of direct products of bipartite graphs. Australasian Journal of Combinatorics 36, 213–221 (2006)
19. Hammack, R.: Minimum cycle bases of direct products of complete graphs. Information Processing Letters 102(5), 214–218 (2007)
20. Stadler, P.F.: Minimum cycle bases of halin graphs. Journal of Graph Theory 43(2), 150–155 (2003)
21. Imrich, W., Stadler, P.F.: Minimum cycle bases of product graphs. Australasian Journal of Combinatorics 26, 233–244 (2002)
22. Jaradat, M.M.: On the basis number and the minimum cycle bases of the wreath product of some graphs. Discussiones Mathematicae Graph Theory 26, 113–134 (2006)
23. Jaradat, M.M., Al-Qeyyam, M.K.: On the basis number and the minimum cycle bases of the wreath product of two wheels. International Journal of Mathematical Combinatorics 1, 52–62 (2008)
24. Hartvigsen, D., Mardon, R.: The all-pairs min cut problem and the minimum cycle basis problem on planar graphs. SIAM Journal on Discrete Mathematics 7(3), 403–418 (1994)
25. Leydold, J., Stadler, P.F.: Minimal cycle bases of outerplanar graphs. Electronic Journal of Combinatorics 5, 209–222 (1998)
26. Frederickson, G.N., Janardan, R.: Designing networks with compact routing tables. Algorithmica 3, 171–190 (1988)

BANDWIDTH on AT-Free Graphs*

Petr Golovach[1], Pinar Heggernes[1], Dieter Kratsch[2], Daniel Lokshtanov[1], Daniel Meister[1], and Saket Saurabh[1]

[1] Department of Informatics, University of Bergen, N-5020 Bergen, Norway
{petr.golovach,pinar.heggernes,daniel.lokshtanov,daniel.meister, saket.saurabh}@ii.uib.no
[2] Laboratoire d'Informatique Théorique et Appliquée, Université Paul Verlaine - Metz, 57045 Metz Cedex 01, France
kratsch@univ-metz.fr

Abstract. We study the classical BANDWIDTH problem from the viewpoint of parameterized algorithms. In the BANDWIDTH problem we are given a graph $G = (V, E)$ together with a positive integer k, and asked whether there is an bijective function $\beta : \{1, \ldots, n\} \to V$ such that for every edge $uv \in E$, $|\beta^{-1}(u) - \beta^{-1}(v)| \leq k$. The problem is notoriously hard, and it is known to be NP-complete even on very restricted subclasses of trees. The best known algorithm for BANDWIDTH for small values of k is the celebrated algorithm by Saxe [*SIAM Journal on Algebraic and Discrete Methods, 1980*], which runs in time $2^{\mathcal{O}(k)}n^{k+1}$. In a seminal paper, Bodlaender, Fellows and Hallet [*STOC 1994*] ruled out the existence of an algorithm with running time of the form $f(k)n^{\mathcal{O}(1)}$ for any function f even for trees, unless the entire W-hierarchy collapses.

We initiate the search for classes of graphs where BANDWIDTH is fixed parameter tractable (FPT), that is, solvable in time $f(k)n^{\mathcal{O}(1)}$ for some function f. In this paper we present an algorithm with running time $2^{\mathcal{O}(k \log k)}n^2$ for BANDWIDTH on AT-free graphs, a well-studied graph class that contains interval, permutation, and cocomparability graphs. Our result is the first non-trivial FPT algorithm for BANDWIDTH on a graph class where the problem remains NP-complete.

1 Introduction

The *bandwidth* of a graph G is the smallest integer b such that there is an bijective function $\beta : \{1, \ldots, n\} \to V$, also called a layout for G, such that for every edge $uv \in E$, $|\beta^{-1}(u) - \beta^{-1}(v)| \leq b$. Given a graph G and an integer k, BANDWIDTH asks whether the bandwidth of G is at most k. The problem arises in sparse matrix computations, where given an $n \times n$ matrix A and an integer k, the goal is to decide whether there is a permutation matrix P such that PAP^T is a matrix whose all non-zero entries lie within the k diagonals on either side of the main diagonal. Standard matrix operations like inversion and multiplication as well as Gaussian elimination can be sped up considerably if the input matrix A can be transformed into a matrix PAP^T of small bandwidth [10].

* This work is supported by the Research Council of Norway.

Y. Dong, D.-Z. Du, and O. Ibarra (Eds.): ISAAC 2009, LNCS 5878, pp. 573–582, 2009.

BANDWIDTH is one of the most well-known and extensively studied graph layout problems [9]. The BANDWIDTH problem is NP-complete [21] and remains NP-complete even on very restricted subclasses of trees, like caterpillars of hair length at most 3 [18]. Furthermore, the bandwidth of a graph is NP-hard to approximate within a constant factor for trees [2]. Polynomial-time algorithms for the exact computation of bandwidth are known for a few graph classes including caterpillars with hair length at most 2 [1], cographs [23], interval graphs [14] and bipartite permutation graphs [12]. A classical algorithm by Saxe [22] solves BANDWIDTH in time $2^{\mathcal{O}(k)}n^{k+1}$, which is polynomial when k is a constant. However, as the value of k grows, the exponent of the polynomial grows with it. A natural question is whether BANDWIDTH can be solved in time $f(k)n^c$ where c is a constant independent of k. This amounts to asking whether BANDWIDTH is *fixed parameter tractable.*

Parameterized complexity is a two-dimensional generalization of "P vs. NP" where, in addition to the overall input size n, one studies how a secondary measurement that captures additional relevant information affects the computational complexity of the problem in question. Parameterized decision problems are defined by specifying the input, the parameter and the question to be answered. The two-dimensional analogue of P is solvability within a time bound of $f(k)n^c$, where n is the total input size, k is the parameter, f is some computable function, and c is a constant that does not depend on k or n. A parameterized problem that can be solved in such time is termed *fixed-parameter tractable* (FPT). There is a hierarchy of intractable parameterized problem classes above FPT, the main ones being:

$$\text{FPT} \subseteq \text{W}[1] \subseteq \text{W}[2] \subseteq \cdots \subseteq \text{W}[P] \subseteq \text{XP}\,.$$

The principal analogue of the classical intractability class NP is W[1], which is a strong analogue, because a fundamental problem complete for W[1] is the k-STEP HALTING PROBLEM FOR NONDETERMINISTIC TURING MACHINES with unlimited nondeterminism and alphabet size — this completeness result provides an analogue of Cook's Theorem in classical complexity. Thus a parameterized problem that is hard for W[1] is unlikely to be fixed parameter tractable. For general background on the theory see the textbooks by Downey and Fellows [7], Flum and Grohe [11] and Niedermeier [19].

In a seminal paper, Bodlaender, Fellows and Hallet showed that BANDWIDTH is hard for W[t] for every $t \geq 1$, even for trees [3]. This rules out the existence of an FPT algorithm for BANDWIDTH unless the entire W-hierarchy collapses. The hardness result in [3] indicates that the tractable cases for BANDWIDTH seem to be few and far between. Here, we initiate the search for classes of graphs where BANDWIDTH is fixed parameter tractable.

For the graph classes for which polynomial time algorithms are known, it has been proved that BANDWIDTH becomes NP-complete (or its complexity remains unknown) on slightly larger graph classes. Therefore it is natural to investigate the *parameterized* complexity of BANDWIDTH on these larger classes of graphs. In this paper we present an algorithm with running time $2^{\mathcal{O}(k \log k)}n^2$ for BANDWIDTH on AT-free graphs. A graph is *AT-free* if for every triple of pairwise

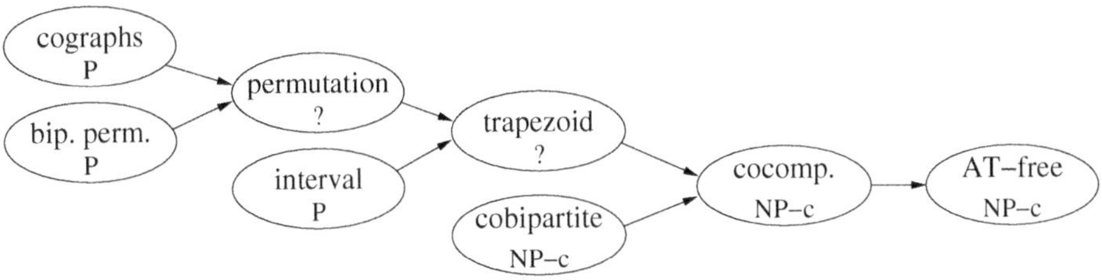

Fig. 1. A graph class inclusion diagram with the classical complexity of BANDWIDTH on these graph classes

non-adjacent vertices, the neighborhood of one of them separates the two others. The class of AT-free graphs contains various well-known graph classes, like interval, permutation, trapezoid, and cocomparability graphs [4]. While BANDWIDTH is polynomial-time solvable on interval graphs [14] and well-studied subclasses of permutation graphs [23,12] it is NP-complete on cocomparability graphs and hence on AT-free graphs [16]. For permutation graphs, the complexity of BANDWIDTH is a well-known open problem. Most natural superclasses of AT-free graphs contain trees, and thus the hardness result in [3] rules out an FPT algorithm for BANDWIDTH on these classes. Thus our FPT algorithm on AT-free graphs essentially settles the parameterized complexity of BANDWIDTH on the chain of natural graph classes above the polynomial cases (Figure 1).

Our algorithm is based on structural properties of AT-free graphs and their relation to interval graphs. The principal idea is to combine layouts for small subgraphs of the input graph to a layout for the whole graph. This approach can be described as sweeping a small window over a layout. For arbitrary graphs, there is no information about the relationship between the vertices in the window. For AT-free graphs, however, we are able to show that we can restrict to windows of vertices that have small distance to a common vertex in the input AT-free graph. This enables us to restrict the number of considered windows and to establish the claimed FPT running time of our algorithm.

2 Definitions and Notation

In this paper, we mainly consider simple finite undirected graphs. However, as an auxiliary structure, we also define a directed graph. By "graph", we always mean undirected graph, and by "digraph", we mean a directed graph.

For a graph $G = (V, E)$, we denote the vertex and edge set of G by respectively $V = V(G)$ and $E = E(G)$, with $n = |V|$. Edges of a graph are denoted as uv, and if uv is an edge of G, we call u and v *adjacent*. The *neighborhood* of a vertex u is the set of vertices that are adjacent to u and is denoted as $N_G(u)$. For two vertices $u, v \in V$, a *u, v-path of G of length r* is a sequence $(u_0, \ldots, u_r)$ of distinct vertices where $u_0 = u$, $u_r = v$ and $u_i u_{i+1} \in E$ for $0 \leq i < r$. The *distance of u and v in G*, denoted by $d_G(u, v)$, is the smallest length of a u, v-path in G. For a vertex u of G and an integer $\ell \geq 1$, the *ball around u of radius ℓ*, $B_G(u, \ell)$, is the set of vertices different from u that are at distance at most ℓ to u in G. Formally, $B_G(u, \ell) = \{x \neq u : \mathrm{d}_G(u, x) \leq \ell\}$. A graph G is *connected* if there is a u, v-path in G for every vertex pair u, v. A set of vertices is a *clique* if its vertices are pairwise adjacent. The *square* of G is $G^2 = (V, \{uv : 1 \leq \mathrm{d}_G(u, v) \leq 2\})$.

For a set $S \subseteq V$, the subgraph of G *induced* by S, denoted by $G[S]$, has vertex set S and all edges of G that have both their endpoints in S. By $G \setminus S$, we denote $G[V \setminus S]$. A graph G is a *subgraph* (not necessarily induced) of a graph H if $V(G) \subseteq V(H)$ and $E(G) \subseteq E(H)$. In this case, we write $G \subseteq H$. A *connected component* of G is a maximal connected induced subgraph of G.

For a graph G, a *layout* β (or *vertex ordering*) is a bijective function from $\{1, \ldots, n\}$ to V. We also write β as $\langle \beta(1), \ldots, \beta(n) \rangle$. For a vertex pair u, v of G, the *distance* between u and v in β is $\mathrm{d}_\beta(u, v) = |\beta^{-1}(u) - \beta^{-1}(v)|$. We write $u \preccurlyeq_\beta v$ if $\beta^{-1}(u) \leq \beta^{-1}(v)$ and $u \prec_\beta v$ if $\beta^{-1}(u) < \beta^{-1}(v)$. The *leftmost* and *rightmost* vertex in β are respectively $\beta(1)$ and $\beta(n)$. For an integer $k \geq 1$, we call β a *k-layout* for G if for every edge uv of G, $d_\beta(u, v) \leq k$.

The *bandwidth* of G, $\mathrm{bw}(G)$, is the smallest integer b such that G has a b-layout. A *minimum bandwidth layout* for G is a k-layout for G with $k = \mathrm{bw}(G)$. Observe that for any two graphs G and H with $G \subseteq H$, $\mathrm{bw}(G) \leq \mathrm{bw}(H)$.

If β is a layout for G and ℓ is between 1 and n, then every i between 1 and $n - \ell + 1$ defines an *ℓ-window of β*: $\langle \beta(i), \ldots, \beta(i + \ell - 1) \rangle$. Informally, an ℓ-window is a portion of β containing exactly ℓ consecutive vertices. Vertices $\beta(i)$ and $\beta(i + \ell - 1)$ are called respectively *left* and *right vertex* of the ℓ-window.

Layouts of sets of vertices are defined analogously to layouts for graphs. Let $U, U' \subseteq V$ and let β and β' be layouts of respectively U and U'. Let $U \cap U' \neq \emptyset$ and let $1 \leq t \leq |U|$ be smallest with $\beta(t) \in U'$. If $\beta(t + i) = \beta'(i + 1)$ for all $0 \leq i \leq |U| - t$ then $\beta \bullet \beta'$ is the layout $\langle \beta(1), \ldots, \beta(|U|), \beta'(|U'| - t + 2), \ldots, \beta'(|U'|) \rangle$; otherwise, if the condition is violated, $\beta \bullet \beta'$ is not defined. Informally, $\beta \bullet \beta'$ is the concatenation of β and β' by overlapping in the common part. Note that the $\bullet$ operator satisfies the associativity law.

In this paper, we study AT-free graphs and subclasses of AT-free graphs. A set $\{u, v, w\}$ of three pairwise non-adjacent vertices of a graph is called *asteroidal triple*, *AT* for short, if for any pair of the vertices there is a path between these two vertices avoiding the neighborhood of the third vertex. A graph that has no asteroidal triple is called *AT-free*. For more information on structural properties of AT-free graphs we refer to [4,5].

Finally, a *cycle* in a digraph G is a sequence $(u_1, \ldots, u_r)$ of distinct vertices where (u_r, u_1) and $(u_1, u_2), \ldots, (u_{r-1}, u_r)$ are arcs of G. A digraph without cycles is called *acyclic*.

3 Preliminary and Auxiliary Results on Bandwidth

The following observation will be important for the running time of our algorithm. The result is easy to establish from the fact that a graph of bandwidth at most k cannot have a vertex of degree more than $2k$.

Lemma 1. *For an arbitrary graph* $G = (V, E)$, $|E| \leq \text{bw}(G) \cdot |V|$.

A graph H is an *interval graph* if its vertices can be assigned closed intervals of the real line such that two vertices of H are adjacent if and only if the assigned intervals have a non-empty intersection. For an arbitrary graph G, an *interval completion* of G is an interval graph H on vertex set $V(G)$ with $E(G) \subseteq E(H)$. If there is no interval completion H' of G with $E(H') \subset E(H)$ then H is a *minimal* interval completion of G.

An interval graph is a *proper interval graph* if there is an interval representation of it where no interval completely contains another interval. For an arbitrary graph G, a *proper interval completion* of G is a proper interval graph H on vertex set $V(G)$ with $E(G) \subseteq E(H)$.

Lemma 2 ([13]). *For any integer* $k \geq 0$, *a graph* G *has a proper interval completion* H *of maximum clique size at most* $k + 1$ *if and only if* $bw(G) \leq k$.

From Lemma 2 it follows that any minimum bandwidth layout β for G defines a proper interval completion of it in the following way. For every edge uv of G, make the set of vertices between (including) u and v in β into a clique by adding the necessary missing edges.

Lemma 3. *For every graph* G, *there is a minimal interval completion* H *of* G *with* $\text{bw}(G) = \text{bw}(H)$.

Proof. Let β be a minimum bandwidth layout for G. Then, β defines a proper interval completion H' of G, where $\text{bw}(G) = \text{bw}(H')$. Since H' is an interval graph, there is a minimal interval completion H of G with $G \subseteq H \subseteq H'$ and $\text{bw}(H) = \text{bw}(G)$. □

An alternative characterization of interval graphs is that a graph G is an interval graph if and only if G has a vertex ordering σ such that for all vertex triples u, v, w of G with $u \prec_\sigma v \prec_\sigma w$, $uw \in E(G)$ implies $vw \in E(G)$ [20]. Such a vertex ordering is called an *interval ordering*.

Theorem 1 ([8]). *Let* H *be an interval graph with interval ordering* σ. *There is a minimum bandwidth layout* β *for* H *such that for every pair* u, v *of non-adjacent vertices of* H, $u \prec_\sigma v$ *implies* $u \prec_\beta v$.

Lemma 4. *Let* H *be an interval graph. There is a minimum bandwidth layout* β *for* H *with the property: for every vertex pair* u, v *of* H, $\text{d}_\beta(u, v) \geq \text{d}_H(u, v) - 2$.

Proof. Let σ be an interval ordering for H. Let β be a minimum bandwidth layout for H with the property of Theorem 1 with respect to σ. We show that β satisfies the lemma. Let u, v be a vertex pair of H. If $\text{d}_H(u, v) \leq 3$, the lemma trivially holds. Let $\text{d}_H(u, v) \geq 4$. Without loss of generality, we can assume $u \prec_\sigma v$. Let $(x_0, \ldots, x_r)$ be a shortest u, v-path of H, where $x_0 = u$ and $x_r = v$. If there is $1 \leq i \leq r - 2$ with $x_{i-1} \prec_\sigma v \prec_\sigma x_i$ then v is adjacent to x_i by the properties of interval orderings, contradicting the choice of the shortest path.

If there is $1 \leq i \leq r-1$ with $x_i \prec_\sigma u \prec_\sigma x_{i+1}$ then u and x_{i+1} are adjacent, which again gives a contradiction. Thus, $u \prec_\sigma x_i \prec_\sigma v$ for all $1 \leq i \leq r-2$. Since $x_2, \ldots, x_{r-2}$ are non-adjacent to u and v, Theorem 1 for β implies that $u \prec_\beta x_i \prec_\beta v$ for all $2 \leq i \leq r-2$. Hence, $\mathrm{d}_\beta(u,v) \geq r-2 = \mathrm{d}_H(u,v) - 2$. □

Lemma 5. *Let G be a connected graph and β a k-layout for G, where $k \geq 1$. Let U be the vertices of a $(k+2)$-window of β with a and b the respectively left and right vertex. Then, $G \setminus U$ has at most $2k$ connected components, and for every connected component C of $G \setminus U$, the vertices of C are either all to the left of a or all to the right of b in β.*

Proof. Since there is no edge uv of G with $u \prec_\beta a \prec_\beta b \prec_\beta v$ by β being a k-layout and $\mathrm{d}_\beta(a,b) = k+1$, no connected component of $G \setminus U$ has vertices to the left of a and to the right of b in β. Since G is a connected graph, every connected component of $G \setminus U$ has a vertex with a neighbor in U. Thus, every connected component of $G \setminus U$ has a vertex to the left of a at distance (with respect to β) at most k to a or to the right of b at distance at most k to b in β. Hence there can be at most $2k$ connected components in $G \setminus U$. □

Lemma 6. *Let G be a graph and let β be a k-layout for G, where $k \geq 1$. For every vertex a of G and every integer $\ell \geq 1$, $|B_G(a,\ell)| \leq 2\ell \cdot k$.*

Proof. Let c and d be the respectively leftmost and rightmost vertex from $B_G(a,\ell)$ in β. Since $\mathrm{d}_\beta(c,a) \leq \ell \cdot k$ and $\mathrm{d}_\beta(a,d) \leq \ell \cdot k$ it follows that $\mathrm{d}_\beta(c,d) \leq 2\ell \cdot k$. Since all vertices from $B_G(a,\ell)$ are between c and d in β, the bound of the lemma follows. Note that by definition $a \notin B_G(a,\ell)$. □

4 Bandwidth of AT-Free Graphs

Combining the results of [16] and [17], we obtain the following:

Theorem 2 ([16,17]). *Let G be an AT-free graph. For every minimal interval completion H of G, $H \subseteq G^2$.*

We use this fact to restrict the number of $(k+2)$-windows to be considered by our algorithm.

Lemma 7. *For a connected AT-free graph G with $\mathrm{bw}(G) \leq k$, there is a k-layout β that satisfies the following. Let U be the vertices of a $(k+2)$-window of β with left vertex a. For every vertex $x \in U$, $\mathrm{d}_G(a,x) \leq 2k+6$.*

Proof. Let H be a minimal interval completion of G with $\mathrm{bw}(H) = \mathrm{bw}(G)$; H exists due to Lemma 3. By Theorem 2 $\mathrm{d}_H(u,v) \geq \frac{1}{2} \cdot \mathrm{d}_G(u,v)$ for every vertex pair u,v of G. Let β be a minimum bandwidth layout for H with the property of Lemma 4. Then, for every vertex pair u,v of G, $\mathrm{d}_\beta(u,v) + 2 \geq \mathrm{d}_H(u,v) \geq \frac{1}{2}\mathrm{d}_G(u,v)$, i.e., $\mathrm{d}_\beta(u,v) \geq \frac{1}{2}\mathrm{d}_G(u,v) - 2$. Let U be the vertices of a $(k+2)$-window of β with left vertex a. Since $\mathrm{d}_\beta(a,x) \leq k+1$ for every $x \in U$, it follows that $\mathrm{d}_G(a,x) \leq 2k+6$. □

We construct a digraph to encode all feasible k-layouts. Let $k \geq 1$ and let $G = (V, E)$ be a connected AT-free graph on $n \geq k+3$ vertices. Note that every graph on at most $k+1$ vertices has bandwidth at most k, and every graph on $k+2$ vertices has bandwidth at most k if and only if the graph has a pair of non-adjacent vertices. For a vertex u of G, we say that *u has few close neighbors* if $|B_G(u, 2k+6)| \leq 2k \cdot (2k+6)$. Let a be a vertex of G that has few close neighbors. An *a-bag* is a tuple $(\mathcal{C}_1, \gamma, \mathcal{C}_2)$ where $\mathcal{C}_1, \mathcal{C}_2 \subseteq V$ such that $U = V \setminus (\mathcal{C}_1 \cup \mathcal{C}_2)$ and $\mathcal{C}_1, \mathcal{C}_2, \gamma$ have the following properties:

- $\{a\} \subseteq U \subseteq B_G(a, 2k+6) \cup \{a\}$ and $|U| = k+2$;
- γ is a layout of U where a is the leftmost vertex;
- let b be the rightmost vertex in γ, a has no neighbor in $\mathcal{C}_2 \cup \{b\}$, and b has no neighbor in $\mathcal{C}_1 \cup \{a\}$;
- $G \setminus U$ has at most $2k$ connected components, and for every connected component C of $G \setminus U$, either $V(C) \subseteq \mathcal{C}_1$ or $V(C) \subseteq \mathcal{C}_2$.

Lemma 8. *For every vertex a of G where a has few close neighbors, the number of a-bags is at most* $\binom{2k\cdot(2k+6)}{k+1} \cdot (k+1)! \cdot 2^{2k} \leq 2^{\mathcal{O}(k \log k)}$.

Proof. Let a be a vertex of G with few close neighbors. Since $|B_G(a, 2k+6)| \leq 2k \cdot (2k+6)$, there are at most $\binom{2k\cdot(2k+6)}{k+1}$ subsets of $B_G(a, 2k+6)$ of size $k+1$. Let U be such a subset. There are $(k+1)!$ possible layouts of U. And since $G \setminus (U \cup \{a\})$ has at most $2k$ connected components according to the definition of an a-bag, there are at most 2^{2k} different partitions of the connected components of $G \setminus (U \cup \{a\})$ into a left and a right set. □

The auxiliary digraph that we will define has a vertex for every bag and arcs between bags representing the fact that one bag can follow another bag in a minimum bandwidth layout. We now describe the construction of the auxiliary digraph $\mathrm{aux}(G, k)$. The digraph $\mathrm{aux}(G, k)$ has a vertex for every a-bag where $a \in V$ has few close neighbors, a source vertex S and a sink vertex T. The vertices of $\mathrm{aux}(G, k)$ corresponding to a-bags are labeled with these a-bags. For two vertices u and u' of $\mathrm{aux}(G, k)$ such that $u \neq u'$ and $u, u' \notin \{S, T\}$ with u labeled with an a-bag $(\mathcal{C}_1, \gamma, \mathcal{C}_2)$ and u' labeled with an a'-bag $(\mathcal{C}'_1, \gamma', \mathcal{C}'_2)$, with $U = V \setminus (\mathcal{C}_1 \cup \mathcal{C}_2)$ and $U' = V \setminus (\mathcal{C}'_1 \cup \mathcal{C}'_2)$, the digraph contains the arc (u, u') if and only if the bags satisfy one of the following two conditions:

1) $\mathcal{C}_1 \subset \mathcal{C}'_1 \subset (\mathcal{C}_1 \cup U)$ and $\mathcal{C}'_2 = \emptyset$ and $\gamma \bullet \gamma'$ is k-layout for $G[U \cup U']$
2) $|U \cap U'| = 1$ and $\mathcal{C}'_1 = \mathcal{C}_1 \cup (U \setminus U')$ and $\gamma \bullet \gamma'$ is k-layout for $G[U \cup U']$.

Note that, by the definition of the $\bullet$ operator for layouts, the end of layout γ is equal to the beginning of layout γ'. For condition 2, this means that the rightmost vertex in γ is equal to the leftmost vertex in γ'. To complete the definition of $\mathrm{aux}(G, k)$, add all arcs (S, u) with u labeled with a bag of the type $(\emptyset, \gamma, \mathcal{C}_2)$ and all arcs (u', T) with u' labeled with a bag of the type $(\mathcal{C}'_1, \gamma', \emptyset)$.

Lemma 9. *The auxiliary digraph* $\mathrm{aux}(G, k)$ *is acyclic.*

Proof. Suppose that $\text{aux}(G, k)$ contains a cycle $(u^{(1)}, \ldots, u^{(r)})$. Let $(\mathcal{C}_1^{(i)}, \gamma^{(i)}, \mathcal{C}_2^{(i)})$ be the bag that $u^{(i)}$ is labeled with, for $1 \leq i \leq r$. According to the definition of $\text{aux}(G, k)$, it holds that $\mathcal{C}_1^{(1)} \subset \cdots \subset \mathcal{C}_1^{(r)} \subset \mathcal{C}_1^{(1)}$. This yields a contradiction. □

The proof of the next lemma will appear in the full version of the paper.

Lemma 10. *The auxiliary digraph* $\text{aux}(G, k)$ *has an* S, T*-path if and only if* $\text{bw}(G) \leq k$. *Furthermore, every* S, T*-path of* $\text{aux}(G, k)$ *defines a* k*-layout for* G.

For the algorithm in the proof of the following theorem, we assume the input graph to be given in adjacency list representation.

Theorem 3. *For given an AT-free graph* G *and an integer* $k \geq 1$, *it can be decided in* $2^{\mathcal{O}(k \log k)} n^2$ *time whether* $\text{bw}(G) \leq k$.

Proof. Let $k \geq 1$. Let $G = (V, E)$ be an AT-free graph. Since $\text{bw}(G) \leq k$ if and only if $\text{bw}(C) \leq k$ for every connected component C of G, the algorithm to be described below is to run on every connected component of G. By these considerations, we can assume that the input graph G to our algorithm is connected. The algorithm is: if $|E| > k \cdot |V|$ or if a vertex has more than $2k$ neighbors then reject, otherwise, compute $\text{aux}(G, k)$, with the distinguished source and sink vertices S and T, and accept if and only if $\text{aux}(G, k)$ contains an S, T-path. Correctness follows from Lemmata 1, 6, and 10.

The running time of the algorithm is mainly determined by the generation of $\text{aux}(G, k)$. Let a be a vertex of G. Since every vertex of G has at most $2k$ neighbors, it can be checked in time $\mathcal{O}(k^3)$ whether vertex a has few neighbors and, if so, to compute $B_G(a, 2k + 6)$. In time $2^{\mathcal{O}(k \log k)} n$, all a-bags can be listed due to Lemma 8. Determining the left side and right side connected components of a single bag takes time $\mathcal{O}(kn)$, and writing down a single a-bag takes time $\mathcal{O}(n)$. In total, the vertices of $\text{aux}(G, k)$ together with the labels can be listed in $2^{\mathcal{O}(k \log k)} n^2$ time. Note that $\text{aux}(G, k)$ has $2^{\mathcal{O}(k \log k)} n$ vertices. Let u be a vertex of $\text{aux}(G, k)$ labeled with an a-bag $(\mathcal{C}_1, \gamma, \mathcal{C}_2)$. If $\mathcal{C}_2 \neq \emptyset$, then all arcs from u go to vertices x that are labeled with a b-bag, where b is the rightmost vertex in γ in case $|\mathcal{C}_2| \geq k + 1$ or is uniquely determined in γ from the size of $\mathcal{C}_2$. Hence, for every vertex of $\text{aux}(G, k)$ there are at most $2^{\mathcal{O}(k \log k)}$ vertices to check, and each check takes $\mathcal{O}(n)$ time because of the left and right side connected components. Hence, in overall $2^{\mathcal{O}(k \log k)} n^2$ time, $\text{aux}(G, k)$ can be generated, and it has $2^{\mathcal{O}(k \log k)} n$ vertices and arcs. Since verifying the existence of an S, T-path in $\text{aux}(G, k)$ takes $2^{\mathcal{O}(k \log k)} n$ time, this concludes the running time analysis for our algorithm. □

5 Concluding Remarks

Our algorithm guesses layouts of small sets of vertices and concatenates them to create a layout for the whole graph. The running time of the algorithm relies on the fact that the considered sets of vertices are of special type, namely that it suffices to consider vertices that are at small distance to a common vertex. Correctness of this restriction follows from the result that distances in minimal interval completions of AT-free graphs provide a constant-factor approximation of the distances in the graph. Thus, we can generalize our algorithm to graph classes with the same property. Which other graph class $\mathcal{C}$ has this property: there is constant c such that for every graph G from $\mathcal{C}$ and every minimal interval completion H of G, $\mathrm{d}_G(u,v) \leq c \cdot \mathrm{d}_H(u,v)$ for all vertex pairs u, v of G?

Classes of graphs of bounded diameter certainly have this property. These are classes of dense graphs. For such classes of dense graphs, the problem becomes trivial, as we show in the following. For a graph G, the *diameter* of G, denoted by $\mathrm{diam}(G)$, is the maximum distance between a vertex pair of G.

Lemma 11. *For an arbitrary connected graph* $G = (V, E)$, $|V| \leq 1 + \mathrm{diam}(G) \cdot \mathrm{bw}(G)$.

Proof. Let β be a minimum bandwidth layout for G. Let a and b be the respectively leftmost and rightmost vertex in β. Then, $|V| - 1 = \mathrm{d}_\beta(a, b) \leq \mathrm{diam}(G) \cdot \mathrm{bw}(G)$. □

In other words, for a class of graphs of bounded diameter, there is only a finite number of graphs of bounded bandwidth. Thus, for such graph classes, deciding whether $\mathrm{bw}(G) \leq k$ for given graph G is trivial when k is fixed. If k is part of the input, we can apply the currently best known exact algorithm for computing bandwidth by Cygan and Pilipczuk, with running time $\mathcal{O}(4.473^n)$ [6], and obtain a $\mathcal{O}(4.473^{dk})$-time algorithm for deciding whether $\mathrm{bw}(G) \leq k$ for a given (connected) graph G with $\mathrm{diam}(G) \leq d$ and integer k. Examples of such graphs are P_{d+1}-free graphs, in particular split graphs and cobipartite graphs, which are well-studied classes of P_5-free graphs. The bandwidth problem is NP-complete when restricted to split and cobipartite graphs [15,16].

We conclude the paper with a few concrete open problems.

- Does there exist an FPT algorithm for BANDWIDTH on AT-free graphs running in time $2^{\mathcal{O}(k)} n^{\mathcal{O}(1)}$? An algorithm with this running time would be interesting even for cocomparability graphs.
- Can bandwidth be FPT-approximated on trees? That is, is there an algorithm that given a tree T and integer k, runs in time $f(k)n^{\mathcal{O}(1)}$ and either correctly answers that $\mathrm{bw}(T) > k$ or outputs a $g(k)$-layout for T for some function g.
- What is the parameterized complexity of BANDWIDTH on caterpillars with hairlength c, for fixed constant $c \geq 3$?

Acknowledgements

We would like to thank Fedor Fomin for insightful discussions around bandwidth.

References

1. Assmann, S.F., Peck, G.W., Sysło, M.M., Zak, J.: The bandwidth of caterpillars with hairs of length 1 and 2. SIAM J. Alg. Disc. Meth. 2, 387–393 (1981)
2. Blache, G., Karpinski, M., Wirtgen, J.: On approximation intractability of the bandwidth problem. Technical report TR98-014, University of Bonn (1997)

3. Bodlaender, H.L., Fellows, M.R., Hallet, M.T.: Beyond NP-completeness for problems of bounded width (extended abstract): hardness for the W hierarchy. In: Proceedings of STOC 1994, pp. 449–458. ACM, New York (1994)
4. Brandstädt, A., Le, V.B., Spinrad, J.: Graph Classes: A Survey. SIAM, Philadelphia (1999)
5. Corneil, D.G., Olariu, S., Stewart, L.: Asteroidal Triple-Free Graphs. SIAM J. Disc. Math. 10, 399–430 (1997)
6. Cygan, M., Pilipczuk, M.: Exact and Approximate Bandwidth. In: Albers, S., et al. (eds.) ICALP 2009. LNCS, vol. 5555, pp. 304–315. Springer, Heidelberg (2009)
7. Downey, R.G., Fellows, M.R.: Parameterized Complexity. Springer, New York (1999)
8. Fishburn, P., Tanenbaum, P., Trenk, A.: Linear discrepancy and bandwidth. Order 18, 237–245 (2001)
9. Garey, M.R., Johnson, D.S.: Computers and Intractability. A Guide to the Theory of NP-Completeness. W.H. Freeman, New York (1979)
10. George, J.A., Liu, J.W.H.: Computer Solution of Large Sparse Positive Definite Systems. Prentice-Hall, New Jersey (1981)
11. Flum, J., Grohe, M.: Parameterized Complexity Theory. Springer, Heidelberg (2006)
12. Heggernes, P., Kratsch, D., Meister, D.: Bandwidth of bipartite permutation graphs in polynomial time. Journal of Disc. Alg. (to appear)
13. Kaplan, H., Shamir, R.: Pathwidth, Bandwidth, and Completion Problems to Proper Interval Graphs with Small Cliques. SIAM J. Computing 25, 540–561 (1996)
14. Kleitman, D.J., Vohra, R.V.: Computing the bandwidth of interval graphs. SIAM J. Disc. Math. 3, 373–375 (1990)
15. Kloks, T., Kratsch, D., Le Borgne, Y., Müller, H.: Bandwidth of Split and Circular Permutation Graphs. In: Brandes, U., Wagner, D. (eds.) WG 2000. LNCS, vol. 1928, pp. 243–254. Springer, Heidelberg (2000)
16. Kloks, T., Kratsch, D., Müller, H.: Approximating the bandwidth for AT-free graphs. J. Alg. 32, 41–57 (1999)
17. Möhring, R.: Triangulating Graphs Without Asteroidal Triples. Discrete Applied Mathematics 64, 281–287 (1996)
18. Monien, B.: The Bandwidth-Minimization Problem for Caterpillars with Hair Length 3 is NP-Complete. SIAM J. Alg. Disc. Meth. 7, 505–512 (1986)
19. Niedermeier, R.: Invitation to Fixed-Parameter Algorithms. Oxford University Press, Oxford (2006)
20. Olariu, S.: An optimal greedy heuristic to color interval graphs. Information Processing Letters 37, 21–25 (1991)
21. Papadimitriou, C.: The NP-completeness of the bandwidth minimization problem. Computing 16, 263–270 (1976)
22. Saxe, J.B.: Dynamic-Programming Algorithms for Recognizing Small-Bandwidth Graphs in Polynomial Time. SIAM Journal on Algebraic and Discrete Methods 1, 363–369 (1980)
23. Yan, J.H.: The bandwidth problem in cographs. Tamsui Oxf. J. Math. Sci. 13, 31–36 (1997)

Editing Graphs into Disjoint Unions of Dense Clusters

Jiong Guo[1,*], Iyad A. Kanj[2,**], Christian Komusiewicz[3,***], and Johannes Uhlmann[3,†]

[1] Universität des Saarlandes, Campus E 1.4, D-66123 Saarbrücken, Germany
jguo@mmci.uni-saarland.de

[2] School of Computing, DePaul University, 243. S. Wabash Avenue, Chicago, IL 60604, USA
ikanj@cs.depaul.edu

[3] Institut für Informatik, Friedrich-Schiller-Universität Jena Ernst-Abbe-Platz 2, D-07743 Jena, Germany
{c.komus,johannes.uhlmann}@uni-jena.de

Abstract. In the Π-Cluster Editing problem, one is given an undirected graph G, a density measure Π, and an integer $k \geq 0$, and needs to decide whether it is possible to transform G by editing (deleting and inserting) at most k edges into a dense cluster graph. Herein, a dense cluster graph is a graph in which every connected component $K = (V_K, E_K)$ satisfies Π. The well-studied Cluster Editing problem is a special case of this problem with $\Pi :=$ "being a clique". In this work, we consider three other density measures that generalize cliques: 1) having at most s missing edges (s-defective cliques), 2) having average degree at least $|V_K| - s$ (average-s-plexes), and 3) having average degree at least $\mu \cdot (|V_K| - 1)$ (μ-cliques), where s and μ are a fixed integer and a fixed rational number, respectively. We first show that the Π-Cluster Editing problem is NP-complete for all three density measures. Then, we study the fixed-parameter tractability of the three clustering problems, showing that the first two problems are fixed-parameter tractable with respect to the parameter (s, k) and that the third problem is W[1]-hard with respect to the parameter k for $0 < \mu < 1$.

1 Introduction

Graph-based data clustering is an important tool for analyzing real-world data, ranging from biological to social network data. In such applications, data items

* Supported by the Excellence Cluster on Multimodal Computing and Interaction (MMCI). Main work was done while the author was with Friedrich-Schiller-Universität Jena.

** Part of this work was done while the author was visiting Friedrich-Schiller-Universität Jena.

*** Supported by a PhD fellowship of the Carl-Zeiss-Stiftung.

† Supported by the DFG, research project PABI, NI 369/7.

Y. Dong, D.-Z. Du, and O. Ibarra (Eds.): ISAAC 2009, LNCS 5878, pp. 583–593, 2009.

are represented as vertices, and there is an edge between two vertices iff the interrelation between the two corresponding data items exceeds some threshold value. A clustering with respect to such a graph means a partition of the vertices into dense subgraphs, also called clusters, such that there are few edges between the clusters. When formulated as a graph modification problem, one thus asks for a minimum-cardinality set of edge modifications, such that the resulting graph is a graph in which every connected component is a cluster. More precisely, the algorithmic task can be formalized as follows:

> Π-Cluster Editing:
> **Input**: An undirected graph $G = (V, E)$, a density measure Π, and an integer $k \geq 0$.
> **Task**: Find a set of at most k edge modifications to transform G into a Π-cluster graph, that is, a graph in which every connected component satisfies Π.

Herein, an edge modification is to insert or delete an edge. Analogously, one defines Π-Cluster Deletion by allowing only edge deletions and Π-Cluster Addition by allowing only edge insertions.

One of the most prominent problems in this context is the NP-hard Cluster Editing problem (also known as Correlation Clustering) [2, 16], where the required density measure is Π :="being a clique". Cluster Editing finds applications in various fields, such as computational biology [3] and machine learning [16], and has been intensively studied from the viewpoints of polynomial-time approximability as well as parameterized algorithmics. In terms of approximability, the currently best known approximation factor is 2.5 [17]. Cluster Editing can be solved in $O(1.83^k + |E|)$ time [3] and several studies concerning provably efficient and effective preprocessing by data reduction have been performed [8, 11]. Successful experimental studies of the parameterized algorithms for Cluster Editing have been conducted mainly in the context of computational biology [3, 6]. The related Cluster Deletion problem is also NP-hard [2].

The density requirement of being cliques has been often criticized for its overly restrictive nature and modeling disadvantages [5, 15]. In this work, we attempt to chart the tractability borderlines of Π-Cluster Editing when the density requirement is relaxed. Therefore, we consider three relaxed density measures, namely, *s-defective cliques*, *average-s-plexes*, and *μ-cliques*. The corresponding modification problems are s-Defective Clique Editing, Average-s-Plex Editing, and μ-Clique Editing. We study the classical and the parameterized complexity of the aforementioned problems. The proposed density measures may provide more realistic models for practical applications and fixed-parameter tractability (FPT) results can serve as a first step in a series of algorithmic improvements, eventually leading to applicability in practice, as it was the case for Cluster Editing [3, 6, 8, 11]. An overview of our results is given in Table 1. Note that for all three density measures the polynomial-time solvability of the addition problem can be easily seen and is included only for the sake of completeness. In the following, we give the exact definitions of the density measures studied in this work, point to related work, and describe our results.

Table 1. The complexity of the problems considered in this work. For s-defective cliques and average-s-plexes, the considered parameter is (s, k), for μ-cliques, the considered parameter is k.

	DELETION	EDITING	ADDITION
s-DEFECTIVE CLIQUE	NP-complete (Thm. 1)	NP-complete (Thm. 1)	$\in P$
	FPT (Thm. 3)	FPT (Thm. 3)	
AVERAGE-s-PLEX	NP-complete (Thm. 4)	NP-complete (Thm. 4)	$\in P$
	FPT (Thm. 5)	FPT (Thm. 5)	
μ-CLIQUE	NP-complete (Thm. 7)	W[1]-hard (Thm. 6)	$\in P$

Defective Cliques. The concept of defective cliques has been used in biological networks to represent a clique with exactly one edge missing [18]. Here, we generalize this notion[1] by allowing up to s missing edges: A graph $G = (V, E)$ is called an s-*defective clique*, if G is connected and $|E| \geq |V| \cdot (|V| - 1)/2 - s$. On the negative side, we prove that s-DEFECTIVE CLIQUE DELETION and EDITING are NP-complete. On the positive side however, we show that s-defective cliques can be characterized by forbidden subgraphs of size at most $2(s+1)$, thus showing the fixed-parameter tractability of s-DEFECTIVE CLIQUE DELETION and EDITING with respect to the parameter (s, k).

Average-s-Plexes. With average-s-plexes, we propose a density measure that concerns the *average* degree of a graph $G = (V, E)$, which is defined as $\bar{d} = 2|E|/|V|$. We call a connected graph $G = (V, E)$ an *average-s-plex* if the average degree $\bar{d}$ of G is at least $|V| - s$ for an integer $1 \leq s \leq |V|$. This density measure is a relaxation of the s-plex notion, which demands that the *minimum* degree of a graph $G = (V, E)$ is $|V| - s$. s-Plexes find applications for example in social network analysis [15]. For s-plexes, the clustering problem s-PLEX EDITING has been previously shown to be NP-hard but fixed-parameter tractable with respect to the parameter (s, k) [12]. Here, we complement this result by showing that AVERAGE-s-PLEX DELETION and EDITING are also NP-hard as well as fixed-parameter tractable with respect to the parameter (s, k). The fixed-parameter tractability result is achieved by a reduction to a more general problem and a subsequent polynomial-time data reduction for the general problem that produces a graph with at most $4k^2 + 8sk$ vertices.

μ-Cliques. With this density measure, we capture the ratio of edges in a graph versus the number of edges in a complete graph of the same size. More precisely, the *density* of a graph $G = (V, E)$ is defined as $2|E|/(|V|(|V| - 1))$. A connected graph $G = (V, E)$ is then called a μ-*clique* for a rational constant $0 \leq \mu \leq 1$ if the density of G is at least μ. We assume that μ is represented by two constant integers a, b such that $\mu = a/b$ (note that a and b are not part of the input). Observe that for $\mu = 0$ every graph is a μ-clique, and that a graph is a 1-clique iff it is a clique. The μ-clique concept was studied for example

[1] Note that Yu et al. [18] introduced a different generalization of defective cliques that is more restrictive than the one considered here.

by Abello et al. [1] and is sometimes also referred to as μ-dense graph [13]. We show that—in contrast to s-Defective Clique Editing and Average-s-Plex Editing—μ-Clique Editing is W[1]-hard and thus presumably fixed-parameter intractable with respect to the parameter k for *any* fixed $0 < \mu < 1$. Note that for $\mu = 1$, the problem is equivalent to Cluster Editing and is thus fixed-parameter tractable. For μ-Clique Deletion we show the NP-hardness, the parameterized complexity remains open.

Preliminaries. We only consider *undirected* graphs $G = (V, E)$, where $n := |V|$ and $m := |E|$. The *(open) neighborhood* $N(v)$ of a vertex $v \in V$ is the set of vertices that are adjacent to v in G. The *degree* of a vertex v, denoted by $\deg(v)$, is the cardinality of $N(v)$. For a set U of vertices, $N(U) := \bigcup_{v\in U} N(v) \setminus U$. We use $N[v]$ to denote the *closed* neighborhood of v, that is, $N[v] := N(v) \cup \{v\}$. For a set of vertices $V' \subseteq V$, the *induced subgraph* $G[V']$ is the graph over the vertex set V' with edge set $\{\{v, w\} \in E \mid v, w \in V'\}$. For $V' \subseteq V$ we use $G - V'$ as an abbreviation for $G[V \setminus V']$ and for a vertex $v \in V$ let $G - v$ denote $G - \{v\}$. A vertex $v \in V(G)$ is called a *cut-vertex* if $G - v$ has more connected components than G. For a graph $G = (V, E)$ let $\overline{G} := (V, \overline{E})$ with $\overline{E} := \{\{u, v\} \mid u, v \in V \wedge u \neq v \wedge \{u, v\} \notin E\}$ denote the *complement graph* of G.

A parameterized problem is *fixed-parameter tractable (FPT)* with respect to a parameter k, if there exists an algorithm solving the problem in time $f(k) \cdot n^{O(1)}$, where n denotes the overall input size and f is a computable function. Downey and Fellows [7] developed a formal framework to show *fixed-parameter intractability*; the basic complexity class for fixed-parameter intractability is called $W[1]$ and there is good reason to believe that $W[1]$-hard problems are not FPT [7, 14].

Due to lack of space, some proofs are deferred to the full version.

2 Defective Cliques

First, we focus on the s-Defective Clique Editing problem. A graph is called an *s-defective clique graph* if every connected component forms an s-defective clique. An edge $\{u, v\} \in \overline{E}$ is called a *missing* edge of G. The following theorem can be obtained by reductions from Cluster Deletion and Editing.

Theorem 1. *s-Defective Clique Deletion and Editing are NP-complete.*

If we delete an arbitrary vertex of an s-defective clique graph, then clearly the resulting graph is still an s-defective clique graph. A graph property that is closed under the operation of deleting vertices (and hence, taking induced subgraphs) is called *hereditary*. It is well known that hereditary graph properties can be described by forbidden induced subgraphs [10]. This means that there exists a set $\mathcal{F}$ of graphs such that a given graph G is an s-defective clique graph iff G is $\mathcal{F}$-free, that is, G does not contain any graph from $\mathcal{F}$ as induced subgraph. A forbidden induced subgraph is *minimal* if each of its proper induced subgraphs is an s-defective clique graph. Clearly, a graph is an s-defective clique graph iff it does not contain any minimal forbidden induced subgraph. Next, we show that, for $s \geq 1$, every minimal forbidden subgraph of s-defective clique graphs contains at most $2(s+1)$ vertices. Note that for $s = 0$ the only forbidden induced subgraph is a path on 3 vertices.

Theorem 2. *For $s \geq 1$, every minimal forbidden induced subgraph of s-defective clique graphs contains at most $2(s+1)$ vertices. Given a graph that is not an s-defective clique graph, a minimal forbidden induced subgraph can be found in $O(nm)$ time.*

Proof. Assume towards a contradiction that there exists a minimal forbidden subgraph $G = (V, E)$ with $|V| > 2(s+1)$. Clearly, we can assume that G is connected, since otherwise we can keep one connected component that is not an s-defective clique and delete all other connected components.

First, we consider the case when G contains a cut-vertex v. Let U denote a set of $s+2$ vertices which together with v induce a connected graph $G' := G[U \cup \{v\}]$ and v remains a cut-vertex in G'. We show that G' is not an s-defective clique graph, a contradiction to the fact that G is minimal (note that $s+3 \leq 2s+2$ for $s \geq 1$). Let $U_1, \ldots, U_\ell$ denote the connected components of $G' - v$. It is not hard to see that there are at least $\frac{1}{2}\sum_{i=1}^{\ell} |U_i| \cdot (|U \setminus U_i|) > s$ edges missing in G', and, hence, G' is not an s-defective clique graph.

In the following, we assume that G does not contain any cut-vertex. Moreover, we can assume that no vertex of G is adjacent to all other vertices of G, since otherwise we can delete it to get a connected graph that has the same number of missing edges as G, thus contradicting the minimality of G. Hence, there are more than $s+1$ missing edges in G, since every vertex is incident to at least one missing edge. Let v be an arbitrary vertex of G and let $A := V \setminus N[v]$. Since the deletion of v results in an s-defective clique graph, it follows that in $G - v$ there are at most s missing edges. Hence, there exists a vertex u that is adjacent to all vertices of $G - v$. Clearly, $u \in A$, since, otherwise, u would be adjacent to all vertices in G. Then, the deletion of u reduces the number of missing edges by one. Thus, $G - u$ is connected and has at least $s+1$ missing edges, and, hence, is not an s-defective clique graph, contradicting the fact that G is minimal.

To find a minimal forbidden induced subgraph proceed as follows. Given a graph $G = (V, E)$ that is not an s-defective clique graph we check for every $v \in V$ whether $G - v$ is an s-defective clique graph in $O(n + m)$ time and delete v if not. It is not hard to observe that we have to consider every vertex at most once. Hence, the overall running time is $O(nm)$. □

The forbidden subgraph characterization given in Theorem 2 directly leads to a search tree algorithm for s-Defective Clique Editing [4].

Theorem 3. *s-*Defective Clique Editing *and* Deletion *are fixed-parameter tractable with respect to the parameter (s, k).*

3 Average-s-Plexes

Here, we consider the Average-s-Plex Editing problem, showing its NP-completeness and fixed-parameter tractability with respect to (s, k). To this end, we need the following problem, called Equal-Size Clique Editing: Given an undirected graph $G = (V, E)$ and two integers $k, d \geq 0$, decide whether it is possible to transform G, by adding and deleting at most k edges, into a vertex-disjoint union of d cliques which have the same size. The edge deletion version of this problem allows only edge deletions. The NP-completeness of both problems can be shown by a reduction from the well-known Clique problem. Reducing from Equal-Size Clique Editing and Deletion we can show the following.

Theorem 4. Average-s-Plex Editing *and* Deletion *are NP-complete.*

In the following, we describe a fixed-parameter algorithm for Average-s-Plex Editing parameterized by (s, k). Our algorithm consists of two main steps. First, we reduce the original problem to a weighted version. Then, we show the fixed-parameter tractability of the weighted version by describing two polynomial-time data reduction rules that yield instances which contain at most $4k^2 + 8sk$ vertices. Note that being an average-s-plex graph is not a hereditary graph property. Hence, the fixed-parameter tractability of Average-s-Plex Editing and Average-s-Plex Deletion cannot be shown by a forbidden subgraph characterization as in the case of s-Defective Clique Deletion and s-Defective Clique Editing.

We begin with describing a weighted version of Average-s-Plex Editing. We introduce three types of weights: two vertex weights and one edge weight. The idea behind these weight types is the following: whenever there are two vertices in G that cannot be separated by at most k edge modifications, we can merge them into a new "super-vertex", since it is clear that they end up in the same connected component of the solution. We say that a super-vertex v "comprises" a vertex u of the input graph, if u is merged into v. When doing so, we must remember for each such super-vertex v:

- How many vertices of the input graph v comprises,
- How many edges there are between the vertices that v comprises, and
- For each vertex w outside v, how many vertices that v comprises are adjacent to w.

The first two aspects can be remembered by introducing two weights for v, $\sigma(v)$ which keeps track of the number of vertices comprised by v, and $\delta(v)$ which keeps track of the number of edges between these vertices. The third aspect can be stored as the edge weight $\omega(e)$ for the edge $e = \{w, v\}$. Herein, we call a vertex pair having no edge between them a non-edge. Then, edges have edge weights at least one and non-edges have edge weight zero.

The "size" of a vertex set S is then simply defined as $\sigma(S) := \sum_{v \in S} \sigma(v)$. The average degree $\bar{d}(V_i)$ of a connected component V_i can be computed as follows:

$$\bar{d}(V_i) = \frac{2\sum_{v \in V_i} \delta(v) + \sum_{v \in V_i} \sum_{u \in N(v)} \omega(\{u,v\})}{\sigma(V_i)}.$$

Similar to the definition of average-s-plex graphs, we say that a graph is a weighted average-s-plex graph, if for each connected component V_i, the average degree $\bar{d}(V_i)$ is at least $\sigma(V_i) - s$. For modifying the weighted graph, we allow the following modifications: increasing $\delta(u)$ by one for some $u \in V$, increasing $\omega(\{u,v\})$ by one for some $\{u,v\} \in E$, decreasing $\omega(\{u,v\})$ by one for some $\{u,v\} \in E$, deleting some $\{u,v\} \in E$ with $\omega(\{u,v\}) = 1$, and adding some edge $\{u,v\}$ to E and setting $\omega(\{u,v\}) := 1$. Each of these operations has cost one, and the overall cost of a modification set S is thus exactly $|S|$. The weighted problem version is then defined as

WEIGHTED AVERAGE-s-PLEX EDITING
Input: A graph $G = (V, E)$, with two vertex-weight functions $\sigma : V \to [1, n]$ and $\delta : V \to [0, n^2]$, an edge weight function $\omega : E \to [1, n^2]$, and a nonnegative integer k.
Question: Is there a set of edge modifications S such that applying S to G yields a weighted average-s-plex graph, and such that $|S| \leq k$?

Observe that we can easily reduce an instance $((V, E), k)$ of AVERAGE-s-PLEX EDITING to an instance of WEIGHTED AVERAGE-s-PLEX EDITING, by setting $\sigma(v) := 1$ and $\delta(v) := 0$ for each $v \in V$, and $\omega(\{u,v\}) := 1$, if $\{u,v\} \in E$; otherwise, $\omega(\{u,v\}) := 0$. Note that this reduction is parameter-preserving, that is, s and k are not changed.

In the following, we present two data reduction rules for WEIGHTED AVERAGE-s-PLEX EDITING which (as we show in Theorem 5) yield instances that contain at most $4k^2 + 8sk$ vertices.

Rule 1. *Remove connected components that are weighted average-s-plexesfrom G.*

The rule is obviously correct, since no optimal solution modifies any edges incident to vertices of such a connected component.

The second reduction rule identifies two vertices that have a large common neighborhood, or a heavy edge between them and "merges" these vertices into a new "super-vertex".

Rule 2. *If G contains two vertices u and v such that $\omega(\{u,v\}) > k$ or u and v have more than k common neighbors, then remove u from G and set*
– $\sigma(v) := \sigma(u) + \sigma(v)$,
– $\delta(v) := \delta(u) + \delta(v) + \omega(\{u,v\})$, *and*
– $\omega(\{v,w\}) := \omega(\{v,w\}) + \omega(\{u,w\})$ *for each $w \in V \setminus \{u,v\}$.*

To see the correctness of the rule, consider the following: we cannot separate u and v using at most k edge modifications, they thus end up in the same connected component. Hence, we can remove one of them, and store the information about its adjacency in the vertex weights and edge weights of the other vertex.

With these two reduction rules we can show our main result of this section.

Theorem 5. (WEIGHTED) AVERAGE-s-PLEX EDITING *and* DELETION *are fixed-parameter tractable with respect to the parameter* (s, k).

Proof. We first show that a yes-instance I of WEIGHTED AVERAGE-s-PLEX EDITING that is reduced with Rules 1 and 2 contains at most $4k^2 + 8sk$ vertices. Let I be such a reduced instance, and let G be the input graph of I. Since I is a yes-instance, there is a weighted average-s-plex graph G' that can be obtained from G by applying at most k edge modifications. We now bound the size of G'. Herein, we call a vertex v "affected" if v is an endpoint of a modified edge.

First, since G is reduced with respect to Rule 1, there is at least one affected vertex in each connected component of G'. Hence, there can be at most $2k$ connected components in G'.

Next, we show that each connected component of G' contains at most $2k + 4s$ vertices. Suppose towards a contradiction that there is a connected component V_i of G' such that $|V_i| > 2k + 4s$. Let $u \in V_i$ be a vertex of maximum degree in V_i. Since G' is a weighted average-s-plex graph, the average vertex degree in $G'[V_i]$ is at least $\sigma(V_i) - s$. Since $|V_i| \leq \sigma(V_i)$, u must be adjacent to at least $|V_i| - s \geq 2k + 3s$ vertices in $G'[V_i]$. We consider two cases for $\sigma(v)$.

Case 1: $\sigma(u) \geq \sigma(V_i)/2$. We show that the average degree of $G'[V_i]$ is less than $\sigma(V_i) - s$, contradicting the assumption that G' is a weighted average-s-plex graph. Since G is reduced with respect to Rule 2, there is no edge that has weight at least $k+1$ in G. In G', the edge weights of edges incident to u have increased by at most k overall. Without loss of generality we can assume that the solution distributes these edge weight increases evenly, since for the average degree of a connected component only the *overall* edge and vertex weights are relevant. The maximum edge weight of edges incident to u in G' is thus at most $k + 2$, since u has $2k + 3s$ neighbors in G' and spreading the edge additions equally leads to a weight increase of at most one for each edge. However, with $\sigma(u) \geq \sigma(V_i \setminus \{u\}) \geq 2k + 3s$, this means that for each edge incident to u, the edge weight is at most $\sigma(u)/2$. This leads to a low average degree. More precisely, we can bound the average degree of V_i as

$$\bar{d}(V_i) \overset{(*)}{<} \frac{2\binom{\sigma(V_i)}{2} - (\sigma(u)/2) \cdot \sigma(V_i \setminus \{u\})}{\sigma(V_i)} \overset{(**)}{<} \frac{2\binom{\sigma(V_i)}{2} - (\sigma(u)/2) \cdot 4s}{\sigma(V_i)}$$
$$\overset{(***)}{\leq} \frac{2\binom{\sigma(V_i)}{2} - (\sigma(V_i)/4) \cdot 4s}{\sigma(V_i)} < \sigma(V_i) - s.$$

Inequality (*) follows from assuming the (maximum-density) case that with exception of the edges incident to u all edges are present and have maximum possible weight, and the fact that the maximum edge weight of edges incident to u is $k+1$, which means that for each such edge, a weight of at least $\sigma(u) - (k+1) > \sigma(u)/2$ is "missing". Inequality (**) follows from $\sigma(V_i \setminus \{u\}) \geq |V_i| - 1 > 4s$ and inequality (***) follows from $\sigma(u) \geq \sigma(V_i)/2$.

Case 2: $\sigma(u) < \sigma(V_i)/2$. First, we show that there must be at least one other vertex $w \in V_i$ that has at least $|V_i| - 2s$ neighbors in $G'[V_i]$. Suppose otherwise. Then it holds for the average degree of $G'[V_i]$ that

$$\bar{d} \overset{(*)}{\leq} \frac{\sigma(u) \cdot (\sigma(V_i) - 1) + (\sigma(V_i) - \sigma(u)) \cdot (\sigma(V_i) - 2s - 1)}{\sigma(V_i)}$$
$$= \frac{\sigma(V_i) \cdot (\sigma(V_i) - 1) - 2s \cdot (\sigma(V_i) - \sigma(u))}{\sigma(V_i)} \overset{(**)}{<} \frac{\sigma(V_i) \cdot (\sigma(V_i) - s)}{\sigma(V_i)}.$$

Inequality (*) follows from assuming the (maximum-density) case that u is adjacent to all vertices in V_i and that all edge and vertex weights have maximum possible values. Inequality (**) follows from $\sigma(V_i) - \sigma(u) > \sigma(V_i)/2$. We have thus shown that there is at least one vertex w that is adjacent to at least $|V_i| - 2s$ vertices in G'. Since u has at least $|V_i| - s$ neighbors in $G'[V_i]$, there must be at least $|V_i| - 3s > 2k + 4s - 3s = 2k + s$ vertices in $G'[V_i]$ that are common neighbors of u and w. Clearly, more than $k + s$ of those vertices are common neighbors of u and w in G. This contradicts that G is reduced with respect to Rule 2.

We have thus shown that a reduced yes-instance contains at most $4k^2 + 8sk$ vertices. We obtain a fixed-parameter algorithm for Weighted Average-s-Plex Editing as follows. First we exhaustively apply the reduction rules, which can clearly be done in polynomial time. If the reduced instance contains more than $4k^2 + 8sk$ vertices, then it is a no-instance. Otherwise, we can solve the problem with running time only depending on s and k, for example by brute-force generation of all possible partitions of the graph. The fixed-parameter tractability of Average-s-Plex Editing, then directly follows from the described reduction to Weighted Average-s-Plex Editing. □

4 μ-Cliques

The main result of this section is that, in contrast to s-Defective Clique Editing and Average-s-Plex Editing, μ-Clique Editing is fixed-parameter intractable with respect to the number k of allowed edge modifications.

Theorem 6. *For any fixed* $0 < \mu < 1$, μ-Clique Editing *is NP-complete and W[1]-hard with respect to the number* k *of allowed edge modifications.*

The reduction used in the proof of Theorem 6 does not work for the edge deletion case. However, we can establish the NP-hardness of μ-Clique Deletion by a reduction from the NP-complete Maximum Triangle Packing problem [9].

Theorem 7. *For any fixed* $0 < \mu < 1$, μ-Clique Deletion *is NP-complete.*

5 Outlook

There are numerous topics for future research, we only point out some of them. For s-Defective Clique Editing and Average-s-Plex Editing clearly further algorithmic improvements are necessary. For instance, s-Defective Clique Editing is still missing a non-trivial kernelization algorithm, whereas for Average-s-Plex Editing a solution algorithm other than the brute-force one that can be applied to the reduced instance is needed. As a next step, experimental studies should then be undertaken regarding the running time of the algorithms and the quality of the produced clusterings. For μ-Clique Editing, other parameterizations should be studied. Finally, the parameterized complexity of μ-Clique Deletion remains open.

References

[1] Abello, J., Resende, M.G.C., Sudarsky, S.: Massive quasi-clique detection. In: Rajsbaum, S. (ed.) LATIN 2002. LNCS, vol. 2286, pp. 598–612. Springer, Heidelberg (2002)

[2] Bansal, N., Blum, A., Chawla, S.: Correlation clustering. Machine Learning 56(1-3), 89–113 (2004)

[3] Böcker, S., Briesemeister, S., Bui, Q.B.A., Truß, A.: Going weighted: Parameterized algorithms for cluster editing. Theor. Comput. Sci. (to appear, 2009)

[4] Cai, L.: Fixed-parameter tractability of graph modification problems for hereditary properties. Inf. Process. Lett. 58(4), 171–176 (1996)

[5] Chesler, E.J., Lu, L., Shou, S., Qu, Y., Gu, J., Wang, J., Hsu, H.C., Mountz, J.D., Baldwin, N.E., Langston, M.A., Threadgill, D.W., Manly, K.F., Williams, R.W.: Complex trait analysis of gene expression uncovers polygenic and pleiotropic networks that modulate nervous system function. Nature Genetics 37(3), 233–242 (2005)

[6] Dehne, F.K.H.A., Langston, M.A., Luo, X., Pitre, S., Shaw, P., Zhang, Y.: The cluster editing problem: Implementations and experiments. In: Bodlaender, H.L., Langston, M.A. (eds.) IWPEC 2006. LNCS, vol. 4169, pp. 13–24. Springer, Heidelberg (2006)

[7] Downey, R.G., Fellows, M.R.: Parameterized Complexity. Springer, Heidelberg (1999)

[8] Fellows, M.R., Langston, M.A., Rosamond, F.A., Shaw, P.: Efficient parameterized preprocessing for Cluster Editing. In: Csuhaj-Varjú, E., Ésik, Z. (eds.) FCT 2007. LNCS, vol. 4639, pp. 312–321. Springer, Heidelberg (2007)

[9] Garey, M.R., Johnson, D.S.: Computers and Intractability: A Guide to the Theory of NP-Completeness. W.H. Freeman, New York (1979)

[10] Greenwell, D.L., Hemminger, R.L., Klerlein, J.B.: Forbidden subgraphs. In: Proceedings of the 4th Southeastern Conference on Combinatorics, Graph Theory and Computing, pp. 389–394 (1973)

[11] Guo, J.: A more effective linear kernelization for Cluster Editing. Theor. Comput. Sci. 410(8-10), 718–726 (2009)

[12] Guo, J., Komusiewicz, C., Niedermeier, R., Uhlmann, J.: A more relaxed model for graph-based data clustering: s-plex editing. In: Goldberg, A., Zhou, Y. (eds.) AAIM 2009. LNCS, vol. 5564, pp. 226–239. Springer, Heidelberg (2009)

[13] Kosub, S.: Local density. In: Brandes, U., Erlebach, T. (eds.) Network Analysis. LNCS, vol. 3418, pp. 112–142. Springer, Heidelberg (2005)

[14] Niedermeier, R.: Invitation to Fixed-Parameter Algorithms. Oxford University Press, Oxford (2006)

[15] Seidman, S.B., Foster, B.L.: A graph-theoretic generalization of the clique concept. Journal of Mathematical Sociology 6, 139–154 (1978)
[16] Shamir, R., Sharan, R., Tsur, D.: Cluster graph modification problems. Discrete Appl. Math. 144(1-2), 173–182 (2004)
[17] van Zuylen, A., Williamson, D.P.: Deterministic algorithms for rank aggregation and other ranking and clustering problems. In: Kaklamanis, C., Skutella, M. (eds.) WAOA 2007. LNCS, vol. 4927, pp. 260–273. Springer, Heidelberg (2008)
[18] Yu, H., Paccanaro, A., Trifonov, V., Gerstein, M.: Predicting interactions in protein networks by completing defective cliques. Bioinformatics 22(7), 823–829 (2006)

A Certifying Algorithm for 3-Colorability of P_5-Free Graphs

Daniel Bruce[1], Chính T. Hoàng[2], and Joe Sawada[3]

[1] Computing and Information Science, University of Guelph, Canada
dbruce01@uoguelph.ca
[2] Physics and Computer Science, Wilfred Laurier University, Canada
Research supported by NSERC
choang@wlu.ca
[3] Computing and Information Science, University of Guelph, Canada
Research supported by NSERC
jsawada@uoguelph.ca

Abstract. We provide a certifying algorithm for the problem of deciding whether a P_5-free graph is 3-colorable by showing there are exactly six finite graphs that are P_5-free and not 3-colorable and minimal with respect to this property.

1 Introduction

An algorithm is *certifying* if it returns with each output a simple and easily verifiable certificate that the particular output is correct. For example, a certifying algorithm for the bipartite graph recognition would return either a 2-coloring of the input graph proving that it is bipartite, or an odd cycle proving it is not bipartite. A certifying algorithm for planarity would return a planar embedding or one of the two Kuratowski subgraphs. The notion of certifying algorithm [8] was developed when researchers noticed that a well known planarity testing program was incorrectly implemented. A certifying algorithm is a desirable tool to guard against incorrect implementation of a particular algorithm. In this paper, we give a certifying algorithm for the problem of deciding whether a P_5-free graph is 3-colorable. We will now discuss the background of this problem.

A class C of graphs is called *hereditary* if for each graph G in C, all induced subgraphs of G are also in $\mathcal{C}$. Every hereditary class of graphs can be described by its *forbidden induced subgraphs*, i.e. the unique set of minimal graphs which do not belong to the class. A comprehensive survey on coloring of graphs in hereditary classes can be found in [11]. An important line of research on colorability of graphs in hereditary classes deals with P_t-free graphs. The induced path on t vertices is called P_t, and a graph is called *P_t-free* if it does not contain P_t as an induced subgraph.

It is known [13] that 5-COLORABILITY is NP-complete for P_8-free graphs and [9] 4-COLORABILITY is NP-complete for P_9-free graphs. On the other hand, the k-COLORABILITY problem can be solved in polynomial time for P_4-free graphs (since they are perfect). In [4] and [5], it is shown that k-COLORABILITY can be solved for the class of P_5-free graphs in polynomial time for every particular value of k. For $t = 6, 7$, the complexity of the problem is generally unknown, except for the case

Y. Dong, D.-Z. Du, and O. Ibarra (Eds.): ISAAC 2009, LNCS 5878, pp. 594–604, 2009.

Table 1. Known complexities for k-colorability of P_t-free graphs

$k\backslash t$	3	4	5	6	7	8	9	10	11	12	...
3	$O(m)$	$O(m)$	$O(n^\alpha)$	$O(mn^\alpha)$	?	?	?	?	?	?	...
4	$O(m)$	$O(m)$	P	?	?	?	NP_c	NP_c	NP_c	NP_c	...
5	$O(m)$	$O(m)$	P	?	?	NP_c	NP_c	NP_c	NP_c	NP_c	...
6	$O(m)$	$O(m)$	P	?	?	NP_c	NP_c	NP_c	NP_c	NP_c	...
7	$O(m)$	$O(m)$	P	?	?	NP_c	NP_c	NP_c	NP_c	NP_c	...
...	...	...	...	...	...	...	...	...	...	...	...

of 3-COLORABILITY of P_6-free graphs [12]. Known results on the k-COLORABILITY problem in P_t-free graphs are summarized in Table 1 (n is the number of vertices in the input graph, m the number of edges, and α is matrix multiplication exponent known to satisfy $2 \leq \alpha < 2.376$ [2]).

In this paper, we study the coloring problem for the class of P_5-free graphs. This class has proved resistant with respect to other graph problems. For instance, P_5-free graphs is the unique minimal class defined by a single forbidden induced subgraph with unknown complexity of the MAXIMUM INDEPENDENT SET and MINIMUM INDEPENDENT DOMINATING SET problems. Many algorithmic problems are known to be NP-hard in the class of P_5-free graphs, for example DOMINATING SET [6] and CHROMATIC NUMBER [7]. In contrast to the NP-hardness of finding the chromatic number of a P_5-free graph, it is known [4] that k-COLORABILITY can be solved in this class in polynomial time for every particular value of k. This algorithm produces a k-coloring if one exists, but does not produce an easily verifiable certificate when such coloring does not exist. We are interested in finding a certificate for non-k-colorability of P_5-free graphs. For this purpose, we start with $k = 3$.

Besides [4], there are several polynomial-time algorithms for 3-coloring a P_5-free graph ([5,10,13]) but none of them is a certifying algorithm. In this paper, we obtain a certifying algorithm for 3-coloring a P_5-free graphs by proving there are a finite number of minimally non-3-colorable P_5-free graphs and each of these graphs is finite.

Theorem 1.1. *A P_5-free graph is 3-colorable if and only if it does not contain any of the six graphs in Fig. 1 as a subgraph.*

It is an easy matter to verify the graphs in Fig. 1 are not 3-colorable, the rest of the paper involves proving the other direction of the theorem. In the last Section, we will discuss open problems arising from our work.

2 Definition and Background

Let k and t be positive integers. An MNkPt is a graph G that (i) is not k-colorable and is P_t-free and (ii) every proper subgraph of G is either k-colorable or has a P_t. We will be interested specifically in the case where $k = 3$ and $t = 5$. We will use the following notations. Let G be a simple undirected graph. A set S of vertices of G is *dominating* if every vertex in $G - S$ has a neighbor in S. A k-clique is a clique on k vertices. $u \sim v$

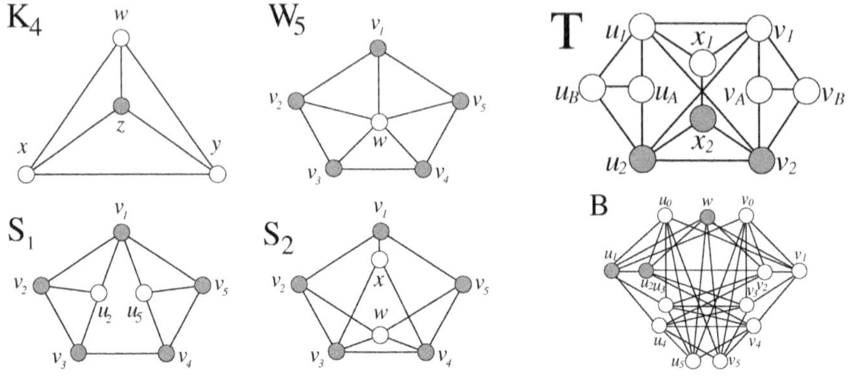

Fig. 1. All 6 MN3P5s

will mean vertex u is adjacent to vertex v. $u \nsim v$ will mean vertex u is not adjacent to vertex v. For any vertex v, $N(v)$ denotes the set of vertices that are adjacent to v. We write $G \cong H$ to mean G is isomorphic to H. The *clique number* of G, denoted by $\omega(G)$, is the number of vertices in a largest clique of G. The *chromatic number* of G, denoted by $\chi(G)$, is the smallest number of colors needed to color the vertices of G. A *hole* is an induced cycle with at least four vertices, and it is odd (or even) if it has odd (or even) length. An *anti-hole* is the complement of a hole. A k-hole (k-anti-hole) is a hole (anti-hole) on k vertices. A graph G is *perfect* if each induced subgraph H of G has $\chi(G) = \omega(G)$.

Theorem 2.1 (The Strong Perfect Graph Theorem [3]). *A graph is perfect if and only if it does not contain an odd hole or odd anti-hole as an induced subgraph.*

Let $\mathfrak{G} = \{K_4, W_5, S_1, S_2, T, B\}$ be the set of graphs in Fig. 1. We will denote these graphs in the following way.

- $P_5(v_1v_2v_3v_4v_5)$ means there is a P_5 being v_1, v_2, v_3, v_4 and v_5.
- $K_4(wxyz)$ means $\{w, x, y, z\}$ form a K_4.
- $W_5(v_1v_2v_3v_4v_5, w)$ means v_1, v_2, v_3, v_4, v_5 and w form a W_5 where $v_1v_2v_3v_4v_5$ form a 5-cycle and w is adjacent to every other vertex.
- $S_1(v_1v_2v_3v_4v_5, u_2, u_5)$ means $v_1, v_2, v_3, v_4, v_5, u_2, u_5$ form an S_1 where v_1 is the only degree 4 vertex and $N(v_1) = \{u_5, u_2, v_5, v_2\}$. Also $N(v_3) = \{v_4, v_2, u_2\}$ and $N(v_4) = \{v_3, v_5, u_5\}$, and $v_1v_2v_3v_4v_5$ form a 5-cycle.
- $S_2(v_1v_2v_3v_4v_5, w, x)$ means $v_1, v_2, v_3, v_4, v_5, w$ and x form an S_2 where $N(w) = \{v_2, v_3, v_4, v_5\}$, $N(x) = \{v_1, v_3, v_4\}$ and $v_1v_2v_3v_4v_5$ form a 5-cycle.
- $T(u_1u_Au_Bu_2, v_1v_Av_Bv_2, x_1, x_2)$ means a T graph is present as shown previously.
- $B(w, u_0u_1u_2u_3u_4u_5, v_0v_1v_2v_3v_4v_5)$ means a B graph is present as shown previously.

We will rely on the following result.

Theorem 2.2 ([1]). *Every connected P_5-free graph has dominating clique or P_3.*

The following lemma is folklore.

Lemma 2.1 (The neighborhood lemma). *Let G be a minimally non k-colorable graph. If u and v are two non-adjacent vertices in G, then $N(u) \nsubseteq N(v)$.*

Proof. Assume $N(u) \subseteq N(v)$. Then the graph $G - v$ admits a k-coloring. By giving u the color of v, we see that G is k-colorable, a contradiction. □

The neighborhood lemma is used predominantly throughout this paper. Writing $\mathbf{N(v,w) \rightarrow u}$ will denote the fact that $N(v) \nsubseteq N(w)$ by the neighborhood lemma so there exists a vertex u where $u \sim v$, but $u \nsim w$.

The following fact is well-known and easy to establish.

Fact 2.1. *In a minimally non k-colorable graph every vertex has degree at least k.* □

3 Intermediate Results

In this section, we establish a number of intermediate results needed for proving the main theorem.

Lemma 3.1. *Let G be an MN3P5 graph with a 5-hole $C = \{v_1, v_2, v_3, v_4, v_5\}$ and a vertex w adjacent to at least 4 vertices of C. Then $G \in \mathfrak{G}$.*

Proof. If w is adjacent to all five vertices of C, then G clearly is isomorphic to W_5. Now, assume $N(w) \cap \{v_1, v_2, v_3, v_4, v_5\} = \{v_2, v_3, v_4, v_5\}$.

We have $\mathbf{N(v_1, w) \rightarrow x}$.

Assume for the moment that $x \nsim v_3, v_4$. We have

$x \sim v_5$, otherwise, we have $P_5(xv_1v_5v_4v_3)$.

$x \sim v_2$, otherwise, we have $P_5(xv_1v_2v_3v_4)$.

But then G contains $S_1(v_1v_2v_3v_4v_5, x, w)$. This means $x \sim v_3$ or $x \sim v_4$. By symmetry, we may assume $x \sim v_3$. We have $x \sim v_2$ or $x \sim v_4$, otherwise, G contains $P_5(xv_1v_2wv_4)$. If $x \sim v_2$ then G properly contains $S_1(v_1v_2v_3v_4v_5, x, w)$, a contradiction. This means $x \sim v_4$; so G contains $S_2(v_1v_2v_3v_4v_5, w, x)$ and $G \cong S_2$. □

Theorem 3.1. *Every MN3P5 graph different from K_4 contains a 5-hole.*

Proof. Let G be an MN3P5 graph different from a K_4. We have $\omega(G) \leq 3$ and $\chi(G) \geq 4$. Thus, G is not perfect. By Theorem 2.1, G contains an odd hole or an odd anti-hole H. H cannot be a hole of size 7 or greater because G is P_5-free. We may assume H is an anti-hole of length at least seven, for otherwise we are done (observe that the hole on five vertices is self-complementary). Let v_1, v_2, v_3, v_4, v_5, v_6, v_7 be the cyclic order of the hole in the complement of G. Then G properly contains $S_1(v_4v_6v_3v_5v_2, v_1, v_7)$, a contradiction. □

Lemma 3.2. *Let G be an MN3P5 graph that has a dominating clique $\{a, b, c\}$. Also assume that there is a vertex $v \notin \{a, b, c\}$ adjacent to two vertices from $\{a, b, c\}$. Then $G \in \mathfrak{G}$.*

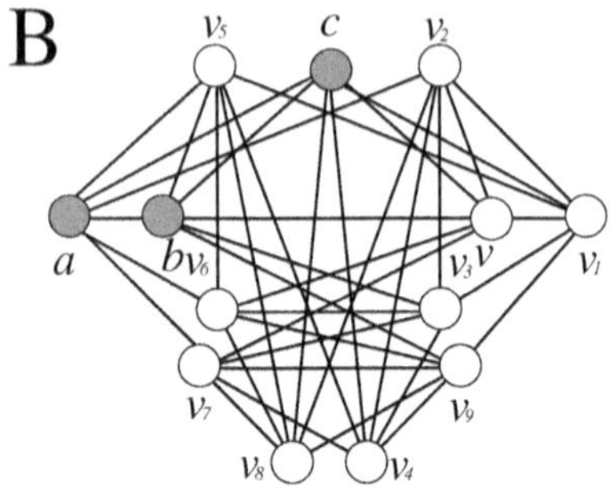

Fig. 2. The graph B obtained in the proof of Lemma 3.2

Proof. The proof is by contradiction. Suppose that $G \notin \mathfrak{G}$. We may assume v is adjacent to b and c. We have $v \nsim a$, otherwise, G contains $K_4(abcv)$. Through repeated applications of the Neighborhood Lemma, we will eventually add nine vertices to G to arrive at a contradiction. In the end, we will obtain the graph B (see Fig. 2 for the order in which vertices are added). Each time we add a vertex we will consider its adjacency to the other vertices of the graph. In every case, the adjacency can be completely determined at each step.

$\mathbf{N(v, a) \rightarrow v_1}$.

- $v_1 \sim c$: since $\{a, b, c\}$ is dominating, v_1 is adjacent to either b or c. Without loss of generality, assume $v_1 \sim c$.
- $v_1 \nsim b$: otherwise, G contains $K_4(bcvv_1)$.

$\mathbf{N(v_1, b) \rightarrow v_2}$.

- $v_2 \sim a$: assume $v_2 \nsim a$. We have $v_2 \sim v$, otherwise, G contains $P_5(v_2v_1vba)$. Also, $v_2 \sim c$ since $\{a, b, c\}$ is a dominating set. But then, G contains $K_4(v_1v_2vc)$.
- $v_2 \nsim c$: otherwise, G contains $W_5(abvv_1v_2, c)$.
- $v_2 \sim v$: otherwise, c has four neighbors in the 5-hole v_2abvv_1 contradicting Lemma 3.1.

$\mathbf{N(v_2, c) \rightarrow v_3}$.

- $v_3 \sim b$: assume $v_3 \nsim b$. We have $v_3 \sim a$ since $\{a, b, c\}$ is a dominating set. We have $v_3 \nsim v_1$, otherwise, G contains $S_1(vbav_3v_2, c, v_1)$. But then G contains $P_5(v_3v_2v_1cb)$.
- $v_3 \nsim v$; otherwise, G contains $W_5(bcv_1v_2v_3, v)$.
- $v_3 \sim v_1$: otherwise, v has four neighbors in the 5-hole $v_3bcv_1v_2$ contradicting Lemma 3.1.
- $v_3 \nsim a$: otherwise G contains $S_1(v_3acvv_1, b, v_2)$.

$\mathbf{N(v_3, v) \rightarrow v_4}$.

- $v_4 \sim c$: assume $v_4 \nsim c$. Then we have $v_4 \sim v_2$, for otherwise G contains $P_5(v_4v_3v_2vc)$; $v_4 \nsim v_1$, for otherwise G contains $K_4(v_1v_2v_3v_4)$; $v_4 \nsim b$, for otherwise G contains $S_1(v_1v_2v_4bc, v_3, v)$; $v_4 \sim a$ because $\{a, b, c\}$ is dominating. But then G contains $P_5(v_4abvv_1)$.

- $v_4 \nsim v_1$: for otherwise G contains $W_5(v_4v_3v_2vc, v_1)$.
- $v_4 \nsim b$: for otherwise, G contains $S_1(vv_1v_3v_4b, v_2, c)$.
- $v_4 \nsim a$: for otherwise, G contains $P_5(v_4abvv_1)$.
- $v_4 \sim v_2$: for otherwise the vertex v_1 has exactly four neighbors in the 5-hole $v_4v_3v_2vc$ contradicting Lemma 3.1.

$\mathbf{N(a, v) \rightarrow v_5}$.

- $v_5 \nsim v_3$: Assume $v_5 \sim v_3$. Then we have $v_5 \sim v_1$, for otherwise G contains $P_5(av_5v_3v_1v)$; $v_5 \nsim v_2$, for otherwise G contains $K_4(v_1v_2v_3v_5)$; $v_5 \sim c$, for otherwise G contains $P_5(v_5v_3v_2vc)$. But now G contains $W_5(v_5cvv_2v_3, v_1)$.
- $v_5 \sim b$: assume $v_5 \nsim b$. Then we have $v_5 \sim v_1$, for otherwise G contains $P_5(v_5abvv_1)$. But then c has four neighbors in the 5-hole v_5abvv_1 contradicting Lemma 3.1.
- $v_5 \nsim c$: for otherwise G contains $K_4(abcv_5)$.
- $v_5 \sim v_1$: for otherwise G contains $P_5(v_5acv_1v_3)$.
- $v_5 \sim v_4$: for otherwise G contains $P_5(v_3v_4cav_5)$.
- $v_5 \nsim v_2$: for otherwise G contains $S_1(cvv_2v_5a, v_1, b)$.

$\mathbf{N(v_5, c) \rightarrow v_6}$.

- $v_6 \sim v$: assume $v_6 \nsim v$. We have $v_6 \sim a$, for otherwise G contains $P_5(v_6v_5acv)$; $v_6 \nsim b$, for otherwise G contains $K_4(abv_5v_6)$; $v_6 \sim v_1$, for otherwise, G contains $P_5(v_6abvv_1)$. But c has four neighbors in the 5-hole v_6abvv_1 contradicting Lemma 3.1.
- $v_6 \nsim b$: for otherwise G contains $W_5(v_5v_6vca, b)$.
- $v_6 \nsim v_2$: for otherwise G contains $P_5(v_2v_6v_5bc)$.
- $v_6 \sim v_3$: for otherwise G contains $P_5(v_5v_6vv_2v_3)$
- $v_6 \sim a$: for otherwise G contains $P_5(v_3v_6v_5ac)$.
- $v_6 \nsim v_1$: for otherwise G contains $S_1(v_6abcv, v_5, v_1)$.
- $v_6 \nsim v_4$: for otherwise G contains $T(v_6av_5b, v_3v_2v_1v, v_4, c)$.

$\mathbf{N(v_4, v_1) \rightarrow v_7}$.

- $v_7 \sim v$: assume $v_7 \nsim v$. Then we have $v_7 \sim v_3$, for otherwise G contains $P_5(v_7v_4v_3v_1v)$; $v_7 \nsim v_2$, for otherwise G contains $K_4(v_2v_3v_4v_7)$; $v_7 \sim c$, for otherwise G contains $P_5(v_7v_3v_2vc)$. Now, G contains $S_1(v_2vcv_7v_3, v_1, v_4)$.
- $v_7 \nsim v_2$: for otherwise G contains $W_5(vv_1v_3v_4v_7, v_2)$.
- $v_7 \nsim v_6$: for otherwise G contains $P_5(v_6v_7v_4v_2v_1)$.
- $v_7 \sim a$: for otherwise G contains $P_5(v_4v_7vv_6a)$.
- $v_7 \sim v_3$: for otherwise G contains $P_5(av_7vv_1v_3)$.
- $v_7 \nsim c$: for otherwise G contains $S_1(v_3v_4cvv_1, v_7, v_2)$.
- $v_7 \nsim b$: for otherwise G contains $P_5(v_7bcv_1v_2)$.
- $v_7 \nsim v_5$: for otherwise G contains $T(av_7v_5v_4, cvv_1v_2, b, v_3)$.

$\mathbf{N(v_6, b) \rightarrow v_8}$.

- $v_8 \sim c$: assume $v_8 \nsim c$. Then we have $v_8 \sim a$ because $\{a, b, c\}$ is a dominating set; $v_8 \sim v_5$, for otherwise G contains $P_5(v_8v_6v_5bc)$. But now, G contains $K_4(av_5v_6v_8)$.
- $v_8 \nsim a$: for otherwise G contains $W_5(v_8v_6v_5bc, a)$.
- $v_8 \nsim v_1$: for otherwise G contains $P_5(bav_6v_8v_1)$.
- $v_8 \sim v_2$: for otherwise G contains $P_5(v_2v_1cv_8v_6)$.

- $v_8 \sim v_5$: for otherwise G contains $P_5(v_8v_2v_1v_5b)$.
- $v_8 \nsim v_4$: for otherwise G contains $P_5(v_4v_8v_6ab)$.
- $v_8 \nsim v$: for otherwise G contains $S_1(bcv_8v_6a, v, v_5)$.
- $v_8 \nsim v_3$: for otherwise G contains $T(v_6av_5b, v_3v_2v_1v, v_8, c)$.
- $v_8 \sim v_7$: for otherwise G contains $P_5(v_8v_5bv_3v_7)$.

$\mathbf{N(v_8, a) \rightarrow v_9}$.

- $v_9 \sim b$: assume $v_9 \nsim b$. We have $v_9 \sim v_2$, for otherwise G contains $P_5(v_9v_8v_2ab)$; $v_9 \sim v_6$, for otherwise G contains $P_5(v_9v_8v_6ab)$. This means G contains $T(v_6v_8v_9v_2, abcv, v_5, v_1)$.
- $v_9 \sim v_1$: assume $v_9 \nsim v_1$. We have $v_9 \sim v_2$, for otherwise G contains $P_5(v_9bav_2v_1)$. This means G contains $T(v_2vv_1c, v_8v_6v_5a, v_9, b)$.
- $v_9 \sim v_6$: for otherwise G contains $P_5(v_1v_9bav_6)$.
- $v_9 \sim v_7$: for otherwise G contains $P_5(v_1v_9bav_7)$.
- $v_9 \sim v_4$: assume $v_9 \nsim v_4$. Then we have $v_9 \sim v_2$, for otherwise G contains $P_5(v_9bav_2v_4)$. This means G contains $T(v_6av_5b, v_9v_2v_1v, v_8, c)$.

But this means G contains $B(c, v_5abv_6v_7v_8, v_2v_1vv_3v_9v_4)$, a contradiction. □

Lemma 3.3. *Let G be an MN3P5 with a dominating clique $\{a, b, c\}$. Let $A = N(a) - \{b, c\}$, $B = N(b) - \{a, c\}$ and $C = N(c) - \{a, b\}$. Suppose A, B and C are pairwise disjoint. Then $G \in \mathfrak{G}$.*

Proof. Some observations are necessary for this proof.

Observation 3.1. *Let X and Y be two distinct elements of $\{A, B, C\}$. Let X' be a component in X with at least two vertices, and y be a vertex in Y. Then either y is adjacent to all vertices of X' or to no vertex of X'.*

Proof. Suppose the Observation is false. Then there are adjacent vertices $v_1, v_2 \in X$ such that y is adjacent to exactly one of v_1, v_2. Without loss of generality, we may assume $X = A$ and $Y = B$. Now, $\{c, b, y, v_2, v_1\}$ induces a P_5, a contradiction. □

Observation 3.2. *Every component in A, B or C is a single edge or one vertex.*

Proof. Assume that one of A, B or C contains a vertex of degree 2. Without loss of generality, assume there is such a vertex $a_0 \in A$ that is adjacent to two other distinct vertices a_1 and a_x in A. Now we have $a_1 \nsim a_x$, for otherwise G contains $K_4(a_1a_xa_0a)$. The Neighborhood Lemma implies $\mathbf{N(a_1, a_x) \rightarrow a_2}$ and $\mathbf{N(a_x, a_1) \rightarrow a_y}$. Observation 3.1 implies $a_2, a_y \in A$. We have $a_y \nsim a_0$, for otherwise G contains $K_4(aa_0a_xa_y)$; $a_2 \nsim a_0$, for otherwise G contains $K_4(aa_0a_1a_2)$; $a_y \sim a_2$, for otherwise G contains $P_5(a_ya_xa_0a_1a_2)$. Then G contains $W_5(a_ya_xa_0a_1a_2, a)$, a contradiction. □

We continue the proof of the Lemma. Assume $G \notin \mathfrak{G}$. Consider the case that two of A, B or C contain an edge. Without loss of generality, assume A contains an edge a_1a_2 and B contains an edge b_1b_2. If a vertex in $\{b_1, b_2\}$ is adjacent to a vertex in $\{a_1, a_2\}$ then by Observation 3.1, G contains $K_4(a_1a_2b_1b_2)$, a contradiction. Suppose some vertex $c_0 \in C$ is adjacent to a vertex in $\{a_1, a_2, b_1, b_2\}$. We may assume $c_0 \sim a_1$. By Observation 3.1, we have $c_0 \sim a_2$. If $c_0 \nsim b_i$ $(i = 1, 2)$ then G contains $P_5(b_ibcc_0a_1)$. So, c_0

is adjacent to all vertices of $\{a_1, a_2, b_1, b_2\}$. But now, G contains $S_1(c_0a_1abb_1, a_2, b_2)$. So, no vertex in C is adjacent to a vertex in $\{a_1, a_2, b_1, b_2\}$. By Fact 2.1 and Observation 3.2, there exists a vertex $a_3 \in A$ with $b_1, b_2 \sim a_3$ and a vertex $b_3 \in B$ with $a_1, a_2 \sim b_3$. Also by Fact 2.1, C contains a vertex c_0. We have $a_3 \sim c_0$, for otherwise G contains $P_5(a_3b_1bcc_0)$; $b_3 \sim c_0$, for otherwise G contains $P_5(b_3a_1acc_0)$; $a_3 \sim b_3$, for otherwise G contains $P_5(b_1a_3c_0b_3a_1)$. But now G contains $T(aa_1a_2b_3, bb_1b_2a_3, c, c_0)$ which is a contradiction. So, at most one of A, B, C contains an edge.

If all of A, B, C is a stable set, then G is obviously 3-colorable. We may assume B, C are stable sets, and A contains an edge. Now there must be one vertex $b_0 \in B$ with $N(b_0)$ contains two adjacent vertices in A. Otherwise, G admits a 3-coloring f as follows. The vertices of C are colored with color 3. Now, for each edge in A, its endpoints are arbitrarily colored with colors 1, 2. The remaining vertices of A are colored with color 1. The vertices of B are colored with color 2 (no vertex of B is adjacent to an endpoint of a edge of A by Observation 3.1), and let $f(a) = 3, f(b) = 1, f(c) = 2$. Thus, f is a 3-coloring which is a contradiction. Therefore, there is a vertex $b_1 \in B$ adjacent to both endpoints in some edge $a_{b1}a_{b2}$ in A. By a similar argument, there is a vertex $c_1 \in C$ adjacent to both endpoints in some edge $a_{c1}a_{c2}$.

Suppose that $a_{b1}a_{b2}$ and $a_{c1}a_{c2}$ are the same edge. For simplicity, write $a_1a_2 = a_{b1}a_{b2} = a_{c1}a_{c2}$. We have $b_1 \nsim c_1$, for otherwise G contains $K_4(a_1a_2b_1c_1)$.

- $\mathbf{N(b_1, a) \to c_2}$. We have $c_2 \in C$ by the fact that B is an independent set.
- $\mathbf{N(c_1, a) \to b_2}$. We have $b_2 \in B$ by the fact that C is an independent set.
- $b_2, c_2 \nsim a_1, a_2$. Otherwise, suppose $b_2 \sim a_1$. Then by Observation 3.1, we have $b_2 \sim a_2$ so G contains $K_4(a_1a_2b_2c_1)$.
- $b_2 \sim c_2$. Otherwise, G contains $P_5(c_1b_2bb_1c_2)$.

Now, G contains $P_5(b_2c_2caa_1)$. Thus, $a_{b1}a_{b2}$ and $a_{c1}a_{c2}$ are distinct edges. We have $b_1 \nsim a_{c1}, a_{c2}$ and $c_1 \nsim a_{b1}, a_{b2}$, for otherwise we are done by the previous case. We have $b_1 \sim c_1$, for otherwise G contains $P_5(b_1a_{b1}aa_{c1}c_1)$. But now G contains $S_1(ab_1a_{b1}b_1c_1a_{c1}, a_{b2}, a_{c2})$, a contradiction. □

Lemma 3.4. *Let G be an MN3P5 with a dominating clique $\{a, b, c\}$. Then $G \in \mathfrak{G}$.*

Proof. If there is a vertex other than a, b and c adjacent to at least two of a, b or c then by Lemma 3.2, $G \in \mathfrak{G}$. Otherwise, the conclusion follows from Lemma 3.3. □

Lemma 3.5. *Let G be an MN3P5 with a dominating clique $\{a, b\}$ of size 2. Then $G \in \mathfrak{G}$.*

Proof. Assume $G \notin \mathfrak{G}$. We may assume G contains no dominating 3-clique, for otherwise we are done by Lemma 3.4. It follows that no vertex v is adjacent to both a, b.

By Theorem 3.1, there is 5-hole $C = v_1v_2v_3v_4v_5$ in G because $G \neq K_4$. Clearly C cannot contain both a and b. WLOG, assume that $|N(a) \cap C| \geq |N(b) \cap C|$. If $b \notin C$ then since $\{a, b\}$ is a dominating clique of G we have $|N(a) \cap C| \geq 3$. If $b \in C$, then a must be adjacent to the 2 vertices in C not adjacent to b. Thus, since $a \sim b$ we also have $|N(a) \cap C| \geq 3$. The case when $|N(a) \cap C| \geq 4$ is handled by Lemma 3.1, so WLOG we may assume either $N(a) \cap C = \{v_1, v_2, v_3\}$ or $N(a) \cap C = \{v_1, v_3, v_4\}$.

Suppose $N(a) \cap C = \{v_1, v_2, v_3\}$. Since $\{a, b\}$ is a dominating clique, we have $b \notin C$ and $b \sim v_4, v_5$. Since no vertex is adjacent to both a and b, G contains $P_5(bv_5v_1v_2v_3)$, a contradiction. Now, we may assume $N(a) \cap C = \{v_1, v_3, v_4\}$. There exists a vertex x with $x \nsim a, v_3, v_4$, for otherwise $\{a, v_3, v_4\}$ is dominating 3-clique. If $x \sim v_5$, then $x \sim v_2$, for otherwise G contains $P_5(xv_5v_4v_3v_2)$; but now G contains $P_5(v_2xv_5v_4a)$. Thus, we have $x \nsim v_5$ and by symmetry $x \nsim v_2$. Since $\{a, b\}$ is a dominating clique, we have $x \sim b$, and $b \sim v_2, v_5$. Recall that no vertex is adjacent to both a, b. Now, G contains $P_5(xbv_5v_4v_3) which is a contradiction.$ □

Theorem 3.2. *If G is an MN3P5 with a dominating clique then* $G \in \mathfrak{G}$.

Proof. If G has a dominating clique of size one or two, then it has a dominating clique of size 2 since G contains no isolated vertices. By Lemma 3.5, $G \in \mathfrak{G}$. If G has a dominating clique of size 3, then Lemma 3.4 implies $G \in \mathfrak{G}$. If G has a dominating clique of size 4 or more, then G contains a K_4 so $G = K_4 \in \mathfrak{G}$ by minimality. □

Lemma 3.6. *Let G be an MN3P5 with a dominating 5-hole. Then G has a dominating* K_3 *or* $G \in \mathfrak{G}$.

Proof. Let $C = v_1v_2v_3v_4v_5$ be an induced 5-hole of G. Assume G does not have a dominating clique. Let X_i be the set of vertices adjacent to v_{i-1} and v_{i+1} and not adjacent to v_{i+2} and v_{i+3} with the subscript taken modulo 5, for $i = 1, 2, 3, 4, 5$. We now prove every vertex of G belongs to exactly one X_i.

Consider a vertex $w \notin C$. By Lemma 3.1, we have $1 \leq |N(w) \cap C| \leq 3$. If w has one neighbor in C, then G obviously contains a P_5. Suppose w has two neighbors a, b in C. If $a \sim b$, then G obviously contains a P_5. Otherwise, a and b have distance two on C and so w belongs to some X_i. We may now assume w has three neighbors on C. If these three neighbors are consecutive on C, then w belongs to some X_i. Now, we may assume $w \sim v_1, v_3, v_4$. There is a vertex x with $x \nsim w, v_4, v_3$, for otherwise $\{w, v_4, v_3\}$ is a dominating clique. Vertex x must have a neighbor in $\{v_1, v_2, v_5\}$ because C is a dominating set. If $x \sim v_5$, then $x \sim v_2$, for otherwise G contains $P_5(xv_5v_4v_3v_2)$; but now G contains $P_5(v_2xv_5v_4w)$. Thus, we have $x \nsim v_5$ and by symmetry $x \nsim v_2$. Now, we have $x \sim v_1$, and G contains $P_5(xv_1v_5v_4v_3)$. Thus, X_1, X_2, X_3, X_4, X_5 is a partition of $V(G)$.

If there are nonadjacent vertices x_1, x_2 with $x_1 \in X_1, x_2 \in X_2$, then G contains $P_5(x_1v_5v_4v_3x_2)$. Thus, there are all possible edges between X_i and X_{i+1} for all i. If every X_i is a stable set, then G is obviously 3-colorable, a contradiction. So we may assume WLOG X_5 contains an edge ab. Then X_1 is a stable set, for otherwise G contains a K_4 with one edge in X_1 and one edge in X_5. Similarly, X_4 is a stable set. If X_2 contains an edge cd, then G contains $S_1(v_1cv_3v_4a, d, b)$. If X_3 contains an edge fg, then G contains $S_1(v_4fv_2v_1a, g, b)$. Thus, X_i is a stable set for $i = 1, 2, 3, 4$. Consider the subgraph H of G induced by X_5. If H contains an odd cycle D, then $D \cup \{v_1\}$ is a K_4 or W_5, or D contains a P_5. Thus H is bipartite. By coloring X_5 with colors 2,3, $X_1 \cup X_4$ with color 1, X_2 with color 2, X_3 with color 3, we see that G is 3-colorable, a contradiction. □

4 Proof of Theorem 1.1

We can now prove the main theorem.

It is a routine matter to verify the "only if" part. We only need prove the "if" part. Suppose G does not contain any of the graphs in Fig. 1 but is not 3-colorable. Then G contains an induced subgraph that is minimally not 3-colorable. It follows that we may assume G is a connected MN3P5 graphs. By Theorem 2.2, G contains a dominating clique or P_3. If G contains a dominating clique, then we are done by Theorem 3.2. So, we may assume G contains no dominating clique and thus contains a dominating P_3 with vertices v_1, v_2, v_3 and edges v_1v_2, v_2v_3. There is a vertex v_4 with $v_4 \sim v_3$ and $v_4 \nsim v_1, v_2$ since v_1v_2 is not a dominating edge. Similarly, there is a vertex v_5 with $v_5 \sim v_1$ and $v_5 \nsim v_2, v_3$. We have $v_5 \sim v_4$, for otherwise G contains a P_5. Thus, $v_1v_2v_3v_4v_5$ is a dominating 5-hole of G, and we are done by Lemma 3.6. □

5 Conclusion and Open Problems

In this paper, we provide a certifying algorithm for the problem of 3-coloring a P_5-graph by showing there are exactly six finite minimally non-3-colorable graphs. Previously known algorithms ([5,10,13]) provide a yes-certificate by constructing a 3-coloring if one exists. Our algorithm provides a no-certificate by finding one of the six graphs of Fig. 1. Since these graphs are finite, our algorithm runs in polynomial time. We do not know if there is a fast algorithm running in, say, $O(n^4)$ to test if a graph contains one of the six graphs of Fig. 1 as a subgraph. We leave this as an open problem.

References

1. Bacsó, G., Tuza, Z.: Dominating cliques in P_5-free graphs. Period. Math. Hungar. 21(4), 303–308 (1990)
2. Coppersmith, D., Winograd, S.: Matrix multiplication via arithmetic progressions. Journal of Symbolic Computation 9(3), 251–280 (1990)
3. Chudnovsky, M., Robertson, N., Seymour, P., Thomas, R.: The strong perfect graph theorem. Annals of Mathematics 164(1), 51–229 (2006)
4. Hoàng, C.T., Kaminśki, M., Lozin, V., Sawada, J., Shu, X.: Deciding k-colorability of P5-free graphs in polynomial time. To appear in Algorithmica
5. Hoàng, C.T., Kaminśki, M., Lozin, V., Sawada, J., Shu, X.: A Note on k-Colorability of P5-Free Graphs. In: Ochmański, E., Tyszkiewicz, J. (eds.) MFCS 2008. LNCS, vol. 5162, pp. 387–394. Springer, Heidelberg (2008)
6. Korobitsyn, D.V.: On the complexity of determining the domination number in monogenic classes of graphs. Diskret. Mat. 2(3), 90–96 (1990) (in Russian); Translation in Discrete Mathematics and Applications 2(2), 191-199 (1992)
7. Kral, D., Kratochvil, J., Tuza, Z., Woeginger, G.J.: Complexity of coloring graphs without forbidden induced subgraphs. In: Brandstädt, A., Le, V.B. (eds.) WG 2001. LNCS, vol. 2204, pp. 254–262. Springer, Heidelberg (2001)
8. Kratsch, D., McConnell, R.M., Mehlhorn, K., Spinrad, J.P.: Certifying algorithms for recognizing interval graphs and permutation graphs. SIAM J. Comput. 36(2), 326–353 (2006)
9. Bang Le, V., Randerath, B., Schiermeyer, I.: On the complexity of 4-coloring graphs without long induced paths. Theoretical Computer Science 389, 330–335 (2007)

10. Mellin, S.: Polynomielle Färbungsalgorithmen für P_k-freie Graphen, Diplomarbeit am Institut für Informatik, Universität zu Köln (2002)
11. Randerath, B., Schiermeyer, I.: Vertex coloring and forbidden subgraphs – a survey. Graphs and Combinatorics 20(1), 1–40 (2004)
12. Randerath, B., Schiermeyer, I.: 3-colorability $\in \mathcal{P}$ for P_6-free graphs. Discrete Applied Mathematics 136, 299–313 (2004)
13. Woeginger, G.J., Sgall, J.: The complexity of coloring graphs without long induced paths. Acta Cybernetica 15(1), 107–117 (2001)

Parameterizing Cut Sets in a Graph by the Number of Their Components

Takehiro Ito[1,⋆], Marcin Kamiński[2], Daniël Paulusma[3,⋆⋆], and Dimitrios M. Thilikos[4,⋆⋆⋆]

[1] Graduate School of Information Sciences, Tohoku University, Aoba-yama 6-6-05, Sendai, 980-8579, Japan
takehiro@ecei.tohoku.ac.jp

[2] Computer Science Department, Université Libre de Bruxelles, Boulevard du Triomphe CP212, B-1050 Brussels, Belgium
marcin.kaminski@ulb.ac.be

[3] Department of Computer Science, University of Durham, Science Laboratories, South Road, Durham DH1 3LE, England
daniel.paulusma@durham.ac.uk

[4] Department of Mathematics, National and Kapodistrian University of Athens, Panepistimioupolis, GR15784 Athens, Greece
sedthilk@math.uoa.gr

Abstract. For a connected graph $G = (V, E)$, a subset $U \subseteq V$ is called a k-cut if U disconnects G, and the subgraph induced by U contains exactly k (≥ 1) components. More specifically, a k-cut U is called a (k, ℓ)-cut if $V \backslash U$ induces a subgraph with exactly ℓ (≥ 2) components. We study two decision problems, called k-Cut and (k, ℓ)-Cut, which determine whether a graph G has a k-cut or (k, ℓ)-cut, respectively. By pinpointing a close relationship to graph contractibility problems we first show that (k, ℓ)-Cut is in P for $k = 1$ and any fixed constant $\ell \geq 2$, while the problem is NP-complete for any fixed pair $k, \ell \geq 2$. We then prove that k-Cut is in P for $k = 1$, and is NP-complete for any fixed $k \geq 2$. On the other hand, we present an FPT algorithm that solves (k, ℓ)-Cut on apex-minor-free graphs when parameterized by $k + \ell$. By modifying this algorithm we can also show that k-Cut is in FPT (with parameter k) and Disconnected Cut is solvable in polynomial time for apex-minor-free graphs. The latter problem asks if a graph has a k-cut for some $k \geq 2$.

1 Introduction

Graph connectivity is a fundamental graph-theoretic property that is well-studied in the context of network robustness. In the literature several measures for graph connectivity are known, such as requiring hamiltonicity, edge-disjoint spanning

⋆ Supported by Grant-in-Aid for Young Scientists (B) 20700003.

⋆⋆ Supported by EPSRC (EP/D053633/1).

⋆⋆⋆ Supported by the project "Kapodistrias" (AΠ 02839/28.07.2008) of the National and Kapodistrian University of Athens (project code: 70/4/8757).

Y. Dong, D.-Z. Du, and O. Ibarra (Eds.): ISAAC 2009, LNCS 5878, pp. 605–615, 2009.

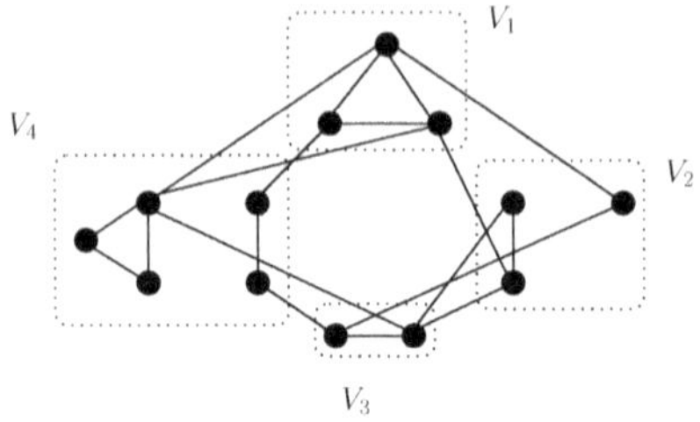

Fig. 1. A graph G with a disconnected cut $V_1 \cup V_3$ that is also a 2-cut and a $(2,4)$-cut and a disconnected cut $V_2 \cup V_4$ that is also a 4-cut and a $(4,2)$-cut

trees, or edge- or vertex-cuts of sufficiently large size. Here, we study the problem of finding a vertex-cut, called a "disconnected cut" of a graph, such that the cut itself is disconnected. As we shall see, this problem is strongly related to several other graph problems such as biclique vertex-covers. We give all further motivation later and first state our problem setting.

Let $G = (V, E)$ be a connected simple graph. For a subset $U \subseteq V$, we denote by $G[U]$ the subgraph of G induced by U. We say that U is a *cut* of G if U disconnects G, that is, $G[V \backslash U]$ contains at least two components. A cut U is *connected* if $G[U]$ contains exactly one component, and *disconnected* if $G[U]$ contains at least two components. We observe that $G[U]$ is a disconnected cut if and only if $G[V \backslash U]$ is a disconnected cut. In Fig.1, the subset $V_1 \cup V_3$ is a disconnected cut, and hence its complement $V_2 \cup V_4$ $\big(= V \setminus (V_1 \cup V_3)\big)$ is also a disconnected cut. This leads to the decision problem Disconnected Cut which asks if a connected graph G has a disconnected cut.

The complexity of Disconnected Cut is open. However, it is known that the problem can be solved in polynomial time for some restricted graph classes, such as graphs with bounded maximum degree, triangle-free graphs, and graphs with a dominating edge (these include cographs) [10]. In particular, we mention that every graph of diameter at least three has a disconnected cut [10].

Besides Disconnected Cut, we study two closely related problems in which we wish to find a cut having a prespecified number of components. For a fixed constant $k \geq 1$, a *k-cut* of a connected graph G is a cut U of G such that $G[U]$ contains exactly k components. More specifically, for a pair (k, ℓ) of fixed constants $k \geq 1$ and $\ell \geq 2$, a k-cut U is called a *(k, ℓ)-cut* of G if $G[V \backslash U]$ consists of exactly ℓ components. Note that a k-cut and a (k, ℓ)-cut are connected cuts if $k = 1$; otherwise (when $k \geq 2$) they are disconnected cuts. It is obvious that, for a fixed pair $k, \ell \geq 2$, a (k, ℓ)-cut U of G corresponds to an (ℓ, k)-cut $V \setminus U$ of G. For example, the disconnected cut $V_1 \cup V_3$ in Fig.1 is a 2-cut and a $(2,4)$-cut, while its complement $V_2 \cup V_4$ is a 4-cut and a $(4,2)$-cut. We study the following two decision problems, where k and ℓ are fixed, *i.e.*, not part of the input. The k-Cut problem asks if a connected graph has a k-cut. The (k, ℓ)-Cut problem asks if a connected graph has a (k, ℓ)-cut.

Our results and the paper organization. Our three main results are as follows. First, we show that DISCONNECTED CUT is strongly related to several other graph problems. This way we determine the computational complexity of (k,ℓ)-CUT. Second, we determine the computational complexity of k-CUT. Third, we give an FPT algorithm that solves (k,ℓ)-CUT for apex-minor-free graphs when parameterized by $k+\ell$. In the following, we explain our results in detail.

In Section 2 we define our terminology. Section 3 contains our first result. We state our motivation for studying these three types of cut problems. We then pinpoint relationships to other cut problems, and to graph homomorphism, biclique vertex-cover and vertex coloring problems. We show a strong connection to graph contractibility problems. This way we prove that (k,ℓ)-CUT is polynomially solvable for $k=1, \ell\geq 2$, and is NP-complete otherwise.

Section 4 gives our second result: we classify the computational complexity of k-CUT. For $k=1$ this is equivalent to asking if a graph G has a connected cut. We show that k-CUT is polynomially solvable for $k=1$, while it becomes NP-complete for every fixed constant $k\geq 2$. Note that the result for (k,ℓ)-CUT in Section 3 does not imply this result, because ℓ is fixed and the subgraph obtained after removing a (k,ℓ)-cut must consist of *exactly* ℓ components.

In Section 5 we present our third result: an FPT algorithm that solves (k,ℓ)-CUT on apex-minor-free graphs when parameterized by $k+\ell$. We also show that k-CUT is FPT in k for apex-minor-free graphs, and that DISCONNECTED CUT is polynomially solvable for this class (which includes the class of planar graphs).

In Section 6 we state some further results and mention a number of open problems that are related to chordal, claw-free and line graphs.

2 Preliminaries

The graphs we consider are undirected and without multiple edges. Unless stated otherwise, they do not contain loops either. Let $G=(V,E)$ be a graph. The *neighborhood* of a vertex $u\in V$ is $N(u)=\{v\,|\,uv\in E\}$. Two disjoint nonempty subsets $U,U'\subset V$ are *adjacent* if there exist $u\in U$ and $u'\in U'$ with $uu'\in E$. The *distance* $d_G(u,v)$ between two vertices u and v in G is the number of edges in a shortest path between them. The *diameter* $\mathrm{diam}(G)$ is defined as $\max\{d_G(u,v)\mid u,v\in V\}$.

Terms in Section 3. A *diagonal coloring* of G is a function $c:V\rightarrow\{1,2,3,4\}$ such that all four *colors* $1,2,3,4$ are used, and no edge has the colors $1,3$ or $2,4$ at its end-vertices. Note that a diagonal coloring does not have to be proper.

A *model graph* is a simple graph with two types of edges: solid and dotted edges. Let H be a fixed model graph with vertex set $\{h_1,\ldots,h_k\}$. An H-partition of a graph G is a partition of V_G into k (nonempty) sets $V_1,\ldots,V_k$ such that for all vertices $u\in V_i$, $v\in V_j$ and for all $1\leq i<j\leq k$ the following two conditions hold. First, if h_ih_j is a solid edge of H, then $uv\in E_G$. Second, if h_ih_j is a dotted edge of H, then $uv\notin E_G$. Let $2K_2$ be the model graph with vertices $h_1,\ldots,h_4$ and solid edges h_1h_3, h_2h_4, and $2S_2$ be the model graph with vertices $h_1,\ldots,h_4$ and dotted edges h_1h_3, h_2, h_4.

A *homomorphism* from a graph G to a graph H is a vertex mapping $f : V_G \to V_H$ satisfying the property that $f(u)f(v) \in E_H$ whenever $uv \in E_G$. It is *vertex-surjective* if $f(V_G) = V_H$. Here we used the shorthand notation $f(S) = \{f(u) \mid u \in S\}$ for a subset $S \subseteq V$. It is called a *compaction* if f is *edge-surjective, i.e.*, for every edge $xy \in E_H$ with $x \neq y$ there exists an edge $uv \in E_G$ with $f(u) = x$ and $f(v) = y$. We then say that G *compacts to* H. For an induced subgraph H of G, a homomorphism f from G to H is called a *retraction* if $f(h) = h$ for all $h \in V_H$. In that case we say that G *retracts to* H.

The *edge contraction* of an edge $e = uv$ in a graph G removes the two end-vertices u and v from G, and replaces them by a new vertex adjacent to precisely those vertices to which u or v were adjacent. If H can be obtained from G by a sequence of edge contractions, then G is *contractible to* H (or H*-contractible*). This is equivalent to saying that G has a so-called H*-witness structure* $\mathcal{W}$, which is a partition of V_G into $|V_H|$ sets $W(h)$, called H*-witness sets*, such that each $W(h)$ induces a connected subgraph of G and for every two $h_i, h_j \in V_H$, witness sets $W(h_i)$ and $W(h_j)$ are adjacent in G if and only if h_i and h_j are adjacent in H. Clearly, by contracting the vertices in the witness sets $W(h)$ to a single vertex for every $h \in V_H$, we obtain the graph H. As an example, viewing each component of each V_i in the graph G in Fig.1 as a witness set shows that G is $K_{2,4}$-contractible. In general, the witness sets $W(h)$ might not be unique.

The cycle and path on n vertices are denoted by C_n and P_n, respectively. A graph G is called *reflexive* if every vertex i in G has a loop ii. We denote the reflexive cycle consisting of n vertices by $\mathcal{C}_n$. A graph $G = (V, E)$ is *complete* p*-partite* if V can be partitioned into p independent sets $V_1, \ldots, V_p$ such that $uv \in E$ if and only if $u \in V_i$ and $v \in V_j$ for some $1 \leq i < j \leq p$. For $p = 2$, $|V_1| = k$, and $|V_2| = \ell$, we speak of a *biclique* $K_{k,\ell}$.

Terms in Section 4. A *hypergraph* H is a pair $(Q, \mathcal{S})$ consisting of a set $Q = \{q_1, \ldots, q_m\}$, called the *vertices* of H, and a set $\mathcal{S} = \{S_1, \ldots, S_n\}$ of nonempty subsets of Q, called the *hyperedges* of H. With a hypergraph $H = (Q, \mathcal{S})$ we associate its *incidence graph* $I = (Q, \mathcal{S}, E_I)$, which is a bipartite graph with partition classes Q and $\mathcal{S}$, where for any $q \in Q, S \in \mathcal{S}$ we have $qS \in E_I$ if and only if $q \in S$. A 2*-coloring* of a hypergraph $H = (Q, \mathcal{S})$ is a partition (Q_1, Q_2) of Q such that $Q_1 \cap S_j \neq \emptyset$ and $Q_2 \cap S_j \neq \emptyset$ for $1 \leq j \leq n$. The Hypergraph 2-Colorability problem asks whether a given hypergraph has a 2-coloring. This problem, also known as Set Splitting, is NP-complete (cf. [11]).

Terms in Section 5. A connected graph G contains H as a *minor* if H can be obtained from G by a sequence of edge deletions and edge contractions; otherwise G is H*-minor-free*. A graph G which is not planar but has a vertex v such that $G \setminus v$ is a planar graph is called an *apex*. The *edge subdivision* of an edge $e = uv$ in a graph G removes e from G and replaces it by a new vertex w adjacent to u and v. A *subdivision* of a graph G is a graph obtained from G after performing a sequence of edge subdivisions. In Figure 2 three examples of an *elementary wall* are given. A *wall* W of height h is a subdivision of an elementary wall of height

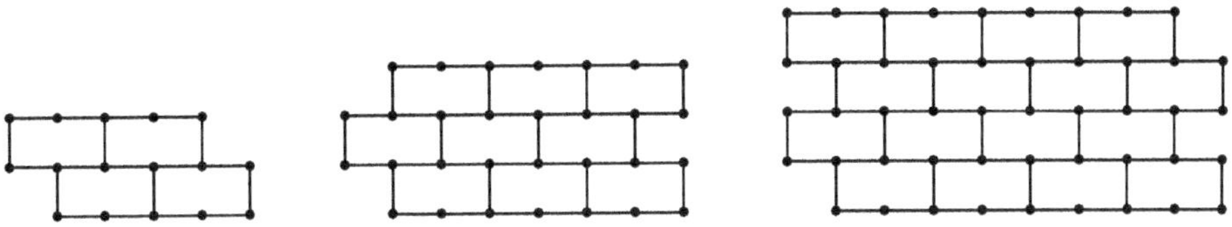

Fig. 2. Elementary walls of height 2, 3, 4 with perimeter of length 14, 22, 30, resp.

h and has a unique planar embedding. A *brick* in W is a face cycle of length 6. The unique face cycle of length greater than 6 is called the *perimeter* P_W of W.

For a graph G with subgraph D, let $\partial_G(D)$ be the set of all vertices of D incident to an edge in $E_G \setminus E_D$. A pair (X, Y) of subsets of V_G such that $G = G[X \cup Y]$ is called a *separation* of G of *order* $|X \cap Y|$. Let G be a graph containing a wall W with perimeter P. Let K' be the unique component of $G \setminus P$ that contains $W \setminus P$. The graph $K = K' \cup P$ is called the *compass* of W in G. A *layout* of K (with respect to the wall W in G) is a family $(C, D_1, \ldots, D_m)$ of connected subgraphs of K that satisfies the following six properties:

1. $K = C \cup D_1 \cup \cdots \cup D_m$ $(m \geq 0)$
2. $W \subseteq C$
3. $Y \subseteq X$ for each separation (X, Y) of C of order ≤ 3 with $V_W \subseteq X$
4. $\partial_G(D_i) \subseteq V_C$ for all $1 \leq i \leq m$
5. $|\partial_G(D_i)| \leq 3$ for all $1 \leq i \leq m$
6. $\partial_G(D_i) \neq \partial_G(D_j)$ for all $1 \leq i < j \leq m$.

We define $\hat{C}$ as the graph obtained from C by adding, for $i \in \{1, \ldots, m\}$, a new vertex d_i and all edges between vertices in $\partial_G(D_i) \cup \{d_i\}$ (so $\partial_G(D_i) \cup \{d_i\}$ becomes a clique). We call $\hat{C}$ the *core* of the layout and $D_1, \ldots, D_m$ its *extensions.* The layout $(C, D_1, \ldots, D_m)$ is *flat* if $\hat{C}$ is a planar graph. We call the wall W *flat* (in G) if the compass of W has a flat layout.

A problem is *fixed parameter tractable* if an instance (I, k) can be solved in time $O(f(k)n^c)$, where f denotes a computable function and c a constant independent of k. The class FPT is the class of all fixed-parameter tractable decision problems.

Terms in Section 6. A *chordal* graph is a graph with no induced cycles of length larger than three. A *claw-free* graph is a graph that does not contain $K_{1,3}$ as an induced subgraph. The *line graph* $L(G)$ of a graph G with edges $e_1, \ldots, e_p$ is the graph $L(G)$ with vertices $u_1, \ldots, u_p$ such that there is an edge between any two vertices u_i and u_j if and only if e_i and e_j share one end-vertex in G. Note that a line graph is claw-free.

3 Relationship to Other Problems

The Disconnected Cut problem can be formulated in several different ways as shown in [10]. We summarize and extend them in the proposition below.

Proposition 1. *Leg G be a connected graph. Then the following statements are equivalent.*

- *G has a disconnected cut.*
- *G has a diagonal coloring.*
- *G has a $2S_2$-partition.*
- *G allows a vertex-surjective homomorphism to C_4.*
- *$\overline{G}$ has a spanning subgraph that consists of two bicliques.*
- *$\overline{G}$ has a $2K_2$-partition.*

If $\mathrm{diam}(G) = 2$, *the above statements are also equivalent to*

- *G allows a compaction to C_4.*
- *G is contractible to some biclique $K_{k,\ell}$ for some $k, \ell \geq 2$.*

Proof. Below we only show that a graph $G = (V, E)$ with $\mathrm{diam}(G) = 2$ has a disconnected cut if and only if G is contractible to $K_{k,\ell}$ for some $k, \ell \geq 2$. The (straightforward) proofs of all other statements can be found in [10].

Let $G = (V, E)$ be a graph with $\mathrm{diam}(G) = 2$. Suppose G has a disconnected cut U. Let $X_1, \ldots, X_k$ be the components of $G[U]$ and let $Y_1, \ldots, Y_\ell$ be the components of $G[V \backslash U]$. Suppose some X_i is not adjacent to some Y_j. Then the distance from a vertex of X_i to a vertex of Y_i is at least three. This is not possible. Hence, G is $K_{k,\ell}$-contractible with as witness sets the sets X_i for $i = 1, \ldots, k$ and Y_j for $j = 1, \ldots, \ell$. Suppose G is $K_{k,\ell}$-contractible with witness structure $\mathcal{W}$. Let the partition classes of $K_{k,\ell}$ be $A = \{a_1, \ldots, a_k\}$ and $B = \{b_1, \ldots, b_\ell\}$. Then $W(a_1) \cup \ldots \cup W(a_k)$ is a disconnected cut of G. □

We now describe the different frameworks related to the equivalent statements in Proposition 1. As we shall see in all these problem settings the DISCONNECTED CUT problem pops up as a missing case (often *the* missing case).

1. Cut sets. In the literature, various kinds of cut sets have been studied. For instance, a cut U of a graph $G = (V, E)$ is a *k-clique cut* if $G[U]$ has a spanning subgraph consisting of k complete graphs, a *strict k-clique cut* if $G[U]$ consists of k components that are complete graphs, a *stable cut* if U is an independent set and a *matching cut* if $E_{G[U]}$ is a matching. The problem that asks whether a graph has a k-clique cut is polynomially solvable for $k = 1$ [16] and $k = 2$ [4]. The latter paper also shows that deciding if a graph has a strict 2-clique cut is polynomially solvable. On the other hand, the problems that asks whether a graph has a stable cut [2] or a matching cut [5], respectively, are NP-complete. Recently, the problem that asks if a graph has a stable cut of size *at most k* has been shown to be in FPT [13].

2. H-partitions. The authors of [7] prove that the problem that asks if a graph allows an H-partition can be solved in polynomial time for all fixed model graphs H with at most four vertices, except for $H = 2K_2$ or $H = 2S_2$. From Proposition 1 it is clear that these two cases correspond exactly to the DISCONNECTED CUT problem. We note that the list version of the $2S_2$-problem (and consequently the $2K_2$-problem) is NP-complete. This follows directly from a result of Hell and Feder [8] who show that the problem of deciding whether a graph G retracts to C_4 is NP-complete. A variant on H-partitions that allows empty blocks V_i in an H-partition can be found in [9].

3. Compactions. We note that any edge-surjective homomorphism from a graph G to a connected graph H is vertex-surjective (whereas the reverse is not necessarily true). Vikas [15] showed that the C_4-COMPACTION problem that asks if there exists a compaction from a graph G to C_4 is NP-complete. By a modification of his proof, one can easily show that the C_4-COMPACTION problem stays NP-complete for the class of graphs of diameter three. Unfortunately, only for graphs of diameter two, the C_4-COMPACTION problem is equivalent to the DISCONNECTED CUT problem (cf. Proposition 1).

4. Contractibility. The H-CONTRACTIBILITY problem asks if a graph G is H-contractible. A slight modification of the proof of Proposition 1 shows that a graph with diameter two has an (k, ℓ)-cut if and only if it is $K_{k,\ell}$-contractible for $k, \ell \geq 2$. Observe that the $(1, \ell)$-CUT problem is equivalent to the $K_{1,\ell}$-CONTRACTIBILITY problem for the class of all connected graphs. Brouwer & Veldman [3] show that $K_{k,\ell}$-CONTRACTIBILITY is polynomially solvable for $k = 1$. For $k, \ell \geq 2$, they show that it is NP-complete. As the gadget in their NP-completeness reduction has diameter two, we obtain the following result.

Proposition 2. *The (k, ℓ)-CUT problem is in P for $k = 1$ and NP-complete, even for the class of connected graphs of diameter two, for $k, \ell \geq 2$.*

5. Vertex-covers. The problem of deciding if a graph has has a spanning subgraph that consists of at most k mutually vertex-disjoint *bicliques* is called the k-BICLIQUE VERTEX-COVER problem. It has applications in data mining, e-commerce, information retrieval and network management. This problem is polynomially solvable if $k = 1$ and NP-complete if $k \geq 3$ [10]. The missing case is $k = 2$ and Proposition 1 shows that this case is equivalent to the DISCONNECTED CUT problem. Due to Proposition 2 we can easily show the following.

Corollary 1. *The problem of deciding if a graph has a spanning subgraph consisting of two vertex-disjoint graphs, one of which is complete k-partite and the other one is complete ℓ-partite, is NP-complete for $k, \ell \geq 2$.*

4 Cuts with a Prespecified Number of Components

We determine the computational complexity of the k CUT problem. Note that we cannot obtain this result from Proposition 2. We omit its proof due to page restrictions.

Theorem 1. *The k-CUT problem is in P for $k = 1$ and NP-complete, even for the class of connected graphs of diameter two, for $k \geq 2$.*

5 Apex-Minor-Free Graphs

As K_5 is an apex and a planar graph is K_5-minor-free, a planar graph is an example of an apex-minor free graph. For these graphs we prove that (k,ℓ)-CUT is fixed parameter tractable in $k+\ell$, that k-CUT is fixed parameter tractable in k, and that DISCONNECTED CUT is polynomially solvable. Here we apply a win-win approach. From Theorem 2 below, which is a variant of the Trinity Lemma [14] for apex-minor-free graphs, an apex-minor-free graph either has small treewidth or a large wall. If the treewidth is small, we show that the problem is polynomially solvable. If the graph has a large wall we show it *always* has a (k,l)-cut (k-cut, or disconnected cut, resp.). We discuss each case separately and refer to [6] for the definitions of tree decomposition and treewidth, as we do not need them here.

Theorem 2 ([12]). *There exists a computable function $f : \mathbb{N}^2 \rightarrow \mathbb{N}$ and an algorithm that, given an apex H, a graph G, and a non-negative integer h, outputs in polynomial time one of the following:*

1. *a tree decomposition of G of width at most $f(|V_H|, h)$, or*
2. *an H-minor of G, or*
3. *a wall of height $\geq h$ in G, and a flat layout $(C, D_1, \ldots, D_m)$ of the compass K of W in G such that the treewidth of each of the extensions $D_1, \ldots, D_m$ is at most $f(|V_H|, h)$.*

First assume $G = (V, E)$ has small treewidth. A seminal result of Courcelle [6] is that in any class of graphs of bounded treewidth, every problem definable in monadic second-order logic can be solved in time *linear* in the number of vertices of the graph. We need the following proposition, whose proof we omitted.

Proposition 3. *Let k, ℓ be two fixed integers. Then the (k,ℓ)-*CUT*, the k-*CUT *and* DISCONNECTED CUT *problem can be defined in monadic second order logic.*

Now suppose we have a "large" wall. We arrange its bricks into walls of height two (a group of four bricks). Each group has at least one middle vertex not adjacent to any vertex in the outerface of the group. The groups of four give a tiling of the wall and the set of middle vertices from different groups forms an independent set, whose removal leaves the graph connected. Hence we have shown Lemma 1 and continue with Lemma 2 before summarizing in Theorem 3.

Lemma 1. *A wall W of height $\geq h$ contains an independent set S of cardinality $(h-1)^2/4$ such that $W \setminus S$ is connected.*

Lemma 2. *Let k, ℓ be two fixed integers, and H an apex. There exists a constant t such that each connected H-minor-free graph G with $tw(G) \geq t$ has a (k,ℓ)-cut.*

Proof. Let $h = 2(\sqrt{\ell-1}+\sqrt{k-1})+3$ and let t be the smallest integer such that $f(|V_H|, h) \leq t-1$ (for the function f from Theorem 2). By Theorem 2, an H-minor-free graph G with $\mathrm{tw}(G) \geq t$ has a flat wall W of height $\geq h$. Then, due to the flat layout of its compass K, there exists a cycle Q in W with $V_Q \cap V_{P_W} = \emptyset$

such that there is a component R of $G \setminus Q$ not containing perimeter P_W. Now let Q' be a subgraph of G containing Q and such that $G \setminus Q'$ has exactly two components: $R' \subset R$ and the component T containing P_W. Note that Q can be chosen in such a way that R' has a wall $W_{R'}$ of height $2\sqrt{\ell - 1} + 1$, and T has a wall W_T of height $2\sqrt{k-1}+1$. From Lemma 1, R' contains an independent set $S_{R'}$ of size $\ell - 1$, and T contains an independent set S_T of size $k - 1$ such that $W_{R'} \backslash S_{R'}$ and $W_T \backslash S_T$ are connected.

For each $v \in S_T$ we define C_v to be v together with the union of all those components of $G \setminus v$ that do not contain P_{W_T}. For each $v \in S_{R'}$ we define C_v analogously. Let $I = \bigcup_{v \in S_T} C_v$ and $J = \bigcup_{v \in S_{R'}} C_v$. Now, $U = Q' \cup (R' \backslash J) \cup S'$ is a cut with exactly k components, and $G \setminus U$ has exactly ℓ components. □

Theorem 3. *Let H be an apex. For the class of connected H-minor-free-graphs, the following statements hold:*

*(i) The (k, ℓ)-*CUT *problem is fixed parameter tractable in $k + \ell$;*
*(ii) The k-*CUT *problem is fixed parameter tractable in k;*
(iii) The DISCONNECTED CUT *problem is solvable in polynomial time.*

Proof. Let t be the constant from Lemma 2 which guarantees that an H-minor-free graph G with $\text{tw}(G) \geq t$ is a YES-instance of the (k, ℓ)-CUT problem. We first check if $\text{tw}(G) < t$. We can do so as recognizing such graphs is fixed parameter tractable in t [1]. So, if $\text{tw}(G) \geq t$ we are done. Suppose $\text{tw}(G) < t$. By Proposition 3, the (k, ℓ)-CUT problem is expressible in monadic second order logic and therefore solvable on graphs of bounded treewidth [6].

For (ii) we take $\ell = 2$ and repeat all arguments. If $\text{tw}(G) \geq t$ then G has a $(k, 2)$-cut, and hence a k-cut. If $\text{tw}(G) < t$ then we apply Proposition 3 on the k-CUT problem. For (iii) we take $k = \ell = 2$ and proceed similarly. □

6 Further Results and Related Open Problems

The main open problem is to determine the computational complexity of the DISCONNECTED CUT problem. The related DISCONNECTED SEPARATOR problem asks whether a connected graph G has a disconnected cut separating two specified vertices s and t of G. This problem is NP-complete. We omit the proof.

Theorem 4. *The* DISCONNECTED SEPARATOR *problem is* NP*-complete, even for the class of connected graphs of diameter 3.*

6.1 Chordal Graphs

We can solve DISCONNECTED CUT in polynomial time for this class. We omit the proof. What about k-CUT or even (k, ℓ)-CUT?

Proposition 4. *The* DISCONNECTED CUT *problem is solvable in polynomial time for connected chordal graphs.*

6.2 Claw-Free Graphs

Is Disconnected Cut polynomially solvable for this class or even for its subclass of line graphs? Every graph of diameter at least three has a disconnected cut [10]. Hence, we only need to consider graphs of diameter two. Then the following connections become relevant. Proofs are omitted.

Proposition 5. *For the class of connected claw-free graphs of diameter two,* Disconnected Cut, (2, 2)-Cut *and* C_4-Contractibility *are equivalent.*

Proposition 6. *Let* $L(G)$ *be the line graph of a connected graph* G *with diameter two. If the diameter of* $L(G)$ *is two, then* G *has a disconnected cut if and only if* $L(G)$ *has a disconnected cut.*

Note that Proposition 6 does not hold if G has diameter at least three: take $G = P_4 = p_1p_2p_3p_4$, which has diameter three and disconnected cut $\{p_1, p_3\}$. However, $L(G) = P_3$ does not have a disconnected cut. The condition that $L(G)$ has diameter two is necessary as well. Take two triangles and merge them in one vertex in order to obtain the graph $(\{u, v, w, x, y\}, \{uv, vw, wu, ux, xy, yu\})$. This graph has diameter two but no disconnected cut. However, its line graph is a K_4 on vertices e_1, e_2, e_3, e_4 extended with two vertices d_1, d_2 and edges d_1e_1, d_1e_2 and d_2e_3, d_2e_4. This graph has diameter three and disconnected cut $\{d_1, e_3, e_4\}$.

References

1. Bodlaender, H.L.: A linear time algorithm for finding tree-decompositions of small treewidth. In: Proceedings of STOC, pp. 226–234 (1993)
2. Brandstädt, A., Dragan, F.F., Le, V.B., Szymczak, T.: On stable cutsets in graphs. Discrete Appl. Math. 105, 39–50 (2000)
3. Brouwer, A.E., Veldman, H.J.: Contractibility and NP-completeness. J. Graph Theory 11, 71–79 (1987)
4. Cameron, K., Eschen, E.M., Hoáng, C.T., Sritharan, R.: The complexity of the list partition problem for graphs. SIAM J. Discrete Math. 21, 900–929 (2007)
5. Chvátal, V.: Recognizing decomposable graphs. J. Graph Theory 8, 51–53 (1984)
6. Courcelle, B.: The Monadic Second-Order Logic of Graphs. I. Recognizable Sets of Finite Graphs. Inf. Comput. 85(1), 12–75 (1990)
7. Dantas, S., de Figueiredo, C.M.H., Gravier, S., Klein, S.: Finding H-partitions efficiently. RAIRO-Theoret. Inf. Appl. 39, 133–144 (2005)
8. Feder, T., Hell, P.: List homomorphisms to reflexive graphs. J. Combin. Theory Ser. B 72, 236–250 (1998)
9. Feder, T., Hell, P., Klein, S., Motwani, R.: List partitions. SIAM J. Discrete Math. 16, 449–478 (2003)
10. Fleischner, H., Mujuni, E., Paulusma, D., Szeider, S.: Covering graphs with few complete bipartite subgraphs. Theoret. Comput. Sci. 410, 2045–2053 (2009)
11. Garey, M.R., Johnson, D.S.: Computers and Intractability. W.H. Freeman and Co., New York (1979)

12. Kamiński, M., Thilikos, D.M.: Induced linkages in apex-minor-free graphs (manuscript)
13. Marx, D., O'Sullivan, B., Razgon, I.: Threewidth reduction for constrained separation and bipartization problems. CoRR abs/0902.3780 (2009) (manuscript)
14. Robertson, N., Seymour, P.D.: Graph minors XIII. The disjoint paths problem. J. Combin. Theory Ser. B 63, 65–110 (1995)
15. Vikas, N.: Computational complexity of compaction to reflexive cycles. SIAM J. Comput. 32, 253–280 (2002)
16. Whitesides, S.H.: An algorithm for finding clique cut-sets. Inform. Process. Lett. 12, 31–32 (1981)

Inapproximability of Maximal Strip Recovery*

Minghui Jiang

Department of Computer Science, Utah State University, Logan, UT 84322-4205, USA
mjiang@cc.usu.edu

Abstract. In comparative genomic, the first step of sequence analysis is usually to decompose two or more genomes into syntenic blocks that are segments of homologous chromosomes. For the reliable recovery of syntenic blocks, noise and ambiguities in the genomic maps need to be removed first. Maximal Strip Recovery (MSR) is an optimization problem recently proposed by Zheng, Zhu, and Sankoff for reliably recovering syntenic blocks from genomic maps in the midst of noise and ambiguities. Given d genomic maps as sequences of gene markers, the objective of MSR-d is to find d subsequences, one subsequence of each genomic map, such that the total length of the syntenic blocks (substrings of consecutive gene markers that appear identically in all d subsequences) is maximized. A polynomial-time $2d$-approximation for MSR-d was previously known. In this paper, we show that even the most basic version of MSR-2, in which all gene markers are distinct and in positive orientation, is APX-hard. Moreover, we provide the first explicit lower bounds on approximating MSR-d for all constants $d \geq 2$.

1 Introduction

In comparative genomic, the first step of sequence analysis is usually to decompose two or more genomes into syntenic blocks that are segments of homologous chromosomes. For the reliable recovery of syntenic blocks, noise and ambiguities in the genomic maps need to be removed first. A genomic map is a sequence of gene markers. A gene marker appears in a genomic map in either positive or negative orientation. Given d genomic maps, *Maximal Strip Recovery* (MSR-d) is the problem of finding d subsequences, one subsequence of each genomic map, such that the total length of strips of these subsequences is maximized [10,6]. Here a *strip* (or syntenic block) is a maximal string of at least two markers that appear consecutively in each of the d subsequences, either with all markers in positive orientation and in the same order, or with all markers in negative orientation and in reverse order. For example, the two genomic maps (the markers in negative orientation are underlined)

$$\begin{array}{cccccccccccc} 1 & 2 & 3 & 4 & 5 & 6 & 7 & 8 & 9 & 10 & 11 & 12 \\ \underline{8} & \underline{5} & \underline{7} & \underline{6} & 4 & 1 & 3 & 2 & \underline{12} & \underline{11} & \underline{10} & 9 \end{array}$$

have two subsequences

$$\begin{array}{ccccccccc} 1 & 3 & & 6 & 7 & 8 & 10 & 11 & 12 \\ \underline{8} & \underline{7} & \underline{6} & & 1 & 3 & \underline{12} & \underline{11} & \underline{10} \end{array}$$

* Supported in part by NSF grant DBI-0743670.

Y. Dong, D.-Z. Du, and O. Ibarra (Eds.): ISAAC 2009, LNCS 5878, pp. 616–625, 2009.

of the maximum total strip length 8. The strip $\langle 1,3\rangle$ is positive and forward in both subsequences; the other two strips $\langle 6,7,8\rangle$ and $\langle 10,11,12\rangle$ are positive and forward in the first subsequence, but are negative and backward in the second subsequence.

The problem MSR-2 was introduced by Zheng, Zhu, and Sankoff [10], and was later generalized to MSR-d by Chen, Fu, Jiang, and Zhu [6]. For MSR-2, Zheng et al. [10] presented a potentially exponential-time heuristic that solves a subproblem of Maximum-Weight Clique. For MSR-d, $d \geq 2$, Chen et al. [6] presented a polynomial-time $2d$-approximation.

On the complexity side, Chen et al. [6] showed that several close variants of the problem MSR-d are intractable. In particular, they showed that (i) MSR-2 is NP-complete if duplicate markers are allowed in each genomic map, and that (ii) MSR-3 is NP-complete even if the markers in each genomic map are distinct. The complexity of MSR-2 with no duplicates, however, was left as an open problem.

In the biological context, a genomic map may contain duplicate markers as a paralogy set [10, p. 516], but such maps are relatively rare. Indeed MSR-2 without duplicates is the most common version of MSR-d in application. Theoretically, MSR-2 without duplicates is the most basic and hence the most interesting version of MSR-d. Also, the previous NP-hardness proofs for both (i) MSR-2 with duplicates and (ii) MSR-3 without duplicates [6] rely on the fact that a marker may appear in a genomic map in either positive or negative orientation. A natural question is whether there is any version of MSR-d that remains NP-hard even if all markers in the genomic maps are in positive orientation.

The following is a precise formulation of the *most basic version* of MSR-d:

INSTANCE: Given d sequences G_i, $1 \leq i \leq d$, where each sequence is a permutation of $\langle 1,\ldots,n\rangle$.

QUESTION: Find a subsequence G'_i of each sequence G_i, $1 \leq i \leq d$, and find a set of strips S_j, where each strip is a sequence of length at least two over the alphabet $\{1,\ldots n\}$, such that each subsequence G'_i is the concatenation of the strips S_j in some order, and the total length of the strips S_j is maximized.

The main result of this paper is the following theorem that settles the computational complexity of the most basic version of Maximal Strip Recovery, and moreover provides the first explicit lower bounds on approximating MSR-d for all constants $d \geq 2$:

Theorem 1. MSR-d *for any constant* $d \geq 2$ *is APX-complete. Moreover, it is NP-hard to approximate* MSR-2, MSR-3, *and* MSR-d *(*$d \geq 4$*) within any constant less than* $\frac{2320}{2319}$ (> 1.0004), $\frac{474}{473}$ (> 1.0021), *and* $\frac{285}{284}$ (> 1.0035), *respectively, even if all markers are distinct and appear in positive orientation in each genomic map.*

For any constant $d \geq 2$, MSR-d admits a polynomial-time $2d$-approximation algorithm [6] and is thus in APX. In the following two sections, we show that MSR-d for any constant $d \geq 2$ is APX-hard, and provide explicit constant factors for the hardness of approximation. For any two constants d and d' such that $2 \leq d < d'$, the problem MSR-d is a special case of the problem MSR-d' with $d' - d$ additional copies of a genomic map. Thus the APX-hardness of MSR-2 implies the APX-hardness of MSR-d for all constants $d \geq 2$. To derive better lower bounds on approximation factors, however, and to present the proof progressively, we first show that MSR-3 and MSR-4 are

APX-hard, by two relatively simple L-reductions, then show that MSR-2 is APX-hard, by a more elaborate L-reduction.

Preliminaries. Given two optimization problems X and Y, an *L-reduction* [8] from X to Y consists of two polynomial-time functions f and g and two positive constants α and β satisfying the following two properties:

1. For every instance x of X, $f(x)$ is an instance of Y such that

$$\text{opt}(f(x)) \leq \alpha \cdot \text{opt}(x), \tag{1}$$

2. For every feasible solution y to $f(x)$, $g(y)$ is a feasible solution to x such that

$$|\text{opt}(x) - \text{val}(g(y))| \leq \beta \cdot |\text{opt}(f(x)) - \text{val}(y)|. \tag{2}$$

Here $\text{val}(y)$ denotes the value of a solution y, and $\text{opt}(x)$ denotes the value of the optimal solution to an instance x. For two maximization problems X and Y, the two properties of L-reduction imply the following inequality on the relative errors of approximation:

$$\frac{\text{opt}(x) - \text{val}(g(y))}{\text{opt}(x)} \leq \alpha\beta \cdot \frac{\text{opt}(f(x)) - \text{val}(y)}{\text{opt}(f(x))}.$$

That is, if there is a polynomial-time $\frac{1}{1-\epsilon}$-approximation for Y, then there is a polynomial-time $\frac{1}{1-\alpha\beta\epsilon}$-approximation for X.

2 MSR-3 and MSR-4 Are APX-Hard

We prove that MSR-3 and MSR-4 are APX-hard by two similar L-reductions from Max-IS-3 and Max-IS-4, respectively. Max-IS-Δ is the problem Maximum Independent Set on graphs of maximum degree Δ. Both Max-IS-3 and Max-IS-4 are APX-hard; see [4]. Moreover, Chlebík and Chlebíková [7] showed that it is NP-hard to approximate Max-IS-3 and Max-IS-4 within any constant less than 1.010661 and 1.0215517, respectively. Before we present the L-reductions, we first show that MSR-3 is NP-hard by a reduction in the classical style, which is perhaps more familiar to most readers.

2.1 NP-Hardness Reduction from Max-IS-3 to MSR-3

We need to introduce some preliminaries. A *linear forest* is a graph in which every connected component is a path. The *linear arboricity* of a graph is the minimum number of linear forests into which the edges of the graph can be decomposed. A famous conjecture on linear arboricity [2] is that any Δ-regular graph has linear arboricity at most $\lceil(\Delta+1)/2\rceil$. This conjecture has been confirmed for graphs of small constant degrees. In particular, the proofs of the conjecture for $\Delta = 3$ and 4 are constructive and lead to polynomial-time algorithms for decomposing a graph of maximum degree $\Delta = 3$ and 4 into at most $\lceil(\Delta+1)/2\rceil = 2$ and 3 linear forests, respectively [2,1,3].

We now present the NP-hardness reduction from Max-IS-3 to MSR-3. Let G be a graph of maximum degree 3. Partition the edges of G into two linear forests E_1 and E_2. Let V_1 and V_2 be the vertices of G that are *not* incident to any edges in E_1 and E_2, respectively. Let n be the number of vertices in G. Construct three genomic maps G_0, G_1, and G_2, where each map is a permutation of $4n$ distinct markers all in positive orientation:

- n pairs of vertex markers $\overset{i}{\subset}$ and $\overset{i}{\supset}$, $1 \le i \le n$;
- n pairs of dummy markers $\overset{i}{\sqsubset}$ and $\overset{i}{\sqsupset}$, $1 \le i \le n$.

G_0 consists of the $2n$ pairs of vertex and dummy markers in an alternating pattern:

$$\overset{1}{\subset}\,\overset{1}{\supset}\quad \overset{1}{\sqsubset}\,\overset{1}{\sqsupset}\quad \cdots \quad \overset{n}{\subset}\,\overset{n}{\supset}\quad \overset{n}{\sqsubset}\,\overset{n}{\sqsupset}$$

G_1 and G_2 are represented schematically as follows:

$$G_1 : \langle V_1 \rangle \; \langle E_1 \rangle \; \langle D \rangle$$
$$G_2 : \langle D \rangle \; \langle E_2 \rangle \; \langle V_2 \rangle$$

1. $\langle E_1 \rangle$ and $\langle E_2 \rangle$ consist of vertex markers of the vertices incident to the edges in E_1 and E_2, respectively. The markers of the vertices in each path $v_1 v_2 \ldots v_k$ are grouped together in an interleaving pattern: for $1 \le i \le k$, the left marker of v_i, the right marker of v_{i-1} (if $i > 1$), the left marker of v_{i+1} (if $i < k$), and the right marker of v_i are consecutive.
2. $\langle V_1 \rangle$ and $\langle V_2 \rangle$ consist of vertex markers of the vertices in V_1 and V_2, respectively. The left marker and the right marker of each pair are consecutive.
3. $\langle D \rangle$ is the reverse permutation of the n pairs of dummy markers:

$$\overset{n}{\sqsubset}\,\overset{n}{\sqsupset}\quad \cdots \quad \overset{1}{\sqsubset}\,\overset{1}{\sqsupset}$$

This completes the construction. We refer to Figure 1 (a) and (b) for an example. Two pairs of markers *intersect* in a genomic map if a marker of one pair appears between the two markers of the other pair. The following property of our construction is obvious:

Proposition 1. *Two vertices are adjacent in the graph G if and only if the corresponding two pairs of vertex markers intersect in one of the two genomic maps G_1, G_2.*

We say that three subsequences of the three genomic maps G_0, G_1, G_2 are *canonical* if each strip of the subsequences is either a pair of vertex markers or a pair of dummy markers. We have the following lemma on canonical subsequences:

Lemma 1. *If the three genomic maps G_0, G_1, G_2 have three subsequences of total strip length l, then they must have three subsequences of total strip length at least l such that each strip of the subsequences is either a pair of vertex markers or a pair of dummy markers.*

Proof. There is an algorithm that transforms the subsequences into canonical form without reducing the total strip length. □

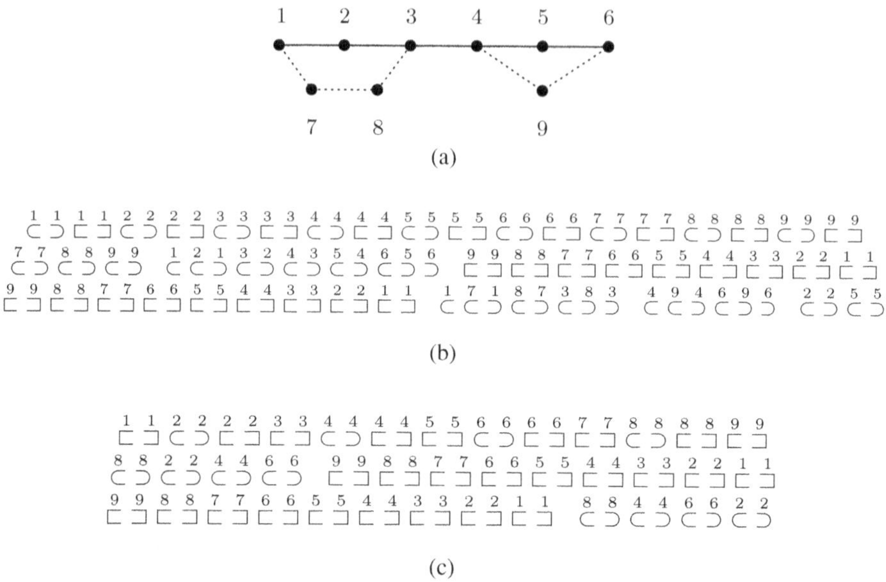

Fig. 1. (a) The graph G: E_1 is a single (solid) path $\langle 1, 2, 3, 4, 5, 6\rangle$, E_2 consists of two (dotted) paths $\langle 1, 7, 8, 3\rangle$ and $\langle 4, 9, 6\rangle$, $V_1 = \{7, 8, 9\}$, $V_2 = \{2, 5\}$. (b) The three genomic maps G_0, G_1, G_2. (c) The three subsequences of the genomic maps G_0, G_1, G_2 corresponding to the independent set $\{2, 4, 6, 8\}$ in the graph G.

The following lemma establishes the NP-hardness of MSR-3:

Lemma 2. *The graph G has an independent set of at least k vertices if and only if the three genomic maps G_0, G_1, G_2 have three subsequences of total strip length at least $2(n + k)$.*

2.2 L-Reductions from Max-IS-3 to MSR-3 and from Max-IS-4 to MSR-4

We present an L-reduction (f, g, α, β) from Max-IS-3 to MSR-3 as follows. The function f, given a graph G of maximum degree 3, constructs the three genomic maps G_0, G_1, G_2 as in our NP-hardness reduction. Let k^* be the number of vertices in a maximum independent set in G, and let l^* be the maximum total strip length of three subsequences of G_0, G_1, G_2. Since a simple greedy algorithm (which repeatedly selects a vertex not adjacent to the previously selected vertices) finds an independent set of at least $n/(3+1)$ vertices in the graph G of maximum degree 3, we have $k^* \geq n/(3+1)$. Then, by Lemma 2, we have

$$l^* = 2(n + k^*) \leq 2((3 + 1)k^* + k^*) = 2(3 + 2)k^* = 10k^*.$$

Choose $\alpha = 10$, then property (1) of L-reduction is satisfied.

The function g, given three subsequences of the three genomic maps G_0, G_1, G_2, transforms the subsequences into canonical form as in the proof of Lemma 1, then returns an independent set of vertices in the graph G corresponding to the pairs of vertex markers that are strips of the subsequences. Let l be the total strip length of the subsequences, and let k be the number of vertices in the independent set returned by the function g. Again, by Lemma 2, we have

$$k^* - k \leq k^* - (l/2 - n) = (n + k^*) - l/2 = (l^* - l)/2.$$

Choose $\beta = 1/2$, then property (2) of L-reduction is also satisfied. The L-reduction from Max-IS-3 to MSR-3 can be generalized:

Lemma 3. *Let $\Delta \geq 3$ and $d \geq 3$ be integer constants. If there is a polynomial-time algorithm for decomposing any graph of maximum degree Δ into $d - 1$ linear forests, then there is an L-reduction from* Max-IS-Δ *to* MSR-d *with constants $\alpha = 2(\Delta + 2)$ and $\beta = 1/2$.*

Recall that there exist polynomial-time algorithms for decomposing a graph of maximum degree 3 and 4 into at most 2 and 3 linear forests, respectively [2,1,3]. Thus, besides the L-reduction from Max-IS-3 to MSR-3 with constants $\alpha = 2(3 + 2)$ and $\beta = 1/2$, we also have by Lemma 3 an L-reduction from Max-IS-4 to MSR-4 with constants $\alpha = 2(4+2)$ and $\beta = 1/2$. Chlebík and Chlebíková [7] showed that it is NP-hard to approximate Max-IS-3 and Max-IS-4 within any constant less than 1.010661 $(> \frac{1}{1-(3+2)/474})$ and 1.0215517 $(> \frac{1}{1-(4+2)/285})$, respectively. Therefore, it is NP-hard to approximate MSR-3 and MSR-4 within any constant less than $\frac{1}{1-1/474} = \frac{474}{473}$ (> 1.0021) and $\frac{1}{1-1/285} = \frac{285}{284}$ (> 1.0035), respectively. The lower bound for MSR-4 extends to MSR-d for constants $d \geq 4$.

3 MSR-2 Is APX-Hard

We prove that MSR-2 is APX-hard by an L-reduction from Ep-Occ-Max-Eq-SAT with $p = 3$ and $q \geq 2$. Given a set X of n variables and a set $\mathcal{C}$ of m clauses, where each variable has exactly p literals (in p different clauses) and each clause is the disjunction of exactly q literals (of q different variables), Ep-Occ-Max-Eq-SAT is the problem of finding an assignment of X that satisfies the maximum number of clauses in $\mathcal{C}$. Note that $np = mq$. Berman and Karpinski [5] showed that it is NP-hard to approximate E3-Occ-Max-E2-SAT within any constant less than $\frac{464}{463}$; we will use this to derive an approximation lower bound for MSR-2. Before we present the L-reduction, we first present an NP-hardness reduction in the classical style.

3.1 NP-Hardness Reduction from Ep-Occ-Max-Eq-SAT to MSR-2

Let $(X, \mathcal{C})$ be an instance of Ep-Occ-Max-Eq-SAT, where X is a set of n variables x_i, $1 \leq i \leq n$, and $\mathcal{C}$ is a set of m clauses C_j, $1 \leq j \leq m$. Without loss of generality, assume that the p literals of each variable are neither all positive nor all negative. Since $p = 3$, it follows that each variable has either 2 positive and 1 negative literals, or 1 positive and 2 negative literals. We construct two genomic maps G_1 and G_2, each map a permutation of $2(5n + m + qm + 2)$ distinct markers all in positive orientation:

- 1 pairs of variable markers $\overset{i}{<}\ \overset{i}{>}$ for each variable x_i, $1 \le i \le n$;
- 2 pairs of true markers $\overset{i,1}{\blacktriangleleft}\ \overset{i,1}{\blacktriangleright}$ and $\overset{i,2}{\blacktriangleleft}\ \overset{i,2}{\blacktriangleright}$ for each variable x_i, $1 \le i \le n$;
- 2 pairs of false markers $\overset{i,1}{\triangleleft}\ \overset{i,1}{\triangleright}$ and $\overset{i,2}{\triangleleft}\ \overset{i,2}{\triangleright}$ for each variable x_i, $1 \le i \le n$;
- 1 pair of clause markers $\overset{j}{\Subset}\ \overset{j}{\Supset}$ for each clause C_j, $1 \le j \le m$;
- q pairs of literal markers $\overset{j,t}{\subset}\ \overset{j,t}{\supset}$, $1 \le t \le q$, for each clause C_j, $1 \le j \le m$;
- 2 pairs of dummy markers $\overset{1}{\sqsubset}\ \overset{1}{\sqsupset}$ and $\overset{2}{\sqsubset}\ \overset{2}{\sqsupset}$.

The construction is done in two steps: first arrange the variable markers, the true/false markers, the clause markers, and the dummy markers into two sequences $\breve{G}_1$ and $\breve{G}_2$, next insert the literal markers at appropriate positions in the two sequences to obtain the two genomic maps G_1 and G_2.

The two sequences $\breve{G}_1$ and $\breve{G}_2$ are represented schematically as follows:

$$\begin{array}{l} \breve{G}_1 :\ \langle x_1 \rangle \ \cdots \ \langle x_n \rangle \quad \overset{1}{\sqsubset}\ \overset{1}{\sqsupset}\ \overset{2}{\sqsubset}\ \overset{2}{\sqsupset} \quad \overset{1}{\Subset}\ \overset{1}{\Supset} \ \cdots \ \overset{m}{\Subset}\ \overset{m}{\Supset} \quad \overset{1}{<}\ \overset{1}{>} \ \cdots \ \overset{n}{<}\ \overset{n}{>} \\ \breve{G}_2 :\ \langle x_n \rangle \ \cdots \ \langle x_1 \rangle \quad \overset{m}{\Subset}\ \overset{m}{\Supset} \ \cdots \ \overset{1}{\Subset}\ \overset{1}{\Supset} \quad \overset{2}{\sqsubset}\ \overset{2}{\sqsupset}\ \overset{1}{\sqsubset}\ \overset{1}{\sqsupset} \end{array}$$

For each variable x_i, $\langle x_i \rangle$ consists of the corresponding four pairs of true/false markers $\overset{i,1}{\blacktriangleleft}\ \overset{i,1}{\blacktriangleright}\ \overset{i,2}{\blacktriangleleft}\ \overset{i,2}{\blacktriangleright}\ \overset{i,1}{\triangleleft}\ \overset{i,1}{\triangleright}\ \overset{i,2}{\triangleleft}\ \overset{i,2}{\triangleright}$ in $\breve{G}_1$ and $\breve{G}_2$, and in addition the pair of variable markers $\overset{i}{<}\ \overset{i}{>}$ in $\breve{G}_2$. These markers are arranged in the two sequences in a special pattern as follows (the indices i are omitted for simpler notations):

$$\begin{array}{c} \overset{1}{\triangleleft}\ \overset{2}{\blacktriangleleft}\ \overset{1}{\triangleright}\ \overset{2}{\blacktriangleright} \qquad \overset{2}{\triangleleft}\ \overset{1}{\blacktriangleleft}\ \overset{2}{\triangleright}\ \overset{1}{\blacktriangleright} \\ \overset{1}{\blacktriangleleft}\ \overset{1}{\triangleleft}\ \overset{1}{\blacktriangleright}\ \overset{1}{\triangleright}\ <\ >\ \overset{2}{\triangleleft}\ \overset{2}{\blacktriangleleft}\ \overset{2}{\triangleright}\ \overset{2}{\blacktriangleright} \end{array}$$

Now insert the literal markers to the two sequences $\breve{G}_1$ and $\breve{G}_2$ to obtain the two genomic maps G_1 and G_2. First, $\breve{G}_1 \rightarrow G_1$. For each positive literal (resp. negative literal) of a variable x_i that occurs in a clause C_j, place a pair of literal markers $\overset{j,t}{\subset}\ \overset{j,t}{\supset}$, $1 \le t \le q$, around a false marker $\overset{i,s}{\triangleleft}$ (resp. true marker $\overset{i,s}{\blacktriangleright}$), $1 \le s \le 2$. The four possible positions of the three pairs of literal markers of each variable x_i are as follows:

$$\begin{array}{c} \subset\overset{1}{\triangleleft}\supset\ \overset{2}{\blacktriangleleft}\ \overset{1}{\triangleright}\ \subset\overset{2}{\blacktriangleright}\supset \qquad \subset\overset{2}{\triangleleft}\supset\ \overset{1}{\blacktriangleleft}\ \overset{2}{\triangleright}\ \subset\overset{1}{\blacktriangleright}\supset \\ \overset{1}{\blacktriangleleft}\ \overset{1}{\triangleleft}\ \overset{1}{\blacktriangleright}\ \overset{1}{\triangleright}\ <\ >\ \overset{2}{\triangleleft}\ \overset{2}{\blacktriangleleft}\ \overset{2}{\triangleright}\ \overset{2}{\blacktriangleright} \end{array}$$

Next, $\breve{G}_2 \rightarrow G_2$. Without loss of generality, assume that the q pairs of literal markers of each clause C_j appear in G_1 with ascending indices:

$$\overset{j,1}{\subset}\ \overset{j,1}{\supset}\ \cdots\ \overset{j,q}{\subset}\ \overset{j,q}{\supset}$$

Insert the q pairs of literal markers in G_2 immediately after the pair of clause markers $\overset{j}{\Subset}\ \overset{j}{\Supset}$, in an interleaving pattern:

$$\overset{j,q}{\subset}\ \cdots\ \overset{j,1}{\subset}\ \overset{j,q}{\supset}\ \cdots\ \overset{j,1}{\supset}$$

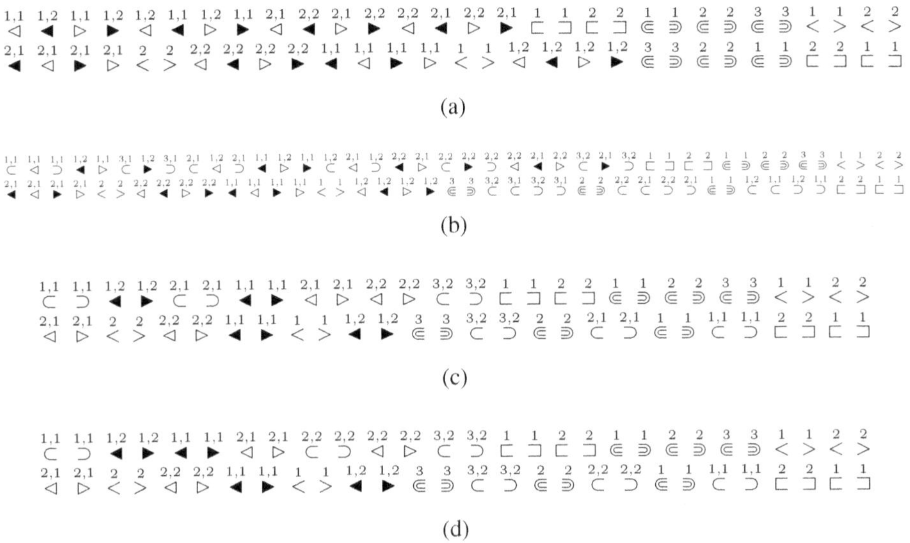

Fig. 2. MSR-2 construction for the E3-Occ-Max-E2-SAT instance $C_1 = x_1 \vee x_2$, $C_2 = x_1 \vee \bar{x}_2$, and $C_3 = \bar{x}_1 \vee \bar{x}_2$. (a) The two sequences $\check{G}_1$ and $\check{G}_2$. (b) The two genomic maps G_1 and G_2. (c) Two canonical subsequences for the assignment $x_1 = \mathit{true}$ and $x_2 = \mathit{false}$. (d) Two other canonical subsequences for the assignment $x_1 = \mathit{true}$ and $x_2 = \mathit{false}$.

This completes the construction. We refer to Figure 2 (a) and (b) for an example of the two steps.

We say that two subsequences of the two genomic maps G_1 and G_2 are *canonical* if each strip of the two subsequences is a pair of markers. Two examples of canonical subsequences are shown in Figure 2 (c) and (d). We have the following lemma on canonical subsequences:

Lemma 4. *If the two genomic maps G_1 and G_2 have two subsequences of total strip length l, then they must have two subsequences of total strip length at least l such that each strip is a pair of markers and, moreover,* (i) *the two pairs of dummy markers are two strips,* (ii) *the m pairs of clause markers and the n pairs of variable markers are $m + n$ strips,* (iii) *at most one pair of literal markers of each clause is a strip,* (iv) *either both pairs of true markers or both pairs of false markers of each variable are two strips.*

Proof. There is an algorithm that transforms the subsequences into canonical form without reducing the total strip length. The algorithm performs incremental operations on the subsequences such that the following eight conditions are satisfied progressively:

1. Each strip that includes a dummy marker is a pair of dummy markers.
2. The two pairs of dummy markers are two strips.
3. Each strip that includes a clause or variable marker is a pair of clause markers or a pair of variable markers.
4. The m pairs of clause markers and the n pairs of variable markers are $m + n$ strips.

5. Each strip that includes a literal marker is a pair of literal markers.
6. At most one pair of literal markers of each clause is a strip.
7. Each strip that includes a true/false marker is a pair of true markers or a pair of false markers.
8. Either both pairs of true markers or both pairs of false markers of each variable are two strips. □

The following lemma establishes the NP-hardness of MSR-2:

Lemma 5. *The variables in X have an assignment that satisfies at least k clauses in $\mathcal{C}$ if and only if the two genomic maps G_1 and G_2 have two subsequences of total strip length at least* $2(3n + m + k + 2)$.

3.2 L-Reduction from Ep-Occ-Max-Eq-SAT to MSR-2

We present an L-reduction (f, g, α, β) from Ep-Occ-Max-Eq-SAT to MSR-3 as follows. The function f, given the Ep-Occ-Max-Eq-SAT instance $(X, \mathcal{C})$, constructs the two genomic maps G_1 and G_2 as in our NP-hardness reduction. Let k^* be the maximum number of clauses in $\mathcal{C}$ that can be satisfied by an assignment of X, and let l^* be the maximum total strip length of two subsequences of G_1 and G_2. Since a random assignment of each variable independently to either true or false with equal probability $\frac{1}{2}$ satisfies each disjunctive clause of q literals with probability $1 - \frac{1}{2^q}$, we have $k^* \geq \frac{2^q-1}{2^q}m$. Then, by Lemma 5, we have

$$l^* = 2(3n+m+k^*+2) = \left(6\,\frac{q}{p} + 2\right) m + 2k^* + 4 \leq \left(\left(6\,\frac{q}{p} + 2\right)\frac{2^q}{2^q - 1} + 2 + \frac{4}{k^*}\right) k^*.$$

The function g, given two subsequences of the two genomic maps G_1 and G_2, transforms the subsequences into canonical form as in the proof of Lemma 4, then returns an assignment of X corresponding to the choices of true or false markers. Let l be the total strip length of the subsequences, and let k be the number of clauses in $\mathcal{C}$ that are satisfied by this assignment. Again, by Lemma 5, we have

$$k^* - k \leq k^* - (l/2 - 3n - m - 2) = (3n + m + k^* + 2) - l/2 = (l^* - l)/2.$$

Let $\epsilon > 0$ be an arbitrary small constant. Note that by brute force we can check whether $k^* < 2/\epsilon$ and, in the affirmative case, compute an optimal assignment of X that satisfies the maximum number of clauses in $\mathcal{C}$, all in $m^{O(1/\epsilon)}$ time, which is polynomial in m for a constant ϵ. Therefore we can assume without loss of generality that $k^* \geq 2/\epsilon$. Then, with the two constants $\alpha = (6\,\frac{q}{p} + 2)\frac{2^q}{2^q-1} + 2 + 2\epsilon$ and $\beta = 1/2$, both properties (1) and (2) of L-reduction are satisfied. In particular, for $p = 3$ and $q = 2$,

$$\alpha\beta = \left(3\frac{q}{p} + 1\right)\frac{2^q}{2^q - 1} + 1 + \epsilon = 5 + \epsilon.$$

Berman and Karpinski [5] showed that it is NP-hard to approximate E3-Occ-Max-E2-SAT within any constant less than $\frac{464}{463} = \frac{1}{1-1/464}$. Thus it is NP-hard to approximate MSR-2 within any constant less than

$$\lim_{\epsilon \to 0} \frac{1}{1 - 1/((5 + \epsilon) \cdot 464)} = \frac{1}{1 - 1/2320} = \frac{2320}{2319} > 1.0004.$$

4 Concluding Remarks

A strip of length l has $l - 1$ *adjacencies* between consecutive markers. In general, k strips of total length l have $l - k$ adjacencies. It can be checked that our L-reductions still work even if the objective function is changed from the total strip length to the total number of adjacencies in the strips. The only effect of this change is that the constant α is halved and correspondingly the constant β is doubled (from $1/2$ to 1). Since the product $\alpha\beta$ is unaffected, Theorem 1 remains valid.

Postscript. This manuscript was written in December 2008. The author was later informed by Binhai Zhu in January 2009 that Lusheng Wang and he [9] had independently and almost simultaneously proved a weaker result that MSR-2 is NP-hard even if all gene markers are distinct in each genomic map.

References

1. Akiyama, J., Chvátal, V.: A short proof of the linear arboricity for cubic graphs. Bull. Liber. Arts & Sci., NMS No. 2, 1–3 (1981)
2. Akiyama, J., Exoo, G., Harary, F.: Covering and packing in graphs III: cyclic and acyclic invariants. Mathematica Slovaca 30, 405–417 (1980)
3. Akiyama, J., Exoo, G., Harary, F.: Covering and packing in graphs IV: linear arboricity. Networks 11, 69–72 (1981)
4. Alimonti, P., Kann, V.: Some APX-completeness results for cubic graphs. Theoretical Computer Science 237, 123–134 (2000)
5. Berman, P., Karpinski, M.: Improved approximation lower bounds on small occurrence optimization. In: Electronic Colloquium on Computational Complexity (2003); Report TR03-008
6. Chen, Z., Fu, B., Jiang, M., Zhu, B.: On recovering syntenic blocks from comparative maps. In: Yang, B., Du, D.-Z., Wang, C.A. (eds.) COCOA 2008. LNCS, vol. 5165, pp. 319–327. Springer, Heidelberg (2008)
7. Chlebík, M., Chlebíková, J.: Complexity of approximating bounded variants of optimization problems. Theoretical Computer Science 354, 320–338 (2006)
8. Papadimitriou, C.H., Yannakakis, M.: Optimization, approximation, and complexity classes. Journal of Computer and System Sciences 43, 425–440 (1991)
9. Wang, L., Zhu, B.: On the tractability of maximal strip recovery. In: Chen, J., Cooper, S.B. (eds.) TAMC 2009. LNCS, vol. 5532, pp. 400–409. Springer, Heidelberg (2009)
10. Zheng, C., Zhu, Q., Sankoff, D.: Removing noise and ambiguities from comparative maps in rearrangement analysis. IEEE/ACM Transactions on Computational Biology and Bioinformatics 4, 515–522 (2007)

The Complexity of Perfect Matching Problems on Dense Hypergraphs

Marek Karpiński[1,⋆], Andrzej Ruciński[2,⋆⋆], and Edyta Szymańska[2,⋆⋆⋆]

[1] Department of Computer Science, University of Bonn
marek@cs.uni-bonn.de
[2] Faculty of Mathematics and Computer Science, Adam Mickiewicz University, Poznań
rucinski,edka@amu.edu.pl

Abstract. In this paper we consider the computational complexity of deciding the existence of a perfect matching in certain classes of dense k-uniform hypergraphs. Some of these problems are known to be notoriously hard. There is also a renewed interest recently in the very special cases of them. One of the goals of this paper is to shed some light on the tractability barriers for those problems.

It has been known that the perfect matching problems are NP-complete for the classes of hypergraphs H with *minimum* $((k-1)-$wise) *vertex degree* δ at least $c|V(H)|$ for $c < \frac{1}{k}$ and trivial for $c \geq \frac{1}{2}$, leaving the status of the problems with c in the interval $[\frac{1}{k}, \frac{1}{2})$ widely open. In this paper we show, somehow surprisingly, that $\frac{1}{2}$, in fact, is not a *threshold* for the tractability of the perfect matching problem, and prove the existence of an $\epsilon > 0$ such that the perfect matching problem for the class of hypergraphs H with δ at least $(\frac{1}{2}-\epsilon)|V(H)|$ is solvable in *polynomial time*. This seems to be the first polynomial time algorithm for the perfect matching problem on hypergraphs for which the existence problem is nontrivial. In addition, we consider parallel complexity of the problem, which could be also of independent interest in view of the known results for graphs.

1 Introduction

In recent years hypergraphs gained a lot of interest as a natural generalization of graphs as well as a model for certain discrete optimization problems. For instance, Asadpour, Feige and Saberi [AFS08] reduced a max-min allocation problem, known as *the Santa Claus Problem*, to finding a perfect matching in a class of bipartite hypergraphs. Since they relied on a rather non-constructive, Hall-type sufficient condition of Haxell [Ha95], they could not solve their problem efficiently.

From the computational point of view, more satisfactory is another, Dirac-type sufficient condition given by Rödl et al. [RRS09]. Recall that the celebrated

⋆ Research supported partly by DFG grants and the Hausdorff Center grant EXC59-1.
⋆⋆ Research supported by grant N201 036 32/2546.
⋆⋆⋆ Research supported by grant N206 017 32/2452.

Y. Dong, D.-Z. Du, and O. Ibarra (Eds.): ISAAC 2009, LNCS 5878, pp. 626–636, 2009.

Dirac theorem for graphs guarantees a Hamilton cycle in every n-vertex graph with minimum degree at least $\frac{1}{2}n$, and thus, a perfect matching when n is even. In [RRS09], the authors used the minimum degree of a $(k-1)$-tuple of vertices in a k-uniform hypergraph and determined the best possible bound on this parameter guaranteeing a perfect matching. If n is sufficiently large and divisible by k, then the threshold values turned out to be close to $\frac{1}{2}n$.

As a consequence, the decision problem asking whether a given k-uniform hypergraph with the minimum $(k-1)$-wise degree above $\frac{1}{2}n$ contains a perfect matching is trivial. Szymańska observed in [Sz09] that the argument presented in [RRS09] can be transformed into a deterministic polynomial time algorithm. Moreover, she also showed that answering the question whether a k-uniform hypergraph with minimum $(k-1)$-wise vertex degree at least $c|V(H)|$, $c < \frac{1}{k}$, contains a perfect matching is NP-complete.

Those results leave a "hardness gap" between $\frac{1}{k}$ and $\frac{1}{2}$. By the counterexamples introduced in [RRS09] it is apparent that in this case there exist hypergraphs of minimum $(k-1)$-wise vertex degree below $\frac{1}{2}|V(H)|$ without a perfect matching. This motivated us to investigate the complexity of the existence problem for hypergraphs in the gap interval. Interestingly, it turned out that at least in the upper end of this interval, when still both answers, *yes* and *no* are possible, the problem is polynomial. Indeed, in this paper we provide a polynomial time algorithm which for every hypergraph of the minimum $(k-1)$-wise vertex degree at least $(\frac{1}{2}-\epsilon)|V(H)|$ constructs a perfect matching if one exists and otherwise exhibits a certificate for non-existence (cf. Theorem 2).

Our second result regards the parallelization of the problem. While the perfect matching problem in graphs can be decided and computed in polynomial time, the parallel complexity of the decision problem remains unknown. Apart from randomized results, only some special classes of graphs have efficient parallel algorithms. This includes dense graphs, in particular Dirac's graphs, that is, graphs with minimum degree $\delta \geq \frac{n}{2}$. Dalhaus, Hajnal and Karpiński in [DKH93] gave an NC^2 parallel algorithm finding a perfect matching in such graphs and showed that for the minimum degree at least cn, $c < \frac{1}{2}$, the problem is as hard as for all graphs. Motivated by the results of [DKH93] and [Sa09], we investigate the parallel complexity of the perfect matching problem in dense hypergraphs. Our Theorem 5 implies that the problem of deciding whether a given $k-$uniform hypergraph H, with minimum $(k-1)$-wise vertex degree at least $c|V(H)|$, $c > \frac{1}{2}$, contains a perfect matching admits an NC algorithm. Along the way, we also design parallel algorithms for constructing *almost* perfect matchings in graphs with restricted $(k-1)$-wise degrees (cf. Theorems 3,4). These algorithms serve as subroutines in the main algorithm.

In the following subsections we introduce our notation, define formally the problems in question and state our results (Theorems 2, 3, 4 and 5, and Proposition 1). Section 2 contains three parallel algorithms and their analysis which proves Theorems 3, 4 and 5. The last section is devoted to an outline of the proof of Theorem 2. A complete proof of Theorem 2, as well as a proof of Proposition 1, will be presented in a full version of the paper.

1.1 Basic Definitions and Notation

Hypergraphs. A *hypergraph* $H = (V, E)$ is a finite set of vertices V together with a family E of distinct, nonempty subsets of V, called edges. In this paper we consider *k-uniform hypergraphs* (or, shortly, *k-graphs*) in which, for a fixed $k \geq 2$, each edge is of size k.

A matching in a hypergraph H is a set $M \subseteq E$ of disjoint edges (we often treat M as a subhypergraph of H and identify M with $E(M)$). The number $|M|$ of edges in a matching M is called *the size* of the matching, while the number of vertices missing from M, that is, the number $|V(H) \setminus V(M)|$ is called *the deficiency* of M in H. Note that the deficiency of any matching in H equals n modulo k. In other words, if $n \equiv q \pmod{k}$, then r-deficient matchings are possible if and only if $r = q + \ell k$ for some $\ell \geq 0$, and such matchings have, of course, size $\lfloor n/k \rfloor - \ell$. A matching is *perfect* if its deficiency is 0, or equivalently if its size is $\frac{1}{k}|V(H)|$. So, a necessary condition for the existence of a perfect matching in H is that $|V(H)| \equiv 0 \pmod{k}$.

For a k-graph H and a set of $k-1$ vertices S, let $N_H(S)$ be the set of edges of H containing S and put $deg_H(S) = |N_S(H)|$. We define $\delta(H) = \min_S deg_H(S)$ and refer to it as the $(k-1)$*-wise, collective minimum degree of* H, or simply, *minimum co-degree*, as we will not consider any other kinds of degrees in hypergraphs.

Furthermore, for all integers $k \geq 2$, $r \geq 0$, and $n \geq k$, denote by $t(k, n, r)$ the smallest integer t such that every k-graph H on n vertices and with $\delta(H) \geq t$ contains an r-deficient matching.

Three classes of computational problems. For $k \geq 2$, by $\mathbf{PM}(k)$ we denote the problem of deciding whether a k-graph $H = (V, E)$ contains a perfect matching. The problem $\mathbf{PM}(2)$ is the classical problem of deciding the existence of a perfect matching in a graph, and is known to be in the polynomial class $\mathcal{P}$ since the paper by Edmonds [Ed65]. For all $k \geq 2$, $\mathbf{PM}(k)$ is equivalent to a decision problem called *exact cover by k-sets*, which is known to be NP-complete for $k \geq 3$ [GJ79].

Given integers $k \geq 3$ and $r \geq 0$, let $\mathbf{PM}(k, r)$ denote the problem of deciding whether a $k-$graph $H = (V, E)$ with $|V(H)| \equiv r \pmod{k}$ contains an r-deficient matching. In particular, when $0 < r < k$, $\mathbf{PM}(k, r)$ asks for a matching in H which, although non-perfect, is as perfect as one can get. Note also that $\mathbf{PM}(k, 0)$=$\mathbf{PM}(k)$.

Given integers $k \geq 3$, $r \geq 0$ and a real $c > 0$, by $\mathbf{PM}(k, r, c)$ we denote the same problem as $\mathbf{PM}(k, r)$ but restricted to $k-$graphs $H = (V, E)$ with minimum co-degree $\delta(H) \geq c|V(H)|$. When $r = 0$, $\mathbf{PM}(k, 0, c)$ can be viewed as the perfect matching problem for dense k-graphs.

1.2 Known Results

Existential Results. Let us begin with the perfect case $r = 0$. For $k = 2$ (graphs), it is very easy to show that, for even n, $t(2, n, 0) = \frac{1}{2}n$. For all integers $k \geq 3$ and sufficiently large $n \geq k$, the value of $t(k, n, 0)$ is exactly determined in

[RRS09]. It is proved there that $t(k,n,0) = \frac{1}{2}n - k + c_{k,n}$, where $c_{k,n}$ is an explicit constant depending on the parities of k, n and $\frac{1}{k}n$, and satisfying $\frac{3}{2} \le c_{k,n} \le 3$. Hence, in particular, $\frac{1}{2}n - k + \frac{3}{2} \le t(k,n,0) \le \frac{1}{2}n - k + 3 \le \frac{1}{2}n$. In [RRSS08] only a slightly weaker upper bound, $t(k,n,0) \le \frac{1}{2}n + \frac{1}{4}k$, but with a simpler proof, was shown.

As for the deficient matchings (case $r > 0$), a striking difference between perfect and almost perfect matchings was observed in [RRS09]. Indeed, it was proved there that for $n \equiv r \pmod{k}$ and $k \ge 3$, $t(k,n,r) = \frac{n-r}{k}$ for $r \ge (k-2)k$, and $\frac{n-r}{k} \le t(k,n,r) \le \frac{n}{k} + O(\log n)$ for $0 < r < (k-2)k$. Thus, in all cases other than the perfect one, the threshold value of $\delta(H)$ for the existence of an r-deficient matching in H is around $\frac{1}{k}n$, while in the perfect case it is around $\frac{1}{2}n$.

Computational Results. An immediate consequence of the results in [RRS09] is that the decision problem $\mathbf{PM}(k,0,c)$ is trivial for every $c \ge \frac{1}{2}$, while $\mathbf{PM}$ (k,r,c), $r > 0$, is trivial already for $c > \frac{1}{k}$. (By trivial we mean that the answer is *yes* for every instance.)

In [Sz09], it was shown by a polynomial reduction of $\mathbf{PM}(k)$ to $\mathbf{PM}(k,r,c)$ that for all $k \ge 3$, $r \ge 0$, and every constant $c < \frac{1}{k}$, $\mathbf{PM}(k,r,c)$ is NP-complete. It follows that $\mathbf{PM}(k,r)$ is NP-complete too, although this can be derived by a direct reduction from $\mathbf{PM}(k)$. Those results have established a "phase transition" at $c = \frac{1}{k}$ for $\mathbf{PM}(k,r,c)$, $r > 0$, but left a hardness gap of $[\frac{1}{k}, \frac{1}{2})$ for $\mathbf{PM}(k,0,c)$.

On the positive side, [Sz09] provided a polynomial time algorithm for the corresponding search problem when $c > \frac{1}{k}$, and $r > 0$. It was also observed in [Sz09] that the existential proof from [RRS09] can be turned into a polynomial time algorithm finding a perfect matching when $c \ge \frac{1}{2}$.

1.3 New Results

One goal of this paper is an attempt to understand the complexity of $\mathbf{PM}(k,0,c)$ in the gap interval $c \in [\frac{1}{k}, \frac{1}{2})$. Theorem 2 below shows that at least in the upper end of the interval the decision problem $\mathbf{PM}(k,0,c)$ is polynomial in time. Another part of this paper is devoted to an alternative, constructive proof of the bound $t(k,n,0) \le \frac{1}{2}n + \frac{1}{4}k$ from [RRSS08]. In fact, we turned that proof into a parallel algorithm (see Theorem 5 below), showing that $\mathbf{PM}(k,0,c)$ is not only in $\mathcal{P}$ but also in the NC class. In the next two subsections we formulate our results in detail.

Hardness Taxonomy. Concerning the problem $\mathbf{PM}(k,0,c)$, the results from [RRS09] and [Sz09] described in Sect. 1.2 have left a hardness gap for $c \in [\frac{1}{k}, \frac{1}{2})$.

Problem 1. What is the computational complexity of $\mathbf{PM}(k,0,c)$ when $c \in [\frac{1}{k}, \frac{1}{2})$?

We present two results which suggest different answers to this problem. To put the first of them into a right context, recall that by [RRS09] we know already that $\mathbf{PM}(k,k,c)$ is trivial for $c > \frac{1}{k}$. In other words, every k-graph H with

$\delta(H) \geq c|V(H)|$, where $c > \frac{1}{k}$ and $|V(H)|$ is divisible by k, has a k-deficient matching.

Proposition 1. *For all $k \geq 3$, $\mathbf{PM}(k)$ is NP-complete even when restricted to k-graphs containing a k-deficient matching.*

It means that knowing that a k-graph has a matching just one edge short from a perfect one, does not help in deciding the existence of the latter. This could suggest that $\mathbf{PM}(k, 0, c)$ is NP-complete for all $c \in [\frac{1}{k}, \frac{1}{2})$. However, it turns out that it is not so. Indeed, in Sect. 3 we describe an algorithm, called PERFECTMATCH, which, for some $c < \frac{1}{2}$ places $\mathbf{PM}(k, 0, c)$ in $\mathcal{P}$.

Theorem 2. *For all $k \geq 3$ there exists $\epsilon > 0$ such that if $c \geq \frac{1}{2} - \epsilon$, then $\mathbf{PM}(k, 0, c)$ and its search version are in $\mathcal{P}$.*

Remark 1. Theorem 2 reveals an interesting feature: it provides a polynomial time algorithm which, unlike the algorithms in [DKH93], [Sa09], [Sz09], or those described in the next section, takes as inputs instances which may not posses a desired matching, and decides whether they indeed have one. If the answer is *yes*, the algorithm, in fact, computes in polynomial time a perfect matching, while when the answer is *no*, it provides an evidence (in a form of a witness).

Parallel Algorithms. As the model of computation we choose the version EREW PRAM. Recall that, as shown in [RRS09], the problem $\mathbf{PM}(k, 0, \frac{1}{2})$ is trivial, that is, for all H with $\delta(H) \geq \frac{1}{2}n$, H has a perfect matching. As observed in [Sz09], the existential proof from [RRS09] can be turned into a polynomial time search algorithm of complexity $O(n^{k^2+2k} \log^4 n)$. Here we present a parallel algorithm which places the search version of $\mathbf{PM}(k, 0, c)$, $c > \frac{1}{2}$, in the class NC. Recall that $NC = \bigcup_i NC^i$, and a problem is in NC^i if it admits an algorithm of running time $O(\log^i n)$, using a polynomial number of processors.

Our algorithm, PAR-PERFECTMATCH, is based on the existential proof in [RRSS08] and uses as subroutines two other parallel algorithms of independent interest, PAR-LARGEDEFMATCH(r) and PAR-SMALLDEFMATCH(r), which find r-deficient matchings for, resp., large and small, positive values of r, under increasingly restrictive conditions on $\delta(H)$.

The properties of these algorithms are presented in the following theorems. The first of them provides a parallel algorithm which finds an r-deficient matching for large r, but relatively small δ.

Theorem 3. *For every $k \geq 3$ and $r \geq (k-2)k$ there exists a constant n_0, and a parallel algorithm, called PAR-LARGEDEFMATCH(r), which in every k-graph H on $n \geq n_0$ vertices with $n \equiv r \pmod{k}$ and $\delta(H) \geq \frac{n-r}{k}$ finds an r-deficient matching in $O(\log^3 n)$ rounds using a polynomial number of processors. It follows that the search version of $\mathbf{PM}(k, r, c)$ is in the class NC^3 for $r \geq (k-2)k$ and $c \geq \frac{1}{k}$.*

If the degree condition is strengthened just a little, we can find in parallel a matching of any smaller, but positive, deficiency r. The algorithm PAR-SMALLDEFMATCH(r), given below, uses the algorithm from Theorem 3 as a subroutine.

Theorem 4. *For every $k \geq 3$ and $0 < r < (k-2)k$ there exist constants n_0 and $C > 0$, and a parallel algorithm, called* PAR-SMALLDEFMATCH(r)*, which in every k-graph on $n \geq n_0$ vertices with $n \equiv r \pmod{k}$ and $\delta(H) \geq \frac{n}{k} + C \log n$ finds an r-deficient matching in a polylogarithmic number of rounds using a polynomial number of processors. It follows that the search version of* $\mathbf{PM}(k, r, c)$ *is in the class NC for $0 < r < (k-2)k$ and $c > \frac{1}{k}$.*

Finally, if $\delta(H)$ exceeds $\frac{1}{2}n$, then we are in position to compute in parallel a perfect matching in H. This is the main result of this section.

Theorem 5. *For every $k \geq 3$ there exists constant n_0, and a parallel algorithm, called* PAR-PERFECTMATCH*, which in every k-uniform hypergraph on $n \geq n_0$ vertices with n divisible by k and such that $\delta(H) \geq \frac{n}{2} + \frac{k}{4}$ finds a perfect matching in a polylogarithmic number of rounds using a polynomial number of processors. It follows that the search version of* $\mathbf{PM}(k, 0, c)$ *is in the class NC for $c > \frac{1}{2}$.*

The above three theorems will be proved in the next section. A summary of all computational results about $\mathbf{PM}(k, r, c)$ is displayed in Table 1.

Table 1. The complexity of $\mathbf{PM}(k, r, c)$ with $k \geq 3$. For every *t=trivial* problem there exists an *NC parallel* algorithm finding an r-deficient matching.

r \ c	$c < \frac{1}{k}$	$\frac{1}{k}$	$(\frac{1}{k}, \frac{1}{2} - \epsilon)$	$(\frac{1}{2} - \epsilon, \frac{1}{2}]$	$c > \frac{1}{2}$
$r \geq (k-2)k$	NP-com	t	t	t	t
$0 < r < (k-2)k$	NP-com	?	t	t	t
$r = 0$	NP-com	?	?	in $\mathcal{P}$	t

2 Description and Analysis of Parallel Algorithms

In this section we prove Theorems 3–5. Each proof consists of a description of the algorithm followed by a proof of its correctness.

2.1 Proof of Theorem 3

The construction below generalizes the ideas from [DKH93] to hypergraphs. *The intersection graph* of a hypergraph H has the edges of H as its vertices, and two vertices are adjacent if the corresponding edges of H intersect. Observe that the matchings in H map one-to-one with the independent sets of the intersection graph. When we refer to *a MIS algorithm*, we always mean a parallel algorithm from [Lu86] which places the maximal independent set problem in NC^2.

ALGORITHM. PAR-LARGEDEFMATCH(r), $r \geq (k-2)k$

In: k-graph H with $n \geq n_0$, $n \equiv r \pmod{k}$ and $\delta(H) \geq \frac{n-r}{k}$
Out: r-deficient matching M_1

1. Compute **in parallel** a maximal matching M_1 in H applying a MIS algorithm to the intersection graph of H. Let $W := V(H) \setminus V(M_1)$.
2. Repeat while $|W| > r$
   (a) Arbitrarily divide W into $t := \left\lfloor \frac{|W|}{(k-1)k} \right\rfloor$ disjoint sets S of size $|S| = (k-1)k$. Call this family of sets $\mathcal{S}$. Define an auxiliary bipartite graph $G = (V_1, V_2, E(G))$ as follows:
   - $V_1 = M_1$ and $V_2 = \mathcal{S}$; thus $|V_2| = t$.
   - For each $e \in V_1$ and $S \in V_2$ put **in parallel** an edge $\{e, S\} \in E(G)$ if and only if there are two vertices $u_e, v_e \in e$, $u_e \neq v_e$ and two disjoint $(k-1)$–element subsets X_S, Y_S of S such that $e'_{e,S} := X_S \cup \{u_e\} \in H$ and $e''_{e,S} := Y_S \cup \{v_e\} \in H,$.

   (b) Compute **in parallel** a maximal matching M_2 in G using a parallel MIS algorithm.
   (c) For every edge $(e, S) \in M_2$ **in parallel** absorb into M_1 the set of vertices $X_S \cup Y_S$, by replacing e with $e'_{e,S}$ and $e''_{e,S}$, at the same time releasing from M_1 the remaining $k-2$ vertices of e, i.e., $M_1 := (M_1 - \{e\}) \cup \{e'_{e,S}, e''_{e,S}\}$. Set $W := V(H) \setminus V(M_1)$.
3. Return M_1.

To show that the above algorithm computes a desired matching we need the following fact.

Fact 6. *Any maximal matching M_2 in the bipartite graph G defined in the algorithm saturates every vertex of V_2, that is, $V(M_2) \supseteq V_2$.*

The algorithm PAR-LARGEDEFMATCH(r) finds an r-deficient matching in $O(\log n)$ iterations and thus its time complexity is $O(\log^3 n)$. Note that in the case of graphs discussed in [DKH93], only one iteration in step 2 was sufficient, saving one logarithmic factor in time complexity.

2.2 Proof of Theorem 4

Let us begin by noting that without loss of generality we may restrict the range of r to $0 < r \leq k$. Indeed, if $r_1 < r_2$ and $r_i \equiv n \pmod{k}$, $i = 1, 2$, then any r_1-deficient matching contains an r_2-deficient matching.

The algorithm PAR-SMALLDEFMATCH presented below uses as subroutine PAR-LARGEDEFMATCH. In addition, following the absorbing technique introduced in [RRS09], we will need another parallel subroutine which computes a so called powerful matching.

Definition 7 (absorbing edge,[RRS09]). *Given a set S of $k+1$ vertices, an edge $e \in H$ is called* S-absorbing *if there are two disjoint edges e' and e'' in H such that $|e' \cap S| = k-1$, $|e' \cap e| = 1$, $|e'' \cap S| = 2$ and $|e'' \cap e| = k-2$.*

The key feature of the absorbing edge is that there are $\Theta(n^k)$ of them for every set S in the input hypergraph H (see Fact 2.2 in [RRS09]).

Definition 8 (powerful matching). *A matching M in a k-graph H is called* powerful *if for every set $S \subset V$ of size $k+1$ the number of S-absorbing edges in M is at least $k-2$.*

To construct a small, powerful matching in H, we first create an auxiliary graph $G = (X \cup Y, E)$, where X is an independent set. The vertices in Y represent all matchings in H of size $k-2$, while the vertices in X represent the families $\mathcal{F}_S$ of all matchings of size $k-2$ consisting of S-absorbing edges, where S runs through all subsets of vertices of size $k+1$. The xy edges of G, where $x \in X$ and $y \in Y$, exhibit the membership of the matchings in the families $\mathcal{F}_S$, while the $y'y''$ edges, where $y', y'' \in Y$, indicate whether the two matchings represented by y' and y'' have a vertex in common. Now, our goal is to construct an independent subset D of Y of size $O(\log n)$ which dominates all vertices of X. Then the union of the $(k-2)$-matchings represented by the vertices of D forms the desired powerful matching in H. This can be done efficiently in parallel if

$$deg_G(x) \geq c|Y| \text{ for all } x \in X \text{ and } \Delta(G[Y]) = o\left(\frac{1}{\log n}|Y|\right). \qquad (1)$$

ALGORITHM. PAR-INDDOMSET

In: graph $G = (X \cup Y, E)$, $G[X] = \emptyset$, satisfying (1)
Out: independent subset $D \subseteq Y$ dominating X, $|D| = O(\log n)$

1. Repeat until $X = \emptyset$:
   (a) For all $y \in Y$ compute **in parallel** $deg_G(y, X)$; set y_0 for the lexicographically first y for which $deg_G(y, X) \geq \frac{c}{2}|X|$;
   (b) Set $D := D \cup \{y_0\}$; $X := X \setminus \{x : xy_0 \in E\}$,
   $Y := Y \setminus (\{y_0\} \cup \{y : yy_0 \in E\})$
2. Return D.

ALGORITHM. PAR-SMALLDEFMATCH(r), $0 < r \leq k$

In: k-graph H with $\delta(H) \geq \frac{n}{k} + C \log n$ and $n \geq n_0$, $n \equiv r \pmod{k}$.
Out: r-deficient matching M

1. Compute a powerful matching M_0 ($|M_0| \leq \frac{1}{k} C \log n$), as in Definition 8, applying PAR-INDDOMSET to the auxiliary graph G described above.
2. $H' := H - V(M_0)$, [notice that $\delta(H') \geq \frac{1}{k}|V(H')|$].
3. Compute a $(k(k-2)+r)$ –deficient matching M_1 using algorithm PAR-LARGEDEFMATCH$(k(k-2)+r)$ in H'.
4. $T := V(H) \setminus (V(M_1) \cup V(M_0))$, [notice that $|T| = k(k-2)+r$].

5. Repeat until $|T| = r$: [$k - 2$ sequential iterations]
   (a) for an arbitrary set $S \subseteq T$, $|S| = k+1$, and an $S-$absorbing edge $e \in M_0$, set $M_0 := M_0 \setminus \{e\} \cup \{e', e''\}$, where e', e'' are as in Definition 7;
   (b) $T := V(H) \setminus V(M_0' \cup M_1)$.
6. Return $M := M_0 \cup M_1$.

It is clear that the above algorithm returns an r-deficient matching in a poly-logarithmic number of steps.

2.3 Proof of Theorem 5

In our construction we will apply an *absorbing configuration* motivated by the proof in [RRSS08].

Definition 9 (absorbing configuration). Given a set $S = \{x_1, x_2, \ldots, x_k\}$ of k vertices, a triplet of vertex disjoint edges $e_1, e_2, e_3 \in H$ is called *S-absorbing configuration* if there are four disjoint edges f_1, f_2, f_3 and f_4 in H such that $f_1 \cap e_1 = \{v\}$, $|f_1 \cap e_2| = k - 1$, $f_2 \cap e_1 = \{w\}$, $f_2 = \{u\} \cup \{x_1, \ldots, x_{k-1}\}$ and $f_3 \cap e_3 = T$ and $f_4 = \{x_k\} \cup (e_3 - T) \cup (e_1 - \{v, w\})$.

ALGORITHM PAR-PERFECTMATCH

In: k-graph H with $\delta(H) \geq \frac{n}{k} + \frac{k}{4}$ and $n \geq n_0$, $n \equiv 0 \pmod{k}$.
Out: perfect matching M

1. Compute a k-deficient matching M_1 using the parallel algorithm. PAR-SMALLDEFMATCH(k) in H.
2. $T := V(H) - V(M_1)$.
3. For every triple of edges $e_1, e_2, e_3 \in M_1$, **in parallel** check if they span an absorbing configuration as in Definition 9.
4. Use an absorbing configuration found in step 3 to absorb the vertices of T and obtain a perfect matching M.
5. Return M.

The existence of an absorbing configuration in a hypergraph H with $\delta(H) \geq \frac{n}{k} + \frac{k}{4}$, searched for in step 3, is guaranteed by the proof in [RRSS08].

3 Toward Understanding the Hardness Gap (the Proof of Theorem 2)

Claim 5.2 in [RRS09] asserts that if H is a k-graph on $n > n_0$ vertices, n divisible by k, $\delta(H) \geq (\frac{1}{2} - \epsilon)n$, and at least one of two conditions, (i) or (ii), holds, then H has a perfect matching.

We say that a partition $V(H) = A \cup B$ is *even-complete* [*odd-complete*] if for all even [odd] r, the subhypergraphs $E_r := E_r(A, B)$ and $K_r(A, B)$ differ only by an ϵ'-fraction of edges, where ϵ' is a function of ϵ which tends to zero

---

Algorithm. PerfectMatch

In: k-graph H with $\delta(H) \geq (\frac{1}{2} - \epsilon)n$ and $n \geq n_0$, $n \equiv 0 \pmod{k}$.
Out: YES if H has a perfect matching, NO otherwise.

1. If (i) or (ii) hold, return YES.
2. Otherwise, let $A := N_H(S_1)$, $B := V \setminus A$, where $(S_1, \ldots, S_k)$ is a k-tuple violating (i).
3. If k is odd and (A, B) is odd-complete, swap A and B around;
4. If (A, B) is even-complete set $k' = k - 1$ if k is odd and $k' = k - 2$ otherwise and do:
   (a) Identify the set S of all 0.3-small vertices of $E_{k'}$ and move them to the other side, that is, reset $A := A \triangle S$ and $B := B \triangle S$.
   (b) If $|A|$ is even or $\bigcup_{r\ odd} E_r \neq \emptyset$, return YES
   (c) Return NO
5. If (A, B) is odd-complete (and so k is even) set $k' = \frac{k}{2} + 1$ if k is divisible by 4 and $k' = \frac{k}{2}$ otherwise and do:
   (a) Identify the set S of all 0.3-small vertices of $E_{k'}$; reset $A := A \triangle S$ and $B := B \triangle S$.
   (b) If $|A| \equiv \frac{n}{k} \pmod{2}$ or $\bigcup_{r\ even} E_r \neq \emptyset$, return YES.
   (c) Return NO.

---

when ϵ does. (For the definitions of $E_r := E_r(A, B)$ and $K_r(A, B)$, as well as of c-small vertices in E_r, see [RRS09], Sect. 4.1)

Because of the lack of space here we give the correctness proof in the full version of the paper.

References

[AFS08] Asadpour, A., Feige, U., Saberi, A.: Santa Claus Meets Hypergraph Matchings. In: Proc. of APPROX-RANDOM 2008, pp. 10–20 (2008)

[DKH93] Dalhaus, E., Hajnal, P., Karpiński, M.: On the parallel complexity of Hamiltonian cycle and matching problem on dense graphs. J. Alg. 15, 367–384 (1993)

[Ed65] Edmonds, J.: Paths, trees and flowers. Canad J. Math. 17, 449–467 (1965)

[GJ79] Garey, M.R., Johnson, D.S.: Computers and intractability. Freeman, New York (1979)

[Ha95] Haxell, P.E.: A Condition for Matchability in Hypergraphs. Graphs and Combinatorics 11, 245–248 (1995)

[KO06] Kühn, D., Osthus, D.: Critical chromatic number and the complexity of perfect packings in graphs. In: 17th ACM-SIAM Symposium on Discrete Algorithms 2006 (SODA), pp. 851–859 (2006)

[Lu86] Luby, M.: A simple parallel algorithm for the maximal independent set problem. SIAM J. Comput. 15(4), 1036–1053 (1986)

[RRS09] Rödl, V., Ruciński, A., Szemerédi, E.: Perfect matchings in large uniform hypergraphs with large minimum collective degree. JCT A 116(3), 613–636 (2009)

[RRSS08] Rödl, V., Ruciński, A., Schacht, M., Szemerédi, E.: A note on perfect matchings in uniform hypergraphs with large minimum collective degree. Commen. Math. Univ. Carol. 49(4), 633–636 (2008)

[Sz09] Szymańska, E.: The Complexity of Almost Perfect Matchings in Uniform Hypergraphs with High Codegree. In: Proc. of IWOCA (2009) (to appear)

[Sa09] Särközy, G.: A fast parallel algorithm for finding Hamiltonian cycles in dense graphs. Discrete Mathematics 309, 1611–1622 (2009)

GPSR Compliance

The European Union's (EU) General Product Safety Regulation (GPSR) is a set of rules that requires consumer products to be safe and our obligations to ensure this.

If you have any concerns about our products, you can contact us on ProductSafety@springernature.com

In case Publisher is established outside the EU, the EU authorized representative is:

Springer Nature Customer Service Center GmbH
Europaplatz 3
69115 Heidelberg, Germany

Batch number: 10761593

Printed by Printforce, the Netherlands

Lecture Notes in Computer Science 5878

Commenced Publication in 1973
Founding and Former Series Editors:
Gerhard Goos, Juris Hartmanis, and Jan van Leeuwen

Yingfei Dong Ding-Zhu Du Oscar Ibarra (Eds.)

Algorithms and Computation

20th International Symposium, ISAAC 2009
Honolulu, Hawaii, USA, December 16-18, 2009
Proceedings

Springer

Volume Editors

Yingfei Dong
University of Hawaii
Deptartment of Electrical and Computer Engineering
Holmes Hall 483
2540 Dole Street
Honolulu, HI 96822, USA
E-mail: yingfei@hawaii.edu

Ding-Zhu Du
University of Texas at Dallas
Department of Computer Science
Richardson, TX 75080, USA
E-mail: dzdu@utdallas.edu

Oscar Ibarra
University of California
Department of Computer Science
Santa Barbara, CA 93106, USA
E-mail: ibarra@cs.ucsb.edu

Library of Congress Control Number: 2009939433

CR Subject Classification (1998): F.2, G.1, G.2, G.3, G.4, I.2.2, I.3, D.2

LNCS Sublibrary: SL 1 – Theoretical Computer Science and General Issues

ISSN 0302-9743
ISBN-10 3-642-10630-7 Springer Berlin Heidelberg New York
ISBN-13 978-3-642-10630-9 Springer Berlin Heidelberg New York

springer.com

Printed in Germany

Typesetting: Camera-ready by author, data conversion by Scientific Publishing Services, Chennai, India
Printed on acid-free paper SPIN: 12805785 06/3180 5 4 3 2 1 0

Preface

The papers in this volume were presented at the 20th Annual International Symposium on Algorithms and Computation, held December 16–18, 2009, in Hawaii, USA. In response to the Call-for-Papers, 279 papers were submitted. Each paper received at least three reviews by either Program Committee members or experts selected by Program Committee members. In all, 120 papers were selected based on the review reports and are included in this volume. We wish to thank all who submitted papers for consideration and all Program Committee members and reviewers for their excellent and hard work. To our invited speakers, we wish to thank you for sharing your expertise. Finally, we wish to thank our colleagues who contributed to the success of the symposium and the sponsors for their assistance and support.

December 2009

Yingfei Dong
Ding-Zhu Du
Oscar Ibarra

Organization

Program Committee Co-chairs

Yingfei Dong	University of Hawaii, USA
Ding-Zhu Du	University of Texas at Dallas, USA
Oscar Ibarra	University of California, Santa Barbara, USA

Advisory Committee Chair

Takeshi Tokuyama	Tohoku University, Japan

Advisory Committee

Tetsuo Asano	Japan Advanced Institute of Science and Technology, Japan
Kyung-Yong Chwa	Korea Advanced Institute of Science and Technology, Republic of Korea
Francis Y.L. Chin	The University of Hong Kong, Hong Kong
Ding-Zhu Du	The University of Texas at Dallas, USA
Peter Eades	The University of Sydney, Australia
Wen Lian Hsu	Academia Sinica, Taiwan
Der Tsai Lee	Academia Sinica, Taiwan
Takao Nishizeki	Tohoku University, Japan

Local Arrangements Chair

Yingfei Dong	University of Hawaii, USA

Publicity, Webmaster, and Online Conference System (OCS) Co-chairs

Xiaofeng Gao	University of Texas at Dallas, USA
Donghyun Kim	University of Texas at Dallas, USA

Financial Co-chairs

Jiaofei Zhong	University of Texas at Dallas, USA
Feng Zou	University of Texas at Dallas, USA

Program Committee

Alberto Apostolico	Georgia Insititute of Technology, USA
Giorgio Ausiello	Università di Roma, Italy
Wolfgang Bein	University of Nevada, USA
Jin-Yi Cai	University of Wisconsin, USA
Zhipeng Cai	Mississippi State University, USA
Zhenfu Cao	Shanghai Jiaotong University, China
Wei Chen	Microsoft Research Asia, China
Maggie Cheng	Missouri University of Science and Technology, USA
Siu-Wing Cheng	Hong Kong University of Science and Technology, Hong Kong
Yongxi Cheng	University of Alberta, Canada
Marek Chrobak	University of California, Riverside, USA
Kyung-Yong Chwa	Korea Advanced Institute of Science and Technology, Republic of Korea
Bhaskar Dasgupta	University of Illinois at Chicago, USA
Narsingh Deo	University of Central Florida, USA
Omer Egecioglu	University of California, Santa Barbara, USA
Juraj Hromkovic	ETH Zentrum, Switzerland
Wenlian Hsu	Academia Sinica, Taiwan
Xiaodong Hu	Chinese Academy of Sciences, China
Kazuo Iwama	Kyoto University, Japan
Ravi Janardan	University of Minnesota, USA
Tao Jiang	University of California, Riverside, USA
Liying Kang	Shanghai University, China
Samir Khuller	University of Maryland, USA
Angsheng Li	Chinese Academy of Sciences, China
Shlomo Moran	Technion, USA
Koji Nakano	Hiroshima University, Japan
Hung Ngo	University of Buffalo, US
Mitsunori Ogihara	University of Miami, USA
E.K. Park	National Science Foundation, USA
Suneeta Ramaswami	Rutgers University, USA
Sanjay Ranka	University of Florida, USA
Sanguthevar Rajasekaran	University of Connecticut, USA
Balasubramanian Ravikumar	Sonoma State University, USA
Paul Spirakis	University of Patras, Greece

Kazuo Sugihara	University of Hawaii, USA
Xiaoming Sun	Tsinghua University, China
Subhash Suri	University of California, Santa Barbara, USA
My T. Thai	University of Florida, USA
Takeaki Uno	NII, Japan
Jie Wang	University of Massachusetts Lowell, USA
Lusheng Wang	City University of Hong Kong, Hong Kong
Wei Wang	Xi'an Jiaotong University, China
Weili Wu	University of Texas at Dallas, USA
Yibo Xue	Tsinghua University, China
Hsu-Chun Yen	National Taiwan University, Taiwan
Wenan Zang	Hong Kong University, Hong Kong
Zhao Zhang	Xingjiang University, China
Hong Zhu	Huadong Normal University, China

Referees

Abdullah Arslan
Sang Won Bae
Gill Barequet
Doina Bein
Said Bettayeb
Hans-J. Boeckenhauer
Vincenzo Bonifaci
Peter Brucker
Tian-Ming Bu
Saverio Caminiti
B. Charron-Bost
Zhi-Zhong Chen
Ho-Lin Chen
Yinjie Chen
Bhadrachalam Chitturi
Ovidiu Daescu
Mirela Damian
Thomas Erlebach
Zheng Fang
Paolo Ferragina
Robin Flatland
Hervé Fournier
Paolo Giulio Franciosa
Rajiv Gandhi
Xiaofeng Gao
Bill Gauvin
Mordecai Golin
Prosenjit Gupta
Xin Han
Rafi Hassin
Tomio Hirata
Junwei Huang
Keiko Imai
Volkan Isler
Naoyuki Kamiyama
Donghyun Kim
Sangjin Kim
Carl Kingsford
Dennis Komm
Richard Kralovic
Zbyněk Křivka
Yokesh Kumar
Luigi Laura
Ronny Lempel
Jian Li
Chun-Cheng Lin
Hsueh-I Lu
Pinyan Lu
Joe Mitchell
Shuichi Miyazaki
Tobias Moemke
Linda Morales
Pat Morin
Hiroki Morizumi
Hiroshi Nagamochi
Harumichi Nishimura
Yoshio Okamoto
Joseph O'Rourke
Sheung-Hung Poon
Iris Reinbacher
Kunihiko Sadakane
Yan Shi
Yi Shi
Charles Shields Jr.
Marcelo Siqueira
Sagi Snir
Bettina Speckmann
Andreas Sprock
William Steiger
Monika Steinova
Hal Sudborough
Hisao Tamaki
Suguru Tamaki
Ramki Thurimella
Takeshi Tokuyama
Shi-Chun Tsai
Ryuhei Uehara
Antoine Vigneron
Xiang Wan
Yajun Wang
Zhong Wang
James Willson
Yulai Xie
Jinhui Xu

Bo Yan
Jie Yang
Sivan Yogev
Neal Young
Tian-Li Yu
Shmuel Zaks
Guochuan Zhang
Wei Zhang
Jiaofei Zhong
Binhai Zhu
Feng Zou

Sponsoring Institutions

The University of Hawaii, USA
The University of Texas at Dallas, USA

On Lower Bounds for Constant Width Arithmetic Circuits

V. Arvind, Pushkar S. Joglekar, and Srikanth Srinivasan

Institute of Mathematical Sciences
C.I.T Campus, Chennai 600 113, India
{arvind,pushkar,srikanth}@imsc.res.in

Abstract. For every $k > 1$, we give an explicit polynomial that is computable by a linear-sized monotone circuit of width $2k$ but has no subexponential-sized monotone circuit of width k. As a consequence we show that the constant-width and the constant-depth hierarchies of monotone arithmetic circuits (both commutative and noncommutative) are infinite. We also prove hardness-randomness tradeoffs for identity testing of constant-width circuits analogous to [6,4].

1 Introduction

Nisan, in a seminal paper [9], showed exponential size lower bounds for noncommutative formulas (and noncommutative Algebraic Branching Programs (ABPs) – see [9,10] for the definition of ABPs) that compute the noncommutative permanent or determinant polynomials in the ring $\mathbb{F}\langle X\rangle$, where $X = \{x_1, \cdots, x_n\}$ are noncommuting variables. By Ben-Or and Cleve's result [3], we know that bounded-width arithmetic circuits (both commutative and noncommutative) are at least as powerful as formulas (indeed, width three is sufficient). Our motivation is whether we can Nisan's result [9] to show size lower bounds for *noncommutative bounded-width* circuits.

An *arithmetic circuit* over a field $\mathbb{F}$ and variables $x_1, x_2, \cdots, x_n$ is a directed acyclic graph with each node of indegree zero labeled by a variable or a scalar constant. Internal nodes of the DAG are of indegree two and are labeled by + or × (indicating a plus or multiply gate). A node of the DAG is designated as the output gate. Each internal node computes a polynomial (by adding or multiplying its input polynomials). The *polynomial computed* by the circuit is the polynomial computed at the output gate. The *size* of a circuit is the number of nodes in it.

A *layered circuit* has its vertex set partitioned into sets $V_1 \cup V_2 \cup \ldots \cup V_t$ such that (i) V_1 contains all the indegree zero nodes, (ii) Each child of an internal node $g \in V_i$, for $i > 1$, is either in V_1 or in V_{i-1}. The *width* of a layered circuit is $\max_{i>1} |V_i|$.

An arithmetic circuit over the field $\mathbb{R}$ is *monotone* if all the scalars used are nonnegative. Finally, a layered arithmetic circuit is *staggered* if, in each layer $i > 1$, all nodes except possibly one is a product gate of the form $g = u \times 1$, for some gate u from the previous layer.

Y. Dong, D.-Z. Du, and O. Ibarra (Eds.): ISAAC 2009, LNCS 5878, pp. 637–646, 2009.

Width-w staggered circuits are essentially the same as straight-line programs with w registers. A width $w-1$ arithmetic circuit can be easily converted to a width-w staggered circuit (for both commutative and noncommutative circuits). More precisely, for any layered arithmetic circuit of width w and size s, there is a staggered arithmetic circuit, computing the same polynomial, of width $w+1$ and size $O(ws)$.

Ben-Or and Cleve's seminal result on bounded-width circuits [3] shows that size s arithmetic formulas (commutative or noncommutative) can be evaluated by staggered arithmetic circuits of width three and size $O(s^2 n)$. Bounded width circuits are studied under various restrictions in [7,8,5]. However, these papers have not considered proving explicit lower bounds.

We first observe that width-2 arithmetic circuits are universal.

Proposition 1. *Any polynomial $f \in \mathbb{F}[x_1, x_2, \cdots, x_n]$ (or in $\mathbb{F}\langle x_1, \cdots, x_n\rangle$) of degree d with s monomials can be computed by a width two arithmetic circuit of size $O(d \cdot s)$. Furthermore, if $\mathbb{F} = \mathbb{R}$ and f is monotone, it can be computed by a width-two size $O(d \cdot s)$ monotone circuit.*

We note that Nisan's rank argument [9] is not useful for proving lower bounds for *noncommutative* bounded width circuits. For the noncommutative "palindromes" polynomial $P(x_0, x_1) = \sum_{w \in \{x_0, x_1\}^n} ww^R$, the communication matrix $M_n(P)$ is of rank 2^n and hence any noncommutative ABP for it is exponentially large [9]. However, we can easily give a width-2 noncommutative arithmetic circuit for $P(x_0, x_1)$ of size $O(n)$. Indeed, we can even ensure that each gate in this circuit is *homogeneous*.

Proposition 2. *The palindromes polynomial $P(x_0, x_1)$ has a width-2 noncommutative arithmetic circuit of size $O(n)$.*

We believe that a good candidate explicit polynomial not computable by width-2 circuits of polynomial size is P_k^ℓ (defined in Section 2) for suitable k. A lower bound argument still eludes us. However, if we consider *monotone* constant-width circuits then even in the commutative case we can show exponential size lower bounds for monotone width-k circuits computing P_k^ℓ. Since P_k^ℓ is computable by depth $2k$ arithmetic circuits (of unbounded fanin), it follows that the constant-width and the constant-depth hierarchies of monotone arithmetic circuits are infinite. We present these results in Section 2.

Remark 1. Regarding the separation of the *constant-depth* hierarchy of monotone circuits, a separation is shown by Raz and Yehudayoff in [11]. They show a superpolynomial separation between depth k *multilinear* circuits and depth $k+1$ monotone circuits for all $k \geq 2$. In contrast, our separation works only for monotone circuits, and only for infinitely many k. Nonetheless, we show stronger separations. More precisely, [11] shows a separation of $2^{(\log s)^{1+\Omega(1/k)}}$ (i.e. there is a polynomial computable by circuits of depth $k+1$ and size s but not by depth k circuits of size $2^{(\log s)^{1+\Omega(1/k)}}$). On the other hand, our separation is at least $2^{(\log s)^c}$ for any $c > 0$ (see the full version [1] for details).

A related question is the comparative power of noncommutative ABPs and noncommutative formulas. Noncommutative formulas have polynomial size noncommutative ABPs. However, $s^{O(\log s)}$ is the best known formula size upper bound for noncommutative ABPs of size s. An interesting question is whether we can prove a separation result. A separation in the *monotone* case can be easily derived from an old result of Snir [12]. Let $\{x_0, x_1\}$ be noncommuting variables. Let L be the set of all monomials of degree $2n$ with an equal number of x_0 and x_1, and consider the polynomial $E = \sum_{w \in L} w$. We can show that E is computable by a monotone homogeneous ABP of size $O(n^2)$, whereas any monotone formula for E has size $n^{\Omega(\log n)}$.

To illustrate again the power of constant-width circuits, there is a width-2 circuit of $n^{O(1)}$ size for computing the polynomial E if the field $\mathbb{F}$ has at least cn^2 distinct elements for a constant c. This is based on the well-known Ben-Or trick [2] for computing the elementary symmetric polynomials in depth 3. These observations are additional motivation for the study of constant-width arithmetic circuits. In this extended abstract several proofs are omitted due to lack of space. The full version of this paper is available on the ECCC [1].

2 Monotone Constant Width Circuits

We show that *monotone* constant-width arithmetic circuits form an infinite hierarchy. Our construction implies that constant-depth monotone arithmetic circuits too form an infinite hierarchy. For positive integers k and ℓ we define a polynomial P_k^ℓ on ℓ^{2k} variables. For each ℓ let $P_1^\ell(x_1, x_2, \ldots, x_{\ell^2}) = \sum_{i=1}^{\ell} \prod_{j=1}^{\ell} x_{(i-1)\ell+j}$. Given P_k^ℓ, define for each ℓ

$$P_{k+1}^\ell(x_1, x_2, \ldots, x_{\ell^{2k+2}}) = \sum_{i=1}^{\ell} \prod_{j=1}^{\ell} P_k^\ell(x_{(i-1)\ell^{2k+1}+(j-1)\ell^{2k}+1}, \ldots, x_{(i-1)\ell^{2k+1}+j\ell^{2k}}).$$

Clearly, P_k^ℓ is a homogeneous polynomial of degree ℓ^k on ℓ^{2k} variables computable by a depth $2k$ monotone formula of size $O(\ell^{2k})$. Furthermore, P_k^ℓ are the "hardest" polynomials for constant-depth circuits.

Proposition 3. *Given a depth k arithmetic circuit C of size s, there is a projection reduction from C to the polynomial P_k^ℓ where $\ell = O(s^{2k})$.*

It is easy to see the following from the fact that a monotone depth $2k$ arithmetic circuit of size s can be simulated by a monotone width $2k$ circuit of size $O(s)$.

Proposition 4. *For any positive integers ℓ and k there is a monotone circuit of width $2k$ and size $O(\ell^{2k})$ that computes P_{2k}^ℓ.*

For a polynomial $f \in \mathbb{F}[X]$, where $X = \{x_1, x_2, \cdots, x_n\}$ let mon(f) denotes the set of nonzero monomials in the polynomial f. Let var(f) be the set of all variables occurring in the monomials in mon(f). If an arithmetic circuit C computes f, we sometimes denote mon(f) by mon(C) and var(f) by var(C).

A layered circuit C is *minimal* if there is no smaller circuit C' of the same width s.t $\text{mon}(C) = \text{mon}(C')$. For any monotone circuit C, there is a minimal circuit C' of the same width s.t $\text{mon}(C') = \text{mon}(C)$ and has the following properties.

- The only constants used in C' are 0 and 1, and no gate is ever multiplied by a constant.
- Given any node g in C' computing a polynomial p, there is a monomial m such that $\text{mon}(m \cdot p) \subseteq \text{mon}(C')$. In particular, this implies that if C' computes a homogeneous multilinear polynomial of degree d, then p must be a homogeneous multilinear polynomial; moreover, $\text{mon}(p) \subseteq \text{mon}(C')$ if $\deg(p) = d$.
- If C' computes a homogeneous multilinear polynomial of degree d, and if a node g in layer i also computes a polynomial p of degree d, then g is a child of a sum gate g' in layer $i+1$, if such a layer exists.

We call a minimal circuit satisfying the above a *good* minimal circuit. We now state a useful property of good minimal circuits C satisfying $\text{mon}(C) \subseteq P_k^\ell$, for any $\ell, k \geq 1$.

Lemma 1. *Fix any $\ell, k \geq 1$. We can write $P_k^\ell = \sum_{i=1}^{\ell} P_i$, where var$(P_i) \cap$ var$(P_j) = \emptyset$ for any $i \neq j$. Suppose C is a good minimal circuit such that mon$(C) \subseteq$ mon(P_k^ℓ). If a gate g in C computes a polynomial p of degree less than $d := \ell^k$, or a product of two polynomials each of degree less than d, then var$(p) \subseteq$ var(P_i) for a* unique *i. Moreover, if p is of degree d, then we in fact have mon$(p) \subseteq$ mon(P_i).*

We will now state and prove the lower bound: P_k^ℓ has no small width-k monotone circuits. In fact, we prove a stronger statement: that P_k^ℓ is even hard to "approximate" by polynomial size width-k monotone circuits.

Theorem 1. *For each $k > 0$ there is $\ell_0 \in \mathbb{Z}^+$ such that for all $\ell > \ell_0$ and any width-k monotone circuit C such that mon$(C) \subseteq$ mon(P_k^ℓ) and $|\text{mon}(C)| \geq \frac{|\text{mon}(P_k^\ell)|}{2}$, the circuit C is of size at least $\frac{2^\ell}{10}$.*

Proof. For $i \in \mathbb{Z}^+$ and $j \in [w]$, denote by $g_{i,j}$ the jth node in layer i of C and by $f_{i,j}$ the polynomial computed by $g_{i,j}$. For a set of monomials M, we say that a circuit C_1 *computes* M if $\text{mon}(C_1) \supseteq M$.

W.l.o.g., we assume throughout that C is a good minimal circuit. The proof is by induction on k. The case $k = 1$ is distinct and easy to handle. Thus, we consider as the induction base case the case $k = 2$. Consider a width two monotone circuit C such that $\text{mon}(C) \subseteq \text{mon}(P_2^\ell)$ and $|\text{mon}(C)| \geq |\text{mon}(P_2^\ell)|/2 = \ell^{\ell+1}/2$. Let f denote the polynomial computed by C. We know that both f and P_2^ℓ are homogeneous polynomials of degree $d = \ell^2$.

We can write $P_2^\ell = \sum_{i=1}^{\ell} P_i$, where $\text{var}(P_i) = \{x_{(i-1)\ell^3+1}, \ldots, x_{i\ell^3}\}$. Note that $\text{var}(P_i) \cap \text{var}(P_j) = \emptyset$ for $i \neq j$. Let $f = \sum_{i=1}^{\ell} P_i'$ where $\text{mon}(P_i') \subseteq \text{mon}(P_i)$ for each i.

Since C is good and f is homogeneous, each gate of C computes only homogeneous polynomials. Consider the lowest layer (say, i_0) when the circuit C computes a degree d monotone polynomial. W.l.o.g. $f_{i_0,1}$ is such a polynomial. We list some crucial properties satisfied by $g_{i_0,1}$ and C.

1. As i_0 is minimal, $g_{i_0,1}$ is a product gate computing the product of two polynomials of degree less than d. Hence, by Lemma 1, $\text{mon}(f_{i_0,1}) \subseteq \text{mon}(P_i)$ for exactly one i. W.l.o.g. we assume $i = 1$.
2. Since $\deg(f_{i_0,1}) = d$ and C is good, there is a sequence of nodes g_{i,j_i}, for $i > i_0$ such that for each i, g_{i,j_i} is a sum gate with $g_{i-1,j_{i-1}}$ as child (here, $j_{i_0} = 1$), and hence $\text{mon}(f_{i_0,1}) \subseteq \text{mon}(f_{i_0+1,j_{i_0+1}}) \subseteq \text{mon}(f_{i_0+2,j_{i_0+2}}) \ldots$, and each f_{i,j_i} is homogeneous of degree d. We assume, w.l.o.g, that $j_i = 1$ for each $i > i_0$.

By the choice of i_0, the node $g_{i_0,2}$ either computes a polynomial of degree less than d or a product of two polynomials of degree less than d. Hence, $\text{var}(f_{i_0,2}) \subseteq \text{var}(P_i)$ for some i. If $i > 1$, we assume w.l.o.g. that $\text{var}(p) \subseteq \text{var}(P_2)$. Consider the circuit C with all variables in $\text{var}(P_1) \cup \text{var}(P_2)$ set to 0. The polynomial computed by the new circuit C' is now $f' = f - P_1' - P_2' = \sum_{i=3}^{\ell} P_i'$. Let $q_{i,j}$ denote the new polynomial computed by the node $g_{i,j}$. Each $q_{i_0,j}$ is now a constant polynomial.

Consider the monotone circuit C'' obtained from C' as follows: we remove all the gates below layer i_0; the gate $g_{i_0,2}$ in layer i_0 is replaced the constant 1, if it computes a nonzero constant in C' and 0 otherwise; from layer i_0 onwards, all nodes of the form $g_{i,1}$ are replaced by the constant 0. Clearly, C'' is a width 1 circuit. We will refer to the nodes of C'' with the same names as the corresponding nodes in C'. For any node $g_{i,2}$ in C'' ($i \geq i_0$), let $q_{i,2}'$ be the polynomial it now computes. Crucially, we observe the following from the above construction.

Claim 2. *For each $i \geq i_0$, $mon(q_{i,2}') \supseteq mon(q_{i,2}) \setminus mon(q_{i,1})$.*

We now finish the proof of the base case. Define a sequence $i_1 < i_2 < \ldots < i_t$ of layers as follows: for each $j \in [t]$, i_j is the least $i > i_{j-1}$ such that $\text{mon}(q_{i,1}) \supsetneq \text{mon}(q_{i_{j-1},1})$, and $\text{mon}(q_{i_t,1}) = \text{mon}(f')$. Clearly, t is at most the size of C. Note that it must be the case that $q_{i_j,1} = q_{i_j-1,1} + q_{i_j-1,2}$. Hence, we have $\text{mon}(q_{i_j,1}) = \text{mon}(q_{i_j-1,1}) \cup \text{mon}(q_{i_j-1,2}) = \text{mon}(q_{i_{j-1},1}) \cup (\text{mon}(q_{i_j-1,2}) \setminus \text{mon}(q_{i_j-1,1}))$. By the above claim, the set $\text{mon}(q_{i_j-1,2}) \setminus \text{mon}(q_{i_j-1,1})$, which we will denote by S_j, can be computed by a width-1 circuit. Thus, $\text{mon}(f') = \text{mon}(q_{i_t,1}) = \text{mon}(q_{i_0,1}) \cup \bigcup_{j=1}^{t} S_j$, where each S_j can be computed by a width-1 circuit. Since $q_{i_0,1}$ is the zero polynomial, we have $\text{mon}(f') = \bigcup_{j=1}^{t} S_j$.

Now, consider any width-1 monotone circuit computing a set $S \subseteq P_2^{\ell}$.

Claim 3. *The set S is of the form $mon(p)$ where $p = (\sum_{i \in X_1} x_i) \prod_{j \in X_2} x_j$, and $X_1 \cap X_2 = \emptyset$.*

Clearly, as each S_j satisfies $S_j \subseteq \text{var}(P_i')$ for some i, it has at most ℓ^3 monomials. Therefore, if the monotone circuit C is of size less than 2^{ℓ}, then it computes a polynomial of the form $P_1' + P_2' + f'$, where f' has at most $2^\ell \ell^3$ monomials. Since $|\text{mon}(P_i')| \leq |\text{mon}(P_i)| = \ell^\ell$ for each i, we have for suitably large ℓ, $|\text{mon}(C)| \leq 2\ell^\ell + 2^\ell \ell^3 < 3\ell^\ell < \frac{\ell^{\ell+1}}{2} = \frac{|\text{mon}(P_2^\ell)|}{2}$, and the base case follows.

The induction step is similar to the base case. As induction hypothesis we assume that any monotone circuit $\hat{C}$ of width $k-1$ such that $\text{mon}(\hat{C}) \subseteq \text{mon}(P_{k-1}^\ell)$ and $|\text{mon}(\hat{C})| \geq |\text{mon}(P_{k-1}^\ell)|/2$ must be of size at least $2^\ell/10$.

We write $P_k^\ell = \sum_{i=1}^\ell P_i$, with $\text{var}(P_i) = \{x_{(i-1)\ell^{2k+1}+1}, \ldots, x_{i\ell^{2k+1}}\}$ and further $P_i = \prod_{j=1}^\ell Q_{ij}$, where each Q_{ij} is of type P_{k-1}^ℓ. We have $\text{var}(Q_{ij}) = \{x_{(i-1)\ell^{2k+1}+(j-1)\ell^{2k}+1}, \ldots, x_{(i-1)\ell^{2k+1}+j\ell^{2k}}\}$. Let d denote $\deg(P_k^\ell) = \ell^k$.

Consider any width $k-1$ circuit $\hat{C}$ of size less than $2^\ell/10$ such that $\text{mon}(\hat{C}) \subseteq \text{mon}(P_k^\ell)$. For any $i, j \in [\ell]$, by fixing all the variables outside $\text{var}(P_i)$ to 0 and all the variables in $\text{var}(P_i) \setminus \text{var}(Q_{ij})$ to 1, we obtain a width $k-1$ circuit $\hat{C}_{ij}$ of the same size s.t $\text{mon}(\hat{C}_{ij}) \subseteq \text{mon}(Q_{ij})$. By the induction hypothesis, we see that $|\text{mon}(\hat{C}_{ij})| \leq |\text{mon}(Q_{ij})|/2$. Clearly $\text{mon}(\hat{C}) \subseteq \bigcup_{i=1}^\ell \text{mon}(\hat{C}_{i1}) \times \text{mon}(\hat{C}_{i2}) \times \ldots \times \text{mon}(\hat{C}_{i\ell})$. Therefore, $|\text{mon}(\hat{C})| \leq \sum_i \prod_j |\text{mon}(\hat{C}_{ij})| \leq \sum_i \prod_j |\text{mon}(Q_{ij})|/2 = |\text{mon}(P_k^\ell)|/2^\ell$. We have established the following claim.

Claim 4. *For any width $k-1$ circuit $\hat{C}$ of size less than $2^\ell/10$ such that $mon(\hat{C}) \subseteq mon(P_k^\ell)$, we have $|mon(\hat{C})| \leq \frac{|mon(P_k^\ell)|}{2^\ell}$.*

Now, consider any monotone width-k circuit C of size at most $2^\ell/10$ such that $\text{mon}(C) \subseteq \text{mon}(P_k^\ell)$. We will show that $|\text{mon}(C)| < |\text{mon}(P_k^\ell)|/2$. W.l.o.g., we can assume that C is a good minimal circuit. Let f denote the polynomial computed by C; we write $f = \sum_{i=1}^\ell P_i'$, where $\text{mon}(P_i') \subseteq \text{mon}(P_i)$ for each i.

Let i_0 be the first layer where a polynomial of degree d is computed. Exactly as in the base case, we fix a sequence of nodes g_{i,j_i} for each $i \geq i_0$ such that g_{i,j_i} is a sum gate with $g_{i-1,j_{i-1}}$ as a child such that $\text{mon}(f_{i_0,j_{i_0}}) \subseteq \text{mon}(f_{i_0+1,j_{i_0+1}}) \subseteq \text{mon}(f_{i_0+2,j_{i_0+2}}) \ldots$, and each f_{i,j_i} computes a homogeneous polynomial of degree d. Renaming nodes if necessary, we assume $j_i = 1$ for all i.

Now consider $f_{i_0,j}$ for $j > 1$. As in the base case, $\text{var}(f_{i_0,j}) \subseteq \text{var}(P_s)$ for some $s \in [\ell]$. Thus, there is a set $S \subseteq [\ell]$ s.t $|S| < k$ such that $\bigcup_{j>1} \text{var}(f_{i_0,j}) \subseteq \bigcup_{s \in S} \text{var}(P_s)$. W.l.o.g., assume that $S \subseteq [k]$.

Consider the circuit C' obtained by setting the variables in $\bigcup_{s\in[k]} \text{var}(P_s)$ to 0. Let $q_{i,j}$ be the polynomial computed by $g_{i,j}$ in C'. The polynomial computed by C' is $f' = f - \sum_{s\in[k]} P_s'$. Each $q_{i_0,j}$ is now simply a constant for each j, and that the size of C' is at most the size of C which by assumption is bounded by $2^\ell/10$. Using this size bound we will argue that C' cannot compute too many monomials.

We now modify C' as follows: we remove all the gates below layer i_0; each gate $g_{i_0,j}$ with $j > 1$ is replaced by 1 if it computes a nonzero polynomial in C' and 0 otherwise; from layer i_0 onwards, all nodes of the form $g_{i,1}$ are replaced by the constant 0. Call this new circuit C''. Clearly, C'' has size at most the size of C and width at most $k-1$. For ease of notation, we will refer to the nodes of C'' with the same names as the corresponding nodes in C'. For any node $g_{i,j}$

in C'' ($i \geq i_0$ and $j > 1$), let $q'_{i,j}$ be the polynomial it now computes. As in the base case, we observe the following from the above construction.

Claim 5. *For each $i \geq i_0$ and each $j > 1$, $mon(q'_{i,j}) \supseteq mon(q_{i,j}) \setminus mon(q_{i,1})$.*

Using Claim 5, we show that the circuit C' was essentially just using the gates $g_{i,1}$ to store the sum of polynomials computed using width $k-1$ circuits.

Construct a sequence of layers $i_1 < i_2 < \ldots < i_t$ in C' as follows: for each $j \in [t]$, i_j is the least $i > i_{j-1}$ such that $\text{mon}(q_{i,1}) \supsetneq \text{mon}(q_{i_{j-1},1})$, and $\text{mon}(q_{i_t,1}) = \text{mon}(f')$. Surely, t is at most the size of C'. Now, fix any i_j for $j \geq 1$. Clearly, it must be the case that $q_{i_j,1} = q_{i_j-1,1} + q_{i_j-1,s}$ for some $s > 1$; therefore, we have $\text{mon}(q_{i_j,1}) \subseteq \text{mon}(q_{i_j-1,1}) \cup (\text{mon}(q_{i_j-1,s}) \setminus \text{mon}(q_{i_j-1,1}))$. Denote the set $\text{mon}(q_{i_j-1,s}) \setminus \text{mon}(q_{i_j-1,1})$ by S_j. Since the above holds for all j, and $\text{mon}(q_{i_j-1,1}) = \text{mon}(q_{i_{j-1},1})$, we see that $\text{mon}(f') = \text{mon}(q_{i_t,1}) \subseteq \text{mon}(q_{i_0,1}) \cup \bigcup_j S_j = \bigcup_j S_j$, since $q_{i_0,1}$ is the zero polynomial.

By Claim 5, for each j, there is a width $k-1$ circuit C'' of size at most the size of C such that $S_j \subseteq \text{mon}(C'') \subseteq P_k^\ell$. If the size of C (and hence that of C' and C'') is at most $2^\ell/10$, it follows from Claim 4 that $|S_j| \leq |\text{mon}(P_k^\ell)|/2^\ell$. Hence, we see that $|\text{mon}(f')| \leq t|\text{mon}(P_k^\ell)|/2^\ell$, which is at most $|\text{mon}(P_k^\ell)|/10$. Since the polynomial f computed by the circuit C is of the form $f' + \sum_{i\in[k]} P'_i$, where $|\text{mon}(P'_i)| \leq |\text{mon}(P_i)| = |\text{mon}(P_k^\ell)|/\ell$. Therefore, $|\text{mon}(f)| \leq \frac{k}{\ell}|\text{mon}(P_k^\ell)| + |\text{mon}(f')| \leq |\text{mon}(P_k^\ell)| \left(\frac{k}{\ell} + \frac{1}{10}\right) < \frac{|\text{mon}(P_k^\ell)|}{2}$ for large enough ℓ. This proves the induction step. ∎

From the remarks following the definition of P_k^ℓ and Proposition 4, we have the following corollary of Theorem 1.

Corollary 1. *For any fixed $k \in \mathbb{Z}^+$ and any $n \in \mathbb{Z}^+$, the explicit polynomial $P_k^{\lfloor n^{1/2k} \rfloor} \in \mathbb{F}[x_1, \ldots, x_n]$ has linear-sized monotone circuits of depth $2k$ and width $2k$, but no subexponential sized monotone circuits of depth k or width k.*

The analogue of Theorem 1 and the above corollary can also be derived for noncommutative circuits [1].

3 Identity Testing for Constant Width Circuits

Impagliazzo and Kabanets [6] showed that derandomizing polynomial identity testing is equivalent to proving arithmetic circuit lower bounds: If there are explicit polynomials that require superpolynomial size arithmetic circuits, they can be used in a Nisan-Wigderson type "arithmetic" pseudorandom generator to derandomize polynomial identity testing. Dvir et al [4] show that if there are explicit polynomials that require superpolynomial size constant-depth arithmetic circuits then polynomial identity testing for constant-depth arithmetic circuits can be derandomized.

We prove a similar result showing that hardness for constant-width arithmetic circuits yields a derandomization of polynomial identity testing for constant-width circuits. A family $\{P_{n,p} \mid n \in \mathbb{Z}^+,\ p \text{ prime}\}$ where $P_{n,p}(\overline{x}) \in \mathbb{F}_p[x_1, \cdots, x_n]$

is a multilinear polynomial is called *explicit* if the coefficient of each monomial m of the polynomial $P_{n,p}$ can be computed in time $2^{(n+p)^{O(1)}}$.

Lemma 2. *Let $f \in \mathbb{F}[x_1, x_2, \cdots, x_n]$ be a polynomial of degree m computed by a staggered arithmetic circuit of size s and width w. Then $H_i(f)$ (the ith homogeneous component of f) is computable by a staggered circuit of size* poly(s, m) *and width $w + O(1)$, provided $\mathbb{F}$ has at least* $\deg(f) + 1$ *many elements.*

Proof. Write $f(x_1z, x_2z, \cdots, x_nz) = \sum_{i=0}^{m} H_i(f)z^i$ where $m = \deg(f)$ and $H_i(f)$ is the ith homogeneous part of f. Let $\{z_0, z_1, \cdots, z_m\}$ be $m+1$ distinct field elements. Consider the matrix $(m+1) \times (m+1)$ matrix M where $M_{ij} = z_{i-1}^{j-1}$ for $i, j \in [m+1]$. Let $g(\overline{x}, z) = f(x_1z, x_2z, \cdots, x_nz)$. We have the system of equations $M(H_0(f), H_1(f), \cdots, H_m(f))^T = (g(\overline{x}, z_0), g(\overline{x}, z_1), \cdots, g(\overline{x}, z_m))^T$.

Since M is invertible, there are $a_{ij} \in \mathbb{F}$ such that $H_i(f) = \sum_{j=0}^{m} a_{ij} g(\overline{x}, z_j)$. Since $f(\overline{x})$ has a width w circuit of size s, $g(\overline{x}, z)$ has a (staggered) circuit of width $w + O(1)$ of size $O(s)$. It follows that each $H_i(f)$ has a circuit of width $w + O(1)$ and size $O(ms)$. ■

Lemma 3. *Let $P(x_1, x_2, \cdots, x_n, y)$ be a polynomial, over a sufficiently large field $\mathbb{F}$, computed by a width-w staggered circuit of size s. Suppose the maximum degree of y in P is r. Then for each j the jth partial derivative $\frac{\partial^j P}{\partial y^j}$ is computable by a staggered circuit of width $w + O(1)$ and size $(rs)^{O(1)}$.*

The proof is similar to that of Lemma 2.

Lemma 4. [4, Lemma 3.2] *For $g \in \mathbb{F}[x_1, x_2, \cdots, x_n]$ let $H_{\leq k}(g) = \sum_{i=0}^{k} H_i(g)$. Let $P \in \mathbb{F}[x_1, x_2, \cdots, x_n, y]$ and $\deg_y(P) = r$. Suppose $f \in \mathbb{F}[x_1, x_2, \cdots, x_n]$ such that $P(\overline{x}, f(\overline{x})) = 0$ and $\frac{\partial P}{\partial y}(\overline{0}, f(\overline{0}))$ is equal to $\xi \neq 0$. Let $P(\overline{x}, y) = \sum_{i=1}^{r} C_i(\overline{x})y^i$. Then for each $k \geq 0$ there is a polynomial $Q_k \in \mathbb{F}[y_0, y_1, \cdots, y_r]$ such that $H_{\leq k}(f) = H_{\leq k}(Q_k(C_0, C_1, \cdots, C_r))$.*

Using the above lemmata, we prove the following theorem.

Theorem 6. *Let $P \in \mathbb{F}[x_1, x_2, \cdots, x_n, y]$ and $\deg_y(P) = r \geq 1$ such that P has a staggered circuit of size s and width w. Suppose that $P(\overline{x}, f(\overline{x})) = 0$ for some polynomial $f \in \mathbb{F}[x_1, x_2, \cdots, x_n]$ with* $\deg(f) = m$. *Then f has a staggered circuit of size* poly$(s, (m+r)^r)$ *and width $w + O(1)$ if $char(\mathbb{F}) > r$ and $\mathbb{F}$ is sufficiently large.*

Proof. We can assume that $\frac{\partial P}{\partial y}(\overline{0}, f(\overline{0})) = \xi \neq 0$. For, if $\frac{\partial P}{\partial y}(x, f(x)) \equiv 0$ we can replace P by $\frac{\partial P}{\partial y}$. Since $char(\mathbb{F}) > r$, there is a $j : 1 \leq j \leq r$ such that $\frac{\partial^j P}{\partial y^j}(x, f(x)) \not\equiv 0$. Hence, we can assume $\frac{\partial P}{\partial y}(x, f(x)) \not\equiv 0$. Therefore, there is an $a \in \mathbb{F}^n$ such that $\frac{\partial P}{\partial y}P(a, f(a)) \neq 0$. We can assume that $a = 0$ by appropriately shifting P as in [4]. Let $P(\overline{x}, y) = \sum_{i=1}^{r} C_i(\overline{x})y^i$.

By Lemma 4, there is $Q_k \in \mathbb{F}[y_0, \cdots, y_r]$ such that $H_{\leq k}(f) = H_{\leq k}(Q_k(C_0, C_1, \cdots, C_r))$ for each $0 \leq k \leq m$. Putting $k = m$ and letting

$Q_m = Q$ we have $f(\overline{x}) = H_{\leq m}(Q(C_0, C_1, \cdots, C_r))$. Let $y^* = (C_0(0), \cdots, C_r(0))$ and $\deg(Q) = M$. Define $I_M = \{(\alpha_0, \alpha_1, \cdots, \alpha_r) \mid \alpha_i \in \mathrm{N}, \sum \alpha_i \leq M\}$. By expanding Q at y^* we get $Q(\overline{y}) = \sum_{\overline{\alpha} \in I_M} Q_\alpha \prod_{i=0}^{r}(y_i - y_i^*)^{\alpha_i}$. Thus, $f(\overline{x}) = H_{\leq m}[\sum_{\overline{\alpha} \in I_M} Q_\alpha \prod_{i=0}^{r}(C_i(\overline{x}) - C_i(0))^{\alpha_i}]$.

As the constant term of $C_i(\overline{x}) - C_i(0)$ is zero, if we consider $\prod_{i=1}^{r}(C_i(\overline{x}) - C_i(0))^{\alpha_i}$ for some $\overline{\alpha}$ with $\sum_i \alpha_i > m$ then we will get monomials of degree more than m whose net contribution to $f(\overline{x})$ must be zero. Hence, $f(\overline{x}) = H_{\leq m}[\sum_{\overline{\alpha} \in I_m} Q_\alpha \prod_{i=0}^{r}(C_i(\overline{x}) - C_i(0))^{\alpha_i}]$, where $I_m = \{(\alpha_0, \alpha_1, \cdots, \alpha_r) \mid \alpha_i \in \mathrm{N}, \sum \alpha_i \leq m\}$. Clearly, $|I_m| \leq (m + r)^r$. Now, the polynomial $\prod_{i=0}^{r}(y_i - y_i^*)^{\alpha_i}$ has a simple $O(1)$-width circuit C'. We can compute $\prod_{i=0}^{r}(C_i(\overline{x}) - C_i(0))^{\alpha_i}$ by plugging in the staggered width $w + O(1)$ circuit for $C_i(\overline{x})$ (as obtained in Lemma 3) where y_i is input to C'. Thus, we obtain a circuit of width $w + O(1)$ for $\sum_{\overline{\alpha} \in I_m} Q_\alpha \prod_{i=0}^{r}(C_i(\overline{x}) - C_i(0))^{\alpha_i}$ that is of size polynomial in s and $(m + r)^r$. By Lemma 2 we can compute its homogeneous components and their partial sums with constant increase in width. Hence, $f(\overline{x})$ is computable by a circuit of width $w + O(1)$ and size polynomial in s and $(m + r)^r$. ■

We apply Theorem 6 to prove the main result of this section.

Theorem 7. *There is a constant $c_1 > 0$ so that the following holds. Suppose there is an explicit family $\{P_{m,q} \mid m \in \mathbb{Z}^+,\ q \text{ prime}\}$ where $P_{m,q}(\overline{x}) \in \mathbb{F}_q[x_1, \cdots, x_m]$ is a multilinear polynomial which can not be computable by arithmetic circuits of width $w + c_1$ and size 2^{m^ϵ}, $\epsilon > 0$. Then for any constant $c_2 > 0$ and prime $p > (\log n)^{c_2}$, there is a deterministic $2^{(\log n)^{O(1)}}$ time algorithm that takes as input a circuit C of size $n^{O(1)}$ and width w computing a polynomial $f \in \mathbb{F}_p[x_1, \ldots, x_n]$, with each variable of individual degree at most $(\log n)^{c_2}$, and checks if the polynomial computed by C is identically zero.*

Proof Sketch. Let $m = (\log n)^{c_3}$ and $\ell = (\log n)^{c_4}$ where c_3 is suitably chosen depending on ϵ and c_2, and c_4 is suitably larger than c_3. Construct the Nisan-Wigderson design $S_1, \cdots, S_n \subset [\ell]$ such that $|S_i| = m$ for each i and $|S_i \cap S_j| \leq \log n$.

Consider $F(y_1, y_2, \cdots, y_\ell) = C(P_m(\overline{y}|S_1), P_m(\overline{y}|S_2), \cdots, P_m(\overline{y}|S_n))$. For any input $\overline{y} \in \mathbb{F}^\ell$ we can evaluate F by evaluating $P_m(\overline{y}|S_i)$ for each i and then evaluating C on the resulting values. Since the P_m are explicit polynomials and $|S_i|$ has $(\log n)^{O(1)}$ size we can evaluate P_m in time $2^{(\log n)^{O(1)}}$. We test if $F(\overline{y}) \equiv 0$ using a brute-force algorithm based on the Schwartz-Zippel lemma. Consider a finite set $S \subseteq \mathbb{F}_p$, such that $|S|$ is more than $\deg(F)$ (go to suitable extension field if necessary). Check if $F(\overline{a}) \equiv 0$ for all $\overline{a} \in S^\ell$ in time $n^{O(\ell)}$. If all the tests returned zero then return $C \equiv 0$ otherwise $C \not\equiv 0$.

Suppose the algorithm fails. After hybridization and fixing variables in C, we get a nonzero polynomial F_2 of the form $F_2(\overline{y}|S_{i+1}, x_{i+1}) = F_1(P_m(\overline{y}|S_1 \cap S_{i+1}), P_m(\overline{y}|S_2 \cap S_{i+1}), \cdots, P_m(\overline{y}|S_i \cap S_{i+1}), x_{i+1})$, where $F_1(x_1, x_2, \ldots, x_{i+1})$ can be computed by a width w circuit of size poly(n) and $F_2(\overline{y}|S_{i+1}, P_m(\overline{y}|S_{i+1})) \equiv 0$. The polynomials $P_m(\overline{y}|S_j \cap S_{i+1})$ depend only on $\log n$ variables and hence are computable by brute force width-2 staggered circuits of size $O(n \log n)$. Clearly,

F_1 is computable by a staggered circuit of size poly(n) and width at most $w+1$. Combining these circuits, F_2 is computable by a staggered circuit C' of size poly(n) and width $w + O(1)$. Applying Theorem 6 to C' we get a circuit of width $w + O(1)$ to compute P_m contradicting the size bound in the hardness assumption. ■

We observe an analogue of [6, Theorem 4.1] for bounded width circuits. If the identity testing problem for bounded-width arithmetic circuits over $\mathbb{Q}$ is in NSUBEXP then *either* NEXP $\not\subseteq$ P/poly *or* the Permanent polynomial is computable by polynomial size bounded-width arithmetic circuit over $\mathbb{Q}$.

Acknowledgements. We thank Amir Yehudayoff for pointing out the separation in [11] and for many valuable comments.

References

1. Arvind, V., Joglekar, P., Srinivasan, S.: On Lower Bounds for Constant Width Arithmetic Circuits. In: Electronic Colloquium of Computational Complexity, TR09-073, `http://eccc.hpi-web.de/report/2009/073/`
2. Ben-Or, M.: Unpublished notes
3. Ben-Or, M., Cleve, R.: Computing Algebraic Formulas Using a Constant Number of Registers. SIAM J. Comput. 21(1), 54–58 (1992)
4. Dvir, Z., Shpilka, A., Yehudayoff, A.: Hardness-randomness tradeoffs for bounded depth arithmetic circuits. In: Proc. Symp. on Theory of Computing, pp. 741–748 (2008)
5. Jansen, M., Raghavendra Rao, B.V.: Simulation of arithmetical circuits by branching programs preserving constant width and syntactic multilinearity. In: CSR (2009)
6. Kabanets, V., Impagliazzo, R.: Derandomization of polynomial identity tests means proving circuit lower bounds. In: Proc. of the thirty-fifth annual ACM Sym. on Theory of computing, pp. 355–364 (2003)
7. Limaye, N., Mahajan, M., Raghavendra Rao, B.V.: Arithmetizing classes around NC1 and L. In: Thomas, W., Weil, P. (eds.) STACS 2007. LNCS, vol. 4393, pp. 477–488. Springer, Heidelberg (2007); Preliminary version in STACS 2007
8. Mahajan, M., Raghavendra Rao, B.V.: Arithmetic circuits, syntactic multilinearity, and the limitations of skew formulae. In: Ochmański, E., Tyszkiewicz, J. (eds.) MFCS 2008. LNCS, vol. 5162, pp. 455–466. Springer, Heidelberg (2008)
9. Nisan, N.: Lower bounds for non-commutative computation. In: Proc. of the 23rd annual ACM Sym. on Theory of computing, pp. 410–418 (1991)
10. Raz, R., Shpilka, A.: Deterministic polynomial identity testing in non commutative models. Computational Complexity 14(1), 1–19 (2005)
11. Raz, R., Yehudayoff, A.: Lower Bounds and Separations for Constant Depth Multilinear Circuits. Computational Complexity 18(2), 171–207 (2009)
12. Snir, M.: On the Size Complexity of Monotone Formulas. In: Proc. 7th Intl. Colloquium on Algorithms Languages and Programming, pp. 621–631 (1980)

Spending Is Not Easier Than Trading: On the Computational Equivalence of Fisher and Arrow-Debreu Equilibria

Xi Chen[1] and Shang-Hua Teng[2]

[1] Princeton University
[2] University of Southern California

Abstract. It is a common belief that computing a market equilibrium in Fisher's spending model is easier than computing a market equilibrium in Arrow-Debreu's exchange model. This belief is built on the fact that we have more algorithmic success in Fisher equilibria than Arrow-Debreu equilibria. For example, a Fisher equilibrium in a Leontief market can be found in polynomial time, while it is PPAD-hard to compute an approximate Arrow-Debreu equilibrium in a Leontief market.

In this paper, we show that even when all the utilities are additively separable, piecewise-linear and concave, computing an approximate equilibrium in Fisher's model is PPAD-hard. Our result solves a long-term open question on the complexity of market equilibria. To the best of our knowledge, this is the first PPAD-hardness result for Fisher's model.[1]

1 Introduction

1.1 Market Equilibria: Fisher's Model vs. Arrow-Debreu's Model

In 1891, Irving Fisher introduced one of the most fundamental exchange market models in his Ph.D. thesis [2]. It considers a market in which there are m buyers and n divisible goods. We denote the amount of good j, $j \in [n]$, in the market by $c_j > 0$. Every buyer i comes to the market with a certain amount of money, denoted by $w_i > 0$. The goal of a buyer is to obtain a bundle of goods, denoted by $\mathbf{a}_i \in \mathbb{R}^n_+$, that maximizes her utility function $u_i : \mathbb{R}^n_+ \to \mathbb{R}_+$.

Fisher showed that if all the utility functions u_i satisfy some mild conditions then there always exists an equilibrium price vector $\mathbf{p} \in \mathbb{R}^n_+$. At this price, one can find a bundle of goods $\mathbf{a}_i$ for each buyer i such that $\mathbf{a}_i$ maximizes her utility under the budget constraint that $\mathbf{a}_i \cdot \mathbf{p} \le w_i$, and at the same time, the market demands equal to the market supply: $\sum_{i\in[m]} a_{i,j} \le c_j$ for all $j \in [n]$.

Fisher's model is a special case of the more general model of exchange economics considered by Arrow and Debreu [3]: In an exchange economy there are

[1] Recently, Vazirani and Yannakakis independently proved that the problem of computing an approximate Fisher equilibrium in a market with additively separable and PLC utility functions is PPAD-hard [1]. They also showed that the problem of finding an exact Arrow-Debreu equilibrium in such markets is in PPAD and thus, both problems are PPAD-complete.

Y. Dong, D.-Z. Du, and O. Ibarra (Eds.): ISAAC 2009, LNCS 5878, pp. 647–656, 2009.

m traders and n divisible goods. Trader i has an *initial endowment* of $w_{i,j} \geq 0$ of good j, and a utility function $u_i : \mathbb{R}_+^n \to \mathbb{R}_+$. The individual goal of a trader is to obtain a new bundle of goods that maximizes her utility.

In a sense, Fisher's model focuses more on spending than trading as in Arrow-Debreu's model. In his model, money can be viewed as a special kind of good. All but one "special" trader only have money as their endowments, and money has no value to their utilities; the special trader, sometime called the "*market*", has all the goods and her interest is to collect all the money.

Over the last two decades we have more algorithmic success in computing a market equilibrium in Fisher's model than computing a market equilibrium in Arrow-Debreu's model.

- For the latter, polynomial-time algorithms are only known for markets with utility functions that are linear [4,5,6,7,8,9,10,11,12] or satisfy weak gross substitutability [13]. These algorithms critically used the fact that the set of equilibria of these markets is convex. Progress on markets with non-convex set of equilibria has been relatively slow. There are only a few algorithms in this case. Devanur and Kannan [14] gave a polynomial-time algorithm for markets with piecewise-linear and concave (PLC) utilities and a constant number of goods. Codenotti, McCune, Penumatcha, and Varadarajan [15] gave a polynomial-time algorithm for CES markets when the elasticity of substitution $s \geq 1/2$.

  For Leontief markets, in which each utility function is of the form $\min_j a_j x_j$, the problem of finding an approximate Arrow-Debreu equilibrium is known to be PPAD-hard [16,17,18]. In [19], Chen *et al.* proved that finding an approximate Arrow-Debreu equilibrium, even if all the utilities are additively separable [2] and PLC, is PPAD-hard. Recently, Vazirani and Yannakakis [1] showed that the problem of computing an *exact* Arrow-Debreu equilibrium in such markets is a member of PPAD and thus, is complete in PPAD.
- For Fisher's model, polynomial-time algorithms are given not only for linear markets but also for Leontief and many other markets, e.g., the hybrid linear Leontief markets [20]. We know that an (approximate) market equilibrium in any Fisher's economy with CES utilities can be found in polynomial time [4,15,12,21,7,22]. In fact, Ye [21] proved that if every utility function is the minimum of a collection of homogeneous linear functions, then one can find a Fisher equilibrium in polynomial time.

1.2 Our Results

It remains open whether there is a family of concave utility functions for which it is PPAD-hard to compute a Fisher equilibrium. The family of utility functions that has drawn most attention is the additively separable, piecewise-linear, and concave (PLC) functions. In [23], Vazirani remarked that obtaining a polynomial-time algorithm for markets with additively separable and concave utilities is

[2] A function $u(x_1, \ldots, x_m)$ from $\mathbb{R}_+^m$ to $\mathbb{R}_+$ is *additively separable* if there exist m real-valued functions $f_1, \ldots, f_m$ such that $u(x_1, \ldots, x_m) = \sum_{j=1}^{m} f_j(x_j)$.

a premier open question today. Although the recent result of Chen *et. al.* shows that finding an Arrow-Debreu equilibrium in markets with additively separable and PLC utility functions is PPAD-hard [19], the complexity of Fisher equilibria remains unclear.

In this paper, we show that the problem of finding a Fisher equilibrium remains to be PPAD-hard when the utility functions are additively separable and PLC. It then follows from the membership result of Vazirani and Yannakakis [1] that this problem is PPAD-complete. Therefore, for this seemingly simple class of utility functions, finding a Fisher equilibrium is as hard as finding an Arrow-Debreu equilibrium. Recently, Vazirani and Yannakakis [1] also independently proved that the problem of finding an approximate Fisher equilibrium in such markets is complete in PPAD.

Proof Sketch: We prove the PPAD-hardness of the problem by giving a reduction from SPARSE BIMATRIX [24]: the problem of finding an approximate Nash equilibrium in a sparse two-player game (see Section 2.1 for definition).

Similar to [19], our reduction starts by constructing a family of markets $\mathcal{M}_n$ for every $n \geq 1$, which we refer to as the *price-regulating* markets. There are $2n$ goods in $\mathcal{M}_n$, and every approximate equilibrium price $\mathbf{p}$ satisfies the following *price-regulation* property: $p_{2k-1} + p_{2k} \approx 3$ and $1/2 \leq p_{2k-1}/p_{2k} \leq 2$, for every $k \in [n]$. This allows us to encode n $[0,1]$-variables $x_1, \ldots, x_n$ using $\mathbf{p}$ as follows

$$x_k = p_{2k} - \big(p_{2k} + p_{2k+1}\big)/3, \qquad \text{for every } k \in [n]. \tag{1}$$

Moreover, the price-regulation property is *stable* with respect to "small perturbations" to $\mathcal{M}_n$: When new buyers are added to $\mathcal{M}_n$ (without introducing new goods), this property remains to hold as long as the total amount of money of these new buyers is small compared to that of the buyers in $\mathcal{M}_n$.

Given an $n \times n$ two-player game $(\mathbf{A}, \mathbf{B})$, we construct a market $\mathcal{M}$ in polynomial time by adding new buyers to $\mathcal{M}_{2n+1}$ (with $4n + 2$ goods). All the new buyers have very little money compared to those buyers in $\mathcal{M}_{2n+1}$ so the price-regulation property still holds. This enables us to encode a pair of probability distributions $(\mathbf{x}, \mathbf{y})$ of n dimensions (with $2n$ $[0,1]$-variables) using the first $4n$ entries of the price vector $\mathbf{p}$, as in (1). By using the price-regulation property, we show how to set the utilities of those new buyers appropriately so that we can control their preferences over the goods and ultimately implement all the Nash equilibrium constraints over $(\mathbf{x}, \mathbf{y})$ through $\mathbf{p}$. As a result, given any (approximate) market equilibrium price vector $\mathbf{p}$ of $\mathcal{M}$, the pair $(\mathbf{x}, \mathbf{y})$ obtained (after normalization) must be an approximate Nash equilibrium of $(\mathbf{A}, \mathbf{B})$.

2 Preliminaries

2.1 Complexity of Nash Equilibria

A two-player game is defined by the payoff matrices $(\mathbf{A}, \mathbf{B})$ of its two players. In this paper, we assume that both players have n choices of actions and thus, both $\mathbf{A}$ and $\mathbf{B}$ are square matrices with n rows and columns. We use $\Delta^n \subset \mathbb{R}^n$ to denote the set of probability distributions of n dimensions.

For $\epsilon \geq 0$, we say a pair of probability distributions $(\mathbf{x}, \mathbf{y})$, where $\mathbf{x}, \mathbf{y} \in \Delta^n$, is an *$\epsilon$-well-supported* Nash equilibrium of $(\mathbf{A}, \mathbf{B})$, if for all $i, j \in [n]$,

$$\mathbf{A}_i \mathbf{y}^T - \mathbf{A}_j \mathbf{y}^T > \epsilon \Rightarrow x_j = 0 \quad \text{and} \quad \mathbf{x}\mathbf{B}_i - \mathbf{x}\mathbf{B}_j > \epsilon \Rightarrow y_j = 0, \tag{2}$$

where we let $\mathbf{A}_i$ denote the ith row of $\mathbf{A}$, and $\mathbf{B}_i$ denote the ith column of $\mathbf{B}$.

A two-player game $(\mathbf{A}, \mathbf{B})$ is said to be *normalized* if every entry of $\mathbf{A}$ and $\mathbf{B}$ is between -1 and 1. We say a two-player game $(\mathbf{A}, \mathbf{B})$ is *sparse* if every row and every column of $\mathbf{A}$ and $\mathbf{B}$ have at most 10 nonzero entries.

Let Sparse Bimatrix denote the following search problem: given an $n \times n$ sparse normalized two-player game, find an n^{-6}-well-supported Nash equilibrium. By [24], we know that Sparse Bimatrix is PPAD-complete.

2.2 Markets with Additively Separable PLC Utilities

Let $\mathcal{G} = \{G_1, \ldots, G_n\}$ denote a set of n divisible goods, and $\mathcal{T} = \{T_1, \ldots, T_m\}$ denote a set of buyers. For each good G_j, we use $c_j > 0$ to denote the amount of G_j in the market. For each buyer T_i, we use $w_i > 0$ to denote her money and $u_i : \mathbb{R}^n_+ \rightarrow \mathbb{R}_+$ to denote her utility function. In this paper, we will mainly focus on markets with additively separable, piecewise-linear and concave utilities.

A continuous function $r(\cdot)$ over $\mathbb{R}_+$ is said to be *t-segment* piecewise linear and concave (PLC) if $r(0) = 0$; and there exists a tuple

$$\left[\theta_0 > \theta_1 > \ldots > \theta_t \geq 0; 0 < a_1 < a_2 < \ldots < a_t\right]$$

of length $2t+1$ such that (letting $a_0 = 0$)

1. $\forall i \in [0 : t-1]$, the restriction of $r(\cdot)$ over $[a_i, a_{i+1}]$ is a segment of slope θ_i;
2. the restriction of $r(\cdot)$ over $[a_t, +\infty)$ is a ray of slope θ_t.

The $(2t+1)$-tuple is called the *representation* of $r(\cdot)$. Also we say $r(\cdot)$ is *strictly monotone* if $\theta_t > 0$, and is *α-bounded* for some $\alpha \geq 1$ if $\alpha \geq \theta_0$ and $\theta_t \geq 1$.

Definition 1. *A function $u(\cdot) : \mathbb{R}^n_+ \rightarrow \mathbb{R}_+$ is said to be an* additively separable PLC *function if there exist* PLC *functions $r_1(\cdot), \ldots, r_n(\cdot)$ from $\mathbb{R}_+$ to $\mathbb{R}_+$ such that $u(\mathbf{a}) = \sum_{j \in [n]} r_j(a_j)$ for all $\mathbf{a} \in \mathbb{R}^n_+$.*

In such a market, we use, for every $T_i \in \mathcal{T}$, $r_{i,j}(\cdot) : \mathbb{R}_+ \rightarrow \mathbb{R}_+$ to denote her PLC function with respect to good $G_j \in \mathcal{G}$ and thus, $u_i(\mathbf{a}) = \sum_{j \in [n]} r_{i,j}(a_j)$.

We use $\mathbf{p} \in \mathbb{R}^n_+$ to denote a price vector, where $\mathbf{p} \neq \mathbf{0}$ and p_j is the price of G_j. Given $\mathbf{p}$, we let $\text{OPT}(i, \mathbf{p})$ denote the set of allocations that maximize u_i:

$$\text{OPT}(i, \mathbf{p}) = \text{argmax}_{\,\mathbf{a} \in \mathbb{R}^n_+,\ \mathbf{a} \cdot \mathbf{p} \leq w_i}\, u_i(\mathbf{a}).$$

We let $\mathcal{X} = \{\mathbf{a}_i \in \mathbb{R}^n_+ : i \in [m]\}$ denote an allocation of the market: for each buyer $T_i \in \mathcal{T}$, $\mathbf{a}_i \in \mathbb{R}^n_+$ is the amount of goods that T_i receives.

Definition 2. *A market equilibrium is a nonzero vector $\mathbf{p} \in \mathbb{R}^n_+$ such that there exists an allocation $\mathcal{X} = \{\mathbf{a}_i : i \in [m]\}$, which has the following two properties:*

1. *Every buyer gets one of the optimal bundles:* $\forall i \in [m]$, $\mathbf{a}_i \in \mathrm{OPT}(i, \mathbf{p})$;
2. *The market clears:* $\forall j \in [n]$, *we have*

$$\textstyle\sum_{i\in[m]} a_{i,j} \le c_j \quad \textit{and} \quad \sum_{i\in[m]} a_{i,j} = c_j \ \ \textit{if } p_j > 0.$$

In general, not every market has such an equilibrium price vector. However, for the additively separable PLC markets, the following condition guarantees the existence of an equilibrium:

> If for every buyer $T_i \in \mathcal{T}$ there exists a good $G_j \in \mathcal{G}$ such that the PLC function $r_{i,j}(\cdot)$ is *strictly monotone*, then a market equilibrium $\mathbf{p}$ exists.

It is a corollary of Maxfield [25]. Moreover, one can show that (e.g., see [14,19,1]), if all parameters of $\mathcal{M}$ are rational numbers, then it must have a rational equilibrium $\mathbf{p}$, and the number of bits needed to describe $\mathbf{p}$ is polynomial in the input size of $\mathcal{M}$ (i.e., the number of bits we need to describe the market $\mathcal{M}$).

We are interested in the problem of finding an approximate market equilibrium in an additively separable PLC market.

Definition 3. *Let $\mathcal{M}$ be an additively separable* PLC *market. Then we say* $\mathbf{p} \in \mathbb{R}^n_+$ *is an* ϵ-approximate equilibrium *for some* $\epsilon \ge 0$, *if there exists an allocation* $\mathcal{X} = \{\mathbf{a}_i \in \mathbb{R}^n_+ : i \in [m]\}$ *such that* $\mathbf{a}_i \in \mathrm{OPT}(i, \mathbf{p})$ *for all* $i \in [m]$; *and*

$$\left|\textstyle\sum_{i\in[m]} a_{i,j} - c_j\right| \le \epsilon \cdot c_j. \qquad \textit{for all } G_j \in \mathcal{G}.$$

We make some further restrictions on the markets we are interested in. We say an additively separable PLC market $\mathcal{M}$ is α-bounded for some $\alpha \ge 1$, if for all T_i and G_j the PLC function $r_{i,j}(\cdot)$ is either the zero function or α-bounded. We call an additively separable PLC market $\mathcal{M}$ a 2-linear market, if for all T_i and G_j, $r_{i,j}(\cdot)$ has at most two segments. Finally we say an additively separable PLC market $\mathcal{M}$ is t-sparse, for some positive integer t, if for any T_i, the number of $j \in [n]$ such that $r_{i,j}(\cdot)$ is not the zero function is at most t. In another word every buyer T_i is interested in at most t goods.

We use FISHER to denote the following search problem: given a 2-linear additively separable PLC market $\mathcal{M}$, which is 81-bounded, 43-sparse and satisfies the condition of Maxfield, find an n^{-21}-approximate market equilibrium, where n denotes the number of goods in the market. The main result of the paper is

Theorem 1 (Main). FISHER *is* PPAD-hard.

3 A Price-Regulating Market

In this section, we construct a family of *price-regulating* markets $\{\mathcal{M}_n : n \ge 1\}$ in Fisher's setting. For every positive integer n, $\mathcal{M}_n$ has n buyers, $2n$ goods and satisfies the following price regulation property.

Property 1. Let $\mathbf{p}$ *be an* ϵ*-approximate equilibrium with* $\epsilon < 1$, *then* $\forall k \in [n]$,

$$3/(1+\epsilon) \le p_{2k-1} + p_{2k} \le 3/(1-\epsilon) \quad \textit{and} \quad 1/2 \le p_{2k-1}/p_{2k} \le 2.$$

Theorem 1 then follows from Theorem 2. Note that if $\mathbf{p}$ is an N^{-21}-approximate equilibrium, then by definition it is also an n^{-21}-approximate equilibrium.

Theorem 2. *If* $\mathbf{p}$ *is an* n^{-21}*-approximate market equilibrium of* $\mathcal{M}$*, then* $(\mathbf{x},\mathbf{y})$ *constructed above must be an* n^{-6}*-well-supported Nash equilibrium of* $(\mathbf{A},\mathbf{B})$.

We prove Theorem 2 in the rest of this section. Let $\mathbf{p} \in \mathbb{R}_+^{4n+2}$ be an n^{-21}-approximate equilibrium of $\mathcal{M}$. It is easy to see that $p_k > 0$ for all k. Let $\mathcal{X}$ be an optimal allocation with respect to $\mathbf{p}$ that clears the market approximately:

$$\mathcal{X} = \big\{\mathbf{a}_i, \mathbf{a}_\mathbf{u}, \mathbf{a}_\mathbf{v} \in \mathbb{R}_+^{4n+2} : i \in [2n+1], \mathbf{u} \in U \text{ and } \mathbf{v} \in V\big\}.$$

We start with some notation. We let

$$\mathcal{T}^* = \big\{T_i : i \in [2n+1]\big\}, \quad \mathcal{T}_U = \big\{T_\mathbf{u} : \mathbf{u} \in U\big\}, \quad \text{and} \quad \mathcal{T}_V = \big\{T_\mathbf{v} : \mathbf{v} \in V\big\}.$$

Let $\mathcal{T}' \subseteq \mathcal{T}$ be a subset of buyers and $k \in [4n+2]$, then we let $a_k[\mathcal{T}']$ denote the amount of good G_k that buyers in $\mathcal{T}'$ receive in the final allocation $\mathcal{X}$. For $\mathcal{T}' \subseteq \mathcal{T}_U \cup \mathcal{T}_V$ and $k \in [4n]$, we let

$$s_k[\mathcal{T}'] = \textstyle\sum_{T_\mathbf{u} \in \mathcal{T}' \cap \mathcal{T}_U} s_{\mathbf{u},k} + \sum_{T_\mathbf{v} \in \mathcal{T}' \cap \mathcal{T}_V} s_{\mathbf{v},k}.$$

By the construction, we have $c_{4n+1} = c_{4n+2} = 1$ and $1 < c_k = 1 + \Theta(1/n^{11}) < 2$ for every $k \in [4n]$. By the definition of ϵ-approximate equilibria, we also have

$$\Big| s_k[\mathcal{T}_\mathbf{u} \cup \mathcal{T}_\mathbf{v}] - a_k[\mathcal{T}_\mathbf{u} \cup \mathcal{T}_\mathbf{v}] + 1 - a_k[\mathcal{T}^*] \Big| < 2/n^{21}, \quad \text{for all } k \in [4n]. \tag{5}$$

To prove Theorem 2, we first show that the price vector $\mathbf{p}$ must satisfy the following price-regulation property. The proof is similar to that of Property 1, which mainly uses the fact that the buyers in $\mathcal{T}^*$ possess almost all the money in $\mathcal{M}$. So we omit it here, which can be found in the full version [26].

Lemma 1 (Price Regulation). *For every* $k \in [2n+1]$*, we have*

$$1/2 \le p_{2k-1}/p_{2k} \le 2 \quad \textit{and} \quad 3 - O\left(1/n^{11}\right) \le p_{2k-1} + p_{2k} \le 3 + O\left(1/n^{10}\right).$$

Using Lemma 1, we analyze the behavior of $T_\mathbf{u}$, $\mathbf{u} \in U$, as follows.

Behavior of $T_\mathbf{u}$: Let $\mathbf{u} = (i,j,1) \in U$ where $1 \le i \ne j \le n$. Let $\mathbf{C} = \mathbf{A}_i - \mathbf{A}_j$, $m \le 20$ be the number of nonzero entries in $\mathbf{C}$ and $C = \sum_{k \in [n]} C_k$. By Lemma 1 and the optimality of $\mathbf{a}_\mathbf{u}$, $T_\mathbf{u}$ first buys the following bundle of goods:

$$\big\{ s_{\mathbf{u},2(n+k)-1} \text{ of } G_{2(n+k)-1} \text{ and } s_{\mathbf{u},2(n+k)} \text{ of } G_{2(n+k)} : k \in [n] \text{ and } C_k \ne 0 \big\}. \tag{6}$$

Using Lemma 1, one can show that her money left is $3/n^{12} - O(1/n^{13}) > 0$.

After buying this bundle, $T_\mathbf{u}$ buys G_{2j-1} up to $1/n^{12}$, and her money left is $\Omega(1/n^{12})$ by Lemma 1. Finally, $T_\mathbf{u}$ buys G_{2j} up to $1/n^{12}$, and then spends all the money left, if any, to buy G_{4n+1}.

The behavior of $T_{\mathbf{v}}$, $\mathbf{v} \in V$, is similar so we omit it here. The analysis above gives us the following lemma. The proof can be found in the full version [26].

Lemma 2. *Let* $\mathbf{p}$ *be an* n^{-21}*-approximate equilibrium. Then for any* $k \in [2n]$,

$$s_{2k}[\mathcal{T}_U \cup \mathcal{T}_V] - a_{2k}[\mathcal{T}_U \cup \mathcal{T}_V] = \Omega(1/n^{19}) \implies p_{2k} = (p_{2k-1} + p_{2k})/3.$$
$$s_{2k}[\mathcal{T}_U \cup \mathcal{T}_V] - a_{2k}[\mathcal{T}_U \cup \mathcal{T}_V] = O(1/n^{21}) \implies p_{2k} = 2(p_{2k-1} + p_{2k})/3.$$

Let $\mathbf{x}', \mathbf{y}'$ denote the two vectors obtained from $\mathbf{p}$ as in (3). By Lemma 1, we have $0 \leq x'_k, y'_k \leq 1 + O(1/n^{10})$, for any $k \in [n]$. Finally, we state the following two lemmas. Both proofs use Lemma 2, which can be found in [26].

Lemma 3. *Let* $\epsilon = n^{-6}$. *Then for all* i, j *with* $1 \leq i \neq j \leq n$, *we have*

$$(\mathbf{A}_i - \mathbf{A}_j)\mathbf{y}'^T > \epsilon/2 \Rightarrow x'_j = 0 \quad \textit{and} \quad \mathbf{x}'(\mathbf{B}_i - \mathbf{B}_j) > \epsilon/2 \Rightarrow y'_j = 0,$$

where $\mathbf{A}_i$ *denotes the* ith *row of* $\mathbf{A}$ *and* $\mathbf{B}_i$ *denotes the* ith *column of* $\mathbf{B}$.

Lemma 4. $\exists i, j \in [n]$ *such that* $x'_i \geq 1 - O(1/n^{11})$ *and* $y'_j \geq 1 - O(1/n^{11})$.

Now assume $\mathbf{x}'$ and $\mathbf{y}'$ satisfy both lemmas. In particular, Lemma 4 implies that $\mathbf{x}', \mathbf{y}' \neq \mathbf{0}$. Therefore, we can normalize them to get two probability distributions $\mathbf{x}$ and $\mathbf{y}$ using (4). Theorem 2 then follows directly from Lemma 3 and 4. The proof can be found in [26].

Acknowledgement. We would like to thank Nikhil Devanur for suggesting us to consider the complexity of Fisher equilibria for additively separable PLC utility functions and for his valuable intuition on this problem.

References

1. Vazirani, V., Yannakakis, M.: PPAD-completeness of Fisher markets under piecewise-linear, concave utility functions (2009) (manuscript)
2. Brainard, W., Scarf, H.: How to compute equilibrium prices in 1891. Cowles Foundation Discussion Papers 1272, Cowles Foundation, Yale University (2000)
3. Arrow, K., Debreu, G.: Existence of an equilibrium for a competitive economy. Econometrica 22, 265–290 (1954)
4. Eisenberg, E., Gale, D.: Consensus of subjective probabilities: The pari-mutuel method. Annals of Mathematical Statistics 30(1), 165–168 (1959)
5. Nenakov, E., Primak, M.: One algorithm for finding solutions of the Arrow-Debreu model. Kibernetica, 127–128 (1983)
6. Deng, X., Papadimitriou, C., Safra, M.: On the complexity of equilibria. In: Proceedings of the 34th annual ACM symposium on Theory of computing, pp. 67–71 (2002)
7. Devanur, N., Papadimitriou, C., Saberi, A., Vazirani, V.: Market equilibrium via a primal-dual algorithm for a convex program. Journal of the ACM 55(5), 1–18 (2008)
8. Jain, K.: A polynomial time algorithm for computing the Arrow-Debreu market equilibrium for linear utilities. In: Proceedings of the 45th Annual IEEE Symposium on Foundations of Computer Science, pp. 286–294 (2004)

9. Garg, R., Kapoor, S.: Auction algorithms for market equilibrium. In: Proceedings of the 36th annual ACM symposium on Theory of computing, pp. 511–518 (2004)
10. Jain, K., Mahdian, M., Saberi, A.: Approximating market equilibria. In: Proceedings of the 6th International Workshop on Approximation Algorithms, pp. 98–108 (2003)
11. Devanur, N.R., Vazirani, V.V.: An improved approximation scheme for computing Arrow-Debreu prices for the linear case. In: Proceedings of the 23rd Conference on Foundations of Software Technology and Theoretical Computer Science, pp. 149–155 (2003)
12. Ye, Y.: A path to the Arrow-Debreu competitive market equilibrium. Mathematical Programming 111, 315–348 (2008)
13. Codenotti, B., Pemmaraju, S., Varadarajan, K.: On the polynomial time computation of equilibria for certain exchange economies. In: Proceedings of the 16th annual ACM-SIAM symposium on Discrete algorithms, pp. 72–81 (2005)
14. Devanur, N.R., Kannan, R.: Market equilibria in polynomial time for fixed number of goods or agents. In: Proceedings of the 49th annual IEEE Symposium on Foundations of Computer Science, pp. 45–53 (2008)
15. Codenotti, B., McCune, B., Penumatcha, S., Varadarajan, K.R.: Market equilibrium for CES exchange economies: Existence, multiplicity, and computation. In: Sarukkai, S., Sen, S. (eds.) FSTTCS 2005. LNCS, vol. 3821, pp. 505–516. Springer, Heidelberg (2005)
16. Codenotti, B., Saberi, A., Varadarajan, K., Ye, Y.: Leontief economies encode nonzero sum two-player games. In: Proceedings of the 17th annual ACM-SIAM Symposium on Discrete Algorithms, pp. 659–667 (2006)
17. Chen, X., Deng, X.: Settling the complexity of two-player Nash equilibrium. In: Proceedings of the 47th Annual IEEE Symposium on Foundations of Computer Science, pp. 261–272 (2006)
18. Huang, L.-S., Teng, S.-H.: On the approximation and smoothed complexity of Leontief market equilibria. In: Preparata, F.P., Fang, Q. (eds.) FAW 2007. LNCS, vol. 4613, pp. 96–107. Springer, Heidelberg (2007)
19. Chen, X., Dai, D., Du, Y., Teng, S.H.: Settling the complexity of Arrow-Debreu equilibria in markets with additively separable utilities. In: Proceedings of the 50th Annual IEEE Symposium on Foundations of Computer Science (2009)
20. Chen, X., Huang, L.S., Teng, S.H.: Market equilibria with hybrid linear-Leontief utilities. In: Spirakis, P.G., Mavronicolas, M., Kontogiannis, S.C. (eds.) WINE 2006. LNCS, vol. 4286, pp. 274–285. Springer, Heidelberg (2006)
21. Ye, Y.: Exchange market equilibria with Leontief's utility: Freedom of pricing leads to rationality. Theoretical Computer Science 378(2), 134–142 (2007)
22. Jain, K., Vazirani, V., Ye, Y.: Market equilibria for homothetic, quasi-concave utilities and economies of scale in production. In: Proceedings of the 16th annual ACM-SIAM symposium on Discrete algorithms, pp. 63–71 (2005)
23. Nisan, N., Roughgarden, T., Tardos, E., Vazirani, V.: Algorithmic Game Theory. Cambridge University Press, Cambridge (2007)
24. Chen, X., Deng, X., Teng, S.H.: Sparse games are hard. In: Spirakis, P.G., Mavronicolas, M., Kontogiannis, S.C. (eds.) WINE 2006. LNCS, vol. 4286, pp. 262–273. Springer, Heidelberg (2006)
25. Maxfield, R.: General equilibrium and the theory of directed graphs. Journal of Mathematical Economics 27(1), 23–51 (1997)
26. Chen, X., Teng, S.H.: Spending is not easier than trading: On the computational equivalence of Fisher and Arrow-Debreu equilibria. arXiv:0907.4130 (2009)

The Identity Correspondence Problem and Its Applications

Paul C. Bell and Igor Potapov

Department of Computer Science, The University of Liverpool
p.c.bell@liverpool.ac.uk, potapov@liverpool.ac.uk

Abstract. In this paper we study several closely related fundamental problems for words and matrices. First, we introduce the Identity Correspondence Problem (ICP): whether a finite set of pairs of words (over a group alphabet) can generate an identity pair by a sequence of concatenations. We prove that ICP is undecidable by a reduction of Post's Correspondence Problem via several new encoding techniques. In the second part of the paper we use ICP to answer a long standing open problem concerning matrix semigroups: "Is it decidable for a finitely generated semigroup S of integral square matrices whether or not the identity matrix belongs to S?". We show that the problem is undecidable starting from dimension four even when the number of matrices in the generator is 48. From this fact, we can immediately derive that the fundamental problem of whether a finite set of matrices generates a group is also undecidable. We also answer several questions for matrices over different number fields.

1 Introduction

Combinatorics on words has many connections to several areas of mathematics and computing. It is well known that words are very suitable objects to formulate fundamental properties of computations. One such property that may be formulated in terms of operations on words is the exceptional concept of undecidability. A problem is called undecidable if there exists no algorithm that can solve it. A famous example is Post's Correspondence Problem (PCP) originally proved undecidable by Post in 1946. It plays a central role in computer science due to its applicability for showing the undecidability of many computational problems in a very natural and simple way.

In the spirit of Post's Correspondence Problem, in this paper, we introduce the Identity Correspondence Problem (ICP): whether a finite set of pairs of words (over a group alphabet) can generate an identity pair by a sequence of concatenations. We prove that ICP is undecidable by a reduction of Post's Correspondence Problem via several new encoding techniques that are used to guarantee the existence of an identity pair only in the case of a correct solution for PCP. We believe that the ICP will be useful in identifying new areas of undecidable problems related to computational questions in abstract algebra, logic and combinatorics on words.

Y. Dong, D.-Z. Du, and O. Ibarra (Eds.): ISAAC 2009, LNCS 5878, pp. 657–667, 2009.

In the second part of the paper, we use the Identity Correspondence Problem to answer several long standing open problems concerning matrix semigroups [6]. Taking products of matrices is one of the fundamental operations in mathematics. However, many computational problems related to the analysis of matrix products are algorithmically hard and even undecidable. Among the oldest results is a remarkable paper by M. Paterson, where he shows that it is undecidable whether the multiplicative semigroup generated by a finite set of 3×3 integer matrices contains the zero matrix (also known as the mortality problem), see [18]. Since then, many results were obtained about checking the freeness, boundedness and finiteness of matrix semigroups and the decidability of different reachability questions such as the membership problem, vector reachability, scalar reachability etc. See [2,3,4,5,7,8,9,13] for example for some related decidability results.

The membership problem asks whether a particular matrix is contained within a given semigroup. The membership problem is undecidable for 3×3 integral matrix semigroups due to Paterson's results and also for the group $\mathrm{SL}(4, \mathbb{Z})$ of 4×4 integer matrices of determinant 1, shown by Mikhailova [16]. Another important problem in matrix semigroups is the identity problem: Decide whether a finitely generated matrix semigroup contains the identity matrix. The identity problem is equivalent to the following problem: Given a finitely generated matrix semigroup $S \subseteq \mathbb{Z}^{n \times n}$, decide whether a subset of the generators of S generates a non-trivial matrix group. In general it is undecidable whether or not the monoid described by a given finite representation is a group. However, this decision problem is reducible to a very restricted form of the uniform word problem and it is not directly applicable to the proof of undecidability of the group problem in matrix semigroups [17].

The question about the membership of the identity matrix for matrix semigroups is a well known open problem and was recently stated in "Unsolved Problems in Mathematical Systems and Control Theory", [6] and also as Problem 5 in [11]. The embedding methods used to show undecidability in other results do not appear to work here [6]. As far as we know, only two decidability results are known for the identity problem. Very recently the first general decidability result for this problem was proved in the case of 2×2 integral matrix semigroups, see [9]. It is also known that in the special case of commutative matrix semigroups [1] the problem is decidable in any dimension.

In this paper we apply ICP to answer the long standing open problem: "Is it decidable for a finitely generated semigroup S of square integral matrices whether or not the identity matrix belongs to S?". We show that the identity problem is undecidable starting from dimension four even when the number of matrices in the generator is fixed. In other words, we can define a class of finite sets $\{M_1, M_2, \ldots, M_k\}$ of four dimensional matrices such that there is no algorithm to determine whether or not the identity matrix can be represented as a product of these matrices. From this fact, we can immediately derive that the fundamental problem of whether a finite set of 4×4 matrices generates a group is also undecidable. In our proofs we use the fact that free groups can be embedded into the multiplicative group of 2×2 integral matrices. This allows us to transfer the undecidability of ICP into undecidability results on matrices.

We also provide a number of other corollaries. In particular, the identity and group problems are undecidable for double quaternions and a set of rotations on the 3-sphere. Therefore, there is no algorithm to check whether a set of linear transformations or a set of rotations in dimension 4 is reversible. Also, the question of whether any diagonal matrix can be generated by a 4×4 integral matrix semigroup is undecidable.

2 Identity Correspondence Problem

Notation: Given an alphabet $\Sigma = \{a, b\}$, we denote the concatenation of two letters $x, y \in \Sigma$ by xy or $x \cdot y$. A *word* over Σ is a concatenation of letters from alphabet Σ, i.e., $w = w_1 w_2 \cdots w_k \in \Sigma^*$. We denote throughout the paper the *empty word* by ε. We shall denote a *pair of words* by either (w_1, w_2) or $\frac{w_1}{w_2}$.

The free group over a generating set H is denoted by $\mathrm{FG}(H)$, i.e., the free group over two elements a and b is denoted as $\mathrm{FG}(\{a, b\})$, we shall also write this as $\mathrm{FG}(a, b)$ by abuse of notation. For example, the elements of $\mathrm{FG}(a, b)$ are all the words over the alphabet $\{a, b, a^{-1}, b^{-1}\}$ that are reduced, i.e., that contain no subword of the form $x \cdot x^{-1}$ or $x^{-1} \cdot x$ (for $x \in \{a, b\}$).

Problem 1. *Identity Correspondence Problem (ICP) - Given a finite set of pairs of words $\Pi = \{(u_1, v_1), (u_2, v_2), \ldots, (u_m, v_m)\} \subset \mathrm{FG}(a, b) \times \mathrm{FG}(a, b)$. Does the equation $u_{s_1} u_{s_2} \cdots u_{s_k} = v_{s_1} v_{s_2} \cdots v_{s_k} = \varepsilon$ hold for any nonempty finite sequence of indices $s = (s_1, s_2, \ldots, s_k)$, where ε is the empty word (identity)?*

We shall reduce a restricted form of Post's Correspondence Problem (PCP) [13] to the Identity Correspondence Problem in a constructive way. We shall require the following theorem:

Theorem 1. *[13,15] Let $\Sigma = \{a, b\}$ be a binary alphabet and $P = \{(r_1, t_1), (r_2, t_2), \ldots \ldots, (r_n, t_n)\} \subseteq \Sigma^* \times \Sigma^*$ be a set of pairs of words. It is undecidable to determine if there exists a finite sequence of indices $2 \leq x_1, x_2, \ldots, x_k \leq n-1$ such that: $r_1 r_{x_1} r_{x_2} \cdots r_{x_k} r_n = t_1 t_{x_1} t_{x_2} \cdots t_{x_k} t_n$. This result holds even for $n = 7$.*

A first step towards the proof of undecidability of Problem 1 was shown in [2] where the following theorem was presented (although in a different form).

Theorem 2. *[2] Let $\Sigma = \{a, b\}$ be a binary alphabet and $X = \{(u_1, v_1), (u_2, v_2), \ldots \ldots, (u_n, v_n)\} \subseteq \mathrm{FG}(\Sigma)^* \times \mathrm{FG}(\Sigma)^*$. It is undecidable to determine if there exists a finite sequence $s_1, s_2, \ldots, s_k$ where $1 \leq s_i \leq n$ and* exactly one *$s_i = n$ such that $u_{s_1} u_{s_2} \cdots u_{s_k} = v_{s_1} v_{s_2} \cdots v_{s_k} = \varepsilon$.*

The reason Theorem 2 does not prove Problem 1 is undecidable is the restriction that the final pair of words (u_n, v_n) is used exactly one time. Despite many attempts, it is not clear how one may remove this restriction in the construction of the proof, since it is essential that this pair be used once to avoid the pathological case of several incorrect solutions canceling with each other.

The main idea of this paper is to show a new non-trivial encoding which contains the encoding used in Theorem 2 but avoids the requirement that a specific element be used one time. The idea is that by encoding the set X *four times* using four different alphabets and adding 'borders' to each pair of words such that for cancelation to occur, each of these alphabets must be used in a specific order, any incorrect solutions using a single alphabet will not be able to be canceled later on.

We shall reduce an instance of the restricted Post's Correspondence Problem of Theorem 1 to an instance of the Identity Correspondence Problem. Let here and throughout $\Sigma = \{a, b\}$ and define new alphabets $\Gamma_i = \{a_i, b_i\}$ for $1 \leq i \leq 4$ and $\Gamma_B = \{x_j | 1 \leq j \leq 8\}$ such that the alphabets are distinct (specifically, the intersection of the free groups generated by any two different alphabets equals $\{\varepsilon\}$). Let us define mappings $\delta_i : \mathrm{FG}(\Sigma) \to \mathrm{FG}(\Gamma_i)$ by $\delta_i(a) = a_i$, $\delta_i(b) = b_i$, $\delta_i(a^{-1}) = a_i^{-1}$, $\delta_i(b^{-1}) = b_i^{-1}$ for $1 \leq i \leq 4$. Note that each δ_i is a homomorphism that may be applied to words over $\mathrm{FG}(\Sigma)$ in the natural way.

Let $\Gamma = \Gamma_1 \cup \Gamma_2 \cup \Gamma_3 \cup \Gamma_4 \cup \Gamma_B$. Define $\phi_i : \mathbb{Z}^+ \to \{a_i, b_i\}^*$ by $\phi_i(j) = a_i^j b_i$. Similarly, let $\psi_i : \mathbb{Z}^+ \to \{a_i^{-1}, b_i^{-1}\}^*$ be defined by $\psi_i(j) = (a_i^{-1})^j b_i^{-1}$. These morphisms will be used to ensure a product is in a specific order.

Let $P = \{(s_1, t_1), (s_2, t_2), \ldots, (s_n, t_n)\} \subseteq \Sigma^* \times \Sigma^*$ be a given restricted PCP instance. An instance of ICP consists of a set of 48 pairs of words for $n = 7$:

$$W = W_0 \cup W_1 \cup \ldots \cup W_{15} \subseteq \mathrm{FG}(\Gamma)^* \times \mathrm{FG}(\Gamma)^*$$

$$\begin{aligned}
W_0 &= \left\{ \tfrac{x_8}{x_8} \cdot \tfrac{v_{11}^{-1} u_{11}}{b_1} \cdot \tfrac{x_1^{-1}}{x_1^{-1}} \right\}, & W_1 &= \left\{ \tfrac{x_1}{x_1} \cdot \tfrac{u_{1j}}{\phi_1(j)} \cdot \tfrac{x_1^{-1}}{x_1^{-1}} | 2 \leq j \leq n-1 \right\}, \\
W_2 &= \left\{ \tfrac{x_1}{x_1} \cdot \tfrac{u_{1n} v_{1n}^{-1}}{b_1^{-1}} \cdot \tfrac{x_2^{-1}}{x_2^{-1}} \right\}, & W_3 &= \left\{ \tfrac{x_2}{x_2} \cdot \tfrac{v_{1j}^{-1}}{\psi_1(j)} \cdot \tfrac{x_2^{-1}}{x_2^{-1}} | 2 \leq j \leq n-1 \right\}, \\
W_4 &= \left\{ \tfrac{x_2}{x_2} \cdot \tfrac{v_{21}^{-1} u_{21}}{b_2} \cdot \tfrac{x_3^{-1}}{x_3^{-1}} \right\}, & W_5 &= \left\{ \tfrac{x_3}{x_3} \cdot \tfrac{u_{2j}}{\phi_2(j)} \cdot \tfrac{x_3^{-1}}{x_3^{-1}} | 2 \leq j \leq n-1 \right\}, \\
W_6 &= \left\{ \tfrac{x_3}{x_3} \cdot \tfrac{u_{2n} v_{2n}^{-1}}{b_2^{-1}} \cdot \tfrac{x_4^{-1}}{x_4^{-1}} \right\}, & W_7 &= \left\{ \tfrac{x_4}{x_4} \cdot \tfrac{v_{2j}^{-1}}{\psi_2(j)} \cdot \tfrac{x_4^{-1}}{x_4^{-1}} | 2 \leq j \leq n-1 \right\}, \\
W_8 &= \left\{ \tfrac{x_4}{x_4} \cdot \tfrac{v_{31}^{-1} u_{31}}{b_3} \cdot \tfrac{x_5^{-1}}{x_5^{-1}} \right\}, & W_9 &= \left\{ \tfrac{x_5}{x_5} \cdot \tfrac{u_{3j}}{\phi_3(j)} \cdot \tfrac{x_5^{-1}}{x_5^{-1}} | 2 \leq j \leq n-1 \right\}, \\
W_{10} &= \left\{ \tfrac{x_5}{x_5} \cdot \tfrac{u_{3n} v_{3n}^{-1}}{b_3^{-1}} \cdot \tfrac{x_6^{-1}}{x_6^{-1}} \right\}, & W_{11} &= \left\{ \tfrac{x_6}{x_6} \cdot \tfrac{v_{3j}^{-1}}{\psi_3(j)} \cdot \tfrac{x_6^{-1}}{x_6^{-1}} | 2 \leq j \leq n-1 \right\}, \\
W_{12} &= \left\{ \tfrac{x_6}{x_6} \cdot \tfrac{v_{41}^{-1} u_{41}}{b_4} \cdot \tfrac{x_7^{-1}}{x_7^{-1}} \right\}, & W_{13} &= \left\{ \tfrac{x_7}{x_7} \cdot \tfrac{u_{4j}}{\phi_4(j)} \cdot \tfrac{x_7^{-1}}{x_7^{-1}} | 2 \leq j \leq n-1 \right\}, \\
W_{14} &= \left\{ \tfrac{x_7}{x_7} \cdot \tfrac{u_{4n} v_{4n}^{-1}}{b_4^{-1}} \cdot \tfrac{x_8^{-1}}{x_8^{-1}} \right\}, & W_{15} &= \left\{ \tfrac{x_8}{x_8} \cdot \tfrac{v_{4j}^{-1}}{\psi_4(j)} \cdot \tfrac{x_8^{-1}}{x_8^{-1}} | 2 \leq j \leq n-1 \right\},
\end{aligned}$$

where $u_{ik} = \delta_i(s_k)$, $v_{ik} = \delta_i(t_k)$ for $1 \leq k \leq n$ and $1 \leq i \leq 4$. Given any two words $w_1, w_2 \in \mathrm{FG}(\Gamma)^*$, recall that we denote by $\frac{w_1}{w_2}$ the pair of words $(w_1, w_2) \in \mathrm{FG}(\Gamma)^* \times \mathrm{FG}(\Gamma)^*$ in the above table.

Note that each word in each pair from W_i has a so called 'border letter' on the left and right from alphabet $\mathrm{FG}(\Gamma_B)$. These are used to restrict the type of sequence[1] that can lead to an identity pair. The central element of each word (i.e. excluding the 'border letters') corresponds to particular words from P and

[1] The only sequences that may lead to identity pair should be of the form of a cycle or insertions of cycles.

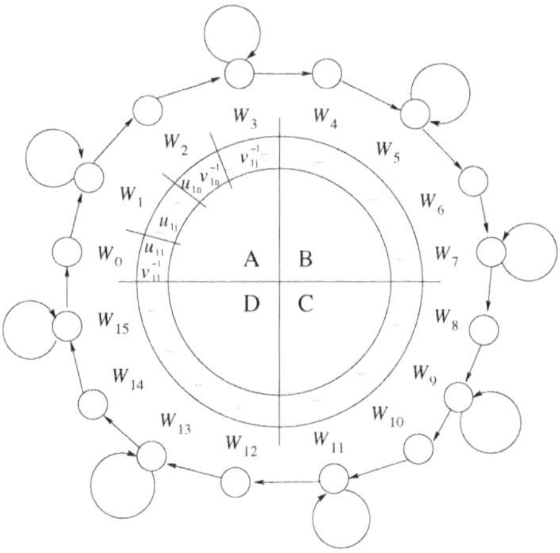

Fig. 1. The structure of a product which forms the identity

we encode instance P four times separately, first in W_0, W_1, W_2, W_3, secondly in W_4, W_5, W_6, W_7 etc. using different alphabets for each encoding [2]. This may be seen in Fig. 1, where A, B, C and D each separately encode instance P.

The second words from each pair in W use an encoding which will ensure that the first words are concatenated in a particular order within each A, B, C and D part. We show in Lemma 1 that this encoding enforces a correct encoding of the instance P within each part if that part gets reduced down to two letters in the second word (the first and last 'border letters'). In particular, we adapt here our recently introduced index encoding technique from [2].

One of the important encoding concepts is a *cycle* of a set W. We see that the first and last letters of any pair of words from set $W_i \subset W$ only cancel with a pair of words from set $W_{i+1 \bmod 16}$ for $0 \leq i \leq 15$ and with elements from W_i itself if $i \bmod 2 = 1$.

Definition 1. *A cycle w of W is a word pair from W^* of the following form:*

$$w = w_i \cdot w_{(i+1) \bmod 16} \cdot \dots \cdot w_{(i+15) \bmod 16} \in W^* \tag{1}$$

for some i: $0 \leq i \leq 15$, where $w_y \in W_y$ if $y \bmod 2 \equiv 0$ and $w_y \in W_y^*$ if $y \bmod 2 \equiv 1$.

For example a cycle could use element W_4 followed by a product of elements from W_5, then element W_6, followed by a product of elements from W_7 etc. As previously mentioned, the idea of the encoding is that a correct solution to the PCP instance P will be encoded *four times* in a correct solution to W, in elements from $\{W_0, \dots, W_3\}, \{W_4, \dots, W_7\}, \{W_8, \dots, W_{11}\}$ and $\{W_{12}, \dots, W_{15}\}$ separately.

2 In the case of an incorrect solution for a PCP instance, the use of different alphabets for the four parts create an ordered sequence of non-empty parts that cannot be trivially canceled from the left or right side.

Proposition 1. *If ICP has a solution it must be constructed by a sequence of pairs of words which form either a single cycle or a pattern generated by cycle insertions.*

Definition 2. *For any product $Y \in W^*$ let us denote by a* decomposition by parts *of Y, the decomposition $Y = Y_1 Y_2 \cdots Y_k$ where for each $1 \leq i \leq k$, if $Y_i \subset \mathrm{FG}(\Gamma_i \cup \Gamma_B)^* \times \mathrm{FG}(\Gamma_i \cup \Gamma_B)^*$ then $Y_{i+1} \subset \mathrm{FG}(\Gamma_j \cup \Gamma_B)^* \times \mathrm{FG}(\Gamma_j \cup \Gamma_B)^*$ where $1 \leq i, j \leq 4$ and $i \neq j$.*

For a cycle Q, the decomposition by parts of Q clearly gives either 4 or 5 parts in the decomposition. For example, we may have $Q = X_1 X_2 X_3 X_4 X_5$ where $X_1 \in \mathrm{FG}(\Gamma_i \cup \Gamma_B)^* \times \mathrm{FG}(\Gamma_i \cup \Gamma_B)^*$ and thus $X_2 \in \mathrm{FG}(\Gamma_{(i \bmod 4)+1} \cup \Gamma_B)^* \times \mathrm{FG}(\Gamma_{(i \bmod 4)+1} \cup \Gamma_B)^*$ etc. X_5 is either empty or uses the same alphabet as X_1.

The first words from each pair in each A, B, C, D part of Fig. 1 will store all words from the instance of restricted PCP, P separately. If we concatenate the first words of one of these parts in the correct order and have the empty word (excluding initial and final 'border letters'), then this corresponds to a solution of P. By a correct order, we mean that if we have $u_{i1} u_{i2} \cdots u_{ik}$ for example, then they should be concatenated with $(v_{i1} v_{i2} \cdots v_{ik})^{-1} = v_{ik}^{-1} \cdots v_{i2}^{-1} v_{i1}^{-1}$. If the concatenation of these words equals ε, then we have a correct solution to P.

The encoding in the second words using ϕ_i, ψ_i and $\{b_i, b_i^{-1} | 1 \leq i \leq 4\}$ is used to ensure that any solution to W *must* use such a correct ordering in each A, B, C, D part. The next lemma formalizes this concept and is a modification of the technique presented in [2]. It also can be seen as a simpler version of Fixed Element PCP, see [4].

Lemma 1. *Given any part $X \in \mathrm{FG}(\Gamma_j \cup \Gamma_B)^* \times \mathrm{FG}(\Gamma_j \cup \Gamma_B)^*$, if the second word of X consists of only the initial and final 'border letters' $x_p x_q^{-1}$ where $(p, q) \in \{(8, 2), (2, 4), (4, 6), (6, 8)\}$ (i.e. it is of length two and over Γ_B), then the second words of X must be of the form*

$$x_p \cdot b_j \phi_j(z_1) \phi_j(z_2) \cdots \phi_j(z_k) \cdot b_j^{-1} \cdot \psi_j(z_k) \cdots \psi_j(z_2) \psi_j(z_1) \cdot x_q^{-1},$$

where $2 \leq z_1, z_2, \ldots, z_k \leq n-1$. (This corresponds to a 'correct' palindromic encoding of the PCP instance P within this part.)

Lemma 2. *If there exists a solution to the PCP instance P, then there exists a solution to the Identity Correspondence Problem instance W.*

Proof. Assume we have a solution to P with indices $2 \leq i_1, i_2, \ldots, i_k \leq n-1$, i.e., $s_1 s_{i_1} \cdots s_{i_k} s_n = t_1 t_{i_1} \cdots t_{i_k} t_n$. If we form a cycle of elements from W using these indices for j in the u, v words, we see that all elements will cancel and we will form the word $\frac{\varepsilon}{\varepsilon}$ as required (see Fig. 1). □

We shall define four 'types' of these parts, A, B, C, D where type A parts use alphabet $\mathrm{FG}(\Gamma_1 \cup \Gamma_B)^* \times \mathrm{FG}(\Gamma_1 \cup \Gamma_B)^*$, type B parts use $\mathrm{FG}(\Gamma_2 \cup \Gamma_B)^* \times \mathrm{FG}(\Gamma_2 \cup \Gamma_B)^*$, type C parts use $\mathrm{FG}(\Gamma_3 \cup \Gamma_B)^* \times \mathrm{FG}(\Gamma_3 \cup \Gamma_B)^*$ and type D parts use $\mathrm{FG}(\Gamma_4 \cup \Gamma_B)^* \times \mathrm{FG}(\Gamma_4 \cup \Gamma_B)^*$ as in Fig. 1. A cycle thus has a decomposition which is a permutation of $ABCD$.

Let us define a function $\zeta : W^* \to \mathbb{N}$. Given any product Y with the decomposition by parts $Y = Y_1 Y_2 \cdots Y_k$, let $\zeta(Y)$ denote the sum of non-identity words from the set of pairs of words $\{Z_1, Z_2, \ldots, Z_k\}$ where Z_i is a pair of words constructed from Y_i where we exclude the first and last letters (from Γ_B) in each pair of words in Y_i. Note that $Z_i \in \mathrm{FG}(\Gamma_j \cup \Gamma_B)^* \times \mathrm{FG}(\Gamma_j \cup \Gamma_B)^*$ for some $1 \leq j \leq 4$.

Thus for a single cycle Q, $0 \leq \zeta(Q) \leq 10$ since it can be decomposed to a maximum of 5 parts. If the first and second words in each decomposed part have a non-reducible word in between the borders, we have that $\zeta(Q)$ is equal to 10. If $\zeta(Q)$ equals 0, it means that all words in between the border elements are reducible to identity.

Lemma 3. *If there exists no solution to the encoded PCP instance P then for any cycle $Q \in W^+$ and its decomposition by parts $Q = X_1 X_2 X_3 X_4 X_5$ the following holds: any $X_r \neq (\varepsilon, \varepsilon)$ where $1 \leq r \leq 5$; $\zeta(Q) \geq 4$ and $Q \neq (\varepsilon, \varepsilon)$, i.e., a single cycle cannot be a solution to the Identity Correspondence Problem.*

Proof. Let Q be a single cycle of the form (1). Since it is a cycle, the 'border letters' of each pair will all cancel with each other and thus we may ignore letters from $\mathrm{FG}(\Gamma_B)$. Let $Q = X_1 X_2 X_3 X_4 X_5$ be its decomposition by parts (thus X_5 can be empty and four of the 'parts' use different alphabets).

Let us consider X_r which is either equal to X_1 X_2, X_3, X_4 or X_5. We will show that X_r cannot be equal to $(\varepsilon, \varepsilon)$. Since Q is a cycle, which has a specific structure, the first word of this element, when concatenated, equals $v_{r1}^{-1} u_{r1} u_{rj_1} \cdots u_{rj_m} u_{rn} v_{rn}^{-1} v_{rk_l}^{-1} \cdots v_{rk_1}^{-1}$ for some $1 \leq r \leq 4$ and $m, l \geq 0$. If $j_i = k_i$ for all $1 \leq i \leq m$ with $m = n$ then this is a correct encoding of the PCP instance P which we have assumed has no solution, thus this word does not equal ε in this case. Therefore the elements must not be in a correct sequence if the first word equals ε. In this case however, the second word will now not equal ε by the choice of the morphisms ϕ_i and ψ_i as shown in Lemma 1. If we have such an incorrect ordering then when we multiply the second set of words (since also each morphism uses a different alphabet) they never equal ε which is not difficult to see. So assuming that there is no solution to the PCP instance P, for any part X_r, $X_r \neq (\varepsilon, \varepsilon)$, i.e., at least one word in the pairs of words of each part does not equal ε. Thus, if there exists no solution to PCP, then $4 \leq \zeta(ABCD) \leq 8$ for a cycle $ABCD$. □

It follows from Lemma 3 that the statements of Prop. 1 can be restricted further. Then in Lemma 4 we show that a concatenation of cycles cannot form a solution either which means that if ICP has a solution it should be a solution in the form of a single cycle.

Proposition 2. *If ICP has a solution it must be given by a concatenation of cycles.*

Lemma 4. *Given an instance of the Identity Correspondence Problem W encoding an instance P of restricted Post's Correspondence Problem, if there exists* *no solution to P then for any product $X \in W^+$, it holds that $X \neq (\varepsilon, \varepsilon)$, i.e., if there is no solution to P, there is no solution to W.*

Proof. Let $X = X_1 X_2 \cdots X_k$ be the decomposition by parts of X. Assume $X = (\varepsilon, \varepsilon)$ is a solution to W, then clearly a permutation of the elements of X is possible so that we may assume without loss of generality that X starts with a part of type A.

Due to the 'border constraints', Proposition 2 gives us a restricted form of sequences that may lead to an identity pair, i.e., a type A pair of words must be followed by a type B pair of words which must be followed by a type C pair of words etc. Clearly then X must be of the form $ABCD \cdot ABCD \cdots ABCD$ if it equals $(\varepsilon, \varepsilon)$ since a single cycle is not a solution to W by Lemma 3.

Assuming that there is no solution to the PCP instance P, for any part, Y_i, we proved in Lemma 3 that $Y_i \neq (\varepsilon, \varepsilon)$, i.e., at least one word in the pairs of words of each part does not equal ε (even excluding initial and final border letters). Thus, crucially, if there exists no solution to PCP, then $4 \leq \zeta(ABCD) \leq 8$ for a cycle $ABCD$.

We have that $\zeta(ABCD) \geq 4$. We shall now prove $\zeta(ABCDABCD) \geq 4$, i.e., by adding another cycle to the existing one, the number of 'empty parts' does not decrease. This means that we cannot reduce such a product to $(\varepsilon, \varepsilon)$ and thus if there exists no solution to instance P, there exists no solution to the Identity Correspondence Problem instance W as required. To see this, consider how many parts can be cancelled by adding a cycle. For example if the first word has an A part which cancels with the A part of the second cycle, then the first word for the B, C, D parts of the first cycle must be ε. But since no part can be equal to $(\varepsilon, \varepsilon)$ we know that in the first $ABCD$ cycle, the second word of the B, C, D parts must not equal ε. The only element that can cancel the second word of $ABCD$ is thus the D part of the *second* cycle. However this implies that the second word of the A, B, C parts of the second cycle all equal ε, thus the first word of the B, C parts of the second cycle cannot be ε and we have at least four non ε parts (the first and second words of the B, C parts).

The same argument holds to cancel any part thus we cannot reduce more than 4 parts by the concatenation of two cycles. The first word can cancel at most two parts and the second words can cancel at most two parts but since we start with eight nonempty parts we remove only four parts at most leaving four remaining parts. Thus $\zeta(ABCDABCD) \geq \zeta(ABCD) + \zeta(ABCD) - 4 \geq 4$ as required. In fact, it is not difficult to see that this argument can be applied iteratively and thus $\zeta((ABCD)^k) \geq 4$ always holds. If there is no solution to ICP then a concatenation of cycles cannot form a solution. □

Theorem 3. *The Identity Correspondence Problem is undecidable for $n = 48$.*

Proof. Given an instance of the Identity Correspondence Problem, $W \subseteq \mathrm{FG}(\Gamma)^* \times \mathrm{FG}(\Gamma)^*$ which encodes an instance of restricted Post's Correspondence Problem P. If there exists a solution to P, Lemma 2 implies that there also exists a solution to W. If there does not exist a solution to P, for a single cycle of W, Lemma 3 shows that there also exists no solution. Finally, Lemma 4 shows that if there is no solution to P, any possible solution to W must be a concatenation of cycles but we then showed that such a concatenation cannot actually be a solution to W by considering the decomposition of the product by parts and the maximum number of cancellations and thus W has a solution if and only if P has a solution proving the undecidability.

It remains to prove that we may define the problem over a binary group alphabet $\{a, b, a^{-1}, b^{-1}\}$. This is not difficult however by a standard technique which we now outline. Given a group alphabet $\Sigma_1 = \{y_1, \ldots, y_k, y_1^{-1}, \ldots, y_k^{-1}\}$ and a binary group alphabet $\Sigma = \{a, b, a^{-1}, b^{-1}\}$. Define $\sigma : \Sigma_1 \to \Sigma^*$ by $\sigma(y_i) = a^i b$ and $\sigma(y_i^{-1}) = (a^{-1})^i b^{-1}$. It is not difficult to see that this is an injective morphism and applying iteratively to each letter in each word of W proves the undecidability of the Identity Correspondence Problem over a binary group alphabet. □

Problem 2. *Group Problem - Is the semigroup generated by a finite set of pairs of words* $P = \{(u_1, v_1), (u_2, v_2), \ldots, (u_n, v_n)\} \subset \mathrm{FG}(a, b) \times \mathrm{FG}(a, b)$ *a group?*

Theorem 4. *The Group Problem is undecidable.*

Proof. Let us assume by contradiction that the group problem is decidable for a semigroup S defined by pairs of words over a group alphabet and the operation of pairwise concatenation. If the identity element can be generated by the concatenation of word pairs $(u_{i_1}, v_{i_1})(u_{i_2}, v_{i_2}) \cdot \ldots \cdot (u_{i_k}, v_{i_k}) = (u_{i_1} u_{i_2} \cdot \ldots \cdot u_{i_k}, v_{i_1} v_{i_2} \cdot \ldots \cdot v_{i_k}) = (\varepsilon, \varepsilon)$ then any cyclic permutation of words in this concatenation is also equal to $(\varepsilon, \varepsilon)$. Thus every element in the set of all pairs used in the generation of identity has an inverse element and this set generates a subgroup. Therefore the identity problem can be solved by checking if any nonempty subset of the original pairs generates a group. If there is a subset of S which generates a group then the identity element is in S. Otherwise the identity element is not generated by S. □

3 Applications of ICP

In this section we will provide a number of new results in matrix semigroups using the undecidability of ICP. It was not previously known whether the identity problem for matrix semigroups was decidable for any dimension greater than two. The decidability of the two dimensional case was recently proved to be decidable in [9].

Theorem 5. *Given a semigroup S generated by a fixed number n of square four dimensional integral matrices, determining whether the identity matrix belongs to S is undecidable. This holds even for $n = 48$.*

Proof. We shall use a standard encoding to embed an instance of the Identity Correspondence Problem into a set of integral matrices. Given an instance of

ICP say $W \subseteq \Sigma^* \times \Sigma^*$ where $\Sigma = \{a, b, a^{-1}, b^{-1}\}$ generates a free group. Define the morphism $\rho : \Sigma^* \to \mathbb{Z}^{2\times 2}$:

$$\rho(a) = \begin{pmatrix} 1 & 2 \\ 0 & 1 \end{pmatrix}, \rho(b) = \begin{pmatrix} 1 & 0 \\ 2 & 1 \end{pmatrix}, \rho(a^{-1}) = \begin{pmatrix} 1 & -2 \\ 0 & 1 \end{pmatrix}, \rho(b^{-1}) = \begin{pmatrix} 1 & 0 \\ -2 & 1 \end{pmatrix}.$$

It is known from the literature that ρ is an injective homomorphism, i.e., the group generated by $\{\rho(a), \rho(b)\}$ is free, see for example [14]. For each pair of words $(w_1, w_2) \in W$, define the matrix $A_{w_1,w_2} = \rho(w_1) \oplus \rho(w_2)$ where $\oplus$ denotes the direct sum of two matrices. Let S be a semigroup generated by $\{A_{w_1,w_2} | (w_1, w_2) \in W\}$. If there exists a solution to ICP, i.e., $(\varepsilon, \varepsilon) \in W^+$, then we see that $\rho(\varepsilon) \oplus \rho(\varepsilon) = I_4 \in S$ where I_4 is the 4×4 identity matrix. Otherwise, since ρ is an injective homomorphism, $I_4 \notin S$. □

It follows from the above construction that another open problem concerning the reachability of any diagonal matrix in a finitely generated integral matrix semigroup stated in [6] and as Open Problem 6 in [11] is also undecidable.

Corollary 1. *Given a finitely generated semigroup of integer matrices S, determining whether there exists any diagonal matrix in S is algorithmically undecidable.*

Proof. This result follows from the proof of Theorem 5. Note that in that theorem, the morphism ρ is injective and thus the only diagonal matrix in the range of ρ is the 2×2 identity matrix I_2 (corresponding to $\rho(\varepsilon)$), since diagonal matrices commute. Clearly then, the only diagonal matrix in the semigroup S of Theorem 5 is given by $\rho(\varepsilon) \oplus \rho(\varepsilon) = I_4$ where I_4 is the 4×4 identity matrix. Since determining if this matrix is in S was shown to be undecidable, it is also undecidable to determine if there exists any diagonal matrix in S. □

Theorem 6. *Given a finite set of rotations on the 3-sphere. Determining whether this set of rotations generates a group is undecidable.*

Proof. For the definitions of quaternions used in this theorem, see [5]. The set of all unit quaternions forms the unit 3-sphere and any pair of unit quaternions a and b can represent a rotation in 4 dimensional space. We can rotate a point $x = (x_1, x_2, x_3, x_4)$ on the 3-sphere, represented by a quaternion $q_x = x_1 + x_2 i + x_3 j + x_4 k$, in the following way: $a q_x b^{-1}$. We can define a morphism ξ from a group alphabet to unitary quaternions: $\xi(a) = (3/5) + (4/5) \cdot i; \xi(b) = (3/5) + (4/5) \cdot j$. It was proven in [5] that ξ is an injective homomorphism. Now we convert pairs of words from an instance of the Identity Correspondence Problem into pairs of quaternions $\{(a_1, b_1), \ldots, (a_n, b_n)\}$. Thus we reduce the group problem for pairs of words over a group alphabet to the question of whether a finite set of rotations, $\{(a_1, b_1), \ldots, (a_n, b_n)\}$, represented by pairs of quaternions, generates a group. □

Acknowledgements. We would like to thank Prof. Tero Harju for useful discussions concerning this problem.

References

1. Babai, L., Beals, R., Cai, J., Ivanyos, G., Luks, E.M.: Multiplicative Equations over Commuting Matrices. In: 7th ACM-SIAM Symp. on Discrete Algorithms, pp. 498–507 (1996)
2. Bell, P., Potapov, I.: On the Membership of Invertible Diagonal and Scalar Matrices. Theoretical Computer Science 372(1), 37–45 (2007)
3. Bell, P., Potapov, I.: On Undecidability Bounds for Matrix Decision Problems. Theoretical Computer Science 391(1-2), 3–13 (2008)
4. Bell, P., Potapov, I.: Periodic and Infinite Traces in Matrix Semigroups. In: Geffert, V., Karhumäki, J., Bertoni, A., Preneel, B., Návrat, P., Bieliková, M. (eds.) SOFSEM 2008. LNCS, vol. 4910, pp. 148–161. Springer, Heidelberg (2008)
5. Bell, P., Potapov, I.: Reachability Problems in Quaternion Matrix and Rotation Semigroups. Information and Computation 206(11), 1353–1361 (2008)
6. Blondel, V.D., Cassaigne, J., Karhumäki, J.: Problem 10.3, Freeness of Multiplicative Matrix Semigroups. In: Blondel, V.D., Megretski, A. (eds.) Unsolved Problems in Mathematical Systems and Control Theory, pp. 309–314. Princeton University Press, Princeton (2004)
7. Blondel, V.D., Tsitsiklis, J.: The Boundedness of All Products of a Pair of Matrices is Undecidable. Systems and Control Letters 41(2), 135–140 (2000)
8. Cassaigne, J., Harju, T., Karhumäki, J.: On the Undecidability of Freeness of Matrix Semigroups. Intern. J. Alg. & Comp. 9, 295–305 (1999)
9. Choffrut, C., Karhumäki, J.: Some Decision Problems on Integer Matrices. Theoretical Informatics and Applications 39, 125–131 (2005)
10. D'Allesandro, F.: Free Groups of Quaternions. IJAC 14(1) (2004)
11. Harju, T.: Post Correspondence Problem and Small Dimensional Matrices. LNCS, vol. 5583, pp. 39–46. Springer, Berlin (2009)
12. Halava, V., Harju, T.: On Markov's Undecidability Theorem for Integer Matrices, TUCS Technical Report, Number 758 (2006)
13. Halava, V., Harju, T., Hirvensalo, M.: Undecidability Bounds for Integer Matrices using Claus Instances. IJFCS 18(5), 931–948 (2007)
14. Lyndon, R.C., Schupp, P.E.: Combinatorial Group Theory. Springer, Heidelberg (1977)
15. Matiyasevich, Y., Senizergues, G.: Decision Problems for Semi-Thue Systems with a Few Rules. Theoretical Computer Science 330(1), 145–169 (2005)
16. Mihailova, K.A.: The Occurrence Problem for a Direct Product of Groups. Dokl. Akad. Nauk 119, 1103–1105 (1958) (in Russian)
17. Otto, F.: On Deciding whether a Monoid is a Free Monoid or a Group. Acta Informatica 23, 99–110 (1986)
18. Paterson, M.: Unsolvability in 3 × 3 Matrices. Studies in Applied Mathematics 49 (1970)

Fast Distributed Approximation Algorithm for the Maximum Matching Problem in Bounded Arboricity Graphs

Andrzej Czygrinow[1], Michał Hańćkowiak[2,⋆], and Edyta Szymańska[2,⋆⋆]

[1] School of Mathematical and Statistical Sciences, Arizona State University, Tempe, AZ, 85287-1804, USA
andrzej@math.la.asu.edu

[2] Faculty of Mathematics and Computer Science, Adam Mickiewicz University, Poznań, Poland
mhanckow, edka@amu.edu.pl

Abstract. We give a deterministic distributed approximation algorithm for the maximum matching problem in graphs of bounded arboricity. Specifically, given $0 < \epsilon < 1$ and a positive integer a, the algorithm finds a matching of size at least $(1-\epsilon)m(G)$, where $m(G)$ is the size of the maximum matching in an n-vertex graph G with arboricity at most a. The algorithm runs in $O(\log^* n)$ rounds in the message passing model and it is the first sublogarithmic algorithm for the problem on such a wide class of graphs. Moreover, the result implies that the known $\Omega(\sqrt{\log n/\log\log n})$ lower bound on the time complexity for a constant or polylogarithmic approximation does not hold for graphs of bounded arboricity.

1 Introduction

Designing distributed approximation algorithms for special families of networks has attracted a lot of interest in recent years and efficient algorithms for many classical graph-theoretic problems in constant-degree graphs, unit-disk graphs, or planar networks have been given ([CH06], [KMNW05a], [KMNW05b], [BE08]). At the same time, most of the problems that admit efficient solutions in the above classes of graphs are provably intractable (see, e.g., [KMW04]) or seem unapproachable in general networks. In this paper, we focus on the maximum matching problem and give a fast, deterministic, distributed approximation algorithm for the problem in graphs of bounded arboricity.

1.1 Model of Computation, Definitions and Notation

We will consider a synchronous, message-passing model of computations (referred to as *LOCAL* in [Pe00]). In this model a graph is used to represent an

⋆ This work was supported by grant N206 017 32/2452 for years 2007-2010.
⋆⋆ This work was supported by grant N206 017 32/2452 for years 2007-2010.

Y. Dong, D.-Z. Du, and O. Ibarra (Eds.): ISAAC 2009, LNCS 5878, pp. 668–678, 2009.

underlying network. The vertices of the graph correspond to computational units, and edges represent communication links. The network is synchronized and in one round a vertex can send and receive messages from all of its neighbors. In addition, in the same round, a vertex can perform some local computations. Neither the amount of local computations nor the size of a message is restricted in any way and the running time of the algorithm is the number of rounds needed to solve a problem. We assume that vertices have unique identifiers from $\{1, \ldots, n\}$, where n is the order of the graph.

Recall that a *matching* in a graph $G = (V, E)$ is a set $M \subseteq E$ of disjoint edges. A matching M is called *maximal* if there is no matching M' such that M is a proper subset of M', and a matching M is called *maximum* if there is no matching M' with $|M'| > |M|$. The size of a maximum matching in G is denoted by $m(G)$. In this paper, we will analyze the distributed complexity of the matching problem in uniformly sparse graphs. The uniform sparsity of a graph can be formalized using the concept of arboricity.

Definition 1. *The arboricity of a graph* $G = (V, E)$ *is defined by* $arb(G) = \max\{\lceil \frac{|E(G')|}{|V(G')|-1} \rceil \mid G' \subseteq G, |V(G')| \geq 2\}$.

It follows from the definition that if a graph G has arboricity at most a then for every subset $U \subseteq V(G)$ the subgraph induced by U has at most $a|U|$ edges. We will use this fact repeatedly in the proofs. In addition, in view of the Nash-Williams Theorem ([Di05]), $E(G)$ is a union of $\lfloor a \rfloor$ forests. Of course, $arb(G) \leq 1$ if and only if G is a forest, and when G is planar then $arb(G) \leq 3$. Moreover, graphs without a forbidden minor have bounded arboricity which depends on the minor. On the other hand, a graph of constant arboricity may contain $K_{\sqrt{n}}$ as a minor.

We will follow standard graph-theoretic notation. In particular, if $G = (V, E)$ is a graph and $U, W \subseteq V$ then $E(U, W)$ denotes the set of edges with one endpoint in U and the other in W. We set $e(U, E) = |E(U, W)|$. If $U = \{u\}$ then we write $E(u, W)$. In addition, we set $deg(u, W) := e(u, W)$, write $deg(u)$ for $deg(u, V)$, and use $\Delta(G)$ for $\max_{v \in V(G)} deg(v)$. For $U \subset V(G)$, $G[U]$ denotes the subgraph of G induced by U. The number of edges in $G[U]$ we denote by $e(U)$.

1.2 Related Work

The theory of distributed algorithms for graph-theoretic problems has seen a dynamic growth in recent years (see the survey by Elkin [El04] for a general overview). The distributed complexity of the sequentially polynomial matching problem has been intensively studied and many questions remain open in the general case. In [HKP01] a poly-logarithmic time (in n), distributed algorithm for the maximal matching problem is given. By virtue of its maximality the returned matching is a $\frac{1}{2}$-approximation of a maximum matching. This was further improved in [CHS02] to $\frac{2}{3}$-approximation in $O(\log^4 n)$ time for general graphs and in [CH03] a $(1 - \epsilon)$-approximation for bipartite graphs in time $(\log n)^{O(1/\epsilon)}$ was given. With the help of randomization it is possible to reduce the expected running time to $O(\log n)$ ([II86]) and the current best approximation ratio, $(1-\epsilon)$ for any $\epsilon > 0$, within this frame is due to Lotker et al. [LPP08].

One of the fundamental interests in the distributed computing is the trade-off between time and approximation ratio. On the negative site, Kuhn et al. [KMW05] proved a lower bound of $\Omega(\sqrt{\log n / \log\log n})$ on the time complexity for any distributed algorithm, deterministic or randomized to compute a constant or polylogarithmic approximation for maximum matching. The lower bound holds even if messages have unbounded size. This, somehow surprising limitation of local algorithms justified the study of the problem in special classes of graphs. It has turned out that if the underlying graph has more structure, much faster algorithms are available. For example, in the case of bounded degree graphs it is possible to find a maximal matching in $O(\log^* n)$ rounds using the algorithm of Cole and Vishkin [CV86] or Panconesi and Rizzi [PR01]. In Sect. 2, building on this fact, we give an algorithm which finds a $(1 - \epsilon)$-approximation of the maximum matching in this class of graphs in $O(\log^* n)$ rounds.

Many problems admit very fast approximations in planar networks. In [CHS06] the authors presented a maximum matching algorithm which yields an asymptotically optimal approximation ratio of $(1+O(1/\log n))$ but the number of rounds is $O(\log^{29} n)$. Later Czygrinow et al. [CHW08] provided an $(1-\epsilon)$-approximation of the maximum matching in this class of graphs in $O(\log^* n)$ rounds. Note, however, that all the above algorithms use planarity extensively, and the fact that contracting a subgraph preserves planarity is a fundamental principle of the approach in [CHW08].

A common generalization of bounded degree graphs and planar graphs is the class of uniformly sparse graphs, that is graphs of bounded arboricity. Recently, these graphs were considered by Barenboim and Elkin ([BE08]), who gave a $O(\log n / \log\log n)$-time algorithm for the maximal independent set problem based on forest-decomposition. Motivated by their result we study distributed approximation for the maximum matching problem in the same family of graphs. Unfortunately, we cannot apply the ideas from [CHW08] or [BE08] and instead must use different techniques. Indeed, the forest decomposition method, suitable for bounded arboricity graphs carries a lower bound, proved in [BE08], and the complexity which makes it insufficient for our task. On the other hand, bounded arboricity is not preserved under subgraph contraction and the technique from [CHW08] fails as well.

Our approach is based on a reduction to the narrower class of bounded degree graphs.

1.3 Main Result

Our main result is summarized in the following theorem.

Theorem 1. *Given $0 < \epsilon < 1$ and a positive integer a, there exits a distributed algorithm (*Matching*) which finds in a graph G with $arb(G) \leq a$ a matching M such that*

$$|M| \geq (1-\epsilon)m(G).$$

The algorithm runs in $c\log^* n$ rounds, where c depends only on ϵ and a.

We assume that the order n of the underlying graph G and the arboricity of G are globally known. According to our knowledge, this is the first $O(\log^* n)$-time distributed approximation for a non-trivial problem in the class of graphs of bounded arboricity. Also, our result shows that the lower bound on the distributed time complexity for a constant or polylogarithmic approximation proved in [KMW05], mentioned in the preceding subsection, does not hold in graphs of bounded arboricity.

In addition, as an application of Theorem 1, we design an algorithm finding an approximate orientation of bounded arboricity graphs (see Theorem 4 in Sect. 5).

1.4 Organization

The rest of the paper is structured as follows. In the next section we give a procedure for finding an approximation of a maximum matching in graphs of bounded degrees. In Sect. 3, we describe the main algorithm and its analysis is presented in Sect. 4. Finally, in Sect. 5 we outline how the matching algorithm can be used to find an orientation of almost all edges in a graph of bounded arboricity.

2 Bounded Degree Graphs

In this section we will give a procedure which approximates a maximum matching in bounded degree graphs. Although many problems are known to admit very fast distributed algorithms in the class of bounded degree graphs, according to our knowledge a $(1-\xi)$-approximation of the maximum matching in $O(\log^* n)$ rounds has not been explicitly discussed. The procedure given here is then used in Sect. 3 in the main matching algorithm. Our method will rely on augmenting paths. Let M be a matching in a graph G. A vertex v of G is called *M-saturated* if $v \in V(M)$ and is called *M-free* if $v \notin V(M)$. A path P in G of length $2l-1$ is called *M-augmenting* if the edges of P alternate between $E(G) \setminus M$ and M, and the endpoints of P are M-free. The following well known result can be applied in the approximation of $m(G)$.

Theorem 2. *[HK73]*

(a) Let M be an arbitrary matching in G. If there are no M-augmenting paths of lengths at most $2k-1$, then $|M| \geq \left(1 - \frac{1}{k+1}\right) m(G)$.

(b) Let $k \geq 2$ and let M be a matching in a graph G such that there are no M-augmenting paths of length at most $2k-3$ (in particular, the matching is maximal). Let $\mathcal{P}$ be a maximal set of vertex-disjoint M-augmenting paths of length $2k-1$ and let $M' = M \div (\bigcup_{P \in \mathcal{P}} E(P))$ be the matching obtained by augmenting M along the paths from $\mathcal{P}$. Then there are no M'-augmenting paths of length at most $2k-1$ in G.

Consequently, together parts (a) and (b) of Theorem 2 assure, that the matching M' satisfies $|M'| \geq \left(1 - \frac{1}{k+1}\right) m(G)$.

The general idea behind our algorithm in this section is to find a maximal matching M and then to improve M iteratively, using a maximal collection of pair-wise vertex-disjoint M-augmenting paths of length $2l - 1$ ($l = 2, \dots, k$), as described in part (b) of the above theorem.

In order to find a maximal set of vertex-disjoint M-augmenting paths of length $2l - 1$ we will compute a maximal independent set of vertices (MIS) in a special *path graph* H_l. Similar approach has been used in [LPP08]. Let the vertices of H_l represent all M-augmenting paths of length $2l - 1$ in G, and there is an edge between $P_u, P_v \in V(H_l)$ whenever the paths P_u and P_v corresponding to u and v share a vertex in G. An ID of $v \in V(H_l)$ can be defined as the sequence of ID's of vertices from the path P_v in G. Note that unique ID's which are bounded by n^p for some constant p, are required by the algorithm from [CV86] to run in $O(\log^* n)$ rounds. In our case p depends only on the error of approximation. We remark that for any constant l, H_l can be obtained in a constant number of rounds and each round in H_l can be simulated by a constant number of rounds in G.

MatchingInBoundedDegree

Input: A graph G with $\Delta(G) \leq L$ (where L is a constant), $\xi > 0$.
Output: A matching M in G with $|M| > (1 - \xi)m(G)$.

1. Find a maximal matching M in G by computing a MIS in the line graph of G using the algorithm from [CV86].
2. For $l := 2$ to $\lceil \frac{1}{\xi} - 1 \rceil$ do
   (a) Construct the path graph H_l for G with respect to M.
   (b) Find a MIS I in H_l using the algorithm from [CV86].
   (c) Augment M along the paths in G that correspond to vertices in I.
3. Return M.

Theorem 3. *Given* $0 < \xi < 1$ *and a positive integer* L, MatchingInBoundedDegree *finds in a graph* G *with* $\Delta(G) \leq L$ *a matching* M *such that* $|M| \geq (1 - \xi)m(G)$ *in* $O(\log^* |G|)$ *rounds.*

Proof. The algorithm proceeds in iterations with $l = 2, \dots, k$; $k = \lceil \frac{1}{\xi} - 1 \rceil$. In the lth iteration the procedure accepts a matching M obtained in the previous iteration and constructs the path graph H_l on all M-augmenting paths of length $2l - 1$. Then it finds a MIS in H_l which gives us an independent set of pairwise disjoint M-augmenting paths of length $2l - 1$ in G. Once a maximal independent set of paths is found, the edges in paths that belong to M and those that are not in M are exchanged. By part (b) of Theorem 2, the new matching does not have augmenting paths of lengths which are less than or equal to $2l - 1$. Consequently, the resulting matching is of cardinality at least $\left(1 - \frac{1}{l+1}\right) m(G)$. Moreover, since $\Delta(H_l)$ is a function of L and l, and the identifiers of vertices from H_l are bounded by n^{2l}, the [CV86] algorithm can be used to find a MIS in H_l in $c \log^* n$ time, where the constant c depends only on L and ξ. □

3 The Algorithm

In this section, we present the main algorithm. The idea behind our algorithm is to reduce the search for a large matching in G to its carefully chosen subgraphs of bounded degree which span a relatively large matching compared to $m(G)$. To solve the latter problem we invoke twice procedure MatchingInBoundedDegree from Sect. 2. Before we describe the algorithm, we need to develop some more notation. The matching algorithm will take as an input a positive integer a, a graph $G = (V, E)$ with $arb(G) \leq a$ and a real number $0 < \epsilon < 1$. Based on these parameters some auxiliary constants and certain subsets of V will be calculated by the procedure. Let

$$C := \frac{20a}{\epsilon^2}, \ C_1 := \epsilon C \tag{1}$$

and consider sets U_1, U_2, and U_3 defined as

$$U_1 := \{v : deg(v) \leq C\} \tag{2}$$

$$U_2 := \{v \notin U_1 : deg(v, U_1) \geq C - C_1\}, \tag{3}$$

$$U_3 := V \setminus (U_1 \cup U_2). \tag{4}$$

Moreover, consider two more subsets of U_2 and their union:

$$L_{2,1} := \{v \in U_2 : deg(v, U_1) \geq 2C\}, \tag{5}$$

$$L_{2,2} := \{v \in U_2 : deg(v, U_2) \geq 2C\}, \tag{6}$$

$$L_2 := L_{2,1} \cup L_{2,2}. \tag{7}$$

In addition to the notation from (1)–(7), set $u_i := |U_i|$ for $i = 1, 2, 3$. Note that the maximum degree of graph $G' := G[U_1 \cup (U_2 \setminus L_2)]$ is at most $4C$ and so, using procedure MatchingInBoundedDegree from the previous section, we are able to approximate a maximum matching in this subgraph. Let M_1 denote the obtained matching. It will follow from the arboricity bound that $L_{2,2}$ is negligible with respect to $m(G)$. If $L_{2,1}$ is so too, we are done. Otherwise, in the next step we disregard the subset $L'_{2,1} \subseteq L_{2,1}$ which contains the vertices with less than C neighbors in $U_1 \setminus V(M_1)$. We will use only the vertex set $L_{2,1} \setminus L'_{2,1}$ to enlarge the matching M_1. To this end, we consider the bipartite graph $G[U_1 \setminus V(M_1), L_{2,1} \setminus L'_{2,1}]$ and construct a subgraph $G'' \subseteq G[U_1 \setminus V(M_1), L_{2,1} \setminus L'_{2,1}]$ with $\Delta(G'') \leq C$. Finally, we invoke MatchingInBoundedDegree again to approximate a maximum matching in graph G'' and find M_2. The algorithm returns $M_1 \cup M_2$. The bottom line is that either $L_{2,1}$ is negligible or $L'_{2,1}$ is negligible with respect to $L_{2,1}$.

Matching

Input: $G = (V, E)$ with $arb(G) \leq a$, $0 < \epsilon < 1$.
Output: A matching M in G such that $|M| \geq (1 - \epsilon)m(G)$.

1. Compute constants and sets defined in (1)–(7).
2. Use procedure MATCHINGINBOUNDEDDEGREE (with $\xi := \epsilon/3$ and $L := 4C$) to compute a matching M_1 in $G' := G[U_1 \cup (U_2 \setminus L_2)]$.
3. Let $L'_{2,1} := \{v \in L_{2,1} : deg(v, U_1 \setminus V(M_1)) < C\}$.
4. For every vertex $v \in L_{2,1} \setminus L'_{2,1}$ delete arbitrarily $deg(v, U_1 \setminus V(M_1)) - C$ edges. Let G'' denote the subgraph of $G(U_1 \setminus V(M_1), L_{2,1} \setminus L'_{2,1})$ obtained in this way.
5. Use procedure MATCHINGINBOUNDEDDEGREE (with $\xi := \epsilon/7$ and $L := C$) to find a matching M_2 in G''.
6. Output $M := M_1 \cup M_2$.

---

4 Analysis

This section is devoted to the proof of Theorem 1. The argument is divided into three sections. First we recall some facts about matchings in bipartite graphs. Next we prove that certain subsets of $V(G)$ can be neglected when estimating $m(G)$. Finally, we give the main proof of Theorem 1. In what follows we assume that ϵ is given and use notation from (1)-(7). Observe that we may assume that $\epsilon < 1/2$.

4.1 Matchings in Bipartite Graphs

Let us start with the following easy consequence of Hall's theorem.

Fact 1. *Let $G = (X \cup Y, E)$ be a bipartite graph, $0 \le \xi \le 1$, and let $L > 0$. If $\max_{x \in X} deg(x) \le L$ and $\min_{y \in Y} deg(y) \ge (1-\xi)L$, then G contains a matching of size at least $(1-\xi)|Y|$.*

Next, we will observe that the graph G contains a matching which constitutes a significant fraction of the vertices of the set U_2.

Corollary 1. $u_2 \le 2m(G)$.

Proof. From Fact 1, the bipartite graph $G[U_1, U_2]$ contains a matching of size $(1-\epsilon)u_2 \ge u_2/2$. □

4.2 Negligible Sets

Recall the notation from (1)–(7) with $u_i := |U_i|$ for $i = 1, 2, 3$ and set $l_{i,j} := |L_{i,j}|$ for $i, j = 1, 2$. Here we show that u_3 and $l_{2,2}$ are both small with respect to $m(G)$.

Fact 2. $u_3 < \epsilon/9 \cdot m(G)$.

Proof. Write $\sum_{v \in U_3} deg(v) = 2e(U_3) + e(U_3, U_2) + e(U_3, U_1)$. By the definition of U_3, we have $\sum_{v \in U_3} deg(v) > Cu_3$, and $e(U_3, U_1) < (C - C_1)u_3$. Moreover, by the arboricity bound, $e(U_3) \le au_3$ and $e(U_3, U_2) \le a(u_3 + u_2)$. Therefore, $Cu_3 < 3au_3 + au_2 + (C - C_1)u_3$, and so, $u_3 < \frac{a}{C_1 - 3a} u_2 \le \frac{\epsilon}{20 - 3\epsilon} u_2$, by the choice of C_1 in (1). By Corollary 1, $u_3 < \epsilon/9 \cdot m(G)$. □

Recall that $L_{2,2} \subseteq L_2 \subseteq U_2$ as defined in (2)-(7). Next, we argue that $L_{2,2}$ is negligible.

Fact 3. $l_{2,2} \leq \epsilon/19 \cdot m(G)$.

Proof. The argument is analogous to the one for Fact 2. Consider $\sum_{v \in L_{2,2}} deg(v, U_2) = 2e(L_{2,2}) + e(L_{2,2}, U_2 \setminus L_{2,2})$ and further, $2Cl_{2,2} \leq 2al_{2,2} + au_2$, which together with our choice of C in (1) gives $l_{2,2} \leq \frac{a}{2(C-a)} u_2 < \frac{\epsilon}{38} u_2 \leq \frac{\epsilon}{19} \cdot m(G)$. □

4.3 Proof of the Main Result

We will now prove Theorem 1 using the observations from previous sections.

Proof. Let M be the matching returned by MATCHING and let M_1 and M_2 denote matchings found in Steps 2 and 5. We will show that

$$|M| \geq (1 - \epsilon) m(G). \tag{8}$$

Let $L'_{2,1}$ be as in Step 3 of MATCHING. We will consider two cases based on the relative size of the set $L'_{2,1}$ in $L_{2,1}$. Set $m_1 := |M_1|$, $m_2 := |M_2|$. In the first case we will show that M_1 already guarantees the approximation ratio claimed in the theorem. The second case regards the situation in which M_2 has to be included in the solution to obtain a desired approximation error.

Case 1: Suppose $|L'_{2,1}| \geq \epsilon/3 \cdot |L_{2,1}|$. Then, by the arboricity bound and definition of $L'_{2,1}$, $\frac{\epsilon}{3} Cl_{2,1} \leq C|L'_{2,1}| \leq e(L'_{2,1}, V(M_1)) \leq a(l_{2,1} + 2m_1)$, which implies

$$l_{2,1} \leq \frac{6a}{\epsilon C - 3a} m_1 \leq \frac{\epsilon}{3} m_1 \leq \frac{\epsilon}{3} m(G). \tag{9}$$

The matching M_1 found in Step 2 of MATCHING is such that

$$|M_1| \geq (1 - \epsilon/3) m(G') \geq (1 - \epsilon/3)(m(G) - u_3 - l_{2,1} - l_{2,2}).$$

Consequently, by Fact 2, Fact 3, and (9)

$$|M_1| \geq (1 - \epsilon/3)(m(G) - u_3 - l_{2,1} - l_{2,2}) \geq (1 - \epsilon) m(G).$$

Case 2: Suppose $|L'_{2,1}| < \epsilon/3 \cdot |L_{2,1}|$. Let $G'' := G[U_1 \setminus V(M_1), L_{2,1} \setminus L'_{2,1}]$. We apply Fact 1 to G'' with $L := C$ and $\xi = 0$. Noting that for every $u \in U_1$, $deg(u) \leq C$ and $\forall v \in L_{2,1} \setminus L'_{2,1}$, $deg(v, U_1 \setminus V(M_1)) \geq C$, we obtain the bound

$$m(G'') = l_{2,1} - l'_{2,1} > (1 - \epsilon/3) l_{2,1}. \tag{10}$$

Let M^* be a maximum matching in G, i.e. $|M^*| = m(G)$. We consider the partition of M^* into three subsets. Let $M_1^* := \{e \in M^* | e \cap (U_3 \cup L_{2,2}) \neq \emptyset\}$. Then, by Fact 2 and Fact 3, $|M_1^*| \leq (u_3 + l_{2,2}) < \epsilon/5 \cdot m(G)$. Let $M_2^* := \{e \in M^* | e \cap L_{2,1} \neq \emptyset\}$. Then, clearly, $|M_2^*| \leq l_{2,1}$ and by (10) the matching M_2 obtained in Step 5 of MATCHING satisfies

$$|M_2| > (1 - \epsilon/7)m(G'') \geq (1 - \epsilon/7)(1 - \epsilon/3)l_{2,1} \geq (1 - \epsilon/2)|M_2^*|.$$

Finally, let $M_3^* := M^* \setminus (M_1^* \cup M_2^*)$. Every $e \in M_3^*$ is contained in $U_1 \cup (U_2 \setminus L_2)$. Thus $M_3^* \leq m(G')$ and M_1, obtained in Step 2 of MATCHING, satisfies $|M_1| \geq (1 - \epsilon/3)|M_3^*|$. Hence, the matching M returned by MATCHING is such that

$$|M| = |M_1| + |M_2| \geq (1 - \epsilon/2)(|M_2^*| + |M_3^*|) = (1 - \epsilon/2)(|M^*| - |M_1^*|)$$
$$> (1 - \epsilon/2 - \epsilon/5 + \epsilon/10)m(G) \geq (1 - \epsilon)m(G).$$

This finishes the proof of (8). Finally, note that for a fixed ϵ, both M_1 and M_2 can be found in $O(\log^* n)$ rounds and so the running time of MATCHING is $c \log^* n$ where c depends only on ϵ and a. □

5 Finding an Approximate Nash-Williams Orientation

In this section, we will briefly describe one application of our main result. As mentioned in the Introduction, every graph of arboricity a can be decomposed into a edge-disjoint forests. The edges of every forest can be oriented so that the maximum out-degree equals one. Consequently, one can obtain an orientation of the entire graph with maximum out-degree at most a. Nonetheless, finding such an orientation efficiently in distributed model is not easy [BE08]. Our algorithm MATCHING can quickly find an approximate orientation having this property. The result is formalized in the following theorem.

Theorem 4. *Given $0 < \epsilon < 1$ and a positive integer a, there is a distributed algorithm which finds in a graph G with $arb(G) \leq a$ an orientation of all but at most ϵn edges with the property that the out-degree in this directed subgraph is at most a. The algorithm runs in $O(\log^* n)$ rounds.*

Proof. Consider an auxiliary bipartite graph $H = (U, W)$ obtained from G in the following way. The vertices of U correspond to the edges of G, that is $U := E(G)$, while W consists of a copies of every vertex of G. We put an edge between $e \in U$ and $v \in W$ if v is a copy of a vertex incident to e. Obviously, $deg(e) = 2a$ for every $e \in U$ and hence, the arboricity of H is at most $2a$.

As explained prior to Theorem 4, the fact that G has arboricity at most a implies the existence of an orientation of the edges of G with maximum out-degree at most a. This orientation in turn yields the existence of a matching in H that saturates U. This matching consists of the edges $((v_1, v_2), v)$, where (v_1, v_2) is an oriented edge of G and v is one of the a copies of v_1 in H. Such a selection is always possible because there are at most a edges which start at v_1. Hence, $m(H) = |U|$.

We construct an approximate orientation of G invoking procedure MATCHING for H. It gives us a matching M such that $|M| \geq (1 - \epsilon)m(H) = (1 - \epsilon)|U|$. Finally, every edge $e = \{u, v\} \in E(G) = U$ which is saturated by M is oriented from u to v if the M-neighbor of e in H is a copy of u, and from v to u otherwise.

The time complexity of this procedure is dominated by MATCHING and equals $O(\log^* n)$. □

We remark that in the subgraph of G induced by the orientation found above (i.e. on $(1-\epsilon)|E(G)|$ edges), it is then possible to compute a vertex coloring and a MIS in $O(\log^* n)$ rounds using the procedure from [CV86].

References

[BE08] Barenboim, L., Elkin, M.: Sublogarithmic distributed MIS algorithm for sparse graphs using Nash-Williams decomposition. In: PODC 2008, pp. 25–34 (2008)

[CV86] Cole, R., Vishkin, U.: Deterministic coin tossing with applications to optimal parallel list ranking. Information and Control 70, 32–53 (1986)

[CH03] Czygrinow, A., Hańćkowiak, M.: Distributed Algorithm for Better Approximation of the Maximum Matching. In: Warnow, T.J., Zhu, B. (eds.) COCOON 2003. LNCS, vol. 2697, pp. 242–251. Springer, Heidelberg (2003)

[CH06] Czygrinow, A., Hańćkowiak, M.: Distributed approximation algorithms in unit-disc graphs. In: Dolev, S. (ed.) DISC 2006. LNCS, vol. 4167, pp. 385–398. Springer, Heidelberg (2006)

[CHS02] Czygrinow, A., Hańćkowiak, M., Szymańska, E.: Distributed algorithm for approximating the maximum matching. Discrete Applied Math. 143(1-3), 62–71 (2004)

[CHS06] Czygrinow, A., Hańćkowiak, M., Szymańska, E.: Distributed Approximation Algorithms for Planar Graphs. In: Calamoneri, T., Finocchi, I., Italiano, G.F. (eds.) CIAC 2006. LNCS, vol. 3998, pp. 296–307. Springer, Heidelberg (2006)

[CHW08] Czygrinow, A., Hańćkowiak, M., Wawrzyniak, W.: Fast distributed approximations in planar graphs. In: Taubenfeld, G. (ed.) DISC 2008. LNCS, vol. 5218, pp. 78–92. Springer, Heidelberg (2008)

[Di05] Diestel, R.: Graph Theory, 3rd edn. Springer, Heidelberg (2005)

[El04] Elkin, M.: An Overview of Distributed Approximation. ACM SIGACT News Distributed Computing Column 35(4-132), 40–57 (2004)

[HKP01] Hańćkowiak, M., Karoński, M., Panconesi, A.: On the distributed complexity of computing maximal matchings. SIAM J. Discrete Math. 15(1), 41–57 (2001)

[HK73] Hopcroft, A., Karp, R.: An $n^{5/2}$ algorithm for maximum matching in bipartite graphs. SIAM J. on Comp. 2, 225–231 (1973)

[II86] Israeli, A., Itai, A.: A fast and simple randomized parallel algorithm for maximal matching. Info. Proc. Lett. 22(2), 77–80 (1986)

[KMW04] Kuhn, F., Moscibroda, T., Wattenhofer, R.: What Cannot Be Computed Locally! In: Proc. PODC, pp. 300–309 (2004)

[KMNW05a] Kuhn, F., Moscibroda, T., Nieberg, T., Wattenhofer, R.: Fast Deterministic Distributed Maximal Independent Set Computation on Growth-Bounded Graphs. In: Fraigniaud, P. (ed.) DISC 2005. LNCS, vol. 3724, pp. 273–287. Springer, Heidelberg (2005)

[KMNW05b] Kuhn, F., Moscibroda, T., Nieberg, T., Wattenhofer, R.: Local Approximation Schemes for Ad Hoc and Sensor Networks. In: 3rd ACM Joint Workshop on Foundations of Mobile Computing (DIALM-POMC), pp. 97–103 (2005)

[KMW05] Kuhn, F., Moscibroda, T., Wattenhofer, R.: The price of being near-sighted. In: Proc. SODA, pp. 980–989 (2005)

[LPP08] Lotker, Z., Patt-Shamir, B., Pettie, S.: Improved distributed approximate matching. In: SPAA 2008, pp. 129–136 (2008)

[PR01] Panconesi, A., Rizzi, R.: Some simple distributed algorithms for sparse networks. Distributed Computing 14(2), 97–100 (2001)

[Pe00] Peleg, D.: Distributed Algorithms, A Locality-Sensitive Approach. SIAM Press, Philadelphia (2000)

An Improved Approximation Algorithm for the Traveling Tournament Problem*

Daisuke Yamaguchi[1], Shinji Imahori[2],
Ryuhei Miyashiro[3], and Tomomi Matsui[1]

[1] Faculty of Science and Engineering, Chuo University,
Kasuga, Bunkyo-ku, Tokyo 112-8551, Japan
[2] Graduate School of Information Science and Technology,
The University of Tokyo, Hongo, Bunkyo-ku, Tokyo 113-8656, Japan
imahori@mist.i.u-tokyo.ac.jp
[3] Institute of Symbiotic Science and Technology,
Tokyo University of Agriculture and Technology,
Naka-cho, Koganei, Tokyo 184-8588, Japan
r-miya@cc.tuat.ac.jp

Abstract. This paper describes the traveling tournament problem, a well-known benchmark problem in the field of tournament timetabling. We propose an approximation algorithm for the traveling tournament problem with the constraints such that both the number of consecutive away games and that of consecutive home games are at most k. When $k \leq 5$, the approximation ratio of the proposed algorithm is bounded by $(2k-1)/k + O(k/n)$ where n denotes the number of teams; when $k > 5$, the ratio is bounded by $(5k-7)/(2k) + O(k/n)$. For $k = 3$, the most investigated case of the traveling tournament problem to date, the approximation ratio of the proposed algorithm is $5/3 + O(1/n)$; this is better than the previous approximation algorithm proposed for $k = 3$, whose approximation ratio is $2 + O(1/n)$.

Keywords: traveling tournament problem, approximation algorithm, lower bound, timetabling, scheduling.

1 Traveling Tournament Problem

In the field of tournament timetabling, the traveling tournament problem (TTP) is a well-known benchmark problem established by Easton, Nemhauser and Trick [2]. The objective of TTP is to make a round-robin tournament that minimizes the total traveling distance of participating teams. The problem TTP includes optimization aspects similar to those of the traveling salesman problem (TSP) and vehicle routing problems. However, TTP is surprisingly harder than TSP: there is a 10-team TTP instance that has not yet been solved exactly [6]. This contrasts starkly to TSP, for which a 10-city instance of TSP is easy. For further discussions related to TTP and its variations, see [4,5].

* This is an extended abstract. Proofs of theorems and lemmas are omitted due to page limitation. They are available in our technical report [7].

Y. Dong, D.-Z. Du, and O. Ibarra (Eds.): ISAAC 2009, LNCS 5878, pp. 679–688, 2009.

In the following, we introduce some terminology and then define TTP. We are given a set T of n teams, where $n \geq 4$ and even. Each team in T has its home venue. A game is specified by an ordered pair of teams. A double round-robin tournament is a set of games in which every team plays every other team once at its home venue and once at away (i.e., at the venue of the opponent); consequently, exactly $2(n-1)$ slots are necessary to complete a double round-robin tournament.

Each team stays at its home venue before a tournament; then it travels to play its games at the chosen venues. After a tournament, each team returns to its home venue if the last game is played at away. When a team plays two consecutive away games, the team goes directly from the venue of the first opponent to the other without returning to its home venue.

Let V be the set of venues satisfying $|V| = n$. For any pair of venues $i, j \in V$, $d_{ij} \geq 0$ denotes the distance between the venues i and j. We denote the distance matrix (d_{ij}) by D, whose rows and columns are indexed by V. Throughout this paper we assume that triangle inequality ($d_{ij} + d_{jk} \geq d_{ik}$), symmetry ($d_{ij} = d_{ji}$), and $d_{ii} = 0$ hold for any $i, j, k \in V$.

Given a constant (positive integer) $k \geq 3$, the traveling tournament problem [2] is defined as follows.

Traveling Tournament Problem (TTP(k))
Input: a set of teams T and a distance matrix $D = (d_{ij})$, indexed by V.
Output: a double round-robin tournament S of n teams such that
C1. no team plays more than k consecutive away games;
C2. no team plays more than k consecutive home games;
C3. game i at j immediately followed by game j at i is prohibited;
C4. the total distance traveled by the teams is minimized.

In this paper, we assume that n is sufficiently larger than a fixed parameter k. Constraints C1 and C2 are called the *atmost* constraints, and Constraint C3 is called the *no-repeater* constraint. In the remainder of this paper, a double round-robin tournament satisfying the above conditions C1–C3/C1–C4 are called a *feasible*/*optimal* tournaments.

Various studies on TTP have been appeared in recent years, and most of them considered TTP(3) [6]. Most of the best upper bounds of TTP instances are obtained using metaheuristic algorithms; on the other hand, few researches have been done to explore lower bounds and exact methods for TTP (see [5] for example). Recently, three of the authors of this paper proposed $(2+(9/4)/(n-1))$-approximation algorithm for TTP(3), which is the first approximation algorithm with a constant ratio [3].

In this paper, we propose an approximation algorithm for TTP(k). When $k \leq 5$, the approximation ratio of our algorithm is bounded by $(2k-1)/k + O(k/n)$; when $k > 5$, the approximation ratio is bounded by $(5k-7)/(2k) + O(k/n)$. For $k = 3$, the approximation ratio of our algorithm is $5/3 + O(1/n)$; that improves the approximation ratio of the previous algorithm for TTP(3), whose ratio is $2 + (9/4)/(n-1) = 2 + O(1/n)$.

2 Algorithm

A key idea of our algorithm is the use of the Kirkman schedule and a shortest Hamilton cycle. The classical Kirkman schedule satisfies the property that the orders of opponents in almost all teams are very similar to a mutual cyclic order of teams. Roughly speaking, our algorithm constructs an almost shortest Hamilton cycle passing all the venues and finds a permutation of teams such that the above cyclic order corresponds to the obtained Hamilton cycle.

In Section 2.1, we introduce how to make a single round-robin schedule with a specific structure. In Section 2.2, we construct double round-robin schedules based on the single round-robin schedule proposed in Section 2.1. In Section 2.3, we consider an assignment of venues to teams in the schedules of Section 2.2.

In the following, "schedule without HA-assignment" means that "round-robin schedule without concepts of home game, away game and venue." In other words, in a schedule without HA-assignment, only a sequence of opponents of each team is decided, but the venues of these games are not specified.

2.1 Single Round-Robin Schedule

Denote the set of n teams by $T = \{0, 1, \dots, n-1\}$ and the set of $n-1$ slots $S = \{0, 1, \dots, n-2\}$. A single round-robin schedule (without HA-assignment) is a matrix K whose (t, s) element $K(t, s)$ denotes the opponent of team t at slot s. More precisely, K is a matrix such that
(1) rows and columns are indexed by T and S, respectively,
(2) every element of K is a team,
(3) a row of K indexed by t consists of teams $T \setminus \{t\}$, and
(4) for any pair $(t, s) \in T \times S$, a team $t' = K(t, s)$ satisfies $K(t', s) = t$.

The *Kirkman schedule* K^* is a matrix defined by

$$K^*(t,s) = \begin{cases} s-t \mod n-1 & (t \neq n-1 \text{ and } [s-t \neq t \mod n-1]), \\ n-1 & (t \neq n-1 \text{ and } [s-t = t \mod n-1]), \\ s/2 & (t = n-1 \text{ and } s \text{ is even}), \\ (s+n-1)/2 & (t = n-1 \text{ and } s \text{ is odd}). \end{cases}$$

Lemma 1. *The Kirkman schedule K^* is a single round-robin schedule.*
Proof. Omitted due to page limitation.

Next, we define HA-assignments of the Kirkman schedule. For constructing variations of HA-assignments, we introduce a function $f : U \to \{\mathrm{H}, \mathrm{A}\}$ where $U = \{i \in \mathbb{Z} \mid i \neq 0 \mod n-1\}$. Given a function f, we define the negated function $\neg f : U \to \{\mathrm{H}, \mathrm{A}\}$ by

$$\neg f(i) = \begin{cases} \mathrm{H}\ (f(i) = \mathrm{A}), \\ \mathrm{A}\ (f(i) = \mathrm{H}). \end{cases}$$

Similarly, we define that $\neg$H is A and $\neg$A is H. We say that the function f is *HA-feasible* if f satisfies

(F1) $\forall i, \forall j \in U$, $[i = j \mod n-1$ implies $f(i) = f(j)]$, and
(F2) $\forall i \in U$, $f(i) = \neg f(-i)$.

From the above definition, an HA-feasible function f is uniquely defined by the sequence $(f(1), f(2), \ldots, f(n/2-1))$. The second condition (F2) implies that the sequence $(f(-n/2+1), \ldots, f(-2), f(-1))$ is the reverse of the sequence $(\neg f(1), \neg f(2), \ldots, \neg f(n/2-1))$. Given an HA-feasible function f, we define that for any pair of teams $t, t' \in T \backslash \{n-1\}$, the game between t and t' is $f(t'-t)$-game of team t and $\neg f(t'-t) = f(t-t')$-game of team t'. (Here we note that H-game means a home game, and A-game means an away game.) For constructing a complete HA-assignment, we need to define an HA-pattern of team $n-1$. We introduce a *root sequence* $(r_0, r_1, \ldots, r_{n-2}) \in \{\mathrm{H}, \mathrm{A}\}^{n-1}$ defined by

$$r_i = \begin{cases} \mathrm{H} & (i \neq n-3 \text{ and } [i \in \{0, 1, \ldots, k-1\} \mod 2k]), \\ \mathrm{A} & (i \neq n-3 \text{ and } [i \in \{k, k+1, \ldots, 2k-1\} \mod 2k]), \\ r_{n-2} & (i = n-3). \end{cases}$$

For example, when $n = 32$ and $k = 5$ the root sequence is

$$(\underbrace{\mathrm{HHHHH}}_{k}\underbrace{\mathrm{AAAAA}}_{k}\underbrace{\mathrm{HHHHH}}_{k}\underbrace{\mathrm{AAAAA}}_{k}\underbrace{\mathrm{HHHHH}}_{k}\underbrace{\mathrm{AAAA\overset{r_{n-3}}{H}}}_{k}\overset{r_{n-2}}{\mathrm{H}}).$$

We define that team $n-1$ plays r_s-game at slot s and the opponent of $n-1$ at slot s plays $\neg r_s$-game (at slot s).

As a consequence, given an HA-feasible function f (and the root sequence), we can construct an HA-assignment of the Kirkman schedule K^* as follows. For any pair $(t, s) \in T \times S$,

$$\text{at slot } s, \text{ team } t \text{ plays } \begin{cases} f(s-2t)\text{-game } (t \neq n-1 \text{ and } [s-t \neq t \mod n-1]), \\ \neg r_s\text{-game } (t \neq n-1 \text{ and } [s-t = t \mod n-1]), \\ r_s\text{-game } (t = n-1). \end{cases}$$

When $t \neq n-1$ and $[s-t \neq t \mod n-1]$, the opponent of team t, denoted by $K^*(t, s)$, is defined by $K^*(t, s) = s-t \mod n-1$ and team t plays $f(K^*(t, s) - t)$-game at slot s. Thus, $K^*(t, s) - t = (s-t) - t = s - 2t \mod n-1$ and Definition (F1) implies $f(K^*(t, s) - t) = f(s-2t)$. The remaining cases are trivial.

Next, we define variations of HA-assignments by introducing k HA-feasible functions $f_1, f_2, \ldots, f_k$. For each $\alpha \in \{1, 2, \ldots, k\}$, we settle a function f_α by a sequence $(f_\alpha(1), f_\alpha(2), \ldots, f_\alpha(n/2-1))$, defined below. First, we consider an infinite sequence that contains k consecutive 'A's and k consecutive 'H's alternately. Next, we clip a sequence of length $n/2-1$ whose top $k-\alpha+1$ elements are $(\overbrace{\mathrm{A}, \mathrm{A}, \ldots, \mathrm{A}}^{k-\alpha}, \mathrm{H})$. Lastly, we set the first and second elements to 'A' and change the penultimate element to the same element as the last (if it is required). When $k = 3$, we additionally set the third and fourth elements to 'H.' For example, when $n = 32$ and $k = 5$, the sequence

$F_\alpha = (f_\alpha(-15), \ldots, f_\alpha(-2), f_\alpha(-1), *, f_\alpha(1), f_\alpha(2), \ldots, f_\alpha(15))$
becomes
$F_1 =$ (AAHHHHAAAAAHHHH $*$ AAAAHHHHHAAAAHH)
$F_2 =$ (AAHHHHHAAAAAHHH $*$ AAAHHHHHAAAAAHH)
$F_3 =$ (AAAHHHHHAAAAAHH $*$ AAHHHHHAAAAAHHH)
$F_4 =$ (AAAAHHHHHAAAAHH $*$ AAHHHHAAAAAHHHH)
$F_5 =$ (AAAAAHHHHHAAAHH $*$ AAHHHAAAAAHHHHH).

In the rest of this paper, X_α ($\alpha \in \{1, 2, \ldots, k\}$) denotes the Kirkman schedule with an HA-assignment induced by an HA-feasible function f_α.

2.2 Feasible Double Round-Robin Schedule

In the previous section, we introduced an HA-feasible function f_α, the sequence

$$F_\alpha = (f_\alpha(-n/2+1), \ldots, f_\alpha(-2), f_\alpha(-1), *, f_\alpha(1), f_\alpha(2), \ldots, f_\alpha(n/2-1))$$

and the Kirkman schedule (with an HA-assignment) X_α for any $\alpha \in \{1, 2, \ldots, k\}$. We set the center element (denoted by $*$) of F_α to A or H, and denote the obtained sequence by F_α^{A} or F_α^{H}, respectively. Here we assume that the first element of F_α^{A} is adjacent with the last element of F_α^{A} (and similarly assume for F_α^{H}). Then an HA-pattern of team $t \in T \setminus \{n-1\}$ in schedule X_α is obtained by a cyclic permutation of sequence F_α^{A} or F_α^{H}. In addition, the HA-pattern of team $n-1$ in schedule X_α is obtained by the root sequence.

From the definition of HA-feasible functions $f_1, f_2, \ldots, f_k$ and the root sequence, the following property holds.

Theorem 1. *For any single round-robin schedule $X_\alpha \in \{X_1, X_2, \ldots, X_k\}$, X_α satisfies the atmost constraints.*
Proof. Omitted due to page limitation.

Next, we show a property of sequences F_α^{A} and F_α^{H}, which plays an important role in constructing a double round-robin schedule.

Lemma 2. *For any $\alpha \in \{1, 2, \ldots, k\}$, (1) every consecutive three elements of the cyclic sequence F_α^{A} is neither* (HAH) *nor* (AHA)*; (2) every consecutive three elements of the cyclic sequence F_α^{H} is neither* (HAH) *nor* (AHA).
Proof. Omitted due to page limitation.

Lemma 2 and the definition of the root sequence imply the following.

Corollary 1. *For any single round-robin schedule $X_\alpha \in \{X_1, X_2, \ldots, X_k\}$, X_α satisfies: (1) the HA-pattern of each team at slots $(n-2, 0, 1)$ is neither* (HAH) *nor* (AHA)*; (2) the HA-pattern of each team at slots $(n-3, n-2, 0)$ is neither* (HAH) *nor* (AHA).

For any $\alpha \in \{1, 2, \ldots, k\}$, given a single round-robin schedule X_α defined above, we construct a double round-robin schedule as follows. First, we construct a single round-robin schedule, denoted by Y_α, by exchanging the first slot for the last slot of X_α. Next, we construct a double round-robin schedule by the ordinary mirroring as follows. Denote $\overline{Y}_\alpha$ the schedule obtained from Y_α by reversing the home and away. We concatenate two single round-robin schedules Y_α and $\overline{Y}_\alpha$ to obtain a double round-robin schedule, denoted by Z_α.

Theorem 2. *For any $\alpha \in \{1, 2, \ldots, k\}$, both of the single round-robin schedules Y_α and $\overline{Y}_\alpha$ satisfy the atmost constraints.*

Proof. Omitted due to page limitation.

Theorem 3. *For any double round-robin schedule $Z_\alpha \in \{Z_1, Z_2, \ldots, Z_k\}$, Z_α is a feasible schedule.*

Proof. Omitted due to page limitation.

2.3 Assignment of Venues

In Section 2.1, we defined that T is a set of teams and described a method for constructing a double round-robin schedule of teams in T. In this section, we say that T is a set of *imaginary* teams (without venues) and each venue in V represents a *real* team. We propose an algorithm for finding a bijection between the set of venues V and the set of imaginary teams T. In this section, 'a team $t \in T$' means 'an imaginary team $t \in T$.'

Herein, we describe our algorithm. First, we choose $\alpha \in \{1, 2, \ldots, k\}$ randomly and construct a double round-robin schedule Z_α with imaginary teams T. Next, we apply Christofides' algorithm for the traveling salesman problem to a complete undirected graph with vertex set (venue set) V and edge length defined by D, and obtain a Hamilton cycle H_C. (Here we note that the length of an undirected edge is well-defined by D, since D satisfies symmetry $d_{ij} = d_{ji}$.) We denote a Hamilton cycle H_C by a sequence $(v_0, v_1, \ldots, v_{n-1})$ of vertices (venues). Lastly, we choose $\beta \in \{0, 1, \ldots, n-1\}$ randomly and construct a bijection $\pi : T \to V$ defined by $\pi(i) = v_j$ where $T = \{0, 1, \ldots, n-1\}$ and $j = i + \beta \mod n$.

To determine an expected value of total traveling distance obtained by the above algorithm, we introduce an undirected graph G_α defined by a double round-robin schedule Z_α. The graph G_α has a vertex set T, and a (multi) edge set $E(\alpha)$ with partition $\{E_t(\alpha) \mid t \in T\}$ where every edge in a multiset $E_t(\alpha)$ corresponds to a move of team $t \in T$ in Z_α. More precisely, multiset $E_t(\alpha)$ consists of following (at most) four types of edges;

(1) when team t plays two consecutive away games, $E_t(\alpha)$ includes an edge between two opponents,
(2) when team t plays consecutive pair of home and away games, $E_t(\alpha)$ includes an edge between t and opponent of the away game,
(3) if t plays away game at first slot, $E_t(\alpha)$ includes an edge between t and opponent in the away game,

(4) if t plays away game at last slot, $E_t(\alpha)$ includes an edge between t and opponent in the away game.

If we have a bijection $\pi : T \rightarrow V$, the corresponding total traveling distance becomes $\sum_{\{i,j\} \in E(\alpha)} d_{\pi(i)\pi(j)}$.

Next, we define a partition of $E(\alpha)$ consists of three subsets, called *irregular edges*, *regular Hamilton edges*, and *regular non-Hamilton edges*. For any $t \in T$, an edge e in $E_t(\alpha)$ is called *irregular* if e satisfies at least one of the following conditions;

(I1) e has at least one parallel edge in $E_t(\alpha)$,
(I2) e connects a pair of vertices in $\{t-5, t-4, \dots, t+5\} \cup \{n-2, n-1, 0\} \cup \{t-n/2+1, t-n/2+2, t-n/2+3\} \cup \{t+n/2-3, t+n/2-2, t+n/2-1\}$, where every integer $t+i$ appearing above corresponds to a vertex (team) $t' \in T$ with $t' = t+i \mod n$,
(I3) e corresponds to a move between a pair of slots in $\{(0,1), (n-3, n-2), (n-2, n-1), (n-1, n), (2n-4, 2n-3)\}$,
(I4) e corresponds to a move caused by an away game at first slot (if it exists),
(I5) e corresponds to a move caused by an away game at last slot (if it exists),
(I6) $e \in E_{n-1}(\alpha)$, i.e., e corresponds to a move of team $n-1$.

Every non-irregular edge in $E(\alpha)$ is called *regular*. If a regular edge in $E(\alpha)$ connects a pair of vertices $\{t, t'\}$ with $t' = t+1 \mod n$, the edge is called *regular Hamilton*. Here we note that every regular Hamilton edge corresponds to an edge in Hamilton cycle H^* defined by cyclic sequence $(0, 1, \dots, n-1)$ of T. The rests of regular edges in $E(\alpha)$ are called *regular non-Hamilton* edges.

In the following, we determine the number of irregular edges in $E(\alpha)$. For any team $t \in T$, undirected graph $(T, E_t(\alpha))$ is an Eulerian graph such that every vertex $t' \in T \setminus \{t\}$ has two incident edges. Thus, every pair of vertices has at most two parallel edges corresponding to consecutive three games with the HA-pattern (H, A, H). From the definition of Z_α, the number of vertex pairs with parallel edges in $E_t(\alpha)$ is bounded by a constant and thus the number of irregular edges in $E_t(\alpha)$ satisfying Condition I1 is bounded by a constant. Obviously, the number of irregular edges in $E_t(\alpha)$ satisfying Conditions I2–I5 is also bounded by a constant. Consequently, for any team $t \in T \setminus \{n-1\}$, $E_t(\alpha)$ contains a constant number of irregular edges. Since $|E_{n-1}(\alpha)| = \mathrm{O}(n)$, the total number of irregular edges in $E(\alpha)$ is $\mathrm{O}(n)$.

Next, we discuss the number of regular Hamilton edges. Every regular Hamilton edge corresponds to a consecutive pair of away games of a team $t \in T \setminus \{n-1\}$. Thus, every team $t \in T \setminus \{n-1\}$ has at most $(2n-2)(k-1)/(2k) \leq n(k-1)/k$ regular Hamilton edges.

Finally, every regular non-Hamilton edge corresponds to a consecutive pair of home game and away game of a team $t \in T \setminus \{n-1\}$. Consequently, every team $t \in T \setminus \{n-1\}$ has at most $(2n-2)2/(2k) \leq 2n/k$ regular non-Hamilton edges.

The below table shows upper bounds of sizes of three sets.

irregular edges	$O(n)$
regular Hamilton edges	$n(n-1)(k-1)/k$
regular non-Hamilton edges	$2n(n-1)/k$

3 Lower Bounds

In the rest of this paper, we consider a complete undirected graph $\overline{G}$ with vertex set (venue set) V and edge length defined by the distance matrix D. We can relate every move of a real team to an undirected edge in $\overline{G}$.

Let Ψ^* be an optimal tournament of a given instance and ψ^* the optimal value. In the tournament Ψ^*, each real team $v \in V$ visits every venue of opponent exactly once, and we call the sequence of moves of v among venues a tour of v. For each real team $v \in V$, ψ^*_v denotes a tour distance of team v in Ψ^*. Obviously, $\sum_{v \in V} \psi^*_v = \psi^*$ holds.

Let η^* be the length of shortest Hamilton cycle of a complete undirected graph $\overline{G}$. Since distance matrix D satisfies triangle inequalities, $\psi^*_v \geq \eta^*$ holds for any $v \in V$. Consequently, we have the following lemma.

Lemma 3. *The length η^* of a shortest Hamilton cycle satisfies that $\psi^* \geq n\eta^*$.*

For each real team $v \in V$, ψ^{home}_v denotes the sum of distances corresponding to moves leaving or returning to its home v. We introduce a ratio $a^* = (\sum_{v \in V} \psi^{\text{home}}_v)/\psi^*$. Let τ^* be the length of minimum spanning tree of a complete undirected graph $\overline{G}$. Then we have the following.

Lemma 4. *The length τ^* of a minimum spanning tree satisfies $(1-\frac{a^*}{2})\psi^* \geq n\tau^*$.*
Proof. Omitted due to page limitation.

We denote the sum total of distances of ordered pairs of venues by Δ, i.e., $\Delta \stackrel{\text{def.}}{=} \sum_{v \in V} \sum_{u \in V} d_{vu}$.

Lemma 5. *The sum of distances Δ satisfies $(a^* + \frac{k-2}{2})\psi^* \geq \Delta$.*
Proof. Omitted due to page limitation.

4 Approximation Ratio

Here we discuss the approximation ratio of our algorithm. Let $(v_0, v_1, \ldots, v_{n-1})$ be a sequence of vertices (venues) corresponding to Hamilton cycle H_C obtained by Christofides' algorithm. Our algorithm chooses $\beta \in \{0, 1, \ldots, n-1\}$ randomly and construct a bijection $\pi : T \to V$ defined by $\pi(i) = v_j$ where $T = \{0, 1, \ldots, n-1\}$ and $j = i + \beta \mod n$.

First, we consider the sum of weights of irregular edges, denoted by a random variable W_{IR}. For each irregular edge (t, t') connecting $t, t' \in T$, $\pi(t)$ (and $\pi(t')$) is randomly assigned to a vertex in V. For any $v \in V$, the triangle inequalities imply that $d_{\pi(t)\pi(t')} \leq d_{\pi(t)v}+d_{v\pi(t')}$, and thus $d_{\pi(t)\pi(t')} \leq (1/n)\sum_{v\in V}(d_{\pi(t)v}+d_{v\pi(t')})$ holds. The expectation of $d_{\pi(t)\pi(t')}$ with respect to random selection of β satisfies

$$\mathrm{E}[d_{\pi(t)\pi(t')}] \leq \left(\frac{1}{n}\right)\sum_{v\in V}(\mathrm{E}[d_{\pi(t)v}]+\mathrm{E}[d_{v\pi(t')}])$$

$$=\left(\frac{1}{n}\right)\sum_{v\in V}\left(\left(\frac{1}{n}\right)\sum_{u\in V}d_{uv}+\left(\frac{1}{n}\right)\sum_{u\in V}d_{vu}\right)=\left(\frac{2}{n^2}\right)\sum_{v\in V}\sum_{u\in V}d_{uv}=\left(\frac{2}{n^2}\right)\Delta.$$

As discussed in Section 2.3, the number of irregular edges is bounded by $\mathrm{O}(n)$ and consequently $\mathrm{E}[W_{\mathrm{IR}}] \leq \mathrm{O}(n)(2/n^2)\Delta = \mathrm{O}(1/n)\Delta$ holds.

Next, we consider the sum of weights of regular Hamilton edges, denoted by W_{RH}. On the length of H_{C}, the following is a well-known theorem.

Lemma 6. [1] *The length of H_C is less than or equal to $\tau^*+(1/2)\eta^*$, where τ^* and η^* denote the length of minimum spanning tree and shortest Hamilton cycle of a complete undirected graph $\overline{G}$.*

Since regular Hamilton edges in H^* are contained in Hamilton cycle H_{C} for any $\beta \in \{0,1,\ldots,n-1\}$, the above randomization implies that the expectation satisfies $\mathrm{E}[W_{\mathrm{RH}}] \leq \frac{n(n-1)(k-1)}{k}\left(\frac{\eta_{\mathrm{C}}}{n}\right) \leq \frac{(n-1)(k-1)}{k}(\tau^*+(1/2)\eta^*)$, where η_{C} denotes the length of H_{C}.

Lastly, we consider the sum of weights of regular non-Hamilton edges, denoted by W_{RnH}. Recall that every regular non-Hamilton edge corresponds to a consecutive pair of a home game and an away game of a team $t \in T\setminus\{n-1\}$. We fix team $t \in T\setminus\{n-1\}$, $\beta \in \{0,1,\ldots,n-1\}$, and permutation $\pi : T \to V$ with respect to β. For any team $t' \in T\setminus\{t,n-1\}$, a regular non-Hamilton edge corresponding to a move of team t (real team $\pi(t)$) between venues $\pi(t)$ and $\pi(t')$ appears with probability at most $2/k$ with respect to random choice of $\alpha \in \{0,1,\ldots,k-1\}$ (and consequently sequence F_α), because our algorithm constructs a double round-robin tournament by mirroring. Thus, we have that

$$\mathrm{E}[W_{\mathrm{RnH}}] \leq \sum_{t\in T\setminus\{n-1\}}\sum_{\beta\in\{0,1,\ldots,n-1\}}\left(\frac{1}{n}\right)\left(\sum_{t'\in T\setminus\{n-1,t\}}\left(\frac{2}{k}\right)d_{\pi(t)\pi(t')}\right)$$

$$\leq \sum_{t\in T}\sum_{\beta\in\{0,1,\ldots,n-1\}}\left(\frac{1}{n}\right)\left(\sum_{t'\in T}\left(\frac{2}{k}\right)d_{\pi(t)\pi(t')}\right)$$

$$=\left(\frac{2}{kn}\right)\sum_{\beta\in\{0,1,\ldots,n-1\}}\left(\sum_{t\in T}\sum_{t'\in T}d_{\pi(t)\pi(t')}\right)=\left(\frac{2}{k}\right)\Delta.$$

Finally, we determine the approximation ratio of our algorithm.

Theorem 4. *When $k \leq 5$, the approximation ratio of our algorithm is bounded by $(2k-1)/k+\mathrm{O}(k/n)$. If $k>5$, the ratio is bounded by $(5k-7)/(2k)+\mathrm{O}(k/n)$.*
Proof. Omitted due to page limitation.

We note that, if we run our algorithm for every pair of $\alpha \in \{1, 2, \ldots, k\}$ and $\beta \in \{0, 1, \ldots, n-1\}$, the above approximation ration can be always attainable in polynomial time.

References

1. Christofides, N.: Worst-case analysis of a new heuristic for the traveling salesman problem. Report 388, Graduate School of Industrial Administration, Carnegie-Mellon University (1976)
2. Easton, K., Nemhauser, G., Trick, M.: The traveling tournament problem: description and benchmarks. In: Walsh, T. (ed.) CP 2001. LNCS, vol. 2239, pp. 580–585. Springer, Heidelberg (2001)
3. Miyashiro, R., Matsui, T., Imahori, S.: An approximation algorithm for the traveling tournament problem. In: The 7th International Conference on the Practice and Theory of Automated Timetabling (PATAT 2008), Université de Montréal, CD-ROM (2008)
4. Rasmussen, R.V., Trick, M.A.: A Benders approach for the constrained minimum break problem. European Journal of Operational Research 177, 198–213 (2007)
5. Rasmussen, R.V., Trick, M.A.: Round robin scheduling — a survey. European Journal of Operational Research 188, 617–636 (2008)
6. Trick, M.: Challenge traveling tournament problems (2009), `http://mat.gsia.cmu.edu/TOURN/`
7. Yamaguchi, D., Imahori, S., Miyashiro, R., Matsui, T.: An improved approximation algorithm for the traveling tournament problem, Mathematical Engineering Technical Report, METR09-42, Department of Mathematical Informatics, Graduate School of Information Science and Technology, The University of Tokyo

The Fault-Tolerant Facility Allocation Problem

Shihong Xu and Hong Shen

School of Computer Science, The University of Adelaide, SA 5005, Australia
{shihong.xu,hong.shen}@adelaide.edu.au

Abstract. We study the problem of *Fault-Tolerant Facility Allocation (FTFA)* which is a relaxation of the classical *Fault-Tolerant Facility Location (FTFL)* problem [1]. Given a set of sites, a set of cities, and corresponding facility operating cost at each site as well as connection cost for each site-city pair, *FTFA* requires to allocate each site a proper number of facilities and further each city a prespecified number of facilities to access. The objective is to find such an allocation that minimizes the total combined cost for facility operating and service accessing. In comparison with the *FTFL* problem which restricts each site to at most one facility, the *FTFA* problem is less constrained and therefore incurs less cost which is desirable in application. In this paper, we consider the metric *FTFA* problem where the given connection costs satisfy triangle inequality and we present a polynomial-time algorithm with approximation factor 1.861 which is better than the best known approximation factor 2.076 for the metric *FTFL* problem [2].

1 Introduction

Facility location problem [3] is one of the classical problems that has been widely studied in operations research. In this problem, we are given a set $\mathcal{F}$ of n_f sites and a set $\mathcal{C}$ of n_c cities: each site $i \in \mathcal{F}$ is associated with a non-negative cost f_i for facility operating and each site-city pair $(i,j), i \in \mathcal{F}, j \in \mathcal{C}$ a connection cost c_{ij} for service accessing. The objective is to find a subset of $\mathcal{F}$ to deploy facilities and further connect each city in $\mathcal{C}$ to one site in the subset in order to access service so that the total combined cost for facility operating and service accessing is minimized.

In this paper, we study a generalization of the facility location problem called *Fault-Tolerant Facility Allocation* (*FTFA*). In this problem, each site allows an unlimited number of facilities and each city requires a prespecified number of connections for fault tolerance purpose. Let the connectivity requirement of city $j \in \mathcal{C}$ be r_j, the objective of *FTFA* is to allocate each site a proper number of facilities and each city the required number of facilities so that the total combined cost is minimized. The *FTFA* problem can be formulated by the following integer linear program.

$$\begin{array}{ll} \text{minimize} & \sum_{i\in\mathcal{F}} f_i y_i + \sum_{i\in\mathcal{F}}\sum_{j\in\mathcal{C}} c_{ij}x_{ij} \\ \text{subjected to} & \forall j \in \mathcal{C}: \sum_{i\in\mathcal{F}} x_{ij} \geq r_j \\ & \forall i \in \mathcal{F}, j \in \mathcal{C}: y_i \geq x_{ij} \\ & \forall i \in \mathcal{F}, j \in \mathcal{C}: x_{ij}, y_i \in \mathbb{Z}^+ \end{array} \qquad (1)$$

Y. Dong, D.-Z. Du, and O. Ibarra (Eds.): ISAAC 2009, LNCS 5878, pp. 689–698, 2009.

In the above formulation, non-negative integer y_i indicates how many facilities are opened at site i and x_{ij} indicates how may connections between site i and j are established. The first constraint is to ensure fault tolerance of connections — each city $j \in \mathcal{C}$ is assigned totally r_j connections and is able to tolerate up to $r_j - 1$ connection failures. The second constraint ensures fault tolerance of facilities — each city $j \in \mathcal{C}$ is assigned distanct facilities for any two connections and is able to tolerate up to $r_j - 1$ facility . In this paper, we consider the metric version of the problem where the given connection costs satisfy triangle inequality.

FTFA is very close to the well studied *Fault-Tolerant Facility Location (FTFL)* problem [1,4,5,2] which has the same objective function and constraints as *FTFA* except the range of variants: For any $i \in \mathcal{F}, j \in \mathcal{C}$, x_{ij} and y_i are any non-negative integers ($\mathbb{Z}^+$) in *FTFA* but binary integers (0 or 1) in *FTFL*. Without the restriction on the maximum number of facilities at each site, *FTFA* is less constrained and therefore incurs less cost which is desirable in application. *FTFA* can be applied in the deployment of ATMs, server farms etc. where multiple facilities can be deployed at one site if necessary and share the duty to service clients together. We also notice that the problem becomes the classical *Uncapacitated Facility Location (UFL)* problem when connectivity requirement $r_j = 1$ for all $j \in \mathcal{C}$. The relation between *UFL*, *FTFA* and *FTFL* is: *UFL* $\subseteq$ *FTFA* $\subseteq$ *FTFL*. The second inclusion relation is implied by a special case of *FTFL* with facility set $\mathcal{F}' = \mathcal{F} \times \{1, 2, ..., R\}$, where $R = \max_{j \in \mathcal{C}} r_j$ is the number of identical facilities in each site.

Considering the relation between *FTFA* and *UFL*, we demonstrate in Section 3 how to adapt an existing *UFL* algorithm to deal with the new problem. We study a *UFL* problem with facility set $\mathcal{F}'$ and city set $\mathcal{C}' = \{(j, p), j \in \mathcal{C}, 1 \leq p \leq r_j\}$ for any given *FTFA* problem, where $\{(j, p), 1 \leq p \leq r_j\}$ presents all cities derived from original city $j \in \mathcal{C}$. We make the derived solution feasible to *FTFA* by ensuring additionally that every two ports of a city are connected with distinct facilities. The proposed algorithm runs through R phases and in each phase employs a subroutine to pick the most *cost-efficient* star iteratively. Instead of employing two types of events as in the algorithms of Jain *et al.* [6,7,8], our algorithm need to process three types of events: One for facilities opened in a previous phase, one for facilities opened in the current phase and another for opening a new facility in the current phase. Combined with the same factor-revealing LP as in [6,8], our analysis in Section 4 shows that the proposed algorithm is also 1.861 approximation for the metric *FTFA* problem (The MMS algorithm [6,8] is 1.861-approximation for the metric *UFL* problem). Running time of the algorithm is $O(mR \log m)$, where $m = n_f n_c$.

2 Related Work

Facility location problem and its variants occupy a central place in operations research [3]. For the simplest version of the problem *UFL*, the first approximation algorithm was built by Cornuejols *et al.* [9]. They obtain $(1-e^{-1})$-approximation algorithm for the maximization variant of the problem. The first approximation algorithm for the minimization variant, a greedy algorithm achieving a guarantee of $O(\log n)$ in the general (nonmetric) case due to [10]. For the metric version of the problem, the first constant-factor approximation algorithm for this problem was due to Shmoys *et al.* [11], who gave a 3.16-approximation algorithm using the filtering technique of Lin and Vitter [12] to round the optimal solution of a linear program; Chudak, Williamson [13] and Sviridenko [14] improved the approximation ratio to 1.736 and 1.582 by rounding an optimal fractional solution to a linear programming relaxation.

Aside from the above results which are achieved through linear programming and rounding technique, primal dual algorithms for facility location problem also have been studied extensively. Charikar and Guha [15] obtained an 1.853-approximation algorithm and an 1.728-approximation algorithm by using primal-dual theory and greedy augmentation; Jain *et al.* [16,6,7,8] presented greedy algorithms based on dual fitting and factor-revealing LP technique, achieving approximation guarantee 1.861 and 1.61. Different from traditional primal dual schema [17,18], dual fitting relaxes the feasibility of the dual solution first in the algorithm and then shrink the solution a factor in the analysis to make the shrunk solution feasible, which is further shown to be the approximation factor of the algorithm. Mahdian *et al.* further improved the approximation ratio to 1.52 [19] by adding a scaling and greedy augmentation procedure to the JMS [7,8] algorithm. Byrka [20] modified the Chudak and Shmoys's algorithm [21] and obtained a new one which touches the approximability limit in the first time. Their new approach give a 1.5-approximation algorithm which is currently the best known for the problem.

Fault Tolerant Facility Location [1] is a generalization of *UFL*, where connectivity at different cities (i. e. the number of distinct facilities that serve a city) are specified to meet fault-tolerant requirements. Guha *et al.* [5] obtained a 3.16-approximation algorithm by rounding the optimal fractional solution to the problem and further improve the result to 2.41 by employing a greedy local improvement step. Recently, Swamy and Shmoys [2] presented a 2.076-approximation by using LP rounding. All these results hold for both uniform connectivity case and nonuniform connectivity case. Guha and Khuller [22] proved that the best approximation factor to *UFL* is not better than 1.463, assuming NP $\not\subseteq$ DTIME$[n^{O(\log\log n)}]$. This result also holds for the fault-tolerant generations of the problem including *FTFL* and *FTFA*.

3 The Algorithm

Without loss of generality, assume the set of connectivity requirements $\mathcal{R} = \{1, 2, 3, ..., R\}$, otherwise add dummy cities with the missing connectivity requirements. Further assume r_j ports at each city, R facilities at each site, all ports of a city must be connected in the order from 1 to r_j and all facilities at a site can be opened, if necessary, in the order from 1 to R. We use vector $\boldsymbol{y}^p$ to denote whether the *p-th* facility is opened for all facilities and $\boldsymbol{x}^p$ whether the p-th port of a city is connected with the site for all site-city pairs. It is clear that $x_{ij}^p = 0$ if $p > r_j$ for any site-city pair, furthermore $\boldsymbol{y} = \sum_{p=1}^{R} \boldsymbol{y}^p$ and $\boldsymbol{x} = \sum_{p=1}^{R} \boldsymbol{x}^p$. For the sake of simplicity, let vector $\boldsymbol{X}^b = \sum_{p=1}^{b} \boldsymbol{x}^p$ and $\boldsymbol{Y}^b = \sum_{p=1}^{b} \boldsymbol{y}^p, 1 \leq b \leq R$, our algorithm evolves the solution from the initial stage (suppose $\boldsymbol{X}^0$ and $\boldsymbol{Y}^0$), through R phases, to $\boldsymbol{X}^R$ and $\boldsymbol{Y}^R$. For the convenience of explanation, we first rewrite program (1) as

$$\begin{array}{ll} \text{minimize} & \sum_{i\in\mathcal{F}}\sum_{p\in\mathcal{R}}(f_i y_i^p + \sum_{j\in\mathcal{C}} c_{ij}x_{ij}^p) \\ \text{subjected to} & \forall j \in \mathcal{C}, 1 \leq p \leq r_j : \sum_{i\in\mathcal{F}} x_{ij}^p \geq 1 \\ & \forall j \in \mathcal{C}, i \in \mathcal{F} : Y_i^R \geq X_{ij}^R \\ & \forall j \in \mathcal{C}, i \in \mathcal{F}, p \in \mathcal{R} : x_{ij}^p, y_i^p \in \{0,1\}. \end{array} \tag{2}$$

The LP-relaxation of the above program can be obtained by allowing x_{ij} and y_i to be non-negative real numbers. The dual problems of the LP relaxation is

$$\begin{array}{ll} \text{maximize} & \sum_{j\in\mathcal{C}}\sum_{p=1}^{r_j} \alpha_j^p \\ \text{subjected to} & \forall i \in \mathcal{F}, p \in \mathcal{R} : \sum_{j\in\mathcal{C}} \beta_{ij}^p \leq f_i \\ & \forall i \in \mathcal{F}, j \in \mathcal{C}, p \in \mathcal{R} : \alpha_j^p - \beta_{ij}^p \leq c_{ij} \\ & \forall i \in \mathcal{F}, j \in \mathcal{C}, p \in \mathcal{R} : \alpha_j^p, \beta_{ij}^p \geq 0. \end{array} \tag{3}$$

We use similar interpretation of dual variables α_j^p and β_{ij}^p as in [6,8] which will be explained in detail later. One of the interesting observation is that we can extract $p \in \mathcal{R}$ in the constraints and the objective function if we have $\alpha_j^p = 0$ when $p > r_j$. We utilize this observation to design a high level algorithm which decomposes the problem into R subproblems and process them in the order. In fact, our algorithm opens facilities through R phases, establishing one connection for each city in each phase if it is *not-fully-connected.* A facility opened in one phase at site i can be used for free by any city j in a later phase, suppose p, under the condition $X_{ij}^p \leq Y_i^p$. In each phase $p \in \mathcal{R}$, the set of *not-fully-connected* cities is denoted by $\mathcal{C}_p = \{j \in \mathcal{C} : r_j \geq p\}$.

The process of our algorithm is presented in Algorithm 1: In the p-th phase, the solution inherited from the last phase is $(\boldsymbol{X}^{p-1}, \boldsymbol{Y}^{p-1})$ which serves as the input of the subroutine together with $\mathcal{F}$ and $\mathcal{C}_p$. Note that cities with $r_j < p$ is already *fully-connected* and therefore not included in $\mathcal{C}_p$. Suppose the new opened facilities and new established connections are denoted by $(\boldsymbol{x}^p, \boldsymbol{y}^p)$, then in the next phase, we have $\boldsymbol{X}^p = \boldsymbol{X}^{p-1} + \boldsymbol{x}^p$ and $\boldsymbol{Y}^p = \boldsymbol{Y}^{p-1} + \boldsymbol{y}^p$. The algorithm ends when all R phases are finished.

In the p-th phase algorithm, we use a notation of star and a definition of *cost efficiency.* A star is composed of a facility and a group of cities that are connected with the facility. Considering the time before the new star is selected, we define the *cost efficiency* of a star to be

$$\text{eff}(i, p, C') = \frac{f_i^p + \sum_{j\in C'} c_{i,j}}{|C'|}, \tag{4}$$

where f_i^p is the cost paid to open a facility at site i in phase p and C' the set of city members in the star. The two items in the numerator represent the total

---

Algorithm 1. 1.861-Approximation Algorithm

Input: f_i, r_j, c_{ij} for any $i \in \mathcal{F}, j \in \mathcal{C}$.

Output: x_{ij}, y_i for any $i \in \mathcal{F}, j \in \mathcal{C}$.

(1) Initially set vector $\boldsymbol{X}^0 \leftarrow \mathbf{0}, \boldsymbol{Y}^0 \leftarrow \mathbf{0}$ and the No. of current phase $p \leftarrow 1$.

(2) While $p \leq R$:

  (a) Invoke the *p-th* phase algorithm with input $(\boldsymbol{X}^{p-1}, \boldsymbol{Y}^{p-1}, \mathcal{F}, \mathcal{C}_p)$, suppose the output is $(\boldsymbol{X}^p, \boldsymbol{Y}^p)$.

  (b) Set $p \leftarrow p + 1$.

(3) Set $\boldsymbol{x} = \boldsymbol{X}^{r_{\max}}$ and $\boldsymbol{y} = \boldsymbol{Y}^{r_{\max}}$

The *p-th* Phase Algorithm

(1) Let $U \subseteq \mathcal{C}_p$ be the set of un-fully-connected cities, initially set $U \leftarrow \mathcal{C}_p$.

(2) While $U \neq \phi$:

  (a) Find the most *cost-efficient* star (i, p, C') according to Formula (4).

  (b) Open the facility i, if it is not already open, and establish a connection to facility i for all cities in C'.

  (c) Set $f_i \leftarrow 0, U \leftarrow U \setminus C'$.

---

cost of the star and therefore the cost efficiency of a star is actually the average payment of all city members to establish the star. Let $U \subseteq \mathcal{C}_p$ be the set of *not-fully-connected* cities in phase p, $C' \subseteq U$ a set of cities to be connected with facility i. As an open facility can be accessed for free under certain condition, the cost paid to the facility is equal to zero if no new facility have to be opened. Formally, we have

$$f_i^p = \begin{cases} f_i & \text{if a new replica of } i \text{ must be opened;} \\ 0 & \text{otherwise.} \end{cases}$$

Now the dual variables α_j^p and β_{ij}^p can be used to find the most *cost-efficient* star. We use similar interpretation as in [6,8]: α_j^p is the total cost (including the connection cost and the contribution to open facilities) paid by the *p-th* port of city j and β_{ij}^p the contribution received by facility i from the *p-th* port of city j. As such, the most *cost-efficient* star in each iteration of the subroutine can be found in this way: if the dual variables of all unconnected cities are raised simultaneously, the most *cost-efficient* star will be the first star (i, p, C') achieving

$$\sum_{j \in C'} \max(t - c_{ij}, 0) = f_i^p,$$

where $\alpha_j^p = t$ and $\beta_{ij}^p = \max(t - c_{ij}, 0)$. The *p-th* phase algorithm open the most *cost-efficient* star repeatedly until all the cities in $\mathcal{C}_p$ are connected with a facility. Once a city is connected, it is removed from U; in contrast, a facility is never removed, instead it can be reused for free under certain condition. The subroutine is very close to the MMS algorithm [8,6] proposed for the *UFL* problem. The difference is, here we need to ensure the feasibility of the solution by maintaining $X_{ij}^p \leq Y_i^p$ for any $i \in \mathcal{F}, j \in \mathcal{C}, p \in \mathcal{R}$. For the sake of simplicity, we say $y_i^p \leftarrow 1$ means a new facility at site i is opened and $x_{ij}^p \leftarrow 1$ a new connection between city j and facility i is established. In order to maintain the feasibility of a solution, we consider three cases for any $j \in \mathcal{C}_p$:

1. $X_{ij}^{p-1} < Y_i^{p-1}$: In this case, feasibility of the solution maintains if we set $x_{ij}^p \leftarrow 1$. There is no need to open a new facility at site i and $f_i^p = 0$. We say the facility i is eligible to be connected by city j.
2. $X_{ij}^{p-1} = Y_i^{p-1}$ and $y_i^p = 0$: In this case, we must open a new facility at site i in order to establish a new connection for city j and therefore $f_i^p = f_i$. The operating cost is shared between cities in C' and any city contributed to open the facility can be connected.
3. $X_{ij}^{p-1} = Y_i^{p-1}$ and $y_i^p = 1$: This case is the result of the second case. Feasibility of the solution maintains if we set $x_{ij}^p \leftarrow 1$. We do not have to open a new facility and $f_i^p = 0$.

In the above three cases, only the second one involves more than one cities. Suppose each facility has a list of cities and these cities are ordered according to their connection costs to the facility, the most *cost-efficient* star will consist of a facility and a set containing the first k cities in this order, for some k. Therefore the algorithm can be finished efficiently in polynomial time. We use three types of events to process the above cases respectively. Note that $f_i^p = 0$ implies that once a city $j \in U$ has enough credit to pay the connection cost, i. e. $c_{ij} = t$, the star is formed. We restate the *p-th* phase algorithm based on these observations.

Remark 1. Algorithm 1 is independent on the order of city sets being processed, e. g., we can also process cities in order $\mathcal{C}_R, \mathcal{C}_{R-1}..., \mathcal{C}_1$.

4 Analysis

In this section, we assume the function of connection cost forms a metric. In order to show the performance of Algorithm 1, we claim that the maximum cost ratio in each phase is bounded by a constant for any instance of the problem. Formally, let $\mathcal{I}$ be an instance of the *FTFA* problem, we define

$$\lambda_{p,\mathcal{I}} = \max_{i \in \mathcal{F}, C' \subseteq \mathcal{C}_p} \frac{\sum_{j \in C'} \alpha_j^p}{f_i + \sum_{j \in C'} c_{ij}}$$

as the maximum cost ratio with respect to any possible star (i, p, C').

Claim. The cost of solution in each phase is equal to $\sum_{j \in \mathcal{C}_p} \alpha_j^p$ and the maximum cost ratio $\lambda_{p,\mathcal{I}}$ is bounded by a constant λ for any phase $p \in \mathcal{R}$ and any instance $\mathcal{I}$.

We use dual fitting technique here to analysis the approximation factor of the high level algorithm by shrinking the dual solution λ times to make it 'fit' to the original problem.

Algorithm 2. Restatement of the *p-th* Phase Algorithm

(1) At the beginning, all cities are *unconnected*, namely, $t \leftarrow 0, U \leftarrow \mathcal{C}_p$. Assume a city $j \in U$ has r_j ports each with some credit which increases from zero simultaneously with time before the port is connected.

(2) While $U \neq \phi$, increase time t until an instance of *Event-1* or *Event-2* or *Event-3* occurs. If two events occur at the same time, process them in arbitrary order.
  - (a) *Event-1*: A city $j \in U$ has enough credit to be connected with an eligible facility, suppose i, i. e. $t = c_{ij}$ and $X_{ij}^{p-1} < Y_i^{p-1}$. Set $X_{ij}^p \leftarrow X_{ij}^{p-1} + 1$ in this case.
  - (b) *Event-2*: A facility $i \in \mathcal{F}$ receives enough payment from cities in U to open its *p-th* facility, i. e. $\sum_{j\in U} \max(t - c_{ij}, 0) = f_i$. In this case, let $C' = \{j \in U : c_{ij} \leq t\}$, set $Y_i^p \leftarrow Y_i^{p-1} + 1$ and $X_{ij}^p \leftarrow X_{ij}^{p-1} + 1$ for any $j \in C'$.
  - (c) *Event-3*: A city $j \in U$ has enough credit to be connected with a new opened facility, i. e. $t = c_{ij}$. Set $X_{ij}^p \leftarrow X_{ij}^{p-1} + 1$ in this case.
  - (d) For any city $j \in U$, set $\alpha_j^p \leftarrow t$ and remove city j from U if it is connected with a facility in phase p.

Theorem 1. *If the p-th phase algorithm fulfills the claim, Algorithm 1 is a* λ-*approximation algorithm to the FTFA problem.*

Proof. It is not hard to see that X_{ij}^p stops increasing when $p > r_j$ because a city j is included in $\mathcal{C}_p$ only when $p \leq r_j$. Therefore, we have $\forall i \in \mathcal{F}, j \in \mathcal{C} : X_{ij}^R \leq Y_i^R$ because Y_i^p is increasing monotonously (we never close a facility). Feasibility of solution is proved.

In order to show the cost ratio, we compose an extra instance of the problem and its feasible dual solution. Let $\beta_{ij}^p = \max(\alpha_j^p/\lambda - c_{ij}, 0)$ for any $i \in \mathcal{F}, j \in \mathcal{C}, p \in \mathcal{R}$ and $C' = \{j \in \mathcal{C} : \alpha_j^p \geq \lambda c_{ij}\}$, we have $\sum_{j\in\mathcal{C}} \beta_{ij}^p = \sum_{j\in C'} \beta_{ij}^p = \sum_{j\in C'}(\alpha_j^p/\lambda - c_{ij})$. According to the claim, we have

$$\sum_{j\in C'} (\alpha_j^p/\lambda - c_{ij}) \leq f_i$$

for any $i \in \mathcal{F}, p \in \mathcal{R}$. That is, there exist dual variable $\beta_{ij}^p \geq 0$ such that

$$\forall i \in \mathcal{F}, p \in \mathcal{R} : \textstyle\sum_{j\in\mathcal{C}} \beta_{ij}^p \leq f_i \tag{5}$$

$$\text{and } \forall i \in \mathcal{F}, j \in \mathcal{C}, p \in \mathcal{R} : \alpha_j^p/\lambda - \beta_{ij}^p \leq c_{ij}. \tag{6}$$

We note that the above inequalities are exactly the constraints of the dual problem (3). Let OPT be the optimal solution to the primal problem of $\mathcal{I}$. From Inequality (5) and (6), we know $(\alpha/\lambda, \beta)$ is a feasible solution to the dual problem of $\mathcal{I}$. Due to the weak duality theorem, which states that the optimum of the dual maximization problem is no more than the optimum of the primal minimization problem, we have

$$\sum_{j \in \mathcal{C}} \sum_{p=1}^{r_j} \alpha_j^p / \lambda \leq OPT. \tag{7}$$

Considering that the cost of solution in each phase is equal to $\sum_{j \in \mathcal{C}_p} \alpha_j^p$, the theorem is established.

Remark 2. Another interesting observation is that the greedy algorithm for the *UFL* problem, like the MMS algorithm [6,8] or JMS Algorithm [7,8], can also be analysed through decomposing the optimal solution into a group of stars. This is true because any solution to *UFL* can be decomposed into stars without overlap or interference. However in the *FTFA* problem, a city is involved in multiple stars and how to assign their costs to achieve the balance between regarding stars is a nontrivial task. Fortunately, by using the dual fitting technique, we get an alternative approach to reveal the approximation factor.

From the algorithm, we can see that all payments are either for the connection cost or facility cost. Therefore the first part of the claim is complied by the algorithm. Now, we only need to find a proper value of $\lambda \geq 1$ such that for any instance $\mathcal{I}$ of the *FTFA* problem and any phase $p \in \mathcal{R}$

$$\max_{i \in \mathcal{F}, \mathcal{C}' \subseteq \mathcal{C}_p} \frac{\sum_{j \in \mathcal{C}'} \alpha_j^p}{f_i + \sum_{j \in \mathcal{C}'} c_{ij}} \leq \lambda.$$

It is clear that we only need to consider cities with $\alpha_j^p \geq \lambda c_{ij}$. Without loss of generality, suppose there are k such cities in $\mathcal{C}_p$ and further $\alpha_1^p \leq \alpha_2^p \leq ... \leq \alpha_k^p$. We consider some important properties of the *p-th* algorithm in order to find a proper value of λ. We have the following lemmas on the contribution received by a facility and the triangle inequality between connection costs respectively.

Lemma 1. Given an instance $\mathcal{I}$ and phase $p \in \mathcal{R}$, $\sum_{j=h}^{k} \max(\alpha_h^p - c_{ij}, 0) \leq f_i$ for any facility $i \in \mathcal{F}$ and any city $h, 1 \leq h \leq k$.

Proof. omitted.

Lemma 2. Given an instance $\mathcal{I}$ and phase $p \in \mathcal{R}$, $\alpha_j^p \leq \alpha_h^p + c_{ij} + c_{ih}$ for any facility $i \in \mathcal{F}$ and city $h, j, 1 \leq h, j \leq k$.

Proof. omitted.

The above lemmas present some important properties of the *p-th* phase algorithm and the following result turns out that they are enough to bound the ratio of the total cost of a derived solution to that of an optimal solution. Let λ_k be the maximum of the following linear program.

$$\begin{aligned} &\text{maximize} \quad \frac{\sum_{j=1}^{k} \alpha_j}{f + \sum_{j=1}^{k} d_j} \\ &\text{subjected to } \forall 1 \leq j < h \leq k: \ \alpha_h \leq \alpha_j + d_j + d_h \\ &\qquad \forall 1 \leq h \leq k: \ \sum_{j=h}^{k} \max(\alpha_h - d_j, 0) \leq f \\ &\qquad \forall 1 \leq j \leq k: \ \alpha_j, d_j, f \geq 0 \end{aligned} \tag{8}$$

If λ_k has an upperbound with respect to any integer k, we are able to choose this upperbound as the value of λ with respect to the claim. Linear programs like Program (8) are also called factor revealing LPs in the literature [23,6,7,8].

Corollary 1. *Algorithm 1 is a 1.861-approximation algorithm for the FTFA problem.*

Proof. omitted.

5 Discussion

The *FTFA* problem is useful in fault tolerant network design. Suppose the usability required by city j is b_j. If downtime of facilities or links is predicable, e. g. a system where each facility need a fraction of time to 'rest', and the downtime ratio is uniformly b for all facilities, the regarding network design problem can be model as a *FTFA* problem with r_j equals $\lceil b_j/(1-b) \rceil$. If downtime is unpredictable, we can solve the problme by setting r_j to be $\lceil \log_b(1-b_j) \rceil$. In both cases, we can apply the proposed algorithm to solve directly. However, if facilities or links have nonuniform downtime, the constraints on connectivity becomes $\sum_{i \in \mathcal{F}}(1-b_{ij})x_{ij} \geq b_j$ for the deterministic model and $\prod_{i \in \mathcal{F}: x_{ij}=1} b_{ij} \leq 1 - b_j$ for the stochastic model and we leave regarding problems for future research. *FTFA* can also be extended into *Fault-Tolerant k-Facility Allocation* which has an extra constraint on the maximum number of open facilities as showed in our another paper [24].

References

1. Jain, K., Vazirani, V.V.: An approximation algorithm for the fault tolerant metric facility location problem. In: Jansen, K., Khuller, S. (eds.) APPROX 2000. LNCS, vol. 1913, pp. 177–182. Springer, Heidelberg (2000)
2. Swamy, C., Shmoys, D.B.: Fault-tolerant facility location. ACM Trans. Algorithms 4(4), 1–27 (2008)
3. Pitu, B., Mirchandani, R.L.F. (eds.): Discrete Location Theory. John Wiley, New York (1990)
4. Guha, S., Meyerson, A., Munagala, K.: Improved algorithms for fault tolerant facility location. In: SODA 2001: Proceedings of the twelfth annual ACM-SIAM symposium on Discrete algorithms, Philadelphia, PA, USA, Society for Industrial and Applied Mathematics, pp. 636–641 (2001)
5. Guha, S., Meyerson, A., Munagala, K.: A constant factor approximation algorithm for the fault-tolerant facility location problem. J. Algorithms 48(2), 429–440 (2003)
6. Mahdian, M., Markakis, E., Saberi, A., Vazirani, V.: A greedy facility location algorithm analyzed using dual fitting. In: Approximation, Randomization, and Combinatorial Optimization: Algorithms and Techniques, pp. 127–137 (2001)
7. Jain, K., Mahdian, M., Saberi, A.: A new greedy approach for facility location problems. In: STOC 2002: Proceedings of the thiry-fourth annual ACM symposium on Theory of computing, pp. 731–740. ACM, New York (2002)

8. Jain, K., Mahdian, M., Markakis, E., Saberi, A., Vazirani, V.V.: Greedy facility location algorithms analyzed using dual fitting with factor-revealing LP. Journal of the ACM 50(6), 795–824 (2003)
9. Cornuejols, G., Fisher, M.L., Nemhauser, G.L.: Location of bank accounts to optimize float: An analytic study of exact and approximate algorithms. Management Science 23(8), 789–810 (1977)
10. Hochbaum, D.S.: Heuristics for the fixed cost median problem. Mathematical Programming 22(1), 148–162 (1982)
11. Shmoys, D.B., Tardos, E., Aardal, K.: Approximation algorithms for facility location problems. In: Proceedings of the 29th Annual ACM Symposium on Theory of Computing, pp. 265–274 (1997)
12. Lin, J.H., Vitter, J.S.: e-approximations with minimum packing constraint violation. In: STOC 1992: Proceedings of the twenty-fourth annual ACM symposium on Theory of computing, pp. 771–782. ACM, New York (1992)
13. Chudak, F.A.: Improved approximation algorithms for uncapacitated facility location. In: Bixby, R.E., Boyd, E.A., Ríos-Mercado, R.Z. (eds.) IPCO 1998. LNCS, vol. 1412, pp. 180–194. Springer, Heidelberg (1998)
14. Sviridenko, M.: An improved approximation algorithm for the metric uncapacitated facility location problem. In: Proceedings of the 9th International IPCO Conference on Integer Programming and Combinatorial Optimization, London, UK, pp. 240–257. Springer, Heidelberg (2002)
15. Charikar, M., Guha, S.: Improved combinatorial algorithms for facility location problems. SIAM J. Comput. 34(4), 803–824 (2005)
16. Jain, K., Vazirani, V.V.: Approximation algorithms for metric facility location and k-median problems using the primal-dual schema and Lagrangian relaxation. Journal of the ACM 48(2), 274–296 (2001)
17. Jain, K., Vazirani, V.V.: Primal-dual approximation algorithms for metric facility location and k-median problems. In: IEEE Symposium on Foundations of Computer Science, pp. 2–13 (1999)
18. Vazirani, V.V.: Approximation Algorithms. Springer, Berlin (2001)
19. Mahdian, M., Ye, Y., Zhang, J.: A 1.52-approximation algorithm for the uncapacitated facility location problem. In: Jansen, K., Leonardi, S., Vazirani, V.V. (eds.) APPROX 2002. LNCS, vol. 2462, pp. 229–242. Springer, Heidelberg (2002)
20. Byrka, J.: An optimal bifactor approximation algorithm for the metric uncapacited facility location problem. In: Charikar, M., Jansen, K., Reingold, O., Rolim, J.D.P. (eds.) RANDOM 2007 and APPROX 2007. LNCS, vol. 4627, pp. 29–43. Springer, Heidelberg (2007)
21. Chudak, F.A., Shmoys, D.B.: Improved approximation algorithms for the uncapacitated facility location problem. SIAM J. Comput. 33(1), 1–25 (2004)
22. Guha, S., Khuller, S.: Greedy strikes back: Improved facility location algorithms. Journal of Algorithms 31, 228–248 (1999)
23. McEliece, R., Rodemich, E., Rumsey, H., Welch, L.: New upper bounds on the rate of a code via the delsarte-macwilliams inequalities. IEEE Transactions on Information Theory 23(2), 157–166 (1977)
24. Xu, S., Shen, H.: Fault-tolerant k-facility allocation (manuscript)

Tighter Approximation Bounds for Minimum CDS in Wireless Ad Hoc Networks*

Minming Li[1], Peng-Jun Wan[2], and Frances Yao[1]

[1] Department of Computer Science, City University of Hong Kong
minmli@cs.cityu.edu.hk, csfyao@cityu.edu.hk
[2] Department of Computer Science, Illinois Institute of Technology
wan@cs.iit.edu

Abstract. Connected dominating set (CDS) has a wide range of applications in wireless ad hoc networks. A number of approximation algorithms for constructing a small CDS in wireless ad hoc networks have been proposed in the literature. The majority of these algorithms follow a general two-phased approach. The first phase constructs a dominating set, and the second phase selects additional nodes to interconnect the nodes in the dominating set. In the performance analyses of these two-phased algorithms, the relation between the independence number α and the connected domination number γ_c of a unit-disk graph plays the key role. The best-known relation between them is $\alpha \leq 3\frac{2}{3}\gamma_c + 1$. In this paper, we prove that $\alpha \leq 3.4306\gamma_c + 4.8185$. This relation leads to tighter upper bounds on the approximation ratios of two approximation algorithms proposed in the literature.

1 Introduction

Connected dominating set (CDS) has a wide range of applications in wireless ad hoc networks (cf. a recent survey [3] and references therein). Consider a wireless ad hoc network with undirected communication topology $G = (V, E)$. A CDS of G is a subset $U \subset V$ satisfying that each node in $V \setminus U$ is adjacent to at least one node in U and the subgraph of G induced by U is connected. A number of distributed algorithms for constructing a small CDS in wireless ad hoc networks have been proposed in the literature. The majority of these distributed algorithms follow a general two-phased approach [1,2,4,8,10,11,12]. The first phase constructs a dominating set, and the nodes in the dominating set are called dominators. The second phase selects additional nodes, called connectors, which together with the dominators induce a connected topology. The algorithms in [1,2,4,8,10,11] differ in how to select the dominators and connectors. For example, the algorithm in [2] selects the dominators using the Chvatal's greedy

* This work was partially supported by Research Grants Council of Hong Kong SAR under Project No. CityU 122807 and No. CityU 117408, and National Basic Research Program of China Grant 2007CB807900 and 2007CB807901. Peng-Jun Wan was supported in part by National Science Foundation of USA under grants CNS-0831831 and CNS-0916666.

Y. Dong, D.-Z. Du, and O. Ibarra (Eds.): ISAAC 2009, LNCS 5878, pp. 699–709, 2009.

algorithm [5] for Set Cover, the algorithms in [1,10] select an arbitrary maximal independent set (MIS) as the dominating set, and all the algorithms in [4,8,11,12] choose a special MIS with 2-hop separation property as the dominating set.

The approximation ratios of these two-phased algorithms [1,2,4,8,10,11] have been analyzed when the communication topology is a unit-disk graph (UDG). For a wireless ad hoc network in which all nodes lie in a plane and have equal maximum transmission radii normalized to one, its communication topology $G = (V, E)$ is often modelled by a UDG in which there is an edge between two nodes if and only if their Euclidean distance is at most one. Except the algorithms in [2,10] which have logarithmic and linear approximations ratios respectively, all other algorithms in [1,4,8,11,12] have constant approximation ratios. The algorithm in [1] targets at distributed construction of CDS in linear time and linear messages. With this objective, it trades the size of the CDS with the time complexity, and thus its approximation ratio is a large constant (but less than 192). The analyses of the algorithms in [4,8,11,12] rely on the relation between the independence number (the size of a maximum independent set) α and the connected domination number (the size of a minimum connected dominating set) γ_c of a connected UDG G. A loose relation $\alpha \leq 4\gamma_c + 1$ was obtained in [11], which implies an upper bound of 8 on the approximation ratios of both algorithms in [4,11]. A refined relation $\alpha \leq 3.8\gamma_c + 1.2$ was discovered in [13]. With such a refined relation, the upper bound on the approximation ratios of both algorithms in [4,11] was reduced from 8 to 7.6, and an upper bound of $5.8 + \ln 5 \approx 7.41$ on the approximation ratio of the algorithms in [8] was derived (the bound $4.8 + \ln 5 \approx 6.41$ in [8] was incorrect). The best-known relation $\alpha \leq 3\frac{2}{3}\gamma_c + 1$ if G has at least two nodes was recently proven in [12]. As a result, the upper bound on the approximation ratio of the algorithm in [11] was further reduced to $7\frac{1}{3}$ in [12]. Another greedy approximation algorithm was also proposed in [12] and its approximation ratio was proven to be bounded by $6\frac{7}{18}$.

In this paper, we first prove a further improved relation $\alpha \leq 3.4306\gamma_c + 4.8185$ in Section 3. The proof for this bound employs an integrated area and length argument, and involves some other interesting extreme geometric problems which are studied in Section 2. Subsequently in Section 4, we provide tighter analyses of the approximation algorithm in [11] and the other greedy algorithm in [12]. We prove that the approximation ratio of the former algorithm is at most 6.862 and the approximation ratio of the latter algorithm is at most 6.075.

We remark that a recent paper [7] claimed that for any connected UDG G,

$$\alpha \leq 3.453\gamma_c + 8.291.$$

However, as discovered in [12], the proof for a key geometric extreme property underlying such claim was missing, and such proof is far from being apparent or easy. Such property is rigorously proved in Lemma 5. Consequently, the bound claimed in [7] can be treated at most as a conjecture at the time of its publication rather than a proven result.

In the remaining of this section, we introduce some terms and notations. For any point u and any $r > 0$, we use $disk_r(u)$ to denote the closed disk of radius r centered at u, and $circle_r(u)$ to denote the boundary circle of $disk_r(u)$. A path or a polygon is said to be inscribed in a circle if all its vertices lie on the circle. The Lebesgue measure (or area) of a measurable set $A \subset \mathbb{R}^2$ is denoted by $|A|$. The topological boundary of a set $A \subset \mathbb{R}^2$ is denoted by ∂A. For the simplicity of presentation, the line segment between two points u and v and its length are both denoted by uv by slightly abusing the notation, but the actual meaning can be clearly told from the context.

2 Canonical Polygons and Inscribed Polygons

Suppose that s, o and t are three points from the left to the right on a horizontal line with $os = 1$ and $ot = 0.5$. For any pair of points u and v on $circle_{1/\sqrt{3}}(o)$, let $\widehat{uv}$ be the arc in $circle_{1/\sqrt{3}}(o)$ from u to v in the counterclockwise manner. Denote by ϕ the radian of $\widehat{uv}$, and let $k = \lceil \phi/(\pi/3) \rceil$. We construct a path Q of k edges from u to v with all vertices on $\widehat{uv}$ as follows: If ϕ is an integer multiple of $\frac{\pi}{3}$, then all edges of Q are tangent to $circle_{0.5}(o)$; otherwise, all edges except the$\lceil k/2 \rceil$-th edge are tangent to $circle_{1/\sqrt{3}}(o)$ (we remark that in this case, the $\lceil k/2 \rceil$-th edge is disjoint from $circle_{1/\sqrt{3}}(o)$). The path Q is referred to as the *canonical path* inscribed in $circle_{1/\sqrt{3}}(o)$ from u to v.

For any point u which lies on the right side the the vertical line through t, we construct a polygon P as follows: let u_1 and u_2 be the two points on $circle_{1/\sqrt{3}}(o)$ such that the two line segments u_1u and u_2u are tangent to $circle_{1/\sqrt{3}}(o)$ and u_1 is above the line st. Then, P is surrounded by u_1u, u_2u and the canonical path from u_1 to u_2. The polygon P is referred to as the *canonical polygon* of u. The point u is called the *base vertex* of P, and the angle $\theta = \arccos \frac{1}{2ou}$ is called the *base angle* of P. Note that if u is on the ray ot, then P is symmetric with respect to the line ot, and the area of $P \cap disk_{1.5}(s)$ is a function of the base angle θ, which is denoted by $f(\theta)$. In this section, we will derive the explicit expression of $f(\theta)$ and explore some useful properties of the function $f(\theta)$. We will also prove that for any canonical polygon P, $|P \cap disk_{1.5}(s)| \geq f(\theta)$ where θ is the base angle of P.

We first introduce a geometric function g on $[0, \pi]$ defined as follows. For any $\theta \in [0, \pi]$, let v be a point on $circle_{1/2}(o)$ satisfying that $\angle tov = \theta$ and v is above st. Let w be the point on $circle_{1.5}(s)$ satisfying that vw is tangent to $circle_{1/2}(o)$ and w lies to the right of v. Then, $g(\theta)$ is defined to be the area of the region surrounded by arc tw and the three line segments ot, ov and vw (see Figure 1). The next lemma presents the explicit expressions of $g(\theta)$ and its first and second order derivatives.

Due to space limit, we omit some of the proofs in this version.

Lemma 1. *Let* $\beta = \theta - \arccos \frac{1+2\cos\theta}{3}$. *Then,*

$$g(\theta) = \frac{9}{8}\beta - \frac{3}{4}\sin\beta + \frac{\sqrt{6}}{4}\sin\frac{\beta}{2},$$

$$g'(\theta) = \frac{13}{8} - 1.5\cos\beta,$$

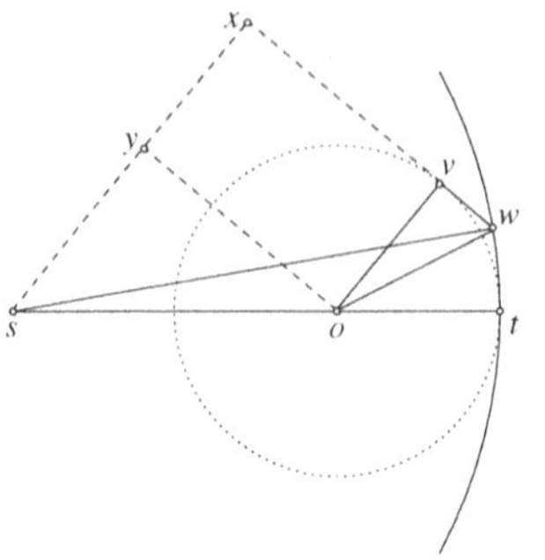

Fig. 1. Calucaltion of $\angle wst$ and $g(\theta)$

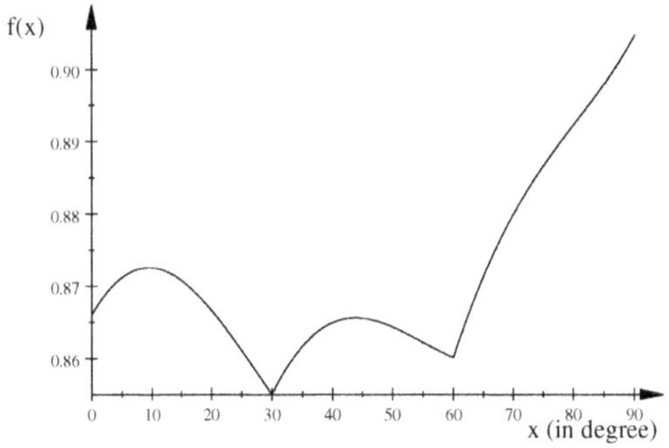

Fig. 2. The curve of f on $[0°, 90°)$

$$g''(\theta) = 1.5\left(1 - \sqrt{1 - \frac{1}{2 + \cos\theta}}\right)\sin\beta.$$

In addition, g is increasing and convex on $[0, \pi]$, while both g' and g'' are increasing on $\left[0, \frac{\pi}{2}\right]$.

It is easy to show that $f(\theta) = 2g(\theta) + h(\theta)$, where

$$h(\theta) = \frac{1}{4\sqrt{3}}\left\lfloor 6 - \frac{\theta}{\frac{\pi}{6}}\right\rfloor + \frac{1}{6}\sin 2\left(\left(6 - \left\lfloor 6 - \frac{\theta}{\frac{\pi}{6}}\right\rfloor\right)\frac{\pi}{6} - \theta\right).$$

Figure 2 is the curve of f on $[0°, 90°)$. We observe and will prove later that f is increasing on $[60°, 90°)$. However, on either of the two intervals $[0°, 30°]$ and $[30°, 60°]$, f is neither monotone, nor concave, and nor convex. Fortunately, on either of these two intervals f has the following weak but still nice *quasi-concave* property: f is said to be quasi-concave on an interval $[a, b] \subset [0°, 90°)$ if for each triple of increasing values $\theta_1, \theta_2, \theta_3$ in $[a, b]$, $f(\theta_2) \geq \min\{f(\theta_1), f(\theta_3)\}$.

Lemma 2. *f is quasi-concave on $\left[0, \frac{\pi}{6}\right]$ and $\left[\frac{\pi}{6}, \frac{\pi}{3}\right]$ respectively, and increasing on $\left[\frac{\pi}{3}, \frac{\pi}{2}\right)$.*

Denote $f\left(\frac{\pi}{6}\right)$ by σ. Then,

$$\sigma = \frac{\sqrt{3}}{6} - \frac{1}{2} + \frac{\sqrt{8+2\sqrt{3}}}{4} + \frac{3\pi}{8} - \frac{9}{4}\arccos\frac{1+\sqrt{3}}{3} \approx 0.855\,053\,28.$$

It is easy to verify that $f(0) = \sqrt{3}/2$ and $f(32^\circ) < f\left(\frac{\pi}{3}\right) < f(34^\circ)$. So, by Lemma 2 we have the following corollary.

Corollary 1. *The minimum of f on the interval $[0^\circ, 90^\circ)$ (respectively, $[32^\circ, 90^\circ)$ and $[34^\circ, 90^\circ)$) is achieved at 30° (respectively, 32° and 60°).*

Finally, we prove the following extreme property of the canonical polygons.

Lemma 3. *For any canonical polygon P with base angle θ, $|P \cap disk_{1.5}(s)| \geq f(\theta)$.*

Next, we prove the following lemma about inscribed polygons.

Lemma 4. *Suppose that P is a polygon inscribed in $circle_{1/\sqrt{3}}(o)$ satisfying that $disk_{0.5}(o) \subseteq P$. Then,*

$$|P| > \sqrt{3}/2,$$
$$|P \cap disk_{1.5}(s)| \geq \sigma.$$

3 Independence Number vs. Connected Domination Number

In this section, we present an improved upper bound on the independence number in terms of the connected domination number.

Theorem 1. *Let α and γ_c be the independence number and connected domination number of a connected UDG G. Then,*

$$\alpha \leq 3.4306\gamma_c + 4.8185.$$

We prove the above theorem by an integrated area and length argument. Let U be a minimum CDS of G, and define

$$\Omega = \bigcup_{u \in U} disk_{1.5}(u).$$

Consider a maximum independent set I of G. We construct the Voronoi diagram defined by I. For each $o \in I$, we use $Vor(o)$ to denote its Voronoi cell and call the set $Vor(o) \cap \Omega$ as the *truncated Voronoi cell* of o. Clearly, $|\Omega|$ is the total area of truncated Voronoi cells of all nodes in I. We partition I into two subsets I_1 and I_2 defined by

$$I_1 = \left\{o \in I : disk_{1/\sqrt{3}}(o) \subset \Omega\right\},$$
$$I_2 = I \setminus I_1.$$

Denote by α_1 and α_2 the size of I_1 and I_2 respectively. The next lemma provides a lower bound on each truncated Voronoi cell.

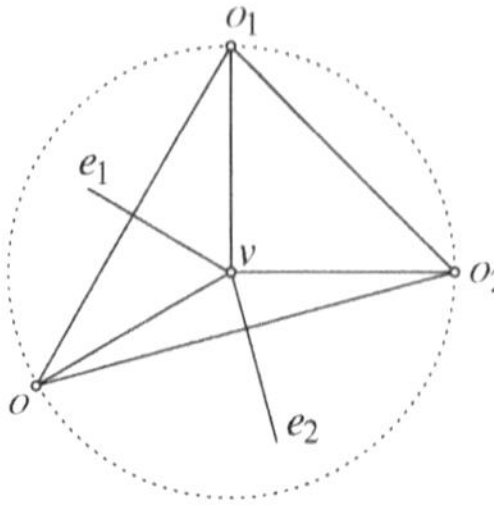

Fig. 3. Any vertex of $Vor\,(o)$ is apart from o by at least $1/\sqrt{3}$

Lemma 5. *For each o in I_1 (respectively, I_2), the area of its truncated Voronoi cell is at least $\sqrt{3}/2$ (respectively, σ).*

Proof. Since the pairwise distances of the points in I are at least one, the distance between o and each side of $Vor\,(o)$ is at least 0.5 and consequently $disk_{0.5}\,(o) \subset Vor\,(o)$. Next, we show that no vertex of $Vor\,(o)$ is inside $disk_{1/\sqrt{3}}\,(o)$. Let v be a vertex of $Vor\,(o)$, and e_1 and e_2 be the two sides of $Vor\,(o)$ incident to v (see figure 3). Let o_1 (respectively, o_2) be the point which is symmetric to o with respect to e_1 (respectively, e_2). Then, both o_1 and o_2 belong to I, and hence the three sides of $\triangle oo_1o_2$ are all at least 2. Clearly, v is the center of $\triangle oo_1o_2$. Since at least one of the three central angles of $\triangle oo_1o_2$ is at most 120°, the circumscribing radius of $\triangle oo_1o_2$ is at least $1/\sqrt{3}$. Thus, $ov \geq 1/\sqrt{3}$.

Let s be the node in the MCDS U closest to o. Then, $o \in disk_1\,(s)$. If $disk_{1/\sqrt{3}}\,(o) \subseteq Vor\,(o)$, then $|Vor\,(o) \cap \Omega| \geq \left|disk_{1/\sqrt{3}}\,(o) \cap disk_{1.5}\,(s)\right| > \sqrt{3}/2$. So, we assume $disk_{1/\sqrt{3}}\,(o)$ is not fully contained in $Vor\,(o)$. Then $Vor\,(o)$ intersects $circle_{1/\sqrt{3}}(o)$. We construct a polygon $P \subseteq Vor\,(o)$ satisfying that P is inscribed in $circle_{1/\sqrt{3}}(o)$ and $disk_{0.5}\,(o) \subseteq P \subseteq Vor\,(o)$. Let Q be the sequence of intersecting points between $Vor\,(o)$ and $circle_{1/\sqrt{3}}\,(o)$ in the counterclockwise order. For each pair of successive u and v in Q, if $\angle uov \leq \frac{\pi}{3}$, we add to P a side between u and v; otherwise, we add to P a path inscribed in the arc from u to v satisfying that each edge in this path is either tangent to or disjoint from $circle_{1/\sqrt{3}}\,(o)$ (see Figure 4). The resulting polygon P meets the requirement. By Lemma 4, $|P| \geq \sqrt{3}/2$.

If $o \in I_1$, then

$$P \subseteq Vor\,(o) \cap disk_{1/\sqrt{3}}\,(o) \subseteq Vor\,(o) \cap \Omega,$$

hence

$$|Vor\,(o) \cap \Omega| \geq |P| \geq \sqrt{3}/2.$$

Now, we assume that $o \in I_2$. Note that $|P \cap disk_{1.5}\,(s)|$ grows when moving o away from s along a fixed radius of $disk_{1.5}\,(s)$. By Lemma 4, $|P \cap disk_{1.5}\,(s)| \geq \sigma$. Since

$$P \cap disk_{1.5}\,(s) \subseteq Vor\,(o) \cap \Omega,$$

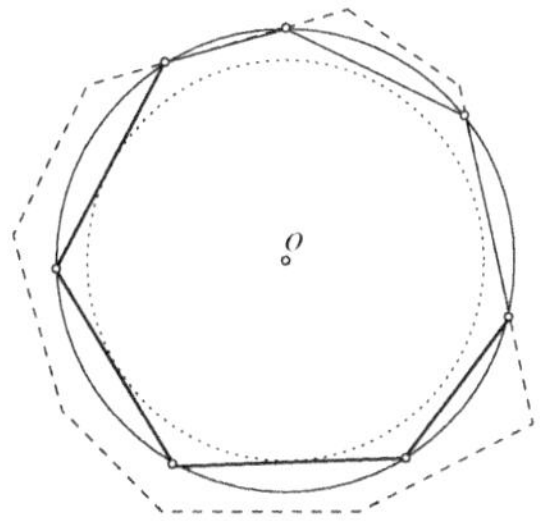

Fig. 4. Inserting operations

we have

$$|Vor\,(o) \cap \Omega| \geq |P \cap disk_{1.5}\,(s)| \geq \sigma.$$

We define

$$\Omega' = \bigcup_{v \in U} disk_{1.5-1/\sqrt{3}}\,(v)\,.$$

The next lemma gives an upper bound on the length of $\partial\Omega'$.

Lemma 6. *The length of* $\partial\Omega'$ *is at most* $2\left(1 - 1/\sqrt{3}\right)\alpha_2$.

By Lemma 5,

$$|\Omega| \geq \frac{\sqrt{3}}{2}\alpha_1 + \sigma\alpha_2 = \frac{\sqrt{3}}{2}\alpha - \left(\frac{\sqrt{3}}{2} - \sigma\right)\alpha_2,$$

which implies

$$\alpha \leq \frac{|\Omega|}{\frac{\sqrt{3}}{2}} + \left(1 - \frac{\sigma}{\frac{\sqrt{3}}{2}}\right)\alpha_2. \tag{1}$$

It is easy to prove by induction on γ_c that

$$|\Omega| \leq \frac{9}{2}\left((\gamma_c - 1)\left(\arcsin\frac{1}{3} + \frac{\sqrt{8}}{9}\right) + \frac{\pi}{2}\right), \tag{2}$$

and the length of of $\partial\Omega'$ is at most

$$2\left(3 - \frac{2}{\sqrt{3}}\right)\left((\gamma_c - 1)\arcsin\frac{1}{3 - \frac{2}{\sqrt{3}}} + \frac{\pi}{2}\right).$$

By Lemma 6,

$$\begin{aligned}\alpha_2 &\leq \frac{2\left(3 \quad \frac{2}{\sqrt{3}}\right)\left((\gamma_c - 1)\arcsin\frac{1}{3 - \frac{2}{\sqrt{3}}} + \frac{\pi}{2}\right)}{2\left(1 - \frac{1}{\sqrt{3}}\right)} \\ &= \frac{\sqrt{3} + 7}{2}\left((\gamma_c - 1)\arcsin\frac{1}{3 - \frac{2}{\sqrt{3}}} + \frac{\pi}{2}\right).\end{aligned} \tag{3}$$

The three inequalities (1), (2) and (3) imply altogether that α is at most

$$(\gamma_c - 1)\left(\begin{array}{c}\sqrt{27}\left(\arcsin\frac{1}{3}+\frac{\sqrt{8}}{9}\right)\\ +\frac{\left(1-\frac{\sigma}{\frac{\sqrt{3}}{2}}\right)(\sqrt{3}+7)}{2}\arcsin\frac{1}{3-\frac{2}{\sqrt{3}}}\end{array}\right)+\frac{\pi}{2}\left(\sqrt{27}+\left(1-\frac{\sigma}{\frac{\sqrt{3}}{2}}\right)\frac{\sqrt{3}+7}{2}\right)$$

$$\approx 3.4305176\gamma_c + 4.8184688.$$

Thus, Theorem 1 follows.

4 Tighter Approximation Ratios

In this section, we derive tighter bounds on the approximation ratio of the distributed algorithm proposed in [11] and the other greedy algorithm proposed in [12]. For the convenience of presentation, we call them **WAF** and **WWY** respectively. Let $G = (V, E)$ be a unit-disk graph. We denote by α and γ_c the independence number and connected domination number of G respectively. For any finite set S, we use $|S|$ to denote the cardinality of S.

The CDS produced by the algorithm **WAF** consists of a maximal independent set I and a set C of connectors. Specifically, let T be an arbitrary rooted spanning tree of G. The set I is selected in the first-fit manner in the breadth-first-search ordering in T. Let s be the neighbor of the root of T which is adjacent to the largest number of nodes in I. Then, C consists of s and the parents (in T) of the nodes in $I \setminus I(s)$. It was proved in [11] that $I \cup C$ is a CDS and $|I \cup C| \leq 8\gamma_c - 1$. Later on, two progressively improved tighter bounds $7.6\gamma_c + 1.4$ and $7\frac{1}{3}\gamma_c$ were obtained in [13] and [12] respectively. The next theorem further improves the bound on $|I \cup C|$.

Theorem 2. *The CDS produced by the algorithm* ***WAF*** *has size at most* $6.862\gamma_c + 8.637$.

Proof. Let I and C be the set of nodes selected by the algorithm **WAF** in the first phase and the second phase respectively. Since $|C| \leq |I| - 1$, we have

$$|I \cup C| \leq 2|I| - 1 \leq 2(3.4306\gamma_c + 4.8185) - 1 \leq 6.862\gamma_c + 8.637.$$

So, the theorem follows.

In the next, we study the algorithm **WWY**. The first phase of this algorithm is the same as the algorithm **WAF**, and we let I be the selected maximal independent set. But the second phase selects the connectors in a more economic way. For any subset $U \subseteq V \setminus I$, let $q(U)$ be the number of connected components in $G[I \cup U]$. For any $U \subseteq V \setminus I$ and any $w \in V \setminus I$, we define

$$\triangle_w q(U) = q(U) - q(U \cup \{x\}).$$

The value $\triangle_w q(U)$ is referred to as the *gain* of w with respect to U. The following lemma was proved in [12].

Lemma 7. *Suppose that $q(U) > 1$ for some $U \subseteq V \setminus I$. Then, there exists a $w \in V \setminus (I \cup U)$ such that $\triangle_w q(U) \geq \max\{1, \lceil q(U)/\gamma_c \rceil - 1\}$.*

The second phase of the algorithm **WWY** runs as follows. We use C to denote the sequence of selected connectors. Initially C is empty. While $q(C) > 1$, choose a node $w \in V \setminus (I \cup C)$ with *maximum* gain with respect to C and add w to C. When $q(C) = 1$, then $I \cup C$ is a CDS. It was proved in [12] that $|I \cup C| \leq 6\frac{7}{18}\gamma_c$. We derive a tighter bound on the output CDS in the theorem below.

Theorem 3. *The CDS produced by the algorithm* ***WWY*** *has size at most $6.075\gamma_c + 5.425$.*

Proof. Let I and C be the set of nodes selected by the algorithm **WWY** in the first phase and the second phase respectively. If $\gamma_c = 1$, then $|I| \leq 5$ and $|C| \leq 1$, hence $|I \cup C| \leq 6$. Thus, the theorem holds trivially if $\gamma_c = 1$. If $|I| \leq 3\gamma_c + 2$, then $|I \cup C| \leq 2|I| - 1 \leq 6\gamma_c + 3$, and the theorem also holds. From now on, we assume that $\gamma_c \geq 2$ and $|I| > 3\gamma_c + 2$.

We break C into three contiguous (and possibly empty) subsequences C_1, C_2 and C_3 as follows. C_1 is the shortest prefix of C satisfying that $q(C_1) \leq 3\gamma_c + 2$, and $C_1 \cup C_2$ is the shortest prefix of C satisfying that $q(C_1 \cup C_2) \leq 2\gamma_c + 1$. We can prove that

$$|C_1| \leq \begin{cases} \frac{|I|}{3} - \gamma_c & \text{if } q(C_1) \leq 3\gamma_c + 1, \\ \frac{|I|-2}{3} - \gamma_c & \text{if } q(C_1) = 3\gamma_c + 2; \end{cases}$$

$$|C_2| \leq \begin{cases} \frac{\gamma_c}{2} & \text{if } q(C_1) \leq 3\gamma_c + 1, \\ \frac{\gamma_c+1}{2} & \text{if } q(C_1) = 3\gamma_c + 2; \end{cases}$$

$$|C_3| \leq 2\gamma_c - 1.$$

From the first two inequalities, we have

$$|C_1 \cup C_2| \leq \frac{|I|}{3} - \frac{\gamma_c}{2}.$$

Using the third inequality, we have

$$|C| \leq \frac{|I|}{3} - \frac{\gamma_c}{2} + 2\gamma_c - 1 = \frac{|I|}{3} + \frac{3}{2}\gamma_c - 1.$$

So,

$$\begin{aligned} |I \cup C| &\leq \frac{4|I|}{3} + \frac{3}{2}\gamma_c - 1 \\ &\leq \frac{4}{3}(3.4306\gamma_c + 4.8185) + \frac{3}{2}\gamma_c - 1 \\ &\leq 6.075\gamma_c + 5.425. \end{aligned}$$

Thus, the theorem follows.

5 Discussions

In this paper, we obtained a tighter relation between the independence number and connected domination number of a connected UDG. We actually proved the following stronger result on packing. Let V be a set of n nodes of a connected dominating set, and Γ be the unions of unit-disks centered at V. Then, we can pack in Γ at most $3.4306n + 4.8185$ points whose pairwise distances are greater than or equal to one. We'd like to emphasize that here we allow two points packed in Γ to have distance equal to one. On the other hand, a packing of $3n + 3$ points in Γ whose pairwise distances are greater than one was presented in [12]. It was also conjectured $3n+3$ is the exact bound. Thus, there is still a gap between the bound $3.4306n + 4.8185$ derived in this paper and the conjectured bound $3n + 3$.

References

1. Alzoubi, K.M., Wan, P.-J., Frieder, O.: Message-Optimal Connected Dominating Sets in Mobile Ad Hoc Networks. In: ACM Mobihoc (2002)
2. Bharghavan, V., Das, B.: Routing in Ad Hoc Networks Using Minimum Connected Dominating Sets. In: IEEE ICC (1997)
3. Blum, J., Ding, M., Cheng, X.: Applications of Connected Dominating Sets in Wireless Networks. In: Du, D.-Z., Pardalos, P. (eds.) Handbook of Combinatorial Optimization, pp. 329–369. Kluwer Academic Publisher, Dordrecht (2004)
4. Cadei, M., Cheng, X., Du, D.-Z.: Connected Domination in Ad Hoc Wireless Networks. In: Proc. 6th International Conference on Computer Science and Informatics (2002)
5. Chvátal, V.: A greedy heuristic for the set-covering problem. Mathematics of Operation Research 4(3), 233–235 (1979)
6. Clark, B.N., Colbourn, C.J., Johnson, D.S.: Unit Disk Graphs. Discrete Mathematics 86, 165–177 (1990)
7. Funke, S., Kesselman, A., Meyer, U., Segal, M.: A simple improved distributed algorithm for minimum CDS in unit disk graphs. ACM Transactions on Sensor Networks 2(3), 444–453 (2006)
8. Li, Y.S., Thai, M.T., Wang, F., Yi, C.-W., Wan, P.-J., Du, D.-Z.: On Greedy Construction of Connected Dominating Sets in Wireless Networks. Journal on Wireless Communications and Mobile Computing 5(8), 927–932 (2005)
9. Lin, F., Yang, X. (eds.): Geometric Measure Theory: An Introduction. International Press (2003)
10. Stojmenovic, I., Seddigh, M., Zunic, J.: Dominating sets and neighbor elimination based broadcasting algorithms in wireless networks. In: Proc. IEEE Hawaii Int. Conf. on System Sciences (January 2001)
11. Wan, P.-J., Alzoubi, K.M., Frieder, O.: Distributed Construction of Connected Dominating Set in Wireless Ad Hoc Network. ACM/Springer Mobile Networks and Applications 9(2), 141–149 (2004); A preliminary version of this paper appeared in IEEE INFOCOM (2002)

12. Wan, P.-J., Wang, L., Yao, F.: Two-Phased Approximation Algorithms for Minimum CDS in Wireless Ad Hoc Networks. In: IEEE ICDCS 2008, pp. 337–344 (2008)
13. Wu, W., Du, H., Jia, X., Li, Y., Huang, S.C.-H.: Minimum connected dominating sets and maximal independent sets in unit disk graphs. Theoretical Computer Science 352(1-3), 1–7 (2006)

Maximal Strip Recovery Problem with Gaps: Hardness and Approximation Algorithms

Laurent Bulteau[1,2], Guillaume Fertin[2], and Irena Rusu[2]

[1] École Normale Supérieure, 45 rue d'Ulm, 75000 Paris, France
[2] Laboratoire d'Informatique de Nantes-Atlantique (LINA), UMR CNRS 6241
Université de Nantes, 2 rue de la Houssinière, 44322 Nantes Cedex 3 - France
Laurent.Bulteau@ens.fr, {Guillaume.Fertin,Irena.Rusu}@univ-nantes.fr

Abstract. Given two comparative maps, that is two sequences of markers each representing a genome, the Maximal Strip Recovery problem (MSR) asks to extract a largest sequence of markers from each map such that the two extracted sequences are decomposable into non-overlapping strips (or synteny blocks). This aims at defining a robust set of synteny blocks between different species, which is a key to understand the evolution process since their last common ancestor. In this paper, we add a fundamental constraint to the initial problem, which expresses the biologically sustained need to bound the number of intermediate (non-selected) markers between two consecutive markers in a strip. We therefore introduce the problem δ-gap-MSR, where δ is a (usually small) non-negative integer that upper bounds the number of non-selected markers between two consecutive markers in a strip. Depending on the nature of the comparative maps (i.e., with or without duplicates), we show that δ-gap-MSR is NP-complete for any $\delta \geq 1$, and even APX-hard for any $\delta \geq 2$. We also provide two approximation algorithms, with ratio 1.8 for $\delta = 1$, and ratio 4 for $\delta \geq 2$.

Keywords: algorithmic complexity, approximation algorithms, comparative maps, genome comparison, synteny blocks.

1 Introduction

In comparative genomics, finding *synteny blocks* (that is, regions with similar content and gene order) of two genomes is a crucial task, as the decomposition of genomes into synteny blocks allows to estimate the nature of genome rearrangement events that hold during the evolution process since the last common ancestor of the genomes.

In addition to the difficulty to define a synteny block precisely, another difficulty is introduced by the quality of genome annotation. Zheng et al. [9] make a list of possible errors and ambiguities introduced by the mapping technology, which is used to obtain a representation of a genome as a sequence of *markers*, called a *genomic map*. Each marker represents a small, specific element which has been identified on the genome, at a specific position which is the *marker's*

Y. Dong, D.-Z. Du, and O. Ibarra (Eds.): ISAAC 2009, LNCS 5878, pp. 710–719, 2009.

position. Comparing two genomes is then possible using their genomic maps, assuming that the pairs of identical markers on the two genomes are known (the maps are then called *comparative maps*). Comparative maps are less precise than genome sequences (either as DNA sequences or as sequences of genes), but still allow the identification of synteny blocks.

The problem that needs to be solved when no error occurs is the following: *Given two comparative maps, decompose them into non-intersecting synteny blocks.* In case of errors or ambiguities, Zheng et al. [9] propose to switch to the following problem: *Given two comparative maps, find a longest (possibly non-contiguous) subsequence of markers in each comparative map, such that the subsequences are decomposable into non-intersecting synteny blocks.* The idea behind this maximization problem is that true synteny is possibly interrupted by erroneous or ambiguous markers, which should be discarded before searching for synteny blocks.

The problem, called MAXIMAL STRIP RECOVERY (MSR), is obtained from this maximization problem using comparative maps with signed, but not duplicated, markers, and a specific definition of synteny blocks. Synteny blocks are defined as *strips*, which are contiguous sequences of *at least two* markers that occur on each genome either in the same order, or in reverse order and with a reversed sign. Zheng et al [9] and Choi et al. [4] propose two heuristics to solve the MSR problem. Chen et al. [3] devise a 4-approximation algorithm for it, propose its extension, called MSR-d, to an arbitrary number $d \geq 2$ of genomes and show that MSR-3 is NP-complete. The NP-completeness of MSR (or equivalently MSR-2) is a result obtained by Wang et al. [8], who also propose FPT algorithms for MSR-d (with arbitrary d) and MSR-DU, the variant of MSR where duplicated markers are allowed in the maps and in different synteny blocks.

The MSR problem takes into account the need to keep as much of the data as possible from the initial comparative maps and the need to have conflict-free synteny blocks. However, it is too permissive as it allows two consecutive elements from one strip to be separated by an arbitrary long gap (in terms of intermediate markers) on the initial comparative maps, and possibly to be very close on one map and very far from each other on the other. As the discarded elements are supposed to be errors and ambiguities (which are rather the exception than the rule), and the elements kept in the subsequences are supposed to be the safe information (which is the major part of the comparative information), it follows that a safe synteny block should not allow arbitrarily long gaps.

We therefore introduce and study in this paper the δ-gap-MSR problem, a restriction of the MSR problem where the allowed gaps along the comparative maps between two consecutive elements in a strip are upper bounded by parameter δ, where δ is a given (usually small) non-negative integer. We investigate the algorithmic complexity of δ-gap-MSR depending on the allowed multiplicity for a marker and prove the results given in Table 1. For the NP-complete or APX-hard cases, we provide two approximation algorithms, whose approximation ratios are given in Table 2.

Table 1. Complexity of variants of MSR

Problem	Without duplicates	With duplicates (-DU variant)
0-gap-MSR	P (Section 4.2)	?
1-gap-MSR	NP-complete (Section 3.1)	NP-complete (Section 3.1)
δ-gap-MSR ($\delta \geq 2$)	APX-complete (Section 3.2)	APX-complete (Section 3.2)
MSR	NP-complete [8]	NP-complete [8]

Table 2. Best approximation ratios of variants of MSR

Problem	Without duplicates	With duplicates (-DU variant)
0-gap-MSR	-	4 (Section 4.2)
1-gap-MSR	1.8 (Section 4.1)	4 (Section 4.2)
δ-gap-MSR ($\delta \geq 2$)	4 (Section 4.2)	4 (Section 4.2)
MSR	4 [3]	4 [3]

The organization of the paper is as follows. In Section 2, we introduce some notations, and we define formally MSR, MSR-DU, δ-gap-MSR and δ-gap-MSR-DU. We prove in Section 3 the hardness results (NP-completeness for $\delta = 1$ in Section 3.1, APX-completeness for $\delta \geq 2$ in Section 3.2). We then give approximation algorithms in Section 4: a 1.8-approximation for 1-gap-MSR in Section 4.1, and a general 4-approximation in Section 4.2. Due to space constraints, most of the proofs are omitted from this paper.

2 Notations and Definitions

A *comparative map* $\mathcal{M}$ is a sequence of signed integers, where the absolute value of each integer represents a specific marker, and the sign represents the orientation of the marker on the chromosome. A marker may appear several times in a comparative map, possibly with different orientations: in this case, we say that the comparative map $\mathcal{M}$ has *duplicates* (the presence of duplicates is useful if we do not want to distinguish paralogs in the comparative map). A *sequence* $\mathcal{M}$ is denoted $\mathcal{M}=\langle m_1, m_2, \ldots, m_l\rangle$, and its ith element m_i is (also) denoted $\mathcal{M}[i]$.

A *subsequence* σ of $\mathcal{M}$ is a sequence $\langle\sigma_1, \ldots, \sigma_h\rangle$ of markers from $\mathcal{M}$ with $h \geq 2$ and positions $i_1 < i_2 < \ldots < i_h$ respectively on $\mathcal{M}$. The vector $(i_1, \ldots, i_h)$ is denoted $idx(\sigma, \mathcal{M})$. The *gap* of σ in $\mathcal{M}$ is $\max\{i_{k+1} - i_k - 1 \;:\; 1 \leq k < h\}$, its *length* $|\sigma|$ is h. Two subsequences σ and τ are *non-overlapping* in $\mathcal{M}$ if one appears strictly before the other (i.e., if the last element of $idx(\sigma, \mathcal{M})$ is strictly smaller than the first element of $idx(\tau, \mathcal{M})$ or vice-versa). The *reversed opposite* of $\langle\sigma_1, \ldots, \sigma_h\rangle$ is $\langle -\sigma_h, -\sigma_{h-1}, \ldots, -\sigma_1\rangle$.

Given two comparative maps $\mathcal{M}_1$ and $\mathcal{M}_2$, a *prestrip* is a subsequence σ of $\mathcal{M}_1$ such that either σ or its reversed opposite is a subsequence of $\mathcal{M}_2$, and such that the markers in σ are pairwise different. The *gap* of a prestrip is the maximum of the gaps of the two corresponding subsequences in $\mathcal{M}_1$ and $\mathcal{M}_2$. Two prestrips are *non-overlapping* if the corresponding subsequences are non-overlapping, both in $\mathcal{M}_1$ and $\mathcal{M}_2$. A *strip* is a prestrip with gap 0. Strips represent synteny blocks between two comparative maps. A prestrip can also be seen as a synteny bock, but only if we consider that there is noise in the comparative maps (false markers appear between two consecutive markers of the "true" synteny block). A set of prestrips $\mathcal{S}$ is said to be *feasible* if it contains pairwise non-overlapping prestrips, and we write $||\mathcal{S}||$ for its *total size*: $||\mathcal{S}|| = \sum_{\sigma \in \mathcal{S}} |\sigma|$.

We finally define some notions of graph theory: a graph $G = (V, E)$ is *cubic* if every vertex $u \in V$ has degree exactly 3. A set $X \subset V$ is said to be *independent* if for every edge $(u, v) \in E$, $u \notin X$ or $v \notin X$. The cardinality of a maximum independent set of G is written $\alpha(G)$.

The problems MSR (for MAXIMAL STRIP RECOVERY, see [9]) and MSR-DU [3] are defined, in their decision formulation, as follows:

Problem: MSR
Input: Two comparative maps $\mathcal{M}_1$ and $\mathcal{M}_2$ without duplicates, $k \in \mathbb{N}$.
Question: Is there a feasible set $\mathcal{S}$ of prestrips of $\mathcal{M}_1$ and $\mathcal{M}_2$, s.t. $||\mathcal{S}|| \geq k$?

Problem: MSR-DU
Input: Two comparative maps $\mathcal{M}_1$ and $\mathcal{M}_2$ (possibly with duplicates), $k \in \mathbb{N}$.
Question: Is there a feasible set $\mathcal{S}$ of prestrips of $\mathcal{M}_1$ and $\mathcal{M}_2$, s.t. $||\mathcal{S}|| \geq k$?

The idea behind both those problems is that, if we find a set of compatible prestrips with maximum total size, the elements appearing in no prestrip are considered as noise: we can remove them to "clean" the data. Indeed, once those elements are removed, the comparative maps can be partitioned into common strips, i.e. we have decomposed both genomes into synteny blocks with the same set of blocks in both genomes. Heuristics for the first problem have been given in [9,4]. They have been improved in [3] into a 4-approximation algorithm. Finally, those problems have been proved NP-complete in [8], where an FPT algorithm is also provided.

The variant we introduce, δ-gap-MSR, takes into account the fact that it is unlikely that long sequences of markers can appear only from noise and errors. If a large number of elements are inserted between two consecutive elements of a prestrip (thus, if it has a large gap), then they are not errors, and the prestrip cannot be considered a synteny block of the original genomes. Thus we define the following two problems:

Problem: δ-gap-MSR
Input: Two comparative maps $\mathcal{M}_1$ and $\mathcal{M}_2$ without duplicates, $k \in \mathbb{N}$.
Question: Is there a feasible set $\mathcal{S}$ of prestrips of $\mathcal{M}_1$ and $\mathcal{M}_2$, such that every $\sigma \in \mathcal{S}$ has gap at most δ, and $||\mathcal{S}|| \geq k$?

Problem: δ-gap-MSR-DU
Input: Two comparative maps $\mathcal{M}_1$ and $\mathcal{M}_2$ (possibly with duplicates), $k \in \mathbb{N}$.

Question: Is there a feasible set $\mathcal{S}$ of prestrips of $\mathcal{M}_1$ and $\mathcal{M}_2$, such that every $\sigma \in \mathcal{S}$ has gap at most δ, and $||\mathcal{S}|| \geq k$?

With the gap constraint we introduce, we keep only prestrips which are nearly contiguous, while tolerating some noise in the input data. Note that those problems are defined for uni-chromosomal genomes. However, algorithms can easily be adapted to handle multi-chromosomal instances.

3 Hardness Results

3.1 NP-Hardness of 1-gap-MSR

In this section, we prove the following theorem.

Theorem 1. *1-gap-MSR and 1-gap-MSR-DU are NP-hard.*

Note that we need to consider only 1-gap-MSR (without duplicates) since NP-hardness of 1-gap-MSR-DU directly follows from NP-hardness of 1-gap-MSR.

The proof uses a reduction from a variant of MAXIMUM INDEPENDENT SET, 3-colored-MIS, which is defined below. A *3-edge-coloring* (also known as Tait Coloring) of a cubic graph $G = (V, E)$ is a partition of its edges in three classes $E = E^{\text{A}} \cup E^{\text{B}} \cup E^{\text{C}}$ such that if two edges $e_1, e_2 \in E$ are incident to a common vertex, they belong to different classes.

Problem: 3-colored-MIS
Input: A cubic graph $G = (V, E)$, a 3-edge-coloring $(E^{\text{A}}, E^{\text{B}}, E^{\text{C}})$ of G, an integer k.
Question: Is $\alpha(G) \geq k$?

Lemma 2. *3-colored-MIS is NP-hard.*

Starting from any instance of 3-colored-MIS, we construct two comparative maps as follows. First, we assign a list of 4 positive integers (or 4 "markers") to each vertex $u \in V$: they are denoted y_u^{A1}, y_u^{A2}, y_u^{B1} and y_u^{B2}. We also assign a list of 10 integers $x_{uv}^1, \ldots, x_{uv}^{10}$ to each edge $(u, v) \in E^{\text{C}}$, in such a way that no integer appears in two different lists. We will also use peg markers: written with the symbol $\times$, they are integers appearing only once, either in $\mathcal{M}_1$ or in $\mathcal{M}_2$ (and thus cannot belong to any prestrip).

We construct the comparative maps with the following iterative procedure. Suppose we have arbitrarily ordered the vertices in V. In that case:

1. For all $(u, v) \in E^{\text{A}}$ such that $u < v$, add $\langle y_u^{\text{A1}}, y_v^{\text{A1}}, y_u^{\text{A2}}, y_v^{\text{A2}}, \times, \times \rangle$ to $\mathcal{M}_1$.
2. For all $(u, v) \in E^{\text{B}}$ such that $u < v$, add $\langle y_u^{\text{B1}}, y_v^{\text{B1}}, y_u^{\text{B2}}, y_v^{\text{B2}}, \times, \times \rangle$ to $\mathcal{M}_2$.
3. For all $(u, v) \in E^{\text{C}}$ such that $u < v$, add $\Gamma_1(u, v)$ to $\mathcal{M}_1$, $\Gamma_2(u, v)$ to $\mathcal{M}_2$, where Γ_1 and Γ_2 are defined as:

$$\begin{aligned}\Gamma_1(u,v) = \langle\, & x_{uv}^1, x_{uv}^5, x_{uv}^2, x_{uv}^6, x_{uv}^3, x_{uv}^7, x_{uv}^4, \times, \times,\\ & y_u^{\text{B1}}, x_{uv}^8, y_u^{\text{B2}}, x_{uv}^9, y_v^{\text{B1}}, x_{uv}^{10}, y_v^{\text{B2}}, \times, \times \rangle;\\ \Gamma_2(u,v) = \langle\, & x_{uv}^1, x_{uv}^8, x_{uv}^2, x_{uv}^9, x_{uv}^3, x_{uv}^{10}, x_{uv}^4, \times, \times,\\ & y_u^{\text{A1}}, x_{uv}^5, y_u^{\text{A2}}, x_{uv}^6, y_v^{\text{A1}}, x_{uv}^7, y_v^{\text{A2}}, \times, \times \rangle.\end{aligned}$$

Property 3. *Let $G = (V, E)$ be an n-vertex cubic graph with a 3-edge-coloring, and let $\mathcal{M}_1$ and $\mathcal{M}_2$ be the two comparative maps obtained by the construction defined above. Then the optimal value of* 1*-gap-MSR over* $(\mathcal{M}_1, \mathcal{M}_2)$ *equals* $4n + 2\alpha(G)$.

Proof (of Theorem 1). The above property directly implies that our construction (which can clearly be done in polynomial time) leads to a reduction from 3-colored-MIS to 1-gap-MSR, which proves Theorem 1. □

3.2 δ-gap-MSR and δ-gap-MSR-DU Are APX-hard

In this section, we prove the following theorem.

Theorem 4. *δ-gap-MSR and δ-gap-MSR-DU are APX-hard for any $\delta \geq 2$.*

As in the previous section, we note that we need to consider only δ-gap-MSR (without duplicates) since APX-hardness of δ-gap-MSR-DU directly follows from APX-hardness of δ-gap-MSR. For this, we use an L-reduction [7] from the variant of Maximum Independent Set restricted to cubic graphs, that we call 3-MIS here. Note that the L-reduction refers to the *optimization* versions of problems δ-gap-MSR and δ-gap-MSR-DU, which are easy to deduce from the decision versions presented here.

Problem: 3-MIS
Input: A cubic graph $G = (V, E)$, an integer k.
Question: Is $\alpha(G) \geq k$?

It is proved in [1] that 3-MIS is APX-hard. Given a cubic graph $G = (V, E)$, our reduction consists in constructing two comparative maps $\mathcal{M}_1$ and $\mathcal{M}_2$, having properties P1, P2 and P3 described below, where Ω denotes the set of all prestrips of $\mathcal{M}_1$ and $\mathcal{M}_2$ having gap at most δ:

P1. There exists a bijection Φ between V and Ω
P2. Every prestrip in Ω has length 2
P3. Two prestrips σ_1 and σ_2 of Ω are overlapping iff $\left(\Phi^{-1}(\sigma_1), \Phi^{-1}(\sigma_2)\right) \in E$

Let P_k denote the path graph with k vertices.

Lemma 5. *Given a cubic graph $G = (V, E)$, one can compute in polynomial time a partition of E into two classes E^B and E^W (for "Black" and "White" edges), such that (1) each connected component of (V, E^B) (called "black component") is isomorphic to a path P_k , and (2) each connected component of (V, E^W) (called "white component") is isomorphic to a path $P_{k'}$, with $k' \leq 4$.*

The first step of the reduction is to compute a partition of E into two classes E^{B} and E^{W} according to Lemma 5. We then construct two comparative maps $\mathcal{M}_1$ and $\mathcal{M}_2$, satisfying properties P1, P2 and P3. Moreover, incompatibilities in $\mathcal{M}_1$ (resp. $\mathcal{M}_2$) will correspond to black (resp. white) edges. We begin by assigning a different pair of integers (x_a, x'_a) to every vertex $a \in V(G)$; we write $\Phi(a) = \langle x_a, x'_a \rangle$.

Then, for every black component B_i of order k, let $V(B_i) = \{a_h : 1 \leq h \leq k\}$ and let $(a_h, a_{h+1}) \in E^{\text{B}}$ for $1 \leq h < k$; we construct the following sequence:

$$I_i = \left\langle x_{a_1}, \times, \times^{\delta-2}, x_{a_2}, x'_{a_1}, \times^{\delta-2}, \ldots, x_{a_h}, x'_{a_{h-1}}, \times^{\delta-2}, \ldots, \times, x'_{a_k} \right\rangle$$

where $\times^l$ represents l consecutive peg markers. The full comparative map $\mathcal{M}_1$ is given by $\mathcal{M}_1 = \left\langle I_1, \times^{\delta+1}, I_2, \times^{\delta+1}, \ldots \right\rangle$.

For $\mathcal{M}_2$, we use a similar construction, but we need to take the reversed opposite of some subsequences to avoid creating undesired prestrips. For a white component W_j having 4 vertices, say a, b, c and d with $(a, b), (b, c), (c, d) \in E^{\text{W}}$, we create the following sequence:

$$J_j = \langle x_a, x_b, x'_a, -x'_c, x'_b, -x'_d, -x_c, -x_d \rangle \ .$$

If W_j is of order three (resp. two), we remove the extra elements from J_j, i.e. $J_j = \langle x_a, x_b, x'_a, -x'_c, x'_b, -x_c \rangle$ (resp. $J_j = \langle x_a, x_b, x'_a, x'_b \rangle$). Finally, $\mathcal{M}_2$ is created in the same way as $\mathcal{M}_1$: $\mathcal{M}_2 = \left\langle J_1, \times^{\delta+1}, J_2, \times^{\delta+1}, \ldots \right\rangle$.

Lemma 6. *The set Ω of the prestrips of $\mathcal{M}_1$ and $\mathcal{M}_2$ with gap less than or equal to δ is exactly $\{\Phi(a) : a \in V\}$. Moreover, $\Phi(a)$ and $\Phi(b)$ overlap in $\mathcal{M}_1$ iff $(a, b) \in E^B$, and $\Phi(a)$ and $\Phi(b)$ overlap in $\mathcal{M}_2$ iff $(a, b) \in E^W$.*

The consequence of this lemma is that $\mathcal{M}_1$ and $\mathcal{M}_2$ satisfy the three properties P1, P2 and P3 defined above. The reduction we have described is an L-reduction from 3-MIS to δ-gap-MSR: indeed, Φ transforms an independent set of size l into a feasible set of prestrips with gap δ of total size $2l$, and Φ^{-1} does the reverse operation. So δ-gap-MSR, like 3-MIS, is APX-hard for $\delta \geq 2$.

4 Approximation Algorithms

4.1 1.8-approximation for 1-gap-MSR

In this section, we present an approximation algorithm for 1-gap-MSR. Our result is the following.

Theorem 7. *There exists a factor-1.8 approximation algorithm for 1-gap-MSR.*

Proof. Our algorithm makes uses of an exact algorithm to solve Maximum Weight Independent Set (MWIS) on claw-free graphs. A *claw* is the 4-vertex graph (V, E) with $V = \{a, b, c, d\}$ and $E = \{(a, b), (a, c), (a, d)\}$. A graph is said to be claw-free if none of its induced subgraphs is isomorphic to a claw. The variant of MWIS on claw-free graphs, Claw-Free-MWIS (which is known to be in P, [6]), is stated as follows:

Problem: Claw-Free-MWIS
Input: A claw-free graph $G = (V, E)$, a weight function $w : V \to \mathbb{R}^+$, $k \in \mathbb{R}^+$
Question: Is there an independent set X of G such that $\sum_{x \in X} w(x) \geq k$?

Our 1.8-approximation algorithm (given in Algorithm 1) works as follows. Given two comparative maps $\mathcal{M}_1$ and $\mathcal{M}_2$, compute the set Ω of all prestrips with length 2 or 3 (and gap at most 1). Longer prestrips are ignored, since they can be split into smaller ones appearing in Ω. Select a subset $V^\lambda \subseteq \Omega$ (according to some parameter λ: see the selection process described below), and create E^λ, the set of all overlapping pairs of prestrips of V^λ. The pair (V^λ, E^λ) forms a graph which is claw-free (see Lemma 8). An independent set for this graph (computable in polynomial time) yields a feasible set of prestrips V^λ_{Ind}.

The selection of V^λ amongst Ω is done as follows: given a prestrip σ of $\mathcal{M}_1$ and $\mathcal{M}_2$, take the values of $\text{idx}(\sigma, \mathcal{M}_2) - \lambda$ modulo 9. This is done by the arithmetic function π_9, which takes the values of a list modulo 9: for example, if σ has indices $(30, 32, 33)$ in $\mathcal{M}_2$, and $\lambda = 5$, then $\text{idx}(\sigma, \mathcal{M}_2) - \lambda = (25, 27, 28)$, and $\pi_9(\text{idx}(\sigma, \mathcal{M}_2) - \lambda) = (7, 9, 1)$. If the result of $\pi_9(\text{idx}(\sigma, \mathcal{M}_2) - \lambda)$ belongs to some list (the list T in Algorithm 1), add σ to V^λ. We only need to test the 9 different values of λ to obtain 9 different feasible sets of prestrips.

Finally, Lemma 9 proves that there exists some λ for which the total size of the corresponding V^λ_{Ind} is at least $5/9^{th}$ of a maximum feasible set of prestrips of $\mathcal{M}_1$ and $\mathcal{M}_2$. Thus, Algorithm 1 is a polynomial-time algorithm giving a 1.8-approximation to 1-gap-MSR, and Theorem 7 is proved. □

Algorithm 1. A factor-1.8 approximation algorithm for 1-gap-MSR

Input: Two comparative maps $\mathcal{M}_1$, $\mathcal{M}_2$ without duplicates.
$T \leftarrow \{(1,2,3), (2,3,4), (3,4,5), (1,3), (2,3), (2,4), (3,4), (3,5),$
$(6,7), (6,8), (7,8), (7,9), (8,9)\}$;
$\Omega \leftarrow$ set of all prestrips of $\mathcal{M}_1$ and $\mathcal{M}_2$ of length 2 or 3, with gap at most 1;
for $\lambda \leftarrow 1$ **to** 9 **do**
  $V^\lambda \leftarrow \{\sigma \,:\, \sigma \in \Omega,\ \pi_9(\text{idx}(\sigma, \mathcal{M}_2) - \lambda) \in T\}$;
  $E^\lambda \leftarrow \{(\sigma_1, \sigma_2) \,:\, \sigma_1, \sigma_2$ overlapping prestrips of $V^\lambda\}$;
  $w(\sigma) \leftarrow |\sigma|$ (for all $\sigma \in V^\lambda$);
  $V^\lambda_{Ind} \leftarrow$ Maximum Weight Independent Set of (V^λ, E^λ) with weight w;
end for
return $\max\{||V^\lambda_{Ind}|| \,:\, 1 \le \lambda \le 9\}$;

Lemma 8. *For each λ, the graph (V^λ, E^λ) created by Algorithm 1 is claw-free.*

Lemma 9. *If $\mathcal{O}$ is a feasible set of prestrips of $\mathcal{M}_1$, $\mathcal{M}_2$ with gap 1, Algorithm 1 provides a solution of total size at least* $5||\mathcal{O}||/9$.

4.2 Reduction to Maximum Weight Independent Set

In this section we consider the variants of Maximum Weight Independent Set on two classes of graphs: interval graphs and 2-interval graphs.

An *interval graph* is a graph $G = (V, E)$, where every vertex in V is seen as an interval I of $\mathbb{R}$, and such that $(I, J) \in E$ iff (1) I and J are distinct intervals from V, and (2) $I \cap J \neq \emptyset$.

A *2-interval graph* is a graph $G = (V, E)$, where every vertex in V is seen as a pair of disjoint intervals (I_1, I_2) of $\mathbb{R}$ (also called a *2-interval*), and such that $((I_1, I_2), (J_1, J_2)) \in E$ iff (1) (I_1, I_2) and (J_1, J_2) are distinct 2-intervals from V, and (2) $(I_1 \cup I_2) \cap (J_1 \cup J_2) \neq \emptyset$.

Problem: Interval-MWIS
Input: An interval graph $G = (V, E)$, a weight function $w : V \to \mathbb{R}^+$, $k \in \mathbb{R}^+$
Question: Is there an independent set X of G such that $\sum_{x \in X} w(x) \geq k$?

Problem: 2-Interval-MWIS
Input: A 2-interval graph $G = (V, E)$, a weight function $w : V \to \mathbb{R}^+$, $k \in \mathbb{R}^+$
Question: Is there an independent set X of G such that $\sum_{x \in X} w(x) \geq k$?

The problem Interval-MWIS is known to be polynomial [5]. On the other hand, 2-Interval-MWIS is APX-hard, and we know a 4-approximation for it [2].

Theorem 10. *There exists a factor-4 approximation algorithm for δ-gap-MSR for all $\delta \geq 2$, and for δ-gap-MSR-DU for all $\delta \geq 0$.*

Proof. In this proof, we describe a reduction from δ-gap-MSR to 2-Interval-MWIS. Given a pair of comparative maps and a maximal gap δ, we construct a set of 2-intervals in the following way. First, compute the set Ω of all prestrips of $\mathcal{M}_1$ and $\mathcal{M}_2$ having gap at most δ. Then, to each prestrip $\sigma \in \Omega$, assign the following 2-interval (where l is $|\mathcal{M}_1| + 1$):

$$\begin{aligned} I_\sigma = (& [\min(\text{idx}(\sigma, \mathcal{M}_1)), \max(\text{idx}(\sigma, \mathcal{M}_1))], \\ & [\min(\text{idx}(\sigma, \mathcal{M}_2)) + l, \max(\text{idx}(\sigma, \mathcal{M}_2)) + l]) \,, \end{aligned}$$

with the weight:

$$w(I_\sigma) = |\sigma| \,.$$

We denote $G_\delta(\mathcal{M}_1, \mathcal{M}_2)$ the weighted 2-interval graph with vertex set $\{I_\sigma : \sigma \in \Omega\}$ and weight w. It has the following property:

Property 11. *The set $\{I_\sigma : \sigma \in \mathcal{S}\}$ is an independent set of $G_\delta(\mathcal{M}_1, \mathcal{M}_2)$ with weight W iff $\mathcal{S}$ is a feasible subset of Ω with total size W.*

The 4-approximation algorithm for δ-gap-MSR and δ-gap-MSR-DU is the following (adapted from the 4-approximation algorithm for MSR and MSR-DU [3]):

1. Compute the weighted 2-interval graph $G_\delta(\mathcal{M}_1, \mathcal{M}_2)$ as described above.
2. Compute X, a 4-approximation to 2-Interval-MWIS($G_\delta(\mathcal{M}_1, \mathcal{M}_2)$).
3. Deduce a feasible set of prestrips $\mathcal{S} = \{\sigma : I_\sigma \in X\}$.

Property 11 tells us that the total size of $\mathcal{S}$ is the weight of X, and that δ-gap-MSR-DU($\mathcal{M}_1, \mathcal{M}_2$) and 2-Interval-MWIS($G_\delta(\mathcal{M}_1, \mathcal{M}_2)$) have the same optimal values: so $\mathcal{S}$ is indeed a 4-approximation of the optimal solution of δ-gap-MSR-DU($\mathcal{M}_1, \mathcal{M}_2$). We have proved Theorem 10. □

Theorem 12. *There exists an exact polynomial-time algorithm for 0-gap-MSR.*

Proof. We consider the case where $\mathcal{M}_1$ has no duplicates and the maximum gap is 0 (we only consider strips instead of prestrips): this is the case for instances of 0-gap-MSR.

We use the same reduction as for Theorem 10, with the difference that now, $G_0(\mathcal{M}_1, \mathcal{M}_2)$ is in fact an interval graph. It can be seen by considering intervals

$$I'_\sigma = [\min(\text{idx}(\sigma, \mathcal{M}_1)), \max(\text{idx}(\sigma, \mathcal{M}_1))].$$

We no longer need to consider the interval coming from $\mathcal{M}_2$ for the following reason. If two strips overlap in $\mathcal{M}_2$, since they have gap zero, they must have a common marker m appearing in $\mathcal{M}_2$. But since m can appear only once in $\mathcal{M}_1$, they also overlap in $\mathcal{M}_1$. Thus I_σ and I_τ intersect iff I'_σ and I'_τ intersect: $G_0(\mathcal{M}_1, \mathcal{M}_2)$ can thus be seen as an interval graph. Hence, we can adapt the previous algorithm to obtain an optimal solution, and complete the proof of Theorem 12:

1. Compute the weighted *interval* graph $G_\delta(\mathcal{M}_1, \mathcal{M}_2)$.
2. Compute X, an *optimal* solution to Interval-MWIS($G_\delta(\mathcal{M}_1, \mathcal{M}_2)$).
3. Deduce a *maximal* feasible set of prestrips $\mathcal{S} = \{\sigma \ : \ I_\sigma \in X\}$.

□

References

1. Alimonti, P., Kann, V.: Hardness of approximating problems on cubic graphs. In: Bongiovanni, G., Bovet, D.P., Di Battista, G. (eds.) CIAC 1997. LNCS, vol. 1203, pp. 288–298. Springer, Heidelberg (1997)
2. Bar-Yehuda, R., Halldórsson, M.M., Naor, J., Shachnai, H., Shapira, I.: Scheduling split intervals. SIAM J. Comput. 36(1), 1–15 (2006)
3. Chen, Z., Fu, B., Jiang, M., Zhu, B.: On recovering syntenic blocks from comparative maps. In: Yang, B., Du, D.-Z., Wang, C.A. (eds.) COCOA 2008. LNCS, vol. 5165, pp. 319–327. Springer, Heidelberg (2008)
4. Choi, V., Zheng, C., Zhu, Q., Sankoff, D.: Algorithms for the extraction of synteny blocks from comparative maps. In: Giancarlo, R., Hannenhalli, S. (eds.) WABI 2007. LNCS (LNBI), vol. 4645, pp. 277–288. Springer, Heidelberg (2007)
5. Golumbic, M.: Algorithmic Graph Theory and Perfect Graphs. Academic Press, New York (1980)
6. Minty, G.J.: On maximal independent sets of vertices in claw-free graphs. J. Comb. Theory, Ser. B 28(3), 284–304 (1980)
7. Papadimitriou, C., Yannakakis, M.: Optimization, approximation, and complexity classes. J. Comput. Syst. Sci. 43(3), 425–440 (1991)
8. Wang, L., Zhu, B.: On the tractability of maximal strip recovery. In: Chen, J., Cooper, S.B. (eds.) TAMC 2009. LNCS, vol. 5532, pp. 400–409. Springer, Heidelberg (2009)
9. Zheng, C., Zhu, Q., Sankoff, D.: Removing noise and ambiguities from comparative maps in rearrangement analysis. IEEE/ACM Trans. Comput. Biology Bioinform. 4(4), 515–522 (2007)

The Directed Hausdorff Distance between Imprecise Point Sets

Christian Knauer[1], Maarten Löffler[2], Marc Scherfenberg[1], and Thomas Wolle[3]

[1] Institute of Computer Science, Freie Universität Berlin, Germany
scherfen@mi.fu-berlin.de
[2] Dep. of Information and Computing Sciences, Utrecht University, The Netherlands
loffler@cs.uu.nl
[3] NICTA Sydney, Australia
thomas.wolle@nicta.com.au

Abstract. We consider the directed Hausdorff distance between point sets in the plane, where one or both point sets consist of imprecise points. An imprecise point is modelled by a disc given by its centre and a radius. The actual position of an imprecise point may be anywhere within its disc. Due to the direction of the Hausdorff Distance and whether its tight upper or lower bound is computed there are several cases to consider. For every case we either show that the computation is NP-hard or we present an algorithm with a polynomial running time. Further we give several approximation algorithms for the hard cases and show that one of them cannot be approximated better than with factor 3, unless P=NP.

1 Introduction

The analysis and comparison of geometric shapes are essential tasks in various application areas within computer science, such as pattern recognition and computer vision. Beyond these fields also other disciplines evaluate the shape of objects such as cartography, molecular biology, medicine, or biometric signal processing. In many cases patterns and shapes are modeled as finite sets of points.

The *Hausdorff* distance is an important tool to measure the similarity between two sets of points (or, more generally, any two subsets of a metric space). It is defined as the largest distance from any point in one of the sets, to the closest point in the other set (see Section 1.3 for a formal definition). This distance is used extensively in pattern matching.

Data imprecision is a phenomenon that has existed as long as data is being collected. In practice, data is often sensed from the real world, and as a result has a certain error region. On the one hand, many application fields of computational geometry use algorithms that take this into account. However, these algorithms are mostly heuristics, and do not benefit from theoretical guarantees. On the other hand, algorithms from computational geometry are provably correct and efficient, often under the assumption that the input data is correct. If we want these algorithms to be used in practice, they need to take imprecision into account in the analysis. Thus not surprisingly, data imprecision in computational geometry is receiving more and more attention.

Y. Dong, D.-Z. Du, and O. Ibarra (Eds.): ISAAC 2009, LNCS 5878, pp. 720–729, 2009.

In this paper, we study several variants of the important and elementary problem of computing the Hausdorff distance under the Euclidean metric between *imprecise* point sets.

1.1 Related Work

The Hausdorff distance is one of the most studied similarity measures. For a survey about similarity measures and shape matching refer to [2]. A straightforward, naive algorithm computes the Hausdorff distance between two point sets A and B consisting of m and n points, respectively, in $O(mn)$ time. Using Voronoi diagrams and a more sophisticated approach the running time can be reduced to $O((m + n) \log n)$, [1].

The study of imprecision within computational geometry started around twenty years ago, when Guibas *et al.* [7] introduced *epsilon geometry* as a way to handle computational imprecision. In this model, each point is assumed to be at most ε away from its given location.

For a given measure on a set of imprecise points, one of the simplest questions to ask in this model is what are the possible output values? Each input point can be anywhere in a given region, and depending on where each point is, the output will have a different value. This leads to the problem of placing the points in their regions such that this value is minimised or maximised. One of the first results of this kind is due to Goodrich and Snoeyink [6], who show how to place a set of points on a set of vertical line segments such that the points are in convex position and the area or perimeter of the convex hull is minimised in $O(n^2)$ time. A similar problem is studied by Mukhopadhyay *et al.* [12], and their result was later generalised to isothetic line segments [11].

Nagai and Tokura [13] thoroughly study the efficient computation of lower and upper bounds for a variety of region shapes and measures; in particular they study the diameter, the width, and the convex hull, and all their algorithms run in $O(n \log n)$ time. However, not all of their bounds are tight. Van Kreveld and Löffler [14] study the same problems and give algorithms to compute tight bounds, though the running times of the algorithms can be much higher and some variants are proven to be NP-hard.

Related work concerning the Hausdorff distance in an imprecise context includes [5] and [8].

1.2 Contribution

In this paper, we assume that an imprecise point is modelled by a disc with a given centre and radius. In general, it is possible that the discs intersect. We assume we have two sets of points, P and Q, and that at least one of them is imprecise. We want to compute the directed Hausdorff distance from P to Q. This includes both the tight lower and upper bound on the possible values, for each combination. This leads to six different cases. Additionally, in some settings the problems become easier if we restrict the model of imprecision to disjoint discs or discs that all have the same radius; we state these results separately. Our results are summarised in Table 1.

Table 1. P and Q are point sets and $\tilde{P}$ and $\tilde{Q}$ are imprecise point sets. Results are shown for the case when all sets have $O(n)$ elements. *can be computed exactly in $O(n^3)$ if the discs are disjoint and the answer is smaller than $r(\sqrt{5-2\sqrt{3}}-1)/2$ where r is the radius of the smallest disc in $\tilde{Q}$.

setting		tight lower bound	tight upper bound
$h(\tilde{P}, Q)$	[general]	$O(n^2)$	$O(n \log n)$
$h(P, \tilde{Q})$	[general]	NP-hard*, 4-APX in $O(n^3 \log^2 n)$	$O(n \log n)$
	[disjoint unit discs]	3-APX-hard, 3-APX in $O(n^{10} \log n)$	$O(n \log n)$
$h(\tilde{P}, \tilde{Q})$	[general]	NP-hard	$O(n^2)$
	[const. depth in $\tilde{P}$]		$O(n \log n)$

In the next section, we review some definitions and structures that we use to obtain our results. After that, we present our three main results. In Section 2, we give a general algorithm for computing the upper bound, which works in all settings in the table, though it can be simplified (conceptually) in some settings. In Section 3, we prove hardness of computing the lower bound in most settings. Finally, in Section 4, we give algorithmic results for computing the lower bound, exactly in some cases and approximately in others. Due to severe space constraints, we are not able to include most of the details of the algorithms. In what follows, we describe the main ideas and structure of our results, but we encourage the interested reader to consult the full version of this paper instead, which can be found in [9].

1.3 Preliminaries

The directed Hausdorff distance h from a point set $P = \{p_1, \ldots, p_m\}$ to a point set $Q = \{q_1, \ldots, q_n\}$ with an underlying Euclidean metric can be computed in $O((n+m)\log n)$ time, see [1], and is defined as (see Fig. 1 for an example):

$$h(P, Q) = \max_{p \in P} \min_{q \in Q} \|p - q\|$$

Let $\tilde{P}$ and $\tilde{Q}$ denote two imprecise point sets consisting of m and n closed discs respectively. We call a set $P = \{p_1, \ldots, p_m\}$ a *precise realisation* of $\tilde{P} = \{\tilde{p}_1, \ldots, \tilde{p}_m\}$ if $p_i \in \tilde{p}_i$ for all i. We also write $P \Subset \tilde{P}$ in this case.

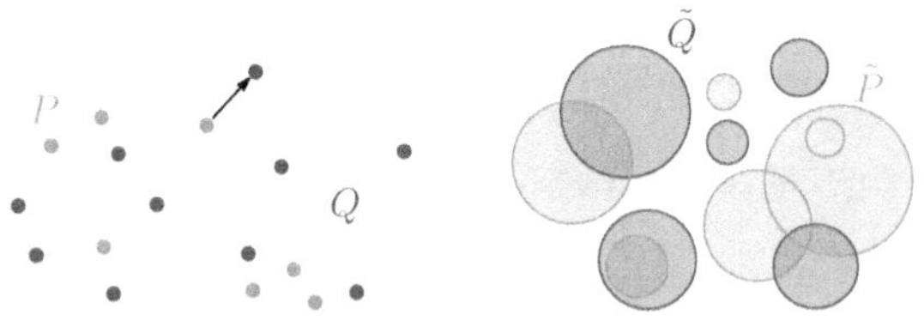

Fig. 1. (a) $h(P, Q)$ is defined by the pair of points indicated by the arrow. (b) An example input of imprecise points.

We define the directed Hausdorff distance between a precise and an imprecise or two imprecise point sets as the interval of all possible outcomes for that distance.

$$h(P,\tilde{Q}) = \{h(P,Q) \mid Q \Subset \tilde{Q}\}, \quad h(\tilde{P},Q) = \{h(P,Q) \mid P \Subset \tilde{P}\}$$
$$h(\tilde{P},\tilde{Q}) = \{h(P,Q) \mid P \Subset \tilde{P}, Q \Subset \tilde{Q}\}$$

Further, we denote the tight upper and lower bounds of this interval by $h_{\max}$ and $h_{\min}$ respectively, for example

$$h_{\max}(P,\tilde{Q}) = \max h(P,\tilde{Q}) \text{ and hence } h(P,\tilde{Q}) = [h_{\min}(P,\tilde{Q}), h_{\max}(P,\tilde{Q})].$$

2 Algorithm for Computing the Tight Upper Bound

In this section, we consider the following problem. Given are two set of discs $\tilde{P}$ and $\tilde{Q}$. The radii may be all different; an example input is shown in Fig. 1(b). We want to place point sets $P \Subset \tilde{P}$ and $Q \Subset \tilde{Q}$ such as to maximise the directed Hausdorff distance $h(P,Q)$. In other words, we want to place the points in P and Q such that one point from P is as far as possible away from all points in Q. The placements of the remaining points of P do not matter. So, we need to identify which point $\hat{p} \in P$ will play this important role. We need to place $\hat{p}$ such that after we placed all points in Q as far away from $\hat{p}$ as possible, this distance is maximised.

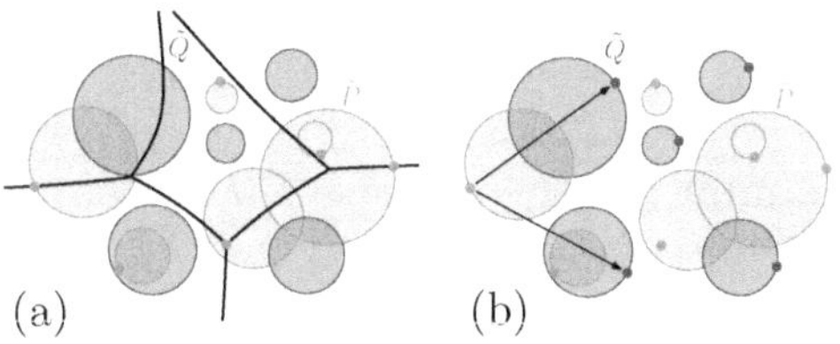

Fig. 2. (a) The inverted additive Voronoi Diagram (iaVD) of $\tilde{Q}$. The point set P placed locally optimal. (b) The points in Q are all placed as far away from $\hat{p}$ as possible.

2.1 Basic Algorithm

We will first compute the *inverted additive Voronoi Diagram* (iaVD) of $\tilde{Q}$. This is a subdivision of the plane into regions where each point x in the plane is associated with the disc in $\tilde{Q}$ whose furthest point is closest to x. See Fig. 2(a) for an example. This diagram can be computed in $O(n \log n)$ time [4], since it corresponds to the additively weighted Voronoi Diagram (also known as Apollonius diagram) of the centres of $\tilde{Q}$, where the weight of a point is minus the radius of the corresponding disc.

Using the iaVD, we can place each point $p \in \tilde{p} \in \tilde{P}$ at a locally optimal position, as if it were $\hat{p}$. We identify three possible placement types for p that are locally optimal, as is illustrated in Fig. 2.

1. A vertex of the iaVD.
2. An intersection point between a Voronoi edge and a disc from $\tilde{P}$.
3. A point on the boundary of $\tilde{p}$ that is furthest away from the iaVD site whose cell contains the centre of $\tilde{p}$

We can now iterate over all points in P and their locally optimal placements, and determine $\hat{p}$ by keeping track of the locally optimal placement $p \in \tilde{p}$ such that the shortest distance between p and (the furthest point on) any disc in $\tilde{Q}$ is maximised. Once $\hat{p}$ is known, we place all points in Q as far away from $\hat{p}$ as possible, and all points in $P \backslash \{\hat{p}\}$ anywhere inside their discs. The result is shown in Fig. 4(b). As it is possible that there are $O(mn)$ locally optimal placements of the second type (namely: an intersection between a disc boundary and a Voronoi edge), we conclude with the following theorem.

Theorem 1. *Given two sets $\tilde{P}$ and $\tilde{Q}$ of imprecise points of size m and n, respectively, we can compute $h_{\max}(\tilde{P}, \tilde{Q})$ and precise realisations $P \Subset \tilde{P}$ and $Q \Subset \tilde{Q}$ with $h(P,Q) = h_{\max}(\tilde{P}, \tilde{Q})$ in $O(nm + n \log n)$ time.*

2.2 Faster Algorithms in Special Cases

In this section we show how the above result can be improved under certain assumptions. To speed up the algorithm, we make some observations about the nature of locally optimal placements.

Lemma 1. *Let $\tilde{p}$ be a disc in $\tilde{P}$, and let $\tilde{q}_1$ and $\tilde{q}_2$ be two discs in $\tilde{Q}$, such that the part of the bisector of $\tilde{q}_1$ and $\tilde{q}_2$ that is in the iaVD slices through $\tilde{p}$ (that is, it is not connected to a vertex of the iaVD inside $\tilde{p}$). Then the optimal placement of p occurs on the same side of this bisector where the centre of $\tilde{p}$ is, regardless of the rest of the iaVD.*

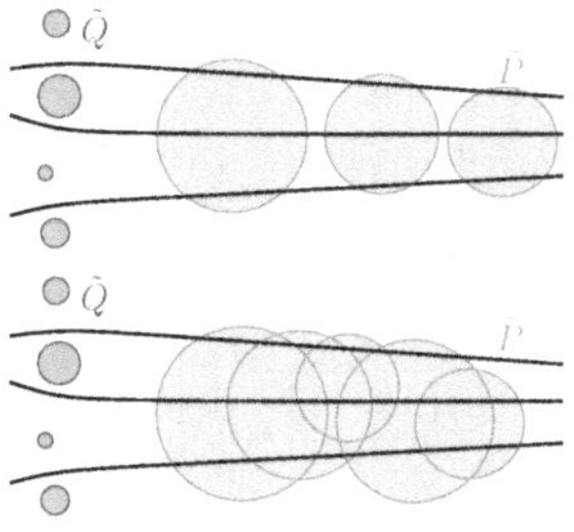

Fig. 3. (a) There could be a quadratic number of intersections between the edges of the iaVD of $\tilde{Q}$ and the discs in $\tilde{P}$. (b) When the discs overlap, the *union* of $\tilde{P}$ has fewer intersections with the iaVD.

Proof. Some notation: let p_c be the centre of $\tilde{p}$, q_{c1} the centre of $\tilde{q}_1$ and q_{c2} the centre of $\tilde{q}_2$. Now let f_1 be the point on the boundary of $\tilde{p}$ that is furthest away from q_{c1} (this would be the type 3 placement if $\tilde{q}_1$ was the only player), and similarly let f_2 be the point furthest away from q_{c2}. Now, suppose w.l.o.g. that p_c is on the same side as q_{c1}. Now, suppose that the optimal placement p is on the other side, that is, on the side of q_{c2}. Then we observe that f_2 must be on the side of q_{c1}, because q_{c2}, p_c and f_2 lie on a line. This means that along the boundary of $\tilde{p}$, the intersection points with the bisector have a better value than any other point on the side of q_{c2}, in particular, better than p, which is a contradiction. (Note that if there are other cells of the iaVD involved, the value of p could only be lower). □

This lemma basically says that if we want to place a certain point p locally optimally, we can start looking by walking from the centre of $\tilde{p}$ and never have to cross edges of the iaVD that slice through $\tilde{p}$, like illustrated in Fig. 3. This makes us arrive at the following conclusion.

Corollary 1. *Let $\tilde{p}$ be a disc in $\tilde{P}$, and suppose that the iaVD has t vertices inside $\tilde{p}$. Then we can find the locally optimal placement for p in $O(t)$ time.*

This immediately implies that if the discs of $\tilde{P}$ do not overlap, we can simply place all points p independently in linear time.

Now, assume that the discs of $\tilde{P}$ are disjoint, or that the intersection depth is at most some constant c. Then, clearly, each vertex of the iaVD can appear in at most c discs of $\tilde{P}$. So, if each disc $\tilde{p}_i$ contains t_i vertices of the iaVD, we have $\sum_i t_i \leq cn$, and we can find all locally optimal placements in $O(n)$ time.

Theorem 2. *Given two sets $\tilde{P}$ and $\tilde{Q}$ of imprecise points of size m and n, respectively, where the discs in $\tilde{P}$ have constant intersection depth, we can compute $h_{\max}(\tilde{P}, \tilde{Q})$ and precise realisations $P \Subset \tilde{P}$ and $Q \Subset \tilde{Q}$ with $h(P, Q) = h_{\max}(\tilde{P}, \tilde{Q})$ in $O((m+n)\log(m+n))$ time.*

The algorithm described in this section works in the most general setting. However, in some more specific settings, the algorithm can be simplified. For example, when the discs of $\tilde{Q}$ are unit discs, the iaVD is simply the normal Voronoi diagram. When P is not imprecise, there are of course only m possible locations for $\hat{p}$, and we do not need to look for all three placement types. This results in the running times as indicated in Table 1.

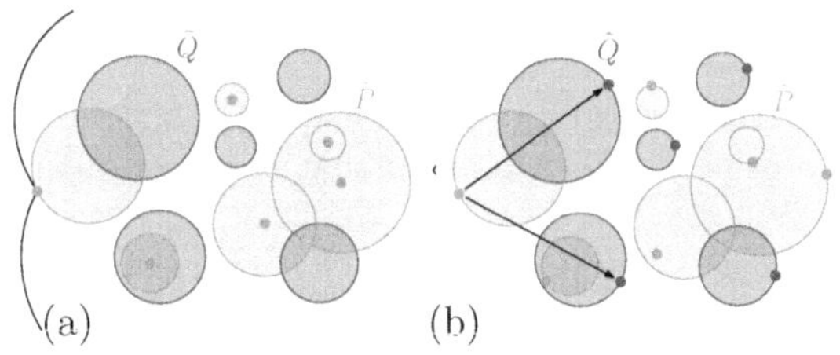

Fig. 4. (a) An example input. (b) The optimal output, shown as a set of circles covering Q.

3 Hardness Results for Tight Lower Bounds

In this section, we consider a transformation from the known NP-complete problem PLANAR 3-SAT [10] to the problem of computing $h_{\min}(P, \tilde{Q})$ for a set P of points and a set $\tilde{Q}$ of discs with radius r. In the PLANAR 3-SAT problem, we are given as input a 3-SAT formula f with the additional property that the graph $G(f)$ is planar, where $G(f)$ has a vertex for each variable and each clause in f, and there is an edge between a variable vertex and a clause vertex if the variable occurs in the clause. Having the boolean formula f and a planar embedding of $G(f)$, the transformation is as follows (see Fig. 5(a,b) for a general overview):

For each variable vertex v in $G(f)$, we construct a *cycle* C of alternating points in P and discs in $\tilde{Q}$. The distance between consecutive points and discs is ϵ, such that $r = 2.5\epsilon$ (see Fig. 5(c)). There may be bends up to a certain angle, and also other geometric features necessary to connect cycles and chains. When looking only at the points P^C and discs $\tilde{Q}^C$ corresponding to a cycle C, we observe that by the construction of C, there are two realisations $Q_0^C, Q_1^C \Subset \tilde{Q}^C$ such that $h(P^C, Q_0^C) = \epsilon$ and $h(P^C, Q_1^C) = \epsilon$. These two realisations represent the two possible boolean values the variable for that cycle can have.

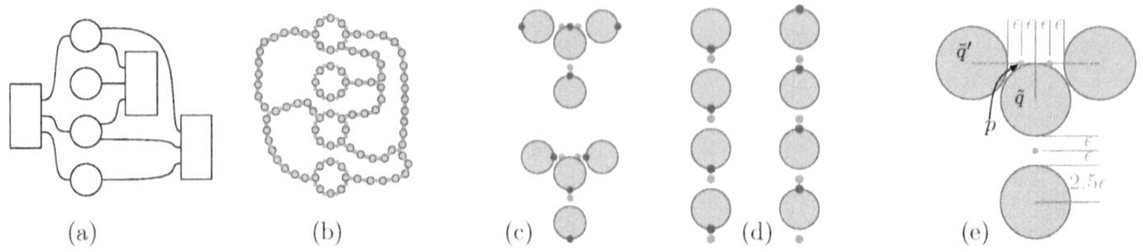

Fig. 5. (a) Planar embedding of $G(f)$, circles represent variables and rectangles represent clauses. (b) Rough overview of how $G(f)$ is transformed into P and $\tilde{Q}$, some details are misrepresented. the chain starts with p followed by $\tilde{q}'$, all other points and discs belong to the cycle. (c,d) Two realisations (representing opposite boolean values) with Hausdorff distance ϵ of chains, cycles and connections. (e) Connection of a chain to a cycle.

For each edge $\{v, c\}$ in $G(f)$, we construct a *chain* of alternating points in P and discs in $\tilde{Q}$ with distance ϵ (see Fig. 5(d)). The chain connects the cycle corresponding to the variable v and the representation of a clause c. One end of this chain is a disc that will be part of a representation of clause c (see Fig. 5(e)), the other end is a point p that is placed near a disc $\tilde{q} \in \tilde{Q}$ of a variable cycle such that p has distance ϵ to either $Q_0^C \cap \tilde{q}$ or $Q_1^C \cap \tilde{q}$ (see Fig. 5(e)).

Each clause vertex in $G(f)$ is represented by three discs and one additional point p^*, such that the disc centres lie on the vertices of an equilateral triangle, and the point has distance ϵ to each of the discs. The three discs are ends of chains that connect to cycles that correspond to the three literals in the clause.

Theorem 3. *Let P be a precise point set and $\tilde{Q}$ be an imprecise point set of pairwise disjunct discs. It is NP-hard to compute a δ-approximation of the directed Hausdorff distance $h_{min}(P, \tilde{Q})$ for $1 \leq \delta < 3$.*

4 Algorithms for Tight Lower Bounds

In this section we present algorithms for computing the minimum of $h(\tilde{P}, Q)$ and $h(P, \tilde{Q})$. As we have seen in the previous section, the latter problem is NP-hard and even hard to approximate in some settings. In the following we give a 4-approximation for the general case, an optimal 3-approximation for disjoint discs and an algorithm for the case which is not NP-hard when the Hausdorff distance is small. Many results in this section rely on similar ideas. Therefore, we will describe several (sub-) algorithms with different approximation factors and running times depending on the value d of the optimal solution. Afterwards, we discuss how to apply them to obtain the results claimed in Table 1.

4.1 Algorithm for Precise Q

In this section, we describe an algorithm for the case where we have an imprecise point set $\tilde{P}$ and a precise point set Q. We place all points in $\tilde{P}$ as close to a point in Q as possible. Fig. 6(a) shows an example. For each pair $(\tilde{p}, q)$ with $\tilde{p} \in P$ and

$q \in Q$ we could simply compute the placement $p \in \tilde{p}$ minimizing the Hausdorff distance and keep track of the longest distance over all pairs. This takes $O(mn)$ time. However, in practice it is probably better to compute the Voronoi diagram of Q first, and locate the discs of $\tilde{P}$ in it. In the worst case, each disc could still intersect linearly many Voronoi cells (although the input needs to be contrived for this). Also, note that as soon as a disc from $\tilde{P}$ is discovered to contain a point from Q, we can stop the computation and just place the point there.

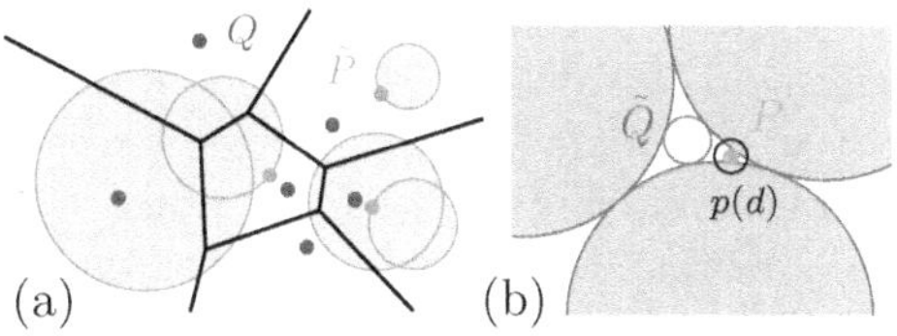

Fig. 6. (a) Placing points in P. (b) Discs of radius at most c can only intersect at most two discs of $\tilde{Q}$.

Theorem 4. *Let $\tilde{P}$ denote an imprecise point set consisting of m discs and Q denote a precise point set consisting of n points. The tight lower bound of $h(P, \tilde{Q})$ can be computed in $O(mn)$ time.*

4.2 Algorithms for Imprecise $\tilde{Q}$

In the case where P is precise and $\tilde{Q}$ is imprecise, we have several simple and more involved algorithms. The algorithms are described in detail in the full version [9]; here we merely mention the resulting theorems.

First, we describe a simple subalgorithm CANDIDATES to establish the following lemma.

Lemma 2. *Let P denote a precise point set consisting of m points and $\tilde{Q}$ denote an imprecise point set consisting of n discs. It is possible to reduce the possible values of $h_{min}(P, \tilde{Q})$ to $O(m^3 + m^2 n)$ many candidates in $O(m^3 + m^2 n)$ time.*

Next, we describe an algorithm INDEPENDENTSETS , which is summarised in the following theorem.

Theorem 5. *Let P denote a precise point set consisting of m points and $\tilde{Q}$ denote an imprecise point set consisting of n* disjoint *discs. Algorithm* INDEPENDENTSETS *computes whether the tight lower bound for $h(P, \tilde{Q})$ is smaller than $r(\sqrt{5 - 2\sqrt{3}} - 1)/2$ where r is the radius of the smallest disc in $\tilde{Q}$. If this is the case, the exact value of $h_{min}(P, \tilde{Q})$ is computed. The running time is $O(m^3 + m^2 n + n \log^2 n)$.*

Finally, we describe another algorithm GROWNDISCS .

Theorem 6. *Let P denote a precise point set consisting of m points and $\tilde{Q}$ denote an imprecise point set consisting of n discs. Given a c-approximation to the geometric k-covering problem that runs in $T(k, m)$ time, we can compute a $(c + 2)$-approximation to the tight lower bound of $h(P, \tilde{Q})$ in $O(m^3 + m^2 n + (n^2 T(k, m) + mn + m\sqrt{m}) \log(m + n))$ time, where $k \le n$ is an internal parameter of the optimal solution.*

We now proceed to describe how to use those results and combine the algorithms to solve various variants of the problem we are interested in.

For the remainder of this section let $r_{\min}$ and $r_{\max}$ denote the radius of the smallest and largest disc in $\tilde{Q}$. When P is precise and $\tilde{Q}$ is imprecise, we note that by Theorem 6 Algorithm GrownDiscs immediately presents a 4-approximation for the case when the discs may have different radii and overlap, which we obtain by plugging in a 2-approximation algorithm for geometric k-covering that runs in $O(m \log k)$ time [3]. The running time of the entire algorithm then becomes $O(m^3 + m^2 n + mn^2 \log(m+n) \log n)$ in the worst case.

We can improve this algorithm by first testing whether $v < c \cdot r_{\min}$ using Algorithm IndependentSets and Theorem 5, without increasing the asymptotic running time. If it is, then we can actually compute the exact solution.

Furthermore, when the discs are disjoint and all have the same size, we can improve this result to a 3-approximation by combining Algorithm GrownDiscs and a trivial algorithm called CentrePoints which simply places every imprecise point at the centre of its disc. First we test whether $v > r/2 = r_{\max}/2$, by applying CentrePoints and checking whether the resulting Hausdorff distance is larger than $3/2r$. If it is, we are done. Otherwise, note that each cell of $\mathcal{A}$ is a subset of the intersection of $k \leq 4$ discs, because $\tilde{Q}$'s discs are disjoint and $v < r/2$. Therefore, by Theorem 6 we can obtain a 3-approximation from Algorithm GrownDiscs by plugging in an exact algorithm to solve the geometric k-covering problem.

We can solve the geometric 4-covering problem exactly by computing the arrangement circles around the points to be covered or radius d. The arrangement has quadratic complexity. Then we need to find out whether there are three cells that are together in all cells. There are $O(m^8)$ such combinations to test, and by keeping track of which discs are already taken care of each can be tested in constant time. So, using this algorithm, we have a $1 + 2 = 3$-approximation to the original problem for disjoint unit discs. The total running time now becomes $O(n^2 m^8 \log(m+n))$.

Theorem 7. *Let P denote a precise point set consisting of m points and $\tilde{Q}$ denote an imprecise point set consisting of n* disjoint *discs of the same radius. The tight lower bound for $h(P, \tilde{Q})$ is 3-approximable in time $O(m^3 + m^2 n + n \log^2 n)$.*

5 Conclusions and Future Work

We studied computing tight lower and upper bounds on the directed Hausdorff distance between two point set, when at least one of the sets has imprecision. We gave efficient exact algorithms for computing the upper bound, prove that computing the lower bound is NP-hard in most settings, and provide approximation algorithms. Furthermore, we show that in one special case, our approximation algorithm is optimal. In other settings, a gap in the factor between the hardness result and approximation still remains. When both sets are imprecise, we don't have any constructive results for the lower bound.

All our results hold for the directed Hausdorff distance. An obvious next step would be to extend them to the undirected Hausdorff distance. We can immediately solve the upper bound problem in that case using our results, since it is just the minimum of the two directed distances. However, computing lower bounds seems to be more complicated, because there one needs to find a single placement of both point sets that minimises the distance in both directions at the same time.

Other directions of future work include looking at other underlying metrics than the Euclidean metric, other similarity measures than the Hausdorff distance, or, as is common in shape matching, allowing some transformation of the point sets.

References

1. Alt, H., Behrends, B., Blömer, J.: Approximate matching of polygonal shapes. Ann. Math. Artif. Intell. 13, 251–266 (1995)
2. Alt, H., Guibas, L.: Discrete Geometric Shapes: Matching, Interpolation, and Approximation - A Survey. In: Handbook on Computational Geometry, pp. 251–265 (1995)
3. Feder, T., Greene, D.H.: Optimal algorithms for approximate clustering. In: Proc. 20th Ann. ACM Symp. on Theory of Comp., pp. 434–444 (1988)
4. Fortune, S.J.: A sweepline algorithm for Voronoi diagrams. Algorithmica 2, 153–174 (1987)
5. Goodrich, M.T., Mitchell, J.S.B., Orletsky, M.W.: Practical methods for approximate geometric pattern matching under rigid motion. In: Proc. 10th Annu. ACM Sympos. Comput. Geom, pp. 103–112 (1994)
6. Goodrich, M.T., Snoeyink, J.: Stabbing parallel segments with a convex polygon. Comput. Vision Graph. Image Process. 49, 152–170 (1990)
7. Guibas, L.J., Salesin, D., Stolfi, J.: Epsilon geometry: building robust algorithms from imprecise computations. In: Proc. 5th Annu. ACM Sympos. Comput. Geom, pp. 208–217 (1989)
8. Heffernan, P.J., Schirra, S.: Approximate decision algorithms for point set congruence. Comput. Geom. Theory Appl. 4, 137–156 (1994)
9. Knauer, C., Löffler, M., Scherfenberg, M., Wolle, T.: The directed Hausdorff distance between imprecise point sets (Preprint, 2009), `http://arXiv.org/abs/0909.4642`
10. Lichtenstein, D.: Planar formulae and their uses. SIAM J. Comput. 11, 329–343 (1982)
11. Mukhopadhyay, A., Greene, E., Rao, S.V.: On intersecting a set of isothetic line segments with a convex polygon of minimum area. In: Gervasi, O., Gavrilova, M.L. (eds.) ICCSA 2007, Part I. LNCS, vol. 4705, pp. 41–54. Springer, Heidelberg (2007)
12. Mukhopadhyay, A., Kumar, C., Greene, E., Bhattacharya, B.: On intersecting a set of parallel line segments with a convex polygon of minimum area. Inf. Proc. Let. 105(2), 58–64 (2008)
13. Nagai, T., Tokura, N.: Tight Error Bounds of Geometric Problems on Convex Objects with Imprecise Coordinates. In: Japanese Conference on Discrete and Computational Geometry, pp. 252–263 (2000)
14. van Kreveld, M., Löffler, M.: Largest bounding box, smallest diameter, and related problems on imprecise points. In: Dehne, F., Sack, J.-R., Zeh, N. (eds.) WADS 2007. LNCS, vol. 4619, pp. 446–457. Springer, Heidelberg (2007)

Computing Multidimensional Persistence*

Gunnar Carlsson[1], Gurjeet Singh[1], and Afra Zomorodian[2]

[1] Department of Mathematics, Stanford University, USA
[2] Department of Computer Science, Dartmouth College, USA

Abstract. The theory of multidimensional persistence captures the topology of a multifiltration – a multiparameter family of increasing spaces. Multifiltrations arise naturally in the topological analysis of scientific data. In this paper, we give a polynomial time algorithm for computing multidimensional persistence.

1 Introduction

In this paper, we give a polynomial time algorithm for computing the persistent homology of a multifiltration. The computed solution is compact and complete, but not an invariant. Theoretically, this is the best one may hope for since complete compact invariants do not exist for multidimensional persistence [1].

1.1 Motivation

Intuitively, a multifiltration models a growing space that is parameterized along multiple dimensions. For example, the complex with coordinate $(3, 2)$ in Figure 1 is filtered along the horizontal and vertical dimensions, giving rise to a bifiltration. Multifiltrations arise naturally in topological analysis of scientific data. Often, scientific data is in the form of a finite set of noisy samples from some underlying topological space. Our goal is to robustly recover the lost connectivity of the underlying space. If the sampling is dense enough, we approximate the space as a union of balls by placing ϵ-balls around each point. As we increase ϵ, we obtain a growing family of spaces, a one-parameter multifiltration also called a filtration. This approximation is the central idea behind many methods for computing the topology of a point set, such as *Čech*, *Rips-Vietoris* [2], or *witness* [3] complexes. Often, the input point set is also filtered through multiple functions. We now have multiple dimensions along which our space is filtered. That is, we have a multifiltration.

1.2 Prior Work

For one-dimensional filtrations, the theory of persistent homology provides a complete invariant called a *barcode*, a multiset of intervals [4]. Each interval in

* The authors were partially supported by the following grants: G. C. by NSF DMS-0354543; A. Z. by DARPA HR 0011-06-1-0038, ONR N 00014-08-1-0908, and NSF CCF-0845716; all by DARPA HR 0011-05-1-0007.

Y. Dong, D.-Z. Du, and O. Ibarra (Eds.): ISAAC 2009, LNCS 5878, pp. 730–739, 2009.

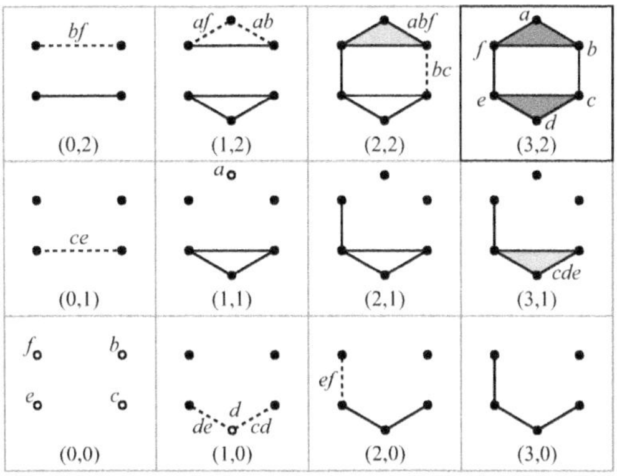

Fig. 1. A bifiltration. The complex with labeled vertices is at coordinate $(3, 2)$. Simplices are highlighted and named at the critical coordinates that they appear.

the barcode corresponds to the lifetime of a single topological attribute within the filtration. Since features have long lives, while noise is short-lived, a quick examination of the intervals gives a robust estimation of the topology. The existence of a complete compact invariant, as well as efficient algorithms and fast implementations have led to successful application of persistence to a variety of problems, such as shape description [5], denoising volumetric density data [6], detecting holes in sensor networks [7], analyzing neural activity in the visual cortex [8], and analyzing the structure of natural images [9], to name a few.

For multifiltrations of dimension higher than one, the situation is much more complicated. The theory of *multidimensional persistence* shows that no complete discrete invariant exists, where *discrete* means that the structure of the target for the invariant does not depend on the structure of the underlying field [1]. Instead, the authors propose an incomplete invariant, the rank invariant, which captures important persistent information. Unfortunately, this invariant is not compact, requiring large storage, so its direct computation using the one-dimensional algorithm is not feasible. A variant of the problem of multidimensional persistence has appeared in computer vision [10]. A partial solution, called *vineyards*, has been offered [11]. A full solution, however, has not been attempted by any prior work.

1.3 Contributions

In this paper, we provide a complete solution to the problem of computing multidimensional persistence. We recast persistence as a problem within computational algebraic geometry, allowing us to utilize powerful algorithms from this area. We then exploit the structure provided by a multifiltration to greatly simplify the algorithms. Finally, we show that the resulting algorithms are polynomial time, unlike their original counterparts, which are EXPSPACE-complete, requiring exponential space and time. We begin with a brief review of necessary concepts in Section 2, and recast the problem into an algebraic geometric framework. Section 3 contains the main contribution of this paper, where we use the structure of multifiltrations to simplify the traditional algorithms.

2 Background and Approach

In this section, we review concepts from algebraic topology and computational algebraic geometry. We then present our approach of computing multidimensional persistence using algorithms from the latter area. Due to lack of space, we omit significant portions of our work, referring the interested reader to our manuscript for a complete description [12]. Our treatment of algebraic geometry and its algorithms follow Chapter 5 of Cox, Little, and O'Shea [13].

Our goal is the computation of the persistent homology of a multifiltration. Let $\mathbb{N} \subseteq \mathbb{Z}$ be the set of non-negative integers. A topological space X is *multifiltered* if we are given a family of subspaces $\{X_u\}_u$, where $u \in \mathbb{N}^n$ and $X_u \subseteq X$ such that for $u, w_1, w_2, v \in \mathbb{N}^n$, the diagrams

$$\begin{array}{ccc} X_u & \longrightarrow & X_{w_1} \\ \downarrow & & \downarrow \\ X_{w_2} & \longrightarrow & X_v \end{array} \tag{1}$$

commute whenever $u \leq w_1, w_2 \leq v$. We call the family of subspaces $\{X_u\}_u$ a *multifiltration*, such as the example in Figure 1. In this paper, we assume our input is a multifiltered simplicial complex that has the following property:

Definition 1 (one-critical). *A multifiltered complex where each cell has a unique minimal critical grade at which it enters the complex is* one-critical.

The bifiltration in Figure 1 is one-critical, as are most multifiltrations that arise in practice [12].

Given a simplicial complex K, we may define *chain groups* C_i as the free Abelian groups on oriented i-simplices. The *boundary operator* $\partial_i \colon C_i \to C_{i-1}$ connects the chain groups into a *chain complex* C_*:

$$\cdots \to C_{i+1} \xrightarrow{\partial_{i+1}} C_i \xrightarrow{\partial_i} C_{i-1} \to \cdots . \tag{2}$$

Given any chain complex, the *ith homology group* is

$$H_i = \ker \partial_i / \operatorname{im} \partial_{i+1}. \tag{3}$$

Given a multifiltration $\{X_u\}_u$, for each pair $u \leq v \in \mathbb{N}^n$, $X_u \subseteq X_v$ by definition, so $X_u \hookrightarrow X_v$, inducing a map $\iota_i(u, v)$ at the homology level $H_i(X_u) \to H_i(X_v)$ that maps a homology class in X_u to the one that contains it in X_v. The *ith persistent homology* is the image of ι_i for all pairs $u \leq v$.

Our work rests on the theory of persistence [4,1]. The key insight is this: Persistent homology of a multifiltration is standard homology of a single multigraded module that encodes the multifiltration using polynomial coefficients. Let

$A^n = k[x_1, \ldots, x_n]$ be the n-graded *polynomial ring*, graded by $A^n_v = kx^v, v \in \mathbb{N}^n$. We define an n-graded module over this ring as follows.

Definition 2 (chain module). *Given a multifiltered simplicial complex $\{K_u\}_u$, the ith* chain module *is the n-graded module over the graded polynomial ring A^n*

$$C_i = \bigoplus_u C_i(K_u), \tag{4}$$

where the k-module structure is the direct sum structure and $x^{v-u}: C_i(K_u) \to C_i(K_v)$ is the inclusion $K_u \hookrightarrow K_v$.

These graded chain modules C_i are finitely generated, and for one-critical filtrations, they are also free, so we may choose bases for them.

Definition 3 (standard basis). *The* standard basis *for the ith chain module C_i is the set of i-simplices in critical grades.*

Given standard bases, we may write the boundary operator $\partial_i : C_i \to C_{i-1}$ explicitly as a matrix with polynomial entries. We now have a new n-graded chain complex (2) that encodes the multifiltration. The homology of this chain complex is the persistent homology of the multifiltration [1]. By definition (3), we may compute homology in three steps:

1. Compute $\operatorname{im} \partial_{i+1}$: This is a submodule of the polynomial module, and its computation is the *submodule membership problem* in computational algebraic geometry. We may solve this problem by computing the *reduced Gröbner basis* using the BUCHBERGER and reduction algorithms, and then dividing using the DIVIDE algorithm.
2. Compute $\ker \partial_i$: The is the *(first) syzygy module*, which we may compute using *Schreyer's algorithm*.
3. Compute H_i: This task is simple, once the above two tasks are complete. We need to test whether the generators of the syzygy submodule are in the boundary submodule, a task which may be completed using the tools above.

While the above algorithms solve the membership problem, they have not been used in practice due to their complexity. The submodule membership problem is a generalization of the *Polynomial Ideal Membership Problem (PIMP)* which is EXPSPACE-complete, requiring exponential space and time [14,15]. Indeed, the Buchberger algorithm, in its original form is doubly-exponential. Therefore, while our reformulation of multidimensional persistence gives us algorithms, we need to make them faster to make this approach feasible.

3 Multigraded Algorithms

In this section, we show that multifiltrations provide additional structure that may be exploited to simplify the algorithms for our three tasks. These simplifications convert these intractable algorithms into polynomial time algorithms.

3.1 Exploiting Homogeneity

The key property that we exploit for simplification is homogeneity.

Definition 4 (homogeneous). *Let M be an $m \times n$ matrix with monomial entries. The matrix M is* homogeneous *iff*

1. *every column $\mathbf{f}$ of M is associated with a coordinate in the multifiltration ($u_\mathbf{f}$) and thus a corresponding monomial $x^{u_\mathbf{f}}$,*
2. *every non-zero element M_{jk} may be expressed as the quotient of the monomials associated with column k and row j, respectively.*

Any vector $\mathbf{f}$ endowed with a coordinate $u_\mathbf{f}$ that may be written as above is homogeneous, *e.g. the columns of M.*

We will show that (1) all boundary matrices ∂_i may be written as homogeneous matrices initially, and (2) the algorithms for computing persistence only produce homogeneous matrices and vectors. That is, we maintain homogeneity as an invariant throughout the computation. We begin with our first task.

Lemma 1. *For a one-critical multifiltration, the matrix of $\partial_i \colon C_i \to C_{i-1}$ written in terms of the standard bases is homogeneous.*

Proof. Recall that we may write the boundary operator $\partial_i \colon C_i \to C_{i-1}$ explicitly as a $m_{i-1} \times m_i$ matrix M in terms of the standard bases for C_i and C_{i-1}, as shown in matrix (Equation 5) for ∂_1. From Definition 3, the standard basis for C_i is the set of i-simplices in critical grades. In a one-critical multifiltration, each simplex σ has a unique critical coordinate u_σ (Definition (1)). In turn, we may represent this coordinate by the monomial x^{u_σ}. For instance, simplex a in Figure 1 has critical grade $(1, 1)$ and monomial $x^{(1,1)} = x_1 x_2$. We order these monomials using $>_{\text{lex}}$ and use this ordering to rewrite the matrix for ∂_i. The matrix entry M_{jk} relates σ_k, the kth basis element for C_i to $\hat{\sigma}_j$, the jth basis element for C_{i-1}. If $\hat{\sigma}_j$ is not a face of σ_k, then $M_{jk} = 0$. Otherwise, $\hat{\sigma}_j$ is a face of σ_k. Since a face must precede a co-face in a multifiltration, $u_{\sigma_k} >_{\text{lex}} u_{\hat{\sigma}_j} \implies x^{u_{\sigma_k}} >_{\text{lex}} x^{u_{\hat{\sigma}_j}}$, and $M_{jk} = x^{u_{\sigma_k}} / x^{u_{\hat{\sigma}_j}} = x^{u_{\sigma_k} - u_{\hat{\sigma}_j}}$. That is, the matrix is homogeneous.

For example, $\hat{\sigma}_1 = a$ is a face of $\sigma_1 = ab$, so $M_{11} = x_1 x_2^2 / x_1 x_2 = x_2$ in the matrix for ∂_1 for the bifiltration in Figure 1.

Corollary 1. *For a one-critical multifiltration, the boundary matrix ∂_i in terms of the standard bases has monomial entries.*

Proof. The result is immediate from the proof of the previous lemma. The matrix entry is either 0, a monomial, or $x^{u(\sigma_k) - u(\hat{\sigma}_j)}$, a monomial.

Below, we show the homogeneous matrix for ∂_1 for the bifiltration in Figure 1, where we augment the matrix with the associated monomials. We assume we are computing over $\mathbb{Z}_2$.

$$\left[\begin{array}{cc|cccccccc} & & ab & bc & cd & de & ef & af & bf & ce \\ & & x_1x_2^2 & x_1^2x_2^2 & x_1 & x_1 & x_1^2 & x_1x_2^2 & x_2^2 & x_2 \\ \hline a & x_1x_2 & x_2 & 0 & 0 & 0 & 0 & x_2 & 0 & 0 \\ d & x_1 & 0 & 0 & 1 & 1 & 0 & 0 & 0 & 0 \\ b & 1 & x_1x_2^2 & x_1^2x_2^2 & 0 & 0 & 0 & 0 & x_2^2 & 0 \\ c & 1 & 0 & x_1^2x_2^2 & x_1 & 0 & 0 & 0 & 0 & x_2 \\ e & 1 & 0 & 0 & 0 & x_1 & x_1^2 & 0 & 0 & x_2 \\ f & 1 & 0 & 0 & 0 & 0 & x_1^2 & x_1x_2^2 & x_2^2 & 0 \end{array}\right] \tag{5}$$

We next focus on our second task, showing that given a homogeneous matrix as input, the algorithms produce homogeneous vectors and matrices. Let F be an $m \times n$ homogeneous matrix. Let $\{\mathbf{e_1}, \ldots, \mathbf{e_m}\}$ and $\{\hat{\mathbf{e}}_\mathbf{1}, \ldots, \hat{\mathbf{e}}_\mathbf{n}\}$ be the standard bases for the graded polynomial rings R^m and R^n, respectively. A homogeneous matrix associates a coordinate and monomial to the row and column basis elements. For example, since x_1 is the monomial for row 2 of matrix (5), we have $u_{\mathbf{e_2}} = (1, 0)$ and $x^{u_{\mathbf{e_2}}} = x_1$. Each column $\mathbf{f}$ in F is homogeneous and may be written in terms of rows:

$$\mathbf{f} = \sum_{i=1}^{m} c_i \frac{x^{u_\mathbf{f}}}{x^{u_{\mathbf{e_i}}}} \mathbf{e_i}, \tag{6}$$

where $c_i \in k$ and we allow $c_i = 0$ when a row is not used. For example, column $\mathbf{g}$ for edge ab in our bifiltration may be written as:

$$\mathbf{g} = x_2\mathbf{e_1} + x_2x_2^2\mathbf{e_3} = \frac{x_2x_2^2}{x_1x_2}\mathbf{e_1} + \frac{x_2x_2^2}{1}\mathbf{e_3} = \frac{x^{u_\mathbf{g}}}{x^{u_{\mathbf{e_1}}}}\mathbf{e_1} + \frac{x^{u_\mathbf{g}}}{x^{u_{\mathbf{e_3}}}}\mathbf{e_3} = \sum_{i \in \{1,3\}} \frac{x^{u_\mathbf{g}}}{x^{u_{\mathbf{e_i}}}}\mathbf{e_i}.$$

Consider the BUCHBERGER algorithm [13]. The algorithm repeatedly computes S-polynomials of homogeneous vectors.

Lemma 2. *The S-polynomial $S(\mathbf{f}, \mathbf{g})$ of homogeneous vectors $\mathbf{f}$ and $\mathbf{g}$ is homogeneous.*

Proof. A zero S-polynomial is trivially homogeneous. A non-zero S-polynomial $S(\mathbf{f}, \mathbf{g})$ implies that $\mathbf{h} = \text{LCM}(\text{LM}(\mathbf{f}), \text{LM}(\mathbf{g}))$ is non-zero. By the definition of LCM, the leading monomials of $\mathbf{f}$ and $\mathbf{g}$ contain the same basis element $\mathbf{e_j}$. We have, $\text{LM}(\mathbf{f}) = \frac{x^{u_\mathbf{f}}}{x^{u_{\mathbf{e_j}}}}\mathbf{e_j}$, $\text{LM}(\mathbf{g}) = \frac{x^{u_\mathbf{g}}}{x^{u_{\mathbf{e_j}}}}\mathbf{e_j}$, and:

$$\mathbf{h} = \text{LCM}(\text{LM}(\mathbf{f}), \text{LM}(\mathbf{g})) = \text{LCM}\left(\frac{x^{u_\mathbf{f}}}{x^{u_{\mathbf{e_j}}}}, \frac{x^{u_\mathbf{g}}}{x^{u_{\mathbf{e_j}}}}\right)\mathbf{e_j} = \frac{\text{LCM}\left(x^{u_\mathbf{f}}, x^{u_\mathbf{g}}\right)}{x^{u_{\mathbf{e_j}}}}\mathbf{e_j}.$$

Let $x^\ell = \text{LCM}(x^{u_\mathbf{f}}, x^{u_\mathbf{g}}) = x^{\text{LCM}(u_\mathbf{f}, u_\mathbf{g})}$, giving us $\mathbf{h} = \frac{x^\ell}{x^{u_{\mathbf{e_j}}}}\mathbf{e_j}$. We now have

$$\frac{\mathbf{h}}{\text{LT}(\mathbf{f})} = \frac{\frac{x^\ell}{x^{u_{\mathbf{e_j}}}}\mathbf{e_j}}{c_\mathbf{f}\frac{x^{u_\mathbf{f}}}{x^{u_{\mathbf{e_j}}}}\mathbf{e_j}} = \frac{x^\ell}{c_\mathbf{f}x^{u_\mathbf{f}}},$$

where $c_\mathbf{f} \neq 0$ is the field constant in the leading term of $\mathbf{f}$. Similarly, we get

$$\frac{\mathbf{h}}{\text{LT}(\mathbf{g})} = \frac{x^\ell}{c_\mathbf{g}x^{u_\mathbf{g}}}, \quad c_\mathbf{g} \neq 0.$$

Putting it together, we have

$$\begin{aligned} S(\mathbf{f},\mathbf{g}) &= \frac{\mathbf{h}}{\mathrm{LT}(\mathbf{f})}\mathbf{f} - \frac{\mathbf{h}}{\mathrm{LT}(\mathbf{g})}\mathbf{g} \\ &= \left(\frac{x^\ell}{c_{\mathbf{f}} x^{u_{\mathbf{f}}}} \sum_{i=1}^{m} c_i \frac{x^{u_{\mathbf{f}}}}{x^{u_{\mathbf{e_i}}}} \mathbf{e_i}\right) - \left(\frac{x^\ell}{c_{\mathbf{g}} x^{u_{\mathbf{g}}}} \sum_{i=1}^{m} c'_i \frac{x^{u_{\mathbf{g}}}}{x^{u_{\mathbf{e_i}}}} \mathbf{e_i}\right) = \sum_{i=1}^{m} d_i \frac{x^\ell}{x^{u_{\mathbf{e_i}}}} \mathbf{e_i}, \end{aligned}$$

where $d_i = c_i/c_{\mathbf{f}} - c'_i/c_{\mathbf{g}}$. Comparing with Equation (6), we see that $S(\mathbf{f},\mathbf{g})$ is homogeneous with $u_{S(\mathbf{f},\mathbf{g})} = \ell$.

Having computed the S-polynomial, BUCHBERGER next divides it by the current homogeneous basis G using a call to the DIVIDE algorithm [13].

Lemma 3. DIVIDE$(\mathbf{f}, (\mathbf{f_1}, \ldots, \mathbf{f_t}))$ *returns a homogeneous remainder vector* $\mathbf{r}$ *for homogeneous vectors* $\mathbf{f}, \mathbf{f_i} \in R^m$.

Proof. Initially, we set $\mathbf{r}$ and $\mathbf{p}$ to be $\mathbf{0}$ and $\mathbf{f}$, respectively, so they are both trivially homogeneous. Since both $\mathbf{f_i}$ and $\mathbf{p}$ are homogeneous, we have $\mathbf{f_i} = \sum_{j=1}^{m} c_{ij} \frac{x^{u_{\mathbf{f_i}}}}{x^{u_{\mathbf{e_j}}}} \mathbf{e_j}$, $\mathbf{p} = \sum_{j=1}^{m} d_j \frac{x^{u_{\mathbf{p}}}}{x^{u_{\mathbf{e_j}}}} \mathbf{e_j}$. Since $\mathrm{LT}(\mathbf{f_i})$ divides $\mathrm{LT}(\mathbf{p})$, the terms must share basis element $\mathbf{e_k}$ and we have $\mathrm{LT}(\mathbf{f_i}) = c_{ik} \frac{x^{u_{\mathbf{f_i}}}}{x^{u_{\mathbf{e_k}}}} \mathbf{e_k}$, $\mathrm{LT}(\mathbf{p}) = d_k \frac{x^{u_{\mathbf{p}}}}{x^{u_{\mathbf{e_k}}}} \mathbf{e_k}$, $\mathrm{LT}(\mathbf{p})/\mathrm{LT}(\mathbf{f_i}) = \frac{d_k}{c_{ik}} \cdot \frac{x^{u_{\mathbf{p}}}}{x^{u_{\mathbf{f_i}}}}$, where $x^{u_{\mathbf{p}}} >_{\mathrm{lex}} x^{u_{\mathbf{f_i}}}$ so that the division makes sense. Then, $\mathbf{p}$ is assigned to

$$\begin{aligned} \mathbf{p} - (\mathrm{LT}(\mathbf{p})/\mathrm{LT}(\mathbf{f_i}))\mathbf{f_i} &= \sum_{j=1}^{m} d_j \frac{x^{u_{\mathbf{p}}}}{x^{u_{\mathbf{e_j}}}} \mathbf{e_j} - \left(\frac{d_k}{c_{ik}} \cdot \frac{x^{u_{\mathbf{p}}}}{x^{u_{\mathbf{f_i}}}}\right) \sum_{j=1}^{m} c_{ij} \frac{x^{u_{\mathbf{f_i}}}}{x^{u_{\mathbf{e_j}}}} \mathbf{e_j} \\ &= \sum_{j=1}^{m} \left(d_j - \frac{d_k \cdot c_{ij}}{c_{ik}}\right) \frac{x^{u_{\mathbf{p}}}}{x^{u_{\mathbf{e_j}}}} \mathbf{e_j} = \sum_{j=1}^{m} d'_j \frac{x^{u_{\mathbf{p}}}}{x^{u_{\mathbf{e_j}}}} \mathbf{e_j}, \end{aligned}$$

where $d'_j = d_j - d_k \cdot c_{ij}/c_{ik}$ and $d'_k = 0$, so the subtraction eliminates the kth term. The final sum means that $\mathbf{p}$ is now a new homogeneous polynomial with the same coordinate $u_{\mathbf{p}}$ as before. Similarly, $\mathrm{LT}(p)$ is added to $\mathbf{r}$ and subtracted from $\mathbf{p}$, and neither action changes the homogeneity of either vector. Both remain homogeneous with coordinate $u^{\mathbf{p}}$.

Theorem 1 (homogeneous Gröbner). *The* BUCHBERGER *algorithm computes a homogeneous Gröbner basis for a homogeneous matrix.*

Proof. Initially, the algorithm sets G to be the set of columns of the input matrix F, so the vectors in G are homogeneous by Lemma 1. The algorithm then computes the S-polynomial of homogeneous vectors $\mathbf{f}, \mathbf{g} \in G$. By Lemma 2, the S-polynomial is homogeneous. It then divides the S-polynomial by G. Since the input is homogeneous, DIVIDE produces a homogeneous remainder $\mathbf{r}$ by Lemma 3. Since only homogeneous vectors are added to G, it remains homogeneous. We may extend this result easily to the reduced Gröbner basis.

Using similar arguments, we may show the following result.

Theorem 2 (homogenous syzygy). *For a homogeneous matrix, all matrices encountered in the computation of the syzygy module are homogeneous.*

3.2 Optimizations

We have shown that the structure inherent in a multifiltration allows us to compute using homogeneous vectors and matrices whose entries are monomials only. We next explore the consequences of this restriction on both the data structures and complexity of the algorithms.

By Definition (4), an $m \times n$ homogeneous matrix naturally associates monomials to the standard bases for R^m and R^n. Moreover, every non-zero entry of the matrix is a quotient of these monomials as the matrix is homogeneous. Therefore, we do not need to store the matrix entries, but simply the field elements of the matrix along with the monomials for the bases. We may modify two standard data structures to represent the matrix.

- *linked list:* Each column stores its monomial as well as a linked-list of its non-zero entries in sorted order. The non-zero entries are represented by the row index and the field element. The matrix is simply a list of these columns in sorted order.
- *matrix:* Each column stores its monomial as well as the column of field coefficients. If we are computing over a finite field, we may pack bits for space efficiency.

The linked-list representation is appropriate for sparse matrices as it is space-efficient at the price of linear access time. This is essentially the representation used for computing in the one-dimensional setting [4]. In contrast, the matrix representation is appropriate for dense matrices as it provides constant access time at the cost of storing all zero entries. The multidimensional setting provides us with denser matrices, as we shall see, so the matrix representation becomes a viable structure.

In addition, the matrix representation is optimally suited to computing over the field $\mathbb{Z}_2$, the field often commonly employed in topological data analysis. The matrix entries each take one bit and the column entries may be packed into machine words. Moreover, the only operation required by the algorithms is *symmetric difference* which may be implemented as a binary XOR operation provided by the chip. This approach gives us bit-level parallelism for free: On a 64-bit machine, we perform symmetric difference 64 times faster than on the list. The combination of these techniques allow the matrix structure to perform better than the linked-list representation in practice.

We may also exploit homogeneity to speed up the computation of new vectors and their insertion into the basis. We demonstrate this briefly using the BUCHBERGER algorithm. We order the columns of input matrix G using the POT rule for vectors [13]. Suppose we have $\mathbf{f}, \mathbf{g} \in G$ with $\mathbf{f} > \mathbf{g}$. If $S(\mathbf{f}, \mathbf{g}) \neq 0$, $\mathrm{LT}(\mathbf{f})$ and $\mathrm{LT}(\mathbf{g})$ contain the same basis, which the S-polynomial eliminates. So, we have $S(\mathbf{f}, \mathbf{g}) < \mathbf{g} < \mathbf{f}$. This implies that when dividing $S(\mathbf{f}, \mathbf{g})$ by the vectors in G, we need only consider vectors that are smaller than $\mathbf{g}$. Since the vectors are in sorted order, we consider each in turn until we can no longer divide. By the POT rule, we may now insert the new remainder column here into the basis G. This gives us a constant time insertion operation for maintaining the ordering, as well as faster computation of the Gröbner basis.

3.3 Complexity

Our optimizations from the last section allow us to give simple polynomial bounds on our multigraded algorithms. These bounds, in turn, imply that we may compute multidimensional persistence in polynomial time.

Lemma 4. *Let F be an $m \times n$ homogeneous matrix of monomials. The Gröbner basis G contains $O(n^2 m)$ vectors in the worst case. We may compute G using* Buchberger *in $O(n^4 m^3)$ worst-case time.*

Proof. In the worst case, F contains nm unique monomials. Each column $\mathbf{f} \in F$ may have any of the nm monomials as its monomial when included in the Gröbner basis G Therefore, the total number of columns in the G is $O(n^2 m)$. In computing the Gröbner Basis, we compare all columns pairwise, so the total number of comparisons is $O(n^4 m^2)$. Dividing the S-polynomial takes $O(m)$ time. Therefore, the worst-case running time is $O(n^4 m^3)$.

We omit the proof of the following due to lack of space and refer the reader to the full manuscript [12].

Lemma 5. *Let F be an $m \times n$ homogeneous matrix of monomials and G be the Gröbner Basis of F. The Syzygy module S for G may be computed using* Schreyer's algorithm *in $O(n^4 m^2)$ worst-case time.*

Theorem 3. *Multidimensional persistence may be computed in polynomial time.*

4 Conclusion

In this paper, we develop polynomial time algorithms for multidimensional persistence by recasting the problem into computational algebraic geometry. Although the recast problem is Expspace-complete, we exploit the multigraded setting to develop practical algorithms. We have implemented all our algorithms and provide statistical experiments to demonstrate their feasibility in the full manuscript [12]. For additional speedup, we plan to parallelize the computation by batching and threading the XOR operations. We also plan to apply our algorithms toward studying scientific data. For instance, for zero-dimensional homology, multidimensional persistence corresponds to *clustering* multiparameterized data, This gives us a fresh perspective, as well as a new arsenal of computational tools, to attack an old and significant problem in data analysis.

References

1. Carlsson, G., Zomorodian, A.: The theory of multidimensional persistence. Discrete & Computational Geometry 42(1), 71–93 (2009)
2. Gromov, M.: Hyperbolic groups. In: Gersten, S. (ed.) Essays in Group Theory, pp. 75–263. Springer, New York (1987)
3. de Silva, V., Carlsson, G.: Topological estimation using witness complexes. In: Proc. Symposium on Point-Based Graphics, pp. 157–166 (2004)

4. Zomorodian, A., Carlsson, G.: Computing persistent homology. Discrete & Computational Geometry 33(2), 249–274 (2005)
5. Collins, A., Zomorodian, A., Carlsson, G., Guibas, L.: A barcode shape descriptor for curve point cloud data. Computers and Graphics 28, 881–894 (2004)
6. Gyulassy, A., Natarajan, V., Pascucci, V., Bremer, P.T., Hamann, B.: Topology-based simplification for feature extraction from 3D scalar fields. In: Proc. IEEE Visualization, pp. 275–280 (2005)
7. de Silva, V., Ghrist, R., Muhammad, A.: Blind swarms for coverage in 2-D. In: Proceedings of Robotics: Science and Systems (2005), `http://www.roboticsproceedings.org/rss01/`
8. Singh, G., Memoli, F., Ishkhanov, T., Sapiro, G., Carlsson, G., Ringach, D.L.: Topological analysis of population activity in visual cortex. Journal of Vision 8(8), 1–18 (2008)
9. Carlsson, G., Ishkhanov, T., de Silva, V., Zomorodian, A.: On the local behavior of spaces of natural images. International Journal of Computer Vision 76(1), 1–12 (2008)
10. Frosini, P., Mulazzani, M.: Size homotopy groups for computation of natural size distances. Bull. Belg. Math. Soc. Simon Stevin 6(3), 455–464 (1999)
11. Cohen-Steiner, D., Edelsbrunner, H., Morozov, D.: Vines and vineyards by updating persistence in linear time. In: Proc. ACM Symposium on Computational Geometry, pp. 119–126 (2006)
12. Carlsson, G., Singh, G., Zomorodian, A.: Computing multidimensional persistence, `http://arxiv.org/abs/0907.2423`
13. Cox, D.A., Little, J., O'Shea, D.: Using algebraic geometry, 2nd edn. Graduate Texts in Mathematics, vol. 185. Springer, New York (2005)
14. Mayr, E.W.: Some complexity results for polynomial ideals. Journal of Complexity 13(3), 303–325 (1997)
15. von zur Gathen, J., Gerhard, J.: Modern Computer Algebra, 2nd edn. Cambridge University Press, Cambridge (2003)

Locating an Obnoxious Line among Planar Objects*

Danny Z. Chen and Haitao Wang**

Department of Computer Science and Engineering
University of Notre Dame, Notre Dame, IN 46556, USA
{dchen, hwang6}@nd.edu

Abstract. Given a set P of n points in the plane such that each point has a positive weight, we study the problem of finding an *obnoxious* line that intersects the convex hull of P and maximizes the minimum weighted Euclidean distance to all points of P. We also consider a variant of this problem whose input is a set of m polygons with totally n vertices in the plane such that each polygon has a positive weight and whose goal is to locate an obnoxious line with respect to the weighted polygons. We improve the previous results for both problems. Our algorithms are based on new geometric observations and interesting algorithmic techniques.

1 Introduction

Determining the locations of undesirable or obnoxious facilities among a set of geometric objects has been an important research topic. In this paper, the target obnoxious facility is a line in the plane. Formally, given a set of n points in the plane, $P = \{p_1, p_2, \ldots, p_n\}$, where each point p_i has a weight $w_i > 0$, a line l is said to be *obnoxious* if there is at least one point of P lying on each side of l (i.e., l intersects the convex hull of P) and the value of $\min\{w_i \cdot d(l, p_i) \mid p_i \in P\}$ is maximized, where $d(l, p_i)$ denotes the Euclidean distance between the line l and the point p_i. The *obnoxious line location* (OLL) problem seeks an obnoxious line with respect to the weighted points of P. When all the point weights are equal, we call it the *unweighted* OLL problem.

In a variant, a set of m (possibly intersecting) polygons, $\mathcal{P} = \{P_1, P_2, \ldots, P_m\}$, with a total of n vertices in the plane is given, where each polygon P_i has a weight $w_i > 0$, and the goal is to compute an obnoxious line l with respect to $\mathcal{P}$ such that there is at least one polygon of $\mathcal{P}$ lying on each side of l and the value of $\min\{w_i \cdot d(P_i, l) \mid P_i \in \mathcal{P}\}$ is maximized, where $d(P_i, l)$ is the minimum distance of any point in P_i to l. We call this variant the *OLL problem among weighted polygons*, denoted by OLLP. When all the polygon weights are equal, we call it the *unweighted* OLLP problem.

* This research was supported in part by NSF under Grants CCF-0515203 and CCF-0916606.

** Corresponding author. The work of this author was also supported in part by a graduate fellowship from the Center for Applied Mathematics, University of Notre Dame.

Y. Dong, D.-Z. Du, and O. Ibarra (Eds.): ISAAC 2009, LNCS 5878, pp. 740–749, 2009.

These problems are motivated by applications of locating a linear route through existing facilities such that the route is noxious or hazardous to its surroundings or of obtaining the maximum clearance of a route with respect to the existing facilities. The weights may represent various importance of the facilities. For examples, one may need to build or find pipelines, roads, railway lines, or sailing routes to transport noxious materials.

A key ingredient to our solution for OLL is the following problem, called *disks separability*, which may be of interest in its own right: Given n disks in the plane, either find a line l in the plane such that l does not intersect the interior of any disk and there is at least one disk on each side of l, or report that there exists no such line. We have not found any previous work on this problem.

The OLL problem was first studied in [8], with an $O(n^3)$ time algorithm. The previously best-known algorithm for the OLLP problem, of $O(mn + n\log^2 n \log m + m^2 \log n \log^2 m)$ time, was given in [7] based on parametric search. By treating each input point as a degenerate polygon and using the method in [7] (with $m = n$), OLL is solved in $O(n^2 \log^3 n)$ time. The previously best solution for the unweighted OLLP takes $O((m^2 + n \log m) \log n)$ time [7].

The unweighted OLL problem is equivalent to the widest empty corridor problem [9]. By using topological sweeping and duality, the problem was solved in $O(n^2)$ time and $O(n)$ space [9]. The dynamic version of the widest empty corridor problem was also studied [11]. Other problem variants include k-dense corridors [2,11,15], L-shaped corridors [4], and curved corridors [1]. The "dual" problem of OLL, which seeks a line minimizing the maximum weighted distance between the line and a given set of points in the plane, has been studied as well. The unweighted version is equivalent to the set width problem which is solvable in $O(n \log n)$ time [10,12]. The weighted version was solved in $O(n^2 \log n)$ time [12]. It turns out that solving the OLL problem resorts to considerably different algorithmic techniques than its "dual" problem.

In this paper, we present an $O(n^2 \log n)$ time algorithm for solving the OLL problem, improving the $O(n^2 \log^3 n)$ time solution in [7]. Our algorithm is based on parametric search where our $O(n^2)$ solution for the disks separability problem works as a decision procedure. For the OLLP problem, we give an $O(mn + n\log^2 n + m^2 \log n)$ time algorithm, improving the $O(mn + n\log^2 n \log m + m^2 \log n \log^2 m)$ time result in [7]. For the unweighted OLLP problem, by a slight modification of the $O((m^2 + n \log m) \log n)$ time algorithm in [7], we reduce its running time to $O(m^2 + n \log m)$.

2 The Disks Separability Problem

Given a set of disks in the plane, $\mathcal{D} = \{D_i \mid 1 \leq i \leq n\}$, where each disk D_i is centered at the point (x_i, y_i) with radius r_i, the disks separability problem seeks to either find a line l in the plane such that l does not intersect the interior of any disk in $\mathcal{D}$ and there is at least one disk of $\mathcal{D}$ on each side of l, or report that there exists no such line. We call a line that satisfies these requirements a *separation line* of $\mathcal{D}$. We say that $\mathcal{D}$ is *separable* if there exists a separation line. In this paper, when we say a line *intersects* a disk, we mean that the line intersects the interior of the disk (and thus the line is not tangent to the disk); similarly, when we say two disks *intersect* each other, we mean that the intersection in their interior is not empty. The main result of this section is as follows.

Theorem 1. *Given a set $\mathcal{D}$ of n disks in the plane, in $O(n^2)$ time, we can either find a separation line for $\mathcal{D}$ or report that no separation line exists.*

To prove the theorem, we begin with an obvious but critical observation.

Observation 1. *Given $\mathcal{D}$, a separation line exists if and only if there is a separation line that is a common tangent of two different disks in $\mathcal{D}$.*

Based on Observation 1, to solve the disks separability problem, it suffices to compute the common tangents of all pairs of different disks in $\mathcal{D}$, and for each tangent line, check whether it is a separation line. Our algorithm follows this idea. Since there are totally $O(n^2)$ common tangents, a straightforward solution takes $O(n^3)$ time. An more efficient algorithm is given below.

First, we remove those disks each of which is completely contained in another disk in $O(n \log n)$ time by a sweeping algorithm. Clearly, the removal of such disks does not affect the solution of the problem. Thus, we assume that in $\mathcal{D}$, no disk contains any other disk. Hence, depending on their positions and radii, every pair of disks in $\mathcal{D}$ has at least two and at most four common tangents. For each disk D_i, denote by L_i the set of common tangents between D_i and all other disks in $\mathcal{D}$. Thus, $|L_i| = O(n)$. Each tangent in L_i determines a tangent point on D_i. Denote by T_i the set of tangent points on D_i determined by the tangents in L_i. If all points of T_i are arranged in a counterclockwise order around D_i, then we say that they are *sorted* or in a *sorted order*. We have the following lemma.

Lemma 1. *Suppose the tangent point set T_i is sorted for each $i = 1, 2, \ldots, n$. Then in $O(n^2)$ time, we can either find a separation line for $\mathcal{D}$ or report that no separation line exists.*

Proof. We show that in $O(n^2)$ time, the following algorithm can check every tangent line in $\cup_{i=1}^{n} L_i$ to determine whether it is a separation line.

Consider an arbitrary L_i. For any D_j with $i \neq j$, if D_i intersects D_j, then there are two common tangents between D_i and D_j in L_i. Let p_1 and p_2 be the two tangent points on D_i determined by these two tangents. The two points p_1 and p_2 cut the bounding circle of D_i into two *open* arcs (p_1 and p_2 do not belong to these two arcs). One of the two open arcs must have the following property: A tangent line of D_i intersects the disk D_j if and only if the corresponding tangent point is on that arc of D_i. We call this open arc the *engaging arc* of D_i with D_j. Further, we call p_1 the *entering point* and p_2 the *exiting point* if we pass the engaging arc by moving from p_1 counterclockwise to p_2 along the boundary of D_i; otherwise, p_1 is the exiting point and p_2 is the entering point (see Fig. 1(a)). If D_i does not intersect D_j and D_i is not tangent to D_j, then they have four common tangents in L_i. Similarly, the corresponding four tangent points on D_i cut the circle of D_i into four open arcs, two of them are engaging arcs, and two of the tangent points are entering points and the other two are exiting points (see Fig. 1(b)). If D_i is tangent to D_j, although they have three common tangents, we still view them as four tangents (two of them coincide) and we handle this case in the same way as for the case with four different common tangents.

Hence, there are $O(n)$ engaging arcs on the circle of D_i determined by the tangent points in T_i. Half of the points in T_i are entering points and the other half are exiting points. Denote by A_i the set of all engaging arcs on D_i. Note that all engaging arcs of A_i, including their entering points and exiting points, can be determined when computing the set L_i. In other words, in $O(n)$ time, we can compute the sets L_i, T_i, and A_i, and also determine which points in T_i are entering points or exiting points.

For each tangent line l of D_i, observe that l does not intersect any disk if and only if its corresponding tangent point on D_i does not belong to any engaging arc in A_i. Further, to determine whether l is a separation line, we also need to check whether there is at least one disk of $\mathcal{D}$ on each side of l (i.e., whether l intersects the interior of the convex hull of all disks in $\mathcal{D}$). To rule out all tangent lines to the convex hull of $\mathcal{D}$, we use a preprocessing procedure, as follows. We first compute the convex hull of $\mathcal{D}$, in $O(n \log n)$ time by the algorithm in [13,14]. For each disk D_i, if there is one (closed) arc of D_i that is part of the convex hull, we call it the *forbidden arc* of D_i. It is shown in [14] that the number of arcs and line segments in the convex hull of $\mathcal{D}$ is at most $2n-1$. Thus, the number of forbidden arcs on all disks is at most $2n-1$. To decide whether any tangent of D_i is a separation line, we only need to determine whether there is a point in T_i (already given in sorted order) that does not belong to any of the forbidden arcs of D_i (if any) or the engaging arcs in A_i. We remove the points on all T_i's that belong to any of the forbidden arcs, which takes $O(n^2)$ time since the number of forbidden arcs on all disks is at most $2n-1$. Then, for each T_i, we use a simple *tangent sweeping* algorithm to solve the problem in $O(n)$ time, as follows.

Since the points in T_i are sorted, starting from the first point of T_i and following the sorted order, our algorithm sweeps the circle of D_i by a sweeping point. Initially, we use $O(n)$ time to compute the number of arcs in A_i that cover the starting point. Denote this number by m. In the sweeping, whenever we encounter an entering point, we increase m by 1, meaning that one more engaging arc covers the sweeping point now; when we encounter an exiting point, we decrease m by 1. During the sweeping, if the value of m ever becomes 0, then the algorithm stops and reports that the tangent line of D_i passing through the current tangent point is a separation line. Otherwise, the algorithm stops at the last point of T_i and reports that no tangent to D_i is a separation line. The correctness of the algorithm is obvious and its running time is $O(n)$.

Since we have at most n T_i's, in $O(n^2)$ time, we either find a separation line or report that no separation line exists. The lemma thus follows.

With the above lemma, to prove Theorem 1, it remains to sort all T_i's in $O(n^2)$ time. This is done in the lemma below with its proof in [3].

Lemma 2. *In $O(n^2)$ time, we can obtain all sorted sets T_i's, for $i = 1, 2, \ldots, n$.*

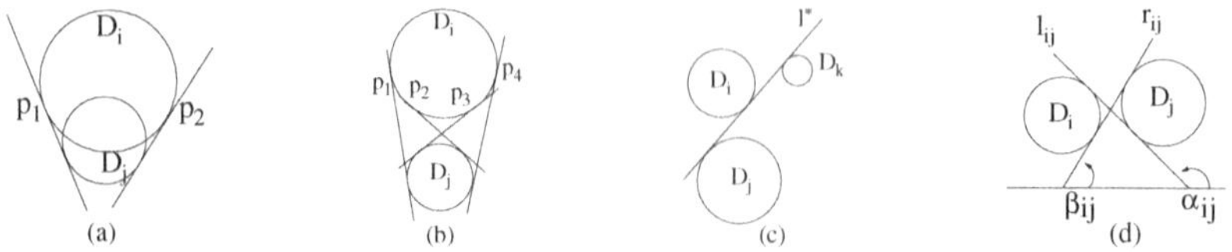

Fig. 1. (a) The case of D_i intersecting D_j: The engaging arc of D_i with D_j is the lower open arc on D_i between points p_1 and p_2; p_1 is its entering point and p_2 is its exiting point. (b) The non-intersection case: There are two engaging arcs on D_i; p_1 and p_3 are two entering points and p_2 and p_4 are two exiting points. (c) The line l^* is the inner common tangent of D_i and D_j (resp., D_i and D_k); the tangent point of D_i on l^* is between those of D_j and D_k on l^*. (d) Illustrating the two inner common tangents.

3 Obnoxious Line Location among Weighted Points

In this section, based on geometric observations (some are from [7,8]), we apply parametric search to solve the OLL problem, which uses the disks separability algorithm as a decision making procedure. Below is the result of this section.

Theorem 2. *Given a set of n weighted points in the plane, an obnoxious line can be determined in $O(n^2 \log n)$ time.*

Given a point set $P = \{p_1, p_2, \ldots, p_n\}$, where each p_i has a weight $w_i > 0$, for any $\epsilon > 0$ and each $1 \leq i \leq n$, define $D_i(\epsilon)$ to be a disk centered at p_i with radius $\frac{\epsilon}{w_i}$. Let $\mathcal{D}(\epsilon)$ be the disk set $\{D_i(\epsilon) \mid 1 \leq i \leq n\}$. Suppose l^* is an obnoxious line for P and ϵ^* is the corresponding error, i.e., $\epsilon^* = \min\{w_i \cdot d(l^*, p_i) \mid p_i \in P\}$. For any $\epsilon > 0$, note that if $\mathcal{D}(\epsilon)$ is separable, then $\epsilon \leq \epsilon^*$ must hold; otherwise, $\epsilon^* < \epsilon$. Based on the results in [8], an obnoxious line l^* has the following (almost self-evident) properties. We say that two disks are *outer tangent* to each other if they are mutually tangent and no disk is contained by the other.

Observation 2. *One of two cases must hold for l^*: (1) There exist two disks in $\mathcal{D}(\epsilon^*)$ outer tangent to each other and l^* passes through their tangent point (l^* is perpendicular to the line connecting the two disk centers); (2) there exist three disks D_i, D_j, and D_k in $\mathcal{D}(\epsilon^*)$ such that l^* is both the inner common tangent of D_i and D_j and the inner common tangent of D_i and D_k, and the tangent point of D_i on l^* is between the tangent points of D_j and D_k on l^* (see Fig. 1(c)).*

Let ϵ_{ij} be the error such that $D_i(\epsilon_{ij})$ and $D_j(\epsilon_{ij})$ are outer tangent to each other, which corresponds to case (1) in Observation 2. Denote by E_1 the set of all such $O(n^2)$ possible ϵ_{ij}'s. Similarly, a corresponding error in case (2) of Observation 2 is determined by three disks, and we denote by E_2 the set of all these errors. Thus, we have $|E_1| = O(n^2)$ and $|E_2| = O(n^3)$. By Observation 2, ϵ^* must be in $E_1 \cup E_2$. By using our disks separability algorithm and the selection algorithm [6], a straightforward approach can find ϵ^* and l^* in $O(n^3)$ time. A more efficient algorithm based on parametric search is given below.

3.1 The Parametric Search

We follow the same high-level framework of the parametric search in [7], but handle a lot of details differently and more efficiently.

We first compute the two elements $\epsilon_1 < \epsilon_2$ in E_1, where ϵ_1 is the largest error such that $\mathcal{D}(\epsilon_1)$ is separable and ϵ_2 is the smallest error such that $\mathcal{D}(\epsilon_2)$ is not separable. Both ϵ_1 and ϵ_2 can be computed in $O(n^2 \log n)$ time by utilizing the disks separability algorithm. Note that it must be $\epsilon_1 \leq \epsilon^* < \epsilon_2$. Thus, if $\epsilon_1 < \epsilon^*$, then ϵ^* is attained in a situation corresponding to case (2) of Observation 2. Furthermore, since the interval (ϵ_1, ϵ_2) contains no value in E_1, for any $\epsilon', \epsilon'' \in (\epsilon_1, \epsilon_2)$ and any $i \neq j$, $D_i(\epsilon') \cap D_j(\epsilon') = \emptyset$ if and only if $D_i(\epsilon'') \cap D_j(\epsilon'') = \emptyset$; in other words, $D_i(\epsilon')$ and $D_j(\epsilon')$ have inner common tangents if and only if $D_i(\epsilon'')$ and $D_j(\epsilon'')$ have inner common tangents. In the following discussion, ϵ is always restricted to be inside the interval (ϵ_1, ϵ_2).

For any two disks $D_i(\epsilon)$ and $D_j(\epsilon)$ that have inner common tangents, denote by l_{ij} (resp., r_{ij}) the inner common tangent that rotates clockwise (resp., counterclockwise) when we increase the value of ϵ. Denote by $\alpha_{ij}(\epsilon)$ (resp., $\beta_{ij}(\epsilon)$) the angle defined by l_{ij} (resp., r_{ij}) and the x-axis (see Fig. 1(d)). To adapt to the forthcoming parametric search, we restrict all the angles $\alpha_{ij}(\epsilon)$ and $\beta_{ij}(\epsilon)$ in the interval $[0, \pi)$. In [7], it was claimed (without a formal proof) that $\alpha_{ij}(\epsilon)$ is decreasing and $\beta_{ij}(\epsilon)$ is increasing when ϵ is increased. However, at least in our problem setting, this is not always the case. Precisely, initially, $\alpha_{ij}(\epsilon)$ is decreasing, but after $\alpha_{ij}(\epsilon) = 0$ (if ever), $\alpha_{ij}(\epsilon)$ starts to decrease from π. Thus, the curve defined by $\alpha_{ij}(\epsilon)$ consists of at most two disjoint continuous decreasing pieces. Fig. 2(a) shows two curves defined by two different α functions. By a slight abuse of notation, we also use $\alpha_{ij}(\epsilon)$ to denote the curve defined by $\alpha_{ij}(\epsilon)$ with $\epsilon \in (\epsilon_1, \epsilon_2)$. For each continuous piece of the curve $\alpha_{ij}(\epsilon)$, define the *curve piece span* as the difference of the largest value of $\alpha_{ij}(\epsilon)$ and the smallest value of $\alpha_{ij}(\epsilon)$ in that curve piece. We then have the next lemma.

Lemma 3. *For any curve $\alpha_{ij}(\epsilon)$, if it has one piece, then its piece span is less than $\pi/2$; if it has two pieces, then the sum of its two piece spans is less than $\pi/2$. Further, there exist no two ϵ' and ϵ'' in (ϵ_1, ϵ_2) such that $\epsilon' \neq \epsilon''$ and $\alpha_{ij}(\epsilon') = \alpha_{ij}(\epsilon'')$.*

Proof. When $\epsilon = 0$, each of the two disks $D_i(\epsilon)$ and $D_j(\epsilon)$ degenerates into a point that is the center of the disk. Denote by l_1 the line containing the two centers, and by α_1 the angle defined by l_1 and the x-axis. Let ϵ_{ij} be the error such that $D_i(\epsilon_{ij})$ and $D_j(\epsilon_{ij})$ are outer tangent to each other and neither of them is contained by the other. Let p' be the tangent point of $D_i(\epsilon_{ij})$ and $D_j(\epsilon_{ij})$. Denote by l_2 the line passing through p' and perpendicular to l_1, and by α_2 the angle defined by l_2 and the x-axis. Refer to Fig. 2(b) for an example. Note that when we increase ϵ from 0 to ϵ_{ij}, l_{ij} will rotate from l_1 to l_2. Thus, l_{ij} rotates exactly $\pi/2$ if we increase ϵ from 0 to ϵ_{ij}. By the definitions of ϵ_1 and ϵ_2, we have $0 \leq \epsilon_1$ and $\epsilon_2 \leq \epsilon_{ij}$. Hence l_{ij} rotates less than $\pi/2$ when $\epsilon \in (\epsilon_1, \epsilon_2)$. Note that the amount of angle rotated by l_{ij} is exactly the piece span if the

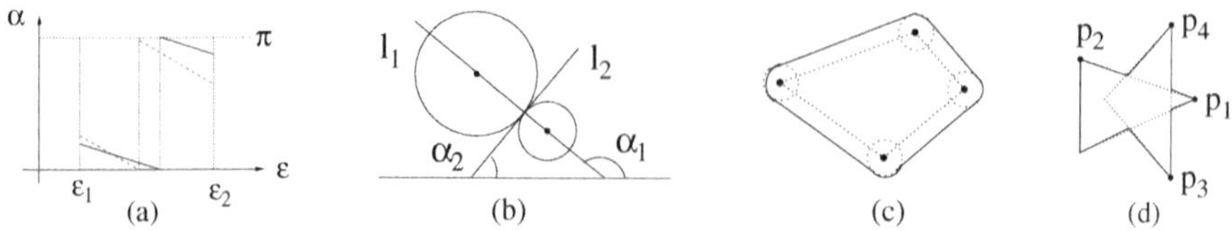

Fig. 2. (a) Illustrating two α curves, one solid (black) and one dotted (red). Each curve has two pieces. The two curves can intersect each other at most once. (b) Illustrating two outer tangent disks and the corresponding l_1, l_2, α_1, and α_2. (c) Illustrating the Minkowski sum (the solid curve) of a polygon (the dashed one) and a disk centered at the origin. (d) Illustrating two convex hulls whose boundaries ∂_i and ∂_j intersect transversally four times. The solid segments are the boundary of their union, i.e., ∂; p_1 and p_2 are on $\partial \setminus \partial_j$ and p_3 and p_4 are on $\partial \setminus \partial_i$.

curve $\alpha_{ij}(\epsilon)$ has one piece or the sum of the two piece spans if the curve has two pieces. The first part of the observation thus follows. The second part is quite straightforward and we omit it due to the space limit.

The curve $\beta_{ij}(\epsilon)$ has similar properties as in Lemma 3.

Note that if all $O(n^2)$ inner common tangent angles α_{ij} and β_{ij} between the $D_i(\epsilon)$'s are sorted at $\epsilon = \epsilon^*$ (ϵ^* is not yet known), then we have $\alpha_{ij}(\epsilon^*) = \beta_{ik}(\epsilon^*)$ if three disks D_i, D_j, and D_k determine l^* as shown in case (2) of Observation 2 (e.g., see Fig. 1(c)). Thus, l^* can be found if we sort all $\alpha_{ij}(\epsilon^*)$'s and $\beta_{ij}(\epsilon^*)$'s. Although we do not know the value of ϵ^*, the sorting can still be done by parametric search, in which the disks separability algorithm is utilized to make decision for each comparison. Further, although each α curve or β curve may not be strictly increasing or decreasing, we can still make our parametric search work by using the special properties in Lemma 3, as explained below.

To compare any two different α angles $\alpha_{ij}(\epsilon^*)$ and $\alpha_{tk}(\epsilon^*)$, we claim that the two curves $\alpha_{ij}(\epsilon)$ and $\alpha_{tk}(\epsilon)$ have at most one intersection if they do not overlap (Lemma 4 in [7] shows a similar result). Intuitively, one curve is decreasing "faster" than the other, and thus, although each of them may have two pieces, they still have at most one intersection (see Fig. 2(a)). We omit the formal proof here. If the two curves overlap (this can be determined in constant time), then we have $\alpha_{ij}(\epsilon^*) = \alpha_{tk}(\epsilon^*)$; otherwise, we need to compute the value ϵ' (if any) such that $\alpha_{ij}(\epsilon') = \alpha_{tk}(\epsilon')$. Although both $\alpha_{ij}(\epsilon)$ and $\alpha_{tk}(\epsilon)$ may have two pieces, we can still compute their intersection in constant time. After obtaining ϵ', we can follow the standard parametric search technique to determine the comparison result of $\alpha_{ij}(\epsilon^*)$ and $\alpha_{tk}(\epsilon^*)$ and possibly shrink the interval (ϵ_1, ϵ_2), which uses the disks separability algorithm. The comparison of any two different β angles $\beta_{ij}(\epsilon^*)$ and $\beta_{tk}(\epsilon^*)$ can be performed similarly. To compare any two values $\alpha_{ij}(\epsilon^*)$ and $\beta_{tk}(\epsilon^*)$, by Lemma 3 and similar observations for the β curves, we can show that the two curves $\alpha_{ij}(\epsilon)$ and $\beta_{tk}(\epsilon)$ have at most one intersection. We omit the proof here. Similarly, their intersection can be computed in constant time. By the standard parametric search technique, we can determine the comparison result of $\alpha_{ij}(\epsilon^*)$ and $\beta_{tk}(\epsilon^*)$ and possibly shrink the interval (ϵ_1, ϵ_2). In summary, we can sort all $O(n^2)$ angles $\alpha_{ij}(\epsilon^*)$ and $\beta_{ij}(\epsilon^*)$ by the standard parametric search for sorting, in which each comparison takes $O(n^2)$ time. By Cole's parametric search [5], we obtain l^* and ϵ^* in $O(n^2 \log n)$ time. Theorem 2 thus follows.

4 Obnoxious Line Location among Weighted Polygons

In this section, we present an improved algorithm for the OLLP problem. It should be noted that by slight modifications on our OLL algorithm, OLLP is solvable in $O(n^2 \log n)$ time. We also give an improved algorithm for the unweighted version. Our results in this section are summarized below.

Theorem 3. *The OLLP problem is solvable in $O(mn+n\log^2 n+m^2 \log n)$ time; its unweighted version is solvable in $O(m^2 + n \log m)$ time.*

We first present an algorithm for the decision version of OLLP which works as a crucial subroutine in the algorithm for OLLP. Our improved algorithm for the unweighted OLLP problem is discussed in [3] due to the space limit.

4.1 The Decision Version of OLLP

Let $\mathcal{P} = \{P_1, P_2, \ldots, P_m\}$ be a set of m polygons with a total of n vertices in the plane such that each polygon P_i has a weight $w_i > 0$. Without loss of generality, we assume that all polygons in $\mathcal{P}$ are convex since a closest point of a polygon P_i to a line not intersecting P_i is always at a vertex of the convex hull of P_i. (If any P_i is not convex, then we simply replace it by its convex hull; this takes $O(n)$ time for the set $\mathcal{P}$.)

For any value $\epsilon > 0$, let $\hat{P}_i(\epsilon)$ be the region that is the Minkowski sum of P_i and the disk centered at the origin with radius $\frac{\epsilon}{w_i}$, for each $1 \leq i \leq m$ (see Fig. 2(c)). Clearly, $\hat{P}_i(\epsilon)$ is convex. Note that the boundary of $\hat{P}_i(\epsilon)$ consists of line segments and arcs, and each arc corresponds to a vertex of P_i. Let $\hat{\mathcal{P}}(\epsilon) = \{\hat{P}_i(\epsilon) \mid 1 \leq i \leq m\}$. We call a line l a *separation line* for $\hat{\mathcal{P}}(\epsilon)$ if l does not intersect the interior of any region in $\hat{\mathcal{P}}(\epsilon)$ and there is at least one region of $\hat{\mathcal{P}}(\epsilon)$ on each side of l. If there exists a separation line for $\hat{\mathcal{P}}(\epsilon)$, then we say that $\hat{\mathcal{P}}(\epsilon)$ is *separable.* Suppose l^* is an obnoxious line for $\mathcal{P}$ and ϵ^* is the corresponding error. Given $\epsilon > 0$, if $\hat{\mathcal{P}}(\epsilon)$ is separable, then it must be $\epsilon \leq \epsilon^*$; otherwise, $\epsilon^* < \epsilon$.

The decision version of OLLP is equivalent to the following problem: Given $\epsilon > 0$, determine whether $\hat{\mathcal{P}}(\epsilon)$ is separable, and if yes, find a separation line. The algorithm in [7] solves this problem in $O((m^2+n\log m)\log n)$ time. We give the following improved result.

Theorem 4. *The decision version of OLLP is solvable in $O(n\log n+m^2)$ time.*

Our algorithm can be viewed as an extension to the disks separability algorithm, with an additional preprocessing procedure. We first give in Lemma 4 an algorithm under the assumption that every two different regions in $\hat{\mathcal{P}}(\epsilon)$ have at most four common tangents, and then show that we can preprocess $\hat{\mathcal{P}}(\epsilon)$ so that this assumption holds while the solution for the original problem is not affected. The algorithm for the following lemma makes use of the techniques of duality, curve arrangement construction and a similar procedure to the tangent sweeping algorithm. Due to the space limit, we leave the details in the full paper [3].

Lemma 4. *If any two different regions in $\hat{\mathcal{P}}(\epsilon)$ have at most four common tangents, then in $O(n \log n + m^2)$ time, we can either find a separation line for $\hat{\mathcal{P}}(\epsilon)$ or report that no separation line exists.*

To satisfy the conditions of Lemma 4, we perform preprocessing on $\hat{\mathcal{P}}(\epsilon)$, which uses a simple sweeping algorithm to find all connected components of $\hat{\mathcal{P}}(\epsilon)$ and then for each connected component, computes its convex hull. It should be noted that in this setting, two regions in $\hat{\mathcal{P}}(\epsilon)$ are considered to be *connected* if and only if they intersect in their interior. Since all regions in $\hat{\mathcal{P}}(\epsilon)$ are convex, this preprocessing can be done in $O(n \log n)$ time by using the algorithm in [13]. Clearly, the number of connected components in $\hat{\mathcal{P}}(\epsilon)$ is no more than m. Let $\hat{\mathcal{P}}'$ be the set of convex hulls thus resulted, with $|\hat{\mathcal{P}}'| \leq m$. The next lemma establishes the fact that $\hat{\mathcal{P}}'$ can be used as the input to the algorithm in Lemma 4 to solve our original problem, which yields the result of Theorem 4.

Lemma 5. *A line is a separation line for $\hat{\mathcal{P}}(\epsilon)$ if and only if it is a separation line for $\hat{\mathcal{P}}'$, and any two different convex hulls in $\hat{\mathcal{P}}'$ have at most four common tangents.*

Proof. First of all, note that a separation line for $\hat{\mathcal{P}}'$ must also be a separation line for $\hat{\mathcal{P}}(\epsilon)$. On the other hand, suppose l is a separation line for $\hat{\mathcal{P}}(\epsilon)$. Denote by $\hat{\mathcal{P}}_{(1)}(\epsilon)$ the set of regions in $\hat{\mathcal{P}}(\epsilon)$ on one side of l and by $\hat{\mathcal{P}}_{(2)}(\epsilon)$ the set of regions in $\hat{\mathcal{P}}(\epsilon)$ on the other side. Since $\hat{\mathcal{P}}_{(1)}(\epsilon)$ and $\hat{\mathcal{P}}_{(2)}(\epsilon)$ are separated by l, any connected component of $\hat{\mathcal{P}}(\epsilon)$ cannot contain both a region in $\hat{\mathcal{P}}_{(1)}(\epsilon)$ and a region in $\hat{\mathcal{P}}_{(2)}(\epsilon)$. Let $\hat{\mathcal{P}}'_{(1)}$ (resp., $\hat{\mathcal{P}}'_{(2)}$) be the set of convex hulls of the connected components of the regions in $\hat{\mathcal{P}}_{(1)}(\epsilon)$ (resp., $\hat{\mathcal{P}}_{(2)}(\epsilon)$). Since l does not intersect the interior of any region in $\hat{\mathcal{P}}_{(1)}(\epsilon)$ (resp., $\hat{\mathcal{P}}_{(2)}(\epsilon)$), l cannot intersect the interior of any convex hull in $\hat{\mathcal{P}}'_{(1)}$ (resp., $\hat{\mathcal{P}}'_{(2)}$). Thus, l is also a separation line for $\hat{\mathcal{P}}'$.

Obviously, two different convex hulls in $\hat{\mathcal{P}}'$ have at most four common tangents if one of the following cases holds: (1) They do not intersect in their interior; (2) one convex hull contains the other. The only remaining case to consider is that the two convex hulls intersect in their interior and neither of them contains the other. Let $\hat{P}'_i$ and $\hat{P}'_j$ be any two such convex hulls in $\hat{\mathcal{P}}'$. Denote by ∂_i and ∂_j their boundaries, respectively. We claim that ∂_i and ∂_j can intersect each other transversally at most twice. (An intersection of two curves is *transversal* if the two curves cross each other at the intersection point.) Let ∂ be the boundary of the union of $\hat{P}'_i$ and $\hat{P}'_j$ (see Fig. 2(d)). Thus, ∂ consists of parts of ∂_i and parts of ∂_j. Suppose ∂_i and ∂_j intersect each other transversally more than twice. Then there must exist two distinct vertices p_1 and p_2 of ∂_i lying on $\partial \setminus \partial_j$ and two distinct vertices p_3 and p_4 of ∂_j lying on $\partial \setminus \partial_i$, such that their cyclical order around ∂ is p_1, p_3, p_2, and p_4 (see Fig. 2(d)). But this implies the two connected components of $\hat{\mathcal{P}}(\epsilon)$ corresponding to $\hat{P}'_i$ and $\hat{P}'_j$ intersect in their interior, a contradiction. Since ∂_i and ∂_j can intersect each other transversally at most twice, $\hat{P}'_i$ and $\hat{P}'_j$ have at most two common tangents. The lemma thus follows.

4.2 Solving OLLP (the Optimization Version)

We follow the high-level algorithmic framework in [7], but use our improved OLLP decision algorithm to make decisions. Further, we show that our parametric search strategy for OLL in Section 3 can be somehow applied. Due to the space limit, all the details are given in the full paper [3].

References

1. Bereg, S., Díaz-Báñez, J.M., Seara, C., Ventura, I.: On finding widest empty curved corridors. Computational Geometry: Theory and Applications 38(3), 154–169 (2007)
2. Chattopadhyay, S., Das, P.: The k-dense corridor problems. Pattern Recognition Letters 11(7), 463–469 (1990)
3. Chen, D.Z., Wang, H.: Locating an obnoxious line among planar objects (2009) (manuscript)
4. Cheng, S.: Widest empty L-shaped corridor. Information Processing Letters 58(6), 277–283 (1996)
5. Cole, R.: Slowing down sorting networks to obtain faster sorting algorithms. Journal of the ACM 34(1), 200–208 (1987)
6. Cormen, T., Leiserson, C., Rivest, R., Stein, C.: Introduction to Algorithms, 2nd edn. MIT Press, Cambridge (2001)
7. Díaz-Báñez, J.M., Ramos, P.A., Sabariego, P.: The maximin line problem with regional demand. European Journal of Operational Research 181(1), 20–29 (2007)
8. Drezner, Z., Wesolowsky, G.O.: Location of an obnoxious route. Journal of Operational Research Society 40(11), 1011–1018 (1989)
9. Houle, M., Maciel, A.: Finding the widest empty corridor through a set of points. In: Tousssaint, G.T. (ed.) Snapshots of Computational and Discrete Geometry. TR SOCS-88.11, Dept. of Computer Science, McGill University, Montreal, Canada (1988)
10. Houle, M.E., Toussaint, G.T.: Computing the width of a set. IEEE Trans. Pattern Anal. Mach. Intell. 10(5), 761–765 (1988)
11. Janardan, R., Preparata, F.P.: Widest-corridor problems. Nordic Journal of Computing 1(2), 231–245 (1994)
12. Lee, D.T., Wu, Y.F.: Geometric complexity of some location problems. Algorithmica 1(1-4), 193–211 (1986)
13. Nielsen, F., Yvinec, M.: Output-sensitive convex hull algorithms of planar convex objects. International Journal of Computational Geometry and Applications 8(1), 39–66 (1998)
14. Rappaport, D.: A convex hull algorithm for discs, and applications. Computational Geometry: Theory and Applications 1(3), 171–187 (1992)
15. Shin, C., Shin, S.Y., Chwa, K.: The widest k-dense corridor problems. Information Processing Letters 68(1), 25–31 (1998)

Finding Fullerene Patches in Polynomial Time

Paul Bonsma[1,⋆] and Felix Breuer[2,⋆⋆]

[1] Humboldt Universität zu Berlin, Computer Science Department,
Unter den Linden 6, 10099 Berlin
bonsma@informatik.hu-berlin.de

[2] Freie Universität Berlin, Mathematics Department, Arnimallee 3, 14195 Berlin
felix.breuer@fu-berlin.de

Abstract. We consider the following question, motivated by the enumeration of fullerenes. A *fullerene patch* is a 2-connected plane graph G in which inner faces have length 5 or 6, non-boundary vertices have degree 3, and boundary vertices have degree 2 or 3. The degree sequence along the boundary is called the *boundary code* of G. We show that the question whether a given sequence S is a boundary code of some fullerene patch can be answered in polynomial time when such patches have at most five 5-faces. We conjecture that our algorithm gives the correct answer for any number of 5-faces, and sketch how to extend the algorithm to the problem of counting the number of different patches with a given boundary code.

1 Introduction

In this paper we consider a graph theoretical problem that is motivated by the generation and enumeration of fullerenes, a problem to which a lot of work has been devoted in mathematical chemistry. A *fullerene* is a molecule consisting only of carbon atoms, which are arranged in a spherical structure, such that every carbon atom is bound to three other carbon atoms in hexagon and pentagon patterns. Since their discovery in 1985, these structures have created an entire new research branch in chemistry, but they have also inspired a lot of research in other fields such as graph theory and algorithm engineering. In this paper we analyze a fascinating question from this area for the first time from a computational complexity viewpoint, and present a strongly improved algorithm. We use basic graph theoretic terminology as in defined [9]. For detailed definitions see also Section 2.

In graph theoretical terms, fullerenes can be modelled by 3-regular plane graphs with only 5-faces and 6-faces (*fullerene graphs*). In the study of how fullerenes are generated and can be enumerated, the following concept is essential [10,6,5]: a fullerene patch can be obtained from a fullerene graph by taking a cycle in the plane graph and removing every vertex and edge outside of the cycle. This motivates the following definition: a *fullerene patch* (or simply *patch*) is a 2-connected plane graph in which every inner face has length 5 or 6, every non-boundary vertex has degree 3, and boundary vertices

⋆ Supported by DFG grant BO 3391/1-1.

⋆⋆ Supported by the Graduate School "Methods for Discrete Structures" in Berlin, DFG grant GRK 1408.

Y. Dong, D.-Z. Du, and O. Ibarra (Eds.): ISAAC 2009, LNCS 5878, pp. 750–759, 2009.

have degree 2 or 3. (*Boundary vertices and edges* are those that are incident with the unique unbounded face, the *outer face*. All other faces are *inner faces*.) The problem we study can informally be posed as follows: given a fullerene patch, is it a subgraph of some fullerene graph? This is the *decision problem*. We will also discuss the important *counting* version of the problem, which asks in how many ways a fullerene patch can be completed to a fullerene graph.

These problems are well-studied in chemistry and combinatorics [12,8,4,7,5], and algorithms have been developed for special cases (see below). Little is however known about the computational complexity of the problem. To illustrate how little is known, and how much this problem differs from the 'usual' problems studied in algorithmics: it is not even known whether the decision problem is decidable[1]!

In this paper we give a polynomial time algorithm for a broad class of instances of the decision problem. We conjecture that our algorithm actually solves the decision problem in polynomial time for all instances (see Section 4). We will also sketch how to extend our approach to solve the counting problem. Our algorithm will be formulated for a slightly more general version of the above problem. We will pose a number of open questions and conjectures on the complexity of this generalization.

A sequence $S = x_0, \ldots, x_{k-1}$ is a *boundary code* of a fullerene patch G if the boundary vertices of G can be labeled $v_0, \ldots, v_{k-1}$ in cyclic order along the boundary (i.e. the cycle $v_0, \ldots, v_{k-1}, v_0$ is the boundary of the outer face of G), and the degree $d(v_i) = x_i$ for all i. It can be seen that a patch G with boundary code S can be completed to a fullerene graph if and only if there exists a fullerene patch with *complementary* boundary code $\overline{S} := 5 - x_0, 5 - x_1, \ldots, 5 - x_{k-1}$; the boundary cycles can be identified such that vertices of degree 3 are identified with vertices of degree 2, to yield a 3-regular planar graph. So to answer the problem, we only need to know whether a patch with a prescribed boundary code exists, and therefore we formulate the problem FULLERENE PATCH - BOUNDARY CODE as follows: Given a sequence S of twos and threes of length n, does there exist a fullerene patch with boundary code S?

This problem is slightly more general because we do not require that S is the complement of a boundary code for some patch. Implications of this generalization are discussed in Section 4. This problem, and the related problems of counting or generating all possible solutions, are also known as the *PentHex Puzzle* in the literature. A fullerene patch G for which S is a boundary code will also be called a *solution* to S. For a sequence S, we use $(S)^x$ to denote the sequence obtained by repeating S x times.

Let $f_5(G)$ denote the number of inner faces of G of length 5. It is well-known that $f_5(G)$ is determined by the degrees on the boundary. Let $d_i(G)$ denote the number of boundary vertices of G with degree i.

Proposition 1. *For a fullerene patch G, $f_5(G) = 6 - d_2(G) + d_3(G)$.*

(This expression follows from Euler's formula by elementary arguments). Note that from this expression, it also follows that fullerene graphs have exactly twelve 5-faces. For a sequence S of twos and threes, define $d_i(S)$ to be the number of times i occurs in S,

[1] Informally speaking, it is possible to exhaustively enumerate all possible solutions, and check whether one of these gives a valid solution, so the problem is Turing recognizable. However when no solution is found, it is not clear when one may terminate and return 'no'.

and $f_5(S) = 6 - d_3(S) + d_2(S)$. It is known that when $f_5(S) \leq 5$, any patch with boundary code S has size $O(n^2)$ (throughout, n denotes the length of S). Bornhöft, Brinkmann and Greinus give precise upper and lower bounds for possible sizes [3]. On the other hand, when $f_5(S) \geq 6$ there may be infinitely many patches with boundary code S. Consider for instance $S = (2,3)^5$. A solution G with six 5-faces and no 6-faces exists, but arbitrarily many 'layers' of 6-faces may be added around G while maintaining the same boundary code.

When $f_5(G) = 0$ for a fullerene patch G, G is called a *hexagonal patch*. Even in this case, the problem is not trivial: Guo, Hansen and Zheng [13] showed that even boundary codes S with $f_5(S) = 0$ may have multiple solutions, although they all have the same size. Their construction can be extended to show that exponentially many solutions are possible. Nevertheless, we have shown in [1] that in this case counting can be done in polynomial time:

Theorem 2. *The number of hexagonal patches that satisfy a boundary code S of length n can be computed in time $O(n^3)$.*

The algorithm is based on a known technique that uses the fact that hexagonal patches can be mapped uniquely to the hexagonal lattice (the infinite 3-regular planar graph where all faces have length 6) with a locally injective homomorphism [13,12]. Previous algorithms for the case $f_5(S) = 0$ focused on the simple special case of patches where the aforementioned mapping is injective [8], or used an algorithm that branches for every face [8,4]. No explicit complexity bounds are given in [4] and [8], but we observe that the worst case complexity is superexponential, a rough bound is $O\left(n^{n^2}\right)$. Other algorithms use variants of this branching approach that apply to sequences S of a special form [7,5], or simply generate all possible patches and categorize them according to boundary codes [6].

Intuitively, the essential new idea in our algorithm is that we have found a way to guess the positions of the 5-faces in advance; we only branch once for each 5-face, instead of for every face. In addition this is done such that the number of possible guesses $O(n^3)$ for one 5-face is not much more than the maximum number of positions of a 5-face; the maximum number of faces of the patch is $O(n^2)$. It can then be checked in polynomial time $O(n^3)$ whether for these guessed positions of the 5-faces a valid solution exists. This way, for any $f_5(S) \leq 5$, we show that the problem can be solved in polynomial time $O(n^{3f_5+3})$ (Section 3), a vast improvement on the complexity of previous algorithms. This is a rather rough bound; in Section 4 we discuss improvements. We start in Section 2 with definitions. Due to space constraints, some details are omitted, see also [2].

2 Preliminaries

For basic graph theoretic notions not defined here we refer to [9]. A *walk* W of *length* k in a graph G is a sequence of vertices $W = v_0, \dots, v_k$ such that $v_i v_{i+1} \in E(G)$ for all i. This is also called a (v_0, v_k)*-walk*. W is *closed* if $v_k = v_0$. It is a *path* if all vertices are distinct, and a *cycle* or k*-cycle* if it is closed and $v_i \neq v_j$ for all distinct $i, j \in \{0, \dots, k-1\}$. A graph is *planar* if it admits a planar embedding or simply *embedding*, which is a drawing in the plane without edge crossings. A *plane graph* is a graph together with a fixed (planar) embedding. The unbounded face is called the *outer face*, all other faces *inner faces*. For every vertex in a plane graph, the clockwise order of edges around every vertex defines a cyclic order on the incident edges. We say that a walk $W = v_0, \ldots, v_k$ *turns left (right) at i* if v_iv_{i+1} follows (precedes) $v_{i-1}v_i$ in this clockwise order around v_i, for $1 \leq i \leq k-1$. If the walk is closed, this is also defined for $i = 0$, as expected. We will mostly consider graphs with maximum degree 3 and walks with $v_{i-1} \neq v_{i+1}$, in which case the walk turns left or right at every $1 \leq i \leq k-1$. A closed walk W in a plane graph is a *facial walk* or simply *face* if it is a minimal closed walk that turns left at every index. If W has length k this is also called a *k-face*. Observe that a graph is 2-connected if and only if every facial walk is a cycle. Throughout, we will only consider plane graphs of which every component is 2-connected, with an embedding such that only the outer face is incident with multiple components. So every component C has a facial cycle that is incident with the outer face. Such a cycle is called a *boundary cycle* of C. Vertices and edges that are part of a boundary cycle are called *boundary vertices and edges*, respectively. For a sequence $\sigma = \sigma_0, \ldots, \sigma_k$, by σ^{-1} we denote the *reversed* sequence $\sigma_k, \ldots, \sigma_0$. It if is a sequence of numbers, $\overline{\sigma}$ denotes the *complementary sequence* $5 - \sigma_0, 5 - \sigma_1, \ldots, 5 - \sigma_k$. $(\sigma)^x$ denotes the sequence consisting of $x \geq 0$ repetitions of σ. We call sequences *lists* when their elements are sequences again. We will use the notation '$|$' to separate sequences in a sequence list, i.e. $2,3,3 \mid 2,2,2$ is a list consisting of two sequences. A patch G is a *solution* to a sequence $S = x_0, \ldots, x_{k-1}$ if G has a boundary cycle $v_0, \ldots, v_{k-1}, v_0$ with $d(v_i) = x_i$ for all $0 \leq i \leq k-1$. A plane graph of which every component is a fullerene patch, embedded such that all components are incident with the outer face, is called a *patch set*. For a sequence list $S = S_1 \mid S_2 \mid \ldots \mid S_k$, a patch set G is called a *solution to* S if the components of G can be numbered $G_1, \ldots, G_k$ such that G_i is a solution to S_i for all $1 \leq i \leq k$. If G is a solution to S, S is called a *boundary code* of G. For a sequence S consisting of d_2 twos and d_3 threes, $f_5(S) = 6 - d_2 + d_3$. For a sequence list $S = S_1 \mid \ldots \mid S_k$, $f_5(S) = \sum_{i=1}^k f_5(S_i)$ and $d_j(S) = \sum_{i=1}^k d_j(S_i)$ (for $j = 2,3$). For a vertex v in a connected plane graph G, the *distance to the boundary* of v is the minimum length over all (v,w)-paths where w is a boundary vertex of G. For a connected plane graph G, $\text{dist}(G)$ denotes the maximum distance to the boundary over all vertices $v \in V(G)$. For a disconnected plane graph G in which every component is incident with the outer face, $\text{dist}(G)$ denotes the maximum of $\text{dist}(C)$ over all components C of G.

3 The Algorithm

Below we will define graph operations on patches G that use use shortest paths $P = u_0, \ldots, u_l$ from the boundary of G to a 5-face f. So u_0 is a boundary vertex of G, u_l is incident with f, and no shorter path with these properties exists. To limit the number of possible operations, we first give an upper bound for $\text{dist}(G)$, which bounds the length of such a path P. The next lemma can be proved using similar techniques as those in [3]. We remark that with more effort, the bound can be improved, but for our purposes this suffices.

Lemma 3. *Let G be a patch with $f_5(G) \leq 5$ with boundary length n. Then dist(G) $\leq$ $n-3$.*

There always exist shortest paths of a very restricted type, called 1-bend paths[2], which ensures that we only have to consider a polynomial number of possible operations.

Definition 4. *A path $P = u_0, \ldots, u_l$ in a patch G with u_0 on the boundary and no other vertices on the boundary is a* 1-bend path of length l with bend at b *if there exists an even $b \in \{0, \ldots, l\}$ such that P turns left at $i \in \{1, \ldots, l-1\}$ if and only if $i \leq b$ and i is odd, or $i > b$ and i is even.*

Note that the choice of l, b and u_0 uniquely determines P, provided that a 1-bend path with these parameters exists.

Lemma 5. *Given a patch G and inner face f of G, there exists a shortest path P from the boundary to f that is a 1-bend path.*

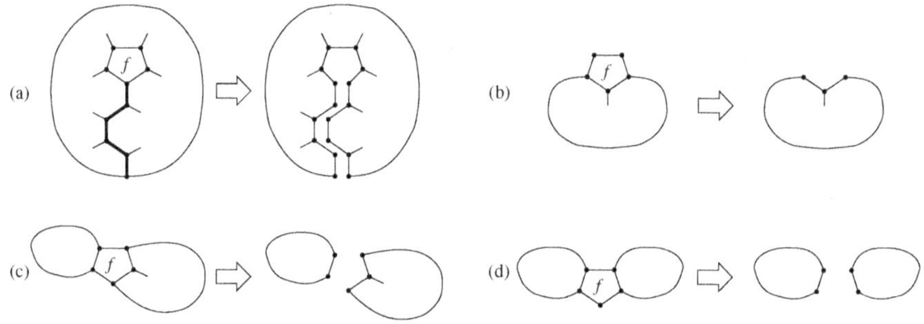

Fig. 1. Cutting a patch using 5-face f and path P

Let G be a patch with 5-face f, and let $P = u_0, \ldots, u_l$ be a 1-bend shortest path in G from the boundary of G to f. For any such path P and 5-face f we define the *cutting operation* using f and P as shown in Figure 1. In the case that P has non-zero length, this operation is defined as follows. See Figure 1(a), where the bold edges indicate P. Every vertex u_i of P is replaced by two vertices v_i and w_i, and every edge u_iu_{i+1} is replaced by two two edges v_iv_{i+1} and w_iw_{i+1}. Edges xu_i with $x \notin V(P)$ are replaced by either xv_i or xw_i, as shown in Figure 1(a), such that a planar embedding is maintained, and every inner face other than f corresponds again to an inner face. Observe that v_0, w_0, v_l and w_l receive degree 2, and for all other i, $d(v_i) + d(w_i) = 5$.

In the case that P has length zero, f must contain a boundary edge, because G has maximum degree 3. In this case, the cutting operation simply consists of deleting all edges of f that are boundary edges of G, and the resulting isolated vertices (see Figure 1(b),(c),(d)). Note that when f contains only boundary edges (G is just this 5-cycle), the resulting graph is the empty graph. We say that an edge set is *connected* if it induces a connected subgraph. When the boundary edges of f are not connected, this

[2] Using 1-bend paths was suggested to us by Gunnar Brinkmann.

operation disconnects the graph (Figure 1(c),(d)). However since f has length 5, at most two components result.

It is easy to see that these operations preserve a plane embedding, and that the vertex degree and face length conditions are maintained. In addition, every component of G' is 2-connected, since every edge lies again on a cycle (corresponding to an inner face).

Proposition 6. *Let G' be obtained by a cutting operation using face f and path P of a patch G. Every component of G' is again a patch.*

Now we will consider how the boundary code changes when applying a cutting operation to a patch G. Formally, since earlier operations may have disconnected the graph, we have to consider the case that G is a patch set. The result G' is again a patch set (which may contain more or fewer components). If G is a solution to a sequence S, then G' is a solution to some S' that can be obtained by one of the *sequence operations* on S that are defined below. Since the cutting operation is only applied to one component of G, we define sequence operations only for the case that G is a single patch; generalizing these definitions to patch sets is straightforward. Let f and P respectively be the 5-face and path used for the cutting operation on G, so P is a 1-bend path of length l with bend at b. Let S be a boundary code of G. We consider four cases.

If $l \geq 1$, then f contains no boundary edges. Then it can be seen that a boundary code of the new patch G' can be obtained from S by replacing a single three by the new sequence shown on the right, see Figure 1(a):

$$\ldots,3,\ldots \Rightarrow \ldots,2,\sigma,2,3,3,3,3,2,\overline{\sigma}^{-1},2,\ldots,$$

where σ is a sequence of twos and threes of length $l-1$. Since P is a 1-bend path of length l with bend at b, $\sigma = \sigma_1,\ldots,\sigma_{l-1}$ is the following sequence: $\sigma_i = 3$ if and only if $i \leq b$ and i is odd, or $i > b$ and i is even. The corresponding operation on sequences of twos and threes is called a *sequence operation of type I of length l*.

Now consider the case that $l = 0$, so f contains at least one boundary edge. In the first case, f contains $x < 5$ boundary edges which are connected. So S is a cyclic permutation of a sequence of the form shown on the left:

$$3,(2)^{x-1},3,y_0,\ldots,y_{n_1} \Rightarrow 2,(3)^{4-x},2,y_0,\ldots,y_{n_1}.$$

Here the numbers indicate degrees of boundary vertices incident with f. The sequence shown on the right is then a boundary code of the resulting patch (see Figure 1(b)). The corresponding sequence operation is called a *sequence operation of type II*, with $1 \leq x \leq 4$. Now suppose that the boundary edges of f are not connected. Since f is a 5-face, this means that f contains either two isolated boundary edges (Figure 1(c)), or one isolated boundary edge and a pair of adjacent boundary edges (Figure 1(d)). In other words, for $a \in \{0,1\}$, the boundary code S is a cyclic permutation of a sequence of the form shown on the left:

$$3,(2)^a,3,y_0,\ldots,y_{n_1},3,3,z_0,\ldots,z_{n_2} \Rightarrow 2,(3)^b,2,y_0,\ldots,y_{n_1} \mid 2,(3)^c,2,z_0,\ldots,z_{n_2}.$$

A cutting operation on G then yields a boundary code list of the form shown on the right (consisting of two sequences), where b and c are non-negative integers satisfying

$a+b+c=1$. Sequence operations that replace a sequence of the first form with two sequences of the second form are called *sequence operations of type III*. In the final case where f contains only boundary edges, the boundary code of G is simply $2,2,2,2,2$, and the resulting empty graph has an empty boundary code. Such an operation is called a *sequence operation of type IV*. To summarize, for every possible cutting operation, we have defined a corresponding sequence operation.

Proposition 7. *Let G' be the patch set obtained from a cutting operation on a patch set G. If S is a boundary code of G, then a boundary code of G' can be obtained by applying a sequence operation of type I, II, III or IV to one of the sequences in S.*

Similarly, it can be checked that we did not define sequence operations that do not correspond to cutting operations. So by considering the corresponding reversed cutting operations, the following proposition can be proved.

Proposition 8. *Let S' be a sequence list obtained from a sequence list S by one of the sequence operations defined above. If S' has a solution, then S has a solution.*

When we use the expression *'all sequence operations of length at most d'*, this includes all sequence operations of type II, III and IV.

Proposition 9. *Let S be a sequence list of twos and d_3 threes. There are less than $d_3^2 + d^2 d_3$ ways to apply a sequence operation of type I, II or III of length at most d to S.*

Algorithm 1.

INPUT: A sequence S of length n, which either has no solution or
a solution G with $\text{dist}(G) \leq n-3$.
OUTPUT: The existence of a solution G to S.

1. **call** TEST($S, n-3$)
2. **If** $S = 2,2,2,2,2$ **then** output('yes') **else** output('no')

Subroutine TEST(S: sequence list, d: integer):

1. **if** $f_5(S) = 0$ **then**
2.   **if** for every sequence S' in list S a hexagonal patch exists, **then**
3.   output('yes'), **halt**
4. **else**
5.   **for** all possible ways to apply a type I, II or III operation of length at most d to S:
6.   Let S' be the resulting sequence list
7.   **while** a type IV operation can be applied to S':
8.   Let S' be the resulting sequence list
9.   **call** TEST(S', d)

Algorithm 1 now shows our algorithm that decides whether for a given sequence S, a patch with boundary code S exists. Line 2 of the subroutine TEST requires some additional explanation: here it is tested whether a sequence list S with $f_5(S) = 0$ admits a solution. Note that if one sequence S' in list S has $f_5(S') > 0$, then another sequence S'' must have $f_5(S'') < 0$, so then the condition is obviously not satisfied. Otherwise, every sequence S' in the list S has $f_5(S') = 0$, and we can use the algorithm from Theorem 2 for every sequence. If S is the empty list (which may occur after applying type IV sequence operations), then the condition is trivially satified. We first prove the correctness of Algorithm 1.

Theorem 10. *Let S be a sequence with length n, such that S either has no solution, or a solution G with $dist(G) \leq n-3$. Then Algorithm 1 returns whether S has a solution.*

Proof: We prove by induction over $f_5(S)$ that if S has a solution G with $\text{dist}(G) \leq d$, then TEST(S,d) returns 'yes', provided that no type IV operation can be applied to S. (Note that if initially a type IV operation can be applied, 'yes' will be returned in Line 2 instead.) If $f_5(S) = 0$ the statement is clear, so assume $f_5(S) \geq 1$. Now there exists a 1-bend path P in G from the boundary to a 5-face f, of length at most d (Lemma 5). Therefore G can be transformed to a patch set G' with $f_5(G') = f_5(G) - 1$ by a cutting operation of length at most d (Proposition 6). Note that cutting operations do not increase the distance to the boundary, so $\text{dist}(G') \leq \text{dist}(G) \leq d$. Proposition 7 shows that a sequence operation on S (of length at most d) exists such that the resulting sequence list S' is a boundary code for G'. By our assumption, this operation is of type I, II or III. Since the algorithm tries all possibilities to do such sequence operations of length at most d on S, in one of the iterations of the for-loop, S' is considered. Applying type IV operations to S' as long as possible (Line 7) does not change the fact that S' has a solution G' with $\text{dist}(G') \leq d$, so by induction on $f_5(S)$, the recursive call TEST(S',d) returns 'yes'.

On the other hand, if 'yes' is returned by the algorithm, then S has a solution: this is again clear if $f_5(S) = 0$ or if $S = 2,2,2,2,2$. Otherwise, let S' be the sequence list obtained from S in the recursion branch in which 'yes' is returned. By induction over $f_5(S)$, S' then has a solution G'. Proposition 8 shows that from G', a solution G to S can be obtained by applying the appropriate reversed cutting operation. □

The complexity of Algorithm 1 can be bounded using the following observations: (i) On input S, the depth of the recursion tree is at most $f_5(S)$. (ii) TEST(S, d) makes at most $d_3^2(S) + d^2 d_3(S)$ recursive calls, when $f_5(S) \geq 1$ (Proposition 9). (iii) A sequence operation of length at most d increases $d_3(S)$ by at most $d+2$. (iv) TEST(S,d) has complexity $O(n^3) = O(d_3^3(S))$ when $f_5(S) = 0$, since then the complexity is determined by the algorithm from Theorem 2. Combining these observations properly yields the following complexity bound.

Theorem 11. *Let S be a sequence with $k = f_5(S)$ and length n. The time complexity of Algorithm 1 on input S is $O\left(k!\, k^3\, n^{2k+3}\, \frac{(n+k)!}{n!}\right)$.*

If $f_5(S) \leq 5$, the condition of Theorem 10 is satisfied by Lemma 3. So combining Theorem 10 and 11 then yields:

Theorem 12. *Let S be a sequence with $k = f_5(S) \leq 5$ and length n. Algorithm 1 returns whether S has a solution, in time $O(n^{3k+3}) \subseteq O(n^{18})$.*

4 Discussion

We gave the first polynomial time algorithm for finding fullerene patches with a given boundary code S, when $f_5(S) \leq 5$. This opens up the way to further studies of computational complexity of this problem, and can be used to as a basis for developing fast practical algorithms for this problem.

Our focus was on proving membership in P and introducing new algorithmic techniques. We remark that with a more detailed topological proof it can be shown that for any patch, there exists an alternating left-right path from some 5-face to the boundary of length $O(n)$. Applying this result in our algorithm would improve the complexity to $O(n^{2f_5+3})$. The exponent can be improved further, but we do not know whether it is possible to entirely remove the parameter f_5 from it (see below). In addition there are many ad-hoc improvements possible to reduce the branching, but that is beyond the scope of this paper.

Observe that there is only one part where we needed the assumption that $f_5(S) \leq 5$, namely in Lemma 3 that bounds dist(G) by $n-3$, for *any* solution G. For $f_5 \geq 6$ such a statement does not hold since in that case, arbitrarily large solutions may exist. However, to answer the question whether at least one solution exists, proving a weaker statement suffices:

Conjecture 1. *For any sequence S of twos and threes of length n, either no solution exists, or at least one solution G with dist$(G) \leq \max\{n-3, 10\}$ exists.*

Note that a proof of Conjecture 1 would show that Algorithm 1 solves the problem for any value of $f_5(S)$, with complexity as stated in Theorem 11. That is, in polynomial time for any fixed f_5. (The small cases with $n < 13$ can be treated correctly by initially setting the parameter $d = 10$ instead of $d = n-3$.)

Algorithm 1 can be extended to the counting problem, by generating solutions, storing them in a list and comparing equivalence (see the full version for details). The only problem is that there may be exponentially many solutions, so in polynomial time one can only decide this way if there are at least $p(n)$ solutions for some polynomial $p(n)$.

Question 2. *Can the number of different solutions to a boundary code S with $f_5(S) \leq 5$ be determined in polynomial time?*

Recall that originally we considered the problem whether a given patch can be completed to a fullerene graph. Expressed in terms of the boundary code problem, this restricts the problem to sequences S such that the complement $\overline{S}$ also has a solution, which we will call *real* sequences. This restriction has some advantages: for instance this implies that $f_5(S) \leq 12$, so proving Conjecture 1 would show that the restricted problem can be solved in polynomial time (without the condition 'for any fixed f_5'). Secondly, we expect that real sequences can only have polynomially many solutions.

Question 3. *It there a polynomial $p(n)$ such that every real sequence S of length n with $f_5(S) \leq 5$ has at most $p(n)$ solutions?*

This would imply that the approach sketched above solves the counting problem in polynomial time, when restricted to real sequences. However, we expect that the general problem cannot be solved in polynomial time without restricting f_5. Note that in general patches may have arbitrarily many 5-faces.

Question 4. *Is Fullerene Patch - Boundary Code NP-hard?*

Finally, considering how the complexity depends on the parameter f_5, this problem is an excellent candidate to be considered from the viewpoint of parameterized complexity [11]. Our algorithm has complexity $n^{O(f_5)}$. An algorithm with complexity $f(f_5)n^{O(1)}$ for some computable function f (a *fixed parameter tractable (FPT) algorithm*) would be preferable, but we do not know whether such an algorithm is possible.

Question 5. *Does there exist an FPT algorithm for the problem Fullerene Patch - Boundary Code parameterized by* $f_5(S)$*, or is this problem* $W[1]$*-hard?*

Acknowledgement. We thank Gunnar Brinkmann for introducing us to this subject and his suggestions.

References

1. Bonsma, P., Breuer, F.: Counting hexagonal patches and independent sets in circle graphs (2009), `http://arxiv.org/abs/0808.3881v2`
2. Bonsma, P., Breuer, F.: Finding fullerene patches in polynomial time (2009), `http://arxiv.org/abs/0907.2627`
3. Bornhöft, J., Brinkmann, G., Greinus, J.: Pentagon–hexagon-patches with short boundaries. European J. Combin. 24(5), 517–529 (2003)
4. Brinkmann, G., Coppens, B.: An efficient algorithm for the generation of planar polycyclic hydrocarbons with a given boundary. MATCH Commun. Math. Comput. Chem. (2009)
5. Brinkmann, G., Dress, A.W.M.: A constructive enumeration of fullerenes. J. Algorithms 23(2), 345–358 (1997)
6. Brinkmann, G., Fowler, P.W.: A catalogue of growth transformations of fullerene polyhedra. J. Chem. Inf. Comput. Sci. 43, 1837–1843 (2003)
7. Brinkmann, G., Nathusius, U.v., Palser, A.H.R.: A constructive enumeration of nanotube caps. Discrete Appl. Math. 116(1-2), 55–71 (2002)
8. Deza, M., Fowler, P.W., Grishukhin, V.: Allowed boundary sequences for fused polycyclic patches and related algorithmic problems. J. Chem. Inf. Comput. Sci. 41, 300–308 (2001)
9. Diestel, R.: Graph theory, 3rd edn. Springer, Berlin (2005)
10. Endo, M., Kroto, H.W.: Formation of carbon nanofibers. J. Phys. Chem. 96, 6941–6944 (1992)
11. Flum, J., Grohe, M.: Parameterized Complexity Theory. Texts in Theoretical Computer Science. An EATCS Series. Springer, Berlin (2006)
12. Graver, J.E.: The (m,k)-patch boundary code problem. MATCH Commun. Math. Comput. Chem. 48, 189–196 (2003)
13. Guo, X., Hansen, P., Zheng, M.: Boundary uniqueness of fusenes. Discrete Appl. Math. 118, 209–222 (2002)

Convex Drawings of Internally Triconnected Plane Graphs on $O(n^2)$ Grids*

Xiao Zhou and Takao Nishizeki

Grad. Sch. of Inf. Sci., Tohoku University, Sendai 980-8579, Japan
zhou@ecei.tohoku.ac.jp, nishi@ecei.tohoku.ac.jp

Abstract. In a convex grid drawing of a plane graph, every edge is drawn as a straight-line segment without any edge-intersection, every vertex is located at a grid point, and every facial cycle is drawn as a convex polygon. A plane graph G has a convex drawing if and only if G is internally triconnected. It has been known that an internally triconnected plane graph G of n vertices has a convex grid drawing on a grid of $O(n^3)$ area if the triconnected component decomposition tree of G has at most four leaves. In this paper, we improve the area bound $O(n^3)$ to $O(n^2)$, which is optimal up to a constant factor. More precisely, we show that G has a convex grid drawing on a $2n \times 4n$ grid. We also present an algorithm to find such a drawing in linear time.

1 Introduction

Recently automatic aesthetic drawing of graphs has created intense interest due to their broad applications, and as a consequence, a number of drawing methods have come out [12]. The most typical drawing of a plane graph is a *straight line drawing*, in which all edges are drawn as straight line segments without any edge-intersection. A straight line drawing is called a *convex drawing* if every facial cycle is drawn as a convex polygon. One can find a convex drawing of a plane graph G in linear time if G has one [3,4,12].

A straight line drawing of a plane graph is called a *grid drawing* if all vertices are put on grid points of integer coordinates. This paper deals with a *convex grid drawing* of a plane graph. Throughout the paper we assume for simplicity that every vertex of a plane graph G has degree three or more. Then G has a convex drawing if and only if G is "internally triconnected" [2,9,10]. One may thus assume that G is internally triconnected. If either G is triconnected [1,2] or the "triconnected component decomposition tree" $T(G)$ of G has two or three leaves [9], then G has a convex grid drawing on an $(n-1) \times (n-1)$ grid, where n is the number of vertices in G. If $T(G)$ has exactly four leaves, then G has a convex grid drawing on a $2n \times n^2$ grid [11]. Thus, G has a convex grid drawing of $O(n^3)$ area if $T(G)$ has at most four leaves.

* This work is supported in part by a Grant-in-Aid for Scientific Research (C) 19500001 from Japan Society for the Promotion of Science (JSPS).

Y. Dong, D.-Z. Du, and O. Ibarra (Eds.): ISAAC 2009, LNCS 5878, pp. 760–770, 2009.

In this paper, we improve the area bound $O(n^3)$ above to $O(n^2)$, which is optimal up to a constant factor because a plane graph of nested triangles needs an $\Omega(n^2)$ area in any straight line drawing [5]. More precisely, we show that an internally triconnected plane graph G has a convex grid drawing on a $2n \times 4n = O(n^2)$ grid if $T(G)$ has exactly four leaves, and present an algorithm to find such a drawing in linear time.

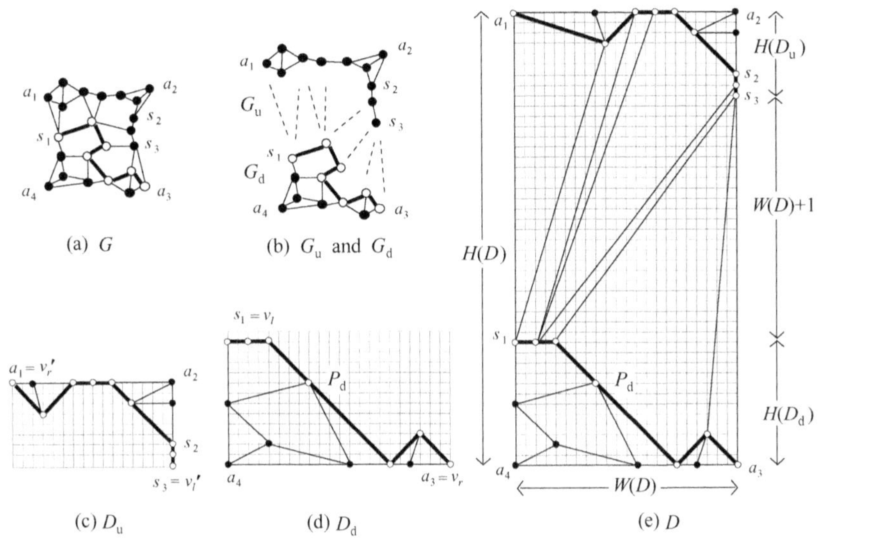

Fig. 1. (a) A plane graph G, (b) subgraphs G_u and G_d, (c) a drawing D_u of G_u, (d) a drawing D_d of G_d, and (e) a convex grid drawing D of G

2 Outline of Our Algorithm

In this section, we outline our algorithm, which is a modification of the algorithm in [11].

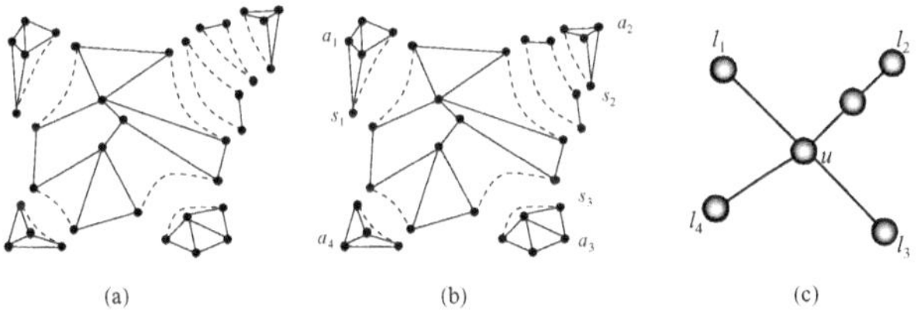

Fig. 2. (a) Split components of the graph G in Fig. 1(a), (b) triconnected components of G, and (c) a decomposition tree $T(G)$

The plane graph G in Fig. 1(a) is internally triconnected, the triconnected components of G are depicted in Fig. 2(b), and the triconnected component decomposition tree $T(G)$ of G having four leaves l_1, l_2, l_3 and l_4 is depicted in Fig. 2(c). We draw G so that the contour of the outer face of G is a rectangle, as illustrated in Fig. 1(e). We first appropriately choose four vertices a_1, a_2, a_3 and a_4 as the four apices of the rectangular contour. We then divide G into an upper subgraph G_u and a lower subgraph G_d, as illustrated in Fig. 1(b), so that G_u contains a_1 and a_2 and G_d contains a_3

and a_4. Using the "pentagon algorithm" in [11], we then obtain "inner convex" grid drawings D_u of G_u and D_d of G_d, both of $O(n^2)$ area, as illustrated in Figs. 1(c) and (d). More precisely, D_u has width $W(D_\mathrm{u}) \leq 2n_\mathrm{u} - 2$ and height $H(D_\mathrm{u}) \leq 2n_\mathrm{u} - 2$, and D_d has width $W(D_\mathrm{d}) \leq 2n_\mathrm{d} - 2$ and height $H(D_\mathrm{d}) \leq 2n_\mathrm{d} - 2$, where n_u and n_d are the numbers of vertices in G_u and G_d, respectively, and hence $n_\mathrm{u} + n_\mathrm{d} = n$. We then shift either vertex a_1 to the left or a_3 to the right so that these two drawings have the same width $\max\{2n_\mathrm{d} - 2, 2n_\mathrm{u} - 2\}$. We next arrange D_d and D_u so that $y(a_3) = y(a_4) = 0$ and $y(a_1) = y(a_2) = H(D_\mathrm{d}) + H(D_\mathrm{u}) + \max\{2n_\mathrm{d} - 2, 2n_\mathrm{u} - 2\} + 1$, as illustrated in Fig. 1(e), where $y(a_1)$, $y(a_2)$, $y(a_3)$ and $y(a_4)$ are the y-coordinates of a_1, a_2, a_3 and a_4, respectively. We finally draw, by straight line segments, all the edges of G that are contained in neither G_u nor G_d. Thus, the width $W(D)$ of the resulting drawing D of G is

$$W(D) \leq \max\{2n_\mathrm{d} - 2, 2n_\mathrm{u} - 2\} < 2n,$$

and the height $H(D)$ of D is

$$H(D) \leq 2n_\mathrm{d} - 2 + 2n_\mathrm{u} - 2 + \max\{2n_\mathrm{d} - 2, 2n_\mathrm{u} - 2\} + 1 < 4n.$$

Hence, the area of the drawing D is $2n \times 4n = O(n^2)$. The selection of apices a_1, a_2, a_3 and a_4, the division of G to G_u and G_d and some others are different from those in [11].

3 Preliminaries

In this section, we give some definitions, and outline the pentagon algorithm in [11] which is used to draw G_d and G_u.

We denote by $G = (V, E)$ an undirected connected simple graph with vertex set V and edge set E. We often denote the set of vertices of G by $V(G)$ and the set of edges by $E(G)$. An edge joining vertices u and v is denoted by (u, v).

A $W \times H$ *integer grid* consists of $W + 1$ vertical grid lines and $H + 1$ horizontal grid lines, and has a rectangular contour. We call W and H the *width* and *height* of the integer grid, respectively. We denote by $W(D)$ the width of the minimum integer grid enclosing a grid drawing D of a graph, and by $H(D)$ the height of D.

A plane graph G divides the plane into connected regions, called *faces*. The boundary of a face is called a *facial cycle*. We denote by $F_\mathrm{o}(G)$ the outer facial cycle of G. A vertex on $F_\mathrm{o}(G)$ is called an *outer vertex*, while a vertex not on $F_\mathrm{o}(G)$ is called an *inner vertex*. In a convex drawing D of a plane graph G, all facial cycles must be drawn as convex polygons. The convex polygonal drawing of $F_\mathrm{o}(G)$ is called the *outer polygon* of D. We call a vertex of a polygon an *apex* in order to avoid the confusion with a vertex of a graph.

We call a vertex v of a connected graph G a *cut vertex* if its removal from G results in a disconnected graph, that is, $G - v$ is not connected. A connected graph G is *biconnected* if G has no cut vertex. We call a pair $\{u, v\}$ of vertices in a biconnected graph G a *separation pair* if its removal from G results in a disconnected graph, that is, $G - \{u, v\}$ is not connected. A biconnected graph G is *triconnected* if G has no separation pair. A biconnected plane graph G is *internally triconnected* if, for any separation pair $\{u, v\}$ of G, both u and v are outer vertices and each connected component of $G - \{u, v\}$ contains an outer vertex.

Let $G = (V, E)$ be a biconnected graph, and let $\{u, v\}$ be a separation pair of G. Then, G has two subgraphs $G'_1 = (V_1, E'_1)$ and $G'_2 = (V_2, E'_2)$ such that

(a) $V = V_1 \bigcup V_2$, $V_1 \bigcap V_2 = \{u, v\}$; and
(b) $E = E'_1 \bigcup E'_2$, $E'_1 \bigcap E'_2 = \emptyset$, $|E'_1| \geq 2$, $|E'_2| \geq 2$.

For a separation pair $\{u, v\}$ of G, $G_1 = (V_1, E'_1 + (u, v))$ and $G_2 = (V_2, E'_2 + (u, v))$ are called the *split graphs* of G with respect to $\{u, v\}$. The new edges (u, v) added to G_1 and G_2 are called the *virtual edges*. Even if G has no multiple edges, G_1 and G_2 may have. Dividing a graph G into two split graphs G_1 and G_2 is called *splitting*. Reassembling the two split graphs G_1 and G_2 into G is called *merging*. Merging is the inverse of splitting. Suppose that a graph G is split, the split graphs are split, and so on, until no more splits are possible, as illustrated in Fig. 2(a) for the graph in Fig. 1(a) where virtual edges are drawn by dotted lines. The graphs constructed in this way are called the *split components* of G. The split components are of three types: triconnected graphs; triple bonds (i.e. a set of three multiple edges); and triangles (i.e. a cycle of length three). The *triconnected components* of G are obtained from the split components of G by merging triple bonds into a bond and triangles into a ring, as far as possible, where a *bond* is a set of multiple edges and a *ring* is a cycle [8]. Thus the triconnected components of G are of three types: (a) triconnected graphs; (b) bonds; and (c) rings. Two triangles in Fig. 2(a) are merged into a single ring, and hence the graph in Fig. 1(a) has six triconnected components as illustrated in Fig. 2(b).

Let $T(G)$ be a tree such that each node corresponds to a triconnected component H_i of G and there is an edge (H_i, H_j), $i \neq j$, in $T(G)$ if and only if H_i and H_j are triconnected components with respect to the same separation pair, as illustrated in Fig. 2(c). We call $T(G)$ a *triconnected component decomposition tree* or simply a *decomposition tree* of G [8]. $T(G)$ has four leaves for the graph G in Fig. 1(a). (See Fig. 2(c).) If G is triconnected, then $T(G)$ consists of a single isolated node and hence $T(G)$ has exactly one leaf.

Let G be an internally triconnected plane graph such that $T(G)$ has exactly four leaves. Then every leaf of $T(G)$ is a triconnected graph and the outer polygon of every convex drawing of G must have four or more apices [10,11]. Our algorithm obtains a convex grid drawing of G whose outer polygon has exactly four apices and is a rectangle in particular, as illustrated in Fig. 1(e).

In Section 4, we will present an algorithm to draw G, which uses the following "canonical decomposition" [2,12]. Let $G = (V, E)$ be an internally triconnected plane graph, and let $V = \{v_1, v_2, \cdots, v_n\}$. Let v_1, v_2 and v_n be three arbitrary outer vertices appearing counterclockwise on $F_o(G)$ in this order. We may assume that v_1 and v_2 are consecutive on $F_o(G)$; otherwise, add a virtual edge (v_1, v_2) to the original graph, and let G be the resulting graph. Let $\Pi = (U_1, U_2, \cdots, U_m)$ be an ordered partition of V into nonempty subsets $U_1, U_2, \cdots, U_m$. We denote by G_k, $1 \leq k \leq m$, the subgraph of G induced by $U_1 \bigcup U_2 \bigcup \cdots \bigcup U_k$, and denote by $\overline{G_k}$, $0 \leq k \leq m - 1$, the subgraph of G induced by $U_{k+1} \bigcup U_{k+2} \bigcup \cdots \bigcup U_m$.

We say that Π is a *canonical decomposition* of G (with respect to vertices v_1, v_2 and v_n) if the following three conditions (cd1)–(cd3) hold:

(cd1) $U_m = \{v_n\}$, and U_1 consists of all the vertices on the inner facial cycle containing edge (v_1, v_2).
(cd2) For each index k, $1 \leq k \leq m$, G_k is internally triconnected.
(cd3) For each index k, $2 \leq k \leq m$, all the vertices in U_k are outer vertices of G_k, and
  (a) if $|U_k| = 1$, then the vertex in U_k has two or more neighbors in G_{k-1} and has one or more neighbors in $\overline{G_k}$ when $k < m$; and
  (b) if $|U_k| \geq 2$, then each vertex in U_k has exactly two neighbors in G_k, and has one or more neighbors in $\overline{G_k}$.

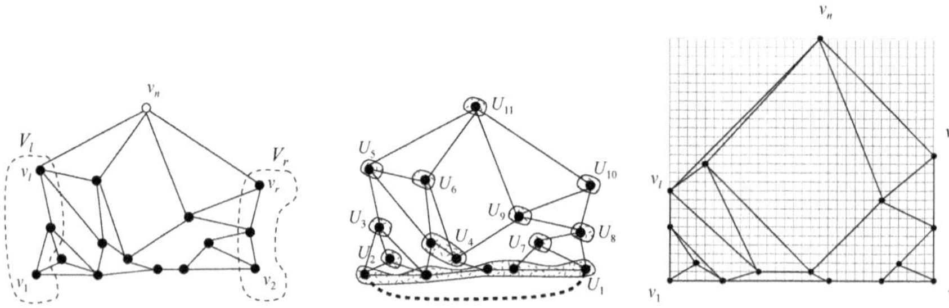

(a) G (b) Π (c) D

Fig. 3. (a) An internally triconnected plane graph G, (b) a canonical decomposition Π of G, and (c) a pentagonal drawing D of G

A canonical decomposition $\Pi = (U_1, U_2, \cdots, U_{11})$ with respect to vertices v_1, v_2 and v_n of the graph in Fig. 3(a) is illustrated in Fig. 3(b). If $T(G)$ has at most three leaves, then G has a canonical decomposition [11].

Let G be a plane graph having a canonical decomposition $\Pi = (U_1, U_2, \cdots, U_m)$ with respect to vertices v_1, v_2 and v_n, as illustrated in Fig. 3(b). Miura *et al.* [11] give a linear-time algorithm, called the *pentagon algorithm*, to find a convex grid drawing of G with a pentagonal outer polygon, as illustrated in Fig. 3(c). The algorithm is based on the so-called shift methods given by Chrobak and Kant [2] and de Fraysseix *et al.* [6], and will be used by our convex grid drawing algorithm in Section 4 to draw G_{u} and G_{d}.

We then outline the pentagon algorithm. Let v_l be an arbitrary outer vertex on the path going from v_1 to v_n clockwise on $F_{\text{o}}(G)$, and let $v_r (\neq v_l)$ be an arbitrary outer vertex on the path going from v_2 to v_n counterclockwise on $F_{\text{o}}(G)$, as illustrated in Fig. 3(a). Let V_l be the set of all vertices on the path going from v_1 to v_l clockwise on $F_{\text{o}}(G)$, and let V_r be the set of all vertices on the path going from v_2 to v_r counterclockwise on $F_{\text{o}}(G)$. The pentagon algorithm in [11] obtains a convex grid drawing of G whose outer polygon is a pentagon with apices v_1, v_2, v_r, v_n and v_l, as illustrated in Fig. 3(c).

More precisely, the pentagon algorithm obtains an "inner convex" grid drawing D_k for each k, $1 \leq k \leq m$, in which all inner facial cycles are convex polygons.

Let $F_o(G_k) = w_1, w_2, \cdots, w_t$, $w_1 = v_1$, and $w_t = v_2$, as illustrated in Fig. 4. Let w_f be the vertex with the maximum index f among all the vertices w_i, $1 \leq i \leq t$, on $F_o(G_k)$ that are contained in V_l. Let w_g be the vertex with the minimum index g among all the vertices w_i that are contained in V_r. Of course, $1 \leq f < g \leq t$. We denote by $\angle w_i$ the interior angle of apex w_i of the outer polygon of D_k. We call w_i a *convex apex* of the polygon if $\angle w_i < \pi$. The drawing D_k of G_k satisfies the following six conditions (sh1)–(sh6). (See Fig. 4.)

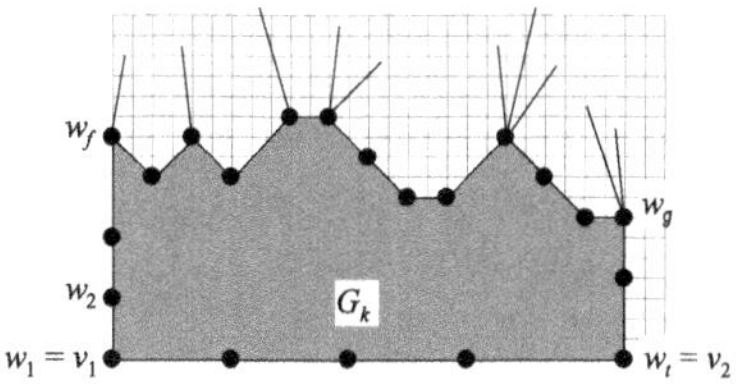

Fig. 4. Drawing D_k of graph G_k

(sh1) w_1 is on the grid point $(0, 0)$, and w_t is on the grid point $(2|V(G_k)| - 2, 0)$.
(sh2) $x(w_1) = x(w_2) = \cdots = x(w_f)$, $x(w_f) < x(w_{f+1}) < \cdots < x(w_g)$, $x(w_g) = x(w_{g+1}) = \cdots = x(w_t)$, where $x(w_i)$ is the x-coordinate of w_i.
(sh3) Every edge (w_i, w_{i+1}), $f \leq i \leq g - 1$, has slope $-1, 0$, or 1.
(sh4) The Manhattan distance between any two grid points w_i and w_j, $f \leq i < j \leq g$, is an even number.
(sh5) Every inner face of G_k is drawn as a convex polygon.
(sh6) Vertex w_i, $f + 1 \leq i \leq g - 1$, has one or more neighbors in $\overline{G_k}$ if w_i is a convex apex.

The pentagon algorithm obtains a convex grid drawing D of $G = G_m$ on a $W \times H$ grid with $W = 2n - 2$ and $H \leq n^2 - n - 2$ in linear time [11]. Thus the area of D is $O(n^3)$. However, we observe that the algorithm obtains a convex grid drawing D of $O(n^2)$ area if one chooses $v_r = v_2$, as follows.

Lemma 1. *For a plane graph G having a canonical decomposition $\Pi = (U_1, U_2, \cdots, U_m)$, the pentagon algorithm obtains an inner convex grid drawing D_k of G_k, $1 \leq k \leq m$, such that $H(D_k) \leq W(D_k) = 2|V(G_k)| - 2$ if one chooses $v_r = v_2$.* (Proof is omitted in this extended abstract.)

Figure 8(d) depicts the convex drawing of the graph in Fig. 8(a) obtained by the pentagon algorithm with choosing $v_r = v_2$. It should be noted that the outer polygon is not a pentagon but is a quadrangle with apices v_1, v_2, $v_n (= w)$ and v_l.

4 Convex Grid Drawing Algorithm

In this section we present a linear algorithm to find a convex grid drawing D of an internally triconnected plane graph G whose decomposition tree $T(G)$ has exactly four leaves. Such a graph G does not have a canonical decomposition, and hence none of the pentagon algorithm and those in [1], [2], [7], [9] can find a convex grid drawing of G. Our algorithm draws the outer facial cycle $F_o(G)$ as a rectangle as illustrated in Fig. 1(e). The algorithm first divides G into an upper subgraph G_u and a lower subgraph G_d as illustrated in Fig. 1(b), then draws G_u and G_d by using the pentagon algorithm [11] with choosing $v_r = v_2$ as illustrated in Figs. 1(c) and (d), and finally combine these two drawings to a convex grid drawing of G as illustrated in Fig. 1(e).

as illustrated in Fig. 8(b), where $U_m = \{w\}$, $v_1 = a_4$ and $v_2 = a_3$. We choose $v_l = s_1$ and $v_r = a_3$, as illustrated in Fig. 8(a). Using the pentagon algorithm in [11], we obtain a convex grid drawing $D_m = D'_{\rm d}$ of $G_m = G'_{\rm d}$, in which the outer polygon of D_m is a quadrangle with apices a_4, a_3, w and s_1, as illustrated in Fig. 8(d). Our drawing $D_{\rm d}$ of $G_{\rm d}$ is an intermediate drawing of D_m, that is, $D_{\rm d}$ is the drawing D_{m-1} of G_{m-1} induced by $U_1 \cup U_2 \cup \cdots \cup U_{m-1}$, as illustrated in Fig. 8(c). Note that $G_{\rm d} = G'_{\rm d} - w = G_{m-1}$. Since we choose $v_r = v_2$, by Lemma 1 we have $H(D_{\rm d}) \le W(D_{\rm d}) \le 2n_{\rm d} - 2$.

4.3 Drawing of $G_{\rm u}$

We now explain how to draw $G_{\rm u}$.

If the degree of s_3 is one in $G_{\rm u}$, then let P_3 be the maximal induced path with an end s_3 such that all the intermediate vertices of P_3 have degree two in $G_{\rm u}$, and let s_2 be the other end of P_3, as illustrated in Fig. 9(a). Otherwise, let P_3 be the trivial path consisting only of s_3 and let $s_2 = s_3$. Let $G'_{\rm u}$ be a graph obtained from G by contracting all the vertices of $G_{\rm d}$ and all the vertices of P_3 except s_2 to a single vertex w, as illustrated in Fig. 9(b) for the graph G in Fig. 1(a). Then clearly $G'_{\rm u}$ is biconnected. Similarly to $G'_{\rm d}$, $G'_{\rm u}$ has a canonical decomposition $\Pi = (U_1, U_2, \cdots, U_m)$ with respect to a_2, a_1 and w, as illustrated in Fig. 9(c). We choose $v_l = s_2$ and $v_r = a_1$, as illustrated in Fig. 9(c). Using the pentagonal algorithm in [11], we obtain a convex grid drawing $D'_u = D_m$ of $G'_{\rm u}$, in which the outer polygon is a quadrangle with apices a_2, a_1, w and s_2, as illustrated in Fig. 9(e). Figure 9(d) depicts the drawing D_{m-1} of $G_{m-1} = G'_u - w$ induced by $U_1 \cup U_2 \cup \cdots \cup U_{m-1}$. We obtain a drawing $D_{\rm u}$ of $G_{\rm u}$ from D_{m-1} by putting the vertices of P_3 on grid points having the same x-coordinate as a_2, as illustrated in Fig. 9(f). Clearly $|V(G_{m-1})| = n_{\rm u} - |V(P_3)| + 1$ and $|V(P_3)| \ge 1$. Therefore, by Lemma 1,

$$W(D_{\rm u}) = W(D_{m-1}) \le 2|V(G_{m-1})| - 2 \le 2n_{\rm u} - 2|V(P_3)| \le 2n_{\rm u} - 2$$

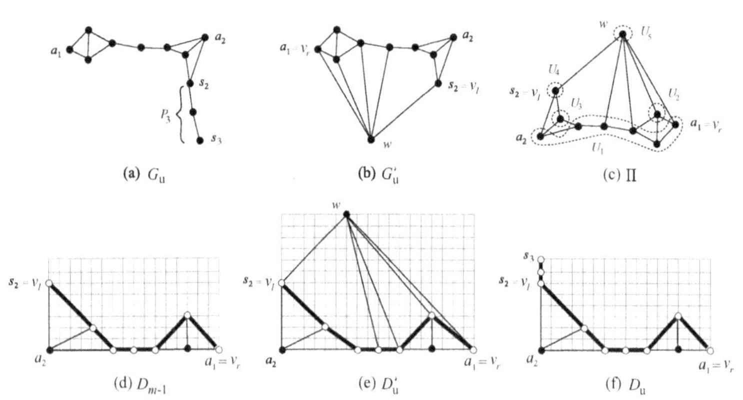

Fig. 9. (a) $G_{\rm u}$, (b) $G'_{\rm u}$, (c) a canonical decomposition of $G'_{\rm u}$, (d) a drawing D_{m-1} of G_{m-1}, (e) a drawing $D'_{\rm u} = D_m$ of $G'_{\rm u}$, and (f) a drawing $D_{\rm u}$ of $G_{\rm u}$

and

$$H(D_u) \leq W(D_{m-1}) + |V(P_3)| - 1 \leq 2n_u - 2|V(P_3)| + |V(P_3)| - 1 \leq 2n_u - 2.$$

4.4 Drawing of G

If $W(D_d) \neq W(D_u)$, then we widen the narrower one of D_d and D_u by the shift method in [2] so that both have the same width, which is at most $\max\{2n_d - 2, 2n_u - 2\}$. Since we combine the two drawings D_d and D_u of the same width to a drawing D of G, we have

$$W(D) \leq \max\{2n_d - 2, 2n_u - 2\} < 2n.$$

We arrange D_d and D_u so that $y(a_3) = y(a_4) = 0$ and $y(a_1) = y(a_2) = H(D_d) + H(D_u) + W(D) + 1$, as illustrated in Fig. 1(e). Since $n_d + n_u = n$, we have

$$H(D) = H(D_d) + H(D_u) + W(D) + 1 < (2n_d - 2) + (2n_u - 2) + 2n + 1 < 4n.$$

We finally draw, by straight line segments, all the edges of G that are contained in neither G_u nor G_d. This completes the grid drawing D of G. (See Fig. 1(e).)

Since the conditions (sh5) and (sh6) hold for D_d and D_u, one can prove similarly as in [11] that the drawing D obtained above is a convex grid drawing of G. Clearly the algorithm takes linear time. We thus have the following theorem.

Theorem 1. *Assume that G is an internally triconnected plane graph, every vertex of G has degree three or more, and the triconnected component decomposition tree $T(G)$ has exactly four leaves. Then our algorithm finds a convex grid drawing of G on a $2n \times 4n$ grid in linear time.*

5 Conclusions

In this paper, we showed that every internally triconnected plane graph G whose decomposition tree $T(G)$ has exactly four leaves has a convex grid drawing on a $2n \times 4n = O(n^2)$ grid, and we present a linear algorithm to find such a drawing. The area bound $O(n^2)$ is optimal up to a constant factor since the nested triangles graph needs $\Omega(n^2)$ area. The remaining problem is to obtain an algorithm for an internally triconnected plane graph whose decomposition tree has five or more leaves.

References

1. Bonichon, N., Felsner, S., Mosbah, M.: Convex drawings of 3-connected plane graphs. Algorithmica 47(4), 399–420 (2007)
2. Chrobak, M., Kant, G.: Convex grid drawings of 3-connected planar graphs. Int. J. Comp. Geometry and Applications 7, 211–223 (1997)
3. Chiba, N., Onoguchi, K., Nishizeki, T.: Drawing planar graphs nicely. Acta Inform. 22, 187–201 (1985)

4. Chiba, N., Yamanouchi, T., Nishizeki, T.: Linear algorithms for convex drawings of planar graphs. In: Bondy, J.A., Murty, U.S.R. (eds.) Progress in Graph Theory, pp. 153–173. Academic Press, London (1984)
5. Dolev, D., Leighton, F.T., Trickey, H.: Planar embedding of planar graphs. In: Advances in Computer Research. VLSI Theory, vol. 2, pp. 147–161 (1984)
6. de Fraysseix, H., Pach, J., Pollack, R.: How to draw a planar graph on a grid. Combinatorica 10, 41–51 (1990)
7. Felsner, S.: Convex drawings of plane graphs and the order of dimension of 3-polytopes. Order 18, 19–37 (2001)
8. Hopcroft, J.E., Tarjan, R.E.: Dividing a graph into triconnected components. SIAM J. Compt. 2(3), 135–138 (1973)
9. Miura, K., Azuma, M., Nishizeki, T.: Canonical decomposition, realizer, Schnyder labeling and orderly spanning trees of plane graphs. Int. J. Found. of Computer Science 16(1), 117–141 (2005)
10. Miura, K., Azuma, M., Nishizeki, T.: Convex drawings of plane graphs of minimum outer apices. Int. J. Found. of Computer Science 17(5), 1115–1127 (2006)
11. Miura, K., Kamada, A., Nishizeki, T.: Convex grid drawings of plane graphs of rectangular contours. J. Graph Algorithms and Applications 12(2), 197–224 (2008)
12. Nishizeki, T., Rahman, M.S.: Planar Graph Drawing. World Scientific, Singapore (2004)

A Self-stabilizing and Local Delaunay Graph Construction*

Riko Jacob[1], Stephan Ritscher[1], Christian Scheideler[2], and Stefan Schmid[2]

[1] Institut für Informatik, Technische Universität München, D-85748 Garching, Germany
jacob@in.tum.de, ritsches@in.tum.de
[2] Department of Computer Science, University of Paderborn, D-33102 Paderborn, Germany
scheideler@upb.de, schmiste@mail.upb.de

Abstract. This paper studies the construction of self-stabilizing topologies for distributed systems. While recent research has focused on chain topologies where nodes need to be linearized with respect to their identifiers, we go a step further and explore a natural 2-dimensional generalization. In particular, we present a local self-stabilizing algorithm that constructs a *Delaunay graph* from any initial connected topology and in a distributed manner. This algorithm terminates in time $O(n^3)$ in the worst-case. We believe that such self-stabilizing Delaunay networks have interesting applications and give insights into the necessary geometric reasoning that is required for higher-dimensional linearization problems.

1 Introduction

Open distributed systems such as peer-to-peer systems are often highly dynamic in the sense that nodes join and leave continuously. In addition to these natural membership changes, a system is sometimes under attack, e.g., a botnet may block entire network fractions by a denial-of-service attack. For these reasons, there is a considerable scientific interest in robust and "self-healing" topologies that can be maintained in a distributed manner even under high churn.

An important concept to build robust networks is *topological self-stabilization*: A self-stabilizing network can provably recover from *any* connected state, that is, eventually the network always returns to a desirable (to be specified) state. Despite its relevance, topological self-stabilization is a relatively new area and today, we still know only very little about the design of self-stabilizing algorithms. In particular, while much existing literature focuses on *eventual* stabilization, the required *convergence times* are still not well understood.

Recently, progress was made in the area of *graph linearization* where nodes need to be arranged in a chain network which respects the node identifiers. In this paper, we go one step further and explore the *2-dimensional* case. We assume nodes are distributed in the Euclidean plane and are arbitrarily connected.

* Research supported by the DFG project SCHE 1592/1-1. Due to space constraints, many proofs and simulation results are only presented in the technical report [9].

Y. Dong, D.-Z. Du, and O. Ibarra (Eds.): ISAAC 2009, LNCS 5878, pp. 771–780, 2009.

A natural 2-dimensional analogon of linearization is the *Delaunay graph*, whose edge set includes all nearest neighbor connections between node pairs. Delaunay graphs are an important graph family in various CS domains, from computational geometry to wireless networking. This is due to their desirable properties such as locality, sparseness or planarity. We find that while insights from graph linearization are useful for self-stabilizing Delaunay graphs as well, the construction and analysis is more involved, requiring a deeper geometric reasoning.

1.1 Related Work

Researchers in the field of self-stabilization study algorithms that provably converge to a desirable system state from *any* initial configuration. In the seminal work by E.W. Dijkstra in 1974 [4], the problem of self-stabilization in a token ring is examined. Subsequently, many aspects of distributed systems have been explored from a self-stabilization point of view, including communication protocols, graph theory problems, termination detection, clock synchronization, and fault containment. Also general techniques for self-stabilization have been considered: In [1], Awerbuch and Varghese showed that every local algorithm can be made self-stabilizing if all nodes keep a log of the state transitions until the current state.

However, much of this work is not applicable to scenarios where faults include changes in the *topology* (e.g., see [6] for an early work on topological self-stabilization): A single fault may require the involvement of all nodes in the system and is hence expensive to repair. To reduce this overhead, researchers have started to study so-called superstabilizing protocols [5]. Topological self-stabilization is still in its infancy. Often, recovery algorithms do not work generally but only from certain degenerate network states (see, e.g., the technical report of the *Chord network)*. A notable recent exception is [8] which describes a truly self-stabilizing algorithm for skip graphs. Unfortunately, however, skip graphs do not maintain locality in the sense that nodes which are close in the metric space are also close with respect to the hop distance, and therefore cannot be used in our context.

In order to shed light onto the fundamental principles enabling provable topological self-stabilization, researchers have started to examine the most simple networks such as *line or ring graphs* (e.g., [3, 7]). Our paper goes one step further and initiates the study of self-stabilizing constructions of 2-dimensional graphs. As a case study, we consider the important family of Delaunay graphs. We assume nodes have (x, y) coordinates and are distributed in the Euclidean plane. As Delaunay graphs include all nearest neighbor edges, our algorithms also involve a kind of 2-dimensional linearization. However, it turns out that the problem is more involved, and the reasoning requires geometric techniques. Still we are able to prove a $O(n^3)$ convergence time in the worst-case.

1.2 Our Contributions

This paper presents the first self-stabilizing algorithm to build a Delaunay graph from *any* weakly connected network. Our algorithm is *local* in the sense that nodes are only allowed to communicate with their topological neighbors. Besides correctness, we are able to derive a $O(n^3)$ worst-case bound on the convergence time (i.e., number of communication rounds). We believe that this result has interesting implications, and that our geometric reasoning can give general insights into the design of higher-dimensional "nearest-neighbor graphs" respecting the closeness of nodes in a self-stabilizing manner. If the initial network contains the Delaunay graph, the convergence time is at most n rounds.

Compared to the trivial strategy to obtain a complete graph in $O(\log n)$ rounds in a first phase and then compute the Delaunay graph "locally" at each node in a second phase, our algorithm provides several advantages. First of all, it is not necessary to distinguish between different execution phases: Each node will perform updates according to the same set of rules at any time; only like this, the algorithm is truly self-stabilizing. Furthermore, our algorithm can deal efficiently with small topology changes: If only a small number of nodes joins or leaves, the topology is repaired *locally*; a complete re-computation is not needed. Finally the simulations show that the maximal degree and the total number of edges remain rather small in general. This keeps the resource requirements at each node small.

2 Model and Preliminaries

This section first introduces some notations and definitions from geometry. Subsequently, the Delaunay graph is introduced together with some important properties. In this paper, we will consider non-degenerate cases only, that is, we assume that no two nodes are at the same location, no three points are on a line, and no four points are on a circle.

2.1 Geometry

We consider the 2-dimensional Euclidean space $\mathbb{R}^2$. The *scalar product* is written as $\langle\cdot,\cdot\rangle$ and the *Euclidean norm* (the distance from the origin) is given by $\|x\| = \sqrt{\langle x, x\rangle}$. We make use of the following notation. Let $B(x,r)$ denote the *disk* (or ball) with center $x \in \mathbb{R}^2$ and radius $r \in \mathbb{R}$, i.e., $B(x,r) := \{y \in \mathbb{R}^2 : \|x-y\| \leq r\}$. Note that the border explicitly belongs to the ball in our model, and hence, a point $y \in B(x,r)$ may lie on the border. $C(x,y) := B(\frac{1}{2}(x+y), \frac{1}{2}\|x-y\|)$ is the disk *between* $x, y \in \mathbb{R}^2$. Similarly, $C(x,y,z) := B(c,r)$ with $r = \|x-c\| = \|y-c\| = \|z-c\|$ is the disk defined by non-collinear $x,y,z \in \mathbb{R}^2$. For a vector $x \neq 0$ we define $0 \neq \perp x \in \mathbb{R}^2$ to be the perpendicular, i.e., $\langle x, \perp x\rangle = 0$. Note that $\perp x$ is unique up to constant factors.

By $\angle xzy$ we denote the area spanned by the vectors x and y attached to z, i.e., the area that can be expressed as a linear combination of the vectors x and y with non-negative factors. In particular, $\angle xzy = \angle yzx$. If a node u is contained in this area, we write $u \in \angle xzy$.

This paper makes use of the following simple geometric facts. For two general points $a, b \in \mathbb{R}^2$, due to the triangle inequality, we have that $\|a+b\| \leq \|a\| + \|b\|$.

Moreover, $\|a+b\| = \|a\| + \|b\| \Leftrightarrow \exists t \geq 0 : a = t \cdot b$. Pythagoras' law says that for any $a, b \in \mathbb{R}^2$ with $\langle a, b \rangle = 0$, it holds that $\|a+b\|^2 = \|a\|^2 + \|b\|^2$. If we know two points on the border of a disk, then their midpoint must be on a specific straight line. Formally, let $u, v, x \in \mathbb{R}^2$. Then $\|u - x\| = \|v - x\|$ if and only if $x = \frac{1}{2}(u+v) + t(u-v)$ for some $t \in \mathbb{R}$. For the Euclidean norm, it holds for $C = C(u, v)$ for $u, v \in \mathbb{R}^2$ that $w \in C$ and $\|w - u\| \geq \|v - u\|$ imply $w = v$.

For some proofs we want to choose a disk $\tilde{C}$ contained in a bigger disk C with at least two points on the border of $\tilde{C}$. We can make the following observations.

Fact 2.1. *Let $C = B(x, r)$ be a disk with $u, v \in C$ and $u \neq v$. Then there is a disk $\tilde{C} = B(\tilde{x}, \tilde{r}) \subseteq C$ with $\|u - \tilde{x}\| = \|v - \tilde{x}\| = \tilde{r}$.*

For the opposite direction, given a set of points, we need a disk containing all of them, with at least three on the border.

Fact 2.2. *Let $V \subset \mathbb{R}^2$ be a finite set of points, not all of them collinear. Then there are three different, not collinear points $u, v, w \in V$ with $C(u, v, w) \supset V$.*

2.2 Delaunay Graphs

We consider graphs with an *embedding* into $\mathbb{R}^2$. Let $V \subset \mathbb{R}^2$ be a finite set and $E \subset \binom{V}{2}$, then $G = (V, E)$ is called *undirected embedded graph* with *nodes* V and *edges* E. Let $n = |V|$ be the cardinality of V. We define $N_G(u) = \{v \in V : \{u, v\} \in E\}$ as the *neighbors* of u. Moreover, let $\overline{N}_G(u) = N_G(u) \cup \{u\}$ denote the *neighbors* of u *including* u.

Usually we speak of a *directed* graph $G = (V, E)$ with $E \subset V^2$. Then a *directed edge* from u to v is denoted by (u, v), the *undirected edge* $\{u, v\}$ represents the two directed edges (u, v) and (v, u) and $N_G(u) = \{v \in V : (u, v) \in E\}$. $\overline{N}_G(u)$ is defined analogously. Note that any undirected graph can be seen as a directed graph with this interpretation of undirected edges. This will be done implicitly throughout the paper. A directed graph is called *strongly connected*, if for every pair (u, v) of nodes $u, v \in V$ there is a directed path from u to v. A direct graph is *weakly connected*, if the graph obtained by replacing all directed edges by undirected edges is connected.

Armed with these definitions, we can now define the Delaunay graph.

Definition 2.3 (Delaunay Graph). *The* Delaunay Graph

$$G_D(V) = (V, E_D(V))$$

of the vertices V is an undirected embedded graph defined by $\{u, v\} \in E_D(V) \Leftrightarrow u \neq v \wedge \exists C = B(x, r) : C \cap V = \{u, v\}$ i.e., u and v are connected, if and only if there is a disk containing only these two points of V.

Recall that we will consider non-degenerate cases, that is, we assume there is no disk $B(x, r)$ with four different points $x_1, \ldots, x_4 \in V$ on its border, i.e. $\forall B(x, r) : |V \cap \{y \in \mathbb{R}^2 : \|x - y\| = r\}| \leq 3$. It is easy to see that the Delaunay graph on a given node set always includes the convex hull edges.

2.3 Properties

We can give several equivalent formulations of Definition 2.3 that will be useful in our analysis. In a Delaunay graph, two nodes u and v are connected if and only if either they are the only two nodes in the disk $C(u, v)$, or if there exists a third node w such that u, v, and w are the only three nodes in $C(u, v, w)$. [2]

Lemma 2.4. *Let $G = (V, E_D(V))$ be a Delaunay graph. Then*

$$\{u,v\} \in E_D(V) \Leftrightarrow u \neq v \wedge (C(u,v) \cap V = \{u,v\} \vee \\ \vee \exists w \in V \setminus \{u,v\} : C(u,v,w) \cap V = \{u,v,w\})$$

The following lemma states that in a Delaunay graph, for each pair of non-adjacent nodes, there must be a "close" neighboring node.

Lemma 2.5. *Let $G = (V, E_D(V))$ be a Delaunay graph and $\{u,v\} \notin E_D(V)$. Then every disk $C = B(x,r)$ containing u and v must contain at least one neighbor $w \in N_G(u)$ with $\|w - x\| < r$.*

We need some properties about restrictions of Delaunay graphs to a subset of nodes $U \subset V$. It is easy to see, that the restriction of the Delaunay graph of V to U is contained in the Delaunay graph on U:

Lemma 2.6

$$U \subset V \Rightarrow E_D(U) \supset E_D(V) \cap (U \times U).$$

Proof. Let $\{u,v\}$ be an edge in $E_D(V) \cap (U \times U)$. Then by Definition 2.3 there is a disk $C = B(x,r)$ such that $C \cap V = \{u,v\}$. Since $U \subset V$, $C \cap U = \{u,v\}$ and thus $\{u,v\} \in E_D(U)$. □

Combining this lemma with the previous one, additional insights can be gained. Let us pick U such that is contains the neighbors $\overline{N}_G(u)$ of a node u. Then u has the same neighbors in the Delaunay graph on U as in the original Delaunay graph.

Lemma 2.7. *Let $G = (V, E_D(V))$ be a Delaunay graph, $u \in V$ and $\overline{N}_G(u) \subset U \subset V$. Then $\overline{N}_{G_D(U)}(u) = \overline{N}_G(u)$.*

Proof. $\overline{N}_{G_D(U)}(u) \supset \overline{N}_G(u)$ is clear by Lemma 2.6. Now let $\{u,v\} \in (U \times U) \setminus E_D(V)$. So, by Lemma 2.5, in each disc $C = B(x,r)$ containing v, w there is a neighbor w of u (i.e. $w \in \overline{N}_G(u) \subset U$). Thus, by Definition 2.3, $\{u,v\} \notin E_D(U)$. □

The next, important characterization of Delaunay graphs also argues about edges that are *not* Delaunay. If and only if two nodes u and v are not connected, there must exist two neighbors x and y of u, such that the disk $C(u,v,x)$ contains only y, and x and y lie on different sides of the line connecting u and v.

Lemma 2.8. *Let $G = (V, E_D(V))$ be a Delaunay graph. Then*

$$\{u,v\} \notin E_D(V) \Leftrightarrow \exists x,y \in V \setminus \{u,v\} : C(u,v,x) \cap V \supset \{u,v,x,y\} \wedge \\ \wedge \langle x-u, \perp(v-u)\rangle \cdot \langle y-u, \perp(v-u)\rangle \leq 0$$

That is, x and y must be on different sides of the line connecting u and v. One can even choose $x, y \in N_G(u)$.

We will later need the existence of special edges in Delaunay graphs. First, we observe that a Delaunay node is always connected to the *closest* node, that is, the Delaunay graph contains the nearest neighbor graph. The following lemma follows directly from the observation that, for two closest neighbors $u, v \in V$, $C(u,v) \cap V = \{u, v\}$.

Lemma 2.9. *Let $G = (V, E_D(V))$ be a Delaunay graph and $u \in V$. Then u is connected to the node $v \in V \setminus \{u\}$ with minimal Euclidean distance to u.*

Another important property of Delaunay graphs is that they are connected.

Lemma 2.10. *Every Delaunay graph $G = (V, E_D(V))$ is connected. [12]*

Moreover, it can be shown that these graphs have a planar embedding.

Lemma 2.11. *Every Delaunay graph $G = (V, E_D(V))$ is planar. [2]*

2.4 Local Algorithms and Self-stabilization

The main objective of this paper is to devise a distributed algorithm—essentially a simple set of rules—which is run by every node all the time. Independently from the initial, weakly connected topology (nodes can be connected to any other nodes from all over the metric space), a self-stabilizing algorithm is required to eventually terminate with a correct Delaunay graph as defined in Definition 2.3. During the execution of this algorithm, each node will add or remove edges to other nodes using *local interactions* only. In order to evaluate the algorithm's performance, a synchronous model is investigated (similarly to [11]) where time is divided into *rounds*. In a round, each node is allowed to perform an update of its neighborhood, that is, remove existing edges and connect to other nodes. We study the *time complexity* of the algorithm and measure the number of rounds (in the worst-case) until a Delaunay graph is formed and the algorithm stops.

3 Self-stabilizing Algorithm

This section presents our algorithm *ALG*. During the execution of *ALG*, all nodes continuously calculate a Delaunay graph on their neighbors, that is, each node u computes the Delaunay graph on the node set $\overline{N}(u)$—a triangulation consisting of circular edges ("convex hull") and radial edges. In the following, we will call the considered node the *active* node and the calculated Delaunay graph its so-called *local Delaunay graph*. Here *active* is *not* referring to an calculation order but emphasizes the local role of the computing node for its local Delaunay graph. Note that the local Delaunay graph of a node u, denoted by $G_L(G, u) = (\overline{N}_G(u), E_D(\overline{N}_G(u))$, also contains edges that are not incident to u, but connect neighbors of u.

The construction of the local Delaunay graph $G_L(G, u)$ is reminiscent of the *1-localized Delaunay graph* $LDEL^{(1)}(\overline{N}_G(u))$ introduced by Li et al. [10]. The major difference is that [10] assumes an underlying unit disk graph to define the neighbors of a node whereas in our construction the current approximation of the Delaunay graph is used (which can be arbitrarily bad initially).

Informally, the active node keeps edges to neighbors in the local Delaunay graph, and forms edges among them in a circular order around it. All other nodes are deferred to some Delaunay neighbor of the active node. The *Delaunay update* $\tilde{G} = (V, \tilde{E})$ of G is the union of these update edges for all nodes in G. Due to the division into rounds, the updates are well-defined and the actions of different nodes in the same round do not interfere.

Definition 3.1 (Stable and Temporary Edges). Stable edges *are undirected and are currently—from a local point of view—consistent with the Delaunay properties.* Temporary edges *on the other hand are directed and will appear, be forwarded, and disappear again (i.e., become stable) during the execution of our algorithm.*

We are now ready to formally define the Delaunay update:

Definition 3.2 (Delaunay Update). *Let $G = (V, E)$ be a directed graph.*

- *The* local Delaunay graph *of u is* $G_L(G, u) = (\overline{N}_G(u), E_D(\overline{N}_G(u))$.
- *Each node u selects the following edges $E_S(G, u)$ from $E_D(\overline{N}_G(u))$, which will be kept for the next round:*

$$E_S(G, u) = E_{stable}(G, u) \cup E_{temp}(G, u)$$

where Rule I*:*

$$\begin{aligned} E_{stable} = &\{\{u, v\} : v \in N_{G_L(G,u)}(u)\} \\ &\quad \textit{(undirected edges from } u \textit{ to its neighbors in } G_L(u)\textit{)} \\ \cup\ &\{\{v, w\} : v, w \in N_{G_L(G,u)}(u) \wedge \\ &\quad \nexists x \in N_{G_L(G,u)}(u) : x \in \angle vuw\} \\ &\quad \textit{(undirected circular edges between } u\textit{'s neighbors)} \end{aligned}$$

and Rule II*:*

$$\begin{aligned} E_{temp}(G, u) = &\{(v, w) : v \in N_{G_L(G,u)}(u), w \in N_G(u) \setminus \overline{N}_{G_L(G,u)}(u) \wedge \\ &\quad \forall x \in N_{G_L(G,u)}(u) : \|x - w\| \geq \|v - w\|\} \\ &\quad \textit{(directed edges from } u\textit{'s non-neighbors to neighbors)} \end{aligned}$$

Rule II keeps directed edges between a node's neighbor and a non-neighbor if there is no closer neighbor to the non-neighbor (a nearest connection strategy).
- *Then the* Delaunay update *is $\tilde{G} = (V, \tilde{E})$ with*

$$\tilde{E} = \bigcup_{u \in V} E_S(G, u),$$

the graph that arises when all nodes have chosen their new neighbors for the next round.

Observe that *ALG* follows a nearest neighbor strategy in the sense that temporary circular edges are only allowed from closest neighbors to non-neighbors of the active node. Moreover, an important property of our algorithm is that temporary edges are forwarded to closer nodes. We will say the edge (u, v) is *passed* to node w, if (u, v) is replaced by (w, v); the node pointed to remains the same.

4 Analysis

We start with two fundamental properties of the Delaunay updates.

Lemma 4.1. *Let $G = (V, E)$ be a directed embedded graph and $\tilde{G} = (V, \tilde{E})$ its Delaunay update. Then Delaunay edges of G will also be in $\tilde{G}$, that is,*

$$(u, v) \in E \cap E_D(V) \quad \Rightarrow \quad \{u, v\} \in \tilde{E}.$$

Proof. Since $(u, v) \in E$, $u, v \in \overline{N}_G(u)$. By Lemma 2.6, $\{u, v\} \in E_D(\overline{N}_G(u))$ and by Definition 3.2, Rule I, $\{u, v\} \in \tilde{E}$. □

Moreover, the following lemma claims that Delaunay updates maintain connectivity.

Lemma 4.2. *Let $G = (V, E)$ be a directed embedded graph and $\tilde{G} = (V, \tilde{E})$ its Delaunay update. If G is (weakly or strongly) connected, then so is $\tilde{G}$.*

Proof. It is enough to show, that for every neighbor w of u in G there is a directed path from u to w in $\tilde{G}$. By Definition 3.2, we have to consider two cases. If $w \in N_{G_L(G,u)}(u)$, then $(u, w) \in \tilde{E}$ is a path from u to w. Otherwise $(v, w) \in E$ for some $v \in N_{G_L(G,u)}(u)$, since directed edges are forwarded between nodes, while the pointed-to node remains the same. Thus (u, v) and (v, w) form a path from u to w. □

Note that Lemma 4.2 proves that all paths are maintained during updates.

4.1 Superfluous Edges

Lemma 4.1 implies that if every Delaunay edge will be created in some round, we end up with a supergraph of $G_D(V)$. Assuming that this happened, this section will show that all non-Delaunay edges will disappear after a few rounds, so that we are left with just the Delaunay graph.

First we need that the circular connections of a node's Delaunay neighbors are Delaunay edges.

Lemma 4.3. *Let $G = (V, E)$ be a directed embedded graph with $N_G(u) \supseteq N_{G_D(V)}(u)$. Then*

$$\{\{v, w\} : v, w \in N_{G_L(G,u)}(u) \land \nexists x \in N_{G_L(G,u)}(u) : x \in \angle vuw\} \subseteq E_D(V).$$

The following helper lemma is crucial for our convergence analysis, as it shows that non-Delaunay edges become shorter over time. The lemma takes into account that *ALG* follows a nearest neighbor strategy.

Lemma 4.4. *Let $G = (V, E)$ be a directed embedded graph with $E \supseteq E_D(V)$ and $\tilde{G} = (V, \tilde{E})$ its Delaunay update. Then for every non-Delaunay edge in $\tilde{G}$ there is a strictly longer non-Delaunay edge in G, formally, $(v, w) \in \tilde{E} \setminus E_D(V) \Rightarrow \exists (u, w) \in E \setminus E_D(V) : \|u - w\| > \|v - w\|$.*

We are now ready to prove that superfluous edges disappear quickly in at most n rounds.

Lemma 4.5. *Let $G = (V, E)$ be a directed embedded graph with $E \supseteq E_D(V)$, i.e., G is a supergraph of the Delaunay graph $G_D(V)$. Then ALG converges to $G_D(V)$ in at most n rounds.*

4.2 Fixpoint and Convergence

We will first show that there is no "dead end", i.e., as long as we do not reach the Delaunay graph, local updates will change the graph.

Lemma 4.6. *Let $V \subset \mathbb{R}^2$ be a finite set of nodes in general positions. Then the Delaunay graph $G = G_D(V) = (V, E_D(V))$ is the only weakly connected stable graph on the nodes V, i.e., the only graph that equals its Delaunay update $\tilde{G} = (V, \tilde{E})$.*

For the convergence proof we need a potential function.

Definition 4.7 (Potential ϕ). *Let $G = (V, E)$ be a directed embedded graph. Then the* potential *$\phi_G(v)$ of a node v is defined as the number of nodes $w \in V$ that are better approximations of the Delaunay neighbors than its current neighbors. This means they would be neighbors of v in the local Delaunay graph containing v, its neighbors and w. Formally*

$$\phi_G(v) = |\{w \in V \setminus \overline{N}_G(v) : \{v, w\} \in E_D(\overline{N}_G(v) \cup \{w\})\}|.$$

The potential of the whole graph is $\phi(G) = \sum_{v \in V} \phi_G(v)$.

We now observe that the potential $\phi(G)$ is *monotone*.

Lemma 4.8. *Let $G = (V, E)$ be a directed embedded graph and $\tilde{G} = (V, \tilde{E})$ its Delaunay update. Then $\phi(G) \geq \phi(\tilde{G})$.*

Combining all our insights, we can now prove our main result.

Theorem 4.9. *Let $G = (V, E)$ be a directed embedded, weakly connected graph. Then ALG requires at most $O(n^3)$ rounds (i.e. Delaunay updates) until the topology converges to the Delaunay graph $G_D(V)$.*

Proof. Consider the sequence of graphs $G_0 = G, G_1, \ldots$, where G_{i+1} is the Delaunay update of $G_i = (V, E_i)$. Due to Lemma 4.2 each graph in this sequence is weakly connected. As soon as $E_D(V) \subseteq E_i$, we know $G_{i+n} = G_D(V)$ from Lemma 4.5. So we just have to consider the case $E_D(V) \not\subseteq E_i$.

From Lemma 4.8 we know that the potential cannot increase. In particular, it holds that once a node leaves the potential set

$$\{w \in V \setminus \overline{N}_G(v) : \{v, w\} \in E_D(\overline{N}_G(v) \cup \{w\})\},$$

it will never be member of the set again. Therefore, it remains to show that after every at most n steps, the cardinality of the set decreases (by a positive integer value): Since the potential is bounded by $n \cdot (n-1)$ and the only graph with potential 0 is the Delaunay graph, this gives the desired bound on the convergence time.

Now assume for the case of contradiction that the potential set has the same cardinality for more than n rounds. This implies that no new Delaunay edge appeared during this time period. Since each temporary edge is forwarded no more than $n-1$ times, the topology must describe a Delaunay fixpoint in the sense of Lemma 4.6. Since the graph is connected, it must be the Delaunay graph. This contradiction proves the claim. □

References

[1] Awerbuch, B., Varghese, G.: Distributed program checking: A paradigm for building self-stabilizing distributed protocols. In: Proc. FOCS, pp. 258–267 (1991)
[2] Berg, M.d., Cheong, O., Kreveld, M.v., Overmars, M.: Computational Geometry: Algorithms and Applications. Springer, Heidelberg (2008)
[3] Clouser, T., Nesterenko, M., Scheideler, C.: Tiara: A self-stabilizing deterministic skip list. In: Proc. SSS (2008)
[4] Dijkstra, E.: Self-stabilization in spite of distributed control. Communications of the ACM 17, 643–644 (1974)
[5] Dolev, S., Herman, T.: Superstabilizing protocols for dynamic distributed systems. Chicago Journal of Theoretical Computer Science 4, 1–40 (1997)
[6] Gafni, E.M., Bertsekas, D.P.: Asymptotic optimality of shortest path routing algorithms. IEEE Trans. Inf. Theor. 33(1), 83–90 (1987)
[7] Gall, D., Jacob, R., Richa, A., Scheideler, C., Schmid, S., Täubig, H.: Brief announcement: On the time complexity of distributed topological self-stabilization. In: Proc. SSS (2009)
[8] Jacob, R., Richa, A., Scheideler, C., Schmid, S., Täubig, H.: A distributed polylogarithmic time algorithm for self-stabilizing skip graphs. In: Proc. PODC (2009)
[9] Jacob, R., Ritscher, S., Scheideler, C., Schmid, S.: A self-stabilizing and local delaunay graph construction. Tech. Report TR-TI-09-307, University of Paderborn (2009)
[10] Li, X.-Y., Calinescu, G., Wan, P.-J.: Distributed construction of planar spanner and routing for ad hoc wireless networks. In: Proc. INFOCOM (2002)
[11] Onus, M., Richa, A., Scheideler, C.: Linearization: Locally self-stabilizing sorting in graphs. In: Proc. ALENEX (2007)
[12] Stojmenovic, I.: Handbook of Wireless Networks and Mobile Computing. Wiley, Chichester (2002)

Succinct Greedy Geometric Routing in the Euclidean Plane*

Michael T. Goodrich and Darren Strash

Computer Science Department, University of California, Irvine, USA

Abstract. We show that greedy geometric routing schemes exist for the Euclidean metric in $\mathbf{R}^2$, for 3-connected planar graphs, with coordinates that can be represented *succinctly*, that is, with $O(\log n)$ bits, where n is the number of vertices in the graph.

1 Introduction

Geometric routing algorithms perform message passing using geometric information stored at the nodes and edges of a network.

Greedy Geometric Routing. Perhaps the simplest routing rule is the *greedy* one:

- If a node v receives a message M intended for a destination $w \neq v$, then v should forward M to a neighbor that is closer to w than v is.

This rule can be applied in any metric space, of course, but simple and natural metric spaces are preferred over cumbersome or artificial ones. Thus, we are interested in greedy routing schemes that assign network nodes to virtual coordinates in a natural metric space.

Interest in greedy geometric routing in fixed-dimensional Euclidean spaces has expanded greatly since the work by Papadimitriou and Ratajczak [8], who showed that any 3-connected planar graph can be embedded in $\mathbf{R}^3$ so as to support greedy geometric routing. Indeed, their conjecture that such embeddings are possible in $\mathbf{R}^2$ spawned a host of additional papers (e.g., see [1,2,3,6,7]). Leighton and Moitra [5] settled this conjecture by giving an algorithm to produce a greedy embedding of any 3-connected planar graph in $\mathbf{R}^2$, and a similar result was independently found by Angelini *et al.* [1]. Greedy embeddings in $\mathbf{R}^2$ were previously known for graphs containing Delaunay triangulations [6], and existentially (but not algorithmically) for triangulations [2].

Succinct Geometric Routing. In spite of their theoretical elegance, these results settling the Papadimitriou-Ratajczak conjecture have an unfortunate drawback, in that the virtual coordinates of nodes in these solutions require $\Omega(n \log n)$ bits each in the worst case. These space inefficiencies reduce the applicability of these results for greedy geometric routing, since one could alternatively keep routing tables of size $O(n \log n)$ bits

* This work was supported by NSF grants 0724806, 0713046, 0830403, and ONR grant N00014-08-1-1015.

Y. Dong, D.-Z. Du, and O. Ibarra (Eds.): ISAAC 2009, LNCS 5878, pp. 781–791, 2009.

at each network node to support message passing. Indeed, such routing tables would allow for network nodes to be identified using labels of only $O(\log n)$ bits each, which would significantly cut down on the space, bandwidth, and packet header size needed to communicate the destination for each packet being routed. Thus, for a solution to be effectively solving the routing problem using a greedy geometric routing scheme, we desire that it be *succinct*, that is, it should use $O(\log n)$ bits per virtual coordinate. Succinct greedy geometric routing schemes are known for fixed-dimensional hyperbolic spaces [3,7], but we are unaware of any prior work on succinct greedy geometric routing in fixed-dimensional Euclidean spaces.

Our Results. We provide a succinct greedy geometric routing scheme for 3-connected planar graphs in $\mathbf{R}^2$. At the heart of our scheme is a new greedy embedding for 3-connected planar graphs in $\mathbf{R}^2$ which exploits the tree-like topology of a spanning (Christmas cactus) subgraph. Our embedding allows us to form a coordinate system which uses $O(\log n)$ bits per vertex, and allows distance comparisons to be done just using our coordinate representations consistently with the Euclidean metric.

2 Finite-Length Coordinate Systems

Let us begin by formally defining what we mean by a coordinate system, and how that differs, for instance, from a simple compression scheme. Let Σ be an alphabet, and let Σ^* a set of finite-length strings over Σ. We define a *coordinate system* f for a space S:

1. f is a map, $f : \Sigma^* \to S$, which assigns character strings to points of S.
2. f may be *parameterized*: the assignment of strings to points may depend on a fixed set of parameters.
3. f is *oblivious*: the value of f on any given $x \in \Sigma^*$ must depend only on f's parameters and x itself. It cannot rely on any other character strings in Σ^*, points in S, or other values of f.

If f is lacking property 3, we prefer to think of f as a *compression scheme*.

3 Greedy Routing in Christmas Cactus Graphs

Our method is a non-trivial adaptation of the Leighton and Moitra scheme [5], so we begin by reviewing some of the ideas from their work.

A graph G is said to be a *Christmas cactus graph* if: (1) each edge of G is in at most one cycle, (2) G is connected, and (3) removing any vertex disconnects G into at most two components. For ease of discussion, we consider any edge in a Christmas cactus graph that is not in a simple cycle to be a simple cycle itself (a 2-cycle); hence, every edge in is in exactly one simple cycle. The *dual tree* of a Christmas cactus graph G is a tree containing a vertex for each simple cycle in G with an edge between two vertices if their corresponding cycles in G share a vertex. Rooting the dual tree at an arbitrary vertex creates what we call a *depth tree*.

Having a depth tree allows us to apply the rooted tree terminology to cycles in G. In particular: *root*, *depth*, *parent*, *child*, *ancestor*, and *descendant* all retain their familiar definitions. We define the *depth* of a node v to be the minimum depth of any cycle containing v. The unique node that a cycle C shares with its parent is called the *primary node* of C. Node v is a *descendant* of a cycle C if v is in a cycle that is a descendant of C and v is not the primary node of C. Node v is a *descendant* of node u if removing neighbors of u with depth less than or equal to u leaves u and v in the same component.

Greedy Routing with a Christmas Cactus Graph Embedding. Working level by level in a depth tree, Leighton and Moitra [5] embed the cycles of a Christmas cactus graph on semi-circles of increasing radii, centered at the origin. Within the embedding we say that vertex u is above vertex v if u is embedded farther from the origin than v, and we say that u is to the left of v if u is embedded in the positive angular direction relative to v. We can define below and right similarly. These comparisons naturally give rise to directions of movement between adjacent vertices in the embedding: up, down, left, and right.

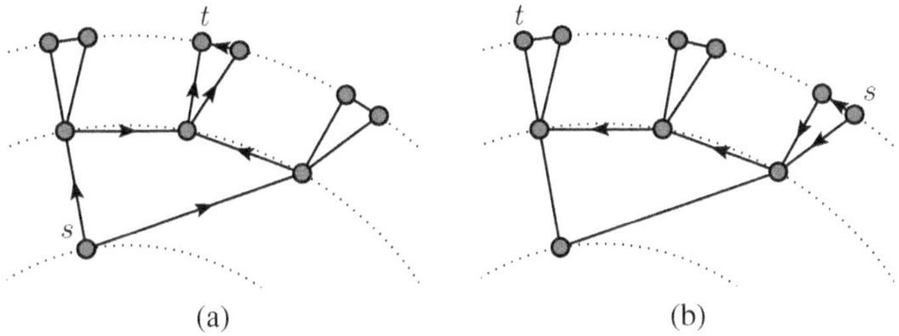

Fig. 1. Arrows indicate valid greedy hops. (a) Descendants of s can be reached by a simple path of up and right hops, up and left hops, or a combination of the two. (b) If t is not a descendant of s, then we route down and (left or right) in the direction of t until we reach an ancestor of t.

Routing from start vertex s to a terminal vertex t in a Christmas cactus graph embedding can be broken down into two cases: (1) t is a descendant of s, and (2) t is not a descendant of s.

1. As shown in Fig. 1(a), if t is a descendant of s, then we can route to t by a simple path of up and right hops, up and left hops, or a combination of the two.
2. As shown in Fig. 1(b), if t is not a descendant of s, then we route to the least common (cycle) ancestor of s and t. Suppose, without loss of generality, that t is to the left of s, then we can reach this cycle by a sequence of down and left hops. Once on the cycle, we can move left until we reach an ancestor of t. Now we are back in case 1.

This routing scheme immediately gives rise to a simple succinct compression scheme for 3-connected planar graphs, which we discuss in the full version of this paper.

4 Toward a Succinct Greedy Embedding

Given a 3-connected planar graph, we can find a spanning Christmas cactus subgraph in polynomial time [5]. Therefore, we restrict our attention to Christmas cactus graphs.

Our results apply to 3-connected planar graphs with little or no modification. In this section, we construct a novel greedy embedding scheme for any Christmas cactus graph in $\mathbf{R}^2$. We then build a coordinate system from our embedding and show that the coordinates can be represented using $O(\log^2 n)$ bits. In the next section, we show how to achieve an optimal $O(\log n)$-bit representation.

Heavy Path Decompositions. We begin by applying the Sleator and Tarjan [9] *heavy path decomposition* to the depth tree T for G.

Definition 1. *Let T be a rooted tree. For each node v in T, let $n_T(v)$ denote the number of descendants of v in T, including v. For each edge $e = (v, \text{parent}(v))$ in T, label e as a heavy edge if $n_T(v) > n_T(\text{parent}(v))/2$. Otherwise, label e as a light edge. Connected components of heavy edges form paths, called heavy paths. Vertices that are incident only to light edges are considered to be zero-length heavy paths. We call this the* ***heavy path decomposition*** *of T.*

For ease of discussion, we again apply the terminology from nodes in T to cycles in G. A cycle in G is on a heavy path H if its dual node in T is on H. Let H be a heavy path in T. We say that $\text{head}(H)$ is the cycle in H that has minimum depth, we define $\text{tail}(H)$ similarly. Let C_1 and C_2 be two cycles such that $C_1 = \text{parent}(C_2)$ and let $\{p\} = V(C_1) \cap V(C_2)$. If C_1 and C_2 are on the same heavy path then we call p a *turnpike*. If C_1 and C_2 are on different heavy paths (where $C_1 = \text{tail}(H_1)$ and $C_2 = \text{head}(H_2)$) then we call p an *off-ramp* for H_1 and the vertices $v \in V(C_2) \setminus \{p\}$ *on-ramps* for H_2.

An Overview of Our Embedding Strategy. Like Leighton and Moitra [5], we lay the cycles from our Christmas cactus graph on concentric semi-circles of radius $1 = R_0 < R_1 < R_2 \ldots$; however, our embedding has the following distinct differences: we have $\Theta(n \log n)$ semi-circles instead of $O(n)$ semi-circles, on-ramps to heavy paths are embedded on special semi-circles which we call *super levels*, turnpikes are placed in a predefined position when cycles are embedded, and the radii of semi-circles can be computed without knowing the topology of the particular Christmas cactus graph being embedded.

To make our embedding scheme amenable to a proof by induction, we modify the input Christmas cactus graph. After constructing a greedy embedding of this modified graph, we use it to prove that we have a greedy embedding for the original graph.

Modifying the Input Christmas Cactus Graph. Given a Christmas cactus graph G on n vertices, we choose a depth tree T of G, and compute the heavy path decomposition of T. For a cycle C on a heavy path H, we define $\text{relativeDepth}(C)$ to be $\text{depth}(C) - \text{depth}(\text{head}(H))$. For each C_1, $C_2 = \text{child}(C_1)$ forming a light edge in T, let $\{p\} = V(C_1) \cap V(C_2)$. Split p into two vertices p_1 and p_2 each on their own cycle, and connect p_1 to p_2 with a path of $n - 1 - \text{relativeDepth}(C_1)$ edges. The new graph G' is also a Christmas cactus graph, and our new depth tree T' looks like T stretched out so that heads of heavy paths (from T) are at depths that are multiples of n. We continue to call the paths copied from T heavy paths (though they do not form a heavy path decomposition of T'), and the newly inserted edges are *dummy* edges.

Embedding the Modified Christmas Cactus Graph in $\mathbb{R}^2$. Given a Christmas cactus graph G on n vertices, run the modification procedure described above and get G' and T'. We embed G' in phases, and prove by induction that at the end of each phase we have a greedy embedding of an induced subgraph of G'.

Lemma 1 (Leighton and Moitra [5]). *If points* $c = (0, 1+z)$, $b = (-\sin\beta, \cos\beta)$, *and* $a = (-(1+\epsilon)\sin(\beta-\alpha), (1+\epsilon)\cos(\beta-\alpha))$ *are subject to the constraints* $0 < \alpha \leq \pi/2$, $0 < \beta \leq \pi/2$, $0 < \epsilon \leq (1-\cos\beta)/6$, $0 \leq z \leq \epsilon$, *and* $\sin\alpha \leq \frac{\epsilon(1-\cos\beta)}{2(1+\epsilon)}$ *then* $d(a,c) - d(b,c) \geq \epsilon^2 > 0$.

We begin by embedding the root cycle, $C = (v_0, \ldots, v_{k-1})$, of T'. We trace out a semi-circle of radius $R_0 = 1$ centered at the origin and divide the perimeter of this semi-circle into $2n+1$ equal arcs. We allow vertices to be placed at the leftmost point of each arc, numbering these positions 0 to $2n$. We place vertices $v_0, \ldots, v_{k-1}$ clockwise into any k distinct positions, reserving position n for C's turnpike. If C does not have a turnpike, as is the case if C is a dummy edge or the tail of a heavy path, then position n remains empty. The embedding of C is greedy (proof omitted here).

Inductive Step: Suppose we have a greedy embedding all cycles in T' up to depth i, call this induced subgraph G'_i. We show that the embedding can be extended to a greedy embedding of G'_{i+1}. Our proof relies on two values derived from the embedding of G'_i.

Definition 2. *Let* s, t *be any two distinct vertices in* G'_i *and fix* $n_{s,t}$ *to be a neighbor of* s *such that* $d(s,t) > d(n_{s,t}, t)$. *We define* $\delta(G'_i) = \min_{s,t}\{d(s,t) - d(n_{s,t}, t)\}$.

We refer to the difference $d(s,t) - d(n_{s,t}, t)$ as the *delta value* for distance-decreasing paths from s to t through $n_{s,t}$.

Definition 3. *Let* $\beta(G'_i)$ *to be the minimum (non-zero) angle that any two vertices in the embedding of* G'_i *form with the origin.*

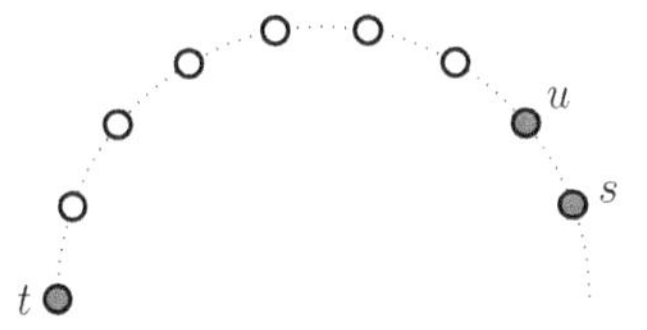

Fig. 2. s, u and t form a lower bound for $\delta(G'_0)$

Since we do not specify exact placement of all vertices, we cannot compute $\delta(G'_0)$ and $\beta(G'_0)$ exactly. We instead compute positive underestimates, δ_0 and β_0, by considering hypothetical vertex placements, and by invoking the following lemma.

Lemma 2. *Let* s *and* u *be two neighboring vertices embedded in the plane. If there exists a vertex* t *that is simultaneously closest to the perpendicular bisector of* su *(on the* u *side), and farthest from the line* su*, then the delta value for* s *to* t *through* u *is the smallest for any choice of* t*.*

Applying the above lemma to all hypothetical s, u, and t placements for the embedding of G'_0 leads to the underestimate $\delta_0 = 2-\sqrt{2+2\cos\frac{\pi}{2n+1}} < d(s,t)-d(u,t) \leq \delta(G'_0)$ where s, u, and t are shown in Fig. 2. Trivially, $\beta_0 = \frac{\pi}{2n+1} \leq \beta(G'_0)$.

We now show how to obtain a greedy embedding of G'_{i+1}, given a greedy embedding of G'_i and values δ_i and β_i.

Let $\epsilon_i = \min\{\delta_i/3, R_i\frac{1-\cos\frac{2}{3}\beta_i}{6}\}$. Trace out a semi-circle of radius $R_{i+1} = R_i + \epsilon_i$ centered at the origin. Each cycle at depth $i+1$ of T' has the form $C = (v, x_1, \ldots, x_m)$ where v, the primary node of C, has already been embedded on the ith semi-circle. We embed vertices x_1 to x_m in two subphases:

Subphase 1 We first embed vertex x_1 from each C. Choose an orientation for C so that x_1 is not a turnpike.[1] We place x_1 where the ray beginning at the origin and passing through v meets semi-circle $i+1$. We now show that distance decreasing paths exist between all pairs of vertices embedded thus far.

Distance decreasing paths between vertices in G'_i are preserved by the induction hypothesis. For t placed during this subphase: t has a neighbor v embedded on semi-circle i. If $s = v$ then s's neighbor t is strictly closer to t. Otherwise if $s \in G'_i$ then since t is within distance $\delta_i/3$ of v, then s's neighbor u that is closer to v is also closer to t. If s was placed during this subphase then s is within distance $R_i\frac{1-\cos\frac{2}{3}\beta_i}{6}$ from its neighbor v, and the perpendicular bisector of sv contains s on one side and every other vertex placed on the other side. Therefore s's neighbor v is closer to t.

The next subphase requires new underestimates, which we call δ_i^1 and β_i^1. By construction, $\beta_i^1 = \beta_i$. No s–t paths within G'_i decrease the delta value. Paths from $s \in G'_i$ to t placed in this subphase have delta value at least $\delta_i/3$ by design. For paths from s placed in this subphase, s's neighbor v is the closest vertex to the perpendicular bisector of sv on the v side. If we translate v along the perpendicular bisector of sv to a distance of R_{i+1} from sv, this hypothetical point allows us to invoke Lemma 2 to get an underestimate for the delta value of all paths beginning with s. Therefore, our new underestimate is: $\delta_i^1 = \min\{\delta_i/3, \sqrt{R_{i+1}^2 + \epsilon_i^2} - R_{i+1}\}$.

Subphase 2 We now finish embedding each cycle $C = (v, x_1, \ldots x_m)$. Let the value $\alpha = \min\{\beta_i^1/3, \delta_i^1/(3R_{i+1})\}$, s.t. $\sin\alpha \leq \frac{\epsilon_i(1-\cos\frac{2}{3}\beta_i^1)}{2(1+\epsilon_i)}$. Trace out an arc of length $R_{i+1}\alpha$ from the embedding of x_1, clockwise along semi-circle $i+1$. We evenly divide this arc into $2n+1$ positions, numbered 0 to $2n$. Position 0 is already filled by x_1. We embed vertices in clockwise order around the arc in $m-1$ distinct positions; reserving position n for C's turnpike. If there is no such node, position n remains empty.

This completes the embedding of G'_{i+1}. We show that the embedding of G'_{i+1} is greedy. We only need to consider distance decreasing paths that involve a vertex placed during this subphase. For t placed during this subphase, t is within distance $\delta_i^1/3$ from an x_1, therefore, all previously placed $s \neq x_1$ have a neighbor u that is closer to t. If

[1] For the case where C is a 2-cycle and x_1 is a turnpike we insert a temporary placeholder vertex p into C with edges to v and x_1, and treat p as the new x_1. We can later remove this placeholder by the triangle inequality.

$s = x_1$ the s's neighbor closer to t is x_2. Finally, for s placed during this subphase, let the cycle that s is on be $C = (v, x_1, \ldots, x_m)$. For $s = x_i \neq x_m$, since $\alpha \leq \beta_i^1/3$, the interior of the sector formed by x_1, x_m and the origin is empty, therefore t is either on the x_{i-1} side of the perpendicular bisector to $x_{i-1}x_i$ or on the x_{i+1} side of the perpendicular bisector to $x_i x_{i+1}$. If $s = x_m$ If t is embedded to the left s, the closer neighbor is x_{m-1}. Otherwise, applying Lemma 1, our choice of $\sin\alpha \leq \frac{\epsilon_i(1-\cos\frac{2}{3}\beta_i^1)}{2(1+\epsilon_i)}$ forces the perpendicular bisector to sv to have s on one side, and all nodes to the right of s on the other side. All cases are considered, so the embedding of G'_{i+1} is greedy.

To complete the inductive proof, we must compute δ_{i+1} and β_{i+1}. Trivially, $\beta_{i+1} = \frac{\alpha}{2n} \leq \beta(G'_{i+1})$. Distance decreasing paths between vertices placed before this subphase will not update the delta value. Therefore, we only evaluate paths with s or t embedded during this subphase. By design, paths from s previously placed to t placed during this subphase have a delta value $\geq \delta_i^1/3$. Distance-decreasing paths from s placed in this subphase to $t \in G'_{i+1}$ take two different directions. If s's neighbor u which is closer to t is on semi-circle $i+1$ then points that are closest to the perpendicular bisector to su are along the perimeter of the sector formed by s, u, and the origin. The point closest to the perpendicular bisector is where the first semi-circle intersects the sector. We translate this point down $R_{i+1}+2$ units along the perpendicular bisector, and we have an underestimate for the delta value for any path beginning with a left/right edge. If s's neighbor that is closer to t is on the ith semi-circle, then a down edge is followed. To finish, we evaluate down edges su added during the second subphase. The closest vertex to the perpendicular bisector to su on the u side is either u, or the vertex placed in the next clockwise position the $i+1$th semi-circle. Translating this point $2R_{i+1}$ units away from su along the perpendicular bisector gives us the an underestimate for paths beginning with su.

This completes the proof for the greedy embedding of G'. To obtain a greedy embedding for G, we repeatedly collapse dummy edges in G' until we get G. When we collapse an edge (p, x_1), where p is the primary node for the 2-cycle, we collapse the edge to vertex p in the embedding. Collapsing in the other direction may break distance decreasing paths through p and neighbors of p embedded on the same semi-circle as p. After collapsing all such dummy edges, we have a greedy embedding of G.

We call the levels where the on-ramps to heavy paths are embedded *super levels*, and all other levels are *baby levels*. There are $n-1$ baby levels between consecutive super levels and, since any path from root to leaf in a depth tree travels through $O(\log n)$ different heavy paths, there are $O(\log n)$ super levels.

Our Coordinate System. Let v be a vertex in G. We define level(v) to be the number of baby levels between v and the previous super level (zero if v is on a super level) and cycle(v) to be the position, 0 to $2n$, where v is placed when its cycle is embedded. These values can be assigned to vertices without performing the embedding procedure.

Let s be v's ancestor on the first super level. The path from s to v passes through $O(\log(n))$ heavy paths, entering each heavy path at an on-ramp, and leaving at an off-ramp. We define v's coordinate to be a $O(\log n)$-tuple consisting of the collection of (level$(\cdot)$, cycle$(\cdot)$) pairs for each off-ramp where a change in heavy paths occurs on the path from s to v, and the pair (level(v), cycle(v)), which is either an off-ramp or a turnpike. Using the coordinate for v and the parameter n, we can compute the Euclidean coordinates for all the turnpikes and off-ramps on the path from s to v, including the coordinate for v. Thus, we have defined a coordinate system for the Euclidean plane.

Using a straightforward encoding scheme, each level-cycle pair is encoded using $O(\log n)$ bits. Since a coordinate contains $O(\log n)$ of these pairs, we encode each coordinate using $O(\log^2 n)$ bits.

Greedy Routing with Coordinate Representations. By design, the routing scheme discussed in Sect. 3 is greedy for our embedding. We develop a comparison rule using the potential number of edges that may be traversed on a specific path from s to t.

Let s_i be the vertex between super levels i and $i+1$, whose level-cycle pair is in position i of s's coordinate. We define t_i similarly. Let superlevel(s) be the position that contains the level-cycle pair for s itself. Let h be the smallest integer such that s_h and t_h differ. Using the level-cycle pairs for s_h and t_h, we can compute the level-cycle pair for the off-ramps on the least common ancestor C that diverge toward s and t, which we call s_C and t_C. That is, if $\text{level}(s_h) = \text{level}(t_h)$ then $s_C = s_h$ and $t_C = t_h$. Otherwise, assume without loss of generality that $\text{level}(s_h) < \text{level}(t_h)$, then s_C's pair is $(\text{level}(s_h), \text{cycle}(s_h))$ and t_C is a turnpike with the pair $(\text{level}(s_h), n)$.

We define l, r, d, u be the potential number of left, right, down, and up edges that may be traversed from s to t. Values d and u are simply the number of semi-circles passed through by down and up hops, respectively. That is,

$$d = (\text{superlevel}(s) \cdot n + \text{level}(s)) - (hn + \text{level}(s_C))$$
$$u = (\text{superlevel}(t) \cdot n + \text{level}(t)) - (hn + \text{level}(t_C)).$$

If $\text{cycle}(t_C) < \text{cycle}(s_C)$, then we count the maximum number left edges on the path from s to t_C, and the maximum number of right edges from t_C to t. That is,

$$l = \begin{cases} \text{cycle}(s) + 2n(d-1) + \text{cycle}(s_C) - \text{cycle}(t_C) & \text{if } s \neq s_C, \\ \text{cycle}(s_C) - \text{cycle}(t_C) & \text{if } s = s_C. \end{cases}$$

$$r = \begin{cases} 2n(u-1) + \text{cycle}(t) & \text{if } t \neq t_C, \\ 0 & \text{if } t = t_C. \end{cases}$$

If $\text{cycle}(t_C) \geq \text{cycle}(s_C)$, then we count the maximum number of right edges on the path from s to t_C, and the maximum number of right edges from t_C to t. That is,

$$l = 0$$
$$r = r_1 + r_2, \text{ where}$$

$$r_1 = \begin{cases} 2n - \text{cycle}(s) + 2n(d-1) + \text{cycle}(t_C) - \text{cycle}(s_C) & \text{if } s \neq s_C, \\ \text{cycle}(t_C) - \text{cycle}(s_C) & \text{if } s = s_C. \end{cases}$$

$$r_2 = \begin{cases} 2n(u-1) + \text{cycle}(t) & \text{if } t \neq t_C, \\ 0 & \text{if } t = t_C. \end{cases}$$

Our comparison rule is:

$$D(s,t) = l + r + (2n+1)u + d.$$

Following the routing scheme from Sect. 3, any move we make toward the goal will decrease $D(\cdot,\cdot)$, and any move away from the goal will increase $D(\cdot,\cdot)$. Therefore, we can use this comparison rule to perform greedy routing in our embedding efficiently, and comparisons made while routing will evaluate consistently with the corresponding Euclidean coordinates under the L_2 metric.

5 An Optimal Succinct Greedy Embedding

Conceptually, the level$(\cdot)$ and cycle$(\cdot)$ values used in the previous section are encoded as integers whose binary representation corresponds to a path from root to a leaf in a full binary tree with n leaves. Instead of encoding with a static $O(\log n)$ bits per integer, we will modify our embedding procedure so we can further exploit the heavy path decomposition of the dual tree T, using *weight-balanced binary trees* [4].

Definition 4. *A* ***weight-balanced binary tree*** *is a binary tree which stores weighted items from a total order in its leaves. If item i has weight w_i, and all items have a combined weight of W then item i is stored at depth $O(\log W/w_i)$. An inorder listing of the leaves outputs the items in order.*

Encoding the Level Values. Suppose we have a depth tree T for G, and a heavy path decomposition of T. Let C be a simple cycle in G on some heavy path H and let C_{next} be the next cycle on the heavy path H, if it exists. Let $n(C)$ be the number of vertex descendants of C in G. We define a weight function $\gamma(\cdot)$ on the cycles in G as follows:

$$\gamma(C) = \begin{cases} n(C) & \text{if } C = \text{tail}(H), \\ n(C) - n(C_{\text{next}}) & \text{if } C \neq \text{tail}(H). \end{cases}$$

For each heavy path H, create a weight-balanced binary tree B_H containing each cycle C in H as an item with weight $\gamma(C)$, and impose a total order so that cycles are in their path order from head(H) to tail(H).

Let v be a vertex whose coordinate we wish to encode, and suppose v is located between super levels l and $l+1$. Let v_i be the vertex whose level-cycle pair is in position i of v's coordinate. Let v_i be contained in cycle C_i (such that v_i is not C_i's primary node) on heavy path H_i. The code for level(v_i) is a bit-string representing the path from root to the leaf for C_i in the weight-balanced binary tree B_{H_i}. Let W_i be the combined weight of the items in B_{H_i}. Since C_i is at a depth of $O(\log W_i/\gamma(C_i))$, this is length of the code. Thus, the level values in v's coordinate are encoded with $O(\sum_{0\le i\le l} \log W_i/\gamma(C_i))$ bits total. By design, this sum telescopes to $O(\log n)$ bits.

Encoding the Cycle Values. For a node v in G we define a weight function $\mu(v)$ to be the number of descendants of v in G.

Let $C = (p, x_1, x_2, \ldots, x_m)$ be a cycle in G, where p is the primary node of C. Let x_h be the turnpike that connects C to the next cycle on the heavy path, if it exists. Let x_i have weight $\mu(x_i)$ and impose a total order so $x_j < x_k$ if $j < k$. For each cycle C, we create a weight-balanced binary tree B_C containing nodes x_1 to x_m as follows. We first create two weight-balanced binary trees B_C^1 and B_C^2 where B_C^1 contains x_j

for $j < h$ and B_C^2 contains x_k for $k > h$. If no such x_h exists, then choose an integer $1 \le k \le m$ and insert items x_j for $j < k$ into B_C^1 and insert the remaining items into B_C^2. We form our single weight-balanced binary tree B_C in two steps: (1) create a tree B_C^3 with B_C^1 as a left subtree and a node for x_h as a right subtree, and (2) form B_C with B_C^3 as a left subtree and B_C^2 as a right subtree. We build B_C in this way to ensure that every turnpike is given the same path within its tree, and hence the same cycle code and value.

The code for cycle(v_i) is a bit-string representing the path from root to the leaf for v_i in the weight-balanced binary tree B_{C_i}. Let W_i be the combined weight of the items in B_{C_i}. Since v_i is at a depth of $O(\log W_i/\mu(v_i))$, this is length of the code. Thus, the cycle values in v's coordinate are encoded with $O(\sum_{0 \le i \le l} \log W_i/\mu(v_i))$ bits total. By design, this sum telescopes to $O(\log n)$ bits.

Interpreting the Codes. Let c be the smallest integer constant such that item i stored in the weight-balanced binary tree is at depth $\le c \log W/w_i$. We can treat the position of i in the weight-balanced binary tree as a position in a full binary tree of height $c \log n$. We interpret this code to be the number of tree nodes preceding i in an in-order traversal of the full binary tree. Using our codes as described, we require $2n^c - 2$ baby levels between each super level and $8n^c - 1$ cycle positions.

An Overview of the Optimal Embedding. Let T be the depth tree for our Christmas cactus graph G. We create weight-balanced binary trees on the heavy paths in T and on each of the cycles in G, giving us the level and cycle codes for every vertex. We adjust the graph modification procedure so that adjacent cycles on heavy paths are spaced out according to the level codes. That is, adjacent cycles on the same heavy path have heavy dummy edges (dummy edges that are considered to be on the heavy path) inserted between them so that they are placed on the appropriate baby levels. For cycles on different heavy paths, we insert dummy edges to pad out to the next superlevel, and heavy dummy edges to pad out to the appropriate baby level.

We embed the modified graph analogously to our $O(\log^2 n)$ embedding, except that the cycle codes dictate vertex placements. We augment our coordinate system to store the level value for elements on the root cycle, otherwise it is not possible to compute the corresponding Euclidean point from our succinct representation. The same comparison rule applies to our new coordinate system, with little change to account for the new range of cycle values. Using this embedding scheme and coordinate system we achieve optimal $O(\log n)$ bits per coordinate.

References

1. Angelini, P., Frati, F., Grilli, L.: An algorithm to construct greedy drawings of triangulations. In: Tollis, I.G., Patrignani, M. (eds.) GD 2008. LNCS, vol. 5417, pp. 26–37. Springer, Heidelberg (2009)
2. Dhandapani, R.: Greedy drawings of triangulations. In: Proceedings of the 19th ACM-SIAM Symposium on Discrete Algorithms (SODA 2008), pp. 102–111. SIAM, Philadelphia (2008)
3. Eppstein, D., Goodrich, M.T.: Succinct greedy graph drawing in the hyperbolic plane. In: Tollis, I.G., Patrignani, M. (eds.) GD 2008. LNCS, vol. 5417, pp. 14–25. Springer, Heidelberg (2009)

4. Knuth, D.E.: Optimum binary search trees. Acta Informatica 1, 14–25 (1971)
5. Leighton, T., Moitra, A.: Some results on greedy embeddings in metric spaces. In: Proceedings of the 49th Annual IEEE Symposium on Foundations of Computer Science (FOCS 2008), pp. 337–346. IEEE Press, Los Alamitos (2008)
6. Lillis, K.M., Pemmaraju, S.V.: On the efficiency of a local iterative algorithm to compute delaunay realizations. In: McGeoch, C.C. (ed.) WEA 2008. LNCS, vol. 5038, pp. 69–86. Springer, Heidelberg (2008)
7. Muhammad, R.B.: A distributed geometric routing algorithm for ad hoc wireless networks. In: Proceedings of the 4th International Conference on Information Technology (ITNG 2007), pp. 961–963. IEEE Press, Los Alamitos (2007)
8. Papadimitriou, C.H., Ratajczak, D.: On a conjecture related to geometric routing. Theoretical Computer Science 344(1), 3–14 (2005)
9. Sleator, D.D., Tarjan, R.E.: A data structure for dynamic trees. Journal of Computer and System Sciences 26(3), 362–391 (1983)

Electric Routing and Concurrent Flow Cutting*

Jonathan Kelner and Petar Maymounkov

Massachusetts Institute of Technology
kelner@mit.edu, petar@csail.mit.edu

Abstract. We investigate an oblivious routing scheme amenable to distributed computation and resilient to graph changes, based on electrical flow. Our main technical contribution is a new rounding method which we use to obtain a bound on the $L_1 \to L_1$ operator norm of the inverse graph Laplacian.

1 Introduction

Overview. We address a vision of the Internet where every participant exchanges messages with their direct friends and no one else. Yet such an Internet should be able to support reliable and efficient routing to remote locations identified by unchanging names in the presence of an ever changing graph of connectivity.

Modestly to this end, this paper investigates the properties of routing along the electric flow in a graph (*electric routing* for short) for intended use in such distributed systems, whose topology changes over time. We focus on the class of expanding graphs which, we believe, gives a good trade-off between applicability and precision in modeling a large class of real-world networks. We address distributed representation and computation. As a measure of performanace, we show that electric routing, being an oblivious routing scheme, achieves minimal maximum edge congestion (as compared to a demand-dependent optimal scheme). Furthermore, we show that electric routing continues to work (on average) in the presence of large edge failures ocurring after the routing scheme has been computed, which attests to its applicability in changing environments. We now proceed to a formal definition of oblivious routing and statement of our results.

Oblivious routing. The object of interest is a graph $G = (V, E)$ (with $V = [n]$ and $|E| = m$) undirected, positively edge-weighted by $w_{u,v} \geqslant 0$, and not necessarily simple. The intention is that higher $w_{u,v}$ signifies stronger connectivity between u and v; in particular, $w_{u,v} = 0$ indicates the absence of edge (u, v). For analysis purposes, we fix an arbitrary orientation "$\to$" on the edges (u, v) of G, i.e. if (u, v) is an edge then exactly one of $u \to v$ or $v \to u$ holds. Two important operators are associated to every G. The *discrete gradient* operator $B \in \mathbf{R}^{E\times V}$, sending functions on V to functions on the *undirected* edge set E, is defined as $\chi^*_{(u,v)}B := \chi_u - \chi_v$ if $u \to v$, and $\chi^*_{(u,v)}B := \chi_v - \chi_u$ otherwise, where χ_y is

* Research partially supported by NSF grant CCF-0843915.

Y. Dong, D.-Z. Du, and O. Ibarra (Eds.): ISAAC 2009, LNCS 5878, pp. 792–801, 2009.

the Kronecker delta function with mass on y. For $e \in E$, we use the shorthand $B_e := (\chi_e B)^*$. The *discrete divergence* operator is defined as B^*.

A *(single-commodity) demand* of amount $\alpha > 0$ between $s \in V$ and $t \in V$ is defined as the vector $d = \alpha(\chi_s - \chi_t) \in \mathbf{R}^V$. A *(single-commodity) flow* on G is defined as a vector $f \in \mathbf{R}^E$, so that $f_{(u,v)}$ equals the flow value from u towards v if $u \to v$, and the negative of this value otherwise. We also use the notation $f_{u \to v} := f_{(u,v)}$ if $u \to v$, and $f_{u \to v} := -f_{(u,v)}$ otherwise. We say that flow f *routes* demand d if $B^* f = d$. This is a linear algebraic way of encoding the fact that f is an (s,t)-flow of amount α. A *multi-commodity demand*, also called a *demand set*, is a matrix whose columns constitute the individual demands' vectors. It is given as the direct product $\oplus_\tau d_\tau$ of its columns. A *multi-commodity flow* is represented as a matrix $\oplus_\tau f_\tau$, given as a direct product of its columns, the single-commodity flows. For clarity, we write $f_{\tau,e}$ for $(f_\tau)_e$. The flow $\oplus_\tau f_\tau$ *routes* the demand set $\oplus_\tau d_\tau$ if $B^* f_\tau = d_\tau$, for all τ, or in matrix notation $B^*(\oplus_\tau f_\tau) = \oplus_\tau d_\tau$. The *congestion* $\|\cdot\|_G$ of a multi-commodity flow measures the load of the most-loaded edge, relative to its capacity. It is given by

$$\| \oplus_\tau f_\tau \|_G := \max_e \sum_\tau |f_{\tau,e}/w_e| = \|(\oplus_\tau f_\tau)^* W^{-1}\|_{1\to 1}, \text{ where } \|A\|_{1\to 1} := \sup_{x \neq 0} \frac{\|Ax\|_1}{\|x\|_1}. \quad (1)$$

An *oblivious routing scheme* is a (not necessarily linear) function $R : \mathbf{R}^V \to \mathbf{R}^E$ which has the property that $R(d)$ routes d when d is a valid single-commodity demand (according to our definition). We extend R to a function over demand sets by defining $R(\oplus_\tau d_\tau) := \oplus_\tau R(d_\tau)$. This says that each demand in a set is routed independently of the others by its corresponding R-flow. We measure the "goodness" of an oblivious routing scheme by the maximum traffic that it incurs on an edge (relative to its capacity) compared to that of the optimal (demand-dependent) routing. This is captured by the *competitive ratio* η_R of the routing scheme R, defined

$$\eta_R := \sup_{\oplus_\tau d_\tau} \sup_{\substack{\oplus_\tau f_\tau \\ B^*(\oplus_\tau f_\tau) = \oplus_\tau d_\tau}} \frac{\|R(\oplus_\tau d_\tau)\|_G}{\| \oplus_\tau f_\tau \|_G}. \quad (2)$$

Let $\mathcal{E}$ denote the (yet undefined) function corresponding to electric routing. Our main theorem states:

Theorem 1. *For every undirected graph G with unit capacity edges, maximum degree $d_{\max}$ and vertex expansion $\alpha := \min_{S \subseteq V} \frac{|E(S,S^{\complement})|}{\min\{|S|,|S^{\complement}|\}}$, one has $\eta_{\mathcal{E}} \leqslant \left(4 \ln \frac{n}{2}\right) \cdot \left(\alpha \ln \frac{2d_{\max}}{2d_{\max} - \alpha}\right)^{-1}$. This is tight up to a factor of $O(\ln \ln n)$.*

The competitive ratio in Theorem 1 is best achievable for any oblivious routing scheme up to a factor of $O(\ln \ln n)$ due to a lower bound for expanders, i.e. the case $\alpha = O(1)$, given in [1]. Theorem 1 can be extended to other definitions of graph expansion, weighted and unbounded-degree graphs. We omit these extensions for brevity. We also give an unconditional, but worse, bound on $\eta_{\mathcal{E}}$:

Theorem 2. *For every unweighted graph on m edges, electrical routing has $\eta_{\mathcal{E}} \leqslant O(m^{1/2})$. Furthermore, there are families of graphs with corresponding demand sets for which $\eta_{\mathcal{E}} = \Omega(m^{1/2})$.*

Electric routing. Let $W = \mathrm{diag}(\ldots, w_e, \ldots) \in \mathbf{R}^{E\times E}$ be the edge weights matrix. We appeal to a known connection between graph Laplacians and electric current [2,3]. Graph edges are viewed as wires of resistance w_e^{-1} and vertices are viewed as connection points. If $\varphi \in \mathbf{R}^V$ is a vector of vertex potentials then, by Ohm's law, the *electric flow* (over the edge set) is given by $f = WB\varphi$ and the corresponding demand is $B^* f = L\varphi$ where the *(un-normalized) Laplacian* L is defined as $L = B^*WB$. Central to the present work will be the vertex potentials that induce a desired (s,t)-flow, given by $\varphi^{[s,t]} = L^\dagger(\chi_s - \chi_t)$, where $L^\dagger$ is the pseudo-inverse of L. Thus, the electric flow corresponding to the demand pair (s,t) is $WB\varphi^{[s,t]} = WBL^\dagger(\chi_s - \chi_t)$. We define the *electric routing* operator as

$$\mathcal{E}(d) = WBL^\dagger d \tag{3}$$

The vector $\mathcal{E}(\chi_s - \chi_t) \in \mathbf{R}^E$ encodes a unit flow from s to t supported on G, where the flow along an edge (u,v) is given by $[\![st, uv]\!] := \mathcal{E}(\chi_s - \chi_t)_{u\to v} = (\varphi_u^{[s,t]} - \varphi_v^{[s,t]})w_{u,v}$.[1] (Our convention is that current flows towards lower potential.) When routing an indivisible message (an IP packet e.g.), we can view the unit flow $\mathcal{E}(\chi_s - \chi_t)$ as a distribution over (s,t)-paths defined recursively as follows: Start at s. At any vertex u, *forward* the message along an edge with positive flow, with probability proportional to the edge flow value. Stop when t is reached. This rule defines the *electric walk* from s to t. It is immediate that the flow value over an edge (u,v) equals the probability that the electric walk traverses that edge.

Let "$\sim$" denote the vertex adjacency relation of G. In order to make a (divisible or indivisible) forwarding decision, a vertex u must be able to compute $[\![st, uv]\!]$ for all neighbors $v \sim u$ and all pairs $(s,t) \in \binom{V}{2}$. We address this next.

Representation. In order to compute $[\![st, uv]\!]$ (for all $s,t \in V$ and all $v \sim u$) at u, it suffices that u stores the vector $\varphi^{[w]} := L^\dagger \chi_w$, for all $w \in \{w : w \sim u\} \cup \{u\}$. This is apparent from writing

$$[\![st, uv]\!] = (\chi_u - \chi_v)L^\dagger(\chi_s - \chi_t) = (\varphi^{[u]} - \varphi^{[v]})^*(\chi_s - \chi_t), \tag{4}$$

where we have (crucially) used the fact that $L^\dagger$ is symmetric. The vectors $\varphi^{[w]}$ stored at u comprise the *(routing) table* of u, which consists of $\deg(u)\cdot n$ real numbers. Thus the per-vertex table sizes of our scheme grow linearly with the vertex degree – a property we call *fair* representation. It seems that fair representation is key for routing in heterogenous sytems consisting of devices with varying capabilities.

Equation (4), written as $[\![st, uv]\!] = (\chi_s - \chi_t)^*(\varphi^{[u]} - \varphi^{[v]})$, shows that in order to compute $[\![st, uv]\!]$ at u, it suffices to know the *indices* of s and t (in the $\varphi^{[w]}$'s). These indices could be represented by $O(\ln n)$-bit opaque vertex ID's and could be carried in the message headers. Routing schemes that support opaque vertex addressing are called *name-independent.* Name independence allows for vertex name persistence across time (i.e. changing graph topology) and across multiple co-existing routing schemes.

[1] The bilinear form $[\![st, uv]\!] = \chi_{s,t} BL^\dagger B^* \chi_{u,v}$ acts like a "representation" of G, hence the custom bracket notation.

Computation. We use an idealized computational model to facilitate this exposition. The vertices of G are viewed as processors, synchronized by a global step counter. During a time step, pairs of processors can exchange messages of arbitrary (finite) size as long as they are connected by an edge. We describe an algorithm for computing approximations $\tilde{\varphi}^{[v]}$ to all $\varphi^{[v]}$ in $O(\ln n/\lambda)$ steps, where λ is the Fiedler eigenvalue of G (the smallest non-zero eigenvalue of L). If G is an expander, then $\lambda = O(1)$. At every step the algorithm sends messages consisting of $O(n)$ real numbers across every edge and performs $O(\deg(v) \cdot n)$ arithmetic operations on each processor v. Using standard techniques, this algorithm can be converted into a relatively easy-to-implement asynchronous one. (We omit this detail from here.) It is assumed that no graph changes occur during the computation of vertex tables.

A vector $\zeta \in \mathbf{R}^V$ is *distributed* if ζ_v is stored at v, for all v. A matrix $M \in \mathbf{R}^{V\times V}$ is *local* (with respect to G) if $M_{u,v} \neq 0$ implies $u \sim v$ or $u = v$. It is straightforward that if ζ is distributed and M is local, then $M\zeta$ can be computed in a single step, resulting in a new distributed vector. Extending this technique shows that for any polynomial $q(\cdot)$, the vector $q(M)\zeta$ can be computed in $\deg(q)$ steps.

The Power Method gives us a matrix polynomial $q(\cdot)$ of degree $O(\ln n/\lambda)$ such that $q(L)$ is a "good" approximation of $L^\dagger$. We compute the distributed vectors $\zeta^{[w]} := q(L)\chi_w$, for all w, in parallel. As a result, each vertex u obtains $\tilde{\varphi}^{[u]} = (\zeta_u^{[1]}, \dots, \zeta_u^{[n]})$, which approximates $\varphi^{[u]}$ according to Theorem 3 and the symmetry of L. In one last step, every processor u sends $\tilde{\varphi}^{[u]}$ to its neighbors. The approximation error n^{-5} is chosen to suffice in accordance with detailed analysis given in [4]. The next theorem is proven in the full version of the paper [4]:

Theorem 3. *Let λ be the Fiedler (smallest non-zero) eigenvalue of G's Laplacian L, and let G be of bounded degree $d_{\max}$. Then $\|\zeta^{[v]} - \varphi^{[v]}\|_2 \leqslant n^{-5}$, where $\zeta^{[v]} = (2d_{\max})^{-1}\sum_{\omega=0}^{k} M^\omega \chi_v$ and $M = I - L/2d_{\max}$, as long as $k \geqslant \Omega(\lambda^{-1} \cdot \ln n)$.*

Robustness and latency. In order to get a handle on the analysis of routing in an ever-changing network we use a simplifying assumption: the graph does not change during the computation phase while it can change afterwards, during the routing phase. This assumption is informally justified because the computation phase in expander graphs (which we consider to be the typical case) is relatively fast, it takes $O(\ln n)$ steps. The routing phase, on the other hand, should be as "long" as possible before we have to recompute the scheme. Roughly, a routing scheme can be used until the graph changes so much from its shape when the scheme was computed that both the probability of reaching destinations and the congestion properties of the scheme deteriorate with respect to the new shape of the graph. We quantify the robustness of electric routing against edge removals in the following two theorems, proven in [4]:

Theorem 4. *Let G be an unweighted graph with Fiedler eigenvalue $\lambda = \Theta(1)$ and maximum degree $d_{\max}$, and let $f^{[s,t]}$ denote the unit electric flow between s*

and t. For any $0 < p \leqslant 1$, let $Q_p = \{e \in E : |f_e^{[s,t]}| \geqslant p\}$ be the set of edges carrying more than p flow. Then, $|Q_p| \leqslant \min\{2/(\lambda p^2), 2d_{\max}\|L^\dagger\|_{1\to1}/p\}$.

Note that part one of this theorem, i.e. $|Q_p| \leqslant 2/(\lambda p^2)$, distinguishes electric routing from simple schemes like shortest-path routing. The next theorem studies how edge removals affect demands when "the entire graph is in use:"

Theorem 5. *Let graph G be unweighted of bounded degree $d_{\max}$ and vertex expansion α. Let f be a routing of the uniform multi-commodity demand set over V (single unit of demand between every pair of vertices), produced by an η-competitive oblivious routing scheme. Then, for any $0 \leqslant x \leqslant 1$, removing a x-fraction of edges from G removes at most $x \cdot (\eta \cdot d_{\max} \cdot \ln n \cdot \alpha^{-1})$-fraction of flow from f.*

The expected number of edges traversed between source and sink reflects the latency of a routing. The next theorem, also proven in [4], bounds the average latency of electric routing in arbitrary graphs:

Theorem 6. *The latency of every electric walk on an undirected graph of bounded degree $d_{\max}$ and vertex expansion α is at most $O(\min\{m^{1/2}, d_{\max}\alpha^{-2}\ln n\})$.*

Analysis. The main hurdle is Theorem 1, which we attack in two steps. First, we show that any linear routing scheme R (i.e. scheme for which the operator $R : \mathbf{R}^V \to \mathbf{R}^E$ is linear) has a distinct worst-case demand set, known as *uniform demands*, consisting of a unit demand between the endpoints of every edge of G. Combinging this with the formulaic expression for electric flow (3) gives us an operator-based geometric bound for $\eta_{\mathcal{E}}$, which in the case of a bounded degree graph is simply $\eta_{\mathcal{E}} \leqslant O(\|L^\dagger\|_{1\to1})$ where the operator norm $\|\cdot\|_{1\to1}$ is defined by $\|A\|_{1\to1} := \sup_{x\neq0}\|Ax\|_1/\|x\|_1$. This is shown in Theorem 7. Second, we give a rounding-type argument that establishes the desired bound on $\|L^\dagger\|_{1\to1}$. This argument relies on a novel technique we dub *concurrent flow cutting* and is our key technical contribution. This is done in Theorem 8. This concludes the analysis of the congestion properties of electric flow.

The computational procedure for the vertex potentials $\varphi^{[v]}$'s (above) only affords us approximate versions $\tilde{\varphi}^{[v]}$ with ℓ_2 error guarantees. We need to ensure that, when using these in place of the exact ones, all properties of the exact electric flow are preserved. For this purpose, it is convenient to view the electric flow as a distribution over paths (i.e. the electric walk, defined above) and measure the total variation distance between the walks induced by exact and approximate vertex potentials. It is then easy to verify that any two multi-commodity flows, whose respective individual flows have sufficiently small variation distance, have essentially identical congestion and robustness properties. This is pursued in detail in [4].

Related work. Two bodies of prior literature concern themselves with oblivious routing. One focuses on approximating the shortest-path metric [5,6,7,8], the other focuses on approximating the minimal congestion universally across all possible demand sets [9,10]. The algorithms in these works are essentially best possible in terms of competitive characteristics, however they are not distributed and do not address (competitive) performance in the presence of churn. It is not obvious how to provide efficient distributed variants for these routing schemes that are additionally resistant to churn. The primary reason for this are the algorithmic primitives used. Common techniques are landmark (a.k.a. beacon) selection [5,6], hierarchical tree-decomposition or tree-packings [9]. These approaches place disproportiantely larger importance on "root" nodes, which makes the resulting schemes vulnerable to individual failures. Furthermore, these algorithms require more than (quasi-)linear time (in the centralized model), which translates to prohibitively slow distributed times.

We are aware of one other work in the theoretical literature by Goyal, et al. [11] that relates to efficient and churn-tolerant distributed routing. Motivated by the proliferation of routing schemes for trees, they show that expanders are well-approximated by the union of $O(1)$ spanning trees. However, they do not provide a routing scheme, since routing over unions of trees is not understood.

Concurrently with this paper, Lawler, et al. [12] study just the congestion of electric flow in isolation from other considerations like computation, representation or tolerance to churn. Their main result is a variant of our graph expansion-based bound on $\|L^{\dagger}\|_{1\to1}$, given by Theorem 8. Our approaches, however, are different. We use a geometric approach, compared to a less direct probabilistic one. Our proof exposes structural information about the electric flow, which makes the fault-tolerance of electric routing against edge removal an easy consequence. This is not the case for the proofs found in [12].

Organization. Section 2 relates the congestion of electric flow to $\|L^{\dagger}\|_{1\to1}$. Section 3 obtains a bound on $\|L^{\dagger}\|_{1\to1}$ by introducing the concurrent flow cutting method. Section 4 contains remarks and open problems.

2 The Geometry of Congestion

Recall that given a multi-commodity demand, electric routing assigns to each demand the corresponding electric flow in G, which we express (3) in operator form $\mathcal{E}(\oplus_\tau d_\tau) := WBL^{\dagger}(\oplus_\tau d_\tau)$. Electric routing is oblivious, since $\mathcal{E}(\oplus_\tau d_\tau) = \oplus_\tau \mathcal{E}(d_\tau)$ ensures that individual demands are routed independently from each other. The first key step in our analysis, Theorem 7, entails bounding $\eta_{\mathcal{E}}$ by the $\|\cdot\|_{1\to1}$ matrix norm of a certain natural graph operator on G. This step hinges on the observation that all linear routing schemes have an easy-to-express worst-case demand set. This theorem is proven in [4]:

Theorem 7. *For every undirected, weighted graph G, let $\Pi = W^{1/2}BL^{\dagger}B^*W^{1/2}$, then $\eta_{\mathcal{E}} \leqslant \|W^{1/2}\Pi W^{-1/2}\|_{1\to1}$.*

Using Theorem 7, the unconditional upper bound in Theorem 2 is simply a consequence of basic norm inequalities. (See [4] for a proof.) Theorem 1 provides a much stronger bound on $\eta_{\mathcal{E}}$ when the underlying graph has high vertex expansion. The lower bound in Theorem 1 is due to Hajiaghayi, et al. [1]. They show that every oblivious routing scheme is bound to incur congestion of at least $\Omega(\ln n/\ln\ln n)$ on a certain family of expander graphs. The upper bound in Theorem 1 follows from Theorem 7, Theorem 8 and using that $\|\Pi\|_{1\to 1} = O(\|L^\dagger\|_{1\to 1})$ for unweighted bounded-degree graphs. Thus in the next section we derive a bound on $\|L^\dagger\|_{1\to 1}$ in terms of vertex expansion.

3 L_1 Operator Inequalities

The main results here are an upper and lower bound on $\|L^\dagger\|_{1\to 1}$, which match for bounded-degree expander graphs. In this section, we present vertex expansion versions of these bounds that assume bounded-degree.

Theorem 8. *Let graph $G = (V, E)$ be unweigthed, of bounded degree $d_{\max}$, and vertex expansion*

$$\alpha = \min_{S\subseteq V} \frac{|E(S, S^{\complement})|}{\min\{|S|, |S^{\complement}|\}}, \quad \textit{then} \quad \|L^\dagger\|_{1\to 1} \leqslant \left(4\ln\frac{n}{2}\right)\cdot\left(\alpha \ln \frac{2d_{\max}}{2d_{\max}-\alpha}\right)^{-1}. \tag{5}$$

The proof of this theorem (given in the next Section) boils down to a structural decomposition of unit (s,t)-electric flows in a graph (not necessarily an expander). We believe that this decomposition is of independent interest. In the case of bounded-degree expanders, one can informally say that the electric walk corresponding to the electric flow between s and t takes every path with probability exponentially inversely proportional to its length. We complement Theorem 8 with a lower bound on $\|L^\dagger\|_{1\to 1}$, proven in [4]:

Theorem 9. *Let graph $G = (V, E)$ be unweighted, of bounded degree $d_{\max}$, with metric diameter D. Then, $\|L^\dagger\|_{1\to 1} \geqslant 2D/d_{\max}$ and, in particular, $\|L^\dagger\|_{1\to 1} \geqslant (2\ln n)\cdot(d_{\max}\ln d_{\max})^{-1}$ for all bounded-degree, unweighted graphs with vertex expansion $\alpha = O(1)$.*

3.1 Proof of Upper Bound on $\|L^\dagger\|_{1\to 1}$ for Expanders

Proof (Proof of Theorem 8).

Reformulation: We start by transforming the problem in a more manageable form,

$$\|L^\dagger\|_{1\to 1} := \sup_{y\neq 0} \frac{\|L^\dagger y\|_1}{\|y\|_1} = \max_w \|L^\dagger \chi_w\|_1 \overset{(*)}{\leqslant} \frac{n-1}{n} \max_{s\neq t} \|L^\dagger(\chi_s - \chi_t)\|_1, \tag{6}$$

where the latter inequality comes from

$$\|L^\dagger \chi_s\|_1 = \|L^\dagger \pi_{\perp\mathbb{1}}\chi_s\|_1 = \|n^{-1}\sum_{t\neq s} L^\dagger(\chi_s - \chi_t)\|_1 \leqslant \frac{n-1}{n}\max_t \|L^\dagger(\chi_s - \chi_t)\|_1.$$

Pick any vertices $s \neq t$ and set $\psi = L^{\dagger}(\chi_s - \chi_t)$. In light of (6) our goal is to bound $\|\psi\|_1$. We think of ψ as the vertex potentials corresponding to electric flow with imbalance $\chi_s - \chi_t$. By an easy perturbation argument we can assume that no two vertex potentials coincide.

Index the vertices in $[n]$ by increasing potential as $\psi_1 < \cdots < \psi_n$. Further, assume that n is even and choose a median c_0 so that $\psi_1 < \cdots < \psi_{n/2} < c_0 < \psi_{n/2+1} < \cdots < \psi_n$. (If n is odd, then set c_0 to equal the pottential of the middle vertex.)

We aim to upper bound $\|\psi\|_1$, given as $\|\psi\|_1 = \sum_v |\psi_v| = \sum_{v:\psi_v>0} \psi_v - \sum_{u:\psi_u<0} \psi_u$. Using that $\sum_w \psi_w = 0$, we get $\|\psi\|_1 = 2\sum_{v:\psi_v>0} \psi_v = -2\sum_{u:\psi_u<0} \psi_u$.

Assume, without loss of generality, that $0 < c_0$, in which case

$$\|\psi\|_1 = -2\sum_{u:\psi_u<0} \psi_u \leqslant 2\sum_{i=1}^{n/2} |\psi_i - c_0| =: 2N \tag{7}$$

In what follows we aim to upper-bound N.

Flow cutting: Define a collection of cuts (S_i, S_i^{C}) of the form $S_i = \{v : \psi_v \leqslant c_i\}$, for integers $i \geqslant 0$, where S_i will be the smaller side of the cut by construction. Let k_i be the number of edges cut by (S_i, S_i^{C}) and p_{ij} be the length of the jth edge across the same cut. The cut points c_i, for $i \geqslant 1$, are defined according to $c_i = c_{i-1} - \Delta_{i-1}$, where $\Delta_{i-1} := 2\sum_j \frac{p_{i-1,j}}{k_{i-1}}$. The last cut, $(S_{r+1}, S_{r+1}^{\mathsf{C}})$, is the first cut in the sequence $c_0, c_1, \ldots, c_{r+1}$ with $k_{r+1} = 0$ or, equivalently, $S_{r+1} = \emptyset$.

Bound on number of cuts: Let $n_i = |S_i|$. The isoperimetric inequality for vertex expansion (5) applied to (S_i, S_i^{C}) and the fact that $n_i \leqslant n/2$, by construction, imply

$$\frac{k_i}{n_i} \geqslant \alpha. \tag{8}$$

Let l_i be the number of edges crossing (S_i, S_i^{C}) that do not extend across c_{i+1}, i.e. edges that are not adjacent to S_{i+1}. The choice $\Delta_i := 2\sum_j p_{ij}/k_i$ ensures that $l_i \geqslant k_i/2$. These edges are supported on at least $l_i/d_{\max}$ vertices in S_i, and therefore $n_{i+1} \leqslant n_i - l_i/d_{\max}$. Thus,

$$n_{i+1} \leqslant n_i - \frac{l_i}{d_{\max}} \leqslant n_i - \frac{k_i}{2d_{\max}} \overset{(8)}{\leqslant} n_i - \frac{\alpha n_i}{2d_{\max}} = n_i\left(1 - \frac{\alpha}{2d_{\max}}\right), \tag{9}$$

Combining inequality (9) with $n_0 = n/2$, we get

$$n_i \leqslant \frac{n}{2}\left(1 - \frac{\alpha}{2d_{\max}}\right)^i \tag{10}$$

The stopping condition implies $S_r \neq \emptyset$, or $n_r \geqslant 1$, and together with (10) this results in

$$r \leqslant \log_{1/\theta} \frac{n}{2}, \text{ with } \theta = 1 - \frac{\alpha}{2d_{\max}}. \tag{11}$$

Amortization argument: Continuing from (7),

$$N = \sum_{i=1}^{n/2} |\psi_i - c_0| \overset{(*)}{\leqslant} \sum_{i=0}^{r} (n_i - n_{i+1}) \sum_{j=0}^{i} \Delta_j, \tag{12}$$

where $(*)$ follows from the fact that for every vertex $v \in S_i - S_{i+1}$ we can write $|\psi_v - c_0| \leqslant \sum_{j=0}^{i} \Delta_j$.

Because $BL^{\dagger}(\chi_s - \chi_t)$ is a unit flow, we have the crucial (and easy to verify) property that, for all i, $\sum_j p_{ij} = 1$. In other words, the total flow of "simulatenous" edges is 1. So,

$$\Delta_i = 2 \sum_j \frac{p_{ij}}{k_i} = \frac{2}{k_i} \overset{(8)}{\leqslant} \frac{2}{\alpha n_i} \tag{13}$$

Now we can use this bound on Δ_j in (12),

$$\sum_{i=0}^{r} (n_i - n_{i+1}) \sum_{j=0}^{i} \Delta_j \overset{(*)}{=} \sum_{i=1}^{r} n_i \Delta_i \leqslant \frac{2}{\alpha} \sum_{i=0}^{r} 1 = \frac{2}{\alpha}(r+1),$$

where to derive $(*)$ we use $n_{r+1} = 0$. Combining the above inequality with (11) concludes the proof.

4 Conclusions

Our main result in Theorem 1 attests to the good congestion properties on graphs of bounded degree and high vertex expansion, i.e. $\alpha = O(1)$. A variation on the proof of this theorem establishes a similar bound on $\eta_{\mathcal{E}}$, however, independent of the degree bound and as a function of the *edge expansion* $\beta = \min_{S \subseteq V} \frac{\mathrm{vol}(E(S,S^{\complement}))}{\min\{\mathrm{vol}(S),\mathrm{vol}(S^{\complement})\}}$, where $\mathrm{vol}(S) := \sum_{v \in S} \sum_{u:u \sim v} w_{u,v}$ and $\mathrm{vol}(E(S,S^{\complement})) := \sum_{(u,v):u \in S, v \in S^{\complement}} w_{u,v}$.

The bounded degree assumption is also implicit in our computational procedure in that all vertices must know an upper bound on $d_{\max}$ in order to apply M in Theorem 3. Using a generous bound on $d_{\max}$ is not desirable because it slows down the mixing of the power polynomial. To avoid this complication, we describe a symmetrization "trick" in [4].

We conclude with a couple of open questions. A central concern, widely-studied in social-networks, are Sybil Attacks [13]. These can be modeled as graph-theoretic noise, as defined in [14]. It is interesting to understand how such noise affects electric routing. We suspect that any $O(\ln n)$-competitive oblivious routing scheme, which outputs its routes in the "next hop" model, must maintain $\Omega(n)$-size routing tables at every vertex. In the *next hop* model, every vertex v must be able to answer the question "What is the flow of the (s,t)-route in the neighborhood of v?" in time $O(\mathrm{polylog}(n))$, using its own routing table alone and for every source-sink pair (s,t).

References

1. Hajiaghayi, M., Kleinberg, R., Leighton, T., Räcke, H.: New lower bounds for oblivious routing in undirected graphs. In: Proceedings of the seventeenth annual ACM-SIAM symposium on Discrete algorithm, pp. 918–927. ACM, New York (2006)
2. Doyle, P.G., Snell, J.L.: Random walks and electric networks. Arxiv preprint math.PR/0001057 (2000)
3. Spielman, D.: Graphs and networks. Lecture Notes
4. Kelner, J., Maymounkov, P.: Electric routing and concurrent flow cutting (full version), http://arxiv.org/abs/0909.28b59
5. Thorup, M., Zwick, U.: Approximate distance oracles. Journal of the ACM (JACM) 52(1), 1–24 (2005)
6. Thorup, M., Zwick, U.: Compact routing schemes. In: Proceedings of the thirteenth annual ACM symposium on Parallel algorithms and architectures, pp. 1–10. ACM, New York (2001)
7. Abraham, I., Gavoille, C., Malkhi, D.: On space-stretch trade-offs: Lower bounds. In: Proceedings of the eighteenth annual ACM symposium on Parallelism in algorithms and architectures, pp. 207–216. ACM, New York (2006)
8. Abraham, I., Gavoille, C., Malkhi, D., Nisan, N., Thorup, M.: Compact name-independent routing with minimum stretch (2008)
9. Räcke, H.: Optimal hierarchical decompositions for congestion minimization in networks. In: Proceedings of the 40th annual ACM symposium on Theory of computing, pp. 255–264. ACM, New York (2008)
10. Harsha, P., Hayes, T., Narayanan, H., Räcke, H., Radhakrishnan, J.: Minimizing average latency in oblivious routing. In: Proceedings of the nineteenth annual ACM-SIAM symposium on Discrete algorithms, Society for Industrial and Applied Mathematics, Philadelphia, PA, USA, pp. 200–207 (2008)
11. Goyal, N., Rademacher, L., Vempala, S.: Expanders via random spanning trees. In: Proceedings of the Nineteenth Annual ACM-SIAM Symposium on Discrete Algorithms, Society for Industrial and Applied Mathematics, Philadelphia, PA, USA, pp. 576–585 (2009)
12. Lawler, G., Narayanan, H.: Mixing times and l_p bounds for Oblivious routing. In: Workshop on Analytic Algorithmics and Combinatorics, ANALCO 2009 (2009)
13. Douceur, J.: The sybil attack. In: Druschel, P., Kaashoek, M.F., Rowstron, A. (eds.) IPTPS 2002. LNCS, vol. 2429, p. 251. Springer, Heidelberg (2002)
14. Kale, S., Peres, Y., Seshadhri, C.: Noise Tolerance of Expanders and Sublinear Expander Reconstruction. In: Proceedings of the 2008 49th Annual IEEE Symposium on Foundations of Computer Science, vol. 00, pp. 719–728. IEEE Computer Society, Washington (2008)

A Polynomial-Time Algorithm for the Universally Quickest Transshipment Problem in a Certain Class of Dynamic Networks with Uniform Path-Lengths

Naoyuki Kamiyama[1] and Naoki Katoh[2]

[1] Department of Information and System Engineering, Chuo University
kamiyama@ise.chuo-u.ac.jp
[2] Department of Architecture and Architectural Engineering, Kyoto University
naoki@archi.kyoto-u.ac.jp

Abstract. In this paper, we consider the universally quickest transshipment problem in a dynamic network where each arc has not only a capacity but also a transit time. The problem asks for minimizing the time when the last supply reaches the sink as well as simultaneously maximizing the amount of supply which has reached the sink at every time step. In this paper, we propose a polynomial-time algorithm for the problem in the class of dynamic networks which is a generalization of grid networks with uniform capacity and uniform transit time.

1 Introduction

Recently, it is very important to establish crisis management systems against large-scale disasters such as big earthquakes, conflagrations and tsunamis to effectively guide residents to a safe place. In the network flow theory, the problem of finding the most effective plan to evacuate people to a safe place has been modelled as the *evacuation problem* by using *dynamic flows*. In the evacuation problem, we are given a *dynamic network* $\mathcal{D} = (D = (V, A), c, \tau, b, s)$, where $D = (V, A)$ is a digraph that consists of a vertex set V and an arc set A, each $x \in V$ has the supply $b(x)$, each $a \in A$ has the capacity $c(a)$ and the transit time $\tau(a)$, and a special vertex $s \in V$ is designated as a sink. If we consider the urban evacuation, vertices model buildings, exits and so on, and arcs model pathways or roads. For each $x \in V$, $b(x)$ represents the number of people which exist at x. For each $a \in A$, $c(a)$ represents the number of people which can enter a per unit time, and $\tau(a)$ represents the time required to traverse a. Then, the evacuation problem asks for finding the minimum time required to send all the supplies to s as well as a *dynamic flow* which attains the optimal evacuation time. Practically, it is not sufficient to minimize the time when the last supply reaches the sink, and it is more desirable that we can simultaneously maximize the amount of supply which has reached the sink at every time step. We call such a dynamic flow a *universally quickest transshipment*. The problem of finding a universally quickest transshipment is called the *universally quickest transshipment problem* (in short UQT).

Previous works. It is known [1] that there always exists a universally quickest transshipment. This result can be derived by reducing the problem to the *lexicographically*

Y. Dong, D.-Z. Du, and O. Ibarra (Eds.): ISAAC 2009, LNCS 5878, pp. 802–811, 2009.

maximum flow problem on a *time-expanded network* (for details, see Sec. 3). Hajek and Ogier [2] presented the first polynomial-time algorithm for the case where the transit time of every arc is equal to zero, which needs $O(|V|)$ maximum flow computations. Fleischer [3] gave an algorithm which can solve the problem for this case in the same time complexity as that of the maximum flow algorithm presented by Goldberg and Tarjan [4]. Baumann and Skutella [5] presented an algorithm whose running time is bounded by a polynomial in the input size and the number of breakpoints of the piece-wise linear function which represents the amount of supply that has reached the sink at each time step in a universally quickest transshipment. Unfortunately, in the worst case the number of the breakpoints is exponential in the input size [5]. Fleischer and Skutella [6] gave an approximation algorithm for this problem.

Our contribution. In this paper, we present a polynomial-time algorithm for UQT in a dynamic network satisfying the following two conditions called the *uniform path-length condition* [7] and the *fully connected condition* [8]. The *length* of a directed path P is defined by the sum of the transit times of all the arcs contained in P. In this paper, we allow a directed path to pass through the same vertex or arc more than once. Then, if for each $x \in V \setminus \{s\}$ the lengths of all the directed paths from x to s are identical, we call $\mathcal{D}$ a *dynamic network with uniform path-lengths*. Notice that we allow that a length of a path from $x \in V$ and that of a path from $y \in V$ with $y \neq x$ are not identical. For each $x \in V$, we denote by $\varrho_D(x)$ the set of $a \in A$ whose head is x. The *connectivity* from x to s in D with the capacity function c is defined by the value of a maximum flow from x with infinite supply to s in D where each $a \in A$ has the capacity $c(a)$ (for the definitions, see Sec. 2). Then, if for every $x \in V \setminus \{s\}$ the connectivity from x to s in D with the capacity function c is equal to the sum of $c(a)$ over $a \in \varrho_D(s)$ whose tail is reachable from x, we say that a triplet (D, c, s) or $\mathcal{D}$ is *fully connected*. The class of fully connected dynamic networks with uniform path-lengths is a generalization of grid networks with uniform capacity and uniform transit time which frequently appear in modelling cities or buildings.

Our technique. Our algorithm uses a *compressed time-expanded network* [7] which is originally introduced for solving the evacuation problem in a dynamic network with uniform path-lengths. Its size is bounded by a polynomial in the input size (for details, see Sec. 4). Thus, since a time-expanded network to which we reduce UQT in a general dynamic network is originally introduced for solving the evacuation problem in a general dynamic network, it seems that we can straightforwardly solve UQT in a fully connected dynamic network with uniform path-lengths by reducing it to the lexicographically maximum flow problem in a compressed time-expanded network. However, since a compressed time-expanded network is obtained from a time-expanded network by merging some arcs and vertices, this is not the case. In order to overcome this difficulty, we first introduce the *hierarchical lexico-max flow problem* (in short HLF) which is a generalization of classical lexicographically maximum flow problems, and we present a polynomial-time algorithm for this problem. Then, we transform a compressed time-expanded network so that a hierarchical lexico-max flow in this network yields a universally quickest transshipment.

Outline. Sec. 2 gives necessary definitions and reviews elementary results. In Sec. 3, we define HLF, and then we show that this problem can be solved in polynomial time. In Sec. 4, we prove the polynomial-time solvability of UQT in a fully connected dynamic network with uniform path-lengths by reducing it to HLF.

2 Preliminaries

Let $\mathbb{R}_{\geq 0}$, $\mathbb{R}_{>0}$ and $\mathbb{Z}_{\geq 0}$ be the sets of nonnegative reals, positive reals and nonnegative integers. We may not distinguish between a singleton $\{x\}$ and its element x. For each $n \in \mathbb{Z}_{\geq 0}$, let $[n] = \{1, \ldots, n\}$. Given two ordered sequences $(x_1, \ldots, x_k)$ and $(y_1, \ldots, y_k)$ of reals, we write $(x_1, \ldots, x_k) \geq_{\mathrm{L}} (y_1, \ldots, y_k)$ if we have $x_i = y_i$ for every $i \in [k]$, or there exists $p \in [k]$ such that $x_i = y_i$ holds for every $i \in [p-1]$ and $x_p > y_p$ holds. Given a set X of k reals, we denote by $\sigma(X)$ the sequence $(x_1, \ldots, x_k)$ of k reals in X satisfying $x_1 \leq \cdots \leq x_k$. Given a function $f\colon X \to \mathbb{R}_{\geq 0}$ on a set X, we use the abbreviation $f(Y) = \sum_{x \in Y} f(x)$ for each $Y \subseteq X$. Given a real M and a set X, we call a function $f\colon X \to \mathbb{R}_{\geq 0}$ a *distribution* of M to X if $f(X) = M$ holds.

Let $D = (V, A)$ be a digraph that consists of a vertex set V and an arc set A. In this paper, we assume without loss of generality that every digraph has no parallel arcs. We denote by $a = xy$ an arc a with a tail x and a head y. It may be simply written as xy. For each $X \subseteq V$, let $\delta_D(X)$ (resp., $\varrho_D(X)$) be the set of arcs $xy \in A$ with $x \in X$ and $y \notin X$ (resp., $x \notin X$ and $y \in X$). A vertex x is said to be *reachable* from a vertex y when there exists a directed path from y to x.

Dynamic networks. Let $\mathcal{D} = (D, c, \tau, b, s)$ be a *dynamic network* which consists of a digraph D, a capacity function $c\colon A \to \mathbb{R}_{>0}$, a transit time function $\tau\colon A \to \mathbb{Z}_{\geq 0}$, a supply function $b\colon V \to \mathbb{R}_{\geq 0}$ and a sink $s \in V$. Since we consider the evacuation to s, we assume without loss of generality that $\delta_D(s) = \varnothing$ and $b(s) = 0$ hold, and s is reachable from every vertex.

We define a *dynamic flow* $f\colon A \times \mathbb{Z}_{\geq 0} \to \mathbb{R}_{\geq 0}$ as follows. For each $a \in A$ and $\theta \in \mathbb{Z}_{\geq 0}$, we denote by $f(a, \theta)$ the flow rate entering the tail of a at time step θ which reaches the head of a at the time step $\theta + \tau(a)$. We call f *feasible* if it satisfies the *capacity constraint* $f(a, \theta) \leq c(a)$ for every $a \in A$ and $\theta \in \mathbb{Z}_{\geq 0}$, the *flow conservation*

$$\sum_{a \in \delta_D(x)} \sum_{\theta=0}^{\Theta} f(a, \theta) \leq \sum_{a \in \varrho_D(x)} \sum_{\theta=0}^{\Theta - \tau(a)} f(a, \theta) + b(x) \quad (x \in V,\ \Theta \in \mathbb{Z}_{\geq 0}),$$

and the *demand constraint*

$$\sum_{a \in \varrho_D(s)} \sum_{\theta=0}^{\Theta - \tau(a)} f(a, \theta) = b(V) \tag{1}$$

for some $\Theta \in \mathbb{Z}_{\geq 0}$. For each feasible dynamic flow f, we define the *evacuation time* $\mathsf{et}(f)$ of f by the minimum time step Θ satisfying (1). Then, the *evacuation problem* asks for finding the minimum evacuation time among all the feasible dynamic flows as well as an optimal dynamic flow which attains the minimum evacuation time. Now we formally define the *universally quickest transshipment problem* (UQT) which we will consider in this paper. This problem is a variant of the evacuation problem in which we are asked to simultaneously maximize the amount of supply which have reached the sink at every time step, i.e., UQT asks for finding a feasible dynamic flow $f^* \in F^*$ satisfying

$$\sum_{a \in \varrho_D(s)} \sum_{\theta=0}^{\Theta-\tau(a)} f^*(a,\theta) \geq \sum_{a \in \varrho_D(s)} \sum_{\theta=0}^{\Theta-\tau(a)} f(a,\theta)$$

for every $f \in F^*$ and $\Theta \in \{0\} \cup [\Theta^*]$, where Θ^* is the optimal objective value for the evacuation problem in $\mathcal{D}$ and F^* is the set of optimal dynamic flows for the evacuation problem in $\mathcal{D}$. We call such f^* a *universally quickest transshipment*. It is known [1] that there always exists a universally quickest transshipment (for details, see Sec. 3).

Static networks. Let $\mathcal{S} = (H, u, v, S, T)$ be a *static network* which consists of a digraph $H = (N, L)$, a capacity function $u\colon L \to \mathbb{R}_{\geq 0}$, a supply function $v\colon S \to \mathbb{R}_{\geq 0}$, a source set $S \subseteq N$ and a sink set $T \subseteq N$ with $S \cap T = \varnothing$. For each flow $\xi\colon L \to \mathbb{R}_{\geq 0}$ and $X \subseteq N$, let $\partial\xi(X) = \xi(\delta_H(X)) - \xi(\varrho_H(X))$. Then, we call ξ *feasible* if it satisfies the *capacity constraint* $\xi(a) \leq u(a)$ for every $a \in L$, and the *flow conservation* in which $\partial\xi(x) = 0$ for $x \in V \setminus (S \cup T)$, $0 \leq \partial\xi(x) \leq v(x)$ for $x \in S$ and $\partial\xi(x) \leq 0$ for $x \in T$. Given a static network $\mathcal{S}$, the *maximum flow problem* asks for finding a feasible flow ξ^* maximizing $-\partial\xi^*(T)$ which is the *value* of ξ^* from S to T. We call such ξ^* a *maximum flow* in $\mathcal{S}$, and we denote by $\mathsf{mv}(\mathcal{S})$ the value of a maximum flow in $\mathcal{S}$. It is known [4] that this problem can be solved in polynomial time and let $\mathsf{MF}(n,m)$ be the time required to solve this problem in a static network with n vertices and m arcs.

For solving the evacuation problem, Ford and Fulkerson [9] introduced the *time-expanded network* $\mathcal{D}(\Theta)$ which is a static network for $\mathcal{D}$ with a time horizon Θ (see Fig. 1). The vertex set of $\mathcal{D}(\Theta)$ consists of vertices $x(\theta)$ ($x \in V$, $\theta \in \{0\} \cup [\Theta]$) and x^* ($x \in V \setminus \{s\}$). The arc set of $\mathcal{D}(\Theta)$ consists of the following three parts. The first part contains an arc $a(\theta) = x(\theta)y(\theta + \tau(a))$ with a capacity $c(a)$ for each $a = xy \in A$ and $\theta \in \{0\} \cup [\Theta - \tau(a)]$, and the second part contains an arc $x(\theta)x(\theta+1)$ with infinite capacity for each $x \in V \setminus \{s\}$ and $\theta \in \{0\} \cup [\Theta - 1]$. The arcs of the second part are called *holdover arcs*. The third part contains an arc $x^*x(\theta)$ with infinite capacity for each $x \in V \setminus \{s\}$ and $\theta \in \{0\} \cup [\Theta]$. We define the source set and the sink set of $\mathcal{D}(\Theta)$ by $\{x^* \mid x \in V \setminus \{s\}\}$ and $\{s(\theta) \mid \theta \in \{0\} \cup [\Theta]\}$, respectively. For each $x \in V \setminus \{s\}$, the supply of x^* is $b(x)$. It is known [9] that there exists a feasible dynamic flow f in $\mathcal{D}$ with $\mathsf{et}(f) \leq \Theta$ if and only if $\mathsf{mv}(\mathcal{D}(\Theta)) = b(V)$. We can construct a feasible dynamic flow f in $\mathcal{D}$ with $\mathsf{et}(f) \leq \Theta$ from a maximum flow ξ in $\mathcal{D}(\Theta)$ whose value is $b(V)$ by setting $f(a,\theta) = \xi(a(\theta))$ for each $a \in A$ and $\theta \in \{0\} \cup [\Theta - \tau(a)]$.

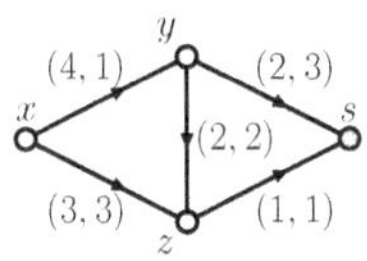

(a)

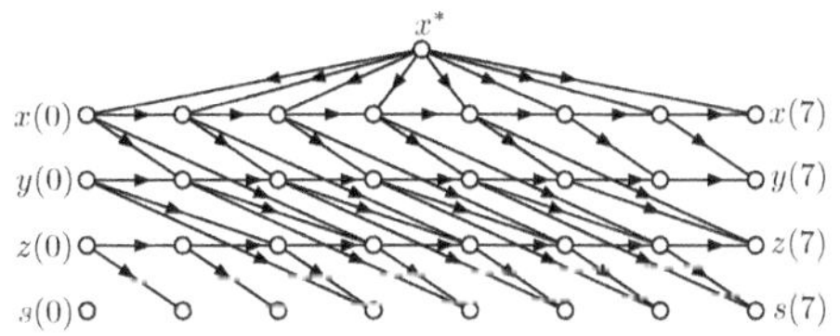

(b)

Fig. 1. (a) A dynamic network $\mathcal{D}$. A pair of numbers attached to an arc shows its capacity and transit time, respectively. (b) $\mathcal{D}(7)$. We omit y^* and z^* as well as the arcs leaving these vertices.

3 The Lexicographically Maximum Flow Problem

In this section, we introduce the *hierarchical lexico-max flow problem* (HLF) which is a generalization of two classical lexicographically maximum flow problems called the *ordered* and *unordered lexico-max flow problem* (OLF and ULF, respectively).

We first define OLF. Suppose that we are given a static network $\mathcal{S}$ with $T = \{x_1, \dots, x_k\}$ and a linear ordering $x_1 \prec \cdots \prec x_k$. For each $i \in [k]$, let $\mathcal{S}_i$ be a static network which is the same as $\mathcal{S}$ except that it has the sink set $\{x_1, \dots, x_i\}$ instead of T. OLF asks for finding a maximum flow ξ^* in $\mathcal{S}$ with $-\partial\xi^*(\{x_1, \dots, x_i\}) = \mathsf{mv}(\mathcal{S}_i)$ for every $i \in [k]$. We call such ξ^* an *ordered lexico-max flow* in $\mathcal{S}$ with respect to $\prec$.

Theorem 1 ([1]). *Given a static network $\mathcal{S}$ with $T = \{x_1, \dots, x_k\}$ and a linear ordering $x_1 \prec \cdots \prec x_k$, there is an ordered lexico-max flow in $\mathcal{S}$ with respect to $\prec$.*

By Theorem 1, we can see the existence of a universally quickest transshipment in $\mathcal{D}$. Consider $\mathcal{D}(\Theta)$ where Θ is the optimal objective value of the evacuation problem in $\mathcal{D}$. Then, $\mathsf{mv}(\mathcal{D}(\Theta))$ is equal to $b(V)$. Thus, UQT in $\mathcal{D}$ is equivalent to OLF in $\mathcal{D}(\Theta)$ with respect to a linear ordering $s(0) \prec \cdots \prec s(\Theta)$. The following theorem is not essential, but it renders our algorithm very concise. We omit the proof.

Theorem 2. *Given a dynamic network $\mathcal{D}$, there exists an ordered lexico-max flow ξ^* in $\mathcal{D}(\Theta)$ with respect to a linear ordering $s(0) \prec \cdots \prec s(\Theta)$ such that ξ^* uses no holdover arc, where Θ is the optimal objective value for the evacuation problem in $\mathcal{D}$.*

Next we define ULF. Suppose that we are given a static network $\mathcal{S}$ and a weight function $w \colon T \to \mathbb{R}_{>0}$. Then, ULF asks for finding a maximum flow ξ^* in $\mathcal{S}$ satisfying

$$\sigma(\{\tfrac{-\partial\xi^*(x)}{w(x)} \mid x \in T\}) \geq_{\mathrm{L}} \sigma(\{\tfrac{-\partial\xi(x)}{w(x)} \mid x \in T\})$$

for every maximum flow ξ in $\mathcal{S}$, where in this inequality we denote by $\{\cdot\}$ a multiset. We call such ξ^* an *unordered lexico-max flow* in $\mathcal{S}$ with respect to w.

Theorem 3 ([10,11]). *Given a static network $\mathcal{S}$ and a weight function $w \colon T \to \mathbb{R}_{>0}$, there always exists an unordered lexico-max flow in $\mathcal{S}$ with respect to w.*

It is known [11] that this problem can be solved in polynomial time. Let $\mathsf{LM}(n, m)$ be the time required to solve ULF in a static network with n vertices and m arcs.

In HLF, we are given a static network $\mathcal{S}$ in which T is decomposed into subsets $T_1, \dots, T_k$, a linear ordering $T_1 \prec \cdots \prec T_k$ and a weight function $w \colon T \to \mathbb{R}_{>0}$. For each $i \in [k]$, we generalize the definition of $\mathcal{S}_i$ given above to a static network which is the same as $\mathcal{S}$ except that it has the sink set $\cup_{t=1}^{i} T_t$. If a feasible flow ξ in $\mathcal{S}$ satisfies $-\partial\xi(\cup_{t=1}^{i} T_t) = \mathsf{mv}(\mathcal{S}_i)$ for every $i \in [k]$, we call ξ *eligible* in $\mathcal{S}$ with respect to $\prec$. Then, HLF asks for finding an eligible flow ξ^* in $\mathcal{S}$ with respect to $\prec$ such that

$$\sigma(\{\tfrac{-\partial\xi^*(x)}{w(x)} \mid x \in T_i\}) \geq_{\mathrm{L}} \sigma(\{\tfrac{-\partial\xi(x)}{w(x)} \mid x \in T_i\}) \tag{2}$$

for every eligible flow ξ in $\mathcal{S}$ with respect to $\prec$ and $i \in [k]$, where in (2) we denote by $\{\cdot\}$ a multiset. We call such ξ^* a *hierarchical lexico-max flow* in $\mathcal{S}$ with respect to $\prec$ and w. In this paper, we show that there always exists a hierarchical lexico-max flow, and we can find it in polynomial time. We omit the proof.

Theorem 4. *Suppose that we are given a static network $\mathcal{S}$ in which T is decomposed into subsets $T_1, \ldots, T_k$, a linear ordering $T_1 \prec \cdots \prec T_k$ and a weight function $w\colon T \to \mathbb{R}_{>0}$. Then, there always exists a hierarchical lexico-max flow in $\mathcal{S}$ with respect to $\prec$ and w, and we can find it in $O(k \cdot \mathsf{LM}(|N|, |L|))$ time.*

4 The Universally Quickest Transshipment Problem

In this section, we prove that we can solve UQT in a fully connected dynamic network with uniform path-lengths by solving HLF. As described in Sec. 3, UQT in a dynamic network is equivalent to OLF in its time-expanded network. In this section, if the input dynamic network satisfies the uniform path-length condition and the fully connected condition, we can find in polynomial time an ordered lexico-max flow in the time-expanded network by solving HLF in *compressed time-expanded network* [7].

Here we introduce notations related to $\mathcal{D}$ with uniform path-lengths. For each $x \in V$, we define $l(x)$ as the length of a directed path from x to s. Notice that since $\mathcal{D}$ satisfies the uniform path-length condition, $l(x)$ is well-defined for each $x \in V$. From here, we fix k as the number of the distinct path-lengths in $\mathcal{D}$, i.e., $k = |\{l(x) \mid x \in V\}|$. Let us arrange the distinct values of $l(x)$ $(x \in V)$ as $l_1, \ldots, l_k$ satisfying $l_1 < \cdots < l_k$. For each $x \in V$, we write $\mathsf{lev}(x) = i$ when $l(x) = l_i$ holds. We assume without loss of generality that there exists $x \in V$ with $\mathsf{lev}(x) = k$ and $b(x) > 0$. Given a time horizon Θ $(\geq l_k)$, we partition the interval $(0, 1, \ldots, \Theta)$ into disjoint subintervals $I_1, \ldots, I_k$ satisfying $I_i = (l_i, l_i + 1, \ldots, l_{i+1} - 1)$ for each $i \in [k]$, where $l_{k+1} = \Theta + 1$.

Here we introduce the *compressed time-expanded network* for $\mathcal{D}$ with uniform path-lengths. For this, we first investigate the structure of $\mathcal{D}(\Theta)$ (see Fig. 2(a)). Every feasible flow in $\mathcal{D}(\Theta)$ does not use vertices from which any sink is not reachable and arcs entering or leaving such vertices. Thus, we assume without loss of generality that such vertices and arcs are removed from $\mathcal{D}(\Theta)$. Moreover, since we consider an ordered lexico-max flow in $\mathcal{D}(\Theta)$ with respect to a linear ordering $s(0) \prec \cdots \prec s(\Theta)$, we assume by Theorem 2 that all holdover arcs are removed from $\mathcal{D}(\Theta)$. For each $\theta \in \{0\} \cup [\Theta]$, let $\mathcal{L}(\theta)$ be a subnetwork of $\mathcal{D}(\Theta)$ induced by the vertex set $\{x(\theta - l(x)) \mid x \in V \text{ with } l(x) \leq \theta\}$. By the uniform path-length condition, the underlying digraph of $\mathcal{L}(\theta)$ with $\theta \in I_i$ is isomorphic to the subgraph of D induced by vertices $x \in V$ with $\mathsf{lev}(x) \leq i$. Furthermore, $\mathcal{D}(\Theta)$ consists of $\Theta + 1$ components $\mathcal{L}(\theta)$ $(\theta \in \{0\} \cup [\Theta])$, sources x^* $(x \in V \setminus \{s\})$ and arcs $x^*x(\theta)$ $(x \in V \setminus \{s\},\ \theta \in \{0\} \cup [\Theta - l(x)])$.

Intuitively, the compressed time-expanded network $\mathcal{C}(\Theta)$ is obtained from $\mathcal{D}(\Theta)$ by merging $\mathcal{L}(\theta)$ for all $\theta \in I_i$ into a single network $\mathcal{L}_i$ for each $i \in [k]$ (see Fig. 2(b)). More precisely, $\mathcal{C}(\Theta)$ has k components $\mathcal{L}_i$ $(i \in [k])$. The underlying digraph of $\mathcal{L}_i$ is isomorphic to the subgraph of D induced by vertices $x \in V$ with $\mathsf{lev}(x) \leq i$, i.e., the underlying digraph of $\mathcal{L}(\theta)$ with $\theta \in I_i$. Let x_i (resp., a_i) denote the vertex (resp., the arc) of $\mathcal{L}_i$ corresponding to $x \in V$ (resp., $a \in A$). The capacity of an arc a_i of $\mathcal{L}_i$ is equal to $|I_i| \cdot c(a)$. Furthermore, $\mathcal{C}(\Theta)$ contains sources x^* $(x \in V \setminus \{s\})$ and arcs x^*x_i $(x \in V \setminus \{s\},\ i \in \{\mathsf{lev}(x), \ldots, k\})$, and the sink set of $\mathcal{C}(\Theta)$ is $\{s_i \mid i \in [k]\}$. For each $x \in V \setminus \{s\}$, the supply of x^* is equal to $b(x)$. It is known [7] that $\mathsf{mv}(\mathcal{D}(\Theta)) = \mathsf{mv}(\mathcal{C}(\Theta))$ holds and we can find the optimal objective value for the evacuation problem for $\mathcal{D}$ in $O(k|A| \cdot \mathsf{MF}(k|V|, k|A|))$ time.

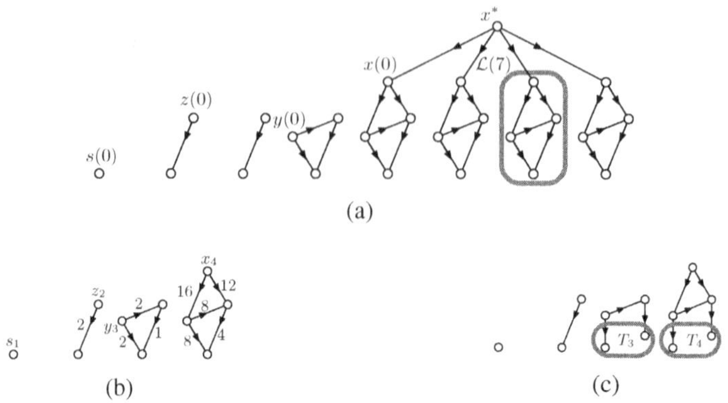

Fig. 2. (a) The layered structure of $\mathcal{D}(7)$ in Fig. 1(b). We omit y^* and z^* as well as the arcs leaving these vertices. (b) $\mathcal{C}(7)$ for $\mathcal{D}(7)$ in (a). We omit x^*, y^* and z^* as well as the arcs leaving these vertices. A number attached to an arc represents its capacity. (c) The modified $\mathcal{C}(7)$.

4.1 Reduction to the Lexicographically Maximum Flow Problem

In the sequel, we assume that $\mathcal{D}$ satisfies the fully connected condition and the uniform path-length condition, and let us fix Θ as an optimal objective value for the evacuation problem in $\mathcal{D}$. In order to show the reduction, we slightly transform $\mathcal{C}(\Theta)$ (see Fig. 2(c)). For each $i \in [k]$, let $[\varrho_D(s)]_{\leq i} = \{xs \in \varrho_D(s) \mid \mathsf{lev}(x) \leq i\}$. Notice that $[\varrho_D(s)]_{\leq i}$ is the set of $a \in \varrho_D(s)$ such that there exists a_i in $\mathcal{L}_i$. Then, for each $i \in [k]$, we replace s_i in $\mathcal{L}_i$ by new sinks s_i^a $(a \in [\varrho_D(s)]_{\leq i})$, and let T_i be the set of these new sinks. Moreover, we change the head of a_i to s_i^a for each $a \in [\varrho_D(s)]_{\leq i}$. Let $\cup_{t=1}^{k} T_t$ be the sink set of $\mathcal{C}(\Theta)$ (denoted by T). From here, we abuse the notation $\mathcal{C}(\Theta)$ to denote this modified network. We show that OLF in $\mathcal{D}(\Theta)$ with respect to a linear ordering $s(0) \prec \cdots \prec s(\Theta)$ can be solved by solving HLF in $\mathcal{C}(\Theta)$ with respect to a linear ordering $T_1 \prec_c \cdots \prec_c T_k$ and a weight function $w \colon T \to \mathbb{R}_{>0}$ defined by $w(s_i^a) = c(a)$ for each $s_i^a \in T$.

Our goal is to prove that an ordered lexico-max flow in $\mathcal{D}(\Theta)$ with respect to $\prec$ can be constructed from a hierarchical lexico-max flow in $\mathcal{C}(\Theta)$ with respect to $\prec_c$ and w. Recall that $\mathcal{C}(\Theta)$ is obtained from $\mathcal{D}(\Theta)$ by merging several components. Hence, in order to construct an ordered lexico-max flow in $\mathcal{D}(\Theta)$ with respect to $\prec$ from a hierarchical lexico-max flow in $\mathcal{C}(\Theta)$ with respect to $\prec_c$ and w, for each maximum flow φ in $\mathcal{C}(\Theta)$ we consider distributions γ_i^a $(s_i^a \in T)$ such that each γ_i^a is a distribution of $-\partial\varphi(s_i^a)$ to I_i with $\gamma_i^a(\theta) \leq c(a)$. We call such distributions γ_i^a $(s_i^a \in T)$ *φ-legal*. By the way of construction of $\mathcal{C}(\Theta)$, it is easy to see that given a maximum flow ξ in $\mathcal{D}(\Theta)$, we can construct a maximum flow φ in $\mathcal{C}(\Theta)$ and φ-legal distributions γ_i^a $(s_i^a \in T)$ with $-\partial\xi(s(\theta)) = \sum\{\gamma_i^a(\theta) \mid s_i^a \in T_i\}$ for every $i \in [k]$ and $\theta \in I_i$ by merging the subflows of ξ in $\mathcal{L}(\theta)$ $(\theta \in I_i)$ for each $i \in [k]$ and setting $\gamma_i^a(\theta) = \xi(a(\theta - \tau(a)))$. We say that such ξ is *generated* by γ_i^a $(s_i^a \in T)$. Furthermore, the following lemma asserts the opposite direction also holds. The proof of this lemma will be given in Sec. 4.2.

Lemma 1. *Suppose that we are given a fully connected dynamic network $\mathcal{D}$ with uniform path-lengths, a maximum flow φ in $\mathcal{C}(\Theta)$ and φ-legal distributions γ_i^a $(s_i^a \in T)$.*

Then, there is a maximum flow ξ in $\mathcal{D}(\Theta)$ with $-\partial\xi(s(\theta)) = \sum\{\gamma_i^a(\theta) \mid s_i^a \in T_i\}$ for every $i \in [k]$ and $\theta \in I_i$, i.e., γ_i^a $(s_i^a \in T)$ generate a maximum flow in $\mathcal{D}(\Theta)$.

From Lemma 1 together with the observation given prior to the lemma, it follows that finding an ordered lexico-max flow ξ^* in $\mathcal{D}(\Theta)$ with respect to $\prec$ reduces to the choice of a maximum flow φ^* in $\mathcal{C}(\Theta)$ and φ^*-legal distributions γ_i^a $(s_i^a \in T)$ that generate ξ^*. Notice that there always exist such φ^* and γ_i^a $(s_i^a \in T)$ by Theorem 1.

Under the constraint that a maximum flow φ in $\mathcal{C}(\Theta)$ is fixed, among all possible φ-legal distributions we consider how to determine φ-legal distributions γ_i^a $(s_i^a \in T)$ which generate an ordered lexico-max flow ξ in $\mathcal{D}(\Theta)$ with respect to $\prec$. We can construct such ξ by distributing $-\partial\varphi(s_i^a)$ greedily to $\gamma_i^a(\theta)$ as much as possible by giving a higher priority to $\gamma_i^a(\theta)$ with smaller θ. More formally, we define $\gamma_i^a\colon I_i \to \mathbb{R}_{\geq 0}$ by

$$\gamma_i^a(\theta) = \begin{cases} c(a), & \text{if } \theta \in \{l_i, \ldots, l_i + \lfloor \frac{-\partial\varphi(s_i^a)}{c(a)} \rfloor - 1\}, \\ -\partial\varphi(s_i^a) - \lfloor \frac{-\partial\varphi(s_i^a)}{c(a)} \rfloor \cdot c(a), & \text{if } \theta = l_i + \lfloor \frac{-\partial\varphi(s_i^a)}{c(a)} \rfloor, \\ 0, & \text{otherwise}, \end{cases} \tag{3}$$

for each $i \in [k]$ and $s_i^a \in T_i$. We call φ-legal distributions γ_i^a $(s_i^a \in T)$ *proper* if γ_i^a is defined by (3), and we denote by ξ_φ a maximum flow in $\mathcal{D}(\Theta)$ which is generated by the proper φ-legal distribution.

In the sequel, let φ^* be a hierarchical lexico-max flow in $\mathcal{C}(\Theta)$ with respect to $\prec_c$ and w. By using the above notions, our goal is equivalent to showing that ξ_{φ^*} is an ordered lexico-max flow in $\mathcal{D}(\Theta)$ with respect to $\prec$. It is not difficult to see that this can be prove the following Facts A and B (we omit the proofs of these facts), and the reduction is done. For each $i \in [k]$, let $[\mathcal{C}(\Theta)]_i$ be the same as $\mathcal{C}(\Theta)$ except that it has the sink set $\cup_{t=1}^i T_t$ instead of T. Let Φ^* be the set of maximum flows φ in $\mathcal{C}(\Theta)$ with $-\partial\varphi(\cup_{t=1}^i T_t) = \mathsf{mv}([\mathcal{C}(\Theta)]_i)$ for every $i \in [k]$. Notice that $\varphi^* \in \Phi^*$.

A. For a maximum flow φ in $\mathcal{C}(\Theta)$ with $\varphi \notin \Phi^*$, ξ_φ is not an ordered lexico-max flow in $\mathcal{D}(\Theta)$ with respect to $\prec$, i.e., we need to choose $\varphi \in \Phi^*$.
B. For every maximum flow φ in $\mathcal{C}(\Theta)$ with $\varphi \in \Phi^*$ and for every $\theta \in \{0\} \cup [\Theta]$, $\sum\{-\partial\xi_{\varphi^*}(s(\theta)) \mid \theta \in \{0\} \cup [\theta]\} \geq \sum\{-\partial\xi_\varphi(s(\theta)) \mid \theta \in \{0\} \cup [\theta]\}$.

4.2 Proof of Lemma 1

In this subsection, we give the proof of Lemma 1. We prove the lemma by giving the way of constructing such ξ. In the proof, we fix $i \in [k]$ and let V_i be the set of the vertices of $\mathcal{L}_i$. Furthermore, let $\sum\{\gamma_i^a(\theta) \mid s_i^a \in T_i\} = \Gamma_i^a(\theta)$, and $S_i = V_i \setminus T_i$.

We first give the sketch of the proof. Recall that $\mathcal{L}_i$ is constructed by merging $\mathcal{L}(\theta)$ for all $\theta \in I_i$. Thus, we determine the value of ξ for the arcs from all x^* to $\mathcal{L}(\theta)$ $(\theta \in I_i)$ in $\mathcal{D}(\Theta)$ by distributing $\varphi(x^* x_i)$ in $\mathcal{C}(\Theta)$ to these arcs. For each $x_i \in S_i$, let λ_{x_i} be a distribution of $\varphi(x^* x_i)$ to I_i. Then, let $\tilde{\mathcal{L}}(\theta)$ be a static network which is the same as $\mathcal{L}(\theta)$ except that each vertex $x(\theta - l(x))$ of $\mathcal{L}(\theta)$ (i.e., the vertex of $\mathcal{L}(\theta)$ corresponding to x_i of $\mathcal{L}_i$) has the supply $\lambda_{x_i}(\theta)$ (i.e., $\lambda_{x_i}(\theta)$ represents the supply of a vertex in $\tilde{\mathcal{L}}(\theta)$ corresponding to x_i in $\mathcal{L}_i$). In order to prove the lemma, we show that there exist distributions λ_{x_i} $(x_i \in S_i)$ with $\lambda_{S_i}(\theta) = \Gamma_i^a(\theta)$ for each $\theta \in I_i$, where

$\lambda_{S_i}(\theta) = \sum\{\lambda_{x_i}(\theta) \mid x_i \in S_i\}$. Note that $\tilde{\mathcal{L}}(\theta)$ has the single sink $s(\theta)$, and that flow values of arcs entering $s(\theta)$ are given as $\gamma_i^a(\theta)$ for each $a \in [\varrho_D(s)]_{\leq i}$. Thus, once $\lambda_{x_i}(\theta)$'s satisfying $\lambda_{S_i}(\theta) = \Gamma_i^a(\theta)$ are given, flow values of arcs leaving all x^* are also given. The remaining task is therefore to find a maximum flow to $s(\theta)$ in $\tilde{\mathcal{L}}(\theta)$ which is consistent with $\lambda_{S_i}(\theta) = \Gamma_i^a(\theta)$. Namely, we will prove that for every $\theta \in I_i$ there exists a maximum flow in $\tilde{\mathcal{L}}(\theta)$ whose value is the sum of the supplies of the vertices of $\tilde{\mathcal{L}}(\theta)$, i.e., $\lambda_{S_i}(\theta)$. This can be done by using the fact that $\mathcal{D}$ is fully connected.

Here we consider how to find such distributions λ_{x_i} $(x_i \in S_i)$. We obtain such λ_{x_i} $(x_i \in S_i)$ by determining $\lambda_{x_i}(\theta; s_i^a)$ which is determined as follows and setting $\lambda_{x_i}(\theta) = \sum\{\lambda_{x_i}(\theta; s_i^a) \mid s_i^a \in T_i\}$. It is known that we can decompose the subflow of φ in $\mathcal{L}_i$ into flows φ_i^a $(s_i^a \in T_i)$ in $\mathcal{L}_i$ satisfying the capacity constraint and for every $s_i^a, s_i^{\bar{a}} \in T_i$ (i) $\partial\varphi_i^a(s_i^{\bar{a}}) = \partial\varphi(s_i^a)$ if $s_i^{\bar{a}} = s_i^a$ and (ii) $\partial\varphi_i^a(s_i^{\bar{a}}) = 0$ otherwise. The value $\partial\varphi_i^a(x_i)$ represents the allocation of $\varphi(x^* x_i)$ to the sink s_i^a. Let $\lambda_{x_i}(\theta; s_i^a) = \partial\varphi_i^a(x_i) \cdot \frac{\gamma_i^a(\theta)}{-\partial\varphi(s_i^a)}$. Furthermore, as mentioned above, let $\lambda_{x_i}(\theta) = \sum\{\lambda_{x_i}(\theta; s_i^a) \mid s_i^a \in T_i\}$. Then, λ_{x_i} is a distribution of $\varphi(x^* x_i)$ to I_i and $\lambda_{S_i}(\theta) = \Gamma_i^a(\theta)$ for every $\theta \in I_i$. We show that for λ_{x_i} $(x_i \in S_i)$ obtained in this way that there exists a maximum flow in $\tilde{\mathcal{L}}(\theta)$ whose value is $\lambda_{S_i}(\theta)$ by the following lemma. We omit the proof.

Lemma 2. *Given a static network $\mathcal{S}$ such that T consists of the single sink t and $S = N \setminus \{t\}$, (H, u, t) is fully connected, and $v(X) \leq u(R(X))$ holds for every $X \subseteq N \setminus \{t\}$, we have $\mathsf{mv}(\mathcal{S}) = v(N \setminus \{t\})$, where let $R(X)$ be the set of $a \in \varrho_H(t)$ whose tail is reachable from at least one vertex of X in H.*

Let $\tilde{D} = (\tilde{V}, \tilde{A})$, $\tilde{u}$ and $\tilde{v}$ be the underlying digraph, the capacity function and the supply function of $\tilde{\mathcal{L}}(\theta)$, respectively. Recall that $s(\theta)$ is the single sink of $\tilde{\mathcal{L}}(\theta)$. By Lemma 2, in order to prove that there exists a maximum flow in $\tilde{\mathcal{L}}(\theta)$ whose value is $\lambda_{S_i}(\theta)$, it suffices to show that $(\tilde{D}, \tilde{u}, s(\theta))$ is fully connected and $\tilde{v}(X) \leq \tilde{u}(\tilde{R}(X))$ holds for every $X \subseteq \tilde{V} \setminus \{s(\theta)\}$, where let $\tilde{R}(X)$ be the set of $a \in \varrho_{\tilde{D}}(s(\theta))$ whose tail is reachable from at least one vertex of X in $\tilde{D}$. We omit the proof that $\tilde{v}(X) \leq \tilde{u}(\tilde{R}(X))$ holds for every $X \subseteq \tilde{V} \setminus \{s(\theta)\}$. We show that $(\tilde{D}, \tilde{u}, s(\theta))$ is fully connected by using that $\mathcal{D}$ (i.e., (D, c, s)) is fully connected. Recall that $\tilde{D}$ is isomorphic to the subgraph $D_{\leq i}$ of D induced by vertices $x \in V$ with $\mathsf{lev}(x) \leq i$. Furthermore, by the definition of the capacities in the $\mathcal{D}(\Theta)$, $(\tilde{D}, \tilde{u}, s(\theta))$ and $(D_{\leq i}, c_{\leq i}, s)$ are equivalent, where $c_{\leq i}$ is the restriction of c on $D_{\leq i}$. Since $\mathsf{lev}(x) \leq i$ implies $\mathsf{lev}(y) \leq i$ for each $xy \in A$ by the nonnegativity of the transit time, the connectivity from x to s in $D_{\leq i}$ with a capacity function $c_{\leq i}$ and that in D with a capacity function c are identical for every vertex x of $D_{\leq i}$. Hence, since $\mathcal{D}$ is fully connected and $(\tilde{D}, \tilde{u}, s(\theta))$ and $(D_{\leq i}, c_{\leq i}, s)$ are equivalent, $(\tilde{D}, \tilde{u}, s(\theta))$ is also fully connected.

4.3 Time Complexity

By the proof of Lemma 1, we can obtain the following algorithm for our problem.

Step 1: Find a hierarchical lexico-max flow φ in $\mathcal{C}(\Theta)$ with respect to a linear ordering $T_1 \prec \cdots \prec T_k$ and a weight function $w\colon T \to \mathbb{R}_{>0}$ defined by $w(s_i^a) = c(a)$.
Step 2: Construct from φ flows φ_i^a $(i \in [k],\ s_i^a \in T_i)$.
Step 3: Calculate $\lambda_{x_i}(\cdot; s_i^a)$ by using the proper φ-legal distributions γ_i^a $(s_i^a \in T)$ and then construct $\lambda_{x_i}(\cdot)$ from $\lambda_{x_i}(\cdot; s_i^a)$.

Step 4: Find a maximum flow in $\tilde{\mathcal{L}}(\theta)$ for each $\theta \in \{0\} \cup [\Theta]$, and construct an ordered lexico-max flow in $\mathcal{D}(\Theta)$ with respect to $s(0) \prec \cdots \prec s(\Theta)$ by combining them.

Step 1 can be executed in polynomial time by Theorem 4, and it is known that Step 2 can be done in $O(k|V| \cdot \mathsf{MF}(|V|, |A|))$ time. However, the time required for Step 3 through 4 is not in general bounded by a polynomial in the input size since we independently find a maximum flow in $\tilde{\mathcal{L}}(\theta)$ for each $\theta \in \{0\} \cup [\Theta]$. In order to reduce the time required for these steps, we partition the interval $(0, \ldots, \Theta)$ into subintervals whose number is bounded by a polynomial in the input size, and we simultaneously find maximum flows in $\tilde{\mathcal{L}}(\theta)$ for each subinterval via one maximum flow computation. More precisely, we do it as follows. Let us fix $i \in [k]$. From (3), we see that every γ_i^a takes the form such that $I_i = (l_i, l_i+1, \ldots, l_{i+1}-1)$ can be partitioned into at most three nonempty subintervals in each of which γ_i^a does not change. Therefore, overlaying these subintervals for all $s_i^a \in T_i$ onto the interval I_i, we can decompose I_i into at most $3|T_i|$ subintervals in each of which every γ_i^a does not change. In each subinterval I, instead of computing a maximum flow in $\tilde{\mathcal{L}}(\theta)$ for all $\theta \in I$, it is sufficient to compute a maximum flow for only one $\theta \in I$ since the supply of each vertex of $\tilde{\mathcal{L}}(\theta)$ is identical for every $\theta \in I$. By these augments, the following theorem holds.

Theorem 5. *Given a fully connected dynamic network $\mathcal{D}$ with uniform path-lengths, we can solve UQT in the time required to find the optimal objective value for the evacuation problem in $\mathcal{D}$ plus $O(k \cdot \mathsf{LM}(k|V|, k|A|) + k|V| \cdot \mathsf{MF}(|V|, |A|))$ time.*

Acknowledgement. This research was supported by a Grant-in-Aid Research (B) from the Ministry of Education, Culture, Sports, Science and Technology of Japan.

References

1. Minieka, E.: Maximal, lexicographic, and dynamic network flows. Operations Research 21, 517–527 (1973)
2. Hajek, B., Ogier, R.G.: Optimal dynamic routing in communication networks with continuous traffic. Networks 14, 457–487 (1984)
3. Fleischer, L.: Faster algorithms for the quickest transshipment problem. SIAM Journal on Optimization 12(1), 18–35 (2001)
4. Goldberg, A.V., Tarjan, R.E.: A new approach to the maximum-flow problem. Journal of the ACM 35(4), 921–940 (1988)
5. Baumann, N., Skutella, M.: Solving evacuation problems efficiently–earliest arrival flows with multiple sources. In: Proceedings of FOCS 2006, pp. 399–410 (2006)
6. Fleischer, L., Skutella, M.: Quickest flows over time. SIAM J. Comput. 36(6), 1600–1630 (2007)
7. Hall, A., Hippler, S., Skutella, M.: Multicommodity flows over time: Efficient algorithms and complexity. Theoretical Computer Science 379(3), 387–404 (2007)
8. Kamiyama, N., Katoh, N., Takizawa, A.: An efficient algorithm for the evacuation problem in a certain class of networks with uniform path-lengths. Disc. Appl. Math. (to appear)
9. Ford Jr., L.R., Fulkerson, D.R.: Flows in Networks. Princeton University Press, Princeton (1962)
10. Megiddo, N.: Optimal flows in networks with multiple sources and sinks. Mathematical Programming 7, 97–107 (1974)
11. Fujishige, S.: Lexicographically optimal base of a polymatroid with respect to a weight vector. Mathematics of Operations Research 5, 186–196 (1980)

Strong Robustness of Randomized Rumor Spreading Protocols

Benjamin Doerr*, Anna Huber, and Ariel Levavi**

Max-Planck-Institut für Informatik
Campus E1 4
66123 Saarbrücken, Germany
{doerr, ahuber, levavi}@mpi-inf.mpg.de

Abstract. Randomized rumor spreading is a classic protocol to disseminate information in a network. At SODA 2008, a quasirandom version of this protocol was proposed and competitive bounds for its run-time were proven. This prompts the question: to what extent does the quasirandom protocol inherit the second principal advantage of randomized rumor spreading, namely robustness against transmission failures?

A tentative solution was proposed at ICALP 2009 where it was demonstrated that if each transmission reaches its destination with a probability of $p \in (0,1]$, the run-time increases by a factor of approximately $4/p$. In this paper, we follow up on this research and provide a result precise up to $(1 \pm o(1))$ factors. We limit ourselves to the network in which every two vertices are connected by a direct link. Run-times accurate to their leading constants are unknown for all other non-trivial networks.

For networks on n nodes, we show that after $(1+\varepsilon)(\log_{1+p} n + \frac{1}{p} \ln n)$ rounds, the quasirandom protocol with probability $1 - n^{-c(\varepsilon,p)}$ has informed all nodes in the network. Note that this is faster than the intuitively natural $1/p$ factor increase over the run-time of approximately $\log_2 n + \ln n$ for the non-corrupted case.

We also provide a corresponding lower bound for the classical model. This demonstrates that the quasirandom model is at least as robust as the fully random model despite the greatly reduced degree of independent randomness.

1 Introduction

Disseminating information in a network, that is, making information known to a single node available to all other nodes, is a classic problem. A simple, yet powerful approach is *randomized rumor spreading*, also known as *random phone calls*. In this setting, each node already informed participates in the dissemination

* Partially supported by the German Science Foundation (DFG) via its priority program "SPP 1307: Algorithm Engineering", grant DO 479/4-1.

** Supported by the Study Scholarship awarded by the German Academic Exchange Service (DAAD).

Y. Dong, D.-Z. Du, and O. Ibarra (Eds.): ISAAC 2009, LNCS 5878, pp. 812–821, 2009.

process by randomly calling neighbors and passing along copies of the information. Besides being self-organized, this approach has two crucial advantages. (i) It is fast. For many important network topologies, $O(\log n)$ rounds suffice to inform all n nodes with high probability. (ii) It is robust against transmission failure. Often, a constant fraction of independently chosen transmission failures does not cause serious problems, but merely increases the time required by a constant factor.

Success of the basic randomized rumor spreading protocol has motivated the study of several variants. At SODA 2008, a quasirandom version of the protocol was proposed. This version is structurally simpler, uses less randomness, and is especially beneficial in that each vertex contacts each neighbor at most once. Nonetheless, most run-time guarantees known for the classical model still hold for the quasirandom version, some in an even stronger form. However, little was previously known about the robustness of this model.

In this paper, we offer a detailed investigation of the robustness of the quasirandom protocol. We use the following model of lossy communication. We assume that independently for all transmissions the message reaches its target with a certain probability $0 < p < 1$. For networks in which every two nodes are connected by a link, we demonstrate that this lossiness only increases the run-time by a small constant factor, which we precisely determine for each value of p. Surprisingly, it is smaller than $1/p$, e.g., 1.828 for $p = 1/2$. The same result also holds for the classical, fully random model. We will not give a formal proof for this, as only slight modifications of our methods are necessary.

In addition, we show that the corresponding slow-down for the classical model is at least this factor. This shows that the quasirandom model is at least as robust as the classical model.

This is the first time, for the classical as well as for the quasirandom model, that results precise up to the leading constant are shown for the robustness.

1.1 Randomized Rumor Spreading

The classic, fully random *randomized rumor spreading* protocol was first investigated by Frieze and Grimmett [1]. They proposed the following model. Given is a network modeled on an undirected graph $G = (V, E)$. At the start of the protocol, a single vertex $s \in V$ knows a piece of information that is to be disseminated to all other vertices. We say that v is *informed*. The protocol proceeds in rounds (hence it assumes a common clock). In each round, every informed vertex v chooses a neighbor $u_v \in N(v) := \{u \in V \mid \{u, v\} \in E\}$ uniformly at random and sends a copy of the information to it. This results in u_v becoming informed, if it is not already, and in u_v participating in the dissemination process in subsequent rounds. This process defines a random variable T_s, which denotes the number of rounds after which all vertices in the network are informed, assuming that the initially informed vertex is s. The *broadcast time* T is then defined as the maximum of all T_s, $s \in V$, where the T_s are defined over independent probability spaces.

Frieze and Grimmett show that if the network is a complete graph on n vertices, the broadcast time satisfies $T = (1 \pm o(1))(\log_2 n + \ln n)$ with probability $1 - o(1)$. For hypercubes and random graphs $G(n,p)$ for $p \geq (1+\varepsilon)\ln(n)/n$, Feige, Peleg, Raghavan and Upfal [2] also determine a broadcast time of $\Theta(\log n)$ with probability $1 - 1/n$, albeit without making the implicit constant precise. They also provide the general bounds of $12n \log n$ and $O(\Delta(G)(\mathrm{diam}(G) + \log n))$ for arbitrary n-vertex graphs.

For reasons of space, we shall not extensively discuss the practical side of the randomized rumor spreading protocol. We refer the interested reader to the aforementioned paper [2] as well as the paper by Karp, Shenker, Schindelhauer and Vöcking [3] for a general discussion, or the works of Demers et al. [4] and Kempe, Dobra and Gehrke [5] for particular applications. What are generally recognized as the three key advantages of randomized rumor spreading are speed (logarithmic broadcast time on important network topologies); self-organization (there is no central authority involved); and robustness against transmission failure. Contrary to broadcast times, significantly less work has been done to quantify the robustness of the randomized rumor spreading protocol.

As far as we are aware, the only results on the robustness of randomized rumor spreading are due to Elsässer and Sauerwald [6]. They consider the model where each transmission does not reach its destination with (independently sampled) failure probability $f < 1$. Denoting the success probability of a transmission by $p := 1 - f$, they assert that the broadcast time for all graphs in this lossy model is at most a factor of $O(1/p)$ larger than in the model without transmission failures.

1.2 Quasirandom Rumor Spreading

The above results show that randomized rumor spreading is a very powerful approach to dissemination problems. However, taking all decisions independently at random also has some unwanted effects. For example, a vertex may contact one of its neighbors twice before contacting all of its other neighbors. This may only be a minor problem for dense graphs like the complete graph, but for sparse graphs, it may increase the broadcast time significantly.

Motivated by the paradigm of quasirandomness, Friedrich, Sauerwald and the first author [7] suggest the following *quasirandom rumor spreading* protocol.

In this model, each vertex is equipped with a cyclic permutation of its neighbors. As before, the protocol proceeds in rounds, and all informed vertices participate in the dissemination process. However, each vertex only directs its first transmission to a random neighbor. Subsequently, it informs the successors of the first addressee on its list. We shall not make any assumptions about the structure of these cyclic lists.

Before analyzing the quasirandom protocol, let us discuss it from an implementation point of view. From a theory perspective, we immediately note that the quasirandom model requires each vertex to store the permutation of its neighbors, which may utilize up to $\Theta(n \log n)$ bits. This is not necessary for the fully random model. However, we may assume that in most networks each vertex already has some list or array of its neighbors, since the information regarding how to contact a neighbor must be stored somewhere. In this scenario, the use of the lists does not increase the complexity. Rather, it appears that the quasirandom protocol needs less resources. In particular, it requires significantly fewer random bits. This is beneficial if we consider randomness costly, and useful if we want to trace an actual run of the protocol.

The core question to be answered is whether the quasirandom protocol works well even if we are not permitted to design the lists. Surprisingly, the answer is positive.

For all lists that can be present at each vertex, $O(\log n)$ rounds suffice with high probability to inform all vertices of a complete graph K_n, a hypercube Q_n, an expander graph on n vertices (some extra conditions are needed here), or a random graph $G(n,p)$ with $p \geq (1+\varepsilon)(\ln n)/n$ [7,8]. Naturally, the lower bound of $\log_2 n$ rounds valid for the fully random model also holds for the quasirandom model. Once again these bounds fall into the right order of magnitude.

Sharper bounds analogous to those defined by Frieze and Grimmett are known for the complete graph. In [9], it is shown that with probability $1-o(1)$, the number of rounds needed to inform all vertices is $(1 \pm o(1))(\log_2 n + \ln n)$.

For some settings, we observe better broadcast times than in the classical model. One example is the random graph with edge probability $p = (\ln n + \omega(1))/n$ only minimally above the connectivity threshold. Nevertheless, with probability $1-o(1)$, the random graph is such that with high probability the quasirandom protocol needs only $O(\log n)$ rounds independent of the starting point. This is a notable advantage over the fully random model. Feige et al. [2] demonstrate that for $p = (\ln n + O(\log \log n))/n$, the random graph with probability $1-o(1)$ is such that $\Theta(\log^2 n)$ rounds are necessary to spread the rumor with high probability.

1.3 Robustness of the Quasirandom Protocol

The above results show that the broadcast time of the quasirandom rumor spreading protocol is quite well understood. Together with the experimental investigation [10], all results indicate that the quasirandom protocol achieves comparable or better broadcast times than the random model. For the equally important aspect of robustness, much less is known. Since it would typically seem that robustness of randomized algorithms is caused by the large number of independent random decisions taken by the algorithm, one may conclude that the quasirandom protocol is less robust.

The following result from [8] is the first to debunk this assertion. Let G be a graph, $T \in \mathbb{N}$ and $\gamma \geq 1$ such that the quasirandom protocol independent of the starting vertex with probability $1-n^{-\gamma}$ succeeds in informing all other nodes within T rounds. Then in the presence of transmission failures (independently chosen with probability $1-p$), independent of the starting vertex $4\gamma(1/p)T$ rounds suffice to inform all vertices with probability $1-2n^{-\gamma}$.

Naturally, this leaves room for constant factor differences between the two models. The only other result pertaining to robustness is the experimental

Data Structures for Range Median Queries

Gerth Stølting Brodal and Allan Grønlund Jørgensen

MADALGO*, Department of Computer Science, Aarhus University, Denmark
{gerth,jallan}@cs.au.dk

Abstract. In this paper we design data structures supporting *range median* queries, i.e. report the median element in a sub-range of an array. We consider static and dynamic data structures and batched queries. Our data structures support *range selection* queries, which are more general, and dominance queries (*range rank*). In the static case our data structure uses linear space and queries are supported in $O(\log n/\log\log n)$ time. Our dynamic data structure uses $O(n\log n/\log\log n)$ space and supports queries and updates in $O((\log n/\log\log n)^2)$ time.

1 Introduction

The median of a set S of size n is an element in S that is larger than $\lfloor\frac{n-1}{2}\rfloor$ other elements from S and smaller than $\lceil\frac{n-1}{2}\rceil$ other elements from S. In the range median problem one must preprocess an input array A of size n into a data structure that given indices i and j, $1 \le i \le j \le n$, a query must return an index i', $i \le i' \le j$, such that $A[i']$ is the median of the elements in the subarray $A[i,j] = [A[i], A[i+1], \ldots, A[j]]$. This problem is considered in [1,2,3,4,5]. In the batched case, the input is an array of size n and a set of k queries, $(i_1, j_1), \ldots, (i_k, j_k)$, and the output is the answer to these k queries [6]. Range median queries are naturally generalized to *range selection*, given indices i, j and s, return the index of the s'th smallest element in $A[i,j]$. A related problem is *range dominance* (or *range rank*) queries, given indices i, j and a value e, return the number of elements from $A[i,j]$ that are less than or equal to e (dominated by e). This corresponds to 3-sided range counting queries for a set of points.

Previous Work. Previously, the best linear space data structure supported range selection queries in $O(\log n)$ time [4,5]. In the dynamic case the only known data structure uses $O(n\log n)$ space and supports updates and queries in $O(\log^2 n)$ time [4]. For dominance queries, linear space data structures supporting queries in $O(\log n/\log\log n)$ time is known, as well as a matching lower bound [7,8,9]. In the dynamic case [10] describes an $O(n)$ space data structure that supports dominance queries in $O((\log n/\log\log n)^2)$ time and updates in $O(\log^{9/2} n/(\log\log n)^2)$ time. A query lower bound of $\Omega((\log n/\log\log n)^2)$ for data structures with $O(\log^{O(1)} n)$ update time is proved in [7].

* Center for Massive Data Algorithmics, a Center of the Danish National Research Foundation.

Y. Dong, D.-Z. Du, and O. Ibarra (Eds.): ISAAC 2009, LNCS 5878, pp. 822–831, 2009.

Our Results. In this paper we use the RAM model of computation with word-size $\Theta(\log n)$. Our data structures use the same basic approach as in [4]. We design a static linear space data structure that supports both range selection and range rank queries in $O(\log n/\log\log n)$ time. This is the best known for range median data structures using $O(n \log^{O(1)} n)$ space, and for range dominance queries this is optimal. Our dynamic data structure uses $O(n \log n/\log\log n)$ space and supports queries and updates in $O((\log n/\log\log n)^2)$ time. For dominance queries this query time is optimal. We prove an $\Omega(\log n/\log\log n)$ time lower bound on range median queries for data structures that can be updated in $O(\log^{O(1)} n)$ time using a reduction from the marked ancestor problem [11], leaving a significant gap to the achieved upper bound. With our static data structure we improve the $O(n \log k + k \log n)$ time bound for the batched range median problem achieved in [4] to $O(n \log k + k \log n/\log\log n)$ time. If $k > \sqrt{n}$ we construct our static data structure in $O(n \log n) = O(n \log k)$ time and perform k queries. This takes $O(n \log n + k \log n/\log\log n) = O(n \log k + k \log n/\log\log n)$ time. If $k < \sqrt{n}$ then $O(n \log k)$ time is already achieved by [6,4].

2 Simple Range Selection Data Structure

In this section we describe the data structure of Gfeller and Sanders [4], which uses linear space and supports queries in $O(\log n)$ time. First, we describe a data structure that uses $O(n \log n)$ space and supports queries in $O(\log n)$ time. Then the space is reduced to $O(n)$ using standard techniques. The main idea is the following. Sort the input elements and place them in the leaves of a binary search tree. Consider a search for the s'th smallest element in $A[i,j]$. If the left subtree of the root contains s or more elements from $A[i,j]$ then it contains the s'th smallest element from $A[i,j]$. If not, it is in the right subtree. We augment each node of the tree with prefix sums such that the number of elements from $A[1,j]$ contained in the left subtree can be determined for any j, and we use fractional cascading [12] to avoid a search for the needed prefix sums in each node.

2.1 Basic Structure

Let $A = [y_1, \dots, y_n]$ be the input array. We sort A and build a complete binary search tree T that stores the n elements in the leaves in sorted order. We introduce the following notation. For a node v in T, let T_v denote the subtree rooted at v, and $|T_v|$ the number of leaves in T_v. The *x-predecessor* of an index (x-coordinate) i in T_v is the largest index i' such that $i' \leq i$ and $y_{i'} \in T_v$. If no such index exists the x-predecessor of i is zero. The *x-rank* of an index i in T_v is the number of elements from $A[1,i]$ contained in T_v. An x-rank is essentially a prefix sum. If we know the x-rank of $i-1$ and j in T_v, we know the number of elements from $A[i,j]$ in T_v since this is the difference between the two. Notice that in T_v, the x-rank of j and the x-rank of the x-predecessor of j are equal.

Each node v of T stores two indices for each element $y_i \in T_v$ in an array A_v of size $|T_v|$. Let $y_i \in T_v$ and r_i be the x-rank of i in T_v. The r_i'th pair of indices stored in A_v is the x-rank of i in the left subtree, and the x-rank of i

in the right subtree of v. These are fractional cascading indices, meaning that the x-rank of i in the left (right) subtree is the position of the indices stored for the x-predecessor of i in the left (right) subtree. The arrays are constructed by scanning A, starting with y_1, and *inserting* the elements, y_i, in increasing i order into T. For each element y_i, the search path to y_i is traversed, and in each visited node v the pair of indices for y_i are appended to A_v. The data structure can be built in $O(n \log n)$ time and uses $O(n \log n)$ words of space.

Range Selection Query. A range selection query is given two indices i and j, and an integer s, where $1 \leq i \leq j \leq n$, and must return the s'th smallest element in $A[i,j]$. In a node v of T the search is guided using the x-ranks stored for the x-predecessor of $i-1$ and j in T_v. In the root this is the $i-1$'th and j'th pair stored in the root's array. By subtracting the x-ranks for the left subtree we learn how many elements, s', from $A[i,j]$ the left subtree of v contains. If $s \leq s'$ the search continues in the left subtree. Otherwise, we set $s = s - s'$ and continue the search in the right subtree. Notice that each step learns the x-ranks of $i-1$ and j in the children nodes, which are needed to lookup the indices stored for the x-predecessors of $i-1$ and j in the subsequent step. A query takes $O(\log n)$ time since each step takes constant time. Given indices i and j in a range median query we return the $s = \lfloor \frac{j-i}{2} \rfloor + 1$'th smallest element in $A[i,j]$. Given indices i, j and a value e in a range rank query, we do a predecessor search for e in T: In each step where the search continues to the right child, the number of elements from $A[i,j]$ in the left subtree is computed as above, and these are added up. When a leaf is reached this sum is the rank of e in $A[i,j]$.

2.2 Getting Linear Space

We reduce the space usage of the data structure to $O(n)$ by replacing the arrays stored in each node by simple rank and select data structures [13] as follows. In each node v, the array A_v is partitioned into chunks of size $\log n$. The last entry of each chunk, i.e. every $\log n$'th entry of A_v, is stored as before. For the remaining entries of a chunk, one bit is stored, indicating whether the corresponding element resides in the left or right subtree. These bits are packed in order into one word, which we denote a direction word. This reduces the space to $O(n)$ bits per level of T. Even though v no longer stores an x-rank for each element in T_v, a needed x-rank is easily computed from the stored chunks in constant time. Let r_j be the x-rank of j in T_v, and let $j' = \lfloor j / \log n \rfloor$. The indices stored in the $j'-1$'th chunk yields the x-rank, r_λ, in the left subtree of $y_\lambda \in T_v$. The first $r_j - j' \log n$ bits in the direction word from the j''th chunk determines how many elements from $A[\lambda+1, j]$ that reside in the left subtree. The sum of these is the x-rank of j in the left subtree. The latter is computed using complete tabulation. The extra table needed for this uses $O(n)$ additional space.

3 Improving Query Time

In this section we generalize the data structure from Section 2 and obtain a linear space data structure that supports range selection queries in

$O(\log n/ \log\log n)$ time. First, we describe a data structure that supports queries in $O(\log n/ \log\log n)$ time but uses slightly more than $O(n)$ space. Then we reduce the space to $O(n)$ by generalizing the ideas from Section 2.2.

3.1 Structure

The data structure is a balanced search tree T storing the n elements from $A = [y_1, \ldots, y_n]$ in the leaves in sorted order. The fan-out of T is $f = \lceil \log^{\varepsilon} n \rceil$ for some constant $0 < \varepsilon < 1$. The height of T is $O(\log n/ \log f) = O(\log n/ \log\log n)$. Each node $v \in T$ contains $f \cdot |T_v|$ prefix sums: For each element, $y_i \in T_v$, and for each child index, $1 \leq \ell \leq f$, v stores the number of elements from $A[1, i]$ that reside in the first ℓ subtrees of T_v. We denote by t_ℓ^i such a prefix sum. These prefix sums are stored in $|T_v|$ bit-matrices, one matrix M_i for each $y_i \in T_v$. The ℓ'th row of bits in M_i is the number t_ℓ^i. The rows form a non-decreasing sequence of numbers by construction. The matrices are stored consecutively in an array A_v as above, i.e. M_i is stored before M_j if $i < j$, and the x-rank of i in T_v defines the position of M_i in A_v. If $y_j \notin T_v$ then v does not store a matrix M_j, but it is still well defined and equal to the matrix $M_{j'}$, that is stored in v, where j' is the x-predecessor of j in T_v. Each matrix is stored in two different ways. In the first copy each row is stored in one word. In the second copy each matrix is divided into sections of $g = \lfloor \log n/f \rfloor$ columns. The first section contains the first g bits of each of the f rows, and these are stored in one word. This is the g most significant bits of each prefix sum stored in the matrix. The second section contains the last three bits of the first section and then the following $g - 3$ bits, and so on. The reason for this overlap of three bits will become clear later. We think of each section as an $f \times g$ bit matrix. For technical reasons, we ensure that the first column of each matrix only contain zero entries by prepending a column of zeroes to all matrices before the division into sections.

3.2 Range Selection Query

A query is given indices i, j and s and locates the s'th smallest element in $A[i, j]$. In each node we consider the matrix $M' = M_j - M_{i-1}$ (row-wise subtraction). The ℓ'th row of M' is $t_\ell^j - t_\ell^{i-1}$, i.e. the number of elements from $A[i, j]$ contained in the first ℓ subtrees. We compute the smallest ℓ such that the ℓ'th row in M' stores a number greater than or equal to s, and this defines the subtree containing the s'th smallest element in T_v. In the following pages we describe how to compute ℓ without explicitly constructing the entire matrix M'.

The intuitive idea to guide a query in a given node, v, is as follows. Let $K = |T_v \cap A[i, j]|$ be the number of elements from $A[i, j]$ contained in T_v. We consider the section from M' containing the $\lceil \log K \rceil$'th least significant bit of each row. All the bits stored in M' before this section are zero and thus not important. Using word-level parallelism we find an interval $[\ell_1, \ell_2] \subseteq [1, f]$, where the g bits of M' match the corresponding g bits of s and the following row. These indices define the subtrees of T_v that can contain the s'th smallest element in T_v. We then try to determine which of these subtrees contain the

s'th smallest element. First, we consider the children of v defined by the endpoints of the interval, ℓ_1 and ℓ_2. If neither of these contain the s'th smallest element in $A[i, j]$, we know that the subtree of T_v containing the s'th smallest element stores approximately a factor of 2^g elements from $A[i, j]$ fewer than T_v, since the g most significant bits of the **prefix** sum of the row corresponding to this subtree are the same as the bits in the preceding row. Stated differently, the number of elements in this subtree does not influence the g most important bits of the prefix sum, and thus it must be small. In this case we determine ℓ in $O(\log \log n)$ time using a standard binary search. The point is that this can only occur $O(\log n/g)$ times, and the total cost of these searches is $O(\log n \log \log n/f) = O(\log^{1-\varepsilon} n \log \log n) = o(\log n/ \log \log n)$. In the remaining nodes we use constant time.

There are several technical issues that must be worked out. The most important is that we cannot actually produce the needed section of M' in constant time. Instead, we compute an approximation where the number stored in the g bits of each row of the section is at most one too large when compared to the g bits of that row in M'. The details are as follows.

In a node $v \in T$ the search is guided using M_{p_i} and M_{p_j} where p_i is the x-predecessor of $i - 1$ in T_v and p_j is the x-predecessor of j in T_v. For clarity we use M_{i-1} and M_j for the description. A query maintains an index c, initially one, defining which section of the bit-matrices that is currently in use i.e. c defines the section of M' containing the $\lceil \log K \rceil$'th least significant bit. We maintain the following invariant regarding the c'th section of M' in the remaining subtree: in M', all bits before the c'th section are zero, i.e. the important bits of M' are stored in the c'th section or to the right of it. For technical reasons, we ensure that the most important bit of the c'th section of M' is zero. This is true before the query starts since the first bit in each row of each stored matrix is zero.

We compute the approximation of the c'th section of M' from the c'th section of M_j and M_i. This approximation we denote $w^{i,j}$ and think of it as a $f \times g$ bit-matrix. Basically, the word containing the c'th section of bits from M_{i-1} is subtracted from the corresponding word in M_j. However, subtracting the c'th section of g bits of t_ℓ^{i-1} from the corresponding g bits of t_ℓ^j does not encompass a potential cascading carry from the lower order bits when comparing the result with the matching g bits of $t_\ell^j - t_\ell^{i-1}$, the ℓ'th row of M'. This means that in the c'th section, the ℓ'th row of M_{i-1} could be larger than ℓ'th row of M_j. To ensure that each pair of rows is subtracted independently in the computation of $w^{i,j}$, we prepend an extra one bit to each row of M_j and an extra zero bit to each row of M_i to deal with cascading carries. Then we subtract the c section of M_{i-1} from the c'th section of M_j, and obtain $w^{i,j}$. After the subtraction we ignore the value of the most significant bit of each row in $w^{i,j}$ (it is masked out). After this computation, each row in $w^{i,j}$ contain a number that either matches the corresponding g bits of M', or a number that is one larger. Since the most important bit of the c'th section of M' is zero, we know that the computation does not *overflow*. If all bits in $w^{i,j}$ are zero the algorithm never needs to consider the current section again, and it is skipped in the remaining subtree by increasing c by one, without breaking the invariant, and $w^{i,j}$ is recomputed.

Searching $w^{i,j}$. Let $s_b = s_1, \ldots, s_g$ be the g bits of s defined by the c'th section, initially the g most important bits of s. If we had actually computed the c'th section of M' then only rows matching s_b and the first row containing an even larger number can define the subtree containing the s'th smallest element. However, since the rows can contain numbers that are one *to large*, we also consider all rows matching $s_b + 1$, and the first row storing a number even larger. Therefore, the algorithm locates the first row of $w^{i,j}$ storing a number greater than or equal to s_b and the first row greater than $s_b + 1$. The indices of these rows we denote ℓ_1 and ℓ_2, and the subtree containing the s'th smallest element corresponds to at row between ℓ_1 and ℓ_2. Subsequently, it is checked whether the ℓ_1'th or ℓ_2'th subtree contains the s'th smallest element in T_v using the first copy of the matrices (where the rows are stored separately). If this is not the case, then the index of the correct subtree is between $\ell_1 + 1$ and $\ell_2 - 1$, and it is determined by a binary search. The binary search uses the first copy of the matrices. In the c'th section of M', the g bits from the $\ell_1 + 1$'th row represents at number that is at least $s_b - 1$, and the $\ell_2 - 1$'th row a number that is at most $s_b + 1$. Therefore, the difference between the numbers stored in row $\ell_1 - 1$ and $\ell_2 - 1$ in M' is at most two. This means that in the remaining subtree, the c'th section of bits from M' $(t^{j}_{\ell} - {}^{i-1}_{\ell}$ for $1 \leq \ell \leq f)$ is a number between zero and two. Since the following section stores the last three bits of the current section, the algorithm safely skips the current section in the remaining subtree, by increasing c by one, without violating the invariant. We need two bits to express a number between zero and two, and the third bit ensures that the most significant bit of the c'th section of M' is zero. After the subtree, T_ℓ, containing the s'th smallest element is located s is updated as before, $s = s - (t^{j}_{\ell-1} - t^{i-1}_{\ell-1})$. Let $r_{i-1} = t^{i-1}_{\ell} - t^{i-1}_{\ell-1}$, be the x-rank of $i-1$ in T_ℓ, and $r_j = t^{j}_{\ell} - t^{j}_{\ell-1}$, the x-rank of j in T_ℓ. In the subsequent node the algorithm uses the r_{i-1}'th and the r_j'th stored matrix to guide the search. This corresponds to the matrix stored for the x-predecessor of $i-1$ and the x-predecessor of j in T_ℓ (fractional cascading).

In the next paragraph we explain how to determine ℓ_1 and ℓ_2 in constant time. Thus, if the search continues in the ℓ_1'th or ℓ_2'th subtree, the algorithm used $O(1)$ time in the node. Otherwise, a binary search is performed, which takes $O(\log f)$ time, but in the remaining subtree an additional section is skipped. An additional section may be skipped at most $\lceil 1 + \log n/(g-3) \rceil = O(f)$ times. When the search is guided using the last section there will not be any problems with cascading carries. This means that the search continues in the subtree corresponding to the first row of $w^{i,j}$ where the number stored is at least as large as s_b, and a binary search is never performed in this case. We conclude that a query takes $O(\log n / \log\log n + f \log f) = O(\log n / \log\log n)$ time.

Given i, j and e in a rank query we use a linear space predecessor data structure (van Emde Boas tree [14]) that in $O(\log\log n)$ time yields the predecessor e_p of e in the sorted order of A. Then, the path from e_p to the root in T is traversed, and during this walk the number of elements from $A[i,j]$ in subtrees hanging of to the left are added up using the first copy of the bit matrices. The data structures uses $O(nf \log n / \log\log n) = O(n \log^{1+\varepsilon} n / \log\log n)$ space.

Determining ℓ_1 and ℓ_2. The remaining issue is compute ℓ_1 and ℓ_2. A query maintains a search word, s_w, that contains f independent blocks of the g bits from s that corresponds to the c'th section. Initially, this is the g most important bits of s. To compute s_w we store a table that maps each g-bit number to a word that contains f copies of these g bits. After updating s we update s_w using a bit-mask and a table look-up. A query knows $w^{i,j} = v_1^1, \ldots, v_g^1, \ldots, v_1^d, \ldots, v_g^d$ and s_w which is $s_b = s_1, \ldots, s_g$ concatenated f times. The g-bit block $v_1^\ell, \ldots, v_g^\ell$ from $w^{i,j}$ we denote $w_\ell^{i,j}$ and the ℓ'th block of $s_1, \ldots, s_g$ from s_w we denote s_w^ℓ. We only describe how to find ℓ_1, ℓ_2 can be found similarly. Remember that ℓ_1 is the index of the first row in $w^{i,j}$ that stores a number greater than or equal to s_b. We make room for an extra bit in each block and make it the most significant. We set the extra bit of each $w_\ell^{i,j}$ to one and the extra bit of each s_w^ℓ to zero. This ensures that $w_\ell^{i,j}$ is larger than s_w^ℓ, for all ℓ, when both are considered $g+1$ bit numbers. s_w is subtracted from $w^{i,j}$ and because of the extra bit, this operation subtracts s_w^ℓ from $w_\ell^{i,j}$, for $1 \le \ell \le f$, independently of the other blocks. Then, all but the most significant (fake) bit of each block are masked out. The first one-bit in this word reveals the index ℓ of the first block where $w_\ell^{i,j}$ is at least as large as s_w^ℓ. This bit is found using complete tabulation.

3.3 Getting Linear Space

In this section we reduce the space usage of our data structure to $O(n)$ words. The previous data structure stores a matrix for each element on each level of the tree, and every matrix uses $O(f \log n)$ bits of space. Instead we only store a matrix for every $t = \lceil f \log n \rceil$'th element. In each node, the sequence of matrices is divided into chunks of size t and only the last matrix of each chunk is explicitly stored. For each of the remaining elements in a chunk, $\lceil \log f \rceil$ bits are used to describe in which subtree it resides. The description for $d = \lfloor \log n / \lceil \log f \rceil \rfloor$ elements are stored in one word, which we denote a direction word. Prefix sums are stored after each direction word summing up all previous directions words in the chunk, i.e. storing how many elements that was inserted in the first ℓ subtrees for $\ell = 1, \ldots, f$. Since each chunk stores the direction of t elements, at most $\lceil f \log t / \log n \rceil = O(1)$ words are needed to store these f prefix sums. We denote it a prefix word. The data structure uses $O(n)$ words of space.

Range Selection Query. The query works similarly to above. The main difference is that we do not use the matrices M_{i-1} and M_j to compute $w^{i,j}$ since they are not necessarily stored. Instead, we use two matrices that are stored which are *close* to M_{i-1} and M_j. The direction and update words enables us to exactly compute any row of M_j and M_{i-1} in constant time. Therefore, the main difference compared to the previous data structure, is that the potential difference between $w^{i,j}$, that we compute, and the c'th section of M' is marginally larger, and for this reason the overlap between blocks is increased to four.

In a node $v \in T$ a query is guided as follows. Let r_i be the x-rank of $i-1$ and r_j the x-rank of j in T_v. Let $i' = \lfloor r_i/t \rfloor$ and $j' = \lfloor r_j/t \rfloor$. The matrices stored in the i''th and j''th chunk respectively are used to guide the search. These matrices we denote M_a and M_b. Since v stores every t'th matrix from above, $t_\ell^j - t_\ell^b \leq t$ for any $1 \leq \ell \leq f$. If we ignore a potential cascading carry, then adding $t_\ell^j - t_\ell^b$ to t_ℓ^b only affects the last $\log t = (1+\varepsilon)\log\log n$ bits of t_ℓ^b. This means that, unless the search is using the last section, each row in the currently considered section of M_b represents a number that is at most one smaller than if we had used the corresponding section from M_j. The same is true for M_a.

We can obtain the value of any row in M_j as follows. From the direction and prefix words from the j''th chunk we compute for each ℓ, $1 \leq \ell \leq f$, how many of the first $r_j - j't$ elements represented in the chunk that reside in the first ℓ children. These are the elements considered in M^j but not in M^b. Formally, the $p = \lfloor (r_j - j't)/d \rfloor$'th prefix word stores how many of the first pd elements from the chunk that reside in the first ℓ children for $1 \leq \ell \leq f$. Using complete tabulation on the following direction word, we obtain a word storing how many of the following $r_j - j't - pd$ elements from the chunk that reside in the first ℓ children for all $1 \leq \ell \leq f$. Adding this to the p'th prefix word, gives for each $1 \leq \ell \leq f$, the difference between the ℓ'th row of M_j and M_b. The difference between M_a and M_{i-1} can be computed similarly. Thus, any row of M_j and M_{i-1}, and the last section of M_j and M_{i-1} can be computed in constant time.

If the last section is used it is computed exactly in constant time and the search is guided as above. Otherwise, we compute the difference between each row in the c'th section of M_a and M_b, yielding $w^{a,b}$. Since the ℓ'th row, for $1 \leq \ell \leq f$, in the current section of M_b might be one to small compared to ℓ'th row in the current section of M_j, the ℓ'th row in $w^{a,b}$ may be one to small compared to the corresponding g bits of M'. Similarly, each row in $w^{a,b}$ might also one to the large since the current section of bits from M_a may be one smaller than in the current section of M_{i-1}. As above, the computation of $w^{a,b}$ does not consider cascading carries from lower order bits and for this reason the ℓ'th row of $w^{a,b}$ may additionally be one to large when compared to the same bits in M'. Therefore, the first row of $w^{a,b}$ that is at least $s_b - 1$ and the first row greater than $s_b + 2$ are located as above. As above, the subtree we are searching for is defined by a row between these two, and if it is not one of these, a binary search is used to determine it. In this case, by the same arguments as earlier, each row in the c'th section of M' in the remaining subtree, represents at number between zero and six. Since we have an overlap of four bits between sections, we safely move to the next section after every binary search.

Dominance queries are supported similarly to above.

4 Dynamic Range Selection

In this section we briefly sketch how our data structure can be made dynamic. Our dynamic data structure uses $O(n \log n / \log\log n)$ space and supports queries and updates in $O((\log n / \log\log n)^2)$ time, worst case and amortized respectively. Details will appear in the full paper.

Our data structure maintains a set of points, $S = \{(x_i, y_i)\}$, under insertions and deletions. A query is given values x_l, x_r and an integer s and returns the point with the s'th smallest y value among the points in S with x-value between x_l and x_r. We store the points from S in a weight-balanced search tree [15,16], ordered by y-coordinate. In each node of the tree we maintain the bit-matrices, defined in the static structure, dynamically using a weight-balanced search tree over the points in the subtree, ordered by x-coordinate. The main issue is efficient generation of the needed sections of the bit-matrices used by queries. The quality of the approximation is worse than in the static data structure, and we increase the overlap between sections to $O(\log\log n)$. Otherwise, a search works as in the static data structure.

5 Lower Bound for Dynamic Data Structures

In this section we describe a reduction from the marked ancestor problem to a dynamic range median data structure. In the marked ancestor problem the input is a complete tree of degree b and height h. An update marks or unmarks a node of the tree, initially all nodes are unmarked. A query is provided a leaf v of the tree and must return whether there exist a marked ancestor of v. Let t_q and t_u be the query and update time for a marked ancestor data structure. Alstrup et al. proved the following lower bound trade-off for the problem, $t_q = \Omega(\frac{\log n}{\log(t_u w \log n)})$ [11], where w is the word size.

Reduction. Let T denote a marked ancestor tree of height h and degree b. For each node v in T we associate two pairs of elements, which we denote *start-mark* and *end-mark*. We translate T into an array of size $4|T|$ by a recursive traversal of T, where we for each node v outputs its start-mark, then recursively visit each of v's children, and then output v's end-mark. Start-marks are used to mark a node, and end-marks ensure that markings only influences the answer for queries in the marked subtree. When a node v is unmarked, start-mark=end-mark=(0,1) and when v is marked, start-mark is set to (1,1) and end-mark to (0,0).

A marked ancestor query for a leaf v is answered by returning yes if and only if the range median from the subarray ranging from the beginning of the array to the start-mark element associated with v is one. If zero nodes are marked, the array is on the form $[0, 1, \ldots, 0, 1]$. Since the median in any range that can be considered by a query is zero, any marked ancestor query returns no. If v or one of its ancestors is marked there will be more ones than zeros in the range for v, and the query answers yes. A node u that is not an ancestor of v has its start-mark and end-mark placed either before v's marks or after v's marks, and independently of whether u is marked or not, it contributes and equal number of zeroes and ones to v's query range. Since the reduction requires an overhead of $O(1)$ for both queries and updates we get the following lower bound.

Theorem 1. *Any data structure that supports updates in* $O(\log^{O(1)} n)$ *time uses* $\Omega(\log n / \log\log n)$ *time to support a range median query.*

6 Main Open Problems

There are two main open problems. First, what is the lower bound on the query time for range selection queries in static $O(n \log^{O(1)} n)$ space data structures? We can prove that any $O(n \log^{O(1)} n)$ space data structure needs $\Theta(\log n / \log \log n)$ time for three-sided range median queries by a reduction from two dimensional rectangle-stabbing [8]. Furthermore, there is a gap between the upper and lower bounds for batched range median problem for $k = \Omega(n^{1+\varepsilon})$, and the lower bound [6] is only valid in the comparison model.

References

1. Krizanc, D., Morin, P., Smid, M.H.M.: Range mode and range median queries on lists and trees. Nord. J. Comput. 12(1), 1–17 (2005)
2. Petersen, H.: Improved bounds for range mode and range median queries. In: Proc. 34th Conference on Current Trends in Theory and Practice of Computer Science, pp. 418–423 (2008)
3. Petersen, H., Grabowski, S.: Range mode and range median queries in constant time and sub-quadratic space. Inf. Process. Lett. 109(4), 225–228 (2009)
4. Gfeller, B., Sanders, P.: Towards optimal range medians. In: Proc. 36th International Colloquium on Automata, Languages and Programming, pp. 475–486 (2009)
5. Gagie, T., Puglisi, S.J., Turpin, A.: Range quantile queries: Another virtue of wavelet trees. In: Proc. 16th String Processing and Information Retrieval Symposium, pp. 1–6 (2009)
6. Har-Peled, S., Muthukrishnan, S.: Range medians. In: Proc. 16th Annual European Symposium on Algorithms, pp. 503–514 (2008)
7. Pătraşcu, M.: Lower bounds for 2-dimensional range counting. In: Proc. 39th ACM Symposium on Theory of Computing, pp. 40–46 (2007)
8. Pătraşcu, M.: (Data) STRUCTURES. In: Proc. 49th Annual IEEE Symposium on Foundations of Computer Science, pp. 434–443 (2008)
9. JáJá, J., Mortensen, C.W., Shi, Q.: Space-efficient and fast algorithms for multidimensional dominance reporting and counting. In: Proc. 15th International Symposium on Algorithms and Computation, pp. 558–568 (2004)
10. Nekrich, Y.: Orthogonal range searching in linear and almost-linear space. In: Proc. 10th International Workshop on Algorithms and Data Structures, pp. 15–26 (2007)
11. Alstrup, S., Husfeldt, T., Rauhe, T.: Marked ancestor problems. In: Proc. 39th Annual Symposium on Foundations of Computer Science, Washington, DC, USA, pp. 534–543. IEEE Computer Society, Los Alamitos (1998)
12. Chazelle, B., Guibas, L.J.: Fractional cascading: I. A data structuring technique. Algorithmica 1(2), 133–162 (1986)
13. Jacobson, G.J.: Succinct static data structures. PhD thesis. Carnegie Mellon University, Pittsburgh, PA, USA (1988)
14. van Emde Boas, P., Kaas, R., Zijlstra, E.: Design and implementation of an efficient priority queue. Mathematical Systems Theory 10, 99–127 (1977)
15. Nievergelt, J., Reingold, E.M.: Binary search trees of bounded balance. In: Proc. 4th Annual ACM symposium on Theory of computing, pp. 137–142 (1972)
16. Arge, L., Vitter, J.S.: Optimal external memory interval management. SIAM Journal on Computing 32(6), 1488–1508 (2003)

Deletion without Rebalancing in Multiway Search Trees*

Siddhartha Sen[1] and Robert E. Tarjan[1,2]

[1] Princeton University
{sssix,ret}@cs.princeton.edu
[2] HP Laboratories, Palo Alto CA 94304

Abstract. Many database systems that use a B^+ tree as the underlying data structure do not do rebalancing on deletion. This means that a bad sequence of deletions can create a very unbalanced tree. Yet such databases perform well in practice. Avoidance of rebalancing on deletion has been justified empirically and by average-case analysis, but to our knowledge no worst-case analysis has been done. We do such an analysis. We show that the tree height remains logarithmic in the number of insertions, independent of the number of deletions. Furthermore the amortized time for an insertion or deletion, excluding the search time, is $O(1)$, and nodes are modified by insertions and deletions with a frequency that is exponentially small in their height. The latter results do not hold for standard B^+ trees. By adding periodic rebuilding of the tree, we obtain a data structure that is theoretically superior to standard B^+ trees in many ways. We conclude that rebalancing on deletion can be considered harmful.

1 Introduction

Deletion in balanced search trees [2–6, 8–10, 14, 15, 18] is a problematic operation. First, if items are stored in the internal nodes of the tree, deletion can require swapping the item to be deleted with its predecessor or successor: this moves the deletion position to the bottom of the tree, where the deletion can be done easily. Second, the rebalancing needed to keep the height of the tree (and the worst-case search time) logarithmic is more complicated than that needed for insertion. Indeed, the original paper on AVL trees [1] did not discuss deletion, and many textbooks neglect it. Third, if operations on the search tree occur in parallel, as in many database systems that use B or B^+ trees, the synchronization necessary to do rebalancing on deletion reduces the available parallelism [7]. Whereas rebalancing *must* be done on insertion into a B or B^+ tree to guarantee correctness (nodes cannot become overfull), it is optional on deletion, since a B or B^+ tree remains a valid search tree even if it has underfilled nodes.

The first problem with deletion can be overcome by storing the items only in the external nodes of the tree, storing keys in the internal nodes to support search. B^+ trees [6]

* Research at Princeton University partially supported by NSF grants CCF-0830676 and CCF-0832797 and US-Israel Binational Science Foundation grant 2006204. The information contained herein does not necessarily reflect the opinion or policy of the federal government and no official endorsement should be inferred.

Y. Dong, D.-Z. Du, and O. Ibarra (Eds.): ISAAC 2009, LNCS 5878, pp. 832–841, 2009.

are an example of such a data structure. This takes extra space, but the space penalty may be worth the benefits. The second and third problems can be addressed by avoiding rebalancing on deletion. But then the tree need no longer have a height logarithmic in the number of items. Nevertheless, this method has been used successfully in Berkeley DB [16, 17], which uses B^+ trees with underfilled nodes, and in other database systems.

Avoiding rebalancing on deletion has been justified empirically [7, 13, 16, 17] and by average-case analysis [11, 12], but to our knowledge no one has studied its worst-case efficiency, perhaps because of the assumption that the worst case, however unlikely, is terrible. Here we undertake such a study. Perhaps surprisingly, our results provide ample theoretical justification for avoiding rebalancing on deletion.

One may wonder how this is possible. It is easy to construct an example showing that the tree height can become arbitrarily large, even if there is only one item left in the tree [10]. Furthermore the idea of deletion without rebalancing has also been used in red-black trees, resulting in unforseen and unfortunate consequences in at least one application [19]. Nevertheless, it is still possible that the height could remain logarithmic in the number of insertions. We show that this is indeed the case. We also show that the amortized time per insertion or deletion is $O(1)$, and that nodes are affected by updates with a frequency exponentially small in their heights. These latter results do not hold for standard B^+ trees. Thus in some ways deletion with rebalancing is not only not helpful but actually harmful. Our results provide theoretical support for the design decision made in Berkeley DB and other database systems to avoid rebalancing on deletion. In a companion paper [19] we present similar results for balanced binary trees. These results require careful design of the deletion method; certain natural choices result in the tree height becoming linear in the number of insertions in the worst case.

The remainder of our paper consists of five sections. In Section 2 we define the B^- tree, a relaxed form of B^+ tree in which deletions are done without rebalancing. B^- trees are essentially those used in Berkeley DB. We describe how to do searches, insertions, and deletions in such trees. Insertions require node-splitting, which can be done either bottom-up or top-down; we describe both methods. Deletions require only deletion of empty nodes. In Section 3 we analyze B^- trees. We show that the height, and hence the search time, is $O(\log_b m)$, where m is the total number of insertions and b is the maximum node degree. This bound is independent of the number of deletions. We also show that an insertion or deletion takes $O(1)$ amortized time in addition to a search, and that nodes are modified by insertions and deletions with a frequency that is exponentially small in their height. (These results require $b > 3$ if node-splitting is top-down.)

In Section 4 we discuss how and when to rebuild the tree. Such rebuilding eliminates two drawbacks of B^- trees: it keeps the space used proportional to the number of items in the tree, and it keeps the height logarithmic in the number of items. If rebuilding is done appropriately often, the amortized rebuilding time is $O(\epsilon)$ per insertion for an arbitrarily small positive constant ϵ. In Section 5 we sketch how to do rebalancing with deletion while retaining our inverse exponential bounds on node updates. Section 6 contains some concluding remarks, including a comparison between the case of multiway trees and that of binary trees.

2 B$^-$ Trees

In our discussion of multiway search trees we denote by m, d, n, and h, respectively, the number of insertions, the number of deletions, the current number of items in the tree, and the tree height. We assume that the initial tree is empty. We measure the time of an operation by counting the number of nodes examined or modified. In B$^+$ trees, the structure of internal nodes differs from that of external nodes: internal nodes contain keys and pointers to children; external nodes contain items (and their keys) but no pointers. To allow for this difference, we define our trees using two parameters, which give upper bounds on the sizes of the internal and external nodes. A *B^- tree of type* $b > 2$, $c > 0$ consists of an ordered tree whose external nodes all have the same depth, each of whose internal nodes has at most b children, and each of whose external nodes contains at least one and at most c items. Generally we think of b and c as large but within a small constant factor of each other (though our results do not require this). Each item has a distinct key selected from a totally ordered universe. (If keys are not distinct we break ties by item identifier.) Increasing key order corresponds to left-to-right node order. That is, if external node y is to the right of external node x, all items in y have larger key than all items in x. In order to facilitate searching, each internal node x with j children contains $j - 1$ keys in increasing order, alternating with pointers to its children; if a pointer to node y immediately follows (precedes) key k, then all items in the subtree rooted at y have key greater than (not greater than) k. We allow $j = 1$; an internal node with one child contains no key.

To search for the item (if any) with a given key k in a B$^-$ tree, start at the root and repeat the following *search step* until reaching an external node: in the current node x, find the largest key in x less than k and replace x by the child indicated by the pointer immediately following k; if there is no such k, replace x by the leftmost child of x. Upon reaching an external node, check whether any of its items have the desired key. The time for a search is $h + 1$.

To insert a new item into a B$^-$ tree, if the tree is empty merely create a new external node containing the item. Otherwise, do a search for the key of the item. Upon reaching an external node, insert the new item into this node. If the node overflows (because it now has $c + 1$ items), split it into two external nodes, one containing the smallest $\lceil (c+1)/2 \rceil = \lfloor c/2 + 1 \rfloor$ items and the other containing the remaining (largest) $\lfloor (c+1)/2 \rfloor = \lceil c/2 \rceil$ items; in the parent, replace the pointer to the original external node by two pointers to the two nodes formed by the split, separated by a copy of the largest key in the new external node containing the smaller keys. If this causes the parent to overflow, because it now has $b+1$ children and b keys, split it, but in a slightly different way: find a median of its keys; put the keys smaller than the median and the child pointers preceding the median in a new node, and put the keys larger than the median and the child pointers following the median in another new node; in the parent, replace the pointer to the old node by two pointers to the new nodes, separated by the median. That is, the median is promoted to the parent, not copied. Walk back up the path toward the root, splitting nodes in this way, until some node does not overflow or the root splits. If the root splits, create a new root containing pointers to the two new nodes formed by the split, separated by the promoted or copied median. (The old root could be an internal or external node.)

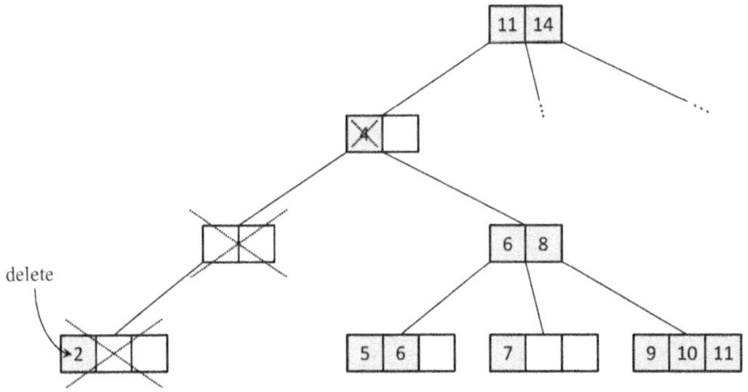

Fig. 1. Deletion in a B^- tree with $b = c = 3$. Deleting item 2 in the above tree causes the additional node and item deletions shown in dotted crosses.

An alternative way to do an insertion is to split full nodes top-down as the search proceeds, rather than splitting overfull nodes bottom-up after the search. This method only gives good bounds if $b > 3$. To do an insertion, do a search on the key of the new item, but if the current node of the search is full (it has b children), find a median of its keys, split the remaining keys and the child pointers into two nodes, containing the keys less than the median and the child pointers preceding the median, and the keys greater than the median and the child pointers following the median, respectively; in the parent, replace the pointer to the old child by pointers to the two new nodes, separated by the median. (The parent cannot be full; if it were, it would have split previously.) If the root splits, create a new root as in the bottom-up method. Continue the search from the appropriate one of the two nodes formed by the split. On reaching an external node, insert the new item into this node and split the node if it is overfull. (Again, the parent cannot be full.)

To delete an item in a B^- tree, find the external node containing it and delete the item from the node. If the node is now empty, delete it, as well as the pointer from its parent and one of the keys next to this pointer (either key will do if there are two). If before the deletion the parent contained no key (and only one child pointer), delete the parent as well, and walk back up the path toward the root deleting each node with no children until reaching a node that still has at least one child or until reaching the root. If the root has no children, delete it; the tree is now empty. (See Figure 1.)

The insertion methods we have described are the standard bottom-up and top-down insertion methods for B^+ trees. In the standard deletion method, a node that becomes underfull is refilled, by fusing it with a sibling (the inverse of splitting) and then possibly resplitting. Also, a root with only one child is deleted. What "underfull" means depends on whether node-splitting during insertion is bottom-up or top-down; in the former case, an internal node is underfull if it has less than $\lceil b/2 \rceil$ children; in the latter case, if it has less than $\lfloor b/2 \rfloor$ children. In either case, an external node is underfull if it has less than $\lceil c/2 \rceil$ items. A B^+ tree has height at most $\log_{\lceil b/2 \rceil}(n/c) + 1$ if splitting is bottom-up, at most $\log_{\lfloor b/2 \rfloor}(n/c) + 1$ if top-down. Avoiding refilling simplifies deletion considerably, but at the expense of having underfull nodes, which worsens space utilization and can increase the time for accesses and updates. Nevertheless, our analysis in the next section provides theoretical support for this method.

3 Analysis of B$^-$ Trees

To analyze B$^-$ trees we use the potential method of amortized analysis [20]. To each state of the data structure we assign a non-negative *potential*, zero for an empty structure. We define the *amortized cost* of an operation to be its actual cost plus the net increase in potential it causes. Then for any sequence of operations starting with an empty structure, the sum of the actual costs is at most the sum of the amortized costs.

We use exponential potential functions similar to those we have used to analyze balanced binary trees [9, 19]. Each node has a non-negative potential; the potential of a tree is the sum of the potentials of its nodes. We begin by analyzing bottom-up splitting; then we modify the analysis to handle top-down splitting. We define the potential of a node of height $h > 0$ with j children to be $\max\{0, j - \lfloor b/2+1 \rfloor\}\lceil b/2 \rceil^h$ and the potential of an external node containing j items to be $\max\{0, j - \lfloor c/2+1 \rfloor\}\lceil b/2 \rceil / \lceil c/2 \rceil$; to cover temporarily overfull nodes, we allow $j = b+1$ for an internal node, $j = c+1$ for an external node.

A deletion cannot increase the potential. Ignoring the effect of splits, an insertion increases the potential by at most $\lceil b/2 \rceil / \lceil c/2 \rceil$, by adding one item to an external node. If an overfull external node (containing $c+1$ items) splits, the potential of both new nodes is zero; the potential of the parent (if any) increases by at most $\lceil b/2 \rceil$. The potential of the old external node was $(c+1-\lfloor c/2+1 \rfloor)\lceil c/2 \rceil / \lceil b/2 \rceil = \lceil b/2 \rceil$, so the split does not increase the total potential. If an overfull internal node (having $b+1$ children) splits, the potential of both new nodes resulting from the split is zero, and the potential of the parent (if any) increases by at most $\lceil b/2 \rceil^{h+1}$. The potential of the old node that split was $(b+1-\lfloor b/2+1 \rfloor)\lceil b/2 \rceil^h = \lceil b/2 \rceil^{h+1}$, so the split does not increase the potential. If the node that splits is the root, the potential decreases by $\lceil b/2 \rceil^{h+1}$. Since the potential can never be negative and increases by a total of at most $m\lceil b/2 \rceil / \lceil c/2 \rceil$, we obtain the following theorem:

Theorem 1. *A B$^-$ tree built by m insertions with bottom-up splitting intermixed with arbitrary deletions has height at most* $\log_{\lceil b/2 \rceil}(m/\lceil c/2 \rceil) + 1$.

Thus as long as m, the total number of insertions, is polynomial in n, the current number of items in the tree, the B$^-$ tree built by an update sequence has height within a multiplicative constant of that of the B$^+$ tree built by the same update sequence; the constant depends on the degree of the polynomial. If m is linear in n, the height of the B$^-$ tree is within an additive constant of that of the B$^+$ tree.

By truncating the potential function, we can obtain an inverse-exponential bound on the number of splits, and the number of nodes, of a given height. For any fixed h let the potential function be as defined above for nodes of height at most h, zero for nodes of height greater than h. Then every split of a node at height h reduces the potential by $\lceil b/2 \rceil^{h+1}$, which gives the following theorem:

Theorem 2. *In a B$^-$ tree built by m insertions with bottom-up splitting intermixed with arbitrary deletions, the number of splits of nodes at height h is at most* $m/(\lceil b/2 \rceil^h \lceil c/2 \rceil)$.

Corollary 1. *In a B$^-$ tree built by m insertions with bottom-up splitting intermixed with arbitrary deletions, the number of deletions of roots of height $h > 0$ is at most*

$m/(\lceil b/2\rceil^{h-1}\lceil c/2\rceil)$. *The number of deletions of non-roots of height* h *is at most* $m/(\lceil b/2\rceil^{h}\lceil c/2\rceil)$.

Proof. The number of nodes deleted at height h is at most the number added. For a node to be added at height h, a split must occur at height $h-1$. Each deletion of a non-root at height h must correspond to a previous split at height h. Thus both parts of the corollary follow from Theorem 2. □

Corollary 2. *With bottom-up splitting, the amortized time for an insertion, excluding the search time, is* $O(1)$. *The amortized time for a deletion is zero. (Each deletion can be charged to the corresponding insertion.)*

Proof. The sum over all heights of the bounds given by Theorem 2 on node splits and by Corollary 1 on node deletions is a geometric series summing to $O(m)$. □

By using a related but different potential function that increases as a node becomes empty rather than as it becomes full, we can obtain a bound on node deletions as a function of d, the number of item deletions, rather than m. We define the potential of a root to be zero, that of a child of height $h > 0$ with j children to be $\max\{0, \lceil b/2\rceil - j\}\lceil b/2\rceil^{h}$, and that of an external child containing j items to be $\max\{0, \lceil c/2\rceil - j\}\lceil b/2\rceil/\lceil c/2\rceil$; to cover underfull nodes, we allow $j = 0$.

An insertion cannot increase the potential: splitting a node creates two nodes of potential zero and adds a child to the parent; if there is no parent, a new root is created, of potential zero. Ignoring node deletions, an item deletion increases the potential by at most $\lceil b/2\rceil/\lceil c/2\rceil$. Deletion of a node of height h decreases the potential by that of the deleted node, namely $\lceil b/2\rceil^{h+1}$, and increases the potential of the parent by at most the same amount, resulting in no net increase. A deletion of a root of height $h > 0$ must be preceded by a deletion of its only child, reducing the potential by $\lceil b/2\rceil^{h}$. If we truncate the potential function by letting the potential of nodes of height greater than h be zero for some fixed h, then a deletion of a node of height h other than the root reduces the potential by $\lceil b/2\rceil^{h+1}$. This gives the following theorem:

Theorem 3. *In a* B^- *tree built by insertions with bottom-up splitting intermixed with* d *arbitrary deletions, the number of deletions of roots of height* $h > 0$ *is at most* $d/(\lceil b/2\rceil^{h-1}\lceil c/2\rceil)$. *The number of deletions of non-roots of height* h *is at most* $d/(\lceil b/2\rceil^{h}\lceil c/2\rceil)$.

For top-down splitting we obtain the same results, but with $\lfloor b/2\rfloor$ in place of $\lceil b/2\rceil$, so the bounds are slightly worse. We assume $b > 3$. Let the potential of a node of height $h > 0$ with j children be $\max\{0, j - \lceil b/2\rceil\}\lfloor b/2\rfloor^{h}$ and that of an external node containing j items be $\max\{0, j - \lfloor b/2 + 1\rfloor\}\lfloor b/2\rfloor/\lceil c/2\rceil$. An argument like that preceding Theorem 1 gives the following:

Theorem 4. *A* B^- *tree built by* m *insertions with top-down splitting intermixed with arbitrary deletions has height at most* $\log_{\lfloor b/2\rfloor}(m/\lceil c/2\rceil) + 1$.

By defining the potential of a node of height exceeding h to be zero for any fixed h, we obtain the following analogues of Theorem 2, Corollary 1, and Corollary 2:

Theorem 5. *In a B^- tree built by m insertions with top-down splitting intermixed with arbitrary deletions, the number of splits of nodes at height h is at most $m/(\lfloor b/2 \rfloor^h \lceil c/2 \rceil)$.*

Corollary 3. *In a B^- tree built by m insertions with top-down splitting intermixed with arbitrary deletions, the number of deletions of roots of height $h > 0$ is at most $m/(\lfloor b/2 \rfloor^{h-1} \lceil c/2 \rceil)$. The number of deletions of non-roots of height h is at most $m/(\lfloor b/2 \rfloor^h \lceil c/2 \rceil)$.*

Corollary 4. *With top-down splitting, the amortized time for an insertion, excluding the search time, is $O(1)$.*

To obtain a bound on node deletions as a function of d, we let the potential of a root be zero, the potential of a child of height $h > 0$ with j children be $\max\{0, \lfloor b/2 \rfloor - j\} \lfloor b/2 \rfloor^h$, and that of an external child containing j items be $\max\{0, \lceil c/2 \rceil - j\} \lfloor b/2 \rfloor / \lceil c/2 \rceil$. An argument like that preceding Theorem 3 gives the following:

Theorem 6. *In a B^- tree built by insertions with top-down splitting intermixed with d arbitrary deletions, the number of deletions of roots of height $h > 0$ is at most $d/(\lfloor b/2 \rfloor^{h-1} \lceil c/2 \rceil)$. The number of deletions of non-roots of height h is at most $d/(\lfloor b/2 \rfloor^h \lceil c/2 \rceil)$.*

4 Rebuilding the Tree

B^- trees have two possible disadvantages: their space usage may be $\omega(n)$, where n is the current number of items, and the search time may exceed $O(\log_b n)$. This can only occur when d/m approaches one. For applications in which insertions significantly outnumber deletions, this is not a concern, but for other applications it may be. We can solve both problems by periodically rebuilding the tree. How to rebuild the tree, and how often, are interesting questions that deserve careful study and experimentation. We describe a simple rebuilding method that takes $O(\epsilon)$ amortized time per insertion for an arbitrarily small positive constant ϵ, analogous to the rebuilding method for binary trees presented in [19].

To rebuild the tree, we initialize a new tree to empty. We traverse the old tree in symmetric order, deleting each item and inserting it into the new tree. To facilitate the rebuilding process, we maintain the right spine (the path from the root to the rightmost external node) of the new tree as a list. The list provides the location of the next insertion: the new item is inserted into the last node on the list. Thus there is no need to search for the insertion location. The list also contains consecutively all the nodes at which splits occur. When a node is split, the new node containing the larger keys replaces the original node on the list. A split at the root additionally creates a new root, which is added to the end of the list. Each insertion into the new tree takes $O(1)$ amortized time, and the entire rebuilding process takes $O(n)$ time.

To decide when to rebuild the tree, we keep track of n and rebuild the tree when the space usage or the tree height (or both) becomes too high. If we rebuild the tree when the space usage becomes greater than an for a suitably large but fixed constant a, then the space usage of the tree is always $O(n)$, and the rebuilding time is $O(1/a)$ per insertion. Similarly, if we rebuild the tree when its height exceeds $\log_{\lceil b/2 \rceil}(n/c) + g$ for a suitably large but fixed constant g, then the height of the tree is always $O(\log_b n)$ and the rebuilding time is $O(1/\lceil b/2 \rceil^g)$ per insertion. The larger the values of a and g, the smaller the overhead for rebuilding, but the worse the space usage and the larger the tree height can become in terms of n.

Rebuilding can also be done incrementally, by maintaining both the old tree and the new tree and doing $O(\epsilon)$ rebuilding work per insertion. For example, we can start the rebuilding process when the tree exceeds the maximum space usage or height, and then move two items from the old tree to the new tree for each subsequent insertion or deletion until the old tree becomes empty. During rebuilding, insertions occur in the new tree if the key of the item is less than the largest key of an item in the old tree moved so far, and in the old tree otherwise. We maintain the left spine of the old tree and the right spine of the new tree as lists, so the next item to be deleted from the old tree and its new location in the new tree can be found in $O(1)$ time. We need to update these lists when insertions and deletions occur, but this takes $O(1)$ amortized time per insertion or deletion. If n is the number of items in the old tree at the start of the rebuilding process, the number of items in the new tree at the end of the rebuilding process is between $n/2$ and $2n$.

Whether the tree is rebuilt incrementally or all at once, the space usage of the tree is always $O(n)$ and the tree height is always $O(\log_b n)$. The inverse-exponential bounds on node updates in Section 3 also continue to hold.

5 Deletion with Weak Rebalancing

As an alternative to rebuilding, one can obtain our inverse-exponential bounds on node updates (with a worse base), while guaranteeing linear space usage and $O(\log_b n)$ search time, by refilling nodes that become too empty. This makes deletion as complicated as in standard B^+ trees, but achieves $O(1)$ rebalancing time per insertion or deletion (not including search time).

To obtain the bounds we seek, we need to allow nodes to become emptier than in standard B^+ trees. Rebalancing during a deletion can be done either bottom-up or top-down. In either case, when an external node contains too few items or an internal node has too few children, we fuse it with a sibling, by combining the contents of both nodes and, if the node is internal, adding the key in the parent that is between the pointers to the two nodes being fused. Whether the node is internal or not, we replace the key in the parent and its adjacent pointers by a pointer to the new node. Then, if the new node is too full, we resplit it evenly, as in an insertion. To rebalance bottom-up, we start at the external node that has just lost an item, fuse it with a sibling if necessary, and walk up toward the root fusing each node that is too empty with a sibling, until reaching a node that is full enough, or resplitting a fused node, or reaching the root. If the root is internal and loses all but one child, we delete it and make its only child the new root. Top-down rebalancing works in the same way, except that we pre-emptively do the fusing top-down during the search; if the root is left with only one child as a result of a fusion of its children, we delete it.

One can easily parameterize the results based on the sizes at which fusions and resplits occur, but here we consider one illustrative special case. Suppose fusion is bottom-up, we fuse an internal node when it has less than $\lfloor b/8 \rfloor$ children, an external node when it has less than $\lfloor c/8 \rfloor$ items, and we resplit an internal node if it has more than $\lceil b/2 \rceil$ children, an external node when it has more than $\lceil c/2 \rceil$ items. We need b and c sufficiently large, say at least 16. Theorems 1, 2, 4, and 5, and Corollaries 2 and 4 still hold, since nodes created by fusion with or without resplitting have zero potential. The tree height is at most $\log_{\lfloor b/8 \rfloor}(n/\lfloor c/8 \rfloor) + 1$. Furthermore we have the following result:

Theorem 7. *Whether splitting during insertion is bottom-up or top-down, the number of node fusings (with or without resplittings) of nodes at height h is at most $d/(\lfloor b/8 \rfloor^h \lfloor c/8 \rfloor)$.*

Proof. Let the potential of a root be zero, that of an internal node of height h with j children be $\max\{0, 2\lfloor b/8 \rfloor - j\}\lfloor b/8 \rfloor^h$, and that of an external node with j items be $\max\{0, 2\lfloor c/8 \rfloor - j\}\lfloor b/8 \rfloor/\lfloor c/8 \rfloor$. An insertion does not increase the potential. Excluding node fusings and resplittings, a deletion increases the potential by at most $\lfloor b/8 \rfloor/\lfloor c/8 \rfloor$. A fusing of two nodes at height h without a resplit does not increase the potential, since the fused node has potential at least $\lfloor b/8 \rfloor^{h+1}$ less than the sum of the potentials of the two nodes that fuse, and the potential of the parent increases by at most this amount. A fusing of two nodes at height h with a resplit does not change the potential of the parent and decreases the total potential of the affected nodes at height h by at least $\lfloor b/8 \rfloor^{h+1}$. If we change the potential of nodes exceeding a fixed height h to zero, then any fusing at height h decreases the potential by at least $\lfloor b/8 \rfloor^{h+1}$. The theorem follows. □

6 Remarks

We have shown that in B^+ trees deletion without rebalancing preserves a logarithmic height bound, and that with this method node updates during insertions and deletions occur exponentially infrequently in the height of the node. If one rebuilds the tree periodically, the space usage remains linear and the height remains logarithmic in the number of items in the tree; the rebuilding time can be made $O(\epsilon)$ per insertion for an arbitrarily small positive constant ϵ. We have obtained similar inverse exponential bounds if rebalancing is done on deletion but nodes are allowed to be less full than in standard B^+ trees. Such "weak" rebalancing takes $O(1)$ amortized time per insertion or deletion, whereas in standard B^+ trees the amortized time per insertion or deletion can be logarithmic in n. Our bounds for weak rebalancing have worse constants than those for rebalancing without deletion, however.

We have obtained similar results for balanced binary trees [9, 19]. In the case of binary trees it is not so simple to do deletion without rebalancing and still obtain good bounds: the deletion method we have analyzed here can produce long chains of unary nodes, but binary trees do not have unary nodes. We could eliminate unary nodes in our solution for multiway trees by making the parent of such a node point to its only child, but this method fails unless one keeps track, for each pointer replacing a path of unary nodes, of the length of the path it replaces. This is what we did to solve the problem for binary trees, and this seems to be required. See [19].

References

1. Adel'son-Vel'skii, G.M., Landis, E.M.: An algorithm for the organization of information. Sov. Math. Dokl. 3, 1259–1262 (1962)
2. Andersson, A.: Balanced search trees made simple. In: Dehne, F., Sack, J.-R., Santoro, N. (eds.) WADS 1993. LNCS, vol. 709, pp. 60–71. Springer, Heidelberg (1993)
3. Bayer, R.: Binary B-trees for virtual memory. In: SIGFIDET, pp. 219–235 (1971)
4. Bayer, R.: Symmetric binary B-trees: Data structure and maintenance algorithms. Acta Inf. 1, 290–306 (1972)
5. Bayer, R., McCreight, E.M.: Organization and maintenance of large ordered indexes. Acta Informatica 1(3), 173–189 (1972)
6. Comer, D.: The ubiquitous B-tree. ACM Computing Surveys 11(2), 121–137 (1979)
7. Gray, J., Reuter, A. (eds.): Transaction Processing: Concepts and Techniques. Morgan Kaufmann, San Mateo (1993)
8. Guibas, L.J., Sedgewick, R.: A dichromatic framework for balanced trees. In: FOCS, pp. 8–21 (1978)
9. Haeupler, B., Sen, S., Tarjan, R.E.: Rank-balanced trees. In: WADS, pp. 351–362 (2009)
10. Jannink, J.: Implementing deletion in B^+-trees. SIGMOD Record 24(1), 33–38 (1995)
11. Johnson, T., Shasha, D.: Utilization of B-trees with inserts, deletes and modifies. In: PODS, pp. 235–246 (1989)
12. Johnson, T., Shasha, D.: B-trees with inserts and deletes: Why free-at-empty is better than merge-at-half. Journal of Computer and System Sciences 47(1), 45–76 (1993)
13. Mohan, C., Levine, F.: ARIES/IM: an efficient and high concurrency index management method using write-ahead logging. SIGMOD Record 21(2), 371–380 (1992)
14. Nievergelt, J., Reingold, E.M.: Binary search trees of bounded balance. SIAM J. on Comput. 2(1), 33–43 (1973)
15. Olivié, H.J.: A new class of balanced search trees: Half balanced binary search trees. ITA 16(1), 51–71 (1982)
16. Olson, M.A., Bostic, K., Seltzer, M.I.: Berkeley DB. In: USENIX Annual, FREENIX Track, pp. 183–191 (1999)
17. Olson, M.A., Bostic, K., Seltzer, M.I.: Berkeley DB (2000)
18. Sedgewick, R.: Left-leaning red-black trees, `www.cs.princeton.edu/~rs/talks/LLRB/LLRB.pdf`
19. Sen, S., Tarjan, R.E.: Deletion without rebalancing in balanced binary trees. In: SODA (to appear, 2010)
20. Tarjan, R.E.: Amortized computational complexity. SIAM J. Algebraic and Disc. Methods 6, 306–318 (1985)

Counting in the Presence of Memory Faults

Gerth Stølting Brodal[1], Allan Grønlund Jørgensen[1], Gabriel Moruz[2,*], and Thomas Mølhave[3,**]

[1] MADALGO***, Department of Computer Science, Aarhus University
[2] MADALGO***, Institut für Informatik, Goethe University Frankfurt am Main
[3] Department of Computer Science, Duke University

Abstract. The faulty memory RAM presented by Finocchi and Italiano [1] is a variant of the RAM model where the content of any memory cell can get corrupted at any time, and corrupted cells cannot be distinguished from uncorrupted cells. An upper bound, δ, on the number of corruptions and $O(1)$ reliable memory cells are provided. In this paper we investigate the fundamental problem of counting in faulty memory. Keeping many reliable counters in the faulty memory is easily done by replicating the value of each counter $\Theta(\delta)$ times and paying $\Theta(\delta)$ time every time a counter is queried or incremented. In this paper we decrease the expensive increment cost to $o(\delta)$ and present upper and lower bound tradeoffs decreasing the increment time at the cost of the accuracy of the counters.

1 Introduction

Modern memory chips are made from increasingly smaller and complicated circuits that work at low voltage levels and offer large storage capacities [2]. Unfortunately, these improvements have increased the likelihood of *soft memory errors*, where arbitrary bits flip, corrupting the contents of the affected memory cells [3]. Soft memory errors are triggered by phenomena such as power failures, cosmic rays, and manufacturing defects. Even though the occurrence rate of these errors in individual memories is quite low they are a serious concern in applications running on clusters, where the frequency of soft memory errors is much larger. The soft memory errors rate is predicted to increase in the future [4]. Since the amount of cosmic rays increases with altitude, soft memory errors are a serious concern in fields like avionics and space research [5].

Corrupted memory cells can have significant consequences for algorithms. For instance, a single corruption in a sorted array can force a standard binary search to end up $\Omega(n)$ cells away from the correct position. Soft memory errors can also be exploited to break the security of software systems. This has been demonstrated in works breaking Java Virtual Machines [6], cryptographic protocols [7,8], and smart-cards [9].

Soft memory errors can be addressed by using replication and error correcting codes at the hardware level, but this approach is not always popular since the increased circuitry requirements is costly with respect to performance, storage capacity, and money.

* Partially supported by the DFG grant ME 3250/1-1.

** Work done while at MADALGO.

*** Center for Massive Data Algorithmics, a Center of the Danish National Research Foundation.

Y. Dong, D.-Z. Du, and O. Ibarra (Eds.): ISAAC 2009, LNCS 5878, pp. 842–851, 2009.

In software, memory errors have been addressed in a variety of settings, with the main focus on ensuring that code runs as expected, anticipating critical errors caused by hardware errors and malicious attacks. Errors are detected using techniques such as algorithm based fault tolerance [10], assertions [11], control flow checking [12], procedure duplication [13], and automatically correcting heap-based memory errors [14].

Most algorithms and data structures assume a perfectly reliable storage, but algorithms dealing with unreliable data were also proposed. These include fault-tolerant pointer-based data structures [15], fault-tolerant sorting networks [16], fault-tolerant parallel models [17], the liar model [18], and locally mendable distributed networks [19].

Faulty Memory RAM. Recently, the *faulty-memory RAM* model was proposed in [1]. This model is a regular RAM with word size w where any memory cell can get corrupted at any time during the execution of an algorithm, and a cell containing corrupted data cannot be distinguished from a cell that does not. Algorithms are provided with an upper bound, δ, on the number of corruptions that may occur during execution. We let $\alpha \leq \delta$ denote the actual number of corruptions that have taken place during the computation. Given that registers in the processor are considered incorruptible, $O(1)$ safe memory locations are provided. It is assumed that reading a word from memory is an atomic operation. An algorithm is *resilient* if it works correctly for all uncorrupted data. For instance, a resilient sorting algorithm outputs a sequence where all uncorrupted elements appear in sorted order and corrupted elements can appear anywhere in the output. The correctness of algorithms is usually proved by assuming that an adaptive adversary (worst-case) performs up to δ corruptions during the execution of an algorithm.

Several problems have been addressed in the faulty-memory RAM, see a recent survey [20] for more information. For instance optimal comparison based sorting algorithms and (static and dynamic) dictionaries [1,21,22,23], and priority queues [24] have been proposed. In [25] it is shown that resilient sorting algorithms are of practical interest. Motivated by the increased soft memory errors frequency on clusters operating with massive data sets, in [26] resilient algorithms are linked to external-memory algorithms, providing the first external-memory algorithms resilient to memory faults.

Our results. We investigate maintaining many counters in the faulty memory RAM:

Definition 1. *A resilient counter with* additive error γ *is a data structure with an* increment *operation and a* query *operation. The query operation returns an integer between* $v - \gamma$ *and* $v + \gamma$ *where* v *is the number of increment operations preceding the query.*

We investigate upper and lower bound tradeoffs between the time needed for n increase operations and the additive error of the counter. We only consider data structures where no information is stored in safe memory between operations, therefore the counters are stored completely in unreliable memory. Our results are summarized in Figure 1.

In Section 2 we prove that any resilient counter with non-trivial additive error must use $\Omega(\delta)$ space, and that a deterministic query operation requires $\Omega(\delta)$ time. Furthermore, we prove a lower bound tradeoff between the increment time and the additive error, stating that if an increment operation takes $t \leq \delta$ time, the additive error is at least $\lfloor \delta/t \rfloor$ in the worst case, i.e. (increment time) $\times$ (additive error) $\geq \delta$. The lower

Time (n increments)	Query time	Additive error γ	Space	Section
$O(\delta n)$	$O(\delta)$	0	$O(\delta)$	-
$O(nt\log(\delta/t) + \alpha\log(\alpha/t))$	$O(\delta)$	α/t	$O(\delta)$	3.1
$O(n + \alpha\log\alpha)$	$O(\delta)$	$\alpha\log\delta$	$O(\delta)$	3.1
$O(n)$	$O(\delta^2)$	$O(\alpha^2)$	$O(\delta)$	3.2
$O(n + \alpha\sqrt{\delta})$	$O(\delta)$	α	$O(\delta)$	3.3
Expected $O(n)$	$O(\delta)$	α	$O(\delta)$	3.4

Fig. 1. Overview of our upper bounds

bounds suggest that an optimal resilient counting data structure is characterized by an $O(\delta)$ space bound, $O(t)$ increment time, $O(\alpha/t)$ additive error and $O(\delta)$ query time.

In Section 3.1 and 3.3 we provide deterministic data structures where both the increment time and the additive error depend on α. The first result in Section 3.1 provides a tradeoff between the increment time and the additive error that does not blow up the space used by the data structure nor the query time. Given any $t \geq 1$ the data structure has additive error α/t and supports n increments in $O(nt\log(\delta/t) + \alpha\log(\alpha/t))$ time. A small change to this data structure gives a data structure with additive error $\alpha\log\delta$ that supports n increments in $O(n + \alpha\log\alpha)$ time. In Section 3.3 we describe a data structure with additive error α that supports n increments in $O(n + \alpha\sqrt{\delta})$ time. This is optimal for $n = \Omega(\alpha\sqrt{\delta})$.

In Section 3.2 we describe a deterministic data structure where the time used by an increment is independent of the number of possible corruptions. The data structure supports increments in $O(1)$ time in the worst case. The additive error of the data structure is $O(\alpha^2)$ and queries are supported in $O(\delta^2)$ time.

Finally, in Section 3.4 we present a randomized data structure with additive error α, that supports n increments in $O(n)$ time in expectation and supports queries in $O(\delta)$ time in the worst case. This is optimal up to constant factors.

The additive error of any of our resilient counters can be reduced by a factor of t by using t counters. Each increment operation increments all t counters and the query operation returns the sum of all t counters divided by t. However, this produces a new tradeoff by increasing the increment time and space by a factor of t. Similarly, any of our resilient counters can be used to create a new counter that supports both decrement and increment operations with the same additive error. This is achieved by using two counters; one to count the number of increment operations and one to count the number of decrement operations.

Preliminaries. Throughout the paper we denote by *reliable value* a value stored in unreliable memory that can be retrieved reliably despite possible corruptions. This is achieved by replicating the given value in $2\delta + 1$ consecutive cells. Since at most δ of the copies can be corrupted, the majority of the $2\delta + 1$ elements are uncorrupted. The value can be retrieved in $O(\delta)$ time with the majority algorithm in [27], which scans the $2\delta + 1$ values keeping a single majority candidate and a counter in safe memory.

2 Lower Bounds and Tradeoffs

We present some simple lower bounds on space and time for resilient counters.

Space. Any resilient counter data structure with non-trivial additive error must use more than δ space. If the data structure uses δ space or less, the adversary can corrupt the entire structure and force a query operation to return any arbitrary value.
Deterministic Query. Any deterministic algorithm uses at least δ probes in the worst case for a query. If a query algorithm reads at most δ memory cells the adversary can simulate any value by corrupting δ cells. This means that the adversary can completely control the value returned by a query, making it impossible to get a non-trivial bound on the additive error.
Deterministic Increment. If an increment takes k time the adversary can roll back the changes to the data structure done by the last $\lfloor \delta/k \rfloor$ increments, or do the changes to the data structure corresponding to $\lfloor \delta/k \rfloor$ increments. Thus, the counter has additive error at least $\lfloor \delta/k \rfloor$ in the worst case.

3 Data Structures

3.1 Replicating Bits

In this section we describe a data structure that is parameterized with an integer t, $1 \leq t \leq \delta$. The data structure uses $O(\delta)$ space and has additive error $\lfloor \alpha/t \rfloor$. The time used for n increments is $O(nt\log(\delta/t) + \alpha\log(\alpha/t))$, and queries take $O(\delta)$ time.
Structure. The data structure maintains the bits of the binary representation of the counter value separately, each bit replicated depending on its significance as follows. For $i = 0, \ldots, \lfloor \log(\delta/t) \rfloor$ the i'th least significant bit is replicated $t2^{i+1}$ times in $t2^{i+1}$ different memory cells. The value of the remaining $w - \lfloor \log(\delta/t) \rfloor$ most significant bits are stored in a reliable variable v. The memory cells are stored in one array of size $O(\delta)$.
Increment. Increments are implemented as binary addition, where we consider the i'th bit to be one if at least $t2^i$ of the $t2^{i+1}$ copies of it are non-zero. The i'th bit is set by writing the value of the bit in all of the $t2^{i+1}$ copies.
Query. The query algorithm reliably retrieves the value of the $w - \lfloor \log(\delta/t) \rfloor$ bits stored in v. For the lower order bits, we add 2^i to the sum, for $i = 0, \ldots, \lfloor \log(\delta/t) \rfloor$, if at least $t2^i$ of the $t2^{i+1}$ copies of the i'th least significant bit are non-zero.
Additive Error. Since the value of the i'th bit is given by the majority value of $t2^{i+1}$ copies, the adversary must use $t2^i$ corruptions to alter the i'th bit. Changing the i'th bit changes the value stored in the data structure by 2^i, yielding an additive error of $\lfloor \alpha/t \rfloor$.
Complexity. If no corruptions occur, we update the i'bit of the counter every 2^i increments, taking $O(t2^i)$ time. Similarly, we update v after $\Theta(\delta/t)$ increments in $O(\delta)$ time. Therefore, if we ignore corruptions, the time used for n increments is $O(nt\log(\delta/t))$.

The only way corruptions can influence the running time of increment operations is by changing the value of a bit. Assume the adversary corrupts the i'th bit, using $t2^i$ corruptions. After a number of increments a cascading carry affects this (corrupted) bit and the increment operation writes the $t2^{i+1}$ copies of the $i+1$'th bit. We charge the work needed to move the $t2^i$ corrupted bits to the corruptions that caused them. These corrupted bits can be charged in $\log(\delta/t) - i$ such cascading carries. However, when k increments have been performed, where $kt > \alpha$, the time used by the increments alone is $O(kt\log\delta/t)$ dwarfing the time needed to deal with corruptions. Otherwise, the number stored in the data structure is at most $k + \alpha/t \leq 2\alpha/t$. Thus, the most significant bit written in an increment operation is the $\lceil \log(\alpha/t) \rceil$ least significant bit. We conclude that the extra time needed to deal with corruptions is $O(\alpha \log(\alpha/t))$.

Theorem 1. *The counter structure uses $O(\delta)$ space and has additive error $\lfloor \alpha/t \rfloor$. The time used for n increments is $O(nt \log(\delta/t) + \alpha \log(\alpha/t))$ and queries take $O(\delta)$ time.*

Trading off Additive Error for Increment Time. We can reduce the time for n increments to $O(n + \alpha \log \alpha)$ by storing the $\lfloor \log \log \delta \rfloor$ least significant bits in the same memory cell. For $i = \lfloor \log \log \delta \rfloor + 1, \ldots, \log \delta$ the i'th least significant bit is replicated in $2^{i+1}/\lfloor \log \delta \rfloor$ memory cells. The remaining bits are stored in a reliable value v as before. One corruption can change the $\lfloor \log \log \delta \rfloor$ least significant bits causing an additive error of at most $\lfloor \log \delta \rfloor$, and $2^i/\lfloor \log \delta \rfloor$ corruptions are needed to corrupt the i'th bit. The increment and the query are basically the same.

Corollary 1. *The counter structure uses $O(\delta)$ space and has additive error $\alpha \log \delta$. The time used for n increments is $O(n + \alpha \log \alpha)$ and queries use $O(\delta)$ time.*

3.2 Round-Robin Counting

In this section we describe a data structure that uses $O(\delta)$ space and has $O(\alpha^2)$ additive error. Increments are supported in constant time, and queries use $O(\delta^2)$ time.
Structure. The data structure consists of an array A of $k = 2\delta + 3$ integers $C_1, \ldots, C_k$ used as counters, and a round-robin index i. The structure is initialized by setting all counters to zero and i to one. We denote by *corrupted counter* a counter that has been changed directly by the adversary.
Increment. If i is not in the range $1, \ldots, k$, it has been corrupted and we reset it to one. Next, we increment first C_i and then i. If i becomes $k + 1$ we set it to one. Note that i could have been corrupted to a value in $1, \ldots, k$, but we do not check if this happened.

Let v_j be the number of times the increment algorithm has incremented C_j, and let $v = \sum_{j=1}^{k} v_j$ denote the correct value of the counter. If no corruption has taken place, then $C_1 = \cdots = C_r = d + 1$ and $C_{r+1} = \cdots = C_k = d$, where $d = \lfloor v/k \rfloor$ and $r = v \bmod k$. Furthermore, if no counter has been corrupted, $v = \sum_{j=1}^{k} C_j$, regardless of corruptions of the round robin index i.
Query. Let α_i be the number of times i has been corrupted. The key observation for the query algorithm is that for any two uncorrupted counters, C_a and C_b, we have $|v_a - v_b| \leq \alpha_i + 1$, which means that $|v/k - v_a| \leq \alpha_i + 1$.

First, we compute a value m larger than or equal to at least one uncorrupted counter, and smaller than or equal to at least one uncorrupted counter. Since the difference between two uncorrupted counters is at most $\alpha_i + 1$, $m \in \{\frac{v}{k} - \alpha_i - 1, \frac{v}{k} + \alpha_i + 1\}$. After computing m, simply returning mk yields an additive error of $O((\alpha + 1)k) = O((\alpha + 1)\delta)$. To improve the additive error we locate $O(\alpha)$ counters which are too far from m and ignore them.

We store m in safe memory and compute it as in [21] as follows. Initially, we set m to $-\infty$. The k counters are scanned $\lceil k/2 \rceil$ times. In each iteration we update m to the minimum counter larger than the current m. Since $k = 2\delta + 3$, after $\lceil k/2 \rceil$ iterations there exist two uncorrupted counters, such that one is smaller and one is larger than m.

Next, we find a bound, x, on the number of the counters that are too far away from m as follows. Initially, we set x to one. Then, the number of counters c outside the range $\{m-x,\ldots,m+x\}$ is counted in a scan. If $c \geq x$ we increment x and recompute c. This process ends when x becomes larger than c. Finally, we scan the k counters maintaining a sum, initially zero, in safe memory. If a counter stores a value in the range $\{m-x, m+x\}$ we add it to the sum. If a counter is outside the range, it is far from m, and we add m to the sum. Finally, we return the computed sum.

Additive Error. Let α_c be the number of times a counter was corrupted by the adversary. By definition, $\alpha_i + \alpha_c = \alpha \leq \delta$. First we recall that for any two uncorrupted counters, C_a and C_b, we have $|v_b - v_a| \leq 1 + \alpha_i$, and that the value of m is in the range $\{\frac{v}{k} - \alpha_i - 1, \frac{v}{k} + \alpha_i + 1\}$. Therefore, if $x \geq \alpha_i + 1$ in the above algorithm, then c, the number of counters that are not in the range $\{m - x, m + x\}$, is at most α_c, the number of counter corruptions. At most α_c corrupted counters can be counted by c, and we conclude that when the algorithm terminates, then $x \leq \alpha_i + \alpha_c + 1$.

Let S be the set of counters not counted by c, i.e. all counters in the range $\{m - x, m + x\}$. All uncorrupted counters in S are unchanged and do not contribute to the error. Let C_j be a corrupted counter in S. By definition of m and x we know that $|v_j - C_j| \leq |v_j - m| + |m - C_j| \leq \alpha_i + 1 + x \leq 2\alpha + 1$. Therefore, each corrupted counter in S can affect the additive error by $O(\alpha)$. We add m to the result for all counters outside the range $\{m - x, m + x\}$. By definition of m, the value for uncorrupted counters not in S differs from m by at most $\alpha_i + 1$. Similarly, for any corrupted counter C_j not in S the difference between m and v_j is at most $\alpha_i + 1$. There are at most $x = O(\alpha)$ counters not in S, and at most α_c corrupted counters in S, leading to an additive error of $O(\alpha^2)$.

Complexity. The increment operation uses $O(1)$ time to update a counter and the round robin index. The query time is given by the time used to compute m and x, that is $O(\delta^2)$.

Theorem 2. *The counter data structure described uses $O(\delta)$ space and has an additive error of $O(\alpha^2)$. Increments are supported in $O(1)$ time and queries in $O(\delta^2)$ time.*

3.3 Counting by Scanning Bits

We describe a counter data structure that uses $O(\delta)$ space with additive error α. It performs n increments in $O(n + \alpha\sqrt{\delta})$ time, and answers queries in $O(\delta)$ time. First, we describe a simpler data structure with an additive error of α that supports n increments in $O(n + \alpha\delta)$ time. Subsequently, we reduce the cost for n increments to $O(n + \alpha\sqrt{\delta})$.

Structure. The data structure stores an array A of δ memory cells, a reliable variable v, and a round-robin index i. Each cell of A is used to store a single bit. We initialize all values in A to zero, v to zero, and i to one.

Increment. If $A[i] = 0$ we set $A[i] = 1$ and set $i = 1 + (i + 1 \mod \delta)$. Otherwise, we count the number of non-zero entries in A. We add this number plus one (for the current increment) to v and set all entries in A to zero.

Query. We count the number v' of non-zero entries in A, retrieve v, and return $v + v'$.

Additive Error. Every time we add a value, k, to the reliable value v in an increment we have seen $k - 1$ non-zero entries in A. The only way a cell in A can be non-zero is if it was set to one by an earlier increment operation, or the adversary corrupted it. Conversely, a cell is set to zero either after updating the reliable value or by a corruption.

Thus, the number returned by a query differs by at most α from the actual number of increments performed.
Complexity. If no corruptions occur, the increment operation takes $O(1)$ amortized time, since setting a value in A to one takes $O(1)$ time and updating v takes $O(\delta)$ time and occurs every $\delta+1$ increments. Every corruption to the round robin index i or an element of A can force us to scan A and reliably add a value to v, and this takes $O(|A| + \delta) = O(\delta)$ time. Therefore, n increments take $O(n + \alpha\delta)$ time.

Improving Increment Time by Packing. We improve the time used for n increments to $O(n + \alpha\sqrt{\delta})$ by packing elements in A to an auxiliary array. In addition to the reliable value v and the array A of size δ, we store an array P of size δ, which is logically divided into $\Theta(\sqrt{\delta})$ blocks of $\sqrt{\delta}$ consecutive memory cells.
Increment. First, we test if i is in the range $\{1, \ldots, \delta\}$. If not then i has been corrupted and we set it to one. Then, we test whether $A[i] = 0$ and if so, we set $A[i] = 1$ and increment i. If i becomes $\delta + 1$ we set i to one. However, unlike the simpler data structure, if $A[i] \neq 0$, a *packing phase* is initiated. In the packing phase we scan A from left to right starting from $A[1]$ until we encounter a zero, or the end of A is reached. During the scan we count the amount, c, of non-zero entries read and set all these entries to zero. After the scan i is set to one. Then, we set c entries in P to one as follows. Let d_j be the index in P of the first element in the j'th logical block. We scan P from d_1. If we see an entry storing a zero, we set it to one, and decrement c. If we see something else we go to the start of the following logical block and continue. We stop the packing phase when c reaches zero or a non-zero element, or the boundary of the last block is found. If $c > 0$ after the packing phase, we count the amount of non-zero elements in A and P in a scan and set all entries to zero. This count summed with c is added to v.
Query. The query operation returns the sum of v and the number of ones in A and P.
Additive Error. Similarly to the simpler data structure, each corruption can only change the value of the data structure by one. It follows that the additive error is α.
Complexity. We analyze the time used between two consecutive updates of v and this time-frame we denote a round. The array A consists of a number of sections of non-zero elements separated by zeros. Note that the packing phase removes at least one section. If no corruptions occur, increments can only extend sections. A corruption, of a cell in A or of the index i, may extend a section, connect two sections, create a section or split an existing section in two. The same things can happen in an increment following a corruption of the index i. Thus, the number of sections created during a round is bounded by one plus the number of corruptions, and a section is moved only once in P.

Moving t non-zero entries from A to P in a packing phase takes $O(t + \sqrt{\delta})$ time, and the clean ending the round takes $O(\delta)$ time. Let c_p be the number of increments and α_p be the number of corruptions in the p'th round. Since the packing phase is called at most $\alpha_p + 1$ times, the time used in the p'th round is $O(c_p + \alpha_p\sqrt{\delta} + \delta)$. We show that the $O(\delta)$ time used for the clean can be payed for by the c_p increments and the α_p corruptions, by charging $O(1)$ per increment and $O(\sqrt{\delta})$ per corruption.

If we copy elements to the i'th logical block in P in a packing phase and encounter a non-zero entry before filling all the $\sqrt{\delta}$ cells, at least one cell in the block is corrupted. Furthermore, we never put elements in the i'th block again unless a new corruption occur, setting a zero in the first entry of the block. This means that the only block that is changed by a packing phase that is not completely filled or has a cell that has been corrupted since the last time it was updated, is the last block considered in the phase.

When an increment performs a clean, ending the round, the first block of all logical blocks contained a non-zero entry during the packing phase. We categorize the $\sqrt{\delta}$ logical blocks as *filled* blocks, *corrupted* blocks, and *last* blocks. A filled block is a logical block which a packing phase has filled with $\sqrt{\delta}$ non-zero entries, a corrupted block contains a cell that has been corrupted during the round and which is not filled, and a last block is a block that does not contain a corrupted cell, but was not completely filled during the packing phase that put a one in the first entry of the block.

There are at most $\alpha_p + 1$ packing phases in a round, thus at most $\alpha_p + 1$ last blocks, and at most α_p corrupted blocks. If there are f filled blocks then we have performed at least $f\sqrt{\delta} - \alpha_p$ increments in the round. This means that there are $\sqrt{\delta} - f$ other blocks (corrupted, last) and since there are $O(\alpha_p)$ blocks that are not filled, $\sqrt{\delta} - f = O(\alpha_p)$. We have charged each increment $\Theta(1)$, which means that the increments have payed at least $f\sqrt{\delta} - \alpha_p$. It remains to charge $\delta - (f\sqrt{\delta} - \alpha_p) = \sqrt{\delta}(\sqrt{\delta} - f) + \alpha_p$ to the α_p corruptions. Since $\sqrt{\delta} - f = O(\alpha_p)$, we have charged enough if each corruption pays $\Theta(\sqrt{\delta})$. We conclude that n increments take $O(n + \alpha\sqrt{\delta})$ time.

Theorem 3. *The counter data structure uses $O(\delta)$ space and has additive error α. The time used for n increments is $O(n + \alpha\sqrt{\delta})$ and queries are answered in $O(\delta)$ time.*

3.4 Using Randomization to Obtain Fast Increments

In this section we describe a randomized data structure that uses $O(\delta)$ space and has additive error α. The expected time used for n increments is $O(n)$, and queries are supported in $O(\delta)$ time in the worst case. The data structure is similar to the data structures in Section 3.3 but randomization is used to find an empty cell fast.

Structure. The data structure stores an array A of size $k = 3\delta$ and a resilient variable v. Initially, v and all entries in A are set to zero.

Increment. We pick a random index $r \in \{1, \ldots, k\}$ and *probe* $A[r]$. If $A[r] = 0$, we set $A[r] = 1$ and return. Otherwise, the probe *failed* and we do one of two things: with probability $\frac{k-1}{k}$ we restart the increment operation and with probability $\frac{1}{k}$ we *clean* the array. The clean operation counts the number of non-zero entries in A and adds this plus one (the current increment) to the reliable value v, then it sets all entries in A to zero.

Query. The query operation is the same as the one in Section 3.3, it simply counts the number of non-zero entries in A and returns the sum of this number and v.

Additive Error. As in Section 3.3 the additive error is α since each unreliable array entry contributes at most one to the result.

Complexity. The query operation simply scans A and retrieves v in $O(\delta)$ time. The expected time analysis of the increment operation is more involved. The sequence of n increments is logically divided into $\lceil n/t \rceil$ rounds of $t = \lceil \delta/2 \rceil$ increments. We prove that the expected cost of each round is $O(t)$, and then the bounds follow from linearity of expectation. We split each full round in two parts, the first part consists of the increments performed before the first clean in the round, and the remaining increments are the second part. If a round does not do a clean, we additionally charge for repeatedly doing failed probes until a clean would be performed. When the first part starts, the state of the array A could be anything. When the second part starts, the array stores only zero values. We divide the cost of the t increments into three.

The cost of successful probes, the cost of failed probes and the cost of doing cleans. The cost of the successful probes is $O(t)$. The cost of failed probes, is divided into two, a cost for the failed probes in the first part and a cost for the failed probes in the second part. The first part ends when the first clean is performed. We charge the first failed probe in each increment to the increment itself. The remaining number of failed probes is upper bounded by the number of times we restart the increment operation before we clean, and a clean is performed with probability $\frac{k-1}{k}$. Thus, the probability of doing exactly f additional failed probes is $(\frac{k-1}{k})^f \frac{1}{k}$. This means that the expected cost of failed probes in the first part is bounded by $t + \sum_{f=0}^{\infty} f(\frac{k-1}{k})^f(\frac{1}{k}) = O(t)$. In the second part we place at most t ones in A and the adversary can at most introduce δ non-zero entries. Therefore, during each increment in the second part, half of the entries in A contains a zero. This means that for each increment in the second part we expect to do one failed probe implying that the expected cost of failed probes in the second part is linear in the number of increments. Each round makes one clean in the first part, and for each increment in the second part, the probability of doing a clean is at most $\frac{1}{2}\sum_{f=1}^{\infty} \frac{1}{2^f}(\frac{k-1}{k})^{f-1}\frac{1}{k} \leq 2/k$. Thus, the expected cost for doing cleans in the second part is $O(1)$ per increment, we conclude that the expected cost of a full round is $O(t)$.

Only the last round remains. If this is the first round, it has no first part, and by the analysis above the cost of this round is linear in the number of increments. If the last round is not the first round, the expected cost is $O(\delta)$ even if zero increments has been performed. We charge this cost to the second to last round.

Theorem 4. *The counter data structure described uses $O(\delta)$ space and has additive error α. The expected time used for n increments is $O(n)$, and queries use $O(\delta)$ time.*

4 Open Problems

The main open problem is whether there exists a data structure that given any $t \geq 1$ has additive error $O(\alpha/t)$, supports increments in $O(t)$ time and queries in $O(\delta)$ time. One resilient counter needs $\Omega(\delta)$ space. It would be interesting too see if one can store k counters using $o(k\delta)$ space with each counter having a non-trivial bound on the additive error. Most of the counters presented in this paper require $\Theta(\delta)$ space for a reliable variable which seems hard to share among several counters. It may be interesting to see if one can use the safe memory to store some state to achieve this and possibly circumventing the lower bound tradeoff between increments and additive error.

References

1. Finocchi, I., Italiano, G.F.: Sorting and searching in the presence of memory faults (without redundancy). In: Proc. 36th Annual ACM Symposium on Theory of Computing, pp. 101–110 (2004)
2. Constantinescu, C.: Trends and challenges in VLSI circuit reliability. IEEE micro 23(4), 14–19 (2003)
3. Tezzaron Semiconductor: Soft errors in electronic memory - a white paper (2004), `http://www.tezzaron.com/about/papers/papers.html`

4. Baumann, R.: Soft errors in advanced computer systems. IEEE Design and Test of Computers 22(3), 258–266 (2005)
5. Taber, A., Normand, E.: Single event upset in avionics. IEEE Transactions on Nuclear Science 40(2), 120–126 (1993)
6. Govindavajhala, S., Appel, A.W.: Using memory errors to attack a virtual machine. In: IEEE Symposium on Security and Privacy, pp. 154–165 (2003)
7. Bar-El, H., Choukri, H., Naccache, D., Tunstall, M., Whelan, C.: The sorcerer's guide to fault attacks. Proceedings of the IEEE 94(2), 370–382 (2006)
8. Boneh, D., DeMillo, R.A., Lipton, R.J.: On the importance of checking cryptographic protocols for faults. In: Fumy, W. (ed.) EUROCRYPT 1997. LNCS, vol. 1233, pp. 37–51. Springer, Heidelberg (1997)
9. Skorobogatov, S.P., Anderson, R.J.: Optical fault induction attacks. In: Kaliski Jr., B.S., Koç, Ç.K., Paar, C. (eds.) CHES 2002. LNCS, vol. 2523, pp. 2–12. Springer, Heidelberg (2003)
10. Huang, K.H., Abraham, J.A.: Algorithm-based fault tolerance for matrix operations. IEEE Transactions on Computers 33, 518–528 (1984)
11. Rela, M.Z., Madeira, H., Silva, J.G.: Experimental evaluation of the fail-silent behaviour in programs with consistency checks. In: Proc. 26th Annual International Symposium on Fault-Tolerant Computing, pp. 394–403 (1996)
12. Abadi, M., Budiu, M., Erlingsson, Ú., Ligatti, J.: Control-flow integrity. In: Proc. 12th ACM conference on Computer and communications security, pp. 340–353 (2005)
13. Pradhan, D.K.: Fault-tolerant computer system design. Prentice-Hall, Inc., Englewood Cliffs (1996)
14. Novark, G., Berger, E.D., Zorn, B.G.: Exterminator: Automatically correcting memory errors with high probability. Communications of the ACM 51(12), 87–95 (2008)
15. Aumann, Y., Bender, M.A.: Fault tolerant data structures. In: Proc. 37th Annual Symposium on Foundations of Computer Science, pp. 580–589 (1996)
16. Leighton, T., Ma, Y.: Tight bounds on the size of fault-tolerant merging and sorting networks with destructive faults. SIAM Journal on Computing 29(1), 258–273 (2000)
17. Chlebus, B.S., Gasieniec, L., Pelc, A.: Deterministic computations on a pram with static processor and memory faults. Fundamenta Informaticae 55(3-4), 285–306 (2003)
18. Ravikumar, B.: A fault-tolerant merge sorting algorithm. In: Proc. 8th Annual International Conference on Computing and Combinatorics, pp. 440–447 (2002)
19. Kutten, S., Peleg, D.: Tight fault locality. SIAM Journal on Computing 30(1), 247–268 (2000)
20. Finocchi, I., Grandoni, F., Italiano, G.F.: Designing reliable algorithms in unreliable memories. Computer Science Review 1(2), 77–87 (2007)
21. Brodal, G.S., Fagerberg, R., Finocchi, I., Grandoni, F., Italiano, G.F., Jørgensen, A.G., Moruz, G., Mølhave, T.: Optimal resilient dynamic dictionaries. In: Arge, L., Hoffmann, M., Welzl, E. (eds.) ESA 2007. LNCS, vol. 4698, pp. 347–358. Springer, Heidelberg (2007)
22. Finocchi, I., Grandoni, F., Italiano, G.F.: Resilient search trees. In: Proc. 18th ACM-SIAM Symposium on Discrete Algorithms, pp. 547–553 (2007)
23. Finocchi, I., Grandoni, F., Italiano, G.F.: Optimal resilient sorting and searching in the presence of dynamic memory faults. Theoretical Computer Science (to appear, 2009)
24. Jørgensen, A.G., Moruz, G., Mølhave, T.: Priority queues resilient to memory faults. In: Proc. 10th International Workshop on Algorithms and Data Structures, pp. 127–138 (2007)
25. Ferraro-Petrillo, U., Finocchi, I., Italiano, G.F.: The price of resiliency: A case study on sorting with memory faults. In: Azar, Y., Erlebach, T. (eds.) ESA 2006. LNCS, vol. 4168, pp. 768–779. Springer, Heidelberg (2006)
26. Brodal, G.S., Jørgensen, A.G., Mølhave, T.: Fault tolerant external memory algorithms. In: Proc. 11th Algorithms and Data Structures Symposium, pp. 411–422 (2009)
27. Boyer, R.S., Moore, J.S.: MJRTY: A fast majority vote algorithm. In: Automated Reasoning: Essays in Honor of Woody Bledsoe, pp. 105–118 (1991)

A Simple, Fast, and Compact Static Dictionary

Scott Schneider and Michael Spertus

Symantec Corporation, Culver City CA 90230, USA

Abstract. We present a new static dictionary that is very fast and compact, while also extremely easy to implement. A combination of properties make this algorithm very attractive for applications requiring large static dictionaries:

1. High performance, with membership queries taking $O(1)$-time with a near-optimal constant.
2. Continued high performance in external memory, with queries requiring only 1-2 disk seeks. If the dictionary has n items in $\{0, ..., m-1\}$ and d is the number of bytes retrieved from disk on each read, then the average number of seeks is $\min\left(1.63, 1 + O\left(\frac{\sqrt{n}\log m}{d}\right)\right)$.
3. Efficient use of space, storing n items from a universe of size m in $n\log m - \frac{1}{2}n\log n + O(n + \log\log m)$ bits. We prove this space bound with a novel application of the Kolmogorov-Smirnov distribution.
4. Simplicity, with a 20-line pseudo-code construction algorithm and 4-line query algorithm.

1 Introduction

A static dictionary is a data structure that stores a subset S of a finite set $U = \{0, ..., m-1\}$ and answers membership queries. It is possible to store these n items with fewer than $n\log m$ bits[1], the space required for a simple list of the items. The theoretical minimum space required is [13]

$$B = \left\lceil \log \binom{m}{n} \right\rceil = n\log\left(\frac{m}{n}e\right) - \Theta\left(\frac{n^2}{m}\right) - O(\log n)$$

To scale well to very large data sizes, a dictionary (or any other data structure) needs not only good time and space complexity, but also good locality. Indeed, according to a standard textbook on large data structures, "We measure a computer system's speed in terms of the number of accesses made to disk, because this access time—which is measured in fractions of a second—dominates the total time involved in a search." [19]

1.1 Previous Work

Much progress has been made in reducing both space usage and membership query time for static dictionaries. Fredman, Komlós, and Szemerédi [8] created

[1] All logarithms in this paper will have base 2.

Y. Dong, D.-Z. Du, and O. Ibarra (Eds.): ISAAC 2009, LNCS 5878, pp. 852–861, 2009.

the first dictionary to run membership queries in $O(1)$ worst case time with low space usage ($n \log m + o(n \log m)$ bits) for a broad range of values of m and n. Brodnik and Munro gave the first $O(B)$ space usage static dictionary [1], and later improved this to $B+o(B)$.[2] Pagh reduced this further to $B+o(n)$ bits.[13] Several recent papers [16,9,15] have the same or even lower space usage, while also supporting rank and select operations in constant time.

All of the papers above, except the first, save space by a technique called *quotienting.* (To our knowledge, quotienting was first described in section 6.4, exercise 13 in [11].) After dividing the dictionary's n elements into buckets, each bucket only needs to store enough information to distinguish its items from the subset of U that could be hashed into that bucket. This information is the item's *quotient.* If the b buckets divide U evenly, then this is only $\lceil \log \frac{m}{b} \rceil$ bits. Thus, dictionaries that use quotienting can save $\lfloor N \log b \rfloor$ bits at the cost of needing to store the number of items in each bucket.

Dynamic dictionaries also support fast insertion and deletion, though at the cost of positive space overhead.[14,7] Jensen's and Pagh's dynamic dictionary performs very well in external memory, requiring only $1 + O\left(\sqrt{\frac{\log m}{d}}\right)$ disk accesses per lookup, insertion, or deletion.[10]

Cleary's bidirectional probing hash table (another dynamic dictionary) is also noteworthy because it reduces space usage by a technique similar to the one used in this paper.[4] His table requires $(1+\epsilon)\, n \log \frac{m}{n} + O(n)$ bits, but has $O\left(\frac{1}{\epsilon^2}\right)$ expected lookup and insertion time. The $\#C[i] - \#V[i]$ values he computes are offsets, with a similar distribution to the offsets used in this paper.

1.2 This Paper

This paper presents a static dictionary that is compact, fast, and extremely simple. It is not succinct and does not support rank or select, but it contributes to the field in several ways. First, its compact size, fast lookup, and high locality make it an excellent practical choice for use in software projects. Queries have average $O(1)$ running time and require only 1-2 disk seeks, on average, when data is stored on disk. Second, our unusual use of the Kolmogorov-Smirnov distribution to prove our space bound is interesting in itself. Finally and most importantly, the extreme simplicity of this dictionary (a few dozen lines of code on top of a bit vector implementation) gives it a unique edge. Though succinct dictionaries save more space, they may be too complex for many real-world projects. At Symantec, this static dictionary is being integrated into our enterprise data loss prevention product. (See `http://vontu.com`.) Thus, we think it is appropriate and valuable for this dictionary to have a proper analysis in the literature.

Our dictionary uses $n \log m - \frac{1}{2} n \log n + O(n + \log \log m)$ bits, in its simplest form. Although larger than other known algorithms, it is still much less than required to store the elements directly and its $O(1)$ average membership query time has a very low constant. In the worst case, membership queries access $\Theta(\log \log n)$ memory locations, with very high locality.

Our static dictionary consists of two arrays. One stores a $\log \frac{2m}{n}$ bit quotient for each item (plus satellite data, if any). The other stores pointers to the start of each bucket in the quotient array. It saves space by compressing these pointers with a technique similar to one in Cleary's hash table.[4] We prove these space savings using the Kolmogorov-Smirnov distribution.

Section 2 describes the simplest form of our data structure and does some simple analysis. Section 3 uses the Kolmogorov-Smirnov distribution to prove the space bound stated above. Section 4 then describes how to reduce the average number of disk seeks to $\min\left(1.63, 1 + O(\frac{\sqrt{n}\log m}{d})\right)$, where d is the amount of data the operating system reads from the disk at once (typically several kilobytes). Section 5 shows empirical results on the number of disk seeks per query and other metrics. Finally, section 6 concludes this paper.

2 Algorithm

The simplest form of our data structure just hashes the n items into n buckets and stores two arrays: the bucket offset array and the quotient array.

The quotient array concatenates items in all buckets together into an array of n items. It stores only item quotients, which are $\log \frac{2m}{n}$ bits each.[2] Section 2.3 will discuss the choice of hash and quotient functions.

In the bucket offset array, each entry stores data that indicates the quotient array index for the first item in each bucket. By looking at two consecutive entries in the bucket offset array, we can find the quotient array range for a bucket and do a binary search.

If the bucket offset array simply stored indexes, it would use $\log n$ bits per item, cancelling any space savings. Instead, we compress indexes by observing that the expected average value of the index for bucket number i is simply i. Each item in this array stores an *offset* – the difference between the quotient array index for the first item in the bucket and the bucket's number. Thus, the number of bits needed per offset is determined by the size of the largest offset.

Intuitively, one would expect this to be $O(\sqrt{n})$. This turns out to be true, though the proof is not trivial. It is easy to show that the expected offset for any particular offset is $O(\sqrt{n})$, but it is not easy to show that the largest of the n offsets, which have high statistical dependence, is $O(\sqrt{n})$. Section 3 will show this using the Kolmogorov-Smirnov distribution.

The total space for the dictionary, in terms of the largest offset, $|\Delta_{max}|$, is

$$n \log \frac{2m}{n} + n(\lceil \log |\Delta_{max}| \rceil + 1) + O(\log \log m) \qquad (1)$$

The $+1$ is for the sign bit in the offset array. The last term is the space for encoding hash function data.

[2] We will assume that m and n are powers of 2. If they are not, the quotient array may store another extra bit per item.

2.1 Dictionary Construction

Algorithm 1 shows the basic construction of our static dictionary. The main elements are the OFFSETS and QUOTIENTS arrays. The algorithm's inputs are S, m, hash function h, and quotient function q.

Construction is quite simple. We sort values in S by $h(x)$ and $q(x)$ and then iterate through it, adding quotients to the QUOTIENTS array. Whenever $h(x)$ increases, we set values in the OFFSETS array, subtracting x's index in the QUOTIENTS array from the OFFSETS array index.

Other than sorting S, all steps are linear. If the largest offset or maximum number of items in one bucket (which affects the worst-case running time) is much higher than its expected value, we will try again with a different hash function.[3] This is an expected $O(1)$ number of tries and section 3 will show that, once we have uniformly distributed items, we are virtually assured that the largest offset will be $O\left(\sqrt{n}\right)$. Thus, the expected construction time is $O(n \log n)$.

Algorithm 1. Dictionary Construction

Sort S, first by $h(x)$ and then by $q(x)$.
Create the $(n+1)$-element TEMPOFFSETS array with $\log n$ bits per item.
Create the n-element QUOTIENTS array with $\log \frac{2m}{n}$ bits per item.
$currentItemNum \leftarrow 0$
$minOffset \leftarrow 0$
$maxOffset \leftarrow 0$
for $bucketNum = 0 \rightarrow n$ **do**
  TEMPOFFSETS[$bucketNum$] $\leftarrow currentItemNum - bucketNum$
  $minOffset \leftarrow \min(minOffset,$ TEMPOFFSETS[$bucketNum$]$)$
  $maxOffset \leftarrow \max(maxOffset,$ TEMPOFFSETS[$bucketNum$]$)$
  while $h(S[currentItemNum]) = bucketNum$ **do**
    QUOTIENTS[$currentItemNum$] $\leftarrow q(S[currentItemNum])$
    $currentItemNum \leftarrow currentItemNum + 1$
  end while
end for
Create the $(n+1)$-element OFFSETS array with
  $\lceil \log \max(|minOffset|, maxOffset) \rceil$ bits per item
Copy items from from TEMPOFFSETS to OFFSETS.

2.2 Dictionary Operations

We test the membership of x in two steps. First, we check the offsets at $h(x)$ and $h(x)+1$ to find indexes for the start and end of x's bucket. Then, we do a binary search in the corresponding range of the quotient array for $q(x)$.

[3] We may try several hash functions to find a lower largest offset – one below $\sqrt{n}$. This is unlikely to save more than 1 bit per item, however. There is only a 3.6% chance that the largest offset will be under $\frac{1}{2}\sqrt{n}$. Repetition is probably only worthwhile when the largest offset is bigger than $\sqrt{n}$.

Algorithm 2. Membership Query

$leftIndex \leftarrow h(x) + \text{OFFSETS}[h(x)]$
$rightIndex \leftarrow (h(x)+1) + \text{OFFSETS}[h(x)+1] - 1$
Binary search QUOTIENTS from *leftIndex* to *rightIndex* for $q(x)$.
Return true if $q(x)$ was found in QUOTIENTS, or false otherwise.

The average number of items in each bucket is 1. Logarithm is concave, so the average time required for membership is $O(1)$.

The worst case time for membership is logarithmic in the size of the largest bucket. In general, this could be as high as n, but section 2.1 showed how we limit the maximum bucket size. We set this limit close to the expected largest bucket size, which is $\Theta\left(\frac{\log n}{\log\log n}\right)$.[6] Thus, the worst case time is $\Theta(\log\log n)$.

1-2 disk seeks in membership queries. When a typical operating system reads from disk, it reads a fixed-size chunk. This is usually one physical disk track – around 8 kilobytes.[18] This chunk is not centered on the requested data, but starts at a predetermined alignment. Thus, if the data chunk size is d bytes and we want to read x bytes, the average number of disk accesses is $1 + \frac{x-1}{d}$. We will consider $O\left(\frac{\log m}{d}\right)$ to be negligibly small, though $O\left(n^{\epsilon}/d\right)$ is worth noting.

The first read in a membership query is to two consecutive offsets. This is $2(\log|\Delta_{max}|+1)$ bits. $|\Delta_{max}|$ cannot be larger than $O(n)$, so the chance that this read spans two chunks is negligibly small.

In addition, there is a $\frac{1}{e}$ chance that this bucket is empty. If so, the membership query is done in one read. Otherwise, the second step is a binary search of a bucket. The average bucket size is 1, so the average number of disk reads in the binary search is 1. Thus, with 1 access for reading offsets and a $1-\frac{1}{e}$ chance of 1 more access, the expected total number of accesses is $2-\frac{1}{e} \approx 1.63$.

The worst case involves a binary search of the largest bucket. It has $\Theta\left(\frac{\log n}{\log\log n}\right)$ items with $\log\frac{m}{n}$ bits each. This requires $\Theta\left(\frac{\log m \log n}{d \log\log n}\right)$ additional accesses, for a total of $2+\Theta\left(\frac{\log m \log n}{d \log\log n}\right)$.

Adding satellite data does not change this much. The quotient array stores this extra data, so the average number of disk accesses does not change. If we have t bits per item, the worst case changes to $2+\Theta\left(\frac{(\log m+t)\log n}{d\log\log n}\right)$ accesses.

2.3 Hash and Quotient Functions

Our algorithm has an especially great need for a hash function that distributes S uniformly, so that offsets are low. Let items in S be values from $0, 1, ...m-1$. We use the well-known hash function, $h(x) = (ax \bmod p) \bmod n$ [11,3], where p is a prime number such that $m \le p < 2m$ and a is a randomly chosen number such that $1 \le a < p$.

The argument that $ax \bmod p$ has a uniform distribution is similar to the well-known argument that it is universal. If $\Pr(ax \bmod p = ay \bmod p) \leq \frac{1}{p}$ for random a and $x \neq y$, then $\Pr(ax \bmod p > ay \bmod p) \leq \frac{ax \bmod p}{p}$.

This $h(x)$ is slightly distorted. Hash values are slightly more likely to be less than the value r, where $r = p \bmod n$. The probability of getting a particular hash value less than r is $\frac{1}{p}\left\lfloor \frac{p}{n}+1 \right\rfloor$, as opposed to the $\frac{1}{p}\left\lfloor \frac{p}{n} \right\rfloor$ probability of getting a particular hash value greater than or equal to r.

This creates a positive expected value for offsets, which is highest at r:

$$\langle |x \in S : h(x) < r| \rangle = n \cdot \Pr(h(x) < r) = nr\frac{1}{p}\lfloor \frac{p}{n} + 1 \rfloor$$

$$\begin{aligned}\langle \text{Offset}(r) \rangle &= nr\frac{1}{p}\lfloor \frac{p}{n} + 1 \rfloor - r = \frac{r}{p}(p - r + n) - r \\ &= \frac{r(n-r)}{p}\end{aligned}$$

Thus, the largest expected offset is $O(\frac{n^2}{m})$. For $n = O(\sqrt{m})$, this hardly matters. For larger n, we can simply adjust our offset calculation to subtract the expected number of items from the bucket's quotient array index.

The corresponding quotient function is $q(x) = \lfloor \frac{ax \bmod p}{n} \rfloor$. Its largest value is $\lfloor \frac{p-1}{n} \rfloor$, so a quotient requires $log \frac{2m}{n}$ bits.

3 The Largest Offset

The amount of space saved in the dictionary depends on the largest offset, which determines how many bits are used for each item in the array of b buckets.

Theorem 1. *For large n and b, the probability that the largest absolute value of an offset in the OFFSETS array is less than $\lambda\sqrt{n}$ approaches*

$$p(\lambda) = \sum_{k=-\infty}^{+\infty} (-1)^k e^{-2k^2\lambda^2}$$

Proof. Let $0 \leq x_{(1)} \leq ... \leq x_{(n)} < m$ be the sorted values for the items. The normalized sum function, $F(i)$, is equal to the largest f such that $x_{(f)} < i\frac{m}{b}$, where $0 \leq i \leq b$. (If $x_{(1)} \geq i\frac{m}{b}$, then $F(i) = 0$.) $F(i)$ is monotonically non-decreasing from 0 to n. The offsets, Δ_i, can in the simplest case[4] be written

$$\Delta_i = F(i) - \left\lfloor i\frac{n}{b} \right\rfloor$$

The absolute value of the maximum offset is therefore approximated by

$$\left|\frac{\Delta_{max}}{n}\right| = \max\left|\frac{F(i)}{n} - \frac{i}{b}\right| \tag{2}$$

[4] In the more complicated case that $n = \omega(\sqrt{m})$, we will replace $\lfloor i\frac{n}{b} \rfloor$ with $< F(i) >$, the expected number of values below i.

The key observation is that the right hand side approximates the Kolmogorov-Smirnov distribution with error bounded by $1/b$.[17,12] $F(i)/n$ is the items' observed distribution and i/b is their expected, theoretical distribution – the uniform distribution. $|\Delta_{max}/n|$ corresponds to λ.

While the Kolmogorov-Smirnov test only applies to continuous distributions, we can get a uniform distribution of integers by choosing a uniform distribution of real numbers and rounding down to integers, which will not change the maximum offset by more than 1. Therefore, the standard tables and computations for the Kolmogorov-Smirnov distribution will estimate the probability distribution for the largest offset.

To complete the proof, we simply note that $p(\lambda)$ is the limit, for large n, of the Kolmogorov-Smirnov distribution, as in [12].

As Smirnov's pre-calculated table of $p(\lambda)$ shows, λ has a high probability of being close to 1.[17] For example, there is a 90% probability that the maximum offset will be less than $1.23\sqrt{n}$ and a 99.9% probability that the largest offset is less than $1.95\sqrt{n}$. The probability that $\lambda = \omega(1)$ is well-approximated by

$$1 - p(\lambda) \approx 1 - \left(1 - 2e^{-2\lambda^2}\right) = 2e^{-2\lambda^2}$$

Even for $\lambda = \sqrt{\log n}$, $1 - p(\lambda) = \frac{2}{n^2}$ is still quite small. This proves the statement in section 2.1 that the largest offset is extremely unlikely to be much larger than $\sqrt{n}$.

Corollary 1. *Our data structure's average space usage is*

$$n \log m - \frac{1}{2} n \log n + O\left(n + \log \log m\right)$$

Proof. To prove this, we simply plug the expected value of the largest offset, from theorem 1, into equation (1).

4 Reducing Disk Seeks Further

We can create locality between a membership query's two reads by interleaving values in the bucket offsets array and the quotient array in the data structure stored on disk. Suppose a bucket has a 0 offset and 1 item. In this case, the query reads offsets for the query item's bucket and the next bucket, and then checks the bucket's item, which is between these two offests. This is only 1 disk access.

An item's bucket's offset is the main factor in the distance between the first and last memory locations that a query will read. The average absolute offset value is $c\sqrt{n}$, with $c = O(1)$. The number of bits required for an (offset, quotient) pair is

$$\log |\Delta_{max}| + 1 + \log \frac{m}{n} \approx \frac{1}{2} \log n + 1 + \log \frac{m}{n} < \log m$$

So the average memory span is $c\sqrt{n}\frac{\log m}{8}$ bytes. This, combined with $\frac{1}{e}$ chance that a bucket is empty, gives the average, total number of disk seeks for a membership query as $1 + \left(1 - \frac{1}{e}\right) \min\left(1, \frac{c\sqrt{n}\log m}{8d}\right) = \min\left(1.63, 1 + O\left(\frac{\sqrt{n}\log m}{d}\right)\right)$. (Again, satellite data does not change this much.) For $d = 8$ kilobytes, we rarely avoid this second lookup for $n \geq 10^7$, but if d rises to 128 kilobytes, we will often avoid the second lookup for $n \approx 10^9$. We achieve this by increasing locality, not by optimizing specifically for this model of memory access.

5 Empirical Results

In empirical testing, we measured the average size of the largest offset and the number of disk seeks per membership query. We did not give overall query times because our tables fit mostly or entirely in memory, so numbers would largely be meaningless due for other memory configurations, disk cache sizes, etc.

We set n to powers of 2 and m to 2^{64}. We generated dictionary members randomly and tested 50 tables at each size. We measured locality by counting the number of distinct pages of various sizes (2k, 8k, and 32k) that were accessed during the query. We did this both for dictionary members and random values (which are mostly non-members), but for the sake of clarity, include the random value results only for the 8k block size. In actual testing, most accesses did not result in a disk seek because much of the dictionary was resident in memory. As a result, practical usage typically results in fewer disk seeks than the already good numbers here.

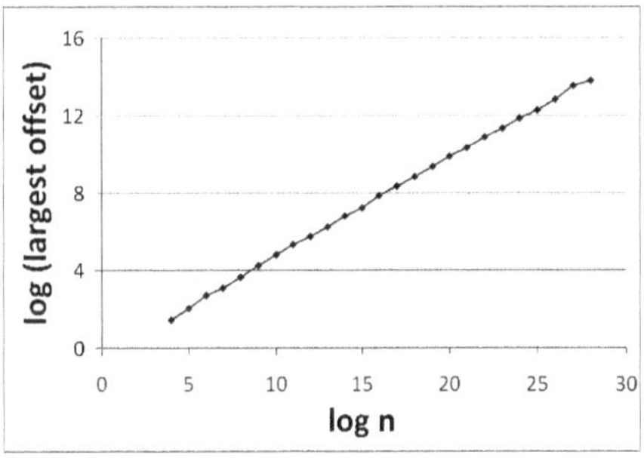

Fig. 1. Largest offset

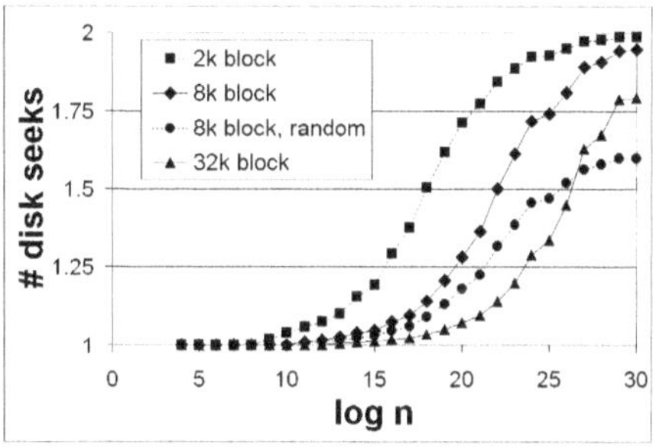

Fig. 2. Disk seeks

As expected, the largest offset is proportional to $\sqrt{n}$. Also, the number of disk seeks tends toward 2 for members and 1.63 for random values. The three curves for positive tests show that, for constant $\frac{\sqrt{n}}{d}$, we have approximately the same number of disk seeks.

Although all of these results merely confirm the theoretical analysis, several useful observations came from writing the empirical tests. First, most of the code for our dictionary dealt with bit-packing and manipulating data that is not byte-aligned. Like other common languages, C++ does not have arrays whose element size is a specified number of bits. Aside from this, the implementation only runs a few dozen lines of code. Second, we noticed that the glibc implementation of the `rand()` function has a rather short cycle of several million items. Tests with larger n required a better source of random numbers.

6 Concluding Remarks

We have described a static dictionary that is simple, compact, and fast. When it is too large to fit in memory, membership queries require only 1-2 disk seeks. These properties make this data structure very practical for real-world applications, such as the enterprise data leakage prevention product for which we developed it.

We saved space by using *quotienting* and by compressing indexes, each of which is $O(n)$, into offsets that are $O\left(\sqrt{n}\right)$. We use a novel application of the Kolmogorov-Smirnov distribution to prove this, as well as our space usage. We presented an optimization that reduced the number of disk seeks to 1, when the number of bytes that the operating system retrieves from the disk on each read is $\Omega\left(\sqrt{n}\right)$.

One major optimization that we have not analyzed is reducing offsets by adding empty slots to the QUOTIENTS array. We can add these empty slots before large negative offsets and after large positive offsets. It is not hard to imagine that adding $O\left(\sqrt{n}\right)$ empty slots ($O\left(\sqrt{n}\right)\log\frac{m}{n}$ bits of space) could reduce the largest offset by a factor of 2, saving n bits. We plan to investigate in the future how far this can reduce the largest offset.

As future work, we also plan to apply our analysis techniques to Cleary's bidirectional probing hash table. Cleary observed that, with a load factor of 95%, 99% of the offsets were between -15 and +15.[4] Kolmogorov-Smirnov theorem 3 in [5] will give the expected number of offsets larger than a specified value, but this and other theorems cannot accommodate a load factor below 1. A theoretical understanding of load factor's impact on offsets would show, for the first time, the space-time trade-offs in Cleary's dictionary.

Acknowledgments

We would like to thank Peter McCullagh for suggesting the relevance of the Kolmogorov-Smirnov distribution. We are also grateful to Janos Simon, Tzicker Chiueh, Kent Griffin, Carey Nachenberg, Tsuen "Johnny" Ngan, Jeffrey Wilhelm, and Matthew Yeo for their help in discussing, implementing, and analyzing this data structure.

References

1. Brodnik, A., Munro, J.I.: Membership in Constant Time and Minimum Space. In: van Leeuwen, J. (ed.) ESA 1994. LNCS, vol. 855, pp. 72–81. Springer, Heidelberg (1994)

2. Brodnik, A., Munro, J.I.: Membership in Constant Time and Almost-Minimum Space. SIAM Journal of Computing 28, 1627–1640 (1999)
3. Carter, J., Wegman, M.: Universal Classes of Hash Functions. Journal of Computer and System Sciences 18, 143–154 (1979)
4. Cleary, J.G.: Compact Hash Tables Using Bidirectional Linear Probing. IEEE Transactions on Computers 33, 828–834 (1984)
5. Feller, W.: On the Kolmogorov-Smirnov Limit Theorems for Empirical Distributions. Annals of Mathematical Statistics 19(2), 177–189 (1948)
6. Vitter, J., Flajolet, P.: Average-case analysis of algorithms and data structures. In: van Leeuwen, J. (ed.) Handbook of Theoretical Computer Science, pp. 431–524. Elsevier, Amsterdam (1990)
7. Fotakis, D., Pagh, R., Sanders, P., Spirakis, P.: Space Efficient Hash Tables With Worst Case Constant Access Time. In: Alt, H., Habib, M. (eds.) STACS 2003. LNCS, vol. 2607, pp. 271–282. Springer, Heidelberg (2003)
8. Fredman, M., Komlós, J., Szemerédi, E.: Storing a Sparse Table with O(1) Worst Case Access Time. Journal of the ACM 31(3), 538–544 (1984)
9. Grossi, R., Orlandi, A., Raman, R., Rao, S.: More Haste, Less Waste: Lowering the Redundancy in Fully Indexable Dictionaries. In: STACS 2009, pp. 517–528 (2009)
10. Jensen, M., Pagh, R.: Optimality in External Memory Hashing. Algorithmica 52(3), 403–411 (2008)
11. Knuth, D.: Sorting and Searching. The Art of Computer Programming, vol. 3. Addison-Wesley Publishing Company, Reading (1973)
12. Kolmogoroff, A.: Confidence limits for an unknown distribution function. Annals of Mathematical Statistics 12, 461–463 (1941)
13. Pagh, R.: Low Redundancy in Static Dictionaries with Constant Query Time. SIAM Journal on Computing 31(2), 353–363 (2001)
14. Pagh, R., Rodler, F.F.: Cuckoo hashing. In: Meyer auf der Heide, F. (ed.) ESA 2001. LNCS, vol. 2161, pp. 121–133. Springer, Heidelberg (2001)
15. Patrascu, M.: Succincter. In: FOCS 2008, pp. 305–313 (2008)
16. Raman, R., Raman, V., Rao, S.: Succinct Indexable Dictionaries with Applications to Encoding k-ary Trees, Prefix Sums and Multisets. ACM Transactions on Algorithms 3(4), Article 43 (2007)
17. Smirnov, N.: On the estimation of the discrepancy between empirical curves of distribution for two independent samples. Bulletin Mathématique de l'Université de Mouscou, 2(fasc. 2) (1939)
18. Vitter, J.: Algorithms and Data Structures for External Memory. Now Publishers, Inc., Hanover (2008)
19. Witten, I.H., Moffat, A., Bell, T.C.: Managing gigabytes: compressing and indexing documents and images. Morgan Kaufmann Publishers, Taylor & Francis, San Francisco, London (1999)

Reconstructing Polygons from Scanner Data

Therese Biedl[1,*], Stephane Durocher[2], and Jack Snoeyink[3]

[1] David R. Cheriton School of Computer Science, University of Waterloo
biedl@cs.uwaterloo.ca
[2] Department of Computer Science, University of Manitoba
durocher@cs.umanitoba.ca
[3] Department of Computer Science, University of North Carolina at Chapel Hill
snoeyink@cs.unc.edu

Abstract. A range-finding scanner can collect information about the shape of an (unknown) polygonal room in which it is placed. Suppose that a set of scanners returns not only a set of points, but also additional information, such as the normal to the plane when a scan beam detects a wall. We consider the problem of reconstructing the floor plan of a room from different types of scan data. In particular, we present algorithmic and hardness results for reconstructing two-dimensional polygons from points, point/normal pairs, and visibility polygons. The polygons may have restrictions on topology (e.g., to be simply connected) or geometry (e.g., to be orthogonal). We show that this reconstruction problem is NP-hard in most models, but for some assumptions allows polynomial-time reconstruction algorithms which we describe.

1 Introduction

Range scanners have been configured in many ways: looking in to capture objects on a platform or in-situ, looking down to capture terrain or urban environments, and looking out to capture rooms or factory floors. Some scanners [3,11] provide – in addition to point coordinates – surface labels, normals, or unobstructed portions of scanned rays.

The problem of reconstructing surfaces from the resulting point clouds has been given both theoretical and practical consideration. Theoretical solutions can provably reconstruct the correct surface when the sample points are sufficiently dense relative to local feature size [1,2]. Applied solutions handle noisy data and often incorporate additional information such as estimated normals [11].

We consider the problem of reconstructing the two-dimensional floor plan of a polygonal room using different types of scanned data. Specifically, we ask whether having additional information about each data point and knowing something about the geometry (monotonicity and/or orthogonality) of the room allows efficient reconstruction from less dense data.

* Supported by NSERC.

Y. Dong, D.-Z. Du, and O. Ibarra (Eds.): ISAAC 2009, LNCS 5878, pp. 862–871, 2009.

1.1 Models and Problem Definition

We consider five models for input data that may be obtained by scanning a room with one or more scanners, as illustrated in Fig. 1.

1. A *point scan* is a set of points, each of which lies on a wall (edge) of the scanned room (polygon).
2. A *point-wall scan* is a point scan for which each point records the line containing the wall on which it lies (i.e., the orientation of the wall).
3. A *point-normal scan* is point-wall scan for which each point-line pair records a normal perpendicular to the line that points towards the room's interior.
4. A *segment scan* is a point-normal scan for which each point records the position of its scanner; this implies that the entire line segment from a scanner to the corresponding scan point must be inside the room.
5. A *visibility-polygon scan* is a set of visibility polygons, i.e., the entire region visible from each scanner.

Given n observations under of one of the five models, the polygon reconstruction problem is to determine whether there exists a polygon that is consistent with the input data. Without additional restrictions, the answer is always "yes" in the point-scan model, as well as in all five models if a solution may include additional edges not encountered by any input point. In this paper, therefore, we assume that each wall has been seen, i.e., that each polygon edge contains at least one scan point.

1.2 Related Work

We have already mentioned results [1,2,11] on reconstructing shapes from densely-spaced samples on the surface in the point-scan model. Because we primarily

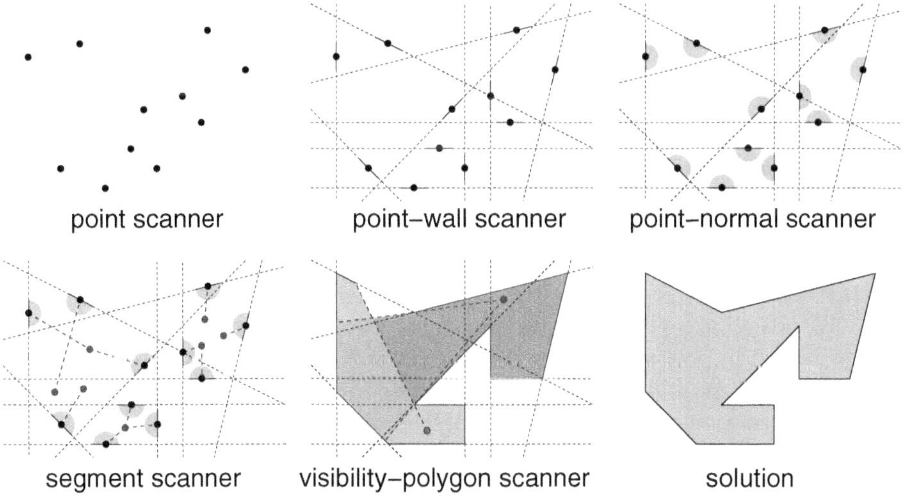

Fig. 1. Input instances of the five scan models and a common solution

consider scan models that provide additional information, sample points need not be closely spaced as long as each edge includes at least one sample point. Thus, our approaches are more closely related to previous work on reconstructing a polygon from a given vertex set. O'Rourke [12] gives an $O(n \log n)$ time algorithm for reconstructing an orthogonal polygon when all vertices must meet edges at a right angle; when a solution exists, it is unique. This problem is NP-hard if edges at a vertex meet either straight or at right angles [14] and also NP-hard if edges must be parallel to one of three (or more) given directions [5].

The reconstruction problem can be formulated as a matching problem with additional restrictions in a graph $G = (V, E)$. Each sample point corresponds to a segment on the polygon's boundary: let V contain two vertices for each sample point, one for each direction out from the segment. Join two vertices by an edge in E if the corresponding rays intersect. The polygon reconstruction problem reduces to finding a spanning subgraph $H \subseteq G$ that has specific properties. In particular, we require that H be a perfect matching that is simple (no matching edges cross each other). Furthermore, if each pair of vertices induced by a sample point is joined by an edge, the resulting subgraph must be connected.

If the constraints of simplicity and connectivity are dropped, the problem is reducible to finding a perfect matching and is solvable in polynomial time [4,6,7,10]. Adding either constraint renders the problem hard: finding a non-crossing 2-factor in a geometric graph was shown to be NP-hard by Jansen and Woeginger [9], and a connected 2-factor is simply a Hamiltonian cycle, which is well known to be NP-hard to find [6], even in grid graphs [8]. Neither of these results, however, directly implies hardness for the polygon reconstruction problems we consider.

1.3 Our Results

We first show that the reconstruction problem is NP-hard (Section 2). Our reduction is to the visibility-polygon scan model, but straightforward modifications can reduce to the point-wall, point-normal, or segment scan models.

For positive results, we consider geometric restrictions to the allowable configurations of polygons. A *star-shaped polygon* is entirely visible from some point in its interior (i.e., there is a single scanner). The interior of a *monotone*[1] *polygon* intersects every vertical line in at most one line segment. The boundary of a monotone polygon splits into two chains, the *upper* and *lower chains*, both of which are monotone. Every edge in an *orthogonal polygon* is either horizontal or vertical.

Our hardness result implies that the reconstruction problem remains NP-hard even for orthogonal polygons. We show that two cases for monotone polygons can be solved in polynomial time: orthogonal monotone polygons for point-wall scans (Section 3), and monotone polygons for point-normal scans (Section 4). The reconstruction problem is also solvable for star-shaped polygons (Section 5). Finally, we present a lower bound showing that the running times of our algorithms are optimal (Section 6).

[1] For simplicity, we use the term *monotonicity* to refer to *x-monotonicity*.

2 Hardness Results

In this section, we prove that reconstructing a simply-connected polygon from a visibility-polygon scan is NP-hard. We use a reduction from ORTHOGONAL NON-CROSSING SPANNING TREE, which was shown to be NP-hard by Jansen and Woeginger [9]. An *orthogonal graph* is a graph drawn in the plane such that every edge is either a horizontal or vertical line segment connecting two vertices and no edge contains any vertex in its interior.

ORTHOGONAL NON-CROSSING SPANNING TREE
Instance. An orthogonal graph G.
Question. Find a graph $H \subseteq G$ that is a spanning tree of G such that no two edges in H cross.

Theorem 1. *Polygon reconstruction under the visibility-polygon scan model is NP-hard.*

Proof. Given any orthogonal graph G, we construct an instance of the visibility-polygon scan problem, $f(G)$, by replacing each vertex v with the vertex gadget $f(v)$ illustrated in Fig. 2. In this gadget, there is a gap in the corresponding polygon edge for every neighbour of the vertex. This allows either connecting to the corresponding neighbouring vertex gadget via the corridor formed by a pair of parallel edges (blue, dashed), or closing off the gap by extending an edge (red, dotted). If a vertex has degree less than four, then the positions of the corresponding scanners can be moved accordingly such that there is no gap (Fig. 2B).

Let d denote the minimum Euclidean distance between the position of any two vertices in the orthogonal drawing of G. Choose any $\epsilon \in (0, d/2)$. Now $f(G)$ consists of a set of vertex gadgets such that a gadget of width and height $d/2 - \epsilon$ is centered at the position of every vertex $v \in V$. See Figs. 2 and 3.

Components $f(v_1)$ and $f(v_2)$ can be joined by a pair of horizontal or vertical parallel edges forming a corridor if and only if vertices v_1 and v_2 are adjacent in G. Each edge in the corridor completes a partial edge in one of the two vertex gadgets. Note that the resulting instance $f(G)$ can be constructed in time proportional to the size of G on a Cartesian grid.

If G has a non-crossing spanning tree, then $f(G)$ has a simple polygonal solution formed by including the corridors that correspond to edges of the spanning tree. On the other hand, if $f(G)$ has a simple polygonal solution, then all vertex gadgets must be joined. Since every edge of the polygon must be seen by a scan, joining edges complete partial edges, and are therefore orthogonal. Since the polygon must be simple, the solution selects both or neither edge in a corridor, and edges from two crossing corridors cannot be selected simultaneously. Therefore, G has a non-crossing spanning tree. □

Since all edges in the reduction are orthogonal, the visibility-polygon scan problem remains NP-hard for orthogonal polygons. The construction can be modified to show hardness for the point-normal scan and segment scan models.

have identical y-coordinates. Detecting multiple data points on a vertical edge is slightly more complicated. We can deduce the position of the substring VV by the absence of the substring HH due to multiple data points; we then delete one of the data points without affecting the solution. Finally, vertices can report VH or HV as they choose; if we find the sequence begins or ends with H, we assign responsibility to a vertex and swap with the adjacent V. □

4 Monotone Polygons

In this section we consider the reconstruction problem for monotone (not necessarily orthogonal) polygons from a point-normal scan. In this case, each input point knows the orientation and interior of the polygon boundary passing through it. In the monotone setting, these half-spaces determine whether each non-vertical edge belongs to the upper or lower chain of the polygon. This leaves the set of vertical edges to be assigned to chains. Nevertheless, the problem is non-trivial; in particular, a solution is not necessarily unique (e.g., see Fig. 5).

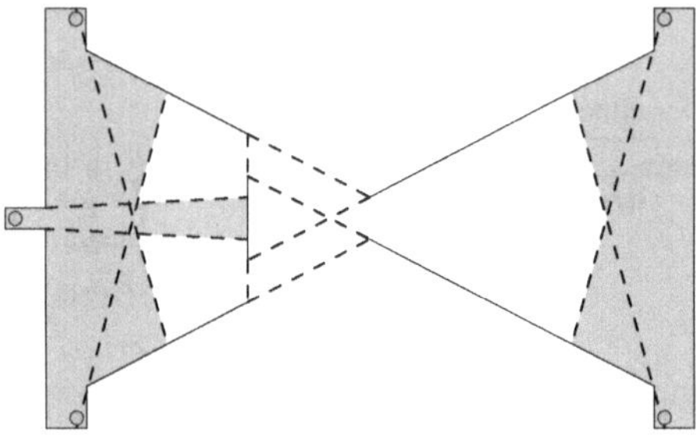

Fig. 5. Even under the visibility-polygon scan model this input instance has two distinct monotone solution polygons; the central vertical edge could belong to the lower or upper chain

Theorem 3. *A monotone polygon can be reconstructed from a point-normal scan in $O(n \log n)$ time.*

We sketch the ideas behind the dynamic programming algorithm that determines the chain to which vertical edges are assigned. Scan all data points from left to right and update a function that stores whether there is a partial solution (in the form of an upper and lower chain) up to the current x-coordinate, with some conditions on where the upper and lower chains end.

Let t denote an x-coordinate that is not the coordinate of any data point. The *upper-left line* of t is the line through the last data point before t for which the edge is not vertical and is in the upper chain (i.e., the associated half-space points downward). The *upper-right line* of t is the line through the first data point after t for which the edge is not vertical and is in the upper chain. We define the *lower-left* and *lower-right* lines of t analogously.

Observe that the vertical line through t intersects the upper chain of any solution necessarily in either the upper-left line or the upper-right line; otherwise one of the corresponding data points could not be used for the upper chain (and could not be used for the lower chain by the given normals). We compute a partial solution and prescribe which of the two lines it uses for the upper chain, and correspondingly for the lower. Thus, define $f(t, u, \ell) \in \{\text{true}, \text{false}\}$, where $u, \ell \in \{L, R\}$, and $f(t, u, \ell) = \text{true}$ if and only if there exist disjoint, monotone, upper and lower chains that may end at t with the upper line $u \in \{L, R\}$ and lower line $\ell \in \{L, R\}$, as in Fig. 6.

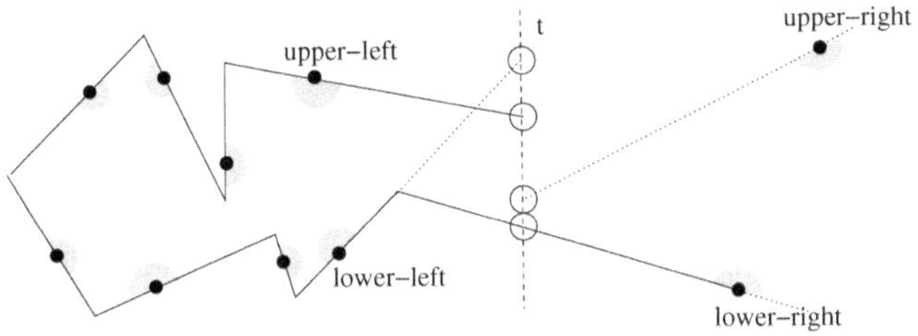

Fig. 6. In this example $f(t, L, R) = \text{true}$. White circles indicate points in which the upper/lower chain might be required to end.

We can initialize $f(t, u, \ell)$ at $t = -\infty$ and update $f(t, u, \ell)$ as t increases. We omit the details, most of which are straightforward; the full paper gives details on one of the more complex cases. Each update can be done in constant time. The time complexity for this algorithm is linear once the data points have been sorted by x-coordinate, resulting in a total running time of $O(n \log n)$. □

5 Star-Shaped Polygons

We briefly describe simple results for reconstructing a star-shaped polygon, i.e., a polygon that is entirely visible from some point in its interior. The region of points that sees all of a star-shaped polygon is its *kernel.*

Reconstructing a star-shaped polygon is straightforward in the point-normal model. Any point in the kernel must be visible to all data points. To compute the kernel, it suffices to compute the intersection I of the set of half-spaces associated with data points; this can be achieved in $O(n \log n)$ time [13]. Either all or none of the points in I are visible to all data points. Thus, to compute the polygon, select any point o in the interior of I, sort all data points in clockwise order around o, and compute the polygon defined by them in this order; either this polygon is star-shaped or there is no solution. Consequently, a solution is unique if it exists.

Theorem 4. *A star-shaped polygon can be reconstructed from a point-normal scan in $O(n \log n)$ time. Moreover, the solution is unique.*

We can also reconstruct a star-shaped polygon from a point-wall scan, but the time complexity increases. Consider the arrangement defined by the set of lines that pass through walls. As for point-normal scans, the kernel of a star-shaped polygon must be one of the cells defined by this arrangement, i.e., one of the maximal connected regions that do not contain a point on a line. There are $O(n^2)$ such cells for n lines. For each cell we can select a point o and attempt to reconstruct a star-shaped polygon with o in its kernel as explained above. The corresponding time complexity is $O(n^3 \log n)$. We suspect that this time can be improved: instead of repeating the $O(n \log n)$ test in every cell, it might be possible to update the intersection of half-planes dynamically each time a half-plane is crossed. Furthermore, we believe that the solution, if one exists, is unique. Both of these questions remain open.

Theorem 5. *A star-shaped polygon can be reconstructed from a point-wall scan in polynomial time.*

6 Lower Bound

We show the following lower bound which follows by a reduction from sorting. Details are omitted due to space constraints.

Theorem 6. *Any algorithm that reconstructs an orthogonal polygon from point-scans, point-wall scans, point-normal scans, or segment scans requires $\Omega(n \log n)$ comparisons. Furthermore, this lower bound also applies to the cases for which a solution must be orthogonal monotone or orthogonal, monotone, and star-shaped.*

7 Discussion and Directions for Future Research

If a solution is not unique, a natural question is to determine the number of additional scanners necessary to reveal the true solution. This question is NP-hard since Theorem 1 shows hardness for an instance of the corresponding decision problem. Approximation algorithms might be interesting to consider.

Several variants of our problem have not yet been considered. In particular:

- Can we reconstruct a monotone polygon from a point-wall scan?
- What other restrictions on a solution make reconstruction feasible in polynomial time? For example, a reasonable assumption could be that a room has four walls, each of which is a polygonal chain: two x-monotone walls and two y-monotone walls.
- Finally, a natural question is to consider the corresponding problems in three or higher dimensions.

Acknowledgements

The authors wish to thank Joseph Mitchell for suggesting that we examine lower bounds (Section 6) as well as a possible technique for improving the running time of our algorithm for star-shaped polygons under the point-wall scan model (Section 5).

References

1. Amenda, N., Choi, S., Kolluri, R.: The power crust, union of balls and the medial axis. Comp. Geom.: Theory & App. 19(2–3), 127–153 (2001)
2. Amenta, N., Choi, S., Kolluri, R.: The power crust. In: Proc. ACM Symp. Solid Model. & App., pp. 249–260 (2001)
3. DeltaSphere, Inc. Deltasphere 3D laser scanner (2001), `http://www.deltasphere.com/DeltaSphere-3000.htm`
4. Edmonds, J.: Paths, trees, and flowers. Can. J. Math. 17, 449–467 (1965)
5. Formann, M., Woeginger, G.J.: On the reconstruction of simple polygons. Bull. Eur. Assoc. Theor. Comp. Sc. 40, 225–230 (1990)
6. Garey, M.R., Johnson, D.S.: Computers and Intractability. W.H. Freeman, New York (1979)
7. Hopcroft, J.E., Karp, R.M.: An $n^{5/2}$ algorithm for maximum matchings in bipartite graphs. SIAM J. Comp. 1(4), 472–484 (1988)
8. Itai, A., Papadimitriou, C.H., Szwarcfiter, J.L.: Hamilton paths in grid graphs. SIAM J. Comp. 11(4), 676–686 (1982)
9. Jansen, K., Woeginger, G.J.: The complexity of detecting crossingfree configurations in the plane. BIT 33(4), 580–595 (1993)
10. Lovász, L., Plummer, M.D.: Matching theory. In: Ann. Disc. Math. North-Holland Mathematics Studies, vol. 29 (1986)
11. Nyland, L., Lastra, A., McAllister, D., Popescu, V., McCue, C.: Capturing, processing and rendering real-world scenes. In: El-Hakim, S.F., Gruen, A. (eds.) Videometrics and Optical Methods for 3D Shape Measurement, Electronic Imaging. SPIE Int. Soc. Optical. Eng. (2001)
12. O'Rourke, J.: Uniqueness of orthogonal connect-the-dots. In: Toussaint, G.T. (ed.) Computational Morphology, pp. 97–104. Elsevier, Amsterdam (1988)
13. Preparata, F., Shamos, M.: Computational Geometry: An Introduction. Springer, Heidelberg (1985)
14. Rappaport, D.: On the complexity of computing orthogonal polygons from a set of points. Technical Report SOCS-86.9, McGill University, Montréal, Canada (1986)

Computing Large Matchings in Planar Graphs with Fixed Minimum Degree

Robert Franke, Ignaz Rutter, and Dorothea Wagner

Universität Karlsruhe (TH), KIT
{franke,rutter,wagner}@ira.uka.de

Abstract. In this paper we present algorithms that compute large matchings in planar graphs with fixed minimum degree. The algorithms give a guarantee on the size of the computed matching and run in linear time. Thus they are faster than the best known algorithm for computing maximum matchings in general graphs and in planar graphs, which run in $O(\sqrt{n}m)$ and $O(n^{1.188})$ time, respectively. For the class of planar graphs with minimum degree 3 the bounds we achieve are known to be best possible. Further, we discuss how minimum degree 5 can be used to obtain stronger bounds on the matching size.

1 Introduction

A *matching* is a set of independent (i.e., pairwise non-adjacent) edges in a graph. A *maximum* matching is a matching of maximum cardinality, and a *maximal* matching cannot be enlarged by adding edges. Finding maximum matchings, or large matchings in general, has many applications, see for example the book on matching theory of Lovász and Plummer [8]. To-date the asymptotically fastest (but rather complicated) algorithm for finding maximum matchings in general graphs runs in $O(\sqrt{n}m)$ time [9], where n and m are the numbers of vertices and edges of the given graph, respectively. Only recently faster algorithms for dense graphs, for planar graphs, for graphs of bounded genus, and for general H-minor free graphs have been suggested. They are all based on fast matrix multiplication (which, as a tool, is not very practical) and run in $O(n^\omega)$ time for dense graphs [11], $O(n^{\omega/2})$ time for planar graphs [12] and for graphs of bounded genus [21], and in $O(n^{3\omega/(\omega+3)}) \subset O(n^{1.326})$ time for H-minor free graphs [21], where $\omega \leq 2.376$ is the exponent in the running time of the best-known matrix-multiplication algorithm [4]. However, for practical purposes often slower, but less complicated algorithms are used. They are based on repeatedly finding augmenting paths and have a running time of $O(nm\,\alpha(n,m))$ [19].

There has been a sequence of more and more general characterizations of graphs with *perfect matchings* [15,7,20], i.e., matchings of size $n/2$. This has also led to algorithms that test the existence of or compute perfect matchings in $o(\sqrt{n}m)$ time in, e.g., bipartite k-regular graphs [18,3] and 3-regular biconnected graphs [1]. Moreover, for planar bipartite graphs a perfect matching can be computed in $O(n \log^3 n)$ time if it exists [10,5].

Y. Dong, D.-Z. Du, and O. Ibarra (Eds.): ISAAC 2009, LNCS 5878, pp. 872–881, 2009.

There are combinatorial results that prove lower bounds on the size of maximum matchings in certain graph classes. Nishizeki and Baybars [13] show that planar graphs with minimum degrees 3,4 and 5 have matchings of size at least $(n+2)/3$, $(2n+3)/5$ and $(5n+6)/11$, respectively. Biedl et al. [2] show that maxdeg-3 graphs have a matching of size $(n-1)/3$, 3-regular graphs have a matching of size $(4n-1)/9$ and 3-connected planar graphs have a matching of size $(n+4)/3$. However, these proofs are not constructive, in particular they do not indicate a way to find such a matching faster than by computation of a maximum matching.

Recently, Rutter and Wolff [17] (a preliminary version appeared in [16]) gave fast algorithms that achieve the tight bounds of Biedl et al. Their algorithms compute matchings of size $(n-1)/3$ in maxdeg-3 graphs in linear time, of size $(4n-1)/9$ in 3-regular graphs in $O(n \log^4 n)$ time and of size $(n+4)/3$ in 3-connected planar graphs in linear time.

However, none of these results can be used to obtain matchings of guaranteed size in planar graphs with fixed minimum degree. In fact the question how fixed minimum degrees can be exploited algorithmically was posed as an open question in [17]. We answer this question and show that the tight bounds of Nishizeki and Baybars [13] for minimum degree 3 can be reached in linear time. We further analyze our algorithm in the context of minimum degree 5 and show that with some small modification it yields a matching of size $(2n+1)/5$ in this case.

The bounds of Nishizeki and Baybars [13] for 1-connected planar graphs with minimum degree 3 were also obtained by Papadimitriou and Yannakakis [14]. They analyze the structure of maximum matchings and show that the structure is such that the free vertices can be balanced against the matching edges. We show that if we construct the matching accordingly, this balancing can be done locally: there is a pairing of free vertices with matching edges such that each free vertex is adjacent to its partner.

The paper is structured as follows. In Section 2 we give a simple algorithm that already gives a non-trivial guarantee on the matching size, yet fails to reach the tight bound of $(n+2)/3$ for planar minimum degree 3 graphs. Section 3 then shows how these structural constraints can be employed to obtain a linear-time algorithm that finds matchings of size $(n+2)/3$ in planar graphs with minimum degree 3 and discuss how our how our approach can be generalized to obtain better bounds for planar graphs with minimum degree 5. We conclude and pose some open questions in Section 4. For full proofs we refer the reader to the long version[6] of this article.

2 Exploiting Minimum Degrees

In this section we describe a simple linear-time matching algorithm that already gives a non-trivial guarantee for planar mindeg 3 graphs. Our tight analysis then shows which aspects of the algorithm need to be improved in order to achieve the tight bounds of Nishizeki and Baybars [13].

We then show that certain additional structural requirements on the matching ensure that for minimum degree $\delta = 3$ we obtain the tight bound of Nishizeki and Baybars [13]. This analysis forms the basis of the algorithm presented in Section 3 where we show that a corresponding matching can be found quickly.

In order to present the algorithms we need some standard notation for graphs and matchings. Let $G = (V, E)$ be a graph and let M be a matching of G. We denote the *degree* of a vertex v by $d(v)$. A vertex in V is *free* (with respect to M) if it is not incident to an edge of M. An *augmenting path* P (with respect to M) is a path that alternates between edges in M and edges in $E \setminus M$ and starts and ends at different free vertices. In this case the symmetric difference of P and M is a matching of size $|M| + 1$. A matching is *k-free* if it does not admit an augmenting path of length at most k.

2.1 Algorithm Based on Short Augmenting Paths

We propose the following two-step algorithm MATCH3AUG: (1) Compute a maximal matching. (2) Iteratively find augmenting paths of length 3.

It is not hard to see that MATCH3AUG can be implemented to run in linear time. In the following we analyze the size of 3-free matchings in planar graphs with minimum degree δ. To this end, we divide the free vertices into two disjoint sets that we bound independently.

Let $G = (V, E)$ be a planar graph with minimum degree δ and let M be a 3-free matching. Let $e \in M$ be an edge such that there is a free vertex $v \in V$ that is incident with both endpoints of e. We say that v *covers* e and that e is *covered*. An edge of the matching that is not covered by a vertex is *open*. Let M_C and M_O denote the set of covered and open edges of M, respectively. Moreover, let F_C denote the set of vertices that cover an edge and let F_O be the set of free vertices that do not cover any edge. Note that by definition M_C and M_O form a partition of M and F_C, F_O form a partition of the free vertices of V. Hence we have that $|M| = |M_C| + |M_O|$ and $n = 2 \cdot |M| + |F_C| + |F_O|$. We now bound the number of free vertices from below by independently bounding $|F_C|$ and $|F_O|$.

Lemma 1. *Let $G = (V, E)$ be a planar graph with minimum degree δ, let M be a 3-free matching and let M_C,M_O, F_C and F_O be defined as above. Then,*

$$|F_C| \leq |M_C| \tag{1}$$

$$|F_O| \leq 2 \cdot \frac{|M_O| - 2}{\delta - 2}. \tag{2}$$

Proof. First note that Equation (1) holds since the vertex covering an edge is unique as there would be an augmenting path of length 3 otherwise.

For the proof of Equation (2) consider the bipartite auxiliary graph $G' = (V', E')$ whose vertices are the vertices in F_O and the open edges of M. We connect a vertex $v \in F_O$ with an edge m of M_O if v is adjacent to an endpoint of m in G. The graph G' is planar as it can be obtained as a minor of G by contracting matching edges and removing edges that are not incident to a free vertex. Since no vertex of F_O covers an edge, each vertex in F_O has degree at least δ in G'. Equation (2) now follows from $|E'| \leq 2 \cdot |V'| - 4 = 2 \cdot (|F_O| + |M_O|) - 4$ (bipartite, planar) and $|E'| \geq \delta |F_O|$ (minimum degree). □

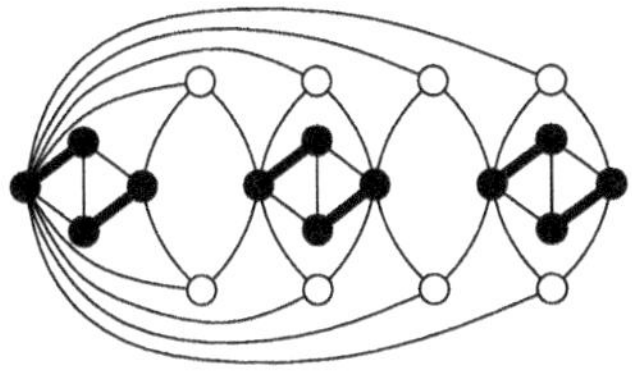

Fig. 1. Planar graph with n vertices, min-deg 3 and a 3-free matching with only $(n+4)/4$ edges

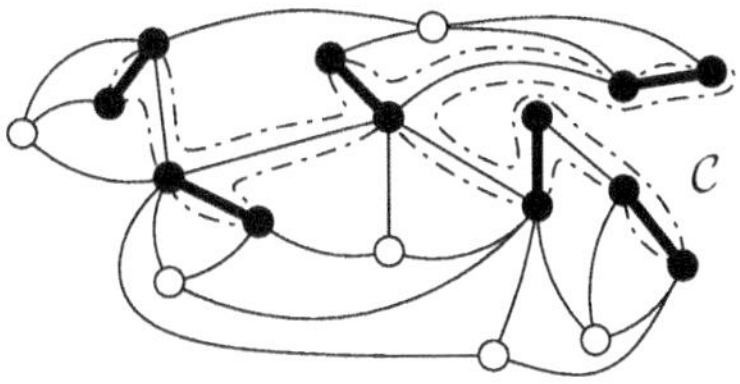

Fig. 2. Construction of curve $\mathcal{C}$ that separates matched and free vertices of a pure tree-like matching

Theorem 1. *Let $G = (V, E)$ be a planar graph with n vertices, minimum degree $\delta \in \{3, 4\}$ and let M be a 3-free matching. Then the following holds:*

$$|M| \geq \frac{(\delta - 2) \cdot n + 4}{2 \cdot (\delta - 1)}. \tag{3}$$

Proof. Follows from Lemma 1 and $|V| = 2 \cdot |M| + |F_O| + |F_C|$. □

Equation (3) does not hold for $\delta = 5$ as in this case the bound on $|F_C|$, which is independent of δ, is too weak. By Theorem 1 MATCH3AUG computes in linear time matchings of size at least $(n+4)/4$ in planar graphs with minimum degree 3 and matchings of size $(n + 2)/3$ in planar graphs with minimum degree 4.

In order to obtain the bound $(n + 2)/3$ for $\delta = 3$ we would like to improve the bound of $|F_O|$ from Equation (2). However, Fig. 1 shows that our analysis is tight. Roughly speaking the problem is that the graph induced by the matching in this example is no connected.

2.2 More Structure via Pure Tree-Like Matchings

Let $G = (V, E)$ be a planar graph with a fixed embedding, i.e., for every vertex v we have a cyclic ordering $\sigma(v)$ of its incident edges, and let M be a matching of G. Let G_M be the graph that is induced by the matched vertices of M. A matched vertex v is *cyclically pure* if its incident edges in G_M form an interval in $\sigma(v)$, further M is called *pure* if all matched vertices are cyclically pure. The matching M is called *tree-like* if G_M is a tree.

Lemma 2. *Let $G = (V, E)$ be a planar embedded graph and let M be a pure tree-like matching in G such that all free vertices have degree at least δ. Let F_M be the set of free vertices that have only matched neighbors. If F_M is not empty then there is a vertex $v \in F_M$ that has matched neighbors $x_1, \ldots, x_{\delta-2}$ and each x_i has no other neighbor in F_M.*

Proof. In this proof we distinguish between *outer* vertices, i.e., matched vertices with free neighbors and *inner* vertices, i.e., matched vertices that are not outer. To prove the lemma we consider the subgraph G' of G that is induced by the edges that have one endpoint in F_M. We show that all outer vertices share a common face in the embedding inherited from G. The main argument is that in a drawing of G with the given embedding there exists a simple closed curve $\mathcal{C}$ that contains all outer vertices, encloses all inner ones and separates the matched vertices from the free vertices (see Fig. 2). Hence, by planarity, the vertices in F_M must have a parenthetical structure, where the most interior ones have the desired property. □

This result on pure tree-like matchings can be used to improve the bound on $|F_O|$ and hence the bound on 3-free matchings.

Lemma 3. *Let $G = (V, E)$ be a planar graph with minimum degree δ, let M be a pure tree-like 3-free matching and let M_O and F_O be defined as above. Then,*

$$|F_O| \leq \frac{|M_O| - 2}{\delta - 2}. \tag{4}$$

With the stronger bound of Equation 4 it follows that a pure tree-like 3-free matching in a planar graph with n vertices and minimum degree 3 has size at least $n/3$. For minimum degrees 4 and 5 the bound on $|F_C|$ is now weaker than the bound on $|F_O|$. Hence to obtain even stronger bounds we would need to improve the bound on the size of F_C.

Unfortunately, it is not easily possible to find a maximal pure tree-like matching in a given graph. Instead we show that we can construct such a matching by carefully removing free vertices when we cannot continue with enlarging the matching. The main part is to show that the number of removed vertices is bounded by the number of matching edges.

3 Algorithm

In this section we describe an algorithm that computes in linear time a matching of size at least $(n + 2)/3$ in planar graphs with minimum degree 3. To show that our algorithm actually finds a matching of this size we use the following argument. In the course of the algorithm we perform a series of steps each of which either increases the size of the matching by 1 or deletes a free vertex. Whenever a vertex is deleted, we make sure that there is an edge in the matching that "remembers" it in such a way that each matching edge "remembers" at most one vertex and no vertex is ever "forgotten". The algorithm finishes when there are no free vertices left. The bound then follows from the observation that there can be at most as many free vertices as matching edges.

The algorithm works as follows. We start by adding an arbitrary edge to the matching, which clearly is both pure and tree-like. We then enlarge the matching and make sure it remains pure and tree-like. To find an adequate spot to try to enlarge the matching we use Lemma 2: If there are only free vertices that also have free neighbors (i.e., $F_M = \emptyset$), we can easily find an edge that can be used to enlarge the matching, see Section 3.1. If F_M is not empty (i.e. there are free vertices which have only matched neighbors) the lemma yields a free vertex v and a matched vertex x such that v is the only neighbor of x in F_M. In this case we try to enlarge the matching by two different strategies: a) If there is an augmenting path $vxyu$ of length 3, we will use this fact to swap xy for two new matching edges (Section 3.2). b) If x has free neighbors which have further free neighbors, we will use one of these and add an edge between two free vertices to the matching (Section 3.1).

In case neither of these strategies can be applied we remove v and show that there is a suitable matching edge that can remember it. The algorithm stops when no free vertices are left. In the following sections we describe these steps in detail and prove that they preserve a pure tree-like matching.

3.1 Enlargement by Adding a Suitable Edge

In this section we discuss how to enlarge a pure tree-like matching M by adding a suitable edge such that the outcome is still pure and tree-like. Consider a matched vertex x that has free neighbors and some of these have further free neighbors. Since the edges that connect x to free vertices form an interval in $\sigma(x)$, there exist a leftmost and a rightmost free neighbor of x (they coincide if x has only one free neighbor). To preserve cyclic purity we need that the leftmost or rightmost free neighbor u of x has a free neighbor u'. This situation occurs if x has at most one free neighbor that belongs to F_M and x is adjacent to a free vertex that is not in F_M, see Fig. 3a.

Lemma 4. *Let $G = (V, E)$ be a planar graph and let M be a pure tree-like matching in G such that each free vertex has degree at least δ. Further let x be a matched vertex such that the leftmost or rightmost free neighbor of x is adjacent to a free vertex. Then there is a graph $G' = (V, E')$ with $E' \subseteq E$ and a pure tree-like matching M' of G' such that $|M'| = |M| + 1$ and each free vertex has degree at least δ in G'.*

Proof. Without loss of generality, we can assume that the leftmost free neighbor u of x has a free neighbor. We now scan $\sigma(u)$ beginning with x until we find the first free neighbor u'. Let M' be $M \cup \{uu'\}$ and let G' be the graph that we obtain by removing all edges between u or u' and another matched vertex except for xu and uu'. We show that G' and M' satisfy the claim.

First, it is obvious that $|M'| = |M| + 1$ holds and each free vertex has the same degree as before since we only deleted edges that have both endpoints matched. It remains to show that M' is pure and tree-like. The vertex x is cyclically pure since u was the leftmost free neighbor of x and a possible edge xu' has been removed. Vertex u is cyclically pure, since it has just two matched neighbors x and u' and the edges ux and uu' are adjacent in $\sigma(u)$. Vertex u' is cyclically pure, because u is its only matched neighbor. The other matched vertices remain also cyclically pure as removing edges never violates cyclical purity. Thus M' is pure. Moreover, M' is tree-like since $G_{M'}$ can be obtained by adding the branch xuu' to G_M (u and u' were free and thus not in G_M). □

Fig. 3. Illustration of the different cases that can occur in the algorithm for the candidate vertex x and its unique neighbor v in F_M

3.2 Exploiting Existence of an Augmenting Path of Length 3

In this section we describe how to make use of an augmenting path of length at most 3 in our context. Let $G = (V, E)$ be a planar embedded graph, let M be a pure tree-like matching in G such that all free vertices have degree at least 3 and let $vxyu$ be an augmenting path of length 3. We show that we can modify G and M such that M is enlarged by 1 and remains pure and tree-like. The problem is that just using the augmenting path to enlarge the matching may violate cyclic purity at the vertices u, v, x and y. Instead we show that there exists a suitable augmenting path of length 3 which leads (after removing some edges whose endpoints are both matched) to an enlarged pure tree-like matching. An example of this situation is shown in Fig. 3b.

Lemma 5. *Let $G = (V, E)$ be a planar graph and let M be a pure tree-like matching in G such that each free vertex has degree at least δ. Let $vxyu$ be an augmenting path of length 3. Then there is a graph $G' = (V, E')$ with $E' \subseteq E$ and a pure tree-like matching M' such that $|M'| = |M| + 1$ and each free vertex has degree at least δ in G'.*

Proof. Let x_ℓ and x_r be the leftmost and rightmost free neighbor of x, respectively, and define y_r,y_ℓ analogously. Choose $v' \in \{x_\ell, x_r\}$, $u' \in \{y_\ell, y_r\}$ such that v' and u' are distinct.

This is always possible, otherwise $x_\ell = x_r = y_\ell = y_r$ and x and y both have only one free neighbor, which is actually shared by x and y, contradicting $v \neq u$.

We set $M' := (M \setminus \{xy\}) \cup \{v'x, yu'\}$ and let G' be the graph obtained from G by removing all edges that connect v' or u' to a matched vertex other than their matching partner. Clearly $|M'| = |M| + 1$ holds. Similar to the proof of Lemma 4 it can be seen that M' is a pure tree-like matching in G' and that every free vertex of G' has degree at least δ. □

3.3 Linear-Time Algorithm

Lemma 4 and Lemma 5 yield together with Lemma 2 the simple algorithm, whose structure was described in the beginning of this section. A pseudo-code description of our approach is shown in Algorithm 1.

Algorithm 1. MATCHMINDEG3

```
Select an arbitrary edge e and set M ← {e}
while there are still free vertices do
   if F_M ≠ ∅ then
      Select a matched vertex x and a free vertex v according to Lemma 2
      if there is an augmenting path vxyu then
         Enlarge the matching according to Lemma 5
      else if x has a free neighbor outside of F_M then
         The leftmost or rightmost free neighbor of x suits to apply Lemma 4
      else
         Remove v (the matching edge that is incident to x remembers v)
   else
      Select a matched vertex that has free neighbors and apply Lemma 4
```

Theorem 2. *Let G be a planar embedded graph with n vertices and minimum degree 3. The algorithm* MATCHMINDEG3 *computes a matching of cardinality at least $(n+1)/3$ in $O(n)$ time.*

Proof. We begin this proof by stating and justifying some loop invariants for the while-loop in MATCHMINDEG3.

(a) Each removed vertex is remembered by an adjacent matching edge
(b) Each matching edge remembers at most one vertex.
(c) The matching is pure and tree-like.
(d) Each free vertex has degree at least 3.

Invariants (a),(b) are needed to prove the correctness of the algorithm while Invariants (c),(d) ensure that the conditions of Lemmas 2, 4 and 5 are satisfied.

The algorithm preserves the invariants: When a free vertex v is removed, it is remembered by a matching edge xy (x and v are adjacent) that is not part of an augmenting path of length 3 and x has no other adjacent free vertices. Hence x and y do not have other free neighbors. Thus xy will not have to remember another vertex and it is never removed from the matching (Fig. 3c,3d). Thus Invariants (a) and (b) hold throughout the algorithm. Invariant (c) holds since we change the matching only by using Lemmas 4 and 5, which preserve the invariant. These lemmas together with the fact that we exclusively remove vertices that have only matched neighbors guarantee Invariant (d).

The size of the computed matching can now be seen as follows. Invariants (a), (b) and the observation that the last removed vertex has an additional remembering edge yields the bound $|F| \leq |M| - 1$ where F is the set of free vertices of G with respect to the output matching M. Using the equation $|F| = n - 2 \cdot |M|$ yields the bound $(n+1)/3 \leq |M|$.

Next, we discuss how to realize MATCHMINDEG3 in linear running time. Each iteration decreases the number of free vertices by at least 1. Thus the algorithm stops after at most n iterations. We show that each iteration of the while-loop runs in amortized $O(1)$ time.

For each vertex we store whether it is matched and if it has free neighbors. When a vertex v becomes matched it requires $O(d(v))$ time to propagate this information to its neighbors such that they can update their number of free neighbors. The overall-time spent in this step is linear since a matched vertex stays matched (although its matching partner may change).

For matched vertices with free neighbors, we additionally store the first and the last edge to a free vertex. With this information we can check whether a given edge is part of an augmenting path of length 3 since this involves only a constant number of vertices. Note that the first and last edge can be updated in constant time when we remove an edge or match a free vertex. Moreover, for each matched vertex we store its *match-degree*, i.e., its number of free neighbors in F_M. When the last free neighbor of a free vertex v gets matched or v is deleted, v notifies its neighbors which can update their match-degree. Both cases occur at most once for each free vertex and thus this notification work needs linear time in total. By keeping a list of vertices with match-degree 1 we can find a candidate vertex x as in Lemma 2 in constant time.

The operation provided by Lemma 5 can be realized in constant time. The total time for all applications of the procedure provided by Lemma 4 is linear. This can be seen by considering an occurrence of this case. Determining the leftmost and rightmost free neighbor of a matched vertex x is possible in constant time. Checking which one of these is not in F_M is also directly possible since we store whether a vertex has free neighbors or not. When we found the suitable free vertex u we have to traverse $\sigma(u)$ in order to find a suitable free neighbor v and delete edges to matched vertices. Afterwards, also $\sigma(v)$ has to be traversed for the same removal issue. This takes $O(d(u) + d(v))$ time and since u and v are matched afterwards, they will not be processed in the same way again. Finally, removing a free vertex v can also be done in $O(d(v))$ time. □

3.4 A Better Bound for Minimum Degree 5

For minimum degree 5 the fact that Lemma 3 yields a candidate vertex v with three neighbors having only v as neighbor in F_M can be used to improve the bound.

Theorem 3. *Let $G = (V, E)$ be a planar graph with n vertices and minimum degree 5. A matching of size at least $(2n + 1)/5$ can be computed in $O(n)$ time.*

4 Conclusion and Future Work

In this paper we have shown that it is possible to exploit minimum degrees in planar graphs algorithmically to compute matchings of guaranteed size quickly. Our algorithms run in linear time and yield matchings of size at least $(n + 2)/3$ and $(2n + 1)/5$ for planar graphs with minimum degrees 3 and 5, respectively.

While $(n+2)/3$ is tight for planar graphs with minimum degree 3, it is known that planar graphs with minimum degree 4 and 5 admit matchings of size $(2n + 3)/5$ and $(5n + 6)/11$, respectively. We leave open the question, whether these tight bounds can be achieved in linear time.

References

1. Biedl, T., Bose, P., Demaine, E., Lubiw, A.: Efficient algorithms for Petersen's theorem. J. Algorithms 38, 110–134 (2001)
2. Biedl, T., Demaine, E.D., Duncan, C.A., Fleischer, R., Kobourov, S.G.: Tight bounds on maximal and maximum matchings. Discrete Math. 285(1–3), 7–15 (2004)
3. Cole, R., Ost, K., Schirra, S.: Edge-coloring bipartite multigraphs in $O(E \log D)$ time. Combinatorica 21, 5–12 (2001)
4. Coppersmith, D., Winograd, S.: Matrix multiplication via arithmetic progressions. In: Proc. 19th Annu. ACM Conf. Theory Comput (STOC 1987), pp. 1–6 (1987)
5. Fakcharoenphol, J., Rao, S.: Planar graphs, negative weight, shortest paths, and near linear time. J. Comput. System Sci. 72, 868–889 (2006)
6. Franke, R., Rutter, I., Wagner, D.: Computing large matchings in planar graphs with fixed minimum degree. Technical Report 2009-18, Fakultät für Informatik, Universität Karlsruhe(2009), `http://digbib.ubka.uni-karlsruhe.de/volltexte/1000012970`
7. Hall, P.: On representatives of subsets. Jour. London Math. Soc. 10, 26–30 (1935)
8. Lovász, L., Plummer, M.D.: Matching Theory. North Holland, Amsterdam (1986)
9. Micali, S., Vazirani, V.V.: An $O(\sqrt{|V|} \cdot |E|)$ algorithm for finding maximum matchings in general graphs. In: Proc. 21st Annu. IEEE Sympos. Found. Comput. Sci (FOCS 1980), pp. 17–27 (1980)
10. Miller, G.L., Naor, J.: Flow in planar graphs with multiple sources and sinks. SIAM J. Comput. 24(5), 1002–1017 (1995)
11. Mucha, M., Sankowski, P.: Maximum matchings via Gaussian elimination. In: Proc. 45th Annu. IEEE Sympos. Foundat. Comput. Sci (FOCS 2004), pp. 248–255 (2004)
12. Mucha, M., Sankowski, P.: Maximum matchings in planar graphs via Gaussian elimination. Algorithmica 45(1), 3–20 (2006)
13. Nishizeki, T., Baybars, I.: Lower bounds on the cardinality of the maximum matchings of planar graphs. Discrete Math. 28(3), 255–267 (1979)
14. Papadimitriou, C.H., Yannakakis, M.: Worst-case ratios for planar graphs and the method of induction on faces. In: Proc. 22nd Annu. IEEE Sympos. Foundat. Comput. Sci (FOCS 1981), pp. 358–363 (1981)
15. Petersen, J.: Die Theorie der regulären Graphs. Acta Mathematica 15, 193–220 (1891)
16. Rutter, I., Wolff, A.: Computing large matchings fast. In: Proc. 19th Annu. ACM-SIAM Sympos. Discr. Algorithms (SODA 2008), pp. 183–192 (2008)
17. Rutter, I., Wolff, A.: Computing large matchings fast. Transactions on Algorithms (to appear)
18. Schrijver, A.: Bipartite edge coloring in $O(\Delta m)$ time. SIAM J. Comput. 28, 841–846 (1999)
19. Tarjan, R.E.: Data structures and network algorithms. SIAM, Philadelphia (1983)
20. Tutte, W.T.: The factorization of linear graphs. J. Lond. Math. Soc. 22, 107–111 (1947)
21. Yuster, R., Zwick, U.: Maximum matching in graphs with an exluded minor. In: Proc. 18th Annu. ACM-SIAM Sympos. Discr. Algorithms (SODA 2007), pp. 108–117 (2007)

Crossing-Free Acyclic Hamiltonian Path Completion for Planar *st*-Digraphs

Tamara Mchedlidze and Antonios Symvonis

Dept. of Mathematics, National Technical University of Athens, Athens, Greece
{mchet,symvonis}@math.ntua.gr

Abstract. In this paper we study the problem of existence of a crossing-free acyclic hamiltonian path completion (for short, HP-completion) set for embedded upward planar digraphs. In the context of book embeddings, this question becomes: given an embedded upward planar digraph G, determine whether there exists an upward 2-page book embedding of G preserving the given planar embedding.

Given an embedded st-digraph G which has a crossing-free HP-completion set, we show that there always exists a crossing-free HP-completion set with at most two edges per face of G. For an embedded N-free upward planar digraph G, we show that there always exists a crossing-free acyclic HP-completion set for G which, moreover, can be computed in linear time. For a width-k embedded planar st-digraph G, we show that it can be efficiently tested whether G admits a crossing-free acyclic HP-completion set.

1 Introduction

A *k-page book* is a structure consisting of a line, referred to as *spine*, and of k half-planes, referred to as *pages*, that have the spine as their common boundary. A *book embedding* of a graph G is a drawing of G on a book such that the vertices are aligned along the spine, each edge is entirely drawn on a single page, and edges do not cross each other. If we are interested only in two-dimensional structures we have to concentrate on 2-page book embeddings and to allow spine crossings. These embeddings are also referred to as 2-page *topological* book embeddings.

For acyclic digraphs, an upward book embedding can be considered to be a book embedding in which the spine is vertical and all edges are drawn monotonically increasing in the upward direction. As a consequence, in an upward book embedding of an acyclic digraph G the vertices of G appear along the spine in topological order. If G is planar upward digraph and an upward embedding of G on the plane is given, we are interested to determine a 2-page upward topological book embedding of G which preserves its plane embedding and has minimum number of spine crossings. Giordano et al. [5] showed that an embedded upward planar digraph always admits an upward topological 2-page book embedding (which preserves its plane embedding) with at most one spine crossing per edge. However, in their work no effort was made to minimize the total number of spine crossings.

Y. Dong, D.-Z. Du, and O. Ibarra (Eds.): ISAAC 2009, LNCS 5878, pp. 882–891, 2009.

The *acyclic hamiltonian path completion with crossing minimization problem* (*Acyclic-HPCCM*) was inspired by its equivalence with the problem of determining an upward 2-page topological book embedding with a minimum number of spine crossings for an embedded planar st-digraph [8].

In the *hamiltonian path completion problem* (*HPC*) we are given a graph[1] G and we are asked to identify a set of edges S (refereed to as an *HP-completion set*) such that, when the edges of S are embedded on G they turn it to a hamiltonian graph, that is, a graph containing a hamiltonian path[2]. The resulting hamiltonian graph G_S is referred to as the *HP-completed graph* of G. When we treat the HP-completion problem as an optimization problem, we are interested in HP-completion sets of minimum size. When the input graph G is an embedded planar digraph, an HP-completion set S for G must be naturally extended to include an embedding of its edges on the plane, yielding to an embedded HP-completed digraph G_S. In general, G_S is not planar, and thus, it is natural to attempt to minimize the number of edge crossings of the embedding of the HP-completed digraph G_S instead of the size of the HP-completion set S. This problem is known as *HP-completion with crossing minimization problem* (*HPCCM*) and was first defined in [8]. When the input digraph G is acyclic, we can insist on HP-completion sets which leave the HP-completed digraph G' also acyclic. We refer to this version of the problem as the *Acyclic-HPC problem*. Analogously, we define the *acyclic-HPCCM* which, as stated above, is equivalent to determining 2-page upward topological book embeddings with minimum number of spine crossings for embedded upward planar digraphs. When dealing with the acyclic-HPCCM problem, it is natural to first examine whether there exists an acyclic HP-completion set for a digraph G of zero crossings, i.e., a *crossing-free acyclic HP-completion set* for G. In terms of an upward 2-page topological book embedding, this question is formulated as follows: given an embedded upward planar digraph G, determine whether there exists an upward 2-page book embedding of G without spine crossings preserving G's embedding.

In this paper we focus on crossing-free hamiltonian path completion sets for embedded upward planar digraphs. Our results include:

1. *Given an embedded st-digraph G which has a crossing-free HP-completion set, we show that there always exists a crossing-free HP-completion set with at most two edges per face of G (Theorem 1).*

This result finds application to upward 2-page book embeddings. The problem of spine crossing minimization in an upward topological book embedding is defined with a scope to improve the visibility of such drawings. For the class of upward planar digraphs that always admit an upward 2-page book embedding (i.e. a topological book embedding without spine crossings) it make sense to define an additional criterion of visibility. When a graph is embedded in a book, its faces are split by the spine into several adjacent parts. It is clear that the

[1] In this paper, we assume that G is directed.

[2] In the literature, a *hamiltonian graph* is traditionally referred to as a graph containing a hamiltonian cycle. In this paper, we refer to a hamiltonian graph as a graph containing a hamiltonian path.

visibility of a drawing improves if each face is split into as few parts as possible. This result implies that the upward planar digraphs which admit an upward 2-page book embedding also admit one such embedding where each face is divided to at most 3 parts by the spine.

2. Given an embedded N-free upward planar digraph G, we show how to construct a crossing-free HP-completion set for G (Theorem 3). The class of embedded N-free upward planar digraphs is the class of embedded upward planar digraphs that does not contain as a subgraph the embedded N-graph of Figure 1.a. N-free upward planar digraphs have been studied in the context of partially ordered sets (posets) and lattices [1]. The class of N-free upward planar digraphs contains the class of series-parallel digraphs which has been thorough studied in the context of book embeddings [4].

3. Given a width-k embedded planar st-digraph G, we show how to determine whether G admits a crossing-free HP-completion set (Theorem 5). It follows that for fixed-width embedded planar st-digraphs, it can be tested in polynomial time whether there exists a crossing-free HP-completion set (and thus, a 2-page upward book embedding). The result is based on a reduction to the *minimum setup scheduling* problem.

For reasons of space, some proofs have been omitted and can be found in [7].

2 Terminology and Notation

Let $G = (V, E)$ be a graph. Throughout the paper, we use the term *"graph"* to refer to both directed and undirected graphs. We use the term *"digraph"* when we want to restrict our attention to directed graphs. We assume familiarity with basic graph theory [6,3]. A *drawing* Γ of graph G maps every vertex v of G to a distinct point $p(v)$ on the plane and each edge $e = (u, v)$ of G to a simple open curve joining $p(u)$ with $p(v)$. A drawing in which every edge (u, v) is a a simple open curve monotonically increasing in the vertical direction is an *upward drawing.* A drawing Γ of graph G is *planar* if no two distinct edges intersect except at their end-vertices. Graph G is called *planar* if it admits a planar drawing Γ. An embedding of a planar graph G is the equivalence class of planar drawings of G that define the same set of faces or, equivalently, of face boundaries. A planar graph together with the description of a set of faces F is called an *embedded planar graph.* Let $G = (V, E)$ be an embedded planar graph, E' be a superset of edges containing E, and $\Gamma(G')$ be a drawing of $G' = (V, E')$. When the deletion from $\Gamma(G')$ of the edges in $E' - E$ induces the embedded planar graph G, we say that $\Gamma(G')$ *preserves the embedded planar graph* G. Let $G = (V, E)$ be a digraph. A vertex of G with in-degree (resp. out-degree) equal to zero (0) is called a *source* (resp., *sink*). An *st-digraph* is an acyclic digraph with exactly one source and exactly one sink. Traditionally, the source and the sink of an st-digraph are denoted by s and t, respectively. An st-digraph which is planar and, in addition, embedded on the plane so that both of its source and sink appear on the boundary of its external face, is referred to as a *planar st-digraph.* In a planar st-digraph G each face f is bounded by two directed paths

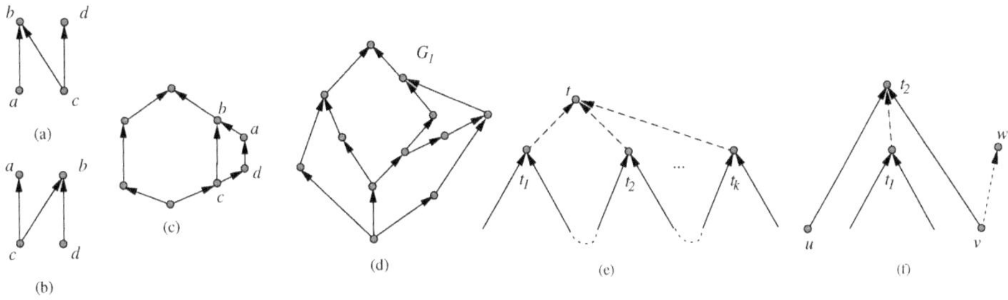

Fig. 1. (a) Embedded N-digraph. (b) Embedded И-digraph. (c) Planar digraph that is N-free if treated as an embedded planar digraph, but not N-free as a planar digraph. (d) An embedded N-free planar st-digraph G_1. (e)-(f) The construction for the proof of Theorem 3.

which have two common end-vertices. The common origin (resp., destination) of these paths is called the *source* (resp., *sink*) of f and is denoted by $source(f)$ (resp., $sink(f)$). The leftmost (resp., rightmost) of these two paths is called a *left border* (resp., *right border*) of face f. The *bottom-left* (rest., *bottom-right*) edge of a face f is the first edge on its left(resp., right) border. Similarly we define the *top-left* and the *top-right* edge of a face border. The *right(left) border* of an st-digraph is the rightmost(leftmost) path from its source s to its sink t. A new edge e that is inserted to a face f of a planar st-digraph G, with its origin and destination on the the left and right border of f, respectively, is called a *left-to-right oriented* edge. Analogously, we define a *right-to-left oriented* edge. Following the terminology of posets, the digraph $G_N = (V_N, E_N)$, where $V_N = \{a, b, c, d\}$ and $E_N = \{(a, b), (c, b), (c, d)\}$ is called an N*-digraph*. Then, any digraph that does not contain G_N as a subgraph is called an N*-free* digraph. This definition can be extended to embedded planar digraphs by insisting on a specific embedding. If we adopt the embedding of Figure 1.a. we refer to an *embedded* N*-digraph* while, if we adopt the embedding Figure 1.b. we refer to an *embedded* И-digraph. An embedded planar digraph G is then called N-free (И-free) if it does not contain any embedded N-digraph (И-digraph) as a subgraph. Figure 1.c shows an embedded N-free digraph. However, when its embedding is ignored, the digraph is not N-free since vertices a, b, c, d comprise a N-digraph.

Let $G = (V, E)$ be an embedded planar st-digraph. The external face is split into two faces, s^* and t^*. s^* is the face to the left of the left border of G while t^* is the face to the right of the right border of G. For each $e = (u, v) \in E$, we denote by $left(e)$ (resp. $right(e)$) the face to the left (resp. right) of edge e as we move from u to v. The *dual* of an st-digraph G, denoted by G^*, is a digraph such that: (i) there is a vertex in G^* for each face of G; (ii) for every edge $e \neq (s, t)$ of G, there is an edge $e^* = (f, g)$ in G^*, where $f = left(e)$ and $g = right(e)$. If G^* after this construction contains multiply edges, we substitute them by single edges. It is a well known fact that the dual graph G^* of any planar st-digraph G, is also a planar st-digraph with source s^* and sink t^*. The following definitions were given in [5] for maximal planar st-digraph. Here we extend them for planar st-digraphs. Let $G = (V, E)$ be a planar st-digraph and G^* be the dual digraph of G. Let $v_1^* = s^*, v_2^*, \ldots, v_m^* = t^*$ be the set of vertices of G^* where the indices are given according to an st-numbering of G^*. By the definition of the dual st-digraph, a vertex v_i^* of G^* $(1 \leq i \leq m)$ corresponds to a face of G. In the following we denote by v_i^* both the vertex of the dual digraph G^* and its corresponding face in digraph G. Face v_k^* is called the *k-th face* of G. Let V_k be the subset of the vertices of G that belong to faces $v_1^*, v_2^*, \ldots, v_k^*$. The subgraph of G induced by vertices in V_k is called the *k-facial subgraph* of G and is denoted by G_k. The next lemma describes how, given an st-digraph G and an st-numbering of its dual, G can be incrementally constructed from its faces. The proof is identical to the proof given in [5] for maximal planar st-digraphs.

Lemma 1. *Assume a planar st-digraph G and let $v_1^* = s^*, v_2^*, \ldots, v_m^* = t^*$ an st-numbering of its dual G^*. Consider the kth-facial subgraph G_k and the $k+1$-th face v_{k+1}^* of G, $(1 \leq k < m)$. Let s_{k+1} be the source of v_{k+1}^*, t_{k+1} be the sink of v_{k+1}^*, s_{k+1}, u_1^l, $u_2^l, \ldots,$ u_i^l, t_{k+1} be its left border, and s_{k+1}, u_1^r, $u_2^r, \ldots,$ u_j^r, t_{k+1} be its right border. Then:*

a. G_k is a planar st-digraph.
b. The vertices $s_{k+1}, u_1^l, u_2^l, \ldots, u_i^l, t_{k+1}$ are vertices of the right border of G_k.
c. G_{k+1} can be built from G_k by an addition of a single directed path s_{k+1}, u_1^r, $u_2^r, \ldots, u_j^r, t_{k+1}$. □

Let $G = (V, E)$ be an embedded planar st-digraph which has an acyclic crossing-free HP-completion set S. By $G_S = (V, E \bigcup S)$ we denote the HP-completed acyclic digraph and by P_{G_S} the resulting hamiltonian path. Note that, as S creates zero crossings with G, each edge of S is drawn within a face of G and, therefore, G_S is a planar st-digraph.

3 Two Edges Per Face Are Enough

In this Section, we prove that an embedded planar st-digraph G which has a crossing-free acyclic HP-completion set, always admits a crossing-free HP-completion set with at most two edges per face of G. This result implies that the upward planar st-digraphs which admit an upward 2-page book embedding also admit one such embedding where each face is divided to at most 3 parts by the spine. This improves the quality of the book embedding drawing.

Theorem 1. *Assume an embedded planar st-digraph G which has an acyclic crossing-free HP-completion set S. Then, there exists another acyclic crossing-free HP-completion set S' for G containing at most two edges per face of G.*

Sketch of proof: The proof is based on the fact that any three consecutive edges of the HP-completion set drawn on the same face can be substituted by a single HP-completion edge (see Figure 2). □

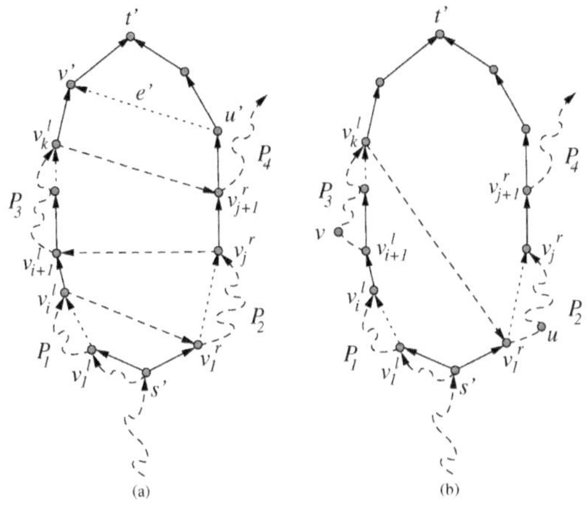

Fig. 2. (a) A crossing-free acyclic HP-completion set S which places at least three edges to a face of an st-digraph. (b) An equivalent crossing-free acyclic HP-completion set S' where the three edges $(v_i^l, v_1^r), (v_j^r, v_{i+1}^l), (v_k^l, v_{j+1}^r)$ were substituted by a single edge (v_k^l, v_1^r).

4 Embedded N-Free Upward Planar Digraphs Always Have Crossing-Free Acyclic HP-Completion Sets

In this Section, we study embedded N-free upward planar digraphs. We establish that any embedded N-free upward planar digraph G has a crossing-free acyclic HP-completion set with at most one edge per face of G. Recall that the class of embedded N-free upward planar digraphs is the class of embedded upward planar digraphs that does not contain as a subgraph the embedded N-graph of Figure 1.a. For the class of N-free upward planar embedded digraphs, which is substantially larger than the class of N-free upward planar digraphs, we show that there is always a crossing-free acyclic HP-completion set that can be computed in linear time, thus improving the results given in [1,4]

Theorem 2. *Any embedded N-free planar st-digraph $G = (V, E)$ has an acyclic crossing-free HP-completion set S which contains exactly one edge per face of G. Moreover, S can be computed in $O(V)$ time.*

Proof. Let G^* be the dual graph of G and let $s^* = v_1^*, \dots, v_m^* = t^*$ be the vertices of G^* ordered according to an st-numbering of G^*. Let G_{k-1} be the $(k-1)$-facial subgraph of G. By Lemma 1, G_k can be constructed from G_{k-1} by adding to the right border of G_{k-1} the directed path forming the right border of v_k^*.

We prove the following stronger statement than the one in the theorem:

Statement 1. *For any G_k $(1 \le k < m)$ there exists an acyclic crossing-free HP-completion set S_k such that the following holds: Let P_k be the resulting hamiltonian paht and let e be an edge of the right border of G_k that is also the bottom-left edge of a face $f \in \{v_{k+1}^*, \dots v_m^*\}$. Then, edge e is traversed by P_k.*

Proof of Statement 1. If $k = 1$, G_1 consist of a single path, that is the left border of G (Figure 3.a). We let $S_1 = \{\emptyset\}$ and set P_1 to G_1. As all the edges of G_1 are traversed by P_1 it is clear that, any edge e on the right border of G_1 that is also a bottom-left edge of any other face t is traversed by P_1. Assume now that the statement is true for any G_{k-1}, $k < m$. We will show that it is true for G_k. Denote by S_{k-1} a crossing-free acyclic HP-completion set of G_{k-1} and by P_{k-1} the produced hamiltonian path. Let e be an edge on the right border of G_{k-1} that is also the bottom-left edge of v_k^*. By the induction hypothesis, P_{k-1} passes through $e = (s_k, v)$ (see Figure 3.b). Denote by s_k and t_k the source and the sink of v_k^* respectively, and by $v_1^r, \ldots, v_{m_k}^r$ the vertices of the right border of v_k^*. By Lemma 1, s_k and t_k are vertices of the right border of G_{k-1} and G_k can be built from G_{k-1} by adding the path $s_k, v_1^r, \ldots, v_{m_k}^r, t_k$ to it. Suppose first that $m_k \neq 0$ (i.e., the right border of v_k^* contains at least one vertex). Set $S_k = S_{k-1} \bigcup \{(v_{m_k}^r, v)\}$, and $P_k = P_{k-1}[s \ldots s_k], v_1^r, \ldots, v_{m_k}^r, P_{k-1}[v \ldots t]$ (see Figure 3.c). It is clear that P_k is a hamiltonian path of G_k. This is because P_{k-1} is hamiltonian path of G_{k-1} and P_k traverses all newly added vertices. It is also easy to see that S_k is acyclic: the edge $(v_{m_k}^r, v)$ which was added to S_{k-1} creates a single directed path: from vertex s_k to the vertex v, which were already connected by the directed edge (s_k, v) in G_{k-1}. We now show that the bottom-left edge e of any $f \in \{v_{k+1}^* \cdots v_m^*\}$, where e is also on the right border of G_k, is traversed by P_k. The only edge that was added to G_{k-1} to create G_k and is not traversed by P_k, is $e' = (v_{m_k}^r, t_k)$, that is, e' is the last edge of the right border of v_k^*. If e' is also the left bottom edge of a f then the graph has an embedded N-digraph as a subgraph (see the subgraph induced by the vertices $u, t_k, v_{m_k}^r, w$ in Figure 3.c), a contradiction. Otherwise, if the bottom-left edge of f coincides with any other edge of the right border of v_k^*, then the statement holds. If f has its bottom-left edge on the right border of G_{k-1} then, by the induction, a bottom left edge of f is traversed by P_{k-1} and, thus, by P_k. Consider now the case where $m_k = 0$, that is, the right border of v_k^* is a single, transitive edge (see Figure 3.d). In this case, no new vertex is added to G_k, so we set $S_{k+1} = S_k$ and $P_{k+1} = P_k$. Consider now a face $f \in \{v_{k+1}^* \cdots v_m^*\}$. If the bottom-left edge e of f is on the right border of G_k and coincides with the transitive edge (s_k, t_k), then u, t_k, s_k, w form an embedded N-digraph (see Figure 3.d), a contradiction. So e is not (s_k, t_k) and, hence, it is an edge of the right border of G_{k-1}. So, by the induction hypothesis, e is traversed by P_{k-1} and hence by P_k. This completes the proof of the statement. Having proved Statement 1, the theorem follows from the observation that $G_m = G$. The bound on the time needed to compute the crossing-free HP-completion set easily follows from the incremental nature of the described constructive proof. □

Corollary 1. *Any И-free embedded planar st-digraph $G = (V, E)$ has an acyclic crossing-free HP-completion set S which contains exactly one edge per face of G. Moreover, S can be computed in $O(V)$ time.*

Proof. Reverse the edges of G^* and repeat the proof of Theorem 2.

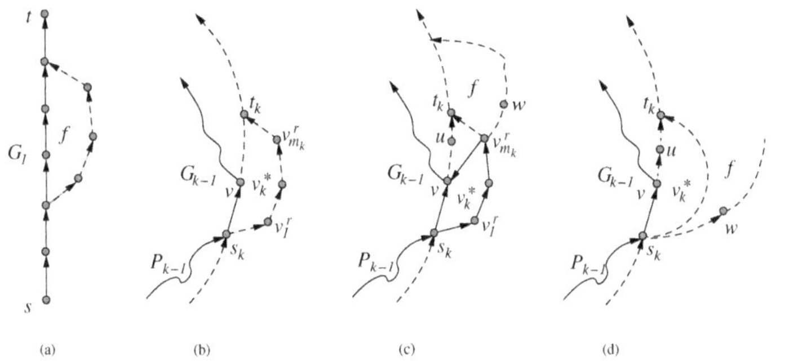

Fig. 3. (a) $G_1 = P_1$ and a face f. The bottom left edge of f is traversed by P_1. (b) G_{k-1} and v_k^*. P_{k-1} is denoted by solid line. (c) A graph G_k for the case that the right border of v_k^* contains at least one vertex. The newly constructed P_k is denoted by solid line. (d) A graph G_k for the case that the right border of v_k^* is a transitive edge.

Theorem 3. *Any embedded N-free upward planar digraph $G = (V, E)$ has an acyclic crossing-free HP-completion set that can be computed in $O(V)$ time.*

Proof. We just prove that, any embedded N-free upward planar digraph G can be transformed to an embedded N-free upward planar st-digraph G' by the addition of few edges. Then, the result follows from Theorem 2. Consider an upward planar embedding Γ of $G = (V, E)$ that is N-free. If the outer face of G contains more than one sink (source), then we add a new super-sink (super-source) vertex. Let $t_1, \ldots, t_k$ be the sinks of G in the outer face. By adding a new vertex t and by joining each t_i to t by an edge, the embedding Γ of G is preserved and remains N-free, because each t_i $1 \leq i \leq k$ has out-degree zero (see Figure 1.e). Let now some sink t_1 be placed in a inner face of G. Let t_2 be the sink of that face. We add edge (t_1, t_2). The addition of the edge (t_1, t_2) creates an embedded N-digraph only if there are edges (v, t_2) and (v, w) in G with (v, w) is the edge following (v, t_2) (in counter clockwise order), out of v. But then, there is already an embedded N-digraph in G (the digraph induced by the vertices u, t_2, v, w in Figure 1.f). A clear contradiction, so (t_1, t_2) can be added to G without creating any embedded N-digraph as a subgraph. The sources are treated similarly. The transformation of G into an st-digraph can be easily completed in linear time. □

5 Crossing-Free Acyclic HP-Completion Sets for Fixed Width st-Digraphs

In this section we establish that for any embedded planar st-digraph G of bounded width, there is a polynomial time algorithm determining whether there exists a crossing-free HP-completion set for G. In the case that such an HP-completion set exists, we can easily construct it. A set Q of vertices of G is called *independent* if the graph incident to Q has no edges. Following the terminology of partially ordered sets, we call *width* of G, and denote it by $width(G)$, the maximum integer r such that G has an independent set of cardinality r. In Minimum Setup Scheduling (MSS) we are given a number of jobs that are to be executed in sequence by a single processor. There are constraints which require that certain jobs be completed before another may start; these constraints are given in the form of *precedence dag*. In addition, for each pair i, j of jobs there is a *setup cost* representing the cost of performing job j immediately after job i, denoted by $cost(i, j)$. The objective is to find a one-processor schedule for all jobs which satisfies all the precedence constraints and minimizes the total setup cost incurred. The main idea of the result presented in this section is a simple application of an algorithm solving the *minimum setup scheduling* problem . Given a precedence dag D and a matrix C of costs, $s(D, C)$ denotes the total setup cost of a minimum cost schedule satisfying the constraints given by D. The next theorem follows from the complexity analysis given in [2].

Theorem 4 ([2]). *Given an n-vertex precedence dag D of width k and a matrix C of setup costs, we can compute in $O(n^k k^2)$ time a setup cost $s(P, C)$ of minimum cost schedule, satisfying the constraints given by P.*

In the rest of this section, we show that given a planar st-digraph G the problem of determining whether there is a crossing-free acyclic HP-completion set for G can be presented as an instance of MSS. Let $G = (V, E)$ is an embedded planar st-digraph. We define the setup cost matrix as follows. Set $C_G[i, j] = 0$ if $(v_i, v_j) \in E$ or v_i and v_j belong to the opposite borders of the same face of G, otherwise set $C_G[i, j] = 1$.

Lemma 2. *Let $G = (V, E)$ be an embedded planar st-digraph. Let also $s(G, C_G)$ be a setup cost of minimum cost schedule satisfying the constraints given by G and setup costs given by C_G. G has an acyclic crossing-free HP-completion set iff $s(G, C_G) = 0$.*

Proof. ($\Rightarrow$) Assume that G has a crossing-free acyclic HP-completion set S and the vertices in the sequence $v_1, v_2, \ldots, v_n$ are enumerated as they appear in the hamiltonian path which is created when S is embedded on G. Then, the sequence $v_1, v_2, \ldots, v_n$ presents a schedule satisfying constraints given by G, otherwise an embedding of S in G would create a cycle. The setup cost for this schedule is $\sum_{i=1}^{n-1} C_G[i, i+1]$. We know that S does not create any crossing with G. Therefore, any two successive vertices v_i and v_{i+1} of the resulting hamiltonian path are either connected by an edge of the graph or belong to the opposite borders of the same face, and thus, $C_G[i, i+1] = 0$. So, we have shown that there is a schedule of setup cost zero and, thus, $s(G, C_G) = 0$.

($\Leftarrow$) Assume now that $s(G, C_G) = 0$, i.e., there exists a one-processor schedule for the jobs represented by the vertices of G which has total setup cost zero and satisfies the precedence constraints given by G. Let v_1, $v_2, \ldots,$ v_n be the jobs as they appear in this schedule. We construct the set of edges S as follows: Consider any two successive jobs v_i and v_{i+1}. If they are not connected by an edge (v_i, v_{i+1}) of G, then we add this edge to S. All the edges added to S

correspond to two jobs with setup cost zero, and hence represent edges which connect two vertices of the opposite borders of the same face. So we have that S creates in G a hamiltonian path which does not cross any edge of G. Finally we note that (*i*) there can be no crossings among the edges of S and, (*ii*) the addition of S to G does not create any cycle. □

Theorem 5. *Let G be a planar st-digraph of width $k \in \mathbb{N}$. Then, in $O(k^2 n^k)$ time we can decide whether G has an crossing-free HP-completion set. In the event that such a set exists, it can be easily computed in the same time bounds.*

References

1. Alzohairi, M.: N-free planar ordered sets that contain no covering four-cycle have pagenumber two. The Arabian Journal for Science and Engineering 31(2A), 213–220 (2005)
2. Colbourn, C.J., Pulleyblank, W.R.: Minimizing setups in ordered sets of fixed width. order 1, 225–229 (1985)
3. Diestel, R.: Graph Theory, 3rd edn. Springer, Heidelberg (2005)
4. Giacomo, E.D., Didimo, W., Liotta, G., Wismath, S.K.: Book embeddability of series-parallel digraphs. Algorithmica 45(4), 531–547 (2006)
5. Giordano, F., Liotta, G., Mchedlidze, T., Symvonis, A.: Computing upward topological book embeddings of upward planar digraphs. In: Tokuyama, T. (ed.) ISAAC 2007. LNCS, vol. 4835, pp. 172–183. Springer, Heidelberg (2007)
6. Harary, F.: Graph Theory. Addison-Wesley, Reading (1972)
7. Mchedlidze, T., Symvonis, A.: Crossing-free acyclic hamiltonian path completion for planar st-digraphs, http://arxiv.org/abs/0909.2787 arXiv:0909.2787
8. Mchedlidze, T., Symvonis, A.: Crossing-optimal acyclic hamiltonian path completion and its application to upward topological book embeddings. In: Das, S., Uehara, R. (eds.) WALCOM 2009. LNCS, vol. 5431, pp. 250–261. Springer, Heidelberg (2009)

Covering a Graph with a Constrained Forest* (Extended Abstract)

Cristina Bazgan[1], Basile Couëtoux[1], and Zsolt Tuza[2]

[1] Université Paris-Dauphine, LAMSADE, France
{bazgan,couetoux}@lamsade.dauphine.fr
[2] Computer and Automation Institute, Hungarian Academy of Sciences, Budapest and Department of Computer Science, University of Veszprém, Hungary
tuza@sztaki.hu

Abstract. Given an undirected graph on n vertices with weights on its edges, Min WCF(p) consists of computing a covering forest of minimum weight such that each of its tree components contains at least p vertices. It has been proved that Min WCF(p) is NP-hard for any $p \geq 4$ (Imielinska *et al.*, 1993) but $(2 - \frac{1}{n})$-approximable (Goemans and Williamson, 1995). While Min WCF(2) is polynomial-time solvable, already the unweighted version of Min WCF(3) is NP-hard even on planar bipartite graphs of maximum degree 3. We prove here that for any $p \geq 4$, the unweighted version is NP-hard, even for planar bipartite graphs of maximum degree 3; moreover, the unweighted version for any $p \geq 3$ has no ptas for bipartite graphs of maximum degree 3. The latter theorem is the first-ever APX-hardness result on this problem. On the other hand, we show that Min WCF(p) is polynomial-time solvable on graphs with bounded treewidth, and for any p bounded by $O(\frac{\log n}{\log \log n})$ it has a ptas on planar graphs.

1 Introduction

Let $G = (V, E)$ be a graph with $|V| = n$ vertices. An edge cover of G is a subset of the edge set E such that every vertex is incident with at least one edge in the covering set. Finding the minimum size, $\rho(G)$, of an edge cover of a graph is a fundamental problem. As proved by Gallai [8], it is strongly related to determining the maximum size, $\nu(G)$, of a matching in G. A famous result of [8] states that any graph G without isolated vertices satisfies the identity $\nu(G)+\rho(G) = n$. As a matter of fact, the relation is much more than quantitative: every maximum matching of a graph can be extended to a minimum edge cover, and also conversely, every minimum edge cover contains a maximum matching. In this way, one can derive a minimum-size edge cover from a maximum matching M just by adding an arbitrary incident edge for each vertex missing from M. Hence, a minimum-size edge cover can be found in polynomial time.

* Research supported in part by the Hungarian Scientific Research Fund, OTKA grant T-049613.

Y. Dong, D.-Z. Du, and O. Ibarra (Eds.): ISAAC 2009, LNCS 5878, pp. 892–901, 2009.

In the case where the graph $G = (V, E)$ has weights on its edges, the minimum-weight edge cover problem can be reduced to the problem of finding a minimum-weight perfect matching; a simple reduction is described e.g. in the first volume of Schrijver's monograph [15, Section 19.2]. As a consequence, an optimal solution can be found in $O(n^3)$ time by the results of Edmonds and Johnson [7]. It should be noted, however, that the relation between maximum matchings and minimum edge covers does not remain valid for weighted graphs, neither for uniform hypergraphs of edge size greater than two [16].

A problem that generalizes the minimum-weight edge cover problem in a very natural way is the Min Weighted Constrained Forest problem, denoted by Min WCF(p) in the sequel. It consists of computing a spanning forest of G of minimum weight such that every tree component contains at least p vertices, for a given integer p. Although traditionally p is assumed to be a constant, the methods proving our positive results will allow us to take p as a function of n, too.

Monnot and Toulouse [14] proved that the unweighted version of Min WCF(3) is *NP*-hard even on planar bipartite graphs of maximum degree three. Imielinska *et al.* [10] showed that Min WCF(p) is *NP*-hard for $p \geq 4$, and that a greedy algorithm achieves a 2-approximation. Interestingly enough, a different algorithm studied by Laszlo and Mukherjee [12] has exactly the same tight worst-case ratio of 2, as well as a common generalization of those two approaches [13]. With the methods of Goemans and Williamson [9], just a slightly better ratio $2 - \frac{1}{n}$ can be achieved.

Let us denote by Min CF(p) the unweighted version of the problem. Until now nothing was known about the complexity of Min CF(p) for $p \geq 4$. We settle this problem by showing that Min CF(p) is *NP*-hard for any $p \geq 4$, already on planar bipartite graphs with maximum degree three.

Moreover, we study non-approximability of these problems for the first time. In this direction we prove that dropping the condition of planarity, Min CF(p) becomes *APX*-hard for any $p \geq 3$ on bipartite graphs, and even on those with maximum degree three. It also turns out that this weakening in the condition necessarily has to appear in non-approximability results, since we can design a polynomial-time approximation scheme for Min WCF(p) on planar graphs. In this result we may allow p to be bounded by $O(\frac{\log n}{\log \log n})$. An important tool in the proof is an algorithm computing an optimal solution for Min WCF(p) on any input graph of treewidth at most k in $O(k^{ck} p^{2k+2} n)$ time, for some constant c. This time bound is valid without any restrictions on the growth of p, hence applicable also in graph classes which cannot be treated with Courcelle's powerful method via monadic second-order logic [4]. (The latter would require to fix p as a constant.)

It is worth noting that the unweighted case admits a much simpler approach than the weighted one, with the following improved approximation ratio:

Remark 1. Contrary to the weighted case, it is very easy to find increasingly better approximations for Min CF(p) as p gets large. Indeed, since every feasible solution has at most n/p connected components, the optimum can never have a value smaller than $n - n/p = \frac{p-1}{p}n$, and hence any spanning tree gives a $(1 + \frac{1}{p-1})$-approximation.

Our results on NP- and APX-hardness are proved in Section 3, while efficient algorithms are presented in Section 4. Due to space limitations, some proofs are not given in the paper.

2 Preliminaries

We begin with some basic definitions, summarized in two groups.

Problems. First, we formally define the problem we will study in the sequel.
Min Weighted Constrained Forest p (Min WCF(p))
Input: An undirected graph $G = (V, E)$ with non-negative weights on its edges.
Output: A spanning forest of minimum weight where every tree is of order at least p (i.e., it contains at least p vertices).

The *unweighted* version of Min WCF(p) will be denoted by Min CF(p).

In our proofs we shall use the following problems:
3-Dimensional Matching (3DM)
Input: Three disjoint sets A, B, C of the same size q, and a set $\mathcal{T} \subseteq A \times B \times C$ of triplets.
Question: Does $\mathcal{T}$ contain a perfect matching, that is a subset $\mathcal{M} \subseteq \mathcal{T}$ such that $|\mathcal{M}| = q$ and no two elements of $\mathcal{M}$ agree in any coordinate (i.e., for any $(a, b, c), (a', b', c') \in \mathcal{M}$ we have $a \neq a'$, $b \neq b'$, and $c \neq c'$)?

We can associate a bipartite graph with any instance of 3DM as follows. We take an 'element-vertex' for each element of $A \cup B \cup C$, and one 'triplet-vertex' for each triplet in $\mathcal{T}$. There is an edge connecting a triplet-vertex to an element-vertex if and only if the element is a member of the triplet. Moreover, we say that the instance is planar if the graph associated with it is planar.

Max 3-Dimensional Matching (Max 3DM)
Input: Three disjoint sets A, B, C of the same size and a set $\mathcal{T} \subseteq A \times B \times C$ of triplets.
Output: A matching (set of mutually disjoint triplets) $\mathcal{M} \subseteq \mathcal{T}$ of maximum size.

The restricted versions of 3DM and of Max 3DM where each element of $A \cup B \cup C$ appears in exactly two triplets will be denoted by $3DM_2$ and Max $3DM_2$, respectively.

Approximability. Given an instance x of an optimization problem A and a feasible solution y of x, we denote by $v(x, y)$ the value of the solution y, and by $opt_A(x)$ the value of an optimum solution of x. The *performance ratio* of y is

$$R(x, y) = \max\left\{\frac{v(x, y)}{opt_A(x)}, \frac{opt_A(x)}{v(x, y)}\right\}.$$

For a constant $c > 1$, an algorithm is a *c-approximation* if for any instance x of the problem it returns a solution y such that $R(x, y) \leq c$. An optimization problem is said to be *constant approximable* if, for some $c > 1$, there exists a polynomial-time c-approximation for it. The class of problems which are constant approximable is denoted by APX. An optimization problem has a *polynomial-time approximation scheme* (a *ptas*, for short) if, for every constant $\varepsilon > 0$, there exists a polynomial-time $(1+\varepsilon)$-approximation for it.

3 Complexity for $p \geq 3$

In this section we prove that the constrained forest problem is intractable for every $p \geq 4$, even on some very restricted classes of problem instances. We first deal with time complexity and then with approximation hardness.

3.1 *NP*-Hardness for $p \geq 4$, Planar Bipartite Unweighted Graphs

Theorem 1. *For any $p \geq 4$,* Min CF(p) *is NP-hard, even on planar bipartite graphs with maximum degree 3.*

Proof. First, we prove *NP*-hardness for $p = 4$. We construct a polynomial reduction from 3DM to Min CF(4). Let $I = (A, B, C, \mathcal{T})$ be an instance of 3DM where $|A| = |B| = |C| = q$ and $\mathcal{T} = \{T_1, \ldots, T_m\}$. Assume, without loss of generality, that each element occurs in at least one triplet (otherwise I admits no perfect matching at all).

The graph instance $G(I) = (V, E)$ of Min CF(4) is constructed as follows; see an illustration in Figure 1 with a particular instance I of 3DM. For each triplet $T_\ell = (a_i, b_j, c_k) \in \mathcal{T}$ we create a copy of a star with 3 branches, that we shall call *star gadget*, with vertices $a_i^\ell, b_j^\ell, c_k^\ell, d^\ell$ and edges $(a_i^\ell, d^\ell), (b_j^\ell, d^\ell), (c_k^\ell, d^\ell)$. These stars are assumed to be vertex-disjoint. For their union we denote

$$V_{\mathcal{T}} = \bigcup_{T_\ell = (a_i, b_j, c_k) \in \mathcal{T}} \{a_i^\ell, b_j^\ell, c_k^\ell, d^\ell\} \qquad \text{and}$$

$$E_{\mathcal{T}} = \bigcup_{T_\ell = (a_i, b_j, c_k) \in \mathcal{T}} \{(a_i^\ell, d^\ell), (b_j^\ell, d^\ell), (c_k^\ell, d^\ell)\}.$$

To the elements of $A \cup B \cup C$ we assign *linking gadgets*; if an element appears in h triplets, then $h - 1$ gadgets will be associated with it.

Consider first the set A. We define the set $J_A \subset A \times \mathbb{N} \times \mathbb{N}$ to be the collection of all (a_i, r, s) with the following properties: $a_i \in A$, and r, s are two consecutive indices of triplets containing a_i, in the sense that $a_i \in T_r$ and $a_i \in T_s$ but $a_i \notin T_\ell$ for any $r < \ell < s$. (In this general setting we allow that some a_i occur in just one member of $\mathcal{T}$ — hence generating no linking gadgets — although in such a situation I would easily be reducible to a smaller instance.) Each $(a_i, r, s) \in J_A$ defines a linking gadget inducing a 'claw' in $G(I)$, with vertices $a_i^r, a_i^s, a_i^{r,s}, \overline{a}_i^{r,s}$ and edges $(a_i^r, a_i^{r,s}), (a_i^{r,s}, a_i^s), (a_i^{r,s}, \overline{a}_i^{r,s})$. For their union we denote

$$V_A = \bigcup_{(a_i,r,s)\in J_A} \{a_i^{r,s}, \overline{a}_i^{r,s}\} \quad \text{and} \quad E_A = \bigcup_{(a_i,r,s)\in J_A} \{(a_i^r, a_i^{r,s}), (a_i^{r,s}, a_i^s), (a_i^{r,s}, \overline{a}_i^{r,s})\}.$$

The sets J_B and J_C are defined in a similar way. Each $(b_j, r, s) \in J_B$ defines a linking gadget inducing a 'claw' in $G(I)$, with vertices $b_j^r, b_j^s, b_j^{r,s}, \overline{b}_j^{r,s}, \underline{b}_j^{r,s}$ and edges $(b_j^r, b_j^{r,s}), (b_j^{r,s}, b_j^s), (b_j^{r,s}, \underline{b}_j^{r,s}), (\underline{b}_j^{r,s}, \overline{b}_j^{r,s})$. For their union we denote

$$V_B = \bigcup_{(b_j,r,s)\in J_B} \{b_j^{r,s}, \underline{b}_j^{r,s}, \underline{b}_j^{r,s}\} \qquad \text{and}$$

$$E_B = \bigcup_{(b_j,r,s)\in J_B} \{(b_j^r, b_j^{r,s}), (b_j^{r,s}, b_j^s), (b_j^{r,s}, \underline{b}_j^{r,s}), (\underline{b}_j^{r,s}, \overline{b}_j^{r,s})\}.$$

The same is done for the set C in complete analogy to B, hence introducing the linking gadgets $(c_k, r, s) \in J_C$ and obtaining vertex set V_C and edge set E_C.

After all these preparations, let the graph $G(I) = (V, E)$ has vertex set $V = V_T \cup V_A \cup V_B \cup V_C$ and edge set $E = E_T \cup E_A \cup E_B \cup E_C$. Since an element of $A \cup B \cup C$ appearing in h triplets has h copies and $h-1$ linking gadgets in $G(I)$, we can see that the total number of vertices is precisely $12|\mathcal{T}| - 8q$. Hence, the transformation is linear with respect to input size.

We are going to show that I contains a perfect matching if and only if $G(I)$ contains a forest of weight at most $9|\mathcal{T}| - 6q$ in which every tree component is of order at least 4.

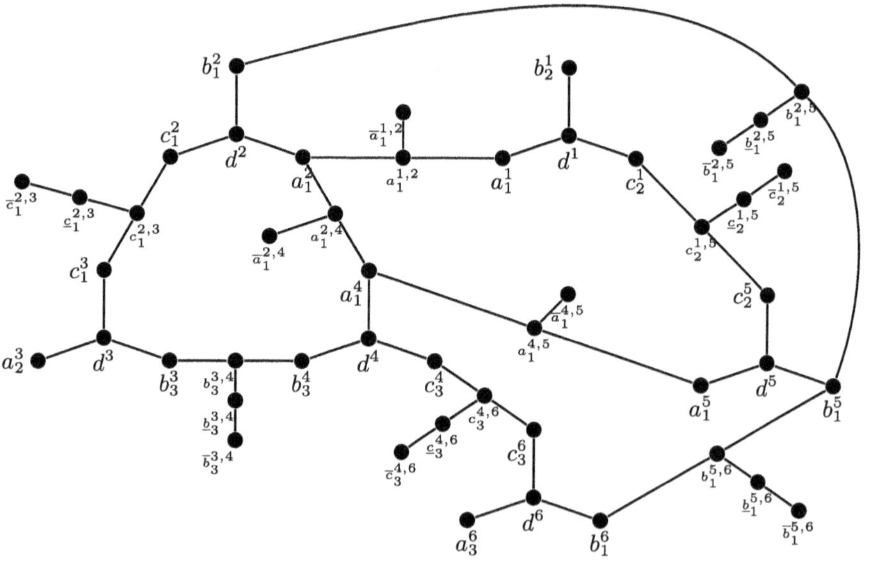

Fig. 1. Graph $G(I)$ derived from instance $I = (A, B, C, \mathcal{T})$ of 3DM with $A = \{a_1, a_2, a_3\}$, $B = \{b_1, b_2, b_3\}$, $C = \{c_1, c_2, c_3\}$, and $\mathcal{T} = \{T_1, \ldots, T_6\}$ with $T_1 = (a_1, b_2, c_2)$, $T_2 = (a_1, b_1, c_1)$, $T_3 = (a_2, b_3, c_1)$, $T_4 = (a_1, b_3, c_3)$, $T_5 = (a_1, b_1, c_2)$, $T_6 = (a_3, b_1, c_3)$

Suppose first that $\mathcal{M}$ is a perfect matching of I. For a $T_\ell = (a_i, b_j, c_k) \in \mathcal{M}$ we write $f_M(a_i) = f_M(b_j) = f_M(c_k) = \ell$, and further denote by F_T the set of edges of the star gadgets corresponding to triplets of $\mathcal{M}$:

$$F_T = \bigcup_{T_\ell=(a_i,b_j,c_k)\in\mathcal{M}} \{(a_i^\ell, d^\ell), (b_j^\ell, d^\ell), (c_k^\ell, d^\ell)\}$$

and

$$F_{a_i} = \bigcup_{\substack{r<f_M(a_i)\\(a_i,r,s)\in J_A}} \{(d^r, a_i^r), (a_i^r, a_i^{r,s}), (a_i^{r,s}, \overline{a}_i^{r,s})\} \cup$$

$$\bigcup_{\substack{s>f_M(a_i)\\(a_i,r,s)\in J_A}} \{(d^s, a_i^s), (a_i^s, a_i^{r,s}), (a_i^{r,s}, \overline{a}_i^{r,s})\},$$

$$F_{b_j} = \bigcup_{\substack{r<f_M(b_j)\\(b_j,r,s)\in J_B}} \{(b_j^r, b_j^{r,s}), (b_j^{r,s}, \underline{b}_j^{r,s}), (\underline{b}_j^{r,s}, \overline{b}_j^{r,s})\} \cup$$

$$\bigcup_{\substack{s>f_M(b_j)\\(b_j,r,s)\in J_B}} \{(b_j^s, b_j^{r,s}), (b_j^{r,s}, \underline{b}_j^{r,s}), (\underline{b}_j^{r,s}, \overline{b}_j^{r,s})\}.$$

The set F_{c_k} is defined like F_{b_j}. Let $F_M = \bigcup_{e\in A\cup B\cup C} F_e \cup F_T$. Forest F_M covers $G(I)$ with trees of order 4. Thus, it is a forest of weight $9|\mathcal{T}| - 6q$.

Conversely, let F be a covering forest of $G(I)$ of weight at most $9|\mathcal{T}| - 6q$, where each tree is of order at least 4. Remark that since the graph $G(I)$ has exactly $12|\mathcal{T}| - 8q$ vertices, a covering forest without isolated edges and isolated vertices is of weight at least $9|\mathcal{T}| - 6q$. Thus, every tree in F is of order exactly 4, and F is of weight $9|\mathcal{T}| - 6q$. All edges $(a_i^{r,s}, \overline{a}_i^{r,s})$ are in F since this is the only edge incident to $\overline{a}_i^{r,s}$. Since F has only trees of order 4, just one of the edges $(a_i^{r,s}, a_i^r)$ and $(a_i^{r,s}, a_i^s)$ is in F. Therefore in the family $\{a_i^\ell\}_\ell$ there exists exactly one vertex that is not incident to a linking gadget in F. Since forest F is of weight $9|\mathcal{T}| - 6q$, it contains $3|\mathcal{T}| - 2q$ trees. Since for any element of A, B, C the number of linking gadgets associated is equal to the number of occurrences of this element in $\mathcal{T}$ minus 1, there are $3|\mathcal{T}| - 3q$ linking gadgets in $G(I)$. It means that there are q trees of F which do not cover any linking gadget. Each of these remaining trees therefore covers a star gadget of the form $a_i^\ell, b_j^\ell, c_k^\ell, d^\ell$. From these star gadgets we extract a collection of q triplets (a_i, b_j, c_k), which is a valid solution in I since every triplet appears in $\mathcal{T}$ by construction and every element of the triplets appears in exactly one occurrence.

Since 3DM is *NP*-hard even on planar instances [5] and the previous reduction preserves planarity, we can restrict MIN CF(4) to planar bipartite graphs of maximum degree 3. This completes the proof for $p = 4$. The construction of $G(I)$ for $p > 4$ consists of replacing some edges of $G(I)$ for $p = 4$ by paths of length $p - 3$. □

3.2 *APX*-Hardness for $p \geq 3$, Unweighted Bipartite Graphs

Applying L-reduction from MAX $3DM_2$, the following result can be obtained.

Theorem 2. *For any $p \geq 3$,* MIN CF(p) *is* APX*-hard. Moreover,* MIN CF(p) *for $p \geq 3$ is not $\frac{95(8p-7)+1}{95(8p-7)}$-approximable on bipartite graphs of maximum degree 3, unless P=NP.*

4 Efficient Algorithms

4.1 Exact Algorithms for Bounded Treewidth Graphs

In this section we present an efficient algorithm solving MIN WCF(p) on graph classes of bounded treewidth. For undefined details on tree decomposition we refer to [11].

Theorem 3. *The problem* MIN WCF(p) *on graphs with treewidth bounded by k can be solved in time $O(k^{ck}p^{2k+2}n)$ for some constant c, where n is the number of vertices.*

Remark 2. The time bound is polynomial in n whenever $k(\log k + \log p) = O(\log n)$ and in particular if $k = O(\frac{\log n}{\log \log n})$ and $p = O(\log n)$.

Proof. Let G be a graph of treewidth k, and $(T_G, \{X \mid x \in V(T_G)\})$ a tree decomposition of G with width k. As a notational convention, the vertex subset of G associated with node x of T_G is denoted by the corresponding upper-case letter X, and we shall use the same subscript for them where necessary. We view T_G as a *rooted* tree. By assumption, the set X associated with any node x of T_G contains at most $k+1$ vertices of $V(G)$. We shall assume further that every node x is of one of the following types:

- a *start* node that has no children (a leaf in T_G),
- a *join* node that has two children x_1, x_2 and $X_1 = X_2 = X$,
- an *introduce* node that has one child x_1 and $X_1 = X \setminus \{v\}$ for some $v \in V(G)$,
- a *forget* node that has one child x_1 and $X_1 = X \cup \{v\}$ for some $v \in V(G)$.

This is called a "nice tree decomposition" in the literature, see pp. 149–150 of [11]. As it is well known, every graph admits a tree decomposition of size $O(n)$ satisfying these conditions. It is also easy to see that any tree decomposition of width k can be modified to one with the properties above in $O(kn)$ time.

For any node x of T_G, let $T_G(x)$ be the rooted subtree of T_G containing exactly the node x and its descendants. The vertices of G which appear in the sets associated with the nodes of $T_G(x)$ define a subgraph of G, denoted by $G(x)$. Remark that $T_G(x)$ is a tree decomposition of $G(x)$.

Consider any node x of T_G. For each spanning forest F (possibly with many vertices of degree zero in F) of the subgraph $G[X]$ induced by X in G, we consider all partitions $\mathcal{P} = (C_1, \dots, C_\ell)$ of X and all ℓ-tuples $I = (i_1, \dots, i_\ell) \in \{1, \dots, p\}^\ell$ such that the following conditions are met:

– if $v_i, v_j \in X$ are in the same tree component of F, then v_i, v_j belong to the same partition class C_r,
– if $|C_r| \leq p$, then $|C_r| \leq i_r \leq p$, and otherwise $i_r = p$.

Let $f(x, F, \mathcal{P}, I)$ be defined as the value of a minimum-weight spanning forest of $G(x)$ in which X induces precisely F, and in which two vertices belong to the same tree component if and only if they are in the same class of $\mathcal{P}$.

Lemma 1. *For any vertex x of T_G, for any forest F over the subgraph $G[X]$, for any partition $\mathcal{P}$ of X, and for any ℓ-tuple I of integers satisfying the conditions above, we can determine $f(x, F, \mathcal{P}, I)$ in $O(k^{ck}p^{2k+2})$ time if the corresponding values f are available for the child(ren) of x.*

Sketch of Proof. One has to consider each type of nodes separately. Due to space limitations, we give the argument for *join* nodes only, which is the most time-consuming case. If x is a *join* node, then $\mathcal{P}$ has to be generated from two finer partitions over X, one for X_1 and the other for X_2. We can do this by artificially completing F with sets E_{blue}, E_{red} of blue and red edges (not necessarily edges of G) to obtain a forest, each tree component of which spans a class of $\mathcal{P}$. We then consider the forests F_{blue} and F_{red} which are respectively the forests defined by $F \cup E_{blue}$ and $F \cup E_{red}$. The tree components of these forests define the partitions $\mathcal{P}_{blue}$ and $\mathcal{P}_{red}$ over the vertices of X. The vectors I_1 and I_2 are such that, for every partition class C_j of $\mathcal{P}$, i_j is equal to the minimum of p and the sum of the values of the blue and red classes included in C_j minus $|C_j|$. Then,

$$f(x, F, \mathcal{P}, I) = \min_{F_{blue}, F_{red}, I_1, I_2} f(x_1, F, \mathcal{P}_{blue}, I_1) + f(x_2, F, \mathcal{P}_{red}, I_2) - w(F).$$

The number of relevant blue-red forests is not larger than the number B_{k+1} of partitions of a $(k+1)$-element set. For each of them, processing all combinations of $O(p^{k+1})$ records at X_1 and the same amount of data at X_2 may need $O(p^{2k+2})$ time. This yields the upper bound $B_{k+1} \cdot O(p^{2k+2})$ altogether. □

To conclude the proof of Theorem 3, we traverse T_G in postorder, and compute $f(x, F, \mathcal{P}, I)$ for all possible choices of node x, spanning forest F of $G[X]$, partition $\mathcal{P}$ of X, and sequence I satisfying the conditions given at the beginning of the proof. Then the solution of MIN WCF(p) on G is equal to the smallest value of f occurring at the root of T_G for a sequence $I = (p, \ldots, p)$ of any length.

The overall upper bound is obtained by the facts that for any k, the number of choices for $\mathcal{P}$ is bounded above by B_{k+1}, the number of vertex-labeled trees of order t is t^{t-2} (this puts a bound on the choices for F), and the number of sequences I at x is at most p^{k+1}. The most time-consuming step is to compute f at a *join* node; it can be organized by taking all combinations of the values stored at x_1 and x_2. □

Remark 3. The same method can be applied to solve the more general problem where, instead of a uniform condition p for the constrained forest, the input graph $G = (V, E)$ on vertex set $V = \{v_1, \ldots, v_n\}$ is given together with a vector

$(p_1, \ldots, p_n)$ of natural numbers $p_i \le p$, and it is required that each v_i be contained in a tree component which has at least p_i vertices in the spanning forest. The steps of the algorithm above can be adjusted for this variant.

4.2 Ptas for Planar Graphs

Theorem 4. *For* $p = O(\frac{\log n}{\log \log n})$, MIN WCF$(p)$ *on planar graphs admits a ptas.*

Proof. Given a planar embedding of an input graph, we consider the set of the vertices which are on the exterior face, they will be called *level 1 vertices*. By induction we define *level* k as the vertices which are on the exterior face when we have removed the vertices of levels smaller than k [1]. A planar embedding is k-*level* if it has no nodes of level greater than k. If a planar graph is k-level, it has a k-outerplanar embedding.

If we want to achieve an approximation within $1 + \epsilon$, let us consider $k = \lceil \frac{4(p-1)}{\epsilon} \rceil$. Let X_t be the set of vertices of level t and let H_i, $0 \le i \le k-1$, be the graph obtained from G by deleting all edges between the vertices X_t and X_{t+1} for all t such that $t \equiv i \pmod{k}$. The set of vertices $\bigcup_{t+1 \le j \le t+k} X_j$, for $t \equiv i \pmod{k}$ is therefore a component in H_i since we have deleted all edges from this set of vertices to the other vertices of the graph. Hence, the subgraph containing exactly $\bigcup_{t+1 \le j \le t+k} X_j$ is k-outerplanar, and so is H_i.

Since H_i is k-outerplanar, it has treewidth at most $3k - 1$ [2]. By applying Theorem 3 and Remark 3, we can efficiently determine an optimal forest S_i on H_i; the condition on p remains true on vertices not incident with edges deleted, and is changed to zero on the boundary vertices of H_i, that means vertices in X_{t+1}, X_{t+k} with $t \equiv i \pmod{k}$. We complete the solution S_i obtained on H_i into a feasible solution F_i on G by choosing the useful edges of smallest cost greedily within the edges of G until every tree is of order at least p.

We are going to prove that the best forest among $F_0, \ldots, F_{k-1}$ is an $(1 + \epsilon)$-approximation of the optimal value on G. We will use two lemmas and some notation for the proof. For any tree T of G, let $w_{max}(T)$ denote the maximum weight of the edges of T, $w_{max}(T) = \max_{e \in E(T)} w(e)$. Let $g : V \to \mathbb{R}$ be defined as $g(v) = \min_{T \text{ tree}, |V(T)|=p, v \in V(T)} w_{max}(T)$.

Lemma 2. *Let F be a spanning forest of G, and $V' = \{v_1, \ldots, v_r\}$ a set of vertices such that any tree T of F which is not of order at least p contains a vertex $v_i \in V'$. Then we can construct a forest of G in which every tree component has order at least p and the total weight is at most $w(F) + \sum_{v \in V'} g(v)$.*

Lemma 3. $\sum_{v \in V} g(v) \le 2(p-1)\, opt(G)$.

Now we are in a position to complete the proof of the theorem. Let $V_i = \bigcup_{t \equiv i \pmod{k}} (X_{t+1} \cup X_{t+k})$. By Lemma 2, starting from S_i we can construct a forest F_i satisfying the property that every tree component is as large as required, moreover $w(S_i) + \sum_{v \in V_i} g(v) \ge w(F_i)$. Since S_i is an optimal solution of a relaxed problem, we have $w(S_i) \le opt(G)$ and so $opt(G) + \sum_{v \in V_i} g(v) \ge w(F_i)$.

Moreover, using Lemma 3 we obtain the following inequality $\sum_{1\le i\le k} g(V_i) = 2\sum_{v\in V} g(v) \le 4(p-1)\,opt(G)$. Hence, there exists an i_0 such that $\sum_{v\in V_{i_0}} g(v) \le \frac{4(p-1)}{k}\,opt(G)$, which implies $\min_{1\le i\le k} w(F_i) \le w(F_{i_0}) \le (1+\frac{4(p-1)}{k})\,opt(G) \le (1+\epsilon)\,opt(G)$. The overall running time of the algorithm is k times what we need for graphs of treewidth at most k, that is $O((\frac{4p}{\varepsilon})^{dp/\varepsilon}n)$, where d is a constant. Hence, since $p = O(\frac{\log n}{\log\log n})$ we have a ptas. □

Remark 4. We can improve the time of the ptas given in Theorem 4 using sphere cut decomposition instead of bounded treewidth decomposition. In this way we obtain a ptas whose running time is $O((4p)^{8p/\varepsilon}n)$

References

1. Baker, B.S.: Approximation algorithms for NP-complete problems on planar graphs. Journal of the Association for Computing Machinery 41(1), 153–180 (1994)
2. Bodlaender, H.L.: A partial k-arboretum of graphs with bounded treewidth. Theoretical Computer Science 209, 1–45 (1998)
3. Chlebik, M., Clebikova, J.: Complexity of approximating bounded variants of optimization problems. Theoretical Computer Science 354, 320–338 (2006)
4. Courcelle, B.: The monadic second-order logic of graphs. III. Tree-decompositions, minors and complexity issues. RAIRO Informatique Théorique Appliquée 26, 257–266 (1992)
5. Dyer, M.E., Frieze, A.M.: Planar 3DM is NP-complete. Journal of Algorithms 7, 174–184 (1986)
6. Dorn, F., Penninkx, E., Bodlaender, H.L., Fomin, F.: Efficient Exact Algorithms on Planar Graphs: Exploiting Sphere Cut Decompositions. Technical Report UU-CS-2006-006
7. Edmonds, J., Johnson, E.L.: Matching, Euler tours and the Chinese postman. Mathematical Programming 5, 88–124 (1973)
8. Gallai, T.: Über extreme Punkt- und Kantenmengen. Annales Universitatis Scientiarum Budapestinensis de Rolando Eötvös Nominatae. Sectio Mathematica 2, 133–138 (1959)
9. Goemans, M.X., Williamson, D.P.: A general approximation technique for constrained forest problems. SIAM Journal of Computing 24, 296–317 (1995)
10. Imielinska, C., Kalantari, B., Khachiyan, L.: A greedy heristic for a minmum-weight forest problem. Operations Research Letters 14, 65–71 (1993)
11. Kloks, T. (ed.): Treewidth. LNCS, vol. 842. Springer, Heidelberg (1994)
12. Laszlo, M., Mukherjee, S.: Another greedy heuristic for the constrained forest problem. Operations Research Letters 33, 629–633 (2005)
13. Laszlo, M., Mukherjee, S.: A class of heuristics for the constrained forest problem. Discrete Applied Mathematics 154, 6–14 (2006)
14. Monnot, J., Toulouse, S.: The path partition problem and related problems in bipartite graphs. Operations Research Letters 35, 677–684 (2007)
15. Schrijver, A.: Combinatorial Optimization. Springer, Heidelberg (2003)
16. Tuza, Z.: Extensions of Gallai's graph covering theorems for uniform hypergraphs. Journal of Combinatorial Theory Series B 52, 92–96 (1991)

Tri-Edge-Connectivity Augmentation for Planar Straight Line Graphs

Marwan Al-Jubeh[1], Mashhood Ishaque[1], Kristóf Rédei[1], Diane L. Souvaine[1], and Csaba D. Tóth[1,2]

[1] Department of Computer Science, Tufts University, Medford, MA, USA *
{maljub01,mishaque,kredei01,dls}@cs.tufts.edu
[2] Department of Mathematics, University of Calgary, AB, Canada **
cdtoth@ucalgary.ca

Abstract. It is shown that if a planar straight line graph (PSLG) with n vertices in general position in the plane can be augmented to a 3-edge-connected PSLG, then $2n-2$ new edges are enough for the augmentation. This bound is tight: there are PSLGs with $n \geq 4$ vertices such that any augmentation to a 3-edge-connected PSLG requires $2n-2$ new edges.

1 Introduction

Connectivity augmentation is a classical problem in graph theory with application in network design. Given a graph $G(V, E)$, with vertex set V and edge set E, and a constant $k \in \mathbb{N}$, find a minimum set E' of *new edges* such that $G'(V, E \cup E')$ is k-connected (respectively, k-edge-connected). For $k = 2$, Eswaran and Tarjan [2] and Plesník [10] showed independently that both edge- and vertex-connectivity augmentation problems can be solved in linear time. Watanabe and Nakamura [16] proved that the edge-connectivity augmentation problem can be solved in polynomial time for any $k \in \mathbb{N}$. The runtime was later improved (using the edge-splitting technique of Lovász [6] and Mader [7]) by Frank [3] and by Nagamochi and Ibaraki [8]. Jackson and Jordán [4] proved that the vertex-connectivity augmentation problem can be solved in polynomial time for any $k \in \mathbb{N}$. For related problems, refer to surveys by Nagamochi and Ibaraki [9], and by Kortsarz and Nutov [5].

The results on connectivity augmentation of *abstract* graphs do not apply if the input is given with a planar embedding, which has to be respected by the new edges (e.g., in case of physical communication or transportation networks). A *planar straight line graph* (PSLG) is a graph $G = (V, E)$, where V is a set of distinct points in the plane, and E is a set of straight line segments between the points in V such that two segments may intersect at their endpoints only. Given a PSLG $G = (V, E)$ and an integer $k \in \mathbb{N}$, the *embedding preserving k-connectivity* (resp., *k-edge-connectivity*) *augmentation* problem asks for a set of new edges E' of minimal cardinality such that $G = (V, E \cup E')$ is a k-connected (resp., k-edge-connected) PSLG. Since every planar graph has at

* Partially supported by NSF grant CCF-0830734.
** Supported by NSERC grant RGPIN 35586. Research done at Tufts University.

Y. Dong, D.-Z. Du, and O. Ibarra (Eds.): ISAAC 2009, LNCS 5878, pp. 902–912, 2009.

least one vertex of degree at most 5, the embedding preserving k-connectivity (k-edge-connectivity) augmentation problems make sense for $1 \leq k \leq 5$ only.

Rutter and Wolff [11] showed that both the embedding preserving vertex- and edge-connectivity augmentation problems are NP-hard for $k = 2, \ldots, 5$. They reduce PLANAR3SAT to a decision problem whether a PSLG with m vertices of degree $k-1$ can be augmented to a k-edge-connected PSLG with $m/2$ new edges. The preservation of the input embedding imposes a severe restriction: for example, a path (as an abstract graph) can be augmented to a 2-connected graph by adding one new edge—however if a path is embedded as a zig-zag path with n vertices in convex position, then it takes $n-2$ (resp., $\lfloor n/2 \rfloor$) new edges to augment it to a 2-connected (2-edge-connected) PSLG [1]. If the vertices of a PSLG G are in convex position, then it *cannot* be augmented to a 3-connected (or 3-edge-connected) PSLG, since any triangulation (which is a *maximal* augmentation) on $n \geq 3$ vertices in convex position has a vertex of degree 2.

A few combinatorial bounds are known on the maximum number of new edges needed for the embedding preserving augmentation of PSLGs. It is easy to see that every PSLG with $n \geq 2$ vertices and p connected components can be augmented to a *connected* PSLG by adding at most $p - 1 \leq n - 1$ new edges. Every connected PSLG G with $n \geq 3$ vertices and $b \geq 2$ distinct 2-connected blocks can be augmented to a *2-connected* PSLG by adding at most $b - 1 \leq n - 2$ new edges [1]. Every connected PSLG with $n \geq 3$ vertices can be augmented to a *2-edge-connected* PSLG by adding at most $\lfloor (2n-2)/3 \rfloor$ new edges [13]. These bounds are tight in the worst case [1].

Valtr and Tóth [14] recently characterized the PSLGs that can be augmented to a 3-connected or a 3-edge-connected PSLG, respectively. In particular, a PSLG $G = (V, E)$ can be augmented to a *3-connected* PSLG if and only if V is not in convex position and E does not contain a chord of the convex hull $\mathrm{ch}(V)$. It can be augmented to a *3-edge-connected* PSLG if and only if V is not in convex position and E does not contain a chord of the convex hull $\mathrm{ch}(V)$ such that all vertices on one side of the chord are incident to $\mathrm{ch}(V)$. A PSLG with these properties is called *3-augmentable* and *3-edge-augmentable*, respectively. They also showed that every 2-edge-connected 3-edge-augmentable PSLG can be augmented to a 3-edge-connected PSLG with at most $n-2$ new edges, and this bound is the best possible [14].

Results. We show that every 3-edge-augmentable PSLG with n vertices can be augmented to 3-edge-connected PSLG with at most $2n-2$ new edges. This bound is the best possible. Our proof is algorithmic, and the new edges can be computed in $O(n \log^2 n)$ time in the real RAM model.

Lower bounds. We present two lower bound constructions. First, consider the empty graph with $n-1$ vertices in convex position, $n \geq 4$, and one vertex in the interior of their convex hull (Fig. 1, left). The only 3-connected or 3-edge-connected augmentation is a wheel graph with $2n-2$ edges. Second, consider a triangulation with m vertices, $2m-5$ bounded faces and the outer face. Put a singleton vertex in each bounded face, and 2 singletons behind each edge in the outer face as in Fig. 1, right. The only 3-edge connected augmentation is obtained by adding 3 new edges for each singleton in a bounded face, and 5 new edges for a pair of singletons behind each edge. A graph with $n = m + (2m-5) + 6 = 3m + 1$ vertices requires $3(2m-5) + 5 \cdot 3 = 2n - 2$ new edges.

Notation for bridges, components, and vertices along the convex hull. We distinguish two types of bridges, edges, singletons, and non-singleton components. In the input graph G, let b_h (resp., g_h, r_h, and s_h) denote the number of bridges (resp., edges, reflex vertices, and singletons) along the convex hull $\mathrm{ch}(G)$. Clearly, we have $b_h \leq g_h$. Let c_h be the number of non-singleton components with at least one vertex incident on the convex hull. Let $b_i = b - b_h$, $c_i = c - c_h$, $r_i = r - r_h$, and $s_i = s - s_h$.

Stage 1. Deformable edges for bounded faces. For every bounded face F in G, add a deformable edge $\tau(F)$ parallel to an arbitrary edge in E adjacent to F. We have created $\boxed{f}$ deformable edges, each of which is parallel to an existing edge of G.

Stage 2. Convex hull edges. Augment G_1 with the edges of the convex hull $\mathrm{ch}(G)$ (if they are not already present in G_1). The number of vertices along the convex hull is $r_h + s_h$, and so we have added at most $\boxed{r_h + s_h - g_h}$ new edges.

Lemma 3. *At the end of stage 2, we have* $\lambda_h(G_2) \leq c_h + s_h + g_h$.

Proof. Since all hull edges are already included in the PSLG G_2, the hull vertices are in one 2-edge-connected component of G_2. If any two hull vertices are connected by a path in the interior of $\mathrm{ch}(G)$, then they are also in the same 3-edge-connected component. Walk around the boundary of $\mathrm{ch}(G)$ in counterclockwise orientation, starting from an arbitrary vertex. We will assume that every vertex along the walk belongs to distinct 3-edge-connected components unless we can show that it is connected by a path in the interior of $\mathrm{ch}(G)$ to a previous vertex along the walk. The walk visits all $c_h + s_h$ (singleton or non-singleton) components of the input graph G incident on the convex hull. There are $c_h + s_h$ edges along which the walk arrives to a component of G for the first time, and the walk traverses g_h edges of G (staying in the same component of G). Let z denote the number of remaining hull edges: these are not edges of G and they lead to a component previously visited by the walk. The convex hull has a total of $c_h + s_h + g_h + z$ edges. Assume that the walk arrives to component $C \subseteq G$ for the kth time, $k \geq 2$, at some vertex v. Let u denote the last vertex of C visited by the walk. Then C contains a path from u to v whose edges lie in the interior of $\mathrm{ch}(G)$, hence v and u are 3-edge-connected in G_2. Hence, we have $\lambda_h(G_2) \leq c_h + s_h + g_h$. □

Stage 3. Connecting non-singleton components. For each non-singleton component of G_2 that lies in the interior of $\mathrm{ch}(G)$, we add two new edges to connect it to the component containing the convex hull.

Repeat the following procedure while there is a non-singleton component in the interior of $\mathrm{ch}(G)$. Refer to Fig. 2. Let G_h denote the connected component that contains the hull edges. Let H be the disjoint union of all non-singleton components in the interior of $\mathrm{ch}(G)$. Let U denote the set of vertices of $\mathrm{ch}(H)$. Each vertex $u \in U$ is a reflex vertex in G_2. Shoot a ray along the bisector of the reflex angle at each $u \in U$, until it hits an edge $v_u w_u$ outside of $\mathrm{ch}(H)$. Pick a vertex $u \in U$ for which the distance between u and the supporting line of $v_u w_u$ is minimal. Compute the

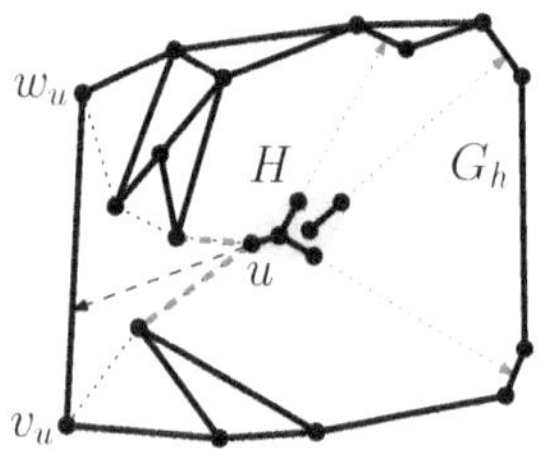

Fig. 2. Connecting a non-singleton component

shortest (i.e., geodesic) paths $\mathrm{path}(u, v_u)$ and $\mathrm{path}(u, w_u)$ with respect to the connected PSLG G_h. Any internal vertex of these paths is closer to the supporting line of $v_u w_u$ than u, and so it must be a vertex of G_h. Augment G_2 with the edges incident to u along each of $\mathrm{path}(u, v_u)$ and $\mathrm{path}(u, w_u)$. The new edges are not bridges, and they partition the reflex angle at u into three convex angles. We have used $\boxed{2c_i}$ new edges. All c_i non-singleton components have been connected, and the number of reflex vertices in the interior of $\mathrm{ch}(G)$ has decreased by at least c_i. The resulting PSLG G_3 has one non-singleton component (which contains all hull edges) and all singletons lie in bounded faces.

Stage 4. Eliminate bridges in the interior of $\mathrm{ch}(G)$**.** It was shown by Abellanas *et al.* [1, Lemma 4] that if $G = (V, E)$ is a connected PSLG with a bridge $e \in E$, then it can be augmented with one new edge e' to a PSLG $(V, E \cup \{e'\})$ in which e and e' are not bridges. In other words, we can decrease the number of bridges by adding one new edge. We use at most $\boxed{b_i}$ new edges to eliminate bridges of G_3, which lie inside $\mathrm{ch}(G)$. We obtain a PSLG G_4, on which we execute Algorithm 1 during the stage 5.

Algorithm 1

Input: A PSLG $G_4 = (V, E_4)$ that consists of some singletons and a 2-edge-connected component such that the boundary of the outer face is a simple polygon P_4; and a function τ that maps a unique edge $\tau(F)$ to every bounded face F of G_4.
Output: $G_5 = (V, E_5)$.
For every bounded face F in G_4, set $C(F) := \mathrm{int}(F)$. Let R be the set of reflex vertices lying in the interior of P_4. Compute W_u and $\overrightarrow{a}_u$ with respect to G_4 for every vertex $u \in R$. Set $E_5 := E_4$.
while $R \neq \emptyset$ **do**
  if there is a vertex $u \in R$ that sees a non-singleton vertex v in W_u (Fig. 3(b)), **then**
    set $E_5 := E_5 \cup \{uv\}$ and $R := R \setminus \{u\}$. Edge uv splits face F_u into two faces, and it also splits region $C(F_u)$ into two regions. If $v \in R$ and uv splits the reflex angle at v into two convex angles, then set $R := R \setminus \{v\}$.
  else if there is a vertex $u \in R$ such that $\overrightarrow{a}_u$ does not hit edge $\tau(F_u)$ (Fig. 3(c)), **then**
    pick a vertex $u \in R$ such that $e_u \neq \tau(F_u)$ and the part of face F_u which lies on one side of $\overrightarrow{a}_u$ incident on $\tau(F_u)$ is minimal. Set $v_1 v_2 = e_u$ such that v_1 is on the same side of $\overrightarrow{a}_u$ as $\tau(F_u)$. Compute the shortest path $\mathrm{path}(u, v_2)$ in F, and denote its vertices by $\mathrm{path}(u, v_2) = (u = w_0, w_1, w_2, \ldots, w_\ell = v_2)$. Set $E_5 := E_5 \cup \{w_j w_{j+1} : 0 \leq j \leq \ell - 1\}$ and $R := R \setminus \{w_j : 0 \leq j \leq \ell - 1\}$. The new edges split F into $\ell + 1$ faces. The rays $\overrightarrow{a}_{w_j}$, $j = 0, 1, \ldots, \ell - 1$ (in this order) split region $C(F_u)$ into $\ell + 1$ regions. Each new edge $w_j w_{j+1}$ lies between two bisectors $\overrightarrow{a}_{w_j}$ and $\overrightarrow{a}_{w_{j+1}}$, and hence in a unique new region, which contains a unique new face incident on $w_j w_{j+1}$.
  else
    for every $u \in R$, ray $\overrightarrow{a}_u$ hits edge $\tau(F_u)$ (Fig. 3(d)). Pick a vertex $u \in R$ such that the distance between u and the supporting line of the edge $\tau(F_u)$ is minimal. Let $vw = \tau(F_u)$. Set $E_5 := (E_5 \setminus \{\tau(F_u)\}) \cup \{uv, uw\}$ and $R := R \setminus \{u\}$. The removal of $\tau(F_u)$ merges face F_u with an adjacent face, and then the two new edges split it into three faces; ray $\overrightarrow{a}_u$ splits region $C(F_u)$ into two regions.
  end if
end while

---

or f only if they are the special edges $\tau(F_1)$ or $\tau(F_2)$, and the bisectors of all reflex angles in those faces hit these edges.

By Lemma 2, there is a vertex u on the convex hull $Q = \text{ch}(H)$ such that the reverse wedge of the exterior angle at u does not intersect e or f. Vertex u is reflex in G_4, it lies in the interior of P_4, and so it is initially in R. Since the edges of G_4 incident on u lie inside $\text{ch}(H)$, the reverse wedge W_u is part of the reverse wedge of $\text{ch}(H)$ at u. Hence $\vec{a}_u$ always hits some edge h in $G_4 - H$, with $h \neq e, f$. It follows that Algorithm 1 does not modify e or f as long as $u \in R$. In the while loop where Algorithm 1 removes u from R, there are three possible cases: (1) some vertex v in W_u is visible from u and we add a new edge uv; (2) ray $\vec{a}_u$ hits an edge $vw \neq \tau(F_u)$ outside of H, and we add all edges along the geodesic $\text{path}(u, v)$; (3) ray $\vec{a}_u$ hits an edge $vw = \tau(F_u)$ outside of H, and we add the edges uv and uw. In all three cases, we add new edges between H and $G_4 - H$. This contradicts the assumption that there are only two edges between H and $G_5 - H$. We conclude that $\{e', f'\}$ is not a 2-bridge in G_5. □

Lemma 5. *At the end of stage 5, we have* $\lambda(G_5) \leq c_h + g_h + s$.

Proof. At the end of stage 2, we have $\lambda_h(G_2) \leq c_h + s_h + g_h$ by Lemma 3. Stages 3-4 did not change the convex hull edges, so at the end of stage 4 we have $\lambda_h(G_4) \leq c_h + s_h + g_h$. Since every 3-edge-connected component of G_5 is either one of the s_i singletons or incident on the outer face P_5, we have $\lambda(G_5) = \lambda_h(G_5) + s_i$. It is enough to show that $\lambda_h(G_5) \leq \lambda_h(G_4)$. Assume that P_4 is the boundary of the outer face in G_4. Algorithm 1 may change P_4 by replacing an edge vw of P_4 by the edges vu and uw for some reflex vertex $u \in \text{int}(P_4)$. In each such step, a new vertex appears along the outer face, however this vertex is connected to another vertex of the outer face by a path that lies in the interior of P_5. None of these steps increases the number of components incident on the outer face, and so we have $\lambda_h(G_5) \leq \lambda_h(G_4)$. □

Stage 6. Connecting singletons. There are s_i singletons in the interior of $\text{ch}(G)$, which lie in convex regions $C_j \in \mathcal{C}$ with property (♡). In each convex region C_j, $j = 1, 2, \ldots, \ell$, we replace the deformable edge $e_j = u_j v_j$ by a path between u_j and v_j that lies entirely in C_j and passes through all singletons in C_j. Let m be the number of singletons in the interior of C_j. Label them as $w_1, w_2, \ldots w_m$ as follows. First label the singletons on the left of $\overrightarrow{u_j v_j}$ in the decreasing order of angles $\angle(v_j, u_j, w_i)$; then label the singletons on the right of $\overrightarrow{u_j v_j}$ in increasing order of angles $\angle(u_j v_j, w_i)$. Replace edge $u_j v_j$ by the simple path $(u_j, w_1, w_2, \ldots, w_m, v_j)$. We obtain a 2-edge-connected PSLG G_6. The number of edges has increased by $\boxed{s_i}$. Each of the s_i singletons of G_5 becomes a 3-edge-connected component in G_6. Hence the number of 3-edge-connected components does not change, and we have $\lambda(G_6) = \lambda(G_5) \leq c_h + g_h + s$.

Stage 7. Eliminating 2-bridges. The input PSLG G was 3-edge-augmentable. In stages 1-6, we have not added any chords of $\text{ch}(G)$, and so at the beginning of stage 6, we have a 2-edge-connected 3-edge-augmentable PSLG G_6. We to obtain a 3-edge-connected PSLG G_7 with at most $\lambda(G_6) - 1 = \boxed{c_h + g_h + s - 1}$ new edges by [14]. This completes our augmentation algorithm.

Theorem 1. *Every 3-edge-augmentable* PSLG G *with* $n \geq 4$ *vertices can be augmented to 3-edge-connected* PSLG *with at most* $2n - 2$ *new edges.*

Proof. In stages 1-7, we have added at most $b_i + c + f + r + 2s - 1$ new edges. If G is not a forest, this is at most $2n - 2$ by Corollary 1. If G is a forest and has an edge (bridge) along $\text{ch}(G)$, then $b_h \geq 1$, and again $b_i + c + f + r + 2s - 1 \leq 2n - 2$.

Let G be a forest with no edges along the convex hull (i.e., $b_h = 0$). We would like to show that our augmentation algorithm used fewer than $b_i + c + f + r + 2s - 1$ new edges. We distinguish four cases.

• *Case 1. A non-singleton component of G has two vertices, u and v, along the convex hull.* By adding all hull edges in stage 2, we eliminate the bridges along the path between u and v, and so we add fewer than b_i new edges in stage 4.

• *Case 2. Every component of G has at most one vertex along the convex hull, and $c_i \geq 1$.* Then in stage 3 we add two new edges to a vertex of each non-singleton component in the interior of $\text{ch}(G)$. The two edges form a circuit with G_h, and so it either decreases $\lambda_h(G_3)$ or eliminates a bridge in the interior of $\text{ch}(G)$.

• *Case 3. Every component of G has at most one vertex along the convex hull, $c_i = 0$, but G_2 has a bridge.* Then in stage 4 we add a new edge for each bridge. The first such new edge creates a circuit, which either contains another bridge of G_3 (eliminating at least two bridges at once), or contains a hull edge, decreasing $\lambda_h(G_3)$ by at least one.

• *Case 4. G consists of n singletons.* Then one can augment G to a plane Hamiltonian circuit (e.g., the Euclidean TSP tour on n vertices), with n new edges. We can augment the edge-connectivity from 2 to 3 with at most $n-2$ new edges [14]. In this case, again, G can be augmented to a 3-edge-connected PSLG with at most $2n - 2$ new edges. □

References

1. Abellanas, M., García, A., Hurtado, F., Tejel, J., Urrutia, J.: Augmenting the connectivity of geometric graphs. Comput. Geom. Theory Appl. 40(3), 220–230 (2008)
2. Eswaran, K.P., Tarjan, R.E.: Augmentation problems. SIAM J. Comput. 5(4), 653–665 (1976)
3. Frank, A.: Augmenting graphs to meet edge-connectivity requirements. SIAM J. Discrete Math. 5(1), 22–53 (1992)
4. Jackson, B., Jordán, T.: Independence free graphs and vertex connectivity augmentation. J. Comb. Theory Ser. B 94, 31–77 (2005)
5. Kortsarz, G., Nutov, Z.: Approximating minimum cost connectivity problems. In: Gonzalez, T.F. (ed.) Handbook of Approximation Algorithms and Metaheuristics, ch. 58. CRC Press, Boca Raton (2007)
6. Lovász, L.: Combinatorial Problems and Exercises. North-Holland, Amsterdam (1979)
7. Mader, W.: A reduction method for edge-connectivity in graphs. Ann. Discr. Math. 3, 145–164 (1978)
8. Nagamochi, H., Ibaraki, T.: Augmenting edge-connectivity over the entire range in $\tilde{O}(nm)$ time. J. Algorithms 30, 253–301 (1999)
9. Nagamochi, H., Ibaraki, T.: Graph connectivity and its augmentation: applications of MA orderings. Discrete Appl. Math. 123, 447–472 (2002)
10. Plesník, J.: Minimum block containing a given graph. Arch. Math. 27(6), 668–672 (1976)
11. Rutter, I., Wolff, A.: Augmenting the connectivity of planar and geometric graphs. In: Proc. Conf. Topological Geom. Graph Theory, Paris, pp. 55–58 (2008)

12. Souvaine, D.L., Tóth, Cs.D.: A vertex-face assignment for plane graphs. Comput. Geom. Theory Appl. 42(5), 388–394 (2009)
13. Tóth, Cs.D.: Connectivity augmentation in planar straight line graphs. In: Proc. Intl. Conf. on Topological and Geometric Graph Theory, Paris, pp. 51–54 (2008)
14. Tóth, Cs.D., Valtr, P.: Augmenting the edge connectivity of planar straight line graphs to three. In: Proc. 13th Spanish Meeting on Comp., Geom., Zaragoza (2009)
15. Tutte, W.T.: Connectivity in Graphs. University of Toronto Press (1966)
16. Watanabe, T., Nakamura, A.: Edge-connectivity augmentation problems. Comp. Sys. Sci. 35(1), 96–144 (1987)

Upward Star-Shaped Polyhedral Graphs

Seok-Hee Hong[1] and Hiroshi Nagamochi[2]

[1] University of Sydney
shhong@it.usyd.edu.au
[2] Kyoto University
nag@amp.i.kyoto-u.ac.jp

Abstract. We introduce a new wider class of polyhedra called *upward (star-shaped) polyhedra*, and present a graph-theoretic characterization. Our proof includes a drawing algorithm which constructs an upward polyhedron with n vertices in $O(n^{1.5})$ time. Moreover, we can test whether a given plane graph is an *upward polyhedral graph* in linear time. Our result is the first graph-theoretic characterization of *non-convex* polyhedra, which solves an open problem posed by Grünbaum [6], and a generalization of the Steinitz' theorem [9].

1 Introduction

There is a rich literature on *convex polytopes* [4,10]. The most well-known Steinitz' theorem [9] characterizes a convex 3-polytope as a triconnected planar graph, and some variations of Steinitz' theorem are available [10]. Many researchers investigated *polytope graphs*, i.e., the graphs of polytopes, of *convex* polytopes [4,6,8]. However, as pointed out by Grünbaum [6], non-convex polyhedra have not been well studied, in particular, the polyhedral graphs of non-convex polyhedra.

We define "polyhedra" in the three dimensional Euclidean space $\Re^3$ mostly according to Grünbaum [5]. A *polyhedron* is a collection of planar, compact, simply-connected polygonal regions; the boundary of such a region is called a *simple polygon* and the region itself is referred as a *face* of the polyhedron. A simple polygon consists of a finite number of line segments of positive length (the *edges* of the polygon); the endpoints of the edges are the *vertices* of the polygon. Each vertex of a polygon belongs to precisely two edges (said to be mutually adjacent), and the edges form a simple circuit (Jordan polygon). We consider *nonsingular* polyhedra without holes in $\Re^3$. We assume that each edge is shared by precisely two faces and that all faces containing a given vertex form a single circuit of at least three faces. A polyhedron is called *acoptic* [5] if the relative interiors of its elements (faces, edges and vertices) are disjoint.

In this paper, we call an acoptic polyhedron a *spherical polyhedron* if it has no singular points and *no holes*, and two faces sharing an edge are *non-coplanar* (i.e., they are in different planes). Spherical polyhedra satisfy Euler's formula which gives a combinatorial relationship between the numbers of vertices n, edges m and faces f: $n - m - f = 2$. A given set of vertices, edges and faces which satisfies this formula forms a topological sphere, a *combinatorial embedding* of a planar

Y. Dong, D.-Z. Du, and O. Ibarra (Eds.): ISAAC 2009, LNCS 5878, pp. 913–922, 2009.

graph. We call such a planar graph of a spherical polyhedron P the *vertex-edge graph* of P, denoted by $G(P)$.

In this paper, as the first step for discovering a new class of polyhedra which admits a graph-theoretic characterization, we introduce a new wider class of spherical polyhedra, called *upward (star-shaped) polyhedra* such that (i) each face is star-shaped, (ii) all the vertices, edges and faces (except the bottom face) are visible from a view point, and (iii) any two faces sharing two vertices are non-coplanar. We present a complete graph-theoretic characterization, *biconnected* planar graphs with special conditions, by presenting a constructive proof.

2 Planar/Plane Graphs, Splitness and Decomposition

Throughout the paper, a graph stands for a simple undirected graph. Let $G = (V, E)$ be a graph. The minimum degree of a vertex in G is denoted by $\delta(G)$. A cycle C is *k-connected* in G if G has a vertex v not in C and k vertices $u_1, u_2, \ldots, u_k$ in C such that G has k internally vertex-disjoint paths P_i, $i = 1, 2, \ldots, k$, each joins v and u_i.

A face is characterized by the cycle of G that surrounds the region, called a *facial cycle*. A set F of facial cycles in a drawing is called a *combinatorial embedding* of a planar graph G. A *plane* graph $G = (V, E, F)$ is a planar graph with a fixed plane embedding F of G, where we always denote the fixed outer facial cycle in F by f_o. A vertex (resp., an edge) in f_o is called an *outer vertex* (resp., an *outer edge*), while a vertex (resp., an edge) not in f_o is called an *inner vertex* (resp., an *inner edge*). The following lemma states that every facial cycle of $G(P)$ of a spherical polyhedron P is triconnected (see [7] for a proof).

Lemma 1. *Let $G = (V, E, F)$ be a biconnected planar graph with fixed embedding. If there is a spherical polyhedron P with $G(P) = G$, then each cycle $f \in F$ is triconnected.* □

We say that two faces f and f' in a biconnected simple planar graph with a fixed embedding are *adjacent* if they share an edge. For a pair $\{u, v\}$ of vertices, two faces f and f' are *joined* at $\{u, v\}$, if they share the two vertices u and v. We say that two faces f and f' are *linearly-joined* if they share an edge or at least three vertices (note that they are not linearly-joined if they share exactly two vertices but no edge). The following lemma describes a forbidden structure for a biconnected planar graph G to be realized as spherical polyhedra (see [7] for a proof).

Lemma 2. *Let $G = (V, E, F)$ be a biconnected planar graph with fixed embedding. If G has a separation pair $\{u, v\}$ and three edges (u, w_i), $i = 1, 2, 3$, incident to u such that each edge (u, w_i) is shared by a pair of linearly-joined faces at $\{u, v\}$, then there is no spherical polyhedron P with $G(P) = G$.* □

Let $G = (V, E, F)$ be a biconnected simple planar graph with the outer face f^o. The *splitness* $\xi(u, v)$ of a pair $\{u, v\}$ of vertices is defined by the number of pairs

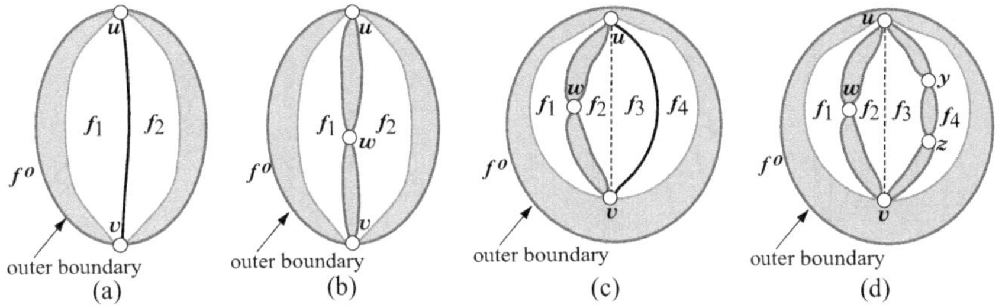

Fig. 1. (a) and (b) show a pair of outer vertices u and v with $\xi(u,v) = 1$, where two inner faces f_1, f_2 and the outer face f^o are joined at $\{u,v\}$, and f_1 and f_2 are linearly-joined at $\{u,v\}$; (c) and (d) show a pair of vertices u and v with $\xi(u,v) = 2$, where inner faces f_1, f_2, f_3 and f_4 are joined at $\{u,v\}$, and f_i and f_{i+1} ($i = 1,3$) are linearly-joined at $\{u,v\}$ ($f_2 \neq f_1 \neq f_4 \neq f_3$ but possibly $f_2 = f_3$)

of linearly-joined inner faces joined at $\{u,v\}$. We define the *splitness* $\xi(G)$ of a plane graph G as the maximum of $\max\{\xi(u,v)+1 \mid u \text{ and } v \text{ are outer vertices}\}$ and $\max\{\xi(u,v) \mid \text{one of } u \text{ and } v \text{ is an inner vertex}\}$. See Fig. 1.

We now briefly review the decomposition of a biconnected graph G into *triconnected components*, based on the *SPQR tree* [2]. Each node ν in the SPQR tree is associated with a graph $\sigma(\nu) = (V_\nu, E_\nu)$ ($V_\nu \subseteq V$), called the *skeleton* of ν, which corresponds to a triconnected component of G. In fact, we use a modified SPQR tree *without* Q-nodes, called SPR tree. Thus, there are only three types of nodes in the SPR tree: *S-node* (the skeleton is a simple cycle), *P-node* (the skeleton consists of two vertices with at least 3 edges), and *R-node* (the skeleton is a triconnected simple graph). The edges of the SPR tree are defined by the *virtual edges* which are introduced by the decomposition process. If two triconnected components have a virtual edge in common, then the nodes that represent the two triconnected components in the SPR tree are joined by an edge that represents the virtual edge. We treat the SPR tree of a graph G as a rooted tree $\mathcal{T}$ by choosing a node ν^* as its root. For a node ν, let $Ch(\nu)$ denote the set of children of ν. We denote the graph formed from $\sigma(\nu)$ by deleting its *parent virtual edge* e' as $\sigma^-(\nu) = (V_\nu, E_\nu^-)$, ($E_\nu^- = E_\nu - \{e'\}$). Let $G^-(\nu)$ denote the subgraph of G which consists of the vertices and real edges in the graphs $\sigma^-(\mu)$ for all descendants μ of ν, including ν itself.

When G is a plane graph, we also treat graphs $\sigma^-(\nu)$ and $G^-(\nu)$ as *plane graphs* induced from the embedding of G. A node is called *outer* if its skeleton has a face whose vertices are all outer vertices of G. We choose the root of SPR tree $\mathcal{T}$ of G as an outer node. We always choose an outer R-node as the root of $\mathcal{T}$ (note that G is assumed to have at least one outer R-node). A real edge in the skeleton of a node is called *outer* (resp., *inner*) if it is an outer (resp., inner) edge of G. Then we have the following property (see [7] for a proof).

Lemma 3. *Let G be a biconnected planar graph with a fixed embedding, and f be a face in F. Then the facial cycle f is triconnected if and only if the SPR tree of G has an R-node ν whose skeleton $\sigma(\nu)$ contains a face f' with $V(f') \subseteq V(f)$.*

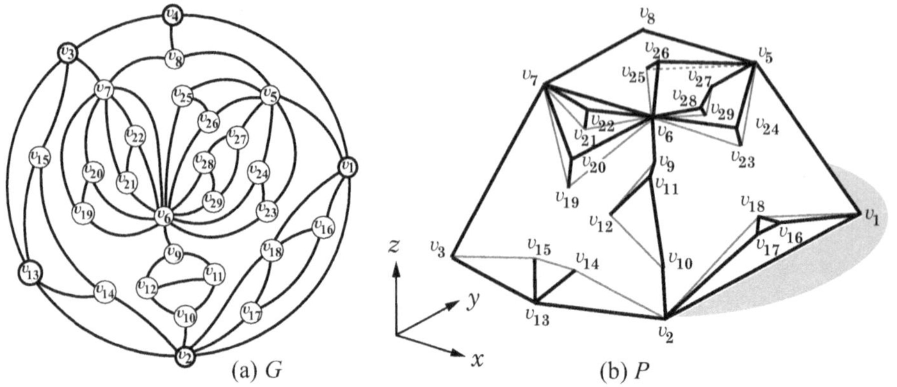

Fig. 2. (a) A biconnected plane graph G which satisfies the necessary condition of Theorem 1: (b) an upward polyhedron P with $G(P) = G$

3 Upward Polyhedral Graphs

Now we present our main theorem, a characterization of *upward polyhedral graphs.*

Theorem 1. *A plane graph G with an outer face f^o is an upward polyhedral graph (i.e., the vertex-edge graph of an upward polyhedron P with the bottom face f^o) if and only if it is a biconnected plane graph with $\delta(G) \geq 3$ and $\xi(G) \leq 1$, and the cycle f^o is triconnected in G.* □

Fig. 2 shows an example of a plane graph G and a proper upward polyhedron P with $G(P) = P$. Based on Theorem 1, we can test whether a given plane graph is an upward polyhedral graph in linear time, since the splitness $\xi(G)$ and triconnectivity of cycle f^o can be computed in linear time after constructing the SPR tree in linear time [2].

Necessity of Theorem 1: Let P be an upward polyhedron. Obviously, $G(P)$ is simple and satisfies $\delta(G(P)) \geq 3$. Also $G(P)$ is biconnected because otherwise $G(P)$ would have a face whose boundary is not a simple cycle, contradicting the assumption on spherical polyhedra. We omit a proof for the reason why $\xi(G(P)) \leq 1$ is necessary (see [7] for a proof).

Lemma 4. *The vertex-edge plane graph $G(P)$ of an upward polyhedron P satisfies $\xi(G(P)) \leq 1$.* □

4 Proof for Sufficiency

This section gives a proof sketch of the sufficiency of Theorem 1. We first prepare a building block of the proof. For three points $p_1, p_2, p_3 \in \Re^3$ which are not collinear, the plane that contains these points is denoted by $H(p_1, p_2, p_3)$.

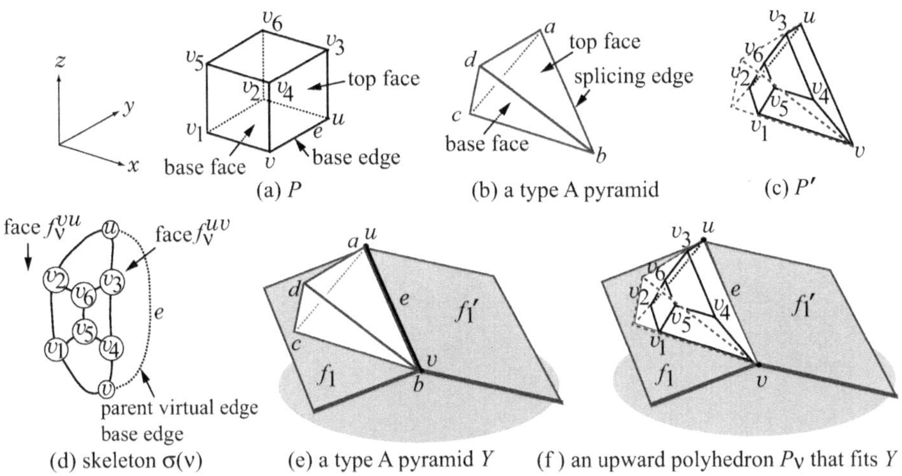

Fig. 3. (a) A convex polyhedron P with base face (u, v, v_1, v_2) and top face (u, v, v_4, v_3); (b) a pyramid $\Delta(a, b, c, d)$ of type A with base face (a, b, c) and top face (a, b, d); (c) a convex polyhedron P' transformed from P into the interior of pyramid $\Delta(a, b, c, d)$ of (b); (d) the skeleton $\sigma(\nu)$ of an R-node ν; (e) a type A pyramid Y; (f) a convex polyhedron P_ν that fits into Y

A triangle (p_1, p_2, p_3) is a polygon with three apices p_1, p_2 and p_3. A convex polygon with exactly four non-coplanar vertices $p_1, p_2, p_3, p_4 \in \Re^3$ is called a *pyramid*, and is denoted by $\Delta(p_1, p_2, p_3, p_4)$.

Consider an ordered sequence of four points $a, b, c, d \in \Re^3$. The pyramid $Y = \Delta(a, b, c, d)$ is called of *type A*, if face (a, b, c) is not visible and face (a, b, d) is visible (see Fig. 3(b)). We call the faces (a, b, c) and (a, b, d) of a type A pyramid the *base face* and the *top face*, respectively. The pyramid $Y = \Delta(a, b, c, d)$ is called of *type B*, if both faces (a, b, c) and (a, b, d) are not visible (see Fig. 4(b)). We call the faces (a, b, c) and (a, b, d) of a type B pyramid the *side faces*. The edge (a, b) of a pyramid of a type A or B is called the *splicing edge*.

Type A Pyramids. Let P be a convex polyhedron, and f_b and f_t be the two adjacent faces, where we call f_b and f_t the base face and the top face of P. Let $e = (u, v)$ be the edge shared by f_b and f_t, which we call the *base edge*, where we assume that u and v appear consecutively in this order when we traverse the top face in the clockwise order (see Fig. 3(a)).

We say that a convex polyhedron P with the base face f_b, the top face f_t and the base edge $e = (u, v)$ *fits into* a type A pyramid $Y = \Delta(a, b, c, d)$ in $\Re^3$, if P can be placed in $\Re^3$ so that (i) the positions of vertices u and v are equal to a and b, respectively; (ii) the region f_b (resp., f_t) is contained in the triangle (a, b, c) (resp., (a, b, d)); and (iii) all faces of P except for the base face f_b are visible.

We next show that a triconnected planar graph with the base and top faces can be realized as a convex polyhedron that fits into any type A pyramid in the following sense (see [7] for a proof).

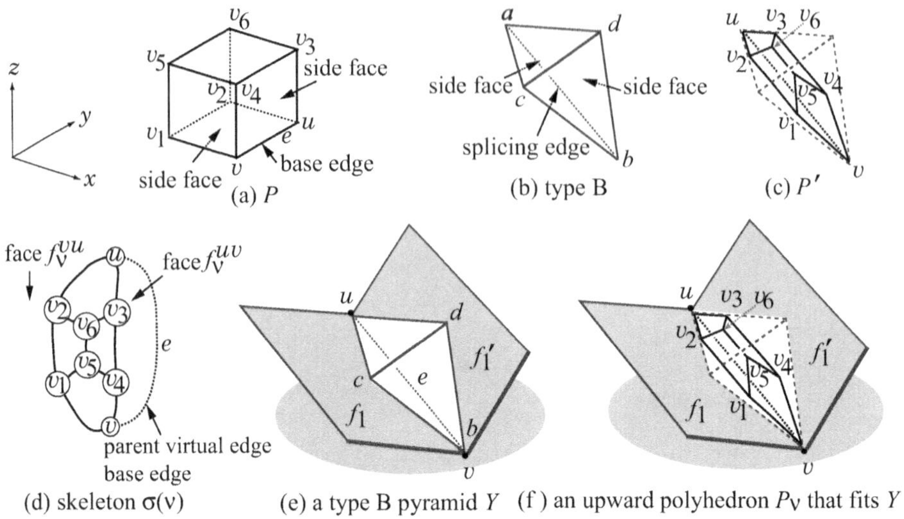

Fig. 4. (a) A convex polyhedron P with side faces (u, v, v_1, v_2) and (u, v, v_4, v_3); (b) a pyramid $\Delta(a, b, c, d)$ of type B with side faces (a, b, c) and (a, b, d); (c) a convex polyhedron P'' transformed from P into the interior of pyramid $\Delta(a, b, c, d)$ of (b); (d) the skeleton $\sigma(\nu)$ of an R-node ν; (e) a type B pyramid Y; (f) a convex polyhedron P_ν that fits into Y

Lemma 5. *Let $G = (V, E, F)$ be a triconnected planar graph, and $Y = \Delta(a, b, c, d)$ be a type A pyramid in $\Re^3$. Then for any two adjacent faces f_b and f_t in G designated as the base and the top faces, there is a convex polyhedron P with $G(P) = G$ that fits into Y.* □

Type B Pyramids. Let P be a convex polyhedron, and f_{s1} and f_{s2} be the two adjacent faces, where we call f_{s1} and f_{s2}, the side faces of P. Let $e = (u, v)$ be the edge shared by f_{s1} and f_{s2}, which we call the *base edge* (see Fig. 4(a)).

We say that a convex polyhedron P with the two side faces f_{s1} and f_{s2} and the base edge $e = (u, v)$ *fits into* a type B pyramid $Y = \Delta(a, b, c, d)$ in $\Re^3$, if P can be placed in $\Re^3$ so that (i) the positions of vertices u and v are equal to a and b, respectively; (ii) the region f_{s1} (resp., f_{s2}) is contained in the triangle (a, b, c) (resp., (a, b, d)); and (iii) all faces of P except for the side faces f_{s1} and f_{s2} are visible. We next show that any triconnected planar graph with the side faces can be realized as a convex polyhedron that fits into a given type B pyramid (see [7] for a proof).

Lemma 6. *Let $G = (V, E, F)$ be a triconnected planar graph, and $Y = \Delta(a, b, c, d)$ be a type B pyramid in $\Re^3$. Then for any two adjacent faces f_{s1} and f_{s2} in G designated as the two side faces, there is a convex polyhedron P with $G(P) = G$ that fits into Y.* □

We now present the main idea of our algorithm that constructs a desired upward polyhedron. Recall that the structure of each R-node in the SPR tree is given by its skeleton $\sigma(\nu)$, which is a triconnected planar graph and admits a convex polyhedra P_ν with $G(P_\nu) = \sigma(\nu)$ by Steinitz' theorem. In fact, our upward polyhedron P consists of these convex polyhedra P_ν with necessary projective transformations using bounding pyramids.

Our algorithm first chooses an R-node ν^* whose skeleton contains at least three outer vertices of G, and designate ν^* as the root of the SPR tree of G. We first realize the skeleton $\sigma(\nu^*)$ of the root R-node ν^* as an upward convex polyhedron P_{ν^*} of the R-node as follows. We choose a type A pyramid $Y = (a, b, c, d)$ on the xy-plane ($H(a, b, c)$ is the xy-plane and d is a point over the xy-plane). By designating an outer edge $e = (u, v)$ of $\sigma(\nu^*)$ as the base edge, and the outer face $f^o(\sigma(\nu^*))$ and the face f sharing e as the base and top faces, respectively, we convert P_{ν^*} into a convex polyhedron P that fits into Y. By Lemma 5, the resulting polyhedron P is upward.

Starting with setting P as an upward convex polyhedron P_{ν^*} for the root R-node ν^*, the algorithm traverses the rooted SPR tree in a DFS manner, and computes an "appropriate" convex polyhedron P_ν whenever it visits an R-node ν. The polyhedron P_ν will be combined with the current polyhedron P to update P as a new upward polyhedron. We construct a type A or type B pyramid Y on a face of P to choose the "appropriate" convex polyhedron for P_ν. Then the convex polyhedron P_ν will be attached to the current polyhedron P in such a way that (i) the bottom face $f^o(P_\nu)$ is completely contained in the region of some face f of P, and (ii) one edge e of $f^o(P_\nu)$ is completely contained in some edge of f without generating any other intersection between $f^o(P_\nu)$ and f. This means that all faces in P are always simple polygons.

Types of nodes in the SPR tree and virtual edges. For a biconnected plane graph G with $\delta(G) \geq 3$ and $\xi(G) \leq 1$ which has a triconnected outer face f^o, the root of the SPR tree of G is now chosen as an R-node ν^* whose skeleton

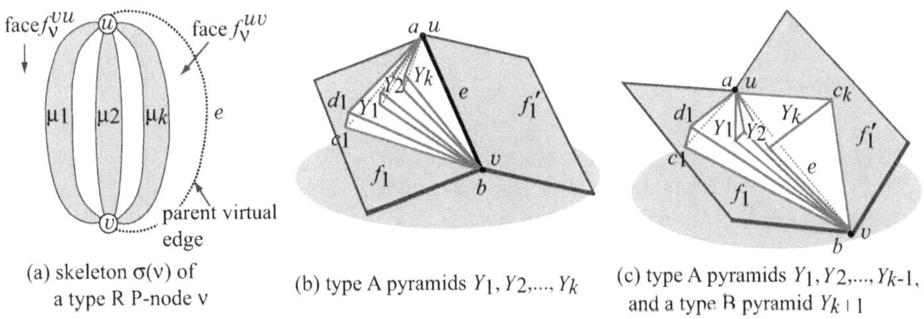

Fig. 5. (a) The skeleton $\sigma(\nu)$ of a type R P-node ν; (b) type A pyramids $Y_1, Y_2, \ldots, Y_k$ for convex edge e; (c) type A pyramids $Y_1, Y_2, \ldots, Y_{k-1}$ and a type B pyramid Y_k for concave edge e

contains at least three outer vertices of G. For each non-root R-node ν, we treat the skeletons $\sigma^-(\nu)$ and $\sigma(\nu)$ as plane graphs induced from the plane embedding of G. Note that $\sigma(\nu)$ contains the virtual parent edge of ν, and is triconnected. For an edge (u, v), we denote the two faces of $\sigma(\nu)$ containing an edge (u, v) by f_ν^{uv} and f_ν^{vu}, where u and v appear in this order when we traverse the cycle f_ν^{uv} in the clockwise order (see Figs. 3(d) and 4(d)).

Then a biconnected simple plane graph G with $\delta(G) \geq 3$ and $\xi(G) \leq 1$ is characterized by the following conditions: (C1) Each outer P-node ν has no inner S-node child and no inner real edge in its skeleton; (C2) Each non-outer P-node ν with an S-node child has no other S-node child and no real edge in its skeleton (ν cannot have an outer S-node or an outer real edge); and (C3) For each S-node, no two real edges are adjacent in the skeleton $\sigma^-(\nu)$.

By condition (C1), for an outer P-node ν, exactly one edge in $\sigma^-(\nu)$ corresponds to an outer R-node, an outer S-node or an outer real edge, and the rest of edges correspond to only R-nodes (note that the virtual parent edge of $\sigma(\nu)$ may be an outer real edge or corresponds to an outer S-node). We call an outer P-node ν *type R* (resp., *type S* or *type E*) if all edges in $\sigma^-(\nu)$ correspond to R-nodes (resp., one edge in $\sigma^-(\nu)$ corresponds to an outer S-node or an outer real edge). By condition (C2), a non-outer P-node ν can have at most one of a non-outer S-node or an inner real edge. We call a non-outer P-node ν *type R* (resp., *type S* or *type E*) if all edges in $\sigma^-(\nu)$ correspond to R-nodes (resp., one edge in $\sigma^-(\nu)$ corresponds to an S-node or a real edge). See Figs. 5 and 6.

We distinguish the following three cases for a virtual edge e in the current polyhedron P: (i) e is not in the boundary of the bottom face $f^o(P)$, and e is a convex edge; (ii) e is not in the boundary of $f^o(P)$, and e is a concave edge; and (iii) e is in the boundary of $f^o(P)$. We always treat an edge e in (iii) as a concave edge which is created by the side face containing e and the plane $H(f^o(P))$.

Replacing virtual edges. We now describe how to process virtual edges in the current upward polyhedron P. In the following, we omit the details on the treatment of visibility of non-bottom faces, and star-shaped faces for simplicity of algorithm description. We focus on how to add a new polyhedron P_ν to the current polyhedron P without losing the upward property.

Let $e = (u, v)$ be the virtual edge to be processed (i.e., the virtual edge with the highest priority in the DFS traversal of the SPR tree). The virtual edge $e = (u, v)$ corresponds to an S-, R-, or P-node. Let f_1, f_1' be the two faces of P that share the edge $e = (u, v)$, where we assume that f_1' denotes the region obtained from the plane $H(f^o(P))$ by removing the region $f^o(P)$ if e is in the boundary of $f^o(P)$. Also assume that u and v appear in this order when we traverse the facial cycle f_1 in the clockwise order. Let ν be the child node to which e corresponds.

We distinguish the three cases: ν is an S-node, R-node, or P-node. Here we consider the case where $e = (u, v)$ corresponds to an R-node ν, which is the most essential part in our construction (see [7] for S- and P-node cases). We first construct a convex polyhedron P_ν of the skeleton $\sigma(\nu)$. Note that $\sigma(\nu)$ contains

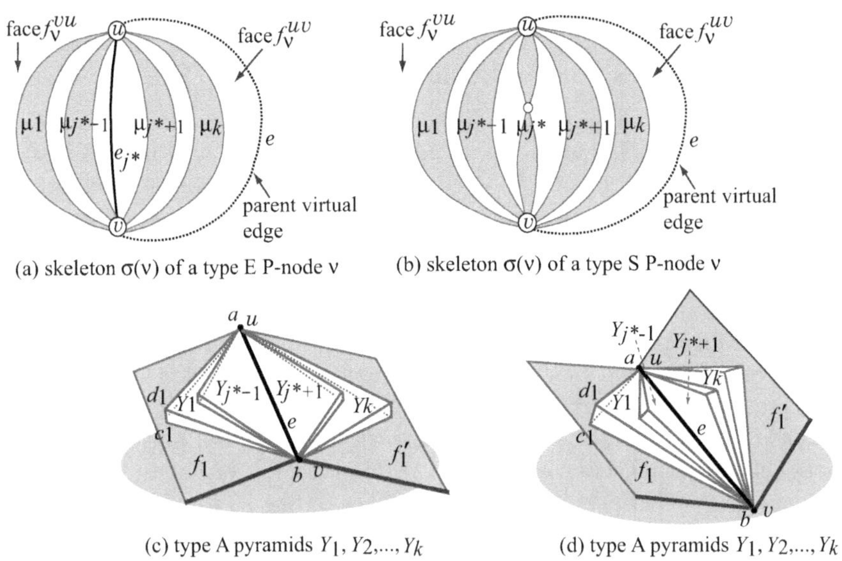

Fig. 6. (a) The skeleton $\sigma(\nu)$ of a type E P-node ν; (b) the skeleton $\sigma(\nu)$ of a type S P-node ν; (c) type A pyramids $Y_1, Y_2, \ldots, Y_k$ for convex edge e; (d) type A pyramids $Y_1, Y_2, \ldots, Y_k$ for concave edge e

the parent *virtual* edge $e = (u, v)$, which is not an element of a polyhedron for the input graph G, and thus we have to eliminate it when we add P_ν to the current polyhedron P. We then have the following two cases.

Case (1). $e = (u, v)$ is a *convex* edge in P: Let us designate $e = (u, v)$ as the base edge, and f_ν^{vu}, f_ν^{uv} as the base and top faces of P_ν. We then choose a type A pyramid $Y = \Delta(a, b, c, d)$ so that a and b are the endpoints of (u, v) in P, c is a point in the face f_1, and d is a point in the plane $H(f'_1)$ and above the face f_1 (note that such d can be chosen so that $\Delta(a, b, c, d)$ is type A, since P is upward and e is convex).

By Lemma 5, we can assume that P_ν fits into Y. By definition of fitting polyhedra, we see that the polyhedron P modified by adding P_ν remains upward, and the parent virtual edge e temporarily introduced to construct P_ν has disappeared in the resulting P, since the top face f_ν^{uv} of P_ν is contained in the plane $H(f'_1)$ (see Fig. 3(f)). Note that by choosing a point c outside $H(f'_1)$, we can realize edge e as an element of the resulting polyhedron, if necessary.

Case (2). $e = (u, v)$ is a *concave* edge in P (including the case where e is in the boundary of $f^o(P)$): Let us designate $e = (u, v)$ as the base edge, and f_ν^{uv}, f_ν^{vu} as the side faces of P_ν. We then choose a type B pyramid $Y = \Delta(a, b, c, d)$ so that a and b are the endpoints of (u, v) in P, c is a point in the face f_1, and d is a point in the face f'_1 (note that such c, d can be chosen so that $\Delta(a, b, c, d)$ is type B, since P is upward and e is concave).

By Lemma 6, we can assume that P_ν fits Y. By definition of fitting polyhedra, we see that the polyhedron P modified by adding P_ν remains upward, and the edge e in P_ν is concealed by the two invisible side faces of P_ν in the resulting P.

Correctness. The algorithm realizes real edges in R-, S- and P-nodes of the current polyhedron P, and the set of edges of the final polyhedron consists only of the real edges of G, i.e., $G(P) = G$. By the way of choosing convex polyhedra P_ν that fit into bounding pyramids on the current polyhedron P, all non-bottom faces of P are always visible. Also, when we place a convex polyhedra P_ν, we keep the convexity of the base edge e, if e was convex in P before adding P_ν to P. We can also easily maintain all the faces as star-shaped as follows.

When we introduce a new face f in the current polyhedron P (i.e., f is a non-bottom face of P_ν), we choose an internal point in the region of f as the *view point* r_f of f. When the face f is modified by placing the bottom face $f^o(P_\mu)$ of a polyhedron P_μ on f (or f is extended with a face f_t of a polyhedron P_μ), we choose the polygonal shape of such face $f^o(P_\mu)$ (or f_t) so that the view point r_f remains as a visible point in the resulting face f. See [7] for details.

Time complexity. We can transform a convex polyhedron with n vertices into a convex polyhedron that fits into a pyramid Y in $O(n)$ time. Hence our algorithm for constructing an upward polyhedron runs in $O(n + T(n))$ time, where $T(n)$ denotes the time complexity of an algorithm for constructing a convex polyhedron of a triconnected planar graph (the current best runs in $O(n^{1.5})$ time [1,3]).

References

1. Chrobak, M., Goodrich, M.T., Tamassia, R.: Convex drawings of graphs in two and three dimensions. In: Proceedings of SoCG 1996, pp. 319–328 (1996)
2. Di Battista, G., Tamassia, R.: On-line maintenance of triconnected components with SPQR-trees. Algorithmica 15, 302–318 (1996)
3. Eades, P., Garvan, P.: Drawing stressed planar graphs in three dimensions. In: Brandenburg, F.J. (ed.) GD 1995. LNCS, vol. 1027, pp. 212–223. Springer, Heidelberg (1996)
4. Grünbaum, B.: Convex Polytopes. Graduate Text in Mathematics 221 (2003)
5. Grünbaum, B.: Acoptic polyhedra. In: Advances in Discrete and Computational Geometry, Contemporary Mathematics, AMS, pp. 163–199 (1998)
6. Grünbaum, B.: Graphs of polyhedra; polyhedra as graphs. Discrete Mathematics 307(3-5), 445–463 (2007)
7. Hong, S.-H., Nagamochi, H.: Extending Steinitz' theorem to non-convex polyhedra, Kyoto University, Technical report 2008-012 (2008)
8. Kalai, G.: Polytope skeletons and paths. In: Handbook of Discrete and Computational Geometry. CRC Press, Boca Raton (1997)
9. Steinitz, E.: Polyeder und Raumeinteilungen, Encyclopädie der mathematischen Wissenschaften, Band 3 (Geometrie), Teil 3AB12, 1-139 (1922)
10. Ziegler, G.M.: Lectures on Polytopes. Springer, Heidelberg (1995)

Conditional Hardness of Approximating Satisfiable Max 3CSP-q*

Linqing Tang

State Key Lab. of Computer Science, Institute of Software,
Chinese Academy of Sciences, Beijing, 100080, China
Graduate University of Chinese Academy of Sciences, Beijing, China
linqing@ios.ac.cn

Abstract. In this article, we study the approximability of satisfiable Max 3CSP-q for $q > 3$ be a prime. We give a $(\frac{1}{q} + \frac{1}{q^2} - \frac{1}{q^3}) + \epsilon$-hardness result for approximate Max 3CSP-q even on satisfiable instances, conditioned on Khot's d-to-1 Conjecture, for any finite constant integer $d < q/2$.

1 Introduction

An Max *3CSP-q* instance $\mathcal{I}$ is specified by a set of variables, which get values from a fix finite domain (say $\mathbb{Z}_q$), and a set of constraints, with each constraint be a predicate applied to 3 variables. The objective of Max *3CSP-q* is to find an assignment to the variables that satisfies as many constraints as possible. When $q = 2$ it is the Boolean Max 3CSPs, which includes Max 3SAT as special case.

It is well-known that for most Max 3CSP-qs it is NP-hard to find the optimal assignment, a natural question asks the performance of finding an approximation solution. Over recent years, this question has been answered successfully for many subcases of 3CSPs. For example, Håstad [5] and Zwick [15] successfully specified the approximability of all Boolean Max 3CSP. They showed that the random assignment can get the best approximation ratio for Max E3-Sat even on satisfiable instances, Håstad [5] also gives tight approximability result for Max E3Lin-q and many other CSPs.

In order to show some hardness results of approximation, Khot introduced the Unique Game Conjecture, based on this conjecture [6], many natural problems' tight hardness of approximation were proved, such as Max Cut [7], Min Vertex Cover [8] and so on. Moreover, based on Unique Game Conjecture, both the algorithm of Charikar et al. [3] for Boolean Max kCSP and the algorithm of Charikar et al. [2] for Unique Games are nearly tight. Recently, Raghavendra [13] even shows that any Max CSP's optimal hardness result is equal to the integrality gap of certain semidefinite programming, conditioned on Unique Game Conjecture.

* This work is supported by Hundred Talent Program of Chinese Academy of Sciences under Angsheng Li.

Y. Dong, D.-Z. Du, and O. Ibarra (Eds.): ISAAC 2009, LNCS 5878, pp. 923–932, 2009.

Conjecture 1 (Unique Game Conjecture). [6] For all constant $\zeta, \delta > 0$, there exists a constant $k = k(\zeta, \delta)$ such that it is NP-hard to determine whether a unique Label Cover instance with label sets of size k has value at least $1 - \zeta$ or at most δ.

An interesting question is how much satisfiable promise can help in approximation algorithm for CSP instances. There are CSP instances do benefit from the satisfiable promise, for example the Max k-Lin-q instances for any $k \geq 3$ and $q \geq 2$, since we can use Gaussian Elimination to get an assignment in polynomial time for satisfiable linear equation systems whereas it hard to approximate within $1/q + \epsilon$ even on almost satisfiable Max k-Lin-q instances. There are also instances that do not benefit, for example Boolean 3SAT, which is still hard to approximate with in $7/8 + \epsilon$ on satisfiable instances. For general Max 3CSP-q instances, it is still an interesting open question.

We remark that all hardness results conditioned on Unique Game Conjecture make no sense for satisfiable Max CSP instances, since there is a hierarchical unsatisfiability ($1 - \zeta$ for completeness) in Unique Game Conjecture. Until recently, a breakthrough result of O'Donnell and Wu [11] shows that satisfiable Max 3NTW is hard to approximate within $5/8 + \epsilon$, conditioned on Khot's d-to-1 Conjecture, i.e., Max 3NTW do not benefit from the satisfiable promise. Their result also solves a long-standing open question which asks the exact approximability of satisfiable Boolean Max 3CSP and the power of PCP system with 3 nonadaptive queries and perfect completeness.

In this paper, we try to extend O'Donnell and Wu's hardness results [11] to approximate satisfiable Max 3CSP-q, here $q > 3$ is a prime. We are able to show

Theorem 1. *If Khot's d-to-1 Conjecture is true for any finite constant integer $q/2 > d \geq 2$, then for any $\epsilon > 0$ it is NP-hard to approximate Max 3CSP-q within $(\frac{1}{q} + \frac{1}{q^2} - \frac{1}{q^3}) + \epsilon$ even on satisfiable instances.*

When $q = 2$, it is exactly O'Donnell and Wu's hardness results [11] for NTW. Actually our CSP instance is a natural extension of 3NTW to large domain Z_q. We construct a special kind of predicate Γ on 3 variables, which accepting q^2+q-1 input patterns of the total q^3, and show it is Approximation Resistant [1] (actually hierarchical approximation resistant [1]) even on satisfiable instances. Here by Approximation Resistant we mean it is NP-hard to approximate the Max 3CSP-q instances defined by predicate Γ better than the performance of random assignment. So the satisfiable promise does not help in approximating Max 3CSP-q instances defined by our predicate Γ, to our best knowledge, Γ is the Predicate on 3 variables with least accepting inputs up to now which satisfies the above property.

By the standard argument of the relations between Max CSPs and the PCP verifier systems, we know that conditioned on d-to-1 Conjecture,

$$PCP_{1,\frac{1}{q}+\frac{1}{q^2}-\frac{1}{q^3}+\epsilon}[O(log), 3] = NP$$

if the proof of the PCP verifier is encoded with q-bit. Before us, Engebretsen and Holmerin [4] had shown

$$PCP_{1,\frac{1}{q}+\frac{1}{q^2}+\epsilon}[O(log),3] = NP$$

using different kind of accepting Predicate in the PCP system.

2 Preliminaries

In this article $q > 3$ is a prime, ω is a primitive q'th root of unity. We use the notations $[q] \triangleq \{0, q, ..., q-1\} \cong \{\omega^i : i \in [q]\}$ and $[q]^* \triangleq [q]\backslash\{0\}$, define $S_{\mathcal{T}}$ be the set

$$\{(\omega^x, \omega^y, \omega^z) \in [q]^3 : x = y = z \ \ or \ \ x + y + z = 1 \mod q\}$$

then $|S_{\mathcal{T}}| = q^2 + q - 1$. For $\mathbf{a} \in [q]^n$, let $N(\mathbf{a}) \triangleq \{i \in [n], a_i \neq 0\}$. For two vectors $\mathbf{a} \in [q]^S$, $\mathbf{b} \in [q]^T$ with $T \subseteq S$, we denote $\mathbf{a}\backslash\mathbf{b} \triangleq \mathbf{a}|_{S\backslash T}$.

2.1 Fourier Analysis and Noise Sensitive Operator

For function $f : [q]^n \to \mathbf{C}$, let

$$f = \sum_{\mathbf{a}\in[q]^n} \widehat{f}_{\mathbf{a}}\chi_{\mathbf{a}}$$

be the Fourier decomposition of f with $\chi_{\mathbf{a}}(\mathbf{x}) = \omega^{\sum_{i=1}^n a_i x_i}$ and $\widehat{f}_{\mathbf{a}} = E_{\mathbf{x}}[f(\mathbf{x})\overline{\chi_{\mathbf{a}}}(\mathbf{x})]$.

Definition 1. *We say a function $f : [q]^n \to \boldsymbol{C}$ is* q-folded, *if $f(\boldsymbol{x} + a \cdot \boldsymbol{1}) = \omega^a f(\boldsymbol{x})$ for any $\boldsymbol{x} \in [q]^n$ and $a \in [q]$, here $\boldsymbol{1} = (1, ..., 1)$. Say a function f* respects exponentiation *if $f(a\boldsymbol{x}) = f(\boldsymbol{x})^a$ for any $a \in [q]^*$ and $\boldsymbol{x} \in [q]^n$.*

It is easy to check that if f is q-folded, then $E_{\mathbf{x}}[f(\mathbf{x})] = 0$ and $\widehat{f}_{\mathbf{a}} = 0$ unless $\sum_i a_i = 1 \mod q$ and if f respects exponentiation, then all $\widehat{f}_{\mathbf{a}}$ are real values.

Definition 2. *For a function $f : [q]^n \to \boldsymbol{C}$ and $1 \leq i \leq n$, define the* Influence of i'th coordinate *on f as*

$$Inf_i(f) = \sum_{a\in[q]^n, a_i\neq 0} |\widehat{f}_a|^2$$

Similarly for any $\emptyset \neq S \subseteq [n]$, define the Influence of S *on f as:*

$$Inf_S(f) = \sum_{a\in[q]^n, S\subseteq N(a)} |\widehat{f}_a|^2$$

Definition 3. *For $0 < \rho < 1$, define the Bonami-Beckner operator T_ρ mapping function $f : [q]^n \to \boldsymbol{C}$ to function $T_\rho f : [q]^n \to \boldsymbol{C}$ as*

$$T_\rho f = \sum_{a \in [q]^n} \rho^{|N(a)|} \widehat{f}_a \chi_a$$

Lemma 1. *For any function $f : [q^n] \to \boldsymbol{C}$ with $|f(x)| \le 1$ and $0 < \gamma < 1$, we have*

$$\sum_{i \in [n]} Inf_i(T_{1-\gamma} f) \le \frac{1}{2e \ln \frac{1}{1-\gamma}}$$

and

$$\sum_{S \subseteq [n], |S| \le d} Inf_S(T_{1-\gamma} f) \le (\frac{d}{2\gamma})^d$$

2.2 Correlation of Distributed Spaces

We need to extend the definition of *correlation* for product probability spaces which is introduced by Mossel [9].

Definition 4. *$(\Omega, \Psi; \mu)$ is a finite correlated probability space. μ is the probability distribution on the product set $\Omega \times \Psi$ with the marginals of μ on Ω and Ψ have full support. Define the* correlation *between Ω and Ψ as:*

$$\rho(\Omega, \Psi; \mu) = \max\{|E_\mu[f(x)\overline{g(y)}]|\}$$

with the maximum restricted on $f : \Omega \to \boldsymbol{C}$ and $g : \Psi \to \boldsymbol{C}$ satisfying $E_\mu[f(x)] = 0$, $E_\mu[g(y)] = 0$, $E_\mu[|f(x)|^2] \le 1$ and $E_\mu[|g(y)|^2] \le 1$.

Lemma 2. *[9] For $1 \le i \le n$, let $(\Omega_i \times \Psi_i, \mu_i)$ be correlated spaces and T_i be the corresponding Markov operator associated with Ω_i and Ψ_i. Define*

$$\Omega = \prod_i \Omega_i, \quad \Psi = \prod_i \Psi_i, \quad \mu = \prod_i \mu_i, \quad T = \prod_i T_i$$

If $\rho(\Omega_i \times \Psi_i, \mu_i) = \rho_i$, then

$$\rho(\Omega, \Psi; \mu) = \max \rho(\Omega_i, \Psi_i; \mu_i)$$

and for all $f : \Omega \to \boldsymbol{C}$ with Efron-Stein Decomposition $f(x) = \sum_{S \subseteq [n]} f_S(x_S)$, it holds that

$$\| T(f_S) \|_2 \le (\prod_{i \in S} \rho_i) \| f_S \|_2$$

Lemma 3. *[9] Let $(\Omega \times \Psi, \mu)$ be correlated space and let T be the corresponding Markov operator associated with Ω and Ψ. Let f be Ω measurable function with $E[f] = 0$ and $E[|f|^2] = 1$, then among all g that are Ψ measurable satisfying $E[g] = 0$ and $E[|g|^2] = 1$, a maximizer of $|E[f\overline{g}]|$ is given by*

$$f = \frac{Tg}{\sqrt{E[|Tg|^2]}}$$

with $|E[f\overline{g}]| = \sqrt{E[|Tg|^2]}$.

We remark that Mossel proved the above lemmas in the case that the functions are real-valued, and our functions are complex-valued. However, these results are still true by a similar way to verify.

2.3 Hypercontractivity Inequality

Definition 5. *For $1 \leq p \leq q \leq \infty$ and $(\boldsymbol{F}, \|\cdot\|)$ is a separable Banach space and let X be a random vector with values in $\boldsymbol{F}$ such that $E[\|X\|^p] < \infty$. Then we say X is* $(2, q, \eta)$-hypercontractive *if*

$$\forall v \in \boldsymbol{F} \quad \|v + \eta X\|_q \leq \|v + X\|_2$$

Conditioned on Oleszkiewicz's result [12], Wolff [14] shows the following

Lemma 4. *Let μ be a discrete measure of random variable X, with the mass of least atom equal to α, then for $2 < p < \infty$ such that $\frac{1}{p-1} \leq \ln \frac{1}{\alpha}$, X is $(2, p, \eta)$-hypercontractive for $\eta \leq (\frac{\beta^{2/p} - \alpha^{2/p}}{\beta \alpha^{2/p-1} - \alpha \beta^{2/p-1}})^{1/2}$, where $\beta = 1 - \alpha$.*

We call a collection of finitely many orthonormal complex random variables, one of which is the constant 1, an *orthonormal ensemble.* Given a sequence of independent complex random variables $X_1, ..., X_n$, we can view them as a sequence of ensembles $\mathcal{X} = \{\mathcal{X}_1, ..., \mathcal{X}_n\}$, with each $\mathcal{X}_i = \{X_{i,0}, X_{i,1}\}$ by setting $X_i = X_{i,1}$ and $X_{i,0} = 1$.

Definition 6. *For $1 \leq p \leq q \leq \infty$ and $0 < \eta < 1$, we say a sequence of ensembles $\mathcal{X}$ is (p, q, η)-hypercontractive if*

$$\|(T_\eta Q)(\mathcal{X})\|_q \leq \|Q(\mathcal{X})\|_p$$

for every multi-linear polynomial Q over $\mathcal{X}$.

Proposition 1. *[9] Let $\mathcal{X}$ and $\mathcal{Y}$ be two independent sequences have n_1 and n_2 ensembles respectively. Assume both $\mathcal{X}$ and $\mathcal{Y}$ are (p, q, η)-hypercontractive, then the sequences of ensembles $\mathcal{X} \cup \mathcal{Y}$ is also (p, q, η)-hypercontractive.*

2.4 The d-to-1 Conjectures

Definition 7. *A d-to-1 Label Cover instance $\mathcal{L}(G, R_1, R_2, \{\pi_e\})$ consists of a bi-regular bipartite graph $G = (U, V; E)$, each edge e associated with a d-to-1 constraint π_e, the goal is to find a labeling of the vertices, i.e., functions $L_U : U \to R_1$ and $L_V : V \to R_2$, that maximizes the fraction of satisfied edges. An edge $e = (u, v)$ is satisfied if $\pi_e(L(v)) = (L(u))$. The value of $\mathcal{L}$ is defined as the maximum fraction of edges satisfied simultaneously. A d-to-1 constraint π from R_2 to R_1 is that for any $i \in R_1$, $\pi^{-1}(i) \subseteq R_2$ satisfies $|\pi^{-1}(i)| \leq d$.*

The following well-known conjecture is the start of our reduction.

Conjecture 2 (d-to-1 Conjecture). [6] For any $\delta > 0$, there exists a constant $k = k(\delta)$ such that it is NP-hard to determine whether a d-to-1 Label Cover instance with label sets of size k has value 1 or at most δ.

3 The PCP Verifier and Conditional Hardness

In this section, we construct a reduction from d-to-1 Label Cover instance to Max 3CSP-q instance, then conditioned on d-to-1 Conjecture [6], we show that for any $\epsilon > 0$, it's NP-hard to approximate the Max 3CSP-q problem within $(q^2 + q - 1)/q^3 + \epsilon$ on satisfiable instance by the standard argument.

3.1 The Test Distribution

Given a d-to-1 Label Cover instance $\mathcal{L}(G(U,V), R_1, R_2, \{\pi_e\})$, our test needs to generate $(\mathbf{x}, \mathbf{y}, \mathbf{z}) \in [q]^{R_1} \times [q]^{R_2} \times [q]^{R_2}$ according to some fixed distributions $\mathcal{T}_{\pi_e}$ and $\mathcal{I}_{\pi_e}$ for each π_e. We first define some basic distributions.

Definition 8. *For an integer $D > 0$, define distributions $\mathcal{T}(D)$, $\mathcal{N}(D)$ and $\mathcal{I}(D)$ generating*

$$(x, y_1, ..., y_D, z_1, ..., z_D) \in [q] \times [q]^D \times [q]^D$$

such that: $\mathcal{T}(D)$ first chooses $x, y_1, ..., y_D$ independently and uniformly from $[q]$, then for each $i \in [D]$, define $z_i = 1 - x - y_i \mod q$. $\mathcal{N}(D)$ first draws from $\mathcal{T}(D)$, then randomly choose $i \in [D]$, let $y_i = z_i = x$. $\mathcal{I}(D)$ first draws from $\mathcal{T}(D)$, then rerandomize x.

Definition 9. *Define the mixed distributions ($0 < \delta < 1$ is a parameter):*

$$\mathcal{T}_\delta(D) = (1-\delta)\mathcal{T}(D) + \delta\mathcal{N}(D) \text{ and } \mathcal{I}_\delta(D) = (1-\delta)\mathcal{I}(D) + \delta\mathcal{N}(D)$$

It is easy to find that $\mathcal{T}(D)$ and $\mathcal{I}(D)$ have the same marginal distributions on $\mathcal{X}$, $\mathcal{Y}$, $\mathcal{Z}$ and on $\mathcal{Y} \times \mathcal{Z}$.

Definition 10. *For a d-to-1 constraint $\pi : [R_2] \to [R_1]$, we define a related distribution $\mathcal{T}_\pi$ as follows: Let $\mathcal{X}^i = [q]^{\{i\}}$ and $\mathcal{Y}^i, \mathcal{Z}^i = [q]^{\pi^{-1}(i)}$ with $|\pi^{-1}(i)| = d_i$, $\mathcal{T}_\delta^i \triangleq \mathcal{T}_\delta(d_i)$ on $X^i \times \mathcal{Y}^i \times \mathcal{Z}^i$ and $\mathcal{T}_\pi \triangleq \otimes_{i=1}^{R_1} \mathcal{T}_\delta^i$ which is on*

$$\prod_{i=1}^{R_1} (\mathcal{X}^i \times \mathcal{Y}^i \times \mathcal{Z}^i) = [q]^{R_1} \times [q]^{R_2} \times [q]^{R_2}$$

Similarly, we can define distribution $\mathcal{I}_\pi \triangleq \otimes_{i=1}^{R_1} \mathcal{I}_\delta^i$ with $\mathcal{I}_\delta^i \triangleq \mathcal{I}_\delta(d_i)$.

Lemma 5. *If function $g : [q]^{R_2} \to \mathbf{C}$ is q-folded and satisfies $\|g\|_2 \leq 1$, let T be the Markov Operator between $[q]^{R_1} \times [q]^{R_2}$ and $[q]^{R_2}$ under the distribution $\mathcal{T}_\pi$ for some d-to-1 constraint π with $2d < q$, then there is some constant $\varrho_0 < 1$ which is independent of R_1 and R_2, such that*

$$\|Tg_S\|_2 \leq \varrho_0^{|S|} \|g_S\|_2$$

where $g_S = \sum_{b \in [q]^{R_2}, \pi(N(b))=S} \widehat{g}_b \chi_b$.

Lemma 5 is verified directly. We can not follow the way of [11] since the based graph of $\mathcal{T}_\delta(D)$ is not connected any more, however we can use the fact that g is q-folded to make this work.

We denote the distribution $\mathcal{T}_\delta^*(\gamma)(D)$ as first generate $(x, \mathbf{y}, \mathbf{z})$ by $\mathcal{T}_\delta(D)$ then rerandomizing each bit in $\mathbf{y}, \mathbf{z}$ independently with probability γ.

Lemma 6. *The correlation between* $[q]^{R_1}$ *and* $[q]^{R_2} \times [q]^{R_2}$ *under* $\mathcal{T}_\pi^*(\gamma) \triangleq \otimes \mathcal{T}_\delta^*(\gamma)(d_i)$ *is bounded above by* $\varrho_1 = 1 - \frac{\gamma^{2d}}{2}$ *for any d-to-1 constraint* π.

3.2 The PCP Verifier

Given a d-to-1 Label Cover instance $\mathcal{L}(G(U, V; E), R, \{\pi_e\})$, we construct an instance of Max 3CSP-q, presented as a $O(logn)$-randomness PCP verifier. As usual, in our PCP system the given proof is supposed to contain a label for each vertex, which is of folded q-ary long code form. An Long code function $f : [q]^n \to \{\omega^0, ..., \omega^{q-1}\}$ is defined as $f(\mathbf{x}) = \omega^{x_i}$ for some $i \in [n]$. We may w.l.o.g. assume the functions given by our proof satisfy the following conditions (which is always satisfied if the proof gives long code functions): q-folded and respects exponentiation. The PCP verifier works as follows:

1. Pick an edge $e = (u, v) \in E$ at random, let π_e be the d-to-1 constraints on edge e. Let f_u, f_v be the supposed q-ary Long codes of the labels for u, v respectively.
2. Generate a triple $(\mathbf{x}, \mathbf{y}, \mathbf{z})$ from the distribution $\mathcal{T}_{\pi_e}$ on $[q]^{R_1} \times [q]^{R_2} \times [q]^{R_2}$.
3. Accept if and only if $(f_u(\mathbf{x}), f_v(\mathbf{y}), f_v(\mathbf{z}))$ is in $S_\mathcal{T}$.

Completeness: The Completeness part is easy and standard.

Soundness: We will show that if the PCP verifier accepts with probability exceeding $(q^2 + q - 1)/q^3 + \epsilon$, then we can find a labeling assignment for the d-to-1 Label Cover instance that satisfies at least θ fraction of edges, here $\theta > 0$ is independent of the label size R_1 and R_2.

We define an indictor function $P : [q]^3 \longrightarrow \mathbb{R}$ as $P(x, y, z) = 1$ if $(\omega^x, \omega^y, \omega^z)$ is in $S_\mathcal{T}$, $P(x, y, z) = 0$ otherwise. Then

$$P(x, y, z) = \sum_{\mathbf{a} \in [q]^3} \hat{P}_\mathbf{a} \chi_\mathbf{a}(x, y, z) = \sum_{\mathbf{a} \in [q]^3} \hat{P}_\mathbf{a} \omega^{a_1 x} \omega^{a_2 y} \omega^{a_3 z}$$

with $\hat{P}_\mathbf{a} = E_{x,y,z}[P(x, y, z)\overline{\chi_\mathbf{a}}(x, y, z)]$ and $|\hat{P}_\mathbf{a}| \leq 1$ since $|P(x, y, z)| \leq 1$ for all $(x, y, z) \in [q]^3$. Now we have

$$Pr[\text{Verifier accepts}] = E_{u,v} E_{\mathbf{x},\mathbf{y},\mathbf{z}}[(f_u(\mathbf{x}), f_v(\mathbf{y}), f_v(\mathbf{z})) \text{ is in } S_\mathcal{T}]$$

$$= E_{u,v} E_{\mathbf{x},\mathbf{y},\mathbf{z}}[\sum_{\mathbf{a} \in [q]^3} \hat{P}_\mathbf{a} f_u(\mathbf{x})^{a_1} f_v(\mathbf{y})^{a_2} f_v(\mathbf{z})^{a_3}]$$

$$= \frac{q^2 + q - 1}{q^3} + E_{u,v} E_{\mathbf{x},\mathbf{y},\mathbf{z}}[\sum_{0 \neq \mathbf{a} \in [q]^3} \hat{P}_\mathbf{a} f_u(\mathbf{x})^{a_1} f_v(\mathbf{y})^{a_2} f_v(\mathbf{z})^{a_3}]$$

If the PCP verifier accepts with probability at least $(q^2+q-1)/q^3+\epsilon$, then for at least $\epsilon/2$ fraction of edges (u,v) (call such edges 'good'), we have

$$E_{\mathbf{x},\mathbf{y},\mathbf{z}}[\sum_{\mathbf{0}\neq\mathbf{a}\in[q]^3}\hat{P}_{\mathbf{a}}f_u(\mathbf{x})^{a_1}f_v(\mathbf{y})^{a_2}f_v(\mathbf{z})^{a_3}]\geq\epsilon/2$$

Fixed a good edge $e=(u,v)$ with d-to-1 constraint π_e, there exists $\mathbf{0}\neq\mathbf{a}\in[q]^3$ such that

$$|E_{\mathbf{x},\mathbf{y},\mathbf{z}}[\hat{P}_{\mathbf{a}}f_u(\mathbf{x})^{a_1}f_v(\mathbf{y})^{a_2}f_v(\mathbf{z})^{a_3}]|\geq\epsilon/2q^3$$

Since $|\hat{P}_{\mathbf{a}}|\leq 1$, then

$$|E_{\mathbf{x},\mathbf{y},\mathbf{z}}[f_u(\mathbf{x})^{a_1}f_v(\mathbf{y})^{a_2}f_v(\mathbf{z})^{a_3}]|\geq\epsilon/2q^3$$

Note the above expectation is over $(\mathbf{x},\mathbf{y},\mathbf{z})\in_{\mathcal{T}_{\pi_e}}[q]^{R_1}\times[q]^{R_2}\times[q]^{R_2}$.

Let $\delta>0$ be a small enough value (say $\delta\leq(\epsilon/6q^3)^2\leq\epsilon/2q^3$) used in the definition of distribution $\mathcal{T}_\pi$, then Lemma 7 and Lemma 8 imply that $a_1a_2a_3\neq 0$. Theorem 2 (since $\delta\leq(\epsilon/6q^3)^2$ then $3\sqrt{\delta}\leq\epsilon/2q^3$) implies that there exists some $\tau=\tau(\epsilon,q,d)>0$, $i\in[R_1]$ and some $\mathbf{a}\in[q]^{R_2}$ with $N(\mathbf{a})\subseteq\pi_e^{-1}(i)$, $|N(\mathbf{a})|>0$ such that

$$\min\{Inf_i(T_{1-\gamma'/2}f_u),Inf_{N(\mathbf{a})}(T_{1-\gamma/2}f_v)\}\geq\tau$$

for small enough $\gamma=\gamma((\epsilon,q,d))$ and $\gamma'=\gamma'(\epsilon,q,d)$.

For each $u\in U$ and $v\in V$, define

$$L_u=\{i\in[R_1]:Inf_i(T_{1-\gamma'/2}f_u)\geq\tau\}$$

$$L_v=\cup S\subseteq[R_2]:|S|\leq d,Inf_S(T_{1-\gamma/2}f_v)\geq\tau$$

Then for each 'good' edge $e=(u,v)$, $L_u\neq\varnothing$ and there exists some $i\in L_u$ such that $\pi^{-1}(i)\cap L_v\neq\emptyset$. Moreover, as $\| T_{1-\gamma'/2}f_u\|_2^2\leq 1$ and $\| T_{1-\gamma/2}f_v\|_2^2\leq 1$, we have $|L_u|\leq\frac{1}{2e\tau\ln\frac{1}{(1-\gamma'/2)}}$ and $|L_v|\leq\frac{d(\frac{d}{\gamma})^d}{\tau}$ by Lemma 1. We define a label assignment for the d-to-1 label cover instance as follows: random choose $i\in L_u$ as assignment of u and random choose $i\in L_v$ as assignment of v, edge (u,v) is satisfied by the assignment with probability at least

$$|L_u|^{-1}\cdot|L_u|^{-1}\geq(2e\tau\ln\frac{1}{(1-\gamma'/2)}\times\frac{\tau}{d(\frac{d}{\gamma})^d})=\frac{2e\tau^2\gamma^d}{d^{d+1}}\ln\frac{1}{(1-\gamma'/2)}$$

Note that there are remarkable fraction of 'good' edges (at least $\epsilon/2$), so the assignment we defined satisfies at least $\theta=\frac{\epsilon e\tau^2\gamma^d}{d^{d+1}}\ln\frac{1}{(1-\gamma'/2)}$ fraction of edges in expectation, which finish our proof by the observation that θ is independent of R_1 and R_2.

3.3 Some Lemmas Used in Soundness Proof

In this section, we present the lemmas we used to prove the soundness of PCP verifier, we omit their proof for space limit, instead we give some proof sketch of these lemmas.

We assume $f : [q]^{R_1} \to \mathbf{C}$, $g : [q]^{R_2} \to \mathbf{C}$ respect exponentiation and satisfy $E[f] = 0$, $E[g] = 0$, $|f(\mathbf{x})| \leq 1$ and $|g(\mathbf{y})| \leq 1$ for all $\mathbf{x} \in [q]^{R_1}$, $\mathbf{y} \in [q]^{R_2}$ in all lemmas of this subsection.

Lemma 7. *For any nonzero vector* $(a, b) \in [q]^2$, *we have*

$$|E_{\mathcal{T}_{\pi_e}}[f(\boldsymbol{x})^a g(\boldsymbol{y})^b]| \leq \delta$$

Lemma 8. *For any* $a, b \in [q]^*$,

$$|E_{\mathcal{T}_{\pi_e}}[g(\boldsymbol{y})^a g(\boldsymbol{z})^b]| \leq \delta$$

Theorem 2. *For any* $d, \delta > 0$ *and* $a, b, c \in [q]^*$, *there exists constant* τ, γ *depending only on* d, δ, q *such that the following holds: if for every* $i \in [R_1]$ *and* $\boldsymbol{a} \in [q]^{R_2}$ *with* $N(\boldsymbol{a}) \subseteq \pi^{-1}(i)$ *and* $\sum_{j \in \pi^{-1}(i)} a_j \neq 0$ *we have*

$$Min\{Inf_i(T_{1-\gamma/2}f), Inf_{S_a}(T_{1-\gamma/2}g)\} \leq \tau$$

then

$$|E_{\mathcal{T}_{\pi_e}}[f(\boldsymbol{x})^a g(\boldsymbol{y})^b g(\boldsymbol{z})^c]| \leq 3\sqrt{\delta}$$

Proof (Proof Sketch of Lemma 7, Lemma 8 and Theorem 2). The Proof of Lemma 7 uses standard Fourier techniques directly. For Lemma 8, we prove it by induction on R_1.

The main idea to prove Theorem 2 is to use Invariance principle argument to transfer bounding $|E_{\mathcal{T}_e}[f(\mathbf{x})^a g(\mathbf{y})^b g(\mathbf{z})^c]|$ to bounding $|E_{\mathcal{I}_e}[f(\mathbf{x})^a g(\mathbf{y})^b g(\mathbf{z})^c]|$ which is much easier. This line of work includes [10], [9] and [11]. All the techniques we used are similar as [11] for proving their Theorem 6.3. However, since our functions are $\mathbf{C}$-valued other than $\mathbf{R}$-valued, we need more careful analysis and more complicated verification, specifically we need to use Lemma 5 instead of to bound the correlation between $\mathcal{X} \times \mathcal{Y}$ and $\mathcal{Z}$ under $\mathcal{T}_e$, and we also need different kind of Hypercontractive argument Lemma 4.

4 Discussion

In this paper, we try to understand the effect of satisfiable promise to approximate Max 3CSP-q. We are able to construct a special kind of Predicate on 3 variables with $q^2 + q - 1$ accepting inputs of the total q^3 and show that Max 3CSP-q instances defined by such Predicate is Approximation Resistant even on satisfiable instances conditioned on d to 1 Conjecture, which means that satisfiable promise does not help in approximating such instances. It is still an interesting open question whether the Predicate we defined is the Predicate with least accepting inputs which satisfies the above mentioned property. However, when $q = 2$, i.e., in the boolean case, it is the case. On the other hands, an better approximation algorithm for satisfiable Max 3CSP-q would also be interesting.

References

1. Austrin, P., Mossel, E.: Approximation Resistant Predicates From Pairwise Independence. ECCC TR08-009 (2008)
2. Charikar, M., Makarychev, K., Makarychev, Y.: Near-Optimal Algorithms for Unique Games. In: Proceedings of the 38-th Annual ACM Symposium on Theory of Computing, pp. 205–214 (2006)
3. Charikar, M., Makarychev, K., Makarychev, Y.: Near-Optimal Algorithms for Maximum Constraint Satisfaction Problems. In: Proceedings of the 18th Annual ACM-SIAM Symposium on Discrete Algorithms, pp. 62–68 (2007)
4. Engebretsen, L., Holmerin, J.: Three-query PCPs with perfect completeness over non-Boolean domains. Random Structures and Algorithms 27(1), 46–75 (2005)
5. Håstad, J.: Some optimal inapproximability reaults. J.ACM 48(4), 798–859 (2001)
6. Khot, S.: On the power of unique 2-pover 1-round games. In: Proceedings of 34th ACM Symposium on Theory of Computating, pp. 767–775 (2002)
7. Khot, S., Kindler, G., Mossel, E., Donnell, R.: Optimal Inapproximability Results for MAX-CUT and Other 2-Variable CSPs? In: Proceedings of 45th Annual IEEE Symposium of Foundations of Computer Science, pp. 146–154 (2004)
8. Khot, S., Regev, O.: Vertex cover might be hard to approximate to within $2-\epsilon$. In: Proceedings of 18th IEEE Conference on Computational Complexity, pp. 379–386 (2003)
9. Mossel, E.: Gaussian bounds for noise correlation of functions. arXiv Report math/0703683v3 (2007)
10. Mossel, E., Donnell, R., Oleszkiewicz, K.: Noise stability of functions with low influences:invariance and optimality. In: Proceedings of 48th Annual IEEE Symposium of Foundations of Computer Science, pp. 307–317 (2007)
11. O'Donnell, R., Wu, Y.: Conditional hardness for satisfiable 3-CSPs. In: Proceedings of the 41-th Annual ACM Symposium on Theory of Computing, pp. 493–502 (2009)
12. Oleszkiewicz, K.: On a nonsymmetric version of the Khinchine-Kahane inequality. Progress in Probability 56, 156–168 (2003)
13. Raghavendra, P.: Optimal Algorithms and Inapproximability Results for Every CSP? In: Proceedings of the 40th Annual ACM Symposium on Theory of Computing (to appear) (2008)
14. Wollf, P.: Hypercontractivity of simple random variables. In: The 46th Annual Symposium on Foundations of Computer Science, pp. 740–746 (2005)
15. Zwick, U.: Approximation algorithms for constraint satisfaction problems involving at most three variables per constraint. In: Proceedings of the 9th Annual ACM-SIAM Symposium on Discrete Algorithms, pp. 201–210 (1998)

The Roles of Advice to One-Tape Linear-Time Turing Machines and Finite Automata (Extended Abstract)

Tomoyuki Yamakami

Department of Information Science, University of Fukui
3-9-1 Bunkyo, Fukui, 910-8507 Japan

Abstract. We discuss the power and limitations of various "advice," when it is given particularly to weak computational models of one-tape two-way linear-time Turing machines and one-way finite (state) automata. Of various advice types, we consider deterministically-chosen advice, which is selected depending only on input size, and randomly-chosen advice, which is chosen according to certain probability distributions. We show that machines can be significantly enhanced in computational power when advice is provided; on the contrary, there are clear limitations on such a power.

1 Introduction

How can we enhance the computational power of an underlying machine? An obvious way is to provide a piece of supplemental information besides original input information so that the machine takes advantages of such extra knowledge to solve a given problem more efficiently. A notion of so-called *advice* in computational complexity refers to such additional information, given to the machine, which depends only on the size of the input. Since Karp and Lipton [4] initiated it in early 1980s, the study of advice has attracted many researchers in various fields of computer science. To grip a better understanding of the roles of such advice, we intend to take a rather intuitive but systematic approach toward an investigation of the strengths and limitations of the advice particularly on weak models of advised computations.

One-tape two-way one-head Turing machines (or 1TMs, in short) running in linear time could be one of the most basic types of computational models ever discussed in computational complexity theory. As were shown in [3,5,6], certain variants of this machine model are closely tied to one-way finite (state) automata with no memory space. Advised computations of one-way deterministic finite automata (or 1dfa's, in short) were initially studied in [1,6] and deterministic linear-time 1TMs with advice were discussed in [6]. Importantly, it was shown in [6] that deterministic linear-time 1TMs that take linear-size advice are no more powerful than 1dfa's together with advice of size equal to input size. This characterization makes it easier for us to handle the one-tape linear-time model. Recently, a series of studies [7,8] revealed an excessive power as well as an unexpected weakness of advice when it is given to various underlying finite

Y. Dong, D.-Z. Du, and O. Ibarra (Eds.): ISAAC 2009, LNCS 5878, pp. 933–942, 2009.

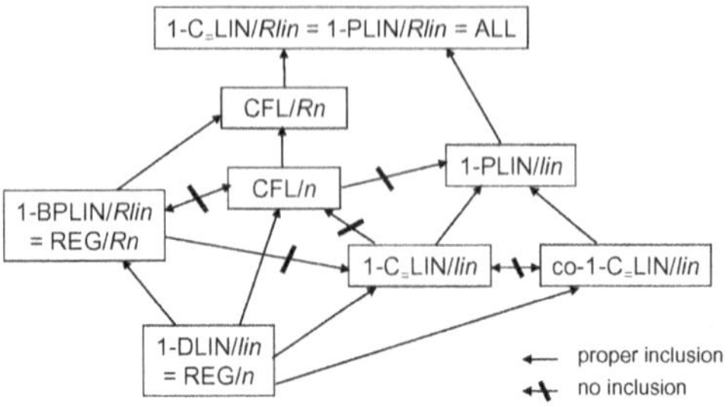

Fig. 1. A hierarchy of advised language families

automata. In addition to standard (deterministic) advice, in this paper, we also study *randomized advice* (in which each advice string is chosen at random according to a certain probability distribution), which allows its underlying machines, possibly incorporated with the machine's coin-tossing mechanism, to err. When the overall error probability is always at most a constant away from 1/2 (less than 1/2, resp.), we use the term "bounded error" ("unbounded error", resp.) to describe that. As we later demonstrate, a piece of such randomized advice gives a significantly high power to the machines.

Based on the aforementioned model of linear-time 1TMs, we are focused only on the following language families: 1-DLIN (deterministic), 1-BPLIN (bounded-error probabilistic), 1-PLIN (unbounded-error probabilistic), and 1-$C_=$LIN (error probability exactly 1/2). These language families can be viewed as "scaled-down" versions of the well-known complexity classes, P, BPP, PP, and $C_=$P. Their advised counterparts are succinctly denoted as 1-DLIN/lin, 1-PLIN/lin, and 1-$C_=$LIN/lin. When randomized advice is provided, we write their corresponding families as 1-BPLIN/$Rlin$, 1-PLIN/$Rlin$, and 1-$C_=$LIN/$Rlin$. Similarly, using the finite-automata models, we define REG/n and CFL/n respectively for regular languages with advice and context-free languages with advice. We further introduce two additional language families with randomized advice: REG/Rn and CFL/Rn.

In this paper, we shall present new class separations among the above-listed advised language families. Our results are summarized in Fig. 1. To obtain these results, we need to develop new characterizations of advised families and further cultivate their new structural properties, which are interesting on their own light.

2 Basic Notions and Notation

Let $\mathbb{N}$ be the set of all *nonnegative integers*. For any pair $m, n \in \mathbb{N}$ with $m \leq n$, $[m,n]_{\mathbb{Z}}$ denotes the *integer interval* $\{m, m+1, \ldots, n\}$. For simplicity, write $[m]$ for $[1,m]_{\mathbb{Z}}$. Let $\mathbb{R}^{\geq 0}$ be the set of all *nonnegative real numbers*. A function $f : \mathbb{N} \to \mathbb{R}^{\geq 0}$ is said to be *negligible* if $f(n) \leq 1/p(n)$ for any non-zero polynomial p and for all but finitely-many numbers $n \in \mathbb{N}$. An *alphabet* Σ is a nonempty finite set and a *string* over Σ is a series of symbols taken from Σ. Let λ denote the *empty string*. The *length* of a string x, denoted $|x|$, is the total number of symbols in x. For any string x over Σ and any symbol σ in Σ, the notation $\#_\sigma(x)$ denotes the number of σ's in x. A *probability ensemble* μ over Σ^* is an infinite series $\{\mu_n\}_{n\in\mathbb{N}}$, in which each μ_n is a probability distribution over Σ^n.

Our basic model of computation is *one-tape (or single-tape) two-way one-head Turing machines* (or 1TMs), each of which can be expressed as a sextuple $(Q, \Sigma, \delta, q_0, Q_{acc}, Q_{rej})$, where Q is a set of *inner states*, Σ is an input alphabet, q_0 ($\in Q$) is the *initial state*, δ is a *transition function*, Q_{acc} ($\subseteq Q$) is a set of *accepting states*, and Q_{rej} ($\subseteq Q$) is a set of *rejecting states*. This machine M is equipped with one input/work tape, on which an input is initially marked by two endmarkers ¢ and \$, and a tape head either moves in both directions or stays still. We say that M runs in *linear time* if the longest computation path (even in a case of probabilistic computation) of M on every input x of length n is bounded from above by a certain linearly-bounded function in n. In other words, any *computation tree* corresponding to the input x has hight at most $O(n)$. (See [6] for a detailed discussion on this so-called *strong definition* for running time.) When δ is deterministic (probabilistic, resp.), we succinctly call M 1DTM (1PTM, resp.). Notice that finite (state) automata are a special case of these linear-time 1TMs. Let REG, CFL, and DCFL denote, respectively, the families of regular languages, of context-free languages, and of deterministic context-free languages.

We say that a 1PTM M has *bounded error probability* if there exists a constant $\varepsilon \in [0, 1/2)$ such that, for every input string x, either $\mathrm{Prob}_M[M(x) = 1] \geq 1-\varepsilon$ or $\mathrm{Prob}_M[M(x) = 0] \geq 1-\varepsilon$. Let 1-DLIN (1-BPLIN, 1-PLIN, and 1-$\mathrm{C_{=}}$LIN, resp.) denote the collection of all languages that are recognized by 1DTMs (1PTMs with bounded error, 1PTMs with unbounded error, and 1PTMs with error probability exactly 1/2, resp.) in linear time using the strong definition.

To feed a piece of supplemental information besides inputs to 1TMs, we use a "track" notation of [6]. For two symbols $\sigma \in \Sigma$ and $\tau \in \Gamma$, the notation $[\begin{smallmatrix}\sigma\\ \tau\end{smallmatrix}]$ expresses a new symbol made from σ and τ. For a 1TM equipped with an input/work tape, this symbol $[\begin{smallmatrix}\sigma\\ \tau\end{smallmatrix}]$ is written in a single cell, which consists of two tracks, whose upper track contains σ and the lower track contains τ. A tape head of the 1TM scans this symbol $[\begin{smallmatrix}\sigma\\ \tau\end{smallmatrix}]$ at once. For any two strings x and y of the same length n, where $x = x_1x_2\cdots x_n$ and $y = y_1y_2\cdots y_n$, let $[\begin{smallmatrix}x\\ y\end{smallmatrix}]$ denote the string $[\begin{smallmatrix}x_1\\ y_1\end{smallmatrix}][\begin{smallmatrix}x_2\\ y_2\end{smallmatrix}]\cdots[\begin{smallmatrix}x_n\\ y_n\end{smallmatrix}]$ of length n. The track notation can be further extended to the case where $|x| \neq |y|$. If $|x| < |y|$ and $y = y_1y_2$ with $|x| = |y_1|$, then $[\begin{smallmatrix}x\\ y\end{smallmatrix}]$ denotes $[\begin{smallmatrix}x\\ y_1\end{smallmatrix}][\begin{smallmatrix}\#^{|x|-|y|}\\ y_2\end{smallmatrix}]$; if $|x| > |y|$ and $x = x_1x_2$ with $|y| = |x_1|$, then $[\begin{smallmatrix}x\\ y\end{smallmatrix}]$ denotes $[\begin{smallmatrix}x_1\\ y\end{smallmatrix}][\begin{smallmatrix}x_2\\ \#^{|x|-|y|}\end{smallmatrix}]$, where # is a special symbol for "blank."

For our later use, we give a brief description of one-way probabilistic finite automata. Here, we assume that all vectors are row vectors. For any matrix M, M^T denotes its *transposed* matrix. A *one-way (rational) probabilistic finite automaton* (or 1pfa, in short) M is a quintuple $(Q, \Sigma, \nu_{ini}, \{M_\sigma\}_{\sigma\in\Sigma}, F)$, where ν_{ini} is an *initial state vector* with rational entries, each M_σ is a $|Q|\times|Q|$ *stochastic matrix* with rational entries, and F ($\subseteq Q$) is a set of *final states*. Without loss of generality, we can assume that ν_{ini} has always value 1 in its q_0-entry and 0 in all the other entries. The set F induces a vector μ_F defined as follows: for each state $q \in Q$, the q-entry of ξ_F has value 1 if $q \in F$, and 0 otherwise. For any string $x = \sigma_1\sigma_2 \cdots \sigma_n$ in Σ^n, let M_x denote $M_{\sigma_1} M_{\sigma_2} \cdots M_{\sigma_n}$. The *accepting* (*rejecting*, resp.) *probability* of M on the input x is defined as $p_{acc}(x) = \nu_{ini} M_x \xi_F^T$ ($p_{rej}(x) = 1 - p_{acc}(x)$, resp.).

3 Deterministic Computation with Advice

We formally define the notion of *advice* and describe how to use it on one-tape Turing machines and finite automata. An *advice function* h is a function mapping $\mathbb{N}$ to Γ^*, where Γ is a certain alphabet (which is referred to as an *advice alphabet*). The advised language family* 1-DLIN/lin (REG/n, resp.) [6] is defined as the collection of all languages S over any alphabet Σ such that there exist an advice alphabet Γ, an advice function $h : \mathbb{N} \to \Gamma^*$, and a linear-time 1DTM (1dfa, resp.) M satisfying the following two conditions: (i) there are two constants $c, d > 0$ such that, for every length $n \in \mathbb{N}$, $|h(n)| \leq cn + d$ ($|h(n)| = n$, resp.) and (ii) for every string $x \in \Sigma^*$, $x \in S$ iff M accepts $\left[{x \atop h(|x|)} \right]$. It is important to note that this scheme of providing advice strings is computationally equivalent to Karp-Lipton's original one [4] if its underlying computation is, for instance, polynomially time-bounded. Naturally, REG/n contains non-regular languages (for instance, $L_{eq} = \{0^n1^n \mid n \in \mathbb{N}\}$). Surprisingly, these two advised families coincide.

Lemma 1. [6] 1-DLIN/lin = REG/n.

In a similar fashion, we can define two more advised language families, 1-PLIN/lin and 1-C$_=$LIN/lin, by expanding the definition of 1-DLIN/lin using 1PTMs in lieu of 1DTMs. It is shown in [6] that CFL $\cap$ 1-C$_=$LIN $\nsubseteq$ REG/n. From this result, since 1-DLIN/lin $\subseteq$ 1-C$_=$LIN/lin, it follows that REG/n is properly included in 1-C$_=$LIN/lin.

For a further analysis of languages in REG/n, it is useful to find their machine-independent characterization. In Theorem 1, we give such a characterization, which naturally yields the so-called *swapping lemma* for regular languages [7]. For our notational convenience, for any language S, we define $S(x) = 1$ if $x \in S$; $S(x) = 0$ if $x \notin S$.

Theorem 1. *For any language S over an alphabet Σ, the following two statements are equivalent. Let $\Delta = \{(x, n) \in \Sigma^* \times \mathbb{N} \mid |x| \leq n\}$.*
1. *S is in* REG/n.
2. *There exists an equivalence relation $\equiv_S$ over Δ such that*
   (a) *the total number of equivalence classes in $\Delta / \equiv_S$ is finite; and*
   (b) *for any number $n \in \mathbb{N}$ and any two strings $x, y \in \Sigma^*$ with $|x| = |y| \leq n$, the following relation holds: $(x, n) \equiv_S (y, n)$ iff, for all z with $|xz| = n$, $S(xz) = S(yz)$.*

---

* The "advice" definition of Damm and Holzer [1] for 1dfa's is quite different from our current definition. However, these two definitions are equivalent for, e.g., polynomial time-bounded computations.

4 Power of Randomized Advice

Our underlying machines so far enable to process a single advice string $h(n)$ per each input length n. It is also possible to give the machines "randomized" advice strings, which are produced according to a certain fixed probability distribution. It turns out that such randomized advice may give an enormous power to the machine's language recognition.

In this paper, *randomized advice* refers to a probability ensemble over advice strings. Let $m, n \in \mathbb{N}$, let x be any input string, and let D_m be any probability distribution over Γ^m, where Γ is an advice alphabet. The notation $\left[\begin{smallmatrix}x\\ D_m\end{smallmatrix}\right]$ indicates a *random variable* that expresses $\left[\begin{smallmatrix}x\\ y\end{smallmatrix}\right]$ over all strings y in Γ^m, where $\left[\begin{smallmatrix}x\\ y\end{smallmatrix}\right]$ is chosen randomly with probability $D_m(y)$. In addition, for a machine M, the notation $M(\left[\begin{smallmatrix}x\\ D_m\end{smallmatrix}\right])$ denotes a random variable expressing the outcome $M(\left[\begin{smallmatrix}x\\ y\end{smallmatrix}\right])$ of M on any input $\left[\begin{smallmatrix}x\\ y\end{smallmatrix}\right]$ when $\left[\begin{smallmatrix}x\\ y\end{smallmatrix}\right]$ is chosen randomly according to D_m.

We write 1-BPLIN/*Rlin* to denote the collection of all languages L such that there exist a linear-time 1PTM M, an error bound $\varepsilon \in [0, 1/2)$, an advice probability ensemble $\{D_n\}_{n\in\mathbb{N}}$, and a linearly-bounded function $\ell : \mathbb{N} \to \mathbb{N}$ satisfying: for every input x, $\mathrm{Prob}_{M,D_m}[M(\left[\begin{smallmatrix}x\\ D_m\end{smallmatrix}\right]) = L(x)] \geq 1 - \varepsilon$, where $m = \ell(|x|)$. Similarly, we can define 1-C$_=$LIN/*Rlin* and 1-PLIN/*Rlin*; however, those two language families are so powerful that they can recognize *all* languages. This can be "roughly" shown for every language A by simply taking the randomized advice D_n that generates all elements in $\Sigma^n - A$ with equal probability.

Proposition 1. *The advised language family* 1-C$_=$LIN/*Rlin (as well as* 1-PLIN/*Rlin) consists of all languages.*

Hereafter, we are focused on the advised language family 1-BPLIN/*Rlin*. Let us first present a useful characterization of 1-BPLIN/*Rlin* using one-way finite automata. A language A over an alphabet Σ is in REG/*Rn* if there exist a 1dfa M, a constant $\varepsilon \in [0, 1/2)$, and an advice probability ensemble $\{D_n\}_{n\in\mathbb{N}}$ that satisfy the following condition: for any $n \in \mathbb{N}$ and $x \in \Sigma^n$, if $x \in A$ then M accepts $\left[\begin{smallmatrix}x\\ D_n\end{smallmatrix}\right]$ with probability $\geq 1 - \varepsilon$; otherwise, M rejects $\left[\begin{smallmatrix}x\\ D_n\end{smallmatrix}\right]$ with probability $\geq 1 - \varepsilon$. In a similar fashion, we can define another advised language family CFL/*Rn* using *one-way nondeterministic pushdown automata* (instead of 1dfa's) with randomized advice. Obviously, REG/*Rn* $\subseteq$ CFL/*Rn*. With regard to REG/*Rn*, we obtain the following lemma, whose proof is in essence similar to [6, Lemma 6.1].

Lemma 2. 1-BPLIN/*Rlin* = REG/*Rn*.

Even for weaker language families, such as REG/*Rn* and REG/n, randomized advice becomes more resourceful than deterministic advice. Let us consider the language $Pal = \{ww^R \mid w \in \{0,1\}^*\}$ (even-length palindromes), which is in DCFL, where w^R is w *in reverse*. It is shown in [7] that Pal is not in REG/n. Since we can show that Pal belongs to REG/*Rn*, we therefore obtain the following proposition.

Proposition 2. DCFL $\cap$ REG/*Rn* $\nsubseteq$ REG/n.

It is not difficult to show that $Dup = \{ww \mid w \in \{0,1\}^*\}$ (duplicated strings) is in REG/Rn. Since $Dup \notin$ CFL/n [7], we obtain that REG/$Rn \nsubseteq$ CFL/n. Moreover, it holds that $Dup \in$ 1-C$_=$LIN/lin. Therefore, we conclude:

Proposition 3. REG/$Rn \cap$ 1-C$_=$LIN/$lin \nsubseteq$ CFL/n.

Notice that Proposition 3 immediately yields another class separation between CFL/n and CFL/Rn since REG/$Rn \subseteq$ CFL/Rn.

Theorem 2. *1.* 1-C$_=$LIN/$lin \neq$ co-1-C$_=$LIN/$lin \neq$ 1-PLIN/lin.
2. REG/$Rn \nsubseteq$ 1-C$_=$LIN/$lin \cup$ co-1-C$_=$LIN/lin.

To prove the theorem, we need a key lemma, which gives new criteria that every language in 1-C$_=$LIN/lin must satisfy. In early 1970s, Dieu [2] showed (in our terminology) that, for any language L in 1-C$_=$LIN, there exists a number $m \in \mathbb{N}$ such that, for any $u, y, v \in \Sigma^*$, if $\{uy^iv \mid i \in [0, m-1]_{\mathbb{Z}}\} \subseteq L$ then $\{uy^iv \mid i \in \mathbb{N}\} \subseteq L$. Unfortunately, his criteria cannot be extended to our advised language family 1-C$_=$LIN/lin, mainly because advice strings may differ for different input sizes. We thus need to seek other criteria for 1-C$_=$LIN/lin. The next lemma provides such criteria.

Lemma 3. *Let A be any language in* 1-C$_=$LIN/lin *over an alphabet Σ. There exists a positive integer m that satisfies the following statement. Let $n, \ell \in \mathbb{N}$ and $z \in \Sigma^*$ satisfy that $n \geq 2$, $\ell \leq n-1$, and $|z| = \ell$. Let $A_{n,z} = \{w \in \Sigma^{n-\ell} \mid wz \in A\}$. There exists a subset $S \subseteq A_{n,z}$ with $|S| \leq m$ such that, for each string $y \in \Sigma^\ell$, if $\{wy \mid w \in S\} \subseteq A$ then $\{xy \mid x \in A_{n,z}\} \subseteq A$.*

We shall prove Theorem 2 using Lemma 3. The proof of the theorem exemplifies usefulness of the criteria given in the lemma.

Proof of Theorem 2. (1) We have already proven that $Dup \in$ 1-C$_=$LIN/lin. Next, we show that $\overline{Dup} \notin$ 1-C$_=$LIN/lin. Let $A = \overline{Dup}$ for simplicity and we wish to prove that $A \notin$ 1-C$_=$LIN/lin. Now, let us assume to the contrary that $A \in$ 1-C$_=$LIN/lin. By Lemma 3, there is a positive integer m that satisfies the conclusion of the lemma. Choose the minimal even integer n for which $2^{n/2} - 1 > m + 1$, and set $z = 1^{n/2}$. Clearly, we have $|A_{n,z}| = 2^{n/2} - 1$. The lemma gives a subset $S \subseteq A_{n,z}$ with $|S| \leq m$. Take any string y in $\Sigma^{n/2} - S \cup \{z\}$. Since $wy \in A$ for any $w \in S$, the lemma concludes that $yy \in A$. This is a contradiction against $A = \overline{Dup}$. Thus, we obtain $A \notin$ 1-C$_=$LIN/lin.

(2) Since Dup is in REG/Rn and REG/Rn is closed under complementation, $\overline{Dup}$ is also in REG/Rn. However, since (1) implies that $\overline{Dup} \notin$ 1-C$_=$LIN/lin, it follows that REG/$Rn \nsubseteq$ 1-C$_=$LIN/lin. From this, we obtain that co-REG/$Rn \nsubseteq$ co-1-C$_=$LIN/lin. Therefore, REG/Rn = co-REG/$Rn \nsubseteq$ co-1-C$_=$LIN/lin. □

Let us return to the proof of Lemma 3.

Proof of Lemma 3. Let $A \in$ 1-C$_=$LIN/lin be any language over an alphabet Σ. Similar to Lemma 1, we can show that there exists a 1pfa

$M = (Q, \Sigma, \nu_{ini}, \{M_\sigma\}_{\sigma\in\Sigma}, F)$ and an advice function h such that, for every string $x \in \Sigma^*$, $x \in A$ iff $\mathrm{Prob}_M[M([{}^{x}_{h(|x|)}]) = 1] = 1/2$. Recall from Section 2 our assumption on ν_{ini} and ξ_F.

We set $m = |Q|$ and choose arbitrarily triplet n, ℓ, and z that satisfy $|z| = \ell$. Note that if $\ell = 0$ then the lemma trivially holds. In what follows, we assume that $\ell > 0$. For simplicity, let $r = \tau_1\tau_2\cdots\tau_{n-\ell}$ and $s = \tau_{n-\ell+1}\tau_{n-\ell+2}\cdots\tau_n$ if $h(n) = \tau_1\tau_2\cdots\tau_n$. Consider $A_{n,z} = \{w \in \Sigma^{n-\ell} \mid wz \in A\}$. Since the lemma is trivially true when $|A_{n,z}| \leq |Q|$, now we assume that $|A_{n,z}| > |Q|$.

For notational convenience, we write $\tilde{w} = [{}^{w}_{r}]$, $\tilde{x} = [{}^{x}_{r}]$, $\tilde{z} = [{}^{z}_{s}]$, and $\tilde{y} = [{}^{y}_{s}]$. Note that the accepting probability $p_{acc}(wz) =_{def} \nu_{ini}M_{\tilde{w}}M_{\tilde{z}}\xi_F^T$ (note that this notation suppresses the advice string $h(n)$) equals $1/2$ for all strings $wz \in A$. Consider the set $T = \{\nu_{ini}M_{\tilde{w}} \mid w \in A_{n,z}\}$ of vectors. Choose a maximal set S' of linearly independent vectors in T (those form a set of *basis vectors*). Clearly, since $\nu_{ini}M_{\tilde{w}}$ has dimension $|Q|$, there are at most $|Q|$ linearly independent vectors. Thus, we have $|S'| \leq |Q| = m$. The desired set S is now defined as $S = \{w \in A_{n,z} \mid \nu_{ini}M_{\tilde{w}} \in S'\}$. Note that, for any vector $\nu_{ini}M_{\tilde{x}}$ in $T - S'$ can be written as a linear combination of basis vectors: (*) $\nu_{ini}M_{\tilde{x}} = \sum_{w\in S}\alpha_w\cdot\nu_{ini}M_{\tilde{w}}$, where $\{\alpha_w\}_{w\in S}$ is a set of certain real numbers. Note that $\sum_{w\in S}\alpha_w = 1$.

At last, we shall show the desired property of S. Let y be any string in Σ^ℓ and assume that $wy \in A$ for all strings w in S. From the above note, since $p_{acc}(wy) = 1/2$, we obtain that $p_{acc}(xy) = \left(\sum_{w\in S}\alpha_w\cdot\nu_{ini}M_{\tilde{w}}\right)M_{\tilde{y}}\xi_F^T = \sum_{w\in S}\alpha_w\cdot p_{acc}(wy) = \frac{1}{2}\sum_{w\in S}\alpha_w = \frac{1}{2}$. Therefore, it follows that $xy \in A$. This completes the proof of the lemma. □

Proposition 3 implies that 1-C$_=$LIN$/lin \not\subseteq$ CFL$/n$. This result can be improved as follows.

Proposition 4. 1-C$_=$LIN $\not\subseteq$ CFL$/n$.

Proof. Consider the language $Equal_6$, which consists of all strings w over the alphabet $\Sigma_6 = \{a_1, a_2, \ldots, a_6, \#\}$ such that, for any pair $a, a' \in \Sigma_6 - \{\#\}$, $\#_a(w) = \#_{a'}(w)$. Note that $Equal_6$ is not in CFL$/n$ [7]; thus, it suffices to show that $Equal_6$ belongs to 1-C$_=$LIN. For any pair $i, j \in [1,6]_{\mathbb{Z}}$, we write $L_{i,j}$ for the language $\{w \in \Sigma_6^* \mid \#_{a_i}(w) = \#_{a_j}(w)\}$. Note that $Equal_6 = \bigcap_{i=2}^{6} L_{1,i}$. Clearly, each language $L_{i,j}$ belongs to 1-C$_=$LIN. Since 1-C$_=$LIN is closed under intersection [6], we conclude that $Equal_6$ is also in 1-C$_=$LIN. □

Next, we show another class separation between CFL$/n$ and 1-PLIN$/lin$.

Theorem 3. CFL $\not\subseteq$ 1-PLIN$/lin$. *Thus,* CFL$/n \not\subseteq$ 1-PLIN$/lin$.

This theorem follows from the next lemma, which gives new criteria for languages in 1-PLIN$/lin$. This lemma can be compared to Lemma 3.

Lemma 4. *Let $A \in$ 1-PLIN$/lin$ over an alphabet Σ. There exists a positive constant m that satisfies the following statement. Let $n, \ell \in \mathbb{N}$ be any numbers with $m \leq |\Sigma|^\ell$ and $\ell \leq n-1$. There exists a set $S = \{w_1, w_2, \ldots, w_m\} \subseteq \Sigma^\ell$ with $|S| = m$ for which the following implication holds: for any set $T \subseteq \Sigma^{n-\ell}$, if*

$|\{A(w_1y)A(w_2y)\cdots A(w_my) \mid y \in T\}| \geq 2^m$, then it follows that, for any string $x \in \Sigma^\ell$, there exists a pair $y, y' \in T$ of strings such that $xy \in A$ and $xy' \notin A$.

From this lemma, Theorem 3 easily follows.

Proof of Theorem 3. Let $\Sigma = \{0,1\}$ for simplicity. Consider the language $IP_* = \{axy \mid a \in \{\lambda, 0, 1\}, x, y \in \Sigma^*, |x| = |y|, x^R \odot y \equiv 0 \pmod 2\}$. Since $IP_* \in$ CFL [8], we need to show that $IP_* \notin$ 1-PLIN/lin.

Let us assume that $IP_* \in$ 1-PLIN/lin. We take a constant m that satisfies Lemma 4. Choose a sufficiently large even number n and let $\ell = n/2$. There is a subset $S = \{w_1, w_2, \ldots, w_m\} \subseteq \Sigma^\ell$ with m distinct elements that satisfy the lemma. To each binary sequence $r = (r_1, r_2, \ldots, r_m)$, we assign a string $y_r \in \Sigma^{n-\ell}$ for which $w_i^R \odot y_r \equiv r_i \pmod 2$ for all indices $i \in [m]$. Define $T = \{y_r \mid r \in \Sigma^m\}$ (here, r is seen as a string). Note that $|T| = 2^m$. Since n is larger than 2^m, there is a string $x \in \Sigma^\ell$ satisfying $x^R \odot y_r \equiv 0 \pmod 2$ for every $r \in \Sigma^m$. For this x, the lemma yields a pair $y, y' \in T$ such that $xy \in IP_*$ and $xy' \notin IP_*$. This is a contradiction against the choice of x. Hence, $IP_* \notin$ 1-PLIN/lin. □

Now, let us present the proof of Lemma 4. This proof relies on an idea, similar to the early part of the proof of Lemma 3, of translating "$A \in$ 1-PLIN/lin" to the existence of a 1pfa M and an advice function h satisfying the property that $\mathrm{Prob}_M[M([\begin{smallmatrix} x \\ h(|x|) \end{smallmatrix}]) = A(x)] > 1/2$ for every input x. This "translation" is not difficult to prove.

Proof of Lemma 4. Let A be any language in $\in$ 1-PLIN/lin over an alphabet Σ. For this A, we can take an appropriate 1pfa M and an advice function h, as described above. To make our argument simple, we make the following assumption: the *success* probability of M on any input string never becomes exactly 1/2. This can be done by an appropriate modification of the given 1pfa M (see, e.g., [6]). Choose n and ℓ arbitrarily. Similar to the proof of Lemma 3, we can define a constant $m > 0$ and a subset S (induced from basis vectors). The only difference here is that we do not need to take a fixed input string z that leads M to accept. Now, let $S = \{w_1, w_2, \ldots, w_m\}$ and fix $x \in \Sigma^\ell$ arbitrarily. Using the same notations as in the proof of Lemma 3, there exist a series $\{\alpha_{w_i}\}_{i \in [1,m]_\mathbb{Z}}$ of real numbers such that, for every string $y \in \Sigma^{n-\ell}$, $\sum_{i=1}^m \alpha_{w_i} = 1$ and $\nu_{ini} M_{\tilde{x}} = \sum_{i=1}^m \alpha_{w_i}(\nu_{ini} M_{\tilde{w}_i})$, where $\tilde{x} = [\begin{smallmatrix} x \\ r \end{smallmatrix}]$ and $\tilde{w}_i = [\begin{smallmatrix} w_i \\ r \end{smallmatrix}]$.

Next, we let $T \subseteq \Sigma^{n-\ell}$ and assume that, for any binary series $r = (r_1, r_2, \ldots, r_m)$, there is a string $y_r \in T$ such that r coincides with the m-bit string $A(w_1y_r)A(w_2y_r)\cdots A(w_my_r)$. For each $y \in \Sigma^{n-\ell}$ and each $i \in [m]$, let $p_{acc}(w_iy) = 1/2 + \beta_{i,y}$ (as before, this notation suppresses the advice string) for a certain real number $\beta_{i,y} \in [-1/2, 1/2] - \{0\}$. For such y, we have $p_{acc}(xy) = \sum_i \alpha_{w_i} p_{acc}(w_iy) = \frac{1}{2} + \sum_i \alpha_{w_i}\beta_{i,y} = \frac{1}{2} + \sum_i (-1)^{1-A(w_iy)}|\beta_{i,y}|\alpha_{w_i}$. We define a new series $r = (r_1, \ldots, r_m)$ as follows: let $r_i = 0$ (or equivalently $A(w_iy_r) = 0$) if $\alpha_{w_i} < 0$, and $r_i = 1$ (or $A(w_iy_r) = 1$) if $\alpha_{w_i} \geq 0$. Note that its corresponding string y_r exists by our assumption. Take this y_r as our desired string y, because $\sum_i (-1)^{1-A(w_i y)} |\beta_{i,y}| \alpha_{w_i} = \sum_i |\beta_{i,y}||\alpha_{w_i}| > 0$, which implies $xy \in A$. Next, we define r' as the bitwise negation of r and let y' be $y_{r'}$. Similar to the previous case, we have $\sum_i (-1)^{1-A(w_i y')} |\beta_{i,y'}| \alpha_{w_i} = \sum_i (-1)|\beta_{i,y'}||\alpha_{w_i}| < 0$. This implies that $xy' \notin A$. This completes the proof of the lemma. □

5 Limitation of Randomized Advice

The previous section has demonstrated a power of randomized advice; for example, we have shown that REG/$Rn \nsubseteq$ CFL/$n \cup$ 1-C$_=$LIN/lin. To the contrary, this section shall discuss a limitation of the randomized advice. In particular, we wish to show that CFL $\nsubseteq$ REG/Rn. This result significantly extends the previously-known separation CFL $\nsubseteq$ REG/n [7]

Theorem 4. CFL $\nsubseteq$ REG/Rn.

From this theorem, we can deduce that CFL/Rn properly contains REG/Rn because, otherwise, CFL is included in REG/Rn, contradicting the theorem.

Hereafter, we shall prove Theorem 4. To do so, we borrow an idea from communication complexity theory because our randomized-advice setting is loosely related to two-party one-way communication with shared randomness. First, we define a new complexity class Aver-REG/n. The class Aver-REG/n consists of all *distributional problems* (A, μ), where A is a language over an alphabet Σ and $\mu = \{\mu_n\}_{n\in\mathbb{N}}$ is a probability ensemble over Σ^*, such that there exist a 1dfa M, an advice function h, and a constant $\varepsilon \in [0, 1/2)$ satisfying the following condition: for every length $n \in \mathbb{N}$, $\mathrm{Prob}_{x\sim\mu_n}[M([\begin{smallmatrix} x \\ h(n) \end{smallmatrix}]) = A(x)] \geq 1 - \varepsilon$, where "$x \sim \mu_n$" means that x is chosen randomly according to μ_n.

Proposition 5. *If $A \in$ REG/Rn, then $(A, \mu) \in$ Aver-REG/n for any probability ensemble μ.*

Recall that our goal is to present a context-free language A that does not belong to REG/Rn. For this purpose, by Proposition 5, it suffices to show that (A, μ) does not belong to Aver-REG/n for a certain probability ensemble μ. We present a simple example of such language, known as REG/n-pseudorandom languages [8]. Formally, a language L over an alphabet Σ is called REG/n-*pseudorandom* if, for every language $A \in$ REG/n over Σ, the function $\ell(n) =_{def} \left| \frac{|(A \triangle L)\cap\Sigma^n|}{|\Sigma^n|} - \frac{1}{2} \right|$ is negligible, where $A \triangle L = (A - L) \cup (L - A)$.

Lemma 5. *If A is REG/n-pseudorandom, then (A, μ_{uni}) is not in Aver-REG/n, where $\mu_{\mathrm{uni}} = \{\mu_{\mathrm{uni},n}\}_{n\in\mathbb{N}}$ is the uniform probability ensemble (i.e., each $\mu_{\mathrm{uni},n}$ is uniform over Σ^n).*

Proof. We prove a contrapositive. Let (A, μ_{uni}) be any distributional problem in Aver-REG/n with an input alphabet Σ. There exist a 1dfa M, a constant $\varepsilon \in [0, 1/2)$, and an advice function h such that, for every $n \in \mathbb{N}$, $\mathrm{Prob}_{x\sim\mu_{\mathrm{uni},n}}[M([\begin{smallmatrix} x \\ h(n) \end{smallmatrix}]) = A(x)] \geq 1 - \varepsilon$. For our convenience, we express ε as $\frac{1}{2} - \varepsilon'$ with $\varepsilon' > 0$. Now, let us define $B = \{x \in \Sigma^* \mid M([\begin{smallmatrix} x \\ h(|x|) \end{smallmatrix}]) = 1\}$ and consider the symmetric difference $B \triangle A$. We then obtain $\mathrm{Prob}_{x\sim\mu_{\mathrm{uni},n}}[x \in B\triangle A] = \mathrm{Prob}_{x\sim\mu_{\mathrm{uni},n}}[M([\,{}^{x}_{h(n)}\,]) \neq A(x)] \leq \frac{1}{2} - \varepsilon'$. From this bound, it thus follows that $\ell(n) = ||(B\triangle A)\cap\Sigma^n|/|\Sigma^n| - 1/2| = \left|\mathrm{Prob}_{x\sim\mu_{\mathrm{uni},n}}[x \in B\triangle A] - 1/2\right| \geq \varepsilon'$. This means that A cannot be REG/n-pseudorandom. □

With Proposition 5 and Lemma 5, it becomes rather an easy task to give the proof of Theorem 4. Since we already know that there exists a context-free language that is also REG/n-pseudorandom [8], we thus conclude by Proposition 5 and Lemma 5 that CFL contains a language not in REG/Rn. This proves Theorem 4.

To close this section, we still need to prove Proposition 5. This follows immediately from a new characterization of REG/Rn. This characterization is an immediate consequence of Yao's principle [9] and it is, to some extent, analogous to a well-known result on one-way communication with public coin.

Lemma 6. *Let A be any language over an alphabet Σ. The following two statements are equivalent.*

1. *A is in* REG/Rn.
2. *There are a 1dfa M, an alphabet Γ, and a constant $\varepsilon \in [0, 1/2)$ that satisfy the following: for every probability ensemble $\{\mu_n\}_{n\in\mathbb{N}}$ over Σ^*, there exists an advice function $h : \mathbb{N} \to \Gamma^*$ such that* $\mathrm{Prob}_{x\sim\mu_n}[M([\,{}^{x}_{h(n)}\,]) = A(x)] \geq 1-\varepsilon$ *for every length $n \in \mathbb{N}$.*

References

1. Damm, C., Holzer, M.: Automata that take advice. In: Hájek, P., Wiedermann, J. (eds.) MFCS 1995. LNCS, vol. 969, pp. 149–152. Springer, Heidelberg (1995)
2. Dieu, P.D.: On a class of stochastic languages. Zietschr. math. Logik und Grundlagen d. Math. Bd. 17, 421–425 (1971)
3. Hennie, F.C.: One-tape, off-line Turing machine computations. Information and Control 8, 553–578 (1965)
4. Karp, R.M., Lipton, R.: Turing machines that take advice. In: L'Enseignement Mathématique. 2nd series, vol. 28, pp. 191–209 (1982)
5. Kobayashi, K.: On the structure of one-tape nondeterministic Turing machine time hierarchy. Theoretical Computer Science 40, 175–193 (1985)
6. Tadaki, K., Yamakami, T., Lin, J.: Theory of one tape linear time Turing machines. To appear in Theoretical Computer Science (2009); A preliminary version appeared in Van Emde Boas, P., Pokorný, J., Bieliková, M., Štuller, J. (eds.): SOFSEM 2004. LNCS, vol. 2932, pp. 335–348. Springer, Heidelberg (2004); See also arXiv:cs/0310046
7. Yamakami, T.: Swapping lemmas for regular and context-free languages (Manuscript) (2008) arXiv:0808.4122
8. Yamakami, T.: Immunity and pseudorandomness of context-free languages (Manuscript) (2009) arXiv:0902.0261
9. Yao, A.C.: Probabilistic complexity: Towards a unified measure of complexity. In: Proc. 18th IEEE Annual Symposium on Foundation of Computer Science, pp. 222–227 (1977)

Of Choices, Failures and Asynchrony: The Many Faces of Set Agreement

Dan Alistarh[1], Seth Gilbert[1], Rachid Guerraoui[1], and Corentin Travers[2,⋆]

[1] Swiss Federal Institute of Technology, Lausanne, Switzerland
[2] Technion, Haifa, Israel

Abstract. Set agreement is a fundamental problem in distributed computing in which processes collectively choose a small subset of values from a larger set of proposals. The impossibility of fault-tolerant set agreement in asynchronous networks is one of the seminal results in distributed computing. The complexity of set agreement in synchronous networks has also been a significant research challenge. Real systems, however, are neither purely synchronous nor purely asynchronous. Rather, they tend to alternate between periods of synchrony and periods of asynchrony.

In this paper, we analyze the complexity of set agreement in a "partially synchronous" setting, presenting the first (asymptotically) tight bound on the complexity of set agreement in such systems. We introduce a novel technique for simulating, in fault-prone asynchronous shared memory, executions of an asynchronous and failure-prone message-passing system in which some fragments appear synchronous to some processes. We use this technique to derive a lower bound on the round complexity of set agreement in a partially synchronous system by a reduction from asynchronous wait-free set agreement. We also present an asymptotically matching algorithm that relies on a distributed asynchrony detection mechanism to decide as soon as possible during periods of synchrony.

By relating environments with differing degrees of synchrony, our simulation technique is of independent interest. In particular, it allows us to obtain a new lower bound on the complexity of early deciding k-set agreement complementary to that of [12], and to re-derive the combinatorial topology lower bound of [13] in an algorithmic way.

1 Introduction

Set agreement was first introduced by Chaudhuri [6] to capture the power of allowing more choices than *consensus* [16], where only a single decision value is permitted. Each process p_i begins with an initial value v_i; eventually, every process outputs one of the initial values as a decision. In k-set agreement, the set of all values output can be of size at most k. When $k = 1$, set agreement reduces to *consensus*. When $k = n$, the problem is trivial, i.e., processes can act entirely independently.

⋆ Supported in part by a Sam & Cecilia Neaman Fellowship.

Y. Dong, D.-Z. Du, and O. Ibarra (Eds.): ISAAC 2009, LNCS 5878, pp. 943–953, 2009.

In a collection of seminal papers, Borowsky, Gafni, Herlihy, Saks, Shavit, and Zaharoglou [14, 17, 5] showed that fault-tolerant asynchronous set agreement is impossible (while at the same time revealing a deep connection between distributed computing and algebraic topology). Chaudhuri et al. [7] further developed these techniques, establishing a tight lower bound on the round complexity of *synchronous* set agreement: in a system with t failures, at least $\lfloor t/k \rfloor + 1$ rounds are necessary. More recently, Gafni et al. [12] and Guerraoui et al. [13] considered the feasibility of reaching an *early decision*: how fast can an algorithm tolerating up to t failures decide in an execution with at most $f < t$ failures? They both show, in different ways, that at least $\lfloor f/k \rfloor + 2$ rounds are needed.

Set agreement has been extensively studied in both synchronous and asynchronous systems. Real world distributed systems, however, are neither purely synchronous nor purely asynchronous. Instead, they tend to exhibit periods of synchrony when the network is well behaved, and periods of asynchrony when the network is poorly behaved. To describe such a system, Dwork et al. [9] introduced the idea of *partial synchrony*. They assume for every execution some (unknown) time GST (*global stabilization time*), after which the system is synchronous. In this paper, we study the *feasibility* and *complexity* of set agreement in the context of partially synchronous systems, determining the minimum-sized window of synchrony in which k-set agreement can be solved. Of course, the lower bounds for synchronous systems [7, 10] imply an immediate lower bound here of $\lfloor \frac{t}{k} \rfloor + 1$ rounds. The question, then, is whether there exists any matching algorithm that terminates in a synchronous window of size $\lfloor \frac{t}{k} \rfloor + 1$, or is there some inherent cost to tolerating asynchrony? Moreover, how does this cost depend on t and k?

We answer these questions by showing that at least $\lfloor \frac{t}{k} \rfloor + 2$ synchronous rounds are required for k-set agreement, and demonstrating an algorithm that terminates in any window of synchrony of size at least $\lfloor \frac{t}{k} \rfloor + 4$ rounds. Together, these results tightly bound the inherent price of tolerating some asynchrony.

Lower Bound By Reduction. The technique for deriving the lower bound is an important contribution in end of itself, as it provides new insights into the complexity of set agreement. Instead of relying on topology, as is typically required for set agreement lower bounds, we derive our result by reducing the feasibility of asynchronous set agreement to the problem of solving set agreement in a window of size $\lfloor \frac{t}{k} \rfloor + 1$. Since asynchronous set agreement is known to be impossible, this reduction immediately implies that at least $\lfloor \frac{t}{k} \rfloor + 2$ synchronous rounds are required for k-set agreement.

Early Deciding Synchronous Set Agreement. Our technique turns out to be of more general interest, as we can re-derive and extend existing lower bounds for synchronous *early deciding* set agreement. It has been previously shown [12, 13] that even in an execution with $f < t$ failures, some process cannot decide prior to round $\lfloor f/k \rfloor + 2$.

Using our simulation technique, we re-derive both previous lower bounds in a simpler and more general manner, in the standard model where t and n are bounded and known. Of note, both lower bounds are corollaries of a single theorem that relates the number of processes which decide early with the worst-case round complexity of an algorithm. Basically, we show that if d processes decide by round $\lfloor f/k \rfloor + 1$ in executions with at most f failures, then in the worst-case, some process takes at least time $\lfloor t/k \rfloor + E(\cdot) + 1$ to decide (where E is a function of t, k and d). Due to space limitations, we leave the presentation of these results to the full version of the paper [2].

Upper Bound for Eventually Synchronous Agreement. We then present the first known algorithm for set agreement that tolerates periods of asynchrony. Our algorithm guarantees correctness, regardless of asynchrony, and terminates as soon as there is a window of synchrony of size $\lfloor t/k \rfloor + O(1)$. For simplicity, we show synchronous round complexity of $\lfloor t/k \rfloor + 4$; a tighter analysis in the full version yields $\lfloor t/k \rfloor + 3$ for $t \geq k^2$. Note that a previous paper [3] presents an algorithm that exactly matches the lower bound, for the special case of *consensus* ($k = 1$). Thus, closing the one round gap between the two bounds remains an intriguing challenge.

Two basic ideas underlie our algorithm. First, processes collectively execute an *asynchrony detection* sub-protocol that determines whether a round appears synchronous or asynchronous. A process can decide when it sees $\lfloor t/k \rfloor + O(1)$ synchronous rounds. Even so, different sets of processes may have different views of the actual execution when the decision occurs, since there are only $\lfloor t/k \rfloor + O(1)$ rounds to exchange information. Second, each process maintains an *estimate*, i.e., a value that it is leaning toward choosing. In each round, each process adopts the minimum estimate that it receives. If a process is about to decide, however, it can *elevate the priority* of its estimate, causing other processes to adopt its value instead.

Implications. Several implications arise from our simulation technique and its usage. First, it provides additional evidence that the impossibility of fault-tolerant asynchronous k-set agreement is a central result in distributed computing, as it implies non-trivial results in both partially synchronous and synchronous models. Second, it highlights close connections between models that have differing levels of synchrony. In particular, our simulation technique takes advantage of structural similarities between *eventually synchronous* set agreement and *early deciding* set agreement to establish lower bounds in two different models of synchrony. The uncertainty regarding asynchrony (found in a partially synchronous execution) turns out to be fundamentally similar to the uncertainty regarding failures (found in an early deciding execution).

2 Model

In this section, we define three basic models of computation. The *partially synchronous model* $ES_{n,t}$ consists of n deterministic processes $\Pi = \{p_1, \ldots, p_n\}$, of which up to $t < n$ may fail by crashing. (Note that the algorithm in Section 4 uses $t < n/2$.) The processes communicate via a message-passing network, modeled much as in [9,8,15]: time is divided into *rounds*; however, there is no assumption that every message broadcast in a round is also delivered in that round. Instead, we assume only that if all non-failed processes broadcast a message in round r, then each process receives at least $n - t$ messages in that round[1]. We assume that the network is *partially synchronous*: there is some round GST after which every message sent by a non-failed process is delivered in the round in which it is sent. The *synchronous model* $S_{n,t}$ is identical to $ES_{n,t}$, except that we assume every process knows, *a priori*, that $GST = 0$, i.e., that every message is delivered in the round that it is sent.

The *asynchronous model* $AS_{n,k}$ consists of n processes Π, up to k of which may crash. The processes communicate via single-writer, multi-readers (SWMR) registers. The memory is organized in arrays $X[1..n]$ of n registers; entry $X[i]$ of an array can be written only by p_i. In addition to read() and write() operations, a process can also invoke X.snapshot() to read all the contents of X in a logically instantaneous single operation. Let x and x' be the result of any two snapshot operations on X. We assume that the following hold: *Containment:* $x \subseteq x' \vee x' \subseteq x$; *Self inclusion:* Let v be the value written by p_i in $X[i]$ prior to invoking X.snapshot(), with no intervening x.write($\cdot$) operations; let x be the result of the snapshot operation; then $x[i] = v$. Implementation of snapshot on top of SWMR registers can be found in [1, 4], and thus they provide no extra power. k-set agreement is impossible in $AS_{n,k}$ [5, 14, 17].

3 Simulating Synchronous Views: A Lower Bound for *k*-Set Agreement

In this section, we present an algorithm for simulating, in the asynchronous model $AS_{n,k}$, executions in $ES_{n,t}$ of a k-set agreement algorithm $\mathcal{A}$. The simulation pseudocode is presented in Figure 1. Each process in $AS_{n,k}$ simulates one process executing $\mathcal{A}$ in $ES_{n,t}$. (We refer to the processes in $AS_{n,k}$ as *simulators.*) Each execution e simulated by our algorithm is a valid execution of $\mathcal{A}$ in model $ES_{n,t}$, and satisfies the following property: at least one process, whose simulator is correct, observes a window of synchrony of length at least $\lfloor t/k \rfloor + 1$. More precisely, for each simulated execution e, there exists a round R and a process p such that p cannot distinguish, by the end of round $R + \lfloor t/k \rfloor + 1$, execution e from some execution e' in which the global stabilization round GST is equal to R. We say that p has a *synchronous view* of length $\lfloor t/k \rfloor + 1$.

Our simulation relies on the ideas introduced in [10]. The goal in [10] is to simulate, in $AS_{n,k}$, executions of the synchronous model $S_{n,t}$. The simulation ensures that (1) whenever a message is not delivered by some process, the sender is simulated as failed in every following round and, (2) in each round, at most k new failures occur. The first property guarantees that the simulated execution is synchronous, while the second property implies that up to $\lfloor t/k \rfloor$ rounds can be simulated without exceeding the maximum number of failures t allowed by the model $S_{n,t}$.

[1] This can be implemented by delaying a round $r+1$ message until at least $n-t$ round r messages have been received.

We observe that in an execution of $\lfloor t/k \rfloor + 1$ rounds simulated with the technique described in [10], although more than t processes might have failed in the simulated execution, at least one process p observes no more than t failures and still perceives the execution as synchronous. Thus, if we assume a k-set agreement protocol for model $ES_{n,t}$ where every process decides by the end of round $GST + \lfloor t/k \rfloor + 1$, process p must obtain a decision. If its associated simulator does not fail, the decision can be propagated to every simulator which allows them to decide. In the other case, we repeat the simulation for another $\lfloor t/k \rfloor + 1$ rounds, again resulting in a process either deciding or its associated simulator failing. Eventually, after $k + 1$ repetitions (which we refer to as "phases"), we argue that some process decides and does not fail.

3.1 Basic Setup

The simulation depends on three parameters: the algorithm $\mathcal{A}$ being simulated, the number of phases $numP$, and an array $R_1, R_2, \ldots, R_{numP+1}$ where each R_i is the first round in the ith phase.

For process p_i, the algorithm $\mathcal{A}$ is described by a function compute(r, rec), where r is a round number and rec a set of messages received by p_i in round r. The compute function returns a pair (d_i, m_i), where m_i is the message to be sent in the next round, and d_i is the decision value or $\perp$. Without loss of generality, we assume that each process sends the same message to all other processes.

3.2 Simulating Synchronous Rounds

Each process in $AS_{n,k}$ simulates one process executing $\mathcal{A}$ in $ES_{n,t}$. The simulation begins with a call to propose(v_i) (line 5), where v_i is p_i's proposal.

The simulation is divided into *phases* (lines 8–13): in each round of a phase in which no simulators fail, there is at least one process that sees the phase as a *window of synchrony* within a (possibly) asynchronous execution.

Round overview. In order to simulate round r (lines 11–13), process p_i invokes simulate(m_i, r) (line 12), where m_i is its message for round r, which was computed previously. The simulate procedure returns pairs $\langle j, m_j \rangle$, where m_j is the round r message "sent" by p_j. The simulator then calls the compute function (line 13), which returns d_i, a possible decision, and m_i, the next message to send. If $d_i \neq \perp$, simulator p_i writes the decision value d_i in the shared array DEC before deciding that value (line 13). Similarly, if a simulator observes that a value $\neq \perp$ has been written in DEC, it decides that value (lines 14–16).

Simulating a round. The simulate function (lines 17–30) carries out the send/receive step. For round r, simulator p_i writes the message m_i into the register $VAL[r][i]$ (line 18), and then performs repeated snapshots of $VAL[r]$ (line 19) to discover the messages of other simulators. Since k simulators may fail in $AS_{n,k}$, the simulator cannot wait for messages from all n simulators. As soon as p_i discovers $n - k$ messages in its snapshot of $VAL[r]$, it continues. The variable M_i stores the set of up to k processes from which some message was

```
Parameters: A, numP, [R_1, ..., R_{numP+1}]
Shared variables:
    AC[1..R_{numP+1}][1..n], array of adopt-commit objects
    DEC[1..n], VAL[1..R_{numP} + 1][1..n], array of SWMR registers.
procedure propose(v_i): start Task T1; start Task T2;
Task T1:
    (_, m_i) ← compute(0, v_i, true) % messages for the first round
    for ρ = 1 to numP do
        % Begin a new phase:
        S_i ← ∅
        for r = R_ρ to R_{ρ+1} − 1 do
            rec_i ← simulate(m_i, r) % Simulate send/receive of round r.
            (d_i, m_i) ← compute(r, rec_i) % Compute message for next round. if
            d_i ≠ ⊥ then DEC[i].write(d_i); stop T2; return d_i
Task T2:
    repeat for j = 1 to n do dec_i[j] ← DEC[i]  until (∃ℓ : dec_i[ℓ] ≠ ⊥)
    stop T1; return dec_i[ℓ]
procedure simulate( m_i, r ) % Simulate round r where p_i sends message m_i.
    rec_i ← ∅; VAL[r][i].write(m_i)
    repeat view_i ← VAL[r].snapshot() until |{j : view_i[j] = ⊥}| ≤ k
    M_i ← {j : view_i[j] = ⊥}
    for j = 1 to n do
        if j ∈ S_i ∪ M_i then state_i[j] ← AC[r][j].propose(suspect)
        else state_i[j] ← AC[r][j].propose(alive)
        if state_i[j] = (commit, suspect ) then S_i ← S_i ∪ {j}
        else if state_i[j] = (adopt, suspect ) then S_i ← S_i ∪ {j};
        rec_i ← rec_i ∪ {⟨j, VAL[r][j]⟩}
        else rec_i ← rec_i ∪ {⟨j, VAL[r][j]⟩}
    % Complete view of round r, if necessary:
    if |rec_i| < n − t then rec_i ← {⟨j, view_i[j]⟩ : view_i[j] ≠ ⊥} ;
    if ⟨i, m_i⟩ ∉ rec_i then rec_i ← rec_i ∪ {⟨i, view_i[i]⟩} ;
    return rec_i
```

Fig. 1. Simulating $\mathcal{A}$ in $AS_{n,k}$, code for simulator p_i

missed. The simulators then need to agree on which messages to deliver in the simulated round and which messages were missed; if a message is missed from some process p_j, then the simulators collectively "fail" process p_j in all future simulated rounds.

Adopt-commit objects. The simulators use *adopt-commit* objects (introduced in [10, 18]) to coordinate which processes have "failed" in the simulation. An adopt-commit object AC is invoked via a propose(v) operation, and returns a decision (dec, v) where $dec \in \{\mathsf{adopt}, \mathsf{commit}\}$. The object satisfies the following properties: *1. Termination:* Each invocation by a correct process terminates. *2. Validity:* If a process decides (dec, v) then some process invoked AC.propose(v).

3. Agreement: If a process decides (commit, v), then every decision is $(\cdot, v)$. *4. Convergence:* If every process proposes the same v, then (commit, v) is the only possible decision. Implementations of adopt-commit objects in $AS_{n,k}$ can be found in [10, 18]. These implementations also satisfy: *5. Commit Validity:* Assume p_j invokes $AC.\mathsf{propose}(v)$; then p_j cannot get back (commit, v') with $v \neq v'$.

Agreeing on failures. After completing the snapshots, the simulators use the adopt-commit objects to agree on which processes have failed. Simulator p_i stores in S_i a set of suspected processes, and it resets S_i to $\emptyset$ at the beginning of each phase. Throughout the phase, processes are added to S_i based on the output of the adopt-commit objects. If a process p_i misses a message from a process p_j in round r (i.e., if $p_j \in M_i$), or if process p_i suspects p_j (i.e., if $p_j \in S_i$), then its simulator proposes suspecting p_j using $AC[r][j]$ (line 22). Otherwise, the simulator proposes that p_j is *alive* (line 23).

There are three possible decisions. First, $(\mathsf{commit}, \mathit{suspect})$ (line 24): in this case, the simulation fails process p_j in round r. By *agreement*, we know that every simulator either adopts or commits to suspecting p_j, and so process i adds p_j to S_i. Second, $(\mathsf{adopt}, \mathit{suspect})$ (line 25): in this case, we cannot determine whether p_j is failed or not in round r; even so, to be safe, simulator p_i adds p_j to S_i. We know, however, by *validity* that some process proposed p_j as alive, and so we know that $VAL[r][j]$ contains the message from p_j, which we add to the set rec_i of messages to deliver. Finally $(\cdot,\ \mathit{alive})$ (line 26): as in the second case, we add the message from $VAL[r][j]$ to rec_i. Notice that if any simulator commits to failing p_j, then every other simulator will either adopt or commit to failing p_j and add p_j to S_i. By *convergence*, in the following round, every simulator commits to failing p_j. By using the adopt-commit objects in this way, we ensure that simulated views remain synchronous.

The end of the phase. This approach results in simulating up to k new failures in each round. Eventually, the number of simulated failures may surpass t, the bound on failures in $ES_{n,t}$. Consequently, for some processes, the set of messages rec_i may no longer contain an appropriate set of messages to deliver. In that case, simulator p_i augments the set rec_i to ensure that it contains enough messages ($|rec_i| \geq n - t$, line 28) and that it contains the round r message of p_i (line 29).

Therefore, not all processes may maintain a synchronous view. However, we can show that if the length of the phase is at most $\lfloor t/k \rfloor + 1$ rounds, at least one process is able to maintain its synchronous view through the end of a phase.

3.3 Lower Bound on Set Agreement in $ES_{n,t}$

We now show how to use the simulation technique to prove a lower bound on set agreement in $ES_{n,t}$. We begin, for the sake of contradiction, by assuming that algorithm $\mathcal{A}$ solves k-set agreement in $ES_{n,t}$ in any window of synchrony of size $\lfloor t/k \rfloor + 1$. The simulation uses $k+1$ phases, each of length $\lfloor t/k \rfloor + 1$, i.e., $R_\rho = (\rho - 1)(\lfloor t/k \rfloor + 1) + 1$. We show that the resulting simulation of $\mathcal{A}$ solves k-set agreement in $AS_{n,k}$, which is known to be impossible, implying that no such algorithm $\mathcal{A}$ exists. This implies that any k-set agreement protocol requires at least $\lfloor t/k \rfloor + 2$ synchronous rounds to decide.

First, we list some of the properties of the simulation, whose proofs can be found in the full version of the paper [2]: *(P1)* the simulated execution of $\mathcal{A}$ is a valid execution of $\mathcal{A}$ in $ES_{n,t}$; *(P2)* each phase ρ appears synchronous: if a process sees $f \leq t$ failures by the end of round r, then it perceives the first r rounds of phase ρ as synchronous; *(P3)* there are at most t simulated failures at the beginning of the last simulated round in phase ρ; *(P4)* some simulated process sees no new failures in the last round. Since each phase is of length $\lfloor t/k \rfloor + 1$, and since $\mathcal{A}$ guarantees a decision in a window of synchrony of size $\lfloor t/k \rfloor + 1$, it follows from properties *(P1)*–*(P4)* that by the end of phase ρ, a process either decides, having seen the entire phase as synchronous, or its associated simulator fails.

Lemma 1. *For every phase ρ, if no process decides and writes its decision to DEC prior to the end of phase ρ, then at least one process that begins phase ρ fails before beginning phase $\rho + 1$.*

We conclude that our simulation of algorithm $\mathcal{A}$ solves k-set agreement in $AS_{n,k}$. *Agreement* follows from the fact that our simulation is a valid simulation of $\mathcal{A}$ in $ES_{n,t}$, and *termination* follows from the fact that if there is no decision, then at least one simulator fails in every phase; since there are only k possible failures in $AS_{n,k}$, by the end of phase $k+1$, some process must decide.

Lemma 2. *The algorithm in Figure 1 simulating $\mathcal{A}$ solves k-set-agreement in $AS_{n,k}$.*

Since k-set agreement is impossible in $AS_{n,k}$, we conclude:

Theorem 1. *There is no algorithm $\mathcal{A}$ for $ES_{n,t}$ that decides by round $GST + \lfloor t/k \rfloor + 1$, i.e., within a window of synchrony of size $\lfloor t/k \rfloor + 1$.*

4 k-Set Agreement Algorithm for $ES_{n,t}$

We present an algorithm named K4 which solves k-set agreement in a window of synchrony of size $\lfloor t/k \rfloor + 4$. The algorithms requires a majority of correct processes, i.e., $t < n/2$. The pseudocode can be found in Figure 2. The proof of correctness for the protocol can be found in the full version of this paper [2].

4.1 Description

K4 is a round-based full-information protocol. Each process maintains a local estimate est_i, representing its preferred decision, and sets $Active_i$ and $Failed_i$, which denote the processes that p_i believes to be alive and failed (respectively). In every round, each process broadcasts its entire state (line 5), and receives all the messages for the current round (line 6), updating its view of which processes have failed and which rounds are synchronous (lines 7–10). A process decides if it receives a message from another process that has already decided (lines 11–12),

```
1 procedure propose(v_i)_i
2     est_i ← v_i; r_i ← 1; msgSet_i ← ∅; sFlag_i ← false
3     Active_i ← []; Failed_i ← []; AsynchRound_i ← [ ]
4     while true do
5         send( est_i, r_i, sFlag_i, Active_i, Failed_i, AsynchRound_i, decide_i ) to all
6         wait until received at least (n − t) messages for round r_i
7         msgSet_i[r_i] ← messages that p_i receives in round r_i
8         Active_i[r_i] ← processes from which p_i gets messages in round r_i
9         Failed_i[r_i] ← Π \ Active_i[r_i]
10        updateSynchDetector() % Update the state of p_i.
11        if ( ∃msg_p ∈ msgSet_i with msg_p.decided_p = true ) then
12            decide_i ← true; est_i ← msg_p.est_p
13        if (sCount_i = ⌊t/k⌋ + 4) then decide_i ← true
14        if (decided_i = false) then
15            flagProcs_i ← { p ∈ Active_i[r_i] | sFlag_p = true }
16            if flagProcs_i ≠ ∅ then est_i ← min_{q∈flagProcs_i}(est_q)
17            else est_i ← min_{q∈Active_i[r_i]}(est_q)
18        r_i ← r_i + 1
19 procedure updateSynchDetector()
20    for every msg_j ∈ msgSet_i[r_i] do
21        for round r from 1 to r_i − 1 do
22            Active_i[r] ← msg_j.Active_j[r] ∪ Active_i[r]
23            Failed_i[r] ← msg_j.Failed_j[r] ∪ Failed_i[r]
24    for round r from 1 to r_i − 1 do
25        AsynchRound_i[r] ← false
26        for round k from r + 1 to r_i do : if (Active_i[k] ∩ Failed_i[r] ≠ ∅) then
          AsynchRound_i[r] ← true
27    sFlag_i ← false
28    sCount_i ← max_ℓ(∀r′ ∈ [r_i − ℓ, r_i], AsynchRound_i[r′] = false)
29    if sCount_i = ⌊t/k⌋ + 3 then sFlag_i ← true
```

Fig. 2. The K4 algorithm, at process p_i

or if it sees $\lfloor t/k \rfloor + 4$ consecutive synchronous rounds (line 13). If no decision is reached, then the estimate est_i is updated in lines 15–17. There are two key components to K4: accurately determining whether rounds are synchronous (which is critical for ensuring liveness), and updating the estimate (which is critical for ensuring agreement).

Detecting Asynchrony. updateSynchDetector() merges information into the *Active* and *Failed* sets; if a process believes that p_ℓ was active in round r, then p_ℓ is added to $Active[r]$; if it believes that p_ℓ was failed during round r, then p_ℓ is added to $Failed[r]$ (see lines 20–23). It then determines based on $Active[r]$ and $Failed[r]$ sets whether round r seems synchronous (lines 24-26). A round r is deemed asynchronous if some process p_ℓ is believed to have failed in round r (i.e., $p_\ell \in Failed[r]$), and yet is also believed to be alive at some later round $k > r$ (i.e., $p_\ell \in Active[k]$). Finally, process p_i sets a flag $sFlag$ to true if it sees the previous $\lfloor t/k \rfloor + 3$ rounds as synchronous (line 29).

Updating the estimate. Each process updates the estimate in every round. Estimates have two levels of priority: if a process has seen $\lfloor t/k \rfloor + 3$ synchronous rounds, i.e., if it is "ready to decide," then its estimate has high priority; other estimates are awarded normal priority. A process adopts the minimum prioritized estimate, if one exists (line 16); otherwise, it adopts the minimum estimate received in the current round (line 17).

5 Conclusion

We have presented a new technique for simulating synchronous and partially synchronous executions in asynchronous shared memory. Our technique allows us to characterize the complexity of set agreement in partially synchronous systems, as well as to refine earlier lower bounds for early-deciding synchronous set agreement. One direction of future work is to extend our lower bound results to other tasks by encapsulating the Extended BG simulation [11]. On the algorithmic side, the solvability of set agreement in partially synchronous systems without a majority of correct processes remains an open question.

References

1. Afek, Y., Attiya, H., Dolev, D., Gafni, E., Merritt, M., Shavit, N.: Atomic snapshots of shared memory. J. ACM 40(4), 873–890 (1993)
2. Alistarh, D., Gilbert, S., Guerraoui, R., Travers, C.: Of choices, failures and asynchrony: The many faces of set agreement. Technical report, `http://lpd.epfl.ch/alistarh/kset-ps/kset-TR-full.pdf`
3. Alistarh, D., Gilbert, S., Guerraoui, R., Travers, C.: How to solve consensus in the smallest window of synchrony. In: Taubenfeld, G. (ed.) DISC 2008. LNCS, vol. 5218, pp. 32–46. Springer, Heidelberg (2008)
4. Attiya, H., Rachman, O.: Atomic snapshots in $o(n \log n)$ operations. SIAM J. Comput. 27(2), 319–340 (1998)
5. Borowsky, E., Gafni, E.: Generalized FLP impossibility result for t-resilient asynchronous computations. In: STOC, pp. 91–100 (1993)
6. Chaudhuri, S.: More choices allow more faults: Set consensus problems in totally asynchronous systems. Inf. Comput. 105(1), 132–158 (1993)
7. Chaudhuri, S., Herlihy, M., Lynch, N.A., Tuttle, M.R.: A tight lower bound for k-set agreement. In: FOCS, pp. 206–215. IEEE, Los Alamitos (1993)
8. Dutta, P., Guerraoui, R.: The inherent price of indulgence. Distributed Computing 18(1), 85–98 (2005)
9. Dwork, C., Lynch, N., Stockmeyer, L.: Consensus in the presence of partial synchrony. J. ACM 35(2), 288–323 (1988)
10. Gafni, E.: Round-by-round fault detectors: Unifying synchrony and asynchrony (extended abstract). In: PODC, pp. 143–152 (1998)
11. Gafni, E.: The Extended BG simulation and the characterization of t-resiliency. In: STOC (2009)
12. Gafni, E., Guerraoui, R., Pochon, B.: From a static impossibility to an adaptive lower bound: the complexity of early deciding set agreement. In: STOC, pp. 714–722 (2005)

13. Guerraoui, R., Herlihy, M.P., Pochon, B.: A topological treatment of early-deciding set-agreement. In: Shvartsman, M.M.A.A. (ed.) OPODIS 2006. LNCS, vol. 4305, pp. 20–35. Springer, Heidelberg (2006)
14. Herlihy, M., Shavit, N.: The topological structure of asynchronous computability. J. ACM 46(6), 858–923 (1999)
15. Keidar, I., Shraer, A.: Timeliness, failure-detectors, and consensus performance. In: PODC, pp. 169–178 (2006)
16. Lamport, L., Shostak, R., Pease, M.: The byzantine generals problem. ToPLaS 4(3), 382–401 (1982)
17. Saks, M.E., Zaharoglou, F.: Wait-free k-set agreement is impossible: The topology of public knowledge. SIAM J. Comput. 29(5), 1449–1483 (2000)
18. Yang, J., Neiger, G., Gafni, E.: Structured derivations of consensus algorithms for failure detectors. In: PODC, pp. 297–308 (1998)

Step-Assembly with a Constant Number of Tile Types

Ján Maňuch[1], Ladislav Stacho[2], and Christine Stoll[2]

[1] Department of Computer Science, University of British Columbia, Canada
[2] Department of Mathematics, Simon Fraser University, Canada

Abstract. Demaine *et. al.* [1] have introduced a model for staged assembly for constructing self-assembled shapes as an extension to the multiple tile model [2]. In their model the assembly proceeds in several bins and several stages with different sets of tiles and supertiles applied on each bin and in each stage. Taking advantage of all these features they showed that a constant number of tile types is sufficient to self-assemble any given shape.

In this paper, we consider a simplified model of staged assembly, called the step assembly model, in which we only have one bin in each step and assembly happens by attaching tiles one by one to the growing structure as in the standard assembly model. We show that in this simplified model a constant number of tile types (24) is sufficient to assemble a large class of shapes. This class includes all shapes obtained from any shape by scaling by a factor of 2. For general shapes, we note that the tile complexity of this model has connections to the monotone connected node search number of a spanning tree of the shape.

1 Introduction

Self-assembly is the process by which simple parts autonomously assemble into larger, more complex objects. Self-assembly occurs in nature, for example, when atoms combine to form molecules, and molecules combine to form crystals. It has been suggested that intricate self-assembly schemes will ultimately be useful for circuit fabrication, nano-robotics, DNA computing, and amorphous computing [3–6]. Current research focuses on self-assembling sieves to remove viruses from serum, and nanomanufacturing drug-delivery and medical-imaging devices [1].

The standard model to study the process of self-assembly is the Tile Assembly Model proposed by [7] which considers the assembly of square blocks called "tiles" and a set of glues called "binding domains". Each of the four sides of a tile can have a glue on it that determines interactions with neighbouring tiles. It is assumed that there is an infinite supply of tiles of each tile type. The process of self-assembly is initiated by a single seed tile and proceeds by attaching tiles one by one. A tile can only bind to the growing complex if it binds strongly enough, as determined by the *temperature* τ.

Branched DNA molecules [8] provide a direct physical motivation for this model. DNA double-crossover molecules, each bearing four "sticky ends"

Y. Dong, D.-Z. Du, and O. Ibarra (Eds.): ISAAC 2009, LNCS 5878, pp. 954–963, 2009.

analogous to the four sides of a tile, have been designed to self-assemble into a periodic two dimensional lattice [9–12]. The binding interactions between double-crossover molecules may be redesigned by changing the base sequence of their sticky ends, thus allowing arbitrary sets of tiles to be investigated in the laboratory. Tiles can also be implemented using protein-based designs, where unit-length nanorods (made of proteins) are joined at right angles at their midpoints to form a plus sign [1]. Protein nanorod structures are, unlike DNA based assemblies, very rigid. It is believed that this rigidity will allow them to be used for the nanomanufacture of macroscale objects.

In this paper we introduce a modification to the standard model of self-assembly which, instead of having a single set of tiles, allows for several sets of tiles. We immerse a seed tile into one of the sets, "filter out" the assembled shape from this set and place this assembled shape into a different set of tiles where it now acts as a seed and assembly continues. We do not restrict the number of times a shape can be filtered out and placed in a new set of tiles, nor do we restrict the number of different sets of tiles. Using this model, we give an upper bound on the number of tile types required to uniquely assemble an arbitrary shape. We note that this bound is related to the monotone connected node search number (cf. [13]) of the underlying spanning tree. Lastly, we exhibit a large class of shapes that have constant tile complexity (24 tile types suffice). This class includes all shapes scaled by a factor of 2. We also show that for these scaled shapes (by a factor of 2), 14 tile types suffice.

A similar model, called the staged assembly model, has been proposed in [1]. The key differences between our step assembly model and this model are that in our model growth occurs by addition of single tiles to the seed configuration similarly as in the standard assembly model, while in the stage assembly model two assemblies are allowed to attach to form a larger assembly. Moreover, our model uses only one bin, while the staged assembly model uses multiples bins at each stage for later mixing of assembled supertiles (assemblies). In [1], a bound of at most 16 tile types for uniquely assembling an arbitrary shape, if we do not require a bond between every two adjacent tiles is given. If scaling by a factor of two is allowed, then the authors show that any arbitrary shape can be assembled using at most 8 binding domains (and hence a constant number of tile types), and this construction ensures that there is a bond between every two adjacent tiles (a full shape is assembled).

The reader may be also interested in other models of self-assembly, for instance, flexible glue model, unique shape model, multiple temperature model, multiple tile model [2].

2 Description of the Step Tile Assembly Model

We will consider the square lattice, i.e., the graph with vertex set $\mathbb{Z}\times\mathbb{Z}$ and edge set $\{(u,v) : |u,v| = 1\}$. The directions $\mathcal{D} = \{N, E, S, W\}$ are used to indicate the natural directions in the lattice. Formally, they are functions from $\mathbb{Z}\times\mathbb{Z}$ to $\mathbb{Z}\times\mathbb{Z}$: $N(x,y) = (x, y+1)$, $E(x,y) = (x+1, y)$, $S(x,y) = (x, y-1)$, and $W(x,y) = (x-1, y)$. Note that $E^{-1} = W$ and $N^{-1} = S$.

A tile t is a square represented by a 4-tuple $(t_N, t_E, t_S, t_W) \in \Sigma^4$ indicating the binding domains on the north, east, south, and west side, respectively. We will use *null* to indicate the lack of a binding domain, and will assume $null \in \Sigma$. The special tile $empty = (null, null, null, null)$ represents an empty space when placed onto the grid. A *configuration* on a set of tiles T is a map $C : \mathbb{Z} \times \mathbb{Z} \to T$. We define the *vertex set* of configuration C as $V(C) = \{(x, y) : C(x, y) \neq empty\}$. A configuration C is *finite* if $V(C)$ is finite. We will refer to $C(x, y)$ as the tile at the vertex (x, y) in C. Given a configuration C and a set of vertices $V \subseteq \mathbb{Z} \times \mathbb{Z}$, a *sub-configuration* of C induced by V is the map $C[V] : \mathbb{Z} \times \mathbb{Z} \to T$ such that $C[V](x, y) = C(x, y)$ for all $(x, y) \in V$, and $C[V](x, y) = empty$, otherwise. If G is any subgraph of the lattice graph, then we sometimes abuse the notation of $C[V(G)]$ to simply $C[G]$. Given two configurations C and D, we define their union to be the following map from $\mathbb{Z} \times \mathbb{Z}$ to $T \cup \{\infty\}$:

$$(C \cup D)(x, y) = \begin{cases} C(x, y) & \text{if } D(x, y) = empty \text{ or } C(x, y) = D(x, y), \\ D(x, y) & \text{if } C(x, y) = empty \text{ or } C(x, y) = D(x, y), \\ \infty & \text{otherwise.} \end{cases}$$

Note that $C \cup D$ is a configuration whenever it is a map to T. Equivalently, $C \cup D$ is not a configuration if there exists $(x, y) \in V(C) \cap V(D)$ such that $C(x, y) \neq D(x, y)$.

A *strength function* $g : \Sigma \times \Sigma \to \mathbb{N} = \{0, 1, 2, ...\}$ measures the interaction strength between binding domains. We denote by s_Σ the strength function satisfying $s_\Sigma(\sigma, \sigma') = 1$, if $\sigma = \sigma' \neq null$, and $s_\Sigma(\sigma, \sigma') = 0$ otherwise. We call s_Σ a *simple* strength function. In this paper we restrict ourselves to simple strength functions.

Given a tile t, a configuration C, and a direction d, we denote the interaction strength in configuration C between tile t at position (x, y) and its respective neighbouring tile by $g_d^C(t, x, y) = g(t_d, C(d(x, y))_{d^{-1}})$. Note that we do not require that $C(x, y) = t$. In particular, if $C(x, y) \neq t$, then $g_d^C(t, x, y)$, $d \in \mathcal{D}$ tells us how t would bind if it were in C. Given $(x, y) \in \mathbb{Z} \times \mathbb{Z}$ and $d \in \mathcal{D}$, we say that there is a bond between positions (x, y) and $d(x, y)$ in C if $g_d^C(C(x, y), x, y) = 1$, i.e. the binding domain on the abutting sides of the two tiles is the same.

Under the Tile Assembly Model a *tile system* is a 5-tuple $\mathbf{T} = (\Sigma, T, \phi, g, \tau)$, where T is a finite set of tiles with binding domains from Σ and contains the tile *empty*, ϕ is a set of configurations on T called *seed configurations*, g is a strength function, and τ is a threshold parameter called *temperature*. We will be working with seed configurations consisting of a single tile; formally a configuration C_t, where $t \in T$, satisfying $C_t(0, 0) = t$, and $C_t(x, y) = empty$ for all $(x, y) \in \mathbb{Z} \times \mathbb{Z} \setminus \{(0, 0)\}$. To distinguish these tile systems from step tile systems which we will define shortly, we will sometimes refer to the above tile systems as *simple* tile systems.

Self-assembly is now defined as a relation between configurations on T. Let C and D be two configurations of T, such that $C = D$ except at position (x, y), where $C(x, y) = empty$, and $D(x, y) = t$, for some $t \in T \setminus \{empty\}$. Then we write $C \to_{\mathbf{T}} D$, if $\sum_{d \in \mathcal{D}} g_d^C(t, x, y) \geq \tau$. For simple strength functions and

$\tau = 1$, which are considered in this paper, $C \rightarrow_{\mathbf{T}} D$ if and only if there exists $d \in \mathcal{D}$ such that $g_d^C(t, x, y) = 1$. The relation $\rightarrow_{\mathbf{T}}^+$ is the transitive closure of $\rightarrow_{\mathbf{T}}$.

A tile system $\mathbf{T}$ and the relation $\rightarrow_{\mathbf{T}}^+$ define the partially ordered set of configurations called *assemblies* of $\mathbf{T}$: $Asmb(\mathbf{T}) = \{A : S \rightarrow_{\mathbf{T}}^+ A$, where $S \in \phi\}$, and the set of *terminal assemblies* of $\mathbf{T}$: $Term(\mathbf{T}) = \{A \in Asmb(\mathbf{T}) : \nexists B$ such that $A \rightarrow_{\mathbf{T}}^+ B\}$. A tile system uniquely produces A if $Term(\mathbf{T}) = \{A\}$.

A *step tile system* is a 5-tuple $\mathbf{T}_{step} = (\Sigma, \{T_i\}_{i=1}^k, \{C_t\}, g, \{\tau_i\}_{i=1}^k)$, where $\{T_i\}_{i=1}^k$ is a sequence of finite sets of tiles (each set including the tile *empty*), C_t is the initial seed configuration (consisting of the single tile t), g is a strength function, and $\{\tau_i\}_{i=1}^k$ is a sequence of temperatures. We define the (simple) tile system at step 1 as $\mathbf{T}_1 = (\Sigma, T_1, \{C_t\}, g, \tau_1)$ and the (simple) tile system at step i as $\mathbf{T}_i = (\Sigma, T_i, Term(\mathbf{T}_{i-1}), g, \tau_i)$ for $2 \le i \le k$. This allows us to view step tile systems as a sequence of simple tile systems.

We define the set of terminal assemblies of $\mathbf{T}_{step}$ as $Term(\mathbf{T}_{step}) = Term(\mathbf{T}_k)$. Given a configuration A, we say that the step tile system $\mathbf{T}_{step}$ uniquely produces A, if $Term(\mathbf{T}_{step}) = \{A\}$. We are interested in the number of tile types used in a tile system and define the *tile complexity* of $\mathbf{T}_{step}$ as $|\bigcup_{i=1}^k T_i| - 1$ (where we are subtracting 1 to exclude the *empty* tile).

3 Tile Complexity of Step-Assembled Shapes

A *shape* S is a connected induced subgraph of the lattice. We say a configuration A *has shape* S, if $V(A) = V(S)$. We say a tile system uniquely produces shape S, if the the terminal assembly of the tile system is unique and has shape S.

A caterpillar K is a graph consisting of a single path, called the *spine*, and additional vertices attached to the internal vertices of the spine by single edges, called *hairs*. We refer to these vertices as *hairtips*. Given a tree F, a caterpillar K with spine P and set of hairtips L is a *natural for* F if it is a subgraph of F and all vertices in L are leaves in F. If it will be clear which tree F we mean, we will omit the reference to F and simply say, K is a natural caterpillar. Moreover, K is *maximal* if it is natural and there is no natural caterpillar properly containing K. If v is an endpoint of the spine P of the caterpillar K, we say that K is *anchored at* v. Moreover, K is *maximal anchored at* v, if K is a natural caterpillar anchored at v and it is not properly contained in any natural caterpillar anchored at v.

Given a tree F, the sequence of sets of caterpillars $\{L_i\}_{i=1}^{\delta}$ forms a *level decomposition* D_F of F if it satisfies all of the following:

(LD1) All caterpillars in $\bigcup_{i=1}^{\delta} L_i$ are edge disjoint, and cover all edges of F.
(LD2) L_1 consists of a single maximal caterpillar of F.
(LD3) For every $2 \le i \le \delta$, caterpillars in L_i are vertex disjoint.
(LD4) For every $2 \le i \le \delta$, every caterpillar K in L_i is a maximal caterpillar anchored at a vertex v on the spine of some caterpillar in L_{i-1}. We call v the *anchor* of K.

Note that the anchor of K is either an internal vertex of the spine of a caterpillar K' or the anchor of K', where $K' \in L_{i-1}$. A vertex v has *level* i in D_F, if i is the smallest index such that v is a vertex of a caterpillar in L_i. The *depth of a level decomposition* $D_F = \{L_i\}_{i=1}^{\delta}$ is δ. The *level-depth of a tree* F, LD-depth(F), is the smallest δ such that F has a level decomposition of depth δ[1]. The *level-depth of a shape* S, LD-depth(S), is the minimum level-depth of all spanning trees of S.

A vertex v is a *single anchor* (*double anchor*) if it is an anchor for exactly one (two) caterpillar(s) in D_F. Note that if F is a spanning tree of a shape S, then v cannot be an anchor for more than two caterpillars, since $deg_F(v) \leq 4$, and every anchor is an internal vertex of the spine of some caterpillar. Note that since for every $1 < i \leq \delta$ caterpillars of L_i are vertex disjoint, any vertex v of level j that is a double anchor is an anchor for one caterpillar of L_{j+1} and one caterpillar of L_{j+2}; moreover, $\deg_F(v) = 4$ and the remaining two edges incident to v belong to a caterpillar of L_j. Note that in this case v cannot be a leaf of the caterpillar of L_j.

Theorem 1. *Any shape S that has a spanning tree F of level-depth δ can be uniquely produced by a step tile system with tile complexity at most $68\delta + 8$. Moreover, if the maximum degree of F is 3, S can be uniquely produced by a step tile system with tile complexity at most $44\delta + 8$, and if the maximum degree of F is 2, only 14 tile types suffice.*

Proof. Let $D_F = \{L_i\}_{i=1}^{\delta}$ be a level decomposition of F of level-depth δ. In the step tile system we are constructing, each tile set will contain only a single non-empty tile type. The binding domains will be taken from the set $\Sigma = \{null, a_1, a_2, \ldots, a_{\delta+2}, b_1, b_2, \ldots, b_\delta\}$. We define a partial function $\overline{x}$ on Σ as follows: $\overline{a_i} = b_i$ and $\overline{b_i} = a_i$, for $1 \leq i \leq \delta$. We also define a *rank* on Σ as follows: rank$(a_i) = i$, rank$(b_i) = i$, and rank$(null) = 0$. Our construction is guided by a depth first search ordering of the vertices of F obtained as follows. Let s_0 be an endpoint of the spine of the caterpillar in L_1 (without loss of generality we assume that $s_0 = (0, 0)$). The sequence $\{s_j\}_{j=0}^{|S|-1}$ is obtained by a depth first search on the vertices of F starting at the vertex s_0. At each branching point the depth first search will choose the next vertex to visit as follows: choose a non-visited neighbour that is a leaf if possible, or otherwise, choose a non-visited neighbour of the highest level. For every vertex s_j of F let $t^j = (t_N^j, t_E^j, t_S^j, t_W^j)$ be the tile that will be placed onto s_j. The tile sets used will be the sets $T_j = \{t^j, empty\}$. Next we specify the tile types t^j for $j = 1, \ldots, |S|$ in this order. Let $d \in \mathcal{D}$ be the unique direction such that $s_1 = d(s_0)$. Set $t_d^0 = a_1$ and $t_\alpha^0 = null$, if $\alpha \neq d$.

For $j > 1$ and $\deg_F(s_j) = 1$, let s_k be the neighbour of s_j such that $k < j$. Let d be the direction such that $s_k = d(s_j)$. Set $t_d^j = t_{d^{-1}}^k$ (this ensures that the tile t^j can attach to the neighbouring tile t^k), and $t_\alpha^j = null$, if $\alpha \neq d$.

For $j > 1$ and $\deg_F(s_j) = 2$, let s_k be the unique neighbour of s_j such that $k < j$. Note that s_{j+1} is also a neighbour of s_j, and, because $\deg_F(s_j) = 2$, s_k, s_j, and s_{j+1} all belong to the same caterpillar in some L_i. Let d_1, d_2 be directions such that $s_k = d_1(s_j)$, and $s_{j+1} = d_2(s_j)$. Set $t_{d_1}^j = t_{d_1^{-1}}^k$ (this ensures

[1] The level-depth of a tree has close connections to the monotone connected node search number of the tree[13] which we will discuss in the full version of this paper.

that the tile t^j can attach to its neighbouring tile t^k), $t^j_{d_2} = \overline{t^j_{d_1}}$, and $t^j_\alpha = null$, if $\alpha \notin \{d_1, d_2\}$. For $j > 1$ and $\deg_T(s_j) = 3$, we distinguish two cases:

Case (i) *s_j is not an anchor:* Then s_j belongs to exactly one caterpillar K in L_i for some i. Note that s_{j+1} is a neighbour of s_j and it is a leaf. Let s_k and s_m be the other two neighbours of s_j such that $k < j$ and $m > j$ (see Figure 1(a)). Note that $s_m = s_{j+2}$ and that s_j and all its neighbours belong to the same caterpillar. Let d_1, d_2, d_3 be directions such that $d_1(s_j) = s_k$, $d_2(s_j) = s_{j+1}$, and $d_3(s_j) = s_m$. Set $t^j_{d_1} = t^k_{d_1^{-1}}$ (this ensures that tile t^j can attach to the neighbouring tile t^k), $t^j_{d_2} = a_{i+1}$ (hence using a new binding domain for hairs of K), $t^j_{d_3} = \overline{t^j_{d_1}}$, and $t^j_\alpha = null$, if $\alpha \notin \{d_1, d_2, d_3\}$.
Case (ii) *s_j is an anchor:* Then s_j must be a single anchor, since only vertices of degree four can be double anchors. Let s_j be an anchor for a caterpillar K' in L_{i+1}. It follows that s_j also belongs to a caterpillar K in L_i (see Figure 1(b)). Let s_k, s_m, d_1, d_2, and d_3 be defined as in case (i). Note that s_{j+1} is a neighbour of s_j and its level is $i + 1$. Set $t^j_{d_1} = t^k_{d_1^{-1}}$ (this ensures that tile t^j can attach to the neighbouring tile t^k), $t^j_{d_2} = a_{i+1}$, (as the caterpillar K' will use a new pair of binding domains a_{i+1}, b_{i+1}), $t^j_{d_3} = \overline{t^j_{d_1}}$, and $t^j_\alpha = null$, if $\alpha \notin \{d_1, d_2, d_3\}$.

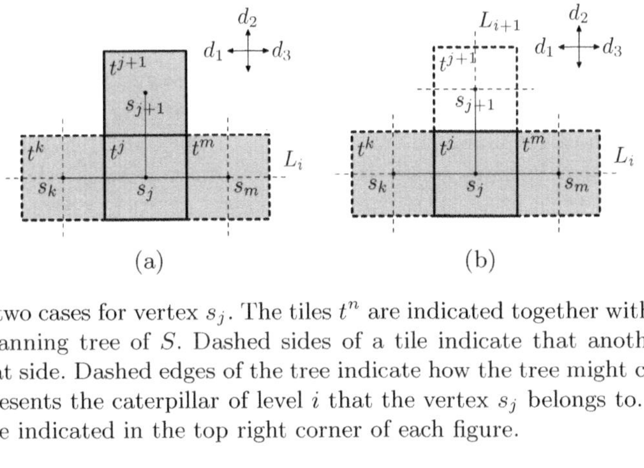

Fig. 1. The two cases for vertex s_j. The tiles t^n are indicated together with the vertices s_n of the spanning tree of S. Dashed sides of a tile indicate that another tile could attach to that side. Dashed edges of the tree indicate how the tree might continue. The shading represents the caterpillar of level i that the vertex s_j belongs to. The relative directions are indicated in the top right corner of each figure.

The case when $j > 1$ and $\deg_F(s_j) = 4$ can be proved similarly.[2]
Consider configuration $C : \mathbb{Z} \times \mathbb{Z} \to \{t^j\}_{j=0}^{|S|-1}$ defined by

$$C(v) = \begin{cases} t^j & \text{if } v = s_j \text{ for some } s_j \in V(T), \\ empty & \text{otherwise.} \end{cases}$$

Clearly, the configuration C has shape S. We will show that the step tile system $\mathbf{T}_{step} = (\Sigma, \{t^j, empty\}_{j=1}^{|S|-1}, \{C_{t^0}\}, s_\Sigma, \{1\}_{j=1}^{|S|-1})$ uniquely produces C. To this

[2] The complete proof will appear in the full version.

end, let $C_j = C[V_j]$ where $V_j = \{s_0, \ldots, s_j\}$, be the sub-configuration of C induced by the first $j+1$ vertices, $(j = 0, \ldots, |S|-1)$. We say that the configuration C_j is *hierarchical*, if all of the following hold:

(H1) All non-*null* binding domains exposed on C_j have distinct ranks.
(H2) Let $x_1, x_2, \ldots, x_m$ be all exposed non-*null* binding domains on C_j ordered increasingly by their ranks. Let $t^{i_1}, t^{i_2}, \ldots, t^{i_m}$ be the corresponding tiles in C_j. Then $i_1 \leq i_2 \leq \ldots \leq i_m$.

We will show by induction that the tile system $\mathbf{T}_j$ of $\mathbf{T}_{step}$ uniquely produces C_j for each $j \geq 1$.

By definition of the step tile system $\mathbf{T}_{step}$, the seed configuration is $C_0 = C_{t^0}$. Since C_0 contains only one tile, condition (H2) is trivially satisfied. Since all non-*null* binding domains exposed on C_0 have distinct rank (as there is only one non-*null* binding domain, namely a_1, exposed), C_0 satisfies condition (H2). Hence, C_0 is hierarchical. Now we show that for every $j \geq 1$, if the set of seed configurations of $\mathbf{T}_j$ is $\{C_{j-1}\}$, and C_{j-1} is hierarchical, then $\mathbf{T}_j$ produces the unique terminal assembly C_j, and C_j is hierarchical. The claim will follow.

By the induction hypothesis, the set of seed configurations of $\mathbf{T}_j$ is $\{C_{j-1}\}$, i.e., $\mathbf{T}_j = \{\Sigma, \{t^j, empty\}, \{C_{j-1}\}, s_\Sigma, 1\}$. Let the vertex s_k be the neighbour of s_j in the tree F such that $k < j$. Since $C_{j-1}(s_j) = empty$, by our construction t^j can attach to C_{j-1} at position s_j in step j. Since t^j is the only non-*empty* tile in this step, it will attach at this position. Next we show that t^j cannot attach anywhere else. Let r be the rank of the binding domain via which tiles t^j and t^k are attached to each other in our construction. Due to our assignment of binding domains, r is a lowest rank of non-*null* binding domains on t^j (there could be two binding domains on t^j with lowest rank), and r is the highest rank of exposed binding domains on t^k. By the depth-first search, either $t^k = t^{j-1}$ or t^{j-1} is a leaf of F and the algorithm backtracked to t^k, in which case the tiles $t^{k+1}, \ldots, t^{j-1}$ do not have any exposed non-*null* binding domains. This implies that the binding domain of highest rank exposed on C_{j-1} is on t^k, since C_{j-1} is hierarchical. As r is the highest rank of exposed binding domains on t^k, it follows that r is the highest rank of exposed binding domains on C_{j-1}. Since all binding domains exposed on C_{j-1} have distinct rank and r is a lowest rank of non-null binding domains on t^j, tile t^j cannot attach anywhere else. Hence, $\mathbf{T}_j$ uniquely produces C_j. It remains to show that C_j is hierarchical. Let $X = x_1, x_2, \ldots, x_m$ be the exposed non-*null* binding domains of C_{j-1} ordered by increasing rank. In C_j, tile t^j is attached to t^k via binding domain x_m which is of rank r. Since C_{j-1} satisfies (H1), and r is the lowest non-*null* rank of the binding domains on t^j, and the exposed non-*null* binding domains on t^j in C_j are distinct, it follows that C_j satisfies (H1). Let $Y = y_1, y_2, \ldots, y_k$ be the exposed non-*null* binding domains of C_j ordered by increasing rank. Since r is the highest rank all non-*null* binding domains exposed on C_{j-1} and r is also the lowest rank of all non-*null* binding domains on tile t^j, in the sequence Y all non-*null* binding domains of t^j are at the end of the sequence. Since all the preceding elements are a subsequence of X, they satisfy (H2). Moreover, since all new (if any) binding domains in Y come from tile t^j, C_j also satisfies (H2). Thus, C_j is hierarchical.

Hence, the step tile system $\mathbf{T}_{step}$ uniquely produces the terminal assembly C, and therefore, $\mathbf{T}_{step}$ uniquely produces shape S.

It is not difficult to check that the tile complexity of $\mathbf{T}_{step}$ is as required in all three cases. We will leave the details for the full version of this paper.

4 Shapes with Constant Tile Complexity

A *rectangle* R is a shape for which there exist integers $N \geq 2$ and $M \geq 2$ and a vertex (x_0, y_0) such that vertex $(x, y) \in R$ if and only if $x_0 \leq x < x_0 + N$ and $y_0 \leq y < y_0 + M$. Given a shape S and an integer $k \geq 1$, we say S has a *rectangle decomposition* $\{R_i\}_{i=1}^m$ *with connectors of size* k if there exist m disjoint rectangles, $R_1, \ldots, R_m$, such that $V(S) = \bigcup_{i=1}^m V(R_i)$ and for each $i > 1$, there exists $j < i$ such that the rectangles R_i and R_j are joined in S by at least k edges. A selected set of k consecutive such edges is called the *connector* of R_i and R_j. Note the the endvertices of this connector in R_i (and R_j, respectively) induce a path of length $k - 1$ which is called an interface of R_i (R_j). Observe that if R_i connects to l other rectangles it will have l interfaces.

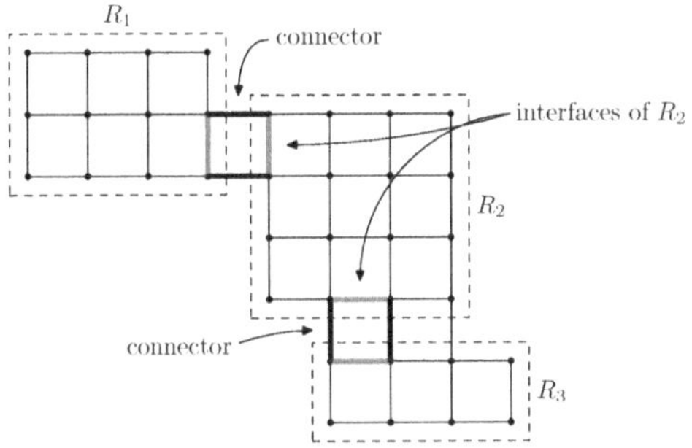

Fig. 2. A rectangle decomposition of a shape with connectors of size 2. Connector edges are depicted in bold and interface edges are depicted in grey.

Theorem 2. *A shape S which has a rectangle decomposition $\mathbf{R} = \{R_i\}_{i=1}^m$ with connectors of size 2, can be assembled using at most 24 non-empty tiles types.*

Proof. We will show that S has a spanning tree F of maximum degree at most 3 and level-depth at most 2. Given this spanning tree, we will obtain a tile system following the construction in the proof of Theorem 1. Due to our choice of F, this tile system will only use 24 non *empty* tile types. Note that direct application of the result of Theorem 1 would give a bound of 96 non-*empty* tile types.

To obtain a suitable spanning tree F of S it can be shown[3] that S has a spanning subgraph G satisfying all of the following conditions:

[3] The proof will appear in the full version.

(i) The maximum degree of G is 3.
(ii) G contains a cycle C that consists of all connector edges and all boundary edges of each rectangle in $\mathbf{R}$ which are not interface edges.
(iii) Every vertex not on the cycle C is on a "horizontal" path whose west endpoint belongs to C.

Given this spanning subgraph G of S, we obtain a spanning tree F of S by deleting from the cycle C an edge that is on the northern boundary of S. Certainly F will still have maximum degree three. It remains to show that F has level-depth at most 2. Let the caterpillar K consist of the path in F that we obtained by deleting an edge from the cycle C of G (this is the spine of K), together with any paths P_w that include only a single edge. Let $L_1 = \{K\}$. If $K = F$, then L_1 trivially forms a level decomposition of F. Otherwise let L_2 be the set consisting of all the paths P_w that are not included in the caterpillar K. Then all caterpillars in $L_1 \cup L_2$ are edge-disjoint and cover all edges of F. Hence, $\{L_1, L_2\}$ satisfies condition (LD1). Let u and v be the endpoints of the spine of K. Since u and v are on the northern boundary of S, they are leafs of F (any branching occurs on the western boundary of a rectangle). Thus K is a maximal caterpillar of F, and hence $\{L_1, L_2\}$ satisfies condition (LD2). By our construction, all paths P_w are vertex disjoint. Thus, $\{L_1, L_2\}$ satisfies condition (LD3). Every path P_w is a maximal caterpillar anchored at w, where w is an internal vertex of the spine of caterpillar K in L_1. Hence, $\{L_1, L_2\}$ satisfies condition (LD4). Therefore, $\{L_1, L_2\}$ is a level decomposition of F, and hence F has depth at most 2.

Now construct a tile system from the spanning tree F of S as described in the proof of Theorem 1. However, due to our choice of F this tile system will use far less than the 96 tile types that Theorem 1 guarantees. Our tile system will use $4 \cdot 3$ tile types that use both a_1 and b_1 (exactly once) and no other non-*null* binding domains. There will be six tile types that use exactly three non-*null* binding domains (a_1, b_1, and a_2) (this is because any vertices of degree 3 are on the western boundary of a rectangle). There are two tile types that use both a_2 and b_2 (exactly once) and no other non-*null* binding domains, because all the paths in L_2 are horizontal. For the same reason, there are only two tile types with only one non-*null* binding domain which is taken from $\{a_2, b_2\}$. Furthermore, there will be two more tile types corresponding to the endpoints of the spine of the

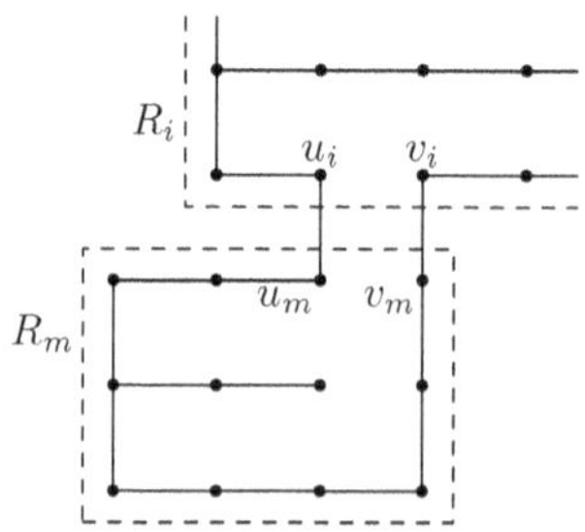

Fig. 3. Extending the spanning subgraph to include R_m

caterpillar of L_1 (both of these tiles have only one non-*null* binding domain - either a_1 or b_1). Therefore, our tile system uses only 24 non-empty tiles.

If we are allowed to scale shapes by a factor of 2, then any shape can be assembled using a constant number of tile types.

Theorem 3. *Given an arbitrary shape S, let S' be the shape obtained by scaling S by a factor of 2. Then S' can be uniquely produced by a step assembly system at temperature 1 using at most 14 non-empty tile types.*

References

1. Demaine, E.D., Demaine, M.L., Fekete, S.P., Ishaque, M., Rafalin, E., Schweller, R.T., Souvaine, D.L.: Staged self-assembly: nanomanufacture of arbitrary shapes with $O(1)$ glues. Natural Computing 7(3), 347–370 (2008)
2. Aggarwal, G., Cheng, Q., Goldwasser, M.H., Kao, M.-Y., Espanes, P.M., Schweller, R.T.: Complexities for generalized models of self-assembly. SIAM J. Comput. 34(6), 1493–1515 (2005)
3. Winfree, E., Yang, X., Seeman, N.: Universal computation via self-assembly of DNA: Some theory and experiments. In: Proceedings of the Second Annual Meeting on DNA Based Computers, pp. 191–214 (1996)
4. Reif, J.H.: Local parallel biomolecular computation. In: Proc. DNA-Based Computers, pp. 217–254 (1997)
5. Gomez-Lopez, M., Preece, J., Stoddart, J.: The art and science of self-assembling molecular machines. Nanotechnology 7, 183–192 (1996)
6. Abelson, H., Allen, D., Coore, D., Hanson, C., Homsy, G., Knight, T.F., Nagpal, R., Rauch, E., Sussman, G.J., Weiss, R.: Amorphous computing. Communications of the ACM 43, 74–82 (2000)
7. Rothemund, P.W.K., Winfree, E.: The program-size complexity of self-assembled squares. In: Proceedings of STOC, pp. 459–468 (2000)
8. Seeman, N.: DNA nanotechnology: novel DNA constructions. Annual Review of Biophysics and Biomolecular Structure 27, 225–248 (1998)
9. Winfree, E., Liu, F., Wenzler, L.A., Seeman, N.C.: Design and self-assembly of two dimensional DNA crystals. Nature 394, 539–544 (1998)
10. Mao, C., LaBean, T.H., Reif, J., Seeman, N.: Logical computation using algorithmic self-assembly of DNA triple-crossover molecules. Nature 407, 493–496 (2000)
11. LaBean, T., Yan, H., Kopatsch, J., Liu, F., Winfree, E., Reif, J.H., Seeman, N.: Construction, analysis, ligation, and self-assembly of DNA triple crossover complexes. Journal of the American Chemical Society 122, 1848–1860 (2000)
12. Rothemund, P., Papadakis, N., Winfree, E.: Algorithmic self-assembly of DNA Sierpinski triangles. PLoS Biology 2, 2041–2053 (2004)
13. Kirousis, L.M., Papadimitriou, C.H.: Interval graphs and searching. Discrete Mathematics 55(1), 181–184 (1985)

Lower Bounds on Fast Searching

Donald Stanley[1] and Boting Yang[2]

[1] Department of Mathematics and Statistics, University of Regina
stanley@math.uregina.ca
[2] Department of Computer Science, University of Regina
boting@cs.uregina.ca

Abstract. Given a graph, suppose that intruders hide on vertices or along edges of the graph. The fast searching problem is to find the minimum number of searchers required to capture all intruders satisfying the constraint that every edge is traversed exactly once and searchers are not allowed to jump. In this paper, we prove lower bounds on the fast search number. We present a linear time algorithm to compute the fast search number of Halin graphs and their extensions. We present a quadratic time algorithm to compute the fast search number of cubic graphs.

1 Introduction

Let $G = (V, E)$ be a graph with vertex set V and edge set E. In the fast search model, every edge of G is initially contaminated and our aim is to clean the whole graph by a sequence of steps. Each intruder can move at a great speed at any time from vertex u to vertex v along a path that contains no searchers. There are two types of actions for searchers, i.e., placing a searcher on a vertex and sliding a searcher along an untraversed edge from one endpoint to the other, and an edge is cleared only by a sliding action in a proper way. First we place all the searchers on a subset of vertices of G (allowing multiple searchers to be placed on any vertex). Then every edge of G will be traversed exactly once by one searcher to clear the edge. A contaminated edge uv can be cleared in one of two ways by one sliding action: (1) sliding a searcher from u to v along uv while at least one searcher is located on u, and (2) sliding a searcher from u to v along uv while all edges incident on u except uv are already cleared. A *fast search strategy* is a sequence of actions such that the final action leaves all edges of G *cleared*. The minimum number of searchers needed to clear G in the fast searching is the *fast search number* of G, denoted by $fs(G)$. A fast search strategy that uses $fs(G)$ searchers to clear G is called an *optimal fast search strategy.*

The edge search model and the node search model are two major models of graph searching problems. In edge searching problem introduced in [6], there are three types of actions for searchers and an edge can be cleared only by sliding actions. In the node searching problem introduced in [4], there are only two types of actions for searchers and an edge is cleared if both endpoints are occupied by searchers. The fast search model was first introduced by Dyer, Yang and Yaşar [2]. In [2], a linear time algorithm is presented for computing the fast search number of trees.

Y. Dong, D.-Z. Du, and O. Ibarra (Eds.): ISAAC 2009, LNCS 5878, pp. 964–973, 2009.

Lower bounds are very important issue in graph searching. It is usually very difficult to show that a graph requires at least certain number of searchers to clear. In this paper, we prove several lower bounds on fast searching. Using one of our lower bounds, we can compute the fast search number of grids, which have been studied in many search models. We also use our lower bounds to compute the fast search number of Halin graphs and cubic graphs. It is an open problem whether there is a polynomial time algorithm to compute the edge search number or the node search number of Halin graphs. We will present a linear time algorithm for computing the fast search number of Halin graphs. It is NP-hard to find the edge search number or the node search number of cubic graphs [5]. However, for the fast searching problem, we will show that the fast search number of cubic graphs can be found in $O(n^2)$ time, where n is the number of vertices in the graph.

2 Definitions and Notation

Unless otherwise stated, all graphs in this paper will be finite, undirected and without loops, multiple edges, or isolated vertices. Throughout this paper, we use $G = (V, E)$ to denote a graph with vertex set V and edge set E, and we also use $V(G)$ and $E(G)$ to denote the vertex set and edge set of G respectively. We use uv to denote an edge with endpoints u and v. Definitions omitted here can be found in [7].

For a graph $G = (V, E)$, the *degree* of a vertex $v \in V$, denoted by $\deg_G(v)$, is the number of edges incident on v. A vertex is *odd* when its degree is odd. Similarly, a vertex is *even* when its degree is even. A vertex with degree 1 is called a *leaf*. Let V_{odd} be the set of all odd vertices in G, and $V_{\text{even}} = V \setminus V_{\text{odd}}$. For trees and forests, a vertex of degree greater than 1 is called an *internal vertex*.

A *component* of a graph G is a maximal connected subgraph of G. A *cut-edge* or *cut-vertex* of a graph is an edge or vertex whose deletion increases the number of components. A *block* of a graph G is a maximal connected subgraph of G that has no cut-vertex. If G itself is connected and has no cut-vertex, then G is a block. It is easy to see that an edge of G is a block if and only if it is a cut-edge. If a block has at least 3 vertices, then it is 2-connected. Thus, the blocks of a graph are its isolated vertices, its cut-edges, and its maximal 2-connected subgraphs. The *block-cutpoint graph* of G is a bipartite graph B in which one partite set consists of the cut-vertices of G and the other partite set has a vertex v_i for each block B_i of G such that uv_i is an edge of B if and only if $u \in V(B_i)$. When a connected graph G is not a single block, its block-cutpoint graph is a tree whose leaves must correspond to blocks of G, which are called *leaf blocks*. Note that every leaf block has exactly one cut-vertex. A leaf block of G is *strong* if the cut-vertex of the leaf block is incident to a cut-edge of G.

A *path* is a list $v_0, e_1, v_1, \ldots, e_k, v_k$ of vertices and edges such that each edge e_i, $1 \le i \le k$, has endpoints v_{i-1} and v_i and each vertex appears exactly once (except that its first vertex might be the same as its last). A *cycle* is a path that begins and ends on the same vertex.

A *Halin graph* is a plane graph constructed from a plane embedding of a tree with at least 4 vertices and with no vertices of degree 2, by connecting all leaves with a cycle in the natural cyclic order defined by the embedding of the tree. An *extended Halin graph* is a connected graph $H = F \cup C$ with no multiple edges, where F is a forest and C is a cycle containing all leaves of F.

In the fast search model, initially, we have a graph G with all edges contaminated (an invisible intruder hides on any edge or vertex of G). We assume that before the first sliding action is carried out, we have placed all the searchers on vertices of G so that these searchers will clear G by a sequence of sliding actions. So, a fast search strategy $\mathcal{S} = (s_0, s_1, \ldots, s_{|E|})$ is a sequence of actions that consists of two parts, the first part s_0 is the placing action that places all the searchers on vertices of G, and the second part $s_1, \ldots, s_{|E|}$ is a sequence of sliding actions such that the final action leaves all edges of G clean. The number of searchers used by $\mathcal{S}$ to clear G is denoted by $fs(G; \mathcal{S})$. If $\mathcal{S}$ is an optimal fast search strategy of G, then $fs(G; \mathcal{S}) = fs(G)$.

We say that a vertex in G is *occupied* at some moment if at least one searcher is located on this vertex at this moment. A sliding action that moves a searcher from a vertex u to vertex v along the edge uv is called a *sliding-out action* on u and a *sliding-in action* on v. Given a graph $G = (V, E)$ and a fast search strategy $\mathcal{S}$ of G, define $\alpha_{\mathcal{S}}(v)$ as the number of searchers on v just before the first sliding action of $\mathcal{S}$ and $\beta_{\mathcal{S}}(v)$ as the number of searchers on v after the final action of $\mathcal{S}$, and define $V^s_{\mathcal{S}} = \{v \in V\colon \deg(v) \geq 2$ and the first clean edge incident on v is cleared by a sliding-out action on $v\}$, and $V^t_{\mathcal{S}} = \{v \in V\colon \deg(v) \geq 2$ and the last clean edge incident on v is cleared by a sliding-in action on $v\}$. In the case with no ambiguity, we delete subscripts from the above notation.

3 Lower Bounds

Lemma 1. *For a graph $G = (V, E)$ and a fast search strategy $\mathcal{S}$ of G, we have the following properties:* (i) $fs(G; \mathcal{S}) = \frac{1}{2}\sum_{v \in V}(\alpha(v) + \beta(v))$; (ii) $\alpha(v) + \beta(v) \equiv \deg(v)$ *(mod 2) for each vertex $v \in V$;* (iii) $\alpha(v) \geq 2$ *for any $v \in V^s$; and* (iv) $\beta(v) \geq 2$ *for any $v \in V^t$.*

Proof. (i) Because all searchers are placed on vertices of G before the first sliding action, and no searcher can be removed from G, we know that every searcher is counted twice in $\sum_{v \in V}(\alpha(v) + \beta(v))$. Thus $fs(G; \mathcal{S})$ is half of this sum.

(ii) For any vertex $v \in V$, there are $\alpha(v)$ searchers on v just before the first sliding action. Since every searcher can only slide along an untraversed edge, when an edge incident on v is cleared, the number of searchers on v increases by one or decreases by one. Thus, if $\deg(v)$ is even, then $\alpha(v)$ has the same parity as $\beta(v)$; and if $\deg(v)$ is odd, then $\alpha(v)$ has the opposite parity from $\beta(v)$. In both cases, $\deg(v)$ has the same parity as $\alpha(v) + \beta(v)$.

(iii) Since $\deg(v) \geq 2$ and the first clean edge incident on v is cleared by a sliding-out action on v, we know there are at least two searchers on v just before this sliding-out action on v.

(iv) Since $\deg(v) \geq 2$, just before the last edge incident on v is cleared, there are both contaminated edge and clean edge incident on v. Thus, there is at least one searcher on v just before the last edge incident on v is cleared. After this edge is cleared by a sliding-in action on v, there are at least two searchers on v.

Lemma 2. *For a graph $G = (V, E)$, a vertex $v \in V$, and a fast search strategy $\mathcal{S}$ of G, we have the following:* (i) *if* $\deg(v)$ *is odd, then* $\alpha(v) + \beta(v) \geq 1$; (ii) *if* $v \in V^s \cup V^t$, *then* $\alpha(v) + \beta(v) \geq 2$; (iii) *if* $v \in V^s \cup V^t$ *and* $\deg(v)$ *is odd, then* $\alpha(v)+\beta(v) \geq 3$; (iv) *if* $v \in V^s \cap V^t$, *then* $\alpha(v)+\beta(v) \geq 4$; *and* (v) *if* $v \in V^s \cap V^t$ *and* $\deg(v)$ *is odd, then* $\alpha(v) + \beta(v) \geq 5$.

Proof. (i) If $\deg(v)$ is odd, it follows from Lemma 1(ii) that $\alpha(v) + \beta(v)$ is also an odd number. Thus $\alpha(v) + \beta(v) \geq 1$.

(ii) It follows from Lemma 1(iii) or (iv) that $\alpha(v) + \beta(v) \geq 2$.

(iii) From Lemma 1(ii), we know that $\alpha(v) + \beta(v)$ is an odd number. From (ii), we have $\alpha(v) + \beta(v) \geq 3$.

(iv) It follows from Lemma 1(iii) and (iv) that $\alpha(v) + \beta(v) \geq 4$.

(v) From Lemma 1(ii), we know that $\alpha(v) + \beta(v)$ is an odd number. From (iv), we have $\alpha(v) + \beta(v) \geq 5$.

We can now show the main result of this section.

Theorem 1. $fs(G) \geq \frac{1}{2}|V_{\text{odd}}| + |(V^s \cup V^t) \setminus (V^s \cap V^t)| + 2|V^s \cap V^t|$.

Proof. For any vertex $v \in V^s \cup V^t$ or v is odd, v must be in exactly one of the following five cases.

1. $v \in V_{\text{odd}} \setminus (V^s \cup V^t)$. From Lemma 2(i), we have $\alpha(v) + \beta(v) \geq 1$.
2. $v \in (V^s \cup V^t) \setminus (V^s \cap V^t)$ and v is even. It follows from Lemma 2(ii) that $\alpha(v) + \beta(v) \geq 2$.
3. $v \in (V^s \cup V^t) \setminus (V^s \cap V^t)$ and v is odd. It follows from Lemma 2(iii) that $\alpha(v) + \beta(v) \geq 3$.
4. $v \in V^s \cap V^t$ and v is even. It follows from Lemma 2(iv) that $\alpha(v)+\beta(v) \geq 4$.
5. $v \in V^s \cap V^t$ and v is odd. It follows from Lemma 2(v) that $\alpha(v)+\beta(v) \geq 5$.

From Lemma 1(i), we have $fs(G) = \frac{1}{2}\sum_{v\in V}(\alpha(v)+\beta(v)) = \frac{1}{2}\sum_{v\in V_{\text{odd}}}(\alpha(v)+\beta(v)) + \frac{1}{2}\sum_{v\in V_{\text{even}}}(\alpha(v)+\beta(v)) \geq \frac{1}{2}|V_{\text{odd}}| + |(V^s \cup V^t) \setminus (V^s \cap V^t)| + 2|V^s \cap V^t|$.

Theorem 2. *For a graph $G = (V, E)$ with no leaves, $fs(G) \geq \frac{1}{2}|V_{\text{odd}}| + 2$.*

An application of Theorem 2 is to find the fast search number of grids. An $n \times m$ *grid*, denoted by $G_{n\times m}$, is the cartesian product of the two paths with n ($n \geq 2$) and m ($m \geq 2$) vertices, respectively.

Theorem 3. *For $m \geq n \geq 2$, $fs(G_{n\times m}) = n + m - 2$.*

Lemma 3. *Let $G = (V, E)$ be a graph with no leaves. For every strong leaf block B_s and any fast search on G, there exists $v \in V(B_s) \cap (V^s \cup V^t)$.*

Proof. Let $c \in V(B_s)$ be a cut-vertex in G. Assume that the cut-edge of G incident on c is cleared by a sliding-in action on c. We consider two cases:

Case 1. Before clearing the cut-edge all edges of B_s have been cleared. This implies that $c \in V^t$ because the cut-edge is the only edge of G incident on c which is not in B_s.

Case 2. After clearing the cut-edge there are still uncleared edges of B_s. In this case consider the last cleared edge of B_s. The searcher that clears this edge ends at some $v \in V(B_s)$ and clearly $v \in V^t$.

So considering these two cases, we have seen that if the cut-edge is cleared by a sliding-in action on c, then there is $v \in V(B_s) \cap V^t$. Similarly, if the cut-edge is cleared by a sliding-out action on c, then there is $v \in V(B_s) \cap V^s$. This completes the proof of the lemma.

Theorem 4. *For a connected graph G containing cut-edges and no leaves, $fs(G) \geq \frac{1}{2}|V_{\text{odd}}| + \tau$, where τ is the number of strong leaf blocks in the block-cutpoint graph of G.*

The theory built on graph minors plays an important role in graph searching. Most of graph searching problems are minor closed. However this is not true for fast searching [2]. If the edge contraction is the only operation allowed in the graph transformation, then we have the following result.

Theorem 5. *Given a graph G, let H be the graph obtained from G by edge contractions. Then $fs(H) \leq fs(G)$.*

4 Halin Graphs

In this section, we show how to compute the fast search number and an optimal fast search strategy for Halin graphs in linear time. Our algorithm can compute the fast search number and an optimal fast search strategy for extended Halin graphs in the same time bound.

The following algorithm SEARCHHALIN(H) describes a fast search strategy that can clear H using $\frac{1}{2}|V_{\text{odd}}(H)| + 2$ searchers. The output of the algorithm is an optimal fast search strategy $\langle V_p, A_s \rangle$, where V_p is a multiset of vertices on which we place searchers and A_s is a sequence of arcs corresponding to sliding actions, that is, an arc (u, v) corresponds to a sliding along the arc from tail u to head v.

Algorithm. SEARCHHALIN(H)
Input: A Halin graph H.
Output: A fast search strategy $\langle V_p, A_s \rangle$ of H.

1. Decompose H into a plane embedding of a tree T and a cycle $C = v_0 v_1 \ldots v_c v_0$ such that $v_0, v_1, \ldots, v_c$ are all leaves of T in the natural cyclic order defined by the embedding of T.
2. Place two searchers λ_1 and λ_2 on v_0, and slide λ_1 along $v_0 v_1$ from v_0 to v_1. λ_1 and λ_2 will slide only on edges of C, and when they meet at a vertex of C, they will stay at this vertex forever.

3. Place a searcher on the vertex of C that is occupied by λ_1, slide the searcher along the contaminated edge in T, and slide λ_1 to the next vertex on C.
4. When a searcher λ slides to a vertex u that has been occupied by another searcher λ', if $\lambda' \notin \{\lambda_1, \lambda_2\}$ and there is at least one contaminated edge incident on u, then keep sliding λ along a contaminated edge; if $\lambda' \in \{\lambda_1, \lambda_2\}$ and there is a contaminated edge incident on u (this edge must be on C), then keep λ on u and slide λ' along this contaminated edge.
5. If there is an even vertex $u \in V(T)$ occupied by one searcher and there is only one contaminated edge incident on u, then slide the searcher along this contaminated edge from u to the other endpoint and go to Step 4.
6. If all edges of H are cleared, then stop and output the multiset of vertices V_p on which searchers are placed and then output the sequence of arcs A_s in the order when searchers slide along them; otherwise, go to Step 3.

Theorem 6. *For any Halin graph H, $fs(H) = \frac{1}{2}|V_{\text{odd}}(H)| + 2$.*

Proof. Since Halin graphs have no leaves, it follows from Theorem 2 that $fs(H) \geq \frac{1}{2}|V_{\text{odd}}(H)|+2$. We will show that the fast search strategy described in Algorithm SEARCHHALIN(H) can clear H using $\frac{1}{2}|V_{\text{odd}}(H)| + 2$ searchers.

From Step 1 of SEARCHHALIN(H), we know all leaves of T have degree 3 in H. Since every internal vertex of T has degree at least 3, we have $|V(C)| =>$ $\frac{1}{2}|V(T)|$. From Step 2, we know that searchers λ_1 and λ_2 slide only on edges of C, and when they meet at a vertex of C, they will stay at this vertex forever. Thus, all edges of T are cleared by other searchers, called T-searchers. From Step 3, we know that every T-searcher is placed on a leaf of T. From Steps 4 and 5, when a searcher (including λ_1 and λ_2) passes through a vertex u that has been occupied by another searcher (neither λ_1 nor λ_2), the parity of the number of contaminated edges incident on u does not change because two contaminated edges incident on u are cleared by this searcher. Thus, if an odd vertex of T is occupied by a searcher (neither λ_1 nor λ_2) who is the first searcher sliding to this odd vertex, then this searcher will stay on this odd vertex forever. If an even vertex of T is occupied by a searcher who is the first searcher sliding to this even vertex, then this searcher will slide out from this even vertex when only one contaminated edge is incident on this even vertex. Notice that $|V(C)| > \frac{1}{2}|V(T)| \geq \frac{1}{2}|V_{\text{odd}}(T)|$ and T is connected. Hence, every searcher can start from a leaf of T. On the other hand, every odd vertex of T is either the start or the end vertex of exactly one searcher except λ_1 and λ_2. Therefore, the total number of searchers used by the fast search strategy is $\frac{1}{2}|V_{\text{odd}}(T)| + 2 = \frac{1}{2}|V_{\text{odd}}(H)| + 2$.

Theorem 7. *Given a connected graph G containing a cut-edge uv, let G_1 and G_2 be the two subgraphs with $u \in V(G_1)$ and $v \in V(G_2)$ after uv is deleted.* (i) *If there is an optimal fast search strategy of G_1 such that there is a searcher starting from u and there is an optimal fast search strategy of G_2 such that there is a searcher starting from v, then $fs(G) = fs(G_1) + fs(G_2) - 1$.* (ii) *If for any optimal fast search strategy of G_1 there is no searcher starting from u and for any optimal fast search strategy of G_2 there is no searcher starting from v, then $fs(G) = fs(G_1) + fs(G_2) + 1$.* (iii) *If there is an optimal fast search strategy of*

G_1 such that there is a searcher starting from u and for any optimal fast search strategy of G_2 there is no searcher starting from v, then $fs(G) = fs(G_1)+fs(G_2)$.

Corollary 1. *Given a connected graph G containing a cut-edge uv, let G_1 and G_2 be the two subgraphs with $u \in V(G_1)$ and $v \in V(G_2)$ after uv is deleted. If both $\deg_{G_1}(u)$ and $\deg_{G_2}(v)$ are odd, then $fs(G) = fs(G_1) + fs(G_2) - 1$.*

Theorem 8. *For any extended Halin graph H, the fast search number and an optimal fast search strategy of H can be computed in linear time.*

5 Cubic Graphs

In this section, we show how to compute the fast search number and an optimal fast search strategy for cubic graphs and their extension in quadratic time.

Algorithm. SEARCHCUBIC(G)
Input: A connected cubic graph $G = (V, E)$ with $|V| > 4$.
Output: A fast search strategy $\langle V_p, A_s \rangle$ of G, where V_p is a multiset of vertices on which we place searchers and A_s is a sequence of arcs corresponding to sliding actions, that is, an arc (u, v) corresponds to a sliding along the arc from tail u to head v.

1. If G has no cut-vertex, then pick two adjacent vertices b_1 and b_2, and let $L_0 \leftarrow \{b_1, b_2\}$, $\tau \leftarrow 2$ and P be the edge b_1b_2; otherwise, compute the block-cutpoint graph of G to find all the leaf blocks $B_1, \ldots, B_\tau$ and their cut-vertices $a_1, \ldots, a_\tau$, and then pick vertices $b_i \in V(B_i) \setminus \{a_i\}$, $1 \le i \le \tau$, and let $L_0 \leftarrow \{b_1, \ldots, b_\tau\}$ and P be a path in G between b_1 and b_2.
2. Totally order L_0 by letting $b_1 \prec b_2 \prec \cdots \prec b_\tau$. Initially, let $L \leftarrow L_0$, $X \leftarrow G$, $\mathcal{P} \leftarrow \emptyset$, and $i \leftarrow 1$.
3. While X contains edges, do the following:
   (a) Let $\mathcal{P} \leftarrow \mathcal{P} \cup \{P\}$. Write $P = p_1p_2 \ldots p_j$, where $p_1, p_j \in L$ satisfying $p_1 \prec p_j$. Totally order $V(P)$ by letting $p_1 \prec p_2 \prec \cdots \prec p_j$. Update L by adding all internal vertices of P, and by extending the order on L so that it is compatible with the order on $V(P)$. For example if $x \in L \setminus V(p)$ and $y \in V(p) \setminus \{p_1, p_j\}$ then we can let $x \prec y$ if $x \prec p_j$ and $y \prec x$ if $p_j \prec x$.
   (b) Let $X \leftarrow X - E(P)$. If $E(X) = \emptyset$, then go to Step 4; otherwise, set $i \leftarrow i + 1$.
   (c) If G has any cut-vertex, then go to Step 3d. If $i = 2$, then find a path P' in X between b_1 and b_2, otherwise, arbitrarily pick two vertices from L which are in the same component in X, and find a path P' in X between them. Let $P \leftarrow P'$ and go to Step 3a.
   (d) If $2 \le i \le \tau - 1$, then pick b_{i+1} and an arbitrary element of $L \setminus L_0$ which is in the same component in X, and find a path P' in X between them; if $\tau \le i \le 2\tau - 1$, then pick $b_{i-\tau+1}$ and an arbitrary element of $L \setminus L_0$ which is in the same component in X, and find a path P' in X between them; otherwise, arbitrarily pick two vertices from L which are in the same component in X, and find a path P' in X between them. Let $P \leftarrow P'$ and go to Step 3a.

4. Each edge uv of G is oriented to (u, v) if $u \prec v$. Let A_s be the sequence of all oriented edges in G listed in the left lexicographical order coming from order of the vertices (i.e., $(u, v) \prec (u', v')$ if $u \prec u'$, or $u = u'$ and $v \prec v'$). Let V_p be the multiset of the smallest vertex of each path in $\mathcal{P}$. Stop and output the fast search strategy $\langle V_p, A_s \rangle$.

The correctness of Algorithm SEARCHCUBIC(G) is shown with the following series of lemmas.

Lemma 4. *Any vertex in $L \setminus L_0$ has degree 0 or 1 in X.*

Lemma 5. *In Step 3b, any vertex incident on a removed edge must be in L. So $L_0 \cup \{u\colon deg_C(u) = 1\} \subset L$.*

Lemma 6. *If G is a block and $i \geq 2$, then after updating X in Step 3b every nonempty component of X has at least one vertex of degree 1. If G has more then one leaf block, then after updating X in each iteration of Step 3b, every nonempty component of X has at least one vertex of degree 1.*

Proof. Note that a nonempty component is a maximal connected subgraph with at least one edge. Consider a nonempty component C of X. Observe that if we can show that C has a vertex of degree 0 or 1 we are done. Observe there must be $v \in V(C)$, such that some edge e incident on v has been removed (in other words e is an edge of some path $P \in \mathcal{P}$) as otherwise C would be all of G. Then by Lemma 5, $v \in L$. If v is not a b_i then by Lemma 4, the degree of v in C is either 0 or 1 and we are done.

So assume $v = b_i$. If we are in the case of G being a block, then, since we have completed at least two iterations of Step 3b, at least two edges incident on b_i have been removed, and so $\deg_X(b_i) = 1$. This completes the proof when there is only one block, so for the rest of the proof we assume there is more then one leaf block in G. Let B_i be the leaf block of $v = b_i$. So an edge e incident on b_i has been removed. Let v' be the other endpoint of e. In Step 1, b_i is not incident on a cut-edge, so v' is on the same leaf block B_i of G as b_i, and so $v' \notin L_0$. By Lemma 5, $v \in L$. If $v' \in V(C)$, we are done since Lemma 4 would then imply that $\deg_C(v') = 0$ or 1.

So now assume that $v' \notin V(C)$. Since B_i has no cut edges, edges other than e must have been removed from B_i. At least one of these must be incident on a vertex $v'' \in V(C)$. If $v'' = b_i$ then at least two edges adjacent to b_i have been removed and so b_i has degree 0 or 1 in C. If $v'' \neq b_i$ then since $v'' \in B_i$ it is not b_j for $i \neq j$ as well. So $v'' \in L \setminus L_0$. Thus the degree of v'' in C is 0 or 1. Since C contains no vertices of degree 0, we are done proving the lemma.

Lemma 7. *Any cut-edge of any nonempty component C of any of the updated graphs X has on either side of it at least one vertex of $L_0 \cup \{u : \deg_C(u) = 1\}$.*

Proof. Consider a cut-edge e of X. If $X = G$, on either side of any cut-edge there is at least one vertex of L_0. So the statement is true for G. Let P be the path that was deleted from the old X to get the updated X. Assume the statement is true for the previous X. We want to show it still holds for the updated X. There are two cases:

Case 1. e is a cut-edge of the previous X. If P is in a different component of the old X from e then C is a component of the old X and so we can use the same vertices as we did for the old X. So assume that P is in the same component C'. Then there is $v \in V(P)$ on one side of e connected to e in the new X. Notice that by Lemma 4 $v \in V(P)$ implies that $v \in L_0 \cup \{u : \deg_C(u) = 1\}$. So we can use that vertex for the one side of e and the vertex from the old X for the other.

Case 2. $e = uu'$ is not a cut-edge of the previous X. Then there was a path Q in the old X between u and u' whose intersection with P is nonempty. So $Q \setminus P$ is a disjoint union of, possibly empty, paths in the new X. Endpoints of the components of the paths containing u and u' give $v, v' \in V(P)$ on different sides of e in the new X. Again by Lemma 4, these are the required points.

Lemma 8. *Step 3 of the algorithm terminates and* $\cup_{P\in\mathcal{P}} E(P) = E(G)$.

Proof. We can always perform the first iteration of Step 3. In case G has a single block then b_1 and b_2 are adjacent so in the first iteration of Step 3(b) we removed a single edge. So the updated X is still connected and there is a path from b_1 to b_2. So we can perform the second iteration of Step 3. For any other iterations if the present X is not empty, then by Lemma 6, each nonempty component of X has a vertex of degree 1. Since any vertex of degree 1 is incident on a cut-edge, we can apply Lemmas 7 and 5 to get that each nonempty component of X has at least two elements of L as vertices. This means we can proceed with Step 3 of the algorithm. Notice that if G has more then one leaf block, none of the paths from a vertex of degree 1 to a b_j can pass through another vertex of L_0, since once a path enters a leaf block it has no way to leave. So for the early steps we can take a b_j and any vertex of degree 1 in the same component of X. Since each step removes edges, after finitely many steps X will be empty. This will mean that $\cup_{P\in\mathcal{P}} E(P) = E(G)$ and also that we go to Step 4.

Lemma 9. *At Step 4 of Algorithm* SEARCHCUBIC*(G),* $|\mathcal{P}| \leq |V|/2 + \tau$.

Lemma 10. *At each iteration of Step 3, for any* $v \in (L \setminus \{b_1\}) \cap (\cup_{P\in\mathcal{P}} V(P))$*, there is* $v' \in L$ *such that* $v' \prec v$ *and an edge* $e \in \cup_{P\in\mathcal{P}} E(P)$*, between* v *and* v'.

Theorem 9. *Algorithm* SEARCHCUBIC*(G) describes a fast search strategy that clears G using at most* $|V|/2 + \tau$ *searchers.*

Proof. Notice that since by Lemma 8 the paths from $\mathcal{P}$ cover G and every edge of G is covered exactly one time, to clear G we need to show that anytime a searcher slides from vertex u to v along edge uv, either:

(1) all the edges incident on u except uv have been cleared, or
(2) another searcher remains on u.

Looking at Step 3 of the algorithm, we see that at least two paths have endpoints at b_1 and this implies that in fact three paths have endpoints at b_1. Since b_1 is the smallest, three searchers start there and move out. This means that (2) occurs when the first two searchers leave and (1) occurs when the last searcher leaves.

By Lemma 10, any vertex $v \neq b_1$ has an adjacent v' such that $v' < v$. Since we are searching lexicographically, this means that a searcher λ goes from v' to v before any searcher leaves v. If λ remains on v for the rest of the search, then any searcher that leaves v satisfies 2).

So assume that λ continues on from v. In this case there is another searcher λ' that either begins its search at v of that ends its search at v. If λ' ends its search at v, then it must come from a lower vertex. Thus it arrives at v before λ leaves and it follows that both (1) and (2) are satisfied when λ leaves clearing the third and final vertex incident on v. If λ' begins its search at v, then when the first of λ and λ' leaves v, (2) is satisfied and when the second leaves, thus clearing the final edge, (1) is satisfied.

So we have proven that any time a searcher leaves a vertex at least one of (1) and (2) is satisfied. Since every edge of G is traversed exactly once, this strategy clears G. If follows from Lemma 9 that this fast search strategy needs at most $|V|/2 + \tau$ searchers.

Theorem 10. *For a connected cubic graph $G = (V, E)$, let τ be the number of leaf blocks in the block-cutpoint graph of G if G contains at least one cut-vertex, and define $\tau = 2$ if G contains no cut-vertex. Then $fs(G) = |V|/2 + \tau$.*

Theorem 11. *For any connected cubic graph $G = (V, E)$, the fast search number and an optimal fast search strategy of G can be computed in $O(|V|^2)$ time.*

References

1. Borie, D., Parker, R., Tovey, C.: Algorithms for recognition of regular properties and decomposition of recursive graph families. Annals of Operations Research 33, 127–149 (1991)
2. Dyer, D., Yang, B., Yaşar, Ö.: On the fast searching problem. In: Fleischer, R., Xu, J. (eds.) AAIM 2008. LNCS, vol. 5034, pp. 143–154. Springer, Heidelberg (2008)
3. Halin, R.: Studies on minimally n–connected graphs. In: Welsh, D.J.A. (ed.) Combinatorial mathematics and its applications, pp. 129–136. Academic Press, New York (1971)
4. Kirousis, L., Papadimitriou, C.: Searching and pebbling. Theoretical Computer Science 47, 205–218 (1996)
5. Makedon, F., Papadimitriou, C., Sudborough, I.: Topological Bandwidth. SIAM Journal on Algebraic and Discrete Methods 6, 418–444 (1985)
6. Megiddo, N., Hakimi, S., Garey, M., Johnson, D., Papadimitriou, C.: The complexity of searching a graph. Journal of ACM 35, 18–44 (1988)
7. West, D.B.: Introduction to Graph Theory. Prentice-Hall, Englewood Cliffs (1996)

Approximation Algorithms for the Firefighter Problem: Cuts over Time and Submodularity

Elliot Anshelevich[1], Deeparnab Chakrabarty[2], Ameya Hate[1], and Chaitanya Swamy[2,⋆]

[1] Department of Computer Science, Rensselaer Polytechnic Institute
[2] Dept. of Combinatorics & Optimization, University of Waterloo

Abstract. We provide approximation algorithms for several variants of the FIREFIGHTER problem on general graphs. The Firefighter problem models the case where an infection or another diffusive process (such as an idea, a computer virus, or a fire) is spreading through a network, and our goal is to stop this infection by using targeted vaccinations. Specifically, we are allowed to vaccinate at most B nodes per time-step (for some budget B), with the goal of minimizing the effect of the infection. The difficulty of this problem comes from its temporal component, since we must choose nodes to vaccinate at every time-step while the infection is spreading through the network, leading to notions of "cuts over time".

We consider two versions of the Firefighter problem: a "non-spreading" model, where vaccinating a node means only that this node cannot be infected; and a "spreading" model where the vaccination itself is an infectious process, such as in the case where the infection is a harmful idea, and the vaccine to it is another infectious idea. We give complexity and approximation results for problems on both models.

1 Introduction

Faced with an epidemic that is spreading through a population, and a limited supply of vaccine (or simply a lack of time to administer it), it is necessary to decide whom to vaccinate. Questions about the spread of disease and epidemics in a social network have often been modeled using graph theory (e.g. [11, 3]), and correspond to fundamental graph-theoretic concepts [22]. Moreover, these graph theoretic principles can be applied to many diffusive network processes, including epidemics in computer networks, the spread of innovations and ideas, and viral marketing [23]. In this paper, we focus specifically on inhibiting the spread of an epidemic or an idea by using vaccination.

Model and the Firefighter problem. We model our network of agents as a directed[1] graph $G = (V, E)$ and a source node s. All nodes in the graph are in one

⋆ Research supported in part by NSERC grant 32760-06 and an Ontario ministry Early Researcher Award.

[1] We use a directed graph since it is more general – an undirected graph is just a directed graph with two arcs per edge.

Y. Dong, D.-Z. Du, and O. Ibarra (Eds.): ISAAC 2009, LNCS 5878, pp. 974–983, 2009.

of three states: they can be *infected*, *vaccinated*, or *vulnerable*, that is neither vaccinated nor infected. At time $\tau = 0$, all nodes are vulnerable, except node s, which is infected. At each $\tau > 0$, any vulnerable vertex v which is connected to an infected node u, such that $(u, v) \in E$, gets infected at time $\tau + 1$, unless it is vaccinated at time step τ. Infected and vaccinated nodes stay infected and vaccinated, respectively. We call a node *saved* if it is either vaccinated or if all paths from any infected node to it contains at least one vaccinated node.

Definition 1. *A vaccination strategy is a set $\Psi \subseteq V \times J$ where V is the set of vertices of graph G and $J = \{1, 2, \ldots |V|\}$. The vertex v is vaccinated at time $\tau \in J$ by the vaccination strategy Ψ if $(v, \tau) \in \Psi$. A vaccination strategy Ψ is valid with respect to budget B, if the following two conditions are satisfied:*

i. if $(v, \tau) \in \Psi$ then v is not infected at time τ,
ii. let $\Psi_\tau = \{(v, \tau) \in \Psi\}$; then $|\Psi_\tau| \leq B$ for $\tau = 1 \ldots |V|$.

The first condition implies we can only vaccinate vulnerable nodes, and the second requires that no more than B nodes are vaccinated at any time-step.

We consider two objectives in this paper. The first objective, which we call MAXSAVE, is to maximize the number of non-infected nodes at the end, given a fixed budget B. The second objective, which we call MINBUDGET, is to minimize the budget B needed per time instant in order to save a given set of nodes, $T \subseteq V$. They can be formally described as follows.

MAXSAVE(G, B, s, T)
INSTANCE: A rooted graph $(G(V, E), s)$, integer $B \geq 1$ and $T \subseteq V$
OBJECTIVE: Find a valid vaccination strategy Ψ such that if s is the only infected node at time 0, then at the end of the above process the number of non-infected nodes that belong to T is maximized.

This problem is also referred to as the FIREFIGHTER PROBLEM in the literature when $T = V$ [20, 16].

MINBUDGET(G, s, T)
INSTANCE: A rooted graph $(G(V, E), s)$, and $T \subseteq V$
OBJECTIVE: Find a valid vaccination strategy Ψ with minimum possible budget B, such that if s is the only infected node at time 0, then at the end of the above process all nodes in T are saved.

We also consider a variant of the above model, where the vaccination is also a process that spreads through the network. In the case of ideas propagating through a social network, this represents the fact that an antidote to a harmful idea is often another idea, which can be just as infectious. In disease propagation, this represents the fact that vaccines can be infectious as well, since they are often an attenuated version of the actual disease. In this *Spreading Vaccination Model*, if at time step $\tau > 0$ a node u is vaccinated and there is a vulnerable node v such that $(u, v) \in E$, then at time $\tau + 1$, the node v also gets vaccinated. Note that a vulnerable node may be adjacent to both an infected node and a vaccinated

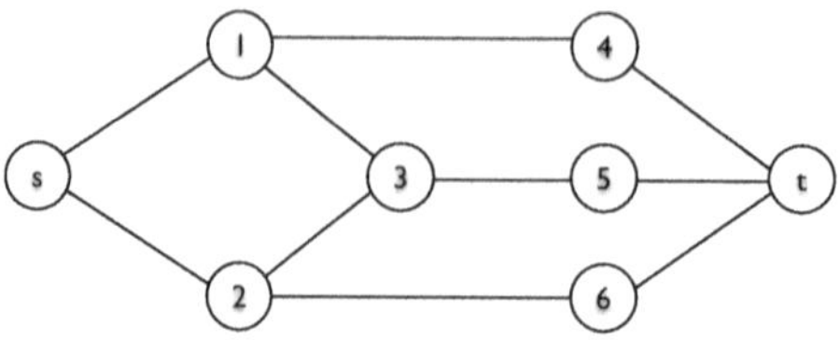

Fig. 1. This example shows that sometimes vaccinating nodes far away from the infection is the only way to save all the required nodes

node, in which case we assume that the vaccine prevails over the infection and the vulnerable node is vaccinated in the subsequent time step (assuming the opposite does not change the quality of our results). In the spreading model, we will say that a node is vaccinated *directly* when it is vaccinated by the vaccination strategy, and it is vaccinated *indirectly* when it is vaccinated by the spread of the vaccine through the network.

Example 1. To gain some intuition about this problem, consider the example shown in Figure 1 using the non-spreading model of vaccination.

Consider the MinBudget objective for this example. The infection begins at node s, and the goal is to find the smallest number B of nodes that need to be vaccinated at every time step so that we can save the node t, which we assume cannot itself be vaccinated. If we were only allowed to cut nodes during the first time-step, this would be equivalent to the minimum s-t node-cut problem. However, unlike previous works, such as [22], which examine the *static* problem of vaccinating a 'cut' before the infection has started spreading, we need to find the "best" *cut over time* (where best depends on the considered objective). This temporal nature of the problem, complicates matters: intuitively, the tradeoff is between vaccinating a small set of nodes close to the infection source early, or spreading out (over time) the vaccination of a larger set of nodes which are farther away from the source.

For instance, in the above example, a minimum s-t node-cut is $\{1, 2\}$, which requires $B = 2$. However, there *is* a solution to the above problem with $B = 1$, but the final set of vaccinated nodes does not form a minimum s-t node-cut. One such solution is to vaccinate vertices 4, 6, and 5 at time steps 1, 2, and 3 respectively, leading to the final set of vaccinated nodes being $\{4, 5, 6\}$ which is not a minimum cut. In fact, it is not hard to come up with examples where the optimal value of B is much smaller than the size of a minimum node s-t cut *and* the final set of vaccinated nodes is much larger than the size of a minimum node s-t cut (e.g., take a graph where s has k neighbors, each of which is connected to t via k long internally node-disjoint paths). Thus, this "cuts over time" problem is quite different from the classical min-cut problem, and in fact is known to be NP-hard (even when the graph is a tree!) [15].

Our Results. In Section 2, we consider the model of spreading vaccinations. In general, our results show that this model is more tractable than the model with non-spreading vaccinations. For MaxSave we show that this problem reduces to maximizing a submodular function with a matroid constraint. Therefore a simple greedy algorithm provides a 2-approximation, and a recent result of [6] lets us prove a $(1 - 1/e)$-factor approximation. For MinBudget we give a $O(\log n)$ approximation algorithm, and show that this approximation ratio is tight, by showing a set-cover hardness.

The non-spreading model, on the other hand, does not yield itself to good approximation algorithms. In fact, we show in Section 3 that it is NP-hard to approximate MaxSave in general graphs by a factor of n^{α}, for any $\alpha < 1$. For MinBudget, we give a $O(\sqrt{n})$ factor approximation algorithm for general graphs based on a natural LP relaxation for the problem. For *directed* layered graphs with ℓ layers, we give a $O(H_{\ell}) = O(\log \ell)$ approximation algorithm. The latter algorithm is combinatorial and requires just one max-flow computation. We show that this result is tight by proving that the integrality gap of the LP is $\Omega(\log \ell)$ for ℓ-layered directed graphs.

Section 3.1 is devoted to vaccination strategies when the underlying graph is a tree. This special case has received a lot of attention [21,26], is computationally difficult [15], and is in fact a generalization of a complex scheduling problem (details in the full version [1]). For this special case we show that both the spreading and the non-spreading models are equivalent, so the stronger results from Section 2 hold for the non-spreading model as well. In addition, our algorithm for layered graphs also implies a $O(\log h)$ approximation algorithm for MinBudget on trees with height h. Note that this is stronger than the $O(\log n)$ algorithm we have for general graphs in the spreading model.

Related Work. Questions about epidemic propagation have been studied in several fields, (e.g., [4,29]), although most of this research models the epidemic as a dynamic system and ignores the effect of the network structure. Recently, a few groups have considered the spread of viruses or ideas on Internet-like topologies, such as small-world networks [32] and preferential attachment models [5, 25]. The papers [10, 13] study targeted vaccinations in this context, and show that they can be used to significantly reduce the effect of an epidemic. These studies assume certain properties of the networks (based on where these networks arise from).

Various recent papers consider modeling vaccination by using graph cuts. For example, the work of Hayrapetyan et al. [22] and others [3, 12] fully utilizes the social-network structure to "cut off" and contain various diffusive processes in a social network. As mentioned earlier, all this work is only concerned with vaccinating a set of nodes before the infection begins, however, and does not have the temporal component of the Firefighter problem. A lot more work has been done on maximizing the spread of an infection (instead of trying to stop it using vaccinations), by selecting the best nodes to infect initially [11,23].

The *Firefighter problem* was first introduced by B. Hartnell [20], and there has been much work on this problem; see, e.g., [16] for a survey. However, much of the work has focused on special graph structures, such as grids [9, 18, 31], and that too usually with the MaxSave objective. The Firefighter problem is

NP-complete even when the underlying graph is a tree [15], although [21] and [26] give approximation algorithms for this case, and [28] shows how to solve the problem in in polynomial time for special cases of trees.

We have recently learnt that, independent of our work, Chalermsook and Chuzhoy [7] obtain some similar results for the non-spreading version of the MinBudget problem. In particular, they also give an $O(\log \ell)$-approximation for ℓ-layered directed graphs, and obtain an improved approximation for trees.

2 Spreading Vaccination Model

We first show a few simple hardness results about this model, and then give approximation algorithms for both our objectives. Due to lack of space, detailed proofs appear in the full version [1] of the paper.

2.1 General Properties

We make certain useful observations about this model. Let $N(v, i)$ be the set of all the nodes that are a distance of at most i from v.

Lemma 1. *At time τ, all nodes in the neighborhood $N(s, \tau)$ will either be vaccinated or infected.*

Now, since all the nodes in the neighborhood $N(s, \tau)$ will be either infected or vaccinated by time τ, any optimal vaccination strategy would not vaccinate any node in this neighborhood at time $> \tau$. Since any valid strategy can vaccinate only B nodes at any time-step, it means that an optimal strategy would vaccinate at most $B \cdot \tau$ nodes directly in the neighborhood $N(s, \tau)$.

We define a set $\Gamma(v)$ for every node $v \in V$ by

$$\Gamma(v) = \{(u, \tau) | u \in V \text{ and } 0 < \tau \leq (d(s, v) - d(u, v))\}$$

The tuple (v, τ) essentially represents the event of vaccinating node v at time τ.

Theorem 1. *A node $v \in V$ is vaccinated by the vaccination strategy Ψ iff $\Psi \cap \Gamma(v) \neq \emptyset$.*

This theorem tells us that vaccinating an element of $\Gamma(v)$ is exactly what is needed to save a node v, and this provides insight into the structure of the problem.

2.2 Approximation for MaxSave

As we explain in the full version of the paper, the MaxSave problem can be modeled as a problem of maximizing a submodular set function on a collection of sets that form a partition matroid. On the basis of this knowledge, techniques like the greedy algorithm [17] can be used to obtain a $\frac{1}{2}$ approximation for MaxSave, while the randomized algorithm of [6] can be used to obtain a $(1-1/e)$ approximation.

Theorem 2. *There is a randomized algorithm which gives with high probability a $(1-1/e)$-approximation for the* MaxSave *problem. Additionally, a simple greedy algorithm gives a $\frac{1}{2}$-approximation.*

The gist of the proof is as follows. A partition matroid consists of disjoint sets $E_1, \ldots, E_k$, and a set S is called independent if $|S \cap E_i| \leq \ell_i$, for some given numbers $\ell_1, \ldots, \ell_k$. We argue that one can actually consider vaccination strategies that satisfy only property (ii) in Definition 1. This set of strategies forms a partition matroid, since we can only choose at most B nodes at every time-step to vaccinate (so $\ell_i = B$ for all i). We next show that the function $f(\Psi)$ defined (suitably) as the number of nodes saved by using the (possibly invalid) vaccination strategy Ψ is submodular, by using Theorem 1, and if Ψ satisfies the budget-constraint then there is a valid vaccination strategy Ψ' such that $f(\Psi) = f(\Psi')$. Thus, we can use the results of [6, 17] on maximizing a submodular function subject to a matroid constraint to obtain the desired approximations.

We note that a $(1-1/e)$-approximation can also be obtained by applying a randomized rounding technique similar to [26] to a modified version of the MaxSave problem. We believe, however, that modeling the problem using partition matroids and submodular functions is fruitful since it provides further insight into the problem, and also yields a rather simple and efficient, combinatorial, deterministic, $\frac{1}{2}$-approximation algorithm.

2.3 Approximation for MinBudget

Consider an instance of MinBudget. First, suppose that we know the optimal budget B that is needed in order to save all nodes of T. Below we give an algorithm that saves all nodes in T using a budget of at most $B \log n$.

By slight adjustments to the proof of Theorem 2, we know that by running the greedy algorithm with budget B, we save at least half of the nodes in T. The greedy algorithm in this case chooses the nodes to vaccinate in each time-step one at a time, always picking the node that saves the most nodes of T. For this purpose the greedy algorithm needs to know exactly which nodes will be saved if a node u is vaccinated at time τ, which we can compute in poly-time. Once finished with the first time-step, the algorithm goes on to the second, and so on.

The complete algorithm is as follows. Repeat the following steps $\log n$ times:

- vaccinate nodes in graph G using the greedy algorithm with budget B.
- Construct graph G_1 from G by removing all the vertices that were vaccinated directly and indirectly in the previous step. Let T_1 be the nodes of T that are in G_1.
- Set $G = G_1$ and $T = T_1$.

It is clear that the new graph G_1 will always contain the original source node s as it is never vaccinated by the greedy algorithm. The $\log n$ applications of the greedy algorithm yield an algorithm that vaccinates $B \log n$ nodes at each step, since at each step, we can simply combine the (at most) B nodes that the greedy algorithm vaccinates at that time step in the $\log n$ runs. We call the resulting algorithm, the *RepGreedy* algorithm.

Theorem 3. *The* RepGreedy *algorithm saves all nodes of T by vaccinating at most $B \log n$ nodes per time-step.*

Finally, to obtain an $O(\log n)$-approximation algorithm without knowing B, we simply do a binary search on B, and run *RepGreedy* for every choice of B.

We complement the above result with the following inapproximability result.

Theorem 4. *The* MinBudget *problem is as hard as set cover, and hence, cannot be approximated in poly-time to a factor better than* $\log n$ *unless P=NP.*

3 Non-spreading Vaccination Model

The non-spreading model is considerably more difficult than the spreading model. One of the main reasons is that Lemma 1 (or any simple modification of it) is no longer true. The MaxSave problem is NP-complete for bipartite graphs [28] and for cubic graphs (3-regular) [24]. The MaxSave problem is NP-complete even when restricted to trees with maximum degree three [15]. We prove the following about the inapproximability of MaxSave

Theorem 5. *The* MaxSave*(G(V,E),s,B,T) problem cannot be approximated in poly-time to the factor of* n^α *where* $n = |V|$ *and* $\alpha < 1$, *unless P=NP.*

We introduce an auxiliary problem, Save-t, which asks whether a specified node t can be saved by vaccinating one node (other than t) at a time. The NP-completeness of this problem follows from known NP-completeness proofs. We then give a gap introducing reduction from the Save-t problem to the MaxSave problem such that if there exists any n^α approximation for the MaxSave problem then we can solve the Save-t problem in polynomial time. The proof appears in full version.

In the remainder of the section, we focus on the MinBudget problem. Note that we need to save *all* the nodes in a set T with the minimum number of vaccinations required per time instant. To simplify notation, we consider the following equivalent problem: we add a new node t with edges from all nodes in T to t, and consider the problem of saving t with minimum budget *under the additional constraint that t itself cannot be vaccinated.* We call s the source and t the sink. Let $\mathcal{P}$ denote the collection of all s-t paths.

$$\begin{aligned} &\text{Minimize} \quad B && \text{(Primal)} \\ &\text{s.t.} \quad \sum_{v \in V} x_v^\tau \le B && \forall \tau = 1, \ldots, n \\ &\sum_{i=1}^{k} \sum_{\tau=1}^{i} x_{v_i}^\tau \ge 1 && \forall (s, v_1, \cdots, v_k, t) \in \mathcal{P} \\ &x \ge 0. \end{aligned}$$

$$\begin{aligned} &\text{Maximize} \quad \sum_{P \in \mathcal{P}} f_P && \text{(Dual)} \\ &\text{s.t.} \quad \sum_{\tau=1}^{n} z_\tau \le 1 \\ &\sum_{P \in \mathcal{P} : v \in P^{(\tau)}} f_P \le z_\tau && \forall v \in V, \tau = 1, \ldots, n \\ &z, f \ge 0. \end{aligned}$$

(Primal) gives the LP relaxation of the problem, and (Dual) is the dual of this LP. The primal LP has a variable x_v^τ that indicates whether vertex v is vaccinated at time τ or not. We consider τ going up to n since it is clear that there is no need to vaccinate any vertex after time n. The first constraint bounds the number of vaccinations at every time step. The second constraint says that for every path $(s, v_1, \cdots, v_k, t)$ to the sink t, one of the nodes, say v_i, must be vaccinated *by* time i. This is a necessary and sufficient condition for this path not to transmit the infection to t. In the dual, we have a flow-variable f_P for every s-t path P. The second constraint in the dual is a bit subtle: it says that for every τ, the total flow through a vertex v via paths such that v lies at a distance τ or more from s on the path, is at most z_τ. We use $P^{(\tau)}$ to denote the portion of the path from the τth vertex to t; that is, if $P = (s, v_1, \ldots, v_\tau, \ldots, v_k, t)$, then $P^{(\tau)} = (v_\tau, \ldots, v_k, t)$.

Although the primal LP above has exponentially many constraints, it can be solved in polynomial time since one can give an efficient separation oracle for the LP. Strictly speaking the LP (Primal) may have an integrality gap of $n = |V|$. However note that if OPT denotes the optimal value of (Primal), then in fact $\lceil OPT \rceil$ is a lower bound on the minimum budget, and by comparing the budget of our solution against this lower bound, we prove the following theorems. Similarly, when we say "integrality gap" below, we mean the worst-case ratio of the (integer) optimum budget and $\lceil OPT \rceil$.

Theorem 6. *In the non-spreading model, there is a* $2\sqrt{n}$*-approximation for the* MINBUDGET *problem in general graphs.*

(Proof Sketch) At a high level, the algorithm recognizes the set of vertices to be vaccinated by time i by looking at the fraction vaccinated by time i. If this fraction is larger than $1/\sqrt{n}$, then the node is vaccinated by day i. We can then show that in the remaining graph, infection can reach t only using paths of length longer than $\sqrt{n}$, and thus there is a cut of size $\sqrt{n}$ which separates s and t. Thus vaccinating this cut as well completes the algorithm. The analysis is slightly subtle and is deferred to the full version.

We now present an improved approximation algorithm for layered graphs. An s-t *directed layered graph* with ℓ layers is one where (i) s has only outgoing edges, t has only incoming edges; (ii) all nodes except t can be partitioned into sets $L_0 := \{s\}, L_1, L_2, \ldots, L_\ell$ such that for every node $v \in L_{i+1}$ (so $v \neq t$) and every incoming edge (u, v) of v, we have $u \in L_i$. Note that now we only need to consider $\tau = 1, \ldots, \ell$ in (Primal) and (Dual). Let $H_r = 1 + 1/2 + \cdots + 1/r$.

Theorem 7. *There is an* H_ℓ*-approximation for the* MINBUDGET *problem in* s-t *directed* ℓ*-layered graphs. Furthermore, there are* ℓ*-layered instances showing that the integrality gap of* (Primal) *is at least* $H_\ell = \Omega(\log \ell)$.

(Proof Sketch) The algorithm sets capacity $1/iH_\ell$ on each vertex of layer i, for all i, and simply computes a minimum s-t vertex cut. It then divides the cut into ℓ pieces, corresponding to the vertices vaccinated on day i. Using the dual LP, we can show that our solution is within H_ℓ of the LP optimum. The integrality gap example is a similar layered graph. We defer the details to the full version.

3.1 Vaccination on Trees

When G is a tree rooted at s, the following observation establishes the equivalence between the spreading model and non-spreading model. For the spreading model on general graphs we defined a function $\Gamma(v)$ as a set of all tuples (u, τ) such that if u is vaccinated directly at time τ then the node v will be saved. For a tree, it is easy to observe that a node v will be saved if any of its ancestors is vaccinated directly before the infection reaches v. Therefore, the optimal strategy will be the same on a given tree irrespective of the vaccination model being spreading or non-spreading. This implies that all the positive results from Section 2 also hold for trees. Since the MinBudget problem on trees with height h yields an instance of MinBudget on an s-t directed graph with h layers, we immediately obtain the following Corollary of Theorem 7.

Corollary 1. *There is an $O(\log h)$-approximation for* MinBudget *on trees, where the set T is the set of leaves and h is the height of the tree.*

Acknowledgments. We thank David Kempe, Sanjeev Khanna, Jon Kleinberg and, Malik Magdon-Ismail for interesting and productive discussions on the Firefighter problem.

References

1. Full version can be found at `www.cs.rpi.edu/~eanshel/pubs.html`
2. Ahuja, R., Magnanti, T., Orlin, J.: Network Flows: Theory, Algorithms, and Applications. Prentice-Hall, Englewood Cliffs (1993)
3. Aspnes, J., Chang, K., Yamposlkiy, A.: Inoculation strategies for victims of viruses and the sum-of-squares partition problem. In: Proc. 16th ACM SODA (2005)
4. Bailey, N.: The Mathematical Theory of Infectious Diseases and its Applications. Hafner Press (1975)
5. Barabasi, A., Albert, R., Jeong, H.: Mean-field theory for scale-free random networks. Physica A 272 (1999)
6. Calinescu, G., Chekuri, C., Pal, M., Vondrak, J.: Maximizing a Monotone Submodular Function subject to a Matroid Constraint. In: Proc. 12th IPCO (2007)
7. Chalermsook, P., Chuzhoy, J.: Resource Minimization for Fire Containment. To appear in Proc. ACM SODA (2010)
8. Crosby, S., Finbow, A., Hartnell, B., Moussi, R., Patterson, K., Wattar, D.: Designing Fire Resistant Graphs. Congr. Numerantium 173 (2005)
9. Develin, M., Hartke, S.G.: Fire Containment in grids of dimension three or higher. Discrete Applied Mathematics 155(17) (2007)
10. Dezső, Z., Barabási, A.: Halting viruses in scale-free networks. Physical Review E 65 (2002)
11. Dreyer Jr., P.A., Roberts, F.S.: Irreversible k-threshold processes: Graph-theoretical threshold models of the spread of disease and of opinion. Discrete Applied Mathematics 157(7) (2009)
12. Engelberg, R., Könemann, J., Leonardi, S., Naor, J(S.): Cut problems in graphs with a budget constraint. In: Correa, J.R., Hevia, A., Kiwi, M. (eds.) LATIN 2006. LNCS, vol. 3887, pp. 435–446. Springer, Heidelberg (2006)

13. Eubank, S., Kumar, V., Marathe, M., Srinivasan, A., Wang, N.: Structural and algorithmic aspects of massive social networks. In: Proc. 15th ACM SODA (2004)
14. Feige, U.: A threshold of ln n for approximating set cover. J. ACM 45 (1998)
15. Finbow, S., King, A.D., MacGillivray, G., Rizzi, R.: The Fire fighter problem on graphs of maximum degree three. Discrete Mathematics 307 (2007)
16. Finbow, S., MacGillivray, G.: The Firefighter Problem: A survey of results, directions and questions (Manuscript) (2007)
17. Fisher, M.L., Nemhauser, G.L., Wolsey, L.A.: An analysis of approximations for maximizing submodular set functions - II. Math. Prog. Study 8 (1978)
18. Fogarty, P.: Catching Fire on Grids, M.Sc. Thesis, Department of Mathematics, University of Vermont (2003)
19. Giakkoupis, G., Gionis, A., Terzi, E., Tsaparas, P.: Models and algorithms for network immunization. Technical Report C-2005-75, Department of Computer Science, University of Helsinki (2005)
20. Hartnell, B.L.: Firefighter! An application of domination. Presentation. In: 25th Manitoba Conference on Combinatorial Mathematics and Computing, University of Manitoba in Winnipeg, Canada (1995)
21. Hartnell, B., Li, Q.: Firefighting on trees: How bad is the greedy algorithm? Congr. Numer. 145 (2000)
22. Hayrapetyan, A., Kempe, D., Pál, M., Svitkina, Z.: Unbalanced graph cuts. In: Brodal, G.S., Leonardi, S. (eds.) ESA 2005. LNCS, vol. 3669, pp. 191–202. Springer, Heidelberg (2005)
23. Kempe, D., Kleinberg, J.M., Tardos, É.: Influential nodes in a diffusion model for social networks. In: Caires, L., Italiano, G.F., Monteiro, L., Palamidessi, C., Yung, M. (eds.) ICALP 2005. LNCS, vol. 3580, pp. 1127–1138. Springer, Heidelberg (2005)
24. King, A., MacGillivray, G.: The Firefighter Problem For Cubic Graphs. Discrete Mathematics 307 (2007)
25. Kumar, R., Raghavan, P., Rajagopalan, S., Sivakumar, D., Tomkins, A., Upfal, E.: Stochastic models for the web graph. In: Proc. 41st IEEE FOCS (2000)
26. Leizhen, C., Verbin, E., Yang, L.: Firefighting on trees $(1 - 1/e)$–approximation, fixed parameter tractability and a subexponential algorithm. In: Hong, S.-H., Nagamochi, H., Fukunaga, T. (eds.) ISAAC 2008. LNCS, vol. 5369, pp. 258–269. Springer, Heidelberg (2008)
27. Leizhen, C., Weifan, W.: The Surviving Rate of a Graph. To appear in SIAM Journal on Discrete Mathematics (2009)
28. MacGillivray, G., Wang, P.: On The Firefighter Problem. JCMCC, 47 (2003)
29. Nowak, M., May, R.: Virus Dynamics: Mathematical Principles of Immunology and Virology. Oxford University Press, Oxford (2000)
30. Raz, R., Safra, S.: A sub-constant error-probability low-degree test, and sub-constant error-probability PCP characterization of NP. In: Proc. 29th ACM STOC (1997)
31. Wang, P., Moeller, S.: Fire Control on graphs. J. Combin. Math. Combin. Comput. 41 (2002)
32. Watts, D., Strogatz, S.: Collective dynamics of 'small-world' networks. Nature 393 (1998)

Constant-Factor Approximations of Branch-Decomposition and Largest Grid Minor of Planar Graphs in $O(n^{1+\epsilon})$ Time

Qian-Ping Gu[1] and Hisao Tamaki[2]

[1] School of Computing Science, Simon Fraser University, Burnaby BC Canada
qgu@cs.sfu.ca
[2] Department of Computer Science, Meiji University, Kawasaki Japan
tamaki@cs.meiji.ac.jp

Abstract. We give constant-factor approximation algorithms for computing the optimal branch-decompositions and largest grid minors of planar graphs. For a planar graph G with n vertices, let $\text{bw}(G)$ be the branchwidth of G and $\text{gm}(G)$ the largest integer g such that G has a $g \times g$ grid as a minor. Let $c \geq 1$ be a fixed integer and α, β be arbitrary constants satisfying $\alpha > c + 1.5$ and $\beta > 2c + 1.5$. We give an algorithm which constructs in $O(n^{1+\frac{1}{c}} \log n)$ time a branch-decomposition of G with width at most $\alpha\, \text{bw}(G)$. We also give an algorithm which constructs a $g \times g$ grid minor of G with $g \geq \frac{\text{gm}(G)}{\beta}$ in $O(n^{1+\frac{1}{c}} \log n)$ time. The constants hidden in the Big-Oh notations are proportional to $\frac{c}{\alpha-(c+1.5)}$ and $\frac{c}{\beta-(2c+1.5)}$, respectively.

Keywords: Graph algorithms, branch-decompositions, graph minors.

1 Introduction

The notions of branchwidth and branch-decompositions are introduced by Robertson and Seymour [19] in relation to the more celebrated notions of treewidth and tree-decompositions [17,18] in the graph minor theory. Grid minors also play an important role in the graph minor theory. All those notions have important algorithmic applications. A graph of small treewidth/branchwidth admits efficient dynamic programming algorithms for a vast class of problems on the graph [2,5]. A tree-/branch-decomposition based dynamic programming algorithm usually runs in exponential time in the width of the tree-/branch-decomposition. Grid minors are fundamental in many algorithms studied in the algorithmic graph minor theory and bidimensionality theory [8,9,10,11]. The ratio of the treewidth or branchwidth of a graph over the largest size of the grid minor of the graph typically appears in the exponent of the running time of those algorithms.

For an arbitrary graph G, the treewidth $\text{tw}(G)$ of G and the branchwidth $\text{bw}(G)$ of G are linearly related by inequalities $\text{bw}(G) \leq \text{tw}(G) + 1 \leq \lfloor \frac{3\text{bw}(G)}{2} \rfloor$ and there are simple translations between tree- and branch-decompositions that prove these inequalities [19]. For general graphs, the problem of deciding if a given

Y. Dong, D.-Z. Du, and O. Ibarra (Eds.): ISAAC 2009, LNCS 5878, pp. 984–993, 2009.

graph has treewidth smaller than k is NP-complete, if k is part of the input [1]. If k is upper-bounded by a constant, then both the decision problem and the optimal decomposition problem can be solved in linear time [6], although the dependency of the time on k is huge. There are exact parallels for branchwidth and branch-decompositions to these results: NP-completeness [22] and a linear-time algorithm for fixed k [7]. For an important subclass of graphs, planar graphs, the decision problem for branchwidth can be solved in $O(n^2)$ time by the well-known rat-catching algorithm of Seymour and Thomas [22] and optimal branch-decompositions can be constructed in $O(n^3)$ time [14,22], while it is not known whether the problem of deciding the treewidth of a planar graph is polynomially solvable or NP-complete (the problem is widely believed NP-hard though).

Approximation algorithms for computing the width and minimum-width decompositions have also been extensively studied (see recent work [3,4] for literature). Because of the relationship between treewidth/decomposition and branchwidth/decomposition stated above, the approximation problems for these two types of width/decomposition are almost equivalent. For general graphs, the best known approximation factor achievable in polynomial time is $O(\sqrt{\log k})$ [12], where k is the optimal width, and constant-factor approximation algorithms take time exponential in the optimal width [3,20]. For planar graphs, the best known approximation result for treewidth is the obvious $O(n^2 \log n)$ time 1.5-approximation algorithm, which uses the rat-catching algorithm of Seymour and Thomas and a binary search. Tree-decompositions take $O(n^3)$ time for 1.5-approximation with a similar approach. Bodlaender, Grigoriev, and Koster give another constant-factor approximation algorithm for treewidth of planar graphs that runs in $O(n^2 \log n)$ time but uses less memory [4].

Computing large grid minors of planar graphs is a key ingredient in the algorithmic graph minor theory and bidimensionality theory [8,9,10,11]. It is not known whether a largest grid minor can be computed in polynomial time for planar graphs. For a graph G, let $\mathrm{gm}(G)$ denote the largest size of a grid minor of G, that is, the largest integer g such that G has a $g \times g$ grid minor. From the definition of branchwidth (see Section 2), $\mathrm{gm}(G) \leq \mathrm{bw}(G)$. The branchwidth $\mathrm{bw}(G)$ is also upper-bounded by some function of $\mathrm{gm}(G)$. For general graphs, the known upper bound is $\mathrm{bw}(G) \leq 20^{2(\mathrm{gm}(G))^5}$, while for planar graphs a linear bound $\mathrm{bw}(G) \leq 4\mathrm{gm}(G)$ is known [21]. This linear bound gives an algorithm which finds a $g \times g$ grid minor with $g \geq \frac{\mathrm{gm}(G)}{4}$ for planar graphs[1]. An $O(n^2 \log n)$ time algorithm which gives the same bound of $g \geq \frac{\mathrm{gm}(G)}{4}$ is also known [4]. The inequalities $\mathrm{tw}(G) \leq \lfloor \frac{3\mathrm{bw}(G)}{2} \rfloor$ and $\mathrm{bw}(G) \leq 4\mathrm{gm}(G)$ give a linear bound $\mathrm{tw}(G) \leq 6\mathrm{gm}(G)$ for planar graphs. This bound is improved to $\mathrm{tw}(G) \leq 5\mathrm{gm}(G)$ [13,24]. These bounds $\mathrm{bw}(G) \leq 4\mathrm{gm}(G)$ and $\mathrm{tw}(G) \leq 5\mathrm{gm}(G)$ for planar graphs has been exploited in many algorithms developed under the bidimensionality theory, which work on a large grid minor if they find one and otherwise use a

---

[1] An $O(n^2 \log n)$ time implementation can be realized by using the rat-catching algorithm of Seymour and Thomas [22] to construct an oracle for the tangles which are required in constructing the grid minor.

tree/branch-decomposition of small width. In those applications, improving the coefficient in the bound is important, as it appears in the exponent of the running time of the algorithms. Recently, the present authors have improved the bound showing that $\mathrm{bw}(G) \leq 3\mathrm{gm}(G) + 2$ for planar graph G [15]. From the relation $\mathrm{tw}(G) \leq \lfloor \frac{3\mathrm{bw}(G)}{2} \rfloor$, this gives $\mathrm{tw}(G) \leq 4.5\mathrm{gm}(G)+2$ for planar graphs, improving the previous $\mathrm{tw}(G) \leq 5\mathrm{gm}(G)$ results of [13,24]. The algorithm implied by their proof, given a planar graph G on n vertices and a positive integer k, runs in $O(n^2)$ time and either finds a $k \times k$ grid minor or certifies that $\mathrm{bw}(G) \leq 3k - 1$. This algorithm yields a 3-approximation algorithm for finding the largest grid minor of planar graphs which is the best known in terms of the approximation ratio but not faster than other known constant-factor approximation algorithms. For branchwidth, the result may not have significant algorithmic consequences, because of the rat-catching algorithm of Seymour and Thomas mentioned earlier.

The purpose of this paper is to build on the ideas in [15] to develop faster constant-factor algorithms both for branch-decompositions and for finding large grid minors of planar graphs. Our results are the fastest known constant-factor approximation algorithms for both of the problems. To gain in speed, we sacrifice the approximation ratio. In fact, our algorithms are parameterized and provide a trade-off between the running time and the approximation ratio. Our algorithm for grid minors actually finds a more general cylinder minor from which a grid minor can be straightforwardly derived. A $k \times k'$ cylinder is a cartesian product of a cycle on k vertices and a path on h vertices. Our main results are expressed in the following theorem.

Theorem 1. *Let $c \geq 1$ be a fixed integer, $\delta > 0$ be a constant, and $\lambda = \frac{1}{2}$ or 1. There is an algorithm which, given a planar graph G with n vertices and an integer k, in $O(n^{1+\frac{1}{c}})$ time constructs either a branch-decomposition of G with width at most $(2\lambda(c+1) + \frac{1}{2} + \delta)k$ or a $k \times \lceil \lambda k \rceil$ cylinder minor of G, where the constant hidden in the Big-Oh notation is proportional to $\frac{c}{\delta}$.*

Because a $k \times \lceil \lambda k \rceil$ cylinder has branchwidth $\min\{2\lceil \lambda k \rceil, k\}$ [15], Theorem 1 with $\lambda = \frac{1}{2}$, together with a binary search, implies the following result.

Theorem 2. *Let $c \geq 1$ be a fixed integer, $\delta > 0$ be an arbitrary constant, and $\alpha = \delta + c + 1.5$. Given a planar graph G with n vertices, we can in $O(n^{1+\frac{1}{c}} \log n)$ time construct a branch-decomposition of G with width at most $\alpha\, \mathrm{bw}(G)$.*

Theorem 2 can be readily extended to planar hypergraphs. This theorem naturally implies an $O(n^{1+\epsilon})$ time constant-factor approximation algorithm for tree-decompositions of planar graphs, with an additional multiplicative factor of 1.5. Since a $k \times \lceil \lambda k \rceil$ cylinder has a $k \times \lceil \lambda k \rceil$ grid minor, taking $\lambda = 1$, the following result can be obtained from Theorem 1 and a binary search.

Theorem 3. *Let $c \geq 1$ be a fixed integer, $\delta > 0$ be an arbitrary constant, and $\beta = \delta + 2c + 1.5$. Given a planar graph G with n vertices, we can in $O(n^{1+\frac{1}{c}} \log n)$ time construct a $g \times g$ grid minor of G with $g \geq \frac{\mathrm{bw}(G)}{\beta}$.*

The next section gives preliminaries of the paper. We describe the basic approach for our algorithm in Section 3. We prove Theorem 1 in Section 4.

2 Preliminaries

A *hypergraph* G consists of a set $V(G)$ of vertices and a set $E(G)$ of edges, where each edge e of $E(G)$ is a subset of $V(G)$ with at least two elements. For a set $E \subseteq E(G)$ of edges let $V(E)$ denote $\cup_{e \in E} e$. A hypergraph G is a graph if $|e| = 2$ for every edge $e \in E(G)$. We say a vertex v and an edge e are incident to each other if $v \in e$. We say that two edges e_1 and e_2 are incident to each other if $e_1 \cap e_2 \neq \emptyset$. A hypergraph H is a subgraph of G if $V(H) \subseteq V(G)$ and $E(H) \subseteq E(G)$. A sub-hypergraph H of G is *induced* by $V \subseteq V(G)$ if $V(H) = V$ and $E(H) = \{e \in E(G) | V_G(e) \subseteq V\}$. A subgraph H of G is *induced* by $E \subseteq E(G)$ if $E(H) = E$ and $V(H) = V(E)$. We denote by $G[E]$ the subgraph induced by E.

For a hypergraph G and a subset $A \subseteq E(G)$ of edges, we denote $E(G) \setminus A$ by $\overline{A}$ when G is clear from the context. A *separation* of hypergraph G is an ordered pair $(A, \overline{A})$ of subsets of $E(G)$. For each $A \subseteq E(G)$, we denote by $\partial(A)$ the vertex set $V(A) \cap V(\overline{A})$. The *order* of separation $(A, \overline{A})$ is $|\partial(A)| = |\partial(\overline{A})|$.

The notions of branchwidth and branch-decomposition are introduced by Robertson and Seymour [19]. A *branch-decomposition* of hypergraph G is a pair (ϕ, T) where T is a tree each internal node of which has degree 3 and ϕ is a bijection from the set of leaves of T to $E(G)$. Consider an edge e' of T and let L_1 and L_2 denote the sets of leaves of T in the two respective subtrees of T obtained by removing e'. We say that the separation (L_1, L_2) is induced by this edge e' of T. We define the width of the branch-decomposition (ϕ, T) to be the largest order of the separations induced by edges of T. The *branchwidth* of G, denoted by $\text{bw}(G)$, is the minimum width of all branch-decompositions of G. In the rest of this paper, we identify a branch-decomposition (ϕ, T) with the tree T, leaving the bijection implicit and regarding each leaf of T as an edge of G.

The contraction of an edge e in a hypergraph G is to remove e from G, identify all vertices of e by a new vertex, and make all edges of G incident to e incident to the new vertex. A hypergraph H is a minor of hypergraph G if H is isomorphic to a hypergraph obtained from G through a (possibly empty) sequence of edge contractions and edge/vertex deletions (which take the subgraphs induced by the remaining sets of edges/vertices).

It suffices to prove Theorem 1 for biconnected planar graphs since if a planar graph G is not biconnected, the problems of finding branch-decompositions and grid minors of G can be solved individually for each biconnected component.

Let Σ be a fixed sphere. A planar embedding of a hypergraph G is a mapping $\rho : V(G) \cup E(G) \rightarrow \Sigma \cup 2^{\Sigma}$ satisfying the following properties.

- For $u \in V(G)$, $\rho(u)$ is a point of Σ, and for distinct $u, v \in V(G)$, $\rho(u) \neq \rho(v)$.
- For each edge $e \in E(G)$, if $e = \{u, v\}$ then $\rho(e)$ is a closed segment with end points $\rho(u)$ and $\rho(v)$, otherwise ($|e| \geq 3$) $\rho(e)$ is a topological disc (a simply connected closed region) of Σ and for each vertex $u \in e$, $\rho(u)$ is on the boundary of $\rho(e)$.
- For distinct $e_1, e_2 \in E(G)$, $\rho(e_1) \cap \rho(e_2) = \{\rho(u) | u \in e_1 \text{ and } u \in e_2\}$.

A hypergraph is planar if it has a planar embedding. A plane hypergraph is a pair (G, ρ), where ρ is a planar embedding of G. We may simply use G to denote the plane hypergraph (G, ρ), leaving the embedding ρ implicit. For a plane hypergraph G, each connected component of $\Sigma \setminus (\cup_{e \in E(G)} \rho(e))$ is a face of G.

Let G be a plane hypergraph. We say that a curve μ on the sphere Σ is *G-normal* if μ intersects with edges of G only at vertices of G. We may use *normal* for G-normal, leaving G implicit. The length of a normal curve μ, denoted by $|\mu|$, is the number of connected components of $\mu \setminus \bigcup_{v \in V(G)} \rho(v)$. For vertices $u, v \in V(G)$, we define the *normal distance* $\mathrm{nd}_G(u, v)$ to be the length of the shortest normal curve between $\rho(u)$ and $\rho(v)$. A *noose* of G is a closed normal curve on Σ that does not intersect with itself. A *minimum noose* satisfying certain properties is a noose with the minimum length satisfying the properties.

Let ν be a noose of G and let R_1 and R_2 be the two open regions of the sphere separated by ν. Then, ν induces a separation $(A, \overline{A})$ of G, with $A = \{e \in E(G) \mid \rho(e) \subseteq R_1 \cup \nu\}$ and $\overline{A} = \{e \in E(G) \mid \rho(e) \subseteq R_2 \cup \nu\}$. We also say that noose ν induces edge-subset A of G if ν induces a separation $(A, \overline{A})$ having A on one side. We call a separation or an edge-subset *noose-induced* if it is induced by some noose. We say a noose *separates* edge sets X and Y if the noose induces a separation $(A, \overline{A})$ with $X \subseteq A$ and $Y \subseteq \overline{A}$.

Let G be a plane hypergraph and let A be a noose-induced edge-subset of G. We denote by $G|A$ a hypergraph defined by $V(G|A) = (V(G) \setminus V(A)) \cup \partial(A)$ and $E(G|A) = (E(G) \setminus E(A)) \cup \{\partial(A)\}$. Note that in $G|A$, we replace all the hyperedges of A by a single hyperedge $\partial(A)$. We also assume that the embedding of $G|A$ is naturally derived from that of G by "painting out" the drawing of the subgraph A by a disc representing hyperedge $\partial(A)$. For a collection $\mathcal{A} = \{A_1, \ldots, A_r\}$ of mutually disjoint edge-subsets of G, we denote $(\ldots(G|A_1)|\ldots)|A_r$ by $G|\mathcal{A}$.

3 Basic Approach

We first give some known results on which our algorithm relies and the basic approach of our algorithm. Let G be a plane hypergraph, $(A, \overline{A})$ a noose-induced separation of G, and T_A and $T_{\overline{A}}$ branch-decompositions of $G|\overline{A}$ and $G|A$, respectively. We define $T_A + T_{\overline{A}}$ to be the tree obtained from T_A and $T_{\overline{A}}$ by first identifying the leaf of T_A and the leaf of $T_{\overline{A}}$ both corresponding to $\partial(A)$, joining the two edges incident to these leaves into one edge and removing the identified leaves. The following lemma is straightforward from the definition of branch-decompositions.

Lemma 1. *Let G be a plane hypergraph, $(A, \overline{A})$ a noose-induced separation of G, and T_A and $T_{\overline{A}}$ branch-decompositions of $G|\overline{A}$ and $G|A$ respectively. Then $T_A + T_{\overline{A}}$ is a branch-decomposition of G with width* $\max\{|\partial(A)|, k_A, k_{\overline{A}}\}$ *where k_A is the width of T_A and $k_{\overline{A}}$ is the width of $T_{\overline{A}}$.*

We use this lemma to recursively construct branch-decompositions of a given plane hypergraph. For each recursive step, we need some known results from the rat-catching characterization of the branchwidth of planar hypergraphs due to Seymour and Thomas [22].

Let G be a plane hypergraph and $l > 0$ an integer. We say that a vertex $u \in V(G)$ *directly captures* with order l an edge $e \in E(G)$ if the following holds:

1. There is a noose ν which induces separation $(\{e\}, E(G) \setminus \{e\})$.
2. There are two vertices $v_1, v_2 \in e$ that separates ν into two normal curves μ_1 and μ_2 such that $\mathrm{nd}_G(u, v_1) + \mathrm{nd}_G(u, v_2) + \max\{|\mu_1|, |\mu_2|\} < l$.

The rat-catching characterization of the branchwidth of a plane hypergraph G with $|e| < k$ for every $e \in E(G)$ implies that if there is a vertex u of G that directly captures with order l every edge of G, then the branchwidth of G is at most $\max\{l-1, k-1\}$ [22]. Further, Tamaki [23] shows that, given such a vertex u, a branch-decomposition of width $\max\{l-1, k-1\}$ can be constructed in $O(n)$ time, where n is the number of vertices of G. The following lemma directly follows from these results.

Lemma 2. *[23] Let G be a plane hypergraph with n vertices and $|e| < k$ for every $e \in E(G)$. If there is a vertex u of G such that $\mathrm{nd}_G(u, v) \leq h$ for every vertex $v \in V(G)$ then a branch-decomposition of G with width at most $\max\{2h + \lceil \frac{k-1}{2} \rceil, k-1\}$ can be constructed in $O(n)$ time.*

The following lemma (an application of Lemma 3.5 of [15]) gives a base for constructing the cylinder minors. A proof for the lemma can be found in [16].

Lemma 3. *Let G be a plane graph and $k, k' > 0$ integers. Let X and Y be edge sets of G satisfying the following conditions.*
1. Each of separations $(X, \overline{X})$ and $(Y, \overline{Y})$ is noose-induced.
2. $G[Y]$ is biconnected.
3. For any vertex $u \in \partial(X)$ and any vertex $v \in \partial(Y)$, $\mathrm{nd}_G(u, v) \geq k'$.
4. There is no noose of G with length $< k$ that separates X and Y.
Then G has a $k \times k'$ cylinder as a minor and such a minor can be constructed in time linear in $|V(\overline{X} \cap \overline{Y})|$.

Let G be a plane hypergraph and u a vertex of G. For any positive integers h and k, a collection $\mathcal{A}$ of noose-induced edge-subsets of G is (k, h)*-shallowing* for (G, u), if it satisfies the following conditions.

1. $u \in V(\overline{A})$ for every $A \in \mathcal{A}$.
2. $A \cap B = \emptyset$ for every pair of distinct elements $A, B \in \mathcal{A}$.
3. $|\partial(A)| < k$ for every $A \in \mathcal{A}$.
4. For each vertex v of $G|\mathcal{A}$, $\mathrm{nd}_G(u, v) \leq h$.

A (k, h)-shallowing collection $\mathcal{A}$ of noose-induced edge-subsets is used to reduce the problem of decomposing G, via Lemma 1, to subproblems of decomposing $G|\overline{A}$ for each $A \in \mathcal{A}$. Based on this notion of (k, h)-shallowing collection, we give a recursive procedure used in our algorithm. In this procedure, the input G and k to the algorithm are global. The precise value of parameter h used in the algorithm shall be specified later. For now, we simply remark that $h = O(k)$. The parameter λ is either $\frac{1}{2}$ or 1.

Procedure Branch-Grid(U, u)
Input: A noose-induced edge-subset U of G s.t. $G|U$ is biconnected, $u \in \partial(U)$.
Output: Either a branch-decomposition of $G|U$ of width at most $2h + \frac{k}{2}$ or a $k \times \lceil \lambda k \rceil$ cylinder minor of G.
Steps:

1. If $\mathrm{nd}_{G|U}(u, v) \leq h$ for each $v \in V(G|U)$ then apply Lemma 2 to find a branch-decomposition of $G|U$. Otherwise, proceed to the next step.
2. Try to find a collection $\mathcal{A}$ of noose-induced edge-subsets of $G|U$ that is (k, h)-shallowing for $(G|U, u)$. When unsuccessful, we are able to apply Lemma 3 to find a cylinder minor of G and terminate the algorithm, as we prove later. If we find such a collection, proceed to the next step.
3. For each $A \in \mathcal{A}$, choose a vertex $v_a \in \partial(\overline{A})$ of graph $G|\overline{A}$ and call Branch-Grid$(\overline{A}, v_a)$ to construct a branch-decomposition T_A or a cylinder minor of $G|\overline{A}$.
   If we find a branch-decomposition for every $A \in \mathcal{A}$, apply Lemma 2 to $(G|U)|\mathcal{A}$ to construct a branch-decomposition T_0 of $(G|U)|\mathcal{A}$ and use Lemma 1 to combine these branch-decompositions T_A, $A \in \mathcal{A}$, and T_0 into a branch-decomposition T of $G|U$ and return T.

To bound the number of recursive calls in which each fixed vertex is involved in the computation of Step 2, we enforce some "progress" when we recurse on each noose-induced edge-subset in $\mathcal{A}$. Let u be a vertex of G and let $d > 0$ be arbitrary. We say that a noose-induced edge-subset A of G is *d-progressive for* (G, u) if it satisfies the following conditions.

1. $u \in \overline{A}$.
2. Let v_a be an arbitrary vertex in $\partial(\overline{A})$ of $H = G|\overline{A}$ and let v be an arbitrary vertex of G. Then the following holds:
   (a) If $\mathrm{nd}_G(u, v) \leq d$ then $v \in V(\overline{A})$.
   (b) If $v \in V(H)$ then $\mathrm{nd}_H(v_a, v) \leq \mathrm{nd}_G(u, v) - d$.

We say a collection of noose-induced edge-subsets is d-progressive for (G, u) if each of its members is d-progressive for (G, u). Informally, if a noose-induced edge-subset A of $G|U$ is d-progressive for $(G|U, u)$ and Branch-Grid(U, u) makes a recursive call Branch-Grid$(\overline{A}, v_a)$, then each vertex of $G|U$ gets closer to v_a in $(G|U)|\overline{A}$ than to u in $G|U$ by the amount of d, as long as it appears in $(G|U)|\overline{A}$. This is how we enforce a progress in the recursion.

4 Algorithm Details

We now give the details of our algorithm, including the precise value of parameter h which depends on a positive integer c and a positive constant δ.

For a plane hypergraph G, a vertex u of G, and a nonnegative integer d, let $\mathrm{reach}_G(u, d) = \bigcup\{v \in V(G) | \mathrm{nd}_G(u, v) \leq d\}$ denote the set of vertices with normal distance d or smaller from u. Let $\lambda = \frac{1}{2}$ or 1. We define $d_1 = (\frac{\delta}{2}+1)\lceil\frac{k-1}{2}\rceil$ and for positive integer $i \geq 2$, $d_i = d_i(k) = d_1 + (i-1)(\lceil\lambda k\rceil - 1)$.

Theorem 1 relies on the following lemma which guarantees that we can find a sufficiently shallowing and progressive collection of noose-induced edge-subsets during the recursion, as long as $\mathrm{bw}(G) < k$. A sketch of the proof for the lemma shall be given later.

Lemma 4. *Let $c \geq 1$ be a fixed integer and δ an arbitrary positive constant. Let G be a biconnected plane graph, k a positive integer, U a noose-induced edge-subset of G with $|\partial(U)| < k$, and u an arbitrary vertex in $\partial(U)$ of $G|U$. Let M denote the number of vertices of $G|U$ in* $\mathrm{reach}_{G|U}(u, d_{c+1})$. *If* $\mathrm{reach}_{G|U}(u, d_{c+1}) \neq V(G|U)$, *we can in $O(M^{1+\frac{1}{c}})$ time either*
(1) find a $\frac{\delta}{2}k$-progressive and (k, d_{c+1})-shallowing collection $\mathcal{A}$ of noose-induced edge-subsets for $(G|U, u)$ such that for each $A \in \mathcal{A}$, $G[A]$ is biconnected, or (2) find a $k \times \lceil\lambda k\rceil$ cylinder minor of $G|U$.

In executing Step 2 of Procedure Branch-Grid, we invoke Lemma 4 and obtain a d-progressive and (k, h)-shallowing collection of noose-induced edge-subsets with $d = \frac{\delta}{2}k$ and $h = d_{c+1}$. When the search for such a collection is unsuccessful, Lemma 4 ensures that a $k \times \lceil\lambda k\rceil$ cylinder minor of G is found. Theorem 1 follows from the following lemma which in turn is proved assuming Lemma 4 is true. A proof for the lemma can be found in [16].

Lemma 5. *Given a biconnected plane graph G and an integer $k \geq 3$, suppose Branch-Grid($\{e\}, u$) is called, where e is an arbitrary edge of G and $u \in e$. The algorithm either gives a branch decomposition of G with width at most $2d_{c+1} + \frac{k}{2}$ or a $k \times \lceil\lambda k\rceil$ cylinder minor of G. The execution time of this call is $O(n^{1+\frac{1}{c}})$, where n is the number of vertices of G.*

Now we give a proof sketch for Lemma 4 to complete the proof of Theorem 1. We first describe the main ideas. Let G, U and u be as in Lemma 4. For each subgraph X of $G|U$ that are at distance h away from u (a precise definition is given later), we try to find a separation of order smaller than k that separates X from u (this can be done in linear time by solving the vertex-disjoint Menger problem on plane hypergraphs, details can be found in [16]). If we fail to find such a separation for any X, then by Lemma 3, we obtain a cylinder minor of G that certifies $\mathrm{bw}(G) \geq k$. If we do obtain separation $(A_X, \overline{A_X})$ of order smaller than k for each X that separates X from u, then we hope that these edge-subsets A_X constitute a (k, h)-shallowing collection for $(G \mid U, u)$. There are two issues to be resolved in this approach.

1. These subsets may not be disjoint with each other as required by the definition of (k, h)-shallowing collections.
2. Even though the algorithm for the vertex-disjoint Menger problem runs in linear time, the computation must be repeated for each X and may result in a quadratic running time.

The first issue is resolved by Lemma 5.2 of [16]. The second issue is resolved by a layered tree approach described below combined with Lemma 5.1 of [16] that helps localizing the graph on which the algorithm for the vertex-disjoint Menger problem is executed.

Let G and u be as above. The *layer tree* for (G, u), denoted by $\mathrm{LT}(G, u)$, is defined as follows. Recall that $d_1 = (\frac{\delta}{2} + 1)\lceil \frac{k-1}{2} \rceil$ and for positive integer $i \geq 2$, $d_i = d_i(k) = d_1 + (i-1)(\lceil \lambda k \rceil - 1)$.

1. The root of the tree is $V(G)$.
2. Each biconnected component of $G[V(G) \setminus \mathrm{reach}_G(u, d_1)]$ is in level 1 of the tree and is a child node of the root.
3. For each i, $2 \leq i \leq c$, each biconnected component X of $G[V(G) \setminus \mathrm{reach}_G(s, d_i)]$ is in level i of the tree and is a child node of the connected component of $G[V(G) \setminus \mathrm{reach}_G(s, d_{i-1})]$ that contains X.

Let X be a non-root node of $\mathrm{LT}(G, u)$ and Y a child node of X. Since each vertex u of G in $\overline{X}$ is of normal distance at least $d_i - d_{i-1} + 1 \geq \frac{k}{2}$ from each vertex v of G in $V(Y)$ and $G[Y]$ is biconnected, we can apply Lemma 3 to obtain a noose-induced edge-subset A with $\partial(A) < k$ separating $\overline{X}$ and Y, assuming $\mathrm{bw}(G) < k$. Our strategy is to find at least one such noose along the path from each leaf in level $c+1$ to the root.

Let m denote the number of leaves in level $c+1$ of $\mathrm{LT}(G, u)$. We classify nodes of $\mathrm{LT}(G, u)$ as *crowded* or *uncrowded* by induction on its tree structure. We classify each leaf in level $c+1$ as crowded. We classify a node in other levels as crowded if it has more than $m^{\frac{1}{c}}$ crowded child nodes. Otherwise it is uncrowded. We call a parent-child pair (X, Y) *processable*, if X is uncrowded, Y is crowded, and no ancestor of X is crowded. In $\mathrm{LT}(G, u)$, for every leaf Z in level $c+1$, the path from the root to Z contains exactly one processable parent-child pair.

To "process" a parent-child pair (X, Y), that is, to find a minimum noose separating $\overline{X}$ and Y, we solve the vertex-disjoint Menger problem for planar hypergraphs. To localize the problem, we define a hypergraph $H(X, Y)$ for each parent-child pair X, Y in $\mathrm{LT}(G, u)$ as follows. Let i be the level of node X in $\mathrm{LT}(G, u)$. Then, $H(X, Y) = (G|\overline{X})|\mathcal{B}$, where $\mathcal{B}$ is the collection of biconnected components of $X \setminus \mathrm{reach}_G(u, d_{i+1})$. Let ν be a minimum noose in G separating $\overline{X}$ and a $Y \in \mathcal{B}$ in G and suppose $|\nu| < k$. By Lemma 5.1 of [16], ν is also a minimum noose separating the edge $\partial(\overline{X})$ and the edge $\partial(Y)$ in $H(X, Y)$. Thus, running the algorithm for the vertex-disjoint Menger problem in $H(X, Y)$ is sufficient for deciding if there is a noose of G of size smaller than k separating $\overline{X}$ and Y and, if there is, finding such a noose.

From the approach described above we get Lemma 4 (the proof details can be found in [16]).

References

1. Arnborg, S., Cornell, D., Proskurowski, A.: Complexity of finding embedding in a k-tree. SIAM J. Disc. Math. 8, 277–284 (1987)
2. Arnborg, S., Lagergren, J., Seese, D.: Easy problems for tree-decomposable graphs. Journal of Algorithms 12, 308–340 (1991)

3. Amir, E.: Approximation Algorithms for Treewidth Algorithmica (April 2008)
4. Bodlaender, H.L., Grigoriev, A., Koster, A.M.C.A.: Treewidth Lower Bounds with Brambles. Algorithmica 51(1), 81–98 (2008)
5. Bodlaender, H.L.: A tourist guide through treewidth. Acta Cybernetica 11, 1–21 (1993)
6. Bodlaender, H.L.: A linear time algorithm for finding tree-decomposition of small treewidth. SIAM J. Comput. 25, 1305–1317 (1996)
7. Bodlaender, H.L., Thilikos, D.: Constructive linear time algorithm for branchwidth. In: Degano, P., Gorrieri, R., Marchetti-Spaccamela, A. (eds.) ICALP 1997. LNCS, vol. 1256, pp. 627–637. Springer, Heidelberg (1997)
8. Dorn, F., Fomin, F.V., Thilikos, D.M.: Catalan Structures and Dynamic Programming in H-minor-free graphs. In: Proc. of the 2008 Symposium on Discrete Algorithms, SODA 2008, pp. 631–640 (2008)
9. Demaine, E.D., Hajiaghayi, M.T.: Graphs excluding a fixed minor have grids as large as treewidth, with combinatorial and algorithmic applications through bidimensionality. In: Proc. of the 2005 Symposium on Discrete Algorithms, SODA 2005, pp. 682–689 (2005)
10. Demaine, E.D., Hajiaghayi, M.T.: Bidimensionality, map graphs, and grid minors, arXiv:Computer Science, DM/052070, v1 (2005)
11. Demaine, E.D., Hajiaghayi, M.T., Kawarabayashi, K.: Algorithmic graph minor theory: decomposition, approximation, and coloring. In: Proc. of the 2005 IEEE Symposium on Foundation of Computer Science, FOCS 2005, pp. 637–646 (2005)
12. Feige, U., Hajiaghayi, M.T., Lee, J.R.: Improved approximation algorithms for minimum weight vertex separators. SIAM J. Comput. 38(2), 629–657 (2008)
13. Grigoriev, A.: Tree-width and large grid minors in planar graphs (2008) (submitted for publication)
14. Gu, Q.P., Tamaki, H.: Optimal branch decomposition of planar graphs in $O(n^3)$ time. ACM Trans. Algorithms 4(3), article No. 30, 1–13 (2008)
15. Gu, Q.P., Tamaki, H.: Improved bound on the planar branchwidth with respect to the largest grid minor size. Technical Report, SFU-CMPT-TR 2009-17 (July 2009)
16. Gu, Q.P., Tamaki, H.: Constant-factor approximations of branch-decomposition and largest grid minor of planar graphs in $O(n^{1+\epsilon})$ time. Technical Report, SFU-CMPT-TR 2009-18 (July 2009)
17. Robertson, N., Seymour, P.D.: Graph minors I. Excluding a forest. Journal of Combinatorial Theory, Series B 35, 39–61 (1983)
18. Robertson, N., Seymour, P.D.: Graph minors II. Algorithmic aspects of tree-width. J. Algorithms 7, 309–322 (1986)
19. Robertson, N., Seymour, P.D.: Graph minors X. Obstructions to tree decomposition. J. of Combinatorial Theory, Series B 52, 153–190 (1991)
20. Robertson, N., Seymour, P.D.: Graph minors XIII. The disjoint paths problem. J. of Combinatorial Theory, Series B 63, 65–110 (1995)
21. Robertson, N., Seymour, P.D., Thomas, R.: Quickly excluding a planar graph. J. of Combinatorial Theory, Series B 62, 323–348 (1994)
22. Seymour, P.D., Thomas, R.: Call routing and the ratcatcher. Combinatorica 14(2), 217–241 (1994)
23. Tamaki, H.: A linear time heuristic for the branch-decomposition of planar graphs. In: Di Battista, G., Zwick, U. (eds.) ESA 2003. LNCS, vol. 2832, pp. 765–775. Springer, Heidelberg (2003)
24. Thomas, R.: Tree decompositions of graphs, `http://www.math.gatech.edu/thomas/SLIDE/slide.ps,p.3`

PTAS for k-Tour Cover Problem on the Plane for Moderately Large Values of $k^\star$

Anna Adamaszek[1], Artur Czumaj[1], and Andrzej Lingas[2]

[1] Centre for Discrete Mathematics and its Applications (DIMAP) and Department of Computer Science, University of Warwick, UK
A.M.Adamaszek@warwick.ac.uk, A.Czumaj@warwick.ac.uk
[2] Department of Computer Science, Lund University, 22100 Lund, Sweden
Andrzej.Lingas@cs.lth.se

Abstract. Let P be a set of n points in the Euclidean plane and let O be the origin point in the plane. In the *k-tour cover problem* (called frequently the *capacitated vehicle routing problem*), the goal is to minimize the total length of tours that cover all points in P, such that each tour starts and ends in O and covers at most k points from P.

The k-tour cover problem is known to be $\mathcal{NP}$-hard. It is also known to admit constant factor approximation algorithms for all values of k and even a polynomial-time approximation scheme (PTAS) for small values of k, $k = \mathcal{O}(\log n / \log \log n)$.

In this paper, we significantly enlarge the set of values of k for which a PTAS is provable. We present a new PTAS for all values of $k \leq 2^{\log^\delta n}$, where $\delta = \delta(\varepsilon)$. The main technical result proved in the paper is a novel reduction of the k-tour cover problem with a set of n points to a small set of instances of the problem, each with $\mathcal{O}((k/\varepsilon)^{\mathcal{O}(1)})$ points.

1 Introduction

The *k-tour cover problem* (k-TC), is a very natural and well known generalization of the traveling salesperson problem (TSP) to include several tours [2,3,8,12]. Namely, we are given a set P of points (sites), a distinguished point O outside P, called the origin as well as a distance function defined on $P \cup \{O\}$. A tour is a cycle whose vertices are in $P \cup \{O\}$. The length of a tour is the sum of distances between the adjacent points on the tour. The objective is to find a set of tours, each including the origin and at most k points in P, which covers all points in P and achieves the minimum total length.

In Operations Research, the k-TC problem is well known as the *capacitated vehicle routing problem* [12]. The name comes from its standard application when the points in P represent customer locations, and the origin O stands for a depot. Then, a fleet of vehicles located at the depot must serve all the customers, so

* Research supported in part by the *Centre for Discrete Mathematics and its Applications (DIMAP)*, EPSRC award EP/D063191/1, and by VR grant 621-2005-408.

Y. Dong, D.-Z. Du, and O. Ibarra (Eds.): ISAAC 2009, LNCS 5878, pp. 994–1003, 2009.

that each vehicle can serve at most k customers. The objective is to minimize the total distance traveled by the fleet. The k-TC problem (capacitated vehicle routing problem) is one of the central special cases of a more general vehicle routing problem, introduced by Dantzig and Ramser [5] fifty years ago, and studied very extensively in the literature ever since (cf. [9,12]).

The k-TC problem contains the TSP problem as a special case and it is known to be $\mathcal{NP}$-hard for all $k \geq 3$. For this reason, the research on k-TC has focused on heuristic algorithms and approximation algorithms. The most extensively studied variants of k-TC are the metric one, when the distance function is symmetric and satisfies the triangle inequality, and in particular the *two-dimensional Euclidean* one, when the points are placed in the plane and the distance is Euclidean.

The general metric case of k-TC for $k \geq 3$ has been shown to be APX-complete [2], i.e., complete for the class of optimization problems admitting constant factor approximations. However, the approximability status of the two-dimensional *Euclidean k-TC* problem, in particular, the problem of the existence of a PTAS, has not been completely settled yet. One of the first studies of two-dimensional Euclidean k-TC has been due to Haimovich and Rinnooy Kan [8], who presented several heuristics for the metric and Euclidean k-TC, including a PTAS for the two-dimensional Euclidean k-TC with $k < c \log\log n$, for some constant c [8, Section 6]. Asano et al. [3] substantially subsumed this result by designing a PTAS for $k = \mathcal{O}(\log n/\log\log n)$. They also observed that Arora's [1] or Mitchell's [10] PTAS for the two-dimensional Euclidean TSP implies a PTAS for the corresponding k-TC where $k = \Omega(n)$. There has not been any significant progress since the paper by Asano et al. [3] until very recently, when Das and Mathieu [6] showed a *quasi-polynomial* time approximation scheme (QPTAS) for the two-dimensional Euclidean k-TC for every k. Their algorithm combines the approach developed by Arora [1] for Euclidean TSP with some new ideas to deal with k-TC and gives a $(1+\varepsilon)$-approximation for the two-dimensional Euclidean k-TC in time $n^{\log^{\mathcal{O}(1/\varepsilon)} n}$ (this bound holds for any k).

In this paper we focus on the two-dimensional Euclidean variant of k-TC. (To simplify the notation, we shall further refer to this variant as to k-TC).

Our **main result** is a new PTAS for k-TC for all values of $k \leq 2^{\log^{\delta} n}$, where $\delta = \delta(\varepsilon)$. This significantly enlarges the set of values of k for which a PTAS is known. Our PTAS relies on a novel reduction of an instance of k-TC with a set of n points to an instance or a small number of independent instances of the problem with a small number of points. Our first reduction takes any instance of k-TC on n points and reduces it to an instance with $\mathcal{O}((k/\varepsilon)^{\mathcal{O}(1)} \log^2(n/\varepsilon))$ points. Then we present a refinement, where the instance of k-TC is reduced to a small set of instances of k-TC, each with $\mathcal{O}((k/\varepsilon)^{\mathcal{O}(1)})$ points. These results, when combined with the recent QPTAS due to Das and Mathieu [6], give the aforementioned PTAS for k-TC for all values $k \leq 2^{\log^{\delta} n}$, where $\delta = \delta(\varepsilon)$.

For simplicity of the presentation, we will present $(1+\mathcal{O}(\varepsilon))$-approximation algorithms; reduction to $(1+\varepsilon)$-approximation is straightforward.

2 Preliminaries

We assume a fixed origin in the plane and denote it by O. For a tour $\mathcal{T}$, its (Euclidean) length is denoted by $|\mathcal{T}|$. For a set U of tours, we set $|U|$ to $\sum_{\mathcal{T} \in U} |\mathcal{T}|$.

For a set P of points in the plane, we denote by $TSP(P)$ the minimum length of a TSP-tour through P and by $\text{OPT}(P)$ the minimum length of a solution to k-TC (i.e., the minimum length of a set of tours, each through the origin and containing at most k points of P, which covers all points in P). When P is clear from the context, we shall simply use the notation OPT.

For a point $p \in P$, we denote by $r(p)$ the distance of p from the origin O.

The following simple lower bound plays a very important role in the previous approaches to k-TC, see [3, Proposition 2] and [8, Lemma 1].

Fact 1. $\text{OPT}(P) \geq \frac{2}{k} \sum_{p \in P} r(p)$.

Following [3], we shall term $\frac{2}{k} \sum_{p \in P} r(p)$ as the *radial cost* of P, and denote by $rad(P)$. Among other things, Haimovich and Kan considered the so called *iterated tour partitioning heuristic* for k-TC in [8]. The heuristic starts from constructing a TSP-tour T through P. Then, it considers all k-tour covers resulting from partitioning T into paths visiting exactly k points (assuming that n is divisible by k), and connecting the endpoints of the paths with O. The heuristic outputs the shortest among these solutions.

Fact 2. [3] *If the iterated tour partitioning heuristic uses a TSP tour U, then it returns a k-tour cover of total length not exceeding $(1 - \frac{1}{k}) \cdot |U| + rad(P)$.*

Note that given a TSP tour, the iterated tour partitioning heuristic can be implemented in time $\mathcal{O}(k\frac{n}{k} + n)$ by repeatedly updating the previous partition and k-tour cover to the next one in time $\mathcal{O}(\frac{n}{k})$. Using the minimum spanning tree heuristic for TSP we can find a 2-approximation of the TSP in time $\mathcal{O}(n \log n)$. Hence, we obtain the following.

Corollary 1. *If the iterated tour partitioning heuristic uses the minimum spanning tree heuristic for TSP then it returns a $(3 - \frac{2}{k})$-approximation of an optimal k-tour cover of an n-point set and it can be implemented in time $\mathcal{O}(n \log n)$.*

3 PTAS for Moderate Values of *k*

In this section we present a reduction that takes as an input any instance of the k-tour cover problem on a set of n points in the Euclidean plane and reduces it to an instance of the problem with $\mathcal{O}((k/\varepsilon)^{\mathcal{O}(1)} \log^2(n/\varepsilon))$ points. Then, we apply this reduction to obtain a PTAS for the k-tour cover problem for all $k \leq 2^{\log^{\delta} n}$, where δ is some positive constant, $\delta = \delta(\varepsilon)$.

Our construction uses a series of transformations that eliminate most of the input points and reduce the input problem instance to one significantly smaller.

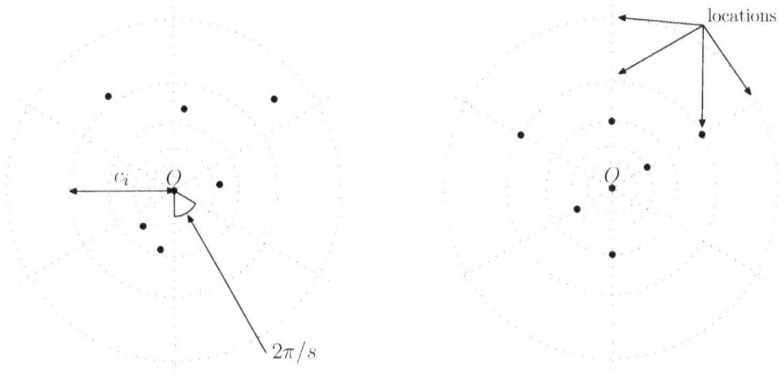

Fig. 1. The structure of circles, rays, and locations. The point labeled O is the origin. Other fat dots represent the points from P. In the right picture each point has been moved to its nearest location.

3.1 Removing Close Points

Let L be the maximum distance from a point in P to the origin O, that is, $L = \max\{p \in P : r(p)\}$. Since $\text{OPT} \geq 2L$, we can ignore any point that is at a distance at most $L\varepsilon/n$ from the origin: covering all such points with 1-tours will give us additional cost not greater than $n \cdot 2\frac{L\varepsilon}{n} \leq \varepsilon \cdot \text{OPT}$. Therefore, from now on, we will consider only the points p with $r(p) \geq L\varepsilon/n$.

3.2 Circles, Rays, and Locations

Let us create *circles* around the origin, the i-th circle with a radius

$$c_i = \frac{L\varepsilon}{n} \cdot \left(1 + \frac{\varepsilon}{k}\right)^i , \qquad \text{for } 0 \leq i \leq \left\lceil \log_{(1+\varepsilon/k)} \frac{n}{\varepsilon} \right\rceil .$$

Let us draw *rays* from the origin with the angle between any pair of neighboring rays equal to $2\pi/s$ (that is, partition the space into s sectors) with $s = \lceil \frac{2\pi k}{\varepsilon} \rceil$.

Define a *location* to be any point on the plane that is the intersection of a circle and a ray. Since

$$\log_{(1+\varepsilon/k)} \frac{n}{\varepsilon} = \frac{\log \frac{n}{\varepsilon}}{\log(1 + \varepsilon/k)} = \Theta\left(\frac{k}{\varepsilon} \cdot \log \frac{n}{\varepsilon}\right) ,$$

there are $\Theta\left(\frac{k}{\varepsilon} \log(n/\varepsilon)\right)$ circles and $\Theta\left(\frac{k}{\varepsilon}\right)$ rays. Therefore we obtain:

Claim 1. *The total number of locations T satisfies $T = \Theta(k^2\varepsilon^{-2} \log(n/\varepsilon))$.*

Now, we modify P by moving each point from P to its nearest location.

Claim 2. *The operation of moving each point to its nearest location can change the cost of a k-tour cover by at most $\varepsilon \cdot \text{OPT}$.*

Proof. Let p be a point in P. Suppose that p lies between the circles with radius c_i and c_{i+1} (the distance between p and the origin is in the interval $[L\varepsilon/n, L]$, so we know such circles exist). The distance between these circles equals $c_{i+1} - c_i = \frac{\varepsilon}{k} \cdot c_i$. The distance between two consecutive locations at the i-th circle is less than $2\pi c_i/s \leq \frac{\varepsilon}{k} \cdot c_i$. Therefore the distance between p and its nearest location is at most $\sqrt{2} \cdot (\frac{1}{2} \cdot \frac{\varepsilon}{k} c_i) < \frac{\varepsilon}{k} \cdot c_i \leq \frac{\varepsilon}{k} \cdot r(p)$.

If we move a point $p \in P$ by a distance at most $\frac{\varepsilon}{k} \cdot r(p)$, the cost of a tour can change by at most $2\frac{\varepsilon}{k} \cdot r(p)$. If we add up the changes of the cost generated by moving all points in P, then this total change is upper bounded by $\sum_{p \in P} 2\frac{\varepsilon}{k} \cdot r(p)$. Next, we use Fact 1 to conclude that the total cost of moving all the points is at most $\varepsilon \cdot \text{OPT}$. □

From a k-tour cover U' for a modified instance of the problem (where all points have been moved to their nearest locations) we can easily get a k-tour cover U for the original version of the problem such that $|U| \leq |U'| + \varepsilon \cdot \text{OPT}$. So a PTAS for the modified version yields a PTAS for the original version. In the rest of this paper we will consider the modified version of the problem.

3.3 Trivial and Nontrivial Tours

We say that a tour *visits* a location if it contains at least one point from that location. (If an edge passes trough a location, but the tour does not contain any point from that location, then the tour does not visit that location.) We call a tour *trivial* if it visits only a single location in P; it is *nontrivial* otherwise.

Theorem 1. *There is an optimal solution in which there are at most T nontrivial tours.*

Proof. We say that a set of tours $t_1, t_2, \ldots, t_m$ ($m \geq 2$) forms a *cycle* if there is a set of locations $\ell_1, \ell_2, \cdots, \ell_m, \ell_{m+1} = \ell_1$ such that each tour t_i visits locations ℓ_i and ℓ_{i+1}. Note that the origin is not considered as a location.

To prove our theorem we will need the following:

Lemma 1. *There is an optimal solution in which there are no cycles.*

Proof. Let U be such an optimal solution which minimizes the sum over all its nontrivial tours of the number of locations visited by that tour.

Let us suppose that U has a cycle, and let $t_1, t_2, \ldots, t_m$ be a minimal cycle (m is minimal). Let $\ell_1, \ell_2, \ldots, \ell_m$ be the locations in which the consecutive tours meet. From the minimality of the cycle we know that both tours and locations are pairwise distinct.

Let $v(t, \ell)$ denote the number of points from a location ℓ visited by a tour t. Let $\min = \min_{i \in \{1,\ldots,m\}}\{v(t_i, \ell_i)\}$. Now we are ready to swap points between the tours: the i-th tour, instead of visiting $v(t_i, \ell_i)$ points in the location ℓ_i and $v(t_i, \ell_{i+1})$ points in the location ℓ_{i+1} will now visit $(v(t_i, \ell_i) - \min)$ points in ℓ_i and $(v(t_i, \ell_{i+1}) + \min)$ points in ℓ_{i+1}. Here ℓ_{m+1} denotes ℓ_1.

Observe that the modification does not change the number of points visited by each tour. It also does not increase the length of any tour. Therefore, we obtain another optimal solution, in which the sum over all nontrivial tours of the number of locations visited by that tour is smaller than in U (we managed to remove one location from each tour t_i for which $v(t_i, \ell_i) = \min$). This is a contradiction with the minimality of that sum in U.

Therefore the optimal solution U has no cycles. □

Consider an optimal solution without cycles. Note that the lack of 2-cycles means that no two tours visit the same pair of locations. To each nontrivial tour we can assign a pair of distinct locations visited by this tour. The chosen pairs are in one-to-one correspondence with the nontrivial tours and they induce an acyclic undirected graph on the locations.

Hence, we can have at most $T - 1$ nontrivial tours in an acyclic solution, so using Lemma 1 we have proved the theorem. □

3.4 Reduction to an Instance of k-TC with $(k \log n/\varepsilon)^{\mathcal{O}(1)}$ Points

Observe that Theorem 1 implies that there is an optimal solution in which at most Tk points are covered by nontrivial tours. Therefore it is enough to consider only solutions which fulfill that property.

If the number of points in a location ℓ is greater than Tk, some of the points will have to be covered by trivial tours. We may assume, without loss of generality, that among all trivial tours visiting a given location there is at most one that visits less than k points. Moreover, if at least one point from some location is visited by a nontrivial tour, we can assume that all trivial tours visiting that location contain exactly k elements. Therefore, for each location ℓ containing c_ℓ points, we only have to consider at most $\min\{c_\ell, c_\ell - k \cdot \lceil \frac{c_\ell - Tk}{k} \rceil\} \leq Tk$ points for nontrivial tours. After finding a $(1+\varepsilon)$-approximation for such reduced case, we will add trivial tours covering all remaining points. That will give us $(1+\varepsilon)-$approximation for the original problem.

Corollary 2. *One can reduce the k-TC problem on n points to one on at most $T^2 k$ points.*

3.5 PTAS for k-TC with $k \leq 2^{\log^\delta n}$

We use Corollary 2 to reduce any instance of k-TC with the input set of n points P to an instance of k-TC with $N = T^2 k = \Theta(k^5 \varepsilon^{-4} \log^2(n/\varepsilon))$ input points. For such input instance, we apply the quasi-polynomial time approximation scheme for k-TC due to Das and Mathieu [6]. The obtained algorithm returns a $(1+\varepsilon)$-approximation in time $N^{\log^{\mathcal{O}(1/\varepsilon)} N}$. This gives polynomial time for all $k \leq 2^{\log^\delta n}$ for some constant $\delta = \delta(\varepsilon) > 0$. Hence, we have the following main theorem.

Theorem 2. *There is a PTAS for the k-TC problem provided that $k \leq 2^{\log^\delta n}$ for some positive constant $\delta = \delta(\varepsilon)$.*

4 Refinement: Reduction to $(k/\varepsilon)^{\mathcal{O}(1)}$ Points

In the preceding section, we have demonstrated that the problem of close approximation of the k-TC problem on the input set of n points in the plane reduces to that for a multi-point-set of size polynomial in k/ε and polylogarithmic in n in the relevant locations. In this section, we shall eliminate the polylogarithmic dependency of n in the reduction. This will have only a relatively small effect on the asymptotics for the size of the largest k in terms of n for which we can attain a PTAS and we will obtain a PTAS for all $k \leq 2^{\log^{\delta'} n}$, where comparing to the bound in Theorem 2, we will have $\delta' > \delta$. However, for small values of k this will lead to a faster PTAS. Hopefully, because it removes completely the dependency on n from the size of the reduced instance, it also might be a step towards a PTAS for arbitrary values of k.

Our approach resembles Baker's method [4] of closely approximating several hard problems on planar graphs. It relies on the following separation lemma.

Lemma 2. *Let P be a set of points situated in the locations and let $\varepsilon > 0$. There is a clustering of the circles into rings of $\lceil \log_{1+\frac{\varepsilon}{k}}(6/\varepsilon) \rceil$ consecutive circles and there are positive integers $a = \mathcal{O}(\varepsilon^{-1})$ and $b \in \{1, \ldots, a\}$ such that if we mark each $(b+ja)$-th ring then any k-tour cover U of P can be transformed to a k-tour cover U' of the points in the unmarked rings such that*

1. *no tour in U' visits two points in P separated by a marked ring, and*
2. *$|U'| \leq (1 + \frac{\varepsilon}{2})|U|$.*

Furthermore, the points in the marked rings can be covered with k-tours of total length at most $\frac{\varepsilon}{2}|U|$ produced by the iterated tour partitioning heuristic from [8] (cf. Section 2).

Proof. Let t denote a tour obtained by removing its edges incident to O. Suppose that t crosses one of the marked rings. Let i be the number of the most inner circle of the ring. Denote the circle by C_i. It follows by straightforward calculation and the definition of the circles that each minimal fragment of t crossing the aforementioned ring is at least $\frac{2}{\varepsilon}$ times longer than the doubled radius of C_i. We can appropriately split the tour t along C_i into smaller ones by connecting pairs of crossing points on C_i with O or just with themselves, see Figure 2.

The total length of the smaller tours is longer than $|t|$ by at most $\frac{\varepsilon}{2}$ of the total length of the aforementioned fragments of t.

We may assume, without loss of generality, that the aforementioned marked ring is the outermost among those crossed by t. We can iterate the elimination of the crossings of the smaller resulting tours but for their edges incident to O with more inner marked rings. Note that then other disjoint fragments of t will be charged with the increase of the length of the union of the resulting smaller tours. Finally, by applying short-cutting, we can drop the points in the marked rings from the resulting tours.

We conclude that we can transform U into a k-tour cover U' of the points in P in the unmarked rings such that no tour in U' crosses any marked ring (but for its edges incident to O) and $|U'| \leq (1 + \frac{\varepsilon}{2})|U|$.

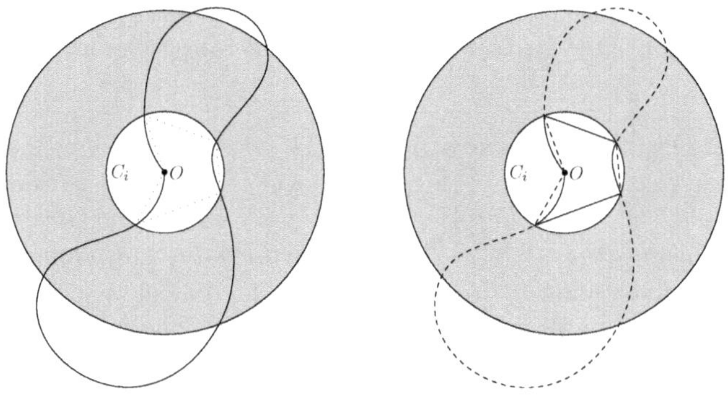

Fig. 2. Splitting t into smaller tours. The grey area is the marked ring. In the left picture dotted lines represent the lines which will be added to our solution. The right picture shows two separate tours obtained from the original tour (one is marked with a dashed line, and the other with a solid one), before the short-cutting.

It remains to show that we can set a and $b \in \{1, \ldots, a\}$ such that one can easily cover the points in P contained in the marked rings with k-tours of total length not exceeding $\frac{\varepsilon|U|}{2}$.

Let R_j denote the set of points from P lying in the j-th ring. Set a to $\lceil \frac{24}{\varepsilon} \rceil$. For each $b \in \{1, \ldots, a\}$, let P_b be the set of points in P in the marked rings, $P_b = \sum_{j \equiv b \bmod a} R_j$. We shall show that there is some $b \in \{1, \ldots, a\}$ such that by applying the k-TC heuristic given in Corollary 1 for P_b, we can cover P_b with k-tours of length at most $\frac{\varepsilon|U|}{2}$. For this purpose, we shall observe that $\sum_j TSP(R_j) \leq 3 \cdot TSP(P)$.

Suppose for the sake of this observation that the tour t considered in the first part of the proof is an n-tour, i.e., an optimal TSP tour of $P \cup \{O\}$. Apply almost the same transformation to the tour t as before with the exception that instead of connecting the outer cut part by two rays to O, we connect the cutting points directly. By the triangle inequality, the total length of the so modified TSP tour t is at most $(1 + \frac{\varepsilon}{2}) \cdot TSP(P)$. The modified TSP tour t can be easily reduced to the non-necessarily optimal TSP tours of the unmarked regions by short-cutting. Assuming first for a moment that the unmarked rings are the even ones, and then conversely, that the unmarked rings are the odd ones, and that $\varepsilon < \frac{1}{2}$, we conclude that $\sum_j TSP(R_j) \leq 3 \cdot TSP(P)$.

Using Fact 2 we get that

$$\sum_{b \in \{1,\ldots,a\}} \text{OPT}(P_b) \leq \sum_{b \in \{1,\ldots,a\}} \sum_{j \equiv b (mod\ a)} \text{OPT}(R_j) = \sum_j \text{OPT}(R_j)$$
$$\leq \sum_j (rad(R_j) + TSP(R_j)) \leq rad(P) + 3 \cdot TSP(P) \leq 4|U|.$$

There must be some $b \in \{1, \ldots, a\}$ such that $\text{OPT}(P_b) \leq \frac{4}{a}|U| \leq \frac{\varepsilon|U|}{6}$. Thus, if we apply the 3-approximation algorithm for the k-tour cover of P_b, which is a composition of the iterated tour partitioning heuristic with the minimum spanning tree heuristic for TSP, we obtain a k-tour cover of P_b of length at most $\frac{\varepsilon|U|}{2}$. □

Theorem 3. *The k-TC problem on a set P of n points on the plane can be reduced to a collection of $\mathcal{O}(\varepsilon^{-1}\log(n/\varepsilon)/\log(1/\varepsilon))$ disjoint k-tour cover problems, each on $\mathcal{O}(k^5\varepsilon^{-6}\log^2(1/\varepsilon))$-point set and each having the maximum distance to the origin at most $(1/\varepsilon)^{\mathcal{O}(1/\varepsilon)}$ larger than the minimum one, such that $(1+\varepsilon)$-approximate solutions to each of the latter problems yield a $(1+\mathcal{O}(\varepsilon))$-approximation to the original k-tour cover problem. The reduction can be done in time $\mathcal{O}(n\log n)$ for a fixed ε.*

Proof. Move the points to the locations and compute the sets R_j of input points lying in the rings for a fixed ε. This all can be easily done in time $\mathcal{O}(n\log n)$ by using standard data structures for point location [11].

Next, compute the value a (the distance between marked rings) and for each $b\in\{1,\dots,a\}$, compute a 3-approximate k-tour cover of the set P_b of points contained in the marked rings. All the a computations take $\mathcal{O}(an\log n)=\mathcal{O}(n\log n)$ time by Corollary 1.

Fix b to that minimizing the length of the aforementioned tour. It follows from Lemma 2 that the produced cover of P_b has length at most $\frac{\varepsilon}{2}\text{OPT}$. Now we will have to compute approximate solutions for each maximal sequence of consecutive not marked rings. Let us denote the number of such sequences by q. We can easily compute that $q=\mathcal{O}(\varepsilon^{-1}\log\frac{n}{\varepsilon}/\log\frac{1}{\varepsilon})$. For $i=1,\dots,q$, let I_i denote the set of points contained in such i-th sequence. Note that these point sets can be also easily computed in time $\mathcal{O}(n\log n)$.

It follows from Lemma 2 that if we compute separately $(1+\varepsilon)$-approximation of the optimal cover with k-tours for each set I_i, then the union of these coverings will have length at most $(1+\mathcal{O}(\varepsilon))\text{OPT}$.

Note that for a given i, the number of locations in I_i is $\mathcal{O}(a\cdot\frac{k}{\varepsilon}\cdot\log_{(1+\frac{\varepsilon}{k})}\frac{1}{\varepsilon})=\mathcal{O}(k^2\varepsilon^{-3}\log\frac{1}{\varepsilon})$. Hence, by the discussion in Section 3, we can account to the intended $(1+\varepsilon)$-approximation of $\text{OPT}(I_i)$ the trivial tours decreasing the point-multiplicity in each location to $\mathcal{O}(k^3\varepsilon^{-3}\log\frac{1}{\varepsilon})$. Thus, for each I_i we can reduce the problem to one with $\mathcal{O}(k^5\varepsilon^{-6}(\log\frac{1}{\varepsilon})^2)$ points.

Each I_i consists of $\mathcal{O}(\varepsilon^{-1})$ consecutive rings and for a point in a ring the maximum distance to the origin is at most $\mathcal{O}(\varepsilon^{-1})$ times larger than the minimum one. Hence, for a point in I_i the maximum distance to the origin is at most $(1/\varepsilon)^{\mathcal{O}(1/\varepsilon)}$ times larger than the minimum one.

The appropriate q sets of points can be computed in time $\mathcal{O}(n\log n)$ and they specify the problems to which we approximately reduce the original k-tour cover problem. □

5 Final Remarks

The central open question left is whether there is a PTAS for the k-TC problem for all values of k. While we have enlarged the set of values of k for which a

PTAS exists, we still do not know how to reach polynomial values for k, even $k = n^{0.001}$. In particular, a PTAS k-TC for $k = \Theta(\sqrt{n})$ is elusive. For arbitrary values of k, the best currently known result is either a quasi-polynomial time approximation scheme by Das and Mathieu [6] that runs in time $n^{\log^{\mathcal{O}(1/\varepsilon)} n}$, or the polynomial-time constant-factor approximation algorithm due to Haimovich and Rinnooy Kan [8]. Similarly as in [3], we believe that the case $k = \Theta(\sqrt{n})$ is the hardcore of the difficulty in obtaining a PTAS for all values of k.

Following [8], let us observe that if we divide the range of k into a logarithmic number of intervals of the form $[\varepsilon^{-2i}, \varepsilon^{-2(i+1)})$, then for k in at most one of the intervals none of the inequalities $TSP(P) \leq \varepsilon \cdot rad(P)$, $rad(P) \leq \varepsilon \cdot TSP(P)$ hold. Note that if any of the inequalities holds then by plugging any PTAS for TSP in the iterated tour partitioning heuristic, we obtain an $(1 + \mathcal{O}(\varepsilon))$-approximation of k-TC. Thus, the aforementioned heuristic is a PTAS for a substantial range of k depending on P: for every set of points P there is k_0 such that there is a polynomial-time $(1 + \mathcal{O}(\varepsilon))$-approximation algorithm for k-TC for every k with $k_0/\varepsilon < k < \varepsilon k_0$. Despite this observation and despite recent progress in [3,6], the problem of designing a PTAS for *all* k remains open: we believe that our paper sheds the light on this problem and is a step towards a PTAS for arbitrary k.

References

1. Arora, S.: Polynomial time approximation schemes for Euclidean traveling salesman and other geometric problems. Journal of the ACM 45(5), 753–782 (1998)
2. Asano, T., Katoh, N., Tamaki, H., Tokuyama, T.: Covering points in the plane by k-tours: a polynomial time approximation scheme for fixed k. IBM Tokyo Research Laboratory Research Report RT0162 (1996)
3. Asano, T., Katoh, N., Tamaki, H., Tokuyama, T.: Covering points in the plane by k-tours: Towards a polynomial time approximation scheme for general k. In: Proc. 29th Annual ACM Symposium on Theory of Computing, pp. 275–283 (1997)
4. Baker, B.S.: Approximation algorithms for NP-complete problems on planar graphs. Journal of the ACM 41(1), 153–180 (1994)
5. Dantzig, G.B., Ramser, R.H.: The truck dispatching problem. Management Science 6(1), 80–91 (1959)
6. Das, A., Mathieu, C.: A quasi-polynomial time approximation scheme for Euclidean capacitated vehicle routing. In: Proc. 21st Annual ACM-SIAM Symposium on Discrete Algorithms (2010)
7. Garey, M.R., Johnson, D.S.: Computers and Intractability. In: A Guide to the Theory of NP-completeness, W.H. Freeman and Company, New York (1979)
8. Haimovich, M., Rinnooy Kan, A.H.G.: Bounds and heuristics for capacitated routing problems. Mathematics of Operation Research 10(4), 527–542 (1985)
9. Laporte, G.: The vehicle routing problem: An overview of exact and approximate algorithms. European Journal of Operational Research 59(3), 345–358 (1992)
10. Mitchell, J.S.B.: Guillotine subdivisions approximate polygonal subdivisions: A simple polynomial-time approximation scheme for geometric TSP, k-MST, and related problems. SIAM Journal on Computing 28(4), 1298–1309 (1999)
11. Preparata, F., Shamos, M.: Computational Geometry – an Introduction. Springer, New York (1985)
12. Toth, P., Vigo, D.: The Vehicle Routing Problem. SIAM, Philadelphia (2001)

Optimal Randomized Algorithm for the Density Selection Problem*

Tien-Ching Lin and D.T. Lee**

Institute of Information Science, Academia Sinica, Taipei, Taiwan
{kero,dtlee}@iis.sinica.edu.tw
**Department of Computer Science and Information Engineering,
National Taiwan University, Taipei, Taiwan

Abstract. In the paper we consider a generalized version of three well-known problems: SELECTION PROBLEM in computer science, SLOPE SELECTION PROBLEM in computational geometry and MAXIMUM-DENSITY SEGMENT PROBLEM in bioinformatics. Given a sequence $A = (a_1, w_1), (a_2, w_2), \ldots, (a_n, w_n)$ of n ordered pairs (a_i, w_i) of real numbers a_i and $w_i > 0$ for each $1 \le i \le n$, two nonnegative real numbers ℓ, u with $\ell \le u$ and a positive integer k, the DENSITY SELECTION PROBLEM is to find the consecutive subsequence $A(i^*, j^*)$ over all $O(n^2)$ consecutive subsequences $A(i, j)$ satisfying width constraint $\ell \le w(i, j) = \sum_{t=i}^{j} w_t \le u$ such that the rank of its density $d(i^*, j^*) = \sum_{t=i^*}^{j^*} a_t / w(i^*, j^*)$ is k. We will give a randomized algorithm for density selection problem that runs in optimal expected $O(n \log n)$ time.

1 Introduction

Let $A = (a_1, w_1), (a_2, w_2), \ldots, (a_n, w_n)$ be a sequence of n ordered pairs (a_i, w_i) of real numbers a_i and width $w_i > 0$ for each $1 \le i \le n$. A *segment* $A(i, j)$ is a consecutive subsequence of A starting with index i and ending with index j. The *width* $w(i, j)$ of segment $A(i, j)$ is $\sum_{t=i}^{j} w_t$. The *density* $d(i, j)$ of segment $A(i, j)$ is $\sum_{t=i}^{j} a_t / w(i, j)$. Given a sequence $A = (a_1, w_1), (a_2, w_2), \ldots, (a_n, w_n)$ of n ordered pairs (a_i, w_i) of real numbers a_i and $w_i > 0$ for each $1 \le i \le n$, two nonnegative real numbers ℓ, u and a positive integer k, the DENSITY SELECTION PROBLEM (DSP) is to find the feasible segment $A(i^*, j^*)$ over all feasible segments such that the rank of its density $d(i^*, j^*)$ is k. We say that a segment $A(i, j)$ is *feasible* if its width satisfies $\ell \le w(i, j) \le u$. A sequence A is called *uniform width* if all w_i's are identical for each i, otherwise it is called *non-uniform width*.

The density selection problem for uniform width such that $\ell = 1$, $u = 1$ is the most well-known selection problem in computer science. Hoare [11] and

* Research supported in part by the National Science Council under the Grants No. NSC-94-2213-E-001-004, NSC-95-2221-E-001-016-MY3, and NSC 94-2752-E-002-005-PAE, and by the Taiwan Information Security Center (TWISC), National Science Council under the Grant No. NSC94-3114-P-001-001-Y.

Y. Dong, D.-Z. Du, and O. Ibarra (Eds.): ISAAC 2009, LNCS 5878, pp. 1004–1013, 2009.

Floyd and Rivest [9] gave an optimal expected $O(n)$ time randomized algorithm respectively. Blum, Floyd, Pratt, Rivest, and Tarjan [2] gave an optimal $O(n)$ time deterministic algorithm. The density selection problem such that k is equal to the total number of feasible segments is exactly the extensively studied maximum-density segment problem [4,10,12,14,15,18,20] which arises from the problem of finding the biologically meaningful region, called the most GC-ratio region, in a DNA sequence. When we let the input sequence $A = (a_1, w_1), (a_2, w_2), \ldots, (a_n, w_n)$ correspond to a given DNA sequence with uniform width such that $a_i = 1$ if the corresponding nucleotide in the DNA sequence is G or C, and $a_i = 0$ if the corresponding nucleotide in the DNA sequence is A or T. It is obvious that the output feasible segment then corresponds to the most GC-ratio region of the given DNA sequence. The density selection problem for fixed $\ell = 0, u = \infty$, also known as the slope selection problem [3,5,8,13,16], has received much attention in computational geometry. Cole et al. [5] first gave an optimal $O(n \log n)$ time deterministic algorithm for the slope selection problem by combining an approximate counting scheme, the AKS sorting network and parametric search technique. Brönnimann and Chazelle [3] modified their approximate counting scheme combining ε-net to obtain another optimal algorithm for this problem. Dillencourt et al. [8] and Matoušek [16] both gave an optimal randomized Monte Carlo algorithm respectively using the random sampling technique. Katz and Sharir [13] gave an optimal deterministic algorithm using expander graph and approximation technique. In this paper we will give an optimal randomized Monte Carlo algorithm for the density selection problem, using the random sampling technique [8,16], that runs in $O(n)$ space and optimal expected $O(n \log n)$ time. Therefore, it can solve the slope selection problem in optimal expected $O(n \log n)$ time as well.

On the other hand, it was observed that the compositional heterogeneity is highly correlated to the GC content of the genomic sequences [18,21]. The GC-ratios of the DNA sequences in all organisms vary from 25% to 75%. The typical GC-ratios of mammalian genomes stay in 45-50% and the GC-ratios of human DNA in 30-60%, but the GC-ratios have the greatest variations among bacteria's DNA sequences. Therefore, we are also interested in finding the range of the GC-ratios of a DNA sequence for a species. We will consider the DENSITY RANGE QUERY PROBLEM (DRQP) as follows. The input consists of a sequence A of n ordered pairs, two width bounds ℓ, u with $\ell \leq u$ and two real numbers d_l, d_r with $d_l \leq d_r$, the reporting mode of the DRQP is to report all feasible segments $A(i, j)$ satisfying $d_l \leq d(i, j) \leq d_r$ and the counting mode is to count the total number of feasible segments $A(i, j)$ satisfying $d_l \leq d(i, j) \leq d_r$. We will show that the reporting mode and counting mode can be solved in optimal $O(n \log m + h)$ and optimal $O(n \log m)$ time respectively, where $m = min\{\frac{u-\ell}{w_{min}}, n\}$ and h is the output size. Clearly, when $u = \ell$, both DSP and DRQP can easily be solved in $O(n)$ time and space. Therefore, from here on we assume $u > \ell$.

The rest of the paper is organized as follows. Section 2 solves the density range query problem. Section 3 gives an algorithm for the density selection problem. Section 4 gives some conclusion.

2 Algorithm for Density Range Query Problem

In this section we consider the density range query problem. Without loss of generality, we may assume $w_i \geq 1$ for each i and $w_{min} = 1$ for DRQP, since the problem for a sequence A of n ordered pairs (a_i, w_i) with respect to width bounds ℓ and u is equivalent to the problem for a sequence B of n ordered pairs $(\frac{a_i}{w_{min}}, \frac{w_i}{w_{min}})$ with respect to width bounds $\frac{\ell}{w_{min}}$ and $\frac{u}{w_{min}}$.

We first transform the DRQP into a geometric slope range query problem in $O(n)$ time as follows. We define the point set $P = \{p_0, p_1, \ldots, p_n\}$ in $\mathbf{R}^2$ according to the prefix sums of the sequence A, where $p_i = (x_i, y_i) = (\sum_{t=1}^{i} w_t, \sum_{t=1}^{i} a_t)$, $i = 1, 2, \ldots, n$ and $p_0 = (0, 0)$. It is easy to see that the slope $m(i, j)$ of the line segment $s(i, j)$ connecting p_i and p_j is equal to the density $d(i+1, j)$ of segment $A(i+1, j)$, so we can define a line segment $s(i, j)$ is feasible if its corresponding segment $A(i+1, j)$ is feasible.

> Given a point set $P = \{p_0, p_1, \ldots, p_n\}$ in $\mathbf{R}^2$, two width bounds ℓ, u and two density bounds d_l, d_r, find all feasible line segments $s(i, j)$ such that $d_l \leq m(i, j) \leq d_r$.

We can further transform this geometric slope range query problem into its dual problem, by transforming points into lines and vice versa. Consider the dual transform that maps the point $p_i = (x_i, y_i)$ into the dual line $l_i : y = x_i x - y_i$. For any two points p_i, p_j, their corresponding dual lines l_i, l_j will intersect at the point with abscissa $x_{ij} = (y_j - y_i)/(x_j - x_i) = m(i, j)$. It means that the abscissa of the intersection point of the two corresponding dual lines l_i, l_j is equal to the slope $m(i, j)$ of line segment $s(i, j)$. Again, we say that an intersection point of two dual lines l_i, l_j is feasible if $\ell \leq x_j - x_i \leq u$.

> Given a set of dual lines $L = \{l_0, l_1, \ldots, l_n\}$ in $\mathbf{R}^2$, where $l_i : y = x_i x - y_i$, two width bounds ℓ, u and two density bounds d_l, d_r, find all feasible intersection points $p_{ij} = (x_{ij}, y_{ij})$ such that their abscissae $x_{ij} \in [d_l, d_r]$.

Let $L_{a,b}$ denote the subset $\{l_a, l_{a+1}, \ldots, l_b\}$ of L starting with *left index* a and ending with *right index* b. For each dual line l_j we have a set of feasible dual lines $L_{c_j,d_j} = \{l_{c_j}, l_{c_j+1}, \ldots, l_{d_j}\}$, such that each $l_i \in L_{c_j,d_j}$ satisfies $\ell \leq x_j - x_i \leq u$. Without confusion we shall for simplicity denote L_{c_j,d_j} as L_j. Since the slope sequence $\{x_j\}_{j=1}^{n}$ of L is monotonically increasing, the left and right index sequences $\{c_j\}_{j=1}^{n}$ and $\{d_j\}_{j=1}^{n}$ are monotonically increasing respectively. Therefore, we can obtain sequences $\{c_j\}_{j=1}^{n}$ and $\{d_j\}_{j=1}^{n}$ by a linear scan of the sequence $\{x_j\}_{j=1}^{n}$. To solve the dual problem, it suffices to iterate on each j finding all feasible intersection points $p_{ij} = (x_{ij}, y_{ij})$ of L_j and l_j such that their abscissae $x_{ij} \in [d_l, d_r]$.

Instead of solving the dual problem directly we will further transform the dual problem into an orthogonal range query problem in computational geometry. For each dual line $l_i : y = x_i x - y_i$ in L, we let $q_i = (u_i, v_i) = (x_i d_l - y_i, x_i d_r - y_i)$ be the point with abscissa u_i defined by the intercept of l_i at $x = d_l$ and ordinate

v_i defined by the intercept of l_i at $x = d_r$. Let $Q = \{q_0, q_1, \ldots, q_n\}$ and $Q_j = \{q_i \in Q \mid \ell \leq x_j - x_i \leq u\}$. By the monotonically increasing property of the slope sequence $\{x_j\}_{j=0}^{n}$, we know that the slope of l_j is larger than the slope of l_i for each $l_i \in L_j$. Therefore, a dual line l_i in L_j will intersect l_j in $[d_l, d_r]$ if and only if $u_i \geq u_j$ and $v_i \leq v_j$. To solve the dual problem, it is now equivalent to making an orthogonal range query of the form $R_j = [u_j, \infty) \times (-\infty, v_j]$ to report all the points of Q_j which lie in R_j for each $j = 1, 2, \ldots, n$.

We first develop a reporting mode algorithm for the DRQP. Our reporting mode algorithm for the DRQP will iterate from $j = 1$ to n. At any iteration j, we will maintain a data structure $\zeta(Q_j)$ in the current window Q_j such that we can make an orthogonal range query of the form $R_j = [u_j, \infty) \times (-\infty, v_j]$, and then we delete points $q_{c_j}, q_{c_j+1}, \ldots, q_{c_{j+1}-1}$ from $\zeta(Q_j)$ and insert points $q_{d_j+1}, q_{d_j+2}, \ldots, q_{d_{j+1}}$ into $\zeta(Q_j)$ to obtain $\zeta(Q_{j+1})$. We will use a data structure called *priority search tree* to support the above orthogonal range query. A priority search tree [17] is a hybrid of a heap and a balanced binary search tree used for orthogonal range query where at least one of sides of the query range is unbounded. We will make the priority search tree $\zeta(Q_j)$ dynamic to support insertion and deletion operations as well. The priority search tree $\zeta(Q_j)$ can be constructed by using any balanced binary search tree and the performance of the priority search tree is summarized in the following lemma.

Lemma 1 ([7, Theorem 10.9, Page 221]). *The priority search tree $\zeta(S)$ for a set S of n points in $\mathbf{R}^2$ can be constructed in $O(n \log n)$ time and $O(n)$ space. Using the priority search tree we can report all points in a query range of the form $R = [u, w] \times (-\infty, v]$ in $O(\log n + h)$ time, where h is the number of reported points that lie in R.*

McCreight [17] shows that a balanced priority search tree can be made dynamic to support both insertion and deletion operations in $O(\log n)$ time if the number of rotations per updating operation can be bounded by a constant. Tarjan [22] shows that a class of balanced binary trees can be updated in $O(1)$ rotations. For example, a red-black tree belongs to the class. Therefore, if we use a red-black tree as our balanced binary search tree to implement dynamic priority search tree, then both insertion and deletion operations can be updated in $O(\log n)$ time. Since the reporting mode algorithm for the DRQP needs to do totally n times range queries, insertions and deletions on the window Q_j with $|Q_j| \leq m$, the overall running time is therefore $O(n \log m + h)$ by Lemma 1, where $m = min\{u - \ell, n\}$ and h is the output size. We can also develop a counting mode algorithm by using the order-statistics tree data structure similarly. Due to page limitation, we omit it here. Thus, we obtain the following theorem.

Theorem 1. *The reporting and counting mode of the density range query problem can be solved in $O(n)$ space and optimal $O(n \log m + h)$ time and optimal $O(n \log m)$ time respectively, where $m = min\{\frac{u-\ell}{w_{min}}, n\}$ and h is the output size.*

Now, we show that both reporting and counting algorithms of the DRQP are optimal in the worst case. It is known that the ELEMENT UNIQUENESS PROBLEM, i.e., to determine if a set of n real numbers $y_1, y_2, \ldots, y_n$ are all distinct, has a lower bound of $\Omega(n \log n)$ time in the algebraic decision tree model of computation [1]. We can transform an instance of element uniqueness problem to an instance of the DRQP with $\ell = 0$, $u = \infty$, $d_l = d_r = 0$ and $w_i = 1$ for each i in $O(n)$ time by letting $a_1 = y_1$, $a_i = y_i - y_{i-1}$ for $i = 2, \ldots, n$. The output of the reporting mode of the DRQP is an empty set (or The output of the counting mode of the DRQP is 0) if and only if $y_1, y_2, ..., y_n$ are all distinct. Therefore, both the reporting and counting mode of the DRQP has a lower bound of $\Omega(n \log n)$ time in the algebraic decision tree model of computation.

3 Algorithm for Density Selection Problem

In this section we give an optimal randomized Monte Carlo algorithm for the DSP based on three subroutines, random sampling subroutine, reporting mode and counting mode algorithms for the DRQP. The DSP is equivalent to the following problem.

> Given a set of lines $L = \{l_0, l_1, \ldots, l_n\}$ in $\mathbf{R}^2$, where $l_i : y = x_i x - y_i$, find the feasible intersection point $p_{i^*j^*} = (x_{i^*j^*}, y_{i^*j^*})$ such that its abscissa $x_{i^*j^*}$ is the k-th smallest among all feasible intersection points.

For convenience we shall without confusion use the intersection point p_{ij} and its abscissa x_{ij} interchangeably. We first develop a random sampling subroutine running in expected $O(n \log n)$ time to randomly generate $\frac{nN_f}{2N}$ to $\frac{3nN_f}{2N}$ feasible intersection points allowing duplicates such that they all lie in a given interval $[d_l, d_r]$, where N and N_f are the total numbers of intersection points and feasible intersection points in $[d_l, d_r]$ respectively. Note that N and N_f can be obtained by the counting algorithm for the DRQP. Dillencourt et al. [8] developed a random sampling subroutine running in $O(n \log n)$ time by merge sort technique to randomly generate n intersection points such that they all lie in a given interval $[d_l, d_r]$. We summarize it in the following lemma.

Lemma 2 (Dillencourt et al. [8]). *Let $L = \{l_0, l_1, \ldots, l_n\}$ be a set of lines in $\mathbf{R}^2$ and $[d_l, d_r]$ be a given interval. We can obtain a random sampling S by randomly generateing n intersection points of L allowing duplicates in $O(n \log n)$ time such that all points of S are in $[d_l, d_r]$.*

We carefully analyze their random sampling subroutine and find that it can be used to randomly generate $\frac{nN_f}{2N}$ to $\frac{3nN_f}{2N}$ feasible intersection points allowing duplicates such that they all lie in a given interval $[d_l, d_r]$ with high probability by using the well-known Chebyshev's inequality in probability theory.

Whenever we select a random intersection point in $[d_l, d_r]$, it has a probability $\frac{N_f}{N}$ such that it is feasible. Consider such an event as a "success" in performing n independent Bernoulli trials, each with a probability $\frac{N_f}{N}$. Let X_i be the random variable, attaining value 1 with probability $p_x = \frac{N_f}{N}$ and value 0 if otherwise. Let $X = X_1 + X_2 + \cdots + X_n$ be the total number of feasible intersection points for a random sampling S obtained by Lemma 2. The expected value of X is

$\mu = np_x = \frac{nN_f}{N}$ and the standard deviation of X is $\sigma = \sqrt{np_x(1-p_x)} = \sqrt{\frac{nN_f}{N}(1-\frac{N_f}{N})} \le \sqrt{\frac{nN_f}{N}}$. By Chebyshev's inequality, for any $\lambda \ge 0$ we have $Pr[|X-\mu| \ge \lambda\sigma] \le \frac{1}{\lambda^2}$, so the probability $Pr[\mu+\lambda\sigma \ge X \ge \mu-\lambda\sigma] \ge 1-\frac{1}{\lambda^2}$. Therefore, if we choose $\frac{nN_f}{2N} \ge 2\lambda^2 = 4$, we have $Pr[\frac{3nN_f}{2N} \ge \frac{nN_f}{N} + \lambda\sqrt{\frac{nN_f}{N}} \ge \mu+\lambda\sigma \ge X \ge \mu-\lambda\sigma \ge \frac{nN_f}{N} - \lambda\sqrt{\frac{nN_f}{N}} \ge \frac{nN_f}{2N}] \ge \frac{1}{2}$. Hence, if $N_f \ge \frac{8N}{n}$, we can obtain a random sampling S of n intersection points in $[d_l, d_r]$ such that it contains $\frac{nN_f}{2N}$ to $\frac{3nN_f}{2N}$ feasible intersection points with probability no less than $\frac{1}{2}$. Otherwise, if $N_f < \frac{8N}{n}$ we can solve the density selection problem directly by using reporting algorithm for DRQP to enumerate N_f feasible intersection points in $O(n\log m + N_f) = O(n\log m + \frac{8N}{n}) = O(n\log m + n) = O(n\log n)$ time and then select the k-th smallest feasible intersection point d^* from those feasible intersection points by using any standard selection algorithm in $O(n)$ time. Thus, we can assume $N_f \ge \frac{8N}{n}$ from now on. Therefore, we have the following random sampling subroutine.

Lemma 3. *Let $L = \{l_0, l_1, \ldots, l_n\}$ be a set of lines in $\mathbf{R}^2$. Let N and N_f (assuming $N_f \ge \frac{8N}{n}$) be the total numbers of intersection points and feasible intersection points of L in a given interval $[d_l, d_r]$ respectively. We can randomly generate in expected $O(n\log n)$ time a set of n intersection points allowing duplicates in $[d_l, d_r]$ such that they contain M to $3M$ feasible points, where $M = \frac{nN_f}{2N}$.*

We now start to solve the DSP. We shall consider a more general problem, called Density Selection Range Query Problem (DSRQP) defined as follows. Given an interval $[d_l, d_r]$ which contains $N = N_f + N_i$ intersection points where N_f and N_i are the total number of feasible and infeasible intersection points in $[d_l, d_r]$ respectively, we would like to find the k-th smallest feasible intersection point among the N_f feasible intersection points in the interval $[d_l, d_r]$. Let d^* denote the k-th smallest feasible intersection point in the interval $[d_l, d_r]$. Note that the DSP is just a special case of this problem such that $N = \frac{n(n-1)}{2}$, $N_f = O((u-\ell)n)$ and $[d_\ell, d_r] = (-\infty, \infty)$.

The randomized algorithm for the DSRQP will contract the interval $[d_l, d_r]$ into a smaller subinterval $[d_{l'}, d_{r'}]$ such that it also contains d^* and the new subinterval $[d_{l'}, d_{r'}]$ contains at most $O(N_f/\sqrt{M})$ feasible intersection points. It will repeat to contract the interval several times until the interval $[d_{l'}, d_{r'}]$ contains not only d^* but also at most $O(n)$ feasible intersection points. It then outputs all the feasible intersection points in $[d_{l'}, d_{r'}]$ by the reporting mode algorithm for the DRQP and finds the feasible intersection point d^* with an appropriate rank by using any standard selection algorithm.

Our randomized algorithm for the DSRQP runs as follows: We first use our random sampling subroutine to randomly generate a set of feasible intersection points $S' = \{s_1, s_2, \ldots, s_F\}$ in $[d_l, d_r]$. If F is smaller than M or greater than $3M$ we repeat our random sampling subroutine again. From Lemma 3 the probability that the set of n randomly generated intersection points contains M to $3M$ feasible intersection points is no less than $1/2$, so we would perform the random sampling subroutine at most twice on average. Assume that we have obtained a random sampling S' which contains M to $3M$ feasible points. We then try to use this random sampling $S^{'}$ to obtain a smaller subinterval $[d_{l'}, d_{r'}]$ as follows. For each of the selected random feasible intersection point in $S^{'}$, it has a probability $\frac{k}{N_f}$ such that it is smaller than or equal to d^*. Consider such an event as a "success" in performing F independent Bernoulli trials, each with a probability $\frac{k}{N_f}$. Let X_i be the random variable, attaining value 1 with probability $p_x = \frac{k}{N_f}$ and value 0 with probability $p_x = 1 - \frac{k}{N_f}$. Let $X = X_1 + X_2 + \cdots + X_F$ be the total number of sample feasible intersection points falling before d^*. The expected value of X is $\mu_x = Fp_x = \frac{Fk}{N_f}$ and the standard deviation of X is $\sigma_x = \sqrt{Fp_x(1-p_x)}$. It means that the average number of feasible intersection points in $S^{'}$ which is smaller than or equal to d^* is $\frac{Fk}{N_f}$. Hence we expect that the w-th smallest element in $S^{'}$, where $w = \lfloor Fp_x \rfloor = \lfloor \frac{Fk}{N_f} \rfloor$ should be a good approximation for the k-th smallest feasible intersection point d^*. Let $l^{'} = \max\{1, \lfloor \frac{Fk}{N_f} - t\frac{\sqrt{F}}{2} \rfloor\}$ and $r^{'} = \min\{F, \lceil \frac{Fk}{N_f} + t\frac{\sqrt{F}}{2} \rceil\}$, for some constant t to be determined later. Therefore, after we get a successful random sampling $S^{'}$, we can find the $l^{'}$-th smallest element $d_{\ell'}$ and the $r^{'}$-th smallest element $d_{r'}$ in $S^{'}$ by any standard selection algorithm in $O(|S^{'}|)$ time to obtain a subinterval $[d_{\ell'}, d_{r'}]$. The key step of our randomized algorithm for the DSRQP is to check whether the subinterval $[d_{\ell'}, d_{r'}]$ satisfies the following two conditions by the counting algorithm for the DRQP:

(1) The density d^* of the k-th smallest feasible intersection point lies in the subinterval $[d_{\ell'}, d_{r'}]$.
(2) The subinterval $[d_{\ell'}, d_{r'}]$ contains at most $\frac{t^2 N_f}{(t-1)\sqrt{M}}$ $(< \frac{2tN_f}{\sqrt{M}})$ feasible intersection points and contains at most $\frac{3t^2 N}{2(t-1)\sqrt{M}}$ intersection points.

If either (1) or (2) is violated, we repeat our randomized algorithm for the DSRQP from scratch again until both (1) and (2) are satisfied: i.e. we need to randomly select F, where $M \le F \le 3M$, feasible intersection points with replacement in the interval $[d_\ell, d_r]$ by running the random sampling algorithm again to obtain a new subinterval $[d_{\ell'}, d_{r'}]$ and then check the above two conditions (1) and (2) for the new subinterval $[d_{\ell'}, d_{r'}]$. Let k_1 and k_2 be the total number of feasible intersection points lying in $[d_\ell, d_{\ell'})$ and $[d_\ell, d_{r'}]$ respectively. Note that d^* lies in the subinterval $[d_{\ell'}, d_{r'}]$ if and only if $k_1 < k$ and $k_2 \ge k$. If both of these conditions hold, we replace the current interval $[d_\ell, d_r]$ by the subinterval $[d_{\ell'}, d_{r'}]$ and let $k^{'} = k - k_1$.

Note that the density selection algorithm starts with $N = \frac{n(n-1)}{2}$ intersection points and $N_f = O((u-\ell)n)$ feasible intersection points in the initial interval $[d_\ell, d_r] = (-\infty, \infty)$. Therefore, after the first successful random sampling which satisfies conditions (1) and (2) we have an interval $[d_{\ell'}, d_{r'}]$ which contains the $k^{'}$-th smallest feasible intersection point d^* and it contains $O(\frac{N_f}{\sqrt{M}})$ feasible intersection points and $O(\frac{N}{\sqrt{M}})$ intersection points. That is in each iteration we try to prune the numbers of intersection points and feasible intersection points roughly by a factor of $O(\sqrt{M}) = O(\sqrt{\frac{nN_f}{N}})$ respectively, so after one successful random sampling we still maintain the ratio of the number of feasible intersection points and the number of intersection points in $[d_{\ell'}, d_{r'}]$ to be $O(\frac{N_f}{N})$. Therefore, we can repeat the same procedure for the next iteration. After the second successful random sampling which satisfies conditions (1) and (2) we have an interval $[d_{\ell''}, d_{r''}]$ which contains the k''-th smallest feasible intersection point d^* and which has $O(\frac{N_f}{\sqrt{M}}/\sqrt{M}) = O(\frac{N_f}{M}) = O(\frac{N}{n}) = O(n)$ feasible intersection points. We can then enumerate all feasible intersection points from this interval $[d_{\ell''}, d_{r''}]$ by the reporting mode algorithm for the DRQP and select the k''-th smallest feasible intersection point d^* from those feasible intersection points by using any standard selection algorithm. We now show that with a high probability the key step of our randomized algorithm for the DSRQP is satisfied.

Lemma 4. *Let N and N_f be the total numbers of intersection points and feasible intersection points in $[d_l, d_r]$ respectively. For a random choice of F independent feasible intersection points with replacement in the interval $[d_\ell, d_r]$ where $M \leq F \leq 3M$, we can find a subinterval $[d_{\ell'}, d_{r'}]$ containing at most $t\sqrt{F}$ sample feasible intersection points such that the probability that it contains at least $\frac{t^2 N_f}{(t-1)\sqrt{M}}$ feasible intersection points is at most $e^{-\sqrt{M}/2(t-1)}$ and the probability that it contains at least $\frac{3t^2 N}{2(t-1)\sqrt{M}}$ intersection points is at most $e^{-\sqrt{M}/2(t-1)}$.*

Proof. To show the subinterval $[d_{\ell'}, d_{r'}]$ contains at most $\frac{t^2 N_f}{(t-1)\sqrt{M}}$ feasible intersection points with high probability $1 - e^{-\sqrt{M}/2(t-1)}$, we just need to show $[d_{\ell'}, d_{r'}]$ contains at most $\frac{t^2 N_f}{(t-1)\sqrt{F}}$ $(\leq \frac{t^2 N_f}{(t-1)\sqrt{M}})$ feasible intersection points with high probability $1 - e^{-\sqrt{M}/2(t-1)}$. Assume that a successful random sampling $S' = \{s_1, s_2, \ldots, s_F\}$ with replacement in $[d_\ell, d_r]$ in the random sampling subroutine for the DSRQP gives a subinterval $[d_{\ell'}, d_{r'}]$ containing at most $t\sqrt{F}$ sample feasible intersection points. Let N' and N'_f be the total numbers of intersection points and feasible intersection points in $[d_{\ell'}, d_{r'}]$ respectively. Assume that $N'_f \geq \frac{t^2 N_f}{(t-1)\sqrt{F}}$. Hence, whenever we select a random feasible intersection point s_i in $[d_\ell, d_r]$, it has probability larger than $\frac{t^2 N_f/(t-1)\sqrt{F}}{N_f} = \frac{t^2}{(t-1)\sqrt{F}}$ such that s_i lies in $[d_{\ell'}, d_{r'}]$. We again think such an event as a "success", each with a probability of success equal to $p \geq \frac{t^2}{(t-1)\sqrt{F}}$. Let X_i be the random variable, attaining value 1 with probability $p \geq \frac{t^2}{(t-1)\sqrt{F}}$ if the i-th selected feasible intersection point falls in $[d_{\ell'}, d_{r'}]$ and value 0 with probability $1-p$ if otherwise. Let $X = X_1 + X_2 + \cdots + X_F$ be the total number of selected feasible intersection points falling in $[d_{\ell'}, d_{r'}]$. The expectation of the random experiment is $\mu = Fp \geq \frac{t^2 F}{(t-1)\sqrt{F}} = \frac{t^2\sqrt{F}}{t-1}$. By the Chernoff bound, we have $Pr[X \leq t\sqrt{F}] \leq Pr[X \leq (1-\frac{1}{t})\mu] \leq e^{-\mu/2t^2} \leq e^{-\sqrt{F}/2(t-1)} \leq e^{-\sqrt{M}/2(t-1)}$. Therefore, we have the joint probability $Pr[(N_f^{'} \geq \frac{t^2 N_f}{(t-1)\sqrt{F}}) \cap (X \leq t\sqrt{F})] \leq e^{-\sqrt{M}/2(t-1)}$. On the other hand, note that $d_{\ell'}$ and $d_{r'}$ are the $l^{'}$-th and $r^{'}$-th smallest elements in the random sampling $S^{'}$ respectively. It means that the random sampling $S^{'}$ contains exactly $r^{'} - \ell^{'}$ $(\leq t\sqrt{F})$ sample feasible intersection points lying in $[d_{\ell'}, d_{r'}]$. Therefore, we have $Pr[(N_f^{'} \geq \frac{t^2 N_f}{(t-1)\sqrt{F}}) \cap (S^{'}$ contains exactly $r^{'} - \ell^{'}$ sample feasible intersection points in $[d_{\ell'}, d_{r'}])] \leq Pr[(N_f^{'} \geq \frac{t^2 N_f}{(t-1)\sqrt{F}}) \cap (X \leq t\sqrt{F})] \leq e^{-\sqrt{M}/2(t-1)}$. The first part of the lemma follows. Due to page limitation we omit the proof of the second part here.

Lemma 5. *Let N and N_f be the total numbers of intersection points and feasible intersection points in $[d_l, d_r]$ respectively. For a random choice of F independent feasible intersection points with replacement in the interval $[d_\ell, d_r]$ where $M \leq F \leq 3M$, we can find a subinterval $[d_{\ell'}, d_{r'}]$ containing at most $t\sqrt{F}$ sample feasible intersection points such that the probability that the k-th smallest feasible intersection point d^* not lying in the subinterval $[d_{\ell'}, d_{r'}]$ is at most $2e^{-t^2/2}$.*

Proof. Let Y_i be the random variable, attaining value 1 with probability $p = \frac{k}{N_f}$ if the i-th sample feasible intersection point is no greater than d^* and value 0 with probability $1-p$ if otherwise. If the $r^{'}$-th smallest feasible intersection point $d_{r'}$ in S' is smaller than d^*, it means that at least $r^{'}$ among the F randomly sample feasible intersection points fall before d^*. Let $Y = Y_1 + Y_2 + \cdots + Y_F$ be the total number of sample feasible intersection points falling before d^*. By the Chernoff bound, we have $Pr[Y \geq r^{'}] = Pr[Y \geq \mu + t\frac{\sqrt{F}}{2}] \leq e^{-t^2/2}$. Similarly, by the Chernoff bound we have $Pr[Y \leq l^{'}] = Pr[Y \leq \mu - t\frac{\sqrt{F}}{2}] \leq e^{-t^2/2}$.

Now, we can choose t large enough such that $2e^{-t^2/2} \leq \frac{1}{4}$ and choose M large enough such that $2e^{-\sqrt{M}/2(t-1)} \leq \frac{1}{4}$, i.e. choose $N_f \geq \frac{2cN}{n}$ for some large enough constant c such that $e^{-\sqrt{M}/2(t-1)} \leq e^{-\sqrt{c}/2(t-1)} \leq \frac{1}{8}$. For example, we can choose $t = 2.1$ and $c = 21$ respectively. Therefore, we just need to repeat the key step at most twice on the average in the randomized algorithm for the DSP, otherwise we can solve the DSP directly by using reporting algorithm for DRQP and any standard selection algorithm.

Theorem 2. *The* Density Selection Problem *can be solved in $O(n)$ space and expected $O(n \log n)$ time.*

4 Conclusion

In the paper we considered an interesting density selection problem. It is a generalization of three well known problems, the maximum density segment problem, slope selection problem and selection problem. We have presented a randomized algorithm for this problem running in expected $O(n \log n)$ time. But whether the density selection problem can be solved by a deterministic algorithm within the same time bound remains to be seen.

References

1. Ben-Or, M.: Lower bounds for algebraic computation trees. In: Proc. 15th Annu. ACM Sympos. Theory Comput., pp. 80–86 (1983)
2. Blum, M., Floyd, R.W., Pratt, V., Rivest, R.L., Tarjan, R.E.: Time bound for selection. Journal of Computer and System Sciences 7(4), 448–461 (1973)
3. Brönnimann, H., Chazelle, B.: Optimal slope selection via cuttings. Computational Geometry — Theory and Applications 10(1), 23–29 (1998)
4. Chung, K.-M., Lu, H.-I.: An optimal algorithm for the maximum-density segment problem. SIAM Journal on Computing 34(2), 373–387 (2004)
5. Cole, R., Salowe, J.S., Steiger, W.L., Szemeredi, E.: An optimal-time algorithm for slope selection. SIAM Journal on Computing 18(4), 792–810 (1989)
6. Cormen, T.H., Leiserson, C.E., Rivest, R.L.: Introdution to algorithms. MIT Press, Cambridge (1998)
7. De Berg, M., van Kreveld, M., Overmars, M., Schwarzkopf, O.: Computational Geometry: Algorithms and Applications. Springer, Berlin (1997)
8. Dillencourt, M.H., Mount, M.H., Netanyahu, N.S.: A randomized algorithm for slope selection. International Journal of Computational Geometry and Applications 2(1), 1–27 (1992)
9. Robert, W.F., Ronald, L.R.: Expected time bounds for selection. Communications of the ACM 18(3), 165–172 (1975)
10. Goldwasser, M.H., Kao, M.-Y., Lu, H.-I.: Linear-time algorithms for computing maximum-density sequence segments with bioinformatics applications. Journal of Computer and System Sciences 70(2), 128–144 (2005)
11. Hoare, C.A.R.: Algorithm 63 (partition) and algorithm 65 (find). Communications of the ACM 4(7), 321–322 (1961)
12. Huang, X.: An algorithm for identifying regions of a DNA sequence that satisfy a content requirement. Comp. Applications in the Biosciences 10(3), 219–225 (1994)
13. Katz, M.J., Sharir, M.: Optimal slope selection via expanders. Information Processing Letters 47(3), 115–122 (1993)
14. Kim, S.K.: Linear-time algorithm for finding a maximum-density segment of a sequence. Information Processing Letters 86(6), 339–342 (2003)
15. Lin, Y.-L., Huang, X., Jiang, T., Chao, K.-M.: Locating non-overlapping maximum average segments in a given sequence. Bioinformatics 19(1), 151–152 (2003)
16. Matoušek, J.: Randomized optimal algorithm for slope selection. Information Processing Letters 39(4), 183–187 (1991)
17. McCreight, E.M.: Priority search trees. SICOMP 14(2), 257–276 (1985)
18. Nekrutenko, A., Li, W.-H.: Assessment of compositional heterogeneity within and between eukaryotic genomes. Genome Research 10, 1986–1995 (2000)
19. Ohler, U., Niemann, H., Liao, G., Rubin, G.M.: Joint modeling of DNA sequence and physical properties to improve eukaryotic promoter recognition. Bioinformatics 17, 199–206 (2001)
20. Rice, P., Longden, I., Bleasby, A.: Emboss: The European molecular biology open software suite. Trends Genet. 16, 276–277 (2000)
21. Stojanovic, N., Florea, L., Riemer, C., Gumucio, D., Slightom, J., Goodman, M., Miller, W., Hardison, R.: Comparison of five methods for finding conserved sequences in multiple alignments of gene regulatory regions. Nucleic Acids Research 27, 3899–3910 (1999)
22. Tarjan, R.E.: Updating a balanced search tree in O(1) rotations. Information Processing Letters 16, 253–257 (1983)

New Results on Simple Stochastic Games*

Decheng Dai[1] and Rong Ge[2]

[1] Tsinghua University
ddc02@mails.tsinghua.edu.cn
[2] Princeton University
rongge@cs.princeton.edu

Abstract. We study the problem of solving simple stochastic games, and give both an interesting new algorithm and a hardness result. We show a reduction from fine approximation of simple stochastic games to coarse approximation of a polynomial sized game, which can be viewed as an evidence showing the hardness to approximate the value of simple stochastic games. We also present a randomized algorithm that runs in $\tilde{O}(\sqrt{|V_{\mathrm{R}}|!})$ time, where $|V_{\mathrm{R}}|$ is the number of RANDOM vertices and $\tilde{O}$ ignores polynomial terms. This algorithm is the fastest known algorithm when $|V_{\mathrm{R}}| = \omega(\log n)$ and $|V_{\mathrm{R}}| = o(\sqrt{\min |V_{\mathrm{min}}|, |V_{\mathrm{max}}|})$ and it works for general (non-stopping) simple stochastic games.

1 Introduction

1.1 Simple Stochastic Games

Simple stochastic games are games played by two players on a graph, it is a restricted version of general stochastic games introduced by Shapley [1]. In a simple stochastic game, two players (MAX and MIN) move a pebble along directed edges in a graph. The vertices in the graph can have one of the three labels: MAX, MIN or RANDOM. If the pebble is on a vertex labeled MAX(or MIN), then MAX(or MIN) player decides through which out going edge the pebble should move; if the pebble is on a vertex labeled RANDOM, then the pebble moves along a randomly chosen edge. The graph also has a special vertex called the "1-sink". The MAX player wins if and only if the pebble is moved to 1-sink.

SSGs have many interesting applications. In complexity theory, SSGs are used in the analysis of space bounded computations with alternations and randomness [2]. In practice, SSGs are used to model reactive systems. In such systems, RANDOM vertices are used to model stochastic environmental changes, MAX vertices are used to model adversary or arbitrary behaviors, MIN vertices are used to model choices of the system. The 1-sink vertex represents a failure. The goal of the system is thus minimizing the probability of failure (reaching 1-sink vertex).

---

* Supported by the National Natural Science Foundation of China Grant 60553001 and the National Basic Research Program of China Grant 2007CB807900, 2007CB807901.

Y. Dong, D.-Z. Du, and O. Ibarra (Eds.): ISAAC 2009, LNCS 5878, pp. 1014–1023, 2009.

Finding the optimal strategies for SSGs has been an interesting open problem for a long time. A lot of algorithms have been purposed. Condon [2] proved the decision version of SSG is in **NP** $\cap$ **coNP**, and later in 1993, she showed several iterative algorithms for SSG in [3], but all of these algorithms require exponential time. She also suggested an approximation version of SSG problem, but there are no polynomial time algorithms known. Our gap amplification result gives an evidence on why the approximation problem is also difficult. Ludwig gave a sub-exponential ($\tilde{O}(2^{\sqrt{n}})$, $\tilde{O}$ hides polynomial terms) time randomized algorithm for SSGs in [4], which uses local search techniques. Somla [5] purposed a new iterative algorithm in 2004 which might be better than previous algorithms, however there's no evidence that shows the algorithm runs in polynomial time. Recently, Gimbert and Horn [6] presented a new non-iterative algorithm that runs in time $\tilde{O}(|V_{\mathrm{R}}!|)$. This highlights one of the main reasons the problem has exponential complexity: the existence of random vertices.

1.2 Our Results

In this paper, we investigate the SSG problem in both hardness and algorithmic aspects. On the hardness side, we show that a coarse approximation of SSGs is as hard as a fine approximation. This is done by constructing a new game G' from a game G, such that G' has polynomial size and a coarse approximation to G' would give a fine approximation of game G. Viewed pessimistically this can be an evidence that shows it is hard to even approximate SSGs; viewed optimistically, this may give hope of deriving a good approximation algorithm for SSGs.

At the algorithmic side, we present an algorithm based on the algorithm of Gimbert and Horn [6]. They considered a set of strategies called **f**-strategies, and showed at least one of **f**-strategies is optimal. However they were not able to distinguish "good" **f**-strategies and "bad" **f**-strategies. By finding a way to evaluate the "correctness" of **f**-strategies, we are able to apply local search algorithms to find the optimal **f**-strategy, and reduce the running time to $\tilde{O}(\sqrt{|V_{\mathrm{R}}|!})$. Our algorithm is the fastest known randomized algorithm for solving SSGs when $|V_{\mathrm{R}}| = \omega(\log n)$ and $|V_{\mathrm{R}}| = o(\sqrt{\min\{|V_{\max}|, |V_{\min}|\}})$.

In Sect.2 we give definitions for Simple Stochastic Games and strategies. Then we describe the reduction from fine approximation to coarse approximation in Sect.3. After that, we give a brief introduction to **f**-strategies and then present our algorithm.

2 Basic Definitions

There are many variations of SSGs, we define the game formally as follows

Definition 1 (Simple Stochastic Games). *A simple stochastic game is specified by a directed graph $G =< V, E >$ and a starting vertex $v_{start} \in V$. Each vertex $v \in V$ has 2 outgoing edges and a label (MAX, MIN or RANDOM).*

V_{min}, V_{max}, V_R are the sets of vertices with label MIN, MAX and RANDOM respectively. There's a special vertex v_1 (the 1-sink) in the graph.

Initially the pebble is at v_{start}. If the pebble is at a MAX/MIN vertex, then the corresponding player moves the pebble along one of the outgoing edges. If the pebble is at a RANDOM vertex, then the pebble moves along a random outgoing edge (both edges are chosen with probability 1/2). If the pebble reaches v_1 then MAX player wins, otherwise MIN player wins.

Solving SSGs means calculating the winning probabilities for the players if they all follow optimal strategy. Informally, the strategy of a player decides which edge should the pebble follow in the game. Although a strategy can decide the edge by considering history or using random coins, it's well known that *positional* optimal strategies exist for simple stochastic games ([1,2]). A positional strategy makes the decision only by the current position of the pebble. Formally, a positional strategy for MAX player α is a function from $V_{\max}$ to V, for any vertex $v \in V$, $(v, \alpha(v))$ is an edge and it is the outgoing edge that the MAX player would choose if the pebble is currently at vertex v. Similarly, a positional strategy for MIN player β is a function from $V_{\min}$ to V. From now on when we mention strategy we mean positional strategy.We define the value of a vertex to be the winning probability of the MAX player if initially the pebble is at this vertex, and denote this by $\mathsf{val}(v)$, the value of the game is $\mathsf{val}(v_{\text{start}})$. When it's not clear which game we are talking about, we use $\mathsf{val}[G](v)$ to specify the value of v in game G.

In simple stochastic games, MIN player wins the game by forcing the pebble to move infinitely many steps without reaching the 1-sink. Sometimes it's easier to consider the situation that the game has two sinks: a 0-sink and a 1-sink. The game guarantees no matter what strategies the players use, with probability 1 the game will reach one of the sinks in finitely many steps. The goal of MAX player is to reach the 1-sink and the goal of MIN player is to reach the 0-sink. This variation of SSG is called stopping simple stochastic games (stopping-SSG). Condon showed in [2] that any SSG can be converted to a stopping-SSG in polynomial time while the change in the value of the game is exponentially small.

3 Coarse Approximation Is as Hard as Fine Approximation

Since no polynomial time algorithms has been discovered for exactly solving SSGs, Condon[2] purposed the following "approximation" version of the problem. Consider the following sets,

$$L_{yes} = \{G : \text{ the value of } G \text{ is at least } \frac{1}{2} + \epsilon\} \ ,$$

$$L_{no} = \{G : \text{ the value of } G \text{ is at most } \frac{1}{2} - \epsilon\} \ .$$

An ϵ-gap SSG decision problem is to determine whether G is in L_{yes} or L_{no} given it is in one of them.Intuitively it might seem for some large enough ϵ this problem is easy to solve. However, we give a gap amplification reduction showing that when enlarging ϵ from $(1/\operatorname{poly}(n))$ to $(1/2 - e^{-n^\rho})$ for any $\rho < 1$, the problem does not become easier. This reduction is analogue to the hardness amplification results for clique and chromatic number problem.

Theorem 1. *For any fixed constant $0 < \rho < 1$ and $c > 0$, if the $(1/2 - e^{-n^\rho})$-gap SSG decision problem is in* **P**, *then the (n^{-c})-gap SSG decision problem is in* **P**.

Proof. First we prove the theorem for stopping SSG.

Now let's assume $G = \langle V, E\rangle$ is a stopping-SSG with n vertices. There are 3 special vertices in a stopping SSG: v_{start}, the starting vertex; v_1, the 1-sink vertex; v_0, the 0-sink vertex. We construct another game $G' = \langle V', E'\rangle$ of size N (which is polynomial in n) such that,

- G' has value larger than $(1 - e^{-N^\rho})$ if G has value larger than $(1/2 + n^{-c})$
- G' has value less than e^{-N^ρ} if G has value less than $(1/2 - n^{-c})$

Let $\{G_0, G_{i,j} | i \in \{0,1\}, j \in \{1, ..., K\}\}$ be $2K + 1$ copies of G. We replace the out-going edges for v_0 and v_1 in each of these games to connect them together in the following way (their two outgoing edges will point to the same location, so it doesn't matter what label they have)

- Connect v_0 in G_0 to v_{start} in $G_{0,1}$, connect v_1 in G_0 to v_{start} in $G_{1,1}$.
- Connect v_1 in $G_{0,j} (1 \le j \le K)$ to v_{start} in G_0, connect v_0 in $G_{0,j} (j < K)$ to v_{start} in $G_{0,j+1}$.
- Connect v_0 in $G_{1,j} (1 \le j \le K)$ to v_{start} in G_0, connect v_1 in $G_{1,j} (j < K)$ to v_{start} in $G_{1,j+1}$.
- The starting vertex in G' is v_{start} in G_0, and the 0-sink vertex is v_0 in $G_{0,K}$ and the 1-sink vertex is v_1 in $G_{1,K}$.

In this constructed game G', the MIN(MAX) player must win G_0 and all $G_{0,j}(G_{1,j})$ to win G'. Let p to be the value in G. By induction, it is easy to prove the probability to reach v_1 in $G_{1,K}$ is $(p^K)/(p^K + (1-p)^K)$.

Let $K = n^d$ where $d = (c+1)/(1-\rho)$, then $N = (2K+1)n = O(n^{d+1})$ when $p \le 1/2 - n^{-c}$ we have,

$$\begin{aligned}\frac{p^K}{p^K + (1-p)^K} &\le \frac{(\frac{1}{2} - n^{-c})^K}{(\frac{1}{2} - n^{-c})^K + (\frac{1}{2} + n^{-c})^K} \\ &\le (1 - 2n^{-c})^K \le e^{-n^{d-c}} \\ &\le e^{-N^\rho} \quad ,\end{aligned}$$

so the value of G' is less than e^{-N^ρ} in this case. Similarly, when $p \ge 1/2 + n^{-c}$, we have the value of G' is larger than $1 - e^{-N^\rho}$. That is,

$$\begin{aligned}\mathsf{val}(G) \le \frac{1}{2} - n^{-c} \Rightarrow \mathsf{val}(G') \le e^{-N^\rho} \quad , \\ \mathsf{val}(G) \ge \frac{1}{2} + n^{-c} \Rightarrow \mathsf{val}(G') \ge 1 - e^{-N^\rho} \quad .\end{aligned}$$

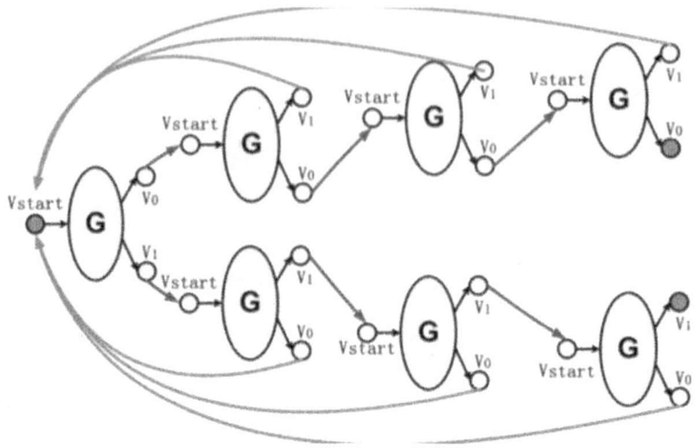

Fig. 1. An example on constructing G' from G, for $K = 3$. Every ellipse is a copy of G. Three solid vertices are v_{start},v_0,v_1. It is easy to check that the probability to reach the v_0 in G' is exactly $p^3/(p^3 + (1-p)^3)$, in which $p = \mathsf{val}[G](v_0)$.

By applying the algorithm for $(1/2 - e^{-n^{\rho}})$-gap SSG decision problem on G', the algorithm would be able to distinguish between $\mathsf{val}(G) > 1/2 + n^{-c}$ and $\mathsf{val}(G) < 1/2 - n^{-c}$.

For general (possibly non-stopping) SSG, we use Condon's reduction in [2] that transforms a SSG G to a stopping SSG G' whose value is arbitrarily close to the value of G. The constructed stopping game G' adopts all the vertices of G and inserts cnm new vertices($m = |E|$ in G). For any vertex v, $|\mathsf{val}[G](v) - \mathsf{val}[G'](v)| \le 2^{(2-c)n}$. By combining these two constructions, we can reduce solving the (n^{-c})-gap decision problem to $(1/2 - e^{-n^{\rho}})$-gap SSG decision problem.

4 Fast Algorithm for SSGs with Few Random Vertices

An interesting case for solving simple stochastic games is when there are a few random vertices. Gimbert and Horn[6] found an algorithm that runs in $\tilde{O}(|V_{\text{R}}|!)$ time. Their algorithm is based on enumerating a special kind of strategies called **f**-strategies. To avoid simple and time consuming enumerations, our algorithm relies on the following Lemma:

Lemma 1 (Main Lemma). *There's a partial order in* **f***-strategies such that the following holds:*

1. *Any maximal* **f** *corresponds to a pair of optimal strategies.*
2. *Two* **f***-strategies can be compared in polynomial time*
3. *If* **f** *is not maximal, then in polynomial time we can find* **g** *which is better than* **f**

The **f**-strategies are first introduced in [6], and they proved a theorem(Lemma 3 in this paper) on testing whether the **f**-strategy is optimal or not. The brute-force idea is to enumerate all possible $O(n!)$ **f**-strategies and use Lemma 3 to find the optimal one. Our major contrubution is this Main Lemma, which reduces the problem to a local maximal searching problem and thus enabled us to design faster algorithms.

4.1 f-Strategies

In this section we'll first briefly describe [6]'s ideas on what are **f**-strategies and how to test their optimality; this is first introduced in [6] and we mention it again for completeness. Then we show how the partial order in the Main Lemma is defined and prove the Main Lemma. Finally we use existing randomized algorithms for local search problems to improve the expected running time to $\tilde{O}(\sqrt{|V_R|!})$.

Let $\mathbf{f} = \langle r_1, \ldots, r_m \rangle$ (for simplicity let $r_0 = v_1$) be a permutation of the random vertices, where $m = |V_R|$ is the the number of the random vertices. A **f**-strategy is a pair of positional strategies associated to **f**.

Let R_i be the first i random vertices in the permutation **f**. The *consuming set* C_i is a set of vertices from which player MAX has a strategy $\sigma_\mathbf{f}$ for moving the pebble to R_i and at the same time avoid touching any other random vertices, no matter what strategy player MIN chooses. Similarly, there's also a strategy $\tau_\mathbf{f}$ for player MIN, such that no matter what player MAX does, vertices outside C_i can never reach a vertex in R_i without touching other random vertices.Obviously $C_0 \subseteq C_1 \subseteq \ldots \subseteq C_m$. This pair of strategies $(\sigma_\mathbf{f}, \tau_\mathbf{f})$ is called the **f**-strategy regarding to the permutation **f**. For any permutation **f**, let $\mathsf{val}_\mathbf{f}(r_i)$ be the probability for player MAX to win if the game starts at vertex r_i, when players follow the **f**-strategies $(\sigma_\mathbf{f}, \tau_\mathbf{f})$. The following lemmas are first proved in [6].

Lemma 2 (f-strategy). *Given any permutation* **f***, the corresponding* $\sigma_\mathbf{f}$ *and* $\tau_\mathbf{f}$ *always exist and can be found in polynomial time.*

Lemma 3. *If* **f** *satisfies the* Consistency *and* Progressive *conditions, then the* **f***-strategy is an optimal strategy for the game.*

Consistency: $\mathsf{val}_\mathbf{f}(r_1) \geq \cdots \geq \mathsf{val}_\mathbf{f}(r_m)$

Progressive: For any random vertex r_i $(i > 0)$ *with* $\mathsf{val}_\mathbf{f}(r_i) > 0$*, at least one of its outgoing edges points to a vertex in* C_{i-1}*.*

There always exists a permutation **f** *that satisfy both conditions.*

For constructing $\{C_i\}$ in polynomial time and more discussions about the *Consistency* and *Progressive* conditions, see [6].

4.2 The Partial Order for f-Strategies

A natural way to improve the algorithm by Gimbert and Horn would be smartly updating **f** when it is not *Consistent* or not *Progressive*. However, it is hard to tell which permutation better by simply looking at the values for vertices.

To estimate whether a particular ordering is good or not, we construct a new SSG with respect to the ordering.

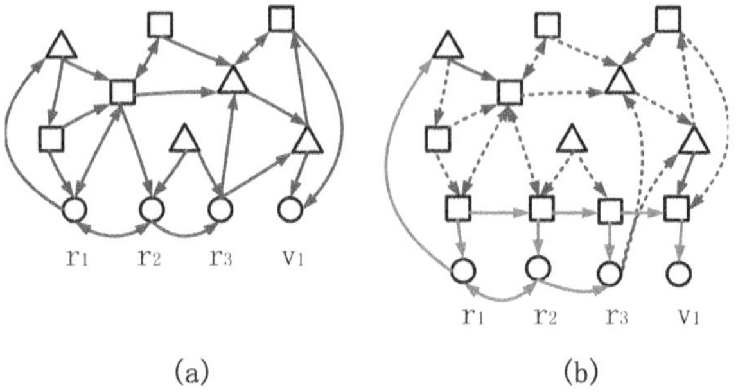

Fig. 2. (a)The game G, Δ are MIN vertices, $\square$ are MAX vertices, $\bigcirc$ are RANDOM vertices. (b)The graph $G_{\mathbf{f}}$, in which $\mathbf{f} = \langle r_1, r_2, r_3, v_1 \rangle$. 4 MAX vertices are added. The dashed lines are the original edges and the solid lines are the added edges.

Definition 2 (value measure H(f)). *Let G be a SSG and $\mathbf{f}$ be an ordering of random vertices, $G_{\mathbf{f}}$ is a new SSG. $G_{\mathbf{f}}$ has all the vertices and edges in G and m new vertices $u_1, u_2, ..., u_m$, all of them are MAX vertices. The two outgoing edges of u_i go to r_i and r_{i+1} (both outgoing edges of u_m go to r_m). All edges of the form (v, r_i) in G are replaced by (v, u_i) in G_r. Let $H(\mathbf{f}) \triangleq \sum_{i=1}^{m} \mathsf{val}[G_{\mathbf{f}}](u_i)$.*

Let $H_{OPT} = \sum_{i=1}^{m} \mathsf{val}[G](r_i)$. An example on how to compute $H(\mathbf{f})$ is showed in Fig 2. In G, the values are $\mathsf{val}(r_1) = 0, \mathsf{val}(r_2) = 0.5, \mathsf{val}(r_3) = \mathsf{val}(v_1) = 1$. In $G_{\mathbf{f}}$, the values are $\mathsf{val}(r_1) = \mathsf{val}(r_2) = \mathsf{val}(r_3) = \mathsf{val}(v_1) = 1$. So $H(\mathbf{f}) = 4 > H_{OPT}$.

Lemma 4. *For any permutation $\mathbf{f}$, $H(\mathbf{f}) \geq H_{OPT}$. When $\mathbf{f}$ is both Consistent and Progressive, $H(\mathbf{f}) = H_{OPT}$*

Proof. Consider a permutation $\mathbf{f}$ and its corresponding $G_{\mathbf{f}}$, assume α, β is a pair of optimal strategies for the original game G. Now we construct a strategy α' for player MAX in Game $G_{\mathbf{f}}$: $\alpha'(v) = \alpha(v)$ for all $v \in G$; $\alpha'(u_i) = r_i$ for all $u_i, 1 \leq i \leq m$. When player MAX takes this strategy, it is easy to check β is the also best response for player MIN in G'. So for every $v \in G$, $\mathsf{val}[G_{\alpha,\beta}](v) = \mathsf{val}[G_{\mathbf{f},\alpha',\beta}](v)$ and $\sum_{i=1}^{m} \mathsf{val}(u_i) = \sum_{i=1}^{m} \mathsf{val}[G_{\mathbf{f},\alpha',\beta}](v) = H_{OPT}$. However, MAX may have better strategies in $G_{\mathbf{f}}$, so $H(\mathbf{f}) \geq H_{OPT}$.

When $\mathbf{f}$ is Consistent and Progressive in G, we first prove that $\mathbf{f}$ is also Consistent and Progressive in $G_{\mathbf{f}}$. Let C_i be the consuming sets in G regarding to $\mathbf{f}$. By analyzing the structure of graph $G_{\mathbf{f}}$, we have $C_i' = C_i \cup \{u_1, u_2, ..., u_i\}$ are consuming sets in $G_{\mathbf{f}}$. Consider the strategies (α', β) as defined in the former case. Using the definition of $\mathbf{f}$-strategy, it is easy to verify that (α', β) are $\mathbf{f}$-strategy for $G_{\mathbf{f}}$. So $\mathsf{val}[G_{\mathbf{f}}](r_i) = \mathsf{val}[G](r_i)$, which means $\mathbf{f}$ is still Consistent and Progressive for $G_{\mathbf{f}}$. Therefore $H(\mathbf{f}) = H_{OPT}$.

To compute the optimal strategy and values for $G_{\mathbf{f}}$, we use the following Lemma.

Lemma 5. *For any permutation $\mathbf{f}$ and the $\mathbf{f}$-strategy (σ, τ) in G, there is an optimal strategy (σ', τ') for $G_{\mathbf{f}}$ such that for all $v \in G$, $\tau(v) = \tau'(v)$ and $\sigma(v) = \sigma'(v)$.*

Proof. Let the permutation $\mathbf{f} = \langle r_0, \dots r_m \rangle$ and the corresponding $\mathbf{f}$-strategy $(\sigma_{\mathbf{f}}, \tau_{\mathbf{f}})$. Now we construct strategy for $G_{\mathbf{f}}$ satisfies the conditions.

Denote the consuming sets for $\mathbf{f}$ as $\{C_i\}$. Let $\mathbf{g} = (r_0, r_{t_1}, r_{t_2}, ..., r_{t_m})$ be a permutation for $G_{\mathbf{f}}$ which is consistent and progressive (this ordering always exists by Lemma 3). Since G and $G_{\mathbf{f}}$ have the same random vertices, $\mathbf{f}$ and $\mathbf{g}$ are permutations over the same set. Denote the consuming sets for $\mathbf{g}$ as $\{C'_i\}$, we have $\bigcup_i C_i = \bigcup_i C'_i$. By the construction of $G_{\mathbf{f}}$, we have $C_i = \cup_{j=1}^{i}(C_j \cup \{u_j\})$ and $C'_i = \cup_{j=1}^{\max t_1, t_2, \dots, t_i}(C_j \cup \{u_j\})$.This is because in the $\mathbf{f}$-strategy in G, player MAX's strategy ensures C_i can reach R_i while MIN strategy ensures that no other vertices outside C_i can reach R_i.

Now the optimal strategies (σ', τ') are defined as follows.

$$\sigma'(v) = \begin{cases} \sigma(v), \text{ if } v \in G \\ r_{t_i}, \text{ if } v = u_{t_i} \text{ and } t_i = \max\limits_{j \le i} t_j \\ u_{t_i+1}, \text{ if } v = u_{t_i} \text{ and } t_i < \max\limits_{j \le i} t_j \end{cases} \tag{1}$$

Since the MIN vertices in $G_{\mathbf{f}}$ are the same with G, we simply let $\tau'(v) = \tau(v)$.

Then for any i, the strategy σ' makes sure that no matter what strategy the MIN player uses, C'_i always reach a vertex in $\{r_0, r_1, ..., r_{t_i}\}$. Similarly, the strategy τ' makes sure that no matter what strategy that the MAX player uses, vertices outside C'_i can never reach a vertex in $\{r_0, r_1, ..., r_{t_i}\}$. So (σ', τ') is a valid $\mathbf{f}$-strategy for the permutation $\mathbf{g}$. Since $\mathbf{g}$ is consistency and progressive, (σ', τ') is therefore optimal by Lemma 3.

By this lemma we can find the optimal strategy for MIN player in $G_{\mathbf{f}}$ in polynomial time, because we know that the strategy for player MIN in $\mathbf{f}$-strategy for G is also an optimal strategy for MIN player in $G_{\mathbf{f}}$. By using linear programming we can find the optimal strategy for player MAX in polynomial time.

Definition 3 (progressiveness measure P(f)). *For an permutation* $\mathbf{f} = \langle r_1, \dots r_m \rangle$, $P(\mathbf{f})$ *is the smallest* $i (i > 0)$ *such that* r_i *does not have an outgoing edge to* C_{i-1}. *If there's no such* i *or* $\mathsf{val}(r_i) = 0$ *then* $P(r) = m + 1$.

Denote the set of all permutations over the random vertices as Π. Searching this space and output the consistent and progressive one takes $\tilde{O}(m!)$ time. But an partial order over Π may help us to find this ordering. We say $\mathbf{f} > \mathbf{g}$ if (1) $H(\mathbf{f}) < H(\mathbf{g})$ or (2) $H(\mathbf{f}) = H(\mathbf{g})$ and $P(\mathbf{f}) > P(\mathbf{g})$.

Any maximal element in $(\Pi, >)$ corresponds to an permutation that is both Consistent and Progressive. Therefore we have proved the first 2 parts of the Main Lemma. To prove part 3 of the Main Lemma, we use the following lemma as a tool to upperbound the Π value.

Lemma 6. *If function* $f : V \to [0, 1]$ *satisfy the following conditions, then* $\mathsf{val}(v) \le f(v)$ *for every vertex* v.

1. *For vertex* v_1, $f(v_1) = 1$;
2. *For vertex* $v \in V_R$, *assume the two outgoing edges are* (v, w_1), (v, w_2), $f(v) \geq (f(w_1) + f(w_2))/2$;
3. *For vertex* $v \in V_{MAX}$, *assume the two outgoing edges are* (v, w_1), (v, w_2), $f(v) \geq \max(f(w_1), f(w_2))$;
4. *For vertex* $v \in V_{MIN}$, *assume the two outgoing edges are* (v, w_1), (v, w_2), $f(v) \geq \min(f(w_1), f(w_2))$.

Due to the limit of space, we defer the proof to the full version of the paper.

Lemma 7. *If an permutation* $\mathbf{f} = \langle r_1, \ldots r_m \rangle$ *is not maximal, then there exists an element* r_i *in* $\mathbf{f}$, *by deleting* r_i *and reinsert it in appropriate place we get a new ordering* $\mathbf{g}$ *such that* $\mathbf{g} > \mathbf{f}$.

Proof. If the ordering $\mathbf{f}$ is not consistent in $G_{\mathbf{f}}$, then there exists some t such that $\mathsf{val}[G_{\mathbf{f}}](r_t) < \mathsf{val}[G_{\mathbf{f}}](r_{t+1})$. Find a place $q > t$ so that $\mathsf{val}[G_{\mathbf{f}}](u_q) < \mathsf{val}[G_{\mathbf{f}}](r_{t+1})$ (if there's no such place then let $q = m + 1$). Delete r_t and reinsert it right before q (if $q = m + 1$ then insert it at the tail). Define $f(v) = \mathsf{val}[G_{\mathbf{f}}](v)$, then for graph $G_{\mathbf{g}}$ f is a valid value function that satisfy the requirements of Lemma 6. Therefore for any vertex v $\mathsf{val}[G_r](v) \geq \mathsf{val}[G_{\mathbf{g}}](v)$. Particularly for the current position of r_t, the corresponding u vertex is u_q in $G_{\mathbf{g}}$, $f(u_q) > \max(f(u_{q+1}), f(r_t))$, so even after reducing $f(u_q)$ to $\max(f(u_{q+1}), f(r_t))$, f is still valid. That is, $H(\mathbf{g}) < H(\mathbf{f})$, $\mathbf{g} > \mathbf{f}$.

If the ordering $\mathbf{f}$ is consistent but not progressive, then assume $P(\mathbf{f}) = t$. Define a graph among the random vertices and r_0 as follows: if an original outgoing edge of r_i goes to a vertex v that is in $C_j \backslash C_{j-1}$, then there's an edge from r_i to r_j. Use breadth first search to find $t' > t$, such that the following holds:

1. There's an edge from $r_{t'}$ to $\{r_0, r_1, ..., r_{t-1}\}$.
2. There's a path from r_t to $r_{t'}$.

Note that such t' must exist because otherwise following the $\mathbf{f}$-strategy, starting from r_t, the pebble will never be able to reach r_0, and therefore the value of r_t is 0, which contradict with the fact that $P(r) \neq m+1$. Also, $\mathsf{val}(r_{t'}) = \mathsf{val}(r_t)$, because if the path from r_t to $r_{t'}$ is $(w_0, w_1, ..., w_k)$ ($w_0 = r_t$ and $w_k = r_{t'}$), then since $\mathsf{val}(w_i) = (\mathsf{val}(w_{i+1}) + \mathsf{val}(r^*))/2$, both w_{i+1} and r^* are ranked lower than t, $\mathsf{val}(w_{i+1}) \leq \mathsf{val}(r_t)$, $\mathsf{val}(r^*) \leq \mathsf{val}(r_t)$. But $\mathsf{val}(w_0) = \mathsf{val}(r_t)$, by induction for all i $\mathsf{val}(w_i) = \mathsf{val}(w_0) = \mathsf{val}(r_t)$.

Now delete $r_{t'}$ and insert it back before r_t to get a new ordering $\mathbf{g}$. Define $f(v) = \mathsf{val}[G_{\mathbf{f}}](v)$, then for graph $G_{\mathbf{g}}$ f is a valid value function that satisfy the requirements of Lemma 6. Therefore for any vertex v $\mathsf{val}[G_{\mathbf{f}}](v) \geq \mathsf{val}[G_{\mathbf{g}}](v)$. Either all values are equal, in this case $H(\mathbf{f}) = H(\mathbf{g})$ but $P(\mathbf{g}) > P(\mathbf{f})$ so $\mathbf{g} > \mathbf{f}$; or some values are different, in this case $H(\mathbf{g}) < H(\mathbf{f})$ so $\mathbf{g} > \mathbf{f}$.

Since there are only polynomially many ways to delete and reinsert an element, a better ordering can always be found in polynomial time.

4.3 The Randomized Algorithm

Now we can use the existed randomized local minimum searching algorithm to solve the simple stochastic game. The algorithm to solve the value of a simple stochastic game $G = (V, E)$:

1. Randomly choose $\sqrt{|V_{\mathrm{R}}|!}\log(|V_{\mathrm{R}}|!)$ permutations, and let $\mathbf{f_0}$ be the maximal permutation among them;
2. Starting from $\mathbf{f_0}$, repeatedly find better permutation until a maximal permutation is found

By Lemma 7, we can always find a better permutations unless $\mathbf{f}$ is maximal, and there are only $|V_{\mathrm{R}}|!$ permutations, the algorithm will eventually find a maximal permutation and thus the optimal strategy.

The first step takes $\tilde{O}(\sqrt{|V_{\mathrm{R}}|!})$ time, after that, each iteration of the loop will take poly$(|V_{\mathrm{R}}|)$ time, so the key is how many iterations step 2 needs.

Lemma 8. *The probability that step 2 needs more than $\sqrt{|V_R|!}$ steps is no more than $1/(|V_R|)!$.*

Proof. Consider any total ordering of the permutations that agrees with the partial ordering we defined. The probability that none of the $\sqrt{|V_{\mathrm{R}}|!}$ largest elements are chosen is at most $(1-\sqrt{|V_{\mathrm{R}}|!}/(|V_{\mathrm{R}}|!))^{\sqrt{|V_{\mathrm{R}}|!}\log(|V_{\mathrm{R}}|!)} = e^{-\log(|V_{\mathrm{R}}|!)} = 1/(|V_{\mathrm{R}}|)!$.

Therefore, the expectation of number of iterations is at most $\sqrt{|V_{\mathrm{R}}|!}$. The running time is $\tilde{O}(\sqrt{|V_{\mathrm{R}}|!})$.

References

1. Shapley, L.: Stochastic games. In: Proceedings of the National Academy of Sciences (1953)
2. Condon, A.: The complexity of stochastic games. Inf. Comput. 96(2), 203–224 (1992)
3. Condon, A.: On algorithms for simple stochastic games. In: Advances in Computational Complexity Theory. DIMACS Series in Discrete Mathematics and Theoretical Computer Science, vol. 13, pp. 51–73. American Mathematical Society, Providence (1993)
4. Ludwig, W.: A subexponential randomized algorithm for the simple stochastic game problem. Inf. Comput. 117(1), 151–155 (1995)
5. Somla, R.: New algorithms for solving simple stochastic games. In: Proceedings of the Workshop on Games in Design and Verification (GDV 2004). Electronic Notes in Theoretical Computer Science, vol. 119, pp. 51–65. Elsevier, Amsterdam (2005)
6. Gimbert, H., Horn, F.: Solving simple stochastic games. In: Beckmann, A., Dimitracopoulos, C., Löwe, B. (eds.) CiE 2008. LNCS, vol. 5028, pp. 206–209. Springer, Heidelberg (2008)

Worst-Case and Smoothed Analysis of k-Means Clustering with Bregman Divergences

Bodo Manthey[1] and Heiko Röglin[2,*]

[1] Department of Applied Mathematics, University of Twente
b.manthey@utwente.nl
[2] Department of Quantitative Economics, Maastricht University
heiko@roeglin.org

Abstract. The k-means algorithm is the method of choice for clustering large-scale data sets and it performs exceedingly well in practice. Most of the theoretical work is restricted to the case that squared Euclidean distances are used as similarity measure. In many applications, however, data is to be clustered with respect to other measures like, e.g., relative entropy, which is commonly used to cluster web pages. In this paper, we analyze the running-time of the k-means method for Bregman divergences, a very general class of similarity measures including squared Euclidean distances and relative entropy. We show that the exponential lower bound known for the Euclidean case carries over to almost every Bregman divergence. To narrow the gap between theory and practice, we also study k-means in the semi-random input model of smoothed analysis. For the case that n data points in $\mathbb{R}^d$ are perturbed by noise with standard deviation σ, we show that for almost arbitrary Bregman divergences the expected running-time is bounded by $\text{poly}(n^{\sqrt{k}}, 1/\sigma)$ and $k^{kd} \cdot \text{poly}(n, 1/\sigma)$.

1 Introduction

Clustering a set of objects into a certain number of classes so as to maximize the similarity of objects in the same class is a fundamental problem with applications in various areas like information retrieval, bioinformatics, and data compression. Usually the objects are represented by points in $\mathbb{R}^d$, and they are to be clustered into k classes $\mathcal{C}_1, \ldots, \mathcal{C}_k$ that can be represented by centers $c_1, \ldots, c_k \in \mathbb{R}^d$ such that the sum $\sum_{i=1}^{k} \sum_{x \in \mathcal{C}_i} d(x, c_i)$ becomes minimal for some distance measure d. A common distance function d are squared Euclidean distances but in many practical applications other distance measures are required. For instance, when clustering text documents like web pages often the *bag-of-words model* [7] is applied, in which the objects to be clustered are probability distributions over the set of all words. A popular distance measure for probability distributions is the *Kullback-Leibler divergence* (KLD, also known as relative entropy). Both

* Supported by a fellowship within the Postdoc-Program of the German Academic Exchange Service (DAAD).

Y. Dong, D.-Z. Du, and O. Ibarra (Eds.): ISAAC 2009, LNCS 5878, pp. 1024–1033, 2009.

squared Euclidean distances and KLD are special cases of *Bregman divergences*, a very general class that contains most practically important distance measures.

Even though a lot of theoretical research has been conducted on clustering algorithms, the by far most successful algorithm in industrial and scientific applications is the seemingly ad hoc *k-means method* [6], a local search algorithm due to Lloyd [12]: Start with an arbitrary set of k centers and repeat the following two steps until the process stabilizes: 1) Assign every data point to its closest center. 2) Readjust the positions of the centers such that they are optimal for the current assignment. The k-means method works very well in practice. One of its distinguished features is its speed: It has been observed that the number of iterations it needs to find a local optimum is much smaller than the number of objects to be clustered [8, Section 10.4.3]. This is in stark contrast to its worst-case running-time: The only upper bound is $n^{O(kd)}$ [11], which is based on the observation that no clustering appears twice in a run of k-means. On the other hand, Vattani [15] showed that k-means can run for $2^{\Omega(n)}$ iterations in the worst case. This lower bounds holds for all $d \geq 2$.

To reconcile theory and practice, Arthur and Vassilvitskii considered the k-means method for squared Euclidean distances in the framework of *smoothed analysis*. This notion has been introduced by Spielman and Teng [14] and it is based on a two-step input model: An adversary specifies an instance, which is then subject to slight random perturbation. The smoothed running-time is defined to be the worst expected running-time the adversary can achieve. If it is small, then (artificial) worst-case instances might still exist, but they are encountered only with very small probability if inputs are subject to some small amount of random noise. In practice, such noise can come, e.g., from measurement errors or numerical imprecision. Unlike worst-case or average-case analyses, smoothed analyses are neither dominated by single worst-case instances nor by completely random instances, and they lead to more realistic conclusions. Arthur and Vassilvitskii [4] showed that for squared Euclidean distances the smoothed running-time of k-means is $\text{poly}(n^k, 1/\sigma)$ if the data points are perturbed by Gaussian noise with standard deviation σ. We improved this bound to $\text{poly}(n^{\sqrt{k}}, 1/\sigma)$ and we additionally obtained a bound of $k^{kd} \cdot \text{poly}(n, 1/\sigma)$ [13]. Recently, Arthur et al. [3] showed that the smoothed running-time of k-means is polynomial in n and $1/\sigma$.

With only a few exceptions [1,2,5], the theoretical knowledge about k-means clustering is limited to the case of squared Euclidean distances. In this paper, we initiate the theoretical study of the k-means method for general Bregman divergences. We show that the lower bound of $2^{\Omega(n)}$ for the worst-case running-time is valid for almost every Bregman divergence, leading, as for squared Euclidean distances, to a huge discrepancy between theory and practice for many commonly used distance measures like Kullback-Leibler divergence or Itakura-Saito divergence. To obtain more realistic theoretical results, we also analyze the smoothed running-time of k-means for general Bregman divergences. We show that for almost arbitrary Bregman divergences, the smoothed running-time of k-means is upper-bounded by $\text{poly}(n^{\sqrt{k}}, 1/\sigma)$ and $k^{kd} \cdot \text{poly}(n, 1/\sigma)$.

1.1 k-Means Method

An instance for k-means clustering is a set $\mathcal{X} \subseteq \mathbb{R}^d$ consisting of n points. The aim is to find a clustering $\mathcal{C}_1, \ldots, \mathcal{C}_k$ of $\mathcal{X}$, i.e., a partition of $\mathcal{X}$, as well as cluster centers $c_1, \ldots, c_k \in \mathbb{R}^d$ such that the potential $\sum_{i=1}^{k} \sum_{x \in \mathcal{C}_i} d_\Phi(x, c_i)$ is minimized, where d_Φ denotes some distance measure on $\mathbb{R}^d$. Given the cluster centers, every data point should be assigned to the cluster whose center is closest to it. The other way round, given the clusters, the centers $c_1, \ldots, c_k$ should be chosen so as to minimize the potential. In the next section, we will see that for Bregman divergences this is the case if the centers are chosen as the centers of mass, i.e., $c_i = \frac{1}{|\mathcal{C}_i|} \sum_{x \in \mathcal{C}_i} x$. The k-means method for Bregman divergences proceeds now as follows (observe that since the potential decreases in every step, no clustering occurs twice, and the algorithm eventually terminates):

1. Select cluster centers $c_1, \ldots, c_k \in \mathbb{R}^d$ arbitrarily.
2. Assign every $x \in \mathcal{X}$ to the cluster $\mathcal{C}_i$ whose cluster center c_i is closest to it. (If the closest center is not unique and a point is already assigned to one of the closest clusters, then do not change its assignment.)
3. Set $c_i = \frac{1}{|\mathcal{C}_i|} \sum_{x \in \mathcal{C}_i} x$.
4. If clusters or centers have changed, goto 2. Otherwise, terminate.

1.2 Bregman Divergences

One of the most commonly used functions is $d_\Phi(x, c) = \|x - c\|^2$, i.e., squared Euclidean distances. But also other distance measures are common, e.g., Kullback-Leibler divergence [7]. Both are special cases of *Bregman divergences* [5].

Definition 1. *Let $X \subseteq \mathbb{R}^d$, and let $\Phi : X \to \mathbb{R}$ be a strictly convex function such that Φ is differentiable on the relative interior* ri(X) *of X. The* Bregman divergence $d_\Phi : X \times \text{ri}(X) \to [0, \infty)$ *is defined as*

$$d_\Phi(x, c) = \Phi(x) - \Phi(c) - (x - c)^T \nabla\Phi(c) .$$

Here, $\nabla\Phi(c)$ is the gradient of Φ at c. The basic intuition behind Bregman divergences is the following: c corresponds to a cluster center and x to a data point. Let $\overline{\Phi}(x) = \Phi(c) + (x - c)^T \nabla\Phi(c)$ be the linear interpolation of $\Phi(x)$ from c. Then $d_\Phi(x, c)$ measures how well this interpolation is: $d_\Phi(x, c) = \Phi(x) - \overline{\Phi}(x)$. Since Φ is strictly convex, we have $\overline{\Phi}(x) \leq \Phi(x)$ with equality only for $x = c$. Thus, d_Φ is non-negative and $d_\Phi(x, c) = 0$ if and only if $x = c$.

For a finite set of points $C \subseteq X$, we denote the center of mass of C by $\text{cm}(C) = \frac{1}{|C|} \sum_{x \in C} x$. For Bregman divergences the potential can be expressed in terms of the center of mass in the following way [5, Proposition 1]: For every c,

$$\sum_{x \in C} d_\Phi(x, c) = \sum_{x \in C} d_\Phi(x, \text{cm}(C)) + |C| \cdot d_\Phi(\text{cm}(C), c) .$$

In particular, this means that the center of mass minimizes the potential for a given cluster C, as it does for squared Euclidean distances.

Another important property of Bregman divergences is that the bisector of two centers c and c', i.e., the set $\{x \in X \mid d_\Phi(x,c) = d_\Phi(x,c')\}$, is a hyperplane, which follows immediately from the definition of d_Φ. The only known worst-case bound for the running-time of k-means on squared Euclidean distances comes from the observation that no clustering can repeat during the execution of k-means. This yields a bound of $W \leq n^{3kd}$ [3,11]. The proof of this bound relies only on the fact that the bisectors are hyperplanes. Hence, also for general Bregman divergences, the worst-case number of iterations cannot exceed W.

In the following, we present some prominent Bregman divergences.

Mahalanobis Distances. Let us assume that we want to cluster objects that are each characterized by d quantities. If these quantities are independent, then clusters should be hyperspherically-shaped and squared Euclidean distances provide a good distance measure. However, if the coordinates are correlated, then clusters are expected to have hyperelliptic shapes and squared Euclidean distances are not the right measure. In that case, let $B \in \mathbb{R}^{d\times d}$ be the covariance matrix of the components of the data points and assume that it is invertible. This means that the matrix B is symmetric and positive definite. Let $A = B^{-1}$, then the right distance measure taking into account the correlations is the *Mahalanobis distance* d_{m_A} for $m_A(x) = x^T Ax$. The gradient of m_A is $\nabla m_A(c) = 2Ac$, which yields $d_{m_A}(x,c) = (x-c)^T A(x-c)$.

Kullback-Leibler Divergence and Generalized I-Divergence. The *Kullback-Leibler divergence* (KLD, relative entropy) is a very popular Bregman divergence. Here, $X = \{x \in \mathbb{R}^d \mid x \geq 0, \sum_{i=1}^d x_i \leq 1\}$ and an element $x = (x_1,\ldots,x_d) \in X$ represents a probability distribution on a discrete set with $d+1$ elements (where $(x_1,\ldots,x_{d+1})$ with $x_{d+1} = 1 - \sum_{i=1}^d x_i$ is the vector of probabilities). For $\mathrm{KLD}(x) = \sum_{i=1}^{d+1} x_i \log(x_i)$, we obtain $d_{\mathrm{KLD}}(x,c) = \sum_{i=1}^{d+1} x_i \log(\frac{x_i}{c_i})$, where $x_{d+1} = 1 - \sum_{i=1}^d x_i$ and $c_{d+1} = 1 - \sum_{i=1}^d c_i$. Intuitively, the Kullback-Leibler divergence is a measure for the expected difference in the number of bits that are required to code samples drawn according to x when, on the one hand, we use an optimal code based on c and, on the other hand, we use an optimal code based on x. KLD plays a crucial role in a variety of applications like clustering text documents and image classification [7].

We will also consider the *generalized I-divergence* (GID), which generalizes KLD to a larger domain: For this, we have $X = \{x \in \mathbb{R}^d \mid x \geq 0\}$, the potential function $\mathrm{GID}(x) = \sum_{i=1}^d x_i \log(x_i)$, and $d_{\mathrm{GID}}(x,c) = \sum_{i=1}^d x_i \log(\frac{x_i}{c_i}) - \sum_{i=1}^d (x_i - c_i)$.

Itakura-Saito Divergence. Another Bregman divergence that is commonly used in signal processing and in particular in speech processing is the *Itakura-Saito divergence* (ISD) [5,10]. We have again $X = \{x \in \mathbb{R}^d \mid x \geq 0\}$, and the potential function is given by the Burg entropy $\mathrm{ISD}(x) = -\sum_{i=1}^d \log(x_i)$. From this, we get the Bregman divergence $d_{\mathrm{ISD}}(x,c) = \sum_{i=1}^d \frac{x_i}{c_i} - \log(\frac{x_i}{c_i}) - 1$.

1.3 Perturbation Models for Bregman Divergences

If the Bregman divergence is defined on the whole space $\mathbb{R}^d$, i.e., if $X = \mathbb{R}^d$, then it is often natural to assume that the points are perturbed by adding Gaussian noise to them. More precisely, we can assume that an adversary is allowed to place initially n points in $[0,1]^d$, and that each of these points is perturbed by adding a Gaussian with standard deviation σ to each of its coordinates.

On the other hand, if X is a proper subset of $\mathbb{R}^d$, as it is the case for KLD or GID, then such a perturbation model cannot be applied as it might yield points outside the feasible region X. For this reason, we decided to consider very general perturbation models that need to satisfy only a couple of properties, which we will summarize in the following. In Section 2, we present concrete perturbation models with these properties for some special Bregman divergences.

We assume that the perturbation model is parameterized by some $\sigma \in (0,1]$ that measures the amount of randomness in the sense that the smaller σ is chosen, the weaker is the perturbation. If every point is perturbed by Gaussian noise, then σ can be chosen as the standard deviation. We assume that the following properties are satisfied for $\sigma \in (0,1]$:

- For any $\varepsilon \geq 0$, any hyperplane H, and any point in $x \in X \cap [0,1]^d$, the probability that x has a distance of at most ε from H after the perturbation is bounded from above by $\sqrt{\varepsilon}/\sigma$.
- For any $x \in X \cap [0,1]^d$, the density of the perturbation of x is bounded from above by $(1/\sigma)^d$ on $\mathbb{R}^d$.

Let us remark two things about our assumptions on the perturbation model: For Gaussian noise, the probability of a point being close to a hyperplane is even bounded by ε/σ. However, to gain some flexibility for choosing other perturbation models, we use the weaker bound of $\sqrt{\varepsilon}/\sigma$. Second, the bound on the density immediately implies that for any $\varepsilon \geq 0$, any $c \in \mathbb{R}^d$, and any $x \in X \cap [0,1]^d$, the perturbed version of x lies in the hyperball with radius ε and center c with probability at most $(2\varepsilon/\sigma)^d$.

Additionally, we need the property that perturbed points cannot be too far away from their initial positions in $X \cap [0,1]^d$. For this, let D be chosen such that with probability at least $1 - W^{-1}$ every point from the perturbed point set $\mathcal{X}$ is contained in the hypercube $\mathcal{D} = [-D, 1+D]^d$, where $W \leq n^{3kd}$ denotes the worst number of steps. The bounds on the smoothed running-time that we obtain depend polynomially on D. For Gaussian random vectors with mean in $[0,1]^d$ and standard deviation $\sigma \leq 1$, D can be chosen polynomially in n.

1.4 Parameterization

In this section, we make precise what we mean by "almost arbitrary Bregman divergences." To do this, we define a couple of parameters of Bregman divergences. For the remainder of the paper we assume that X, the domain of the distance measure, is a convex set.

For $\varepsilon \geq 0$, let $\mathcal{I}(\varepsilon)$ be the interior of $X \cap \mathcal{D}$ that has a distance of at least ε to the boundary: $\mathcal{I}(\varepsilon) = \{x \in X \cap \mathcal{D} \mid \operatorname{dist}(x, \partial(X \cap \mathcal{D})) \geq \varepsilon\}$.

For a given perturbation model, we choose ε^* such that $\Pr[x \notin \mathcal{I}(\varepsilon^*)] \leq n^{-13}$, where x denotes the perturbed version of an arbitrary point in $X \cap [0,1]^d$. In the following, we use the notations $\mathcal{I} = \mathcal{I}(\varepsilon^*)$ and $\mathcal{I}' = \mathcal{I}(\varepsilon^*/(2n))$. An important property of this definition is the following: If $A \subseteq \mathcal{X}$ is a subset of the data points, and A contains a point from $\mathcal{I}$, then $\operatorname{cm}(A) \in \mathcal{I}(\varepsilon^*/n) \subseteq \mathcal{I}'$, i.e., the center of mass of A is also at a distance of at least ε^*/n from the boundary.

To relate the Bregman divergence d_Φ to squared Euclidean distances, we introduce two parameters ξ and ξ' such that

$$\forall x, y \in X \cap \mathcal{D}\colon d_\Phi(x,y) \geq \xi \cdot \|x-y\|^2 \quad \text{and} \quad \forall x, y \in \mathcal{I}'\colon d_\Phi(x,y) \leq \xi' \cdot \|x-y\|^2 .$$

Observe that for the definition of ξ', only the interior of $X \cap D$ is relevant. This is important as otherwise ξ' would be unbounded for many Bregman divergences. The ratio ξ'/ξ is closely related to the μ in the notion of μ-similarity introduced by Ackermann et al. [2]. However, Bregman divergences like KLD, GID, or ISD are not μ-similar for any μ on their whole domain. To make them μ-similar, their domains have been restricted such that all data points must be sufficiently far away from the singularities. We emphasize that no such restrictions are necessary for our smoothed analysis. There may be points close to the boundary of the domain, but we can take special care of those points. This technical challenge is the reason for the definition of $\mathcal{I}$ and $\mathcal{I}'$ above.

We also need the following lower bound on the "second derivative" of Φ, which follows by a simple calculation from the previous definition: $\frac{\|\nabla\Phi(x)-\nabla\Phi(y)\|}{\|x-y\|} \geq 2\xi$ for all $x, y \in X \cap \mathcal{D}$ with $x \neq y$. Similarly, we need an upper bound (only for the interior): $Q' := \sup_{x,y \in \mathcal{I}', x \neq y} \frac{\|\nabla\Phi(x)-\nabla\Phi(y)\|}{\|x-y\|}$.

1.5 Our Results

In the following, we assume $d \leq n$, $k \leq n$, and $d \geq 4$, which are no severe restrictions from a practical point of view. Let P be the maximal potential after the first iteration of k-means, provided that all points of $\mathcal{X}$ lie in $\mathcal{D}$.

Theorem 2. *Let d_Φ be a Bregman divergence. Then the smoothed running-time of k-means is bounded from above by $\frac{P}{\xi} \cdot \operatorname{poly}(n^{\sqrt{k}}, \frac{1}{\sigma})$ and by $P \cdot k^{kd} \cdot \frac{Q'^2 \xi'^3}{4\xi^5 \varepsilon^{*2}} \cdot \operatorname{poly}(n, \frac{1}{\sigma})$, where the polynomials are independent of d, k, and the parameters.*

The second bound in the theorem yields a polynomial smoothed running-time if $k, d \in O(\sqrt{\log n / \log\log n})$. Indeed, k and d are usually much smaller than n in practice.

On the negative side, in Section 3, we transfer the lower bound of $2^{\Omega(n)}$ for squared Euclidean distances to all good-natured Bregman divergences, where "good-natured" means that all third order derivatives exist and are bounded in a small region, which includes Mahalanobis distances, KLD, GID, and ISD.

1.6 Technical Contribution

In an earlier analysis [13], we presented two different approaches for analyzing the smoothed running-time, leading to upper bounds of $k^{kd} \cdot \text{poly}(n, \sigma^{-1})$ and $\text{poly}(n^{\sqrt{k}}, \sigma^{-1})$ for squared Euclidean distances. Both of these approaches are based on a novel lemma about perturbed point sets, stating that, given any Voronoi partition of the point set, it is unlikely that many points are close to the bisectors [13, Lemma 2.1]. Clearly, the structure of the smoothed analysis presented in this paper is similar to the earlier one [13]. However, we had to tackle several severe problems when transferring the results from squared Euclidean distances to general Bregman divergences. First of all, the proof of the aforementioned lemma about perturbed point sets cannot be generalized directly to Bregman divergences. In the course of finding a generalization, we found a shorter and simpler proof of the lemma. Given this result, the bound of $k^{kd} \cdot \text{poly}(n, \sigma^{-1})$ follows roughly in the same way as in the Euclidean case, but some additional technical problems have to be addressed. Let us describe the main problem by way of example: For KLD, the parameters ξ' and Q' can become arbitrarily large for points close to the boundary of X. Even after the perturbation, some of the points might still be too close to the boundary to obtain reasonable upper bounds for ξ' and Q'. Essentially, we show that the kd points that are closest to the boundary can be handled separately and that all other points are sufficiently far away from the boundary (i.e., they lie in $\mathcal{I}$) to bound ξ' and Q' in a reasonable way.

An obvious question is whether the smoothed polynomial bound [3] carries over to Bregman divergences. The problem with adapting the proof of this bound is that it exploits specific properties of Gaussian perturbations. It uses, in particular, the property that the projection of a Gaussian random vector onto a lower-dimensional subspace is still a Gaussian with the same standard deviation. It would be very interesting to see if it is possible to relax some of these requirements or if it is possible to design more general perturbation models that still meet the requirements needed for the smoothed polynomial bound.

In order to prove the lower bound, we first observe that all Mahalanobis distances (in particular squared Euclidean distances) exhibit the same worst-case behavior. Then we show that all "good-natured" Bregman divergences (including all commonly considered examples like KLD, GID, or ISD) behave locally like some Mahalanobis distance, which makes a transfer of the known lower bound for the Euclidean case possible.

Due to lack of space, all proofs are deferred to the full version of this paper. In the following section, we will only apply Theorem 3 to four common Bregman divergences.

2 Applying the Smoothed Analysis

2.1 Mahalanobis Distances

For Mahalanobis distances, we use the same perturbation model that has been used for squared Euclidean distances [4,13]: The adversary chooses n points in

$[0,1]^d$. Then the d coordinates are perturbed by independent Gaussian perturbations of standard deviation σ. We can choose $D = \text{poly}(n)$. Then $\mathcal{X} \subseteq \mathcal{D} = [-D, D+1]^d$ with a probability of at least $1 - W^{-1}$ since Gaussians are concentrated around their mean, which is in $[0,1]^d$. After one iteration of k-means, every point is assigned to a cluster center within a distance of at most poly(n).

Let $A \in \mathbb{R}^{d \times d}$ be an arbitrary symmetric positive definite matrix, and consider k-means using m_A. Scaling the matrix does not change the behavior of k-means. Thus, we assume that A is scaled such that the smallest eigenvalue, which is positive, equals 1. Let $\lambda_{\max}$ be the largest eigenvalue of A. A short calculation shows that the parameters can be chosen such that Theorem 2 yields the following bound:

Theorem 3. *The smoothed running-time of k-means using m_A is bounded from above by $\lambda_{\max} \cdot \text{poly}(n^{\sqrt{k}}, \frac{1}{\sigma})$ and $k^{kd} \cdot \lambda_{\max}^6 \cdot \text{poly}(n, \frac{1}{\sigma})$.*

2.2 Kullback-Leibler Divergence

We have to be more careful when choosing a perturbation model for Kullback-Leibler divergence. KLD is defined on a simplex. Thus, we cannot use Gaussian perturbations since these might result in points outside of the domain of KLD.

To get a perturbation model, we take into account that a point represents a probability distribution on a finite set $\{1, 2, \ldots, d+1\}$. For instance, assume that we want to classify web pages based on a list $w_1, \ldots, w_{d+1}$ of words (the so-called *bag-of-words model* [7]). For a specific web page, let n_i be the number of occurrences of w_i. Then $x_i = \frac{n_i}{\sum_{j=1}^{d+1} n_j}$ is the relative frequency of w_i. Based on the vectors x, web pages can be clustered according to their topics since pages about similar topics are likely to contain similar words. To perturb instances, the idea is to add a random number of each word to the web page.

Let us make this more precise. For a point $x \in X$, we obtain $x' \in \mathbb{R}^{d+1}$ by adding the component $x_{d+1} = 1 - \sum_{i=1}^{d} x_i$. Then we draw random numbers $y_1, \ldots, y_{d+1}$ independently according to some probability distribution to be specified in a moment. Let $S = \sum_{i=1}^{d+1} x_i + y_i = 1 + \sum_{i=1}^{d+1} y_i$. Then we obtain the perturbed point $z \in \mathbb{R}^d$ by setting $z_i = \frac{x_i + y_i}{S}$. By construction, $z \geq 0$ and $\sum_{i=1}^{d} z_i \leq 1$. We use the exponential distribution [9], whose density is $\frac{1}{\theta} \cdot \exp\left(-\frac{x}{\theta}\right)$ for a positive parameter θ.

In the full version of this paper, we show that this perturbation model satisfies the requirements of Section 1.3 for $\theta = 8d\sigma^{d/(d+1)}$. Analyzing the parameters ξ, ξ', Q', and ε^* as well as the potential P after the first iteration yields the following theorem.

Theorem 4. *The smoothed running-time of k-means using* KLD *is bounded from above by* $\text{poly}(n^{\sqrt{k}}, \frac{1}{\sigma})$. *and* $k^{kd} \cdot \text{poly}(n, \frac{1}{\sigma})$.

2.3 Generalized I-Divergence

For generalized I-divergence and Itakura-Saito divergence, we use the same perturbation model, except for rescaling. Since we do not have to rescale, this allows us to let the adversary choose any density function f bounded by $\frac{1}{2\sqrt{d}\sigma}$ whose tail bounds are sufficiently small: The probability of a number greater than poly(n) must be bounded by $\frac{1}{ndW}$. Then we perturb a point by adding independent random numbers drawn according to f. For this perturbation model, Theorem 4 carries over to GID and ISD. Details can be found in the full version.

3 Lower Bound

In this section, we transfer the exponential lower bound proved by Vattani [15] to almost arbitrary Bregman divergences.

Theorem 5 (Vattani [15]). *For squared Euclidean distances, there exist sets $\mathcal{X} \subseteq \mathbb{R}^d$ of n points on which the k-means method requires $2^{\Omega(n)}$ iterations when initialized with a particular set of cluster centers. Here, k depends on n and $d \geq 2$ is arbitrary.*

First, we show that all Mahalanobis distances are equivalent in terms of the worst-case number of iterations. Squared Euclidean distances are a special case of Mahalanobis distances. Thus, we get an exponential lower bound for all Mahalanobis distances. Let $W_{d_\Phi}^{k,d}(n)$ be the maximum number of iterations of k-means on any instance of n points in $\mathbb{R}^d$ using d_Φ as the distance measure.

Lemma 6. *For every symmetric positive definite matrix $A \in \mathbb{R}^{d \times d}$, we have $W_{m_A}^{k,d}(n) = W_{m_I}^{k,d}(n)$ for all $n, k, d \in \mathbb{N}$.*

Now we transfer worst-case instances for Mahalanobis distances to instances for arbitrary good-natured Bregman divergences. For this, we use the observation that any good-natured Bregman divergence d_Φ behaves locally at some point z_0 like the Mahalanobis distance d_{m_H}, where H is the Hessian matrix of Φ at z_0. Hence, essentially we only need to scale down the worst-case instance for d_{m_H} and embed it locally into a small space around z_0.

Lemma 7. *Let $\Phi : X \to \mathbb{R}$ be a strictly convex function with $X \subseteq \mathbb{R}^d$ and the following properties: There exist a $z_0 \in X$ and a $\zeta > 0$ such that*

- *$Z = \{z \in \mathbb{R}^d \mid \|z - z_0\|_\infty \leq \zeta\} \subseteq X$,*
- *all third-order derivatives of Φ exist on Z and their absolute values are bounded, and*
- *the Hessian matrix of Φ at z_0 is positive definite.*

Then $W_{d_\Phi}^{k,d}(n) \geq W_{m_I}^{k,d}(n)$.

Combining Vattani's lower bound with Lemma 6 and Lemma 7, we obtain the main result of this section.

Theorem 8. *The worst-case number of iterations of k-means for the following Bregman divergences is at least $\exp(\Omega(n))$ for n points and $d \geq 2$: Mahalanobis distances for any symmetric positive definite matrix A, Kullback-Leibler divergence (KLD), generalized I-divergence (GID), Itakura-Saito divergence (ISD).*

References

1. Ackermann, M.R., Blömer, J.: Coresets and approximate clustering for Bregman divergences. In: Proc. of the 20th ACM-SIAM Symp. on Discrete Algorithms (SODA), pp. 1088–1097 (2009)
2. Ackermann, M.R., Blömer, J., Sohler, C.: Clustering for metric and non-metric distance measures. In: Proc. of the 19th ACM-SIAM Symp. on Discrete Algorithms (SODA), pp. 799–808 (2008)
3. Arthur, D., Manthey, B., Röglin, H.: k-means has polynomial smoothed complexity. In: Proc. of the 50th Ann. IEEE Symp. on Found. of Computer Science, FOCS (to appear, 2009)
4. Arthur, D., Vassilvitskii, S.: Worst-case and smoothed analysis of the ICP algorithm, with an application to the k-means method. SIAM Journal on Computing 39(2), 766–782 (2009)
5. Banerjee, A., Merugu, S., Dhillon, I.S., Ghosh, J.: Clustering with Bregman divergences. Journal of Machine Learning Research 6, 1705–1749 (2005)
6. Berkhin, P.: Survey of clustering data mining techniques. Technical report, Accrue Software, San Jose, CA, USA (2002)
7. Dhillon, I.S., Mallela, S., Kumar, R.: A divisive information-theoretic feature clustering algorithm for text classification. Journal of Machine Learning Research 3, 1265–1287 (2003)
8. Duda, R.O., Hart, P.E., Stork, D.G.: Pattern Classification. John Wiley & Sons, Chichester (2000)
9. Feller, W.: An Introduction to Probability Theory and Its Applications, vol. II. John Wiley & Sons, Chichester (1971)
10. Gray, R.M., Buzo, A., Gray Jr., A.H., Matsuyama, Y.: Distortion measures for speech processing. IEEE Transactions on Acoustics, Speech, and Signal Processing 28(4), 367–376 (1980)
11. Inaba, M., Katoh, N., Imai, H.: Variance-based k-clustering algorithms by Voronoi diagrams and randomization. IEICE Transactions on Information and Systems E83-D(6), 1199–1206 (2000)
12. Lloyd, S.P.: Least squares quantization in PCM. IEEE Transactions on Information Theory 28(2), 129–137 (1982)
13. Manthey, B., Röglin, H.: Improved smoothed analysis of the k-means method. In: Proc. of the 20th ACM-SIAM Symp. on Discrete Algorithms (SODA), pp. 461–470 (2009)
14. Spielman, D.A., Teng, S.-H.: Smoothed analysis of algorithms: Why the simplex algorithm usually takes polynomial time. Journal of the ACM 51(3), 385–463 (2004)
15. Vattani, A.: k-means requires exponentially many iterations even in the plane. In: Proc. of the 25th ACM Symp. on Computational Geometry (SoCG), pp. 324–332 (2009)

Succinct Index for Dynamic Dictionary Matching*

Wing-Kai Hon[1], Tak-Wah Lam[2], Rahul Shah[3],
Siu-Lung Tam[2], and Jeffrey Scott Vitter[4]

[1] National Tsing Hua University, Taiwan
wkhon@cs.nthu.edu.tw
[2] University of Hong Kong, Hong Kong
{twlam,sltam}@cs.hku.hk
[3] Louisiana State University, Louisiana, USA
rahul@csc.lsu.edu
[4] Texas A&M University, Texas, USA
jsv@tamu.edu

Abstract. In this paper we revisit the dynamic dictionary matching problem, which asks for an index for a set of patterns $P_1, P_2, \ldots, P_k$ that can support the following query and update operations efficiently. Given a query text T, we want to find all the occurrences of of these patterns; furthermore, as the set of patterns may change over time, we also want to insert or delete a pattern. The major contribution of this paper is the first succinct index for dynamic dictionary matching. Prior to our work, the most compact index is given by Chan *et al.* (2007), which is based on the compressed suffix arrays (Grossi and Vitter (2005) and Sadakane (2003)) and the FM-index (Ferragina and Manzini (2005)), and it requires $O(n\sigma)$ bits where n is the total length of patterns and σ is the alphabet size. We develop a dynamic succinct index using a different (and simpler) paradigm based on suffix sampling. The new index not only improves the space complexity to $(1 + o(1))n \log \sigma + O(k \log n)$ bits, but also the time complexity of the query and update operations. Specifically, the query and update operations respectively take $O(|T| \log n + occ)$ and $O(|P| \log \sigma + \log n)$ times, where occ is the number of occurrences.

1 Introduction

Given a pattern P and a text T finding all the occurrences of P in T has been a fundamental problem and has developed into a very mature research field. The earliest work involved developing algorithms to solve the problem in $O(|P| + |T|)$ time [14]. When the text remains relatively static but patterns keep changing, one would like to build an index on the text T and treat the patterns as queries. Data structures like suffix trees [16,20] and suffix arrays [15] achieve optimal

* This work is supported in part by Taiwan NSC Grant 96-2221-E-007-082-MY3 (W. Hon), Hong Kong RGC Grant HKU 7140/06E (T. Lam), and US NSF Grant CCF-0621457 (R. Shah and J. S. Vitter).

Y. Dong, D.-Z. Du, and O. Ibarra (Eds.): ISAAC 2009, LNCS 5878, pp. 1034–1043, 2009.

query performance. The space for these structures was considered to be "linear" but this was only when measured in terms of number of words and in asymptotic sense. In the stricter information-theoretic sense (which measures space in bits), this could be $\Theta(\log n)$ times more than the optimal. Furthermore, the hidden constants in the asymptotic notions often make these indexes about 20 to 60 times bigger than the original text.

Recently, Ferragina and Manzini [10] and Grossi and Vitter [11] presented text indexes based on the concept of Burrows-Wheeler transform (BWT) [6] whose space bounds are very close to the size of the compressed text. This has evolved into a thriving research field with many new application-specific compressed/succinct indexes developed.

The *dictionary matching* problem is an orthogonal problem to the text indexing problem. Here, some number of patterns are given beforehand and then a text comes in as the query. We need to find which patterns appear in this query text and at which locations. Hence, the index is built on the set of patterns. More formally, the problem is defined as follows.

Index:	A set of patterns $P_1, P_2, ..., P_k$ with total n characters.
Query:	A text T of size $\|T\|$ characters.
Output:	For each P_i occurring in T, all locations ℓ where P_i matches T beginning at position ℓ.

In the dynamic version of the problem, we support two update operations, namely *insert*(P) and *delete*(P). These operations, respectively, insert a new pattern to the set and delete any of the existing patterns from the set.

Dictionary matching problem has a long history starting with Aho and Corasick in 1975 [1] who solved the problem optimally for the case of static patterns. Amir *et al.* [3] gave a solution for dynamic case where inserts and deletes of patterns are allowed. Their approach consists of constructing a generalized suffix tree of the patterns with suffix links. In particular, suffix links are exploited to avoid repeatedly matching the characters of T when different positions of T are examined for pattern occurrences.

However, a major problem with all the above solutions was that the index takes too much space. With the advent of the field of compressed data structures, it remained to be shown that a space-efficient index can be designed for dictionary matching. Also, the issue of dynamism was somewhat hard to achieve with some of the earlier compressed indexing solutions. Chan *et al.* [7] were the first to present $O(n\sigma)$ bit index to solve this problem. Their solution mainly relied on Compressed Suffix Arrays (CSA) [11,18] and the subsequent Compressed Suffix Tree (CST) [19] with ingenious extension of suffix link operations. However, this solution remained from optimal in space (it only achieves big-O term) usage.

In this paper, we take a different approach than CSA or BWT based indexes. Our approach is based on directly sparsifying suffix links and using only sampled suffixes. A similar approach, but with a rather different sampling criteria, was considered by Kärkkäinen and Ukkonen [13] to solve the text indexing problem.

1.1 Comparisons with Previous Results

The solution by Aho and Corasick for the static case required $O(n \log n)$ bits and answered the queries in optimal $O(|T| + occ)$ time. When the dynamic case was addressed by Amir *et al.*, they achieved $O(n \log n)$-bit index but their query and update times had an extra multiplicative factor of $O(\log n / \log \log n)$.

The first attempt to reduce index space for the dynamic dictionary matching problem was given by Chan *et al.* [7]. Their approach builds on compressed suffix arrays (CSA) and compressed suffix trees. They extend the compressed suffix tree representation to use suffix links (using LCA queries). However, their approach in a way uses CSA as a black-box tool and hence it not only remains complicated but also that it does not attain the best possible bounds. For the case of constant alphabet size (i.e., $\sigma = O(1)$), they achieve $O(n)$ bits index. However, their search and update times have an extra multiplicative factor of $O(\log^2 n)$ which comes from attempting to dynamize some of more sophisticated data structures underlying the CSA based approach.

Hon *et al.* [12] introduced the suffix sampling technique to obtain stronger bounds for the static version of this problem. Namely, they achieved $O(n \log \sigma)$ bits index with $O(|T| \log \log n + occ)$ time. They also showed that space bounds like $nH_h + o(n \log \sigma) + O(k \log n)$ are achievable if the $\log \log n$ multiplicative factor can be increased to $\log n$; here, H_h denotes the hth-order empirical entropy of the set of patterns.

In this paper, we build on suffix sampling technique of [12] to achieve the best known bounds for the dynamic dictionary matching problem. One of the consequences of suffix sampling is that it results in sparsification of suffix links which in turn results in the space savings. We also show that in our approach we can split the data structure into two parts: one part is where the compressed text can be stored separately and the other part is the indexing overhead. By choosing the sampling rate appropriately, we can arbitrarily reduce the second part of the data structure, achieving space very close to the one required for the compressed representation of the text. Our space requirement and query times are as given in Table 1. We also note that we can achieve the entropy-compressed bound like $nH_h + o(n \log \sigma) + O(k \log n)$ under the assumption that the generative model from which the dynamic pattern statistics are taken remains the same. Not only our method is simpler to understand (and implement) but it may also greatly enhance the understanding of the nature of this problem.

Table 1. Summary of Results

Result	Space (bits)	Query Time	Update Time
[3]	$O(n \log n)$	$O((\|T\| + occ) \log n / \log \log n)$	$O(\|P\| \log n / \log \log n)$
[7]	$O(n\sigma)$	$O((\|T\| + occ) \log^2 n)$	$O(\|P\| \log^2 n)$
this	$O(n \log \sigma)$	$O(\|T\| \log n + occ)$	$O(\|P\| \log \sigma + \log n)$
this	$(1 + o(1))n \log \sigma + O(k \log n)$	$O(\|T\| \log n + occ)$	$O(\|P\| \log \sigma + \log n)$

2 Preliminaries

2.1 Basic Notation

Let $\mathcal{S} = \{S_1, S_2, \ldots, S_r\}$ be a set of r strings over an alphabet Σ of size σ. Let \$ and # be two characters not in Σ, whose alphabetic orders are, respectively, smaller than and larger than any character in Σ. Let $\mathcal{C}$ be a compact trie such that each string $S_i\$$ or $S_i\#$ corresponds to a distinct leaf in $\mathcal{C}$; also, each edge is labeled by a sequence of characters, such that for each leaf representing some string $S_i\$$ (or $S_i\#$), the concatenation of the edge labels along the root-to-leaf path is exactly $S_i\$$ (or $S_i\#$). For each node v, we use $path(v)$ to denote the concatenation of edge labels along the path from root to v. Note that for each S_i, there must be some internal node v_i such that $path(v_i) = S_i$.

Definition 1. *For any string Q, the* locus *of Q in $\mathcal{C}$ is defined to be the lowest node v (i.e., farthest from the root) such that $path(v)$ is a prefix of Q.*

2.2 Suffix Tree

The *suffix tree* [16,20] for a set $\mathcal{S}$ of strings $\{S_1, S_2, \ldots, S_r\}$ is a compact trie storing all suffixes of each $S_i\$$ and each $S_i\#$. It can be stored in $O(m \log m)$-bit space where $m = |\mathcal{S}|$ denote the total number of characters in the strings of $\mathcal{S}$. For each internal node v in the suffix tree, it is shown that there exists a unique internal node u in the tree, such that $path(u)$ is equal to the string obtained from removing the first character of $path(v)$. Usually, a pointer is stored from v to such a u; this pointer is known as the *suffix link* of v.

By utilizing the suffix links, the suffix tree can be updated according to the insertion or deletion of S_i in the set $\mathcal{S}$ with $O(|S_i| \log \sigma)$ time [9].[1] In addition, we can efficiently find the loci of all suffixes of any text T within the suffix tree in $O(|T| \log \sigma)$ time [3].

2.3 Review: Dictionary Matching with Suffix Trees

Let $\Delta = \{P_1, P_2, \ldots, P_k\}$ be the set of patterns that are currently stored in the collection. Let Σ be the alphabet, and σ be its size. Let $n = \sum |P_i|$ be the total characters of the patterns in Δ. Suppose that we store the suffix tree for Δ; also for each i, we mark the node v_i with $path(v_i) = P_i$. Then we have the following:

Lemma 1. *Let $T(j)$ denote the jth suffix of a text T and let u be the locus of $T(j)$ in the suffix tree of Δ. Then, P_i appears at position j in T if and only if the marked node v_i is an ancestor of u.*

In case the set of patterns is static, we can store a pointer in each node of the suffix tree, pointing to the nearest marked ancestor. Then by the previous lemma, we can answer the dictionary matching query in $O(|T| \log \sigma + occ)$ time,

[1] That is, we insert or delete all suffixes of S_i in the suffix tree.

since finding all loci of all suffixes of T can be done in $O(|T| \log \sigma)$ time. In case the set of patterns is dynamic, the above scheme of storing pointers does not work well, as in the worst case a single pattern update can cause many nodes to change their nearest marked ancestors. Nevertheless, Amir *et al.* [3] showed that with suitable maintenance of the marked ancestors, we can answer the dictionary matching query in $O((|T| + occ) \log n / \log \log n)$ time and we can update a pattern P in $O(|P| \log n / \log \log n)$ time.[2]

3 Towards Succinctness with Sparse Suffix Tree

A major problem with the existing suffix-tree-based solutions is the index space, requiring $O(n \log n)$ bits which can be $\Theta(\log n)$ times more than the storage of the patterns in the plain form. To achieve space reduction, our idea is to *selectively* sample one suffix per every d suffixes, and maintain a compact trie $\mathcal{C}$ from these sampled suffixes. Intuitively, the resulting trie is a suffix tree for the original patterns, when we imagine every d characters of a pattern are merged into a single meta-character.

Our query is answered analogously as in the original suffix tree scheme. Basically, when a text T is given, we shall locate T positions, say π_i for $i = 1$ to $|T|$, in the compact trie $\mathcal{C}$ which respectively represents the locus of $T[i..|T|]$. Because of the similarity of $\mathcal{C}$ and an ordinary suffix tree, finding the loci can be done efficiently by exploiting "suffix links". However, since each meta-character represents d original characters, the computation of loci will be done by d separate traversals in $\mathcal{C}$, where the jth traversal computes the loci of those suffixes $T[i..|T|]$ with $i (\text{mod } d) = j$. Afterwards, we report the occurrences by finding the marked ancestors of each locus, using the data structure in Section 4.

3.1 Implementation Details

Various performance tradeoffs can be obtained by varying the sampling factor d. We first consider the simple case where d is set to $0.5 \log_\sigma n$. In this case, the number of suffixes is reduced from n to at most $\lceil n/d \rceil + k$. Consequently, the compact trie $\mathcal{C}$ has $O(n/d + k)$ nodes, so that its space is $O((n/d + k) \times \log n) = O(n \log \sigma + k \log n)$ bits.

Recall that $\mathcal{C}$ is similar to an ordinary suffix tree; indeed, we can analogously define suffix link for each internal node v in $\mathcal{C}$, which is the node u such that $path(u)$ is the same as the string obtained by removal of first d characters of $path(v)$ (i.e., removal of its first meta-character). However, due to the effect of merging characters, the alphabet size has increased from σ to $\sigma^d = \sqrt{n}$.

When a query text T is given, our target is to obtain the locus of each suffix of T in $\mathcal{C}$. We may first treat T as a meta-text T' by blocking every d characters. Then, we can utilize the suffix links and find the loci of each suffix of T' in $O((|T|/d+1) \log n)$ time, since there are $O(|T|/d+1)$ meta-characters, each from

[2] When σ is not a constant, an additive $O(|T| \log \sigma)$ and $O(|P| \log \sigma)$ term will be added to the query time and update time, respectively.

an alphabet of $\sqrt{n}$. Note that these loci may not be the same as the loci of those $T[i..|T|]$ with $i (\text{mod } d) = 1$, but they are closely related. For instance, the locus of T can be at most d nodes further from the locus of T'. In general, the locus of each $T[i..|T|]$ with $i (\text{mod } d) = 1$ can be obtained in an extra $O(d \log \sigma) = O(\log n)$ time through traversal in $\mathcal{C}$. As a result, the loci of roughly $1/d$ of all suffixes of T are obtained. To find the other loci, we can repeat the procedure for $d-1$ times, where at the jth time we search $\mathcal{C}$ with the meta-text formed by blocking $T[j+1..|T|]$. This gives the following lemma.

Lemma 2. *When $d = 0.5 \log_\sigma n$, the compact trie $\mathcal{C}$ requires $O(n \log \sigma + k \log n)$ bits of space. On any input text T, the loci of all suffixes of T in $\mathcal{C}$ can be obtained in $O(|T| \log n)$ time.*

Next, we briefly discuss two ideas of further reducing the space terms. The first one is to reduce the $O(k \log n)$ terms, under tha natural assumption that all patterns in the set Δ are distinct. For this case, we shall classify patterns into two groups, one for those longer than d, the other for those with length at most d. The number of patterns, k_1, in the first group is at most n/d, and these patterns will be indexed by a compact trie $\mathcal{C}'$ using Lemma 2. The number of patterns, k_2, in the second group is at most $\Theta(\sqrt{n} \log_\sigma n)$, whose total length is at most $\Theta(\sqrt{n} (\log_\sigma n)^2)$; these patterns will be stored in an ordinary suffix tree $\mathcal{R}$, and requires only $o(n)$ bits of space. Once the loci of all suffixes of T are located in both trees, we can proceed as before to output the marked ancestors of these loci. We summarize the above discussion as follows:

Lemma 3. *Assuming patterns in Δ are distinct. When $d = 0.5 \log_\sigma n$, we can store the compact trie $\mathcal{C}'$ and the $o(n)$-bit suffix tree $\mathcal{R}$, in total $O(n \log \sigma)$ bits of space, such that on any input text T, the loci of all suffixes of T in $\mathcal{C}'$ and in $\mathcal{R}$ can be obtained in $O(|T| \log n)$ time.*

The second idea to reduce space is by raising the sampling factor d. In particular, we set $d = \log n \log_\sigma n$.[3] Then, we can immediately obtain a lemma similar to Lemma 2, such that the space of $\mathcal{C}$ is reduced to $o(n \log \sigma) + O(k \log n)$ bits and finding all loci is done in $O(|T| \log^2 n)$ time. The increased in time to find loci is due to the inefficiency in extending each of the "approximate" locus (obtained from searching T' in $\mathcal{C}$) to the true locus. In fact, each such extension can be reduced to the *prefix matching* problem in [12], which can be solved more efficiently using $O(d/\log_\sigma n + \log k) = O(\log n)$ time (see Lemma 4 of [12]).[4] The extra space required to support the reduction is $O(k \log n)$ bits in total. Thus, we have the following lemma:

[3] Due to the increase in d, we can no longer combine this idea with the first one; as a result, we do not classify short and long patterns, and the $O(k \log n)$ term reappears.

[4] The idea is to maintain an extra data structure, called String B-tree [9], to manage the marked nodes so that once we obtain an approximate locus, we can easily jump to the nearest marked ancestor of the true locus. Due to space limitation, we defer the details to the full paper.

Lemma 4. *When* $d = \log n \log_\sigma n$*, the compact trie* $\mathcal{C}$ *requires* $o(n \log \sigma) + O(k \log n)$ *bits of space. On any input text* T*, the loci of all suffixes of* T *in* $\mathcal{C}$ *can be obtained in* $O(|T| \log n)$ *time.*

4 New Approach for Dynamic Marked Ancestors

Let $\mathcal{C}$ be a rooted tree with m nodes, where some k nodes are marked. The dynamic marked ancestor problem is to index $\mathcal{C}$ so that on given any node v, we can report all the ancestors of v which are marked; in addition, the tree can be updated by insertion or deletion of nodes, and by marking or unmarking nodes. Existing solutions [3,2] are achieved by the reduction to parentheses maintenance problem. In the following, we use an alternative approach where we solve the problem via management of one-dimensional intervals.

4.1 Reduction for Semi-static Case: Intervals Management

When the structure of the tree is static, and the set of marked nodes is fixed, the marked ancestor problem can be easily and optimally solved, simply by maintaining a pointer in each node to its nearest marked ancestor. Nevertheless, we shall show a non-optimal solution, which acts as a stepping stone towards an efficient solution for the dynamic case.

First, we perform a pre-order traversal of the tree. Each node is assigned the order in which it is first visited as its label. For instance, the root has label 1 and its leftmost child has label 2. For each marked node v, let v' denote the last node visited in the subtree rooted at v; also, let L_v and $L_{v'}$ be their labels, respectively. It is easy to check that v is a marked ancestor of a node u if and only if the label of u falls in the interval $[L_v, L_{v'}]$.

Using the *interval tree*, we can maintain the k intervals corresponding to the k marked nodes in $O(k \log m)$ bits, such that for any node u with label L_u, we can report all *occ* intervals containing L_u in $O(\log k + occ)$ time; that is, we can find all marked ancestors of u in $O(\log k + occ)$ time.

In fact, if the tree structure is static, the above scheme can also handle marking or unmarking of a tree node. Each such operation simply corresponds to inserting or deleting an interval in the interval tree. For this semi-static case, we can apply the dynamic interval tree by Arge and Vitter [4], where each update can be done in $O(\log k)$ time, while the query time and the space requirement remain unchanged.

4.2 Reduction for Dynamic Case: Elastic Intervals Management

Note that the interval tree scheme cannot be directly used to handle the fully dynamic case. In particular, when a node is inserted or deleted in the tree,[5] it

[5] Here, node insertion includes the case where a node is inserted into the middle of an existing edge, thus splitting one edge into two edges. On the other hand, when a degree-1 internal node is deleted, we reverse the process so that its parent edge and its child edge will be merged to a single edge.

can cause the pre-order label of many nodes to change, which in turn can cause the intervals of many marked nodes to change.

However, observe that the relative order of the pre-order label of the existing nodes, before and after the updates, are not changed. This motivates us to represent each marked node v by an "elastic" interval (instead of a fixed interval when v is marked), where endpoints are represented by pointers to v and v', so that its interval can be flexibly changed according to the current ranks of v and v' in the tree.

Now, suppose that the *relative* rank of two nodes can be compared online in $f(m)$ time, where m is the number of nodes in the tree. Then the dynamic interval tree of Arge and Vitter can easily be adapted to support each update in $O(f(m) \log k)$ time and each query in $O(f(m)(\log k + occ))$ time. One simple solution is to overlay a balanced binary tree for the nodes so that the exact rank of any node can be computed in $O(\log m)$ time, thus comparison can be made in $O(\log m)$ time. A more complicated solution is by Dietz and Sleator [8] or by Bender *el.* [5], which is an $O(m \log m)$-bit data structure for maintaining order in a list of items. In this order-maintenance data structure, an item can be inserted into the list in $O(1)$ time when either its predecessor or its successor is given, while it can be deleted (freely) in $O(1)$ time; given two items, we can compare their rank in the list in $O(1)$ time. Thus, we can obtain a solution of dynamic marked ancestor by interval tree without any sacrifice in query efficiency.

Yet, there are two important points to note for using the final scheme. First, the insertion of a node v in a tree will require the knowledge of which node is v's predecessor or successor. This can be immediately done when v is the first child of its parent (so that its predecessor is known), or v is inserted in the middle of an existing edge (whose successor is known). However, it will be time-consuming in case v is the *last* child of its parent, in which case we may need to find its successor by traversing to the root and finding the first branch to the right. Thus, the position of where a node is inserted will greatly affect the time in updating.

Second, as the endpoints of the interval for a marked node v is now replaced by pointers to v and v', it will cause a serious problem if v' can be deleted while v is marked (in that case, the endpoint becomes undefined). To avoid this problem, whenever we mark a node v, we will create a *dummy* node $\hat{v}$ and insert it as the rightmost child of v; on the other hand, $\hat{v}$ will be deleted only when v becomes unmarked. As $\hat{v}$ will always be the last node visited in the subtree rooted at v, $\hat{v} = v'$ by definition, so that the interval of each marked nodes will always be well-defined.

5 All in a Nutshell

We are now ready to combine the sparse suffix tree (Section 3) and the dynamic marked ancestor data structures (Section 4) to see their overall performance.

When $d = 0.5 \log_\sigma n$ and assuming the patterns are distinct, we can solve the dictionary matching query as follows. Recall that we maintain a compact trie $\mathcal{C}'$ for long patterns (length longer than d) and a suffix tree $\mathcal{R}$ for short patterns (length at most d).

1. We locate the loci of all suffixes of T in $\mathcal{C}'$ in $O(|T|\log n)$ time. (Lemma 3)
2. Then, we apply the dynamic interval tree to report all marked ancestors of these $|T|$ loci in a total of $O(|T|\log n + occ_\ell)$ time, where occ_ℓ denote the number of occurrences of long patterns.
3. Next, we traverse the suffix tree in $O(|T|\log\sigma)$ time to locate the $|T|$ loci of all suffixes of T in $\mathcal{R}$.
4. Then, we use a brute-force method to report all marked ancestors of these $|T|$ loci in a total of $O(|T| \times d) = O(|T|\log n)$ time.

Thus, in total, $O(|T|\log n + occ)$ time is required.

To support the update when a pattern P is inserted, we perform the following. Firstly, when P is shorter than d, we add P and its suffixes into the suffix tree $\mathcal{R}$, using $O(|P|\log\sigma)$ time. After that, we mark the node v with $path(v) = P$, using $O(1)$ time. Otherwise, when P is long, we shall update the compact trie $\mathcal{C}'$ and the dynamic marked ancestor data structures as follows:

1. We first insert the $\lceil |P|/d \rceil$ suffixes of P into $\mathcal{C}'$, using $O((|P|/d + 1)\log n)$ time, by exploiting the suffix links. In addition, we will ensure that for each node inserted to the tree $\mathcal{C}$, if it is not inserted into the middle of some existing edge, then it will be inserted as the *first* child of its parent.
2. Then, for each node inserted, we find either its predecessor or its successor in the pre-order traversal in $O(1)$ time. Then, we make the corresponding change in the Dietz-Sleator order-maintenance data structure, using an extra $O(1)$ time per node. In total, this takes $O(|P|/d + 1)$ time.
3. Next, we mark the node v with $path(v) = P$ in $\mathcal{C}'$. This involves adding a dummy node $\hat{v}$ as the rightmost child of v. For this step, we find the successor of $\hat{v}$ in $\mathcal{C}'$ by traversing from $\hat{v}$ to the root, and finding the first branch to the right. This takes $O(|P|)$ time. After that, we update the order-maintenance data structure in $O(1)$ time. In total, adding $\hat{v}$ takes $O(|P|)$ time.
4. After that, we add the elastic interval corresponding to the marked node v to the dynamic interval tree. This step takes $O(\log k)$ time.

As the most time-consuming step is Step 1, pattern insertion can be supported in $O((|P|/d+1)\log n) = O(|P|\log\sigma + \log n)$ time. To support pattern deletion, it can be done similarly (and more easily) with the above steps, using the same time bound. This gives the following theorem.

Theorem 1. *Suppose that the patterns in Δ are distinct. Then we can maintain an $O(n\log\sigma)$-bit index for Δ, such that on any given text T, a dictionary matching query can be answered in $O(|T|\log n + occ)$ time. Also, the index supports insertion or deletion of a pattern in Δ in $O(|P|\log\sigma + \log n)$ time.*

When $d = \log n \log_\sigma n$, answering a dictionary matching query will only involve the search in the compact trie $\mathcal{C}$ for $|T|$ loci, and subsequently finding the marked ancestors of each locus using the data structures of Section 4. In total, this can be done in $O(|T|\log n + occ)$ time. For updates due to pattern insertion or deletion, it can be done in similar time as the above.[6] This gives the following theorem.

[6] Though we will need to handle a single update in the String B-tree data structure, this can easily be done in $O(|P|)$ time. Details are deferred in the full paper.

Theorem 2. *Suppose that patterns in Δ are stored separately in $n \log \sigma$ bits. Then we can maintain an $o(n \log \sigma) + O(k \log n)$-bit index for Δ, such that dictionary matching query can be answered in $O(|T| \log n + occ)$ time. The index supports insertion or deletion of a pattern in $O(|P| \log \sigma + \log n)$ time.*

References

1. Aho, A., Corasick, M.: Efficient String Matching: An Aid to Bibliographic Search. Communications of the ACM 18(6), 333–340 (1975)
2. Alstrup, S., Husfeldt, T., Rauhe, T.: Marked Ancestor Problems. In: Proceedings of Symposium on Foundations of Computer Science, pp. 534–544 (1998)
3. Amir, A., Farach, M., Idury, R., La Poutre, A., Schaffer, A.: Improved Dynamic Dictionary Matching. Information and Computation 119(2), 258–282 (1995)
4. Arge, L., Vitter, J.S.: Optimal External Memory Interval Management. SIAM Journal on Computing, 1488–1508 (2003)
5. Bender, M.A., Cole, R., Demaine, E.D., Farach-Colton, M., Zito, J.: Two Simplified Algorithms for Maintaining Order in a List. In: Proceedings of European Symposium on Algorithms, pp. 152–164 (2002)
6. Burrows, M., Wheeler, D.J.: A Block-sorting Lossless Data Compression Algorithm. Tech Report 124, Digital Equipment Corporation, CA, USA (1994)
7. Chan, H.L., Hon, W.K., Lam, T.W., Sadakane, K.: Compressed Indexes for Dynamic Text Collections. ACM Transactions on Algorithms 3(2) (2007)
8. Dietz, P.F., Sleator, D.D.: Two Algorithms for Maintaining Order in a List. In: Proceedings of Symposium on Theory of Computing, pp. 365–372 (1987)
9. Ferragina, P., Grossi, R.: The String B-tree: A New Data Structure for String Searching in External Memory and Its Application. Journal of the ACM 46(2), 236–280 (1999)
10. Ferragina, P., Manzini, G.: Indexing Compressed Text. Journal of the ACM 52(4), 552–581 (2005)
11. Grossi, R., Vitter, J.S.: Compressed Suffix Arrays and Suffix Trees with Applications to Text Indexing and String Matching. SIAM Journal on Computing 35(2), 378–407 (2005)
12. Hon, W.-K., Lam, T.-W., Shah, R., Tam, S.-L., Vitter, J.S.: Compressed Index for Dictionary Matching. In: DCC 2008, pp. 23–32 (2008)
13. Kärkkäinen, J., Ukkonen, E.: Sparse Suffix Trees. In: Proceedings of International Conference on Computing and Combinatorics, pp. 219–230 (1996)
14. Knuth, D.E., Morris, J.H., Pratt, V.B.: Fast Pattern Matching in Strings. SIAM Journal on Computing 6(2), 323–350 (1977)
15. Manber, U., Myers, G.: Suffix Arrays: A New Method for On-Line String Searches. SIAM Journal on Computing 22(5), 935–948 (1993)
16. McCreight, E.M.: A Space-economical Suffix Tree Construction Algorithm. Journal of the ACM 23(2), 262–272 (1976)
17. Overmars, M.H.: The Design of Dynamic Data Structures. LNCS, vol. 156. Springer, Heidelberg (1983)
18. Sadakane, K.: New text indexing functionalities of the compressed suffix arrays. Journal of Algorithms 48(2), 294–313 (2003)
19. Sadakane, K.: Compressed Suffix Trees with Full Functionality. Theory of Computing Systems, 589–607 (2007)
20. Weiner, P.: Linear Pattern Matching Algorithms. In: Proceedings of Symposium on Switching and Automata Theory, pp. 1–11 (1973)

Range Non-overlapping Indexing

Hagai Cohen and Ely Porat

Department of Computer Science, Bar-Ilan University, 52900 Ramat-Gan, Israel
{cohenh5,porately}@cs.biu.ac.il

Abstract. We study the *non-overlapping indexing* problem: Given a text T, preprocess it in order to answer queries of the form: given a pattern P, report the maximal set of non-overlapping occurrences of P in T. A generalization of this problem is the *range non-overlapping indexing* where in addition we are given two indexes i, j to report the maximal set of non-overlapping occurrences between these two indexes. We suggest new solutions for these problems. For the non-overlapping problem our solution uses $O(n)$ space with query time of $O(m+occ_{NO})$. For the range non-overlapping problem we propose a solution with $O(n \log^{\epsilon} n)$ space for some $0 < \epsilon < 1$ and $O(m + \log\log n + occ_{ij,NO})$ query time.

1 Introduction and Related Work

Given a text T of length n over an alphabet Σ, the *text indexing* problem is to build an index on T which can answer pattern matching queries efficiently: Given a pattern P of length m, we want to report all its occurrences in T. There are some known solutions for this problem. For instance, the *suffix tree*, proposed by Weiner [1], which is a compacted trie storing all suffixes of the text. A suffix tree for text T of length n requires $O(n)$ space and can be built in $O(n)$ preprocessing time. It has query time of $O(m + occ)$ where occ is the number of occurrences of P in T.

Range text indexing, also known as *position restricted substring searching*, is the problem of finding a pattern P in a substring of the text T between two given positions i, j. A solution for this problem was presented by Makinen and Navarro [2]. It uses $O(n \log^{\epsilon} n)$ space and has query time of $O(m + \log\log n + occ)$. Their solution is based on another problem - the range searching problem.

The *range searching* problem is to preprocess a set of points in a d-dimensional space for answering queries about the set of points which are contained within a specific range. Alstrup et al [3] proposed a solution for the orthogonal two dimensional range searching problem when all the points are from $n \times n$ grid, which costs $O(n \log^{\epsilon} n)$ space and has $O(\log\log n + k)$ query time, where k is the number of points inside the range. Grossi and Iwona in [4] have shown how to use Alstrup's data structure to get all the points inside a particular range in a specific order using some kind of rank function.

In text indexing we are sometimes interested in reporting only the non-overlapping occurrences of P in T. There is such interest in fields such as pattern recognition, computational linguistics, speech recognition, data compression, etc. For instance, we might want to compress a text by replacing each

Y. Dong, D.-Z. Du, and O. Ibarra (Eds.): ISAAC 2009, LNCS 5878, pp. 1044–1053, 2009.

non-overlapping occurrence of a substring of it with a pointer to a single copy of the substring.

Another problem is the *string statistics* problem [5,6] which consists of preprocessing a text T such that when given a query pattern P, the maximum number of non-overlapping occurrences of P in T can be reported efficiently. However, in the string statistics problem we only return the number of non-overlapping occurrences not the actual occurrences. In this paper, we present the first non-trivial solution for the *non-overlapping indexing* problem where we want to report the maximal sequence of non-overlapping occurrences of P in T.

Keller et al [7] proposed a solution for a generalization of this problem called the *range non-overlapping indexing* where we want to report the non-overlapping occurrences in a substring of T. Their solution has query time of $O(m + occ_{ij,NO} \log\log n)$ and uses $O(n \log n)$ space, where $occ_{ij,NO}$ is the number of the maximal non-overlapping occurrences in the substring $T[i:j]$.

Crochemore et al [8] suggested another solution for the *range non-overlapping indexing* problem. Their solution has optimal query time of $O(m + occ_{ij,NO})$ but requires $O(n^{1+\epsilon})$ space.

In this paper, we present new solutions for the *non-overlapping indexing* problem, which use the periodicity of the text and pattern in order to minimize the query time. Our solution for *non-overlapping indexing* problem uses $O(n)$ space with optimal query time of $O(m + occ_{NO})$. For the *range non-overlapping indexing* problem we present a solution of $O(n \log^{\epsilon} n)$ space for some $0 < \epsilon < 1$ with $O(m + \log\log n + occ_{ij,NO})$ query time.

2 Preliminaries

Let n be the length of the text T. And let m be the length of the pattern P. For two integers i, j $(i \leq j)$, $T[i:j]$ is the substring of T from i to j.

We will use the Suffix Tree as our main data structure. Each leaf in the Suffix Tree represents a suffix in the text. In each leaf we save two values: y - the start location of its suffix in the text and x - the location of the leaf in a left to right order of all the leaves of the Suffix Tree (lexicographic order, for example). We have two orders on the leaves and therefore on the suffixes as well: x-order and y-order. The y-order is the text order and the x-order is the suffix tree leaves order.

When we search for a pattern P in a Suffix Tree ST of T, we finish searching at some node v. The subtree rooted by v has all the occurrences of P in T in its leaves. We denote by l and r the x-value of the leftmost leaf and the x-value of the rightmost leaf of that subtree respectively. Therefore, the occurrences of P in T are all the leaves with x-value between l and r.

We denote the number of occurrences of P in T by occ. The non-overlapping occurrences will be denoted as occ_{NO}. The occurrences of P in $T[i:j]$ and the non-overlapping occurrences of P in $T[i:j]$ will be denoted by occ_{ij} and $occ_{ij,NO}$ respectively.

3 A Solution for Non-overlapping Indexing

We use a new approach for solving this problem. Our solution uses the periodicity of the text and the pattern. We divide patterns for two types: **periodic** and **aperiodic**. A different strategy will be used for each type.

Definition 1. *A pattern that can appear more than twice overlapping is called a* ***periodic pattern****. A pattern that can appear at most twice overlapping is called an* ***aperiodic pattern****.*

3.1 Aperiodic Pattern

In the aperiodic case we use the periodicity of the pattern to answer a query. We use the familiar Suffix Tree to get all the leaves that correspond to the given pattern. After we have all the leaves, we need to remove the overlapping occurrences. This can be done by sorting the leaves in y-order, going over the sorted list and filtering the overlapping occurrences. However, sorting occ items costs $O(occ \log occ)$ which is greater than the optimal $O(occ)$. In order to solve this sorting part we use the following theorem.

Theorem 1. *All occurrences of a pattern can be reported and sorted in text order in $O(m + occ)$ time using $O(n)$ space.*

Proof. We use a Suffix Tree to get all the occurrences in $O(m + occ)$ time. For sorting all the occurrences we will use a renaming method on the Suffix Tree.

Each leaf has its location, i.e., its y-value index, in the whole tree. Saving this location for a leaf costs $\log n$ bits because the whole tree has n leaves. Hence, the domain for the location is n. Nevertheless, we are interested in the order of the leaves in a subtree of the occurrences and not in the whole tree. Thus, we would like to save the location of a leaf for a subtree with less leaves. If we save for each leaf its location in a subtree with less leaves, for example $\sqrt{n}$ leaves, it will cost us only $\log \sqrt{n}$. Therefore, for each leaf, aside from keeping its location in the whole suffix tree, we save its location in a subtree of size $\sqrt{n}$, its location in a subtree of size $\sqrt[4]{n}$, and so on for all subtrees of size $\sqrt[2^i]{n}$ for $i \geq 1$ until we reach a constant size.

We use Radix Sort which can sort n numbers in a domain of n^2 in $O(n)$ time for sorting the leaves by their locations. Given a subtree whose leaves we wish to sort in y-order, we can sort them by the locations of the subtree of size at most $O(occ^2)$, this will cost us only $O(occ)$ by using Radix Sort because we sort occ items in a domain of at most occ^2.

In each leaf we save:

$\log n$ bits for its location in the ST.
$\log \sqrt{n}$ bits for its location in the subtree of size $\sqrt{n}$.
$\log \sqrt[4]{n}$ bits for its location in the subtree of size $\sqrt[4]{n}$.
etc.

This sums as following: $\log n + \log \sqrt{n} + \log \sqrt[4]{n} + \log \sqrt[8]{n} ... = \log n + \frac{1}{2}\log n + \frac{1}{4}\log n + \frac{1}{8}\log n + ... \leq 2\log n$.

Therefore, we save only $2\log n$ bits per leaf. We have n leaves summing up to $n \cdot 2\log n = O(n \log n)$ bits which is $O(n)$ space. □

Theorem 1 provides us a sorted list of all occurrences in $O(m + occ)$ query time and $O(n)$ space. By filtering the overlapping occurrences which costs $O(occ)$ time, we are through. Because in an aperiodic pattern, $O(occ) = O(occ_{NO})$, the query time equals to $O(m + occ_{NO})$.

3.2 Periodic Pattern

The periodic case is more complex. In this case we use the periodicity of the text in order to answer a query.

Definition 2. *A node in the Suffix Tree which represents a suffix which is a periodic pattern is called* ***period node****.*

Definition 3. *Let s be a string. We define a* ***period*** *of s to be a string p, such that $s = p^t \acute{p}$, for $t \geq 1$ where $\acute{p}$ is prefix of p.*

Lemma 1. *A period node has only one son which can also be a period node.*

Proof. Let a be a period node. Therefore, the string represented by a has a period p. For a son of a to be a period node too it must continue the period p. If a ends with a character c than the node which has the next character in p that is after that c is the period node. There can be only one child of a which can start with this character. Thus, a period node can have only one son which is also period node. □

Note that by the period definition a string can have more than one period, by taking $p_{new} = pp$ for example. Nevertheless, each period must overlap with a shorter period of the same string. Thus, there can't be more than one such character c.

Definition 4. *The path that starts from the a period node and goes through all nodes which continue that period in the Suffix Tree is called a* ***period path****. We denote the number of nodes in a period path to be the period path length.*

Lemma 2. *Let pp_2 be a period path on a path PT to the root in the Suffix Tree. Than pp_2 must be at least twice as long as the previous period path pp_1 on PT.*

Proof. On PT, between pp_1 and pp_2 there must be at least one node which is not a period node. For pp_2 to be a period path is must represent a period suffix which must be started with the period of pp_1. Moreover, pp_2 period suffix should be continued by the character of the next node after pp_1 which is not a period node. After it there must be the period of pp_1 again. Therefore, pp_2 length is at least twice longer than pp_1. □

Lemma 3. *The largest number of different period paths contained in the path from the root to a period node is* $\log n$.

Proof. Let PT be the path from the root to some period node in the Suffix Tree. According to Lemma 2 each period path on PT must be at least twice as long as the previous period path on PT. Therefore, if we have more than $\log n$ period paths on PT, than the length of the last period path must be greater than n which is the text length. Hence, the number of period paths on the same path can't be more than $\log n$. □

Definition 5. *A* ***period sequence*** *is the maximum substring in the text of some period which is repeated more than twice. We will mark it as* $[s, e]$*, where* s *and* e *are the start index and the end index of the period sequence accordingly. The period sequences of a period pattern are all the period sequences which start with that periodic pattern. The* ***period length*** *of a period sequence is the length of the period inside repeated the sequence.*

Example 1. Lets T be the text "abababcabababcabababc". The period sequences are: For the period "ab", the period sequences are [1,6], [8,13], [15,20] which have period length 2. For the period "abababc", the period sequence is [1,21] which has period length 7.

Lemma 4. *Given a list* L *of all the period sequences of some periodic pattern in the text. All the non-overlapping occurrences of that periodic pattern can be retrieved from this list in* $O(occ_{NO})$ *time.*

Proof. For a period sequence $[s, e]$ with period length pl, the non-overlapping occurrences of a periodic pattern P with length m are the group: $s+i*step|step = \lceil \frac{m}{pl} \rceil * pl, 0 \le i \le \frac{e-s-1}{step}$ which can be easily calculated.

We report all the occurrences in each period sequence in L. The number of all occurrences we report is $O(occ_{NO})$, so the total time for reporting all the non-overlapping occurrences from L is $O(occ_{NO})$. □

For answering the non-overlapping indexing we use the following data structure. We build a data structure for each period path on the Suffix Tree, saving a list of all period sequences sorted by their length for each period path. Each period node in the Suffix Tree is on a period path. We save a pointer from each period node on the Suffix Tree to the period sequence list of its period path. This pointer will point to the period node appropriate length on the period sequences list.

Theorem 2. *Using the data structure described above, all period sequences of a period pattern can be retrieved in* $O(m + occ_{NO})$ *time using* $O(n \log n)$ *space.*

Proof. On a query we go to that data structure, get all the period sequences and by Lemma 4 we calculate all the non-overlapping occurrences. This will take $O(occ_{NO})$ time, because the number of the period sequences that we will get is less than the number of the non-overlapping occurrences of the pattern. If we come up with a long pattern we won't get shorter period sequences which don't fit the pattern so we won't get unnecessary period sequences.

The space for this data structure is $O(n \log n)$. This is because there are n nodes where each one can be at most in $\log n$ period paths. For each node, we save all its period paths so it needs $O(n \log n)$. □

Now, we will show how to reduce the space needed for this data structure.

Definition 6. *Let's define a* ***degree of a period sequence*** *to be the maximum degree of a period sequence included in it plus one. A period sequence without any period sequences in it will has the degree* 0.

Lemma 5. *The maximum degree of a period sequence can be at most* $\log n$.

Proof. Let ps be the period sequence with the maximum degree in the text. In ps there is a period sequence with a degree decreased by one. In that period sequence there is another period sequence with a degree decreased by one. And so on until we receive a period sequence ps_0 with the degree 0. The length of each period sequence from ps_0 to ps is at least twice the length of the period sequence in it. The maximum length of ps can be at most n, therefore, its degree can be at most $\log n$. □

Lemma 6. *There are at most* $O(n)$ *period sequences.*

Proof. We will count the number of period sequences in each degree:

The number of period sequences of degree 0 can be at most n.
The number of period sequences of degree 1 can be at most $\frac{n}{2}$.
The number of period sequences of degree 2 can be at most $\frac{n}{4}$.
...
The number of period sequences of degree $\log n$ can be at most 1.

Summery: $n + \frac{n}{2} + .. + 1 \leq 2n = O(n)$ □

Theorem 3. *The data structure in Theorem 2 can be saved using only* $O(n)$ *space.*

Proof. We save all the period paths in a data structure. Each one with its own period sequences. Each period sequence appears in only one period path. From Lemma 6 we have $O(n)$ period sequences. Therefore we save at most $O(n)$ space for all the period sequences. Thus, we need only $O(n)$ space for this data structure. □

Corollary 1. *Using these two different strategies for each type of pattern we can solve the non-overlapping indexing problem in* $O(n)$ *space with* $O(m + occ_{NO})$ *query time.*

4 A Solution for Range Non-overlapping Indexing

We propose a better solution for this problem. Our solution costs $O(n \log^{\epsilon} n)$ space for some $0 < \epsilon < 1$ and has query time of $O(m + \log\log n + occ_{ij,NO})$.

to a period sequence $[s, e]$. We get only points which have x values between l and r. The x value of a point is the index of the suffix of s in the left to right order of all the ST leaves. So if we get a point (x, y) with x between l and r it means that the corresponding period sequence $[s, e]$ has an occurrence of P. This happens because all the leaves in the ST between l and r are occurrences of P.

Now, we need to prove two more things. The first is that all the period sequences we get from the RSDS are suitable for us and that we haven't got unnecessary period sequences, which don't fit to P in the range i, j. The second thing is that we didn't miss any period sequence which can have some suitable occurrences.

We start by proving that we get all the occurrences of P in the range $[i, j]$ from the period sequences we get in the three queries. Period sequences of P in T can be in some cases. Let $[s, e]$ be our period sequence.

The first case is that $[s, e]$ is out of the range $[i, j]$, $s < e \leq i < j$ or $i < j \leq s < e$. In this case we wouldn't like to get this period sequences at all. The first two queries will not resolve these period sequences because we do a query on $[i.j]x[l, r]$ but s is out of range and the points corresponding to this period sequence have y value equals s. Nevertheless, we can get this period sequence in the third query. However, it will be at most one period sequence which can be checked in $O(1)$ time.

The second case, which is the simplest, is that $[s, e]$ is fully inside the range $[i, j]$, $i \leq s < e \leq j$. In this case we get all the suitable period sequences from the first query. Nevertheless, we can get the same period sequence twice, first from the first query and again from the second query. Therefore, we will have to check any period sequence that we get in order to prevent duplicate occurrence reporting.

The third case is when only e or s is out of range but not both, $s < i < e \leq j$ or $i \leq s < j < e$. This time if $i \leq s < j < e$ we will get the period sequence from the first query. Otherwise, If $s < i < e \leq j$ we will get the period sequence from the second query.

The fourth case is when the range $[i, j]$ is fully inside the period sequence $[s, e]$, $s < i < j < e$. In order to solve this case we have the third query which will resolve the last start of a period sequence which is before index i. This period sequence can be checked in $O(1)$ time. □

Corollary 2. *Using these two different strategies for each type of pattern, the range non-overlapping text indexing problem can be solved in* $O(n \log^{\epsilon} n)$ *space for some* $0 < \epsilon < 1$ *and query time of* $O(m + \log\log n + occ_{ij,NO})$.

5 Conclusion

We have studied the problem of non-overlapping indexing. In this paper, we provide the first non-trivial solution for this problem. In addition we proposed a better solution for a generalization of this problem, the range non-overlapping problem.

References

1. Weiner, P.: Linear pattern matching algorithms. In: FOCS, pp. 1–11. IEEE, Los Alamitos (1973)
2. Mäkinen, V., Navarro, G.: Position-restricted substring searching. In: Correa, J.R., Hevia, A., Kiwi, M. (eds.) LATIN 2006. LNCS, vol. 3887, pp. 703–714. Springer, Heidelberg (2006)
3. Alstrup, S., Brodal, G.S., Rauhe, T.: New data structures for orthogonal range searching. In: 41st Symposium on Foundations of Computer Science (FOCS 2000), pp. 198–207 (2000)
4. Bialynicka-Birula, I., Grossi, R.: Rank-sensitive data structures. In: Consens, M.P., Navarro, G. (eds.) SPIRE 2005. LNCS, vol. 3772, pp. 79–90. Springer, Heidelberg (2005)
5. Apostolico, A., Preparata, F.P.: Data structures and algorithms for the string statistics problem. Algorithmica 15(5), 481–494 (1996)
6. Brodal, G.S., Lyngsø, R.B., Östlin, A., Pedersen, C.N.S.: Solving the string statistics problem in time o(n log n). In: Widmayer, P., Triguero, F., Morales, R., Hennessy, M., Eidenbenz, S., Conejo, R. (eds.) ICALP 2002. LNCS, vol. 2380, pp. 728–739. Springer, Heidelberg (2002)
7. Keller, O., Kopelowitz, T., Lewenstein, M.: Range non-overlapping indexing and successive list indexing. In: Dehne, F., Sack, J.-R., Zeh, N. (eds.) WADS 2007. LNCS, vol. 4619, pp. 625–636. Springer, Heidelberg (2007)
8. Crochemore, M., Iliopoulos, C.S., Kubica, M., Rahman, M.S., Walen, T.: Improved algorithms for the range next value problem and applications. In: Albers, S., Weil, P. (eds.) STACS. Dagstuhl Seminar Proceedings, Internationales Begegnungs- und Forschungszentrum fuer Informatik (IBFI), Schloss Dagstuhl, Germany, vol. 08001, pp. 205–216 (2008)

Querying Two Boundary Points for Shortest Paths in a Polygonal Domain*
(Extended Abstract)

Sang Won Bae[1,**] and Yoshio Okamoto[2,***]

[1] Department of Computer Science and Engineering, POSTECH, Pohang, Korea
swbae@postech.ac.kr
[2] Graduate School of Information Science and Engineering,
Tokyo Institute of Technology, Tokyo, Japan
okamoto@is.titech.ac.jp

Abstract. We consider a variant of two-point Euclidean shortest path query problem: given a polygonal domain, build a data structure for two-point shortest path query, provided that query points always lie on the boundary of the domain. As a main result, we show that a logarithmic-time query for shortest paths between boundary points can be performed using $\tilde{O}(n^5)$ preprocessing time and $\tilde{O}(n^5)$ space where n is the number of corners of the polygonal domain and the $\tilde{O}$-notation suppresses the polylogarithmic factor. This is realized by observing a connection between Davenport-Schinzel sequences and our problem in the parameterized space. We also provide a tradeoff between space and query time; a sublinear time query is possible using $O(n^{3+\epsilon})$ space. Our approach also extends to the case where query points should lie on a given set of line segments.

1 Introduction

A polygonal domains $\mathcal{P}$ with n corners and h holes is a polygonal region of genus h whose boundary consists of n line segments. The holes and the outer boundary of $\mathcal{P}$ are regarded as *obstacles*. Then, the geodesic distance between any two points p, q in a given polygonal domain $\mathcal{P}$ is defined to be the length of a shortest obstacle-avoiding path between p and q.

The Euclidean shortest path problem in a polygonal domain has drawn much attention in the history of computational geometry [7]. In the *two-point shortest path query* problem, we preprocess $\mathcal{P}$ so that we can determine a shortest path (or its length) quickly for a given pair of query points $p, q \in \mathcal{P}$. While we can compute a shortest path in $O(n \log n)$ time from scratch [5], known structures for logarithmic time query require significantly large storage [3]. Chiang and Mitchell [3] developed several data structures that can answer a two-point query quickly with tradeoffs between storage

* The authors would like to thank Matias Korman for fruitful discussion.
** Supported by the Brain Korea 21 Project.
*** Supported by Global COE Program "Computationism as a Foundation for the Sciences" and Grant-in-Aid for Scientific Research from Ministry of Education, Science and Culture, Japan, and Japan Society for the Promotion of Science.

Y. Dong, D.-Z. Du, and O. Ibarra (Eds.): ISAAC 2009, LNCS 5878, pp. 1054–1063, 2009.

Table 1. Summary of new and known results on exact two-point shortest path queries, where $\epsilon > 0$ is arbitrary and $0 < \delta \leq 1$ is a parameter. [new] denotes our results

Query domain	Preprocessing time	Space	Query time	Ref.
$\mathcal{P}$	$O(n^{11})$	$O(n^{11})$	$O(\log n)$	[3]
$\mathcal{P}$	$O(n^{10} \log n)$	$O(n^{10} \log n)$	$O(\log^2 n)$	[3]
$\mathcal{P}$	$O(n^{5+10\delta+\epsilon})$	$O(n^{5+10\delta+\epsilon})$	$O(n^{1-\delta} \log n)$	[3]
$\mathcal{P}$	$O(n^5)$	$O(n^5)$	$O(\log n + h)$	[3]
$\mathcal{P}$	$O(n + h^5)$	$O(n + h^5)$	$O(h \log n)$	[3]
$\partial\mathcal{P}$	$O(n^4 \lambda_{65}(n) \log n)$	$O(n^4 \lambda_{66}(n))$	$O(\log n)$	[new]
$\partial\mathcal{P}$	$O(n^{3+\delta} \lambda_{65}(n^\delta) \log n)$	$O(n^{3+\delta} \lambda_{66}(n^\delta))$	$O(n^{1-\delta} \log n)$	[new]
m segments	$O(m^2 n^{3+\delta} \lambda_{65}(n^\delta) \log n)$	$O(m^2 n^{3+\delta} \lambda_{66}(n^\delta))$	$O(n^{1-\delta} \log(m+n))$	[new]

usage and query time. Most notably, $O(\log n)$ query time can be achieved by using $O(n^{11})$ space and preprocessing time; sublinear query time by $O(n^{5+\epsilon})$ space and preprocessing time. Their results are summarized in Table 1. For more results on shortest paths in a polygonal domain, we refer to a survey by Mitchell [7].

In this paper, we focus on a variant of the problem, in which possible query points are restricted to a subset of $\mathcal{P}$; the boundary $\partial\mathcal{P}$ of the domain $\mathcal{P}$ or a set of line segments within $\mathcal{P}$. In many applications, possible pairs of source and destination do not span the whole domain $\mathcal{P}$ but a specified subset of $\mathcal{P}$. For example, in an urban planning problem, the obstacles correspond to the residential areas and the free space corresponds to the walking corridors. Then, the query points are restricted to the spots where people depart and arrive, which are on the boundary of obstacles.

Therefore, our goal is to design a data structure using much less resources than structures of Chiang and Mitchell [3] when the query domain is restricted to the boundary of a given polygonal domain $\mathcal{P}$ or to a set of segments in $\mathcal{P}$. To our best knowledge, no prior work seems to investigate this variation. As a main result, in Section 3, we present a data structure of size $O(n^4 \lambda_{66}(n))$ that can be constructed in $O(n^4 \lambda_{65}(n) \log n)$ time and can answer a $\partial\mathcal{P}$-restricted two-point shortest path query in $O(\log n)$ time. Here, $\lambda_m(n)$ stands for the maximum length of a Davenport-Schinzel sequence of order m on n symbols [10]. It is good to note that $\lambda_m(n) = O(n \log^* n)$ for any constant m as a convenient intuition, while tighter bounds are known [10,8]. We also provide a tradeoff between space and query time in Section 4. In particular, we show that one can achieve sublinear query time using $O(n^{3+\epsilon})$ space and preprocessing time. New results in this paper are also summarized in Table 1.

Our data structure is a subdivision of two-dimensional domain parameterized in a certain way. The domain is divided into a number of grid cells in which a set of constrained shortest paths between query points have the same structure. Each grid cell is divided according to the projection of the lower envelope of functions stemming from the constrained shortest paths. With careful investigation into this lower envelope, we show the claimed upper bounds.

Also, our approach readily extends to the variant where query points are restricted to lie on a given segment or a given set of segments in $\mathcal{P}$. We discuss this extension in Section 5.

2 Preliminaries

We are given as input a polygonal domain $\mathcal{P}$ with h holes and n corners. More precisely, $\mathcal{P}$ consists of an outer simple polygon in the plane $\mathbb{R}^2$ and a set of h (≥ 0) disjoint simple polygons inside P. As a set, $\mathcal{P}$ is the region contained in its outer polygon *excluding* the holes, also called the *free space*. The complement of $\mathcal{P}$ in the plane is regarded as *obstacles* so that any feasible path does not cross the boundary $\partial\mathcal{P}$ and lies inside $\mathcal{P}$. It is well known from earlier works that there exists a *shortest (obstacle-avoiding) path* between any two points $p, q \in \mathcal{P}$ [6].

Letting V be the set of all corners of $\mathcal{P}$, any shortest path from $p \in \mathcal{P}$ to $q \in \mathcal{P}$ is a simple polygonal path and can be represented by a sequence of line segments connecting points in $V \cup \{p, q\}$ [6]. The *length* of a shortest path is the sum of the Euclidean lengths of its segments. The geodesic distance, denoted by $d(p, q)$, is the length of a shortest path between p and q. Also, we denote by $|\overline{pq}|$ the Euclidean length of segment $\overline{pq}$.

A *two-point shortest path query* is given as a pair of points (p, q) with $p, q \in \mathcal{P}$ and asks to find a shortest path between p and q. In this paper, we deal with a restriction where the queried points p and q lie on the boundary $\partial\mathcal{P}$.

A *shortest path tree* $SPT(p)$ for a given source point $p \in \mathcal{P}$ is a spanning tree of the corners V plus the source p such that the unique path to any corner $v \in V$ from the source p in $SPT(p)$ is a shortest path between p and v. The complexity of $SPT(p)$ for any $p \in \mathcal{P}$ is at most linear in n. A shortest path map $SPM(p)$ for the source p is a decomposition of the free space $\mathcal{P}$ into cells in which any point x has a shortest path to p through the same sequence of corners in V. Once $SPT(p)$ is obtained, $SPM(p)$ can be computed as an additively weighted Voronoi diagram of $V \cup \{p\}$ with weight assigned by the geodesic distance to p [6]; thus, the combinatorial complexity of $SPM(p)$ is linear. A cell of $SPM(p)$ containing a point $q \in \mathcal{P}$ has the common last corner $v \in V$ along the shortest path from p to q; we call such a corner v the *root* of the cell or of q with respect to p. An $O(n \log n)$ time algorithm, using $O(n \log n)$ working space, to construct $SPT(p)$ and $SPM(p)$ is known by Hershberger and Suri [5].

An *SPT-equivalence decomposition* $\mathcal{A}^{SPT}$ of $\mathcal{P}$ is the subdivision of $\mathcal{P}$ into cells in which every point has topologically equivalent shortest path tree. An $\mathcal{A}^{SPT}$ can be obtained by overlaying n shortest path maps $SPM(v)$ for every corner $v \in V$ [3]. Hence, the complexity of $\mathcal{A}^{SPT}$ is $O(n^4)$. Note that $\mathcal{A}^{SPT} \cap \partial\mathcal{P}$ consists of at most $O(n^2)$ points; they are intersection points between any edge of $SPM(v)$ for any $v \in V$ and the boundary $\partial\mathcal{P}$. We call those intersection points, including the corners V, the *breakpoints*. The breakpoints induce $O(n^2)$ intervals along $\partial\mathcal{P}$. We shall say that a breakpoint is *induced by* $SPM(v)$ if it is an intersection of an edge of $SPM(v)$ and $\partial\mathcal{P}$.

Given a set Γ of algebraic surfaces and surface patches in $\mathbb{R}^d$, the *lower envelope* $\mathcal{L}(\Gamma)$ of Γ is the pointwise minimum of all given surfaces or patches in the d-th coordinate. The *minimization diagram* $\mathcal{M}(\Gamma)$ of Γ is a decomposition of $\mathbb{R}^{d-1}$ into faces, which are maximally connected region over which $\mathcal{L}(\Gamma)$ is attained by the same set of functions. In particular, when $d = 3$, the minimization diagram $\mathcal{M}(\Gamma)$ is simply a projection of the lower envelope onto the xy-plane. Analogously, we can define the *upper envelope* and the *maximization diagram*.

As we intensively exploit known algorithms on algebraic surfaces or surface patches and their lower envelopes, we assume a model of computation in which several primitive operations dealing with a constant number of given surfaces can be performed in constant time: testing if a point lies above, on or below a given surface, computing the intersection of two or three given surfaces, projecting down a given surface, and so on. Such a model of computation has been adopted in many research papers; see [9,1,10,2].

3 Structures for Logarithmic Time Query

In this section, we present a data structure that answers a two-point query restricted on $\partial\mathcal{P}$ in $O(\log n)$ time. To ease discussion, we parameterize the boundary $\partial\mathcal{P}$. Since $\partial\mathcal{P}$ is a union of $h+1$ closed curves, it can be done by parameterizing each curve by arc length and merging them into one. Thus, we have a bijection $p\colon [0,|\partial\mathcal{P}|) \to \partial\mathcal{P}$ that maps a one-dimensional interval into $\partial\mathcal{P}$, where $|\partial\mathcal{P}|$ denotes the total lengths of the $h+1$ closed curves forming $\partial\mathcal{P}$.

A shortest path between two points $p, q \in \mathcal{P}$ is either the segment $\overline{pq}$ or a polygonal chain through corners in V. Thus, unless $d(p,q) = |\overline{pq}|$, the geodesic distance is taken as the minimum of the following functions $f_{u,v}\colon [0,|\partial\mathcal{P}|) \times [0,|\partial\mathcal{P}|) \to \mathbb{R}$ over all $u, v \in V$, which are defined as follows:

$$f_{u,v}(s,t) := \begin{cases} |\overline{p(s)u}| + d(u,v) + |\overline{vp(t)}| & \text{if } u \in VP(p(s)) \text{ and } v \in VP(p(t)), \\ \infty & \text{otherwise,} \end{cases}$$

where $VP(x)$, for any point $x \in \mathcal{P}$, denotes the *visibility profile* of x, defined as the set of all points $y \in \mathcal{P}$ that are *visible from* x; that is, $\overline{xy}$ lies inside $\mathcal{P}$. The symbol ∞ can be replaced by an upper bound of $\max_{s,t} d(p(s),p(t))$; for example, the total length $|\partial\mathcal{P}|$ of the boundary of the polygonal domain $\mathcal{P}$.

Since the case where $p(s)$ is visible from $p(t)$, so the shortest path between them is just the segment $\overline{p(s)p(t)}$, can be checked in $O(\log n)$ time using $O(n^2 \log n)$ space [3], we assume from now on that $p(s) \notin VP(p(t))$. Hence, our task is to efficiently compute the lower envelope of the $O(n^2)$ functions $f_{u,v}$ on a 2-dimensional domain $\mathcal{D} := [0,|\partial\mathcal{P}|) \times [0,|\partial\mathcal{P}|)$.

3.1 Simple Lifting to 3-Dimension

Using known results on the lower envelope of the algebraic surfaces in 3-dimension, we can show that a data structure of size $O(n^{6+\epsilon})$ for $O(\log n)$ query can be built in $O(n^{6+\epsilon})$ time as follows.

Fix a pair of intervals I_s and I_t induced by the breakpoints. Since I_s belongs to a cell of an SPT-equivalence decomposition, it fixes the set $V_s := V \cap VP(p(s))$ of corners visible from $p(s)$ for any $s \in I_s$ and further, for a fixed $u \in V_s$, a unique $v \in V$ that minimizes $f_{u,v}(s,t)$ for any $(s,t) \in I_s \times I_t$ over all $v \in V$ [3]. This implies that for each such subdomain $I_s \times I_t \subset \mathcal{D}$ we extract at most n functions, possibly appearing at the lower envelope. Moreover, in $I_s \times I_t$, such a function is represented explicitly; for $u \in V_s$ and $v \in V_t$,

$$f_{u,v}(s,t) = \sqrt{(x(s)-x_u)^2 + (y(s)-y_u)^2} + d(u,v) + \sqrt{(x(t)-x_v)^2 + (y(t)-y_v)^2},$$

where $x(s)$ and $y(s)$ are the x- and the y-coordinates of $p(s)$, and x_u and y_u are the x- and the y-coordinates of a point $u \in \mathbb{R}^2$. Note that $x(s)$ and $y(s)$ are linear functions in s by our parametrization.

For each $u \in V$, let $g_u(s,t) := f_{u,v}(s,t)$ be a function defined on $I_s \times I_t$, where $v \in V_t$ minimizes $f_{u,v'}$ in $I_s \times I_t$ over all $v' \in V_t$. Observe that the graph of g_u is an algebraic surface with degree at most 4 in 3-dimensional space. Applying any efficient algorithm that computes the lower envelope of algebraic surfaces in $\mathbb{R}^3$, we can compute the lower envelope of the functions g_u in $O(n^{2+\epsilon})$ time [9]. Repeating this for every such subdomain $I_s \times I_t$ yields $O(n^{6+\epsilon})$ space and preprocessing time.

Since we would like to provide a point location structure in domain $\mathcal{D}$, we need to find the minimization diagram $\mathcal{M}$ of the computed lower envelope. Fortunately, our domain is 2-dimensional, so we can easily project it down on $\mathcal{D}$ and build a point location structure with an additional logarithmic factor.

In another way around, one could try to deal with *surface patches* on the whole domain $\mathcal{D}$. Consider a fixed corner $u \in V$ and its shortest path map $SPM(u)$. The number of breakpoints induced by $SPM(u)$ is at most $O(n)$, including the corners V themselves. This implies at most an $O(n^2)$ number of combinatorially different paths between any two boundary points $p(s)$ and $p(t)$ via u. That is, for a pair of intervals I_s and I_t, we have a unique path via u and its length is represented by a partial function of (s,t) defined on a rectangular subdomain $I_s \times I_t \subset \mathcal{D}$. Hence, we have $O(n^2)$ such partial functions for each $u \in V$, and thus $O(n^3)$ in total. Each of them defines an algebraic surface patch of constant degree on a rectangular subdomain. Consequently, we can apply the same algorithm as above to compute the lower envelope of those patches in $O((n^3)^{2+\epsilon}) = O(n^{6+\epsilon})$ time and space.

3.2 $O(n^{5+\epsilon})$-Space Structure

Now, we present a way of proper grouping of subdomains to reduce the time/space bound by a factor of n. We call a subdomain $I_s \times I_t \subset \mathcal{D}$, where both I_s and I_t are intervals induced by breakpoints, a *grid cell*. Thus, $\mathcal{D}$ consists of $O(n^4)$ grid cells. We will decompose $\mathcal{D}$ into $O(n^3)$ *blocks* of $O(n)$ grid cells.

Consider a pair of boundary edges $S, T \subset \partial\mathcal{P}$ and let b_S and b_T be the number of breakpoints on S and on T, respectively. Let $s_0, \ldots, s_{b_S}$ and $t_0, \ldots, t_{b_T}$ be the breakpoints on S and on T, respectively, in order $s_0 < s_1 < \cdots < s_{b_S}$ and $t_0 < t_1 < \cdots < t_{b_T}$. Take b_T grid cells with $s \in [s_0, s_1)$ and let $\mathcal{C} := [s_0, s_1) \times [t_0, t_{b_T}) \subseteq [s_0, s_{b_S}) \times [t_0, t_{b_T})$ be their union. We redefine the functions $f_{u,v}$ on domain $\mathcal{C}$. As discussed above, for any $s \in [s_0, s_1)$, we have a common subset V_s of corners visible from $p(s)$.

For any $u \in V_s$, let $g_u(s,t) := \min_{v \in V} f_{u,v}(s,t)$ be a function defined on $\mathcal{C}$ and b_T^u be the number of breakpoints on T induced by $SPM(u)$. The following is our key observation.

Lemma 1. *The graph of $g_u(s,t)$ on $\mathcal{C}$ consists of at most $b_T^u + 1$ algebraic surface patches with constant maximum degree.*

Proof. If $g_u(s,t) = f_{u,v}(s,t)$ for any $(s,t) \in \mathcal{C}$ and some $v \in V$, then $p(t)$ lies in a cell of $SPM(u)$ with root v; by the definition of g_u, the involved path goes directly from $p(s)$ to u and follows a shortest path from u to $p(t)$. On the other hand, when we walk along T as t increases from t_1 to t_{b_T}, we encounter b_T^u breakpoints induced by $SPM(u)$; thus, $b_T^u + 1$ cells of $SPM(u)$. Hence, the lemma is shown. □

Moreover, the partial function corresponding to each patch of γ_u is defined on a rectangular subdomain $[s_0, s_1) \times [t_i, t_j)$ for some $1 \leq i < j \leq b_T$. This implies that the lower envelope of g_u on $\mathcal{C}$ is represented by that of at most $\sum_u (b_T^u + 1) = n + b_T$ surface patches.

Though this envelope can be computed in $O((n + b_T)^{2+\epsilon})$ time, we do further decompose $\mathcal{C}$ into $\lceil \frac{b_T}{n} \rceil$ blocks of at most n grid cells. This can be simply done by cutting $\mathcal{C}$ at $t = t_{in}$ for each $i = 1, \ldots, \lfloor \frac{b_T}{n} \rfloor$. For each such block of grid cells, we have at most $2n$ surface patches and thus their lower envelope can be computed in $O(n^{2+\epsilon})$ time. Hence, we obtain the following consequence.

Theorem 1. *One can preprocess a given polygonal domain $\mathcal{P}$ in $O(n^{5+\epsilon})$ time into a data structure of size $O(n^{5+\epsilon})$ for $O(\log n)$-time two-point shortest path queries restricted to the boundary $\partial\mathcal{P}$, where ϵ is an arbitrarily small positive number.*

Proof. Recall that $\sum_S b_S = \sum_T b_T = O(n^2)$. For a pair of boundary edges S and T, we can compute the lower envelope of the functions $f_{u,v}$ in $O(b_S \lceil \frac{b_T}{n} \rceil n^{2+\epsilon})$. Summing this over every pair of boundary edges, we have

$$\sum_{S,T} O(b_S \lceil \frac{b_T}{n} \rceil n^{2+\epsilon}) = O(n^{4+\epsilon}) \cdot \sum_T (\frac{b_T}{n} + 1) = O(n^{5+\epsilon}).$$

A point location structure on the minimization diagram can be built with additional logarithmic factor, which is subdued by $O(n^{\epsilon})$. □

3.3 Further Improvement

The algorithms we described so far compute the lower envelope of surface patches in 3-space in order to obtain the minimization diagram $\mathcal{M}$ of the functions $f_{u,v}$. In this subsection, we introduce a way to compute $\mathcal{M}$ rather directly on $\mathcal{D}$, based on more careful analysis.

Basically, we make use of the same scheme of partitioning the domain $\mathcal{D}$ into blocks of (at most) n grid cells as in Section 3.2. Let $\mathcal{C}$ be such a block defined as $[s_0, s_1) \times [t_0, t_{b_T})$ such that $[s_0, s_1)$ is an interval induced by the breakpoints and $[t_0, t_{b_T})$ is a union of $b_T \leq n$ consecutive intervals in which we have $b_T - 1$ breakpoints $t_1, \ldots, t_{b_T - 1}$.

By Lemma 1, the functions $g_u = \min_{v \in V} f_{u,v}$ restricted to $\mathcal{C}$ can be split into at most $2n$ partial functions h_i with $1 \leq i \leq 2n$ defined on a subdomain $\mathcal{C}_i \subseteq \mathcal{C}$. Each $h_i(s,t)$ is represented explicitly as $h_i(s,t) = |p(s)u_i| + d(u_i, v_i) + |v_i p(t)|$ in $\mathcal{C}_i$, so that we have $h_i(s,t) = f_{u_i,v_i}(s,t)$ for any $(s,t) \in \mathcal{C}_i$ and some $u_i, v_i \in V$. Note that it may happen that $u_i = u_j$ or $v_i = v_j$ for some i and j; in particular, if $u_i = u_j$, we have $\mathcal{C}_i \cap \mathcal{C}_j = \emptyset$. Also, as discussed in Section 3.2, $\mathcal{C}_i$ is represented as $[s_0, s_1) \times [t_k, t_{k'})$ for some $0 \leq k < k' \leq b_T$.

In this section, we take the partial functions h_i into account, and thus the goal is to compute the minimization diagram $\mathcal{M}$ of surface patches defined by the h_i. We start with an ordering on the set V_s of corners visible from $p(s)$ for any $s \in [s_0, s_1)$ based on the following observation.

Lemma 2. *The angular order of corners in V_s seen at s is constant if s varies within $[s_0, s_1)$.*

Without loss of generality, we assume that as s increases, $p(s)$ moves along $\partial\mathcal{P}$ in direction that the obstacle lies to the *right*; that is, $p(s)$ moves *clockwise* around each hole and *counter-clockwise* around the outer boundary of $\mathcal{P}$. By Lemma 2, we order the corners in V_s in counter-clockwise order at $p(s)$ for any $s \in [s_0, s_1)$; let $\prec$ be a total order on V_s such that $u \prec u'$ if and only if $\angle p(s_0)p(s)u < \angle p(s_0)p(s)u'$.

From now on, we investigate the set

$$B(i,j) := \{(s,t) \in \mathcal{C}_i \cap \mathcal{C}_j \mid h_i(s,t) = h_j(s,t)\},$$

which is a projection of the intersection of two surface patches defined by h_i and h_j. One can easily check that $B(i,j)$ is a subset of an algebraic curve of degree at most 8.

Lemma 3. *The set $B(i,j)$ is t-monotone. That is, for fixed t, there is at most one $s \in [s_0, s_1)$ such that $(s,t) \in B(i,j)$.*

Lemma 4. *The set $B(i,j)$ is either an empty set or an open curve whose endpoints lie on the boundary of $\mathcal{C}_i \cap \mathcal{C}_j$. Moreover, $B(i,j)$ is either a linear segment parallel to the t-axis or s-monotone.*

Proof Sketch. Let I_s and I_t be intervals such that $\mathcal{C}_i \cap \mathcal{C}_j = I_s \times I_t$. Note that $I_s = [s_0, s_1)$ and $I_t = [t_k, t_{k'})$ for some $0 \le k < k' \le b_T$.

First, note that if $u_i = u_j$, $\mathcal{C}_i \cap \mathcal{C}_j = \emptyset$ and thus $B(i,j) = \emptyset$, so the lemma is true. Thus, we assume that $u_i \ne u_j$. Regarding v_i and v_j, there are two cases: $v_i = v_j$ or $v_i \ne v_j$. In the former case, we get $|p(s)u_i| - |p(s)u_j| = d(u_j, v_j) - d(u_i, v_i)$ from equation $h_i(s,t) = h_j(s,t)$. Observe that variable t is readily eliminated from the equation, and thus if there exists $(s',t') \in \mathcal{C}_i \cap \mathcal{C}_j$ with $(s',t') \in B(i,j)$, we have $(s',t) \in B(i,j)$ for every other $t \in I_t$. Hence, by Lemma 3, $B(i,j)$ is empty or a straight line segment in $\mathcal{C}_i \cap \mathcal{C}_j$ which is parallel to t-axis, and thus the lemma is shown.

Now, we consider the latter case where $v_i \ne v_j$. Without loss of generality, we assume that $u_i \prec u_j$. Recall that if $u_i \prec u_j$, then $\angle p(s_0)p(s)u_i < \angle p(s_0)p(s)u_j$ for any s in the interior of I_s. We denote $\theta_i(s) := \angle p(s_0)p(s)u_i$ and $\theta_j(s) := \angle p(s_0)p(s)u_j$. On the other hand, we also have a similar relation for v_i and v_j. Let $\phi_i(t) := \angle p(t_0)p(t)v_i$ and $\phi_j(t) := \angle p(t_0)p(t)v_j$. Observe that $\phi_i(t)$ and $\phi_j(t)$ are continuous functions of t, and if $\phi_i(t') = \phi_j(t')$ at $t = t'$, then $p(t')$ is a breakpoint induced by $SPM(u_i)$ or $SPM(u_j)$. Since I_t contains no such breakpoint induced by $SPM(u_i)$ or $SPM(u_j)$ in its interior, either $\phi_i(t) < \phi_j(t)$ or $\phi_i(t) > \phi_j(t)$ for all t in the interior of I_t; that is, the sign of $\phi_j(t) - \phi_i(t)$ is constant.

Since for any $s, s' \in I_s$ with $s' > s$ we have $|p(s')p(s)| = s' - s$ by our parametrization, we can represent $|p(s)u_i| = \sqrt{(s+a_i)^2 + b_i^2}$ and $|v_i p(t)| = \sqrt{(t+c_i)^2 + d_i^2}$, where a_i, b_i, c_i and d_i are constants depending on u_i, v_i, and parametrization p. More specifically, $s + a_i$ denotes a signed distance between $p(s)$ and the perpendicular foot of

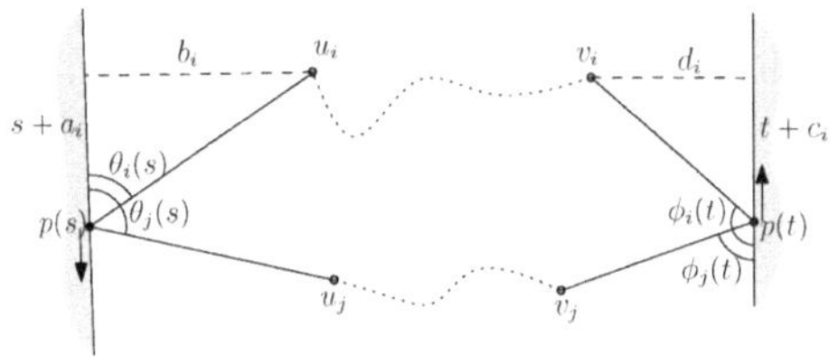

Fig. 1. Illustration to the proof of Lemma 4

u_i onto the line supporting $p(I_s)$, and b_i is the distance between u_i and the line supporting $p(I_s)$. See Figure 1. Thus, $h_i(s,t)$ can represented as $h_i(s,t) = \sqrt{(s+a_i)^2 + b_i^2} + \sqrt{(t+c_i)^2 + d_i^2} + d(u_i, v_i)$.

The last step of the proof is done by analysis on derivative $\frac{ds}{dt}$: we have $\frac{ds}{dt} = \frac{-\cos(\phi_i(t))+\cos(\phi_j(t))}{\cos(\theta_i(s))-\cos(\theta_j(s))}$, and the sign of $\frac{ds}{dt}$ is constant while $t \in I_t$. □

Now, we know that $B(i,j)$ can be seen as the graph of a partial function $\{s = \gamma(t)\}$. Also, Lemma 4 implies that $B(i,j)$ *bisects* $\mathcal{C}_i \cap \mathcal{C}_j$ into two connected regions $R(i,j)$ and $R(j,i)$, where $R(i,j) := \{(s,t) \in \mathcal{C}_i \cap \mathcal{C}_j \mid h_i(s,t) < h_j(s,t)\}$ and $R(j,i) := \{(s,t) \in \mathcal{C}_i \cap \mathcal{C}_j \mid h_i(s,t) > h_j(s,t)\}$. Let $\mathcal{M}(i)$ be the set of points (s,t) where the minimum of $h_j(s,t)$ over all j is attained by $h_i(s,t)$. We then have $\mathcal{M}(i) = \mathcal{C}_i \setminus \bigcup_j R(j,i)$ for each i.

For easy explanation, from now on, we regard the s-axis as the *vertical axis* in $\mathcal{D}$ so that we can say a point lies *above* or *below* a curve in this sense.

The idea of computing $\mathcal{M}(i)$ is using the lower and the upper envelopes of the bisecting curves $B(i,j)$. In order to do so, we extend $B(i,j)$ to cover the whole t-interval $I_t = [t_k, t_{k'})$ in $\mathcal{C}_i \cap \mathcal{C}_j$ by following operation: For each endpoint of $B(i,j)$, if it does not lie on the vertical line $\{t = t_k\}$ or $\{t = t_{k'}\}$, attach a horizontal segment to reach the vertical line as shown in Figure 2(a). We denote the resulting curve by $\beta(i,j)$; if $B(i,j) = \emptyset$, define $\beta(i,j)$ as the horizontal segment connecting two points (s_0, t_k) and $(s_0, t_{k'})$ in $\mathcal{C}_i \cap \mathcal{C}_j$. Observe now that $\beta(i,j)$ bisects $\mathcal{C}_i \cap \mathcal{C}_j$ into regions $R(i,j)$ and $R(j,i)$, which lie *above* and *below* $\beta(i,j)$, respectively.

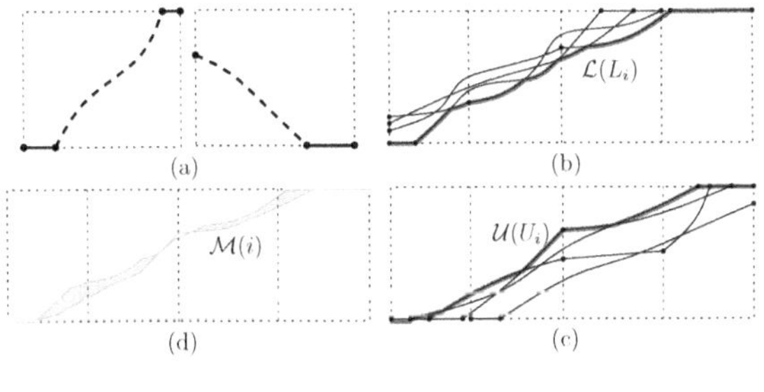

Fig. 2. (a) How to extend $B(i,j)$ (dashed line) to $\beta(i,j)$ by attaching horizontal segments (solid line); (b) $\mathcal{L}(L_i)$, (c) $\mathcal{U}(U_i)$, and (d) the region $\mathcal{M}(i)$ between them. Here, the dotted boxes are grid cells whose union is $\mathcal{C}_i$ and the s-axis appears vertical.

Let $\beta(i,j)^+ \subseteq \mathcal{C}_i \cap \mathcal{C}_j$ be the region above $\beta(i,j)$ and $\beta(i,j)^-$ be the region below $\beta(i,j)$. For a fixed i with $1 \leq i \leq 2n$, we classify the $\beta(i,j)$ into two sets L_i and U_i such that $\beta(i,j) \in L_i$ if $R(j,i) = \beta(i,j)^+$ or $\beta(i,j) \in U_i$ if $R(j,i) = \beta(i,j)^-$. Recall that $\mathcal{M}(i) = \mathcal{C}_i \setminus \bigcup_j R(j,i)$. Thus, $\mathcal{M}(i)$ is equal to $\mathcal{C}_i \setminus \left(\bigcup_{\beta \in L_i} \beta^+ \cup \bigcup_{\beta \in U_i} \beta^- \right)$. The boundary of $\bigcup_{\beta \in L_i} \beta^+$ is the lower envelope $\mathcal{L}(L_i)$ of L_i; symmetrically, the boundary of $\bigcup_{\beta \in U_i} \beta^-$ is the upper envelope $\mathcal{U}(U_i)$ of U_i. Therefore, $\mathcal{M}(i) = \mathcal{L}(L_i)^- \cap \mathcal{U}(U_i)^+$, the region below the lower envelope $\mathcal{L}(L_i)$ of L_i and above the upper envelope $\mathcal{U}(U_i)$ of U_i, and it can be obtained by computing the overlay of two envelopes $\mathcal{L}(L_i)$ and $\mathcal{U}(U_i)$. See Figure 2(b)–(d). We exploit known results on the Davenport-Schinzel sequences to obtain the following lemma [10,4]. Proof is omitted due to lack of space.

Lemma 5. *The set $\mathcal{M}(i)$ is of combinatorial complexity $O(\lambda_{66}(n))$ and can be computed in $O(\lambda_{65}(n) \log n)$ time, where $\lambda_m(n)$ is the maximum length of a Davenport-Schinzel sequence of order m on n symbols.*

We can compute the minimization diagram $\mathcal{M}$ by computing each $\mathcal{M}(i)$ in $O(n\lambda_{65}(n) \log n)$ time. In the same time bound, we can build a point location structure on $\mathcal{M}$. Finally, we conclude our main theorem.

Theorem 2. *One can preprocess a given polygonal domain $\mathcal{P}$ in $O(n^4\lambda_{65}(n) \log n)$ time into a data structure of size $O(n^4\lambda_{66}(n))$ for $O(\log n)$-time two-point shortest path queries restricted on the boundary $\partial\mathcal{P}$.*

4 Tradeoffs between Space and Query Time

In this section, we provide a space/query-time tradeoff. We use the technique of partitioning V, which has been used in Chiang and Mitchell [3].

Let δ be a positive number with $0 < \delta \leq 1$. We partition the corner set V into $m = n^{1-\delta}$ subsets $V_1, \ldots, V_m$ of near equal size $O(n^\delta)$. For each such subset V_i of corners, we run the algorithm described above with little modification: We build the shortest path maps $SPM(u)$ only for $u \in V_i$ and care about only $O(n^{1+\delta})$ breakpoints induced by such $SPM(u)$. Thus, we consider only the paths from $p(s)$ via $u \in V_i$ and $v \in V$ to $p(t)$, and thus $O(n^{1+\delta})$ functions $f_{u,v}$ for $u \in V_i$ and $v \in V$.

Since we deal with less number of functions, the cost of preprocessing reduces from $O(n)$ to $O(n^\delta)$ at several places. We take blocks of $O(n^\delta)$ grid cells contained in $\mathcal{D}$ and the number of such blocks is $O(n^{2+\delta})$. For each such block, we spend $O(n^\delta\lambda_{65}(n^\delta) \log n)$ time to construct a point location structure for the minimization map $\mathcal{M}_i$ of the functions. Iterating all such blocks, we get running time $O(n^{2+2\delta}\lambda_{65}(n^\delta) \log n)$ for a part V_i of V. Repeating this for all such subsets V_i yields $O(n^{3+\delta}\lambda_{65}(n^\delta) \log n)$ construction time.

Each query is processed by a series of m point locations on every $\mathcal{M}_i$, taking $O(m \log n) = O(n^{1-\delta} \log n)$ time.

Theorem 3. *Let δ be a fixed parameter with $0 < \delta \leq 1$. Using $O(n^{3+\delta}\lambda_{65}(n^\delta) \log n)$ time and $O(n^{3+\delta}\lambda_{66}(n^\delta))$ space, one can compute a data structure for $O(n^{1-\delta} \log n)$-time two-point shortest path queries restricted on the boundary $\partial\mathcal{P}$.*

Remark that when $\delta = 1$, we obtain Theorem 2, and that $O(n^{3+\epsilon})$ time and space is enough for sublinear time query. Note that if $O(n)$ time is allowed for processing each query, $O(n^2)$ space and $O(n^2 \log n)$ preprocessing time is sufficient.

5 Extensions to Segments-Restricted Queries

Let $\mathcal{S}_s$ and $\mathcal{S}_t$ be two sets of m_s and m_t line segments, respectively, within $\mathcal{P}$. In this section, we restrict a query pair (p, q) of points to lie on $\mathcal{S}_s$ and $\mathcal{S}_t$ each. We will refer to this type of two-point query as a *$(\mathcal{S}_s, \mathcal{S}_t)$-restricted two-point query*. As we did above, we take two segments $S \in \mathcal{S}_s$ and $T \in \mathcal{S}_t$ and let b_S and b_T be the number of breakpoints — the intersection points with an edge of $SPM(v)$ for some $v \in V$ — on S and T, respectively. Also, parameterize S and T as above so that we have two bijections $p : [0, |S|] \to S$ and $q : [0, |T|] \to T$.

Any path from a point on S leaves to one of the two sides of S. Thus, the idea of handling such a segment within the free space $\mathcal{P}$ is to consider two cases separately. Here, we regard S and T as directed segments in direction of movement of $p(s)$ and $q(t)$ as s and t increases. Details can be found in a full version of the paper.

Theorem 4. *Let $\mathcal{S}_s$ and $\mathcal{S}_t$ be two sets of m_s and m_t (possibly crossing) line segments, respectively, within $\mathcal{P}$, and δ be a fixed parameter with $0 < \delta \leq 1$. Then, using $O(m_s m_t n^{3+\delta} \lambda_{65}(n^\delta) \log n)$ time and $O(m_s m_t n^{3+\delta} \lambda_{66}(n^\delta))$ space, one can compute a data structure for $O(n^{1-\delta} \log(n + m_s + m_t))$-time $(\mathcal{S}_s, \mathcal{S}_t)$-restricted two-point queries.*

References

1. Agarwal, P.K., Aronov, B., Sharir, M.: Computing envelopes in four dimensions with applications. SIAM J. Comput. 26(6), 1714–1732 (1997)
2. Agarwal, P.K., Sharir, M.: Arrangements and their applications. In: Sack, J.-R., Urrutia, J. (eds.) Handbook of Computationaal Geometry, pp. 49–119. Elsevier Science Publishers B.V, Amsterdam (2000)
3. Chiang, Y.-J., Mitchell, J.S.B.: Two-point Euclidean shortest path queries in the plane. In: Proc. 10th ACM-SIAM Sympos. Discrete Algorithms (SODA), pp. 215–224 (1999)
4. Hershberger, J.: Finding the upper envelope of n line segments in $O(n \log n)$ time. Inf. Process. Lett. 33(4), 169–174 (1989)
5. Hershberger, J., Suri, S.: An optimal algorithm for Euclidean shortest paths in the plane. SIAM J. Comput. 28(6), 2215–2256 (1999)
6. Mitchell, J.S.B.: Shortest paths among obstacles in the plane. Internat. J. Comput. Geom. Appl. 6(3), 309–331 (1996)
7. Mitchell, J.S.B.: Shortest paths and networks. In: Handbook of Discrete and Computational Geometry, 2nd edn., ch. 27, pp. 607–641. CRC Press, Inc., Boca Raton (2004)
8. Nivasch, G.: Improved bounds and new techniques for Davenport-Schinzel sequences and their generatlizations. In: Proc. 20th ACM-SIAM Sympos. Discrete Algorithms, pp. 1–10 (2009)
9. Sharir, M.: Almost tight upper bounds for lower envelopes in higher dimensions. Discrete Comput. Geom. 12, 327–345 (1994)
10. Sharir, M., Agarwal, P.K.: Davenport-Schinzel Sequences and Their Geometric Applications. Cambridge University Press, New York (1995)

Pattern Matching for 321-Avoiding Permutations

Sylvain Guillemot and Stéphane Vialette

LIGM, CNRS UMR 8049, Université Paris-Est
{guillemo,vialette}@univ-mlv.fr

Abstract. Given two permutations π and σ, the NP-complete Permutation Pattern problem is to decide whether π contains σ as a pattern. In case both π and σ are 321-avoiding, we prove the Permutation Pattern problem to be solvable in $O(k^2n^6)$ time, where $k = |\sigma|$ and $n = |\pi|$, and give a $O(kn^{4\sqrt{k}+12})$ time algorithm if only σ is 321-avoiding. Finally, we show W[1]-hardness of a 2-colored version of this latter problem.

1 Introduction

A permutation π is said to contain the pattern (shorter permutation) σ, in symbols $\sigma \preceq \pi$, if there exists a subsequence of entries of π that has the same relative order as σ (alternatively, σ is involved in π). Otherwise, π avoids σ. For example, 3215674 contains the pattern 132 since the subsequence 154 is ordered in the same way as 132. Pattern involvement in permutations has become a very active area of research. For one, pattern containment restrictions are often used to describe classes of permutations that are sortable under various conditions [8]. For another, a great deal of study has been devoted to counting pattern-avoiding permutations [3].

We consider here the Permutation Pattern problem. Given two permutations σ and π, this problem is to decide whether $\sigma \preceq \pi$ (the problem is ascribed to H. Wilf in [4]). The Permutation Pattern problem is NP-hard [4], but is solvable in $O(n^k)$ time if σ has size k and π has size n. Improvements were presented in [2] and [1], the latter describing a $O(n^{0.47k+o(k)})$ time algorithm. Also, the problem is known to be polynomial-time solvable (in k and n) if σ is *separable*, *i.e.*, σ contains neither the pattern 2413 nor 3142 [4]. In case σ is monotone, a $O(n \log\log n)$ time algorithm is known [7].

We focus in this paper on the Permutation Pattern problem in case σ (possibly σ and π) avoids a pattern of length 3. Knuth proved in [9] that for all six of the patterns of length 3 it is true that the number of permutations of size n that avoid the pattern is the Catalan number. First, it is easy to see that the Permutation Pattern problem is polynomial-time solvable if the pattern σ avoids 132, 312, 213 or 231 since σ is clearly separable in this case. Monotone patterns, *i.e.*, 123 and 321, however, deserve separate consideration (we focus here on 321-avoiding permutations but if a permutation avoids 123 then its reverse avoids 321). In case both π and σ are 321-avoiding, we prove the

Y. Dong, D.-Z. Du, and O. Ibarra (Eds.): ISAAC 2009, LNCS 5878, pp. 1064–1073, 2009.

PERMUTATION PATTERN problem to be solvable in $O(k^2n^6)$ time, where $k = |\sigma|$ and $n = |\pi|$, and give a $O(kn^{4\sqrt{k}+12})$ time algorithm if only σ is 321-avoiding. Finally, we show W[1]-hardness of a 2-colored version of this latter problem.

This paper is organized as follows. Section 2 briefly reviews the needed material and some basic properties are derived. We consider in Section 3 the PERMUTATION PATTERN problem in case both σ and π are 321-avoiding whereas Section 4 is devoted to the case σ only is 321-avoiding. Due to space constraints, complete proofs are deferred to the full paper.

2 Generalities

2.1 Permutation Patterns

We will use two different representations of permutations. The usual *array representation* of π is simply the sequence $\pi(1) \dots \pi(n)$. A second representation describes a permutation π by its graph, *i.e.*, a set of points in the plane where no two two points are aligned horizontally or vertically. We will write $p \in \pi$ to mean that p is a point in the graph of π, and the x- and y-coordinates of p will be denoted by $x(p)$ and $y(p)$, respectively.

Definition 1 (Embedding). *Given two permutations σ and π, an* embedding *of σ into π is an injective mapping $\phi : \sigma \rightarrow \pi$ which is order-preserving for x-coordinates and y-coordinates,* i.e., *for any two points $p, p' \in \sigma$, it holds that: (i) $x(p) < x(p') \Rightarrow x(\phi(p)) < x(\phi(p'))$, and (ii) $y(p) < y(p') \Rightarrow y(\phi(p')) < y(\phi(p'))$.*

When there exists an embedding of σ into π, we say that σ *occurs* in π, denoted by $\sigma \preceq \pi$. We also say that π *contains* σ; when this does not hold, we say that π *avoids* σ. A permutation is *k-increasing* iff it can be partitioned in k increasing subsequences. It is well-known that a permutation π is k-increasing iff its longest decreasing subsequence has length at most k, or equivalently if π avoids $k+1 \dots 1$. We focus here on 2-increasing permutations, which are also characterized as 321-avoiding permutations. These permutations were introduced by [8] as *queue-sortable permutations*. This characterization yields a linear-time recognition algorithm for this class, which can be turned into a certifying algorithm.

Proposition 1. *Given a permutation π of size n, in $O(n)$ time we can decide if π is 2-increasing, and (i) output a partition in two increasing subsequences (as a positive certificate), and (ii) or output an occurrence of* 321 *into π (as a negative certificate).*

2.2 Stair-Decompositions

We introduce here the notion of *stair-decomposition* that will play an important role in our study of 2-increasing permutations.

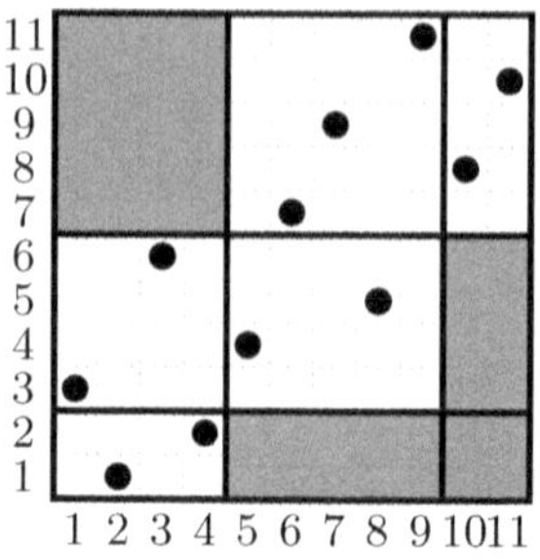

Fig. 1. A stair-decomposition of the permutation $\pi = 3\ 1\ 6\ 2\ 4\ 7\ 9\ 5\ 11\ 8\ 10$. The white blocks on the two diagonals contain increasing subsequences, while the grey blocks contain no point.

Definition 2 (Stair-decomposition). *A* stair-decomposition *of a permutation π consists of: (i) a partition of the horizontal axis in intervals $I_1, \ldots, I_k$, and (ii) a partition of the vertical axis in intervals $J_1, \ldots, J_k$, such that: if we let $S_{i,j}$ be the square at the intersection of intervals I_i and J_j, then the graph of π contains points only in the squares $S_{i,i}$ or $S_{i,i+1}$, and the points of π inside a nonempty square form an increasing subsequence.*

A stair-decomposition $\mathcal{D}$ can also be described by its *blocks* $B_1, \ldots, B_k$, where for each i, B_{2i-1} is the points inside $S_{i,i}$, and B_{2i} is the points inside $S_{i,i+1}$. Figure 1 illustrates the definition.

Lemma 1. *π is 2-increasing iff π has a stair-decomposition.*

3 Both σ and π Are 2-Increasing

3.1 Preliminaries

Definition 3 (Ordered preforest). *An* ordered preforest *consists of (i) a forest F with node set $N(F)$, (ii) a* depth *$d_F(u) \in \mathbb{N}^*$ assigned to each node $u \in N(F)$, (iii) a total order $<_F$ on $N(F)$ satisfying the following conditions. For each i, the* level *L_i of F is the set of nodes of depth i. We then require that: (i) for each $v \in N(F)$ with parent u, it holds that $d_F(v) = d_F(u) + 1$; (ii) $<_F$ ranks nodes by increasing depth, i.e. if $d_F(u) < d_F(v)$ then $u <_F v$; and (iii) for each $u \in N(F)$, the children of u form an interval of $<_F$.*

Condition 3 amounts to say that there is a plane drawing of F such that (i) all nodes of L_i have their y-coordinate equal to i, (ii) in this drawing, $<_F$ orders the nodes of L_i from left to right. The total order $<_F$ can be seen as a breadth-first traversal of F. Given $u \in N(F)$, we denote by $parent_F(u)$ its parent node, and by $children_F(u)$ its set of child nodes. If all roots of F have depth 1, then F is called an *ordered forest*. If F is an ordered preforest and if $i \in \mathbb{Z}$, $F \uparrow i$ denotes the ordered preforest which contains the nodes of F of depth $\geq i+1$, and where each such node has new depth $d_{F'}(u) = d_F(u) - i$.

We will also need the notion of *split* of an ordered forest. Suppose that F is an ordered forest with n nodes. Let $x = x_1 \dots x_n$ be the depth-first traversal of F (defined in the standard way). A *split* of F is a partition A, B of $N(F)$ s.t. A forms a prefix of x, and B forms a suffix of x. Note that the splits of F are in number $n+1$. From the ordered forest F, we can then define $F[A], F[B]$ which are ordered preforests.

3.2 Representation of 2-Increasing Permutations

We describe here a way to represent 2-increasing permutations by ordered forests. Let F be an ordered forest with n nodes. Let $L_1, \dots, L_h$ be the levels of F. The *anchor* of F is the leftmost node of L_1. An *odd-alternating traversal* of F is a traversal of F which proceeds as follows: start with $i = 1$; while $i \leq h$, enumerate the elements of $L_i \cup L_{i+1}$ by a depth-first traversal, and increment i by 2. An *even-alternating traversal* of F is defined in a similar way, except that it starts at $i = 0$ ($L_0 = \emptyset$ by convention).

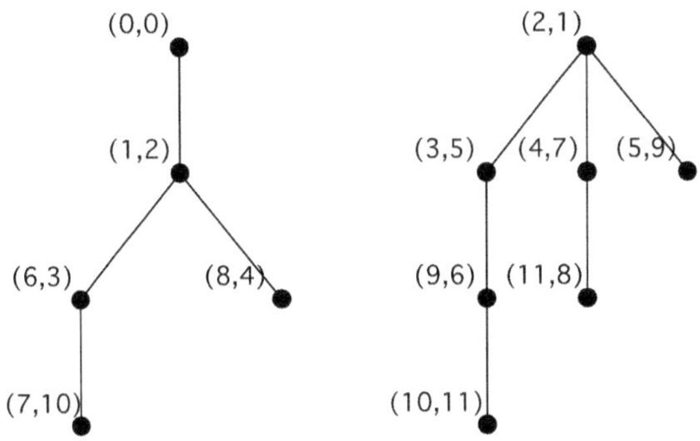

Fig. 2. An ordered forest F; the labeling of its nodes defines the permutation $\pi_F =$ 2 1 5 7 9 3 10 4 6 11 8

To the ordered forest F, we associate a permutation π_F of size $n-1$, defined as follows: (i) label the nodes of F with increasing x-coordinates from 0 to $n-1$, by an odd-alternating traversal of F; (ii) label the nodes of F with increasing y-coordinates from 0 to $n-1$, by an even-alternating traversal of F; (iii) ignore the anchor of F, and consider the $n-1$ remaining points as the graph of a permutation π_F. This process is illustrated in Figure 2. The following Lemma shows that a stair-decomposition of π_F can be obtained from the ordered forest F; it implies that the construction always produces a 2-increasing permutation.

Lemma 2. *The sets $L_1, \dots, L_h$ form a stair-decomposition of π_F.*

The above result implies that π_F is 2-increasing. The following proposition shows that every 2-increasing permutation can be represented in this way.

Proposition 2. *For any 2-increasing permutation π of size n, there exists an ordered forest F with $n+1$ nodes such that $\pi = \pi_F$.*

3.3 Embeddings

We introduce here notions of embedding for ordered forests, and we characterize embeddings of 2-increasing permutations in terms of embeddings of their associated forests (Proposition 3).

Definition 4 (Embedding). *Let F be an ordered preforest with levels $L_1, \ldots, L_h$, and let F' be an ordered preforest with levels $L'_1, \ldots, L'_{h'}$. An* embedding *of F into F' is an injective mapping $\phi : N(F) \to N(F')$ such that for each $1 \leq i \leq h$: (i) $\phi(L_i) \subseteq L'_i$; (ii) if $p, p' \in L_i$, then $p <_F p' \Leftrightarrow \phi(p) <_{F'} \phi(p')$; and (iii) if $p \in L_i$ and $p' \in L_{i+1}$, then $p \leq_F parent_F(p') \Leftrightarrow \phi(p) \leq_{F'} parent_{F'}(\phi(p'))$.*

In words, the definition requires that ϕ maps points of F to points of F' on the same level (Point 1), preserves the ordering between points on a same level (Point 2), and preserves the parent-child ordering between points on consecutive levels (Point 3). We write $F \preceq F'$ iff there exists an embedding of F into F'. We now define a second notion of embedding called *twist embedding.*

Definition 5 (Twist embedding). *We say that there exists a* twist embedding *of F into F' (denoted by $F \trianglelefteq F'$) iff there exists A, B split of F and A', B' split of F' s.t. (i) $F[A] \preceq F'[B']$ and (ii) $F[B] \preceq F'[A'] \uparrow 2$.*

Given two permutations π_1, π_2, their sum $\pi_1 \oplus \pi_2$ is the permutation π whose graph can be partitioned in two rectangles R_1, R_2, with R_2 above and at the right of R_1, s.t. $\pi|R_1 = \pi_1, \pi|R_2 = \pi_2$. We say that π is *simple* iff it cannot be written as $\pi_1 \oplus \pi_2$ (where π_1, π_2 are not empty). The following proposition characterizes embeddings of 2-increasing permutations in terms of twist embeddings of their associated forests.

Proposition 3. *Let F, F' be two ordered forests, and let h be the height of F'. Suppose that π_F and $\pi_{F'}$ are simple. Then: $\pi_F \preceq \pi_{F'}$ iff there exists $0 \leq i \leq h$ s.t. $F \trianglelefteq F' \uparrow i$.*

3.4 Finding an Embedding

We now describe a polynomial-time algorithm to decide if $F \trianglelefteq F'$ (Proposition 5). By Proposition 3, this will yield a polynomial-time algorithm for the Permutation Pattern problem for 2-increasing permutations (Theorem 1). For the purpose of deciding the embedding relation between forests, we need a result on *labeled dags* (Proposition 4). Before stating this result, we introduce the following definitions.

Definition 6 (Labeled dag). *Let Σ be an alphabet, and let D be a symmetric reflexive relation over Σ. A Σ, D-labeled dag consists in a dag $G = (V, A)$, together with a labeling $\lambda : V \to \Sigma$, having the following property: for each $u, v \in V$ distinct, u, v are adjacent in G iff $(\lambda(u), \lambda(v)) \in D$.*

Definition 7 (Morphism). *Let G, G' be two Σ, D-labeled dags. A* morphism *of G into G' is an injective mapping $\phi : V(G) \to V(G')$ such that: (i) ϕ is arc-preserving: for each $u \in V(G)$, $(u,v) \in A(G) \Leftrightarrow (\phi(u), \phi(v)) \in A(G')$; (ii) ϕ is label-preserving: for each $u \in V(G)$, u and $\phi(u)$ have the same label.*

We denote $G \preceq G'$ iff there exists a morphism of G into G'. It turns out that this property can be efficiently decided, thanks to an algorithm based on topological sorting.

Proposition 4. *Given two Σ, D-labeled dags G and G', we can decide in polynomial time if $G \preceq G'$.*

Proof (Sketch). We first perform a topological sorting of G', which gives us a linear extension of G'. Let $y = y_1 \dots y_n$ be the labeling of this linear extension. We now perform a topological sort of G as follows. We maintain Min the set of minimal elements in G, and j a position in y; initially $j = 0$. At each step of the topological sort, we need to pick the next element in Min. We proceed as follows: (i) for each $x \in Min$ labeled by a, we compute m_x as the first position $i > j$ s.t. $y_j = a$ (if it exists), or ∞ otherwise; (ii) if all values m_x are ∞, we stop and we answer negatively. Otherwise, we choose the element $x \in Min$ with minimum m_x, and we update $j \leftarrow m_x$. Note that there is a unique choice for x at each step, since no two elements of Min can have the same label. When the algorithm completes the topological sort, it answers positively. The algorithm clearly runs in polynomial-time and correctness is proved in the full version. □

In fact, a more involved algorithm can solve the problem in linear $O(|G| + |G'|)$ time (see journal version). The following Proposition shows how these results allow us to find embeddings (Point 1) and twist-embeddings (Point 2) between two forests.

Proposition 5. *Let F and F' be two ordered forests, with $|F| = k$ and $|F'| = n$. Then: (i) We can decide in $O(n)$ time if $F \preceq F'$, and (ii) we can decide in $O(kn^2)$ time if $F \trianglelefteq F'$.*

Proof. Point 1. Suppose that F has levels $L_1, \dots, L_h$ and that F' has levels $L'_1, \dots, L'_h$. Let $\Sigma = \{1, \dots, h\}$ and let D be the set of pairs $(i,i), (i, i+1)$. We represent F by a labeled dag G_F as follows: (i) G_F has vertex set $N(F)$, (ii) each vertex of G_F is labeled by its depth in F, (iii) for each $u, v \in L_i$, G_F contains an arc (u,v) iff $u <_F v$, (iv) for each $u \in L_i, v \in L_{i+1}$, G_F contains an arc (u,v) iff $u \leq_F parent_F(v)$. It is readily seen from Definitions 4 and 7 that embeddings of F into F' correspond to morphisms of G_F into $G_{F'}$. It follows that $F \preceq F'$ iff $G_F \preceq G_{F'}$, which can be decided in $O(|G_F| + |G_{F'}|)$ time by the above algorithm. Note that G_F may have size $O(k^2)$ and that $G_{F'}$ may have size $O(n^2)$. However, we can apply the above algorithms to their *transitive reductions*, whose size is now $O(k)$ and $O(n)$, and which can be constructed in linear time from F and F'. This is correct since G_F and its transitive reduction has the same linear extensions. We therefore obtain a $O(k+n) = O(n)$ time algorithm to decide if $F \preceq F'$.

Point 2. It follows from Definition 5 that to decide if $F \trianglelefteq F'$, we need to examine every split $S = A, B$ of F, every split $S' = A', B'$ of F', and in each case to test if $F[A] \preceq F'[B']$ and $F[B] \preceq F'[A'] \uparrow 2$. For a given S, S', the two tests are carried out in $O(n)$ time by Point 1. Now, the possible S are in number $k+1$ and the possible S' are in number $n+1$, leading to the claimed $O(kn^2)$ running time. □

Propositions 3 and 5 yield a polynomial-time algorithm for the Permutation Pattern problem for 2-increasing permutations.

Theorem 1. *Let σ, π be two 2-increasing permutations, with $|\sigma| = k$ and $|\pi| = n$. We can decide in $O(k^2n^6)$ time if $\sigma \preceq \pi$.*

4 Only σ Is 2-Increasing

We now consider the restriction of the Permutation Pattern problem to the case when σ is 2-increasing but π is arbitrary. For convenience, we define the problem so that a stair-decomposition of σ is also given as input.

Name: 2-Increasing Permutation Pattern (2IPP)
Input: A 2-increasing permutation σ of size k, a stair-decomposition $\mathcal{D}$ of σ with r blocks, each of size at most s, and a permutation π of size n.
Question: Decide if $\sigma \preceq \pi$.

4.1 Algorithms

We show that the 2IPP problem can be solved in $n^{O(\sqrt{k})}$ time. The algorithm is by combining two different strategies for solving the problem. We will describe a first strategy which solves the problem in $n^{O(s)}$ time, and a second strategy which solves the problem in $n^{O(r)}$ time. Recall that s is the maximum size of the blocks of $\mathcal{D}$, and that r is the number of blocks of $\mathcal{D}$. A combination of both strategies will yield an algorithm with the claimed $n^{O(\sqrt{k})}$ running time.

The first strategy uses a simple dynamic-programming approach. It stems from the observation that to find an embedding of σ into π, it suffices to find embeddings ϕ_i of B_i into π (for every i), and to check the consistency between pairs of consecutive embeddings ϕ_i, ϕ_{i+1}. To formalize this idea, we need the following definitions. Given a block B_i, a *partial solution* for B_i is a pair $S = (R, \psi)$, where $R = (x_1, y_1, x_2, y_2)$ is a rectangle in the graph of π, and ψ is an embedding of B_i into π whose image is included into R (*i.e.*, for each $p \in B_i$, it holds that $x_1 \leq x(\psi(p)) < x_2$ and $y_1 \leq y(\psi(p)) < y_2$). Intuitively, a partial solution specifies the restriction of an embedding to B_i, as well as the region to which the points of B_j ($j \geq i$) are mapped. Let $\mathcal{S}_i$ denote the set of partial solutions for B_i.

Definition 8. *Given a partial solution $S = (R, \psi) \in \mathcal{S}_i$ with $R = (x_1, y_1, x_2, y_2)$, we define the predicate $\Pi(i, S)$ to hold iff there exists an embedding ϕ of*

$B_i \cup \ldots \cup B_r$ into π such that: (i) $\phi|B_i = \psi$; (ii) if $p \in B_j$ with $j > i$ then: (i) if $j > i+1$: then $x(\phi(p)) \geq x_2$ and $y(\phi(p)) \geq y_2$; (ii) if $j = i+1$ and i odd: then $x(\phi(p)) \geq x_1$ and $y(\phi(p)) \geq y_2$; and (iii) if $j = i+1$ and i even: then $x(\phi(p)) \geq x_2$ and $y(\phi(p)) \geq y_1$.

The predicates $\Pi(i, S)$ can be computed by dynamic programming thanks to the following Lemma. Given partial solutions $S = (R, \psi) \in \mathcal{S}_i$ and $S' = (R', \psi') \in \mathcal{S}_{i+1}$, with $R = (x_1, y_1, x_2, y_2)$, and $R' = (x'_1, y'_1, x'_2, y'_2)$, say that S and S' are *compatible* iff (i) either B_i, B_{i+1} are on the same column (i odd), and it holds that: $x'_1 = x_1$, $x'_2 = x_2$, $y'_1 = y_2$, and for any $p \in B_i$, $p' \in B_{i+1}$, $x(p) < x(p') \Leftrightarrow x(\psi(p)) < x(\psi'(p'))$, or (ii) B_i, B_{i+1} are on the same row (i even), and it holds that: $y'_1 = y_1$, $y'_2 = y_2$, $x'_1 = x_2$, and for any $p \in B_i$, $p' \in B_{i+1}$, $y(p) < y(p') \Leftrightarrow y(\psi(p)) < y(\psi'(p'))$.

Lemma 3. *Suppose that $i < r$, and let $S \in \mathcal{S}_i$. Then: $\Pi(i, S)$ holds iff there exists $S' \in \mathcal{S}_{i+1}$ compatible with S such that $\Pi(i+1, S')$ holds.*

The above lemma yields a polynomial-time algorithm for the 2IPP problem when s is bounded.

Proposition 6. 2IPP *is solvable in $O(rn^{2s+5})$ time.*

Proof. By dynamic-programming, we compute the predicates $\Pi(i, S)$ for each $1 \leq i \leq r$ and each $S \in \mathcal{S}_i$, using Lemma 3. An element of $\mathcal{S}_i$ consists of a rectangle R and of an embedding ψ; there are $O(n^4)$ choices for R and $O(n^s)$ choices for ψ (since $|B_i| \leq s$). It follows that each $\mathcal{S}_i$ has size $O(n^{s+4})$. Consider now the running time needed to compute a value $\Pi(i, S)$ assuming that the values $\Pi(i+1, S')$ are available. By Lemma 3, we need to examine every $S' = (R', \psi') \in \mathcal{S}_{i+1}$ compatible with S. There are $O(n)$ choices for R' (since three of the four coordinates are determined by R), and $O(n^s)$ choices for ψ'. It follows that computing any $\Pi(i, S)$ is $O(n^{s+1})$ time. Therefore, the algorithm, as a whole, runs in $O(rn^{s+4}n^{s+1}) = O(rn^{2s+5})$ time. □

The second strategy solves the problem in $n^{O(r)}$ time, where r is the number of blocks in the stair-decomposition $\mathcal{D}$. For the sake of clarity, we formulate our approach as a non-deterministic algorithm. This is an extension of the following idea. If σ is increasing, then we can find an embedding of σ into π by the following simple non-deterministic algorithm: (i) choose a point p_1 in π; (ii) for $i = 2$ to $|\sigma|$, choose a point p_i in π such that $x(p_i) > x(p_{i-1})$ and $y(p_i) > y(p_{i-1})$. The key idea in this algorithm is that we only need to memorize the image of the last point of σ examined so far. We will extend this idea to the case where σ has a stair-decomposition with r blocks: in this case, in each block B_i the points will be examined from left to right, and we will memorize the image of the *first* and *last* point in each row and each column of $\mathcal{D}$. At the time of considering a new point in B_i, we will choose an image for this point in π, and check that this image is well-positioned with respect to the last points of the current row and column, and with respect to the first points of the next row and column. We need to specify the order in which the points of σ will be examined. This leads us to the following definition.

Definition 9. *An* insertion order *for σ with respect to $\mathcal{D}$ is an enumeration of the points of σ such that: (i) on each row, the points are ordered by increasing y-coordinate, and (ii) on each column, the points are ordered by increasing x-coordinate.*

Lemma 4. *There exists an insertion order for σ with respect to $\mathcal{D}$.*

Let us show how this notion of insertion order allows us to solve the problem. Suppose that $s = p_1 \dots p_k$ is an insertion order for σ with respect to $\mathcal{D}$. Define a *configuration* to be a tuple C consisting of: (i) two values firstr_i and lastr_i (which may be equal to $\perp$) for each row i, and (ii) two values $\mathrm{firstc}_i, \mathrm{lastc}_i$ for each column i. We then find an embedding of σ into π using (non-deterministic) Algorithm 1. This yields a polynomial-time algorithm for the 2IPP problem when the number of blocks r is bounded as shown in the following proposition.

Algorithm 1. A non-deterministic algorithm for the 2IPP problem

1: choose values $\mathrm{firstr}_i, \mathrm{firstc}_i$ in $\{1, \dots, n\}$ such that for each i,
2:     $\mathrm{firstr}_i \leq \mathrm{firstr}_{i+1}$, $\mathrm{firstc}_i \leq \mathrm{firstc}_{i+1}$;
3: initialize the values lastr_i and lastc_i to $\perp$;
4: **for** $h = 1$ to k **do**
5:     if p_h belongs to row i, column j of $\mathcal{D}$, then:
6:     choose p point of π;
7:     check that $x(p) < \mathrm{firstc}_{j+1}$ and $y(p) < \mathrm{firstr}_{i+1}$;
8:     if $\mathrm{lastr}_i = \perp$, check that $y(p) = \mathrm{firstr}_i$; if $\mathrm{lastr}_i \neq \perp$, check that $y(p) > \mathrm{lastr}_i$;
9:     if $\mathrm{lastc}_j = \perp$, check that $x(p) = \mathrm{firstc}_j$; if $\mathrm{lastc}_j \neq \perp$, check that $x(p) > \mathrm{lastc}_j$;
10:     update $\mathrm{lastc}_j \leftarrow x(p)$, $\mathrm{lastr}_i \leftarrow y(p)$.
11: **end for**
12: accept at the end of the algorithm.

Proposition 7. 2IPP *is solvable in $O(kn^{2r+3})$ time.*

Proof. We perform a deterministic simulation of the non-deterministic Algorithm 1. For each configuration C and each $0 \leq i \leq k$, we compute a predicate $P(i, C)$ which holds iff there exists of an execution of the algorithm which reaches configuration C at the beginning of loop $h = i + 1$. The predicates $P(i, C)$ are then computed by induction on i; (i) obtaining $P(0, C)$ takes $O(r)$ time; (ii) for $0 \leq i < k$, if the values $P(i, C)$ have been previously computed then we can deduce the values $P(i + 1, C')$, where each value $P(i, C)$ gives rise to n values $P(i + 1, C')$ (since we need to examine every possible choice of p in Line 6). Finally, we accept iff $P(k, C)$ holds for at least one C. The claimed running time follows by observing that, if $\mathcal{D}$ has p rows and q columns, then the number of configurations is bounded by $n^{2(p+q)}$ and that $p+q \leq r+1$. Hence, the predicates $P(i, C)$ are computed in $O(kn^{2r+3})$ time. □

Combining the two above strategies in a nontrivial way, we achieve a $n^{O(\sqrt{k})}$ running time for solving 2IPP, as stated in the following Theorem.

Theorem 2. 2IPP *can be solved in $O(kn^{4\sqrt{k}+12})$ time.*

4.2 Hardness Results

While we are still unable to settle the parameterized complexity of the 2IPP problem, we can show hardness results for a colored version of the problem. The c-COLORED 2IPP problem is defined as follows: given σ, $\mathcal{D}$ and π where σ and π are c-colored permutations (*i.e.*, a color in $[c]$ is associated to each point of the permutation), find a color-preserving embedding of σ into π. It is not difficult to see that the algorithmic results of the previous section can be adapted to this colored version. When $c \geq 2$, we are able to prove the asymptotic optimality of these algorithms, conditioned by the Exponential-Time Hypothesis (ETH). Recall that ETH is the assumption that 3-SAT cannot be solved in $2^{o(n)}$ time (where n is the number of variables). It was shown in [5] that CLIQUE cannot be solved in $n^{o(k)}$ time assuming ETH.

Theorem 3. *(i)* 2-COLORED 2IPP *parameterized by k is* W[1]*-hard and cannot be solved in $n^{o(\sqrt{k})}$ time assuming ETH. (ii)* 2-COLORED 2IPP *parameterized by s is* WNL*-hard and cannot be solved in $n^{o(s)}$ time assuming ETH.*

The parameterized class WNL was introduced in [6] to capture the parameterized complexity of problems solvable by k-dimensional dynamic programming.

References

1. Ahal, S., Rabinovich, Y.: On Complexity of the Subpattern Problem. SIAM Journal on Discrete Mathematics 22(2), 629–649 (2008)
2. Albert, M.H., Aldred, R.E.L., Atkinson, M.D., Holton, D.A.: Algorithms for Pattern involvement in Permutations. In: Eades, P., Takaoka, T. (eds.) ISAAC 2001. LNCS, vol. 2223, pp. 355–366. Springer, Heidelberg (2001)
3. Bona, M.: Combinatorics of permutations. Discrete Mathematics and Its Applications. Chapman-Hall, CRC Press, Boca Raton (2004)
4. Bose, P., Buss, J.F., Lubiw, A.: Pattern Matching for Permutations. Information Processing Letters 65(5), 277–283 (1998)
5. Chen, J., Chor, B., Fellows, M., Huang, X., Juedes, D., Kanj, I., Xia, G.: Tight lower bounds for certain parameterized NP-hard problems. In: Proceedings of 19th Annual IEEE Conference on Computational Complexity (CCC), pp. 150–160 (2004)
6. Guillemot, S.: Parameterized complexity and approximability of the SLCS problem. In: Grohe, M., Niedermeier, R. (eds.) IWPEC 2008. LNCS, vol. 5018, pp. 115–128. Springer, Heidelberg (2008)
7. Hunt, J., Szymanski, T.: A fast algorithm for computing longest common subsequences. Communications of the ACM 20, 350–353 (1977)
8. Knuth, D.E.: Fundamental Algorithms, The Art of Computer Programming, 2nd edn., vol. 1. Addison-Wesley, Reading (1973)
9. Knuth, D.E.: Sorting and Searching. The Art of Computer Programming, 2nd edn., vol. 3. Addison-Wesley, Reading (1998)
10. Marcus, A., Tardos, G.: Excluded permutation matrices and the Stanley-Wilf conjecture. Journal of Combinatorial Series A 107(1), 153–160 (2004)

Folding a Better Checkerboard

Erik D. Demaine[1,*], Martin L. Demaine[1],
Goran Konjevod[2], and Robert J. Lang[3]

[1] MIT Computer Science and Artificial Intelligence Laboratory, 32 Vassar St., Cambridge, MA 02139, USA
{edemaine,mdemaine}@mit.edu
[2] Department of Computer Science and Engineering, Arizona State University, Tempe, AZ 85287, USA
goran@asu.edu
[3] http://langorigami.com

Abstract. Folding an $n \times n$ checkerboard pattern from a square of paper that is white on one side and black on the other has been thought for several years to require a paper square of semiperimeter n^2. Indeed, within a restricted class of foldings that match all previous origami models of this flavor, one can prove a lower bound of n^2 (though a matching upper bound was not known). We show how to break through this barrier and fold an $n \times n$ checkerboard from a paper square of semiperimeter $\frac{1}{2}n^2 + O(n)$. In particular, our construction strictly beats semiperimeter n^2 for (even) $n > 16$, and for $n = 8$, we improve on the best seamless folding.

1 Introduction

Within the world of origami, the use of two-colored paper (white on one side, colored on the other) has been widespread for many years, leading to creation of origami figures in which both colors are used for artistic effect. In the early days of western origami, two-colored figures tended to have relatively simple patterns (the penguin [11] has been a perennial favorite), but with the 1993 publication of John Montroll's *Origami Inside-Out* [10], the genre received a significant boost, as this book displayed much more complex patterns: striped tigers, spotted cows, and most notably, an 8×8 checkerboard with squares alternating in color—each folded from a single square sheet with no cuts.

The checkerboard, in fact, has a long history as an origami subject. The smaller sizes—2×2 and 3×3—make interesting puzzles, accessible even to relative beginners at folding. Larger checkerboards, and in particular the 8×8 used for chess, are significant challenges to the origami designer. Even so, there are several solutions in the origami literature. To the best of our knowledge, the earliest chessboard from a single uncut square was folded by Max Hulme [7] in 1977, but several other designs exist by now [1,8,2,5].

* Partially supported by NSF CAREER award CCF-0347776.

Y. Dong, D.-Z. Du, and O. Ibarra (Eds.): ISAAC 2009, LNCS 5878, pp. 1074–1083, 2009.

One of the most important attributes of complex origami designs such as these is their *efficiency*, which is simply a measure of the size of the finished figure relative to the original sheet of paper. For the design of a checkerboard with unit squares, a measure of the efficiency is the size, in the same units, of the square (or other shape) of paper from which it is folded. For example, among 8×8 boards, Hulme's [7] is folded from a 64×64 square, Casey's [1] and Kirschenbaum's [8] are folded from 40×40 squares, Montroll's [10] is folded from a 36×36 square, and Dureisseix's [5] and Chen's [2] are folded from 32×32 squares.

A secondary question, which plays more of an aesthetic role, is whether any square of the folded board is crossed by a folded edge. The most aesthetically pleasing checkerboards have *seamless* squares, in which each square is an unbroken surface. If one or more squares is crossed by a folded edge, we call such a solution *unconstrained.* The most common form of unconstrained square has a folded edge crossing its diagonal.

We may then ask a general question: what is the smallest square of paper required to fold an $n \times n$ checkerboard with unconstrained or seamless squares?

Origamists have experimented with this problem for specific values of n other than just 8. Figure 1 shows sample foldings for $n \in \{2, 3, 4\}$. The 2×2 and 3×3 solutions are elegant and have been plausibly conjectured, although not yet proved, to be optimal. Foldings become increasingly difficult as n grows. Although the three examples in Figure 1 are all seamless, in the larger sizes both seamless and unconstrained square solutions exist, with unconstrained solutions often turning out to be slightly more efficient. In particular, an unconstrained 4×4 checkerboard can be folded from an 8×8 square [5,7].

Perimeter limit. An interesting property of many examples is that the perimeter of the square in the folded form follows the edges of the colored/white boundary (or exterior boundary) in the pattern. Indeed, finding a way to map the boundary of the square to the colored/white boundary of the surface pattern has been a powerful and fruitful design approach for checkerboards (as well as many other two-colored origami figures). However, in many of the folded examples not all of the boundary of the square is used to create colored/white boundary in the pattern; there are often small bits of boundary that double back on themselves or that remain hidden from view as part of the folding design. The significance of the square perimeter in the construction of checkerboards was first explicitly identified by Dureisseix [5] in his construction of an unconstrained-square 8×8 from a 32×32 square. See also [4, sec. 15.4.2, pp. 238–239] for a discussion.

For an $n \times n$ checkerboard of unit squares viewed as a pattern of white squares on a colored field, the total boundary of the $n^2/2$ white squares (both white/colored and white/boundary) is $4(n^2/2) = 2n^2$, which leads directly to the following conjectured relationship between the size of a checkerboard and the minimum size square (or, in general, any other convex shape) required to fold it:

Conjecture 1 (Checkerboard Perimeter Limit). The minimum scaling of a convex polygon required to fold an $n \times n$ checkerboard of unit squares has semiperimeter at least n^2.

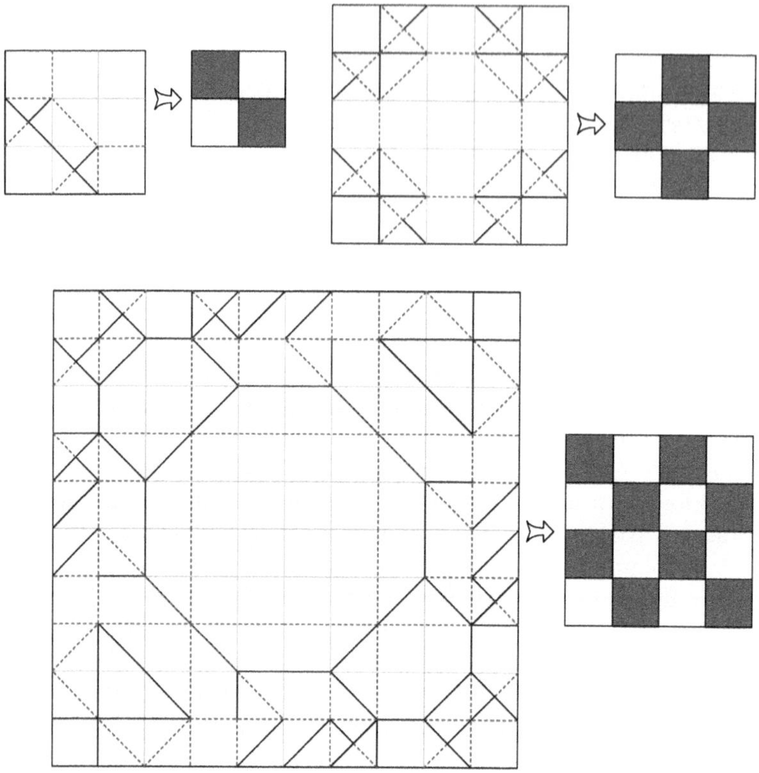

Fig. 1. Top left: Seamless 2×2 checkerboard from a 3×3 square, crease pattern (left) and folded form (right). Top right: Seamless 3×3 checkerboard from a 5×5 square. Bottom: Seamless 4×4 checkerboard from a 10×10 square. Dashed lines are valley folds; solid lines are mountain folds.

For folding from a square, the minimum size square would have a side length of $L_{\min} = \lceil n^2/2 \rceil$. Thus, for example, the paper required to fold an 8×8 checkerboard would have a semiperimeter of 64, and so the minimum size square would be at least 32×32. This conjectured limit would seem to be borne out by the published examples of 8×8 checkerboards. In particular the designs of Chen and Dureisseix achieve this bound. Also, while Montroll's published board is folded from a 36×36 grid, he has reported in personal communication with the authors that he has an unpublished design for a checkerboard from a 32×32 grid. On the other hand, the best known seamless 8×8 boards [1,8] use 40×40 unit squares.

Table 1 lists the conjectured lower bounds for n from 2 to 8 for square paper, along with the best known unconstrained and seamless solutions.

Our results. After a certain amount of experimentation, we found that these limits seemed plausible, and a challenge to even meet. To our knowledge, the only general construction for $n \times n$ checkerboards for arbitrary n is what follows from a general construction of two-color patterns [3]. This construction starts

Table 1. Checkerboard size, conjectured limit on square size, and sizes of best previous known unconstrained and seamless solutions

n	conjectured limit	best previous example	
		unconstrained	seamless
2	2	3	3
3	5	5	5
4	8	8	10
5	13	?	?
6	18	?	?
7	25	?	?
8	32	32 [2,5]	40 [1,8]

by accordion-folding a paper square into a long rectangular strip, and "flips" the strip at each color reversal. This method requires a square of side length $2n^2 + O(n)$, or semiperimeter $4n^2 + O(n)$, though it is at least seamless.

An improvement we developed, shown in Figure 2, asymptotically approaches the conjectured lower bound for unconstrained squares. In this approach, the perimeter of the checkerboard is mapped precisely to the checkerboard, with a bit effectively "wasted" in the turns at the end. However, this excess material increases only linearly with n. For an $n \times n$ checkerboard (n even), the amount of "square diagonal" needed, including turns, can be shown to be $\frac{1}{2}n^2 + 4n - 5$, which leads to a semiperimeter of

$$n^2 + 8n - 10,$$

which, indeed, asymptotically approaches the limit of n^2 for large n (though exceeds n^2 for all $n \geq 2$). To our knowledge, this is the first construction for general n other than the straightforward folding from a rectangular strip that follows from a general construction of two-color patterns [3].

Based on these examples, algorithms, and many empirical folding tests, the perimeter conjecture seemed to all of the present authors likely to be the true limit for an $n \times n$ checkerboard. Thus, it was with some surprise that we discovered a new approach to folding a checkerboard that beats the conjectured limit, at least for sufficiently large n. This paper presents an algorithmic implementation of this approach and shows that it leads to a semiperimeter asymptotically shorter than the conjectured limit by a factor of 2:

Theorem 1. *A seamless $n \times n$ checkerboard can be folded from a square of semiperimeter $\frac{1}{2}n^2 + O(n)$.*

Our construction strictly beats semiperimeter n^2 for all (even) $n > 16$. Along the way, we present a seamless checkerboard construction from a rectangle of semiperimeter $\frac{1}{2}n^2 + O(n)$ but aspect ratio $\Theta(n)$. This construction achieves better lower-order terms, in particular improving the best seamless 8×8 checkerboard folding from 40×40 to $34 + \varepsilon \times 34 + \varepsilon$ for any $\varepsilon > 0$.

Finally, our approach suggests a general technique for folding an arbitrary two-color pattern. This technique is based on solving a TSP-with-neighborhoods

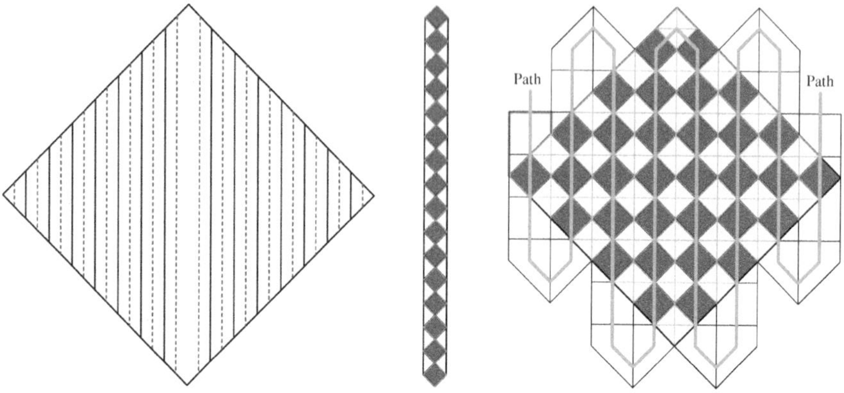

Fig. 2. Top: A square is pleated diagonally to realize a strip of squares and half-squares. Bottom: The diagonal is wound back and forth across the desired checkerboard, with folded turns at the ends.

instance, which has the potential to beat the natural limit determined by perimeter length (an Euler tour). However, fully optimizing the construction does not appear to be easy, and the TSP-with-neighborhoods solution provides only a lower bound for the construction.

2 Construction

In the full paper, we give two efficient foldings of a checkerboard from paper of semiperimeter $\frac{1}{2}n^2 + O(n)$. The first construction uses a rectangle of paper of aspect ratio $\Theta(n)$. The second construction is from a square of paper. We focus here on details of the square construction, but touch on the rectangle construction to build intuition and for its smaller lower-order terms.

2.1 Rectangular Paper

Figure 3 shows the idea of our efficient checkerboard folding from a rectangular piece of paper. We fold the paper into an $n \times n$ square with n square tabs sticking up in alternate rows, and with strips of length $n/2$ hanging off the sides of each row. The back sides of the strips are colored, while the $n \times n$ square and tabs are white. Thus folding the strips to cover the $n \times n$ square turns it colored, and folding the white tabs alternating up and down makes a checkerboard pattern.

The effect of this approach is that, instead of tracing the boundary of the white (or colored) regions with the paper perimeter, we use the paper perimeter to create strips of color that can visit all the squares in the checkerboard. The tabs can be constructed from the pleats introduced by strip folding, without increasing the demand on perimeter.

The folding uses a $\frac{1}{2}n^2 + 2 + \varepsilon \times 4n + \varepsilon$ rectangle, for a semiperimeter of $\frac{1}{2}n^2 + 4n + 2 + \varepsilon$ for any desired $\varepsilon > 0$. This bound beats the conjectured limit

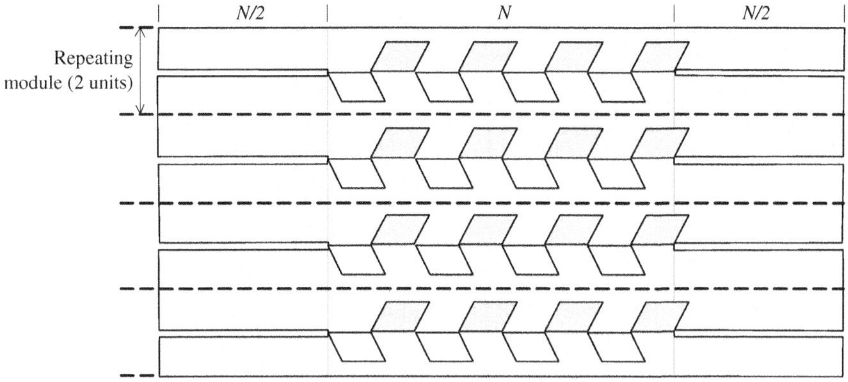

Fig. 3. Efficient construction of a checkerboard from rectangular paper

for all $n > 8$. An 8×8 checkerboard would use a $34 + \varepsilon \times 32 + \varepsilon$ rectangle, better than the 40×40 of the best previous seamless checkerboards [1,8] and respectably close to the 32×32 of the best unconstrained checkerboards [2,5].

2.2 Square Paper

Our construction from a square piece of paper uses the same overall approach, but rebalanced using a different arrangement of strips shown in Figure 4. The arrangement is parameterized by an integer m which we will set to minimize the aspect ratio of the paper. Along the left and right, we place m strips of width 2 and length $n/2$, which folded inward cover the middle of the square from left to right. Along the top and bottom, we place $n/2$ strips of width 2 and length $n/2 - m$, which folded inward cover the remaining area.

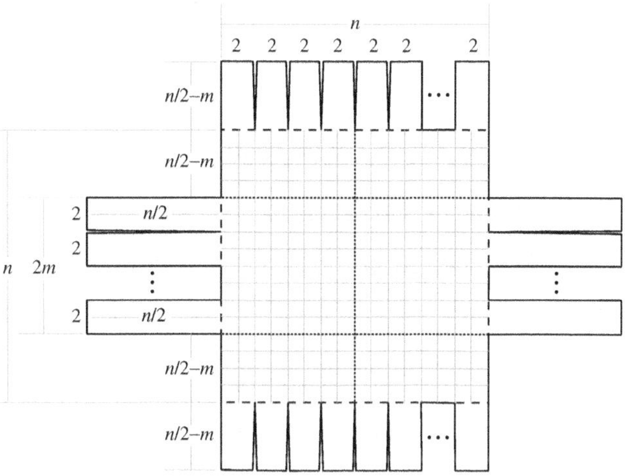

Fig. 4. Dissection pattern for an $n \times n$ square (n even)

Figure 5 shows the necessary gadgets to fold this pattern of strips. All pleats fall on the half-integer grid. Each depth-L slot between two strips consumes a perimeter segment of length $2L + 2$. Turning a corner between a horizontal strip of length L_w and a vertical strip of length L_h happens at a corner of the paper, and consumes $L_w + L_h + 1$ of the vertical perimeter and L_h of the horizontal perimeter. (This corner gadget can be flipped to consume more horizontal than

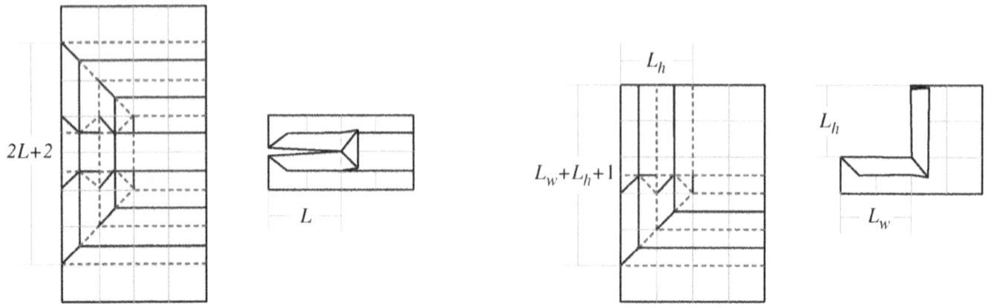

Fig. 5. Left: crease pattern and folded form for a slot (here with $L = 2$). Right: crease pattern and folded form for a corner (here with $L_w = L_h = 2$).

vertical perimeter, but we use only the one orientation for simplicity; tuning here would only improve by an additive constant.)

Although the construction is not yet complete, we compute the semiperimeter used so far. The vertical edge consists of m strips of width 2, $m-1$ slots of depth $L = n/2$, and two contributions from the upper and lower reflex corners with $L_w = n/2$ and $L_h = n - 2m$, giving a total of

$$V_1 = 2m + (m-1)\,(2(n/2)+2) + 2\,(n/2 + (n-2m) + 1) = mn + 2n.$$

The horizontal edge consists of $n/2$ strips of width 2, $n/2 - 1$ slots of depth $L = n/2 - m$, and two contributions from the left and right reflex corners with $L_w = n/2$ and $L_h = n - 2m$, giving a total of

$$H_1 = 2(n/2) + (n/2-1)\,(2\,(n/2-m)+2) + 2(n-2m) = \tfrac{1}{2}n^2 - mn + 3n - 2m - 2.$$

In the middle of the square, we fold an $n \times n$ array of *universal tabs* as shown in Figure 6. The universal tab creates a tab joined along a desired edge of the square that it covers, with the additional feature that the crease pattern's interface is identical on all four sides, consisting of four pleats, two on each side of the symmetry line, each pleat $\frac{1}{2}$ unit wide. Thus, a universal tab, rotated into any of the four possible orientations, can be folded from a 5×5 square (outlined by the heavy dashed line in Figure 6) whose creases will mate with the creases of adjacent universal tabs, no matter their orientations. Thus we can choose the orientation of each universal tab independently.

We orient universal tabs covered by horizontal strips to join along a horizontal edge (top or bottom), and we orient universal tabs covered by vertical strips to join along a vertical edge (left or right). This choice enables the strips to thread alternately above and below each universal tab it visits, forming the desired checkerboard color pattern.

It may seem wasteful to fold n^2 universal tabs when we need only half that many, but the perimeter consumed by these tabs is only linear in n (in particular, at most $5n$). We also obtain the feature that the same model can form any $n \times n$ bitmap of colored and white pixels: without changing the crease pattern, and just changing the overlap order, the strips can go either over or under each universal

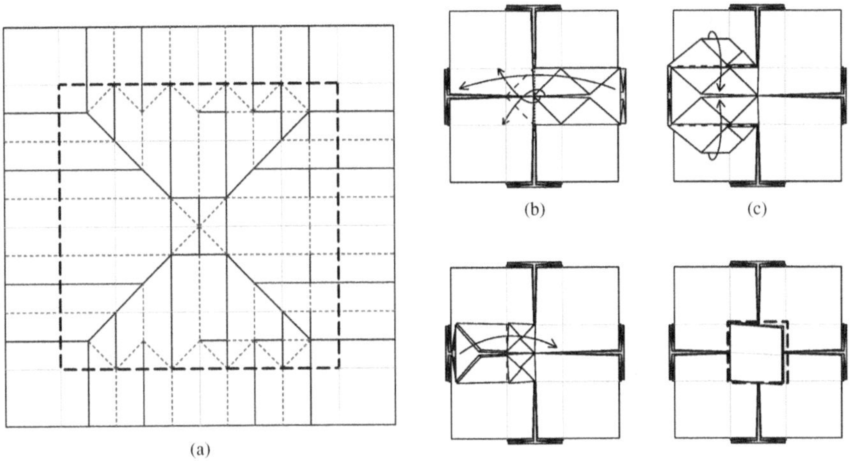

Fig. 6. (a) Crease pattern for the left-oriented universal tab. (b) Collapse the crease pattern to this, then spread two layers and fold the flap to the left. (c) Fold two layers to the center line. (d) Fold the flap back to the right. (e) The completed universal tab with hinge along the left side. The heavy dashed line in the crease pattern maps to the boundary of the tab in the last figure.

tab, resulting in the two possible colors. This idea of a universal pixel display was introduced to us by Masashi Tamaka[1] who designed a 4×4 model.

We can optimize the construction by re-using the pleats from the slots and reflex corners in the folding of the tabs. Each slot creates at least two pleats on either side of the slot, with the important property that the pleats have the correct parity to join up with the pleats in the universal tabs. Figure 7 shows where additional pleats must be added; each heavy dashed line indicates a single pair of pleats at that location, contributing two units to the perimeter along that side. Looking first at the vertical direction, we see that each of the m strips requires four additional pleats running down the middle of the strip, adding $4m$ to the vertical dimension. There are

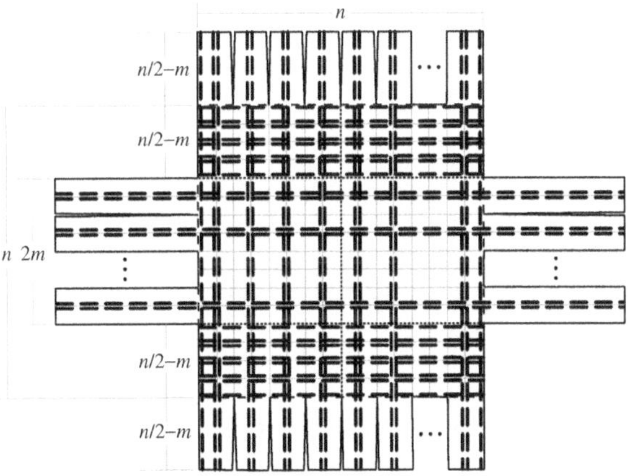

Fig. 7. Where pleats must be added to create universal tabs. Each heavy dashed line indicates a pair of pleats added at that location.

[1] Personal communication, June 2009.

also $4(n/2 - m)$ horizontal pleats in each of the sections above and below the horizontal strips, for a total of $V_2 = 2n$ units added to the vertical edge length. In the other dimension, we must add four vertical pleats running down the middle of each of the $n/2$ strips, plus two pleats along the left and right boundary, for a total of $H_2 = 2n + 2$ units added to the horizontal edge length.

One loose end to tie up: each universal tab requires exactly four pleats in each direction, two on each side of the vertical and horizontal symmetry lines. For the sets of pleats that come from slots or reflex corners, there will in general be more than four pleats running in one, the other, or both directions. This issue does not create any problems: we can put into place all but two pleats, then overlay the universal tab crease pattern on the "sandwich" of layers to complete the tab.

Therefore the entire crease pattern requires a rectangle of size

$$\begin{aligned} V &= V_1 + V_2 &&= mn + 2n + 4m \\ H &= H_1 + H_2 &&= \tfrac{1}{2}n^2 - mn + 3n - 2m - 2 + 2n + 2 \\ & &&= \tfrac{1}{2}n^2 - mn + 5n - 2m \end{aligned}$$

Setting $V = H$, we obtain

$$m = \frac{n(n+6)}{4(n+3)} = \tfrac{1}{4}n + \tfrac{3}{4} - O(1/n)$$

But m must be an integer, so we use either $m = \lfloor \tfrac{1}{4}(n+3) \rfloor$ or $m = \lceil \tfrac{1}{4}(n+3) \rceil$. In the first case, H dominates, while in the second case, V dominates. The best construction is the minimum of these two options, which (for n even) works out to a side length of

$$\tfrac{1}{4}n^2 + 4n + 4 - \tfrac{5}{2}(n \bmod 4).$$

The resulting semiperimeter is

$$\tfrac{1}{2}n^2 + 8n + 8 - 5(n \bmod 4),$$

which beats n^2 for all (even) $n > 16$. For $n = 8$, this construction uses a 52×52 square, worse than our rectangle construction, but for $n > 16$ the square construction wins.

3 More General Patterns

Our construction can be generalized to arbitrary two-color patterns. We fold the perimeter of the sheet into strips of lengths sufficient to reach every colored region of the final design. In case the pleats formed during the strip folding suffice to generate the tabs, we are done. Otherwise, we must make additional pleats first, before folding the strips.

There are multiple new ingredients that would be necessary for an optimal solution here. One issue is that these initial pleats and the strips together determine the width of the strips. Another is that with an irregular design, it may be necessary to run a complex optimization procedure to determine an optimal solution. A first attempt at formulating this optimization might be to notice that every colored region in the design must be reached by a strip, and therefore also by a point on the perimeter of the original sheet. Thus, in a solution, the perimeter will hit every colored region. In other words, the perimeter will form a solution to an instance of *TSP with neighborhoods* (also known as *one-of-a-set TSP* and *group TSP*).

However, finding such a tour will not be enough in general. Namely, colored regions of the folded design must be covered by the strips, not just touched by them. Thus larger colored regions in the design may require special wider strips, or multiple strips. We leave a more rigorous formulation of this optimization problem as a topic for future research.

Acknowledgments

We thank Marc Kirschenbaum and John Montroll for helpful early discussions.

References

1. Casey, S.: Chess board. In: Proceedings of the West Coast Origami Guild, vol. 19, pp. 3–12 (August 1989)
2. Chen, S.Y.: Checker board. In: Proceedings of the Annual OUSA Convention, pp. 72–75 (2001), `http://www.origami-usa.org/files/CheckerBoard.PDF`
3. Demaine, E.D., Demaine, M.L., Mitchell, J.S.B.: Folding flat silhouettes and wrapping polyhedral packages: New results in computational origami. Computational Geometry: Theory and Applications 16(1), 3–21 (2000)
4. Demaine, E.D., O'Rourke, J.: Geometric Folding Algorithms: Linkages, Origami, Polyhedra. Cambridge University Press, Cambridge (July 2007)
5. Dureisseix, D.: Chessboard. British Origami 201, 20–24 (2000)
6. Fujimoto, S.: Crowding butterflies. In: BOS Autumn Convention, p. 27 (1989)
7. Guy, M., Venables, D.: The Chess Sets of Martin Wall, Max Hulme and Neal Elias. British Origami Society Booklet 7 (1979)
8. Kirschenbaum, M.: Chess board. The Paper 61, 24–30 (1998), `http://dev.origami.com/images_pdf/chessboard.pdf`
9. Lang, R.J.: Origami Design Secrets: Mathematical Methods for an Ancient Art. AK Peters, Wellesley (2003)
10. Montroll, J.: Origami Inside-Out. Dover, Newyork (1993)
11. Yoshizawa, A.: Origami Museum I. Kamakura Shobo Publishing Co., Tokyo (1987)

Finding All Approximate Gapped Palindromes*

Ping-Hui Hsu[1], Kuan-Yu Chen[1], and Kun-Mao Chao[1,2,3]

[1] Department of Computer Science and Information Engineering
[2] Graduate Institute of Biomedical Electronics and Bioinformatics
[3] Graduate Institute of Networking and Multimedia
National Taiwan University, Taipei, Taiwan 106
kmchao@csie.ntu.edu.tw

Abstract. We study the problem of finding all *maximal* approximate gapped palindromes in a string. More specifically, given a string S of length n, a parameter $q \geq 0$ and a threshold $k > 0$, the problem is to identify all substrings in S of the form uvw such that (1) the Levenshtein distance between u and w^r is at most k, where w^r is the reverse of w and (2) v is a string of length q. The best previous work requires $O(k^2n)$ time. In this paper, we propose an $O(kn)$-time algorithm for this problem by utilizing an incremental string comparison technique. It turns out that the core technique actually solves a more general incremental string comparison problem that allows the insertion, deletion, and substitution of multiple symbols.

Keywords: palindrome, incremental string comparison, string matching.

1 Introduction

A word is called a palindrome if it reads the same both forward and backward. In other words, a palindrome is a word of the form uau^r, where u is a string, a is a symbol (or an empty word), and u^r is the reverse of u. A palindrome in a string is *maximal* if it can not be extended outward while preserving a palindromic structure. The recognition of palindromes in a string has always been an intriguing question both in theory and practice. For example, finding all maximal palindromes in a string was studied and well solved in linear time with the help of suffix tree [15] and constant-time LCA (least common ancestor) queries [12]. In [4,5], the authors studied the problems of palindromes in Sturmian words or in ternary square-free words. Moreover, the problem of identifying palindromes in compressed texts were investigated in [10].

In DNA and RNA sequences, there is a similar structure called quasi-palindrome which plays an important role in genome research [3,8,16]. A quasi-palindrome can be seen as a pair of reverse complementary repeats in a string that are separated by a number of characters. The complementarity relation on

* This research was supported in part by NSC grants NSC 95-2221-E-002-126-MY3 and NSC 97-2221-E-002-097-MY3 from the National Science Council, Taiwan.

Y. Dong, D.-Z. Du, and O. Ibarra (Eds.): ISAAC 2009, LNCS 5878, pp. 1084–1093, 2009.

nucleotides means that A is complementary to T (U) and C is complementary to G. Several functions of the quasi-palindrome in genomes have been discovered. For example, the quasi-palindromic structure is a sign of replication origins in the nucleotide sequence [3]. By estimating the appearance of replication origins in advance, the biologists can avoid much labor-intensive work. A study of quasi-palindromes also shows that they may control male germ-line gene expression [16]. Thus, the recognition of quasi-palindromic structure draws much attention and raises interesting computational problems. For example, the *gapped palindrome* problem is to recognize a word structure of the form uvu^r, where strings u and u^r are called *arms* and string v is called *gap*. In [8], the computation of gapped palindromes in a string, with some length constraints on the arms and the gap, are done in linear time.

Since mutations may occur in DNA and RNA sequences during evolution, looking for the *approximate* quasi-palindromes (or gapped palindromes) is in a sense biologically more meaningful. A tool called Inverted Repeats Finder (https://tandem.bu.edu/cgi-bin/irdb/irdb.exe) identifies approximate gapped palindromes heuristically. The tool uses a set of statistically based criteria to detect candidates of approximate gapped palindromes. By applying some alignment techniques, it then conforms whether these candidates are palindromes.

In this paper, we study the problem of finding all approximate gapped palindromes in a string. More specifically, we allow the Levenshtein distance between two arms to be at most k and the length of gap to be a fixed q. It should be noted that we identify all such palindromes while a previous work [1] can only identify a set of them. We show that solving our problem is essentially solving the incremental string comparison problem. The research of incremental string comparison was initiated by Landau *et al.*. The incremental approach proposed in [9] allows one to append a single symbol to one string at a time. Throughout the paper we aim to solve a more general incremental problem in which we allow multiple symbols to be appended to and deleted from the strings. As a result, our algorithm finds all maximal approximate gapped palindromes in $O(kn)$ time while the best previous work requires $O(k^2n)$ time [11], where n is the string size.

2 Preliminaries

The Levenshtein distance (edit distance) between two strings is the minimum number of editing steps that convert one string into another. Given two string $A[1..m]$ and $B[1..n]$, one can calculate their edit distance by dynamic programming. We refer to matrix $D[0..m, 0..n]$ as the *edit-distance table* of strings A and B. Initially, $D[i, 0] = i$ for $0 \le i \le m$ and $D[0, j] = j$ for $1 \le j \le n$. Then the cell $D[i, j]$, where $i, j > 0$, stores the edit distance of strings $A[1..i]$ and $B[1..j]$. We also say that the cells $D[i, j]$ where $j - i = d$ are on diagonal d of D. Below we introduce some important properties of the edit-distance table that are constantly used in later discussion.

Lemma 1 ([13]). *Given an edit-distance table D, we have* $D[i, j] - D[i-1, j-1] \in \{0, 1\}$.

In other words, Lemma 1 implies that $D[i, j] \geq D[i-1, j-1]$, i.e. the values of the cells on the same diagonal are monotonically increasing. To compute table D, as noticed by Ukkonen [13], it suffices to find the last cell that has value h, for all h, on each diagonal of table D. In other words, all values in D are determined when those last cells that have value h, for all h, are determined. To specify the last cells with value h, the terms $L_d^D(h)$ and *h-cell* are defined as follows.

Definition 1. *Given an edit-distance table D, we define* $L_d^D(h)$ *to be the last cell on diagonal d that has value h. In other words,* $L_d^D(h)$ *is the cell* $D[i, j]$ *where* $i = \max\{i' : D[i', i'+d] = h\}$ *and* $j = i + d$. *Furthermore, we refer to such cell* $L_d^D(h)$ *as the h-cell on diagonal d of D.*

As mentioned in [9,13], one can compute the index (i, j) of $L_d^D(h)$ for all $h > 0$ by the following recurrence:

$$i = Slide_d\left(\max\left\{\begin{matrix} i_1 \\ i_2 + 1, \\ i_3 + 1 \end{matrix}\right\}\right), j = i + d,$$

where $Slide_d(i) = \max\{i' : A[i..i'] = B[i+d, i'+d]\}$ and i_1, i_2, and i_3 are the row indices of $L_{d-1}^D(h-1), L_d^D(h-1)$ and $L_{d+1}^D(h-1)$. Notice that, for brevity we do not explicitly specify the bounds of the value of h. However, all the h-cells are assumed to be valid while mentioned. The following lemmas demonstrate the relationship between the values of adjacent cells of D.

Lemma 2 ([14]). *Given an edit-distance table D, we have* $D[i, j] - D[i-1, j], D[i, j] - D[i, j-1] \in \{-1, 0, 1\}$.

Lemma 3. *Given an edit-distance table D, we have* $D[i+1, j], D[i, j+1] \in \{h, h+1\}$ *for any h-cell* $D[i, j]$ *in D.*

Proof. Lemma 3 follows Lemmas 1, 2 immediately. □

3 Incremental String Comparison

Recall that given a string S of size n, a parameter $q \geq 0$ and a threshold $k > 0$, our problem is to find all maximal approximate gapped palindromes in S that are of gap size q and the edit distance between its two arms is at most k. Here we sketch a basic algorithm for finding all approximate gapped palindromes.

Note that the *maximal* property implies that each pair of prefixes found in the basic algorithm cannot be further extended. At iteration i, to find all pairwise prefixes one can compute the edit-distance table of strings $S[1..i]^r$ and $S[i+q+1..n]$. At iteration $i+1$, we need to compute the edit-distance table of $Cat(S[i+1], S[1..i])^r)$ and $Del(1, S[i+q+1..n])$. Here $Cat(\hat{S}, S)$ denotes the string obtained by concatenating string $\hat{S}$ and string S, and $Del(c, S)$ denotes

```
procedure BASIC ALGORITHM
for i = 1 to n − q − 1 do
    find all possible pairs of prefixes of (S[1..i])^r and S[i + q + 1..n] such that
    the edit distance of pairwise prefixes is at most k.
end for
```

Fig. 1. A basic algorithm for finding all approximate gapped palindromes

the string obtained by deleting the first c symbols from S. Thus the challenge lies in, given two strings A and B how to compare $Cat(a, A)$ and $Del(1, B)$, provided the comparison result of A and B, where a is an additional symbol. That is we aim to solve an incremental string comparison problem. In what follows we formulate this problem in a more general form.

Problem 1. THE $(+t_1, -t_2)$-INCREMENTAL-STRING-COMPARISON PROBLEM (abbr. $(+t_1, -t_2)$-ISC). *Given the edit-distance table of strings A and B, how to efficiently compute the edit-distance table of strings $Cat(\hat{A}, A)$ and $Del(t_2, B)$, where $\hat{A}$ is an arbitrarily appended string with size t_1 ($t_1 = |\hat{A}|$) and t_2 is the number of deleted symbols from B?*

The (+1,0)-ISC problem has been studied by Landau *et al.*. In [9], they show that there exist nice properties between the given edit-distance table and the desired edit-distance table. Based on the properties observed, they further devise a clever incremental algorithm to compute the *h-cells* for all $h \geq 0$ on each diagonal of the desired table. In this paper we extend their ideas and show that the same properties hold even for the more general $(+t_1, -t_2)$-ISC problem, where $t_1 \geq 0$ and $0 \leq t_2 \leq n$.

Let D denote the edit-distance table of strings $A[1..m]$ and $B[1..n]$, and let D' denote the edit-distance table of $Cat(\hat{A}, A)$ and $Del(t_2, B)$, where $|\hat{A}| = t_1$. The $(+t_1, -t_2)$-ISC problem is to compute table D' from table D. For convenience of our discussion, we label the indices of table D' by an offset. We assume that the first row of D' is of index $-t_1$ and the first column of D' is of index t_2, i.e. $D'[-t_1..m, t_2..n]$. Notice that $D[i, j]$ stores the edit distance of $A[1..i]$ and $B[1..j]$, and $D'[i, j]$ stores the edit distance of $Cat(\hat{A}, A[1..i])$ and $Del(t_2, B[1..j])$. We define the *difference table* C of D and D' as follows.

Definition 2. *The difference table of D and D' is table $C[0..m, t_2..n]$ whose entry $C[i, j] = D'[i, j] - D[i, j]$ for $0 \leq i \leq m$ and $t_2 \leq j \leq n$.*

In other words, table C can be obtained by (1) putting table D' on the top of table D such that the cells of the same indices between the two tables are overlapped, and (2) storing the difference of values of two overlapped cells in a cell of C. Figure 2 shows an example where $t_1 = 1$ and $t_2 = 2$.

Kim and Park [7] showed that for the case where either $t_1 = 1$ or $t_2 = 1$, the value range of the cells in C is constrained. Besides, there exist some monotonic properties in table C. We shall demonstrate that these properties hold for the

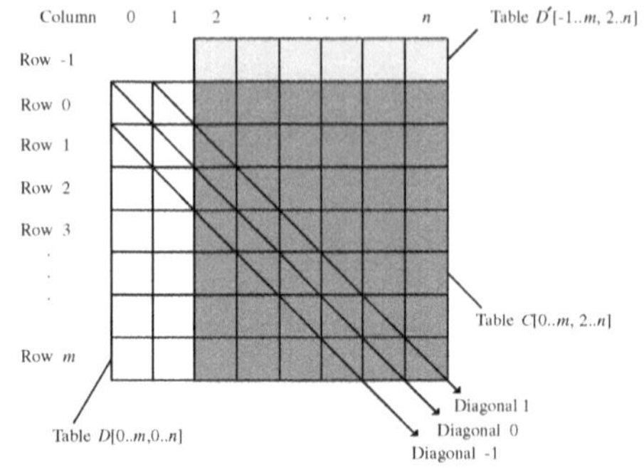

Fig. 2. Table $D'[-1..m, 2..n]$ (gray cells) is placed on the top of table $D[0..m, 0..n]$ (white cells) and the overlapped region forms the difference table $C[0..m, 2..n]$ (dark cells)

more general cases where $t_1 \geq 0$ and $0 \leq t_2 \leq n$. Due to space limitations, the proofs of the following lemmas are omitted.

Lemma 4. *For* $1 \leq i \leq m$ *and* $t_2 + 1 \leq j \leq n$, $\min\{C[i-1, j], C[i-1, j-1], C[i, j-1]\} \leq C[i, j] \leq \max\{C[i-1, j], C[i-1, j-1], C[i, j-1]\}$.

Lemma 5. *For any column* j *of* C*, we have* $C[0, j] \leq C[1, j] \leq \ldots \leq C[m, j]$.

Lemma 6. *For any row* i *of* C*, we have* $C[i, t_2] \geq C[i, t_2 + 1] \geq \ldots \geq C[i, n]$.

We introduce the following notations for comparing the order of indices of cells in D and D'.

Definition 3. *Given cell* $D[i, j]$ *and cell* $D'[i', j']$*, we define* $D[i, j] \succeq D'[i', j']$ *iff* $i \geq i'$*. Similarly, we define* $D[i, j] \succ D'[i', j']$ *iff* $i > i'$*. Moreover, we define* $D[i, j] \sim D'[i', j']$ *iff* $i = i'$ *and* $j = j'$*. We say cell* $D[i, j]$ *coincides with cell* $D'[i', j']$ *iff* $D[i, j] \sim D'[i', j']$

Suppose that table D' is placed on the top of D as before. For those *h-cells* $L_d^{D'}(h)$ lying in the overlapped region, we define function f, analogous to the notion of *key values* mentioned in [9].

Definition 4. *For an h-cell* $L_d^{D'}(h)$*, we let* $f(L_d^{D'}(h)) = g$ *if* $L_d^{D'}(h) \sim L_d^D(g)$ *for some integer* g*. Otherwise, we let* $f(L_d^{D'}(h)) = g - \frac{1}{2}$ *such that* $g = \min\{g' : L_d^D(g') \succ L_d^{D'}(h)\}$.

Now we prove the central theorem of our algorithm. Again, we only consider the *h-cells* $L_d^{D'}(h)$ lying in the overlapped region.

Theorem 1. *For h-cells* $L_d^{D'}(h)$ *and* $L_{d+1}^{D'}(h)$*, we have* $\lfloor f(L_{d+1}^{D'}(h)) \rfloor \geq f(L_d^{D'}(h))$.

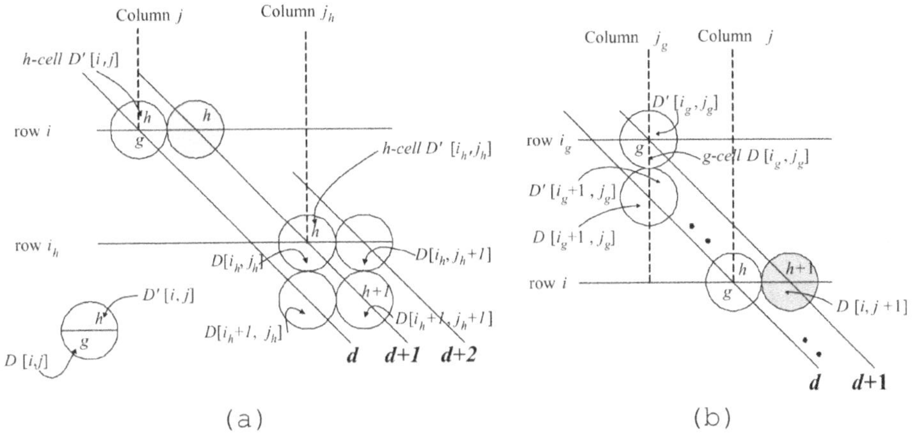

Fig. 3. (a) Case 1: $D'[i,j+1] = h$ (the gray cell). In this case, we show $D[i_h+1,j_h+1] = \min\{D[i_h+1,j_h]+1, D[i_h,j_h]+\delta(i_h+1,j_h+1), D[i_h,j_h+1]+1\} > g$. Note that each circle represents two overlapped cells when table D' is placed on the top of table D. The top-right value in a circle indicates the value of the cell from D' and the bottom-left value in a circle indicates the value of the cell from D. (b) Case 2: $D[i,i+d+1] = h+1$ (gray cell). In this case, we show that $D'[i_g,j_g] \leq h$ by first proving that $D[i_g+1,j_g] = g$ and $D'[i_g+1,j_g] \leq h$.

Proof. Let (i,j) be the position of $L_d^{D'}(h)$ and $D[i,j] = g$. To prove $\lfloor f(L_{d+1}^{D'}(h)) \rfloor \geq f(L_d^{D'}(h))$, it must be shown that $f(L_{d+1}^{D'}(h)) \geq g$. By Lemma 1, $D'[i,j+1] \in \{h, h+1\}$. Depending on the value of cell $D'[i,j+1]$, we divide the proof into two cases.

Case 1: $D'[i,j+1] = h$. (See Figure 3(a) for an illustration.)
Observe that, by definition, $f(L_{d+1}^{D'}(h)) \geq g$ *iff* $L_{d+1}^{D'}(h) \succeq L_{d+1}^{D}(g)$. Let (i_h, j_h) be the position of $L_{d+1}^{D'}(h)$. We have $L_{d+1}^{D'}(h) \succeq L_{d+1}^{D}(g)$ *iff* $D[i_h+1,j_h+1] > g$. By the recursive relation, we have

$$D[i_h+1,j_h+1] = \min\{D[i_h+1,j_h]+1, D[i_h,j_h]+\delta(i_h+1,j_h+1), D[i_h,j_h+1]+1\}.$$

In the following, we show that the three parameters of function min are greater than g, implying $D[i_h+1,j_h+1] > g$.

By Lemma 1, $D[i_h+1,j_h] \geq D[i,j] = g$ since they both are cells on diagonal d and i_h+1 is greater than i. Thus, $D[i_h+1,j_h]+1 > g$. We next show that the second parameter, $D[i_h,j_h]+\delta(i_h+1,j_h+1)$, is greater than g. By Lemma 6, $C[i,j] \geq C[i,j+1]$. That is, $D'[i,j]-D[i,j] \geq D'[i,j+1]-D[i,j+1]$. Because $D'[i,j] = h$, $D[i,j] = g$ and $D'[i,j+1] = h$, we have $D[i,j+1] \geq g$. According to Lemma 1, $D[i_h,j_h] \geq D[i,j+1] \geq g$ (because they are cells on diagonal $d+1$ and $i_h \geq i$). Besides, since $L_{d+1}^{D'}(h) =$ cell $D'[i_h,j_h]$, which implies $A[i_h+1] \neq B[i_j+1]$, we have $\delta(i_h+1,j_h+1) = 1$. Hence we obtain $D[i_h,j_h]+\delta(i_h+1,j_h+1) > g$.

Finally, we show that the third parameter, $D[i_h, j_h+1]+1$, is greater than g. According to Lemma 6, $C[i_h, j_h] \geq C[i_h, j_h+1]$. That is, $D'[i_h, j_h] - D[i_h, j_h] \geq D'[i_h, j_h+1] - D[i_h, j_h+1]$. Together with Lemma 3, we have (1) $D'[i_h, j_h] = h$, (2) $D[i_h, j_h] \geq g$ and (3) $D'[i_h, j_h+1] \in \{h, h+1\}$. Thus, we conclude that $D[i_h, j_h+1] \geq g$, implying $D[i_h, j_h+1]+1 > g$.

Case 2: $D'[i, j+1] = h+1$. (See Figure 3(b) for an illustration.)
If $L^D_{d+1}(g)$ does not exist, i.e. the smallest value of the cells on diagonal $d+1$ of D is at least $g+1$, we have $f(L^{D'}_{d+1}(h)) \geq g$ immediately. Thus, we assume $L^D_{d+1}(g)$ exists. By definition, $f(L^{D'}_{d+1}(h)) \geq g$ *iff* $L^{D'}_{d+1}(h) \succeq L^D_{d+1}(g)$. Let (i_g, j_g) be the position of $L^D_{d+1}(g)$. We have $L^{D'}_{d+1}(h) \succeq L^D_{d+1}(g)$ *iff* $h \geq D'[i_g, j_g]$. Hence, in the following we show that $h \geq D'[i_g, j_g]$.

First we show that $i_g < i$. By Lemma 6, $D'[i,j] - D[i,j] \geq D'[i, j+1] - D[i, j+1]$. Besides, we have (1) $D'[i,j] = h$, (2) $D[i,j] = g$ and (3) $D'[i, j+1] = h+1$. It follows that $D[i, j+1] \geq g+1$, implying $i > i_g$. Combining $i > i_g$ with Lemma 1, we further evaluate $D'[i_g+1, j_g]$ and $D[i_g+1, j_g]$ as follows.

- Since $D'[i_g+1, j_g] \leq D'[i,j]$ by Lemma 1 and $D'[i,j] = h$, we derive $D'[i_g+1, j_g] \leq h$.
- Since $D[i_g+1, j_g] \leq D[i,j]$ by Lemma 1 and $D[i,j] = g$, we derive $D[i_g+1, j_g] \leq g$. Moreover, since $D[i_g+1, j_g] \in \{g, g+1\}$ by Lemma 3, it follows that $D[i_g+1, j_g] = g$.

By Lemma 5, $D'[i_g+1, j_g] - D[i_g+1, j_g] \geq D'[i_g, j_g] - D[i_g, j_g]$. Since $D'[i_g+1, j_g] \leq h$, $D[i_g+1, j_g] = g$, and $D[i_g, j_g] = g$, we conclude that $h \geq D'[i_g, j_g]$. □

4 The Algorithm

In this section, inspired by [9], we describe an algorithm for the $(+\mathbf{t_1}, -\mathbf{t_2})$-ISC problem as follows.

Given an edit-distance table D, for a fixed h we refer to the list of all *h-cells* of D as the *h-wave* of D. A wave is implemented as a doubly-linked list, in which each node stores the table index of an *h-cell* in D. Thus, we can access the *h-cells* on diagonals $d-1$ and $d+1$ in constant time when provided the pointer to the *h-cell* on diagonal d. Besides, we also construct links across the waves to connect those cells on the same diagonal. Therefore, we can also access the $(h\text{-}1)$-*cell* and $(h{+}1)$-*cell* on diagonal d in constant time when given the pointer to the *h-cell* on diagonal d.

As mentioned before, let D denote the edit-distance table of strings A and B, and let D' denote the edit-distance table of $Cat(\hat{A}, A)$ and $Del(t_2, B)$, where $|\hat{A}| = t_1$. Note that we label the indices of D' by an offset, i.e. the first row is labeled $-t_1$ and the first column is labeled t_2. The algorithm takes the waves of D as input, and aims to compute the 0-*wave*, 1-*wave*, ..., k-*wave* of D'. Below we show how the scheme works.

We partition a wave of D' into a series of blocks according to their f's values, i.e. the cells having the identical f's value are in the same block. By Theorem 1, each block contains either a single cell or several consecutive cells of the wave. For those cells in the same block, if their f's values are integer g, we know each of those cells coincides with a certain *g-cell* of D. Observe that if two nodes of the doubly-linked lists represent a cell x of D' and x's coincided cell in D respectively, these two nodes store the same table index. That is, we can imitate one node by the other node if they represent the coincided cells. Thus, in the implementation we can construct this block by "cutting" a piece from the *g-wave*. If the f's value of a cell is not an integer, i.e. $g - \frac{1}{2}$, we compute this cell by the recurrence in Section 2.

Observe that the waves of D' may include cells lying outside the overlapping region of D and D'. We can not obtain those cells (which are outside the overlapping region) of D' by cutting the waves of D; however, they can be computed by the recurrence in Section 2.

Definition 5. *For $h \geq 0$, the partition of the h-wave can be parameterized as:*

- *$s(h)$: the number of blocks;*
- *$b^i(h)$: the f's value of cells in the ith (for $1 \leq i \leq s(h)$) block;*
- *$l^i(h)$: the left boundary of the ith (for $1 \leq i \leq s(h)$) block (formally, $l^i(h)$ is the smallest d such that $f(L_d^{D'}(h)) = b^i(h)$);*
- *$r^i(h)$: the right boundary of the ith (for $1 \leq i \leq s(h)$) block (formally, $r^i(h)$ is the largest d such that $f(L_d^{D'}(h)) = b^i(h)$).*

The number of blocks, $s(h)$, depends on the possible f's values of cells. Lemma 7 shows that the value of $s(h)$ is bounded by $O(t_1 + t_2)$. Besides, with Lemma 8 and Corollary 1 we can obtain the f's values of *h-cells* by checking the f's values of *(h-1)-cells*. As space is limited, the proofs of these lemmas are omitted.

Lemma 7. *For $h \geq 0$, we have $s(h) = O(t_1 + t_2)$.*

Lemma 8. *For diagonals d and d', $d < d'$, if $f(L_d^{D'}(h-1)) = f(L_{d'}^{D'}(h-1)) = g$, we have $f(L_{d''}^{D'}(h)) = g + 1$ for $d < d'' < d'$.*

Corollary 1. *For boundaries $l^i(h-1)$ and $r^i(h-1)$ of the ith block of the $(h-1)$-wave, we have $f(L_d^{D'}(h)) = b^i(h-1) + 1$ for $l^i(h-1) < d < r^i(h-1)$.*

By Corollary 1 we know how to obtain the f's values of *h-cells* lying in between the boundaries of each block of the *(h-1)-wave*. These *h-cells* are called *trivial* since we can immediately retrieve their f's values by examining the partition. As for the remaining *h-cells*, we compute them by the recurrence in Section 2 and then obtain their f's values. As for the remaining *h-cells*, their f's values can be obtained once we have their table indices, which can be computed by the recurrence in Section 2. Moreover, those *h-cells* are called *non-trivial.*

Now we have the f's value (if exists) of each *h-cell.* The partition of the *h-wave* is easily determined by checking the f's values of all the non-trivial *h-cells.*

We then cut the corresponding pieces of waves of D to assemble the *h-wave* of D'. Combining those pieces with the non-trivial *h-cells* computed, we derive the whole *h-wave*.

Theorem 2. *There exists an algorithm that computes the* 0*-wave,* $\ldots$*, and* k*-wave of* D' *and their partitions in* $O((t_1 + t_2) \times k)$ *time.*

Proof. We analyze the time of computing the non-trivial cells and the time of doing the cutting operations.

For an *h-cell*, the (h-1)*-cell* on the same diagonal may (1) lie in the overlapped region, (2) lie outside the overlapped region, or (3) not exist. Therefore, the non-trivial *h-cells* can be divided into three groups. In the following we show that there are in total $O((t_1 + t_2) \times k)$ non-trivial *h-cells* for $h = 0, 1, \ldots,$ and k.

The first group are those cells on the diagonals $l^1(h-1)$, $r^1(h-1)$, $l^2(h-1)$, $r^2(h-1), \ldots, l^{s(h-1)}(h-1)$, and $r^{s(h-1)}(h-1)$. Thus, by Lemma 7 we know that there are $O(t_1 + t_2)$ cells for each fixed h, implying that there are in total $O((t_1 + t_2) \times k)$ cells for $h = 0, 1, \ldots,$ and k. To calculate the size of the second group, we count the number of (h-1)*-cell* lying outside the overlapped region. We have that the number of 0*-cells*,1*-cells*, $\ldots$, and (k-1)*-cells* lying outside the overlapped region is $O(t_1 \times k)$. The reasons are that (1) all the 0*-cells*,1*-cells*, $\ldots$, and (k-1)*-cells* lie on $O(k)$ different diagonals and (2) each diagonal of D' contains at most t_1 cells lying outside the overlapped region. The third group only contains the leftmost and rightmost cells of each wave of D', leading to a total of $O(k)$ cells for $h = 1, 2, \ldots,$ and k. With the help of suffix tree (constructed in advance) and constant-time LCA queries, each non-trivial cell can be computed in constant time. Thus, the computation time of those non-trivial *h-cells* is $O((t_1 + t_2) \times k)$ time.

Due to the doubly-linked list structure, each cutting operation can be done in constant time. Besides, for each wave of D' there are $O(t_1 + t_2)$ blocks by Lemma 7. Therefore, we can do all the cutting operations in $O((t_1 + t_2) \times k)$ time. □

Theorem 3. *Given a string* S *of size* n *and a threshold* k*, where* k *specifies the edit distance between two arms, the problem of finding all maximal approximate gapped palindromes (with fixed gap length) can be solved in* $O(kn)$ *time.*

Proof. We have shown that to solve the problem is essentially to solve the (+1,-1)-ISC problem. We first construct the generalized suffix tree for strings S and S^r (the reverse of S) in linear time. Thus, the $Slide_d$ function in Section 2 can be done in constant time. It immediately implies that during the process of solving the $(+1, -1)$-ISC problem, we can compute the *h-cell* in constant time. Thus, by Theorem 2 each iteration (except the first one) in the basic algorithm of Figure 1 spends $O(k)$ time. By using the approach of [14], the first iteration can be done in $O(k^2)$ time. Hence the total time is $O(kn)$. □

5 Concluding Remarks

The core technique used in this paper is called incremental string comparison. We generalize the previous result of Landau *et al.* [9] by allowing multiple symbols appended to and removed from the heads of two strings simultaneously. In fact, this technique can be extended to handle the *substitute* operation, since a *substitute* operation can be seen as a *remove* operation followed by an *append* operation. More specifically, we allow a prefix part of the string to be substituted by another string.

References

1. Allison, L.: Finding Approximate Palindromes in Strings Quickly and Simply. In: CoRR, vol. abs/cs/0412004, informal publication (2004)
2. Chao, K.M., Zhang, L.: Sequence Comparison: Theory and Methods. Springer, Heidelberg (2009)
3. Chew, D.S.H., Choi, K.P., Leung, M.Y.: Scoring schemes of palindrome clusters for more sensitive prediction of replication origins in herpesviruses. Nucleic Acids Research 33(15), e134 (2005)
4. Currie, J.D.: Palindrome positions in ternary square-free words. Theoretical Computer Science 396(1-3), 254–257 (2008)
5. Glen, A.: Occurrences of palindromes in characteristic Sturmian words. Theoretical Computer Science 352(1-3), 31–46 (2006)
6. Gusfield, D.: Algorithms on strings, trees, and sequences: computer science and computational biology. Cambridge University Press, New York (1997)
7. Kim, S.R., Park, K.: A dynamic edit distance table. In: Giancarlo, R., Sankoff, D. (eds.) CPM 2000, vol. 1848, pp. 60–68. Springer, Heidelberg (2000)
8. Kolpakov, R., Kucherov, G.: Searching for gapped palindromes. In: Ferragina, P., Landau, G.M. (eds.) CPM 2008. LNCS, vol. 5029, pp. 18–30. Springer, Heidelberg (2008)
9. Landau, G.M., Myers, E.W., Schmidt, J.P.: Incremental string comparison. SIAM Journal on Computing 27(2), 557–582 (1998)
10. Matsubara, W., Inenaga, S., Ishino, A., Shinohara, A., Nakamura, T., Hashimoto, K.: Efficient algorithms to compute compressed longest common substrings and compressed palindromes. Theoretical Computer Science 410(8-10), 900–913 (2009)
11. Porto, A.H.L., Barbosa, V.C.: Finding approximate palindromes in strings. Pattern Recognition 35(11), 2581–2591 (2002)
12. Schieber, B., Vishkin, U.: On finding lowest common ancestors: simplification and parallelization. SIAM Journal on Computing 17(6), 1253–1262 (1988)
13. Ukkonen, E.: Algorithms for approximate string matching. Information and Control 64, 100–118 (1985)
14. Ukkonen, E.: Finding approximate patterns in strings. Journal of Algorithms 6(1), 132–137 (1985)
15. Ukkonen, E.: On-line construction of suffix trees. Algorithmica 14, 249–260 (1995)
16. Warburton, P.E., Giordano, J., Cheung, F., Gelfand, Y., Benson, G.: Inverted repeat structure of the human genome: the X-chromosome contains a preponderance of large, highly homologous inverted repeats that contain testes genes. Genome Research 14, 1861–1869 (2004)

General Pseudo-random Generators from Weaker Models of Computation*

George Karakostas

McMaster University, Dept. of Computing and Software, 1280 Main St. West, Hamilton, Ontario L8S 4K1, Canada
karakos@mcmaster.ca

Abstract. The construction of pseudo-random generators (PRGs) has been based on strong assumptions like the existence of one-way functions or exponential lower bounds for the circuit complexity of Boolean functions. Given our current lack of satisfactory progress towards proving these assumptions, we study the implications of constructing PRGs for weaker models of computation to the derandomization of general classes like *BPP*. More specifically, we show how PRGs that fool monotone circuits could lead to derandomization for general complexity classes, and how the Nisan-Wigderson construction could use hardness results for monotone circuits to produce pseudo-random strings.

Keywords: Pseudo-random generators, circuit complexity, monotone circuit complexity.

1 Introduction

One of the central issues in computational complexity is the understanding of the the additional (if any) power randomization can offer to solving problems efficiently. The investigation of the exact relations between randomized computational classes like *BPP* and non-randomized ones (like *P, NP, EXP*) has been the subject of intense research for the last two decades. One of the major steps towards a better understanding of randomness in computation was the realization [15] that hardness results can lead to the construction of good pseudo-random generators. This observation led to the construction of pseudo-random generators based on the (assumed) intractability of the discrete logarithm function [3] or the existence of one-way functions [18], [7]. The seminal paper by Nisan and Wigderson [11] was the first to connect the existence of predicates with high circuit complexity to the existence of good pseudo-random generators. The hardness assumption of [11] was still very strong (although not as strong as previous assumptions, e.g., the existence of one-way functions): the predicate used by the generator had to be extremely hard *on average*. A series of results based on hardness amplification techniques like Yao's XOR lemma [18] or error-correcting codes [16] can be used in order to relax this requirement

* Research supported by an NSERC Discovery grant.

Y. Dong, D.-Z. Du, and O. Ibarra (Eds.): ISAAC 2009, LNCS 5878, pp. 1094–1103, 2009.

for an extremely hard function to constant hardness [9], to mildly hard [5], to worst-case hard functions [2].

Unfortunately, this spectacular progress towards weaker hardness assumptions for generators of a given power has not brought us close to resolving the still open issue of the power of randomness in computation. This is due to the bleak current state-of-the-art in lower bounds for general circuits. We are not even close to proving any of the hardness assumptions above, and if the past is an indication of the future, we shouldn't expect their proof any time soon. Motivated by these difficulties, we ask whether looking for explicit hard functions in the general circuit model in order to construct pseudo-random generators is really an overkill. Indeed, let's assume that we can show the following 'theorem': if there is a (general) circuit of a certain size that can distinguish a truly random string to a string produced by a generator that uses only a random seed with some probability, then there is a construction in a restricted model of computation (e.g., a monotone circuit) of a certain size that can distinguish the two strings with a somewhat smaller probability. The contrapositive of the above 'theorem' would imply that if we can construct a generator that fools *all* restricted constructions (e.g., monotone circuits) of a certain size, then this generator would fool *all* general circuits of a certain size with a somewhat smaller probability. We apply this general framework to the restricted model of monotone circuits (in fact circuits that compute slice functions). There has been great progress in proving strong unconditional lower bounds for monotone circuits, for example [14], [10], [13], [4]. Razborov's approximation technique has even been extended to general circuits with only a few NOT gates [1]. In Section 2 we formalize the intuition above as Theorem 1 and give its (almost trivial) proof. Then we study the Nisan-Wigderson construction as a generator for monotone circuits. The hardness requirements of this generator are much stronger than just worst-case hardness, and there are no strong enough hardness amplification techniques for monotone circuits yet (indeed, it may be the case that no such techniques exist). Hence the generator still isn't unconditionally pseudo-random. Two facts though allow us to be optimistic about this approach: The first is the tremendous success of proving lower bounds for monotone circuits. The second fact is that the Nisan-Wigderson construction was built with fooling general circuits in mind. Once we set a more modest goal (fooling monotone circuits only, for example), other constructions may do better in terms of hardness requirements. We discuss these issues together with other possible directions in what we consider to be the most important part of this work, Section 4.

2 Monotone Circuit Tests from General Circuit Tests

In what follows, $\mathcal{C}$ and $\mathcal{C}_M$ are the classes of (general) circuits and monotone circuits respectively.

Definition 1. *A function* $G : \{0,1\}^d \rightarrow \{0,1\}^n$ *is an* (s, ε) pseudo-random generator *if no circuit of size* s *can distinguish* G *from the uniform distribution* U_n *with advantage greater than* ε*. That is, for every circuit* C *of size at most* s*,*

$$\left|Pr\left[C(U_n) = 1\right] - Pr\left[C(G(U_d)) = 1\right]\right| \leq \varepsilon$$

The circuit C in the definition above can be thought as a test that the alleged PRG has to pass in order to be pseudo-random.

Lemma 1. *If there is circuit $C \in \mathcal{C}$ of size s such that*

$$\left|Pr_{y\in\{0,1\}^n}\left[C(y) = 1\right] - Pr_{x\in\{0,1\}^d}\left[C(G(x)) = 1\right]\right| > \varepsilon$$

then there is a monotone circuit $C_M \in \mathcal{C}_M$ of size $|C_M| = O(s + n\log^2 n)$ such that

$$\left|Pr_{y\in\{0,1\}^n}\left[C_M(y) = 1\right] - Pr_{x\in\{0,1\}^d}\left[C_M(G(x)) = 1\right]\right| > \frac{\varepsilon}{2(n+1)}$$

Proof: W.l.o.g. we will assume that

$$\mathrm{Pr}_{y\in\{0,1\}^n}\left[C(y) = 1\right] - \mathrm{Pr}_{x\in\{0,1\}^d}\left[C(G(x)) = 1\right] > \varepsilon$$

Let $y_i, G_i(x)$ be the ith bit of $y, G(x)$. Then

$$\begin{aligned}
&\mathrm{Pr}_{y\in\{0,1\}^n}\left[C(y) = 1\right] - \mathrm{Pr}_{x\in\{0,1\}^d}\left[C(G(x)) = 1\right] = \\
&\sum_{k=0}^{n}\Bigg(\mathrm{Pr}_y\Big[\sum_i y_i = k\Big]\mathrm{Pr}_y\Big[C(y) = 1 | \sum_i y_i = k\Big] - \\
&-\mathrm{Pr}_x\Big[\sum_i G_i(x) = k\Big]\mathrm{Pr}_x\Big[C(G(x)) = 1 | \sum_i G_i(x) = k\Big]\Bigg) > \varepsilon
\end{aligned}$$

Therefore there is $0 \leq k_0 \leq n$ such that

$$\begin{aligned}
&\mathrm{Pr}_y\Big[\sum_i y_i = k_0\Big]\mathrm{Pr}_y\Big[C(y) = 1 | \sum_i y_i = k_0\Big] - \\
&-\mathrm{Pr}_x\Big[\sum_i G_i(x) = k_0\Big]\mathrm{Pr}_x\Big[C(G(x)) = 1 | \sum_i G_i(x) = k_0\Big] > \frac{\varepsilon}{n+1}
\end{aligned} \quad (1)$$

Let C_M be the following slice function of monotone circuit complexity $O(s + n\log^2 n)$ (see [17]):

$$C_M(y) = \begin{cases} 0, & \text{if } \sum_i y_i < k_0 \\ 1, & \text{if } \sum_i y_i > k_0 \\ C(y), & \text{if } \sum_i y_i = k_0 \end{cases}$$

Then we have

$$\begin{aligned}
&\Pr_{y\in\{0,1\}^n}\big[C_M(y)=1\big]-\Pr_{x\in\{0,1\}^d}\big[C_M(G(x))=1\big]=\\
&=\Pr_y\Big[\sum_i y_i<k_0\Big]\Pr_y\Big[C_M(y)=1|\sum_i y_i<k_0\Big]-\\
&-\Pr_x\Big[\sum_i G_i(x)<k_0\Big]\Pr_x\Big[C_M(G(x))=1|\sum_i G_i(x)<k_0\Big]+\\
&+\Pr_y\Big[\sum_i y_i>k_0\Big]\Pr_y\Big[C_M(y)=1|\sum_i y_i>k_0\Big]-\\
&-\Pr_x\Big[\sum_i G_i(x)>k_0\Big]\Pr_x\Big[C_M(G(x))=1|\sum_i G_i(x)>k_0\Big]+\\
&+\Pr_y\Big[\sum_i y_i=k_0\Big]\Pr_y\Big[C_M(y)=1|\sum_i y_i=k_0\Big]-\\
&-\Pr_x\Big[\sum_i G_i(x)=k_0\Big]\Pr_x\Big[C_M(G(x))=1|\sum_i G_i(x)=k_0\Big]
\end{aligned}$$

From the definition of C_M we know that

$$\Pr_y\Big[C_M(y)=1|\sum_i y_i<k_0\Big]=\Pr_x\Big[C_M(G(x))=1|\sum_i G_i(x)<k_0\Big]=0$$

$$\Pr_y\Big[C_M(y)=1|\sum_i y_i>k_0\Big]=\Pr_x\Big[C_M(G(x))=1|\sum_i G_i(x)>k_0\Big]=1$$

hence

$$\begin{aligned}
&\Pr_{y\in\{0,1\}^n}\big[C_M(y)=1\big]-\Pr_{x\in\{0,1\}^d}\big[C_M(G(x))=1\big]=\Pr_y\Big[\sum_i y_i>k_0\Big]-\\
&-\Pr_x\Big[\sum_i G_i(x)>k_0\Big]+\Pr_y\Big[\sum_i y_i=k_0\Big]\Pr_y\Big[C_M(y)=1|\sum_i y_i=k_0\Big]-\\
&-\Pr_x\Big[\sum_i G_i(x)=k_0\Big]\Pr_x\Big[C_M(G(x))=1|\sum_i G_i(x)=k_0\Big]
\end{aligned}$$

Let

$$\begin{aligned}
A&=\Pr_y\Big[\sum_i y_i>k_0\Big]-\Pr_x\Big[\sum_i G_i(x)>k_0\Big]\\
B&=\Pr_y\Big[\sum_i y_i=k_0\Big]\Pr_y\Big[C_M(y)=1|\sum_i y_i=k_0\Big]-\\
&\quad-\Pr_x\Big[\sum_i G_i(x)=k_0\Big]\Pr_x\Big[C_M(G(x))-1|\sum_i G_i(x)=k_0\Big]
\end{aligned}$$

If $|A|>\frac{\varepsilon}{2(n+1)}$ then the threshold function

$$T_{k_0+1}(y)=\begin{cases}0, & \text{if } \sum_i y_i\le k_0\\ 1, & \text{if } \sum_i y_i>k_0\end{cases}$$

has the desired properties, and the lemma holds. So assume that $|A| \leq \frac{\varepsilon}{2(n+1)}$. Also, from (1) we know that $B > \frac{\varepsilon}{n+1}$, therefore

$$\Pr_{y\in\{0,1\}^n}\big[C_M(y)=1\big] - \Pr_{x\in\{0,1\}^d}\big[C_M(G(x))=1\big] = A+B > \frac{\varepsilon}{2(n+1)}$$

The proof is the same in case

$$\Pr_{x\in\{0,1\}^d}\big[C(G(x))=1\big] - \Pr_{y\in\{0,1\}^n}\big[C(y)=1\big] > \varepsilon \qquad \square$$

From the lemma above we get the following theorem:

Theorem 1. *There is a constant $\gamma > 0$ such that if*

$$\Big|Pr_{y\in\{0,1\}^n}\big[C_M(y)=1\big] - Pr_{x\in\{0,1\}}\big[C_M(G(x))=1\big]\Big| \leq \frac{\varepsilon}{2(n+1)}$$

for all monotone circuits C_M with size $|C_M| \leq \gamma s + \gamma n^2 \log n$ for some $s > 0$, then

$$\Big|Pr_{y\in\{0,1\}^n}\big[C(y)=1\big] - Pr_{x\in\{0,1\}^d}\big[C(G(x))=1\big]\Big| \leq \varepsilon$$

for all circuits $C \in \mathcal{C}$ of size $|C| \leq s$.

Notice that nowhere in the above did we demand that G be a special kind of ciruit (e.g. monotone). Also notice that we can strengthen Lemma 1 and Theorem 1 by replacing the monotone circuit C_M by a monotone *slice* function C_M. Therefore in order to construct PRG's for general circuits, it is enough to construct PRG's for monotone (slice) circuits.

3 The Nisan-Wigderson Generator for Monotone Circuits

As an illustrative example, we demonstrate how one can use the Nisan-Wigderson PRG for monotone circuits. We cannot stretch enough the fact that this PRG was constructed in order to fool *general* Boolean circuits, and therefore it may very well *not* be the appropriate way to go. Nevertheless, at this point this is one of the most successful constructions (in terms of derandomization power) and a very good example for pointing out some of the different issues one may encounter during the design of a PRG for *monotone* circuits. Again, we emphasize that in what follows we make assumptions some of which may possibly be proven false, but we make them nevertheless, since they help us in illustrating our ideas or they can be transformed to other assumptions which are not so easily proven/disproved.

The main difficulty for applying the Nisan-Wigderson(NW) construction as is, is our inability to use the standard conversion of a circuit that is a good distinguisher between a truly random sequence and the output of the NW generator, to a circuit that approximates a hard function. This standard conversion uses the XOR of the output of such a circuit with a random bit, to produce the approximator, but in a monotone setting we cannot simulate the XOR. Therefore we need to modify the hardness assumption used by the NW generator:

Assumption 1. *There is a monotone predicate* $P : \{0,1\}^l \to \{0,1\}$ *such that no* monotone *circuit of size* s *can compute* P *or* $\bar{P}$ *correctly on more than a fraction* $\frac{1}{2} + \frac{\varepsilon}{2n(n+1)}$ *of the* 2^l *inputs (*n *is the number of (pseudo)-random bits we need).*[1]

This assumption differs from the usual hardness assumptions associated with the NW generator in two major points: it is an assumption about the *monotone* complexity of a (monotone) predicate P, and it is also an assumption on the *monotone* complexity of the *non-monotone* complement of P. Also notice that the hardness assumption is stronger as far as the fraction of correct answers is concerned: it is $\frac{1}{2} + \frac{\varepsilon}{2n(n+1)}$ instead of $\frac{1}{2} + \frac{\varepsilon}{n}$. The need for these modifications will become apparent when we try to prove that the NW construction is indeed a PRG. First we describe the Nisan-Wigderson construction.

Initially, this construction produces a collection of sets with small intersections (called a *design*):

Lemma 2. *[11] For every* $l, n \in \mathbb{N}$ *there exists a family of sets* $S_1, \ldots, S_n \subset \{1, \ldots, d\}$ *such that*

1. $d = O(\frac{l^2}{\log n})$
2. $|S_i| = l, \ \forall i$
3. $|S_i \cap S_j| \le \log n, \ \forall i \neq j$

Moreover, such a family can be computed by a deterministic TM in time $poly(n, 2^d)$*. In case* $l = C \log n$ *for some constant* $C > 0$*, there exists such a design with* $d = O(C^2 \log n)$ *that can be computed by a deterministic TM in time* $poly(n)$*.*

The generator in [11] uses the fact that if we have a uniformly distributed string x of length d, we can produce a family of substrings x_{S_i} with the above properties that will behave as independent when they are used as inputs to a hard predicate. The NW construction will output the string

$$NW^P_{l,n}(x) = P(x_{S_1})P(x_{S_2}) \ldots P(x_{S_n}).$$

Following the approach of [11] we show that under our Assumption 1 the NW construction is a PRG.

Theorem 2. *Under Assumption 1 the NW construction is a* $(\gamma s - O(n^2 \log n), \varepsilon)$ *PRG for some constant* $\gamma > 0$*.*

Proof: The proof of the theorem is virtually the same as in [11], but we repeat it here in order to illustrate new directions and open problems that arise from the use of a weaker model (monotone circuits).

---

[1] As stated, this assumption refers to monotone functions that are unbiased over *all inputs*. In fact, we can relax it by just requiring them to be unbiased over a particular *distribution* on all inputs that can be easily sampled. For example, we may talk about Razborov's CLIQUE function and the inputs to it are picked from only the "good" and "bad" inputs of Razborov's proof with equal probability.

We will assume that the NW construction is not a PRG with the required size-error parameters and will arrive at a contradiction. Since we assume that the NW construction is not an $(\gamma s - O(n^2 \log n), \varepsilon)$ PRG (for some constant $\gamma > 0$ to be picked later), there is a (possibly non-monotone) circuit C of size $\gamma s - O(n^2 \log n)$ such that

$$\left| \Pr_{y \in \{0,1\}^n} \left[C(y) = 1 \right] - \Pr_{x \in \{0,1\}^d} \left[C(NW^P_{l,n}(x)) = 1 \right] \right| > \varepsilon$$

From Lemma 1 and after an appropriate choice of γ above, we get that there is a *monotone* circuit C_M of size $s - O(n^2 \log n)$ such that

$$\left| \Pr_{y \in \{0,1\}^n} \left[C_M(y) = 1 \right] - \Pr_{x \in \{0,1\}^d} \left[C_M(NW^P_{l,n}(x)) = 1 \right] \right| > \frac{\varepsilon}{2(n+1)}$$

We assume that

$$\Pr_{y \in \{0,1\}^n} \left[C_M(y) = 1 \right] - \Pr_{x \in \{0,1\}^d} \left[C_M(NW^P_{l,n}(x)) = 1 \right] > \frac{\varepsilon}{2(n+1)} \tag{2}$$

If this is not the case, we *complement* C_M and we work with this complemented monotone circuit (which now is *non-monotone*). Later we will see that this will not affect our proof. Following closely previous proofs related to the NW construction, we define the following hybrid distribution on $\{0,1\}^n$:

Distribution D_i**:** The first i bits are the first i bits of $NW^P_{l,n}(x)$, where x is chosen uniformly over d-bit strings, and the rest $n-i$ bits $r_{i+1}, r_{i+2}, \ldots, r_n$ are chosen uniformly and independently at random.

Since D_0 is U_n and D_n is $NW^P_{l,n}(x)$, we have that there is an i such that

$$\Pr\left[C_M(D_i) = 1\right] - \Pr\left[C_M(D_{i-1}) = 1\right] > \frac{\varepsilon}{2n(n+1)}$$

or if we expand it

$$\begin{aligned} &\Pr_{x, r_{i+1}, \ldots, r_n} \left[C_M(P(x_{S_1}) \ldots P(x_{S_{i-1}}) P(x_{S_i}) r_{i+1} \ldots r_n) = 1 \right] - \\ &\quad - \Pr_{x, r_i, \ldots, r_n} \left[C_M(P(x_{S_1}) \ldots P(x_{S_{i-1}}) r_i r_{i+1} \ldots r_n) = 1 \right] > \frac{\varepsilon}{2n(n+1)} \end{aligned} \tag{3}$$

At this point, the proof of the pseudorandomness properties of the NW construction when the hardness of general circuits is used, utilizes a standard transformation of a distinguisher satisfying inequality (3) to a predictor [18]. By renaming r_i to b we get that

$$\Pr_{x, b, r_{i+1}, \ldots, r_n} \left[C_M(P(x_{S_1}) \ldots P(x_{S_{i-1}}) b r_{i+1} \ldots r_n) \oplus b = P(x_{S_i}) \right] > \frac{1}{2} + \frac{\varepsilon}{2n(n+1)}$$

Using standard averaging arguments, we can fix $b, r_{i+1}, \ldots, r_n$ as well as the bits of x not in S_i while preserving this prediction probability. Unfortunately, if b is fixed to 1, this transformation doesn't work in our case if we had just assumed the hardness of P (and not $\bar{P}$ as well), because it results in a *non-monotone* circuit approximating P (destroying our contradiction argument).

We rename x_{S_i} to z and we notice that the values of P calculated by the NW construction depend only on $|S_i \cap S_j| \leq \log n, j \neq i$ bits of z. Therefore these values are *monotone* functions P_j of z. So, depending on the (fixed) value of b, we have that either

$$\Pr_z\left[C'_M(P_1(z)\dots P_{i-1}(z)) = P(z)\right] > \frac{1}{2} + \frac{\varepsilon}{2n(n+1)}$$

or

$$\Pr_z\left[\,\overline{C'_M}(P_1(z)\dots P_{i-1}(z)) = P(z)\right] > \frac{1}{2} + \frac{\varepsilon}{2n(n+1)}$$

for some monotone circuit C'_M. Notice that we will also arrive at this point in the case (2) doesn't hold.

Each P_j can be computed by a DNF of size $O(n \log n)$, and there are at most n such functions. If we plug in C'_M the monotone circuits for each P_j, then we get a monotone circuit C'_M of size at most $s - O(n^2 \log n) + O(n^2 \log n) = s$ (for an appropriate choice of the constant in the first big-O), and such that either

$$\Pr_z\left[C'_M(z) = P(z)\right] > \frac{1}{2} + \frac{\varepsilon}{2n(n+1)}$$

or

$$\Pr_z\left[C'_M(z) = \overline{P}(z)\right] > \frac{1}{2} + \frac{\varepsilon}{2n(n+1)}$$

In either case, this contradicts Assumption 1. □

4 Discussion and Open Problems

In this section we explore some new directions and open problems suggested by the use of hard monotone functions for the construction of PRGs for general circuits.

4.1 Hardness Assumptions

The first (obvious) open question is whether there is indeed a monotone predicate P with the hardness properties of Assumption 1. It seems plausible that if a monotone function is very hard to approximate, then its *non-monotone* complement is even harder to approximate with *monotone* circuits. Razborov's approximation technique has been extended to general circuits with a limited number of NOT gates [1] quite naturally. Hence it is conceivable that techniques which would prove hardness of approximation for monotone circuits can be extended to prove hardness of approximation for circuits with, say, only one NOT gate before their output. Assumption 1 can be reformulated as follows:

Assumption 2. *There is a predicate $P : \{0,1\}^l \to \{0,1\}$ such that no monotone circuit or circuit with exactly one NOT gate right before its output of size s can compute P correctly on more than a fraction $\frac{1}{2} + \frac{\varepsilon}{2n(n+1)}$ of the 2^l inputs (n is the number of (pseudo)-random bits we need).*

The monotonicity of P was necessary so that the circuit we get after plugging in the DNF for each P_j is still monotone. But we can modify Assumption 1 as follows:

Assumption 3. *There is a (general) predicate* $P : \{0,1\}^l \to \{0,1\}$ *such that no monotone circuit of size* s *can compute* P *or* $\bar{P}$ *correctly on more than a fraction* $\frac{1}{2} + \frac{\varepsilon}{2n(n+1)}$ *of the* 2^l *inputs, using as advice the output of oracles for* P *that have* $l - \log n$ *input bits fixed (*n *is the number of (pseudo)-random bits we need).*

Assumption 3 is a version of strong non-self-reducibility that trades the generalization of the circuit family for which we need strong lower bounds (monotone with advice instead of just monotone) for the generalization of the hard predicate used in the NW construction.

How hard is it to design PRGs for monotone circuits? Impagliazzo, Shaltiel and Wigderson [8] showed that derandomizing BPP using a pseudo-random generator implies that EXP $\not\subset$ P/poly. Impagliazzo and Kabanets [6] also show that derandomization of RP or BPP would imply superpolynomial lower bounds for Boolean or arithmetic circuits. Since the assumptions used above imply the construction of pseudo-random generators, their proof would immediately imply strong circuit lower bounds. Therefore proving any of our assumptions (provided any of them is true) is *as difficult as* proving such lower bounds. Notice though that our assumptions are about a *weaker* model than general boolean circuits, and in this model the development of (worst case) lower bounds has been very successful so far.

4.2 Hardness Amplification for Monotone Predicates

Unfortunately the XOR function is not monotone, therefore Yao's XOR lemma cannot be applied directly in order to amplify the hardness of a function in a way useful for Theorem 2. Nevertheless, there are already some (weak) hardness amplification results for monotone functions. The recent work by O'Donnell [12] is in fact an amplification that applies to monotone circuits and their complements, in just the way we need it for Assumption 1. Instead of the XOR of several copies of a function, [12] uses two monotone functions: first it uses the $REC - MAJ - 3_l$ function (l is the depth of a ternary tree of majority-of-3's) of several copies of P to go from an $(1 - 1/poly(n))$-hard predicate P for *general* polynomial-size circuits to a new $(1/2 + o(1))$-hard predicate for polynomial circuits, and then it uses the Tribes function of Ben-Or and Linial, to transform the new predicate to a new $(1/2 + 1/n^{-1/2+\varepsilon})$-hard predicate for polynomial circuits. The definitions of the $REC - MAJ - 3_l$ and Tribes functions can be found in [12]. An important property of these functions is that $REC - MAJ - 3_l(\bar{P}, \ldots, \bar{P}) = \overline{REC - MAJ - 3_l}(P, \ldots, P)$ and $\text{Tribes}(\bar{P}, \ldots, \bar{P}) = \overline{\text{Tribes}}(P, \ldots, P)$, therefore the construction can use a mild version of Assumption 1 to produce a harder version of this same assumption. The starting point of this amplification is Impagliazzo's hard-core set theorem [5], which holds with respect to any model of computation closed under majority (therefore it also holds for our model of computation with respect to which we assume hardness of some predicate, namely monotone circuits). But O'Donnell's construction analysis seems to be an overkill for our case, since we only need hardness against *monotone* circuits of some size. Nevertheless, this amplification method probably doesn't achieve the power of Yao's XOR lemma. The existence of powerful amplification techniques for monotone predicates remains an interesting open problem.

Acknowledgements. I am thankful to A. Viglas, I. Tourlakis, and N. Galesi for helpful discussions, and V. Kabanets for pointing out [8],[6].

References

1. Amano, K., Maruoka, A.: Potential of the approximation method. In: 37th FOCS, pp. 431–440 (1996)
2. Babai, L., Fortnow, L., Nisan, N., Wigderson, A.: BPP has subexponential time simulations unless EXPTIME has publishable proofs. Computational Complexity 3(4), 307–318 (1993)
3. Blum, M., Micali, S.: How to generate cryptographically strong sequences of pseudo-random bits. SIAM J. on computing 13(4), 850–864 (1984)
4. Harnik, D., Raz, R.: Higher Lower Bounds for Monotone Size. In: 32nd STOC, pp. 191-201 (2000)
5. Impagliazzo, R.: Hard-core distributions for somewhat hard problems. In: 36th FOCS (1995)
6. Impagliazzo, R., Kabanets, V.: Derandomizing Polynomial Identity Tests means proving circuit lower bounds. In: 35th STOC (2003)
7. Impagliazzo, R., Levin, L., Luby, M.: Pseudorandom generators from any one-way function. In: 21st STOC (1989)
8. Impagliazzo, R., Shaltiel, R., Wigderson, A.: Near-optimal conversion of hardness into pseudo-randomness. In: 40th FOCS, pp. 181–190 (1999)
9. Impagliazzo, R., Wigderson, A.: P=BPP if E requires exponential circuits. In: 29th STOC, pp. 220–229 (1998)
10. Karchmer, M., Wigderson, A.: Monotone circuits for connectivity require super-logarithmic depth. SIAM J. on Discrete Mathematics 3(2), 255–265 (1990)
11. Nisan, N., Wigderson, A.: Hardness vs. Randomness. JCSS 49(2), 149–167 (1994)
12. O'Donnell, R.: Hardness Amplification Within NP. In: 34th STOC, pp. 751–760 (2002)
13. Raz, R., McKenzie, P.: Separation of the Monotone NC Hierarchy. Combinatorica 19(3), 403–435 (1999)
14. Razborov, A.: Lower bounds on the monotone complexity of some Boolean functions. Dokl. Akad. Nauk. SSSR 281(4), 598–607 (1985) (In Russian)
15. Shamir, A.: On the generation of cryptographically strong pseudo-random sequences. In: Even, S., Kariv, O. (eds.) ICALP 1981. LNCS, vol. 115. Springer, Heidelberg (1981)
16. Sudan, M., Trevisan, L., Vadhan, S.: Pseudorandom generators without the XOR lemma. In: 31st STOC, pp. 537–546 (1999)
17. Wegener, I.: The complexity of Boolean functions. John Wiley, Chichester (1987)
18. Yao, A.C.: Theory and applications of trapdoor functions. In: 23rd FOCS, pp. 80–91 (1982)

Random Generation and Enumeration of Bipartite Permutation Graphs

Toshiki Saitoh[1], Yota Otachi[2], Katsuhisa Yamanaka[3], and Ryuhei Uehara[1]

[1] School of Information Science, JAIST, Asahidai 1-1, Nomi, Ishikawa 923-1292, Japan
{toshikis,uehara}@jaist.ac.jp
[2] Department of Computer Science, Gunma University, 1-5-1 Tenjin-cho, Kiryu, Gunma 376-8515, Japan
otachi@comp.cs.gunma-u.ac.jp
[3] Graduate School of Information Systems, The University of Electro-Communications, Chofugaoka 1-5-1, Chofu, Tokyo 182-8585, Japan
yamanaka@is.uec.ac.jp

Abstract. Connected bipartite permutation graphs without vertex labels are investigated. First, the number of connected bipartite permutation graphs of n vertices is given. Based on the number, a simple algorithm that generates a connected bipartite permutation graph uniformly at random up to isomorphism is presented. Finally an enumeration algorithm of connected bipartite permutation graphs is proposed. The algorithm is based on the reverse search, and it outputs each connected bipartite permutation graph in O(1) time.

Keywords: Bipartite permutation graph, counting, Dyck path, enumeration, Motzkin path, random generation.

1 Introduction

Recently we have to process huge amounts of data in the area of data mining, bioinformatics, etc. In most cases, we have to use some certain structure to solve problems efficiently. We need three efficiencies to deal with the complex structure; it has to be represented efficiently, essentially different instances have to be enumerated efficiently, and its properties have to be checked efficiently. From the viewpoint of graph classes, the previously studied structures are relatively primitive. Although trees are widely investigated as a model of such structured data [6,10,12,14], there are few results for more complex graph classes. Recently, distance-hereditary graphs [11] and proper interval graphs [16] are investigated from this viewpoint.

In this paper, we investigate counting, random generation, and enumeration of a graph class called *bipartite permutation graphs*. More precisely, we aim to count, generate, and enumerate unlabeled connected bipartite permutation graphs. From the practical point of view, "unlabeled" and "connected" are reasonable properties to avoid redundancy. On the other hand, however, they are also challenges to developing efficient algorithms. Especially, unlabeled property requires us to avoid generating isomorphic graphs. In other words, we have to recognize isomorphic graphs and suppress generating/counting/enumerating them twice or more. Roughly speaking, the graph isomorphism problem has to be solved efficiently for our target graph classes in this context.

Y. Dong, D.-Z. Du, and O. Ibarra (Eds.): ISAAC 2009, LNCS 5878, pp. 1104–1113, 2009.

The graph isomorphism problem is one of well-known basic problems, and it is still hard on restricted graph classes [20]. There are two well known graph classes that the graph isomorphism problem can be solved in polynomial time; interval graphs [13] and permutation graphs [3]. Hence, they are the final goal in this framework. We mention that these graph classes have been widely investigated since they are very basic graph classes from the viewpoint of graph theory. Therefore many useful properties have been revealed, and many efficient algorithms have been developed for them (see, e.g., [2,7,17]). From the practical point of view, when an efficient algorithm for a graph class is developed and implemented, we need many graphs belonging to the class to check the reliability of the algorithm. Hence, for such popular graph classes, efficient random generator and enumerator are required. On the other hand, the counting of such graphs is rather mathematical. From the viewpoint of combinatorics, the counting of graphs having a certain structure is an important issue. In combinatorics, the notion of Dyck path is one of basic tools, and it appears in a number of areas [18,19]. One natural extension of the notion of Dyck path is known as Motzkin path; while a Dyck path is a sequence of $+1$ and -1, a Motzkin path is a sequence of $+1$, -1, and 0. We will show that an unlabeled connected bipartite permutation graph is strongly related to an extension of a Motzkin path, which is known as a 2-Motzkin path [5], that consists of $+1$, -1, $+0$, and -0. Our counting result also gives a new insight of this area.

Saitoh et al. have obtained such results for proper interval graphs which form a subclass of interval graphs [16]. We turn to bipartite permutation graphs that form a subclass of permutation graphs, and show the similar results for them. As we will see, bipartite permutation graphs have a certain structure, which can be seen as a generalization of the structure appearing in proper interval graphs implicitly. That is, developing some new nontrivial techniques based on the results in proper interval graphs, we advance the results in [16] to bipartite permutation graphs.

Due to space limitation, all proofs are omitted.

2 Preliminaries

Interval graph: A graph $G = (V, E)$ with $V = \{v_1, v_2, \ldots, v_n\}$ is an *interval graph* if there is a finite set of intervals $\mathcal{I} = \{I_{v_1}, I_{v_2}, \ldots, I_{v_n}\}$ on the real line such that $\{v_i, v_j\} \in E$ iff $I_{v_i} \cap I_{v_j} \neq \emptyset$ for each i and j with $0 < i, j \leq n$. We call the interval set $\mathcal{I}$ an *interval representation* of G. For each interval I, we denote by $L(I)$ and $R(I)$ the left and right endpoints of the interval, respectively. An interval representation is *proper* if no two distinct intervals I and J exist such that I properly contains J or vice versa. An interval graph is *proper* if it has a proper interval representation. If an interval graph G has an interval representation $\mathcal{I}$ s. t. every interval in $\mathcal{I}$ has the same length, G is said to be a *unit interval graph*. Such interval representation is called a *unit interval representation*. It is well known that proper interval graphs coincide with unit interval graphs [15]. That is, given a proper interval representation, we can transform it to a unit interval representation. A simple constructive way of the transformation can be found in [1]. We can assume without loss of generality that $L(I) \neq L(J)$ (and hence $R(I) \neq R(J)$), and $R(I) \neq L(J)$ for any two distinct intervals I and J in a unit interval representation $\mathcal{I}$.

Let Σ be an alphabet {'[', ']'}. We encode a unit interval representation $\mathcal{I}$ of a unit interval graph G by a string $s(\mathcal{I})$ in Σ^* as follows; we sweep the interval representation from left to right, and for each $I \in \mathcal{I}$ encode $L(I)$ and $R(I)$ by '[' and ']', respectively. We call the encoded string a *string representation* of G. We say that string x in Σ^* is *balanced* if the number of '['s in x equals that of ']'s. Clearly $s(\mathcal{I})$ is a balanced string of $2n$ letters. Using the construction in [1], $s(\mathcal{I})$ can be constructed from a proper interval representation $\mathcal{I}$ in O(n) time and vice versa since the ith '[' and the ith ']' give the left and right endpoints of the ith interval, respectively. (We assume that each interval representation is given by a list of the endpoints of intervals from left to right.)

We define '$\bar{[}$' = ']' and '$\bar{]}$' = '[' respectively. For two strings $x = x_1x_2 \cdots x_n$ and $y = y_1y_2 \cdots y_m$ in Σ^*, we say that x is *smaller* than y if (1) $n < m$, or (2) $n = m$ and there exists an index $i \in \{1, \ldots, n\}$ such that $x_{i'} = y_{i'}$ for all $i' < i$ and $x_i =$ '[' and $y_i =$ ']'. If x is smaller than y, we denote $x < y$. (This is so called "lexicographical order with length preferred.") For a string $x = x_1x_2 \cdots x_n$ we define the *reverse* $\bar{x}$ of x by $\bar{x} = \bar{x}_n\bar{x}_{n-1} \cdots \bar{x}_1$. A string x is *reversible* if $x = \bar{x}$. A connected proper interval graph G is said to be *reversible* if its string representation is reversible.

Lemma 1 (See, e.g., [4, Corollary 2.5]). *Let G be a connected proper interval graph, and $\mathcal{I}$ and $\mathcal{I}'$ be any two unit interval representations of G. Then either $s(\mathcal{I}) = s(\mathcal{I}')$ or $s(\mathcal{I}) = \overline{s(\mathcal{I}')}$ holds. That is, the string representation of a proper interval graph is unique up to isomorphism.*

Permutation graph: A graph $G = (V, E)$ with $V = \{1, 2, \ldots, n\}$ is said to be a *permutation graph* iff there is a permutation π over V such that $\{i, j\} \in E$ iff $(i-j)(\pi(i)-\pi(j)) < 0$. Intuitively, each vertex i in a permutation graph corresponds to a line ℓ_i joining two points on two parallel lines L_1 and L_2. Then two vertices i and j are adjacent iff the corresponding lines ℓ_i and ℓ_j intersect. The ordering of vertices gives the ordering of the points on L_1, and the ordering by permutation π over V gives the ordering of the points on L_2. We call the intersection model a *line representation* of the permutation graph. For two line representations $\mathcal{L}$ and $\mathcal{L}'$, suppose $\mathcal{L}$ contains (i, j) iff $\mathcal{L}'$ contains (i, j). Then we call them *isomorphic* and denote by $\mathcal{L} = \mathcal{L}'$.

When a permutation graph is bipartite, it is said to be a *bipartite permutation graph* (see Fig. 1). Then the following lemma holds:

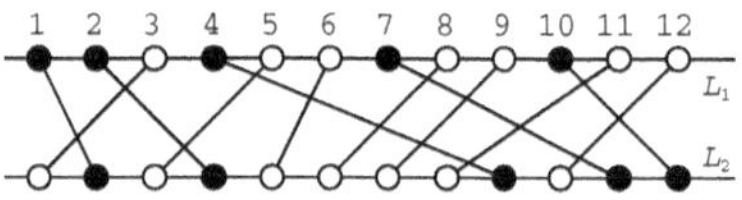

Fig. 1. Bipartite permutation graph

Lemma 2. *Let $G = (X, Y, E)$ be a connected bipartite permutation graph with $|X|, |Y| > 0$ and $\mathcal{L} = (L_1, L_2)$ its line representation. Without loss of generality, we assume that $v_1 \in X$ corresponds to $(1, i)$ for some i with $1 \le i \le n$. Then X and Y satisfy that $X = \{v_i \mid v_i$ corresponds to (i, j) with $i < j\}$ and $Y = \{v_i \mid v_i$ corresponds to (i, j) with $i > j\}$.*

Let $\mathcal{L} = (L_1, L_2)$ be a line representation of a bipartite permutation graph $G = (X, Y, E)$. For a connected bipartite permutation graph G, we can construct essentially equivalent representations by flipping $\mathcal{L}$. There are three operations that play important roles in this paper. On a *horizontal flip* $\mathcal{L}^H$ (H-flip for short) of $\mathcal{L}$, each line (i, j) on $\mathcal{L}$ is mapped to the line $(n-i+1, n-j+1)$. On a *vertical flip* $\mathcal{L}^V$ (V-flip for short) of $\mathcal{L}$, each line (i, j) on $\mathcal{L}$ is mapped to the line (j, i). For a line representation $\mathcal{L}$, $(\mathcal{L}^H)^V = (\mathcal{L}^V)^H$ gives us a *rotation* of $\mathcal{L}$. Hence we denote the line representation by $\mathcal{L}^R$ after this operation.

One important property is that they are unique up to isomorphism like Lemma 1:

Lemma 3. *Let $G = (V, E)$ be a connected bipartite permutation graph, and $\mathcal{L}$ and $\mathcal{L}'$ any two line representations of G. Then one of $\mathcal{L} = \mathcal{L}'$, $\mathcal{L} = \mathcal{L}'^H$, $\mathcal{L} = \mathcal{L}'^V$, and $\mathcal{L} = \mathcal{L}'^R$ holds. That is, the line representation of G is unique up to isomorphism.*

Let $G = (V, E)$ be a connected bipartite permutation graph, and $\mathcal{L}, \mathcal{L}^H, \mathcal{L}^V, \mathcal{L}^R$ its four line representations. Then some of them can be isomorphic; G is *H-symmetric*, *V-symmetric*, and *R-symmetric* if $\mathcal{L} = \mathcal{L}^H$, $\mathcal{L} = \mathcal{L}^V$, and $\mathcal{L} = \mathcal{L}^R$, respectively.

Here, we map each representation $\mathcal{L}$ to a string $s(\mathcal{L})$ in Σ^* as follows. We first sweep the endpoints from left to right on L_1, and construct a string $s_1(\mathcal{L})$ by adding '[' when the endpoint is in X, and ']' when the endpoint is in Y (e.g., $s_1(\mathcal{L})$ = [[][]][]][]] in Fig. 1). Next we sweep the endpoints from left to right on L_2, and construct a string $s_2(\mathcal{L})$ by adding '[' when the endpoint is in Y, and ']' when the endpoint is in X (e.g., $s_2(\mathcal{L})$ = [][][[[[][]] in Fig. 1). Finally, we concatenate $s_2(\mathcal{L})$ after $s_1(\mathcal{L})$ and obtain the resultant string (e.g., $s(\mathcal{L})$ = [[][]][]][]][][][[[[][]] in Fig. 1).

Using the string, we define a *canonical* representation of G as follows. We first suppose that all strings $s(\mathcal{L}), s(\mathcal{L}^H), s(\mathcal{L}^V), s(\mathcal{L}^R)$ are distinct. Then the canonical representation is the one corresponding to the smallest string. When G satisfies exactly one symmetricalness with respect to H-flip, V-flip, or rotation, then four possible representations gives two distinct strings. Then the canonical representation is the one corresponding to the smaller string. If G satisfies two symmetricalnesses, the last symmetricalness is also satisfied. Hence, in the case, four representations are isomorphic and this gives the unique canonical representation. By Lemma 3, this rule gives us a one-to-one mapping between bipartite permutation graphs and canonical representations.

Dyck path, Motzkin path, and 2-Motzkin path: A path in the (x, y) plane from $(0, 0)$ to $(2n, 0)$ with steps $(1, 1)$ and $(1, -1)$ is called a *Dyck path* of length $2n$ if it never pass below the x-axis. It is well known that the number of Dyck paths of length n is given by the nth *Catalan number* $C(n) := \frac{1}{n+1}\binom{2n}{n}$ (see [19, Corollary 6.2.3] for further details). We will use one of the generalized notions of Catalan number; $C(n, k) := \frac{k+1}{n+1}\binom{n+1}{(n-k)/2}$, which gives us the number of subpaths of Dyck paths from $(0, 0)$ to (n, k). This can be obtained by a generalized Raney's lemma about m-Raney sequences with letting $m = 2$; see [8, Equation (7.69), p. 349] for further details. A path in the (x, y) plane from $(0, 0)$ to $(n, 0)$ with steps $(1, 0)$, $(1, 1)$, and $(1, -1)$ is called a *Motzkin path* of length n if it never go below the x-axis (see [19, Exercise 6.38] for further details). The number of Motzkin paths of length n is called *Motzkin number* $\mathcal{M}(n)$; e.g., $\mathcal{M}(1) = 1, \mathcal{M}(2) = 2, \mathcal{M}(3) = 4, \mathcal{M}(4) = 9, \mathcal{M}(5) = 21, \mathcal{M}(6) = 51$. A *2-Motzkin* path is a Motzkin path that has two kinds of step $(1, 0)$. We distinguish them by $(1, +0)$ and $(1, -0)$. Deutsch and Shapiro show that 2-Motzkin paths have correspondences to ordered trees and others [5].

In paths above, each step consists of $(1, x)$ for some x in $\{\pm 1, \pm 0\}$. Hence we will denote a path by a sequence of such integers x in $\{\pm 1, \pm 0\}$.

Machine Model: Time complexity is measured by the number of arithmetic operations. Especially we assume that each binomial coefficient and each (generalized) Catalan number can be computed in O(1) time. Moreover we assume that the basic arithmetic operations of these numbers can be done in O(1) time. This assumption is out of the standard RAM model. We have to multiply the time complexity of calculation of these numbers to the complexities we show to obtain the time complexity in the standard RAM model. We employ the assumption only in Section 3 to simplify the discussion. The enumeration algorithm in Section 4 does not require the assumption, and all the results are valid on the standard RAM model.

3 Counting and Random Generation

Let $P(n)$ be the set of permutations corresponding to connected bipartite permutation graphs of n vertices, and $\mathcal{B}_n$ the set of distinct (up to isomorphism) connected bipartite permutation graphs of n vertices. We denote a (not necessarily canonical) line representation of a permutation π by $\mathcal{L}_\pi = (L_1, L_2)$, and the graph of π by $G_\pi = (X, Y, E)$. Without loss of generality, we assume that X contains the vertex corresponding to $(1, \pi(1))$ in $\mathcal{L}_\pi$ for $\pi(1) > 1$. Now, we construct a 2-Motzkin path as follows. For each i with $1 \le i \le n$, we see the endpoints at i on L_1 and L_2. Let p_i and q_i be the endpoints on L_1 and L_2, respectively. We say that p_i is in X (and Y) if p_i is the endpoint of a vertex corresponding to $(i, \pi(i))$ in X (and Y, respectively). Similarly, we say that q_i is in X (and Y) if q_i is the endpoint of a vertex corresponding to $(\pi^{-1}(i), i)$ in X (and Y, respectively). If G_π is not connected, in each connected component, we assume that the vertex corresponding to the leftmost point on L_1 belongs to X. Then the value z_i is defined as follows;

$$z_i = \begin{cases} +1 & \text{if } p_i \text{ is in } X \text{ and } q_i \text{ is in } Y, \\ -1 & \text{if } p_i \text{ is in } Y \text{ and } q_i \text{ is in } X, \\ +0 & \text{if } p_i \text{ and } q_i \text{ are in } X, \\ -0 & \text{if } p_i \text{ and } q_i \text{ are in } Y. \end{cases}$$

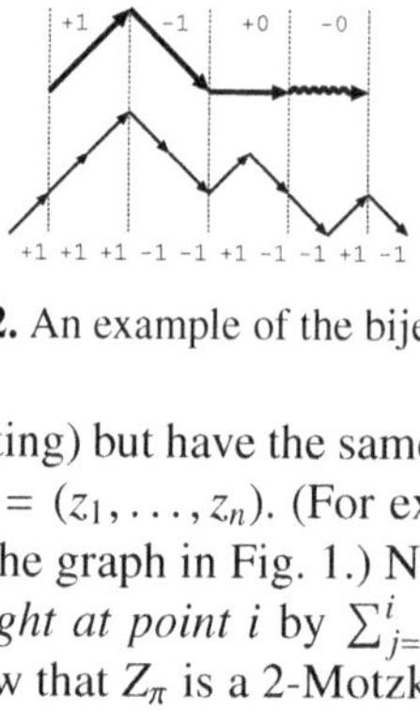

Fig. 2. An example of the bijection

That is, two values +0 and −0 are distinguished (for counting) but have the same value. From the sequence $z_1, \ldots, z_n$, we can consider a path $Z_\pi = (z_1, \ldots, z_n)$. (For example, $Z_\pi = (+1, +0, -0, +0, -0, -0, +1, -0, -1, +1, -1, -1)$ for the graph in Fig. 1.) Note that $\pi = \pi'$ iff $Z_\pi = Z_{\pi'}$. For the path Z_π, we define its *height at point* i by $\sum_{j=1}^{i} z_j$. To simplify, we define that the height at point 0 is 0. We show that Z_π is a 2-Motzkin path that has positive height at point i, $1 < i < n$, iff $\pi \in P(n)$. To this end, we need a property of connected permutation graphs.

Lemma 4 ([9, Lemma 3.2]). *Let π be a permutation on $\{1, \ldots, n\}$. Then G_π is disconnected iff there exists $k < n$ such that $\{\pi(1), \pi(2), \ldots, \pi(k)\} = \{1, 2, \ldots, k\}$.*

Then we have the following lemma.

Lemma 5. *A sequence $Z = (z_1, \ldots, z_n)$ on the alphabet $\{+1, -1, +0, -0\}$ is constructed from $\pi \in P(n)$ in the above way iff Z is a 2-Motzkin path such that Z has height 0 at point 0 and n, and positive height at point i with $0 < i < n$.*

From the above characterization, we can count the number of elements in $P(n)$. Deutsch and Shapiro [5] have shown the following bijection between 2-Motzkin paths of length

n and Dyck paths of length $2(n+1)$: In a 2-Motzkin path, we replace $+1$ by $(+1,+1)$, -1 by $(-1,-1)$, $+0$ by $(+1,-1)$, and -0 by $(-1,+1)$; Then add $+1$ before the obtained sequence, and add -1 after the sequence. Fig. 2 shows an example. Note that a 2-Motzkin path has height k at point i iff the corresponding Dyck path has height $2k+1$ at point $2i+1$. The bijection gives the following lemma, which yields $|P(n)| = C(n-1)$.

Lemma 6 **([5]).** *The number of 2-Motzkin paths of length n is $C(n+1)$.*

We can show that the bijection is also a bijection for restricted paths. For $z \in \{+1,-1,+0,-0\}$, we define $-z$ naturally; $-z = \pm b$ iff $z = \mp b$ for $b \in \{0,1\}$. A Dyck path $D = (d_1, \ldots, d_{2n})$ is *symmetric* if $z_i = -z_{n-i+1}$ for $1 \le i \le n$.

Lemma 7 **([16]).** *The number of symmetric Dyck paths of length $2n$ is $\binom{n}{\lfloor n/2 \rfloor}$.*

A 2-Motzkin path $Z = (z_1, \ldots, z_n)$ is *semi-symmetric* if $z_i = -z_{n-i+1}$ for $1 \le i \le n$, and Z is *symmetric* if $z_i = -z_{n-i+1}$ for $z_i \in \{+1,-1\}$ and $z_i = z_{n-i+1}$ for $z_i \in \{+0,-0\}$. A 2-Motzkin path can be semi-symmetric only if its length is even. Obviously, the bijection is also a bijection between symmetric 2-Motzkin paths of length n and symmetric Dyck paths of length $2(n+1)$. Furthermore, if n is even, there is a bijection between semi-symmetric 2-Motzkin paths of length n and symmetric Dyck paths of length $2(n+1)$. From the above observation and Lemma 7, we have the following corollary.

Corollary 1. *The number of symmetric 2-Motzkin paths of length n is $\binom{n+1}{\lfloor (n+1)/2 \rfloor}$. If n is even, the number of semi-symmetric 2-Motzkin paths of length n is also $\binom{n+1}{\lfloor (n+1)/2 \rfloor}$.*

Any given $\pi \in P(n)$, Lemma 3 implies that there exist at most four line representations $\mathcal{L}_\pi$, $\mathcal{L}_\pi^H$, $\mathcal{L}_\pi^V$, and $\mathcal{L}_\pi^R$ for a graph G_π. We define four subsets of $P(n)$ as follows: (1) $P^H(n) = \{\pi \in P(n) \mid \mathcal{L}_\pi \text{ is H-symmetric}\}$, (2) $P^V(n) = \{\pi \in P(n) \mid \mathcal{L}_\pi \text{ is V-symmetric}\}$, (3) $P^R(n) = \{\pi \in P(n) \mid \mathcal{L}_\pi \text{ is R-symmetric}\}$, and (4) $P^F(n) = P^H(n) \cap P^R(n) \cap P^V(n)$.

Proposition 1. *(1) If n is odd, $P^H(n)$ and $P^V(n)$ are empty. (2) $P^F(n) = P^H(n) \cap P^V(n) = P^V(n) \cap P^R(n) = P^R(n) \cap P^H(n)$.*

Lemma 8. $|\mathcal{B}_n| = \frac{1}{4}\left(|P(n)| + |P^H(n)| + |P^V(n)| + |P^R(n)|\right)$.

Lemma 8 implies that it suffices to count the elements of $P(n)$, $P^H(n)$, $P^V(n)$, and $P^R(n)$ to show the size of $\mathcal{B}_n$. For the random generation, $|P^F(n)|$ is also necessary.

Lemma 9. *(1) $|P^V(n)| = C(n/2-1)$ for even n. (2) $|P^R(n)| = \binom{n-1}{\lfloor (n-1)/2 \rfloor}$. (3) $|P^H(n)| = \binom{n-1}{\lfloor (n-1)/2 \rfloor}$ for even n. (4) $|P^F(n)| = \binom{(n-2)/2}{\lfloor (n-2)/4 \rfloor}$ for even n.*

Lemmas 8, 9, and Proposition 1 together show the number of elements of $\mathcal{B}_n$. We use a well-known relation $2\binom{2m-1}{m-1} = \binom{2m}{m}$ for the even case.

Theorem 1. *For $n \ge 2$, the number of connected bipartite permutation graphs of n vertices is*

$$|\mathcal{B}_n| = \begin{cases} \frac{1}{4}\left(C(n-1) + C(n/2-1) + \binom{n}{n/2}\right) & \text{if } n \text{ is even,} \\ \frac{1}{4}\left(C(n-1) + \binom{n\ 1}{(n-1)/2}\right) & \text{if } n \text{ is odd.} \end{cases}$$

Theorem 2. *For any given positive integer n, a connected bipartite permutation graph with n vertices can be generated uniformly at random in $O(n)$ time and $O(n)$ space.*

The complexity in Theorem 2 is measured by the number of operations. However, the algorithm runs in $O(n)$ expected time even in the standard RAM model.

4 Enumeration

In this section we give an efficient algorithm to enumerate all bipartite permutation graphs of n vertices. Our algorithm can enumerate such graphs in O(1) time for each.

Our approach is to repeatedly enumerate all bipartite permutation graphs of the specified number of vertices. If we can enumerate all bipartite permutation graphs with $p = |X|$ and $q = |Y|$, such graphs of n vertices can be enumerated by repeating the method for each pair of $(p, q) = (\lceil \frac{n}{2} \rceil, \lfloor \frac{n}{2} \rfloor), (\lceil \frac{n}{2} \rceil + 1, \lfloor \frac{n}{2} \rfloor - 1), \ldots, (n-1, 1)$. By the above observation and Lemma 3, it is sufficient to enumerate all canonical representations of bipartite permutation graphs with $p = |X|$ and $q = |Y|$.

We first define a tree structure, *family tree*, among the set of canonical representations. The algorithm traverses the family tree efficiently. We need some definitions. Let $S_{p,q}$ be the set of canonical representations of bipartite permutation graphs of p vertices in X and q vertices in Y. We assume $p \geq q$ without loss of generality. The *root* $R_{p,q}$ in $S_{p,q}$ is the smallest representation in $S_{p,q}$; $s(R_{p,q}) = [[\cdots[]]\cdots][[\cdots[]]\cdots]$ (Fig. 3). As we will see, the root corresponds to the root vertex in a tree structure among $S_{p,q}$.

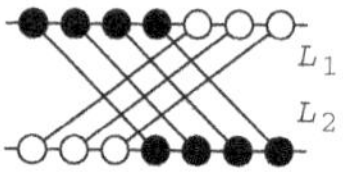

Fig. 3. $R_{4,3}$ in $S_{4,3}$

Let $\mathcal{L} = (L_1, L_2)$ be a representation in $S_{p,q} \setminus \{R_{p,q}\}$. Let $s(\mathcal{L}) = x_1 x_2 \cdots x_{2n}$, $s_1(\mathcal{L}) = x_1 x_2 \cdots x_n$, and $s_2(\mathcal{L}) = x_{n+1} x_{n+2} \cdots x_{2n}$. Now we define "the parent" $P(\mathcal{L})$ of the representation $\mathcal{L}$ in $S_{p,q}$ as follows. We have two cases.

Case (a): $s_1(\mathcal{L}) \neq s_1(R_{p,q})$. Let i be the index of $s_1(\mathcal{L})$ s. t. x_i = ']' and $x_{i'}$ = '[' for all $i' < i$, and j be the index of $s_1(\mathcal{L})$ s. t. x_j = '[' and $x_{j'}$ = ']' for all $i \leq j' < j$. Then j is the *swappable point* of $\mathcal{L}$. $P(\mathcal{L})$ is the representation obtained from $\mathcal{L}$ by swapping two endpoints at $j-1$ and j on L_1 (Fig. 4(a)).

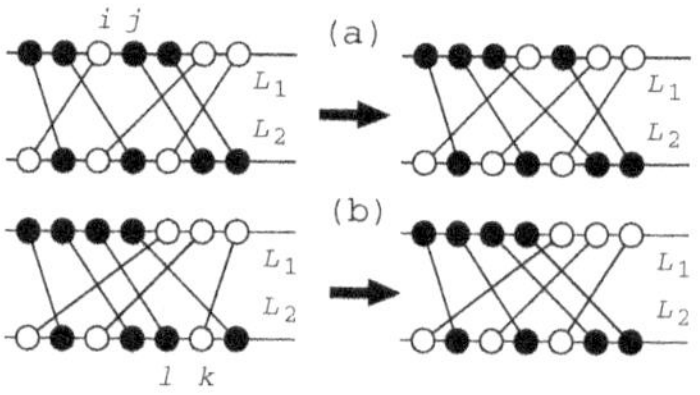

Fig. 4. Examples of the parents

Case (b): $s_1(\mathcal{L}) = s_1(R_{p,q})$. In this case we define $P(\mathcal{L})$ by swapping two points on L_2 at the *swappable point* defined similar to the case (a) (Fig. 4(b)).

$P(\mathcal{L})$ is called the *parent* of $\mathcal{L}$ and $\mathcal{L}$ is called a *child* of $P(\mathcal{L})$. We can observe that $s(P(\mathcal{L}))$ is smaller than $s(\mathcal{L})$, and the parent $P(\mathcal{L})$ of $\mathcal{L}$ in $S_{p,q} \setminus \{R_{p,q}\}$ is always defined. Since $\mathcal{L}$ is canonical, so is $P(\mathcal{L})$. The next lemma shows we finally obtain the root in $S_{p,q}$ by repeatedly finding the parent.

Lemma 10. *Let $\mathcal{L}$ be a representation in $S_{p,q} \setminus \{R_{p,q}\}$. The sequence obtained by repeatedly finding the parent ends up with the root $R_{p,q}$.*

By merging all these sequences we have the *family tree* $T_{p,q}$ of $S_{p,q}$; the root of $T_{p,q}$ corresponds to $R_{p,q}$, the vertices of $T_{p,q}$ correspond to representations in $S_{p,q}$, and each edge corresponds to a relation between a representation in $S_{p,q} \setminus \{R_{p,q}\}$ and its parent.

Now we give an algorithm that enumerates all representations in $S_{p,q}$. The algorithm traverses a family tree and enumerates canonical representations corresponding to the vertices of the family tree. To traverse a family tree, we design finding all children of

Algorithm 1. find-all-children($\mathcal{L}$)

```
1 begin
2   for each nominated point i on L1 do
3     if L1[i] is connected and canonical then find-all-children(L1[i])
4   end
5   for each nominated point i on L2 do
6     if L2[i] is connected and canonical then find-all-children(L2[i])
7   end
8 end
```

a given canonical representation. We need some definitions. $\mathcal{L}_1[i]$ is the line representation obtained from $\mathcal{L}$ by swapping two endpoints at i and $i+1$ on L_1, and similarly $\mathcal{L}_2[i]$ is the line representation obtained from $\mathcal{L}$ by swapping two endpoints at $i-1$ and i on L_2. If $\mathcal{L} = P(\mathcal{L}_1[i])$ (and $\mathcal{L} = P(\mathcal{L}_2[i])$), we say i is a *nominated point* on L_1 (and L_2, respectively). $\mathcal{L}_1[i]$ (and $\mathcal{L}_2[i]$) is a child of $\mathcal{L}$ only if i is a nominated point on L_1 (and L_2) and $\mathcal{L}_1[i]$ (and $\mathcal{L}_2[i]$, respectively) is connected and canonical.

For a string $s(\mathcal{L}) = x_1x_2\cdots x_{2n}$, we define the *connectivity value* $c(i)$ for $i = 0, 1, \ldots, 2n$ as follows: $c(0) = c(n) = 0$, and

$$c(i) = \begin{cases} c(i-1)+1 & \text{if } (x_i = \text{`['} \text{ and } i < n) \text{ or } (x_i = \text{`]'} \text{ and } i > n) \\ c(i-1)-1 & \text{if } (x_i = \text{`]'} \text{ and } i < n) \text{ or } (x_i = \text{`['} \text{ and } i > n) \end{cases}$$

Intuitively, $c(i)$ for $i < n$ is the number of '['s minus the number of ']'s in $x_1x_2\cdots x_i$, and $c(i)$ for $i > n$ is the number of ']'s minus the number of '['s in $x_{n+1}x_{n+2}\cdots x_i$. A bipartite permutation graph is connected iff we have $c(i) \neq c(n+i)$ for each $i = 1, 2, \ldots, n-1$. We say $\mathcal{L}$ is *connected* if $c(i) \neq c(n+i)$ for each $i = 1, 2, \ldots, n-1$.

All children can be enumerated as follows. We construct $\mathcal{L}_1[i]$ for each $i = 1, 2, \ldots, n-1$, then check whether or not (1) i is a nominated point on L_1, (2) $\mathcal{L}_1[i]$ is connected and (3) $\mathcal{L}_1[i]$ is canonical. If all conditions are satisfied, $\mathcal{L}_1[i]$ is a child. Similarly, we check whether or not $\mathcal{L}_2[i]$ is a child for each $i = 2, 3, \ldots, n$. We show that (1) the list of nominated points can be maintained efficiently, and (2) efficient way to check if a representation is connected and canonical.

Lemma 11. *(1) Let $\mathcal{L} = (L_1, L_2)$ be a representation in $S_{p,q}$. There exist at most 3 nominated points on L_1 and L_2. (2) Given $\mathcal{L}$ and its nominated points, we can construct the list of nominated points of each child in $O(1)$ time.*

Now we have **Algorithm** 1, that generates all children of a given representation $\mathcal{L}$. For each nominated point i on L_1 (and L_2), it first checks if $\mathcal{L}_1[i]$ (and $\mathcal{L}_2[i]$) is connected and canonical, and next recursively calls it for $\mathcal{L}_1[i]$ (and $\mathcal{L}_2[i]$, respectively) if it satisfies the conditions. By calling the algorithm recursively at $R_{p,q}$ in $S_{p,q}$, we can traverse the family tree $T_{p,q}$ and enumerate all representations in $S_{p,q}$.

By Lemma 11, steps 2 and 5 can be done in O(1) time in each recursive call. The remaining task is checking whether or not $\mathcal{L}$ is connected and canonical efficiently.

We first consider the check of connectivity of a representation. By symmetry we only consider $\mathcal{L}_1[i]$ without loss of generality. Assume $\mathcal{L}$ is connected. Then $\mathcal{L}_1[i]$ is connected only if $c(i) \neq c(n+i)$ and $c(i+1) \neq c(n+i+1)$. We can check such conditions in O(1) time using an array of size $2n$ to maintain the sequences of connectivity values of $\mathcal{L}_1[i]$. Update of the array also can be done in O(1) time. Therefore, the connectivity of $\mathcal{L}_1[i]$ can be checked in O(1) time.

Next we check whether or not $\mathcal{L}$ is canonical. When $p \neq q$, $s(\mathcal{L})$ is canonical if $s(\mathcal{L})$ is the smallest string among $s(\mathcal{L}^V)$, $s(\mathcal{L}^H)$ and $s(\mathcal{L}^R)$. If $p = q$, we need more discussions. Let $\mathcal{L}$ be a representation in $S_{p,q}$ and G be the bipartite permutation graph corresponding to $\mathcal{L}$. Then there exists a line representation $\mathcal{L}'$ obtained from $\mathcal{L}$ by swapping lines corresponding to vertices in X and ones in Y. Similarly, we denote by $\mathcal{L}^{V'}$, $\mathcal{L}^{H'}$, $\mathcal{L}^{R'}$ the representations obtained from $\mathcal{L}^V$, $\mathcal{L}^H$, $\mathcal{L}^R$ by swapping lines corresponding to vertices in X and ones in Y, respectively. Then $\mathcal{L}$ is canonical iff $s(\mathcal{L})$ is the smallest string among $s(\mathcal{L}^V)$, $s(\mathcal{L}^H)$, $s(\mathcal{L}^R)$, $s(\mathcal{L}')$, $s(\mathcal{L}^{V'})$, $s(\mathcal{L}^{H'})$ and $s(\mathcal{L}^{R'})$. Using a similar idea in [16], we have the following lemma.

Lemma 12. *One can determine whether or not $\mathcal{L} = (L_1, L_2)$ is canonical in $O(1)$ time.*

Therefore steps 3 and 6 in **Algorithm** 1 can be computed in O(1) time.

Lemma 13. *Our algorithm uses $O(n)$ space and runs in $O(|S_{p,q}|)$ time.*

By Lemma 13, our algorithm generates each representation in O(1) time "on average". **Algorithm** 1 may return from the deep recursive calls without outputting any representation after generating a representation corresponding to the leaf of a large subtree in the family tree. This delay can be canceled by outputting the representations in the "prepostorder" in which representation are outputted in the preorder (and postorder) at the vertices of odd (and even, respectively) depth of a family tree (see [12] for the further details). Thus we have the following lemma.

Lemma 14. *After outputting the root in $O(n)$ time, our algorithm enumerates every representation in $S_{p,q}$ in $O(1)$ time in worst case.*

Now we turn to enumerate all canonical representations corresponding to bipartite permutation graphs of n vertices. By applying Lemma 14 for each $(p,q) = (\lceil \frac{n}{2} \rceil, \lfloor \frac{n}{2} \rfloor), (\lceil \frac{n}{2} \rceil + 1, \lfloor \frac{n}{2} \rfloor - 1), \ldots, (n-1, 1)$ in this order, we can enumerate all representations; every non-root representation is generated in O(1) time. However, $R_{p,q}$ in $S_{p,q}$ is not constructed from the last outputted representation in $S_{p-1,q+1}$ in O(1) time.

This delay can be canceled as follows. Let $\mathcal{L} = (L_1, L_2)$ be a representation in $S_{p,q}$. Then $\mathcal{L}$ is *jump representation* if $s_1(\mathcal{L}) = s_1(R_{p,q})$ and $s_2(\mathcal{L}) = []]\cdots][[\cdots[]$ (see Fig. 5). When jump representation in $S_{p,q}$ is generated, we construct a representation $\mathcal{K}$ in $S_{p+1,q-1}$ by swapping the three lines (p, n), $(n-1, n-2)$, $(n, n-1)$ to $(p, n-1)$, $(n-1, n)$, $(n, n-2)$, respectively. We note that the line $(n-1, n-2)$ is switched to a line corresponding to a vertex in X, and $\mathcal{K}$ can be generated from $\mathcal{L}$ in O(1) time. Then we enumerate all representations in $S_{p+1,q-1}$ by traversing $T_{p+1,q-1}$ as follows. After $\mathcal{K}$ is generated, the descendants of $\mathcal{K}$ in $T_{p+1,q-1}$ are enumerated by

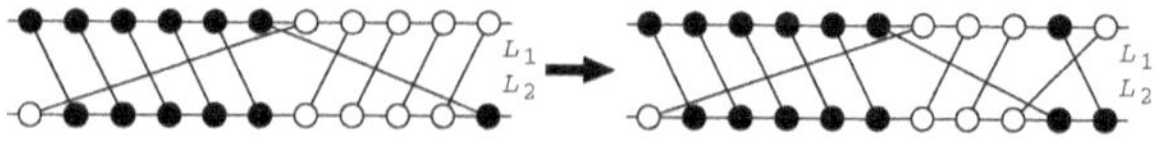

Fig. 5. Construction of a representation in $S_{7,4}$ from the jump representation in $S_{6,5}$

Algorithm 1, and we construct $P(\mathcal{K})$. Then we traverse the descendants of $P(\mathcal{K})$ except the subtree rooted at $\mathcal{K}$ and construct $P(P(\mathcal{K}))$. We repeat this process until the root is generated. We note that $P(\mathcal{K})$ can be generated in O(1) time by maintaining the swappable point and its data structure can be updated in O(1) time.

We note that (1) swapping two endpoints of a canonical representation corresponds to adding or removing one edge in the corresponding graph and (2) a graph can be constructed from the graph corresponding to a jump representation by a constant number of operations to add and remove edges. Therefore we have the following theorem.

Theorem 3. *(1) After outputting the root in* $S_{\lceil \frac{n}{2} \rceil, \lfloor \frac{n}{2} \rfloor}$, *one can enumerate every canonical representation of a bipartite permutation graph of n vertices in* $O(1)$ *time. (2) The algorithm enumerates every connected bipartite permutation graph of n vertices in* $O(1)$ *time.*

References

1. Bogart, K.P., West, D.B.: A short proof that 'proper=unit'. Disc. Math. 201, 21–23 (1999)
2. Brandstädt, A., Le, V.B., Spinrad, J.P.: Graph Classes: A Survey. SIAM, Philadelphia (1999)
3. Colbourn, C.J.: On Testing Isomorphism of Permutation Graphs. Networks 11, 13–21 (1981)
4. Deng, X., Hell, P., Huang, J.: Linear-time Representation Algorithms for Proper Circular-arc Graphs and Proper Interval Graphs. SIAM J. on Comp. 25(2), 390–403 (1996)
5. Deutsch, E., Shapiro, L.W.: A bijection between ordered trees and 2-Motzkin paths and its many consequences. Disc. Math. 256(3), 655–670 (2002)
6. Geary, R., Rahman, N., Raman, R., Raman, V.: A Simple Optimal Representation for Balanced Parentheses. In: Sahinalp, S.C., Muthukrishnan, S.M., Dogrusoz, U. (eds.) CPM 2004, vol. 3109, pp. 159–172. Springer, Heidelberg (2004)
7. Golumbic, M.C.: Algorithmic Graph Theory and Perfect Graphs. Annals of Disc. Math. 57 (2004)
8. Graham, R.L., Knuth, D.E., Patashnik, O.: Concrete Mathematics. Addison-Wesley Publishing Company, Reading (1989)
9. Koh, Y., Ree, S.: Connected permutation graphs. Disc. Math. 307(21), 2628–2635 (2007)
10. Nakano, S.-i.: Efficient generation of plane trees. IPL 84(3), 167–172 (2002)
11. Nakano, S.-i., Uehara, R., Uno, T.:: A new approach to graph recognition and applications to distance-hereditary graphs. J. of Computer Science and Technology 24(3), 517–533 (2009)
12. Knuth, D.E.: Generating All Trees, History of Combinatorial Generation. Fascicle 4 of The Art of Computer Programming, vol. 4. Addison-Wesley Publishing Company, Reading (2005)
13. Lueker, G.S., Booth, K.S.: A Linear Time Algorithm for Deciding Interval Graph Isomorphism. J. of the ACM 26(2), 183–195 (1979)
14. Munro, J.I., Raman, V.: Succinct Representation of Balanced Parentheses and Static Trees. SIAM J. on Comp. 31, 762–776 (2001)
15. Roberts, F.S.: Indifference graphs. In: Harary, F. (ed.) Proof Techniques in Graph Theory, pp. 139–146. Academic Press, London (1969)
16. Saitoh, T., Yamanaka, K., Kiyomi, M., Uehara, R.: Random Generation and Enumeration of Proper Interval Graphs. In: Das, S., Uehara, R. (eds.) WALCOM 2009. LNCS, vol. 5431, pp. 177–189. Springer, Heidelberg (2009)
17. Spinrad, J.P.: Efficient Graph Representations. AMS (2003)
18. Stanley, R.P.: Enumerative Combinatorics, Cambridge, vol. 1 (1997)
19. Stanley, R.P.: Enumerative Combinatorics, Cambridge, vol. 2 (1999)
20. Uehara, R., Toda, S., Nagoya, T.: Graph Isomorphism Completeness for Chordal Bipartite Graphs and Strongly Chordal Graphs. Disc. App. Math. 145(3), 479–482 (2004)

A Combinatorial Algorithm for Horn Programs*

R. Chandrasekaran[1] and K. Subramani[2]

[1] Department of Computer Science,
University of Texas at Dallas,
Dallas, TX
chandra@utdallas.edu
[2] LDCSEE,
West Virginia University,
Morgantown, WV
ksmani@csee.wvu.edu

Abstract. In this paper, we design and analyze a simple, greedy algorithm for a class of linear programs called Horn programs. This algorithm, which we term as the "Lifting Algorithm", is a variant of the Stressing Algorithm proposed for Difference Constraint systems in [5] and runs in time $O(m \cdot n^2)$ on a Horn system with m constraints and n variables. Inasmuch as Horn constraints subsume difference constraints, and all known algorithms for the problem of checking feasibility in Difference Constraint Systems run in time $\Omega(m \cdot n)$, the running time of our algorithm is only a factor n worse than the best known running time for checking feasibility in Difference Constraint Systems. Horn programs arise in a number of application areas including econometrics and program verification; consequently, their study is well-motivated. An important feature of our algorithm is that it uses only one operator, viz., addition. We also show that our algorithm can identify the linear and lattice point feasibility of Extended Horn Systems in $O(m \cdot n^2)$ time.

1 Introduction

In this paper, we analyze the "Lifting Algorithm" for solving a class of linear programs called Horn programs. Horn programs are a specialized class of linear programs that generalize three classes of constraints, viz., propositional Horn formulas, renamable Horn formulas and Difference constraint systems [2]. Horn programs arise in a number of application domains such as Program Verification and Econometrics; hence, algorithmic improvements in feasibility checking of these programs provide immediate impact. To the best of our knowledge, this problem was solved through linear programming techniques to date [6].

* This research was supported in part by the Air-Force Office of Scientific Research under contract FA9550-06-1-0050 and in part by the National Science Foundation through Award CCF-0827397.

Y. Dong, D.-Z. Du, and O. Ibarra (Eds.): ISAAC 2009, LNCS 5878, pp. 1114–1123, 2009.

2 Statement of Problem and Preliminaries

Let

$$\mathbf{A} \cdot \boldsymbol{x} \geq \boldsymbol{b}, \ \boldsymbol{x} \geq \mathbf{0} \tag{1}$$

denote a polyhedral system.

In System (1), $\mathbf{A}$ is an $m \times n$ integral matrix, $\boldsymbol{b}$ is an integral m-vector and $\boldsymbol{x} = [x_1, \ x_2, \ \ldots x_n]^T$ is a variable n-vector.

Definition 1. *System (1) is said to be a Difference Constraint System (DCS), if each row of* $\mathbf{A}$ *contains at most two non-zero entries with one of these entries being* 1 *and the other being* -1.

Definition 2. *System (1) is said to be a Horn system or a Horn polytope if*

(i) The entries in $\mathbf{A}$ *belong to the set* $\{0, 1, -1\}$.
(ii) Each row contains at most one positive entry.

The matrix $\mathbf{A}$ *is said to satisfy the Horn structure.*

Note that a Horn system could include *absolute* constraints, i.e., constraints of the form: $x_1 \geq 5$ and $-x_2 \geq -6$. However, observe that a constraint of the form: $x_1 \geq 5$ can be replaced by a constraint of the form: $x_1' \geq 0$, where $x_1' = x_1 - 5$. The idea is that the resultant constraint system is feasible if and only if the original system is. We will show later that any constraint with no positive coefficient can be set aside until a solution to the remaining system is found. Accordingly, we can assume that the Horn system under consideration does not have any absolute constraints.

Let System (1) denote a Horn system. Assume that some column of $\mathbf{A}$ (say $\mathbf{A}_{.1}$) has no negative entry. We can then rewrite System (1) as:

$$\mathbf{A} = \begin{array}{|c|c|} \hline \vec{1} & \mathbf{C} \\ \hline \vec{0} & \hat{\mathbf{A}} \\ \hline \end{array} ; \ \vec{\mathbf{b}} = \begin{array}{|c|} \hline \vec{\tilde{\mathbf{b}}} \\ \hline \vec{\hat{\mathbf{b}}} \\ \hline \end{array}$$

Lemma 1. *System (1) is feasible if and only if the system*

$$\hat{\mathbf{A}} \cdot \vec{\hat{\mathbf{x}}} \geq \vec{\hat{\mathbf{b}}}, \ \vec{\hat{\mathbf{x}}} \geq \mathbf{0}$$

is feasible.

Proof: Clearly

$$\hat{\mathbf{A}} \cdot \vec{\hat{\mathbf{x}}} \geq \vec{\hat{\mathbf{b}}}, \ \vec{\hat{\mathbf{x}}} \geq \mathbf{0}$$

is a necessary condition for System (1) to be feasible since these constraints are included in System (1). To show that it is also sufficient, let

$$x_1 = \max\{0, \max_{j: a_{1,j}=1} [\tilde{\mathbf{b}}_j - \sum_{k=2}^{n} a_{j,k} \hat{\mathbf{x}}_k]\}$$

It is easy to see that $\vec{\mathbf{x}} = [x_1, \vec{\hat{\mathbf{x}}}]$ is a feasible solution of System (1). □

Lemma 2. *If any column of* **A** *has no +1, and the system is feasible, then there exists a solution for System (1) in which the corresponding variable is set to* 0.

Proof: Let x_a denote the variable corresponding to this column in **A**. Let $\boldsymbol{t}$ denote a solution to System (1); assume that $t_a \neq 0$. Now consider the assignment $\boldsymbol{t'}$, which is constructed as follows: $t'_i = t_i$, $i \neq a$, $t_a = 0$. Constraints not involving x_a are trivially satisfied. Consider a constraint of the form: $-x_a - g() \geq b$; since this constraint is satisfied by $t_a > 0$, it is clearly satisfied by decreasing the value of x_a from t_a to 0. Since the constraint was chosen arbitrarily, we can reason similarly about every constraint in which x_a appears. □

Definition 3. *A Horn system is said to be standardized if every row and every column of the defining matrix* **A** *has at least one positive entry and one negative entry.*

The justification for disregarding rows in which all entries are negative (including absolute constraints of this type) will be provided in Section 4. For the rest of this paper, we assume that the Horn polytope under consideration is standardized.

3 The Lifting Algorithm

Algorithm 3.1 describes the details of the Lifting Algorithm; the algorithm operates in a series of $(n-1)$ phases. In each phase, every variable is processed in two stages: In the first stage, the constraints in which the variable appears positively are isolated. In the second stage, we determine the constraint with the largest Right-Hand Side (RHS) from the isolated set. If (and only if) the largest RHS is positive, we LIFT() the variable. The *lifting* of a variable, say x_i, involves two steps: Assume that the variable being processed is x_i and assume that the largest RHS of a constraint in which it appears positively is r. If r is non-positive, we proceed to process the next variable. Otherwise, the constraint system is changed to reflect the following substitution: $x'_i = x_i - r$. Note that under this substitution, the RHS of any constraint in which x_i appears positively will decrease, whereas the RHS of any constraint in which x_i appears negatively will *increase*. Thus, if $r = 7$, a constraint of the form $x_i - x_j \geq 7$ becomes $x'_i - x_j \geq 0$, whereas a constraint of the form $x_k - x_i - x_j \geq 3$ becomes $x_k - x'_i - x_j \geq 10$. Essentially, the origin of the Horn system has been lifted to a new value and hence the name "Lifting Algorithm."

Assume that the constraint matrix **A** is standardized. The coefficients of each variable (i.e., each column of **A**) are stored in two lists, viz., the *in-list* which contains the row indices in which the variable appears positively and the *out-list* which contains the row indices in which the variable appears positively. We assume that the RHS $\boldsymbol{b}$ is stored in column format for RAM access.

To process a variable, say x_i, we only need to process the corresponding *outlist*, *inlist*, and the components of $\boldsymbol{b}$ indexed by both lists. Clearly, this processing can be accomplished in $O(m)$ time; thus all variables can be processed in $O(m \cdot n)$ time and the algorithm itself runs in $O(m \cdot n^2)$ time, since there are $O(n)$ iterations.

Function Horn-Check ($\mathbf{A}$, $\boldsymbol{b}$, $\boldsymbol{o}$)

{Note that the constraint system that we are trying to solve is $\mathbf{A} \cdot \boldsymbol{x} \geq \boldsymbol{b}$, $\boldsymbol{x} \geq \mathbf{0}$. Initially, $\boldsymbol{o} = \mathbf{0}$. }
if ($\boldsymbol{b} \leq \mathbf{0}$) **then**
Set $\boldsymbol{x} = \boldsymbol{o}$
return ($\boldsymbol{x}$)
end if
for ($r = 1$ **to** $(n-1)$) **do**
for (each variable x_i, $i = 1, 2, \ldots, n$) **do**
Let F_i denote the set of constraints of the form: $x_i - x_k \geq c$, $c > 0$.
Let $l_i' : x_i - x_j \geq r$ denote the constraint in F_i with the largest Right Hand Side (RHS).
{l_i' is called the pivot constraint.}
if ($r > 0$) **then**
Lift($\boldsymbol{b}$, $\boldsymbol{o}$, i, r)
end if
end for
end for
if (any constraint has a positive RHS) **then**
assert("System is infeasible.")
Set $\boldsymbol{x} = -\mathbf{1}$
return ($\boldsymbol{x}$)
else
assert("System is feasible")
Set $\boldsymbol{x} = \boldsymbol{o}$
return ($\boldsymbol{x}$)
end if

Algorithm 3.1. A fast algorithm for checking non-emptiness of Horn polytopes

Function Lift ($\boldsymbol{b}$, $\boldsymbol{o}$, i, r)

Replace the variable x_i by the variable $x_i' = x_i - r$.
In each constraint that x_i appears positively, the corresponding b value is decremented by r.
In each constraint that x_i appears negatively, the corresponding b value is incremented by r.
The replacement will only change $\boldsymbol{b}$; $\mathbf{A}$ will remain unaltered.
Set $o_i = o_i + r$

Algorithm 3.2. The Lifting Procedure

4 Correctness

We need the following concepts from linear algebra to establish the correctness of Algorithm 3.1.

Definition 4. *Given a matrix* $\mathbf{A}$*, a minor of* $\mathbf{A}$ *is the determinant of any square sub-matrix obtained by deleting one or more rows and columns.*

Definition 5. *A linear diophantine equation is an equation of the form:* $\sum_{i=1}^{n} \alpha_i x_i = n$*, where the variables are all integral. A conjunction of such equations is termed a simultaneous linear diophantine system and can be represented in matrix form as:* $\mathbf{A} \cdot \boldsymbol{x} = \boldsymbol{b}$.

The problem of solving a simultaneous linear diophantine system is in P, but determining non-negative solutions to such a system is identical to Integer Programming and hence strongly NP-complete [4].

Borosh and Treybig proved the following lemma about non-negative solutions of simultaneous linear diophantine systems [1].

Theorem 1. *Let* $\mathbf{A} \cdot \boldsymbol{x} = \boldsymbol{b}$ *denote a polyhedral system with* $\mathbf{A}$ *having dimensions* $m \times n$*. Let* Δ *denote the maximum of the absolute value of all minors of the augmented matrix* $[\mathbf{A} \mid \boldsymbol{b}]$*. Then the polyhedral system has an integral solution if and only if it has one with all entries bounded by* $(n+2) \cdot \Delta$.

We now establish the solution-preserving nature of the LIFT() procedure.

Lemma 3. *Let System (1) represent a Horn system. Let* $\mathbf{A} \cdot \boldsymbol{x}' \geq \boldsymbol{b}'$ *denote the resultant system after variable* x_i *has been lifted. Then System (1) has a solution if and only if the resultant system has one.*

Proof: As per the hypothesis, x_i is the variable that was lifted. Assume that the pivot constraint is: $x_i - f(x_1, x_2, \ldots, x_n) \geq r$, $r > 0$, where $f()$ denotes a sum of some of the other variables in the system. This means that in every constraint of the constraint system in which x_i appears, either x_i appears negatively or it appears positively with a RHS not greater than r. From the constraint, $x_i - f() \geq r$, we can conclude that $x_i \geq f() + r$. Now observe that $f()$ is a sum of *positive* variables and its value is at least 0 is *any satisfying solution.* It therefore follows that x_i is at least r in any satisfying solution. Accordingly, replacing the variable x_i by the variable $x_i' = x_i - r$, $x_i \geq 0$ does not alter the solution space at all. The LIFT() procedure merely implements this change of origin by incrementing o_i by r and adjusting the b values to reflect this substitution. □

We can extend the above argument inductively to conclude that Algorithm 3.1 produces a sequence of polyhedra each of which is feasible if and only if System (1) is. Alternatively, we can say that the LIFT() procedure is *sound*, i.e., it preserves the solution space of the original system. Inasmuch as Algorithm 3.1 is constituted exclusively of calls to the LIFT() procedure, we can conclude that Algorithm 3.1 is sound.

Lemma 4. *Let System (1) represent a Horn system enclosing a lattice point (integral vector). Then at most* $(m+n)\cdot(m+n+2)\cdot\Delta$ *calls to the* LIFT() *procedure can be made before the* $\boldsymbol{b}$ *vector becomes non-positive, where* Δ *is the maximum of the absolute value of all minors of the augmented matrix* $[\mathbf{A},\ -\mathbf{I} \mid \boldsymbol{b}]$.

Proof: We can rewrite System (1) in the form:

$$\begin{aligned} \mathbf{A}\cdot\boldsymbol{x} - \mathbf{I}\cdot\boldsymbol{z} &= \boldsymbol{b} \\ \boldsymbol{x},\ \boldsymbol{z} &\geq \mathbf{0} \end{aligned} \tag{2}$$

Note that the matrix $[\mathbf{A},\ -\mathbf{I}]$ has $m+n$ columns; applying Theorem 1, we conclude that if System (2) encloses a lattice point, then it must have an integral solution in which all components are bounded by $(m+n+2)\cdot\Delta$.

Observe that each call to the LIFT() procedure results in one component increasing by at least unity. Therefore, if more than $(m+n)\cdot(m+n+2)\cdot\Delta$ calls are made to the LIFT() procedure then the soundness of the LIFT() procedure assures us that in *any* solution to System (1), at least one component is greater than $(m+n)\cdot(m+n+2)\cdot\Delta$ and this contradicts Theorem 1. □

Essentially, Lemma 4 establishes that if executed sufficiently many times, the LIFT() procedure is *complete* from the perspective of lattice point feasibility in Horn systems.

Lemma 5. *Let System (1) denote a Horn system. Then it is feasible if and only if it encloses a lattice point.*

Proof: Clearly, if System (1) encloses a lattice point then it is feasible. Assume that it does not enclose a lattice point, but does enclose a fractional solution. From the theory of linear programming, we know that System (1) has a basic feasible solution and that all basic feasible solutions have components that are bounded by an exponential function of m and n and the elements of $\mathbf{A}$ and $\boldsymbol{b}$ [3]; let us call this bound Ξ. Now consider what happens when more than Ξ calls are made to the LIFT() procedure; the soundness of lifting implies that every solution has to have at least one component greater than Ξ, and that contradicts the linear programming bound in [3]. It follows that a fractional solution cannot exist either. □

We have now established the following two facts:

(i) A Horn system of the form: $\mathbf{A}\cdot\boldsymbol{x}\geq\boldsymbol{b},\ \boldsymbol{x}\geq\mathbf{0}$, with $\boldsymbol{b}$ integral, is feasible if and only if it contains a lattice point. The set of lattice points satisfying the above Horn system is bounded below by $\mathbf{0}$.

(ii) The LIFT() procedure is both sound and (if carried out sufficiently many times) a complete procedure for the purpose of checking feasibility in Horn systems.

Definition 6. *Given a non-empty set of vectors* $\mathcal{S}$, *a vector* $\boldsymbol{y}\in\mathcal{S}$, *is said to be a minimal element, if* $\boldsymbol{x}\in S\Rightarrow\boldsymbol{y}\leq\boldsymbol{x}$, *where the* $\leq$ *relation holds componentwise.*

It is not hard to see that if a set contains a minimal element, then this element is unique. Two elements $\boldsymbol{u}$ and $\boldsymbol{v}$ in S are incomparable, if neither $\boldsymbol{u} \leq \boldsymbol{v}$ nor $\boldsymbol{v} \leq \boldsymbol{u}$.

We remark that our definition of minimal element is different from the standard definition of minimal element; in the standard definition, an element of a set $\boldsymbol{y}$ is declared to be minimal, as long as there is no element $\boldsymbol{z}$, such that $\boldsymbol{z} \leq \boldsymbol{y}$ and $z_i < y_i$, for at least one $i = 1, 2, \ldots, n$. In other words, as per the standard definition, a set could have multiple minimal elements, which are mutually incomparable. We will be using our definition for the rest of the paper.

Lemma 6. *Given a non-empty set S of vectors, which is closed and bounded below, and a partial order "$\leq$" defined on the elements of S, either S has a minimal element $\boldsymbol{z}$, or there exists a pair of elements $\boldsymbol{u}$, $\boldsymbol{v} \in S$, such that $\boldsymbol{u}$ and $\boldsymbol{v}$ are incomparable and there is no element $\boldsymbol{z} \in S$, such that $\boldsymbol{z} \leq \boldsymbol{u}$ and $\boldsymbol{z} \leq \boldsymbol{v}$.*

Proof: Suppose that S does not have a minimal element. Then, as per our definition, there must be at least two elements, say $\boldsymbol{u}$ and $\boldsymbol{v}$ which are incomparable, since if every pair of elements is comparable, then the elements would form a chain under the "$\leq$" relationship and every chain which is bounded below has a minimal element. Now, consider the case in which corresponding to every pair of incomparable elements (say $(\boldsymbol{u}, \boldsymbol{v})$), there is an element $\boldsymbol{z} \in S$, such that $\boldsymbol{z} \leq \boldsymbol{u}$ and $\boldsymbol{z} \leq \boldsymbol{v}$. We call $\boldsymbol{z}$ the dominator of $\boldsymbol{u}$ and $\boldsymbol{v}$. Observe that we can create a set of dominators of all pairs of elements that are mutually incomparable; either the elements of this set form a chain or we can create the set of their dominators. As this process repeats, we will be left with a single element, since S is closed and bounded above or we will have two elements that are mutually incomparable and which are not dominated by another element in S. If the process results in a single element then this element is clearly the minimal element of S, violating the hypothesis that S did not have a minimal element. □

The proof of the following lemma is available in the full paper.

Lemma 7. *The solution set S of the Horn system $\{\mathbf{A} \cdot \boldsymbol{x} \geq \boldsymbol{b},\ \boldsymbol{x} \geq \mathbf{0}\}$ contains a minimal element if $S \neq \emptyset$.*

Theorem 2. *If the solution set S of the Horn system $\{\mathbf{A} \cdot \boldsymbol{x} \geq \boldsymbol{b},\ \boldsymbol{x} \geq \mathbf{0}\}$ is non-empty and if Algorithm 3.1 asserts that the Horn system is feasible, then the solution returned is the minimal element of S.*

Proof: The proof is available in the full version of the paper. □

Lemma 8. *Suppose at some step of the algorithm, we have a vector $\overrightarrow{\mathbf{o}}$ and the RHS is $\overrightarrow{\mathbf{b}'}$. Let the minimal element of the polyhedron $\{\mathbf{A} \cdot \boldsymbol{x} \geq \overrightarrow{\mathbf{b}'},\ \boldsymbol{x} \geq \mathbf{0}\}$ be $\overrightarrow{\hat{\mathbf{x}}}$. The the minimal element of the system $\{\mathbf{A} \cdot \boldsymbol{x} \geq \boldsymbol{b},\ \boldsymbol{x} \geq \mathbf{0}\}$ is $\boldsymbol{x} = \overrightarrow{\mathbf{o}} + \overrightarrow{\hat{\mathbf{x}}}$.*

Proof: Follows readily from the previous lemma. □

Corollary 1. *The minimal element of a feasible Horn system is integral.*

Proof: The only operation in Algorithm 3.1 is integer addition; given that we start with $\boldsymbol{o} = \mathbf{0}$ and that $\boldsymbol{b}$ is integral, it must be the case that the point returned by Algorithm 3.1, which is minimal by the previous theorem, is integral. □

Observe that if a polyhedron $\mathbf{P}$ has a unique minimal element, then this element is obtained by minimizing the linear function $\boldsymbol{p} \cdot \boldsymbol{x}$ over $\mathbf{P}$, where $\boldsymbol{p} > \mathbf{0}$ is an arbitrary positive vector. Therefore, without loss of generality, we can assume that Algorithm 3.1 is, in essence, solving the following linear program:

$$\begin{aligned} \Psi : \min \quad & \sum_{i=1}^{n} x_i \\ \mathbf{A} \cdot \boldsymbol{x} &\geq \boldsymbol{b} \\ \boldsymbol{x} &\geq \mathbf{0} \end{aligned} \tag{3}$$

We are now going to argue that if $\mathbf{A}$ in System (3) has a row with no positives, then this row can be discarded. Assume that there is a row in $\mathbf{A}$ which has no positives; let this row be $l_i : -f(x_1, x_2, \ldots, x_n) \geq b_i$, where $f()$ is a sum function of some of the variables. First observe that b_i cannot be positive since the sum of a collection of positive variables is at least 0 and hence the negation of this sum is at most 0. Thus, if b_i is positive, the Horn system is infeasible. Likewise, if b_i is 0, then each variable in l_i must be 0 since the sum of non-negative variables can be 0 if and only if each variable is individually 0. Thus b_i must be negative; the constraint is therefore imposing the condition $f() \leq -b_i$, where $-b_i > 0$. Now if we delete the constraint l_i and solve System (3), we get the minimal element of the resultant Horn system, say $\boldsymbol{u}$. If $\boldsymbol{u}$ does not satisfy l_i, then the original Horn system is infeasible. Even if this row is not deleted and LIFT() is performed, if the value of the RHS b'_i becomes positive, it will continue to stay positive, and the sytem will be declared infeasible by the algorithm when it is infeasible. Accordingly, we can focus our attention on standardized Horn systems only.

The only issue that we have to establish is that if we organize the LIFT() procedure as described in Algorithm 3.1, then $(n-1)$ phases, with each phase examining each variable, are sufficient.

Lemma 9. *If System (3) is feasible, then in the optimal basis $x_i = 0$ for at least one i.*

Proof: Let $\boldsymbol{z}$ denote the solution to System (3), and let $z_i > 0$, $\forall i$. We use Ψ_z to denote the value of the objective function $\sum_{i=1}^{n} x_i$ at this point. Let $k = \min_{i=1}^{n} z_i$. Observe that $\boldsymbol{u} = (\boldsymbol{z} - \boldsymbol{k})$ is also a solution to System (3), and $\boldsymbol{u} < \boldsymbol{z}$. Hence $\Psi_u < \Psi_z$, thereby contradicting the optimality of $\boldsymbol{z}$. The lemma follows. □

Claim. For each $i = 1, 2, \ldots, n$, the o_i value in Algorithm 3.1 increases monotonically with each recursive call.

Proof: Observe that the only operation performed on the o_i values is the addition of a positive number in the LIFT() procedure. □

We use $\mathbf{L_i}$ to denote the state of the outermost **for** loop of Algorithm 3.1 when the loop index r is i. We use S_i to denote the set of variables that are lifted during $\mathbf{L_i}$ of Algorithm 3.1. By convention, $S_0 = \mathbf{X} = \{x_1, x_2, \ldots, x_n\}$, i.e., we say that all the variables of the system are lifted during $\mathbf{L_0}$. Thus, if a variable is lifted only during $\mathbf{L_0}$, it means that the variable is not lifted at all.

Definition 7. *A variable in the Horn system (3) is said to be saturated at level* i*, if it is lifted during* $\mathbf{L_i}$*, but never afterwards during Algorithm 3.1.*

We use Z_i to denote the set of variables that are saturated at level i. It is important to note that there could exist a variable $x_a \in \mathbf{X}$, such that $x_a \in S_i$, but $x_a \notin Z_i$. In other words, a variable which is lifted during $\mathbf{L_i}$ need not necessarily be saturated at Level i. However, if $x_a \in Z_i$, then x_a is necessarily part of S_i since x_a is lifted during $\mathbf{L_i}$. We thus have $Z_i \subseteq S_i$, $\forall i = 0, 1, \ldots, (n-1)$.

Lemma 10. *If System (3) is feasible, then there exists at least one variable* x_i*, which is never lifted by Algorithm 3.1.*

Proof: Observe that when variable x_i is lifted, o_i increases in value. From Claim 4, we know that o_i can never decrease for any variable x_i. Thus, if System (3) is feasible and all the variables of $\mathbf{X}$ are lifted at least once, by Algorithm 3.1, then on its termination, $o_i > 0, \forall i$, thereby contradicting Lemma 9. □

Lemma 10 establishes that there is at least one variable which is saturated at level 0.

Lemma 11. *If* $S_i = \emptyset$*, then* $S_j = \emptyset$*,* $j = (i+1), (i+2), \ldots, (n-1)$*. Furthermore, System (3) is feasible and the unique minimal element is returned by Algorithm 3.1.*

Proof: If no variable was lifted during $\mathbf{L_i}$, then no variable occurred positively in a a constraint with a positive RHS. This situation will not change at the commencement of $\mathbf{L_{i+1}}$, and hence S_j will be empty as well, for $j = (i+1), \ldots, (n-1)$. Let $\mathbf{T}: \ \mathbf{A} \cdot \boldsymbol{x} \geq \boldsymbol{b'}, \ \boldsymbol{x} \geq \mathbf{0}$ denote the constraint system that results after $\mathbf{L_i}$ completes execution. Since all the entries in $\boldsymbol{b'}$ are non-positive, $\mathbf{T}$ is feasible, and as discussed in Theorem 2, Algorithm 3.1 returns the minimal element of the Horn system. □

Theorem 3. *Assume that System (3) has a feasible basis. Then,*

$$(S_i \neq \emptyset) \rightarrow (Z_i \neq \emptyset), \ i = 0, 1, 2, \ldots n.$$

Proof: The proof of this theorem is provided in the full paper. □

The correctness of Algorithm 3.1 follows, since we have shown that if System (3) has a feasible basis, then Algorithm 3.1 will detect this basis in $(n-1)$ iteration of the outermost **for** loop.

5 Conclusions

This paper presented a strongly polynomial algorithm called the "Lifting Algorithm" for checking both linear feasibility and integer feasibility in Horn polytopes. Our algorithm is a variant of the Stressing Algorithm outlined in [5] for detecting negative cost cycles and runs in $O(m \cdot n^2)$ time, which is worse than the best known time for checking the feasibility of difference constraint systems, by a factor of n. Two features of the algorithm that bear mention are its greediness and its simplicity (the algorithm uses only the addition operation). From the polyhedral perspective, we showed that every feasible (extended) Horn system had a minimal element and that if the RHS of a feasible Horn system is integral, then it must enclose a lattice point. The Lifting Algorithm actually produces the minimal element of the feasible Horn system. It must be noted that while minimizing a positive function over a Horn system is in P, optimizing an arbitrary linear function over a Horn system is NP-hard.

References

1. Borosh, I., Treybig, L.B.: Bounds on positive integral solutions to linear diophantine equations. American Mathematical Society 55(2), 299–304 (1976)
2. Chandru, V., Rao, M.R.: Linear programming. In: Algorithms and Theory of Computation Handbook. CRC Press, Boca Raton (1999)
3. Papadimitriou, C.H., Steiglitz, K.: Combinatorial Optimization. Prentice Hall, Englewood Cliffs (1982)
4. Schrijver, A.: Theory of Linear and Integer Programming. John Wiley and Sons, New York (1987)
5. Subramani, K.: A Zero-Space algorithm for negative cost cycle detection in networks. Journal of Discrete Algorithms 5(3), 408–421 (2007)
6. Truemper, K.: Personal communication (2003)

Online Maximum Directed Cut

Amotz Bar-Noy and Michael Lampis

Doctoral Program in Computer Science,
Graduate Center, City University of New York
amotz@sci.brooklyn.cuny.edu, mlampis@gc.cuny.edu

Abstract. We investigate a natural online version of the well-known MAXIMUM DIRECTED CUT problem on DAGs. We propose a deterministic algorithm and show that it achieves a competitive ratio of $\frac{3\sqrt{3}}{2} \approx 2.5981$. We then give a lower bound argument to show that no deterministic algorithm can achieve a ratio of $\frac{3\sqrt{3}}{2} - \epsilon$ for any $\epsilon > 0$ thus showing that our algorithm is essentially optimal. Then, we extend our technique to improve upon the analysis of an old result: we show that greedily derandomizing the trivial randomized algorithm for MAXDICUT in general graphs improves the competitive ratio from 4 to 3, and also provide a tight example.

1 Introduction

The MAXIMUM CUT and MAXIMUM DIRECTED CUT problems are among the most famous and widely studied problems in computer science and during the past few decades countless papers have been devoted to them. Their objective is to partition the vertices of an edge-weighted graph so that the weight of the edges going from one side of the partition to the other is maximized.

MAXIMUM CUT was one of the 21 original problems shown to be NP-complete in Karp's seminal paper [7]. Since then, many results have appeared in the literature dealing with restrictions of the problem; for example in [10] it is shown that the problem remains NP-hard for graphs of maximum degree 3, while in [5] it is shown that it is solvable in polynomial time for planar graphs. The approximability of both problems has also been well-studied. It has long been known that MAXCUT and MAXDICUT can be approximated within factors of 2 and 4 respectively with a trivial randomized algorithm which randomly places each vertex on either side of the partition with equal probability (see e.g. [1]). The celebrated paper of Goemans and Williamson [4] achieves an approximation ratio of 1.1383 for MAXCUT using semi-definite programming and improving on its results a ratio of 1.165 is achieved for MAXDICUT in [3]. The ratio for MAXCUT is shown to be best possible under the Unique Games Conjecture in [8]. In addition to these results, several other results using combinatorial algorithms are known for special cases of MAXCUT (see for example [2]).

MAXDICUT is the less studied of the two problems, especially from the point of view of combinatorial algorithms. One exception is the paper by Halperin and

Y. Dong, D.-Z. Du, and O. Ibarra (Eds.): ISAAC 2009, LNCS 5878, pp. 1124–1133, 2009.

Zwick [6] which takes a combinatorial approach to the problem and gives a 2-approximation algorithm via a reduction to matching and a 2.22-approximation algorithm which runs in linear time. Few other combinatorial results are known for this problem. In fact, the restriction of MaxDiCut to DAGs was only recently shown to be NP and APX-hard in [9].

To the best of our knowledge, neither problem has been studied before in an online setting and the only algorithms applicable in this case are the folklore trivial randomized algorithms. In this paper we propose the study of MaxDiCut as an online problem motivated by several reasons: First, studying a problem in an online setting can often lead us to discover some unknown combinatorial structure which we can then exploit, either in an online or an offline setting. This is especially true in the case of MaxDiCut since the vast majority of known results rely on complex tools such as semi-definite programming, while ignoring the more combinatorial aspect of the problem. Second, studying hard offline problems in an online setting makes sense because often in practice some heuristic is used which makes a single pass over the input in an attempt to find an approximate solution. Therefore, it would be useful to have some results regarding the quality of the solution produced.

It should be clarified at this point what we mean by an online setting, since one could potentially define countless online variations of MaxDiCut. The model we propose is one where the vertices are revealed one by one. Along with each vertex, all edges incident to it whose other endpoint has already been revealed are also revealed. This description is very natural. However, if this is all the information the adversary has to reveal, it is easy to force any deterministic algorithm to have unbounded competitive ratio thus making the problem uninteresting (more on this in Section 2). This is why we add an extra restriction: we force the adversary to reveal along with each vertex the *total* weight of its incoming and its outgoing edges. We believe that this model strikes a good balance: it only gives the algorithm a little information about the future, just enough to make the problem reasonable. It also fits well with our motivation from single-pass algorithms, since in cases like that one would likely make decisions based on past decisions and local information, such as the degree of a vertex.

Furthermore, motivated by the result of [9] mentioned above, we investigate the problem restricted to DAGs. Motivated by the problem's own structure we impose one further restriction on the adversary: we require the vertices to be revealed in an order consistent with the partial ordering implied by the DAG itself. This can be an interresting restriction consistent with our above motivation since this is probably the most natural order in which a single-pass algorithm would examine the vertices of the DAG.

Our main results regarding the problem on DAGs are an algorithm with competitive ratio $\frac{3\sqrt{3}}{2} \approx 2.5981$ and an essentially matching lower bound, which also applies in the case of general graphs. Thus, it follows from our results that $\frac{3\sqrt{3}}{2}$ is the best ratio achievable for this problem. What is also interesting is that the intuition gained from our analysis on DAGs is then applied in the general case of the problem to give a deterministic 3-competitive algorithm. In fact, this can be interpreted as an improvement in the analysis of an old result: our algorithm can be seen as a derandomization of the trivial randomized algorithm and we show that this derandomization improves the competitive ratio from 4 to 3. To the best of our knowledge, this result was not known before, even concerning the analysis of this derandomized algorithm in the offline case (for example in [4] it is mentioned that the best previously known algorithm for MaxDiCut has a ratio of 4), and it is interesting to consider its contrast with the undirected case where derandomizing the trivial algorithm does not improve its guarantee.

2 Definitions and Preliminaries

In the remainder we consider only directed graphs, so we will use the terms graph and digraph interchangeably. Let us give the definition of MaxDiCut.

Definition 1. *Given a digraph* $G(V,E)$ *and a weight function on the edges* $w : E \to \mathbb{N}$, MaxDiCut *is the problem of finding a partition of* V *into two sets* V_0 *and* V_1 *so that the weight of the edge set* $C = \{(u,v) \mid u \in V_0, v \in V_1\}$ *is maximized. That is, the objective is to maximize* $\sum_{e \in C} w(e)$.

The restriction of the problem where all edges have the same weight is often called cardinality or unweighted MaxDiCut. In this paper we focus on the weighted version. Because we focus on the weighted problem when we refer to the in-degree (resp. out-degree) of a vertex, we mean the total weight of its incoming (resp. outgoing) edges. Also when we refer to the size of a cut, we mean the total weight of the edges in the cut.

When a vertex u is placed in V_0 (resp. V_1) we will often say that 0 (resp. 1) was assigned to u. We denote by $w_{\text{in}}(u)$ the total weight of incoming edges at vertex u and by $w_{\text{out}}(u)$ the total weight of outgoing edges. By $w_{\text{in}}^0(u)$ (resp. $w_{\text{in}}^1(u)$) we denote the incoming edges of u whose other endpoint has been assigned 0 (resp. 1). Similarly for $w_{\text{out}}^0(u)$ and $w_{\text{out}}^1(u)$. In a context where the solution produced by an algorithm is compared with another (optimal) solution, when we use this notation we refer to the algorthm's choices, unless otherwise specified. That is, $w_{\text{in}}^0(u)$ means the total weight of edges coming to u from vertices to which *the algorithm* assigned 0. Some extra shorthand notation relevant in the case of our algorithm for general graphs is defined in Section 5.

The online model

Definition 2. *In the* online MaxDiCut *problem, an adversary chooses a graph* $G(V,E)$ *and an ordering of the vertices of* V. *Then, for each vertex* u *in this order the adversary reveals to the algorithm* $w_{\text{in}}(u)$, $w_{\text{out}}(u)$ *and the weights of edges connecting* u *to previously revealed vertices. The algorithm then makes an irrevocable decision to assign* 0 *or* 1 *to* u *and the process continues, until all of* G *has been revealed.*

Perhaps the detail of this definition which needs justification is why we demand that the adversary reveal the total in-degree and out-degree each vertex will have in the *final* graph G, thus revealing some information about the future. Aside from practical considerations (it could be argued that this is a more realistic model for some applications) the main reason is that not revealing any such information makes the problem impossible. Consider for example the setting where the adversary does not reveal such information. So, when the first vertex is given the algorithm must assign a value. If it picks 0 the adversary is free to claim later that this vertex is a sink in G with many heavy edges going into it, which are lost from the solution. Similarly, the adversary can respond to a 1 by revealing that the vertex is a source. This simple trick would be enough to make the competitive ratio of any deterministic algorithm tend to ∞. That is why revealing $w_{\text{in}}(u)$ and $w_{\text{out}}(u)$ are necessary.

In the case of DAGs we restrict the setting a little bit more. Recall that the edges of a DAG imply a partial ordering on its vertices. We demand that the adversary must reveal the vertices of G in a way consistent with this partial ordering. This eliminates the need to reveal $w_{\text{in}}(u)$, since when this restriction is followed the tails of all edges incoming to u must be revealed before u, therefore the algorithm already has enough information to calculate $w_{\text{in}}(u)$.

The motivation behind this restriction is once again single-pass algorithms. In the case of DAGs it is quite likely that one would scan through the graph in the order dictated by the directed edges. Moreover, this restriction seems natural and may be interesting in its own right. Of course it should also be noted that, as our results point out, this restriction does not have a huge impact on the problem since we show a lower bound for this special case which is not very far from the competitiveness of our algorithm even in the general case.

3 An Online Algorithm for MaxDiCut on DAGs

We present an algorithm (Algorithm 1 shown below) based on a weighted comparisons idea, leaving a parameter $c > 1$ to be fixed later to a value that will achieve the best competitive ratio. After this optimization we will show a ratio of $\frac{3\sqrt{3}}{2} < 2.5981$.

Algorithm 1. Weighted Comparison Online Algorithm

When a vertex u is revealed, compare $w_{\text{out}}(u)$ to $cw_{\text{in}}^0(u)$.

- If $w_{\text{out}}(u) > cw_{\text{in}}^0(u)$ then assign 0 to u.
- Otherwise, assign 1 to u.

It is not hard to understand the motivation behind this algorithm, if one considers the naive greedy algorithm which at every vertex compares $w_{\text{in}}^0(u)$, which is the payoff that would be obtained by assigning 1, with $w_{\text{out}}(u)$, which is the potential payoff of assigning 0. An adversary could easily fool this naive algorithm into assigning a long string of consecutive 0s in a path, making its competitive ratio tend to ∞. However, in the algorithm we propose, we only choose to assign 0 to a vertex when tempted by a payoff much larger (c times larger) than what could be obtained by assigning 1. Even though the adversary can still fool the algorithm into assigning a long string of consecutive 0s, it will have to offer exponentially increasing edge weights to do so. Eventually, a vertex will be encountered where the algorithm assigns 1, and the profit obtained by doing so will (at least partially) outweigh the loss of edges that occured up to that point.

Let us now provide a formal analysis of the algorithm's performance to establish its competitive ratio. Suppose we are given a graph $G(V,E)$, and let (V_0,V_1) be the partition of V decided by the algorithm. Let $E_{ij} = \{(u,v) \in E \mid u \in V_i, v \in V_j\}$. Thus, the cut produced by the algorithm has weight $\text{SOL} = \sum_{e \in E_{01}} w(e)$.

Consider the two subgraphs of G, $G_0(V_0, E_{00})$ and $G_1(V, E_{01} \cup E_{10} \cup E_{11})$. Let OPT be the size of an optimal cut of G and OPT_0 (resp. OPT_1) the size of an optimal cut of G_0 (resp. G_1).

Lemma 1. $\text{OPT} \leq \text{OPT}_0 + \text{OPT}_1$

We will compare SOL independently to OPT_0 and OPT_1. We use the following Lemmas as intermediate steps (the proofs have been omitted due to space constraints).

Lemma 2. $c\text{SOL} \geq \text{OPT}_1$

Lemma 3. $\text{SOL} \geq (c-1)\sum_{e \in E_{00}} w(e)$

Lemma 4. $\text{OPT}_0 \leq \frac{c}{c^2-1}\text{SOL}$

Theorem 1. *Algorithm 1 is $(c + \frac{c}{c^2-1})$-competitive.*

Taking the derivative of $\frac{c}{c^2-1} + c$ we find that a minimum is obtained for $c = \sqrt{3}$. In that case the competitive ratio achieved is $\frac{3\sqrt{3}}{2}$.

The fact that the above analysis is tight for $c = \sqrt{3}$ will be confirmed in Section 4 where it will be shown that this is in fact the best ratio achievable by any algorithm. But before that, we could also give a simple example to show that our analysis is tight for any c. First, let us point out that when looking for such an example we may assume without loss of generality that in cases where $w^0_{\text{in}}(u) = cw_{\text{out}}(u)$ the algorithm assigns to u an arbitrary value chosen by the adversary. This is because the adversary could easily adjust $w_{\text{out}}(u)$ by $\pm\epsilon$ without affecting the size of the optimal solution significantly.

Consider a directed path with $2n$ edges and the vertices labeled $0, 1, \ldots, 2n$. We give the edges weights $w((i, i+1)) = c^i$. The algorithm assigns 0 to vertex 0 and then without loss of generality assigns 0 to all vertices until $2n-2$. Then it assigns 1 to $2n-1$ and $2n$. The produced solution has weight $w((2n-2, 2n-1)) = c^{2n-2} = (c^2)^{n-1}$. The optimal solution has weight $c^{2n-1} + c^{2n-3} + c^{2n-5} + \ldots + c = c\frac{(c^2)^n - 1}{c^2-1} = \frac{c^3}{c^2-1}(c^2)^{n-1} - \frac{c}{c^2-1}$.

4 A $3\sqrt{3}/2 - \epsilon$ Lower Bound

In this Section we show that the competitive ratio of $3\sqrt{3}/2$ achieved by the algorithm of Section 3 is essentially the best possible. We show this by giving a strategy which the adversary can follow to force any deterministic algorithm's competitive ratio arbitrarily close to $3\sqrt{3}/2$.

The construction we use is very simple: it is a directed path, similar to the tight examples of the previous Section. However, now the edge weights are picked so that any algorithm can be defeated. The adversary's strategy is still to fool the algorithm into assigning a long string of 0s thus worsening the competitive ratio. But now the edge weights will eventually converge. We make use of the following simple observation:

Lemma 5. *If for some vertex $w^0_{\text{in}}(u) \geq w_{\text{out}}(u)$ then it is optimal to assign* 1 *to that vertex.*

The adversary must find a sequence of edge weights for the path, call them $w_0, w_1, \ldots$, such that assigning a long string of 0s is inevitable for any algorithm with a good competitive ratio. If the adversary can also make the sequence of weights decrease at some point without violating this principle the previous observation will complete the proof of a lower bound.

The above methodology is illustrated in the following Lemma.

Lemma 6. *For a given number $\lambda > 1$, if the sequence defined by $w_0 = 1, w_1 = \lambda, w_2 = \lambda^2 - 1$ and $w_i = \lambda(w_{i-1} - w_{i-3})$ is not always increasing, then no deterministic online algorithm can achieve a ratio less than λ for* MaxDiCut *on DAGs.*

Proof. Suppose that for some i we have $w_i \leq w_{i-1}$. The adversary will follow this strategy: the graph used is a directed path with edge weights $w_0, w_1, \ldots$. While the algorithm has not yet assigned a 1 keep revealing vertices and set edge weights according to the sequence w_i. When the algorithm assigns 1 or when the edge with weight w_i is revealed, terminate the instance, that is reveal only the other endpoint of the single outstanding edge and inform the algorithm that this is the last vertex.

For this strategy to demonstrate a lower bound of λ it is sufficient to have that for all k $w_k + w_{k-2} + \ldots \geq \lambda w_{k-1}$. The left-hand side of this equation is the optimal solution if we terminate after edge k (we will denote this by OPT_k), while w_{k-1} is the algorithm's solution in that case (observe that we terminate on the algorithm's first 1 so the algorithm always picks exactly one edge in the cut). Therefore, if this inequality holds for all k the competitive ratio will be at least λ independent of the point where the instance was terminated.

Clearly, $\text{OPT}_1 = w_1 = \lambda w_0$ and $\text{OPT}_2 = w_2 + w_0 = \lambda w_1$. Now, we can use induction and we have $\text{OPT}_k = w_k + \text{OPT}_{k-2} = \lambda(w_{k-1} - w_{k-3}) + \text{OPT}_{k-2} \geq \lambda(w_{k-1} - w_{k-3}) + \lambda w_{k-3} = \lambda w_{k-1}$. Therefore, if the instance terminates at any point, the competitive ratio is at least λ. Lemma 5 guarantees that the instance will be terminated at i at the latest for any reasonable algorithm. □

The question is for which λ the sequence w_i decreases at some point and for which it is always increasing. In order to answer this we would like to obtain a closed form of w_i. One way to obtain such a form is to solve the corresponding equation $x^3 = \lambda(x^2 - 1)$. If the roots of this equation are distinct, say r_1, r_2, r_3 we get that $w_i = A_1 r_1^i + A_2 r_2^i + A_3 r_3^i$ for some constants A_1, A_2, A_3 determined by w_0, w_1, w_2.

Lemma 7. *If $\lambda < \frac{3\sqrt{3}}{2}$ the equation $x^3 = \lambda(x^2 - 1)$ has three distinct roots. One of them is real and belongs in $(-1, 0)$ and the other two are complex.*

Lemma 8. *If $\lambda < \frac{3\sqrt{3}}{2}$ there exists an i such that for the sequence w defined in Lemma 6 we have $w_i \leq w_{i-1}$.*

Using Lemmata 6 and 8 we have the following result.

Theorem 2. *No deterministic algorithm can achieve a competitive ratio of $\frac{3\sqrt{3}}{2} - \epsilon$ for any $\epsilon > 0$ for the* MaxDiCut *problem on DAGs.*

5 General Graphs

In this Section we show that a natural extension of the online algorithm for DAGs of Section 3 results in a 3-competitive online algorithm for the case of general digraphs.

Before we go on, let us first introduce some additional notation, which will be needed in this case. Recall that in Section 3 we used the notation $w_{\text{in}}(u), w_{\text{out}}(u)$ to refer to the total weight of edges incoming and outgoing from u respectively. However, the online model we assumed in that case guaranteed that at the time when a vertex u was revealed, the tails of its incoming edges had already been revealed (and therefore assigned a value). Similarly, we knew that none of the heads of edges coming out of u had been revealed yet at the time u was given.

In the online model for general graphs this is not necessarily the case. However, we now make the assumption that whenever the adversary reveals a vertex, both its total in-degree and its total out-degree are revealed. Therefore, we introduce a shorter notation as follows: for a given vertex u and a specific assignment to all the vertices revealed before u we will denote by $C_1(u)$ the certain payoff which can be achieved by assigning 1 to this vertex. In other words, $C_1(u)$ is the total weight of edges incoming to u with their tails already revealed and assigned 0. Similarly, we denote by $C_0(u)$ the certain payoff of assigning 0, which is the total weight of edges coming out of u whose heads have already been assigned 1. We denote by $P_0(u), P_1(u)$ the maximum additional payoff which can be achieved by assigning 0 or 1 respectively to u. That is, $P_0(u)$ is the total weight of edges outgoing from u to vertices not yet revealed, and $P_1(u)$ is the total weight of edges incoming to u from such vertices.

Our proposed algorithm is Algorithm 2 (shown below). It is clear that using this notation we would have in the online model for DAGs of Section 3 that for all u, $P_0(u) = w_{\text{out}}(u)$, $P_1(u) = 0$, $C_1(u) = w^0_{\text{in}}(u)$, $C_0(u) = 0$. This new notation will ease the analysis of our algorithm for general graphs.

Algorithm 2. Doubling online algorithm for general graphs

When u is revealed calculate $C_0(u), C_1(u), P_0(u), P_1(u)$.

- If $C_0(u) + \frac{P_0(u)}{2} > C_1(u) + \frac{P_1(u)}{2}$ assign 0 to u.
- Otherwise, assign 1 to u.

Following our previous remark that in the model for DAGs $\forall u, C_0(u) = P_1(u) = 0$ it is easy to see that, if restricted to DAGs, Algorithm 2 is exactly Algorithm 1 with $c = 2$ (which implies that it would have a ratio of $8/3$ for that special case). Moreover, the intuition is essentially the same: give twice as much weight to a certain payoff as you give to an uncertain one. The question is, what is this algorithm's competitive ratio for general graphs?

First, let us point out that Algorithm 2 is at most 4-competitive. This can be seen if we compare its performance to the trivial randomized algorithm. (The proof has been omitted due to space constraints)

Theorem 3. *Let* SOL *be the solution produced by Algorithm 2 for a graph* $G(V, E)$. *Then* $\text{SOL} \geq \frac{|E|}{4}$.

A competitive ratio of 4 immediately follows from Theorem 3. We will now show that this can actually be improved to a ratio of 3.

Theorem 4. *Algorithm 2 is 3-competitive.*

Proof. Let SOL be the cut produced by the algorithm and OPT an optimal cut. Once again we will gradually change OPT to SOL by bribing the adversary to change the assignment for vertices on which OPT and SOL disagree. Then we will bound the amount of bribing needed.

Number the vertices $1, 2, \ldots, n$ in the order in which they were revealed. Let $\text{OPT}_0, \text{OPT}_1, \ldots, \text{OPT}_n$ be a sequence of cuts defined as follows: OPT_i gives the same assignment as SOL to the first i vertices and the same assignment as OPT to the others. Therefore, $\text{OPT}_0 \equiv \text{OPT}$ and $\text{OPT}_n \equiv \text{SOL}$.

Suppose that OPT and SOL agree on the assignment of some vertex i. Then $\text{OPT}_{i-1} \equiv \text{OPT}_i$.

Suppose that for some vertex i the optimal cut assigns 0 while the algorithm assigned 1. Consider then the cut OPT_{i-1}. From the edges incident on i the cut OPT_{i-1} gets at most $C_0(i) + P_0(i)$. The cut OPT_i gets at least $C_1(i)$. The cuts OPT_{i-1} and OPT_i differ only in vertex i therefore we have $\text{OPT}_{i-1} \leq \text{OPT}_i + C_0(i) + P_0(i) - C_1(i)$. But the algorithm assigned 1 to i therefore, $C_0(i) + \frac{P_0(i)}{2} \leq C_1(i) + \frac{P_1(i)}{2} \Rightarrow C_0(i) - C_1(i) \leq \frac{P_1(i) - P_0(i)}{2}$. Using this we get that $\text{OPT}_{i-1} \leq \text{OPT}_i + \frac{P_0(i) + P_1(i)}{2}$.

Using similar arguments we can prove that in the symmetric case where the optimal cut assigns 1 and the algorithm assigns 0 we again have $\text{OPT}_{i-1} \leq \text{OPT}_i + \frac{P_0(i) + P_1(i)}{2}$. Therefore, we have $\text{OPT} = \text{OPT}_0 \leq \text{OPT}_n + \sum_{i=1}^{n} \frac{P_0(i) + P_1(i)}{2}$

However, for a vertex i we have that $P_0(i) + P_1(i)$ is the total weight of edges whose first endpoint to be revealed is i. Therefore, taking the sum of $P_0(i) + P_1(i)$ for all vertices gives us exactly $|E|$, since then every edge's weight is counted exactly once (on its endpoint which was revealed first). Using also Theorem 3 we have $\text{OPT} \leq \text{SOL} + \frac{|E|}{2} \leq 3\text{SOL}$ □

Even though it has been known for decades that the trivial randomized algorithm is 4-competitive, to the best of our knowledge this is the first time that it has been examined whether its greedy derandomization actually offers an improved competitive ratio. We prove that it does, even in an online setting.

It should also be clear that, even though Algorithm 2 is a derandomization of the trivial algorithm, Theorem 4 does not imply that the randomized algorithm is also 3-competitive (it can be seen that it is in fact tightly 4-competitive by taking a bipartite graph $G(V_1, V_2, E)$ with all edges oriented from V_1 to V_2, so that $\text{OPT} = |E|$). Quite interestingly, this is a case where derandomizing really helps improve the guarantee on the algorithm's performance.

Finally, let us point out that there is a simple example which demonstrates that the analysis of Algorithm 2 is tight.

Theorem 5. *There exists a graph $G(V, E)$ for which the results of Theorems 3 and 4 are tight.*

Proof. First, observe that without loss of generality we can assume that when for some vertex $C_0(u) + \frac{P_0(u)}{2} = C_1(u) + \frac{P_1(u)}{2}$ then the algorithm assigns to u some arbitrary value chosen by the adversary. This is because simply by adding a small ϵ to P_0 or P_1 the adversary can make the algorithm behave in either way.

Now, the graph G we will use is a directed path on four vertices P_4. We label the vertices $1, 2, 3, 4$ and set $w((1, 2)) = w((2, 3)) = 1$ and $w((3, 4)) = 2$. The order in which the vertices are presented is $2, 1, 3, 4$. When 2 is revealed, without loss of generality, the algorithm assigns 0. So, the edge $(1, 2)$ is lost independent of the choice for 1. When 3 is revealed we have $C_0(3) = P_1(3) = 0$ and $C_1(3) = \frac{P_0(3)}{2}$ so again without loss of generality the algorithm assigns 1 and the edge $(3, 4)$ is lost. Now $\text{SOL} = 1$ but $\text{OPT} = 3$ and $|E| = 4$. □

6 Conclusions and Further Work

In this paper we introduced a natural online setting for the study of MaxDiCut and its restriction to DAGs. We completely solved the problem in DAGs for deterministic algorithms by providing an algorithm and an essentially matching lower bound. In addition, and perhaps more interestingly, we showed that the intuition gained from this problem can help in the general case, by improving a folklore result concerning the approximation ratio of the basic greedy algorithm for general graphs.

Many other topics are worth considering. For the specific problem on DAGs it would be interesting to consider randomized algorithms against an oblivious adversary. This would likely defeat our lower bound but it is not clear what would be a better algorithm. For the general case, there is a small gap between the competitiveness of our algorithm and the lower bound for DAGs. Could this gap be closed? Finally, it would be interesting to see if and how any of the ideas of this paper can be applied in the undirected case of the problem.

References

1. Alon, N., Spencer, J.: The Probabilistic Method. Wiley-Interscience, Hoboken (2004)
2. Bazgan, C., Tuza, Z.: Combinatorial 5/6-approximation of max cut in graphs of maximum degree 3. J. Discrete Algorithms 6(3), 510–519 (2008)
3. Feige, U., Goemans, M.: Approximating the value of two power proof systems, with applications to MAX 2SAT and MAX DICUT. In: Proceedings of the Third Israel Symposium on the Theory of Computing and Systems, 1995, pp. 182–189 (1995)
4. Goemans, M.X., Williamson, D.P.: Improved approximation algorithms for maximum cut and satisfiability problems using semidefinite programming. J. ACM 42(6), 1115–1145 (1995)
5. Hadlock, F.: Finding a Maximum Cut of a Planar Graph in Polynomial Time. SIAM Journal on Computing 4, 221 (1975)
6. Halperin, E., Zwick, U.: Combinatorial approximation algorithms for the maximum directed cut problem. In: Proceedings of the twelfth annual ACM-SIAM symposium on Discrete algorithms, PA, USA, pp. 1–7. Society for Industrial and Applied Mathematics Philadelphia (2001)
7. Karp, R.M.: Reducibility among combinatorial problems. In: Miller, R.E., Thatcher, J.W. (eds.) Complexity of Computer Computations, pp. 85–103. Plenum Press, Newyork (1972)
8. Khot, S., Kindler, G., Mossel, E., O'Donnell, R.: Optimal inapproximability results for MAX-CUT and other 2-variable CSPs? In: Proceedings. 45th Annual IEEE Symposium on Foundations of Computer Science, 2004, pp. 146–154 (2004)
9. Lampis, M., Kaouri, G., Mitsou, V.: On the algorithmic effectiveness of di-graph decompositions and complexity measures. In: Hong, S.-H., Nagamochi, H., Fukunaga, T. (eds.) ISAAC 2008. LNCS, vol. 5369, pp. 220–231. Springer, Heidelberg (2008)
10. Papadimitriou, C.H., Yannakakis, M.: Optimization, approximation, and complexity classes. J. Comput. Syst. Sci. 43(3), 425–440 (1991)

Maintaining Nets and Net Trees under Incremental Motion*

Minkyoung Cho, David M. Mount, and Eunhui Park

Department of Computer Science
University of Maryland
College Park, Maryland
{minkcho,mount,ehpark}@cs.umd.edu

Abstract. The problem of maintaining geometric structures for points in motion has been well studied over the years. Much theoretical work to date has been based on the assumption that point motion is continuous and predictable, but in practice, motion is typically presented incrementally in discrete time steps and may not be predictable. We consider the problem of maintaining a data structure for a set of points undergoing such incremental motion. We present a simple online model in which two agents cooperate to maintain the structure. One defines the data structure and provides a collection of certificates, which guarantee the structure's correctness. The other checks that the motion over time satisfies these certificates and notifies the first agent of any violations.

We present efficient online algorithms for maintaining both nets and net trees for a point set undergoing incremental motion in a space of constant dimension. We analyze our algorithms' efficiencies by bounding their competitive ratios relative to an optimal algorithm. We prove a constant factor competitive ratio for maintaining a slack form of nets, and our competitive ratio for net trees is proportional to the square of the tree's height.

1 Introduction

Motion is a pervasive concept in geometric computing. The problem of maintaining discrete geometric structures for points in motion has been well studied over the years. The vast majority of theoretical work in this area falls under the category of kinetic data structures (KDS) [7]. KDS is based on the assumption that points move continuously over time, where the motion is specified by algebraic functions of time. This makes it possible to predict the time of future events, and so to predict the precise time in the future at which the structure will undergo its next discrete change. In practice, however, motion is typically presented incrementally over a series of discrete time steps by a *black-box*, that is, a function that specifies the locations of the points at each time step. For example, this black-box function may be the output of a physics integrator, which determines

* This work has been supported by the National Science Foundation under grant CCR-0635099 and the Office of Naval Research under grant N00014-08-1-1015.

Y. Dong, D.-Z. Du, and O. Ibarra (Eds.): ISAAC 2009, LNCS 5878, pp. 1134–1143, 2009.

the current positions of the points based on the numerical solution of a system of differential equations [1,12]. Another example arises in the use of Markov-chain Monte-Carlo (MCMC) algorithms such as the Metropolis-Hastings algorithm [2] and related techniques such as simulated annealing [10].

In this paper we consider the maintenance of two well known structures, nets and net trees, for a set of moving points. Let P denote a finite set of points in some (continuous or discrete) metric space $\mathcal{M}$. Given $r > 0$, an *r-net* for P is a subset $X \subseteq P$ such that every point of P lies within distance r of some point X, and no two points of X are closer than r. (We will actually work with a generalization of this definition, which will present in Section 2.) Each point of P can be associated with a covering point of X that lies within distance r. This point is called its *representative*. We can easily derive a tree structure, by building a series of nets with exponentially increasing radius values, and associating each point at level $i-1$ with its representative as parent at level i.

The net tree has a number of advantages over coordinate-based decompositions such as quadtrees. The first is that the net tree is intrinsic to the point set, and thus the structure is invariant under rigid motions of the set. This is an important consideration with kinetic point sets. Another advantage is that the net tree can be defined in general metric spaces, because it is defined purely in terms of distances. A number of papers have been written about improvements to and applications of the above net-tree structure in metric spaces of constant doubling dimension. (See, for example [11,8,4,6].) Note that the net tree is a flexible structure in that there may be many possible choices for the points that form the nets at each level of the tree and the assignment of points to parents.

Although there has been much research on maintaining geometric structures in continuous contexts, such as KDS, there has been comparatively little theoretical work involving efficiency of algorithms for incremental black-box motion. Gao *et al.* [5] observe that their data structure (which is very similar to our net-tree structure) can be updated efficiently in the black-box context, but they do not consider the issue of global efficiency. A major issue here is the computational model within which efficiency is to be evaluated. In the absence of any *a priori* assumptions about the point motions, the time required to update the point locations and verify the correctness of the data structure is already $\Omega(|P|)$. With each time step the points could move to entirely new locations, thus necessitating that the data structure be rebuilt from scratch. This need not always be the case, however. It has been widely observed (see, e.g., [5,9,13]) that when the underlying motion is continuous, and/or the time steps are small, the relative point positions are unlikely to change significantly. Hence, the number of discrete structural changes per time step is likely to be small. We desire a computational model that allows us to exploit any underlying continuity in the motion to enhance efficiency, without the strong assumptions of KDS.

We introduce such a computational model for the online maintenance of geometric structures under incremental black-box motion. Our approach is similar other models for incremental motion, such as the observer-tracker model of Yi and Zhang [14] and the IM-MP model of Mount *et al.* [13]. Our model involves the interaction of two agents, an *observer* and a *builder*. The observer monitors the motions of the points over time, and the builder is responsible for maintaining the data structure. These two agents communicate through a set of boolean conditions, called *certificates*, which effectively "prove" the correctness of the current structure (exactly as they do in KDS). Based on the initial point positions, the builder constructs the initial structure and the initial certificates, and communicates these certificates to the observer. The observer monitors the point motion and, whenever it detects that a certificate has been violated, it informs the builder which certificates have been violated. The builder then queries the new locations the points, updates the data structure, and informs the observer of any updates to the certificate set. An algorithm for maintaining a data structure in this model is essentially a communication protocol between the observer and the builder. The total computational cost of an algorithm is defined to be the communication complexity between these two agents. One advantage of this model is that it divorces low-level motion issues from the principal algorithmic issues involving the design of the net structure itself.

Our main results are efficient online algorithms for maintaining both nets and net trees for a point set undergoing incremental motion. In each case, our algorithm is allowed some additional slackness in the properties of the net to be maintained. For example, while the optimal algorithm is required to maintain all points within distance r of each net point, we allow our algorithm for nets to maintain a covering distance of $2r$ for nets and $4r$ for net trees. (See Section 2 for the exact slackness conditions.) Because our principal motivation is in maintaining net trees under motion, we impose the assumption that the input points to our r-net algorithm arise from an $(r/2)$-net.

We establish the efficiency of our online algorithms by proving an upper bound on the *competitive ratio* on the communication cost of our algorithm, that is, the worst-case ratio between the communication costs of our algorithm (subject to the slackness conditions) and any other algorithm (without the slackness). The exact results are presented in Sections 3 and 4. Assuming that the points are in a space of constant doubling dimension (e.g., Euclidean of constant dimension), we achieve a competitive ratio of $O(1)$ for the maintenance of a net and $O(\log^2 \Phi)$ for the net tree, where Φ is the aspect ratio of the point set (the ratio between the maximum and minimum interpoint distances). Our online algorithm makes no *a priori* assumptions about the motion of the points. The competitive ratio applies even if the optimal algorithm has full knowledge of point motion, and it may even have access to unlimited computational resources. The constant factors hidden by the asymptotic notation grow exponentially in the doubling dimension of the underlying metric space.

2 Preliminaries

We begin with some basic definitions, which will be used throughout the paper. Let $\mathcal{M}$ denote a metric space, with associated distance function dist: $\mathcal{M} \times \mathcal{M} \rightarrow \mathbb{R}$. (This means that dist is symmetric, positive definite, and satisfies the triangle inequality.) Throughout, we let P be a finite subset of points in $\mathcal{M}$. For a point $p \in \mathcal{M}$ and a real $r \in \mathbb{R}^+$, let $b(p,r) = \{q \in \mathcal{M} : \mathrm{dist}(p,q) < r\}$ denote the open ball of radius r centered at p. The *doubling constant* of the metric space is defined to be the minimum value λ such that every ball b in $\mathcal{M}$ can be covered by at most λ balls of at most half the radius. The *doubling dimension* of the metric space is defined as $d = \log_2 \lambda$. Throughout, we assume that $\mathcal{M}$ is a space of constant doubling dimension.

Consider a point set P in $\mathcal{M}$. Given $r \in \mathbb{R}^+$, an *r-net* for P [8] is a subset $X \subseteq P$ such that,

$$\max_{p \in P} \min_{x \in X} \mathrm{dist}(p,x) \; < \; r \qquad \text{and} \qquad \min_{\substack{x,y \in X \\ x \neq y}} \mathrm{dist}(x,y) \; \geq \; r.$$

The first constraint is called the *covering constraint*, and the second is called the *packing constraint*. Each point $p \in P$ may be associated with a *representative* $x \in X$ (denoted by $\mathrm{rep}(p)$) lying within distance r. No two representatives are closer than r. Note that the choice of representative is not necessarily unique.

In order to establish our competitive ratio, we will need to relax the r-net definition slightly. Given constants $\alpha, \beta \geq 1$, an (α,β)*-slack r-net* is a subset $X \subseteq P$ of points such that,

$$\max_{p \in P} \min_{x \in X} \mathrm{dist}(p,x) \; < \; \alpha r \qquad \text{and} \qquad \forall x \in X, \; |\{X \cap b(x,r)\}| \; \leq \; \beta.$$

Thus, we allow each point to be farther from the closest net point by a factor of α, and we allow net points to be arbitrarily close to each other, but there cannot be more than β points within distance r of any net point. Clearly, an (α,β)-slack r-net is an (α',β')-slack r-net for $\alpha' \geq \alpha$ and $\beta' \geq \beta$. When we wish to make the distinction clearer, we will use the term *strict r-net* to denote the standard definition, which arises as a special case when $\alpha = \beta = 1$.

Before introducing net trees, we first introduce the concept of the *aspect ratio* (or *spread*) of a kinetic point set P. Consider two positive reals δ and Δ, which denote the minimum and maximum distances, respectively, between any two points of P throughout the course of the motion. We define $\Phi(P)$ to be Δ/δ. By scaling distances, we may assume that $\delta = 1$.

A *net tree* of P is defined as follows. The leaves of the tree consists of the points of P. Note that by our assumption that $\delta = 1$, these form a 1-net of P itself, which we denote by $P^{(0)} = P$. The tree is based on a series of nets, $P^{(1)}, P^{(2)}, \ldots, P^{(m)}$, where $m = \lceil \log_2 \Phi \rceil$, and $P^{(i)}$ is a (2^i)-net for $P^{(i-1)}$. Observe that $|P^{(m)}| = 1$. Recall that, for each $p \in P^{(i-1)}$, there is a point $x \in P^{(i)}$, called its representative, such that $\mathrm{dist}(p,x) \leq 2^i$. We declare this point to be a *parent* of p, which (together with the fact that $|P^{(m)}| = 1$) implies that the resulting structure is a rooted tree. An easy consequence of the packing and covering constraints is that the number of children of any node of this tree is a constant (depending on the doubling constant of the containing metric space). Our definition is based on the simplest forms of net tree [5,11], in contrast to more sophisticated forms given elsewhere [8,4,6].

This definition can be easily generalized to assume that the nets forming each level of the tree are (α, β)-slack nets. We refer to such a tree as an (α, β)-slack net tree. Assuming that α and β are constants, this relaxation will affect only the constant factors in the asymptotic complexity bounds.

Because our ultimate interest is in maintaining net trees under incremental motion, it will be convenient to impose an additional constraint on the points P. In a net tree, the input to the ith level of the tree is a (2^{i-1})-net, from which we are to compute a 2^i net. Thus, in our computation of an r-net, we will assume that the point set P is an $(r/2)$-net.

Recall that we interested in maintaining points under incremental black-box motion. More formally, we assume that the points change locations synchronously at discrete time steps $T = \{0, 1, \ldots, t_{\max}\}$. Given a point $p \in P$ and $t \in T$, we use p to refer to the point in the symbolic sense, and (when time is significant) we use p_t to denote its position at time t.

Let us now consider the certificates used in the maintenance of an r-net. In order to maintain an (α, β)-slack r-net, the observer must be provided enough information to verify that the covering and packing constraints are satisfied. At any time t, let P_t denote the current point set, and let X_t denote the current slack net. We assume the incremental maintenance of any net is based on the following two types of certificates, where the former validates the covering constraint and the latter validates the packing constraint.

Assignment Certificate(p, x): For $p \in P$ and $x \in X$, $\mathrm{rep}(p) = x$, and therefore at each time t, $\mathrm{dist}(p_t, x_t) < \alpha r$.
Packing Certificate(x): For $x \in X$, at each time t we have $|X_t \cap b(x_t, r)| \leq \beta$.

The first condition requires constant time to verify. The second condition involves answering a spherical range counting query. Observe that $O(|P|)$ certificates suffice to maintain the net.

3 Incremental Maintenance of a Slack Net

In this section we present an online algorithm for maintaining an r-net for a set of points undergoing incremental motion, and we provide an analysis of its competitive ratio. Given our metric space $\mathcal{M}$, let $\beta = \beta(\mathcal{M})$ denote the maximum number of balls of radius r that can overlap an arbitrary ball of radius r, such that the centers of these balls are at distance at least r from each other. We shall show in Lemma 1(iii) below that $\beta \leq \lambda^2 = 4^d$, where λ is the doubling constant of $\mathcal{M}$, and d is the doubling dimension of $\mathcal{M}$. The main result of this section is as follows.

Theorem 1. *Consider any metric space $\mathcal{M}$ of constant doubling dimension, and let $\beta = \beta(\mathcal{M})$ be as defined above. There exists an incremental online algorithm, which for any real $r > 0$, maintains a $(2, \beta)$-slack r-net for any point set P under incremental motion in $\mathcal{M}$. Under the assumption that P is a $(2, \beta)$-slack $(r/2)$-net, the algorithm achieves a competitive ratio of at most $(\beta\lambda^3 + 2)(\beta + 2) = O(\lambda^7) = O(1)$.*

The algorithm begins by inputting the initial placements of the points, and it communicates an initial set of certificates to the observer. (We will discuss how this is done below.) Recall that the observer then monitors the point motions over time, until first arriving at a time step t when one or more of these certificates is violated or when a point of P is explicitly inserted or deleted. It then wakes up the builder and informs it of the current event.

Our algorithm maintains the points of the slack r-net, which we denote by X, and the assignment of each point $p \in P$ to its representative $\text{rep}(p) \in X$. For each point $p \in P$, we maintain a subset $\text{cand}(p) \subseteq X$, called the *candidate list.* Before describing the update process, we present the following utility operations.

Reassign(p): If $\text{cand}(p) \neq \emptyset$, repeatedly extract elements from $\text{cand}(p)$ until finding a candidate $x \in X$ that lies within distance $2r$ of p. If such a candidate is found, set $\text{rep}(p) \leftarrow x$, and create a new assignment certificate involving p and x. If no such candidate exists, invoke Create-net-point(p).

Create-net-point(p): We assume the precondition that $\text{cand}(p) = \emptyset$. Add p to the current net X. Set $\text{rep}(p) \leftarrow p$. For each point $p' \in P \setminus \{p\}$ such that $\text{dist}(p', p) < 2r$, add p to $\text{cand}(p')$. Finally, create a packing certificate for p.

Remove-net-point(x): First, x is removed from both X and all the candidate lists that contain it. Remove any packing certificate involving x. For all $p \in P$ such that $\text{rep}(p) = x$, invoke Reassign(p).

Note that the reassignment operation generates one new assignment certificate (for p) and may create one packing certificate if p is added to X. Let us now consider the possible actions of the builder, in response to any event (point insertion or deletion) or a certificate violation (assignment or packing).

Insert-point(p): Set $\text{cand}(p)$ to be the set of net points $x \in X$ such that $\text{dist}(p, x) < 2r$. Then invoke Reassign(p).

Delete-point(p): All certificates involving p are removed. If $p \in X$, then invoke Remove-net-point(p). Finally, remove p from P.

Assignment-certificate-violation(p): Let $x = \text{rep}(p)$. Remove x from p's candidate list and invoke Reassign(p).

Packing-certificate-violation(x): Invoke Remove-net-point(x), for each $x \in X \cap b(x, r)$. (This results in at least $\beta + 1$ removals.)

Observe that the processing of any of the above events results in a constant number of changes to the certificate set, and hence in order to account for the total communication complexity, it suffices to count the number of operations performed.

Initially, X is the empty set, and we start the process off by invoking the insertion operation for each $p \in P$, placing it at its starting location. Observe that after the processing of each assignment- and packing-certificate violation, the condition causing the violation has been eliminated, and therefore it is easy to see that the above protocol maintains a $(2, \beta)$-slack r-net for P.

Recall that P denotes the point set, which is in motion of some finite time period. For any time t, let $N_t(o)$ denote the *optimal neighborhood* consisting of the points of P that have o assigned as their representative by the optimal algorithm. We will show that each of the operations performed by our algorithm can be charged to some operation of the optimal algorithm, in such a manner that each optimal operation is charged a constant number of times.

We first present some geometric preliminaries, which will be useful later in the analysis. Recall that λ denotes the doubling constant of the metric space. Due to space limitations, the proofs of the lemmas appear in the full version of this paper [3].

Lemma 1. *(i) Given any (α, β)-slack r-net X, at most $\beta\lambda^{\lg\lceil 2R/r\rceil}$ points of X can lie within any ball of radius R.*
(ii) Let P be an (α, β)-slack $(r/2)$-net, and let X be an (α, β)-slack r-net for P. Then the number of points of X (respectively, P) that lie within a ball of radius $2^k r$ is at most $\beta\lambda^{k+1}$ (respectively, $\beta\lambda^{k+2}$).
(iii) Let Z be a set of balls of radius r whose centers are taken from a (strict) r-net. Then any ball b of radius r (not necessarily in Z) can have a nonempty intersection with at most λ^2 balls of Z.

Given the motion sequence for the point set P, let n denote the total number of operations performed by our online slack-net algorithm, and let n^* denote the total number of operations processed by any correct (e.g., the optimal) algorithm. In order to establish the competitive ratio, it suffices to show

$$n \leq (\beta\lambda^3 + 2)(\beta + 2)n^*. \tag{1}$$

Let n_A^*, n_C^*, n_R^*, n_I^*, and n_D^*, and denote, respectively, the total number of assignments, net point creations, net-point removals, point insertions, and point deletions performed by the optimal algorithm. Let n_A, n_C, n_R, n_I, and n_D denote corresponding quantities for our slack-net algorithm. Thus, we have $n = n_A + n_C + n_R + n_I + n_D$.

First, we bound the total number of assignments in terms of the number of point insertions and slack-net creations. Changes in assignment in our algorithm occur as a result of running of the reassignment operator. Since the assignment is made to some point of the candidate list, it suffices to bound the total number of insertions into candidate lists. This occurs when points are inserted and when net points are created.

Lemma 2. $n_A \leq \beta\lambda^3 n_C + \beta\lambda^2 n_I$.

Since point insertions and deletions must be handled by any correct algorithm, we have $n_I = n_I^*$ and $n_D = n_D^*$. The total number of net point removals (n_R) cannot exceed the total number of net point creations (n_C). Thus, it suffices to bound n_C, the total number of slack-net point creations.

Before bounding n_C, we make a useful observation. Whenever a net point x is created, it adds itself to the candidate list of all points that lie within distance $2r$. Since (by strictness) the diameter of any optimal neighborhood is at most $2r$, it follows that, in the absence of other events, the creation of net-point x within an optimal neighborhood inhibits the creation of any other net points within this neighborhood. Given $t \leq t'$, let $(t, t']$ denote the interval $[t + 1, t']$.

Lemma 3. *Let o denote an optimal net point. Suppose that no optimal assignments occur to the points of $N(o)$ during the time interval $(t, t']$, $x \in N(o)$ is added to X at time t, and x is not removed from X throughout $(t, t']$. Then, for any time in $(t, t']$ no point $p \in N(o)$ will be added to X.*

This implies that, without optimal assignments, each optimal net point o can have at most one corresponding slack net point $x \in N(o)$. Furthermore, if x is removed from X (e.g., as the result of a packing-certificate violation), only one of the points of $N(o)$ may replace it as a slack-net point. When a net point within an optimal neighborhood is created to replace a removed net point, we call this a *recovery*. Whenever a packing-certification violation occurs, at least $\beta+1$ slack-net points are removed. The following lemma implies that the number of recovered net points is strictly smaller.

Lemma 4. *The number of net points recovered as a result of the processing of a packing-certificate violation is at most β.*

We say that an optimal neighborhood $N(o)$ is *crowded* (at some time t) if $|N_t(o) \cap X_t| \geq 2$. Our next lemma states that whenever a packing-certificate violation occurs, at least two of removed net points lie in the same crowded neighborhood.

Lemma 5. *Consider a packing-certificate violation which occurs in the slack net but not within the optimal net, and let $X' \subseteq X$ denote the net points that have been removed as a result of its handling. Let O' denote a subset of O of overlapping neighborhoods, that is, $O' = \{o \mid N(o) \cap X' \neq \emptyset\}$. Then, there exists $o \in O'$ such that $|N(o) \cap X'| \geq 2$.*

The handling of the packing-certificate violation removes the points of X', and by Lemma 3, this optimal neighborhood will recover at most one net point. Thus, in the absence of other effects (optimal reassignment in particular) the overall crowdedness of the system strictly decreases after processing each packing-certificate violation. Intuitively, crowdedness increases whenever a point of the slack net is moved from one optimal neighborhood to another. But such an event implies that the optimal algorithm has modified an existing assignment. We can therefore charge each slack-net point creation to such an optimal reassignment.

We summarize that above analysis to obtain the following bound.

Lemma 6. $n_C \leq (\beta + 2)\, n_A^* + n_C^* + n_D^*$.

The proof of Theorem 1 follows by combining the observations of this section with Lemmas 2 and 6.

4 Incremental Maintenance Algorithm for Net Tree

The main result of this section is presented below. In contrast to the results of the previous section, the slackness parameter α in the net increases from 2 to 4 and the competitive ratio increases by a factor of $O(\lambda h^2)$. Recall from Section 3 that $\beta = \lambda^2$.

Theorem 2. *Consider any metric space $\mathcal{M}$ of constant doubling dimension. There exists an online algorithm, which maintains a $(4, \beta)$-slack net tree for any point set P under incremental motion in $\mathcal{M}$. The algorithm achieves a competitive ratio of at most $(\beta\lambda^4 + \beta\lambda^3 + 4)(\beta + 3)h^2 = O(\lambda^8 h^2) = O(h^2)$.*

Intuitively, our algorithm for maintaining the net tree is based on applying the algorithm of the previous section to maintain the net defining each level of the tree. Creation and removal of net points at level will result in point insertions and deletions at other levels of the tree. Let $O^{(i)}$ and $X^{(i)}$ denote nets at level i of the tree generated by (strict) optimal algorithm and our slack-net algorithm, respectively. It will be convenient to use $P^{(i)}$ as a pseudonym for $X^{(i-1)}$.

The competitive analysis of the previous section was based on the relationship between slack-net points and neighborhoods of the optimal net. However, except at the leaf level, the points of $P^{(i)}$ need not reside in level i of the optimal net tree. Consider an optimal net point $o \in O^{(i)}$. We define the optimal neighborhood, still denoted by $N(o)$, to be the set of points of P lying in the leaves of the tree that are descended from o. Because of the exponential decrease in the radius values, it is easy to see that the descendants of o lie within distance of o of $2^i + 2^{i-1} + \cdots + 1 < 2^{i+1} = 2r_i$. Thus, the diameter of $N(o)$ is at most $4r_i$. This implies that, if we choose any point $x \in N(o)$ to be in our $(4, \beta)$-slack net, it can be used to cover all the points of the optimal neighborhood.

Let us now consider the operations performed by our algorithm at level i. Each operation is defined in terms of the analogous single-net operation of Section 3 applied to the points at this level of the net tree. In each case the operation may cause events to propagate to higher levels of the tree. We begin by describing a few utility operations. Throughout i denotes the tree level at which the operation is being applied.

Tree-reassign(p, i)**:** Invoke Reassign(p) on $X^{(i)}$, but use Tree-create-net-point at level-i (rather than Create-net-point).

Tree-create-net-point(p, i)**:** Invoke Create-net-point(p) on $X^{(i)}$, with one difference. The point p is added to the candidate lists of points within distance $4r_i$ (rather than $2r_i$). Invoke Tree-insert-point$(p, i + 1)$.

Tree-remove-net-point(x, i)**:** First, invoke Remove-net-point(x) on $X^{(i)}$, and then invoke Tree-delete-point$(x, i + 1)$.

With the aid of these utility operations, we now present the actions taken by the builder in response to the various events.

Tree-insert-point(p, i)**:** Invoke Insert-point(p) on $P^{(i)}$, but with the following change. Set cand(p) to be the set of net points $x \in X^{(i)}$ such that dist$(p, x) < 4r_i$ (rather than $2r_i$).

Tree-delete-point(p, i)**:** Invoke Delete-point(p) on $P^{(i)}$. If $p \in X^{(i)}$, invoke Tree-remove-net-point(p, i).

Tree-assignment-certificate violation(p, i)**:** Invoke the single-net assignment certificate violation for p on $P^{(i)}$, but use Tree-reassign(p, i) (rather than Reassign(p)).

Tree-packing-certificate-violation(x, i): Invoke the single-net packing certificate violation on $X^{(i)}$, but use Tree-remove-net-point(x, i) (rather than Remove-net-point(x)).

As with single-net operations, each operation may induce addition certificate violations. These violations are stored in a priority queue, and violations are first applied to the lower levels of the tree and then propagate upwards.

Due to space limitations, we have omitted the competitive analysis for above the net tree algorithm. The analysis appears in the full version of the paper [3]. The analysis involves showing that each operation performed by our algorithm can be charged to some operation performed by the optimal algorithm, such that each optimal operation is charged at most $O(h^2)$ times, where h is the height of the tree.

References

1. Adams, B., Pauly, M., Keiser, R., Guibas, L.J.: Adaptively sampled particle fluids. In: ACM SIGGRAPH, p. 48 (2007)
2. Chib, S., Greenberg, E.: Understanding the metropolis-hastings algorithm. The American Statistician 49, 327–335 (1995)
3. Cho, M., Mount, D.M., Park, E.: Maintaining nets and net trees under incremental motion. Technical Report CS-TR-4941, UMIACS-TR-2009-10, University of Maryland, College Park (2009)
4. Cole, R., Gottlieb, L.-A.: Searching dynamic point sets in spaces with bounded doubling dimension. In: Proc. 38th Annu. ACM Sympos. Theory Comput., pp. 574–583 (2006)
5. Gao, J., Guibas, L.J., Nguyen, A.: Deformable spanners and applications. Comput. Geom. Theory Appl. 35, 2–19 (2006)
6. Gottlieb, L.-A., Roditty, L.: An optimal dynamic spanner for doubling metric spaces. In: Halperin, D., Mehlhorn, K. (eds.) Esa 2008. LNCS, vol. 5193, pp. 478–489. Springer, Heidelberg (2008)
7. Guibas, L.: Kinetic data structures. In: Mehta, D., Sahni, S. (eds.) Handbook of Data Structures and App., pp. 23–1–23–18. Chapman and Hall/CRC, Boca Raton (2004)
8. Har-Peled, S., Mendel, M.: Fast construction of nets in low dimensional metrics, and their applications. SIAM J. Comput. 35, 1148–1184 (2006)
9. Kahan, S.: A model for data in motion. Proc. 23rd Annu. ACM Sympos. Theory Comput., 265–277 (1991)
10. Kirkpatrick, S., Gelatt, C.D., Vecchi, M.P.: Optimization by simulated annealing. Science 220(4598), 671–680 (1983)
11. Krauthgamer, R., Lee, J.R.: Navigating nets: Simple algorithms for proximity search. In: Proc. 15th Annu. ACM-SIAM Sympos. Discrete Algorithms, pp. 798–807 (2004)
12. Monaghan, J.J.: Smoothed particle hydrodynamics. Reports on Progress in Physics 68, 1703–1759 (2005)
13. Mount, D.M., Netanyahu, N.S., Piatko, C., Silverman, R., Wu, A.Y.: A computational framework for incremental motion. In: Proc. 20th Annu. ACM Sympos. Comput. Geom., pp. 200–209 (2004)
14. Yi, K., Zhang, Q.: Multi-dimensional online tracking. In: Proc. 20th Annu. ACM-SIAM Sympos. Discrete Algorithms, pp. 1098–1107 (2009)

Distributed Scheduling of Parallel Hybrid Computations

Shivali Agarwal[1], Ankur Narang[1], and Rudrapatna K. Shyamasundar[2]

[1] IBM India Research Laboratory, New Delhi
{shivaaga,annarang}@in.ibm.com
[2] Tata Institute of Fundamental Research, Mumbai
shyam@tifr.res.in

Abstract. Multicore computing is fast becoming the norm. Improving parallel programming productivity without compromising performance on multicores is a serious challenge facing research community and systems vendors. Towards this end, efficient run-time scheduling of parallel programs helps programmer by dynamically mapping tasks onto processors and scheduling them in appropriate order. Distributed scheduling of parallel computations on multiple *places*[1] while ensuring low time and message complexity in bounded space is a very challenging problem. We attempt to address this challenge for *hybrid* parallel computations which contain tasks that have pre-specified affinity to a place and also tasks that can be mapped to any place in the system. This paper presents online distributed scheduling algorithms for *hybrid* parallel computations assuming both unconstrained and bounded space per place. We also present the time and message complexity for distributed scheduling of *hybrid* computations. To the best of our knowledge, this is the first time distributed scheduling algorithms for *hybrid* parallel computations have been presented and analyzed for time and message bounds under both unconstrained space and bounded space.

Keywords: Work Stealing; Scheduling; Multithreaded Computation; Algorithm.

1 Introduction

Multicore computing is fast becoming the norm. Improving parallel programming productivity without compromising performance on multicores is a serious challenge facing research community and systems vendors. Towards this end, efficient run-time scheduling of parallel programs helps programmer by dynamically mapping tasks onto processors and scheduling them in appropriate order. The DARPA HPCS[2] program is geared towards developing languages and run-time systems that increase programmer productivity. These include X10 [1], Chapel[3] and Fortress[4] which are based on partitioned global address space (PGAS[5]) paradigm. These languages have in-built support for initial placement (mapping of tasks to places) of parallel programs and therefore

[1] Place is a group of processors with shared memory.
[2] www.highproductivity.org
[3] http://chapel.cs.washington.edu
[4] http://research.sun.com/projects/plrg
[5] http://x10-lang.org/

Y. Dong, D.-Z. Du, and O. Ibarra (Eds.): ISAAC 2009, LNCS 5878, pp. 1144–1154, 2009.

data locality comes implicitly with the programs. We refer to activities(threads) that have pre-specified placement as *affinity annotated* activities. Further, there are activities (threads) in the parallel program that can be run on any place. We call such activities *anyplace activities*. Parallel computations that have both affinity annotated activities and anyplace activities are referred to as *hybrid* parallel computations. The run-time system needs to provide online distributed scheduling of large hybrid parallel computations on many-core and massively parallel architectures.

The two distributed scheduling problems we address are as follows: Given, **(a)** An input hybrid computation DAG that represents a parallel computation with fine to medium grained parallelism. It is assumed to have no logical deadlocks due to dependencies; **(b)** A cluster of n SMPs (each SMP[6] also referred to as *place*, has fixed number(m) of processors and memory) as the target architecture on which to schedule the computation DAG. For both problems one needs to map the anyplace activities onto places and generate a schedule for all the nodes of the computation DAG in an online and distributed fashion. Specifically, for the first problem we assume that the input is a *strict* (section 2) computation DAG and there is unconstrained(sufficient) space per place. Here, we need to design a distributed scheduling algorithm that minimizes the time and message complexity. For the second problem we assume that the input is a *terminally strict*(section 2) parallel computation DAG and the space per place is bounded. Here, the aim is to ensure *physical* (due to cyclic resource dependency) deadlock free execution while keeping low time and message complexity for execution.

Scheduling of dynamically created tasks for shared memory multi-processors has been a well studied problem. The work on Cilk [2] promoted the strategy of *randomized work stealing*. Here, a processor that has no work (*thief*) randomly steals work from another processor (*victim*) in the system. [2] proved efficient bounds on space ($O(P \cdot S_1)$) and time ($O(T_1/P + T_\infty)$) for scheduling of *fully-strict* (section 2) computations in an SMP platform; where P is the number of processors, T_1 and S_1 are the time and space for sequential execution respectively, and T_∞ is execution time on infinite processors. [3] analyzed time complexity ($O(T_1/P + T_\infty)$) for scheduling *general* parallel computations on SMP platform but does not consider space or message complexity bounds. [4] considers data locality oriented work stealing in an SMP but does not provide time complexity analysis.

[5] considers work-stealing algorithms in a distributed-memory environment, with adaptive parallelism and fault-tolerance. Here task migration was entirely pull-based (via a randomized work stealing algorithm) hence it ignored affinity and also didn't provide any formal proof for the deadlock-freedom or resource utilization properties. The work in [6] described a *multi-place*(distributed) deployment for parallel computations for which initial placement based scheduling strategy is appropriate. A *multi-place* deployment has multiple places connected by an interconnection network where each *place* has multiple processors connected as in an SMP platform. [6] also provided a scheduling strategy based on initial placement and proved space bounds for physical deadlock free execution of *terminally strict* (section 2) computations. It resorted to a degenerate mode called Doppelgänger mode to prevent physical deadlocks due to cyclic resource dependency. The computation did not respect affinity in this mode and no time

[6] Symmetric MultiProcessor: group of processors with shared memory.

or communication bounds were provided. Also, the aspect of load balancing was not addressed. [7] considers affinity driven distributed scheduling of multi-place parallel computations with physical deadlock freedom (using distributed deadlock avoidance strategy). It provides time and message complexity lower and upper bounds and deadlock freedom proof but the algorithms and proofs are limited to affinity driven activities and it assumes same distances between places on the cluster. Here, the assumption of all activities to be affinity driven forces the programmer to ensure load-balance across places which reduces productivity and performance especially when dynamic load-balancing is needed.

In this paper, we relax this assumption by additionally allowing anyplace activities in the input *hybrid* computation DAG. This generalization allows more parallel applications to be expressed easily by the programmer. Also, we design novel distributed scheduling algorithms that incorporate inter-place prioritized random work stealing to provide *automatic dynamic load balancing* across places. It is proved that with suitable choice of probability distribution, the prioritized random work stealing across places is efficient. Further, it leads to low average communication cost when the distances between the places are different (e.g. 3D torus interconnect). This paper leverages the distributed deadlock avoidance strategy for deadlock free execution and time and message complexity proofs in [7] for efficient scheduling of hybrid parallel computations. To the best of our knowledge this is the first work on distributed scheduling algorithms for **hybrid** parallel computations in a *multi-place* setup for both unconstrained and bounded space. Our main contributions are:

- We present an online multi-place distributed scheduling algorithm for *strict* multi-place hybrid parallel computations assuming unconstrained (sufficient) space per place. This algorithm incorporates (a) intra-place work stealing, (b) remote place work pushing for affinity annotated activities and (c) prioritized random work stealing across places for anyplace activities. We show that prioritized random stealing across places is efficient. We also present the time and message complexity bounds of the scheduling algorithm.
- For bounded space per place, we present a novel distributed scheduling algorithm for terminally strict multi-place hybrid computations with provable physical deadlock free execution.

2 System and Computation Model

The system on which the *computation DAG* is scheduled is assumed to be cluster of *SMPs* connected by an *Active Message Network*. Each *SMP* is a group of processors with shared memory. Each SMP is also referred to as *place* in the paper. Active Messages (*(AM)*[7] is a low-level lightweight RPC(remote procedure call) mechanism that supports unordered, reliable delivery of matched request/reply messages. We assume that there are n places and each place has m processors(also referred to as workers). Each place also has one processor called *interface processor* that deals with interaction between the places during remote pushing or remote stealing of activities. The

[7] Active Messages defined by the AM-2: http://www.lrr.in.tum.de/ weissc/am.html

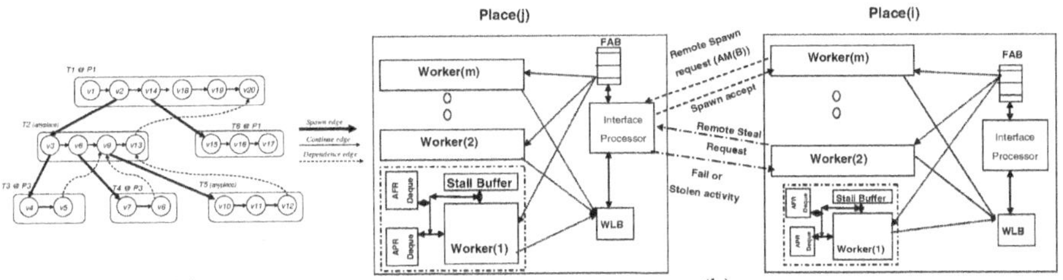

(a) (b)

Fig. 1. (a) Hybrid Multi-Place Computation Dag. (b) Distributed Scheduling.

interconnection network connecting the places(SMPs) can in general provide different distances (latencies) between the places such as 3D torus interconnect. Our scheduling algorithm considers this asymmetry in distances between places to optimize performance.

The parallel computation, to be dynamically scheduled on the system, is assumed to be specified by the programmer in languages such as X10 and Chapel. To describe our distributed scheduling algorithms, we assume that the parallel computation has a *DAG*(directed acyclic graph) structure and consists of nodes that represent basic operations like *and*, *or*, *not*, *add* and others. There are edges between the nodes in the computation DAG (Fig. 1(a)) that represent creation of new activities (*spawn* edge), sequential execution flow between nodes within a thread/activity (*continue* edge) and synchronization dependencies (*dependence* edge) between the nodes. In the paper we refer to the parallel computation to be scheduled as the *computation DAG*. At a higher level the parallel computation can also be viewed as a computation tree of *activities*. Each *activity* is a *thread* (as in multi-threaded programs) of execution and consists of a set of nodes (basic operations). Each activity is either assigned to a specific place (called *AF activity*) or, can be executed on any place (called *AP activity*). An AF activity shall be executed on the assigned place only. Such computation is called *hybrid multi-place* computation and DAG is referred to as *hybrid multi-place* computation DAG (Fig. 1(a): v1..v20 denote nodes, T1..T6 denote activities and P1..P3 denote places).

The structure of dependencies between the nodes can vary depending on the input parallel computation. In *fully-strict* and *strict* computations the dependencies can go from a node to its immediate parent and to any of its ancestors in the computation DAG, respectively. In a *terminally strict* computation, introduced in [6] and shown in Fig. 1(a), the dependencies arise due to an activity waiting for the completion of its descendants. Every dependency edge, therefore, goes from the last instruction of an activity to one of its ancestor activities with the following restriction: In a subtree rooted at an activity called Γ_r, if there exists a dependence edge from any activity in the subtree to the root activity Γ_r, then there cannot exist any dependence edge from the activities in the subtree to the ancestors of Γ_r. A *terminally strict multi place* computation is defined as a terminally strict computation where each activity has an affinity to a place.

2.1 Useful Notations

The set of places is denoted by $P = \{P_1, \cdots, P_n\}$. The set of workers at place P_i, is denoted by $\{W_i^1, W_i^2..W_i^m\}$. S_1 denotes the space required by a single processor execution schedule. The size in bytes of the largest activation frame in the computation is denoted by S_{max}. If node u enables node v then we place an edge, referred as *enable* edge from u to v. The tree formed over all nodes with enable edges is referred to as *enabling tree* [3]. $depth(u)$ denotes the distance of node u from the root in the enabling tree. The root node is assumed to be at depth 0. $T_{\infty,n}$ denotes the execution time of the computation DAG over n places with infinite processors at each place. T_{∞}^{k} denotes the execution time for activities at place P_k using infinite processors. Note that, $T_{\infty,n} \leq \sum_{1 \leq k \leq n} T_{\infty}^{k}$. T_1^k denotes the minimum time taken by a single processor for the activities assigned to place k. D_{max} denotes the maximum depth of the computation tree in terms of number of activities. The *depth* of an activity is defined as the distance from the root activity in the computation tree.

3 Distributed Scheduling in Unconstrained Space

Consider a *strict* multi-place hybrid computation DAG. During execution, the computation DAG unfolds in an online fashion in a breadth-first manner across places when the AF activities are pushed onto their respective remote places and when AP activities are stolen from remote places. Stealing work from remote places helps in load balancing across places. Within a place, the online unfolding of the computation DAG happens in a depth-first manner to enable efficient space and time execution. Within a place randomized work-stealing (referred as *local stealing*) of AF or AP activities, is enabled to allow load-balanced execution. If the worker fails to obtain activities using local steals then it tries to steal AP activities from other places in a distance prioritized random fashion (referred as *remote stealing*). Here, the remote place is selected randomly using a probability distribution that prefers closer places over farther places. This helps in reducing the communication cost associated with remote stealing on a cluster where the distances between the places are different (such as 3D torus interconnect). Since sufficient space is guaranteed to exist at each place, physical deadlocks due to lack of space cannot happen in this algorithm.

3.1 Algorithm Design

Each place maintains one *Fresh Activity Buffer* (*FAB*) which is managed by the *interface* processor at that place. An activity that has affinity for a remote place is pushed into the FAB at that place. Each worker at a place has: (a) an *APR Deque* that contains anyplace ready activities, (b) an *AFR Deque* that contains affinity annotated ready activities and (c) *Stall Buffer* that contains stalled activities (refer Fig. 1(b)). The FAB at each place as well as the AFR Deque and APR Deque at each worker are implemented using concurrent deque data-structure. Each place also maintains a *Worker List Buffer* (*WLB*) that is a list of workers that have anyplace activities ready to be stolen. WLB is implemented as a concurrent linked list and is maintained by the interface processor. WLB aids in *remote stealing* where the remote workers which attempt to steal activities from this place get information about available workers for stealing from WLB. The distributed scheduling algorithm is given in Fig. 2.

At any step, an activity A at the rth worker (at place i) W_i^r, may perform following actions:

1. **Spawn**:
   (a) A spawns activity B at place, P_j, $i \neq j$: A sends *AM(B)* (active message for B) to the interface processor at the remote place. Since, the remote place, P_j, is guaranteed to have memory for activity B, it is successfully inserted in the FAB at P_j and A continues execution (Fig. 1(b)).
   (b) A spawns B locally: B is successfully created and starts execution whereas A is pushed into the bottom of the APR Deque if A is an anyplace activity else push A into the bottom of the AFR Deque.
2. **Terminates** (A terminates): The worker at place P_i, W_i^r, where A terminated, picks an activity from the bottom of the AFR Deque for execution. If none available in its AFR Deque, then it picks activity from the bottom of APR Deque. If none obtained then same as *Empty Deque* (case 3) below.
3. **Empty Deque**: Worker W_i^r has both AFR Deque and APR Deque as empty.
   (a) *Local Stealing:* Worker attempts stealing from the top of other local workers' AFR Deque or APR Deque. If successful, start execution of stolen activity and update WLB, else attempt pick from FAB.
   (b) *Pick from FAB:* Pick from FAB at place P_i. If successful start execution of new activity and update FAB, else attempt remote stealing.
   (c) *Remote Stealing:* Worker attempts to steal from remote place chosen randomly, P_j. W_i^r send remote steal request to interface processor at P_j. Interface processor looks at the WLB at place P_j and attempts local steal of anyplace activity from available worker, W_j^s. If this succeeds it passes this stolen activity to the worker, W_i^r, which requested remote steal and updates WLB at P_j. W_i^r starts execution of stolen anyplace activity (Fig. 1(b)). If remote steal fails (no available anyplace activity) then repeat from start *Empty Deque* (case 3).
4. **Stalls** (A stalls): An activity may stall due to dependencies in which case it is put in the stall buffer in a stalled state. Then same as *Terminates* (case 2) above.
5. **Enables** (A enables B): The termination of an activity A may enable a stalled activity B in which case the state of B changes to enabled and it is pushed onto the top of the APR Deque if it is an anyplace activity else it is pushed on top of the AFR Deque.

Fig. 2. Distributed Scheduling Algorithm

3.2 Time Complexity Analysis

The detailed time complexity analysis using potential function on ready nodes in the system follows as in [3] [7] . In this section, we give a brief intuitive explanation of time and message complexity. Our unique contributions are (a) proof that prioritized random inter-place work stealing is efficient using suitable probability density function, (b) proof of the lower and upper bounds of time complexity and message complexity for the **multi-place** distributed scheduling algorithm presented in section 3.1 that includes (1) *intra-place work stealing*, (2) *remote-place work stealing* and (3) *remote place affinity driven work pushing*.

Below, *throw* represents an attempt by a worker(*thief*) to steal an activity. It can be an *intra-place* throw when the activity is stolen from another local worker(*victim*), or *remote-place* throw when it is stolen from a remote place. For potential function based analysis, each ready node u is assigned a potential $3^{2w(u)-1}$ or $3^{2w(u)}$ depending upon whether it is assigned for execution or not ($w(u) = T_{\infty,n} - depth(u)$). The total potential of the system at step i is denoted by ϕ_i and $\phi_i(D_i)$ denotes potential of all APR Deques and AFR Deques that have some ready nodes.

Prioritized Random Inter-Place Work Stealing. We prove that distance-prioritized inter-place work stealing works efficiently with suitable choice of probability distribution across places. Consider a 2D torus interconnect across places. Let the place where a processor attempts to steal be denoted by the *start* place. The places around the start place can be viewed as rings. The rings increase in size as we move to rings at increasing distance from the start place, i.e. there are more places in a ring farther away from the start place than the ring closer to the start place. (refer Fig. 3). In a remote steal

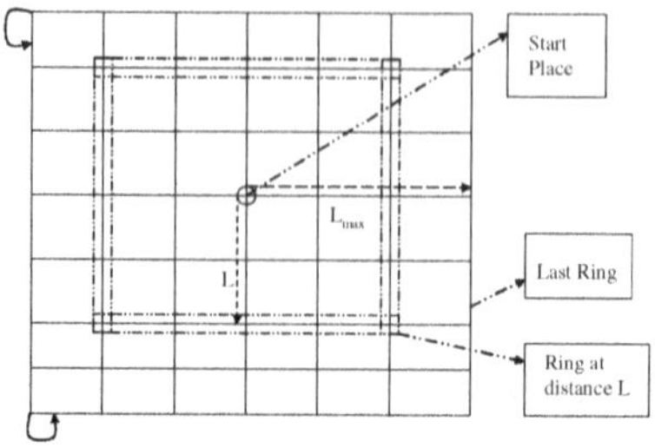

Fig. 3. 2D Torus with rings of places

attempt from the start place, the places on the same ring are chosen with equal probability. This probability decreases with increasing ring distance from the start place but the total probability of choosing a processor over all processors across all places should be equal to 1. In order to model this scenario, consider a generalized Balls and Weighted Bins game where P balls are thrown independently but *non-uniformly* at random into P bins. We derive an upper bound on the probability of the un-successful steal attempts using Markov's inequality.

Lemma 1. Prioritized Balls and Weighted Bins Game: *Let there be n places arranged in a 2D torus topology. Suppose that P balls are thrown independently but non-uniformly at random into P bins, where for i = 1,...,P, bin i has a weight $W(i)$. The total weight is $W = \sum_{1 \leq i \leq P} W(i)$. For each bin i, define a random variable $X(i)$ as: (1) $X(i) = W(i)$, if some ball lands in bin(i)*
(2) $X(i) = 0$, otherwise
Let, l_{max} be the distance of the start place from the last ring. Define the probability distribution of choosing rings as follows. Let, γ/l_{max} be the probability of choosing the last ring at distance l_{max} from the source of the steal request, where $0 < \gamma < 1$. The probability of selecting other rings is chosen appropriately so that the sum of choosing processor across all processors equals 1. (For example, let $\gamma = 3/4$. Here, we assign a probability of $5/4l_{max}$ to each of the first $l_{max}/2$ rings and probability of $3/4l_{max}$ to each of the last $l_{max}/2$ rings.)
If $X = \sum_{1 \leq i \leq p} X(i)$, then for β in the range $0 < \beta < 1$, we have:
$PrX \geq \beta.W > 1 - 1/((1-\beta)e^{\gamma/2}$.

Proof. A ring at distance l from the start place has $8l$ places. Since, each place has m processors, the ring at distance l has $8lm$ processors and each of the processors have equal probability that a ball will land in that processor(bin).

Now, for each bin i, consider the random variable, $W(i) - X(i)$. It takes on a value $W(i)$ when no ball lands on bin(i) otherwise is take value 0. Thus, we have,

$$E[W(i) - X(i)] = W(i) * \text{Probability that no ball lands in bin(i)} \quad (3.1a)$$
$$\leq W(i) * [1 - \text{Min. prob. that any ball lands in bin(i)}]^P \quad (3.1b)$$
$$\leq W(i) * [1 - \gamma/(l_{max} \cdot 8l_{max}m)]^{mn} \quad (3.1c)$$
$$\leq W(i)/e^{(l_{max}+1).\gamma/(2.l_{max})} \quad (3.1d)$$

$$\because n = 4l_{max}(l_{max} + 1); (1 - 1/x)^x \leq 1/e$$
$$\leq W(i)/e^{\gamma/2}, \text{for large } l_{max} \quad (3.1e)$$

It follows that: $E[W - X] \leq W/e^{\gamma/2}$.
From Markov's inequality we have:
$Pr\{(W - X) > (1 - \beta)W\} < E[W - X]/((1 - \beta).W)$
$\Rightarrow Pr\{X < \beta.W\} \leq 1/((1 - \beta).e^{\gamma/2})$
$\Rightarrow Pr\{X \geq \beta.W\} > 1 - 1/((1 - \beta)e^{\gamma/2}$

We can see that due to skewed probability of balls choosing which bin to go, the probability of successful attempts goes down compared to the case of uniform probability [7]. Even though we chose ring distance based probability variation, actual processor distance based probability variation can be similarly analyzed with suitable probability distribution. By choosing $\beta = 1/5$, $\gamma = 3/4$ one can show that after $O(mn)$ remote place throws across the system, the potential of anyplace ready activities in $\phi_i(D_i)$ decreases by at least $1/16$. The time and message complexity lower and upper bounds are given by theorems below. Detailed proofs follow by extending the analysis in [7].

Theorem 1. *Consider a strict multi-place hybrid computation DAG with work for place P_k, denoted by T_1^k, being executed by the distributed scheduling algorithm (section 3.1). Let the critical-path length for the computation be $T_{\infty,n}$. The lower bound on the expected execution time is $O(\max_k T_1^k/m + T_{\infty,n})$ and the upper bound is $O(\sum_k (T_1^k/m + T_\infty^k))$. Moreover, for any $\epsilon > 0$, the lower bound for the execution time is $O(\max_k T_1^k/m + T_{\infty,n} + \log(1/\epsilon))$ with probability at least $1 - \epsilon$. Similar probabilistic upper bound exists.*

Theorem 2. *Consider the execution of a strict hybrid multi-place computation DAG with critical path-length $T_{\infty,n}$ by the Distributed Scheduling Algorithm (section 3.1). Then, the total number of bytes communicated across places has the expectation $O(I \cdot S_{max} \cdot n_d) + m \cdot T_{\infty,n} \cdot S_{max} \cdot n_d)$. Further, the lower bound on number of bytes communicated within a place has the expectation $O(m \cdot T_{\infty,n} \cdot S_{max} \cdot n_d)$, where n_d is the maximum number of dependence edges from the descendants to a parent and I is the number of remote spawns from one place to a remote place. Moreover, for any $\epsilon > 0$, the probability is at least $(1 - \epsilon)$ that the lower bound on the intra-place communication overhead per place is $O(m \cdot (T_{\infty,n} + \log(1/\epsilon)) \cdot n_d \cdot S_{max})$. Similarly message upper bounds exist.*

4 Distributed Scheduling of Hybrid Computation in Bounded Space

Due to limited space on real systems, the distributed scheduling algorithm has to limit online breadth first expansion of the computation DAG while minimizing the impact on execution time and simultaneously providing deadlock freedom guarantee. Due to bounded space constraints this distributed online scheduling algorithm has guaranteed deadlock free execution for *terminally strict* multi-place hybrid computations. Due to space constraints at each place in the system, the algorithm needs to keep track of space availability at each worker and place to ensure physical deadlock freedom. It does so by ensuring that remote activity pushing, inter-place stealing and intra-place stealing happen only when there is sufficient space to execute the remaining path to the leaf in the current path. This tracking of available space and using depth based ordering of activities for execution from FAB help in ensuring distributed deadlock avoidance [7]. An activity can be in one of the following stalled states: (a) *local-stalled* due to lack of space at a worker, (b) *remote-stalled* due to failed spawn onto a remote place, (c) *depend-stalled* due to synchronization dependencies.

We assume that maximum depth of the computation tree (in terms of number of activities), D_{max}, can be estimated fairly accurately prior to the execution from the parameters used in the input parallel computation [7]. D_{max} value is used in our distributed scheduling algorithm to ensure physical deadlock free execution. The assumption on knowledge of D_{max} prior to execution holds true for the kernels and large applications of the **Java Grande Benchmark suite** [8].

4.1 Distributed Data-Structures and Algorithm Design

The data structures used for bounded space scheduling algorithm are described in Fig. 4.

Let *AM(T)* denote the active message for spawning the activity T. The activities in *remote-stalled* state are tracked using a linked list using activity IDs with the head and

Each worker has the following data-structures (Fig. 5(a)):

- ***AF PrQ***, ***AP PrQ*** and ***StallBuffer***: *AF PrQ* and *AP PrQ* are priority queues that contain AF activities and AP activities respectively. These activities can be either in ready state or *local-stalled* state. The *StallBuffer* contains activities in *depend-stalled* and *remote-stalled* states. The total current size of all these three data-structures is denoted by B_i^r and is kept bounded by $O(D_{max} \cdot S_{max})$ bytes.
- ***AFR Deque***: This contains AF activities (affinity annotated activities) in the current executing path on this worker. This has total space of $O(S_1)$ bytes.
- ***APR Deque***: This contains AP activities (anyplace activities) in the current executing path on this worker. This has total space of $O(S_1)$ bytes.
- ***AMRejectMap***: This is a one-to-one map from a place-id,say P_j, to the tuple $[U,\ AM(V)$, head, tail]. This tuple contains, $AM(V)$, the active message rejected in a remote-spawn attempt at place P_j; U, the activity stalled due to the rejected active message and head and tail of the linked list of activities in *remote-stalled* state due to lack of space on the place, P_j. This map occupies $O(n \cdot S_{max})$ space per worker.

Each place P_i, has the following data-structures (Fig. 5(a)):

- ***WLB***: Work List Buffer is a map from activity depth to list of workers. This can be implemented using a concurrent red-black tree and is managed by the interface processor. WLB is used to respond efficiently to remote steal requests coming from other places by providing instant list of available workers that have at least one AP activity to steal of the given or higher depth. From each worker the depth of the highest AP activity available to be stolen is kept at the WLB. It needs to be kept updated with updates onto *AP PrQ* and *AP Deque* of the workers at that place. It occupies $O(D_{max} + m)$ bytes in space.
- ***FAB***: Fresh Activity Buffer is a concurrent priority queue that is managed by the interface processor. It contains the fresh AF activities spawned by remote places onto this place. The current size of FAB is denoted by F_i and is bounded by $O(D_{max} \cdot S_{max})$ bytes.
- ***WorkRejectMap***: This is a one-to-many map from depth to list of workers. For each depth this map contains the list of workers whose spawns were rejected from this place. It occupies $O(m \cdot n + D_{max})$ space.

The priority queue (used for *AF PrQ*, *AP PrQ* and FAB) uses the *depth* of an activity as the priority with higher depth denoting higher priority. The *depth* of an activity is defined as the distance from the root activity in the computation tree.

Fig. 4. Multi-place Distributed Data-Structures

[8] http://www.epcc.ed.ac.uk/research/activities/java-grande

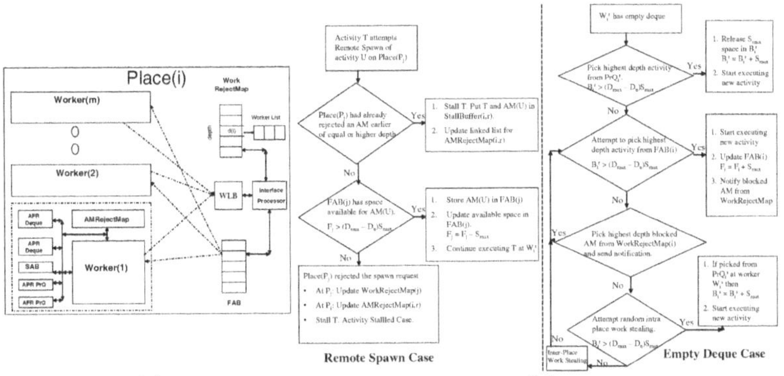

(a) (b)

Fig. 5. (a) Distributed Data Structures For Bounded Space Scheduling. (b) Remote Spawn and Empty Deque Case In Bounded Space Scheduling Algorithm.

At any time, a worker W_i^r takes the following actions. It might be executing an activity T(@depth D_t).

1. ***Local Spawn*: T spawns activity U locally**. T is pushed to the bottom of the *AF Deque* or *AP Deque* depending upon whether T is an AF activity or AP activity respectively. U starts executing.
2. ***Remote Spawn*: T attempts remote spawn of U(@depth D_u) at a remote place P_j, $i \neq j$**
// Refer Remote Spawn Case Flow Chart in Fig. 5(b)
3. ***Receives Notification*: W_i^r receives notification from place P_j on available space for spawn**
   (a) Get pair $< P_j, < R, AM(V), head(U), tail(S) >>$ from *AMRejectMap(i,r)*.
   (b) Send the tuple, $< AM(V), U >$, to Place P_j. Put R in $AFPrQ(i,r)$ or $APPrQ(i,r)$ depending upon whether R is an AF activity or AP activity.
   (c) Update head in the tuple for the pair with key as P_j as: $head = U \rightarrow$Next().
4. ***Termination*: T terminates**
   - if(*AF Deque(i,r)* is non-empty) then pick the bottommost activity from *AF Deque(i,r)*
   - else if(*AP Deque(i,r)* is non-empty) then pick the bottommost activity from *AP Deque(i,r)*
   - else { Same as case *Empty Deque*.}
5. ***Empty Deque*: W_i^r has both *AF Deque* and *AP Deque* as empty**
// Refer Empty Deque Case Flow Chart in Fig. 5(b)
6. ***Activity Enabled*: activity U gets enabled**
   (a) Set state of U to enabled.
   (b) Insert U in $AFPrQ(i,r)$ or $APPrQ(i,r)$ depending upon whether U is an AF activity or AP activity.
   (c) if(activity enabled was in state *remote-stalled*) then perform other actions as in the case *Receives Notification*.
7. ***Activity Stalled*: T (@depth D_t) stalls**
Let U(@depth D_u) be the next bottommost activity in *Deque(i,r)*.
   - State of T is changed to an appropriate stalled state. T is removed from *AF Deque / AP Deque*. If T is in *local-stalled* state (due to lack of space at worker) then it is moved into *AF PrQ* or *AP PrQ* else it is moved into *StallBuffer*. Update B_i^r. $B_i^r \leftarrow B_i^r - S_{max}$.
   - if($B_i^r > ((D_{max} - D_u) \cdot S_{max})$) { Execute U. }
   - else { Activity Stalled case for U.}

Fig. 6. Multi-place Distributed Scheduling Algorithm

tail of the list available at the tuple corresponding to the place in the map *AMRejectMap*. For notation purpose, the suffix *(i)* and *(i,r)* denote that data-structure is located at place P_i and worker W_i^r respectively.

Computation starts with root of the computation DAG which is at depth 1. The computation starts at a worker W_0^s, at the default place P_0. At any point of time a worker at a place, W_i^r, can either be executing an activity, T, or be idle. The detailed algorithm is presented in Fig. 6. The actions taken by the interface processor have been kept implicit in the description for sake of brevity.

Distributed deadlock freedom can be proved by induction as in [7] and has been left for brevity. The essence lies in showing that when an activity gets rejected then a higher depth activity must be executing at that place and then using induction one can show that all activities eventually become leaf and get executed starting from maximum depth activities and going backwards to lower depth activities as the space gets released by completed activities. The following theorem gives the space bound.

Theorem 3. *A terminally strict computation scheduled using algorithm in Fig 6 uses* $O(m \cdot (D_{max} \cdot S_{max} + n \cdot S_{max} + S_1))$ *bytes as space per place.*

The inter-place message complexity is same as theorem 2 (assuming similar order of number of throws for inter-place work stealing) as there is constant amount of work for handling rejected remote spawns and notification of space availability. For intra-place work stealing again the message complexity is same as theorem 2.

5 Conclusions and Future Work

We have presented the design of novel distributed scheduling algorithms for hybrid parallel computations for both bounded and unconstrained space along with time and message complexity bounds. This is the first work for distributed scheduling of hybrid parallel computations. In future, we plan to look into time complexity, space-time trade offs and multi-core implementations of the bounded space scheduling algorithm.

References

1. Charles, P., Donawa, C., Ebcioglu, K., Grothoff, C., Kielstra, A., von Praun, C., Saraswat, V., Sarkar, V.: X10: An object-oriented approach to non-uniform cluster computing. In: OOPSLA 2005 Onward! Track, San Diego, California, pp. 519–538 (2005)
2. Blumofe, R.D., Leiserson, C.E.: Scheduling multithreaded computations by work stealing. J. ACM 46(5), 720–748 (1999)
3. Arora, N.S., Blumofe, R.D., Plaxton, C.G.: Thread scheduling for multiprogrammed multiprocessors. In: SPAA, Puerto Vallarta, Mexico, pp. 119–129 (1998)
4. Acar, U.A., Blelloch, G.E., Blumofe, R.D.: The data locality of work stealing. In: SPAA, New York, NY, USA, December 2000, pp. 1–12 (2000)
5. Blumofe, R.D., Lisiecki, P.A.: Adaptive and reliable parallel computing on networks of workstations. In: USENIX Annual Technical Conference, Anaheim, California (1997)
6. Agarwal, S., Barik, R., Bonachea, D., Sarkar, V., Shyamasundar, R.K., Yellick, K.: Deadlock-free scheduling of x10 computations with bounded resources. In: SPAA, San Diego, CA, USA, December 2007, pp. 229–240 (2007)
7. Agarwal: S., Narang, A., Shyamasundar, R.K.: Affinity driven distributed scheduling algorithms for parallel computations. Technical Report RI09010, IBM India Research Labs, New Delhi (July 2009)

I/O-Efficient Contour Tree Simplification*

Lars Arge and Morten Revsbæk

MADALGO**, Department of Computer Science
University of Aarhus, Denmark
{large,mrevs}@madalgo.au.dk

Abstract. We present an I/O-efficient version of an algorithm for simplifying contour trees of two- and three-dimensional scalar fields described by Carr et al. [7]. Our algorithm uses optimal $\mathcal{O}(Sort(N)) = \mathcal{O}(\frac{N}{B} \log_{M/B} \frac{N}{B})$ I/Os, where N is the size of the contour tree, M the size of main memory and B the disk block size.

1 Introduction

Many modern applications involve processing and analysis of massive data representing two- and three-dimensional scalar fields. Since the data is often too large to fit in main memory of even large machines, this highlights the need for I/O-efficient algorithms, that is, algorithms which minimize movement of data between fast main memory and slower external memory. Furthermore, the availability of large and detailed data naturally leads to focus on data simplification. In this paper we develop I/O-efficient algorithms for simplifying contour trees of two- and three-dimensional scalar fields. A key part of this algorithm is an I/O-efficient algorithm for a batched version of a variant of the union-find problem, which we believe is of independent interest.

Preliminaries. Algorithms designed to minimize I/O are normally designed in the *external-memory* or *I/O-model* of computation [2]. In this model, the machine consists of an infinite size external memory (disk) and a main memory of size M. A block of B consecutive elements can be transferred between main memory and disk in one *I/O operation* (or simply *I/O*). Computation can only occur on elements in main memory, and the complexity of an algorithm is measured in terms of the number of I/Os it uses to solve a problem. Many fundamental problems have been considered in the I/O-model. For example, sorting N elements takes $\Theta(Sort(N)) = \Theta(\frac{N}{B} \log_{M/B} \frac{N}{B})$ I/Os [2].Refer to recent surveys for further results [10].

A d-dimensional *scalar field* $f : \mathbb{R}^d \rightarrow \mathbb{R}$ is a function that associates a real-value with every point in the d-dimensional Euclidean space. A *terrain* is simply

* Supported in part by the Danish National Research Foundation, the Danish Strategic Research Council, and by the US Army Research office.

** Center for Massive Data Algorithmics—a center of the Danish National Research Foundation.

Y. Dong, D.-Z. Du, and O. Ibarra (Eds.): ISAAC 2009, LNCS 5878, pp. 1155–1165, 2009.

a two-dimensional scalar field. A terrain is often represented by a planar triangulation $\mathbb{M} = (V, E)$, where a height is associated with each vertex of V. By linearly interpolating heights in each triangle based on the heights associated with the vertices of the triangles, $\mathbb{M}$ defines a piecewise-linear scalar field h (height function). For a given height $l \in \mathbb{R}$, the *l-level set* of a two-dimensional scalar field h is defined as $\{(x, y) \in \mathbb{R} | h(x, y) = l\}$. A *contour* of h at height l is then simply a connected component of the l-level set. In general a contour is a closed polygonal curve, but a contour through a local minima or maxima of h reduces to a single point, and at saddle points of h contours merge or split. The *Contour tree* of h is a tree representation of the topological changes in the l-level sets (the contours) of h as l increases; it encodes how contours are created or destroyed at minima and maxima, and merge or split at saddle points. See Section 3 for a formal definition. The definitions of l-level sets, contours and the contour tree naturally extends to three-dimensional scalar fields [6,9].

Previous related results. The union-find problem is the problem of maintaining a partition Π of a set $\mathbb{U} = \{x_1, x_2, \dots\}$ under UNION and FIND operations, where a UNION(x_i, x_j) joins the set containing x_i and the set containing x_j, and FIND(x_i) determines the set in Π containing x_i. Recently, Agarwal et al. [1] described an $\mathcal{O}(Sort(N))$ I/O algorithm for the batched version of the union-find problem, provided that none of the union operations are *redundant*, that is, for each UNION(x_i, x_j), x_i and x_j are in different sets. This is optimal [1]. Since the developed algorithms are quite complicated, Agarwal et al. [1] also described and implemented a simple $\mathcal{O}(Sort(N) \log(\frac{N}{M}))$ I/O algorithm.

Carr et al. [7] described an $\mathcal{O}(N \log N)$ time internal-memory algorithm for simplifying contour trees of two- and three-dimensional scalar fields. The simplification algorithm uses the fact that removal of a leaf in a contour tree naturally corresponds to modifying a region around a local minima or maxima in the corresponding scalar field. Thus various geometric measures for this region, such as e.g. height, area or volume, can be used to determine the significance of the leaf. The simplification algorithm first computes the relevant geometric measures, and then it iteratively removes the least significant leaf (and superfluous internal vertices), while updating the measures.

Our Results. In Section 3 we present an I/O-efficient version of the algorithm for simplifying contour trees of two- and three-dimensional scalar fields described by Carr et al. [7]. Our algorithm uses optimal $\mathcal{O}(Sort(N))$ I/Os. A key to obtaining the I/O-efficient contour tree simplification algorithm is an $\mathcal{O}(Sort(N))$ I/O algorithm for the batched (off-line) version of a problem we call *union-find with set properties* We present this algorithm in Section 2.

We believe our solution to the batched union-find with set properties problem is of independent interest. Recently, Agarwal et al. [1] described an $\mathcal{O}(Sort(N))$ I/O algorithm for removing small height depressions in terrain models using their efficient batched union-find algorithm. In the full version of this paper we show how to use batched union-find with set properties to extend this result such that depressions can be removed based on any of a number of local geometric measures, rather than only on the depth of the depression.

2 I/O-Efficient Batched Union-Find with Set Properties

Usually the find operation FIND(x_i) is implemented such that it returns an arbitrary (but unique) element in the set containing x_i. Sometimes, like in our contour tree simplification and flooding algorithms, one would like to maintain a certain property of each set in Π, such that it is returned by a find operation. This lead us to the following *union-find with set properties* problem.

Definition 1. *A union-find with set properties problem is given by a* universe $\mathbb{U} = \{x_1, x_2, \ldots\}$, *a property set* $\mathbb{P}$, *a* property function $\omega : \mathbb{P} \times \mathbb{P} \rightarrow \mathbb{P}$ *and a* sequence $\Sigma = \langle \sigma_1, \sigma_2, \ldots, \sigma_N \rangle$ *of* UNION($\mathrm{x_i}, \mathrm{x_j}$) *and* FIND($\mathrm{x_i}$) *operations. Furthermore, each element* $x_i \in \mathbb{U}$ *is associated with a* base property $p_i \in \mathbb{P}$.

The problem is to maintain a partition Π *of the elements in* $\mathbb{U}$ *together with a set property* $p \in \mathbb{P}$ *for each set in* Π *under the sequence of operations in* Σ*; if* x_i *is in set* $S_i \in \Pi$ *with set property* p_i *and* x_j *in set* $S_j \in \Pi$ *with set property* p_j *then a* UNION(x_i, x_j) *operation in* Σ *creates* $S_k = S_i \cup S_j$ *and assigns* S_k *the set property* $p_k = \omega(p_i, p_j)$*; a* FIND(x_i) *operation in* Σ *returns* p_i.

Below we show how to solve the batched union-find with set properties problem in $\mathcal{O}(Sort(N))$ I/Os. Our algorithm first constructs what we call a *set tree forest* encoding the sequence Σ. To do so we use the batched union-find algorithm of Agarwal et al. [1] (and thus we require that Σ does not contain any redundant union operations). After that the problem is solved by applying an I/O-efficient graph traversal techniques to the set tree forest.

Set Tree Forest F_Σ**.** In the following, we will say that the i'th operation σ_i in a batched union-find with set properties sequence $\Sigma = \langle \sigma_1, \sigma_2, \ldots, \sigma_N \rangle$ has *timestamp* i. Let S be a set in the partition Π resulting from performing all union operations in Σ, and consider the subset $\Sigma_S = \langle \sigma_i, \sigma_j, \ldots, \sigma_k \rangle$ of Σ consisting of all operations that operate on an element in S, ordered by timestamp. Below we define the *set tree* T_{Σ_S} to be a tree with a leaf for each element in S and an internal vertex v for each operation σ_l in Σ_S. The *set tree forest* F_Σ is then simply the forest containing a set tree for every set in Π.

To define the edges of set tree T_{Σ_S}, we first assume that all operations in Σ_S are union operations. Let σ_l be a union operation UNION(x', x'') where x' and x'' belong to sets S' and S'' before σ_l is performed, respectively; we say that σ_l is the *creator* of the set $S' \cup S''$ and the *destructor* of both S' and S''. Then the node v in T_{Σ_S} corresponding to σ_l has edges to the nodes that are roots in the set trees $T_{\Sigma_{S'_l}}$ and $T_{\Sigma_{S''_l}}$, where $\Sigma_{S'_l}$ and $\Sigma_{S''_l}$ are the operations in Σ_S with time stamp less than l operating on elements in S' and S'', respectively. In the general case where some of the operations in Σ_S are find operations, we replace some of the edges defined above with two edges. More precisely, let σ_l be a find operation FIND(x') where x' belong to set S' before σ_l is performed (that is, after all union operations in Σ_S with timestamp less than l have been performed), and let v_c and v_d be the nodes corresponding to the creator and destructor of S', respectively. Then the edge between v_c and v_d is replaced with two edges between v in T_{Σ_S} corresponding to σ_l and v_c and v_d, respectively.

Constructing F_Σ. Obviously, constructing the edges is the hard part of constructing F_Σ. The main idea in our solution is to encode each vertex in F_Σ using the unique elements returned by the find operations when using the standard batched union-find algorithm [1] on a sequence Σ' obtained from Σ. Our algorithm for constructing F_Σ works as follows. First we construct the union-find sequence Σ' by replacing each UNION(x', x'') operation in Σ with a sequence of union and find operations $\langle \text{FIND}(x'), \text{FIND}(x''), \text{UNION}(x', x''), \text{FIND}(x')\rangle$. Then we solve the batched union-find problem on Σ', obtaining for each union operation $\sigma_t = \text{UNION}(x', x'')$ in Σ the unique elements $r_{S'}$, $r_{S''}$ for the sets S',S'' destroyed by σ_t, and the unique element $r_{S' \cup S''}$ for the set $S' \cup S''$ created by σ_t. Similarly, for each $\sigma_t = \text{FIND}(x')$ operation in Σ we obtain a unique element $r_{S'}$ for the set S' containing x'. We then encode the vertex v in F_Σ corresponding to the union operation σ_t as the tuple $(r_{S' \cup S''}, t)$, and the vertex v corresponding to a find operation σ_t as $(r_{S'}, t)$. Finally, the leaf corresponding to an element $x_i \in \mathbb{U}$ is encoded as $(x_i, 0)$.

The encoding of the nodes in F_Σ allows us to construct the edges relatively easily. The key is that the children of a node $(r_{S' \cup S''}, t)$ corresponding to the union operation σ_t are the nodes of the form $(r_{S'}, t')$ and $(r_{S''}, t'')$ with maximal t' and t'' smaller than t. Similarly, the child of a node $(r_{S'}, t)$ corresponding to find operation σ_t is the node $(r_{S'}, t')$ with maximal t' smaller than t. Thus to construct the child edges $((r_{S'}, t'), (r_{S' \cup S''}, t))$ and $((r_{S''}, t''), (r_{S' \cup S''}, t))$ for each union operation σ_t and $((r_{S'}, t'), (r_{S'}, t))$ for each find operation in Σ, we first construct edges $((r_{S'}, t), (r_{S' \cup S''}, t))$, $((r_{S''}, t), (r_{S' \cup S''}, t))$ and $((r_{S'}, t), (r_{S'}, t))$ with incorrect timestamps. Then we sort the vertices lexicographically and the edges lexicographically according to their first vertex. Finally we simultaneously process the two sorted lists in order; this way we will meet one of the edges $((r_{S'}, t), (r_{S' \cup S''}, t))$ corresponding to union operation σ_t at the same time as we meet node $(r_{S'}, t')$, and we can thus update the first t in the edge to t'. The other edge for union operation σ_t, as well as the edge for a find operation, are updated in the same way.

Besides performing a constant number of scans and sorts, our algorithm uses $\mathcal{O}(Sort(N))$ I/Os to solve the batched union-find problem on Σ', so we construct F_Σ in $\mathcal{O}(Sort(N))$ I/Os.

Traversing F_Σ. After having constructed F_Σ, we can solve our batched union-find with set properties problem using a simple traversal of F_Σ from the leaves towards the roots, while applying the property function at each union node to the properties already computed at the child nodes. Utilizing that the timestamps of the nodes in F_Σ define a topologically order on the nodes of the DAG obtained from F_Σ by directing each edge from the child to the parent vertex, the traversal can be performed in $\mathcal{O}(Sort(N))$ using a standard technique called *time-forward processing* [8]. Details are deferred to the full paper.

Theorem 1. *The batched union-find with set properties problem can be solved in $O(sort(N))$ I/O operations, provided that all union operations are non-redundant.*

REMARK. Since the $\mathcal{O}(Sort(N))$ I/O batched union-find algorithm of Agarwal et al. [1] is quite complicated, they also described a simpler and practical $\mathcal{O}(Sort(N)\log(N/M))$ algorithm. This algorithms can relatively easily be modified to solve the batched union-find with set properties problem.

3 I/O-Efficient Contour Tree Simplification

In this section we describe our I/O-efficient version of the contour tree simplification algorithm of Carr et al. [7]. Before describing the algorithm, we first define the contour tree and discuss how several geometric measures can be associated with the edges of the tree. For simplicity and due to lack of space, we will only consider terrains (two-dimensional scalar fields).

3.1 Preliminaries

Contour tree. Consider the l-level set of a two-dimensional scalar field (terrain) h defined by a planar triangulation $\mathbb{M}$; we denote this set h_l. Recall that the contours at level l is simply the connected components of h_l. As l increases, new contours are *created* in h_l at local minima of h and *destroyed* at local maxima of h; at saddle points of h contours will split or merge (in effect destroying contours and creating new ones). Thus we call local minima, local maxima and saddle points *critical points* of h. At all other *regular points* of h the topology of h_l does not change. Two contours $c \in h_l$ and $c' \in h_{l'}$ are said to be *equivalent* if they are created and destroyed at the same critical points of h. We use c_v^w to denote the equivalence class of all contours created at critical point v and destroyed at critical point w and refer to the specific contour of c_v^w in the l-level set as $c_v^w(l)$.

We define the *contour tree* $\mathcal{C}$ of h as the tree with a vertex for each critical vertex in $\mathbb{M}$, and an edge $e = (v, w)$ for each contour equivalence class c_v^w of h; refer to Figure 1. Let V_v^w be the set of regular vertices in $\mathbb{M}$ contained in a contour of c_v^w. The *augmented contour tree* $\mathcal{C}_{aug}$ is then obtained by augmenting each edge $e = (v, w)$ of $\mathcal{C}$ with the vertices in V_v^w as follows: if $V_v^w = \{a_1, a_2, \ldots, a_k\}$ is sorted in order of increasing height, then e is split into edges $(v, a_1), (a_1, a_2) \ldots (a_k, w)$; refer again to Figure 1. In the following, when referring to an edge $e = (v, w)$ of $\mathcal{C}$ or $\mathcal{C}_{aug}$ we will always have $h(v) < h(w)$, w is called an *upper neighbor* of v and v a *lower neighbor* of w. We say that e is an *upper leaf edge* of $\mathcal{C}$ and $\mathcal{C}_{aug}$ if e is the only incident edge of w; similarly, e is *lower leaf edge* if e is the only incident edge of v.

Associating geometric measures with the edges of $\mathcal{C}$. Carr et al. [7] define the *upstart region* $up(c)$ of a contour c to be the region of h reachable

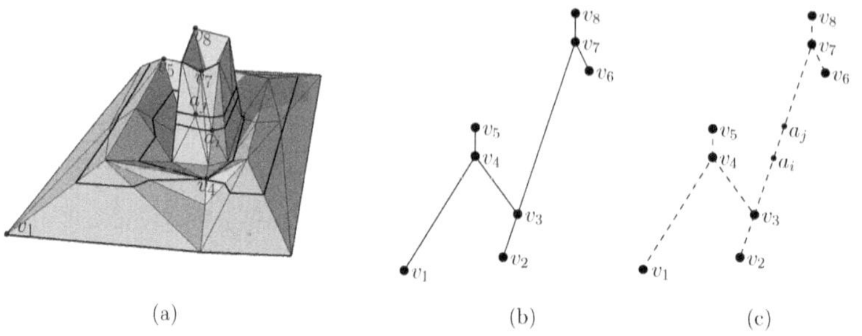

Fig. 1. (a) A terrain where the visible critical vertices have been marked, and where the contours through the regular vertices a_i and a_j are shown, along with the merging of two contours into a new contour at the saddle vertex v_4. (b) Contour tree of the terrain. (c) Augmented contour tree of the terrain.

from a point on c by a path that initially *ascends* from c and never returns to c. Similarly, the *downstart region* $down(c)$ of c is the region of h reachable from a point on c by a path that initially *descends* from c and never returns to c. Refer to Figure 2(a). An upstart region $h_e^{up} = \lim_{l \to h(v)}(up(c_v^w(l)))$ and downstart region $h_e^{down} = \lim_{l \to h(w)}(down(c_v^w(l)))$ can also naturally be associated with each edge $e = (v, w)$ in $\mathcal{C}$ [5]. Refer to Figure 2. Since h_e^{up} and h_e^{down} are regions of the two dimensional scalar field h it makes sense to consider geometric measures (such as height, area or volume [7]) of these regions. In general we will denote the geometric measure of h_e^{up} and h_e^{down} as $\sigma(h_e^{up})$ and $\sigma(h_e^{down})$ respectively.

3.2 Simplification Algorithm

In this section we describe the contour tree simplification algorithm of Carr et al. [7] and how it can be made I/O-efficient. The simplification algorithm consist of two phases: In the first phase $\sigma(h_e^{down})$ and $\sigma(h_e^{up})$ are computed for each edge e of the contour tree $\mathcal{C}$, and in the second phase $\mathcal{C}$ is simplified by iteratively removing edges and vertices guided by the computed geometric measures. Below we describe the two phases and how to perform them I/O-efficiently.

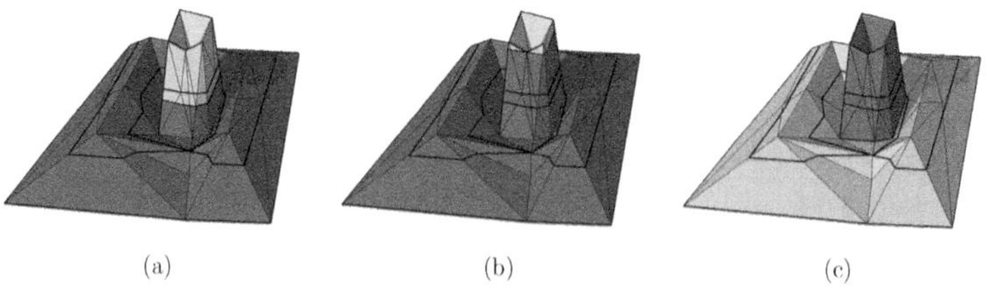

Fig. 2. (a) The downstart (dark) and upstart (light) regions of a contour through a regular vertex. (b) The region h_e^{down} (dark) of $e = (v_3, v_7)$ in the contour tree in Figure 1(b). (c) The region h_e^{up} (dark) of $e = (v_3, v_7)$ in the contour tree in Figure 1(b).

Computing geometric measures. Let $e = (v, w)$ be an edge in $\mathcal{C}$ and assume that the regular vertices augmented to e in $\mathcal{C}_{aug}$ are $\{a_1, a_2, \ldots, a_k\}$ in order of increasing height. Consider the set V_e^{up} of vertices in $\mathcal{C}_{aug}$ reachable from v by paths that initially contain (v, a_1); similarly consider the set V_e^{down} of vertices in $\mathcal{C}_{aug}$ reachable from w by paths that initially contain (a_k, w). To compute geometric measures of h_e^{down} and h_e^{up} for each edge e of $\mathcal{C}$, Carr et al. [7] utilizes that for a range of measures $\sigma(h_e^{down})$ (and $\sigma(h_e^{up})$) can be expressed as the sum of polynomial functions associated with each vertex in V_e^{down} (V_e^{up}) [4] [7]. Their algorithm repeatedly visits and (conceptually) removes a leaf from $\mathcal{C}_{aug}$ until all vertices have been visited (removed). At each visited vertex v it computes two polynomial functions $p_v^{up}(l)$ and $p_v^{down}(l)$ from the polynomial functions computed for the already visited neighbors of v in $\mathcal{C}_{aug}$. Once the algorithm has visited all vertices in $\mathcal{C}_{aug}$, it is easy to use the computed functions to obtain the geometric measures associated with edges in $\mathcal{C}$. The polynomial function $p_{a_i}^{up}(l)$ for $1 \leq i \leq k$ expresses the relevant geometric measure of the upstart region for contours $c_v^w(l)$ with $h(a_{i-1}) \leq l < h(a_i)$; similar $p_{a_i}^{down}(l)$ expresses the relevant geometric measure of the downstart region for contours $c_v^w(l)$ with $h(a_i) < l \leq h(a_{i+1})$. Therefore $\sigma(h_e^{down})$ is simply $p_{a_k}^{down}(h(w))$ and $\sigma(h_e^{up})$ is $p_{a_1}^{up}(h(v))$. Refer to [7] and [5] for the details of the algorithm.

In terms of I/O, the main issues with the above algorithm are that the vertices are visited in a somewhat unpredictable order, and that when visiting a vertex v the functions for the already visited neighbors of v need to be obtained. Both issues may result in the algorithm using $\Omega(N)$ I/Os. To obtain a predictable order, we consider an Euler tour τ of $\mathcal{C}_{aug}$ (rooted in an arbitrary vertex), where all but the last occurrence of any vertex have been removed. By definition, v appears in τ after each child of v and before the parent of v. Thus τ is equal to one of the possible vertex orders one can obtain in the algorithm by Carr et al. [7] when visiting the vertices in $\mathcal{C}_{aug}$ by repeatedly visiting and removing a leaf. Utilizing that τ defines a topological order on the vertices in the DAG obtained by directing each edge in $\mathcal{C}_{aug}$ from the vertex that occur first in τ to the vertex that occur last in τ, the last issue (the traversal of $\mathcal{C}_{aug}$) can be solved using the standard technique *time-forward processing* [8]. Since Euler tour computation and time-forward processing can be performed in $\mathcal{O}(sort(N))$ I/Os [8], we can obtain an algorithm for computing $\sigma(h_e^{down})$ and $\sigma(h_e^{up})$ for each edge e of $\mathcal{C}$ in $\mathcal{O}(sort(N))$ I/Os. Details will appear in the full paper.

Simplification using geometric measures. Having computed geometric measures, the simplification algorithm of Carr et al. [7] simplifies the contour tree $\mathcal{C}$ of a terrain h, by iteratively removing leaf edges e of $\mathcal{C}$. It is easily seen that a leaf edge in $\mathcal{C}$ corresponds to an extremum (local minimum or maximum) of h, and the main observation made by Carr et al. [7] is that removing a leaf edge e corresponds to modifying h in the region h_e^{up} (if e is an upper leaf edge) or the region h_e^{down} (if e is a lower leaf edge). Thus it is natural to remove leaf edges based on a *significance* associated with each leaf edge, equal to the geometric measure of the region affected by removing the leaf edge ($\sigma(h_e^{up})$ for an upper leaf edge e and $\sigma(h_e^{down})$ for a lower leaf edge e).

More precisely, the iterative algorithm works as follows [7]: In each iteration the lowest significance leaf edge $e = (v, w)$ is removed among all upper leaf edges where v has more than one upper neighbor and all lower leaf edges where w has more than one lower neighbor. The edge is removed by applying a *leaf prune operation* on e. If the removal of e means that an internal vertex v only has one upper neighbor w' and one lower neighbor w'', a *vertex reduction operation* is then performed on v. The vertex reduction operation removes v by merging the edges $e'' = (w'', v)$ and $e' = (v, w')$ into $e''' = (w'', w')$. If either e' or e'' were already leaf edges then the vertex reduction operation creates a new leaf e'''. Note that the regions associated with e''' are given by $h^{up}_{e'''} = h^{up}_{e''}$ and $h^{down}_{e'''} = h^{down}_{e'}$ and that by definition $h^{up}_{e'}$ is included in $h^{up}_{e''}$ and $h^{down}_{e''}$ is included in $h^{down}_{e'}$. The iteration continues until all leafs with significance less than a given significance threshold τ have been removed from $\mathcal{C}$. We denote the obtained tree $\mathcal{C}(\tau)$.

In terms of I/O, the problem with the above algorithm is not only to predict the order in which leaf prune and vertex reduction operations are performed. Actually performing the operations and obtaining the simplified contour tree $\mathcal{C}(\tau)$ also seems to be difficult to do in less than $\Omega(N)$ I/Os. Below we show how to address the two above problems and obtain an $\mathcal{O}(sort(N))$ algorithm. We first show how to predict the sequence $\mathcal{S}_\tau$ of leaf prune and vertex reduction operations performed by the algorithm of Carr et al. [7] to obtain $\mathcal{C}(\tau)$. Then we describe how to perform the operations in $\mathcal{S}_\tau$ to obtain $\mathcal{C}(\tau)$ using our batched union-find with set properties algorithm from section 2.

Constructing $\mathcal{S}_\tau$. Consider the contour tree $\mathcal{C}(t)$ for $0 \leq t \leq \tau$. Each edge in $\mathcal{C}(t)$ is either an edge in $\mathcal{C}$ or it represents a number of edges in $\mathcal{C}$ that have been joined by vertex reduction operations. In the following we will explicitly consider an edge e' in $\mathcal{C}(t)$ as a set of edges in $\mathcal{C}$, and use the notation $e \in e'$ to denote that the edge e in $\mathcal{C}$ is a member of e' in $\mathcal{C}(t)$. Assume that σ is monotonic increasing, any edge e''' created by a vertex reduction operation that merges the edges e' and e'' will then have $\sigma(h^{up}_{e'''}) \geq \sigma(h^{up}_{e'})$ and $\sigma(h^{up}_{e'''}) \geq \sigma(h^{up}_{e''})$. Using this we can proof the following lemma (proof will appear in full paper).

Lemma 1. *Let e be an edge with $\sigma(h^{up}_e) < \sigma(h^{down}_e)$ in the contour tree $\mathcal{C}$. There is an upper leaf edge $e' = (v', w')$ in $\mathcal{C}(\sigma(h^{up}_e))$ such that $e \in e'$ and such that the significance of e' is greater than or equal to $\sigma(h^{up}_e)$.*

Lemma 1 has a dual version which states that for an edge e in $\mathcal{C}$ where $\sigma(h^{down}_e) < \sigma(h^{up}_e)$, there is a lower leaf edge e' in $\mathcal{C}(\sigma(h^{down}_e))$ such that $e \in e'$ and such that the significance of e' is greater than or equal to $\sigma(h^{down}_e)$. Using the edge property shown in lemma 1, we can now prove a property of the vertices in $\mathcal{C}$ that will allow us to construct $\mathcal{S}_\tau$.

Lemma 2. *Consider an internal vertex v of $\mathcal{C}$ with r incident edges $e_1 \ldots e_r$ leading to upper neighbors and s incident edges $\hat{e}_1 \ldots \hat{e}_s$ leading to lower neighbors. Assume that the edges are sorted such that $\sigma(h^{up}_{e_1}) < \ldots < \sigma(h^{up}_{e_r})$ and $\sigma(h^{down}_{\hat{e}_1}) < \ldots < \sigma(h^{down}_{\hat{e}_s})$.*

When simplifying $\mathcal{C}$ to obtain $\mathcal{C}(\tau)$, a leaf prune operation is performed on an upper leaf with significance $\sigma(h^{up}_{e_i})$, for each edge e_i with $\sigma(h^{up}_{e_i}) < \tau$, where

$1 \le i \le r-1$. *Similarly, a leaf prune operation is performed on a lower leaf edge with significance* $\sigma(h^{down}_{\hat{e}_i})$ *for each edge* $\hat{e}_i$ *with* $\sigma(h^{down}_{\hat{e}_i}) < \tau$, *where* $1 \le i \le s-1$.

Proof (sketch). Consider the edge e_i where $1 \le i \le r-1$. Since the region $h^{down}_{e_i}$ includes the region $h^{up}_{e_{i+1}}$, we know that $\sigma(h^{up}_{e_i}) < \sigma(h^{down}_{e_i})$ due to the monotonicity of σ. Using lemma 1 we get that $\mathcal{C}(\sigma(h^{up}_{e_i}))$ has an upper leaf edge e'_i such that $e_i \in e'_i$ with significance greater than or equal to $\sigma(h^{up}_{e_i})$. Applying lemma 1 to e_{i+1} gives that $\mathcal{C}(\sigma(h^{up}_{e_i}))$ also contains an edge e'_{i+1} such that $e_{i+1} \in e'_{i+1}$. Which essentially proves the first part of the lemma. Using the dual version of lemma 1 we can prove the second part of the lemma. □

Lemma 2 allows us to compute all leaf prune operations in $\mathcal{S}_\tau$. To compute the vertex reductions we use that a vertex reduction operation on a vertex v is enabled by a leaf prune operation that causes v to have exactly one upper and one lower neighbor. From lemma 2 we get that the leaf prune operation that enables a vertex reduction on v is the operation that removes a leaf with significance equal to the largest of $\sigma(h^{up}_{e_{r-1}})$ and $\sigma(h^{down}_{\hat{e}_{s-1}})$. Thus we can construct $\mathcal{S}_\tau$ such that each leaf prune operation is represented (implicitly) by an edge of $\mathcal{C}$ with a *timestamp* equal to the significance of the leaf edge being removed, and such that a vertex reduction is represented by two edges of $\mathcal{C}$ (the two being removed) with a *timestamp* equal to the timestamp of the leaf prune operation that enables the vertex reduction. The construction can easily be done in $\mathcal{O}(sort(N))$ I/Os by sorting and scanning the edges of $\mathcal{C}$ a constant number of times.

Computing $\mathcal{C}(\tau)$ from $\mathcal{S}_\tau$. Having computed the sequence of leaf prune and vertex reduction operations $\mathcal{S}_\tau$ used to obtain $\mathcal{C}(\tau)$ from $\mathcal{C}$, we can now compute $\mathcal{C}(\tau)$ using an instance of batched union-find with set properties. From $\mathcal{S}_\tau$ we construct a sequence Σ_τ of union and find operations. Σ_τ contains a union operation $\sigma_t = \text{UNION}(e'', e')$ operation for each vertex reduction operation in $\mathcal{S}_\tau$ represented by the timestamp t and edges e'' and e'. At the end of Σ_τ a find operation $\sigma_{\tau+1} = \text{FIND}(e)$ is appended for each edge e in $\mathcal{C}$. The leaf prune operations (which we only implicitly represented) in $\mathcal{S}_\tau$ are ignored. The partition Π_0 consists of a singleton set with set property $(v, w, \sigma(h^{up}_e), \sigma(h^{down}_e))$ for each edge $e = (v, w)$ in $\mathcal{C}$. The set property function ψ is defined such that when unioning two sets with set properties $(w'', v, \sigma(h^{up}_{e''}), \sigma(h^{down}_{e''}))$ and $(v, w', \sigma(h^{up}_{e'}), \sigma(h^{down}_{e'}))$, the new set property is $(w'', w', \sigma(h^{up}_{e''}), \sigma(h^{down}_{e'}))$.

Let Π_t be the partition resulting from performing the union operations in Σ_τ with timestamp less than or equal to t. We say that a set in Π_t with set property $(v', w', \sigma(h^{up}_{e'}), \sigma(h^{down}_{e'}))$ *represents* an edge $e' = (v', w')$ with associated geometric measures $\sigma(h^{up}_{e'})$ and $\sigma(h^{down}_{e'})$, if and only if $\sigma(h^{up}_{e'}) > t$ and $\sigma(h^{down}_{e'}) > t$. If the edges represented by sets in Π_t are exactly the edges in $\mathcal{C}(t)$, we say that Π_t *represents* $\mathcal{C}(t)$. Note that the set property function is defined such that if two sets representing edges (w'', v) and (v, w') are unioned then the resulting set represents the edge (w'', w') resulting from performing a vertex reduction on v. A simple induction proof on $\mathcal{S}_\tau$ proves the following.

Lemma 3. *The partition* Π_τ *represents the simplified contour tree* $\mathcal{C}(\tau)$.

Since Π_τ represents $\mathcal{C}(\tau)$ we can obtain the actual edges from the find operations at the end of Σ_τ sequence: We simply collect the edges represented by the set properties returned by all the find operations and remove duplicate edges.

Constructing Σ_τ from $\mathcal{S}_\tau$ can be done in a simple scan using $\mathcal{O}(\frac{N}{B})$ I/Os. After that the union-find with set properties problem is solved in $\mathcal{O}(sort(N))$ I/Os (Theorem 1), and finally duplicate edges can also be removed in a sorting step using $\mathcal{O}(sort(N))$ I/Os. Overall we have computed $\mathcal{C}(\tau)$ from $\mathcal{S}_\tau$ in $\mathcal{O}(sort(N))$ I/Os. Thus we have our main Theorem.

Theorem 2. *Given a significance threshold τ and a contour tree $\mathcal{C}$ of a 2d scalar field, we can simplify $\mathcal{C}$ as described in [7] using $\mathcal{O}(sort(N))$ I/O operations.*

REMARK 1. Our simplification algorithm (Lemma 1 and 2) requires that σ is monotonically increasing. This is indeed the case for most interesting geometric measures, but not all. For example, consider using height as geometric measure, such that for $e = (v, w)$ the measure $\sigma(h_e^{up})$ is given by the height difference $h(z) - h(v)$, where z is the highest leaf that can be reached from v by a path in $\mathcal{C}_{aug}$ that visit vertices of monotonically increasing height. In this case σ is not monotonically increasing. However, we can relatively easy define another monotone function σ', such that simplifying $\mathcal{C}$ with the algorithm by Carr et al [7] yields the same result for both σ and σ'. This way we can just use σ' in our algorithm.

REMARK 2. Theorem 2 extends to 3d scalar fields.

REMARK 3. Let t_e^c be the time at which an edge e is created and t_e^d the time at which e is destroyed. By slightly changing the way we construct Σ_τ from $\mathcal{S}_\tau$ (utilizing the implicit representation of leaf prune operations), we can compute t_e^c and t_e^d for every edge e in a contour tree $\mathcal{C}(t)$ for $t \geq 0$ using $O(sort(N))$ I/Os. By building an I/O-efficient interval tree [3] on the intervals $[t_e^c; t_e^d]$, we obtain a data structure that for any query threshold τ, can return $\mathcal{C}(\tau)$ using $\mathcal{O}(\log_B(N) + \frac{T}{B})$ I/Os, where T is the size of $\mathcal{C}(\tau)$.

Acknowledgements

The authors wish to thank Hamish Carr for discovering a bug in an earlier draft.

References

1. Agarwal, P., Arge, L., Yi, K.: I/O-efficient batched union-find and its applications to terrain analysis. In: Proc. ACM Symposium on Computational Geometry (2006)
2. Aggarwal, A., Vitter, J.S.: The Input/Output complexity of sorting and related problems. Communications of the ACM 31(9), 1116–1127 (1988)
3. Arge, L., Vitter, J.S.: Optimal external memory interval management. SIAM Journal on Computing 32(6), 1488–1508 (2003)
4. Bajaj, C.L., Pascucci, V., Schikore, D.R.: The contour spectrum. In: IEEE Symposium on Volume Visualization, pp. 167–175 (1998)

5. Carr, H.: Topological Manipulation of Isosurfaces. PhD thesis, University of British Columbia (2004)
6. Carr, H., Snoeyink, J., Axen, U.: Computing contour trees in all dimesions. Computational Geometry: Theory and Applications 24, 75–94 (2003)
7. Carr, H., Snoeyink, J., van de Panne, M.: Simplifying flexible isosurfaces using local geometric measures. In: IEEE Visualization 2004 (2004)
8. Chiang, Y.-J., Goodrich, M.T., Grove, E.F., Tamassia, R., Vengroff, D.E., Vitter, J.S.: External-memory graph algorithms. In: Proc. ACM-SIAM Symposium on Discrete Algorithms, pp. 139–149 (1995)
9. van Kreveld, M., van Oostrum, R., Bajaj, C., Pascucci, V., Schikore, D.: Contour trees and small seed sets for isosurface traversal. In: Proc. 13th Annu. ACM Sympos. Comput. Geom., pp. 212–220 (1997)
10. Vitter, J.S.: External memory algorithms and data structures: Dealing with MASSIVE data. ACM Computing Surveys 33(2), 209–271 (2001)

Algorithms for Computing the Maximum Weight Region Decomposable into Elementary Shapes

Jinhee Chun, Ryosei Kasai, Matias Korman, and Takeshi Tokuyama

Graduate School of Information Sciences, Tohoku University, Japan
{jinhee,ryosei,mati,tokuyama}@dais.is.tohoku.ac.jp

Abstract. Motivated from the image segmentation problem, we consider the problem of finding the maximum weight region with a shape decomposable into elementary shapes in $n \times n$ pixel grid where each pixel has a real valued weight. We give efficient algorithms for several interesting cases. This shows string constrast to the NP-hardness results to find the maximum weight union for the corresponding cases.

1 Introduction

Let $\mathbf{P}$ be an $n \times n$ pixel plane, and consider a family $\mathcal{F} \subset 2^{\mathbf{P}}$ of *pixel regions*. A pixel of $\mathbf{P}$ is the unit square $p(i,j) = [i-1, i] \times [j-1, j]$ where $1 \leq i \leq n$ and $1 \leq j \leq n$. The pixel p at the (i,j) position in the grid has a real value $W(p) = W(i,j)$ called the weight of p. We can regard the array $(W(p))_{p \in \mathbf{P}}$ as a real-valued matrix $W = (W(i,j))(1 \leq i,j \leq n)$ where we count the indices of rows from bottom to top. For conveniences' sake, we define $W(0,j) = W(n+1,j) = W(i,0) = W(i,n+1) = 0$ for each i and j. We consider the following *maximum-weight region* problem:

Find a region $R \in \mathcal{F}$ maximizing $W(R) = \sum_{p \in R} W(p)$.

The maximum-weight region problem is considered in several applications such as image processing [1], data mining [2,3], surface approximation [4,5,7], and radiation therapy [5]. The difficulty of the problem depends on the family $\mathcal{F}$. If $\mathcal{F} = 2^{\mathbf{P}}$, the problem is trivial, since R is obtained as the set of all pixels with positive weights. On the other hand, the problem is NP-hard if $\mathcal{F}$ is the set of all connected regions in $\mathbf{P}$ (in the usual 4-neighbor topology) [1].

The following is a list of previously known families for which the maximum-weight regions can be computed efficiently (definitions will be given later): x-monotone regions, based monotone regions, rectilinear convex regions, staircase convex regions centered at a pixel r (called stabbed union of rectangles in [4,7]), and digital star-shaped regions [6]. More generally, we can solve the problem if $\mathcal{F}$ can be represented as the family of closures in a graph defined on $\mathbf{P}$ (see [5,9]). All of the above families can be treated in this framework (although this approach might not lead to the most efficient algorithm).

A natural question is to solve the maximum weight region problem for more general families of regions. In this paper, we consider the problem of finding the maximum weight region constructed as an aggregation of more than one basic region. Ideally, we

Y. Dong, D.-Z. Du, and O. Ibarra (Eds.): ISAAC 2009, LNCS 5878, pp. 1166–1174, 2009.

would like to compute the maximum weight region represented as a union of basic regions given in the above list. Unfortunately, we show a negative result even on the simplest possible case: it is NP-hard to compute the maximum weight region represented as a union of a based x-monotone region (based monotone region with the x-axis as its base line) and a based y-monotone region. Moreover, it is NP-hard to have any finite ratio approximation algorithm for computing the maximum weight region represented as a union of two digital star-shaped regions with given two centers.

In order to have tractable computational problems, we consider a different formulation: a region decomposable into pairwise disjoint basic regions (i.e., represented as a disjoint union of basic regions). We give a series of novel algorithms listed as follows:

(1). Given k axis parallel base lines, the maximum-weight region decomposable into base monotone regions corresponding to the base lines can be computed in $O(N^{1.5})$ time, where $N = n^2$ is the number of pixels.
(2). If we consider k base segments instead of lines, we give a FPT algorithm for a special case and an $n^{O(k)}$ algorithm for the general case.
(3). The maximum weight region decomposable into two digital star shaped regions with given different centers can be computed in $O(N^3)$ time.

We also show that the maximum-weight region decomposable into k staircase convex regions or k rectilinear convex regions can be computed in polynomial time (if k is a constant). We also show that the union problem can be solved in a similar fashion for these regions.

1.1 Motivation from Image Segmentation Applications

Separating an object in an image from its background is a central problem in pattern recognition and computer vision. This operation is commonly called *image segmentation* and many practical methods are proposed in the literature. Consider a pixel plane $\mathbf{P}$ representing a (say, monochromatic) picture, where each pixel p has a real value $f(p)$ represents the brightness level of the pixel p. The segmented image should be a pixel region with a nice geometric property. Clearly, the quality of the segmentation depends on the separation of brightness levels in the image and background. However, it is non-trivial to formulate the image segmentation problem in the form of a nice optimization problem (so as to have a trade off between the output quality and computational complexity).

Asano et al. [1] proposed an *optimization-based* image segmentation method that gives a robust solution with theoretical guarantee. The general mathematical framework defines a family $\mathcal{F}$ of grid regions, and finds the region $R \in \mathcal{F}$ maximizing an objective function $\Phi(R)$. The function Φ needs a kind of convexity (Asano et al. [1] particularly considered the *intraclass variance*), and solved the problem via a parametric optimization framework.

Asano's framework gives a method to compute the segmentation problem provided that the following key problem can be solved efficiently: compute the region $R \in \mathcal{F}$ maximizing $\sum_{p \in R}(f(p)-\theta)$. Once the parameter value θ is fixed, we assign the weight $W(p) = f(p) - \theta$ to each pixel p, transforming the problem into the maximum weight region problem. The optimal value of the parameter θ (among the $O(N)$ possible values) is found using parametrization techniques.

Segmentation in color pictures can be also reduced to the maximum weight region problem by using a three-dimensional parameter space.

Our positive results imply that shapes that are decomposable into two or more pairwise disjoint fundamental shapes can be treated in Asano et al.'s framework. This allows a variety of objects to be handled in a robust fashion, and the authors believe that it gives significant advancement to the theoretical aspect of image segmentation problem. On the other hand, the NP-hardness says that it is difficult to segment an object that is an overlay of two highly intersecting basic objects. Thus, the first picture of Figure 1 is difficult to segment, while the second picture is easier since it can be decomposed into two star shaped regions as shown in the rightmost picture.

2 Regions Decomposable to Base Monotone Regions

A base line of the pixel grid $\mathbf{P}$ is a vertical line $x = i$ or horizontal line $y = i$ where $0 \leq i \leq n$. For a given horizontal base line $\ell : y = i$, its *based monotone region* is the region $\{p(s, j) : 1 \leq s \leq n, g(s) < j \leq f(s)\}$ for a suitable pair of functions satisfying $g(j) \leq i \leq f(j)$ for each j. In other words, it is a union of segments of columns intersecting the base line. It is a special case of *x-monotone region*, where the intersection with each column can be any connected segment. Note that the j-th column part is empty if $g(j) = f(j)$, thus we do not assume connectivity of the region. The vertical base line case is analogously defined. A based monotone region with the base line $x = 0$ (resp. $y = 0$) is often called *based x-monotone* (resp. *based y-monotone*) region.

A based monotone region with a horizontal base line ℓ is subdivided into the based monotone regions of the upper and lower halfplanes of ℓ: They are the pixel region defined by $\{p(s, j) : 1 \leq s \leq n, i \leq j \leq f(s)\}$ and $\{p(s, j) : 1 \leq s \leq n, g(s) < j \leq i\}$, respectively. The family of base monotone region of the upper and lower half planes of ℓ are denoted by $\mathcal{U}(\ell)$ and $\mathcal{D}(\ell)$, respectively (meaning that each column grows upward and downward from the base line, respectively). Similarly, for a vertical base line m, we define families $\mathcal{L}(m)$ and $\mathcal{R}(m)$ of base monotone region in the left and right halfplanes of m, respectively.

Given a set of k base lines, a region R is called a feasible region if it can be decomposed into pairwise disjoint base monotone regions with respect to the given base lines. Figure 2 shows a feasible region of given six base lines, and Figure 3 gives intuition of a segmentation using four grid boundary lines as base lines.

Fig. 1. Union and decomposition

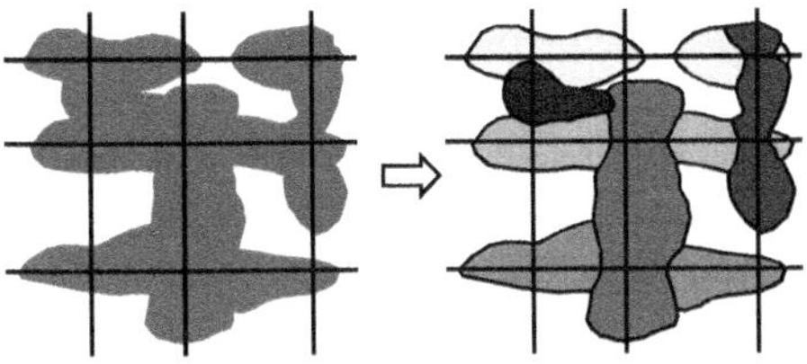

Fig. 2. A feasible region and its decomposition

Fig. 3. A segmented flower

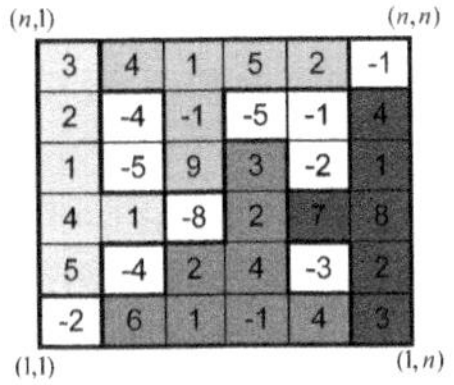

Fig. 4. Room-edge problem

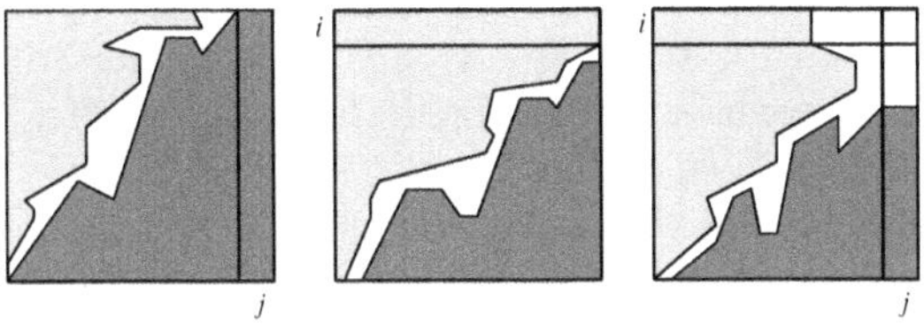

Fig. 5. Computing $UR(i, j)$

2.1 Room-Edge Problem

First, we consider a special case (called room-edge problem) where we are given the four (or less) boundary lines of $\mathbf{P}$ as the set of base lines. See Figures 3 and 4. We abbreviate $\mathcal{U}$, $\mathcal{D}$, $\mathcal{L}$ and $\mathcal{R}$ for the families of base monotone regions of upper, lower, left, and right halfplanes with respect to the bottom, top, right, and left boundary edges of $\mathbf{P}$. (note that $\mathcal{R}$ grows rightwards from the *left* edge, etc.) A feasible region R of the room-edge problem can be decomposed into disjoint regions $\mathbf{U}, \mathbf{D}, \mathbf{L}, \mathbf{R}$ such that $\mathbf{U} \in \mathcal{U}$, $\mathbf{D} \in \mathcal{D}$, $\mathbf{L} \in \mathcal{L}$ and $\mathbf{R} \in \mathcal{R}$. Imagine that we paint $\mathbf{U}$, $\mathbf{D}$, $\mathbf{R}$, and $\mathbf{L}$ by blue, green, yellow and red in a way that each color starts painting from a boundary edge in the direction perpendicular to the edge without breaking or mixing of colors. Then, the feasibility means that we can paint the region with the four colors. We would like to maximize $W(R) = W(\mathbf{U}) + W(\mathbf{D}) + W(\mathbf{L}) + W(\mathbf{R})$.

2.2 Algorithms for the Room-Edge Problem

Painting with only one color is easy (and well known) to be solvable in linear time: Suppose that we would like to maximize $w(\mathbf{U})$ for $\mathbf{U} \in \mathcal{U}$. Now, it suffices to consider the prefix sums $\text{prefix}(s; j) = \sum_{k=0}^{s} W(k, j)$ for each column j, and compute the array $U(i, j) = \max_{0 \le s \le i} \text{prefix}(s; j)$, which is the maximum prefix in j-th column up to the i-th row. Then, $W(\mathbf{U}) = \sum_{1 \le j \le n} U(n, j)$, and the region $\mathbf{U}$ is obtained as $\{(i, j) : U(i-1, j) \ne U(n, j)\}$. This computation can be done in $O(N) = O(n^2)$ time. Similarly to $U(i, j)$, we compute tables of $D(i, j)$, $R(i, j)$, and $L(i, j)$ to be utilized later (definitions are rotated suitably).

Next, let us discuss painting with two colors. There are basically two cases: Painting from opposite edges (e.g., $\mathbf{D}$ and $\mathbf{U}$), and painting from adjacent edges (e.g., $\mathbf{U}$ and $\mathbf{R}$). It is easy to paint from opposite edges, since we can consider each column (or row) separately. Indeed, if we negate all the weights, the complement of the solution region for this case is nothing but the maximum weight x-monotone region.

Thus, we consider the adjacent case using **U** (blue) and **R** (yellow). Let $UR(i, j)$ be the maximum weight if we color the pixel grid up to (i, j) by blue and yellow; that is, we consider the submatrix of W with respect to the columns up to the j-th column and the rows up to the i-th row.

Theorem 1. *We can compute the matrix UR in $O(N)$ time.*

Proof. We compute the table $UR(*, *)$ by dynamic programming classifying the optimal painting of $UR(i, j)$ into two cases (as shown in Figure 5): if $(i, j) \in \mathbf{U}$ we have $UR(i, j) = UR(i, j-1) + U(i, j)$ (Figure 5, left). Otherwise (Figure 5, center or right), we have $UR(i, j) = UR(i-1, j) + R(i, j)$. Thus, using the recursive formula $UR(i, j) = \max\{UR(i, j-1) + U(i, j), UR(i-1, j) + R(i, j)\}$ and the precomputed tables U and R, we can compute $UR(*, *)$ in linear time. Also note that the optimal region attaining $UR(n, n)$ is given by backtracking. □

Next, let us consider painting with three colors. Suppose we paint from bottom, top and left edges to maximize $W(\mathbf{U} \cup \mathbf{D} \cup \mathbf{R})$. We precompute tables UR, UD and UL in $O(N)$ time and define $UDR(j)$ as the maximum weight of coloring the pixels in the first j columns using three colors.

Theorem 2. *We can compute $UDR(j)$ for all $1 \leq j \leq n$ in $O(N)$ time.*

Proof. If the j-th column of the region attaining $UDR(j)$ does not intersect the left region **R**, we have $UDR(j) = UDR(j-1) + UD(j)$. where $UD(j)$ is the maximum weight painting of the j-to column from top and bottom. Otherwise, the picture is divided into upper half and lower half by the intersecting row of **R**, and each half is painted by two colors. Thus, we have $UDR(j) = \max_{0 \leq i \leq n}\{DR(i, j) + UR(i+1, j)\}$ for this case. Since we prepared the tables UD, DR, UR, the above formula can be computed in $O(n)$ time for each j. Hence, each new entry of UDR can be computed in $O(n)$ time by taking the maximum of the above two cases, and the table $UDR(*)$ is computed in $O(n^2) = O(N)$ time. □

Finally, we consider painting by four colors from four edges. If there exists a vertical line $x = j$ separating **R** and **L**, we can decompose the problem into two instances of the three-color paintings. Thus, the optimal value for this case is computed as $\max_{0 \leq j \leq n}\{UDR(j) + UDL(j+1)\}$, where $UDL(j+1)$ is the optimal region of UDL painting of the region to the right of the partition line $x = j$. This type of the solution region can be computed from the arrays UDR and UDL for the three-color paintings in $O(N)$ time. Similarly, we can solve in $O(N)$ time if there exists a separating horizontal line.

Thus, we need to consider the case where the solution does not allow such a partition line. We guess the longest column/row length of each colored region, and decompose the pixel plane into five rectangular parts as shown in the left picture of Figure 6. The center rectangle cannot be painted by any color, since each color is blocked by another region to paint. Moreover, in each other rectangular part, the painting is done by two colors, and can be done independently of the painting of other regions. Therefore, we

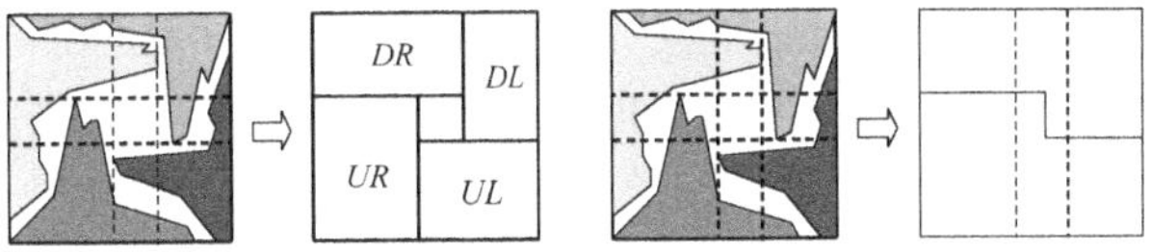

Fig. 6. Decomposition into two-colored rectangles and L-shaped regions

can obtain the optimal four-color painting by combining the two-color paintings of these four rectangular parts. The optimal value is obtained by combining optimal solution of each solution in $O(1)$ time (since tables DL, DR, UL, UR are precomputed) for each possible partitions. Since there are $O(n^4) = O(N^2)$ possible partition patterns, we can solve compute the maximum weight region for the room-edge problem in $O(N^2)$ time with $O(N)$ space.

The time complexity can be improved to $O(N^{1.5})$ if we can use $O(N^{1.5})$ space. We use another decomposition as shown in the right picture of Figure 6, where the pixel grid is decomposed into two L-shaped regions in which each region is painted by using three colors. There are four types of L-shaped regions, each of which is union of two rectangles containing a corner of the grid. Let us focus on the L-shaped region $\mathcal{L}(s,t,u)$ that is a union of two rectangles containing the left-lower corner: the tall rectangle has height s and width u, and the short rectangle has a height t. Other types can be handled analogously.

We define $F(s,t,u)$ to be the optimal three color painting (from both sides and the bottom edge). We have the following recursive formula

$$F(s,t,u) = \max\{F(s-1,t,u) + R(s,u), UR(s,u) + UL(t,u+1), F(s,t,u-1)\}.$$

The first case is where the s-th row is not penetrated by **U**; thus, the row can only be reached from the left, and painted in the color of **R**. The second case is where the u-th column is not penetrated by the **L**, thus we can cut the shape at the u-th column to have two 2-colored rectangles; one has height s, and the other has t. The third case is where the s-th row is penetrated by **U**, and the u-th column is penetrated by **L**; thus, the u-th column beyond the t-th row is blocked, and cannot be colored.

We can compute each new entry of the table $F(*,*,*)$ in $O(1)$ time using the precomputed tables, and hence the time complexity for computing all entries of the three-dimensional array F is $O(n^3) = O(N^{1.5})$. We need $O(N^{1.5})$ space to store it. Analogously, we can compute 3D-tables for other types of L-shaped regions. Then, we can compute the weight $UDLR$ of the maximum-weight four-color painting from them in $O(N^{1.5})$ additional time. Thus, we have the following:

Theorem 3. *The room-edge problem can be solved either $O(N^{1.5})$ time and space or $O(N^2)$ time using $O(N)$ space. If we use only three colors, the problem can be solved in $O(N)$ time.*

2.3 Allowing k Base Lines

Now, let us consider the case where we are given k base (either vertical or horizontal) lines. We solve the problem of finding the maximum weight (possibly non-connected) region that can be decomposed into base monotone polygons corresponding to the separating lines. The arrangement of k separating lines decompose the pixel grid into $O(k^2)$ cells that are rectangular subgrids.

Lemma 1. *If we consider the union of the optimal solution of room-edge problem of all cells of the arrangement, it is decomposable into pairwise disjoint base monotone regions of the separating lines, and attains the maximum weight.*

Proof. The key observation is taht each cell can be solved independently: suppose that a monotone region $R(\ell)$ based by a line ℓ is cut by another base line ℓ' (parallel to ℓ) into $R_1 \cup R_2$ where R_2 is separated from ℓ by ℓ'. Then, we can replace $R(\ell)$ by R_1 and $R(\ell')$ by $R(\ell') \cup R_2$ to obtain a new pair of based monotone regions, where the new $R(\ell)$ does not intersect ℓ'. In this way, we can reform the decomposition such that its each component do not intersect other parallel base lines. Cutting with partition lines orthogonal to base lines into subregions does not affect because of definition of base monotone regions. Thus, the union of the solutions of the room-edge problem gives the global solution of the problem with k base lines. □

For each cell G_t, if it has $O(N_t)$ pixels, the room-edge problem inside the cell can be solved in $O(N_t^{1.5})$ time. Since the summation of N_t over all cells is $O(N)$, the total time complexity of solving the room-edge problem insider all cells is $O(N^{1.5})$. Thus, we have the following theorem.

Theorem 4. *The maximum weight region decomposable into based monotone regions of given k base lines can be computed in $O(N^{1.5})$ time.*

3 Other Types of Regions

Due to 10 pages limitation of this proceedings, proofs of theorems in this section will be given in the full version.

Allowing k base segments

Let us consider a segment s of the base line ℓ, and suppose that we only consider the monotone regions R such that the intersection of the closure of R and ℓ is contained in s. Then, we call R has s as its base segment. Given a set of base segments, we consider an analogous problem. Without loss of generality, we assume that the segments are mutually non-intersecting in their interior, since we can refine segments if they intersect. The algorithm given in the previous subsection does not work, since we cannot easily decompose the problem into room problems. We also consider the restricted problem in which the region can be build only on the right side of each vertical base segment and on the upper side of each horizontal base segment.

Theorem 5. *If we are given k non-intersecting base segments and consider based monotone regions in upside (for horizontal segments) and right side (for vertical segments), the optimal region decomposable to these monotone regions can be computed in $O(k^{O(k)}N^2)$ time. If we use four directions, the optimal region is computed in $O(N^{O(k)})$ time.*

Digital star-shaped regions
$\mathbf{G}$ is the graph representing the adjacent relation of pixels of $\mathbf{P}$. A digital ray system is a rooted spanning tree T of $\mathbf{G}$ such that all leaves are located on the boundary of the grid. A digital star-shaped region R associated with T is a rooted subtree of the tree T. The path from root to a pixel is called the digital ray, and the digital star-shaped region is characterized a region such that for any pixel in the region the digital ray to the pixel is also in the region. A construction of T such that any digital star-shaped region is an approximation of a real star-shaped region within $O(\log n)$ Hausdorff distance is known [6]. For simplifying the presentation of time complexity, we consider the case in which the diameter of T is $O(n)$. Given two root positions r_1 and r_2 and digital ray system T_1 and T_2, an admissible decomposition of a region R is a nonintersecting pair of rooted subtrees of T_1 and T_2 such that the union of corresponding regions is R. We say R is an admissible region if it has an admissible decomposition.

Theorem 6. *The maximum weight admissible region can be computed in $O(N^3)$ time.*

Union versus decomposition
We discuss the difference of complexities of the problems for *union* and *decomposition*. The following theorem shows a strong contrast to the positive results in Theorems 1 and 6. An outline of the proof is implicitly given in [8], and is omitted in this version.

Theorem 7. *It is NP-hard to compute the maximum weight union $R = R_1 \cup R_2$ of regions $R_1 \in \mathcal{U}$ and $R_2 \in \mathcal{R}$. Moreover, the maximum weight union of two star-shaped regions is hard to approximate within any given finite ratio.*

Staircase/rectilinear convex regions
A region R is a *staircase convex region* centered at p if R is represented as a union of rectangles containing p, in other words, for each $q \in R$, every L_1 shortest path between p and q is contained in R. A region R is a *rectilinear convex region* if it is both x-monotone and y-monotone; in other words, the intersection of any column or row and R is an interval (or empty). By definition, a staircase convex region is a rectilinear convex region. It is known that the maximum weight staircase convex region for a given center p can be computed in $O(N)$ time, and the maximum weight rectilinear convex region can be computed in $O(N^{1.5})$ time [7,10].

Theorem 8. *For any constant k, the maximum weight region decomposable into k staircase pairwise disjoint convex (or rectilinear convex) regions can be computed in $O(N^{k+1})$ time. Moreover, the problem can be solved in the same time complexity even without the pairwise disjoint condition.*

Note that the difference of computational complexities of the decomposition problem and the union problem has not been revealed for these cases. We conjecture that there is a fixed parameter tractable algorithm for the pairwise disjoint case (when the k center points are given).

4 Concluding Remarks

Study of complexity of combinatorial algorithms of segmenting a figure decomposable into k basic objects has just started by this work. Especially, development of FPT algorithms is important and unsolved in many cases. The three dimensional extension has a potential application to a variation of *open-pit mining* problem [9]. The authors thank Martin Nöllenburg for fruitful discussions. This work is partially supported by MEXT grant on basic research (B) 18300001 and young researcher grant (B) 21700004.

References

1. Asano, T., Chen, D.Z., Katoh, N., Tokuyama, T.: Efficient Algorithms for Optimization-Based Image Segmentation. Int. J. Comput. Geometry Appl. 11(2), 145–166 (2001)
2. Fukuda, T., Morimoto, Y., Morishita, S., Tokuyama, T.: Data Mining Using Two-Dimensional Optimized Association Rules: Scheme, Algorithms, and Visualization. In: SIGMOD Conference 1996, pp. 13–23 (1996)
3. Fukuda, T., Morimoto, Y., Morishita, S., Tokuyama, T.: Data Mining with optimized two-dimensional association rules. ACM Trans. Database Syst. 26(2), 179–213 (2001)
4. Chen, D.Z., Chun, J., Katoh, N., Tokuyama, T.: Efficient algorithms for approximating a multi-dimensional voxel terrain by a unimodal terrain. In: Chwa, K.-Y., Munro, J.I.J. (eds.) COCOON 2004. LNCS, vol. 3106, pp. 238–248. Springer, Heidelberg (2004)
5. Chen, D.Z., Hu, X.S., Luan, S., Wu, X., Yu, C.X.: Optimal Terrain Construction Problems and Applications in Intensity-Modulated Radiation Therapy. Algorithmica 42(3-4), 265–288 (2005)
6. Chun, J., Korman, M., Nöllenburg, M., Tokuyama, T.: Consistent Digital Rays. In: Proc. 24th ACM SOCG, pp. 355–364 (2008)
7. Chun, J., Sadakane, K., Tokuyama, T.: Efficient Algorithms for Constructing a Pyramid from a Terrain. In: Akiyama, J., Kano, M. (eds.) JCDCG 2002. LNCS, vol. 2866, pp. 108–117. Springer, Heidelberg (2003)
8. Chun, J., Korman, M., Nöllenburg, M., Tokuyama, T.: Hardness of the Maximum Union-of-Closures Problem, Preprint (preliminarlily reported in WAAC2008)., `http://www.dais.is.tohoku.ac.jp/~mati/2peakNP.pdf`
9. Hochbaum, D.: A New-Old Algorithm for Minimum-cut and Maximum-flow in Closure Graphs. Network 37(4), 171–193 (2001)
10. Yoda, K., Fukuda, T., Morimoto, Y., Morishita, S., Tokuyama, T.: Computing Optimized Rectilinear Regions for Association Rules. In: Proc. KDD 1997, pp. 96–103 (1997)

I/O and Space-Efficient Path Traversal in Planar Graphs

Craig Dillabaugh[1], Meng He[2], Anil Maheshwari[1], and Norbert Zeh[3]

[1] School of Computer Science, Carleton University, Canada
[2] Cheriton School of Computer Science, University of Waterloo, Canada
[3] Faculty of Computer Science, Dalhousie University, Canada

Abstract. We present a technique for representing bounded-degree planar graphs succinctly while permitting I/O-efficient path traversal. To represent a graph G on N vertices, each with an associated key of $q = O(\lg N)$ bits[1], we use $Nq + O(N) + o(Nq)$ bits. Using this representation, a path of length K can be traversed with $O(K/\lg B)$ I/Os, where B is the disk block size. Our structure may be adapted to represent, with similar space bounds, a terrain modeled as a triangular-irregular network to support traversal of a path that visits K triangles using $O(K/\lg B)$ I/Os. This structure can be used to answer a number of useful queries efficiently, such as reporting terrain profiles, trickle paths and connected components.

1 Introduction

External-memory (EM) data structures and succinct data structures both address the problem of representing very large data sets. In the EM model, the amount of data required to solve a given problem exceeds internal memory. Data structures and algorithms are designed to solve the problem, while minimizing the transfer of data between internal and external memory. For succinct data structures, the aim is to encode the structural component of the data structure using as little space as is theoretically possible, while still permitting efficient navigation. In addition to our own research on traversal in trees [1], the only other research that merges these techniques is on succinct EM data structures for text indexing [2,3].

Here we develop data structures for path traversal in planar graphs. Given a bounded-degree planar graph G, we wish to report a path composed of K vertices in G using a small number of I/O operations. We demonstrate practical applications of our structure by showing how it can be applied to answering queries on triangular irregular network (TIN) models.

1.1 Background

In the External Memory (EM) model [4], the number of elements in the problem instance is denoted by N. Memory is divided into a two-level hierarchy, external

[1] We use $\lg N$ to denote $\log_2 N$.

Y. Dong, D.-Z. Du, and O. Ibarra (Eds.): ISAAC 2009, LNCS 5878, pp. 1175–1184, 2009.

and internal memory. The external memory is assumed to have effectively infinite capacity, but accessing data elements in external memory is slow. The internal memory permits efficient operations, but its capacity is limited to $M < N$ elements. Data are transferred between internal and external memory in blocks of size B, where $1 < B < M/2$. In this paper, we assume that $B = \Omega(\lg N)$, which is true for all realistic values of N and B in an EM setting. We refer to such a transfer as an I/O operation. The efficiency of algorithms in the EM model is evaluated with respect to the number of I/O operations they require.

Nodine *et al.* [5] first considered the problem of blocking graphs in external memory for efficient path traversal. Path traversal is measured in terms of *blocking speed-up*—the worst-case number of vertices (path length) that can be traversed before an I/O is required. They identified the optimal bounds for several classes of graphs. An I/O-efficient algorithm that blocks a bounded-degree planar graph so that any path of length K can be traversed in $O(K/\lg B)$ I/Os was proposed by Agarwal *et al.* [6].

Succinct data structures for trees and graphs were originally proposed by Jacobson [7]; the goal is to represent data structures using space as near the information-theoretic lower bounds as possible, while still permitting efficient navigation. Numerous efforts have been made to improve this for graphs [8,9]. Recently, Chiang *et al.* [10] obtained a succinct data structure for planar graphs that uses $2m + 2n + o(n)$ bits, where m and n are the number of edges and vertices, respectively, in the graph.

2 Preliminaries

A key data structure used in our graph representation is a bit vector $C[1..N]$ that supports `rank` and `select` operations efficiently. The operations $\texttt{rank}_1(C, i)$ and $\texttt{rank}_0(C, i)$ return the number of 1s and 0s in $C[1..i]$, respectively. The operations $\texttt{select}_1(C, r)$ and $\texttt{select}_0(C, r)$ return the position of the rth occurrence of 1 and 0 in C, respectively. The problem of representing a bit vector succinctly to support `rank` and `select` operations in constant time under the word RAM model with word size $\Theta(\lg N)$ bits has been considered in [7,2,11], and these results can be directly applied to the external memory model:

Lemma 1. *A bit vector C of length N containing R 1s can be represented using (a) $N + o(N)$ bits [2] or (b) $\lg \binom{N}{R} + O(N \lg\lg N/\lg N)$ bits [11] to support the access to the bits of C, as well as* `rank` *and* `select` *operations on C, in $O(1)$ time (or $O(1)$ I/Os in external memory).*

Frederickson [12] developed a technique for decomposing planar graphs. A planar graph is divided into overlapping regions which contain two types of vertices, *interior* and *boundary* vertices. Interior vertices occur in a single region and are adjacent only to vertices within that region. Boundary vertices are shared among two or more regions. Lemma 2 summarizes Frederickson's result.

Lemma 2 ([12]). *A planar graph with N vertices can be subdivided into $\Theta(N/r)$ regions of no more than r vertices and with a total of $O(N/\sqrt{r})$ boundary vertices.*

3 Graph Representation

Let G be a planar graph of bounded degree d. Each vertex in G stores a q-bit key. We assume that $q = O(\lg N)$. We perform a two-level partitioning of G inspired by the approach in [13]. This results in a subdivision of G into *regions* of fixed maximum size, which are subdivided into *sub-regions* of smaller fixed maximum size. Within the regions and sub-regions, vertices fall into one of two categories, *interior* vertices, and *boundary* vertices. A vertex is interior to a region if all of its neighbouring vertices in G belong to the same region. A vertex is a region boundary vertex if it has neighbouring vertices in G that belong to different regions. Sub-region vertices are labeled as interior or boundary in the same fashion. Due to the two-level partitioning, a sub-region boundary vertex may also be a region boundary vertex.

Consider some vertex $v \in G$. Based on [6], we define the *α-neighbourhood* of v as follows. Beginning with v, we perform a breadth-first search in G and select the first α vertices encountered. The α-neighbourhood of v is the subgraph of G induced by these vertices. Analogous to the interior and boundary vertices in a region or sub-region, we define *internal* and *terminal* vertices for α-neighbourhoods. The neighbours of an internal vertex belong to the same α-neighbourhood, while terminal vertices have neighbours external to the α-neighbourhood.

In our representation of G, we store each sub-region and the α-neighbourhood of each boundary vertex. When constructing the α-neighbourhood of a sub-region boundary vertex that is not a region boundary vertex, we add an additional constraint that its α-neighbourhood cannot be extended beyond the region it is interior to. Collectively, we refer to the sub-regions and α-neighbourhoods as *components* of the graph. The regions are not explicitly stored, but rather are a collection of their sub-region components. Each component is stored using a succinct representation that allows efficient traversal of the component. To enable traversal of G, each vertex is assigned a unique *graph label.*

3.1 Graph Labeling

In this section, we describe the labeling scheme that enables traversal across the components of the graph. The scheme is based on Bose *et al.* [13] but uses the technique of Frederickson [12] for graph decomposition.

Each vertex v of G is assigned a unique *graph label*, in addition to possibly multiple (in the case of boundary vertices) *region labels* and *sub-region labels.* G is partitioned into a set of regions. We denote the ith region by R_i. Each region R_i is subdivided into sub-regions. We denote by q_i the number of sub-regions in R_i, and denote the jth sub-region of R_i as $R_{i,j}$. In partitioning G, the vertices on the boundary of a sub-region (or region) appear in more than one sub-region (region). Consider a boundary vertex v of $R_{i,j}$. We say that the instance of v appearing in $R_{i,j}$ *defines* v in R_i if there is no sub-region $R_{i,h}$ in R_i, such that $v \in R_{i,h}$ and $h < j$. All subsequent instances of v in any other sub-region are referred to as *duplicates*. Likewise, for a region boundary vertex $v \in R_i$, v is a defining vertex if there is no region R_h containing v, where $h < i$. In our labeling at the region and graph levels, our strategy is to assign defining vertices a unique label, while duplicate vertices are assigned the labels of the defining vertex.

The encoding of a sub-region $R_{i,j}$ induces a permutation of the vertex set within the sub-region. We let the position of each vertex within this permutation serve as its sub-region label. Now we discuss the assignment of region labels to the vertices within a region R_i with q_i sub-regions. There are, including duplicates, a total of $\sum_{j=1}^{q_i} |R_{i,j}|$ vertices in R_i. Let n_i^b be the number of defining sub-region boundary vertices in R_i. We visit each sub-region $R_{i,j}$ for $j = 1, 2, \ldots, q_i$, and assign each defining vertex the next available region label from the set $\{1, \ldots, n_i^b\}$. This process is then repeated and the interior vertices are assigned labels from the set $\{n_i^b + 1, \ldots, |R_i|\}$. Duplicate vertices are assigned the labels of their defining vertices.

The assignment of graph labels mirrors that of region labels. Let n^b be the total number of region boundary vertices over all regions in G. We visit each region R_i, for $i = 1, 2, \ldots, t$, and assign each defining boundary vertex the next available graph label from the set $\{1, \ldots, n^b\}$. As with the region labeling, we then repeat this process and assign each interior vertex the next available label from the set $\{n^b + 1, \ldots |G|\}$. This completes the labeling procedure.

Based on this labeling scheme, observe that the graph labels assigned to all interior vertices of a region are consecutive. Likewise, the region and graph labels assigned to all interior vertices of a sub-region are consecutive.

3.2 Data Structures

We wish to have each sub-region fit in a single disk block. Denote by A the maximum number of vertices that may be stored in a block, and this becomes our maximum sub-region size. Using Lemma 2, we first divide G into regions of size $A \lg^3 N$ by setting $r = A \lg^3 N$. We further divide each region into sub-regions of size A by setting $r = A$. The sub-region $R_{i,j}$ is encoded using:

1. A compact encoding of the graph structure of $R_{i,j}$. This involves a permutation of the vertices in $R_{i,j}$.
2. A bit vector, $\mathcal{B}$ of length $|R_{i,j}|$ for which $\mathcal{B}[i] = 1$ if and only if the corresponding vertex in the encoding's permutation is a boundary vertex.
3. An array of length $|R_{i,j}|$ that stores the q-bit key for each vertex in $R_{i,j}$.

We store two arrays $\mathcal{L}_S$ and $\mathcal{L}_R$, which record for each sub-region and region, respectively, the graph label of the sub-region's (region's) first interior vertex.

We select A such that a sub-region of size A will fit in exactly one block in memory. However, we are only guaranteed that sub-regions will have at most A vertices, therefore some sub-regions will occupy less than a full block. We do not want to waste any bits by storing sub-regions in partially full blocks. We store sub-regions to disk in the following fashion: Let $\preceq_{SR}$ be a total order of the sub-regions of G. Sub-region $R_{j,k}$ comes before $R_{l,m}$ in $\preceq_{SR}$ if either $j < k$, or, if $j = k$ and $k < m$. We write the sub-regions to disk in this order. Prior to writing a sub-region we write its *sub-region offset* value, which is its size in vertices. When we finish writing one sub-region to disk, we immediately start writing the next sub-region. If we come to a block boundary, we skip a pre-defined number of bits at the start of the next block, which we term the *block offset*, and continue writing the bits for the current sub-region. When we overrun a block in this fashion, we record the length of the overrun (in bits) in the block offset. If a sub-region happens to end at a block boundary, we write 0 to the block offset of the next block. Since a sub-region is no larger than a single block, a sub-region will never span portions of more than two blocks. We continue this process until all sub-regions have been written to disk.

Denote by Q the total number of sub-regions used to store G. To facilitate efficient lookup of sub-regions within the disk blocks, we store two bit vectors:

1. Bit vector $\mathcal{B}_R[1..Q]$, where $\mathcal{B}_R[i] = 1$ iff the ith sub-region in $\preceq_{SR}$ is the first sub-region in its region.
2. Bit vector $\mathcal{B}_S[1..Q]$, where $\mathcal{B}_S[i] = 1$ iff the block in which the ith sub-region starts differs from sub-region $i-1$.

We store the α-neighbourhood for each region and sub-region boundary vertex v by encoding the subgraph G_v that comprises v's α-neighbourhood using:

1. An encoding of the graph structure of G_v. This encoding involves a permutation of the vertices in G_v.
2. A bit array of length $|G_v|$ that marks each vertex in G_v as internal or terminal.
3. A variable that records the position of v within the permutation of the vertices of G_v.
4. An array of length $|G_v|$ that stores the key associated with each vertex.
5. An array, $\mathcal{L}_\alpha$, of length $|G_v|$ that stores for each vertex w:
   (a) its graph label if v is a region boundary vertex. Otherwise,
   (b) a region offset of $\lg(A \lg^3 N)$ bits if w is interior to v's region. This offset is calculated as the difference between the graph labels of w and the first interior vertex in v's region (stored in $\mathcal{L}_R$). If w is a boundary vertex of v's region, we store w's region label in this region using $\lg(A \lg^3 N)$ bits.

The α-neighbourhoods of all sub-region boundary vertices are stored together in an array, $\mathcal{SR}_\alpha$. Since the size of the compact encoding of the subgraph may vary between α-neighbourhoods, we pad extra bits where necessary to ensure that the elements of $\mathcal{SR}_\alpha$ are of fixed size. This array is created by visiting each region in turn and appending the α-neighbourhoods of all sub-region boundary vertices to $\mathcal{SR}_\alpha$, ordered by region label. Additionally, we store a bit vector $\mathcal{D}$ of length N where $\mathcal{D}[i] = 1$ iff the vertex with graph label i is a sub-region boundary vertex that is interior to its region.

Lemma 3. *The data structures described above store a bounded-degree planar graph G on N vertices, each with an associated $q = O(\lg N)$-bit key, in $O(N) + Nq + o(Nq)$ bits.*

Proof (Sketch). Let c denote the number of bits per vertex required to store the sub-graph and boundary bit-vector (this applies to both sub-region and α-neighbourhood components). The exact value of c depends on the chosen succinct representation for the graph used to store components. We have assumed blocks of size $B \lg N$ bits, and for the sake of simplicity, we assume that $c+q$ divides $B \lg N$. Thus $(c+q)A = B \lg N$, and $A = \frac{B \lg N}{c+q}$.

When splitting G into regions and sub-regions, we let $r = A \lg^3 N$ for regions, and $r = A$ for sub-regions. By Lemma 2, this results in $\Theta\left(\frac{N}{A \lg^3 N}\right)$ regions with $O\left(\frac{N}{\sqrt{A \lg^3 N}}\right)$ region boundary vertices, and $\Theta(\frac{N}{A})$ sub-regions with $O\left(\frac{N}{\sqrt{A}}\right)$ sub-region boundary vertices. Regions are not explicitly stored, so consider the space required to store the sub-regions. To store a single copy of each vertex, we require $N(c+q)$ bits, which accounts for all internal vertices plus the defining copies of all boundary vertices. Additionally, we need to store the $O\left(N/\sqrt{A}\right)$ duplicate boundary vertices, which requires $O\left(\frac{N}{\sqrt{A}}\right)(c+q) = o(Nq)$ bits. Storing the sub-regions also requires space for the bit arrays $\mathcal{B}_R$ and $\mathcal{B}_S$, plus the block and sub-region offsets when packing the sub-regions to disk. $\mathcal{B}_R$ and $\mathcal{B}_S$ are both of length bounded by the number of sub-regions, of which there are $O(N/A)$, and as such require $o(N)$ bits by Lemma 1. Each block offset requires $\lg B$ bits and there are $O(N/B)$ blocks, so block offsets use $o(N)$ bits. Likewise, the $O(N/A)$ sub-region offsets use $o(N)$ bits. Thus, the total cost for storing the sub-regions is $N(c+q) + o(Nq)$ bits.

We let $\alpha = A^{\frac{1}{3}}$ be the size of the α-neighbourhoods for both region and sub-region boundary vertices (in fact $\alpha = A^{\frac{1}{2}-\epsilon}$, for $\epsilon > 0$, suffices for our analysis). Cumulatively, the number of bits required to store the α-neighbourhoods of all region boundary vertices is $\Theta(\frac{N}{\sqrt{A \lg^3 N}}) \cdot A^{\frac{1}{3}} \cdot (\lg N + q + c) = \Theta(\frac{N}{A^{\frac{1}{6}} \lg^{\frac{3}{2}} N}) \cdot \Theta(\lg N) = o(N)$. The space required by $\mathcal{L}_\alpha$ to store the labels associated with each vertex in the α-neighbourhood of a sub-region boundary vertex is $\lg (A \lg^3 N)$ bits. The space requirement for all α-neighbourhoods of all sub-region boundary vertices is thus $\Theta(\frac{N}{\sqrt{A}}) \cdot A^{\frac{1}{3}} \cdot (\lg (A \lg^3 N) + q + c) = o(Nq)$ bits. It is easy to show that the remaining auxiliary data structures, $\mathcal{D}$, $\mathcal{L}_R$ and $\mathcal{L}_S$, occupy $o(N)$ bits. □

3.3 Navigation

The traversal algorithm operates by loading either a sub-region or the α-neighbourhood of a boundary vertex and traversing that component until a boundary vertex (in the case of a sub-region) or a terminal vertex (in the case of a α-neighbourhood) is encountered, at which time the next component is loaded from memory and traversal continues. Traversal assumes that we have a function `step` available, which, given a vertex v in G and the key value of v, determines where to proceeded in the traversal. A call to the `step` function can have one of three possible outcomes: termination of the traversal, selection of a neighbour of the current vertex, or loading a new component from memory if not all of the current vertices' neighbours are in the currently loaded component.

Now we analyze the I/O complexity. First we show that within each component, labels can be reported at no additional I/O cost (Lemma 4). We then describe how we ensure that components can be identified and loaded in $O(1)$ I/Os (Lemma 5). Finally, we demonstrate that visiting a constant number of components guarantees a progress of $O(\lg A)$ steps along the path (Lemma 6). We omit the proofs in the rest of this paper due to space constraints.

Lemma 4. *Given a sub-region or α-neighbourhood, the graph labels of all interior (sub-regions) and internal (α-neighbourhoods) vertices can be reported without incurring any additional I/Os beyond what is required when the component is loaded into main memory.*

When we arrive at a boundary/terminal vertex, conversions between labels are necessary in order to locate the next component to load. Our labeling scheme is derived from [13], and by Lemma 3.4 in their paper, conversion between these labels can be performed in $O(1)$ time. We can extend this to external memory, and prove the following lemma:

Lemma 5. *When the traversal algorithm encounters a terminal or boundary vertex v, the next component containing v in which the traversal may be resumed can be loaded in $O(1)$ I/O operations.*

Lemma 6. *Using the data structures and navigation scheme described above, a path of length K in graph G can be traversed in $O\left(\frac{K}{\lg A}\right)$ I/O operations.*

Combining Lemmas 3 and 6, we have (note that $A = \Omega(B)$):

Theorem 1. *A bounded-degree planar graph G on N vertices, where each vertex stores a key of q bits, can be represented using $Nq+O(N)+o(Nq)$ bits to support the traversal of a path of length K with $O\left(\frac{K}{\lg B}\right)$ I/O operations.*

The only comparable work to ours is that of Agarwal *et al.* [6]. In addition to storing keys, they use $O(N \lg N)$ bits to store the graph structure. In contrast, we use only $O(N)$ bits to store the graph structure, and we manage to show that the path traversal and other operations can still be performed I/O-efficiently. For very large data sets, this is an enormous saving in space.

4 Representing Triangulated Terrains

Let Σ be a terrain in $\mathbb{R}^3$. Let P be a set of points on the terrain Σ with coordinates x, y and z, where z is the elevation of the point. The triangulation T of the point set P is a model of Σ. By projecting T onto the x, y-plane, we can view T as a planar graph with vertices being the point set P. Each triangle of T is defined by three points from P. If a point is one of the defining points for a triangle, we say that it is *adjacent* to that triangle. Two triangles are adjacent if they share a common edge (and consequently two adjacent points).

We can represent T in a compact fashion as follows. Let $G = (V, E)$ be the dual graph of T. G is a connected planar graph of bounded degree $d = 3$, with a vertex corresponding to each triangle in T. There is no vertex corresponding to the outer face, so edges along the perimeter of T do not have a corresponding edge in the dual G. Starting with G we generate an augmented planar graph $G' = (V', E')$, by adding the point set P to G so that $V' = V \cup P$. We form the edge set E' by adding an edge to E for each vertex pair (v, p) where $v \in V$ and $p \in P$ and where p is adjacent to the face of T that corresponds to v. The new graph G' remains planar but is no longer of bounded degree. However, all the vertices from the original vertex set V are still of degree at most six. In the augmented graph, we refer to the vertices corresponding to triangles as *triangle vertices*, and to the vertices corresponding to points as *point vertices*. In a similar fashion, we refer to the edges of G as *dual edges*, and to those edges added to connect the point and triangle vertices as *point edges*. We have the following lemma that bounds the size of G':

Lemma 7. *Given the dual graph G of triangulation T with N vertices, the augmented graph G' has at most $2N + 2$ vertices.*

For the purpose of quantifying the bit cost of our data structures, we denote by $\varphi = O(\lg N)$ the number of bits required to encode the coordinates of a point. We encode G' using a succinct planar graph data structure. The encoding involves a permutation of the vertices of G'. Let the *augmented graph label*, $\ell(v)$, of the vertex v, be the position of v in this permutation. We store the point set P in an array $\mathcal{P}$ ordered by the augmented graph labels of the points. Finally, we create a bit vector π of length $|V'|$, where $\pi[v] = 0$ if v is a triangle vertex and $\pi[v] = 1$ if v is a point vertex. To summarize, we have:

Lemma 8. *The data structures described above can represent a terrain T composed of N triangles with φ-bit point coordinates, where $\varphi = O(\lg N)$, using $N\varphi + O(N) + o(N\varphi)$ bits, so that, given the label of a triangle, the adjacent triangles and points can be reported in $O(1)$ time.*

4.1 Compact External-Memory TIN Representation

In this section, we extend our data structures for I/O-efficient traversal in bounded-degree planar graphs (Section 3) to terrains. We represent T by its dual graph G. Since the dual graph of the the terrain (and subsequently each component) is a bounded-degree planar graph, it can be represented with the data structures described in Section 3. Each component is a subgraph of G for which we generate the augmented subgraph, as described above. To represent a terrain we must store the augmented subgraph which includes points from the point set P adjacent to the triangles represented by the vertices in the sub-regions and α-neighbourhoods.

Lemma 9. *The space requirement, in bits, to store a component (α-neighbourhood or sub-region) representing a terrain is within a constant factor of the space required to store the terrain's dual graph.*

We have the following theorem due to Theorem 1 and Lemma 9:

Theorem 2. *A terrain T modeled as a TIN on N nodes, where the coordinates of each point can be stored in φ bits, can be represented using $N\varphi+O(N)+o(N\varphi)$ bits to support the traversal of a path which crosses K faces in T with $O\left(\frac{K}{\lg B}\right)$ I/O operations.*

For the case in which we wish to associate a q bit key with each triangle, we can easily extend our approach to represent a terrain using $N(\varphi+q)+O(N)+o(N(\varphi+q))$ bits to provide the same support for path traversals.

4.2 Applications on TIN Models

We now present a number of applications on TIN models represented using our data structures. In this section, we assume that we are given a starting triangle, $t \in T$, as a query parameter. We remove this assumption in Section 4.3.

Terrain Profiles and Trickle Paths: Terrain profiles are a common tool for GIS visualization. The input is a line segment, or chain of line segments possibly forming a polygon, and the output is a profile of the elevation along the line segment(s). The trickle path, or path of steepest descent, from a point p is the path on T that begins at p and follows the direction of steepest descent until it reaches a local minimum or the boundary of T [14]. In analyzing these algorithms, we measure the complexity of a path based on the number, K, of triangles it intersects. When a path intersects a vertex, we consider all triangles adjacent to that vertex to have been intersected. Given this definition, we have:

Lemma 10. *Let T be a terrain stored using our representation. Then:*

(a) Given a chain of line segments, S, the profile of the intersection of S with T can be reported with $O(\frac{K}{\lg B})$ I/Os.
(b) Given a point p, the trickle path from p can be reported with $O(\frac{K}{\lg B})$ I/Os.

Connected Component Queries: In a connected component query, we are given a convex terrain T and a triangle $t \in T$, and wish to report all triangles in the connected component $T' \subset T$ that share a common attribute or property $\mathcal{P}(t)$ (the property of t). A triangle t' is in T' iff $\mathcal{P}(t)=\mathcal{P}(t')$ and there exists a path in G from t to t' consisting only of triangles that share this property.

Theorem 3. *A triangulation T with a q-bit key per triangle can be stored using $Nq+O(N)+o(Nq)$ bits such that a connected component, T', may be reported:*

(a) Using $O\left(\frac{|T'|}{\lg B}\right)$ I/Os if T' is convex.
(b) Using $O\left(\frac{|T'|}{\lg B}+h\log_B h\right)$ I/Os, and $O(h\cdot(q+\lg h))$ bits of temporary storage, if T' is non-convex and/or may contain holes. The value h denotes the number of boundary edges around all the holes and the perimeter of T'.
(c) Using $O\left(\frac{|T'|}{\lg B}+h'\log_B h'\right)$ I/Os and no additional storage if T' is non-convex and/or may contain holes. The value h' is the total number of triangles touching all the holes and the perimeter of T'.

4.3 Terrain Representation with Point Location

For the various applications that we have described, we assumed that a starting triangle in T is given as an input parameter. This is problematic because in real applications, we will typically need to locate the triangle from which we will begin reporting the result. To address this problem, we combine the succinct data structure of Bose *et al.* [13] that supports point location in internal memory with our data structures. This allows us to use $o(N\varphi)$ bits of additional storage to perform planar point location on a terrain, T, in $O(\log_B N)$ I/Os. Asymptotically this does not change the space requirement for T.

Theorem 4. *A terrain T modeled as a TIN on N nodes, where each point coordinate may be stored in φ bits, can be represented T using $N\varphi+O(N)+o(N\varphi)$ bits to support the traversal of a path crossing K faces in T with $O\left(\frac{K}{\lg B}\right)$ I/Os, and to support point location queries incurring $O(\log_B N)$ I/Os.*

References

1. Dillabaugh, C., He, M., Maheshwari, A.: Succinct and I/O efficient data structures for traversal in trees. In: Hong, S.-H., Nagamochi, H., Fukunaga, T. (eds.) ISAAC 2008. LNCS, vol. 5369, pp. 112–123. Springer, Heidelberg (2008)
2. Clark, D.R., Munro, J.I.: Efficient suffix trees on secondary storage. In: SODA, pp. 383–391 (1996)
3. Chien, Y.F., Hon, W.K., Shah, R., Vitter, J.S.: Geometric Burrows-Wheeler transform: Linking range searching and text indexing. In: DCC, pp. 252–261 (2008)
4. Aggarwal, A., Jeffrey, S.V.: The input/output complexity of sorting and related problems. Commun. ACM 31(9), 1116–1127 (1988)
5. Nodine, M.H., Goodrich, M.T., Vitter, J.S.: Blocking for external graph searching. Algorithmica 16(2), 181–214 (1996)
6. Agarwal, P.K., Arge, L., Murali, T.M., Varadarajan, K.R., Vitter, J.S.: I/O-efficient algorithms for contour-line extraction and planar graph blocking (extended abstract). In: SODA, pp. 117–126 (1998)
7. Jacobson, G.: Space-efficient static trees and graphs. FOCS 42, 549–554 (1989)
8. Munro, J.I., Raman, V.: Succinct representation of balanced parentheses, static trees and planar graphs. In: FOCS, pp. 118–126 (1997)
9. Chuang, R.C.N., Garg, A., He, X., Kao, M.Y., Lu, H.I.: Compact encodings of planar graphs via canonical orderings and multiple parentheses. CoRR cs.DS/0102005 (2001)
10. Chiang, Y.T., Lin, C.C., Lu, H.I.: Orderly spanning trees with applications. SIAM J. Comput. 34(4), 924–945 (2005)
11. Raman, R., Raman, V., Rao, S.S.: Succinct indexable dictionaries with applications to encoding k-ary trees and multisets. In: SODA, pp. 233–242 (2002)
12. Frederickson, G.N.: Fast algorithms for shortest paths in planar graphs, with applications. SIAM J. Comput. 16(6), 1004–1022 (1987)
13. Bose, P., Chen, E.Y., He, M., Maheshwari, A., Morin, P.: Succinct geometric indexes supporting point location queries. In: SODA, pp. 635–644 (2009)
14. de Berg, M., Bose, P., Dobrindt, K., van Kreveld, M.J., Overmars, M.H., de Groot, M., Roos, T., Snoeyink, J., Yu, S.: The complexity of rivers in triangulated terrains. In: CCCG, pp. 325–330 (1996)

Improved Algorithms for Finding Consistent Superstrings Based on a New Graph Model*

Jin Wook Kim[1], Siwon Choi[1], Joong Chae Na[2], and Jeong Seop Sim[1,**]

[1] School of Computer and Information Engineering, Inha University
Incheon 402-751, Korea
{gnugi,jssim}@inha.ac.kr, siwonred@gmail.com

[2] Department of Computer Science and Engineering, Sejong University
Seoul 143-747, Korea
jcna@sejong.ac.kr

Abstract. The problems that are related to string inclusion and non-inclusion have been vigorously studied in such diverse fields as data compression, molecular biology, and computer security. Given a finite set of negative strings $\mathbb{N}$ and a finite set of positive strings $\mathbb{P}$, a string α is a consistent superstring if every positive string is a substring of α and no negative string is a substring of α. The shortest (resp. longest) consistent superstring problem is finding a string α that is the shortest (resp. longest) among all the consistent superstrings for the given sets of strings.

In this paper, we first propose a new graph model based on the Aho-Corasick algorithm to represent the consistent superstrings for the given sets. Then, we propose improved algorithms for the problems of the shortest consistent superstring and the longest consistent superstring using the graph model.

Keywords: consistent superstring, Aho-Corasick algorithm, string inclusion, string non-inclusion.

1 Introduction

The problems that are related to string inclusion and non-inclusion have been vigorously studied in such diverse fields as data compression [12,13], molecular biology [2,10], and computer security [7].

The followings are some examples of well-known inclusion or non-inclusion related strings; consider a set of strings $\mathbb{W} = \{w_1, \cdots, w_m\}$ and a string α over a constant size alphabet.

- *longest common substrings*: If α is a substring of w_i for all $1 \leq i \leq m$, then α is called a common substring of $\mathbb{W}$. Among such α's, the longest one

* This research was supported by Basic Science Research Program through the National Research Foundation of Korea(NRF) funded by the Ministry of Education, Science and Technology (2009-0090441).

** Corresponding author.

Y. Dong, D.-Z. Du, and O. Ibarra (Eds.): ISAAC 2009, LNCS 5878, pp. 1185–1194, 2009.

is called the longest common substring of $\mathbb{W}$. Finding the longest common substring of $\mathbb{W}$ is solvable in polynomial time by dynamic programming or using generalized suffix trees [4].

- *shortest common superstrings*: If α is a superstring of w_i for all $1 \leq i \leq m$, then α is called a common superstring of $\mathbb{W}$. Among such α's, the shortest one is called the shortest common superstring of $\mathbb{W}$. The problem of finding the shortest common superstring is NP-hard [8,3].
- *shortest common nonsubstrings*: If α is not a substring of any w_i for all $1 \leq i \leq m$, then α is called a common nonsubstring of $\mathbb{W}$. Among such α's, the shortest one is called the shortest common nonsubstring of $\mathbb{W}$. The problem of finding the shortest common nonsubstring can be solved in polynomial time [11].
- *longest common nonsuperstrings*: If α is not a superstring of any w_i for all $1 \leq i \leq m$, then α is called a common nonsuperstring of $\mathbb{W}$. Among such α's, the longest one is called the longest common nonsuperstring of $\mathbb{W}$. The problem of finding the longest common nonsuperstring is solvable in polynomial time [11,6,9].

Meanwhile, there are some notions that consider string inclusion and non-inclusion at the same time. Given a finite set of negative strings $\mathbb{N}$ and a finite set of positive strings $\mathbb{P}$, a string α is a *consistent superstring* for $\mathbb{N}$ and $\mathbb{P}$ if α is both a common superstring of $\mathbb{P}$ and a common nonsuperstring of $\mathbb{N}$.

- *shortest consistent superstrings*: Among the consistent superstrings for $\mathbb{N}$ and $\mathbb{P}$, the shortest one is called the shortest consistent superstring for $\mathbb{N}$ and $\mathbb{P}$. Jiang and Li [5] introduced the notion of the consistent superstring and they provided approximation algorithms for finding the shortest consistent superstring when $|\mathbb{N}|$ is bounded or when inclusion free holds, i.e., no string in the set includes another as a substring. Jiang and Timkovsky [6] proposed an algorithm for solving this problem using the directed graph model proposed in [11]. There are two nontrivial assumptions in [6] for finding the shortest consistent superstring for $\mathbb{N}$ and $\mathbb{P}$. The first assumption is that every symbol of the alphabet appears at the end of some negative string. The second assumption is that $\mathbb{N} \cup \mathbb{P}$ is inclusion free. Jiang and Timkovsky's algorithm runs in polynomial time when the graph is acyclic or when $|\mathbb{P}|$ is bounded by a constant.
- *longest consistent superstrings*: Among the consistent superstrings for $\mathbb{N}$ and $\mathbb{P}$, the longest one is called the longest consistent superstring for $\mathbb{N}$ and $\mathbb{P}$. Under the same assumptions as above, Jiang and Timkovsky [6] proposed a polynomial time algorithm for finding the longest consistent superstring when the graph is acyclic.

In this paper, we first propose a new graph model based on the Aho-Corasick algorithm to represent consistent superstrings for $\mathbb{N}$ and $\mathbb{P}$. Then, we present improved algorithms for finding the shortest consistent superstring and the longest consistent superstring using the graph model.

Our contributions are as follows:

- We propose a new model that represents consistent superstrings for $\mathbb{N}$ and $\mathbb{P}$ without any of the nontrivial assumptions that were used in [6].
- We propose an algorithm for finding the shortest (or longest) consistent superstring running in time linear to the sum of the lengths of the strings in $\mathbb{N}$ and $\mathbb{P}$ when the graph is acyclic and either when $|\mathbb{N}|$ and $|\mathbb{P}|$ are bounded by a constant or when $\mathbb{N} \cup \mathbb{P}$ is inclusion free.

The remainder of the paper is organized as follows. In Section 2, we describe the various notations and definitions that we use in this paper. In Section 3, we present a new graph model for consistent superstrings. In Section 4, we give algorithms for finding the shortest and the longest consistent superstrings for the given sets of strings.

2 Preliminaries

Let α be a string over a constant size alphabet Σ. We denote the length of α by $|\alpha|$ and the ith ($1 \le i \le |\alpha|$) character of α by $\alpha[i]$. When a string β is $\alpha[i]\alpha[i+1]\cdots\alpha[j]$, we denote β by $\alpha[i..j]$ and we call β a *substring* of α. Conversely, α is called a *superstring* of β. Moreover, $\alpha[1..j]$ ($1 \le j \le |\alpha|$) is called a *prefix* of α and when $j \ne |\alpha|$, $\alpha[1..j]$ is called a *proper prefix* of α. Similarly, we can define a *suffix* of α and a *proper suffix* of α, respectively. For convenience, we assume the empty string λ is both a prefix of α and a suffix of α.

Consider two given sets of strings $\mathbb{N} = \{x_1, x_2, \ldots, x_t\}$ and $\mathbb{P} = \{y_1, y_2, \ldots, y_r\}$ over a constant size alphabet Σ. We call each x_i ($1 \le i \le t$) a negative string, and we call each y_j ($1 \le j \le r$) a positive string. We assume that $\mathbb{N}$ does not contain the empty string λ. Let n denote the sum of the lengths of all strings in $\mathbb{N}$ and let p denote the sum of the lengths of all strings in $\mathbb{P}$. We denote the number of elements in a set $\mathbb{X}$ by $|\mathbb{X}|$. Then, $|\mathbb{N}| = t$ and $|\mathbb{P}| = r$.

When $\mathbb{N}$ and $\mathbb{P}$ are given as above, a consistent superstring (*CSS*) for $\mathbb{N}$ and $\mathbb{P}$ is a string α that includes every positive string y_i ($1 \le i \le r$) and no negative string x_j ($1 \le j \le t$) as a substring. We denote the set of all CSSs for $\mathbb{N}$ and $\mathbb{P}$ by $\mathbb{CS}$. The shortest consistent superstring (*SCSS*) for $\mathbb{N}$ and $\mathbb{P}$ is the shortest string in $\mathbb{CS}$. Similarly, if $\mathbb{CS}$ is a finite set, then the longest consistent superstring (*LCSS*) for $\mathbb{N}$ and $\mathbb{P}$ is the longest string in $\mathbb{CS}$. If we can make an arbitrarily long CSS, then $\mathbb{CS}$ is an infinite set and thus the LCSS for $\mathbb{N}$ and $\mathbb{P}$ does not exist. Note that when $\mathbb{N} = \phi$, α is called a common superstring for $\mathbb{P}$, and when $\mathbb{P} = \phi$, α is called a common non-superstring (*CNSS*) for $\mathbb{N}$. We denote the set of all CNSSs for $\mathbb{N}$ by $\mathbb{CN}$. Note that $\mathbb{CS} \subseteq \mathbb{CN}$. If $\mathbb{CN}$ is a finite set, then the longest common non-superstring (*LCNSS*) is the longest string in $\mathbb{CN}$.

Now, we define the shortest consistent superstring problem and the longest consistent superstring problem:

Problem 1. The shortest consistent superstring problem (the SCSS problem).
Input: *A finite set of negative strings* $\mathbb{N} = \{x_1, x_2, \ldots, x_t\}$ *and a finite set of positive strings* $\mathbb{P} = \{y_1, y_2, \ldots, y_r\}$ *over a constant size alphabet* Σ.

Output: *If* $\mathbb{CS} = \phi$*, the output is 'No SCSS exists.' Otherwise, the output is the SCSS for* $\mathbb{N}$ *and* $\mathbb{P}$*.*

Problem 2. The longest consistent superstring problem (the LCSS problem).
Input: *A finite set of negative strings* $\mathbb{N} = \{x_1, x_2, \ldots, x_t\}$ *and a finite set of positive strings* $\mathbb{P} = \{y_1, y_2, \ldots, y_r\}$ *over a constant size alphabet* Σ*.*
Output: *If* $\mathbb{CS} = \phi$ *or an arbitrarily long CSS can be made, the output is 'No LCSS exists.' Otherwise, the output is the LCSS for* $\mathbb{N}$ *and* $\mathbb{P}$*.*

To avoid some trivial cases, we assume that $\mathbb{N}$ and $\mathbb{P}$ satisfy the following three conditions:

(1) For all x_i and x_j $(i \neq j)$, x_i is not a substring of x_j.
(2) For all y_i and y_j, y_i is not a substring of y_j.
(3) For all x_i and y_j $(i \neq j)$, x_i is not a substring of y_j.

Note that if x_i is a substring of x_j, then any non-superstring of x_i is a non-superstring of x_j, and that if y_i is a substring of y_j, then any superstring of y_j is a superstring of y_i. If x_i is a substring of y_j, then any superstring of y_j cannot be a non-superstring of x_i, and thus any CSS for $\mathbb{N}$ and $\mathbb{P}$ cannot exist. However, Jiang and Timkovsky [6] assumed two more conditions that are not trivial:

(4) For all y_i and x_j, y_i is not a substring of x_j.
(5) For every symbol $a \in \Sigma$, there exists $x \in \mathbb{N}$ such that $a = x[|x|]$.

We say that $\mathbb{N} \cup \mathbb{P}$ is *inclusion free* when $\mathbb{N}$ and $\mathbb{P}$ satisfy the conditions (1)–(4).

To solve the CSS problems, we utilize the well-known Aho-Corasick algorithm. The Aho-Corasick algorithm (the *AC algorithm* for short) is an efficient algorithm for finding all occurrences of any of a finite number of pattern strings in a text string [1]. The AC algorithm consists of constructing a finite state pattern matching machine (the *AC machine* for short) from the pattern strings, and then finding the pattern strings from the text string in a single pass using the AC machine.

The AC machine has three functions: the goto function, the failure function, and the output function. The *goto* function can be represented by a trie of given pattern strings. The *failure* function for a vertex (state) is defined when the goto function for the vertex does not exist. The *output* function for a vertex shows the pattern strings that have been found. Figure 1.(a) shows an example of the machine that can find any string of $\{aa, abba, aba, bb\}$ in any text string. Note that v_2, v_4, v_5, v_6, and v_8 are *output* vertices that indicate the given pattern strings $aa, aba, bb, abba$, and bb, respectively.

Another representation of the AC machine is a deterministic finite automaton that avoids all failure transitions. Figure 1.(b) shows the deterministic finite automaton version of Figure 1.(a).

3 The Graph Model

In this section, we present our graph model, called G_{CSS}, for consistent superstrings for $\mathbb{N}$ and $\mathbb{P}$. We first introduce the intermediate graph G_{AC} and then we define G_{CSS} using G_{AC}.

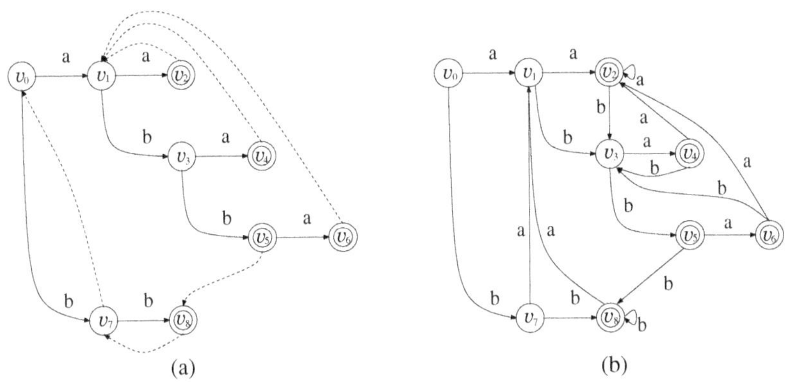

Fig. 1. (a) The AC machine for $\{aa, abba, aba, bb\}$. (b) The deterministic finite automaton version of the AC machine for $\{aa, abba, aba, bb\}$.

Let $\mathbb{T}$ denote a set of prefixes of all strings in $\mathbb{N}$ and $\mathbb{P}$. For example, assume $\mathbb{N} = \{aa, abba\}$ and $\mathbb{P} = \{aba, bb\}$, then $\mathbb{T} = \{\lambda, a, aa, ab, abb, abba, aba, b, bb\}$. Given two sets $\mathbb{N}$ and $\mathbb{P}$ over a constant size alphabet Σ, we define a directed graph $G_{AC} = (V, E)$ with the vertex set V and the edge set E as follows:

(1) For each string $\alpha \in \mathbb{T}$, we define a vertex v, and we do not define any other vertices. Note that there is a one-to-one correspondence between each vertex in V and each string in $\mathbb{T}$. We denote a corresponding vertex of a string α by $ver(\alpha)$ and a corresponding string of a vertex v by $str(v)$. It is clear that $str(ver(\alpha)) = \alpha$. We denote $ver(\lambda)$ by v_0 as a special vertex.
(2) For $\alpha \in \mathbb{T}$ and $\sigma \in \Sigma$, we define a function $\delta : \mathbb{T} \times \Sigma \to \mathbb{T}$. $\delta(\alpha, \sigma)$ is the longest suffix β of $\alpha\sigma$ such that $\beta \in \mathbb{T}$. Then, a directed edge from $ver(\alpha)$ to $ver(\beta)$ labeled with σ is defined if and only if $\beta = \delta(\alpha, \sigma)$. Note that for each string $\alpha \in \mathbb{T}$ and $\sigma \in \Sigma$, only one edge is defined, and thus there are exactly $|\Sigma|$ outgoing edges for each vertex.

Note that G_{AC} is the deterministic finite automaton version of the AC machine for $\mathbb{N} \cup \mathbb{P}$.

We define $V_S(\alpha)$ as a set of $ver(\gamma)$ such that α is a suffix of γ and $\gamma \in \mathbb{T}$. Let $V_S(\mathbb{P})$ denote the union of $V_S(y)$ for every $y \in \mathbb{P}$ and let $V_S(\mathbb{N})$ denote the union of $V_S(x)$ for every $x \in \mathbb{N}$. Note that $V_S(y_i) \cap V_S(y_j) = \phi$ for all $i \neq j$ $(1 \leq i, j \leq r)$. Figure 2 shows G_{AC} for $\mathbb{N} = \{aa, abba\}$ and $\mathbb{P} = \{aba, bb\}$. Since $V_S(aba) = \{v_4\}$, $V_S(bb) = \{v_5, v_8\}$, $V_S(aa) = \{v_2\}$ and $V_S(abba) = \{v_6\}$, $V_S(\mathbb{P}) = \{v_4, v_5, v_8\}$ and $V_S(\mathbb{N}) = \{v_2, v_6\}$.

The number of vertices $|V|$ is $O(n+p)$ since the number of prefixes of a string is the same as the length of the string. The number of edges $|E|$ is also $O(n+p)$ because every vertex has at most $|\Sigma|$ edges that is a constant.

Now we give some notations for G_{AC}. A *path* in a graph is denoted by the sequence $A = (u_1, u_2, \ldots, u_k)$ of vertices such that for all $i = 1, 2, \ldots, k-1$, (u_i, u_{i+1}) is an edge of the graph. The *length* of A is the number of edges in

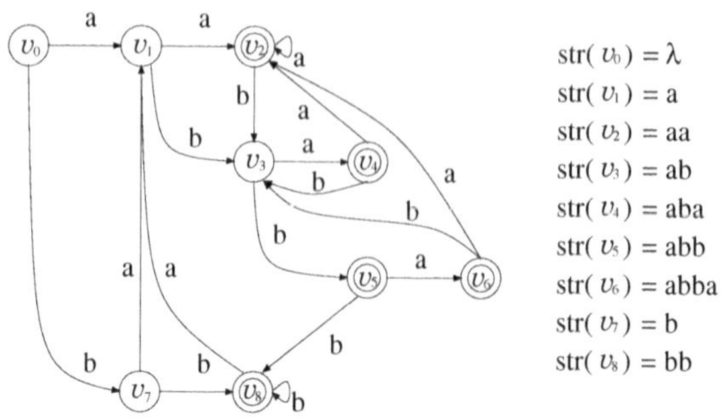

Fig. 2. G_{AC} for $\mathbb{N} = \{aa, abba\}$ and $\mathbb{P} = \{aba, bb\}$, when $\Sigma = \{a, b\}$

A, denoted by $|A|$. For two paths $A_1 = (u_1, \ldots, u_k)$ and $A_2 = (u_k, \ldots, u_l)$ such that the last vertex of A_1 is equal to the first vertex of A_2, let $A_1A_2 = (u_1, \ldots, u_k, \ldots, u_l)$ denote the merged path of A_1 and A_2. Let *λ-path* denote a path that starts at v_0. Let *ℙ-path* denote a λ-path that passes at least one vertex of $V_S(y)$ for every $y \in \mathbb{P}$. Let *ℕ-path* denote a λ-path that does not pass any vertex in $V_S(\mathbb{N})$.

Given a path $A = (u_1, \ldots, u_k)$ in G_{AC}, let $pstr(A)$ denote a string $l_1 \ldots l_{k-1}$ such that each l_i is the label of the edge (u_i, u_{i+1}) for $1 \le i \le k-1$. Conversely, given a string $\alpha = l_1 \ldots l_{k-1}$, let $path(\alpha)$ denote a path $(u_1, \ldots, u_k)$ such that each (u_i, u_{i+1}) is an edge labeled with l_i for $1 \le i \le k-1$. Let λ-$path(\alpha)$ denote $path(\alpha)$ starting at v_0. Note that $path(\alpha)$ as well as λ-$path(\alpha)$ always exists in G_{AC} for every $\alpha \in \Sigma^*$ because for every $v \in V$, there exists an edge labeled with σ for every $\sigma \in \Sigma$.

For a string α and λ-$path(\alpha)$ in G_{AC}, the following lemmas show some properties of the graph G_{AC}.

Lemma 1. *For a string $\beta \in \mathbb{N} \cup \mathbb{P}$, α includes β as a substring if and only if λ-path(α) passes a vertex $v \in V_S(\beta)$.*

Proof. It is clear by the definition of G_{AC}.

Lemma 2. *$\alpha \in \mathbb{CN}$ if and only if λ-path(α) is an ℕ-path in G_{AC}.*

Proof. (if) If λ-$path(\alpha)$ is an ℕ-path in G_{AC}, then λ-$path(\alpha)$ does not pass any vertex in $V_S(\mathbb{N})$, which means $pstr(\lambda\text{-}path(\alpha))$ does not include any string in ℕ by Lemma 1. Thus, $\alpha \in \mathbb{CN}$.

(only if) For λ-$path(\alpha)$ to be an ℕ-path, it should not pass any vertex in $V_S(\mathbb{N})$. Assume λ-$path(\alpha)$ passes $ver(x)$ where $x \in \mathbb{N}$. Then α includes x as a substring by Lemma 1. It contradicts $\alpha \in \mathbb{CN}$. Thus λ-$path(\alpha)$ is an ℕ-path in G_{AC}.

Lemma 3. *$\alpha \in \mathbb{CS}$ if and only if λ-path(α) is both an ℕ-path and a ℙ-path in G_{AC}.*

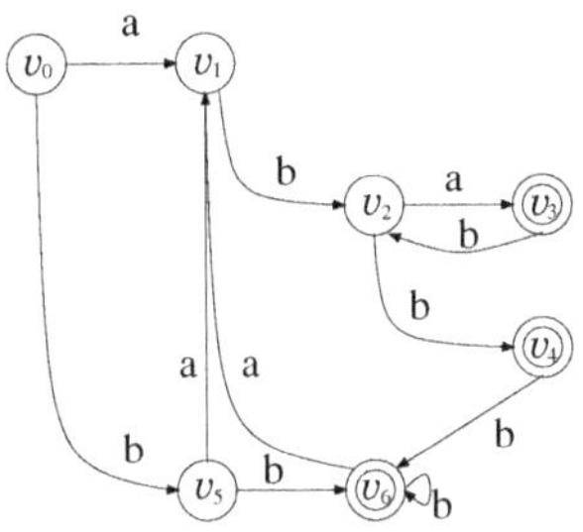

Fig. 3. G_{CSS} for $\mathbb{N} = \{aa, abba\}$ and $\mathbb{P} = \{aba, bb\}$

Proof. (if) If λ-$path(\alpha)$ is an $\mathbb{N}$-path in G_{AC}, then $\alpha \in \mathbb{CN}$ by Lemma 2. If λ-$path(\alpha)$ is a $\mathbb{P}$-path in G_{AC}, i.e., λ-$path(\alpha)$ passes at least one vertex of $V_S(y)$ for every $y \in \mathbb{P}$, then α includes every y as a substring by Lemma 1. Thus $\alpha \in \mathbb{CS}$.

(only if) Since $\alpha \in \mathbb{CS}$, $\alpha \in \mathbb{CN}$ and thus by Lemma 2, λ-$path(\alpha)$ is an $\mathbb{N}$-path in G_{AC}. Also, α includes every positive string $y \in \mathbb{P}$ as a substring. By Lemma 1 λ-$path(\alpha)$ passes one vertex of $V_S(y)$ for every y. Thus λ-$path(\alpha)$ is a $\mathbb{P}$-path in G_{AC}.

Now we define a directed graph $G_{CSS} = (V', E')$ such that $V' = V - V_S(\mathbb{N})$ and $E' = E - \{(v_a, v_b) | v_a \in V_S(\mathbb{N}) \text{ or } v_b \in V_S(\mathbb{N})\}$. That is, G_{CSS} is made from G_{AC} by removing all the corresponding vertices of negative strings and all their adjacent edges. Figure 3 shows an example of G_{CSS}.

Corollary 1. *$\alpha \in \mathbb{CS}$ if and only if λ-path(α) is a $\mathbb{P}$-path in G_{CSS}*

Proof. (if) If λ-$path(\alpha)$ is a $\mathbb{P}$-*path* in G_{CSS}, then λ-$path(\alpha)$ is also a $\mathbb{P}$-*path* in G_{AC} since G_{AC} includes every vertex and every edge in G_{CSS}. Since λ-$path(\alpha)$ is a path in G_{CSS}, λ-$path(\alpha)$ does not pass any vertex in $V_S(\mathbb{N})$, i.e., λ-$path(\alpha)$ is an $\mathbb{N}$-*path* in G_{AC}. Thus, by Lemma 3, $\alpha \in \mathbb{CS}$.

(only if) Since λ-$path(\alpha)$ for $\alpha \in \mathbb{CS}$ is an $\mathbb{N}$-path in G_{AC} by Lemma 3, λ-$path(\alpha)$ never passes any vertex in $V_S(\mathbb{N})$. Since G_{CSS} is the same as G_{AC} except $V_S(\mathbb{N})$ and all their adjacent edges, λ-$path(\alpha)$ exists in G_{CSS}. In addition, since α includes every positive string $y \in \mathbb{P}$ as a substring, λ-$path(\alpha)$ passes at least one vertex of $V_S(y)$ for every y. Thus, λ-$path(\alpha)$ is a $\mathbb{P}$-*path* in G_{CSS}.

4 CSS Problems and Algorithms

In this section, we present algorithms for the CSS problems. We can find CSSs for $\mathbb{N}$ and $\mathbb{P}$ by finding $\mathbb{P}$-paths in G_{CSS} by Corollary 1. We will find $\mathbb{P}$-paths using the vertices in $V_S(\mathbb{P})$. Let us choose a vertex from each $V_S(y_i)$ for $1 \leq i \leq r$ and denote it by w_i. Then, a $\mathbb{P}$-*sequence* is a permutation of $w_1, w_2, \cdots, w_r$. Let $\pi = \langle u_1, u_2, \cdots, u_r \rangle$ be a $\mathbb{P}$-sequence of G_{CSS}. When a $\mathbb{P}$-path A passes all vertices of π in order, i.e., A is the concatenation of paths from v_0 to u_1, from u_1 to u_2, $\ldots$, and from u_{r-1} to u_r, we say that the $\mathbb{P}$-path A *embeds* the $\mathbb{P}$-sequence π. If there exists a $\mathbb{P}$-path embedding a $\mathbb{P}$-sequence π, we call π a *valid*

$\mathbb{P}$-sequence. Otherwise, we call π an *invalid* $\mathbb{P}$-sequence. Note that, for each $\mathbb{P}$-path, there exists at least one $\mathbb{P}$-sequence embedded by the $\mathbb{P}$-path, which means we can find $\mathbb{P}$-paths by considering $\mathbb{P}$-sequences.

4.1 Finding the SCSS

Our algorithm for the SCSS problem consists of three phases.

1. Construct G_{CSS} using the Aho-Corasick algorithm.
2. Find the shortest $\mathbb{P}$-path using $\mathbb{P}$-sequences.
   For each $\mathbb{P}$-sequence $\pi = \langle u_1, \cdots, u_r \rangle$, we find the shortest $\mathbb{P}$-path embedding π by concatenating the shortest paths from v_0 to u_1, from u_1 to u_2, ..., and from u_{r-1} to u_r.
3. Compute the SCSS using the shortest $\mathbb{P}$-path found in Phase 2.
   Concatenate the label of each edge on the shortest $\mathbb{P}$-path while traversing it.

In Phase 2, we might have to consider all $\mathbb{P}$-sequences in the worst case. Let us consider the number of $\mathbb{P}$-sequences. Since there are $|V_S(y_i)|$ vertices for each y_i $(1 \le i \le r)$, the number of possible $\mathbb{P}$-sequences is $r! \prod_{i=1}^{r} |V_S(y_i)|$, which does not exceed $r! \left(\frac{n+r}{r}\right)^r$ due to the property of G_{CSS}.

Consider a set $\mathbb{Z} = \mathbb{P} \cup \{z | z$ is the longest proper prefix of $x \in \mathbb{N}\}$. Assume $\alpha \in \mathbb{Z}$. For each i $(1 \le i \le r)$, if a vertex $u_i \in V_S(y_i)$ is the first vertex on λ-*path*(α) among all the vertices in $V_S(y_i)$, we call u_i a τ-vertex. If every vertex in a $\mathbb{P}$-sequence π is a τ-vertex, we call π a *heading* $\mathbb{P}$-sequence.

Lemma 4. *Every* $\mathbb{P}$*-path embeds a heading* $\mathbb{P}$*-sequence in* G_{CSS}.

Proof. Let A be a $\mathbb{P}$-path in G_{CSS}. For each $y_i \in \mathbb{P}$, let u_i denote the first appeared vertex in A among all the vertices in $V_S(y_i)$. If λ-*path*$(str(u_i))$ is the subpath of A from v_0 to u_i for $1 \le i \le r$, then u_i is a τ-vertex and we are done since the $\mathbb{P}$-sequence $\langle u_1, \cdots, u_r \rangle$ is the heading $\mathbb{P}$-sequence. But λ-*path*$(str(u_i))$ is not necessarily the subpath of A from v_0 to u_i. We will show that u_i is a τ-vertex even when λ-*path*$(str(u_i))$ is not the subpath of A from v_0 to u_i. Let B be the subpath of A from v_0 to u_i. Since $str(u_i)$ is a substring of $pstr(B)$ by the definition of G_{CSS}, $str(u_i)$ includes y_i only once as a suffix, i.e., u_i is the first vertex to appear in λ-*path*$(str(u_i))$ among $V_S(y_i)$. Therefore, u_i is a τ-vertex. Since for each y_i there exists a τ-vertex in A, we can make a heading $\mathbb{P}$-sequence embedded by A. Thus every $\mathbb{P}$-path embeds a heading $\mathbb{P}$-sequence.

By Lemma 4, we can find a $\mathbb{P}$-path by considering heading $\mathbb{P}$-sequences. There are at most $t+1$ τ-vertices in each $V_S(y_i)$ for $1 \le i \le r$. Thus, the number of possible $\mathbb{P}$-sequences is at most $\min\left\{r!(t+1)^r, r!\left(\frac{n+r}{r}\right)^r\right\}$.

When G_{CSS} is acyclic, the number of $\mathbb{P}$-sequences to consider may be drastically reduced due to the following property of acyclic graphs. Consider a path $L = (z_1, \cdots, z_q)$ in a directed acyclic graph (DAG). Then, in any topological order of the DAG, z_i must appear before z_{i+1} for all $1 \le i \le q-1$. Let A be a $\mathbb{P}$-path embedding a valid $\mathbb{P}$-sequence in G_{CSS}, and let TP be a topological order of G_{CSS}. Then, by the above property of a DAG, the vertices in A appear in the same order as in TP and so do the vertices in π. Thus, to find a $\mathbb{P}$-path, it suffices to check only the $\mathbb{P}$-sequences appeared in the same order as in TP. Therefore, the number of possible $\mathbb{P}$-sequences is at most $\min\left\{(t+1)^r, \left(\frac{n+r}{r}\right)^r\right\}$.

Now we analyze the time complexity of our algorithm for the SCSS problem. Phase 1 takes $O(n+p)$ time. G_{AC} can be constructed in $O(n+p)$ time since the constructing part of the Aho-Corasick algorithm runs in time linear to the sum of the lengths of the pattern strings. Removing $ver(x)$ for every $x \in \mathbb{N}$ and its incoming and outgoing edges also takes $O(n+p)$ time since $t \leq n$ and the number of edges is $O(n+p)$.

Making τ-vertices also takes $O(n+p)$ time. For each string α in $\mathbb{Z} = \mathbb{P} \cup \{z|z$ is the longest proper prefix of $x \in \mathbb{N}\}$, we search α in G_{CSS} and check the output function for the vertices during searching α. If $y \in \mathbb{P}$ is included in the output function for a vertex v, then v is in $V_S(y)$, and if v is the first vertex having y as output, then v is a τ-vertex by the definition. Since any positive string is not a substring of another positive string, each vertex has at most one positive string as output, and thus checking the output function for one vertex needs only a constant time. Since the traversing takes $O(n+p)$ time and the checking takes $O(1)$ time, making τ-vertices takes $O(n+p)$ time.

Phase 2 takes $O(r! \min\left\{(t+1)^r, \left(\frac{n+r}{r}\right)^r\right\} \times r(n+p)\log(n+p))$ time in general if we use Dijkstra's algorithm to find the shortest path. Moreover, if G_{CSS} is acyclic, Phase 2 takes $O(\min\left\{(t+1)^r, \left(\frac{n+r}{r}\right)^r\right\} \times (n+p))$ time since the shortest path in a DAG can be found in time linear to the number of vertices.

In the final phase, traversing the path takes $O(r(n+p))$ time in general and if G_{CSS} is acyclic, it takes $O(n+p)$ time.

Theorem 1. *Given two sets of negative strings $\mathbb{N}$ and positive strings $\mathbb{P}$ over a constant size alphabet Σ, the SCSS for $\mathbb{N}$ and $\mathbb{P}$ can be found in $O(r! \min \left\{(t+1)^r, \left(\frac{n+1}{r}\right)^r\right\} r(n+p)\log(n+p))$ time when G_{CSS} is cyclic, and in $O((\min \left\{t+1, \frac{n+r}{r}\right\})^r(n+p))$ time when G_{CSS} is acyclic.*

Corollary 2. *The SCSS for $\mathbb{N}$ and $\mathbb{P}$ can be found in polynomial time when $|\mathbb{P}|$ is bounded by a constant.*

Corollary 3. *The SCSS for $\mathbb{N}$ and $\mathbb{P}$ can be found in $O(n+p)$ time when G_{CSS} is acyclic and either when $|\mathbb{N}|$ and $|\mathbb{P}|$ are bounded by a constant or when $\mathbb{N} \cup \mathbb{P}$ is inclusion free.*

Proof. When $|\mathbb{N}|$ and $|\mathbb{P}|$ are bounded by a constant, t and r are constants, and thus our algorithm runs in $O(n+p)$ time. When $\mathbb{N} \cup \mathbb{P}$ is inclusion free, y_i $(1 \leq i \leq r)$ is not a substring of any $x \in \mathbb{N}$, and thus $|V_S(y_i)| = 1$ for all $1 \leq i \leq r$. Therefore, our algorithm runs in $O(n+p)$ time.

4.2 Finding the LCSS

To find the LCSS, our algorithm finds the longest path in G_{CSS}, and it is almost same as the previous algorithm that was used for the SCSS problem. Instead of finding the shortest path between every pair of vertices in $\mathbb{P}$-sequences, we find the longest path. In general, finding the longest path is NP-hard, and thus we assume G_{CSS} is acyclic.

1. Construct G_{CSS} using the Aho-Corasick algorithm.
2. Find the longest $\mathbb{P}$-path using $\mathbb{P}$-sequences.
   For each $\mathbb{P}$-sequence $\pi = \langle u_1, \cdots, u_r \rangle$, we find the longest $\mathbb{P}$-path embedding π by concatenating the longest paths from v_0 to u_1, from u_1 to u_2, and so on.
3. Compute the LCSS using the longest $\mathbb{P}$-path found in Phase 2.

Theorem 2. *Given two sets of negative strings $\mathbb{N}$ and positive strings $\mathbb{P}$ over a constant size alphabet Σ, the LCSS for $\mathbb{N}$ and $\mathbb{P}$ can be found in $O((\min\{t+1, \frac{n+r}{r}\})^r(n+p))$ time when G_{CSS} is acyclic.*

Corollary 4. *The LCSS for $\mathbb{N}$ and $\mathbb{P}$ can be found in polynomial time when G_{CSS} is acyclic and when $|\mathbb{P}|$ is bounded by a constant.*

Corollary 5. *The LCSS for $\mathbb{N}$ and $\mathbb{P}$ can be found in $O(n+p)$ time when G_{CSS} is acyclic and either when $|\mathbb{N}|$ and $|\mathbb{P}|$ are bounded by a constant or when $\mathbb{N} \cup \mathbb{P}$ is inclusion free.*

References

1. Aho, A.V., Corasick, M.J.: Efficient string matching: an aid to bibliographic search. Communications of the ACM 18(6), 333–340 (1975)
2. Drmanac, R., Crkvenjakov, C.: Sequencing by hybridization (SBH) with oligonucleotide probes as an integral approach for the analysis of complex genomes. International Journal of Genomic Research 1(1), 59–79 (1992)
3. Garey, M.R., Johnson, D.S.: Computers and Intractability. W. H. Freeman and Company, New York (1979)
4. Gusfield, D.: Algorithms on Strings, Trees, and Sequences. Cambridge University Press, Cambridge (1997)
5. Jiang, T., Li, M.: Approximating shortest superstrings with constraints. Theoretical Computer Science 134(2), 473–491 (1994)
6. Jiang, T., Timkovsky, V.G.: Shortest consistent superstrings computable in polynomial time. Theoretical Computer Science 143(1), 113–122 (1995)
7. Kerschbaum, F.: A new way to think about computation: Language-based secure computation. In: 5th International Workshop on Security in Information Systems, pp. 33–42 (2007)
8. Maier, D., Storer, A.: A note on the complexity of superstring problem, Computer Science Laboratory, Report No. 233. Princeton University, Princeton
9. Na, J.C., Kim, D.K., Sim, J.S.: Finding the longest common nonsuperstring in linear time. Information Processing Letter 109(18), 1066–1070 (2009)
10. Pevzner, P., Lipshutz, R.: Towards DNA sequencing chips. In: Privara, I., Ružička, P., Rovan, B. (eds.) MFCS 1994. LNCS, vol. 841, pp. 143–158. Springer, Heidelberg (1994)
11. Rubinov, A.R., Timkovsky, V.G.: String noninclusion optimization problems. SIAM Journal on Discrete Mathematics 11(3), 456–467 (1998)
12. Store, J.: Data Compression: Methods and Theory. Computer Science Press, Rockville (1998)
13. Storer, J., Szymanski, T.: Data compression via textual substitution. Journal of ACM 29, 928–961 (1982)

Two-Vertex Connectivity Augmentations for Graphs with a Partition Constraint* (Extended Abstract)

Pei-Chi Huang[1], Hsin-Wen Wei[2,**], Yen-Chiu Chen[1], Ming-Yang Kao[3], Wei-Kuan Shih[1], and Tsan-sheng Hsu[2]

[1] Department of Computer Science, National Tsing-Hua University, Taiwan
{peggy,ycchen,wshih}@rtlab.cs.nthu.edu.tw
[2] Institute of Information Science, Academia Sinica, Taiwan
{hwwei,tshsu}@iis.sinica.edu.tw
[3] Department of Electrical Engineering and Computer Science, Northwestern University, U.S.A.
kao@northwestern.edu

Abstract. In this paper, we study the two-vertex connectivity augmentation problem in an undirected graph whose vertices are partitioned into k sets. Our objective is to add the smallest number of edges to the graph such that the resulting graph is 2-vertex connected under the constraint that each new edge is between two different sets in the partition. We propose an algorithm to solve the above augmentation problem that runs in linear time in the size of the input graph.

1 Introduction

A graph is called be *x-vertex connected* if it remains connected after the removal of any set of at most $x - 1$ vertices. The notion of *x-vertex connectivity* can be similarly defined. Many algorithms have been developed to solve the problem of making general graphs x-vertex or x-edge connected for various values of x [1, 4–9, 11, 13]. Eswaran and Tarjan [1] presented a linear-time algorithm for the smallest bridge connectivity augmentation problem on general graphs without a partition constraint. Huang et al. [10] introduced a linear-time algorithm for bridge connectivity augmentation with a bipartite constraint. Hsu and Kao [8] designed a linear-time algorithm for 2-vertex connectivity with a bipartite constraint.

In this paper, we focus on augmenting graphs with a partition constraint, which requires that the vertex set of an input graph must be partitioned into k disjoint vertex subsets. Moreover, each edge in the augmentation must be added between two different vertex subsets. We propose a linear-time algorithm that adds the smallest number of edges to a graph G with a given partition

* Supported in part by National Science Council (Taiwan) Grants NSC 97-2221-E-001-011-MY3.

** Corresponding author.

Y. Dong, D.-Z. Du, and O. Ibarra (Eds.): ISAAC 2009, LNCS 5878, pp. 1195–1204, 2009.

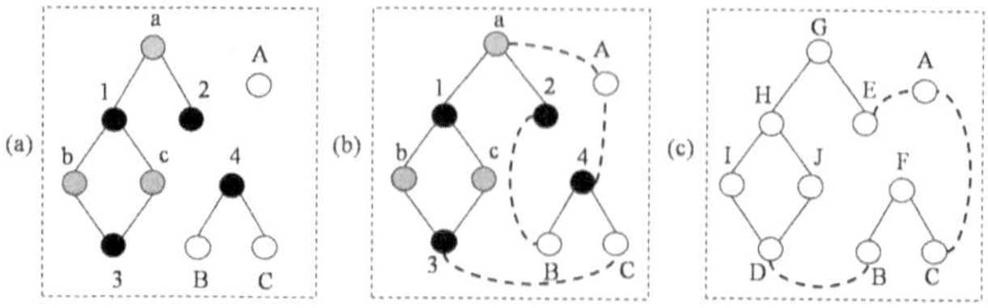

Fig. 1. (a) A graph with a partition of 3 vertex subsets. (b) A smallest 2-vertex connectivity augmentation for the graph and partition in (a) with the added edges marked by dashed lines. (c) A smallest 2-vertex connectivity augmentation for the graph in (a) without the partition constraint where the added edges are marked by dashed lines.

constraint to make G 2-vertex connected, while maintaining the constraint. A *smallest 2-vertex connectivity augmentation* of G, denoted by aug2v(G), is a set of edges with the minimum cardinality whose addition makes G 2-vertex connected. Figure 1(a) shows an example of a graph with a partition of 3 vertex subsets. A smallest 2-vertex connectivity augmentation of Figure 1(a) is shown in Figure 1(b); and a smallest 2-vertex connectivity augmentation of Figure 1(a) without a partition constraint is shown in Figure 1(c). Note that the set of added edges in Figure 1(c) is not valid for the partition since vertices A and C are in the same partition.

2 Preliminaries

2.1 Graph-Theoretical Definitions

Let $G = (V, E)$ be a graph. A vertex with degree one of a graph is called a *leaf*, and a vertex with degree zero of a graph is called an *isolated vertex*. An edge whose endpoints are a vertex u and a vertex v is denoted as $\{u, v\}$. For an edge set E', $G - E'$ denotes G without edges in E', and $G \cup E'$ denotes G with edges in E' added to it. Most of our graph-theoretical definitions can be found in [2, 3, 8, 12].

In this paper, all graphs are undirected and have neither self-loops nor multiple edges. It is assume that a vertex set of an input graph is partitioned into k disjoint subsets. Let P_i denote the ith vertex subset in the given partition of an input graph; i.e., $V = P_1 \cup P_2 \cup \ldots \cup P_k$, and for all P_i, P_j with $i \neq j$, $P_i \cap P_j = \emptyset$.

Our algorithmic problem is to add the smallest number of edges such that the resulting graph is 2-vertex connected, and the two endpoints of each added edge are in different vertex subsets of the given partition unless the two leaves are the same partition.

We use a *block forest* BT(G) [1, 3, 8] to represent the graph obtained by transforming the input graph G. Let BT(G) be the block forest of G, and let each block in G be a vertex of BT(G). Let $\mathcal{F}'$ be a function that maps a new edge set added to BT(G) into a corresponding edge set added to G. Therefore, the solution of the problem can found by solving the problem of a smallest 2-vertex-connectivity augmentation of BT(G), denoted by aug2v(BT(G)). Let E'

be aug2v(BT(G)). If E' is the edge set added to BT(G), then $\mathcal{F}'(E')$ is the corresponding edge set added to G, i.e., aug2v(G) = $\mathcal{F}'(E')$.

To improve the readability of the paper, we assign a distinct color to a vertex subset in the given partition of G; let C_i denote the color for the vertices in P_i. The vertices and the leaves in BT(G) are classified according to the colors as follows. A *mono leaf* (respectively, *mono vertex*) of color C_i in BT(G) is a leaf (respectively, vertex) in BT(G), whose corresponding block consists only of vertices in G of color C_i. A *hybrid leaf* in BT(G) is a leaf in BT(G), whose corresponding block consists of at least two vertices in G of different colors. An *isolated vertex* of color C_i in BT(G) is an isolated block of in G, whose vertices are all of color C_i. A *hybrid isolated vertex* in BT(G) is an isolated block of G that contains at least two vertices of different colors in G. A *legal edge* is added to BT(G) according to the following rules. A edge can be added between C_i and C_j, C_i and a hybrid vertex, or two hybrid vertices, where $i \neq j$.

2.2 Lower Bound on |aug2v(BT(G))|

First, we compute a lower bound for |aug2v(BT(G))|. Let L_i and H denote the set of mono leaves of color C_i and the set of hybrid leaves of BT(G), $1 \leq i \leq k$, respectively. In addition, let L_i^* and H^* be the set of isolated vertices of color C_i and the set of hybrid isolated vertices of BT(G), $1 \leq i \leq k$, respectively.

Let $\hat{L}_{max} = \max\{|L_i| + 2|L_i^*||1 \leq i \leq k\}$. In addition, let $\ell = \sum_{i=1}^{k} |L_i| + |H|$, $I^* = \sum_{i=1}^{k} |L_i^*| + |H^*|$. Let $\mathrm{LOW_{f2v}}(\mathrm{BT}(G)) = \max\{(q - 1) + \max_{i=1}^{n}(D(v_i, \mathrm{BT}(G)) - 1), \hat{L}_{\max}, \lfloor(\ell + 2I^*)/2\rfloor\}$, where q is the number of connected components in BT(G), n is the number of vertices in BT(G), and $D(v_i, \mathrm{BT}(G))$ is the degree of v_i in BT(G).

Lemma 1. $|\mathrm{aug2v}(\mathrm{BT}(G))| \geq \mathrm{LOW_{f2v}}(\mathrm{BT}(G))$.

3 Recolor Vertices of Degree < 2 with Three Colors

In the section, we describes a method that recolors vertices with degree < 2 in a forest so that only 3 colors among vertices of degree < 2 are left after recoloring. First, we use only BT(G) to simplify the paper. Let F = BT(G) be a forest with k different colors among vertices of Degree < 2, $C_1, C_2, \ldots, C_k$, $k \geq 3$. We consider that C_i in F is *saturated* if $|L_i| + 2|L_i^*| > \lfloor(\ell + 2I^*)/2\rfloor$. If at least one color is saturated, F is *color-saturated*; if no color is saturated, F is *color-safe*. Note that at most one color is saturated. Therefore, we consider two cases: (1) BT(G) is color-saturated; this case is discussed in Section 3.1, and (2) BT(G) is color-safe; this case is discussed in Section 3.2.

3.1 BT(G) Is Color-Saturated

In this section, BT(G) is color-saturated. Without loss of generality, we assume that the number of leaves and isolated vertices with the black color in BT(G) is more than the total number of leaves and isolated vertices with the other colors in BT(G). We recolor the leaves and isolated vertices with the non-black colors by the white color.

As shown Lemma 2 below, after this recoloring, the $\mathrm{LOW_{f2v}}$ bound is unchanged. Let F' be a bipartite graph obtained by recoloring F. By Theorem 1, F' is the case when two colors among vertices of degree < 2 is solved.

Lemma 2. *Assume that* BT(G) *is color-saturated. Let F' be the graph obtained by recoloring* BT(G)*. Then,* $\mathrm{LOW_{f2v}}(F') = \mathrm{LOW_{f2v}}(\mathrm{BT}(G))$.

Theorem 1 ([8]). *Let F be a forest with two colors among vertices of degree < 2. We can add the smallest number of edges, i.e.,* $\mathrm{LOW_{f2v}}(F)$*, to make F 2-vertex connected while preserving its bipartiteness in linear time.*

3.2 BT(G) Is Color-Safe

We now present a recoloring algorithm that transforms a tree with k colors, $k > 3$, on vertices of degree < 2 to exactly 3 colors. The recoloring process involves three phases.

3.2.1 Phase 1

To obtain F_1 from F, we assign colors to vertices of degree < 2 in BT(G) as follows: We keep the original color of the internal nodes in F, as well as the original color of mono leaves and isolated vertices in F; however, each hybrid leaf and each hybrid isolated vertex is viewed as one distinct color. Therefore, we assign one distinct color to each hybrid leaf and each hybrid isolated vertex.

3.2.2 Phase 2

We recolor F_1 to obtain F_1' so that the total number of colors among vertices of degree < 2 in F_1' is one less than in F_1. Assume that the k different colors among vertices of degree < 2 in F_1 is $C_1, C_2, \ldots, C_k$, where $k \geq 4$. Let the colors with the least and second least number of leaves and isolated vertices be denoted by C_i and C_j, respectively. We replace all leaves and isolated vertices colored C_i or C_j with a new color. Then, no color is saturated in F_1', and the number of colors in F_1' is one less than in F_1 in Lemma 3.

Lemma 3. *The total number of colors among vertices of degree < 2 in F_1' is one less than in F_1.*

3.2.3 Phase 3

In the following Lemma 4, we show that we can simplify a problem instance of more than 3 colors among vertices of degree < 2 in F_1 by transforming it into a problem instance of 3 colors among vertices of degree < 2. We can apply Lemma 3 iteratively until there are only 3 colors left, denoted as F_2. In Lemma 4, we prove that this iterative transformation only requires linear time. It is straightforward to prove that a lower bound of F_1 is equal to a lower bound of F_2. By Lemma 5, if $F_2 \cup E$ is 2-vertex connected, $F \cup \mathcal{F}'(E)$ is also 2-vertex connected, where E is a set of edges added to F_2, and $\mathcal{F}'(E)$ is the corresponding set of edges added to F.

Lemma 4. *An iterative transformation of vertices of degree < 2 from $k \geq 4$ colors for* BT(G) *to three colors by repeatedly applying the methods described before Lemma 3 can be accomplished in $O(n)$ time.*

Lemma 5. *(1)* $\mathrm{LOW}_{\mathrm{f2v}}(\mathrm{BT}(G))=\mathrm{LOW}_{\mathrm{f2v}}(F_2)$. *(2) Let E be a set of edges. If $F_2 \cup E$ is 2-vertex connected, then $F \cup E'$ is also 2-vertex connected, where $E' = \mathcal{F}'(E)$.*

Since the given forest F can always be transformed into F_2, hereafter, we assume that the given forest BT(G), i.e., F has three colors among vertices of degree < 2, and F is color-safe.

4 BT(G) Is a Color-Safe Tree with Three Colors among Vertices of Degree < 2

In this section, we focus on how to add edges to make BT(G) 2-vertex connected when BT(G) is a color-safe tree with three colors among ℓ leaves.

We define the following notations. Let $T = \mathrm{BT}(G)$ be a tree rooted at v_r, and $\Gamma(T) = \{\tau_i \mid \tau_i$ is a v_r*-branch*, i.e., the subtree of T rooted at a child of v_r.$\}$. A *chain leaf* of T, is a v_r-branch that contains exactly one leaf in the subtree. Let $\mathcal{A} = \{S_1, S_2, \ldots, S_s\}$, $s \geq 2$, and S_i denote the set of the leaves in a subtree τ_i rooted at a child of the root v_r of T, $1 \leq i \leq s$. If S_i is a chain leaf, $|S_i| = 1$.

A set S_i in T is *saturated* if $|S_i|$ is greater than $\lfloor \ell/2 \rfloor$, and we say that T is *set-saturated*, i.e., $1 \leq i \leq s$. If no subset is saturated, T is *set-safe*. If $2(s-1) > \ell$, T is *degree-saturated*; otherwise, T is *degree-safe*. Furthermore, if T is degree-safe, color-safe and set-safe, we define that T is *all-safe*.

In the following, we consider two cases: (1) BT(G) is set-saturated; this case is discussed in Section 4.1, and (2) BT(G) is set-safe; this case is discussed in Section 4.2.

4.1 BT(G) Is Set-Saturated

Given a rooted tree T, a *center* v_c is a vertex whose branches have at most $\lfloor \ell/2 \rfloor$ leaves each. If the root is a center, we also call it a *center root*. $D(v_c, T)$ equals the degree of v_c in T [1, 8].

Let $T = \mathrm{BT}(G)$ with ℓ leaves be the input graph. Note that BT(G) is a tree if and only if T is connected. Here, let the colors among vertices of degree < 2 in T be C_1, C_2 and C_3, respectively. Assume that $|S_i|$ is greater than $\lfloor \ell/2 \rfloor$, that is, S_i in T is set-saturated, where $1 \leq i \leq s$.

Lemma 6. *After rerooting T at a center v_c, the resulting T is set-safe and the rerooting can be done in linear time.*

4.2 BT(G) Is Set-Safe

Let $T = \mathrm{BT}(G)$ be a tree rooted at a center v_c with ℓ leaves and there are only three colors among vertices of degree < 2, where the number of leaves with each color is at most $\lfloor \ell/2 \rfloor$. Assume that T has $|s|$ sets, $s \geq 2$. Here, the case $s = 2$ is trivial. Since the root v_c of T is a center of T, the number of leaves for each $|S_1|, |S_2|, \ldots, |S_s|$ is at most $\lfloor \ell/2 \rfloor$, $s \geq 2$, then no set is saturated.

In this section, T is not only set-safe but also color-safe. Thus, we consider two cases. (1) T is degree-saturated; this case is discussed in Section 4.2.1, and (2) T is degree-safe; this case is discussed in Section 4.2.2.

4.2.1 *T* Is Degree-Saturated

In this case, T is color-safe and set-safe, but degree-saturated. Let $T' = T \cup E_d$, where E_d is the set of edges added to T such that T' is all-safe.

By Lemma 7, we always select and remove two chain leaves, a_j and b_j, of T at a time such that T' is color-safe and set-safe. We continue adding edges in this way until T' is degree-safe, $\tau_i \in \Gamma(T)$. After removing a_j and b_j, $D(v_c, T') = D(v_c, T) - 1$. $\mathrm{LOW}_{\mathrm{f2v}}(T') = \mathrm{LOW}_{\mathrm{f2v}}(T) - 1$.

Lemma 7. *Let T be a color-safe, set-safe but degree-saturated tree. Let E_d be the set of added edges. In addition, let $T' = T \cup E_d$. $\mathrm{LOW}_{\mathrm{f2v}}(T') = \mathrm{LOW}_{\mathrm{f2v}}(T) - |E_d|$ and T' is all-safe.*

4.2.2 *T* Is Degree-Safe

In this section, we discuss the scenario where T is degree-safe. Since T is set-safe and color-safe, T is all-safe.

4.2.2.1 High-level Description of the Proposed Algorithm. Here, we assume that T is all-safe. Our goal is to add edges between the leaves of T such that no added edge connects leaves of the same color or in the same set. To add an edge between the leaves, we define a pair of leaves (a_i, b_i) as *valid* if it satisfies the condition that a_i and b_i are different colors and in different sets. We remove a valid pair of leaves from T each time. After connecting and removing each pair of leaves, the resulting tree, denoted as T', is still all-safe.

We call a chain leaf of T a *singular set*, and a v_c-branch that contains at least two leaves a *non-singular set.* The leaf in a singular set is called a *singular leaf.*

Our algorithm is a greedy recursive one. We process one set at a time. To present our algorithm, we create a ready queue, denoted as RQ. Let T be a tree with $|s|$ sets; and let $\mathcal{A}' = \mathcal{A}$, initially. First, we partition all sets into two pools, denoted as X and Y, respectively. All non-singular sets are placed in pool X and all singular sets are placed in pool Y. Then, we find the smallest set, denoted as S_x, in T, and add all the leaves in S_x to the ready queue. Next, we find a valid pair to connect one leaf in the ready queue with one leaf in Pool X. If any set in X becomes a singular set, we moved it to Pool Y. Therefore, after the ready queue is empty, let $\mathcal{A}' = \mathcal{A}' - S_x$ to remove S_x from the pool, so the number of sets in $\mathcal{A}'$ is reduced by one. We apply the above process to $\mathcal{A}'$ repeatedly until the remaining leaves have been processed. The steps of algorithm are detailed in Algorithm 1. Note that if the number of leaves is odd, we remove one leaf before applying the Algorithm 1.

Algorithm 1. Pairing

Input: A tree T rooted at a center v_c with ℓ leaves, where ℓ is even, the colors among vertices of degree < 2, C_1, C_2 and C_3 respectively, and let $\mathcal{A} = \{S_1, S_2, \ldots, S_s\}$, where $s \geq 2$, and let S_i denote the set of the leaves in a subtree τ_i rooted at a child of the root v_r in T, where $1 \leq i \leq s$, such that T is all-safe;
Output: A set of added edges $M = \{\langle a_i, b_i \rangle \mid 1 \leq i \leq \ell/2,\ a_i \in S_p \text{ and } b_i \in S_q, p \neq q;\}$;
Create a ready queue RQ;
Partition $\mathcal{A}$ into non-singular sets in Pool X and singular sets in Pool Y;
Construct one linked-list data structure for X;
$M = \emptyset$; $RQ = \emptyset$;
while $\mathcal{A} \neq \emptyset$ **do**
  Pick the smallest set S_x form the pools, and put all leaves of S_x into the ready queue;
  while $RQ \neq \emptyset$ **do**
    Pick a leaf a_i from the ready queue;
    Pick a leaf b_i from Pool X such that a_i and b_i are different colors and no color is saturated after removing (a_i, b_i);
    $M = M \cup \{\langle a_i, b_i \rangle\}$;
    Remove (a_i, b_i) from T and $\mathcal{A}$ respectively;
    if Some non-singular sets become singular sets **then**
      Move these sets from Pool X to Pool Y;
    end if
    if Only two singular sets exist in T **then** {* Assume that only a_j and b_j are left. *}
      $M = M \cup \{\langle a_j, b_j \rangle\}$;
    end if
  end while
end while
return M;

4.2.2.2 Finding a Valid Pair. Next, we explain how to find a valid pair of leaves. If singular leaves exist, there are two cases. First, we pick a leaf a_i from the ready queue. Without loss of generality, we assume that $a_i \in C_1$. We also choose a leaf $b_j \in C_2$ and a leaf $b_k \in C_3$ from X by Lemma 8. After removing (a_i, b_j) or (a_i, b_k), we check whether T' is color-saturated in Lemma 9 and select the pair that satisfies the above constraint. After this procedure is completed, we remove this pair from T. Whenever any set in X becomes singular, we move it from X to Y, and update all related counters accordingly.

Lemma 8. *Let T be an all-safe tree. We can always find a valid pair (a, b) in T, where a is removed from the ready queue and b is removed from Pool X.*

Proof. According to Algorithm 1, we always select a pair's components, a and b, from different sets, where $a \in S_i$ and $b \notin S_i$. To prove the lemma by contradiction, we assume a valid pair does not exist. Then, a and b must be the same color and all leaves whose color is different to a must be in S_i. Without loss of generality, let $a, b \in C_1$, all leaves with colors C_2 and C_3 be in S_i. Then, $|S_i| >$ the number of leave with C_2 and $C_3 > \lfloor \ell/2 \rfloor$. However, since T is all-safe, this is a contradiction. Therefore, the lemma is proven. □

Lemma 9. *If T is all-safe, after we pick and remove a valid pair in Algorithm 1, the resulting tree T' remains to be all-safe.*

We implement a dynamic data structure to realize Algorithm 1 so that finding a valid pairing from an all-safe T can be done in constant time. Due to space limitation, we omit the details of the dynamic data structure in this extended abstract.

Theorem 2. *Let T be an input graph. T is an all-safe tree with three colors among vertices of degree < 2. Algorithm 1 finds a smallest 2-vertex connectivity augmentation of T in linear time.*

5 Adding Edges to Make a Forest into a Tree

In this section, we present an algorithm for finding added edges to transform an input forest F into a tree T in linear time. Let E_x be the set of added edges. We prove that $\mathrm{LOW}_{\mathrm{f2v}}(F) = \mathrm{LOW}_{\mathrm{f2v}}(T) + |E_x|$.

Recall that $F = \mathrm{BT}(G)$ is an input graph with three colors among vertices of degree < 2, denoted as C_1, C_2, and C_3 respectively, and F is color-safe. We assume that $\mathrm{BT}(G)$ is a forest F. F has $m + n$ connected components, where m is the number of trees $\{T_1, T_2, \ldots, T_m\}$ that are not isolated vertices in F, and n is the number of isolated vertices $\{I_1^*, I_2^*, \ldots, I_n^*\}$ in F. Here, an isolated vertex is viewed as a tree with two leaves. Let $\hat{\mathcal{A}} = \{\hat{S}_1, \hat{S}_2, \ldots, \hat{S}_s\}$, where $s = (m + n) \geq 3$, be the set of the leaves of connected components in F, where $\hat{S}_i$ (respectively, $\hat{S}_{m+j}$) denotes the set of the leaves of a component T_i (respectively, I_j^*) in F, $1 \leq i \leq m$ and $1 \leq j \leq n$. If $\hat{S}_i$ is an isolated vertex, $|\hat{S}_i| = 2$, $1 \leq i \leq s$.

Our goal is to add edges between all connected components in F such that all the components will be connected to form one connected component. Any edge connects the leaves with different colors or in different sets. To add an edge between these leaves, we define a pair of leaves (a_i, b_i) as *valid* if it satisfies the condition that a_i and b_i are with different colors and in different sets $\hat{S}_j$, $1 \leq j \leq s$. We pick a valid pair of leaves in F each time. After connecting the leaves, the two trees associated with the leaves will be combined to form one new tree and two leaves will be removed from the forest. The new tree will be added to the forest. We denote the remaining forest as F'. After this process, we connect two connected components in F to obtain F' so that the total number of connected components in F' is one less that of F.

Algorithm 2. 2VC(G) Finding a smallest 2-vertex connectivity augmentation with a partition constraint of a graph (G)

Input: A graph G with a partition of k vertex subtrees $P_1, P_2, \ldots, P_k$ with $k \geq 2$;
Output: E is a set of added edges, and an edge is added between two different colors;
Let $F = \mathrm{BT}(G)$;
$E = \emptyset$;
repeat
  switch (F);
  Case 1: F is color-saturated.
    Use the method in Section 3.1 to find E';
  Case 2: F is a tree. {* F is color-safe. *}
    Use the method in Section 3.2 to recolor;
      if F is set-saturated **then**
        Use the method in Section 4.1 to reroot;
      end if
        Case 2.1: F is degree-saturated.
          Use the method in Section 4.2.1 to find E';
        Case 2.2: F is degree-safe. {* Algorithm 1 *}
          Use the method in Section 4.2.2 to find E';
  Case 3: F is a forest. {* F is color-safe. *}
    Transforming F into a tree by finding E' using the method in Section 5;
  Let $E = E \cup E'$;
  Let $F = \mathrm{BT}(F \cup E')$;
until Case 1 or 2 is executed;
return E

Next, we explain how to find a valid pair of leaves. First, we pick a leaf a_i from the smallest set $\hat{S}_x$, and without loss of generality, assume that $a_i \in C_1$. We also choose a leaf $b_j \in C_2$ and a leaf $b_k \in C_3$ from $\hat{\mathcal{A}}$, $b_j \notin \hat{S}_x, b_k \notin \hat{S}_x$. Then, we check whether F' is color-safe after removing (a_i, b_j) or (a_i, b_k), and pick the pair that satisfies the above constraint. In the final step, we remove the selected pair from F, and update all related counters.

Our algorithm is a greedy recursive one by finding a valid pair of leaves. By using appropriate data structures, each such finding can be done in constant time. Due to space limitation, the detailed algorithmic description is omitted.

Theorem 3. *Let F be a forest, and we can find a set of edges E_x such that* $\mathrm{LOW}_{\mathrm{f2v}}(F) = \mathrm{LOW}_{\mathrm{f2v}}(F \cup E_x) + |E_x|$ *in linear time.*

6 Finding an Optimal 2-Vertex Connectivity Augmentation for a Graph

Let F be a forest as an input. We use F and $\mathrm{BT}(G)$ interchangeably to denote the block forest of G. The steps of Algorithm 2 for finding an optimal augmentation for 2-vertex connectivity of a graph $\mathrm{BT}(G)$ with a partition constraint are as follows.

Theorem 4. *Let* BT(G) *be an input graph. The number of added edges returned by Algorithm 2, which finds a smallest 2-vertex connectivity augmentation with a partition constraint of a graph* G*, is equal to* $\mathrm{LOW}_{\mathrm{f2v}}(\mathrm{BT}(G))$*, and it can be computed in linear time.*

References

1. Eswaran, K.P., Tarjan, R.E.: Augmentation problems. SIAM Journal on Computing 5, 653–665 (1976)
2. Gibbons., A.M.: Algorithmic Graph Theory. Cambridge University Press, Cambridge (1985)
3. Harary., F.: Graph Theory. Addison-Wesley, Reading (1969)
4. Hsu, T.-s.: Undirected vertex-connectivity structure and smallest four-vertex-connectivity augmentation (extended abstract). In: Staples, J., Katoh, N., Eades, P., Moffat, A. (eds.) ISAAC 1995. LNCS, vol. 1004, pp. 274–283. Springer, Heidelberg (1995)
5. Hsu, T.-s.: On four-connecting a triconnected graph. Journal of Algorithms 35, 202–234 (2000)
6. Hsu, T.-s.: Simpler and faster vertex-connectivity augmentation algorithms (extended abstract). In: Paterson, M. (ed.) ESA 2000. LNCS, vol. 1879, pp. 278–289. Springer, Heidelberg (2000)
7. Hsu, T.-s.: Simpler and faster biconnectivity augmentation. Journal of Algorithms 45(1), 55–71 (2002)
8. Hsu, T.-s., Kao, M.Y.: Optimal augmentation for bipartite componentwise biconnectivity in linear time. SIAM Journal on Discrete Mathematics 19(2), 345–362 (2005)
9. Hsu, T.-s., Ramachandran, V.: On finding a smallest augmentation to biconnect a graph. SIAM Journal on Computing 22, 889–912 (1993)
10. Huang, P.C., Wei, H.W., Lu, W.C., Shih, W.K., Hsu, T.-s.: Smallest bipartite bridge-connectivity augmentation. Algorithmica 54(3), 353–378 (2009)
11. Rosenthal, A., Goldner, A.: Smallest augmentations to biconnect a graph. SIAM Journal on Computing 6, 55–66 (1977)
12. Tarjan, R.E.: Depth-first search and linear graph algorithms. SIAM Journal on Computing 1, 146–160 (1972)
13. Watanabe, T., Nakamura, A.: A minimum 3-connectivity augmentation of a graph. Journal of Computer and System Sciences 46, 91–128 (1993)

Computing a Smallest Multi-labeled Phylogenetic Tree from Rooted Triplets

Sylvain Guillemot[1], Jesper Jansson[2,⋆,⋆⋆], and Wing-Kin Sung[3,4]

[1] Institut Gaspard Monge - Université Paris-Est, 5 boulevard Descartes, Champs-sur-Marne, 77454 Marne-la-Vallée, France
Sylvain.Guillemot@univ-mlv.fr

[2] Ochanomizu University, 2-1-1 Otsuka, Bunkyo-ku, Tokyo 112-8610, Japan
Jesper.Jansson@ocha.ac.jp

[3] School of Computing, National University of Singapore, 3 Science Drive 2, Singapore 117543
ksung@comp.nus.edu.sg

[4] Genome Institute of Singapore, 60 Biopolis Street, Genome, Singapore 138672

Abstract. We investigate the computational complexity of a new combinatorial problem of inferring a smallest possible multi-labeled phylogenetic tree (MUL tree) which is consistent with each of the rooted triplets in a given set. We prove that even the restricted case of determining if there exists a MUL tree consistent with the input and having just one leaf duplication is NP-hard. Furthermore, we show that the general minimization problem is NP-hard to approximate within a ratio of $n^{1-\epsilon}$ for any constant $0 < \epsilon \leq 1$, where n denotes the number of distinct leaf labels in the input set, although a simple polynomial-time approximation algorithm achieves the approximation ratio n. We also provide an exact algorithm for the problem running in $O^*(7^n)$ time and $O^*(3^n)$ space.

1 Introduction

1.1 Problem Definitions

A *phylogenetic tree* is a rooted, unordered tree in which every internal node has at least two children and where each leaf is labeled by an element from a set of leaf labels. A phylogenetic tree where each leaf label occurs at most once is called a *single-labeled phylogenetic tree*; a phylogenetic tree where each leaf label may occur more than once is called a *multi-labeled phylogenetic tree*, or *MUL tree* for short [6,8,11].[1] For any MUL tree M, denote the set of all leaf labels that occur in M by $\mathcal{L}(M)$. For any leaf label $x \in \mathcal{L}(M)$, the number of *duplications of* x is equal to the number of occurrences of x in M minus 1. The number of *leaf duplications in* M, denoted by $d(M)$, is the total number of duplications

⋆ Funded by the Special Coordination Funds for Promoting Science and Technology.

⋆⋆ Corresponding author.

[1] MUL trees are called *rl-trees* in [6].

Y. Dong, D.-Z. Du, and O. Ibarra (Eds.): ISAAC 2009, LNCS 5878, pp. 1205–1214, 2009.

of all leaf labels in $\mathcal{L}(M)$. Define $m(M)$ as the number of leaves in M. Then, $d(M) = m(M) - |\mathcal{L}(M)|$.

For any two nodes u, v in a rooted tree, the notation $u \prec v$ means that u is a proper descendant of v, and $\mathsf{lca}(u, v)$ denotes the lowest common ancestor (lca) of u and v. (For convenience, any node is considered to be an ancestor of itself.) A *rooted triplet* is a binary phylogenetic tree with exactly three distinctly labeled leaves. The unique rooted triplet on leaf label set $\{x, y, z\}$ where $\mathsf{lca}(x, y) \prec \mathsf{lca}(x, z) = \mathsf{lca}(y, z)$ is denoted by $xy|z$. If $xy|z$ is an embedded subtree of a MUL tree M, i.e., if there exist three leaves ℓ_x, ℓ_y, ℓ_z in M labeled by x, y, and z, respectively, such that $\mathsf{lca}(\ell_x, \ell_y) \prec \mathsf{lca}(\ell_x, \ell_z) = \mathsf{lca}(\ell_y, \ell_z)$ then $xy|z$ and M are said to be *consistent* with each other; otherwise, $xy|z$ and M are *inconsistent.* A set $\mathcal{R}$ of rooted triplets and a MUL tree M are *consistent* with each other if every $xy|z \in \mathcal{R}$ is consistent with M. See Fig. 1 for an example.

In this paper, we consider the following problem, named *the smallest MUL tree from rooted triplets problem* (SMRT): Given a set $\mathcal{R}$ of rooted triplets over a leaf label set L, output a MUL tree M with $\mathcal{L}(M) = L$ which is consistent with $\mathcal{R}$ and which minimizes $d(M)$. We also consider the following decision problem for any positive integer d, termed d-SMRT: Given a set $\mathcal{R}$ of rooted triplets over a leaf label set L, does there exist a MUL tree M with $\mathcal{L}(M) = L$ which is consistent with $\mathcal{R}$ and which satisfies $d(M) \leq d$? In the rest of this paper, we define $k = |\mathcal{R}|$ and $n = |L|$ for any given instance of SMRT or d-SMRT.

1.2 Motivation and Previous Work

The problem of determining whether there exists a *single-labeled* tree consistent with all of the rooted triplets in a given set, and if so, constructing such a tree, can be solved efficiently by a classical algorithm of Aho *et al.* [1]. When no such tree exists because of conflicts in the branching information, one may try to select a largest possible subset of the triplets which is consistent with some tree (*the maximum rooted triplets consistency problem* (MRTC)), find a largest possible subset of the leaves such that the restriction of the input triplets to those leaves is consistent with some tree (*the maximum agreement supertree problem* (MASP)), or build a *phylogenetic network* (a generalization of a phylogenetic tree in which internal nodes may have more than a single parent) which contains all of the rooted triplets. See [3] for a recent survey of related results and many references. In this paper, we consider a new approach: Allow leaf labels to be repeated, but try to minimize the number of such repetitions.

The main application of phylogenetic trees is to describe tree-like evolution for a set of objects; leaves represent the objects while internal nodes correspond to their common ancestors. In the study of evolutionary history, MUL trees arise from the modeling of biological processes where it is necessary to use certain leaf labels more than once. For example, a gene tree can contain several leaves labeled by the same species due to gene duplication events [6,8,11]. As another example, area cladograms, where the names of geographical areas are used to label the leaves, may apply the same label to more than a single leaf (see, e.g., [2,8]). MUL trees can also be useful for studying host-parasite cospeciation [8,10].

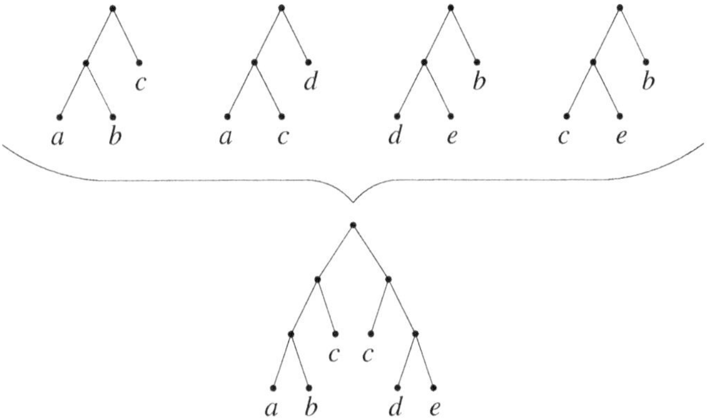

Fig. 1. The set of rooted triplets $\{ab|c,\ ac|d,\ de|b,\ ce|b\}$ is consistent with a MUL tree containing one leaf duplication

Although the problem of inferring a MUL tree from an input set of single-labeled phylogenetic trees that minimizes the number of leaf duplications has not been studied before, several algorithms for manipulating already known MUL trees have been published in the literature. Huber *et al.* [8] presented a method for constructing a phylogenetic network from an input MUL tree. The network output by their method is binary and has the fewest possible reticulation nodes among all binary networks which exhibit the structural information of the input MUL tree. Scornavacca *et al.* [11] considered some computational problems involving the extraction of the unambiguous parts of an input MUL tree. More precisely, [11] proposed linear-time algorithms to identify every so-called observed duplication node (odn) in a MUL tree, testing if two MUL trees are isomorphic, and computing a largest duplication-free rooted subtree of a MUL tree. They also showed that it is an NP-hard problem to prune each of the MUL trees in a given set to a single-labeled tree at odns in such a way that the obtained set of trees can be merged without conflicts into a single-labeled tree.

1.3 Our Results and Organization of the Paper

We present several negative and positive results regarding the computational complexity and polynomial-time approximability of SMRT. Below, we say that an algorithm $\mathcal{A}$ for SMRT is an *α-approximation algorithm* (and that the *approximation ratio of $\mathcal{A}$* is at most α) if, for every input $\mathcal{R}$, the MUL tree output by $\mathcal{A}$ is consistent with $\mathcal{R}$ and contains at most $\alpha \cdot d(M^*)$ leaf duplications, where M^* is an optimal MUL tree (i.e., having the fewest possible number of leaf duplications) consistent with $\mathcal{R}$.

The rest of the paper is organized as follows. Section 2 presents a simple polynomial-time n-approximation algorithm for SMRT. On the negative side, Section 3 proves that d-SMRT is NP-hard even if $d = 1$, and also that SMRT cannot be approximated within a ratio of $n^{1-\epsilon}$ for any constant $0 < \epsilon \leq 1$ in polynomial time, unless P = NP. Finally, Section 4 presents an exact algorithm for SMRT which runs in $O^*(7^n)$ time and $O^*(3^n)$ space.

2 Straightforward n-Approximation of SMRT

We start with the following simple observation.

Lemma 1. *For any set $\mathcal{R}$ of rooted triplets over a leaf label set L with $|L| = n$, there exists a MUL tree with $2n$ leaves which is consistent with $\mathcal{R}$.*

Proof. Let T be an arbitrary single-labeled phylogenetic tree with n leaves bijectively labeled by L. Let M be the MUL tree obtained by taking two copies T_1, T_2 of T and joining the roots of T_1 and T_2 to a new parent root node. Clearly, M has $2n$ leaves and any rooted triplet $xy|z$ over L is consistent with M since T_1 contains leaves labeled by x, y and T_2 contains a leaf labeled by z. □

Consequently, SMRT admits a trivial polynomial-time n-approximation algorithm: Using the algorithm of Aho *et al.* [1], determine if there exists a single-labeled tree consistent with $\mathcal{R}$. If the answer is yes then output this tree, otherwise output the MUL tree from Lemma 1 which has exactly n leaf duplications.

Theorem 1. SMRT *can be approximated within a ratio of n in polynomial time.*

3 Hardness Results for SMRT

This section demonstrates that SMRT is computationally intractable. It is shown that d-SMRT is NP-hard already for $d = 1$ and that SMRT is NP-hard to approximate within a ratio of $n^{1-\epsilon}$ for any constant $0 < \epsilon \leq 1$. (Recall that n denotes the number of distinct leaf labels in the input set $\mathcal{R}$.) To obtain our hardness results, we first prove strong inapproximability bounds for a problem on directed graphs named Acyclic Tree-Partition (defined below) and then give a measure-preserving reduction from Acyclic Tree-Partition to SMRT.

3.1 Hardness of Acyclic Partition and Acyclic Tree-Partition

Definition 1. *Let $D = (V, A)$ be a directed graph. An* acyclic partition *of D is a partition of V into subsets $V_1, \ldots, V_r$ called* classes *such that each class induces an acyclic subgraph of D.*

Definition 2. *Let $D = (V, A)$ be a directed graph. An* acyclic tree-partition *of D consists of a binary rooted tree T with a node set N along with a partition $\{V(x) : x \in N\}$ of V (i.e., a subset $V(x)$ of V is associated to each node x of the tree T) such that:*

1. *for every $x \in N$, $V(x)$ induces an acyclic subgraph of D,*
2. *for any $x, y \in N$ with $x \prec y$, D has no arc from $V(y)$ to $V(x)$.*

Definitions 1 and 2 lead to the following natural problems. The Acyclic Partition problem takes as input a directed graph D and seeks an acyclic partition of D with the smallest possible number of classes; this number is denoted by $ap(D)$.[2] Similarly, the Acyclic Tree-Partition problem seeks an acyclic tree-partition of an input directed graph D with the minimum number of internal nodes, denoted by $atp(D)$. For any positive integer r, the two decision problems r-Acyclic Partition and r-Acyclic Tree-Partition ask if an input directed graph D satisfies $ap(D) \leq r$ and $atp(D) \leq r$, respectively.

Acyclic partitions and acyclic tree-partitions have some useful properties:

Lemma 2. *Let D be a directed graph and let $(T, \{V(x) : x \in N\})$ be an acyclic tree-partition of D. For any set X of ancestors of a leaf in T, the union $\bigcup_{x \in X} V(x)$ induces an acyclic subgraph of D.*

Lemma 3. *For every directed graph D, $atp(D) = ap(D) - 1$.*

Theorem 2. *(i) r-Acyclic Partition is NP-hard for $r = 2$.*
(ii) Acyclic Partition cannot be approximated within $n^{1-\epsilon}$ for any constant $0 < \epsilon \leq 1$ in polynomial time unless P = NP, where n is the number of vertices in the input graph.

Proof. (i) Reduce from Not-all-equal 3SAT, which is known to be NP-hard [7]. Let I be a given instance of Not-all-equal 3SAT with m clauses and construct a directed graph D with $3m$ vertices as follows. For each clause C in I, D contains three vertices C_1, C_2, C_3 forming a directed cycle in D that represent the literals of C. In addition, for each pair of conflicting literals $C_i = x$ and $C'_j = \overline{x}$, D contains the two arcs (C_i, C'_j) and (C'_j, C_i). It is easy to see that there is a one-to-one correspondence between the valid truth assignments for I and the acyclic bipartitions of D: given a truth assignment ϕ, define a bipartition V_t, V_f of D by letting V_t (resp. V_f) contain all literals which are assigned the value *true* (resp. *false*) under ϕ.

(ii) Follows by giving a measure-preserving reduction from Chromatic Number and applying known inapproximability results for this problem [5,13]. The reduction maps a given undirected graph $G = (V, E)$ to a directed graph $D = (V, A)$ by replacing each edge $\{u, v\}$ of G by two arcs $(u, v), (v, u)$. Observe that for any $V' \subseteq V$, V' is an independent set of G if and only if V' induces an acyclic subgraph of D. Therefore, colorings of G correspond to acyclic partitions of D. □

Corollary 1. *(i) r-Acyclic Tree-Partition is NP-hard for $r = 1$.*
(ii) Acyclic Tree-Partition cannot be approximated within $n^{1-\epsilon}$ for any constant $0 < \epsilon \leq 1$ in polynomial time unless P = NP, where n is the number of vertices in the input graph.

3.2 Hardness of SMRT

We first reduce Acyclic Tree-Partition to a *constrained* variant of SMRT that forbids duplications of certain labels (Proposition 1). We then reduce the

[2] $ap(D)$ is also referred to in the literature as *the dichromatic number of D* [9].

constrained variant to the unconstrained SMRT problem (Proposition 2). When combined, these reductions yield the desired hardness results for SMRT, as summarized in Theorem 3. The constrained variant of SMRT is defined as follows.

Definition 3. *Let $\mathcal{R}$ be a set of rooted triplets over a leaf label set L and $U \subseteq L$ a set of* unique labels. *A MUL tree M is* consistent *with the pair $(\mathcal{R}, U)$ if: (i) M is consistent with $\mathcal{R}$; (ii) M has only one occurrence of each label in U.*

The CONSTRAINED-SMRT problem (C-SMRT) takes as input a pair $(\mathcal{R}, U)$ and seeks a MUL tree consistent with $(\mathcal{R}, U)$ containing the fewest duplications.

Proposition 1. *There exists a measure-preserving reduction from* ACYCLIC TREE-PARTITION *to* C-SMRT.

Proof. Given an instance $D = (V, A)$ of ACYCLIC TREE-PARTITION, construct a new instance $(\mathcal{R}, U)$ of C-SMRT with label set $L := V \cup \{z\}$, where z is a new label not belonging to V. The set $\mathcal{R}$ contains exactly the following triplets: for each arc $(u, v) \in A$, let $zu|v \in \mathcal{R}$. The set of unique labels is $U = V$, meaning that only z is allowed to be duplicated. To prove that the reduction is measure-preserving, we show that for every $r \leq |V|$, the following are equivalent:

1. D admits an acyclic tree-partition with r internal nodes;
2. $(\mathcal{R}, U)$ admits a consistent MUL tree with r duplications.

(1) $\Rightarrow$ (2): Suppose D has an acyclic tree-partition consisting of a tree $T = (N, E)$ with r internal nodes and a partition $\{V_x : x \in N\}$ of V. We construct a MUL tree M from T by labeling each leaf by z, and then, above each node x of T, attaching the elements of V_x in the order given by a topological ordering of $D[V_x]$ (where $D[V_x]$ denotes the subgraph of D induced by vertices of V_x).

We introduce the following additional notation: given a MUL tree M, and a sequence of labels $s = x_1 \ldots x_n$, let $R(M, s)$ be the tree obtained by starting with a caterpillar with $n + 1$ leaves $l_0, \ldots, l_n$ (with l_0, l_1 being farthest from the root), substituting l_0 with M, and labeling each leaf $l_i, i \geq 1$ by x_i. We inductively define two MUL trees M_x, M'_x for each node x of T: (i) if x is a leaf then M_x consists of a single leaf labeled by z; (ii) if x is an internal node with two children y, y' then $M_x := (M'_y, M'_{y'})$; (iii) for any node x of T, let s_x be a topological ordering of $D[V_x]$ (which is acyclic by Point 1 of Definition 2), and let $M'_x := R(M_x, s_x)$. Finally, define $M := M'_t$, where t is the root of T.

We now examine the constructed MUL tree M. Clearly, only z is duplicated in M; since $\{V_x : x \in N\}$ is a partition of V, a label $u \in V_x$ appears only once in M (in the subtree between the root of M_x and the root of M'_x). Moreover, since the leaves of M labeled by z correspond to the leaves of T, their number is $r + 1$, hence M has r duplications. Next, we show by a case analysis that M is consistent with $\mathcal{R}$. Consider $zu|v \in \mathcal{R}$, then $(u, v) \in A$ by the construction of $\mathcal{R}$. Let x, y be the nodes of T such that $u \in V_x, v \in V_y$. Four cases are possible:

- if $x = y$: Since $(u, v) \in A$, and since s_x is a topological ordering of $D[V_x]$, it follows that $u <_{s_x} v$. Consider $M'_x = R(M_x, s_x)$. M_x contains a leaf labeled by z, and u, v appear in s_x with $u <_{s_x} v$, so M'_x (and thus M) contains $zu|v$.

- if $x \prec y$ in T: Consider $M'_y = R(M_y, s_y)$. M_y contains leaves labeled by z, u and v appears in s_y, therefore M'_y (and thus M) contains $zu|v$.
- if $y \prec x$ in T: This is impossible according to Point 2 of Definition 2.
- if both $x \not\prec y$ and $y \not\prec x$ in T: Let $c = \mathsf{lca}(x, y)$ in T and let c_x, c_y be the two (distinct) children of c such that $x \preceq c_x, y \preceq c_y$. Consider $M_c = (M'_{c_x}, M'_{c_y})$, then M'_{c_x} contains leaves labeled by z, u and M'_{c_y} contains a leaf labeled by v, hence M_c (and M) contains $zu|v$.

To conclude, M is a MUL tree with r duplications that is consistent with $(\mathcal{R}, U)$.

(2) $\Rightarrow$ (1): Let M be a MUL tree with r duplications which is consistent with $(\mathcal{R}, U)$. We may assume w.l.o.g. that M is binary. By definition, only the label z is duplicated in M. Let $T = (N, E)$ be the subtree of M which connects the leaves labeled by z. For each node x of T, let P_x be the path in M joining x to its parent node in T (or to the root of M if x is the root of T). Then, define V_x as the set of labels appearing in a subtree along the path P_x. It is straightforward to verify that $(T, \{V_x : x \in N\})$ is an acyclic tree-partition of D with r internal nodes (see the full version of this paper for a complete proof). □

Proposition 2. *There exists a measure-preserving reduction from* C-SMRT *to* SMRT.

Proof. Let $(\mathcal{R}, U)$ be any given instance of C-SMRT, where $\mathcal{R}$ is a triplet set over a set L of n leaf labels and $U \subseteq L$ is a set of unique labels. We construct an instance $\mathcal{R}'$ of SMRT by replacing each element of U by $n+1$ copies. Formally, $\mathcal{R}'$ has a leaf label set L' consisting of: (i) for each $x \in U$, labels x_i $(1 \leq i \leq n{+}1)$; (ii) for each $x \in L \backslash U$, a single element x_1. The set $\mathcal{R}'$ consists of the following triplets: for each $xy|z \in \mathcal{R}$ and each i, j, k, let $x_i y_j | z_k \in \mathcal{R}'$. Assume w.l.o.g. that $r \leq n$. We show that $(\mathcal{R}, U)$ has a consistent MUL tree M with r duplications if and only if $\mathcal{R}'$ has a consistent MUL tree M' with r duplications.

($\Rightarrow$): Let M be a MUL tree with r duplications consistent with $(\mathcal{R}, U)$. Construct a MUL tree M' from M by substituting each leaf u having label x by an arbitrary single-labeled binary tree T_u over $\{x_1, \ldots, x_i\}$, where i equals either 1 or $n+1$. Observe that: (i) for each $x \in U$, each label x_i occurs exactly once in M'; (ii) for each $x \in L \backslash U$, the number of occurrences of x in M equals the number of occurrences of x_1 in M'. It follows that $d(M') = d(M) = r$. In addition, for any triplet $x_i y_j | z_k \in \mathcal{R}'$, the corresponding $xy|z \in \mathcal{R}$ is obtained from leaves u, v, w of M; hence $x_i y_j | z_k$ is obtained by selecting the corresponding leaves of T_u, T_v, T_w in M'. This proves that M' is consistent with every triplet in $\mathcal{R}'$.

($\Leftarrow$): Omitted due to space constraints. See the full version of this paper for a complete proof. □

Propositions 1 and 2 together with our hardness results for Acyclic Tree-Partition in Corollary 1 give us the next theorem.

Theorem 3. *(i)* d-SMRT *is* NP*-hard for* $d = 1$*;*
(ii) SMRT *cannot be approximated within* $n^{1-\epsilon}$ *for any constant* $0 < \epsilon \leq 1$ *in polynomial time, unless* $\mathsf{P} = \mathsf{NP}$.

We remark that the analogous Minimum Duplication Supersequence problem [6] for strings behaves quite differently: it is equivalent to the Directed Feedback Vertex Set problem, and as such it is FPT with respect to r (by a result of [4]) and approximable within $O(\log n \log\log n)$ in polynomial time [12].

4 An Exact Algorithm for SMRT

Here, we present an exact exponential-time algorithm for SMRT.

We use a dynamic programming approach, exploiting the recursive structure of the problem. More precisely, we consider pairs of subsets of L of the form (A, B) such that $B \subseteq A \subseteq L$. Subproblems in our dynamic programming approach will correspond to pairs (A, B). For a given pair, we will restrict our attention to specific MUL trees given by the following definition.

Definition 4. *Let (A, B) be a pair of subsets of L with $B \subseteq A \subseteq L$. A binary MUL tree M leaf-labeled by A* complies with (A, B) *if and only if for each $uv|w \in \mathcal{R}$ with $u, v, w \in A$ and $w \notin B$, it holds that $uv|w$ is consistent with M.*

For a given pair (A, B), let $n(A, B)$ denote the minimum value of $d(M)$ taken over every binary MUL tree M leaf-labeled by A which complies with (A, B). We compute the values $n(A, B)$ by dynamic programming, and obtain $n(L, \emptyset)$ as the desired value at the end of the computation. To compute a value $n(A, B)$, we break the computation into two subproblems of the form $(A_1, \_), (A_2, \_)$, where A_1, A_2 are the label sets of the two child subtrees. In order to explain this in detail, we will need the following definitions. A *split* of (A, B) is a pair (A_1, A_2) of subsets of A such that $A_1 \cup A_2 = A$ (observe here that A_1, A_2 are not necessarily disjoint, and that the definition does not depend on B).

Definition 5. *Let (A_1, A_2) be a split of (A, B). We say that (A_1, A_2) is a* nice split *of (A, B) if and only if the following holds: for each $u, v, w \in A$, if $u \in A_i \backslash A_j$, $v \in A_j \backslash A_i$, $w \notin B$ with $i \neq j$ then $\mathcal{R}$ does not contain the rooted triplet $uv|w$.*

From here on, we will denote by B_i the intersection of B with A_i. Also, we define $B' = A_1 \cap A_2$. The next property describes the recursive structure of the problem, characterizing the fact that M complies with (A, B) by conditions on its child subtrees.

Lemma 4. *Let (A, B) be a pair such that $B \subseteq A \subseteq L$ with $|A| \geq 2$ and let M be a binary MUL tree over A, consisting of two MUL trees M_1, M_2 joined by a parent root node. Write $A_1 = \mathcal{L}(M_1)$ and $A_2 = \mathcal{L}(M_2)$, $B' = A_1 \cap A_2$ and $B_i = B \cap A_i$. Then the following are equivalent:*

1. *M complies with (A, B);*
2. *(A_1, A_2) is a nice split of (A, B), and for $i \in \{1, 2\}$, M_i complies with $(A_i, B_i \cup B')$.*

Proof. (1) $\Rightarrow$ (2): We first show that M_i complies with $(A_i, B_i \cup B')$. Suppose that $uv|w \in \mathcal{R}$ with $u, v, w \in A_i$ and $w \notin B_i \cup B'$. Then we also have $u, v, w \in A$ and $w \notin B$, which implies that $uv|w$ is consistent with M (since M complies with (A, B)). Therefore, M has leaves ℓ_u, ℓ_v, ℓ_w labeled by u, v, w such that $\mathsf{lca}(\ell_u, \ell_v) \prec \mathsf{lca}(\ell_u, \ell_w) = \mathsf{lca}(\ell_v, \ell_w)$. What we need to show is that these three leaves all appear in M_i. If this was not the case, we would have ℓ_w appearing in M_j $(j \neq i)$, which would imply that $w \in B'$, contradicting the hypothesis. It follows that ℓ_u, ℓ_v, ℓ_w all appear in M_i, thus $uv|w$ is consistent with M_i.

Next, we show that (A_1, A_2) is a nice split of (A, B). Let $u, v, w \in A$. Suppose $u \in A_i \backslash A_j$, $v \in A_j \backslash A_i$, $w \notin B$ with $i \neq j$. If $\mathcal{R}$ contained the rooted triplet $uv|w$, then $uv|w$ would be consistent with M since M complies with (A, B) and $w \notin B$. But this is impossible since u only appears in M_i and v only appears in M_j.

(2) $\Rightarrow$ (1): To prove that M complies with (A, B), consider any $uv|w \in \mathcal{R}$ with $u, v, w \in A$ and $w \notin B$ and show that $uv|w$ is always consistent with M. There are four (partially overlapping) cases:

1. $u, v, w \in A_i$ and $w \notin B'$: Then $w \notin B_i \cup B'$. Since M_i complies with $(A_i, B_i \cup B')$, we conclude that $uv|w$ is consistent with M_i, and thus with M.
2. $u, v \in A_i, w \in A_j$ with $i \neq j$: Then u, v appear in M_i and w appears in M_j, hence $uv|w$ is consistent with M.
3. $u, w \in A_i, v \in A_j$ with $i \neq j$: We have three mutually exclusive subcases.
   - $u, v \notin B'$: Then $\mathcal{R}$ contains $uv|w$ with $u \in A_i \backslash A_j, v \in A_j \backslash A_i$ and $w \notin B$. This contradicts the assumption that (A_1, A_2) is a nice split of (A, B).
   - $v \in B'$: Then $u, v, w \in A_i$, and we are in Case 1.
   - $u \in B'$: Then $u, v \in A_j, w \in A_i$, and we are in Case 2.
4. $v, w \in A_i, u \in A_j$ with $i \neq j$: This case is symmetric to the previous case. □

Lemma 4 yields recurrence relations for $n(A, B)$ as stated in Lemma 5 below. Say that a split (A_1, A_2) of (A, B) is *proper* if and only if either: (i) A_1, A_2 are proper subsets of A; or (ii) $A_i = A$ and $B' \not\subseteq B$, where $B' = A_1 \cap A_2$.

Lemma 5. *The following recurrence relations for $n(A, B)$ hold:*

1. *Let (A, B) be a pair with $|A| \leq 2$. Then $n(A, B) = 0$.*
2. *Let (A, B) be a pair with $B = A$. Then $n(A, B) = 0$.*
3. *Let (A, B) be a pair with $|A| \geq 3$ and $B \subset A$. Given a split $S = (A_1, A_2)$ of (A, B), let $B' = A_1 \cap A_2, B_i = B \cap A_i$, and define $m(S) = |B'| + n(A_1, B_1 \cup B') + n(A_2, B_2 \cup B')$. Then $n(A, B)$ equals the minimum of the values $m(S)$ over all nice splits S of (A, B) which are proper.*

Lemma 5 allows us to compute $n(A, B)$ by dynamic programming on the pairs ordered by: $(A, B) \leq (A', B')$ if and only if $|A| < |A'|$ or ($|A| = |A'|$ and $|B| \geq |B'|$). This yields a dynamic programming algorithm for solving SMRT. At the end of the algorithm, $n(L, \emptyset)$ gives the value of an optimal solution, and a corresponding optimal MUL tree can be obtained by performing a traceback.

Theorem 4. SMRT *can be solved using $O^*(7^n)$ time and $O(3^n)$ space.*

Proof. To prove the correctness of the algorithm, we can verify that the definition of the relation $\leq$ on pairs is compatible with the above relations. Indeed, when computing $n(A, B)$ in point 3, we recursively call $n(A_i, B_i \cup B')$. Then: (i) either $A_i \subset A$, in which case we have $(A_i, B_i \cup B') < (A, B)$; (ii) or $A_i = A$, then $B' \nsubseteq B$ since the split is proper, therefore $B \subset B_i \cup B'$ and $(A_i, B_i \cup B') < (A, B)$.

We now analyze the complexity of the algorithm. Fix an integer $p \leq n$. For any $A \subseteq L$ of size p, there are 2^p pairs (A, B), so the number of pairs (A, B) with $|A| = p$ is $\binom{n}{p}2^p$. It follows that the total number of pairs considered is $\sum_{p=0}^{n} \binom{n}{p} 2^p = 3^n$, giving the claimed space complexity. Next, for any pair (A, B) with $|A| = p$, there are 3^p splits to consider, and each split is processed in $O(n^3)$ time (i.e., the time required to check that the split is nice and to perform the set operations). Hence, the time complexity is $O(\sum_{p=0}^{n} \binom{n}{p} 2^p 3^p n^3) = O(7^n n^3)$. □

References

1. Aho, A.V., Sagiv, Y., Szymanski, T.G., Ullman, J.D.: Inferring a tree from lowest common ancestors with an application to the optimization of relational expressions. SIAM Journal on Computing 10(3), 405–421 (1981)
2. Brown, G.K., Nelson, G., Ladiges, P.Y.: Historical biogeography of *Rhododendron* section *Vireya* and the Malesian Archipelago. Journal of Biogeography 33, 1929–1944 (2006)
3. Byrka, J., Guillemot, S., Jansson, J.: New results on optimizing rooted triplets consistency. In: Hong, S.-H., Nagamochi, H., Fukunaga, T. (eds.) ISAAC 2008. LNCS, vol. 5369, pp. 484–495. Springer, Heidelberg (2008)
4. Chen, J., Liu, Y., Lu, S., O'Sullivan, B., Razgon, I.: A fixed-parameter algorithm for the directed feedback vertex set problem (Article 21). Journal of the ACM 55(5) (2008)
5. Feige, U., Kilian, J.: Zero knowledge and the chromatic number. Journal of Computer and System Sciences 57(2), 187–199 (1998)
6. Fellows, M., Hallett, M., Stege, U.: Analogs & duals of the MAST problem for sequences & trees. Journal of Algorithms 49(1), 192–216 (2003)
7. Garey, M., Johnson, D.: Computers and Intractability – A Guide to the Theory of NP-Completeness. W. H. Freeman and Company, New York (1979)
8. Huber, K.T., Oxelman, B., Lott, M., Moulton, V.: Reconstructing the evolutionary history of polyploids from multilabeled trees. Molecular Biology and Evolution 23(9), 1784–1791 (2006)
9. Neumann-Lara, V.: The dichromatic number of a digraph. Journal of Combinatorial Theory, Series B 33(3), 265–270 (1982)
10. Page, R.D.M., Charleston, M.A.: Trees within trees: phylogeny and historical associations. Trends in Ecology & Evolution 13(9), 356–359 (1998)
11. Scornavacca, C., Berry, V., Ranwez, V.: From gene trees to species trees through a supertree approach. In: Proc. of the 3rd Int. Conference on Language and Automata Theory and Applications (LATA 2009). LNCS, vol. 5457, pp. 702–714. Springer, Heidelberg (2009)
12. Seymour, P.D.: Packing directed circuits fractionally. Combinatorica 15(2), 281–288 (1995)
13. Zuckerman, D.: Linear degree extractors and the inapproximability of Max Clique and Chromatic Number. Theory of Computing 3(1), 103–128 (2007)

On Partitioning a Graph into Two Connected Subgraphs*

Daniël Paulusma[1] and Johan M.M. van Rooij[2]

[1] Department of Computer Science, University of Durham, Science Laboratories, South Road, Durham DH1 3LE, England
daniel.paulusma@durham.ac.uk

[2] Department of Information and Computing Sciences, Universiteit Utrecht, PO Box 80.089, 3508TB Utrecht, The Netherlands
jmmrooij@cs.uu.nl

Abstract. Suppose a graph G is given with two vertex-disjoint sets of vertices Z_1 and Z_2. Can we partition the remaining vertices of G such that we obtain two connected vertex-disjoint subgraphs of G that contain Z_1 and Z_2, respectively? This problem is known as the 2-Disjoint Connected Subgraphs problem. It is already NP-complete for the class of n-vertex graphs $G = (V, E)$ in which Z_1 and Z_2 each contain a connected set that dominates all vertices in $V \backslash (Z_1 \cup Z_2)$. We present an $\mathcal{O}^*(1.2051^n)$ time algorithm that solves it for this graph class. As a consequence, we can also solve this problem in $\mathcal{O}^*(1.2051^n)$ time for the classes of n-vertex P_6-free graphs and split graphs. This is an improvement upon a recent $\mathcal{O}^*(1.5790^n)$ time algorithm for these two classes. Our approach translates the problem to a generalized version of hypergraph 2-coloring and combines inclusion/exclusion with measure and conquer.

1 Introduction

There are several natural and elementary algorithmic problems that check if the structure of some fixed graph H shows up as a pattern within the structure of some input graph G. One of the most well-known problems is the H-Minor Containment problem that asks whether a given graph G contains H as a minor. A celebrated result by Robertson and Seymour [11] states that the H-Minor Containment problem can be solved in polynomial time for every fixed pattern graph H. They obtain this result by designing an algorithm that solves the following problem in polynomial time for any fixed input parameter k.

Disjoint Connected Subgraphs
Instance: A graph $G = (V, E)$ and mutually disjoint nonempty sets $Z_1, \ldots, Z_t \subseteq V$ such that $\sum_{i=1}^{t} |Z_i| \leq k$.
Question: Do there exist mutually vertex-disjoint connected subgraphs $G_1, \ldots, G_t$ of G such that $Z_i \subseteq V_{G_i}$ for $1 \leq i \leq t$?

* This work has been supported by EPSRC (EP/D053633/1).

Y. Dong, D.-Z. Du, and O. Ibarra (Eds.): ISAAC 2009, LNCS 5878, pp. 1215–1224, 2009.

Our Focus. We are interested in finding a fast *exact* algorithm that solves the 2-Disjoint Connected Subgraphs problem, which is a restriction of the above problem to $t = 2$, and in which Z_1 and Z_2 may have arbitrary size. This problem is already NP-complete however, if one of the sets Z_1 or Z_2 has cardinality 2, as shown in a very recent paper [8]. A trivial algorithm solves the 2-Disjoint Connected Subgraphs problem in $\mathcal{O}^*(2^n)$ time. (The $\mathcal{O}^*$-notation, used throughout the paper, suppresses factors of polynomial order.) However, since connectivity is a "global" property, this problem is an example of a "non-local" problem. Such a problem is typically hard to solve exactly (see e.g. [6]). Arguably the most well-known non-local problem is the Travelling Salesman problem, for which no exact algorithm with better time complexity than $\mathcal{O}^*(2^n)$ is known. Another example of a non-local problem is the Connected Dominating Set problem. The fastest known exact algorithm for the Connected Dominating Set problem runs in $\mathcal{O}^*(1.9407^n)$ time [6], whereas for the general (unconnected) version of the Dominating Set problem an $\mathcal{O}^*(1.5048^n)$ exact algorithm is known [12].

Existing Results. In an attempt to design fast exact algorithms for non-local problems, one can focus on restrictions of the problem to certain graph classes. In [8] it has been shown that 2-Disjoint Connected Subgraphs is already NP-complete for the class of P_5-free graphs, whereas it is polynomially solvable for P_4-free graphs. There, it is also shown that this problem is NP-complete for the class of split graphs. Let $\mathcal{G}^{k,r}$ denote the class of graphs all connected induced subgraphs of which have a connected r-dominating set of size at most k. Somewhat surprisingly, for any fixed k, the 2-Disjoint Connected Subgraphs problem for $\mathcal{G}^{k,r}$ can be solved in polynomial time if $r = 1$, or if one of the given sets Z_1 or Z_2 of vertices has fixed size [8]. However, for any fixed k and $r \geq 2$, the 2-Disjoint Connected Subgraphs problem is NP-complete and the authors of [8] present an algorithm that solves it for n-vertex graphs in the class $\mathcal{G}^{k,r}$ in $\mathcal{O}^*((f(r))^n)$ time, where

$$f(r) = \min_{0<c\leq 0.5}\Big\{\max\Big\{\frac{1}{c^c(1-c)^{1-c}}, 2^{1-\frac{2c}{r-1}}\Big\}\Big\}.$$

In particular, their algorithm solves the 2-Disjoint Connected Subgraphs problem faster than $\mathcal{O}^*(2^n)$ for any n-vertex P_ℓ-free graph. For example, for a P_6-free graph on n vertices it uses $\mathcal{O}^*(1.5790^n)$ time and for a P_7-free graph it uses $\mathcal{O}^*(1.7737^n)$ time. Also, for an n-vertex split graph, this algorithm runs in $\mathcal{O}^*(1.5790^n)$ time.

Our Results and Paper Organization. After explaining our notations and terminology in Section 2, we propose to study the class of graphs G in which Z_1 and Z_2 both have a connected set that dominates $V_G \backslash (Z_1 \cup Z_2)$. In Section 3 we show that the 2-Disjoint Connected Subgraphs stays NP-complete under this restriction and we present an algorithm that solves it in $\mathcal{O}^*(1.2051^n)$ time for this class of graphs. We also show how to use this algorithm to solve the problem in $\mathcal{O}^*(1.2051^n)$ time for the classes of P_6-free graphs and split graphs. Hence, we improved the $\mathcal{O}^*(1.5790^n)$ time algorithm of [8] for these graph classes. Our approach translates the problem to a generalized hypergraph 2-coloring problem, for which we design an exact algorithm with the above running time in Section 4. It uses the recently introduced combined approach of [12] of inclusion/exclusion [1,2,10] with fast measure and conquer based running times [5] for solving the DOMINATING SET problem. Hence, our algorithm shows that this approach is not restricted to DOMINATING SET only but has a larger applicability within the field of covering and partitioning algorithms. Section 5 contains the conclusions and mentions relevant open problems.

2 Preliminaries

All graphs in this paper are undirected, finite, and without multiple edges. Unless explicitly stated otherwise, they do not contain loops. We write P_k to denote the path on k vertices. Let $G = (V, E)$ be a graph. For a subset $S \subseteq V$ we write $G[S]$ to denote the the subgraph of G induced by S. Two subsets $S, T \subseteq V$ are *adjacent* if there is an edge between a vertex in S and a vertex in T. The *distance* $d_G(u, v)$ between two vertices u and v in a graph G is the *length* $|V_P| - 1$ of a shortest path P between them. For any vertex $v \in V$ and set $S \subseteq V$, we write $d_G(v, S)$ to denote the length of a shortest path from v to S, i.e., $d_G(v, S) := \min_{w \in S} d_G(v, w)$. The *neighborhood* of a vertex $u \in V$ is the set $N_G(u) := \{v \in V \mid uw \in E\}$. The set $N_G^r(S) := \{u \in V \mid d_G(u, S) \leq r\}$ is called the *r-neighborhood* of a set S. A set S *r-dominates* a set S' if $S' \backslash S \subseteq N_G^r(S)$. We also say that S r-dominates $G[S']$. A subgraph H of G is an *r-dominating subgraph* of G if V_H r-dominates G. In case $r = 1$, we use "dominating" instead of "1-dominating". A set $S \subseteq V$ is called a *(k, r)-center* of G if $|S| \leq k$ and $N_G^r(S) = V$. A set S is called *connected* if $G[S]$ is connected. The class of graphs all connected induced subgraphs of which have a connected (k, r)-center is denoted by $\mathcal{G}^{k,r}$. The graph G is called a *split graph* if V can be partitioned into a clique and an independent set. A graph G is called *H-free* for some graph H if G does not contain an induced subgraph isomorphic to H.

Observation 1 ([8]). *The class of split graphs and the class of P_ℓ-free graphs for $\ell \in \{5, 6\}$ belong to $\mathcal{G}^{4,2}$, whereas the class of P_ℓ-free graphs for $\ell \geq 7$ belongs to $\mathcal{G}^{1,\ell-3}$.*

Let $G = (V, E)$ be a graph. Let $V' \subset V$ and $p, q \in V \backslash V'$. The *edge contraction* of edge $e = uv$ in G removes the two end-vertices u and v from G, and replaces them by a new vertex that is adjacent to precisely those vertices to which u or v were adjacent. Let $d(v)$ denote the degree of $v \in V$.

A *hypergraph* $H = (Q, \mathcal{S})$ is a set $Q = \{q_1, \ldots, q_m\}$ of *elements* together with a set $\mathcal{S} = \{S_1, \ldots, S_n\}$ of subsets of Q called *hyperedges*. A *2-coloring* of H is a partition of Q into Q_1, Q_2 such that $Q_1 \cap S \neq \emptyset$ and $Q_2 \cap S \neq \emptyset$ for each $S \in \mathcal{S}$. These notions can be generalized as follows. A *2-hypergraph* $H = (Q, \mathcal{L}, \mathcal{R})$ is a set $Q = \{q_1, \ldots, q_m\}$ together with two (not necessarily disjoint) sets $\mathcal{L} = \{L_1, \ldots, L_s\}$ and $\mathcal{R} = \{R_1, \ldots, R_t\}$ of subsets of Q. We call $\mathcal{L}$ and $\mathcal{R}$ the *partition classes* of H. With a 2-hypergraph $H = (Q, \mathcal{L}, \mathcal{R})$ we associate its *incidence graph* I, which is a bipartite graph on $Q \cup \mathcal{L} \cup \mathcal{R}$, where $qS \in E_I$ if and only if $q \in S$ for $q \in Q$ and $S \in \mathcal{L} \cup \mathcal{R}$. Let the *dimension* of a 2-hypergraph $H = (Q, \mathcal{L}, \mathcal{R})$ be $d = |Q| + |\mathcal{L}| + |\mathcal{R}|$. A *2-coloring* of H is a partition of Q into Q_1, Q_2 such that $Q_1 \cap L \neq \emptyset$ for each $L \in \mathcal{L}$ and $Q_2 \cap R \neq \emptyset$ for each $R \in \mathcal{R}$. This leads to the following decision problem.

2-HYPERGRAPH 2-COLORING
Input: a 2-hypergraph H.
Question: does H have a 2-coloring?

Note that a 2-hypergraph $H = (Q, \mathcal{S}, \mathcal{S})$ is 2-colorable if and only if hypergraph $H' = (Q, \mathcal{S})$ is 2-colorable. The HYPERGRAPH 2-COLORABILITY problem asks if a hypergraph is 2-colorable and is NP-complete (cf. [9]).

Observation 2. *The 2-HYPERGRAPH 2-COLORING problem is NP-complete.*

A *path decomposition* of a graph G is a sequence of bags (sets of vertices) $\mathcal{X} = \langle X_1, \ldots, X_r \rangle$ with the following three properties. First, $\bigcup_{i=1}^{r} X_i = V$. Second, for each $uv \in E$, there exists a bag X_i such that $\{u, v\} \subseteq X_i$. Third, if $v \in X_i$ and $v \in X_k$ then $v \in X_j$ for all $i \leq j \leq k$. The *width* of $\mathcal{X}$ is $\max_{1 \leq i \leq r} |X_i| - 1$ and the *pathwidth* $\text{pw}(G)$ of G is the minimum over all widths.

3 The 2-DISJOINT CONNECTED SUBGRAPHS Problem

Let (G, Z_1, Z_2) be an instance of the 2-DISJOINT CONNECTED SUBGRAPHS problem. Let $U = V_G \backslash (Z_1 \cup Z_2)$. We say that G is *semi-connected* with respect to Z_1 and Z_2 if Z_1 and Z_2 each contain a connected set that dominates U. We note that the 2-HYPERGRAPH 2-COLORING problem stays NP-complete for the class of 2-hypergraphs $H = (Q, \mathcal{L}, \mathcal{R})$ with $L_s = R_t = Q$. These 2-hypergraphs have an incidence graph I that is semi-connected with respect to $\mathcal{L}$ and $\mathcal{R}$, because L_s and R_t both dominate $Q = V_I \backslash (\mathcal{L} \cup \mathcal{R})$. Because such a 2-hypergraph H has a 2-coloring if and only if $(I, \mathcal{L}, \mathcal{R})$ is a YES-instance of the 2-DISJOINT CONNECTED SUBGRAPHS problem, the following observation immediately follows from Observation 2.

Observation 3. *The 2-DISJOINT CONNECTED SUBGRAPHS problem is even NP-complete for semi-connected graphs.*

For our main theorem we need the following result. We prove it in Section 4.

Theorem 1. *The 2-HYPERGRAPH 2-COLORING problem can be solved in $\mathcal{O}^*(1.2051^d)$ time for 2-hypergraphs of dimension d.*

Here is our main theorem.

Theorem 2. *The 2-DISJOINT CONNECTED SUBGRAPHS problem can be solved in $\mathcal{O}^*(1.2051^n)$ time for the class of semi-connected graphs.*

Proof. Let (G, Z_1, Z_2) be an instance of the 2-DISJOINT CONNECTED SUBGRAPHS problem with G semi-connected and $U = V_G \backslash (Z_1 \cup Z_2)$. For $i = 1, 2$, let D_i be the connected set in Z_i that dominates U. Transform each D_i to a single vertex by performing edge contractions. Do the same for any other component of $G[Z_1]$ and $G[Z_2]$. This results in two new sets Z'_1 and Z'_2 that are both independent and contain a vertex that is adjacent to all vertices in U. This means we may remove all edges in $G[U]$. Then we obtain a 2-hypergraph (U, Z'_1, Z'_2) of dimension at most n, on which we apply Theorem 1. □

As a consequence of Theorem 2 we find the following.

Corollary 1. *For any fixed $k \geq 1$, the* 2-DISJOINT CONNECTED SUBGRAPHS *problem can be solved in $\mathcal{O}^*(1.2051^n)$ time for any n-vertex graph in $\mathcal{G}^{k,2}$. In particular, this is true for P_6-free graphs and split graphs.*

Proof. The latter statement immediately follows from Observation 1. Let $G = (V, E)$ be a graph in $\mathcal{G}^{k,2}$ for some fixed $k \geq 1$ with two vertex-disjoint sets Z_1 and Z_2. Let $U = V \backslash (Z_1 \cup Z_2)$. We guess a set D_1 of up to k vertices in $G[Z_1 \cup U]$ and a set D_2 of up to k vertices in $G[Z_2 \cup U]$ such that $D_1 \cap D_2 = \emptyset$. We check if $G[D_1]$ and $G[D_2]$ are both connected. If not, we guess other sets instead. Otherwise, we form a new instance (G', Z'_1, Z'_2), where

- G' is the subgraph of G obtained after removing all vertices from U that are neither adjacent to D_1 nor to D_2. The reason we may remove these vertices is that they are redundant in any possible solution (G_1, G_2) in which D_i is a $(k, 2)$-center of G_i for $i = 1, 2$.
- $Z'_1 = Z_1 \cup D_1 \cup \{u \in U \mid u \text{ is adjacent to } D_1 \text{ but not to } D_2.\}$. We may define Z'_1 like this for a similar reason as above.
- $Z'_2 = Z_2 \cup D_2 \cup \{u \in U \mid u \text{ is adjacent to } D_2 \text{ but not to } D_1.\}$.

Then G' is semi-connected with respect to Z'_1 and Z'_2 and we can use Theorem 2. Since the total number of guesses is bounded by a polynomial in n, the result follows. □

4 The Proof of Theorem 1

We first sketch our algorithm for the 2-HYPERGRAPH 2-COLORING problem that runs in time $\mathcal{O}^*(1.2051^d)$ for d-dimensional hypergraphs.

Phase 1. We exhaustively apply a series of reduction rules and afterwards branch on the elements $q \in Q$: either give q color 1 or color 2. In both cases we remove q and all hyperedges that are colored appropriately (color 1 for $\mathcal{L}$, and color 2 for $\mathcal{R}$). We go to Phase 2 if every remaining element appears in at most three hyperedges in $\mathcal{L} \cup \mathcal{R}$.

Phase 2. The algorithm switches to the counting variant of our problem. It now uses a different set of reduction rules. If no such rule is applicable it applies inclusion/exclusion based branching to the hyperedges in $\mathcal{L}$ and $\mathcal{R}$. We go to Phase 3 when we have sufficiently reduced the size of H.

Phase 3. The algorithm solves the remaining problem by dynamic programming over a path decomposition of the incidence graph of the remaining hypergraph.

4.1 The Algorithm in Detail

Throughout the description of the algorithm, we denote the 2-hypergraph under consideration by $H = (Q, \mathcal{L}, \mathcal{R})$ and its incidence graph by I. So, H represents the elements that have not yet a color and the hyperedges that still do not have an element with the *right* color (color 1 for $L \in \mathcal{L}$ and color 2 for $R \in \mathcal{R}$). All other elements and hyperedges have been removed. If we say that an element in Q or a hyperedge in $\mathcal{L} \cup \mathcal{R}$ has a certain *degree*, we refer to its degree in I.

Phase 1: reduce and branch
We first introduce two reduction rules and exhaustively apply them to H.

Rule 1. Deal with elements q appearing in hyperedges of at most one partition class
If q has degree zero, remove q. If q occurs in only one partition class, color it with 1 if it belongs to an hyperedge in $\mathcal{L}$ and with 2 otherwise. Remove q and all hyperedges containing q. If H becomes empty this way, return YES.

Rule 2. Deal with hyperedges S of degree at most one
Let $S = \{q\}$. If $S \in \mathcal{L}$ color q with 1 and with 2 otherwise. Remove S and q. If $\emptyset \in \mathcal{L} \cup \mathcal{R}$, then return NO.

When Rule 1 and 2 cannot be applied anymore, select an element q of maximum degree. If q has degree at most three, go to Phase 2. Otherwise, branch on q. In one branch, color q with 1 and remove q and all hyperedges in $\mathcal{L}$ containing q. In the other branch, color q with 2 and remove q and all hyperedges in $\mathcal{R}$ containing q. After each branching, apply Rule 1 and 2 exhaustively. If in the end the algorithm has returned output YES then we are done. Otherwise, we go to Phase 2 with each 2-hypergraph created after the branching has finished and for which the algorithm has not returned NO.

Phase 2: branch based on inclusion/exclusion
Note that all elements in Q now have degree two or three. Switch to the counting variant: how many 2-colorings does H have? Apply the two new reduction rules below exhaustively.

Rule 3. Deal with hyperedges S of degree at most one
If $S = \{q\}$, then q must get the right color for S in any 2-coloring for H. Suppose w.l.o.g. that $S \in \mathcal{L}$. Hence, the number of 2-colorings for H equals the number of 2-colorings for $H' = (Q \backslash q, \mathcal{L}', \mathcal{R}')$, where $\mathcal{L}'$, $\mathcal{R}'$ denote the sets $\mathcal{L}$, $\mathcal{R}$ minus all hyperedges containing q, respectively. If $S = \emptyset$, return 0; no solutions exist.

Rule 4. Deal with elements q of degree one
Let q belong to $S \in \mathcal{L} \cup \mathcal{R}$. Note that we can not just simply remove q and S. The reason is that S may contain more than one element and these elements can be colored in two ways. This might lead to different 2-colorings (recall that

we want to determine this number). We circumvent this as follows. In Phase 3, we compute the number of 2-colorings by dynamic programming over a path decomposition of I. Adding a set of trees, each connected to only one vertex of the graph, increases the pathwidth by at most a logarithmic factor [3]. This does not influence the exponential running time of Phase 3 as we shall see. Hence, we do remove q and store it in a special set $\mathcal{C}$. If S then gets degree one, we can not apply Rule 3, as S may get its right color from an element in $\mathcal{C}$. Instead, we put S in $\mathcal{C}$ as well, and so on. In Phase 3 we put all elements and hyperedges in $\mathcal{C}$ back into I. These will correspond to trees adjacent to a single vertex in I. Throughout the remainder of Phase 2, our algorithm updates $\mathcal{C}$ whenever it takes some decision on H. If necessary, it removes elements and hyperedges from $\mathcal{C}$ (e.g., after applying Rule 3).

When Rules 3 and 4 cannot be applied anymore, branch on hyperedges. Pick a hyperedge of maximum degree if the maximum degree is at least six. Otherwise, let $d_i(S)$ be the number of elements in hyperedge S of degree i and $o(S)$ be the total number of appearances that elements in S have in the partition class not containing S. Then let $\mathcal{S} \subseteq \mathcal{L} \cup \mathcal{R}$ be the set of hyperedges S with either $d(S) = 5$ and $d_3(S) \geq 3$ or $d(S) = d_3(S) = 4$. Pick an $S \in \mathcal{S}$ with $o(S)$ maximum over all $S \in \mathcal{S}$. If $\mathcal{S} = \emptyset$ go to Phase 3; otherwise branch on the selected S as below.

The *optional* branch computes the number of S*-indifferent 2-colorings*, i.e., in which S may not have received its right color. In this branch, we only remove S. The *forbidden* branch computes the number of S*-incorrect 2-colorings*, i.e., in which S did not receive its right color. Here, we remove S, all its elements, and all hyperedges that are in the partition class not containing S and that contain an element of S, as these hyperedges have received their right color. After each branching, apply Rule 3 and 4 exhaustively. We now compute the number of 2-colorings that correctly coloring S by taking the difference between the two numbers from the two branches. We check if the final difference α at the root of the branching tree is strictly positive. If so return YES, otherwise return NO. Note that the exact value of α has no meaning because Rule 1 does not preserve counting properties. To solve the subproblems obtained from Phase 2, the algorithm requires results from Phase 3. Hence, all generated subproblems are given to Phase 3 after which the described subtractions and checks are performed.

Phase 3: dynamic programming over path decompositions
Note that all elements have degree two or three and all hyperedges have degree at most five with some extra constraints on their vertices in case their degree is four or five. We restore $\mathcal{C}$ and compute a path decomposition of I (which will have small enough width as we shall see). Using this path decomposition we can count the number of 2-colorings and with these numbers we perform the computations described in Phase 2.

4.2 Running Time Analysis

We analyze our algorithm using measure and conquer [5]. We start with the following lemma, the proof of which we only sketch due to page restrictions.

Lemma 4. *Phase 3 starts with $O(1.20509^{d-h})$ subproblems of complexity $h \leq d$.*

Proof Sketch. We assign a weight $0 \leq w(i) \leq 1$ to an element q of degree i and a weight $0 \leq v(i) \leq 1$ to an hyperedge of degree i. This way we define the complexity measure: $k(Q, \mathcal{L}, \mathcal{R}) = \sum_{q \in Q} w(d(q)) + \sum_{S \in (\mathcal{L} \cup \mathcal{R})} v(d(S))$. Let $\Delta v(i) = v(i) - v(i-1)$ and $\Delta w(i) = w(i) - w(i-1)$. We use the following constraints on the weights:

1. $v(0) = v(1) = w(0) = w(1) = 0$
2. $\Delta v(i) \geq 0$ and $\Delta w(i) \geq 0$ for all $i \geq 1$
3. $\Delta v(i) \geq \Delta v(i+1)$ and $\Delta w(i) \geq \Delta w(i+1)$ for all $i \geq 2$
4. $v(2) \geq 2\Delta v(5)$.

Using these constraints we obtain a set of recurrence relations describing the complexity reductions. We need to choose weights that respect the constraints and that minimize the solution to the set of recurrence relations (cf. [12]). It turns out that the following weights give us the desired upper bound in the statement of the lemma:

i	1	2	3	4	5	≥ 6
$v(i)$	0	0.809607	0.963013	0.996566	1	1
$w(i)$	0	0.448902	0.767484	0.934782	0.992583	1

□

Besides Proposition 1 from [4], the proof of Lemma 5 uses Proposition 2 which is obtained by dynamic programming and whose proof we omit.

Proposition 1 ([4]). *For any $\epsilon > 0$, there exists an integer n^*_ϵ such that for every graph G with $n > n^*_\epsilon$ vertices, its pathwidth* $\mathrm{pw}(G)$ *satisfies:*

$$\mathrm{pw}(G) \leq \frac{1}{6}n_3 + \frac{1}{3}n_4 + \frac{13}{30}n_5 + \frac{23}{45}n_6 + n_{\geq 7} + \epsilon n$$

where n_i is the number of vertices of degree i in G for $i \in \{3, \ldots, 6\}$ and $n_{\geq 7}$ is the number of vertices of degree at least 7. Moreover, a path decomposition of the corresponding width can be constructed in polynomial time.

Proposition 2. *Let p be the width of a path decomposition of the incidence graph I of a 2-hypergraph $H = (Q, \mathcal{L}, \mathcal{R})$. Then the number of 2-colorings of H can be counted in $\mathcal{O}^*(2^p)$ time.*

Lemma 5. *The number of 2-colorings of each 2-hypergraph H of complexity h in Phase 3 can be computed in $\mathcal{O}^*(1.1904^h)$ time.*

Proof. By Proposition 2, we can count the number of 2-colorings of H in $\mathcal{O}^*(2^p)$ time. Here, p denotes the width of a path decomposition from Proposition 1. Hence, we need to prove an upper bound on p expressed in k. To this end, we formulate the linear program:

$$\max \quad z = \frac{1}{6}(x_3 + y_3) + \frac{1}{3}y_4 + \frac{13}{30}y_5 + \epsilon \quad \text{such that:}$$

$$1 = \sum_{i=2}^{3} w(i)x_i + \sum_{i=2}^{5} v(i)y_i \qquad \sum_{i=2}^{3} ix_i = \sum_{i=2}^{5} iy_i \qquad x_2 \geq \frac{1}{2}y_4 + \frac{3}{2}y_5$$

In this linear program, x_i and y_i represent the number of elements and hyperedges, respectively, that are of degree i per unit of complexity (unit of k as in Lemma 4) in a worst case instance. Recall that all elements are of degree 2 and 3 and all hyperedges are of degree 2,3,4, or 5 in Phase 3. The objective function comes directly from Proposition 1, now formulated in x_i and y_i. This function gives an upper bound of the pathwidth per unit of complexity and we need the worst case.

The first constraint guarantees that the variables use exactly one unit of complexity. The second constraint counts the number of edges in I in two different ways and demands that equality must hold. To get a good upper bound we add the third constraint. The reason why we may do this is as follows. In Phase 3, every hyperedge of degree four contains at least one element of degree two, and every hyperedge of degree five contains at least three elements of degree two. So, there must exist at least $x_2 \geq \frac{1}{2}y_4 + \frac{3}{2}y_5$ elements of degree two.

The solution to this linear program is $z = 0.251446$ with $x_2 = 0.251446$, $x_3 = 0.502892$, $y_4 = 0.502892$ and $y_2 = y_3 = y_5 = 0$. As a result, $p \leq 0.251446h + \epsilon h$. We choose ϵ sufficiently small such that $2^{\epsilon h}$ may be neglected due to the rounding. If $h \leq n_\epsilon^*$, the pathwidth of H is bounded by a constant and otherwise by $0.251447h$. Hence, the algorithm runs in the desired time. □

Combining Lemma 4 and Lemma 5 proves Theorem 1.

Theorem 1. *2-*Hypergraph 2-Coloring *can be solved in* $\mathcal{O}^*(1.2051^d)$ *time for 2-hypergraphs of dimension d.*

Proof. Let $T(d)$ denote the running time of our algorithm on a 2-hypergraph H of dimension d. Let J be the set of all possible complexities of the subproblems that exist at the start of Phase 3. After all these subproblems have been processed in Phase 3, the algorithm must compute the number of 2-colorings for each hypergraph created after the end of Phase 1. These computations follow the structure of the branching tree, and hence $T(d)$ is dominated by the time spent in Phase 3. This together with Lemma 4 and 5 implies that

$$T(d) \leq \sum_{h \in J} \mathcal{O}(1.2051^{d-h}) \cdot \mathcal{O}^*(1.1904^h) \leq \sum_{h \in J} \mathcal{O}^*(1.2051^d) = \mathcal{O}^*(1.2051^d),$$

since $|J|$ is polynomially bounded as we only use a finite number of weights. □

5 Conclusions and Open Problems

We presented an $\mathcal{O}^*(1.2051^n)$ time algorithm for the 2-Disjoint Connected Subgraphs problem restricted to semi-connected graphs. We also showed how to use this algorithm to solve this problem within the same time for graphs in the class $\mathcal{G}^{k,2}$ for any fixed $k \geq 1$ and, in particular, for split graphs and

P_6-free graphs. We leave it as an open question how to obtain a faster algorithm for graphs in a class $\mathcal{G}^{k,r}$ with $r \geq 3$. Another natural question is to study the class of instances (G, Z_1, Z_2) where only one of the subsets, say Z_1, contains a connected set that dominates $U = V \backslash (Z_1 \cup Z_2)$. For solving this, a similar approach as in [8] can be followed (where brute force techniques are applied depending on the size of $|Z_1|$ and $|Z_2|$). Another approach would be to apply an algorithm that lists all minimal set covers (similar to [7]). By using such an approach one can enumerate all sets $U' \subseteq U$ that are minimal with respect to dominating Z_1. For each choice of U' one can check in polynomial time if $G[Z_2 + (U' \backslash U)]$ is connected. We note that this approach also works for semi-connected instances. However, this seems to lead to much worse running times.

We did not explore the above two questions in detail, as the *main* open question is to find an exact algorithm for the 2-Disjoint Connected Subgraphs problem for general graphs that is faster than the trivial $\mathcal{O}^*(2^n)$ algorithm. For solving this, new techniques that deal with the connectivity issue are necessary. This will be future research.

References

1. Bax, E.T.: Inclusion and exclusion algorithm for the hamiltonian path problem. Inform. Process. Lett. 47, 203–207 (1993)
2. Björklund, A., Husfeldt, T., Koivisto, M.: Set partitioning via inclusion-exclusion. SIAM J. Comput. 39, 546–563 (2009)
3. Ellis, J.A., Sudborough, I.H., Turner, J.: The vertex separation and search number of a graph. Inform. and Comput. 113, 50–79 (1994)
4. Fomin, F.V., Gaspers, S., Saurabh, S., Stepanov, A.A.: On two techniques of combining branching and treewidth. Algorithmica 54, 181–207 (2009)
5. Fomin, F.V., Grandoni, F., Kratsch, D.: Measure and conquer: Domination - a case study. In: Caires, L., Italiano, G.F., Monteiro, L., Palamidessi, C., Yung, M. (eds.) ICALP 2005. LNCS, vol. 3580, pp. 191–203. Springer, Heidelberg (2005)
6. Fomin, F.V., Grandoni, F., Kratsch, D.: Solving connected dominating set faster than 2^n. Algorithmica 52, 153–166 (2008)
7. Fomin, F.V., Grandoni, F., Pyatkin, A.V., Stepanov, A.A.: Combinatorial bounds via measure and conquer: Bounding minimal dominating sets and applications. ACM Trans. Algorithms 5(1) (2008)
8. van 't Hof, P., Paulusma, D., Woeginger, G.J.: Partitioning graphs in connected parts. Theoret. Comput. Sci. (to appear) doi:doi:10.1016/j.tcs.2009.06.028
9. Garey, M.R., Johnson, D.S.: Computers and Intractability. W. H. Freeman and Co., New York (1979)
10. Karp, R.M.: Dynamic programming meets the principle of inclusion-exclusion. Oper. Res. Lett. 1, 49–51 (1982)
11. Robertson, N., Seymour, P.D.: Graph minors. XIII. The disjoint paths problem. J. Combin. Theory Ser. B63, 65–110 (1995)
12. van Rooij, J.M.M., Nederlof, J., van Dijk, T.C.: Inclusion/exclusion meets measure and conquer: Exact algorithms for counting dominating set. In: Fiat, A., Sanders, P. (eds.) ESA 2009. LNCS, vol. 5757, pp. 554–565. Springer, Heidelberg (2009)

Author Index

GPSR Compliance

The European Union's (EU) General Product Safety Regulation (GPSR) is a set of rules that requires consumer products to be safe and our obligations to ensure this.

If you have any concerns about our products, you can contact us on ProductSafety@springernature.com

In case Publisher is established outside the EU, the EU authorized representative is:

Springer Nature Customer Service Center GmbH
Europaplatz 3
69115 Heidelberg, Germany

Batch number: 10761599

Printed by Printforce, the Netherlands